Sturkie's Avian Physiology

Sturkie's Avian Physiology

Sixth Edition

Edited by

Colin G. Scanes
Department of Biological Sciences, University of Wisconsin,
Milwaukee, WI, USA

ELSEVIER

AMSTERDAM • BOSTON • HEIDELBERG • LONDON • NEW YORK • OXFORD • PARIS
SAN DIEGO • SAN FRANCISCO • SINGAPORE • SYDNEY • TOKYO

Academic Press is an imprint of Elsevier

AP

Academic Press is an imprint of Elsevier
32 Jamestown Road, London NW1 7BY, UK
225 Wyman Street, Waltham, MA 02451, USA
525 B Street, Suite 1800, San Diego, CA 92101-4495, USA

Sixth edition

British Library Cataloguing-in-Publication Data
A catalogue record for this book is available from the British Library

Library of Congress Cataloging-in-Publication Data
A catalog record for this book is available from the Library of Congress

ISBN: 978-0-12-407160-5

For information on all Academic Press publications
visit our website at elsevierdirect.com

Typeset by TNQ Book and Journals
www.tnq.co.in

Printed and bound in China

15 16 17 18 19 10 9 8 7 6 5 4 3 2 1

Dedication

**To all who have inspired me—my wife,
my parents, my children, my mentors, my colleagues,
and my students and to Paul Sturkie,
who I had the privilege of knowing.**

Contents

14. Gastrointestinal Anatomy and Physiology

D. Michael Denbow

15. Poultry Bone Development and Bone Disorders

M. Pines and R. Reshef

16. Skeletal Muscle

Sandra G. Velleman and Douglas C. McFarland

17. The Avian Immune System

Pete Kaiser and Adam Balic

Part IV
Metabolism Theme

18. Carbohydrate Metabolism

Colin G. Scanes

19. Adipose Tissue and Lipid Metabolism

Johan Buyse and Eddy Decuypere

20. Protein Metabolism

Colin G. Scanes

24. Thyroids

F.M. Anne McNabb and Veerle M. Darras

25. The Role of Hormones in the Regulation of Bone Turnover and Eggshell Calcification

Christopher G. Dacke, Toshie Sugiyama and Carol V. Gay

Part VI
Reproductive Theme

28. Reproduction in the Female

Alan L. Johnson

29. Reproduction in Male Birds

Jorge Vizcarra, Rebecca Alan and John Kirby

30. Reproductive Behavior

Pierre Deviche

Preface

The new edition is staying true to the vision of Paul Sturkie with the two foci of avian physiology—domesticated birds (mainly chickens) and wild birds. The volume has a cohort of returning authors who have extensively revised their chapters. In addition, there are multiple new chapters and new authors. Some of the more recent research approaches (e.g., genomics, transcriptomics, and proteomics) are covered in the initial chapters. Moreover, new chapters address recent work including the control of feed intake, endocrine disruptors, the metabolic challenges of migration together with magnetoreception, and other senses in birds. The volume also returns to its roots in earlier editions with chapters on blood, as well as carbohydrate, lipid, and protein metabolism.

The professionalism and support of Pat Gonzalez at Elsevier are gratefully acknowledged.

Colin G. Scanes
Department of Biological Science, University of Wisconsin, Milwaukee, Milwaukee, WI, USA

Contributors

Numbers in parenthesis indicate the pages on which the authors' contributions begin.

Rebecca Alan (667), College of the Environment and Life Sciences, University of Rhode Island, Kingston, RI, USA

Adam Balic (403), The Roslin Institute & R(D)SVS, University of Edinburgh, Easter Bush, Midlothian, EH25 9RG, UK

C.M. Bishop (919), School of Biological Sciences, Bangor University, Bangor, Gwynedd, UK

Julio Blas (769), Estación Biológica de Doñana, Consejo Superior de Investigaciones Científicas (CSIC), Seville, Spain

Meredith Bohannon (979), Department of Animal and Avian Sciences, University of Maryland, College Park, MD, USA

Walter Bottje (39), Department of Poultry Science, Division of Agriculture, University of Arkansas, Fayetteville, AR, USA

Eldon J. Braun (285, 975), Department of Physiology, College of Medicine, University of Arizona, Tucson, AZ, USA

Kathleen R. Brazeal (847), Department of Neurobiology, Physiology and Behavior, University of California, Davis, CA, USA

Shane C. Burgess (25), Vice Provost and Dean, Agriculture & Life Sciences; Director Arizona Experiment Station; The University of Arizona, Tucson, AZ, USA

Warren W. Burggren (739), Developmental and Integrative Biology, Department of Biological Science, University of North Texas, Denton, TX, USA

P.J. Butler (919), School of Biosciences, University of Birmingham, Edgbaston, Birmingham, UK

Johan Buyse (443), Laboratory of Livestock Physiology, Department of Biosystems, Faculty of Bioscience Engineering, KU Leuven, Leuven, Belgium

Leah Carpenter (979), Department of Animal and Avian Sciences, University of Maryland, College Park, MD, USA

Tiffany Carro (979), Department of Animal and Avian Sciences, University of Maryland, College Park, MD, USA

Rocco V. Carsia (577), Department of Cell Biology, Rowan University School of Osteopathic Medicine, Stratford, NJ, USA

Vincent M. Cassone (811, 829), Department of Biology, University of Kentucky, Lexington, KY, USA

Yupaporn Chaiseha (717), School of Biology, Institute of Science, Suranaree University of Technology, Thailand

Helen E. Chmura (847), Department of Neurobiology, Physiology and Behavior, University of California, Davis, CA, USA

Larry Clark (89), United States Department of Agriculture, Animal and Plant Health Inspection Service, Wildlife Services, National Wildlife Research Center, Fort Collins, CO, USA

Mark A. Cline (469), Department of Animal and Poultry Sciences, Virginia Tech, Blacksburg, VA, USA

Jamie M. Cornelius (847), Department of Neurobiology, Physiology and Behavior, University of California, Davis, CA, USA

Dane A. Crossley II (193), Developmental Integrative Biology Research Cluster, Department of Biological Sciences, University of North Texas, Denton, TX, USA

Christopher G. Dacke (549), Pharmacology Division, School of Pharmacy and Biomedical Science, University of Portsmouth, Portsmouth, UK

Veerle M. Darras (535), Department of Biological Sciences, Virginia Tech, Blacksburg, VA, USA; Department of Biology, Katholieke Universiteit Leuven, Leuven, Belgium

Alistair Dawson (907), NERC Centre for Ecology & Hydrology, Midlothian, Edinburgh, UK

Karen M. Dean (979), University of Lethbridge, Lethbridge, Canada

Eddy Decuypere (443), Laboratory of Livestock Physiology, Department of Biosystems, Faculty of Bioscience Engineering, KU Leuven, Leuven, Belgium

D. Michael Denbow (337, 469), Department of Animal and Poultry Sciences, Virginia Tech, Blacksburg, VA, USA

Pierre Deviche (695), School of Life Sciences, Arizona State University, Tempe, AZ, USA

Jerry B. Dodgson (3), Department of Microbiology and Molecular Genetics, Michigan State University, East Lansing, MI, USA

Joëlle Dupont (613), Unité de Physiologie de la Reproduction et des Comportements, Institut National de la Recherche Agronomique, 37380 Nouzilly, France

Edward M. Dzialowski (193), Developmental Integrative Biology Research Cluster, Department of Biological Sciences, University of North Texas, Denton, TX, USA

Mohamed E. El Halawani (717), Department of Animal Science, University of Minnesota, St. Paul, MN, USA

Carol V. Gay (549), Department of Biochemistry and Molecular Biology, Penn State University, University Park, PA, USA

Julie Hagelin (89), Institute of Arctic Biology, University of Alaska Fairbanks, Fairbanks, AK, USA; Alaska Department of Fish and Game, Fairbanks, AK, USA

Thomas P. Hahn (847), Department of Neurobiology, Physiology and Behavior, University of California, Davis, CA, USA

Alan L. Johnson (635), Center for Reproductive Biology and Health, The Pennsylvania State University, University Park, PA, USA

Pete Kaiser (403), The Roslin Institute & R(D)SVS, University of Edinburgh, Easter Bush, Midlothian, EH25 9RG, UK

John Kirby (667), College of the Environment and Life Sciences, University of Rhode Island, Kingston, RI, USA

Christine Köppl (71), Cluster of Excellence "Hearing4all", Carl von Ossietzky University, Oldenburg, Germany; Research Center Neurosensory Science, Carl von Ossietzky University, Oldenburg, Germany; Department of Neuroscience, School of Medicine and Health Science, Carl von Ossietzky University, Oldenburg, Germany

Wayne J. Kuenzel (135), Poultry Science Center, University of Arkansas, Fayetteville, AR, USA

Vinod Kumar (811), Department of Zoology, University of Delhi, Delhi, India

Dusan Kunec (25), Institut für Virologie, Zentrum für Infektionsmedizin, Freie Universität Berlin, Robert-von-Ostertag-Str. 7, Berlin, Germany

Scott A. MacDougall-Shackleton (847), Departments of Psychology and Biology, University of Western Ontario, Canada

Douglas C. McFarland (379), The Ohio State University/OARDC, Wooster, OH, USA, South Dakota State University, Brookings, SD, USA

F.M. Anne McNabb (535), Department of Biological Sciences, Virginia Tech, Blacksburg, VA, USA; Department of Biology, Katholieke Universiteit Leuven, Leuven, Belgium

Henrik Mouritsen (113), Institut für Biologie und Umweltwissenschaften, Universität Oldenburg, Oldenburg, Germany; Research Centre for Neurosensory Sciences, University of Oldenburg, Oldenburg, Germany

Casey A. Mueller (739), Developmental and Integrative Biology, Department of Biological Science, University of North Texas, Denton, TX, USA

Mary Ann Ottinger (979), Department of Biology and Biochemistry, University of Houston, Houston, TX, USA, Department of Animal and Avian Sciences, University of Maryland, College Park, MD, USA

M. Pines (367), Institute of Animal Sciences, Volcani Center, Bet Dagan, Israel

Tom E. Porter (15), Department of Animal and Avian Sciences, University of Maryland, College Park, MD, USA

Frank L. Powell (301), Division of Physiology, Department of Medicine, University of California, San Diego, CA, USA

R. Reshef (367), Department of Biology and Department of Evolutionary and Environmental Biology, University of Haifa, Haifa, Israel

Nicole Rideau (613), Unité de Recherches Avicoles, Institut National de la Recherche Agronomique, 37380 Nouzilly, France

Johanna R. Rochester (979), The Endocrine Disruption Exchange, Paonia, CO, USA

Colin G. Scanes (167, 421, 455, 489, 497), Department of Biological Sciences, University of Wisconsin, Milwaukee, WI, USA

Elizabeth M. Schultz (847), Department of Neurobiology, Physiology and Behavior, University of California, Davis, CA, USA

Jean Simon (613), Unité de Recherches Avicoles, Institut National de la Recherche Agronomique, 37380 Nouzilly, France

Toshie Sugiyama (549), Department of Agrobiology, Niigata University, Niigata, Japan

Hiroshi Tazawa (739), Developmental and Integrative Biology, Department of Biological Science, University of North Texas, Denton, TX, USA

Sandra G. Velleman (379), The Ohio State University/OARDC, Wooster, OH, USA, South Dakota State University, Brookings, SD, USA

Jorge Vizcarra (667), Department of Food and Animal Sciences, Alabama A&M University, Huntsville, AL, USA

Heather E. Watts (847), Department of Biology, Loyola Marymount University, Los Angeles, CA, USA

Scott Werner (89), United States Department of Agriculture, Animal and Plant Health Inspection Service, Wildlife Services, National Wildlife Research Center, Fort Collins, CO, USA

J. Martin Wild (55), Department of Anatomy with Radiology, Faculty of Medical and Health Sciences, University of Auckland, Auckland, New Zealand

Shlomo Yahav (869), Department of Poultry and Aquaculture Sciences, Institute of Animal Sciences, ARO, The Volcani Center, Bet-Dagan, Israel

Takashi Yoshimura (829), Laboratory of Animal Physiology, Graduate School of Bioagricultural Sciences, Nagoya University, Furo-cho, Chikusa-ku, Nagoya, Japan, Avian Bioscience Research Center, Graduate School of Bioagricultural Sciences, Nagoya University, Furo-cho, Chikusa-ku, Nagoya, Japan

Part I

Undergirding Themes

Chapter 1

Avian Genomics

Jerry B. Dodgson
Department of Microbiology and Molecular Genetics, Michigan State University, East Lansing, MI, USA

1.1 INTRODUCTION

The fifth edition of *Sturkie's* contained neither an avian genomics chapter, nor any of the subsequent three chapters in this edition. Their inclusion here reflects the fact that all aspects of physiology have become intertwined with our understanding of genes and genomes. The early history of this transition is discussed elsewhere (Siegel et al., 2006), but the keystone event was the sequencing of the chicken genome (International Chicken Genome Sequencing Consortium, 2004). Soon, we will have genome sequences for thousands of avian species (Genome 10K Community of Scientists, 2009), but the fundamental challenge will remain: learning how to read the fascinating stories of avian physiological adaptations and evolution from a long string of a billion or so A, T, G, and C nucleotides per bird.

1.2 GENOME SIZE

Haploid avian genomes are generally the smallest among amniotes (www.genomesize.com), averaging 1.35 Gb (billion base pairs). A narrow range separates the smallest (black-chinned hummingbird, 0.9 Gb) and largest (ostrich, 2.1 Gb) species. Their compactness reflects the low frequency of repetitive elements that derive from transposons and their descendent sequences (International Chicken Genome Sequencing Consortium, 2004). Avian genome size correlates with physiological measures, such as with red cell size and (inversely) with metabolic rate (Gregory, 2002). It was proposed that small genomes were selected during the evolution of flight (Hughes and Hughes, 1995). However, Organ et al. (2007) suggested that contraction in genome size preceded the acquisition of flight, and nonadaptive and neutral explanations for small bird genomes also have support (Lynch and Conery, 2003; Nam and Ellegren, 2012).

1.3 CHROMOSOMES

1.3.1 Karyotypes

Avian karyotypes have been unusually stable during evolution (Burt et al., 1999; Ellegren, 2010). The ancestral avian karyotype is predicted to have $2n=80$ chromosomes, with the only subsequent change in chicken ($2n=78$) being a fusion between ancestral chromosomes 4 and 10 (Shibusawa et al., 2004; Griffin et al., 2007). However, there are exceptions, with avian chromosome numbers ranging from 40 to 126 (Griffin et al., 2007). A particular feature of avian karyotypes is that most species have numerous "microchromosomes", a trait they share with some, but not all, nonavian reptiles (Janes et al., 2010). The definition of a microchromosome is somewhat arbitrary (Masabanda et al., 2004), but, generally, microchromosomes are too small to discriminate by size in standard karyotypes.

In those birds with fewer chromosomes (falcons, Nishida et al., 2008; hawks and eagles, de Oliveira et al., 2005; stone curlew, Nie et al., 2009), some, but not all, microchromosomes have fused to ancestral macrochromosomes or to each other. It remains difficult to determine orthologous relationships because sequences derived from one species' microchromosomes often fail to hybridize to those of another species (e.g., Nie et al., 2009), suggestive of high content of rapidly evolving repetitive DNA. However, in general, translocations appear to have been very rare during avian evolution (Griffin et al., 2007), in comparison to the somewhat more common frequency of chromosome inversions (Warren et al., 2010; Zhang et al., 2011; Skinner and Griffin, 2012). Interestingly, in turkeys there appears to be a predominance of acrocentric (centromere at or near one telomere) chromosomes (Zhang et al., 2011), whereas in falcons and hawks the trend is towards metacentric (centromere near the middle) chromosomes (Nishida et al., 2008).

Sturkie's Avian Physiology.

1.3.2 Sex Chromosomes

Another characteristic that all birds share with some non-avian reptiles is the use of a ZW sex chromosome arrangement in which males are homogametic (ZZ) and females are heterogametic (ZW). However, sex determination has evolved independently several times within the vertebrates, although common genes or a common set of autosomes may be reused (Marshall Graves and Peichel, 2010; Ellegren, 2010). The ratite W is minimally diverged from the Z (and presumably the ancestral autosome), whereas in other birds, W is smaller, gene-poor, and repeat rich (Marshall Graves and Shetty, 2001). The Z-specific gene, *DMRT1*, appears to play a major role in masculinization (Smith et al., 2009), although it appears that both cell autonomous and hormonal sex determination pathways exist, with the interplay between the two yet to be fully elucidated (Zhao et al., 2010). Further aspects of sexual differentiation are discussed in later chapters.

1.3.3 Telomeres and Centromeres

Birds share the canonical TTAGGG telomere repeat with all other vertebrates. However, chickens, turkeys, and other birds possess variable numbers of unusually large telomere repeat blocks, even up to 3–4 Mb (million base pairs) in length (Delany et al., 2000; O'Hare and Delany, 2009). Although the purpose of these mega-telomeres remains unknown, they map preferentially, but not obligatorily, to specific chromosomes (Delany et al., 2007; O'Hare and Delany, 2009). Chicken centromeres also merit special mention. Although most contain typical long (>100 kb pairs) arrays of chromosome-specific simple repeats, the centromeres of GGA5, GGA27, and GGAZ are remarkably short (~30 kb) and lack the usual repeat structure (Shang et al., 2010). Being able to clone and manipulate these centromeres by homologous recombination (Shang et al., 2013) promises to make the chicken the primary model system for the study of vertebrate centromeres. A final point is that the zebra finch and probably other birds possess a germline restricted chromosome, with a function that remains obscure (Itoh et al., 2009).

1.4 GENOME SEQUENCES

1.4.1 Approach

All bird genomes sequenced to date have employed a whole genome shotgun method, in which overlaps between millions of random reads are used to assemble contiguous blocks of sequence (i.e., contigs) along the genome. Due to their relatively low repeat content, avian genomes are ideal for shotgun sequencing. Contigs are then assembled into scaffolds (i.e., aligned groups of contigs containing size-calibrated gaps), using mate-pair reads in which both ends are sequenced from DNA fragments within a selected size range. Even for genomes with deep coverage, this generates hundreds to thousands of scaffolds that, ideally, are ordered and aligned using physical (based on mapping of recombinant clones in bacterial artificial chromosome (BAC) vectors) and/or linkage maps (Table 1.1).

The chicken (International Chicken Genome Sequencing Consortium, 2004) and zebra finch (Warren et al., 2010) were sequenced by the Sanger method, in which reads are derived one-by-one from recombinant clone libraries. This currently remains the gold standard for genome sequencing but no longer is cost-effective with the advent of next-generation sequencing (NGS) methods, which directly sequence collections of (uncloned) DNA fragments in a multiparallel manner. NGS read lengths often are shorter and sometimes more error-prone than Sanger reads, but NGS compensates by much higher coverage, such that the consensus sequence is at least as accurate. Various NGS methods have been developed (Metzker, 2010). The first avian genome to be sequenced via NGS was that of the turkey (Table 1.1), and we can anticipate an onslaught of new bird genomes soon (Genome 10K Community of Scientists, 2009).

1.4.2 Coverage

Most current avian genome sequence assemblies contain about 90–95% of their respective euchromatic genomes (typically 1.1–1.2 Gb; Table 1.1). Coverage is usually estimated by the fraction of different mRNA transcripts that can be found within the assembly. Highly repetitive heterochromatic sequences, especially when repeated in tandem, are nearly impossible to assemble and are missing from all vertebrate genomes, but these contain few genes. For example, centromeres (however, see Shang et al., 2010), telomeres, and rDNA (tandem repeats that encode ribosomal RNA, on GGA16) are generally missing altogether or shown as gaps, and very little of the repeat-rich/gene-poor W chromosome is usually assembled. Sequence scaffolds are ordered and aligned along chromosomes for birds that have dense linkage maps and/or BAC contig physical maps, sometimes assuming a common local order with closely related genomes (comparative maps); however, most NGS-derived avian genomes currently are unordered (Table 1.1). Sequence scaffolds that cannot be placed are arbitrarily clustered on chrUn (chromosome unknown) or, for example, chr1_random if the chromosome but not the location is known, or simply provided as a list of unplaced scaffolds. Even for the chicken, it has been impossible to align sequence scaffolds with specific smaller microchromosomes (GGA29–31, GGA33–38, and most of GGA16 and 32), so any such sequence is on chrUn. In part, this is due to a paucity of aligning markers; however, more generally, microchromosomal DNA is poorly represented in sequence reads. The reasons remain unclear, but they likely relate to microchromosomes being rich in

TABLE 1.1 Avian Reference Genome Sequence Assemblies

Species/ WGS Project[1]	Assembly[1]	Method[2]	Fold Coverage	Sequenced Bases (Gb)[3]	Aligned to Chromosome	Scaffold N50[4] (Mb)	Contig N50[4] (kb)	Approximate Coverage[5]	References
Chicken AADN03	Gallus_gallus-4.0 November 2011	Sanger	6.6×	1.047	Yes	12.9	280	96%	International Chicken Genome Sequencing Consortium, 2004[6]
Turkey ADDD01	Turkey_2.01 February 2011	Roche Illumina	30×	1.062	Yes	0.86	12.5	89%	Dalloul et al., 2010
Zebra finch ABQF01	Taeniopygia_ guttata-3.2.4 February 2013	Sanger	6×	1.232	Yes	8.24	38.6	96%	Warren et al., 2010
Budgerigar AGAI01	Melopsittacus_ undulatus_6.3 February 2012	Roche Illumina	23×	1.117	No	10.6	55.6	NR	[7]
Budgerigar	Koren et al. July 2012	Roche Illumina Pacific Biosciences	63×	1.07	No	NR	100	NR	[8]
Collared flycatcher AGTO01	FicAlb_1.4 November 2012	Illumina	85×	1.116	Yes	7.3	450	NR	Ellegren et al., 2012
Medium ground finch AKZB01	GeoFor_1.0 June 2012	Illumina	115×	1.065	No	5.3	30.5	NR	[9]
Large ground finch	Rands et al. February 2013	Roche	6.5×	0.96	No	0.38	30.5	89%	Rands et al., 2013
Rock pigeon AKCR01	Cliv_1.0 February 2013	Illumina	63×	1.108	No	3.15	26.6	88%	Shapiro et al., 2013
Puerto Rican parrot AOCU01	AV1 January 2013	Illumina	27×	1.175	No	0.019	6.9	76%	[10]

(Continued)

TABLE 1.1 Avian Reference Genome Sequence Assemblies—cont'd

Species/ WGS Project[1]	Assembly[1]	Method[2]	Fold Coverage	Sequenced Bases (Gb)[3]	Aligned to Chromosome	Scaffold N50[4] (Mb)	Contig N50[4] (kb)	Approximate Coverage[5]	References
Peregrine falcon AKMT01	F_peregrinus_v1.0 February 2013	Illumina	107×	1.172	No	3.9	28.6	99%	Zhan et al., 2013
Saker falcon AKMU01	F_cherrug_v1.0 February 2013	Illumina	114×	1.175	No	4.2	31.3	97%	Zhan et al., 2013
Tibetan ground-tit ANZD01	PseHum1.0 January 2013	Illumina	96×	1.030	No	16.3	165	NR	[11]
Mallard duck ADON01	BGI_duck_1.0 April 2013	Illumina	60×	1.107	Yes	1.23	26.1	95%	[12]
White-throated spar-row	ASM38545v1 April 2013	Illumina	95×	1.053	No	4.9	113	NR	[13]

[1]*Whole Genome Shotgun (WGS) project numbers and assembly names and dates from National Center for Biotechnology Information (NCBI) Assembly (http://www.ncbi.nlm.nih.gov/assembly, accessed May 14, 2013). The most recent builds and dates are listed. In some cases, these are more recent updates of those described in references. References are listed for those assemblies not curated in NCBI Assembly.*
[2]*Initial method employed (see Metzker, 2010), although supplemented later by alternative approaches in some cases.*
[3]*Sequenced base total generally includes gaps within scaffolds. Aligned to chromosome indicates whether scaffolds were ordered and aligned to chromosomes, typically using linkage maps (all indicated), bacterial artificial chromosome contig physical maps (chicken, turkey, zebra finch), comparative maps (turkey, flycatcher, duck), and/or radiation hybrid maps (chicken).*
[4]*N50 is the size of a scaffold or contig such that half the sequenced genome is contained in scaffolds or contigs that size or larger.*
[5]*Approximate genome coverage estimates are calculated relative to the euchromatic genome. NR=not reported. Falcon coverage shown is likely an overestimate (see text).*
[6]*International Chicken Genome Sequencing Consortium, 2004. The sequenced individual was a red jungle fowl, the primary wild progenitor of domestic chickens.*
[7]*Ganapathy, G., Howard, J., Jarvis, E.D., Phillippy, A., Warren, W., 2012. Draft genome of Melopsittacus undulates budgerigar version 6.3. Direct submission to NCBI Genbank.*
[8]*Koren, S., Schatz, M.C., Walenz, B.P., Martin, J., Howard, J.T., Ganapathy, G., Wang, Z., Rasko, D.A., McCombie, W.R., Jarvis, E.D., Phillippy, A.M., 2012. Hybrid error correction and de novo assembly of single-molecule sequencing reads. Nat. Biotechnol. 30, 693–700.*
[9]*Zhang, G., Parker, P., Li, B., Li, H., Wang, J., 2012. The genome of Darwin's Finch (Geospiza fortis). Gigascience. Available from: http://dx.doi.org/10.5524/100040.*
[10]*Oleksyk, T.K., Pombert, J.F., Siu, D., Mazo-Vargas, A., Ramos, B., Guiblet, W., Afanador, Y., Ruiz-Rodriguez, C.T., Nickerson, M.L., Logue, D.M., Dean, M., Figueroa, L., Valentin, R., Martinez-Cruzado, J.C., 2012. A locally funded Puerto Rican parrot (Amazona vittata) genome sequencing project increases avian data and advances young researcher education. Gigascience 1, 14.*
[11]*Cai, Q., Lang, Y., Li, Y., Wang, J., 2013. The genome sequence and adaptation to high land of Hume's groundpecker Pseudopodoces humilis. Direct submission to NCBI Genbank.*
[12]*Huang, Y., Li, Y., Burt, D.W., Chen, H., Zhang, Y., Qian, W., Kim, H., Gan, S., Zhao, Y., Li, J., Yi, K., Feng, H., Zhu, P., Li, B., Liu, Q., Fairley, S., Magor, K.E., Du, Z., Hu, X., Goodman, L., Tafer, H., Vignal, A., Lee, T., Kim, K.W., Sheng, Z., An, Y., Searle, S., Herrero, J., Groenen, M.A., Crooijmans, R.P., Faraut, T., Cai, Q., Webster, R.G., Aldridge, J.R., Warren, W.C., Bartschat, S., Kehr, S., Marz, M., Stadler, P.F., Smith, J., Kraus, R.H., Zhao, Y., Ren, L., Fei, J., Morisson, M., Kaiser, P., Griffin, D.K., Rao, M., Pitel, F., Wang, J., and Li, N., 2013. The duck genome and transcriptome provide insight into an avian influenza virus reservoir species. Nat. Genet. 45, 776–783.*
[13]*White-throated sparrow consortium, 2013. Zonotrichia albicollis genome sequencing. Direct submission to NCBI Genbank.*

repetitive sequences and high in GC content. It was initially thought that this made microchromosomal DNA refractile to recombinant DNA cloning (and, indeed, it is rare in clone libraries), but these reads remain underrepresented even in uncloned NGS sequences. The smallest chicken chromosome with reasonable sequence representation is GGA25 (~2.2 Mb), but the sequence assembly is problematic for this and at least two other small chromosomes (GGA28, Gordon et al., 2007; GGA16, Shiina et al., 2007), in part due to repeated sequences. Even though they may be rich in repeats, for the most part, microchromosomes are also gene-rich (International Chicken Genome Sequencing Consortium, 2004), although one cannot be certain about GGA29–38. It seems likely that much of the missing 5–10% of current assemblies (Table 1.1) lies on microchromosomes and W chromosomes. (Falcon assemblies claim 97–99% coverage (Zhan et al., 2013), but this probably is not due to the fact that these genomes contain fewer microchromosomes, but rather because the authors measured coverage by the frequency with which cloned sequences are found, so their test set is biased away from microchromosomes.)

1.5 ANNOTATION

Much of the value of the reference genome sequence depends on annotation (Yandell and Ence, 2012), which links the DNA sequence to all the information available on component genes, mRNAs, proteins, etc. Once the genome is sequenced, there are two broad classes of annotation: (1) evidenced-based, which uses RNA or proteomic data (see Chapters 3 and 4), as well as homology to genes in other species; and (2) ab initio annotation, employing computer searches for open reading frames, likely initiation and stop codons, splice junctions, and other sequence-based characteristics to predict the existence of genes for which experimental evidence is lacking. Transposable elements are annotated based on their repetition in the genome, relatedness to transposons in other species, and their characteristic end structures (Jurka et al., 2005). Annotating regulatory sequences (such as transcription factor binding sites) is more problematic; it also relies on both comparisons to other genomes and evidence from genome-wide DNA methylation and chromatin immunoprecipitation (ChIP) analyses (see Chapter 2). This is exemplified by the human ENCODE project (The ENCODE Project Consortium, 2012), but it will be some time before that level of data is available for any bird. Much of the annotation of avian genome sequences has relied on comparisons to other genomes and has not been manually curated. Thus, the annotations are frequently inaccurate, especially for those genes and other elements whose functions are lineage-specific (i.e., only found in a given species or only in birds). Thus, one should be hesitant to accept conclusions based solely on computer analysis of avian genome sequences in their current state.

1.6 GENOME BROWSERS

Most of the user community depends on one or more genome browsers to utilize sequence data. There are three major browsers: University of California at Santa Cruz (UCSC) Genome Bioinformatics (www.genome.ucsc.edu), Ensembl (www.ensembl.org), and the National Center for Biotechnology Information (NCBI) Map Viewer (www.ncbi.nlm.nih.gov/genome); there are also avian-focused sites such as Avian Genomes (aviangenomes.org) and Bird Base (birdbase.arizona.edu/birdbase). The browsers all employ the same reference sequence information as a series of chromosomes, scaffolds, or both. Any property that is sequence-specific (genes, ChIP binding sites, RNA sequences, homology with other sequences, etc.) can be displayed as a track on the genome (Figure 1.1). Genome browsers are only as good as the underlying sequence assembly and annotation. Not all avian genomes are available at every browser site, and not all annotation tracks are available for each build (i.e., updated assemblies based on new data). The various options are in constant flux.

1.7 GENES

All bird genomes evolved via two whole genome duplication events that preceded the ancestral vertebrate genome (Van de Peer et al., 2009). A commonly cited outcome are the four clusters of *HOX* homeobox developmental transcription factor genes found in most vertebrates (e.g., chicken *HOXA* cluster on GGA2; *HOXB* on GGA27, Figure 1.1; *HOXD* on GGA7; and *HOXC* on chrUn, probably on a microchromosome). In most instances, one or more of the potential four ancestral genes or clusters has been lost during subsequent evolution or, as in the case of the *HOX* clusters, has diverged to perform different functions, thereby providing a selective force leading to its retention. Another major force in gene evolution has been the (usually local) expansion and contraction of gene families. For example, the γ-c clade of olfactory receptor genes (always among the most rapidly diverging gene families) is highly expanded in the chicken and zebra finch, but falcon genomes have only one or two copies (Zhan et al., 2013).

Depending on the methods employed and the available evidence, avian genomes are estimated to contain 15–20,000 protein-coding genes, but keep in mind that each gene locus may generate multiple transcripts and proteins due to alternative splicing, transcriptional start sites, and polyadenylation sites. This number may end up being slightly low once additional transcriptome and proteome data accumulate (Chapters 3 and 4). There is some evidence of a greater rate of gene loss versus gene gain during avian evolution (International Chicken Genome Sequencing Consortium, 2004), but this must be viewed cautiously, given the less fully annotated state of bird genomes. The most reliably identified genes are

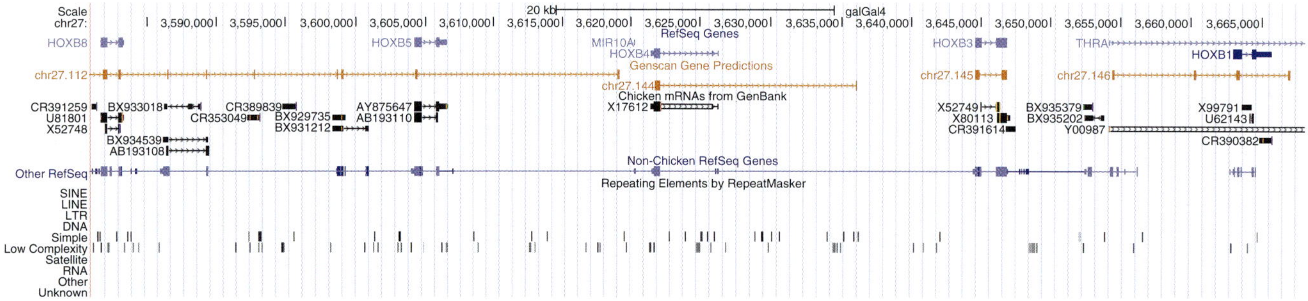

FIGURE 1.1 UCSC genome browser view of the chicken *HOXB* cluster on GGA27. Sequence coordinates chr27:3,581,000–3,668,000 (shown at top) from the November 2011 International Chicken Genome Sequencing Consortium Gallus_gallus-4.0 assembly are shown. In descending order, tracks are appended for the following: (1) RefSeq genes (blue): five *HOXB* genes, a microRNA *MIR10A* locus, and part of the overlapping *THRA* gene; (2) Genscan ab initio gene models (light brown, note numerous incorrect exons); (3) sequences mapping to chicken mRNAs (black); (4) homology to RefSeq genes from other species (blue); and (5) repeated sequences (gray-black boxes) of classes designated at left (interestingly, long interspersed nuclear transposable elements, which are common in most of the genome, are absent here). In the first three tracks, exons are shown as filled boxes with coding regions thicker than untranslated regions, and introns are depicted as narrow lines with arrowheads in the direction of transcription. Chicken mRNAs are complementary DNA clone sequences estimated to be full length, but often (as shown) they are not, and those lacking introns should be considered as likely artifacts (genomic DNA fragments contaminating mRNA). This view was generated with the following track settings: Base Position and RepeatMasker set to full; RefSeq Genes, Genscan Genes, and Chicken mRNAs set to pack; Other RefSeq genes set to dense; and all other tracks set to hide. Although only five *HOXB* genes are shown as RefSeq annotated, using the chicken mRNAs (BX931212, BX934539, BX935202) and homology to Other RefSeq genes (shown only in dense mode here for sake of scale) and expanding the initial coordinate to chr27:3,530,000 also reveals homologues to *HOXB2, HOXB6, HOXB7, HOXB9* and *HOXB13*. The UCSC Genome Browser Gateway at http://genome.ucsc.edu was accessed on May 28, 2013 (Kent et al., 2002).

the RefSeq (http://www.ncbi.nlm.nih.gov/projects/RefSeq/) genes that have been manually annotated, but the RefSeq set is conservative (low false-positive rate and higher false-negative rate). For example, only five of at least 10 likely *HOXB* genes are official chicken RefSeq genes (Figure 1.1). However, by using mRNA information and sequence homology to genes and/or proteins in other species, it often is straightforward to identify a gene of interest, even when it is not a RefSeq gene. This is particularly critical for birds other than chickens, whose genomes are less well annotated.

1.8 TRANSPOSONS

As noted previously, avian genomes contain comparatively low levels of transposable element-derived repeats (less than 10% of the assembled sequence), although these numbers also can diverge widely depending on the estimation methods employed (compare Zhan et al., 2013 to Warren et al., 2010; Dalloul et al., 2010 for chicken, zebra finch, and turkey). The predominant avian transposon is the chicken repeat 1 long interspersed nuclear transposable element, although it appears these comprise a smaller portion of passerine genomes (Warren et al., 2010; Ellegren et al., 2012), and the zebra finch genome is comparatively rich in long terminal repeat transposons. Short interspersed nuclear element (SINE) transposons are extremely rare (less than 0.1% of all avian genome sequences), suggesting that the ability to transpose SINEs died out long ago (International Chicken Genome Sequencing Consortium, 2004). DNA transposons constitute close to 1% of the turkey and chicken genomes but appear quite rarely in other avians. For most of these transposon families, there appear to be few, if any, copies that are still active (Wicker et al., 2005), with the caveat that ~5–10% of the genome remains missing, especially in repeat-rich regions. Overall, it appears that transposon copies are being deleted 3–5 times faster than new ones are being created in avian evolution (Nam and Ellegren, 2012).

1.9 GENOME DIVERSITY

1.9.1 SNP Discovery

At least for chickens and turkeys, genome maps, especially linkage maps, predated the genome sequence (Siegel et al., 2006) and were important complements in aligning sequence scaffolds to chromosomes. However, NGS allows one to sequence a genome first and then use that sequence for high-resolution mapping, both to improve the assembly and to locate trait-encoding loci. Sequencing provides the critical component for linkage analysis: DNA polymorphisms, mostly single-nucleotide polymorphisms (SNPs) and copy number variations. Polymorphism is an enemy of accurate reference genome assembly, so the ideal is to sequence a single (preferentially inbred and genetically monomorphic) individual. This was feasible for chickens, but parallel sample sequencing of three other chickens generated nearly 3 million SNPs, which provided the initial basis for high-density genotyping (International Chicken Polymorphism Map Consortium, 2004). For most other birds, the sequenced individual was, at best, only slightly inbred, thus immediately providing extensive SNP variation between the two copies of each chromosome (usually a ZZ male was sequenced). Additional SNPs can be obtained by NGS sequencing of other individuals (resequencing) or by NGS RNA sequencing (RNA-seq).

1.9.2 SNP Diversity

Avian genomes exhibit high levels of diversity with typical average pairwise heterozygosity rates (π) of 2–10 SNPs per kilobase (International Chicken Polymorphism Map Consortium, 2004; Balakrishnan and Edwards, 2009; Ellegren et al., 2012; Shapiro et al., 2013). These considerably exceed the rate in humans (except for falcons; Zhan et al., 2013), which is presumably a reflection of larger effective population sizes during the evolution of the respective birds. NCBI dbSNP (www.ncbi.nlm.nih.gov/snp) currently lists over 9.4 million reference chicken SNPs. Although commercial breeding has reduced SNP numbers in chickens (Muir et al., 2008), both broilers and (to a lesser extent) layers retain relatively high genetic diversity. This explains why commercial breeders continue to make progress in enhancing economically desirable traits even after 50 years of intense selection; it also testifies to their ability to avoid excessive inbreeding. It should be noted that large populations have not been deeply sequenced from any bird, so the above discussion considers relatively common (and therefore ancient, having had time to spread through the population) SNPs. Indeed, given the enormous worldwide numbers of commercial chickens, one expects that extremely rare SNPs exist at nearly every base pair, but these have extremely low likelihoods of long-term persistence.

1.9.3 Recombination

High-density SNP genotyping arrays have been developed for chickens (Kranis et al., 2013), which allow for both linkage mapping and association analysis. The former relies on meiotic recombination in genotyped family pedigrees, whereas the latter relies on historical linkage disequilibrium (LD, the nonrandom correlation in the co-segregation or association of linked alleles) within a broader population. The greater the local recombination rate, the lower the level of LD, so high recombination rates increase map resolution but require denser marker panels. The chicken genome exhibits a high average recombination rate per Mb of DNA compared to mammals (International Chicken Genome Sequencing Consortium, 2004), along with much greater variation in that rate between chromosomes. This is to be expected, given that proper segregation of microchromosomes should require at least one crossover per meiosis, therefore making them >50 cM. Thus, a 4 Mb microchromosome (e.g., GGA22) should average >12.5 cM/Mb, which is more than 10 times the typical mammalian rate. This should allow for higher resolution in mapping the microchromosomes (which is good because they typically are gene rich), but they need to be much more densely sampled in genotyping panels. The same trend occurs in the zebra finch (and presumably most, if not all, birds), but, interestingly, recombination along the length of individual chromosomes is more variable (Backström et al., 2010). Although there is a clear tendency for higher crossover density within 10 Mb of a chicken telomere, this is much more dramatic in zebra finch (the difference can only be observed on macrochromosomes >20 Mb). As a result, LD should be much greater near the center of larger zebra finch chromosomes, making identification of the specific genes/alleles involved in traits more difficult. (See Ellegren, 2005 and Backström et al., 2010 for more discussion of recombination effects on avian genomes.)

1.10 CONNECTING SEQUENCE TO PHENOTYPE

1.10.1 Avian-Specific Genes

Beyond generating lists of SNPs, genes, transcripts, noncoding RNAs, etc., a major goal for sequence analysis is to increase the understanding of avian phenotypes. For agricultural species, traits that have economic impact tend to receive the greatest attention, and many of these trait alleles have been selected in the last 6000 years since domestication or during commercial breeding. For wild birds, ecological and evolutionary questions predominate, and some of the traits of interest date back millions of years to the time of speciation (Ellegren et al., 2012).

In general, little is known about what genes or alleles make avian physiology unique. As noted above, it is much more difficult to annotate and identify functions for lineage-specific genes/alleles unique to birds. It is known that birds have greatly expanded their repertoire of keratin genes (feather, scale, and claw keratins), and, of course, they retain genes used in egg production in common with most nonavian reptiles (International Chicken Genome Sequencing Consortium, 2004; Warren et al., 2010), but there is obviously much more to learn. In addition to the presence or absence of certain genes, conclusions are also derived from rates of gene evolution, often by comparing the rates of amino acid altering nucleotide substitutions to synonymous changes (K_A/K_S ratios). High ratios suggest positive/diversifying selection of derived alleles that enhance fitness, whereas low ratios imply negative/purifying selection that eliminates diverse alleles in evolutionarily conserved genes. For example, Warren et al. (2010) found evidence for positive selection within zebra finch genes that exhibit differential expression in the auditory forebrain, suggesting a possible role in the evolution of singing behavior. Similar evidence of positive selection (as well as gene duplication) identified candidate genes involved in beak morphology in both Darwin's finches (Rands et al., 2013) and falcons (Zhan et al., 2013).

1.10.2 Mapping Mutations and QTL

The availability of a reference chicken sequence, along with dense SNP maps and genotyping arrays, has facilitated identification of causal alleles for several Mendelian (monogenic) traits (Davey et al., 2006; Gunnarsson et al., 2007; Eriksson et al., 2008; Wright et al., 2009; Dorshorst et al., 2010; Hellström et al., 2010; Dorshorst et al., 2011; Robb et al., 2011; Imsland et al., 2012; Ng et al., 2012; Wang et al., 2012; Wells et al., 2012). A host of chicken lines exist with specific mutant, physiological, or immunological characteristics (Delany, 2004; Robb et al., 2011) that remain open to this sort of analysis. Chickens have also been widely employed for the mapping of numerous quantitative trait loci (QTL). Currently, there are over 3800 QTL in ChickenQTLdb (www.animalgenome.org/cgi-bin/QTLdb/GG/index). Recently, QTL analysis also has been performed in turkey and zebra finch (Aslam et al., 2011; Schielzeth et al., 2012a,b).

1.10.3 Resequencing

With the decreasing costs of NGS, it is now feasible to resequence rather than use SNP genotypes at least for reasonably small numbers of individual birds. In this case, one chooses (1) to sequence whole genomes (often along with parents and/or siblings as controls) or (2) to hope that the allele(s) of interest is coding, therefore sequencing the "exome" (i.e., DNA enriched for exons by hybridization to a synthetic array designed for this purpose; Ng et al., 2009). The challenge then becomes to identify potential candidate causal mutations in a background of unrelated polymorphisms and sequencing errors. Various filters can be applied to reduce the background due to errors, and common polymorphisms can be eliminated from consideration because these are unlikely to generate a typically deleterious mutant trait. The remaining candidate alleles can be searched for those most likely to result in a dramatic phenotype based on evolutionary conservation and predicted effect on the protein product (for coding mutations). As one example, resequencing of pigeon genomes (Shapiro et al., 2013) demonstrated an *EphB2* allele that gives rise to a derived head crest trait. At a broader level of resolution, resequencing can identify genome segments, such as selection signatures, that show unusual absence of diversification in the selected populations. Because the signal detected derives from many SNPs across an LD block, it can be more readily identified relative to the background. In the first application of this approach in birds, Rubin et al. (2010) resequenced nine pooled samples from domestic broiler, layer, and red jungle fowl breeds to identify more than 70 genome regions likely to have been involved in domestication and/or subsequent commercial selection. More recently, Ellegren et al. (2012) resequenced collared and pied flycatchers and found approximately 50 large (~400 kb) divergence islands characterized by high interspecies (and low intraspecies) diversity, at least some of which were likely involved in the split of the two species over the last 2 million years. We are just beginning to see the first of many fascinating stories linking genome sequence to avian physiology, behavior, and evolution. Additional powerful tools relating gene expression to phenotypic traits derive from transcriptomic and proteomic analyses, which are discussed in Chapters 3 and 4.

1.11 CONCLUSIONS AND SUMMARY

The sequencing of the chicken genome was a watershed moment in avian biology. Zebra finch and turkey genome sequences were completed in 2010, and we now stand at the leading edge of a wave of avian genomes. Beyond being a table of contents of genes, transposons, and other elements, the genome sequence is the central foundation for transcriptomics, proteomics, linkage maps, and other tools. In particular, avian genome sequences form the foundation upon which tools such as resequencing and SNP arrays (above), ChIP-seq and methyl-seq (Chapter 2), RNA-seq and microarrays (Chapter 3), and proteomics (Chapter 4) depend. Together, these domains provide a critical genetic reference text for all aspects of avian physiology.

All avian genomes currently are incomplete drafts, with particular deficiencies on microchromosomes. A major challenge remains to fill the gaps in these assemblies and properly align them along chromosomes. An even greater challenge is to accurately annotate all the components that contribute to gene expression and its regulation. To date, most annotation of avian genomes derives from information from other species, and this necessarily misses lineage-specific characteristics that define what it is to be a bird. The coming era of avian genomics will focus on elucidating the function of the various sequence elements. This is where genomics and physiology must join forces to ultimately marry genotype to phenotype.

Although the reference genome sequence is a critical first step, next-generation methods allow for the sequencing of many individuals within any species. This also provides an avenue to address questions of ecology and evolution, such as diversity and speciation, as well as traits of commercial interest in domestic species such as muscle growth, disease resistance, and reproduction. We will soon have genome sequences from thousands of different birds and, at least for some of these, the sequences of hundreds to thousands of individuals. We will also be hearing about the epigenomes, transcriptomes, and proteomes of many of these species. How will we integrate all this data and derive a more thorough understanding of the nearly 200 million years of separate avian evolution and the approximately 10,000 extant birds? This is the challenge for the next generation of avian physiologists.

REFERENCES

Aslam, M.L., Bastiaansen, J.W., Crooijmans, R.P., Vereijken, A., Groenen, M.A., 2011. Whole genome QTL mapping for growth, meat quality and breast meat yield traits in turkey. BMC Genet. 12, 61.

Backström, N., Forstmeier, W., Schielzeth, H., Mellenius, H., Nam, K., Bolund, E., Webster, M.T., Ost, T., Schneider, M., Kempenaers, B., Ellegren, H., 2010. The recombination landscape of the zebra finch *Taeniopygia guttata* genome. Genome Res. 20, 485–495.

Balakrishnan, C.N., Edwards, S.V., 2009. Nucleotide variation, linkage disequilibrium and founder-facilitated speciation in wild populations of the zebra finch (*Taeniopygia guttata*). Genetics 181, 645–660.

Burt, D.W., Bruley, C., Dunn, I.C., Jones, C.T., Ramage, A., Law, A.S., Morrice, D.R., Paton, I.R., Smith, J., Windsor, D., Sazanov, A., Fries, R., Waddington, D., 1999. The dynamics of chromosome evolution in birds and mammals. Nature 402, 411–413.

Dalloul, R.A., Long, J.A., Zimin, A.V., Aslam, L., Beal, K., Blomberg, L.A., Bouffard, P., Burt, D.W., Crasta, O., Crooijmans, R.P., Cooper, K., Coulombe, R.A., De, S., Delany, M.E., Dodgson, J.B., Dong, J.J., Evans, C., Frederickson, K.M., Flicek, P., Florea, L., Folkerts, O., Groenen, M.A., Harkins, T.T., Herrero, J., Hoffmann, S., Megens, H.J., Jiang, A., de Jong, P., Kaiser, P., Kim, H., Kim, K.W., Kim, S., Langenberger, D., Lee, M.K., Lee, T., Mane, S., Marcais, G., Marz, M., McElroy, A.P., Modise, T., Nefedov, M., Notredame, C., Paton, I.R., Payne, W.S., Pertea, G., Prickett, D., Puiu, D., Qioa, D., Raineri, E., Ruffier, M., Salzberg, S.L., Schatz, M.C., Scheuring, C., Schmidt, C.J., Schroeder, S., Searle, S.M., Smith, E.J., Smith, J., Sonstegard, T.S., Stadler, P.F., Tafer, H., Tu, Z.J., Van Tassell, C.P., Vilella, A.J., Williams, K.P., Yorke, J.A., Zhang, L., Zhang, H.B., Zhang, X., Zhang, Y., Reed, K.M., 2010. Multi-platform next-generation sequencing of the domestic turkey (*Meleagris gallopavo*): genome assembly and analysis. PLoS Biol. 8, e1000475.

Davey, M.G., Paton, I.R., Yin, Y., Schmidt, M., Bangs, F.K., Morrice, D.R., Smith, T.G., Buxton, P., Stamataki, D., Tanaka, M., Münsterberg, A.E., Briscoe, J., Tickle, C., Burt, D.W., 2006. The chicken *talpid3* gene encodes a novel protein essential for Hedgehog signaling. Genes Dev. 20, 1365–1377.

Delany, M.E., Krupkin, A.B., Miller, M.M., 2000. Organization of telomere sequences in birds: evidence for arrays of extreme length and for in vivo shortening. Cytogenet. Cell Genet. 90, 139–145.

Delany, M.E., 2004. Genetic variants for chick biology research: from breeds to mutants. Mech. Dev. 121, 1169–1177.

Delany, M.E., Gessaro, T.M., Rodrigue, K.L., Daniels, L.M., 2007. Chromosomal mapping of chicken mega-telomere arrays to GGA9, 16, 28 and W using a cytogenomic approach. Cytogenet. Genome Res. 117, 54–63.

Dorshorst, B., Okimoto, R., Ashwell, C., 2010. Genomic regions associated with dermal hyperpigmentation, polydactyly and other morphological traits in the Silkie chicken. J. Hered. 101, 339–350.

Dorshorst, B., Molin, A.M., Rubin, C.J., Johansson, A.M., Strömstedt, L., Pham, M.H., Chen, C.F., Hallböök, F., Ashwell, C., Andersson, L., 2011. A complex genomic rearrangement involving the endothelin 3 locus causes dermal hyperpigmentation in the chicken. PLoS Genet. 7, e1002412.

Ellegren, H., 2005. The avian genome uncovered. Trends Ecol. Evol. 20, 180–186.

Ellegren, H., 2010. Evolutionary stasis: the stable chromosomes of birds. Trends Ecol. Evol. 25, 283–291.

Ellegren, H., Smeds, L., Burri, R., Olason, P.I., Backström, N., Kawakami, T., Künstner, A., Mäkinen, H., Nadachowska-Brzyska, K., Qvarnström, A., Uebbing, S., Wolf, J.B., 2012. The genomic landscape of species divergence in Ficedula flycatchers. Nature 491, 756–760.

Eriksson, J., Larson, G., Gunnarsson, U., Bed'hom, B., Tixier-Boichard, M., Strömstedt, L., Wright, D., Jungerius, A., Vereijken, A., Randi, E., Jensen, P., Andersson, L., 2008. Identification of the yellow skin gene reveals a hybrid origin of the domestic chicken. PLoS Genet. 4, e1000010.

Genome 10K Community of Scientists, 2009. Genome 10K: a proposal to obtain whole-genome sequence for 10,000 vertebrate species. J. Hered. 100, 659–674.

Gordon, L., Yang, S., Tran-Gyamfi, M., Baggott, D., Christensen, M., Hamilton, A., Crooijmans, R., Groenen, M., Lucas, S., Ovcharenko, I., Stubbs, L., 2007. Comparative analysis of chicken chromosome 28 provides new clues to the evolutionary fragility of gene-rich vertebrate regions. Genome Res. 17, 1603–1613.

Gregory, T.R., 2002. A bird's-eye view of the C-value enigma: genome size, cell size, and metabolic rate in the class aves. Evolution 56, 121–130.

Griffin, D.K., Robertson, L.B.W., Tempest, H.G., Skinner, B.M., 2007. The evolution of the avian genome as revealed by comparative molecular cytogenetics. Cytogenet. Genome Res. 117, 64–77.

Gunnarsson, U., Hellström, A.R., Tixier-Boichard, M., Minvielle, F., Bed'hom, B., Ito, S., Jensen, P., Rattink, A., Vereijken, A., Andersson, L., 2007. Mutations in *SLC45A2* cause plumage color variation in chicken and Japanese quail. Genetics 175, 867–877.

Hellström, A.R., Sundström, E., Gunnarsson, U., Bed'Hom, B., Tixier-Boichard, M., Honaker, C.F., Sahlqvist, A.S., Jensen, P., Kämpe, O., Siegel, P.B., Kerje, S., Andersson, L., 2010. Sex-linked barring in chickens is controlled by the *CDKN2A/B* tumour suppressor locus. Pigment Cell Melanoma Res. 23, 521–530.

Hughes, A.L., Hughes, M.K., 1995. Small genomes for better fliers. Nature 377, 391.

Imsland, F., Feng, C., Boije, H., Bed'hom, B., Fillon, V., Dorshorst, B., Rubin, C.J., Liu, R., Gao, Y., Gu, X., Wang, Y., Gourichon, D., Zody, M.C., Zecchin, W., Vieaud, A., Tixier-Boichard, M., Hu, X., Hallböök, F., Li, N., Andersson, L., 2012. The Rose-comb mutation in chickens constitutes a structural rearrangement causing both altered comb morphology and defective sperm motility. PLoS Genet. 8, e1002775.

International Chicken Genome Sequencing Consortium, 2004. Sequence and comparative analysis of the chicken genome provide unique perspectives on vertebrate evolution. Nature 432, 695–716.

International Chicken Polymorphism Map Consortium, 2004. A genetic variation map for chicken with 2.8 million single-nucleotide polymorphisms. Nature 432, 717–722.

Itoh, Y., Kampf, K., Pigozzi, M.I., Arnold, A.P., 2009. Molecular cloning and characterization of the germline-restricted chromosome sequence in the zebra finch. Chromosoma 118, 527–536.

Janes, D.E., Organ, C.L., Fujita, M.K., Shedlock, A.M., Edwards, S.V., 2010. Genome evolution in Reptilia, the sister group of mammals. Annu. Rev. Genomics Hum. Genet. 11, 239–264.

Jurka, J., Kapitonov, V.V., Pavlicek, A., Klonowski, P., Kohany, O., Walichiewicz, J., 2005. Repbase Update, a database of eukaryotic repetitive elements. Cytogenet. Genome Res. 110, 462–467.

Kent, W.J., Sugnet, C.W., Furey, T.S., Roskin, K.M., Pringle, T.H., Zahler, A.M., Haussler, D., 2002. The human genome browser at UCSC. Genome Res. 12, 996–1006.

Kranis, A., Gheyas, A.A., Boschiero, C., Turner, F., Yu, L., Smith, S., Talbot, R., Pirani, A., Brew, F., Kaiser, P., Hocking, P.M., Fife, M., Salmon, N., Fulton, J., Strom, T.M., Haberer, G., Weigend, S., Preisinger, R., Gholami, M., Qanbari, S., Simianer, H., Watson, K.A., Woolliams, J.A., Burt, D.W., 2013. Development of a high density 600K SNP genotyping array for chicken. BMC Genomics 14, 59.

Lynch, M., Conery, J.S., 2003. The origins of genome complexity. Science 302, 1401–1404.

Marshall Graves, J.A., Peichel, C.L., 2010. Are homologies in vertebrate sex determination due to shared ancestry or to limited options? Genome Biol. 11, 205.

Marshall Graves, J.A., Shetty, S., 2001. Sex from W to Z: evolution of vertebrate sex chromosomes and sex determining genes. J. Exp. Zool. 290, 449–462.

Masabanda, J.S., Burt, D.W., O'Brien, P.C., Vignal, A., Fillon, V., Walsh, P.S., Cox, H., Tempest, H.G., Smith, J., Habermann, F., Schmid, M., Matsuda, Y., Ferguson-Smith, M.A., Crooijmans, R.P., Groenen, M.A., Griffin, D.K., 2004. Molecular cytogenetic definition of the chicken genome: the first complete avian karyotype. Genetics 166, 1367–1373.

Metzker, M.L., 2010. Sequencing technologies – the next generation. Nat. Rev. Genet. 11, 31–46.

Muir, W.M., Wong, G.K., Zhang, Y., Wang, J., Groenen, M.A., Crooijmans, R.P., Megens, H.J., Zhang, H., Okimoto, R., Vereijken, A., Jungerius, A., Albers, G.A., Lawley, C.T., Delany, M.E., MacEachern, S., Cheng, H.H., 2008. Genome-wide assessment of worldwide chicken SNP genetic diversity indicates significant absence of rare alleles in commercial breeds. Proc. Natl. Acad. Sci. U. S. A. 105, 17312–17317.

Nam, K., Ellegren, H., 2012. Recombination drives vertebrate genome contraction. PLoS Genet. 8, e1002680.

Ng, C.S., Wu, P., Foley, J., Foley, A., McDonald, M.L., Juan, W.T., Huang, C.J., Lai, Y.T., Lo, W.S., Chen, C.F., Leal, S.M., Zhang, H., Widelitz, R.B., Patel, P.I., Li, W.H., Chuong, C.M., 2012. The chicken frizzle feather is due to an α-keratin (KRT75) mutation that causes a defective rachis. PLoS Genet. 8, e1002748.

Ng, S.B., Turner, E.H., Robertson, P.D., Flygare, S.D., Bigham, A.W., Lee, C., Shaffer, T., Wong, M., Bhattacharjee, A., Eichler, E.E., Bamshad, M., Nickerson, D.A., Shendure, J., 2009. Targeted capture and massively parallel sequencing of 12 human exomes. Nature 461, 272–276.

Nie, W., O'Brien, P.C., Ng, B.L., Fu, B., Volobouev, V., Carter, N.P., Ferguson-Smith, M.A., Yang, F., 2009. Avian comparative genomics: reciprocal chromosome painting between domestic chicken (*Gallus gallus*) and the stone curlew (*Burhinus oedicnemus*, Charadriiformes)– an atypical species with low diploid number. Chromosome Res. 17, 99–113.

Nishida, C., Ishijima, J., Kosaka, A., Tanabe, H., Habermann, F.A., Griffin, D.K., Matsuda, Y., 2008. Characterization of chromosome structures of Falconinae (Falconidae, Falconiformes, Aves) by chromosome painting and delineation of chromosome rearrangements during their differentiation. Chromosome Res. 16, 171–181.

O'Hare, T.H., Delany, M.E., 2009. Genetic variation exists for telomeric array organization within and among the genomes of normal, immortalized, and transformed chicken systems. Chromosome Res. 17, 947–964.

Organ, C.L., Shedlock, A.M., Meade, A., Pagel, M., Edwards, S.V., 2007. Origin of avian genome size and structure in non-avian dinosaurs. Nature 446, 180–184.

de Oliveira, E.H.C., Habermann, F.A., Lacerda, O., Sbalqueiro, I.J., Wienberg, J., Müller, S., 2005. Chromosome reshuffling in birds of prey: the karyotype of the world's largest eagle (Harpy eagle, *Harpia Harpyja*) compared to that of the chicken (*Gallus gallus*). Chromosoma 114, 338–343.

Rands, C.M., Darling, A., Fujita, M., Kong, L., Webster, M.T., Clabaut, C., Emes, R.D., Heger, A., Meader, S., Hawkins, M.B., Eisen, M.B., Teiling, C., Affourtit, J., Boese, B., Grant, P.R., Grant, B.R., Eisen, J.A., Abzhanov, A., Ponting, C.P., 2013. Insights into the evolution of Darwin's finches from comparative analysis of the *Geospiza magnirostris* genome sequence. BMC Genomics 14, 95.

Robb, E.A., Gitter, C.L., Cheng, H.H., Delany, M.E., 2011. Chromosomal mapping and candidate gene discovery of chicken developmental mutants and genome-wide variation analysis of MHC congenics. J. Hered. 102, 141–156.

Rubin, C.J., Zody, M.C., Eriksson, J., Meadows, J.R., Sherwood, E., Webster, M.T., Jiang, L., Ingman, M., Sharpe, T., Ka, S., Hallböök, F., Besnier, F., Carlborg, O., Bed'hom, B., Tixier-Boichard, M., Jensen, P., Siegel, P., Lindblad-Toh, K., Andersson, L., 2010. Whole-genome resequencing reveals loci under selection during chicken domestication. Nature 464, 587–591.

Shang, W.H., Hori, T., Toyoda, A., Kato, J., Popendorf, K., Sakakibara, Y., Fujiyama, A., Fukagawa, T., 2010. Chickens possess centromeres with both extended tandem repeats and short non-tandem-repetitive sequences. Genome Res. 20, 1219–1228.

Shang, W.H., Hori, T., Martins, N.M., Toyoda, A., Misu, S., Monma, N., Hiratani, I., Maeshima, K., Ikeo, K., Fujiyama, A., Kimura, H., Earnshaw, W.C., Fukagawa, T., 2013. Chromosome engineering allows the efficient isolation of vertebrate neocentromeres. Dev. Cell 24, 635–648.

Shapiro, M.D., Kronenberg, Z., Li, C., Domyan, E.T., Pan, H., Campbell, M., Tan, H., Huff, C.D., Hu, H., Vickrey, A.I., Nielsen, S.C., Stringham, S.A., Hu, H., Willerslev, E., Gilbert, M.T., Yandell, M., Zhang, G., Wang, J., 2013. Genomic diversity and evolution of the head crest in the rock pigeon. Science 339, 1063–1067.

Shibusawa, M., Nishibori, M., Nishida-Umehara, C., Tsudzuki, M., Masabanda, J., Griffin, D.K., Matsuda, Y., 2004. Karyotypic evolution in the Galliformes: an examination of the process of karyotypic evolution by comparison of the molecular cytogenetic findings with the molecular phylogeny. Cytogenet. Genome Res. 106, 111–119.

Schielzeth, H., Kempenaers, B., Ellegren, H., Forstmeier, W., 2012a. QTL linkage mapping of zebra finch beak color shows an oligogenic control of a sexually selected trait. Evolution 66, 18–30.

Schielzeth, H., Forstmeier, W., Kempenaers, B., Ellegren, H., 2012b. QTL linkage mapping of wing length in zebra finch using genome-wide single nucleotide polymorphisms markers. Mol. Ecol. 21, 329–339.

Shiina, T., Briles, W.E., Goto, R.M., Hosomichi, K., Yanagiya, K., Shimizu, S., Inoko, H., Miller, M.M., 2007. Extended gene map reveals tripartite motif, C-type lectin, and Ig superfamily type genes within a subregion of the chicken MHC-B affecting infectious disease. J. Immunol. 178, 7162–7172.

Siegel, P.B., Dodgson, J.B., Andersson, L., 2006. Progress from chicken genetics to the chicken genome. Poult. Sci. 85, 2050–2060.

Skinner, B.M., Griffin, D.K., 2012. Intrachromosomal rearrangements in avian genome evolution: evidence for regions prone to breakpoints. Heredity 108, 37–41.

Smith, C.A., Roeszler, K.N., Ohnesorg, T., Cummins, D.M., Farlie, P.G., Doran, T.J., Sinclair, A.H., 2009. The avian Z-linked gene *DMRT1* is required for male sex determination in the chicken. Nature 461, 267–271.

The ENCODE Project Consortium, 2012. An integrated encyclopedia of DNA elements in the human genome. Nature 489, 57–74.

Van de Peer, Y., Maere, S., Meyer, A., 2009. The evolutionary significance of ancient genome duplications. Nat. Rev. Genet. 10, 725–732.

Wang, Y., Gao, Y., Imsland, F., Gu, X., Feng, C., Liu, R., Song, C., Tixier-Boichard, M., Gourichon, D., Li, Q., Chen, K., Li, H., Andersson, L., Hu, X., Li, N., 2012. The crest phenotype in chicken is associated with ectopic expression of HOXC8 in cranial skin. PLoS One 7, e34012.

Warren, W.C., Clayton, D.F., Ellegren, H., Arnold, A.P., Hillier, L.W., Künstner, A., Searle, S., White, S., Vilella, A.J., Fairley, S., Heger, A., Kong, L., Ponting, C.P., Jarvis, E.D., Mello, C.V., Minx, P., Lovell, P., Velho, T.A., Ferris, M., Balakrishnan, C.N., Sinha, S., Blatti, C., London, S.E., Li, Y., Lin, Y.C., George, J., Sweedler, J., Southey, B., Gunaratne, P., Watson, M., Nam, K., Backström, N., Smeds, L., Nabholz, B., Itoh, Y., Whitney, O., Pfenning, A.R., Howard, J., Völker, M., Skinner, B.M., Griffin, D.K., Ye, L., McLaren, W.M., Flicek, P., Quesada, V., Velasco, G., Lopez-Otin, C., Puente, X.S., Olender, T., Lancet, D., Smit, A.F., Hubley, R., Konkel, M.K., Walker, J.A., Batzer, M.A., Gu, W., Pollock, D.D., Chen, L., Cheng, Z., Eichler, E.E., Stapley, J., Slate, J., Ekblom, R., Birkhead, T., Burke, T., Burt, D., Scharff, C., Adam, I., Richard, H., Sultan, M., Soldatov, A., Lehrach, H., Edwards, S.V., Yang, S.P., Li, X., Graves, T., Fulton, L., Nelson, J., Chinwalla, A., Hou, S., Mardis, E.R., Wilson, R.K., 2010. The genome of a songbird. Nature 464, 757–762.

Wells, K.L., Hadad, Y., Ben-Avraham, D., Hillel, J., Cahaner, A., Headon, D.J., 2012. Genome-wide SNP scan of pooled DNA reveals nonsense mutation in *FGF20* in the scaleless line of featherless chickens. BMC Genomics 13, 257.

Wicker, T., Robertson, J.S., Schulze, S.R., Feltus, F.A., Magrini, V., Morrison, J.A., Mardis, E.R., Wilson, R.K., Peterson, D.G., Paterson, A.H., Ivarie, R., 2005. The repetitive landscape of the chicken genome. Genome Res. 15, 126–136.

Wright, D., Boije, H., Meadows, J.R., Bed'hom, B., Gourichon, D., Vieaud, A., Tixier-Boichard, M., Rubin, C.J., Imsland, F., Hallböök, F., Andersson, L., 2009. Copy number variation in intron 1 of *SOX5* causes the Pea-comb phenotype in chickens. PLoS Genet. 5, e1000512.

Yandell, M., Ence, D., 2012. A beginner's guide to eukaryotic genome annotation. Nat. Rev. Genet. 13, 329–342.

Zhan, X., Pan, S., Wang, J., Dixon, A., He, J., Muller, M.G., Ni, P., Hu, L., Liu, Y., Hou, H., Chen, Y., Xia, J., Luo, Q., Xu, P., Chen, Y., Liao, S., Cao, C., Gao, S., Wang, Z., Yue, Z., Li, G., Yin, Y., Fox, N.C., Wang, J., Bruford, M.W., 2013. Peregrine and saker falcon genome sequences provide insights into evolution of a predatory lifestyle. Nat. Genet. 45, 563–566.

Zhang, Y., Zhang, X., O'Hare, T.H., Payne, W.S., Dong, J.J., Scheuring, C.F., Zhang, M., Huang, J.J., Lee, M.K., Delany, M.E., Zhang, H.B., Dodgson, J.B., 2011. A comparative physical map reveals the pattern of chromosomal evolution between the turkey (*Meleagris gallopavo*) and chicken (*Gallus gallus*) genomes. BMC Genomics 12, 447.

Zhao, D., McBride, D., Nandi, S., McQueen, H.A., McGrew, M.J., Hocking, P.M., Lewis, P.D., Sang, H.M., Clinton, M., 2010. Nature 464, 237–242.

Chapter 2

Transcriptomics of Physiological Systems

Tom E. Porter

Department of Animal and Avian Sciences, University of Maryland, College Park, MD, USA

ABBREVIATIONS

ACTH Adrenocorticotropic hormone gene
ADRB2 Beta 2 adrenergic receptor gene
ALD Anterior latissimus dorsi
cDNA Complementary DNA
DIO2 Thyroid hormone deiodinase 2 gene
DNA Deoxyribonucleic acid
CGA Alpha subunit of thyroid-stimulating hormone gene
ESTs Expressed sequence tags
GLUT1 Glucose transporter 1 gene
GRM8 Type 8 glutamate receptor gene
LPS Lipopolysacharide
MAPK Mitogen-activated protein kinase
miRNA Micro-RNA
mRNA Messenger RNA
NFκB Nuclear factor kappa B gene
NPYR5 Neuropeptide Y receptor type 5 gene
PLD Posterior latissimus dorsi
POMC Pro-opiomelanocortin gene
PPARγ Peroxisome proliferator-activated receptor gamma gene
RNA Ribonucleic acid
RNAseq Massively parallel sequencing of RNA
TTR Transthyretin gene

2.1 INTRODUCTION

How does the brain integrate environmental cues to control behavior, vocalization, and reproduction? How do the endocrine tissues respond to internal and external signals to coordinate physiological responses? What governs immune cell responses to pathogens? What are the mechanisms underlying muscle differentiation and development? How do hepatic, gastrointestinal, and adipose tissues mediate nutrient uptake and metabolism? How does the cardiovascular system respond to systemic and environmental needs? These questions have been asked by physiologists for decades. Multiple approaches have been taken to advance our knowledge in these areas. With the sequencing of the genome for multiple avian species discussed in Chapter 1 and the development of genome-wide tools for analysis of mRNA levels, physiologists have begun to address these still-open questions on a genomic scale. Transcriptomics, also known as transcriptional profiling and functional genomics, involves large-scale and in many cases genome-wide analysis of mRNA levels in samples using DNA microarrays and, more recently, massively parallel sequencing of RNA (RNAseq). The discussion in this chapter highlights some of the efforts by avian physiologists to apply transcriptomics to the open questions noted in this introduction.

2.2 EARLY EFFORTS

The first major efforts in transcriptional profiling in avian species occurred with the chicken. A dramatic increase in the number of expressed sequence tags (ESTs) for the chicken submitted to GenBank occurred between 1999 and 2002, prior to the release of the chicken genome sequence in 2004. During that 4-year period, the number of ESTs for the chicken increased from a few hundred to more than 400,000. Currently, 688,203 ESTs have been deposited in GenBank. Assembly of these ESTs allowed for the production of more than a dozen microarray platforms. Details on the production of these microarrays have been reviewed in detail (Cogburn et al., 2007; Gheyas and Burt, 2013). In 2001, three papers were published that marked the beginning of transcriptomics in birds (Liu et al., 2001; Morgan et al., 2001; Neiman et al., 2001). These first avian cDNA microarrays contained cDNAs representing approximately 2000 genes expressed in lymphoid tissues that were printed on nylon membranes. Several other tissue-specific cDNA microarrays were developed in the following few years (Bailey et al., 2003; Carre et al., 2006; Cogburn et al., 2003, 2004; Ellestad et al., 2006), and two groups produced cDNA microarrays representing more than half the expressed genes in the chicken (Burnside et al., 2005; Cogburn et al., 2004). These system-wide cDNA microarrays were followed by oligonucleotide arrays representing the majority of expressed genes in the chicken (Affymetrix, ARK-Genomics/Operon, Agilent). Within 5 years of the foundation of functional genomics in the chicken and shortly after the release of the chicken genome sequence, tools were established for near genome-wide analysis of mRNA levels in chicken samples. Subsequent to these early efforts in the chicken, similar projects established resources for genomics studies in the zebra finch (Li et al., 2007) and Northern bobwhite quail (Rawat et al., 2010). These genomics tools were rapidly put to use in studies of physiological systems, which will be discussed in this chapter.

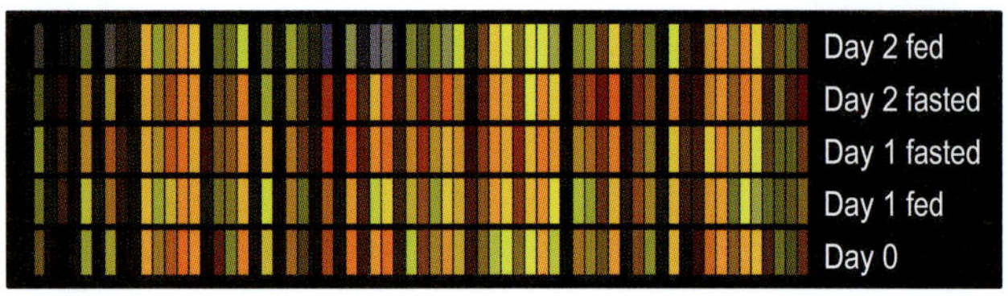

FIGURE 2.1 Heat map illustrating the effects of fasting or feeding of newly hatched chickens on mRNA levels in the hypothalamus. Hypothalamic mRNA samples were analyzed using cDNA microarrays, and the results were clustered. Shown is a cluster of genes for which mRNA levels were increased (red) in response to fasting. Among these were *DIO2* and neuropeptide Y receptor type 5. *The data presented have been published previously (Higgins et al., 2010), but not in this format.*

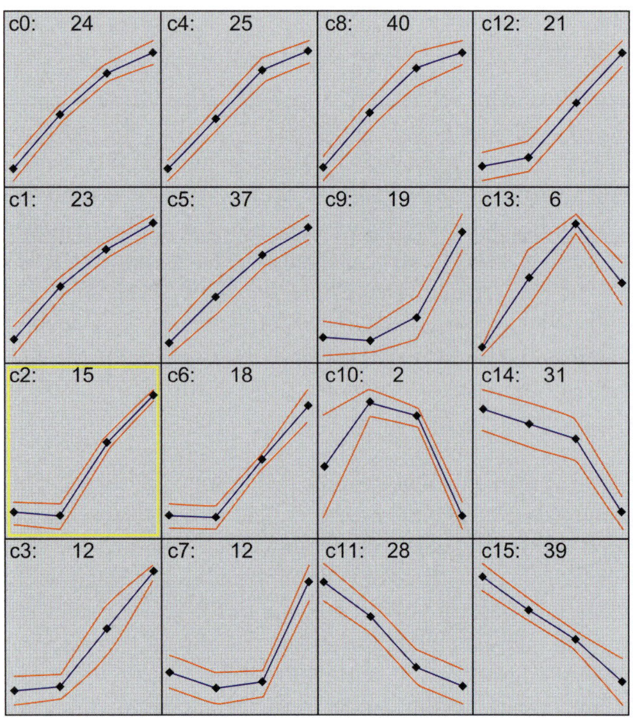

FIGURE 2.2 Self-organizing maps (SOMS) clustering of mRNA levels in the anterior pituitary gland during embryonic development. Pituitary gland mRNA samples from embryonic days 12, 14, 16, and 18 were analyzed using cDNA microarrays, and the results were clustered into SOMS based on mRNA level profiles during development. Cluster 2 (C2, outlined in yellow) contains 15 genes whose expression increased on embryonic day 16. Growth hormone was among the genes in this cluster. *These results have been published previously in a different format (Ellestad et al., 2006).*

Early studies on transcriptomics in birds frequently resulted in long lists of genes that were either upregulated or downregulated in one set of samples relative to another. Although these lists represented candidate genes for potential involvement in physiological responses, they did not add substantially to our comprehensive understanding of the gene interactions mediating those responses. Many studies used hierarchical clustering, heat maps, or self-organizing maps to group genes based on their expression patterns in responses to treatments or during time-course studies. Examples of heat map and SOMS clustering are presented in Figures 2.1 and 2.2, respectively. The rationale used was that genes that respond with similar patterns are likely to be regulated by shared mechanisms or even common transcription factors. More recent transcriptomics studies have placed differentially expressed genes into known or predicted gene networks and pathways based on reports in the literature (see example in Figure 2.3). One commonly used pathway analysis software is Ingenuity Pathway Analysis. Taking multiple approaches to analyze transcriptomics results can lead to a broader understanding of the cellular pathways, gene networks, and transcriptional regulation of gene expression involved in physiological processes.

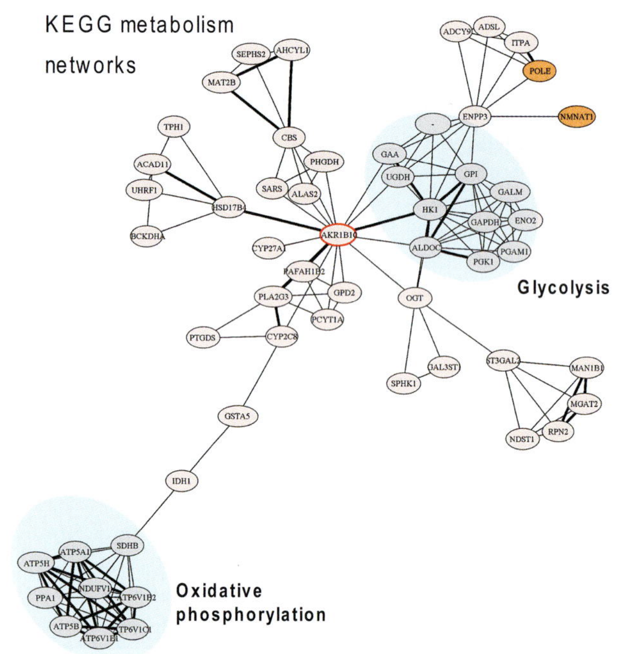

FIGURE 2.3 Network of genes that were differentially expressed in the hypothalamus of chickens genetically selected for high or low body fat. Hypothalamic RNA samples from the two genetic lines were analyzed using cDNA microarrays, and the results were further analyzed using Ingenuity Pathway Analysis software. Genes involved in glycolysis and oxidative phosphorylation were among those that were differentially expressed in the hypothalamus of the fat and lean chickens. Additional details on these findings can be found in Byerly et al. (2010).

2.3 NERVOUS SYSTEM

Vocalization in songbirds is part of the courtship ritual and an important neuroscience model for speech in humans and for sexual dimorphism in the central nervous system. However, the genes involved in establishing differences in vocalization between male and female songbirds were not known. Using microarray technology, differences in hypothalamic gene expression in response to territorial intrusion were found between the spring and autumn in free-living song sparrows (Mukai et al., 2009). Among the genes that were differentially expressed between the seasons were genes involved in thyroid hormone regulation and action, including genes encoding for the alpha subunit of thyroid-stimulating hormone (*CGA*) and transthyretin (*TTR*), supporting a role for thyroid hormones in modulating hypothalamic control of territorial aggression during seasonal reproduction. Gene expression in the high vocal center (HVC) of zebra finches and canaries was similarly characterized (Li et al., 2007). Relative to a whole-brain reference RNA sample, expression of 190 genes was greater in the HVC of both zebra finch and canaries, suggesting that these genes might function in controlling vocalization. Genes expressed specifically within the song control nucleus HVC were also identified in the Bengalese finch using microarrays (Kato and Okanoya, 2010).

In a more comprehensive study, microarray analysis of the basal ganglia identified thousands of genes differentially expressed in area X of singing zebra finches (Hilliard et al., 2012). Microarray analysis was also used to identify genes located on the Z chromosome that are expressed within the song control nucleus of the male zebra finch that are involved in cell survival (Tomaszycki et al., 2009), supporting their role in formation of the sexually dimorphic nucleus involved in masculinization of song. A similar microarray analysis of gene expression in the telencephalon of zebra finch and whitethroat indicated that most of the genes differentially expressed in males were linked to the Z chromosome (Naurin et al., 2011). However, only half of these were differentially expressed in males of both passerine species. Transcriptional profiling of the auditory lobe of zebra finches identified genes whose expression changes with the introduction of a novel song and reverts upon habituation of the birds to the introduced song (Dong et al., 2009; London et al., 2009). Interestingly, RNAseq revealed micro-RNAs (miRNA) in the auditory forebrain responsive to song that target genes whose expression changes with song (Gunaratne et al., 2011), indicating that expression of miRNAs likely contributes to song in birds.

The central nervous system plays an essential role in regulating metabolism, growth, and body composition. However, the full complement of genes expressed within the central nervous system that function in regulating these processes is not known. Transcriptomics has been used to identify genes and gene networks involved in these processes. Transcriptional profiling of the telencephalon of the white-crowned sparrow during the migratory and nonmigratory seasons revealed differences in expression of genes involved in glucose transport, including glucose transporter 1 (Jones et al., 2008). These findings support an increased need by the nervous system for glucose during the migratory season. In a study in which the metabolism of newly hatched chickens was perturbed by fasting, microarray analysis revealed that fasting altered expression in the hypothalamus of genes involved in the regulation of metabolic rate, including thyroid hormone deiodinase 2 (*DIO2*) and pro-opiomelanocortin (*POMC*; Figure 2.1), suggesting hypothalamic modulation of metabolic rate in order to compensate for decreased feed intake (Higgins et al., 2010). Genes not previously associated with hypothalamic regulation of feed intake and metabolism were also identified, including the beta 2 adrenergic receptor (*ADRB2*) and the type 8 glutamate receptor (*GRM8*). Functional relationships between *ADRB2*, *GRM8*, and *POMC* were confirmed in cultures of hypothalamic neurons, and effects were dependent on whether the neurons were derived from chicks that were previously fed or fasted.

In two other reports, hypothalamic gene expression was profiled in genetic lines of chickens divergently selected for high or low body fat (Byerly et al., 2010) or high or low body weight (Ka et al., 2011). Differences in expression of genes associated with glucose sensing, transport, and metabolism were detected in the hypothalamus of birds selected for low or high body fat, suggesting that differences in hypothalamic regulation of body fat in birds might involve the capacity of the hypothalamus to sense and metabolize glucose (Figure 2.3). In contrast, selection for body weight did not affect hypothalamic expression of genes known to regulate feed intake and metabolism, even though the high body weight birds are hyperphagic. These studies demonstrated how involvement of novel gene pathways within the central nervous system in physiological processes can be identified using transcriptomics.

2.4 ENDOCRINE SYSTEM

Transcriptomics has been used to characterize gene expression within endocrine tissues and the responses to hormonal treatments in birds. In one of the first microarray analyses in birds reported, transcriptional profiles were defined within the pineal gland of chicks during a light–dark circadian cycle (Bailey et al., 2003). Expression of hundreds of genes in the pineal gland was found to oscillate during the light–dark cycle, including genes involved in melatonin synthesis. Importantly, this transcriptomics analysis revealed many genes associated with immune responses, stress responses, and hormone binding, suggesting other roles for these genes within the pineal gland or for the pineal gland within these other physiological systems. A similar analysis was subsequently performed in the chick retina (Bailey et al., 2004). Although some overlap in oscillating gene expression was found between the retina and the pineal gland, distinct differences were also noted, suggesting that differences might exist in circadian regulation within the two tissues. Effects of thyroid hormone and growth hormone on hepatic gene expression have been characterized in the chicken using microarrays (Wang et al., 2007b). Dozens of thyroid hormone- and growth hormone-regulated genes were identified. Interestingly, crosstalk between the two systems was noted, as thyroid hormone status affected levels mRNA levels for growth hormone receptor and insulin-like growth factor binding protein 1.

Microarrays have been used to characterize the response of the avian adrenal gland to adrenocorticotropic hormone (ACTH). Injection of ACTH increased mRNA levels for several steroidogenic genes as well as genes with other functions, such as transcription, cell division, and electron transfer (Bureau et al., 2009). The effects of insulin immunoneutralization through administration of antiserum to insulin were evaluated in the chicken (Simon et al., 2012). Microarray analysis of mRNA samples from liver and muscle revealed that expression levels for more than 1000 genes were affected by decreased insulin levels or the elevated glucose levels associated with insulin immunoneutralization. The results demonstrated the wide range of the effects of insulin on the two target tissues. Microarrays were used to characterize changes in gene expression in the pituitary gland during chicken embryonic development (Ellestad et al., 2006). Numerous genes were identified with expression profiles that suggested involvement in differentiation of pituitary thyrotrophs, somatotrophs, and lactotrophs (Figure 2.2). Effects of glucocorticoids on pituitary gene expression were also identified using cDNA microarrays. Treatment with corticosterone of chicken embryonic pituitary cells affected mRNA levels for hundreds of genes, and these were placed into networks of affected genes (Jenkins et al., 2013). The results were also used to identify putative glucocorticoid receptor targets in the chicken, demonstrating the power of pathway and network analysis of transcriptional profiles.

2.5 REPRODUCTIVE SYSTEM

Development and function of the reproductive system involves gonadal differentiation and hormonal effects on reproductive tissues, including the testes, ovary, and oviduct. However, the full extent of the genetic mechanisms underlying these processes is not known. A number of studies have been reported in which investigators used transcriptomics to shed light on the underlying mechanisms controlling development and function of the reproductive system in birds. A comparison of gene expression in the shell glands of juvenile and laying chicken hens using cDNA microarrays identified hundreds of genes that are differentially expressed in the mature shell gland (Dunn et al., 2009). Similar transcriptional profiling of the uterus of the chicken using cDNA microarrays revealed genes expressed specifically in the magnum or the isthmus (Jonchère et al., 2010). Among these were genes encoding for antimicrobial proteins and ion transporters, respectively, supporting their role in antibacterial properties of the egg albumen and for egg shell formation. An analysis of the effects of a synthetic estrogen on gene expression in the oviduct demonstrated that estrogen affects expression of genes associated with epithelial differentiation and tissue remodeling (Song et al., 2011), consistent with the dramatic effects of estrogens on oviduct size and glandular development.

Microarrays were used to identify genes expressed specifically within the germinal disk of the developing oocyte (Elis et al., 2008). These genes are likely to play a role in oocyte maturation or early embryonic development. Other genes found to be expressed in the granulosa cells are more likely involved in follicular maturation. Within the developing gonad of the chicken, miRNAs that were specific to the testes or ovary were identified using microarrays (Bannister et al., 2009), suggesting a role for miRNA expression in gonadal differentiation. One reproductive organ that is often overlooked is the pigeon crop, which produces a "milk" for the nutritional supply of lactating parents' offspring. A comparison of gene expression in nonlactating and lactating pigeon crop using oligonucleotide arrays identified genes that are differentially expressed in the lactating crop (Gillespie et al., 2011). Genes associated with extracellular matrix receptors, adherens tight junctions, and Wnt signaling were found. This finding supported hyperplasia and cellular release into the crop lumen in the formation of pigeon crop milk.

2.6 IMMUNE SYSTEM

Differences exist in immunological responses to pathogens among individuals in a species. However, the genes expressed within cells of the immune system that account for responses to pathogens and differences in responses among individuals are not entirely known. Microarray analysis was used to study gene expression in one lymphoid organ—the spleen—of susceptible and resistant lines of chickens in response to *Campylobacter jejuni* infection (Li et al., 2012b). Not surprisingly, expression of genes for lymphocyte activation and humoral responses, including immunoglobulin heavy and light chains, was increased following infection in the resistant line. Surprisingly, expression of genes related to erythropoiesis and apoptosis was affected in the susceptible line. These differences in genetic responses within the spleen to *C. jejuni* infection could contribute to susceptibility or resistance of individual birds to infection.

In a similar study, transcriptional responses of the spleen to *Escherichia coli* infection were profiled (Sandford et al., 2011). Immunological pathways including cytokine signaling and Toll-like receptors were affected by *E. coli* challenge, and the magnitude of transcriptional changes was correlated with the severity of infection. Surprisingly, immunization prior to *E. coli* challenge had no significant effects on transcriptional profiles in response to *E. coli*. Similarly, transcriptional profiling of macrophage responses to *Salmonella*-derived endotoxins revealed effects on expression of genes for multiple cytokines and Toll-like receptors (Ciraci et al., 2010). In an earlier study of macrophage responses to lipopolysaccharide or *E. coli*, downstream targets of the Toll-like receptor pathway were affected (Bliss et al., 2005). Transcriptional profiling of cecal gene expression in *Salmonella*-challenged neonatal chicks followed by pathway analysis revealed that expression of genes associated with the nuclear factor kappa B complex and apoptosis were affected by *Salmonella* administration (Higgins et al., 2011). In each of these studies, other genes not previously associated with immune responses were identified, which might play a role in immunological responses to pathogens. In an analysis of miRNA expression within the spleen and the bursa of Fabricius of embryonic chicks, divergent expression of numerous miRNAs was noted, suggesting that these miRNAs might play diverse roles in the functions of the various tissues of the immune system (Hicks et al., 2009).

2.7 MUSCLE, LIVER, ADIPOSE, AND GASTROINTESTINAL TISSUES

Multiple tissues are involved in nutrient absorption, metabolism, and partitioning into tissues for animal growth or energy storage. Among these are the intestine, liver, skeletal muscle, and adipose tissues. However, all of the genes expressed in these tissues to regulate growth and nutrient partitioning are not known. In a comparison of skeletal muscle from slow-growing layer and fast-growing broiler chickens, transcriptional profiling using microarrays revealed differences in expression of genes encoding for muscle fiber proteins and regulators of satellite cell proliferation and differentiation (Zheng et al., 2009). Genes associated with slow muscle fibers were expressed at a greater level in the breast muscle of layer chickens than in broiler chickens, whereas mRNA levels for genes associated with satellite cell growth were greater in muscle of broiler chickens than in layer chickens.

In a similar analysis of gene expression in muscle types, microarray analysis was used to identify differentially expressed genes between the anterior latissimus dorsi (ALD) and posterior latissimus dorsi (PLD) muscles of turkeys (Nierobisz et al., 2011). Expression of genes encoding for extracellular matrix proteins was greater in the slow-twitch, red ALD muscle than in the fast-twitch, white PLD. In contrast, expression of genes involved in glycolysis was greater in the PLD than in the ALD. In a comparison of a random-bred turkey line with a line selected for increased body weight, microarray analysis of mRNA levels in breast muscle revealed alterations in expression of genes associated with extracellular matrix, apoptosis, Ca^{2+} signaling, and muscle function.

Transcriptomics has been used to identify genes associated with meat quality of chicken breast muscle. Genes associated with lipid and carbohydrate metabolism were associated with meat quality (Sibut et al., 2011). In a similar study aimed at identifying genes involved in deposition of intramuscular fat in chickens, transcriptional profiling of breast muscle from broiler chickens and a slow-growing Chinese breed using DNA microarrays revealed differential expression of genes involved in lipid metabolism and muscle development (Cui et al., 2012). DNA microarrays for chicken were also used to study pectoralis gene expression in juvenile and sea acclimated king penguins (Teulier et al., 2012). Genes associated with lipid metabolism were upregulated, whereas genes associated with carbohydrate metabolism were downregulated, in older, sea-acclimated penguins. High environmental temperatures impair growth performance in chickens, but the mechanisms involved have not been elucidated. Microarray analysis of gene expression in breast muscle of chickens exposed to chronic heat stress revealed changes in expression of genes involved in protein turnover, tumor necrosis factor signaling, and mitogen-activated protein kinase signaling (Li et al., 2011). In studies aimed at identifying mechanisms responsible for differences in feed efficiency in broiler chickens, transcriptional profiling of breast muscle mRNA levels using microarrays indicated that an upregulation of genes involved in anabolic processes and energy sensing and a downregulation of genes involved in muscle fiber development and function are associated with high feed efficiency (Bottje et al., 2012; Kong et al., 2011).

Excessive deposition of body fat in commercial poultry leads to decreased conversion of feed into muscle for meat. However, the genetic mechanisms involved in accumulation of body fat in poultry are not known. Transcriptional profiling of adipose tissue from chicken lines divergently selected for low or high body fat revealed that genes involved in lipid metabolism and endocrine function were differentially expressed between the genetic lines (Wang et al., 2007a). Genes identified included lipoprotein lipase (*LPL*), fatty acid binding protein, thyroid hormone-responsive protein (*Spot14*), and leptin receptor. In a more comprehensive study of a different pair of chicken lines genetically selected for low or high abdominal fat, microarray analysis of adipose tissue was used to identify genes and gene networks involved in the observed differences in adiposity (Resnyk et al., 2013). Many genes involved in adipogenesis and lipogenesis were upregulated in the fat line. Again, many genes involved in endocrine signaling were also differentially expressed between the two genetic lines, including *TTR*, *DIO1*, *DIO3*, *Spot14*, and chemerin. These findings suggest that differences in adiposity among individual birds might be related to differences in endocrine regulation of adipocyte differentiation and growth and lipid metabolism.

Lipogenesis in birds occurs primarily in the liver, and this process is regulated by the energy needs of the animal. Transcriptional profiling of newly hatched chicks that were fed or fasted revealed that the metabolic perturbation of fasting delayed the upregulation of lipogenic genes in the liver (Richards et al., 2010). Expression of one gene that encodes for a transcription factor that regulates expression of the lipogenic genes, peroxisome proliferator-activated receptor gamma (*PPAR*γ), was also delayed, supporting coordinated regulation of lipogenesis. A similar analysis of transcriptional responses of the liver to fasting of older chickens has also been reported (Désert et al., 2008). In this study, fasting resulted in upregulation of genes involved in ketogenesis, gluconeogenesis, and fatty acid beta-oxidation, whereas genes involved in fatty acid synthesis were downregulated. These findings demonstrated the coordinated regulation of genes involved in nutrient partitioning by the liver in response to the metabolic perturbance of fasting.

Development of the intestine has also been studied using transcriptomics. Transcriptional profiles of the duodenum during embryonic development of the turkey were characterized using microarrays (de Oliveira et al., 2009). Results indicated that expression of peptidase and lipase genes (*LPL*) decreased toward hatch, whereas expression of genes encoding for peptide and glucose transporters (e.g., *PEPT1* and *SLC5AP*) increased toward hatch. Transcriptional profiles of the chicken jejunum were characterized during the first 3 weeks after hatch (Schokker et al., 2009). Microarray analysis indicated that genes involved in morphological and functional development were highly expressed immediately after hatch, while expression declined during later juvenile development.

2.8 CARDIOVASCULAR SYSTEM

Chicken embryos are a widely used model for studies of cardiac development. As such, much is known about development of the heart in chickens. However, the complex relationships among genes necessary for cardiac development are not known. Gene expression profiling of early embryonic heart tissues was used to identify genes associated with cardiac development (Buermans et al., 2010). Results from this analysis included components of Wnt signaling. In another microarray analysis of heart development, differences in gene expression between the left and right ventricle were defined (Krejčí et al., 2012). Not surprisingly, the set of differentially expressed genes included genes associated with cardiac cell differentiation, heart development, and morphogenesis. However, many other genes not associated with these processes were also identified, providing a novel list of candidate genes for future research on the mechanisms underlying cardiac development. The cardiovascular system acclimates to life at higher altitudes and the associated hypoxia. However, the genes involved in cardiac acclimation in birds are not known. Microarray analysis of gene expression in the heart of the Tibetan chicken and chickens not adapted to life at high altitudes has been performed (Li and Zhao, 2009). Results provided a list of candidate genes that might function in chronic acclimation to high altitudes. Although not part of the cardiovascular system, microarray analysis of pectoral muscle was performed on rufous-collared sparrows sampled at 2000 and 4000 m above sea level in the Andes Mountains (Cheviron et al., 2008). Differentially expressed genes included those involved in oxidative phosphorylation and oxidative stress. Interestingly, none of the genes identified remained differentially expressed when high-altitude and low-altitude birds were allowed to acclimate to life at sea level, supporting functional involvement of the candidate genes identified in acclimation of the birds to high altitudes.

2.9 HURDLES AND FUTURE DEVELOPMENTS

One hurdle facing comparative physiologists in performing transcriptomics on nonmodel species is a lack of genomics resources. These include an assembled and annotated genome sequence and genomics tools such as DNA microarrays. However, the number of avian genomes sequenced continues to grow, making functional genomics possible for investigators who are interested in nonmodel and wild species. Furthermore, the advent of RNAseq makes it possible to perform transcriptional profiling in species for which no DNA microarrays exist. With declining costs of RNAseq, this transcriptomics approach is now preferred over microarray analysis for most studies of gene expression in physiological systems. A number of investigators have used RNAseq for characterization of mRNA and miRNA levels in the chicken (Goher et al., 2013; Hicks et al., 2010; Kang et al., 2013; Nie et al., 2012; Wang et al., 2011). Importantly, RNAseq has been used in nonmodel species to sequence and annotate the transcriptome in the dark-eyed junco (Peterson et al., 2012) and song sparrow (Srivastava et al., 2012), identify genes involved plumage coloring in ducks (Li et al., 2012a), study gene dosage compensation in the European crow (Wolf and Bryk, 2011), compare gene expression between the black carrion crow and the gray coated crow (Wolf et al., 2010), and identify genes differentially expressed between the ovaries of laying and broody geese (Xu et al., 2013). These studies using RNA-seq are among the first transcriptomics studies in nonmodel avian species, and they demonstrate the utility of RNAseq for transcriptional profiling in physiological systems of any avian species.

REFERENCES

Bailey, M., Beremand, P., Hammer, R., Reidel, E., Thomas, T., Cassone, V., 2004. Transcriptional profiling of circadian patterns of mRNA expression in the chick retina. J. Biol. Chem. 279, 52247–52254.

Bailey, M.J., Beremand, P.D., Hammer, R., Bell-Pedersen, D., Thomas, T.L., Cassone, V.M., 2003. Transcriptional profiling of the chick pineal gland, a photoreceptive circadian oscillator and pacemaker. Mol. Endocrinol. 17, 2084–2095.

Bannister, S.C., Tizard, M.L.V., Doran, T.J., Sinclair, A.H., Smith, C., 2009. Sexually dimorphic microRNA expression during chicken embryonic gonadal development. Biol. Reprod. 81, 165–176.

Bliss, T.W., Dohms, J.E., Emara, M.G., Keeler, C.L., 2005. Gene expression profiling of avian macrophage activation. Vet. Immunol. Immunopathol. 105, 289–299.

Bottje, W.G., Kong, B.-W., Song, J.J., Lee, J.Y., Hargis, B.M., Lassiter, K., Wing, T., Hardiman, J., 2012. Gene expression in breast muscle associated with feed efficiency in a single male broiler line using a chicken 44K microarray. II. Differentially expressed focus genes. Poult. Sci. 91, 2576–2587.

Buermans, H.P.J., vanWijk, B., Hulsker, M.A., Smit, N.C.H., den Dunnen, J.T., van Ommen, G.B., Moorman, A.F., van denHoff, M.J., 'tHoen, P.A.C., 2010. Comprehensive gene-expression survey identifies Wif1 as a modulator of cardiomyocyte differentiation. PLoS One 5, e15504.

Bureau, C., Hennequet-Antier, C., Couty, M., Guemene, D., 2009. Gene array analysis of adrenal glands in broiler chickens following ACTH treatment. BMC Genomics 10, 430.

Burnside, J., Neiman, P., Tang, J., Basom, R., Talbot, R., Aronszajn, M., Burt, D., Delrow, J., 2005. Development of a cDNA array for chicken gene expression analysis. BMC Genomics 6, 13.

Byerly, M.S., Simon, J., Cogburn, L.A., Le Bihan-Duval, E., Duclos, M.J., Aggrey, S.E., Porter, T.E., 2010. Transcriptional profiling of the hypothalamus during development of adiposity in genetically selected fat and lean chickens. Physiol. Genomics 42, 157–167.

Carre, W., Wang, X., Porter, T., Nys, Y., Tang, J., Bernberg, E., Morgan, R., Burnside, J., Aggrey, S., Simon, J., Cogburn, L., 2006. Chicken genomics resource: sequencing and annotation of 37,557 ESTs from single and multiple tissue cDNA libraries and CAP3 assembly of a chicken gene index. Physiol. Genomics 25, 514–524.

Cheviron, Z.A., Whitehead, A., Brumfield, R.T., 2008. Transcriptomic variation and plasticity in rufous-collared sparrows (*Zonotrichia capensis*) along an altitudinal gradient. Mol. Ecol. 17, 4556–4569.

Ciraci, C., Tuggle, C.K., Wannemuehler, M.J., Nettleton, D., Lamont, S.J., 2010. Unique genome-wide transcriptome profiles of chicken macrophages exposed to Salmonella derived-endotoxin. BMC Genomics 11, 545.

Cogburn, L.A., Porter, T.E., Duclos, M.J., Simon, J., Burgess, S.C., Zhu, J.J., Cheng, H.H., Dodgson, J.B., Burnside, J., 2007. Functional genomics of the chicken – a model organism. Poult. Sci. 86, 2059–2094.

Cogburn, L.A., Wang, X., Carre, W., Rejto, L., Aggrey, S.E., Duclos, M.J., Simon, J., Porter, T.E., 2004. Functional genomics in chickens: development of integrated-systems microarrays for transcriptional profiling and discovery of regulatory pathways. Comp. Funct. Genom. 5, 253–261.

Cogburn, L.A., Wang, X., Carre, W., Rejto, L., Porter, T.E., Aggrey, S.E., Simon, J., 2003. Systems-wide chicken DNA microarrays, gene expression profiling, and discovery of functional genes. Poult. Sci. 82, 939–951.

Cui, H.-X., Liu, R.-R., Zhao, G.-P., Zheng, M.-Q., Chen, J.-L., Wen, J., 2012. Identification of differentially expressed genes and pathways for intramuscular fat deposition in pectoralis major tissues of fast-and slow-growing chickens. BMC Genomics 13, 213.

de Oliveira, J.E., Druyan, S., Uni, Z., Ashwell, C.M., Ferket, P.R., 2009. Prehatch intestinal maturation of turkey embryos demonstrated through gene expression patterns. Poult. Sci. 88, 2600–2609.

Désert, C., Duclos, M.J., Blavy, P., Lecerf, F., Moreews, F., Klopp, C., Aubry, M., Herault, F., Roy, P.L., Berri, C., Douaire, M., Diot, C., Lagarrigue, S., 2008. Transcriptome profiling of the feeding-to-fasting transition in chicken liver. BMC Genomics 9, 611.

Dong, S., Replogle, K.L., Hasadsri, L., Imai, B.S., Yau, P.M., Rodriguez-Zas, S., Southey, B.R., Sweedler, J.V., Clayton, D.F., 2009. Discrete molecular states in the brain accompany changing responses to a vocal signal. Proc. Natl. Acad. Sci. U. S. A. 106, 11364–11369.

Dunn, I.C., Wilson, P.W., Lua, Z., Bain, M.M., Crossan, C.L., Talbot, R.T., Waddington, D., 2009. New hypotheses on the function of the avian shell gland derived from microarray analysis comparing tissue from juvenile and sexually mature hens. Gen. Comp. Endocrinol. 163, 225–232.

Elis, S., Batellier, F., Couty, I., Balzergue, S., Martin-Magniette, M.-L., Monget, P., Blesbois, E., Govoroun, M.S., 2008. Search for the genes involved in oocyte maturation and early embryo development in the hen. BMC Genomics 9, 110.

Ellestad, L., Carre, W., Muchow, M., Jenkins, S., Wang, X., Cogburn, L., Porter, T., 2006. Gene expression profiling during cellular differentiation in the embryonic pituitary gland using cDNA microarrays. Physiol. Genomics 25, 414–425.

Gheyas, A.A., Burt, D.W., 2013. Microarray resources for genetic and genomic studies in chicken: a review. Genesis 51, 337–356.

Gillespie, M.J., Haring, V.R., McColl, K.A., Monaghan, P., Donald, J.A., Nicholas, K.R., Moore, R.J., Crowley, T.M., 2011. Histological and global gene expression analysis of the 'lactating' pigeon crop. BMC Genomics 12, 452.

Goher, M., Hicks, J.A., Liu, H.-C., 2013. The Interplay between MDV and HVT affects viral miRNA expression. Avian Dis. 57, 372–379.

Gunaratne, P.H., Lin, Y.-C., Benham, A.L., Drnevich, J., Coarfa, C., Tennakoon, J.B., Creighton, C.J., Kim, J.H., Milosavljevic, A., Watson, M., Griffiths-Jones, S., Clayton, D.F., 2011. Song exposure regulates known and novel microRNAs in the zebra finch auditory forebrain. BMC Genomics 12, 277.

Hicks, J.A., Tembhurne, P.A., Liu, H.-C., 2009. Identification of microRNA in the developing chick immune organs. Immunogenetics 61, 231–240.

Hicks, J.A., Trakooljul, N., Liu, H.-C., 2010. Discovery of chicken microRNAs associated with lipogenesis and cell proliferation. Physiol. Genomics 41, 185–193.

Higgins, S.E., Ellestad, L.E., Trakooljul, N., McCarthy, F., Saliba, J., Cogburn, L.A., Porter, T.E., 2010. Transcriptional and pathway analysis in the hypothalamus of newly hatched chicks during fasting and delayed feeding. BMC Genomics 11, 162.

Higgins, S.E., Wolfenden, A.D., Tellez, G., Hargis, B.M., Porter, T.E., 2011. Transcriptional profiling of cecal gene expression in probiotic- and Salmonella-challenged neonatal chicks. Poult. Sci. 90, 901–913.

Hilliard, A.T., Miller, J.E., Fraley, E., Horvath, S., White, S.A., 2012. Molecular microcircuitry underlies functional specification in a basal ganglia circuit dedicated to vocal learning. Neuron 73, 537–552.

Jenkins, S.A., Ellestad, L.E., Mukherjee, M., Narayana, J., Cogburn, L.A., Porter, T.E., 2013. Glucocorticoid-induced changes in gene expression in embryonic anterior pituitary cells. Physiol. Genomics 45, 422–433.

Jonchère, V., Réhault-Godbert, S., Hennequet-Antier, C., Cabau, C., Sibut, V., Cogburn, L.A., Nys, Y., Gautron, J., 2010. Gene expression profiling to identify eggshell proteins involved in physical defense of the chicken egg. BMC Genomics 11, 57.

Jones, S., Pfister-Genskow, M., Cirelli, C., Benca, R.M., 2008. Changes in brain gene expression during migration in the white-crowned sparrow. Brain Res. Bull. 76, 536–544.

Ka, S., Albert, F.W., Denbow, D.M., Pääbo, S., Siegel, P.B., Andersson, L., Hallböök, F., 2011. Differentially expressed genes in hypothalamus in relation to genomic regions under selection in two chicken lines resulting from divergent selection for high or low body weight. Neurogenetics 12, 211–221.

Kang, L., Cui, X., Zhang, Y., Yang, C., Jiang, Y., 2013. Identification of miRNAs associated with sexual maturity in chicken ovary by illumina small RNA deep sequencing. BMC Genomics 14, 352.

Kato, M., Okanoya, K., 2010. Molecular characterization of the song control nucleus HVC in Bengalese finch brain. Brain Res. 1360, 56–76.

Kong, B.-W., Song, J.J., Lee, J.Y., Hargis, B.M., Wing, T., Lassiter, K., Bottje, W., 2011. Gene expression in breast muscle associated with feed efficiency in a single male broiler line using a chicken 44K oligo microarray. I. Top differentially expressed genes. Poult. Sci. 90, 2535–2547.

Krejčí, E., Pesevski, Z., DeAlmeida, A.C., Mrug, M., Fresco, V.M., Argraves, W.S., Barth, J.L., Cui, X., Sedmera, D., 2012. Microarray analysis of normal and abnormal chick ventricular myocardial development. Physiol. Res. 61 (Suppl. 1), S137–S144.

Li, C., Wang, X., Wang, G., Li, N., Wu, C., 2011. Expression analysis of global gene response to chronic heat exposure in broiler chickens (*Gallus gallus*) reveals new reactive genes. Poult. Sci. 90, 1028–1036.

Li, M., Zhao, C., 2009. Study on Tibetan chicken embryonic adaptability to chronic hypoxia by revealing differential gene expression in heart tissue. Sci. China, C, Life Sci. 52, 284–295.

Li, S., Wang, C., Yu, W., Zhao, S., Gong, Y., 2012a. Identification of genes related to white and black plumage formation by RNA-seq from white and black feather bulbs in ducks. PLoS One 7, e36592.

Li, X., Swaggerty, C.L., Kogut, M.H., Chiang, H.-I., Wang, Y., Genovese, K.J., He, H., McCarthy, F.M., Burgess, S.C., Pevzner, I.Y., Zhou, H., 2012b. Systemic response to *Campylobacter jejuni* infection by profiling gene transcription in the spleens of two genetic lines of chickens. Immunogenetics 64, 59–69.

Li, X., Wang, X.-J., Tannenhauser, J., Podell, S., Mukherjee, P., Hertel, M., Biane, J., Masuda, S., Nottebohm, F., Gaasterland, T., 2007. Genomic resources for songbird research and their use in characterizing gene expression during brain development. Proc. Natl. Acad. Sci. U. S. A. 104, 6834–6839.

Liu, H.-C., Cheng, H.H., Tirunagaru, V., Sofer, L., Burnside, J., 2001. A strategy to identify positional candidate genes conferring Marek's disease resistance by integrating DNA microarrays and genetic mapping. Anim. Genet. 32, 351–359.

London, S.E., Dong, S., Replogle, K., Clayton, D.F., 2009. Developmental shifts in gene expression in the auditory forebrain during the sensitive period for song learning. Dev. Neurobiol. 69, 437–450.

Morgan, R.W., Sofer, L., Anderson, A.S., Bernberg, E.L., Cui, J., Burnside, J., 2001. Induction of host gene expression following infection of chicken embryo fibroblasts with oncogenic Marek's disease virus. J. Virol. 75, 533–539.

Mukai, M., Replogle, K., Drnevich, J., Wang, G., Wacker, D., Band, M., Clayton, D.F., Wingfield, J.C., 2009. Seasonal differences of gene expression profiles in song sparrow (*Melospiza melodia*) hypothalamus in relation to territorial aggression. PLoS One 4, e8182.

Naurin, S., Hansson, B., Hasselquist, D., Kim, Y.-H., Bensch, S., 2011. The sex-biased brain: sexual dimorphism in gene expression in two species of songbirds. BMC Genomics 12, 37.

Neiman, P.E., Ruddell, A., Jasoni, C., Loring, G., Thomas, S.J., Brandvold, K.A., Lee, R., Burnside, J., Delrow, J., 2001. Analysis of gene expression during myc oncogene-induced lymphomagenesis in the bursa of Fabricius. Proc. Natl. Acad. Sci. U. S. A. 98, 6378–6383.

Nie, Q., Sandford, E., Zhang, X., Nolan, L., Lamont, S., 2012. Deep sequencing-based transcriptome analysis of chicken spleen in response to avian pathogenic *Escherichia coli* (APEC) infection. PLoS One 7, e41645.

Nierobisz, L.S., Sporer, K.R.B., Strasburg, G.M., Reed, K.M., Velleman, S.G., Ashwell, C.M., Felts, J.V., Mozdziak, P.E., 2011. Differential expression of genes characterizing myofibre phenotype. Anim. Genet. 43, 298–308.

Peterson, M.P., Whittaker, D.J., Ambreth, S., Sureshchandra, S., Buechlein, A., Podicheti, R., Choi, J.-H., Lai, Z., Mockatis, K., Colbourne, J., Tang, H., Ketterson, E.D., 2012. De novo transcriptome sequencing in a songbird, the dark-eyed junco (*Junco hyemalis*): genomic tools for an ecological model system. BMC Genomics 13, 305.

Rawat, A., Gust, K.A., Elasri, M.O., Perkins, E.J., 2010. Quail genomics: a knowledgebase for northern bobwhite. BMC Bioinform. 11, S13.

Resnyk, C.W., Carré, W., Wang, X., Porter, T.E., Simon, J., Le Bihan-Duval, E., Duclos, M.J., Aggrey, S.E., Cogburn, L.A., 2013. Transcriptional analysis of abdominal fat in genetically fat and lean chickens reveals adipokines, lipogenic genes and a link between hemostasis and leanness. BMC Genomics 14, 557.

Richards, M.P., Proszkowiec-Weglarz, M., Rosebrough, R.W., McMurtry, J.P., Angel, R., 2010. Effects of early neonatal development and delayed feeding immediately post-hatch on the hepatic lipogenic program in broiler chicks. Comp. Biochem. Physiol. Part B 157, 374–388.

Sandford, E.E., Orr, M., Balfanz, E., Bowerman, N., Li, X., Zhou, H., Johnson, T.J., Kariyawasam, S., Liu, P., Nolan, L.K., Lamont, S.J., 2011. Spleen transcriptome response to infection with avian pathogenic *Escherichia coli* in broiler chickens. BMC Genomics 12, 469.

Schokker, D., Hoekman, A.J.W., Smits, M.A., Rebel, J.M.J., 2009. Gene expression patterns associated with chicken jejunal development. Dev. Comp. Immunol. 33, 1156–1164.

Sibut, V., Hennequet-Antier, C., LeBihan-Duval, E., Marthey, S., Duclos, M.J., Berri, C., 2011. Identification of differentially expressed genes in chickens differing in muscle glycogen content and meat quality. BMC Genomics 12, 112.

Simon, J., Milenkovic, D., Godet, E., Cabau, C., Collin, A., Métayer-Coustard, S., Rideau, N., Tesseraud, S., Derouet, M., Crochet, S., Cailleau-Audouin, E., Hennequet-Antier, C., Gespach, C., Porter, T.E., Duclos, M.J., Dupont, J., Cogburn, L.A., 2012. Insulin immunoneutralization in fed chickens: effects on liver and muscle transcriptome. Physiol. Genomics 44, 283–292.

Song, G., Seo, H.W., Choi, J.W., Rengaraj, D., Kim, T.M., Lee, B.R., Kim, Y.M., Yun, T.W., Jeong, J.-W., Han, J.Y., 2011. Discovery of candidate genes and pathways regulating oviduct development in chickens. Biol. Reprod. 85, 306–314.

Srivastava, A.J., Winker, K., Shaw, T.I., Jones, K.L., Glenn, T.C., 2012. Transcriptome analysis of a North American songbird, *Melospiza melodia*. DNA Res. 19, 325–333.

Teulier, L., Dégletagne, C., Rey, B., Tornos, J., Keime, C., deDinechin, M., Raccurt, M., Rouanet, J.-L., Roussel, D., Duchamp, C., 2012. Selective upregulation of lipid metabolism in skeletal muscle of foraging juvenile king penguins: an integrative study. Proc. R. Soc. Lond., B, Biol. Sci. 279, 2464–2472.

Tomaszycki, M.L., Peabody, C., Replogle, K., Clayton, D.F., Tempelman, R.J., Wade, J., 2009. Sexual differentiation of the zebra finch song system: potential roles for sex chromosome genes. BMC Neurosci. 10, 24.

Wang, H.-B., Li, H., Wang, Q.-G., Zhang, X.-Y., Wang, S.-Z., Wang, Y.-X., Wang, X.-P., 2007a. Profiling of chicken adipose tissue gene expression by genome array. BMC Genomics 8, 193.

Wang, X., Carre, W., Saxton, A.M., Cogburn, L.A., 2007b. Manipulation of thyroid status and/or GH injection alters hepatic gene expression in the juvenile chicken. Cytogenet. Genome Res. 117, 174–188.

Wang, Y., Ghaffari, N., Johnson, C.D., Braga-Neto, U.M., Wang, H., Chen, R., Zhou, H., 2011. Evaluation of the coverage and depth of transcriptome by RNA-seq in chickens. BMC Bioinform. 12 (Suppl. 10), S5.

Wolf, J.B.W., Bayer, T., Haubold, B., Schilhabel, M., Rosenstiel, P., Tautz, D., 2010. Nucleotide divergence vs. gene expression differentiation: comparative transcriptome sequencing in natural isolates from the carrion crow and its hybrid zone with the hooded crow. Mol. Ecol. 19 (Suppl. 1), 162–175.

Wolf, J.B.W., Bryk, J., 2011. General lack of global dosage compensation in ZZ/ZW systems? Broadening the perspective with RNA-seq. BMC Genomics 12, 91.

Xu, Q., Zhao, W.M., Chen, Y., Tong, Y.Y., Rong, G.H., Huang, Z.Y., Zhang, Y., Chang, G.B., Wu, X.S., Chen, G.H., 2013. Transcriptome profiling of the goose (*Anser cygnoides*) ovaries identify laying and broodiness phenotypes. PLoS One 8, e55496.

Zheng, Q., Zhang, Y., Chen, Y., Yang, N., Wang, X.-J., Zhu, D., 2009. Systematic identification of genes involved in divergent skeletal muscle growth rates of broiler and layer chickens. BMC Genomics 10, 87.

Chapter 3

Avian Proteomics

Dusan Kunec
Institut für Virologie, Zentrum für Infektionsmedizin, Freie Universität Berlin, Robert-von-Ostertag-Str. 7, Berlin, Germany

Shane C. Burgess
Vice Provost and Dean, Agriculture & Life Sciences; Director Arizona Experiment Station; The University of Arizona, Tucson, AZ, USA

3.1 INTRODUCTION

Whereas genomics provides and analyzes the entire set of functional elements encoded in a genome, and transcriptomics studies gene expression by measuring RNA levels, proteomics analyzes protein expression, modification, structure, localization, interaction, and function. Having a completed genome sequence of an organism is a key step towards understanding how that organism is built and maintained, and thus its complex biology. This information is stored in the genome in the form of genes, which are transcribed into RNA, and RNA is translated into proteins. The entire set of RNA transcripts and proteins encoded by the genome is called transcriptome and proteome, respectively (Velculescu et al., 1997; Wilkins et al., 1996). Although genes provide instructions, proteins are the functional units of almost all biological processes and the principal structural building blocks of all living organisms. Systems-level understanding of cell physiology is thus inevitably based on understanding the multifaceted interplay of gene expression and protein functional networks.

An individual's genome sequence (with the exception of some regions dedicated to the adaptive immune system) is static, its epigenome (the methylation patterns on DNA) less so, and the transcriptome and proteome are extremely dynamic. These latter two differ from cell to cell, and change dramatically according to conditions that cells are exposed to. The transcriptome is more complicated than the genome because of both frame-shifting and alternative splicing. The proteome is even more complex, because most proteins are co- and posttranslationally modified (Walsh, 2006). More than 200 different types of protein modifications are documented in vertebrates, and more than one of these modifications routinely occurs on most proteins (Walsh, 2006). Measurement of the proteome is also more challenging than that of the transcriptome because the dynamic range of proteins in tissues is higher than of transcripts, and can span over 11 orders of magnitude in body fluids (Anderson and Anderson, 2002), and most importantly from a technical perspective, there is no equivalent of the polymerase chain reaction (PCR) for proteins—we must use very expensive machinery to directly identify proteins.

Although mRNA quantities measured by quantitative (q)PCR, microarrays, or sequencing are often used as surrogates for protein quantities, and indirectly, protein activity, there is no or little correlation between mRNA and levels of its corresponding protein (Gygi et al., 1999a; Cullen et al., 2004; Nagaraj et al., 2011; Marguerat et al., 2012). This means that the presence or quantities of proteins in biological samples cannot be satisfactorily estimated solely through their mRNA levels. In addition, posttranslational modifications often profoundly affect protein activities. Though arguably less sensitive, proteomics methods are more specific for determining what is happening at the protein level—they can identify and quantify protein amounts and posttranslational modifications. Proteomics thus provides a direct measure of the predominant functional units responsible for cellular behavior.

Although not absolutely essential, a sequenced and structurally annotated genome greatly facilitates proteomics. The more accurate the genome assembly and annotation, the more accurate the proteomics methodologies can be; this extends to the individual—the most accurate proteomics experiments will be done using the individual's own genome sequence and, to be more accurate still, the transcriptome that corresponds to the proteome. Conversely, proteomics can be used to improve the structural annotation of genomes (Nanduri et al., 2010; Jaffe et al., 2004).

Red junglefowl (*Gallus gallus*), the major wild ancestor of the domestic chicken, was the first avian and nonmammalian amniote to have its genome sequenced (International Chicken Genome Sequencing C, 2004). The chicken is the principal nonmamamlian vertebrate animal model for studying development, infectious disease, immunology, oncogenesis, and behavior. It is also one of the most important agricultural species for production of meat and eggs. Until additional avian complete genomes became available, the chicken genome served as de facto model bird genome, and most of the proteomics studies have utilized this model to study various aspects of bird biology.

Recently, complete or draft genomes of several other avian species have become available, including several lines of domestic chicken (*Gallus gallus domesticus*), zebra finch (*Taeniopygia guttata*), domestic turkey (*Meleagris gallopavo*), collared flycatcher (*Ficedula albicollis*), pied flycatcher (*Ficedula hypoleuca*), large ground finch (*Geospiza magnirostris*), scarlet macaw (*Ara macao*), mallard duck (*Anas platyrhynchos*), ground tit (*Pseudopodoces humilis*), Puerto Rican parrot (*Amazona vittata*), and budgerigar (*Melopsittacus undulatus*) (Ellegren et al., 2012; Warren et al., 2010; Oleksyk et al., 2012; Dalloul et al., 2010; Rands et al., 2013; Rubin et al., 2010; Huang et al., 2013; Cai et al., 2013; http://aviangenomes.org/, 2013). The turkey, duck, and domestic chicken were sequenced because they are economically important (Dalloul et al., 2010; Rubin et al., 2010; Rao et al., 2012). These species are also used as biomedical models, which is the reason that the zebra finch, scarlet macaw, and Puerto Rican amazon were sequenced—they are important in neuroscience for studying their behavioral, cognitive, and speech abilities (Warren et al., 2010; Oleksyk et al., 2012; Seabury et al., 2013). Darwin's finches are model organisms to study of various aspects of evolution and development (Rands et al., 2013). Flycatchers are important models for speciation (Ellegren et al., 2012), and the genome of the ground tit provides new opportunities to study adaptation mechanisms to extreme conditions (Cai et al., 2013). Collectively, the availability of these additional genomes is opening up the avenues for genome-wide studies of various aspects of bird biology both on the RNA and protein levels. It is expected that large-scale analysis of the avian genome, transcriptome, and proteome will increase our understanding of complex molecular processes that determine phenotype.

3.2 PROTEIN IDENTIFICATION AND ANALYSIS

Although traditional protein biochemistry focuses on studying properties of individual proteins, proteomics encompasses nearly any type of technology that enables studying proteins on a large scale. Among the most commonly used tools are two-hybrid systems (Fields and Song, 1989), protein/peptide microarrays (Haab, 2003; Panse et al., 2004), and mass spectrometry (MS)-based approaches (Fenn et al., 1989). Although the two-hybrid systems and protein/peptide microarrays have only limited applicability—identification of protein–protein interactions—the analytical capabilities of MS make it an ideal tool for a broad range of applications. In addition to its versatility, MS has also the ability to handle the difficulties associated with the complex and dynamic nature of proteomes (Han et al., 2008). MS simplifies and accelerates the analysis and characterization of proteins. Thus, not surprisingly, MS plays a prime and increasingly indispensable role in current large-scale proteomic research. MS-based proteomics refers to approaches that use MS for identifying, characterizing, and/or quantifying proteins in biological samples (Figure 3.1).

Mass spectrometers are used to detect, identify, and quantify small molecules based on their mass and charge (m/z) ratios with high precision, sensitivity, and speed. Many excellent reviews have been written that cover instrumentation and principles of protein identification by MS (Yates et al., 2009; Yates, 1998; Steen and Mann, 2004), and we will not replicate this information. Rather, we will introduce the key methods.

A typical biological sample contains extremely complex proteomes. Because mass spectrometers can analyze only a limited number of different peptides at a time, the sample complexity must be reduced before MS. This is usually done on the protein level by gel electrophoresis, or on the peptide level by various chromatography techniques. Also mass spectrometry-based approaches are often divided into two major groups depending whether sample fractionation includes gel electrophoresis (gel-based approaches) or not (gel-free approaches).

3.2.1 Two-Dimensional Gel Electrophoresis-Based Proteomics

Two-dimensional (2-D) polyacrylamide gel electrophoresis (PAGE) was the first technique that allowed truly complex proteomic analysis and was instrumental for the development of proteomics (O'Farrell, 1975; Rabilloud et al., 2010). 2-D PAGE is the most widely used technique in gel-based proteomics; it simply deconvolutes a protein mixture in two dimensions. Proteins are first separated in the first dimension by their isoelectric point (pI), and then in the second dimension according to their electrophoretic mobility (which is a function of molecular weight and charge of a protein) in a polyacrylamide gel. Separated proteins are then stained and appear as spots in the gel. The amount of protein in a spot is determined by measuring the spot volume.

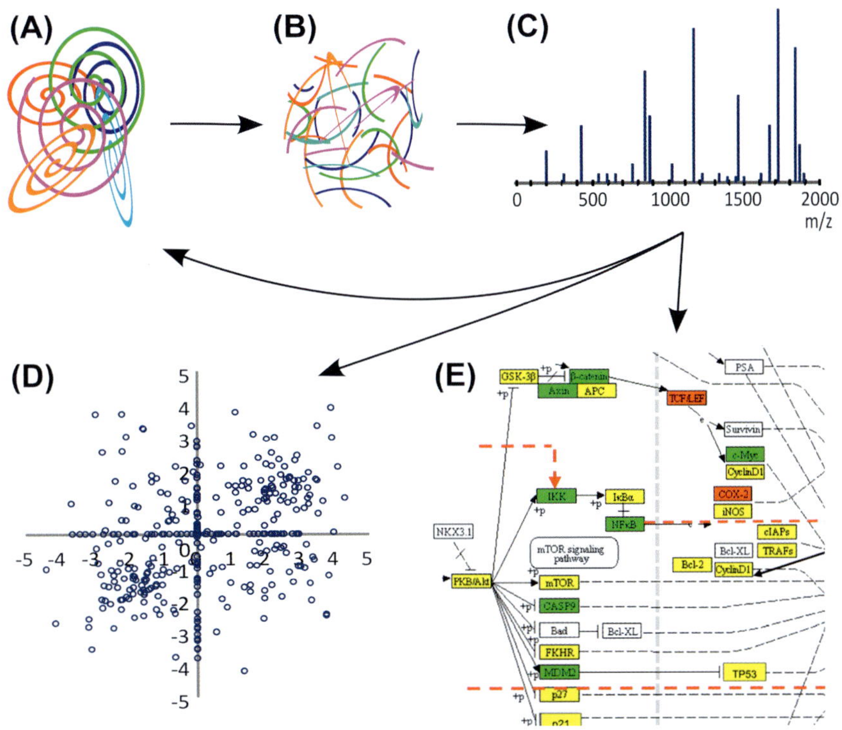

FIGURE 3.1 The workflow of mass spectrometry-based proteomics. The goal of proteomics is physical (protein identity, modifications, structure, localization) and functional (protein interactions, composition of protein complexes) characterization of the proteome. Proteomics denotes the collection of diverse technologies that enable the study of proteins on a large scale. In current proteomic research mass spectrometry (MS) plays prime and indispensable role, because several thousand proteins can be rapidly analyzed, correctly identified and accurately quantified in a single MS-based experiment. Proteins (A) are typically too large for accurate mass determination by mass spectrometry, therefore, they are first digested into peptides (B), which are then analyzed by a mass spectrometer. The obtained peptide masses, peptide (tandem) mass spectra (C), are then searched against predicted peptide masses derived from DNA or protein sequence databases by using one or several different mass spectral search algorithms and their amino acid sequence is determined. The amino acid sequence of an identified peptide is often unique to its parental protein, and such peptides are used for unambiguous identification of proteins that were present in the analyzed sample. In addition to protein identification, MS enables accurate relative or absolute quantification of proteins. As a result, MS is commonly used to identify proteins that are qualitatively and/or quantitatively differentially expressed between studied conditions. The obtained proteomic information is integrated with existing knowledge and/or data from other large-scale studies (mRNA profiling, siRNA screens) to better understand cell or tissue biology at the system level. For example, protein expression levels are compared with mRNA levels (D), and differences in protein expression are analyzed in the context of biological pathways or interaction networks (E) in effort to identify the underlying causalities and mechanistic principles that give rise to a studied phenomenon or phenotype.

This quantification is used as a screening process to select a limited number of corresponding spots that contain different amounts of protein on related gels. The identity of the protein(s) in a selected spot is next identified by MS.

This technique is very useful for comparing two samples that have similar protein expression profiles in order to find proteins that differ between the samples in their expression levels or posttranslational modifications (Rabilloud et al., 2010). The major advantage of this method is that it is intrinsically quantitative. The major drawbacks are the limited capacity of protein separation; poor reproducibility of 2-D gels; and low sensitivity, dynamic range, and throughput. Although efforts to overcome some of the shortcomings inherent to gel-based approaches resulted in the development of improved 2-D gel methods, such as 2-D fluorescence gel electrophoresis (2-D DIGE) (Unlu et al., 1997), gel-free chromatographic approaches, which offer a number of advantages over gel-based ones, now completely dominate the field of proteomics.

3.2.2 Gel-Free Based Proteomics

Experimentation with various gel-free separation techniques resulted in the emergence of multidimensional high-performance liquid chromatography (HPLC) as a viable alternative to gel-based separation techniques. Multidimensional HPLC quickly became the technique of choice for large-scale proteomic studies. Here, in contrast to gel-based approaches, the entire analyzed proteome is first digested and the resulting peptides are then separated by multidimensional HPLC and further analyzed by MS. Digestion of a protein mixture generates a highly complex mixture of peptides, in which the connection between the peptide and the originating protein is lost. Peptides detected and identified by MS are then used to infer the presence of all original proteins in the sample, which is the principle of the "shotgun" proteomics named by an analogy to shotgun DNA sequencing.

Multidimensional HPLC combines several separation steps to improve resolution of complex mixtures of peptides. One of the most popular multidimensional separation methods utilizes strong cation exchange (SCX) and reversed-phase (RP) chromatography to separate peptides in two dimensions: first peptides are separated based on charge, and then on hydrophobicity (Washburn et al., 2001). This separation method is the basis of the shotgun proteomic strategy known as multidimensional protein identification technology (MudPIT) (Washburn et al., 2001).

Another important chromatographic technique that is often used in gel-free approaches is affinity purification. Selective enrichment affinity materials are either used to enrich for peptides that contain certain posttranslational modifications, such as phosphorylation (Ficarro et al., 2002; Cao and Stults, 1999) or glycosylation (Geng et al., 2000; Durham and Regnier, 2006), or peptides that contain a specific selectable amino acid residue, such as cysteine (Wang and Regnier, 2001) or histidine (Wang et al., 2002). Affinity-enriched peptide mixtures are usually directly (online) or offline transferred to RP HPLC columns and further analyzed by MS.

Gel-free approaches overcome many of the drawbacks that are inherent to gel-based methods (for example, proteins with extreme size, pI, or hydrophobicity are amenable for analysis); but more importantly, they allow a large number of proteins to be identified and quantified in a high throughput manner and short time. The major disadvantage of shotgun proteomics is that peptides derived from very low abundance proteins are often undetected (Yates et al., 2009). An additional drawback relates to the loss of link between a peptide and a parental protein after digestion, which can lead to the incorrect identification of parental proteins.

3.3 QUANTITATIVE PROTEOMICS

In addition to protein identification, mass spectrometry can quantify proteins in complex biological samples. The proteome of a cell is highly dynamic and expressed proteins often change their locations, interactions, and modifications in response to different stimuli. The goal of quantitative proteomics is to obtain a snapshot of a proteome at a particular time. Accurate quantification of proteins is important for understanding of physiological or pathological phenomena, and for identification and modeling of functional networks.

Traditionally, protein quantification has been based on the 2-D PAGE approaches, but several gel-free methods allow accurate protein quantification solely by MS. Methods of quantitative proteomics are classified into two major categories: those that use stable isotopes and those that do not. The most popular stable isotope-labeling techniques either label peptides with isobaric tags for relative and absolute quantitation (iTRAQ) (Ross et al., 2004), or label proteins metabolically by incorporation of stable isotope labels with amino acids in cell culture (SILAC) (Ong et al., 2002) or enzymatically with isotope-coded affinity tags (ICATs) (Gygi et al., 1999b).

Protein quantification by label-free approaches is based on the observation that the chromatographic peak area for any given peptide in an LC run (Bondarenko et al., 2002; Chelius and Bondarenko, 2002) and the number of tandem MS spectra of a given peptide (Liu et al., 2004) are proportional to peptide concentration in the analyzed sample. Thus, relative quantification by label-free approaches can be done by measuring and comparing the intensities of precursor ions, or by counting and comparing the number of tandem MS spectra derived from a particular protein in different experiments.

3.4 STRUCTURAL PROTEOMICS

A protein's function is determined by its structure. The major goals of structural proteomics are to elucidate the three-dimensional (3-D) protein structures and to determine the relationship between protein structure and function. Traditionally, static 3-D protein structures are determined by X-ray crystallography (Sherwood et al., 2011) or nuclear magnetic resonance spectroscopy (Wuthrich, 1990). However, experimental determination of protein structure by these methods remains a difficult and laborious task (Sherwood et al., 2011; Hyung and Ruotolo, 2012). Alternatively, protein structures can be predicted computationally by homology modeling, or *ab initio* (Flock et al., 2012). However, and despite decades of intensive research, these approaches do not always produce reliable models (Flock et al., 2012).

Recently, hydrogen-deuterium exchange (Wales and Engen, 2006), covalent labeling (Chance, 2001), or chemical cross-linking (Young et al., 2000; Petrotchenko and Borchers, 2010) have been coupled with MS to emerge as viable methods to probe 3-D protein structure. Protein footprinting methods modify the surface of the protein that is exposed to the solvent by exchanging amide protons with heavier deuterium atoms (Wales and Engen, 2006), or by different covalent modifications (Stocks and Konermann, 2009). The labeling changes the molecular weight of a protein, and this enables MS to identify the modified sites. Protein footprinting methods are used to investigate protein conformation in solution. Chemical cross-linking covalently couples parts of a protein(s) that are close in space under native conditions. Subsequent MS analysis identifies the location and identity of the cross-linked sites, which provides important clues about the structural topology of the protein or protein complexes (Young et al., 2000; Petrotchenko and Borchers, 2010).

3.5 APPLICATION OF PROTEOMICS IN AVIAN RESEARCH

Until recently, red junglefowl, the ancestor of domestic chickens, was the only avian species with a sequenced genome (2004) (International Chicken Genome Sequencing C, 2004). Because proteomics greatly depends on a complete and well-annotated genome sequence, most of the proteomics studies in birds have been done on this animal model. Since then, the complete genomes of zebra finch (2010) (Warren et al., 2010), turkey (2011) (Dalloul et al., 2010), and two flycatchers (2012) (Ellegren et al., 2012) have been sequenced and assembled and many more are underway (Oleksyk et al., 2012; Rands et al., 2013; Rubin et al., 2010; Huang et al., 2013; Cai et al., 2013; Seabury et al., 2013). The increasing number of sequenced avian genomes expands the range of unique bird phenomena that can be studied on a global protein level, but more importantly, makes it possible to study avian proteomes directly through their genomes, obviating the need for cross-species peptide matching.

A number of proteomic studies have been done to study various aspects of bird biology including egg production, embryogenesis, development, metabolism, behavior, cognition, immunity, cancer, disease, and infection.

3.5.1 Organ and Tissue Proteomics

A number of studies have focused on initial description and functional characterization of proteomes of different avian tissues, anatomical structures, or entire organs.

The avian egg is a reproductive cell and a highly elaborate biological structure that protects and nourishes the developing embryo. The major components of the egg—the crystalline shell (Mann et al., 2006, 2007), albumen (egg white) (Mann and Mann, 2011; D'Ambrosio et al., 2008; Mann, 2007), yolk (Farinazzo et al., 2009; Mann and Mann, 2008), and the vitelline membrane (Mann, 2008)—have been extensively characterized by various proteomic approaches.

McCarthy et al. (2006) analyzed the proteomes of the supporting stromal and B cells isolated from the chicken bursa of Fabricius, a unique bird organ and a common experimental system for B-cell development (McCarthy et al., 2006). Proteins were isolated from the two major functional cell types of bursa by a sequential detergent extraction procedure that increased proteome coverage and helped to localize known and previously unknown proteins to different cellular compartments. The analysis identified 5198 proteins in bursa, and of these, 1753 were B-cell specific, 1972 were stroma specific, and 1473 proteins were identified in both cell types. Functional modeling of the identified proteins provided insights about signaling pathways involved in programmed cell death, proliferation, and differentiation. In a similar study, van den Berg et al. (2007) applied whole organ proteomics to study frozen spleen (van den Berg et al., 2007).

To gain a better understanding of B-cell development in the bursa of Fabricius, Korte et al. used a quantitative 2D PAGE approach to study bursal proteomes from the embryonic and posthatch developmental stages. They showed that enzymes of the retinoic acid metabolism play a crucial role in the early development of the primary avian B-cell organ (Korte et al., 2013). Similar observations were done in mammals, where vitamin A plays a similarly important role in the development of secondary lymphoid organs (van de Pavert et al., 2009).

Proteomic analysis of the Harderian gland showed that Harderian gland is a site of active mucosal immunity also due to expression of hematopoietic prostaglandin D synthase (Scott et al., 2005), which is necessary for production of prostaglandin D_2, the potent activator of inflammatory responses (Serhan et al., 2008).

The chicken embryo is one of the most useful and investigated comparative and biomedical models for studying development, physiology, and pathogenesis. Several proteomic studies have used chicken embryos to study embryonic development of retina (Lam et al., 2006; Mizukami et al., 2008; Finnegan et al., 2008, 2010), face (Mangum et al., 2005), cerebrospinal fluid (Parada et al., 2005, 2006), liver (Jianzhen et al., 2007), cardiovascular system (Bon et al., 2010), and vasculature (Soulet et al., 2013).

Lam et al. (2006), Mizukami et al. (2008), and Finnegan et al. (2008) used 2D PAGE to catalog the most abundant proteins in young chicken retina and to identify those that were differentially expressed between different stages of retina development (Lam et al., 2006; Mizukami et al., 2008; Finnegan et al., 2008). These studies identified known and novel proteins that play roles in early ocular growth and neural development. The retinal dysplasia and degeneration (rdd) chick was used as a model to identify proteins that are differentially expressed during the onset of degeneration of retina (Finnegan et al., 2010).

Mangum et al. (2005) studied the development of the first pharyngeal arch, an embryonic structure that is crucial for the formation of the face, as a model for the craniofacial defects in humans (Mangum et al., 2005). This study showed that expression of molecular chaperones, cytoskeletal proteins and plasma proteins associated with vascularization, was altered the most between the different stages of craniofacial development.

Proteomics has been used to characterize the proteome of a chicken embryonic cerebrospinal fluid (CSF) (Parada et al., 2006). This study identified, among others, 14 proteins that are also present in human CSF, and 12 of them are altered in neurodegenerative diseases and/or neurological disorders.

Bon et al. (2010) identified and quantified selected proteomes of three different heart tissues and studied them at three different developmental stages (Bon et al., 2010). By comparing the identified proteomes, it was possible study the changes in proteome expression and to identify proteins that were specific for particular heart structures or developmental stages. Grey et al. (2010) used an alternative approach, based on MALDI tissue imaging MS, to study spatial distribution of proteins in chicken heart structures such as vessels, valves, endocardium, myocardium, and septa.

Initial characterization of the zebra finch retina and optic tectum, a major structure of the midbrain, proteomes have been done using the 1D PAGE approach coupled with MS (Sloley et al., 2007a, 2007b). Because these studies were done before the complete zebra finch genome was available, potential zebra finch proteins had to be identified by cross-species matching using the nonredundant NCBI, Ensemble, and Swissprot protein databases.

3.5.2 Proteomics of Cell Metabolism

Biomineralization is a complex and not well understood process, which starts with the release of matrix vesicles by mineralization competent cells, such as osteoblasts or odontoblasts, into the extracellular matrix (Golub, 2009). Matrix vesicles are essential for the formation of hydroxyapatite, the main primary inorganic component of bones, but their exact role in this process is unknown. To better understand embryonic chicken bone formation, Balcerzak et al. (2008) applied proteomics to identify protein machinery of matrix vesicles, which is essential for the formation of hydroxyapatite (Balcerzak et al., 2008). Functional analysis of the matrix vesicle constituents suggested what roles these protein might have in the mineralization process.

2D PAGE MS was used to compare expression profiles of proteins in the oviduct in chicken hens of different ages during the egg-laying period (Kim et al., 2007). The analysis reveled that anterior gradient homolog 2 (AGR2) protein was among the most differentially expressed proteins. Analysis of the mRNA showed that expression of AGR-2 was limited to the magnum and isthmus of the oviduct, and that this expression was approximately 900-fold higher in the mature oviduct in comparison to the premature one. Because AGR-2 is a secreted protein that shows estrogen-dependent expression, and egg laying is strongly affected by estrogen, it was suggested that AGR-2 might be important for the development of the epithelial cells in the oviduct during the egg-laying period in chickens.

To better understand how birds are prepared for transition from a fat-rich diet *in ovo* to saccharide- and protein-based diet after hatch, Gilbert et al. (2010) analyzed the proteome of chicken small intestine at hatch and during the early posthatch period in two different broiler lines (Gilbert et al., 2010). This study identified differences in expression of digestion and absorption-related proteins between different genetic lines.

Recently, 2D PAGE and MudPIT have been used to discover genetic and molecular mechanisms that compromise sperm mobility in chickens (Froman et al., 2011). Analysis of the sperm proteome from chicken lines of low or high sperm mobility allowed deduction of a proteome-based model that explained well the differences in sperm mobility between lines, and confirmed the initial hypothesis that defects in ATP metabolism and glycolysis are responsible for premature mitochondrial failure, which results in sperm immobility.

The zebra finch is the dominant animal model for studying molecular mechanisms underlying learning, memory, vocalization, and social behavior. A natural perceptual experience, such as a sound of another bird singing, triggers rapid changes in expression of specific genes in the auditory region of the zebra finch brain (Mello et al., 1992). Repeated exposure to the same song leads to stimulus-specific habituation of the original response (Petrinovich and Patterson, 1979). To understand the process of habituation better, Dong et al. (2009) used DNA microarrays and 2D PAGE MS approaches to analyze global changes of gene expression at different stages in the development of habituation. This study showed that exposure to a song induces massive changes in gene expression, and that song response habituation is not a simple loss of the original responses but rather a change of neuronal responses underpinned by a novel and different gene expression profile. Analysis of protein expression showed that habituation is accompanied by a decrease in expression of cellular and mitochondrial proteins that are involved in biosynthesis and energy metabolism.

Neuropeptides are signaling peptides found in neural tissue that modulate a wide range of physiological and behavioral processes including metabolism, reproduction, learning, and memory. Xie et al. (2010) used a combination of bioinformatics, MS, and biochemistry for prediction, identification, and localization of neuropeptidome of the zebra finch. Computational analysis of the zebra finch genome predicted 70 putative pro-hormones and 90 peptides derived from 24 putative pheromones identified in the zebra finch brain by two different MS approaches. The power of MS was further used for localization of a subset of peptides in the major song control nuclei of the zebra finch brain. Furthermore, gene expression of a subset of pheromone genes was anatomically mapped in selected zebra finch brain sections by in situ hybridization.

Birds display an enormous range of plumage colors, and this diversity rivals or exceeds that of plants (Stoddard and Prum, 2011). Still, bird plumage occupies only about 30% of the possible colors that birds can see (Stoddard and Prum, 2011). The molecular mechanisms that determine and drive the development of this diversity are largely unknown. The breeding plumage of male pied flycatchers varies from a brown to dark black. Leskinen et al. (2012) characterized and quantified the proteome of developing pied flycatcher feathers to advance understanding of physiological processes that underlay the variation in pigmentation. In total, 294 proteins were identified in the developing feathers. Sixty-five proteins were linked with epidermal development and/or pigmentation in the developing feathers, and 23 proteins were associated with pigment-containing organelles—melanosomes. The comparison of the brown- and black-specific proteomes revealed several proteins and functional networks that differed in expression between the two phenotypes and that are candidates for further studies. The most pronounced differences were detected in immunological signaling, oxidative stress, energy balance, and protein synthesis networks, and these differences might be responsible for differential feather growth and color pigmentation.

3.5.3 Production Proteomics

Food products derived from farm animals, birds, and fish represent a significant part of the human diet. Understanding the nutrient metabolism, muscle accretion, and fat deposition in food birds provides practical knowledge that can be used to improve feed conversion efficiency, food quality, and the health and welfare of animals.

Animal feed represents the major cost of poultry production. A balanced poultry diet is reflected in optimal growth and production at minimal nutrient expense. Because of the composition of poultry diet, where corn and soybeans are used as major sources of energy and proteins, respectively, some amino acids become more limited than others. Corzo et al. (2005) utilized the power of MS to understand amino acid requirements in chickens. Blood plasma proteome from chickens fed an adequate or lysine-deficient diet was analyzed to identify potential biomarkers of dietary lysine deficiency. The analysis revealed that lysine deficiency might not result in a simple overall reduction of protein synthesis in chickens fed with a lysine-deficient diet, but rather in reduced anabolism of specific proteins. Corzo et al. (2006) and Zhai et al. (2012) also evaluated the effect of dietary methionine on breast muscle accretion in broiler chickens (Corzo et al., 2006; Zhai et al., 2012). This study showed that four canonical pathways related to muscle development (citrate cycle, calcium signaling, actin cytoskeleton signaling, and clathrin-mediated endocytosis signaling) were differentially regulated between chickens that were fed with low- and high-methionine diets (Zhai et al., 2012). In addition, this study suggested that a methionine-rich diet preferably induced muscle accretion by sarcoplasmic over myofibrillar hypertrophy.

The blood plasma is an extremely complex tissue that contains thousands of distinct proteins. It is also the most common tissue used in diagnosis of disease and nutritional status. Several authors analyzed blood plasma protein composition to gain better understanding of protein dynamics during chicken development (Huang et al., 2006), or for discovery of plasma biomarkers that reflect nutritional conditions (Corzo et al., 2004, 2006).

Muscle meat food products derived from birds, and in particular chickens, are important sources of essential nutrients and energy intake in the human diet. Proteomics has an obvious potential to study a broad range of aspects related to the meat production including nutrition, muscle formation, breed differentiation, meat quality, and meat contamination (Paredi et al., 2013).

Chicken strains selected for meat production show dramatic growth rates and accelerated accretion of the pectoralis (breast) major and minor muscles. The proteome of the chicken pectoralis muscle has been extensively profiled (Corzo et al., 2006; Zhai et al., 2012; Doherty et al., 2004; Teltathum and Mekchay, 2009), and the most recent study identified over 5000 unique proteins in the pectoralis muscle of studied birds (Zhai et al., 2012).

A complementary study identified the proteome of the pipping muscle, which is primarily used for breaking the egg's surface during hatching (Sokale et al., 2011). The identified proteins, 676 in all, were analyzed using the assigned Gene Ontology categories for molecular function, biological process, or cellular component. This analysis revealed which protein functions and cellular activities are important for rapid development of pipping muscle during embryogenesis.

A proteomic approach was also used to identify hypothalamic biomarkers associated with high egg production in chickens (Kuo et al., 2005). Comparison of the hypothalamic proteomes from two related chicken lines selected for meat and high egg production resulted in identification of six proteins that differed in their expression between the lines, and some of these proteins are involved in regulation of gene expression, signal transduction, and lipid metabolism. The heterogeneous nuclear ribonucleoprotein H3 was suggested as novel biomarker for high egg production.

A global analysis of the chicken embryo liver proteome resulted in identification of proteins that were differentially expressed between two chicken lines that differed in hepatic lipid metabolism and fat deposition (Huang et al., 2010). Comparative analysis of the identified liver proteins showed that proteins involved in gluconeogenesis, cholesterol metabolism, and fatty acid oxidation were expressed earlier and more abundantly in the liver of lean-line of chickens. In a similar study, to better understand duck liver physiology at the protein level, liver proteome of a domesticated lean

Pekin duck (*Anas platyrhynchos domestica*) was analyzed at different developmental stages (Zheng et al., 2012). Comparison of the identified proteomes showed that proteins involved in transportation were more abundantly expressed in newborn ducks, whereas adult duck liver proteome contained more abundantly expressed proteins associated with carbohydrate and protein metabolism, immune defense, and antioxidation.

3.5.4 Proteomics of Disease and Infection

The recent explosion of animal and pathogen genomes has not only enabled identification of genes involved in the etiology and pathology of diseases (such as mutant gene variants or virulence factors), but also has opened up the door for proteomics to probe the pathogenesis and pathogen–host interactions on a global protein level. Proteomics has greatly improved understanding of diseases; it has been very valuable in diagnostic marker discovery and it has a great potential in drug discovery. Despite its great value, disease proteomics remains to be one of the least-developed areas in avian research. Nevertheless, proteomics has been used to study the pathogenesis, etiology, and pathology of several avian (infectious) diseases. In addition, proteomics has been used to study various human diseases on chicken experimental models (Andrews Kingon et al., 2013).

3.5.4.1 Disease Proteomics

The chicken is an ideal and unique animal model to probe the etiology and progression of spontaneous human epithelial ovarian cancer (EOC), largely because the domestic chicken has a high prevalence of spontaneous ovarian carcinomas. EOC remains the most lethal gynecologic malignancy in part because early detection and therapeutic strategies have been largely unsuccessful (Kurman and Shih, 2010). Hawkridge et al. (2010) used this model to study the onset and progression of EOC by a large-scale biomarker discovery effort involving longitudinal sample collection and protein analysis by MS. Inter- and intra-individual measurement of proteins identified ovomacroglobulin (ovostatin) as a potential EOC biomarker because its levels in plasma were undetected in a healthy individual and significantly higher during later stages in an EOC bird.

Pulmonary hypertension syndrome, or ascites syndrome, is a metabolic pathogenesis in meat-type chickens that is manifested by the formation of ascites. Ascites syndrome is one of the major problems in the chicken industry, and is caused by cardiopulmonary insufficiency during high oxygen demands spurred by a rapid tissue growth (Currie, 1999). Proteomic analysis of the cardiac mitochondrial matrix proteomes of the ascites-resistant and ascites-susceptible line broilers suggested that the mitochondria of susceptible chickens may respond inappropriately to hypoxia (Cisar et al., 2005). A complementary proteomic analysis of the hepatic proteomes of healthy birds and those with ascites suggested that insufficient energy generation in the liver is responsible for development of pulmonary hypertension syndrome (Wang et al., 2012).

3.5.4.2 Proteomics of Infections

Some strains of the highly pathogenic avian influenza A subtype H5N1 cause severe acute encephalopathy and neurodegeneration in poultry and migratory birds. To reveal the mechanisms that cause the observed neuropathogenesis, Zou et al. (2010) used comparative proteomics to identify the proteins that were expressed differently in the brains of healthy and H5N1-infected chickens (Zou et al., 2010). Among the differentially expressed proteins were septin 5 and collapsin response mediator protein 2 (CRMP2). Septin 5 is dysregulated in Parkinson's disease and CRMP2 in Alzheimer's disease, Down syndrome, and human T-cell lymphotropic virus type I associated myelopathy (Vincent et al., 2005; Lubec et al., 1999; Son et al., 2005). This suggests that these proteins might also have a role in neurodegenerative pathologies associated with influenza H5N1 infection.

Gallid herpesvirus 2 (GaHV-2) is an avian oncogenic herpesvirus that causes a highly infectious and rapidly progressive lymphomatous disease of chickens, Marek's disease (MD). GaHV-2 infects and transforms cells in all chicken genotypes, but some chickens are genetically resistant to gross lymphoma formation (Burgess and Davison, 2002). To better understand the molecular mechanisms of differential susceptibility to MD, spleen proteomes of MD-susceptible and MD-resistant chickens were analyzed using the 2D PAGE MS approach (Thanthrige-Don et al., 2010). Among the differentially expressed proteins identified in this study were antioxidants; molecular chaperones; and proteins involved in the activation and migration of T lymphocytes, formation of cytoskeleton, protein degradation, and antigen presentation; and some of these were implicated as potential factors in MD resistance.

In a similar study, analysis of the changes in the chicken spleen proteome induced by the GaHV-2 infections revealed that protein expression was the most altered during early stages of infection. Comparative analysis showed that proteins that were differentially expressed at different timepoints postinfection were involved in a variety of cellular processes that are crucial for the host response to GaHV-2 infection and pathogenesis (Thanthrige-Don et al., 2009).

To better understand how GaHV-2 infection changes the host protein expression, several proteomic profiling studies have been done to determine the proteome of GaHV-2-lytically infected chicken embryo cells (CECs) (Liu et al., 2006; Chien et al., 2011; Ramaroson et al.,

2008; Chien et al., 2012) compare the phosphoproteomes of mock- and GaHV-2-infected CECs (Chien et al., 2011; Ramaroson et al., 2008); and quantify protein expression changes caused by GaHV-2 infection (Chien et al., 2012). Collectively, these studies revealed GaHV-2 infection dramatically changes the protein expression profile of infected cells. Overlaying quantitative and phosphorylation data revealed that GaHV-2 infection altered both protein expression and phosphorylation of proteins from several cellular pathways, and among the most affected processes were RNA transport, signal transduction, initiation of translation, and protein degradation. Perhaps the most interesting discovery of these studies is that GaHV-2 causes unique phosphorylation of the translation initiation factor 4E-binding protein 1 (4E BP1), which is important for the assembly of the protein translation initiation complex after virus infection.

In a complementary study Buza and Burgess (2007) used MudPIT to profile the proteome of the GaHV-2-transformed lymphoblastoid cell line UA01. Functional modeling the UA01 proteome showed that cells had a typical cancer phenotype. UA01 cells were activated, differentiated, and proliferative, but antagonistic to apoptosis, anergy, quiescence, and senescence. Identified cytokines, cytokine receptors, and related proteins suggested that the UA01 proteome had a T-cell regulatory (T-reg) rather than T-helper (Th)-2 phenotype.

Marek's disease, a CD4+ T cell lymphoma of chickens, and many human lymphomas overexpress the Hodgkin's disease antigen CD30 ($CD30^{hi}$). Marek's disease lymphomas, like its human homologs, are formed by a minority of transformed ($CD30^{hi}$) and a majority of non-transformed ($CD30^{lo}$) cells (Burgess and Davison, 2002; Shack et al., 2008). Although the GaHV-2 gene meq is the principal oncogene, which acts as a transcription factor and transforms via the Jun pathway (Levy et al., 2005), the exact mechanism of neoplastic transformation and transition from $CD30^{lo}$ to $CD30^{hi}$ neoplastic phenotype is unknown. As described in this review, most of the proteomics work that has been done in aves has been descriptive, based on differential expression. Kumar et al. (2012), though, compared microRNA, mRNA and protein levels and from this data imputed functional models. As described by many others in many different systems, there was poor overall correlation between mRNA and protein expression (Gygi et al., 1999; Cullen et al., 2004). However, to identify the key regulatory proteins responsible for neoplastic transformation, all gene products which were differentially expressed in the same direction at both mRNA and protein levels (i.e. concordant) were selected for further analysis and these did have an overall positive correlation. Those gene products with the greatest mRNA and protein correlation are known to be involved in human CD30-over expressing lymphomas. When ranked into pentiles based on protein expression levels, the Gene Ontology Biological Processes of cell cycle and proliferation to programmed cell death ratios were greatest in pentile 1. The authors then identified the numbers of putative canonical MDV Meq (the virus oncogene) binding sites in each of the 88 concordantly-expressed genes' promoters; and genes in pentile 1 had the most Meq binding sites. Of the five concordant genes previously implicated in lymphomagenesis in other species most were in pentile 1 suggesting direct transcriptional regulation by Meq. In contrast, one gene product, CST3 was likely regulated by a micro-RNA.

Plasmodium gallinaceum is a protozoal avian malaria parasite and the most relevant animal model of the human parasite *Plasmodium falciparum* sexual stages zygote and ookinete. The early stages of the *P. gallinaceum* life-cycle occur in its definitive host, the mosquito, but this process is largely unknown. To better understand the initial molecular mechanisms of *P. falciparum* vector interaction, Patra et al. (2008) used high-throughput proteomics to identify 966 orthologous proteins of *P. falcipa*rum present in the zygote and ookinete proteomes (Patra et al., 2008). Forty percent of the identified proteins had hypothetical status and the majority of these were transmembrane or secreted proteins. This suggests that these proteins might play important roles in parasite–host interactions.

Coccidiosis of fowl, an intestinal disease caused by a protozoal parasite Eimeria, causes significant losses for the poultry industry. Recently, plasma proteome profiles of two different chicken lines infected with one of three common Eimeria species were compared by the 2D PAGE MS (Gilbert et al., 2011). Fourty-six proteins displayed significantly changed expression in response to Eimeria infection. The differentially expressed proteins were found to participate in innate immunity, blood clotting, and iron and mitochondrial metabolism, and these processes fit well within the host acute-phase responses that are initiated when a tissue is invaded by a microorganism. Some of the identified proteins were suggested as candidate biomarkers for early diagnosis of Eimeria infection.

Salmonella enterica subspecies enterica is a Gram-negative enterobacterium and an important pathogen that infects a broad range of vertebrate species including chicken and man. The subspecies is routinely divided into more than 2500 serotypes (serovars) based on antigenic epitopes (Franklin et al., 2011). *Salmonella* serovars differ in host range and pathogenic potential, but molecular mechanisms underlying these differences are not well understood. Several proteomic studies compared proteomes of different avian *Salmonella* serovars and discovered possible molecular mechanisms responsible for the observed phenotypic differences (Encheva et al., 2005; Osman et al., 2009; Sun and Hahn, 2012).

3.6 CONCLUSIONS

One of the primary goals of biology is to understand how organisms function on a molecular level. Because proteins are the "makers of life", understanding their functions is central to understanding biology. During the last decade, proteomics has emerged as an extremely versatile and comprehensive platform to study proteins on a large scale. It offers a broad range of tools that can be used to determine the identity, structure, quantity, and quality of expressed proteins in biological systems. Thus, while scientific questions in avian research remain essentially the same, proteomics has the potential to transform the form of biological inquiry.

REFERENCES

Anderson, N.L., Anderson, N.G., 2002. The human plasma proteome: history, character, and diagnostic prospects. Mol. Cell. Proteomics 1, 845–867.

Andrews Kingon, G.L., Petitte, J.N., Muddiman, D.C., Hawkridge, A.M., 2013. Multi-peptide nLC-PC-IDMS-SRM-based assay for the quantification of biomarkers in the chicken ovarian cancer model. Methods 61, 323–330.

Balcerzak, M., Malinowska, A., Thouverey, C., Sekrecka, A., Dadlez, M., Buchet, R., et al., 2008. Proteome analysis of matrix vesicles isolated from femurs of chicken embryo. Proteomics 8, 192–205.

van den Berg, B.H.J., Harris, T., McCarthy, F.M., Lamont, S.J., Burgess, S.C., 2007. Non-electrophoretic differential detergent fractionation proteomics using frozen whole organs. Rapid Commun. Mass Spectrom. 21, 3905–3909.

Bon, E., Steegers, R., Steegers, E.A.P., Ursem, N., Charif, H., Burgers, P.C., et al., 2010. Proteomic analyses of the developing chicken cardiovascular system. J. Proteome Res. 9, 268–274.

Bondarenko, P.V., Chelius, D., Shaler, T.A., 2002. Identification and relative quantitation of protein mixtures by enzymatic digestion followed by capillary reversed-phase liquid chromatography-tandem mass spectrometry. Anal. Chem. 74, 4741–4749.

Burgess, S.C., Davison, T.F., 2002. Identification of the neoplastically transformed cells in Marek's disease herpesvirus-induced lymphomas: recognition by the monoclonal antibody AV37. J. Virol. 76, 7276–7292.

Buza, J.J., Burgess, S.C., 2007. Modeling the proteome of a Marek's disease transformed cell line: a natural animal model for CD30 overexpressing lymphomas. Proteomics 7, 1316–1326.

Cai, Q., Qian, X., Lang, Y., Luo, Y., Pan, S., Hui, Y., et al., 2013. The genome sequence of the ground tit *Pseudopodoces humilis* provides insights into its adaptation to high altitude. Genome Biol. 14, R29.

Cao, P., Stults, J.T., 1999. Phosphopeptide analysis by on-line immobilized metal-ion affinity chromatography-capillary electrophoresis-electrospray ionization mass spectrometry. J. Chromatogr. A 853, 225–235.

Chance, M.R., 2001. Unfolding of apomyoglobin examined by synchrotron footprinting. Biochem. Biophys. Res. Commun. 287, 614–621.

Chelius, D., Bondarenko, P.V., 2002. Quantitative profiling of proteins in complex mixtures using liquid chromatography and mass spectrometry. J. Proteome Res. 1, 317–323.

Chien, K.Y., Liu, H.C., Goshe, M.B., 2011. Development and application of a phosphoproteomic method using electrostatic repulsion-hydrophilic interaction chromatography (ERLIC), IMAC, and LC-MS/MS analysis to study Marek's Disease Virus infection. J. Proteome Res. 10, 4041–4053.

Chien, K.Y., Blackburn, K., Liu, H.C., Goshe, M.B., 2012. Proteomic and phosphoproteomic analysis of chicken embryo fibroblasts infected with cell culture-attenuated and vaccine strains of Marek's disease virus. J. Proteome Res. 11, 5663–5677.

Cisar, C.R., Balog, J.M., Anthony, N.B., Donoghue, A.M., 2005. Differential expression of cardiac muscle mitochondrial matrix proteins in broilers from ascites-resistant and susceptible lines. Poult. Sci. 84, 704–708.

Corzo, A., Kidd, M.T., Pharr, G.T., Burgess, S.C., 2004. Initial mapping of the chicken blood plasma proteome. Int. J. Poult. Sci. 3, 157–162.

Corzo, A., Kidd, M.T., Koter, M.D., Burgess, S.C., 2005. Assessment of dietary amino acid scarcity on growth and blood plasma proteome status of broiler chickens. Poult. Sci. 84, 419–425.

Corzo, A., Kidd, M.T., Dozier 3rd, W.A., Shack, L.A., Burgess, S.C., 2006. Protein expression of pectoralis major muscle in chickens in response to dietary methionine status. Br. J. Nutr. 95, 703–708.

Cullen, P., Lorkowski, S., Kratz, M., Werner, M., Marschall, C., Klein, H.-G., et al., 2004. Basic Concepts of Gene Expression. Analysing Gene Expression. Wiley-VCH Verlag GmbH & Co. KGaA, Darmstadt, Germany. pp. 1–95.

Currie, R.J., 1999. Ascites in poultry: recent investigations. Avian Pathol. 28, 313–326.

D'Ambrosio, C., Arena, S., Scaloni, A., Guerrier, L., Boschetti, E., Mendieta, M.E., et al., 2008. Exploring the chicken egg white proteome with combinatorial peptide ligand libraries. J. Proteome Res. 7, 3461–3474.

Dalloul, R.A., Long, J.A., Zimin, A.V., Aslam, L., Beal, K., Blomberg Le, A., et al., 2010. Multi-platform next-generation sequencing of the domestic turkey (*Meleagris gallopavo*): genome assembly and analysis. PLoS Biol. 8.

Doherty, M.K., McLean, L., Hayter, J.R., Pratt, J.M., Robertson, D.H., El-Shafei, A., et al., 2004. The proteome of chicken skeletal muscle: changes in soluble protein expression during growth in a layer strain. Proteomics 4, 2082–2093.

Dong, S., Replogle, K.L., Hasadsri, L., Imai, B.S., Yau, P.M., Rodriguez-Zas, S., et al., 2009. Discrete molecular states in the brain accompany changing responses to a vocal signal. Proc. Natl. Acad. Sci. U.S.A. 106, 11364–11369.

Durham, M., Regnier, F.E., 2006. Targeted glycoproteomics: serial lectin affinity chromatography in the selection of O-glycosylation sites on proteins from the human blood proteome. J. Chromatogr. A 1132, 165–173.

Ellegren, H., Smeds, L., Burri, R., Olason, P.I., Backstrom, N., Kawakami, T., et al., 2012. The genomic landscape of species divergence in *Ficedula* flycatchers. Nature 491, 756–760.

Encheva, V., Wait, R., Gharbia, S.E., Begum, S., Shah, H.N., 2005. Proteome analysis of serovars Typhimurium and Pullorum of *Salmonella enterica* subspecies I. BMC Microbiol. 5, 42.

Farinazzo, A., Restuccia, U., Bachi, A., Guerrier, L., Fortis, F., Boschetti, E., et al., 2009. Chicken egg yolk cytoplasmic proteome, mined via combinatorial peptide ligand libraries. J. Chromatogr. A 1216, 1241–1252.

Fenn, J.B., Mann, M., Meng, C.K., Wong, S.F., Whitehouse, C.M., 1989. Electrospray ionization for mass spectrometry of large biomolecules. Science 246, 64–71.

Ficarro, S.B., McCleland, M.L., Stukenberg, P.T., Burke, D.J., Ross, M.M., Shabanowitz, J., et al., 2002. Phosphoproteome analysis by mass spectrometry and its application to *Saccharomyces cerevisiae*. Nat. Biotechnol. 20, 301–305.

Fields, S., Song, O., 1989. A novel genetic system to detect protein-protein interactions. Nature 340, 245–246.

Finnegan, S., Robson, J.L., Wylie, M., Healy, A., Stitt, A.W., Curry, W.J., 2008. Protein expression profiling during chick retinal maturation: a proteomics-based approach. Proteome Sci. 6, 34.

Finnegan, S., Robson, J., Hocking, P.M., Ali, M., Inglehearn, C.F., Stitt, A., et al., 2010. Proteomic profiling of the retinal dysplasia and degeneration chick retina. Mol. Vis. 16, 7–17.

Flock, T., Venkatakrishnan, A., Vinothkumar, K., Babu, M.M., 2012. Deciphering membrane protein structures from protein sequences. Genome Biol. 13, 160.

Franklin, K., Lingohr, E.J., Yoshida, C., Anjum, M., Bodrossy, L., Clark, C.G., et al., 2011. Rapid genoserotyping tool for classification of *Salmonella* serovars. J. Clin. Microbiol. 49, 2954–2965.

Froman, D.P., Feltmann, A.J., Pendarvis, K., Cooksey, A.M., Burgess, S.C., Rhoads, D.D., 2011. Physiology and endocrinology symposium: a proteome-based model for sperm mobility phenotype. J. Anim. Sci. 89, 1330–1337.

Geng, M., Ji, J., Regnier, F.E., 2000. Signature-peptide approach to detecting proteins in complex mixtures. J. Chromatogr. A 870, 295–313.

Gilbert, E.R., Williams, P.M., Ray, W.K., Li, H.F., Emmerson, D.A., Wong, E.A., et al., 2010. Proteomic evaluation of chicken Brush-Border membrane during the early posthatch period. J. Proteome Res. 9, 4628–4639.

Gilbert, E.R., Cox, C.M., Williams, P.M., McElroy, A.P., Dalloul, R.A., Ray, W.K., et al., 2011. Eimeria species and genetic background influence the serum protein profile of broilers with coccidiosis. PLoS One 6, e14636.

Golub, E.E., 2009. Role of matrix vesicles in biomineralization. Biochim. Biophys. Acta 1790, 1592–1598.

Grey, A.C., Gelasco, A.K., Section, J., Moreno-Rodriguez, R.A., Krug, E.L., Schey, K.L., 2010. Molecular morphology of the chick heart visualized by MALDI imaging mass spectrometry. Anat. Rec. 293, 821–828.

Gygi, S.P., Rochon, Y., Franza, B.R., Aebersold, R., 1999a. Correlation between protein and mRNA abundance in yeast. Mol. Cell. Biol. 19, 1720–1730.

Gygi, S.P., Rist, B., Gerber, S.A., Turecek, F., Gelb, M.H., Aebersold, R., 1999b. Quantitative analysis of complex protein mixtures using isotope-coded affinity tags. Nat. Biotechnol. 17, 994–999.

Haab, B.B., 2003. Methods and applications of antibody microarrays in cancer research. Proteomics 3, 2K116–2122.

Han, X., Aslanian, A., Yates 3rd, J.R., 2008. Mass spectrometry for proteomics. Curr. Opin. Chem. Biol. 12, 483–490.

Hawkridge, A.M., Wysocky, R.B., Petitte, J.N., Anderson, K.E., Mozdziak, P.E., Fletcher, O.J., Horowitz, J.M., Muddiman, D.C., 2010. Measuring the intra-individual variability of the plasma proteome in the chicken model of spontaneous ovarian adenocarcinoma. Anal. Bioanal. Chem. 398, 737–749.

Huang, S.Y., Lin, J.H., Chen, Y.H., Chuang, C.K., Chiu, Y.F., Chen, M.Y., et al., 2006. Analysis of chicken serum proteome and differential protein expression during development in single-comb White Leghorn hens. Proteomics 6, 2217–2224.

Huang, J., Tang, X., Ruan, J., Ma, H., Zou, S., 2010. Use of comparative proteomics to identify key proteins related to hepatic lipid metabolism in broiler chickens: evidence accounting for differential fat deposition between strains. Lipids 45, 81–89.

Huang, Y., Li, Y., Burt, D.W., Chen, H., Zhang, Y., Qian, W., et al., 2013. The duck genome and transcriptome provide insight into an avian influenza virus reservoir species. Nat. Genet. 45, 776–783.

Hyung, S.J., Ruotolo, B.T., 2012. Integrating mass spectrometry of intact protein complexes into structural proteomics. Proteomics 12, 1547–1564.

International Chicken Genome Sequencing C, 2004. Sequence and comparative analysis of the chicken genome provide unique perspectives on vertebrate evolution. Nature 432, 695–716.

Jaffe, J.D., Berg, H.C., Church, G.M., 2004. Proteogenomic mapping as a complementary method to perform genome annotation. Proteomics 4, 59–77.

Jianzhen, H., Haitian, M., Liming, Y., Sixiang, Z., 2007. Developmental changes of protein profiles in the embryonic Sanhuang chicken liver. J. Vet. Med. A Physiol. Pathol. Clin. Med. 54, 464–469.

Kim, N.S., Shen, Y.N., Kim, T.Y., Byun, S.J., Jeon, I.S., Kim, S.H., 2007. Expression of AGR-2 in chicken oviduct during laying period. J. Biochem. Mol. Biol. 40, 212–217.

Korte, J., Frohlich, T., Kohn, M., Kaspers, B., Arnold, G.J., Hartle, S., 2013. 2D DIGE analysis of the bursa of Fabricius reveals characteristic proteome profiles for different stages of chicken B-cell development. Proteomics 13, 119–133.

Kumar, S., Kunec, D., Buza, J.J., Chiang, H.I., Zhou, H., Subramaniam, S., et al., 2012. Nuclear Factor kappa B is central to Marek's disease herpesvirus induced neoplastic transformation of CD30 expressing lymphocytes in-vivo. BMC Syst. Biol. 6, 123.

Kuo, Y.M., Shiue, Y.L., Chen, C.F., Tang, P.C., Lee, Y.P., 2005. Proteomic analysis of hypothalamic proteins of high and low egg production strains of chickens. Theriogenology 64, 1490–1502.

Kurman, R.J., Shih, I.M., 2010. The origin and pathogenesis of epithelial ovarian cancer: a proposed unifying theory. Am. J. Surg. Pathol. 34, 433–443.

Lam, T.C., Li, K.K., Lo, S.C., Guggenheim, J.A., To, C.H., 2006. A chick retinal proteome database and differential retinal protein expressions during early ocular development. J. Proteome Res. 5, 771–784.

Leskinen, P.K., Laaksonen, T., Ruuskanen, S., Primmer, C.R., Leder, E.H., 2012. The proteomics of feather development in pied flycatchers (*Ficedula hypoleuca*) with different plumage coloration. Mol. Ecol. 21, 5762–5777.

Levy, A.M., Gilad, O., Xia, L., Izumiya, Y., Choi, J., Tsalenko, A., et al., 2005. Marek's disease virus Meq transforms chicken cells via the v-Jun transcriptional cascade: a converging transforming pathway for avian oncoviruses. Proc. Natl. Acad. Sci. U.S.A. 102, 14831–14836.

Liu, H.B., Sadygov, R.G., Yates, J.R., 2004. A model for random sampling and estimation of relative protein abundance in shotgun proteomics. Anal. Chem. 76, 4193–4201.

Liu, H.C., Soderblom, E.J., Goshe, M.B., 2006. A mass spectrometry-based proteomic approach to study Marek's Disease Virus gene expression. J. Virol. Methods 135, 66–75.

Lubec, G., Nonaka, M., Krapfenbauer, K., Gratzer, M., Cairns, N., Fountoulakis, M., 1999. Expression of the dihydropyrimidinase related protein 2 (DRP-2) in Down syndrome and Alzheimer's disease brain is downregulated at the mRNA and dysregulated at the protein level. J. Neural. Transm. Suppl. 57, 161–177.

Mangum, J.E., Farlie, P.G., Hubbard, M.J., 2005. Proteomic profiling of facial development in chick embryos. Proteomics 5, 2542–2550.

Mann, K., 2007. The chicken egg white proteome. Proteomics 7, 3558–3568.

Mann, K., 2008. Proteomic analysis of the chicken egg vitelline membrane. Proteomics 8, 2322–2332.

Mann, K., Mann, M., 2008. The chicken egg yolk plasma and granule proteomes. Proteomics 8, 178–191.

Mann, K., Mann, M., 2011. In-depth analysis of the chicken egg white proteome using an LTQ Orbitrap Velos. Proteome Sci. 9, 7.

Mann, K., Macek, B., Olsen, J.V., 2006. Proteomic analysis of the acid-soluble organic matrix of the chicken calcified eggshell layer. Proteomics 6, 3801–3810.

Mann, K., Olsen, J.V., Macek, B., Gnad, F., Mann, M., 2007. Phosphoproteins of the chicken eggshell calcified layer. Proteomics 7, 106–115.

Marguerat, S., Schmidt, A., Codlin, S., Chen, W., Aebersold, R., Bahler, J., 2012. Quantitative analysis of fission yeast transcriptomes and proteomes in proliferating and quiescent cells. Cell 151, 671–683.

McCarthy, F.M., Cooksey, A.M., Wang, N., Bridges, S.M., Pharr, G.T., Burgess, S.C., 2006. Modeling a whole organ using proteomics: the avian bursa of Fabricius. Proteomics 6, 2759–2771.

Mello, C.V., Vicario, D.S., Clayton, D.F., 1992. Song presentation induces gene expression in the songbird forebrain. Proc. Natl. Acad. Sci. U.S.A. 89, 6818–6822.

Mizukami, M., Kanamoto, T., Souchelnytskyi, N., Kiuchi, Y., 2008. Proteome profiling of embryo chick retina. Proteome Sci. 6, 3.

Nagaraj, N., Wisniewski, J.R., Geiger, T., Cox, J., Kircher, M., Kelso, J., et al., 2011. Deep proteome and transcriptome mapping of a human cancer cell line. Mol. Syst. Biol. 7, 548.

Nanduri, B., Wang, N., Lawrence, M.L., Bridges, S.M., Burgess, S.C., 2010. Gene model detection using mass spectrometry. Methods Mol. Biol. 604, 137–144.

O'Farrell, P.H., 1975. High resolution two-dimensional electrophoresis of proteins. J. Biol. Chem. 250, 4007–4021.

Oleksyk, T.K., Pombert, J.F., Siu, D., Mazo-Vargas, A., Ramos, B., Guiblet, W., et al., 2012. A locally funded Puerto Rican parrot (*Amazona vittata*) genome sequencing project increases avian data and advances young researcher education. Gigascience 1, 14.

Ong, S.E., Blagoev, B., Kratchmarova, I., Kristensen, D.B., Steen, H., Pandey, A., et al., 2002. Stable isotope labeling by amino acids in cell culture, SILAC, as a simple and accurate approach to expression proteomics. Mol. Cell. Proteomics 1, 376–386.

Osman, K.M., Ali, M.M., Radwan, M.I., Kim, H.K., Han, J., 2009. Comparative proteomic analysis on *Salmonella* Gallinarum and *Salmonella* Enteritidis exploring proteins that may incorporate host adaptation in poultry. J. Proteomics 72, 815–821.

Panse, S., Dong, L., Burian, A., Carus, R., Schutkowski, M., Reimer, U., et al., 2004. Profiling of generic anti-phosphopeptide antibodies and kinases with peptide microarrays using radioactive and fluorescence-based assays. Mol. Divers. 8, 291–299.

Parada, C., Gato, A., Bueno, D., 2005. Mammalian embryonic cerebrospinal fluid proteome has greater apolipoprotein and enzyme pattern complexity than the avian proteome. J. Proteome Res. 4, 2420–2428.

Parada, C., Gato, A., Aparicio, M., Bueno, D., 2006. Proteome analysis of chick embryonic cerebrospinal fluid. Proteomics 6, 312–320.

Paredi, G., Sentandreu, M.A., Mozzarelli, A., Fadda, S., Hollung, K., de Almeida, A.M., 2013. Muscle and meat: new horizons and applications for proteomics on a farm to fork perspective. J. Proteomics 88, 58–82.

Patra, K.P., Johnson, J.R., Cantin, G.T., Yates 3rd, J.R., Vinetz, J.M., 2008. Proteomic analysis of zygote and ookinete stages of the avian malaria parasite *Plasmodium gallinaceum* delineates the homologous proteomes of the lethal human malaria parasite *Plasmodium falciparum*. Proteomics 8, 2492–2499.

van de Pavert, S.A., Olivier, B.J., Goverse, G., Vondenhoff, M.F., Greuter, M., Beke, P., et al., 2009. Chemokine CXCL13 is essential for lymph node initiation and is induced by retinoic acid and neuronal stimulation. Nat. Immunol. 10. 1193-1U78.

Petrinovich, L., Patterson, T.L., 1979. Field studies of habituation .1. Effect of reproductive condition, number of trials, and different delay intervals on responses of the white-crowned sparrow. J. Comp. Physiol. Psychol. 93, 337–350.

Petrotchenko, E.V., Borchers, C.H., 2010. Crosslinking combined with mass spectrometry for structural proteomics. Mass Spectrom. Rev. 29, 862–876.

Rabilloud, T., Chevallet, M., Luche, S., Lelong, C., 2010. Two-dimensional gel electrophoresis in proteomics: past, present and future. J. Proteomics 73, 2064–2077.

Ramaroson, M.F., Ruby, J., Goshe, M.B., Liu, H.C., 2008. Changes in the *Gallus gallus* proteome induced by Marek's disease virus. J. Proteome Res. 7, 4346–4358.

Rands, C.M., Darling, A., Fujita, M., Kong, L., Webster, M.T., Clabaut, C., et al., 2013. Insights into the evolution of Darwin's finches from comparative analysis of the *Geospiza magnirostris* genome sequence. BMC Genomics 14, 95.

Rao, M., Morisson, M., Faraut, T., Bardes, S., Feve, K., Labarthe, E., et al., 2012. A duck RH panel and its potential for assisting NGS genome assembly. BMC Genomics 13, 513.

Ross, P.L., Huang, Y.L.N., Marchese, J.N., Williamson, B., Parker, K., Hattan, S., et al., 2004. Multiplexed protein quantitation in *Saccharomyces cerevisiae* using amine-reactive isobaric tagging reagents. Mol. Cell. Proteomics 3, 1154–1169.

Rubin, C.J., Zody, M.C., Eriksson, J., Meadows, J.R., Sherwood, E., Webster, M.T., et al., 2010. Whole-genome resequencing reveals loci under selection during chicken domestication. Nature 464, 587–591.

Scott, T.R., Messersmith, A.R., McCrary, W.J., Herlong, J.L., Burgess, S.C., 2005. Hematopoietic prostaglandin D2 synthase in the chicken Harderian gland. Vet. Immunol. Immunopathol. 108, 295–306.

Seabury, C.M., Dowd, S.E., Seabury, P.M., Raudsepp, T., Brightsmith, D.J., Liboriussen, P., et al., 2013. A multi-platform draft de novo genome assembly and comparative analysis for the Scarlet Macaw (*Ara macao*). PLoS One 8, e62415.

Serhan, C.N., Chiang, N., Van Dyke, T.E., 2008. Resolving inflammation: dual anti-inflammatory and pro-resolution lipid mediators. Nat. Rev. Immunol. 8, 349–361.

Shack, L.A., Buza, J.J., Burgess, S.C., 2008. The neoplastically transformed (CD30hi) Marek's disease lymphoma cell phenotype most closely resembles T-regulatory cells. Cancer Immunol. Immunother. 57, 1253–1262.

Sherwood, D., Cooper, J., Sherwood, D., 2011. Crystals, X-rays, and Proteins: Comprehensive Protein Crystallography. Oxford University Press, New York.

Sloley, S., Smith, S., Algeciras, M., Cavett, V., Busby, J.A., London, S., et al., 2007a. Proteomic analyses of songbird (Zebra finch; *Taeniopygia guttata*) retina. J. Proteome Res. 6, 1093–1100.

Sloley, S., Smith, S., Gandhi, S., Busby, J.A., London, S., Luksch, H., et al., 2007b. Proteomic analyses of zebra finch optic tectum and comparative histochemistry. J. Proteome Res. 6, 2341–2350.

Sokale, A., Peebles, E.D., Zhai, W., Pendarvis, K., Burgess, S., Pechan, T., 2011. Proteome profile of the pipping muscle in broiler embryos. Proteomics 11, 4262–4265.

Son, J.H., Kawamata, H., Yoo, M.S., Kim, D.J., Lee, Y.K., Kim, S., et al., 2005. Neurotoxicity and behavioral deficits associated with Septin 5 accumulation in dopaminergic neurons. J. Neurochem. 94, 1040–1053.

Soulet, F., Kilarski, W.W., Roux-Dalvai, F., Herbert, J.M., Sacewicz, I., Mouton-Barbosa, E., et al., 2013. Mapping the extracellular and membrane proteome associated with the vasculature and the stroma in the embryo. Mol. Cell. Proteomics.

Steen, H., Mann, M., 2004. The ABC's (and XYZ's) of peptide sequencing. Nat. Rev. Mol. Cell Biol. 5, 699–711.

Stocks, B.B., Konermann, L., 2009. Structural characterization of short-lived protein unfolding intermediates by laser-induced oxidative labeling and mass spectrometry. Anal. Chem. 81, 20–27.

Stoddard, M.C., Prum, R.O., 2011. How colorful are birds? Evolution of the avian plumage color gamut. Behav. Ecol. 22, 1042–1052.

Sun, J.S., Hahn, T.W., 2012. Comparative proteomic analysis of *Salmonella enterica* serovars Enteritidis, Typhimurium and Gallinarum. J. Vet. Med. Sci. 74, 285–291.

Teltathum, T., Mekchay, S., 2009. Proteome changes in Thai indigenous chicken muscle during growth period. Int. J. Biol. Sci. 5, 679–685.

Thanthrige-Don, N., Abdul-Careem, M.F., Shack, L.A., Burgess, S.C., Sharif, S., 2009. Analyses of the spleen proteome of chickens infected with Marek's disease virus. Virology 390, 356–367.

Thanthrige-Don, N., Parvizi, P., Sarson, A.J., Shack, L.A., Burgess, S.C., Sharif, S., 2010. Proteomic analysis of host responses to Marek's disease virus infection in spleens of genetically resistant and susceptible chickens. Dev. Comp. Immunol. 34, 699–704.

Unlu, M., Morgan, M.E., Minden, J.S., 1997. Difference gel electrophoresis: a single gel method for detecting changes in protein extracts. Electrophoresis 18, 2071–2077.

Velculescu, V.E., Zhang, L., Zhou, W., Vogelstein, J., Basrai, M.A., Bassett Jr., D.E., et al., 1997. Characterization of the yeast transcriptome. Cell 88, 243–251.

Vincent, P., Collette, Y., Marignier, R., Vuaillat, C., Rogemond, V., Davoust, N., et al., 2005. A role for the neuronal protein collapsin response mediator protein 2 in T lymphocyte polarization and migration. J. Immunol. 175, 7650–7660.

Wales, T.E., Engen, J.R., 2006. Hydrogen exchange mass spectrometry for the analysis of protein dynamics. Mass Spectrom. Rev. 25, 158–170.

Walsh, C.T., 2006. Posttranslational Modification of Proteins: Expanding Nature's Inventory. Roberts and Company Publishers, Greenwood Village, CO.

Wang, S., Regnier, F.E., 2001. Proteomics based on selecting and quantifying cysteine containing peptides by covalent chromatography. J. Chromatogr. A 924, 345–357.

Wang, S., Zhang, X., Regnier, F.E., 2002. Quantitative proteomics strategy involving the selection of peptides containing both cysteine and histidine from tryptic digests of cell lysates. J. Chromatogr. A 949, 153–162.

Wang, Y., Guo, Y., Ning, D., Peng, Y., Cai, H., Tan, J., et al., 2012. Changes of hepatic biochemical parameters and proteomics in broilers with cold-induced ascites. J. Anim. Sci. Biotechnol. 3, 41.

Warren, W.C., Clayton, D.F., Ellegren, H., Arnold, A.P., Hillier, L.W., Kunstner, A., et al., 2010. The genome of a songbird. Nature 464, 757–762.

Washburn, M.P., Wolters, D., Yates 3rd, J.R., 2001. Large-scale analysis of the yeast proteome by multidimensional protein identification technology. Nat. Biotechnol. 19, 242–247.

Wilkins, M.R., Pasquali, C., Appel, R.D., Ou, K., Golaz, O., Sanchez, J.C., et al., 1996. From proteins to proteomes: large scale protein identification by two-dimensional electrophoresis and amino acid analysis. Biotechnology (N Y) 14, 61–65.

Wuthrich, K., 1990. Protein structure determination in solution by NMR spectroscopy. J. Biol. Chem. 265, 22059–22062.

Xie, F., London, S.E., Southey, B.R., Annangudi, S.P., Amare, A., Rodriguez-Zas, S.L., et al., 2010. The zebra finch neuropeptidome: prediction, detection and expression. BMC Biol. 8, 28.

Yates 3rd, J.R., 1998. Mass spectrometry and the age of the proteome. J. Mass Spectrom. 33, 1–19.

Yates, J.R., Ruse, C.I., Nakorchevsky, A., 2009. Proteomics by mass spectrometry: approaches, advances, and applications. Annu. Rev. Biomed. Eng. 11, 49–79.

Young, M.M., Tang, N., Hempel, J.C., Oshiro, C.M., Taylor, E.W., Kuntz, I.D., et al., 2000. High throughput protein fold identification by using experimental constraints derived from intramolecular cross-links and mass spectrometry. Proc. Natl. Acad. Sci. U.S.A. 97, 5802–5806.

Zhai, W., Araujo, L.F., Burgess, S.C., Cooksey, A.M., Pendarvis, K., Mercier, Y., et al., 2012. Protein expression in pectoral skeletal muscle of chickens as influenced by dietary methionine. Poult. Sci. 91, 2548–2555.

Zheng, A., Liu, G., Zhang, Y., Hou, S., Chang, W., Zhang, S., et al., 2012. Proteomic analysis of liver development of lean Pekin duck (*Anas platyrhynchos domestica*). J. Proteomics 75, 5396–5413.

Zou, W., Ke, J., Zhang, A., Zhou, M., Liao, Y., Zhu, J., et al., 2010. Proteomics analysis of differential expression of chicken brain tissue proteins in response to the neurovirulent H5N1 avian influenza virus infection. J. Proteome Res. 9, 3789–3798.

Chapter 4

Mitochondrial Physiology

Walter Bottje

Department of Poultry Science, Division of Agriculture, University of Arkansas, Fayetteville, AR, USA

4.1 MITOCHONDRIA: AN INTRODUCTION

4.1.1 Overview

Mitochondria generate 90% of the energy within a cell in the form of adenosine triphosphate (ATP) by oxidative phosphorylation, thus earning them the title of "powerhouse of the cell." The process of oxidative phosphorylation from the respiratory or electron transport chain (ETC) activity was first reported by Kennedy and Lehninger (1949). Energy production, however, is just one of many roles orchestrated by mitochondria. Mitochondria are the only organelle outside the nucleus with a discrete pool of DNA (mitochondrial DNA, or mtDNA). This distinction lead to the accepted theory of an endosymbiotic origin of mitochondria, in which an α-proteobacteria took up a commensal residence within a eukaryotic cell, with more recent evidence of co-evolution of an extant eukaryotic cell (Gray et al., 1999).

According to Lehninger (1965), Rudolf Albert van Kolliker, a Swiss cytologist, first described mitochondria in 1857 and gave them the name of sarcosomes, as they have a distinct granular structure surrounded by a membrane. Later, Benda (1898) renamed the structure mitochondrion—a derivation from Greek for thread (*mitos*) and grain (*chondrion*); this has been the standard name for the organelle since the 1930s (Lehninger, 1965). The synthesis and import of nuclear (n) encoded proteins, which represent 98% of all mitochondrial protein, is tightly coordinated with the synthesis of mtDNA-encoded proteins; this is followed by the coordinated assembly needed for a fully functional mitochondria. Mitochondria play a vital role in programmed cell death (apoptosis), and mitochondrial-generated reactive oxygen species (ROS) make it a major site of endogenous oxidative stress. Mitochondria are dynamic organelles that change morphology and composition in response to physiological signals, such as variations in nutrition, oxygen levels, and metabolic demand (Aw and Jones, 1989). Mitochondria also undergo fission processes associated with mitochondrial biogenesis when additional energy is needed.

4.1.2 Physical Description

Under electron microscopy, mitochondria appear bean-shaped with striations that are visible due to folding of the inner mitochondrial membrane, called cristae, which is where the ETC is located. Electron tomography revealed mitochondria as long tube-like structures that weave throughout the cytosol (Mannella, 2000). Parts of the mitochondrial membrane are contiguous with the endoplasmic and sarcoplasmic reticula, that facilitate the shuttling of molecules such as ATP and adenosine diphosphate (ADP) between the mitochondria and cytosol (Scheffler, 1999; Sharma et al., 2000). Mitochondria have an inner membrane that surrounds the mitochondrial matrix and an outer membrane that encloses an intramembranous space (Figures 4.1 and 4.2). Cytochrome *c*, located in the intramembranous space, is critical for cellular respiration and for initiating normal cell turnover (apoptosis). Most mitochondrial proteins are found in the matrix and are associated with the Krebs cycle, β-oxidation of fats, synthesis of heme proteins, and the iron-sulfur proteins prevalent in the ETC. Mitochondrial DNA is also present in the matrix. The outer mitochondrial membrane is a simple phospholipid bilayer, whereas the inner mitochondrial membrane is highly convoluted, forming cristae that greatly increase its surface area. Due to the presence of the multiprotein complexes of the ETC, the inner mitochondrial membrane contains 70% protein and 30% lipid, compared to the equal ratio of proteins to lipids that is typically found in membranes. The inner mitochondrial membrane also contains cardiolipin, a unique lipid found primarily in mitochondria (Hatefi, 1985).

4.1.3 Mitochondrial and Nuclear DNA Interaction for Assembly and Function

Mitochondrial DNA, a circular molecule with over 16,000 base pairs, contains roughly 37 genes that code for two ribosomal RNAs, 22 transfer RNAs, and 13 proteins, that combine with over 70 other n-encoded proteins to form the respiratory chain (Anderson et al., 1981). Transcription, translation, and mtDNA replication, including synthesis of ribosomal proteins,

Sturkie's Avian Physiology.

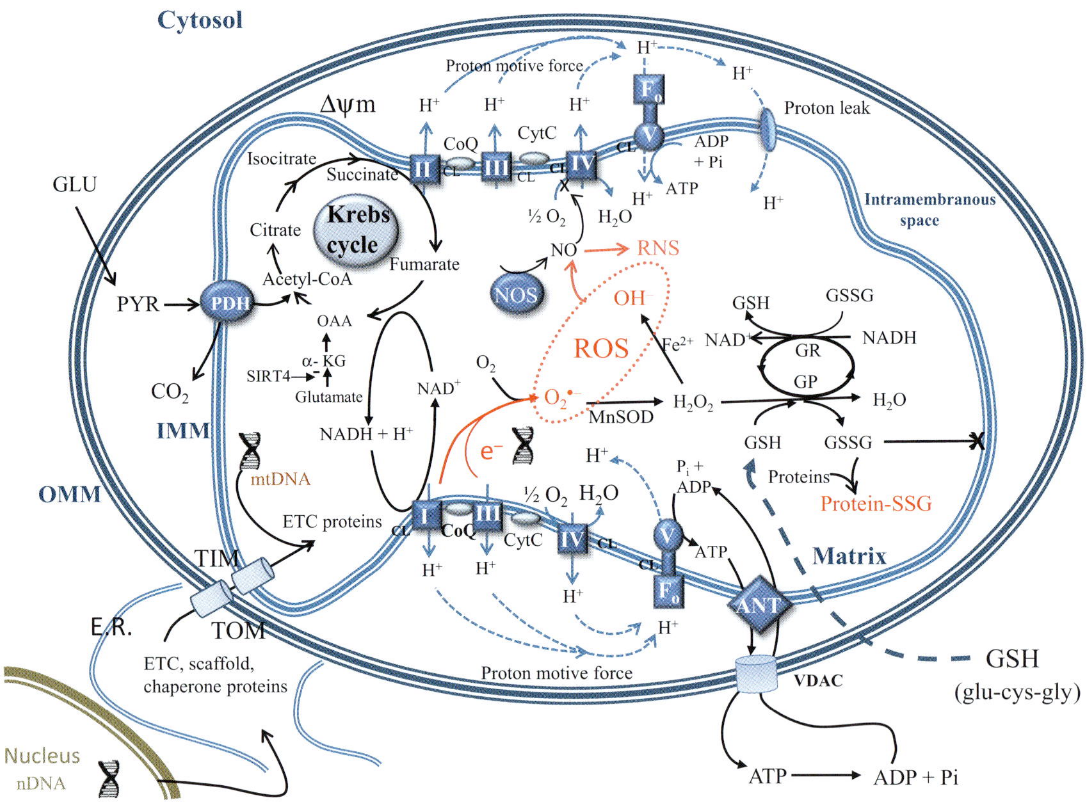

FIGURE 4.1 **Overview of mitochondria.** A stylized mitochondria showing some mitochondrial processes discussed in the text. The mitochondria have both an outer (OMM) and inner (IMM) mitochondrial membrane with the electron transport chain (ETC) (I, II, III, IV, V) located on the IMM. The ETC towards the top shows electrons moving from succinate, an intermediate of the Krebs cycle, to complexes II–IV. The ETC on the bottom shows NADH-linked energy substrates with electrons entering the ETC at complex I. Electrons are passed between complex II and III, and complex I and III, by coenzyme Q (CoQ). Cytochrome *c* (cyt *c*) shuttles electrons from complex III to complex IV. The movement of electrons down the respiratory chain is accompanied by pumping of protons (H^+) in the intramembranous space, sets up a proton motive force that drives adenosine triphosphate (ATP) synthesis when protons flow through ATP synthase (complex V). ATP is transported out of the mitochondria for use by the cell through the adenine nucleotide translocase (ANT) on the inner membrane and the voltage-dependent anion channel (VDAC) on the outer membrane. A mitochondrial nitric oxide synthase (NOS) produces nitric oxide (NO), that competes with oxygen for the active site on cytochrome *c* oxidase. Protons may also move through the membrane at sites other than the ATP synthase in a process called proton leak. Proton leak dissipates the proton motive force without synthesis of ATP but can also attenuate formation of reactive oxygen species (ROS). Electrons (e^-) that leak from the ETC can react with oxygen to form superoxide (O_2^-), which is normally converted to hydrogen peroxide (H_2O_2) by manganese superoxide dismutase (MnSOD). In the presence of free metal ions, H_2O_2 can be converted to the highly reactive hydroxyl radical (OH^-). Collectively, superoxide, H_2O_2, and OH^- are ROS and can cause oxidative damage to cellular structures (e.g., proteins, lipids, DNA). ROS can react with NO to produce reactive nitrogen species (RNS), that also can damage these structures. Glutathione (GSH) is an important endogenous antioxidant that is imported from the cytosol into the mitochondria. The active thiol in GSH is used to reduce lipid peroxides or H_2O_2 to water or lipid alcohols with the concomitant formation of oxidized glutathione (GSSG); it can be recycled to GSH by glutathione reductase (GR), which utilizes reducing equivalents from NADH. Unlike cells, mitochondria cannot export GSSG, and elevations in GSSG in mitochondria that lead to protein disulfides (protein-SSG) formation. This can be particularly detrimental to ETC activity due to the presence of reactive thiol groups in these proteins. The ETC is comprised of nuclear and mitochondrially encoded (mtDNA) proteins. The nuclear-encoded proteins must be transported into the mitochondria, which is facilitated by both outer membrane translocase (TOM) and inner membrane translocase (TIM) proteins. *This figure was adapted from Wallace (1999).*

are all under nuclear regulation. Consequently, mitochondrial function depends upon the tightly coordinated interaction between nDNA and mtDNA-encoded proteins, protein assembly factors, and chaperone proteins involved in protein folding, protein scaffolding, and structural support (Nijtmans et al., 2002; Rabilloud et al., 2002; Ryan and Hoogenraad, 2007).

Nuclear-encoded proteins destined for the mitochondria must be unfolded prior to transport through the outer and inner membrane translocase protein channels (Ryan and Hoogenraad, 2007). After transit through these channels, the proteins are refolded within the mitochondria by chaperone proteins, such as heat shock protein (Hsp) 70, Hsp 60/10 (also called chaperonin 60/10), Hsp 78, and a number of proteases. Chaperone protein expression increases during stress (e.g., heat, oxidative, toxin-mediated), when they are particularly important in repairing damaged proteins.

The D loop of mtDNA contains regulators of mitochondrial transcription and replication. Differences in D-loop

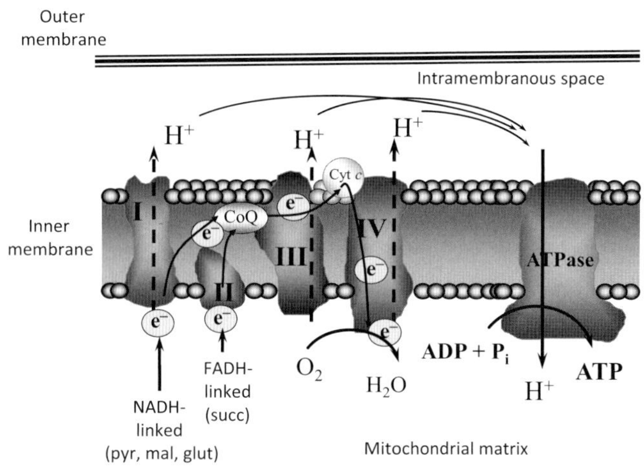

FIGURE 4.2 **Diagrammatic representation of the mitochondrial electron transport chain (ETC) (based on Lehninger et al., 1993).** The respiratory chain consists of five multiprotein complexes (complex I, II, III, IV, and V). Electrons (e^-) enter the ETC either through complex I from NADH-linked energy substrates (e.g., pyruvate, malate, glutamate), or at complex II from FADH-linked substrates (e.g., succinate). The electrons are passed down the ETC (solid arrows) to the terminal electron acceptor, oxygen that is reduced to water. Coenzyme Q (CoQ, ubiquinone) is responsible for the transfer of electrons from complex I and II to complex III. Associated with the movement of electrons along the ETC is the movement of protons (H^+, dashed arrows) from the mitochondrial matrix into the intramembranous space, setting up a proton motive force. The movement of protons through the adenosine triphosphate (ATP) synthase (complex V) provides the energy to support ATP synthesis.

mtDNA have been used to identify animals with different mitochondrial types. Some proteins are part of the mitochondrial import machinery, whereas others are needed for expression of the mitochondrial genome and metabolism. Other proteins are required for mitochondrial roles in apoptosis (Liu and Kitsis, 1996), redox cell signaling, and homeostasis (Bogoyevitch et al., 2000; Levonen et al., 2001; Droge, 2002). Rabilloud et al. (2002) indicated that "mitochondrial function in general, and mitochondrial protein synthesis in particular, depend on the conjugated and coordinated expression of both mitochondrial and nuclear genomes." A complex communication network between mitochondria and the nucleus also exists to coordinate mitochondrial biogenesis and function (Poyton and McEwen, 1996).

4.1.4 Mitochondrial Fusion and Fission

Mitochondria are not discrete or static structures. Rather, they are part of a network that constantly undergoes complex fission and fusion processes, thus enabling mitochondria to communicate and "form into local and widespread mitochondrial syncytia within cells" (Hoppins et al., 2007). Mitochondrial fission and fusion processes are controlled by highly conserved dynamin-related proteins—large GTPase proteins that regulate many membrane-associated activities (Praefcke and McMahon, 2004). Mitochondrial fusion ensures the distribution of mtDNA to maintain functionally competent mitochondria throughout a cell. In contrast, mitochondrial fission ensures that competent mitochondria are distributed equally during cell division and mitochondrial biogenesis. Mitochondrial fission and fusion processes may also play roles in promoting or delaying apoptosis, respectively (Hoppins et al., 2007).

4.1.5 The Respiratory Chain and ATP Synthesis

The ETC, first described by Kennedy and Lehninger (1949), consists of five multiprotein enzyme complexes: complex I (nicotinamide adenine dinucleotide (NADH): ubiquinone oxido-reductase), complex II (succinate: ubiquinone reductase), complex III (ubiquinol: cytochrome *c* oxido-reductase), complex IV (cytochrome *c* oxidase), and the F_1F_0 ATP synthase or ATPase (complex V), and two mobile electron carriers, ubiquinone (Q) and cytochrome *c* (cyt *c*) (Figures 4.1 and 4.2). Electrons enter the respiratory chain at complex I for NADH-linked substrates (e.g., malate and pyruvate) or at complex II for succinate, an $FADH_2$-linked substrate (Figure 4.2). Ubiquinone carries electrons from complex I and II to complex III, whereas cyt *c* shuttles electrons from complex III to complex IV. Electron transfer to O_2 (the terminal acceptor) results in full reduction of O_2 to water.

Electron movement coincides with proton pumping to establish a proton motive force consisting of a membrane potential ($\Delta\psi_m$) and pH (proton) gradient, that provide energy for ATP synthesis as protons flow back into the matrix through ATP synthase (complex V). Protons can also cross at sites other than the ATP synthase due to anion carrier proteins (e.g., adenine nucleotide transporter, glutamate transporter), uncoupling proteins (UCPs), and intrinsic membrane characteristics (Brown and Brand, 1991; Brand et al., 1994, 2005; Rolfe and Brand, 1997; Brookes et al., 1997). Proton leak therefore consumes O_2 and dissipates the membrane potential without ATP synthesis. Uncoupling represents an inefficiency of mitochondrial function, but it minimizes ROS production (see below). Although mitochondrial uncoupling is important in heat generation in brown adipose fat tissue in mammals, brown adipose tissue has not been reported in any avian species; it appears to have been lost early in evolution of the avian lineage from a common ancestor of birds and mammals (Mezentseva et al., 2008).

4.1.5.1 *Ubiquinone (Coenzyme Q)*

Electron transfer from complex I to complex III and from complex II to complex III is carried out by ubiquinone (coenzyme Q; e.g., CoQ_9 and CoQ_{10}). Auto-oxidation of CoQ is a major source of mitochondrial ROS production (Chance et al., 1979; Turrens et al., 1985; Turrens and Boveris, 1980). Animals with relatively more CoQ_{10} had lower mitochondrial ROS production than those with higher CoQ_9 levels (Lass and Sohal, 1999), and CoQ content is highly correlated with complex I and II activities (Ernster and Forsmark-Andree, 1993; Forsmark-Andree et al., 1997).

4.1.5.2 *Cardiolipin*

Cardiolipin (tetra-acyl-diphophatidyl-glycerol) is a unique phosphoglyceride with four long-chain fatty acids (compared to two side chains in typical phospholipids), and is essential for membranes involved in coupled (oxidative) phosphorylation (Hoch, 1992). Full activity requires the interaction of each complex (I–V) with cardiolipin. Yeast lacking cardiolipin exhibited impaired mitochondrial function (Koshkin and Greenberg, 2000). Exogenously added cardiolipin depressed respiratory chain coupling but increased ATP synthase activity in isolated liver mitochondria (Bobyleva et al., 1997).

4.1.6 Assessing Mitochondrial Function

4.1.6.1 *Polarographic Method*

A standard method of assessing mitochondrial function uses an O_2 electrode in a Warburg apparatus to measure O_2 consumption with different respiratory states in freshly isolated mitochondria (Estabrook, 1967). In the presence of NADH- and FADH-linked energy substrates, mitochondria exhibit an initial slow rate of O_2 consumption (state 2 respiration). The addition of ADP stimulates ETC activity and initiates rapid O_2 consumption, that is followed by a slower rate of O_2 consumption (state 4 respiration) when ADP levels decline (i.e., ADP is limiting) due to oxidative phosphorylation and synthesis of ATP (from ADP and inorganic phosphorus). Functional indices calculated from these data include the respiratory control ratio (RCR) and ADP:O ratio (Estabrook, 1967). The RCR represents the degree of coupling or efficiency of respiratory chain activity and is calculated as state 3 (active respiration) divided by state 4 (resting) respiration rate. The ADP:O ratio is the amount of ADP phosphorylated per nanoatom of monomeric oxygen consumed during state 3 respiration and is an index of oxidative phosphorylation. Electron movement down the transport chain is coupled to proton pumping, setting the proton motive force that synthesizes ATP as protons flow through the F_1F_0 ATPase. The ADP:O ratio, an index of oxidative phosphorylation, is determined as ADP added per nmol monomeric oxygen consumed in state 3 respiration. Theoretical ADP:O ratios are 2 (for succinate) and 3 (for malate), which enter at complex I and II, respectively.

ATP synthesis is not 100% efficient, due in part to electron and proton leak. Decreases in the ADP:O ratio (increased O_2 use that is uncoupled from ATP synthesis) occur by proton leakage across the inner mitochondrial membrane at sites other than the F_1F_0 ATPase (Brand et al., 1994) or by electron leakage from the respiratory chain that reacts with O_2 to form ROS, such as superoxide and H_2O_2 (Chance et al., 1979; Boveris and Chance, 1973).

4.1.6.2 *Measurement of Oxygen Flux (Flux Analysis) in Intact Cells*

A relatively new approach to assess mitochondrial function in intact cells using flux analysis of the oxygen consumption rate (OCR) was reported by Wu et al. (2007). The advantage of this method is that mitochondrial function can be assessed within intact cells, which eliminates any artifacts (e.g., shear stress) that might be introduced from the stirring that is required in the polarographic method (Estabrook, 1967). Assessment of glycolytic activity can also be determined simultaneously by assessment of extracellular acidification rate.

4.1.7 Mitochondrial Role in Apoptosis

Mitochondria are critical in initiating programmed cell death or apoptosis (Wallace, 1999). Sandwiched between the inner and outer mitochondrial membranes are cytochrome *c*, apoptosis-inducing factor, and caspases (proteases) that contribute to apoptosis. Apoptosis is initiated by formation of the mitochondrial permeability transition pore (mtPTP) on the inner membrane. The mtPTP forms by the coalescing of the voltage-dependent anion channel (VDAC), adenine nucleotide translocator (ANT), BCL-2-associated X protein, and cyclophilin D. When the mtPTP is formed, the mitochondrial membrane potential is dissipated, followed shortly thereafter by mitochondrial swelling and release of caspases and apoptosis-initiating factor. The proteolytic caspases released into the cytoplasm degrade the cytoskeletal architecture. Events that trigger the opening of the mtPTP include a decrease in energetic capacity of the mitochondria, excessive influx of ionic calcium, and increased ROS generation.

4.2 MITOCHONDRIAL INEFFICIENCIES

4.2.1 Electron Transport Defects and Oxidative Stress

4.2.1.1 Reactive Oxygen Species

Mitochondria are a major source of endogenous oxidative stress. Approximately 2–4% of O_2 used by mitochondria may be converted to ROS by univalent reduction of O_2 to form superoxide ($O_2^{\bullet -}$) following electron (e^-) leak from the respiratory chain (Chance et al., 1979; Turrens and Boveris, 1980; Boveris and Chance, 1973). Superoxide dismutase (SOD) converts $O_2^{\bullet -}$ to H_2O_2 that is reduced to H_2O by glutathione peroxidase (GPx). The relatively nonreactive H_2O_2 is converted to the highly reactive hydroxyl radical ($^{\bullet}OH$) in the presence of Fe^{2+} and Cu^{2+}; due to its lipid solubility, it is able to cross membranes and oxidize proteins, DNA, lipids, and carbohydrates throughout the cell (Yu, 1994) (Figures 4.1 and 4.3).

4.2.1.2 Identification of Site-Specific Defects in Electron Transport

Figure 4.3 shows sites of e-leak that are accentuated by treatment with various chemical inhibitors. The use of these chemicals for identifying e^- transport defects were pioneered by Boveris and co-workers (Chance et al., 1979; Turrens and Boveris, 1980; Boveris and Chance, 1973). Numerous reports followed that confirmed complex I and III as predominant sites of mitochondrial e^- leak, which is associated with numerous metabolic conditions in humans, including Alzheimer disease, cancer, diabetes, and aging (Yu, 1994).

4.2.1.3 Mitochondrial ROS Generation in Avian Species

Chemical inhibitors (Figure 4.3) have been used to assess mitochondrial ROS production in birds. In broilers, site-specific defects in e^- leak at complex I and III were identified in liver, lung, skeletal, and cardiac mitochondria obtained from birds exhibiting fulminant pulmonary hypertension syndrome (Cawthon et al., 2001; Iqbal et al., 2001; Tang et al., 2002). Greater ROS production due to site-specific defects in e^- transport was also identified in complex I and III skeletal muscle, liver, and duodenal mitochondria associated with low feed efficiency in broilers (Bottje et al., 2002; Iqbal et al., 2004;

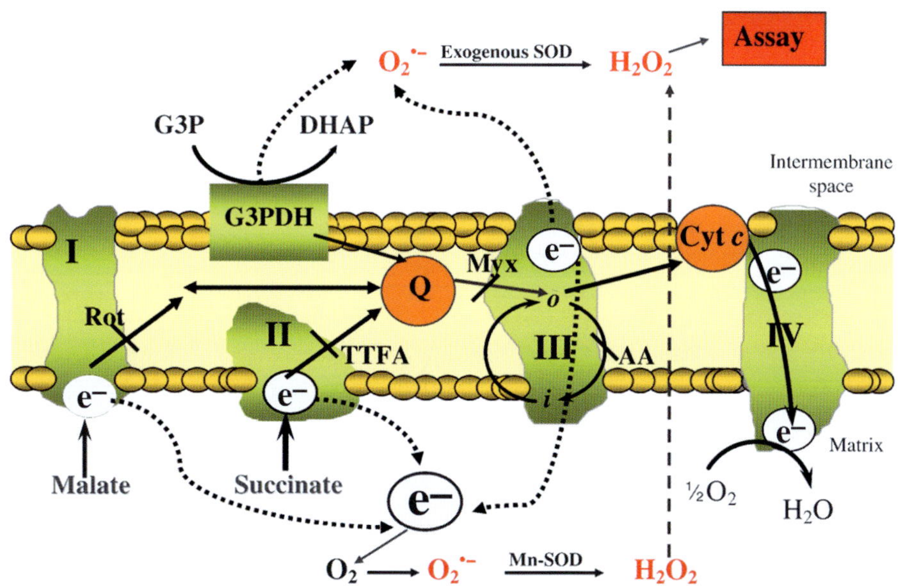

FIGURE 4.3 Diagrammatic representation of identification of site-specific defects in the electron transport chain using chemical inhibitors. The movement of electrons (e^-) along the respiratory chain are shown by solid arrows from complex I or complex II to complex III by the e^- carrier, coenzyme Q (Q) and to complex IV by cytochrome *c* (cyt *c*). The terminal step of electron transport is the full reduction of oxygen (O_2) to water by cytochrome *c* oxidase. Chemical inhibitors of to block e^- transport and identify site-specific defects are rotenone (Rot) at complex I, thenotrifluoroacetone (TTFA) at complex II, and myxothiazol (Myx) (at the outer membrane (*o*)) and antimycin A(AA) (at the inner membrane (*i*)) within complex III. If a site-specific defect exists at any of these sites following chemical inhibition, e^- leak (dotted arrows) results in the univalent reduction of O_2 to superoxide ($O^{\bullet -}$), which is reduced to hydrogen peroxide (H_2O_2) by superoxide dismutase (SOD). Topology of H_2O_2 formation (as in index of reactive oxygen species, ROS) can be determined by adding exogenous SOD to the media in isolated mitochondrial preparations that distinguishes it from endogenous ROS, which is by the mitochondrial located MnSOD (see Miwa et al., 2003). Extramitochondrial e^- leak can also occur when glyceraldehyde 3-phosphate (GSP) is converted by G3P dehydrogenase (G3PDH) to dihyroxacetone phosphate (DHAP). *Reprinted with permission from Ojano-Dirain et al. (2007a).*

Ojano-Dirain et al., 2004, 2007a) as well as complex II in duodenal tissue (Ojano-Dirain et al., 2004). Greater ROS production was likely responsible for higher oxidative stress and lower respiratory chain complex activities consistently observed in animals exhibiting a low feed efficiency phenotype (Ojano-Dirain et al., 2007b; Bottje and Carstens, 2009) and may have been involved in differential gene expression in feed efficiency in broilers (Bottje and Kong, 2013). Mitochondrial ROS production has also been shown to play a role in heat stress (Abe et al., 2006; Mujahid et al., 2006, 2007a,b).

4.2.1.4 Mitochondrial ROS in Normal Cell Function

Although high levels of mitochondrial ROS are detrimental, low levels of mitochondrial ROS are recognized as being vital for normal cell function, acting as second messengers in signal transduction processes (Giulivi and Oursler, 2003; Crawford et al., 1997; Carper et al., 1999; Greiber et al., 2002; Li et al., 2002; Kemp et al., 2003). Baughman and Mootha (2006) hypothesized that there is "a homeostatic role for ROS in maintaining stable respiratory phenotypes across genetic variants of the mitochondrial genome."

4.2.1.5 Mitochondrial ROS and Longevity

Mitochondrial ROS production is generally lower in avian species compared to comparable-sized mammalian species, with an inverse relationship between longevity and mitochondrial ROS production (Herrero and Barja, 1997, 1998). Figure 4.4 presents data of heart mitochondrial ROS production in comparable sized mammals and birds (Herrero and Barja, 1997, 1998) in the presence or absence of ETC inhibitors (Figure 4.3). This species difference is even more remarkable because birds have a number of characteristics that should favor heightened mitochondrial radical production, such as higher body temperatures, metabolic rates, and blood glucose concentrations (Holmes and Austad, 1995; Holmes et al., 2001). The fact that birds have much lower mitochondrial ROS production has been hypothesized to explain lower mitochondrial DNA diversity in avian species compared to mammalian species (Hickey, 2008).

4.2.1.6 Nitric Oxide and Reactive Nitrogen Species

Nitric oxide (NO) produced by mitochondrial nitric oxide synthase (NOS) near the site of the ETC (Giulivi et al., 1998; Giulivi and Oursler, 2003) competitively inhibits cytochrome oxidase, thus regulating mitochondrial O_2 consumption. The release of NO in the presence of ROS can produce a large number of reactive nitrogen species (e.g., peroxynitrite) that damage proteins by nitrosylation. Peroxynitrite was reported to be responsible for decreased activities of complex I and II in mitochondria (Riobo et al., 2001).

4.2.1.7 Mitochondrial ROS: DNA Damage and Respiratory Chain Complex Activities

A balance of mtDNA- and nDNA-encoded proteins is needed for the functional integrity of mitochondria (Nijtmans et al., 2002). Due to its proximity to the respiratory chain and a lack of protective histones, mtDNA is more susceptible to mitochondrial ROS-mediated oxidation than nDNA, and mtDNA oxidation can lead to mitochondrial dysfunction (Kristal et al., 1994; Wei, 1998). Oxidant-medicated repression of

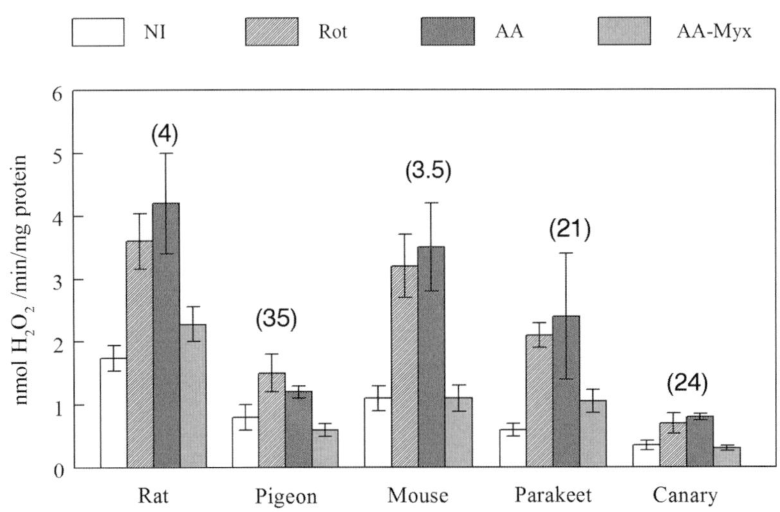

FIGURE 4.4 Hydrogen peroxide (H_2O_2) production rates (expressed as nmol/min/mg protein) in heart mitochondria isolated from mammals and birds of comparable body weights (rat versus pigeon; mouse versus parakeet and canary). Rates of H_2O_2 production are shown for mitochondria treated with no-inhibitor (NI, basal rate), and for mitochondria treated with inhibitors of complex I (rotenone, Rot), and two inhibitors of complex III (antimycin A, AA and myxothiazol, Myx, alone and in combination). The maximum life span for each animal species is shown in parentheses. *Data was obtained from Herrero and Barja (1997, 1998).*

mitochondrial transcription exacerbates dysfunction by inhibiting respiratory protein synthesis (Kristal et al., 1997). Restricted availability of mt-encoded subunits or damaged proteins can lead to diminished complex activities and cell respiration (Wallace, 1999). There are also specific thiols in proteins of complexes I, II, and IV that are particularly susceptible to oxidation and their oxidation leads to decreased complex activity upon exposure to oxidants (Lin et al., 2002b). Inverse relationships between oxidative stress and complex activity were noted in animals with low feed efficiency (Ojano-Dirain et al., 2005; Bottje and Carstens, 2009).

4.2.2 Antioxidants

It is difficult to discuss oxidation without including antioxidant protection. Oxidative stress is unavoidable in eukaryotic organisms and occurs during normal metabolism, primarily within mitochondria. Oxidative stress develops when the generation of ROS overwhelms antioxidant protection (Yu, 1994). Repair of damaged structures (e.g., lipids, proteins) is energetically expensive, requiring considerable input of ATP to either repair or recycle materials within the cell. Mitochondrial ROS are normally metabolized by the enzymatic antioxidants SOD and glutathione peroxidase, as well as by nonenzymatic antioxidants GSH and α-tocopherol (Yu, 1994).

Glutathione is the major endogenous antioxidant system both in the cytosol and within mitochondria (Meister, 1984; Griffith and Meister, 1985; Martensson et al., 1993). Glutathione exists in either a reduced (GSH) or oxidized (GSSG; glutathione disulfide) form; the GSSG:GSH ratio is used as an indicator of oxidative stress. The glutathione reduction-oxidation (redox) system consists of GSH and the GSH recycling enzymes, GSH peroxidase (GPx), and GSH reductase (GR) (Meister, 1984). The enzyme GPx metabolizes peroxides (e.g., H_2O_2), using reducing equivalents from GSH, and catalyzes the reaction shown in Eqn (4.1) (see Figure 4.1 also):

$$2GSH + H_2O_2 \rightarrow GSSG + 2H_2O \qquad (4.1)$$

Low levels of GSSG are maintained by GR, which uses nicotinamide adenine dinucleotide phosphate ($NADPH_2$) to reduce GSSG to GSH, as shown in Eqn (4.2):

$$GSSG + NADPH_2 \rightarrow 2GSH + NADP^+ \qquad (4.2)$$

The glutathione redox system is a vital defense mechanism of mitochondria against free radical damage because mitochondria lack catalase (Martensson et al., 1990), γ-glutamyl synthetase (the rate-limiting enzyme in GSH synthesis) (Meister, 1984), and the ability to export GSSG (Olafsdottir and Reed, 1988). Martensson et al. (1990) noted that the mitochondrial GSH transport system might be designed to efficiently conserve mitochondrial GSH at the expense of cytosolic GSH.

During oxidative stress, toxic levels of GSSG can accumulate within mitochondria (Olafsdottir and Reed, 1988; Cawthon et al., 1999). Unlike cells, mitochondria are unable to export GSSG from the mitochondria, leading to thiolation of critical proteins in the ETC and diminished activities for the respiratory chain complexes. Augustin et al. (1997) reported that mitochondria are not damaged by ROS as long as the mitochondria are in an energized state in which ROS production is minimized. Other research indicates that GSH levels are critical in maintaining or protecting respiratory chain complex activity from oxidation, and positive correlations between GSH and respiratory chain activity have been reported (Bolanos et al., 1996; Cardoso et al., 1999; Ojano-Dirain et al., 2005).

4.2.3 Mitochondrial Uncoupling and Attenuation of Oxidative Stress

Proton motive force provides the power that drives ATP synthesis when protons flow back into the matrix through ATP synthase (complex V). However, protons that flow into the mitochondrial matrix at sites other than the ATP synthase (proton leak) dissipate proton motive force, diminish mitochondrial membrane potential, and short-circuit ATP synthesis (Brand et al., 1994). Oxygen consumption from proton leak can represent 25% of total basal metabolic rate in animals (Rolfe and Brand, 1997). Basal proton leak is facilitated by the intrinsic characteristics of membranes and the presence of intramembranous proteins (e.g., UCPs, ANT) (Dilger et al., 1979; Brown and Brand, 1991; Rolfe and Brand, 1997; Brookes et al., 1997, 1998; Brand et al., 2005). Free fatty acids enhance proton-translocating activities of ANT, UCPs, and phosphate and glutamate carrier proteins (Andreyev et al., 1988, 1989; Echtay et al., 2001; Samartsev et al., 1997; Jaburek et al., 1999). Proton leak is also stimulated or diminished by hyperthyroidism and hypothyroidism, respectively (Hafner et al., 1990).

Proton leak represents an energetic inefficiency, but it plays an important role in attenuating mitochondrial ROS production and endogenous oxidative stress. A self-limiting feedback of superoxide on mitochondrial ROS production (Skulachev, 1996, 1997) is due to increased expression and activity of UCPs and ANT (Echtay et al., 2002; Murphy et al., 2003; Brand et al., 2004). This mechanism is depicted in Figure 4.5, which is a composite of Figures 1 and 2 from Brand et al. (2004).

Uncoupling to reduce mitochondrial ROS generation has been clearly demonstrated in several avian species. In birds, the initial sequencing of the avian uncoupling protein (avUCP) was reported by Raimbault et al. (2001) and Toyomizu et al. (2002). Uncoupling of mitochondria represents an important physiological response to attenuate oxidative stress during both cold and heat stress

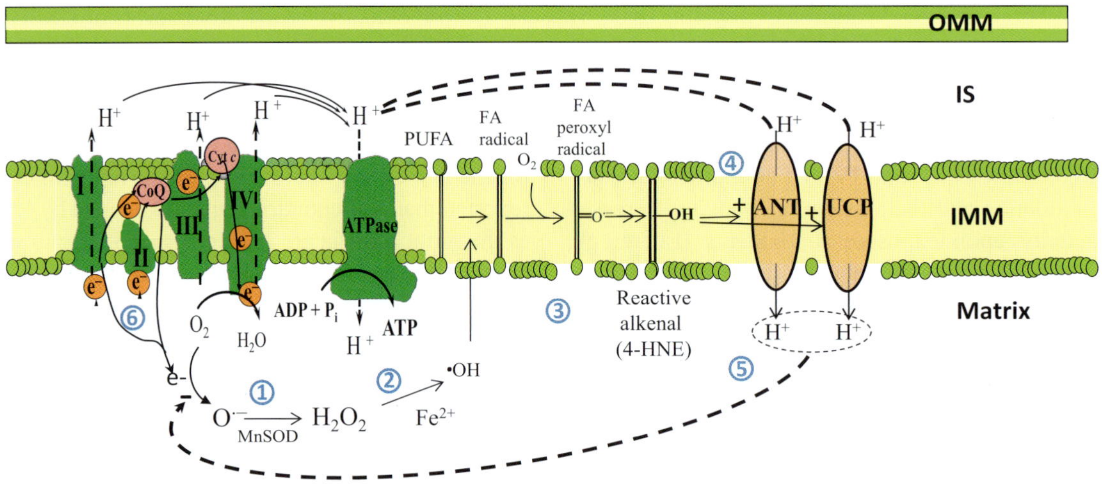

ROS-mediated uncoupling

① e^- leak increases superoxide ($O^{\cdot-}$) that is reduced to H_2O_2 MnSOD
② In the presence of Fe^{2+}, H_2O_2 is converted to hydroxyl radical (·OH)
③ OH initiates lipid peroxidation leading to formation of a stable reactive alkenal (4-HNE)
④ 4-HNE stimulates ANT and UCP activity and expression which
⑤ Increases proton leak that dissipates proton motive force that
⑥ Reduces e^- leak and ROS formation.

FIGURE 4.5 Attenuation of mitochondrial reactive oxygen species formation by increased uncoupling activity of adenine nucleotide translocator (ANT) and uncoupling protein (UCP). Shown in the figure are the outer and inner mitochondrial membranes (OMM and IMM) and intramembranous space (IS), within which proton motive force is generated by pumping of protons from the matrix into the IS. Proton leak occurs when protons move across the IMM at sites other than the adenosine triphosphate synthase. Electron leak from flavin and iron-sulfur centers in proteins of complex I and from coenzyme Q into the mitochondrial matrix cause univalent reduction of oxygen to superoxide ($O^{\cdot-}$). The presence of free iron (Fe^{2+}) that can be released from oxidative damage of matrix proteins (e.g., aconitase) results in the formation of a hydroxyl radical (·OH), which abstracts an electron from polyunsaturated fatty acid to form a carbon-centered fatty acid radical (FA radical) and an FA peroxyl radical in the presence of oxygen. This leads to the formation of a stable reactive alkenal, 4-hydroxy 2-nonenal (4-HNE), which stimulates uncoupling activity (proton leak) by ANT and UCP. Increased proton leak in turn dissipates proton motive force and mitochondrial membrane potential, which reduces electron leak and mitochondrial ROS formation. *Based on Echtay et al. (2002), Murphy et al. (2003), and Brand et al. (2004).*

conditions. Toyomizu et al. (2002) reported upregulation of both ANT and UCP mRNA in cold-stressed chicken skeletal muscle. Increased UCP and ANT expression has been observed in skeletal muscle of chickens (Toyomizu et al., 2002), king penguins (Talbot et al., 2003, 2004), and cold-acclimated ducks (Rey et al., 2010) that attenuates cold-induced increases in mitochondrial ROS production. Increased mitochondrial ROS and oxidative stress has also been observed in heat-stressed chicken skeletal muscle (Mujahid et al., 2007a,b). It was determined that olive oil attenuated mitochondrial ROS production during heat stress (Mujahid et al., 2009). The increased ROS production during heat stress was due to a combination of downregulation of avUCP expression and an increase in inner mitochondrial membrane potential (Mujahid et al., 2006; Kikusato and Toyomizu, 2013).

Differences in proton leak and mitochondrial membrane potential have also been linked to the phenotypic expression of feed efficiency in broilers (Ojano-Dirain et al., 2007a; Bottje et al., 2009). Enhanced ROS production was observed in muscle, liver, and duodenal mitochondria in broilers with low feed efficiency compared with broilers with high feed efficiency due to site-specific defects in electron transport (Ojano-Dirain et al., 2004; Bottje et al., 2004; Iqbal et al., 2005). Using a number of different chemical treatments, broilers with a high feed efficiency phenotype exhibited proton leak that was either lower or equal to, but never higher than, proton leak in broilers exhibiting a high feed efficiency phenotype (Bottje et al., 2009). Observations of increased (<0.06) expression of avUCP and lower membrane potential in low feed efficiency mitochondria (Ojano-Dirain et al., 2004, 2007c) are consistent with the model presented in Figure 4.5 to minimize or attenuate mitochondrial ROS formation.

4.3 MATCHING ENERGY PRODUCTION TO ENERGY NEED

4.3.1 Mitochondrial Biogenesis

Although mitochondria divide during mitosis to ensure daughter cells are provided a full complement of functional mitochondria, mitochondrial biogenesis is stimulated in response to increased energy demand. An early demonstration of mitochondrial biogenesis was provided by Paul and Sperling (1952), who observed that there were more mitochondria in breast muscle in pigeons that are more active than the relatively sedentary commercial chicken. One of the first factors identified to increase mitochondrial biogenesis was PGC-1α (Puigserver et al., 1998; Wu et al., 1999). Cold exposure increased the mRNA expression of PGC-1α, which in turn increased the expression of several mitochondrial proteins, including ATP synthase and cytochrome *c*-oxidases II and IV. PGC-1α and PGC-1β stimulate nuclear respiratory factors (NRF-1 and NRF-2) and mitochondrial transcription factor A (Nisoli et al., 2003, 2004) that upregulate mitochondrial transcription factor A (mtTFA) by independent mechanisms (Meirhaeghe et al., 2003; Lin et al., 2002a). NRF-1 and NRF-2 stimulate mitochondrial protein synthesis, such as ETC proteins, whereas mtTFA stimulates mitochondrial DNA transcription, that is instrumental in synthesis of mitochondrial proteins during mitochondrial biogenesis. Because of its role in mitochondrial biogenesis, PGC-1α has been termed the master regulator of mitochondrial protein synthesis (Nisoli et al., 2003, 2004).

4.3.2 AMP-Activated Protein Kinase

AMP-activated protein kinase (AMPK) is critical for sensing cellular energy (AMP/ATP) status and stimulating mitochondrial biogenesis (Zhou et al., 2001; Hardie et al., 2003; Hardie, 2007; Carling, 2005), as well as regulating animal food intake and overall energy balance (Minokoshi et al., 2004). Once AMPK is phosphorylated by serine–threonine kinase 11 (LKB1) (Hardie, 2005), the activated AMPK phosphorylates several proteins involved in carbohydrate, lipid, and protein metabolism (Kemp et al., 2003; Hardie, 2004, 2007). In general, AMPK reduces ATP-utilizing (anabolic) pathways (e.g., fatty acid synthesis) and increases ATP-generating (catabolic) pathways (e.g., fatty acid oxidation, glycolysis). AMPK is required for stimulating glucose uptake and glycolysis in skeletal muscle cells and astrocytes (Zhou et al., 2001; Almeida et al., 2004). AMPK also upregulates PGC-1α expression (Ojuka, 2004) and therefore presumably plays a role in mitochondrial biogenesis. In conjunction with thyroid hormone receptor activation, PGC-1α upregulates ANT and UCP3 expression, that uncouple mitochondrial oxidative phosphorylation (Masatoshi et al., 2005). Choi et al. (2001) presented evidence that AMPK and the AMPK cascade mechanisms are sensitive to ROS, particularly H_2O_2. Colombo and Moncada (2009) provided evidence that mitochondrial ROS-mediated upregulation of AMPK was associated with an increase of several cellular antioxidants. Thus, AMPK is very important in sensing energy status in cells and could be a pivotal component in growth and development; also, it is responsive to mitochondrial-generated ROS production. The expression of AMPK was higher in species exhibiting a high feed efficiency phenotype (Bottje and Kong, 2013).

4.3.3 Sirtuins

The sirtuins are a family of conserved NAD-dependent deacetylases that regulate many cellular activities including stress response and energy metabolism (Haigis and Sinclair, 2010). SIRT4 is located mainly in mitochondria (Haigis et al., 2006), where it inactivates glutamate dehydrogenase by ADP-ribosylation. The inactivation of glutamate dehydrogenase that converts glutamate to α-ketoglutarate indicates that SIRT4 regulates entry of energy substrates into the Krebs cycle. Knockdown of SIRT4 increased gene expression of mitochondrial and fatty acid metabolism enzymes in hepatocytes and myocytes, and changes in gene expression were associated with SIRT1-dependent fatty acid oxidation (Nasrin et al., 2010). Chau et al. (2010) reported that fibroblast growth factor 21 (FGF21) regulates energy homeostasis in adipocytes through phosphorylation and activation of AMPK by increasing cellular NAD^+ levels as well by deacetylation and activation of PGC-1α and histone 3. Activation of AMPK maintains energy balance by enhancing mitochondrial biogenesis and oxidative metabolism (Hardie, 2007). AMPK increased SIRT1 (NAD^+-dependent type III deacetylase sirtuin 1) by increasing NAD^+ levels that modulate several downstream SIRT1 targets (Canto et al., 2009). AMPK and SIRT1 act in concert with PGC-1α to regulate energy homeostasis in response to differences in nutritional and environmental factors (Reznick and Shulman, 2006; Hardie, 2007).

REFERENCES

Abe, T., Mujahid, A., Sato, K., Akiba, Y., Toyomizu, M., 2006. Possible role of avian uncoupling protein in down-regulating mitochondrial superoxide production in skeletal muscle of fasted chickens. FEBS Lett. 580, 4815–4822.

Almeida, A., Moncada, S., Bolanos, J.P., 2004. Nitric oxide switches on glycolysis through the AMP protein kinase and 6-phosphofructo-2-kinase pathway. Nat. Cell. Biol. 6, 45–51.

Anderson, S., Bankier, A.T., Barrell, B.G., Debruijn, M.L., Coulson, A.R., Drouin, J., Eperon, I.C., Nierlich, D.P., Roe, B.A., Sanger, F., Schreier, P.H., Smith, A.J.H., Staden, R., Young, I.G., 1981. Sequence and organization of the human mitochondrial genome. Nature 290, 457–465.

Andreyev, A., Bondareva, T.O., Dedukhova, V.I., Mokhova, E.N., Skulachev, V.P., Tsofina, L.M., Volkov, N.I., Vygondina, T.V., 1989. The ATP/ADP-antiporter is involved in the uncoupling effect of fatty acids on mitochondria. Eur. J. Biochem. 182, 585–592.

Andreyev, A., Bondareva, T.O., Dedukhova, V.I., Mokhova, E.N., Skulachev, V.P., Volkov, N.I., 1988. Carboxyatractylate inhibits the uncoupling effect of free fatty acids. FEBS Lett. 226, 265–269.

Augustin, W., Wiswedel, I., Noack, J., Reinheckel, T., Reichelt, O., 1997. Role of endogenous antioxidants in defense against functional damage and lipid peroxidation in rat liver mitochondria. Mol. Cell Biochem. 174, 199–205.

Aw, T.Y., Jones, D.P., 1989. Nutrient supply and mitochondrial function. Annu. Rev. Nutr. 9, 229–251.

Baughman, J.M., Mootha, V.K., 2006. Buffering mitochondrial DNA variation. Nat. Genet. 38, 1232–1233.

Benda, C., 1898. Ueber die Spermatogenese der Vertebraten und höherer Evertebraten, II. Theil: Die Histiogenese der Spermien. Arch. Anat. Physiol. 73, 393–398.

Bobyleva, V., Bellei, M., Pazienza, T.L., Muscatello, U., 1997. Effect of cardiolipin on functional properties of isolated rat liver mitochondria. Biochem. Mol. Biol. Int. 41, 469–480.

Bogoyevitch, M.A., Ng, D.C., Court, N.W., Draper, K.A., Dhillon, A., Abas, L., 2000. Intact mitochondrial electron transport function is essential for signalling by hydrogen peroxide in cardiac myocytes. J. Mol. Cell. Cardiol. 32, 1469–1480.

Bolanos, J.P., Heales, J.R., Peuchen, S., Barker, J.E., Land, J.M., Clark, J.B., 1996. NO-mediated mitochondrial damage: a potential neuroprotective role for glutathione. Free Rad. Biol. Med. 21, 995–1001.

Bottje, W.G., Brand, M.D., Ojano-Dirain, C., Lassiter, K., Toyomizu, M., Wing, T., 2009. Mitochondrial proton leak kinetics and relationship with feed efficiency within a single genetic line of male broilers. Poult. Sci. 88, 1683–1693.

Bottje, W.G., Iqbal, M., Tang, Z., Cawthon, D.C., Okimoto, R., Wing, T., Cooper, M., 2002. Association of mitochondrial function with feed efficiency within a single genetic line of male broilers. Poult. Sci. 81, 546–555.

Bottje, W., Carstens, G.E., 2009. Association of mitochondrial function and feed efficiency in poultry and livestock species. J. Anim. Sci. 87, E48–E63.

Bottje, W., Kong, B.W., 2013. Feed efficiency: mitochondrial function to global gene expression. J. Anim. Sci. 91, 1582–1593. http://dx.doi.org/10.2527/jas.2012-5787.

Bottje, W., Iqbal, M., Pumford, N.R., Ojano-Dirain, C., Lassiter, K., 2004. Role of mitochondria in phenotypic expression of feed efficiency. J. Appl. Poult. Res. 13, 1–12.

Boveris, A., Chance, B., 1973. The mitochondrial generation of hydrogen peroxide. General properties and effect of hyperbaric oxygen. Biochem. J. 134 (3), 707–716.

Brand, M.D., Chien, L.F., Ainscow, E.K., Rolfe, D.F.S., Porter, R.K., 1994. The causes and functions of mitochondrial proton leak. Biochem. Biophys. Acta 1187, 132–139.

Brand, M.D., Pakay, J.L., Ocloo, A., Kokoszka, Z., Wallace, D.C., Brookes, P.S., Cornwall, E.J., 2005. The basal proton conductance of mitochondria depends on adenine nucleotide translocase content. Biochem. J. 392, 353–362.

Brand, M.D., Affourtit, C., Esteves, T.C., Green, K., Lambert, A.J., Miwa, S., Pakay, J.L., Parker, N., 2004. Mitochondrial superoxide: production, biological effects, and activation of uncoupling proteins. Free Rad. Biol. Med. 37, 755–767.

Brookes, P.S., Buckingham, J.A., Tenreiro, A.M., Hulbert, A.J., Brand, M.D., 1998. The proton permeability of the inner membrane of liver mitochondria from ectothermic and endothermic vertebrates and from obese rats: correlation with standard metabolic rate and phospholipid fatty acid composition. Comp. Biochem. Physiol. 119B (2), 325–334.

Brookes, P.S., Rofle, D.F.S., Brand, M.D., 1997. The proton permeability of liposomes made from mitochondrial inner membrane phospholipids: comparison to isolated mitochondria. J. Membr. Biol. 155, 167–174.

Brown, G.C., Brand, M.D., 1991. On the nature of the mitochondrial proton leak. Biochem. Biophys. Acta 1059, 55–62.

Canto, C., Gerhart-Hines, Z., Feige, J.N., Lagouge, M., Noriega, L., Milne, J.C., Elliott, P.J., Puigserver, P., Auwerx, J., 2009. AMPK regulates energy expenditure by modulating NAD^+ metabolism and SIRT1 activity. Nature 458, 1056–1060.

Cardoso, S.M., Pereira, C., Oliveira, C.R., 1999. Mitochondrial function is differentially affected upon oxidative stress. Free Rad. Biol. Med. 26, 3–13.

Carling, D., 2004. The AMP-activated protein kinase cascade – a unifying system for energy control. Trends Biochem. Sci. 29, 18–24.

Carling, D., 2005. New roles for the LKB1-> AMPK pathway. Curr. Opin. Cell Biol. 17, 167–173.

Carper, D.A., Sun, J.K., Iwata, T., Sigler, J.S.J., Ibaraki, N., Lin, L.R., Reddy, V., 1999. Oxidative stress induces differential gene expression in human lens epithelial cell line. Invest. Opthalmol. Vis. Sci. 40, 400–406.

Cawthon, D.C., McNew, R., Beers, K.W., Bottje, W.G., 2001. Electron transport chain defect and inefficient respiration may both underlie pulmonary hypertension syndrome (PHS)-associated mitochondrial dysfunction in broilers. Poult. Sci. 80, 474–484.

Cawthon, D., McNew, R., Beers, K.W., Bottje, W.G., 1999. Evidence of mitochondrial dysfunction in broilers with pulmonary hypertension syndrome (Ascites): effect of t-butyl hydroperoxide on function, glutathione and related thiols. Poult. Sci. 78, 114–125.

Chance, B., Sies, H., Boveris, A., 1979. Hydroperoxide metabolism in mammalian organs. Physiol. Rev. 59, 527–605.

Chau, M.D.L., Gap, K., Yang, Q., Wu, Z., Gromada, J., 2010. Fibroblast growth factor 21 regulates energy metabolism by activating the AMPK-SIRT1-PGC-1alpha pathway. Proc. Natl. Acad. Sci. U.S.A. 107, 12553–12558.

Choi, S.L., Kim, S., Lee, K., Kim, J., Mu, J., Birnbaum, M.J., Kim, S., Ha, J., 2001. The regulation of AMP-activated protein kinase by H_2O_2. Biochem. Biophys. Res. Comm. 287, 92–97.

Colombo, S.L., Moncada, S., 2009. AMPKα1 regulates the antioxidant status of vascular endothelial cells. Biochem. J. 421, 163–169.

Crawford, D.R., Wang, Y., Schools, G.P., Kochheise, J., Davies, K.J., 1997. Down-regulation of mammalian mitochondrial RNAs during oxidative stress. Free Rad. Biol. Med. 22, 551–559.

Dilger, J.P., McLaughlin, S.G., McIntosh, T.J., Simon, A.S., 1979. The dielectric constant of phospholipid bilayers and the permeability of membranes to ions. Science 206, 1196–1198.

Droge, W., 2002. Free radicals in the physiological control of cell function. Physiol. Rev. 82, 47–95.

Echtay, K.S., Winkler, E., Fischmuth, K., Slingenberg, M., 2001. Uncoupling proteins 2 and 3 are highly active H^+ transporters and highly nucleotide sensitive when activated by coenzyme Q (ubiquinone). Proc. Natl. Acad. Sci. U.S.A. 98, 1416–1421.

Echtay, K., Roussel, D., St. Pierre, J., Jekabsons, M.B., Cadenas, S., Stuart, J.A., Harper, A., Roebuck, S.J., Morrison, A., Pickering, S., Clapham, J.C., Brand, M.D., 2002. Superoxide activates mitochondrial uncoupling proteins. Nature 415, 96–99.

Ernster, L., Forsmark-Andree, P., 1993. Ubiquinol: an endogenous antioxidant in aerobic organisms. Clin. Invest. 71, S60–S65.

Estabrook, R.W., 1967. Mitochondrial respiratory control and polarographic measurement of ADP:O ratios. Methods Enzymol. 10, 41–47.

Forsmark-Andree, P., Lee, C.P., Dallner, G., Ernster, L., 1997. Lipid peroxidation and changes in the ubiquinone content and the respiratory chain enzymes of submitochondrial particles. Free Rad. Biol. Med. 19, 749–757.

Giulivi, C., Oursler, M.J., 2003. Role of mitochondrial oxygen and nitrogen reactive species in signaling. In: Forman, H.J., Fukuto, J., Torres, M. (Eds.), Signal Transduction by Reactive Oxygen and Nitrogen Species: Pathways and Chemical Principles. Kluwer Academic Publishers, The Netherlands, pp. 311–332.

Giulivi, C., Poderoso, J.J., Boveris, A., 1998. Production of nitric oxide by mitochondria. J. Biol. Chem. 273, 11038–11048.

Gray, M.W., Burger, G., Lang, B.F., 1999. Mitochondrial evolution. Science 5407, 1476–1481.

Greiber, S., Muller, B., Daemisch, P., Pavenstadt, H., 2002. Reactive oxygen species alter gene expression in podocyte: induction of granulocyte macrophage-colony stimulating factor. J. Am. Nephrol. 13, 86–95.

Griffith, O.W., Meister, A., 1985. Origin and turnover of mitochondrial glutathione in the isolated hepatocyte. Proc. Natl. Acad. Sci. U.S.A. 82, 4668–4672.

Hafner, R.P., Brown, g. C., Brand, M.D., 1990. Analysis of the control of respiration rate, phosphorylation rate, proton leak rate and protonmotive force in isolated mitochondria using the 'top-down' approach of metabolic control theory. Eur. J. Biochem. 188, 313–319.

Haigis, M.C., Mostoslavsky, R., Haigis, K.M., Fahie, K., Christodoulou, D.C., Murphy, A.J., Valenzuela, D.M., Yancopoulos, G.D., Karow, M., Blander, G., Wolberger, C., Prolla, T.A., Weindruch, R., Alt, F.W., Guarente, L., 2006. SIRT4 inhibits glutamate dehydrogenase and opposes the effects of calorie restriction in pancreatic beta cells. Cell 126, 941–954.

Haigis, M.C., Sinclair, D.A., 2010. Mammalian sirtuins: biological insights and disease relevance. Ann. Rev. Pathol. 5, 253–295.

Hardie, D.G., 2007. AMP-activated/SNF1 protein kinases: conserved guardians of cellular energy. Nat. Rev. Mol. Cell Biol. 8, 774–785.

Hardie, D.G., 2005. New roles for the LKB1-AMPK pathway. Curr. Opin. Cell Biol. 17, 167–173.

Hardie, D.G., 2004. The AMP-activated protein kinase pathway-new players upstream, and downstream. J. Cell Sci. 117, 5479–5487.

Hardie, D.G., Scott, J.W., Pan, D.A., Hudson, E.R., 2003. Management of cellular energy by the AMP-activated protein kinase system. FEBS Lett. 546, 113–120.

Hatefi, Y., 1985. The mitochondrial electron transport and oxidative phosphorylation system. Annu. Rev. Biochem. 54, 1015–1069.

Herrero, A., Barja, G., 1998. Hydrogen peroxide production of heart mitochondria and aging rate are slower in canaries and parakeets than in mice: sites of free radical generation and mechanisms involved. Mech. Aging Dev. 103, 133–146.

Herrero, A., Barja, G., 1997. Sites and mechanisms responsible for the low rate of free radical production of heart mitochondria in the long lived pigeon. Mech. Aging Dev. 98, 95–111.

Hickey, A.J.R., 2008. An alternate explanation for low mtDNA diversity: an age-old solution? Heredity 100, 443.

Hoch, F., 1992. Cardiolipins and biomembrane function. Biochem. Biophys. Acta 1113, 71–133.

Holmes, D.J., Austad, S.N., 1995. Birds as models for the comparative biology of ageing: a prospectus. J. Gerontol. Biol. Sci. 50A, B59–B66.

Holmes, D.J., Fluckiger, R., Austad, S.N., 2001. Comparative biology of ageing in birds: an update. Exp. Gerentol. 36, 869–883.

Hoppins, S., Lackner, L., Nunnari, J., 2007. The machines that divide and fuse mitochondria. Annu. Rev. Biochem. 76, 751–780.

Iqbal, M., Cawthon, D., Wideman Jr., R.F., Bottje, W.G., 2001. Lung mitochondrial dysfunction in pulmonary hypertension syndrome. I. Site specific defects in electron transport chain. Poult. Sci. 80, 485–495.

Iqbal, M., Pumford, N.R., Tang, Z.X., Lassiter, K., Wing, T., Cooper, M., Bottje, W.G., 2004. Low feed efficient broilers within a single genetic line exhibit higher oxidative stress and protein expression in breast muscle with lower mitochondrial complex activity. Poult. Sci. 83, 474–484.

Iqbal, M., Pumford, N.R., Lassiter, K., Tang, Z., Wing, T., Cooper, M., Bottje, W.G., 2005. Compromised liver mitochondrial function and complex activity in low feed efficient broilers within a single genetic line associated with higher oxidative stress and differential protein expression. Poult. Sci. 84, 933–941.

Jaburek, M., Varecha, M., Gimeno, M., Dembski, M., Jezek, P., Zhang, M., Burn, P., Tartaglia, L.A., Garlid, K.D., 1999. Transport function and regulation of mitochondrial uncoupling proteins 2 and 3. J. Biol. Chem. 274, 26003–26007.

Kemp, B.E., Stapleton, D., Campbell, D.J., Chen, Z.P., Murthy, S., Walter, M., Gupta, A., Adams, J.J., Katsis, F., van Denderen, B., Jennings, I.G., Iseli, T., Michell, B.J., Witters, A.L., 2003. AMP-activated protein kinase, super metabolic regulator. Biochem. Soc. Trans. 31, 162–168.

Kennedy, E.P., Lehninger, A.L., 1949. Oxidation of fatty acids and tricarboxylic acid cycle intermediates by isolated rat liver mitochondria. J. Biol. Chem. 179, 957–963.

Kikusato, M., Toyomizu, M., 2013. Crucial role of membrane potential in heat stress-induced overproduction of reactive oxygen species in avian skeletal muscle. PLoS One 8 (5), 1–9. http://dx.doi.org/10.1371/journal.pone.0064412.

Koshkin, V., Greenberg, M.L., 2000. Oxidative phosphorylation in cardiolipin-lacking yeast mitochondria. Biochem. J. 347, 687–691.

Kristal, B., Park, B., Yu, B.P., 1994. Antioxidants reduce peroxyl-mediated inhibition of mitochondrial transcription. Free Rad. Biol. Med. 16, 653–660.

Kristal, B., Koopmans, S., Jackson, C.T., Ikeno, Y., Par, B., Yu, B.P., 1997. Oxidant-mediated repression of mitochondrial transcription in diabetic rats. Free Rad. Biol. Med. 22, 813–822.

Lass, A., Sohal, R.S., 1999. Comparisons of coenzyme Q bound to mitochondrial membrane proteins among different mammalian species. Free Rad. Biol. Med. 27, 220–226.

Lehninger, A.L., 1965. The mitochondrion. W.A. Benjamin, Inc. New York, NY.

Lehninger, A.L., Nelson, D.L., Cox, M.M., 1993. Principals of Biochemistry. Worth Publishing Co., New York, NY.

Levonen, A.L., Patel, R.P., Brookes, P., Go, Y.M., Jo, H., Parthasarathy, S., Anderson, P.G., Darley-Usmar, V., 2001. Mechanisms of cell signaling by nitric oxide and peroxynitrite: from mitochondria to MAP kinases. Antioxid. Redox Signal. 3, 215–229.

Li, J., Cai, Q., Zhou, H., Xiao, G.U., 2002. Effects of hydrogen peroxide on mitochondrial gene expression of intestinal epithelial cells. World J. Gastroenterol. 8, 1117–1122.

Lin, J., Puigserver, P., Donovan, J., Tarr, P., Spiegelman, B.M., 2002a. Peroxisome proliferator-activated receptor gamma coactivator 1beta (PGC-1beta), a novel PGC-1 related transcription coactivator associated with host cell factor. J. Biol. Chem. 277, 1645–1648.

Lin, T.K., Hughes, G., Muratovska, A., Blaikie, F.H., Brookes, P.S., Darley-Usmar, V., Smith, R.A., Murphy, M.P., 2002b. Specific modification of mitochondrial protein thiols in response to oxidative stress: a proteomics approach. J. Biol. Chem. 277, 17048–17056.

Liu, Y., Kitsis, R.N., 1996. Induction of DNA synthesis and apoptosis in cardiac myocytes by E1A oncoprotein. J. Cell Biol. 133, 325–334.

Mannella, C.A., 2000. Our changing views of mitochondria. J. Bioeng. Biomembr. 32, 1–4.

Martensson, J., Han, J., Griffith, O.W., Meister, A., 1993. Glutathione ester delays the onset of scurvy in ascorbate-deficient guinea pigs. Proc. Natl. Acad. Sci. U.S.A. 90, 317–321.

Martensson, J., Lai, J., Meister, A., 1990. High affinity transport of glutathione is part of a multicomponent system essential for mitochondrial function. Proc. Natl. Acad. Sci. U.S.A. 87, 7185–7189.

Masatoshi, U., Watanabe, K., Sato, K., Akiba, Y., Toyomizu, M., 2005. Possible role for avPGC-1alpha in the control of expression of fiber type, along with avUCP and avANT mRNA's in skeletal muscles of cold-exposed chickens. FEBS Lett. 579, 11–17.

Meirhaeghe, A., Crowley, V., Lenaghan, C., Lelliott, C., Green, K., Stewart, A., Hart, K., Schinner, S., Sethi, J.K., Yeo, G., Brand, M.D., Cortright, R.N., O'Rahilly, S., Montague, C., Vidal-Puig, A.J., 2003. Characterization of the human, mouse and rat PGC1Beta (peroxisome-proliferator-activated receptor-gamma co-activator 1β) gene in vitro and in vivo. Biochem. J. 373, 155–165.

Meister, A., 1984. New aspects of glutathione biochemistry and transport: selective alteration of glutathione metabolism. Fed. Proc. 43, 3031–3042.

Mezentseva, N.V., Kumaratilake, J.S., Newman, S.A., 2008. The brown adipocyte differentiation pathway in birds: an evolutionary road not taken. BMC Biol. 6, 17. http://dx.doi.org/10.1186/1741-7007-6-17.

Minokoshi, Y., Alquier, T., Furukawa, N., Kim, Y., Lee, A., Xue, B., Mu, J., Foufelle, F., Ferre, P., Birnbaum, M.J., Stuck, B.J., Kahn, B.B., 2004. AMP-kinase regulates food intake by responding to hormonal and nutrient signals in the hypothalamus. Nature 428, 569–574.

Miwa, S., St-Pierre, J., Partridge, L., Brand, M.D., 2003. Superoxide and hydrogen peroxide production by drosophila mitochondria. Free Rad. Biol. Med. 35, 938–948.

Mujahid, A., Akiba, Y., Toyomizu, M., 2007a. Acute heat stress induces oxidative stress and decreases adaptation in young white leghorn cockerels by down regulation of avian uncoupling protein. Poult. Sci. 86, 364–371.

Mujahid, A., Sato, K., Akiba, Y., Toyomizu, M., 2006. Acute heat stress stimulates mitochondrial superoxide production in broiler skeletal muscle, possibly via down-regulation of uncoupling protein content. Poult. Sci. 85, 1259–1265.

Mujahid, A., Pumford, N.P., Bottje, W., Kiotaka, K., Miyazawa, T., Akiba, Y., Toyomizu, M., 2007b. Mitochondrial oxidative damage in chicken skeletal muscle induced by acute heat stress. J. Poult. Sci. 44, 439–445.

Mujahid, M., Akiba, Y., Toyomizu, M., 2009. Olive-oil supplemented diet alleviates acute heat stress-induced mitochondrial ROS production in chicken skeletal muscle. Am. J. Physiol. Regul. Integr. Comp. Physiol. 297, R690–R698.

Murphy, M.P., Echtay, K.S., Blaikie, F.H., Asin-Cayuela, J., Cocheme, H.M., Green, K., Buckingham, J.A., Taylor, E.R., Hurrell, F., Hughes, G., Miwa, S., Cooper, C.E., Svistunenko, D.A., Smith, R.A., Brand, M.D., 2003. Superoxide activates uncoupling proteins by generating carbon-centered radicals and initiating lipid peroxidation: studies using a mitochondria-targeted spin trap derived from a-phenyl-*N-tert*-butylnitrone. J. Biol. Chem. 278, 48534–48545.

Nasrin, N., Wu, X., Fortier, E., Feng, Y., Bare, O.C., Chen, S., Ren, X., Wu, Z., Streeper, R.S., Bordone, L., 2010. SIRT4 regulates fatty acid oxidation and mitochondrial gene expression in liver and muscle cells. J. Biol. Chem. 285, 31995–32002.

Nijtmans, L.G.J., Sanz, A.M., Grivell, L.A., Coates, P.J., 2002. The mitochondrial PHB complex: roles in mitochondrial respiratory complex assembly, ageing and degenerative disease. Cell. Mol. Life Sci. 59, 143–155.

Nisoli, E., Clementi, E., Moncada, S., Carruba, M.O., 2004. Mitochondrial biogenesis as a cellular signaling framework. Biochem. Pharmacol. 67, 1–15.

Nisoli, E., Clementi, E., Paolucci, C., Cozzi, V., Tonello, C., Sciorati, C., Bracale, R., Valerio, A., Francolini, M., Moncada, S., Carruba, M.O., 2003. Mitochondrial biogenesis in mammals: the role of endogenous nitric oxide. Science 299, 896–899.

Ojano-Dirain, C., Tinsley, N.B., Wing, T., Cooper, M., Bottje, W.G., 2007a. Membrane potential and hydrogen peroxide production in duodenal mitochondria in broilers chicken (*Gallus gallus*) with low and high feed efficiency. Comp. Biochem. Physiol. 147, 934–941.

Ojano-Dirain, C., Bottje, W., Wing, T., Cooper, M., 2005. Glutathione and respiratory chain complex activities in duodenal mitochondria from broilers with low and high feed efficiency. Poult. Sci. 84, 782–788.

Ojano-Dirain, C., Iqbal, M., Cawthon, D., Swonger, S., Wing, T., Cooper, M., Bottje, W.G., 2004. Site-specific effects in electron transport in duodenal mitochondria is associated with low feed efficiency in broiler breeder males. Poult. Sci. 83, 1394–1403.

Ojano-Dirain, C., Pumford, N.R., Toyomizu, M., Bottje, W.G., 2007b. Association of mitochondrial function and feed efficiency. J. Poult. Sci. 44, 221–237.

Ojano-Dirain, C., Toyomizu, M., Wing, T., Cooper, M., Bottje, W.G., 2007c. Gene expression in breast muscle and duodenum from low and high feed efficient broilers. Poult. Sci. 86, 372–381.

Ojuka, E.O., 2004. Role of calcium and AMP kinase in the regulation of mitochondrial biogenesis and GLUT4 levels in muscle. Proc. Nutr. Soc. 63, 275–278.

Olafsdottir, K., Reed, D.J., 1988. Retention of oxidized glutathione by isolated rat liver mitochondria during hydroperoxide treatment. Biochem. Biophys. Acta 964, 377–382.

Paul, M.H., Sperling, E.D., 1952. Cyclophorus system-XXIII. Correlation of cyclophorase activity and mitochondrial density in striated muscle. Proc. Soc. Exp. Biol. Med. 79, 352–354.

Poyton, R.O., McEwen, J.E., 1996. Crosstalk between nuclear and mitochondrial genomes. Annu. Rev. Biochem. 65, 563–607.

Praefcke, G.J., McMahon, H.T., 2004. The dynamin superfamily: universal membrane tubulation and fission molecules. Nat. Rev. Mol. Cell Biol. 5, 133–147.

Puigserver, P., Wu, Z., Park, C.W., Graves, R., Wright, M., Spiegelman, B.M., 1998. A cold-inducible coactivator of nuclear receptors linked to adaptive thermogenesis. Cell 92, 829–839.

Rabilloud, T., Strub, J.M., Carte, N., Luche, S., van Dorsselaer, A., Lunardi, J., Giege, R., Florentz, C., 2002. Comparative proteomics as a new tool for exploring human mitochondrial tRNA disorders. Biochemistry 41, 144–150.

Raimbault, S., Dridi, S., Denjean, F., Lachuer, J., Couplan, E., Bouillaud, F., Bordas, A., Duchamp, C., Taouis, M., Ricquier, D., 2001. An uncoupling protein homologue putatively involved in facultative muscle thermogenesis in birds. Biochem. J. 353, 441–444.

Rey, B., Roussel, D., Romestaing, C., Belouze, M., Rouanet, J.L., Desplanches, D., Sibille, B., 2010. Up-regulation of avian uncoupling protein in cold-acclimated and hyperthyroid ducklings prevents reactive oxygen species production by skeletal muscle mitochondria. BMC Physiol. 10 (5), 1–12.

Reznick, R.M., Shulman, G.I., 2006. The role of AMP-activated protein kinase in mitochondrial biogenesis. J. Physiol. 574, 33–39.

Riobo, N.A., Clement, E., Melani, M., Boveris, A., Cadenas, E., Moncoda, S., Poderoso, J.J., 2001. Nitric oxide inhibits mitochondrial NADH-ubiquinone reductase activity through the formation of peroxynitrite. Biochem. J. 359, 139–145.

Rolfe, D.F.S., Brand, M.D., 1997. The physiological significance of mitochondrial proton leak in animal cells and tissues. Biosci. Rep. 17, 9–16.

Ryan, M.T., Hoogenraad, N.J., 2007. Mitochondrial-nuclear communications. Annu. Rev. Biochem. 76, 701–722.

Samartsev, V.N., Smirnov, A.V., Zeldi, I.P., Markova, E.N., Mokova, E.N., Skulachev, V.P., 1997. Involvement of the aspartate/glutamate antiporter in fatty acid-induced uncoupling of liver mitochondria. Biochim. Biophys. Acta 1319, 251–257.

Scheffler, I., 1999. Mitochondria. Wiley-Liss Inc., New York, NY.

Sharma, V.K., Ramesh, V., Franzini-Armstrong, C., Sheu, S., 2000. Transport of Ca^{2+} from sarcoplasmic reticulum to mitochondria in rat ventricular myocytes. J. Bioenerg. Biomembr. 32, 97–104.

Skulachev, V.P., 1997. Membrane linked systems preventing superoxide formation. Biosci. Rep. 17, 347–366.

Skulachev, V.P., 1996. Role of uncoupled and non-coupled oxidations in maintenance of safely low levels of oxygen and its one-electron reductants. Q. Rev. Biophys. 29, 169–202.

Talbot, D.A., Duchamp, C., Rey, B., Hanuise, N., Rouanet, J.L., Sibille, B., Brand, M.D., 2004. Uncoupling protein and ATP/ADP carrier increase mitochondrial proton conductance after cold adaptation of king penguins. J. Physiol. 558, 123–135.

Talbot, D.A., Hanuise, N., Rey, B., Rouanet, J., Duchamp, C., Brand, M.D., 2003. Superoxide activates a GDP-sensitive proton conductance in skeletal muscle mitochondria of king penguin (*Aptenodytes patagonicus*). Biochem. Biophys. Acta 312, 983–988.

Tang, Z., Iqbal, M., Cawthon, D., Bottje, W.G., 2002. Heart and breast muscle mitochondrial dysfunction in pulmonary hypertension syndrome in broilers (*Gallus domesticus*). Comp. Biochem. Physiol. A132, 527–540.

Toyomizu, M., Ueda, M., Sato, S., Seki, Y., Sato, K., Akiba, Y., 2002. Cold-induced mitochondrial uncoupling and expression of chicken UCP and ANT mRNA in chicken skeletal muscle. FEBS Lett. 529, 313–319.

Turrens, J.F., Alexander, A., Lehninger, A.L., 1985. Ubisemiquinone is the electron donor for superoxide formation by complex III of heart mitochondria. Arch. Biochem. Biophys. 237, 408–413.

Turrens, J.F., Boveris, A., 1980. Generation of superoxide anion by the NADH dehydrogenase of bovine heart mitochondria. Biochem. J. 191, 421–427.

Wallace, D., 1999. Mitochondrial diseases in man and mouse. Science 283, 1482–1488.

Wei, Y., 1998. Oxidative stress and mitochondrial DNA mutations in human aging. Exp. Biol. Med. 217, 53–63.

Wu, M., Neilson, A., Swift, A.L., Moran, R., Tamagnine, J., Parslow, D., Armistead, S., Lemire, K., Orrell, J., Teich, J., Chomicz, S., Ferrick, D.A., 2007. Multiparameter metabolic analysis reveals a close link between attenuated mitochondrial bioenergetic function and enhanced glycolysis dependency in human tumor cells. Am. J. Physiol. Cell Physiol. 292, C125–C136.

Wu, Z., Puigserver, P., Andersson, U., Zhang, C., Adelmant, G., Mootha, V., Troy, A., Cinti, S., Lowell, B., Scarpulla, R.C., Spiegelman, B.M., 1999. Mechanisms controlling mitochondrial biogenesis and respiration through the thermogenic coactivator PGC-1. Cell 98, 115–124.

Yu, B.P., 1994. Cellular defenses against damage from reactive oxygen species. Physiol. Rev. 74, 139–162.

Zhou, G., Myers, R., Li, Y., Chen, Y., Shen, X., Fenyk-Melody, J., Wu, M., Ventre, J., Doebber, T., Fujii, N., Musi, N., Hirshman, M.F., Goodyear, L.J., Moller, D.E., 2001. Role of AMP-activated protein kinase in mechanism of metformin action. J. Clin. Invest 108, 1167–1174.

Part II

Sensory Biology and Nervous System Theme

Chapter 5

The Avian Somatosensory System: A Comparative View

J. Martin Wild
Department of Anatomy with Radiology, Faculty of Medical and Health Sciences, University of Auckland, Auckland, New Zealand

ABBREVIATIONS

A Arcopallium
Bas Nucleus basorostralis
CN IX Glossopharyngeal nerve
CN XII Hypoglossal nerve
CTB Cholera toxin B-chain
DCN Dorsal column nuclei: nucleus gracilis, nucleus cuneatus et nucleus cuneatus externus
DIVA Nucleus dorsalis intermedius ventralis anterior
DLP Nucleus dorsolateralis posterior thalami
DLPc Nucleus dorsolateralis posterior thalami, pars caudalis
DLPr Nucleus dorsolateralis posterior thalami, pars rostralis
DRG Dorsal root ganglion
EM Electron microscopy
FLM Fasciculus longitudinalis medialis
HA Hyperpallium apicale
HRP Horseradish peroxidase
HVC HVC (proper name)
ICc Central nucleus of the inferior colliculus
ICo Nucleus intercollicularis
IHA Interstitial hyperstristum accessorium
LLDa Anterior division of the dorsal lateral lemniscal nucleus
LLDp Posterior division of the dorsal lateral lemniscal nucleus
LLI Intermediate nucleus of the lateral lemniscus
LLIc Caudal part of the intermediate nucleus of the lateral lemniscus
LLIr Rostral part of the intermediate nucleus of the lateral lemniscus
LLV Ventral nucleus of the lateral lemniscus
LPS Lamina pallio-subpallialis
M Mesopallium
MLd Nucleus mesencephalicus lateralis, pars dorsalis
N Nidopallium
NCL Nidopallium caudolaterale
NI/Nc Neostriatum intermedium/neostriatum caudale
NIf Nucleus interface
nTTD Nucleus tractus descendens nervi trigemini
nVI Nucleus nervi abducentis
NVI Nervus abducens
nVII Nucleus nervi facialis
nXIIts Nucleus nervi hypoglossi, pars tracheosyringealis
OI Nucleus olivaris inferior
Ov Nucleus ovoidalis
PE Nucleus pontis externus
PrV Nucleus sensorius principalis nervi trigemini
RPcvm Ventromedial part of the parvocellular reticular formation
Rt Nucleus rotundus
sP Nucleus subprincipalis
SCi Intermediate part of the core nucleus of the pre-isthmic region (Puelles et al. (1994)
SI Primary somatosensory cortex
SII Secondary somatosensory cortex
SS Synsacral segment
SSp Nucleus supraspinalis
St Striatum
Uva Nucleus unaeformis
VB Ventrobasal complex

5.1 INTRODUCTION

Despite the lengthy interim between this and the previous volume, there has been only a limited amount of new information gathered on the avian somatosensory system. This is largely due to the fact that two protagonists in this area of research, Reinhold Necker and Jaap Dubbeldam, have long since retired. The reader will hopefully understand, therefore, the reasons for the relatively few recent publications in this area and the admittedly biased nature of this author's presentation. New recruits to the study of avian somatosensation are clearly needed!

Necker's (2000a) chapter on the avian somatosensory system was a concise overview of its basic organization, from the types, distribution, and physiology of mechanoreceptors and other sensory receptors found in avian skin, to the synaptic and functional connectivity of the somatosensory pathways ascending throughout the neuraxis (Figure 5.1). He divided the avian somatosensory system into two parts, spinal and trigeminal, noting that the former innervates the body surface and extremities and the latter mainly the beak. However, with regard to ascending systems, the word "spinal" usually implies an origin in the spinal cord itself, rather than in the periphery,

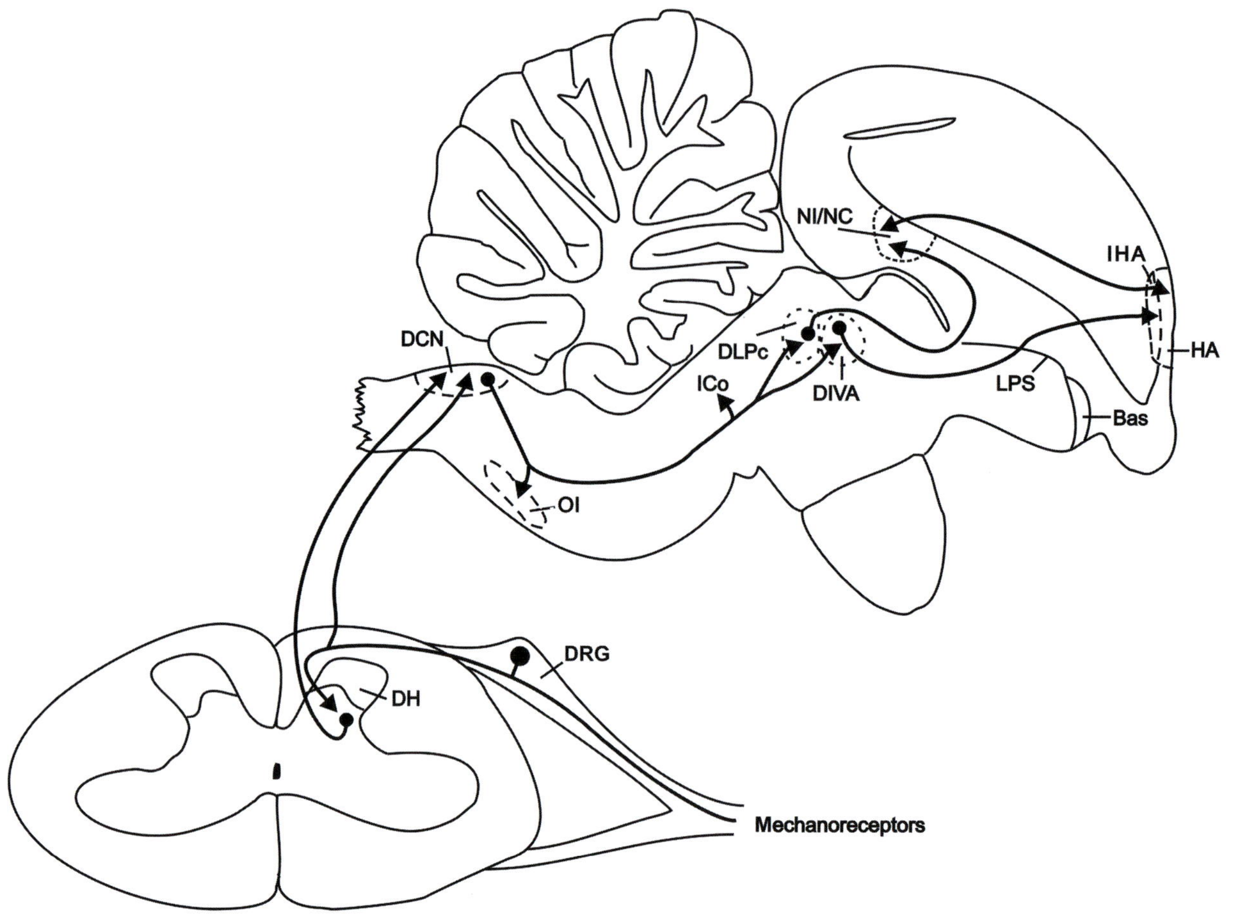

FIGURE 5.1 **Schematic drawing of the brain of a pigeon showing the main somatosensory pathways.** Filled circles denote cell bodies, arrow heads denote terminations. Note that there are substantial deviations from this plan in other species (see text and Figure 5.2). *Adapted from Necker (2000a).*

and "trigeminal" seems to exclude other primary afferent projections from the head, such as the tongue, which is not innervated by the trigeminal nerve in birds. In this chapter, therefore, a division of somatosensory origins into the body (including wings, legs, and claws) and beak and tongue is preferred. Nevertheless, to better appreciate the following material, the reader should first consult Necker's chapters on the somatosensory system (Necker, 2000a) and the spinal cord (Necker, 2000b)—both in the fifth edition of this book.

5.2 BODY SOMATOSENSORY PRIMARY AFFERENT PROJECTIONS IN DIFFERENT SPECIES

5.2.1 Spinal Cord

Our knowledge of primary afferent projections to the bird's spinal cord is based on a variety of techniques, such as tracing degenerating fibers following dorsal rhizotomy (van den Akker, 1970; Leonard and Cohen, 1975a), injections of tracer into dorsal root ganglia or application of tracer to the cut ends of dorsal roots (Necker, 2001), injections of tracer into feather follicles (Wild, 1985), injections or applications of tracer to whole wing and leg nerves (Wild, 1985; Necker and Schermuly, 1985; Schulte and Necker, 1994), cutaneous nerves and skin (Woodbury and Scott, 1991), the vibration sensitive interosseus nerve (Ohmori and Necker, 1995), ankle joint receptors (Gentle et al., 1995), and muscles (Wild, 1985) in pigeons and chickens. Physiological analyses of pigeon dorsal horn responses have been provided by Necker (1985a,b, 1990) for pigeons and by Woodbury (1992) for chickens.

Although the different techniques provide different patterns of projections to the cord and brainstem, it is clear that primary afferent fibers enter and terminate in the cord differently depending on their diameter, as in mammals. Necker (2001) found that at the entry level of C5, for instance, large-diameter fibers entered the dorsal horn from its medial aspect, between laminae IV and V. These fibers continued into the ventral horn, where they terminated densely, and some continued laterally to terminate in the marginal nuclei. Fine-diameter fibers entered the dorsal horn laterally and terminated in various parts of the horn at the level of entry, whereas other fine fibers extended as far as C1 rostrally and at least as far as C8 caudally. Intermediate diameter fibers entered the dorsal horn from dorsal and medial positions to terminate from C1 to C8 in lamina IV and medial lamina V. Some fibers crossed in the dorsal commissure to terminate in similar areas on the contralateral side.

That the primary afferent projections are somatotopically organized is indicated by the fact that the follicles of primary flight feathers and their coverts of the wing are represented in lamina I and medial lamina II, whereas the follicles of secondary flight feathers and their coverts are represented at the lateral edge of the dorsal horn (Wild, 1985; Necker, 1990). Terminal fields of chicken ankle afferents are found in laminae I–III and VI, with a few terminals in deeper laminae (Gentle et al., 1995), which is similar to the projections from the cutaneous nerves of the leg in chicks (Woodbury and Scott, 1991). It should be remembered, however, that in birds as well as in mammals, different types of tracer tend to produce labeling in different laminae, with wheat germ agglutinin tending to produce labeling in laminae I and II, and cholera toxin B-chain producing labeling predominantly in lamina III and IV (Robertson and Grant, 1985).

Primary afferent terminations in the dorsal horn can extend over several segments and are not necessarily heaviest at the level of entry of the dorsal root fibers. Gentle et al. (1995), for instance, found that although ankle joint afferents entered the cord predominantly over synsacral level SS5–7, terminations were split such that there was a small field in SS7 and 8 and a larger field in SS3 and 4. Split terminal fields were also found for whole nerve inputs by Woodbury and Scott (1991). Terminations of wing nerve fibers extend well beyond the rostrocaudal extent of incoming rootlets, although the density of terminations declines with increasing distance from the entry zone (Leonard and Cohen, 1975a; Wild, 1985). These terminations are likely collaterals of fibers ascending or descending to more distant levels. In the case of wing nerve afferents, at least, fibers ascending in the dorsal column to the medulla provide collaterals to the dorsal horn of most, if not all, of the dozen or so cervical segments (Wild, unpublished observations in pigeons). In the greenfinch wing, nerve afferent terminations in the cervical cord are concentrated in medial lamina V (Wild, 1997), as are primary afferent fibers in budgerigars (Wild et al., 1997). Neurons at similar locations and in the nucleus of Bischoff in the upper cervical spinal cord of pigeons can be retrogradely labeled from injections of tracer into either a dorsal thalamic somatosensory nucleus (DIVA; see below) or a multimodal thalamic nucleus (DLP; see below) (Wild, 1989), which could suggest the presence of a spinothalamic projection from medial lamina V of upper cervical spinal cord segments. These findings should be contrasted with those of Schneider and Necker (1989), who found that DIVA injections retrogradely labeled very few cells in the brachial spinal cord intermediate region, but many cells at lumbar levels, suggesting a spinothalamic projection mediating lower limb somatosensory input, but the virtual absence of one from the wing.

5.2.2 Brainstem

Primary afferent projections from the body to the brainstem were first visualized in pigeons by van den Akker (1970) using degeneration techniques following dorsal rhizotomies, and later using the transganglionic horseradish peroxidase (HRP) technique by Wild (1985). Neck primary afferents were also visualized in pigeons by Necker (2001) and limb primary afferents have been visualized in pigeons, chickens, greenfinches, and barn owls (Schulte and Necker, 1994; Necker and Shermuly, 1985; Gentle et al., 1995; Wild, 1997; Wild et al., 2001).

At caudal levels of the dorsal column nuclei (DCN; i.e., gracile, cuneate, and external cuneate nuclei[1]) in pigeons, leg afferent terminations in the gracile nucleus lie medial to those of the wing in the cuneate nucleus, but more rostrally there is substantial overlap of leg and wing inputs throughout the DCN, although wing inputs extend further laterally around the periphery of the medulla than leg inputs. A picture of the representation of the wings can be gained by imagining them spread out over the dorsal and dorsolateral periphery of the medulla. Only in the barn owl have terminations from the wing and leg been seen to be confined to clearly separate gracile and cuneate nuclei (Wild et al., 2001). In pigeons, neck primary afferents were found by Necker (2001) to project rostrally and ventrolaterally in the DCN complex, where they terminated in the external cuneate nucleus (CuE), in a nucleus that he compared with the intermediate nucleus of Cajal, as well as in another nucleus he thought comparable to nucleus *x* located lateral to the nucleus of the descending trigeminal tract (nTTD) and ventral to the descending vestibular nucleus, well rostral to the obex.

As in mammals, primary afferent projections to the DCN are supplemented by a dorsal column postsynaptic system (Figure 5.1), which in pigeons takes its origin from lamina IV of the brachial spinal cord, where mechanosensitive neurons have been located (Necker, 1985a,b, 1991).

In pigeons, finches and barn owls primary afferent projections extend ipsilaterally throughout the medulla and into the pons, where they terminate sparsely in proximity to PrV (Wild, 1985, 1997). In the Australian budgerigar

[1] The avian external cuneate nucleus is not the equivalent of the same named nucleus in mammals, because its inputs are not confined to upper limb proprioceptors and, at least in pigeons, it does not project to the cerebellum (Wild, 1985; Reinke and Necker, 1996).

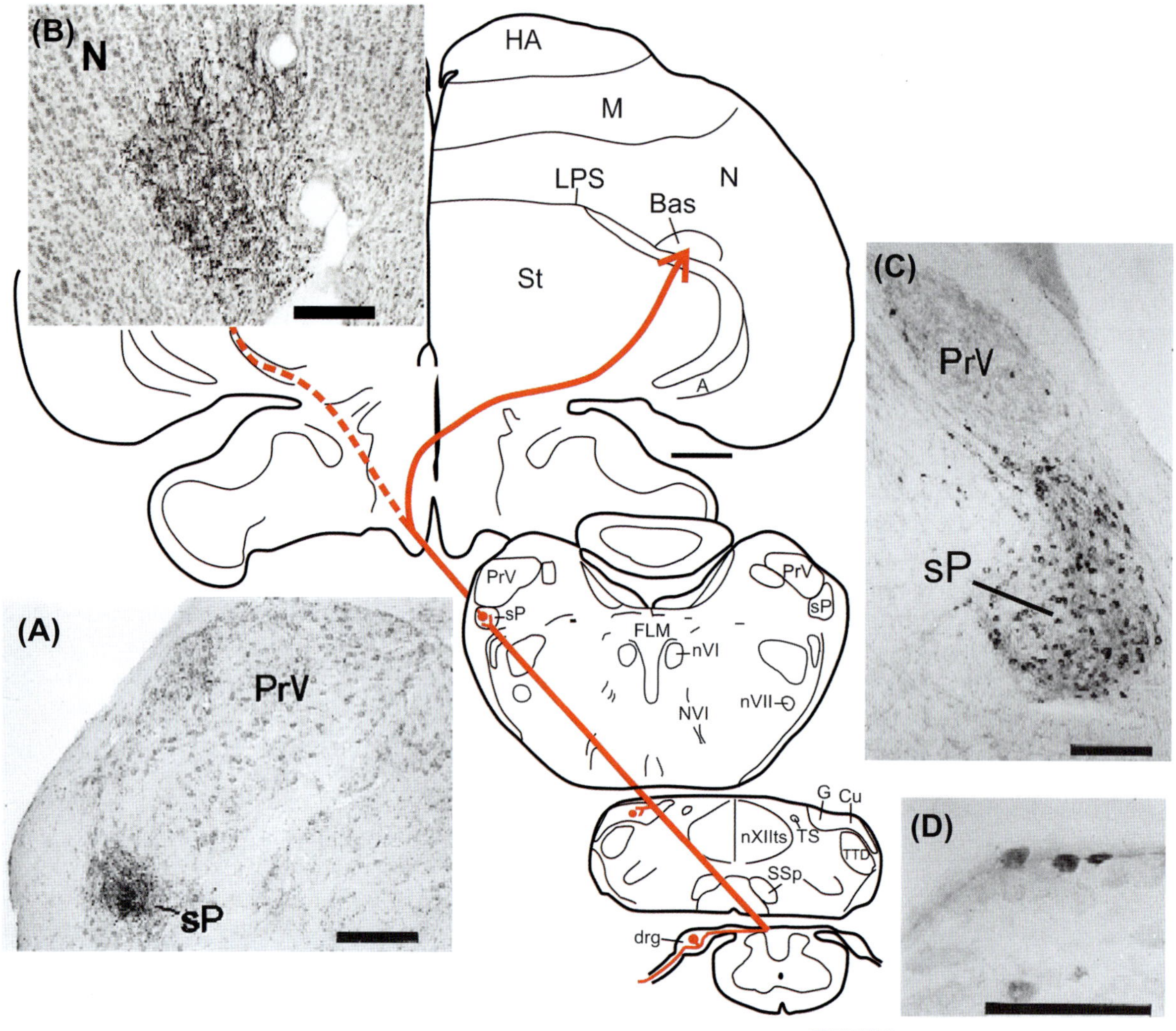

FIGURE 5.2 **Depiction of the somatosensory pathway in the budgerigar.** Primary afferent fibers project to and terminate in the ipsilateral nucleus subprincipalis (sP), cell bodies of which project their axons to the body regions of Bas, predominantly contralaterally (solid lines). Scale bars for schematics: 1 mm. (A) Terminal field in sP following an injection of horseradish peroxidase-conjugated cholera toxin B-chain (CTB) in the ipsilateral radial nerve. The terminal field of sciatic fibers (not shown) is located ventrolateral to the wing terminal field. (B) Terminal field (note its precise borders) in the body part of Bas following an injection of biotinylated dextran amine (BDA) in the contralateral sP. (C) Retrogradely labeled cells in sP following an injection of CTB in the contralateral body part of Bas. (D) Retrogradely labeled cells in a dorsal root ganglion of a spinal segment from the brachial enlargement, following an injection of BDA in the ipsilateral sP. Scale bars for photomicrographs: 200 mm *Adapted from Wild et al. (1997).*

(*Melopsittacus undulatus*) (Wild et al., 1997) and possibly other psittaciforms (Wild, 1981), primary afferents from both wing and leg also reach pontine levels, where they terminate densely and topographically in a previously undescribed nucleus immediately subjacent to PrV, hence called subprincipalis (sP). One implication of these findings in the budgerigar is that here at pontine levels there is a striking instance of a complete, somatotopic representation of the whole body, with the beak and tongue and possibly some other parts of the head being represented massively in PrV (see below) and, ventral to PrV, a smaller representation of the rest of the body in sP (Figure 5.2).

5.3 ASCENDING PROJECTIONS OF THE DORSAL COLUMN NUCLEI

As in mammals (Berkley et al., 1986), the DCN in birds project predominantly contralaterally to several more rostral nuclei via a medial lemniscus, en route to their final targets in the thalamus (Wild, 1989, 1997). The inferior olive (OI) is the first of these, which then projects to the cerebellum via the inferior cerebellar peduncle. Because in pigeons, at least, the DCN do not project directly to the cerebellum, the route to the cerebellum via the OI may be one way that somatosensory inputs serve sensorimotor control.

Another significant target of the DCN is the midbrain, where there are terminations not only in intercollicular regions surrounding the auditory nucleus mesencephalicus lateralis, pars dorsalis (MLd), but also within it (Karten, 1967; Leibler, 1975; Wild, 1989, 1995, 1997). Units recorded electrophysiologically in the region between the dorsal border of MLd and the tectal ventricle respond to both somatosensory and visual stimuli (Ballam, 1982). The region ventral to more caudomedial regions of MLd forms a distinct nucleus called the intermediate part of the core nucleus of the pre-isthmic region (SCi; Puelles et al., 1994). Its neurons are exquisitely sensitive to tactile stimuli applied to feathers and body surface and respond with large-amplitude action potentials (Wild, 1995). Its ascending projections require further definition, but the location of terminations around the ventral periphery of the thalamic auditory nucleus ovoidalis (Ov), which is the main target of MLd projections (Karten, 1967), closely mirrors the location of SCi with respect to MLd (Wild and Williams, 2000). In pigeons and finches, it appears that the DCN also project directly to MLd, where diffuse terminations overlap the much denser terminations of brainstem auditory nuclei (Wild, 1989, 1995, 1997). Speculations concerning the functions of the somatosensory projection to MLd can be found in Wild (1995). In the barn owl, there is a similar but more restricted DCN projection to rostrodorsal parts of MLd (known in barn owls as the central nucleus of the inferior colliculus; Wild et al., 2008). These findings represent but one of several instances in the avian brain of somatosensory and auditory proximity or overlap.

At caudal thalamic levels, the DCN target the medial spiriform nucleus (SpM), which then projects to the cerebellum (Wild, 1992), and a nucleus called dorsolateral posterior thalami (DLP) in pigeons or unaeformis (Uva) in finches (Wild, 1987, 1994; Funke, 1989a). DLP and Uva tend to be multimodal (Korzeniewski, 1987; Korzeniewski and Güntürkün, 1990; Wild, 1994) and in this respect are perhaps similar to the posterior group of thalamic nuclei of mammals (Gamlin and Cohen, 1986). According to Korzeniewski (1987), the whole body, including the beak and head, is represented in DLP in pigeons. However, the origin of the somatosensory inputs that might account for the beak responses requires further specification, despite the fact that Korzeniewski and Güntürkün (1990) found retrogradely labeled cells in nTTD (to which beak afferents project) following tracer injections in DLP (see also Wild, 1989). In the mallard, projections to the diencephalon from nTTD were not observed (Arends, 1981; Arends and Dubbeldam, 1984).

Rostral to DLP, the DCN target the nucleus DIVA, which lies immediately dorsal and lateral to the auditory nucleus Ov and the visual nucleus rotundus (Wild, 1989, 1997). Unlike DLP, DIVA is specifically somatosensory; although it contains a representation of most body parts (including toes, but not the beak), it is weakly somatotopically organized (Schneider and Necker, 1996). DIVA is probably homologous to part of the ventrobasal complex (VB) of mammals.

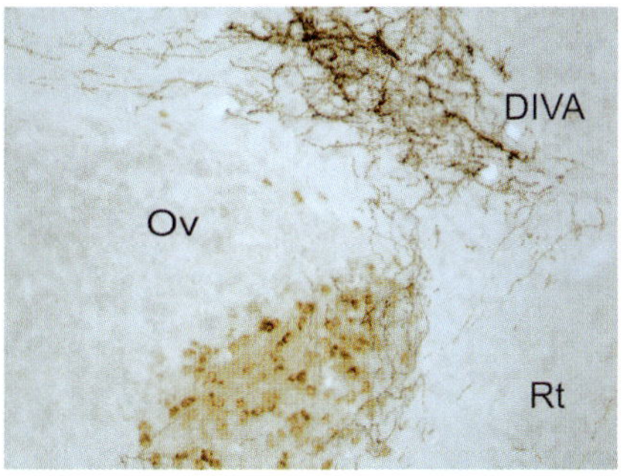

FIGURE 5.3 Labeled fibers and terminations (black) in the lateral part of ovoidalis (Ov) and the overlying dorsal thalamic somatosensory nucleus following an iontophoretic injection of BDA in the dorsal column nucleus (section midline is to the left). Neurons in the lateral part of Ov (brown) were retrogradely labeled by an injection of cholera toxin B-chain in the lateral part of Field L, the primary thalamorecipient auditory field in the telencephalon.

In zebra finches, there is a small but distinct projection from the DCN to the ventrolateral part of Ov, as fibers pass to the overlying DIVA (Figure 5.3). This is yet another instance in the avian brain of somatosensory and auditory proximity or overlap.

Unlike the case for mammals (Berkley et al., 1986), there is some evidence in finches, at least, that single DCN neurons project to more than one more rostral nucleus (e.g., intercollicular nucleus and DIVA) via branched axons (Wild, 1997).

To complete this section, I again note the curious case of the budgerigar, in which the DCN project ipsilaterally to sP in the pons (Wild et al., 1997), as do primary afferents from the wings and legs. Whether the DCN also project to the thalamus in this species is not known. Another variation on the DCN projections occurs in the barn owl, in which there is a projection from both gracile and cuneate nuclei to a large nucleus at the lateral edge of the pons called pontis externus (PE) (Wild et al., 2001). Such a nucleus, which in the barn owl does not receive primary afferent projections directly from the body surface, has thus far not been described for any other species. PE may, however, be the functional equivalent of sP in budgerigar because, like sP, it provides a major body somatosensory and topographically organized input to nucleus basorostralis pallii (Bas) (see below).

The various ascending somatosensory pathways are depicted for four species in Figure 5.4.

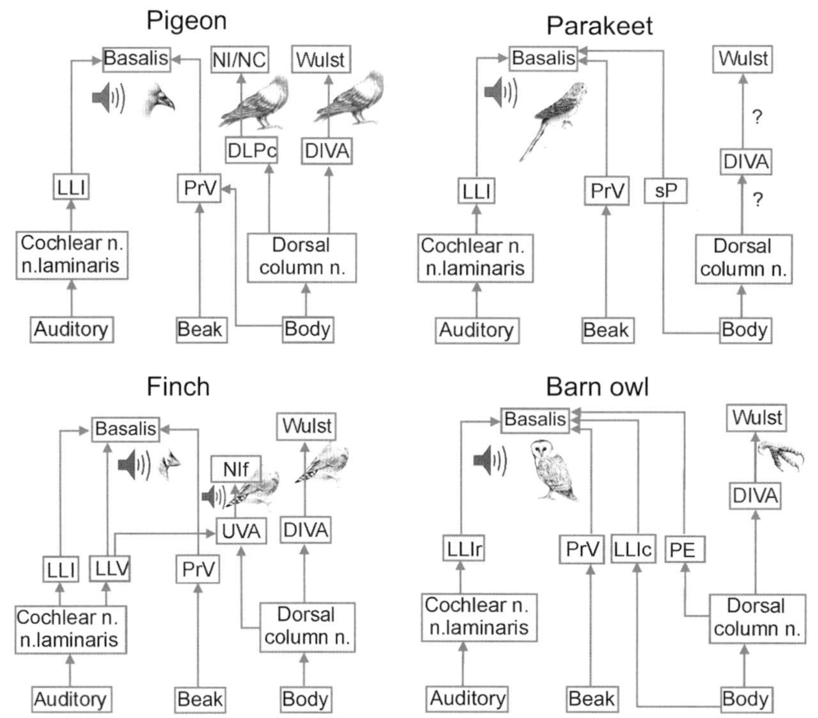

FIGURE 5.4 **Comparative schematic of ascending somatosensory and auditory projections to the telencephalon via thalamic and nonthalamic pathways in different avian groups.** Note the "beakless" representation of the body in the rostral Wulst of pigeons and finches and the complete body + beak + auditory representation in Bas of parakeets (budgerigars) and barn owls (auditory representations are symbolized by speakers). Note also the auditory projection to NIf of finches, which is relayed via projections of the ventral nucleus of the lateral lemniscus to nucleus unaeformis (Coleman et al., 2007), and perhaps by projections of ovoidalis to the nucleus interface (not shown; Wild, unpublished observations). *From Wild et al. (2008).*

5.4 TELENCEPHALIC PROJECTIONS OF THALAMIC NUCLEI RECEIVING SOMATOSENSORY INPUT

The ascending projections of DLP/Uva and DIVA are quite separate. In pigeons, DLP projects to the intermediate and caudal parts of the nidopallium (NI/NC: Gamlin and Cohen, 1986; Funke, 1989a; Wild, 1987). The terminal field of more rostral parts of DLP (DLPr) lies medially adjacent to the visual entopallium, whereas the terminal field of more caudal parts of DLP (DLPc) is more caudal, closer to the thalamorecipient auditory field L, and is predominantly somatosensory (Wild, 1987; Funke, 1989b). In finches, which are songbirds, the equivalent region of the nidopallium to which Uva projects is called nucleus interface (NIf), which is usually regarded as belonging to the song system (Nottebohm et al., 1982). It is possible, however, that following a sufficiently fine-grained electophysiological and anatomical analysis, this region of the nidopallium may be subdivided into visual, somatosensory, and auditory components (Wild, unpublished observations); perhaps only the latter provide an important input to HVC of the song control system in songbirds (Nottebohm et al., 1982; Wild, 1994; Vates, 1996).

In both pigeons and finches, the ascending projections of DIVA are specifically to the interstitial hyperstristum accessorium (IHA) of the rostral, somatosensory Wulst (Wild, 1987, 1997; Funke, 1989a). In finches, the input to IHA is distinctly and regularly patchy, but a somatotopic organization has not been examined. In pigeons, there is a very weak somatotopic organization of this region (Funke, 1989b).

It has been suggested that the somatosensory area in the rostral Wulst of birds may be equivalent to SI in mammals, whereas the more caudal somatosensory area in NI/NC is equivalent to SII (Wild, 1987). The two regions are reciprocally connected (Wild and Williams, 1999). Unlike the somatosensory inputs to SI in mammals, however, the avian classical three-neuron sequence of somatosensory projections from the body surface to the telencephalon—via primary afferent projections to the DCN, contralateral DCN projections to the dorsal thalamus, and DIVA projections to the rostral Wulst—mediates a decreasingly specific somatotopic organization as it ascends. In the spinal dorsal horn, the somatotopy is both anatomically and electrophysiologically clear cut (see above; Necker, 1990; Woodbury, 1992); in the caudal parts of the DCN, it is reasonably clear, but much less clear in more rostral parts of the DCN (Wild, 1985). In DIVA, the somatotopy is unimpressive (Schneider and Necker, 1996), as it is in the rostral Wulst (Funke, 1989b). Furthermore, the beak appears not to be represented in DIVA, but the toes and foot joints are (Schneider and Necker, 1996).

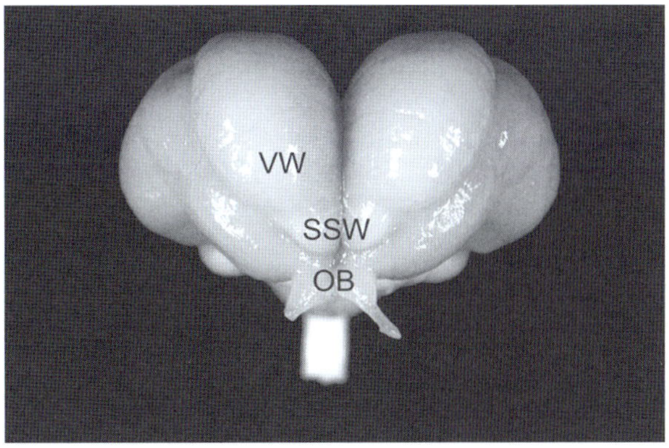

FIGURE 5.5 Photograph of the brain of a barn owl viewed from in front. VW, visual Wulst; SSW, somatosensory Wulst (claw area); OB, olfactory bulb. *From Wild et al. (2008).*

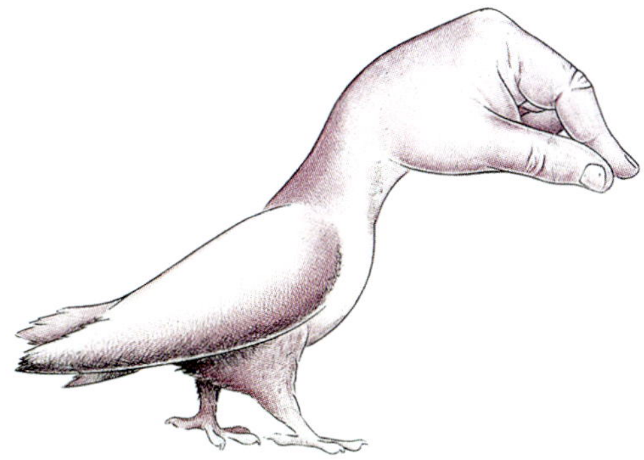

FIGURE 5.6 Illustration of the pigeon body and human hand to illustrate the functional nature of the bird's beak as a combined thumb-and-forefinger grasper and mouth.

In the barn owl, rostral to the rostral Wulst, there is an apparently unique bulge at the frontal pole of the brain that contains a detailed somatotopic representation of the contralateral claw (Manger et al., 2002; Wild et al., 2008) (Figure 5.5). Curiously, a representation of more proximal parts of the lower limb, or of any other body part, was not found in or near this bulge, but the foot and other parts of the body are also represented in what is now called Bas (Reiner et al., 1996, and see below). Apparently, sensory input from the claw is all important for the predatory barn owl. It would be most interesting to determine the representation of the claw in other predatory avian species that strike and capture their prey in a similar way to barn owls, such as eagles.

5.5 SOMATOSENSORY PRIMARY AFFERENT PROJECTIONS FROM THE BEAK AND TONGUE TO THE TRIGEMINAL COLUMN

5.5.1 Principal Sensory Trigeminal Nucleus

The bird's beak or bill varies hugely in shape and size between different species (e.g., compare the beaks of pelican, spoonbill, tucan, kiwi, flamingo, duck, cockatoo, wren)—a variation that is generally correlated with the different feeding habits and preferred foods, on the one hand, and with the size and morphological complexity of the principal sensory trigeminal nucleus (PrV; to which trigeminal beak afferents project)—on the other (Stingelin, 1961). As an instance of microevolutionary processes at work, the size and shape of the beaks of Galapagos finches may even vary according to the annual availability of preferred types of food (Weiner, 1995).

In many species, there are specializations of the beak for various feeding strategies or for probing the substrate. For instance, ducks have an elaborate bill tip organ packed with mechanorecptors (Berkhoudt, 1980), which enable the discrimination of food items through dabbling and sifting. Many shorebirds, such as sandpipers (Scolopacidae), possess Herbst corpuscles densely packed into honeycomb-like cavities in the bill, which enable the detection of food items even remote from the bill (Gerritsen and Meiboom, 1986). This is similar in kiwi, for probing the substrate during their nightime foraging (Cunningham et al., 2007; Martin et al., 2007). In many other species, such as parrots and finches, however, the mechanoreceptors are not in the bone but lie more superficially, nearer the bill surface for the detection, manipulation, and guidance of food items within the mouth (Demery et al., 2011; Krulis, 1978). During feeding, these processes are aided by tongue movements, but differently in different species. Again in parrots and finches, the tongue is loaded with mechanoreceptors (Herbst and Grandry corpuscles) that enhance the efficiency of the cracking and husking of seeds in conjunction with the beak (Krulis, 1978; Wild, 1990; Demery et al., 2011). In contrast, in pigeons the beak functions as a simple grasper, like the forefinger and thumb of humans (Figure 5.6). Once the food object is grasped, the tongue aids the transport of food objects, such as peas or grain, from the beak tip to the back of the mouth, and these items are then swallowed whole (Zweers, 1982).

PrV receives topographic projections from the beak by way of ophthalmic (upper beak), maxillary (palate, lower eyelid), and mandibular (lower beak) branches (Dubbeldam, 1980; Dubbeldam and Karten, 1978; Wild and Zeigler, 1996). Depending on the species, it may also receive projections from the tongue, but not by way of the trigeminal nerve in birds. In pigeons, there does not appear to be a representation of the tongue in PrV (Wild, unpublished observations), whereas in finches, parrots and ducks, there is a substantial lingual representation in PrV.

In ducks, lingual afferents are carried by the glossopharyngeal nerve and terminate in a dorsomedial portion of PrV (Dubbeldam et al., 1979; Dubbeldam, 1980). In finches and parrots, they are carried by the hypoglossal nerve and terminate in a dorsolateral portion of PrV (Wild, 1981, 1990) (Figure 5.7). These afferents are unlikely to be trigeminal afferents that hitchhike via glossopharyngeal nerves (CNs) IX or XII, because their cell bodies are not located in the trigeminal ganglion, but in a combined "jugular" ganglion (Dubbeldam et al., 1979; Wild, 1981). The innervation of the orderly arrays of sensory receptors (Herbst and Grandry corpuscles: Berkhoudt, 1980; Gottschaldt, 1985) in the tongue can be visualized by injecting the hypoglossal nerve with cholera toxin B-chain (CTB), or staining tongue sections with a trichrome stain (Figure 5.7). These receptors have also been visualized with electron microscopy (Toyoshima and Shimamura, 1991), although the source of their innervation was apparently not known to these authors.

5.5.2 Nucleus of the Descending Trigeminal Tract (nTTD)

Because PrV is the principal origin of projections to the telencephalon, it has overshadowed interest in the descending component of the trigeminal column. In fact, there is still only one systematic anatomical analysis of the combined afferent and efferent projections of the spinal trigeminal nucleus in birds, namely that in the mallard duck (Arends and Dubbeldam, 1984; Arends et al., 1984), and few electrophysiological analyses (Silver and Witkovsky, 1973). With Karten, Dubbeldam went on to study the descending trigeminal projections in the pigeon (Dubbeldam and Karten, 1978), using lesions of the trigeminal ganglion to induce axonal degeneration in the tract and nuclei, as had been done previously in ducks. A more recent study of these projections in pigeons was carried out using injections of HRP-conjugated CTB into the three nerve branches or their innervated territories (Wild and Zeigler, 1996), and also CTB in chickens (Wild and Krützfeldt, 2012). The various studies show a fairly consistent pattern of descending projections throughout the three subdivisions of the spinal nucleus, there being a roughly inverted representation of mandibular, maxillary, and ophthalmic afferents throughout pars interpolaris and parts of caudalis, but a clear-cut mediolateral representation at upper cervical spinal levels. Corneal afferents terminate specifically in a ventral portion of the ophthalmic representation in pars caudalis and in the laterally adjacent external cuneate nucleus (Wild, 1999). Dubbeldam and Karten (1978) also described a lateral descending tract in the pigeon, which is also present in certain snakes possessing infra-red detection and hence in birds might carry thermosensitive afferents (see Wild and Zeigler, 1996).

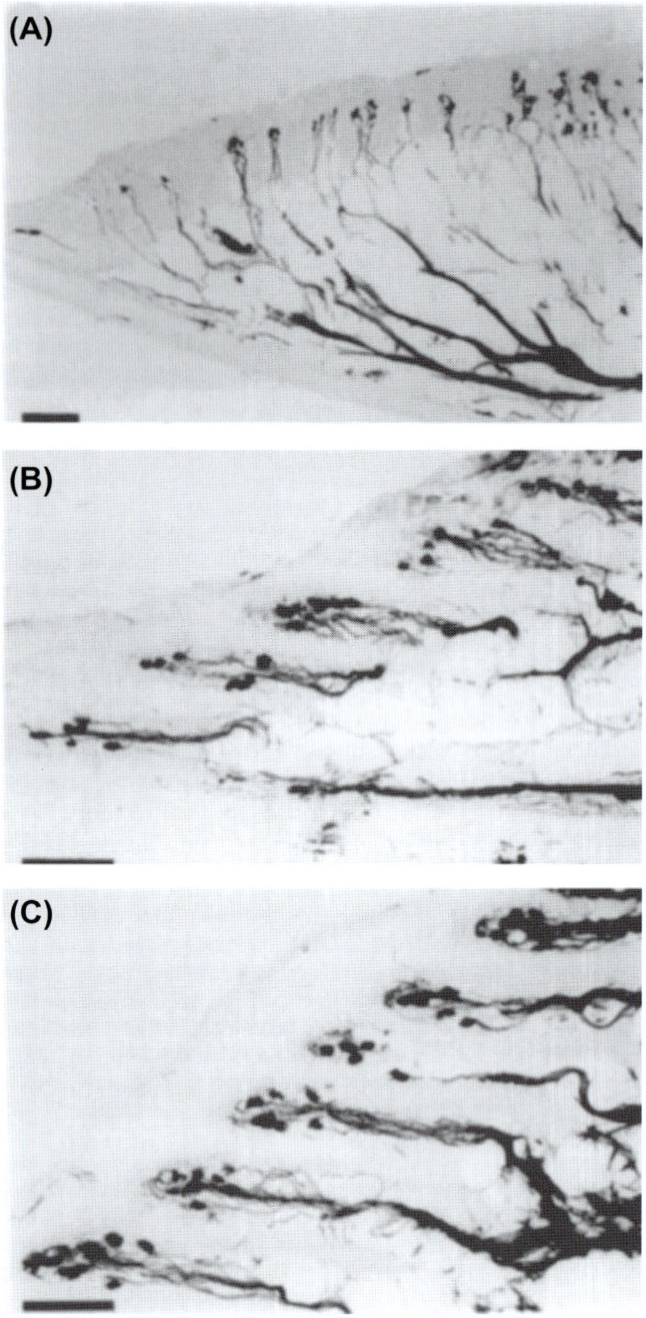

FIGURE 5.7 Photomicrographs of horseradish peroxidase-conjugated cholera toxin B-chain labeled hypoglossal nerve branches innervating the papillae of the tongues of a zebra finch (A), a greenfinch (B), and a gold finch (C), anterior to the left. The terminal "blobs" visible at this magnification (bars = 200 μm) are terminal cell receptors at the sides and apices of the bowls of flask-shaped papillae. Note how the papillae of the zebra finch tongue are perpendicular to the surface, whereas those of the other two species lie in the horizontal plane and are hence at approximately 45° to the surface. *From Wild (1990).*

As in mammals, mechanosensitive afferents that terminate in nTTD in birds arrive over other nerves in addition to the trigeminal. In the mallard tongue, afferents traveling in the glossopharyngeal nerve terminate dorsomedially in nTTD (Dubbeldam et al., 1979). In finch tongue, afferents that travel in the lingual branch of the hypoglossal nerve terminate dorsolaterally in pars interpolaris and pars caudalis of nTTD, but most medially at upper cervical spinal cord levels. Interestingly, this suggests that although the tongue makes contact with both upper and lower beaks during feeding, in the brainstem lingual afferents are aligned with mandibular afferents (which innervate the lower beak) rather than with ophthalmic ones (which innervate the upper beak).

5.6 NUCLEUS BASOROSTRALIS

Anatomical knowledge of trigeminal components of the avian somatosensory system has existed for much longer than knowledge of its spinal components, Wallenberg (1903) described a direct projection from PrV to a nucleus in the rostrobasal part of the avian forebrain via the quintofrontal tract in 1903. Although considerable anatomical and physiological attention has been directed to Bas in several species in the last 40 years, the nucleus remains enigmatic from both an anatomical and to some extent from a functional point of view. The absence of a thalamic relay in the PrV projection remains puzzling, at least in comparison to the organization of the somatosensory system in mammals (Cohen and Karten, 1974). Although Bas has been known for decades to play a major function in the sensory control of feeding in pigeons and ducks, it is now known to be more than a forebrain nucleus dedicated solely to the somatosensory representation of the beak (Witkovsky et al., 1973).

Dubbeldam and colleagues in the 1980s used the mallard to investigate the functional morphology of Bas in relation to the duck's feeding apparatus and its various mechanoreceptors involved in the different sensory phases of feeding (Berkhoudt, 1980). The projections of PrV subnuclei were traced to Bas (Dubbeldam et al., 1981) and the nucleus was mapped electrophysiologically to reveal a distinct functional topography based on the representations of the glossopharyngeal and trigeminal nerve branches innervating the tongue and beak, respectively (Berkhoudt et al., 1981). Projections of PrV to Bas were also traced in pigeons (Wild et al., 1985) and in zebra finches (Wild and Farabaugh, 1996), but Bas is much smaller in these species than in ducks and does not lend itself so readily to detailed electrophysiological mapping. In budgerigars, as in ducks, however, Bas lies atop the pallial–subpallial border rather that lateral to it, so a complete mapping of the representations of beak and tongue has been performed (Wild et al., 1997) (Figure 5.8). As in ducks, a large lingual representation is present rostrally in Bas, but this is hypoglossal territory rather than glossopharyngeal. Behind is an even larger representation of the beak, followed by a smaller one of the head. On the lateral aspect of the head, representation is an indentation that includes a representation of the cochlea and possibly of the semicircular canals; these representations seemingly mirror the position on the lateral aspect of the skull of the cochlear and vestibular apparatus at the medial end of an external auditory meatus. Progressively more

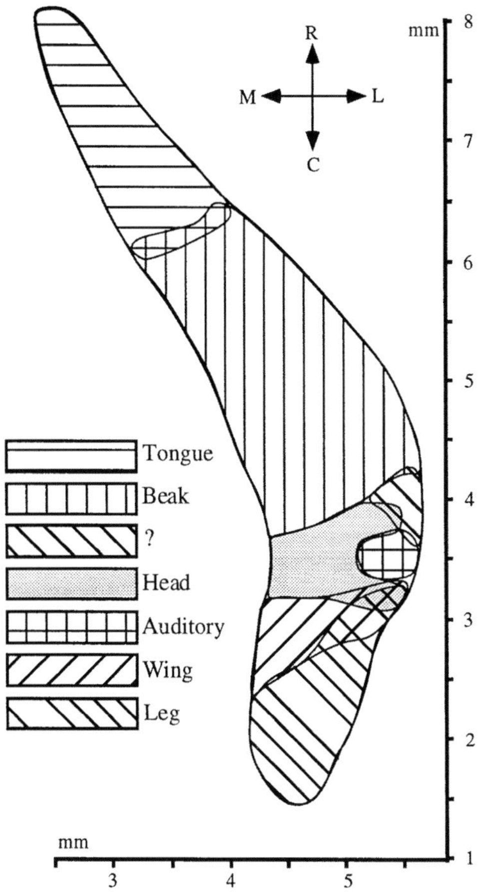

FIGURE 5.8 A two-dimensional map of Bas on the right side of the budgerigar brain. The map is based primarily on systematic recordings of evoked multiple-unit activity in Bas. The area rostral to the auditory area possibly receives inputs from a vestibular nucleus. *From Wild et al. (1997).*

caudal to the representation of the head are representations of the wings, body, and legs and feet. These body inputs are relayed predominantly by the contralateral nucleus sP, lying ventral to PrV, which supplies the beak and tongue inputs. Thus, in the budgerigar, there is a complete representation of the body in Bas, much as there is at pontine levels (see above).

The auditory representation in Bas of all species thus far examined reflects its input from the intermediate nucleus of the lateral lemniscus (LLI, which lies close to PrV), a projection first defined in pigeons (Arends and Zeigler, 1986) and later in zebra finches (Wild and Farabaugh, 1996). In these last two species, however, complete body representations have not been found in Bas, although in finches, as in parrots, there is a large lingual (hypoglossal) representation rostrally in the nucleus.

The largest Bas thus far encountered is that of the barn owl, in which there also appears to be a complete body representation, including a tonotopically organized auditory component (Wild et al., 2001). Injections of tracers in Bas retrogradely label neurons in the massive PE nucleus that forms a bulge on the lateral aspect of the pons and lies ventrolateral to the equally large anterior and posterior divisions of the dorsal nucleus of the lateral lemniscus (LLda and LLDp). In turn, injections in PE retrogradely label neurons in the DCN, thereby completing a three-neuron sequence of somatosensory projections from the body periphery, but one that is quite different from the classic sequence projecting via the thalamus. Medially adjacent to PE is a smaller LLI that presumably provides most of the auditory input to Bas, although Bas seems also to receive a small auditory projection from LLDa.

5.7 THE MEETING OF THE SPINAL AND TRIGEMINAL SYSTEMS

In his chapter on the spinal cord, Necker (2000b) briefly mentioned a difference between pigeons and chickens in the laminar organization of the dorsal horn—a difference that warranted further examination (Wild et al., 2010). In pigeons, the dorsal horn laminae are arranged in mammalian fashion, with lamina II lying dorsal to lamina III; however, in chickens the latter lies medial, not ventral, to the former (cf. Brinkman and Martin, 1974; Leonard and Cohen, 1975b). In fact, as Woodbury (1998) showed, the majority of bird species resemble chickens, rather than pigeons, in the bifid, side by side arrangement of laminae II and III. Nevertheless, the functional organization of these laminae in chickens and their kin is not dissimilar to that in pigeons and their kin, in that smaller diameter and unmyelinated primary afferents terminate in lamina I and II, whereas larger diameter, myelinated primary afferents terminate in lamina III (Woodbury and Scott, 1991; Wild, 1985).

Many years ago, Woodbury asked a related, interesting question about the organization of the trigeminal dorsal horn: Since this was generally considered to be a direct continuation of the spinal dorsal horn, and to have a similar organization, did this organization reflect the organization present at spinal levels in species in which II and III were side by side rather than arranged in mammalian-type laminar fashion? In other words, was the trigeminal dorsal horn also split in chickens, etc., with a side by side arrangement of superficial and deeper laminae? This question has been recently answered, and the findings compared with those in pigeons (Wild and Zeigler, 1996; Wild et al., 2010). It was found that the trigeminal dorsal horn in chickens was organized in a laminar fashion, as it is in pigeons, and remained laminar down to about the level of C3. Thereafter, however, lamina III gradually, over a couple of segments, shifted medially, until from C5 caudally laminae II and III came to lie side by side. It was also shown that, although the three branches of the trigeminal nerve in the chicken terminated throughout the dorsal horn in a similar fashion to that in pigeons—with ophthalmic afferents laterally, mandibular afferents medially and maxillary afferents in between—they did not extend further caudally than about C3, in contrast to C6 or C7 in pigeons. Thus, it appears that trigeminal primary afferents in the chicken do not terminate at cervical spinal levels, having a side by side arrangement of dorsal horn laminae. These findings suggest a discontinuity of spinal and trigeminal dorsal horn organization, the reason for which remains obscure. What is clear, however, is that Woodbury's (1998) hypothesis of a side by side arrangement of laminae II and II defining a novel clade of birds is not supported, for when an appropriate phylogenetic analysis is performed, a reversion to the mammalian type of arrangement of dorsal horn lamina is seen to have occurred several times in avian evolution (Wild et al., 2010).

5.8 THE SOMATOSENSORIMOTOR SYSTEM IN BIRDS

In mammals, somatosensory inputs to the cortex initiate somatomotor responses that reach brainstem and spinal targets via corticobulbar and corticospinal components of the pyramidal tract. Such a tract in birds originating in the rostral part of the Wulst was proposed by Zecha (1962) and shown with contemporary tracing techniques in pigeons and zebra finches by Wild and Williams (2000). The pyramidal tract equivalent originates in the hyperpallium apicale (HA) of the rostral Wulst, implying rather direct connections between somatosensory inputs to IHA and somatomotor outputs from HA. The tract and its terminations are far from meager. As in mammals, the red nucleus is a prime target, which then projects to all levels of the spinal cord (Wild et al., 1979). Further caudally, the tract terminates profusely throughout wide regions of the brainstem reticular formation and secondary sensory nuclei, such as the external cuneate nucleus, but does not proceed further caudally than the upper cervical levels.

An interesting but little-known somatosensorimotor link in the brainstem reflects control of the lower eyelid, which relaxes during sleep to cover the eye, in contrast to the descent of the upper lid in mammals. During waking hours, the lower eyelid appears to be under tonic control to keep the eye open, until a corneal stimulus initiates a defensive reaction. These actions are mediated via corneal afferent input to the ophthalmic part of nTTD and projections of nTTD to the dorsal motor nucleus of the trigeminal motor complex that innervates the two muscles of the lower eyelid (Wild, 1999).

In contrast to motor output from rostral HA via the pyramidal tract, outputs of Bas exit the telencephalon from the arcopallium at the caudal pole of the brain. They reach the arcopallium indirectly via an overlying part of the frontal nidopallium and the caudolateral part of the nidopallium (NCL) (Wild et al., 1985; Dubbeldam and Visser, 1987). In finches, beak and auditory components of Bas were shown to follow this route independently (Wild and Farabaugh,

1996). The descending projections of different major parts of the arcopallium were originally described by Zeier and Karten (1971) in pigeons and shown to include secondary sensory nuclei in the brainstem. This was specifically confirmed using contemporary tracing techniques in zebra finches, in which Bas-recipient components of the lateral arcopallium were shown to target the spinal trigeminal nucleus (Wild and Farabaugh, 1996) and the ventromedial part of the parvocellular reticular formation (RPcvm), which is a major nexus of premotor projections to the jaw and other upper vocal tract motor nuclei (Wild and Krützfeldt, 2012).

5.9 SOMATOSENSORY PROJECTIONS TO THE CEREBELLUM

In the pigeon cerebellum, there are two somatosensory areas—one rostrally (lobules I–VI) and another caudally (lobule IX) (Necker, 2000a; see also Whitlock, 1952). Somatosensory projections to these regions are relayed from the spinal cord (Necker, 1992, 2000a; Okada et al., 1987). Spinal inputs to the anterior cerebellum arise predominantly from neurons in Clarke's column and hence are proprioceptive. Spinal inputs to the posterior cerebellum arise primarily from cervical lamina IV neurons, which are mechanosensitive (Necker, 1992). Another source of putative mechanosensitive spinocerebellar fibers is the paragriseal cells, which in the lumbosacral spinal cord are contacted by the axons of neurons in the contralateral accessory lobes of Lachi (marginal nuclei) (Necker, 1997). These lobes comprise a sensory component of a lumbosacral specialization involved in a sense of equilibrium (Necker, 1999; Necker et al., 2000; Rosenberg and Necker, 2000).

5.10 MAGNETORECEPTION AND THE TRIGEMINAL SYSTEM

The link between magnetoreception and the trigeminal system is highly controversial, which is fully discussed by Mouritsen in the present volume (Chapter 8). My own involvement in this area of research has been fascinating but has also produced some apparently inconsistent results. Initially, I was able to contribute to a careful laboratory conditioning study of magnetic intensity discrimination in homing pigeons by sectioning either the ophthalmic branch of the trigeminal nerve (which innervates the upper beak where the magnetoreceptor was suspected to lie) or the olfactory nerve (as a control)—first in different birds and then in the same birds (Mora et al., 2004). The results showed that a learned discrimination between the presence versus the absence of a magnetic anomaly was reduced to chance by sectioning the ophthalmic nerve but not by sectioning the olfactory nerve. At about the same time, our group showed what appeared to be approximations between cholera toxin B chain-labeled ophthalmic afferents and iron-containing structures in the base of the upper beak—structures that we thought might be some sort of mechanoreceptor that could perhaps signal a magnetic stimulus (Williams and Wild, 2001). Together, these studies seemed to support the work of Beason and Semm (1987) and Semm and Beason (1990), who reported electrophysiological responses of trigeminal nerve cells to magnetic stimulation in the transequatorial migratory bobolink.

Then, together with Anna Gagliardo and colleagues in Pisa, Italy, we performed the same surgery (ophthalmic or olfactory nerve sections) in two groups of homing pigeons prior to releasing them from unfamiliar territory approximately ~50 km from their home loft. The results were exactly the opposite of those gathered in the laboratory (Gagliardo et al., 2006). That is, trigeminal deafferentation had no effect whatsoever on measures of either orientation or homing performance, whereas olfactory nerve deafferentation produced severe effects in both, with the majority of birds never to be seen again. Variations on this theme were performed in subsequent experiments, all with the same effect (Gagliardo et al., 2008, 2009). The role of olfaction in pigeon homing, although fascinating (Gagliardo, 2013), is not relevant here. What is relevant is that, despite the results of these behavioral, real-life homing experiments, other anatomists have produced evidence either in favor of beak (trigeminal) mediated magnetoreception (Fleissner et al., 2003) or completely against it (Treiber et al., 2012).

Then, working with Mouritsen, we showed that the trigeminal ophthalmic nerve was not involved in the detection of magnetic compass information in a European migrant, the European robin (Zapka et al., 2009), but it did appear to be involved in mediating the detection of magnetic field changes, as indicated by stimulus-dependent zenk protein activation in PrV and in parts of the spinal trigeminal nucleus (Heyers et al., 2010).

What the outcome of these results will be is not clear, especially in the absence of unequivocal evidence of a magnetoreceptor in the beak of either pigeons (Treiber et al., 2012) or in any species thought to use the Earth's magnetic field to guide migration. Furthermore, electrophysiological recordings of responses to magnetic stimuli in the trigeminal brainstem complex have not thus far been replicated, although recordings in other brainstem regions have. Technical difficulties in ruling out artifactual responses to electrical and metal components of the recording apparatus should also be appreciated in studies of magnetoreception. Perhaps there are magnetoreceptors in other peripheral locations, such as the lagenar or vestibular apparatus (Wu and Dickman, 2012), but it should be noted that removal of the lagenar was shown to have no effect on homing in pigeons (Walraff, 1972). These and other issues, and the evidence in favor of light-dependent magnetoreception, are discussed by Mouritsen (2013).

5.11 SUMMARY AND CONCLUSIONS

It is hoped that this chapter will lead to the realization and appreciation of considerable diversity in the organization of the somatosensory system in birds. Even within the few species so far examined, there seem to be major deviations from the commonly accepted mammalian-type sequence of somatosensory projections throughout the brain. Of course, within mammals also, there are huge differences in the species-specific representation of the integument and of parts of the head—representations that, as in birds, reflect the functional importance of the body part in the everyday behavior of the animal. However, these distorted representations in mammals are the end stations of a common three-neuron sequence of somatosensory projections from the periphery to the cortex, via the DCN and the thalamus. In contrast, in some species of birds, even the DCN are not an obligatory relay in the sequence of ascending projections from the periphery. Furthermore, in parrots, at least, there appears to be a complete representation of the body and head at pontine levels—a phenomenon not encountered in mammals, or probably in many other species of birds.

Another striking and equally puzzling feature of the somatosensory projections in birds is the complete or partial separation of the representations of the beak and body in Bas or the rostral Wulst, respectively. The great majority of beak inputs are relayed via PrV to Bas. In barn owls and budgerigars, the representation of the beak in Bas is but one part, albeit a major part, of a complete representation of the whole body. In ducks, pigeons, and finches, a representation of the rest of the body is apparently absent in Bas (or has not yet been found), except for that of the cochlea.

In pigeons, the somatosensory, rostral part of the Wulst comprises a weakly somatotopic representation of the body minus the beak, with a second nonsomatotopic body representation further caudally in the nidopallium (Wild, 1987; Funke, 1989a). Whether there is a somatosensory representation of the body in the rostral Wulst of ducks and parrots remains to be determined. In barn owls, the claw—and only the claw—is represented in what may be a specialization of the rostral Wulst. In the dunlin (Pettigrew and Frost, 1985), the finding of a tactile fovea in the rostral part of the telencephalon, representing the specialized probing beak tips, is intriguing, but whether the recording electrodes were in Bas or a body representation in another part of the telencephalon is not clear.

In the final analysis, we should remind ourselves—obvious though it may be—that one of the main functions of the body somatosensory system, at least for most birds, is to enable flight. Once thought to be a meager system (Ariëns Kappers et al., 1936), the complexity of its organization can now be appreciated with the help of the tools of modern neuroanatomy and electrophysiology. With the help of cameras fixed to flying birds (jdp.co.uk/programmes/earthflight), we can marvel at the feedback from feather follicles this system must provide to guide the birds through their aerial worlds (Bilo and Bilo, 1983; Necker, 1985c).

REFERENCES

Arends, J.J.A., 1981. Sensory and Motor Aspects of the Trigeminal System in the Mallard (*Anas platyrhynchos* L.) (Ph.D. thesis). Leiden University, Netherlands.

Arends, J.J.A., Dubbeldam, J.L., 1984. The subnuclei and primary afferents of the descending trigeminal system in the mallard. Neuroscience 13, 781–795.

Arends, J.J.A., Zeigler, H.P., 1986. Anatomical identification of an auditory pathway from a nucleus of the lateral lemniscus to the frontal telencephalon (nucleus basalis) of the pigeon. Brain Res. 398, 375–381.

Arends, J.J.A., Woelders-Blok, A., Dubbeldam, J.L., 1984. The efferent connections of the nuclei of the descending trigeminal tract in the mallard (*Anas platyrhyncos* L.). Neuroscience 13, 797–817.

Ariëns Kappers, C.U., Huber, G.C., Crosby, E.C., 1936. The Comparative Anatomy of the Nervous System of Vertebrates, Including Man. The Macmillan Company, New York.

Ballam, G.O., 1982. Bilateral and multimodal sensory interactions of single cells in the pigeon's midbrain. Brain Res. 245, 27–35.

Beason, R.C., Semm, P., 1987. Magnetic responses of the trigeminal nerve system of the bobolink (*Dolichonyx oryzivorus*). Neurosci. Lett. 80, 229–234.

Berkhoudt, H., 1980. The morphology and distribution f cutaneous mechanoreceptors (Herbst and Grandry corpuscles) in the bill and tongue of the mallard (*Anas platyrhynchos* L.). Netherl. J. Zool. 30, 1–34.

Berkhoudt, H., Dubbeldam, L., Zeilstra, S., 1981. Studies on the somatotopy of the trigeminal system in the mallard, *Anas platyrhynchos* L. IV. Tactile representation in the nucleus basalis. An electrophysiological study. J. Comp. Neurol. 196, 407–420.

Berkley, K.J., Budell, R.J., Blomqvist, A., Bull, M., 1986. Output systems of the dorsal column nuclei in the cat. Brain Res. Rev. 11, 199–225.

Bilo, D., Bilo, A., 1983. Neck flexion related activity of flight control muscles in the flow-stimulated pigeon. J. Comp. Physiol. 153, 111–122.

Brinkman, R.W., Martin, A.H., 1974. A cytoarchitectonic study of the spinal cord of the domestic fowl *Gallus gallus domesticus*. I. Brachial region. Brain Res. 56, 43–62.

Cohen, D.H., Karten, H.J., 1974. The structural organization of avian brain: an overview. In: Goodman, I.J., Schein, M. (Eds.), Birds: Brain and Behavior. Academic Press, New York, pp. 29–73.

Coleman, M.J., Roy, A., Wild, J.M., Mooney, R., 2007. Thalamic gating of auditory responses in telencephalic song control nuclei. J. Neurosci. 27, 10024–10036.

Cunningham, S., Castro, I., Alley, M., 2007. A new prey-detection mechanism for kiwi (*Apteryx* spp.) suggests convergent evolution between paleognathous and neognathous birds. J. Anat. 211, 493–502.

Demery, Z.P., Chappell, J., Martin, G.R., 2011. Vision, touch and object manipulation in Senegal parrots *Poicephalus senegalus*. Proc. R Soc Lond. B. Biol. Sci. 278, 3687–3693.

Dubbeldam, J.L., 1980. Studies on the somatotopy of the trigeminal system in the mallard, *Anas platyrhynchos* L. II. Morphology of the principal sensory nucleus. J. Comp. Neurol. 191, 557–571.

Dubbeldam, J.L., Karten, H.J., 1978. The trigeminal system in the pigeon (*Columba livia*). I. Projections of the Gasserian ganglion. J. Comp. Neurol. 180, 661–678.

Dubbeldam, J.L., Visser, A.M., 1987. The organization of the nucleus basalis-neostriatum complex of the mallard (*Anas platyrhynchos* L.) and its connections with the archistriatum and the paleostriatum complex. Neuroscience 21, 487–517.

Dubbeldam, J.L., Brus, E.R., Menken, S.B.J., Zeilstra, S., 1979. The central projections of the glossopharyngeal and vagus ganglia in the mallard, *Anas platyrhynchos* L. J. Comp. Neurol. 183, 149–168.

Dubbeldam, J.L., Brauch, C.S.M., Don, A., 1981. Studies on the somatotopy of the trigeminal system in the mallard, *Anas platyrhynchos* L. III. Afferents and organization of the nucleus basalis. J. Comp. Neurol. 196, 391–405.

Fleissner, G., Holtkamp-Rötzler, E., Hanzlik, M., Winklhofer, M., Fleissner, G., Petersen, N., Wiltschko, W., 2003. Ultrastructural analysis of a putative magnetoreceptor in the beak of homing pigeons. J. Comp. Neurol. 458, 350–360.

Funke, K., 1989a. Somatosensory areas in the telencephalon of the pigeon. II. Spinal pathways and afferent connections. Exp. Brain Res. 76, 620–638.

Funke, K., 1989b. Somatosensory areas in the telencephalon of the pigeon. I. Response characteristics. Exp. Brain Res. 76, 603–619.

Gagliardo, A., 2013. Forty years of olfactory navigation in birds. J. Exp. Biol. 216, 2165–2171.

Gagliardo, A., Ioale, P., Savini, M., Wild, J.M., 2006. Having the nerve to home: olfactory versus magnetoreceptor mediation of homing in pigeons. J. Exp. Biol. 209, 2888–2892.

Gagliardo, A., Ioalè, P., Savini, M., Wild, J.M., 2008. Navigational abilities of homing pigeons deprived of olfactory or trigeminally mediated magnetic information when young. J. Exp. Biol. 211, 2046–2051.

Gagliardo, A., Ioalè, P., Savini, M., Wild, J.M., 2009. Navigational abilities of adult and experienced homing pigeons deprived of olfactory or trigeminally mediated magnetic information. J. Exp. Biol. 212, 3119–3124.

Gamlin, P.D.R., Cohen, D.H., 1986. A second ascending visual pathway from the optic tectum to the telencephalon in the pigeon (*Columba livia*). J. Comp. Neurol. 250, 296–310.

Gentle, M.J., Hunter, L.N., Sterling, R.J., 1995. Projections of ankle joint afferents to the spinal cord and brainstem of the chicken (*Gallus g. domesticus*). J. Comp. Neurol. 361, 669–680.

Gerritsen, A.F.C., Meiboom, A., 1986. The role of touch in prey density estimation by *Calidris alba*. Neth. J. Zool. 36, 530–562.

Gottschaldt, K.-M., 1985. Structure and function of avian somatosensory receptors. In: In: King, A.S., McLelland, J. (Eds.), Form and Function in Birds, vol. 3. Academic Press, London, pp. 375–461.

Heyers, D., Zapka, M., Hoffmeister, M., Wild, J.M., Mouritsen, H., 2010. Magnetic field changes activate the trigeminal brainstem complex in a migratory bird. PNAS 107, 9394–9399.

Karten, H.J., 1967. The organization of the ascending auditory pathway in the pigeon (*Columba livia*). I. Diencephalic projections of the inferior colliculus (nucleus mesencehalicus lateralis, pars dorsalis). Brain Res. 6, 409–427.

Korzeniewski, E., 1987. Multisensory convergence in the thalamus of the pigeon (*Columba livia*). Neurosci. Lett. 80, 55–60.

Korzeniewski, E., Güntürkün, O., 1990. Sensory properties and afferents of the N. dorsolateralis posterior thalami of the pigeon. J. Comp. Neurol. 292, 457–479.

Krulis, V., 1978. Struktur und Verteilung von Tastrezeptoren im Schnabel-Zungenbereich von Singvögeln, im besonderen der Fringillidae. Rev. Suisse Zool. 85, 385–447.

Leibler, L., 1975. Ascending Binaural and Monaural Pathways to Mesencephalic and Diencephalic Auditory Nuclei in the Pigeon, *Columba livia* (Ph.D. thesis). Massachusetts Institute of Technology, Cambridge, Mass., USA.

Leonard, R.B., Cohen, D.H., 1975a. Spinal terminal fields of dorsal root fibers in the pigeon (*Columba livia*). J. Comp. Neurol. 163, 181–192.

Leonard, R.B., Cohen, D.H., 1975b. A cytoarchitectonic analysis of the spinal cord of the pigeon (*Columba livia*). J. Comp. Neurol. 163, 159–180.

Manger, P.R., Elston, G.N., Pettigrew, J.D., 2002. Multiple maps and activity-dependent representational plasticity in the anterior Wulst of the adult barn owl (*Tyto alba*). Eur. J. Neurosci. 16, 743–750.

Martin, G.R., Wilson, K.-J., Wild, J.M., Parsons, S., Kubke, M.F., Corfield, J., 2007. Kiwi forego vision in the guidance of their nocturnal activities. PLoS One 2 (2), e198. http://dx.doi.org/10.1371/journal.pone.0000198.

Mora, C.V., Davison, M., Wild, J.M., Walker, M.M., 2004. Magnetoreception and its trigeminal mediation in the homing pigeon. Nature 432, 508–511.

Mouritsen, H., 2014. Magnetoreception in birds and its use for long-distance migration. In: Sturkie's Avian Physiology, sixth ed., pp. 113–133.

Necker, R., 1985a. Projection of a cutaneous nerve to the spinal cord of the pigeon. I. Evoked field potentials. Exp. Brain Res. 59, 338–343.

Necker, R., 1985b. Projection of a cutaneous nerve to the spinal cord of the pigeon. II. Responses of dorsal horn neurons. Exp. Brain Res. 59, 344–352.

Necker, R., 1985c. Observations on the function of a slowly-adapting mechanoreceptor associated with filoplumes in the feathered skin of pigeons. J. Comp. Physiol. A 156, 391–394.

Necker, R., 1990. Sensory representation of the wing in the spinal dorsal horn of the pigeon. Exp. Brain Res. 81, 403–412.

Necker, R., 1991. Cells of origin of avian postsynaptic dorsal column pathways. Neurosci. Lett. 126, 91–93.

Necker, R., 1992. Spinal neurons projecting to anterior or posterior cerebellum in the pigeon. Anat. Embryol. 185, 325–334.

Necker, R., 1997. Projections of the marginal nuclei in the spinal cord of the pigeon. J. Comp. Neurol. 377, 95–104.

Necker, R., 1999. Specializations in the lumbosacral spinal cord of birds: morphological and behavioural evidence for a sense of equilibrium. Europ. J. Morphol. 37, 211–214.

Necker, R., 2000a. The somatosensory system. In: Sturkie's Avian Physiology, fifth ed., pp. 57–69.

Necker, 2000b. Functional organization of the spinal cord. In: Sturkie's Avian Physiology, fifth ed. pp. 71–81.

Necker, R., 2001. Spinocerebellar projections in the pigeon with special reference to the neck region of the body. J. Comp. Neurol. 429, 403–418.

Necker, R., Schermuly, C., 1985. Central projections of the radial nerve and one of its cutaneous branches in the pigeon. Neurosci. Lett. 58, 271–276.

Necker, R., Janssen, A., Beissenhirtz, T., 2000. Behavioural evidence of the role of lumbosacral anatomical specializations in pigeons in maintaining balance during terrestrial locomotion. J. Comp. Physiol. A 186, 409–412.

Nottebohm, F., Kelley, D.B., Paton, J.A., 1982. Connections of vocal control nuclei in the canary telencephalon. J. Comp. Neurol. 207, 344–357.
Ohmori, Y., Necker, R., 1995. Central projections of primary afferents from the interosseus nerve in the pigeon. Brain Res. Bull. 38, 269–274.
Okado, N., Ito, R., Homma, S., 1987. The terminal distribution pattern of spinocerebellar fibers: an anterograde labelling study in the posthatching chick. Anat. Embryol. (Berl) 176, 175–182.
Pettigrew, J.D., Frost, B.J., 1985. A tactile fovea in the Scolopacidae? Brain Behav. Evol. 26, 185–195.
Puelles, L., Rrobles, M., Martínez-de-la-Torre, M., Martínez, S., 1994. New subdivision schema for the avian torus semicircularis: neurochemical maps in the chick. J. Comp. Neurol. 340, 98–125.
Reiner, A., Bruce, L., Butler, A., Csillag, A., Kuenzel, W., Medina, L., Paxinos, G., Perkel, D., Shimizu, T., Striedter, G., Wild, J.M., Ball, G., Durand, S., Gunturkun, O., Lee, D., Reinke, H., Necker, R., 1996. Coding of vibration by neurons of the dorsal column nuclei in the pigeon. J. Comp. Physiol. A 179, 263–276.
Reinke, H., Necker, R., 1996. Coding of vibration by neurones of the dorsal column nuclei in the pigeon. J. Comp. Physiol. A 179, 263–276.
Robertson, B., Grant, G., 1985. A comparison between wheat germ agglutinin and choleragenoid-horseradish peroxidise as anterogradely transported markers in central branches of primary sensory neurons in rat with some observations in the cat. Neuroscience 14, 895–905.
Rosenberg, J., Necker, R., 2000. Fine structural evidence of mechanoreception in spinal lumbosacral accessory lobes of pigeons. Neurosci. Lett. 285, 13–16.
Reiner, A., Perkel, D.J., Bruce, L.L., Butler, A.B., Csillag, A., Kuenzel, W., Medina, L., Paxinos, G., Shimizu, T., Striedter, G., Wild, M., Ball, G.F., Durand, S., Güntürkün, O., Lee, D.W., Mello, C.V., Powers, A., White, S.A., Hough, G., Kubikova, L., Smulders, T.V., Wada, K., Dugas-Ford, J., Husband, S., Yamamoto, K., Yu, J., Siang, C., Jarvis, E.D., 2004. Revised nomenclature for avian telencephalon and some related brainstem nuclei. J. Comp. Neurol. 473, 377–414.
Schneider, A., Necker, R., 1989. Spinothalamic projections in the pigeon. Brain Res. 484, 139–149.
Schneider, A., Necker, R., 1996. Electrophysiological investigations of the somatosensory thalamus of the pigeon. Exp. Brain Res. 109, 377–383.
Schulte, M., Necker, R., 1994. Projection of wing nerves to spinal cord and brain stem of he pigeon a studies by transganglionic transport of Fast Blue. J. Brain Res. 35, 313–325.
Semm, P., Beason, R.C., 1990. Responses to small magnetic variations by the trigeminal system of the bobolink. Brain Res. Bull. 25, 735–740.
Silver, R., Witkovsky, P., 1973. Functional characteristics of single units in the spinal trigeminal nucleus of the pigeon. Brain Behav. Evol. 8, 287–303.
Stingelin, W., 1961. Grossenunterschiede des sensiblen Trigeminuskerns bei verschiedenen Vögeln. Rev. Suisse Zool. 68, 247–251.
Toyoshima, K., Shimamura, A., 1991. Ultrastructure of Merkel corpuscles in the tongue of the finch, Lonchura striata. Cell Tiss. Res. 264, 427–436.
Treiber, C.D., Salzer, M.C., Riegler, J., Edelman, N., Sugar, C., Breuss, M., Pichler, P., Cadiou, H., Saunders, M., Lythgoe, M., Shaw, J., Keays, D.A., 2012. Clusters of iron-rich cells in the upper beak of pigeons are macrophages not magnetosensitive neurons. Nature. http://dx.doi.org/10.1038/nature11046.
van den Akker, L.M., 1970. An Anatomical Outline of the Spinal Cord of the Pigeon. Assen: The Netherlands: Royal Van Gorcum Ltd.
Vates, E.G., 1996. Auditory pathways of caudal telencephalon and their relation to the song system of adult male zebra finches (*Taenopygia guttata*). J. Comp. Neurol. 366, 613–642.
Wallenberg, A., 1903. Der Ursprung des Traktus isthmo-striatus (oder bulbo-striatus) der Taube. Neurol. Zentralblatt 22, 98–101.
Wallraff, H.G., 1972. Homing of pigeons after extirpation of their cochleae and lagenae. Nat. New Biol. 236, 223–224.
Weiner, J., 1995. The Beak of the Finch. Vintage.
Whitlock, D.G., 1952. A neurohistological and neurophysiological study of afferent fiber tracts and receptive areas of the avian cerebellum. J. Comp. Neurol. 97, 567–635.
Wild, J.M., 1981. Identification and localization of the motor nuclei and sensory projections of the glossopharyngeal, vagus and hypoglossal nerves in the cockatoo (*Cacatua roseicapilla*), Cacatuidae. J. Comp. Neurol. 203, 352–378.
Wild, J.M., 1985. The avian somatosensory system. I. Primary spinal afferent input to the spinal cord and brainstem in the pigeon (*Columba livia*). J. Comp. Neurol. 240, 377–395.
Wild, J.M., 1987. The avian somatosensory system: connections of regions of body representation in the forebrain of the pigeon. Brain Res. 412, 205–223.
Wild, J.M., 1989. The avian somatosensory system. II. Ascending projections of the dorsal column and external cuneate nuclei in the pigeon. J. Comp. Neurol. 287, 1–18.
Wild, J.M., 1990. Peripheral and central terminations of hypoglossal afferents innervating lingual tactile mechanoreceptor complexes in *Fringillidae*. J. Comp. Neurol. 298, 157–171.
Wild, J.M., 1992. Direct and indirect "cortico"-rubral and rubrocerebellar cortical projections in the pigeon. J. Comp. Neurol. 326, 623–636.
Wild, J.M., 1994. Visual and somatosensory projections to the avian song system via nucleus uvaeformis (Uva) and a comparison with the projections of a similar thalamic nucleus in a non-songbird (*Columba livia*). J. Comp. Neurol. 349, 512–535.
Wild, J.M., 1995. Convergence of somatosensory and auditory projections in the avian torus semicircularis, including the central auditory nucleus. J. Comp. Neurol. 358, 465–486.
Wild, J.M., 1997. The avian somatosensory system: the pathway from wing to Wulst in a passerine (*Chloris chloris*). Brain Res. 759, 122–134.
Wild, J.M., 1999. Trigeminal disynaptic circuit mediating corneal afferent input to M. depressor palpebrae inferioris motoneurons in the pigeon (*Columba livia*). J. Comp. Neurol. 403, 391–406.
Wild, J.M., Farabaugh, S.M., 1996. Organization of afferent and efferent projections of nucleus basalis prosencephali in a passerine (*Taeneopygia guttata*). J. Comp. Neurol. 365, 306–328.
Wild, J.M., Krützfeldt, N.O.E., 2012. Trigeminal and telencephalic projections to jaw and other upper vocal tract premotor neurons in songbirds: sensorimotor circuitry for beak movements during singing. J. Comp. Neurol. 520, 590–605.
Wild, J.M., Williams, M.N., 1999. Rostral wulst in passerine birds. II. Intratelencephalic projections to nuclei associated with the auditory and song systems. J. Comp. Neurol. 413, 520–534.
Wild, J.M., Williams, M.N., 2000. Rostral wulst in passerine birds. I. Origin, course and terminations of an avian 'pyramidal tract'. J. Comp. Neurol. 416, 429–450.
Wild, J.M., Zeigler, H.P., 1996. Central projections and somatotopic organization of trigeminal primary afferents in the pigeon (*Columba livia*). J. Comp. Neurol. 368, 136–152.

Wild, J.M., Cabot, J.B., Cohen, D.H., Karten, H.J., 1979. Origin, course and terminations of the rubrospinal tract in the pigeon (*Columba livia*). J. Comp. Neurol. 197, 639–654.

Wild, J.M., Arends, J.J.A., Zeigler, H.P., 1985. Telencephalic connections of the trigeminal system in the pigeon (*Columba livia*): a trigeminal sensorimotor circuit. J. Comp. Neurol. 234, 441–464.

Wild, J.M., Reinke, H., Farabaugh, S.M., 1997. A non-thalamic pathway contributes to a whole body map in the brain of the budgerigar. Brain Res. 755, 137–141.

Wild, J.M., Kubke, M.F., Carr, C.E., 2001. Tonotopic and somatotopic representation in nucleus basalis of the barn owl, *Tyto alba*. Brain Behav. Evol. 57, 39–62.

Wild, J.M., Kubke, M.F., Pena, J.L., 2008. A pathway for predation in the brain of the barn owl: projections of the gracile nucleus to the 'claw area' of the rostral wulst, via the dorsal thalamus. J. Comp. Neurol. 509, 156–166.

Wild, J.M., Krützfeldt, N.O.E., Altshuler, D., 2010. Trigeminal and spinal dorsal horn (dis)continuity and avian evolution. Brain Behav. Evol. 76, 11–19.

Witkovsky, P., Zeigler, H.P., Silver, R., 1973. The nucleus basalis of the pigeon: a single unit analysis. J. Comp. Neurol. 147, 119–128.

Williams, M.N., Wild, J.M., 2001. Trigeminally innervated iron-containing structures in the beak of homing pigeons, and other birds. Brain Res. 889, 243–246.

Woodbury, C.J., 1992. Physiological studies of cutaneous inputs to dorsal horn laminae I–IV of adult chickens. J. Neurophysiol. 67, 241–254.

Woodbury, C.J., 1998. Two spinal cords in birds: novel insights into early avian evolution. Proc. R. Soc. Lond. B 265, 1721–1729.

Woodbury, C.J., Scott, S.A., 1991. Somatotopic organization of hindlimb skin sensory inputs to the dorsal horn of hatchling chicks (*Gallus g. domesticus*). J. Comp. Neurol. 314, 237–256.

Wu, L.-Q., Dickman, J.D., 2012. Neural correlates of a magnetic sense. Science 336, 1054–1057.

Zapka, M., Heyers, D., Hein, C.M., Engels, S., Schneider, N.-L., Hans, J., Weiler, S., Dryer, D., Kishkinev, D., Wild, M., Mouritsen, H., 2009. Visual, but not trigeminal, mediation of magnetic compass information in a migratory bird. Nature 461, 1274–1278.

Zecha, A., 1962. The 'pyramidal tract' and other telencephalic efferents in birds. Acta Morphol. Neerl. Scand. 5, 194–195.

Zeier, H., Karten, H.J., 1971. The archistriatum of the pigeon: organization of afferent and efferent connections. Brain Res. 31, 313–326.

Zweers, G., 1982. The feeding system of the pigeon (*Columba livia*). Adv. Anat. Embryol. Cell Biol. 73, 1–108.

Chapter 6

Avian Hearing

Christine Köppl

Cluster of Excellence "Hearing4all", Carl von Ossietzky University, Oldenburg, Germany

Research Center Neurosensory Science, Carl von Ossietzky University, Oldenburg, Germany

Department of Neuroscience, School of Medicine and Health Science, Carl von Ossietzky University, Oldenburg, Germany

ABBREVIATIONS

Aivm Ventromedial portion of the intermediate archopallium
CM Caudal nidopallium
ICc Central nucleus of the inferior colliculus
ICx External nucleus of the inferior colliculus
ILD Interaural level difference
ITD Interaural time difference
LLDp Posterior part of the ventral nucleus of the lateral lemniscus
MLd Nucleus mesencephalicus lateralis pars dorsalis
NA Nucleus angularis
NCM Caudal medial nidopallium
Nd Dorsal nidopallium
NL Nucleus laminaris
NM Nucleus magnocellularis

6.1 INTRODUCTION: WHAT DO BIRDS HEAR?

Birds are very vocal. They communicate by a multitude of vocalizations, ranging from simple calls and cries to extremely varied songs, which are often also beautiful to our ears. Unlike simple calls and like human speech, birdsong must be learned and can serve to recognize other birds individually. Even beyond communication, the sense of hearing has special meaning to many birds. Owls that hunt hidden prey or at night rely on their keen sense of hearing. Some birds use echolocation to find their way in dark caves. However, some myths about the hearing of birds also abound. Perhaps the most notorious is the assertion that they can hear ultrasound—that is, frequencies too high for humans to hear. Bird scare devices based on broadcasting ultrasound might scare a lot of mammals, but they are inaudible to birds. Although some birds' songs may contain ultrasonic components (Pytte et al., 2004), they do not hear those themselves. Avian hearing typically remains restricted to below 10kHz (Figure 6.1), somewhat lower than human hearing. As a rule of thumb, small birds hear better at high frequencies than larger birds and vice versa (Dooling et al., 2000; Gleich and Langemann, 2011). Some birds, most notably owls, are exquisitely sensitive and able to hear fainter sounds than we can (Figure 6.1).

In many standard behavioral auditory tests, such as frequency or intensity discrimination, birds do just as well as a typical mammal, including humans (Dooling et al., 2000). On some measures of temporal resolution, especially tasks that require the perception of fast-fluctuating fine structures, birds often outperform mammals (Dooling et al., 2000). Localizing sounds is more of a problem for birds because of their generally small head size which, in combination with their restricted upper hearing range, offers only small interaural cues (Klump, 2000; Köppl, 2009). As a rule, birds can localize sounds in azimuth as well as any mammal of comparable size. However, they cannot localize in elevation. The only exceptions to this are several owl species, which show extraordinary localization performance and prominent specializations associated with that (Klump, 2000). In more naturalistic auditory tasks, birds have been shown to group complex sounds into perceptual categories and patterns, much like humans do with speech syllables, for example (Dooling et al., 2000). Birds also form "auditory objects"—that is, they group concurrent sounds in a complex scene into perceptual objects, such as different individual birds singing or tonal patterns forming galloping rhythms (Dooling et al., 2000).

Obviously, the auditory world of birds is rich, and hearing plays an important part in their lives. This chapter summarizes the sensory aspects of hearing in birds. How do birds' ears work and how is sound processed in the brain? It should be pointed out that, although not the focus of this chapter, much of what is discussed below also applies to crocodilians, which, together with birds (and a lot of extinct forms), are classified as archosaurs (Carroll, 1988). Like birds, crocodilians are quite vocal and their auditory system is very similar (Vergne et al., 2009; Young et al., 2014).

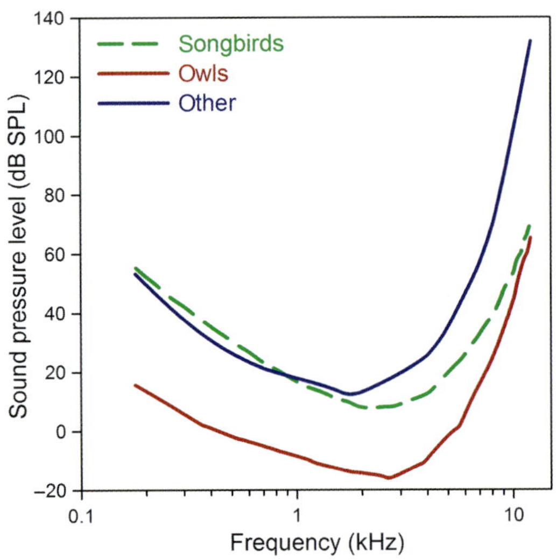

FIGURE 6.1 **Median behavioral audiograms for three groups of birds.** Note that owls hear sounds below 0 dB sound pressure level, which is below the best human thresholds. *Modified from Dooling et al. (2000), with permission from Springer Science + Business Media B.V.*

6.2 OUTER AND MIDDLE EAR

6.2.1 No Specialized Outer Ear Structures, Except in Owls

Birds, like all nonmammals, do not have external ears or pinnae. This is easily explained by their generally small body (and consequently head) size in combination with a range of hearing limited to frequencies below about 10 kHz. The mammalian pinna is a sound reflection and filtering device that provides cues about the direction of a sound source relative to the head (Pickles, 2008). For this to be effective, sound wavelength, as a rule of thumb, must be smaller than head diameter—a condition that is generally not met in birds within their limits of hearing. Some very specialized owls, however, exploit similar principles by bearing asymmetrical outer ears consisting of a facial ruff, skin flaps, and/or even asymmetrical bony ear canals (Norberg, 2002). Effectively, one of the outer ears is pointing slightly upwards and the other downwards, which differentially reflects frequencies above 4 kHz or so (Keller et al., 1998). If the inputs from both ears are compared, this conveys a sensitivity for sound source elevation (see Section 6.4.2). Remarkably, such asymmetrical outer ears arose several times independently, in different owl genera, as an adaptation to hunting by auditory cues (Norberg, 2002).

6.2.2 The Single-Ossicle Middle Ear

Middle ears arose independently in a number of land vertebrate lineages during the Triassic period (Clack and Allin, 2004). In all nonmammalian lineages, the middle ear consists of a simple, piston-like device. It is for the most part homologous with the mammalian stapes (Manley and Sienknecht,

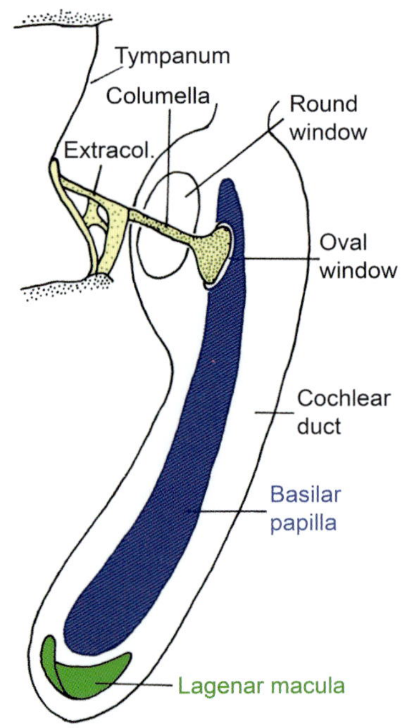

FIGURE 6.2 **Schematic drawing of the cochlear duct of a typical bird, and its connection to the eardrum (tympanum) via the columella and extracolumella (yellow) of the middle ear.** The dorsal, vestibular part of the inner ear is not shown. Two sensory hair-cell organs are located in the cochlear duct: the auditory basilar papilla (blue) and the vestibular lagenar macula (green).

2013). The middle ear of birds consists of an elongated, mostly bony columella that bears a footplate on its medial end that abuts the bony inner-ear capsule. At its peripheral end, the columella grades into the mainly cartilaginous extracolumella that, with one of its several projections, makes close contact to the eardrum (Figure 6.2). The most important part of the extracolumella is the inferior process, which connects the edge of the eardrum to somewhere close to its center. About halfway along its length, the extracolumella is connected almost at a right angle to the columella; the flexible cartilage connection permits the columella to move in a piston-like fashion. Eardrum motion causes the inferior process to tilt on its fulcrum (at the edge of the eardrum), which exerts a force on the columella (Figure 6.3). The force at the columellar footplate greatly exceeds that of the sound arriving at the eardrum because (1) the columella inserts at (roughly) half the length of the extracolumella and (2) the surface area of the eardrum is much larger than that of the footplate (Saunders et al., 2000). By these means, the middle ear acts as an impedance-matching device allowing sound from the low-impedance medium air to very effectively change the pressure in the high-impedance fluids of the inner ear (Figure 6.3). Without it, the inner ear would be about 40 dB less sensitive (Gummer et al., 1989).

An enigma remains the paratympanic organ, which is a receptor patch almost exclusively found in birds and crocodiles, situated on the wall of the tympanic cavity

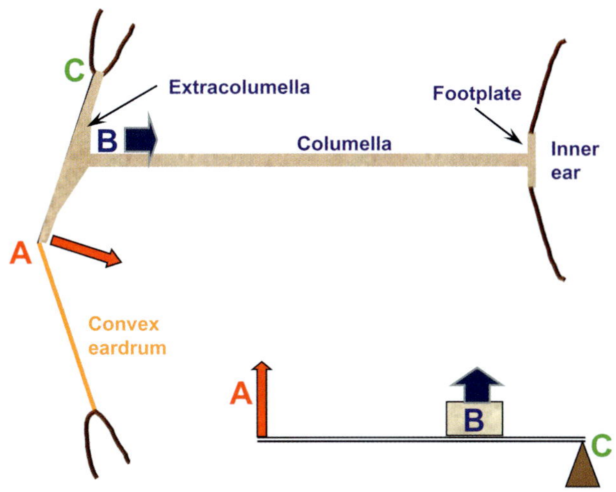

FIGURE 6.3 A schematic diagram of the middle ear of birds and its function. The extracolumella is attached flexibly to the columella, such that movement of the eardrum inwards results in a piston-like movement of the columellar footplate. The illustration at the lower right shows that these structures together build a lever of second order. The arrow length is proportional to the displacement amplitude, whereas the arrow width is proportional to the force behind the movement. Thus, at point B, the displacement is smaller than at A, but the force proportionately larger. C is the fulcrum, the axis of rotation of the lever.

(von Bartheld and Gianessi, 2011). It is made up of hair cells that are innervated by the facial nerve. However, its function is unknown. It has been speculated to act as a barometric device.

6.2.3 Coupled Middle Ears?

The middle ears of birds are not enclosed in bullae as they are in mammals, but are acoustically connected through skull spaces collectively termed the interaural canal. Part of the interaural canal is formed by wide Eustachian tubes that open into the buccal cavity (Christensen-Dalsgaard, 2011). However, there are very likely multiple routes through the highly trabeculated avian skull that have been difficult to visualize. The functional implication of interaural connections is that the ears function as pressure difference receivers, with sound reaching each eardrum from both sides. Depending on the physical dimensions of the head, the wavelength, and the attenuation across the interaural canal, significant interactions between the sounds reaching the eardrum from both sides may result in increased directional cues. Although agreed in principle, the precise extent of those effects in different species of birds is still controversial and measurements by different laboratories vary (reviews in Christensen-Dalsgaard, 2011; Klump, 2000). Small birds probably experience more of such acoustic cross-talk, whereas larger birds experience less and are limited to lower frequencies. Indeed, in the barn owl, interaural canal transmission appears to play no role in the high-frequency range that the owl uses for prey localization (Calford and Piddington, 1988; Moiseff and Konishi, 1981b).

6.3 BASILAR PAPILLA (COCHLEA)

6.3.1 General Morphology and Physiology

The inner ear or labyrinth houses both the vestibular organs and the auditory organ, the basilar papilla. The hearing part is commonly referred to as cochlea or cochlear duct, which is correct in the sense that it is homologous to the mammalian cochlea (Manley and Clack, 2004). However, the avian version is not coiled and the term cochlea ("snail") is thus not entirely appropriate. The cochlear duct houses the auditory basilar papilla and the vestibular lagenar macula, which forms the apical tip of the labyrinth (Figure 6.2). This is another salient difference to mammals, which have lost the lagenar macula. In spite of earlier evidence to the contrary, the avian lagenar macula does not respond to sound (Manley et al., 1991) and does not send afferent fibers to the cochlear nuclei (Kaiser and Manley, 1996).

The avian basilar papilla is typically only slightly curved and a few millimeters long; it tapers from being wide near the apical end to narrow at the basal end (Figures 6.2 and 6.4(A)) (Gleich and Manley, 2000). It is composed of several thousand sensory hair cells, plus supporting cells surrounding each hair cell in a roughly hexagonal pattern (Figure 6.4(C) and (D)). The epithelium sits on a basilar membrane that is largely suspended over the fluid of scala tympani. At both edges, the structure is anchored to the cartilaginous limbus and at the inner (or neural) edge; a varying proportion of hair cells actually sits atop this solid tissue, not on the basilar membrane (Figure 6.5). There are two types of hair cells, tall and short (see Section 6.3.2). All hair cells are covered by the tectorial membrane, an acellular, proteinaceous structure (Goodyear and Richardson, 2002). Sound sets the basilar papilla in motion, which ultimately deflects the mechanosensitive bundles of its hair cells through a shear motion between them and the tectorial membrane. However, the precise modes of mechanical excitation in the avian papilla are still unclear. The basilar papilla is tonotopically organized, such that hair cells at apical locations are maximally sensitive to low frequencies, typically down to about 100 Hz, and those at basal locations are maximally sensitive to high frequencies, typically 5–8 kHz (Gleich et al., 2004). A traveling wave has been observed at the level of the basilar membrane (Gummer et al., 1987). However, unlike in the mammalian cochlea, its frequency resolution does not approximate the known selectivity at the neural level. Furthermore, the hair cells conveying the most sensitive responses appear to sit on or at the edge of the solid limbus (Gleich, 1989; Smolders et al., 1995), suggesting that basilar-membrane motion would not be their direct input. Hair cells in the avian basilar papilla show a very unique and complex pattern of bundle orientation (Gleich and Manley, 2000) (Figure 6.4(B)), which has prompted speculation about excitation waves moving obliquely across the tectorial membrane (Tilney et al., 1987); however, this remains experimentally untested.

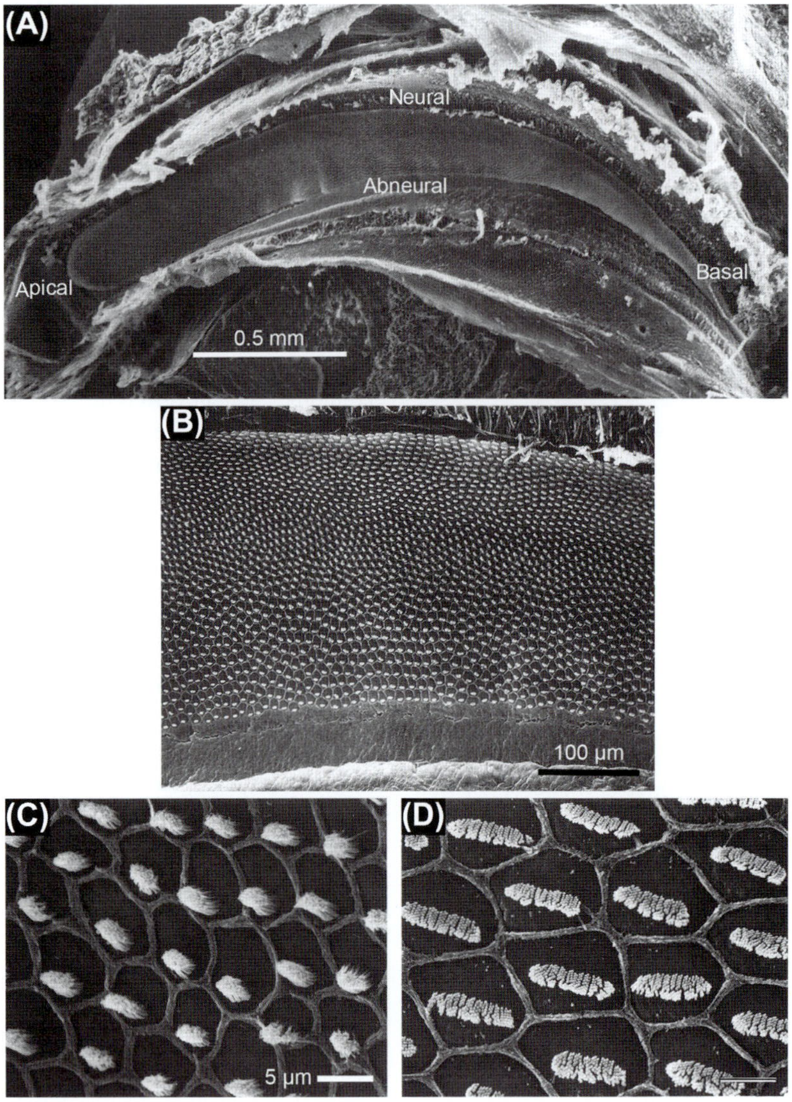

FIGURE 6.4 Surface morphology of the basilar papilla. (A) Low-magnification scanning electron micrograph of the exposed papilla of a chicken, after removal of the overlying tegmentum vasculosum and tectorial membrane. (B) View onto a segment of basilar papilla. Individual hair bundles are identifiable as white structures whose orientation changes across the epithelium. (C) and (D) High-magnification view onto a small group of hair cells. Note the hexagonal fringe of supporting cells surrounding each hair cell and the mechanosensitive hair bundles composed of tightly packed stereovilli. Note also that hair cells from apical regions (C) have smaller but taller bundles than those from basal regions (D). *Reproduced from Cotanche et al. (1994), with permission from Springer Science + Business Media.*

Transduction by the hair cells can be assumed to work according to the "gating-spring" principle established in other hair-cell systems (Fettiplace and Ricci, 2006). Deflection of the mechanosensitive hair bundle directly modulates the open probability of transduction channels associated with the tip links in the bundle. Opening of the transduction channels leads to the depolarizing influx of cations, chiefly K^+ and Ca^{2+}. The movement of K^+ into the cell is due to the unusual ionic composition and electrical potential in the endolymphatic environment facing the apical surfaces of the hair cells. As is typical for the inner ear of all vertebrates, a high K^+concentration and low Na^+ concentration, close to intracellular conditions, are maintained in endolymph, together with a slight to moderate positive potential, the endocochlear potential (Runhaar et al., 1991; Schmidt, 1963; Wangemann, 2002). In the cochlear duct of birds, this metabolically demanding task is carried out by the tegmentum vasculosum, the tissue separating scala vestibuli and scala media and thus directly overhanging the basilar papilla (Figure 6.5). An endocochlear potential of about +15 mV is maintained in avian endolymph (Necker, 1970), much lower than in the mammalian cochlea but higher than in the vestibular parts of the inner ear (Wangemann, 1995).

Hair-bundle morphology varies characteristically along the basilar papilla and thus along the tonotopic gradient.

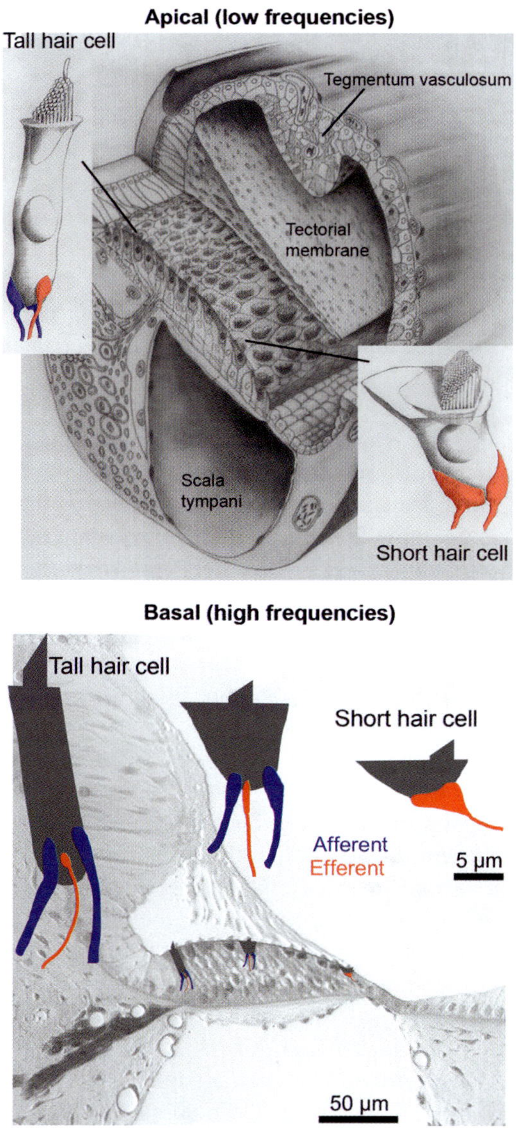

FIGURE 6.5 Overview of basilar-papilla morphology and innervation. The top panel shows a three-dimensional (3D) rendering of a typical cross-section through the apical, low-frequency part. The papilla is wide, with many hair cells across, most of which are tall hair cells. Insets are 3D drawings of two representative hair cells, with afferent (blue) and efferent (red) nerve terminals attached The lower panel shows a light-microscopical cross-section through the basal, high-frequency part, with some hair cells schematically highlighted and also shown enlarged. The lower scale bar applies to the histological image, the upper one to the drawings. The schematic drawings also show the typical innervation pattern, with afferent terminals drawn in blue and efferent terminals in red. *Top panel reproduced from Takasaka and Smith (1971), with permission from Academic Press. Insets reproduced from Manley and Ladher (2008), with permission from the artist, Johanna Kraus. Bottom panel reproduced from Köppl (2011a), with permission of Elsevier BV.*

Hair cells at apical locations have relatively tall bundles composed of relatively few stereovilli, whereas hair cells at basal locations have shorter bundles with many stereovilli (Gleich and Manley, 2000) (Figure 6.4(C) and (D)). This morphology is a major determinant of bundle stiffness and, as in all vertebrates, it is believed that this contributes to tune a hair cell's mechanical frequency response. In addition, avian hair cells are electrically tuned to respond preferentially to different frequencies. This is achieved by varying the number and/or kinetics of two principal ion channels in the cell membrane: voltage-gated Ca^{2+}-channels and Ca^{2+}-activated K^{+}-channels (Art and Fettiplace, 2006; Tan et al., 2013). However, electrical tuning is likely to be less effective or even absent at the highest frequencies (Wu et al., 1995).

One of the most fascinating aspects of hair-cell function is the ability for reverse transduction—that is, the ability to generate mechanical forces and thus to actively amplify low-level stimuli. Active movements of hair bundles have indeed been observed in isolated hair cells in vitro. It is currently believed that this ability is inherent to the hair cells' transduction mechanism, although the molecular components remain unidentified (see review in Martin, 2008). In mammalian cochlear outer hair cells, an additional mechanism evolved, conveying somatic motility through the voltage-sensitive protein prestin in the cell membrane (see review in Russell, 2014). This appears to dominate amplification in the high-frequency range of several kilohertz and above in mammals. Hair cells of the avian basilar papilla have been suggested to represent an intermediate case in which reverse transduction and somatic motility may literally join forces (Beurg et al., 2013). However, this is still controversial and much remains to be learned about the precise mechanisms of active amplification in the bird inner ear. Nevertheless, the presence of amplification is undisputed. One of its indirect manifestations is the occurrence of otoacoustic emissions, which are faint sounds emitted by the inner ear that are believed to be an epiphenomenon of the active processes. Otoacoustic emissions are only measurable under shielded, very quiet laboratory conditions. They may be present spontaneously or can be evoked by a range of stimulation protocols, and they have been observed in a wide range of vertebrates, including birds (Manley and van Dijk, 2008; Taschenberger and Manley, 1997).

Hair cells are secondary sensory cells; they thus do not form an axon. They are contacted by two basic types of neurons, afferent and efferent to the central nervous system (Figure 6.5). Afferent neurons have their cell bodies located in a compact ganglion close to the basilar papilla. Their peripheral fibers enter the basilar papilla at its inner or neural edge and contact typically only one hair cell each. The afferent synapses are of the specialized ribbon synapse type, which occurs only in vertebrate hair cells and photoreceptors and in retinal bipolar cells. They are believed to enable high, sustained rates of transmitter release (see review in Matthews and Fuchs, 2010). The central axons of the afferents collectively form the auditory nerve and terminate on the brainstem neurons comprising the cochlear nucleus. A truly unique feature of the avian auditory papilla is that typically 20–25% of the hair cells receive no afferent contacts (see also Section 6.3.2). In contrast, all hair cells receive efferent input from neurons located in the ventral brainstem (see review in Köppl, 2011b). The predominant effect of evoked efferent activity on the hair cells is inhibitory. However, very little is known about the behavioral contexts in which efferents are active and, consequently, about their broader role in hearing.

6.3.2 Hair-Cell Types: A Remarkable Example of Evolutionary Convergence in Birds and Mammals

When the ancestors of birds and mammals separated, their common heritage included a dedicated auditory hair-cell field, the basilar papilla. This ancestral basilar papilla is likely to have been small, with a uniform hair-cell type, and sensitive to low frequencies only (Manley and Köppl, 1998). Thus, although the basilar papilla itself is homologous between the groups, its specializations are not. The auditory hair-cell types in particular represent remarkable convergent evolutionary developments.

Cross-sections of avian basilar papillae, at first glance, do not show any of the prominent and uniquely mammalian features, such as the tunnel of Corti or the strict arrangement of one row of inner and three rows of outer hair cells. However, looking more closely at the fine structure of the hair cells and especially their innervation, a very salient analogy emerges. Like mammals, birds have two types of hair cells, here called tall and short hair cells. Unlike in mammals, those types are not sharply separated. The extremes are clearly different and well defined, but the two types grade into each other (Gleich and Manley, 2000). Tall hair cells are found on the inner (neural) side of the epithelium and gradually transition into short hair cells towards the outer (abneural) edge (Figure 6.5). Furthermore, the relative numbers of tall and short hair cells are not fixed in the manner of inner and outer hair cells in mammals, but change along the tonotopic gradient. Basal, high-frequency regions show a predominance of short hair cells and vice versa. Most strikingly, short hair cells receive large efferent terminals but are devoid of afferent innervation (Fischer, 1994a) (Figure 6.5). This may be the only example of a sensory cell entirely losing its primary function. It is even more extreme than the innervation of mammalian outer hair cells, which also receive large efferent terminals but retain a sparse afferent innervation (Raphael and Altschuler, 2003). Clearly, avian short hair cells are not the principal sensory cells for hearing. The analogy to the mammalian cochlea, in which inner hair cells take most of the classical sensory role and outer hair cells are specialized for mechanical amplification (Dallos, 1996), is palpable. However, much remains to be learned about the assumed amplificatory role of short hair cells, as there is currently little direct evidence for it. Although short hair cells do exhibit bundle motility in vitro, tall hair cells do as well (Beurg et al., 2013), so it remains unclear what the specific role(s) of short hair cells may be.

6.3.3 Hair-Cell Regeneration: Birds Never Lose Their Hearing

Because human hearing loss often involves damage to and ultimately death of hair cells through noise or ototoxic medication, similar treatments were used on birds in order to find a good preparation to study the underlying phenomena. This led to the remarkable discovery that adult birds are able to regenerate functional hair cells (see review in Rubel et al., 2013). It very quickly became apparent that, in contrast to humans, birds did not necessarily suffer permanent hearing loss. Instead, and unexpectedly, hearing thresholds returned to normal or near-normal within a matter of weeks following damage. This discovery initiated an excited search for the underlying cellular and genetic processes. It is now known that after insult and hair-cell death, new hair cells are regenerated from the surviving supporting cells, either by rekindled cell division or by direct transdifferentiation into hair cells (see review in Stone and Cotanche, 2007). The regeneration process is extensive and detailed. Specific, local hair-cell characteristics, such as hair-cell shape, bundle shape, and orientation, are so well matched to the surviving adjacent areas that it can be difficult to distinguish an undamaged cochlea from a recovered one. In addition, nerve fibers reconnect with the new hair cells and the overlying tectorial membrane is at least partly replaced (Cotanche, 1999), in total leading to an impressive functional recovery (Ryals et al., 2013; Smolders, 1999). Regeneration reaches its limits, however, if the damage extends over very large areas of the basilar papilla and includes supporting cells (Cotanche, 1999; Smolders, 1999).

Even without severely damaging experience, the avian basilar papilla appears to continuously regenerate hair cells at a low level (Ryals and Westbrook, 1990). Thus, there is probably no such thing as age-related hearing loss in birds. Indeed, starlings in aviaries often reach ages of more than 10 years (five times their expected ages in the wild) and they can still have close-to-normal hearing thresholds at that age (Langemann et al., 1999). It is not without irony that human breeding of certain races of songbirds for their loud songs unknowingly involved these mechanisms. Belgian Waterslager canaries have an inherited high-frequency hearing loss and sing a louder and lower-pitched song than normal canaries (Okanoya and Dooling, 1985). In the basilar papilla of those birds, hair cells continuously die at abnormal rates—a process that is incompletely offset by ongoing regeneration of new hair cells (Gleich et al., 1997; Wilkins et al., 2001). Unfortunately, the hope of initiating the same kind of regeneration of hair cells in mammalian cochleae has so far not been fulfilled (Brigande and Heller, 2009).

6.3.4 Cochlear Specializations: Auditory Foveae, Infrasound Hearing

As mentioned above, the basilar papilla is tonotopically organized. Apical regions respond most sensitively to low frequencies, and the characteristic frequency gradually increases towards the basal end. In a typical avian basilar papilla, this tonotopic map is well described by a near-logarithmic function—that is, each doubling of frequency (equivalent to one octave) corresponds to an approximately equal length of papilla (Figure 6.6) (Gleich et al., 2004). However, there are interesting exceptions to this rule. In analogy to visual foveae, regions of enhanced frequency representation, where some frequencies occupy disproportionally more space along the basilar papilla, have been termed *auditory foveae*. In birds, the best known example of such an auditory fovea is the barn owl, and a similar case has been suggested for the kiwi. In the barn owl, the frequencies of the highest octave perceived (5–10 kHz) fully occupy the basal half of the basilar papilla (Köppl et al., 1993), as much space as the lower five or six octaves together (Figure 6.6, red curve). These are the frequencies shown behaviorally to be the most useful for prey localization (Payne, 1971). The anatomical gradients of the owl cochlea, which at least partially determine the frequency response of the hair cells, hardly change along the basal half of the cochlea, concordant with a large expansion of the area processing higher frequencies (Fischer et al., 1988). Because a visual fovea is associated with improved acuity, a larger cochlear representation is often assumed to correlate with enhanced frequency resolution. Interestingly, this is not the case in the barn owl (Köppl et al., 1998), suggesting that the role of the fovea instead may be to lay the foundation for massive parallel processing by increasing the numbers of afferent fibers from this behaviorally important frequency region (Fischer, 1994b; Köppl, 1997b). Although there are no physiological data from the ear of the kiwi, studies of hair-cell morphology also suggested a foveal expansion of frequency representation in the upper range of kiwi hearing, estimated at 4–6 kHz (Corfield et al., 2011).

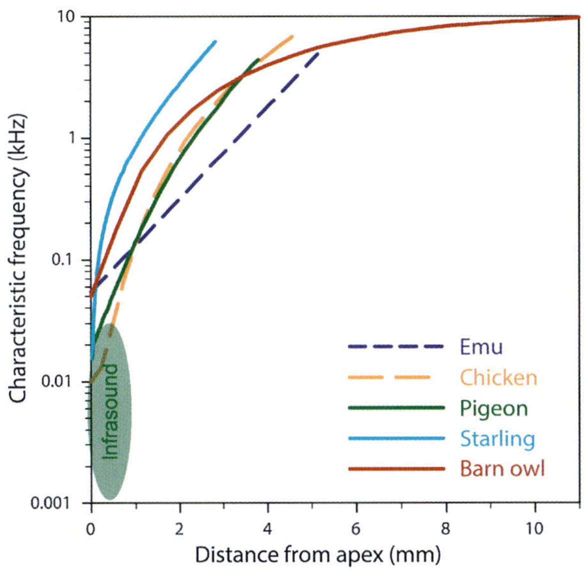

FIGURE 6.6 Tonotopic frequency representation along the avian basilar papilla. The differently colored curves show the mapping functions for several species of birds, as determined from labeling physiologically characterized auditory-nerve fibers to reveal their innervation sites in the papilla. Note the exceptional length of the barn owl basilar papilla and the shallow slope of its frequency map over most of its extent, indicating an auditory fovea (i.e., expanded spatial representation of a narrow, high-frequency band). Infrasound sensitivity in the pigeon was found over the most apical millimeter but it may not be strictly tonotopic and is therefore represented by the green area. *Data from Chen et al. (1994), Gleich et al. (2004), Köppl et al. (1993), Köppl and Manley (1997), and Smolders et al. (1995).*

An equally amazing specialization for extremely low-frequency hearing has been found in pigeons. For some time, it had been known from behavioral experiments that pigeons are sensitive to infrasound, defined as frequencies below those audible to humans (i.e., <15 Hz; Kreithen and Quine, 1979). Schermuly and Klinke (1990b) traced the origin of this sensitivity to hair cells located in the extreme apical-abneural regions of the pigeon basilar papilla. The afferent fibers connecting to these hair cells did not respond to normal acoustic frequencies, but showed responses to infrasound at levels comparable to the behavioral thresholds shown previously (Schermuly and Klinke, 1990a). Quite in contrast to the high-frequency foveae discussed above, the infrasound receptive region appears to co-exist over the most apical 1 mm of the basilar papilla, alongside a conventional logarithmic frequency representation (Figure 6.6, green) (Smolders et al., 1995). The behavioral significance of infrasound hearing remains unknown. Although it has been suggested that pigeons use it as a navigational cue (Hagstrum, 2000; Kreithen and Quine, 1979), conclusive evidence is lacking. Infrasound sensitivity may also not be unique to pigeons. Morphological evidence for an unusual hair-cell area at the apex of the chicken cochlea has also been described (Lavigne-Rebillard et al., 1985) and some neurons in the cochlear nucleus of the chicken have best response frequencies at least as low as 10 Hz (Warchol and Dallos, 1989).

6.3.5 Auditory Nerve: What the Ear Conveys to the Brain

All information that the brain receives about the outside acoustic world is encoded in the activity of the auditory-nerve fibers. Beginning with Sachs et al. (1974), who recorded single-unit activity in the pigeon auditory nerve, there have been numerous studies on the responses of afferent fibers in various bird species to simple, well-controlled stimuli. Here, I briefly summarize the encoding of frequency, sound level, and temporal fine structure.

All avian auditory nerve fibers are spontaneously active. Mean spontaneous rates vary between 45 and 90 spikes/s, depending on species, in adult birds (Köppl, 1997a). Unlike in mammals, there is no consistent, strong correlation in birds between spontaneous rate and sensitivity to tonal stimulation and there is no population of fibers with very low spontaneous activity (<1 spike). Spontaneous activity is irregular and the interspike interval distribution is typically Poisson-like, modified by the refractory period (Gleich and Manley, 2000; Neubauer et al., 2009). Interestingly, a significant number of fibers show preferred intervals, with periods at or close to their characteristic frequency for tonal stimulation. Although some of this activity for very sensitive fibers may be attributable to inadequate sound shielding during experiments, the bulk of evidence points to this being a genuine phenomenon that reflects spontaneous oscillations of the hair-cell membrane potential due to electrical tuning and/or active mechanical amplification processes in the avian cochlea (Gleich and Manley, 2000; Taschenberger and Manley, 1997).

Sound frequency is, of course, collectively coded according to the place principle, where the tonotopic gradient along the basilar papilla is preserved in the projections to the brain. Upon tonal stimulation, each auditory-nerve fiber responds most sensitively to one particular frequency, its characteristic frequency. At increasingly lower and higher frequencies, the fiber's sensitivity decreases rapidly, so that its response area forms a narrow, V-shaped frequency tuning curve (Figure 6.7(A)). In birds, these curves are very sharply tuned, with higher mean quality factors than mammalian responses at the same frequencies, and typically almost symmetrical (Gleich and Manley, 2000). However, the best thresholds of fibers within any narrow range of characteristic frequencies in an individual ear vary in birds more widely than in mammals, by as much as 50 dB (Figure 6.7(B)).

Sound level is encoded in the discharge rates of auditory-nerve fibers. The dynamic ranges over which individual fibers are able to encode have been studied in depth in response to short tonal stimuli of 50–100 ms and vary characteristically. There are (1) saturating fibers, whose discharge rate rises rapidly above threshold, then also levels off abruptly at a break point; (2) sloping-saturating fibers, whose rate does not show an abrupt saturation but continues to rise more slowly above the break point (Figure 6.7(C)); and (3) straight fibers, whose discharge rate slowly rises with a nearly uniform, shallow slope over a broad range of sound levels (Gleich and Manley, 2000; Köppl, 2011a). This shows remarkable parallels to mammals, where the same three types were originally described. However, in contrast to mammals, there are no correlations with other fiber properties in birds. It is believed that in both cases, the discharge behavior of auditory-nerve fibers with sound level is shaped by the underlying processes of active amplification, effecting a compressive nonlinearity. However, in birds, the nonlinearity is suggested to develop and vary locally, perhaps at the level of individual hair cells, while in mammals stronger coupling leads to a shared nonlinear response of larger groups of hair cells (Köppl, 2011a). The wide range of thresholds and the different dynamic ranges together ensure that the avian auditory nerve is able to encode a very wide range of sound levels indeed.

Finally, temporal information is encoded by phase locking of auditory-nerve fibers; that is, spikes are discharged preferentially at a specific phase of a sinusoidal or near-sinusoidal stimulus. It is important to point out that a spike need not occur in every cycle of the stimulus (Figure 6.7(D), inset). Indeed, at frequencies above 300–400 Hz, which is the upper range of sustained discharge rates, many cycles may be skipped, but phase locking still persists into the kilohertz range. Phase locking critically depends on the ability to control spike timing within the temporal window of one stimulus cycle. Thus, the requirement for temporal precision rises with increasing frequency (and shorter cycles) and phase locking invariably fails above a species-specific frequency limit. Avian auditory-nerve fibers typically phase lock up to 3–4 kHz (Figure 6.7(D)), which compares favorably to other vertebrates, including mammals. A famous exception is the barn owl, for whom phase locking persists to frequencies near 10 kHz, corresponding to a precision of spiking, or temporal jitter, of only 20–30 μs (Figure 6.7(D), red curve) (Köppl, 1997c). Although it remains unknown what cellular mechanisms enable this extraordinary precision, it is clearly an adaptation and prerequisite for the superb sound localization abilities of the barn owl (see Section 6.4.2).

6.4 THE AUDITORY BRAIN

6.4.1 Basic Organization of Auditory Pathways

The principal pathways of the avian auditory brain follow a general layout shared by all land vertebrates (Figure 6.8) (Carr and Code, 2000; Grothe et al., 2004). The major afferent pathways flow from the cochlear nucleus, either directly or indirectly, through the brainstem areas of the superior olive and lateral lemniscus, to the midbrain inferior colliculus (also termed torus semicircularis, or nucleus mesencephalicus lateralis pars dorsalis). From there, projections go to multimodal layers of the optic tectum (or superior colliculus) and to the auditory thalamus and from the latter on to the auditory forebrain. Beyond this very general pattern, however, the specific nuclei or subdivisions of nuclei can differ substantially between the major vertebrate groups, and the evolutionary origin and homology of many auditory areas beyond the broad categories outlined above is quite controversial. In general, differences in the central auditory pathways mirror the major types of basilar-papilla specializations and thus are now believed to also be the result of

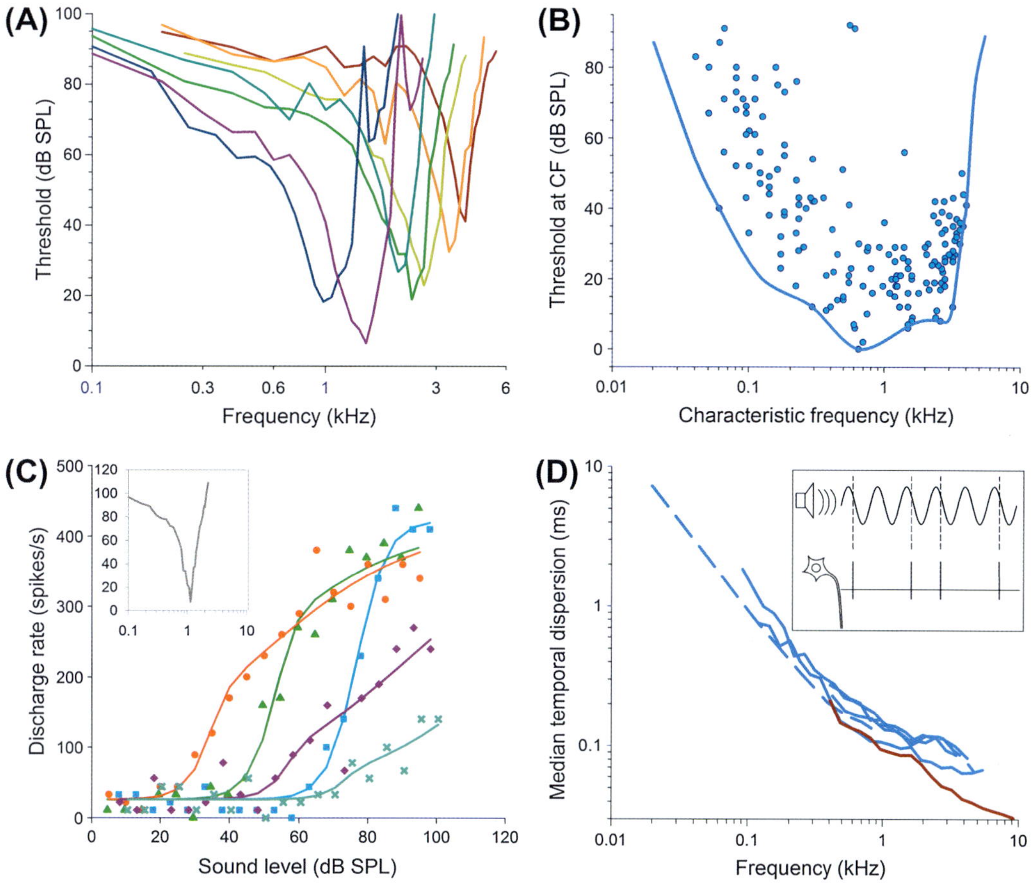

FIGURE 6.7 **Salient characteristics of single auditory-nerve fiber activity.** (A) Examples of frequency tuning curves from the emu (Manley et al., 1997). (B) Distribution of the most sensitive thresholds of a population of fibers from the emu (Manley et al., 1997), as a function of characteristic frequency. Note the large range of thresholds at any one frequency, which is typical and not due to damage. The solid line joins data from the most sensitive fibers, which gives an approximation of the emu's audiogram. (C) Rate-level functions for one particular auditory-nerve fiber in the emu, in response to several different frequencies within the fiber's response area shown in the inset. Note the pronounced sloping-saturating behavior at the most sensitive, characteristic frequency (shown in red). *(Data from Yates et al. (2000).)* (D) The inset illustrates schematically the temporal relationship between a sinusoidal stimulus and phase-locked action potentials. In real auditory-nerve fibers, individual spikes vary in their timing around a mean preferred phase. Median values for this temporal jitter or dispersion are shown in the main graph, as a function of frequency, for several bird species: barn owl (red; Köppl, 1997c), emu (dashed blue; Manley et al., 1997) and, all in blue, chicken (Salvi et al., 1992), pigeon (Hill et al., 1989), starling (Gleich and Narins, 1988), and redwing blackbird (Sachs et al., 1980).

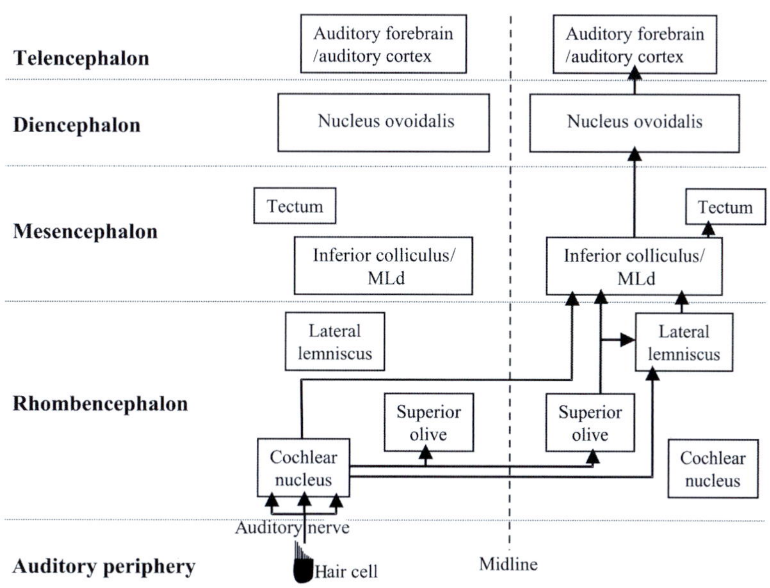

FIGURE 6.8 **Principal wiring diagram of the central ascending auditory pathway of birds.** Each box represents a major center that may contain multiple subareas. *Modified from Grothe et al. (2004), with permission from Springer Science + Business Media B.V.*

independent evolution (Grothe et al., 2004). Birds follow an archosaurian pattern that they share with crocodilians; it is also most similar to that of other, more distantly related, "reptiles".

The cochlear nucleus of birds has two major subdivisions, nucleus magnocellularis (NM) and nucleus angularis (NA), both of which receive a shared input by collaterals of each auditory-nerve fiber. NM and NA probably form the starting points of several auditory processing streams that specialize in different aspects. NM is the relatively more simple of the two, containing one predominant neuron type that is clearly specialized for preserving or even enhancing the temporal information conveyed through phase-locking from the auditory nerve (Carr and Boudreau, 1993; Sullivan and Konishi, 1984). NM is thus the starting point of a "time pathway". The best-known function of this time pathway is the extraction of interaural time differences used in sound localization (see Section 6.4.2). NA then appears to be the origin for everything else, although it is commonly referred to simply as the "intensity pathway" (which does not do justice to the complexity already seen at this level). NA contains a range of cell types having distinct anatomical and physiological properties (see reviews in Grothe et al., 2004; MacLeod and Carr, 2007). Although cell-specific connection patterns have not yet been investigated, NA neurons collectively have several projection targets in the superior olive and lemniscal nuclei, as well as direct projections to the inferior colliculus. Thus, the potential for different functional processing streams is present. In the barn owl, a specific involvement in deriving interaural level differences for sound localization has been demonstrated (see Section 6.4.2).

Although the cochlear nuclei are still strictly monaural, all higher levels show more or less pronounced binaural interaction. A tonotopic organization is, however, retained in the brainstem nuclei. All afferent auditory information then converges again in the midbrain nucleus mesencephalicus lateralis pars dorsalis (MLd), which is considered homologous to the mammalian inferior colliculus (Carr and Code, 2000). The definition of MLd subdivisions differs between authors but, in general, a "core and belt"-type organization is recognized; the common terms used for those are central nucleus (ICc) and external nucleus (ICx). Brainstem projections terminate in the ICc regions which, in turn, project to ICx. The physiology of the avian auditory midbrain has mostly been studied in the context of sound localization (see Section 6.4.2). In the barn owl, successive processing steps culminate in the formation of a neural map of auditory space in the ICx (see review in Konishi, 2003). It is currently unclear to what extent this generalizes to other birds. In a songbird, the zebra finch, responses of MLd neurons cluster into functional groups that were suggested to represent cues for fundamental acoustic percepts, such as pitch, timbre, and rhythm (Woolley et al., 2009).

As a general principle in vertebrates, although a tonotopic representation is maintained in parts of the auditory midbrain, pure tones are often not the most effective stimuli. Instead, selectivity for other derived sound features, such as location or temporal modulation, begins to predominate.

In parallel to the connections from the ICc to the ICx (and on to the superior colliculus), the ICc also initiates the forebrain ascending auditory pathway by projecting to the thalamic nucleus ovoidalis (Figure 6.8) (Carr and Code, 2000). The morphology and connections of nucleus ovoidalis are quite well characterized, but its functional role remains unclear and few studies have probed it physiologically (Carr and Code, 2000; Ondracek and Hahnloser, 2014; Proctor and Konishi, 1997). Nucleus ovoidalis projects on to the primary auditory forebrain area in birds. It is important to point out that the nomenclature used in the avian forebrain has seen major revision in recent years, driven by growing evidence that early interpretations of avian forebrain organization had been erroneous and that the classic nomenclature derived from it was seriously misleading (Jarvis et al., 2005). According to the revised nomenclature, the avian auditory forebrain consists of Field L in the caudal nidopallium, the caudal mesopallium (CM), the dorsal nidopallium (Nd), and the ventromedial portion of the intermediate arcopallium (Aivm) (Wang et al., 2010). Field L is further divided into subregions, with only L2 being the primary thalamorecipient layer. It has been argued that all four avian auditory forebrain areas together form a circuit highly similar in both its columnar and laminar organization to the mammalian auditory cortex (Figure 6.9) (Wang et al., 2010). The physiology of the avian auditory forebrain has been studied most extensively in songbirds (see also Section 6.4.4). Here, a hierarchy of responses is apparent, with a clear tonotopic organization and spectrally defined responses in thalamorecipient neurons, and striking specificities to

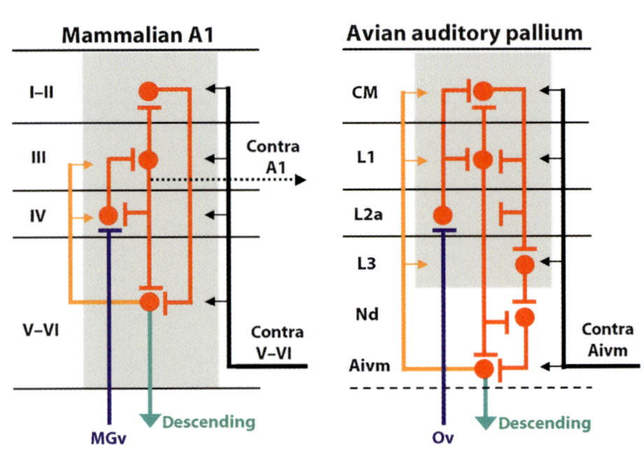

FIGURE 6.9 Suggested analogy in the circuitry between the different layers of mammalian auditory cortex and the principal nuclei of the avian auditory pallium. Thalamic inputs, intrinsic connections, and descending projections are shown in blue, red, and turquoise, respectively. Orange lines and arrows indicate recurrent projections from the deep layers to the more superficial layers. Reentrant projections from the other side of the brain are drawn in black. *Reproduced from Wang et al. (2010).*

species-specific vocalizations emerging in higher-order neurons. Also, correlates of perceptual categories of auditory streaming, such as galloping rhythms, have been found in the responses of Field L neurons (Bee and Klump, 2005; Itatani and Klump, 2011). In the barn owl, the forebrain ascending auditory pathway appears to independently generate a second representation of auditory space that resides in the arcopallium. However, the forebrain representation is different in several salient aspects and probably serves different functions to the midbrain topographical map of space (Cohen and Knudsen, 1999; Vonderschen and Wagner, 2009).

6.4.2 The Generation of an Auditory Space Map in the Barn Owl

Beginning with the classic studies of Roger Payne (1971), who first demonstrated that barn owls use acoustic cues to precisely locate and strike their prey in the dark, many general principles about how the location of sounds in the environment is reconstructed by the brain have been learned from this bird. Unlike vision, the sense of hearing cannot rely on a spatial image of the external world being projected onto the primary receptor surface and relayed to the brain. Sound localization requires central auditory computation, using indirect cues. For birds, these cues are largely binaural in nature—that is, the inputs from both ears are compared for the minute differences in timing and level that arise from different path lengths of the sound to both ears and from sound shadowing effects of the head and body (Klump, 2000). For most birds, both interaural time differences (ITDs) and interaural level differences (ILD) are cues for sound source azimuth; however, the asymmetrical facial ruff of the barn owl (see Section 6.2.1) generates ILDs as a function of sound source elevation. This changes the meaning of ILDs but not necessarily the neural processing steps to derive them. Much of what has been learned from the barn owl is believed to reflect general principles of the neural computation of sound location in birds, if not vertebrates (Grothe et al., 2010; Konishi, 2003; Takahashi, 2010). In the barn owl, the binaural comparisons of time and level are carried out in parallel processing streams in brainstem nuclei that are specialized for each task. At the midbrain level, this information is then combined and used to create a "space map"—that is, a two-dimensional representation of auditory space in azimuth and elevation.

Contrary to intuition, the computation of ITD is not based on a comparison of sound onset at the two ears, but largely relies on ongoing temporal information from neural phase locking in the auditory nerve (see Section 6.3.5). This was shown by clever behavioral experiments with owls wearing headphones that allow for the presentation of stimuli that artificially dissociate onset and ongoing ITDs (Moiseff and Konishi, 1981a). The phase locking of auditory-nerve fibers is preserved by the monaural cochlear NM, which then projects to the binaural nucleus laminaris (NL). NL neurons perform arguably one of the most challenging and extreme tasks in the nervous system: they receive the phase-locked inputs from each side and fire selectively if their inputs coincide within a very narrow, submillisecond time window (Funabiki et al., 2011; Kuba, 2007). Of course, such coincidence naturally occurs only when sounds originate directly in front (or back) of the animal. To encode more lateral sound locations, the incoming NM axons contact several NL neurons serially, thus forming functional delay lines that match and compensate for the range of natural acoustic interaural delays (Figure 6.10). This principle of an array of coincidence detectors, receiving temporal information by delay lines from both sides to create a topographical map of auditory azimuth, was originally formulated by Jeffress (1948). It appears beautifully implemented in the NM–NL circuit of birds in general (Burger and Rubel, 2008; Kubke and Carr, 2006), with the owl displaying a hypertrophied form and achieving the highest temporal resolution (Carr and Boudreau, 1993; Funabiki et al., 2011). The initial topographic map of auditory azimuth is created many times over in NL, separately in each frequency band of the tonotopically organised nucleus. An important task of subsequent processing steps within the ICc of the midbrain is to converge this information across frequencies and thus remove inherent ambiguities due to the cyclic nature of the phase code underlying it (Konishi, 2003).

The first binaural comparison of sound level occurs in the posterior part of the ventral nucleus of the lateral lemniscus (LLDp, formerly also called VLVp). Here, excitatory inputs from the contralateral ear (coming in via the cochlear NA) compete with inhibitory inputs from the ipsilateral ear (coming in via NA and the contralateral LLDp). Depending on the relative strengths of excitation and inhibition, the responses of individual LLDp neurons decrease at different relative levels of ipsi- and contralateral sound,

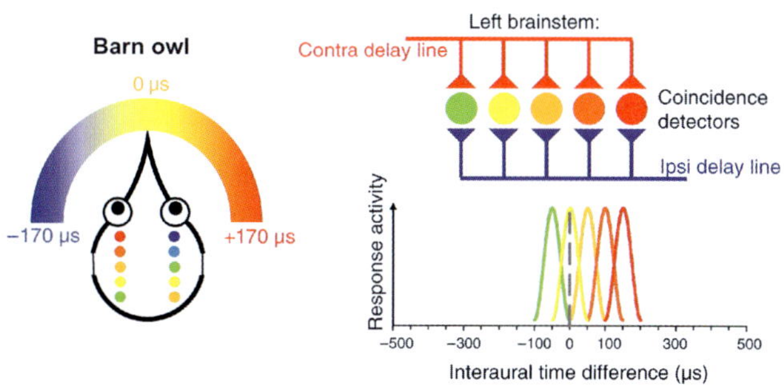

FIGURE 6.10 Schematic illustration of the coding of interaural time differences according to the Jeffress (1948) model and as seen in the barn owl. Each half of the brainstem contains a representation of interaural time differences, corresponding to sound sources mostly in the contralateral acoustic hemifield. This map is created by the basic circuit of delay lines and coincidence detectors illustrated for the left side of the brainstem. Typical responses of barn owl coincidence detector neurons in nucleus laminaris are shown below the circuit diagram, as a function of interaural time difference. Together they cover the interaural time difference range experienced by the owl and form a topographic representation of the auditory azimuth. *Reproduced from Köppl (2009), with permission from Cell Press.*

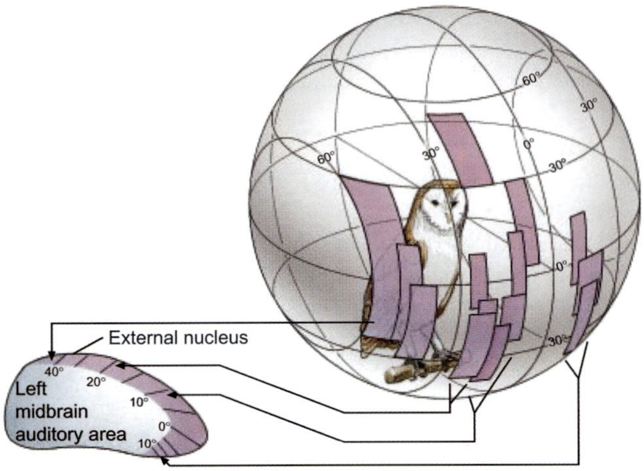

FIGURE 6.11 Auditory space map in the midbrain of barn owls. In the external nucleus of the inferior colliculus (purple), the auditory spatial receptive fields of neurons change systematically with position along the nucleus. Some examples of receptive fields are illustrated as purple rectangles on a sphere surrounding the owl. This spatial map is synthesized from auditory responses selective for specific interaural time and level differences. *Reproduced from Konishi (1993), with permission from Jana Brennings.*

or ILD. Inhibitory strength systematically decreases along the dorsoventral axis of the nucleus and thus establishes a first topographic representation of ILD (see review in Konishi, 2003). At the level of the ICc, the sigmoidal response functions of inputs from LLDp neurons sharpen into bell-like curves, only showing responses to a restricted range of ILDs. However, it remains unknown exactly how this is achieved (see review in Konishi, 2003).

Finally, in the ICx, a two-dimensional map of space-specific neurons with receptive fields bounded both in azimuth and elevation is synthesized (Figure 6.11) (Knudsen and Konishi, 1978), where the azimuthal axis is based on the selectivity for ITD and the elevational axis on the selectivity for ILD (Konishi, 2003; Takahashi et al., 2003). At the level of the individual space-specific neurons, both inputs are combined by a multiplicative process that further sharpens their resulting receptive field (Peña and Konishi, 2001). This auditory space map is relayed from ICx to bimodal neurons of the optic tectum (or superior colliculus), forming a combined auditory-visual spatial representation that contributes to stereotypical orienting behaviors (Knudsen, 2002).

6.4.3 Developmental Plasticity: Auditory Space is Calibrated by Vision

As just outlined, the auditory space representation in the barn owl's midbrain is a computational map generated from indirect cues (ITD and ILD) to sound location. These cues vary with details of head size and shape and are thus to some extent individual (Keller et al., 1998). How does the bird know which ITD and ILD values to associate with precise locations in space? Experiments manipulating the auditory cues of young owls by plugging one ear showed that owls learned, over some weeks, to again precisely localize sounds under these conditions. If instead their visual world was shifted by fitting prism spectacles, it was still the sound localization behavior that slowly adapted to match the changed visual environment (Knudsen, 2002). Thus, owls use vision to learn to associate certain ITDs and ILDs with the correct locations in space. The visual system, with its direct projection map of the retinal image, provides the objective spatial representation of the outside world against which the computational map of the auditory system is calibrated. The site of this plasticity has been identified as the ICx, the site where the auditory space map is first synthesized (Figure 6.12). Additional anatomical connections from the ICc can be induced here, whose ITD selectivity matches the corresponding visual receptive field in the optic tectum; those new connections come to dominate the response of ICx neurons over previously formed, now mismatched, inputs (Knudsen, 2002). The required error signal appears to be provided by reciprocal projections from the optic tectum to the ICx, an area traditionally thought to be exclusively auditory. Like many forms of experience-dependent plasticity, this learning of the meaning of auditory cues is limited to young owls and gradually diminishes in early adult life (Knudsen, 2002). Many of the principles shown for the learning of auditory localization in the owl also apply to the mammalian auditory system (King et al., 2000).

6.4.4 The Special Processing of Birdsong

Much of the research on higher-order auditory processing in birds has concentrated on songbirds for two reasons. First, the behaviorally most relevant acoustic signals are clearly defined and can guide the design of stimuli to test. Second, birdsong is a learned vocalization—a rare trait shared only with human speech and few other avian and mammalian vocalizations (Bolhuis et al., 2010; Jarvis, 2004). Together, this makes the songbird auditory system a rich source for unraveling how vocalizations are processed and learned. Not surprisingly, songbirds have very well-developed auditory forebrain areas (Figure 6.13(A)). However, in addition to that, a whole set of sensorimotor "song nuclei" is present, which are involved in vocal learning and plasticity and which have no clear equivalent in vocal nonlearning birds (Figure 6.13(B)) (see review in Farries, 2004). Of central importance is the nucleus HVC, which appears to be the pivotal interface between the auditory input and the motor output for singing. It is also the nucleus where neurogenesis in adult animals was first shown to produce functional neurons, in this case seasonally (Nottebohm, 2004). In HVC, extremely selective

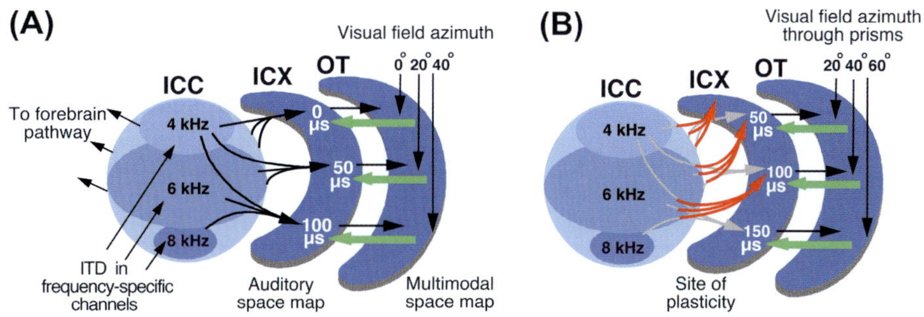

FIGURE 6.12 Schematic summary of adaptive plasticity in the auditory space map of the barn owl, after modified visual experience in early life. (A) Normally, different tonotopic regions from the central nucleus (ICc) that share selectivity for a common interaural time difference converge in the external nucleus (ICx). Together, many such convergent projections contribute the azimuthal axis of the space map that is relayed and combined with matching visual receptive fields in the optic tectum (OT). (B) When visual receptive fields are artificially shifted by fitting prims goggles to the owl, the auditory space map shifts accordingly, by forming new and matching connections between ICc and ICx (red arrows). Green arrows represent an instructive feedback projection from the OT to ICx. *Reprinted from Knudsen (2002), with permission from Macmillan Publishers Ltd.*

auditory responses, not just for species-specific vocalizations but for the individual bird's own song, appear for the first time (Konishi, 2004). HVC is then the starting point of two major motor pathways (Figure 6.13(B)): (1) a fairly direct, posterior pathway projects to the robust nucleus of the arcopallium (RA) and from there to various brainstem motor nuclei controlling the syrinx and respiration (Farries, 2004); and (2) the so-called anterior forebrain pathway, which eventually also leads to RA but via a whole set of intermediate nuclei that are believed to be homologous to the mammalian basal ganglia (Perkel, 2004). The anterior forebrain pathway has been shown to be critical for song learning in young birds (Brainard, 2004). In addition, there is evidence for an involvement in the plastic control of song, even in species that do not normally learn new songs as adults (Konishi, 2004). Therefore, it is believed that the anterior forebrain pathway, although not strictly necessary to produce song in an adult, continuously monitors what the bird actually sings and initiates corrections when necessary.

The purely auditory ascending forebrain pathway (Figure 6.13(A)), which eventually feeds into the specialized sensorimotor song nuclei system, comprises the primary field L and the secondary CM, both shared with other birds, and the secondary caudal medial nidopallium (NCM), an area unique to songbirds. Responses in these auditory areas never show the particular selectivity for the individual bird's own song (Theunissen et al., 2004). However, selectivity for conspecific song arises. As a rule, neurons in the primary auditory area field L are poorly driven by simple tones or noise but respond selectively to spectrotemporal features that characterize conspecific song and other natural sounds (Theunissen and Shaevitz, 2006). Many individual neurons appear to specialize in either spectral or temporal modulation selectivity and cluster accordingly in specific subregions (Nagel et al., 2011). The secondary auditory forebrain areas NCM and CM may add selectivity for familiar songs versus novel songs and have thus been implicated in auditory memory formation (Ondracek and Hahnloser, 2013; Theunissen and Shaevitz, 2006).

6.4.5 Echolocating Birds

Among the fascinating curiosities of the bird world are some species that use echolocation. These are the neotropical oilbird, *Steatornis caripensis*, and the paleotropical swiftlets of the genera *Aerodramus* and *Collocalia*. Both groups have independently evolved echolocation as a means of navigating in the dark caves where they roost and nest (see review in Brinkløv et al., 2013). This lifestyle has many superficial similarities with bats. However, the birds' echolocation calls are not ultrasonic but fall well within the typical avian hearing range. Due to this lower frequency range of the calls, avian echolocation cannot approach the spatial resolution of bat ultrasonic navigation, but few rigorous tests have been conducted (Griffin and Thompson, 1982; Konishi and Knudsen, 1979). Furthermore, the birds were reported to use echolocation only under conditions where they were unable to navigate by vision (Griffin, 1953; Novick, 1959), although anecdotal observations of foraging birds suggest this may be worth re-examining (Brinkløv et al., 2013).

Both the oilbird and the cave swiftlets use brief clicks that are produced by the syrinx (Suthers and Hector, 1982, 1985) and typically emitted in pairs or short trains (Konishi and Knudsen, 1979; Thomassen and Povel, 2006). The dominant frequencies in the clicks are in the range of several kilohertz and appear to match the birds' sensitive hearing range. However, due to their remote geographical distributions, very few species and individuals have been tested. From the studies available, it appears that the hearing capabilities of echolocating birds are unremarkable (Coles et al., 1987; Konishi and Knudsen, 1979). Anatomical data on the relative size of the auditory midbrain suggest a slight enlargement, which may indicate moderate specializations for echolocation processing (Cobb, 1968; Iwaniuk et al., 2006).

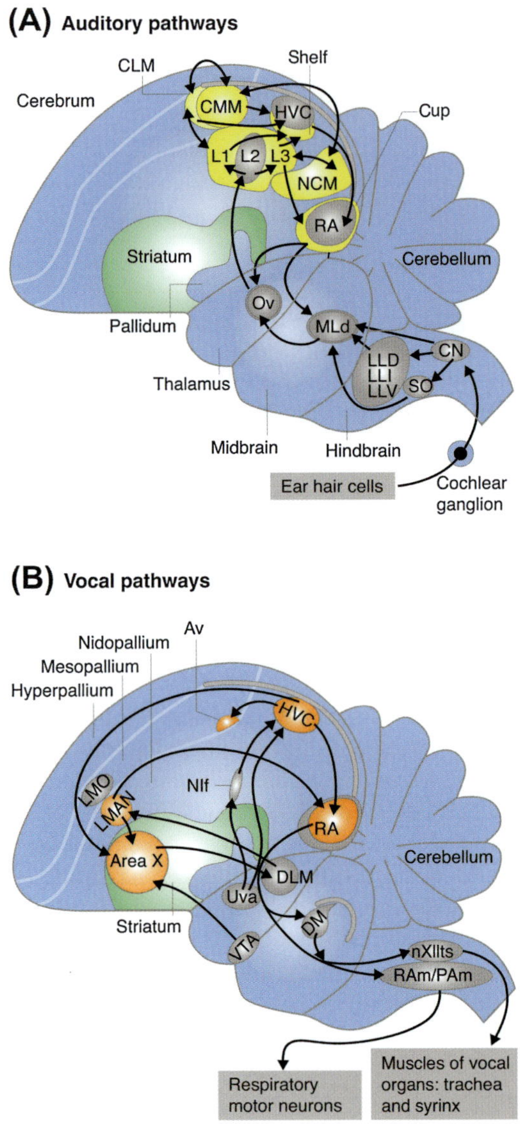

FIGURE 6.13 Schematic diagram of a composite view of parasagittal sections of a songbird brain, giving approximate positions of nuclei and brain regions. (A) Auditory ascending pathways, with brain regions that show increased neuronal activation when the bird hears song highlighted in yellow. (B) Vocal motor pathways. Depicted are connections between the nuclei, of both the direct posterior pathway and the anterior forebrain pathway. Both pathways together form the so-called song system. The orange nuclei in the song system show substantially enhanced neuronal activation when the bird itself is singing. Area X, Area X of the striatum; Av, avalanche; CLM, caudolateral mesopallium; CN, cochlear nucleus; DLM, medial subdivision of the dorsolateral nucleus of the anterior thalamus; DM, dorsomedial subdivision of nucleus intercollicularis of the mesencephalon; HVC, a letter-based name; L1, L2 and L3 are subdivisions of Field L; LLD, lateral lemniscus, dorsal nucleus; LLI, lateral lemniscus, intermediate nucleus; LLV, lateral lemniscus, ventral nucleus; LMAN, lateral magnocellular nucleus of the anterior nidopallium; LMO, lateral oval nucleus of the mesopallium; MLd, dorsal part of the lateral nucleus of the mesencephalon; NIf, interfacial nucleus of the nidopallium; nXIIts, tracheosyringeal portion of the nucleus hypoglossus (nucleus XII); Ov, nucleus ovoidalis; PAm, nucleus paraambiguus medullaris; RA, robust nucleus of the arcopallium; RAm, nucleus retroambiguus medullaris; SO, superior olive; Uva, nucleus uvaeformis; VTA, ventral tegmental area. *Reprinted from Moorman et al. (2011), updated after Bolhuis et al. (2010), reprinted with permission from John Wiley and Sons and from Macmillan Publishers.*

6.5 SUMMARY

The sense of hearing has special meaning to many birds, be it for communication, hunting, or orienting in the dark. Avian hearing typically remains restricted to below 10 kHz, somewhat lower than human hearing and much lower than in a typical mammal. However, within that range, avian hearing is just as sensitive and discriminative. Important insights into auditory physiology have come from studies on birds, such as the mechanisms of regeneration of sensory hair cells after damage, the neural computations underlying sound localization, or the neural processing and learning of vocalizations. Some fascinating auditory specializations in birds are infrasound hearing in pigeons, asymmetric ears in owls, and echolocation in oilbirds and swiftlets.

REFERENCES

Art, J.J., Fettiplace, R., 2006. Contribution of ionic currents to tuning in auditory hair cells. In: Eatock, R.A., Fay, R.R., Popper, A.N. (Eds.), Vertebrate Hair Cells. Springer Science + Business Media, Inc., New York, pp. 204–248.

Bee, M.A., Klump, G.M., 2005. Auditory stream segregation in the songbird forebrain: effects of time intervals on responses to interleaved tone sequences. Brain Behav. Evol. 66, 197–214.

Beurg, M., Tan, X., Fettiplace, R., 2013. A prestin motor in chicken auditory hair cells: active force generation in a nonmammalian species. Neuron 79, 69–81.

Bolhuis, J.J., Okanoya, K., Scharff, C., 2010. Twitter evolution: converging mechanisms in birdsong and human speech. Nat. Rev. Neurosci. 11, 747–759.

Brainard, M.S., 2004. Contributions of the anterior forebrain pathway to vocal plasticity. Ann. N. Y. Acad. Sci. 1016, 377–394.

Brigande, J.V., Heller, S., 2009. Quo vadis, hair cell regeneration? Nat. Neurosci. 12, 679–685.

Brinkløv, S., Fenton, M.B., Ratcliffe, J.M., 2013. Echolocation in oilbirds and swiftlets. Front. Physiol. 4, 123.

Burger, R.M., Rubel, E.W., 2008. Encoding of interaural timing for binaural hearing. In: Basbaum, A.I., Kaneko, A., Shepherd, G.M., Westheimer, G. (Eds.), The Senses: a Comprehensive Reference. Academic Press, San Diego, pp. 613–630.

Calford, M.B., Piddington, R.W., 1988. Avian interaural canal enhances interaural delay. J. Comp. Physiol., A 162, 503–510.

Carr, C.E., Boudreau, R.E., 1993. Organization of the nucleus magnocellularis and the nucleus laminaris in the barn owl: encoding and measuring interaural time differences. J. Comp. Neurol. 334, 337–355.

Carr, C.E., Code, R.A., 2000. The central auditory system of reptiles and birds. In: Dooling, R.J., Fay, R.R., Popper, A.N. (Eds.), Comparative Hearing: Birds and Reptiles. Springer Verlag, New York, pp. 197–248.

Carroll, R.L., 1988. Vertebrate Paleontology and Evolution. Freeman, New York.

Chen, L., Salvi, R., Shero, M., 1994. Cochlear frequency-place map in adult chickens: intracellular biocytin labeling. Hear. Res. 81, 130–136.

Christensen-Dalsgaard, J., 2011. Vertebrate pressure-gradient receivers. Hear. Res. 273, 37–45.

Clack, J.A., Allin, E., 2004. The evolution of single- and multiple-ossicle ears in fishes and tetrapods. In: Manley, G.A., Popper, A., Fay, R.R.

(Eds.), Evolution of the Vertebrate Auditory System. Springer Verlag, New York, pp. 128–163.

Cobb, S., 1968. On the size of the auditory nuclei in some Apodiformes and Caprimulgiformes. Auk 85, 132–133.

Cohen, Y.E., Knudsen, E.I., 1999. Maps versus clusters: different representations of auditory space in the midbrain and forebrain. Trends Neurosci. 22, 128–135.

Coles, R.B., Konishi, M., Pettigrew, J.D., 1987. Hearing and echolocation in the Australian grey swiftlet, *Collocalia spodiopygia*. J. Exp. Biol. 129, 365–371.

Corfield, J., Kubke, M.F., Parsons, S., Wild, J.M., Köppl, C., 2011. Evidence for an auditory fovea in the New Zealand kiwi (*Apteryx mantellii*). PLoS One 6, e23771.

Cotanche, D.A., 1999. Structural recovery from sound and aminoglycoside damage in the avian cochlea. Audiol. Neurootol. 4, 271–285.

Cotanche, D.A., Lee, K.H., Stone, J.S., Picard, D.A., 1994. Hair cell regeneration in the bird cochlea following noise damage or ototoxic drug damage. Anat. Embryol. (Berl) 189, 1–18.

Dallos, P., 1996. Overview: cochlear neurobiology. In: Dallos, P., Popper, A.N., Fay, R.R. (Eds.), The Cochlea. Springer Verlag, New York, pp. 1–43.

Dooling, R.J., Lohr, B., Dent, M.L., 2000. Hearing in birds and reptiles. In: Dooling, R.J., Fay, R.R., Popper, A.N. (Eds.), Comparative Hearing: Birds and Reptiles. Springer Verlag, New York, pp. 308–359.

Farries, M.A., 2004. The avian song system in comparative perspective. Ann. N. Y. Acad. Sci. 1016, 61–76.

Fettiplace, R., Ricci, A.J., 2006. Mechanoelectrical transduction in auditory hair cells. Vertebr. Hair Cell., 154–203.

Fischer, F.P., 1994a. General pattern and morphological specializations of the avian cochlea. Scan. Microsc. 8, 351–364.

Fischer, F.P., 1994b. Quantitative TEM analysis of the barn owl basilar papilla. Hear. Res. 73, 1–15.

Fischer, F.P., Köppl, C., Manley, G.A., 1988. The basilar papilla of the barn owl *Tyto alba*: a quantitative morphological SEM analysis. Hear. Res. 34, 87–102.

Funabiki, K., Ashida, G., Konishi, M., 2011. Computation of interaural time difference in the Owl's coincidence detector neurons. J. Neurosci. 31, 15245–15256.

Gleich, O., 1989. Auditory primary afferents in the starling: correlation of function and morphology. Hear. Res. 37, 255–267.

Gleich, O., Dooling, R.J., Presson, J.C., 1997. Evidence for supporting cell proliferation and hair cell differentiation in the basilar papilla of adult Belgian Waterslager canaries (*Serinus canarius*). J. Comp. Neurol. 377, 5–14.

Gleich, O., Fischer, F.P., Köppl, C., Manley, G.A., 2004. Hearing organ evolution and specialization: Archosaurs. In: Manley, G.A., Popper, A., Fay, R.R. (Eds.), Evolution of the Vertebrate Auditory System. Springer Verlag, New York, pp. 224–255.

Gleich, O., Langemann, U., 2011. Auditory capabilities of birds in relation to the structural diversity of the basilar papilla. Hear. Res. 273, 80–88.

Gleich, O., Manley, G.A., 2000. The hearing organ of birds and crocodilia. In: Dooling, R.J., Fay, R.R., Popper, A.N. (Eds.), Comparative Hearing: Birds and Reptiles. Springer Verlag, New York, pp. 70–138.

Gleich, O., Narins, P.M., 1988. The phase response of primary auditory afferents in a songbird ((*Sturnus vulgaris*) L.). Hear. Res. 32, 81–92.

Goodyear, R.J., Richardson, G.P., 2002. Extracellular matrices associated with the apical surfaces of sensory epithelia in the inner ear: molecular and structural diversity. J. Neurobiol. 53, 212–227.

Griffin, D.R., 1953. Acoustic orientation in the oil bird, *Steatornis*. Proc. Natl. Acad. Sci. U. S. A. 39, 884–893.

Griffin, D.R., Thompson, D., 1982. Echolocation by cave swiftlets. Behav. Ecol. Sociobiol. 10, 119–123.

Grothe, B., Carr, C.E., Cassedy, J.H., Fritzsch, B., Köppl, C., 2004. The evolution of central pathways and their neural processing patterns. In: Manley, G.A., Popper, A., Fay, R.R. (Eds.), Evolution of the Vertebrate Auditory System. Springer Verlag, New York, pp. 289–359.

Grothe, B., Pecka, M., McAlpine, D., 2010. Mechanisms of sound localization in mammals. Physiol. Rev. 90, 983–1012.

Gummer, A.W., Smolders, J.W.T., Klinke, R., 1987. Basilar membrane motion in the pigeon measured with the Mössbauer technique. Hear. Res. 29, 63–92.

Gummer, A.W., Smolders, J.W.T., Klinke, R., 1989. Mechanics of a single-ossicle ear: I. The extra-stapedius of the pigeon. Hear. Res. 39, 1–14.

Hagstrum, J., 2000. Infrasound and the avian navigational map. J. Exp. Biol. 203, 1103–1111.

Hill, K.G., Stange, G., Mo, J., 1989. Temporal synchronization in the primary auditory response in the pigeon. Hear. Res. 39, 63–74.

Itatani, N., Klump, G.M., 2011. Neural correlates of auditory streaming of harmonic complex sounds with different phase relations in the songbird forebrain. J. Neurophysiol. 105, 188–199.

Iwaniuk, A.N., Clayton, D.H., Wylie, D.R.W., 2006. Echolocation, vocal learning, auditory localization and the relative size of the avian auditory midbrain nucleus (MLd). Behav. Brain Res. 167, 305–317.

Jarvis, E., Gunturkun, O., Bruce, L., Csillag, A., Karten, H., Kuenzel, W., Medina, L., Paxinos, G., Perkel, D.J., Shimizu, T., Striedter, G., Wild, J.M., Ball, G.F., Dugas-Ford, J., Durand, S.E., Hough, G.E., Husband, S., Kubikova, L., Lee, D.W., Mello, C.V., Powers, A., Siang, C., Smulders, T.V., Wada, K., White, S.A., Yamamoto, K., Yu, J., Reiner, A., Butler, A.B., Avian Brain Nomenclature Consortium, 2005. Avian brains and a new understanding of vertebrate brain evolution. Nat. Rev. Neurosci. 6, 151–159.

Jarvis, E.D., 2004. Learned birdsong and the neurobiology of human language. Ann. N. Y. Acad. Sci. 1016, 749–777.

Jeffress, L.A., 1948. A place theory of sound localization. J. Comp. Physiol. Psychol. 41, 35–39.

Kaiser, A., Manley, G.A., 1996. Brainstem connections of the macula lagenae in the chicken. J. Comp. Neurol. 374, 108–117.

Keller, C.H., Hartung, K., Takahashi, T.T., 1998. Head-related transfer functions of the barn owl: measurement and neural responses. Hear. Res. 118, 13–34.

King, A.J., Parsons, C.H., Moore, D.R., 2000. Plasticity in the neural coding of auditory space in the mammalian brain. Proc. Natl. Acad. Sci. U. S. A. 97, 11821–11828.

Klump, G.M., 2000. Sound localization in birds. In: Dooling, R.J., Fay, R.R., Popper, A.N. (Eds.), Comparative Hearing: Birds and Reptiles. Springer Verlag, New York, pp. 249–307.

Knudsen, E.I., 2002. Instructed learning in the auditory localization pathway of the barn owl. Nature 417, 322–328.

Knudsen, E.I., Konishi, M., 1978. A neural map of auditory space in the owl. Science 200, 795–797.

Konishi, M., 1993. Listening with two ears. Sci. Am. 268, 66–73.

Konishi, M., 2003. Coding of auditory space. Annu. Rev. Neurosci. 26, 31–55.

Konishi, M., 2004. The role of auditory feedback in birdsong. Ann. N. Y. Acad. Sci. 1016, 463–475.

Konishi, M., Knudsen, E.I., 1979. The oilbird: hearing and echolocation. Science 204, 425–427.

Köppl, C., 1997a. Frequency tuning and spontaneous activity in the auditory nerve and cochlear nucleus magnocellularis of the barn owl, *Tyto alba*. J. Neurophysiol. 77, 364–377.

Köppl, C., 1997b. Number and axon calibres of cochlear afferents in the barn owl. Aud. Neurosci. 3, 313–334.

Köppl, C., 1997c. Phase locking to high frequencies in the auditory nerve and cochlear nucleus magnocellularis of the barn owl, *Tyto alba*. J. Neurosci. 17, 3312–3321.

Köppl, C., 2009. Evolution of sound localisation in land vertebrates. Curr. Biol. 19, R635–R639.

Köppl, C., 2011a. Birds – same thing, but different? Convergent evolution in the avian and mammalian auditory systems provides informative comparative models. Hear. Res. 273, 65–71.

Köppl, C., 2011b. Evolution of the octavolateral efferent system. In: Ryugo, D., Fay, R.R., Popper, A.N. (Eds.), Auditory and Vestibular Efferents. Springer Science + Business Media, LLC, New York, pp. 217–259.

Köppl, C., Gleich, O., Manley, G.A., 1993. An auditory fovea in the barn owl cochlea. J. Comp. Physiol., A 171, 695–704.

Köppl, C., Klump, G.M., Taschenberger, G., Dyson, M., Manley, G.A., 1998. The auditory fovea of the barn owl – no correlation with enhanced frequency resolution. In: Palmer, A.R., Rees, A., Summerfield, A.Q., Meddis, R. (Eds.), Psychophysical and Physiological Advances in Hearing. Whurr Publishers Ltd, London, pp. 153–161.

Köppl, C., Manley, G.A., 1997. Frequency representation in the emu basilar papilla. J. Acoust. Soc. Am. 101, 1574–1584.

Kreithen, M.L., Quine, D.B., 1979. Infrasound detection by the homing pigeon: a behavioral audiogram. J. Comp. Physiol. 129, 1–4.

Kuba, H., 2007. Cellular and molecular mechanisms of avian auditory coincidence detection. Neurosci. Res. 59, 370–376.

Kubke, M.F., Carr, C.E., 2006. Morphological variation in the nucleus laminaris of birds. Int. J. Comp. Psychol. 19, 83–97.

Langemann, U., Hamann, I., Friebe, A., 1999. A behavioral test of presbycusis in the bird auditory system. Hear. Res. 137, 68–76.

Lavigne-Rebillard, M., Cousillas, H., Pujol, R., 1985. The very distal part of the basilar papilla in the chicken: a morphological approach. J. Comp. Neurol. 238, 340–347.

MacLeod, K.M., Carr, C.E., 2007. Beyond timing in the auditory brainstem: intensity coding in the avian cochlear nucleus angularis. Prog. Brain Res. 165, 123–133.

Manley, G.A., Clack, J.A., 2004. An outline of the evolution of vertebrate hearing organs. In: Manley, G.A., Popper, A., Fay, R.R. (Eds.), Evolution of the Vertebrate Auditory System. Springer Verlag, New York, pp. 1–26.

Manley, G.A., Haeseler, C., Brix, J., 1991. Innervation patterns and spontaneous activity of afferent fibres to the lagenar macula and apical basilar papilla of the chick's cochlea. Hear. Res. 56, 211–226.

Manley, G.A., Köppl, C., 1998. Phylogenetic development of the cochlea and its innervation. Curr. Opin. Neurobiol. 8, 468–474.

Manley, G.A., Köppl, C., Yates, G.K., 1997. Activity of primary auditory neurones in the cochlear ganglion of the emu *Dromaius novaehollandiae* I: spontaneous discharge, frequency tuning and phase locking. J. Acoust. Soc. Am. 101, 1560–1573.

Manley, G.A., Ladher, R., 2008. Phylogeny and evolution of ciliated mechanoreceptor cells. In: Dallos, P., Oertel, D. (Eds.), Audition. Academic Press, San Diego, pp. 1–34.

Manley, G.A., Sienknecht, U., 2013. The evolution and development of middle ears in land vertebrates. In: Puria, S., Popper, A.N., Fay, R.R. (Eds.), The Middle Ear. Science, Otosurgery, and Technology. Springer Science + Business Media, LLC, New York, pp. 7–30.

Manley, G.A., van Dijk, P., 2008. Otoacoustic emissions in amphibians, lepidosaurs, and archosaurs. In: Manley, G.A., Fay, R.R., Popper, A.N. (Eds.), Active Processes and Otoacoustic Emissions in Hearing. Springer Science + Business Media, LLC, New York, pp. 211–260.

Martin, P., 2008. Active hair-bundle motility of the hair cells of vestibular and auditory organs. In: Manley, G.A., Fay, R.R., Popper, A.N. (Eds.), Active Processes and Otoacoustic Emissions in Hearing. Springer Science + Business Media, LLC, New York, pp. 93–144.

Matthews, G., Fuchs, P., 2010. The diverse roles of ribbon synapses in sensory neurotransmission. Nat. Rev. Neurosci. 11, 812–822.

Moiseff, A., Konishi, M., 1981a. Neuronal and behavioral sensitivity to binaural time differences in the owl. J. Neurosci. 1, 40–48.

Moiseff, A., Konishi, M., 1981b. The owl's interaural pathway is not involved in sound localization. J. Comp. Physiol., A 144, 299–304.

Moorman, S., Mello, C.V., Bolhuis, J.J., 2011. From songs to synapses: molecular mechanisms of birdsong memory. Molecular mechanisms of auditory learning in songbirds involve immediate early genes, including zenk and arc, the ERK/MAPK pathway and synapsins. Bioessays 33, 377–385.

Nagel, K., Kim, G., McLendon, H., Doupe, A., 2011. A bird brain's view of auditory processing and perception. Hear. Res. 273, 123–133.

Necker, R., 1970. Zur Entstehung der Cochleapotentiale von Vögeln: Verhalten bei O2-Mangel, Cyanidvergiftung und Unterkühlung sowie Beobachtungen über die räumliche Verteilung. Z. Vgl. Physiol. 69, 367–425.

Neubauer, H., Köppl, C., Heil, P., 2009. Spontaneous activity of auditory nerve fibers in the barn owl (*Tyto alba*): analyses of interspike interval distributions. J. Neurophysiol. 101, 3169–3191.

Norberg, R.A., 2002. Independent evolution of outer ear asymmetry among five owl lineages: morphology, function and selection. In: Newton, I., Kavanagh, R., Olsen, J., Taylor, I. (Eds.), Ecology and Conservation of Owls. CSIRO Publishing, Collingwood, Victoria, Australia, pp. 329–342.

Nottebohm, F., 2004. The road we travelled: discovery, choreography, and significance of brain replaceable neurons. Ann. N. Y. Acad. Sci. 1016, 628–658.

Novick, A., 1959. Acoustic orientation in the cave swiftlet. Biol. Bull. Mar. Biol. Lab. Woods Hole 117, 497–503.

Okanoya, K., Dooling, R.J., 1985. Colony differences in auditory thresholds in the canary (*Serinus canarius*). J. Acoust. Soc. Am. 78, 1170–1176.

Ondracek, J.M., Hahnloser, R.H.R., 2014. Advances in understanding the auditory brain of songbirds. In: Köppl, C., Manley, G.A., Popper, A.N., Fay, R.R. (Eds.), Insights from Comparative Hearing Research. Springer Science + Business Media, LLC, New York http://dx.doi.org/10.1007/2506-2013-31.

Payne, R.S., 1971. Acoustic location of prey by barn owls (*Tyto alba*). J. Exp. Biol. 54, 535–573.

Peña, J.L., Konishi, M., 2001. Auditory spatial receptive fields created by multiplication. Science 292, 249–252.

Perkel, D.J., 2004. Origin of the anterior forebrain pathway. Ann. N. Y. Acad. Sci. 1016, 736–748.

Pickles, J.O., 2008. An Introduction to the Physiology of Hearing, third ed. Emerald Group Publishing Ltd, Bingley, UK.

Proctor, L., Konishi, M., 1997. Representation of sound localization cues in the auditory thalamus of the barn owl. Proc. Natl. Acad. Sci. U. S. A. 94, 10421–10425.

Pytte, C.L., Ficken, M.S., Moiseff, A., 2004. Ultrasonic singing by the blue-throated hummingbird: a comparison between production and perception. J. Comp. Physiol., A 190, 665–673.

Raphael, Y., Altschuler, R.A., 2003. Structure and innervation of the cochlea. Brain Res. Bull. 60, 397–422.

Rubel, E.W., Furrer, S.A., Stone, J.S., 2013. A brief history of hair cell regeneration research and speculations on the future. Hear. Res. 297, 42–51.

Runhaar, G., Schedler, J., Manley, G.A., 1991. The potassium concentration in the cochlear fluids of the embryonic and post-hatching chick. Hear. Res. 56, 227–238.

Russell, I.J., 2014. Roles for prestin in harnessing the basilar membrane to the organ of Corti. In: Köppl, C., Manley, G.A., Popper, A.N., Fay, R.R. (Eds.), Insights from Comparative Hearing Research. Springer Science + Business Media, LLC, New York http://dx.doi.org/10.1007/2506-2013-23.

Ryals, B.M., Dent, M.L., Dooling, R.J., 2013. Return of function after hair cell regeneration. Hear. Res. 297, 113–120.

Ryals, B.M., Westbrook, E.W., 1990. Hair cell regeneration in senescent quail. Hear. Res. 50, 87–96.

Sachs, M.B., Woolf, N.K., Sinnott, J.M., 1980. Response properties of neurons in the avian auditory system: comparisons with mammalian homologues and consideration of the neural encoding of complex stimuli. In: Popper, A.N., Fay, R.R. (Eds.), Comparative Studies of Hearing in Vertebrates. Springer-Verlag, New York, Heidelberg, Berlin, pp. 323–353.

Sachs, M.B., Young, E.D., Lewis, R.H., 1974. Discharge patterns of single fibers in the pigeon auditory nerve. Brain Res. 70, 431–447.

Salvi, R.J., Saunders, S.S., Powers, N.L., Boettcher, F.A., 1992. Discharge patterns of cochlear ganglion neurons in the chicken. J. Comp. Physiol., A 170, 227–241.

Saunders, J.C., Duncan, R.K., Doan, D.E., Werner, Y.L., 2000. The middle ear of reptiles and birds. In: Dooling, R.J., Fay, R.R., Popper, A.N. (Eds.), Comparative Hearing: Birds and Reptiles. Springer Verlag, New York, pp. 13–69.

Schermuly, L., Klinke, R., 1990a. Infrasound sensitive neurones in the pigeon cochlear ganglion. J. Comp. Physiol., A 166, 355–363.

Schermuly, L., Klinke, R., 1990b. Origin of infrasound sensitive neurones in the papilla basilaris of the pigeon: an HRP study. Hear. Res. 48, 69–78.

Schmidt, R.S., 1963. Types of endolymphatic potentials. Comp. Biochem. Physiol. 10, 83–87.

Smolders, J.W.T., 1999. Functional recovery in the avian ear after hair cell regeneration. Audiol. Neurootol. 4, 286–302.

Smolders, J.W.T., Ding-Pfennigdorff, D., Klinke, R., 1995. A functional map of the pigeon basilar papilla: correlation of the properties of single auditory nerve fibres and their peripheral origin. Hear. Res. 92, 151–169.

Stone, J.S., Cotanche, D.A., 2007. Hair cell regeneration in the avian auditory epithelium. Int. J. Dev. Biol. 51, 633–647.

Sullivan, W.E., Konishi, M., 1984. Segregation of stimulus phase and intensity coding in the cochlear nucleus of the barn owl. J. Neurosci. 4, 1787–1799.

Suthers, R.A., Hector, D.H., 1982. Mechanism for the production of echolocating clicks by the grey swiftlet, *Collocalia spodiopygia*. J. Comp. Physiol., A 148, 457–470.

Suthers, R.A., Hector, D.H., 1985. The physiology of vocalization by the echolocating oilbird, *Steatornis caripensis*. J. Comp. Physiol., A 156, 243–266.

Takahashi, T.T., 2010. How the owl tracks its prey – II. J. Exp. Biol. 213, 3399–3408.

Takahashi, T.T., Bala, A.D.S., Spitzer, M.W., Euston, D.R., Spezio, M.L., Keller, C.H., 2003. The synthesis and use of the owl's auditory space map. Biol. Cybern. 89, 378–387.

Takasaka, T., Smith, C.A., 1971. The structure and innervation of the pigeon's basilar papilla. J. Ultrastruct. Res. 35, 20–65.

Tan, X., Beurg, M., Hackney, C., Mahendrasingam, S., Fettiplace, R., 2013. Electrical tuning and transduction in short hair cells of the chicken auditory papilla. J. Neurophysiol. 109, 2007–2020.

Taschenberger, G., Manley, G.A., 1997. Spontaneous otoacoustic emissions in the barn owl. Hear. Res. 110, 61–76.

Theunissen, F.E., Amin, N., Shaevitz, S.S., Woolley, S.M., Fremouw, T., Hauber, M.E., 2004. Song selectivity in the song system and in the auditory forebrain. Ann. N. Y. Acad. Sci. 1016, 222–245.

Theunissen, F.E., Shaevitz, S.S., 2006. Auditory processing of vocal sounds in birds. Curr. Opin. Neurobiol. 16, 400–407.

Thomassen, H.A., Povel, G.D.E., 2006. Comparative and phylogenetic analysis of the echo clicks and social vocalizations of swiftlets (Aves: Apodidae). Biol. J. Linn. Soc. 88, 631–643.

Tilney, M.S., Tilney, L.G., DeRosier, D.J., 1987. The distribution of hair cell bundle lengths and orientations suggests an unexpected pattern of hair cell stimulation in the chick cochlea. Hear. Res. 25, 141–151.

Vergne, A.L., Pritz, M.B., Mathevon, N., 2009. Acoustic communication in crocodilians: from behaviour to brain. Biol. Rev. 84, 391–411.

von Bartheld, C.S., Gianessi, F., 2011. The paratympanic organ: a barometer and altimeter in the middle ear of birds? J. Exp. Zool. B Mol. Dev. Evol. 316, 402–408.

Vonderschen, K., Wagner, H., 2009. Tuning to interaural time difference and frequency differs between the auditory arcopallium and the external nucleus of the inferior colliculus. J. Neurophysiol. 101, 2348–2361.

Wang, Y., Brzozowska-Prechtl, A., Karten, H., 2010. Laminar and columnar auditory cortex in avian brain. Proc. Natl. Acad. Sci. U. S. A. 107, 12676–12681.

Wangemann, P., 1995. Comparison of ion transport mechanisms between vestibular dark cells and strial marginal cells. Hear. Res. 90, 149–157.

Wangemann, P., 2002. K^+ cycling and its regulation in the cochlea and the vestibular labyrinth. Audiol. Neurootol. 7, 199–205.

Warchol, M.E., Dallos, P., 1989. Neural response to very low-frequency sound in the avian cochlear nucleus. J. Comp. Physiol., A 166, 83–95.

Wilkins, H.R., Presson, J.C., Popper, A.N., Ryals, B.M., Dooling, R.J., 2001. Hair cell death in a hearing-deficient canary. J. Assoc. Res. Otolaryngol. 2, 79–86.

Woolley, S.M., Gill, P.R., Fremouw, T., Theunissen, F.E., 2009. Functional groups in the avian auditory system. J. Neurosci. 29, 2780–2793.

Wu, Y.-C., Art, J.J., Goodman, M.B., Fettiplace, R., 1995. A kinetic description of the calcium-activated potassium channel and its application to electrical tuning of hair cells. Prog. Biophys. Mol. Biol. 63, 131–158.

Yates, G.K., Manley, G.A., Köppl, C., 2000. Rate-intensity functions in the emu auditory nerve. J. Acoust. Soc. Am. 107, 2143–2154.

Young, B.A., Mathevon, N., Tang, Y., 2014. Reptile auditory neuroethology: what do reptiles do with their hearing? In: Köppl, C., Manley, G.A., Popper, A.N., Fay, R.R. (Eds.), Insights from Comparative Hearing Research. Springer Science + Business Media, LLC, New York-http://dx.doi.org/10.1007/2506-2013-30.

Chapter 7

The Chemical Senses in Birds

Larry Clark

United States Department of Agriculture, Animal and Plant Health Inspection Service, Wildlife Services, National Wildlife Research Center, Fort Collins, CO, USA

Julie Hagelin

Institute of Arctic Biology, University of Alaska Fairbanks, Fairbanks, AK, USA; Alaska Department of Fish and Game, Fairbanks, AK, USA

Scott Werner

United States Department of Agriculture, Animal and Plant Health Inspection Service, Wildlife Services, National Wildlife Research Center, Fort Collins, CO, USA

7.1 CHEMICAL SENSES

The chemical senses generally fall into three categories: chemesthesis (irritation and pain), olfaction (smell), and gustation (taste). Traditionally, the emphasis in describing responsiveness to chemical stimuli has been placed on taste and smell. The reality is more complex. For example, the sensory afferents for chemesthetic perception are in close proximity with olfactory receptors in the nasal cavity and with gustatory receptors in the oral cavity. Because external chemical stimuli can be processed by multiple sensory systems, there has been a great deal of confusion in the literature on the importance of individual sensory modalities. Generally, the principal mediating sensory modality may be related to stimulus type, concentration, and presentation. However, when perception of external chemical stimuli occurs via the integrated perception across modalities, the combined perceptual quality is commonly referred to as flavor.

7.2 CHEMESTHESIS

Chemesthesis is the perception of chemically induced pain. The first neural mediator of noxious stimuli is the nociceptor (Woolf and Ma, 2007). These primary sensory neurons are the interface between the internal and external environments. Nociceptors have cell bodies located in the dorsal root ganglion, a peripheral axon that innervates tissues, and a central axon that enters the spinal cord to transfer information to the central nervous system. Nociceptors have three functions: (1) detection of potentially damaging external noxious stimuli, which is useful in warning an animal to the risk of injury; (2) detection of endogenous inflammatory stimuli, which is useful in initiating and promoting behaviors conducive to healing and repair; and (3) detection of neural damage and ectopic firing. This latter function is a pathological condition of chronic pain. Nociceptors have high thresholds for exogenous stimuli, presumably because it would be maladaptive to defensively respond to every external assault. Nociceptors have low thresholds for endogenous stimuli. This is an adaptive response to promote healing once damage has occurred (Patapoutian et al., 2009).

A major component of the chemesthetic system is the trigeminal nerve (TN). The TN is the principal somatic sensory nerve of the head, and its primary function is the coding of mechanical and thermal stimuli. However, the TN also contains chemoreceptive fibers that mediate the detection of chemical irritants (Silver and Maruniak, 1981). The somatosensory system is the primary somatic sensory system of the rest of the body. Like the TN, the somatosensory system primarily codes for mechanical and thermal stimuli, but it does have sensory afferents that are chemosensory (Gentle, 2011; Necker, 2000; Wild, 1985).

7.2.1 Trigeminal and Somatosensory Nerves

The morphological organization of the peripheral TN in birds is not very different from that found in mammals (Dubbeldam and Karten, 1978; Dubbeldam and Veenman, 1978; Gottschaldt, 1985). The TN is the fifth cranial nerve in birds, arising from the rostrolateral medulla near the caudal surface of the optic lobe (Getty, 1975; Schrader, 1970).

Sturkie's Avian Physiology.

The TN travels along the trochlear nerve (IV), entering a fossa in the floor of the cranial cavity where the trigeminal ganglion (TG) is found. The TG is subdivided into a smaller medial ophthalmic region and a larger lateral maxillomandibular region, from which the nerve splits into three branches. In the chicken (*Gallus gallus domesticus*) the ophthalmic branch innervates the frontal region, the eyeball, upper eyelid, conjunctiva, glands in the orbit, the rostrodorsal part of the nasal cavity, and the tip of the upper jaw. The ophthalmic branch as a communicating ramus with the trochlear nerve serves for motor control of the eye region. This aspect can provide for reflexive response to irritating stimuli to the ocular region. The larger medial ramus accompanies the olfactory nerve into the nasal fossa via the medial orbitonasal foramen. The maxillary branch provides sensory input from the integument of the crown, temporal region, rostral part of the external ear, upper and lower eyelids, the region between the nostrils and eye, conjunctival mucosa, the mucosal part of the palate, and the floor of the medial wall of the nasal cavity. The mandibular branch provides sensory input from the skin and rhamphotheca of the lower jaw, intermandibular skin, wattles, oral mucosa of rostral floor of the mouth, and the palate near the angle of the mouth (Getty, 1975; Schrader, 1970).

7.2.2 Performance Characteristics of Nociceptors

Pain and irritation perception begin with activation of primary sensory nociceptors. In birds, chemosensitive fibers in the TN and somatosensory nerves are similar to mammalian afferents. Most are unmyelinated C-type polymodal nociceptors with conduction velocities of 0.3–1 m/s. However, some myelinated A-delta high-threshold mechanoreceptors with conduction velocities of 5–40 m/s also respond to chemical stimuli. The discharge patterns and conduction velocities for the chicken, mallard (*Anas platyrhyncos*), and pigeon (*Columba livia*) are similar to those observed in mammals (Gentle, 1989; Necker, 1974).

Although birds have slightly different neural architecture relative to mammals, the underlying functions of neural connections have been evolutionarily preserved (Butler and Cotterill, 2006; Dugas-Ford et al., 2012; Güntürkün, 2012). This also applies to the underlying physiological and biochemical processes of chemically induced pain. Generally, birds have the same classes of neuropeptides as mammals, but their structures are not totally homologous. Avian endogenous pain-promoting substances such as substance P, 5-HT, histamine, bradykinin, and acetylcholine evoke inflammation and pain-related behaviors in chickens, pigeons, rats, dogs, and guinea pigs (Szolcsanyi et al., 1986; Gentle and Hill, 1987; Gentle and Hunter, 1993; Koda et al., 1996; Hu et al., 2002; Ohta et al., 2006). Prostaglandins that modulate the pain response in mammals also serve this function in birds, and their effects can be abolished by prostaglandin biosynthase inhibitors such as aspirin-like analgesics (Clark, 1995).

Despite these physiologically mediated similarities, there are profound differences in how birds and mammals respond to exogenous chemical stimuli. In mammals, chemicals such as capsaicin are potent trigeminal irritants. These irritants deplete substance P from afferent terminals and the dorsal root ganglion, producing an initial sensitization followed by desensitization to further chemical stimulation (Szolcsanyi, 1982). In contrast, birds are insensitive to capsaicin (Mason and Maruniak, 1983; Szolcsanyi et al., 1986). Peripheral presentation of capsaicin to pigeons and chickens does not cause release of substance P in avian sensory afferents (Pierau et al., 1986; Szolcsanyi et al., 1986; Sann et al., 1987). These taxon-specific responses to exogenous chemical stimuli underscore taxonomic differences in both endogenous neuropeptides and receptors, whose significance has been implicated in the evolutionary ecology of the taxa (Mason et al., 1991; Clark, 1998; Tewksbury and Nabhan, 2001).

7.2.3 Receptor Mechanisms

Nonselective transient receptor potential (TRP) cation channels are involved in sensory neuron activation events, neurotransmitter release, release of inflammatory mediators, and other aspects of pain transduction (Cortright et al., 2007; Figure 7.1). Most of what is known about TRP channels is derived from work done on mammals (Holzer, 2011). However, increasingly more comparative evolutionary similarities and differences are being characterized for other taxa (Saito and Shingai, 2006; Saito et al., 2011). TRPV1 (initially called VR1) was first cloned in mammals and found to respond to the exogenous vanilloid, capsaicin (Caterina et al., 1997), as well as endogenous agonists, anandamide, and 12-HPETE, which are structurally similar to capsaicin (Zygmunt et al., 1999; Hwang et al., 2000). TRPV1 is also activated by heat (>43 °C) and acid ($pH \leq 6$). The sensation that TRPV1 activation evokes in humans via these polymodal nociceptors is one of tingling and burning, like the sensation produced by capsaicin found in chili peppers. Like its mammalian counterpart, the TRP receptor in birds (cTRPV1, chick dorsal root ganglion) responds to high temperatures (≥45 °C) and extracellular acid solution ($pH \leq 4$). However, cTRPV1 is different, showing a 68% identity and 79% similarity to rat TRPV1. These differences in receptor composition manifest as a poor response to capsaicin (Jordt and Julius, 2002) and explain the behavioral differences in capsaicin sensitivity between birds and mammals; mammals are behaviorally sensitive to capsaicin and birds are not (Mason et al., 1991; Norman et al., 1992).

Currently, 28 TRP channels, grouped into six functional subfamilies, have been characterized. The subfamilies are

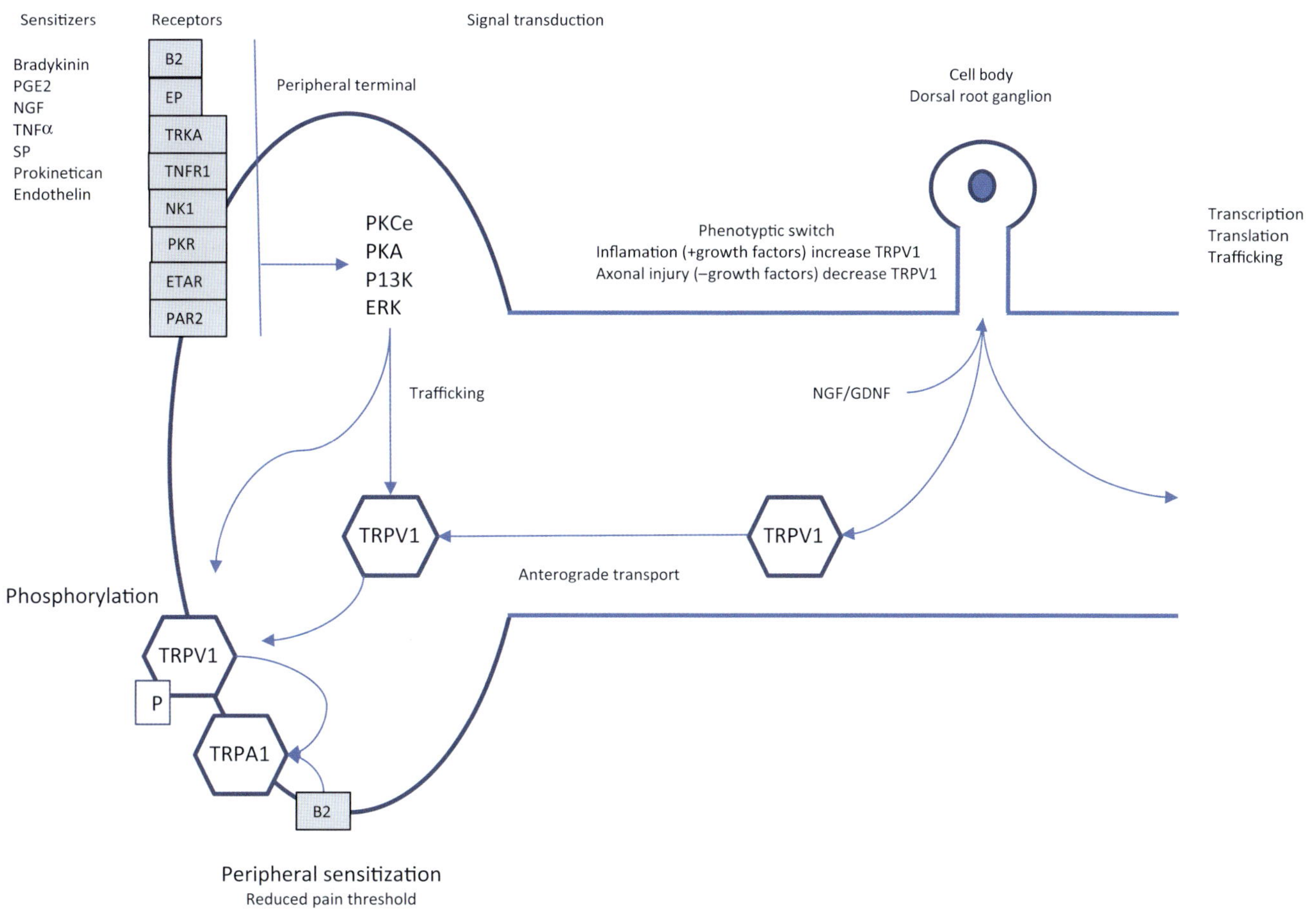

FIGURE 7.1 **Changes in transient receptor potential (TRP) channels produced by inflammation.** Endogenous sensitizers act on receptors expressed by nociceptors to activate intracellular signal transduction pathways. Pathways phosphorylate TRP channels, altering trafficking to the membrane, thresholds, and kinetics. Growth factors, such as nerve growth factor (NGF), are retrogradely transported to the cell body of the nociceptors. Through intracellular signaling pathways, expression of TRP channels is increased and they are transported to the peripheral terminal. Changes in transcription and translation of TRP channels and other proteins can switch the chemical phenotype of the neurons from their state in naive conditions to an altered state during inflammation. B2, bradykinin receptor; ERK, extracellular signal-regulated kinase; ETAR, endothelin receptor type A; GDNF, glial-cell-derived neurotrophic factor; NK1, neurokinin receptor 1; PAR2, protease-activated receptor 2; PGE2, prostaglandin E2; PI3K, phosphoinositide 3-kinase; PK, protein kinase; PKR, prokineticin receptor; TNFα, tumor necrosis factor α; TNFR1, TNF receptor 1; TRKA, tyrosine kinase receptor A. *Adapted from Patapoutian et al. (2009).*

responsive to exogenous compounds that code for qualitative perceptual similarities (e.g., the "hotness" of capsaicin, the "burn" of cinnamon oil, the "coolness" of menthol, the irritation of mustard oil; Holzer, 2011). Although the specific homologies for other TRP channels in birds are generally not known, based on behavioral responsiveness to a variety of mammalian irritants, it is anticipated that TRP channel receptor molecules in birds would be structurally similar and/or have similar expression in nociceptors to that found in mammals for cinnamon oil, allicin (garlic/onion), and menthol and divergent for mustard oil and anthranilate (grape) compounds (Clark, 1998; Stucky et al., 2009).

Digital fluorescence imaging of intracellular calcium $[Ca^{2+}]_I$ in vitro preparations of chicken and rat trigeminal dorsal root ganglia show that there are separate and overlapping populations of neurons that are sensitive to the well-described avian irritant, methyl anthranilate, and capsaicin (Kirifides et al., 2004). In the chicken, 48% of neurons responded to methyl anthranilate, whereas only 16% responded to capsaicin. Moreover, there was a greater change in $[Ca^{2+}]_I$ to equimolar concentrations of methyl anthranilate (78%) relative to capsaicin (43%). Increases in $[Ca^{2+}]_I$ were dependent upon extracellular calcium for both methyl anthranilate and capsaicin. However, responses to methyl anthranilate, but not capsaicin, were dependent on extracellular sodium. This suggests different transduction mechanisms for the two compounds. Together, these observations provide further rationale for the observed behavioral differences in birds to these two compounds. Starlings (*Sturnus vulgaris*) demonstrate congenital avoidance to methyl anthranilate but not capsaicin, although they could be trained to avoid capsaicin in conditioned

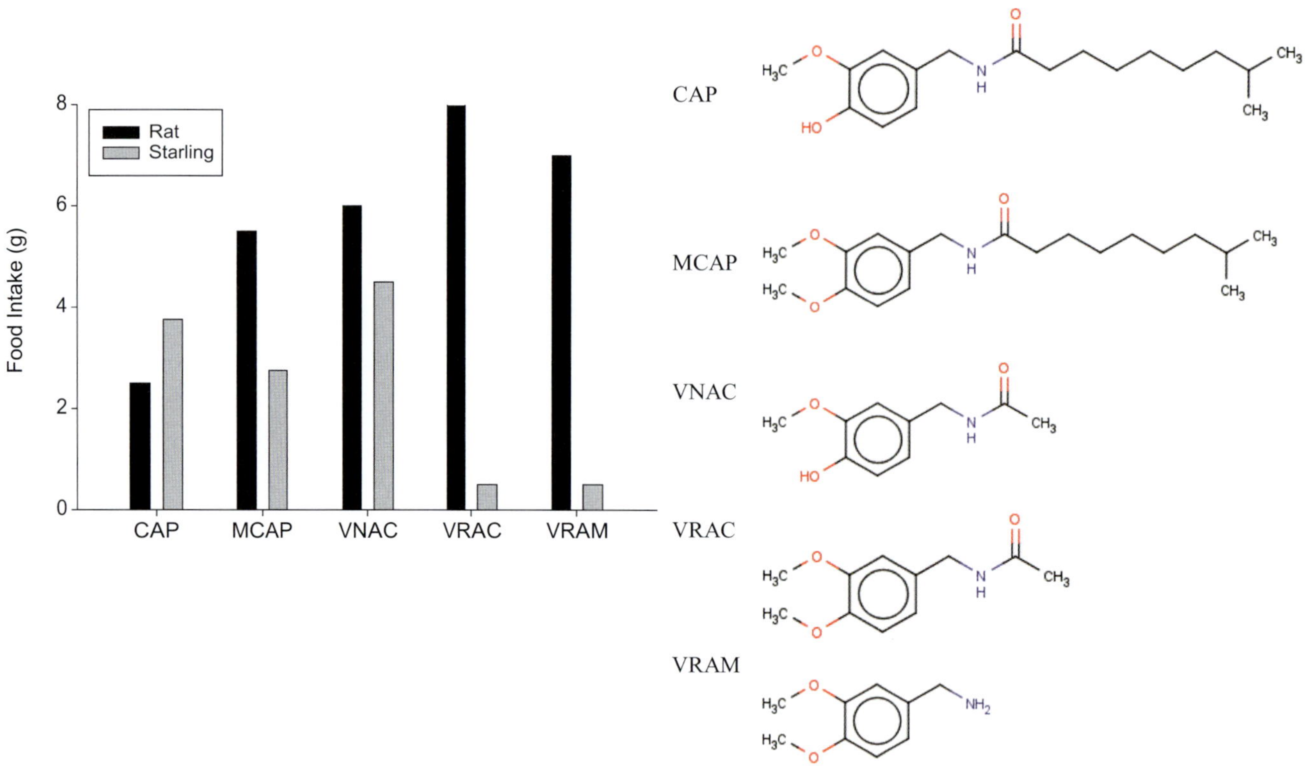

FIGURE 7.2 **Consumption of food treated with capsaicin (CAP), methyl capsaicin (MCAP), vanillyl acetamide (VNAC), veratryl acetamide (VRAC), and veratryl amine (VRAM) in rats and starlings.** Note the general inverse relationship of consumption as structure changes, suggesting functional receptor differences in the two taxa. Bird repellents are more basic and rigid (planar) than mammal aversive compounds. Concentration applied: 1000 ppm. Consumption of 4 g of untreated food is control baseline intake for both species. *Data adapted from Mason et al. (1991).*

avoidance paradigms, and that avoidance was contingent upon an intact ophthalmic branch of the TN (Mason and Clark, 1995). These observations also suggest that while birds can perceive capsaicin, although somewhat poorly, it is not coded as pain, highlighting the importance of central processing in the perceptual interpretation of peripheral signals.

7.2.4 Chemical Structure–Activity Relationships to Irritants

Despite the apparent insensitivity of birds to capsaicin, they can respond to other vanilloid compounds (Figure 7.2). Aromatic compounds that are considered aversive by birds are qualitatively characterized as having an aromatic heterocyclic core, high degree of basicity, high degree of lipophilicity, and a high degree of electronegativity (Figure 7.3). The core aromatic heterocycle of a repellent compound is enhanced by substitutions that affect electron donation: amino > methoxy > methyl > hydroxyl groups. Resonance of lone pairs of electrons enhances repellency as a function of substituent position: ortho > para > meta. Acidic substituents in the electron withdrawing group detract from aversive qualities of the compound. Steric effects and extreme delocalization of lone pairs of electrons, as might occur in meta isomers and aromatic structures with multiple substituted electron donating groups, tend to interfere with repellency (Mason et al., 1989; Clark, 1991a; Clark and Shah, 1991, 1994; Clark et al., 1991; Shah et al., 1991).

Quantitative structure–activity relationships of aromatic compounds and repellency are consistent with earlier qualitative studies. The aversive properties of 14 derivatives of cinnamic acid compounds are characterized by heat of formation (DH(f)), polarizability (XY and YY), and superdelocalizability (Sr). All of these descriptors are electronic (Watkins et al., 1999). These findings generally align with a reanalysis of the quantitative structure–activity relationships of the 117 compounds described above (Clark, 1997). Canonical analysis of the relationship of physicochemical, topological, and electrostatic descriptors and the response shape of the four-parameter fluid intake curve showed that 94% of variance in the response profile could be accounted for by five parameters: polarizability, ES2, ANC, KAPPA2, and CHI2. Polarizability is the relative susceptibility of the electron cloud of a molecule to be distorted by presence of an external electric field. Owing to distortion, an induced electric dipole moment appears. Temporary dipoles induce dipoles in other molecules, resulting in van der

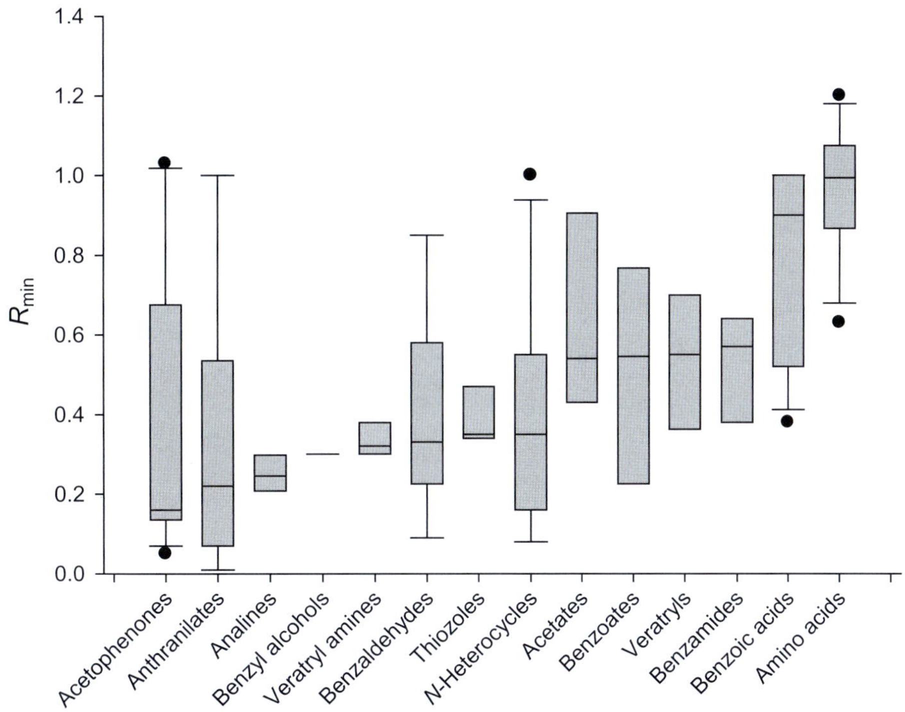

FIGURE 7.3 **The relative reduction of fluid intake for solutions as a function of chemical class, which assumes a benzene parent structure with the nomenclatural taxonomy defined by the principal electron withdrawing group.** Fluid intake is the asymptotic minimum intake in one-bottle 6-h drinking trials (R_{min}). Strongly aversive solutions (where R_{min} is not statistically distinguishable from zero) have $R_{min} \leq 0.2$. Moderately aversive solutions have $0.2 < R_{min} \leq 0.4$, weakly aversive solutions have 0.4 $R_{min} \leq 0.6$, and solutions with $R_{min} > 0.6$ are not aversive at all (not statistically different from water controls). Median R_{min} (solid bars), R_{min} 25–75th percentile (shaded box), R_{min} 5–95th percentile (capped line), and the range of R_{min} (open symbols). *Adapted from Clark (1997).*

Waals intermolecular forces by orienting the temporary and induced dipoles with each other. ES2 is an electrotopological descriptor that describes electronic interactions between molecules. ANC is a partial negative electronic charge descriptor of electrostatic potential that influences molecular interactions. CHI2 and KAPP2 are valence connectivity and shape descriptors that may describe the rigidity of the molecule and accessibility of the molecule to receptor systems. The importance of electronic features of molecules is consistent with studies of TRPA1 channel modulation and activation of cysteine-reactive chemicals. TRP channel activation was found to be more dependent on chemical reactivity relative to molecular shape (Hinman et al., 2006; Macpherson et al., 2007). However, the importance of gaining access to proximity of the TRP channels owing to influences of molecular flexibility and shape still remains to be more fully explored.

7.2.5 Responses to Respiratory Stimuli

Changes in carbon dioxide concentration in the nasopharynx region can cause species-specific changes in reflexive breathing in birds (Hiestand and Randall, 1941). However, concentrations of carbon dioxide that are sufficiently high to be irritating to mammals have no effect on blood pressure, heart rate, tidal volume, breathing frequency, upper airway resistance, or lower airway resistance in geese (*Anser anser*) and chickens. Geese and chickens respond differently than mammals to exposure to sulfur dioxide, but in a similar manner when exposed to ammonia and phenyl diguanide (Callanan et al., 1974; McKeegan et al., 2005).

7.2.6 Nasal and Respiratory Irritation and Interaction of Olfaction and Chemesthesis

The TN is important in the perception of odors (Tucker, 1971; Silver and Maruniak, 1981; Keverne et al., 1986). Electrophysiological evidence shows that the TN responds to odors, although it is generally less sensitive than the olfactory nerve (Tucker, 1963). Behavioral assays yield similar results. Pigeons trained to respond to odors fail to respond after olfactory nerve transections. However, odor responding can be reinstated if the odor concentration is increased (Michelsen, 1959; Henton, 1969; Henton et al., 1966). Odor sensitivity of pigeons decreased by 2–4 log units (vapor saturation) after olfactory nerve transaction (Walker et al., 1979).

Although olfaction can modulate responding to chemical irritants, it is relatively unimportant (Clark, 1995). In European starlings, avoidance of anthranilate compounds was partially a consequence of olfactory cues. When the olfactory nerves were transected, avoidance was only mildly diminished. When the ophthalmic branches of the TN were transected, the starlings became insensitive to the aversive properties of the anthranilates (Mason et al., 1989).

7.2.7 Behavioral Responses to Irritants

Many aromatic molecules are aversive to birds (Kare, 1961; Mason et al., 1989; Crocker and Perry, 1990; Clark and Shah, 1991, 1993; Crocker et al., 1993). Several lines of evidence suggest that a variety of compounds have intrinsic properties that cause them to be aversive on a purely sensory basis. First, the aversive quality is unlearned; that is, avoidance occurs upon initial contact (Clark and Shah, 1991). Second, there is no evidence that consumption is altered by gastrointestinal feedback; intake of fluid treated with those sensory stimuli is constant over time (Clark and Mason, 1993). Third, unlike mammals, birds seem unable to associate the aversive quality of the stimulus with other chemosensory cues, suggesting that conditioned flavor avoidance learning does not occur (Clark, 1996; Clark and Avery, 2013). Fourth, birds do not habituate to the stimulus; avoidance persists in the absence of reinforcement (Clark and Shah, 1994).

7.2.8 Applications

Current interest in chemesthetic function and properties in birds is largely focused in four areas: (1) the evolutionary phylogenetic relationships of receptor mediated perception of noxious stimuli and its consequence to the foraging ecology of birds (Clark, 1998; Tewksbury and Nabhan, 2001); (2) the applicability of using aversive compounds in modulating feeding behavior of birds to develop repellents for prevention of crop damage or otherwise mitigating against damage caused by birds (Mason and Clark, 1997; Clark and Avery, 2013); (3) efforts to gain a better understanding of pathologic pain caused by "debeaking" and promotion of animal welfare in domestic chicken production through better management methods or development of appropriate analgesics (Kuenzel, 2007; Gentle, 2011); and (4) discovery of better analgesics for management of pain in veterinary clinical settings.

7.3 OLFACTION

7.3.1 Morphology of Olfactory System

Air entering a bird's nasal cavity passes through a series of mucous-covered, invaginated chambers called nasal conchae. Nasal conchae influence air flow dynamics and direct odors to the caudal-most chamber, which contains the chemically sensitive olfactory epithelium (reviewed in Roper, 1999; see also: Bang, 1960, 1961, 1963, 1964, 1965, 1966; Bang and Cobb, 1968). The surface of the olfactory epithelium is composed of receptor cells, which detect odorous compounds and occur at the ends of olfactory nerve dendrites. Each receptor cell is surrounded by a cluster of supporting cells and ends in a knob bristling with 6–15 cilia that extend into the lumen. The length of cilia varies by species. Black vultures, for example, have cilia of 40–50 μm, whereas domestic fowl have cilia of 7–10 μm (Shibuya and Tucker, 1967). To gain access to the cilia of receptor cells, odor molecules must diffuse through a mucous membrane. Cilia themselves provide no transport function. Rather, secretions covering cilia provide rapid flow for odor molecules. Olfactory gland secretions must be removed and replaced to maintain diffusion and avoid receptor habituation to odorant molecules. Traction of nearby respiratory cilia facilitates removal of secretions.

The extent of scrolling of caudal conchae correlates with the surface area of olfactory epithelium and the relative size of the olfactory bulb, which is the region of the brain that processes odor input (Bang and Cobb, 1968; Bang, 1971; Bang and Wenzel, 1985; reviewed in Roper, 1999; Hagelin, 2007a). Avian orders with relatively larger olfactory bulbs have lower detection thresholds, indicating they are more sensitive to certain odorous compounds than those with relatively small olfactory bulbs (Clark et al., 1993; Table 7.1, Figure 7.4). Elaborated olfactory systems typically belong to species with demonstrated reliance on odor cues in the field (Stager, 1964; Hutchison and Wenzel, 1980; Hagelin, 2004) and, in some species, correlate positively with the number of olfactory receptor genes (Steiger et al., 2008). Fossil evidence also indicates olfactory bulb size was relatively large early in bird evolution, revealing a previously unrecognized emphasis on smell (Zelenitsky et al., 2011).

Although a larger olfactory bulb size or greater scrolling of receptor epithelium likely indicates greater functional capacity (e.g., more cells and neural circuits; Meisami, 1991), it is important not to dismiss avian species with relatively "unelaborate" olfactory systems (Hagelin, 2007b). Both field and laboratory tests indicate that several taxa with relatively small olfactory bulbs can discriminate between and/or adaptively employ certain odors, such as those related to breeding and nesting (e.g., crested auklets (*Aethia cristatella*) Hagelin et al., 2003; European starlings Clark and Mason, 1985; Gwinner and Berger, 2008; Corsican Blue Tit (*Parus caeruleus ogliastrae*) Petit et al., 2002).

7.3.2 Innervation of Olfactory Receptors

Olfactory receptor cells from each nasal cavity transmit information via the olfactory nerve to the olfactory bulb, located in the anterior region of each brain hemisphere. Each olfactory bulb is composed of concentric cell layers.

TABLE 7.1 Summary of Mean Ratios of Ipsilateral Olfactory Bulb Diameter to Cerebral Hemisphere Diameter and Their Standard Errors (SE) for Several Orders of Birds

Order	*N*	Ratio	SE	Order	*N*	Ratio	SE
Anseriformes	4	19.4	1.5	Psittaciformes	2	8.0	1.4
Apodiformes	8	12.3	1.9	Falconiformes	5	17.4	2.6
Apterygiformes	1	34.0	0.0	Charadriiformes	9	16.4	0.9
Caprimulgiformes	3	23.3	0.7	Galliformes	3	14.2	1.4
Columbiformes	2	20.0	1.4	Piciformes	5	11.4	1.3
Cuculiformes	4	19.5	0.6	Passeriformes	25	13.3	0.7
Gruiformes	14	22.2	0.9	Pelecaniformes	4	12.1	1.6
Gaviformes	1	20.0	0.0	Coraciiformes	5	14.5	1.6
Podicipediformes	2	24.5	1.8	Sphenisciformes	1	17.0	0.0
Procellariiformes	10	29.1	1.4				

Sample sizes indicate the number of species (*N*).
Source: Data adapted from Bang and Cobb (1968).

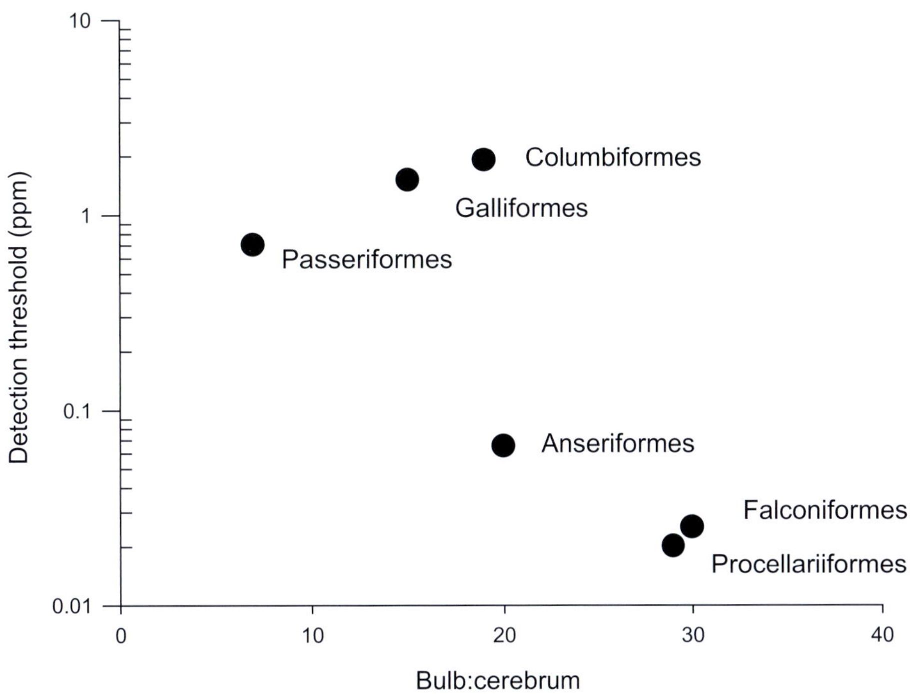

FIGURE 7.4 **Relationship between olfactory detection threshold and relative size of the olfactory bulb for different orders of birds.** *Adapted from Clark and Shah (1993).*

Incoming olfactory nerve fibers constitute the outer layer. Branching nerve terminals penetrate into the adjacent, glomerular layer, where they connect with dendrites of mitral and tufted cells in spherical arborizations called glomeruli. The perikarya of these cells are in the deeper mitral cell layer, where their axons leave to project to many areas of the forebrain.

Like other vertebrates, the olfactory bulbs of birds are bilaterally symmetrical; each is associated with its own (ipsilateral) brain hemisphere. The layering of different cell types within avian olfactory bulbs is qualitatively similar to reptiles, in that well-defined cell layers (like those of mammals) are lacking (Allison, 1953; Andres, 1970). However, there are many interneuron connections in the cell layers between the mitral and glomerular regions. There are no direct connections between the two (contralateral) olfactory bulbs (Rieke and Wenzel, 1978).

Although birds clearly have olfactory bulbs, they appear to lack an accessory olfactory system (Rieke and Wenzel, 1974, 1978). Both olfactory and accessory olfactory structures commonly occur in other vertebrates. The accessory olfactory system is frequently linked to conspecific scent stimuli that modulate social behavior (e.g., reproduction, aggression). However, there is good evidence for mammals that both the main olfactory and accessory olfactory systems can detect and process overlapping sets of odor stimuli (Keller et al., 2009). Accessory olfactory structures include the vomeronasal organ and accessory olfactory bulb. It is possible that accessory olfactory bulbs in birds occur during early embryonic development only, but are lost later on (Matthes, 1934). This idea, however, has received little scientific attention.

7.3.3 Olfactory Neuronal Response

Electrophysiological responses to odor stimuli are taken as definitive evidence of olfactory capacity. These can be recorded from a single "unit" (neuron) or multiunit nerve fibers. Recordings of black vultures indicate that the electro-olfactogram appears primarily during inspiration, which coincides with peak spike activity (Shibuya and Tucker, 1967). Electrophysiological recordings of mammals, amphibians, reptiles, and birds all show similar responses, irrespective of the size of a species' olfactory apparatus (Tucker, 1965; Shibuya and Tonosaki, 1972).

Single-unit responses from within the olfactory bulb of domestic chickens show widely variable rates of spontaneous firing (mean 4.9 spikes/s, range 0.1–32.4 spikes/s) prior to odor exposure (McKeegan, 2002). Odor stimulation modifies spontaneous firing via excitation or inhibition. Avian firing rates appear to fall in between rates reported for mammals and reptiles (McKeegan, 2002, 2009). Single units of chickens responded to two or more odors and revealed surprising sensitivity to biologically relevant scents associated with captivity (e.g., hydrogen sulfide). Responses to extremely low (<0.5 ppm) stepwise changes in concentration to hydrogen sulfide revealed a level of fine-tuning not previously reported for other vertebrates (McKeegan et al., 2002). Continuous presentation of a stimulus can result in physiological adaptation of both single-unit (McKeegan and Lippens, 2003) and nerve-unit recordings, like mammals. Recovery can be achieved within a few minutes of rest.

Olfactory nerve fibers are unmyelinated, which produces slow conduction velocities of about 1.5 mJ/s (Macadar et al., 1980). Interestingly, transected olfactory nerves (which experimentally inhibit olfaction) can repair and recover full physiological capacity within 30 days (Tucker et al., 1974). Although healed nerves are scarred and smaller, recordings and autonomic reflexes to odorants did not differ between controls and nerves that had been cut at least 6 months earlier (Tucker, 1971; Tucker et al., 1974).

Another means of quantifying olfactory neural responses involves calcium imaging (Restrepo et al., 1995). This method uses fluorescence to quantify changes in the flux of calcium ions associated with neural activation (i.e., signal transduction) of a single olfactory receptor neuron (ORN). Jung et al. (2005) tested responses of acutely dissociated ORNs from olfactory epithelium of embryonic domestic chicks. Avian ORNs were placed in Ringer's solution containing liquid solutions of odorants. The fluorescence patterns, which correspond to increases or decreases in Ca^{2+} concentration, were remarkably similar to those of other vertebrates (mammals and fish) that had been tested with the same set of odorants (Jung et al., 2005).

7.3.4 Laboratory Detection Thresholds, Discrimination, and Seasonal Change

Physiological responses (e.g., change in respiration or heart rate) to novel odor stimuli have been observed (Wenzel and Sieck, 1972). Habituation to the stimulus under this paradigm, however, is problematic. Operant and classical conditioning paradigms that use positive or negative reinforcement (Michelsen, 1959; Henton et al., 1966; Henton, 1969) are usually poor at determining olfactory thresholds or discrimination (Calvin et al., 1957). However, two process learning paradigms, such as cardiac conditioning, have proven to be a successful technique for detection, discrimination, and threshold testing (Rescorla and Solomon, 1967; Walker et al., 1986; Clark and Mason, 1989; Clark and Smeraski, 1990; Clark, 1991a; Clark et al., 1993). During cardiac conditioning, an odor (the conditional stimulus) is paired with an aversive experience, such as a shock (the unconditional stimulus). Heart rate is compared before and after stimulus presentation during training until a level of cardiac acceleration is reliably achieved, indicating a bird has learned to associate the odor in anticipation of a shock. Thereafter, tests of detection or odor discrimination can proceed. Most birds tested with this paradigm have shown olfactory capabilities comparable to mammals (Davis, 1973). Even passerines, with the least developed olfactory system, demonstrate behavioral responsiveness to odors (Clark and Mason, 1987; Clark and Smeraski, 1990; Clark, 1991a; Clark et al., 1993) (Table 7.2).

European starlings offer an interesting case study of olfactory structure, function, and seasonality. Male starlings incorporate green plants that are rich in aromatic volatiles into nests, some of which act as a fumigant against parasites and pathogens (Clark and Mason, 1985, 1987, 1988; Clark, 1991b; Gwinner, 1997; Gwinner et al., 2000; Gwinner and Berger, 2005). Starlings are most sensitive to, and can discriminate between, plant odors during spring only, rather than

TABLE 7.2 Summary of Selected Behavioral Olfactory Threshold Data for Different Species of Birds

			Threshold (ppm)		
Species	Ratio[1]	Stimulus	Min	Max	Source
Rock dove (*Columba livia*)	18.0	*n*-Amyl acetate	0.31	29.8	Henton (1969), Henton et al. (1966), Walker et al. (1979), Walker et al. (1986)
		Benzaldehyde	0.47	00.75	Walker et al. (1986)
		Butanethiol	13,820	–	Snyder and Peterson (1979)
		Butanol	0.17	–	Walker et al. (1986)
		n-Butyl acetate	0.11	2.59	Henton (1969), Walker et al. (1986)
		Butyric acid	2.59	–	Henton (1969)
		Ethanethiol	10,080	–	Snyder and Peterson (1979)
		Heptane	0.29	0.38	Stattelman et al. (1975)
		Hexane	1.53	2.98	Stattelman et al. (1975)
		Pentane	16.45	20.76	Stattelman et al. (1975)
Chicken (*Gallus gallus*)	15.0	Heptane	0.31	0.57	Stattelman et al. (1975)
		Hexane	0.64	1.00	Stattelman et al. (1975)
		Pentane	1.58	2.22	Stattelman et al. (1975)
Northern bobwhite (*Colinus virginianus*)	–	Heptane	2.14	3.49	Stattelman et al. (1975)
		Hexane	3.15	4.02	Stattelman et al. (1975)
		Pentane	7.18	10.92	Stattelman et al. (1975)
Black-billed magpie (*Pica pica*)	–	Butanethiol	13,416	–	Snyder and Peterson (1979)
		Ethanethiol	8400	–	Snyder and Peterson (1979)
European starling (*Sturnus vulgaris*)	9.7	Cyclohexane	2.50	–	Clark and Smeraski (1990)
Cedar waxwing (*Bombycilla cedrorum*)	–	Cyclohexane	6.80	86.46	Clark (1991a)
Tree swallow (*Tachycineta bicolor*)	15.0	Cyclohexane	73.42	–	Clark (1991a)
Brown-headed cowbird (*Molothrus ater*)	7.0	Ethyl butyrate	0.76	–	Clark and Mason (1989)
Catbird (*Dumetella carolinensis*)	–	Cyclohexane	35.14	–	Clark et al. (1993)
Eastern phoebe (*Sayornis phoebe*)	–	Cyclohexane	35.61	–	Clark et al. (1993)
European goldfinch (*Carduelis carduelis*)	–	Cyclohexane	13.05	–	Clark et al. (1993)
Great tit (*Parus major*)	–	Cyclohexane	34.10	–	Clark et al. (1993)
Black-capped chickadee (*Parus atricapillus*)	3.0	Cyclohexane	59.95	–	Henton (1969)

[1]*The ratio of the longest axis of the olfactory bulb to that of the ispsilateral cerebral hemisphere.*

in summer and fall. Spring is coincident with nest building and suggests a hormonal influence (Clark and Smeraski, 1990).

Birds treated with testosterone (T), a hormone that enlarges song-learning nuclei of the brain and alters behavior, exhibited enlarged olfactory bulbs year-round, indicating a proximate effect on bulb structure. However, perception of plant odor in T-implanted males was greatest during spring only, indicating that perception was independent of T-treatment and olfactory bulb volume. One hypothesized but untested mechanism is that an increase in receptor cell density in starling olfactory epithelium occurs in spring (DeGroof et al., 2010).

7.3.5 Development

Volatile compounds diffuse through avian eggshell (Rahn et al., 1979), providing an opportunity for odor exposure within the egg (Tolhurst and Vince, 1976; Sneddon et al., 1998). Many vertebrates, including birds, detect and learn chemical information as embryos (e.g., humans: Schaal et al., 2000; Mennella et al., 2001; other mammals: Hepper, 1988; Bilko et al., 1994; amphibians: Mathis et al., 2008; birds: Porter and Picard, 1998; Bertin et al., 2012). Early exposure can cause changes in neuroanatomy, which alters chemosensory perception in a way that can adaptively shape responses later in life (e.g., to food, mates, etc.) (Todrank et al., 2011).

Studies of domestic chickens, the avian model for development, indicate that odor detection can occur before or after young pierce the egg's air sac and begin breathing air (Tolhurst and Vince, 1976; Bertin et al., 2012; Hagelin et al., 2013). ORNs are functional 6 days prior to air-breathing (on embryonic developmental day 13; Lalloué et al., 2003), when nasal passages are full of amniotic fluid. Embryos at this stage swallow frequently, facilitating fluid movement, similar to mammals in utero (Sneddon et al., 1998). Airbreathing begins approximately 2 days prior to hatching, on embryonic developmental day 19 (Tolhurst and Vince, 1976).

The magnitude of embryonic response varies relative to stimulus concentration and timing of exposure (Bertin et al., 2010). Later developmental stages show relatively greater responses to odors (Gomez and Celli, 2008; Bertin et al., 2012). Detectable stimuli include artificial odors (Sneddon et al., 1998), as well as naturally occurring scents, such as nest materials (Gwinner and Berger, 2008), food-related odors (Burne and Rogers, 1999; Cunningham and Nevitt, 2011), and compounds found in plumage scent of at least one alcid species (Hagelin et al., 2013).

7.3.6 Field Studies and Behavioral Ecology

Like other vertebrates, birds detect and respond adaptively to odors (reviewed in Roper, 1999; Hagelin, 2007a; Balthazart and Taziaux, 2009; Caro and Balthazart, 2010). Hagelin (2007a) made a distinction between environmentally derived odors (e.g., food, predators) and those produced by birds themselves (e.g., body odors, fecal odor, preen gland secretions). The latter can have social and reproductive implications. This section considers examples of adaptive olfactory responses to environmental odors as well as bird-derived scents.

The use of olfactory cues for locating food has been documented for numerous species, such as procellariids, vultures, corvids, hummingbirds, honeyguides, parrots, and kiwis (Roper, 1999). Turkey vultures (*Cathartes aura*), for example, are attracted to ethyl-mercaptan, a volatile associated with decomposed carcasses (Stager, 1964, 1967), and locate food without visual cues (Houston, 1986). Procellariiforms also forage over considerable distances. Black-footed albatrosses (*Diomedea nigripes*) respond to bacon grease over 31 km away (20 miles; Miller, 1942), whereas Leach's storm petrel (*Oceanodroma leucorhoa*) home to scent targets at a distance of 1–12 km (Clark and Shah, 1992). Some procellariiformes also respond to a compound that is correlated with prey called dimethyl-sulfide (DMS) (Nevitt et al., 1995). DMS smells like rotten seaweed and results from the breakdown of metabolic products of marine algae (phytoplankton). Petrels, however, do not feed on phytoplankton. Rather, DMS concentrates in locations where a bird's prey (zooplankton, such as krill) is actively grazing on phytoplankton. Grazing by zooplankton lyses phytoplankton cells and thereby creates a DMS odor plume, which some birds follow to locate food (Nevitt, 2011).

With regard to predators, the scent of urine and/or feces has an aversive effect on some avian species (blue tits (*Cyanistes caeruleus*), Amo et al., 2008; house finches (*Carpodacus mexicanus*), Roth et al., 2008; red junglefowl (*Gallus gallus*), Zidar and Løvlie, 2012), but not all (eastern blue birds (*Sialia sialis*), Godard et al., 2007; house wren (*Troglodytes aedon*), Johnson et al., 2011). Application of predator odor can also deter breeding ducks and songbirds (Eicholz et al., 2012; Forsman et al., 2013). Responses appear to be innate rather than learned (Amo et al., 2011b), although sleeping birds are unreactive (Amo et al., 2011a).

Odors are also germane to avian orientation and navigation (reviewed in Wallraff, 2005; Gagliardo, 2013). Homing pigeons, for example, exhibit larger olfactory bulbs than nonhoming breeds (Rehkämper et al., 1988, 2008). Investigators have also altered pigeon homing behavior via experimental disruption of the olfactory system. Manipulations include olfactory nerve transection (Papi et al., 1971; Gagliardo et al., 2006, 2009), anesthesia of olfactory mucosa (Wallraff, 1988), ablating the central piriform cortex of the brain (Papi and Casini, 1990), and nostril plugging. The last of these manipulations indicates that pigeons rely more on their right nostril for olfactory information (Gagliardo et al., 2007, 2011). *ZENK*, an immediate early gene expressed in olfactory neurons, also implicates the use of olfaction during the process of homing (Patzke et al., 2010).

Emerging evidence for passerine species further supports olfaction during migration. For example, adult gray catbirds (*Dumetella carolinensis*) rendered temporarily anosmic (by washing the olfactory tissues with zinc sulfate) oriented differently from adult controls but similarly to juvenile birds, which were migrating for the first time and therefore unable to navigate (Holland et al., 2009). With regard to cellular mechanisms, black-headed buntings (*Emberiza melanocephala*) increase activation of olfactory tissues (as measured by c-fos immunoreactivity) during migration. These birds exhibit a seasonally enhanced emphasis on olfaction while migrating, compared to visual systems (Rastogi et al., 2011).

Many birds produce a variety of odorous compounds (Table 7.3; reviewed in Campagna et al., 2011). For example, a seabird colony, with its dense numbers of birds, burrows, and feces, makes for a potent chemosensory experience. Pioneering work by Grubb (1974) on Leach's storm petrel showed differential return rates to nest sites after surgical manipulation, indicative of olfactory-based homing: 91% for controls, 74% for sham surgery, and 0% for olfactory nerve section. Several petrel species have since been shown to discriminate between the odor of their own nest and conspecific burrows (Mínguez, 1997; De León et al., 2003; Bonadonna et al., 2003a,b). Attraction to home nest odor is also reported for passerines (Caspers and Krause, 2010; Krause and Caspers, 2012).

TABLE 7.3 Some Avian Orders Considered To Be Very Odorous by Ornithologists

Order	Common Name	Number of Species[1]
Procellariiformes	Petrels, shearwaters, diving petrels	16
Ciconiiformes	Herons, storks, new world vultures	12
Anseriformes	Ducks, geese, swans, screamers	49
Charadriiformes	Sandpipers, gulls, auks	23
Psittaciformes	Parrots	14
Cuculiformes	Cuckoos	16
Coraciiformes	Kingfishers, rollers, hoopoes, woodhoopoes	14
Piciformes	Woodpeckers, barbets, tucans	33
Passeriformes	Grackles, starlings, ravens, finches, honeycreepers	46

[1]*Data compiled from Weldon and Rappole, 1997.*

Avian chemical substances are linked with a variety of social contexts (reviewed in Hagelin, 2007a; Hagelin and Jones, 2007; Balthazart and Taziaux, 2009; Caro and Balthazart, 2010). Uropygial gland secretions, for example, show some level of hormonal control and exhibit individual, sex, and age-specific patterns (e.g., Procellariiformes: Mardon et al., 2010, 2011; Anseriformes: Kolattukudy et al., 1987; Galliformes: Karlsson et al., 2010; passerines: Whittaker et al., 2010; Whelan et al., 2010; Shaw et al., 2011; Amo et al., 2012a). Pioneering work by Balthazart and Schoffeniels (1979) indicated male mallards decreased social displays and sexual behavior toward females when their olfactory nerves were sectioned, suggesting that intact olfactory system is critical to courtship and mating. Crested auklets produce a seasonally elevated scent associated with a stereotyped behavior that focuses on the scented region of the body (the nape). Auklets are attracted to natural feather odor, a chemical cocktail of odor compounds, and scented decoys, which suggests odor has a social function (Hagelin et al., 2003; Jones et al., 2004; Hagelin, 2007a). Odorous compounds of crested auklets can also negatively impact ectoparasites in experimental tests (Douglas, 2008, 2013).

Procellariiform seabirds show a surprising level of body odor discrimination, in that they are attracted to mate odors and avoid self-odor (Antarctic petrel (*Pachyptila desolata*) Bonadonna and Nevitt, 2004; blue petrels (*Halobaena caerulea*), Mardon and Bonadonna, 2009). Furthermore, preference for the odor of unrelated individuals over those of kin was recently discovered (European storm petrel (*Hydrobates pelagicus*), Bonadonna and Sanz-Aguilar, 2012). Such results suggest that body odors could provide a mechanism for inbreeding avoidance, known as self-referent phenotype matching (Mateo and Johnston, 2000). This may be particularly important in petrels which are a long-lived philopatric species that mates for life. Petrels are also likely to encounter kin on their natal breeding grounds that they have never met before (Bonadonna and Nevitt, 2004; Bonadonna and Sanz-Aguilar, 2012). Recent evidence for passerines suggests that conspecific odor may provide relevant social information. Bird responses to scent correlated with social rank (house finch, Amo et al., 2012b), sex (European starling, Amo et al., 2012a), and body size (dark-eyed junco (*Junco hyemalis*), Whittaker et al., 2011).

7.3.7 Summary

Every bird tested has exhibited a functional sense of smell (Bang and Wenzel, 1985). The extent of olfactory development also is on par with that found in mammals. However, ornithologists have largely overlooked the role of olfaction in avian biology. Many birds adaptively employ environmental odors; they also produce and respond to conspecific scents. Although passerines have a relatively poorly developed olfactory anatomy, they nonetheless show some degree of olfactory acuity. Other species, such as procellariiformes, have olfactory systems that are acutely sensitive to odor cues and capable of a surprisingly detailed level of conspecific odor discrimination. Given the broad range of contexts that implicate avian olfaction, future interdisciplinary research that compares olfactory mechanisms in birds to better-known vertebrate systems, such as mammals and fish, holds exciting promise.

7.4 GUSTATION

7.4.1 Taste Receptors

Relative to other vertebrates, birds have fewer taste receptors and taste receptor genes (Berkhoudt, 1985; Shi and Zhang, 2005) (Table 7.4). Notwithstanding these observations, birds have a well-developed system for gustation with functional significance for their behavior, ecology, and evolution. Taste receptors are located in taste buds throughout the oral cavity. The greatest concentration of avian taste receptors is found around salivary glands in the soft epithelium of the palate, the posterior tongue, and the oropharynx (Bath, 1906; Lindenmaier and Kare, 1959; Saito, 1966; Ganchrow and Ganchrow, 1985). Afferent taste signals in birds are carried in the glossopharyngeal nerve (cranial nerve IX; Duncan, 1960). The glossopharyngeal nerve innervates the posterior buccal and pharyngeal areas (Kare and Mason, 1986). Unlike mammals, the facial nerve (VII) does not innervate the avian tongue (Wenzel, 1973). Rather, glossopharyngeal afferents in birds enter the medulla and join fibers from the facial (including chorda tympani) and vagus nerves (X) to form a well-developed *fasciculus solitarius* (Lindenmaier and Kare, 1959). The chorda tympani innervates taste buds adjacent to the anterior mandibular salivary glands, situated in the buccal epithelium of the lower jaw (Kare and Mason, 1986).

TABLE 7.4 Abundance of Taste Buds among Vertebrate Species[1]

Species	Taste Buds	Source
Domestic chick (day-old)	5–12	Lindenmaier and Kare (1959)
Domestic chicken (3 months)	24	Lindenmaier and Kare (1959)
Blue tit	24	Gentle (1975)
Bullfinch	41–42	Duncan (1960)
Pigeon	59	Moore and Elliot (1946)
Japanese quail	62	Warner et al. (1967)
European starling	200	Bath (1906)
Parrot	300–400	Bath (1906)
Domestic cat (juvenile)	473	Elliot (1937)
Lizard	550	Schwenk (1985)
Bat	800	Moncrieff (1946)
Domestic cat (adult)	2755	Robinson and Winkles (1990)
Human	6974	Miller and Reedy (1990)
Rabbit	17,000	Moncrieff (1946)
Pig	19,904	Chamorro et al. (1993)
Ox	35,000	Moncrieff (1946)
Catfish	100,000	Hyman (1942)

[1]*Modified from Kare and Mason (1986) and Mason and Clark (2000).*

7.4.2 Response to Sweet

Birds have a well-developed sense of taste that generally corresponds to their feeding habits. Frugivorous and omnivorous birds tend to perceive and prefer sweet more so than species in other foraging guilds. For example, European starlings prefer 0.5–5% D-fructose solutions (w/v) to distilled water (Espaillat and Mason, 1990). Sugar detection thresholds of cockatiels (*Nymphicus hollandicus*) is 0.36 M sucrose, 0.40 M fructose and 0.16 M glucose (Matson et al., 2000, 2001). The sugar detection thresholds of broad-billed hummingbirds (*Cynanthus latirostris*) is between 1.31 and 1.54 mM sucrose, 0.87–1.31 mM fructose, 1.54–1.75 mM glucose and 1.75–3.5 mM of a 1:1 mixture of fructose and glucose (Medina-Tapia et al., 2012). Interestingly, the sweet taste receptor gene Tas1r2 is absent in all bird genomes sequenced thus far, irrespective of their diet (Zhao and Zhang, 2012), suggesting that additional avian receptors may exist for sweet.

The order of preference among nectivorous passerines is sucrose = glucose + fructose = fructose > glucose > xylose (Lotz and Nicolson, 1996). Lesser double-collared sunbirds (*Nectarinia chalybea*) and Cape sugarbirds (*Promerops cafer*) absorb sucrose, glucose, and fructose from ingested food at nearly 100% efficiency, but xylose was excreted (Lotz and Nicolson, 1996; Jackson et al., 1998a,b). Although nectar composition and concentration are often considered independently, these characteristics may have a synergistic effect on the sugar preferences of nectar-feeding birds (Schondube and Martinez del Rio, 2003).

Sugar preferences among nectarivorous and frugivorous birds are concentration-dependent. Although nectarivorous birds in Africa prefer sucrose when offered a choice of 0.25 M solutions of glucose, fructose, and sucrose, no preference among these sugars was observed when their concentration was increased to 0.73 M; the dietary choices in these species indicate the birds had either reached a limit where they had sufficient energy intake or they were affected by postingestion constraints (Downs and Perrin, 1996; Downs, 1997). House finches demonstrated no preference for equicaloric, 2% solutions of hexoses (1:1 mixture of fructose and glucose) and sucrose, and strong preference manifest for hexoses but not sucrose at 4, 6, and 10% concentrations; energetics, rather than sucrase deficiency, may determine finches' sugar preferences (Avery et al., 1999).

Studies of unrelated, nectarivorous birds (including a generalist, nonpasserine nectarivore) have demonstrated a distinct switch from hexose preference at low concentrations to sucrose preference at higher concentrations (Lotz and Schondube, 2006; Fleming et al., 2008; Brown et al., 2010a,c). Sucrose preference at higher concentrations may possibly be explained by taste perception due to differences in solution osmolality or a degree of imprinting due to experience with natural nectar compositions. Village weavers (i.e., generalist passerine nectarivores; *Ploceus cucullatus*) preferred hexose solutions at 5% and 10% sucrose equivalents (SE), yet no sugar preference was observed at 15, 20, and 25% SE (Odendaal et al., 2010). In contrast, dark-capped bulbuls (*Pycnonotus tricolor*), an opportunistic nectarivore, significantly preferred hexose solutions, irrespective of concentration (5–25%), when given a choice between equicaloric hexose and sucrose solutions (Brown et al., 2010b). Interestingly, malachite sunbirds (*Nectarinia famosa*) demonstrated either sucrose preference, no preference, or hexose preference when offered equimolar, equiweight, or equicaloric paired solutions of sucrose and hexose, respectively (Brown et al., 2008).

The bananaquit (*Coereba flaveola*) strongly prefers the most concentrated sucrose solution when the lowest concentration ranged from 276 to 522 mM. From 522 to 1120 mM sucrose concentrations, bananaquits adjust their volumetric food intake to maintain constant energy intake. At a sucrose concentration of 276 mM, however, bananaquits did not maintain their rate of energy intake by increasing food consumption (Mata and Bosque, 2004). Although nectarivorous birds generally prefer concentrated over dilute sugar solutions, the concentration difference that they can discriminate is smaller at low concentrations relative to high concentrations; this pattern may be a consequence of the functional form of intake responses that often results in decelerating sugar intakes with increasing sugar concentration (Martinez del Rio et al. 2001; Leseigneur and Nicolson, 2009). With regard to gender-specific food intake among nectarivorous birds, males take longer to digest than females when fed on sucrose-rich nectars as opposed to hexose-rich nectars; therefore, they can allow themselves a relatively lower digestive capacity (Markman et al., 2006). The digestive transit rates of Cape white-eyes (*Zosterops virens*) fed artificial fruit were faster for glucose- than sucrose-based diets, irrespective of concentration; increased food intake with decreasing glucose concentration and no significant differences in food intake with differing sucrose concentrations were observed (Wellmann and Downs, 2009). Indeed, nectar ingestion rate is determined by viscosity, and total food intake is primarily modulated by sugar concentration (Köhler et al., 2010).

Sugar preference and selection among nectarivorous and frugivorous birds are likely to have coevolutionary effects on flowering and fruit-bearing plants. Among 58 wild fruits studied in Hong Kong, all fruit species contained glucose, all but one contained fructose, and only 11 species contained sucrose; birds are known to eat 29 of these species without detectable sucrose and four with sucrose (Ko, 1996). From a comparative analysis of glucose, fructose, and sucrose in the nectar and fruit juice of 525 tropical and subtropical plant species, passerine nectars and fruits had low sucrose and high hexose content, respectively; the nectar of hummingbird flowers had very high sucrose content; microchiroptera nectars showed hexose richness and microchiropteran fruits had a sucrose content similar to passerine fruits; and megachiroptera nectars and fruits were sucrose-rich (Baker et al., 1998). The dichotomy between sucrose-rich nectars in hummingbird-pollinated plants and predominantly hexose-rich nectars in sunbird-pollinated plants appears to have little to do with bird physiology and may rather reflect patterns of nectar secretion or plant physiology and opportunist nectar feeders (Nicolson and Fleming, 2003; Fleming et al., 2004).

The hummingbird-passerine dichotomy was strongly emphasized until the discovery of South African plants with sucrose-dominant nectars, which are pollinated by passerines that demonstrate sucrose digestion and preference (Lotz and Schondube, 2006). Flowers adapted for specialized passerine nectarivores have nectar similar to that of hummingbird flowers in terms of volume (approx. 10–30 mL), concentration (15–25% w/w) and sucrose content (40–60% of total sugar). In contrast, flowers adapted to generalized bird pollinators are characterized by large volumes (approximately 40–100 mL) of extremely dilute (8–12%) nectar with minimal sucrose (0–5%; Johnson and Nicolson, 2008).

Rufous hummingbirds (*Selasphorus rufus*) preferred 50% sucrose to higher and lower concentrations, and they could distinguish solutions differing by only 1% sucrose (Blem et al., 2000). Sucrase activity is 10 times higher in hummingbirds than in passerines (Schondube and Martinez del Rio, 2004). Neither sex nor temperature affected sugar preferences among green-backed firecrown hummingbirds (*Sephanoides sephaniodes*; Chalcoff et al., 2008). Patterns of hummingbird sugar preference can be affected by different mechanisms, both pre- and postingestive. At low concentrations, gustatory thresholds may play an important role in sugar selection. At intermediate and high concentrations, however, sugar selection can be explained by sugar assimilation rates and velocity of food processing generated by osmotic constraints (Medina-Tapia et al., 2012).

Species belonging to the Sturnidae–Muscicapidae lineage do not express intestinal sucrase, despite having generalist diets comprising fruits with sugars of diverse kinds (Gatica et al., 2006). Members of the Sturnidae-Muscicapidae lineage are intolerant of solutions or fruit above 11–15% sucrose (Brown et al., 2012). Considering the phylogenetic constraint hypothesis for sucrose digestion in the Muscicapoidea superfamily, the lack of sucrase activity is a shared, derived character only for the Cinclidae–Sturnidae–Turdinae lineage (Gatica et al., 2006).

Within an experimental meal with varying sucrose concentration, captive whitebellied sunbirds (*Cinnyris talatala*) demonstrated a measurable increase in feeding frequency and food intake within 10 min after a decrease in sucrose concentration (Köhler et al., 2008). Similarly, Knysna turacos (*Tauraco corythaix*) preferred an artificial sucrose diet to an equicaloric glucose diet at low concentrations, whereas purple-crested turacos (*Gallirex porphyreolophus*) showed no preference for either diet. Both turacos species preferred a sucrose diet to an equimolar glucose diet at low concentrations. At high concentrations, neither species showed a preference for either equicaloric or equimolar diets; thus, energy requirements influence food preferences more than sugar type and birds will select fruit that is higher in energy irrespective of sugar type (Wilson and Downs, 2011).

7.4.3 Response to Salt

A comparison of the sodium chloride rejection thresholds among 58 bird species illustrated rejection thresholds ranging from 0.35% NaCl in a parrot to 37.5% NaCl in the pine siskin (*Carduelis pinus*; Rensch and Neunzig, 1925). Red-winged blackbirds (*Agelaius phoeniceus*) and European starlings preferred 0.1–1% NaCl solutions (w/v) to distilled water (Espaillat and Mason, 1990). The salt detection threshold of cockatiels is 0.16 M NaCl (Matson et al., 2000) and 0.16 M potassium chloride (Matson et al., 2001). With regard to the mechanism of salt perception, sodium in the oral cavity can cross the taste sensory cell membrane through the epithelial Na^+ channel (ENaC), thus triggering an action potential (Roura et al., 2012). Pigeons (*C. livia domestica*) learned to discriminate a safe 0.06 M NaCl solution and a toxic equimolar LiCl solution. Because the pigeons avoided the LiCl solution within a short presentation period of 5 minutes, it is unlikely that the birds were using an interoceptive stimulus of faint, postingestive malaise as a conditioned cue; thus, the pigeons' discrimination performance between the two chloride solutions was attributed to gustation (Nakajima and Onimaru, 2006).

7.4.4 Response to Sour

Sourness is related to the acidity of food, which is often caused by bacterial fermentation and typically evokes a rejection response. With regard to the mechanism of sour perception, the receptors for sour taste are thought to be transmembrane channels that are selective for hydrogen ions (Roura et al., 2012). Red-winged blackbirds and female starlings preferred distilled water to 0.01–0.1 M citric acid solutions (Espaillat and Mason, 1990). For the purpose of investigating sour detection thresholds, Matson et al. (2000) defined sourness as a pH, and they achieved sourness by varying the pH of a 0.05 M citrate buffer system. The sour detection threshold of cockatiels is pH 5.5 citric acid.

7.4.5 Response to Bitter

Bitter taste perception likely evolved as a protective mechanism against the ingestion of harmful compounds in food (Davis et al., 2010). Red-winged blackbirds and European starlings preferred distilled water to 0.5–5% tannic acid solutions (w/v; Espaillat and Mason, 1990). The bitter detection thresholds of cockatiels is 100 μM quinine, 1000 μM gramine, 500 μM hydrolysable tannin and 10,000 μM condensed tannin (Matson et al., 2004). Compared with pigs, chickens showed a lower sensitivity to glucosinolates (i.e., bitter plant metabolites); compared to ruminants, however, chickens showed a higher aversion to glucosinolates (Roura et al., 2012).

Bitter detection thresholds indicate that a birds' rejection of quinine occurs at lower concentrations than phytophagic mammals (Matson et al., 2004). White Leghorn and Rhode Island Red chickens were able to detect 2.0 mM quinine hydrochloride; broiler chickens detected 0.5 mM quinine hydrochloride (Kudo et al., 2010). Domestic chicks (14 days old) can discriminate between an untreated diet and a diet treated with 0.2% quinine hydrochloride (Ueda and Kainou, 2005).

Johnson et al. (2006) explored the functional significance of the phenolic compounds that impart a dark brown color to the nectar of the South African succulent shrub, *Aloe vryheidensis*. Dark-capped bulbuls were more likely to probe model flowers containing dark nectar than those containing clear nectar, suggesting a potential signaling function of dark nectar. The main effect of the phenolics, however, appears to be repellency of ''unwanted'' nectarivores that find their bitter taste unpalatable. Nectar-feeding honey bees and sunbirds are morphologically mismatched for pollinating *A. vryheidensis* flowers and strongly reject its nectar. Thus, the dark phenolic component of the nectar appears to function as a floral filter by attracting some animals visually and deterring others by its taste (Johnson et al., 2006).

The taste receptor type 2 (Tas2r) gene family encodes the chemoreceptors that are directly responsible for the detection of bitter compounds. The Tas2r cluster encodes up to 18 functional bitter taste receptors in the white-throated sparrow (*Zonotrichia albicollis*; Davis et al., 2010). Although the tens to hundreds of taste buds observed among birds pales in comparison to the hundreds to thousands of taste buds found in other vertebrates (Table 7.4), this relative deficit does not preclude birds from detecting bitter compounds as effectively as those species with more taste buds. Future biochemical and genetic studies will be needed to identify the natural ligands for avian Tas2r gene clusters, and the intra- and inter-specific differences in these genes with variation in bitter taste perception (Davis et al., 2010).

7.4.6 Response to Umami

Male starlings preferred 0.7–1% L-alanine solutions to distilled water (Espaillat and Mason, 1990). The T1R1 umami receptor gene and the T1R3 sweet/umami receptor gene have been identified in chickens (Shi and Zhang, 2005). Moreover, the expression of T1R1 has been reported in hypothalamus, liver, and abdominal fat (Byerly et al., 2010). Thus, avian taste receptors and umami receptor genes may be involved in the orchestration of postingestive and metabolic events (Roura et al., 2012). Further research is needed to comparatively investigate avian feeding responses to umami tastants.

7.4.7 Response to Calcium

Calcium-deprived chickens preferred calcium-rich diets when offered a choice (Wood-Gush and Kare, 1966; Hughes and Wood-Gush, 1971). Similarly, consumption of supplementary calcium was inversely related to chicken's dietary calcium content (Taher et al., 1984). Further research is needed to distinguish the behavioral responses of birds to calcium as a tastant (i.e., sensory cue) versus the pre- and postingestive attributes of calcium-rich supplements. Although it is clear that animals can detect calcium in micromolar or low millimolar concentrations, it is less clear what they detect or how they detect it (Tordoff, 2001). The notion that calcium is a distinct taste quality is an anathema to many psychophysicists, who argue that there are very few basic taste qualities (sweet, sour, salty, bitter, and umami). To them, calcium taste is a complex of basic tastes, such as bitterness, sourness, and saltiness (Tordoff, 2001).

Calcium taste varies with both the form and the concentration of salt tested, but it nearly always includes sour and bitter components (Tordoff, 2001). The extracellular calcium-sensing receptor (CaR) is a multimodal sensor for several key nutrients, notably Ca^{2+} and L-amino acids, and is expressed abundantly throughout the gastrointestinal tract in humans (Conigrave and Brown, 2006). Although the T1r3 receptor gene in mice (Tordoff et al., 2008) and the CaR have been identified as calcium sensors, it is yet uncertain if they mediate calcium appetite or taste (Roura et al., 2012) in birds.

7.4.8 Taste Behavior and Applications

Deterrents based merely on offensive flavors are not likely to be effective in the absence of aversive postingestive effects (Provenza, 1995). In this context, flavor is the perceptual integration of chemesthetic, olfactory, and gustatory stimuli. Red-winged blackbirds conditioned with sodium chloride paired with an intraperitoneal injection of a gastrointestinal toxin (lithium chloride) or a free choice of a postingestive, cathartic purgative (anthraquinone) or a postingestive, cholinesterase inhibitor (methiocarb) subsequently avoided the flavor (NaCl; Figure 7.5) and color of food experienced during conditioning. In contrast, blackbirds conditioned with sodium chloride paired with an intraperitoneal injection of an opioid antagonist (i.e., chemesthetic; naloxone hydrochloride) or a free choice of a preingestive, trigeminal irritant (methyl anthranilate) subsequently avoided only the color (not flavor; Figure 7.5) of food experienced during conditioning. Thus, red-winged blackbirds reliably integrate gustatory (and visual) experience with postingestive consequences to procure nutrients and avoid toxins (Werner and Provenza, 2011).

Avian taste behavior has been investigated in context of agricultural production, chemical defenses of insects and plants, coevolution in predator-prey and pollination systems, chemical ecology, conservation biology, and comparative physiology and taxonomy. For example, although avian feeding responses to secondary metabolites are species-specific (Saxton et al., 2011; Rios et al., 2012), increased sugar concentrations (not decreasing acid concentrations) are a functional cue for the onset of bird damage to ripening grapes (Saxton et al., 2009). Although increased sucrose content may deter sucrase-deficient birds from damaging commercial fruit (Brugger and Nelms, 1991), increased sucrose may also lead to increased crop damage by other species obligated to consume more of the less-digestible fruit to meet their energy requirements (Lane, 1997). This compensatory feeding hypothesis notwithstanding, McWhorter and Martinez del Rio (2000) observed a physiological constraint on sugar consumption among nectarivorous hummingbirds; the rate of intestinal sucrose hydrolysis can limit sugar assimilation and reduce sucrose preference. Indeed, the intake responses of nectar-feeding birds manifest from the integration of a behavioral response with the physiological processes that shape it (Martinez del Rio et al., 2001).

Several tastants have been used to condition aversions among birds associated with agricultural production. The risk of accidental poisoning of birds may be reduced by adding an aversive tastant (e.g., D-pulegone, quinine hydrochloride) to granular pesticides (Mastrota and Mench, 1995; Clapperton et al., 2012). Garlic oil was identified as an effective chemical repellent for European starlings (Hile et al., 2004) and quinine sulfate (bitterant) was used to

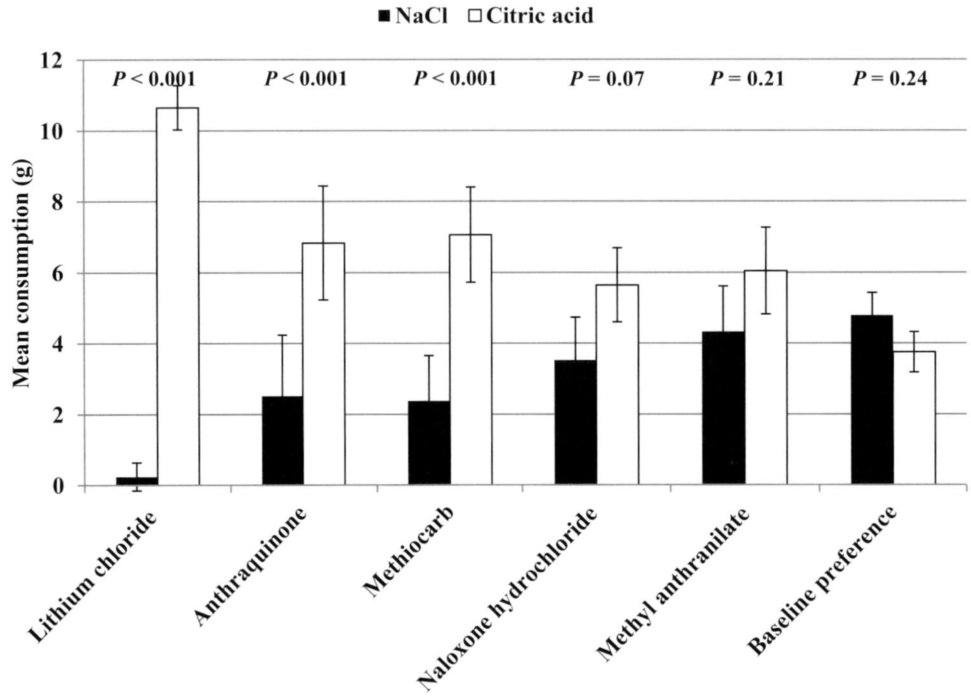

FIGURE 7.5 Mean consumption (±2 SEM) of sodium chloride and citric acid subsequent to NaCl conditioning with: an intraperitoneal injection of a gastrointestinal toxin (lithium chloride), or a free-choice of a postingestive, cathartic purgative (anthraquinone) or a postingestive, cholinesterase inhibitor (methiocarb); or an intraperitoneal injection of an opioid antagonist (naloxone hydrochloride) or a free-choice of a pre-ingestive, trigeminal irritant (methyl anthranilate) in red-winged blackbirds (*Agelaius phoeniceus*). *From Werner and Provenza (2011), baseline preference data from Werner et al. (2008); with permission.*

condition taste aversions and thus reduce destructive feather pecking among laying hens (Harlander-Matauschek et al., 2009, 2010).

Relative preference for specific tastants has been used to enhance feeding for poultry production. The preference of chickens for oily diets (i.e., long-chain versus medium-chain triacylglycerol) is mediated by gustation (Furuse et al., 1996; Mabayo et al., 1996), not satiety (Vermaut et al., 1997). In contrast, the avoidance of a saponin-rich diet is not mediated by taste in domestic chicks (Ueda and Shigemizu, 2001); rather, crop distension causes decreased feed intake associated with tea saponin (Ueda et al., 2002).

Domestic chicks can use unpalatable taste (e.g., quinine) to adapt their visual foraging decisions (Rowe and Skelhorn, 2005; Skelhorn et al., 2008). Moreover, European starlings and domestic chicks can learn to use bitter taste cues to regulate consumption of toxic prey (Skelhorn and Rowe, 2010; Barnett et al., 2011). Similarly, red-winged blackbirds use affective processes (flavor-feedback relationships) to shift preference for both novel and familiar flavors (Werner et al., 2008).

7.4.9 Summary

The conventional notion regarding the "limited ability of birds to taste" (Kassarov, 2001) was shaped by a historic paradigm of taste research (i.e., elementary structure and function). Avian taste perception is currently investigated in context of ontogenetic and phylogenetic relationships within ever-changing environments. Birds use taste cues to select nutrients and avoid toxins; thereby, they affect the distribution, diversity, and coevolution of their prey. Thus, taste cues and postingestive consequences have behavioral, ecological, and evolutionary implications for domestic and wild birds. Future avian gustation research will develop our understanding of comparative biochemistry, molecular biology, and ethology—from an emphasis on anatomical structure to the physiological bases of behavior and performance.

REFERENCES

Allison, A.C., 1953. The morphology of the olfactory system in the vertebrates. Biol. Rev. 28, 195–244.

Amo, L., Caro, S.P., Visser, M.E., 2011a. Sleeping birds do not respond to predator odour. PLoS One 6, e27576.

Amo, L., Visser, M.E., van Oers, K., 2011b. Smelling out predators is innate in birds. Ardea 99, 177–184.

Amo, L., Galvan, I., Tomas, G., Sanz, J.J., 2008. Predator odour recognition and avoidance in a songbird. Funct. Ecol. 22, 289–293.

Amo, L., Avilés, J.M., Parejo, D., Pena, A., Rodríguez, J., Tomas, G., 2012a. Sex recognition by odour and variation in the uropygial gland secretion in starlings. J. Anim. Ecol. 81, 605–613.

Amo, L., López-Rull, I., Pagán, I., Garcia, C.M., 2012b. Male quality and conspecific scent preferences in the house finch, *Carpodacus mexicanus*. Anim. Behav. 84, 1483–1489.

Andres, K.H., 1970. Anatomy and ultrastructure of the olfactory bulb in fish, amphibia, reptiles, birds and mammals. In: Wolstenhome, G.E.W., Knight, J. (Eds.), Taste and Smell in Vertebrates. J and A Churchill, London, pp. 177–196.

Avery, M.L., Schreiber, C.L., Decker, D.G., 1999. Fruit sugar preferences of house finches. Wilson Bull. 111, 84–88.

Baker, H.G., Baker, I., Hodges, S.A., 1998. Sugar composition of nectars and fruits consumed by birds and bats in the tropics and subtropics. Biotropica 30, 559–586.

Balthazart, J., Schoffeniels, E., 1979. Pheromones are involved in the control of sexual behaviour in birds. Naturwissenschaften 66, 55–56.

Balthazart, J., Taziaux, M., 2009. The underestimated role of olfaction in avian reproduction? Behav. Brain Res. 200, 248–259.

Bang, B.G., 1960. Anatomical evidence for olfactory function in some species of birds. Nature 188, 547–549.

Bang, B.G., 1961. The surface pattern of the nasal mucosa and its relation to mucous flow—a study of chicken and herring gull nasal mucosae. J. Morphol. 109, 57–71.

Bang, B.G., 1963. Comparative studies of the nasal organs of birds: a study of 28 species of birds of West Bengal. PAVO 1, 79–89.

Bang, B.G., 1964. The nasal organs of the black and Turkey vultures: a comparative study of the cathartid species *Coragyps atratus atratus* and *Canhartes aura septentrionalis* (with notes on *Cathartes aura falklandica, Pseudogyps bengalensis*, and *Neophron percnopterus*). J. Morphol. 115, 153–184.

Bang, B.G., 1965. Anatomical adaptations for olfaction in the snow petrel. Nature 205, 513–515.

Bang, B.G., 1966. The olfactory apparatus of tube-nosed birds (Procellariiformes). Acta Anat. 65, 391–415.

Bang, B.G., 1971. Functional anatomy of the olfactory system in 23 orders of birds. Acta Anat. (Suppl. 79), S1–S76.

Bang, B.G., Cobb, S., 1968. The size of the olfactory bulb in 108 species of birds. Auk 85, 55–61.

Bang, B., Wenzel, B.M., 1985. Nasal cavity and olfactory system. In: King, A.S., McLelland, J. (Eds.), Form and Function in Birds III. Academic Press, London, pp. 195–225.

Barnett, C.A., Skelhorn, J., Bateson, M., Rowe, C., 2011. Educated predators make strategic decisions to eat defended prey according to their toxin content. Behav. Ecol. 23, 418–424.

Bath, W., 1906. Die Geschmacksorgane der Vögel und Krokodile. Arch. Biontol. 1, 1–47.

Berkhoudt, H., 1985. Structure and function of avian taste receptors. In: Levy, A.S., McLelland, J. (Eds.), Form and Function in Birds III. Academic Press, New York, pp. 463–496.

Bertin, A., Calandreau, L., Arnould, C., Lèvy, F., 2012. The developmental stage of chicken embryos modulates impact of in ovo olfactory stimulation on food preferences. Chem. Senses 37, 253–261.

Bertin, A., Calandreau, L., Arnould, C., Nowak, R., Lèvy, F., Noirot, V., Bouvarel, I., Leterrier, C., 2010. *In ovo* olfactory experience influences post-hatch feeding behaviour in young chickens. Ethology 116, 1027–1037.

Bilko, A., Altbacker, V., Hudson, R., 1994. Transmission of food preference in the rabbit—the means of information-transfer. Physiol. Behav. 56, 907–912.

Blem, C.R., Blem, L.B., Felix, J., van Gelder, J., 2000. Rufous hummingbird sucrose preference: precision of selection varies with concentration. Condor 102, 235–238.

Bonadonna, F., Nevitt, G.A., 2004. Partner-specific odor recognition in an Antarctic seabird. Science 306, 835.

Bonadonna, F., Cunningham, G.B., Jouventin, P., Hesters, F., Nevitt, G.A., 2003a. Evidence for nest-odour recognition in two species of diving petrel. J. Exp. Biol. 206, 3719–3722.

Bonadonna, F., Hesters, F., Jouventin, P., 2003b. Scent of a nest: discrimination of own-nest odours in Antarctic prions, *Pachyptila desolata*. Behav. Ecol. Sociobiol. 54, 174–178.

Bonadonna, F., Sanz-Aguilar, A., 2012. Kin recognition and inbreeding avoidance in wild birds: the first evidence for individual kin-related odour recognition. Anim. Behav. 84, 509–513.

Brown, M., Downs, C.T., Johnson, S.D., 2008. Sugar preferences of nectar feeding birds – a comparison of experimental techniques. J. Avian Biol. 39, 479–483.

Brown, M., Downs, C.T., Johnson, S.D., 2010a. Concentration-dependent sugar preferences of the Malachite sunbird (*Nectarinia famosa*). Auk 127, 151–155.

Brown, M., Downs, C.T., Johnson, S.D., 2010b. Sugar preferences and digestive efficiency in an opportunistic avian nectarivore, the dark-capped bulbul *Pycnonotus tricolor*. J. Ornithol 151, 637–643.

Brown, M., Downs, C.T., Johnson, S.D., 2010c. Sugar preferences of a generalist nonpasserine flower visitor, the African speckled mousebird (*Colius striatus*). Auk 127, 781–786.

Brown, M., Downs, C.T., Johnson, S.D., 2012. African red-winged starlings prefer hexose sugar solutions, but do not like them too sweet. J. Ornithol. 153, 265–272.

Brugger, K.E., Nelms, C.O., 1991. Sucrose avoidance by American robins (*Turdus migratorius*): implications for control of bird damage in fruit crops. Crop Prot. 10, 455–460.

Burne, T.H.J., Rogers, L.J., 1999. Changes in olfactory responsiveness by the domestic chick after early exposure to odorants. Anim. Behav. 58, 329–336.

Butler, A.B., Cotterill, R.M.J., 2006. Mammalian and avian neuroanatomy and the question of consciousness in birds. Biol. Bull. 211, 106–127.

Byerly, M.S., Simon, J., Cogburn, L.A., Bihan-Duval, E.L., Duclos, M.J., Aggrey, S.E., Porter, T.E., 2010. Transcriptional profiling of hypothalamus during development of adiposity in genetically selected fat and lean chickens. Physiol. Genomics 42, 157-167.

Callanan, D., Dixon, M., Widdicombe, J.G., Wise, J.C.M., 1974. Responses of geese to inhalation of irritant gases and injection of phenyl diguanide. Resp. Physiol. 22, 157–166.

Calvin, A.D., Williams, C.M., Westmoreland, N., 1957. Olfactory sensitivity in the domestic pigeon. Am. J. Psychol. 188, 255–256.

Campagna, S., Mardon, J., Celerier, A., Bonadonna, F., 2011. Potential semiochemical molecules from birds: a practical and comprehensive compilation of the last 20 years studies. Chem. Senses 37, 3–25.

Caro, S.P., Balthazart, J., 2010. Pheromones in birds: myth or reality? J. Comp. Physiol. A 196, 751–766.

Caspers, B.A., Krause, E.T., 2010. Odour-based natal nest recognition in the zebra finch (*Taeniopygia guttata*), a colony-breeding songbird. Biol. Lett. 7, 184–186.

Caterina, M.J., Schumacher, M.A., Tominaga, M., Rosen, T.A., Levine, J.D., Julius, D., 1997. The capsaicin receptor: a heat-activated ion channel in the pain pathway. Nature 389, 816–824.

Chalcoff, V.R., Aizen, M.A., Galetto, L., 2008. Sugar preferences of the green-backed firecrown hummingbird (*Sephanoides sephaniodes*): a field experiment. Auk 125, 60–66.

Chamorro, C.A., de Paz, P., Fernández, J.G., Anel, L., 1993. Fungiform papillae of the pig and the wild boar analysed by scanning electron microscopy. Scanning Microsc. 7, 313–322.

Clapperton, B.K., Porter, R.E.R., Day, T.D., Waas, J.R., Matthews, L.R., 2012. Designer repellents: combining olfactory, visual or taste cues with a secondary repellent to deter free-ranging house sparrows from feeding. Pest Manage. Sci. 68, 870–877.

Clark, L., 1991a. Odor detection thresholds in tree swallows and cedar waxwings. Auk 108, 177–180.

Clark, L., 1991b. The nest protection hypothesis: the adaptive use of plant secondary compounds by European starlings. In: Loye, J.E., Zuk, M. (Eds.), Bird-Parasite Interactions: Ecology Evolution and Behaviour. Oxford University Press, Oxford, pp. 205–221.

Clark, L., 1995. Modulation of avian responsiveness to chemical irritants: effects of prostaglandin E1 and analgesics. J. Exp. Zool. 271, 432–440.

Clark, L., 1996. Trigeminal repellents do not promote conditioned odor avoidance in European starlings. Wilson Bull. 108, 36–52.

Clark, L., 1997. A review of the bird repellent effects of 117 carbocylic compounds. In: Mason, J.R. (Ed.), Repellents in Wildlife Management. National Wildlife Research Center, Fort Collins, CO, pp. 343–352.

Clark, L., 1998. Physiological, ecological, and evolutionary bases for the avoidance of chemical irritants by birds. In: Nolan, V. Jr., Ketterson, E.D. (Eds.), Current Ornithology, vol. 14. Plenum Press, New York, pp. 1–37.

Clark, L., Avery, M.L., 2013. Factors influencing the effectiveness of chemical repellents in management of birds on airports. In: DeVault, T.L., Blackwell, B.F., Belant, J.L. (Eds.), Wildlife in Airport Environments. Johns Hopkins University Press, Baltimore.

Clark, L., Mason, J.R., 1985. Use of nest material as insecticidal and anti-pathogenic agents by the European starling. Oecologia 67, 169–176.

Clark, L., Mason, J.R., 1987. Olfactory discrimination of plant volatiles by the European starling. Anim. Behav. 35, 227–235.

Clark, L., Mason, J.R., 1988. Effect of biologically active plants used as nest material and the derived benefit to starling nestlings. Oecologia 77, 174–180.

Clark, L., Mason, J.R., 1989. Sensitivity of brown-headed cowbirds to volatiles. Condor 91, 922–932.

Clark, L., Mason, J.R., 1993. Interactions between sensory and postingestional repellents in starlings: methyl anthranilate and sucrose. Ecol. Appl. 3, 262–270.

Clark, L., Shah, P.S., 1991. Nonlethal bird repellents: in search of a general model relating repellency and chemical structure. J. Wildl. Manage. 55, 538–545.

Clark, L., Shah, P.S., 1992. Information content of prey odor plumes: what do foraging Leach's storm petrels know? In: Doty, R.L., Müller-Schwartze, D. (Eds.), Chemical Signals in Vertebrates. Plenum Press, New York, pp. 421–427.

Clark, L., Shah, P.S., 1993. Chemical bird repellents: possible use in cyanide ponds. J. Wildl. Manage. 57, 657–664.

Clark, L., Shah, P., 1994. Tests and refinements of a general structure-activity model for avian repellents. J. Chem. Ecol. 20, 321–339.

Clark, L., Smeraski, C.A., 1990. Seasonal shifts in odor acuity by starlings. J. Exp. Zool. 255, 22–29.

Clark, L., Avilova, K.V., Bean, N.J., 1993. Odor thresholds in passerines. Comp. Biochem. Physiol. A 104, 305–312.

Clark, L., Shah, P.S., Mason, J.R., 1991. Chemical repellency in birds: relationship between chemical structure and avoidance response. J. Exp. Zool. 260, 310–322.

Conigrave, A.D., Brown, E.M., 2006. Taste receptors in the gastrointestinal tract II. L-amino acid sensing by calcium-sensing receptors: implications for GI physiology. Am. J. Physiol. Gastrointest. Liver Physiol. 291, G753–G761.

Cortright, D.N., Krause, J.E., Broom, D.C., 2007. TRP channels and pain. Biochim. Biophys. Acta 1772, 978–988.

Crocker, D.R., Perry, S.M., 1990. Plant chemistry and bird repellents. Ibis 132, 300–308.

Crocker, D.R., Perry, S.M., Wilson, M., Bishop, J.D., Scanlon, C.B., 1993. Repellency of cinnamic acid derivatives to captive rock doves. J. Wildl. Manage. 57, 113–122.

Cunningham, G.B., Nevitt, G.A., 2011. Evidence for olfactory learning in procellariiform seabird chicks. J. Avian Biol. 42, 85–88.

Davis, R.G., 1973. Olfactory psychophysical parameters in man, rat, dog and pigeon. J. Comp. Physiol. Psychol. 85, 221–232.

Davis, J.K., Lowman, J.J., Thomas, P.J., ten Hallers, B.F.H., Koriabine, M., Huynh, L.Y., Maney, D.L., de Jong, P.J., Martin, C.L., Thomas, J.W., 2010. Evolution of a bitter taste receptor gene cluster in a new world sparrow. Genome Biol. Evol. 2, 358–370.

De León, A., Mínguez, E., Belliure, B., 2003. Self-odour recognition in European storm-petrel chicks. Behaviour 140, 925–933.

DeGroof, G., Gwinner, H., Steiger, S., Kempenaers, B., van der Linden, A., 2010. Neural correlates of behavioural olfactory sensitivity changes seasonally in European Starlings. PLoS One 5, e14337.

Douglas III, H.D., 2008. Prenuptial perfume: alloanointing in the social rituals of the crested auklet (*Aethia cristatella*) and the transfer of arthropod deterrents. Naturwissenschaften 95, 45–53.

Douglas III, H.D., 2013. Colonial seabird's paralytic perfume slows lice down: an opportunity for parasite-mediated selection? Int. J. Parasitol. 43, 399–407.

Downs, C.T., 1997. Sugar preference and apparent sugar assimilation in the red lory. Austral. J. Zool. 45, 613–619.

Downs, C.T., Perrin, M.R., 1996. Sugar preferences of some southern African nectarivorous birds. Ibis 138, 455–459.

Dubbeldam, J.L., Karten, H.J., 1978. The trigeminal system in the pigeon (*Columba livia*). I. Projections of the gasserian ganglion. J. Comp. Neurol. 180, 661–678.

Dubbeldam, J.L., Veenman, C.L., 1978. Studies on the somatotopy of the trigeminal system in the mallard, *Anas platyrhynchos* L: I. The ganglion trigeminale. Neth. J. Zool. 28, 150–160.

Dugas-Ford, J., Rowell, J.J., Ragsdale, C.W., 2012. Cell-type homologies and the origins of the neocortex. PNAS 109, 16974–16979. http://dx.doi.org/10.1073/pnas.1204773109.

Duncan, C.J., 1960. The sense of taste in birds. Ann. Appl. Biol. 48, 409–414.

Eicholz, M.W., Dassow, J.A., Stafford, J.D., Weatherhead, P.J., 2012. Experimental evidence that nesting ducks use mammalian urine to assess predator abundance. Auk 129, 638–644.

Elliot, B., 1937. Total distribution of taste buds on the tongue of the kitten at birth. J. Comp. Neurol. 66, 361–373.

Espaillat, J.E., Mason, J.R., 1990. Differences in taste preference between red-winged blackbirds and European starlings. Wilson Bull. 102, 292–299.

Fleming, P.A., Bakken, B.H., Lotz, C.N., Nicolson, S.W., 2004. Concentration and temperature effects on sugar intake and preferences in a sunbird and a hummingbird. Funct. Ecol. 18, 223–232.

Fleming, P.A., Xie, S., Napier, K., McWhorter, T.J., Nicolson, S.W., 2008. Nectar concentration affects sugar preferences in two Australian honeyeaters and a lorikeet. Funct. Ecol. 22, 599–605.

Forsman, J.T., Monkkonen, M., Korpimaki, E., Thomson, R.L., 2013. Mammalian nest predator feces as a cue in avian habitat selection decisions. Behav. Ecol. 24, 262–266.

Furuse, M., Mabayo, R.T., Okumura, J.-I., 1996. The role of gustation in oil preference in the chicken. Jpn. Poult. Sci. 33, 256–260.

Gagliardo, A., 2013. Forty years of olfactory navigation in birds. J. Exp. Biol. 216, 2165–2171.

Gagliardo, A., Filannino, C., Ioalè, P., Pecchia, T., Wikelski, M., Vallortigara, G., 2011. Olfactory lateralization in homing pigeons: a GPS study on birds released with unilateral olfactory inputs. J. Exp. Biol. 214, 593–598.

Gagliardo, A., Ioalè, P., Savini, M., Wild, J.M., 2006. Having the nerve to home: trigeminal magnetoreceptor versus olfactory mediation of homing in pigeons. J. Exp. Biol. 209, 2888–2892.

Gagliardo, A., Ioalè, P., Savini, M., Wild, J.M., 2009. Navigational abilities of adult and experienced homing pigeons deprived of olfactory or trigeminally mediated magnetic information. J. Exp. Biol. 212, 3119–3124.

Gagliardo, A., Pecchia, T., Savini, M., Odetti, F., Ioalè, P., Vallortigara, G., 2007. Olfactory lateralization in homing pigeons: initial orientation of birds receiving a unilateral olfactory input. Eur. J. Neurosci. 25, 1511–1516.

Ganchrow, D., Ganchrow, J.R., 1985. Number and distribution of taste buds in the oral cavity of hatchling chicks. Physiol. Behav. 34, 889–894.

Gatica, C.D.L., Gonzalez, S.P., Vasquez, R.A., Sabat, P., 2006. On the relationship between sugar digestion and diet preference in two Chilean avian species belonging to the Muscicapoidea superfamily. Rev. Chilena Hist. Nat. 79, 287–294.

Gentle, M.J., 1975. Gustatory behavior of the chicken and other birds. In: Wright, P., Caryl, P.E., Vowles, D.M. (Eds.), Neural and Endocrine Aspects of Behaviors in Birds. Elsevier Scientific, Amsterdam, pp. 305–308.

Gentle, M.J., 1989. Cutaneous sensory afferents recorded from the *nervus intramandibularis* of *Galllus gallus* var *domesticus*. J. Comp. Physiol. A 164, 763–774.

Gentle, M.J., 2011. Pain issues in poultry. Appl. Anim. Behav. Sci. 135, 252–258.

Gentle, M.J., Hill, F.L., 1987. Oral lesions in the chicken: behavioural responses following nociceptive stimulation. Physiol. Behav. 40, 781–783.

Gentle, M.J., Hunter, L.N., 1993. Neurogenic inflammation in the chicken (*Gallus gallus* var *domesticus*). Comp. Biochem. Physiol. C 105, 459–462.

Getty, R., 1975. Sisson and Grossman's the Anatomy of the Domestic Animals. W.B. Saunders Company, Philadelphia.

Godard, R.D., Bowers, B.B., Wilson, C.M., 2007. Eastern bluebirds *Sialia sialis* do not avoid nest boxes with chemical cues from two common nest predators. J. Avian Biol. 38, 128–131.

Gomez, G., Celli, A., 2008. The peripheral olfactory system of the domestic chicken: physiology and development. Brain Res. Bull. 76, 208–216.

Grubb, T.C., 1974. Olfactory navigation to the nesting burrow in Leach's petrel (*Oceanodroma leucorrhoa)*. Anim. Behav. 22, 192–202.

Gottschaldt, K.M., 1985. Structure and function of avian somatosensory receptors. In: King, A.S., McLelland, J. (Eds.), Form and Function in Birds, vol. 3. Academic Press, London, pp. 375–461.

Güntürkün, O., 2012. The convergent evolution of neural substrates for cognition. Psychol. Res. 2, 212–219.

Gwinner, H., 1997. The function of green plants in nests of European starlings (*Sturnis vulgaris*). Behaviour 134, 337–351.

Gwinner, H., Oltrogge, M., Trost, L., Nienaber, U., 2000. Green plants in starling nests: effects on nestlings. Anim. Behav. 59, 301–309.

Gwinner, H., Berger, S., 2005. European starlings: nestling condition, parasites and green nest material during the breeding season. J. Ornithol. 146, 365–371.

Gwinner, H., Berger, S., 2008. Starling males select green nest material by olfaction using experience-independent and experience-dependent cues. Anim. Behav. 75, 971–976.

Hagelin, J.C., 2004. Observations on the olfactory ability of the Kakapo Strigops habroptilus, the critically endangered parrot of New Zealand. Ibis 146, 161–164.

Hagelin, J.C., 2007a. Odors and chemical signaling. In: Jamieson, B.G.M. (Ed.), Reproductive Biology and Phylogeny of Birds, vol. 6B. Science Publishers, Enfield, NH, pp. 75–119.

Hagelin, J.C., 2007b. The citrus-like scent of crested auklets: reviewing the evidence for an avian olfactory ornament. J. Ornithol. 148, S195–S201.

Hagelin, J.C., Jones, I.L., 2007. Bird odors and other chemical substances: a defense mechanism or overlooked mode of intraspecific communication? Auk 124, 741–761.

Hagelin, J.C., Jones, I.L., Rasmussen, L.E.L., 2003. A tangerine-scented social odor in a monogamous seabird. Proc. R. Soc. Ser. B 270, 1323–1329.

Hagelin, J.C., Simonet, J.C., Lyson, T.R., 2013. Embryonic domestic chickens can detect compounds in an avian chemosignal before breathing air. In: East, M.L., Dehnhard, M. (Eds.), Chemical Signals in Vertebrates, vol. 12. Springer, New York, pp. 363–377.

Harlander-Matauschek, A., Beck, P., Piepho, H.-P., 2009. Taste aversion learning to eliminate feather pecking in laying hens, *Gallus gallus domesticus*. Anim. Behav. 78, 485–490.

Harlander-Matauschek, A., Beck, P., Rodenburg, T.B., 2010. Effect of an early bitter taste experience on subsequent feather-pecking behaviour in laying hens. Appl. Anim. Behav. Sci. 127, 108–114.

Henton, W.W., 1969. Conditioned suppression to odorous stimuli in pigeons. J. Exp. Anal. Behav. 12, 175–185.

Henton, W.W., Smith, J.C., Tucker, D., 1966. Odor discrimination in pigeons. Science 153, 1138–1139.

Hepper, P.G., 1988. Adaptive fetal learning: prenatal exposure to garlic affects postnatal preferences. Anim. Behav. 36, 935–936.

Hiestand, W.A., Randall, W.C., 1941. Species differentiation in the respiration of birds following carbon dioxide administration and the location of inhibitory receptors in the upper respiratory tract. J. Cell. Comp. Physiol. 17, 333–340.

Hile, A.G., Shan, Z., Zhang, S.-Z., Block, E., 2004. Aversion of European starlings (*Sturnus vulgaris*) to garlic oil treated granules: garlic oil as an avian repellent. Garlic oil analysis by nuclear magnetic resonance spectroscopy. J. Agr. Food Chem. 52, 2192–2196.

Hinman, A., Chuang, H., Bautista, D.M., Julius, D., 2006. TRP channel activation by reversible covalent modification. PNAS 103, 19564–19568.

Holland, R.A., Thorup, K., Gagliardo, A., Bisson, I.-A., Knecht, E., Mizrahi, D., Wikelski, M., 2009. Testing the role of sensory systems in the migratory heading of a songbird. J. Exp. Biol. 212, 4065–4071.

Holzer, P., 2011. Transient receptor potential (TRP) channels as drug targets for diseases of the digestive system. Pharmacol. Ther. 131, 142–170.

Houston, D.C., 1986. Scavenging efficiency of turkey vultures in tropical forests. Condor 88, 318–323.

Hu, H.J., Bhave, G., Gereau, R.W., 2002. Prostaglandin and protein kinase A-dependent modulation of vanilloid receptor function by metabotropic glutamate receptor 5: potential mechanism for thermal hyperalgesia. J. Neurosci. 22, 7444–7452.

Hughes, B.O., Wood-Gush, D.G.M., 1971. A specific appetite for calcium in domestic chickens. Anim. Behav. 19, 490–499.

Hutchison, L.V., Wenzel, B.M., 1980. Olfactory guidance in foraging by procellariiforms. Condor 82, 314–319.

Hwang, S.W., Cho, H., Kwak, J., Lee, S.-Y., Kang, C.-J., Jung, J., Cho, S., Min, K.H., Suh, Y.-G., Kim, D., Oh, U., 2000. Direct activation of capsaicin receptors by products of lipoxygenases: endogenous capsaicin-like substances. PNAS 97, 6155–6160.

Hyman, L.H., 1942. Comparative Vertebrate Anatomy, second ed. University of Chicago Press, Chicago.

Jackson, S., Nicolson, S.W., Lotz, C.N., 1998a. Sugar preferences and "side bias" in cape sugarbirds and lesser double-collared sunbirds. Auk 115, 156–165.

Jackson, S., Nicolson, S.W., van Wyk, B.-E., 1998b. Apparent absorption efficiencies of nectar sugars in the cape sugarbird, with a comparison of methods. Physiol. Zool. 71, 106–115.

Johnson, L.S., Murphy, S.M., Parrish, G.W., 2011. Lack of predator-odor detection and avoidance in a songbird, the House Wren. J. Field Ornithol. 82, 150–157.

Johnson, S.D., Hargreaves, A.L., Brown, M., 2006. Dark, bitter-tasting nectar function as a filter of flower visitors in a bird-pollinated plant. Ecology 87, 2709–2716.

Johnson, S.D., Nicolson, S.W., 2008. Evolutionary associations between nectar properties and specificity in bird pollination systems. Biol. Lett. 4, 49–52.

Jones, I.L., Hagelin, J.C., Major, H.L., Rasmussen, L.E.L., 2004. An experimental field study of the function of crested auklet feather odor. Condor 106, 71–78.

Jordt, S.-E., Julius, D., 2002. Molecular basis for species-specific sensitivity to "hot" chili peppers. Cell 108, 421–430.

Jung, Y., Wirkus, D., Amendola, D., Gomez, G., 2005. Characteristics of odorant elicited calcium fluxes in acutely-isolated chick olfactory neurons. J. Comp. Physiol. A 191, 511–520.

Kare, M.R., 1961. Comparative aspects of taste. In: Kare, M.R., Halpern, B.P. (Eds.), Physiological and Behavioral Aspects of Taste. University of Chicago Press, Chicago, pp. 13–23.

Kare, M.R., Mason, J.R., 1986. The chemical senses in birds. In: Sturkie, P.D. (Ed.), Avian Physiology. Springer-Verlag, New York.

Karlsson, A.C., Jensen, P., Elgland, M., Laur, K., Fyrner, T., Konradsson, P., Laska, M., 2010. Red junglefowl have individual body odors. J. Exp. Biol. 213, 1619–1624.

Kassarov, L., 2001. Do cyanogenic glycosides and pyrrolizidine alkaloids provide some butterflies with a chemical defense against their bird predators? A different point of view. Behaviour 138, 45–67.

Keller, M., Baum, M.J., Brock, O., Brennan, P.A., Bakker, J., 2009. The main and the accessory olfactory systems interact in the control of mate recognition and sexual behavior. Behav. Brain Res. 200, 268–276.

Keverne, E.B., Murphy, C.L., Silver, W.L., Wysocki, C.J., Meredith, M., 1986. Non-olfactory chemoreceptors of the nose: recent advances in understanding the vomeronasal and trigeminal systems. Chem. Senses 11, 119–133.

Kirifides, M.L., Kurnellas, M.P., Clark, L., Bryant, B.P., 2004. Calcium responses of chicken trigeminal ganglion neurons to methyl anthranilate and capsaicin. J. Exp. Biol. 207, 715–722.

Ko, I.W.P., 1996. Sugar preferences of fruit-eating birds in Hong Kong. Hong Kong Bird Rep., 165–169.

Koda, H., Minagawa, M., Si-Hong, L., Mizumura, K., Kumazawa, T., 1996. H1-receptor-mediated excitation and facilitation of the heat response by histamine in canine visceral polymodal receptors studied in vitro. J. Neurophysiol. 76, 1396–1404.

Köhler, A., Leseigneur, C.D.C., Verburgt, L., Nicolson, S.W., 2010. Dilute bird nectars: viscosity constrains food intake by licking in a sunbird. Am. J. Physiol. Regul. Integr. Comp. Physiol. 299, R1068–R1074.

Köhler, A., Verburgt, L., Fleming, P.A., Nicolson, S.W., 2008. Changes in nectar concentration: how quickly do whitebellied sunbirds (*Cinnyris talatala*) adjust feeding patterns and food intake? J. Comp. Physiol. B 178, 785–793.

Kolattukudy, P.E., Bohnet, S., Rogers, L., 1987. Diesters of 3-hydroxy fatty acids produced by the uropygial glands of female mallards uniquely during the mating season. J. Lipid Res. 28, 582–588.

Krause, E.T., Caspers, B.A., 2012. Are olfactory cues involved in nest recognition in two social species of estrildid finches? PLoS One 7, e36615.

Kudo, K.-i., Shiraishi, J.-i., Nishimura, S., Bungo, T., Tabata, S., 2010. The number of taste buds is related to bitter taste sensitivity in layer and broiler chickens. Anim. Sci. J. 81, 240–244.

Kuenzel, W.J., 2007. Neurobiological basis of sensory perception: welfare implications of beak trimming. Poult. Sci. 86, 1273–1282.

Lalloué, F.L., Ayer-Le-Lièvre, C.S., Sicard, G., 2003. Analysis of the functional maturation of olfactory neurons in chicks before and after birth. Chem. Senses 28, 729–737.

Lane, S.J., 1997. Preferences and apparent digestibilities of sugars by fruit damaging birds in Japan. Ann. Appl. Biol. 130, 361–370.

Leseigneur, C.D.C., Nicolson, S.W., 2009. Nectar concentration preferences and sugar intake in the white-bellied sunbird, *Cinnyris talatala* (Nectariniidae). J. Comp. Physiol. B 179, 673–679.

Lindenmaier, P., Kare, M.R., 1959. The taste end-organs of the chicken. Poult. Sci. 38, 545–550.

Lotz, C.N., Nicolson, S.W., 1996. Sugar preferences of a nectarivorous passerine bird, the lesser double-collared sunbird (*Nectarinia chalybea*). Funct. Ecol. 10, 360–365.

Lotz, C.N., Schondube, J.E., 2006. Sugar preferences in nectar- and fruit-eating birds: behavioral patterns and physiological causes. Biotropica 38, 3–15.

Mabayo, R.T., Okumura, J.-I., Furuse, M., 1996. Dietary flavor modifies oil preferences in the chicken. Appl. Anim. Behav. Sci. 49, 213–221.

Macadar, A.W., Rausch, L.J., Wenzel, B.M., Hutchison, L.V., 1980. Electrophysiology of the olfactory pathway in the pigeon. J. Comp. Physiol. 137, 39–46.

Macpherson, L.J., Bailong, X., Kwan, K.Y., Petrus, M.J., Dubin, A.E., Hwang, S., Cravatt, B., Corey, D.P., and Patapoutian, A., 2007. An ion channel essential for sensing chemical damage. J. Neurosci. 27, 11412–11415.

Mardon, J., Bonadonna, F., 2009. Atypical homing or self-odour avoidance? Blue petrels (*Halobaena caerulea*) are attracted to their mate's odour but avoid their own. Behav. Ecol. Sociobiol. 63, 537–542.

Mardon, J., Saunders, S.M., Bonadonna, F., 2011. From preen secretions to plumage: the chemical trajectory of blue petrels' *Halobaena caerulea* social scent. J. Avian Biol. 42, 29–38.

Mardon, J., Saunders, S.M., Anderson, M.J., Couchoux, C., Bonadonna, F., 2010. Species, gender, and identity: cracking petrels' sociochemical code. Chem. Senses 35, 309–321.

Markman, S., Tadmor-Melamed, H., Arieli, A., Izhaki, I., 2006. Sex differences in food intake and digestive constraints in a nectarivorous bird. J. Exp. Biol. 209, 1058–1063.

Martinez del Rio, C., Schondube, J.E., McWhorter, T.J., Herrera, L.G., 2001. Intake responses in nectar feeding birds: digestive and metabolic causes, osmoregulatoory consequences, and coevolutionary effects. Am. Zool. 41, 902–915.

Mason, J.R., Clark, L., 1995. Capsaicin detection in trained European starlings: the importance of olfaction and trigeminal chemoreception. Wilson Bull. 107, 165–169.

Mason, J.R., Clark, L., 1997. Avian repellents: options, modes of action, and economic considerations. In: Mason, J.R. (Ed.), Repellents in Wildlife Management. USDA, National Wildlife Research Center, Fort Collins, CO, pp. 371–391.

Mason, J.R., Clark, L., 2000. The chemical senses in birds. In: Whittow, G.C. (Ed.), Sturkie's Avian Physiology, fifth ed. Academic Press, San Diego, pp. 39–56.

Mason, J.R., Maruniak, J.A., 1983. Behavioral and physiological effects of capsaicin in red-winged blackbirds. Pharmacol. Biochem. Behav. 19, 857–862.

Mason, J.R., Adams, M.A., Clark, L., 1989. Anthranilate repellency to starlings: chemical correlates and sensory perception. J. Wildl. Manage. 53, 55–64.

Mason, J.R., Bean, N.J., Shah, P.S., Clark, L., 1991. Taxon-specific differences in responsiveness to capsaicin and several analogues: correlates between chemical structure and behavioral aversivenesss. J. Chem. Ecol. 17, 2539–2551.

Mastrota, F.N., Mench, J.A., 1995. Evaluation of taste repellents with northern bobwhites for deterring ingestion of granular pesticides. Environ. Toxicol. Chem. 14, 631–638.

Mata, A., Bosque, C., 2004. Sugar preferences, absorption efficiency and water influx in a neotropical nectarivorous passerine, the bananaquit (*Coereba flaveola*). Comp. Biochem. Physiol. A 139, 395–404.

Mathis, A., Ferrari, M.C.O., Windel, N., Messier, F., Chivers, D.P., 2008. Learning by embryos and the ghost of predation future. Proc. R. Soc. Ser. B 275, 2603–2607.

Matson, K.D., Millam, J.R., Klasing, K.C., 2000. Taste threshold determination and side-preference in captive cockatiels (*Nymphicus hollandicus*). Appl. Anim. Behav. Sci. 69, 313–326.

Matson, K.D., Millam, J.R., Klasing, K.C., 2001. Thresholds for sweet, salt, and sour taste stimuli in cockatiels (*Nymphicus hollandicus*). Zoo Biol. 20, 1–13.

Matson, K.D., Millam, J.R., Klasing, K.C., 2004. Cockatiels (*Nymphicus hollandicus*) reject very low levels of plant secondary compounds. Appl. Anim. Behav. Sci. 85, 141–156.

Mateo, J.M., Johnston, R.E., 2000. Kin recognition and the 'armpit effect': evidence of self-referent phenotype matching. Proc. R. Soc. Ser. B. 267, 695–700.

Matthes, E., 1934. Geruchsorgan, Lubosch Handbuch der vergleichenden Anatomie der Wirbeltiere, Groppert, Kallius, vol. 11, Urban and Schwarzenbeig, Berlin.

McKeegan, D.E.F., 2002. Spontaneous and odour evoked activity in single avian olfactory bulb neurones. Brain Res. 929, 48–58.

McKeegan, D.E.F., 2009. Avian chemoreception: an electrophysiological approach. international symposium on olfaction and taste. Ann. N. Y. Acad. Sci. 1170, 438–441.

McKeegan, D.E.F., Lippens, N., 2003. Adaptation responses of single avian olfactory bulb neurones. Neurosci. Lett. 344, 83–86.

McKeegan, D.E.F., Demmers, T.G.M., Wathes, C.M., Jones, R.B., Gentle, M.J., 2002. Stimulus-response functions of single avian olfactory bulb neurones. Brain Res. 953, 101–111.

McKeegan, D.E.F., Smith, F.S., Demmers, T.G.M., Wathes, C.M., Jones, R.B., 2005. Behavioral correlates of olfactory and trigeminal gaseous stimulation in chickens, *Gallus domesticus*. Physiol. Behav. 84, 761–768.

McWhorter, T.J., Martinez del Rio, C., 2000. Does gut function limit hummingbird food intake? Physiol. Biochem. Zool. 73, 313–324.

Medina-Tapia, N., Ayala-Berdon, J., Morales-Pérez, L., Melo, L.M., Schondube, J.E., 2012. Do hummingbirds have a sweet-tooth? Gustatory sugar thresholds and sugar selection in the broad-billed hummingbird *Cynanthus latirostris*. Comp. Biochem. Physiol. A 161, 307–314.

Meisami, E., 1991. Chemoreception. In: Prosser, C.L. (Ed.), Neural and Integrative Animal Physiology. John Wiley and Sons, New York, NY, pp. 335–362.

Mennella, J.A., Jagnow, C.P., Beauchamp, G.K., 2001. Prenatal and postnatal flavor learning by human infants. Pediatrics 107, e88. http://dx.doi.org/10.1542/peds.107.6.e88.

Michelsen, W.J., 1959. Procedure for studying olfactory discrimination in pigeons. Science 130, 630–631.

Miller, L., 1942. Some tagging experiments with black-footed albatrosses. Condor 44, 3–9.

Miller, Jr., I.J., Reedy, Jr., F.E., 1990. Variations in human taste bud density and taste intensity perception. Physiol. Behav. 47, 1213–1219.

Mínguez, E., 1997. Olfactory nest recognition by British storm-petrel chicks. Anim. Behav. 53, 701–707.

Moncrieff, R.W., 1946. The Chemical Senses. Wiley, New York.

Moore, C.A., Elliot, R., 1946. Numerical and regional distribution of taste buds on the tongue of the bird. J. Comp. Neurol. 84, 119–131.

Nakajima, S., Onimaru, S., 2006. Salt discrimination in domestic pigeons (*Columba livia domestica*): poisonous LiCl solution versus equimolar safe NaCl solution. J. Ethol. 24, 59–65.

Nevitt, G.A., 2011. The neuroecology of dimethyl sulfide: a global-climate regulator turned marine infochemical. Integr. Comp. Biol. 51, 819–825.

Nevitt, G.A., Veit, R.R., Kareiva, P., 1995. Dimethyl sulphide as a foraging cue for Antarctic procellariiform seabirds. Nature 376, 680–682.

Necker, R., 1974. Dependence of mechanoreceptor activity on skin temperature in sauropsids. II. Pigeon and duck. J. Comp. Physiol. 92, 75–83.

Necker, R., 2000. The somatosensory system. In: Whittow, G.C. (Ed.), Sturkie's Avian Physiology, fifth ed. , pp. 57–69.

Nicolson, S.W., Fleming, P.A., 2003. Nectar as food for birds: the physiological consequences of drinking dilute sugar solutions. Plants Syst. Evol. 238, 139–153.

Norman, D.L., Mason, J.R., Clark, L., 1992. Capsaicin effects on consumption of food by cedar waxwings and house finches. Wilson Bull. 104, 549–551.

Odendaal, T.C., Brown, M., Downs, C.T., Johnson, S.D., 2010. Sugar preferences and digestive efficiency of the village weaver: a generalist avian pollinator of African plants. J. Exp. Biol. 213, 2531–2535.

Ohta, T., Ikemi, Y., Murakami, M., Imagawa, T., Otsuguro, K., Ito, S., 2006. Potentiation of transient receptor potential V1 functions by the activation of metabotropic 5-HT receptors in rat primary sensory neurons. J. Physiol. 576, 809–822.

Papi, F., Casini, G., 1990. Pigeons with ablated pyriform cortex home from familiar but not from unfamiliar sites. PNAS 87, 3783–3787.

Papi, F., Fiore, L., Fiaschi, V., Benvenuti, S., 1971. The influence of olfactory nerve section on the homing capacity of carrier pigeons. Monit. Zool. Ital. 5, 265–267.

Patapoutian, A., Tate, S., Woolf, C.J., 2009. Transient receptor potential channels: targeting pain at the source. Nat. Rev. Drug Discov. 8, 55–68.

Patzke, N., Manns, M., Güntürkün, O., Ioalè, P., Gagliardo, A., 2010. Navigation-induced ZENK expression in the olfactory system of pigeons (*Columba livia*). Eur. J. Neurosci. 31, 2062–2072.

Petit, C., Hossaert-McKey, M., Perret, P., Blondel, J., Lambrechts, M.M., 2002. Blue tits use selected plants and olfaction to maintain an aromatic environment for nestlings. Ecol. Lett. 5, 585–589.

Pierau, F.-K., Sann, H., Harti, G., 1986. Resistance of birds to capsaicin and differences in their substance P (SP) system. Proc. Int. Union Physiol. Sci. 16, 207–211.

Porter, R.H., Picard, M., 1998. Effects of early odor exposure in domestic chicks. Reprod. Nutr. Dev. 38, 441–448.

Provenza, F.D., 1995. Origins of food preference in herbivores. In: Mason, J.R. (Ed.), Repellents in Wildlife Management. National Wildlife Research Center, Fort Collins, CO, pp. 81–90.

Rahn, H., Ar, A., Paganelli, C.V., 1979. How bird eggs breathe. Sci. Am. 240, 46–55.

Rastogi, A., Kumari, Y., Rani, S., Kumar, V., 2011. Phase inversion of neural activity in the olfactory and visual systems of a night-migratory bird during migration. Eur. J. Neurosci. 34, 99–109.

Rehkämper, G., Frahm, H.D., Cnotka, J., 2008. Mosaic evolution and adaptive brain component alteration under domestication seen on the background of evolutionary theory. Brain Behav. Evol. 71, 115–126.

Rehkämper, G., Haase, E., Frahm, H.D., 1988. Allometric comparison of brain weight and brain structure volumes in different breeds of the domestic pigeon, *Columba livia* f.d. (fantails, homing pigeons, strassers). Brain Behav. Evol. 31, 141–149.

Rensch, B., Neunzig, R., 1925. Experimentelle untersuchungen über den geschmackssinn der vogel II. J. Ornithol. 73, 633–646.

Rescorla, R.A., Solomon, R.L., 1967. Two-process learning theory: relationships between Pavlovian conditioning and instrumental learning. Psychol. Rev. 74, 151–182.

Restrepo, D., Zviman, M.M., Rawson, N.E., 1995. Imaging of intracellular calcium in chemosensory receptor cells. In: Spielman, A.I., Brand, J.G. (Eds.), Experimental Cell Biology of Taste and Olfaction. CRC Press, Boca Raton, pp. 387–398.

Rieke, G.K., Wenzel, B.M., 1974. The ipsilateral and olfactory projection field in the pigeon. In: Denton, D.A., Coghlan, J.P. (Eds.), Olfaction and Taste, vol. 5. Academic Press, New York, pp. 361–368.

Rieke, G.K., Wenzel, B.M., 1978. Forebrain projections of the pigeon olfactory bulb. J. Morph. 158, 41–55.

Rios, J.M., Mangione, A.M., Marone, L., 2012. Tolerance to dietary phenolics and diet breadth in three seed-eating birds: implications for graminivory. J. Exp. Zool. A 317, 425–433.

Robinson, P.P., Winkles, P.A., 1990. Quantitative study of fungiform papillae and taste buds on the cat's tongue. Anat. Rec. 226, 108–111.

Roper, T.J., 1999. Olfaction in birds. In: In: Snowden, C.T., Roper, T.J. (Eds.), Advances in the Study of Behaviour, vol. 28. Academic Press, Boston, MA, pp. 247–332.

Roth, T.C., Cox, J.G., Lima, S.L., 2008. Can foraging birds assess predation risk by scent? Anim. Behav. 76, 2021–2027.

Roura, E., Baldwin, M.W., Klasing, K.C., 2012. The Avian Taste System: an Update Proceedings of the Australian Poultry Science Symposium, Sydney, New South Wales, Australia 23, 97–104

Rowe, C., Skelhorn, J., 2005. Colour biases are a question of taste. Anim. Behav. 69, 587–594.

Saito, I., 1966. Comparative anatomical studies of the oral organs of the poultry. V. Structures and distribution of taste buds of the fowl. Bull. Fac. Agric. Univ. Miyazaki 13, 95–102.

Saito, S., Shingai, R., 2006. Evolution of thermo TRP ion channel homologs in vertebrates. Physiol. Genomics 27, 219–230.

Saito, S., Fukuta, N., Shingai, R., Tominaga, M., 2011. Evolution of vertebrate transient receptor potential vanilloid 3 channels: opposite temperature sensitivity between mammals and western clawed frogs. PloS Genet. 7, e1002041.

Sann, H., Harti, G., Pierau, F.-K., Simon, E., 1987. Effect of capsaicin upon afferent and efferent mechanisms of nociception and temperature regulation in birds. Can. J. Physiol. Pharmacol. 65, 1347–1354.

Saxton, V.P., Creasy, G.L., Paterson, A.M., Trought, M.C.T., 2009. Behavioral responses of European blackbirds and Australasian silvereyes to varying acid and sugar levels in artificial grapes. Am. J. Enol. Vit. 60, 82–86.

Saxton, V.P., Mulder, I., Creasy, G.L., Paterson, A.M., Ross, J.G., Trought, M.C.T., 2011. Comparative behavioural responses of silvereyes (*Zosterops lateralis*) and European blackbirds (*Turdus merula*) to secondary metabolites in grapes. Austral. Ecol. 36, 233–239.

Schaal, B., Marlier, L., Soussignan, R., 2000. Human foetuses learn odours from their pregnant mother's diet. Chem. Senses 25, 729–737.

Schondube, J.E., Martinez del Rio, C., 2003. Concentration-dependent sugar preferences in nectar-feeding birds: mechanisms and consequences. Funct. Ecol. 17, 445–453.

Schondube, J.E., Martinez del Rio, C., 2004. Sugar and protein digestion in flowerpiercers and hummingbirds: a comparative test of adaptive convergence. J. Comp. Physiol. B 174, 263–273.

Schrader, E., 1970. Die Topographie der Kopfnerven vom Huhn (Ph.D. Dissertation). Freie Univ., Berlin.

Schwenk, K., 1985. Occurrence, distribution and functional significance of taste buds in lizards. Copeia 1, 91–101.

Shah, P., Clark, L., Mason, J.R., 1991. Prediction of avian repellency from chemical structure: the aversiveness of vanillin, vanillyl alcohol and veratryl alcohol. Pestic. Biochem. Physiol. 40, 169–175.

Shaw, C.L., Rutter, J.E., Austin, A.L., Garvin, M.C., Whelan, R.J., 2011. Volatile and semivolatile compounds in gray catbird uropygial secretions vary with age and between breeding and wintering grounds. J. Chem. Ecol. 37, 329–339.

Shi, P., Zhang, J., 2005. Contrasting modes of evolution between vertebrate sweet/umami receptor genes and bitter receptor genes. Mol. Biol. Evol. 23 (2), 292–300.

Shibuya, T., Tucker, D., 1967. Single unit responses of olfactory receptors in vultures. In: Hayashi, T. (Ed.), Olfaction-and Taste. Pergamon Press, Oxford, pp. 219–220.

Shibuya, T., Tonosaki, K., 1972. Electrical responses of single olfactory receptor cells in some vertebrates. In: Schneider, D. (Ed.), Olfaction and Taste. Wissenschaftliche Verlagsgesellschaft MBH, Stuttgart, pp. 102–108.

Silver, W.L., Maruniak, J.A., 1981. Trigeminal chemoreception in the nasal and oral cavities. Chem. Senses 6, 295–305.

Skelhorn, J., Griksaitis, D., Rowe, C., 2008. Colour biases are more than a question of taste. Anim. Behav. 75, 827–835.

Skelhorn, J., Rowe, C., 2010. Birds learn to use distastefulness as a signal of toxicity. Proc. R. Soc. B 277, 1729–1734.

Sneddon, H., Hadden, R., Hepper, P.G., 1998. Chemosensory learning in the chicken embryo. Phys. Behav. 64, 133–139.

Snyder, G.K., Peterson, T.T., 1979. Olfactory sensitivity in the black-billed magpie and in the pigeon. Comp. Biochem. Physiol. A 62, 921–925.

Stager, K.E., 1964. The role of olfaction in food location by the Turkey vulture (*Cathartes aura*). Los Ang. Cnty. Mus. Contrib. Sci. 81, 1–63.

Stager, K.E., 1967. Avian olfaction. Am. Zool. 7, 415–420.

Stattelman, A.J., Talbot, R.B., Coulter, D.B., 1975. Olfactory thresholds of pigeons (*Columba liva*), quail (*Colinus virginianus*) and chickens (*Gallus gallus*). Comp. Biochem. Physiol. A 50, 807–809.

Steiger, S.S., Fidler, A.E., Valcu, M., Kempenaers, B., 2008. Avian olfactory receptor gene repertoires: evidence for a well-developed sense of smell in birds? Proc. R. Soc. Ser. B 275, 2309–2317.

Stucky, C.L., Dubin, A.E., Jeske, N.A., Malin, S.A., McKerny, D.D., Story, G.M., 2009. Roles of transient receptor potential channels in pain. Brain Res. Rev. 60, 2–23.

Szolcsanyi, J., 1982. Capsaicin type pungent agents producing pyrexia. In: Milton, A.S. (Ed.), Pyretics and Antipyretics. Springer-Berlin Heidelberg, pp. 437–478.

Szolcsanyi, J., Sann, H., Pierau, F.-K., 1986. Nociception in pigeons is not impaired by capsaicin. Pain 27, 247–260.

Taher, A.I., Gleaves, E.W., Beck, M., 1984. Special calcium appetite in laying hens. Poult. Sci. 63, 2261–2267.

Tewksbury, J.J., Nabhan, G.P., 2001. Seed dispersal. Directed deterrence by capsaicin in chilies. Nature 412, 403–404.

Todrank, J., Heth, G., Restrepo, D., 2011. Effects of *in utero* odorant exposure on neuroanatomical development of the olfactory bulb and odour preferences. Proc. R. Soc. Ser. B. 278, 1949–1955.

Tolhurst, B.E., Vince, M.A., 1976. Sensitivity to odours in the embryo of the domestic fowl. Anim. Behav. 24, 772–779.

Tordoff, M.G., 2001. Calcium: taste, intake, and appetite. Physiol. Rev. 81, 1567–1597.

Tordoff, M.G., Shao, H., Alarcón, L.K., Margolskee, R.F., Mosinger, B., Bachmanov, A.A., Reed, D.R., McCaughey, S., 2008. Involvement of T1R3 in calcium–magnesium taste. Physiol. Genomics 34, 338–348.

Tucker, D., 1963. Olfactory, vomeronasal and trigeminal receptor responses to odorants. In: Zotterman, Y. (Ed.), Olfaction and Taste. Pergamon, New York, pp. 45–69.

Tucker, D., 1965. Electrophysiological evidence for olfactory function in birds. Nature 207, 34–36.

Tucker, D., 1971. Nonolfactory responses from the nasal cavity: Jacobson's organ and the trigeminal system. In: Beidler, L.M. (Ed.), Handbook of Sensory Physiology IV: Chemical Senses Olfaction. Springer-Verlag, Berlin, pp. 151–181.

Tucker, D., Graziadei, P.C., Smith, J.C., 1974. Recovery of olfactory function in pigeons after bilateral transection of the olfactory nerves. In: Denton, D.A., Coghlan, J.P. (Eds.), Olfaction and Taste, vol. 5. Academic Press, New York, pp. 369–374.

Ueda, H., Kainou, S., 2005. Aversion to quinine is associated with taste sensation in chicks. J. Poult. Sci. 42, 254–262.

Ueda, H., Shigemizu, G., 2001. Feeding response to tea saponin in chicks given diet selection. J. Poult. Sci. 38, 333–342.

Ueda, H., Takagi, A., Katou, K., Matsumoto, S., 2002. Feeding behavior in chicks fed tea saponin and quinine sulfate. J. Poult. Sci. 39, 34–41.

Vermaut, S., De Coninck, K., Flo, G., Cokelaere, M., Onagbesan, M., Decuypere, E., 1997. Effect of deoiled jojoba meal on feed intake in chickens: satiating or taste effect? J. Agr. Food Chem. 45, 3158–3163.

Walker, J.C., Tucker, D., Smith, J.C., 1979. Odor sensitivity mediated by trigeminal nerve in the pigeon. Chem. Senses 4, 107–116.

Walker, J.C., Walker, D.B., Tambiah, C.R., Gilmore, K.S., 1986. Olfactory and nonolfactory odor detection in pigeons: elucidation by a cardiac acceleration paradigm. Physiol. Behav. 38, 575–580.

Wallraff, H.G., 1988. Olfactory deprivation in pigeons: examination of methods applied in homing experiments. Comp. Biochem. Physiol. A Comp. Physiol. 89, 621–629.

Wallraff, H., 2005. Beyond familiar landmarks and integrated routes: goal-oriented navigation by birds. Connect. Sci. 17, 91–106.

Warner, R.L., McFarland, L.Z., Wilson, W.O., 1967. Microanatomy of the upper digestive tract of the Japanese quail. Am. J. Vet. Res. 28, 1537–1548.

Watkins, R.W., Lumley, J.A., Gill, E.L., Bishop, J.D., Langton, S.D., MacNicoll, A.D., Price, N.R., Drew, M.G.B., 1999. Quantitative structure-activity relationships (QSAR) of cinnamic acid bird repellents. J. Chem. Ecol. 25, 2825–2845.

Weldon, P.J., Rappole, J.H., 1997. J. Chem. Ecol. 23, 2609–2632.

Wellmann, A.E., Downs, C.T., 2009. Sugar preferences and digestion by cape white-eyes, *Zosterops virens*, fed artificial fruit diets. Afr. Zool. 44, 106–116.

Wenzel, B.M., 1973. Chemoreception. In: Farner, D.S., King, J.R., Parkes, K.C. (Eds.), Avian Biology, vol. 3. Academic Press, New York, pp. 389–415.

Wenzel, B.M., Sieck, M.H., 1972. Olfactory perception and bulbar electrical activity in several avian species. Physiol. Behav. 9, 287–293.

Werner, S.J., Kimball, B.A., Provenza, F.D., 2008. Food color, flavor, and conditioned avoidance among red-winged blackbirds. Physiol. Behav. 93, 110–117.

Werner, S.J., Provenza, F.D., 2011. Reconciling sensory cues and varied consequences of avian repellents. Physiol. Behav. 102, 158–163.

Whelan, R.J., Levin, T.C., Owen, J.C., Garvin, M.C., 2010. Short-chain carboxylic acids from gray catbird (*Dumetella carolinensis*) uropygial secretions vary with testosterone levels and photoperiod. Comp. Biochem. Physiol. B 156, 183–188.

Whittaker, D.J., Soini, H.A., Atwell, J.W., Hollars, C., Novotny, M.V., Ketterson, E.D., 2010. Songbird chemosignals: volatile compounds in preen gland secretions vary among individuals, sexes and populations. Behav. Ecol. 21, 608–614.

Whittaker, D.J., Richmond, K.M., Miller, A.K., Kiley, R., Burns, C.B., Atwell, J.W., Ketterson, E.D., 2011. Intraspecific preen oil preferences in dark-eyed juncos (*Junco hyemalis*). Behav. Ecol. 22, 1256–1263.

Wild, J.M., 1985. The avian somatosensory system. I. Primary spinal afferent input to the spinal cord and brainstem in the pigeon (*Columba livia*). J. Comp. Neurol. 240, 377–395.

Wilson, A.L., Downs, C.T., 2011. Food preferences of Knysna and purple-crested turacos fed varying concentrations of equicaloric and equimolar artificial fruit. J. Exp. Biol. 214, 607–612.

Woolf, C.J., Ma, Q., 2007. Nociceptors-noxious stimulus detectors. Neuron 55, 353–364.

Wood-Gush, D.G.M., Kare, M.R., 1966. The behaviour of calcium–deficient chickens. Brit. Poult. Sci. 7, 285–290.

Zelenitsky, D.K., Therrien, F.O., Ridgely, R.C., McGee, A.R., Witmer, L.M., 2011. Evolution of olfaction in non-avian theropod dinosaurs and birds. Proc. R. Soc. Ser. B 278, 3625–3634.

Zidar, J., Løvlie, H., 2012. Scent of the enemy: behavioral responses to predator faecal odour in the fowl. Anim. Behav. 84, 547–554.

Zhao, H., Zhang, J., 2012. Mismatches between feeding ecology and taste receptor evolution: an inconvenient truth. PNAS 109, E1464.

Zygmunt, P.M., Petersson, J., Andersson, D.A., Chuang, H., Sørgård, M., Di Marzo, V., Julius, D., Högestätt, E.D., 1999. Vanilloid receptors on sensory nerves mediate the vasodilator action of anandamide. Nature 400, 452–457.

Chapter 8

Magnetoreception in Birds and Its Use for Long-Distance Migration*

Henrik Mouritsen
Institut für Biologie und Umweltwissenschaften, Universität Oldenburg, Oldenburg, Germany; Research Centre for Neurosensory Sciences, University of Oldenburg, Oldenburg, Germany

8.1 INTRODUCTION

The Earth's magnetic field provides potentially useful information, which birds could use for directional and/or positional information. It has been clearly demonstrated that birds are able to sense the compass direction of the Earth's magnetic field and that they can use this information as part of a compass sense. Magnetic information could also be useful as part of a map sense, and there is a growing body of evidence that birds are able to determine their approximate position on the Earth on the basis of geomagnetic cues. In addition to direct uses for orientation and navigation, magnetic information also seems to be able to influence other physiological processes, such as fattening and migratory motivation, as a trigger for changes in behavior. Although the behavioral responses to geomagnetic cues are relatively well understood, the physiological mechanisms enabling birds to sense the Earth's magnetic field are only starting to be understood, and understanding the magnetic sense(s) of animals, including birds, remains one of the most significant unsolved problems in biology. It is very challenging to sense magnetic fields as weak as that of the Earth using only biologically available materials. Only two basic mechanisms are considered theoretically viable in terrestrial animals: iron-mineral-based magnetoreception and radical-pair based magnetoreception. On the basis of current scientific evidence, iron-mineral-based magnetoreception and radical-pair-based magnetoreception mechanisms seem to exist in birds, but they seem to be used for different purposes. Plausible primary sensory molecules and a few brain areas involved in processing magnetic information have been identified in birds for each of these two types of magnetic senses. Nevertheless, we are still far away from understanding the detailed function of any of the at least two different magnetic senses existing in some if not all bird species, and, at present, no primary sensory structure has been identified beyond reasonable doubt to be the source of avian magnetoreception. This is an exciting but challenging field in which several major discoveries are likely to be made in the next 1–2 decades.

8.2 MAGNETIC FIELDS

Moving electric charges such as electrons produce magnetic fields. On the microscopic scale, electron (and nuclear) spins can generate magnetic fields. On the macroscopic scale, a magnetic field, **B**, is, for instance, generated around a wire when current runs through it. The magnetic field at a given location can be described as a three-dimensional (3D) vector for which the strength, **B**, is measured as magnetic flux density using the unit "Tesla" (T), $1\,T = 1\,(V^*s)/m^2 = 1\,(N^*s)/(C^*m) = 10{,}000$ Gauss (V = Volt, s = second, m = meter, N = Newton, C = Coulomb). Some materials, which are called "ferromagnetic," can be permanently magnetized by a magnetic field, and this magnetization remains after the magnetizing field has been removed. Magnetite (Fe_3O_4), an iron oxide, is a well-known example of a ferromagnetic mineral (Mouritsen, 2013).

* Because our knowledge of magnetoreception did not change dramatically over the last few months, there is significant text and content overlap between the present chapter and a chapter focusing on magnetoreception in all kinds of organisms and titled "The Magnetic Senses," which I recently wrote for the textbook *Neurosciences*: Mouritsen, H., 2013. The magnetic senses. In: Galizia, C.G., Lledo, P.M. (Eds), *Neurosciences—From Molecule to Behavior: A University Textbook*. © Springer-Verlag Berlin Heidelberg, pp. 427–443, doi: 10.1007/978-3-642-10,769-6_20. Springer Verlag has permitted the reuse of significant parts of the *Neurosciences* textbook chapter text and figures in the present chapter. Specific references to this text are not given at every location where text is reused because such references would compromise readability and could be misunderstood to be referring to primary research findings.

8.3 THE EARTH'S MAGNETIC FIELD

The Earth generates its own magnetic field (the geomagnetic field), which is mostly caused by electric currents in the liquid outer core of the Earth (the "dynamo effect"). The magnetic field measured at the Earth's surface is similar to the magnetic field one would expect to see if a large dipole magnet was placed in the center of the Earth (see Figure 8.1). The Earth's magnetic field currently has a magnetic field South Pole near the Earth's geographic North Pole (referred to as "Magnetic North" or "Magnetic North Pole" in biology). Throughout this chapter, I will follow the convention used in the bird orientation research literature and use the term "Magnetic North" or "Magnetic North Pole" to refer, not to the physical magnetic North Pole, but to the magnetic pole located closest to the geographic North Pole. Likewise, the magnetic field North Pole near the Earth's geographic South Pole will be referred to as "Magnetic South" or "Magnetic South Pole" (Mouritsen, 2013).

The magnetic field lines leave the Magnetic South Pole and re-enter the Magnetic North Pole. The polarity of the magnetic field lines always points toward Magnetic North; therefore, they can provide a highly reliable directional reference that can be used as the basis for a magnetic compass anywhere on planet Earth except at the magnetic poles. At the magnetic poles, the field lines point directly into the sky (at the Magnetic South Pole) or directly into the Earth (at the Magnetic North Pole). At the magnetic equator, the magnetic field lines are parallel to the Earth's surface. The angle between the magnetic field lines and the Earth's surface is called "magnetic inclination." Thus, magnetic inclination changes gradually from −90° at the Magnetic South Pole to 0° at the magnetic equator to +90° at the Magnetic North Pole (see Figure 8.1). The Earth's magnetic field intensity ranges from c. 30,000 nT (nanoTesla = 10^{-9} T; 1 T = 1 V s m^{-2}; 1 nT = 10^{-5} Gauss) near the magnetic equator to c. 60,000 nT at the magnetic poles. Earth-strength magnetic fields are usually measured with a calibrated three-axial flux-gate magnetometer. In theory, magnetic inclination and magnetic intensity can be useful for determining one's position, but, on most parts of the Earth, magnetic inclination and intensity changes predominantly from North to South but not much from East to

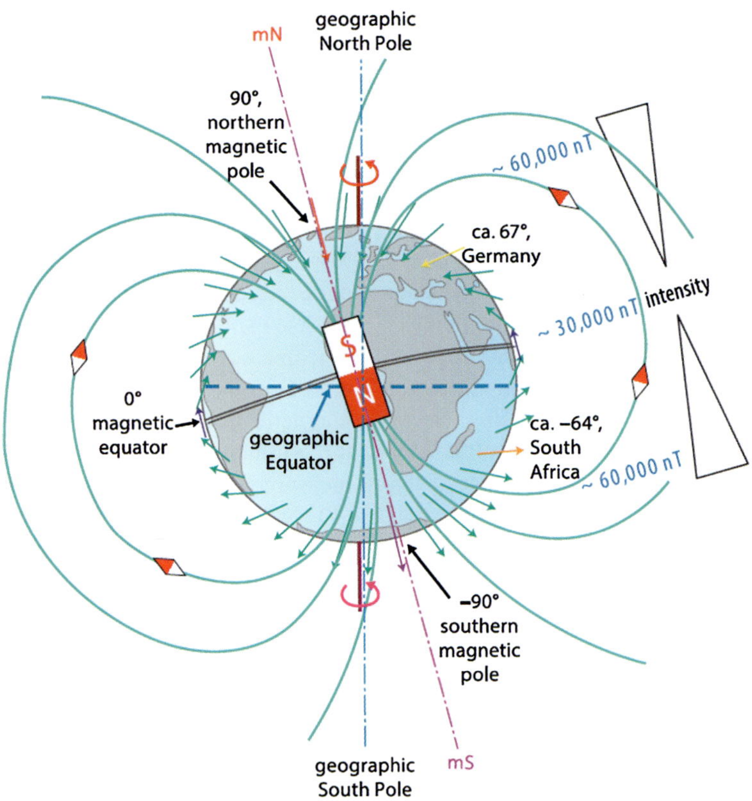

FIGURE 8.1 The Earth's magnetic field (the geomagnetic field). Notice that the southern and northern magnetic poles and the magnetic equator do not coincide with the geographical poles and the geographic equator. Also notice that the magnetic field lines intersect the Earth's surface at different angles depending on the magnetic latitude (blue-green lines and vectors). The intersection angle is called the magnetic inclination. Magnetic inclination is +90° at the Magnetic North Pole (red vector), c. +67° at the latitude of Germany (yellow vector), 0° at the magnetic equator (dark blue vectors), c. −64° at the latitude of South Africa (orange vector), and −90° at the Magnetic South Pole (magenta vector) *(Adapted with permission after Wiltschko and Wiltschko (1996) and Mouritsen (2013).)* The magnetic intensity varies from c. 60,000 nT near the magnetic poles to c. 30,000 nT along the magnetic equator.

West; therefore, it seems easier to determine latitude than longitude from geomagnetic field information (Mouritsen, 2013).

The Magnetic North Pole is currently located in northern Canada, and the Magnetic South Pole is currently located south of Australia. Consequently, the geographic and magnetic poles do not coincide (see Figure 8.1). The deviation between geographic and Magnetic North is called the "magnetic declination." Magnetic declination is the angle between Magnetic North (i.e., the direction in which the north end of a compass needle points in) and Geographic North. The declination is positive when Magnetic North is east of Geographic North and negative when Magnetic North is west of Geographic North. Declination is mostly small, but near the magnetic poles declination can pose a serious problem for navigating birds using a magnetic compass unless they find a way to compensate for it. On the other hand, magnetic declination could, in theory, be a useful parameter to determine, for example, East-West position if it would be combined with other map cues (Mouritsen, 2013).

8.4 CHANGING MAGNETIC FIELDS FOR EXPERIMENTAL PURPOSES

The direction of the magnetic field around a wire can be determined by the "right hand rule": If you grasp around the wire with your right hand so that your thumb is pointing in the direction of the current, then the magnetic field around the wire runs in the direction in which your fingers are pointing. The magnetic field decreases with distance as you move away from the wire. If you create a coil of wire, then the magnetic field created is much stronger inside of the coil than on the outside of the coil because many parallel magnetic field lines created by different parts of the wire coincide and thus add up in the center of the coil. This is the reason why coil constructions are typically used to produce and alter magnetic fields (Mouritsen, 2013).

The typical coil constructions, which are used to produce Earth-strength magnetic fields for scientific experiments, are so-called "Helmholtz coils"—a pair of parallel coils placed one radius apart from each other (Kirschvink, 1991). In a pair of Helmholtz coils, the magnetic field is very homogeneous within a central space of c. 60% of the radius of the coils (Kirschvink, 1991). The magnetic field generated in the center of a pair of Helmholtz coils is $\mathbf{B} = (0.9*10^{-6}\,\mathrm{T\,m/A}*n*I)/R$, where T is the unit Tesla, n is the number of turns in each coil, I is the current flowing through the coils measured in ampere (A), and R is the radius of the coils measured in meters (m) (Kirschvink, 1991). One pair of Helmholtz coils can only alter the magnetic field along one axis. To make any desired 3D magnetic field, three pairs of Helmholtz coils oriented perpendicular to each other are ideally needed. If one adds an artificially created field to an existing field (such as that of the Earth), then the resultant field is calculated by simple vector addition of the two fields (see Figure 8.2; Kirschvink, 1991). Therefore, it is also possible to use a single pair of Helmholtz coils to make any 3D magnetic field, but in that case, this single pair of coils must be oriented very precisely in 3D space (see Figure 8.2; Mouritsen, 2013).

Although the Helmholtz arrangement is easy to calculate and construct, the central homogeneous space can be increased to c. 110% of the radius of the coils by using more elaborate coil designs such as the Merritt-4-coil system (Kirschvink, 1991; Zapka et al., 2009, Figure 20.2 in Mouritsen, 2013). To control for artefacts, one would—independent of the coil design chosen—expect the coils to be "double wrapped" (Kirschvink, 1991; Kirschvink et al., 2010). This means that during construction of the coils, each coil contains two separate but identically wrapped wires, each with separate connectors, so that one can either run current through both halves of the windings in the same direction (then the magnetic field in the center of the coil will change), or one can run the current through one half of the coils in one direction but in the opposite direction through the second half of the windings. In that case, the current running through one half of the windings will create a magnetic field, which exactly cancels the magnetic field produced by the other half of the windings, and the background field is not changed. By using double-wrapped coils, exactly the same amount of current is sent through the coils whether the magnetic field is being changed or not. Double-wrapped coils also allow for truly double-blinded experiments (Kirschvink, 1991; Zapka et al., 2009; Harris et al., 2009; Hein et al., 2010, 2011; Engels et al., 2012). An excellent presentation of the theoretical background and practical instructions on how to construct various coil designs for changing Earth-strength magnetic fields can be found in Kirschvink (1991).

8.5 BIRDS USE INFORMATION FROM THE EARTH'S MAGNETIC FIELD FOR ORIENTATION AND NAVIGATION

Orientation and navigation skills are essential for the survival of all migratory birds. All first-time migrants are faced with the challenge of finding an unfamiliar wintering area, often thousands of kilometers away (Berthold, 1991; Mouritsen and Mouritsen, 2000; Mouritsen, 2003). Many bigger birds are day migratory and travel in groups, which means that young birds of these species might simply follow experienced birds that know the way. However, most small songbirds are night-migratory and travel alone without contact with their parents. Consequently, all of their navigational skills must be based on inherited sensory capabilities and strategies (Mouritsen, 2003). No cues requiring previous

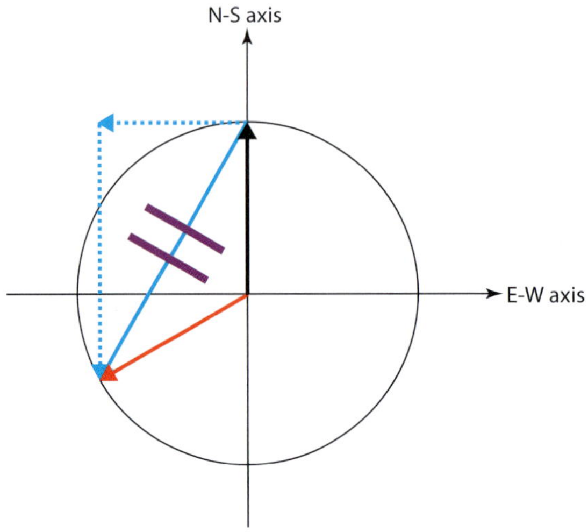

FIGURE 8.2 Magnetic fields are vector fields and can easily be turned with pairs of coils. If we consider the geomagnetic field, it will point toward North (0°) and have a vertical and a horizontal component. Let's say that we want to create a magnetic field with the same strength and inclination as the geomagnetic field but that is turned horizontally 120° counterclockwise. In that case, the vertical component of the field should remain unchanged, and we only have to consider the two dimensions in the horizontal plane. Let's say that the Earth's magnetic field at the relevant location has a horizontal field component of 18,000 nT pointing toward Magnetic North (black vector). If we want to turn the field to point toward 240° (a 120° counterclockwise turn, red vector), then we need to produce a magnetic field vector (the blue vector) that connects the tip of the black vector to the tip of the red vector. The needed field can be produced by a single pair of Helmholtz coils (symbolized by the violet lines) if the coil frames are oriented on the axis defined by half of the wanted angular turn (in this case 120°/−60°) given that the final wanted intensity should remain unchanged. Simple trigonometry can be used to calculate the needed field strength, B_{blue}, of the blue vector. In this case, $B_{blue}=(((\cos(\alpha_{black})*B_{black})-(\cos(\alpha_{red})*B_{red}))^2+((\sin(\alpha_{black})*B_{black})-(\sin(\alpha_{red})*B_{red}))^2)^{1/2}$, where $\alpha_{black}=360°$, $\alpha_{red}=240°$, and $B_{black}=B_{red}=18{,}000\,nT \Rightarrow B_{blue}=((\cos(360°)*18{,}000\,nT-\cos(240°)*18{,}000\,nT)^2+(\sin(360°)*18{,}000\,nT-\sin(240°)*18{,}000\,nT)^2)^{1/2}=(((27{,}000\,nT)^2+(-15{,}588\,nT)^2)^{1/2})=31{,}177\,nT$. If the strength of the final vector should have a different intensity than the original vector, or if the vertical component also needs to be changed, then again a single pair of coils can, in principle, do the job (the needed calculations are 3D), but accurately orienting this pair of coils is very difficult in real life. Therefore, if excellent control of static magnetic fields is required, usually 3D systems of perpendicularly oriented coils are used. Because magnetic fields are vector fields, which all need to be added up to get the total resultant field, instead of producing the direct vector (the blue vector) that connects the tip of the black vector to the tip of the red vector, we can produce two vectors (the dashed blue vectors) along the two coil axes, which in total connect the tip of the black vector to the tip of the red vector. With such systems, each of the needed vectors is much easier to calculate. The needed N–S component is $\cos(\alpha_{black})*B_{black}-\cos(\alpha_{red})*B_{red}$, and the needed E–W component is $\sin(\alpha_{black})*B_{black}-\sin(\alpha)*B_{red}$. If one uses a 3D magnetometer oriented with the x-axis toward North and the y-axis toward East, and one just wants to calculate the values that should be on the display when the wanted field is present, X should read $\cos(\alpha_{red})*B_{red}$ and Y should read $\sin(\alpha)*B_{red}$. Thus, in the case of a 120° counterclockwise turn of the above-mentioned field, X should read $\cos(240°)*18{,}000\,nT=-9000\,nT$ and Y should read $\sin(240°)*18{,}000\,nT=-15{,}588\,nT$. All formulas presented here are valid for geographical angles (North = 0° = 360°, East = 90°, South = 180°, and West = 270°) but have to be modified if mathematical angles are used (East = 0°, North = 90°, West = 180°, and South = 270°). What should the same magnetometer read on X and Y if the same geomagnetic field is turned horizontally to 165°?#

$X=-17{,}387\,nT$; $Y=+4659\,nT$.

experience with the goal can be involved in the orientation strategies of solitary, first-year migrants. These considerations strongly limit the number of possible orientation cues to a few classes of globally or at least regionally consistent cues (Mouritsen, 2003):

1. Celestial cues, including the Sun, the stars, and maybe the polarized light pattern of the sky
2. Geomagnetic cues

In addition to these cues, some authors have suggested that chemical cues, including odors, (Wallraff and Andreae, 2000; Wallraff, 2005; Gagliardo et al., 2006, 2008, 2009) infrasound (sound with frequency below c. 20 Hz; Hagstrum, 2013 but see Wallraff, 1972; Holland, in press), and/or Coriolis forces (the phenomenon that moving liquids and moving air are deflected slightly to the right on the Northern Hemisphere and slightly to the left on the Southern Hemisphere because of the Earth's rotation; Coriolis, 1835) might also be used for orientation and navigation.

However, there seems to be no physiological structure inside of birds that would enable them to detect the Coriolis effect with a reasonable signal-to-noise ratio (Rosenblum et al., 1985; Adair, 1991; Kirschvink et al., 2010). Likewise, it is difficult to imagine how an inexperienced migrant could know, in advance, what the infrasound or odor "landscapes" along its migratory path looks like, and it is difficult to imagine that the infrasound and/or odor landscapes would be simple and consistent enough to be used by inexperienced birds as a primary map cue over thousands of kilometers (but see Wallraff and Andreae, 2000). Furthermore, because the width of the bird's head is much smaller than the wavelength of infrasound, it would be challenging for a

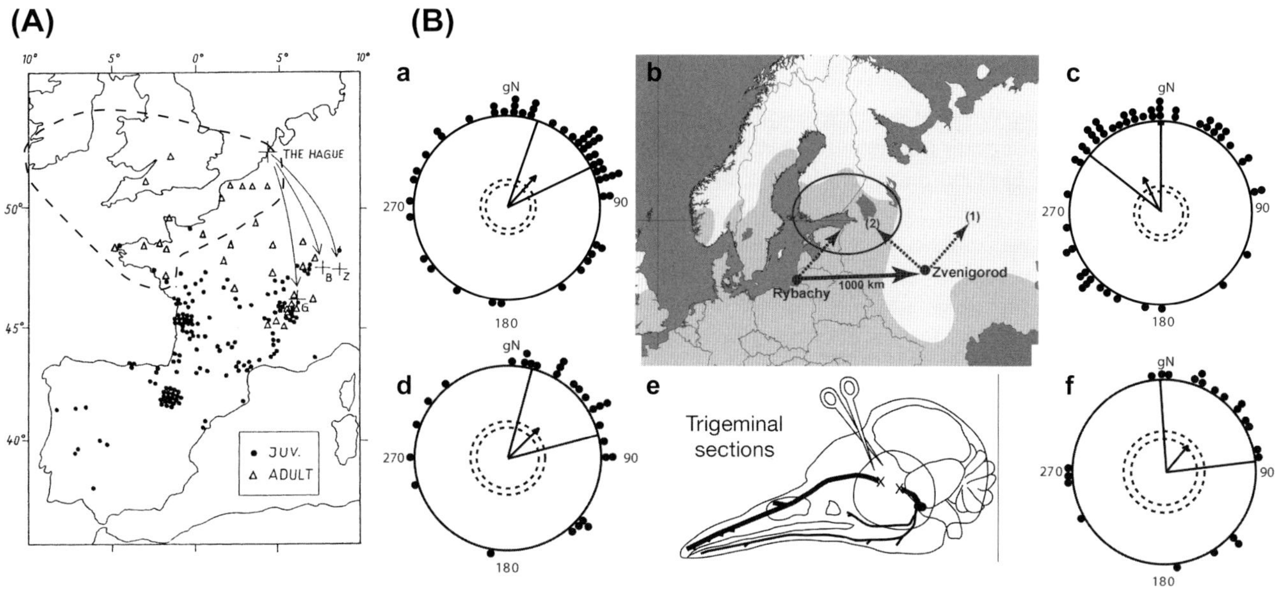

FIGURE 8.3 Displacement experiments provide key evidence for understanding the spatiotemporal orientation strategies of migratory birds. (A) Perdeck's classical experiments in which he displaced >10,000 starlings from Holland S/SSE (→) to Switzerland during autumn migration showed that young starlings (•) on their first autumn migration were unable to correct for the displacement. The young birds show a parallel displaced migration pattern relative to the wintering area of nondisplaced controls (dashed area) whereas adult starlings (Δ) orient directly back to the normal population-specific wintering area of nondisplaced controls *(From Mouritsen (2003) after Schmidt-Koenig (1965); Perdeck (1958))*. (B) Displacement experiments, in which Eurasian Reed Warblers (*Acrocephalus scirpaceus*) were tested in Emlen funnels (Emlen and Emlen, 1966; Mouritsen et al., 2009) before and after displacement, showed that young birds on their first spring migration are already able to correct for 1000-km eastward displacements to a location where they have certainly never been before (b). (a) Orientation of birds at the capture site (Rybachy). (c) Orientation of the same birds after the 1000 km eastward translocation to Zvenigorod. Each dot at the circular diagram periphery indicates the mean orientation of one individual bird. The arrows show group mean directions and vector lengths. The dashed circles indicate the length of the group mean vector needed for significance according to the Rayleigh test (5% and 1% level for inner and outer dashed circles, respectively). The lines flanking group mean vectors indicate the 95% confidence intervals for the mean direction. gN = geographic North. On (b), a map of the displacement region is shown. The shaded light-gray zone represents the breeding range of Eurasian Reed Warblers and the dashed arrows show the expected results in case of (1) no compensation for the displacement or (2) compensation toward the eastern part of the breeding range. Notice that intact birds compensate for the displacement whereas birds that had the ophthalmic branch of the trigeminal nerve cut (d–f) could no longer compensate for the displacement. *Re-assembled after Chernetsov et al. (2008); Kishkinev et al. (2013).*

small bird with a 2 cm wide head to determine from which direction infrasound originates (Mouritsen, 2013).

Thus, the primary orientation system of young inexperienced migrants on their first autumn migration is likely to be based primarily on celestial and magnetic cues, and it is known that when first-time autumn migrants are displaced away from their migration route, they are unable to correct for displacements (Drost, 1938; Perdeck, 1958; Mouritsen and Larsen, 1998; Mouritsen, 2003; Thorup et al., 2007; Holland, in press). Instead, they choose a migration route parallel to their normal route; thus, they do not seem to possess a map sense (see Figure 8.3). The migratory program of first-season solitary migrants can be described as a "clock-and-compass," "calendar-and-compass," or "vector navigation" strategy (Mayr, 1952; Perdeck, 1958; Schmidt-Koenig, 1965; Rabøl, 1978; Berthold, 1991; Mouritsen, 1998b; Mouritsen and Mouritsen, 2000; Mouritsen, 2003) in which the birds fly in a specific direction for a given amount of time independent of their present location. Because the system includes little (see below under magnetic signposts) or no location-related feedback, the orientation strategy of first-time solitary migrants can be mathematically described as a directed random walk: The birds choose their flight direction randomly from a normal-like distribution pointing in their mean migratory direction each evening independent of previous events (Mouritsen, 1998b; Mouritsen and Mouritsen, 2000; Mouritsen et al., 2013). This strategy predicts that the statistical distribution of first-time migrants should be parabolic, and it has been shown that this prediction fits very well with the actual distribution of ringing recoveries of free-flying, first-time migrants in Western Europe (Mouritsen, 1998b; Mouritsen and Mouritsen, 2000).

The orientation task facing adult migrants and young migrants on their first spring migration is fundamentally different from the task faced during their first autumn migration (Kramer, 1957; Rabøl, 1978; Berthold, 1991; Mouritsen, 2003; Holland, in press). Adult migrants and young migrants on their first spring migration are migrating back toward a region with which they have had previous experience; therefore, their orientation system is likely to include local (map) information gained through previous migration experience. Birds that would use sensory information from all useful senses, which would improve their ability to find their way (e.g., magnetic sense, olfaction, vision, and hearing), should have an evolutionary advantage over birds that would only use a single cue or sense. Thus, it is likely that the orientation strategies of experienced migrants are multisensory and involve learned maps (Mouritsen, 2003, 2013; Holland, in press). Indeed, in contrast to first-time migrants, experienced migrants are able to correct for displacements (Perdeck, 1958; Mewaldt, 1964; Thorup et al., 2007; Chernetsov et al., 2008; Kishkinev et al., 2010, 2013) and thus have added a learned map to their orientation program (see Figure 8.3). Interestingly, this map is also functional at locations that have not been visited previously: Birds can appropriately correct their orientation when they are experimentally displaced to far-away locations where they have certainly never been before (Perdeck, 1958; Mewaldt, 1964; Thorup et al., 2007; Chernetsov et al., 2008; Kishkinev et al., 2010, 2013). If their learned map would have been based exclusively on previously experienced local landmarks, then it should not have worked at unfamiliar locations. Although the functional basis of this map sense is not yet understood (Holland, in press), it is almost certainly based on multiple cues and it must involve the detection of larger scale gradients, which can be extrapolated and thus enable birds to return from unfamiliar locations.

8.6 THE MAGNETIC COMPASS OF BIRDS

Friedrich W. Merkel and Wolfgang Wiltschko discovered that birds have a magnetic compass sense in the mid-1960s (Merkel and Wiltschko, 1965; Wiltschko, 1968). When birds are placed in a round cage at night, they show migratory restlessness (or *Zugunruhe* in German, Kramer, 1949): The birds primarily jump/flutter in their migratory direction, and when the magnetic field is turned horizontally in the absence of celestial cues, the birds turn their orientation with the magnetic field (see Figure 8.4). This is the behavioral evidence required to show that a migratory bird species possesses and is able to use a magnetic compass (Wiltschko and Wiltschko, 1995). A magnetic compass has been found in more or less every migratory bird species properly tested for it (Wiltschko and Wiltschko, 1995); therefore, it is quite safe to presume that all migratory birds and potentially birds in general possess a magnetic compass.

It is important to note that there are at least two different magnetic field properties that could potentially

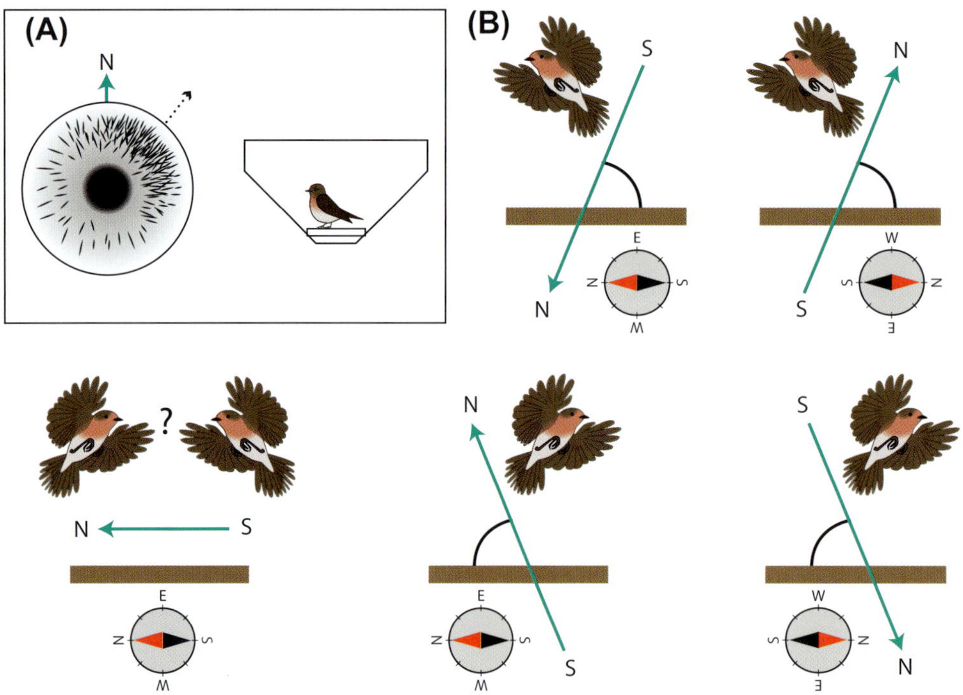

FIGURE 8.4 The Emlen funnel and the inclination compass. (A) The so-called Emlen funnel is the most commonly used orientation cage (Emlen and Emlen, 1966). The mean jumping direction of the birds are recorded on scratch-sensitive paper lining the inclined wall of the funnel (Mouritsen et al., 2009). (B) Early experiments by Wiltschko and Wiltschko (1972) have shown that birds have an inclination compass, which means that the birds measure the angle between the magnetic field lines and the Earth's surface or gravity; thereby, the birds separate between poleward and equatorward, not between North and South like a polarity compass would do (if birds use a polarity compass, then they should have oriented in the direction indicated by the red end of the inserted technical compass). Birds are disoriented in a horizontal magnetic field like the one occurring at the magnetic equator. The flight direction of the inserted bird indicates the springtime mean direction chosen by all bird species tested so far in the given magnetic field (Wiltschko and Wiltschko, 1995). The red arrows indicate the direction if the magnetic field lines. Brown bar = the Earth's surface, N = geographic North, S = geographic South. *Figure and legend is reused from Mouritsen (2013).*

be used as input for a magnetic compass sense. A magnetic polarity compass (e.g., the human ship compass) uses only the horizontal component of the field lines, which points toward Magnetic North anywhere on Earth except at the magnetic poles. On the other hand, a magnetic inclination compass detects only the angle between the geomagnetic field lines and the Earth's surface or gravity—not the polarity of the field lines. The smallest angle between the Earth's surface and the geomagnetic field lines indicates the direction "toward the magnetic equator" whereas the greatest angle indicates "toward the magnetic pole." Because the inclination is opposite on the Northern and Southern Hemisphere, respectively, this holds on both hemispheres. All bird species properly tested so far have a magnetic inclination compass (Wiltschko and Wiltschko, 1972, 1995; see Figure 8.4). Thus, the magnetic compass of night-migratory birds does not separate between North and South like our ship compass, but it distinguishes between "toward the magnetic equator" and "toward the magnetic pole" (the Magnetic North Pole in the Northern Hemisphere and the Magnetic South Pole in the Southern Hemisphere). Furthermore, the birds' magnetic compass sense seems to have a rather narrow functional intensity window, but this window seems to be extendable to new intensities after a few hours of adaptation to a changed magnetic field intensity (Wiltschko, 1978).

8.7 DO BIRDS POSSESS A MAGNETIC MAP?

Many studies have reported that magnetic cues play an important role in birds' sense of position (i.e., that birds have a "magnetic map"). However, the existence of a magnetic map is heavily debated, and the views among researchers range from a magnetic map with a precision of a few kilometers being an established fact (Walcott, 1991; Wiltschko and Wiltschko, 1995; Wiltschko et al., 2010a) to a magnetic map sense being an evergreen phantom (Wallraff, 2001; Gagliardo et al., 2009). One thing is for sure; the natural map sense of birds is multifactorial. It relies on input from olfaction (Papi, 1991; Wallraff, 2001, 2005; Gagliardo et al., 2006, 2008, 2009) and vision (Guilford et al., 2004) and possibly also from magnetic sensing (Dennis et al., 2007; Holland, 2010; Kishkinev et al., 2013) and maybe even from hearing (Hagstrum, 2013 but see Wallraff, 1972; Holland, in press).

Pigeons with opaque lenses that prevented them from detecting any local visual landmarks can return to within c. 5 km of their loft (Schmidt-Koenig and Walcott, 1978). Thus, the precision of nonlandmark-based navigation seems to be a few kilometers, but which cue(s) enable pigeons to home to within c. 5 km of their loft without being able to use visual landmarks? At present, informed neutral observers of the olfaction contra magnetic map controversy agree (Able, 1996; Mouritsen, 2013; both are orientation researchers who have never performed a pigeon release; thus, they have no vested interest in any of the camps) that the evidence suggesting that chemical cues (odors) play an important role in the nonlandmark-based part of pigeons' map sense (Papi, 1991; Wallraff and Andreae, 2000; Wallraff, 2001, 2005; Gagliardo et al., 2006, 2008, 2009) is much more convincing than the evidence supporting an important role of magnetic cues in the pigeons' map (e.g., Walcott, 1991; Wiltschko and Wiltschko, 1995; Dennis et al., 2007; Wiltschko et al., 2010a; Holland, 2010). However, remember that in the end, both camps may be right. The map will be multifactorial because a bird using all available input from all of its senses will have an evolutionary advantage over a bird using only a single cue for a task so essential for the survival of a bird. In any case, it remains very difficult to understand how a magnetic-field-based map sense should be able to function on a scale less than 10 km. Why is that?

The problem for a bird wanting to use a magnetic map is that the average change in magnetic field intensity is only c. 3 nT/km on the North-South axis: The geomagnetic field changes c. 30,000 nT from one of the magnetic poles to the magnetic equator, which are c. 10,000 km apart. Likewise, magnetic inclination changes only c. 0.009°/km along the North-South axis (a 90° change over 10,000 km). On the East-West axis, there is generally very little change in magnetic field intensity and magnetic inclination. Thus, any magnetic-field-based input to a reasonably precise map would require a very accurate magnetic sensory system and a very accurate sense of gravity. However, even if birds have such a system, daily, partly stochastic, natural variations in the geomagnetic field on the order of 30–100 nT in more or less random directions mean that it is very difficult to imagine how a magnetic-field-based map sense could have a precision less than 10–30 km (during magnetic storms generated primarily by the Sun, the geomagnetic field variability can reach 1000 nT; Courtillot and Le Mouël, 1988). Therefore, it is possible that magnetic parameters may only help determine position where the expected differences in the magnetic field parameters are consistently larger than the daily magnetic variations (Mouritsen, 2013).

Other cues such as odors and familiar landmarks may be more significant map parameters at shorter distances. It is much easier to imagine that a magnetic map could be relevant on a much larger spatial scale, and intriguing data exist that suggest that some songbirds can use magnetic cues as an approximate geographic "signpost", which, for example, tells the birds when to increase their fat reserves before crossing the Sahara Desert (Fransson et al., 2001) or when to change their migratory heading (Henshaw et al., 2010). This case, in which specific magnetic parameters trigger a change in behavior, is referred to as a magnetic signpost (Mouritsen, 2013).

8.8 INTERACTIONS WITH OTHER CUES

In most orientation-related contexts, magnetic cues interact with several other sources of similar and/or conflicting information. For instance, night-migratory songbirds not only have a magnetic compass, they also have a Sun compass and a star compass (Emlen, 1975; Schmidt-König et al., 1991; Mouritsen and Larsen, 2001; Cochran et al., 2004; Zapka et al., 2009). They only need information from any one of these compasses to orient in the appropriate direction (Mouritsen, 1998a; Muheim et al., 2006b; Chernetsov et al., 2011; Liu and Chernetsov, 2012). If the three compasses provide conflicting information, then it is not consistent which compass the birds prefer. The preference is probably dependent on the ecological context, the details of the experimental setup, and the conditions under which the birds were housed and tested, and it is likely that various calibrations are taking place in nature (see Figure 8.5,

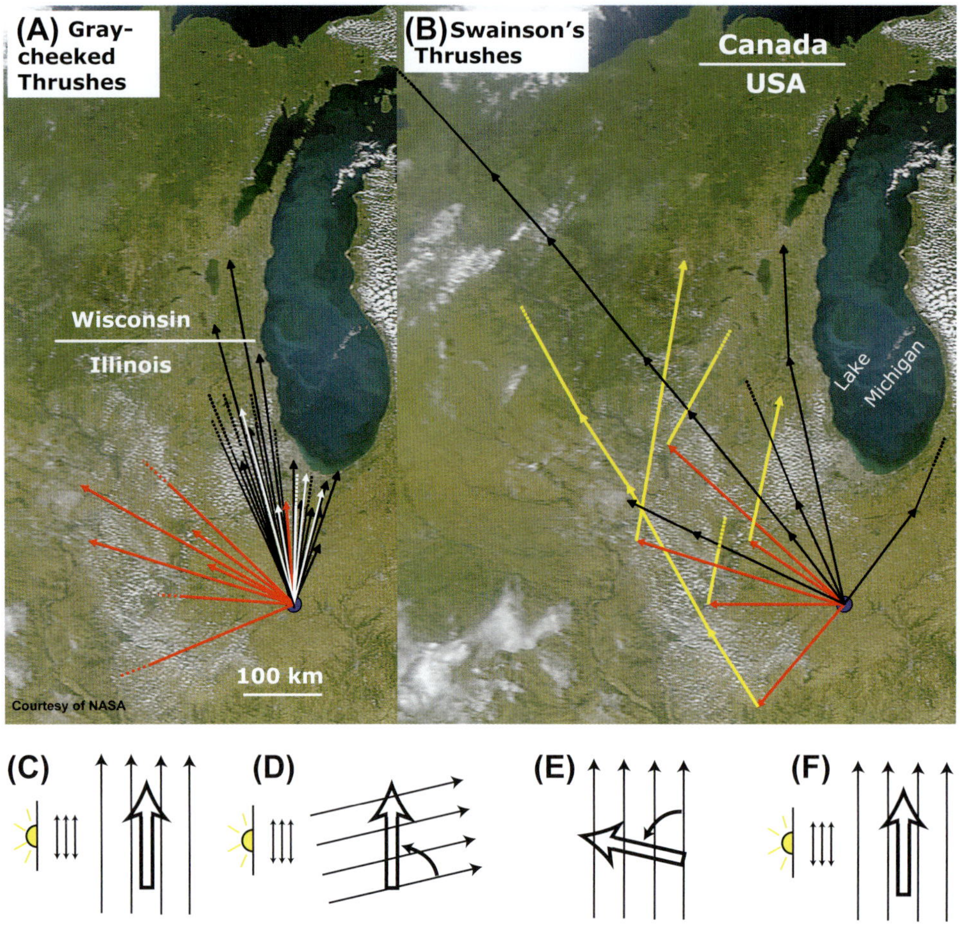

FIGURE 8.5 Some birds calibrate their magnetic compass from celestial cues around sunset. Tracks of free-flying Gray-Cheeked Thrushes (A) and Swainson's Thrushes (B) released from Champaign, Illinois, are shown. The arrows indicate the direction and ground tracks of migratory flights when wind effects were discarded. Black arrows indicate migratory flights of nonmanipulated individuals. Red arrows indicate migratory flights of experimental birds that had experienced a magnetic field turned to 80° East before takeoff, and yellow arrows indicate the migratory flights of the experimental birds during subsequent nights. White arrows indicate the migratory flight paths of experimental birds that did not migrate on the night of magnetic treatment, but they did so 1–6 days later. Connected arrows show flights of the same individual during successive nights. Data are depicted differently in (A) and (B) because for Grey-Cheeked Thrushes, experimental and control birds are different individuals whereas in Swainson's Thrushes, the same experimental individuals were followed for at least two successive nocturnal migrations (because of the large spread in natural headings). Broken lines indicate that birds were lost during tracking at the site where the broken lines start. Notice that the birds that experienced a magnetic field turned 80° to the East during sunset and that were released after all light from the Sun had disappeared migrated toward the West when they embarked on migration on the same night. On later nights, they migrated in the appropriate northerly spring migratory direction. These results mean that the birds had calibrated their magnetic compass from their Sun compass before takeoff and that this calibration happens daily. The reasons are illustrated in (C–F): (C) For control birds all cues gives the same information. (D) If experimental birds calibrate their magnetic compass from sunset-related cues, they will calibrate their magnetic compass so that 80° counter-clockwise to the magnetic field lines will be their "North" for the coming night. (E) After release, the experimental birds experience the natural field lines. Because all light from the Sun has disappeared, no new calibration is possible at time of release and their wrongly sunset-calibrated magnetic compass makes them fly 80° counter-clockwise relative to the natural field lines, which is towards the west for the rest of the first night. (F) On then second night after release, Sun and magnetic cues are in agreement and the birds will reorient into their intended migratory direction. The four thin parallel arrows (C–F) indicate the horizontal direction of the magnetic field lines experienced by the birds. The thick arrow indicates the expected orientation of the birds. The setting Sun and the three lines with double arrowheads indicate whether Sun and polarized light cues were available for calibration. *Figure and parts of the legend from Cochran et al. (2004).*

Cochran et al., 2004; Muheim et al., 2006a; Liu and Chernetsov, 2012). The only experiment with truly free-flying birds performed so far suggested that two species of North-American songbirds used the magnetic compass as their primary compass in midair during spring migration, but that they calibrate this compass on the basis of celestial cues during the sunset period (Cochran et al., 2004); however, this mechanism is not universal (Chernetsov et al., 2011). Other evidence suggests that polarized light cues might be crucial for this calibration (Muheim et al., 2006a); however, so far, it is not understood how a bird's eyes can detect polarized light.

In pigeon homing or any other map-related task, the interactions between different cues seem to be even more complicated. Homing pigeons have been shown to use olfactory cues (Papi, 1991; Wallraff and Andreae, 2000; Wallraff, 2001, 2005; Gagliardo et al., 2006, 2008, 2009), visual landmarks of various kinds (Guilford et al., 2004), outward journey information (reviews in Wiltschko and Wiltschko, 1995; Wallraff, 2005), and maybe magnetic cues (Walcott, 1991; Wiltschko and Wiltschko, 1995; Dennis et al., 2007; Wiltschko et al., 2010a) to estimate their position relative to home when released from a previously unknown location. The relative importance of these map cues is hotly debated, with many apparently contradictory results occurring in the literature. One reason may very well be that in one location, one type of cue may be particularly reliable whereas another cue is more reliable in a different location; therefore, the animals predominantly rely on different cues in different locations. The most convincing experiments performed to date involved cutting of the olfactory nerves or cutting of the ophthalmic branch of the trigeminal nerves in experienced and inexperienced pigeons. The experiments showed that homing pigeons tested around Pisa in Italy need intact olfactory nerves but not intact "magnetic" nerves (see Section 8.9) to home (Gagliardo et al., 2006, 2008, 2009). In procellariiform seabirds such as albatrosses and shearwaters, olfactory cues also seem to be much more important for navigation than magnetic cues (Mouritsen et al., 2003; Nevitt and Bonadonna, 2005; Bonadonna et al., 2005; Gagliardo et al., 2013).

8.9 HOW DO BIRDS SENSE THE EARTH'S MAGNETIC FIELD?

It is challenging to detect the weak geomagnetic field with biological materials. Considering the anatomical constraints and known structures found within small birds, careful models of putative sensory mechanisms often find it hard to explain how a 50,000 nT magnetic field can result in reliable signals in the presence of thermal fluctuations (kT) and other sources of noise. In fact, any biological mechanism that can, in principle, allow detection of 50,000 nT fields is noteworthy (Ritz et al., 2010; Mouritsen, 2013). Only three basic mechanisms are currently considered to be physically viable: (1) induction in highly sensitive electric sensors, (2) iron-mineral-based magnetoreception, and (3) radical-pair based magnetoreception.

8.10 THE INDUCTION HYPOTHESIS

Electromagnetic induction is the production of voltage across an electric conductor situated in a changing magnetic field or a conductor moving through a stationary magnetic field. Thus, in practical terms, if one has an electric wire and one moves this through a magnetic field, then a current will be generated in the wire. If this wire is ring or coil-shaped, directional sensitivity can be achieved. In biological tissues, one would need conductive, liquid-filled, ring-like structures of sufficient size and diameter to generate measurable electrical signals that can be picked up by an electrically sensitive receptor cell. For electromagnetic induction, Lorenzini ampullae are a concrete realization of an electrically sensitive cell operating in saltwater fish (von der Emde, 2013). Their function uses the fact that saltwater is electrically conductive, and aquatic animals could potentially use induction to sense the geomagnetic field (Kalmijn, 1981; Molteno and Kennedy, 2009). However, no strong evidence currently exists that fish actually use their electric sense to deduce information from the geomagnetic field (Kirschvink et al., 2010; Mouritsen, 2013). In land-based animals, it is difficult to imagine how induction could be used to sense the geomagnetic field because air has low conductivity; therefore, the needed structures would have to be realized inside of the animals themselves. In fact, biophysical considerations effectively eliminate induction as a potential source of magnetodetection in terrestrial animals: The required physiological structures filled with conductive liquid would be large and easily detectable, but no such structures have been reported (Kirschvink et al., 2010; Mouritsen, 2013). Thus, for terrestrial animals such as birds, another mechanism must be responsible for magnetoreception.

8.11 THE IRON-MINERAL-BASED HYPOTHESIS

When human beings want to use the direction of the geomagnetic field for orientation, we use a technical compass that is based on a needle made of magnetized iron or a magnetic iron compound that moves in the horizontal plane. Therefore, the first suggestion almost any human thinks of when one asks them how birds may detect the geomagnetic field is "Maybe they have little compass needles in their head." It is not surprising that this suggestion was also the first suggestion scientists came up with. A compass needle-like structure has been realized inside of magnetotactic bacteria (see Figure 8.6(A), Blakemore, 1975; Frankel and Blakemore, 1989). A chain of single-domain magnetite crystals (Blakemore, 1975; Frankel and Blakemore, 1989; Kirschvink et al., 2010) or other very similar iron oxides (Falkenberg

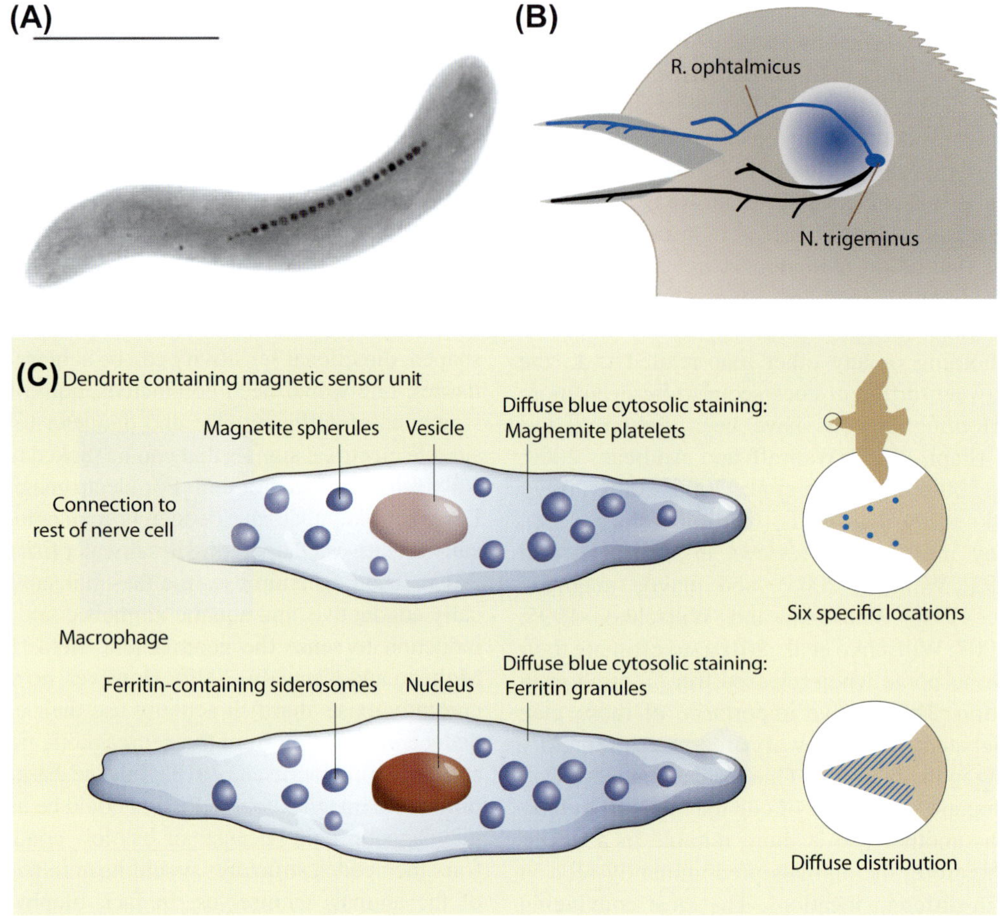

FIGURE 8.6 Iron-mineral structures in birds. (A) Transmission electron micrograph of the magnetotactic bacteria, *Magnetospirillum magnetotacticum*, showing the magnetosome chain inside of the cell. Scale bar: 1 μm *(Photograph © Richard B. Frankel)*. Magnetosomes would be the most straightforward solution for a magnetic field sensor in the nervous system of a bird, but, so far, magnetosomes have not been proven to occur in any bird. (B) Bird head schematically illustrating the anatomical location of the three branches of the trigeminal nerve. (C) Schematic drawing of iron mineral-containing structures in birds' upper beak illustrating the opposing interpretations of Fleissner et al. (2003) and Treiber et al. (2012). *Part (C) reproduced with permission from Mouritsen (2012), (B) and panel composition is reproduced from Mouritsen (2013).*

et al., 2010) would be the easiest realization of a small compass-needle-like structure inside of a bird, but other arrangements of iron-mineral crystals could also work as a magnetic field detector (Solov'yov and Greiner, 2009; Kirschvink et al., 2010). The iron-mineral crystals are expected to transduce the magnetic signal by opening or closing pressure-sensitive ion channels (Johnsen and Lohmann, 2005).

Many studies have documented the presence of magnetite or some other kind of iron mineral crystals in almost any animal, in which researchers have seriously looked for such crystals (e.g., in *Caenorhabditis elegans*, mollusks, insects, crustaceans, and various vertebrates; Mouritsen, 2013). However, the mere existence of iron mineral crystals or even magnetite does not represent significant evidence by itself that such structures have any relevance to magnetoreception (Mouritsen, 2013). Iron is an important element required for proper function of most organisms. Consequently, iron homeostasis is important, and iron mineral deposits may just be a way for an organism to get rid of excess iron. Therefore, only if iron mineral structures are found at consistent, specific locations and are associated with the nervous system do the iron mineral structures qualify as serious magnetosensory candidate structures (Mouritsen, 2013). The existence of magnetite crystal chains, which lead to a magnetically oriented swimming behavior in so-called "magnetotactic bacteria" (Blakemore, 1975; Frankel and Blakemore, 1989; Bazylinski and Frankel, 2004), unequivocally proves that living cells can, in principle, synthesize magnetite that will align with the geomagnetic field. However, the magnetite crystals in these bacteria are not part of an active sensory system; they only lead to passive alignment (Wiltschko and Wiltschko, 1995; Mouritsen, 2013).

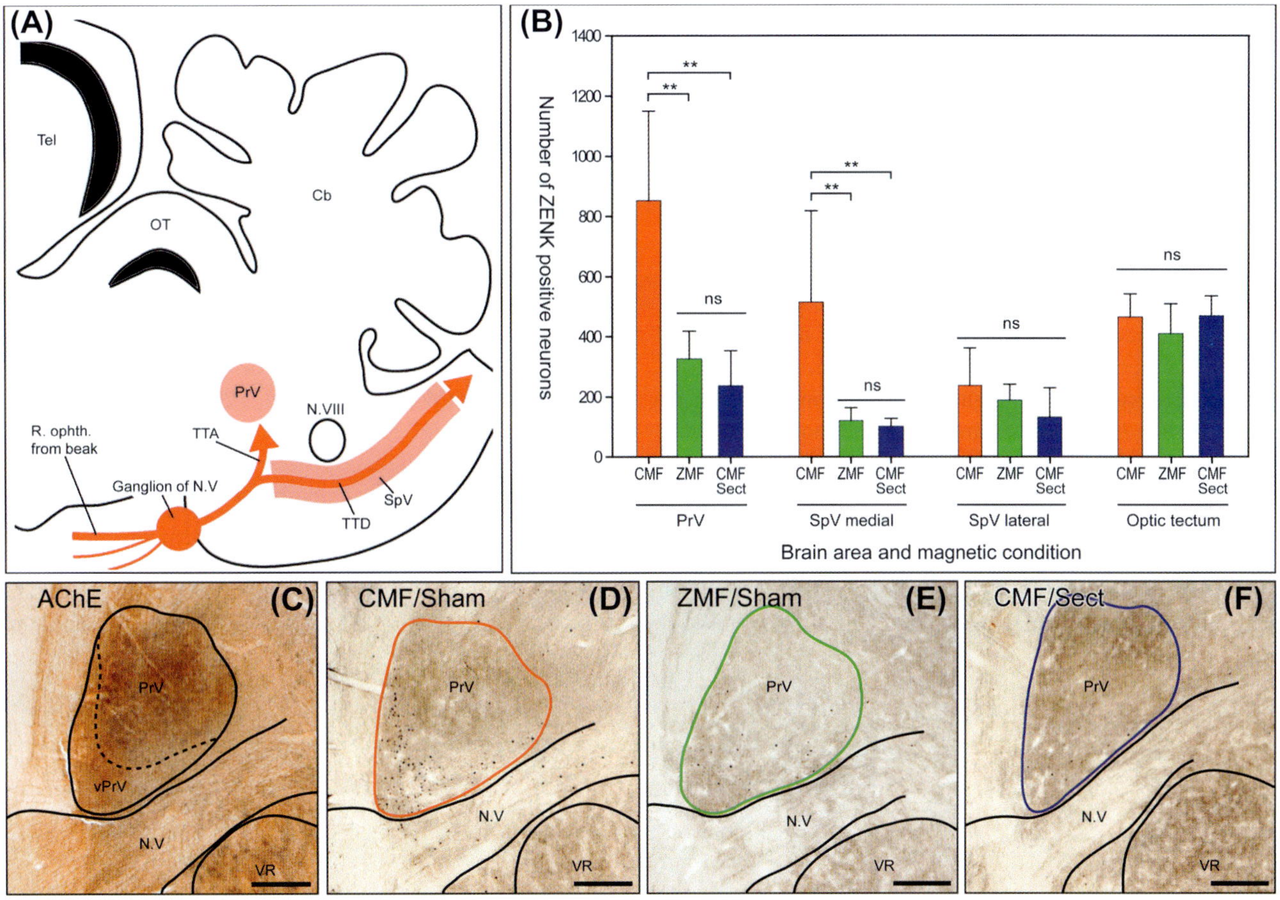

FIGURE 8.7 **There are magnetically activated neurons in the two hindbrain regions, PrV and SpV, which receive sensory input from the ophthalmic branch of the trigeminal nerve (V1) in birds.** (A) Anatomical overview of the termination of the trigeminal nerve in the avian hindbrain. (B–F) In birds with intact V1 (sham-sectioned nerves; (D)), stimulation with a changing magnetic field (CMF) leads to increased expression of the neuronal activity-dependent gene, *ZENK* (black dots in (D–F) are activated neuronal nuclei), in the hindbrain regions (PrV, shown in (C–F)) and SpV, that receive their primary input from the trigeminal nerve. This activation disappears when the magnetic field stimulus is absent (ZMF=zero magnetic field; (E)). The activation also disappears when the CMF is present but the birds had V1 bilaterally sectioned (CMF Sect; (F)). No magnetic field-dependent activation is seen in control regions such as the optic tectum (B). (C) Acetyl cholinesterase (AChE) is a good anatomical marker for identification of the borders of PrV. *From Heyers et al. (2010).*

The currently most promising, but not proven, active, iron-mineral-based magnetoreceptor candidate structures are those reported from the olfactory epithelium of fish (Walker et al., 1997; Eder et al., 2012). Elaborate iron-mineral-based structures thought to be magnetoreceptors have also been reported in the upper beak of birds (Fleissner et al., 2003; Falkenberg et al., 2010). However, recent findings suggest that these structures are macrophages involved in iron homeostasis (Treiber et al., 2012). Thus, at present, no convincingly documented, iron-mineral-based, magnetoreceptive candidate structures are known from birds (see Figure 8.6(C); Mouritsen, 2012).

Conditioning of birds to magnetic stimuli has also proven to be very difficult, and independent replication is rare. It has been reported that homing pigeons, *Colomba livia*, can be conditioned to respond to strong magnetic fields (Mora et al., 2004; Mora and Bingman, 2013). The conditioned response to a very strong magnetic field (c. 2 times the strength of the geomagnetic field) required intact trigeminal nerves (Mora et al., 2004). Consequently, pigeons seem to, in principle, be able to detect strong magnetic field changes via the ophthalmic branch of the trigeminal nerve. However, to use geomagnetic information for a map, animals must be sensitive to changes in the geomagnetic field, which are 3–5 orders of magnitude smaller than the anomalies used in the successful conditioning experiments. Using a very similar paradigm adapted for European robins and using weaker fields resulted in nicely conditioned responses to auditory stimuli but failed to produce a conditioned response to magnetic field stimuli (Kishkinev et al., 2012).

The ophthalmic branch of the trigeminal nerve (see Figure 8.6(B)) terminates in the principal (PrV) and spinal tract (SpV) nuclei of the trigeminal brainstem complex (Williams and Wild, 2001; Heyers et al., 2010; see Figure 8.7). Recent evidence shows that subpopulations of neurons in PrV and SpV in European robins (*Erithacus rubecula*), a night-migrating songbird, are activated by changing magnetic field stimuli but not by a zero magnetic field (Heyers et al., 2010). Furthermore, the activation in the changing magnetic field disappears when the ophthalmic branch of the trigeminal nerve is severed (see Figure 8.7(B–F)). These findings suggest that the ophthalmic branches of the trigeminal nerves carry magnetic information in birds (Heyers et al., 2010). However, the sensory origin (most likely iron-mineral-based) and the biological significance of the trigeminally mediated magnetic information are unclear at present (Mouritsen, 2012, see Figure 8.6(C)). Information from the ophthalmic branch of the trigeminal nerve is neither required nor sufficient for magnetic compass orientation in several night-migrating songbird species (Zapka et al., 2009, see Figure 8.8). Homing experiments with pigeons have shown that pigeons tested around Pisa, Italy, need intact olfactory nerves but not intact trigeminal nerves to home (Gagliardo et al., 2006, 2008, 2009).

The most likely function of the trigeminal nerve-related magnetic sense is to detect large-scale changes in magnetic field strength and/or magnetic inclination, which could be used to determine approximate position. A very recent set of experiments has shown that Eurasian Reed Warblers, *Acrocephalus scirpaceus*, are capable of correcting for a 1000 km eastward displacement (Chernetsov et al., 2008), but that this ability disappears when the ophthalmic branch of the trigeminal nerve is cut (Kishkinev et al., 2013, see Figure 8.3). In addition, experiments with night-migratory songbirds exposed to strong magnetic pulses, which are thought to disturb any magnetite-based magnetic sense for days to weeks but should not have any effect after the treatment itself on a light-dependent magnetoreception mechanism, support the idea that the magnetic map or signpost sense is iron mineral-based (Wiltschko et al., 2009; Holland, 2010; Holland and Helm, 2013).

It has also very recently been suggested that the avian lagena (a part of the birds' vestibular system) plays a role in magnetodetection (Wu and Dickman, 2011, 2012). Whether the lagena provides the primary magnetic information or gravity information to the magnetic sense is not yet clear, but if the seemingly very convincing electrophysiological data (Wu and Dickman, 2012) can be independently replicated,

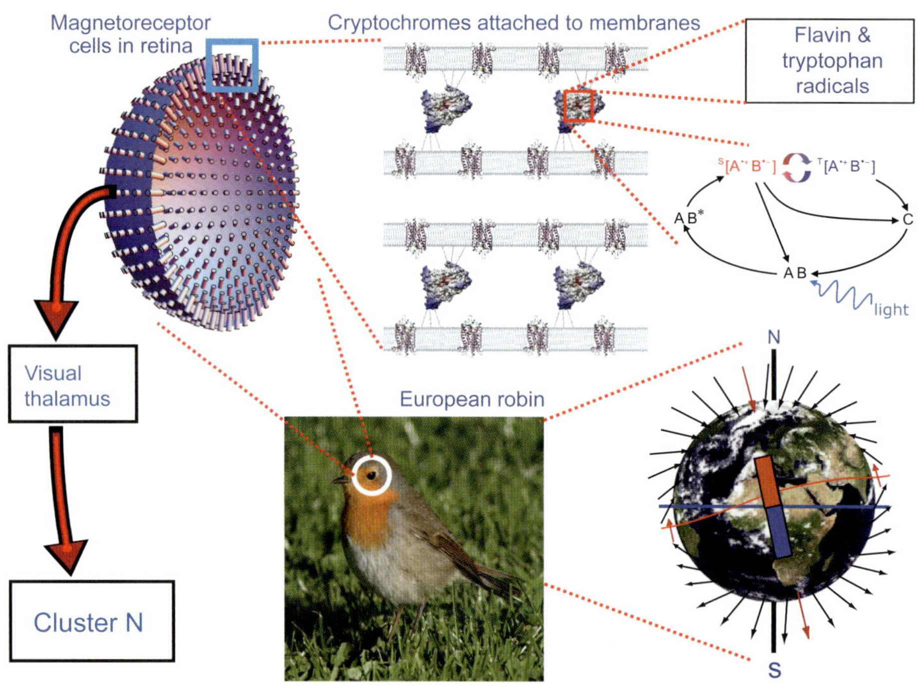

FIGURE 8.8 Summary of the proposed light-dependent magnetic compass sensing hypothesis in birds. Most of the experiments reviewed here were performed on European robins, *Erithacus rubecula (Photo © Henrik Mouritsen)*. The reference direction provided by the Earth's magnetic field is detected in the birds' eyes, where cryptochrome proteins are the most likely light-dependent magnetic sensory molecules. Light absorption is thought to generate long-lived flavin-tryptophan radical pairs within cryptochromes in the retina, the reaction yields of which are determined by the orientation of the molecule with respect to the geomagnetic field vector. If the cryptochromes were associated e.g. with the membrane disks of the outer segments of the photoreceptors, then an ordered structure could result, and different reaction yields in different parts of the retina could be compared to provide a visual impression of the compass bearing (see Figure 8.9). Light-dependent magnetic compass information is transmitted from the retina through the optic nerve to the visual thalamus and from there to Cluster N in the forebrain via the thalamofugal visual pathway (Figure 8.11). If Cluster N is destroyed, European robins can no longer use their magnetic compass (Figure 8.12). The illustration of the cryptochromes bound to photoreceptor membranes is modified from Solov'yov et al. (2010). The reaction scheme is modified from Rodgers and Hore (2009). *Figure and parts of the legend from Mouritsen and Hore (2012).*

then the role of the lagena in magnetoreception is very significant indeed, and the vestibular brainstem nuclei would be a very important processing station for magnetic field information.

8.12 THE LIGHT-DEPENDENT HYPOTHESIS

The magnetic compass behavior of newts (Phillips and Borland, 1992) and birds (Wiltschko et al., 1993; Muheim et al., 2002; Wiltschko et al., 2010b) is dependent on the wavelengths of light being available during behavioral tests. Already in the late 1970s, theoretical considerations led Klaus Schulten to suggest that chemical reactions in photosensitive molecules could form the basis of a magnetic compass sense (Schulten et al., 1978).

The principles of the suggested light-dependent magnetic sensing mechanism are illustrated in Figure 8.8. A light-sensitive molecule (D) absorbs light and uses the light energy to transfer an electron to an acceptor (A); thereby, a radical pair is produced. If this radical pair is long-lived (>1 μs), then it can, depending on the spin of the electrons, exist in one of two states—a singlet state (spins antiparallel) or a triplet state (spins parallel). It is known from chemistry that singlet and triplet states have different chemical properties; thus, they often result in different chemical end products. Earth-strength magnetic fields can theoretically affect this statistical equilibrium and thereby modulate a presently unknown biochemical pathway (Schulten et al., 1978; Ritz et al., 2000, 2010; Rodgers and Hore, 2009; Hore, 2012).

How can we imagine that a bird using a light-induced, radical-pair mechanism would detect the magnetic field? It is possible that a virtual visual image would literally enable birds to "see" the direction of the magnetic field lines (e.g., Ritz et al., 2000, 2010; Solov'yov et al., 2010). If one makes the simple assumption that the sensory molecules are oriented perpendicularly to the eyeball (see Figure 8.9), then the half ball shape of the retina would mean that molecules oriented in all axial directions would occur (Ritz et al., 2000; Mouritsen, 2013). If a bird looks in the direction of the magnetic field lines, then in the line of sight, the retinal molecules would be parallel to the magnetic field and this could lead to a light pixel. At the edge of the eye, the molecules would be perpendicular to the magnetic field, and this could lead to darker pixels. In between, the molecules would be oriented at different angles relative to the magnetic field, and various shades of gray pixels could appear. Altogether, this could lead to a virtual image looking somewhat like the one shown to the right in Figure 8.9 (Ritz et al., 2000; Mouritsen, 2013). This pattern is only for illustrative purposes of the principle. We have much too little information available at present to know what an actual magnetically modulated light pattern seen by a bird would look like (Mouritsen, 2013). The patterns might become easier to see when they move across the retina during so-called head scan behavior (Mouritsen et al., 2004b; Ritz et al., 2010).

If the radical-pair mechanism is responsible for magnetoreception, then it means that it is based on a quantum mechanical effect (Rodgers and Hore, 2009; Ball, 2011; Mouritsen and Hore, 2012; Hore, 2012; Hogben et al., 2012; Solov'yov et al., 2014; Engels et al., 2014). In fact, it might be the only sensory mechanism in biology to be inherently quantum in nature. Critics of the radical-pair mechanism have pointed out that the interaction energy between a geomagnetic field and a radical is typically several orders of magnitude below the background thermal energy, k_BT (Kirschvink et al., 2010). At first glance this might seem like a problem, but because

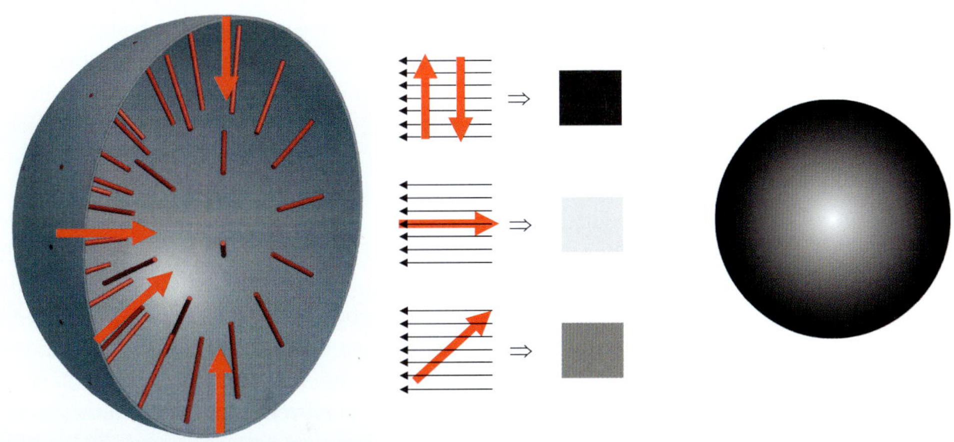

FIGURE 8.9 **How the light-dependent, radical pair-based, magnetic sensing mechanism could lead to perception of a visual image.** Principal illustration suggesting how birds could, in principle, convert a magnetic stimulus into a putative visual image. Left: 3D illustration of the half-sphere of an eyeball. Red pins simulate cryptochrome orientation all pointing toward the center of the eyeball. If a bird would be looking in the direction of the magnetic field lines, one could imagine that the bird would see a pattern similar to the one illustrated on the right because the light sensitivity of one or more cryptochromes would depend on their orientation relative to the axis of the magnetic field lines. *Redrawn after Mouritsen (2013) inspired by Ritz et al. (2000).*

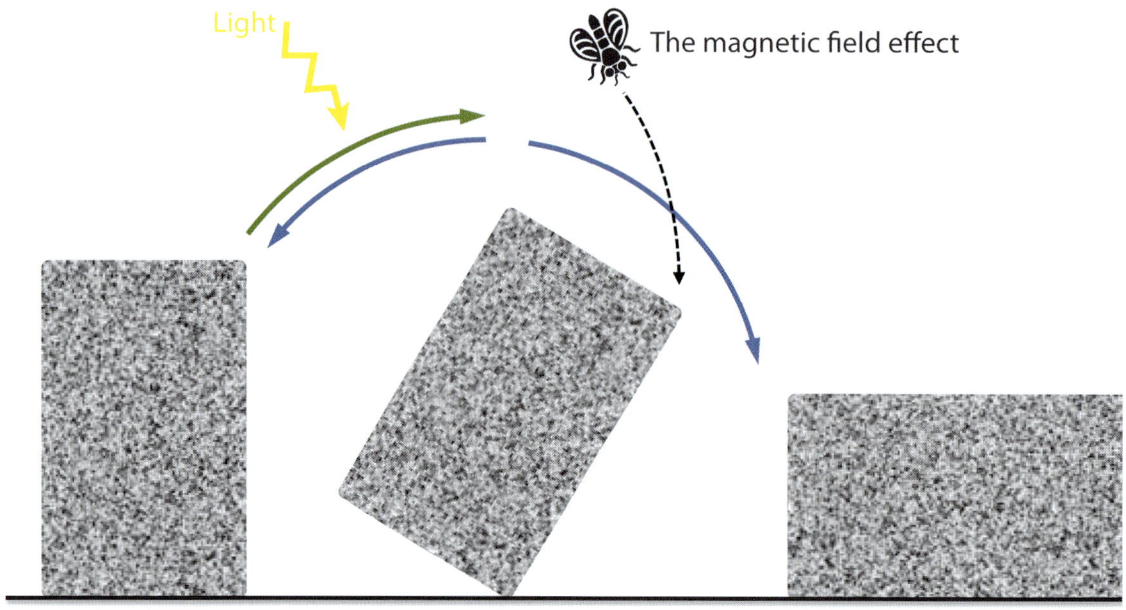

FIGURE 8.10 Granite block analogy of light-dependent magnetoreception. A granite block analogy can help understand how a radical pair-based mechanism can theoretically be used to sense Earth-strength magnetic fields, although the energy exerted by the magnetic field on the radical is much lower than thermal energy, k_BT. Imagine a granite block standing on one of its sides. If a fly lands on the granite block while it is in this position, then there is no way the fly can make the granite block flip over. In fact, quite a lot of energy is needed to move the granite block onto one of its corners, but once it is on one of its corners, a very small amount of energy can influence which way the granite block falls. Now, even a fly landing on the granite block may make it flip over to one or the other side. In the light-dependent, radical pair-based magnetoreception hypothesis, light has much more energy than the magnetic field. It is the absorption of light that brings a photoreceptor molecule (probably a cryptochrome) into an exited state (moves the granite block onto its corner in the analogy), which is then highly sensitive to even very small magnetic field effects (the fly landing onto the granite block in the analogy). *After Hore (2011).*

the spins are not in equilibrium, this is not a fundamental problem. Very weak magnetic interactions can essentially affect radical pair reactions for three reasons (Hore, 2011): (1) radical pair chemistry is controlled by the electron spins in the radicals, (2) the electron spins are not at thermal equilibrium, and (3) the electron spins behave quantum mechanically. Weak magnetic fields affect the coherent behavior of the electron spins in a fundamentally quantum manner in which k_BT plays no role. Instead of comparing interaction energies to k_BT, one must compare the time required for a magnetic interaction to have an effect with the time required for the system to reach thermal equilibrium (Hore, 2011). If the former is shorter than the latter, then the magnetic field can have an effect (Hore, 2011). As an excellent analogy (Hore, 2011), one can imagine a fly and a rectangular granite block. If the granite block is standing on one of its sides (conventional physics, equilibrium), then the fly has no chance to flip the granite block, but if the granite block is balancing on one of its corners (quantum mechanics, nonequilibrium), then, depending on where the fly lands, it might flip the granite block to one or the other side (Figure 8.10).

Which molecule can be responsible for light-dependent magnetoreception? Opsins cannot function as radical-pair-based magnetoreceptors because opsins use the light energy to change a chemical bond, not to transfer an electron. The only currently known photoreceptor molecules found in vertebrates that can use light energy to form long-lived radical pairs are the cryptochromes (Ahmad and Cashmore, 1993; Cashmore et al., 1999; Ritz et al., 2000, 2010; Giovani et al., 2003; Liedvogel et al., 2007a; Biskup et al., 2009; Rodgers and Hore, 2009; Liedvogel and Mouritsen, 2010). Some cryptochromes are known to be involved in circadian clocks (Cashmore et al., 1999; Sancar, 2003). However, in birds, more cryptochromes than the ones thought to be involved in the clock occur; therefore, it is easy to imagine that they can play a role in other biochemical processes (Liedvogel and Mouritsen, 2010). Cryptochromes are related to the DNA repair enzymes called photolyases (Cashmore et al., 1999; Sancar, 2003) and consist of a photolyase homology region and a C-terminal end, which varies greatly between different cryptochromes (Cashmore et al., 1999; Sancar, 2003; Müller and Carell, 2009; Liedvogel and Mouritsen, 2010). The C-terminal is thought to be involved in binding cryptochromes to currently unknown interaction partners (Sancar, 2003; Liedvogel and Mouritsen, 2010; Mouritsen and Hore, 2012). Cryptochromes noncovalently bind the cofactor flavin. The light-induced electron transfer is thought to take place between the flavin and three tryptophane residues within the cryptochrome protein (Gindt et al., 1999; Biskup et al., 2009; Rodgers and Hore, 2009; Solov'yov et al., 2012).

Cryptochromes are predominantly found within photoreceptor cells and ganglion cells in the eyes of birds (Mouritsen et al., 2004a; Möller et al., 2004; Niessner et al., 2011), and cryptochromes are currently the only seriously considered candidate molecules for radical-pair-based magnetoreception in birds (Mouritsen and Ritz, 2005; Rodgers and Hore, 2009; Ritz et al., 2010; Mouritsen and Hore, 2012).

After the suggestion of Klaus Schulten (Schulten et al., 1978), it was shown that the compass orientation behavior of night-migrating songbirds is influenced by the color (i.e., wavelengths) of the light available in the room where the orientation tests are performed (Wiltschko et al., 1993, 2010b). This wavelength dependence is difficult to explain if the eyes and/or pineal organ are not somehow involved in the magnetic compass. In birds, the pineal organ is not needed for magnetic compass orientation (Schneider et al., 1994), but photoreceptor molecules in the pineal organ seem to be essential for magnetic compass orientation in newts (Phillips et al., 2001).

It has been reported that oscillating magnetic fields at specific resonance frequencies in the low-megahertz range disrupt the magnetic compass orientation capabilities of night-migratory songbirds (Ritz et al., 2004, 2009). A recent double-blinded set of experiments has indicated that the disruptive effects of low-megahertz range electromagnetic fields are real, but that they are not limited to specific resonance frequencies (Engels et al., 2014). Furthermore, Engels et al. (2014) could show that anthropogenic electromagnetic noise, omnipresent in most urban environments where birds and humans live, disrupts magnetic compass orientation in migratory European Robins. The intensities of the disruptive fields are ca. 1000 times lower than the current WHO guideline limits for human exposure (Engels et al., 2014). Such effects are still difficult to understand from a theoretical perspective, but they are likely to be diagnostic for the involvement of a fundamentally quantum mechanical mechanism in the birds' magnetic compass sense (Ritz et al., 2009; Mouritsen & Hore, 2012; Engels et al., 2014).

On the molecular level, it has been shown that putatively magnetosensitive cryptochrome molecules exist in the retina of many vertebrates including migratory birds (Mouritsen et al., 2004a; Möller et al., 2004; Liedvogel and Mouritsen, 2010; Niessner et al., 2011). Furthermore, cryptochromes from migratory Garden Warblers (*Sylvia borin*) have been shown to form long-lived radical pairs upon light excitation (Liedvogel et al., 2007a), and effects of Earth-strength magnetic fields on a radical pair reaction in an artificially produced molecule mimicking the reaction principle thought to take place in cryptochromes have supported the theoretical feasibility of the suggested mechanism (Maeda et al., 2008; reviewed in Mouritsen and Hore, 2012).

On the neuroanatomical level, a region named Cluster N (Figure 8.11) is by far the most active forebrain region when night-migrating birds perform magnetic compass orientation, and this activation disappears when the birds' eyes are covered (Mouritsen et al., 2005; Feenders et al., 2008; Zapka et al., 2010). Cluster N consists of parts of the hyperpallium and the dorsal mesopallium (Jarvis et al., 2013) and is the lateral-most part of the visual Wulst in European robins because Cluster N receives its neuronal input from the eyes via the thalamofugal visual pathway (Heyers et al., 2007). The presence and activation of Cluster N at night has been independently replicated in another night-migratory songbird, the Black-Headed Bunting (*Emberiza melanocephala*) (Rastogi et al., 2011). Could Cluster N be a processing center of light-dependent magnetic compass information?

Double-blind experiments with European robins have shown that birds with bilateral Cluster N lesions were unable to orient using their magnetic compass (Zapka et al., 2009, see Figure 8.12). In contrast, sham Cluster N lesions or bilateral sections of the ophthalmic branch of the trigeminal nerves did not influence the robins' ability to use their magnetic compass for orientation (Zapka et al., 2009, see Figure 8.12). Cluster N lesions only affect the magnetic compass because Cluster N lesioned robins orient well using their Sun and star compasses (Zapka et al., 2009, see Figure 8.12). These data (1) show that Cluster N is required for magnetic compass orientation in this species, (2) indicate that Cluster N may be specifically involved in processing magnetic compass information, (3) strongly suggest that a vision-mediated mechanism underlies the magnetic compass in this migratory songbird, (4) indicate that input from the lagena is not sufficient for magnetic compass orientation in robins, and (5) show that the proposed magnetic input to the brain transmitted via the trigeminal nerve is neither necessary nor sufficient for magnetic compass orientation of European robins tested in an orientation cage (Zapka et al., 2009; Mouritsen, 2013). The exact role of Cluster N within the magnetic compass information processing circuit has not been determined, but the existing results raise the distinct possibility that this small part of the visual system enables birds to "see" magnetic compass information (Mouritsen, 2013).

Do these results exclude the possibility that iron-mineral-based and/or trigeminally mediated and/or lagena-mediated magnetoreception exists? Absolutely not! In birds, iron-mineral-based magnetoreception may very well exist, and magnetic field-dependent neuronal activation in trigemino-recipient and lagena-recipient regions has been documented (see previous section). Trigeminally and lagena-mediated magnetoreception just does not seem to be the primary mechanism for the magnetic compass of night-migratory songbirds (Zapka et al., 2009), but it could be a primary source for magnetic positional information (Mora et al., 2004; Kishkinev et al., 2013). In fact, it is likely that light-mediated, radical pair-based magnetoreception and iron-mineral-based magnetoreception

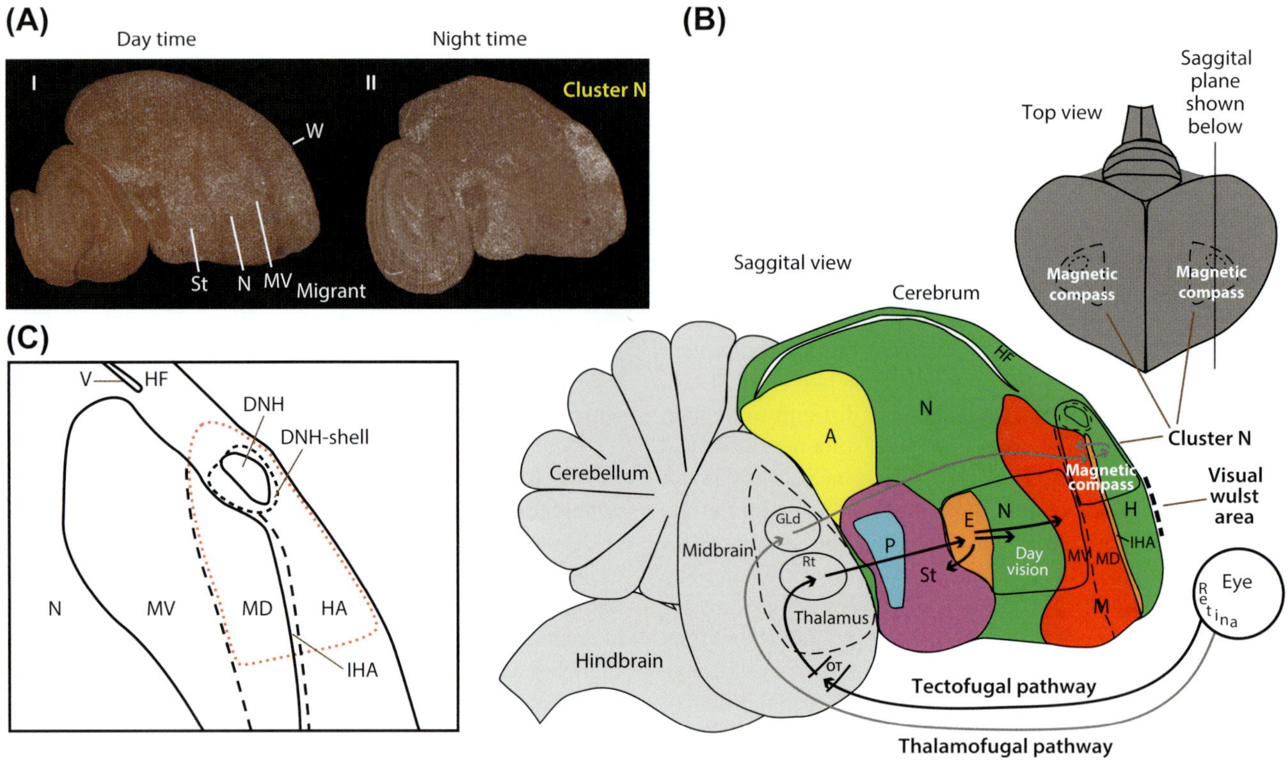

FIGURE 8.11 **Cluster N.** (A) Cluster N is the most active brain area when migratory birds perform magnetic sensing and/or compass orientation at night, and Cluster N is required for magnetic compass orientation (see Figure 8.12). (B) Cluster N is a part of the visual Wulst and receives input from the eyes via the thalamofugal visual pathway (Heyers et al., 2007). Top view of the brain in gray indicates the medial-lateral and the frontal-caudal extent of Cluster N and the DNH and DNH-shell. (C) Cluster N is a functional unit consisting of a part of the hyperpallium, a part of the dorsal mesopallium (Jarvis et al., 2013), and a nucleus embedded within the hyperpallium and named DNH with a shell of cells around the DNH. Anatomy: A = arcopallium, P = pallidum, E = entopallium, St = striatum, N = nidopallium, M = mesopallium, MD = mesopallium dorsale, MV = mesopallium ventrale, H = hyperpallium, v = ventricle, OT = optic tectum, HF = hippocampal formation, IHA = HI = interstitial region of the hyperpallium intercalatum, DNH = dorsal nucleus of the hyperpallium, DNH-shell = shell around the DNH, W = visual Wulst, LGd = Lateral geniculate nucleus, dorsal part, Rt = nucleus rotundus. Scale bar = 0.5 mm. *From Mouritsen (2013) after Mouritsen et al. (2005).*

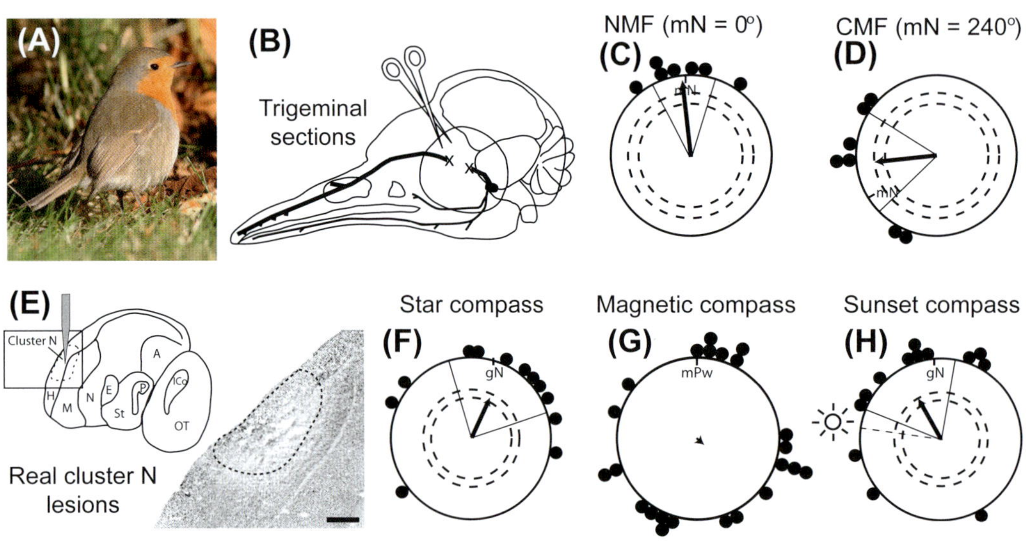

FIGURE 8.12 **The brain region Cluster N is necessary for magnetic compass orientation behavior, but not for star and Sun compass orientation. The trigeminal nerve is neither necessary nor sufficient for magnetic compass orientation in European robins.** (A) The European robin *(Photo © Henrik Mouritsen)*. (B–D) Bilateral sectioning of the ophthalmic branch of the trigeminal nerve (B) does not affect the birds' magnetic compass orientation capabilities (C, D; mN = magnetic North). (E–H) Bilateral chemical lesions of Cluster N (E) destroyed the magnetic compass capabilities of the birds (G), whereas star compass orientation in a planetarium (F) and Sun compass orientation outdoors with view of the setting Sun (H) was unaffected by Cluster N lesions. The circular diagrams are explained in the legend of Figure 8.3. (E) A schematic drawing and an example part of a brain section saggitally cut through the center of Cluster N and stained with a neuronal marker. Scale bar = 500 μm. Rostral is left, caudal is right. Note that the tissue where Cluster N should have been in the lesioned bird (E) is destroyed. Anatomy: A = arcopallium, E = entopallium, H = hyperpallium, ICo = intercollicular complex, M = mesopallium, MD = mesopallium dorsale, MV = mesopallium ventrale, N = nidopallium, OT = optic tectum, P = pallidum. *From Mouritsen (2013) after Zapka et al. (2009).*

mechanisms exist side by side in several animal species and that they may provide the animals with different types of magnetic information (Wiltschko and Wiltschko, 2007; Mouritsen and Hore, 2012).

As so often in biology, when there are two hypotheses how something works, it often turns out that both of them are correct to a certain degree. Furthermore, seemingly unnecessary redundancy seems to be a very common occurrence in biology, probably because organisms that can perform a function in several ways will be more robust to changes and thus be favored by evolution (Mouritsen, 2013).

8.13 IRREPRODUCIBLE RESULTS AND THE URGENT NEED FOR INDEPENDENT REPLICATION

Magnetic sense research is strongly influenced by several claims that nobody has ever been able to independently replicate. This is particularly true for electrophysiological evidence (Semm and Demaine, 1986; Beason and Semm, 1987), but there are also many other examples of contradicting or irreproducible results in the literature, including all claims that humans have a magnetic sense (Westby and Partridge, 1986; Baker, 1989) and the claim that the magnetic compass of birds should only be located in their right eye (Wiltschko et al., 2002, 2003; Liedvogel et al., 2007b; Stapput et al., 2010; Hein et al., 2010, 2011; Engels et al., 2012). These problems with reproducibility do not necessarily mean that the original claims were wrong. However, it means that any result in magnetoreception—and in any other field for that matter—should be treated with caution until a given finding has been independently replicated.

This lack of reproducibility in magnetic sense-related research is unfortunately accompanied by an almost complete lack of double-blind procedures. Considering its history and the fact that humans have no intuitive feel for magnetic stimuli (and therefore are less likely to detect even obvious artifacts), double-blind procedures should become the standard, and studies representing the first independent, double-blind replication of key findings in magnetoreception are as important as the original finding (Mouritsen, 2013).

8.14 WHERE DO WE GO FROM HERE?

Although the magnetic senses are still not completely understood, many studies from different fields support the iron-mineral-based and the light-dependent magnetoreception hypotheses. However, fundamental questions remain in all relevant fields.

For example, functional understanding of any particular iron-mineral-based structure proven to be involved in an active sensory system is lacking (Mouritsen, 2012; Mouritsen and Hore, 2012). Likewise, we yet have to understand biophysically how nature designed radical-pair receptors so that they can be sensitive to Earth-strength magnetic fields at physiological temperatures, a feat that has been approximated, but not yet fully accomplished, in man-made radical-pair reactions (Maeda et al., 2008; Rodgers and Hore, 2009). Furthermore, studies at the protein level suggest that bird cryptochromes have properties optimal for magnetic sensing, such as formation of long-lived radical pairs (Liedvogel et al., 2007a). However, we yet have to demonstrate Earth-strength magnetic field effects on bird cryptochromes at the protein level and *in vivo*.

On the neuroanatomical level, we have just begun to explore the brain circuits processing magnetic information, but we are still far from understanding how a bird gets from the detection of magnetic information to a directional choice, which is made based on integration of information from multiple sensory systems. In addition, so far, none of the reported responses of single neurons to magnetic stimuli have been independently replicated. Even at the behavioral level, where most studies about magnetic senses have been published, a clear separation of experimental parameters has proven difficult, and many behaviors appear to be multimodal, or at least modulated by other modalities, such as vision and olfaction (Mouritsen, 2013).

In conclusion, magnetoreception is an important part of life for birds and a wide variety of other animals, and there are still many opportunities to perform new, groundbreaking research on the molecules, cells, and neural processes underlying any kind of magnetoreception.

REFERENCES

Able, K.P., 1996. The debate over olfactory navigation by homing pigeons. Journal of Experimental Biology 199, 121–124.

Adair, R.K., 1991. Constraints on biological effects of weak extremely-low frequency electromagnetic fields. Physical Review A 43, 1039–1048.

Ahmad, M., Cashmore, A.R., 1993. HY4 gene of *A. thaliana* encodes a protein with characteristics of a blue-light photoreceptor. Nature 366, 162–166.

Baker, R.R., 1989. Human Navigation and Magnetoreception. Manchester University Press, Manchester.

Ball, P., 2011. Physics of life: the dawn of quantum biology. Nature 474, 272–274.

Bazylinski, D.A., Frankel, R.B., 2004. Magnetosome formation in prokaryotes. Nature Reviews Microbiology 2, 217–230.

Beason, R.C., Semm, P., 1987. Magnetic responses of the trigeminal nerve system of the bobolink (*Dolichonyx oryzivorus*). Neuroscience Letters 80, 229–234.

Berthold, P., 1991. Spatiotemporal programmes and genetics of orientation. In: Berthold, P. (Ed.), Orientation in Birds. Birkhäuser Verlag, Basel, pp. 86–105.

Biskup, T., Schleicher, E., Okafuji, A., Link, G., Hitomi, K., Getzoff, E.D., Weber, S., 2009. Direct observation of a photoinduced radical-pair in a cryptochrome blue-light photoreceptor. Angewandte Chemie International Edition 48, 404–407.

Blakemore, R., 1975. Magnetotactic bacteria. Science 190, 377–379.

Bonadonna, F., Bajzak, C., Benhamou, S., Igloi, K., Jouventin, P., Lipp, H.P., Dell'Omo, G., 2005. Orientation in the wandering albatross: interfering with magnetic perception does not affect orientation performance. Proceedings of the Royal Society London B 272 (1562), 489–495.

Cashmore, A.R., Jarillo, J.A., Wu, Y.-J., Liu, D., 1999. Cryptochromes: blue light receptors for plants and animals. Science 284, 760–765.

Chernetsov, N., Kishkinev, D., Mouritsen, H., 2008. A long-distance avian migrant compensates for longitudinal displacement during spring migration. Current Biology 18, 188–190.

Chernetsov, N., Kishkinev, D., Kosarev, V., Bolshakov, C.V., 2011. Not all songbirds calibrate their magnetic compass from twilight cues: a telemetry study. Journal of Experimental Biology 214, 2540–2543.

Cochran, W.W., Mouritsen, H., Wikelski, M., 2004. Migrating songbirds recalibrate their magnetic compass daily from twilight cues. Science 304, 405–408.

Coriolis, G., 1835. Sur les équations du mouvement relatif des systèmes de corps. Journal de l'École polytechnique 15, 142–154.

Courtillot, V., Le Mouël, J.L., 1988. Time variations of the Earth's magnetic field: from daily to secular. Annual Review of Earth and Planetary Science 16, 389–476.

Dennis, T.E., Rayner, M.J., Walker, M.M., 2007. Evidence that pigeons orient to geomagnetic intensity during homing. Proceedings of the Royal Society B 274, 1153–1158.

Drost, R., 1938. Über den Einfluss von Verfrachtungen zur Herbstzugzeit auf den Sperber, *Accipiter nisus* (L.). Zugleich ein Beitrag zur Frage nach der Orientierung der Vögel auf dem Zuge ins Winterquartier (Rouen 1938) Proceedings of the International Ornithology Congress 9, 502–521.

Eder, S.H.K., Cadiou, H., Muhamad, A., McNaughton, P.A., Kirschvink, J.L., Winklhofer, M., 2012. Magnetic characterization of isolated candidate vertebrate magnetoreceptor cells. Proceedings of the National Academy of Sciences of the United States of America 109, 12022–12027.

Emlen, S.T., 1975. The stellar-orientation system of a migratory bird. Scientific American 233, 102–111.

Emlen, S.T., Emlen, J.T., 1966. A technique for recording migratory orientation of captive birds. The Auk 83, 361–367.

Engels, S., Hein, C.M., Lefeldt, N., Prior, H., Mouritsen, H., 2012. Night-migratory songbirds possess a magnetic compass in both eyes. PLoS One 7 (9), e43271.

Engels, S., Schneider, N.-L., Lefeldt, N., Hein, C.M., Zapka, M., Michalik, A., Elbers, D., Kittel, A., Hore, P.J., Mouritsen, H., 2014. Anthropogenic electromagnetic noise disrupts magnetic compass orientation in a migratory bird. Nature 509, 353–356.

Falkenberg, G., Fleissner, G., Schuchardt, K., Kuehbacher, M., Thalau, P., Mouritsen, H., Heyers, D., Wellenreuther, G., Fleissner, G., 2010. Avian magnetoreception: elaborate iron mineral containing dendrites in the upper beak seem to be a common feature of birds. PLoS One 5 (2), e9231.

Feenders, G., Liedvogel, M., Rivas, M., Zapka, M., Horita, H., Hara, E., Wada, K., Mouritsen, H., Jarvis, E.D., 2008. Molecular mapping of movement-associated areas in the avian brain: a motor theory for vocal learning origin. PLoS One 3 (3), e1768.

Fleissner, G., Holtkamp-Rötzler, E., Hanzlik, M., Winklhofer, M., Fleissner, G., Petersen, N., Wiltschko, W., 2003. Ultrastructural analysis of a putative magnetoreceptor in the beak of homing pigeons. Journal of Comparative Neurology 458, 350–360.

Frankel, R.B., Blakemore, R.P., 1989. Magnetite and magnetotaxis in microorganisms. Bioelectromagnetics 10, 223–237.

Fransson, T., Jakobsson, S., Johansson, P., Kullberg, C., Lind, J., Vallin, A., 2001. Bird migration – magnetic cues trigger extensive refuelling. Nature 414, 35–36.

Gagliardo, A., Ioalè, P., Savini, M., Wild, J.M., 2006. Having the nerve to home: trigeminal magnetoreceptor versus olfactory mediation of homing in pigeons. Journal of Experimental Biology 209, 2888–2892.

Gagliardo, A., Ioalè, P., Savini, M., Wild, M., 2008. Navigational abilities of homing pigeons deprived of olfactory or trigeminally mediated magnetic information when young. Journal of Experimental Biology 211, 2046–2051.

Gagliardo, A., Ioalé, P., Savini, M., Wild, M., 2009. Navigational abilities of adult and experienced homing pigeons deprived of olfactory or trigeminally mediated magnetic information. Journal of Experimental Biology 212, 3119–3124.

Gagliardo, A., Bried, J., Lambardi, P., Luschi, P., Wikelski, M., Bonadonna, F., 2013. Oceanic navigation in Cory's shearwaters: evidence for a crucial role of olfactory cues for homing after displacement. Journal of Experimental Biology 216, 2798–2805.

Gindt, Y.M., Vollenbroek, E., Westphal, K., Sackett, H., Sancar, A., Babcock, G.T., 1999. Origin of the transient electron paramagnetic resonance signals in DNA photolyase. Biochemistry 38, 3857–3866.

Giovani, B., Byrdin, M., Ahmad, M., Brettel, K., 2003. Light-induced electron transfer in a cryptochrome blue-light photoreceptor. Nature Structural Biology 10, 489–490.

Guilford, T., Roberts, S., Biro, D., Rezek, I., 2004. Positional entropy during pigeon homing II: navigational interpretation of Bayesian latent state models. Journal of Theoretical Biology 227, 25–38.

Hagstrum, J.T., 2013. Atmospheric propagation modeling indicates homing pigeons use loft-specific infrasonic 'map' cues. Journal of Experimental Biology 216, 687–699.

Harris, S.-R., Henbest, K.B., Maeda, K., Pannell, J.R., Timmel, C.R., Hore, P.J., Okamoto, H., 2009. Effect of magnetic fields on cryptochrome-dependent responses in *Arabidopsis thaliana*. Journal of the Royal Society Interface 6, 1193–1205.

Hein, C.M., Zapka, M., Heyers, D., Kutzschbauch, S., Schneider, N.-L., Mouritsen, H., 2010. Night-migratory garden warblers can orient with their magnetic compass using the left, the right or both eyes. Journal of the Royal Society Interface 7, S227–S233.

Hein, C.M., Engels, S., Kishkinev, D., Mouritsen, H., 2011. Robins have a magnetic compass in both eyes. Nature 471, E11. http://dx.doi.org/10.1038/nature09875.

Henshaw, I., Fransson, T., Jakobsson, S., Kullberg, C., 2010. Geomagnetic field affects spring migratory direction in a long distance migrant. Behavioral Ecology and Sociobiology 64, 1317–1323.

Heyers, D., Manns, M., Luksch, H., Güntürkün, O., Mouritsen, H., 2007. A visual pathway links brain structures active during magnetic compass orientation in migratory birds. PLoS One 2 (9), e937.

Heyers, D., Zapka, M., Hoffmeister, M., Wild, J.M., Mouritsen, H., 2010. Magnetic field changes activate the trigeminal brainstem complex in a migratory bird. Proceedings of the National Academy of Sciences of the United States of America 107, 9394–9399.

Hogben, H.J., Biskup, T., Hore, P.J., 2012. Entanglement and sources of magnetic anisotropy in radical pair-based avian magnetoreceptors. Physical Review Letters 109, 220501.

Holland, R.A., 2010. Differential effects of magnetic pulses on the orientation of naturally migrating birds. Journal of the Royal Society Interface 7, 1617–1625.

Holland, R.A., 2014. True navigation in birds: from quantum physics to global migration. Journal of Zoology 293, 1–15.

Holland, R.A., Helm, B., 2013. A strong magnetic pulse affects the precision of departure direction of naturally migrating adult but not juvenile birds. Journal of the Royal Society Interface 10, 20121047.

Hore, P.J., 2011. The quantum robin. Navigation News 2011 (10), 19–21.

Hore, P.J., 2012. Are biochemical reactions affected by weak magnetic fields? Proceedings of the National Academy of Sciences of the United States of America 109, 1357–1358.

Jarvis, E.D., Yu, J., Rivas, M.V., Horita, H., Feenders, G., Whitney, O., Jarvis, S.C., Jarvis, E.R., Kubikova, L., Puck, A.E.P., Siang-Bakshi, C., Martin, S., McElroy, M., Hara, E., Howard, J., Pfenning, A., Mouritsen, H., Chen, C.-C., Wada, K., 2013. Global view of the functional molecular organization of the avian cerebrum: mirror images and functional columns. Journal of Comparative Neurology 521, 3614–3665.

Johnsen, S., Lohmann, K.J., 2005. The physics and neurobiology of magnetoreception. Nature Reviews Neuroscience 6, 703–712.

Kalmijn, A.J., 1981. Biophysics of geomagnetic field detection. IEEE Transactions on Magnetics 17, 1113–1124.

Kirschvink, J.L., 1991. Uniform magnetic fields and double-wrapped coil systems: improved techniques for the design of bioelectromagnetic experiments. Bioelectromagnetics 13, 401–411.

Kirschvink, J.L., Winklhofer, M., Walker, M.M., 2010. Biophysics of magnetic orientation: strengthening the interface between theory and experimental design. Journal of the Royal Society Interface 7, 179–191.

Kishkinev, D., Chernetsov, N., Mouritsen, H., 2010. A double clock or jetlag mechanism is unlikely to be involved in detection of east-west displacements in a long-distance avian migrant. The Auk 127, 773–780.

Kishkinev, D., Mouritsen, H., Mora, C.V., 2012. An attempt to develop an operant conditioning paradigm to test for magnetic discrimination behaviour in a migratory songbird. Journal of Ornithology 153, 1165–1177.

Kishkinev, D., Chernetsov, N., Heyers, D., Mouritsen, H., 2013. Migratory reed warblers need intact trigeminal nerves to correct for a 1,000 km eastward displacement. PLoS One 8 (6), e65847.

Kramer, G., 1949. Über Richtungstendenzen bei der nächtlichen Zugunruhe gekäfigter Vögel. In: Mayr, E., Schüz, E. (Eds.), Ornithologie als Biologische Wissenschaft, Heidelberg, pp. 269–283.

Kramer, G., 1957. Experiments on bird orientation and their interpretation. Ibis 99, 196–227.

Liedvogel, M., Mouritsen, H., 2010. Cryptochromes – a potential magnetoreceptor: what do we know and what do we want to know? Journal of the Royal Society Interface 7, S147–S162.

Liedvogel, M., Maeda, K., Henbest, K., Schleicher, E., Simon, T., Timmel, C.R., Hore, P.J., Mouritsen, H., 2007a. Chemical magnetoreception: bird cryptochrome 1a is excited by blue light and forms long-lived radical-pairs. PLoS One 2 (10), e1106.

Liedvogel, M., Feenders, G., Wada, K., Troje, N.F., Jarvis, E.D., Mouritsen, H., 2007b. Lateralized activation of Cluster N in the brains of migratory songbirds. European Journal of Neuroscience 25, 1166–1173.

Liu, X., Chernetsov, N., 2012. Avian orientation: multi-cue integration and calibration of compass systems. Chinese Birds 3, 1–8.

Maeda, K., Henbest, K.B., Cintolesi, F., Kuprov, I., Rodgers, C.T., Liddell, P.A., Gust, D., Timmel, C.R., Hore, P.J., 2008. Chemical compass model of avian magnetoreception. Nature 453, 387–390.

Mayr, E., 1952. German experiments on orientation of migrating birds. Biological Reviews Cambridge Philosophical Society 27, 394–400.

Merkel, F.W., Wiltschko, W., 1965. Magnetismus und Richtungsfinden zugunruhiger Rotkehlchen (*Erithacus rubecula*). Die Vogelwarte 23, 71–77.

Mewaldt, R., 1964. California sparrows return from displacement to Maryland. Science 146, 941–942.

Möller, A., Sagasser, S., Wiltschko, W., Schierwater, B., 2004. Retinal cryptochrome in a migratory passerine bird: a possible transducer for the avian magnetic compass. Naturwissenschaften 91, 585–588.

Molteno, T.C.A., Kennedy, W.L., 2009. Navigation by induction-based magnetoreception in elasmobranch fishes. Journal of Biophysics. 380976.

Mora, C.V., Bingman, V.P., 2013. Detection of magnetic field intensity gradient by homing pigeons (*Columba livia*) in a novel "Virtual magnetic map" conditioning paradigm. PLoS One 8 (9), e72869.

Mora, C.V., Davison, M., Wild, J.M., Walker, M.M., 2004. Magnetoreception and its trigeminal mediation in the homing pigeon. Nature 432, 508–511.

Mouritsen, H., 1998a. Redstarts, *Phoenicurus phoenicurus*, can orient in a true-zero magnetic field. Animal Behaviour 55, 1311–1324.

Mouritsen, H., 1998b. Modelling migration: the clock-and-compass model can explain the distribution of ringing recoveries. Animal Behaviour 56, 899–907.

Mouritsen, H., 2003. Spatiotemporal orientation strategies of long-distance migrants. In: Berthold, P., Gwinner, E., Sonnenschein, E. (Eds.), Avian Migration. Springer, Berlin, pp. 493–513.

Mouritsen, H., 2012. Sensory biology: search for the compass needles. Nature 484, 320–321.

Mouritsen, H., 2013. The magnetic senses. In: Galizia, C.G., Lledo, P.-M. (Eds.), Neurosciences – from Molecule to Behavior: A University Textbook. Springer-Verlag Berlin, Heidelberg, pp. 427–443. http://dx.doi.org/10.1007/978-3-642-10769-6_20.

Mouritsen, H., Hore, P.J., 2012. The magnetic retina: light-dependent and trigeminal magnetoreception in migratory birds. Current Opinion in Neurobiology 22, 343–352.

Mouritsen, H., Larsen, O.N., 1998. Migrating young pied flycatchers *Ficedula hypoleuca* do not compensate for geographical displacements. Journal of Experimental Biology 201, 2927–2934.

Mouritsen, H., Larsen, O.N., 2001. Migrating songbirds tested in computer-controlled Emlen funnels use stellar cues for a time-independent compass. Journal of Experimental Biology 204, 3855–3865.

Mouritsen, H., Mouritsen, O., 2000. A mathematical expectation model for bird navigation based on the clock-and-compass strategy. Journal of Theoretical Biology 207, 283–291.

Mouritsen, H., Ritz, T., 2005. Magnetoreception and its use in bird navigation (Rouen 1938) Proceedings of the International Ornithology Congress 915, 406–414.

Mouritsen, H., Huyvaert, K.P., Frost, B.J., Andersson, D.J., 2003. Waved albatrosses can navigate with strong magnets attached to their head. Journal of Experimental Biology 206, 4155–4166.

Mouritsen, H., Janssen-Bienhold, U., Liedvogel, M., Feenders, G., Stalleicken, J., Dirks, P., Weiler, R., 2004a. Cryptochromes and neuronal-activity markers colocalize in the retina of migratory birds during magnetic orientation. Proceedings of the National Academy of Sciences of the United States of America 101, 14294–14299.

Mouritsen, H., Feenders, G., Liedvogel, M., Kropp, W., 2004b. Migratory birds use head scans to detect the direction of the earth's magnetic field. Current Biology 14, 1946–1949.

Mouritsen, H., Feenders, G., Liedvogel, M., Wada, K., Jarvis, E.D., 2005. Night-vision brain area in migratory songbirds. Proceedings of the National Academy of Sciences of the United States of America 102, 8339–8344.

Mouritsen, H., Feenders, G., Hegemann, A., Liedvogel, M., 2009. Thermal paper can replace typewriter correction paper in Emlen funnels. Journal of Ornithology 150, 713–715.

Mouritsen, H., Derbyshire, R., Stalleicken, J., Mouritsen, O.Ø., Frost, B.J., Norris, R.D., 2013. An experimental displacement and over 50 years of tag-recoveries show that monarch butterflies are not true navigators. Proceedings of the National Academy of Sciences of the United States of America 110, 7348–7353.

Muheim, R., Bäckman, J., Akesson, S., 2002. Magnetic compass orientation in European robins is dependent on both wavelength and intensity of light. Journal of Experimental Biology 205, 3845–3856.

Muheim, R., Phillips, J.B., Åkesson, S., 2006a. Polarized light cues underlie compass calibration in migratory songbirds. Science 313, 837–839.

Muheim, R., Moore, F.R., Phillips, J.B., 2006b. Calibration of magnetic and celestial compass cues in migratory birds—a review of cue-conflict experiments. Journal of Experimental Biology 209, 2–17.

Müller, M., Carell, T., 2009. Structural biology of DNA photolyases and cryptochromes. Current Opinion in Structural Biology 19, 277–285.

Nevitt, G.A., Bonadonna, F., 2005. Sensitivity to dimethyl sulphide suggests a mechanism for olfactory navigation by seabirds. Biology Letters 1, 303–305.

Niessner, C., Denzau, S., Gross, J.C., Peichl, L., Bischof, H.J., Fleissner, G., Wiltschko, W., Wiltschko, R., 2011. Avian ultraviolet/violet cones identified as probable magnetoreceptors. PLoS One 6 (5), e20091.

Papi, F., 1991. Olfactory navigation. In: Berthold, P. (Ed.), Orientation in Birds. Birkhäuser, Basel, pp. 52–85.

Perdeck, A.C., 1958. Two types of orientation in migrating starlings, *Sturnus vulgaris* L., and chaffinches, *Fringilla coelebs* L., as revealed by displacement experiments. Ardea 46, 1–37.

Phillips, J.B., Borland, S.C., 1992. Behavioural evidence for use of a light-dependent magnetoreception mechanism by a vertebrate. Nature 359, 142–144.

Phillips, J.B., Deutschlander, M.E., Freake, M.J., Borland, S.C., 2001. The role of extraocular photoreceptors in newt magnetic compass orientation: parallels between light-dependent magnetoreception and polarized light detection in vertebrates. Journal of Expperimental Biology 204, 2543–2552.

Rabøl, J., 1978. One-direction orientation versus goal area navigation in migratory birds. Oikos 30, 216–223.

Rastogi, A., Kumari, Y., Rani, S., Kumar, V., 2011. Phase inversion of neural activity in the olfactory and visual systems of a night-migratory bird during migration. European Journal of Neuroscience 34, 99–109.

Ritz, T., Adem, S., Schulten, K., 2000. A model for photoreceptor-based magnetoreception in birds. Biophysical Journal 78, 707–718.

Ritz, T., Thalau, P., Phillips, J.B., Wiltschko, R., Wiltschko, W., 2004. Resonance effects indicate a radical-pair mechanism for avian magnetic compass. Nature 429, 177–180.

Ritz, T., Wiltschko, R., Hore, P.J., Rodgers, C.T., Stapput, K., Thalau, P., Timmel, C.R., Wiltschko, W., 2009. Magnetic compass of birds is based on a molecule with optimal directional sensitivity. Biophysical Journal 96, 3451–3457.

Ritz, T., Ahmad, M., Mouritsen, H., Wiltschko, R., Wiltschko, W., 2010. Photoreceptor-based magnetoreception: optimal design of receptor molecules, cells, and neuronal processing. Journal of the Royal Society Interface 7, S135–S146.

Rodgers, C.T., Hore, P.J., 2009. Chemical magnetoreception in birds: the radical pair mechanism. Proceedings of the National Academy of Sciences the United States of America 106, 353–360.

Rosenblum, B., Jungerman, R.L., Longfellow, L., 1985. Limits to induction-based magnetoreception. In: Kirschvink, J.L., Jones, D.S., MacFadden, B.J. (Eds.), Magnetite Biomineralization and Magnetoreception in Organisms: A New Biomagnetism. Plenum Press, New York, NY, pp. 223–232.

Sancar, A., 2003. Structure and function of DNA photolyases and cryptochrome blue-light photoreceptors. Chemical Reviews 103, 2203–2237.

Schmidt-Koenig, K., 1965. Current problems in bird orientation. Advances in the Study of Behaviour 1, 217–272.

Schmidt-Koenig, K., Walcott, C., 1978. Tracks of pigeons homing with frosted lenses. Animal Behavior 26, 480–486.

Schmidt-Koenig, K., Ganzhorn, J.U., Ranvaud, R., 1991. The sun compass. In: Berthold, P. (Ed.), Orientation in Birds. Birkhäuser, Basel, pp. 1–15.

Schneider, T., Thalau, H.P., Semm, P., Wiltschko, W., 1994. Melatonin is crucial for the migratory orientation of pied flycatchers (*Ficedula-hypoleuca pallas*). Journal of Experimental Biology 194, 255–262.

Schulten, K., Swenberg, C.E., Weller, A., 1978. A biomagnetic sensory mechanism based on magnetic field modulated coherent electron spin motion. Zeitschrift für Physikalische Chemie, Neue Folge 111, 1–5.

Semm, P., Demaine, C., 1986. Neurophysiological properties of magnetic cells in the pigeon's visual system. Journal of Comparative Physiology A 159, 619–625.

Solov'yov, I.A., Greiner, W., 2009. Micromagnetic insight into a magnetoreceptor in birds: existence of magnetic field amplifiers in the beak. Physical Review. E 80, 041919.

Solov'yov, I.A., Mouritsen, H., Schulten, K., 2010. Acuity of a cryptochrome and vision-based magnetoreception system in birds. Biophysical Journal 99, 40–49.

Solov'yov, I.A., Domratcheva, T., Shahi, A.R.M., Schulten, K., 2012. Decrypting cryptochrome: revealing the molecular identity of the photoactivation reaction. Journal of the American Chemical Society 134, 18046–18052.

Solov'yov, I.A., Ritz, T., Schulten, K., Hore, P.J., 2014. A chemical compass for bird navigation. In: Mohseni, M., Omar, Y., Engel, G., Plenio, M. (Eds), Quantum Effects in Biology. Cambridge University Press, pp. 216–236, in press.

Stapput, K., Güntürkün, O., Hoffmann, K.-P., Wiltschko, R., Wiltschko, W., 2010. Magnetoreception of directional information in birds requires nondegraded vision. Current Biology 20, 1259–1262.

Thorup, K., Bisson, I.-A., Bowlin, M.S., Holland, R.A., Wingfield, J.C., Ramenofsky, M., Wikelski, M., 2007. Evidence for a navigational map stretching across the continental U.S. in a migratory songbird. Proceedings of the National Academy of Sciences of the United States of America 104, 18115–18119.

Treiber, C.D., Salzer, M.C., Riegler, J., Edelman, N., Sugar, C., Breuss, M., Pichler, P., Cadiou, H., Saunders, M., Lythgoe, M., Shaw, J., Keays, D.A., 2012. Clusters of iron-rich cells in the upper beak of pigeons are macrophages not magnetosensitive neurons. Nature 484, 367–370.

von der Emde, G., 2013. Electroreception. In: Galizia, C.G., Lledo, P.-M. (Eds.), Neurosciences – from Molecule to Behavior: A University Textbook. Springer-Verlag, Berlin Heidelberg, pp. 409–425.

Walcott, C., 1991. Magnetic maps in pigeons. In: Berthold, P. (Ed.), Orientation in Birds. Birkhäuser, Basel, pp. 38–51.

Walker, M.M., Diebel, C.E., Haugh, C.V., Pankhurst, P.M., Montgomery, J.C., Green, C.R., 1997. Structure and function of the vertebrate magnetic sense. Nature 390, 371–376.

Wallraff, H.G., 1972. Homing of pigeons after extirpation of their cochleae and lagenae. Nature New Biology 236, 223–224.
Wallraff, H.G., 2001. Navigation by homing pigeons: updated perspective. Ethology Ecology and Evolution 13, 1–48.
Wallraff, H.G., 2005. Avian Navigation: Pigeon Homing as a Paradigm. Springer Verlag, Berlin.
Wallraff, H.G., Andreae, M.O., 2000. Spatial gradients in ratios of atmospheric trace gases: a study stimulated by experiments on bird navigation. Tellus 52B, 1138–1157.
Westby, G.W.M., Partridge, K.J., 1986. Human homing: still no evidence despite geomagnetic controls. Journal of. Experimental Biology 120, 325–331.
Williams, M.N., Wild, J.M., 2001. Trigeminally innervated iron-containing structures in the beak of homing pigeons, and other birds. Brain Research 889, 243–246.
Wiltschko, W., 1968. Über den Einfluss statischer Magnetfelder auf die Zugorientierung der Rotkehlchen (*Erithacus rubecula*). Zeitschrift für Tierpsychology 25, 537–558.
Wiltschko, W., 1978. Further analysis of the magnetic compass of migratory birds. In: Schmidt-Koenig, K., Keeton, W.T. (Eds.), Animal Migration, Navigation and Homing. Springer, Berlin Heidelberg, New York, pp. 302–310.
Wiltschko, W., Wiltschko, R., 2007. Magnetoreception in birds: two receptors for two different tasks. Journal of Ornithology 148, S61–S76.
Wiltschko, W., Wiltschko, R., 1972. Magnetic compass of European Robins. Science 176, 62–64.
Wiltschko, R., Wiltschko, W., 1995. Magnetic Orientation in Animals. Springer Verlag 978-3-642-79751-4. 298p.
Wiltschko, W., Wiltschko, R., 1996. Magnetic orientation in birds. Journal of Experimental Biology 199, 29–38.
Wiltschko, W., Munro, U., Ford, H., Wiltschko, R., 1993. Red light disrupts magnetic orientation of migratory birds. Nature 364, 525–527.
Wiltschko, W., Traudt, J., Güntürkün, O., Prior, H., Wiltschko, R., 2002. Lateralization of magnetic compass orientation in a migratory bird. Nature 419, 467–470.
Wiltschko, W., Munro, U., Ford, H., Wiltschko, R., 2003. Lateralisation of magnetic compass orientation in silvereyes, Zosterops lateralis. Australian Journal of Zoology 51, 597–602.
Wiltschko, W., Munro, U., Ford, H., Wiltschko, R., 2009. Avian orientation: the pulse effect is mediated by the magnetite receptors in the upper beak. Proceedings of the Royal Society London B 276, 2227–2232.
Wiltschko, R., Schiffner, I., Fuhrmann, P., Wiltschko, W., 2010a. The role of magnetite-based receptors in the beak in pigeon homing. Current Biology 20, 1534–1538.
Wiltschko, R., Stapput, K., Thalau, P., Wiltschko, W., 2010b. Directional orientation of birds by the magnetic field under different light conditions. Journal of the Royal Society Interface 7, 163–177.
Wu, L.Q., Dickman, J.D., 2011. Magnetoreception in an avian brain in part mediated by inner ear lagena. Current Biology 21, 418–423.
Wu, L.Q., Dickman, J.D., 2012. Neural correlates of a magnetic sense. Science 336, 1054–1057.
Zapka, M., Heyers, D., Hein, C.M., Engels, S., Schneider, N.-L., Hans, J., Weiler, S., Dreyer, D., Kishkinev, D., Wild, J.M., Mouritsen, H., 2009. Visual but not trigeminal mediation of magnetic compass information in a migratory bird. Nature 461, 1274–1277.
Zapka, M., Heyers, D., Liedvogel, M., Jarvis, E.D., Mouritsen, H., 2010. Night-time neuronal activation of Cluster N in a day- and night-migrating songbird. European Journal of Neuroscience 32, 619–624.

Chapter 9

The Avian Subpallium and Autonomic Nervous System

Wayne J. Kuenzel
Poultry Science Center, University of Arkansas, Fayetteville, AR, USA

9.1 INTRODUCTION

Since the last edition of *Sturkie's Avian Physiology*, significant progress has been made in avian physiology due to the publication of the chicken genome (Hillier et al., 2004) and genomes of other avian species, including the zebra finch (Warren et al., 2010), turkey (Aslam et al., 2012), Japanese quail (Kawahara-Miki et al., 2013), and scarlet macaw (Seabury et al., 2013). In addition to progress in molecular biology, important events and advances have occurred in avian neurobiology, specifically neuroanatomy. In the late 1990s, Dr Tony Reiner formed a small group of avian neuroanatomists called The Thinktank to address a serious problem with the current nomenclature of the avian brain, particularly the forebrain. The persistent, inappropriate terminology of many structures was based upon an outdated and erroneous assumption of homology to mammalian brain structures. In the fall of 2000, Dr Erich Jarvis initiated a proposal for federal funding to organize an avian brain nomenclature forum, in which avian and comparative neuroanatomists would meet to develop a new terminology for the avian forebrain that was more in line with that developed for the mammalian brain. The result was a 2-year in-depth discussion via email of all specific terms that needed to be changed, based upon an evaluation of the current data. In addition, a voting procedure was developed and agreed upon where each structure would be discussed.

In July 2002, a 3-day meeting supported by the National Institutes of Health and the National Science Foundation was held at Duke University in Durham, North Carolina. Twenty-nine comparative neuroanatomists participated. Individual structures in the forebrain were discussed and new names were adopted. Importantly, the revised nomenclature, for the first time, contained the appropriate terms to eliminate the perception that the avian telencephalon was composed mostly of basal ganglia (as once thought). A number of structures residing in the outer region or cortical areas of the telencephalon now contain the suffix *pallium*. This change immediately showed that an extensive pallium exists in the cortical area of the telencephalon surrounding the basal ganglia of birds. The term *pallium* is used because, in birds and other nonmammalian vertebrates, a laminated structure resembling that of a mammalian neocortex is not evident. Hence, the term *cortex* is inappropriate because, by definition, a cortex comprises six layers of cells. The new terms, their rationale, and images showing the revised terms of all pallial, striatal, and pallidal structures can be found in Reiner et al. (2004) and Jarvis et al. (2005).

Since that nomenclature meeting, new data and advancements have occurred, particularly for another critical neuroanatomical region deep within the forebrain called the *subpallium*. Structural and functional subdivisions of that region were detailed and summarized (Kuenzel et al., 2011). The subpallium comprises the basal ganglia and other anatomical components critical to the survival of a particular species. Specifically, it includes structures that help regulate motor movements and ingestive, reproductive, and defensive behaviors. Indeed, the functions listed are quite similar to the neural functions associated with the autonomic nervous system (ANS). Importantly, some structures residing in the subpallium project directly to central structures that serve as premotor nuclei of the ANS. Therefore, the overall purpose of this review is to provide a summary of the structures and functions of the subpallium and indicate where that basal forebrain region connects with components of the ANS. The last section of the chapter includes specific examples and hypotheses about how the subpallial nervous system, ANS, and neuroendocrine systems are interconnected and integrated to regulate the reproductive system, ingestive/feeding behavior, and selected circannual rhythmic behaviors. By focusing on central neural structures, it is hoped that readers will be directed to brain regions that regulate functions or behaviors of interest and will recognize areas where research is greatly needed due to large gaps in our knowledge.

9.2 COMPONENTS OF THE SUBPALLIUM

The subpallium comprises five major groups: (1) dorsal somatomotor basal ganglia; (2) ventral viscerolimbic basal ganglia; (3) extended amygdala; (4) basal telencephalic cholinergic and noncholinergic corticopetal systems; and (5) septum and septal neuroendocrine systems (Figure 9.1). The five groups occupy the mediobasal region of the telencephalon and are surrounded dorsally and ventrolaterally by the pallium.

Significant advances have been made in understanding the embryonic development of the subpallial region of the chick brain from known proliferative zones based upon gene expression and fate mapping data. Details about the histogenetic zones can be obtained in the following references: Puelles et al. (2000, 2007), Cobos et al. (2001a,b), Marín and Rubenstein (2001), Redies et al. (2001), Flames et al. (2007), Abellán and Medina (2008, 2009), García-López et al. (2008), Abellán et al. (2010), and Kuenzel et al. (2011). Each of the five subpallial groups in posthatching and adult birds is discussed in the following sections.

9.2.1 Dorsal Somatomotor Basal Ganglia

9.2.1.1 Structures

The dorsal somatomotor and ventral viscerolimbic components of the basal ganglia comprise the basal ganglia (BG). Discovery of the abundance of dopamine and acetylcholinesterase in the avian subpallium in the 1960s (Spooner and Winters, 1966; Juorio and Vogt, 1967; Karten, 1969; Nauta and Karten, 1970) began a period that reshaped an understanding of the actual location and extent of the avian BG that occupied only the ventromedial portion of the forebrain (Figure 9.2) and not its extent throughout the entire forebrain as previously thought. A structure that clearly marks the boundary between the BG and pallium is the pallial-subpallial lamina (LPS; Figure 9.2(C)). Importantly, the use of immunohistochemistry and an antibody to tyrosine hydroxylase (an indicator of dopamine) and choline acetyltransferase (synthesizes acetylcholine) clearly showed how similar the avian basal ganglia was to that of mammals regarding its location and relative size compared with the rest of the telencephalon.

Key structures of the dorsal somatomotor basal ganglia include the dorsal and ventral parts of the medial striatum (MSt), lateral striatum (LSt), nucleus intrapeduncularis, and globus pallidus (Figure 9.2). The MSt and LSt are generally considered homologous to the mammalian caudate putamen (Reiner et al., 2004). For details of comparisons among birds, mammals, reptiles, and other vertebrate classes, see Reiner et al. (1998).

9.2.1.2 Functions

Overall, both structurally and functionally there are marked similarities between mammals and birds in the dorsal somatomotor basal ganglia. Specifically, models of the functional organization of the somatic basal ganglia, first developed in mammals, comprised two parallel output circuits that had opposing functions in motor control as well as interactions between them (Albin et al., 1989; DeLong, 1990; Gerfen, 1992). Birds have been shown to have similar direct and indirect pathways (Reiner et al., 1998). A modified representation of the direct and indirect motor pathways involving the avian dorsal somatomotor basal ganglia is shown in Figure 9.3. The direct pathway (Figure 9.3(A)) promotes movement of birds (head, wings, legs). Input into the avian striatum (homologous to mammalian caudate putamen) occurs via the corticostriatal projections (from pallium), shown as glutamatergic excitatory input to the lateral and medial striatum (Veenman and Reiner, 1996; Csillag et al., 1997; Reiner et al., 2001; Ding et al., 2003; Ding and Perkel, 2004; Farries et al., 2005). Modulatory dopaminergic input to the striatum likewise occurs from the substantia nigra, pars compacta (SNc) [A9 catecholaminergic (CA) cell group], ventral tegmental area (VTA; A10 CA group), and retrorubral field (A8 CA group) (Brauth et al., 1978; Kitt and Brauth, 1986b; Bailhache and Balthazart, 1993; Moons et al., 1994; Reiner et al., 1994; Wynne and Güntürkün, 1995) (Figure 9.4). Both the lateral and medial striatum receive excitatory inputs from the pallium. The striatum, in turn, projects inhibitory neurons containing both GABA and substance P (SP) to the medial (interna) region of globus pallidus (Figure 9.3(A)), indicated by GPi, (Anderson and Reiner, 1990, 1991; Reiner, 1986; Reiner and Anderson, 1990; Veenman and Reiner, 1994). The medial area of the globus pallidus projects inhibitory GABAergic neurons to a small thalamic region called the ventrointermediate thalamic area (VIA; Karten and Dubbeldam, 1973; Kitt and Brauth, 1982; Medina et al., 1997). The VIA contains glutamatergic neurons that are excitatory and project to the Wulst in the dorsal region of the pallium, which appears comparable to the mammalian primary somatosensory/somatomotor cortex (Wild, 1987; Korzeniewska and Güntürkün, 1990; Medina and Reiner, 2000).

In addition to the previously described direct pathway facilitating body movement, there is an antagonistic pathway that inhibits bird movements (Figure 9.3(B)). This parallel pathway, distinct from the direct pathway, is called the indirect pathway. It occurs from the striatum, comprises neurons co-localized with GABA and enkephalin (ENK) and functions to inhibit unwanted body movements. Nonetheless it has the same excitatory inputs shown previously for the direct motor pathway (Figure 9.3(B)). Those striatal projection neurons are inhibitory and innervate the lateral (externa) portion of the globus pallidus (GPe, Figure 9.3(B); Brauth, 1984; Reiner et al., 1984a,b; Reiner, 1987; Anderson and Reiner, 1990; Veenman and Reiner, 1994; Veenman et al., 1994). Inhibitory GABAergic neurons project to the avian subthalamic nucleus (STN) which contains excitatory

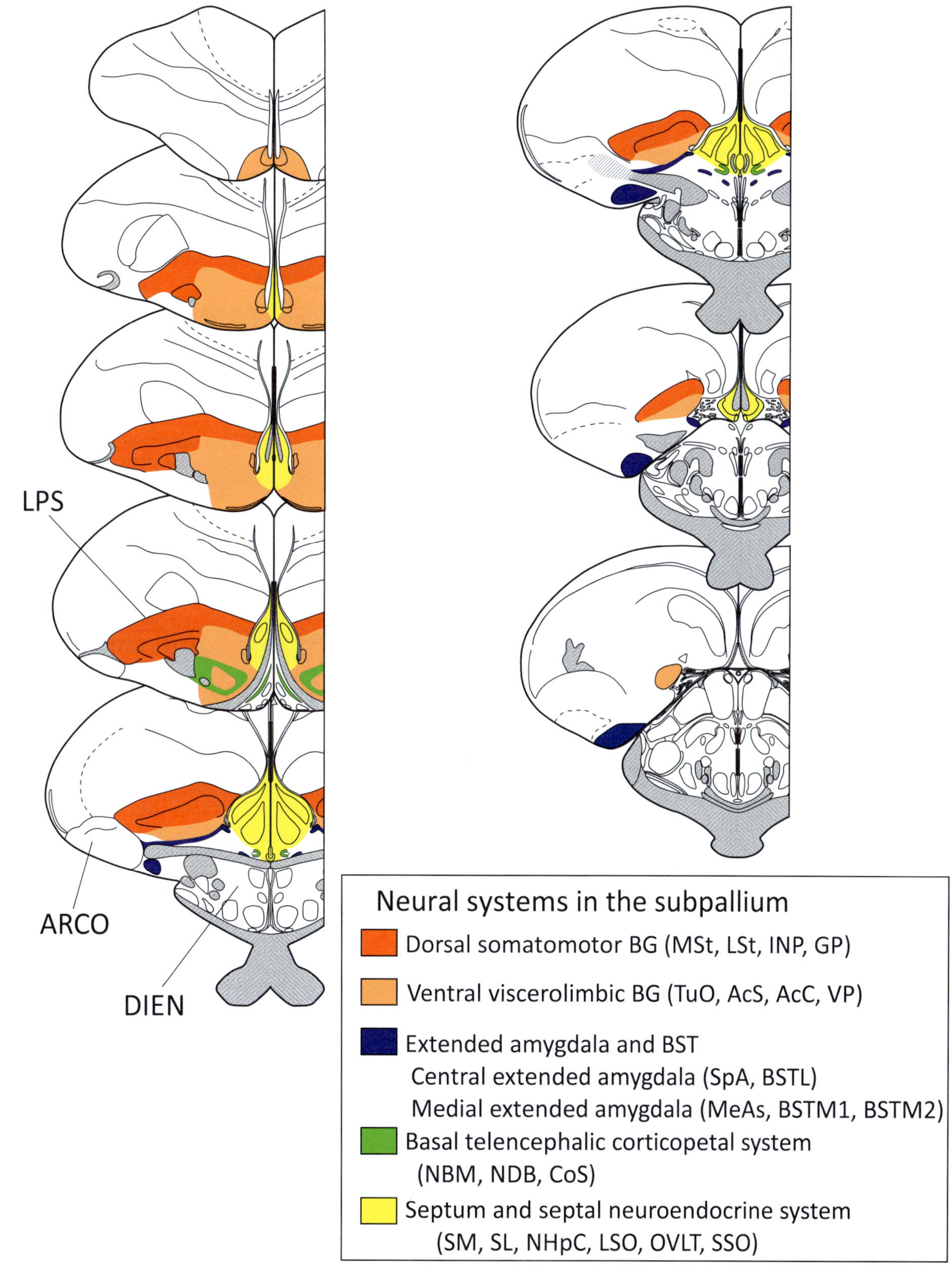

FIGURE 9.1 Five neural systems occupy the avian subpallium: (1) *Dorsal somatomotor basal ganglia* (shown in red) includes the lateral striatum (LSt), medial striatum (MSt), globus pallidus (GP), intrapeduncular nucleus (INP). The LPS marks the dorsal border of the avian subpallium. (2) *Ventral viscerolimbic basal ganglia* (shown in tan) includes the olfactory tubercle (TuO), nucleus accumbens (shell and core, AcS, AcC), ventral pallidum (VP). (3) *Extended amygdala and bed nuclei of the stria terminalis* (shown in blue) include the central extended amygdala and medial extended amygdala. The former is composed of the lateral bed nucleus of the stria terminalis (BSTL) and central extended amygdala (CEA, formerly the dorsal portion of the subpallial amygdala (SpA; see Figure 9.7)); the identification of the central nucleus of the avian amygdala is currently being investigated. The latter comprises the medial bed nucleus of the stria terminalis 1 and 2, (BSTM1, BSTM2), medial extended amygdala (MEA, formerly the ventral portion of the SpA (see Figure 9.7)), and medial amygdala (MeA, formerly the nucleus taeniae amygdala, TnA). (4) *Basal telencephalic corticopetal system* (shown in green) includes the basal magnocellular nucleus (NBM), diagonal band nucleus (NDB) and commissural septal nucleus (CoS). (5) *Septum and septal neuroendocrine system* (shown in yellow) comprise the medial septum (SM), lateral septum (LS), nucleus of the hippocampal commissure (NHpC, formerly known as the bed nucleus of the pallial commissure), and three circumventricular organs: the lateral septal organ (LSO), organum vasculosum of the lamina terminalis (OVLT), and subseptal organ (SSO). Other abbreviations: ARCO = arcopallium, DIEN = diencephalon (hypothalamus, thalamus and epithalamus). *From Kuenzel et al. (2011).*

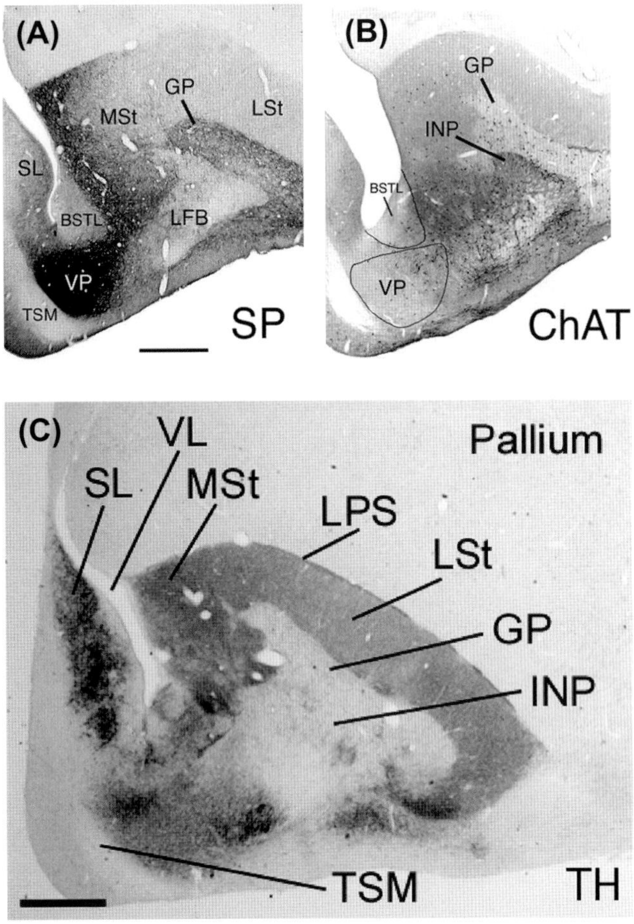

FIGURE 9.2 Components of the dorsal somatomotor basal ganglia. Images of transverse sections of pigeon brain showing immunoreactivity for (A) substance P (SP) and (B) choline acetyltransferase (ChAT). A transverse section of chicken brain shows immunolabeling for (C) tyrosine hydroxylase (TH). The use of TH, an indicator of dopaminergic input, identifies striatal structures of the dorsal somatomotor basal ganglia, particularly the lateral (LSt) and medial striatum (MSt). Note also the increased immunoreactivity of SP fibers in the ventral pallidum (VP), as well as portions of the VP showing increased levels of ChAT neurons and TH fibers. In (B), the field of cholinergic neurons spanning the VP and lateral forebrain bundle (LFB) represents the basal magnocellular cholinergic cell group (NBM). There exists a paucity of ChAT neurons and SP immunoreactivity in the bed neucleus of the stria terminalis (BSTL). Other abbreviations: GP = globus pallidus, INP = intrapeduncular nucleus, LPS = pallial-subpallial lamina, SL = lateral septum, TSM = tractus septopallio-mesencephalicus, VL = lateral ventricle. Scale bar = 1 mm. *From Kuenzel et al. (2011).*

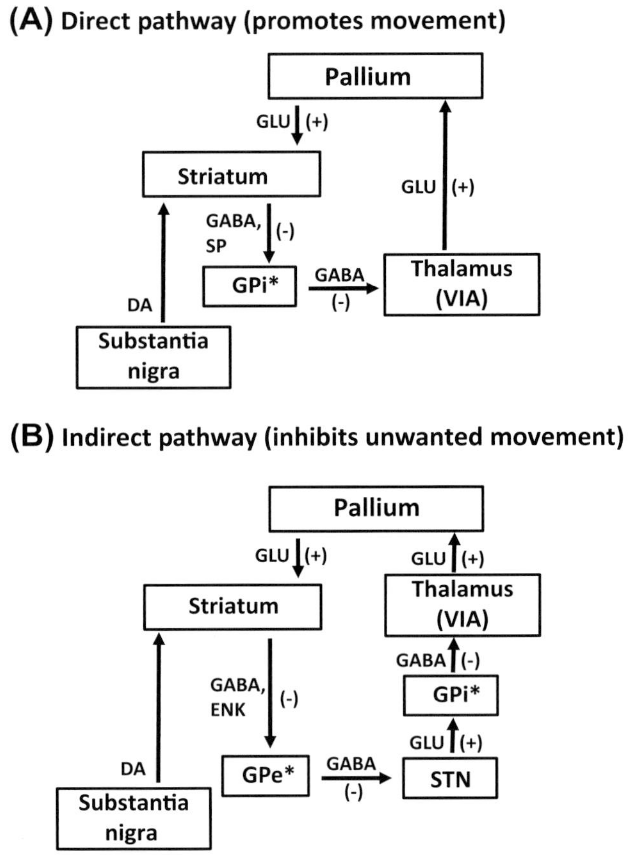

FIGURE 9.3 (A) Schematic diagram showing a direct pathway where excitatory inputs (+) from the pallium and modulatory inputs from the substantia nigra promote skeletal muscle movements in birds. (B) In parallel, an indirect pathway functions to inhibit unwanted movement thereby facilitating the movement initiated by the direct pathway. Abbreviations: DA = dopamine, ENK = enkephalin, GABA = gamma aminobutyric acid, GLU = glutamic acid, GPe* = globus pallidus externa (lateral portion of GP), GPi* = globus pallidus internus (medial portion of GP), STN = subthalamic nucleus, SP = substance P, VIA = ventrointermediate thalamic area.

glutamatergic neurons that in turn project back to the globus pallidus interna (Jiao et al., 2000). The medial globus pallidus comprises predominantly GABAergic inhibitory neurons that project to the VIA. The VIA contains excitatory glutamatergic neurons that project to the pallium (Figure 9.3(B)). In summary, activation of ENK, GABA-containing neurons in the striatum produces an inhibition of neurons in the globus pallidus. The result is a disinhibition of glutatmate containing neurons in the STN. The enhanced excitatory output from the STN causes an increase in GABA release from globus pallidus neurons that project to the thalamic VIA, thereby decreasing glutamate release into specific pallial projection sites. The overall effect of this pathway is to promote suppression of unwanted movements that potentially would conflict with the movements being promoted by the direct pathway (Figure 9.3(A)). Behavioral data supporting the function of the indirect pathway were obtained in rats (Kafetzopoulos and Papadopoulos, 1983; Piallat et al., 1996) and later in pigeons following surgical lesions directed to the STN that produced hyperkinesia and rotation.

Data obtained in lampreys (jawless vertebrates) have shown that this species likewise contains all the major components of the basal ganglia including the striatum, globus pallidus, and STN; thus, it is suggested that this phylogenetic old group of vertebrates had similar, functional direct and indirect pathways forming a core network regulating motor function (Stephenson-Jones et al., 2011). Data

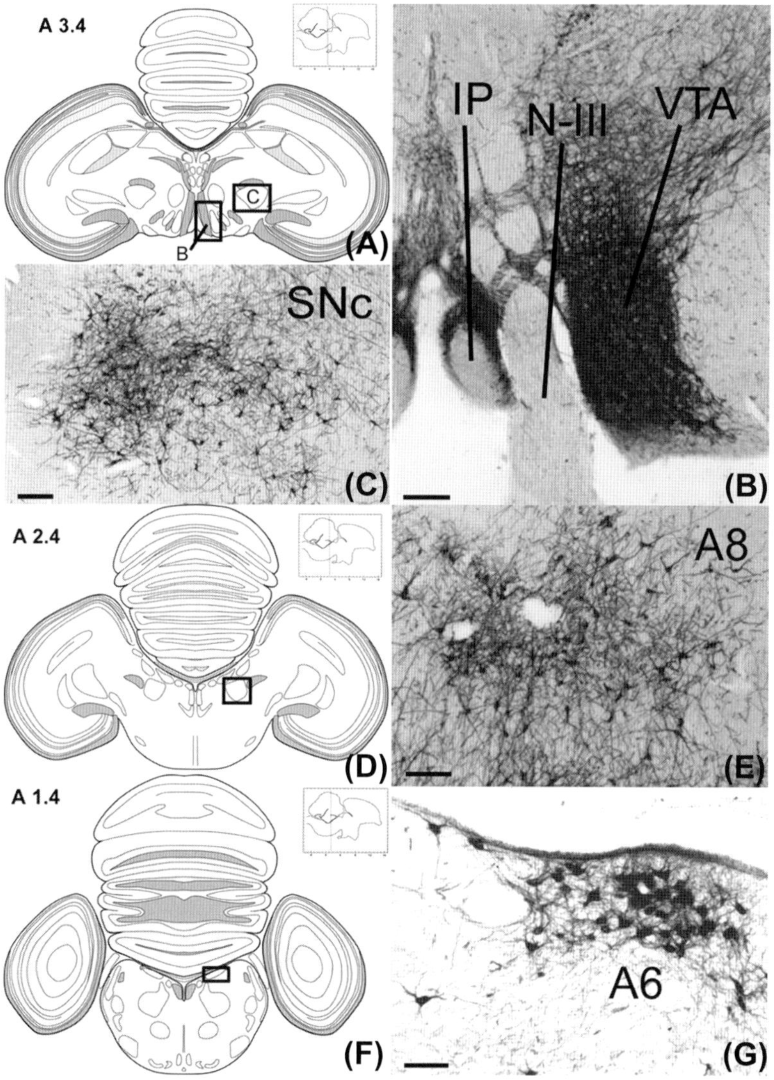

FIGURE 9.4 Sources of modulatory afferent inputs to the basal ganglia in chick brain. (A) Brain atlas plate A3.4 (Kuenzel and Masson, 1988). Boxed in areas show location of images for (B) and (C). (B) Ventral tegmental area (VTA) or A10 dopaminergic cell group. (C) Substantia nigra pars compacta (SNc) or A9 dopaminergic cell group. (D) Chick brain atlas plate A2.4; boxed area showns location of digital image (E) A8 dopaminergic cell group. (F) Chick brain atlas plate A1.4; boxed area shows location of image 4G. (G) Locus coeruleus or A6 noradrenergic cell group. Scale bar for (B, C, E) = 200 μm and for (G) = 100 μm. Other abbreviations: N-III = oculomotor nerve, IP = interpeduncular nucleus. *From Kuenzel et al. (2011).*

support the general hypothesis that the basal ganglia circuits evolved to support selection of motor actions among vertebrates (Redgrave et al., 1999). Further research has shown the marked conservation of the dual-output pathways throughout vertebrate phylogeny (Stephenson-Jones et al., 2012). Additionally, a neural structure in lampreys found to be homologous to the mammalian pedunculopontine nucleus (PPN), located in the midbrain near the substantia nigra and ventral tegmental area, has been identified (Stephenson-Jones et al., 2012). The PPN is thought to be involved in the initiation and modulation of gait and other stereotyped movements in mammals (Pahapill and Lozano, 2000). In birds, the ventral mesencephalic profundus nucleus identified in avian brain atlases (Karten and Hodos, 1967; Kuenzel and Masson, 1988) was renamed the pedunculopontine tegmental nucleus (PPT or PPN; Reiner et al., 2004) due to the presence of cholinergic neurons and its association with components comprising the basal ganglia. The PPN has been proposed to be an extended part of the basal ganglia among vertebrates due to its connectivity with basal ganglia structures (Mena-Segovia et al., 2004).

9.2.2 Ventral Viscerolimbic Basal Ganglia

9.2.2.1 Structures

The ventral viscerolimbic basal ganglia (VVBG) are organized structurally similar to the dorsal somatomotor basal ganglia (DSBG) in that there is a dorsal striatal and a ventral

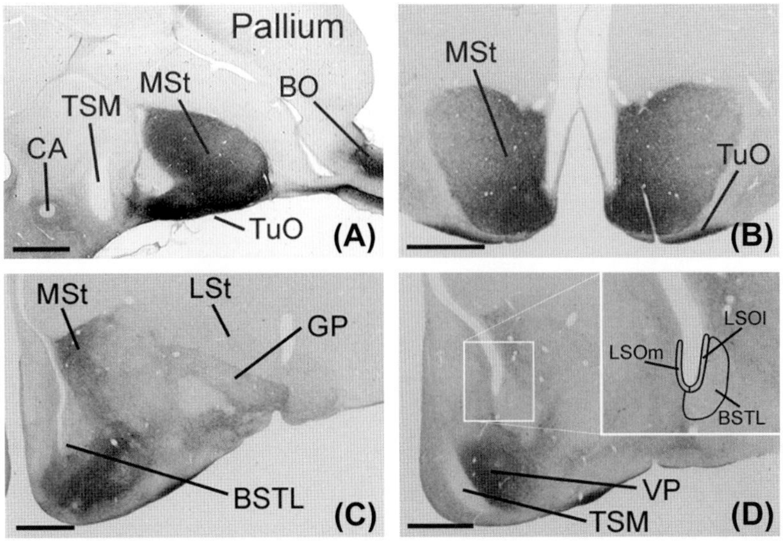

FIGURE 9.5 Sections of chick brain showing components of the viscerolimbic basal ganglia (BG) immunolabeled with an antibody to substance P. (A) Sagittal section near midline showing the ventral portion of medial striatum (MSt) and tuberculum olfactorium (TuO). Note projections from the olfactory bulb (BO) reaching the TuO. (B, C, D) Transverse sections of BG depicting the MSt, TuO and VP. The insert in (D) shows the location of one of the circumventricular organs, the lateral septal organ (LSO), which is associated with the fifth neural system of the subpallium, the septum, and septal neuroendocrine system. Other abbreviations: BSTL = lateral bed nucleus of the stria terminalis, CA = anterior commissure, GP = globus pallidus, TSM = tractus septopallio-mesencephalicus. Scale bars = 1.0 mm. *From Kuenzel et al. (2011).*

pallidal subdivision. The VVBG comprises the ventral portion of the medial striatum, nucleus accumbens (core and shell), olfactory tubercle, and ventral pallidum (Figures 9.5 and 9.6). The striatal subdivision consists of the ventral portion of the medial striatum, nucleus accumbens shell, and core and superficial part of the olfactory tubercle. Striatal components, particularly nucleus accumbens, receive excitatory pallial inputs, most likely glutamatergic (Veenman and Reiner, 1996; Csillag et al., 1997; Reiner et al., 2001; Ding et al., 2003; Ding and Perkel, 2004; Farries et al., 2005). Pallial inputs to the olfactory tubercle come from the olfactory bulb (Figure 9.5), piriform cortex (Bingman et al., 1994), and dorsomedial hippocampus (Atoji and Wild, 2004). Dopaminergic striatal, modulatory input largely originates from the ventral tegmental area A10 cell group (Figure 9.4(B)) and also from the substantia nigra (Figure 9.4(C); Kitt and Brauth, 1986b; Moons et al., 1994; Panzica et al., 1994, 1996). Striatal structures have been shown to have interneurons containing NADPH-diaphorase and nitric oxide synthase (NOS; Vincent et al., 1983; Brüning, 1993; Brüning et al., 1994; Panzica et al., 1994), as well as calcitonin gene-related peptide fibers (Lanuza et al., 2000; Roberts et al., 2002) and thyroid hormone releasing hormone-containing fibers (Jozsa et al., 1988). As stated previously, the striatal subdivision of the ventral viscerolimbic BG, regarding its inputs and outputs, is structurally similar to the DSBG as it receives excitatory glutamatergic input from the pallium and modulatory dopaminergic input largely from the ventral tegmental area (VTA) or A10 cell group and some input from the SNc. Additionally, the medial striatum, nucleus accumbens, and olfactory tubercle of the striatal subdivision, similar to the DSBG, have GABAergic projection neurons co-localized with either substance P or ENK. The viscerolimbic pallidal subdivision includes the ventral pallidum and deep olfactory tubercle (Figure 9.5).

A prominent major structure of the VVBG is the nucleus accumbens. The mammalian nucleus accumbens comprises a central core (AcC) surrounded on its medial, ventral, and lateral sides by a shell (AcS; Herkenham et al., 1984; Záborszky et al., 1985). A third subterritory of the mammalian accumbens has also been identified (Zahm, 2000; Zahm and Brog, 1992; Zahm and Heimer, 1993). Similarly, in avian species, three subterritories of the Ac have been described using antibodies to substance P, neuropeptide Y (NPY), dopamine and cAMP-regulated phosphoprotein (DARPP-32), and calcium-binding proteins (Figure 9.6) (Roberts et al., 2002; Bálint and Csillag, 2007; Abellán and Medina, 2009; Bálint et al., 2011; Husband and Shimizu, 2011). Controversies, however, remain on the boundaries of components of the Ac, the ventral medial striatum, as well as their anatomical organization.

The complex accumbens structure (Figure 9.6), nonetheless, is important to define due to its diverse reciprocal connections with limbic and visceral structures including the hippocampal formation, amygdala, ventral pallidum, lateral hypothalamus, and ventral tegmental area (reviewed in Husband and Shimizu, 2011). It also receives input from central components of the autonomic nervous system, including the lateral hypothalamus (Berk and Hawkin, 1985), lateral bed nucleus of the stria terminalis (Atoji et al., 2006), parabrachial nucleus (Wild et al., 1990), nucleus tractus solitarius (Arends et al., 1988; Bálint and Csillag, 2007), and

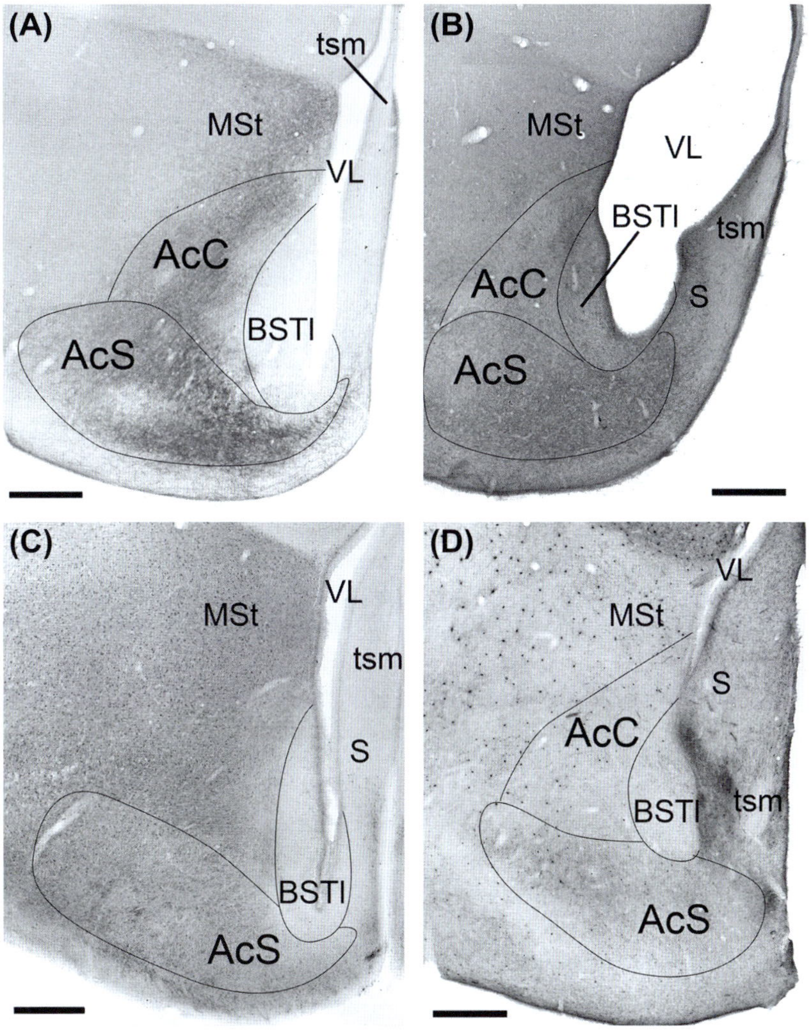

FIGURE 9.6 **Transverse sections of chick brain showing the nucleus accumbens (Ac) core (AcC) and shell (AcS) subdivisions immunolabeled with antibodies to the following neuropeptides/proteins.** (A) Substance P. (B) Neuropeptide Y (NPY). (C) Dopamine and cAMP-regulated phosphoprotein (DARPP-32). (D) Calbindin. The NPY immunoreactivity is useful for demarcation of the shell. Other abbreviations: BSTl = lateral bed nucleus of the stria terminalis, MSt = medial striatum, S = septum, tsm = tractus septopallio-mesencephalicus, VL = lateral ventricle. Scale bars = 500 μm. *From Kuenzel et al. (2011).*

dorsal motor nucleus of the vagus (Arends et al., 1988). As noted previously, striatal subdivisions have dopaminergic input from the ventral tegmental area and substantia nigra. In addition, the nucleus accumbens receives dense noradrenergic input from the A6 area (locus coeruleus; Figure 9.4(F,G)), particularly in the AcS (Kitt and Brauth, 1986a; von Bartheld and Bothwell, 1992; Bailhache and Balthazart, 1993; Reiner et al., 1994; Moons et al., 1995; Mello et al., 1998). Output of the striatal accumbens nucleus as well as that from some neurons of the striatal olfactory tubercle is to the ventral pallidum (VP). The avian VP in turn projects to several diencephalic sites, including the subthalamic nucleus, paraventricular nucleus, lateral hypothalamic area, thalamic reticular nucleus, dorsomedial thalamus, and habenular nucleus (Kitt and Brauth, 1981; Berk and Hawkin, 1985; Veenman et al., 1995; Medina and Reiner, 1997).

9.2.2.2 Functions

A function most widely associated with the VVBG is reward (reward-seeking effects). A useful definition of reward in animal experiments is a demonstration that an unconditioned stimulus can initiate appetitive behavior and evoke approach behavioral effects (Ikemoto and Panksepp, 1999). It is based largely on the extensive data in mammals showing that rats will self-administer dopamine, amphetamine, or electrical stimulation (each of these is initially an unconditioned stimulus) in the region of the nAc (Broekkamp et al., 1975; Cador et al., 1991). The nAc, especially the shell region, is also responsive to stress (Kalivas and Duffy, 1995; King et al., 1997) and the general accumbens region is associated with passive avoidance learning in chicks (Stewart et al., 1996).

A number of anatomical studies have been conducted in mammals and birds regarding the VVBG such that structural details of each component have been described. Nonetheless, a comprehensive neural circuit model similar to that developed for the dorsal somatomotor BG, describing the function of voluntary muscles via the direct-indirect pathway, has not been forthcoming. Perhaps some of the loop circuits between pallium and basal ganglia type structures, such as the anterior forebrain pathway in song birds (reviewed in Doupe et al., 2005), the proposed direct and indirect pathways involving the vocal learning circuit of song birds (Farries et al., 2005; Gale et al., 2008), or the limbic loop associated with the nAc (Husband and Shimizu, 2011) that have been proposed to date may lead to a promising model for the VVBG, which could be tested for regulating reward-motivated behavior.

9.2.3 Extended Amygdaloid Complex: Central Extended Amygdala and Medial Extended Amygdala

Heimer and coworkers recognized that the mammalian amygdaloid complex comprised two major groups, each of which had neuronal structures that extended from them forming a neuronal corridor. The two corridors were named the extended amygdala, with one originating from the central nucleus of the amygdala, while the second was continuous with the medial amygdaloid nucleus. Both passed through a territory ventral to the globus pallidus (anatomically referred to as sublenticular) and the neuronal corridors ended in different components of the bed nuclei of the stria terminalis (Alheid and Heimer, 1988; Alheid et al., 1995). Each corridor possessed similar neurochemical characteristics of the amygdaloid nucleus from which they were confluent. Functionally, the central and medial amygdaloid nuclei and their respective subnuclei of the complex bed nuclei of the stria terminalis represent the major output nuclei of the amygdala of mammals (Swanson and Petrovich, 1998; Swanson, 2000; Paré et al., 2004).

Developmental, neurohistochemical, hodological, and behavioral data suggest strongly that comparable structures forming a subpallial amygdaloid complex exist in birds (Jurkevich et al., 1997, 1999; Aste et al., 1998; Cheng et al., 1999; Panzica et al., 1999; Absil et al., 2002a,b; Roberts et al., 2002; Reiner et al., 2004; Yamamoto et al., 2005; Abellán and Medina, 2009; Xie et al., 2010, 2011). For clarification of the structures and functions of the subpallial amygdaloid complex, the following sections discuss each of its two basic components, first in mammals and then in birds.

9.2.3.1 Central Extended Amygdala

9.2.3.1.1 Structures

The central nucleus of the amygdala—cells within the corridor of the central extended amygdala and the lateral bed nucleus of the stria terminalis (BSTL)—comprise structures of the central extended amygdaloid complex in mammals. Associated with the central nucleus of the amygdala (CeM) are groups of intercalated cells positioned between the lateral amygdala and the CeM (Paré et al., 2004). The major groups of projection neurons from the central extended amygdaloid complex are GABAergic, similar to the dorsal somatomotor and ventral viscerolimbic basal ganglia discussed previously. A difference is that the GABAergic neurons are typically enriched with one of several neuropeptides, including corticotropin-releasing hormone (CRH), neurotensin (NT), somatostatin (SOM) or ENK (Moga and Gray, 1985; Swanson and Petrovich, 1998; Paré and Smith, 1994; Alheid et al., 1995; Poulin and Timofeeva, 2008; Panguluri et al., 2009). The major descending projections of the central extended amygdaloid complex are to the lateral hypothalamus, central gray, parabrachial nucleus, and nucleus of the solitary tract, all of which are structures in the diencephalon and brainstem that are part of the ANS. Functionally the structures and their connections to the ANS regulate physiological processes and behaviors related to ingestion, and fear/anxiety/stress (Alheid and Heimer, 1988; Alheid et al., 1995; Swanson, 2000; de Olmos et al., 2004).

In birds, a major component of the central extended amygdaloid complex, positioned directly below the globus pallidus, was previously termed the subpallial amygdaloid area (SpA; Reiner et al., 2004). It has been proposed that the SpA and its lateral continuation directly ventral to the lateral striatum (LSt) be renamed the central extended amygdala (CEA, Figure 9.7) (Abellán and Medina, 2009). It was also suggested that part of the caudolateral striatum and lateral continuation of the SpA may be the avian structure homologous to the mammalian central nucleus of the amygdala (Abellán and Medina, 2009). Another avian structure that has been suggested to be homologous to the mammalian central amygdaloid nucleus is the compact division of the posterior pallial amygdaloid nucleus (Atoji and Wild, 2006). A structure termed the striatal capsule (composed of a thin group of subpallial neurons), which is located between the nidopallium and lateral striatum (Puelles et al., 2007), has been proposed to be comparable to the intercalated cell masses of the mammalian amygdala (Abellán and Medina, 2009) that occur juxtapositioned to central amygdaloid nucleus of mammals (Paré et al., 2004). The final structure of the complex is the lateral bed nucleus of the stria terminalis (BSTL; Figure 9.7) and it is regarded as comparable to the mammalian bed nucleus of the stria terminalis due to the presence of CRH (Panzica et al., 1986; Ball et al., 1989; Richard et al., 2004), neurotensinergic (Reiner and Carraway, 1987; Atoji et al., 1996; Reiner et al., 2004) and enkephalinergic (Molnar et al., 1994) neurons. The avian BSTL has been shown to have direct connections to components of the autonomic nervous system including hypothalamic structures, parabrachial nucleus, nucleus of the solitary tract, and dorsal motor nucleus of the vagus (Berk, 1987; Arends et al., 1988; Wild et al., 1990; Atoji et al., 2006; Bálint et al., 2011).

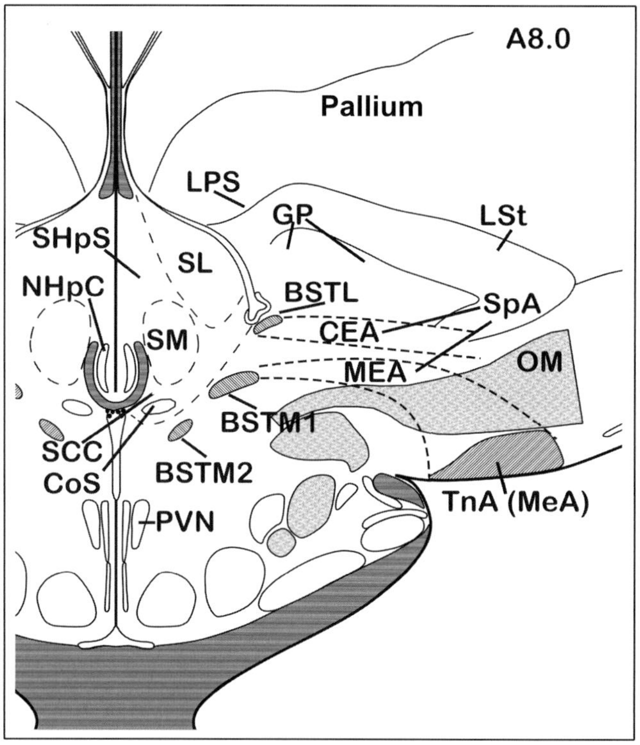

FIGURE 9.7 **Components of the subpallial amygdaloid complex and the septum and septal neuroendocrine systems.** The central extended amygdaloid complex comprises the lateral bed nucleus of the stria terminalis (BSTL), central extended amygdala (CEA), and central nucleus of the amygdala. Boundaries of the CEA in mature birds and location of the central nucleus of the amygdala require additional studies. The medial extended amygdaloid complex includes the medial bed nucleus of the stria terminalis, which has two subnuclei in chickens (BSTM1 and BSTM2), the medial extended amygdala (MEA) and the medial amygdala (formerly termed the nucleus taeniae amygdala (TnA)). Similar to the CEA, the MEA corridor requires refinement. Formerly, they were termed the subpallial amygdala (SpA). The fifth subpallial system, the septum and septal neuroendocrine system, was shown to consist of four divisions including the lateral septum (SL), medial septum (SM), septohippocampal septum (SHpS), and caudocentral septum (SCC) in birds (Goodson et al., 2004). The four septal divisions extend over a rostral-caudal length of at least 3.0 mm in the 2-week-old chick brain; however, for the purpose of this review, they have been compacted into one plane; therefore, their relative locations within the septum are approximate. Other abbreviations: CoS = commissural septal nucleus, GP = globus pallidus, LPS = pallial-subpallial lamina, LSt = lateral striatum, OM = occipitao-mesencephalic tract, PVN = paraventricular nucleus.

9.2.3.1.2 Functions

In mammals, the central extended amygdaloid complex is thought to be involved in food intake and fear/anxiety/stress behavior due to its connectivity to central components of the autonomic nervous system located in the hypothalamus and caudal brainstem (van der Kooy et al., 1984; Luiten et al., 1987; Paré et al., 2004). In particular, CRH neurons in the bed nucleus of the stria terminalis can affect the regulation of appetite (Heimer and Alheid, 1991; Clark and Kaiyala, 2003; Gallagher et al., 2008; Krogh et al., 2008). The stria terminalis is known to have connections with the lateral hypothalamic area, parabrachial nucleus, nucleus of the solitary tract, and dorsal motor nucleus of the vagus; all of the previously listed structures have connectivity to the ANS. In birds, similar central connections with the ANS have been demonstrated showing that ingestive behavior is regulated by similar neural systems in mammals and birds (reviewed in Kuenzel and Blähser, 1993; Kuenzel, 1994, 2000). The same neuropeptide, CRH, has been shown to be important for the expression of stress and anxiety in mammals (Paré et al., 2004) and CRH administered intracerebroventricularly causes distress vocalizations in chicks (Zhang et al., 2004).

9.2.3.2 Medial Extended Amygdala

9.2.3.2.1 Structures

The medial extended amygdaloid complex in mammals comprises the medial amygdala, the neuronal corridor emerging from the medial amygdala, and the medial nuclear complex of the bed nucleus of the stria terminalis. The main input to the medial amygdala originates in the olfactory bulb as well as in the accessory olfactory system and its vomeronasal organ. It is rich in gonadal steroid receptors and projects to various hypothalamic structures involved with reproductive behavior, aggression, and defense (Alheid et al., 1994; Swanson, 2000). Similar to other projection neurons comprising the subpallial structures, the medial extended amygdala has predominantly GABAergic neurons (Alheid et al., 1994, 1995; Swanson, 2000). Glutamatergic neurons have also been identified (Choi et al., 2005).

Avian species have comparable structures. The nucleus taeniae was proposed for some time to be comparable to the mammalian medial amygdala (Cheng et al., 1999) and was officially renamed the nucleus taeniae amygdala (Reiner et al., 2004). Additional data have supported the concept that birds have a medial amygdala and an extended amygdala (Yamamoto et al., 2005). Birds also have the medial bed nucleus of the stria terminalis (BSTM, Figure 9.7) (Jurkevich et al., 1997, 1999; Aste et al., 1998). In addition to arginine vasotocin, neurons of the BSTM of chickens contain galanin (Klein et al., 2006). Embryological studies showed that birds have a subpallial medial amygdala with a corridor of neurons that extends to the BSTM; therefore, it was proposed that birds have a medial extended amygdala (MEA, Figure 9.7) complex (Abellán and Medina, 2009). In addition to GABAergic and glutamatergic neurons, the complex likewise has nitrergic neurons (Panzica et al., 1994; Balthazart et al., 2003) resembling comparable findings of GABAergic and nitrergic neurons in the mammalian medial amygdalar nucleus (Tanaka et al., 1997; Swanson, 2000). In addition to inputs from the olfactory bulbs, the avian subpallial amygdala receives projections from the piriform cortex (Bingman et al., 1994; Veenman et al., 1995). Vasotocin neurons of the BSTM component of the medial extended amygdaloid complex project to the medial preoptic nucleus (Absil et al., 2002a). The avian medial amygdala is reciprocally connected to the hippocampal formation (Atoji et al., 2002; Atoji and Wild, 2004). Importantly, the medial extended amygdaloid complex contains high concentrations of androgen and estrogen receptors as well as the enzyme aromatase, which converts testosterone to estradiol (Balthazart et al., 1998b; Foidart et al., 1999). The abundance of sex steroid receptors is greater in males in this region (Watson and Adkins-Regan, 1989).

9.2.3.2.2 Functions

As stated previously, in mammals the medial extended amygdaloid complex plays a key role in mating, sexual, defensive, and aggressive behaviors. Olfactory input from the main and accessory olfactory systems are important signals to the complex (Swanson, 2000; Choi et al., 2005). Hodological and behavioral data suggest that the BSTM and medial amygdala play a similar role in birds (Panzica et al., 1998; Thompson et al., 1998; Absil et al., 2002a; Xie et al., 2010, 2011). Due to a lack of a defined accessory olfactory system and small olfactory bulbs in most avian species, it has been thought that pheromones and odors do not serve a critical function in this vertebrate class. Nonetheless, it has been shown that pheromones play a role in avian social and sexual behavior (Balthazart and Schoffeniels, 1979; Caro and Balthazart, 2010) and electrophysiological responses in the avian olfactory bulb to odorants are comparable to those in mammals (McKeegan, 2002). Importantly, structures in the medial extended amygdaloid complex have been shown to be sexually dimorphic in avian species. Specifically, the BSTM has been shown to be larger in males, is steroid-responsive, and plays a role in copulatory behavior in mammals (Del Abril et al., 1987; Guillamón and Segovia, 1997) and birds (Kiss et al., 1987; Voorhuis et al., 1988; Panzica et al., 1998; Viglietti-Panzica et al., 1992; Aste et al., 1998; Jurkevich et al., 1999). A subnucleus of the BSTM in chickens, the BSTM2 (ventromedial BSTM), functions in male appetitive sexual behavior (Xie et al., 2010, 2011).

9.2.4 Basal Telencephalic Cholinergic and Noncholinergic Corticopetal System

9.2.4.1 Structures

The basal corticopetal system, sets of neurons at the base of the forebrain that project to the cortex in mammals or pallium in birds and other vertebrates, is composed mostly of cholinergic neurons that in mammals are found overlapping other neuronal types located in the globus pallidus and ventral pallidum of the basal ganglia, the medial septum-diagonal band nucleus, and the magnocellular preoptic nucleus (Gritti et al., 1993, 2003). In birds, three nuclei have been grouped together that comprise this telencephalic corticopetal system: nucleus basalis magnocellularis (NBM), nucleus of the diagonal band, horizontal limb (NDBh), and commissural septal nucleus (CoS; Figure 9.1). All three avian nuclei have in common the presence of cholinergic neurons (Reiner et al., 2004). In the first nucleus listed, the NBM, the cholinergic neurons are found in and about the ventral globus pallidus, ventral pallidum (VP, Figure 9.2), intrapeduncular nucleus, and in and about the lateral and medial forebrain bundle (Medina and Reiner, 1994; Reiner et al., 2004). Of interest is that the NBM is considered equivalent to the mammalian nucleus basalis of Meynert, which contains cholinergic neurons that project to the neocortex, hippocampus, and amygdala (Záborszky et al., 1999). In birds, the NBM receives input from the ventral viscerolimbic basal ganglia, including the nucleus accumbens, as well as the subpallial amygdaloid complex including the arcopallium (Veenman et al., 1995; Medina and Reiner, 1997). The second avian cholinergic nucleus, the NDBh, is similar to cholinergic neurons located in the nucleus of the diagonal band in the mammalian basal forebrain region (Woolf, 1991). The avian NDB projects heavily to the hippocampal and parahippocampal areas (Benowitz and Karten, 1976; Casini et al., 1986; Atoji et al., 2002; Montagnese et al., 2004). The third nucleus, the CoS (Figure 9.7), occurs immediately dorsal to the anterior commissure and just lateral to the midline nucleus, the nucleus of the hippocampal commissure (NHpC, Figure 9.7). The CoS has a major projection to the hippocampus and parahippocampal areas, similar the projections of the NDB (Benowitz and Karten, 1976; Casini et al., 1986; Atoji et al., 2002; Montagnese et al., 2004).

9.2.4.2 Functions

In mammals, the basal forebrain, particularly the nucleus basalis of Meynert, has been implicated in learning and memory. Specifically, corticopetal cholinergic neurons have been shown to play a role in modulating cortical activity and in attentional and arousal processes (Záborszky et al., 1999) that affect learning and memory (Metherate et al., 1988, 1992; Cape and Jones, 2000; Cape et al., 2000). There exists a high correlation between the loss of memory and loss of cholinergic neurons in humans (Auld et al., 2002). Little is known about the role of basal forebrain cholinergic neurons in cognitive processes in birds. Nonetheless, pharmacological blockage of muscarinic cholinergic receptors results in impaired learning and memory in diverse avian species (Patterson et al., 1990; Mineau et al., 1994; Savage et al., 1994; Kohler et al., 1996; Zhao et al., 1997). Additionally, beta-amyloid toxicity, known to damage the basal forebrain cholinergic system in mammals, is known to impair memory in chicks (Gibbs et al., 2010).

9.2.5 Septum and Septal Neuroendocrine Systems

9.2.5.1 Divisions and Structures

The avian septum lies dorsal and anterior to the diencephalon and comprises all neural structures medial to the ventral horns of the lateral ventricle (Figures 9.1 and 9.7). It first appears in rostral coronal sections of forebrain as the ventral horns of the lateral ventricle begin to move apart laterally and the tractus septopallio-mesencephalicus (tsm, Figure 9.6) begins to descend from the hippocampal region into the septum (S, Figure 9.6(B,C,D)). A useful marker of the septal ventral boundary is the anterior commissure, when the mid-septal region appears dorsal to the hypothalamus. Moving caudally, the septum gradually decreases in size. Before the septum disappears, the cortico-habenular and cortico-septal tract that passes through the septum becomes more visible. The septum of mammals has been parcellated into numerous chemoarchitectonic zones that can be grouped into four major divisions (Jakab and Leranth, 1995; Risold and Swanson, 1997a,b). A comparative study was completed using a variety of songbird species and antibodies against 10 neuropeptides and enzymes to develop a nomenclature and framework to establish homologous septal zones among tetrapods (Goodson et al., 2004). Results suggested that the avian septum, similar to that of mammals can be divided into four major divisions: lateral septum (SL), medial septum (SM), septohippocampal septum (SHpS), and caudocentral septum (SCC, Figure 9.7). Coupled with developmental (Puelles et al., 2007; Abellán and Medina, 2009; Abellán et al., 2010) and hodological (Atoji and Wild, 2004; Montagnese et al., 2004, 2008) studies, homologous divisions and structures have been proposed, which will be detailed below.

9.2.5.1.1 Lateral Septum

The developing avian lateral septum (Figure 9.7) comprises three nuclei, the dorsal (SLd) and intermediate (SLi) are striatal derivatives based upon expression of Pax6 and LIM-only gene *Lmo4,* whereas the ventral one (SLv) is a pallidal derivative due to expression of the *Nkx2.1* gene (García-López et al., 2008; Abellán and Medina, 2009; Medina and Abellán, 2009). Hodological studies have shown four major inputs to the lateral septum, including the hippocampal formation, arcopallium/amygdala, diencephalon, and brainstem. Different hippocampal domains project preferentially to different SL regions of birds (Atoji and Wild, 2004; Montagnese et al., 2004, 2008) as well as mammals (Risold and Swanson, 1997b). In addition, in mammals major glutamatergic inputs to the SL have been documented from pyramidal neurons in the hippocampus (Jakab and Leranth, 1995; Risold and Swanson, 1997b). The caudal intermediate arcopallium and nucleus taeniae of the amygdala have been shown to project to the avian SL (Atoji and Wild, 2004; Montagnese et al., 2004, 2008). The mammalian medial amgydala, proposed to be homologous to the avian nucleus taeniae of the amygdala (Thompson et al., 1998; Cheng et al., 1999; Yamamoto et al., 2005), sends projections to the SL (Canteras et al., 1995). Extensive input to the SL originates from the rostral-caudal extent of the ventral diencephalon including the preoptic area (medial preoptic nucleus) (Berk and Butler, 1981; Panzica et al., 1992; Balthazart et al., 1994; Balthazart and Absil, 1997; Atoji and Wild, 2004; Montagnese et al., 2008), hypothalamus (anterior and lateral hypothalamic area), tuberal, and mammillary region (Atoji and Wild, 2004, 2006). Similar projections have been demonstrated in the rat (Canteras et al., 1992, 1994; Risold and Swanson, 1997b). A major output from the mammalian SL descends via the medial forebrain bundle and innervates the medial septal nucleus, nucleus of the diagonal band, lateral hypothalamic area, medial preoptic, anterior hypothalamic, ventral premammillary, and mammillary nuclei (Risold and Swanson, 1997b). The avian SL sends similar projections to the medial preoptic nucleus, anterior and ventromedial hypothalamic nucleus, lateral hypothalamic area, and lateral mammillary nucleus (Atoji and Wild, 2004), thereby showing numerous reciprocal projections between hypothalamic nuclei and SL in birds and mammals. Finally, a fourth extensive set of projections to the mammalian and avian SL originate from the brainstem, specifically from the ventral tegmental area (Kitt and Brauth, 1986b), locus coeruleus (Kitt and Brauth, 1986a), and nucleus raphe/linear caudalis (Cozzi et al., 1991; Atoji and Wild, 2004). Those projections to the SL have been reported to be dopaminergic (Bailhache and Balthazart, 1993; Bottjer, 1993; Moons et al., 1994; Wynne and Güntürkün, 1995), noradrenergic (Bailhache and Balthazart, 1993; Moons et al., 1995; Mello et al., 1998), and serotonergic (Challet et al., 1996), respectively.

Additional chemoarchitectonic studies have shown arginine vasotocin (homologous to mammalian vasopressin) containing dense fiber terminals in the caudal and ventrolateral region of the SL among various vertebrate taxa. Expression of the density of the neuropeptide in vertebrates is testosterone dependent and sexually dimorphic, particularly in mammals (Moore and Lowry, 1998; Goodson and Bass, 2001). The sexual dimorphic expression has been demonstrated in at least three avian species (Voorhuis et al., 1988; Viglietti-Panzica et al., 1992; Jurkevich et al., 1997). The SL is also rich in androgen-concentrating cells (Arnold et al., 1976; Barfield et al., 1978) and in estrogen receptors (Panzica et al., 2001). Other chemoarchitectonic studies have shown that galanin fibers occur in the lateral wall of the septum in Japanese quail (Azumaya and Tsutsui, 1996), collared dove (Dubbeldam et al., 1999), songbirds

(Goodson et al., 2004), chickens (Klein et al., 2006), anuran amphibians (Lazar et al., 1991), and mammals (Melander et al., 1986; Risold and Swanson, 1997a; Chaillou et al., 1999). Of interest is that the distribution of galanin fibers and perikarya, particularly in the lateral septum and medial bed nucleus of the stria terminalis, have been shown to be sexually dimorphic in chickens (Klein et al., 2006), as previously shown for arginine vasotocin. Past studies in mammals have documented galanin and vasopressin co-expression in brain regions known to display sex differences (Miller et al., 1993).

9.2.5.1.2 Medial Septum

The medial septal division in the developing mouse and chicken expresses Tbr1 (Puelles et al., 2000), Lhx6, Lhx7/8, GAD67, and VGLUT2 (Flames et al., 2007; Abellán and Medina, 2009; Abellán et al., 2010) and consists of glutamatergic, GABAergic, cholinergic neurons, and some neurons from pallial septal and pallido-preoptic subdivisions; the latter in the ventral region of the SM expresses some GABAergic neurons (Abellán et al., 2010). Similar to the lateral septum, the medial septum (Figure 9.7) has inputs from the hippocampal complex, the entire rostral-caudal extent of the ventral diencephalon including nuclei from the preoptic, anterior hypothalamus, mammillary and tuberal regions, as well as receives projections from the ventral tegmental area, locus coeruleus, and linear caudalis (Kitt and Brauth, 1986a,b; Atoji and Wild, 2004; Montagnese et al., 2008). Other inputs documented include the nucleus of the diagonal band (Atoji and Wild, 2004; Montagnese et al., 2008), medial division of the bed nucleus of the stria terminalis (Montagnese et al., 2008), and the dorsal lateral nucleus of the anterior medial thalamus (Atoji and Wild, 2004). The SM sends projections to the hippocampal complex (Atoji and Wild, 2004), olfactory tubercle, ventral pallidum, lateral septum, nucleus of the diagonal band, and commissural septal nucleus (Atoji and Wild, 2004; Montagnese et al., 2008). Chemoarchitectonic studies show that most peptidergic and aminergic neurons and fibers show less immunoreactivity compared to the SL. The one exception appears to be the presence of cholinergic neurons, which show their greatest concentration in the medial septum, particularly at caudal levels—not only in a variety of bird species but also the ventromedial septum of most tetrapod groups (Goodson et al., 2004).

9.2.5.1.3 Septohippocampal Septum

A septohippocampal nucleus occurs in mammals (Jakab and Leranth, 1995; Risold and Swanson, 1997a,b) and a third division, the septohippocampal septum (Figure 9.7), has been proposed to exist in birds comprising a small, thin vertical area located above the anterior commissure and extending to the dorsal region of the septum (Goodson et al., 2004). Developmental studies using the pallial marker Emx-1 initially suggested that a discrete, dorsal area of the septohippocampal septum was pallial (Puelles et al., 2007). More recent data using a gene *Lhx5* encoding a transcription factor showed that the entire medial portion of the septohippocampal division from its most dorsal extent ventral to the nucleus of the hippocampal commissure comprises a pallial portion of the septum (Abellán et al., 2010). Chemoarchitectonic data displayed VGLUT2 transporter surrounding the nucleus of the hippocampal commissure, suggesting that the neurons were glutamatergic providing additional support that the septohippocampal septum is a pallial derivative (Abellán et al., 2010). Another source of evidence for its pallial origin comes from the well-known migration of gonadotropin-releasing hormone, type 1 (GnRH-1) neurons from the embryonic olfactory placode into the pallium behind the olfactory bulb and their subsequent migration toward the preoptic, hypothalamic, and septal regions (Figure 9.8(A)). The unusual neuronal migration has been demonstrated in amphibians (Muske and Moore, 1988), mammals (Schwanzel-Fukuda and Pfaff, 1989; Wray et al., 1989), and birds (Murakami et al., 1991; Norgren and Lehman, 1991). Of interest is that for mammalian species, a major portion of the GnRH-1 neurons finally reside in the preoptic area and arcuate nucleus of the hypothalamus (Silverman et al., 1994); in the avian brain such as the chick, 73% of the total GnRH-1 neurons remain in the septum between the medial and lateral septal divisions and surrounding the nucleus of the hippocampal commissure (NHpC, Figure 9.8(B,C); Kuenzel and Golden, 2006). Within the core region of the nucleus of the hippocampal commissure is a dense plexus of corticotropin-releasing hormone fibers (Richard et al., 2004). The avian septohippocampal septum is also characterized by substance P, NPY, and VIP fibers (Goodson et al., 2004). The latter two peptides (NPY and VIP) are associated with reproductive function (refer to Section 9.4.1). Similarly, the septohippocampal n. of rats has been shown to express neuropeptide Y mRNA in cell perikarya and substance P in fibers (Risold and Swanson, 1997a).

9.2.5.1.4 Caudocentral Septum, Cortico-Habenularis, and Cortico-Septal Tract

The fourth major septal area of mammals is the posterior septum, which contains the septofimbrial and triangular nuclei (Jakab and Leranth, 1995; Risold and Swanson, 1997a,b). It has been proposed that birds have a caudocentral septal (Figure 9.7) division that is topographically comparable to the mammalian septofimbrial nucleus and appears to be a caudal expansion of the ventrolateral SL (Goodson et al., 2004). A fiber tract occurs in the caudocentral septum, called the cortico-habenularis and cortico-septal tract. In the rat, a fiber tract that connects the septofibrial nucleus to the hippocampus is the fimbria of hippocampus (Risold and Swanson, 1997b). It has therefore been proposed that the cortico-habenularis and cortico-septal tract be renamed the fimbria

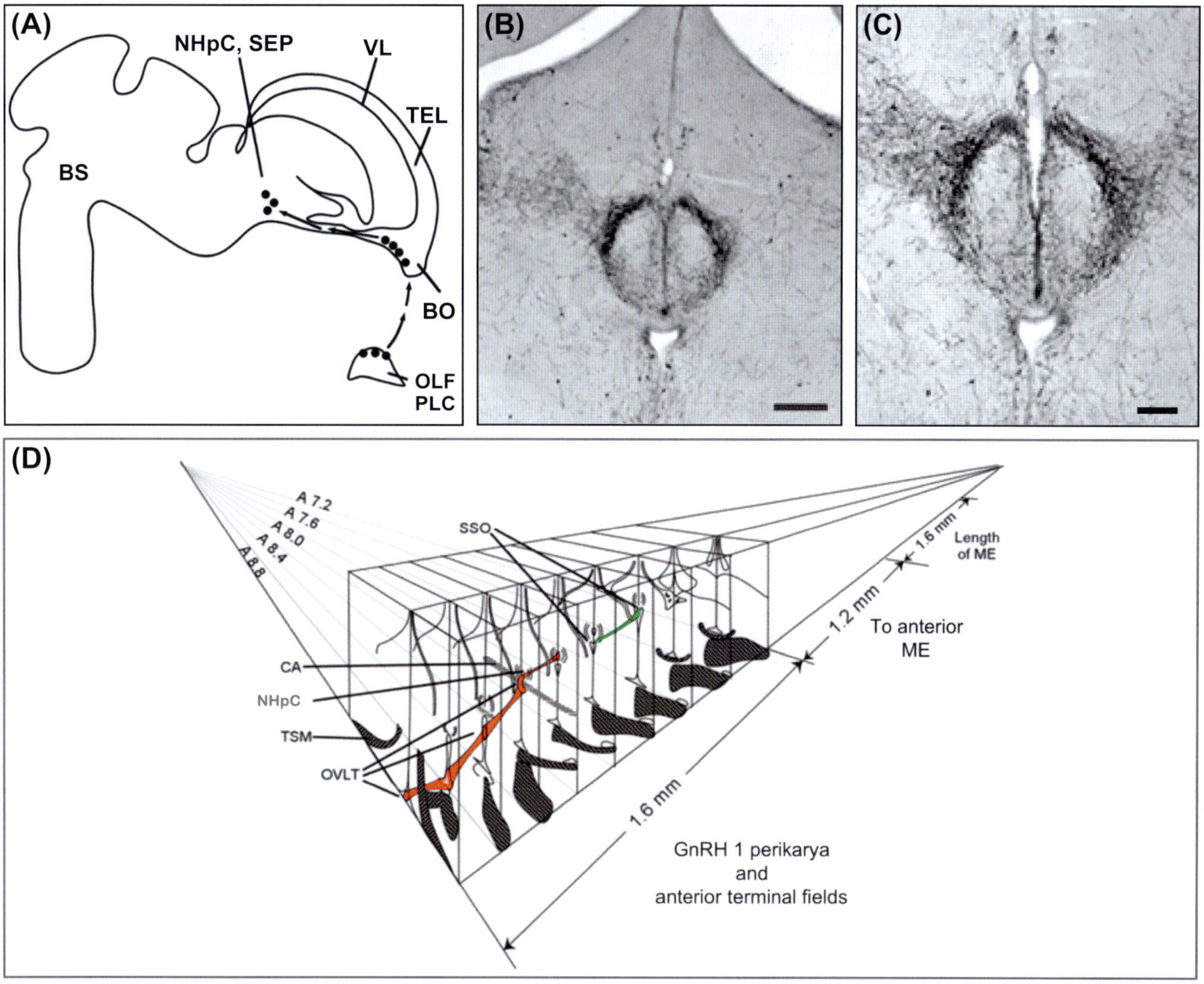

FIGURE 9.8 Origin and distribution of gonadotropin-releasing hormone type one (GnRH-1) neurons in the avian brain. (A) GnRH-1 neurons originate outside the brain in the olfactory placode (OLF PLC) and migrate into the brain to the septum and preoptic area. (B and C) A major site where GnRH-1 neurons migrate during embryonic development occurs surrounding the nucleus of the hippocampal commissure (NHpC) in the septal region. (D) Approximately 73% of GnRH-1 neurons reside in the septum while terminal fields of GnRH-1 neurons occur in the organum vasculosum of the lamina terminalis (OVLT, shown in red) and subseptal organ (SSO, shown in green) in addition to their traditional terminal field in the external zone of the median eminence (ME). Scale bars = 300 μm. Other abbreviations: BO = olfactory bulb, BS = brainstem, CA = anterior commissure, TEL = telencephalon, TSM = tractus septopallio-mesencephalicus, VL = lateral ventricle. *Parts (B), (C), and (D) from Kuenzel and Golden (2006).*

of hippocampus (Goodson et al., 2004; Puelles et al., 2007). In the ventral region of the avian caudocentral septal division occur GnRH-1 neurons that are a continuation of the major group of GnRH-1 found in and about the nucleus of the hippocampal commissure (Kuenzel and Golden, 2006).

9.2.5.1.5 Circumventricular Organs Associated with the Septum: Lateral Septal Organ, Organum Vasculosum of the Lamina Terminalis, and Subseptal Organ

Circumventricular organs (CVOs) are found in the brain of all vertebrate classes. As the name implies, they occur adjacent to the ventricles of the brain in specific regions. Three CVOs are associated with the avian septum including the lateral septal organ (LSO), organum vasculosum of the lamina terminalis (OVLT), and subseptal organ (SSO). There are general characteristics that many CVOs possess, including specialized ependymal cells, a vascular area that has an incomplete blood–brain barrier and cerebrospinal fluid-contacting neurons (Vigh, 1971). Due to their known neuroendocrine function for sustained and integrated activities of vital importance to the survival of a given vertebrate species (Weindl, 1973), three CVOs that border the avian septum will be briefly discussed.

The avian LSO was first named (Kuenzel and van Tienhoven, 1982) and later characterized as having two components, a lateral and medial part (Kuenzel and Blähser, 1994;

Kuenzel et al., 1997). The medial LSO (LSOm) is located at the base of the lateral ventricle in the ventral portion of the lateral septum. It has cerebrospinal fluid-contacting neurons that contain and synthesize VIP, other neuron types, and specialized ependyma. The lateral component (LSOl) occupies the lateral aspect of the base of the lateral ventricle (Figure 9.5(D)), has layers of modified ependymal cells, vasoactive intestinal peptide receptors and lacks a complete blood–brain barrier (Kuenzel and Blähser, 1994; Kuenzel et al., 1997). Vasoactive intestinal peptide containing neurons in the avian (Silver et al., 1988; Wada et al., 2000) and reptilian (Foster et al., 1994; Foster and Soni, 1998) lateral septum have also been shown to contain opsin-like compounds, suggesting that those neurons or others in that organ may serve as encephalic photoreceptors involved in monitoring seasonal time by sensing exposure to daylight.

The OVLT is located near the rostral wall of the third ventricle, termed the lamina terminalis. On both sides of the beginning of the third ventricle is the OVLT, a highly vascular structure. In rodents, the OVLT is restricted to the anterior, ventral hypothalamic region and it is found near the suprachiasmatic nucleus (Duvernoy and Risold, 2007). In contrast, the OVLT of birds is more extensive and comprises a prechiasmatic segment located between the optic chiasma and base of the brain. The vascular OVLT continues dorsal on both sides of the anterior region of the third ventricle continues as a subcommissural segment and passes in front of and dorsal to the anterior commissure to form a precommisssural and supracommissural segment, respectively (Dellmann, 1964; Mikami, 1976). The avian OVLT therefore occupies a strategic position between the anterior roof of the diencephalon and floor of the septum along midline. In mammals, gonadotropin-releasing hormone (GnRH) neurons are found in the preoptic region adjacent to the OVLT and GnRH fibers are found in and about the OVLT (Foster and Younglai, 1991; Silverman et al., 1994). In the chick, a terminal field of GnRH-1 neurons has been observed throughout the entire extent of the OVLT (shown in red) and continues through the length of another CVO, the subseptal organ (shown in green; Figure 9.8D). Data suggest a large terminal field of GnRH-1 neurons occurs along the midline floor of the septal region where GnRH-1 hormone can be secreted into the cerebrospinal fluid of the third ventricle and/or into the capillaries of the CVOs (Kuenzel and Golden, 2006).

The subfornical organ (SFO) of mammals is situated at the meeting point of the two lateral and third ventricles and is suspended from the fornix (Miselis et al., 1979). A subseptal organ was described in birds and the name was suggested in place of the term subfornical organ due to lack of evidence that birds have a fornix (Legait and Legait, 1958). The subseptal organ has a finger-like projection into the dorsal region of the third ventricle and has a dense terminal field of GnRH-1 fibers (Figure 9.2; Walsh and Kuenzel, 1997) as well as a rostral and caudal extension along midline that continues following the caudal end of the OVLT as described previously. In mammals, the SFO is highly vascular and is an extremely sensitive brain site for the induction of water intake by angiotensin II (Miselis et al., 1979) due to the presence of angiotensin receptors in that organ (Mendelsohn et al., 1983). Similarly, the subseptal organ in birds is sensitive to angiotensin II (Takei, 1977; Massi et al., 1986) and has angiotensin II receptor subtypes in its structure (Schäfer et al., 1996).

9.2.5.1.6 Functional Considerations for the Septum

The septum has major connections with the hippocampus, amygdala, diencephalon, and brainstem. Functions attributed to this brain subpallial region include ingestive, defensive, and reproductive behaviors that are fundamental to sustainability of vertebrate species (Swanson, 2000). With respect to inputs from the hippocampus, food caching has been associated with the septum. Specifically, black-capped chickadees are known to display food-storing behavior, a seasonally related activity. A significant increase in septal as well as hippocampal volume occurred during the time of year storing behavior was at its peak (Shiflett et al., 2002). Another seasonal activity displayed by birds is reproductive behavior that usually occurs in the spring when day length increases. The septum plays a major role in courtship and aggressive behavior in a variety of avian species (Ramirez et al., 1988; Goodson, 1998; Goodson and Adkins-Regan, 1999; Panzica et al., 2001). Lesions of the septal region produced a facilitation of male courtship behavior and vocalization in ring doves (Cooper and Erickson, 1976) and facilitated both overt aggression and the number of simple, multipurpose songs in territorial field sparrows (Goodson et al., 1999). When brain sections are immunostained for arginine vasotocin (AVT), the lateral septum of male Japanese quail shows significantly greater immunostaining of AVT fibers when compared with females (Panzica et al., 2001). Similarly, AVT and galanin fibers in male chickens (Jurkevich et al., 1997; Klein et al., 2006) display increased density within the lateral septum. During both courtship and mating behavior between roosters and hens as well as agonistic behavior between two males, significantly greater amounts of Fos protein produced by the immediate early gene *c-fos* occurred in the lateral septum of males (Xie et al., 2010). Increased expression of the gene *c-fos* in neurons is a marker of activation of those cells.

The majority of neurons, GnRH-1 neurons, responsible for development and function of the gonads occur in the septum of birds (Kuenzel and Golden, 2006). It has been hypothesized for a number of years that activation of GnRH-1 neurons is thought to occur by neurons called encephalic photoreceptors that can sense photoperiod (Follett and Davies, 1975; Silver et al., 1988; Kuenzel,

1993; Sharp and Ciccone, 2005). The location of neurons serving as encephalic photoreceptors remains controversial. One possible site, as previously discussed, is in the LSO (Li and Kuenzel, 2008; Li et al., 2009). A second site is the mediobasal hypothalamus, where either neurons associated with thyroid hormone metabolism/opsin five (Yoshimura et al., 2003; Yasuo et al., 2003; Ono et al., 2009; Nakane et al., 2010) or neurons containing melatonin, dopamine, and melanopsin serve as the encephalic photoreceptors (Thayananuphat et al., 2007; Kang et al., 2007; El Halawani et al., 2009). Another candidate site is in and around the paraventricular nucleus, where neurons containing vertebrate ancient opsin occur (Halford et al., 2009).

Hodological studies in birds involving the septum have shown that not only is there a descending neuroendocrine system affecting hypothalamic function but also ascending pathways from key diencephalic structures back to the septum, suggesting that the downstream motor pathways have a number of feedback loops (Berk and Butler, 1981; Berk and Finkelstein, 1983; Atoji and Wild, 2004; Montagnese et al., 2004, 2008). Several hypothalamic structures with direct connections to the septum have been shown in previous studies to be involved with food intake and the regulation of the autonomic nervous system. The reciprocal connections of the septum to hypothalamic structures as well as the presence of the subseptal organ at the floor of the septum, shown to contain angiotensin II receptors, provides evidence that the septum is involved in the regulation of food and water intake, respectively. Indeed, there exists structural and functional evidence that the septum possesses a descending neuroendocrine and integration system involved with basic social and ingestive behaviors and functions.

9.3 COMPONENTS OF THE AUTONOMIC NERVOUS SYSTEM

The basic components of the ANS, including efferent pre- and postganglionic neuronal configurations of the sympathetic nervous system (SNS) and parasympathetic nervous system (PNS), neurotransmitters, receptors, and specific sites in the brain and spinal cord where preganglionic neurons originate can be found in the previous chapter addressing the avian ANS (Kuenzel, 2000). In addition, there are excellent reviews on this subject (Akester, 1979; Bubien-Waluszewska, 1981; Yasuda, 2002). This section includes solely the overall function and central structural components of the ANS in order to show sites where connectivity from subpallial structures occurs to affect the regulation of the ANS. Overall, the ANS is responsible for motor functions and secretions that occur automatically on a daily, seasonal, and annual basis. It involves the integration of sensory information such that smooth muscle, cardiac muscle, and secretory cells located in visceral organs provide the appropriate output to maintain homeostasis of bodily functions, such as heart and respiratory rate, blood pressure, osmotic and pH balance, body temperature particularly in homoiotherms, gastrointestinal movements, reproductive processes, circadian rhythms, and other essential circannual rhythmic functions. To regulate the output, two antagonistic systems are in place—the sympathetic and parasympathetic arms of the ANS.

9.3.1 Sympathetic Nervous System

Similar to mammals, a chain of ganglia exists on both sides of the vertebral column in the thoracolumbar region of the spinal cord of birds. A marked difference, however, exists in avian species. Most of the vertebrae are fused; therefore, it requires experience to dissect the entire spinal cord, spinal nerves, and ganglia associated with each segment of the spinal cord. More than one system of numbering the spinal nerves and ganglia exist due to the variable number of vertebrae among various avian species, as well as different methods that either number the spinal segments and nerves sequentially beginning with the cervical region and ending with the last coccygeal vertebra (Dubbeldam, 1993; Yasuda, 2002) or number the vertebrae, spinal segments, and their associated spinal nerves, dorsal root ganglia, and the chain of autonomic ganglia to a regional group of vertebrae (Nickel et al., 1977; Landmesser, 1978). The regional terms cervical, thoracic, lumbar, and sacral (or synsacral and coccygeal) nerves (and ganglia) have been used, but the transitional zones may be problematic to ascribe a ganglion or spinal nerve to the appropriate region when the synsacral bone is dissected to reveal the caudal spinal cord (Dubbeldam, 1993). Therefore, documentation with images is most helpful, particularly when referencing a specific segment, ganglion, or spinal nerve in a study.

Within each segment of the thoracolumbar region of the spinal cord is a nucleus that identifies the central nervous system origin of the SNS. The intermediomedial nucleus, previously termed the column of Terni, contains the preganglionic motor neurons of the SNS. The nucleus is located at midline of a cross-section of spinal cord, directly dorsal to the central canal (see Figure 9.5 in Kuenzel, 2000). The axonal processes of the neurons move laterally a short distance and innervate one of the autonomic ganglia. A major neurotransmitter released is acetylcholine (ACh), which binds to nicotinic receptors present on postganglionic neurons. The long postganglionic axons project to specific visceral organs and release primarily the neurotransmitter norepinephrine, which binds to subtypes of alpha- or beta-adrenergic receptors.

An autonomic ganglionated nerve unique to birds is Remak's nerve. It appears at the end of the duodenum, running parallel to the intestinal tract and ending at the cloaca. Remak's nerve plays a role in the regulation of gut motility

(Hodgkiss, 1984a,b). Microinjections into individual nerves within ganglia of Remak's nerve revealed four distinct morphological types of neurons. The function of each neuronal type remains unknown (Lunam and Smith, 1996).

9.3.2 Parasympathetic Nervous System

The PNS has a system divided into two sets of preganglionic neurons, one originating in the brain and a second set located in the sacral region of the spinal cord. Preganglionic neurons in the brain are found in motor nuclei of the oculomotor (III), facial (VII), glossopharyngeal (IX), and vagus (X) nerves (Akester, 1979; Dubbeldam, 1993). The cranial nerve that has a major impact on a diverse number of visceral organs regarding parasympathetic regulation of function is the vagus. Those in the sacral region have processes that form a pelvic plexus. In contrast to the SNS, the preganglionic neurons of the PNS project long processes to specific ganglia located within or adjacent to visceral organs or tissues. Similar to the SNS, the preganglionic neurons secrete the neurotransmitter ACh. Nicotinic receptors on postganglionic neurons bind ACh and, in turn, the major neurotransmitter released from postganglionic neurons is ACh, which binds to muscarinic receptors found within the targeted visceral organ.

9.4 FUNCTIONAL NEURAL PATHWAYS INVOLVING THE SUBPALLIUM AND ANS

Two behaviors exhibited by birds that are important to the survival of their species and their welfare are reproductive and ingestive behavior. Each of them to be expressed depends upon structures found in the subpallium and autonomic nervous system. This section attempts to show how each of the behaviors requires the subpallial, autonomic nervous, and neuroendocrine systems in order to regulate the essential ethophysiological processes required on a daily and/or seasonal basis.

9.4.1 Regulation of the Reproductive System

The proper function of the reproductive system depends upon an unusual developmental process. Unlike most neurons that develop and differentiate within the brain, gonadotropin-releasing hormone neurons type 1 (GnRH-1) originate outside of the brain and migrate into the developing organ. It was originally discovered in amphibians (Muske and Moore, 1988) and detailed data showed that GnRH-1 neurons in mammals migrate from the olfactory to preoptic and hypothalamic brain regions (Schwanzel-Fukuda and Pfaff, 1989; Wray et al., 1989). Similar data have been obtained for the embryonic chick (Murakami et al., 1991; Norgren and Lehman, 1991). It has been confirmed that the olfactory placode, outside of the brain, serves as the origin of GnRH-1 neurons in the chick (Figure 9.8(A)) and GnRH-1 precursors are specified through a mutual antagonism between retinoic acid and fibroblast growth factor (FGF), where FGF induces specification of GnRH-1 neurons (Sabado et al., 2012). Once inside of the brain in the olfactory region, the neurons migrate caudally toward the preoptic and hypothalamic area. In contrast to mammals, where the majority of GnRH-1 neurons take up residence in the preoptic area and anterior hypothalamus, including a population in the arcuate nucleus within the caudal mediobasal hypothalamus (Silverman et al., 1994), in the chick brain the majority of GnRH-1 neurons (73%) settle in the septal region of the subpallium (Kuenzel and Golden, 2006). A major group of GnRH-1 neurons occur surrounding the bed nucleus of the pallial commissure, currently renamed the nucleus of the hippocampal commissure (NHpC; Figure 9.8(B,C)). Indeed, developmental biologists have reported that within the septal region, a medium strip of the septum on either side of midline from the dorsal region down to the NHpC, called the septohippocampal septum, appears to have pallial-type neurons rather than the expected striatal and pallidal types using histogenic cell markers (Puelles et al., 2007; Abellán et al., 2010). The migrating GnRH-1 neurons from the pallial, olfactory bulb region could account for some of the pallial neurons found in this part of the septum (see Section 9.2.5).

The GnRH-1 neurons found in the septal and preoptic regions comprise the primary neuroendocrine system responsible for development and maintenance of the reproductive system and influence gender-specific reproductive behavior. A major portion of the neurons form a large terminal field in the median eminence, and their secretions stimulate the anterior pituitary gland to secrete luteinizing hormone (LH) and follicle-stimulating hormone (FSH). In turn, the gonadotropins (LH and FSH) transported to the gonads cause the release of testosterone from the testes or estrogens from the ovaries.

In addition to the major population of GnRH-1 neurons that reside in the septal division of subpallium, another subpallial component, the extended amygdala, likewise impacts reproductive function. The medial extended amygdaloid complex (see Section 9.2.3 for more details) includes the medial bed nucleus of the stria terminalis (BSTM). The BSTM is sexually dimorphic, shown to be larger in males, is steroid-responsive and affects copulatory behavior in birds (Kiss et al., 1987; Voorhuis et al., 1988; Panzica et al., 1998; Viglietti-Panzica et al., 1992; Aste et al., 1998; Jurkevich et al., 1999). A subnucleus of the BSTM in chickens, the BSTM2 (ventromedial BSTM), functions in rooster appetitive sexual behavior (Xie et al., 2010, 2011). Importantly, vasotocin neurons in the BSTM project to the medial preoptic nucleus (Absil et al., 2002b). The medial preoptic nucleus (POM), similar to the BSTM, is sexually dimorphic (Viglietti-Panzica et al., 1986; Adkins-Regan and Watson, 1990). In male quail, the caudal POM was shown to be associated with consummatory sexual behavior, whereas the rostral area affected appetitive sexual behavior (Balthazart et al., 1998a; Balthazart and Ball, 2007).

A study in Japanese quail has included the mapping of an entire reproductive pathway from the POM down to the primary muscle in the cloaca responsible for the production of foam and transfer of semen from the cloaca of the male to that of the female during copulatory behavior (Wild and Balthazart, 2013). The pathway includes a projection from the POM to the intercollicular complex (ICo; Wild and Balthazart, 2013). Within the ICo was found a distinct, elongated nucleus, the dorsomedial nucleus (DM). The DM was shown to project to the caudal brainstem nucleus, called the nucleus retroambigualis (RAm; Wild et al., 1997; Wild and Balthazart, 2013). Anterograde tracing from the RAm showed a continuous projection from that caudal brainstem nucleus through the entire length of the spinal cord to the sacral region of spinal cord. The motor neurons in the ventral horn at that spinal cord level innervate the large cloacal sphincter muscle (mSC; Wild and Balthazart, 2013). The cloacal sphincter muscle has been shown responsible for the production of foam, which is transferred along with semen to the female during copulation to enhance male fertilization (Seiwert and Adkins-Regan, 1998).

Another pathway involving the autonomic nervous system has also been proposed, affecting avian reproductive function as well as dominance and defensive behavior. The key structure involved is the ICo, including a second structure enclosed within the ICo, the DM (described in the previous paragraph; Wild and Balthazart, 2013). An intriguing, earlier proposal was that the dorsomedial part of the ICo and the central gray (GCt) of avian species were comparable to the mammalian periaqueductal gray (PAG; Dubbeldam and den Boer-Visser, 2002). Of interest is that the mammalian PAG has been shown to be part of an emotional motor system (Holstege, 1992). The PAG concept of Dubbeldam and den Boer-Visser (2002) has been expanded and supported by additional data. It has been proposed that the elongated avian ICo containing the enclosed DM plus the dorsolateral nucleus (DL) that joins the GCt are equivalent to the mammalian PAG (Kingsbury et al., 2011). Within the elongated ICo and GCt of birds are distinct parts related to reproduction, particularly copulation and others related to dominance and defensive behavior. The GCt in birds has been shown to project to neurons in the column of Terni (the intermediomedial nucleus, pars commissurae dorsalis) that houses the preganglionic neurons of the sympathetic nervous system (Breazile and Hartwig, 1989). Therefore, the revived avian PAG concept (Kingsbury et al., 2011) provides a fertile ground for additional studies to refine and characterize this very interesting midbrain structure.

In summary, the septum, medial bed nucleus of the stria terminalis, medial preoptic nucleus, and their known connectivity with the neuroendocrine and autonomic nervous systems result in coordinated neural and endocrine hormonal output to help regulate this complex reproductive function of birds.

9.4.2 Regulation of Food Intake

Significant progress has been made in the neural regulation of food intake in mammals. A similar neural system appears to exist in avian species as well. A major signaling molecule, leptin, produced in adipose tissue was shown to play a major role in regulating body weight of mammals (Friedman and Halaas, 1998). The concentration of leptin found in the blood was shown to be proportional to the extent of adipose stores located in the periphery. Leptin appeared to be a metabolic signal that directly affected neural circuits in the central nervous system, particularly in the hypothalamus, that regulated food intake and energy homeostasis. A group of neuropeptides and receptors were found in the hypothalamus of mammals that played a key role in regulating appetite as well as energy expenditure in order to maintain body weight (Woods et al., 1998; Berthoud, 2002). Of interest is that some of the genes responsible for producing those critical neuropeptides and their receptors have been cloned in birds, suggesting strongly that the neural system is evolutionarily conserved.

A selected list of neuropeptides that have been shown to affect food intake and/or energy expenditure in birds is found in Table 9.1. Curiously, some of the peptides had different effects on feeding behavior in birds compared to mammals (indicated by footnotes). Even though the genes producing

TABLE 9.1 Neuropeptides Injected Centrally and Their Effect on Food Intake in Birds[1]

Orexigenic	No Effect	Anorexigenic
Neuropeptide Y	Melanin concentrating hormone[3]	α-Melanocyte stimulating hormone
Agouti-related peptide	Orexins (A and B)[3]	Cocaine and amphetamine regulated transcript
Peptide YY[2]	Motilin[3]	Corticotropin-releasing hormone
Pancreatic polypeptide[2]		Ghrelin[3]
		Cholecystokinin
		Bombesin
		Glucagon-like peptide 1
		Gastrin
		Urotensin I/Urocortin Neuromedin U/S

[1]*Data reported in Furuse (2002) and Richards and Proszkowiec-Weglarz (2007).*
[2]*In mammals, the peptides decrease food intake.*
[3]*In mammals, the peptides increase food intake.*

leptin receptors have been cloned and sequenced in chickens and turkeys (Horev et al., 2000; Ohkubo et al., 2000; Richards and Poch, 2003; Liu et al., 2007), a controversy exists regarding the published sequence of its actual ligand. The consensus is that the leptin-like gene has yet to be cloned for any avian species. Regardless, there exists compelling evidence that an avian-like leptin molecule exists. A leptin gene has been cloned in amphibians; even though it has low amino acid sequence homology to mammalian leptin, a common feature among known leptin proteins is a highly conserved tertiary structure (Crespi and Denver, 2006). Administration of recombinant chicken or mammalian leptin proteins to birds reduced food intake, similar to leptin's effect in rats (Denbow et al., 2000; Lohmus et al., 2003; Dridi et al., 2005). Importantly, immunization against leptin in chickens resulted in increased food intake (Shi et al., 2006).

Key neuropeptides that have been identified in mammals for regulating feeding behavior and energy expenditure (Woods et al., 1998) have likewise been shown to play equivalent roles in birds. Figure 9.9 is a working model regulating food intake in avian species. The model is adapted from current data and concepts developed in mammals. Note that the schematic diagram, two transverse sections of chick brain (Figure 9.9), should be viewed as two parts. Figure 9.9(A) (left side of the brain) shows activity of the paraventricular nucleus (PVN), lateral hypothalamic area (LHy), infundibular nucleus (IN), and SNS when leptin-like signals in the blood are high. The right side of the brain (Figure 9.9(B)) shows activity of the PVN, LHy, IN, and PNS when plasma leptin-like signals are low. The following gives data supporting the concept that in mammals and birds high leptin-like signals (Figure 9.9(A), left side) result in decreased food intake and increased energy expenditure, whereas low leptin-like signals (Figure 9.9(B), right side) increase food intake and reduce energy expenditure.

Specifically, neuropeptide Y (NPY) has been shown to be one of the most potent orexigenic neuropeptides (Kuenzel et al., 1987). One of its locations in the diencephalon is the inferior hypothalamic (IH) and infundibular nucleus (Kuenzel and McMurtry, 1988; Walsh and Kuenzel, 1997); the latter nucleus is considered homologous to the arcuate nucleus, the major hypothalamic nucleus in mammals containing NPY neurons (Schwartz et al., 2000). Additionally, a melanocortin system comprising agouti-related peptide (AgRP) and α-melanocyte-stimulating hormone (α-MSH, a product of pro-opiomelanocortin [POMC] gene) are two other key peptides found within neurons of the mammalian arcuate nucleus regulating food intake (Woods et al., 1998; Cone, 1999; Schwartz et al., 2000). The melanocortin system has likewise been shown to occur in the infundibular nucleus of birds (Phillips-Singh et al., 2003). Similar to mammals, neurons in the avian infundibular nucleus co-expressed NPY and AgRP mRNA, and both NPY and AgRP mRNA were significantly increased by fasting (Boswell

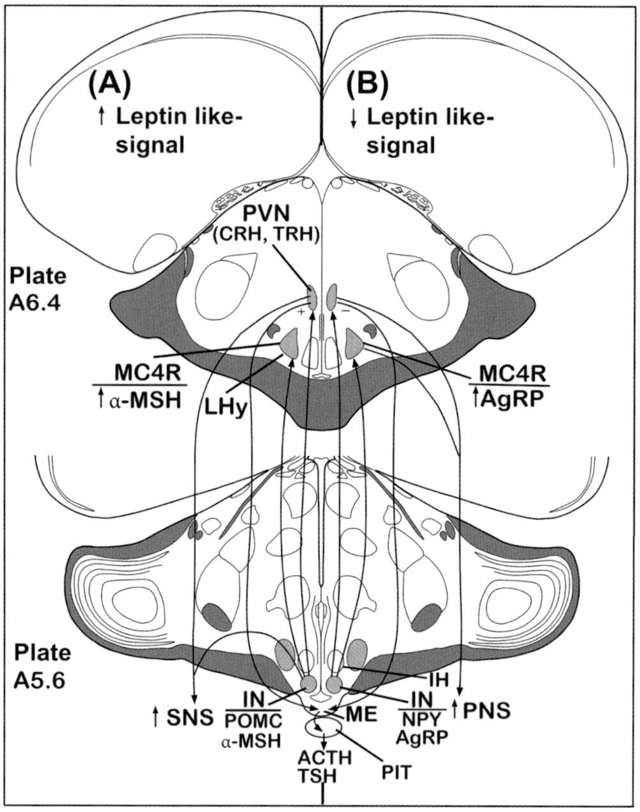

FIGURE 9.9 Two transverse sections of chick brain showing key hypothalamic structures regulating food intake and energy expenditure. (A) Left side of brain sections show activity of three hypothalamic structures, the infundibular (IN), lateral hypothalamic (LHy) and paraventricular (PVN) nuclei, when blood levels of leptin-like signals are high. Specific neurons found in the IN (homologous to the arcuate nucleus of mammals) containing pro-opiomelanocortin (POMC) and α-melanocyte stimulating hormone (α-MSH) are activated. The neurons in turn stimulate paraventricular (PVN) neurons containing corticotropin-releasing hormone (CRH) and thyroid-releasing hormone (TRH) to release their respective neuropeptides into the median eminence (ME). The peptides are transported to the anterior pituitary (PIT), causing the release of the hormones adrenocorticotropic hormone (ACTH) and thyroid-stimulating hormone (TSH) into the vasculature, effecting increased peripheral levels of corticosterone and thyroid hormones. The sympathetic nervous system (SNS) is activated. The increased α-MSH released into the LHy enable the peptide to effectively bind to melanocortin 4 receptors (MC4R). The overall effect is decreased food intake. (B) The right side of brain sections shows the effects of low levels of blood leptin-like signals. Neuropeptide Y (NPY) and agouti-related peptide (AgRP) containing neurons in the IN are activated that stimulate food intake. Additionally, they inhibit activity of CRH- and TRH-containing neurons in the PVN. The AgRP outcompetes α-MSH for MC4R binding sites in the LHy, thereby blocking α-MSH from inhibiting food intake. Additionally, the parasympathetic nervous system (PNS) is activated contributing to an enhanced food intake and decreased energy expenditure.

et al., 2002; Phillips-Singh et al., 2003). Importantly, AgRP administered centrally in birds increases food intake (Kawakami et al., 2000; Tachibana et al., 2001; Strader et al., 2003), whereas α-MSH inhibits feeding (Kawakami et al., 2000; Tachibana et al., 2001; refer to Table 9.1), similar to what has been found in mammals.

One additional component of the melanocortin system that is essential for the actions or signaling of its two key peptides, AgRP and α-MSH, is the identification of their respective receptors. In mammals, the main melanocortin receptor involved is the melanocortin 4 receptor (MC4R; Forbes et al., 2001). Two peptides, α-MSH and AgRP, compete for the same MC4R located on neurons in the lateral hypothalamic area. The peptide α-MSH is an agonist for the MC4R, whereas AgRP serves as an antagonist. When excess α-MSH is produced, the peptide out competes AgRP for binding sites, resulting in decreased food intake and increased catabolism. In contrast, excess AgRP will result in blocking available sites of the MC4R, resulting in elevated food intake and decreased energy expenditure (Cone, 2005). Importantly, birds likewise express the MC4R, which appears to be the key receptor regulating feeding behavior and energy balance based upon the use of an MC4R antagonist and a nonselective MC3-/MC4R agonist in ring doves (Strader et al., 2003). In chickens, the major melanocortin receptor type present in the brain implicated in the regulation of feeding is the MC4R (Takeuchi and Takahashi, 1998, 1999). The working model in birds (Figure 9.9) suggests that the regulation of food intake and energy expenditure in the two classes of vertebrates has similarities. Note, however, that it is a highly simplified model and requires experimental documentation, particularly for identification of the avian leptin-like molecule.

9.4.2.1 An Autonomic Pathway Regulating Food Intake and Energy Balance

The hypothalamic model regulating food intake and energy balance (Figure 9.9) no longer has a structure, the ventromedial hypothalamic nucleus (VMN)—previously termed the "satiety center" in publications for at least 60 years because lesions to that structure resulted in hyperphagia and obesity (Hetherington and Ranson, 1940). The arcuate nucleus has replaced it due to its sets of neuropeptide-containing neurons that either stimulate or inhibit feeding depending upon signals from peripheral organs, particularly lipid stores. In the rat as well as the chick brain, the VMN is positioned just anterior to the IN (arcuate nucleus) in the rostral-caudal plane. The VMN electrolytic lesions placed in experimental animal brains in past publications were large and most likely encroached upon the arcuate nucleus.

The model (Figure 9.9) has retained two neural structures, specifically the paraventricular nucleus and lateral hypothalamic area, which have been in feeding models over the past 50 years. In mammals, both show connectivity to the autonomic nervous system. The PVN projects to the central gray (GCt) and then to the intermediolateral (IML) cell column, where the preganglionic neurons of the SNS originate (Luiten et al., 1987). The PVN as well as the LHy project descending neuronal axons to the nucleus tractus solitarius (nTS) and to the dorsal motor nucleus of the vagus, which are key nuclei of the PNS (Luiten et al., 1987). Hence, a change in activity of SNS compared to the PNS, which are antagonistic, has significant effects on energy expenditure and therefore significantly impacts feeding behavior and homeostatic mechanisms regulating body weight. An energy sensor, adenosine monophosphate-activated protein kinase (AMPK), which is recognized as an important enzyme of a kinase-signaling cascade at the cellular level, stimulates energy-producing (catabolic) pathways (Hardie, 2004). The end result of AMPK activation is the stimulation of NPY/AgRP hypothalamic neurons, resulting in increased food intake and decreased energy expenditure. Coupled with AMPK is another type of sensor, a serine–threonine kinase called mammalian target of rapamycin (mTOR), which has the opposite effect of AMPK. When activated, the mTOR kinase pathway results in the stimulation of POMC-expressing neurons, causing the release of α-MSH and the subsequent reduction in food intake and an increase in energy expenditure (Cota et al., 2006). Utilizing antibodies to mTOR showed that this protein was found in both the PVN and arcuate nucleus with high abundance in arcuate NPY/AgRP neurons (Cota et al., 2006). An excellent review of possible mechanisms regulating appetite and energy expenditure in chickens, which incorporates AMPK and TOR signaling in a model that includes hypothalamic neurons impacting body weight, can be found in Richards and Proszkowiec-Weglarz (2007).

9.4.2.2 Poikilostasis or Shifts in Homeostasis: An Autonomic Hypothesis that May Explain the Regulation of Annual Cycles of Birds (and Perhaps a Consequence of Genetic Selection for Rapid Growth in Broilers and Turkeys)

In the last edition of *Sturkie's Avian Physiology*, the hypothesis of poikilostasis was introduced (Kuenzel, 2000). It was based upon the peptide distribution (Kuenzel and Blähser, 1994) and gene expression (Kuenzel et al., 1997) of the neuropeptide VIP mapped throughout the chick brain and a hypothesis related to a possible neural system regulating the annual cycle of domestic as well as wild avian species (Kuenzel and Blähser, 1993). The unique VIP distribution identified an avian circumventricular organ, the lateral septal organ, subpallial structures, and central components of the autonomic nervous system—a majority of which were previously grouped together and proposed to function as a visceral forebrain system in mammals (van der Kooy et al., 1984). The mammalian visceral forebrain system comprised the medial and lateral prefrontal cortex, paraventricular, arcuate, and posterolateral hypothalamic nuclei, bed nucleus of the stria terminalis, central nucleus of the amygdala, and nucleus tractus solitarius. It was hypothesized that the system influenced autonomic functions and could override brainstem homeostatic mechanisms during periods of stress or emotional activity (van der Kooy et al., 1984). The poikilostasis hypothesis remains a viable one that perhaps can help explain the annual cycle of migratory birds, as well as the behavior and physiology of the modern broiler and turkey, which have been largely selected for growth rate and feed conversion.

Migratory birds, particularly those in temperate zones that migrate north each spring, need to shift from an anabolic state where rapid body weight gains are essential in order to deposit the necessary fat reserves to make the demanding migratory journey to the breeding grounds. Thereafter, a shift to a catabolic state occurs as birds enter a new phase of their annual cycle. Birds become territorial in order to attract and find a mate during the breeding season. The physiological and behavioral processes required for successful reproduction and the raising of young are metabolically demanding. Thereafter, birds undergo a major molt that continues their catabolic state, resulting in the utilization of considerable lipid reserves. A shift to an anabolic state then follows, resulting in increased food intake to prepare birds for a rigorous return migratory flight to the wintering grounds. Due to the marked changes in metabolic states that birds undergo, coupled with short time periods for each particular phase of their annual cycles, birds would unlikely be able to complete their seasonal life history requirements if set points for energy homeostasis were fixed throughout the year. Hence, shifts in homeostasis (poikilostasis) would allow the parasympathetic nervous system to dominate when anabolic processes were critical for rapid body weight gains; conversely, a shift to greater activity of the sympathetic arm of the ANS could occur when energy demands were high and sustained. The hypothesis of poikilostasis suggests that there exists in birds, and perhaps other vertebrate species, the capacity for dynamic shifts in the balance or set points of the ANS in order to successfully complete their annual cycles. The recent molecular models involving neural pathways, energy sensors, and the ability to integrate nutrient and hormonal signaling at the level of the hypothalamus may provide key markers for testing the validity of this hypothesis by focusing upon signals and processes associated with shifts in the balance of the ANS.

With the continued use of genetic selection programs in broilers as well as turkeys for increased growth rate and better feed conversion, the result has been a modern bird showing a high efficiency for the conversion of feed to a quality meat product. A cost, however, for managers of broiler and turkey breeders has been a feed restriction program that requires careful monitoring throughout the life span of birds in order to ensure that parent stock remain healthy and continue to produce an acceptable rate of fertilized eggs (Richards et al., 2010). Data over many years have shown that broilers compared to layers display increased food intake, less overall activity, and a lower basal metabolic rate. Data suggest an imbalance of the ANS, pointing to a physiological system with high parasympathetic and low sympathetic activity in meat-type birds. Due to the current advances in the understanding of the neural regulation of food intake, metabolic signals, energy homeostasis, and genes involved (Richards and Proszkowiec-Weglarz, 2007; Byerly et al., 2009; Yuan et al., 2009), the time may be appropriate to include an examination of genes affecting the ANS in order to develop an objective PNS/SNS ratio indicative of an effective balance between acceptable growth rate, lifetime welfare, and sustained reproductive output for birds selected as parent stock.

9.5 SUMMARY AND CONCLUSIONS

This chapter examined neural structures and functions that comprise five distinct regions in the subpallium of the avian forebrain (Figure 9.1), as well as the basic components of the ANS. Noteworthy is the evidence that the regions are evolutionarily conserved among vertebrates. Therefore, data obtained in mammalian species can be utilized to help discover the function of systems that have not been examined extensively in birds. Subpallial structures and regions are important as several have been shown to have connectivity to pallial (cortical-like) brain regions as well as to key hypothalamic and brainstem structures that are part of the neuroendocrine and autonomic nervous systems. Importantly, the latter two systems involve the production of hormones that can affect the physiology and behavior of organisms for hours, days, and even weeks—indicating processes regulating seasonal activities of birds.

Some of the functions associated with subpallial and ANS components include the neural regulation of feeding, reproductive, voluntary muscle, agonistic, and stress behaviors. Additional functions associated with the subpallium and ANS include reward, memory, and learning. It is hoped that the review will stimulate further reading and examination of the neural systems discussed as well as help suggest future experiments that can advance the understanding of the avian central nervous system, its neural pathways, and functions among various wild and domestic avian species.

ACKNOWLEDGMENTS

Supported in part by National Science Foundation Grant #IOS-0842937, a grant from the Arkansas Biosciences Institute (ABI), the Arkansas Division of Agriculture, and a past grant no. 2005-35203-15850 from the National Research Initiative Competitive Grants Program, USDA/AFRI/National Institute of Food and Agriculture (NIFA). The author wishes to thank Dr Seong Kang for reading the manuscript, helping to organize and format the references, and completing Figures 9.3 and 9.8, as well as Rajamani Selvam for making the final layout of Figures 9.7 and 9.9.

REFERENCES

Abellán, A., Medina, L., 2008. Expression of *cLhx6* and *cLhx7/8* suggests a pallido-pedunculo-preoptic origin for the lateral and medial parts of the avian bed nucleus of the striaterminalis. Brain Res. Bull. 75, 299–304.

Abellán, A., Medina, L., 2009. Subdivisions and derivatives of the chicken subpallium based on expression of LIM regulatory genes and markers of neuron subpopulations during development. J. Comp. Neurol. 515, 465–501.

Abellán, A., Vernier, B., Rétaux, S., Medina, L., 2010. Similarities and differences in the forebrain expression of *Lhx1* and *Lhx5* between chicken and mouse: insights for understanding telencephalic development and evolution. J. Comp. Neurol. 518, 3512–3528.

Absil, P., Braquenier, J.B., Balthazart, J., Ball, G.F., 2002a. Effects of lesions of nucleus taeniae on appetitive and consummatory aspects of male sexual behavior in Japanese quail. Brain Behav. Evol. 60, 13–35.

Absil, P., Papello, M., Viglietti-Panzica, C., Balthazart, J., Panzica, G.C., 2002b. The medial preoptic nucleus receives vasotocinergic inputs in male quail: a tract-tracing and immunocytochemical study. J. Chem. Neuroanat. 24, 27–39.

Adkins-Regan, E., Watson, J.T., 1990. Sexual dimorphism in the avian brain is not limited to the song system of songbirds: a morphometric analysis of the brain of the quail (*Coturnix japonica*). Brain Res. 514, 320–326.

Akester, A.R., 1979. The autonomic nervous system. In: In: King, A.S., McLelland, J. (Eds.), Form and Function in Birds, vol. 1. Academic Press, London, pp. 381–441.

Albin, R.L., Young, A.B., Penney, J.B., 1989. The functional anatomy of basal ganglia disorders. Trends Neurosci. 12, 366–375.

Alheid, G.F., Heimer, L., 1988. New perspectives in basal forebrain organization of special relevance for neuropsychiatric disorders: the striatopallidal, amygdaloid, and corticopetal components of substantia innominata. Neuroscience 27, 1–39.

Alheid, G.F., Beltramino, C., Braun, A., Miselis, R.R., Francois, C., de Olmos, J.S., 1994. Transition areas of the striatopallidal system with the extended amygdala in the rat and primate: observations from histochemistry and experiments with mono- and trans-synaptic tracers. In: In: Percheron, G., McKenzie, J.S., Feger, J. (Eds.), The Basal Ganglia IV. New Ideas and Data on Structure and Function, Advances in Behavioral Biology, vol. 41. Plenum Press, N.Y, pp. 95–107.

Alheid, G.F., de Olmos, J.S., Beltramino, C.A., 1995. Amygdala and extended amygdala. In: Paxinos, G. (Ed.), The Rat Nervous System. Academic Press, San Diego, pp. 495–578.

Anderson, K.D., Reiner, A., 1990. Extensive co-occurrence of substance P and dynorphin in striatal projection neurons: an evolutionarily conserved feature of basal ganglia organization. J. Comp. Neurol. 295, 339–369.

Anderson, K.D., Reiner, A., 1991. Striatonigral projection neurons: a retrograde labeling study of the percentages that contain substance P or enkephalin in pigeons. J. Comp. Neurol. 303, 658–673.

Arends, J.J., Wild, J.M., Zeigler, H.P., 1988. Projections of the nucleus of the tractus solitarius in the pigeon (*Columba livia*). J. Comp. Neurol. 278, 405–429.

Arnold, A.P., Nottebohm, F., Pfaff, D.W., 1976. Hormone concentrating cells in vocal control and other areas of the brain of the zebra finch (*Poephila guttata*). J. Comp. Neurol. 165, 487–511.

Aslam, M.L., Bastiaansen, J.W., Elferink, M.G., Megens, H.J., Crooijmans, R.P., Blomberg le, A., Fleischer, R.C., Van Tassell, C.P., Sonstegard, T.S., Schroeder, S.G., Groenen, M.A., Long, J.A., 2012. Whole genome SNP discovery and analysis of genetic diversity in Turkey (*Meleagris gallopavo*). BMC Genomics 13, 391.

Aste, N., Balthazart, J., Absil, P., Grossmann, R., Mülhbauer, E., Viglietti-Panzica, C., Panzica, G.C., 1998. Anatomical and neurochemical definition of the nucleus of the stria terminalis in Japanese quail (*Coturnix japonica*). J. Comp. Neurol. 396, 141–157.

Atoji, Y., Wild, J.M., 2004. Fiber connections of the hippocampal formation and septum and subdivisions of the hippocampal formation in the pigeon as revealed by tract tracing and kainic acid lesions. J. Comp. Neurol. 475, 426–461.

Atoji, Y., Wild, J.M., 2006. Anatomy of the hippocampal formation. Rev. Neurosci. 17, 3–15.

Atoji, Y., Shibata, N., Yamamoto, Y., Suzuki, Y., 1996. Distribution of neurotensin-containing neurons in the central nervous system of the pigeon and the chicken. J. Comp. Neurol. 375, 187–211.

Atoji, Y., Wild, J.M., Yamamoto, Y., Suzuki, Y., 2002. Intratelencephalic connections of the hippocampus in pigeons (*Columba livia*). J. Comp. Neurol. 447, 177–199.

Atoji, Y., Saito, S., Wild, J.M., 2006. Fiber connections of the compact division of the posterior pallial amygdala and lateral part of the bed nucleus of the stria terminalis in the pigeon (*Columba livia*). J. Comp. Neurol. 499, 161–182.

Auld, D.S., Kornecook, T.J., Bastianetto, S., Quirion, R., 2002. Alzheimer's disease and the basal forebrain cholinergic system: relations to beta-amyloid peptides, cognition, and treatment strategies. Prog. Neurobiol. 68, 209–245.

Azumaya, Y., Tsutsui, K., 1996. Localization of galanin and its binding sites in the quail brain. Brain Res. 727, 187–195.

Bailhache, T., Balthazart, J., 1993. The catecholaminergic system of the quail brain: immunocytochemical studies of dopamine beta-hydroxylase and tyrosine hydroxylase. J. Comp. Neurol. 329, 230–256.

Bálint, E., Csillag, A., 2007. Nucleus accumbens subregions: hodological and immunohistochemical study in the domestic chick (*Gallus domesticus*). Cell Tissue Res. 327, 221–230.

Bálint, E., Mezey, S., Csillag, A., 2011. Efferent connections of nucleus accumbens subdivisions of the domestic chicken (*Gallus domesticus*): an anterograde pathway tracing study. J. Comp. Neurol. 519, 2922–2953.

Ball, G.F., Nock, B., McEwen, B.S., Balthazart, J., 1989. Distribution of alpha 2-adrenergic receptors in the brain of the Japanese quail as determined by quantitative autoradiography: implications for the control of sexually dimorphic reproductive processes. Brain Res. 491, 68–79.

Balthazart, J., Absil, P., 1997. Identification of catecholaminergic inputs to and outputs from aromatase-containing brain areas of the Japanese quail by tract tracing combined with tyrosine hydroxylase immunocytochemistry. J. Comp. Neurol. 382, 401–428.

Balthazart, J., Ball, G.F., 2007. Topography in the preoptic region: differential regulation of appetitive and consummatory male sexual behaviors. Front. Neuroendocrinol. 28, 161–178.

Balthazart, J., Schoffeniels, E., 1979. Pheromones are involved in the control of sexual behaviour in birds. Naturwissenschaften 66, 55–56.

Balthazart, J., Absil, P., Fiasse, V., Ball, G.F., 1994. Effects of the aromatase inhibitor R76713 on sexual differentiation of brain and behavior in zebra finches. Behaviour 131, 225–260.

Balthazart, J., Absil, P., Gerard, M., Appeltants, D., Ball, G.F., 1998a. Appetitive and consummatory male sexual behavior in Japanese quail are differentially regulated by subregions of the preoptic medial nucleus. J. Neurosci. 18, 6512–6527.

Balthazart, J., Foidart, A., Baillien, M., Harada, N., Ball, G.F., 1998b. Anatomical relationships between aromatase and tyrosine hydroxylase in the quail brain: double-label immunocytochemical studies. J. Comp. Neurol. 391, 214–226.

Balthazart, J., Panzica, G.C., Krohmer, R.W., 2003. Anatomical relationships between aromatase-immunoreactive neurons and nitric oxide synthase as evidenced by NOS immunohistochemistry or NADPH diaphorase histochemistry in the quail forebrain. J. Chem. Neuroanat. 25 (1), 39–51.

Barfield, R.J., Ronay, G., Pfaff, D.W., 1978. Autoradiographic localization of androgen-concentrating cells in the brain of the male domestic fowl. Neuroendocrinology 26, 297–311.

Benowitz, L.I., Karten, H.J., 1976. Organization of the tectofugal visual pathway in the pigeon: a retrograde transport study. J. Comp. Neurol. 167, 503–520.

Berk, M.L., 1987. Projections of the lateral hypothalamus and bed nucleus of the stria terminalis to the dorsal vagal complex in the pigeon. J. Comp. Neurol. 260, 140–156.

Berk, M.L., Butler, A.B., 1981. Efferent projections of the medial preoptic nucleus and medial hypothalamus in the pigeon. J. Comp. Neurol. 203, 379–399.

Berk, M.L., Finkelstein, J.A., 1983. Long descending projections of the hypothalamus in the pigeon, *Columba livia*. J. Comp. Neurol. 220, 127–136.

Berk, M.L., Hawkin, R.F., 1985. Ascending projections of the mammillary region in the pigeon: emphasis on telencephalic connections. J. Comp. Neurol. 239, 330–340.

Berthoud, H.R., 2002. Multiple neural systems controlling food intake and body weight. Neurosci. Biobehav. Rev. 26, 393–428.

Bingman, V.P., Casini, G., Nocjar, C., Jones, T.J., 1994. Connections of the piriform cortex in homing pigeons (*Columba livia*) studied with fast blue and WGA-HRP. Brain Behav. Evol. 43, 206–218.

Boswell, T., Li, Q., Takeuchi, S., 2002. Neurons expressing neuropeptide Y mRNA in the infundibular hypothalamus of Japanese quail are activated by fasting and co-express agouti-related protein mRNA. Brain Res. Mol. Brain Res. 100, 31–42.

Bottjer, S.W., 1993. The distribution of tyrosine hydroxylase immunoreactivity in the brains of male and female zebra finches. J. Neurobiol. 24, 51–69.

Brauth, S.E., 1984. Enkephalin-like immunoreactivity within the telencephalon of the reptile *Caiman crocodilus*. Neuroscience 11, 345–358.

Brauth, S.E., Ferguson, J.L., Kitt, C.A., 1978. Prosencephalic pathways related to the paleostriatum of the pigeon (*Columba livia*). Brain Res. 147, 205–221.

Breazile, J.E., Hartwig, H.-G., 1989. Central nervous system. In: In: King, A.S., McLelland, J. (Eds.), Form and Function in Birds, vol. 4. Academic Press, London, pp. 485–566.

Broekkamp, C.L.E., Pijnenburg, A.J.J., Cools, A.R., Van Rossum, J.M., 1975. The effect of microinjections of amphetamine into the neostriatum and the nucleus accumbens on self-stimulation behavior. Psycopharmacologia 42, 179–183.

Brüning, G., 1993. Localization of NADPH-diaphorase in the brain of the chicken. J. Comp. Neurol. 334, 192–208.

Brüning, G., Funk, U., Mayer, B., 1994. Immunocytochemical localization of nitric oxide synthase in the brain of the chicken. Neuroreport 5, 2425–2428.

Bubien-Waluszewska, A., 1981. The cranialnerves. In: In: King, A.S., McLelland, J. (Eds.), Form and Function in Birds, vol. 2. Academic Press, London, pp. 385–438.

Byerly, M.S., Simon, J., Lebihan-Duval, E., Duclos, M.J., Cogburn, L.A., Porter, T.E., 2009. Effects of BDNF, T3, and corticosterone on expression of the hypothalamic obesity gene network in vivo and in vitro. Am. J. Physiol. Regul. Integr. Comp. Physiol. 296, R1180–R1189.

Cador, M., Taylor, J.R., Robbins, T.W., 1991. Potentiation of the effects of reward-related stimuli by dopaminergic-dependent mechanisms in the nucleus accumbens. Psychopharmacology (Berl.) 104, 377–385.

Canteras, N.S., Simerly, R.B., Swanson, L.W., 1992. Connections of the posterior nucleus of the amygdala. J. Comp. Neurol. 324, 143–179.

Canteras, N.S., Simerly, R.B., Swanson, L.W., 1994. Organization of projections from the ventromedial nucleus of the hypothalamus: a *Phaseolus vulgaris*-leucoagglutinin study in the rat. J. Comp. Neurol. 348, 41–79.

Canteras, N.S., Simerly, R.B., Swanson, L.W., 1995. Organization of projections from the medial nucleus of the amygdala: a PHAL study in the rat. J. Comp. Neurol. 360, 213–245.

Cape, E.G., Jones, B.E., 2000. Effects of glutamate agonist versus procaine microinjections into the basal forebrain cholinergic cell area upon gamma and theta EEG activity and sleep-wake state. Eur. J. Neurosci. 12, 2166–2184.

Cape, E.G., Manns, I.D., Alonso, A., Beaudet, A., Jones, B.E., 2000. Neurotensin-induced bursting of cholinergic basal forebrain neurons promotes gamma and theta cortical activity together with waking and paradoxical sleep. J. Neurosci. 20, 8452–8461.

Caro, S.P., Balthazart, J., 2010. Pheromones in birds: myth or reality? J. Comp. Phys. A 196, 751–766.

Casini, G., Bingman, V.P., Bagnoli, P., 1986. Connections of the pigeon dorsomedial forebrain studied with WGA-HRP and 3H-proline. J. Comp. Neurol. 245, 454–470.

Chaillou, E., Tramu, G., Tillet, Y., 1999. Distribution of galanin immunoreactivity in the sheep diencephalon. J. Chem. Neuroanat. 17, 129–146.

Challet, E., Miceli, D., Pierre, J., Repérant, J., Masicotte, G., Herbin, M., Vesselkin, N.P., 1996. Distribution of serotonin-immunoreactivity in the brain of the pigeon (*Columba livia*). Anat. Embryol. (Berl.) 193, 209–227.

Cheng, M., Chaiken, M., Zuo, M., Miller, H., 1999. Nucleus taenia of the amygdala of birds: anatomical and functional studies in ring doves (*Streptopelia risoria*) and European starlings (*Sturnus vulgaris*). Brain Behav. Evol. 53, 243–270.

Choi, G.B., Dong, H.-W., Murphy, A.J., Valenzuela, D.M., Yancopoulos, G.D., Swanson, L.W., Anderson, D.J., 2005. Lhx6 delineates a pathway mediating innate reproductive behaviors from the amygdala to the hypothalamus. Neuron 46, 647–660.

Clark, M.S., Kaiyala, K.J., 2003. Role of corticotropin-releasing factor family peptides and receptors in stress-related psychiatric disorders. Semin. Clin. Neuropsychiatry 8, 119–136.

Cobos, I., Puelles, L., Martinez, S., 2001a. The avian telencephalic subpallium originates inhibitory neurons that invade tangentially the pallium (dorsal ventricular ridge and cortical areas). Dev. Biol. 239, 30–45.

Cobos, I., Shimamura, K., Rubenstein, J.L., Martinez, S., Puelles, L., 2001b. Fate map of the avian anterior forebrain at the four-somite stage, based on the analysis of quail-chick chimeras. Dev. Biol. 239, 46–67.

Cone, R.D., 1999. The central melanocortin system and energy homeostasis. Trends Endocrinol. Metab. 10, 211–216.

Cone, R.D., 2005. Anatomy and regulation of the central melanocortin system. Nat. Neurosci. 8, 571–578.

Cooper, R.L., Erickson, C.J., 1976. Effects of septal lesions on the courtship behavior of male ring doves (*Streptopelia risoria*). Horm. Behav. 7, 441–450.

Cota, D., Barrera, J.G., Seeley, R.J., 2006. Leptin in energy balance and reward: two faces of the same coin? Neuron. 51, 678–680.

Cozzi, B., Viglietti-Panzica, C., Aste, N., Panzica, G.C., 1991. The serotoninergic system in the brain of the Japanese quail. An immunohistochemical study. Cell Tissue Res. 263, 271–284.

Crespi, E.J., Denver, R.J., 2006. Leptin (ob gene) of the South African clawed frog *Xenopus laevis*. Proc. Natl. Acad. Sci. U.S.A. 103, 10092–10097.

Csillag, A., Székely, A.D., Stewart, M.G., 1997. Synaptic terminals immunolabeled against glutamate in the lobus parolfactorius of domestic chicks (*Gallus domesticus*) in relation to afferents from the archistriatum. Brain Res. 750, 171–179.

de Olmos, J.S., Beltramino, C.A., Alheid, G., 2004. Amygdala and extended amygdala of the rat: a cytoarchitectonical, fibroarchitectonical, and chemoarchitectonical survey. In: Paxinos, G. (Ed.), The Rat Nervous System, third ed. Elsevier Academic Press, San Diego, pp. 509–603.

Del Abril, A., Segovia, S., Guillamón, A., 1987. The bed nucleus of the stria terminalis in the rat: regional sex differences controlled by gonadal steroids early after birth. Brain Res. 429, 295–300.

Dellmann, H.D., 1964. On the structure of the vascular organ of the terminal lamina in poultry. Anat. Anz. 115, 174–183.

DeLong, M.R., 1990. Primate models of movement disorders of basal ganglia origin. Trends Neurosci. 13, 281–285.

Denbow, D.M., Meade, S., Robertson, A., McMurtry, J.P., Richards, M., Ashwell, C., 2000. Leptin-induced decrease in food intake in chickens. Physiol. Behav. 69, 359–362.

Ding, L., Perkel, D.J., 2004. Long-term potentiation in an avian basal ganglia nucleus essential for vocal learning. J. Neurosci. 24, 488–494.

Ding, L., Perkel, D.J., Farries, M.A., 2003. Presynaptic depression of glutamatergic synaptic transmission by D1-like dopamine receptor activation in the avian basal ganglia. J. Neurosci. 23, 6086–6095.

Doupe, A.J., Perkel, D.J., Reiner, A., Stern, E.A., 2005. Birdbrains could teach basal ganglia research a new song. Trends Neurosci. 28, 353–363.

Dridi, S., Swennen, Q., Decuypere, E., Buyse, J., 2005. Mode of leptin action in chicken hypothalamus. Brain Res. 1047, 214–223.

Dubbeldam, J.L., 1993. Systema nervosum periphericum. In: Handbook of Avian Anatomy: Nomina Anatomica Avian. Harvard Univ., Cambridge, MA, pp. 555–584 (Nuttall Ornithological Club, Mus. Comp. Zool.).

Dubbeldam, J.L., den Boer-Visser, A.M., 2002. The central mesencephalic grey in birds: nucleus intercollicularis and substantia grisea centralis. Brain Res. Bull. 57, 349–352.

Dubbeldam, J.L., van Ommen, M.H., den Boer-Visser, A.M., 1999. Immunohistochemical characterization of forebrain areas in the collared dove (*streptopelia decaocto*). Eur. J. Morphol. 37, 134–138.

Duvernoy, H.M., Risold, P.Y., 2007. The circumventricular organs: an atlas of comparative anatomy and vascularization. Brain Res. Rev. 56, 119–147.

Farries, M.A., Ding, L., Perkel, D.J., 2005. Evidence for "direct" and "indirect" pathways through the song system basal ganglia. J. Comp. Neurol. 484, 93–104.

Flames, N., Pla, R., Gelman, D.M., Rubenstein, J.L., Puelles, L., Marín, O., 2007. Delineation of multiple subpallial progenitor domains by the combinatorial expression of transcriptional codes. J. Neurosci. 27, 9682–9695.

Foidart, A., Lakaye, B., Grisar, T., Ball, G.F., Balthazart, J., 1999. Estrogen receptor-beta in quail: cloning, tissue expression and neuroanatomical distribution. J. Neurobiol. 40, 327–342.

Follett, B.K., Davies, D.T., 1975. Photoperiodicity and the neuroendocrine control of reproduction in birds. Symp. Zool. Soc. London 35, 199–224.

Forbes, S., Bui, S., Robinson, B.R., Hochgeschwender, U., Brennan, M.B., 2001. Integrated control of appetite and fat metabolism by the leptin-proopiomelanocortin pathway. Proc. Natl. Acad. Sci. U.S.A. 98, 4233–4237.

Foster, R.G., Soni, B.G., 1998. Extraretinal photoreceptors and their regulation of temporal physiology. Rev. Reprod. 3, 145–150.

Foster, W.G., Younglai, E.V., 1991. An immunohistochemical study of the GnRH neuron morphology and topography in the adult female rabbit hypothalamus. Am. J. Anat. 191, 293–300.

Foster, R.G., Grace, M.S., Provencio, I., Degrip, W.J., Garcia-Fernandez, J.M., 1994. Identification of vertebrate deep brain photoreceptors. Neurosci. Biobehav. Rev. 18, 541–546.

Friedman, J.M., Halaas, J.L., 1998. Leptin and the regulation of body weight in mammals. Nature 395, 763–770.

Furuse, M., 2002. Central regulation of food intake in the neonatal chick. Anim. Sci. J. 73, 83–94.

Gale, S.D., Person, A.L., Perkel, D.J., 2008. A novel basal ganglia pathway forms a loop linking a vocal learning circuit and its dopaminergic input. J. Comp. Neurol. 508, 824–839.

Gallagher, J.P., Orozco-Cabal, L.F., Liu, J., Shinnick-Gallagher, P., 2008. Synaptic physiology of central CRH system. Eur. J. Pharmacol. 583, 215–225.

García-López, M., Abellán, A., Legaz, I., Rubenstein, J.L.R., Puelles, L., Medina, L., 2008. Histogenetic compartments of the mouse centromedial and extended amygdala based on gene expression patterns during development. J. Comp. Neurol. 506, 46–74.

Gerfen, C.R., 1992. The neostriatal mosaic: multiple levels of compartmental organization. Trends Neurosci. 15, 133–139.

Gibbs, M.E., Maksel, D., Gibbs, Z., Hou, X., Summers, R.J., Small, D.H., 2010. Memory loss caused by β-amyloid protein is rescued by a β_3-adrenoceptor agonist. Neurobiol. Aging 31, 614–624.

Goodson, J.L., 1998. Vasotocin and vasoactive intestinal polypeptide modulate aggression in a territorial songbird, the violet-eared waxbill (*Estrildidae: Uraeginthus granatina*). Gen. Comp. Endocrinol. 111, 233–244.

Goodson, J.L., Adkins-Regan, E., 1999. Effect of intraseptal vasotocin and vasoactive intestinal polypeptide infusions on courtship song and aggression in the male zebra finch (*Taeniopygia guttata*). J. Neuroendocrinol. 11, 19–25.

Goodson, J.L., Bass, A.H., 2001. Social behavior functions and related anatomical characteristics of vasotocin/vasopressin systems in vertebrates. Brain Res. Brain Res. Rev. 35, 246–265.

Goodson, J.L., Eibach, R., Sakata, J., Adkins-Regan, E., 1999. Effect of septal lesions on male song and aggression in the colonial zebra finch (*Taeniopygia guttata*) and the territorial field sparrow (*Spizella pusilla*). Behav. Brain Res. 98, 167–180.

Goodson, J.L., Evans, A.K., Lindberg, L., 2004. Chemoarchitectonic subdivisions of the songbird septum and a comparative overview of septum chemical anatomy in jawed vertebrates. J. Comp. Neurol. 473, 293–314.

Gritti, I., Mainville, L., Jones, B.E., 1993. Codistribution of GABA- with acetylcholine-synthesizing neurons in the basal forebrain of the rat. J. Comp. Neurol. 329, 438–457.

Gritti, I., Manns, I.D., Mainville, L., Jones, B.E., 2003. Parvalbumin, calbindin, or calretinin in cortically projecting and GABAergic, cholinergic, or glutamatergic basal forebrain neurons of the rat. J. Comp. Neurol. 458, 11–31.

Guillamón, A., Segovia, S., 1997. Sex differences in the vomeronasal system. Brain Res. Bull. 44, 377–382.

El Halawani, M.E., Kang, S.W., Leclerc, B., Kosonsiriluk, S., Chaiseha, Y., 2009. Dopamine-melatonin neurons in the avian hypothalamus and their role as photoperiodic clocks. Gen. Comp. Endocrinol. 163, 123–127.

Halford, S., Pires, S.S., Turton, M., Zheng, L., González-Menéndez, I., Davies, W.L., Peirson, S.N., García-Fernández, J.M., Hankins, M.W., Foster, R.G., 2009. VA opsin-based photoreceptors in the hypothalamus of birds. Curr. Biol. 19, 1396–1402.

Hardie, D.G., 2004. The AMP-activated protein kinase pathway–new players upstream and downstream. J. Cell Sci. 117, 5479–5487.

Heimer, L., Alheid, G.F., 1991. Piecing together the puzzle of basal forebrain anatomy. In: Napier, T.C., Kalivas, P.W., Hanin, I. (Eds.), The Basal Forebrain: Anatomy to Function. Plenum Press, New York, pp. 1–42.

Herkenham, M., Moon-Edley, S., Stuart, J., 1984. Cell clusters in the nucleus accumbens of the rat and the mosaic relationship of opiate receptors, acetylcholinesterase and subcortical afferent terminations. Neuroscience 11, 561–593.

Hetherington, A., Ranson, S., 1940. Hypothalamic lesions and adiposity in the rat. Anat. Rec. 78, 149–172.

Hillier, L.W., Miller, W., Birney, E., Warren, W., Hardison, R.C., Ponting, C.P., Bork, P., Burt, S.W., Groenen, M.A.M., Delany, M.E., Dodgson, J.B., et al., 2004. Sequence and comparative analysis of the chicken genome provide unique perspectives on vertebrate evolution. Nature 432, 695–716.

Hodgkiss, J.P., 1984a. Evidence that enteric cholinergic neurons project orally in the intestinal nerve of the chicken. Q. J. Exp. Physiol. 69, 797–807.

Hodgkiss, J.P., 1984b. Peristalsis and antiperistalsis in the chicken caecum are myogenic. Q. J. Exp. Physiol. 69, 161–170.

Holstege, G., 1992. The emotional motor system. Eur. J. Morphol. 30, 67–79.

Horev, G., Einat, P., Aharoni, T., Eshdat, Y., Friedman-Einat, M., 2000. Molecular cloning and properties of the chicken leptin-receptor (CLEPR) gene. Mol. Cell. Endocrinol. 162, 95–106.

Husband, S.A., Shimizu, T., 2011. Calcium-binding protein distributions and fiber connections of the nucleus accumbens in the pigeon (*Columba livia*). J. Comp. Neurol. 519, 1371–1394.

Ikemoto, S., Panksepp, J., 1999. The role of nucleus accumbens dopamine in motivated behavior: a unifying interpretation with special reference to reward-seeking. Brain Res. Brain Res. Rev. 31, 6–41.

Jakab, R.L., Leranth, C., 1995. Septum. In: Paxinos, G. (Ed.), The Rat Nervous System, second ed. Academic Press, San Diego, pp. 405–442.

Jarvis, E.D., Gunturkun, O., Bruce, L., Csillag, A., Karten, H., Kuenzel, W., Medina, L., Paxinos, G., Perkel, D.J., Shimizu, T., Striedter, G., Wild, J.M., Ball, G.F., Dugas-Ford, J., Durand, S.E., Hough, G.E., Husband, S., Kubikova, L., Lee, D.W., Mello, C.V., Powers, A., Siang, C., Smulders, T.V., Wada, K., White, S.A., Yamamoto, K., Yu, J., Reiner, A., Butler, A.B., 2005. Avian brains and a new understanding of vertebrate brain evolution. Nat. Rev. Neurosci. 6, 151–159.

Jiao, Y., Medina, L., Veenman, C.L., Toledo, C., Puelles, L., Reiner, A., 2000. Identification of the anterior nucleus of the ansa lenticularis in birds as the homolog of the mammalian subthalamic nucleus. J. Neurosci. 20, 6998–7010.

Jozsa, R., Korf, H.W., Csernus, V., Mess, B., 1988. Thyrotropin-releasing hormone (TRH)-immunoreactive structures in the brain of the domestic mallard. Cell Tissue Res. 251, 441–449.

Juorio, A.V., Vogt, M., 1967. Monoamines and their metabolites in the avian brain. J. Physiol. 189, 489–518.

Jurkevich, A., Barth, S.W., Grossmann, R., 1997. Sexual dimorphism of arg-vasotocin gene expressing neurons in the telencephalon and dorsal diencephalon of the domestic fowl. An immunocytochemical and *in situ* hybridization study. Cell Tissue Res. 287, 69–77.

Jurkevich, A., Barth, S.W., Kuenzel, W.J., Kohler, A., Grossman, R., 1999. Development of sexually dimorphic vasotocinergic system in the bed nucleus of stria terminalis in chickens. J. Comp. Neurol. 408, 46–60.

Kafetzopoulos, E., Papadopoulos, G., 1983. Turning behavior after unilateral lesion of the subthalamic nucleus in the rat. Behav. Brain Res. 8, 217–223.

Kalivas, P.W., Duffy, P., 1995. Selective activation of dopamine transmission in the shell of the nucleus accumbens by stress. Brain Res. 675, 325–328.

Kang, S.W., Thayananuphat, A., Bakken, T., El Halawani, M.E., 2007. Dopamine-melatonin neurons in the avian hypothalamus controlling seasonal reproduction. Neuroscience 150, 223–233.

Karten, H.J., 1969. The organization of the avian telencephalon and some speculations on the phylogeny of the amniote telencephalon. Ann. N. Y. Acad. Sci. 167, 146–179.

Karten, H.J., Dubbeldam, J.L., 1973. The organization and projections of the paleostriatal complex in the pigeon (*Columba livia*). J. Comp. Neurol. 148, 61–89.

Karten, H.J., Hodos, W., 1967. A Stereotaxic Atlas of the Brain of the Pigeon (*Columba livia*). The Johns Hopkins University Press, Baltimore, MD.

Kawahara-Miki, R., Sano, S., Nunome, M., Shimmura, T., Kuwayama, T., Takahashi, S., Kawashima, T., Matsuda, Y., Yoshimura, T., Kono, T., 2013. Next-generation sequencing reveals genomic features in the Japanese quail. Genomics 101, 345–353.

Kawakami, S., Bungo, T., Ando, R., Ohgushi, A., Shimojo, M., Masuda, Y., Furuse, M., 2000. Central administration of alpha-melanocyte stimulating hormone inhibits fasting- and neuropeptide Y-induced feeding in neonatal chicks. Eur. J. Pharmacol. 398, 361–364.

King, D., Zigmond, M.J., Finlay, J.M., 1997. Effects of dopamine depletion in the medial prefrontal cortex on the stress-induced incease in extracellular dopamine in the nucleus accumbens core and shell. Neuroscience 77, 141–153.

Kingsbury, M.A., Kelly, A.M., Schrock, S.E., Goodson, J.L., 2011. Mammal-like organization of the avian midbrain central gray and a reappraisal of the intercollicular nucleus. PLoS One. 6, e20720.

Kiss, J.Z., Voorhuis, T.A., van Eekelen, J.A., de Kloet, E.R., de Wied, D., 1987. Organization of vasotocin-immunoreactive cells and fibers in the canary brain. J. Comp. Neurol. 263, 347–364.

Kitt, C.A., Brauth, S.E., 1981. Projections of the paleostriatum upon the midbrain tegmentum in the pigeon. Neuroscience 6, 1551–1566.

Kitt, C.A., Brauth, S.E., 1982. A paleostriatal-thalamic-telencephalic path in pigeons. Neuroscience 7, 2735–2751.

Kitt, C.A., Brauth, S.E., 1986a. Telencephalic projections from midbrain and isthmal cell groups in the pigeon. I. Locus coeruleus and subcoeruleus. J. Comp. Neurol. 247, 69–91.

Kitt, C.A., Brauth, S.E., 1986b. Telencephalic projections from midbrain and isthmal cell groups in the pigeon. II. The nigral complex. J. Comp. Neurol. 247, 92–110.

Klein, S., Jurkevich, A., Grossmann, R., 2006. Sexually dimorphic immunoreactivity of galanin and colocalization with arginine vasotocin in the chicken brain (*Gallus gallus domesticus*). J. Comp. Neurol. 499, 828–839.

Kohler, E.C., Riters, L.V., Chaves, L., Bingman, V.P., 1996. The muscarinic acetylcholine receptor antagonist scopoloamine impairs short-distance homing pigeon navigation. Physiol. Behav. 60, 1057–1061.

Korzeniewska, E., Güntürkün, O., 1990. Sensory properties and afferents of the N. dorsolateralis posterior thalami of the pigeon. J. Comp. Neurol. 292, 457–479.

Krogh, K., Ostergaard, K., Sabroe, S., Laurberg, S., 2008. Clinical aspects of bowel symptoms in Parkinson's disease. Acta Neurol. Scand. 117, 60–64.

Kuenzel, W.J., 1993. The search for deep encephalic photoreceptors within the avian brain, using gonadal development as a primary indicator. Poult. Sci. 72, 959–967.

Kuenzel, W.J., 1994. Central neuroanatomical systems involved in the regulation of food intake in birds and mammals. J. Nutr. 124, 1355S–1370S.

Kuenzel, W.J., 2000. The autonomic nervous system of birds. In: Whittow, G.C. (Ed.), Sturkie's Avian Physiology. Academic Press, San Diego, pp. 101–122.

Kuenzel, W.J., Blähser, S., 1993. The visceral forebrain system in birds: its proposed anatomical components and functions. Poult. Avian Biol. Rev. 5, 29–36.

Kuenzel, W.J., Blähser, S., 1994. Vasoactive intestinal polypeptide (VIP)-containing neurons: distribution throughout the brain of the chick (*Gallus domesticus*) with focus upon the lateral septal organ. Cell Tissue Res. 275, 91–107.

Kuenzel, W.J., Golden, C.D., 2006. Distribution and change in number of gonadotropin-releasing hormone-1 neurons following activation of the photoneuroendocrine system in the chick, *Gallus gallus*. Cell Tissue Res. 325, 501–512.

Kuenzel, W.J., Masson, M., 1988. A Stereotaxic Atlas of the Brain of the Chick (*Gallus domesticus*). Johns Hopkins University Press, Baltimore.

Kuenzel, W.J., McMurtry, J., 1988. Neuropeptide Y: brain localization and central effects on plasma insulin levels in chicks. Physiol. Behav. 44, 669–678.

Kuenzel, W.J., van Tienhoven, A., 1982. Nomenclature and location of avian hypothalamic nuclei and associated circumventricular organs. J. Comp. Neurol. 206, 293–313.

Kuenzel, W.J., Douglass, L.W., Davison, B.A., 1987. Robust feeding following central administration of neuropeptide Y or peptide YY in chicks, *Gallus domesticus*. Peptides. 8, 823–828.

Kuenzel, W.J., Mccune, S.K., Talbot, R.T., Sharp, P.J., Hill, J.M., 1997. Sites of gene expression for vasoactive intestinal polypeptide throughout the brain of the chick (*Gallus domesticus*). J. Comp. Neurol. 381, 101–118.

Kuenzel, W.J., Medina, L., Csillag, A., Perkel, D.J., Reiner, A., 2011. The avian subpallium: new insights into structural and functional subdivisions occupying the lateral subpallial wall and their embryological origins. Brain Res. 1424, 67–101.

Landmesser, L., 1978. The development of motor projection patterns in the chick hind limb. J. Physiol. 284, 391–414.

Lanuza, E., Davies, D.C., Landete, J.M., Novejarque, A., Martínez-García, F., 2000. Distribution of CGRP-like immunoreactivity in the chick and quail brain. J. Comp. Neurol. 421, 515–532.

Lazar, G.Y., Liposits, Z.S., Toth, P., Trasti, S.L., Maderdrut, J.L., Merchenthaler, I., 1991. Distribution of galanin-like immunoreactivity in the brain of *Rana esculenta* and *Xenopus laevis*. J. Comp. Neurol. 310, 45–67.

Legait, H., Legait, E., 1958. Paraphyse et organe subfornical dans la série des vértebrés. C. R. Assoc. Anat. 99, 427–435.

Li, H., Kuenzel, W.J., 2008. A possible neural cascade involving the photoneuroendocrine system (PNES) responsible for regulating gonadal development in an avian species, *Gallus gallus*. Brain Res. Bull. 76, 586–596.

Li, H., Proudman, J., Kuenzel, W.J., 2009. Differential regulation of gene expression and release of FSH and prolactin by long day and sulfamethazine in chicks. Gen. Comp. Endocrinol. 161, 262–266.

Liu, X., Dunn, I.C., Sharp, P.J., Boswell, T., 2007. Molecular cloning and tissue distribution of a short form chicken leptin receptor mRNA. Domest. Anim. Endocrinol. 32, 155–166.

Lohmus, M., Sundstrom, L.F., El Halawani, M., Silverin, B., 2003. Leptin depresses food intake in great tits (*Parus major*). Gen. Comp. Endocrinol. 131, 57–61.

Luiten, P.G., Ter, Horst, G.J., Steffens, A.B., 1987. The hypothalamus, intrinsic connections and outflow pathways to the endocrine system in relation to the control of feeding and metabolism. Prog. Neurobiol. 28, 1–54.

Lunam, C.A., Smith, T.K., 1996. Morphology and projections of neurons in Remak's nerve of the domestic fowl revealed by intracellular injection of biocytin. Cell Tissue Res. 284, 215–222.

Marín, O., Rubenstein, J.L., 2001. A long, remarkable journey: tangential migration in the telencephalon. Nat. Rev. Neurosci. 2, 780–790.

Massi, M., De Caro, G., Mazzarella, L., Epstein, A.N., 1986. The role of the subfornical organ in the drinking behavior of the pigeon. Brain Res. 381, 289–299.

McKeegan, D.E., 2002. Spontaneous and odour evoked activity in single avian olfactory bulb neurones. Brain Res. 929, 48–58.

Medina, L., Abellán, A., 2009. Development and evolution of the pallium. Semin. Cell Dev. Biol. 20, 698–711.

Medina, L., Reiner, A., 1994. Distribution of choline acetyltransferase immunoreactivity in the pigeon brain. J. Comp. Neurol. 342, 497–537.

Medina, L., Reiner, A., 1997. The efferent projections of the dorsal and ventral pallidal parts of the pigeon basal ganglia, studied with biotinylated dextran amine. Neuroscience 81, 773–802.

Medina, L., Reiner, A., 2000. Do birds possess homologues of mammalian primary visual, somatosensory and motor cortices? Trends Neurosci. 23, 1–12.

Medina, L., Veenman, C.L., Reiner, A., 1997. Evidence for a possible avian dorsal thalamic region comparable to the mammalian ventral anterior, ventral lateral, and oral ventroposterolateral nuclei. J. Comp. Neurol. 384, 86–108.

Melander, T., Hokfelt, T., Rokaeus, A., 1986. Distribution of galaninlike immunoreactivity in the rat central nervous system. J. Comp. Neurol. 248, 475–517.

Mello, C.V., Pinaud, R., Ribeiro, S., 1998. Noradrenergic system of the zebra finch brain: immunocytochemical study of dopamine-beta-hydroxylase. J. Comp. Neurol. 400, 207–228.

Mena-Segovia, J., Bolam, J.P., Magill, P.J., 2004. Pedunculopontine nucleus and basal ganglia: distant relatives or part of the same family? Trends Neurosci. 27, 585–588.

Mendelsohn, F.A., Aguilera, G., Saavedra, J.M., Quirion, R., Catt, K.J., 1983. Characteristics and regulation of angiotensin II receptors in pituitary, circumventricular organs and kidney. Clin. Exp. Hypertens. A 5, 1081–1097.

Metherate, R., Tremblay, N., Dykes, R.W., 1988. Transient and prolonged effects of acetylcholine on responsiveness of cat somatosensory cortical neurons. J. Neurophysiol. 59, 1253–1276.

Metherate, R., Cox, C.L., Ashe, J.H., 1992. Cellular bases of neocortical activation: modulation of neural oscillations by the nucleus basalis and endogenous acetylcholine. J. Neurosci. 12, 4701–4711.

Mikami, S., 1976. Ultrastructure of the organum vasculosum of the lamina terminalis of the Japanese quail, Coturnix coturnix japonica. Cell Tissue Res. 172, 227–243.

Miller, M.A., Kolb, P.E., Raskind, M.A., 1993. Extra-hypothalamic vasopressin neurons coexpress galanin messenger RNA as shown by double in situ hybridization histochemistry. J. Comp. Neurol. 329, 378–384.

Mineau, P., Boag, P.T., Beninger, R.J., 1994. The effects of physostigmine and scopolamine on memory for food caches in the black-capped chickadee. Pharmacol. Biochem. Behav. 49, 363–370.

Miselis, R.R., Shapiro, R.E., Hand, P.J., 1979. Subfornical organ efferents to neural systems for control of body water. Science 205, 1022–1025.

Moga, M.M., Gray, T.S., 1985. Peptidergic efferents from the intercalated nuclei of the amygdala to the parabrachial nucleus in the rat. Neurosci. Lett. 61, 13–18.

Molnar, M., Casini, G., Davis, B.M., Bagnoli, P., Brecha, N., 1994. Distribution of proenkephalin mRNA in the chicken and pigeon telencephalon. J. Comp. Neurol. 348, 419–432.

Montagnese, C.M., Szekely, A.D., Adam, A., Csillag, A., 2004. Efferent connections of septal nuclei of the domestic chick (*Gallus domesticus*): an anterograde pathway tracing study with a bearing on functional circuits. J. Comp. Neurol. 469, 437–456.

Montagnese, C.M., Zachar, G., Balint, E., Csillag, A., 2008. Afferent connections of septal nuclei of the domestic chick (*Gallus domesticus*): a retrograde pathway tracing study. J. Comp. Neurol. 511, 109–150.

Moons, L., Van, G.J., Ghijsels, E., Vandesande, F., 1994. Immunocytochemical localization of L-dopa and dopamine in the brain of the chicken (*Gallus domesticus*). J. Comp. Neurol. 346, 97–118.

Moons, L., D'Hondt, E., Pijcke, K., Vandesande, F., 1995. Noradrenergic system in the chicken brain: immunocytochemical study with antibodies to noradrenaline and dopamine-beta-hydroxylase. J. Comp. Neurol. 360, 331–348.

Moore, F.L., Lowry, C.A., 1998. Comparative neuroanatomy of vasotocin and vasopressin in amphibians and other vertebrates. Comp. Biochem. Physiol. C Pharmacol. Toxicol. Endocrinol. 119, 251–260.

Murakami, S., Seki, T., Wakabayashi, K., Arai, Y., 1991. The ontogeny of luteinizing hormone-releasing hormone (LHRH) producing neurons in the chick embryo: possible evidence for migrating LHRH neurons from the olfactory epithelium expressing a highly polysialylated neural cell adhesion molecule. Neurosci. Res. 12, 421–431.

Muske, L.E., Moore, F.L., 1988. The nervus terminalis in amphibians: anatomy, chemistry and relationship with the hypothalamic gonadotropin-releasing hormone system. Brain Behav. Evol. 32, 141–150.

Nakane, Y., Ikegami, K., Ono, H., Yamamoto, N., Yoshida, S., Hirunagi, K., Ebihara, S., Kubo, Y., Yoshimura, T., 2010. A mammalian neural tissue opsin (Opsin 5) is a deep brain photoreceptor in birds. Proc. Natl. Acad. Sci. U. S. A 107, 15264–15268.

Nauta, W.J.H., Karten, H.J., 1970. A general profile of the vertebrate brain, with sidelights on the ancestry of the cerebral cortex. In: Schmitt, F.O. (Ed.), The Neurosciences, Second Study Program. Rockefeller University Press, New York, pp. 7–26.

Nickel, R., Schummer, A., Seiferle, E., Siller, W.G., Wight, P.A.L., 1977. Anatomy of the Domestic Birds. Springer-Verlag, New York.

Norgren Jr, R.B., Lehman, M.N., 1991. Neurons that migrate from the olfactory epithelium in the chick express luteinizing hormone-releasing hormone. Endocrinology 128, 1676–1678.

Ohkubo, T., Tanaka, M., Nakashima, K., 2000. Structure and tissue distribution of chicken leptin receptor (cOb-R) mRNA. Biochim. Biophys. Acta 1491, 303–308.

Ono, H., Nakao, N., Yoshimura, T., 2009. Identification of the photoperiodic signaling pathway regulating seasonal reproduction using the functional genomics approach. Gen. Comp. Endocrinol. 163, 2–6.

Pahapill, P.A., Lozano, A.M., 2000. The pedunculopontine nucleus and Parkinson's disease. Brain 123, 1767–1783.

Panguluri, S., Saggu, S., Lundy, R., 2009. Comparison of somatostatin and corticotrophin-releasing hormone immunoreactivity in forebrain neurons projecting to taste-responsive and non-responsive regions of the parabrachial nucleus in rat. Brain Res. 1298, 57–69.

Panzica, G.C., Viglietti-Panzica, C., Fasolo, A., Vandesande, F., 1986. CRF-like immunoreactive system in the quail brain. J. Hirnforsch 27, 539–547.

Panzica, G.C., Aste, N., Viglietti-Panzica, C., Fasolo, A., 1992. Neuronal circuits controlling quail sexual behavior. Chemical neuroanatomy of the septo-preoptic region. Poul. Sci. Rev. 4, 249–259.

Panzica, G.C., Arévalo, R., Sánchez, F., Alonso, J.R., Aste, N., Viglietti-Panzica, C., Aijón, J., Vázquez, R., 1994. Topographical distribution of reduced nicotinamide adenine dinucleotidde phosphate-diaphorase in the brain of the Japanese quail. J. Comp. Neurol. 342, 97–114.

Panzica, G.C., Garzino, A., García-Ojeda, E., 1996. Coexistence of NADPH-diaphorase and tyrosine hydroxylase in the mesencephalic catecholaminergic system of the Japanese quail. J. Chem. Neuroanat. 11, 37–47.

Panzica, G.C., Castagna, C., Viglietti-Panzica, C., Russo, C., Tlemçani, O., Balthazart, J., 1998. Organizational effects of estrogens on brain vasotocin and sexual behavior in quail. J. Neurobiol. 37, 684–699.

Panzica, G., Pessatti, M., Viglietti-Panzica, C., Grossmann, R., Balthazart, J., 1999. Effects of testosterone on sexually dimorphic parvocellular neurons expressing vasotocin mRNA in the male quail brain. Brain Res. 850, 55–62.

Panzica, G.C., Aste, N., Castagna, C., Viglietti-Panzica, C., Balthazart, J., 2001. Steroid-induced plasticity in the sexually dimorphic vasotocinergic innervation of the avian brain: behavioral implications. Brain Res. Brain Res. Rev. 37, 178–200.

Paré, D., Quirk, G.J., Ledoux, J.E., 2004. New vistas on amygdala networks in conditioned fear. J. Neurophysiol. 92, 1–9.

Paré, D., Smith, Y., 1994. GABAergic projection from the intercalated cell masses of the amygdala to the basal forebrain in cats. J. Comp. Neurol. 344, 33–49.

Patterson, T.A., Lipton, J.R., Bennett, E.L., Rozenzweig, M.R., 1990. Cholinergic receptor antagonists impair formation of intermediate-term memory in the chick. Behav. Neural Biol. 54, 63–74.

Phillips-Singh, D., Li, Q., Takeuchi, S., Ohkubo, T., Sharp, P.J., Boswell, T., 2003. Fasting differentially regulates expression of agouti-related peptide, pro-opiomelanocortin, prepro-orexin, and vasoactive intestinal polypeptide mRNAs in the hypothalamus of Japanese quail. Cell Tissue Res. 313, 217–225.

Piallat, B., Benazzouz, A., Benabid, A.L., 1996. Subthalamic nucleus lesion in rats prevents dopaminergic nigral neuron degeneration after striatal 6-OHDA injection: behavioural and immunohistochemical studies. Eur. J. Neurosci. 8, 1408–1414.

Poulin, A.-M., Timofeeva, E., 2008. The dynamics of neuronal activation during food anticipation and feeding in the brain of food-entrained rats. Brain Res. 1227, 128–141.

Puelles, L., Kuwana, E., Puelles, E., Bulfone, A., Shimamura, K., Keleher, J., Smiga, S., Rubenstein, J.L., 2000. Pallial and subpallial derivatives in the embryonic chick and mouse telencephalon, traced by the expression of the genes Dlx-2, Emx-1, Nkx-2,1, Pax-6, and Tbr-1. J. Comp. Neurol. 424, 409–438.

Puelles, L., Martínez-de-la-Torre, M., Paxinos, G., Watson, C., Martínez, S., 2007. The Chick Brain in Stereotaxic Coordinates. Academic Press, San Diego.

Ramirez, J.M., Salas, C., Portavella, M., 1988. Offense and defense after lateral septal lesions in *Columba livia*. Internat. J. Neurosci. 41, 241–250.

Redgrave, P., Prescott, T.J., Gurney, K., 1999. The basal ganglia: a vertebrate solution to the selection problem? Neuroscience 89, 1009–1023.

Redies, C., Medina, L., Puelles, L., 2001. Cadherin expression by embryonic subdivisions and derived gray matter structures in the telencephalon of the chicken. J. Comp. Neurol. 438, 253–285.

Reiner, A., 1986. The co-occurrence of substance P-like immunoreactivity and dynorphin-like immunoreactivity in striatopallidal and striatonigral projection neurons in birds and reptiles. Brain Res. 371, 155–161.

Reiner, A., 1987. The distribution of proenkephalin-derived peptides in the central nervous system of turtles. J. Comp. Neurol. 259, 65–91.

Reiner, A., Anderson, K.D., 1990. The patterns of neurotransmitter and neuropeptide co-occurrence among striatal projection neurons: conclusions based on recent findings. Brain Res. Rev. 15 (3), 251–265.

Reiner, A., Carraway, R.E., 1987. Immunohistochemical and biochemical studies on Lys8-Asn9-neurotensin 8-13 (LANT6)-related peptides in the basal ganglia of pigeons, turtles, and hamsters. J. Comp. Neurol. 257, 453–476.

Reiner, A., Brauth, S.E., Karten, H.J., 1984a. Evolution of the amniote basal ganglia. Treands Neurosci. 7, 320–325.

Reiner, A., Davis, B.M., Brecha, N.C., Karten, H.J., 1984b. The distribution of enkephalinlike immunoreactivity in the telencephalon of the adult and developing domestic chicken. J. Comp. Neurol. 228, 245–262.

Reiner, A., Karle, E.J., Anderson, K.D., Medina, L., 1994. Catecholaminergic perikarya and fibers in the avian nervous system. In: Smeets, W.J.A.J., Reiner, A. (Eds.), Phylogeny and Development of Catecholaminergic Systems in the CNS of Vertebrates. Cambridge University Press, Cambridge, pp. 135–181.

Reiner, A., Medina, L., Veenman, C.L., 1998. Structural and functional evolution of the basal ganglia in vertebrates. Brain Res. Rev. 28, 235–285.

Reiner, A., Stern, E.A., Wilson, C.J., 2001. Physiology and morphology of intratelencephalically projecting corticostriatal-type neurons in pigeons as revealed by intracellular recording and cell filling. Brain Behav. Evol. 58, 101–114.

Reiner, A., Perkel, D.J., Bruce, L.L., Butler, A.B., Csillag, A., Kuenzel, W., Medina, L., Paxinos, G., Shimizu, T., Striedter, G., Wild, M., Ball, G.F., Durand, S., Güntürkün, O., Lee, D.W., Mello, C.V., Powers, A., White, S.A., Hough, G., Kubikova, L., Smulders, T.V., Wada, K., Dugas-Ford, J., Husband, S., Yamamoto, K., Yu, J., Siang, C., Jarvis, E.D., 2004. Revised nomenclature for avian telencephalon and some related brainstem nuclei. J. Comp. Neurol. 473, 377–414.

Richard, S., Martinez-Garcia, F., Lanuza, E., Davies, D.C., 2004. Distribution of corticotropin-releasing factor-immunoreactive neurons in the central nervous system of the domestic chicken and Japanese quail. J. Comp. Neurol. 469, 559–580.

Richards, M.P., Poch, S.M., 2003. Molecular cloning and expression of the turkey leptin receptor gene. Comp. Biochem. Physiol. B Biochem. Mol. Biol. 136, 833–847.

Richards, M.P., Proszkowiec-Weglarz, M., 2007. Mechanisms regulating feed intake, energy expenditure, and body weight in poultry. Poult. Sci. 86, 1478–1490.

Richards, M.P., Rosebrough, R.W., Coon, C.N., McMurtry, J.P., 2010. Feed intake regulation for the female broiler breeder: In theory and in practice. J. Appl. Poult. Res. 19, 182–193.

Risold, P.Y., Swanson, L.W., 1997a. Chemoarchitecture of the rat lateral septal nucleus. Brain Res. Brain Res. Rev. 24, 91–113.

Risold, P.Y., Swanson, L.W., 1997b. Connections of the rat lateral septal complex. Brain Res. Brain Res. Rev. 24, 115–195.

Roberts, T.F., Hall, W.S., Brauth, S.E., 2002. Organization of the avian basal forebrain: chemical anatomy in the parrot (*Melopsittacus undulatus*). J. Comp. Neurol. 454, 383–408.

Sabado, V., Barraud, P., Baker, C.V., Streit, A., 2012. Specification of GnRH-1 neurons by antagonistic FGF and retinoic acid signaling. Dev. Biol. 362, 254–262.

Savage, L.M., Stanchfield, M.A., Overmeier, J.B., 1994. The effects of scopolamine, diazepam, and lorazepam on working memory in pigeons: an analysis of reinforcement procedures and sample problem type. Pharmacol. Biochem. Behav. 48, 183–191.

Schäfer, F., Müller, A.R., Schmid, H.A., Gerstberger, R., Simon, E., 1996. Angiotensin II receptor subtypes in the duck subfornical organ: an electrophysiological and receptor autoradiographic investigation. Brain Res. 711, 118–124.

Schwanzel-Fukuda, M., Pfaff, D.W., 1989. Origin of luteinizing hormone-releasing hormone neurons. Nature 338, 161–164.

Schwartz, M.W., Woods, S.C., Porte Jr, D., Seeley, R.J., Baskin, D.G., 2000. Central nervous system control of food intake. Nature 404, 661–671.

Seabury, C.M., Dowd, S.E., Seabury, P.M., Raudsepp, T., Brightsmith, D.J., Liboriussen, P., Halley, Y., Fisher, C.A., Owens, E., Viswanathan, G., Tizard, I.R., 2013. A multi-platform draft de novo genome assembly and comparative analysis for the Scarlet Macaw (*Ara macao*). PLoS One 8, e62415.

Seiwert, C.M., Adkins-Regan, E., 1998. The foam production system of the male Japanese quail: characterization of structure and function. Brain Behav. Evol. 52, 61–80.

Sharp, P.J., Ciccone, N.A., 2005. The gonadotrophin releasing hormone neurone: key to avian reproductive function. In: Dawson, A., Sharp, P.J. (Eds.), Functional Avian Endocrinology. Narosa Publishing House Pvt Ltd, India, pp. 59–72.

Shi, Z.D., Shao, X.B., Chen, N., Yu, Y.C., Bi, Y.Z., Liang, S.D., Williams, J.B., Taouis, M., 2006. Effects of immunisation against leptin on feed intake, weight gain, fat deposition and laying performance in chickens. Br. Poult. Sci. 47, 88–94.

Shiflett, M.W., Gould, K.L., Smulders, T.V., DeVoogd, T.J., 2002. Septum volume and food-storing behavior are related in parids. J. Neurobiol. 51, 215–222.

Silver, R., Witkovsky, P., Horvath, P., Alones, V., Barnstable, C.J., Lehman, M.N., 1988. Coexpression of opsin- and VIP-like-immunoreactivity in CSF-contacting neurons of the avian brain. Cell Tissue Res. 253, 189–198.

Silverman, A.J., Livne, I., Witkin, J.W., 1994. The gonadotropin-releasing hormone (GnRH), neuronal systems: immunocytochemistry and in situ hybridization. In: Knobil, E., Neill, J.D. (Eds.), The Physiology of Reproduction, second ed. Raven Press Ltd, New York, pp. 1683–1709.

Spooner, C.E., Winters, W.D., 1966. Neuropharmacological profile of the young chick. Int. J. Neuropharmacol. 5, 217–236.

Stephenson-Jones, M., Samuelsson, E., Ericsson, J., Robertson, B., Grillner, S., 2011. Evolutionary conservation of the basal ganglia as a common vertebrate mechanism for action selection. Curr. Biol. 21, 1081–1091.

Stephenson-Jones, M., Ericsson, J., Robertson, B., Grillner, S., 2012. Evolution of the basal ganglia: dual-output pathways conserved throughout vertebrate phylogeny. J. Comp. Neurol. 520, 2957–2973.

Stewart, M.G., Kabai, P., Harrison, E., Steele, R.J., Kossut, M., Gierdalski, M., Csillag, A., 1996. The involvement of dopamine in the striatum in passive avoidance training in the chick. Neuroscience 70, 7–14.

Strader, A.D., Schioth, H.B., Buntin, J.D., 2003. The role of the melanocortin system and the melanocortin-4 receptor in ring dove (*Streptopelia risoria*) feeding behavior. Brain Res. 960, 112–121.

Swanson, L.W., 2000. Cerebral hemisphere regulation of motivated behavior. Brain Res. 886, 113–164.

Swanson, L.W., Petrovich, G.D., 1998. What is the amygdala? Trends Neurosci. 21, 323–331.

Tachibana, T., Sugahara, K., Ohgushi, A., Ando, R., Kawakami, S., Yoshimatsu, T., Furuse, M., 2001. Intracerebroventricular injection of agouti-related protein attenuates the anorexigenic effect of alpha-melanocyte stimulating hormone in neonatal chicks. Neurosci. Lett. 305, 131–134.

Takei, Y., 1977. The role of the subfornical organ in drinking induced by angiotension in the Japanese quail, Coturnix coturnix japonica. Cell Tissue Res. 185, 175–181.

Takeuchi, S., Takahashi, S., 1998. Melanocortin receptor genes in the chicken–tissue distributions. Gen. Comp. Endocrinol. 112, 220–231.

Takeuchi, S., Takahashi, S., 1999. A possible involvement of melanocortin 3 receptor in the regulation of adrenal gland function in the chicken. Biochim. Biophys. Acta 1448, 512–518.

Tanaka, M., Ikeda, T., Hayashi, S., Iijima, N., Amaya, F., Hisa, Y., Ibata, Y., 1997. Nitrergic neurons in the medial amygdala project to the hypothalamic paraventricular nucleus of the rat. Brain Res. 777, 13–21.

Thayananuphat, A., Kang, S.W., Bakken, T., Millam, J.R., El Halawani, M.E., 2007. Rhythmic dependent light induction of gonadotrophin-releasing hormone-I expression and activation of dopaminergic neurones within the premammillary nucleus of the turkey hypothalamus. J. Neuroendocrinol. 19, 399–406.

Thompson, R.R., Goodson, J.L., Ruscio, M.G., Adkins-Regan, E., 1998. Role of the archistriatal nucleus taeniae in the sexual behavior of male Japanese quail (*Coturnix japonica*): a comparison of function with the medial nucleus of the amygdala in mammals. Brain Behav. Evol. 51, 215–229.

van der Kooy, D., Koda, L.Y., McGinty, J.F., Gerfen, C.R., Bloom, F.E., 1984. The organization of projections from the cortex, amygdala, and hypothalamus to the nucleus of the solitary tract in rat. J. Comp. Neurol. 224, 1–24.

Veenman, C.L., Reiner, A., 1994. The distribution of GABA-containing perikarya, fibers, and terminals in the forebrain and midbrain of pigeons, with particular reference to the basal ganglia and its projection targets. J. Comp. Neurol. 339, 209–250.

Veenman, C.L., Reiner, A., 1996. Ultrastructural morphology of synapses formed by corticostriatal terminals in the avian striatum. Brain Res. 707, 1–12.

Veenman, C.L., Albin, R.L., Richfield, E.K., Reiner, A., 1994. Distributions of $GABA_A$, $GABA_B$, and benzodiazepine receptors in the forebrain and midbrain of pigeons. J. Comp. Neurol. 344, 161–189.

Veenman, C.L., Wild, J.M., Reiner, A., 1995. Organization of the avian "corticostriatal" projection system: a retrograde and anterograde pathway tracing study in pigeons. J. Comp. Neurol. 354, 87–126.

Vigh, B., 1971. Das Paraventrikularorgan und das zirkumventrikuläre System des Gehirns. Stud. Biol. Hung. 10 (Akad Kiadó, Budapest).

Viglietti-Panzica, C., Panzica, G.C., Fiori, M.G., Calcagni, M., Anselmetti, G.C., Balthazart, J., 1986. A sexually dimorphic nucleus in the quail preoptic area. Neurosci. Lett. 64, 129–134.

Viglietti-Panzica, C., Anselmetti, G.C., Balthazart, J., Aste, N., Panzica, G.C., 1992. Vasotocinergic innervation of the septal region in the Japanese quail: sexual differences and the influence of testosterone. Cell Tissue Res. 267, 261–265.

Vincent, S.R., Johansson, O., Hökfelt, T., Skirboll, L., Elde, R.P., Terenius, L., Kimmel, J., Goldstein, M., 1983. NADPH-diaphorase: a selective histochemical marker for striatal neurons containing both somatostatin- and avian pancreatic polypeptide (APP)-like immunoreactivities. J. Comp. Neurol. 217, 252–263.

von Bartheld, C.S., Bothwell, M., 1992. Development and distribution of noradrenergic and cholinergic neurons and their trophic phenotypes in the avian ceruleus complex and midbrain tegmentum. J. Comp. Neurol. 320, 479–500.

Voorhuis, T.A.M., Kiss, J.Z., de Kloet, E.R., de Wied, D., 1988. Testosterone-sensitive vasotocin-immunoreactive cells and fibers in the canary brain. Brain Res. 442, 139–146.

Wada, Y., Okano, T., Fukada, Y., 2000. Phototransduction molecules in the pigeon deep brain. J. Comp. Neurol. 428, 138–144.

Walsh, K.M., Kuenzel, W.J., 1997. Effect of sulfamethazine on sexual precocity and neuropeptide Y neurons within the tuberoinfundibular region of the chick brain. Brain Res. Bull. 44, 707–713.

Warren, W.C., Clayton, D.F., Ellegren, H., Arnold, A.P., Hillier, L.W., Kunstner, A., Searle, S., White, S., Vilella, A.J., Fairley, S., Heger, A., Kong, L., Ponting, C.P., Jarvis, E.D., Mello, C.V., Minx, P., Lovell, P., Velho, T.A., Ferris, M., Balakrishnan, C.N., Sinha, S., Blatti, C., London, S.E., Li, Y., Lin, Y.C., George, J., Sweedler, J., Southey, B., Gunaratne, P., Watson, M., Nam, K., Backstrom, N., Smeds, L., Nabholz, B., Itoh, Y., Whitney, O., Pfenning, A.R., Howard, J., Volker, M., Skinner, B.M., Griffin, D.K., Ye, L., McLaren, W.M., Flicek, P., Quesada, V., Velasco, G., Lopez-Otin, C., Puente, X.S., Olender, T., Lancet, D., Smit, A.F., Hubley, R., Konkel, M.K., Walker, J.A., Batzer, M.A., Gu, W., Pollock, D.D., Chen, L., Cheng, Z., Eichler, E.E., Stapley, J., Slate, J., Ekblom, R., Birkhead, T., Burke, T., Burt, D., Scharff, C., Adam, I., Richard, H., Sultan, M., Soldatov, A., Lehrach, H., Edwards, S.V., Yang, S.P., Li, X., Graves, T., Fulton, L., Nelson, J., Chinwalla, A., Hou, S., Mardis, E.R., Wilson, R.K., 2010. The genome of a songbird. Nature 464, 757–762.

Watson, J.T., Adkins-Regan, E., 1989. Neuroanatomical localization of sex steroid concentrating cells in the Japanese quail (*Coturnix japonica*): autorediography with (^{3}H)-estradiol, (^{3}H)-testosterone and (^{3}H)-dihydrotestosterone. Neuroendocrinology 49, 51–64.

Weindl, A., 1973. Neuroendocrine aspects of circumventricular organs. In: Ganong, W.E., Martini, L. (Eds.), Frontiers in Neuroendocrinology. Oxford Univ. Press, London and New York, pp. 3–32.

Wild, J.M., 1987. Thalamic projections to the paleostriatum and neostriatum in the pigeon (*Columba livia*). Neuroscience 20, 305–327.

Wild, J.M., Balthazart, J., 2013. Neural pathways mediating control of reproductive behavior in male Japanese quail. J. Comp. Neurol. 521, 2067–2087.

Wild, J.M., Arends, J.J., Zeigler, H.P., 1990. Projections of the parabrachial nucleus in the pigeon (*Columba livia*). J. Comp. Neurol. 293, 499–523.

Wild, J.M., Li, D., Eagleton, C., 1997. Projections of the dorsomedial nucleus of the intercollicular complex (DM) in relation to respiratory-vocal nuclei in the brainstem of pigeon (*Columba livia*) and zebra finch (*Taeniopygia guttata*). J. Comp. Neurol. 377, 392–413.

Woods, S.C., Seeley, R.J., Porte Jr, D., Schwartz, M.W., 1998. Signals that regulate food intake and energy homeostasis. Science 280, 1378–1383.

Woolf, N.J., 1991. Cholinergic systems in mammalian brain and spinal cord. Prog. Neurobiol. 37, 475–524.

Wray, S., Nieburgs, A., Elkabes, S., 1989. Spatiotemporal cell expression of luteinizing hormone-releasing hormone in the prenatal mouse: evidence for an embryonic origin in the olfactory placode. Brain Res. Dev. Brain Res. 46, 309–318.

Wynne, B., Güntürkün, O., 1995. Dopaminergic innervation of the telencephalon of the pigeon (*Columba livia*): a study with antibodies against tyrosine hydroxylase and dopamine. J. Comp. Neurol. 357, 446–464.

Xie, J., Kuenzel, W.J., Anthony, N.B., Jurkevich, A., 2010. Subpallial and hypothalamic areas activated following sexual and agonistic encounters in male chickens. Physiol. Behav. 101, 344–359.

Xie, J., Kuenzel, W.J., Sharp, P.J., Jurkevich, A., 2011. Appetitive and consummatory sexual and agonistic behavior elicits FOS expression in aromatase and vasotocin neurons within the preoptic area and bed nucleus of the stria terminalis of male domestic chickens. J. Neuroendocrinol. 23, 232–243.

Yamamoto, K., Sun, Z., Wang, H.B., Reiner, A., 2005. Subpallial amygdala and nucleus taeniae in birds resemble extended amygdala and medial amygdala in mammals in their expression of markers of regional identity. Brain Res. Bull. 66, 341–347.

Yasuda, M., 2002. The Anatomical Atlas of Gallus. University of Tokyo Press, Tokyo.

Yasuo, S., Watanabe, M., Okabayashi, N., Ebihara, S., Yoshimura, T., 2003. Circadian clock genes and photoperiodism: comprehensive analysis of clock gene expression in the mediobasal hypothalamus, the suprachiasmatic nucleus, and the pineal gland of Japanese quail under various light schedules. Endocrinology 144, 3742–3748.

Yoshimura, T., Yasuo, S., Watanabe, M., Iigo, M., Yamamura, T., Hirunagi, K., Ebihara, S., 2003. Light-induced hormone conversion of T4 to T3 regulates photoperiodic response of gonads in birds. Nature 426, 178–181.

Yuan, L., Ni, Y., Barth, S., Wang, Y., Grossmann, R., Zhao, R., 2009. Layer and broiler chicks exhibit similar hypothalamic expression of orexigenic neuropeptides but distinct expression of genes related to energy homeostasis and obesity. Brain Res. 1273, 18–28.

Záborszky, L., Alheid, G.F., Beinfeld, M.C., Eiden, L.E., Heimer, L., Palkovits, M., 1985. Cholecystokinin innervation of the ventral striatum: a morphological and radioimmunological study. Neuroscience 14, 427–453.

Záborszky, L., Pang, K., Somogyi, J., Nadasdy, Z., Kallo, I., 1999. The basal forebrain corticopetal system revisited. Ann. N. Y. Acad. Sci. 877, 339–367.

Zahm, D.S., 2000. An integrative neuroanatomical perspective on some subcortical substrates of adaptive responding with emphasis on the nucleus accumbens. Neurosci. Biobehav. Rev. 24, 85–105.

Zahm, D.S., Brog, J.S., 1992. On the significance of subterritories in the "accumbens" part of the rat ventral striatum. Neuroscience 50, 751–767.

Zahm, D.S., Heimer, L., 1993. Specificity in the efferent projections of the nucleus accumbens in the rat: comparison of the rostral pole projection patterns with those of the core and shell. J. Comp. Neurol. 327, 220–232.

Zhang, R., Tachibana, T., Takagi, T., Koutoku, T., Denbow, D.M., Furuse, M., 2004. Serotonin modifies corticotropin-releasing factor-induced behaviors of chicks. Behav. Brain Res. 151, 47–52.

Zhao, W.Q., Feng, H., Bennett, P., Ng, K.T., 1997. Inhibition of intermediate-term memory following passive avoidance training in neonate chicks by a presynaptic cholinergic blocker. Neurobiol. Learn. Mem. 67, 207–213.

Part III

Organ Systems Theme

Chapter 10

Blood

Colin G. Scanes
Department of Biological Sciences, University of Wisconsin, Milwaukee, WI, USA

10.1 INTRODUCTION

Blood consists of the following components:

- Plasma
- Formed elements, all containing nuclei
  - Erythrocytes
  - Leukocytes
  - Thrombocytes

Blood is critically important to the physiology of birds. The functions of blood include the following:

- Transportation of respiratory gases (oxygen and carbon dioxide), electrolytes, nutrients (e.g., glucose, amino acids, fatty acids), metabolites (e.g., lactate), waste compounds, hormones (from endocrine cell to target organ or tissue), heat, as well as some pathogens and toxicants are also transported via the blood stream
- Protection via antibodies, leukocytes, and thrombocytes
- Water and electrolyte homeostasis
- Clotting in the event of injury of blood vessels

Blood is more viscous than water, with the viscosity of chicken blood being 3.1 times greater than the viscosity of water; the viscosity of blood from domestic ducks and geese is 4.2 times greater than the viscosity of water (Sturkie, 1986). Moreover, the viscosity avian plasma is approximately 1.4 times greater than the viscosity of water (calculated from Sturkie, 1986 for chickens, ostriches, domestic ducks, and geese).

10.2 PLASMA

Table 10.1 summarizes mean concentrations of the constituents of plasma in birds. The osmotic pressure of avian plasma comes from the following (Peltonen and Sankari, 2011):

- Electrolytes: 95% (see Table 10.1)
- Glucose, amino acids, urea, and other organic molecules: 5%
- Proteins: 0.5% of colloid osmotic pressure

The circulating concentrations of electrolytes vary with physiological state. Venous plasma concentrations of sodium and plasma osmolarity are elevated by either dehydration or hyperthermia in chickens (Arad et al., 1983). Hyperthermia also results in elevated concentrations of chloride ions, blood urea nitrogen, and uric acid in chickens (Arad et al., 1983).

10.2.1 Circulating Electrolytes

The electrolytes are maintained within strict limits in view of their critical role in blood osmotic pressure and polarization/depolarization of nerve and muscle cells. The mean circulating concentrations of electrolytes such as sodium, potassium, calcium, and chloride across a large number of avian species are the summarized in Table 10.1 (additional materials can be found on the companion website http://booksite.elsevier.com/9780124071605/). There are shifts of the circulating concentrations of electrolytes with physiological state. For instance, acute heat stress is followed by decreases in plasma concentrations of sodium and potassium ions (chickens: Borges et al., 2004). Moreover, plasma concentrations of sodium are reduced with fasting and the concomitant reductions in metabolism (chickens: Christensen et al., 2012). There are differences in venous plasma concentrations of sodium and chloride between migratory and nonmigratory species of passeriform birds (Heatley et al., 2013).

Circulating concentrations of bicarbonate change with pCO_2 (see Section 10.4). For instance, there are decreases in plasma concentrations of bicarbonate ions due to panting associated with acute heat stress (chickens: Borges et al., 2004) and the hyperventilation with exercise (chickens: Gleeson and Brackenbury, 1984). Differences have been reported for venous plasma concentrations of bicarbonate between migratory and nonmigratory species of passeriform birds (Heatley et al., 2013).

10.2.2 Circulating Nutrients and Other Small Organic Molecules

The mean circulating concentrations of glucose, uric acid, and urea across a large number of avian species are summarized in Table 10.1.

Sturkie's Avian Physiology.

10.2.2.1 Plasma Concentrations of Glucose

The plasma concentrations of glucose are more than double those in mammals (Table 10.1). Plasma concentrations of glucose are relatively refractory to physiological changes such as insulin administration or fasting. Plasma concentrations of lactate are increased in exercising birds, for example, increasing from 2.1 to 8.4 mM in chickens on a treadmill (Gleeson and Brackenbury, 1984).

10.2.2.2 Uric Acid and Urea

Nitrogenous waste is carried in the circulation in birds as both uric acid and urea. Plasma concentrations of uric acid and urea in birds are summarized in Table 10.1. It is widely assumed that the circulating concentrations of uric acid and urea are very similar in birds (Sturkie, 1986). However, this is only the case when expressed as mg/dL, not on a millimolar (mmol/L) basis. Circulating concentrations of nitrogenous waste are influenced by physiological state. For example, heat stress reduces circulating concentrations of urate (chickens: Lin et al., 2006).

TABLE 10.1 Mean Circulating Concentrations of Constituents in Avian Plasma

Parameter	Plasma Concentration
Protein (g/L) (100 species)	39.6 ± 0.74
Albumin (g/L) (63 species)	15.9 ± 0.55
Globulin (g/L) (55 species)	18.8 ± 0.96
Glucose (mM or mmol/L) (139 species)	15.4 ± 0.32
Sodium (mEquiv/L) (47 species)	152.5 ± 1.13
Chloride (mEquiv/L) (43 species)	112.6 ± 1.28
Calcium (mEquiv/L) (49 species)	2.56 ± 0.10
Potassium (mEquiv/L) (46 species)	3.21 ± 0.19
Uric acid (mM or mmol/L) (60 species)	0.50 ± 0.04
Blood urea nitrogen (mM or mmol/L) (43 species)	1.11 ± 0.09

For details of circulating concentration of serum/plasma constituents in individual avian species see Table 1 in the companion website (http://booksite.elsevier.com/9780124071605/).

10.2.2.3 Circulating Antioxidants

Antioxidants have a protective role, reducing damage from free radicals and the immune related respiratory burst. The major antioxidant in the plasma is uric acid with vitamin E and four carotenoids also being antioxidants. There are changes in circulating antioxidant capacity with stress and diet (Cohen et al., 2007, 2009).

10.2.3 Plasma Proteins

The mean circulating concentrations of protein, albumin, and globulin across a large number of avian species are summarized in Table 10.1. Table 10.2 summarizes the concentrations of prealbumin, albumin, α-globulin, β-globulin, and γ-globulin in representative avian species.

The colloidal osmotic pressure due to plasma proteins is much lower that exerted by electrolytes and small organic molecules, such as glucose, and is lower than that in mammals (Peltonen and Sankari, 2011). Based on electrophoretic analysis, the major proteins in serum are the following:

- Prealbumin (with thyroid hormone binding capacity and also known as transthyretin)
- Albumin (binding fatty acids, lipophilic hormones, etc.)
- Globulins (α, β, and γ)

In the avian embryo, the major embryonic plasma protein is α-fetoprotein. This α-fetoprotein binds metals. The α-fetoprotein has been isolated from chick embryo serum

TABLE 10.2 Concentrations of Plasma Proteins in Chickens and Wild Birds

Species	Prealbumin	Albumin	α-Globulin	β-Globulin	γ-Globulin	Reference
Black-fronted piping-guan (*Aburria jacutinga*)	3.4	27.3	4.8	6.8	1.5	Motta et al., 2013
Chicken[1]	Not reported but identified	15 17	4 8	8 12	15 9	Hasegawa et al., 2002 Peltonen and Sankari, 2011
St Vincent parrot (*Amazona guildingii*)	6	17.4	2.5	4.6	3.0	Deem et al., 2008
Red grouse (*Lagopus lagopus scoticus*)	1.7	19.7	17.9			Wilson and Wilson, 1978

[1]*Broiler breeder adults.*
Source: Peltonen and Sankari (2011).

but does not appear to bind steroid hormones, unlike its mammalian counterpart (Ido and Matsuno, 1982).

10.2.3.1 Extracellular Fluid Protein

Based on studies with suction blister fluids, the concentrations of proteins in extracellular fluids are about half those in plasma (Peltonen and Sankari, 2011).

10.2.3.2 Albumin

Albumin is synthesized in the liver (chicken: Fujii et al., 1996). The functions of albumin include providing a colloidal osmotic effect and facilitating transport of nutrients and lipophilic hormones. For instance, albumin binds to, transports, and distributes lipids, including fatty acids. In the presence of albumin, there is increased uptake of saturated fatty acids, such as palmitate, by chick embryo cardiac cells (Paris et al., 1978).

There are some changes in the plasma concentration of albumin with physiological state. For instance, induction of early molt by reduced photoperiod and feed withdrawal for 48 h results in reductions in albumin concentrations in the plasma but increases in globulin (chickens: Gildersleeve et al., 1983). There are marked diurnal rhythms of plasma concentrations of albumin (chickens: Gildersleeve et al., 1983). Moreover, plasma concentrations of albumin are decreased in laying females (Morgan, 1975).

10.2.3.3 Globulins

There are multiple proteins in the globulin grouping. Broadly, globulins can be divided into three categories based on electrophoretic mobility:

- Alpha globulins
  - α_1-globulins, including apolipoprotein A-I, which binds lipids (chicken: Roman et al., 2009) containing transcortin (discussed below) and retinol binding protein
  - α_2-globulins, including angiotensinogen (precursor for angiotensin I and II) and ceruloplasmin
- Beta-globulins (including plasminogen)
  - β_1-globulins
  - β_2-globulins, including transferrin in mammals but not in birds
- Gamma-globulin fraction, which is reported to contain transferrin in at least some birds (Torres-Medina et al., 1971) and the immunoglobulins (Ig), IgA, IgM, and IgY (equivalent to IgG in mammals).

Globulin concentrations are influenced by physiological state. For instance, induction of early molting, by reduced photoperiod and feed withdrawal, is accompanied by increases in plasma concentrations of total globulin (chickens: Gildersleeve et al., 1983). Serum concentrations of α_1-globulin decline during growth in male chickens (Peltonen and Sankari, 2011).

10.2.3.4 Specific Transporter Proteins

10.2.3.4.1 Ceruloplasmin

The principal circulating transporter protein that binds copper ions is ceruloplasmin. This glycoprotein is produced by the liver. Avian ceruloplasmin has been purified and partially characterized (chicken: Disilvestro and Harris, 1985; Calabrese et al., 1988; goose: Hilewicz-Grabska et al., 1988). Chicken ceruloplasmin binds five atoms of copper per molecule (Calabrese et al., 1988).

Plasma concentrations of ceruloplasmin are influenced by physiological and nutritional status. Copper status influences circulating concentrations of ceruloplasmin; these being depressed by copper deficiency (chicken: Baumgartner et al., 1978; Kaya et al., 2006) and elevated by dietary copper supplementation (chicken: Koh et al., 1996). Plasma concentrations of ceruloplasmin are also elevated acutely following injection of *Escherichia coli* endotoxin (chicken: Curtis and Butler, 1980; Lin et al., 2006) and by glucocorticoids (chicken: Lin et al., 2004). Circulating concentrations of ceruloplasmin are depressed in birds receiving dietary supplementation with 5-aminolevulinic acid, a precursor for porphyrin and then heme (chicken: Sato et al., 2012).

10.2.3.4.2 Insulin-Like Growth Factor Binding Proteins

Insulin-like growth factor 1 (IGF1) is transported in the plasma, predominantly bound to insulin-like growth factor binding proteins (IGFBPs). There are at least five IGFBPs in birds:

- IGFBP 1 (chicken: NCBI Reference Sequence: NM_001001294.1; turkey: NCBI Reference Sequence: XM_003204673.1; zebra finch (*Taeniopygia guttata*): NCBI Reference Sequence: XM_002192038.2)
- IGFBP 2 (chicken: GenBank: AJ544105.1; zebra finch (*Taeniopygia guttata*): NCBI Reference Sequence: XM_002191294.2) (Schoen et al., 1994)
- IGFBP 3 (chicken: NCBI Reference Sequence: NM_001101034.1)
- IGFBP 4 (chicken NCBI Reference Sequence: NM_204353.1)
- IGFBP 5 (chicken NCBI Reference Sequence: XM_422069.3).

10.2.3.4.3 Retinol-Binding Protein

Vitamin A is transported bound to retinol-binding protein (RBP), which can complex with prealbumin (chicken: Abe et al., 1975). Avian RBP has been purified and its crystal structure characterized (Zanotti et al., 2001). Transthyretin also binds RBP (Eguchi et al., 2008). Synthesis of retinol binding protein by the liver is inhibited *in vivo* by vitamin E in the feed or *in vitro* by α-tocopherol in the incubation media (chicken: Zhou et al., 2012).

10.2.3.4.4 Transferrin

Transferrin transports iron (two molecules of iron for each molecule of transferrin) in the blood (reviewed in Lambert et al., 2005); in turn, at least some of the transferrin is bound to transferrin receptor. The circulating concentrations of iron are reported as ~2 μg/mL (chicken: Morgan, 1975; turkey: Huff et al., 2010). Iron that is bound to transferrin can be transferred to the other major iron binding protein, the yolk precursor phosvitin (Morgan, 1975).

The structure of avian transferrin has been reported (chicken: Guha Thakurta et al., 2003; Lambert et al., 2005). In addition to transporting iron, transferrin facilitates the uptake of ferrous iron during development of erythrocytes (chick embryo blood cells: van Bockxmeer and Morgan, 1982). Transferrin also binds a circulating enzyme, butyrylcholinesterase (chicken: Weitnauer et al., 1999).

Transferrin is synthesized by liver and secreted into the blood (chicken: Fujii et al., 1996). In birds, expression of the transferrin gene is restricted to the liver and oviduct (McKnight et al., 1980). The chicken transferrin receptor has been characterized based on the cDNA (chicken: Gerhardt et al., 1991). The transferrin receptor can induce programmed cell death or apoptosis (chicken: Ohno et al., 2008).

The circulating concentrations of transferrin and the transferrin receptor are reported as, respectively, 2.4 mg/mL (Morgan, 1975) and 1.2 μg/mL (chicken: Wiwanitkit et al., 2007). Plasma transferrin receptor concentrations reflect erythropoietic activity.

There are effects of physiology on circulating concentrations of transferrin. Circulating concentrations of transferrin are elevated by either estrogen or iron deficiency (chicken: McKnight et al., 1980). Hepatic expression of transferrin increased by estrogen and glucocorticoids (chicken: McKnight et al., 1980). Plasma concentrations of transferrin are also increased in chickens with a *Staphylococcus aureus* infection (Chamanza et al., 1999). Similarly, there is a strong tendency for plasma concentrations of transferrin receptor to be increased in birds infected with *Plasmodium gallinaceum*, the protozoan that causes malaria in poultry (chicken: Wiwanitkit et al., 2007).

10.2.3.4.5 Hormone Transport

Lipophilic hormones, including the steroid hormones, are transported in the plasma bound to proteins, including albumin and specific binding proteins. The rationale for lipophilic hormones having binding proteins in the circulation is to facilitate the distribution of the hormones throughout their target organ(s) (Richardson et al., 2005).

10.2.3.4.5.1 Sex Steroids Although a specific sex hormone-binding protein (SHBP) is found in amphibians, reptiles, and mammals, SHBP is missing from birds (Wingfield et al., 1984). This is confirmed in later reports (Breuner and Orchinik, 2009; Malisch and Breuner, 2010).

10.2.3.4.5.2 Thyroid Hormones Homologues of mammalian thyroxine-binding globulin are not present in birds (Schreiber and Richardson, 1997). Thyroid hormones bind to both albumin (Schreiber and Richardson, 1997) and transthyretin in birds as in all other vertebrate classes (Schreiber and Richardson, 1997; Richardson et al., 2005). Avian transthyretin has been purified as a 15 kDa protein (chicken: Eguchi et al., 2008). Transthyretin also binds retinol binding protein (Eguchi et al., 2008).

10.2.3.4.5.3 Corticosteroids Birds have a corticosteroid-binding globulin (CBG) or transcortin. This α_1-globulin binds and transports corticosterone in the plasma, for example, in the American kestrel (Whitman et al., 2011), chicken (Murakami, 1991), white crowned sparrow (Breuner and Orchinik, 2009), and zebra finch (Schmidt et al., 2010). CBG also binds cortisol, a major glucocorticoid in avian embryos (Schmidt et al., 2010). The concentration of CBG is decreased by stress, such as handling, in the American kestrel (Whitman et al., 2011).

10.2.3.5 Gamma Globulins

Gamma globulins include IgA, IgM, and IgY (equivalent to both IgE and IgG in mammals). In growing chickens, the concentrations of immunoglobulins are as follows (based on Klasing, 1998):

- Immunoglobulin A: 0.3 g/L
- Immunoglobulin M: 2.7 g/L
- Immunoglobulin Y: 5.5 g/L

The concentrations of IgY are reported to be markedly lower in laying hens (0.8 g/L) (adult female chickens: Cetin et al., 2010). This may be due to the transportation of circulating IgY into the yolk of the ovum.

10.2.3.6 Enzymes

There are multiple enzymes in the plasma of birds, including the following: alkaline phosphatase asparate aminotransferase, cholinesterase, creatine phosphotransferase, glutamic oxaloacetic transaminase (GOT), and lactic acid dehydrogenase (LDH). The activities of asparate aminotransferase and creatine phosphotransferase are most frequently determined in hematological studies.

The activity of enzymes is reported to be influenced by both physiology and the external environment. For instance, venous plasma alanine transaminase (SGPT or ALAT or SGPT) activity is elevated by either dehydration or hyperthermia (Arad et al., 1983). Induction of early molt by reduced photoperiod and feed withdrawal for 48h results in shifts in plasma components with reductions in glutamic pyruvic transaminase (GPT) and increases in alkaline phosphatase (Alk P), LDH, and GOT. Moreover, depressed plasma cholinesterase activity has been used as biomarker for exposure to organophosphates and carbamate and hence wildlife risk assessment. However, it is critically important to have baseline activity for the individual species of bird being investigated (Santos et al., 2012).

10.3 ERYTHROCYTES

Table 10.3 summarizes multiple erythrocyte characteristics including number, hematocrit, and erythrocyte size in wild birds and adult chicks.

10.3.1 Structure of the Erythrocyte

Avian erythrocytes are ovoid cells (Figure 10.1), with an average volume of 145 fL across multiple avian species with consistent sizes (see Table 10.3):

- Long diameter: approximately 12.5 μm
- Short diameter: 6.8 μm
- Thickness: 3.2 μm

The volume of erythrocytes declines during embryonic development from 170 fL at day 10 to 140 fL at day 18 (chicken: Tazawa et al., 2011). There is marked asymmetry, with a ratio of erythrocyte length to width of 1.7 μm in chickens and 1.8 μm across wild birds (Table 10.3).

There is a centrally located nucleus (long diameter approximately 6.5 μm, short diameter 2.8 μm). The avian erythrocyte differs from that in mammals by the presence of a nucleus and mitochondria and by being larger. The most abundant protein in erythrocytes is hemoglobin (Figure 10.1). Erythrocytes of wild birds contain more hemoglobin than those of chickens (Table 10.3). The nuclei of avian erythrocytes also contain hemoglobin (Davis, 1961). This is contiguous with cytoplasmic hemoglobin via pores in the nuclei membrane (Davis, 1961).

Microtubules stabilize the asymmetric structure of erythrocytes (Winckler and Solomon, 1991). These are composed of tubulin, with 95% of the tubulin from chicken erythrocytes being tubulin-βVI (Sharma et al., 2010). The erythrocyte anion transport protein (erythrocyte band 3) is the major transmembrane protein in the erythrocyte. It has two functions:

- An anchor for the intracellular cytoskeleton
- The key anion transporter exchanging bicarbonate (HCO_3^-) for chloride (Cl^-) ions

Another plasma membrane protein in the turkey erythrocyte is goblin, which is phosphorylated in the presence of β-adrenergic agonists (Alper et al., 1980). The fatty acid composition of the cell membrane of avian erythrocytes is modified by fatty acids in the diet (e.g., cockatiel; Heinze et al., 2012).

TABLE 10.3 Characteristics of Erythrocytes in Wild Birds (Excluding Poultry Species) and Chickens

Parameter	Wild Birds (Mean)	Chickens[1] (Mean)
Erythrocytes ($10^6/\mu L$)	3.33 (164 species)	3.2
Packed cell volume/ hematocrit (%)	44.0 (158 species)	44.0
Hemoglobin (%)	14.5 (134 species)	10.1
Erythrocyte volume (fL)	149.4 (81 species)	149.4
Erythrocyte length (μm)	12.5 (364 species)	12.2
Erythrocyte width (μm)	6.8 (362 species)	7.1
Ratio of erythrocyte length to width	1.844 (362 species)	1.718
Erythrocyte cross-sectional area[2] (μm^2)	68.3 (362 species)	68.0
Erythrocyte hemoglobin (fg)	50.8 (23 species)	32

[1]*Summarized early data from mature/adult chickens with thickness 3.6 μm (Sturkie, 1986).*
[2]*Area = $\pi \times$ (½ length × ½ width).*
Additional materials can be found on the companion website: http://booksite.elsevier.com/9780124071605

10.3.2 Erythrocyte Chromatin and Transcription

Chicken erythrocytes exhibit transcription (e.g., enolase; Toll-like receptors 3, 9, and 21; myxovirus resistance 1; Morera et al., 2011). Chromatin structure is critical in the control of DNA replication, repair, and transcription. It is reasonable to conclude that the chromatin is in a decondensed state (Morera et al., 2011). The 30 nm fiber is the first hierarchical level of chromatin folding, allowing transcription and access by regulatory factors but not DNA replication in chicken erythrocytes (Scheffer et al., 2011). There are also strong associations between linker histones and chromatin in chicken erythrocyte nuclei (Koutzamani et al., 2002).

The ratio of RNA to hemoglobin decreases rapidly in a linear manner, from approximately 240 μg per mg on day 3 to 35 μg per mg on day 6 (chicken: Baumann et al., 2003). Subsequently, the ratio of RNA to hemoglobin declines at a much slower rate, to less than 10 μg per mg at day 17 (chicken: Baumann et al., 2003). There are large reductions in the erythrocyte adenosine triphosphate (ATP) content, decreasing from a molar ratio of RNA to Hb from 6.8 (day 3) to 0.5 (day 17) (chicken: Baumann et al., 2003).

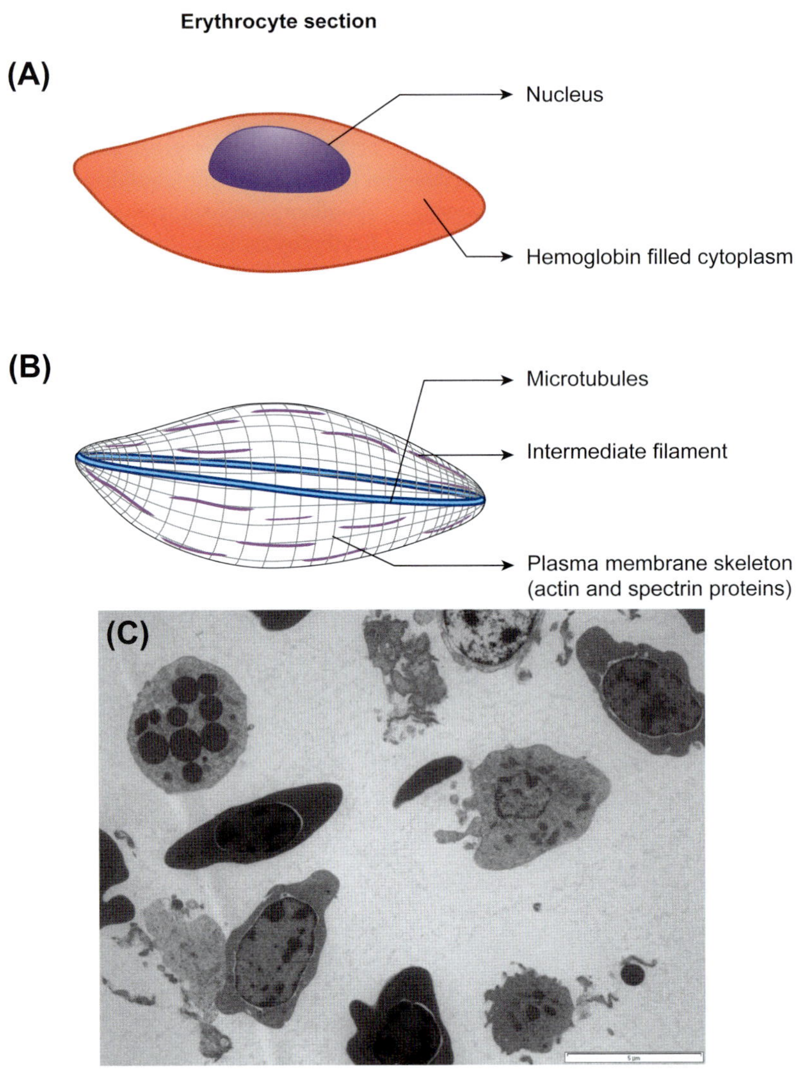

FIGURE 10.1 **Schematic images of avian erythrocytes.** (A) Section through avian erythrocyte showing nucleus and hemoglobin filled cytoplasm. (B) Structure of avian erythrocyte showing cytoskeleton maintaining shape of the cell and supporting the nucleus. Marginal band of microtubules (blue) and membrane cytoskeleton (shown as cross-hatch) maintaining shape of the cell. Intermediate filaments (purple) support the nucleus. *(Based on Joseph-Silverstein and Cohen (1984); also see http://hermes.mbl.edu/BiologicalBulletin/MMER/COH/CohFig1.html (accessed 23.07.13.).)* (C) Electron microscopy of chicken bone marrow cells showing developing erythrocytes with electron dense hemoglobin (bar 5 μm). *Reproduced from Shini et al. (2008), with permission from Elsevier.*

10.3.3 Metabolism of Erythrocytes

Surprisingly, metabolism and expression of metabolic enzymes in avian erythrocyte has received relatively little attention. It has been assumed that glucose is the major substrate for energy needs of the erythrocyte. However, glucose fails to maintain erythrocyte ATP *in vitro* (chicken: Mathew et al., 1993). This is not unexpected in view of the very low active transport of glucose into avian erythrocytes. In contrast, glutamine was effective in maintaining erythrocyte ATP *in vitro* (chicken: Mathew et al., 1993). There is evidence that the avian erythrocyte employs the citric acid cycle based on older literature (chicken: Dajani and Orten, 1958). Moreover, chicken erythrocytes can use glycine as a substrate (Dajani and Orten, 1959). However, there was no detectable activity of TCA cycle enzymes in the pigeon erythrocyte (Kalomenopoulou and Beis, 1990). Metabolic enzyme activities (including hexokinase, phosphofructokinase, and pyruvate kinase) have been reported in pigeon erythrocytes (Kalomenopoulou and Beis, 1990). High activities for the following enzymes have been reported for erythrocytes from the little penguin (*Eudyptula minor*): triphosphate isomerase, lactate dehydrogenase, glyceraldehyde phosphate dehydrogenase, and phosphoglycerate kinase (Nicol et al., 1988). In the little penguin (*E. minor*) erythrocytes, about half of the glucose is metabolized to lactate (Nicol et al., 1988). There is a need to see studies on the expression and its control of key metabolic genes by avian erythrocytes.

Erythrocytes have a high concentration of ATP in early embryonic development (e.g., 15 mmoles per liter at day 3 of incubation; chicken: Baumann et al., 2003). Erythrocytes from chick embryos also have significant levels of CTP and UTP (chicken: Baumann et al., 2003). In addition, erythrocytes from chick embryos contain another high energy compound, 2,3-bisphosphoglycerate (2,3-BPG) (chicken: Baumann et al., 2003). *In vitro*, ATP concentrations are depressed by hypoxia in erythrocytes from 10 day embryos (chicken: Baumann et al., 2003).

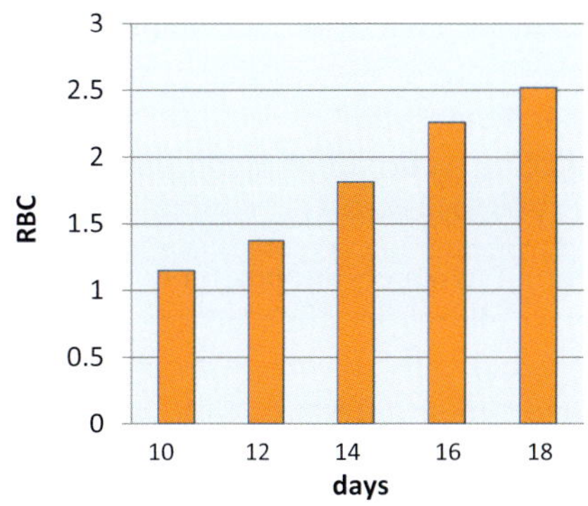

FIGURE 10.2 Changes in erythrocyte numbers (10^6 per µL) during embryonic development of the chicken. *Data from Tazawa et al. (2011).*

10.3.4 Number of Erythrocytes and Packed Cell Volume (Hematocrit)

Table 10.3 summarizes mean the packed cell volume (hematocrit) and erythrocyte concentrations across multiple avian species and a comparison with earlier data in chickens. Across avian species, hematocrit is reported to be inversely proportional to $\log_{10}$ body weight; declining from 52% on average in birds at 10 g to 40% in birds at 1 kg (Bishop and Butler, 1995). When erythrocytes are removed, the hematocrit declines as the plasma volume is more rapidly restored. Following removal of 30% of the blood, there is an almost immediate decrease in hematocrit and erythrocyte concentration (Japanese quail: Schindler et al., 1987).

When oxygen partial pressure is low, there is an increase in erythrocyte concentrations and hematocrit. For instance, hypoxia increases hematocrit (chickens and Japanese quail: Rosse and Waldmann, 1966). Similarly, following removal of 30% of the blood and the immediate decrease in erythrocyte concentration, there is then a rapid recovery with 72 h (Japanese quail: Schindler et al., 1987). Indicative of the acceleration of erythropoiesis, there are increases in the proportion of immature red blood cells, namely reticulocytes, in the blood and bone marrow (Japanese quail: Gildersleeve et al., 1985a,b; Schindler et al., 1987).

There are marked increases in the concentration of erythrocytes during embryonic development (chicken: Tazawa et al., 2011). These changes are illustrated in Figure 10.2. As might be expected, the concentration of erythrocytes is increased in response to high altitude (e.g., turkey embryos: Bagley et al., 1990). Similarly and not unexpectedly, hematocrit is reduced with copper deficiency (chicken: Baumgartner et al., 1978). In chickens, there appears to be a sex difference in hematocrit and erythrocyte number between breeding males and females (Table 10.4), perhaps due to the high concentrations of yolk precursors in the plasma of sexually adult females and/or direct effects of estrogens and/or androgens.

TABLE 10.4 Comparison of Erythrocyte Concentrations and Hematocrit in Male and Female Adult Chickens

Species	Erythrocyte Concentration (10^6/µL)	Hematocrit (%)
Male	3.6	42.9
Female	2.8	29.2

Source: Calculated from data summarized in Sturkie (1986).

There are also decreases in erythrocyte concentrations caused by some toxicants, such as after *in ovo* administration of 1-(2-chlorophenyl)-1-(4-chlorophenyl)-2,2,2-trichloroethane o,p'-DDT (Japanese quail: Bryan et al., 1989). Chemicals such as acetophenylhydrazine and the organophosphorus pesticide foschlor reduce the circulating concentration of erythrocyte, thereby inducing anemia in birds (chickens: Coll and Ingram, 1978; Japanese quail: Gromysz-Kałkowska et al., 1985). Moreover, petroleum oil decreases packed cell volume (mallard ducks: Lee et al., 2012). In a meta-analysis of data from 37,000 broiler chickens, mycotoxin was demonstrated to depress hematocrit (Andretta et al., 2012).

10.3.5 Production of Erythrocytes

Avian erythrocytes are produced in the bone marrow. *In vitro*, embryonic marrow cells have been demonstrated as capable of developing into erythrocytes (e.g., Japanese quail: Brandon et al., 2000). The avian erythrocyte series can be summarized as follows (based on Williams, 1972):

1. Erythroblasts (cell proliferation) → Polychromatic erythroblasts (reduce cell volume with hemoglobin synthesis initiated)
2. Polychromatic erythroblasts → Reticulocytes (with further reduced cell volume but with hemoglobin synthesis)
3. Reticulocytes → mature erythrocytes (asymmetric elongated)

During the morphogenesis of mature erythrocytes, microtubules play a critical role in the development of the elongated asymmetric shape (Winckler and Solomon, 1991). Thus, more rounded cells in the circulation represent immature erythrocytes/reticulocytes.

10.3.5.1 Erythropoietin

Evidence for an avian erythropoietin comes from the ability of chicken plasma to stimulate heme accumulation in chicken erythroid cells *in vitro*, with plasma from anemic chickens being more effective (Coll and Ingram, 1978). Moreover, *in vivo* the rate of erythropoiesis changes in a manner consistent with control by erythropoietin, being increased in anemia (chicken: Rosse and Waldmann, 1966) and depressed by polycythemia (Japanese quail: Rosse and Waldmann, 1966). During development of bone air sacs in the bones of posthatching pigeons, there are decreases in hemopoietic bone marrow as a percentage of body weight and increased distribution in the ulna, radius, femur, tibiotarsus, scapula, furcula, and the caudal vertebrae (Schepelmann, 1990).

Erythropoietin (EPO) acts by binding to the EPO receptor (EpoR). There is increased proliferation of avian erythroblasts with over-expression of EpoR (Mikulits et al., 2000). Moreover, iron repression of ferritin heavy chain (ferH) expression is prevented by overexpression of both EpoR and c-Kit, another cytokine growth factor receptor (Mikulits et al., 2000).

10.3.6 Lifespan of Erythrocytes

Despite having a nucleus, avian erythrocytes do not divide, failing to enter a new S phase (Williams, 1972). Bird erythrocytes have a very limited lifespan that is much less than in mammals (Brace and Atland, 1956; reviewed Beuchat and Chong, 1998), with all labeled cells disappearing in 42 days (ducks), 35 days (chickens, *Gallus gallus*) and 48 days (pigeons, *Columba livia*) (Rodman et al., 1957) and a mean survival time of 34 days in the (Japanese quail *Coturnix coturnix*: Nirmallan and Robinson, 1973).

The mean lifespan (± the standard error of mean) for four species of avian erythrocytes was calculated as 39.7 ± 3.3 days. In contrast, the longevity of mammalian erythrocytes is longer: 85.6 ± 10.5 days for 11 species (calculated from Röhme, 1981).

10.3.7 Hemoglobin

Hemoglobin binds to oxygen at the lungs and releases oxygen at the tissues. Avian hemoglobin has been purified (chicken: Matsuda and Takae, 1963), crystallized, and characterized, including x-ray analysis (e.g., bar-headed goose: greylag goose: Liang et al., 2001; ostrich: Sundaresan et al., 2009). As with mammals, hemoglobin is a tetrameric protein with four protein subunits, each attached to a ferrous iron-containing heme unit. The molecular basis of the greater oxygen affinity and a lower Bohr effect of early embryonic hemoglobin has been established (Chapman et al., 1980). ATP and other nucleotides are the major determinants of oxygen saturation (chicken: Baumann et al., 2003).

There are multiple forms of avian hemoglobin, with six forms in the chicken (four embryonic and two adult):

- Embryo major forms:
  - Hb P with π α globin and ϱ β
  - Hb P′
- Embryo minor forms:
  - Hb M α^D globin
  - Hb E α^A globin
- Adult major form: Hb A α^A globin
- Adult minor form: Hb D α^D globin

In the chicken, the relative contribution of two major embryonic hemoglobins peaks at 4 days of embryonic development and then decreases gradually. They are no longer detectable from 15 days. The two minor ones increase up to 6–7 days and then decrease, but they are still present in the blood at hatching (Cirotto et al., 1975). The shifts in composition of α-globin in hemoglobin in chick embryo blood are summarized in Figure 10.3.

The binding of oxygen to hemoglobin is markedly affected by organic phosphates in the avian erythrocyte. In most birds, it is thought that the major regulator is inositol pentaphosphate; however, in ostriches there is evidence that inositol tetrakisphosphate (inositol-P_4) plays a critical role (Sundaresan et al., 2009). Inositol hexakisphosphate (inositol-P_6) decreases the affinity of chicken or pigeon hemoglobin for oxygen (Vandecasserie et al., 1971). Similarly, inositol-P_5 lowers the oxygen affinity of hemoglobin (Isaacks et al., 1977).

Arteriovenous differences in the blood concentration of oxygen are constant in either resting or exercising birds, being independent of body mass (Bishop and Butler, 1995).

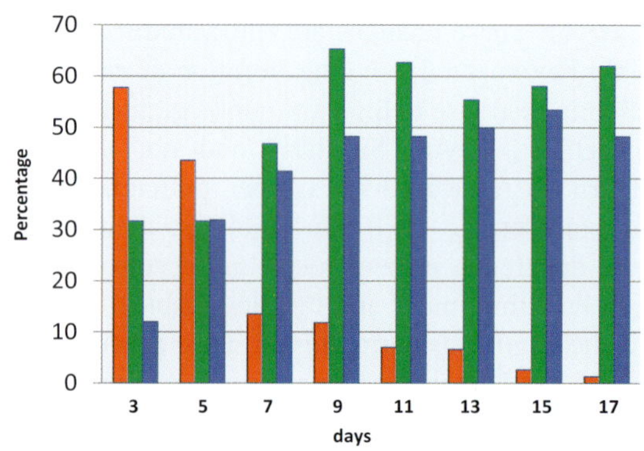

FIGURE 10.3 **Shifts in the proportion of α-globins during embryonic development of the chicken. Red: π-α-globin. Green: α-A-globin. Blue: α-D-globin.**

10.3.7.1 Hemoglobin Genes

In birds as in mammals, there are multiple hemoglobin genes: seven in the chicken (Reitman et al., 1993) and six in the zebra finch (Alev et al., 2009). These represent a small family of genes encoding α- and β-globins. The avian hemoglobin genes can be summarized as follows:

- Chicken π (Pi) α: homologous to duck π and orthologous to α1 in zebra finch
- Chicken αD: homologous to turkey, ostrich, and quail αD and orthologous to α2 in zebra finch
- Chicken αA: homologous to turkey, ostrich, and quail αA and orthologous to α3 in zebra finch
- Chicken ϱ (rho or P) β: possibly orthologous to β1 in zebra finch
- Chicken βH: possibly orthologous to β2 in zebra finch
- Chicken βA: homologous to turkey, duck, flamingo, condor, macaw, pigeon, and ostrich β major and orthologous to β3 in zebra finch
- Chicken β ε (epsilon or E): marked homologies to chicken ϱ

In adult zebra finches, hemoglobin consists of 74% α3 and 26% α2 and all β3 globin. In the embryo, hemoglobin consists of 28% α1, 55% α2, and 16% α3 and 44% β1, 33% β2, and 24% β3 (Alev et al., 2009).

The α-globin genes are located on chromosome 14 in the chicken. In embryonic erythrocytes, the region containing the multiple α-globin genes is located in a crown of DNA loops surrounding the nuclear matrix (Iarovaia et al., 2009).

10.3.7.2 Adaptations of Hemoglobin to Flight at High Altitudes

The structures of hemoglobin in birds that fly at high altitudes have increased affinity for oxygen (see Figure 10.4). For instance, there are subtle changes to the protein structure (single substitutions in α-globin) in both the bar-headed goose, a species that migrates across the Himalayas, and the Andean goose (Jessen et al., 1991). Similarly, the oxygen affinity of hemoglobin in driving birds, such as emperor penguins, is increased (Meir and Ponganis, 2009). This can be readily seen from the left shift in the O_2~Hb dissociation curve for the bar-headed goose and the emperor penguin compared with the domestic duck (and most other birds; see Figure 10.4).

10.3.7.3 Adaptations of Hemoglobin to Embryonic Development

There are marked changes in the affinity of hemoglobin for oxygen, with higher affinities earlier in development (see Figure 10.5). The molecular basis of the greater oxygen affinity and a lower Bohr effect of early embryonic hemoglobin has been established (Chapman et al., 1980).

10.3.7.4 Glycation of Hemoglobin

Despite elevated circulating concentrations of glucose, the percentage hemoglobin that is glycated is lower in most birds examined than in mammals (e.g., <2% in rooks (*Corvus frugilegus*) and mute swans (*Cygnus olor*) (Mikšik and Hodný, 1992; also reviewed in Beuchat and Chong, 1998). Even in hummingbirds, where plasma concentrations of glucose can be more than 40 mmol/L, glycated hemoglobin is within the mammalian range (3.5–5.0; Beuchat and Chong, 1998).

10.3.7.5 Hemoglobin and Nutrition

In birds, as might be expected considering their low iron availability, circulating concentrations of hemoglobin are reduced (Tako et al., 2010).

10.3.8 Carbonic Anhydrase

Carbonic anhydrase in erythrocytes catalyzes the reaction between carbon dioxide and water to yield carbonic acid.

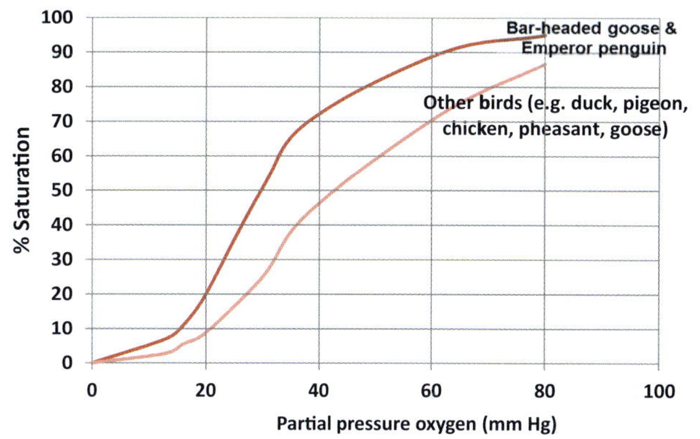

FIGURE 10.4 Oxygen dissociation curves for hemoglobin in avian erythrocytes. *Data calculated for pigeons, domestic ducks and geese, Mucovy ducks, chickens and red-necked pheasants from Christensen and Dill (1935) and for domestic goose, bar-headed geese, and emperor penguins from Meir and Ponganis (2009).*

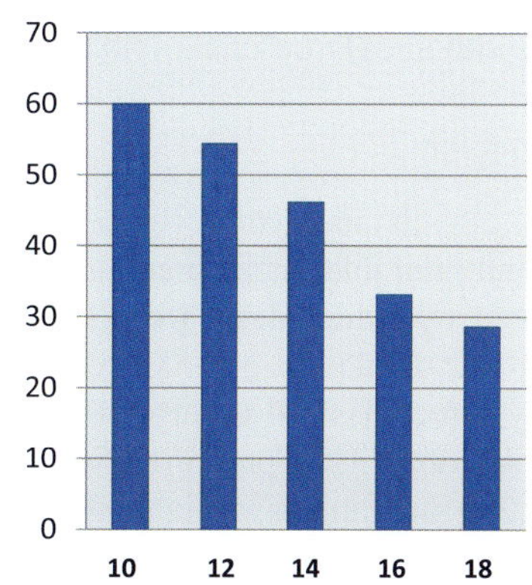

FIGURE 10.5 Changes in the oxygen affinity (p50) of chicken hemoglobin (y-axis) during the development (x-axis in days) of the chick embryo. *Data from Tazawa et al. (1976).*

This in turn forms bicarbonate (HCO_3^-) and protons (H^+) reversibly.

$$CO_2 + H_2O \leftrightarrow H_2CO_3 \leftrightarrow HCO_3^- + H^+$$

Carbonic anhydrate (CA, EC 4.2.1.1) has been purified from avian erythrocytes (e.g., chicken: Bernstein and Schraer, 1972; ostrich: Ozensoy et al., 2005; pigeon: Ozensoy et al., 2011). Carbonic anhydrase activity is increased by β-adrenergic agonists based on studies with Japanese quail erythrocytes (Igbo et al., 1994). Moreover, hypoxia increases carbonic anhydrase activity in avian erythrocytes (Glombitza et al., 1996).

There are increases in the amount of carbonic anhydrase II (CAII) and the isoenzyme CAIII relative to hemoglobin as female chickens commences egg production (Nishita et al., 2011a,b). Ferrous iron increases the activity of carbonic anhydrase from avian erythrocytes (Wu et al., 2007).

10.3.9 Transporters

10.3.9.1 Anion Transporter

The key anion transporter exchanges bicarbonate (HCO_3^-) for chloride (Cl^-) ions. The transcripts for the anion transporter have been characterized in the chicken, with chicken erythroid cells expressing anion transport proteins (band 3) (Kim et al., 1988). The v-erbA oncogene suppresses expression of anion transport proteins (band 3) by chicken erythroid cells (Zenke et al., 1988).

10.3.9.2 Sodium and Potassium Transport

There is considerable movement of sodium and potassium across the erythrocyte plasma membrane. The flux of sodium ions across the plasma membrane of erythrocytes from turkeys is two- to three-fold that in humans (Palfrey and Greengard, 1981). The $Na^+/K^+/2Cl^-$ cotransporter in avian erythrocytes exhibits sensitivity to β-adrenergic agonists and to oxygen concentrations (reviewed in Gibson et al., 2000). Transport of potassium into chicken erythrocytes declines with age (Drew et al., 2002).

10.3.9.3 Glucose

Glucose entry into avian erythrocytes is low, being both nonsaturable and saturable carrier mediated; it can be inhibited by inhibited by cytochalasin (Simons, 1983a,b). The 3-0-methylglucose uptake by pigeon erythrocytes that can be inhibited by cytochalasin B is very low, as there are only about 200 copies of GLUT1 protein per erythrocyte (compared to 300,000 in humans) (Diamond and Carruthers, 1993). Chicken erythroblasts express two glucose transporters (GLUT), GLUT 1 and GLUT 3 (Mathew et al., 1994). During differentiation to erythrocytes, expression of GLUT 1 and GLUT 3 declines rapidly to very low levels (Mathew et al., 1994). Simple diffusion of monosaccharides is markedly greater in erythrocytes from chicken embryos than older birds (Ingermann et al., 1985).

10.3.9.4 Amino Acids and Urea

Amino acids are transported across the plasma membrane of the avian erythrocyte with, for instance, uptake by chicken erythrocytes of glycine, leucine, and lysine being sodium dependent (Somes et al., 1981; Lerner et al., 1984). Avian erythrocytes are reported to have little expression of urea transporter-B (Liu et al., 2011).

10.3.10 Hormonal Effects on Erythrocytes

Catecholamines enhance sodium transport via β-adrenergic receptors, activation of adenylate cyclase, and cyclic adenosine 3′,5′-monophosphate (cyclic 3′,5′-AMP) in turkey erythrocytes (Gardner et al., 1973; reviewed in Palfrey and Greengard, 1981). β-adrenergic agonists stimulate potassium influx into turkey erythrocytes (Furukawa et al., 1980).

There are at least three β- adrenergic receptors in the cell membrane of the avian erythrocyte—βtrunc, β3C, and β4C-receptors (turkey: Chen et al., 1994; Baker, 2010). The turkey has been used as a model for β-adrenergic receptors. These are functional, with β-adrenergic receptor agonists increasing adenylate cyclase activity (Stadel et al., 1982; Peters et al., 1984). Avian β-adrenergic receptors are desensitized by β-adrenergic receptor agonists and sensitized by β-adrenergic receptor antagonists due to structural changes (Stadel et al., 1982; Peters et al., 1984). Chick embryo erythrocytes have A2α adenosine receptors that also activate adenylate cyclase (Glombitza et al., 1996). There are progressive reductions in the erythrocyte responses to beta-adrenergic and A2 receptor agonists during embryonic development due to cGMP-inhibited phosphodiesterase 3 (Baumann et al., 1999).

There is also evidence for the effects of thyroid hormones on avian erythrocytes as erythrocytes from hypothyroid turkeys show lower sensitivity to β-adrenergic agonists (Furukawa et al., 1980).

10.3.11 Effect of Stressors

There are effects of stressors on the avian erythrocyte. For instance, exposure to crude oil is followed by degeneration of the erythrocyte mitochondria in herring gulls (*Larus argentatus*) and Atlantic puffins (*Fratercula arctica*) (Leighton, 1985). Avian erythrocytes exhibit marked increases in production of both hydrogen peroxide and superoxide *in vitro* in response to glucose or glucose 6-phosphate together with shifts in the oxygen dissociation curve for hemoglobin (Zhang et al., 2011). In contrast, the effects on either reactive oxygen species (ROS) or hemoglobin are not observed with mammalian red blood cells (Zhang et al., 2011). Similarly, ROS are produced by mitochondria from avian erythrocytes, such as from quail or parrots (Montgomery et al., 2012). These effects may explain the shorter lifespan of avian erythrocytes.

10.3.12 Other Roles for the Avian Erythrocyte

It has been suggested that avian erythrocytes may have immune-like roles in addition to gas exchange (Morera et al., 2011).

10.4 BLOOD GASES

Table 10.5 summarizes arterial and venous pO_2, pCO_2, and pH in multiple species of resting birds. There is no change in arterial pO_2 during short-term flight (Butler et al., 1977), presumably due to the increased rate of respiration. Moreover, in chickens exercising on a treadmill, arterial pO_2 paradoxically increases (Gleeson and Brackenbury, 1984), again presumably due to increased respiration. The changes in venous pO_2 and pCO_2 during rapid growth and adult meat-type chickens are shown in Table 10.6. There is a decline in venous pO_2 during growth, presumably reflecting the requirements of metabolism.

Arterial pCO_2 is influenced by respiration, whereas venous pCO_2 reflects metabolism. For instance, pCO_2 is decreased in heat-stressed panting emus (Jones et al., 1983) and chickens running on a treadmill (Gleeson and Brackenbury, 1984). However, pCO_2 in venous plasma is increased during growth (Table 10.6) and reduced with fasting (young rapidly growing chickens: Christensen et al., 2012). These changes are presumed to reflect tissue metabolism.

10.5 LEUKOCYTES

There are five types of leukocytes: lymphocytes, heterophils (equivalent to neutrophils in mammals), basophils, eosinophils, and monocytes. Figure 10.6 shows the structure of avian leukocytes. Table 10.7 summarizes the concentrations of leukocytes and differential counts in both wild bird species and the domestic chicken. Klasing (1998) estimated that chickens produce 0.76 g leukocytes per kilogram per day. Leukocytes are found in the "buffy" layer when the hematocrit of avian blood is being determined.

10.5.1 Populations

Leukocyte numbers per unit volume of blood vary. Table 10.7 summarizes circulating concentrations of leukocytes in wild birds and chickens. Blood leukocyte concentrations and populations vary in birds, with a high ratio of heterophils to lymphocytes in *Accipitriformes* (63.1%:30.5%) and *Struthioniformes* (65.4%:26.0%) and low ratio in *Passeriformes* (19.6%:67.9%) (C.G. Scanes, unpublished observations).

A reciprocal relationship frequently exists between heterophil and lymphocyte concentrations. The stress hormone corticosterone increases heterophils and decreases lymphocytes in young chickens (see Figure 10.7; Gross and Siegel, 1983). This leads to large shifts in the heterophil/lymphocyte ratio (H/L) (Gross and Siegel, 1983). Similarly, stimulation of the adrenal cortical cells with adrenocorticotropic hormone is accompanied by changes in both increases in heterophils and decreases in lymphocytes in young chickens (Davison and Flack, 1981). There is a concomitant reduction in volume of heterophils with corticosterone administration in young chickens (Shini et al., 2008). The relationship between stress and leukocyte populations is further supported by the greater H/L ratio in injured Adélie penguins (Vleck et al., 2000); in young chickens exposed to stressors, such as atmospheric ammonia, intermittent electric shock (McFarlane and Curtis, 1989), and environmental heat stress (Borges et al., 2004); and in young turkeys after *E. coli* challenge together with transportation stress (Huff et al., 2008, 2010).

Marked changes in leukocyte populations have been reported in some birds in captivity, with an almost tripled proportion of heterophils with a concomitant decline in lymphocytes (Rufous-collared sparrows, *Zonotrichia capensis*: Ruiz et al., 2002). This may reflect an effect of stress.

There are also independent effects of environment on populations of different leukocytes. Circulating concentrations of leukocytes are depressed by 71.4% in blue-fronted Amazon parrots (*Amazona aestiva*) maintained in captivity compared to wild birds (Deem et al., 2005). This is due to decreases in both the percentage and number of heterophils (Deem et al., 2005). Although the percentage of lymphocytes is increased in blue-fronted Amazon parrots (*A. aestiva*) maintained in captivity, there is no change in the concentration (Deem et al., 2005). There are marked increases in total leukocyte numbers with botulism in black-faced spoonbills (*Platalea minor*; Chou et al., 2008). In a meta-analysis of studies on the effects of mycotoxins on chickens, Andretta et al. (2012) reported reductions in the concentrations of total leukocytes together with that of both heterophils and lymphocytes.

10.5.1.1 Number of Leukocytes

There are physiological shifts in both the total concentrations of leukocytes and in the relative proportion of heterophils and lymphocytes. Challenging chickens with the parasite *Plasmodium juxtanucleare* is accompanied by increases in the leukocyte concentration in the blood (Silveira et al., 2009).

TABLE 10.5 Examples of Arterial and Venous pO_2 and pCO_2 in Resting Birds with pH as a Comparison

Species	Arterial	Venous	Reference
pO_2 (mm Hg or Torr)			
Order *Anseriformes*			
Domestic duck	82		Kawashiro and Scheid, 1975
Order *Columbiformes*			
Pigeon (*Columbia livea*)	87	57	Butler et al., 1977
Mourning doves (*Zenaida macroura*)		49	Harms and Harms, 2012
Order *Falconiformes*			
Southern crested caracaras (*Caracara plancus*)	99	–	Escobar et al., 2011
Order *Galliformes*			
Chicken (*Gallus gallus*) adult hen	93		Gleeson and Brackenbury, 1984
Rapidly growing chicken		44	Mean from Olanrewaju et al., 2010; van As et al., 2010 and Christensen et al., 2012
Order *Passeriformes*			
Boat-tailed grackles (*Quiscalus major*)		47	Harms and Harms, 2012
House sparrows (*Passer domesticus*)		40	Harms and Harms, 2012
Order *Psittaciformes*			
Amazon parrot (*Amazona aestiva*).	98.1		Valéria et al., 2008
pCO_2 (mm Hg or Torr)			
Order *Anseriformes*			
Domestic duck	38		Kawashiro and Scheid, 1975
Order *Columbiformes*			
Pigeon (*Columbia livea*)	27	35	Butler et al., 1977
Mourning doves (*Zenaida macroura*)		29	Harms and Harms, 2012
Order *Falconiformes*			
Southern crested caracaras (*Caracara plancus*)	25	–	Escobar et al., 2011
Order *Galliformes*			
Chicken (*Gallus gallus*)	25		Gleeson and Brackenbury, 1984
Rapidly growing meat type	–	56	Mean from Olanrewaju et al., 2010; van As et al., 2010; Christensen et al., 2012
Order *Passeriformes*		–	
Boat-tailed grackles (*Quiscalus major*)		29	Harms and Harms, 2012
House sparrows (*Passer domesticus*)		38	Harms and Harms, 2012
Order *Psittaciformes*			
Amazon parrot (*Amazona aestiva*).	22		Valéria et al., 2008
pH			
Order *Columbiformes*			
Pigeon (*Columbia livea*)	7.43	7.36	Butler et al., 1977

TABLE 10.5 Examples of Arterial and Venous pO_2 and pCO_2 in Resting Birds with pH as a Comparison—cont'd

Species	Arterial	Venous	Reference
Order *Galliformes*			
Rapidly growing chicken		7.35	Mean from Olanrewaju et al., 2010; van As et al., 2010
Adult chicken		7.42	Martin et al., 2010
Order *Falconiformes*			
Southern crested caracaras (*Caracara plancus*)	7.54		Escobar et al., 2011

TABLE 10.6 Changes in Venous pO_2 and pCO_2 during Growth in Chickens

	pO_2 (mm Hg)	pCO_2 (mm Hg)	Reference
Young broiler chicks			
11 day old	58	48	van As et al., 2010
33 day old	48	59	van As et al., 2010
47 day old	35	69	van As et al., 2010
Adult meat-type female chicken (broiler breeder)	46	38	Martin et al., 2010

Additional materials can be found on the companion website: http://booksite.elsevier.com/9780124071605

Blood leukocyte populations are markedly influenced by stress. There are increases in heterophils and decreases in lymphocytes in young chickens receiving administration of exogenous corticosterone, the stress hormone (see Figure 10.7; Gross and Siegel, 1983). This leads to concomitant large shifts in the H/L ratio (Gross and Siegel, 1983). Similarly, stimulation of the adrenal cortical cells with adrenocorticotropic hormone is accompanied by changes in both increases in heterophils and decreases in lymphocytes in young chickens (Davison and Flack, 1981).

There is a concomitant reduction in volume of heterophils with corticosterone administration in young chickens (Shini et al., 2008). The relationship between stress and leukocyte populations is further supported by the greater H/L ratio in injured Adélie penguins (Vleck et al., 2000) and in young chickens exposed to stressors such as atmospheric ammonia, intermittent electric shock, environmental heat stress (McFarlane and Curtis, 1989) and lipopolysaccharide (LPS) (Shini et al., 2008). In response to intravenous cellulose microparticles, there are decreases in the percentage of heterophils and increases in eosinophils (chickens: Wang et al., 2003).

Changes in the leukocyte system were seen after poisoning with both chlorphenvinphos and foschlor, in the form of distinct neutrophilic leukocytosis and eosinopenia. A marked increase in the number of basophils was found after ingesting chlorphenvinphos, and a distinct increase in monocytes after ingesting both pesticides (Japanese quail: Gromysz-Kałkowska et al., 1985).

10.5.2 Heterophils

Heterophils are the major phagocytic leukocytes. They are presumed to be equivalent to the neutrophils in mammals.

10.5.2.1 Structure

Heterophils are the major polymorphonuclear leukocytes in birds (Figure 10.6). They have a diameter of 10–15 μm (Sturkie, 1986). Heterophils are weakly basophilic and pseudoeosinophic, staining brilliant red with Wright's stain (Sturkie, 1986). Electron microscopy studies of heterophils (duck and goose: Maxwell, 1973; chicken: Shini et al., 2008) reveal that avian heterophils have the following characteristics: The nuclei have two or three lobes. Mitochondria are sparse. There are two types of intracellular granules: dense and less dense small granules. The dense granules can be either oval or spherical, with diameters between 0.4 and 1.8 μm (ducks and geese: Maxwell, 1973).

Stressors, such as corticosterone or LPS, rapidly influence circulating heterophils; reducing cell size as indicated by cross-sectional area and increasing secretory granule diameter (chickens: Shini et al., 2008).

10.5.2.2 Function

Heterophils are the major phagocytic leukocytes in birds, with an analogous role to neutrophils in mammals (Figure 10.6; reviewed in Harmon, 1998). They are thought to have an important role in innate immune system, including mediating the acute inflammation response. Heterophils exhibit superior ability to monocytes, both to phagocytize and kill bacteria (chicken and turkey: Stabler et al., 1994).

Opsonization of bacteria enhances phagocytosis heterophils (Stabler et al., 1994). Heterophils release superoxide

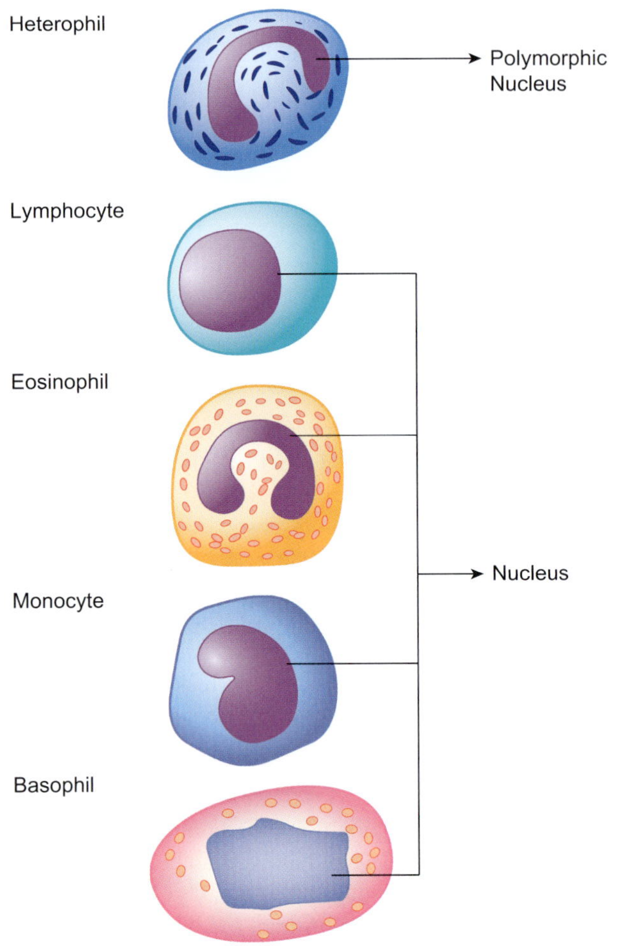

FIGURE 10.6 **Avian leukocytes.** (A) Schematic microscopic images of leukocytes.

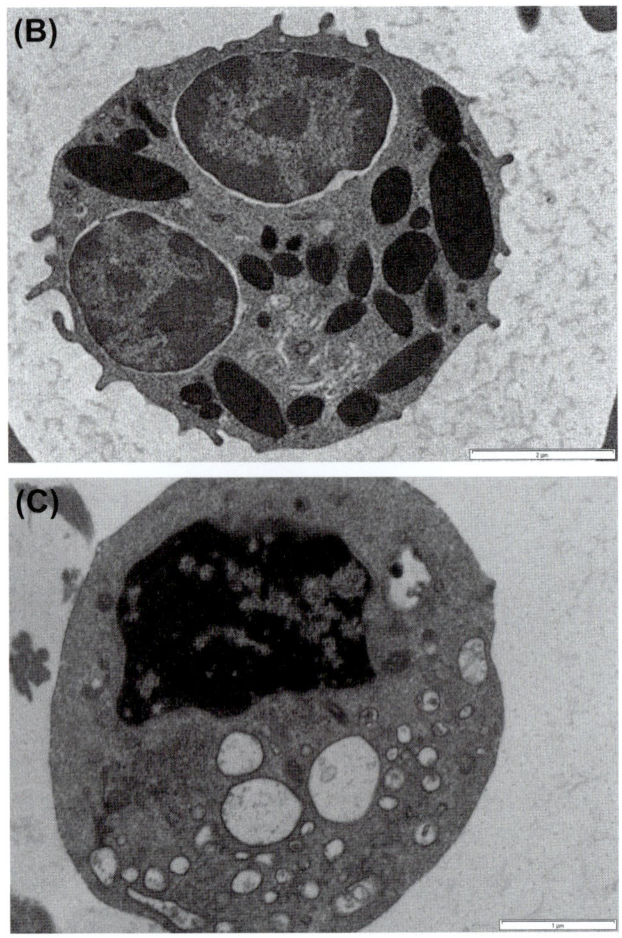

FIGURE 10.6—Cont'd, (B) Electron microscopy of heterophils from peripheral blood of chickens with large cytoplasmic granules (bar 2 μm). *(Reproduced from Shini et al. (2008), with permission from Elsevier.).* (C) Electron microscopy of agranular lymphocytes from peripheral blood of chickens (bar 1 μm). *Reproduced from Shini et al. (2008), with permission from Elsevier.*

(chicken and turkey: Stabler et al., 1994). Chicken heterophils that are infected with bacteria, such as *Mycoplasma* gallisepticum, attract lymphocytes (Lam, 2002). Heterocytes have also been shown to phagocytize fungi, such as *Candida albicans* (ring doves (*Streptopelia risoria*); Terrón et al., 2003). The phagocytic activity of heterophils is increased following priming with interferon-gamma (chicken: Kogut et al., 2005). Interleukin-2 (IL-2) has marked effects on phagocytizing heterophils; increasing phagocytosis, bacteriocidal activity, and the expression of the inflammatory cytokine, IL-8 (chicken: Kogut et al., 2002, 2003). Avian heterocytes show decreases in phagocytic activity with age (ring doves (*S. risoria*): Terrón et al., 2004). Melatonin increases both phagocytic activity by heterophils and the percentage of ingested *C. albicans* killed while depressing superoxide anion levels (O_2^-; ring doves (*S. risoria*): Terrón et al., 2003). Aflatoxins decrease both the phagocytic and bactericidal activity of heterophils (chicken: Chang and Hamilton, 1979a).

Another aspect of the innate immune system are the granules in heterophils that contain β-defensins: gallinacins

TABLE 10.7 Mean Concentration of Thrombocytes and Leukocytes, Together with Differential Count, in Wild Birds (Excluding Poultry Species) and Chickens

Parameter	Wilds Birds (Mean)	Chicken (Mean)[1]
Thrombocytes ($10^3/\mu L$)	30.4 (14 species)	34.4
Leukocytes ($10^3/\mu L$)	16.1 (97 species)	25.5
Heterophils (%)	45.3 (80 species)	25.9
Lymphocytes (%)	45.5 (79 species)	57.6
Monocytes (%)	3.0 (77 species)	5.6
Eosinophils (%)	4.1 (77 species)	1.7
Basophils (%)	1.6 (77 species)	2.4

[1]*Mean of meta-analysis of studies with 37,000 broiler birds (Andretta et al., 2012) and early data summarized in Sturkie (1986).*

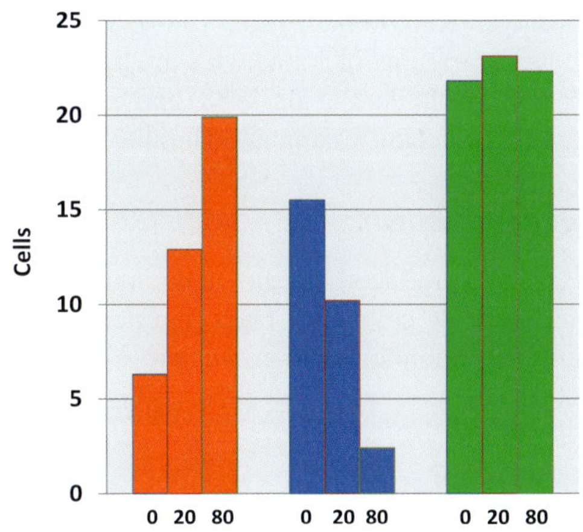

FIGURE 10.7 **Effect of corticosterone (in diet in ppm) on heterophils (red), lymphocytes (blue), and total leukocytes (green) in young chickens.** *Data from Gross and Siegel (1983).*

(Gal 1, Gal 1-α, and Gal 2), cathepsin, lysozyme, acid phosphatase, β-glucuronidase, and α-glucosidase (Harwig et al., 1994; also reviewed Harmon, 1998).

10.5.2.3 Number

Circulating concentrations of heterophils are increased by stresses (Figure 10.7) and by toxicants such as the herbicide paraquat (Japanese quail: Clark et al., 1988) and administration of organophosphate insecticides, either chlorphenvinphos or foschlor (Japanese quail: Gromysz-Kałkowska et al., 1985). Increases in the heterophil percentage have been also reported after removal of 30% of the blood volume (Japanese quail: Schindler et al., 1987). Chicken heterophils can be killed by some pathogens, such as *S. aureus* (Lowder et al., 2009).

10.5.2.4 Production

Heterophils are produced in the bone marrow. *In vitro*, embryonic marrow cells have been demonstrated to be capable of developing into heterophils (e.g., Japanese quail: Brandon et al., 2000).

10.5.3 Lymphocytes

Avian lymphocytes can be isolated using specific monoclonal antibodies, such as K55 monoclonal antibody, which binds chicken lymphocytes (Schmaier et al., 2011). There are B and T lymphocytes. T lymphocytes include the following (chicken: Gehad et al., 2002; Kushima et al., 2004):

- ($CD3^+$) T cells, T-helper cells ($CD4^+$)
- T-cytotoxic/suppressor cells ($CD8^+$)
- $CD3^+$ CD4- CD8- T cells

Chowdhury et al. (2005) reported that 17.5% of the lymphocytes in the blood of domestic turkeys are B lymphocytes, whereas 16% are $CD4^+$ and 16% are $CD8^+$.

10.5.3.1 Structure

Avian lymphocytes are weakly basophilic. Circulating avian lymphocytes have been characterized by electron microscopy studies (ducks and geese together with pigeons, turkeys, Japanese quail: Maxwell, 1974; also see Figure 10.6). The lymphocytes have a nonlobed nucleus with a high nuclear:cytoplasmic ratio.

Two types of lymphocytes have been reported in the circulation of birds: small and medium sized (Maxwell, 1974). In the small lymphocytes, the nuclei are pleomorphic in shape. The cytoplasm contains multiple mitochondria, Golgi apparatus, some endoplasmic reticulum, and a few granules. The small-sized lymphocytes have extensive pseudopodia. The medium-sized lymphocytes are oval with a spherical nucleus. They have relatively few intracellular organelles (mitochondria and endoplasmic reticulum) and a few pseudopodia.

There are shifts in ultrastructure with physiology. The stressor LPS rapidly reduces the cell size of circulating lymphocytes, as indicated by cell diameter or cross-sectional area (chickens: Shini et al., 2008).

10.5.3.2 Function

Lymphocytes play important roles in both humoral and cell mediate immunity. The IL-2 receptor alpha chain (CD25) is found on cell membranes of both $CD4^+$ and $CD8^+$ lymphocytes (chicken: Teng et al., 2006).

10.5.3.3 Number

Circulating concentrations of lymphocytes are decreased by stresses (as previously described for leukocyte populations) and toxicants, such as the herbicide paraquat (Clark et al., 1988).

10.5.4 Eosinophils

10.5.4.1 Structure

Avian eosinophils are oval granulocytes with considerable cytoplasm. The round granules stain red after the uptake of the stain or dye eosin. The nucleus has two lobes (i.e., bi-lobed) (Sturkie, 1986). Electron microscopy studies reveal that avian eosinophils have multiple dense granules (0.1–1.6 µm diameter) with crystalline cores (Maxwell and Siller, 1972).

10.5.4.2 Function

The function of avian eosinophils is not fully understood (Campbell, 1997; Campbell and Ellis, 2007).

10.5.4.3 Number

In a differential count, the percentage of eosinophils appears to be impacted by some stressors. Eosinophils are reported to be increased in some birds in captivity (Rufous-collared sparrows, *Z. capensis*: Ruiz et al., 2002). In contrast, the pesticides chlorphenvinphos and foschlor decrease eosinophil numbers (Japanese quail: Gromysz-Kałkowska et al., 1985).

10.5.5 Monocytes

10.5.5.1 Structure

Avian monocytes are oval cells with considerable cytoplasm. Electron microscopy reveals that avian monocytes have multiple small dense granules (0.1–1.6 μm diameter) with crystalline cores (Maxwell and Siller, 1972). They have kidney-shaped nuclei and extensive cytoplasm containing a well-developed Golgi apparatus, rough endoplasmic reticulum, microtubules, vesicles, and mitochondria (ducks and geese together with pigeons, turkeys, Japanese quail: Maxwell, 1974).

10.5.5.2 Function

The function of avian monocytes is not fully known. Monocytes produce nitric oxide (NO) in response to LPS challenge (chickens: Bowen et al., 2009) with, in turn, NO-inducing vasodilation. Expression of inducible NOS (iNOS) and hence the production of NO is stimulated by lipopolysaccharide in both monocytes and macrophages (chickens: Bowen et al., 2007).

The IL-2 receptor alpha chain (CD25) is found on cell membranes of monocytes (chicken: Teng et al., 2006). Chicken monocytes can exert a chemotaxic effect on heterophils and lymphocytes due to release of macrophage inflammatory protein (MIP)-1β (Lam, 2002). Opsonization of bacteria enhances phagocytosis by monocytes (Stabler et al., 1994). Aflatoxins decrease the phagocytic activity of monocytes (chicken: Chang and Hamilton, 1979b). The transformation of avian monocytes into very powerful phagocytic cells, the macrophage, is well documented (Grecchi et al., 1980).

10.5.5.3 Number

The effects of stress have been reported on monocytes as a percentage of total leukocytes, with increases after *E coli* challenge together with transportation stress (turkeys: Huff et al., 2010). Monocyte numbers are reduced initially after challenge with Gram-negative bacteria cell wall LPS, but then recover and increase (chickens: Gehad et al., 2002; Bowen et al., 2009). Pesticides, such as chlorphenvinphos and foschlor, increase monocyte numbers (Japanese quail: Gromysz-Kałkowska et al., 1985). In contrast, circulating concentrations of monocytes are decreased by the herbicide paraquat (Clark et al., 1988) and by the mycotoxicant aflatoxin (chicken: Chang and Hamilton, 1979b).

10.5.5.4 Production

Monocytes are produced in the bone marrow. *In vitro*, embryonic marrow cells have been reported as capable of developing into monocytes (e.g., Japanese quail: Brandon et al., 2000).

10.5.6 Basophils

10.5.6.1 Structure

The nuclei and cytoplasmic granules of basophils take up basic stains, such as hematoxylin, giving a pale purple appearance with a clear cytoplasm (Sturkie, 1986). The ultrastructure of avian basophils has been reported (ducks and geese: Maxwell, 1973). Basophils contain cytoplasmic granules (see Figure 10.6). These vary in diameter between 0.1 and 0.8 μm in ducks and geese and up to 1.0 μm in turkeys (Maxwell, 1973). The nuclei of basophils are reported as polymorphic (Sturkie, 1986) or nonlobulated (ducks and geese: Maxwell, 1973).

10.5.6.2 Function

The functions of avian basophils are not fully established.

10.5.6.3 Numbers

Parasites and toxicants influence basophil numbers. Challenging chickens with the parasite *P. juxtanucleare* is accompanied by increases in the basophil concentration in the blood (Silveira et al., 2009). The pesticide chlorphenvinphos increases basophil numbers (Japanese quail: Gromysz-Kałkowska et al., 1985).

10.6 THROMBOCYTES

Thrombocytes are the functional equivalent of platelets in mammals. Avian thrombocytes can be recognized by specific monoclonal antibodies for surface antigens (e.g., chicken: Horiuchi et al., 2004; guinea fowl: Bódi et al., 2009). Avian thrombocytes can be isolated using specific monoclonal antibodies such against chicken α_{2b} integrins 14 (Schmaier et al., 2011). Chicken thrombocytes express many of the genes specific for mammalian platelet formation (Schmaier et al., 2011). In view of their functions, some researchers suggest that thrombocytes should be considered as leukocytes (Seliger et al., 2012). Thrombocytes are found in the "buffy" layer when the hematocrit of avian blood is being determined.

10.6.1 Structure

Avian thrombocytes have the following characteristics (Japanese quail: Belleville et al., 1982):

- Cell diameter: 5.4 μm
- Nuclear diameter: 4.2 μm.

Avian thrombocytes can be confused with lymphocytes but are readily recognized by specific monoclonal antibodies for specific surface antigens (e.g., chicken: Horiuchi et al., 2004; guinea fowl: Bódi et al., 2009). The ultrastructure of the avian thrombocyte has been characterized (ducks, geese, pigeons, turkeys, Japanese quail: Maxwell, 1974). Avian thrombocytes have an irregular shape with multiple pseudopodia. Similarly, the nucleus is irregular in shape. There is more cytoplasm than is seen in small lymphocytes. There are multiple small mitochondria, an endoplasmic reticulum, prominent microtubules, and a few dense granules.

10.6.2 Function

Chicken thrombocytes, like mammalian platelets, release serotonin in response to either thrombin or collagen, but appear much less sensitive to bovine thrombin (Schmaier et al., 2011). Avian thrombocytes differ from mammalian platelets by their inability to form shear-resistant aggregates and their much lower levels of $\alpha(_2b)\beta_3$ integrin, a protein required for aggregate formation (Schmaier et al., 2011). Avian thrombocytes are activated by thrombin and adhere to fibrinogen as part of the process of blood clotting (Lacoste-Eleaume et al., 1994). Avian thrombocytes aggregate and induce hemostasis, a temporary stopping of blood flow (Grant and Zucker, 1973). Aggregation occurs in response to serotonin (Stiller et al., 1975). Avian thrombocytes differ from mammalian platelets by their inability to form shear-resistant aggregates and their much lower levels of $\alpha(_2b)\beta_3$ integrin, a protein required for aggregate formation (Schmaier et al., 2011).

There are avian homologs of the mammalian platelet proteins. For example, integrin GPIIb-IIIa is present on the surface of chicken thrombocytes (Lacoste-Eleaume et al., 1994).

Avian thrombocytes produce platelet-derived growth factor (PDGF). Expression of PDGF A and B-chain mRNA are, respectively, low and high in thrombocytes, but expression of both are increased further following exposure to type 1 collagen (Horiuchi et al., 2001, 2002). The PDGF is presumed to play a role in tissue repair. PDGF B-chain has been characterized from cDNA of the transcript from chicken thrombocytes (Horiuchi et al., 2002).

Avian thrombocytes, like mammalian platelets, release serotonin from intracellular granules in avian thrombocytes (Stiller et al., 1975). Serotonin is released in response to either thrombin or collagen but appear much less sensitive to bovine thrombin (chicken: Schmaier et al., 2011). The serotonin in turn induces vasoconstriction, aggregation of thrombocytes and blood clotting (Stiller et al., 1975; Lacoste-Eleaume et al., 1994; Chapman et al., 2008). Moreover, as with mammalian platelets, chicken thrombocytes release pro-inflammatory cytokines in response to lipopolysaccharide, a surrogate for pathogens. There are large increases in expression of IL-1β, IL-6, IL-8, and IL-12 together with interferon-α and cyclooxygenase in response to lipopolysaccharide, which is a surrogate for pathogens (Ferdous et al., 2008; Scott and Owens, 2008; St Paul et al., 2012). Toll-like receptors (TLR) recognize pathogens and are expressed by chicken thrombocytes. For example, TLR 2, 3, and 4, are expressed in response to lipopolysaccharide (St Paul et al., 2012).

Thrombocytes should also be considered as part of the innate immune system. For instance, avian thrombocytes are phagocytotic (Carlson et al., 1968) with bacteria but probably not protozoa (chicken: Chang and Hamilton, 1979c; DaMatta et al., 1998). Chicken thrombocytes have been reported exhibit increased phagocytosis in response to lipopolysaccharide (St Paul et al., 2012). Thrombocytes express the IgY Fc receptor (chicken: Viertlboeck et al., 2009). The IL-2 Receptor alpha chain (CD25) is found on cell membranes of thrombocytes, like monocytes, together with both $CD4^+$ and $CD8^+$ lymphocytes (chicken: Teng et al., 2006). There is increased release of nitrite from chicken thrombocytes in response to lipopolysaccharide and high expression of inducible nitric oxide synthase (St Paul et al., 2012).

10.6.3 Number

Table 10.7 summarizes thrombocyte concentrations in wild birds and poultry. The number of thrombocytes in the blood can be influenced by environmental factors. The circulating concentration of thrombocytes (thrombocytopenia) decreases with viral infection, such as by bursal disease virus in chickens (Lima et al., 2005). Moreover, there are marked changes in thrombocytes in chickens challenged with *P. juxtanucleare* to experimentally induce chicken malaria with infected and reduced numbers of thrombocytes (Silveira et al., 2009). Similarly, there are decreases in the number of thrombocytes following intravenous administration of cellulose microparticles (Wang et al., 2003). Thrombocyte numbers increase after challenge with LPS (chicken: Gehad et al., 2002). Pesticides can influence thrombocyte concentrations, being for instance increased by foschlor but decreased by chlorphenvinphos (Japanese quail: Gromysz-Kałkowska et al., 1985).

10.6.4 Production

Thrombocytes are produced from thromboblasts in the blood marrow. *In vitro*, embryonic marrow cells have been demonstrated to be capable of developing into thrombocytes (e.g., Japanese quail: Brandon et al., 2000).

10.7 CLOTTING

Avian blood appears to clot more slowly than does mammalian blood (Japanese quail: Belleville et al., 1982). For instance, the whole blood clotting time is 38 min for Japanese quail compared 8 min for human blood (Belleville et al., 1982). Avian blood does not clot in responses to glass activation (Japanese quail: Belleville et al., 1982). Table 10.8 provides examples of clotting times in birds.

The overall scheme for clotting in birds is presumed to be the same as that in mammals (Figure 10.8). Fibrinogen is cleaved by the protease, thrombin, to fibrin, which then polymerized. Avian thrombin has been isolated (Japanese quail: Belleville et al., 1982). Prothrombin is activated to thrombin by a cascade of clotting factors.

10.8 AVIAN BLOOD MODELS

10.8.1 β-adrenergic Receptors

Erythrocytes from both turkeys and chickens have been employed as models for β-adrenergic receptor and adenylate cyclase for many years (turkey: Schramm et al., 1972; Bilezikian and Aurbach, 1973a,b; Oye and Sutherland, 1996). The fluxes of sodium ions across the plasma membrane of erythrocytes from turkeys is two- to three-fold that in humans (Palfrey and Greengard, 1981). Lines of chickens have been selected for high and low leucine uptake (Somes et al., 1981). These also have elevated glycine, lysine uptake, and the Na^+-K^+-ATPase activity

TABLE 10.8 Examples of Blood Coagulation Times

Test	Chicken[1,2]	Turkey[1]	Japanese Quail[3]	Hispaniolan Parrots (*Amazona ventralis*)[2]	Umbrella Cockatoos (*Cacatua alba*)[2]
Lee-White whole blood clotting (min)	52[1]	75	38	–	–
Capillary tube test (min)	78[1]	124	–	–	–
One-stage prothrombin test (sec)	15.9[1] (10.0[2])	16.9	10	10.0	10.0
Whole-blood thromboplastin test (sec)	16.9[1] (9.1[2])	19.5	(78[4])	10.3	–

[1]*Bigland and Starr (1965).*
[2]*Morrisey et al. (2003).*
[3]*Belleville et al. (1982).*
[4]*Partial thromboplastin time.*

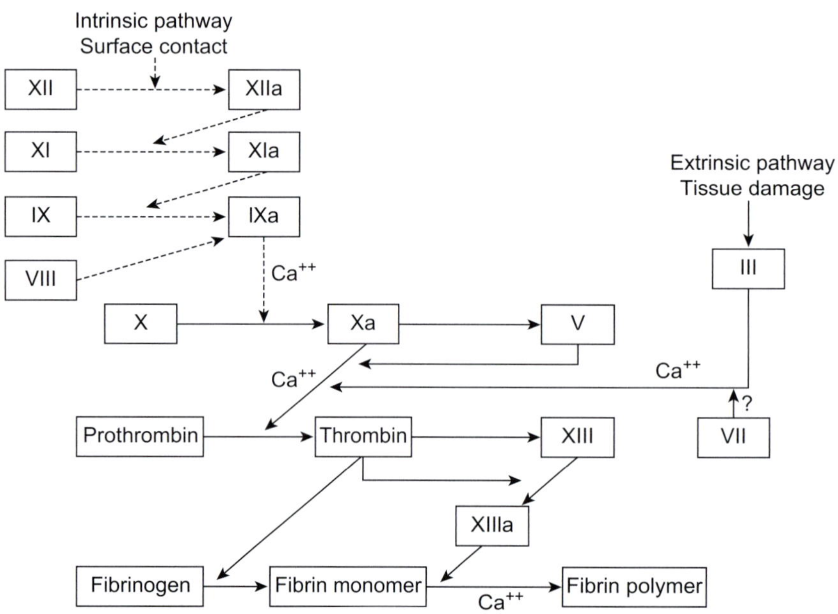

FIGURE 10.8 Cascade of coagulation of avian blood with a series of clotting factors.

(Somes et al., 1981; Lerner et al., 1982). Using a lysate of turkey erythrocytes, roles of protein kinases in the desensitization of β-adrenergic receptors has been demonstrated (Nambi et al., 1985). Adenylate cyclase is inhibited by calcium ions in turkey erythrocyte membranes (Demirel et al., 1998).

10.8.2 Transgenic Chickens

Transgenic chickens can produce desired proteins, such as human erythropoietin. Chickens producing human erythropoietin have been developed (Koo et al., 2010; Penno et al., 2010).

10.8.3 Avian IgY Antibodies

Chicken IgY antibodies are being used for immunoaffinity fractionation of human plasma due to their high avidity and specificity, low nonspecific binding, and ready source in eggs (Huang, and Fang, 2008).

10.8.4 Nutritional Models

10.8.4.1 Vitamin K

Historically, vitamin K was discovered based on nutritional studies in chicken (McCullem, 1957). This fat-soluble vitamin is critically important for the synthesis of proteins in the blood coagulation cascade.

10.8.4.2 Iron Availability

A model for nutritional studies for iron availability has been proposed using chickens together with the stable isotope ^{58}Fe (Tako et al., 2010).

REFERENCES

Abe, T., Muto, Y., Hosoya, N., 1975. Vitamin A transport in chicken plasma: isolation and characterization of retinol-binding protein (RBP), prealbumin (PA), and RBP-PA complex. J. Lipid Res. 16, 200–210.

Alev, C., Shinmyozu, K., McIntyre, B.A., Sheng, G., 2009. Genomic organization of zebra finch alpha and beta globin genes and their expression in primitive and definitive blood in comparison with globins in chicken. Dev. Genes Evol. 219, 353–360.

Alper, S.L., Palfrey, H.C., DeRiemer, S.A., Greengard, P., 1980. Hormonal control of protein phosphorylation in turkey erythrocytes. Phosphorylation by cAMP-dependent and Ca2+-dependent protein kinases of distinct sites in goblin, a high molecular weight protein of the plasma membrane. J. Biol. Chem. 255, 11029–11039.

Andretta, I., Kipper, M., Lehnen, C.R., Lovatto, P.A., 2012. Meta-analysis of the relationship of mycotoxins with biochemical and hematological parameters in broilers. Poult. Sci. 91, 376–382.

Arad, Z., Marder, J., Eylath, U., 1983. Serum electrolyte and enzyme responses to heat stress and dehydration in the fowl (*Gallus domesticus*). Comp. Biochem. Physiol. A Comp. Physiol. 74A, 448–453.

Bagley, L.G., Christensen, V.L., Gildersleeve, R.P., 1990. Hematological indices of turkey embryos incubated at high altitude as affected by oxygen and shell permeability. Poult. Sci. 69, 2035–2039.

Baker, J.G., 2010. A full pharmacological analysis of the three turkey β-adrenoceptors and comparison with the human β-adrenoceptors. PLoS One 5, e15487.

Baumann, R., Blass, C., Götz, R., Dragon, S., 1999. Ontogeny of catecholamine and adenosine receptor-mediated cAMP signaling of embryonic red blood cells: role of cGMP-inhibited phosphodiesterase 3 and hemoglobin. Blood 94, 4314–4320.

Baumann, R., Gotz, R., Dragon, S., 2003. NTP pattern of avian embryonic red cells: role of RNA degradation and AMP deaminase/5'-nucleotidase activity. Am. J. Physiol. 284, R771–R779.

Baumgartner, S., Brown, D.J., Salevsky Jr., E., Leach Jr., R.M., 1978. Copper deficiency in the laying hen. J. Nutr. 108, 804–811.

Belleville, J., Cornillon, B., Paul, J., Baguet, J., Clendinnen, G., Eloy, R., 1982. Haemostasis, blood coagulation and fibrinolysis in the Japanese quail. Comp. Biochem. Physiol. A Comp. Physiol. 71, 219–230.

Bernstein, R.S., Schraer, R., 1972. Purification and properties of an avian carbonic anhydrase from the erythrocytes of *Gallus domesticus*. J. Biol. Chem. 247, 1306–1322.

Beuchat, C.A., Chong, C.R., 1998. Hyperglycemia in hummingbirds and its consequences for hemoglobin glycation. Comp. Biochem. Physiol. A Mol. Integr. Physiol. 120, 409–416.

Bigland, C.H., Starr, R.M., 1965. Comparison of simple blood coagulation tests in birds. Can. Vet. J. 6, 233–236.

Bilezikian, J.P., Aurbach, G.D., 1973a. A beta-adrenergic receptor of the turkey erythrocyte. I. Binding of catecholamine and relationship to adenylate cyclase activity. J. Biol. Chem. 248, 5577–5583.

Bilezikian, J.P., Aurbach, G.D., 1973b. A beta-adrenergic receptor of the turkey erythrocyte. II. Characterization and solubilization of the receptor. J. Biol. Chem. 248, 5584–5589.

Bishop, C., Butler, P., 1995. Physiological modelling of oxygen consumption in birds during flight. J. Exp. Biol. 198, 2153–2163.

Bódi, I., Nagy, N., Sinka, L., Igyártó, B.Z., Oláh, I., 2009. Novel monoclonal antibodies recognise guinea fowl thrombocytes. Acta Vet. Hung. 57, 239–246.

Borges, S.A., Fischer da Silva, A.V., Majorka, A., Hooge, D.M., Cummings, K.R., 2004. Physiological responses of broiler chickens to heat stress and dietary electrolyte balance (sodium plus potassium minus chloride, milliequivalents per kilogram). Poult. Sci. 83, 1551–1558.

Bowen, O.T., Erf, G.F., Chapman, M.E., Wideman Jr., R.F., 2007. Plasma nitric oxide concentrations in broilers after intravenous injections of lipopolysaccharide or microparticles. Poult. Sci. 86, 2550–2554.

Bowen, O.T., Dienglewicz, R.L., Wideman, R.F., Erf, G.F., 2009. Altered monocyte and macrophage numbers in blood and organs of chickens injected i.v. with lipopolysaccharide. Vet. Immunol. Immunopathol. 131, 200–210.

Brace, K., Atland, P.D., 1956. Lifespan of the duck and chicken erythrocytes as determined with C-14. Proc. Soc. Exp. Biol. Med. 92, 615–617.

Brandon, C., Eisenberg, L.M., Eisenberg, C.A., 2000. WNT signaling modulates the diversification of hematopoietic cells. Blood 96, 4132–4141.

Breuner, C.W., Orchinik, M., 2009. Pharmacological characterization of intracellular, membrane, and plasma binding sites for corticosterone in house sparrows. Gen. Comp. Endocrinol. 163, 214–224.

Bryan, T.E., Gildersleeve, R.P., Wiard, R.P., 1989. Exposure of Japanese quail embryos to o,p'-DDT has long-term effects on reproductive behaviors, hematology, and feather morphology. Teratology 39, 525–535.

Butler, P.J., West, N.H., Jones, D.R., 1977. Respiratory and cardiovascular responses of the pigeon to sustained, level flight in a wind-tunnel. J. Exp. Biol. 71, 7–26.

Calabrese, L., Carbonaro, M., Musci, G., 1988. Chicken ceruloplasmin. Evidence in support of a trinuclear cluster involving type 2 and 3 copper centers. J. Biol. Chem. 263, 6480–6483.

Campbell, T.W., 1997. Hematology. In: Ritchie, W.B., Harrison, G.J., Harrison, L.R. (Eds.), Avian Medicine: Principles and Application. Wingers Publishing, Lake Worth, Florida, pp. 176–198.

Campbell, T.W., Ellis, C.K., 2007. Avian and Exotic Animal Hematology and Cytology 3rd edition. Blackwell, Ames, IA.

Carlson, H.C., Sweeny, P.R., Tokaryk, J.M., 1968. Demonstration of phagocytic and trephocytic activities of chicken thrombocytes by microscopy and vital staining techniques. Avian Dis. 12, 700–715.

Cetin, E., Silici, S., Cetin, N., Güçlü, B.K., 2010. Effects of diets containing different concentrations of propolis on hematological and immunological variables in laying hens. Poult. Sci. 89, 1703–1708.

Chamanza, R., Toussaint, M.J., van Ederen, A.M., van Veen, L., Hulskamp-Koch, C., Fabri, T.H., 1999. Serum amyloid A and transferrin in chicken. A preliminary investigation of using acute-phase variables to assess diseases in chickens. Vet. Q. 21, 158–162.

Chang, C.F., Hamilton, P.B., 1979a. Impaired phagocytosis by heterophils from chickens during aflatoxicosis. Toxicol. Appl. Pharmacol. 48, 459–466.

Chang, C.F., Hamilton, P.B., 1979b. Impairment of phagocytosis in chicken monocytes during aflatoxicosis. Poult. Sci. 58, 562–566.

Chang, C.F., Hamilton, P.B., 1979c. The thrombocyte as the primary circulating phagocyte in chickens. J. Reticuloendothel. Soc. 25, 585–590.

Chapman, B.S., Tobin, A.J., Hood, L.E., 1980. Complete amino acid sequences of the major early embryonic alpha-like globins of the chicken. J. Biol. Chem. 255, 9051–9059.

Chapman, M.E., Taylor, R.L., Wideman Jr., R.F., 2008. Analysis of plasma serotonin levels and hemodynamic responses following chronic serotonin infusion in broilers challenged with bacterial lipopolysaccharide and microparticles. Poult. Sci. 87, 116–124.

Chen, X.H., Harden, T.K., Nicholas, R.A., 1994. Molecular cloning and characterization of a novel beta-adrenergic receptor. J. Biol. Chem. 269, 24810–24819.

Chou, S.J., Shieh, Y.C., Yu, C.Y., 2008. Hematologic and biochemistry values for black-faced spoonbills (*Platalea minor*) with and recovering from botulism. J. Wildl. Dis. 44, 781–784.

Chowdhury, S.R., Smith, T.K., Boermans, H.J., Woodward, B., 2005. Effects of feed-borne *Fusarium* mycotoxins on hematology and immunology of turkeys. Poult. Sci. 84, 1698–1706.

Christensen, E.H., Dill, D.B., 1935. Oxygen dissociation curves of bird blood. J. Biol. Chem. 109, 443–448.

Christensen, K., Vizzier Thaxton, Y., Thaxton, J.P., Scanes, C.G., 2012. Changes in body temperature during growth and in response to fasting in growing modern meat type chickens. Brit. Poult. Sci. 53, 531–537.

Cirotto, C., Di Tella, A.S., Geraci, G., 1975. The hemoglobins of the developing chicken embryos. Fractionation and globin composition of the individual component of total erythrocytes and of a single erythrocyte type. Cell Differ. 4, 87–99.

Clark, M.W., Gildersleeve, R.P., Thaxton, J.P., Parkhurst, C.R., McRee, D.I., 1988. Hematological effects of ethyl methanesulfonate, paraquat and phenylhydrazine in Japanese quail. Comp. Biochem. Physiol. C 89, 15–30.

Cohen, A., Klasing, K., Ricklefs, R., 2007. Measuring circulating antioxidants in wild birds. Comp. Biochem. Physiol. B Biochem. Mol. Biol. 147, 110–121.

Cohen, A.A., McGraw, K.J., Robinson, W.D., 2009. Serum antioxidant levels in wild birds vary in relation to diet, season, life history strategy, and species. Oecologia 161, 673–683.

Coll, J., Ingram, V.M., 1978. Stimulation of heme accumulation and erythroid colony formation in cultures of chick bone marrow cells by chicken plasma. J. Cell Biol. 76, 184–190.

Curtis, M.J., Butler, E.J., 1980. Response of caeruloplasmin to *Escherichia coli* endotoxins and adrenal hormones in the domestic fowl. Res. Vet. Sci. 28, 217–222.

Dajani, R., Orten, J.M., 1958. A study of the citric acid cycle in erythrocytes. J. Biol. Chem. 231, 913–924.

Dajani, R., Orten, J.M., 1959. The utilization of glycine by the nucleated erythrocyte. J. Biol. Chem. 234, 877–879.

DaMatta, R.A., Seabra, S.H., de Souza, W., 1998. Further studies on the phagocytic capacity of chicken thrombocytes. Submicrosc. Cytol. Pathol. 30, 271–277.

Davis, H.G., 1961. Structure in nucleated erythrocytes. J. Biophys. Biochem. Cytol. 9, 671–687.

Davison, T.F., Flack, I.H., 1981. Changes in the peripheral blood leucocyte populations following an injection of corticotrophin in the immature chicken. Res. Vet. Sci. 30, 79–82.

Deem, S.L., Noss, A.J., Cuéllar, R.L., Karesh, W.B., 2005. Health Evaluation of free-ranging and captive blue-fronted Amazon parrots (*Amazona aestiva*) in the Gran Chaco, Bolivia. J. Zoo. Wildl. Med. 36, 598–605.

Deem, S.L., Ladwig, E., Cray, C., Karesh, W.B., Amato, G., 2008. Health assessment of the ex situ population of St Vincent parrots (*Amazona guildingii*) in St Vincent and the Grenadines. J. Avian Med. Surg. 22, 114–122.

Demirel, E., Ugur, O., Onaran, H.O., 1998. Ca2+-induced inhibition of adenylyl cyclase in turkey erythrocyte membranes. Pharmacology 57, 222–228.

Diamond, D.L., Carruthers, A., 1993. Metabolic control of sugar transport by derepression of cell surface glucose transporters. An insulin-independent recruitment-independent mechanism of regulation. J. Biol. Chem. 268, 6437–6444.

Disilvestro, R.A., Harris, E.D., 1985. Purification and partial characterization of ceruloplasmin from chicken serum. Arch. Biochem. Biophys. 241, 438–446.

Drew, C., Lapaix, F., Egee, S., Thomas, S., Ellory, J.C., Staines, H.M., 2002. Age-dependent changes in cation transport in the chicken erythrocyte. Comp. Biochem. Physiol. A Mol. Integr. Physiol. 133, 169–178.

Eguchi, R., Ishihara, A., Yamauchi, K., 2008. Interaction of diethylstilbestrol and ioxynil with transthyretin in chicken serum. Comp. Biochem. Physiol. C Toxicol. Pharmacol. 147 C, 345–350.

Escobar, A., Thiesen, R., Vitaliano, S.N., Belmonte, E.A., Werther, K., Valadão, C.A., 2011. Cardiorespiratory effects of isoflurane anesthesia in crested caracaras (*Caracara plancus*). J. Zoo Wildl. Med. 42, 12–17.

Ferdous, F., Maurice, D., Scott, T., 2008. Broiler chick thrombocyte response to lipopolysaccharide. Poult. Sci. 87, 61–63.

Fujii, M., Yoshino, I., Suzuki, M., Higuchi, T., Mukai, S., Aoki, T., Fukunaga, T., Sugimoto, Y., Inoue, Y., Kusuda, J., Saheki, T., Sato, M., Hayashi, S., Tamaki, M., Sugano, T., 1996. Primary culture of chicken hepatocytes in serum-free medium (pH 7.8) secreted albumin and transferrin for a long period in free gas exchange with atmosphere. Int. J. Biochem. Cell. Biol. 28, 1381–1391.

Furukawa, H., Loeb, J.N., Bilezikian, J.P., 1980. Beta-adrenergic receptors and isoproterenol-stimulated potassium transport in erythrocytes from normal and hypothyroid turkeys. Quantitative relation between receptor occupancy and physiologic responsiveness. J. Clin. Invest. 66, 1057–1064.

Gardner, J.D., Klaeveman, H.L., Bilezikian, J.P., Aurbach, G.D., 1973. Effect of beta-adrenergic catecholamines on sodium transport in turkey erythrocytes. J. Biol. Chem. 248, 5590–5597.

Gehad, A.E., Lillehoj, H.S., Hendricks 3rd, G.L., Mashaly, M.M., 2002. Initiation of humoral immunity. II. The effects of T-independent and T-dependent antigens on the distribution of lymphocyte populations. Dev. Comp. Immunol. 26, 761–771.

Gerhardt, E.M., Chan, L.N., Jing, S.Q., Qi, M.Y., Trowbridge, I.S., 1991. The cDNA sequence and primary structure of the chicken transferrin receptor. Gene 102, 249–254.

Gibson, J.S., Cossins, A.R., Ellory, J.C., 2000. Oxygen-sensitive membrane transporters in vertebrate red cells. J. Exp. Biol. 203, 1395–1407.

Gildersleeve, R.P., Satterlee, D.G., Johnson, W.A., Scott, T.R., 1983. The effects of forced molt treatment on blood biochemicals in hens. Poult. Sci. 62, 755–762.

Gildersleeve, R.P., Galvin, M.J., Thaxton, J.P., McRee, D.I., 1985a. Hematological response of Japanese quail to acute hemorrhagic stress. Comp. Biochem. Physiol. A Comp. Physiol. 81, 403–409.

Gildersleeve, R.P., Phelps, P.V., Thaxton, J.P., McRee, D.I., 1985b. Effect of phlebotomy on reticulocyte numbers in Japanese Quail. Poult. Sci. 64, 1990–1995.

Gleeson, M., Brackenbury, J.H., 1984. Effects of body temperature on ventilation, blood gases and acid-base balance in exercising fowl. Q. J. Exp. Physiol. 69, 61–72.

Glombitza, S., Dragon, S., Berghammer, M., Pannermayr, M., Baumann, R., 1996. Adenosine causes cAMP-dependent activation of chick embryo red cell carbonic anhydrase and 2,3-DPG synthesis. Am. J. Physiol. 271, R973–R981.

Grant, R.A., Zucker, M.B., 1973. Avian thrombocyte aggregation and shape change in vitro. Am. J. Physiol. 225, 340–343.

Grecchi, R., Saliba, A.M., Mariano, M., 1980. Morphological changes, surface receptors and phagocytic potential of fowl mono-nuclear phagocytes and thrombocytes in vivo and in vitro. J. Pathol. 130, 23–31.

Gromysz-Kałkowska, K., Szubartowska, E., Kaczanowska, E., 1985. Peripheral blood in the Japanese quail (*Coturnix coturnix japonica*) in acute poisoning by different insecticides. Comp. Biochem. Physiol. C 81, 209–212.

Gross, W.B., Siegel, H.S., 1983. Evaluation of heterophil/lymphocyte ratio as a measure of stress in chickens. Avian Dis. 27, 972–979.

Guha Thakurta, P., Choudhury, D., Dasgupta, R., Dattagupta, J.K., 2003. Structure of diferric hen serum transferrin at 2.8 A resolution. Acta Crystallogr. D Biol. Crystallogr. 59, 1773–1781.

Harmon, B.G., 1998. Avian heterophils in inflammation and disease resistance. Poult. Sci. 77, 972–977.

Harms, C.A., Harms, R.V., 2012. Venous blood gas and lactate values of mourning doves (*Zenaida macroura*), boat-tailed grackles (*Quiscalus major*), and house sparrows (*Passer domesticus*) after capture by mist net, banding, and venipuncture. J. Zoo Wildl. Med. 43, 77–84.

Harwig, S.L., Swiderek, K.M., Kokryakov, J.N., Tan, L., Lee, T.D., Panyutich, E.A., Aleshina, G.M., Shamova, O.V., Lehrer, R.I., 1994. Gallinacins: cysteine-rich antimicrobial peptides of chicken leukocytes. FEBS Lett. 342, 281–285.

Hasegawa, M.Y., Fonteque, J.H., Kohayagawa, A., Boretti, L.P., 2002. Avaliação do perfil eletroforético das proteínas séricas em matrizes pesadas (*Gallus gallus domesticus*) da linhagem avian farm. Revista Brasileira de Ciência Avícola 4, 203–207.

Heatley, J.J., Cary, J., Russell, K.E., Voelker, G., 2013. Clinicopathologic analysis of passeriform venous blood reflects transitions in elevation and habitat. Vet. Med. Res. Rep. 4, 21–29.

Heinze, C.R., Hawkins, M.G., Gillies, L.A., Wu, X., Walzem, R.L., German, J.B., Klasing, K.C., 2012. Effect of dietary omega-3 fatty acids on red blood cell lipid composition and plasma metabolites in the cockatiel, *Nymphicus hollandicus*. J. Anim. Sci. 90, 3068–3079.

Hilewicz-Grabska, M., Zgirski, A., Krajewski, T., Plonka, A., 1988. Purification and partial characterization of goose ceruloplasmin. Arch. Biochem. Biophys. 260, 18–27.

Horiuchi, H., Inoue, T., Furusawa, S., Matsuda, H., 2001. Characterization and expression of three forms of cDNA encoding chicken platelet-derived growth factor-A chain. Gene 272, 181–190.

Horiuchi, H., Inoue, T., Furusawa, S., Matsuda, H., 2002. Cloning and characterization of a chicken platelet-derived growth factor B-chain cDNA. Dev. Comp. Immunol. 26, 73–83.

Horiuchi, H., Tanaka, K., Shigeta, A., Yoshida, K., Kushima, K., Ohta, H., Furusawa, S., Matsuda, H., 2004. A monoclonal antibody against chicken thrombocytes reacts with the cells of thrombocyte lineage. J. Vet. Med. Sci. 66, 243–250.

Huang, L., Fang, X., 2008. Immunoaffinity fractionation of plasma proteins by chicken IgY antibodies. Methods Mol. Biol. 425, 41–51.

Huff, G.R., Huff, W.E., Rath, N.C., Anthony, N.B., Nestor, K.E., 2008. Effects of *Escherichia coli* challenge and transport stress on hematology and serum chemistry values of three genetic lines of turkeys. Poult. Sci. 87, 2234–2241.

Huff, G.R., Huff, W.E., Farnell, M.B., Rath, N.C., Solis de Los Santos, F., Donoghue, A.M., 2010. Bacterial clearance, heterophil function, and hematological parameters of transport-stressed turkey poults supplemented with dietary yeast extract. Poult. Sci. 89, 447–456.

Iarovaia, O.V., Borounova, V.V., Philonenko, E.S., Kantidze, O.L., Vassetzky, Y.S., Razin, S.V., 2009. In embryonic chicken erythrocytes actively transcribed alpha globin genes are not associated with the nuclear matrix. J. Cell. Biochem. 106, 170–178.

Ido, E., Matsuno, T., 1982. Purification and physicochemical and immunological analysis of chicken alpha-fetoprotein. Jpn. J. Med. Sci. Biol. 35, 87–96.

Igbo, I.N., Reigel Jr., C.E., Greene, I.M., Kenny, A.D., 1994. Effect of reserpine pretreatment on avian erythrocyte carbonic anhydrase activation by isoproterenol. Pharmacology 49, 112–120.

Ingermann, R.L., Stock, M.K., Metcalfe, J., Bissonnette, J.M., 1985. Monosaccharide uptake by erythrocytes of the embryonic and adult chicken. Comp. Biochem. Physiol. A Comp. Physiol. 80, 369–372.

Isaacks, R., Harkness, D., Sampsell, R., Adler, J., Roth, S., Kim, C., Goldman, P., 1977. Studies on avian erythrocyte metabolism. Inositol tetrakisphosphate: the major phosphate compound in the erythrocytes of the ostrich (*Struthio camelus camelus*). Eur. J. Biochem. 77, 567–574.

Jessen, T.H., Weber, R.E., Fermi, G., Tame, J., Braunitzer, G., 1991. Adaptation of bird hemoglobins to high altitudes: demonstration of molecular mechanism by protein engineering. Proc. Natl. Acad. Sci. U.S.A. 88, 6519–6522.

Jones, J.H., Grubb, B., Schmidt-Nielsen, K., 1983. Panting in the emu causes arterial hypoxemia. Respir. Physiol. 54, 189–195.

Joseph-Silverstein, J., Cohen, W.D., 1984. The cytoskeletal system of nucleated erythrocytes. III. Marginal band function in mature cells. J. Cell Biol. 98, 2118–2125.

Kalomenopoulou, M., Beis, I., 1990. Studies on the pigeon red blood cell metabolism. Comp. Biochem. Physiol. B 95, 677–684.

Kawashiro, T., Scheid, P., 1975. Arterial blood gases in undisturbed resting birds: measurements in chicken and duck. Resp. Physiol. 23, 337–342.

Kaya, A., Altiner, A., Ozpinar, A., 2006. Effect of copper deficiency on blood lipid profile and haematological parameters in broilers. J. Vet. Med. A 53, 399–404.

Kim, H.R., Yew, N.S., Ansorge, W., Voss, H., Schwager, C., Vennström, B., Zenke, M., Engel, J.D., 1988. Two different mRNAs are transcribed from a single genomic locus encoding the chicken erythrocyte anion transport proteins (band 3). Mol. Cell Biol. 8, 4416–4424.

Klasing, K.C., 1998. Nutritional modulation of resistance to infectious diseases. Poult. Sci. 77, 1119–1125.

Kogut, M., Rothwell, L., Kaiser, P., 2002. Differential effects of age on chicken heterophil functional activation by recombinant chicken interleukin-2. Dev. Comp. Immunol. 26, 817–830.

Kogut, M.H., Rothwell, L., Kaiser, P., 2003. Priming by recombinant chicken interleukin-2 induces selective expression of IL-8 and IL-18 mRNA in chicken heterophils during receptor-mediated phagocytosis of opsonized and nonopsonized *Salmonella enterica* serovar *enteritidis*. Mol. Immunol. 40, 603–610.

Kogut, M., Rothwell, L., Kaiser, P., 2005. IFN-gamma priming of chicken heterophils upregulates the expression of proinflammatory and Th1 cytokine mRNA following receptor-mediated phagocytosis of *Salmonella enterica* serovar *enteritidis*. J. Interferon. Cytokine Res. 25, 73–81.

Koh, T.S., Peng, R.K., Klasing, K.C., 1996. Dietary copper level affects copper metabolism during lipopolysaccharide-induced immunological stress in chicks. Poult. Sci. 75, 867–872.

Koo, B.C., Kwon, M.S., Lee, H., Kim, M., Kim, D., Roh, J.Y., Park, Y.Y., Cui, X.S., Kim, N.H., Byun, S.J., Kim, T., 2010. Tetracycline-dependent expression of the human erythropoietin gene in transgenic chickens. Transgenic Res. 19, 437–447.

Koutzamani, E., Loborg, H., Sarg, B., Lindner, H.H., Rundquist, I., 2002. Linker histone subtype composition and affinity for chromatin in situ in nucleated mature erythrocytes. J. Biol. Chem. 277, 44688–44694.

Kushima, K., Yoshida, K., Fujita, M., Shigeta, A., Horiuchi, H., Matsuda, H., Furusawa, S., 2004. Chicken peripheral blood CD3+CD4-CD8- cells are regulated by endocrine and nerve systems. J. Vet. Med. Sci. 66, 143–148.

Lacoste-Eleaume, A.S., Bleux, C., Quere, P., Coudert, F., Corbel, C., Kanellopoulos-Langevin, C., 1994. Biochemical and functional characterization of an avian homolog of the integrin GPIIb-IIIa present on chicken thrombocytes. Exp. Cell Res. 213, 198–209.

Lam, K.M., 2002. The macrophage inflammatory protein-1beta in the supernatants of *Mycoplasma gallisepticum*-infected chicken leukocytes attracts the migration of chicken heterophils and lymphocytes. Dev. Comp. Immunol. 26, 85–93.

Lambert, L.A., Perri, H., Halbrooks, P.J., Mason, A.B., 2005. Evolution of the transferrin family: conservation of residues associated with iron and anion binding. Comp. Biochem. Physiol. B Biochem. Mol. Biol. 142, 129–141.

Lee, K.A., Tell, L.A., Mohr, F.C., 2012. Inflammatory markers following acute fuel oil exposure or bacterial lipopolysaccharide in mallard ducks (*Anas platyrhynchos*). Avian Dis. 56, 704–710.

Leighton, F.A., 1985. Morphological lesions in red blood cells from herring gulls and Atlantic puffins ingesting Prudhoe Bay crude oil. Vet. Pathol. 22, 393–402.

Lerner, J., Smagula, R.M., Hilchey, S.E., Somes Jr., R.G., 1982. Amino acid transport and intracellular Na+ and K+ content of chicken erythrocytes genetically selected for high and low leucine transport activity. Comp. Biochem. Physiol. A Comp. Physiol. 73, 243–248.

Lerner, J., Smagula, R.M., Somes Jr., R.G., 1984. Sodium-ion dependence of glycine and lysine transport in chicken erythrocytes genetically selected for high and low leucine transport activity. Comp. Biochem. Physiol. A Comp. Physiol. 78, 277–278.

Liang, Y.H., Liu, X.Z., Liu, S.H., Lu, G.Y., 2001. The structure of greylag goose oxy haemoglobin: the roles of four mutations compared with bar-headed goose haemoglobin. Acta Crystallogr. D Biol. Crystallogr. 57, 1850–1856.

Lima, A., Fehervari, T., Paasch, L.H., Calderón, N.L., 2005. Haematological and histological findings in Leghorn chickens infected with infectious bursal disease virus strain 73688. Acta Vet. Hung. 53, 501–506.

Lin, H., Decuypere, E., Buyse, J., 2004. Oxidative stress induced by corticosterone administration in broiler chickens (*Gallus gallus domesticus*) 1. Chronic exposure. Comp. Biochem. Physiol. B Biochem. Mol. Biol. 139, 737–744.

Lin, H., Decuypere, E., Buyse, J., 2006. Acute heat stress induces oxidative stress in broiler chickens. Comp. Biochem. Physiol. A Mol. Integr. Physiol. 144, 11–17.

Liu, L., Lei, T., Bankir, L., Zhao, D., Gai, X., Zhao, X., Yang, B., 2011. Erythrocyte permeability to urea and water: comparative study in rodents, ruminants, carnivores, humans, and birds. J. Comp. Physiol. B. 181, 65–72.

Malisch, J.L., Breuner, C.W., 2010. Steroid-binding proteins and free steroids in birds. Mol. Cell. Endocrinol. 316, 42–52.

Martin, M.P., Wineland, M., Barnes, H.J., 2010. Selected blood chemistry and gas reference ranges for broiler breeders using the i-STAT handheld clinical analyzer. Avian Dis. 54, 1016–1020.

Mathew, A., Grdisa, M., Johnstone, R.M., 1993. Nucleosides and glutamine are primary energy substrates for embryonic and adult chicken red cells. Biochem. Cell. Biol. 71, 288–295.

Mathew, A., Grdisa, M., Robbins, P.J., White, M.K., Johnstone, R.M., 1994. Loss of glucose transporters is an early event in differentiation of HD3 cells. Am. J. Physiol. 266, C1222–C1230.

Matsuda, G., Takae, H., 1963. The studies on the structure of chicken hemoglobin: 1. Chromatographic purification of chicken hemoglobin. J. Biochem. 54, 156–160.

Maxwell, M.H., 1973. Comparison of heterophil and basophil ultrastructure in six species of domestic birds. J. Anat. 115, 187–202.

Maxwell, M.H., 1974. An ultrastructural comparison of the mononuclear leucocytes and thrombocytes in six species of domestic bird. J. Anat. 117, 69–80.

Maxwell, M.H., Siller, W.G., 1972. The ultrastructural characteristics of the eosinophil granules in six species of domestic bird. J. Anat. 112, 289–303.

McCullem, E.V., 1957. A History of Nutrition. Houghton-Mifflin, Boston.

McFarlane, J.M., Curtis, S.E., 1989. Multiple concurrent stressors in chicks. 3. Effects on plasma corticosterone and the heterophil:lymphocyte ratio. Poult. Sci. 68, 522–527.

McKnight, G.S., Lee, D.C., Palmiter, R.D., 1980. Transferrin gene expression: regulation of mRNA transcription in chick liver by steroid hormones and iron deficiency. J. Biol.Chem. 255, 148–153.

Meir, J.U., Ponganis, P.J., 2009. High-affinity hemoglobin and blood oxygen saturation in diving emperor penguins. J. Exp. Biol. 212, 3330–3338.

Mikulits, W., Schranzhofer, M., Deiner, E.M., Beug, H., Müllner, E.W., 2000. Regulation of ferritin mRNA translation in primary erythroblasts: exogenous c-Kit plus EpoR signaling mimics v-ErbA oncoprotein activity. Biochem. Biophys. Res. Commun. 275, 292–294.

Mikšik, I., Hodný, Z., 1992. Glycated hemoglobin in mute swan (*Cygnus olor*) and rook (*Corvus frugilegus*). Comp. Biochem. Physiol. B 103, 553–559.

Montgomery, M.K., Hulbert, A.J., Buttemer, W.A., 2012. Does the oxidative stress theory of aging explain longevity differences in birds? I. Mitochondrial ROS production. Exp. Gerontol. 47, 203–210.

Morera, D., Roher, N., Ribas, L., Balasch, J.C., Doñate, C., Callol, A., Boltaña, S., Roberts, S., Goetz, G., Goetz, F.W., MacKenzie, S.A., 2011. RNA-Seq reveals an integrated immune response in nucleated erythrocytes. PLoS One 6, e26998.

Morgan, E.H., 1975. Plasma iron transport during egg laying and after oestrogen administration in the domestic fowl (*Gallus domesticus*). Q. J. Exp. Physiol. Cogn. Med. Sci. 60, 233–247.

Morrisey, J.K., Paul-Murphy, J., Fialkowski, J.P., Hart, D., Darien, B., 2003. Estimation of prothrombin times of hispaniolan Amazon parrots (*Amazona ventralis*) and umbrella cockatoos (*Cacatua alba*). J. Avian Med. Surg. 17, 72–77.

Motta, R.O., Romero Marques, M.V., Ferreira Junior, F.C., Andery Dde, A., Horta, R.S., Peixoto, R.B., Lacorte, G.A., Moreira Pde, A., Paes Leme Fde, O., Melo, M.M., Martins, N.R., Braga, E.M., 2013. Does haemosporidian infection affect hematological and biochemical profiles of the endangered Black-fronted piping-guan (Aburria jacutinga)? Peer J. 1, e45.

Murakami, T., 1991. Positive cooperativity of [3H]dexamethasone binding to chick corticosteroid-binding globulin. Comp. Biochem. Physiol. A Comp. Physiol. 100, 361–364.

Nambi, P., Peters, J.R., Sibley, D.R., Lefkowitz, R.J., 1985. Desensitization of the turkey erythrocyte beta-adrenergic receptor in a cell-free system. Evidence that multiple protein kinases can phosphorylate and desensitize the receptor. J. Biol. Chem. 260, 2165–2171.

Nicol, S.C., Melrose, W., Stahel, C.D., 1988. Haematology and metabolism of the blood of the little penguin, *Eudyptula minor*. Comp. Biochem. Physiol. A Comp. Physiol. 89A, 383–386.

Nirmallan, G.P., Robinson, G.A., 1973. The survival time of erythrocytes (DF^{32} P-label) in the Japanese quail. Poult. Sci. 52, 355–359.

Nishita, T., Tomita, Y., Imanari, T., Ichihara, N., Orito, K., Arishima, K., 2011a. Biochemical and developmental characterization of carbonic anhydrase II from chicken erythrocytes. Acta Vet. Scand. 53, 16.

Nishita, T., Tomita, Y., Yorifuji, D., Orito, K., Ochiai, H., Arishima, K., 2011b. Purification of chicken carbonic anhydrase isozyme-III (CA-III) and its measurement in White Leghorn chickens. Acta Vet. Scand. 53, 63.

Ohno, Y., Yagi, H., Nakamura, M., Masuko, K., Hashimoto, Y., Masuko, T., 2008. Cell-death-inducing monoclonal antibodies raised against DT40 tumor cells: identification of chicken transferrin receptor as a novel cell-death receptor. Cancer Sci. 99, 894–900.

Olanrewaju, H.A., Purswell, J.L., Collier, S.D., Branton, S.L., 2010. Effect of ambient temperature and light intensity on physiological reactions of heavy broiler chickens. Poult. Sci. 89, 2668–2677.

Oye, I., Sutherland, E.W., 1996. The effect of epinephrine and other agents on adenyl cyclase in the cell membrane of avian erythrocytes. Biochim. Biophys. Acta 127, 347–354.

Ozensoy, O., Isik, S., Arslan, O., Arslan, M., Scozzafava, A., Supuran, C.T., 2005. Carbonic anhydrase inhibitors. Inhibition of red blood cell ostrich (*Struthio camelus*) carbonic anhydrase with a series of aromatic and heterocyclic sulfonamides. J. Enzyme Inhib. Med. Chem. 20, 383–387.

Ozensoy, O., Arslan, M., Supuran, C.T., 2011. Carbonic anhydrase inhibitors: purification and inhibition studies of pigeon (*Columba livia* var. *domestica*) red blood cell carbonic anhydrase with sulfonamides. J. Enzyme Inhib. Med. Chem. 26, 749–753.

Palfrey, H.C., Greengard, P., 1981. Hormone-sensitive ion transport systems in erythrocytes as models for epithelial ion pathways. Ann. N.Y. Acad. Sci. 372, 291–308.

Paris, S., Samuel, D., Jacques, Y., Gache, C., Franchi, A., Ailhaud, G., 1978. The role of serum albumin in the uptake of fatty acids by cultured cardiac cells from chick embryo. Eur. J. Biochem. 83, 235–243.

Peltonen, L.M., Sankari, S., 2011. Ott's protein osmotic pressure of serum and interstitial fluid in chickens (*Gallus gallus*): effect of age and gender. J. Exp. Biol. 214, 599–606.

Penno, C.A., Kawabe, Y., Ito, A., Kamihira, M., 2010. Production of recombinant human erythropoietin/Fc fusion protein by genetically manipulated chickens. Transgenic Res. 19, 187–195.

Peters, J.R., Nambi, P., Sibley, D.R., Lefkowitz, R.J., 1984. Enhanced adenylate cyclase activity of turkey erythrocytes following treatment with beta-adrenergic receptor antagonists. Eur. J. Pharmacol. 107, 43–52.

Reitman, M., Grasso, J.A., Blumenthal, R., Lewit, P., 1993. Primary sequence, evolution, and repetitive elements of the *Gallus gallus* (chicken) beta-globin cluster. Genomics 18, 616–626.

Richardson, S.J., Monk, J.A., Shepherdley, C.A., Ebbesson, L.O., Sin, F., Power, D.M., Frappell, P.B., Köhrle, J., Renfree, M.B., 2005. Developmentally regulated thyroid hormone distributor proteins in marsupials, a reptile, and fish. Am. J. Physiol. 288, R1264–R1272.

Rodman, G.P., Ebaugh Jr., F.G., Fox, M.R.S., 1957. Life span of red blood cells of the chicken, pigeon, and duck as estimated by use of $Na_2Cr^{51}O_4$ with observations of red cell turnover in mammal, birds and reptile blood. J. Hematol. 12, 355–366.

Röhme, D., 1981. Evidence for a relationship between longevity of mammalian species and life spans of normal fibroblasts *in vitro* and erythrocytes in vivo. Proc. Natl. Acad. Sci. U.S.A. 78, 5009–5013.

Roman, Y., Bed'hom, B., Guillot, A., Levrier, J., Chaste-Duvernoy, D., Bomsel-Demontoy, M.C., Jalme, M.S., 2009. Identification of apolipoprotein A-I in the alpha-globulin fraction of avian plasma. Vet. Clin. Pathol. 38, 206–212.

Rosse, W.F., Waldmann, T.A., 1966. Factors controlling erythropoiesis in birds. Blood 27, 654–661.

Ruiz, G., Rosenmann, M., Fernando Novoa, F., Sabat, P., 2002. Hematological parameters and stress index in Rufous-collared sparrows dwelling in urban environments. Condor 104, 162–166.

Santos, C.S., Monteiro, M.S., Soares, A.M., Loureiro, S., 2012. Characterization of cholinesterases in plasma of three Portuguese native bird species: application to biomonitoring. PLoS One 7, e33975.

Sato, K., Matsushita, K., Takahashi, K., Aoki, M., Fuziwara, J., Miyanari, S., Kamada, T., 2012. Dietary supplementation with 5-aminolevulinic acid modulates growth performance and inflammatory responses in broiler chickens. Poult. Sci. 91, 1582–1589.

Scheffer, M.P., Eltsov, M., Frangakis, A.S., 2011. Evidence for short-range helical order in the 30-nm chromatin fibers of erythrocyte nuclei. Proc. Natl. Acad. Sci. U.S.A. 108, 16992–16997.

Schepelmann, K., 1990. Erythropoietic bone marrow in the pigeon: development of its distribution and volume during growth and pneumatization of bones. J. Morph. 203, 21–34.

Schindler, S.L., Gildersleeve, R.P., Thaxton, J.P., McRee, D.I., 1987. Hematological response of hemorrhaged Japanese quail after blood volume replacement with saline. Comp. Biochem. Physiol. A Comp. Physiol. 87, 933–945.

Schmaier, A.A., Stalker, T.J., Runge, J.J., Lee, D., Nagaswami, C., Mericko, P., Chen, M., Cliché, S., Gariépy, C., Brass, L.F., Hammer, D.A., Weisel, J.W., Rosenthal, K., Kahn, M.L., 2011. Occlusive thrombi arise in mammals but not birds in response to arterial injury: evolutionary insight into human cardiovascular disease. Blood 118, 3661–3669.

Schmidt, K.L., Malisch, J.L., Breuner, C.W., Soma, K.K., 2010. Corticosterone and cortisol binding sites in plasma, immune organs and brain of developing zebra finches: intracellular and membrane-associated receptors. Brain Behav. Immun. 24, 908–918.

Schoen,T.J.,Mazuruk,K.,Waldbillig,R.J.,Potts,J.,Beebe,D.C.,Chader,G.J., Rodriguez, I.R., 1995. Cloning and characterization of a chick embryo cDNA and gene for IGF-binding protein-2. J. Mol. Endocrinol. 15, 49–59.

Schramm, M., Feinstein, H., Naim, E., Lang, E., Lasser, M., 1972. Epinephrine binding to the catecholamine receptor and activation of the adenylate cyclase in erythrocyte membranes (hormone receptor- -adrenergic receptor-cyclic AMP-turkey). Proc. Natl. Acad. Sci. U.S.A. 69, 523–527.

Schreiber, G., Richardson, S.J., 1997. The evolution of gene expression, structure and function of transthyretin. Comp. Biochem. Physiol. B Biochem. Mol. Biol. 116, 137–160.

Scott, T., Owens, M.D., 2008. Thrombocytes respond to lipopolysaccharide through Toll-like receptor-4, and MAP kinase and NF-kappaB pathways leading to expression of interleukin-6 and cyclooxygenase-2 with production of prostaglandin E2. Mol. Immunol. 45, 1001–1008.

Seliger, C., Schaerer, B., Kohn, M., Pendl, H., Weigend, S., Kaspers, B., Härtle, S., 2012. A rapid high-precision flow cytometry based technique for total white blood cell counting in chickens. Vet. Immunol. Immunopathol. 145, 86–99.

Sharma, S., Poliks, B., Chiauzzi, C., Ravindra, R., Blanden, A.R., Bane, S., 2010. Characterization of the colchicine binding site on avian tubulin isotype betaVI. Biochemistry 49, 2932–2942.

Shini, S., Kaiser, P., Shini, A., Bryden, W.L., 2008. Differential alterations in ultrastructural morphology of chicken heterophils and lymphocytes induced by corticosterone and lipopolysaccharide. Vet. Immunol. Immunopathol. 122, 83–93.

Silveira, P., Damatta, R.A., Dagosto, M., 2009. Hematological changes of chickens experimentally infected with *Plasmodium (Bennettinia) juxtanucleare*. Vet. Parasitol. 162, 257–262.

Simons, T.J., 1983a. Characterization of sugar transport in the pigeon red blood cell. J. Physiol. 338, 477–499.

Simons, T.J., 1983b. The role of calcium in the regulation of sugar transport in the pigeon red blood cell. J. Physiol. (Lond.) 338, 501–526.

Somes Jr., R.G., Smagula, R.M., Lerner, J., 1981. Selective breeding of chickens for erythrocytes with high and low leucine transport activity. Am. J. Physiol. 241, C233–C242.

St Paul, M., Paolucci, S., Barjesteh, N., Wood, R.D., Schat, K.A., Sharif, S., 2012. Characterization of chicken thrombocyte responses to Toll-like receptor ligands. PLoS One 7, e43381.

Stabler, J.G., McCormick, T.W., Powell, K.C., Kogut, M.H., 1994. Avian heterophils and monocytes: phagocytic and bactericidal activities against *Salmonella enteritidis*. Vet. Microbiol. 38, 293–305.

Stadel, J.M., Nambi, P., Lavin, T.N., Heald, S.L., Caron, M.G., Lefkowitz, R.J., 1982. Catecholamine-induced desensitization of turkey erythrocyte adenylate cyclase. Structural alterations in the beta-adrenergic receptor revealed by photoaffinity labeling. J. Biol. Chem. 257, 9242–9245.

Stiller, R.A., Belamarich, F.A., Shepro, D., 1975. Aggregation and release in thrombocytes of the duck. Am. J. Physiol. 229, 206–210.

Sturkie, P.D., 1986. Avian Physiology. Springer Verlag, New York.

Sundaresan, S.S., Ramesh, P., Sivakumar, K., Ponnuswamy, M.N., 2009. Purification, crystallization and preliminary X-ray analysis of haemoglobin from ostrich (*Struthio camelus*). Acta Crystallogr. Sect. F Struct. Biol. Cryst. Commun. 65, 681–683.

Tako, E., Rutzke, M.A., Glahn, R.P., 2010. Using the domestic chicken (*Gallus gallus*) as an in vivo model for iron bioavailability. Poult. Sci. 89, 514–521.

Tazawa, H., Ono, T., Mochizuki, M., 1976. Oxygen dissociation curve for chorioallantoic capillary blood of chicken embryo. J. Appl. Physiol. 40, 393–398.

Tazawa, H., Andrewartha, S.J., Burggren, W.W., 2011. Development of hematological respiratory variables in late chicken embryos: the relative importance of incubation time and embryo mass. Comp. Biochem. Physiol. A Mol. Integr. Physiol. 159, 225–233.

Teng, Q.Y., Zhou, J.Y., Wu, J.J., Guo, J.Q., Shen, H.G., 2006. Characterization of chicken interleukin 2 receptor alpha chain, a homolog to mammalian CD25. FEBS Lett. 580, 4274–4281.

Terrón, M.P., Cubero, J., Barriga, C., Ortega, E., Rodríguez, A.B., 2003. Phagocytosis of *Candida albicans* and superoxide anion levels in ring dove (*Streptopelia risoria*) heterophils: effect of melatonin. J. Neuroendocrinol. 15, 1111–1115.

Terrón, M.P., Paredes, S.D., Barriga, C., Ortega, E., Rodríguez, A.B., 2004. Comparative study of the heterophil phagocytic function in young and old ring doves (*Streptopelia risoria*) and its relationship with melatonin levels. J. Comp. Physiol. B 174, 421–427.

Torres-Medina, A., Rhodes, M.B., Mussman, H.C., 1971. Chicken serum proteins: comparison of electrophoretic techniques and localization of transferrin. Poult. Sci. 50, 1115–1121.

Valéria, V.P., Fantoni, D.T., Otsuk, D.A., Auler Jr., J.O.C., 2008. Blood-gas and electrolyte values for Amazon parrots (*Amazona aestiva*). Pesq. Vet. Bras. 28, 108–112.

van As, P., Elferink, M.G., Closter, A.M., Vereijken, A., Bovenhuis, H., Crooijmans, R.P., Decuypere, E., Groenen, M.A., 2010. The use of blood gas parameters to predict ascites susceptibility in juvenile broilers. Poult. Sci. 89, 1684–1691.

van Bockxmeer, F.M., Morgan, E.H., 1982. Comparative aspects of transferrin-reticulocyte interactions: membrane receptors and iron uptake. Comp. Biochem. Physiol. A Comp. Physiol. 71, 211–218.

Vandecasserie, C., Schnek, A.G., Léonis, J., 1971. Oxygen-affinity studies of avian hemoglobins. Chicken and pigeon. Eur. J. Biochem. 24, 284–287.

Viertlboeck, B.C., Schmitt, R., Hanczaruk, M.A., Crooijmans, R.P., Groenen, M.A., Göbel, T.W., 2009. A novel activating chicken IgY FcR is related to leukocyte receptor complex (LRC) genes but is located on a chromosomal region distinct from the LRC and FcR gene clusters. J. Immunol. 182, 1533–1540.

Vleck, C.M., Vertalino, N., Vleck, D., Bucher, T.L., 2000. Stress, corticosterone, and heterophil to lymphocyte ratios in free-living Adélie penguins. Condor 102, 392–400.

Wang, W., Wideman Jr., R.F., Bersi, T.K., Erf, G.F., 2003. Pulmonary and hematological inflammatory responses to intravenous cellulose microparticles in broilers. Poult. Sci. 82, 771–780.

Weitnauer, E., Ebert, C., Hucho, F., Robitzki, A., Weise, C., Layer, P.G., 1999. Butyrylcholinesterase is complexed with transferrin in chicken serum. J. Protein Chem. 18, 205–214.

Whitman, B.A., Breuner, C.W., Dufty Jr., A.M., 2011. The effects of neonatal handling on adrenocortical responsiveness, morphological development and corticosterone binding globulinin nestling American kestrels (*Falco sparverius*). Gen. Comp. Endocrinol. 172, 260–267.

Williams, A.F., 1972. DNA synthesis in purified populations of avian erythroid cells. J. Cell Sci. 10, 27–46.

Wilson, G.R., Wilson, L.P., 1978. Haematology, weight and condition of captive red grouse (Lagopus lagopus scoticus) infected with caecal threadworm (Trichostrongylus tenuis). Res. Vet. Sci. 25, 331–336.

Winckler, B., Solomon, F., 1991. A role for microtubule bundles in the morphogenesis of chicken erythrocytes. Proc. Natl. Acad. Sci. U.S.A. 88, 6033–6037.

Wingfield, J.C., Matt, K.S., Farner, D.S., 1984. Physiologic properties of steroid hormone-binding proteins in avian blood. Gen. Comp. Endocrinol. 53, 281–292.

Wiwanitkit, V., Paritpokee, N., Nithiuthai, S., Boonchalermvichian, C., Bhokaisawan, N., 2007. Change of serum transferrin receptor due to malarial infection, an experiment in *Plasmodium gallinaceum* infected chicken model. J. Vector Borne Dis. 44, 255–258.

Wu, Y., Zhao, X., Li, P., Huang, H., 2007. Impact of Zn, Cu, and Fe on the activity of carbonic anhydrase of erythrocytes in ducks. Biol. Trace Elem. Res. 118, 227–232.

Zanotti, G., Calderone, V., Beda, M., Malpeli, G., Folli, C., Berni, R., 2001. Structure of chicken plasma retinol-binding protein. Biochim. Biophys. Acta 1550, 64–69.

Zenke, M., Kahn, P., Disela, C., Vennström, B., Leutz, A., Keegan, K., Hayman, M.J., Choi, H.R., Yew, N., Engel, J.D., Beug, H., 1988. v-erbA specifically suppresses transcription of the avian erythrocyte anion transporter (band 3) gene. Cell 52, 107–119.

Zhang, Z.W., Cheng, J., Xu, F., Chen, Y.E., Du, J.B., Yuan, M., Zhu, F., Xu, X.C., Yuan, S., 2011. Red blood cell extrudes nucleus and mitochondria against oxidative stress. IUBMB Life 63, 560–565.

Zhou, X.D., Dong, X.F., Tong, J.M., Xu, P., Wang, Z.M., 2012. High levels of vitamin E affect retinol binding protein but not CYP26A1 in liver and hepatocytes from laying hens. Poult. Sci. 91, 1135–1141.

Chapter 11

The Cardiovascular System

Edward M. Dzialowski and Dane A. Crossley II
Developmental Integrative Biology Research Cluster, Department of Biological Sciences, University of North Texas, Denton, TX, USA

11.1 INTRODUCTION

Birds have evolved a high-performance cardiovascular system to meet the rigorous demands of running, flying, swimming, or diving in a variety of environments, some of them extreme. Sustained high levels of activity in these environments place severe demands on the cardiovascular system to provide adequate delivery of oxygen to working vascular beds and to provide efficient removal of metabolic products. Furthermore, birds are endothermic organisms and the cardiovascular system plays a major role in conserving or removing body heat. The descriptions of the component parts of the circulatory system in this chapter illustrate that these transport requirements are met in a variety of ways in birds inhabiting particular environmental niches. This chapter describes the morphological and functional aspects of the avian heart (Section 11.2), circulatory hemodynamics (Section 11.3), and the vascular tree (Section 11.4). A common thread running through this discussion is that the component parts of the circulation must function in an integrated fashion to ensure tissue oxygen delivery matches tissue demands. This is accomplished through the integrative control of circulation by autoregulatory, humoral, and neural mechanisms (Section 11.5). Since the last edition of this book, a significant number of studies have examined the development of cardiovascular control in avian embryos (Section 11.5) which has expanded our understanding of this system throughout ontogeny. Finally, this chapter examines how the cardiovascular system functions within a complex animal interactsing in complex environments (Section 11.6).

Modern birds are derived from theropod dinosaurs (Padian and de Ricqles, 2009), whereas mammals have descended from a group of carnivorous reptiles, the cynodonts. These ancestral lines originated in the Triassic period more than 200 million years ago, so in evolutionary terms avian and mammalian stocks have been separated for a substantial period of time. As one might expect, significant differences in cardiovascular structure and function have arisen in the two groups since their separation, yet a number of similarities in their circulatory systems are also evident. Such similarities probably represent both the conservation of characteristics common to organisms ancestral to the two groups, a shared endothermic phenotype, and the results of convergent evolution once the stocks had divided.

However, our knowledge of cardiovascular structure and function is far more limited in birds than in mammals. In comparing the characteristics of the avian cardiovascular system with those of the mammalian circulation throughout this chapter, we have attempted to clarify the nature of the divergent and convergent features of this system in the two groups.

This review of the avian cardiovascular system encompasses the chapter from the previous edition of this volume (Whittow, 2000), which drew from a number of excellent previous reviews, including major works by Akester (1971), Jones and Johansen (1972), Bennett (1974), Baumel (1975), Akester (1979), Cabot and Cohen (1980), West et al. (1981), and Benzo (1986). This chapter has updated and extended these works and summarized recent contributions in additional areas, such as developments not previously covered.

11.2 HEART

11.2.1 Gross Structure and Function

11.2.1.1 Functional Anatomy

The avian heart, like that of the mammal, is a four-chambered, muscular fluid pump that intermittently pressurizes the central arteries, inducing blood flow to the capillary beds of both the systemic and pulmonary circulations. Functionally, these circuits lie in series with each other, and blood returns to the heart to be pressurized before entering either circuit. As in mammals, the right ventricle pressurizes the pulmonary circulation and the left ventricle pressurizes the systemic circulation. In each case, the pressure differential between the central mean arterial pressure and the central venous pressure drives blood flow (the cardiac output (CO)) through the resistance to flow offered by the microvessels of the circulation. The left and right atria receive blood at central venous pressure before it enters the ventricles. In common with the atria of mammals, these chambers probably function more as blood reservoirs for their respective ventricles than as important "superchargers" for ventricular pressure. The resistance to blood flow, the peripheral resistance, is lower in the pulmonary than in the systemic circuit, so the right ventricle is required to generate less pressure than the left ventricle to produce the same volume flow rate. This difference in ventricular pressure is reflected in the gross anatomy of the ventricles, with the myocardium of the right ventricle being thinner than that of the more powerful left ventricle.

In birds, the heart is located in the cranial part of the common thoracoabdominal cavity, with its long axis slightly to the right of the midline. It is partly enclosed dorsally and laterally by the lobes of the liver. A very thin, but tough, fibrous pericardial sac encloses the heart. This sac contains a small volume of serous fluid that provides lubrication for the rhythmic motion of the cardiac contraction cycle. The pericardium is loosely attached to the dorsal surface of the sternum and the surrounding air sacs and more firmly to the liver. It is also attached, via the peritoneum of the hepatic peritoneal cavities, to the vertebral column. These attachments secure the apex of the heart within the median incisura of the liver and in the caudoventral axis of the thoracoabdominal cavity. The outer fibrous layer of the pericardial sac is continuous with the outer adventitial layer of the large central blood vessels. The pericardial membrane is relatively noncompliant and therefore strongly resists large, rapid increases in cardiac size that might be caused by volume overload of a heart chamber. The noncompliant nature of the pericardial sac may result in some degree of mechanical coupling between the ventricles via the contained incompressible lubricating fluid. For example, an increase in diastolic pressure in one ventricle may be transmitted to the other, increasing pressure and decreasing compliance.

11.2.1.2 Heart Size

In birds, heart mass scales with respect to body mass as $M_h = 0.014 M_b^{0.91}$ (Bishop and Butler, 1995). In mammals, the relationship is $M_h = 0.0058 M_b^{0.98}$ (Prothero, 1979), where M_h is heart mass and M_b is body mass. When compared with mammals, birds of a given body mass have a significantly heavier heart. This may be due to the high aerobic power input needed to sustain flapping flight. Furthermore, unlike mammals, in which heart mass is almost directly proportional to body mass, in birds the exponent denoting proportionality is significantly less than 1. This means that larger birds like swans, ducks, and geese tend to have proportionally smaller hearts in relation to their body mass than do smaller birds. Thus, heart mass represents about 1.1% of body mass for a bird such as the racing pigeon (421 g), compared with 0.8% for the 2.95 kg Pekin duck; this relationship is shown in Figure 11.1 (Grubb, 1983; Bishop and Butler, 1995). In a number of migrating species, the heart becomes hypertrophic before migration (see Section 11.6.1.2). Therefore, migrating birds may have the genetic potential to increase heart size, and therefore CO, through either seasonal humoral mechanisms or in the long term through natural selection. Hummingbirds have proportionally larger hearts than all other birds (shown separately in Figure 11.1), probably reflecting the high aerobic demands of hovering flight. For 25 species of hummingbirds, $M_h = 0.025 M_b^{0.95}$ (Hartman, 1961). In contrast, the ornate tinamou (*Nothoprocta ornata*) and Chilean tinamou (*Nothoprocta perdicaria*) have heart masses that represents 0.24% and 0.28% of body mass, respectively (Altimiras et al., 2013).

Heart mass as a fraction of body mass may also be age dependent. Prior to hatching, the ventricle of Pekin ducks (*Anas platyrhynchos*) represents about 0.4% of body mass and increases significantly upon hatching to 0.75% of body mass (Ream et al., 2013). This increase in heart mass coincides with an increase in aerobic capacity of the hatchling. Similar increases have been observed in the chicken and northern bobwhite quail (*Colinus virginianus*) after hatching (Dzialowski, unpublished data).

11.2.1.3 Cardiac Chambers

The avian heart has two completely divided atria and ventricles. These chambers are functionally equivalent to those of the mammalian heart, serving to distribute CO both to the systemic circulation and to the lungs. In life, the atria are rounded chambers, distended with blood during atrial diastole. In excised hearts, they may collapse, causing auricles to appear. The right atrium tends to be much larger than the left. The wall of the avian atria and ventricles, as in mammals, consists of endocardial, myocardial, and epicardial layers. The atrial walls are generally thin, although atrial muscle is arranged in thick bundles forming muscular arches. The right and left transverse arches are arranged at right angles to the dorsal longitudinal arch and the interatrial septum. The transverse arches branch into smaller bundles, which fuse with a circular muscle band (muscularis basianularis atrii) at the ventral limits of the atria. Contraction of atrial muscle nearly empties the atria. In many species, the atria lack functional inflow valves, so that the importance of atrial contraction for ventricular filling may be slight.

The muscular architecture of the ventricles is more complex than that of the atria and includes a superficial layer, longitudinal muscle of the right ventricle, and sinuspiral and bulbospiral muscles. The left ventricle is cone-shaped and extends to the apex of the heart. Its right wall forms the interventricular septum. The free wall of the right ventricle is continuous with the outer portion of the wall of the

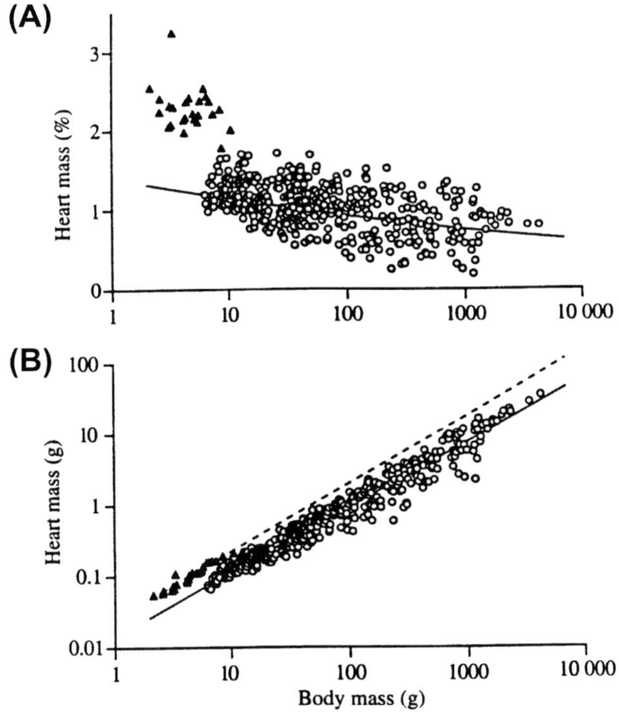

FIGURE 11.1 (A) Heart mass as a percentage of body mass. (B) Heart mass in grams, plotted against body mass (g) for 488 avian species, including 25 species of hummingbird. Hummingbird data are represented by the filled triangles and dashed line; all other species are represented by open circles and solid line. *From Bishop and Butler (1995).*

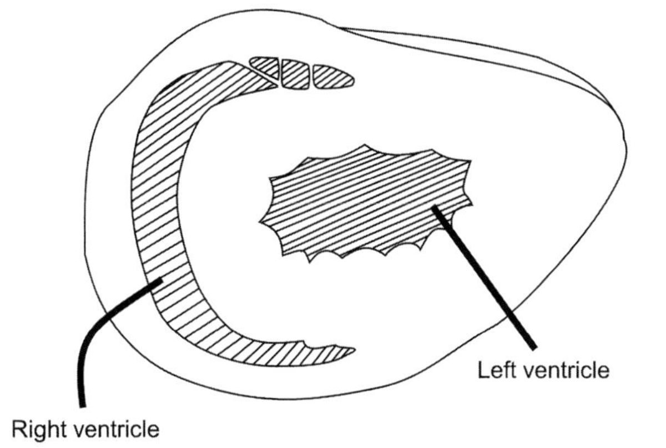

FIGURE 11.2 Transverse section through the ventricles of the avian heart. The lumen of each ventricle is shaded. *Reprinted from West et al. (1981).*

left ventricle and wraps around the right side of the heart to enclose a crescent-shaped cavity, which does not reach the apex of the heart. The muscular walls of the two ventricles are differentially developed, with the wall of the left ventricle being two to three times thicker than that of the right. In addition, the radius of curvature of the wall of the left ventricle is smaller than that of the right (Figure 11.2). This implies both a greater mechanical advantage for pressure generation in the left than in the right ventricle and, according to LaPlace's law, a smaller wall tension for a given left ventricular pressure increment. Therefore, contraction of the myocardial layers of the thick, smallradius wall of the left ventricle enables it to generate systolic pressures four to five times higher than those produced by the right ventricle, without rupturing. The larger radius of curvature and thinner free ventricular wall of the right ventricle reflects the lower systolic pressures generated by this chamber, made possible by the low vascular resistance of the avian lungs. Another consequence of this geometry is that relatively large changes in stroke volume can be made by small changes in the degree of shortening of right ventricular muscle fibers.

11.2.1.4 Valves

Blood entering the left ventricle from the left atrium on atrial systole passes through an orifice guarded by a membranous atrioventricular (AV) valve, similar in general structure to a mammalian AV valve. The valve forms a continuous membrane around the aperture. The valve is tricuspid, not bicuspid as it is in mammals, but in the avian heart the cusps of this valve are poorly defined. The anterior and posterior leaflets are small. The large aortic (medial) leaflet is connected to the bases of the left and noncoronary cusps of the adjacent aortic outflow valve by fibrous tissue. The free margin of the valve is well secured to the left ventricular endocardium by numerous inextensible chordae tendineae. This arrangement prevents valve eversion during ventricular systole.

Blood passing from the right atrium to the right ventricle enters through an orifice guarded by an AV valve that is structurally unique to birds. In pronounced contrast to the fibrous structure characteristic of the mammalian tricuspid valve, in birds the right AV valve consists of a single spiral flap of myocardium attached obliquely to the free wall of the right ventricle (Figure 11.3; Lu et al., 1993a). This spiral flap is apposed to a downward extension of the free wall of the right atrium. The atrial component of the valve extends toward the apex of the ventricle for a shorter distance than does the right ventricular flap. Most of the valve is made of ventricular myocardium and is bilaminar only in its upper portion. The mechanism of valve closure at the start of ventricular systole is unknown. It could be active, by contraction of the muscular flaps, or passive, by deflection of the ventricular flap by a brief backflow of blood at the start of ventricular systole.

The idea of valve closure depending at least partially on active muscular contraction is supported by evidence that at the cellular level both AV valves are closely approached by the electrical conducting system of the myocardium. A complete ring of Purkinje fibers encircles the right AV orifice and connects to the muscular AV valve (Lu et al., 1993a,b). An exception to this anatomical arrangement may be the penguin, in which Adams (1937) did not find a Purkinje

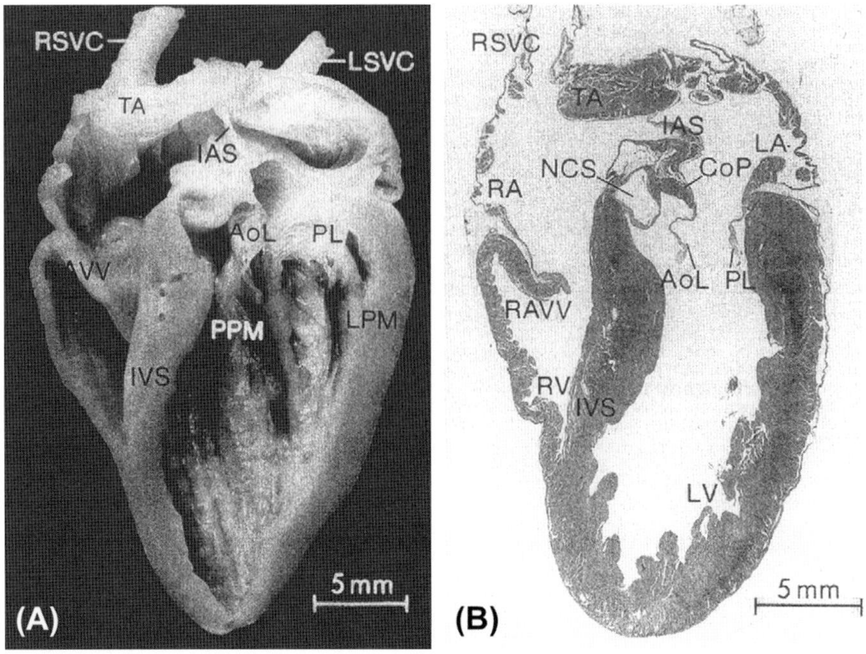

FIGURE 11.3 **Anterior frontal views through the atria and ventricles of a chicken heart, showing both the right and left atrioventricular (AV) valves.** (A) Anterior view of heart dissected in the frontal plane. (B) Frontal histological section of 8 μm thickness, Goldner trichrome stain. AOL, aortic leaflet of left AV valve; COP, connecting part of muscle arch; IAS, interatrial septum; IVS, interventricular septum; LA, left atrium; LPM, left papillary muscle; LSVC, left superior vena cava; LV, left ventricle; NCS, noncoronary sinus of the aorta; PL, posterior leaflet of left AV valve; PPM, posterior papillary muscle; RA, right atrium; RAVV, right AV valve; RSVC, right superior vena cava; RV, right ventricle. *From Lu et al. (1993a).*

ring. The majority of studies, as summarized by Lu et al. (1993a,b), support the idea that both the atrial and ventricular muscular components of the right AV valve are excited to contract via the Purkinje system. However, Szabo et al. (1986) thought that there was an insulating layer of connective tissue between the Purkinje AV ring system and the left ventricular myocardium, suggesting that the muscle flap derived from the left ventricle may work passively. Definitive physiological experiments have not been performed to resolve this issue, but the balance of current evidence suggests that both portions of the valve are electrically activated before the ventricular myocardium to dynamically contract and close the AV orifice at the start of ventricular systole (Lu et al., 1993b). This is clearly very different from the closure mechanism in mammals, in which the leaflets of the tricuspid valve float up into the right AV orifice, moved by the AV pressure differential generated during ventricular systole itself.

The outflow valves from the right and left ventricles are, at first glance, more conventional (mammalian) in nature. The pulmonary outflow valve consists of three semilunar cusps. It prevents regurgitation from the pulmonary artery into the right ventricle, with the valvules closing as pressure in the ventricle falls below that in the pulmonary trunk on ventricular diastole. There are also three semilunar cusps in the aortic outflow valve, but they are much more rigid than those of the pulmonary outflow valve and are firmly attached to underlying myocardium. The cusps are linked by a ring of fibrous tissue that lies within a complete ring of underlying, circumferentially arranged, myocardial cells. The ring is completed by an arch of cardiac muscle that lies between the left coronary cusp of the aortic outflow valve and the aortic leaflet of the left AV valve, as shown in Figure 11.4 (Lu et al., 1993a). This anatomical arrangement contrasts with that in the mammalian heart, in which there is only connective tissue, not myocardium, between that part of the muscular ring lying between the aortic wall and the adjacent mitral valve; in mammals, the myocardial ring is incomplete. In the bird, however, this sphincter-like myocardial cylinder is potentially capable, on contraction, of constricting the left ventricular outflow tract. Lu et al. (1993a) proposed that the muscular ring could act as a sphincter controlling the rate of left ventricular outflow by modulating outflow resistance. Another attractive possibility is that muscular contraction of the myocardial ring could close the relatively rigid cusps of the aortic outflow valve. The middle bundle branch of the Purkinje system is connected to the arch of muscle, so its contraction may start relatively early in the cardiac cycle. Obviously, physiological studies are urgently needed to determine whether either of these intriguing mechanisms operates in the avian heart.

11.2.1.5 Coronary Circulation

Oxygenated blood destined to supply the avian myocardium via the right and left coronary arteries enters the right ventral

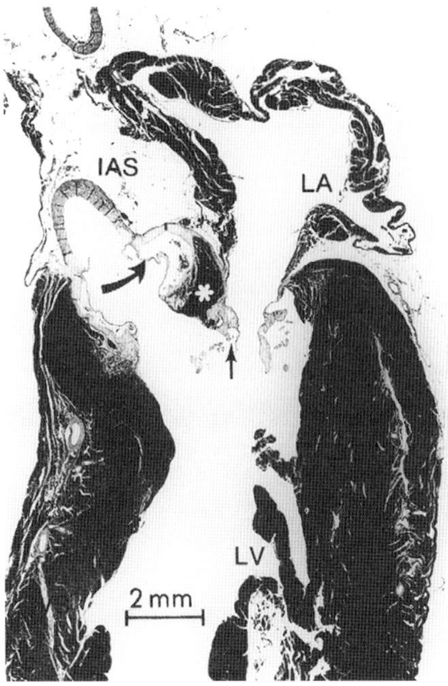

FIGURE 11.4 Photomicrograph of a histological section of the chicken heart, taken in a modified sagittal plane through the anterior column of the muscle arch, indicated by an asterisk. The muscle arch lies between the left coronary aortic valve cusp (curved arrow) and the aortic leaflet of the left AV valve (straight arrow). IAS, interatrial septum; LA, left atrium; LV, left ventricle. *From Lu et al. (1993a).*

and left aortic sinuses, which lie immediately downstream from the cusps of the aortic outflow valve. Most birds have two entrances to the coronary circulation, although there is individual variation so that up to four openings have been observed. In chickens, the right ventral sinus leads into the right coronary artery, which then divides immediately into a superficial and a deep branch (Figure 11.5). The superficial branch follows the groove (coronary sulcus) between the right ventricle and atrium and supplies the cardiac muscle of both chambers. The larger deep branch supplies the ventral wall of the right ventricle, the dorsal walls of both atria, and the muscular right AV valve. In most species, the right coronary artery is dominant and also supplies the ventricular septum, the heart apex, and much of the left ventricular myocardium. The left coronary artery arises from the left aortic sinus and also has a superficial branch that follows the left coronary sulcus. Another superficial branch gives off atrial and ventricular tributaries, and a deep branch supplies the ventral myocardium of the left ventricle. It is not uncommon in chickens for the left coronary artery to be dominant, in which case it supplies almost all of the left ventricular myocardium and the heart apex. There are frequent anastomoses between the branches of the coronary arteries, particularly near the coronary sulcus.

Five groups of cardiac veins, with frequently anastomosing small tributaries, return venous blood from the myocardium into the right atrium via a coronary sinus. Small cardiac veins open directly into the atria and the right ventricle. This basic anatomical pattern of coronary circulation is seen in birds ranging from the chicken, duck, and pigeon (West et al., 1981) to the ostrich (Bezuidenhout, 1984).

The rate of perfusion of avian myocardium is high compared with perfusion rates of most other avian tissues, as shown in Figure 11.6 (see Section 11.4.2.3; Johansen, 1964; Jones et al., 1979; Ellerby et al., 2005). The constantly active cardiac muscle is perfused at a higher rate than resting skeletal muscle. During exercise in the guinea fowl, flow to the ventricles more than doubled (Ellerby et al., 2005). Presumably, as in mammals, the majority of avian coronary blood flow occurs in diastole, so coronary flow might be expected to increase if the diastolic interval is prolonged, provided arterial driving pressure does not fall.

Turkeys sometimes show a congestive cardiomyopathy that is presumed to be of viral origin, the so-called round heart disease. In this condition, systemic hypotension and low CO, caused by reduced left ventricular myocardial shortening, are probably the result of reduced subendocardial coronary perfusion rate (Einzig et al., 1980). Turkeys with round heart disease also show an altered electrocardiography pattern (see Section 11.2.3.4).

11.2.2 Cardiac Variables

The avian cardiovascular system is not merely a replica of the arrangement in mammals, despite similarities in performance between the two systems. Birds have larger hearts, bigger stroke volumes, lower heart rates, and higher COs than mammals of corresponding body mass (Grubb, 1983). In addition, in many avian species mean arterial pressure is higher than that found in mammals of comparable body mass (see Smith, 1994). CO, the product of stroke volume and heart rate, is of particular interest because it is a major determinant of the rate of oxygen delivery to tissues.

In resting birds, left ventricular stroke volume (V_s) was found to be almost directly proportional to body mass (M_b) (Grubb, 1983; Seymour and Blaylock, 2000). For nine species of birds ranging in body mass from 0.035 kg (budgerigar) to 37.5 kg (emu), Grubb found that $V_s = 1.72M_b^{0.97}$, where V_s is in milliliters and M_b is in kilograms. Seymour and Blaylock (2000) found a similar relationship in birds and $V_s = 0.99M_b^{1.03}$ in mammals. Heart rate (f_H, beats/min) at rest was found to be slower in larger birds: $f_H = 178.5M_b^{-0.282}$. CO (mL/kg/min) at rest, the product of stroke volume and heart rate, therefore scaled with the mass of the bird as $CO = 307.0M_b^{0.69}$. The corresponding relationship for mammalian cardiac output is $CO = 166M_b^{0.79}$ (Holt et al., 1968). These results show that birds have a proportionally larger CO compared with a mammal of the same body mass. In larger birds, resting heart rate is slower than in smaller birds. Bishop and Butler (1995) found that for 49 species the allometric

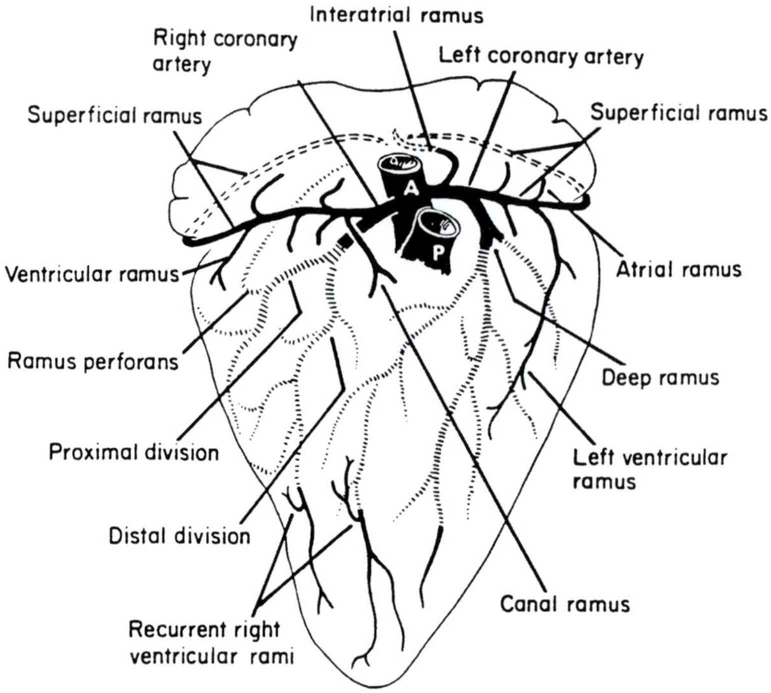

FIGURE 11.5 **Arrangement of the coronary arteries of the chicken, *Gallus*, drawn from the cranioventral aspect.** Solid black and dashed lines represent superficial portions of arteries. Cross-hatched lines represent deep arteries embedded in the myocardium of the ventral and right side of the interventricular septum. A, aorta; P, pulmonary trunk. *From West et al. (1981) with permission.*

relationship for heart rate at rest was $f_H = 125M_b^{-0.37}$, whereas for birds in flight it was $f_H = 480M_b^{-0.19}$. It is interesting that the heart rate–body mass relationship during flight has a shallower slope than in resting animals, indicating that larger species show a greater increase in heart rate in absolute terms in the transition from rest to flight (Figure 11.7). Bishop and Butler (1995) suggested that the body mass exponent of stroke volume in flight should be similar to that at rest ($M_b^{0.96}$), even though the absolute value of stroke volume may increase during flight. Therefore, in flight, it is predicted that CO will scale to body mass as the sum of the exponents for stroke volume and heart rate ($M_b^{0.77}$).

In flight, as at rest, larger species show lower coronary perfusion rates per mass of body tissue. It is likely that this reflects an optimization of the arterial oxygen supply at the tissue level, with the body mass exponent for CO being very similar to the exponent for mass-specific oxygen consumption. Thus, natural selection probably acts on CO to maintain the arteriovenous O_2 difference at a similar level across different avian species. Stroke volume is constrained by cardiac geometry such that, on theoretical grounds, V_s should be closely proportional to M_h and M_b (Schmidt-Nielsen, 1984; Astrand and Rodahl, 1986). Therefore, the lower mass-specific COs of larger avian species, matching their lower mass-specific $\dot{V}_{O_2}$ levels, are reflected in their lower heart rates.

11.2.3 Fine Structure and Cardiac Electrophysiology

11.2.3.1 Fine Structure

Histologically, the atria and ventricles are quite similar, consisting of an external layer, the epicardium, which is separated from an inner endocardium by the mass of heart muscle, and the myocardium. Ventricles are much thicker than the atria due to extensive proliferation of the myocardial layer. The epicardial and endocardial layers are morphologically similar, consisting of loose connective tissue and elastic fibers bordered by a single layer of squamous epithelial or endothelial cells, respectively. The atrial and ventricular septa have endocardial layers facing the lumens of their respective cavities, with myocardial cells between them. In the sparrow and stork, the atrial septum is very thin, in some regions consisting only of two apposed layers of endocardium.

The atrial and ventricular myocardia consist of striated muscle fibers, differing from those of mammals in three notable respects (Sommer and Johnson, 1969, 1970; Hirakow, 1970). First, striated muscle bands are prominent in mammalian cardiac muscle and are also present in bird hearts, with the exception of the M-band. In mammalian cardiac muscle, the M-band is a line of protein molecules connecting adjacent myosin filaments. The significance of the lack of an M-band on the contractile properties of avian

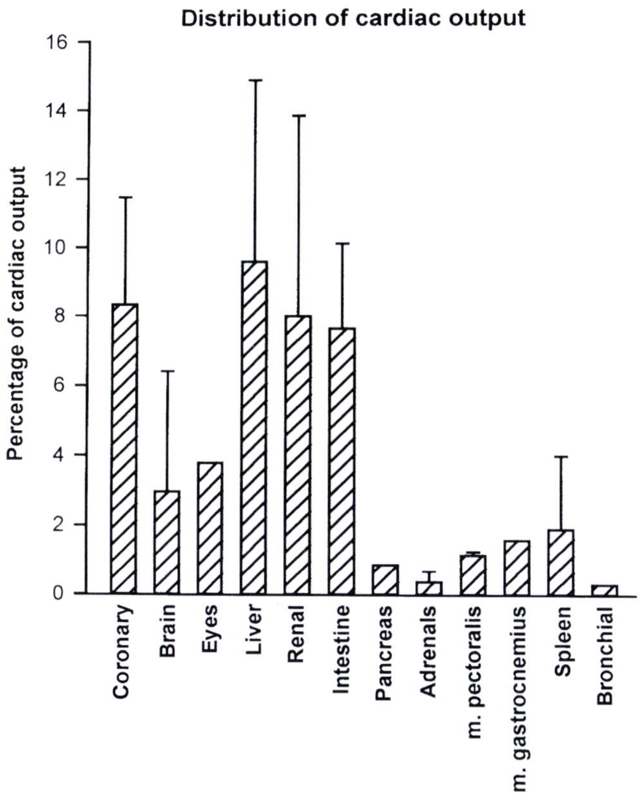

FIGURE 11.6 Organ blood flow plotted as a percentage of cardiac output in birds at rest. Standard error bars are shown where appropriate. *Data from Boelkins et al. (1973) (chicken), Duchamp and Barre (1993) (muscovy duckling), Jones et al. (1979) (Pekin and mallard), Sapirstein and Hartman (1959) (chicken), Stephenson et al. (1994) (Pekin duck) and Wolfenson et al. (1978) (chicken).*

myocytes is unknown. Second, avian cardiac muscle fibers are much smaller in diameter than mammalian fibers and hence there are many more of them in similarly sized hearts. Avian myocardial cells are typically 2–7 μm in diameter compared with the 10–15 μm diameter of mammalian cells. Third, myocardial cells of all fish and reptiles, including avian species, lack transverse tubules (T-tubules), which are prominent in mammalian cardiac muscle. The membrane surrounding the muscle fibers (sarcolemma) consists of two parts—a cell membrane (plasmalemma) and an external layer interconnected with an interstitial network of collagenous fibers. In mammalian cardiac muscle, T-tubules form as invaginations of the plasmalemma, perpendicular to the long axis of the myofilaments. T-tubules lie next to, and form junctions with, sections of the sarcoplasmic reticulum (SR) (diads or triads, Figure 11.8). In mammals, the T-tubule system increases the surface area of the myocardial cells to the extent that the surface-to-volume ratio of a mouse cardiac cell (15 μm diameter) is the same as that of a finch (8 μm diameter). The finch and the mouse have similar cardiac frequencies (Bossen et al., 1978).

The connection between the SR and the plasmalemma occurs through "couplings". In birds, lacking a T-tubule

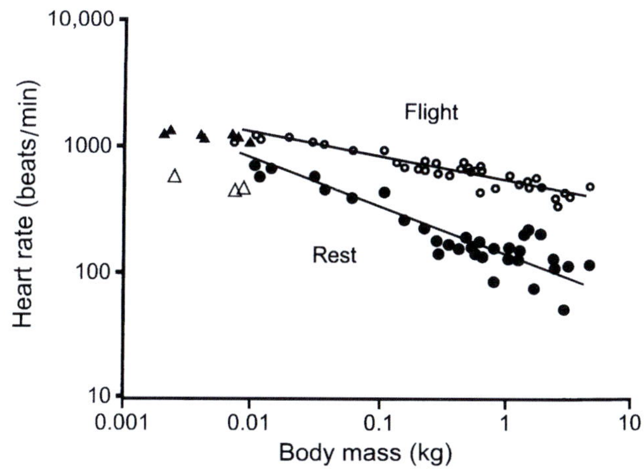

FIGURE 11.7 Heart rate (beats/min) plotted against body mass (kg) for 49 species of birds, including seven species of hummingbirds, at rest and in flight. Hummingbirds, open triangles (rest) and filled triangles (flight). Other species, filled circles (rest) and open circles (flight). *From Bishop and Butler (1995).*

system, these couplings occur at the surface of the cell (Figure 11.8). Couplings are affected by junctional processes that extend from the cytoplasmic face of the SR (junctional sarcoplasmic reticulum (JSR)) that are very closely apposed to the inner surface of the plasmalemma. Birds also possess an extended junctional sarcoplasmic reticulum (EJSR), which occurs in the region of the Z-bands, but it is anomalous. Although it resembles the JSR in most respects, the actual junctional processes may be separated from the plasmalemma by a cleft of several microns. EJSR is much less developed in chicken than in passerines (Sommer et al., 1991). Interestingly, the volume of JSR in mouse hearts and the total volume of JSR (20%) and EJSR (80%) in the finch are virtually identical (Bossen et al., 1978).

11.2.3.2 Excitation–Contraction Coupling

Excitation–contraction coupling describes how an electrical signal, the action potential (AP), traveling along the plasmalemma evokes calcium release from the SR in the region of the myofibrils, causing a change in actin–myosin interactions, which leads to muscle contraction. In cardiomyocytes, the transduction between the electrical signal and Ca^{2+} release from the JSR is effected by a transmitter, which is, in fact, calcium itself. In the first step of this process, the AP causes voltage-dependent Ca^{2+} channels to open in the sarcolemmal membrane through a conformational change in the channels. Ca^{2+} then enters the cell and diffuses to receptors on the junctional processes, where it acts as a transmitter, opening Ca^{2+}-dependent Ca^{2+} channels, which in turn release Ca^{2+} sequestered in the SR. This Ca^{2+}-induced Ca^{2+} release (CICR) is crucial for the physiological function of the bird heart, in which the EJSR, which is the majority of the junctional SR, is separated from the plasmalemma by several microns. Furthermore, CICR allows exquisitely fine

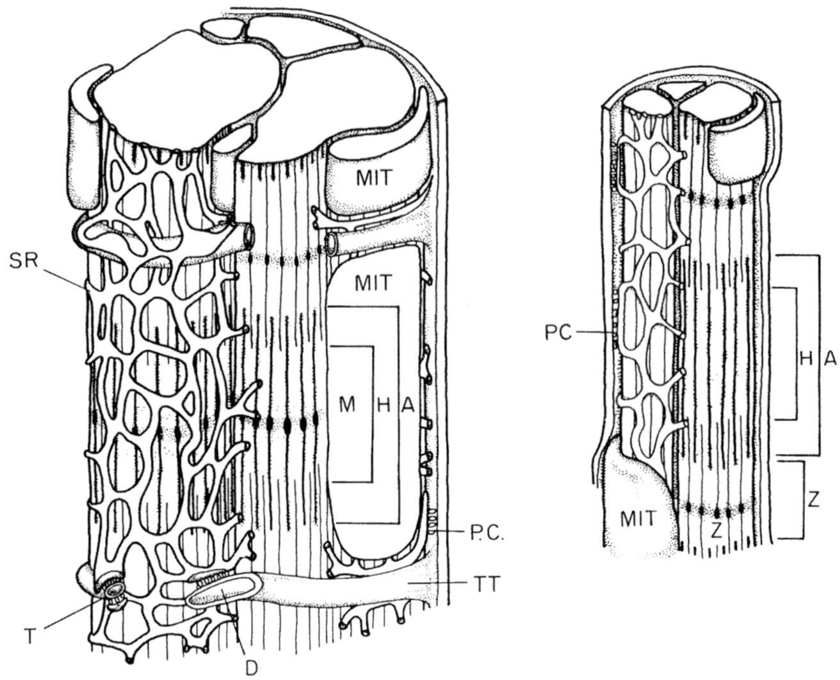

FIGURE 11.8 Comparison of mammalian (left) and avian (right) myocardial cells. The major distinguishing features of avian fibers are smaller cell diameters and absence of M bands and transverse (T)-tubules. SR, sarcoplasmic reticulum; T, triad junction; D, diad junction; PC, peripheral coupling site; TT, transverse tubule; MIT, mitochondria; M, H, A, I, Z, bands of striated muscle. *From Sommer and Johnson (1969).*

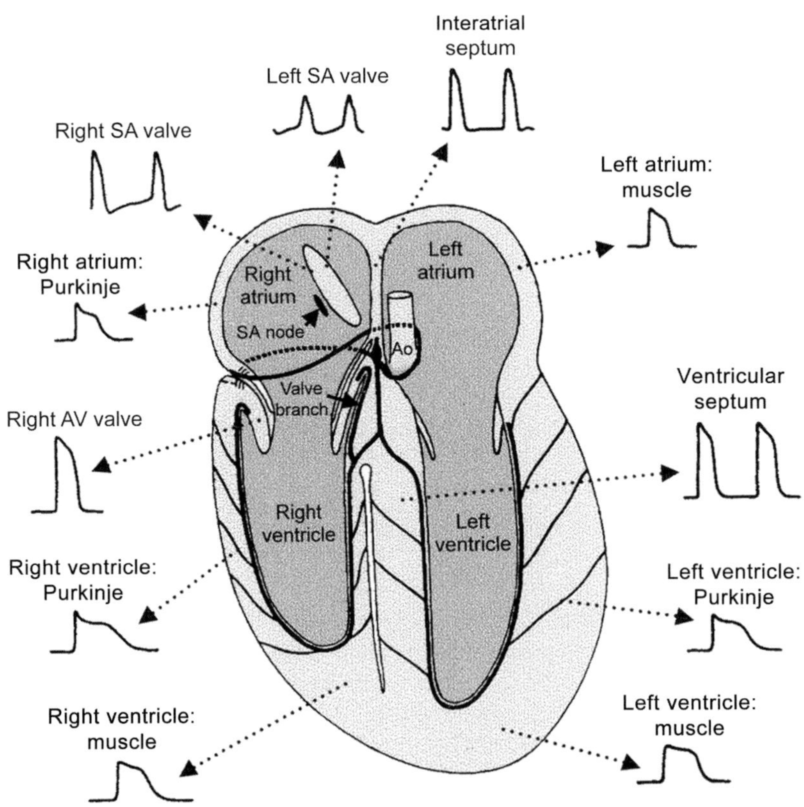

FIGURE 11.9 The Purkinje system of the bird heart (Davies, 1930) and transmembrane action potentials recorded from cells at the indicated sites in chicken and turkey. *Modified from Moore (1965) and Jones and Johansen (1972).*

regulation of force generation in myocytes. Relaxation of the myocytes is due to the uptake of Ca^{2+} into the SR and extrusion by Na^+/Ca^{2+} exchange. In cardiac muscle, contraction is "all or none". Force modulation must be done at the cellular level by regulating not only the amount of Ca^{2+} entering through the sarcolemma, but also its effect on Ca^{2+} release from the SR.

11.2.3.3 Conduction System

The cardiac conduction system of the avian heart consists of the sinoatrial (SA) node, AV node, AV Purkinje ring, His bundle, and three bundle branches (Figure 11.9). Histologically, three types of cells are associated with the conducting system. The first type, pacemaker cells (P-cells), are small and spherical in shape and are found in both the SA and AV nodes. P-cells have the property of repetitive spontaneous depolarization. Transitional cells (T-cells) are much smaller and have fewer microfibrils than cardiac muscle cells; their structure is intermediate between normal cardiac muscle cells and Purkinje fibers. The third type, Purkinje fibers, are large, elongated, brick-shaped cells containing few myofibrils. Many Purkinje cells, however, contain longitudinal fibers called intermediate filaments; these are part of the cytoskeleton, serving to maintain cell shape as the myocardium contracts. Purkinje cells may be up to five times the diameter of myocardial cells.

The SA node is located close to the opening of the venae cavae into the right atrium, although there is considerable species variation in birds. The SA node consists of P-cells and many T-cells and is enclosed in a loosely organized connective tissue sheath. The T-cells transmit impulses from the pacemaker to atrial muscle cells. The SA node is morphologically, and perhaps physiologically, diffuse in birds. The primary electrical pacemaker region appears to change position spontaneously within the node (Hill and Goldberg, 1980).

For normal cardiac function, all cardiac muscle fibers within a given cardiac chamber should contract more or less simultaneously, although it is essential that the atria contract before the ventricles. The wave of excitation initiated in the SA node is delayed at the AV node, allowing the atria to empty before ventricular contraction begins. The electrical impulse that initiates contraction spreads through the atria and ventricles at rates in excess of 1 m/s, whereas conduction through the AV node is two to three orders of magnitude slower. Spread of excitation through myocardial muscle occurs from one muscle cell to the next, as well as along specialized conducting pathways. Individual muscle cells are discrete entities, but they behave electrically as if they were all joined together to form a syncytium. This property results from low electrical resistance in parts of the cell membrane where cell apposition is very close. Junctional complexes, the intercalated disks, commonly join myocytes end-to-end in avian myocardium, occurring at right angles to the long axis of the myofibrils. Intercalated disks consist of two components, desmosomes, which mechanically couple the cells together, and nexuses, which couple cells electrically. A nexus may be viewed as an array of unit electrical resistors, with their number being inversely proportional to the electrical resistance between the cells (Sommer, 1983). Interestingly, there are few nexuses along the longitudinal axes of myocardial cells.

The speed at which the wave of electrical excitation propagates through the ventricles is enhanced by a specialized conducting system of Purkinje fibers, but whether a specialized conducting system also exists in the atria of birds is controversial. In the atrium, waves of electrical and contractile activity proceed in the same direction, from the SA to AV nodes, which may mean that a specialized conducting system is unnecessary here. However, both atria contain Purkinje cells, and these have been described both morphologically and physiologically as being organized to preferentially direct the wave of activation toward the AV node (Davies, 1930; Hill and Goldberg, 1980). The contrarian view is that because the Purkinje cells in the atria are mixed diffusely among the normal myocardial cells, then they may represent the remnants of an embryological anlage left over from the time when that anlage was building the ventricular conducting system (Sommer, 1983).

The atrial wave of excitation crosses to the ventricle through the AV node, which, in birds, is a somewhat controversial structure because many morphological investigations have failed to locate it, although its presence has been established in functional studies. In the chicken, the AV node is located in the right side of the base of the interatrial septum (Davies, 1930; Lu et al., 1993b; Ying et al., 1993), although in Indian fowl, *Pycnonotus cafer*, house sparrow, *Passer domesticus*, and in *Bubo bengalensis*, it is located in the left AV junction (Mathur, 1973). The His bundle and its three bundle branches of Purkinje cells arise from the AV node. The right and left bundle branches emerge from the septum to form a network in the subendocardium of the right and left ventricles, respectively, penetrating the myocardium along the tracts of the coronary arteries (periarterial Purkinje fibers).

An indication of the theropod ancestry of birds can be inferred from the conducting system of the heart. Birds, unlike mammals, possess an AV ring of Purkinje fibers on the right side of the heart, which runs up and around the right AV valve (Figure 11.10; also see Section 11.2.1.4). The middle bundle branch, after separating from the others, runs around the aorta and connects to the AV ring, forming a figure-eight (Lu et al., 1993b).

Purkinje cells conduct electrical impulses much faster than cardiac myocytes. In mammals, part of the reason for this high conduction speed is that Purkinje cells lack a T-tubule system. T-tubules increase the surface area of

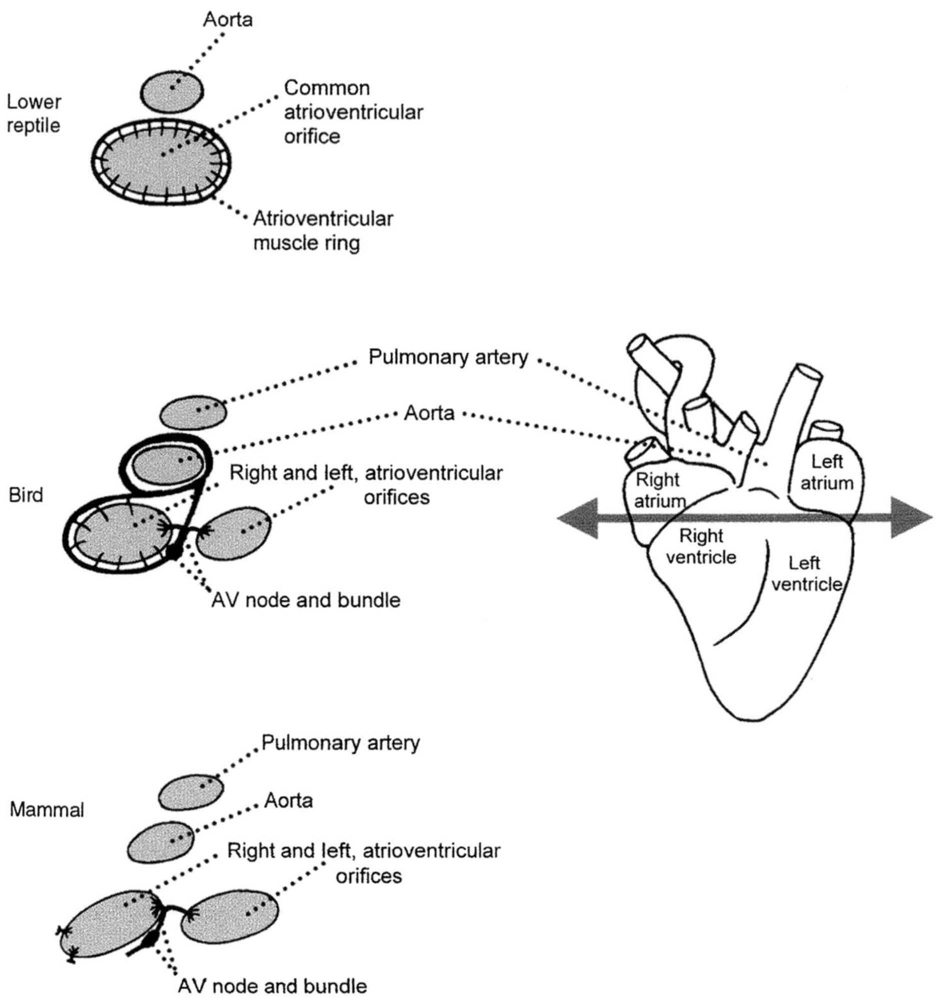

FIGURE 11.10 Diagram of specialized atrioventricular connections in lower reptiles, birds, and mammals as seen when looking into the ventricles after making a transverse section through the heart, as indicated on the diagram of the chicken heart on the righthand side. *From Davies (1930).*

the cell and therefore membrane capacitance (increasing the length of time a given amount of electrical charge will take to alter membrane potential); a high capacitance thus slows conduction velocity. However, in birds, there is no T-tubule system associated with either myocardial or Purkinje cells, yet the latter still conduct impulses at a faster rate. This is because conduction velocity varies directly with cell diameter and avian Purkinje cells are much larger than myocardial cells. Furthermore, a higher conduction velocity may be fostered in Purkinje cells by intermediate filaments, which serve to keep the cell round. Also, the electrical resistance between Purkinje cells is lower than that between myocardial cells because nexus size also increases with cell diameter, acting to further increase conduction velocity. Finally, in mammals with extremely large hearts (i.e., ungulates), the Purkinje cells within a bundle are tightly packed together and surrounded by an insulating membrane so that they behave electrically as a single fiber of a diameter equal to that of the whole bundle. Purkinje cells are likewise bundled in the avian heart, but whether this enhances conduction velocity is uncertain because the bundles lack a connective tissue sheath and are therefore not insulated from surrounding tissues.

11.2.3.4 Electrophysiology

The Purkinje fibers follow the coronary arteries and therefore take a relatively short course through the thick left myocardium. This accounts for the rapidity of arrival of the wave of excitation at a given point on the surface of the left ventricular wall in the avian heart (Lewis, 1916). The sequence of depolarization is, according to Kisch (1951): right ventricle apex, right ventricle base, left ventricle base, left ventricle apex. Moore (1965) mapped epicardial activation in the turkey and suggested that the apical third of the right ventricular epicardium is activated earliest, the upper basilar third is intermediate, and the pulmonary outflow tract is the last region activated in the whole heart. The anterior one-third of the septal region and the middle region of the left ventricle are activated before the basilar regions, with the whole left ventricular epicardium being activated in 12.5 ms. Others have found somewhat different sequences of activation (Lewis, 1916; Mangold, 1919). Kisch's suggestion that the conducting system stimulates heart muscle only at places of direct contact between its terminal fibers and heart muscle and not along the entire course of the conducting system receives support from his own work (Kisch, 1951), showing that subendocardial muscle is activated about 20 ms later than the earliest activated subepicardial muscle, which in turn suggests short cuts of the conductive system to subepicardial muscle. However, Davies (1930), Lu et al. (1993b), and Kharin (2004) suggested that because the bundle branches lack a fibrous sheath, there will be early and widespread propagation of the impulse in the septal region and thence along the bundle to all parts of the ventricles. Kharin (2004) mapped activation patterns with 64 electrodes in the chicken ventricle and found earliest activation in both the right and left ventricle in a mosaic-like pattern. Repolarization of the ventricle follows a different pattern from that observed during depolarization (Kharin, 2004). Repolarization starts at the right ventricle apex and moves down to the left ventricle apex. This is then followed by repolarization of the left ventricle base and finally the right ventricle base.

The conducting system of the bird heart has been investigated by recording transmembrane potentials from cells in the heart of the chicken and turkey (Moore, 1965, 1967). The pacemaker cells of the SA node, in the absence of any extrinsic influences, set the heart rate. Cells that function as pacemakers show a characteristic slow depolarization during diastole, the steepness of the depolarization being related to the degree of automaticity inherent in the cell (the fastest cells to depolarize drive the slower cells), whereas cells not spontaneously active show a steady membrane potential during diastole (Figure 11.9). APs recorded from the junction of the left SA valve with the sinus venosus show diastolic depolarization (prepotentials) with a slow transition to the ascending phase of the actual AP (Figure 11.9), in contrast to the relatively more rapid rise of the AP recorded in the right SA valve itself, indicating that cells in the right SA valve are triggered by the pacemaker cells. The duration of APs recorded from ventricular muscle cells is longer than those recorded from atrial muscle cells (Figure 11.9). Purkinje fibers display a prominent sharp peak to their APs, which is followed by a distinct plateau, a feature not seen in APs from atrial or ventricular muscle cells. The duration of depolarization is also much longer in Purkinje fibers, although diastolic depolarizations have not been recorded from avian Purkinje fibers. The longer duration of the Purkinje APs as compared with those of ventricular myocytes indicates a long refractory period, which would tend to prevent extrasystole and possible fibrillation by assuring a concerted depolarization of the ventricular muscle (Moore, 1965).

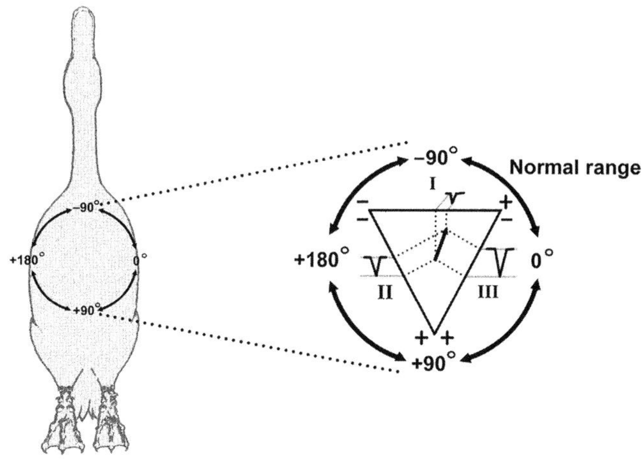

FIGURE 11.11 Relationship of the electrical axis of the heart to potentials recorded from the three standard leads (I, II, and III) as conceived by Einthoven. The duck is presented in ventral view on the left side of the figure. The electrical axis is upward and close to the midline. The arrow in the middle of Einthoven's triangle shows the direction of the axis, while the length represents its magnitude and degrees its orientation. The voltage changes seen at each lead (I, II, and III) are shown.

An electrocardiogram (ECG) shows the summed electrical activity of the heart. An ECG is usually recorded indirectly with electrodes placed on the body surface or just under the skin. (A direct ECG recording would be made by dividing the sternum and placing recording electrodes directly on the surface of the heart.) In birds, as in mammals, three standard leads (I–III) are used, following the model first conceived by Einthoven about 100 years ago. The body is a volume conductor of electricity and the waves of depolarization and repolarization that sweep across the heart can be reduced to a single electrical dipole. The dipole has magnitude (volts), direction, and sense (positive or negative), so it is a vector quantity. In Einthoven's concept, the cardiac vector is situated at the center of an equilateral triangle (Figure 11.11) formed by the bipolar lead connections. Lead I connects across the thorax from the right (negative electrode) to the left (positive electrode) wing bases. Lead II is recorded between right wing base (negative electrode) and left thigh (positive electrode). Lead III is connected between the left wing base (negative electrode) and the left thigh (positive electrode). The arrangement of these leads is such that, in humans and many other mammals, the polarity of the recorded signals is positive. In contrast, in birds, the polarity of the major component of the ECG (ventricular contraction) is negative (Figure 11.12).

There has been some controversy over what the typical avian ECG looks like (Figure 11.12). For instance, Kisch (1951) reported the presence of P, QRS, and T waves, whereas Mangold (1919) reported that the electrocardiogram of birds has no R component, but instead a deep S wave. Sturkie (1986) reported the presence of P, a dominant S, and T waves, a small R, but no Q wave. The status

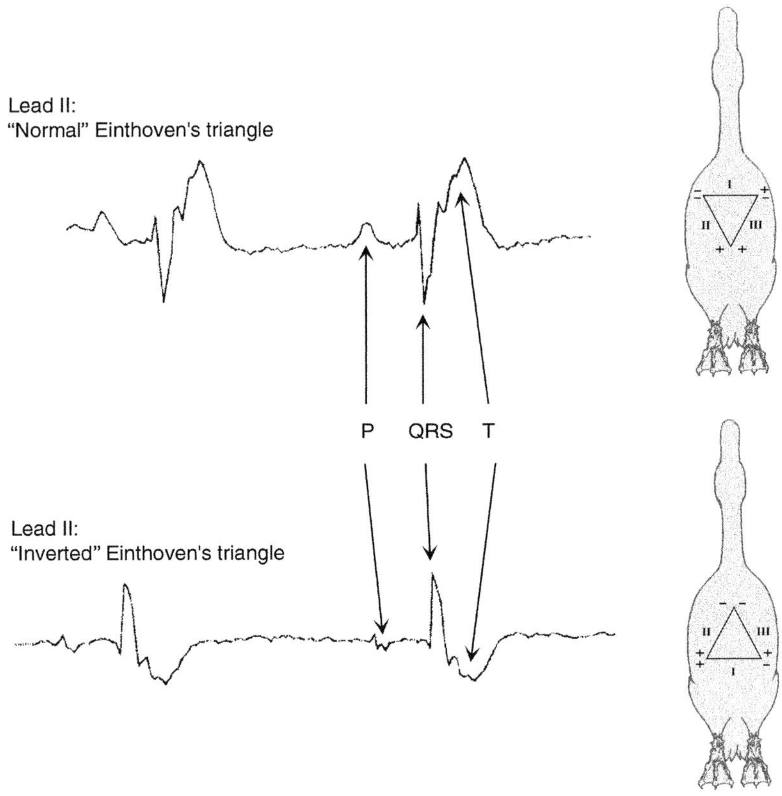

FIGURE 11.12 Upper electrocardiogram (ECG) trace: voltage recorded from lead II when the recording electrodes are placed in the conventional mammalian manner, on the right shoulder and left leg. Einthoven's triangle is shown on the duck on the upper right in ventral view. Lower ECG trace: voltage recorded from lead II when the recording electrodes are placed on the left shoulder and right leg. Einthoven's triangle is inverted as shown on the duck on the lower right in ventral view. The components of the ECG waveform are shown. P indicates atrial contraction, QRS is ventricular activation, and T represents ventricular depolarization. *From Liu et al. (unpublished data).*

of the Q wave has not been clarified by more recent work. There is general agreement that it is absent in the chicken (Goldberg and Bolnick, 1980; Liu and Li, 2005), small in the turkey (McKenzie et al., 1971), and prominent in some duck ECGs (Figure 11.12; Cinar et al., 1996). Aside from being useful in monitoring heart rate, the ECG can also be used to access the timing of the various phases of the cardiac cycle, because components of the ECG can be identified with atrial (P wave) and ventricular (QRS or RS) depolarization as well as repolarization of the ventricles (T wave). The duration of the P wave is the period of atrial depolarization and repolarization, whereas the P–R duration includes the conduction delay at the AV node. QRS or RS represents ventricular activation and the T wave is the period in which the heart is completely depolarized. QT or RT duration represents the duration of a complete cycle of activation and relaxation of the ventricle (Figure 11.12). Most of these intervals are fixed, so it is primarily the period between the T and P waves (i.e., interbeat interval), which shortens and lengthens with increases and decreases in heart rate, respectively. In fact, at very high heart rates, the T wave of one beat may come to overlie the P wave of the next.

Bipolar recording of the standard limb leads means that the cardiac vector is projected along the line between the two electrodes (Figure 11.11). When all leads are used, then it is possible to reconstruct the orientation of the cardiac vector or mean electrical axis (MEA) for any of the events of the cardiac cycle with respect to the plane of orientation of the leads. Standard leads lie in the frontal plane and, in mammals, the MEA of the QRS is oriented downward (inferiorly) and to the left (+60°). In contrast, the MEA of the QRS wave in the bird heart is close to −90°, being oriented along the long axis of the body and superior to the frontal plane. Hence, the QRS or RS wave is of negative polarity and barely represented in lead I, with the highest voltages being recorded by leads II or III (Figure 11.11). The mean electrical axes of the P and T waves can be calculated in a similar fashion. By using other cardiac leads in addition to the standard limb lead I, Szabuniewicz and McCrady (1967) were able to determine the MEAs of the QRS or RS, P, and T waves of the heart in the chicken, not only in the frontal plane (−77.1°) but also in the horizontal (+72.4°) and sagittal (−55.4°) planes.

The MEA of the ventricular depolarization phase is negative, whereas that for the repolarization phase is positive.

Therefore, the QRS or RS component deflects downward or negatively while the T wave is upright or positive (Figure 11.12). Obviously, the heart does not markedly change its position in the chest with each beat. The waves deflect in opposite directions because depolarization causes the ventricular myocardium to become negative and repolarization drives the myocardium positive. Also, the time courses of these waves are different. Repolarization is slower than depolarization, so the T wave is more spread out than the QRS or RS wave. In this context, it should be noted that when the ECG is recorded just for purposes of monitoring heart rate, using a single pair of leads (usually II), the RS wave is often presented as deflecting positively. This is achieved by reversing the polarity of the lead II bipolar electrodes (inverting Einthoven's triangle, Figure 11.12) and is done for artistic reasons.

Further interpretation of the ECG of birds is complicated by variability in electrode recording sites, anatomical differences between species, and the absence of a large bank of data such as has been accumulated for humans and, to a lesser extent, other mammals. In fact, it seems most unlikely that rigorous, detailed investigation of the ECG of birds will ever be used for clinical diagnosis. Boulianne et al. (1992) suggested that only diseases that cause a shift in position of the heart in the torso, and therefore alter the MEA, can be successfully diagnosed by two-dimensional electrocardiography. Round heart disease in chickens and turkeys produces such a shift; the mean RS axis in the frontal plane averages +70° compared with −85° in the normal bird (Hunsaker et al., 1971; also see Section 11.2.1.5).

11.3 GENERAL CIRCULATORY HEMODYNAMICS

The three major constituents of the pulmonary and systemic circulations are (1) the arteries or distributing vessels, (2) the capillaries or exchange vessels, and (3) the veins, which are storage vessels. Arterioles and venules are muscular vessels located upstream and downstream of the capillary beds, respectively. They are regulatory vessels, directly controlling blood flow distribution and indirectly controlling exchange of materials across capillary walls by adjustment of capillary pressure.

The major arteries bifurcate many times before the capillary beds; at each bifurcation, vascular resistance increases (McDonald, 1974). Volume flow in the parent and daughter vessels remains the same in the steady state, but flow velocity in the daughters falls to about 80% of that in the parent vessel. Therefore, the sum of the cross-sectional areas (πr^2, where r is internal radius) of both daughter vessels is greater than that of the parent vessel by about 25%. Hence, as the vessels divide, flow velocity falls; in the capillary circulation, flow velocity is exceptionally low. This allows adequate time for exchange of blood gases, nutrients, and metabolites with the surrounding cells.

Pressure, generated by cardiac contraction, drives blood flow around the circulation. Poiseuille's law relates volume flow ($\dot{Q}$) to the pressure drop ($P_1 - P_2$) along a tube of radius (r) and length (L) during steady flow, as follows:

$$\dot{Q} = (P_1 - P_2) \times \frac{\pi r^4}{8\mu L} \tag{11.1}$$

where μ is blood viscosity. Rearrangement of Eqn (11.1) hints at a somewhat more familiar form,

$$\frac{\pi r^4}{8\mu L} = \frac{P_1 - P_2}{\dot{Q}} \tag{11.2}$$

because the term on the left hand side of Eqn (11.2) is vascular resistance (R). Consequently,

$$R = \frac{P_1 - P_2}{\dot{Q}} \tag{11.3}$$

or, for the whole body, total peripheral resistance (TPR, kPa s/m^3),

$$\text{TPR} = \frac{\text{MAP} - \text{MVP}}{\text{CO}} \tag{11.4}$$

where MAP is mean arterial pressure (kPa), MVP is mean venous pressure (kPa), and CO is cardiac output (m^3/s). In order to compare animals of different sizes, it is common to express CO on a unit weight basis (i.e., m^3/s/kg).

Because the length (L) of any vascular channel is anatomically fixed while blood viscosity will only vary by 2–3 times, then vascular resistance is dominated by the radius of the vessels (Eqn (11.2)). Consequently, with a given pressure drop, halving vessel radius will reduce flow to one sixteenth, as shown in Figure 11.13, for a change in vessel radius from, for instance, two units (center profile) to one (left profile). This has important implications for the control of blood flow distribution.

Poiseuille's Law (Eqn (11.1)) applies to steady flow. However, in the major arteries, flow is highly pulsatile. In pulsatile flow, due to the inertia of the blood and high heartbeat frequencies, flow amplitude may no longer vary linearly with the pressure gradient. Nevertheless, the extent of the deviation from Poiseuille's law can be assessed from a nondimensional constant α (Womersley, 1957)

$$\alpha = r\sqrt{\frac{2\pi f \rho}{\mu}} \tag{11.5}$$

where r is radius, f is heart rate, ρ is blood density, and μ is blood viscosity.

When $\alpha < 0.5$ for the fundamental frequency (i.e., heartbeat frequency), the phase lag is negligible and flow conforms approximately with that predicted by Poiseuille's equation. Calculations for the aorta of a duck give a value

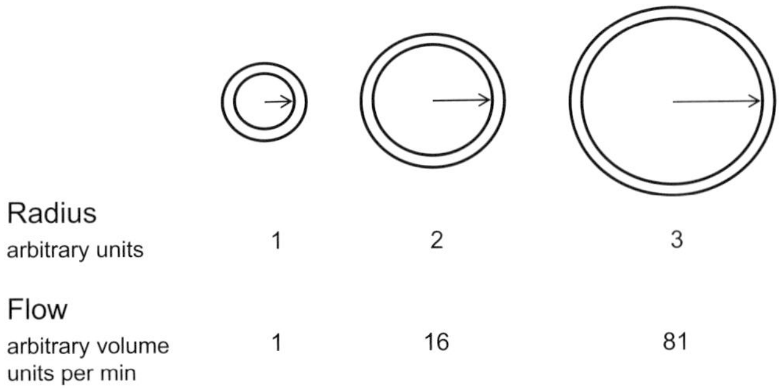

FIGURE 11.13 Effect of change in vessel radius as shown by the length of the arrow, on fluid flow with a constant driving pressure. Flow decreases in proportion to the fourth power of the decrease in radius. Radius dimensions and relative flow volumes are in arbitrary units.

for α of 6.0–7.0 for the fundamental frequency (about 3/s), so that estimation of flow by the Poiseuille formula is not reliable. However, it is obvious that for any given heart frequency, the value of α is directly dependent on the size of the vessel. Hence, in the femoral artery, α is certainly below 1, and flow will vary approximately linearly with the pressure gradient in this vessel. The vessels of the capillary circulation are small and α will be likewise, so application of Poiseuille's formula to this vascular bed would appear, superficially, to be most appropriate. Unfortunately blood viscosity, which can be regarded as a constant in larger vessels, may vary unpredictably in vessels of capillary size (Section 11.4.2). Therefore, caution is the watchword when applying steady flow formulations to flow driven by an oscillating pump. Even so, pulsatile flow has a mean, steady flow component and Pouseuille's law can be applied to this component, as in calculations of TPR (Eqn (11.4)).

Fourier analysis of pulsatile flow (and pressure) waveforms can be used to isolate the mean from the oscillatory components. The latter are resolved as a series of sinusoidal waves at the harmonics of the original waveform (the first harmonic is fundamental frequency (f); second harmonic is $2f$, …, nth harmonic is nf). If pressure and flow waveforms are recorded simultaneously, then dividing the oscillatory component of pressure by flow, at each harmonic, gives the vascular impedance. Consequently, vascular resistance (mean pressure divided by mean flow) can be considered as the impedance at zero frequency. The vascular impedance of any region of the circulatory system is thus determined by relating the corresponding frequency components of pressure and flow waves recorded simultaneously at that region. If pressure and flow are recorded at the input to the aorta, then aortic impedance is an expression not only of the characteristics of the whole systemic circulation but also of the afterload against which the left ventricle must work.

Pressure and flow recorded at the input to the arterial system start synchronously, yet peak flow velocity is reached before peak pressure. This rather anomalous situation is caused by the fact that the pulses travel through the arterial system. Therefore, for the pressure pulse, a positive gradient between an upstream and downstream point in an artery, established when the crest of the pressure wave traverses the upstream point, will reverse when the crest reaches the downstream point. Hence, the pressure gradient oscillates about a mean in all arterial vessels and flow will rise or fall or even reverse with these oscillations in the pressure gradient, although the presence of valves on the outflow tracts of the ventricles may limit the extent of reversal at the root of the aorta. Nevertheless, it is obvious that flow and pressure waves are not "in phase" and the extent of the phase difference, at each harmonic frequency, can be resolved by Fourier analysis.

The input impedance is presented (usually graphically, see Figures 11.14 and 11.15 in Section 11.4) as a set of terms of the values of modulus (|Z|) and phase (Φ) at each frequency obtained by Fourier analysis. Thus:

$$(|Z|) = \frac{\text{Pulsatile pressure}}{\text{Pulsatile flow}} \tag{11.6}$$

and

$$(\phi) = \text{Pressure phase} - \text{flow phase} \tag{11.7}$$

The phase of the impedance will be negative when the flow leads the pressure and positive when the pressure leads the flow.

11.4 THE VASCULAR TREE

11.4.1 Arterial System

11.4.1.1 Gross Anatomy

At least six pairs of aortic arches appear in the embryos of all vertebrates, recapitulating their aquatic ancestry. In birds, not all arches are present at one time and some are extremely transitory, such as the fifth pair of aortic arches, which make their appearance last. Only three arches persist in the adult, represented by the carotid artery (third arch), the aorta (fourth arch), and the pulmonary artery (sixth arch). In terrestrial vertebrates other than birds and mammals, both left and right branches of the fourth aortic arch are retained, whereas only the right persists in birds and the left in mammals. In some avian species, a remnant of the left aortic arch may remain as a solid core of cells; in a few others, such as the belted kingfisher (*Megaceryle alcyon*), the left arch remains patent and functional, although it loses its connection with the root of the aorta (Glenny, 1940). In an interesting series of experiments Stéphan (1949) demonstrated that ligation of the right aortic arch in the embryo causes the left to develop, as in mammals. This finding suggests that the retention or disappearance of aortic arches is dependent upon hemodynamic conditions, and it may be that the persistence of the right arch simply results from the unique development of the ventricular outflow tract in birds.

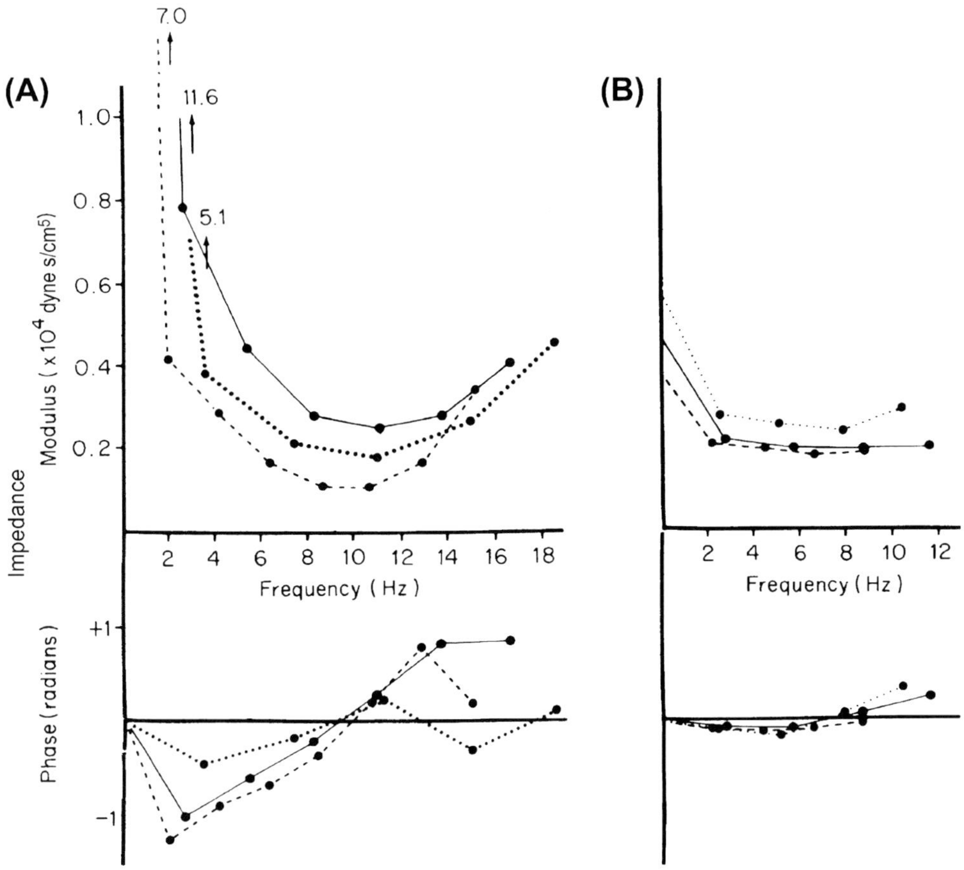

FIGURE 11.14 Impedance modulus and phase versus frequency graphs for aortic (A) and pulmonary (B) circulations in a duck. *From Langille and Jones (1975).*

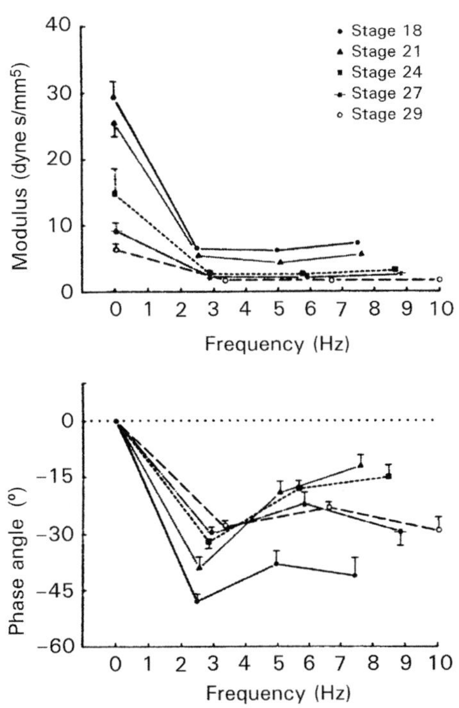

FIGURE 11.15 Impedance modulus (upper) and phase angle (lower) for chick embryos from stages 18–29. The identification key inset in the upper panel also applies to the lower panel. *From Zahka et al. (1989).*

The first major vascular bed supplied by the aorta is that of the heart. The coronary arteries, supplying the nutritional and respiratory circulation of heart muscle, arise from the ascending aorta close to the heart (Section 11.2.1.5). The ascending aorta then gives rise to two very large brachiocephalic trunks (Figure 11.16), supplying blood to the head, wings, and flight muscles. Each brachiocephalic vessel is usually larger in diameter than the continuation of the aorta, reflecting the higher blood flow rates in the brachiocephalics.

All arteries of the head and neck are branches of the carotid arteries. Surprising variation exists in the pattern of the carotid arteries close to the heart. The most common arrangement is two vessels of equal size running side by side (Figure 11.16). Other patterns include a single artery formed by fusion of both carotids (herons, bitterns, and kingfishers); a single vessel due to loss of the right (passerines) or left (plovers) carotid; or two arteries of unequal size (flamingos, sulfur-crested cockatoo).

Blood flow to the brain must not be interrupted, or impairment of brain function rapidly ensues. The carotid arteries lie in a groove in the base of the neck vertebrae close to the axis of rotation and are therefore protected from possible obstruction due to compression from neck movements. Other safety measures are provided by anastomoses between the carotid and vertebral arteries and, at the base of the brain, by either an X-, I-, or H-shaped junction between the carotids. This intercarotid anastomosis has been found in all species of bird except those of the suborder Tyranni (Baumel and Gerchman, 1968). Birds do not possess a cerebral arterial circle of Willis comparable to that of mammals but, because the intercarotid anastomosis is relatively large in many cases, it may represent a more effective collateral circulation than the mammalian arterial circle. Blood to the wings and flight muscles is supplied by the subclavian arteries. Each subclavian divides into two branches, the brachial (wing) and pectoral (flight muscles).

The descending aorta runs caudally, ventral to the vertebral column, giving off paired intercostal and lumbar arteries. Blood is supplied to organs within the abdomen and legs by the following vessels, originating from the descending aorta (see Figure 11.16):

Celiac artery	Liver, spleen, glandular stomach, gizzard, intestine, pancreas
Cranial mesenteric artery	Most of intestine, pancreas
Renal arteries	Kidneys (anterior portion), testes
Femoral arteries	Legs
Ischiatic arteries	Middle and posterior portions of kidney and legs; uterine region of oviduct
Caudal mesenteric artery	Rectum and cloaca
Internal iliac arteries	Walls of pelvis; oviduct
Caudal artery	Tail, terminal branch of aorta

There are, in effect, three pairs of renal arteries in birds, one pair arising from the aorta and two from the ischiatic arteries. However, in the gray heron (*Ardea cinerea*), one pair of renal arteries arises from the femoral arteries instead of the ischiatics.

The ischiatic artery is the major vessel supplying the leg. At the level of the knee, it meets and joins the femoral artery to form the popliteal artery. This artery, passing into the lower leg, divides to form the anterior and posterior tibial arteries. In the tarsal region of the leg and in the axillary region of many birds, there are arteriovenous networks of vessels referred to as *rete mirabile*, particularly prominent in wading and aquatic birds (Midtgård, 1981). These structures serve as heat exchangers, because warm arterial blood is brought into close proximity to venous blood that has traversed the distal parts of the limbs and is therefore colder. The countercurrent arrangement of blood flow ensures that heat can be transferred from arterial to venous blood all along the length of the artery and vein apposed in the *rete*, thereby reducing heat loss to the environment by reducing the temperature of arterial blood flowing through peripheral thinner sections, such as the web of the foot or the wing.

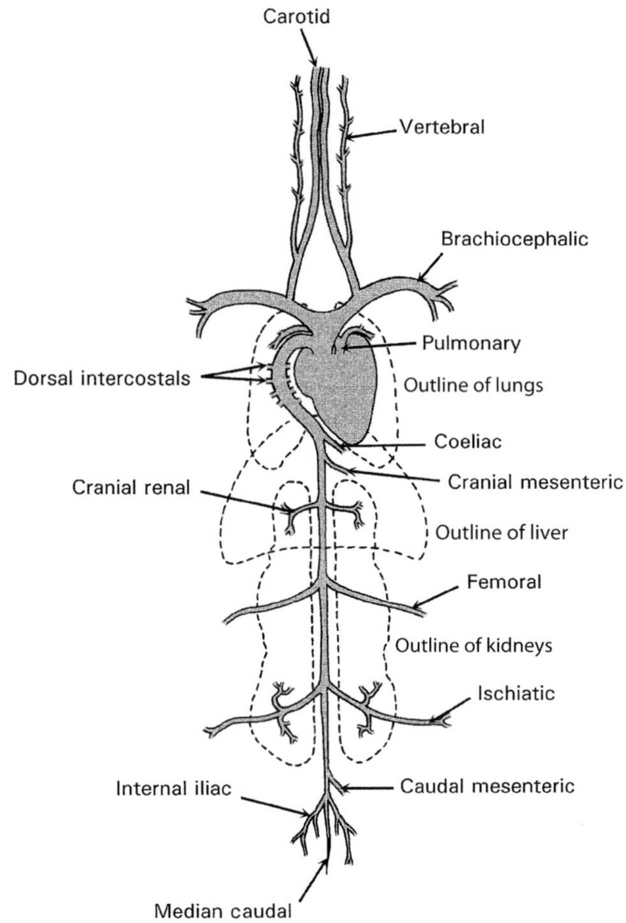

FIGURE 11.16 **The major systemic arteries in the bird.**

11.4.1.2 Functional Morphology of the Arterial Wall

The large arteries have two main functions. First, they serve as low-resistance conduits carrying blood to the arterioles for distribution to the peripheral vascular beds. Second, the whole arterial system serves as a pressure reservoir or *Windkessel*, accepting the volume of blood ejected by the heart and converting the highly pulsatile input into a steady flow of blood through the capillary beds. The *Windkessel* results from wall elasticity, particularly of those vessels close to the heart.

The central arterial vessels are elastic, whereas the more peripheral ones, certainly distal to the second order of branching, are muscular. In elastic arteries, the vast majority of the wall is made up of layers of smooth muscle embedded in elastin fibers, alternating with layers of collagen. One layer, composed of a combination of muscular, elastic, and collagen fibers, forms a single lamellar unit within the wall. Large numbers of concentric lamellar units make up the bulk of the wall of elastic arteries of pigeon, chicken, and the mute swan (*Cygnus olor*), as shown in Figure 11.17 (Bussow, 1973). Interestingly, the lamellar units do not

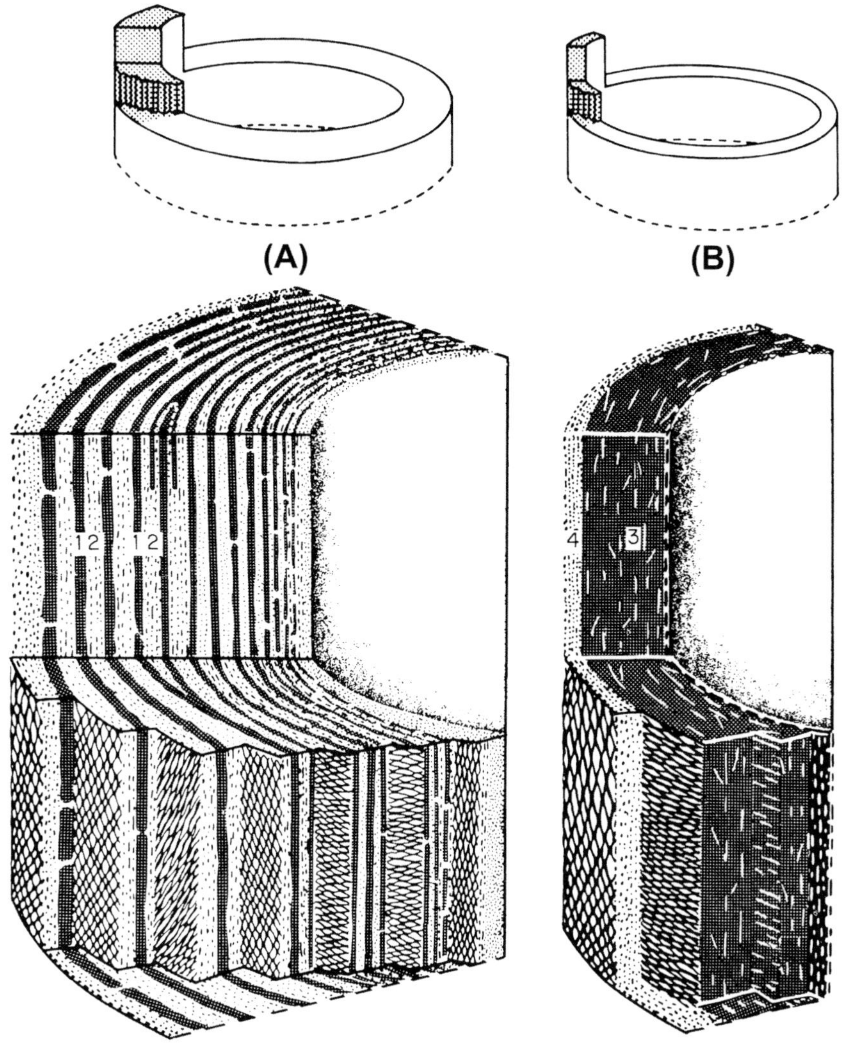

FIGURE 11.17 Schematic depiction of an elastic (A) and muscular (B) artery wall. (A) Wall structure of an avian elastic artery. The wall is composed of fragmented smooth muscle layers embedded in a fine network of elastic fibers (1) alternating with layers of collagen (2). (B) Wall structure of an avian muscular artery consisting of a thin intima, a media made up of smooth muscle cells and elastic fibers (3), and a thick adventitia with a well-defined elastica externa (4). Upper diagrams compare wall thickness of elastic and muscular arteries of similar lumen diameter. The wall of the elastic vessel is about three times as thick as that of a muscular vessel. *From Bussow (1973).*

form complete cylinders around the vessel. This is particularly obvious in vessels very close to the heart where an individual lamella may extend around only one quarter, at most, of the vessel circumference.

Wall structure of the muscular arteries is very different. Muscular vessels consist largely of circumferentially arranged smooth muscle cells with elastic fibers distributed, either singly or in bundles, as a wide-meshed plexus between the muscle cells (Hodges, 1974). The collagenous components are transferred to the outer layer of the wall. An interesting embellishment of the normal structure of muscular arteries occurs in the cranial mesenteric artery of the chicken and turkey (Ball et al., 1963). This vessel is invested by longitudinally arranged smooth muscle fibers, the thickness of this layer being approximately the same as that of the circumferentially oriented smooth muscle within the wall proper. The functional significance of the external muscle layer is unclear, but it may serve to shorten the vessel to accommodate changes in its position brought about by gut movements.

The elastic arteries include the aortic arch and its major branches, the thoracic aorta up to about the level of the celiac artery, and the extrapulmonary portions of the pulmonary arteries, whereas all branches of the abdominal aorta as well as the caudal portion of the aorta itself are muscular. In most regions of the arterial tree, the change from an elastic to a muscular wall occurs rather abruptly, usually at a branch site. An exception to this is the aorta itself. In the aorta, the elastic and muscular portions are separated by a segment of the vessel extending from the coeliac artery to the ischiatic arteries, whose wall structure fits neither description very well. Furthermore, in both the pigeon and turkey the wall in this region is transversely asymmetric with a thick, muscular ventral wall and a thin elastic dorsal wall.

The arteries *in vivo* expand and recoil with every heartbeat, although this behavior is seldom mimicked *in vitro* during experiments designed to study vessel mechanics. Instead, the static rather than dynamic elastic behavior of the arterial wall is usually investigated. Essentially, a short length of excised blood vessel is inflated from a syringe and the pressure change induced by a given volume change is noted. These pressure–volume loops give an immediate and compelling view of how arterial elasticity changes with the degree of inflation.

Furthermore, by using blood vessels from different areas of the body, regional variations in elasticity are revealed. Pressure–volume loops are usually J-shaped, showing that the more a vessel is stretched the more resistant it becomes to further stretch (Figure 11.18; Speckmann and Ringer, 1966). The collagen fibers in the vessel wall inhibit expansion at high pressures, whereas the properties of elastin dominate the lower pressure limb of the curves. It is the compliance of elastin and the stiffness of collagen working in concert that allows uniform and smooth expansion of the vessel wall over a range of distending pressures without formation of aneurysms. In contrast, a wall in which the properties of extension remain constant across the range of distending pressures would be prone to aneurysm formation. Rubber has a straight rather than J-shaped response to distension and aneurysms always occur in the wall of cylindrical balloons when they are inflated.

When a blood vessel is inflated, more pressure is required to expand it than is recovered during recoil of the elastic walls. The ratio between the energy recovered in deflation to that expended in inflation is a measure of the resilience of that vessel (Figure 11.18). Surprisingly, resilience is similar for both thoracic (elastic) and abdominal (muscular) aortae of the turkey. Over the range of arterial pressures encountered in turkeys, resilience lies between 85% and 87%—values that are well above those of most mammals and approaching those obtained for invertebrate blood vessels. In one sense, the higher the resilience, the better because most of the energy cost of stretching elastic vessels with each cardiac ejection will be returned by elastic recoil on deflation. Unfortunately, if the resilience is too high, then the vessels may go into uncontrolled oscillations (resonance), particularly at the high repetition frequencies (heart rates) seen in many birds. Obviously, vascular engineering in birds is close to the edge.

Pressure–volume loops reveal characteristics specific to particular segments of whole vessels, whereas the properties of the materials making up a vessel wall are revealed by stress–strain curves (Figure 11.19(A)). Stress is the deforming force divided by the area of the vessel wall over which it is applied while strain is the ratio of the stretched radius to the unstretched radius of the vessel. Stress–strain curves for the ascending and descending aortae, the brachiocephalic arteries, and the thoracic aorta of the duck (*Anas*

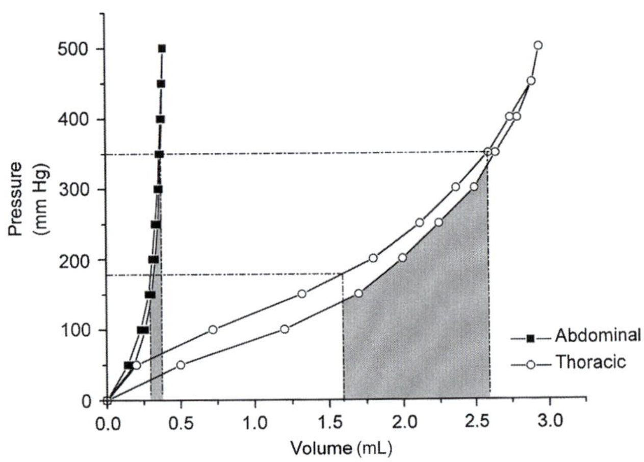

FIGURE 11.18 Typical pressure–volume loops for the abdominal and thoracic regions of the aorta in a turkey. For each loop, the upper curve is the inflation sequence and the lower curve the deflection sequence. Resilience of the vessel is obtained by dividing the area under the deflation sequence (shaded) by the area under the inflation sequence over the range of blood pressures recorded in turkeys. *From Speckmann and Ringer (1966).*

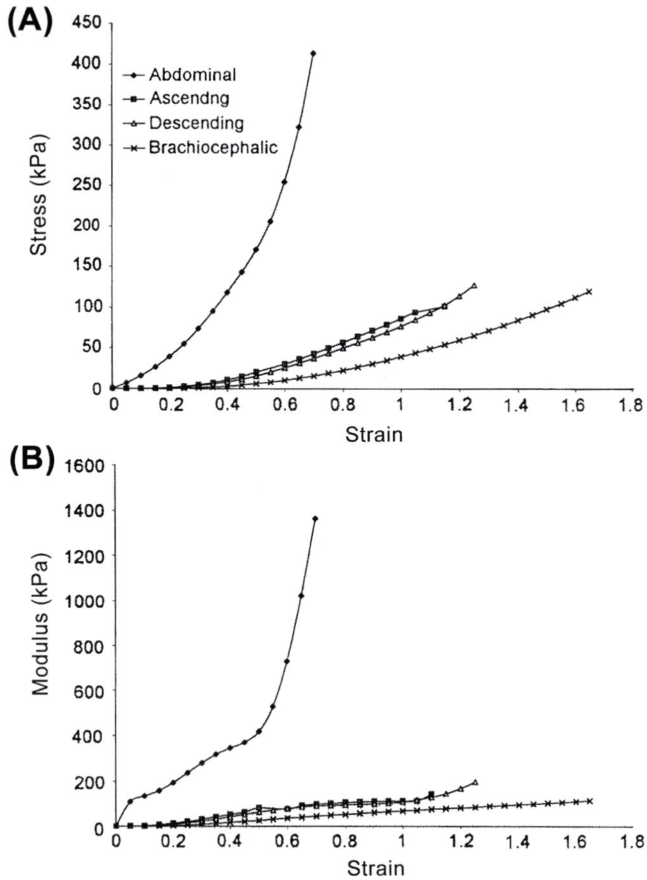

FIGURE 11.19 (A) Stress–strain curves for the blood vessel wall from four regions of the aorta in a duck (*A.platyrhynchos*). (B) The static elastic modulus of these vessels derived from data in A. *From Braun and Jones (unpublished data).*

platyrhynchos) are shown in Figure 11.19(A). The stress–strain relationship for the abdominal aorta lies to the left of that for the other, more central, vessels indicating that, as in the turkey (Figure 11.18), the abdominal aorta is stiffer than the other vessels—a finding that would be expected given that the abdominal portions of the aortae of duck and turkey have more collagen than do more central segments of this vessel.

The slope of the line of a stress–strain curve is the elastic modulus of the material making up the blood vessel wall, but, since the slopes for blood vessels are not linear, the elastic modulus is continually changing. The incremental elastic modulus describes the elastic modulus for a small increment in strain (Bergel, 1961). Incremental elastic moduli for the duck aorta, which are similar to those for the turkey aorta (Speckmann and Ringer, 1964), are shown in Figure 11.19(B). The moduli for the abdominal aorta are well above that for the thoracic, confirming that the former is stiffer than the latter. Nevertheless, these moduli are one to two orders of magnitude below those obtained from arteries in corresponding parts of the mammalian vascular tree. In mammals, the lamellar units in the wall form complete cylinders so that those laminae containing primarily collagen must be stretched to the same extent as the more distensible muscular laminae. Consequently, such rigid laminae are more important in determining the degree of extension of the wall. However, in birds, arterial lamellar units do not form complete cylinders, so there is some "series" coupling between the rigid and elastic components of the wall that allows more distensible laminae to be extended somewhat independently. Avian blood vessels have much thicker walls than mammalian vessels of the same diameter, thus compensating for their lower elastic modulus (Bussow, 1973).

11.4.1.3 Relationship between Arterial Pressure and Flow

Each heartbeat sends a pulse through the arterial system, which arrives later at sites more distal to the heart. The velocity at which the pulse wave travels is lowest in the most distensible vessels and increases in the stiffer peripheral vessels. In ducks, pulse wave velocity increases from 4.4 ± 0.8 m/s in the aortic arch to 11.7 ± 1.2 m/s in the abdominal aorta, with the major increase in velocity occurring in the thoracic aorta (Langille and Jones, 1975).

When the pulse transit time occupies a considerable proportion of each cardiac cycle, then significant phase changes occur between the pressure and flow pulses at different arterial sites. In ducks, the time taken for the pulse to travel from the heart to the distal end of the abdominal aorta is around 20 ms, which is about 5–10% of the cardiac cycle (Langille and Jones, 1975). Consequently, in ducks, there are marked changes in the waveform of the systemic pressure pulse as it travels through the arterial system (Figure 11.20). Both pulse amplitude and the contour of the pulse waveform are altered, with pulse pressure increasing by about 30%. This peaking of the pressure pulse results from a marked increase in the systolic portion, with little change in the diastolic portion (Figure 11.20).

Peaking of the pressure pulse is due to wave transmission effects, primarily wave reflections. All forms of wave motion can be reflected by physical changes in the system they are traveling through. When such changes occur within the arterial system, incident pressure and flow waves will be reflected back toward the heart. These physical changes can be discrete discontinuities, such as those due to arterial branching (McDonald, 1974) or continuous variations in wall compliance due to an increase in arterial stiffening toward the periphery (Langille and Jones, 1975, 1976). However, the major reflecting site seems to be the terminal vascular bed. From this site, pressure and flow pulse waves are reflected back toward the heart to interfere, destructively or constructively, with the incident wave generated by cardiac contraction. This interference means that pressure and flow waves recorded simultaneously at any one site in the arterial system will be quite unlike those recorded at another. In a reflectionless system, pressure and flow pulses sampled at any given site should look similar to those recorded anywhere else in the system.

An essential question concerns the nature of the termination that the peripheral vascular beds present to outgoing pressure and flow waves. Peripheral beds are closed terminations if they present a relatively large impedance to pulsatile flow; they are open if they present a relatively low impedance. In higher vertebrates, reflections produce large oscillations in peripheral pressures that drive small oscillatory flows through the terminal vascular beds (Figure 11.20), indicating a high terminal impedance (i.e., of the closed type). Hence, the pressure should be reflected at the closed end without a phase shift while, to satisfy the condition that high-pressure oscillations are required to drive low oscillatory flows through a high terminal impedance, the reflected flow wave should be inverted. That is, the reflected wave should be 180° out of phase with the incident wave. However, how much of the incident wave reaches the termination (because the incident pulse is attenuated, especially in the smaller vessels, as it propagates through the system) and how much of the reflected wave gets back to the heart (because its amplitude is also reduced by damping) is still a matter of speculation.

The reflection coefficient (that portion of the incident wave reflected by the terminal vascular beds) is extremely sensitive to the state of the peripheral vasculature. Under resting conditions, up to 80% of the incident wave may be reflected. Intense vasoconstriction occurring in ducks

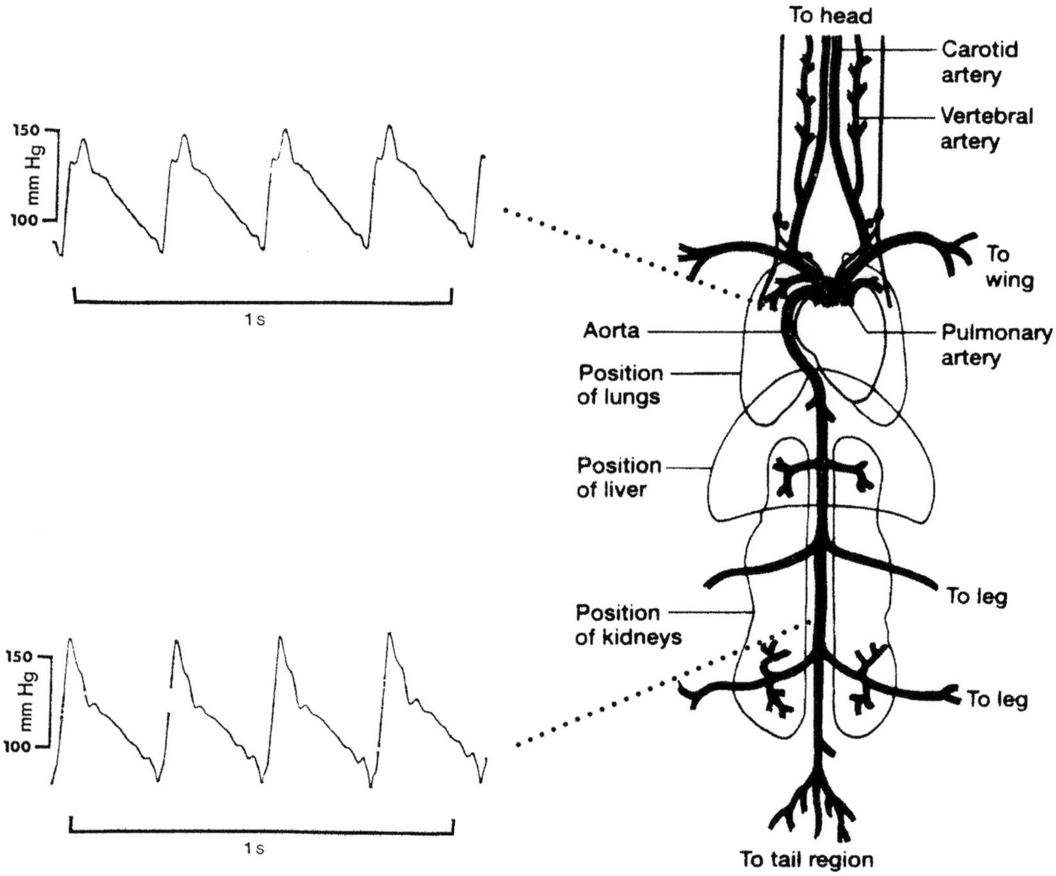

FIGURE 11.20 Diagrammatic representation of pressure waves recorded simultaneously in the proximal and distal aorta of a duck. Amplification and distortion of the pressure wave occurs during propagation along the aorta. *Modified from Langille and Jones (1975).*

during forced diving or when voluntarily diving birds are trapped underwater and unable to surface causes 100% of the incident wave to be reflected. In contrast, vasodilatation of peripheral vascular beds, occurring during exercise or hemorrhage, may reduce the reflection coefficient to zero.

Evaluation of wave propagation through the arterial system is complex, requiring harmonic analyses of pressure and flow waveforms. Fortunately, a simple conceptual analysis of the interaction of incident and reflected waves in the arterial tree is sufficient for the present purpose. For simplicity, consider a pressure wave displaying simple harmonic motion, as illustrated in Figure 11.21 (first harmonic). At the closed end of the system (terminal vascular beds), both incident and reflected pressure waves will be in phase and the waves will sum giving an antinode, evident as an enlarged pressure oscillation. The reflected wave is shown as 40% of the incident wave in Figure 11.21 so that the amplitude of the resultant compound wave will be 140% of the incident wave, as is shown by the wave envelope. Now, consider a point one-quarter wavelength back from the termination. The incident wave left here one-quarter of a cycle before it reached the closed end and the reflected wave takes another one-quarter cycle to return to this point,

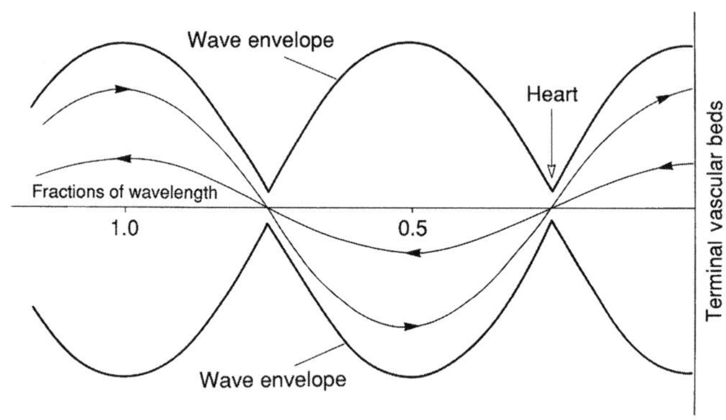

FIGURE 11.21 Interactions between incident and reflected pressure waves at a closed end (terminal vascular beds). For clarity, it is assumed that only 40% of the wave is reflected from the closed end (i.e., reflection coefficient = 0.4). The abscissa is marked in fractions of a wavelength. At the point of reflection, both waves are in phase and sum together. With reference to a point one-quarter wavelength away, the incident wave is 90° earlier and the reflected wave 90° later so that they are 180° out of phase and cancel. This point is a node and the only oscillation is the difference between the maximum amplitudes of the incident and reflected waves. For maximum benefits in terms of promoting cardiac efficiency the heart should be located at this point. The total excursion throughout the cycle is represented by the wave envelope (heavy outer lines). *Modified from McDonald (1974).*

so the incident and reflected waves are now 180° out of phase and destructive interference produces a node, evident as a decrease in pressure. Hence, the amplitude of the resultant wave is some 40% smaller (as shown by the wave envelope in Figure 11.21) than it would be in the absence of reflections.

The flow wave will also be reflected such that the incident and reflected waves are 180° out of phase at the termination but are in phase one-quarter cycle away from the terminal impedance. Hence, the flow waves will cancel one another at the termination and will sum at the heart. Therefore, highly oscillatory cardiac outflow is generated for less pulsatile pressure than would be the case in a nonreflecting system; the advantage of this is that more external cardiac work is available for any given level of cardiac oxygen consumption (Milnor, 1979).

The benefits of such reflections will be maximized when the heart is one-quarter wavelength upstream of the major reflecting site. In ducks, resting heart rate is 2–3 beats/s; therefore, even at the lowest pulse wave velocity (4 m/s), the major reflecting site would have to be located one-third or one-half meter from the heart—a most unlikely possibility. During exercise, however, heart frequency may double and the major reflecting site would now be 16–25 cm from the heart. Unfortunately, exercise is associated with vasodilation, which will reduce the reflection coefficient. In contrast, vasoconstriction during forced or voluntary diving, especially if the bird is trapped underwater, would accentuate reflections but heart frequency is now about 0.33 beats/s (or one beat every 3 s).

Consequently, obtaining the necessary balance between heart rate and reflection coefficient to maximize beneficial reflection effects seems unlikely. What then are the consequences of a mismatch between pulse wavelength and distance between the heart and major reflecting sites? Let us assume that Figure 11.21 describes the second and not the first harmonic of the pressure wave. In this case, the first harmonic will be 90°, not 180°, out of phase at the heart. This represents an antinode for the first harmonic which will add to systolic pressure, causing the heart to expend more energy. The heart will be located at a node for the second harmonic but at an antinode for the third (270° out of phase). As the first three harmonics contribute about 80–90% to the original pulse amplitude, then the net effect will be an increase in systolic pressure. In the duck, this is seen as a significant early systolic shoulder in the pressure pulse recorded in the aortic arch (Figure 11.20).

If the transit time of either pressure or flow waves through the arterial system becomes less than 5% of the cardiac cycle, then reflection effects on the shapes of these waves are not obvious. Nevertheless, reflections occur, but they are diffuse; the pressure wave bounces back and forth between the heart and periphery, until damped to extinction. In humans, atherosclerosis causes a loss of pressure pulse amplification as the pulse travels through the arterial system (O'Rourke et al., 1968); similar observations have been made in the turkey, *Meleagris* (Taylor, 1964), where atherosclerosis is very common (Ball et al., 1972; Manning and Middleton, 1972). Loss of pulse amplification in atherosclerosis results from a generalized stiffening of the major arteries, which acts to speed pulse wave propagation and thereby minimize wave transmission phenomena.

The hummingbird is among the smallest homeothermic vertebrates. Is it possible to predict the hemodynamics of the hummingbird from our knowledge of hemodynamics in the duck? Assuming pulse wave velocity is unaltered, then for similar conditions to hold in both hummingbird and duck, heart frequency (f_H) must increase in the same proportion as the linear dimension (L) of the animal decreases with reduction in body mass (M_b). For birds, the allometric equation relating f_H and M_b is $f_H = k_1 \cdot M_b^{-0.28}$ (Grubb, 1983; see Section 11.2.2), whreas $L = k_2 \cdot M_b^{-0.33}$ where k_1 and k_2 are constants.

According to this analysis, a hummingbird 400 times smaller than a duck will have a f_H 5.3 times higher but L will decrease nearly seven times. Even at the highest f_H reported for the giant hummingbird (*Patagona gigas*) of 1020 beats/min (Lasiewski et al., 1967), pulse transit time as a proportion of the cardiac interval will be considerably less than that in the duck and reflection effects on pulse wave shapes will probably not be obvious (Jones, 1991).

11.4.1.4 Vascular Impedance

While the complex pressure and flow waves recorded in avian arteries (Figure 11.22) are not directly comparable they, like all periodic signals, can be expressed as a sum of sinusoidal signals of ascending frequency (harmonics). These individual harmonics of pressure and flow are directly comparable (see McDonald, 1974). Comparison can be done most conveniently by determining vascular impedance versus frequency. Impedance modulus (amplitude of a pressure harmonic divided by amplitude of the flow harmonic of the same order) is the analog of vascular resistance that is applicable to pulsatile flows. Impedance phase is simply a measure of the degree to which pressure and flow oscillations are out of synchrony.

Figure 11.14(A) illustrates impedance versus frequency curves for the circulation supplied by the descending thoracic aorta of *Anas*, while Figure 11.14(B) illustrates the vascular impedance of the pulmonary circulation. Aortic impedance falls from the value at zero frequency, the peripheral resistance (Z_t), to settle at a steady value at high frequencies; this is an indication of the characteristic impedance (Z_o); (Figure 11.14(B)). Peripheral resistance and characteristic impedance are measures of arteriolar caliber and aortic distensibility respectively, uninfluenced by wave reflection effects. Often, the modulus and phase are not constant but fluctuate;

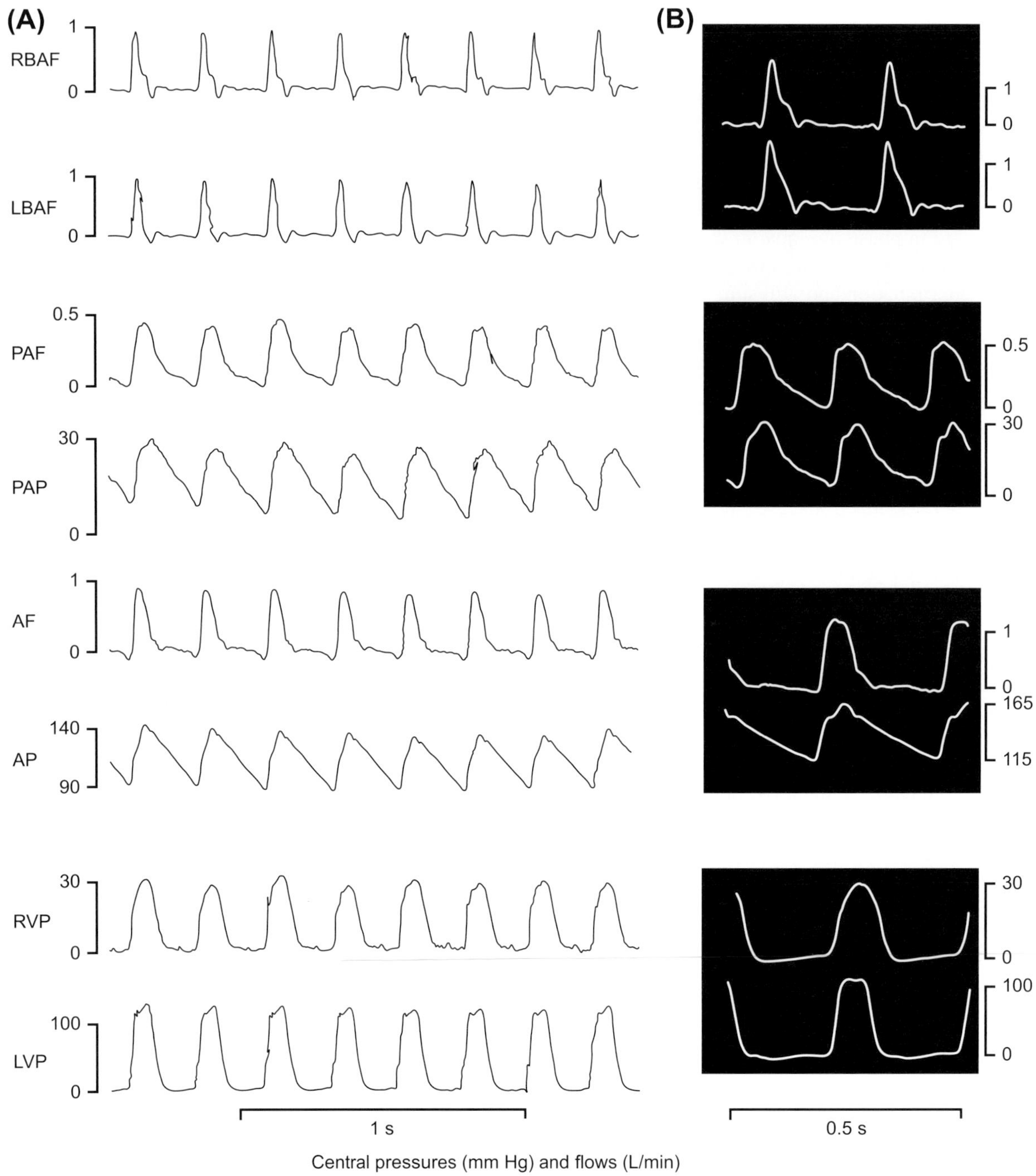

FIGURE 11.22 **Pressures (mm Hg) and flows (L/min) recorded simultaneously in the arterial system of a duck.** Traces, from top to bottom: RBAF, flow in right brachiocephalic artery; LBAF, flow in left brachiocephalic artery; PAF, pulmonary arterial flow; PAP, pulmonary arterial pressure; AF, aortic flow; AP, aortic pressure; RVP, right ventricular pressure; LVP, left ventricular pressure. Panels on right represent paired oscilloscope records of central pressures and flows. Traces match those on the left but each pair was recorded from a different animal. *From Langille and Jones (1975).*

these fluctuations are caused by wave reflection effects from peripheral vascular sites. A well-defined minimum in the impedance modulus of the aortic circulation, and a coincident rise in impedance phase from negative values (pressure lagging flow oscillations) to positive values (pressure leading flow oscillations), are characteristics of a wave reflecting system (Figure 11.14(A)). At the frequency of the impedance minimum (10 Hz), the circulation imposes minimal load for pulsatile flow on the heart (Figure 11.14(A)).

The reflection coefficient depends on the impedance mismatch between the terminal arteriolar bed and the supply artery and is given by the ratio

$$(Z_t - Z_o) / (Z_t + Z_o) \times 100$$

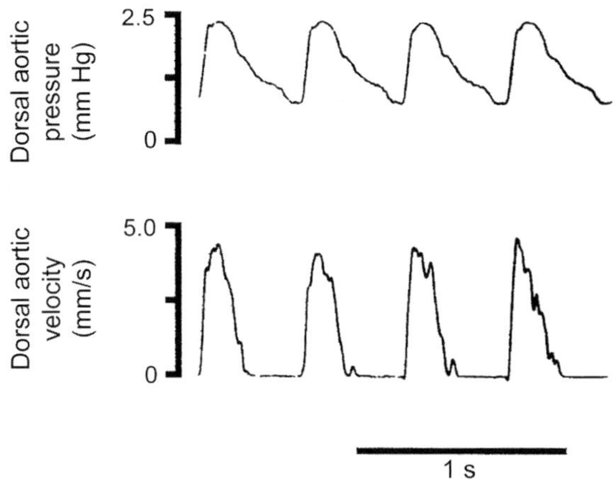

FIGURE 11.23 **Dorsal aortic pressure and flow velocity recorded simultaneously from a stage 24 chick embryo.** The pressure and flow waveforms are similar to those recorded from mature animals (see Figure 11.14) despite the absence of a semilunar valve apparatus. *From Zahka et al. (1989).*

For the aortic circulation, the reflection coefficient is high, being over 80%. However, in the pulmonary circulation, the absence of clear impedance minima suggests that this low-resistance circulation does not give rise to major reflections of the pulse wave. In fact, the reflection coefficient for the pulmonary circuit is only 25%.

Pulsatile pressure and flow are generated by the first, embryonic cardiac contractions. In stage 24, chick embryos pressure and flow waveforms resemble those recorded from mature animals, despite the absence of the semilunar valve apparatus in the heart (Figure 11.23; Zahka et al., 1989). Both peripheral resistance and vascular impedance modulus decrease with development through stages 18–29 (3–6 days, development; Figure 11.15). Over this period, mean dorsal aortic pressure and flow increase 13 and 10 times, respectively, with the increased flow being accommodated by the rapidly expanding arterial bed. The velocity of pulse wave propagation increases with developmental stage, from around 0.5 m/s at stage 18 to nearly 1 m/s at stage 24 (Yoshigi et al., 1997). Hence, pulse transit time will be a negligible fraction of the cardiac cycle (0.35–0.45 s) so that reflection effects will be unimportant. Even if this were not the case, reflection effects would be minimized in the later stages of embryonic development because of the marked decline in vascular resistance. Consequently, the fall in vascular impedance during embryonic development is due to the growth of a larger, more distensible, dorsal aorta in which elastic fibers first appear at stage 29 (Hughes, 1942).

11.4.2 Capillary Beds

11.4.2.1 Gas Exchange

Systemic capillaries form a vital functional interface between the blood and systemic tissues of birds. The pathway between erythrocytes in capillary blood and mitochondria in the surrounding tissue represents the last in a series of resistances in the oxygen transport pathway from the lungs. Oxygen and carbon dioxide move between the systemic capillary blood and the surrounding tissue mitochondria by simple diffusion. Therefore, the diffusion distance from erythrocyte to mitochondrion and the partitioning of diffusion resistance along this route is of immense physiological interest. The concept of a capillary domain, a volume of tissue whose oxygen demands could potentially be satisfied by diffusion from one capillary, was first expounded by August Krogh in 1914 and remains a valuable concept in understanding gas exchange in muscle tissue (Krogh, 1919). A simple explanation of the factors important in capillary blood–tissue gas exchange, based on this model, can be found in West (2008).

Studies on the systemic capillaries of birds, particularly those of flight muscle, have been dominated by two interesting themes: (1) the high workload of the avian pectoralis major muscle during flight, reflected in an increase in oxygen consumption of about five times that at rest (Butler et al., 1977), suggests that the functional anatomy of the pectoral muscle capillaries may reveal adaptations to both high tissue oxygen demand and mechanical tissue deformation during sarcomere shortening; and (2) some species fly at high altitudes (see Section 11.6). Therefore, there may be specific adaptations in capillary density or geometric arrangement that facilitates the delivery of oxygen to the working pectoralis muscle in the face of a relatively low P_{aO_2} (the pressure head for diffusion at the arterial end of systemic capillaries) caused by reduced atmospheric partial pressure of oxygen.

Three parameters may be considered in determining the capillarity of muscle. These are the number of capillaries per muscle fiber, the cross-sectional area of muscle fibers, and the geometrical arrangement of capillaries around each fiber (Snyder, 1990). Gray et al. (1983) found that sections of pure slow red muscle fibers from the anterior latissimus dorsi of chicken had 25% more capillaries per square millimeter than did sections of fast white fibers from the posterior part of this muscle. However, in six other species of birds ranging in mass from 11 g to 6.2 kg, there was no significant correlation between fiber diameter and capillary number per fiber in slow red fibers and fast white fibers of the anterior and posterior latissimus dorsi respectively. Thus, tissue capillary density decreased in muscle with larger fibers. Snyder (1990) estimated the maximum diffusion distance from capillary to mitochondrion in slow red fibers and fast white fibers of the gastrocnemius muscle to be on average 32.4 and 36.5 μm respectively. Torrella et al. (1999) found maximum diffusion distances ranging from 16.8 μm in the oxidative fibers of the mallard pectoralis to 34.4 μm in the glycolitic fibers of the mallard gastrocnemius. These are similar to the values obtained in mammals, suggesting that diffusion distance has been highly conserved in vertebrate evolution. Increasing the number of capillaries per fiber appears to produce diminishing returns, such that beyond two capillaries per fiber there do not appear to be further measurable reductions in diffusion distance. In contrast to the situation in neonatal mammals, the capillary-to-fiber ratio in bird muscle appears to be fixed around hatching. Because the fibers hypertrophy during development, diffusion distances are shortest in newly hatched chicks—some 18 μm (Byers and Snyder, 1984).

The benefits conferred by reducing erythrocyte-to-mitochondrion diffusion distance depend on assumptions made about how the resistance to oxygen diffusion is distributed between source (blood) and sink (mitochondrion). It has been argued that the capillary-to-tissue interface represents the major resistance (Gayeski and Honig, 1986), in which case reducing the overall diffusion distance would be of limited effectiveness compared with increasing the area of this interface (see below).

Compared with the geometric arrangement in mammalian hind limb muscle, there are a larger number of capillary branches running perpendicular to the long axis of muscle fibers in pigeon pectoralis muscle (Mathieu-Costello, 1991, Figure 11.24). These branch from capillaries running parallel to the long axes of the muscle fibers. The branch points move closer together as the pectoralis muscle shortens on the power stroke, but the perpendicular orientation of the branches to the long axis of the muscle fibers does not change appreciably during contraction. The short segments of capillary parallel to the muscle fiber between these branches bow as the muscle shortens, but do not become particularly tortuous. The branches perpendicular to the long axis of the muscle fibers run around the circumference of the fibers; this arrangement, together with their high density, ensures that there is an effective envelope of capillary blood surrounding portions of the fibers. This results in

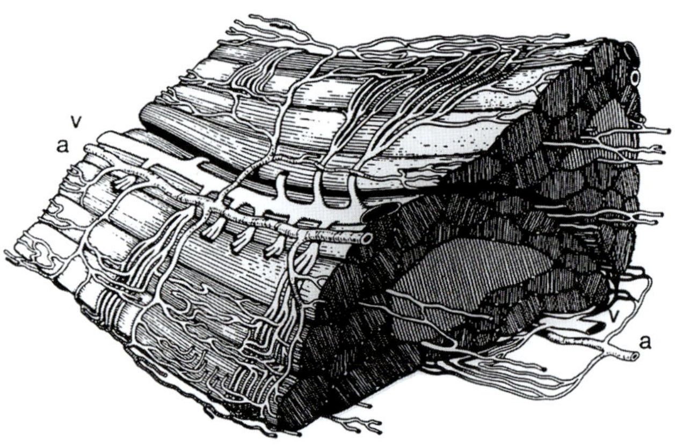

FIGURE 11.24 Schematic diagram illustrating the microvascular geometry in the pectoralis muscle of the pigeon. For clarity, the width and number of capillary branches running perpendicular to the long axis of the muscle fibers have been reduced. a, artery; v, vein. *Reprinted from Mathieu-Costello et al. (1994).*

very effective blood–tissue O_2 transfer (Ellis et al., 1983). Such an arrangement of capillary branches may compensate for the unfavorable rheological properties of avian blood compared with mammalian blood; these properties include relatively low red cell deformability and a low capillary hematocrit.

Flying hummingbirds have the highest mass-specific metabolic rate of any vertebrate and hummingbird flight muscles have the highest oxygen demand of any vertebrate skeletal muscle. It would be expected, therefore, that the adaptations for effective gas exchange at the capillary level would be most obvious in these birds. These could include a reduced diffusion distance from the capillary to the mitochondria or possibly an increase in the capillary-to-fiber contact area. Increases in either of these factors would increase the flux of respiratory gases for a given drop in P_{O_2} across the capillary–mitochondrial diffusion distance. The ratio of capillary surface area to muscle fiber surface area is about twice as large in hummingbird flight muscle as in the rat soleus muscle, with similar mitochondrial density in the two muscles (Mathieu-Costello et al., 1992). Therefore, according to Fick's law, the rate of O_2 diffusion into the avian fiber would be about double that in the rat, all other factors being equal. This supports the idea that the large area of the capillary–muscle fiber interface in avian flight muscle plays an important role in enabling these muscles to maintain an extremely high oxygen usage during flight. The respiration rates of muscle mitochondria in working hummingbird flight muscle are about double those in the locomotor muscles of mammals. Interestingly, the size of this interface in the pectoralis muscle of the only actively flying mammal, the bat, is similar to that in hummingbird flight muscle.

The basic structure of the capillary network in the hummingbird is similar to that in pigeon flight muscle, although capillary density tends to be higher in the hummingbird. This results from the smaller cross-sectional area of the muscle fibers (one-half for aerobic and one-tenth for glycolytic) rather than from a greater number of capillaries surrounding each fiber (Mathieu-Costello et al., 1992). The smaller cross-sectional area of fibers may be an adaptation to reduce the diffusion distance from capillary blood to the mitochondria. However, some experimental and theoretical evidence shows that the main drop in P_{O_2} occurs across the resistance offered by the capillary–fiber interface and P_{O_2} then declines more slowly, largely because of myoglobin facilitated diffusion (Honig et al., 1991). Thus, a large contact area between capillary and fiber is probably a more important factor in effective oxygen delivery than a short path between capillary and mitochondrion. This is particularly true for flight muscle in both birds and bats where fiber myoglobin content, and therefore the potential for facilitated diffusion, is high. Interestingly, a comparison of actively flying and sedentary pigeons showed little effect of flight conditioning on capillary–fiber relationships. Wild pigeons had a greater aerobic capacity than sedentary birds, achieved by a 30% greater cross-sectional area of aerobic fibers in the pectoralis together with a higher density of mitochondria. However, the capillary–fiber ratio was similar, as was capillary length–fiber volume at a given mitochondrial density (Mathieu-Costello et al., 1994). This finding is consistent with the results of Snyder and Coelho (1989), who caused hypertrophy of the right anterior latissimus dorsi of chickens by taping weights to the humerus. They found that the increased number of capillaries per fiber just matched fiber hypertrophy and concluded that muscle growth was the primary determinant of capillarity.

11.4.2.2 Microvascular Fluid Exchange

Capillary fluid balance is maintained by the dynamic interaction between hydrostatic and osmotic forces acting across the capillary wall, as first described by Starling over 100 years ago (Starling, 1896). Starling's original formulation has been modified and refined by Landis (1927) and Kedem and Katchalsky (1958), yielding the following equation to describe microvascular fluid exchange:

$$J_V = K_{FC}\left[(P_c - P_T) - \sigma_d(\pi_P - \pi_T)\right],$$

where J_V is net volume flow across the vascular wall, K_{FC} is capillary filtration coefficient, P_c and P_T are capillary and tissue fluid pressures, respectively, $(\pi_P - \pi_T)$ is colloid osmotic pressure (COP) difference between plasma (P) and tissue (T), and σ_d is the osmotic reflection coefficient.

In the steady state, the capillary blood pressure opposes the blood COP to maintain tissue fluid balance. Blood pressure exceeds COP at the arteriolar end of the capillary and is usually below COP at the venous end. Fluids are secreted at the arteriolar and absorbed at the venous ends of the capillary. Hence, for adequate fluid exchange, COP pressure must offset capillary pressure, the latter being a reflection of the arterial blood pressure (Landis and Pappenheimer, 1963). The value of the COP is determined by the concentrations and species of blood proteins as well as by cations held in the plasma by the Donnan effect of the proteins (Guyton et al., 1975). However, it is now clear that microvascular fluid exchange is a dynamic process in which extravascular forces, such as tissue fluid pressure, tissue COP, and the actual flow of lymph, can influence transcapillary fluid movement (Taylor and Townsley, 1987). In addition to heterogeneity of Starling forces in different areas of the microvascular beds, there is also the possibility that heterogeneity of capillary membrane permeability will also contribute to differences between global and local values of the Starling pressures (Michel, 1997). In the light of recent knowledge, the simplistic steady state view of secretion at the arteriolar end of the capillary and absorption at the venous end of the capillary can only be regarded as a transient phenomenon at best.

Studies on avian species have contributed little to this discussion although microvascular fluid exchange in birds presents some unique and interesting features. For instance, in turkey and duck, the ratio of protein concentration in the interstitial fluid to that in the blood is much lower than the ratio in mammals (Hargens et al., 1974). Hargens et al. (1974) have pointed out that the lower ratio in birds is correlated with a higher arterial blood pressure. Also, birds as a group seem to be highly resistant to hemorrhage, tolerating blood loss much better than mammals. Kovách and Balint (1969) have shown that increased hemorrhage tolerance becomes apparent only during prolonged bleeding because hemodilution continues in the pigeon through the period of blood loss, whereas in the rat no further hemodilution occurs after about 15–20 min of bleeding. Hemodilution is achieved by inflow of isotonic fluid with low protein content.

The restoration of blood volume results from absorption of tissue fluid across the capillary walls due to reduced capillary pressure. This fall in capillary pressure could be brought about by an increase in the ratio between precapillary and postcapillary resistances, as well as by changes in arterial and venous pressures during hemorrhage. Resistance changes across the capillaries seem to be the most important factor in rapid restoration of blood volume in ducks. Blockade of α-adrenergic receptors eliminates vasoconstriction in the skeletal muscle, which forms the major reserve of tissue fluid and leads to a greatly retarded restoration of blood volume (Djojosugito et al., 1968). Djojosugito et al. (1968) attribute the difference in ability to restore blood volume after hemorrhage in ducks and cats to a very pronounced reflex vasoconstriction in duck skeletal musculature and to a capillary surface area in ducks three to five times that in the cat, a condition that increases the rate of absorption of fluid into the vascular system (Folkow et al., 1966).

Many birds, such as the emu and ostrich, have extremely long necks, with the head being held a meter or more above the heart. Yet, the emu's resting mean arterial blood pressure (Grubb et al., 1983) does not differ from that of a pigeon (Butler et al., 1977). An exceptionally high blood pressure to overcome gravitational effects on circulation and to ensure flow to the head is not needed in these birds. In a fluid-filled system, the gravitational pressure of blood in the veins will counterbalance the gravitational pressure in the arteries of the neck, much like the loop of a siphon. In other words, it is no more difficult for blood to flow uphill than downhill in a system of closed tubes like the circulation. Overall circulatory flow around the body occurs due to a pressure difference between the aorta and right atrium, and it matters little what actual route the blood follows.

If pressure is measured in the cerebral circulation of a long-necked bird, this will be lower than that recorded in the aorta just outside the aortic valves by an amount sufficient to cause the required blood flow along the neck artery (a small difference) and by the gravitational effect due to the height the head is held above the heart. If the head is held 1 m above the heart, pressure in the arteries of the head will be about 75 mm Hg less than at the heart. Consequently, in the cranial capillary beds, it is now possible that the hydrostatic pressure will be lower than the COP of the blood and fluid will be continuously removed from the interstitial spaces, with similar consequences to those following excessive alcohol consumption. However, because drunken ostriches are not a common sight, countermeasures such as markedly increased arterial blood pressure, decreased arteriolar resistance, or reduced blood COP must be in effect. It is now possible to obtain a ready supply of long-necked ostriches and emus, and it is hoped that these countermeasures can be subjected to empirical investigation.

11.4.2.3 Distribution of Blood Flow at Rest

The percentage of CO distributed to different organs is closely related to their aerobic metabolic activity and their size. The distribution of CO is ultimately determined by the relative resistance of systemic vascular beds that are arranged in parallel throughout the body. Vascular resistance, in turn, is determined by a variety of hormonal, autoregulatory, and neural control mechanisms (see Section 11.5). Values obtained by different investigators for relative blood flow (% CO) to various organs differ widely, probably reflecting differences in experimental technique, species, and resting conditions. Nevertheless, it is apparent from the limited data available that the heart, liver, kidneys, and intestines receive relatively large percentages of the total CO (Figure 11.6). The avian brain appears to receive about 3% of CO, similar to the proportion of CO going to the brain in a small mammal, such as the rat, at rest (Ollenberger and West, 1998).

By far the highest resting blood flows thus far measured in any avian organ (about 16 mL/min/g of wet tissue weight) are found in the spleen (Figure 11.25). This organ receives a disproportionately large percentage of total CO despite its small size (Figure 11.6). High tissue flow rates also have been found in the mammalian spleen; a flow of some 12 mL/min/g has been reported in conscious dogs (Grindlay et al., 1939). Such high rates are almost certainly related to the dual function of this organ: as a filter for aging erythrocytes, which are eliminated by the process of diapedesis; and as an organ of the reticuloendothelial system, in which the blood is cleaned by phagocytic reticuloendothelial cells as it passes through the splenic sinuses and pulp. Obviously, a high flow rate is needed for this dual filtration role to be effective. In contrast to splenic blood flow, the rate of cerebral blood flow in birds is an order of magnitude less. Blood flow to the whole brain and to individual cerebral regions ranges from 0.43 to just over 2 mL/min/g under normoxic conditions, as shown in Figure 11.25 (Bickler and Julian, 1992; Butler et al., 1988; Faraci and

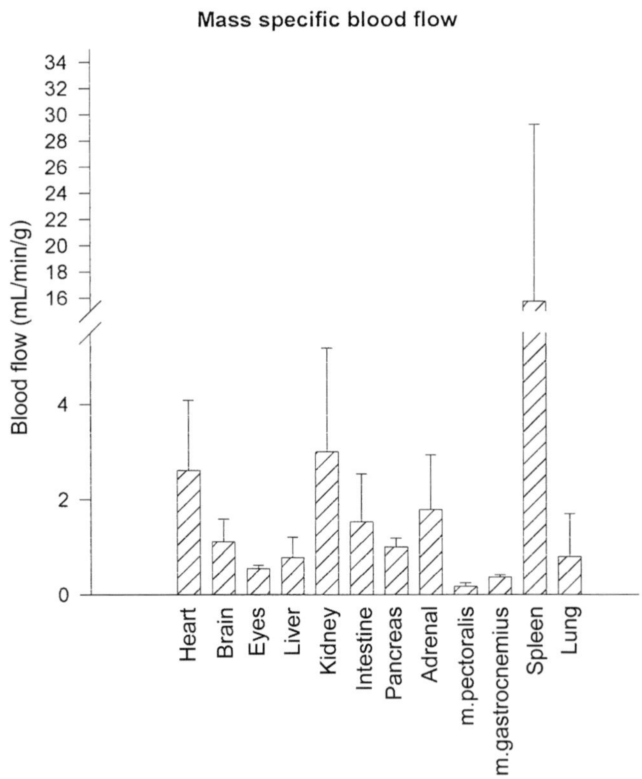

FIGURE 11.25 Mass-specific organ blood flow (mL/min/g) in birds at rest. Standard error bars are shown where appropriate. *Data from Butler et al. (1988) (tufted duck), Duchamp and Barre (1993) (muscovy duckling), Faraci et al. (1985) (Pekin duck), Jones et al. (1979) (Pekin and mallard), Stephenson et al. (1994) (Pekin duck), and Wolfenson et al. (1978) (chicken).*

Fedde, 1986; Faraci et al., 1984; Grubb et al., 1977; Jones et al., 1979; Stephenson et al., 1994; Wolfenson et al., 1982; Ellerby et al., 2005). Both the heart and kidneys have relatively high rates of mass specific blood flow, reflecting the high oxygen demand of the contracting cardiac muscle and the activity of energy-dependent membrane pumps in renal tissue, respectively. Pectoral and gastrocnemius muscle, on the other hand, show relatively low perfusion rates at rest (Figure 11.25).

11.4.3 Venous System

11.4.3.1 Functional Development of Venous System

The embryological development of the avian venous system follows a typical vertebrate pattern. At about the 15-somite stage of the avian embryo, paired cranial and caudal common cardinal veins develop from a vascular plexus in somatic mesoderm (Lillie, 1908; Sabin, 1917). The heart shifts caudally during embryonic development, and the cranial cardinal veins elongate to become the jugular veins. Subclavian veins, returning blood to the right side of the heart from the pectoral region and the wings, arise as tributaries of the caudal cardinal veins (Ede, 1964). The adult avian venous system cranial to the heart differs in detail from the mammalian pattern in that there are two cranial (superior) venae cavae. The right jugular vein is much larger in diameter than the left, and there is an anastomosis between the jugular veins at the base of the head, allowing some blood draining from the left side of the head and neck to return to the heart in the larger right jugular vein.

Development of the venous system caudal to the heart is primarily concerned with the formation of the physiologically important renal and hepatic portal venous circulations (see below). Subcardinal veins, which develop along with the embryonic kidney (mesonephros), initially provide renal drainage into the caudal cardinal veins. Eventually, the anterior portions of the caudal cardinal veins disappear and the subcardinals form a connection with the ductus venosus; this connection becomes the caudal vena cava. The liver develops around the ductus venosus, which subdivides into a capillary bed, forming the hepatic portal circulation. The cranial portion of the ductus venosus becomes the hepatic vein and the caudal portion becomes the hepatic portal vein. Finally, the renal portal veins form a junction with the caudal vena cava via the common iliac veins and the caudal subcardinal veins are replaced by the caudal renal veins.

11.4.3.2 Capacitance Function

The walls of the veins in birds are, as in mammals, thinner than those of arteries; therefore, venous distension depends on a positive transmural pressure gradient. The three basic components of blood vessel walls—tunica intima, media, and externa—are present. The tunica media is composed of circumferential smooth muscle fibers. In larger veins, elastic laminae appear in the tunica externa, which makes up most of the wall tissue. Veins near the heart are frequently invested with cardiac muscle fibers that are apparently functional. The caudal vena cava of the mallard can occasionally be seen to contract at the same frequency as the sinus venosus.

As in mammals, the avian venous system does not necessarily represent a passive conduit returning blood to the heart. The thin, distensible walls of veins mean that these vessels are relatively compliant compared with arteries. As applied to blood vessels, compliance is the ratio of the change in vessel volume (ΔV) resulting from a change in transmural distending pressure (ΔP):

$$\text{Compliance} = \Delta V / \Delta P.$$

In mammals, the entire vascular system has a compliance of about 3 mL (blood) kg (body mass)/mm Hg. The compliance of the arterial vascular segment is only about 3% of that of the venous segment (Rothe, 1983). Therefore, despite the smaller vascular pressures in the venous side of the circulation, at any one time about 60–80% of blood volume is contained in the veins. They are therefore referred to as capacitance vessels. The capacity of the venous circulation can change either passively by changes in transmural pressure or actively by changes in the contractile state of venous smooth muscle. A reduced transmural pressure and therefore a passive elastic recoil of compliant veins, or a reduction in venous compliance by active contraction of smooth muscle in the venous walls, mediated by α-adrenergic receptors (Section 11.5.2.4), would both serve to reduce venous capacitance. This transfers blood toward the heart. All other things being equal, this would tend to increase atrial filling and therefore CO. The large veins of the domestic fowl are well innervated with adrenergic motor fibers (Bennett et al., 1974; Bennett and Malmfors, 1975b), and the density of this innervation suggests that there is active control of venous capacitance (Section 11.5.2.4), although to date there are no physiological studies in birds analogous to those of adrenergic effects on venous capacitance function in mammals (Vanhoutte and Leusen, 1969).

11.4.3.3 Physiological Role of Veins in Exercise and Submersion

Venous pressure in pigeons flying in a low-speed wind tunnel increased to 2.5 mm Hg from the resting value of 1.2 mm Hg (Butler et al., 1977). CO increased 4.4 times during flight, mainly accomplished by an increase in heart rate at a constant stroke volume. In flight, the increased pressure gradient from the venous end of capillaries to the right ventricle increases venous return, right heart filling, and CO via the Frank–Starling relationship. Venous pressure is determined by the relationship between venous volume and compliance, which is reduced as venous smooth muscle contracts and the vein walls stiffen. In mammals, the venous beds of the liver, spleen, and skin act as blood reservoirs that can actively reduce capacitance during exercise, thereby increasing venous pressure and volume of venous return to the heart (Rothe, 1983). Whether these venous vascular beds also constrict during flight exercise in birds is currently unknown.

Active venoconstriction may also be important in the cardiovascular adjustments to diving in birds. Djojosugito et al. (1969) provided indirect evidence that venous pressure increased during diving in ducks and this was confirmed by Langille (1983), whose results suggested that this was due to active venoconstriction. In the latter study, cardiac stroke volume fell if central venous pressure was held constant during diving. This suggests that reduced ventricular contractility, caused by an increase in vagal motor nerve activity to the heart, is normally counteracted during diving by increased venoconstriction-induced filling via the Frank–Starling mechanism (see Section 11.5.3.2.3).

11.4.3.4 Renal Portal System

In common with most other vertebrate groups, birds possess a renal portal circulation. Venous blood making its way back to the heart from the legs and the lower intestine of birds enters the kidneys through a renal portal system. Within the kidneys this blood mixes with postglomerular efferent arteriolar blood in peritubular sinuses that surround all nonmedullary nephron segments and eventually flows toward the renal veins. The physiological significance of the renal portal system is currently poorly understood. About 50–70% of total renal blood flow is contributed by the renal portal vein. However, this percentage is highly variable between animals and can change rapidly in the same individual for no apparent reason (Odlind, 1978). Part of the variability in flow can be attributed to the status of active, innervated renal portal valves within the iliac veins (Figures 11.26 and 11.27; Glahn et al., 1993). The valves receive dense reciprocal motor innervation from the parasympathetic and sympathetic divisions of the autonomic nervous system (see Section 11.5.2.3). Adrenergic stimulation produces relaxation of the smooth muscle of the valve, and cholinergic stimulation produces contraction. In contrast, smooth muscle of the renal portal vein itself shows a predominantly adrenergic contractile response typical of most vascular smooth muscle (Burrows et al., 1983). The distribution of vascular resistance within the portal system, governed by both the portal valve and the alternate, parallel venous pathways for blood returning to the heart (Figure 11.28) determine the volume flow of renal portal blood.

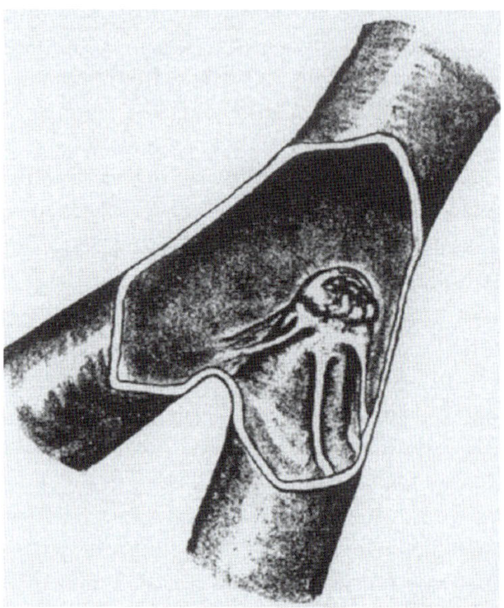

FIGURE 11.26 **Illustration of a renal portal valve.** The valve, composed of smooth muscle, is situated in the external iliac vein at the point where the efferent renal vein joins the iliac vein. The valve is anchored to the vein wall by muscular "tethers". *From Burrows et al. (1983).*

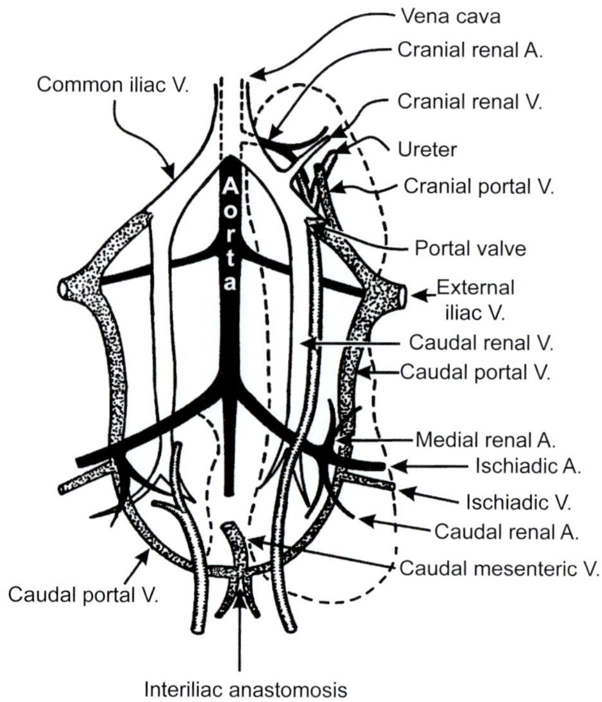

FIGURE 11.27 **Ventral view of the avian kidneys with a simplified representation of the avian renal portal circulation and its connections to the systemic venous system.** *From Wideman et al. (1992).*

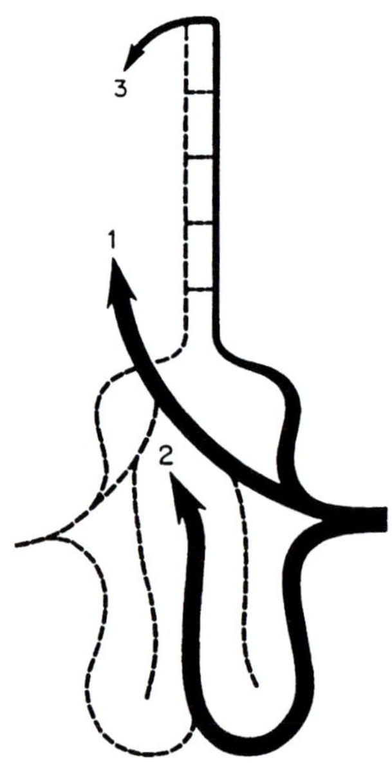

FIGURE 11.28 **Three potential parallel shunt pathways in the renal portal circulation.** (1) The renal portal valve is open under influence of its sympathetic motor innervation—venous blood flows from the external iliac vein through the patent valve into the common iliac vein, bypassing the kidney. (2 and 3) The renal portal valve is partially closed and resistance at the valve is high. Blood returning from the legs in the external iliac veins enters the cranial and caudal portal systems. *From Akester (1967).*

There are several options for a "packet" of venous blood returning from the legs in the external iliac veins (Figures 11.27 and 11.28): if the renal portal valve is open under the influence of sympathetic nervous system activation, blood can flow through the patent valve into the common iliac vein leading to the vena cava and directly back to the right side of the heart, bypassing the kidney. If the valve is partially closed due to parasympathetic stimulation and resistance at the valve is high, blood can alternatively enter the renal portal system by flowing into the cranial and caudal portal veins. Thus the portal veins are arranged functionally in parallel with the direct venous route provided by the common iliac vein and the vena cava. Blood entering the cranial and caudal portal systems eventually drains into the internal vertebral venous sinuses and caudal mesenteric veins respectively and makes its way back to the right side of the heart by this route. Anteriorly the caudal mesenteric vein connects with the portal system of the liver, providing another option for blood flow. The cranial and caudal portal veins are arranged in parallel with each other, and a "packet" of blood from the external iliac vein can enter one or the other, but not both. Therefore, variable fractions of venous blood derived from the legs, tail, and lower digestive tract can return directly to the right side of the heart (via the common iliac vein and vena cava) or enter the renal (cranial portal vein) or renal and hepatic portal systems (caudal portal vein) (Figure 11.28). The overall pattern of venous flow will depend on the distribution of vascular resistance within the portal system.

Several workers have tried to ascribe a functional significance to the renal portal system. This is a difficult task in the face of the possibility of neural, humoral, and local metabolic control all influencing the relative resistances offered by the portal veins and valve. Recent experimental evidence has resulted in the development of the portal compensation hypothesis, in which the parallel interconnecting veins described above are viewed as an anastomosing network, with the peritubular sinuses located in the renal cortex at its center. Therefore, the amount of portal blood flowing to the kidneys will depend on both the resistances and pressures in the parallel shunt pathways described above and resistance and pressure within the peritubular sinuses.

Blood flowing from the renal glomeruli also enters the sinuses and contributes to pressure within them (Wideman et al., 1991), so any reduction in renal arterial pressure below the autoregulatory range should promote inflow to the sinuses from the portal veins. It is known that if the portal system is intact, birds can maintain total renal blood flow in the face of a fall in renal arterial pressure to 40–50 mm Hg. Should arterial pressure fall, the glomerular vessels themselves can, by autoregulation, maintain glomerular filtration rate constant but only down to a minimum pressure of 70 mm Hg (Wideman et al., 1992). The wider autoregulatory range for total renal blood flow may be due partially to an autoregulatory buffering effect by the portal system. Experimentally reducing portal blood flow leads to a narrowing of the range of arterial pressures over which total renal blood flow is maintained constant (Wideman et al., 1992). There are regional differences in renal blood flow, with the anterior part of the kidney apparently receiving a greater contribution from the portal veins. This suggests that renal arterial flow is normally lower in the anterior kidney.

The functional significance of the portal system in the environmental physiology of birds may be related to salt loading and dehydration. Dantzler (1989) has proposed that under these conditions the adaptive response of the preglomerular arterial vessels of reptiliantype nephrons in the superficial renal cortex is to constrict, causing sustained cessation of filtration. Renal blood flow could be maintained under these conditions by a compensatory increase in portal flow, maintaining a nutritional blood supply to the cells of cortical nephrons (Wideman and Gregg, 1988).

11.4.4 Embryonic Shunts

The circulatory system of the developing avian embryo has a number of vascular shunts found only during the embryonic stages (Dzialowski et al., 2011). These vascular shunts allow systemic venous return entering the right atria to bypass the nonventilated lungs, providing a right-to-left shunt. The first major embryonic shunt in embryonic birds is through a pair of ductus arteriosi. In birds, these two embryonic vessels branch from the right and left pulmonary arteries and connect with the descending aorta (Figure 11.29). This arrangement is in contrast to the mammalian fetus that has a single ductus artertiosus connecting the pulmonary artery with the aorta in a position much closer to the heart and origin of these great vessels. During later stages of chicken embryonic development, White (1974) found that a large portion of venous return from the right anterior vena cava was shunted to the dorsal aorta through the paired ductus arteriosi. Tazawa and Takenaka (1985) estimated that the ductus shunt 84% of right ventricular output to the dorsal aorta. A large fraction of dorsal aorta blood flow goes to the umbilical arteries and the embryonic gas exchanger, the chorioallantoic membrane. Patency of the mammalian ductus during the fetal stage is maintained in part by the relaxing influence of prostaglandins. Interestingly, there is no vasodilatory response to prostaglandins in the avian ductus (Dzialowski and Greyner, 2008; Greyner and Dzialowski, 2008). This difference could be due to the fact that the maternal placenta is the source of circulating prostaglandins in the mammalian fetus.

A second shunt exists between the two atria of the heart which functions like the mammalian fetal foramen ovale. The avian right and left atria are connected by perforations or foramina in the wall separating the atria. The right atrium receives posterior venous return, which includes oxygenated blood from the embryonic gas exchanger, the chorioallantoic membrane (Figure 11.29; White, 1974). Tazawa and

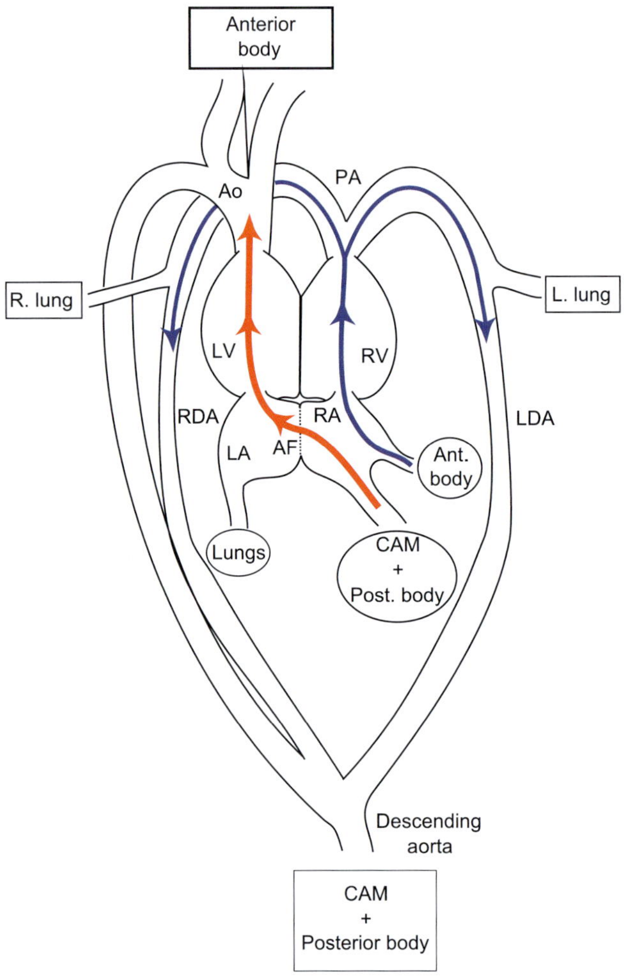

FIGURE 11.29 Embryonic circulation of a developing chicken embryo. The two right-to-left shunts are the left and right ductus arteriosi and the atrial foramina. PA, pulmonary artery; Ao, aorta; LDA, left ductus arteriosus; RDA, right ductus arteriosus; AF, atrial foramina; CAM, chorioallanotic membrane; RV, right ventricle; LV, left ventricle; LA, left atrium; RA, right atrium. *From Dzialowski et al. (2011).*

Takenaka (1985) estimate that 60% of this posterior venous return flows through the atrial foramina resulting in a right-to-left shunt. This shunt ensures oxygenated blood flows to the anterior portion of the embryo through the carotid and subclavian arteries.

Both vascular shunts discussed above must close upon hatching to ensure adequate oxygenation of blood in the lungs. The ductus arteriosus is an oxygen sensitive blood vessel that constricts in response to increases in oxygen (Greyner and Dzialowski, 2008; Agren et al., 2007). In the embryonic chicken, the strength of oxygen induced vasoconstriction increases during hatching. Using *in vitro* myography, the contractile strength of the ductus has been documented to triple between day 18 of incubation and external piping (Belanger et al., 2008). At the same time, arterial P_{O_2} in the hatchling increases from around 30–80 mm Hg (Tazawa et al., 2002). By the time of hatching, the chicken ductus arteriosi are functionally closed (Belanger et al., 2008). Closure of the atrial foramina appears to be caused by pressure changes in the left atrium that occur with hatching (Jaffee, 1965). The perforations appear to have in a valve-like manner, closing when the pressure in the left atrium is greater than the right atrium.

11.5 CONTROL OF THE CARDIOVASCULAR SYSTEM

11.5.1 Control Systems

Both the output of the heart and the resistance to blood flow in the vascular beds are subject to wide variations, depending on the type of activity in which animals are engaged and the intensity of that activity. CO varies in proportion to total body metabolic requirements while vascular resistance varies on a regional basis according to the blood flow requirements of different parts of the body. CO and resistance are both under the control of autoregulatory, humoral, hormonal, and neural influences. Arterial blood pressure is the product of CO and TPR and it is this pressure that produces the driving force to ensure adequate blood flow to the vascular beds. Blood pressure varies but over a proportionally narrower range than either CO or peripheral resistance. Investigations of the regulation of cardiovascular function have been driven by the broad assumption that blood pressure is maintained within sensible limits to ensure adequate tissue perfusion in the face of the variations in CO and peripheral resistance imposed by changes in the external environment or in activity levels.

The cardiovascular system is controlled by several integrated mechanisms operating over time scales ranging from less than a second to months or longer. The most rapid adjustments in CO and peripheral resistance, which may occur within the span of a few heartbeats, are reflexogenic and primarily function to maintain short-term homeostasis and to effect rapid cardiovascular responses to changes in the internal or external environments. Autoregulatory mechanisms acting within vascular beds to modify blood flow as a result of local changes in metabolites or other influences may operate on a time scale of seconds to minutes. Changes in humoral factors, such as levels of oxygen and carbon dioxide, pH, and metabolic products, can affect cardiovascular receptors or circulatory elements directly, producing changes in cardiac and vascular function directed toward correcting these disturbances on a time scale which may extend over long periods. An example of this is in birds undertaking extended migratory flights at high altitudes where inspired oxygen levels, and therefore blood oxygen levels, are much lower than at sea level. Circulating hormones can also affect both the peripheral circulation and the heart, and levels of many of these hormones in the blood may change depending on the activity state of the animal or on the time of year, as during molting or mating.

11.5.2 Control of Peripheral Blood Flow

Contraction of smooth muscle in the walls of arteries provides the means for varying vessel caliber, and the most effective location for altering blood flow by this means will be where the ratio of wall cross-sectional area to lumen area is maximal. The smallest arteries and arterioles possess the highest wall-to-lumen ratio, and it is here that the contraction of individual smooth muscle fibers, coordinated over the whole cross-section of the wall, will give the greatest change in resistance. The largest component of vascular resistance in the circulation is therefore set by the tone of the smooth muscles in the walls of these vessels. Arterial smooth muscle tone is subject to modulation by several mechanisms: (1) intraluminal pressure changes, leading to mechanical autoregulation of blood flow; (2) humoral factors including oxygen tension, levels of local metabolites, extracellular ion concentrations, locally released vasoactive agents, and circulating hormones and vasoactive agents; as well as (3) transmitters released from autonomic nerve terminals.

Smooth muscle fibers in the walls of veins are also influenced by these factors, providing mechanisms for adjusting compliance of the venous walls and thus some control of the rate of return to the heart of blood in the central venous pool. The influences of these regulatory factors on vascular function in birds are discussed in the following sections.

11.5.2.1 Mechanism of Vascular Reactivity

Regulation of smooth muscle contraction differs from skeletal muscle and involves activation of myosin light chain. Vascular smooth muscle contraction and relaxation are regulated by the phosphorylation and dephosphorylation state of the 20-kDa myosin light chain at Ser^{19} (MLC_{20}). In the phosphorylated state, smooth muscle contracts and produces vasoconstriction. Phosphorylation of MLC_{20} is mediated by Ca^{2+}-calmodulin-dependent MLC kinase (MLCK) that are activated by increased cytosolic Ca^{2+} due to Ca^{2+} release from sacroplasmic reticulum stores and extracellular Ca^{2+} entry through membrane bound Ca^{2+} channels (Ganitkevich et al., 2002; Somlyo and Somlyo, 2003; Webb, 2003). Relaxation occurs when myosin light chain phosphatase (MLCP) dephosphorylates MLC_{20}. Several G-protein coupled receptor agonists inhibit MLCP leading to an increase in MLC phosphorylation and contraction without changes in cytoplasmic Ca^{2+} concentration. The result is an increase in Ca^{2+} sensitization (Ganitkevich et al., 2002; Somlyo and Somlyo, 2003; Webb, 2003).

Inhibition of MLCP involves a number of pathways in mammals. MLCP activity is inhibited by the RhoA/Rho-kinase pathway by inactivation of MLCP through phosphorylation of the MLCP regulatory subunit, MYPT1. The second mechanism of MLCP inhibition in mammals is through phosphorylation of the smooth muscle-specific MLCP inhibitor protein CPI-17 (PKC potentiated inhibitor protein-17 kDa; Webb, 2003). Under physiological situations, Ca^{2+} release by the SR, Ca^{2+} influx, and Ca^{2+} sensitization act together to regulate vascular smooth muscle contraction. Their role in regulating avian smooth muscle contraction has only begun to be examined. The protein CPI-17 may not be involved in avian smooth muscle contraction because it was not detected in chicken smooth muscles from aorta, mesenteric artery, gizzard or small intestine (Kitazawa et al., 2004). Additionally, adult chicken arteries contract in response to PKC activation (Kitazawa et al., 2004). In contrast, the Rho kinase inhibitors Y-27632 and hydroxyfasudil produced marked impairment of vessel contractions in femoral artery and ductus arteriosus of chicken embryos and emu (Zoer et al., 2010; Greyner and Dzialowski, 2008; Dzialowski and Greyner, 2008). The role of these pathways in avian vascular smooth muscle contraction needs further study.

11.5.2.2 Autoregulation

Vascular tone within a region of the circulation can be defined as the average level of contraction of smooth muscle fibers in the blood vessel walls within that region. At a steady intraluminal pressure and in the absence of extrinsic influences, spontaneous contractions of individual smooth muscles occur at an intrinsic rate. In resistance vessels, the smooth muscle cells are arranged at a right angle to, or on a shallow helix around, the axis of flow, and the time-averaged tension generated by their contraction, in balance with the intraluminal pressure, will set vessel caliber and thus blood flow. An increase in arterial pressure will distend the vessel wall, increasing the caliber of the vessel and therefore reducing resistance to flow. As a result, blood flow through the vessel will increase. This wall distension also increases the frequency of contraction of the smooth muscles, and this increased vasomotion will then act to reduce the caliber of the vessel, increasing its resistance and restoring blood flow toward its original level. Conversely, a reduction in pressure will reduce wall tension, resulting in a decrease in the rate of smooth muscle spontaneous activity. This leads to vasodilation and an increase in blood flow to offset the effects of reduced perfusion pressure. These myogenic changes in vascular caliber thus provide a mechanism to autoregulate blood flow around a preferred level in the face of variations in tissue perfusion pressure. In most vascular beds, autoregulatory mechanisms are probably limited to the modulation of local blood flow to ensure even blood distribution, as, for example, in the kidney where this mechanism is important in maintaining glomerular flow rate in the face of alterations in arterial blood pressure. Throughout the body, local pressure-induced autoregulation will interact with locally released vasoactive agents and with neurogenically mediated vasomotion resulting from activation of autonomic reflexes to provide balanced adjustments in regional peripheral blood flow.

11.5.2.3 Humoral Factors

Three broad classes of humoral factors affect blood flow in the peripheral vasculature. One class includes chemical factors such as P_{O_2}, P_{CO_2}, lactic acid and other metabolic byproducts, electrolyte concentrations, and pH, which act directly on myocytes. The second class of factors consists of vasoactive agents typically released from local vascular endothelial cells. This group includes nitric oxide, H_2S, prostaglandins, and endothelin; these factors act on smooth muscle cells by receptor-mediated mechanisms. The third class includes circulating vasoactive agents, also acting via receptor-coupled mechanisms to modify smooth muscle contraction.

11.5.2.3.1 Chemical Factors

If the metabolic rate of a tissue increases, as, for example, in skeletal muscle during exercise, regional blood flow will increase due partly to local vasodilation of resistance vessels and precapillary sphincters induced by an increase in the concentration of lactic acid and CO_2 and a fall in pH. Vasodilation under these circumstances is produced by alterations in Ca^{2+} flux and handling at the level of Ca^{2+} channels (Wray and Smith, 2004) and secondarily by direct chemical effects on the contractile apparatus of the vascular myocytes (Mellander and Johansson, 1968). Vasodilation may be further enhanced if local P_{O_2} falls or concentrations of extracellular ions such as K^+ increase. The resulting increase in blood flow is called functional hyperemia and serves to accelerate oxygen delivery to the muscle and to increase the clearance rate of tissue metabolites. During exercise, central arterial pressure may also rise (Butler, 1991; Saunders and Fedde, 1994) and local myogenic autoregulation would, as outlined in Section 11.5.2.2, attempt to limit the rise in blood flow; however in tissue operating at a high metabolic rate, this mechanism is to a large extent overridden by local chemical vasodilatory influences. Vasodilation due to tissue hypoxia and buildup of metabolites also occurs during periods of ischemia, for example, when blood flow to a vascular bed is occluded as shown in Figure 11.30. Upon release of the occlusion, blood flow rises transiently to several times the preocclusion rate, with the increase in flow proportional to the duration of the occlusion. This reactive hyperemia acts to restore tissue oxygen levels and to remove metabolic products accumulating during the period of occlusion. Local hypoxia also reduces the vasoconstrictor effects of norepinephrine applied exogenously or released from sympathetic nerve terminals in avian arteries *in vitro* (Gooden, 1980; see Section 11.5.2.3.2), and might therefore be expected to reduce the *in vivo* effectiveness of vasoconstriction mediated neurogenically or by circulating catecholamines in support of the hyperemic response. Local hypercapnia also produces vasodilation and, in the hind limb vascular bed of the duck, has an even stronger effect than hypoxia in inhibiting neurogenic or catecholamine-induced vasoconstriction (Lacombe and Jones, 1990).

11.5.2.3.2 Locally Released Vasoactive Agents

The endothelium-derived relaxing factor nitric oxide (NO) is a small, rapidly diffusible molecule released by enzymatic cleavage of L-arginine from vascular endothelial cells of mammals (see Umans and Levi, 1995 for review) and birds (Hasegawa et al., 1993). In both of these

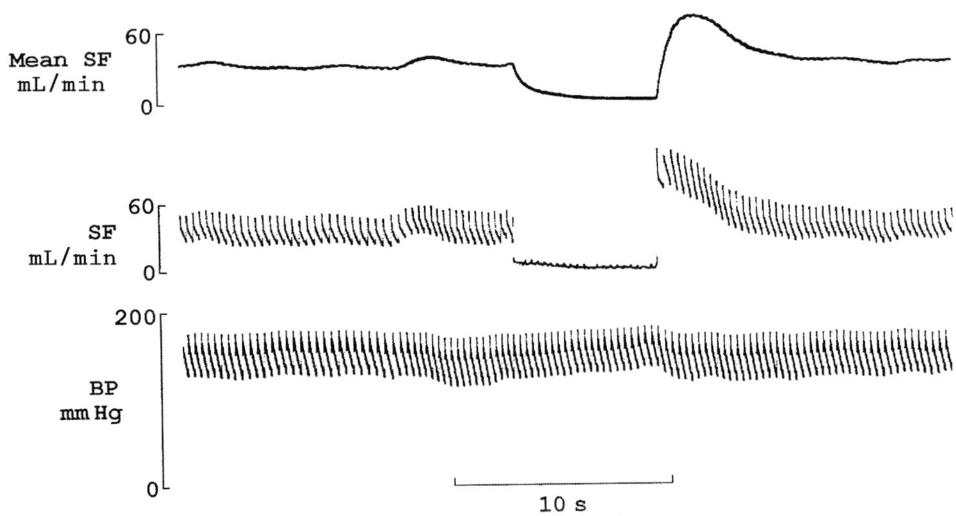

FIGURE 11.30 Change in pulsatile (SF) and mean (mean SF) blood flow in the sciatic artery of a duck in response to arterial occlusion. The period of occlusion is marked by zero flow. BP, arterial blood pressure. Note the hyperemia after the period of occlusion and the subsequent rapid return to the preocclusion flow rate with minimal change in pressure. *From Jones and Johansen (1972).*

vertebrate classes this molecule exerts a powerful vasodilatory effect by relaxing precontracted vascular smooth muscle. The release of NO as a result of acetylcholine (ACh) stimulation of endothelial cells thus provides part of the explanation for the vasodilatory effects of ACh in the circulation (Furchgott and Zawadzki, 1980). In the *in vitro* aorta of the fowl, ACh acting at muscarinic receptors on endothelial cells provokes release of NO which then produces local vasodilation (Hasegawa and Nishimura, 1991; Jarrett et al., 2013). Nitric oxide can produce relaxation through activation of soluble guanylate cyclase which increases levels of cGMP. This in turn activates a cGMP-dependent protein kinase which decreases intracellular Ca^{2+}.

Another factor provoking release of NO from avian vascular endothelial cells is angiotensin II (AII); the vasodilation produced by NO release has been proposed to cause the transient depressor effects on arterial blood pressure observed in some birds immediately after systemic injection of AII (Stallone et al., 1990; Hasegawa et al., 1993; Takei and Hasegawa, 1990). Mechanical stimuli such as flow-induced shear stress (see Section 11.5.2) may also elicit release of NO from endothelial cells in the vasculature of birds. This effect has been reported for mammalian vasculature (Umans and Levi, 1995); given that blood flow rates in birds are generally higher than those in mammals of comparable body size, such a mechanism could provide additional local adjustment of the degree of vasodilation within a vascular bed to match the immediate flow requirements of that bed.

Other locally released vasoactive agents, such as endothelin and H_2S, are important in adjusting regional blood flow in birds. Endothelin is the most potent vasoconstrictor known in mammals (Inagami et al., 1995). In chickens, endothelin has been shown to produce contractions of femoral, mesenteric, and pulmonary arteries in the embryonic stages (Moonen and Villamor, 2011; Villamor et al., 2002) and pulmonary arteries from adults (Martinez-Lemus et al., 2003). In hatchling chickens, endothelin produced contraction at low doses and relaxation at higher doses (Moonen and Villamor, 2011). Moonen and Villamor (2011) suggest that contraction may be mediated by ET_A receptors and at higher doses ET_B receptors. Receptors for this peptide have been demonstrated on avian cardiac myocytes and their activation causes an increase in contractile force (Kohmoto et al., 1993; Hassanpour et al., 2010). Recently, locally released H_2S has been found to be potentially important in regulating vessel tone. It produces both vasorelaxation and vasoconstriction in systemic vessels. Vasorelaxation by H_2S is proposed to act through K_{ATP} channels and cGMP levels in a fashion similar to NO (biological roles are reviewed in Kolluru et al., 2013). H_2S mediated vasoconstriction in the mammalian aorta is mediated by downregulation of cAMP (Lim et al., 2008). In the one study looking at the Pekin duck aorta, H_2S constricted the unstimulated aorta and produced a weak relaxation followed by constriction in the preconstricted vessel (Dombkowski et al., 2005). As with many of these pathways, further research is needed to determine the potential role of H_2S in regulating vessel tone in the avian system.

11.5.2.3.3 Circulating Agents

Circulating catecholamines have powerful effects on all elements of the circulation in birds. The catecholamines epinephrine (EPI) and norepinephrine (NE) are released into the circulation from adrenal chromaffin cells and have direct effects on vascular smooth muscle. Significant amounts of NE are also released into the circulation by activation of the sympathetic nervous system (discussed in Section 11.5.2.4 below). A number of peptides which have vasoactive effects on mammalian vasculature are also present in avian plasma, but the specific actions of most of these peptides on the avian vasculature have not been investigated in detail. Of these peptides, the most extensively studied in birds with respect to vasomotion are AII and avian antidiuretic hormone.

Circulating NE levels in conscious ducks and fowl at rest are in the range of 3–5 nM (Lacombe and Jones, 1990; Kamimura et al., 1995), whereas the resting plasma level of EPI is about half the value for NE (Lacombe and Jones, 1990), as illustrated in Figure 11.31(A). EPI appears to be released solely from the adrenal glands since removal of these glands eliminates EPI from the plasma (Lacombe and Jones, 1990). The loss of adrenal glands does not, however, markedly affect the resting level of circulating NE, which must therefore be due to spill over into the plasma of NE released from sympathetic nerve terminals by autonomic efferent activity. Circulating levels of both EPI and NE vary under different physiological conditions. For example, plasma catecholamines can increase by factors ranging from 2 to >1000 in ducks during involuntary submersion, with end dive levels being proportional to dive length as shown in Figure 11.31(B) (Huang et al., 1974; Hudson and Jones, 1982; Lacombe and Jones, 1990).

Both NE and EPI produce vasomotion in avian vascular smooth muscle, acting via α- and β-adrenergic receptors (Bolton and Bowman, 1969). NE, injected intravenously into conscious ducks, produces vasoconstriction throughout the body. This vasoconstriction causes an increase in blood pressure produced by the collective effect of increases in resistance to flow in individual vascular beds in the body, illustrated in Figure 11.32, by increased resistance in the hind limb vascular bed. These vascular responses to NE are primarily mediated by α-adrenergic receptors (Butler et al., 1986; Wilson and West, 1986; Bolton and Bowman, 1969). Activation of α-adrenoceptors on vascular smooth

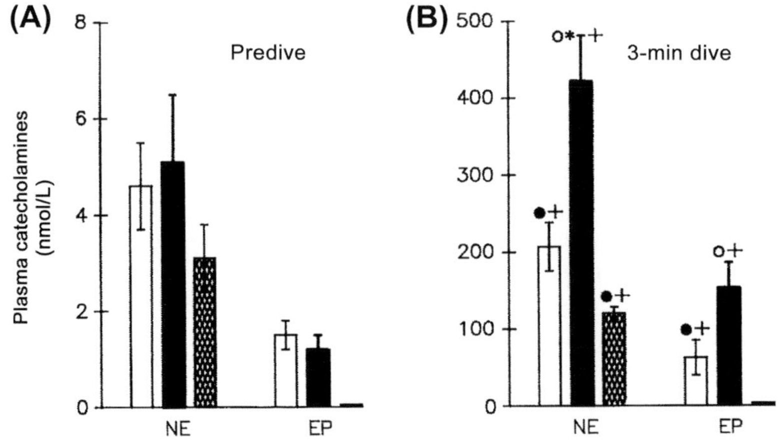

FIGURE 11.31 Plasma levels of norepinephrine (NE) and epinephrine (EP) in intact (open bars), sham-operated (filled bars), and adrenalectomized (cross-hatched bars) ducks (A) before and (B) at the 3 min point during forced submergence. The open circles indicate significant differences from intact ducks; closed circles, differences from sham-operated animals; asterisks, differences from adrenalectomized animals; plus signs, differences from predive value. Adrenalectomy eliminated EP but not NE from the plasma. Note ordinate scale change in B. *From Lacombe and Jones (1990).*

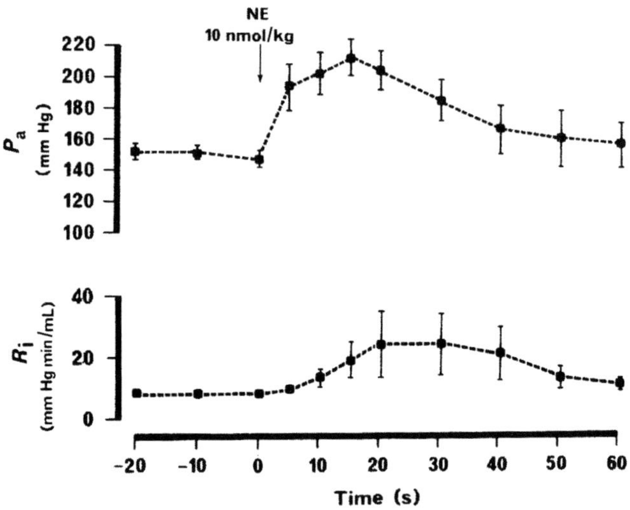

FIGURE 11.32 Responses of mean arterial blood pressure (P_a) and resistance to blood flow in the ischiatic artery (R_i) of an adult duck to a bolus intravenous injection of norepinephrine (NE, at arrow). The R_i values for each data point were calculated from the corresponding P_a and ischiatic blood flow values. *Modified from Wilson and West (1986).*

muscle acts through second-messenger systems to mobilize internal Ca^{2+} stores and to open membrane calcium channels, thus increasing intracellular Ca^{2+} concentration and activating the actin–myosin contractile apparatus (see Webb, 2003; Section 11.5.2.1). Norepinephrine can also have a vasodilatory effect on the vasculature of birds, mediated through β-adrenergic receptors; however, this effect is only apparent systemically after pharmacological blockade of α-adrenoceptors (Butler et al., 1986). β-adrenergic vasodilation in the avian vasculature also works by activating intracellular second messengers, converting the actin–myosin complex to an inactive form to promote relaxation. Combined α- and β-adrenergic blockade appears to eliminate all direct effects of NE on bird vascular smooth muscle. The overall effects of NE on peripheral resistance therefore depend on the relative abundance of α- and β-receptors in individual vascular beds. Epinephrine also acts on adrenoceptors of vascular smooth muscle but binds to α-receptors with a higher affinity than to the β subtype, so exerts a stronger vasoconstrictive effect for the same receptor density than does NE.

Avian AII is similar to the corresponding mammalian peptide in its structure and in the biochemical pathway of its production (see reviews by Wilson, 1989; Henderson and Deacon, 1993). Renin, released from the juxtaglomerular cells lining glomerular afferent arterioles in the kidney, produces the peptide angiotensin I by hydrolysis of angiotensinogen, a plasma α-globulin. Angiotensin-converting enzyme, which has been identified in circulating plasma and fixed in the walls of blood vessels in birds (Henderson and Deacon, 1993), then cleaves angiotensin I to produce AII. A number of stimuli such as systemic hypotension, hypovolemia, decreased plasma or distal tubule ion concentrations (particularly Na^+), or the activation of juxtaglomerular β-receptors causes renin to be secreted into the plasma. This promotes an increase in circulating angiotensin I, making it available for conversion to AII. AII affects circulatory function at several levels, evoking responses in both the central nervous system and the peripheral vasculature. These responses are aimed at the conservation of water and electrolytes in order to counter the original renin secreting stimulus.

Within the central nervous system, AII acts on receptors of some hypothalamic neurons to promote drinking behavior (Evered and Fitzsimons, 1981). In the periphery, exogenous or endogenous AII produces either an increase in systemic arterial blood pressure or a biphasic hypotensive–hypertensive response, depending on species. In ducks (Wilson and West, 1986; Butler et al., 1986) and pigeons

(Evered and Fitzsimons, 1981), systemic injections of AII produced dose-dependent increases in arterial blood pressure. Butler et al. (1986) and Wilson and West (1986) proposed that this response was due to vasoconstriction resulting from AII mediated release of NE from sympathetic nerve terminals and enhanced EPI and NE release from the adrenal glands; in their experiments, α- and β-adrenergic blockade eliminated the pressor effects of AII injection. Indeed, Moore et al. (1981) maintained that AII has no direct vasoconstrictor effects on arterial smooth muscle in the fowl. Wilson (1989), in a review of the renin–angiotensin system in birds, ascribed AII-induced vasoconstriction entirely to the effects of elevated catecholamine secretion.

In fowl and quail systemic circuits, AII injections produce a rapid, transient hypotension followed by a prolonged rise in arterial blood pressure (Nakamura et al., 1982; Takei and Hasegawa, 1990). The hypertensive phase of this response is mediated by adrenergic mechanisms, as in the duck and pigeon, but the transient hypotensive phase appears to be an indirect AII effect on vascular endothelial cells, working via the local release of NO from these cells, as described in Section 11.5.2.3.2. *In vitro* isometric tension studies on the abdominal aorta of embryonic and adult chicken have shown AII induces a vasorelaxation that is inhibited by L-NAME and eliminated by removal of endothelium (Nishimura et al., 2003).

Arginine vasotocin (AVT), released from the posterior pituitary into the circulation under conditions of osmotic or hypovolemic challenge, is the avian homolog of the mammalian antidiuretic hormone arginine vasopressin. However, the effects on avian vasculature of increased endogenous AVT levels after salt loading or hemorrhage are not well understood, nor are the vasomotor effects of systemic AVT injections. In mammals, arginine vasopressin produces vasoconstriction in systemic arterioles and in glomerular afferent arterioles, both serving to facilitate antidiuresis. In birds, AVT also induces vasoconstriction of afferent glomerular arterioles (see Braun, 1982 for review), but its effects on the rest of the circulation are controversial. Several studies have reported no cardiovascular consequences of AVT injection, maintaining that the avian vasculature is not sensitive to this peptide even at doses many times greater than "physiological" levels (Simon-Oppermann et al., 1988; Robinzon et al., 1988). In contrast, Wilson and West (1986) in ducks and chickens and Brummermann and Simon (1990) in ducks found that systemic injections of AVT produced hypotension accompanied by tachycardia. The latter authors proposed that AVT directly relaxes vascular smooth muscle, producing a fall in arterial blood pressure which then evokes a baroreflex-mediated tachycardia. However, Robinzon et al. (1993) found in fowl that the direction of AVT-mediated vascular responses depended on the dose and method of application. Low doses given slowly by intravenous infusion produced hypertension similar to the mammalian response to arginine vasopressin, whereas bolus intravenous doses produced the hypotensive responses reported in other avian studies. Robinzon et al. (1993) therefore proposed that AVT acts primarily via modulation of vascular caliber but that the degree and direction of vasomotion was not uniform in beds throughout the body. Although there appears to be consensus that AVT does have direct vascular effects in birds, the understanding of its specific actions must await further studies on isolated vascular beds *in situ* and on blood vessels *in vitro*.

11.5.2.4 Neural Control

All parts of the systemic and pulmonary vascular trees, except capillary beds, are innervated by the autonomic nervous system. This innervation constitutes the final common pathway for rapid and flexible control by the central nervous system of regional distribution of CO. Autonomic outflow to the vasculature is governed by a variety of reflexogenic inputs from visceral or somatic receptors relayed through brainstem and spinal cord pathways, and is also subject to influences originating at suprabulbar levels of the central nervous system. However, the degree of vasomotion produced in any region of the vasculature for a given intensity of autonomic drive will depend on the densities of effector terminals and postjunctional receptors in that region.

The location of autonomic terminals within the vascular wall is different in arteries and veins. Nerve fibers and terminals in the walls of arteries are, with some exceptions, limited to the tunica adventitia, extending as far as the outer elastic lamina marking the border between the adventitia and the tunica media (Bennett and Malmfors, 1970). In this respect, the innervation pattern of avian arteries is similar to that in mammals (see Hirst and Edwards, 1989 for a review of mammalian arterial innervation). In contrast to the arterial innervation pattern, nerve fibers and terminals in avian systemic veins are commonly apposed to smooth muscle in the tunica media, as well as being located in the adventitia (Bennett and Malmfors, 1970), and in this regard also birds are similar to mammals (see Shepherd and Vanhoutte, 1975 for a review of the innervation of mammalian veins).

In addition to intramural nerve fibers and terminals, large and small nerves course over the outer surfaces of both arteries and veins, and some vessels are completely surrounded by plexi of nerve fibers. Furthermore, throughout the body nerves generally accompany blood vessels, running parallel with the vessels to form neurovascular bundles. These close associations between blood vessels and nerves have an important consequence for experimental investigations of neurogenic vasomotion. In assessing the effectiveness of neural control of vascular resistance in particular beds, electrical stimulation of autonomic nerves is commonly employed. However, some neural pathways to these beds may be interrupted if supply vessels to the beds are manipulated or sectioned (for example, to insert a blood flow probe) between the stimulus site and the expected site of vasomotion.

11.5.2.4.1 Systemic Arterial Innervation

The aorta from its origin to the junction with the celiac artery, the proximal brachiocephalic trunks, and the most proximal parts of the common carotid arteries are elastic vessels (see Section 11.4.1.2), having relatively little smooth muscle and thus a correspondingly sparse vasomotor innervation (Bennett and Malmfors, 1970; Bennett, 1971). The aorta, especially close to its root and adjacent to the root of the pulmonary trunk, has numerous small cells in the adventitia that contain catecholamines (Bennett, 1971). These cells occur singly or in clusters, appear to be similar to amine-containing cells of the carotid body, and are innervated by branches of the vagus nerve. Although these cells may be a source of locally released catecholamines, it has also been suggested that these cells may constitute chemoreceptive aortic bodies (Bennett, 1971; Tcheng and Fu, 1962) similar to those found at the homologous site in mammals. In addition, the adventitia of the aortic arch is innervated by afferent vagal fibers with nerve terminals transducing wall stretch and thus signaling an index of central arterial blood pressure; these constitute the only systemic arterial baroreceptors in birds (Jones, 1973).

The transition from elastic to muscular wall structure occurs in the arterial tree distal to the branching points of major distribution arteries from the great vessels. This transition also marks an increase in the density of innervation of the arterial wall, as shown in the photograph of a branch of the posterior mesenteric artery of the fowl in Figure 11.33.

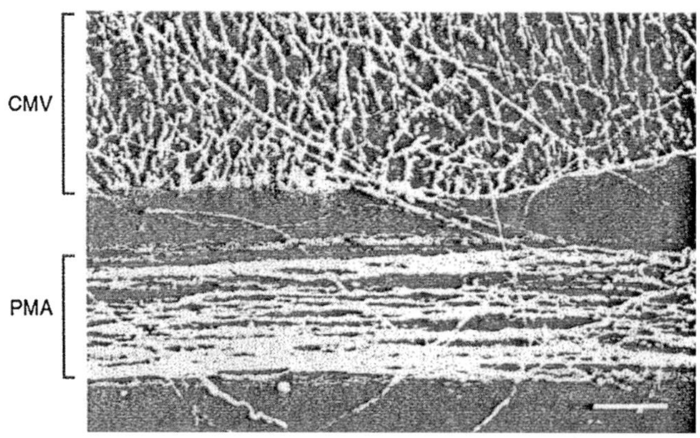

FIGURE 11.33 Adrenergic innervation patterns of a distal branch of the posterior mesenteric artery (PMA) and the coccygeomesenteric vein (CMV) of the chicken. The vessels run side by side with their long axes aligned from left to right in the photograph. Aldehyde fluorescence histochemistry shows that adrenergic nerve fibers and terminals are present in both vessels; in the artery, they run parallel to the vessel axis and are more dense than in the vein. The horizontal bar represents 100 μm. *Modified from Bennett and Malmfors (1970).*

In some arteries such as the ischiatic, varicose fibers can be seen in the media as well as at the medial–adventitial border (Bennett and Malmfors, 1970). Of the large muscular arteries, the common carotids appear to be the most heavily innervated (Bennett and Malmfors, 1970). Given that these arteries convey the majority of blood flow to the avian brain, this density of innervation possibly reflects a requirement for greater autonomic control of cephalic blood flow than in other beds. Vasomotor innervation of the smaller branches of arteries within individual vascular beds supplied by the large arteries has not been systematically described but in all beds muscular arterioles, which constitute the resistance vessels responsible for 70–80% of TPR, are densely innervated (Folkow et al., 1966; Bennett and Malmfors, 1970).

The density of innervation of different regions of the arterial tree is variable among bird species; this variation has important consequences for differences in regional neurogenic control of blood flow among species. Large arteries supplying hind limb muscles in ducks, for instance, are more densely innervated than those in turkeys, and this difference is correlated functionally with an enhanced capability in the duck to generate and maintain neurogenically mediated intense peripheral vasoconstriction (Folkow et al., 1966). These authors proposed that this was a general physiological adaptation in diving birds, enabling the redistribution of blood flow away from those peripheral vascular beds able to withstand periods of ischemia and toward the central circulation, thus conserving blood oxygen for ischemia-sensitive heart and brain tissue.

Electrical stimulation of the sympathetic innervation of a vascular bed evokes increases in vascular resistance in that bed, mimicking the effect of elevated vasoconstrictor outflow from the central nervous system. As illustrated in Figure 11.34, graded increases in stimulation frequency evoke proportionally larger increases in peripheral resistance in the hind limb vascular bed of the duck, producing

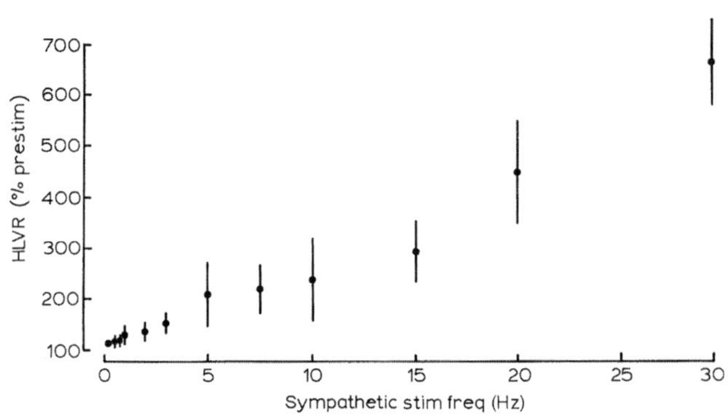

FIGURE 11.34 Relationship between frequency of electrical stimulation of the sympathetic innervation of the hind limb and vascular resistance of this bed (HLVR) in a duck. HLVR for each data point was calculated from the corresponding arterial pressure and ischiatic artery flow. Increases in HLVR are expressed in percentages relative to the prestimulus value. *From Smith (unpublished data).*

a maximal increase of up to 7 times the prestimulus resistance value at a stimulation frequency of 30 Hz. In contrast, stimulus frequencies of less than 10 Hz are required to produce a maximal increase in resistance in the cat hind limb (Folkow, 1952). The hind limb vascular bed in the duck thus requires a greater degree of sympathetic drive than that of the cat to achieve the same order of increase in resistance to flow. This may reflect differences in the distribution of sympathetic terminals to the arteries in this bed in the two species. In support of this, Bennett and Malmfors (1970) noted that the pattern of adrenergic innervation of small intramuscular arteries in birds was not markedly different from that in mammals, but observed a higher terminal density in larger arteries of avians relative to mammals.

Some regions of the avian arterial vasculature have specialized wall structures and unusual innervation patterns, possibly reflecting enhanced capabilities for regional control of blood flow. The anterior mesenteric artery in several species of birds has, in addition to the normal circular smooth muscle in the media, an outer layer of longitudinal smooth muscle in the adventitia (Bolton, 1969; Bennett and Malmfors, 1970; Bell, 1969; Ball et al., 1963; see Section 11.4.1.2). Adrenergic fibers and varicose terminals are found at the adventitial–medial border in this as in other arteries, but these elements also extend into the longitudinal muscle itself, aligned in the direction of the muscle fibers. Longitudinal muscle of this vessel also receives cholinergic innervation (Bolton, 1967; Bell, 1969; Bennett and Malmfors, 1970), although this does not extend to the inner circular muscle (Bell, 1969). Adrenergic nerve stimulation produces contraction of the circular muscle which is mimicked and blocked by α-adrenergic agonists and antagonists, respectively (Bolton, 1969; Bell, 1969; Gooden, 1980). The longitudinal muscle is, however, relaxed by adrenergic nerve stimulation, acting through β-adrenergic receptors (Bolton, 1969; Bell, 1969). Stimulation of cholinergic nerve fibers in an *in vitro* preparation of a segment of the vessel caused the longitudinal muscle to contract, shortening the whole segment. This did not, however, markedly affect resistance to flow through the vessel but in the shortened state adrenergic vasoconstrictor responses of the circular muscle were found to be exaggerated (Bell, 1969). The cholinergic innervation of longitudinal muscle has thus been proposed to potentiate adrenergically mediated control of blood flow, possibly as an adaptation for rapid adjustment of intestinal blood flow during stress (Bell, 1969). The fact that increased sympathetic drive to this vessel has a relaxing effect on the longitudinal muscle as well as a constricting effect on the circular muscle would facilitate the effects of cholinergic input in shortening the vessel segment. Cholinergic input may also help adjust blood flow more precisely when the length of the anterior mesenteric artery changes during gross intestinal movements associated with digestion.

Coronary arteries also possess an outer coat of longitudinal muscle, but adrenergic innervation of this muscle layer is very sparse. In these arteries, the majority of adrenergic nerve fibers and terminals are found at the adventitiomedial border, as in other arteries (Bennett and Malmfors, 1970). There are no reports of cholinergic innervation of the coronary arteries of birds. Cerebral arteries are, however, dually innervated by adrenergic and cholinergic fibers (Tagawa et al., 1979). Adrenergic varicosities are located in the adventitia near the border with the media throughout the cerebral circulation, in common with most other arteries of the body. Cholinergic fibers as well are located in the tunica adventitia but occur less frequently than adrenergic terminals (Tagawa et al., 1979). It is presumed that, since cerebral arteries have no smooth muscle in the adventitia, neurotransmitters released from both types of terminals diffuse into the media to act on muscle fibers there. Tagawa et al. (1979) also reported that, as in mammals, some of the cholinergic innervation of avian cerebral arteries appears to originate from central neurons, especially in the diencephalon, as well as from peripheral neurons. Studies in mammals have shown that the functional significance of reflexogenic vasomotion in overall control of cerebral blood flow is relatively minor compared with neurogenic control of blood flow in other vascular beds in the body; intracerebral blood distribution is influenced mainly by local and possibly circulating humoral factors (Kontos, 1981). This is likely to be true also in birds; Stephenson et al. (1994) estimated that neurogenic contributions to changes in cerebral blood flow in ducks during diving were minor.

11.5.2.4.2 Systemic Venous Innervation

Large veins in birds are more densely innervated than those in mammals. The density of noradrenergic innervation of the chicken caudal vena cava is graded, increasing in portions of the vessel more distal to the heart (Bennett, 1974; Bennett et al., 1974). The caudal vena cava has an outer coat of longitudinal muscle as well as an inner circular layer and both layers receive adrenergic innervation, with the orientation of the terminals and fibers following the direction of the muscle fibers in each layer (Bennett, 1974). There is also a sparse cholinergic innervation of this vessel. The superior venae cavae also have two muscle layers, with a similar adrenergic innervation pattern to the caudal vena cava (Bennett and Malmfors, 1970). The walls of the pectoral, subclavian, celiac, and jugular veins contain both circular and longitudinal smooth muscle, and the innervation of these vessels is through a plexus of varicose fibers and terminals located primarily between the inner and outer muscle layers. Other veins, such as the coccygeomesenteric vein, have primarily circular muscle, innervated in the pattern illustrated in Figure 11.33.

Functionally, adrenergic vasomotion appears to be pre-eminent in veins. Vasomotion of the caudal vena cava and other major veins is mediated primarily by α-adrenergic receptors producing vasoconstriction when activated; no functional β-adrenergic receptors are present (Bennett and Malmfors, 1974). Responses of venous smooth muscle to cholinergic nerve stimulation are weak and variable and probably do not contribute directly to neurogenic control of wall compliance; however, ACh released from cholinergic nerve terminals can modulate the release of NE from local adrenergic nerve terminals and thus may provide fine adjustments of adrenergically generated venomotor tone.

Sympathetically mediated contraction of smooth muscle in the major veins serves to decrease wall compliance and vessel diameter, thus providing a reflexogenic mechanism for reducing the volume of the central venous pool and increasing return of blood to the heart, as discussed in Section 11.4.3.2. Langille (1983) proposed that reflexogenic venoconstriction was responsible for the increased central venous pressure observed in ducks during involuntary submersion. This response may function to aid venous return to the heart to help maintain stroke volume in the face of a reduction in cardiac contractility, which may develop during diving (Djojosugito et al., 1969; Langille, 1983).

The renal portal valves, located bilaterally where the external iliac veins join the junction of the caudal vena cava with the caudal renal veins, are unique to birds. These valves (Figure 11.26) are in the form of sphincters of smooth muscle which can close off the direct route for blood flow from the external iliac veins to the caudal vena cava. Closure increases renal portal blood flow, venous blood from the external iliac vein being then partially redirected into the renal capillaries and thence to the caudal vena cava, as discussed in Section 11.4.3.4. The renal portal valve is heavily innervated by both adrenergic and cholinergic nerve fibers (Akester and Mann, 1969; Bennett and Malmfors, 1970) with reciprocal effects on the smooth muscle of the valve. Adrenergic nerve stimulation and sympathomimetic agents induce relaxation, mediated by β-adrenergic receptors, while cholinergic nerve stimulation and cholinomimetic agents produce contraction via muscarinic receptors (Bennett and Malmfors, 1975a; Sturkie et al., 1978). The importance of this valve in the control of venous blood distribution in the renal portal system has been the subject of some debate (Akester, 1967; Section 11.4.3.4). Functional studies of the avian kidney *in situ* by Glahn et al. (1993) have shown that the status of the renal portal valve affects total renal blood flow by altering the amount of flow in the renal portal system. In conditions of lowered arterial blood pressure in which renal arterial perfusion is below the range of autoregulation of glomerular blood flow, closure of the renal portal valve raises renal portal blood flow to compensate total renal blood flow (Glahn et al., 1993). Neural control of the renal portal valve may thus form a component of the suite of reflexogenic responses to hypotension or hypovolemia.

11.5.2.4.3 Pulmonary Vessel Innervation

Adrenergic innervation of the pulmonary artery in the fowl consists mostly of fibers with few varicose nerve terminals. Fibers and nerve terminals form a plexus at the adventitiomedial border of the artery, with some projections into the circular smooth muscle of the media. There is also smooth muscle oriented longitudinally in the adventitia, but innervation of this is very sparse (Bennett and Malmfors, 1970; Bennett, 1971). Some intrapulmonary arterial branches have dense adrenergic plexi occurring in short segments along their length (Bennett and Malmfors, 1970), which may help to redistribute blood flow to selected gas exchange areas within the lung to optimize local ventilation–perfusion ratios (Hebb, 1969).

The pulmonary veins proximal to the left atrium are very densely innervated with adrenergic nerve fibers and terminals, the density of this innervation being markedly greater than in the pulmonary arteries (Bennett, 1971). Abundant terminal varicosities and nerve fibers are located in a plexus at the adventitiomedial border, with some penetration into the media (Bennett, 1971; Bennett and Malmfors, 1970). There is also longitudinal smooth muscle present in the adventitia, with adrenergic terminals between the muscle fibers (Bennett, 1971; Bennett and Malmfors, 1970). The density of innervation of the longitudinal smooth muscle is reduced close to the left atrium, increasing distally along the vessel until the bifurcation to the lungs. At this junction, there is an abrupt decrease in density of innervation of the entire vessel wall, and within the lungs the major branches of the pulmonary veins are very sparsely innervated. Smaller branches of these vessels appear to have no innervation (Bennett, 1971).

11.5.2.4.4 Autonomic Pathways

The cell bodies of adrenergic postganglionic vasoconstrictor neurons innervating the vasculature are located in paired paravertebral ganglion chains, in prevertebral ganglia and, in some cases, in small ganglia scattered throughout the viscera. Cells in the superior cervical ganglion (representing a fusion of the two most cranial cervical ganglia) innervate blood vessels of the head, including those of the salt and salivary glands, via cephalic extensions anastomosing with several cranial nerves (Bennett, 1974). In birds, pairs of sympathetic ganglia are associated with the cervical vertebrae and cells in these ganglia innervate blood vessels of the neck. The presence of cervical paravertebral ganglia in birds constitutes a major difference in the organization of the sympathetic nervous system between this vertebrate group and mammals. In the caudal part of the neck, in the thorax, and in the wings of birds the vasculature is innervated by neurons in the thoracic paravertebral ganglion chain. Sympathetic fibers reach wing vessels primarily through the brachial plexus. Some thoracic postganglionic neurons also contribute sympathetic fibers to the greater splanchnic nerves innervating anterior abdominal viscera via the celiac plexus (Bennett, 1971).

Vertebrae of the lumbar, sacral, and coccygeal spine are fused to form the synsacrum in birds. The paravertebral ganglion chains from each side in this region are fused in the midline at about the level of the sixth coccygeal segment in avian species so far examined, and this combined sympathetic trunk continues caudally to the pygostyle (Pick, 1970; Akester, 1979; Benzo, 1986). Axons from postganglionic neurons in this part of the sympathetic nervous system innervate abdominal and pelvic viscera via the lesser splanchnic nerves and aortic plexus, the hypogastric plexus, the pelvic plexus, and the cloacal plexus. Some sympathetic postganglionic somata are also located in prevertebral ganglia within these plexi. In addition, lumbosacral sympathetic neurons contribute vasoconstrictor axons to the hind limbs via the lumbosacral plexus (Benzo, 1986; Bennett, 1974).

Sympathetic preganglionic cell bodies that synapse on the postganglionic vasoconstrictor neurons are located in and near a bilateral column of neurons, the column of Terni, in the gray matter near the central canal of the spinal cord (see Section 11.5.3.2). All preganglionic axons innervating postganglionic neurons in the cervical sympathetic chain exit the spinal cord through ventral nerve roots of cranial thoracic segments; there are apparently no connections between the cervical spinal nerve roots and the sympathetic ganglia in the neck (Bennett, 1974; Akester, 1979; Pick, 1970). Although no detailed analysis has been done of the segmental locations of spinal preganglionic neurons projecting to postganglionic vasoconstrictor neurons innervating the thoracic, abdominal, pelvic, and limb regions, it is likely that the locations of these neurons follows the mammalian plan. That is, axons of preganglionic neurons in a particular spinal segment emerge from the cord in the ventral root of that segment to innervate postganglionic neurons in ganglia at that level or within one or two segments rostral or caudal to the exit site (see Gabella, 1976 for review).

Parasympathetic postganglionic neurons innervating the vascular smooth muscle of such organs as salt glands and salivary glands in the head of birds are located in autonomic ganglia near or embedded in the organs (Ash et al., 1969). Cell bodies of preganglionic neurons innervating parasympathetic postganglionic vasodilator neurons associated with structures in the head are located in the oculomotor, facial, glossopharyngeal, and vagal nuclei in the brainstem and course to peripheral ganglia via the respective cranial nerves associated with these nuclei (Akester, 1979). Postganglionic cholinergic nerve fibers innervating arteries and veins of the body originate from the somata of neurons located in plexuses associated with blood vessels themselves or, for vessels of the abdominal viscera, in prevertebral ganglia containing a mixture of adrenergic and cholinergic postganglionic neurons (Bennett and Malmfors, 1970; Bennett, 1974). The location of cells of origin of preganglionic axons innervating the postganglionic neurons producing vasodilation in the viscera and skeletal muscle has not been determined in detail. However, in birds the parasympathetic outflow is organized in cranial and sacral tracts as it is in mammals. It would therefore be expected that preganglionic neuronal somata responsible for vasodilation in thoracic and anterior abdominal viscera and skeletal muscle in the upper part of the body are located in medullary vagal motor nuclei, with their axons running to postganglionic neurons via the vagus nerves. Similarly, vasodilation of posterior abdominal and pelvic viscera and skeletal muscle of the lower part of the body is likely to be mediated by preganglionic neurons with their somata located at sacral levels of the spinal cord and their axons innervating postganglionic neurons via the abdominal and pelvic autonomic nerves.

11.5.3 Control of the Heart

11.5.3.1 Catecholamine Effects on the Heart

Both NE and EPI are present in circulating plasma of birds (see Section 11.5.2.3.3 for discussion), and these amines have cardiac effects which include an increase in the rate of pacemaker depolarization (DeSantis et al., 1975; Bolton and Bowman, 1969) and augmented force of myocardial contraction (DeSantis et al., 1975; Bennett and Malmfors, 1974; Bolton, 1967; Bolton and Bowman, 1969). In isolated heart preparations and in myocardial strips *in vitro*, NE has been found to be as effective as, or more potent than, equimolar EPI in producing inotropic and chronotropic augmentation. This is in marked contrast to the condition in the mammalian heart, in which EPI is the more potent stimulant (Gilman et al., 1990). In whole-animal experiments on birds, intravenously injected boluses of NE and EPI augment CO by transiently increasing both rate and force of contraction, contributing along with peripheral vasoconstriction to catecholamine-mediated hypertension (Wilson and Butler, 1983; Wilson and West, 1986; Bolton and Bowman, 1969; Butler et al., 1986). In the avian as in the mammalian heart, catecholamines appear to act primarily via β-adrenergic receptors on myocardial and pacemaker cells (Bolton and Bowman, 1969; Butler et al., 1986; Bolton, 1967).

11.5.3.2 Neural Control

The anatomical organization of the innervation of the heart in birds has been studied for more than 100 years, first by gross dissection and then with a variety of nerve-specific stains (see for example Ábrahám, 1969; Hirsch, 1970; Pick, 1970), but these techniques have not allowed the patterns of cardiac sympathetic and parasympathetic innervation to be anatomically differentiated. This is a factor of major importance in determining the mechanisms involved in neural control of the heart. It is only recently, with the advent of histochemical techniques specific for adrenergic and cholinergic neurotransmitters or enzymes in the catabolic and anabolic pathways of these transmitters, that the patterns of dual, function-specific cardiac innervation have begun to be explored. The most well-established histochemical method for determining the peripheral distribution of adrenergic nerves is that developed by Falck (1962) to render catecholamines brightly fluorescent under ultraviolet illumination in the light microscope. This sensitive technique shows cell bodies, axons, and nerve terminal varicosities, and most of the descriptions of adrenergic innervation of the heart are based on the use of this technique in whole-mounts or sections of atrial and ventricular tissue. Histochemical assays for the presence of acetylcholinesterase (AChE), based on those developed by Koelle and others (reviewed by Koelle, 1963), have been used to determine the distribution of parasympathetic innervation of the heart. There are also more recent immunohistochemical techniques available for identifying enzymes in the biochemical pathways synthesizing neurotransmitters; specifically, antibodies directed against dopamine β-hydroxylase and tyrosine hydroxylase (NE synthesis) and choline acetyltransferase (ACh synthesis) are commercially available. These techniques, however, have yet to be applied to the avian heart.

Interpretation of the results of studies using transmitter-specific histological techniques to determine patterns of autonomic innervation of the cardiovascular system rests on two major assumptions: (1) that the primary neurotransmitter released by the terminals of sympathetic postganglionic neurons is NE, and (2) that released by parasympathetic terminals is ACh. Functional studies in bird hearts largely support this assumption. However, work on peripheral autonomic anatomy and function in a wide range of vertebrates has emphasized the diversity of autonomic neurotransmitters and neuromodulators in addition to the classic transmitters utilized by this system (see Nilsson and Holmgren, 1994; Armour and Ardell, 1994; Furness and Costa, 1987), so some caution must be exercised in this area. In the following discussion of the innervation of the avian heart, results from studies using neurotransmitter-specific techniques are emphasized.

11.5.3.2.1 Sympathetic Innervation

11.5.3.2.1.1 Anatomy Postganglionic sympathetic nerve fibers arising from neuronal somata extrinsic to the heart form part of an intracardiac nerve plexus distributed throughout all four cardiac chambers. The adrenergic innervation of the proximal part of the venae cavae appears to be continuous with the intracardiac plexus associated with the right atrium (Bennett and Malmfors, 1970). Sympathetic nerve fibers form a network over the epicardial surface of the right atrium, with some fibers penetrating into the thin atrial wall to lie adjacent to bundles of myocardial cells and others passing through the wall to the subendocardium (Smith, 1971a). The overall appearance of the plexus is that of a three dimensional latticework of fibers extending from the epicardium through the wall to the subendocardium, with varying concentrations of smooth (nonvaricose) and varicose nerve fibers and nerve endings in different regions of the atrial myocardium.

Bennett and Malmfors (1970) reported that the most densely innervated region of the heart was the external wall of the right atrium; furthermore, within this area the region adjacent to the confluence of the venae cavae with the wall contained the highest density of fibers and terminals. These authors referred to this area as the "sinu-atrial node", presumably by analogy with the corresponding sharply defined SA node of the mammalian heart. There is, however, some evidence that the cells in the primary pacemaker site in the right atrium of the bird heart may have a functional organization different from that in the mammalian heart, in the area of the junction of the sinus venosus with the SA valves (Moore, 1965; see Section 11.2.3.3). These valves are present in birds but are only represented by a vestigial flap in mammals. A large number of the nerve fibers in this area of the avian heart were varicose, running among the myocardial cells and aligned with the longitudinal axes of these cells. Another area with a high density of adrenergic terminals was in the right atrial wall near the AV border, an area corresponding to the AV node (Bennett and Malmfors, 1970; Section 11.2.3.3). Many nerve varicosities were also associated with blood vessels in the right atrial wall.

Bogusch (1974), (osmium stain) reported in the fowl that the specialized cells of the pacemaker areas and conducting system in the right atrium are well innervated by nerve fibers with frequent varicosities and some bare nerve endings. In contrast, Yousuf (1965), (silver stain) found no investment of nerve fibers into the SA node region and few in the AV node in the sparrow heart. This disparity could be due to a species difference or to a relative lack of sensitivity of the silver stain; in any case, neither technique is neurotransmitter specific, nor as sensitive as the amine fluorescence technique. Full details of the sympathetic innervation of the conducting tissues of the right atrium are still not clear. A combination of immunohistochemical localization of dopamine hydroxylase or tyrosine hydroxylase in nerve terminals and standard histological processing for visualizing pacemaker cells and atrial Purkinje fibers would address this problem.

A small number of strongly fluorescent cell bodies are present in the right atrium and in the walls of some parts of the vasculature after processing for amine related fluorescence. These cells are, however, not associated with intracardiac ganglia (Bennett and Malmfors, 1970) and probably represent small, intensely fluorescent cells or so-called paraganglion cells (Eranko and Eranko, 1977) of as yet uncertain function. There are, however, some fluorescent nerve endings around the cell bodies of nonfluorescent ganglion cells (Bennett and Malmfors, 1970), suggesting some form of sympathetic modulation of the activity of intrinsic cardiac neurons.

The right AV valve has a central layer of connective tissue between two layers of cardiac muscle (Section 11.2.1.4), and nerve bundles have been observed in association with this connective tissue (Smith, 1971a). In this study, several general nerve stains as well as a cholinesterase-specific stain were used, but unfortunately no details are given regarding the specificity of staining of the valvular innervation. However, Bennett and Malmfors (1970) have shown that this valve is innervated by adrenergic nerve fibers with few terminals, arranged in a loose plexus in the leaves of the valve. These observations suggest the existence of some degree of neural control of this valve, possibly by both sympathetic and parasympathetic limbs of the autonomic nervous system.

Adrenergic innervation in the left atrium is less dense than in the right atrium, but more dense than in the ventricles. Adrenergic fibers and terminals of the left atrial cardiac plexus, while distributed in a three-dimensional pattern similar to that in the right atrium, appear to be more evenly spread throughout the left atrial wall with little variation in density. In addition, the interatrial septum also receives adrenergic innervation as an extension of the epicardial plexus (Akester, 1971; Akester et al., 1969). The distal portion of the left AV valve has a fibroelastic structure, while more proximally cardiac muscle is associated with the fibrous skeleton (Section 11.2.1.4). Smith (1971a) has described a fine network of nerve fibers, continuous with the left atrial subendocardial plexus, investing the fibroelastic portion of this valve; no mention of an innervation of the muscular portion of the valve was made in this study. The staining technique used was again not transmitter specific, so the left AV valve may be under sympathetic or parasympathetic influence, or both, or the innervation observed may be afferent. However, regarding the latter possibility, Smith (1971a) stated that no simple or specialized nerve endings typical of sensory receptors were observed in or near the left AV valve.

The avian ventricles are relatively sparsely innervated compared with the atria, but even so are more densely innervated than are the ventricles of the hearts of mammals (Smith, 1971b). Akester et al. (1969) and Akester (1971) reported adrenergic innervation of the interventricular septum, and Bennett and Malmfors (1970) observed that innervation of this region was denser than that in the rest of the ventricle walls. Although these authors did not determine the intraseptal targets of innervation, it is possible that some of these axons may innervate intraseptal Purkinje cells since bare nerve endings have been observed close to these cells (Akester, 1971). Regarding the innervation of the pulmonary and aortic valves, Smith (1971a) reported that they were sparsely innervated in comparison with the left AV valve. Fibers in the pulmonary and aortic valves comprise a plexus arranged to form a widely spaced lattice in the basal parts of the valve leaflets.

In birds, the sympathetic cardiac nerves arise from the most rostral ganglia of the thorax and the most caudal cervical ganglia, but there is some variation in detail among different accounts (see Cabot and Cohen, 1980 for summary). Macdonald and Cohen (1970) and Cabot and Cohen (1980) describe the right sympathetic cardiac nerve in the pigeon as a single trunk formed by the anastomosis of postganglionic nerves arising from the three most caudal cervical ganglia. These ganglia are associated with spinal nerves contributing to the brachial plexus; the most caudal ganglion of this group is thus associated with the last spinal segment contributing to the brachial plexus. This description closely follows that of Malinovsky (1962) in the pigeon. In the chicken, however, the origin of the right cardiac sympathetic nerve is limited to the first thoracic paravertebral ganglion (as defined by its location caudal to the head of the first rib; Pick, 1970; Tummons and Sturkie, 1969; see Figure 11.35).

There has been some controversy over the nomenclature of avian sympathetic ganglia. Birds have a sympathetic ganglion associated with each cervical spinal segment (Gabella, 1976; Pick, 1970) and this observation, as well as variations in opinion on numeration of the ribs in the bird, has made it difficult to be precise about the terminology for the most caudal ganglion contributing to the cardiac nerve.

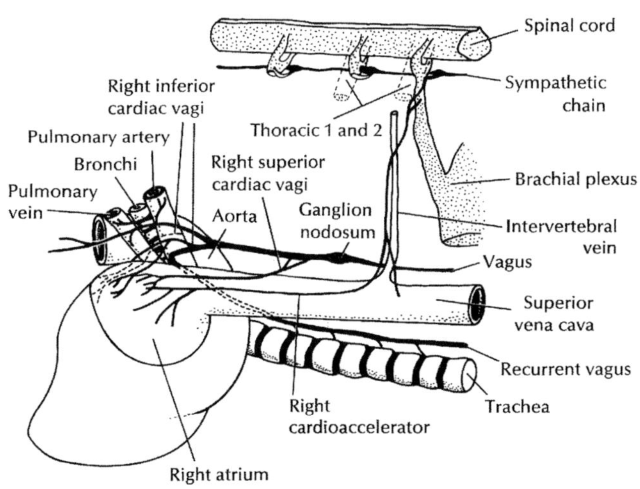

FIGURE 11.35 **Schematic representation of the sympathetic innervation of the chicken heart on the right side.** In this species, the cardiac sympathetic nerve originates from the first thoracic paravertebral ganglion. *Modified from Tummons and Sturkie (1969).*

Malinovsky (1962) holds that this is the first thoracic ganglion, because it lies between the heads of the first and second ribs. However, Macdonald and Cohen (1970) have sided with earlier authors in noting that the first rib should not be considered to mark the first thoracic segment because this rib is reduced in size and does not form part of the ribcage proper. As Cabot and Cohen (1980) have pointed out, the issue of terminology is not critical when considering functional aspects of cardiac sympathetic outflow to the bird heart, but may become important when attempting to draw conclusions about the homology of this pathway with the sympathetic innervation of the mammalian heart. Mammals have no cervical paravertebral ganglion chain, and postganglionic neurons efferent to the heart in this vertebrate class are located in the middle cervical ganglion and the stellate ganglion, with the latter being formed by the condensation of the caudal cervical and first thoracic ganglia. This arrangement has no parallel in birds.

Regardless of the number of ganglia contributing to the right cardiac sympathetic nerve, the course of this nerve to the heart appears to follow the same general pattern in all avian species so far examined. The nerve courses toward the heart in conjunction with a small vertebral vein arising from the lateral aspect of the vertebral column between the last cervical and first thoracic spinal segments. The vessel and nerve merge with the apical pleura of the lung and run together between the pleural fascia. At the ventral surface of the pleura close to the junction of the vertebral vein with the superior vena cava, the nerve turns caudally along the vena cava toward the heart, there forming two rami. The medial ramus divides further and its fascicles enter the cardiac plexus to distribute within the right atrial wall. The lateral ramus joins the right vagus nerve in the vicinity of the right pulmonary artery (Pick, 1970; Cabot and Cohen, 1980; Tummons and Sturkie, 1969; Cabot and Cohen, 1977a; Macdonald and Cohen, 1970; Malinovsky, 1962). The ganglionic origin of the left sympathetic cardiac nerve is similar to that of the right, arising in the chicken from paravertebral ganglion 14 (first thoracic) and in the pigeon by anastomosis of postganglionic branches from ganglia 12–14 (Cabot and Cohen, 1980). In the pigeon, the largest branch contributing to the left sympathetic cardiac nerve arises from ganglion 14, as on the right side. In its path to the heart, the left cardiac nerve in the pigeon divides into two or more fascicles that run in parallel for a short distance then recombine before reaching the superior vena cava. This nerve ramifies again as it runs caudally along the vena cava toward the heart, and the individual rami merge with the cardiac plexus of the left atrium.

The general locations of the cells of origin of the sympathetic postganglionic axons to the avian heart and the intraspinal locations of the cardiac sympathetic preganglionic neurons have been worked out in greatest detail in the pigeon, by Cabot and Cohen and their coworkers. These workers used a combination of degeneration, neuroanatomical tracing, and electrophysiological stimulation and recording techniques. Macdonald and Cohen (1970) undertook a series of neuronal degeneration studies in order to determine the ganglionic distribution of the somata of postganglionic neurons supplying axons to the heart. After section of the right or left cardiac sympathetic nerve in the thorax, the greatest number of degenerating neurons was found in ganglion 14, with lesser numbers in ganglia 12 and 13; no degenerating neurons were observed in ganglia rostral or caudal to this level. Postganglionic cells of origin of fibers in the cardiac sympathetic nerves were thus located bilaterally in the same sympathetic ganglia, which give rise to the postganglionic nerves constituting the cardiac nerve. In experiments in which the ganglia themselves or the interganglionic sympathetic trunks were stimulated electrically, Macdonald and Cohen (1970) obtained positive cardiac chronotropic responses of short latency from ganglia 12 to 16, occasional longer-latency responses from ganglia 17 and 18, and none from ganglia caudal to 18. Experiments in which stimulation of ganglia was combined with sections of the sympathetic trunk above and below the stimulation site confirmed these results. These data were interpreted to mean that preganglionic axons originating in the spinal cord as caudally as the 16th segment converged on postganglionic cardiac neurons in ganglia 12–14. The longer latency cardiac responses to stimulation of ganglia 17 and 18 were attributed to sympathetic activation of the adrenal glands and consequent release of catecholamines into the bloodstream; nerves arising from ganglia 17 and 18 were observed to join the splanchnic nerves, branches of which innervate the adrenal medulla.

Avian sympathetic preganglionic neurons innervating the heart are located in the same area of the spinal gray matter as those innervating the blood vessels, near the midline dorsal and lateral to the central canal, and extending rostrocaudally throughout segments 14–21. Most of the preganglionic neurons are confined to a distinct cell column in the midline, the column of Terni, a nucleus peculiar to the avian spinal cord (Huber, 1936; Cabot and Cohen, 1980). This cell column is the probable homolog of the mammalian intermediolateral cell column. Leonard and Cohen (1975), in a study of the cytoarchitecture of the pigeon spinal gray, reported that the rostral and caudal extents of this nucleus were indistinct due to small clusters of cells which extended into the regions between segments 13 and 14 and caudal to segment 21. Neurotracer studies using retrograde transport of horseradish peroxidase have shown that spinal preganglionic neurons efferent to the postganglionic cells in ganglion 14 of the pigeon are present from the caudal portion of segment 14 to the rostral part of segment 17 (Cabot and Cohen, 1977a). This finding provides strong anatomical support for the earlier conclusion of Macdonald and Cohen (1970), reached on the basis of functional and degeneration studies, that spinal preganglionic inputs to cardiac postganglionic neurons originate from segments 14 to 16, with some inputs possibly coming from segment 17.

The protocol used by Cabot and Cohen (1977a) did not label cardiac preganglionic neurons specifically. In a more recent study of the intraspinal circuitry involved in sympathetic control of the heart, the region of the spinal cord containing cardiac preganglionic motor neurons has been more precisely mapped by Cabot et al. (1991b). In this study, preganglionic neurons associated with ganglion 14 in the pigeon were labeled using fragment C of tetanus toxin (a nontoxic moiety which is retrogradely transported by axons), injected into sympathetic ganglion 14. These experiments confirmed the location of preganglionic neurons in the column of Terni and also demonstrated labeling of neurons lateral to this nucleus (Figure 11.36) in an area which, in the mammalian spinal cord, is occupied by sympathetic preganglionic neurons of the nucleus intercalatus spinalis (Petras and Cummings, 1972), a term that Cabot et al. (1991b) have applied to the corresponding area in the pigeon spinal cord. Their results show that cardiac neurons in the spinal cord are not confined to the column of Terni. No studies have yet been done to label cardiac postganglionic sympathetic neurons in the bird which innervate specific regions of the heart, but this will be a necessary further step in order to determine the intraspinal locations of groups of neurons controlling cardiac functions such as rate and contractility. A potentially useful approach to this problem would be localized injections of a retrograde neuroanatomical tracer such as fragment C of tetanus toxin or pseudorabies virus into specific regions of the heart. This tracer would then be transported retrogradely and transsynaptically so that both cardiac postganglionic neurons and the spinal preganglionic neurons afferent to them could be visualized. This approach has proven advantageous in studies of cardiac autonomic pathways in mammals (Strack et al., 1988).

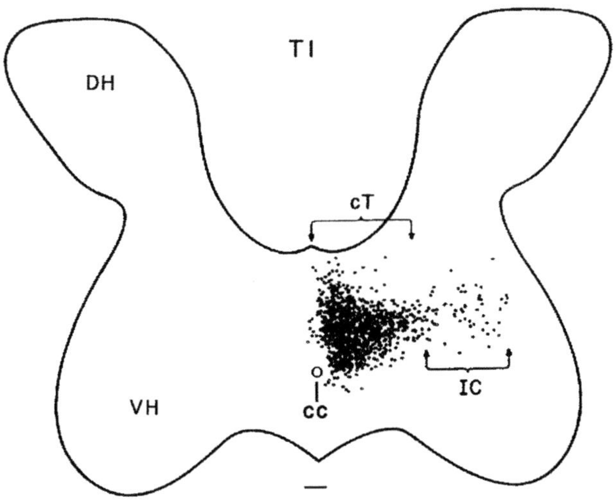

FIGURE 11.36 Mediolateral distribution of cell bodies of spinal preganglionic neurons labeled in the first thoracic segment (T1) of the spinal cord by an injection of a retrograde neurotracer (fragment C of tetanus toxin) into the right paravertebral ganglion 14 of the pigeon. The largest root of the cardiac sympathetic nerve in the pigeon arises from this ganglion. The diagram shows labeled neurons concentrated in the column of Terni (cT, large cluster of dots closest to the central canal (cc)), along with a lesser concentration of cells located more laterally in the nucleus intercalatus spinalis (IC). Although preganglionic neurons innervating the heart were not labeled specifically in this experiment, their cell bodies will be among the labeled population. Abbreviations: DH, dorsal horn; VH, ventral horn. Horizontal scale bar represents 50 μm. *Reprinted from Cabot et al. (1991b).*

The descending projections from higher centers in the central nervous system to sympathetic preganglionic neurons controlling the heart have not in general been delineated in birds. However Cabot et al. (1982), in a series of anatomical studies in the pigeon, determined that the overall pattern of projections to spinal preganglionic neurons is very similar to that found in mammals, and this anatomical data corroborates the results of earlier brain stimulation studies in the bird. In particular, Cabot et al. (1982) have shown anatomically that avian preganglionic neurons receive direct projections from diencephalic and medullary areas which, when electrically stimulated, provoke cardioaugmentation (Macdonald and Cohen, 1973; Folkow and Rubenstein, 1965; Kotilainen and Putkonen, 1974; Feigl and Folkow, 1963).

11.5.3.2.1.2 Sympathetic Control Electrical activation of intramural adrenergic nerves in the avian heart produces augmented force of contraction and cardioacceleration. Bolton and Raper (1966) and Bolton (1967) first demonstrated sympathetically mediated augmentation of force of contraction of electrically paced strips of *in vitro* left ventricle from the fowl heart. In these studies, field stimulation excited both cholinergic and adrenergic nerves, and sympathetically mediated augmentatory effects were then pharmacologically isolated by the application of atropine to eliminate parasympathetic inhibitory effects. After muscarinic blockade, the increased contractile force produced by field stimulation was attributed to activation of adrenergic nerves since this effect could then be blocked by β-adrenergic antagonists. These data provided the first evidence in birds that the sympathetic nervous system can have a positive inotropic effect directly on ventricular myocardial cells.

Similar field stimulation experiments on left and right atria *in vitro* have shown that activation of intramural sympathetic nerves has powerful effects on these chambers. In the left atrium of the fowl heart, increased force of contraction of myocytes resulted from sympathetic nerve activation. These positive inotropic effects were blocked by β-adrenergic antagonists (Koch-Weser, 1971; Bennett and Malmfors, 1974, 1975b). The right atrium *in vitro* responds to field stimulation of intramural sympathetic nerves with an increase in force of contraction and rate of pacemaker discharge, with both effects mediated by β-adrenergic receptors (Pappano and Loffelholz, 1974).

In the isolated chicken heart perfused *in vitro* by Langendorff's method, stimulation of the attached right cardiac sympathetic nerve produces positive chronotropic effects (heart rate increased from 186 to 292 beats/min; Sturkie and Poorvin, 1973). The basal heart rate of this isolated preparation, supplied with (presumably) adequate oxygen in the perfusate at 40 °C, might be expected to be similar to that of the *in situ* chicken heart after bilateral section of the vagus and cardiac sympathetic nerves ("decentralized" state), but this was not the case. The mean rate of *in situ* decentralized hearts was in the range of 235–285 beats/min (Tummons and Sturkie, 1968, 1969). Under these conditions, stimulation of the peripheral stump of the right cardiac sympathetic nerve increased heart rate to 345 beats/min, an increase of 48%. By comparison, stimulation of the right cardiac nerve stump attached to isolated hearts *in vitro* produced a mean heart rate of 292 beats/min. That is, maximal sympathetic stimulation in this preparation could only raise heart rate to a level equivalent to the basal rate of the decentralized heart *in situ*. However, even though basal heart rate was different in these preparations, the proportional increase in rate during sympathetic stimulation was the same in both cases. Thus sympathetically mediated chronotropic effects in the isolated heart may, in relative terms, reflect the capabilities of this control system in the *in vivo* heart.

In a beating heart *in vivo*, stimulation of the cardiac sympathetic nerves produces cardioacceleration. Tummons and Sturkie (1968), in the unanesthetized chicken, showed that either cardiac sympathetic nerve could produce this effect when stimulated: activation of the right nerve increased heart rate by 48% above the prestimulation value, while activation of the nerve on the left side increased heart rate by 32%. In the pigeon, however, Macdonald and Cohen (1970) found that only the right cardiac nerve mediated cardioacceleration when stimulated (Figure 11.37, top panel), whereas stimulation of the left cardiac nerve altered the appearance of the T wave of the electrocardiogram without a chronotropic effect (bottom panel, Figure 11.37). Such functional asymmetry in cardiac control has also been reported for the mammalian heart (Randall, 1994).

In the right atrium the role of the dense adrenergic innervation of the SA area (see Section 11.5.3.2.1) in controlling heart rate is obvious and a number of studies have shown that adjustments of heart rate *in vivo* are made by the sympathetic nervous system under a variety of physiological conditions. Furthermore, the anatomical evidence for widespread cardiac innervation and the data cited above for sympathetic influences on myocardial contractility in the atria and ventricles implies that the sympathetic nervous system can also produce both global and regional enhancement of contractility. Two important functional consequence of this arrangement are that (1) different patterns of sympathetic outflow from the central nervous system to individual chambers of the heart provide the means for augmenting regional contractility differentially to match the chambers' pumping actions to the hydraulic impedances into which they are working; and (2) the output of each chamber can be controlled independently of pumping rate. However, to understand the function of this control system more thoroughly it is necessary to establish whether different populations of pre and postganglionic sympathetic neurons do in fact innervate different cardiac regions and whether such subpopulations may be differentially activated through reflexes driven by receptors in specific cardiac or vascular reflexogenic zones. Such an analysis is complicated by the location, remote from the heart, of cells of origin of the postganglionic sympathetic axons innervating the cardiac chambers, in contrast to the intracardiac locations of the postganglionic parasympathetic neurons.

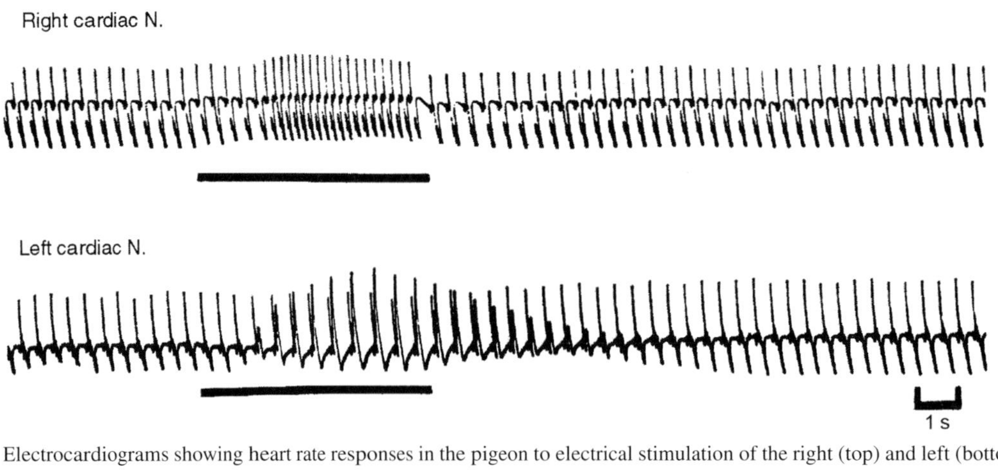

FIGURE 11.37 Electrocardiograms showing heart rate responses in the pigeon to electrical stimulation of the right (top) and left (bottom) cardiac sympathetic nerves. The duration of the stimulus train delivered to each nerve is indicated by the length of the solid bars under the traces. *From Macdonald and Cohen (1970).*

Most studies of sympathetic control of cardiac function have affirmed the role of NE as the transmitter released by avian postganglionic terminals on the myocardium. However, the bird heart contains EPI as well as NE (Sturkie and Poorvin, 1973; DeSantis et al., 1975; data summarized in Holzbauer and Sharman, 1972). It has been suggested that, in other organs such as the rectum of the fowl, adrenergic terminals may release EPI (Komori et al., 1979). In the isolated chicken heart, DeSantis et al. (1975) proposed that both EPI and NE may act as sympathetic neurotransmitters, on the basis of two main lines of evidence. First, sympathetic nerve stimulation, infusion of tyramine (a compound provoking the release of endogenous amines from sympathetic nerve terminals) or depolarization of intracardiac nerve terminals with a potassium-enriched perfusate all produced elevated NE and EPI efflux from the heart. Second, exogenously applied NE and EPI were equipotent in augmenting cardiac rate and strength of contraction. These authors also treated hearts with 6-hydroxydopamine to destroy sympathetic nerve endings and found that this reduced the intracardiac concentrations of both EPI and NE to very low levels. But DeSantis et al. (1975) did not perform the critical experiment of determining the effect of chemical sympathectomy on release of catecholamines during cardiac nerve stimulation. Sturkie and Poorvin (1973), on the other hand, concluded that even though they and other workers had identified stores of both EPI and NE in the heart, only NE appeared to be released during sympathetic nerve stimulation. These authors concluded that EPI was sequestered in nonneuronal stores and would not, therefore, be involved in neurogenic control of the myocardium. Currently, the most widely accepted view is that NE is the primary sympathetic neurotransmitter in the avian heart, as in the mammalian heart.

Autonomic tone is usually taken to mean the level of spontaneous and ongoing activity in autonomic nerves to the heart under basal or resting conditions (i.e., when the animal is not actively moving or engaged in major physiological responses to its environment). Because monitoring spontaneous nerve activity is technically difficult in any preparation other than the acutely anesthetized animal instrumented for nerve recording, the level of basal heart rate is usually taken as the major indicator of cardiac autonomic tone as rate is easily measured under a variety of conditions in the whole animal. An added complication in analyzing tonic autonomic drive to the heart is that both autonomic limbs can strongly affect rate. Because rate is driven in opposite directions by parasympathetic and sympathetic inputs, the chronotropic effects of tonic activity in one limb can only be accurately assessed in the absence of influences from the other limb. Autonomic inputs to the heart can be selectively ablated by a variety of means including surgical section of the vagus or cardiac sympathetic nerves, chemical sympathectomy (by pretreatment with agents which destroy adrenergic nerve terminals or by adrenergic blockade), blockade of cardiac vagal effects with atropine, or a combination of these methods.

The level of sympathetic tone to the heart can be quantified by determining the change from basal heart rate produced by any of the above methods for functional sympathectomy after vagal influences on the heart are removed. Widely varying levels of sympathetic tone have been reported among bird species and even among different studies of the same species. A portion of this variability is likely due to the use of anesthetics. Baseline heart rate itself will change as a consequence of general anesthesia, and anesthetics will also have a blunting effect on autonomic control of the heart (Vatner and Braunwald, 1975; Brill and Jones, 1981; Lumb and Jones, 1984). Therefore, the most accurate assessment of cardiac sympathetic tone would be made in awake, spontaneously breathing animals in a quiescent state after vagal influences on the heart have been eliminated.

Johansen and Reite (1964) investigated the level of autonomic tone to the heart in both awake ducks and those under general anesthesia; the authors did not differentiate between these states in reporting their data, claiming that this made no difference to the outcome of the experiments. In vagotomized ducks in this study, β-adrenergic blockade produced a large fall in heart rate, implying the existence of strong resting sympathetic tone. Tummons and Sturkie (1969) reported that, in awake chickens at 6 days after recovery from surgical division of the cardiac sympathetic nerves, heart rate was about 16% less than that in intact animals. In anaesthetized ducks, Kobinger and Oda (1969), using pharmacological agents to inhibit sympathetic function at central and peripheral levels of the nervous system, also found evidence for significant sympathetic tone to the heart. On the central side, they showed that clonidine mediated depression of vasopressor areas in the medulla, including the sympathetic cardiomotor area, led to a reduction in heart rate. On the peripheral side, depletion of NE from peripheral sympathetic terminals with reserpine, or prevention of NE release from these terminals, also significantly reduced cardiac rate. None of these experiments was done with accompanying vagal blockade. In contrast to the results of Kobinger and Oda (1969), Folkow et al. (1967) found no significant change in heart rate after β-adrenergic blockade in the same species. However, a complicating factor in the latter study was that the agent used (an experimental β-blocker then under development) was acknowledged by the authors to have a partial β-agonist effect, which may have offset any effects of β-blockade on basal heart rate. Butler and Jones (1968, 1971) reported no significant change in heart rate in unanesthetized ducks after β-blockade with propranolol, an agent free of intrinsic β-agonist effects. In awake chickens, Butler (1967) found that β-blockade with the vagi intact produced a significant fall in heart rate to 75% of the control rate. β-blockade after vagotomy produced less of an effect, reducing heart rate to 82% of the level in animals with intact vagi. These results show that significant sympathetic tone is present in the chicken, reinforcing the contention that the level of this tone can only be accurately assessed in the absence of parasympathetic input to the heart. However, the results of Butler (1967) contrast with those of Tummons and Sturkie (1970), who found that sectioning the sympathetic nerve produced a fall of about 16% in heart rate from the level before nerve section, while vagotomy produced a rise of about the same proportion. Combined vagotomy and sympathetic nerve section resulted in a heart rate not significantly different from that in intact animals. The authors therefore concluded that, in resting intact animals, the balance between tonic sympathetic and parasympathetic inputs to the heart maintained rate at the same level as the intrinsic rate in animals after cardiac decentralization. It is clear from the foregoing discussion that further studies of tonic sympathetic drive to the heart must be rigorous in taking both the state of anesthesia and the level of concomitant parasympathetic drive into account.

An electrophysiological analysis of the compound AP of the right cardiac nerve in the pigeon has shown that axons in this nerve can be categorized into two groups separable by conduction velocity, as shown in Figure 11.38 (Cabot and Cohen, 1977a). Fibers of the more slowly conducting group (range 0.4–2 m/s) were shown to mediate sympathetic cardioacceleration. The range of conduction velocities of these axons lies within that of unmyelinated sympathetic postganglionic fibers known to innervate the viscera (Gabella, 1976), and morphological analysis of the pigeon right cardiac nerve confirms that 67% of axons in this nerve are unmyelinated (Macdonald and Cohen, 1970). The faster-conducting group of fibers in this nerve (range 2–5.6 m/s) are likely to be myelinated axons. These make up the remaining 33% of the total number of axons and probably represent so-called sympathetic afferent fibers (Malliani et al., 1979) with receptor endings in the heart, great vessels, or lungs. Afferent fibers in cardiac nerves have been shown to participate in cardiopressor reflexes in the pigeon (Cabot and Cohen, 1977b).

11.5.3.2.2 Parasympathetic Innervation

11.5.3.2.2.1 Anatomy Efferent neurons with their cell bodies in the heart form the final common pathway for parasympathetic control of cardiac function. These postganglionic neurons receive synaptic inputs from terminals of preganglionic neurons with their somata in the brainstem and their axons coursing to the heart in the vagus nerves. It is generally accepted that parasympathetic efferent neurons in the heart are cholinergic, releasing ACh at their myocardial terminals; this neurotransmitter acts to modify myocardial function through postjunctional muscarinic receptors. Indeed, most of the anatomical studies of parasympathetic

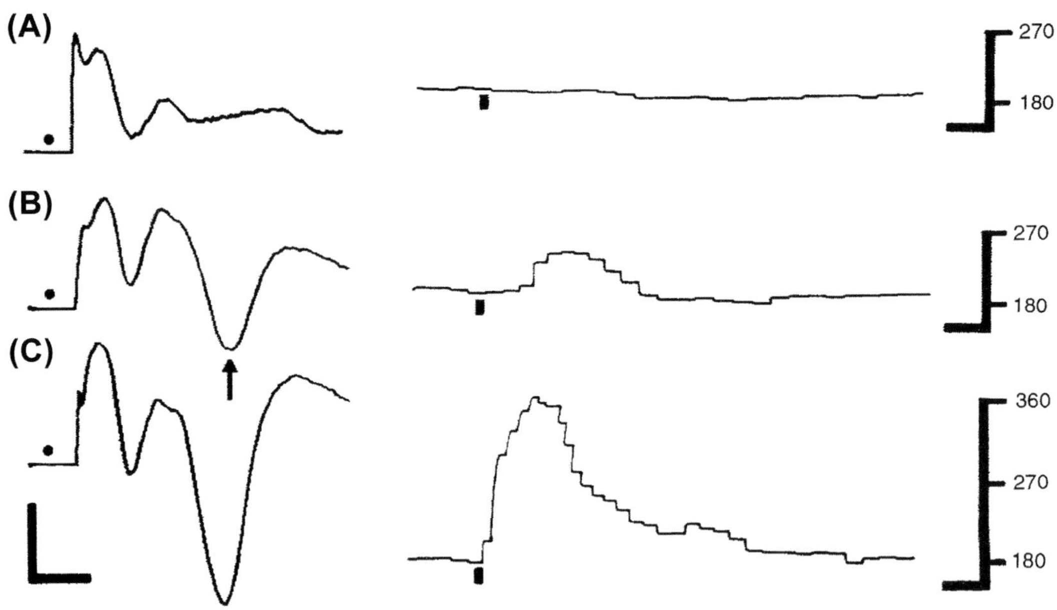

FIGURE 11.38 Correlation between components of the compound action potentials evoked by graded electrical stimulation of the right cardiac sympathetic nerve and chronotropic responses of the heart in the pigeon. The traces on the left represent the compound action potential at three intensities of stimulation increasing from A to C (round dots associated with each trace indicate the start of the stimulation, which consisted of a 200 ms train of 50 Hz pulses). For these traces the vertical calibration bar represents 100 μV and the horizontal bar represents 5 ms. The traces on the right represent beat-by-beat ratemeter recordings showing heart rate responses produced by the same stimuli (filled squares under each trace) which evoke the nerve responses on the left. The calibration bars to the right of each trace indicate heart rate in beats/min. For these traces the horizontal bar represents 1 s. The onset of the component of the compound action potential indicating conduction of cardioaccelerator fibers is indicated by the arrow in trace B; this component strengthens with increased stimulus intensity in trace C, as does the degree of cardioacceleration. *Reprinted from Cabot and Cohen (1977a).*

innervation of the heart have used a histochemical reaction indicating the presence of AChE as a marker to determine the distribution of cholinergic fibers and terminals of the cardiac plexus, as well as the locations of intracardiac neurons. Cabot and Cohen (1980) have extensively reviewed the cholinergic innervation of the heart, so a brief synopsis of that review and the contributions of later workers are combined below.

In all avian species so far examined, all four cardiac chambers receive AChE-positive innervation (Hirsch, 1963; Yousuf, 1965; Smith, 1971a,b; Akester and Akester, 1971; Mathur and Mathur, 1974; Rickenbacher and Müller, 1979; Kirby et al., 1987). Smith (1971a,b), in studies of the innervation pattern in the chicken heart, determined that cholinergic nerves, nerve fibers, and terminals formed a subepicardial ground plexus throughout the atria and ventricles, penetrating into the myocardium. Some fibers were observed to run all the way to the endocardial region, where they contributed to a subendocardial plexus that was also distributed throughout the atria and ventricles. In many avian species, the overall intracardiac distribution of cholinergic innervation parallels the distribution of adrenergic fibers and terminals (Bennett and Malmfors, 1970). The right atrium has been a particular focus for anatomical studies of cholinergic innervation in view of the location of the primary pacemaker in this chamber and the vagally mediated inhibition of pacemaker node discharge rate. Yousuf (1965) noted in the sparrow heart that the region around the SA node was the first part of the heart to receive extrinsic cholinergic innervation during embryological development and observed that, in later developmental stages, this area and the right atrial wall near the AV node were heavily innervated. However, no nerve fibers were observed to enter the SA node area proper, and only a few fibers were present among the AV nodal cells. This pattern has also been observed in the pigeon heart (Mathur and Mathur, 1974). However, Gossrau (as summarized in Akester, 1971) reported that in pigeon, chaffinch, and canary hearts, the pacemaker region had a dense AChE-positive innervation, while in duck and chicken this area was sparsely innervated. Differences among these studies may be partly species dependent, although the contrasting results of Mathur and Mathur (1974) and Gossrau in the pigeon heart may be due to differences in the techniques used. Bogusch (1974) observed a dense cholinergic innervation pattern around subepicardial Purkinje fibers of the fowl right atrium but noted that the density of this innervation decreased as the conducting fibers approached the AV border. In this study, multiple nerve terminals were only loosely associated with Purkinje fibers, leaving some doubt as to the nature of the neuroeffector–tissue relationship in cholinergic control of conduction in the atrium. SA valve remnants have also been reported to be the targets of cholinergic innervation in the bird heart (Akester, 1971; Mathur and Mathur, 1974).

Cholinergic innervation of the left atrium and the ventricles has been less extensively studied than that of the right atrium, but the general pattern of subepicardial and subendocardial plexi with individual nerve fibers and terminals extending into the myocardium, as described by Smith (1971b), appears to hold. Extensive cholinergic innervation has been described in the interatrial and interventricular septa (Akester, 1971), the right AV valve leaflets (Akester, 1971; Mathur and Mathur, 1974), and the left AV valve. Here, the AChE-positive nerve fibers enter the basal half of the valve from the subendocardial plexus (Smith, 1971a). The cholinergic innervation of the avian chordate tendineae is controversial: Smith (1971b) reported no nerve fibers present in chordae tendineae of the chicken heart, whereas Mathur and Mathur (1974) found in the pigeon heart that these structures were innervated. The function of such innervation is obscure because the chordae tendineae contain no actively contracting tissue, serving only to anchor the papillary muscles to the aortic valve leaflets. It is most likely that the nerve fibers observed in these structures are efferents en route to either the valves or the papillary muscle, but some fibers may subserve an afferent function.

The adventitia of coronary arteries associated with all chambers of the bird heart is innervated by cholinergic nerve fibers (Hirsch, 1963; Akester, 1971; Smith, 1971a) but the origin of this innervation is not known. Cholinergic fibers innervating coronary arteries may originate from postganglionic parasympathetic neurons within the heart or may possibly be extrinsic sympathetic cholinergic fibers; in either case, this innervation probably functions to increase coronary blood flow by promoting vasodilation.

In development of the chick heart, Rickenbacher and Müller (1979) determined that cell bodies of intracardiac neurons, clustered into ganglia, first developed in the left ventricle, then a large group of ganglia became evident around the coronary sulcus and the ventral surface of the ventricles and, lastly, ganglia developed in association with the dorsal atrial walls.

In the adult avian heart, ganglia are located primarily in the subepicardial plexus, usually in association with plexus nerves and frequently near the branch points of these nerves. The somata of intracardiac neurons have been characterized as multipolar (possessing more than two processes; Smith, 1971b; Yousuf, 1965), but these observations are limited to the hearts of only two species (chicken and sparrow, respectively). No morphological data exists on the variation of somatic dimensions or on the length or specific projection targets of the processes of intracardiac neurons in any avian species.

Histochemical reactions for AChE have been the primary tool used in analyzing the distribution of intracardiac neurons in the bird heart, and there seems little doubt that these techniques allow visualization of most if not all of these neurons. Such anatomical data, along with evidence from the functional studies cited below, supports the contention that the phenotype of avian intracardiac neurons is cholinergic. However, AChE has been detected in some nonneuronal elements associated with the nervous system (see review by Fibiger, 1982) and demonstration of the presence of this enzyme in neuronal somata in the heart, although necessary, may not be a sufficient criterion for designating these cells cholinergic. In the central nervous system the most widely accepted indicator of cholinergic function is the presence of choline acetyltransferase (ChAT), an enzyme in the pathway for ACh synthesis (Fibiger, 1982). Reliable antibodies directed against ChAT are now commercially available and have begun to be applied to the mammalian peripheral autonomic nervous system. A similar application of ChAT immunohistochemical techniques to the avian heart would help verify the assumption that intracardiac neurons in this vertebrate group are in fact of cholinergic phenotype.

In their analysis of the distribution of the intracardiac ganglia during development, Rickenbacher and Müller (1979) found that the right ventricle wall contained about half of the total number of ganglia, the right atrium about one-fifth, the left ventricle about one-sixth, and the left atrium possessed the fewest ganglia. In the adult bird heart, no quantitative studies of regional neuron distribution have been done to date, so the absolute number of neurons associated with each chamber is not known. However, the general pattern of distribution of intracardiac neurons in the adult heart has been established (see Cabot and Cohen, 1980 for review). Ganglia containing variable numbers of neurons are present on both dorsal and ventral aspects of the left and right atria and ventricles (Yousuf, 1965; Smith, 1971a,b; Rickenbacher and Müller, 1979; Kirby et al., 1987; reviewed by Cabot and Cohen, 1980). Smith (1971b) has reported that a larger proportion of the total number of intracardiac ganglia is found in the ventricles of bird hearts than is the case in mammalian ventricles. Ganglia have been observed near but not within the SA node region (Yousuf, 1965; Smith, 1971b). Yousuf (1965) reported that some neurons in the sulcus terminalis appeared to send projections in the direction of the nodal tissue. The AV nodal region was also reported to be devoid of ganglia (Yousuf, 1965; Smith, 1971b; Mathur and Mathur, 1974) but this region and the AV bundle appeared to be innervated by axons from ganglion neurons in the nearby AV sulcus (Yousuf, 1965). Smith (1971b) described high concentrations of ganglia within the dorsal right atrial wall near the ostia of the superior and inferior venae cavae, near the roots of the pulmonary veins on the dorsal aspect of the left atrium, within the dorsal portion of the AV groove, and clustered around the roots of the pulmonary artery and aorta. In the ventricles, neurons are located in a scattered pattern reaching from the AV groove to the apex on the ventral surface (Smith, 1971b; Rickenbacher and Müller, 1979). There are also numerous ganglia associated with nerves accompanying atrial and ventricular coronary arteries (Mathur and Mathur, 1974; Smith, 1971b).

Studies in the mammalian heart have shown that a number of neuropeptides are colocalized in axons, terminals, and somata of cholinergic intracardiac neurons, as well as in preganglionic terminals contacting these neurons (Steele et al., 1994, 1996). Peptides constitute an important class of neuromodulators in the peripheral autonomic nervous system, and their presence in specific combinations in some peripheral neurons has been proposed to chemically code subpopulations of these neurons for specific functions such as vasomotion or control of muscle cell contractility. In the bird heart, substance P and vasoactive intestinal peptide have been found in intracardiac neurons and their terminals, while somatostatin is present in intracardiac terminals but not in cell bodies (Corvetti et al., 1988). These neuropeptides have been shown to exert powerful modulatory effects on mammalian intracardiac neuronal activity and cardiodynamics (Armour et al., 1993) and their presence in the bird heart indicates that they may play a prominent role in modulation of ganglionic and neuroeffector transmission in this vertebrate group. This constitutes a promising but as yet unexplored area for the comparative study of mechanisms of neural control of the heart.

The course of the avian vagus nerve and its cardiac branches has been described for a number of species (see Pick, 1970; Jones and Johansen, 1972; Cabot and Cohen, 1980 for reviews). The latter review points out that descriptions of the course and major branches of this nerve are consistent among species so a general summary of the avian vagal cardiac innervation will be given here, based on the comprehensive reports of Malinovsky (1962) and Cohen et al. (1970) in the pigeon and the reviews cited above. Inside the cranium, the peripheral trunks of the vagus and glossopharyngeal nerves originate bilaterally from large ganglia composed of a fusion of the proximal ganglion of the glossopharyngeal nerve and the jugular ganglion of the vagus nerve. The trunks of these nerves emerge together from the skull through the jugular foramen, and immediately outside the foramen an anastomosis (of Staderini) connects the vagal trunk to the petrosal ganglion of the glossopharyngeal nerve. The vagal trunk continues caudad in the neck along the dorsomedial aspect of the internal jugular vein, passing over the cervical spinal nerves on their ventral sides. No major vagal branches arise from the trunk along its cervical portion, although Malinovsky (1962) described occasional small anastomoses with the nearby cervical sympathetic trunk. At the level of the thoracic inlet the nodose (alternatively termed distal vagal) ganglia are present as spindle-shaped enlargements of the vagal trunks, as shown in Figure 11.39. Several afferent nerves carrying the axons of receptors important in the control of cardiovascular and respiratory functions arise from each nodose ganglion. Along the length of this

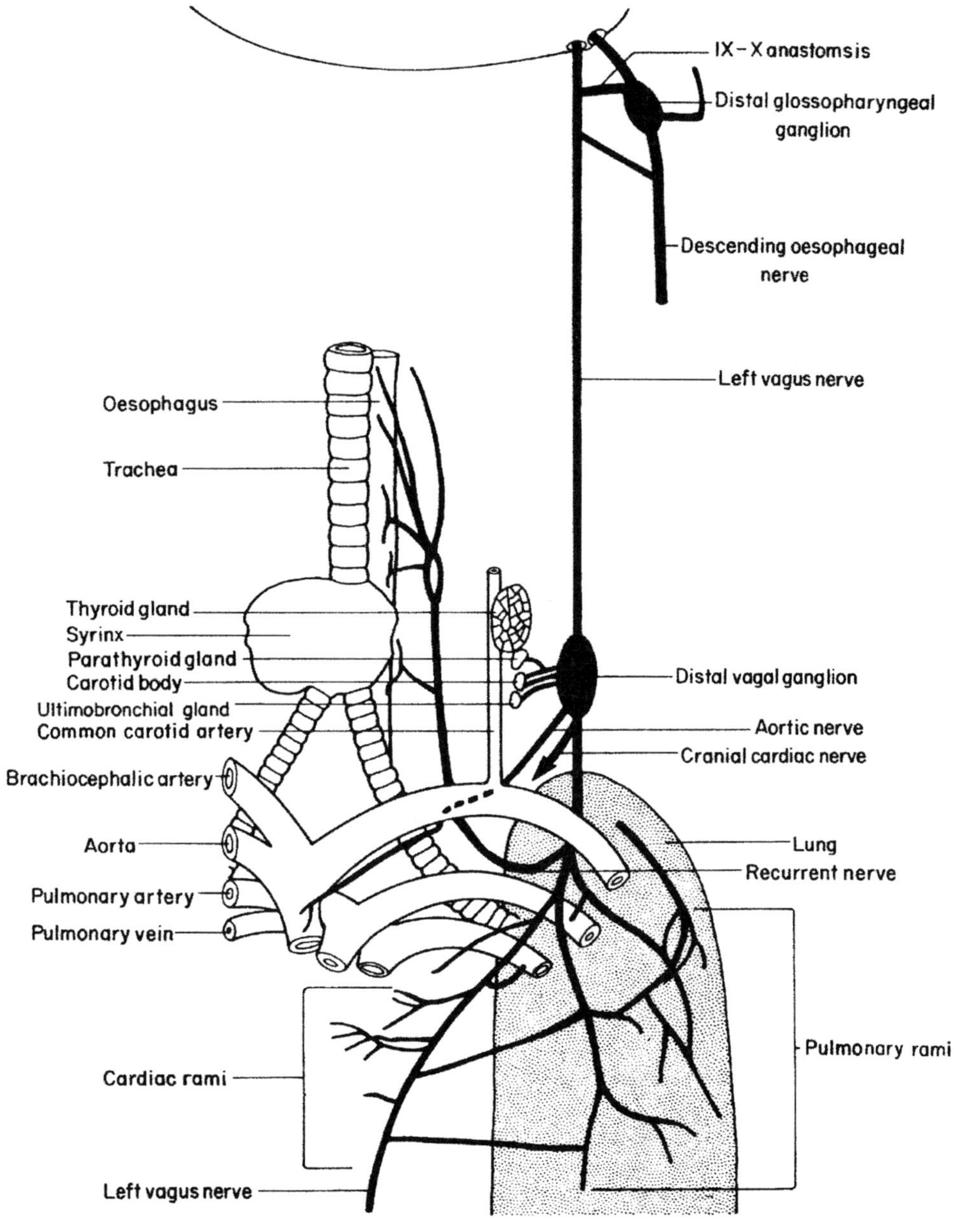

FIGURE 11.39 Schematic diagram depicting a ventral view of the pathway of the left vagus nerve in the area of the upper thorax of the duck. Details of the vagal innervation of the carotid body, ultimobranchial, thyroid, and parathyroid glands, and the aorta are illustrated. *From Jones and Johansen (1972).*

ganglion, branches arise which extend medially to innervate the thyroid, parathyroid and ultimobranchial glands, and the carotid body (Figure 11.39).

The latter structure constitutes the primary locus for peripheral chemoreceptors sensing arterial oxygen and carbon dioxide tensions and pH in the bird (Jones and Purves, 1970; see Section 11.5.4.1). Branches exiting from the caudal portion of the nodose ganglion course to the root of the aorta. On the right side, Nonidez (1935) reported two such branches in the chick, which he designated depressor and accessory depressor nerves by analogy with the mammalian condition. Nonidez (1935) reported no equivalent nerve on the left side, but in other species nerves from the nodose ganglion coursing to the aortic root (designated aortic nerves) have been reported to be present bilaterally (Jones and Purves, 1970; Jones, 1973; Cohen et al., 1970; Jones et al., 1983; Smith and Jones, 1990, 1992; summarized by Smith, 1994; see Figure 11.39). These nerves ramify into a plexus in the aortic wall. Jones (1973) first demonstrated that the aortic nerves carried arterial blood pressure information centripetally and that aortic baroreceptors were involved in regulating and maintaining arterial blood pressure in the duck.

A few millimeters caudal to the nodose ganglion the vagal trunk splits, as shown in Figure 11.40, to form several major divisions as it approaches the pulmonary artery. Two of these circumscribe the pulmonary artery, rejoin, and continue as the main vagal trunk en route to the abdominal cavity while a third forms the recurrent laryngeal nerve, coursing craniad along the trachea; no cardiac vagal branches arise from this nerve. Another major vagal branch courses to the heart to enter the dorsal cardiac plexus (Jones and Johansen, 1972; shown in Figure 11.39). The remaining vagal branches form part of the pulmonary innervation, running to the lungs along the pulmonary arteries. The main vagal trunk, after reforming caudal to the pulmonary artery, passes ventral to the ipsilateral bronchus and over the dorsal surface of the heart. From this portion of the trunk a variable number of smaller branches arise and enter the cardiac plexus. On the right side, these branches enter the heart near the SA and AV nodes and at the caval ostia; on both the left and right sides of the heart vagal branches also enter the cardiac plexus in the vicinity of the AV groove. That the cardiac nerves described here constitute the major efferent vagal innervation of the heart has been confirmed by functional studies in which electrical stimulation of the vagal trunk was combined with surgical section of the trunk and the various cardiac branches (Cohen et al., 1970). Caudal to the origin of the most inferior cardiac branches, the left and right vagal trunks pass ventral to the pulmonary veins where both turn medially, coming to lie in close approximation as they course to the abdominal viscera (Figure 11.40).

The location in the central nervous system of cells of origin of vagal cardioinhibitory fibers in birds has been investigated by a variety of anatomical and physiological techniques. The extent of the medullary regions containing vagal preganglionic motor neurons innervating pharyngeal structures and thoracic and abdominal viscera was originally defined anatomically by determining the extent of

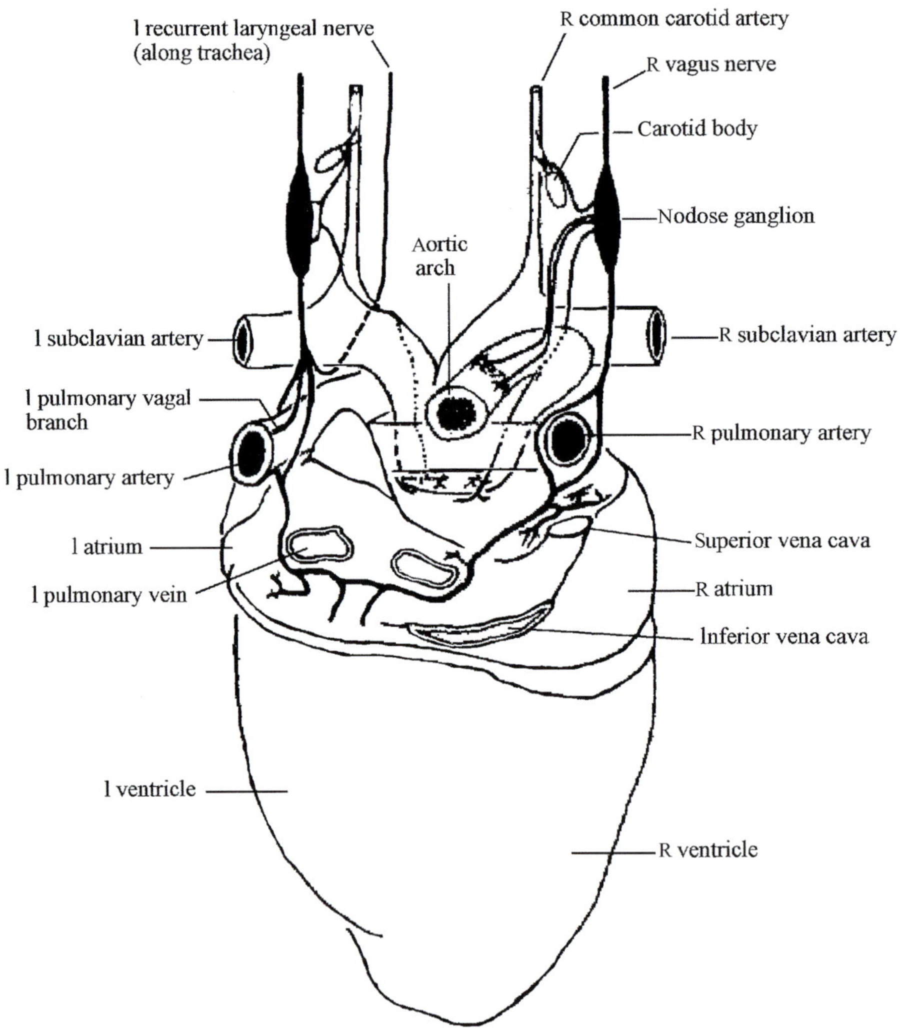

FIGURE 11.40 Schematic diagram to illustrate a dorsal view of the generalized vagal innervation of the avian heart. Note the pathways of the vagal branches as these nerves split close to the pulmonary arteries. For clarity the right pulmonary and right recurrent laryngeal branches are not shown and the right pulmonary vein, depicted next to the left, is not labeled (r, right; l, left). *Modified from Cabot and Cohen (1980).*

retrograde degeneration of neuronal somata after section of the cervical vagus nerve (Cohen et al., 1970) and by retrograde labeling of vagal neurons with neurotracers applied to the peripheral vagus nerve (Katz and Karten, 1979, 1983a,b, 1985; Cabot et al., 1991a). Cohen et al. (1970), working in the pigeon, described three major subdivisions of the dorsal motor nucleus of the vagus nerve (DMV) based on cytoarchitectonic and morphological criteria. The principal medullary location of neurons showing signs of degeneration after section of the cardiac vagal branches was in a region extending from the obex to the rostral pole of the DMV, with the highest density of degenerating cells in an area between 0.6 and 0.8 mm rostral to the obex. At this rostrocaudal level, these cells were primarily located in the most ventral region of the DMV. The results of these experiments suggested that the central arrangement of vagal cardioinhibitory neurons in birds was different than the mammalian condition; in mammals the primary locus for these neurons is the nucleus ambiguus, ventrolateral to the DMV (reviewed by Hopkins, 1987).

In the pigeon, focal electrical stimulation of the regions of the DMV shown anatomically to contain cell bodies of putative cardioinhibitory neurons produced short-latency decreases in heart rate; this response is shown in Figure 11.41 (Cohen and Schnall, 1970). The response was rapid, occurring within one or two cardiac cycles after the start of stimulation, suggesting that cardioinhibitory preganglionic cell bodies were being directly stimulated. If stimulation was continued, complete AV blockade could be produced in some animals; a depressor response invariably occurred secondary to all negative chronotropic responses (Figure 11.41). There was no lateral asymmetry in this response: stimulating in the DMV on either side produced similar cardioinhibitory responses. The cardiac effects produced by central stimulation could be mimicked by stimulating the vagal trunks in the neck or at the thoracic inlet (Cohen and Schnall, 1970). Field potential and single-unit recordings made in the pigeon DMV during stimulation of the vagal trunk provided further confirmation that the cell bodies of a large number of cardioinhibitory neurons were located in the central zone of the DMV rostral to the obex (Schwaber and Cohen, 1978b).

Schwaber and Cohen (1978a), in an electrophysiological study of the vagus nerves, found that when the cervical vagus was stimulated at progressively greater intensities the onset of bradycardia coincided with the elicitation of a specific component of the compound AP (B1 wave, Figure 11.42). This component was generated by the activation of a group of vagal axons conducting in the velocity range of 8–14 m/s. When the vagus nerve was stimulated with sufficient intensity to evoke both the A and B1 components (trace B of Figure 11.42) and a polarizing voltage was applied to the nerve to block the A but not the B1 component, the cardioinhibitory response was maintained. This led the authors to conclude that fibers in the vagus nerve responsible for generating the B1 component of the compound AP were responsible for cardioinhibition. This was confirmed in experiments in which field potential and single-unit activity were mapped in the DMV in correlation with synchronous compound APs recorded from the vagus nerve during stimulation of this nerve (Schwaber

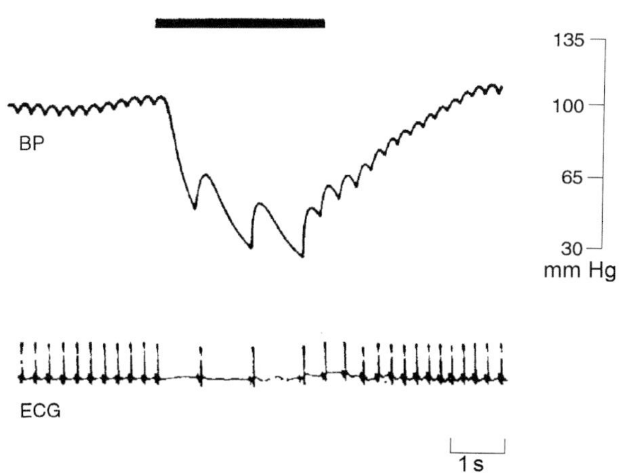

FIGURE 11.41 Arterial blood pressure (BP) and cardiac chronotropic (ECG) responses to focal electrical stimulation in the area of the dorsal motor nucleus of the vagus nerve in the pigeon. The duration of the stimulus train (50 Hz pulses) is shown by the solid horizontal bar above the traces. *From Cohen and Schnall (1970).*

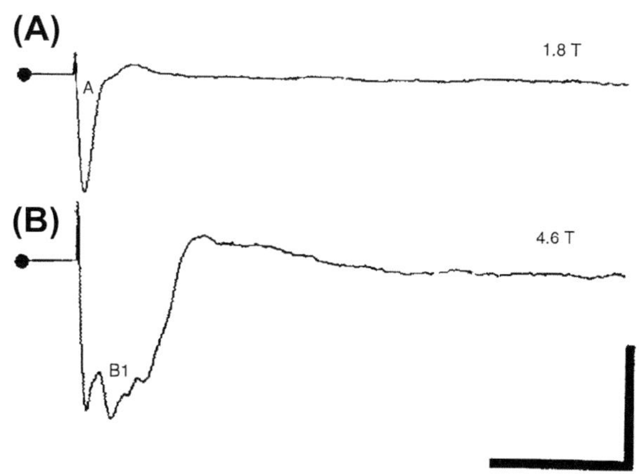

FIGURE 11.42 Compound action potentials evoked in the right vagus nerve by electrical stimulation at the midcervical level in the pigeon. Responses to stimuli of two intensities are shown. (A) Stimulation at 1.8 times the intensity which just evokes a response (threshold intensity, T) produces a short-latency component labeled the A wave, representing the fastest conducting fibers in the vagus (start of stimulus is indicated by the dot at the left side of trace). No change in heart rate is associated with the activation of this group of fibers. (B) Stimulation at 4.6 times the threshold intensity evokes responses in an additional, more slowly conducting group of fibers; this component of the compound action potential is labeled the B1 wave. The fibers responsible for this component conduct in the range of 8–14 m/s and when activated produce bradycardia. The vertical bar represents 250 μV; the horizontal bar represents 5 ms. *From Schwaber and Cohen (1978a).*

and Cohen, 1978b). In these experiments, stimulus-evoked activation of the B1 component in the vagus nerve produced the shortest latency, highest amplitude responses in the region of the DMV, which had been shown in previous studies to contain the somata of putative cardioinhibitory neurons. In addition to the evidence from stimulus-evoked potentials, recordings of spontaneously active single units in the DMV rostral to the obex in awake, paralyzed pigeons demonstrated rhythmic discharge patterns phase-locked to mechanical events in the cardiac cycle (Gold and Cohen, 1984). These authors also showed that single-unit neuronal activity in this area was decreased or eliminated by external conditioning stimuli (light flash, foot shock) which caused heart rate to increase (see Section 11.5.5).

The above anatomical and physiological evidence, taken together, indicates that in the avian brain preganglionic vagal cardioinhibitory neurons are located in the ventrolateral region of the DMV rostral to the obex. However, a more recent reexamination of the question of the location of these neurons was undertaken by Cabot et al. (1991a), using a new and more sensitive method for retrograde neuroanatomical tracing. These authors injected small volumes of the binding fragment of tetanus toxin into selected regions of the pigeon heart. This neurotracer was taken up by local nerve fibers and terminals at the injection site and transported retrogradely to label the somata of vagal preganglionic cardiac neurons in the medulla by two possible routes, both giving similar end results. The first of these was via direct uptake of neurotracer by fibers or terminals of the preganglionic neurons running through or close to the injection sites in the heart; in this case, the tracer would be transported directly back to the cell bodies. The second route was via transsynaptic transport. Fibers and terminals of postganglionic intracardiac neurons took up the neurotracer from the injection sites, and upon traveling to the somata and other processes of these neurons, the tracer would then cross synaptic clefts to the preganglionic terminals contacting the postganglionic cells. From these terminals, the tracer was transported to the somata of the preganglionic neurons. After intracardiac injection of the tracer, labeling of preganglionic vagal neurons was found in two locations in the medulla (Figure 11.43). The majority of label was found in neurons located ventrolateral to the DMV, in a site homologous to the nucleus ambiguus of the mammalian brainstem; these neurons were clustered within 0.5 mm of the obex as shown in the bottom panel of Figure 11.43. A smaller number of labeled neurons were found in more rostral sections in an area bordering the ventrolateral margin of the DMV (top panel of Figure 11.43), in close proximity to the region delineated in the degeneration studies of Cohen et al. (1970) and just ventral to the area described in the functional studies of Schwaber and Cohen (1978a,b). The results of Cabot et al. (1991a) have forced a reevaluation of the central organization of neurons controlling the avian heart, indicating that this organization has much more in common with the mammalian condition than was previously believed. However, these anatomical data have not been confirmed by physiological studies. In addition, it still remains for the membrane and firing properties of vagal cardioinhibitory preganglionic neurons to be investigated to determine if there are functionally discrete subpopulations within this group of neurons and, if so, whether there is any correlation between the functional properties of neurons and their potential roles in controlling specific aspects of cardiodynamic function.

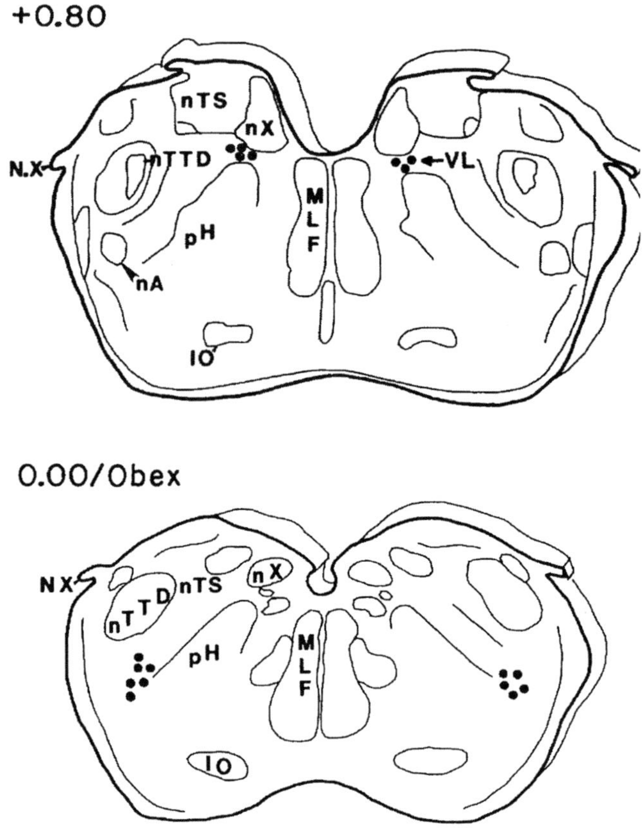

FIGURE 11.43 Outline drawings of transverse sections of the pigeon medulla in the region of the obex, depicting the location of vagal preganglionic neurons labeled by retrograde transport of tetanus toxin binding fragment C injected into the heart. See text for explanation. The top panel shows a section cut 0.8 mm rostral to the obex, with labeled neurons (filled circles) located just ventral to the dorsal motor nucleus of the vagus (nX) in the nucleus ventrolateralis (VL). The bottom panel, representing a section taken at the level of the obex, shows additional labeled neurons in an area of the medulla which may be the avian homolog of the mammalian nucleus ambiguus (nA), ventrolateral to nX. Abbreviations: IO, inferior olivary nucleus; MLF, medial longitudinal fasciculus; nTS, nucleus and tractus solitarius; nTTD, nucleus and tractus trigemini descendens; NX, vagal nerve rootlet; pH, plexus of Horsley. *Reprinted from Cabot et al. (1991a).*

Little is known of the nature and origins of inputs to medullary vagal cardiomotor neurons in birds. Berk and Smith (1994) have shown in the pigeon that peptide containing projections to these neurons arise from the area of the nucleus of the tractus solitarius (NTS). The NTS is one target for afferent information from visceral receptors carried in the vagus, glossopharyngeal, and other cranial nerves, and such peptidergic projections from the NTS to cardiomotor neurons may represent an important viscero-visceral reflex pathway, as has been described in mammals (Loewy and Spyer, 1990). In addition to inputs of peripheral origin, there is an extensive pattern of projections to the DMV from structures located more rostrally in the brain. Berk and Finkelstein (1983) and Berk (1987) demonstrated projections from the bed nucleus of the stria terminalis, the ventral paleostriatum, and the medial and lateral hypothalamus to the DMV. These forebrain inputs thus represent potential pathways through which central nervous control of cardiac function may be exerted in the interests of homeostasis, as well as providing pathways for neutrally mediated alterations in cardiac function which may be required during exercise, feeding or other behaviors, or in response to changes in the external environment. In addition, Cohen and coworkers have explored the central anatomical pathways mediating conditioned responses which target medullary cardiomotor neurons in birds (see Section 11.5.5). These cardiomotor neurons therefore integrate information from visceral and other receptors and from higher levels of the central nervous system to control cardiodynamics, but our knowledge of the integrative mechanisms involved is scant at present.

11.5.3.2.2.2 Parasympathetic Control Acetylcholine acts in the bird heart to depress atrial and ventricular myocyte contractility, rate of discharge of pacemaker tissue, and rate of conduction through the specialized conductive tissues. ACh, released from preganglionic terminals, activates excitatory nicotinic receptors on the membranes of postganglionic neurons in the heart and these neurons in turn release ACh from their effector terminals to inhibit cardiac functions. Intrinsic postganglionic parasympathetic neurons release the majority of ACh which overflows from the isolated heart during vagal nerve stimulation. Only a small fraction of the total amount of ACh recovered in these experiments is released from vagal preganglionic terminals, as shown by a large reduction in vagally evoked ACh release after treatment of the isolated heart preparation to prevent release of the neurotransmitter from postganglionic terminals (Loffelholz et al., 1984).

ACh has different effects on the cell membrane conductances and thus on contractile properties of myocytes in the avian atria and ventricles. Inoue et al. (1983), using intracellular electrode techniques *in vitro*, investigated the effects of ACh on membrane ion conductances of atrial and ventricular myocytes to determine how these might differ. They found that, in ventricular muscle cells, ACh reduced the force of contraction, diminishing both amplitude and time course of the AP, but did not change either resting membrane potential or whole-cell resistance. On the other hand, in atrial myocytes, ACh hyperpolarized the membrane and reduced whole-cell resistance (implying an increase in steady-state ionic conductance) as well as causing a reduction in amplitude and time course of the AP. ACh binds to muscarinic receptors on myocyte membranes, and the authors found that this induced similar decreases in calcium-dependent sodium currents in the two types of cells. However, atrial myocytes exhibited in addition a muscarinically mediated increase in an outward potassium current, which accounted for the hyperpolarization and reduction in resistance induced by ACh; this mechanism was not present in ventricular myocytes. These differences in response to ACh imply that the same neurotransmitter can differentially control atrial and ventricular contractility.

There are several subtypes of muscarinic receptor present in the mammalian heart (see Deighton et al., 1990 for review) and some of these receptor subtypes are also present in the avian heart. The majority of muscarinic receptors on mammalian myocardial cells are of the M_2 subtype (Deighton et al., 1990; Jeck et al., 1988), and this subtype is believed to mediate complex intracellular mechanisms, leading to the inhibition of myocyte functions, which are ultimately responsible for the parasympathetic control of the heart. In a comparison of muscarinic receptor types in the chicken and guinea pig hearts, Jeck et al. (1988) found that receptors of the M_1 subtype predominated in the myocardium of the chicken while the most prevalent type in the guinea pig heart was the M_2 subtype, as found in other studies of the mammalian heart. Detailed analyses of the muscarinic receptor subtypes present in the hearts of other avian species have not been conducted, but if the results of Jeck et al. (1988) in the chicken represent the general avian situation, there are likely to be major differences in receptor-mediated intracellular mechanisms of mucarinic inhibition of myocyte function between birds and mammals.

In avian atrial tissue *in vitro*, and in atria in whole *in situ* or isolated hearts, it is widely accepted that ACh has a strong negative inotropic effect (Jeck et al., 1988; review by Sturkie, 1986), but studies of specific cholinergic effects on ventricular inotropy have been few in the bird. Avian ventricular tissue has a higher density of cholinergic innervation than does that of mammals as outlined above and a higher proportion of the total number of intracardiac neurons is associated with the ventricles of the avian than the mammalian heart (see above, and Smith, 1971b). On the basis of early anatomical evidence suggesting a high density of cholinergic innervation of the ventricular myocytes in birds, Bolton and Raper (1966) compared responses of strips of the right ventricle of fowl and guinea pig hearts *in vitro* to endogenously released and exogenously applied ACh. Field stimulation of electrically paced ventricular strips produced a strong decrease in force of contraction of fowl ventricular tissue, while guinea pig ventricular tissue responded with an increase in force. Atropine, blocking muscarinic receptors, eliminated the negative inotropic response of the fowl ventricle to stimulation, and a strongly positive response was then observed; however, atropine had no effect on the response of the guinea pig ventricle to stimulation. The authors proposed that the fowl heart has a capacity for effective parasympathetic inhibition of ventricular inotropy, mediated by intracardiac release of ACh. Furthermore, blockade of this response unmasked a stimulus-evoked increase in force of contraction which the authors determined was the result of release of NE from sympathetic nerve terminals. They concluded that, in contrast with the mammalian condition, the fowl right ventricle was innervated by both the sympathetic and parasympathetic limbs of the autonomic nervous system. Subsequent *in vitro* work has confirmed these observations (Bolton, 1967; Biegon et al., 1980; Biegon and Pappano, 1980).

The first experimental approach used to explore general questions of parasympathetic efferent control of the avian heart, dating from the early part of the 1900s, was to examine the effects of exogenously applied ACh on the activity of pacemaker cells in the right atrium (reviewed by Jones and Johansen, 1972; Cabot and Cohen, 1980). More detailed characterization of this system has resulted from investigations carried out by Loffelholz and co-workers, and others. In the *in vitro* right atrium, ACh release provoked by field stimulation and by stimulation of the attached right vagus nerve produced a fall in pacemaker discharge rate (Pappano and Loffelholz, 1974; Pappano, 1976; Brehm et al., 1992). This chronotropic effect was blocked by atropine but not by the ganglionic blocker hexamethonium, so it must have been mediated by direct release of ACh from postganglionic parasympathetic neurons in the atrial wall (Pappano and Loffelholz, 1974). On the other hand, hexamethonium in the absence of atropine blocked all effects of electrical stimulation of the attached vagal stump. Vagally mediated bradycardia therefore occurred as a result of the synaptic activation of intracardiac postganglionic neurons by preganglionic terminals. As with cholinergic control of the inotropic state of the atrial and ventricular myocardium discussed above, control of pacemaker rate operates via the synaptic relay of impulses from pre- to postganglionic neurons within intracardiac ganglia.

11.5.3.2.2.2.1 Dromotropic Effects There have been no studies of the direct effects of ACh on the rate of impulse conduction through the specialized conducting cells of the avian heart and few studies of the overall parasympathetic control of this function. There are technical difficulties in identifying the locations of conducting tissues in a viable *in vitro* preparation of cardiac tissue, so work on this intriguing problem in birds has largely been conducted on hearts *in situ*. In the chicken, Goldberg et al. (1983) showed that AV conduction time could be significantly prolonged by stimulation of either vagus nerve; there was no bilateral asymmetry in this response. However, in order to unmask this dromotropic effect the heart was paced through electrodes attached to the SA node, both to control heart rate and to prevent shifts in pacemaker position during vagal nerve stimulation. Bogusch (1974), using anatomical techniques, identified cholinergic fibers and terminals in the region of conducting cells near the AV border and proposed a functional role for this innervation but the study of Goldberg et al. (1983) appears to be the only physiological confirmation of this role in the bird (also see Section 11.2.3.3).

11.5.3.2.2.2.2 Chronotropic Effects The study of vagal control of heart rate in birds has a long history going back to the recognition in the last eighteenth century that activation of the vagus nerves could produce large reductions in heart rate and, in some cases, arrest the heart (see Cabot and Cohen, 1980 for a summary of early work). As these reviewers have pointed out, the nature of vagally mediated bradycardia has been intensively reinvestigated using recently developed techniques in a variety of avian species (Johansen and Reite, 1964; Bopelet, 1974; Peterson and Nightingale, 1976; Langille, 1983; Lindmar et al., 1983; Goldberg et al., 1983; Lang and Levy, 1989; Butler and Jones, 1968; Cohen and Schnall, 1970; Jones and Purves, 1970; and others). The negative chronotropic effects of ACh in the pharmacological studies cited above are paralleled by the effects of electrical stimulation of the vagus nerves. Several preparations have been used to assess the chronotropic consequences of vagal stimulation. These include *in vitro* atrial tissue, isolated beating hearts with attached vagal stumps, open-thorax anesthetized preparations, and anesthetized or awake animals in which only the vagi in the cervical region were exposed. Vagal control of the heart is apparently very robust in all of these preparations and each type of preparation yields results which complement the findings obtained from the others.

Examples of stimulus-induced bradycardia obtained in pigeon are shown in Figure 11.44. The peripheral cut ends of the right (top panel) and left (bottom panel) cervical vagus nerves were stimulated with trains of pulses at a frequency of 50 Hz. In each case the contralateral vagus nerve was intact. In these examples the intensity of stimulation used was capable of arresting the heart. A close parallel to the negative chronotropic effect of peripheral vagal nerve stimulation can be produced by focal electrical stimulation in the medulla (Cohen and Schnall, 1970) where anatomical studies (see above) have shown that cardiac vagal preganglionic neurons are located.

The effectiveness of the left and right vagi in controlling heart rate has been shown by some workers to be equivalent, while other workers have found strong bilateral asymmetry in this system. Bopelet (1974), Peterson and Nightingale (1976), and Goldberg et al. (1983) reported no difference in the chronotropic response of the fowl heart to electrical stimulation of left and right vagus nerves. In the same

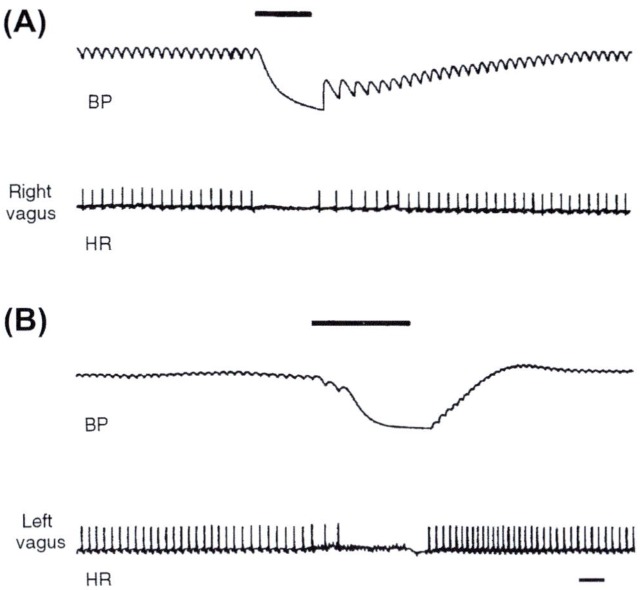

FIGURE 11.44 Effects of stimulating the right (A) and left (B) vagus nerves on heart rate (HR) and arterial blood pressure (BP) in two pigeons. The nerves were exposed in the neck and stimulated with 50 Hz trains of pulses (stimulus duration indicated by the horizontal bars above the BP traces). Horizontal bar below the bottom trace represents 1 s. *From Cohen and Schnall (1970).*

species, however, Sturkie (1986) and Lang and Levy (1989) reported that the right vagus was more effective in altering heart rate than was the left. Johansen and Reite (1964), in ducks and seagulls, Jones and Purves (1970) in ducks, and Cohen and Schnall (1970) in pigeons, all reported a similar asymmetrical response to electrical stimulation of the vagus nerves. Furthermore, in a systematic study of vagal control of heart rate in the duck, Butler and Jones (1968) showed by means of cold blockade of the vagi in the neck and by unilateral and bilateral section of these nerves that vagal dominance could change sides over time. However, these authors did not stimulate the peripheral cut ends of the vagi after nerve section, so it is not known whether the origin of this bilateral asymmetry was within the central nervous system or at the heart.

Lindmar et al. (1983) have quantified the amounts of ACh released in the isolated chicken heart by stimulation of the attached left and right vagal stumps in an attempt to determine if there were bilateral differences in the intracardiac connections of these nerves. If fibers from each vagus innervated approximately the same number of postganglionic intracardiac neurons, the amount of ACh released by stimulation of either nerve alone would be expected to be similar, and this was in fact found to be the case. If there were no overlap in the populations of intracardiac neurons innervated by each vagus, then the sum of ACh released by unilateral stimulation of each nerve should be the same as that released by simultaneous bilateral nerve stimulation. However, Lindmar et al. (1983) found that the sum of ACh released by separate nerve stimulation was significantly greater than that released by bilateral stimulation, indicating that a large proportion of intracardiac neurons must have been bilaterally innervated. In these experiments it was not possible to separate the responses of postganglionic neurons innervating the pacemaker tissue from the responses of neurons subserving other functions, but if it is assumed that the overall pattern of bilateral innervation found by Lindmar et al. (1983) can be applied to the specific subpopulation of neurons controlling heart rate, then this pool of neurons may be controlled by preganglionic fibers running in either nerve. There would thus be little reason to expect that side-to-side shifts in vagal dominance result from factors acting to change the relative intracardiac influences of these nerves. This line of reasoning supports the notion that changes in the pattern of activity of vagal preganglionic neurons in the medulla may be responsible for spontaneous shifts in the dominant vagus nerve. Such central nervous factors might be investigated using the approach of Gold and Cohen (1984) to simultaneously record spontaneous activity from neurons in medullary vagal complexes on both sides.

11.5.3.2.2.2.3 Inotropic Effects Despite the evidence cited above for strong negative inotropic effects of endogenously released or exogenously applied ACh in the avian heart *in vitro*, the inotropic effects of vagal stimulation on *in situ* hearts is controversial. Folkow and Yonce (1967) unmasked a strong reduction in one index of left ventricular contractility, that of peak left ventricular chamber pressure, in the duck heart in response to vagal nerve stimulation when heart rate was kept constant by electrical pacing. These authors also found that CO declined with vagal stimulation during pacing. Because vagal stimulation was shown not to affect peripheral resistance in this study, the fall in CO must have been caused by a decrease in stroke volume. Although contractile force was not measured directly, the authors attributed both the fall in left ventricular pressure and in CO to a vagally mediated decrease in ventricular contractility. That this effect was neurally mediated was demonstrated by the elimination of the effects of vagal stimulation after atropine was administered intravenously. Furnival et al. (1973), using another index of ventricular contractility, that of maximum rate of change of left ventricular pressure, reported results contrary to those of Folkow and Yonce (1967). In their comparative study of the responses of ventricular contractility to vagal stimulation in the dog, duck, and toad, Furnival et al. (1973) reported that only the amphibian heart displayed a significant reduction in contractility. These authors proposed that the ventricles of the ducks in the experiments of Folkow and Yonce (1967) had been subject to very high end-diastolic pressures as a consequence of the experimental protocol, were probably in failure as a result of this treatment, and thus responded abnormally to vagal stimulation. Yet Lang and Levy (1989), using the same index of ventricular contractility in the chicken as that employed by Furnival et al. (1973) in the duck, concluded that vagal stimulation could produce decreases of more than 50% in contractile force (Figure 11.45). Furthermore, the inotropic responses to vagal stimulation in the chicken heart reported by Lang and Levy (1989) were considerably more powerful than those that could be obtained in mammalian hearts.

The controversy surrounding this issue likely hinges on problems inherent in the methods used to evaluate ventricular contractility. Resolution of this issue will only be possible when more direct indices of contractility, such as direct attachment of Walton-Brodie type force gauges to the ventricular walls, estimation of beat-by-beat ejection fraction, or the measurement of CO in combination with ventricular and systemic pressures for calculating stroke work, are employed.

11.5.3.2.2.2.4 Tonic Parasympathetic Activity Studies of tonic parasympathetic restraint of heart rate in several bird species have shown that this, like sympathetic tone, varies over a wide range. The factors discussed in Section 11.5.3.2.1, which affect the evaluation of sympathetic tone (state of anesthesia, presence or absence of tone from the other autonomic limb), apply equally to the analysis of vagal tone, and the protocols used in studies of the extent of vagal restraint will therefore influence the way in which the results of these studies are interpreted. In a study by Johansen and Reite (1964) in awake or anesthetized ducks with intact sympathetic cardiac innervation, section of one vagus nerve (right or left) produced no change in heart rate; the chronotropic response to subsequent section of the remaining vagus was

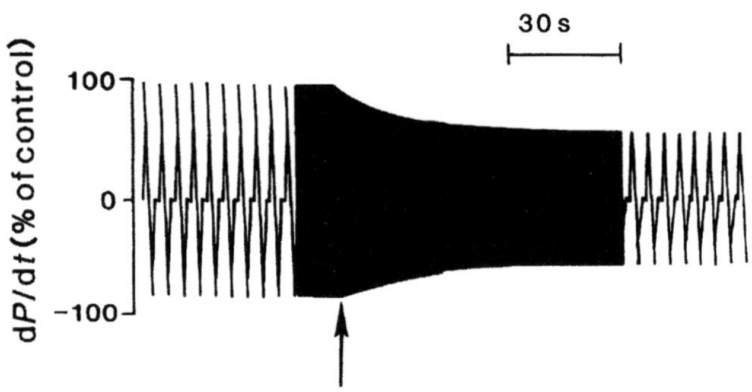

FIGURE 11.45 Effect of stimulation of the right vagus nerve on the rate of change of left ventricular pressure (d*P*/d*t*) in the chicken heart. Vagally induced changes in this variable represent an index of changes in ventricular contractility. The trace depicts the electrically analyzed first derivative of left ventricular chamber pressure; the amplitude of each peak thus represents the maximum rate of change of the corresponding ventricular pressure pulse. The trace was obtained at two speeds; the segment on the left side was recorded at a high chart speed to show the rate of pressure change of individual pulses prior to vagal stimulation. Just before stimulation was started (at the arrow) the chart speed was reduced to display the response to the first part of an 80s stimulus train (frequency, 20 Hz); the 30s time bar over the trace applies to this segment. On the right the recorder was returned to a high speed to display the rate of change of individual pressure pulses once the response to vagal stimulation had reached a plateau. *From Lang and Levy (1989).*

an increase in heart rate up to 65% above that in intact or unilaterally vagotomized animals. These results imply the presence of strong tonic vagal restraint of the heart; however, the authors found that after bilateral vagotomy, β-blockade revealed a high degree of sympathetic tone to the heart. Since the authors did not perform the converse experiment of β-blockade prior to bilateral vagotomy in this study, neither the balance between parasympathetic and sympathetic tone nor the actual degree of parasympathetic restraint on the heart could be ascertained. In unanesthetized ducks, Butler and Jones (1968), using a combination of cold block and section of the cervical vagi, showed that mean resting heart rate rose about 180% over the rate prior to these manipulations. In another study by the same authors, pharmacological blockade of the parasympathetic nervous system with atropine caused mean heart rate to increase to about 150% over the control value in awake ducks (Butler and Jones, 1971). In both of these studies the sympathetic nervous system was functional during evaluations of parasympathetic tone. In chickens Butler (1967) found that heart rate increased to 138% over the control value after bilateral vagal nerve section; in contrast to this, Bopelet (1974) reported an increase of only 8%, and Peterson and Nightingale (1976) found no change in rate in chickens after bilateral vagotomy. Sympathetic influence on the heart had not been eliminated in any of these studies. Sturkie and co-workers (reviewed by Sturkie, 1986) addressed the problem of evaluating parasympathetic tone to the heart by examining the effects of pharmacological or surgical vagotomy in the absence of sympathetic cardiac influences in the chicken using the rationale discussed in Section 11.5.3.2.1. They found that the net restraining effect of tonic parasympathetic activity on the chicken heart was the equivalent of a 20% reduction in heart rate from the rate of the completely decentralized heart. This estimate of vagal tone is substantially lower than the estimates of other workers in chickens (Butler, 1967) or in ducks (Johansen and Reite, 1964; Butler and Jones, 1968, 1971) in which the sympathetic cardiac innervation was functional. The degree of parasympathetic restraint on the heart in the latter studies may have been exaggerated by the presence of ongoing sympathetic drive after the lifting of vagal influence.

11.5.3.2.3 Control of CO

The volume of blood pumped by either the left or right ventricle per unit time is termed CO. The total volume of blood pumped by the heart per unit time is therefore twice the CO, since the outputs of both sides of the heart must be exactly matched over time. CO, usually expressed in units of mL/min, is the product of the rate of contraction of the heart (beats/min) and the volume pumped during each beat, or stroke volume (mL). Rate and stroke volume are determined by factors which may be intrinsic or extrinsic to the heart. Intrinsic factors include atrial pacemaker cell activity and contractile properties of the cardiac muscle fibers. Extrinsic factors affecting rate and stroke volume include autonomic nervous activity and levels of circulating cardiotropic hormones. CO over a given time period will thus be determined by the complex interplay of these intrinsic and extrinsic factors. Some of these have been discussed in previous sections, and references will be made to them as necessary.

11.5.3.2.3.1 Role of Heart Rate in Control of CO The basal level of heart rate in the absence of external influences is primarily determined by the inherent membrane properties of the pacemaker cells of the SA node (Section 11.2.3). The rate of depolarization, and thus the rate of discharge of APs in these cells, is set by the ion conductances through their membranes (particularly K^+), but this process is itself partly dependent on physical factors such as the concentrations of extracellular ions and temperature. Under most circumstances these physical factors are kept within fairly narrow limits by homeostatic mechanisms and should therefore not affect heart rate substantially.

The interaction of the sympathetic and parasympathetic branches of the autonomic nervous system controls heart rate in a complex, nonlinear manner. Part of this complexity arises from the fact that the full chronotropic effects of vagal stimulation on heart rate occur within a few heartbeats whereas it may take up to 30 s for the full expression of the cardiac response to sympathetic stimulation (Figure 11.46). Consequently, the most rapid changes in heart rate appear to be dominated by the parasympathetic system. The major reason for the nonlinearity is that the degree of parasympathetic–sympathetic interaction at the heart may be changed by increasing output of one of these limbs.

The relationship of heart rate to bilateral stimulation of the distal cut ends of the vagus and cardiac sympathetic nerves of the duck *A. platyrhynchos* is illustrated in Figure 11.47 (Furilla and Jones, 1987b). In this figure, 100% represents the frequency of stimulation above which no further changes in heart rate occurred. The heart rate resulting from a given level of vagal and sympathetic stimulation was plotted on perspective graph paper and the surface was drawn, by eye, to encompass all heart rates obtained in the stimulation experiments. Area B represents complete cardiac denervation and area E maximal vagosympathetic stimulation. The effects of varying the intensity of sympathetic stimulation at minimal and maximal vagal activity are represented by the lines A–B and D–E, respectively. Similarly, the effects of varying the intensity of vagal stimulation at minimal and maximal sympathetic activity are represented by the lines B–D and A–E, respectively.

Increasing sympathetic activation at zero vagal activity causes heart rate to rise over 200 beats/min (B to A) while increasing sympathetic drive at maximal vagal activation only increases heart rate by 50 beats/min (D to E). Similarly, increasing vagal activation causes heart rate to fall by more than 200 beats/min at zero sympathetic activity (B to D) and by nearly 400 beats/min at maximal sympathetic activity (A to E). Obviously, the greater the vagal drive, the more this input is able to occlude sympathetic effects on the heart, accounting for the nonlinearity in the interplay between the two branches of the autonomic nervous system at the cardiac pacemaker. This increased parasympathetic

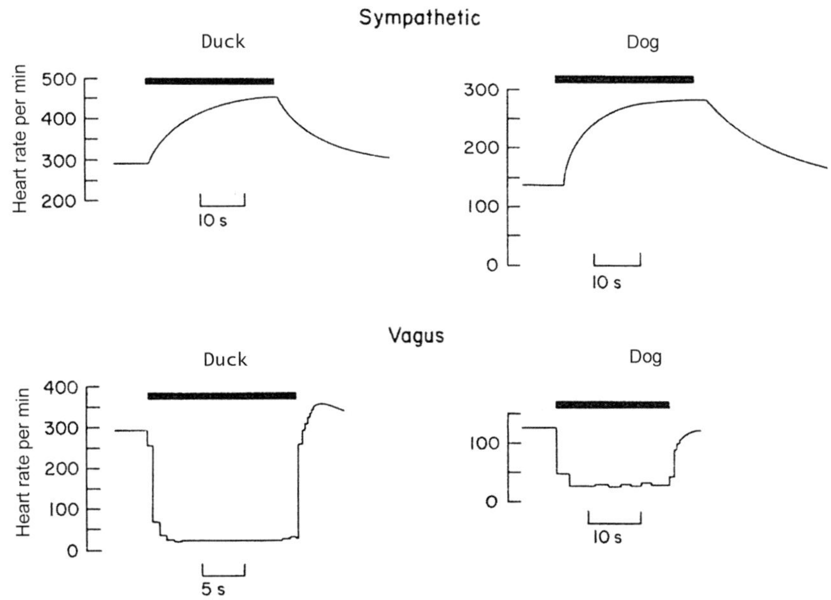

FIGURE 11.46 Comparison of time courses of changes in heart rate evoked by bilateral stimulation of cardiac sympathetic nerves (top panels) and vagus nerves (bottom panels) in a duck (left-hand traces) and dog (right-hand traces). In each panel the thick horizontal bar represents the duration of the stimulus. *Modified from Furilla and Jones (1987a).*

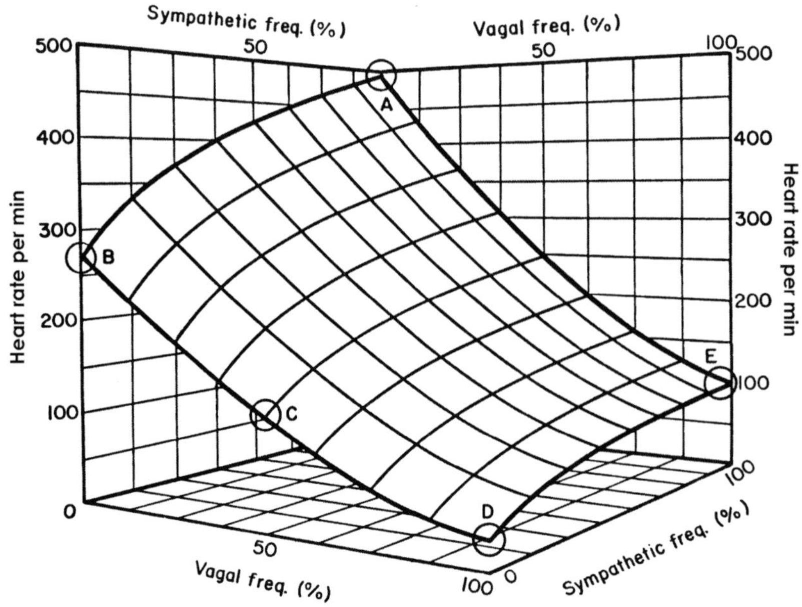

FIGURE 11.47 The relationship of heart rate to bilateral stimulation of the distal cut ends of the vagus and cardiac sympathetic nerves of the Pekin duck. "One-hundred percent" represents the frequency of stimulation above which no further changes in heart rate occurred with increases in stimulation frequency. The heart rate caused by a given level of vagal and sympathetic stimulation was plotted on "perspective" graph paper and the surface was drawn, by eye, to encompass all heart rates obtained in the stimulation experiments. See the text for an explanation of points A–E. *From Furilla and Jones (1987a).*

effectiveness in cardiac control is termed accentuated antagonism and may be mediated through two mechanisms. First, in response to sympathetic stimulation above a certain threshold, there is a cholinergically mediated reduction in prejunctional release of NE. Second, the magnitude of the postjunctional response to a given level of sympathetic stimulation is attenuated by ACh (Levy, 1971).

Short-term heart rate fluctuations are caused by the continued tug-of-war between the two branches of the autonomic nervous system at the pacemaker. This interplay may overlie long-term changes in heart rate caused, for instance, by changes in levels of circulating hormones. The normal homeostatic processes tend to reduce variability and maintain constancy of internal physiological functions, and short-term fluctuations in heart rate are usually seen as perturbations away from the norm. If a series of cardiac intervals is recorded from an animal, the duration of intervals within the series appears quite irregular with apparently random fluctuations occurring all the time. Whether these fluctuations are truly random or patterned, in the latter case providing evidence for chaotic control of heart rate (Denton et al., 1990; Goldberger et al., 1990), is a matter of some controversy. The strongest evidence for chaotic control of heart rate may lie in the morphology of the nerves innervating the heart. The nerves divide repeatedly, like the branching of a tree which is an intrinsically fractal structure. Hence, if the anatomy is fractal, then why should the day-to-day workings of the system not be fractal as well? (Goldberger, cited in Pool, 1989; Goldberger, 1991).

In any event, it is clear that, rather than maintaining a homeostatic steady state, heart rate fluctuates considerably, even when recorded over short time periods (Figure 11.48(A)). The interbeat intervals form a time series which can be transformed into the frequency domain by Fourier analysis, revealing the presence of periodic components within the series. The square of the absolute value of the Fourier transform yields the power spectrum of the heart rate variability (PS/HRV) (Kamath and Fallen, 1993).

The PS/HRV of an intact, resting, duck (*Aythya affinis*) is shown in Figure 11.48(B)(i). This plot reveals a respiratory sinus arrhythmia as a single major peak at the respiratory frequency. This is a manifestation of respiratory modulation of cardiac parasympathetic activity (Shah et al., 2010; the response to sympathetic heart stimulation is too slow to significantly affect heart rate at a high frequency). Blockade of the sympathetic nervous system with a β-antagonist tends to increase heart rate variability although the amplitude of the high-frequency components are reduced (Figure 11.48(B)(ii)), while parasympathetic blockade with atropine gives a regular, unvarying heart rate (Figure 11.48(B)(iii)). This confirms that short-term heart rate control is dominated by the parasympathetic nervous system in *A. affinis*. Similar control of heart rate variability has been found in hatchling emu (*Dromaius novaehollandiae*) (Shah et al., 2010).

The PS/HRV is a quantifier of autonomic responsiveness (Saul, 1990) and allows evaluation of cardiovascular regulation in birds over long time courses and during many types of activities. Also, the influence of other periodic functions

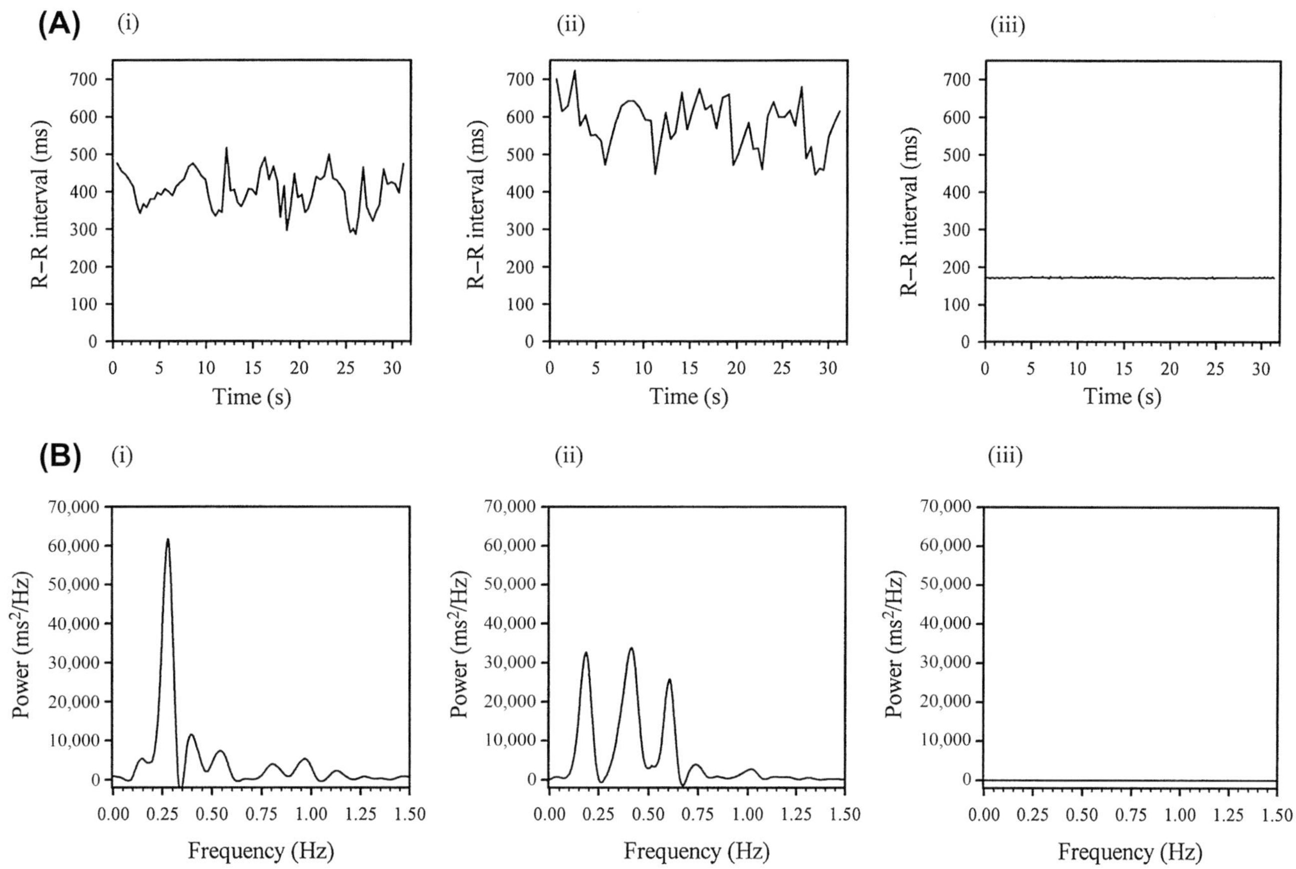

FIGURE 11.48 (A) 30s duration time series of interbeat intervals (R–R intervals) recorded by radiotelemetry from a duck (*Aythya affinis*) while the animal was resting quietly on the water surface. (i) Control; (ii) after sympathetic nervous system blockade with Nadolol (a β-adrenergic antagonist); and (iii) after parasympathetic nervous system blockade with the muscarinic antagonist atropine. (B) Power spectra of heart rate variability derived from the time series shown in (A). (i) Control; (ii) after Nadolol blockade; and (iii) after atropine blockade. Because the time series were too short to adequately display extremely low-frequency components, these components were removed with a high-pass filter from the power spectra. *From McPhail et al. (unpublished data).*

such as arterial blood pressure and vasomotor fluctuations on PS/HRV can be evaluated using this technique. Telemetric recording of heart rate, combined with frequency analysis of cardiac function, will open new doors for studying control of physiological processes in unrestrained, active birds.

11.5.3.2.3.2 Role of Stroke Volume in Control of CO Stroke volume, like heart rate, is dependent upon factors intrinsic and extrinsic to the heart. As all myocytes within the heart contract during each beat, the primary intrinsic factors which determine stroke volume are the inherent contractile properties of each muscle fiber and the resting lengths of all the fibers. The amount of force developed during contraction by a cardiac muscle fiber at a specified precontraction length is properly termed *contractility*, but this term has also been used more loosely to describe the collective contractile properties of all muscle fibers associated with one chamber of the heart. A major problem in quantifying contractility is that the force developed by a single cardiac muscle fiber is difficult to measure in working hearts. Consequently a number of indirect indices have been developed to estimate this variable. These include measuring cardiac outflow volume over time to calculate stroke volume; recording the ventricular peak systolic pressure developed against a fixed afterload or into a constant arterial pressure; and measuring the rate of rise of ventricular pressure during systole. The major assumption in all of these methods is that the measured variable reflects the contractility of all muscle fibers integrated over the dimensions of the whole chamber. However, the variety of indices of contractility used by investigators under different experimental conditions has made it difficult to compare estimates across studies. The direct measurement of volume flow from the ventricle would appear to give the most reliable index of cardiac contractility, being independent of the complicating effects of changing arterial or ventricular pressures. This measurement is also among the most difficult to make, requiring highly invasive procedures to place the appropriate instrumentation.

By analogy with the contraction of skeletal muscle, the amount of force developed by a contracting cardiac muscle fiber depends upon its precontraction length ("preload"). This principle was first applied to the heart by Otto Frank (summarized in Rushmer, 1976), who showed that, within limits, the greater the preload on ventricular muscle in diastole, the more tension was developed during the next systole. This length–tension relationship was further investigated by Ernest Starling and coworkers, who demonstrated that the amount of blood ejected by the left ventricle during systole was proportional to the volume of blood in the ventricle at the end of the diastolic filling phase of the cardiac cycle. These concepts have been combined into the Frank–Starling relationship to describe the intrinsic responses of ventricular stroke volume to changes in cardiac venous return, expressed graphically in Figure 11.49. Elevated contractility of each muscle fiber in the ventricle is evoked by increasing the preload on all of the fibers by increasing the volume of blood filling the ventricle before each beat; this is reflected in an overall increase in ventricular stroke volume. The curve designated A in Figure 11.49 is termed a ventricular function curve. The relationship embodied in this curve demonstrates an autoregulatory feature of cardiac function known as heterometric regulation: a change in the resting fiber length (heterometry) results in a change in contractility in the same direction. This regulatory mechanism is an intrinsic property of cardiac muscle. The consequence of this mechanism for the overall function of the heart is that, if all other conditions remain constant, CO will be determined by venous return. An increase in venous return to the left ventricle via the left atrium will result in greater end-diastolic stretch of the ventricle walls and an increase in stroke volume at the next beat; conversely, stroke volume will be reduced if cardiac return falls. In short, the heart "pumps what it gets" if all other factors are unchanging.

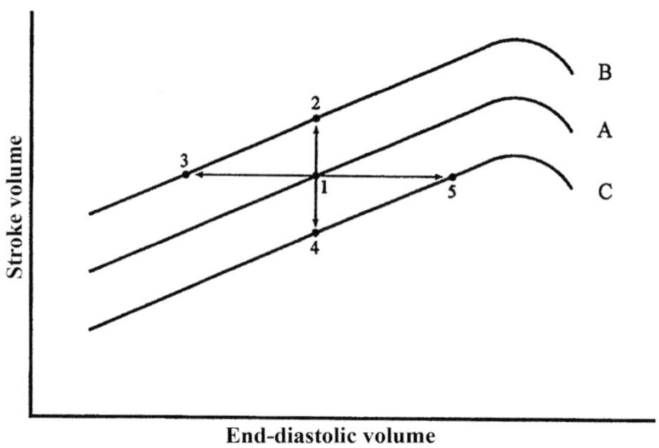

FIGURE 11.49 Idealized graphical representation of the Frank–Starling relationship for cardiac ventricular muscle. (A) Intrinsic ventricular function curve depicting the relationship between end-diastolic volume (representing degree of stretch of muscle fibers) and stroke volume (index of contractility) in the absence of extrinsic influences. The curve peaks and begins to decline at high end-diastolic volumes because resting sarcomere length is maximal here. (B and C) Factors extrinsic to the heart that alter inotropic function of cardiac muscle reset the ventricular function curve to operate over different ranges of stroke volume, independent of end-diastolic volume or initial fiber length. (B) Elevated cardiac sympathetic drive or circulating catecholamines have positive inotropic effects, resetting the curve toward higher stroke volumes. (C) Elevated vagal drive has negative inotropic effects, resetting the curve toward lower stroke volumes. Points 1 through 5 are the operating points assumed for the text discussion of the effects of extrinsic factors on ventricular function.

Cardiac contractility is influenced by extrinsic factors in addition to the intrinsic Frank–Starling mechanism. Circulating hormones such as EPI and the autonomic neurotransmitters NE and ACh (see Sections 11.5.3.1 and 11.5.3.2) directly affect the contractility of cardiac muscle fibers. These extrinsic factors are superimposed on the intrinsic autoregulatory factors governing stroke volume and can shift the whole ventricular function curve (curve A in Figure 11.49) toward higher (curve B) or lower (curve C) stroke volumes at the same resting muscle fiber length or degree of ventricular filling. This type of regulation of stroke volume is referred to as homeometric regulation, to emphasize the fact that changes in contractility can occur independent of resting fiber length. An increase in sympathetic drive to the heart or an increase in the level of circulating catecholamines will increase ventricular inotropic function homeometrically; thus a greater stroke volume will result from the same degree of cardiac filling, as indicated in Figure 11.49 by the arrow from point 1 on curve A to point 2 on curve B. Another way to consider this is that after such a shift in the curve a much smaller end-diastolic volume will give the same stroke volume (arrow from point 1 to point 3). On the other hand, elevated vagal drive to the heart of birds can decrease the contractility of ventricular muscle and will shift the ventricular function curve toward a lower stroke volume (curve C in Figure 11.49). At the new operating point, the same degree of preload will result in a lower stroke volume (arrow from point 1 to point 4); alternatively, a much larger end-diastolic volume will be required to maintain the same stroke volume (arrow from point 1 to point 5).

The arterial pressure against which the ventricle pumps ("afterload") is a major extrinsic factor in determining the magnitude of stroke volume. The pressure generated during the isometric phase of ventricular contraction is a function of the contractility of the muscle fibers, and when chamber pressure exceeds that in the aorta the valves open and blood is ejected from the ventricle during the isotonic phase. If the preload on the ventricle is increased by elevating the arterial blood pressure without a change in contractility or end diastolic volume, stroke volume of subsequent beats will be reduced because more energy will be required to raise chamber pressure above the new level of arterial pressure. Initially, this will leave a larger fraction of the previous end-diastolic volume still in the chamber at the end of systole, resulting in an increased level of resting tension on the muscle fibers during the next filling phase. This increased tension, according to the Frank–Starling mechanism, will quickly result in increased contractility during subsequent beats, restoring stroke volume by heterometric regulation in the face of the increased arterial pressure.

In many species of birds, CO is adjusted to match perfusion requirements of the tissues in a variety of conditions, such as during exercise, hypoxia, or submersion (see Section 11.6). These adjustments appear to be made primarily through alterations in heart rate, with stroke volume remaining relatively unchanged. Changes in CO during exercise are driven by increased heart rate in ducks (Bech and Nomoto, 1982; Kiley et al., 1985), geese (Fedde et al., 1989), and turkeys (Boulianne et al., 1993a,b). However, in the emu (Grubb et al., 1983) and the chicken (Barnas et al., 1985) stroke volume may increase by up to 100% during exercise, contributing significantly to elevated CO. Reflex changes in CO mediated by systemic arterial baroreceptor input also appear to operate via alterations in heart rate, leaving stroke volume relatively unchanged (Section 11.5.4.2). In summary, during exercise, hypoxia, or submersion, birds display significant changes in heart rate, arterial blood pressure, and venous return from the resting condition. In the transition from the resting condition to these altered states, stroke volume also varies. However, in most of the species examined so far, stroke volume returns to values close to those at rest after a short period of initial adjustment. This indicates that intrinsic autoregulation of CO has the potential to play an important role in the maintenance of stroke volume in the face of large-scale circulatory adjustments.

11.5.4 Reflexes Controlling the Circulation

11.5.4.1 Chemoreflexes

In birds, reflex adjustments of CO and vascular caliber are generated by chemoreflexes in response to changes in levels of oxygen, carbon dioxide, and pH in the cerebrospinal fluid and in arterial blood. Receptors sensitive to CO_2 in cerebrospinal fluid are present in the avian central nervous system (Jones et al., 1982), but no detailed studies of the location or transduction properties of these receptors have been done in birds. However, if these receptors are similar to those found in mammals, they may be located at or near the surface of the ventrolateral medulla (reviewed by Schlaefke, 1981). Arterial chemoreceptors in birds are located primarily in the carotid bodies, bilateral structures lying caudal to the thyroid gland, and close to the ultimobranchial gland and nodose ganglia of the vagus nerves and the carotid artery (from which they are supplied with blood) (Figure 11.39; Adams, 1958; Jones and Purves, 1970; Kameda, 2002). The carotid bodies are innervated by the vagus and the recurrent laryngeal nerves (Kameda, 2002). Chemoreceptors have also been reported in aortic bodies associated with the roots of the great vessels in several species of birds (Tcheng et al., 1963; Ito et al., 1999).

The discharge characteristics of arterial chemoreceptors in response to changes in arterial P_{CO_2} (P_{aCO_2}), P_{O_2} (P_{aO_2}) and arterial pH (pH_a) have been studied in the duck and chicken. Receptor discharge rate increases proportionally with P_{aCO_2} and in inverse proportion to P_{aO_2} (Bouverot and Leitner, 1972; Bamford and Jones, 1976; Nye and Powell, 1984; Hempleman et al., 1992). Discharge sensitivities to changes in P_{aCO_2} and P_{aO_2} have been quantified by Hempleman et al. (1992) for carotid body chemoreceptors in the duck. When P_{aO_2} was maintained at a normoxic level (near 100 mm Hg) mean chemoreceptor sensitivity to step changes in P_{aCO_2} (produced by changing the fraction of CO_2 in the air breathed by the bird) was +0.20 impulses/s mm Hg/P_{aCO_2}. Hypoxia (P_{aO_2} 56 mm Hg) potentiated chemoreceptor discharge in response to altered P_{aCO_2} (Figure 11.50). In this condition, the same step changes in P_{aCO_2} as given in normoxia resulted in a sensitivity of +0.32 impulses/s mm Hg/P_{aCO_2}. In the absence of CO_2 in the inspired air, step changes in P_{aO_2} from the normoxic level to about half this level resulted in a mean sensitivity of about −0.10 impulses/s mm Hg/P_{aO_2}. Carotid body chemoreceptors are thus about twice as sensitive to changes in absolute levels of arterial CO_2 as to O_2. Some of the chemoreceptors sampled in this study were also sensitive to the rate of change of O_2 and CO_2 in the blood, indicating that proportional and rate-related information on arterial blood gas status are both transmitted to the central nervous system. However, the maximum rate of change of discharge of most rate-sensitive chemoreceptors was in the lower frequency range of physiologically occurring blood gas oscillations. Therefore, only relatively low

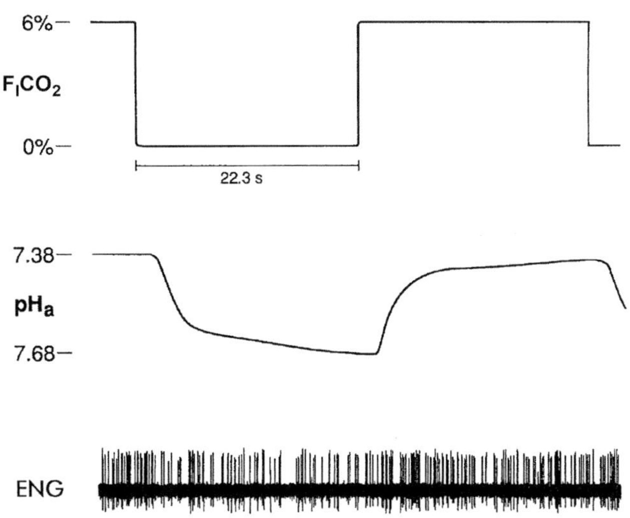

FIGURE 11.50 Single-fiber arterial chemoreceptor response recorded from vagal slip (bottom trace, electroneurogram (ENG)) to step changes in inspired CO_2 level (top trace, fraction of CO_2 in inspired gas F_{ICO_2}) during hypoxia (10% O_2 in inspired gas) in a duck. Arterial pH (middle trace (pH_a)) also changes in step with inspired CO_2. Chemoreceptor discharge is sensitive to the level of CO_2 and pH_a and to rate of change of these variables. *Reprinted from Hempleman et al. (1992).*

frequencies of blood gas oscillation, such as those occurring at rest or during low-intensity activities, will be faithfully transduced. At higher oscillation frequencies, such as those occurring during panting or high-intensity exercise, chemoreceptor inputs to the central nervous system probably represent mean blood gas levels averaged over several oscillatory cycles.

Arterial chemoreceptors are spontaneously active at normoxic and normocapnic blood gas levels in birds and these receptors have been proposed to play an important role in setting the level of eupneic ventilation under these conditions (Bouverot and Leitner, 1972). However, the role of arterial chemoreceptors in reflex control of the avian circulation is less clear. Analyses of the circulatory effects of chemoreceptor activation are complicated by parallel changes in ventilation. During apneic asphyxia in ducks, blood oxygen tension falls, CO_2 tension rises, and carotid body chemoreceptors become progressively more strongly stimulated; this input plays a major role in initiating and maintaining an intense bradycardia (Jones and Purves, 1970; Butler and Taylor, 1973). However, if blood gases in spontaneously breathing ducks are artificially adjusted to mimic the hypoxic and hypercapnic levels achieved during apneic asphyxia, chemoreflex drive acts to increase ventilation, leading to elevated drive from pulmonary receptors. In this state, there is little or no change in heart rate (Butler and Taylor, 1973, 1983). When the rise in ventilation (and thus the elevation in pulmonary receptor input) is prevented by controlling tidal volume and respiratory frequency during hypoxic hypercapnia in paralyzed but unanesthetized animals, heart rate falls to a level midway between that in normoxic, normocapnic animals and that obtained at end-dive (Butler and Taylor, 1973, 1983). This effect is illustrated in Figure 11.51. Subsequent cessation of respiration by stopping the ventilator pump then results in the full expression of chemoreceptor mediated bradycardia: heart rate falls to the same level as at end-dive (Butler and Taylor, 1973, 1983). These experiments show that in ducks, as in mammals, the cardiovascular responses to strong arterial chemoreceptor stimulation during spontaneous breathing are masked by elevated ventilatory drive. Butler and Taylor (1983) have suggested that pulmonary receptors activated by increased ventilation contribute to occlusion of the cardiac chemoreflex.

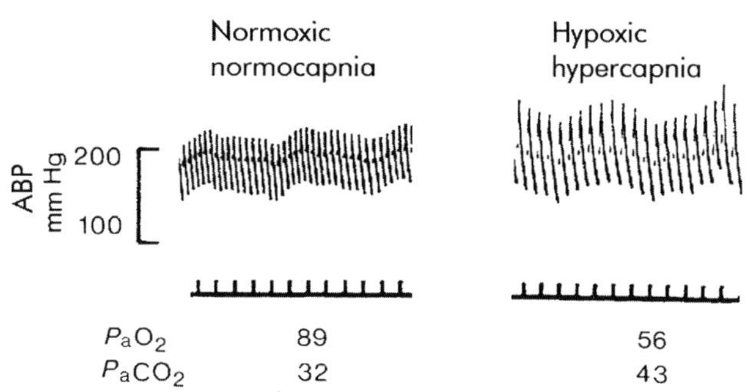

FIGURE 11.51 Effect of systemic hypoxic hypercapnia on heart rate and blood pressure in mallard duck. The animal was artificially ventilated after spontaneous respiratory movements were suppressed with intravenous pancuronium bromide. (Left panel, upper trace) Arterial blood pressure and heart rate during ventilation with air; systemic arterial blood gas values are shown at bottom. (Right panel) When blood gas values were adjusted to match those at end of 1 min submersion by altering CO_2 and O_2 levels in inspired gas, heart rate decreased as a result of arterial chemoreflex activation. In both panels the ticks on the bars under the pressure traces indicate 1 s intervals. *Reprinted from Butler and Taylor (1973).*

Input from peripheral arterial chemoreceptors can reflexly alter peripheral vascular resistance but the general role of chemoreceptors in control of the vasculature has not been established. Indirect evidence for chemoreflex effects on the vasculature comes from a study by Bouverot et al. (1979), who showed a trend toward increased peripheral resistance in carotid body-intact ducks subjected to arterial hypoxia while breathing spontaneously. The same stimulus in animals after denervation of the carotid bodies led to a 40% fall in peripheral resistance, indicating a potential role for these chemoreceptors in generating the vascular response to hypoxia. As with cardiac responses to carotid body stimulation, elevated ventilation during hypoxia may mask the full extent of reflexogenic vasoconstriction. In the study by Bouverot et al. (1979), no attempt was made to evaluate this interaction by controlling ventilation.

It is clear that the full role of arterial chemoreflexes in circulatory control has yet to be defined in birds. Elevated carotid body input can result in both ventilatory and cardiac responses, and both of these responses are important in maintaining oxygen delivery to and CO_2 washout from working tissues. Further experiments are necessary to establish the relative importance of these limbs of the chemoreflex in matching ventilation with perfusion to cope with changes in internal and external levels of these gases.

11.5.4.2 Baroreflexes

Arterial blood pressure provides the driving force for perfusion of the systemic vascular beds and must therefore be maintained within limits that ensure optimal tissue blood flow under a variety of physiological conditions. Blood pressure in birds is maintained by the baroreflex, a mechanism employing negative feedback (see reviews by Bagshaw, 1985; Smith, 1994). Adjustments in blood pressure produced by the baroreflex are driven by afferent signals from arterial baroreceptors. These are mechanoreceptors with their receptor endings embedded in connective tissue of the arterial wall, where they sense changes in arterial pressure as variations in wall tension. An increase in intraarterial pressure results in an increase in circumference of the vessel wall, which in turn stretches baroreceptor endings to increase their frequency of discharge of APs. Arterial baroreceptors in birds are located primarily in the walls of the aorta close to the left ventricular valves (Jones, 1973). Their axons course to the brainstem via aortic nerves which arise from the nodose ganglia of the vagus (Nonidez, 1935). In the few studies of avian baroreceptor function done to date, discharge characteristics in response to changes in blood pressure appear to be similar to those of mammalian high threshold, slowly adapting baroreceptors (Jones, 1969, 1973). Spontaneous baroreceptor impulse generation is phase locked to mechanical events in systole of the cardiac cycle, as shown in Figure 11.52. Baroreceptors are also sensitive to the rate of rise of the pressure pulse. It thus appears that baroreceptors are capable of transmitting information on cardiac rate, peak systolic pressure, and the slope of the aortic pressure waveform (which may in turn reflect cardiac contractility) to the central nervous system.

Mean arterial blood pressures of resting birds are higher than those recorded in mammals of equivalent body weight at rest; indeed, mean resting pressures may exceed

FIGURE 11.52 Relationship between baroreceptor discharge (upper trace, electoneurogram (ENG)), recorded from peripheral cut end of aortic nerve, and blood pressure (lower trace (ABP)), recorded from a brachiocephalic artery in an anesthetized duck. Baroreceptor discharge is synchronous with systolic peak pressure. ABP scale bar represents a pressure span from 100 to 200 mm Hg. *From Jones (1973).*

150 mm Hg in some avian species (Altman and Dittmer, 1971). Avian baroreceptor discharge occurs at resting levels of blood pressure in birds (Jones, 1969, 1973), so it is likely that the baroreflex is tonically active at these pressures. Changes in blood pressure sensed at the receptors are represented to the baroreflex circuitry in the central nervous system by changes in baroreceptor afferent discharge frequency. The baroreflex acts to adjust pressure in a direction opposite to that of the initial pressure change. This reflex operates through both peripheral vascular and cardiac effectors to return blood pressure toward a set level after disturbances. Pressure is thus maintained within fairly narrow limits over time to ensure a constant pressure head for tissue perfusion. Little is known of the central nervous mechanisms involved in blood pressure regulation in birds. Inhibition of NMDA receptors by ketamine attenuated the pressor response in the pigeon suggesting a role for the vagal nerve as in mammals (Lucitti and Hedrick, 2006). It is assumed that the basic organization of central components of the baroreflex is similar to the arrangement in mammals (Lucitti and Hedrick, 2006).

The primary cardiovascular response to changes in blood pressure is a baroreflex-mediated change in CO, seen in Figure 11.53(A) as a rapid fall in heart rate in response to a pharmacologically induced pressure increase. Such reflex responses are completely abolished by bilateral section of the aortic nerves (Figure 11.53(B)). The sensitivity of the blood pressure–heart rate relationship ranges from

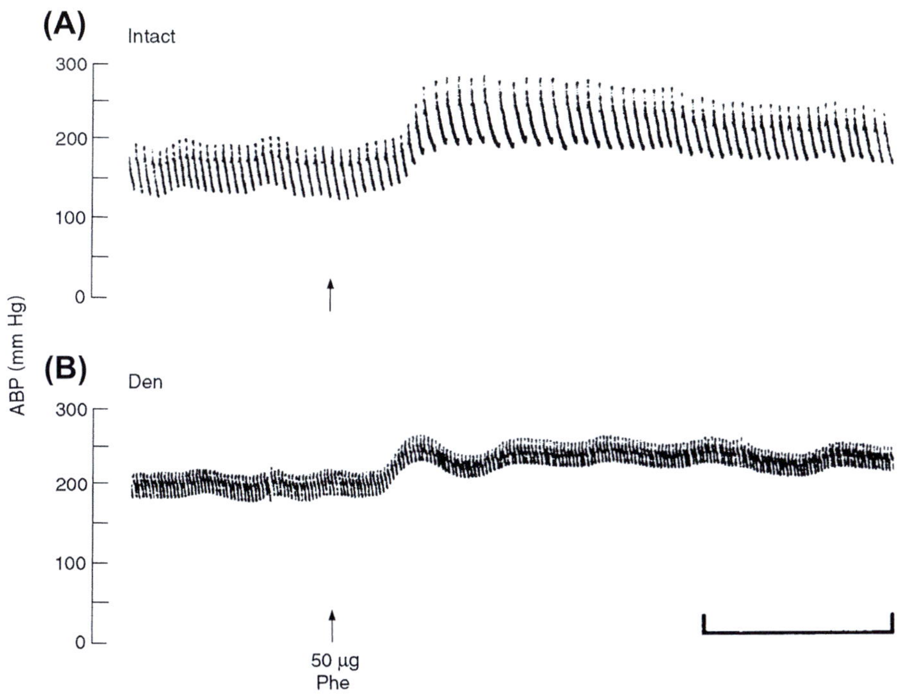

FIGURE 11.53 Heart rate responses to increased arterial blood pressure (ABP) induced by bolus intravenous doses of phenylephrine (Phe, injected at arrows) in the same duck (A) before and (B) after barodenervation by section of aortic nerves. Baroreflex-mediated bradycardia was eliminated by denervation of baroreceptors. Note increased preinjection blood pressure and heart rate in barodenervated animals, indicating a degree of baroreflex-mediated restraint on cardiovascular system in baroreceptor intact animal. *From Smith (unpublished data).*

−0.5 to −3.13 beats/min/mm Hg, depending on species and method of evaluation (chickens, Bagshaw and Cox, 1986; ducks, Smith and Jones, 1990, 1992; Millard, 1980; pigeons, Lucitti and Hedrick, 2006; see Smith, 1994 for further discussion). The baroreflex-mediated effects of changing pressure at the receptors can be mimicked by electrical stimulation of the central cut end of an aortic nerve in an animal in which both aortic nerves are sectioned. This method was used by Smith and Jones (1992) to explore the dynamic role of baroreceptors in controlling the circulation. Stimulation of the aortic nerve in barodenervated animals evokes a decrease in arterial blood pressure in proportion to the stimulus frequency when stimulus current is set just above threshold for a response (Figure 11.54, closed circles). This response works primarily through a fall in CO, mediated by decreased heart rate with no change in stroke volume (Smith and Jones, 1992). This response follows the same pattern as the baroreflex-mediated response to pharmacologically induced pressure changes in baroreceptor-intact birds; that is, in both intact and denervated animal's baroreflex activation was expressed primarily through changes in heart rate. However, the baroreflex can engage peripheral vasomotion as well as cardiac responses: in barodenervated animals, stimulation of the aortic nerve with a current intensity several times the threshold level produced decreases in peripheral resistance as well as in heart rate (Smith and Jones, 1992). These data suggest that when relatively small disturbances in pressure occur they will be compensated by adjusting CO, but larger changes in pressure will be corrected by a combination of cardiac and peripheral vascular adjustments.

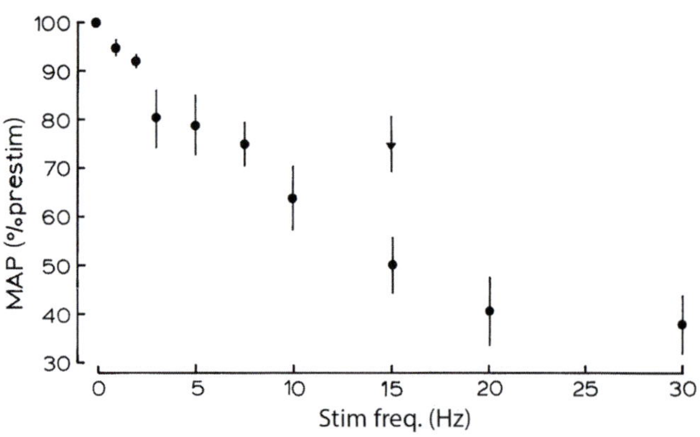

FIGURE 11.54 Normalized responses of mean arterial blood pressure (MAP) to electrical stimulation of aortic nerve in awake ducks spontaneously breathing air (P_{aO_2}: 88 mm Hg, P_{aCO_2}: 26 mm Hg; closed circles) or a hypoxic hypercapnic gas mixture (P_{aO_2}: 62 mm Hg, P_{aCO_2}: 44 mm Hg; triangle). Pressure response at each frequency of nerve stimulation is expressed as a percentage change in MAP relative to MAP just prior to stimulation. Loading the chemoreceptors attenuated the depressor effect of aortic nerve stimulation at 15 Hz. Error bars represent: 1 S.E.M. for five trials in normoxic conditions and three trials during hypoxic hypercapnia in three animals. Stimulus pulse duration and amplitude were constant within each trial. *From Smith (unpublished data).*

The effectiveness of baroreflex control of the circulation can be modified by interaction with other reflexes, such as the chemoreflex, which may be concurrently engaged. In an effort to determine the cause of an apparent reduction in baroreflex function observed during submersion in ducks (Jones et al., 1983; Millard, 1980), Smith and Jones (1992) stimulated the aortic nerve in barodenervated ducks before and during periods of elevated chemoreceptor drive. Stimulation of chemoreceptors was accomplished by ventilating animals with a gas mixture which simulated the hypoxic and hypercapnic blood gas values observed at the end of 2 min of submersion; this significantly decreased the capability of aortic nerve stimulation to affect mean arterial pressure. This is shown in Figure 11.54 by a 50% decrease in the pressure response to aortic nerve stimulation at a frequency of 15 Hz during hypoxic hypercapnia (triangle), compared with the responses during air breathing (closed circles). That this occlusive response was due to chemoreceptor activation was further demonstrated by the attenuation of baroreflex function during perfusion of one vascularly isolated carotid body with venous blood in otherwise normoxic, normocapnic animals which were spontaneously breathing (Smith and Jones, 1992). It therefore appears that the chemoreceptor drive that develops after the first minute of submersion (Jones and Purves, 1970) is at least partly responsible for the attenuation of baroreflex control of cardiovascular variables which occurs during this period.

11.5.4.3 Reflexes from Cardiac Receptors

Small nerve terminals in the shape of simple knobs, plates or rings as well as straight or spirally wound nerve endings have been observed in anatomical studies of avian atria and ventricles (Ábrahám, 1969; also see review by Jones and Milsom, 1982), but bird hearts do not appear to have the complex and highly developed sensory receptor endings present in large numbers in mammalian hearts. Few functional studies of avian cardiac receptors and their reflexogenic effects have been done so far. Jones (1969) established in the duck that some afferent fibers in the cervical vagus had receptor endings associated with the heart, responding to punctate stimulation near the AV junction and discharging spontaneously in patterns which were phase-locked to mechanical events in the cardiac cycle. In the chicken, Estavillo and Burger (1973a,b) found that a majority of cardiac receptors with their cell bodies in the nodose ganglia had receptive fields located in the left ventricle near the aortic valves. Discharge patterns of these receptors were either phase-locked to the cardiac cycle or were irregular and apparently unrelated to mechanical events in the cycle. The discharge of receptors of both types could be modulated by varying inspired CO_2 and pH_a independently or together. In this study, the discharge rate of phasically firing cardiac receptors was proportional to arterial blood pressure over a wide pressure range, and this relationship was reset toward lower discharge rates at increased CO_2 levels, as shown in Figure 11.55. Bilateral section of the middle cardiac nerve, carrying the axons of cardiac receptors in the chicken, produced an immediate rise in arterial blood pressure (Estavillo, 1978; Estavillo et al., 1990) reminiscent of that produced in ducks after section of the aortic nerve (Figure 11.53).

Avian cardiac receptors have been proposed to contribute to blood pressure regulation and to the control of ventilation, providing sensory feedback on intracardiac pressures and volumes. Such feedback may be modulated by changes in the levels of P_{aCO_2} and pH_a (Estavillo and Burger, 1973b; Estavillo et al., 1990). In the latter study, bilateral section of the middle cardiac nerve considerably blunted the increase in ventilation elicited by systemic hypercapnia, in addition to promoting elevated blood pressure.

Birds exhibit a Bezold–Jarisch reflex similar to mammals. It manifests in ducks as a fall in heart rate and arterial blood pressure when cardiac receptors are stimulated (Blix et al., 1976; Jones et al., 1980). Blix et al. (1976) proposed, on the basis of pharmacological stimulation of cardiac receptors, that this reflex contributed to the generation and maintenance of the cardiac chronotropic response to submersion. In a reexamination of this issue, Jones et al. (1980) loaded and unloaded the cardiac receptors by altering left ventricular pressure to provide more realistic physiological stimulation of these receptors before and during submersion. The results of this study failed to confirm a link between cardiac receptor activation and diving bradycardia.

There is as yet insufficient evidence to determine the overall function of reflexes driven by inputs from cardiac receptors. Although these receptors can influence blood pressure and ventilation, they do not appear to have a primary role in pressure or ventilatory regulation. However, they may mediate some of the dynamic interactions between respiratory and circulatory systems, thus helping to correct ventilation–perfusion mismatches that can develop during exercise or in adverse environmental conditions.

11.5.4.4 Reflex Cardiovascular Effects from Skeletal Muscle Afferents

Reflex mediated changes in arterial blood pressure, heart rate, and other cardiovascular variables accompany exercise, hypoxia, and hypoxic hypercapnia in birds. In several studies designed to deduce the respective roles of peripheral arterial chemoreceptors, chemosensitive areas of the central nervous system and arterial baroreceptors in these cardiovascular responses, it has been suggested that inputs from these receptor groups do not account for all of the changes observed. Thus inputs from other receptor type must also be involved. In mammals there is a significant reflexogenic increase in mean arterial blood pressure induced by skeletal muscle activity; receptors for this response appear to be intramuscular terminals of group III afferent fibers (small myelinated fibers) and group IV fibers (unmyelinated or C-fibers) coursing in somatic nerves (see Coote, 1975 for review). Kiley et al. (1979) implicated muscle afferents in the cardiovascular responses to exercise in ducks, and Lillo and Jones (1983) proposed that somatic muscle afferents in ducks were at least partly responsible for those portions of the cardiac and vascular responses in hypoxic hypercapnia and ischemia which were independent of chemoreceptor activation. Furthermore, Solomon and Adamson (1997) demonstrated that ducks express an "exercise pressor" reflex similar to that in mammals. This effect consists of an increase in mean arterial pressure (largely due to elevation of diastolic pressure) induced by and sustained during static contraction of a large hind limb muscle, the gastrocnemius (Figure 11.56(A)). The authors concluded that this was a reflex effect since section of the sciatic nerve carrying

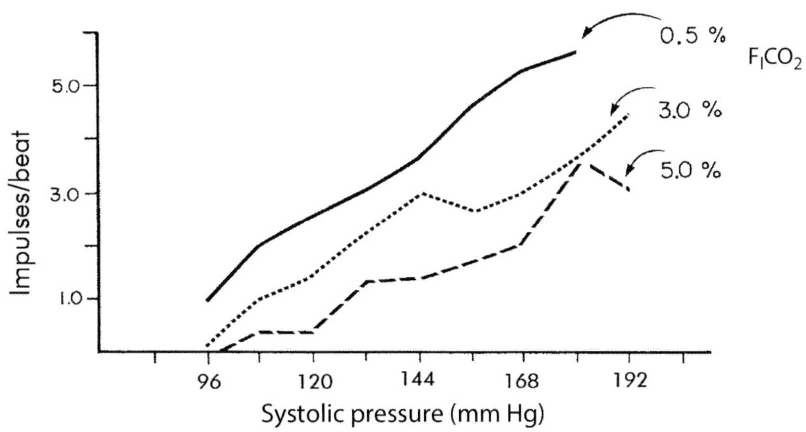

FIGURE 11.55 Modulatory effect of changes in inspired CO_2 (F_{ICO_2}) on the relationship between discharge rate of cardiac mechanoreceptors (impulses per heartbeat) and arterial blood pressure (systolic pressure) in chicken. Curves were obtained by plotting receptor discharge rate at a given blood pressure over a range of pressures produced by bolus intravenous injections of mecholylchloride or epinephrine. Progressive increases in F_{ICO_2} displaced the receptor discharge–blood pressure relationship to progressively lower discharge rates. *From Estavillo and Burger (1973b).*

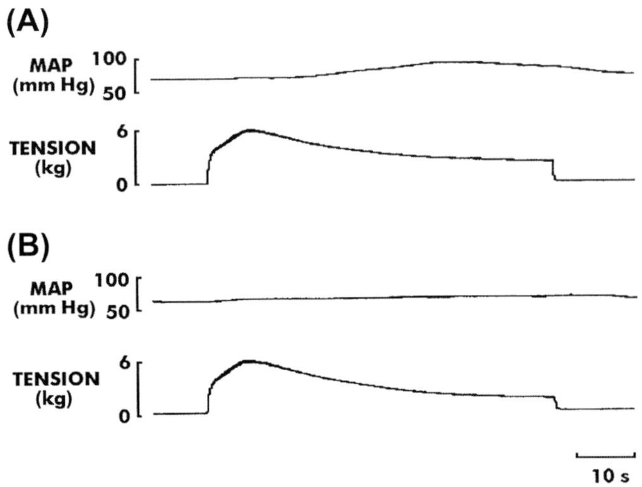

FIGURE 11.56 Pressor effect (upper traces in each panel, mean arterial blood pressure (MAP)) of static contraction of gastrocnemius muscle (lower traces in each panel (tension)) induced by electrical stimulation of sciatic nerve in anesthetized chicken. (A) MAP increases during static muscle contraction when sciatic nerve–spinal cord connection is intact. (B) No change in MAP during static muscle contraction after section of sciatic nerve. Afferent limb of pressoreflex originates within gastrocnemius muscle. *From Solomon and Adamson (1997).*

afferent fibers from the muscle to the spinal cord eliminated the pressor response to muscle contraction (Figure 11.56(B)). Whether the intramuscular receptors involved in these responses were sensing mechanical events related to muscle contraction or chemical signals resulting from intramuscular metabolite buildup as the primary physiological stimulus for this reflex was not clear from this study. The authors acknowledged that receptors sensing either modality, or both, could be driving reflex changes in pressure. The studies done to date on reflex effects of activating skeletal muscle afferents suggest that direct sensory feedback from exercising or hypoxically stressed muscle to the neural circuitry controlling the circulation could be involved in initiating or intensifying cardiovascular responses to muscle activity. Such a mechanism would serve to increase the perfusion of working muscle, but the full contribution of these reflexes remains to be determined.

11.5.5 Integrative Neural Control

As discussed in the previous sections, the major reflexes controlling cardiovascular function are the chemoreflex, the baroreflex, and reflexes driven by receptors within the heart and skeletal muscle. Under some conditions these reflexes may interact, as in the case of chemoreflex occlusion of the baroreflex detailed in Section 11.5.4.2. Furthermore, there are a number of other reflexes which can affect the circulation as part of more global homeostatic control systems, such as those regulating temperature and ventilation. All reflexes influencing the circulation do so through the final common pathways of the autonomic nervous system innervating the heart and vasculature, and several of these reflexes may be engaged at the same time during physiological challenges to the animal. Integrative control of the circulation by the nervous system therefore operates through complex interactions among cardiovascular and other reflexes, but the nature of these interactions is still incompletely understood in any vertebrate group. This large-scale integration makes investigation of the properties of any one reflex a considerable challenge, particularly in unanesthetized animals. One approach to this problem is to ensure that only afferents specific to the reflex in question are stimulated, while activation of afferents associated with other reflexes is prevented. However even this may not be enough to prevent complications in analysis since efferent activity associated with one reflex, and the consequences of this activity, can produce secondary activation of other reflexes. A case in point is the activation of pulmonary reflexes consequent to an increase in ventilation driven by primary stimulation of the arterial chemoreflex. The design of further studies of avian cardiovascular reflexes embedded in complex control systems must therefore take into account the integrative nature of these systems.

A single neurally mediated response, such as a change in heart rate, involving the stimulation of one or more afferent or efferent pathways by a frequently induced or naturally performed behavior, may be subject to habituation or conditioning. Habitation refers to a decrement in the response (not due to sensory adaptation or motor fatigue) resulting from repeated presentations of a single triggering stimulus. In contrast, conditioning involves the animal learning a relationship between two different stimuli. In classical conditioning (Pavlov, 1927) a conditioned and an unconditioned stimulus are presented with little temporal dissociation between them.

The animal learns the relation between these stimuli so that after the initial trials the reflex response can then be evoked by presenting only the conditioned stimulus. A reflex, once conditioned, will anticipate, augment, and have similar effects as the unconditioned reflex.

Cohen and his collaborators have developed a model of conditioned learning in the pigeon for exploring cellular neurophysiological mechanisms of long-term associative learning. In this model animals are trained to respond with a transient and quantifiable increase in heart rate (the conditioned response, CR) to a conditioning stimulus (CS, a 6 s whole-field retinal illumination) by pairing the CS with an unconditional stimulus (US) consisting of a 0.5 s foot shock. After an initial training period of 30–50 paired-stimulus trials, further CS without US reliably evokes conditioned responses. The properties of this CR in the pigeon are very robust, remaining stable for long periods (up to weeks) without habituating (Cohen, 1980, 1984). Such longevity facilitates electrophysiological investigations of changes in properties of neurons involved in the development of associative learning in this system. Using this model as a platform, Cohen and co-workers have anatomically and physiologically characterized the central visual pathways for the CS, the somatosensory pathways for the US, and the descending tracts converging on the autonomic motor neurons of the final common pathway efferent to the heart in the pigeon. Much of this group's work on the efferent components of this system has already been cited in the description of the neuronal circuitry controlling cardiac function appearing earlier in this chapter.

The central and peripheral nervous pathways involved in the CR have been reviewed in detail by Cohen (1980, 1984) and are briefly summarized here. Ganglion cells throughout the retina respond with a phasic burst of APs at the start of a 6-s period of whole-field retinal illumination. This phasic wave of excitation is transmitted synchronously along multistage pathways in the central nervous system to preganglionic sympathetic and parasympathetic motor neurons in the spinal cord and brainstem, respectively. These preganglionic neurons in turn control pools of cardiac postganglionic autonomic neurons in a synergistic manner, acting to potentiate a transient sympathetic cardioacceleration and to facilitate concurrent withdrawal of parasympathetic cardioinhibition.

Studies of the development of conditioned responses after lesioning selected components of the central visual pathways and electrophysiological studies during the conditioning process in pharmacologically immobilized pigeons have shown that the CS is conveyed in parallel through multiple pathways to the visual area of the telencephalon. These include (1) a thalamofugal pathway involving the principal optic nucleus (the avian homolog of the mammalian lateral geniculate nucleus) and (2) a tectofugal pathway projecting through the optic tectum and the nucleus rotundus (Cohen, 1980). In addition, there may be a third visual input pathway implicated in the conditioned response, projecting through the pretectal area and the thalamus. While retinal responses to repeated CS are not modified during the development of conditioning, responses of second and higher-order neurons in the CS pathway are facilitated during this process. The time course of these changes parallels the time course of the development of the conditioned heart rate response, so neurons or their impinging synaptic fields in successive stages in the CS pathways are likely sites for modulation of cardiovascular control during associative learning.

The descending pathways involved in this conditioned cardiac response have been well established. Neurons in the medial region of the hypothalamus, along with those located in the ventral brainstem, project to preganglionic sympathetic and parasympathetic cardiomotor neurons. The medial hypothalamus in turn receives inputs from the avian homolog of the mammalian amygdala, an area which, in both groups of animals, evokes marked cardioactive effects when stimulated.

Furthermore, lesions of the avian amygdala or its hypothalamic projection either produce deficits in conditioning or can prevent development of the conditioned response (Cohen, 1980). The telencephalic projections conveying visual information to the avian amygdala have, however, not as yet been established in detail. Convergence of the CS and US has been demonstrated at each of the established stages in the central pathway of this response. It thus appears that training-induced modification of the CS works through long-term heterosynaptic facilitation in this pathway, and such facilitation constitutes an important element in the constellation of neurophysiological mechanisms for associative learning.

The broad outlines of the central and peripheral neural pathways involved in efferent control of the circulation in birds have been established, but relatively little is known about details of specific afferent and intermediary connections within the central pathways of any of the cardiovascular reflexes in birds. The working assumption guiding studies of these reflexes is that their pathways are similar to those in mammals, but given the differences in cardiovascular reflexogenic zones between birds and mammals, this is not necessarily a valid assumption. Birds have, for instance, only one major arterial baroreceptor site while mammals have two. In addition, most of the input from avian arterial chemoreceptors originates from the carotid bodies while in mammals both aortic and carotid chemoreceptors contribute to cardiovascular chemoreflexes. The organization of the neural circuitry for cardiovascular control in birds has not been investigated in detail, so the similarity of this circuitry to that of mammals remains an open question.

11.5.6 Development of Cardiovascular Control

Recent advancements in general knowledge of cardiovascular control mechanisms in birds have been predominately focused on the embryonic phase of avian life history. Seminal works have extensively documented anatomical maturation of the heart and vasculature. Here, we provide an overview of cardiovascular control during development in birds. It is important to recognize that the majority of studies investigating avian development have focused on the domestic chicken and the maturation of cardiovascular control is no exception.

11.5.6.1 Ontogeny of Autonomic Nervous System Control of the Heart

11.5.6.1.1 Cardiac Autonomic Innervation

The ontogeny of parasympathetic and sympathetic fiber innervation of the developing avian embryo heart differs. Branches of the vagus nerve (X) perforate the embryonic chicken ventricle at 15% of incubation and innervate all cardiac chambers by 35% of incubation (Pappano, 1975). Cardiac and sinal branches reach the truncus and atria around 20% of incubation (Kuratani and Tanaka, 1990). In contrast, sympathetic cardiac nerves from the sympathetic ganglia do not reach the heart until 50% of incubation (Higgins and Pappano, 1979; Kirby et al., 1980) and penetrate the myocardium at 75% of incubation (Verberne et al., 1999). The sympathetic fiber origin is either the first pair of thoracic ganglia (Kirby et al., 1980) or the cervical ganglia (Verberne et al., 1999).

Field stimulation studies have found that function lags behind the anatomical appearance of innervation. A cholinergic dependent negative chronotropic response appeared as early as 60% of incubation in response to autonomic receptor drugs and atrial field stimulation. The negative chronotropic response appeared earlier, at 50% of incubation, when pretreated with the cholinesterase inhibitor physostigmine (Pappano and Löffelholz, 1974). In contrast, an adrenergic dependent, positive chronotropic response to field stimulation was not observed until hatching (Pappano and Löffelholz, 1974). An adrenergic dependent response was produced as early as 50–60% of incubation by potentiating the release of catecholamines from postganglionic neurons with tyramine (Pappano, 1975; Crossley, 1999). This potentiating effect of tyramine increased with age until 90% of incubation (Crossley et al., 2003b). Collectively the data suggest functional control by the autonomic nervous system lags behind the innervation of the heart.

11.5.6.1.2 Cardiac Cholinergic and Adrenergic Receptors

It has long been established that embryonic chicken pacemaker cells express muscarinic cholinergic receptors as early as 10% of incubation (Cullis and Lucas, 1936; Coraboeuf et al., 1970; Pappano and Löffelholz, 1974). Stimulation of these receptors produces an acetylcholine dependent negative chronotropic response that is blocked by the cholinergic antagonist atropine. There are two stages to the development of acetylcholine's negative chronotropic action (Dufour and Posternak, 1960). Between 15% and 30% of incubation, increased acetylcholine sensitivity is due to changes in membrane permeability. This is followed by functional parasympathetic innervation of the heart, which occurs between 40% and 55% of incubation (Pappano, 1977).

As early as 10% of incubation, pacemaker cells exhibit β-adrenergic receptor mediated chronotropic responses with cells expressing both β_1 and β_2 receptors (Berry, 1950; Fingl et al., 1952; McCarty et al., 1960; Lenselink et al., 1994). Pacemaker cell sensitivity to epinephrine is constant from 60% to 85% of incubation and decreases during the later 85–95% of incubation (Löffelholz and Pappano, 1974). This decrease in sensitivity is associated with increased circulating catecholamine levels during the later stage of embryonic development (Crossley et al., 2003b). The embryonic β-adrenergic receptors at 20% of incubation that respond to β-adrenergic stimulation with inotropic and chronotropic cardiac effects (McCarty et al., 1960; Shigenobu and Sperelakis, 1972; Frieswick et al., 1979; Higgins and Pappano, 1981). Maximal adrenergic stimulation produces a constant twitch force but the isoproterenol sensitivity of embryonic White Leghorn ventricular tissue increases from 75% to 90% of incubation. This is followed by a decrease prior to hatching suggesting development of a reduced adrenergic sensitivity (Higgins and Pappano, 1981). Tyramine (10 mg/kg) produces a similar inotropic and chronotropic effect on the embryonic chicken heart during this developmental period (Crossley, 1999). An age dependent decrease in sensitivity to epinephrine over this incubation window has also been shown in a broiler strain of embryonic chickens due in part to internalization of β-adrenergic receptors (Lindgren and Altimiras, 2009). Overall, developing chickens exhibit β-adrenergic mediated inotropic and chronotropic cardiovascular responses early in incubation.

11.5.6.2 Ontogeny of Vascular Contractility

11.5.6.2.1 Vascular Reactivity Regulation

Embryonic circulation must meet the demands of a rapidly growing embryo while also developing into their adult phenotype. Adult vascular smooth muscle and endothelial cells must regulate vascular tone, blood pressure, and blood flow distribution. During vascular morphogenesis developing vascular smooth muscle and endothelial cells must proliferate, migrate, and produce the extracellular matrix, while at the same time acquiring the physiological capacity to regulate vascular tone though smooth muscle activity (Owens et al., 2004; Rzucidlo et al., 2007).

As noted in Section 11.5.2.1, smooth muscle contraction relies on phosphorylation of MLC20 by the Ca^{2+} dependent MLCK and the Rho-kinase pathway. The expression of MLC20, MLCK, and MLCP in chicken embryo vascular smooth muscle occurs as early as 50% of incubation (Ogut and Brozovich, 2000). The embryonic aorta tonically contracts in response to increased cytosolic Ca^{2+} as early as 40% of incubation. The levels of MLC20 phosphorylation plateau at 75% of incubation (Ogut and Brozovich, 2000). The Rho-kinase pathway, providing Ca^{2+} sensitization, appears to mature during embryonic development. Contractions of the femoral artery and ductus arteriosus are inhibited by the Rho-kinase pathway inhibitors Y-27632 and hydroxyfasudil (Greyner and Dzialowski, 2008; Zoer et al., 2010). The effect of Rho Kinase inhibitors increased with incubation age, suggesting a developmental augmentation in the RhoA/Rho kinase-mediated increase in calcium sensitivity of the contractile apparatus.

11.5.6.2.2 Vascular Adrenergic Receptors

The vasculature expresses adrenergic receptors early in embryonic development. Epinephrine increases arterial pressure in the embryo after 15% of incubation demonstrating that adrenoceptors are present and functional (Hoffman and van Mierop, 1971; Girard, 1973). Pharmacological manipulation to identify receptor types with specific adrenergic receptor agonists and antagonists indicate that α-adrenergic and β-adrenergic receptors are present as early as 30% of incubation (Saint-Petery and van Mierop, 1974; Koide and Tuan, 1989). Further studies to isolate differences between vascular beds of the embryo indicate that as early as 60% incubation, the mesenteric circulation expresses α-adrenergic receptors (Rouwet et al., 2000). Chicken femoral and carotid arteries increase α_1-adrenergic receptor and receptor-independent stimulation contractile reactivity from 70% to 90% incubation (Le Noble et al., 2000). In contrast, pulmonary arteries lack α-adrenergic mediated contraction during all of embryonic development (Villamor et al., 2002). Stimulation of β-adrenergic receptors produces a relaxation in some vessels including the femoral artery. Both the sensitivity and responsiveness of the vasculature to β-adrenergic simulation increased with incubation age.

Although adrenoceptors appear functional in the arterial vasculature as early as 15% of incubation, there is a delay in sympathetic nervous control. Arterial vascular constriction to exogenous NE is present before any neurogenic responses (Le Noble et al., 2000). Sympathetic control of arterial vascular resistance seems to be restricted to the later stages of embryonic development in chickens. The femoral artery contracts in response to perivascular nerve stimulation in late-stage embryonic chickens. In contrast, neither the carotid nor pulmonary arteries respond to nerve stimulation. In summary, there is an increase in adrenergic control of the vasculature that initially comes from circulating catecholamines and later sympathetic nervous control.

11.5.6.2.3 Vascular Cholinergic Receptors and Endothelial Control

Cholinergic control of vessel reactivity appears early in the developing avian embryo. Muscarinic receptor stimulation relaxes the aorta, femoral artery, carotid artery, mesenteric arteries, pulmonary arteries, and ductus arteriosus of chicken embryos in a nitric oxide (NO)-mediated endothelium-dependent manner (Greyner and Dzialowski, 2008; Le Noble et al., 2000; Rouwet et al., 2000; Martinez-Lemus et al., 2003; Nishimura et al., 2003; Villamor et al., 2002). This endothelium-dependent, NO-mediated relaxation appears to be critical for regulating embryonic circulation during early development. There is early expression of NO synthase mRNA in multiple tissues of the developing chicken cardiovascular system as early as 15% of incubation (Groenendijk et al., 2005). As early as 50% of incubation, isolated cardiomyocytes respond to sodium nitroprusside, an NO donor, and L-arginine, the NO precursor, suggesting the NO/cGMP pathway is already functional (Ungureanu-longrois et al., 1997). Isolated chorioallantoic membrane (CAM) arteries contract in respond to acetylcholine (Lindgren et al., 2010) and NO produces increased CAM blood flow as early as 50% of incubation (Dunn et al., 2005). Early in incubation, NO donors decrease ventricular preload without changing arterial resistance, which may occur through venodilation (Bowers et al., 1996). At 45% of incubation, sodium nitroprusside produces a vasodilation, suggesting early maturation of the cellular pathway involving cGMP activation (Altimiras and Crossley, 2000). In addition, NO plays a role in maintaining systemic vascular dilation at 70 and 90% of incubation in chickens (Iversen, et al., accepted article). Finally, the endothelium-dependent relaxation of pulmonary and systemic arteries is constant during internal and external pipping (Villamor et al., 2002; Le Noble et al., 2000). The hatching chicken lacks the depression of pulmonary endothelial function observed in neonatal mammal.

Other mediators of endothelium-dependent relaxation have received limited attention in embryonic chickens. Adenosine is involved in the CAM angiogenic response to hypoxia at 50 and 65% of incubation. Embryonic whole body vascular resistance decreases in response to adenosine at 50–70% of incubation (Adair et al., 1989). Adenosine receptors have been found on embryonic red blood cells at 50% of incubation (Glombitza et al., 1996). These studies indicate that adenosine receptors are present in the developing chicken cardiovascular system.

11.5.6.3 Developmental Integration of Autonomic Cardiovascular Regulation

11.5.6.3.1 Afferent Pathways

The carotid bodies and special mechanosensory nerve endings in the aorta provide the main cardiovascular reflexes in birds. The carotid bodies constitute the primary loci for peripheral chemoreceptors sensing arterial oxygen and carbon dioxide tensions and pH (Section 11.5.4.1). Initially, carotid bodies consisting of mesenchyme-like cells appear around 25% of incubation and by 40% of incubation have migrated to the adult location (Murillo-Ferrol, 1967). At 60% of incubation, the parenchyma has a large number of dispersed granule-containing cells, glomus cells or Type I cell (Kameda, 1994). The appearance of these cells coincides with peak serotonin immunoreactivity (Kameda,

1990) and appearance of long axon synaptic junctions with glomus cells (Kameda, 1994). At 70% of incubation, the developing glomus cells already express features of mature glomus cells (Kameda, 1994). Hypoxic-induced glomus cell catecholamine secretion increases after hatching and is coupled to decreased constitutive catecholamine release (Donnelly, 2005).

Reflexogenic sensory afferents of the cardiovascular system originate from the nodose ganglion. The nodose ganglion bifurcate after emerging from the cell body with one branch to the heart and viscera and the other centrally located, making connections within the central nervous system and solitary tract nucleus. The primordium of the ganglion appears at 15% of incubation and is followed by proliferation of ganglion, peaking at 30% of incubation. An apoptosis-induced decrease in cell number occurs by the time of hatching (Harrison et al., 1994). Early neuron development is independent of neurotrophins and becomes dependent on brain-derived neurotrophic factor as the neurons develop towards their target tissues. The development of the neurons is regulated by other trophic factors such as nerve growth factor during embryonic develpoment (Hedlund and Ebendal, 1980).

Overall, the development of cardiovascular reflexogenic afferent pathway connections with the central nervous system occurs during the first half of incubation. The sensitivity of the carotid bodies increases when the animal hatches, but the changes occurring in baroreceptive areas are unknown.

11.5.6.3.2 Tonic Heart Regulation

The role of cholinergic tonic activity does not appear to be involved in maintaining baseline cardiovascular function during early embryonic development. The cholinergic muscarinic receptor antagonist atropine does not alter heart rate in White Leghorn chicken until just prior to or after hatching (Pickering, 1895; Saint-Petery and van Mierop, 1974; Tazawa et al., 1992; Crossley and Altimiras, 2000) and has no effect on arterial pressure (Tazawa et al., 1992; Crossley and Altimiras, 2000). However, this is not universal for embryonic birds, as different species differ in terms of the presence of a heart rate response to atropine administration (Crossley et al., 2003b; Swart et al., 2014). Embryonic chicken strains including broilers, black Sumatran Bantams, and red Jungle fowl vary in terms of the presence of functional cholinergic tone on heart rate, appearing as early as early as 60% to as late as 90% of incubation (Crossley et al., 2003b; Chiba et al., 2004; Crossley and Altimiras, 2012). An absence of cholinergic tone does not exclude possible recruitment and intermittent activity of the parasympathetic nervous system. Continuous recordings of instantaneous heart rate in embryonic chickens reveal heart rate decelerations likely due to transient increase parasympathetic activity (Kato et al., 2002). Although cholinergic receptors are expressed early in development and the parasympathetic efferent pathway is functional under some conditions at 60% of incubation (Pappano et al., 1973), cardiovascular function and development continues without a tonic cholinergic stimulation throughout at least the first half of incubation.

In contrast, there is clear adrenergic tone present during the majority of embryonic bird development. α-adrenergic and β-adrenergic receptor mediated stimulatory tones on both heart rate and arterial pressure appear early in development (Saint-Petery and van Mierop, 1974; Koide and Tuan, 1989; Tazawa et al., 1992; Crossley and Altimiras, 2000). As early as 30% of incubation, a positive β-adrenergic chronotropic tone on heart rate and arterial pressure appears (Saint-Petery and van Mierop, 1974; Girard, 1973). This tone is critical to basal baseline function (Tazawa et al., 1992; Crossley, 1999; Crossley and Altimiras, 2000). In chickens, the magnitude of the β-adrenergic chronotropic tone increases in strength from 10% to 20% as the embryos progress from 40% to 95% of incubation, respectively (Crossley, 1999). Baseline heart rate is unchanged after pharmacological blockade of sympathetic nervous terminals with 6-hydroxydopamine or ganglionic blockade (Tazawa et al., 1992; Crossley, 1999; Crossley et al., 2003b). This suggests that adrenergic heart rate tone in chicken embryos originates from circulating catecholamines. A similar change in the β-adrenergic heart rate tone during the final 30% of incubation has been observed in emus, domestic geese, and Canada geese (Crossley et al., 2003a; Swart et al., 2014).

Although the direct effects of β-adrenergic inhibition on the heart is mediated through direct actions on the heart and pacemaker tissue, the mechanisms regulating the depressive actions of α-adrenergic receptor antagonists on heart rate present from 40% of incubation through hatching is unclear (Koide and Tuan, 1989; Tazawa et al., 1992; Crossley, 1999; Crossley and Altimiras, 2000). While this tone is present and the magnitude is maximal between 60% and 95% of incubation, α-adrenergic receptors have been reported to be absent in the chicken heart (Crossley, 1999; Chess-Williams et al., 1991). Therefore, the bradycardic response to α-adrenergic receptor blockade may be secondary to the pronounced vasodilation after α-adrenergic receptor antagonist treatment (Crossley and Altimiras, 2000). Vasodilation may produce blood pooling in the CAM vasculature, reduced venous return, or decreasing heart rate and CO.

11.5.6.3.3 Tonic Vasculature Regulation

Total and regional vascular resistance and capacitance is regulated by the sympathetic nervous system in adult vertebrates. This vascular control is mediated by changes in catecholamine release (adrenaline and noradrenaline) from sympathetic nerve terminals as well as the adrenal medulla.

These ligands bind to adrenoceptors (AR) types that are heterogeneously distributed with the dominant AR type dependent on the vascular bed. This allows differential hemodynamic responses to systemic catecholamine release. The vascular tone resulting from activation of different receptors is dependent on the vascular areas affected and if stimulation is arising from sympathetic nerve endings or circulating catecholamines (Guimaraes and Moura, 2001).

A strong α-AR tone maintaining vasculature constriction is active in some vascular beds and is functional from 30% to 95% of incubation (Girard, 1973; Saint-Petery and van Mierop, 1974; Koide and Tuan, 1989; Tazawa et al., 1992; Crossley, 1999; Crossley and Altimiras, 2000). A α-AR mediated vascular tone prevails in the skeletal muscles and has limited effects on the heart, intestines, and yolk sac (Mulder et al., 2001). Further α-ARs are absent from the CAM vascular tree illustrating the regionalization of AR types (Lindgren et al., 2010). Although the time course has not been elucidated, a similar strong α-AR vascular tone has been observed in the black Sumatran bantam, red jungle fowl and broiler chickens (Crossley, unpublished observation; Crossley and Altimiras, 2012). In white leghorns, the receptor subtype responsible for maintaining embryonic vascular tone are likely α_1-adrenergic receptors given similar responses following a nonspecific α-AR antagonist (phentolamine) and an α_1-adrenergic receptors specific antagonist (prazosin) (Crossley and Altimiras, 2000). In white leghorn chicken embryos, there is an increase in the dependence on α-AR mediated vasoconstriction from a 10% contribution to resting arterial pressure to over 55% just prior to hatching (Crossley, 1999). Limited numbers of studies conducted in other species show that embryonic emus, domestic geese, and Canada geese possess a pronounced α-AR tone during the final 30% of incubation (Crossley et al., 2003a; Swart et al., 2014). Overall, it is clear that embryonic chickens depend on α-mediated vasoconstriction during the majority of development, which is a feature shared by the other avian species investigated.

A β-AR vasodilator tone opposes the α-AR mediated vascular tone during embryonic development. The β-AR vasodilator tone in white leghorn chickens appears at 30% of incubation, increases in magnitude from 35% to 60% of incubation where it stabilizes until the final 5% of incubation when it disappears (Saint-Petery and van Mierop, 1974; Crossley, 1999; Girard, 1973). An active β-tone depressing arterial pressure is present in other chicken strains from 60% to 90% of incubation (Koide and Tuan, 1989; Tazawa et al., 1992; Crossley, unpublished; Crossley and Altimiras, 2012). This tone reaches maximal strength at 90% of incubation in other strains of embryonic chickens and embryonic emus (Crossley, unpublished data; Crossley et al., 2003b; Crossley and Altimiras, 2012). Unlike the chickens and emus, domestic and Canada geese embryos have a β-AR vasodilator tone at 70% of incubation that is absent at 90% (Swart et al., 2014), so clearly regulatory mechanisms may differ between species.

As with adrenergic tone on heart rate, the adrenergic vascular tone during development is due to levels of circulating plasma catecholamines. This is supported by measurements following sympathectomy with 6-hydroxydopamine or hexamethonium ganglionic blockade, illustrating that resting arterial pressure in embryonic chickens is unaltered by the treatment (Crossley, 1999; Crossley and Altimiras, 2000; Crossley et al., 2003b).

11.5.6.3.4 Baroreflex Regulation

Investigations of developing baroreflex function have been limited to domestic chicken breeds and have focused on characterizing reflex parameters of the cardiac limb or heart rate responses to changes in arterial pressure only (Altimiras and Crossley, 2000; Elfwing et al., 2011; Mueller et al., 2013). Two methods have been successfully utilized to quantify changes in baroreflex gain (sensitivity) and set point during chicken development; pharmacological assessment or Oxford method and spontaneous baroreflex sensitivity assessments (Altimiras and Crossley, 2000; Elfwing et al., 2011; Mueller et al., 2013). Each approach has merits and deficiencies however collectively baroreflex function in chickens is present as early as 80% of 21day incubation period, dependent in part on the breed studied (Altimiras and Crossley, 2000; Elfwing et al., 2011). In addition to the timing of operation during chicken development, changes in baroreflex function have also been reported. Gain of the reflex is a key parameter used as an index of baroreflex maturation during chicken incubation; however, components such as set point and threshold are also informative. In white leghorn chicken embryos, the baroreflex is functional at 85% of incubation (Altimiras and Crossley, 2000), but only 17% of the embryos showed a change in heart rate after experimental manipulation of blood pressure. This rapidly increases to 33% at 90% of incubation, with a fivefold increase in baroreflex gain upon hatching (Altimiras and Crossley, 2000). Using the spontaneous assessment method in which the correlation between spontaneous fluctuations in arterial pressure and heart rate are analyzed, the reflex is functional by 80% of incubation, while the gain remains constant from this point until 95% of incubation in broiler chickens (Elfwing et al., 2011). Interestingly, spontaneous gain in embryonic broiler chickens is relatively constant from 80% to 95%, averaging 59.8 k/Pa/min. This is a higher gain than that determined with the Oxford method in the same breed, which increased progressively from 10.9 to 30 k/Pa/min over the same developmental period (Elfwing, 2007). Previous arguments regarding the utility of each methodological approach have been made (Di Rienzo, 2001; Elfwing et al., 2011) and may account for these reported differences in ontogeny of chicken baroreflex function. Overall, while previous studies have been novel and informative about the ontogeny of avian baroreflex function, the domestic chicken has been the lone model organism. Investigations of multiple species from different avian lineages with different developmental strategies (precocial versus altricial) are crucial to identify the commonalities of the ontogeny of baroreflex function in birds.

Equally informative will be investigations of the interactions between baroreflex function and other regulatory mechanisms during ontogeny; however, modulation of baroreflex of function via systemic hormones and central neuropeptides has been unexplored. Angiotensin II (AT) is an important regulatory peptide that is crucial for blood volume homeostasis in adults and is a contributor to an integrative series of signaling pathways that impacts baroreflex function. In embryonic chickens, AT also modulates baroreflex function, decreasing both baroreflex gain and operating point at 90% of incubation (Mueller et al., 2013). This change was attributed, in part, to a decrease in vagal inhibition that may be the result of the known elevated plasma levels of AT in embryonic chickens (Crossley et al., 2010). Thus, to deepen the understanding of the development of this reflex, the ontogeny of baroreflex function must be assessed in concert with the maturation of other regulatory components of the embryonic cardiovascular system.

11.5.6.3.5 Cardiovascular Response to Hypoxia

The embryonic cardiovascular response to hypoxic exposure has been as useful tool in assessing the capacity of developing regulatory mechanisms to respond to an environmental insult in addition to exploring the maturation of a typical adult response to the stress. Hypoxia has been suggested to be a relevant nature occurrence during avian ontogeny (Andrewartha et al., 2011). Investigators have utilized two methods to characterize the cardiovascular response to reductions in oxygen: prolonged and acute exposures. In domestic chickens, prolonged bouts (>15 min) of hypoxic exposure result in a negligible heart rate response in early embryos (15–25%), a depression in heart rate during the middle of incubation (67–76%), and an increase in heart rate late (95–100%) in incubation (Khandoker et al., 2003; Andrewartha et al., 2011). Changes in arterial pressure and blood flow distributions have been yet to be investigated under prolonged periods of hypoxic exposure.

Arterial pressure responses to hypoxic exposure have been conducted primarily using acute hypoxia (<10 min). In general, hypoxia causes hypotension in embryonic chickens, which is dependent on the level of hypoxia throughout embryonic development (Tazawa, 1981; Crossley et al., 2003b). Embryonic emus respond similarly at 60% of incubation; however, they transition to a hypertensive response prior to hatching (Crossley et al., 2003a). Acute hypoxia also depresses heart rate in embryonic chickens, between 43% to 100% of incubation, with the intensity of the response dependent on the level of oxygen, unlike the response seen during prolonged exposures (Tazawa, 1981; Tazawa et al., 1985; Crossley et al., 2003a). In contrast, embryonic emus either maintain or increase heart rate during hypoxic exposures during the final 30% of incubation (Crossley et al., 2003a). Blood flow measurements during hypoxic exposures beyond 30% of incubation are limited. Tazawa et al. (1985) measured blood flow in the allantoic artery, the major artery supplying the CAM, of an embryonic chicken at ~70% of incubation and reported a slight decrease during hypoxia from the baseline value of 4.1–4.4 mL/min (Tazawa et al., 1985). Later work reported that severe hypoxia (0–5% O_2) reduces allantoic artery blood flow as much as 0.7 mL/min in embryonic chickens from 40% to 76% of incubation (Van Golde et al., 1997). Given that under normoxic conditions the CAM vascular receives between 52% and 41% of CO in embryos ranging from 50% to 90% of incubation, the depression caused by hypoxia represent an important redistribution of flow (Mulder et al., 1997). The pattern of redistribution favored the increased perfusion to the heart and brain while decreasing it to the liver, yolk and carcass (Mulder et al., 1998). Direct measures of femoral arterial blood flow also illustrate a reduction in limb perfusion during hypoxic exposures (Iversen et al., accepted article). Overall, the embryonic chicken response to acute hypoxic exposure is a depression of overall cardiovascular function.

The mechanisms that underlie the documented changes in cardiovascular function have been investigated in the embryonic chickens. In the white leghorn breed, hypoxia decreases arterial pressure in embryos from 43% to 100% of incubation (Crossley, 1999). Initially, this response was partially attributed to a direct effect of reduced O_2 on the vasculature inducing a relative dilation (Crossley et al., 2003b). From 70% to 90% of incubation, the hypoxic dilation is limited by an α-adrenergic and cholinergic receptor stimulation, with a dilatory β-adrenergic stimulation contributing in the final ~20% of incubation (Crossley et al., 2003b). Although the increased vascular cholinergic effects may resulted from a reflexive mechanisms, the adrenergic stimulations can be attributed to the elevation in plasma catecholamines reported to result from hypoxic exposures during the final 30–40% of incubation (Mulder et al., 2000; Crossley et al., 2003b). Recent investigations have identified a nitric oxide mediated vasodilation at 70 and 90% of incubation as an additional contributing factor to the hypoxic hypotension in embryonic chickens during this period (Iversen et al., accepted article).

Heart rate responses to hypoxia have been suggested to be partially due to the direct actions of low oxygen on cardiac muscle and pacemaker tissue of the embryo (van Golde et al., 1997; Crossley et al., 2003b). During the final 10% of incubation, the hypoxic bradycardia may also be affected by an α-adrenergic stimulation and a cholinergic inhibition induced by the direct effects of low oxygen on autonomic nerve terminals as well as chromaffin tissue (Crossley et al., 2003b). It should be noted that α-adrenergic receptor-mediated redistribution of CO, including preferential perfusion of the heart, is an important component of the embryonic chicken response and inhibition may contribute to the reported chronotropic actions of α-adrenergic during hypoxia (Mulder et al., 1998, 2001).

Clearly, additional regulatory systems in multiple species must be investigated to further develop the model of how the embryonic cardiovascular system responds to periods of hypoxia however the embryo may rely on peripheral mechanisms, such as NO, with limited action from the autonomic nervous system.

11.5.6.4 Development of Humoral and Local Effectors of Cardiovascular Function

A number of hormones play an active role in cardiovascular regulation of the developing embryo. The best studied hormones include angiotensin II (AII), endothelin (ET), and natriuretic peptide (NP), and their roles in the adult have been outlined in Section 11.5.2.3. All of these hormones have been found relatively early in development of the chicken embryo.

The components of the renin-angiotensin system are present during early embryonic chicken development. The egg-laying female provisions AII converting enzyme (ACE) at measurable levels in freshly laid eggs. Over the first 2 days of development, whole embryo levels of ACE-mRNA increase (Savary et al., 2005) and pathways for angiotensin synthesis and signal transduction are present. Between 15% and 20% of incubation, cardiac tissue, brachial arch tissue, and mesonepric tissue all express AII receptor mRNA (Kempf and Corvol, 2001). The CAM expresses AII receptors during at least the last half of incubation (Moellera et al., 1996) and responds to AII at 35% of incubation (Le Noble et al., 1991, 1993).

Angiotensin II appears to be a tonic regulator of cardiovascular function in the developing chicken. *In vivo* AII produces a dose dependent increase in arterial pressure that increases in intensity from 60% of incubation until hatching (Crossley et al., 2010). In addition, embryonic chickens possess high levels of circulating AII over this same time frame and blockade of ACE results in a relative hypotension at 90% of incubation (Crossley et al., 2010; Mueller et al., 2013). Although the embryonic animal responds to AT with an increase in blood pressure similar to adult birds, the embryo lacks the initial AII induced hypotension reported in adults (Crossley et al., 2010). Embryonic chickens also do not depend entirely on α-adrenergic receptors stimulation for the AII-induced hypertension and antagonists for the AII type 1 receptor (AII1R) are ineffective in embryonic chickens until hatching (Crossley et al., 2010). The ACE is active in chicken aortas as early as 50% of incubation and it increases in activity with development (Topouzis et al., 1992). Angiotensin also relaxes isolated aortic rings of embryonic chickens at 90% of incubation (Nishimura et al., 2003). These isolated vessels are lacking the sympathetic nerve inputs that would provide a constrictive response. Collectively, the data illustrate that AII is an important component of embryonic chicken cardiovascular regulation with similarities and differences from the cardiovascular response in adult birds.

Both AII type 1 and type 2 receptors are present on the developing heart (Rabkin, 1996). Angiotensin II can act on the developing chicken heart to produce a positive inotropic effect at 85% of incubation (Freer et al., 1976). Exposure of AII during 35–90% of incubation produces cardiac hypertrophy through activation of the AII type 1 receptors (AII1R) and upregulation of myosin light chain (Mathew et al., 2004; Baker and Aceto, 1990; Aceto and Baker, 1990). The observed cardiac hypertrophy may be due to direct action of AII on the heart and/or in response to changes in embryonic vasculature function. It is clear there is a role for angiotensin in chicken cardiovascular development and regulation and *in vivo* studies are necessary to further clarify the significance of the role.

The locally produced vasoconstrictor, endothelin, is expressed ubiquitously throughout the developing chicken cardiovascular system (Kempf et al. 1998). The myocardium and the outflow tract express endothelin receptor subtype mRNA transcripts at 15% of incubation (Groenendijk et al., 2008). Endothelin converting enzyme 1 activates endothelin and is first detected at 20% of incubation (Ballard and Mikawa, 2002; Hall et al., 2004). Endothelin produces changes in cardiovascular function in both *in vivo* and *in vitro* studies. *In vivo* administration of endothelin antagonists at 20% of incubation reduce cardiac function (Groenendijk et al., 2008). An *in vitro* endothelin-mediated positive inotropic effect was observed in cultured cardiomyocytes from 50% of incubation embryos (Bézie et al., 1996). Myograph studies on isolated aorta, pulmonary arteries, and ductus arteriosus all show a vasoconstriction in response to endothelin (Wingard and Godt, 2002; Villamor et al., 2004, 2002; Martinez-Lemus et al., 2003; Agren et al., 2007). Active wall tension in response to endothelin-1 increases before hatching, which could be critical for the transition to *ex ovo* life (Martinez-Lemus et al., 2003; Villamor et al., 2004).

Natriuretic peptides help regulate sodium and water balance and maintain cardiovascular homeostasis (Takei, 2000; Toop and Donald, 2004; Trajanovska et al., 2007). Mammals express three natriuretic peptide subtypes: atrial natriuretic peptide (ANP), B-type natriuretic peptide

(BNP), and C-type natriuretic peptide (CNP). The subtypes ANP and BNP are produced and released from cardiac myocytes in response to increases in ventricular volume (Takei, 2000; Toop and Donald, 2004; Trajanovska et al., 2007). The chicken genome has four potential natriuretic peptide genes, including BNP and CNP (Houweling et al., 2005; Trajanovska et al., 2007). Natriutetic peptides produce a relaxation of the vasculature that appears as early as 20% of embryonic development (Nakazawa et al., 1990). Studies on isolated cardiomyocytes reveal a response to natriuretic peptides beginning at 50% of incubation (Bézie et al., 1996; Koide et al., 1996). These studies suggest that the receptors are present both on the heart and in the vasculature of the developing embryo. There is also some indication that endothelin and natriuretic peptide interact to regulate cardiomyocyte contractility in embryonic chickens (Bézie et al., 1996). The functional role during development remains to be fully characterized.

11.6 ENVIRONMENTAL CARDIOVASCULAR PHYSIOLOGY

11.6.1 Flight

Avian flight requires adaptations of the respiratory and cardiovascular systems to supply sufficient O_2 to the working muscles. Given the inherent complexity of measuring these parameters in active birds it is not surprising that few studies have examined cardiovascular adjustments that support flight. The two most complete studies examined cardiovascular and respiratory function in flying domestic pigeons (*Columba livia domestica*; Butler et al., 1977; Peters et al., 2005). During flight at wind tunnel speeds of 10 m/s (Butler et al., 1977) or 18.4 m/s (Peters et al., 2005) O_2 consumption increased between 10- and 17-fold. CO increased 7.4-fold and blood oxygen extraction 2.4-fold during flight, increasing convective O_2 transport to meet metabolic demands of flight muscle. During these flying bouts, heart rate (which is the primary mechanism for elevating CO in these animals) increased from a resting rate of 110 beats/min to flying rate of 663 beats/min (Peters et al., 2005). The exercise-induced increase in pigeon heart rate during flight is of a similar magnitude to that of flying bats and larger than those observed in running mammals (reviewed in Peters et al., 2005).

The magnitude of change in CO and O_2 consumption is flight mode dependent in birds. For example, during soaring and gliding, flight heart rate is similar to that of resting birds, while flapping flight increases heart rate 2.2–7.4 times, with the greatest acceleration occurring during take-off (Peters et al., 2005; Sapir et al., 2010; Sakamoto et al., 2013). There is a strong correlation between heart rate, flight mode, and time spent flapping in the black-browed Albatross (*Thalassarche melanophrys*; Sakamoto et al., 2013). During the transition from gliding to flapping, heart rate changes almost instantaneously in the Cape gannet (*Morus capensis*; Ropert-Coudert et al., 2006). Although the method of flight is correlated with changes in heart rate, flight speed does not appear to alter this cardiovascular parameter (Ward et al., 2002; Ros et al., unpublished data). At wind speeds ranging from 0 to 16 m/s, the heart rate of flying cockatiels (*Nymphicus hollandicus*) is relatively constant, averaging 817 beats/min, which is an increase of 2.3-fold from resting levels (Ros et al., unpublished data).

Tissue perfusion is presumably modified during flight as well with an increased distribution of CO to flight muscles during activity; however this parameter has not been quantified to date. The data available quantifies skeletal muscle perfusion during walking and running. Ellerby et al. (2005) found that, during bipedal locomotion in birds, locomotory muscle perfusion increased significantly. In guinea fowl, (*Numida meleagris*) as walking speed increased to 90% of V_{O_2max}, CO increased by 4.9-fold through increased heart rate and stroke volume (Ellerby et al., 2005). The majority of this increase in CO perfused the leg muscles, while blood flow to the brain, spleen, stomach, pancreas, intestines, and kidneys did not change. Bech and Nomoto (1982) reported that sciatic artery blood flow increased 3.7 times in Pekin ducks (*Anas domesticus*) running on a treadmill, suggesting similar CO modifications occur in this species. Tissue perfusion to leg muscles in ducks increased some 5-fold in ducks swimming at close to maximum sustainable rates (Butler et al., 1988). Therefore, by extension, one could assume that similar increases in blood perfusion to flight muscle occur during moderate to energetic flapping flight. In contrast, gliding should result in limited changes in muscle flow rates, especially given the lower expected CO during these types of flying.

11.6.1.1 Altitude

Avian species, such as the Bar-headed goose (*Anser indicus*), are unique among vertebrates in that they make annual migrations across mountain ranges at altitudes between 5500 and 7250 m (Hawkes et al., 2012). At these high elevations, the inspired partial pressure of oxygen (P_{IO_2}) falls to approximately 50 mmHg, a sea level equivalent of 7% O_2. In the O_2 cascade, maximum aerobic capacity is limited by convective movement of O_2 by the cardiovascular system (Hillman et al., 2013). Therefore, to achieve the feats of strenuous activity in a low O_2 environment at high altitudes, modifications to the cardiovascular system should be anticipated, allowing high aerobic activity. A number of studies from Dr William Milsom's group have examined cardiovascular functional adjustments of Bar-headed geese (*A. indicus*) and lowland species to decreasing levels of O_2 both at rest and during flight. Scott and Milsom (unpublished data) compared cardiovascular parameters of resting Bar-headed geese (*A. indicus*) and Peking ducks (*A. domesticus*) while facing decreasing O_2 levels. Resting Bar-headed geese maintain mean arterial pressure at inspired O_2 levels as low as 4%, while MAP falls in the duck at 7% O_2. They found that in response to decreased inspired O_2 levels, Bar-headed geese increased heart rate from resting levels of 160 beats/min to over 300 beats/min. In contrast, Peking duck increased heart rate from 130 to 190 beats/min in response to these same hypoxic challenges. Similar differences in capacity are evident when comparing the response of the Bar-headed goose and the Barnacle goose (*Branta leucopsis*; Figure 11.57; Lague et al., unpublished data). When O_2 is decreased from 12% to 7%, the Bar-headed goose increased heart rate by over 40% and CO by over 100%, while maintaining oxygen consumption. In contrast, oxygen consumption of the Barnacle goose decreased by 18% due to a 4% reduction in heart rate and a relatively minor 10%, increase in CO.

During activity, the capacity to elevated heart rate in low O_2 environments is paramount and has been investigated in the Bar-headed goose. During flight at speeds between 45 and 55 km/h, heart rate increased roughly 2.5- to 3-fold and oxygen consumption increased 16-fold over resting values in these geese. Preliminary analysis of heart rate data for flights at 21, 10.5, and 7% O_2 suggest that these animals are able to maintain an elevated heart rate with the associated CO under hypoxic flight conditions that mirror the oxygen levels experienced during migratory behavior (Meir et al., 2013). Therefore, these animals are able to extend the cardiovascular changes observed at rest to an active hypoxic flight condition.

Coronary vasodilation is critical to maintaining O_2 delivery during bouts of reduced O_2 availability or an increase in myocardial oxygen demand. Both of these factors presumably come into play during high-altitude flight in birds. In Pekin ducks and Bar-headed geese, coronary perfusion has been reported to be 3.5 mL/min/g wet heart mass at sea level when the P_{IO_2} was 142 mm Hg (Faraci et al., 1984). Exposure to severe hypoxia (28 mm Hg P_{IO_2}) increases coronary blood perfusion 5.5 and 2.7 times, respectively, in ducks and geese (Faraci et al., 1984) via a hypoxic coronary vasodilation (Figure 11.58). Although Bar-headed geese are accomplished high-altitude fliers, in contrast to ducks, counterintuitively these animals did not increase coronary blood perfusion to the same extent under hypoxic conditions. This may be compensated for by greater ventricle capillary density in the bar-head goose compared to other species (Scott et al., 2011).

Carbon dioxide is also a potent coronary vasodilator in mammals, coupling increased aerobic metabolism in the myocardium to an increased rate of oxygen delivery, while hypocapnia increases coronary resistance and decreased coronary perfusion. This would obviously be deleterious during high-altitude migratory flight. Interestingly, the relationship between coronary blood flow and P_{aCO_2} in Bar-headed geese appears to be quite different than mammals. Over a range of P_{aCO_2} from about 30 to 60 mm Hg, there is a linear increase in coronary flow with P_{aCO_2}, as in the mammal. However, in the hypocapnic condition, when P_{aCO_2} is 30 mm Hg or lower, there appears to be no effect of P_{aCO_2} on coronary resistance (Faraci and Fedde, 1986). Whether this represents a mechanism to ensure myocardial oxygen delivery during high-altitude flight remains to be determined.

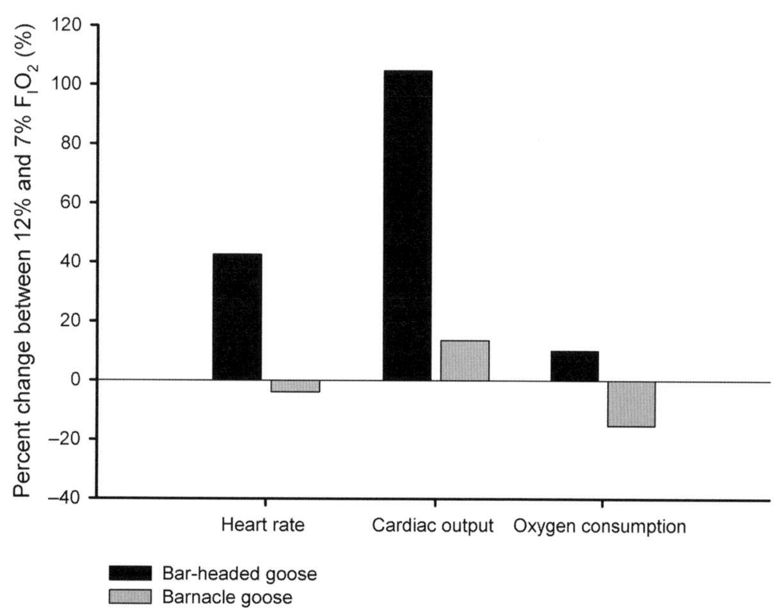

FIGURE 11.57 Percent change in heart rate, cardiac output, and oxygen consumption in response to a decrease in inspired F_{IO_2} from 12% to 7% in the resting Bar-headed goose and resting barnacle goose. *From Lague et al. (unpublished data).*

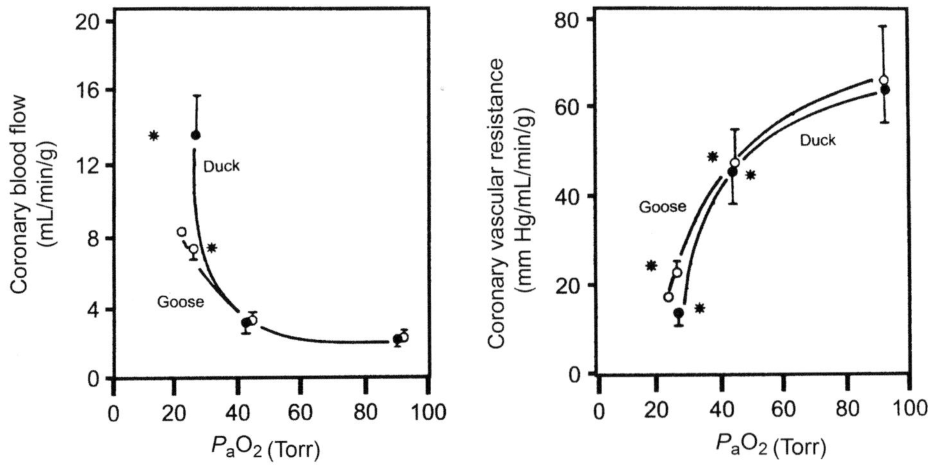

FIGURE 11.58 Responses of the coronary circulation to hypoxia in Pekin ducks and Bar-headed geese. (Left) coronary blood flow (mL/min/g); (right) coronary vascular resistance (mm Hg mL/min/g) as a function of the arterial O_2 partial pressure (P_{aO_2}). All values are means ± S.E.; $n=5$, except at 25 Torr, where means for two geese were plotted. An asterisk represents significant difference from normoxia (highest P_{aO_2} level) at $p \leq 0.05$. *Reprinted from Faraci et al. (1985).*

Activity at high elevation subjects tissue to reduced P_{IO_2} and hypocapnia. The hypocapnia and subsequent respiratory alkalosis results from increased CO_2 washout from greater hypoxemia driven ventilation (Faraci et al., 1985; Scott and Milsom, 2007). This condition could dramatically influence tissue metabolic function with the greatest impact on highly aerobic tissues of the central nervous system. In domestic geese, cerebral blood flow decreased as P_{aCO_2} falls from 50 to 20 mm Hg but then reached a plateau; however, P_{O_2} of cerebral tissue continued to fall as P_{aO_2} fell below 20 mm Hg, suggesting differences in cerebral vascular sensitivity to these two conditions (Bickler and Julian, 1992). Carbon dioxide is normally a potent vasodilator within the cerebral circulation (Faraci and Fedde, 1986; Grubb et al., 1977); however, the cerebral vasculature is fairly insensitive in domestic geese (*Anser anser*) and Pekin ducks, while Bar-headed geese are very insensitive to severe hypocapnia (Bickler and Julian, 1992; Faraci and Fedde, 1986; Grubb et al., 1977).

Given the range of altitude that bird species inhabit and transiently experience during seasonal migration, differing degrees in flight muscle capillarity may also correlate with flight at high elevation. It could be predicted that high-elevation species would exhibit a relatively greater capillary density to muscle fiber density, increasing the total surface area of contact between capillaries and fibers. This phenotype is evident in a number of species that fly at high elevation and that have relatively higher muscle capillarity and smaller fiber sizes. In pectorals muscle from Bar-headed geese and Andean coots (*Fulica ardesiaca*) native to 4200 m, as well as some limb muscles in the coot, the number of capillaries per muscle fiber is increased compared with lowland species (Scott et al. 2009; Leon-Velarde et al., 1993). Further, in the Bar-headed goose, mitochondria of aerobic fibers are redistributed so they are adjacent to capillaries (Scott et al., 2009). These compensatory changes can be induced during ontogeny as evident in increased capillary-to-fiber ratio and possible changes in diffusion distances in Canada goose goslings (*Branta canadensis*) hatching from eggs raised under hypoxic conditions (Snyder et al., 1984; Snyder, 1987). In contrast, wild adult pigeons actively flying at 3800 m maintain similar capillary geometry and density compared to controls at sea level (Mathieu-Costello et al., 1996). Equally important are factors such as changes in $P_{50,}$ the oxygen-carrying capacity of blood or the myoglobin content of muscle fibers, which are also significant adaptations involved in the response. Increases in carrying capacity and high levels of tissue oxygen extraction have been demonstrated in unexercised pigeons acclimatized to high altitude (Weinstein et al., 1985) and Bar-headed geese (Scott, 2011). In short, while the current understanding of cardiovascular and respiratory adjustments that must accompany flight at elevation has progressed, further extensive investigations must be undertaken to explain the capacity of birds to sustain activity in low-oxygen environments.

11.6.1.2 Migration

Extensive migrations involving continuous fasting flight for long periods of time are undertaken annually by many bird species. Migrating birds exhibit extraordinary phenotypic plasticity in organ size and body mass. This organ mass plasticity manifesting as hypertrophy of the heart and other organs is akin to that observed in infrequent feeding reptiles such as snakes (Secor and Diamond, 1998; Andersen et al., 2005). For example, the barnacle goose heart hypertrophies before migration composing approximately 1.1% of body mass (Bishop et al., 1995). In contrast, the

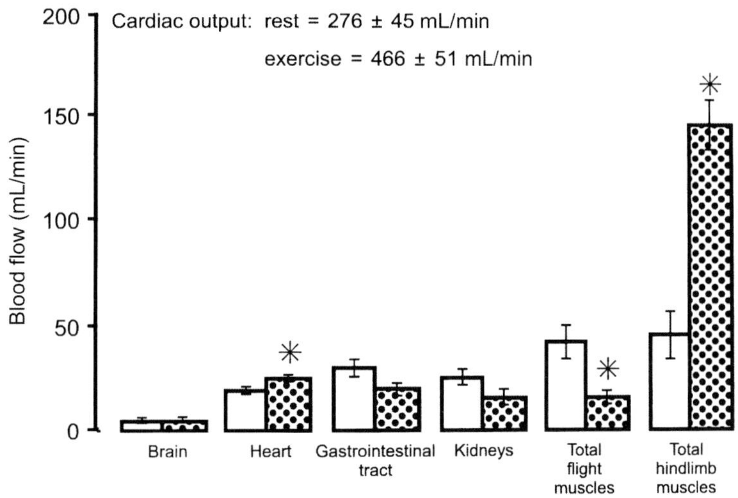

FIGURE 11.59 Histograms showing mean blood flow to selected vascular beds in six tufted ducks before (open bars) and during swimming at a mean velocity of 0.69:0.01 m/s (shaded bars). Asterisks indicate significant differences between preexercise and swimming values ($p<0.05$). *From Butler et al. (1988).*

heart of a typical bird is roughly 0.8% of body mass (Section 11.2.1.2). Heart hypertrophy of the red knot (*Calidris canutus islandica*) and bar-tailed godwit (*Limosa lapponica taymyrensis*) is correlated with increased power requirements needed for migration (Piersma et al., 1999; Landys-Ciannelli et al., 2003). The garden warbler (*Sylvia borin*) migrating across the Sahara exhibited a decrease in heart mass and flight muscle mass that was restored within 9 days of recovery feeding (Bauchinger et al., 2005). Few studies have examined the cardiovascular physiology associated with these long migrations and associated hypertrophy of the heart. During migration of Svalbard Barnacle geese (*B. leucopsis*), heart rate falls on successive days from 317 to 226 beats/min (Butler et al., 1998). Butler et al. (1998) suggested that progressive bradycardia was due to decreased O_2 demand as body mass fell while heart mass remains fixed during the migration. Thus, changes in CO appear to be influenced most heavily by heart rate changes. It remains to be seen how these changes in heart mass before, during, and after migration influence cardiovascular function and performance in these species.

11.6.2 Swimming and Diving

Bird species engage in surface swimming and submerged or diving swimming. In Tufted ducks (*Aythya fuligula*), surface swimming at maximal sustainable speeds is associated with a 70% increase in CO from 276 to 466 mL/min (Bevan and Butler, 1992). As with flight, swimming is associated with changes in regional vascular perfusion with myocardium and active leg musculature blood flow increasing by 30 and 300%, respectively, while flow to other regions remained the same or decreased (Bevan and Butler, 1992). Similar patterns are evident in the redhead duck (*Aythya americana*) during short periods of underwater swimming which results in a 200–500%, increase in hind legs perfusion (Stephenson and Jones, 1992). Clearly, selective perfusion takes place during swimming likely via an interaction between vasodilation in active muscle and vasoconstriction in visceral organs and inactive muscle (Figure 11.59; Butler et al., 1988; Bevan and Butler, 1992).

During a dive, animals must function with the limited O_2 stores carried within the body during the dive duration. The majority of dives are shorter than the aerobic, dive limit (ADL). This is the length of time that a dive can be aerobic relying on the O_2 stores in the body. A large number of diving species, such as ducks and cormorants, perform multiple short dives of less than 60 s in duration. Others species, such as the emperor penguin (*Aptenodytes forsteri*), dive for longer periods of more than 20 min (Meir et al. 2008). These different diving behaviors should by necessity result in species having different physiological capacities to ensure effective O_2 transport; however, regardless of dive duration, the pattern of cardiovascular responses is similar.

The phases of the cardiovascular response to a voluntary dive include an initial tachycardia during the immediate predive followed by a diving bradycardia. After surfacing, there is usually a postdive tachycardia. The predive and postdive tachycardia is suggested to aid in loading and replenishing O_2 stores and elimination of CO_2 (Butler and Jones, 1997). In many species, the diving bradycardia is not a true bradycardia in that the diving heart rate remains elevated in comparison to the resting heart rate.

In diving ducks, Jones and Holeton (1972), Lillo and Jones (1982), Jones et al. (1983), Smith and Jones (1992), and Bevan and Butler (1992) showed that stroke volume was maintained during the large decreases in CO generated during submersion, suggesting the primary factor dictating

CO during these events is heart rate. In addition, there is a negative correlation between minimum heart rate and dive duration as well as depth (Bevan et al. 1997; Meir et al. 2008). The extent of the diving bradycardia may be correlated with changes in blood oxygen levels, regulated by variation in parasympathetic output in ducks (McPhail and Jones, 1999). In diving cormorants (*Phalacrocorax auritus*), heart rate during a shallow dive was dependent upon the level of inspired oxygen (Enstipp et al., 2001). Predive inspiration of hypoxic gas resulted in a stronger bradycardia and predive inspiration of hyperoxic gas resulted in the opposite response. Similar changes in response to inspired oxygen levels and diving have been observed in the lesser scaup duck (*A. affinis*; Borg et al., 2004).

Penguins have been extensively studied to understand their capacity for extended active diving bouts. During dives that are greater than the ADL Emperor penguins become relatively bradycardiac just prior to returning to the surface (Meir et al., 2008). Lesser scaup ducks decrease heart rate below resting levels during forced dives only (Borg et al., 2004). During long voluntary dives of 30 min or if this species is exposed to 9% O_2, heart rate fells below resting values during the later portions of the dive suggesting that a true diving bradycardia may only occur when blood P_{O_2} levels become severely reduced (Borg et al., 2004). Changes occur in tissue perfusion have been suggested to accompany heart rate responses to diving. During long dives Emperor penguins have blood lactate levels that suggest changes occur in blood flow to the skeletal muscles during the dive (Ponganis et al., 2009). The ADL is the duration at which the dive becomes anaerobic and is indicated by blood lactate levels during postdive recovery. The majority of birds carry out multiple short dives that fall within their ADL; however, blood lactate levels remain low in animals that have passed their ADL, with levels increasing during postdive period. Data also suggest a potential arteriovenous shunt through the extremities in the emperor penguin (Ponganis et al., 2009). Ponganis et al. (2009) proposed that this shunt may be important during the initial phases of the dive by allowing enhanced transport of O_2 from the gas exchange surface to the venous system.

Regulation of the cardiovascular response to diving has been explored is some species. Baroreceptor input has been shown to have no direct role in generating or maintaining heart rate responses to voluntary submersion in diving ducks, since these animals display the same degree of bradycardia after denervation of arterial baroreceptors as before (Furilla and Jones, 1987a). However, baroreceptors may play a role in control of heart rate in dabbling ducks (*A. platyrhynchos*), which have been trained to dive voluntarily. Predive heart rate in dabblers ranged from 100 to about 500 beats/min but, regardless of the rate preceding any particular dive, rates during the dive tended to a value of approximately 250 beats/min (Furilla and Jones, 1987b). This response implies that the heart rate was being regulated at a set value during voluntary dives. Removal of baroreceptor input by bilateral section of the aortic nerves eliminated the tendency of dive heart rate to approach the

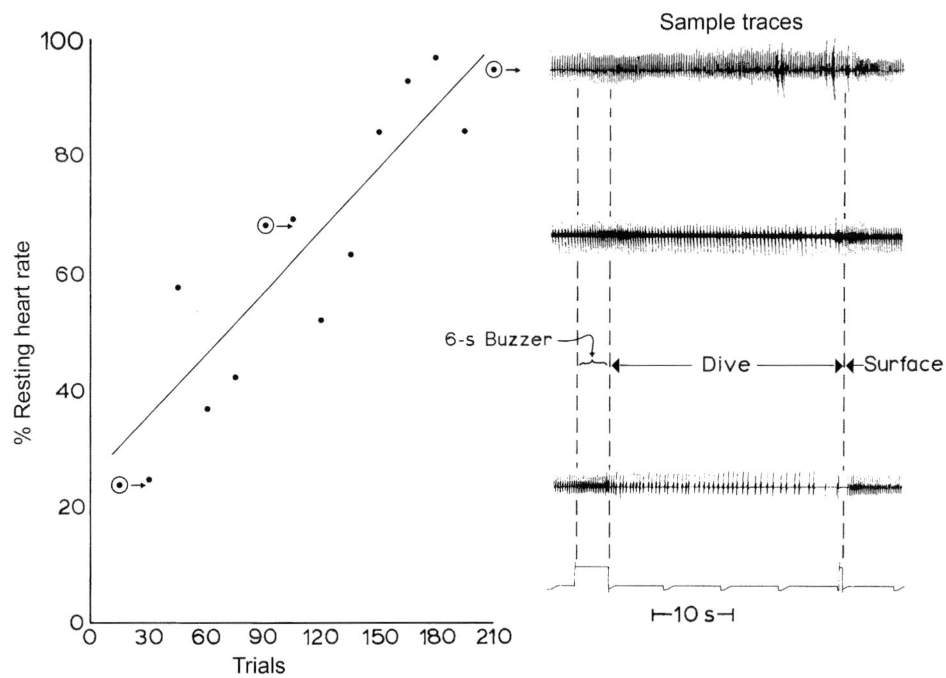

FIGURE 11.60 Reduction in degree of bradycardia attained in 40-s dives by a Pekin duck during repeated trials, each consisting of a 40-s dive immediately preceded by a 6-s buzzer. The graph is a plot of percentage of resting heart rate achieved at the end of a trial, against the number of trials. Sample electrocardiogram traces are taken from the first trial of the series (bottom trace), the 90th trial (middle trace), and the final dive (top trace). *From Gabbott and Jones (1987).*

"set point" value; after barodenervation dive heart rate varied little from the prevailing predive rate (Furilla and Jones, 1987b). Given that the baroreflex normally regulates blood pressure by adjusting CO and vascular resistance, the physiological value of a baroreceptor-dependent set point for heart rate is uncertain. The behavior of arterial blood pressure during voluntary diving in dabbling ducks has not been established. However, there may be some inherent benefit to regulating heart rate under these conditions. If this is true, then strong phasic baroreceptor input to the central nervous system during systole would represent the primary afferent feedback route for heart rate related information.

Stimulation of nasal receptors in diving ducks and chemoreceptors in dabbling ducks are the proximate causes of the development of diving bradycardia (Furilla and Jones, 1986, 1987a,b). Repeatedly submerging the head of a diving or dabbling duck in a laboratory situation causes the bradycardic response to habituate after 100 to 200 dives (Figure 11.60; Gabbott and Jones, 1987). In dabbling ducks, however, extending the period of submergence beyond 40 s virtually eliminates any attenuation of the cardiac response to submergence. Obviously, in dabbling ducks input from the carotid body chemoreceptors is too intense for habitation after 60 s submergence. Similarly, exposing habituated animals to 10 or 15% oxygen in air before submergence causes prominent bradycardia although the very next trial, after breathing room air, evokes the habituated cardiac response. Interestingly, the heart rate response to diving after breathing air with low levels of oxygen is unaffected by training. Consequently, chemoreceptor input, which will be the same in naive and habituated ducks because blood gas levels are the same after 40 s submergence, can be habituated. Habituation of the response occurs within the central nervous system, below the thalamic level. Animals with their higher brain centers surgically removed can be trained as easily as intact ducks (Gabbott and Jones, unpublished data).

Even though CO is reduced during diving, the rate of perfusion of cerebral vasculature and thoracoabdominal cavity is maintain or elevated above predive levels (Heieis and Jones, 1988; Jones et al., 1979). In mallard and Pekin ducks myocardial flow was, on average, 0.73 mL/min/g presubmersion and 0.88 mL/min/g after 144–250 s of submergence. Cerebral flow increased from 0.43 to 3.68 mL/min/g over the same time period. In Pekin ducks forcibly submerged until P_{aO_2} fell to 50 mm Hg, cerebral blood flow increased from 1.58 to 3.2 mL/min/g. Clearly, regardless of the wide range of absolute values measured, cerebral blood flow increases in forced submersion asphyxia, maintaining oxygen delivery to brain tissue. A redistribution of blood flow away from more hypoxia-tolerant regions toward more sensitive regions within the brain itself does not seem to occur in the Pekin duck (Stephenson et al., 1994). However, such heterogeneous regional changes in cerebral blood flow in response to asphyxia have been proposed to occur in neonatal mammals (Goplerud et al., 1989).

ACKNOWLEDGMENTS

This chapter drew heavily from The Cardiovascular System from the 5th edition written by Frank M. Smith, Nigel H. West, and David R. Jones.

REFERENCES

Ábrahám, A., 1969. Microscopic Innervation of the Heart and Blood Vessels in Vertebrates Including Man. Pergamon Press, Oxford, UK.

Aceto, J.F., Baker, K.M., 1990. [Sar1] angiotensin II receptor-mediated stimulation of protein synthesis in chick heart cells. Am. J. Physiol. 258, H806–H813.

Adair, T.H., Montani, J.P., Strick, D.M., Guyton, A.C., 1989. Vascular development in chick embryos, a possible role for adenosine. Am. J. Physiol. 256, H240–H246.

Adams, W.E., 1937. A contribution to the anatomy of the avian heart as seen in the kiwi (*Apterix australis*) and the yellow-crested penguin (*Megadypte antipodum*). J. Zool. 107, 417–441.

Adams, W.E., 1958. Morphology of the Carotid Body and Carotid Sinus. Charles C. Thomas, Springfield, IL.

Agren, P., Cogolludo, A.L., Kessels, C.G., Perez-Vizcaino, F., DeMey, J.G., Blanco, C.E., Villamor, E., 2007. Ontogeny of chicken ductus arteriosus response to oxygen and vasoconstrictors. Am. J. Physiol. 292, R485–R496.

Akester, A.R., 1967. Renal portal shunts in the kidney of the domestic fowl. J. Anat. 101, 569–594.

Akester, A.R., 1971. The heart. In: Bell, D.J., Freeman, B.M. (Eds.), Physiology and Biochemistry of the Domestic Fowl, vol. 2. Academic Press, New York, pp. 745–781.

Akester, A.R., 1979. The autonomic nervous system. In: King, A.S., McLelland, J. (Eds.), Form and Function in Birds, vol. 1. Academic Press, New York, pp. 381–441.

Akester, A.R., Akester, B., 1971. Double innervation of the avian cardiovascular system. J. Anat. 108, 618–619.

Akester, A.R., Mann, S.P., 1969. Adrenergic and cholinergic innervation of the renal portal valve in the domestic fowl. J. Anat. 104, 241–252.

Akester, A.R., Akester, B., Mann, S.P., 1969. Catecholamines in the avian heart. J. Anat. 104, 591.

Altimiras, J., Crossley II, D.A., 2000. Control of blood pressure mediated by baroreflex changes of heart rate in the chicken embryo (*Gallus gallus*). Am. J. Physiol. 278, R980–R986.

Altimiras, J., Deck, L.M.G., Garitano-Zavala, A., 2013. The small heart of the Ornate Tinamou is compatible with endothermy and flight but compromises aerobic metabolism and thermoregulation during recovery from exhaustive activity. FASEB J. 27, 1149.18.

Altman, P.L., Dittmer, D.S., 1971. Respiration and Circulation. Federation of American Societies for Experimental Biology, Bethesda, MD.

Andersen, J.B., Rourke, B.C., Caiozzo, V.J., Bennett, A.F., Hicks, J.W., 2005. Physiology: postprandial cardiac hypertrophy in pythons. Nature 434, 37–38.

Andrewartha, S.J., Tazawa, H., and Burggren, W.W., 2011. Embryonic control of heart rate: examining developmental patterns and temperature and oxygenation influences using embryonic avian models. Respir. Physiol. Neurobiol. 178, 84–95.

Armour, J.A., Ardell, J.L., 1994. Neurocardiology. Oxford University Press, New York.

Armour, J.A., Huang, M.H., Smith, F.M., 1993. Peptidergic modulation of *in situ* canine intrinsic cardiac neurons. Peptides 14, 191–202.

Ash, R.W., Pearce, J.W., Silver, A., 1969. An investigation of the nerve supply to the salt gland of the duck. Q. J. Exp. Physiol. 54, 281–295.

Astrand, P.O., Rodahl, K., 1986. Textbook of Work Physiology. McGraw-Hill, New York.

Bagshaw, R.J., 1985. Evolution of cardiovascular baroreceptor control. Biol. Rev. 60, 121–162.

Bagshaw, R.J., Cox, R.H., 1986. Baroreceptor control of heart rate in chickens (*Gallus domesticus*). Am. J. Vet. Res. 47, 293–295.

Baker, K.M., Aceto, J.F., 1990. Angiotensin II stimulation of protein synthesis and cell growth in chick heart cells. Am. J. Physiol. 259, H610–H618.

Ball, R.A., Sautter, J.H., Katter, M.S., 1963. Morphological characteristics of the anterior mesenteric artery of the fowl. Anat. Rec. 146, 251–256.

Ball, R.A., Sautter, J.H., Waibel, P.E., 1972. Adaptive features in the turkey aorta which precede plaque formation. Atherosclerosis 15, 241–247.

Ballard, V.L.T., Mikawa, T., 2002. Constitutive expression of preproendothelin in the cardiac neural crest selectively promotes expansion of the adventitia of the great vessels in vivo. Dev. Biol. 251, 167–177.

Berry, A., 1950. The effects of epinephrine on the myocardium of the embryonic chick. Circulation 1, 1362–1368.

Bamford, O.S., Jones, D.R., 1976. The effects of asphyxia on afferent activity recorded from the cervical vagus in the duck. Pflugers Arch. 366, 95–99.

Barnas, G.M., Gleeson, M., Rautenberg, W., 1985. Respiratory and cardiovascular responses of the exercising chicken to spinal cord cooling at different ambient temperatures. I. Cardiovascular responses and blood gases. J. Exp. Biol. 114, 415–426.

Bauchinger, U., Wohlmann, A., Biebach, H., 2005. Flexible remodeling of organ size during spring migration of the garden warbler (*Sylvia borin*). Zoology (Jena) 108, 97–106.

Baumel, J.J., 1975. *Aves* heart and blood vessels. In: Getty, R. (Ed.), Sisson and Grossman's the Anatomy of the Domestic Animals. Saunders, Philadelphia.

Baumel, J.J., Gerchman, L., 1968. The avian intercarotid anastomosis and its homologue in other vertebrates. Am. J. Anat. 122, 1–18.

Bech, C., Nomoto, S., 1982. Cardiovascular changes associated with treadmill running in the Pekin duck. J. Exp. Biol. 97, 345–358.

Belanger, C., Copeland, J., Muirhead, D., Heinz, D., Dzialowski, E.M., 2008. Morphological changes in the chicken ductus arteriosi during closure at hatching. Anat. Rec. 291, 1007–1015.

Bell, C., 1969. Indirect cholinergic vasomotor control of intestinal blood flow in the domestic chicken. J. Physiol. 205, 317–327.

Bennett, T., 1971. The adrenergic innervation of the pulmonary vasculature, the lung and the thoracic aorta, and on the presence of aortic bodies in the domestic fowl (*Gallus gallus domesticus* L.). Z. Zellforsch. 114, 117–134.

Bennett, T., 1974. Peripheral and autonomic nervous systems. In: Farner, D.S., King, J.R., Parkes, K.C. (Eds.), Avian Biology, vol. 4. Academic Press, New York, pp. 1–77.

Bennett, T., Malmfors, T., 1970. The adrenergic nervous system of the domestic fowl (*Gallus domesticus* (L.)). Z. Zellforsch. 106, 22–50.

Bennett, T., Malmfors, T., 1974. Regeneration of the noradrenergic innervation of the cardiovascular system of the chick following treatment with 6-hydroxydopamine. J. Physiol. 242, 517–532.

Bennett, T., Malmfors, T., 1975a. Autonomic control of renal portal blood flow in the domestic fowl. Experientia 31, 1177–1178.

Bennett, T., Malmfors, T., 1975b. Characteristics of the noradrenergic innervation of the left atrium in the chick (*Gallus gallus domesticus*, L.). Comp. Biochem. Physiol. C 52, 47–49.

Bennett, T., Cobb, J.L.S., Malmfors, T., 1974. The vasomotor innervation of the inferior vena cava of the domestic fowl (*Gallus gallus domesticus* L.): I. Structural observations. Cell Tissue Res. 148, 521–533.

Benzo, C.A., 1986. Nervous system. In: Sturkie, P.D. (Ed.), Avian Physiology, fourth ed. Springer-Verlag, Berlin, pp. 1–36.

Bergel, D.H., 1961. The static properties of the arterial wall. J. Physiol. 156, 445–457.

Berk, M.L., 1987. Projections of the lateral hypothalamus and bed nucleus of the stria terminalis to the dorsal vagal complex in the pigeon. J. Comp. Neurol. 260, 140–156.

Berk, M.L., Finkelstein, J.A., 1983. Long descending projections of the hypothalamus in the pigeon, *Columba livia*. J. Comp. Neurol. 220, 127–136.

Berk, M.L., Smith, S.E., 1994. Local and commissural neuropeptide-containing projections of the nucleus of the solitary tract to the dorsal vagal complex in the pigeon. J. Comp. Neurol. 347, 369–396.

Bevan, R.M., Boyd, I.L., Butler, P.J., Reid, K., Woakes, A.J., Croxall, J.P., 1997. Heart rates and abdominal temperatures of free-ranging south Georgian shags, *Phalacrocorax georgianus*. J. Exp. Biol. 200, 661–675.

Bevan, R.M., Butler, P.J., 1992. Cardiac output and blood flow distribution during swimming and voluntary diving of the tufted duck (*Aythya fuligula*). J. Exp. Biol. 168, 199–217.

Bézie, Y., Mesnard, L., Longrois, D., Samson, F., Perret, C., Mercadier, J.J., Laurent, S., 1996. Interactions between endothelin-1 and atrial natriuretic peptide influence cultured chick cardiac myocyte contractility. Eur. J. Pharmacol. 311, 241–248.

Bezuidenhout, A.J., 1984. The coronary circulation of the heart of the ostrich. J. Anat. 138, 385–397.

Bickler, P.E., Julian, D., 1992. Regional cerebral blood flow and tissue oxygenation during hypocarbia in geese. Am. J. Physiol. 263, R221–R225.

Biegon, R.L., Epstein, P.M., Pappano, A.J., 1980. Muscarinic antagonism of the effects of phosphodiesterase inhibitor (methylisobutylxanthine) in embryonic chick ventricle. J. Pharmacol. Exp. Ther. 215, 348–356.

Biegon, R.L., Pappano, A.J., 1980. Dual mechanism for inhibition of calcium-dependent action potentials by acetylcholine in avian ventricular muscle. Relationship to cyclic AMP. Circ. Res. 46, 353–362.

Bishop, C.M., Butler, P.J., 1995. Physiological modelling of oxygen consumption in birds during flight. J. Exp. Biol. 198, 2153–2163.

Bishop, C.M., Butler, P.J., Egginton, S., el-Haj, A.J., Gabrielsen, G.W., 1995. Development of metabolic enzyme activity in locomotor and cardiac muscles in the migratory barnacle goose. Am. J. Physiol. 269, R64–R72.

Blix, A.S., Wennergren, G., Folkow, B., 1976. Cardiac receptors in ducks—a link between vasoconstriction and bradycardia during diving. Acta Physiol. Scand. 97, 13–19.

Boelkins, J.N., Mueller, W.J., Hall, K.L., 1973. Cardiac output distribution in the laying hen during shell formation. Comp. Biochem. Physiol. A 46, 735–743.

Bogusch, G., 1974. The innervation of Purkinje fibres in the atrium of the avian heart. Cell Tissue Res. 150, 57–66.

Bolton, T.B., 1967. Intramural nerves in the ventricular myocardium of the domestic fowl and other animals. Br. J. Pharmacol. Chemother. 31, 253–268.

Bolton, T.B., 1969. Spontaneous and evoked release of neurotransmitter substances in the longitudinal muscle of the anterior mesenteric artery of the domestic fowl. Br. J. Pharmacol. 35, 112–120.

Bolton, T.B., Bowman, W.C., 1969. Adrenoreceptors in the cardiovascular system of the domestic fowl. Eur. J. Pharmacol. 5, 121–132.

Bolton, T.B., Raper, C., 1966. Innervation of domestic fowl and guinea-pig ventricles. J. Pharm. Pharmacol. 18, 192–193.

Bopelet, M., 1974. Normal electrocardiogram of the chicken: its variations during vagal stimulation and following vagotomies. Comp. Biochem. Physiol. A 47, 361–369.

Borg, K.A., Milsom, W.K., Jones, D.R., 2004. The effect of O_2 and CO_2 on the dive behavior and heart rate of lesser scaup ducks (*Aythya affinis*): quantification of the critical PaO_2 that initiates a diving bradycardia. Respir. Physiol. Neurobiol. 144, 263–279.

Bossen, E., Sommer, J.R., Waugh, R.A., 1978. Comparative stereology of the mouse and finch left ventricle. Tissue Cell 10, 773–784.

Boulianne, M., Hunter, D.B., Julian, R.J., O'Grady, M.R., Physick Sheard, P.W., 1992. Cardiac muscle mass distribution in domestic turkey and relationship to electrocardiogram. Avian Dis. 36, 582–589.

Boulianne, M., Hunter, D.B., Physick-Sheard, P.W., Viel, L., Julian, R.J., 1993a. Effect of exercise on cardiac output and other cardiovascular parameters of heavy turkeys and relevance to the sudden death syndrome. Avian Dis. 37, 98–106.

Boulianne, M., Hunter, D.B., Viel, L., Physick-Sheard, P.W., Julian, R.J., 1993b. Effect of exercise on the cardiovascular and respiratory systems of heavy turkeys and relevance to sudden death syndrome. Avian Dis. 37, 83–97.

Bouverot, P., Leitner, L.M., 1972. Arterial chemoreceptors in the domestic fowl. Resp. Physiol. 15, 310–320.

Bouverot, P., Douguet, D., Sébert, P., 1979. Role of the arterial chemoreceptors in ventilatory and circulatory adjustments to hypoxia in awake Pekin ducks. J. Comp. Physiol. B 133, 177–186.

Bowers, P.N., Tinney, J.P., Keller, B.B., 1996. Nitroprusside selectively reduces ventricular preload in the stage 21 chick embryo. Cardiovas. Res. 31, E132–E138.

Braun, E.J., 1982. Glomerular filtration in birds—its control. Fed. Proc. 41, 2377–2381.

Brehm, G., Lindmar, R., Loffelholz, K., 1992. Inhibitory and excitatory muscarinic receptors modulating the release of acetylcholine from the postganglionic parasympathetic neuron of the chicken heart. Naunyn-Schmiedeberg's Arch. Pharmacol. 346, 375–382.

Brill, R.W., Jones, D.R., 1981. On the suitability of Innovar, a neuroleptic analgesic, for cardiovascular experiments. Can. J. Physiol. Pharmacol. 59, 1184–1189.

Brummermann, M., Simon, E., 1990. Arterial hypotension in ducks adapted to high salt intake. J. Comp. Physiol. B 160, 127–136.

Burrows, M.E., Braun, E.J., Duckles, S.P., 1983. Avian renal portal valve: a reexamination of its innervation. Am. J. Physiol. 245, H628–H634.

Bussow, H., 1973. Zar wandstruktur der grojen arterien der vogel. Z. Zellforsch. 142, 263–288.

Butler, D.G., Wilson, J.X., Graves, L.E., 1986. α- and β-adrenergic mechanisms mediate blood pressure control by norepinephrine and angiotensin in ducks. Gen. Comp. Endocrinol. 61, 323–329.

Butler, P.J., 1967. The effect of progressive hypoxia on the respiratory and cardiovascular systems of the chicken. J. Physiol. 191, 309–324.

Butler, P.J., 1991. Exercise in birds. J. Exp. Biol. 160, 233–262.

Butler, P.J., Jones, D.R., 1968. Onset of and recovery from diving bradycardia in ducks. J. Physiol. 196, 255–272.

Butler, P.J., Jones, D.R., 1971. The effect of variations in heart rate and regional distribution of blood flow on the normal pressor response to diving in ducks. J. Physiol. 214, 457–479.

Butler, P.J., Jones, D.R., 1997. The physiology of diving of birds and mammals. Physiol. Rev. 77, 837–899.

Butler, P.J., Taylor, E.W., 1973. The effect of hyperoxic hypoxia, accompanied by different levels of lung ventilation, on heart rate in the duck. Resp. Physiol. 19, 176–187.

Butler, P.J., Taylor, E.W., 1983. Factors affecting the respiratory and cardiovascular responses to hypercapnic hypoxia, in Mallard ducks. Resp. Physiol. 53, 109–127.

Butler, P.J., Turner, D.L., Al-Wassia, A., Bevan, R.M., 1988. Regional distribution of blood flow during swimming in the tufted duck (*Aythya fuligula*). J. Exp. Biol. 135, 461–472.

Butler, P.J., West, N.H., Jones, D.R., 1977. Respiratory and cardiovascular responses of the pigeon to sustained, level flight in a wind-tunnel. J. Exp. Biol. 71, 7–26.

Butler, P.J., Woakes, A.J., Bishop, C.M., 1998. Behaviour and physiology of Svalbard barnacle geese, *Branta leucopsis*, during their autumn migration. J. Avian Biol. 29, 536–545.

Byers, R.L., Snyder, G.K., 1984. Effects of maturation on tissue capillarity in chickens. Resp. Physiol. 58, 137–150.

Cabot, J.B., Cohen, D.H., 1977a. Avian sympathetic cardiac fibers and their cells of origin: anatomical and electrophysiological characteristics. Brain Res. 131, 73–87.

Cabot, J.B., Cohen, D.H., 1977b. Anatomical and physiological characterization of avian sympathetic cardiac afferents. Brain Res. 131, 89–101.

Cabot, J.B., Cohen, D.H., 1980. Neural control of the avian heart. In: Bourne, G.B. (Ed.), Hearts and Heart-like Organs, vol. 1. Academic Press, NY, pp. 199–258.

Cabot, J.B., Carroll, J., Bogan, N., 1991a. Localization of cardiac parasympathetic preganglionic neurons in the medulla oblongata of pigeon, *Columba livia*: a study using fragment C of tetanus toxin. Brain Res. 544, 162–168.

Cabot, J.B., Mennone, A., Bogan, N., Carroll, J., Evinger, C., Erichsen, J.T., 1991b. Retrograde, trans-synaptic and transneuronal transport of fragment C of tetanus toxin by sympathetic preganglionic neurons. Neuroscience 40, 805–823.

Cabot, J.B., Reiner, A., Bogan, N., 1982. Avian bulbospinal pathways: anterograde and retrograde studies of cells of origin, funicular trajectories and laminar terminations. Prog. Brain Res. 57, 79–108.

Chess-Williams, R., Austin, C.E., O'Brien, H.L., 1991. α-adrenoceptors do not contribute to the chronotropic or inotropic responses of the avian heart to noradrenaline. J. Auton. Pharmacol. 11, 27–35.

Chiba, Y., Fukuoka, S., Niiya, A., Akiyama, R., Tazawa, H., 2004. Development of cholinergic chronotropic control in chick (*Gallus gallus domesticus*) embryos. Comp. Biochem. Physiol. A. 137, 65–73.

Cinar, A., Bagci, C., Belge, F., Uzun, M., 1996. The electrocardiogram of the Pekin duck. Avian Dis. 40, 919–923.

Cohen, D.H., 1980. The functional neuroanatomy of a conditioned response. In: Thompson, R.F., Hicks, L.H., Shvyrkov, V.B. (Eds.), Neural Mechanisms of Goal-Directed Behavior and Learning. Academic Press, New York, pp. 283–302.

Cohen, D.H., 1984. Identification of vertebrate neurons modified during learning: analysis of sensory pathways. In: Alkon, D.L., Farley, J. (Eds.), Primary Neural Substrates of Learning and Behavioral Change. Cambridge University Press, Cambridge, UK, pp. 129–154.

Cohen, D.H., Schnall, A.M., 1970. Medullary cells of origin of vagal cardioinhibitory fibers in the pigeon. II. Electrical stimulation of the dorsal motor nucleus. J. Comp. Neurol. 140, 321–342.

Cohen, D.H., Schnall, A.M., Macdonald, R.L., Pitts, L.H., 1970. Medullary cells of origin of vagal cardioinhibitory fibers in the pigeon. I. Anatomical studies of peripheral vagus nerve and the dorsal motor nucleus. J. Comp. Neurol. 140, 299–320.

Coote, J.H., 1975. Physiological significance of somatic afferent pathways from skeletal muscle and joints with reflex effects on the heart and circulation. Brain Res. 87, 139–144.

Coraboeuf, E., Obrecht-Coutris, G., Le-Douarin, G., 1970. Acetylcholine and the embryonic heart. Am. J. Cardiol. 25, 285–291.

Corvetti, G., Andreotti, L., Sisto-Daneo, L., 1988. Chick heart peptidergic innervation: localization and development. Basic Appl. Histochem. 32, 485–493.

Crossley II, D.A., 1999. Development of Cardiovascular Regulation in Embryos of the Domestic Fowl (*Gallus gallus*) with Partial Comparison to Embryos of the Desert Tortoise (*Gopherus agassizii*). University of North Texas, Denton. Ph.D. dissertation.

Crossley II, D.A., Altimiras, J., 2000. Ontogeny of autonomic control of cardiovascular function in the domestic chicken *Gallus gallus*. Am. J. Physiol. 279, R1091–R1098.

Crossley II, D.A., Altimiras, J., 2012. Effect of selection for commercially productive traits on the plasticity of cardiovascular regulation in chicken breeds during embryonic development. Poult. Sci. 91, 2628–2636.

Crossley II, D.A., Bagatto, B.P., Dzialowski, E.M., Burggren, W.W., 2003a. Maturation of cardiovascular control mechanisms in the embryonic emu (*Dromiceius novaehollandiae*). J. Exp. Biol. 206, 2703–2710.

Crossley II, D.A., Burggren, W.W., Altimiras, J., 2003b. Cardiovascular regulation during hypoxia in embryos of the domestic chicken *Gallus gallus*. Am. J. Physiol. 284, R219–R226.

Crossley II, D.A., Jonker, S.S., Hicks, J.W., Thornburg, K.L., 2010. Maturation of the angiotensin II cardiovascular response in the embryonic White Leghorn chicken (*Gallus gallus*). J. Comp. Physiol. B. 180, 1057–1065.

Cullis, W.C., Lucas, C.L.T., 1936. Action of acetylcholine on the aneural chick heart. J. Physiol. Suppl. 86, 53–55.

Dantzler, W.H., 1989. Comparative Physiology of the Vertebrate Kidney. Springer-Verlag, New York.

Davies, F., 1930. The conducting system of the bird's heart. J. Anat. 64, 129–146.

Deighton, N.M., Motomura, S., Borquez, D., Zerkowski, H.R., Doetsch, N., Brodde, O.E., 1990. Muscarinic cholinoceptors in the human heart: demonstration, subclassification, and distribution. Naunyn Schmiedebergs Arch. Pharmacol. 341, 14–21.

Denton, T., Diamond, G.A., Helfant, R.H., Khan, S., Karagueuzian, H., 1990. Fascinating rhythm: a primer on chaos theory and its application to cardiology. Am. Heart J. 120, 1419–1440.

DeSantis, V.P., Lindmar, L.R., Loffelholz, K., 1975. Evidence for noradrenaline and adrenaline as sympathetic transmitters in the chicken. Br. J. Pharmacol. 55, 343–350.

Di Rienzo, M., Parati, G., Castiglioni, P., Tordi, R., Mancia, G., Pedotti, A., 2001. Baroreflex effectiveness index: an additional measure of baroreflex control of heart rate in daily life. Am. J. Physiol. 280, R744–R751.

Djojosugito, A.M., Folkow, B., Kovách, A.G.B., 1968. The mechanisms behind the rapid blood volume restoration after hemorrhage in birds. Acta Physiol. Scand. 74, 114–122.

Djojosugito, A.M., Folkow, B., Yonce, L.R., 1969. Neurogenic adjustments of muscle blood flow, cutaneous A–V shunt flow and of venous tone during "diving" in ducks. Acta Physiol. Scand. 75, 377–386.

Dombkowski, R.A., Russell, M.J., Schulman, A.A., Doellman, M.M., Olson, K.R., 2005. Vertebrate phylogeny of hydrogen sulfide vasoactivity. Am. J. Physiol. Regul. Integr. Comp. Physiol. 288, R243–R252.

Donnelly, D.F., 2005. Development of carotid body/petrosal ganglion response to hypoxia. Respir. Physiol. Neurobiol. 149, 191–199.

Duchamp, C., Barre, H., 1993. Skeletal muscle as the major site of nonshivering thermogenesis in cold-acclimated ducklings. Am. J. Physiol. 265, R1076–R1083.

Dufour, J.J., Posternak, J.M., 1960. Effets chronotropes de l'acetylcholine sur le coeur d'embryon de poulet. Helv. Physiol. Pharmacol. Acta 18, 563–580.

Dunn, L.K., Gruenloh, S.K., Dunn, B.E., Reddy, D.S., Falck, J.R., Jacobs, E.R., Medhora, M., 2005. Chick chorioallantoic membrane as an in vivo model to study vasoreactivity: characterization of development-dependent hyperemia induced by epoxyeicosatrienoic acids (EETs). Anat. Rec. 285, 771–780.

Dzialowski, E.M., Greyner, H., 2008. Maturation of the contractile response of the Emu ductus arteriosus. J. Comp. Physiol. B 178, 401–412.

Dzialowski, E.M., Sirsat, T., van der Sterren, S., Villamor, E., 2011. Prenatal cardiovascular shunts in amniotic vertebrates. Respir. Physiol. Neurobiol. 178, 66–74.

Ede, D.A., 1964. Bird Structure. Hutchinson Educational, London.

Elfwing, M., 2007. The Ontogeny of the Baroreflex in Domestic Broiler Chickens (*Gallus Gallus domesticus*). Linköpings Universitet, Linköping. MS thesis.

Elfwing, M., Lundengård, K., Altimiras, J., 2011. Fetal development of baroreflex sensitivity: the chicken embryo as a case model. Respir. Physiol. Neurobiol. 178, 75–83.

Ellerby, D.J., Henry, H.T., Carr, J.A., Buchanan, C.I., Marsh, R.L., 2005. Blood flow in guinea fowl *Numida meleagris* as an indicator of energy expenditure by individual muscles during walking and running. J. Physiol. 564, 631–648.

Einzig, S., Staley, N.A., Mettler, E., Nicoloff, D.M., Noren, G.R., 1980. Regional myocardial blood flow and cardiac function in a naturally occurring congestive cardiomyopathy of turkeys. Cardiovas. Res. 14, 396–407.

Ellis, C.G., Potter, R.F., Groom, A.C., 1983. The krogh cylinder geometry is not appropriate for modelling O_2 transport in contracted skeletal muscle. Adv. Exp. Med. Biol. 159, 253–268.

Eranko, O., Eranko, L., 1977. Morphological indications of SIF cell functions. Adv. Biochem. Psychopharmacol. 16, 525–531.

Estavillo, J.A., 1978. Fiber size and sensory endings of the middle cardiac nerve of the domestic fowl (*Gallus domesticus*). Acta Anat. 101, 104–109.

Enstipp, M.R., Andrews, R.D., Jones, D.R., 2001. The effects of depth on the cardiac and behavioural responses of double-crested cormorants (*Phalacrocorax auritus*) during voluntary diving. J. Exp. Biol. 204, 4081–4092.

Estavillo, J., Burger, R.E., 1973a. Cardiac afferent activity in depressor nerve of the chicken. Am. J. Physiol. 225, 1063–1066.

Estavillo, J., Burger, R.E., 1973b. Avian cardiac receptors: activity changes by blood pressure, carbon dioxide, and pH. Am. J. Physiol. 225, 1067–1071.

Estavillo, J.A., Adamson, T.P., Burger, R.E., 1990. Middle cardiac nerve section alters ventilatory response to $PaCO_2$ in the cockerel. Resp. Physiol. 81, 349–358.

Evered, M.D., Fitzsimons, J.T., 1981. Drinking and changes in blood pressure in response to angiotensin II in the pigeon *Columba livia*. J. Physiol. 310, 337–352.

Falck, B., 1962. Observations on the possibilities of the cellular localization of monoamines by a fluorescence method. Acta Physiol. Scand. 197 (Suppl.).

Faraci, F.M., Fedde, M.R., 1986. Regional circulatory responses to hypocapnia and hypercapnia in Bar-headed geese. Am. J. Physiol. 250, R499–R504.

Faraci, F.M., Kilgore, D.L., Fedde, M.R., 1984. Oxygen delivery to the heart and brain during hypoxia: Pekin duck vs Bar-headed goose. Am. J. Physiol. 247, R69–R75.

Faraci, F.M., Kilgore, D.L., Fedde, M.R., 1985. Blood flow distribution during hypocapnic hypoxia in Pekin ducks and Bar-headed geese. Resp. Physiol. 61, 21–30.

Fedde, M.R., Orr, J.A., Shams, S., Scheid, P., 1989. Cardiopulmonary function in exercising Bar-headed geese during normoxia and hypoxia. Resp. Physiol. 77, 239–252.

Feigl, E., Folkow, B., 1963. Cardiovascular responses in "diving" and during brain stimulation in ducks. Acta Physiol. Scand. 57, 99–110.

Fibiger, H.C., 1982. The organization and some projections of cholinergic neurons of the mammalian forebrain. Brain Res. Rev. 4, 327–388.

Fingl, E., Woodbury, L.A., Hecht, M.H., 1952. Effects of innervation and drugs upon direct membrane potentials of embryonic chick myocardium. J. Pharmacol. Exp. Ther. 104, 103–114.

Folkow, B., 1952. Impulse frequency in sympathetic vasomotor fibres correlated to the release and elimination of the transmitter. Acta Physiol. Scand. 25, 49–76.

Folkow, B., Rubenstein, E.H., 1965. Effect of brain stimulation on "diving" in ducks. Hvalradets Skr. 48, 30–41.

Folkow, B., Yonce, L.R., 1967. The negative inotropic effect of vagal stimulation on the heart ventricles of the duck. Acta Physiol. Scand. 71, 77–84.

Folkow, B., Fuxe, K., Sonnenschein, R.R., 1966. Responses of skeletal musculature and its vasculature during "diving" in the duck: peculiarities of the adrenergic vasoconstrictor innervation. Acta Physiol. Scand. 67, 327–342.

Folkow, B., Nilsson, N.J., Yonce, L.R., 1967. Effects of "diving" on cardiac output in ducks. Acta Physiol. Scand. 70, 347–361.

Freer, R.J., Pappano, A.J., Peach, M.J., 1976. Mechanism for the positive inotropic effect of angiotensin II on isolated cardiac muscle. Circ. Res. 39, 178–183.

Frieswick, K.G.M., Danielson, T., Shideman, F.E., 1979. Adrenergic inotropic responsiveness of embryonic chick and rat hearts. Dev. Neurosci. 2, 276–285.

Furchgott, R.F., Zawadzki, J.V., 1980. The obligatory role of endothelial cells in the relaxation of arterial smooth muscle by acetylcholine. Nature 288, 373–376.

Furilla, R.A., Jones, D.R., 1986. The contribution of nasal receptors to the cardiac response to diving in restrained and unrestrained redhead ducks (*Aythya americana*). J. Exp. Biol. 121, 227–238.

Furilla, R.A., Jones, D.R., 1987a. The relationship between dive and predive heart rates in restrained and free dives by diving ducks. J. Exp. Biol. 127, 333–348.

Furilla, R.A., Jones, D.R., 1987b. Cardiac responses to dabbling and diving the mallard, *Anas platyrhynchos*. Physiol. Zool. 60, 406–412.

Furness, J.B., Costa, M., 1987. The Enteric Nervous System. Churchill Livingston, Edinburgh.

Furnival, C.M., Linden, R.J., Snow, H.M., 1973. The inotropic effect on the heart of stimulating the vagus in the dog, duck and toad. J. Physiol. 230, 155–170.

Gabbott, G.R.J., Jones, D.R., 1987. Habituation of the cardiac response to involuntary diving in diving and dabbling ducks. J. Exp. Biol. 131, 403–415.

Gabella, G., 1976. Structure of the Autonomic Nervous System. J Wiley, New York.

Ganitkevich, V., Hasse, V., Pfitzer, G., 2002. Ca^{2+}-dependent and Ca^{2+}-independent regulation of smooth muscle contraction. J. Muscle Res. Cell. Motil. 23, 47–52.

Gayeski, T.E.J., Honig, C.R., 1986. O_2 gradients from sarcolemma to cell interior in red muscle at maximal VO_2. Am. J. Physiol. 251, H789–H799.

Gilman, A.G., Rall, T.W., Nies, A.S., Taylor, P., 1990. The Pharmacological Basis of Therapeutics, eighth ed. Pergamon, Oxford, UK.

Girard, H., 1973. Adrenergic sensitivity of circulation in the chick embryo. Am. J. Physiol. 224, 461–469.

Glahn, R.P., Bottje, W.G., Maynard, P., Wideman, R.F., 1993. Response of the avian kidney to acute changes in arterial perfusion pressure and portal blood supply. Am. J. Physiol. 264, R428–R434.

Glenny, F.H., 1940. A systematic study of the main arteries in the region of the heart—aves. Anat. Rec. 76, 371–380.

Glombitza, S., Dragon, S., Berghammer, M., Pannermayr, M., Baumann, R., 1996. Adenosine causes cAMP-dependent activation of chick embryo red cell carbonic anhydrase and 2,3-DPG synthesis. Am. J. Physiol. 271, R973–R981.

Gold, M.R., Cohen, D.H., 1984. The discharge characteristics of vagal cardiac neurons during classically conditioned heart rate change. J. Neurosci. 4, 2963–2971.

van Golde, J., Mulder, T., Blanco, C.E., 1997. Changes in mean chorioallantoic artery blood flow and heart rate produced by hypoxia in the developing chick embryo. Pediatr. Res. 42, 293–298.

Goldberg, T.M., Bolnick, D.A., 1980. Electrocardiograms from the chicken, emu, red-tailed hawk and Chilean tinamou. Comp. Biochem. Physiol. A 67, 15–19.

Goldberg, J.M., Johnson, M.H., Whitelaw, K.D., 1983. Effect of cervical vagal stimulation on chicken heart rate and atrioventricular conduction. Am. J. Physiol. 244, R235–R243.

Goldberger, A.L., 1991. Is the normal heartbeat chaotic or homeostatic? NIPS 6, 87–91.

Goldberger, A.L., Rigney, D.R., West, B.J., 1990. Chaos and fractals in human physiology. Sci. Am. 262, 42–49.

Gooden, B.A., 1980. The effect of hypoxia on vasoconstrictor responses of isolated mesenteric arterial vasculature from chicken and duckling. Comp. Biochem. Physiol. C 67, 219–222.

Goplerud, J.M., Wagerle, L.C., Delivoria-Papadopoulos, M., 1989. Regional cerebral blood flow response during and after acute asphyxia in newborn piglets. J. Appl. Physiol. 66, 2827–2832.

Gray, S.D., McDonagh, P.F., Gore, R.W., 1983. Comparison of functional and total capillary densities in fast and slow muscles of the chicken. Pflügers Arch. 397, 209–213.

Greyner, H., Dzialowski, E.M., 2008. Mechanisms mediating the oxygen-induced vasoreactivity of the ductus arteriosus in the chicken embryo. Am. J. Physiol. 295, R1647–R1659.

Grindlay, J.H., Herrick, J.F., Mann, F.C., 1939. Measurement of the blood flow of the spleen. Am. J. Physiol. 127, 106–118.

Groenendijk, B.C.W., Hierck, B.P., Vrolijk, J., Baiker, M., Pourquie, M.J.B.M., Gittenberger-de Groot, A.C., Poelmann, R.E., 2005. Changes in shear stress-related gene expression after experimentally altered venous return in the chicken embryo. Circ. Res. 96, 1291–1298.

Groenendijk, B.C.W., Stekelenburg-De Vos, S., Vennemann, P., Wladimiroff, J.W., Nieuwstadt, F.T.M., Lindken, R., Westerweel, J., Hierck, B.P., Ursem, N.T.C., Poelmann, R.E., 2008. The endothelin-1 pathway and the development of cardiovascular defects in the haemodynamically challenged chicken embryo. J. Vasc. Res. 45, 54–68.

Grubb, B.R., 1983. Allometric relations of cardiovascular function in birds. Am. J. Physiol. 245, H567–H572.

Grubb, B.R., Jorgensen, D.D., Conner, M., 1983. Cardiovascular changes in the exercising emu. J. Exp. Biol. 104, 193–201.

Grubb, B.R., Mills, C.D., Colacino, J.M., Schmidt-Neilsen, K., 1977. Effect of arterial carbon dioxide on cerebral blood flow in ducks. Am. J. Physiol. 232, H596–H601.

Guimaraes, S., Moura, D., 2001. Vascular adrenoceptors: an update. Pharmacol. Rev. 53, 319–356.

Guyton, A.C., Young, D.B., DeClue, J.W., Trippodo, N., Hall, J.E., 1975. Fluid balance, renal function and blood pressure. Clin. Nephrol. 4, 122–126.

Hall, C.E., Hurtado, R., Hewett, K.W., Shulimovich, M., Poma, C.P., Reckova, M., Justus, C., Pennisi, D.J., Tobita, K., Sedmera, D., Gourdie, R.G., Mikawa, T., 2004. Hemodynamic-dependent patterning of endothelin converting enzyme 1 expression and differentiation of impulse-conducting Purkinje fibers in the embryonic heart. Development 131, 581–592.

Hargens, A.R., Millard, R.W., Johansen, K., 1974. High capillary permeability in fishes. Comp. Biochem. Physiol. A 48, 675–680.

Harrison, T.A., Stadt, H.A., Kirby, M.L., 1994. Developmental characteristics of the chick nodose ganglion. Dev. Neurosci. 16, 67–73.

Hartman, F.A., 1961. Smithsonian Miscellaneous Collections, vol. 143, pp. 1–91.

Hasegawa, K., Nishimura, H., 1991. Humoral factor mediates acetylcholine-induced endothelium-dependent relaxation of chicken aorta. Gen. Comp. Endocrinol. 84, 164–169.

Hasegawa, K., Nishimura, H., Khosla, M.C., 1993. Angiotensin II-induced endothelium-dependent relaxation of fowl aorta. Am. J. Physiol. 264, R903–R911.

Hassanpour, H., Teshfam, M., Momtaz, H., Brujeni, G.N., Shahgholian, L., 2010. Up-regulation of endothelin-1 and endothelin type A receptor gene expression in the heart of broiler chickens versus layer chickens. Res. Vet. Sci. 89, 352–357.

Hawkes, L.A., Balachandan, S., Batbayar, N., Butler, P.J., Chua, B., Douglas, D.C., Frappell, P.B., Hou, Y., Milsom, M.K., Newman, S.H., Prosser, D.J., Sathiyaselvam, P., Scott, G.R., Takekawa, J.Y., Natsagdorj, T., Wikelski, M., Witt, M.J., Yan, B., Bishop, C.M., 2012. The paradox of extreme high-altitude migration in Bar-headed geese *Anser indicus*. Proc. R. Soc. B 108, 9516–9519.

Hebb, C., 1969. Motor innervation of pulmonary blood vessels. Part 3. In: Fishman, A.P., Hecht, H.H. (Eds.), The Pulmonary Circulation and Interstitial Space. University of Chicago Press, Chicago.

Hedlund, K.O., Ebendal, T., 1980. The chick embryo nodose ganglion: effects of nerve growth factor in culture. J. Neurocytol. 9, 665–682.

Heieis, M., Jones, D.R., 1988. Blood flow and volume distribution during forced submergence in Pekin ducks (*Anas platyrhynchos*). Can. J. Zool. 66, 1589–1596.

Hempleman, S.C., Powell, F.L., Prisk, G.K., 1992. Avian arterial chemoreceptor responses to steps of CO_2 and O_2. Resp. Physiol. 90, 325–340.

Henderson, I.W., Deacon, C.F., 1993. Phylogeny and comparative physiology of the renin-angiotensin system. In: Robertson, J.I.S., Nicholls, M.G. (Eds.), The Renin-Angiotensin System: Biochemistry and Physiology, vol. 1. Mosby, New York, pp. 2.1–2.28.

Higgins, D., Pappano, A.J., 1979. A histochemical study of the ontogeny of catecholamine-containing axons in the chick embryo heart. J. Mol. Cell. Cardiol. 11, 661–668.

Higgins, D., Pappano, A.J., 1981. Developmental changes in the sensitivity of the chick embryo ventricle to beta-adrenergic agonist during adrenergic innervation. Circ. Res. 48, 245–253.

Hill, J.R., Goldberg, J.M., 1980. P-wave morphology and atrial activation in the domestic fowl. Am. J. Physiol. 239, R483–R488.

Hillman, S.S., Hancock, T.V., Hedrick, M.S., 2013. A comparative meta-analysis of maximal aerobic metabolism of vertebrates: implications for respiratory and cardiovascular limits to gas exchange. J. Comp. Physiol. B 183, 167–179.

Hirakow, R., 1970. Ultrastructural characteristics of the mammalian and sauropsidian heart. Am. J. Cardiol. 25, 195.

Hirsch, E.F., 1963. The innervation of the human heart. V. A comparative study of the intrinsic innervation of the heart in vertebrates. Exp. Mol. Pathol. 2, 384–401.

Hirsch, E.F., 1970. The Innervation of the Vertebrate Heart. Charles C. Thomas, Springfield, IL.

Hirst, G.D.S., Edwards, F.R., 1989. Sympathetic neuroeffector transmission in arteries and arterioles. Physiol. Rev. 69, 546–604.

Hodges, R.D., 1974. The Histology of the Domestic Fowl. Academic Press, London.

Hoffman, L.E., van Mierop, L.H.S., 1971. Effect of epinephrine on heart rate and arterial blood pressure of the developing chick embryo. Pediatr. Res. 5, 472–477.

Holt, J.P., Rhode, E.A., Kines, H., 1968. Ventricular volumes and body weight in mammals. Am. J. Physiol. 215, 704–715.

Holzbauer, M., Sharman, D.F., 1972. The distribution of catecholamines in vertebrates. In: Blaschko, H., Muscholl, E. (Eds.), Handbook of Experimental Pharmacology: Catecholamines, vol. 33. Springer-Verlag, Berlin, pp. 110–185.

Honig, C.R., Gayeski, T.E.J., Groebe, K., 1991. Myoglobin and oxygen gradients. In: Crystal, R.G., West, J.B., Barnes, B.J., Cherniak, N.S., Weibel, E.R. (Eds.), The Lung: Scientific Foundations. Raven, New York.

Hopkins, D.A., 1987. The dorsal motor nucleus of the vagus nerve and the nucleus ambigus: structure and connections. In: Hainsworth, R., McWilliam, P.N., Mary, D.A.S.G. (Eds.), Cardiogenic Reflexes. Oxford University Press, Oxford, UK, pp. 185–203.

Houweling, A.C., Somi, S., Massink, M.P., Groenen, M.A., Moorman, A.F., Christoffels, V.M., 2005. Comparative analysis of the natriuretic peptide precursor gene cluster in vertebrates reveals loss of ANF and retention of CNP-3 in chicken. Dev. Dyn. 233, 1076–1082.

Huang, H.C., Sung, P.K., Huang, T.F., 1974. Blood volume, lactic acid and catecholamines in diving response in ducks. Taiwan Yi Xue Hui Za Zhi 73, 203–210.

Huber, J.F., 1936. Nerve roots and nuclear groups in the spinal cord of the pigeon. J. Comp. Neurol. 65, 43–91.

Hudson, D.M., Jones, D.R., 1982. Remarkable blood catecholamine levels in forced dived ducks. J. Exp. Zool. 224, 451–456.

Hughes, A.F.W., 1942. The histogenesis of the arteries of the chick embryo. J. Anat. 77, 266–287.

Hunsaker, W.G., Robertson, A., Magwood, S.E., 1971. The effect of round heart disease on the electrocardiogram and heart weight of turkey poults. Poult. Sci. 50, 1712–1720.

Inagami, R., Naruse, M., Hoover, R., 1995. Endothelium as an endocrine organ. Annu. Rev. Physiol. 57, 171–189.

Inoue, D., Hachisu, M., Pappano, A.J., 1983. Acetylcholine increases resting membrane potassium conductance in atrial but not in ventricular muscle during muscarinic inhibition of Ca^{++}-dependent action potentials in chick heart. Circ. Res. 53, 158–167.

Ito, S., Ohta, T., Nakazato, Y., 1999. Characteristics of 5-HT-containing chemoreceptor cells of the chicken aortic body. J. Physiol. 515, 49–59.

Iversen, N.K., Wang, T., Baatrup, E., and Crossley, D.A. II, 2014. The role of nitric oxide in the cardiovascular response to chronic and acute hypoxia in white leghorn Chicken (*Gallus domesticus*). Acta Physiol, http://dx.doi.org/10.111/apha.12286.

Jaffee, O.C., 1965. Hemodynamic factors in the development of the chick embryo heart. Anat. Rec. 151, 69–76.

Jarrett, C., Lekic, M., Smith, C.L., Pusec, C.M., Sweazea, K.L., 2013. Mechanisms of acetylcholine-mediated vasodilation in systemic arteries from mourning doves (*Zenaida macroura*). J. Comp. Physiol. B 183, 959–967.

Jeck, D., Lindmar, R., Löffelholz, K., Wanke, M., 1988. Subtypes of muscarinic receptor on cholinergic nerves and atrial cells of chicken and guinea-pig hearts. Br. J. Pharmacol. 93, 357–366.

Johansen, K., 1964. Regional distribution of circulating blood during submersion asphyxia in the duck. Acta Physiol. Scand. 62, 1–9.

Johansen, K., Reite, O.B., 1964. Cardiovascular responses to vagal stimulation and cardioaccelerator nerve blockade in birds. Comp. Biochem. Physiol. 12, 479–487.

Jones, D.R., 1969. Avian afferent vagal activity related to respiratory and cardiac cycles. Comp. Biochem. Physiol. A 28, 961–965.

Jones, D.R., 1973. Systemic arterial baroreceptors in ducks and the consequences of their denervation on some cardiovascular responses to diving. J. Physiol. 234, 499–518.

Jones, D.R., 1991. Cardiac energetics and design of arterial systems. In: Blake, R.W. (Ed.), Efficiency and Economy in Animal Physiology. Cambridge University Press, Cambridge, UK, pp. 159–168.

Jones, D.R., Holeton, G.F., 1972. Cardiac output of ducks during diving. Comp. Biochem. Physiol. A 41, 639–645.

Jones, D.R., Johansen, K., 1972. The blood vascular system of birds. In: Farner, D.S., King, J.E. (Eds.), Avian Biology, vol. 2. Academic Press, New York, pp. 157–285.

Jones, D.R., Milsom, W.K., 1982. Peripheral receptors affecting breathing and cardiovascular function in non-mammalian vertebrates. J. Exp. Biol. 100, 59–91.

Jones, D.R., Milsom, W.K., Gabbott, G.R.J., 1982. Role of central and peripheral chemoreceptors in diving responses of ducks. Am. J. Physiol. 243, R537–R545.

Jones, D.R., Milsom, W.K., West, N.H., 1980. Cardiac receptors in ducks: the effect of their stimulation and blockade on diving bradycardia. Am. J. Physiol. 238, R50–R56.

Jones, D.R., Purves, M.J., 1970. The carotid body in the duck and the consequences of its denervation upon the cardiac responses to immersion. J. Physiol. 211, 279–294.

Jones, D.R., Bryan, R.M., West, N.H., Lord, R.H., Clark, B., 1979. Regional distribution of blood flow during diving in the duck (*Anas platyrhynchos*). Can. J. Zool. 57, 995–1002.

Jones, D.R., Milsom, W.K., Smith, F.M., West, N.H., Bamford, O.S., 1983. Diving responses in ducks after acute barodenervation. Am. J. Physiol. 245, R222–R229.

Kamath, M.V., Fallen, E.L., 1993. Power spectral analysis of heart rate variability: a noninvasive signature of cardiac autonomic function. Crit. Rev. Biomed. Eng. 21, 245–311.

Kameda, Y., 1990. Ontogeny of the carotid body and glomus cells distributed in the wall of the common carotid artery and its branches in the chicken. Cell Tissue Res. 261, 525–537.

Kameda, Y., 1994. Electron microscopic study on the development of the carotid body and glomus cell groups distributed in the wall of the common carotid artery and its branches in the chicken. J. Comp. Neurol. 348, 544–555.

Kameda, Y., 2002. Carotid body and glomus cells distributed in the wall of the common carotid artery in the bird. Microsc. Res. Tech. 59, 196–206.

Kamimura, K., Nishimura, H., Bailey, J.R., 1995. Blockade of beta-adrenoceptor in control of blood pressure in fowl. Am. J. Physiol. 269, R914–R922.

Kato, K., Moriya, K., Dzialowski, E., Burggren, W.W., Tazawa, H., 2002. Cardiac rhythms in prenatal and perinatal emu embryos. Comp. Biochem. Physiol. A 131, 775–785.

Katz, D.M., Karten, H.J., 1979. The discrete anatomical localization of vagal aortic afferents within a catecholamine-containing cell group in the nucleus solitarius. Brain Res. 171, 187–195.

Katz, D.M., Karten, H.J., 1983a. Subnuclear organization of the dorsal motor nucleus of the vagus nerve in the pigeon, *Columba livia*. J. Comp. Neurol. 217, 31–46.

Katz, D.M., Karten, H.J., 1983b. Visceral representation within the nucleus of the tractus solitarius in the pigeon, *Columba livia*. J. Comp. Neurol. 218, 42–73.

Katz, D.M., Karten, H.J., 1985. Topographic representation of visceral target organs within the dorsal motor nucleus of the vagus nerve of the pigeon *Columba livia*. J. Comp. Neurol. 242, 397–414.

Kedem, O., Katchalsky, A., 1958. Thermodynamic analysis of the permeabililty of biological membranes to non-electrolytes. Biochim. Biophys. Acta 27, 229–246.

Kempf, H., Corvol, P., 2001. Angiotensin receptor(s) in fowl. Comp. Biochem. Physiol. A. 128, 77–88.

Kempf, H., Linares, C., Corvol, P., Gasc, J.M., 1998. Pharmacological inactivation of the endothelin type A receptor in the early chick embryo: a model of mispatterning of the branchial arch derivatives. Development 125, 4931–4941.

Kharin, S.N., 2004. Depolarisation and repolarisation sequences of ventricular epicardium in chickens (*Gallus gallus domesticus*). Comp. Biochem. Physiol. A 137, 237–244.

Khandoker, A.H., Dzialowski, E.M., Burggren, W.W., Tazawa, H., 2003. Cardiac rhythms of late pre-pipped and pipped chick embryos exposed to altered oxygen environments. Comp. Biochem. Physiol. A 136, 289–299.

Kiley, J.P., Faraci, F.M., Fedde, M.R., 1985. Gas exchange during exercise in hypoxic ducks. Resp. Physiol. 59, 105–115.

Kiley, J.P., Kuhlmann, W.D., Fedde, M.R., 1979. Respiratory and cardiovascular responses to exercise in the duck. J. Appl. Physiol. 47, 827–833.

Kirby, M.L., Conrad, D.C., Stewart, D.E., 1987. Increase in the cholinergic cardiac plexus in sympathetically aneural chick hearts. Cell Tissue Res. 247, 489–496.

Kirby, M.L., McKenzie, J.W., Weidman, T.A., 1980. Developing innervation of the chick heart: a histoflourescence and light microscopic study of sympathetic innervation. Anat. Rec. 196, 333–340.

Kisch, B., 1951. The electrocardiogram of birds: chicken, duck, pigeon. Exp. Med. Surg. 9, 103–124.
Kitazawa, T., Polzin, A.N., Eto, M., 2004. CPI-17-deficient smooth muscle of chicken. J. Physiol. 557, 515–528.
Kobinger, W., Oda, M., 1969. Effects of sympathetic blocking substances on the diving reflex of ducks. Eur. J. Pharmacol. 7, 289–295.
Koch-Weser, J., 1971. Beta-receptor blockade and myocardial effects of cardiac glycosides. Circ. Res. 28, 109–118.
Koelle, G.B., 1963. Cytological distributions and physiological functions of cholinesterases. In: Eichler, O., Farah, A. (Eds.), Handbook of Experimental Pharmacology. Springer-Verlag, Heidelberg.
Kohmoto, O., Ikenouchi, H., Hirata, Y., Momomura, S., Serizawa, T., Barry, W.H., 1993. Variable effects of endothelin-1 on [Ca^{2+}]i transients, pHi, and contraction in ventricular myocytes. Am. J. Physiol. 265, H793–H800.
Koide, M., Akins, R.E., Harayama, H., Yasui, K., Yokota, M., Tuan, R.S., 1996. Atrial natriuretic peptide accelerates proliferation of chick embryonic cardiomyocytes in vitro. Differentiation 61, 1–11.
Koide, M., Tuan, R., 1989. Adrenergic regulation of calcium-deficient hypertension in chick embryos. Am. J. Physiol. 257, H1900–H1909.
Kolluru, G.K., Shen, X., Bir, S.C., Kevil, C.G., 2013. Hydrogen sulfide chemical biology: pahtophysiological roles and detection. Nitric Oxide 35, 5–20.
Komori, S., Ohashi, H., Okada, T., Takewaki, T., 1979. Evidence that adrenaline is released from adrenergic neurons in the rectum of the fowl. Br. J. Pharmacol. 65, 261–269.
Kontos, H.A., 1981. Regulation of the cerebral circulation. Annu. Rev. Physiol. 43, 397–407.
Kotilainen, P.V., Putkonen, P.T.S., 1974. Respiratory and cardiovascular responses to electrical stimulation of the avian brain with emphasis on inhibitory mechanisms. Acta Physiol. Scand. 90, 358–369.
Kovách, A.G.B., Balint, T., 1969. Comparative study of haemodilation after haemorrhage in the pigeon and the rat. Acta Physiol. Acad. Sci. Hung. 35, 231–243.
Krogh, A., 1919. The number and distribution of capillaries in muscles with calculations of the oxygen pressure head necessary for supplying the tissue. J. Physiol. 52, 409–415.
Kuratani, S., Tanaka, S., 1990. Peripheral development of the avian vagus nerve with special reference to the morphological innervation of heart and lung. Anat. Embryol. 182, 435–446.
Lacombe, A.M.A., Jones, D.R., 1990. The source of circulating catecholamines in forced dived ducks. Gen. Comp. Endocrinol. 80, 41–47.
Landis, E.M., 1927. Micro-injection studies of capillary permeability. II. The relation between capillary pressure and the rate at which fluid passes through the walls of single capillaries. Am. J. Physiol. 82, 217–238.
Landis, E.M., Pappenheimer, J.R., 1963. Exchange of substances through the capillary walls. In: Hamilton, W.F., Dow, P. (Eds.), Handbook of Physiology, Section 2: Circulation, vol. II. American Physiological Society, Washington, DC, pp. 961–1034.
Landys-Ciannelli, M.M., Piersma, T., Jukema, J., 2003. Strategic size changes of internal organs and muscle tissue in the bar-tailed Godwit during fat storage on a spring stopover site. Func. Ecol. 17, 151–159.
Lang, S.A., Levy, M.N., 1989. Effects of vagus nerve on heart rate and ventricular contractility in chicken. Am. J. Physiol. 256, H1295–H1302.
Langille, B.L., 1983. Role of venoconstriction in the cardiovascular responses of ducks to head immersion. Am. J. Physiol. 244, R292–R298.
Langille, B.L., Jones, D.R., 1975. Central cardiovascular dynamics of ducks. Am. J. Physiol. 228, 1856–1861.
Langille, B.L., Jones, D.R., 1976. Examination of elastic nonuniformity in the arterial system using a hydraulic model. J. Biomech. 9, 755–761.
Lasiewski, R.C., Weathers, W.W., Bernstein, M.H., 1967. Physiological responses of the giant hummingbird, *Patagona gigas*. Comp. Biochem. Physiol. 23, 797–813.
Le Noble, F.A.C., Hekking, J.W.M., Van Straaten, H.W.M., Slaaf, D.W., Boudier, H.A.J.S., 1991. Angiotensin II stimulates angiogenesis in the chorio-allantoic membrane of the chick embryo. Eur. J. Pharmacol. 195, 305–306.
Le Noble, F.A.C., Ruijtenbeek, K., Gommers, S., De Mey, J.G.R., Blanco, C.E., 2000. Contractile and relaxing reactivity in carotid and femoral arteries of chicken embryos. Am. J. Physiol. 278, H1261–H1268.
Le Noble, F.A.C., Schreurs, N.H.J.S., Van Straaten, H.W.M., Slaaf, D.W., Smits, J.F.M., Rogg, H., Struijker-Boudier, H.A.J., 1993. Evidence for a novel angiogensin II receptor involved in angiogenesis in chick embryo chorioallantoic membrane. Am. J. Physiol. 264, R460–R465.
Lenselink, D.R., Kuhlmann, R.S., Lowrence, J.M., Kolesari, G.L., 1994. Cardiovascular teratogenicity of terbutaline and ritodrine in the chick embryo. Am. J. Obstet. Gynecol. 171, 501–506.
Leon-Velarde, F., Sanchez, J., Bigard, A.X., Brunet, A., Lesty, C., Monge, C.C., 1993. High altitude tissue adaptation in Andean coots: capillarity, fiber area, fiber type and enzymatic activities of skeletal muscle. J. Comp. Physiol. B 163, 52–58.
Leonard, R.B., Cohen, D.H., 1975. Responses of sympathetic postganglionic neurons to peripheral nerve stimulation in the pigeon (*Columba livia*). Exp. Neurol. 49, 466–486.
Levy, M.N., 1971. Sympathetic–parasympathetic interactions in the heart. Circ. Res. 29, 437–445.
Lewis, T., 1916. The spread of the excitatory process in the vertebrate heart. V. The bird's heart. Philos. Trans. R. Soc. Lond., B 207, 298–311.
Lillie, F.R., 1908. Development of the Chick. Holt, New York.
Lillo, R.S., Jones, D.R., 1982. Effect of cardiovascular variables on hyperpnea during recovery from diving in ducks. J. Appl. Physiol. 52, 206–215.
Lillo, R.S., Jones, D.R., 1983. Influence of ischemia and hyperoxia on breathing in ducks. J. Appl. Physiol. 55, 400–408.
Lim, J.J., Liu, Y., Khin, E.S.W., Bian, J., 2008. Vasoconstrictive effect of hydrogen sulfide involves downregulation of cAMP in vascular smooth muscle cells. Am. J. Physiol. 295, C1261–C1270.
Lindgren, I., Altimiras, J., 2009. Chronic prenatal hypoxia sensitizes beta-adrenoceptors in the embryonic heart but causes postnatal desensitization. Am. J. Physiol. 297, R258–R264.
Lindgren, I., Zoer, B., Altimiras, J., Villamor, E., 2010. Reactivity of chicken chorioallantoic arteries, avian homologue of human fetoplacental arteries. J. Physiol. Pharmacol. 61, 619–628.
Lindmar, R., Loffelholz, K., Weide, W., Weis, S., 1983. Evidence for bilateral vagal innervation of postganglionic parasympathetic neurons in chicken heart. J. Neur. Transm. 56, 239–247.
Liu, C., Li, R., 2005. Electrocardiogram and heart rate in response to temperature acclimation in three representative vertebrates. Comp. Biochem. Physiol. A 142, 416–421.
Loewy, A.D., Spyer, K.M., 1990. Central Regulation of Autonomic Functions. Oxford University Press, Oxford, UK.
Löffelholz, K., Brehm, R., Lindmar, R., 1984. Hydrolysis, synthesis, and release of acetylcholine in the isolated heart. Fed. Proc. 43, 2603–2606.

Mulder, A.L.M., van Golde, J.M.C.G., van Goor, A.A.C., Giussani, D.A., Blanco, C.E., 2000. Developmental changes in plasma catecholamine concentrations during normoxia and acute hypoxia in the chick embryo. J. Physiol. 527, 593–599.

Löffelholz, K., Pappano, A.J., 1974. Increased sensitivity of sinoatrial pacemaker to acetylcholine and to catecholamines at the onset of autonomic neuroeffector transmission in chick embryo heart. J. Pharmacol. Exp. Ther. 191, 479–486.

Lu, Y., James, T.N., Bootsma, M., Terasaki, F., 1993a. Histological organization of the right and left atrioventricular valves of the chicken heart and their relationship to the atrioventricular Purkinje ring and the middle bundle branch. Anat. Rec. 235, 74–86.

Lu, Y., James, T.N., Yamamoto, S., Terasaki, F., 1993b. Cardiac conduction in the chicken: gross anatomy plus light and electron microscopy. Anat. Rec. 236, 493–510.

Lumb, W.V., Jones, E.W., 1984. Veterinary Anesthesia, second ed. Lea and Febiger, Philadelphia, PA.

Lucitti, J.L., Hedrick, M.S., 2006. Characterization of baroreflex gain in the domestic pigeon (*Columba livia*). Comp. Biochem. Physiol. A 143, 103–111.

Macdonald, R.L., Cohen, D.H., 1970. Cells of origin of sympathetic pre- and postganglionic cardioacceleratory fibers in the pigeon. J. Comp. Neurol. 140, 343–358.

Macdonald, R.L., Cohen, D.H., 1973. Heart rate and blood pressure responses to electrical stimulation of the central nervous system in the pigeon (*Columba livia*). J. Comp. Neurol. 150, 109–136.

Malinovsky, L., 1962. Contribution to the anatomy of the vegetative nervous system in the neck and thorax of the domestic pigeon. Acta Anat. 50, 326–347.

Malliani, A., Pagani, M., Bergamaschi, M., 1979. Positive feedback sympathetic reflexes and hypertension. Am. J. Cardiol. 44, 860–865.

Mangold, E., 1919. Elektrographischel Untersuchungen del Erregungsverlaufes im Vogelherzen. Arch. Ges. Physiol. (Pflügers) 175, 327–354.

Manning, P.J., Middleton, C.C., 1972. Atherosclerosis in wild turkeys: morphological features of lesions and lipids in serum and aorta. Am. J. Vet. Res. 33, 1237–1246.

Martinez-Lemus, L.A., Hester, R.K., Becker, E.J., Ramirez, G.A., Odom, T.W., 2003. Pulmonary artery vasoactivity in broiler and leghorn chickens: an age profile. Poult. Sci. 82, 1957–1964.

Mathew, S., Mascareno, E., Siddiqui, M.A.Q., 2004. A ternary complex of transcription factors, nishéd and NFATc4, and co-activator p300 bound to an intronic xequence, intronic regulatory element, is pivotal for the up-regulation of myosin light chain-2v gene in cardiac hypertrophy. J. Biol. Chem. 279, 41018–41027.

Mathieu-Costello, O., 1991. Morphometric analysis of capillary geometry in pigeon pectoralis muscle. Am. J. Anat. 191, 74–84.

Mathieu-Costello, O., Agey, P.J., Normand, H., 1996. Fiber capillarization in flight muscle of pigeons native and flying at high altitude. Resp. Physiol. 103, 187–194.

Mathieu-Costello, O., Suarez, R.K., Hochachka, P.W., 1992. Capillary-to-fiber geometry and mitochondrial density in hummingbird flight muscle. Resp. Physiol. 89, 113–132.

Mathieu-Costello, O., Agey, P.J., Logemann, R.B., Florez-Duquett, M., Bernstein, M.H., 1994. Effect of flying activity on capillary-fiber geometry in pigeon flight muscle. Tissue Cell 26, 57–73.

Mathur, P.N., 1973. Distribution of the specialized conducting tissue in the avian heart. Ind. J. Zool. 1, 17–27.

Mathur, R., Mathur, A., 1974. Nerves and nerve terminations in the heart of *Columba livia*. Anat. Anz. 136, 40–47.

McCarty, L.P., Lee, W.C., Shideman, F.E., 1960. Measurement of the inotropic effects of drugs on the innervated and noninnervated embryonic chick heart. J. Pharmacol. Exp. Ther. 129, 315–321.

McDonald, D.A., 1974. Blood Flow in Arteries. Williams and Wilkins, Baltimore, MD.

McPhail, L.T., Jones, D.R., 1999. The autonomic nervous control of heart rate in ducks during voluntary diving. Physiol. Biochem. Zool. 72, 164–169.

McKenzie, B.E., Will, J.A., Hardie, A., 1971. The electrocardiogram of the turkey. Avian Dis. 15, 737–744.

Meir, J.U., Jardine, W., York, J., Chua, B., Milsom, W.K., 2013. Heart rate and metabolic rate of Bar-headed geese flying in hypoxia. FASEB J. 27, 1149.16.

Meir, J.U., Stockard, T.K., Williams, C.L., Ponganis, K.V., Ponganis, P.J., 2008. Heart rate regulation and extreme bradycardia in diving emperor penguins. J. Exp. Biol. 211, 1169–1179.

Mellander, S., Johansson, B., 1968. Control of resistance, exchange, and capacitance functions in the peripheral circulation. Pharmacol. Rev. 20, 117–196.

Michel, C.C., 1997. Starling: the formulation of his hypothesis of microvascular fluid exchange and its significance after 100 years. Exp. Physiol. 82, 1–30.

Midtgård, U., 1981. The Rete tibiotarsale and arterio-venous association in the hind limb of birds: a comparative morphological study on countercurrent heat exchange systems. Acta Zool. 62, 67–87.

Millard, R.W., 1980. Depressed baroreceptor-cardiac reflex sensitivity during simulated diving in ducks. Comp. Biochem. Physiol. A 65, 247–249.

Milnor, W.R., 1979. Aortic wavelength as a determinant of the relation between heart rate and body size in mammals. Am. J. Physiol. 237, R3–R6.

Moellera, I., Small, D.H., Reed, G., Harding, J.W., Mendelsohn, F.A.O., Chaia, S.Y., 1996. Angiotensin IV inhibits neurite outgrowth in cultured embryonic chicken sympathetic neurones. Brain Res. 725, 61–66.

Moonen, R.M., Villamor, E., 2011. Developmental changes in mesenteric artery reactivity in embryonic and newly hatched chicks. J. Comp. Physiol. B. 181, 1063–1073.

Moore, A.F., Strong, J.H., Buckley, J.P., 1981. Cardiovascular actions of angiotensin in the fowl (*Gallus domesticus*). I. Analysis. Res. Commun. Chem. Path. Pharmacol. 32, 423–445.

Moore, E.N., 1965. Experimental electrophysiological studies on avian hearts. Ann. N. Y. Acad. Sci. 127, 127–144.

Moore, E.N., 1967. Phylogenetic observations on specialized cardiac tissues. Bull. N. Y. Acad. Med. 43, 1138–1159.

Mueller, C.A., Burggren, W.W., Crossley II, D.A., 2013. Angiotensin II and baroreflex control of heart rate in embryonic chickens (*Gallus gallus domesticus*). Am. J. Physiol. 305, R855–863.

Mulder, A.L.M., Van Goor, C.A., Giussani, D.A., Blanco, C.E., 2001. Alpha-adrenergic contribution to the cardiovascular response to acute hypoxemia in the chick embryo. Am. J. Physiol. 281, R2004–R2010.

Mulder, T.L., van Golde, J.C., Prinzen, F.W., Blanco, C.E., 1997. Cardiac output distribution in the chick embryo from stage 36 to 45. Cardiovasc. Res. 34, 525–528.

Mulder, A.L.M., van Golde, J.C., Prinzen, F.W., Blanco, C.E., 1998. Cardiac output distribution in response to hypoxia in the chick embryo in the second half of the incubation time. J. Physiol. 508, 281–287.

Murillo-Ferrol, N.L., 1967. The development of the carotid body in *Gallus domesticus*. Acta Anat. 68, 102–126.

Nakamura, Y., Nishimura, H., Khosla, M.C., 1982. Vasodepressor action of angiotensin in conscious chickens. Am. J. Physiol. 243, H456–H462.

Nakazawa, M., Kajio, F., Ikeda, K., Takao, A., 1990. Effect of atrial natriuretic peptide on hemodynamics of the stage 21 chick embryo. Pediatr. Res. 27, 557–560.

Nilsson, S., Holmgren, S., 1994. Comparative Physiology and Evolution of the Autonomic Nervous System. Harwood Academic Publishers, London, UK.

Nishimura, H., Yang, y., Hubert, C., Gasc, J.M., Ruijtenbeek, K., DeMey, J., Struijker Boudier, H.A.J., Corvol, P., 2003. Maturation-dependent changes of angiotensis receptor expression in fowl. Am. J. Physiol. 285, R231–R242.

Nonidez, J.F., 1935. The presence of depressor nerves in the aorta and carotid of birds. Anat. Rec. 62, 47–73.

Nye, P.C.G., Powell, F.L., 1984. Steady-state discharge and bursting of arterial chemoreceptors in the duck. Resp. Physiol. 56, 369–384.

O'Rourke, M.F., Blazek, J.V., Morrells, C.L., Krovetz, L.J., 1968. Pressure wave transmission along the human aorta. Circ. Res. 23, 567–579.

Odlind, B., 1978. Blood flow distribution in the renal portal system of the intact hen: a study of a venous system using microspheres. Acta Physiol. Scand. 102, 342–356.

Ogut, O., Brozovich, F.V., 2000. Determinants of the contractile properties in the embryonic chicken gizzard and aorta. Am. J. Physiol. 279, C1722–C1732.

Ollenberger, G.P., West, N.H., 1998. Distribution of regional cerebral blood flow in voluntarily diving rats. J. Exp. Biol. 201, 549–558.

Owens, G.K., Kumar, M.S., Wamhoff, B.R., 2004. Molecular regulation of vascular smooth muscle cell differentiation in development and disease. Physiol. Rev. 84, 767–801.

Padian, K., de Ricqles, A., 2009. The orgin and evolution of birds: 35 years of progress. Comptes Rendus. Palevol. 8, 257–280.

Pappano, A.J., 1975. Development of autonomic neuroeffector transmission in the chick embryo heart. In: Lieberman, M., Sano, T. (Eds.), Developmental and Physiological Correlates of Cardiac Muscle. Raven Press, New York, pp. 235–248.

Pappano, A.J., 1976. Onset of chronotropic effects of nicotinic drugs and tyramine on the sino-atrial pacemaker in chick embryo heart: relationship to the development of autonomic neuroeffector transmission. J. Pharmacol. Exp. Ther. 196, 676–684.

Pappano, A.J., 1977. Ontogenetic development of autonomic neuroeffector transmission and transmitter reactivity in embryonic and fetal hearts. Pharmacol. Rev. 29, 3–33.

Pappano, A.J., Löffelholz, K., 1974. Ontogenesis of adrenergic and cholinergic neuroeffector transmission in chick embryo heart. J. Pharmacol. Exp. Ther. 191, 468–478.

Pappano, A., Löffelholz, K., Skowronek, C., 1973. Onset of cholinergic neuroeffector transmission in chick embryo heart. Pharmacologist 15, 198.

Pavlov, I.P., 1927. Conditioned Reflexes: an Investigation of the Physiological Activity of the Cerebral Cortex (G.V. Anrep, Trans.). Oxford University Press, Oxford, UK.

Peters, G.W., Steiner, D.A., Rigoni, J.A., Mascilli, A.D., Schnepp, R.W., Thomas, S.P., 2005. Cardiorespiratory adjustments of homing pigeons to steady wind tunnel flight. J. Exp. Biol. 208, 3109–3120.

Peterson, D.F., Nightingale, T.E., 1976. Functional significance of thoracic vagal branches in the chicken. Resp. Physiol. 27, 267–275.

Petras, J.M., Cummings, J.F., 1972. Autonomic neurons in the spinal cord of the rhesus monkey: a correlation of the findings of cytoarchitectonics and sympathectomy with fiber degeneration following dorsal rhizotomy. J. Comp. Neurol. 146, 189–218.

Pick, J., 1970. The Autonomic Nervous System. Lippincott, Philadelphia, PA.

Pickering, J.W., 1895. Further experiments on the embryonic heart. J. Physiol. 18, 470–483.

Piersma, T., Gudmundsson, G.A., Lilliendahl, K., 1999. Rapid changes in the size of different functional organ and muscle groups during refueling in a long-distance migrating shorebird. Physiol. Biochem. Zool. 72, 405–415.

Ponganis, P.J., Stockard, T.K., Meir, J.U., Williams, C.L., Ponganis, K.V., Howard, R., 2009. O_2 store management in diving emperor penguins. J. Exp. Biol. 212, 21–224.

Pool, R., 1989. Is it healthy to be chaotic? Science 243, 604–607.

Prothero, J., 1979. Heart weight as a function of body weight in mammals. Growth 43, 139–150.

Rabkin, S.W., 1996. The angiotensin II subtype 2 (AT2) receptor is linked to protein kinase C but not cAMP-dependent pathways in the cardiomyocyte. Can. J. Physiol. Pharmacol. 74, 125–131.

Randall, W.C., 1994. Efferent sympathetic innervation of the heart. In: Armour, J.A., Ardell, J.L. (Eds.), Neurocardiology. Oxford University Press, Oxford, UK, pp. 77–94.

Ream, S., Duquaine, A., Goy Sirsat, S.K., Sirsat, T.S., Dzialowski, E.M., 2013. Development of endothermy in the Pekin duck (*Anas pekin*). FASEB J. 27, 1149.17.

Rickenbacher, J., Müller, E., 1979. The development of cholinergic ganglia in the chick embryo heart. Anat. Embryol. 155, 253–258.

Robinzon, B., Koike, T.I., Marks, P.A., 1993. At low dose, arginine vasotocin has vasopressor rather than vosodepressor effect in chickens. Gen. Comp. Endocrinol. 91, 105–112.

Robinzon, B., Koike, T.I., Neldon, H.L., Hendry, I.R., el Halawani, M.E., 1988. Physiological effects of arginine vasotocin and mesotocin in cockerels. Br. Poult. Sci. 29, 639–652.

Ropert-Coudert, Y., Wilson, R.P., Grémillet, D., Kato, A., Lewis, S., Ryan, P.G., 2006. Electrocardiogram recordings in free-ranging gannets reveal minimum difference in heart rate during flapping versus gliding flight. Mar. Ecol. Prog. Ser. 328, 275–284.

Rothe, C.F., 1983. Venous system: physiology of the capacitance vessels. In: Shepherd, J.T., Abboud, F.M. (Eds.), Handbook of Physiology: The Cardiovascular System. Circulation and Organ Blood Flow, Part 1, vol. III. American Physiological Society, Bethesda, MD.

Rouwet, E.V., De Mey, J.G.R., Slaaf, D.W., Heineman, E., Ramsay, G., Le Noble, F.A.C., 2000. Development of vasomotor responses in fetal mesenteric arteries. Am. J. Physiol. 279, H1097–H1105.

Rushmer, R.F., 1976. Cardiovascular Dynamics, fourth ed. W. B. Saunders, Philadelphia.

Rzucidlo, E.M., Martin, K.A., Powell, R.J., 2007. Regulation of vascular smooth muscle cell differentiation. J. Vasc. Surg. 45 (Suppl. A), A25–A32.

Sabin, F.R., 1917. Origin and development of the primitive vessels of the chick and of the pig. Contr. Embryol. 6, 63.

Saint-Petery, L.B., van Mierop, L.H.S., 1974. Evidence for presence of adrenergic receptors in the 6-day chick embryo. Am. J. Physiol. 227, 1406–1410.

Sakamoto, K.Q., Takahashi, A., Iwata, T., Yamamoto, T., Yamamoto, M., Trathan, P.N., 2013. Heart rate and estimated energy expenditure of flapping and gliding in black-browed albatrosses. J. Exp. Biol. 216, 3175–3182.

Sapir, N., Wikelski, M., McCue, M.D., Pinshow, B., Nathan, R., 2010. Flight modes in migrating European bee-eaters: heart rate may indicate low metabolic rate during soaring and gliding. PLoS One 5, e13956.

Sapirstein, L.A., Hartman, F.A., 1959. Cardiac output and its distribution in the chicken. Am. J. Physiol. 196, 751.

Saul, P.J., 1990. Beat-to-beat variations of heart rate reflect modulation of cardiac autonomic outflow. NIPS 5, 32–37.

Saunders, D.K., Fedde, M.R., 1994. Exercise performance of birds. In: Jones, J.H. (Ed.), Advances in Veterinary Science and Comparative Medicine: Comparative Vertebrate Exercise Physiology; Phyletic Adaptations, vol. 38B. Academic Press, New York, pp. 139–190.

Savary, K., Michaud, A., Favier, J., Larger, E., Corvol, P., Gasc, J.M., 2005. Role of the renin-angiotensin system in primitive erythropoiesis in the chick embryo. Blood 105, 103–110.

Schlaefke, M.E., 1981. Central chemosensitivity: a respiratory drive. Rev. Physiol. Biochem. Pharmacol. 90, 171–244.

Schmidt-Nielsen, K., 1984. Scaling: Why is Animal Size so Important? Cambridge University Press, Cambridge, UK.

Schwaber, J.S., Cohen, D.H., 1978a. Electrophysiological and electron microscopic analysis of the vagus nerve of the pigeon, with particular reference to the cardiac innervation. Brain Res. 147, 65–78.

Schwaber, J.S., Cohen, D.H., 1978b. Field potential and single unit analyses of the avian dorsal motor nucleus of the vagus and criteria for identifying vagal cardiac cells of origin. Brain Res. 147, 79–90.

Scott, G.R., 2011. Elevated performance: the unique physiology of birds that fly at high altitudes. J. Exp. Biol. 214, 2455–2462.

Scott, G.R., Egginton, S., Richards, J.G., Milsom, W.K., 2009. Evolution of muscle phenotype for extreme high altitude flight in the Bar-headed goose. Proc. R. Soc. Lond. B 276, 3645–3653.

Scott, G.R., Milsom, W.K., 2007. Control of breathing and adaptation to high altitude in the Bar-headed goose. Am. J. Physiol. Regul. Integr. Comp. Physiol. 293, R379–R391.

Scott, G.R., Schulte, P.M., Egginton, S., Scott, A.L.M., Richards, J.G., Milsom, W.K., 2011. Molecular evolution of cytochrome c oxidase underlies high-altitude adaptation in the Bar-headed goose. Mol. Biol. Evol. 28, 351–363.

Secor, S.M., Diamond, J., 1998. A vertebrate model of extreme physiological regulation. Nature 395, 659–662.

Seymour, R.S., Blaylock, A.J., 2000. The principle of Laplace and scaling of ventricular wall stress and blood pressure in mammals and birds. Physiol. Biochem. Zool. 73, 389–405.

Shah, R., Greyner, H., Dzialowski, E.M., 2010. Autonomic control of heart rate and its variability during normoxia and hypoxia in emu (*Dromaius novaehollandiae*) hatchlings. Poult. Sci. 89, 135–144.

Shepherd, J.T., Vanhoutte, P.M., 1975. Veins and Their Control. W. B. Saunders, London.

Shigenobu, K., Sperelakis, N., 1972. Calcium current channels induced by catecholamines in chick embryonic potassium hearts whose fast sodium channels are blocked by tetrodotoxin or elevated potassium. Circ. Res. 31, 932–952.

Simon-Oppermann, C., Simon, E., Gray, D.A., 1988. Central and systemic antidiuretic hormone and angiotensin II in salt and fluid balance of birds as compared to mammals. Comp. Biochem. Physiol. A 90, 789–803.

Smith, F.M., 1994. Blood pressure regulation by aortic baroreceptors in birds. Physiol. Zool. 67, 1402–1425.

Smith, F.M., Jones, D.R., 1990. Effects of acute and chronic baroreceptor denervation on diving responses in ducks. Am. J. Physiol. 258, R895–R902.

Smith, F.M., Jones, D.R., 1992. Baroreflex control of arterial blood pressure during involuntary diving in ducks (*Anas platyrhynchos* var.). Am. J. Physiol. 263, R693–R702.

Smith, R.B., 1971a. Intrinsic innervation of the avian heart. Acta Anat. 79, 112–119.

Smith, R.B., 1971b. Observations on nerve cells in human, mammalian and avian cardiac ventricles. Anat. Anz. 129, 436–444.

Snyder, G.K., 1987. Muscle capillarity in chicks following hypoxia. Comp. Biochem. Physiol. A 87, 819–822.

Snyder, G.K., 1990. Capillarity and diffusion distances in skeletal muscles in birds. J. Comp. Physiol. B 160, 583–591.

Snyder, G.K., Coelho, J.R., 1989. Microvascular development in chick anterior latissimus dorsi following hypertrophy. J. Anat. 162, 215–224.

Snyder, G.K., Byers, R.L., Kayar, S.R., 1984. Effects of hypoxia on tissue capillarity in geese. Resp. Physiol. 58, 151–160.

Solomon, I.C., Adamson, T.P., 1997. Static muscular contraction elicits a pressor reflex in the chicken. Am. J. Physiol. 272, R759–R765.

Somlyo, A.P., Somlyo, A.V., 2003. Ca^{2+} sensitivity of smooth muscle and nonmuscle myosin II: modulated by G proteins, kinases, and myosin phosphatase. Physiol. Rev. 83, 1325–1358.

Sommer, J.R., 1983. The implications of structure and geometry on cardiac electrical activity. Ann. Biomed. Eng. 11, 149–157.

Sommer, J.R., Johnson, E.A., 1969. Cardiac muscle: a comparative ultrastructural study with special reference to frog and chicken hearts. Z. Zellforsch. Mikrosk. Anat. 98, 437–468.

Sommer, J.R., Johnson, E.A., 1970. Comparative ultrastructure of cardiac cell membrane specialization: a review. Am. J. Cardiol. 25, 184–194.

Sommer, J.R., Bossen, E., Dalen, H., Dolber, P., High, T., Jewett, P., Johnson, E.A., Junker, J., Leonard, S., Nassar, R., Scherer, B., Spach, M., Spray, T., Taylor, I., Wallace, N.R., Waugh, R., 1991. To excite a heart: a bird's view. Acta Physiol. Scand. S599, 5–21.

Speckmann, E.W., Ringer, R.K., 1964. Static elastic modulus of the turkey aorta. Can. J. Physiol. Pharmacol. 42, 553–561.

Speckmann, E.W., Ringer, R.K., 1966. Volume pressure relationships in the turkey aorta. Can. J. Physiol. Pharmacol. 44, 901–907.

Stallone, J.N., Nishimura, H., Nasjletti, A., 1990. Angiotensin II binding sites in aortic endothelium of domestic fowl. Am. J. Physiol. 258, E777–E782.

Starling, E.H., 1896. On the absorption of fluids from connective tissue spaces. J. Physiol. 19, 312–326.

Steele, P.A., Gibbins, I.L., Morris, J.L., 1996. Projections of intrinsic cardiac neurons to different targets in the guinea-pig heart. J. Auton. Nerv. Syst. 47, 177–187.

Steele, P.A., Gibbins, I.L., Morris, J.L., Mayer, B., 1994. Multiple populations of neuropoptide-containing intrinsic neurons in the guinea-pig heart. Neuroscience 62, 241–250.

Stéphan, F., 1949. Les suppliances obtenues experimentalement dans le systeme des arcs aortiques de l'embryon d'oiseau. C. R. Soc. Anat. 36, 647.

Stephenson, R., Jones, D.R., 1992. Blood flow distribution in submerged and surface-swimming ducks. J. Exp. Biol. 166, 285–296.

Stephenson, R., Jones, D.R., Bryan, R.M., 1994. Regional cerebral blood flow during submergence asphyxia in Pekin duck. Am. J. Physiol. 266, R1162–R1168.

Strack, A.M., Sawyer, W.B., Marubio, L.M., Loewy, A.D., 1988. Spinal origin of sympathetic preganglionic neurons in the rat. Brain Res. 455, 187–191.

Sturkie, P.D., 1986. Heart: contraction, conduction, and electrocardiography. In: Sturkie, P.D. (Ed.), Avian Physiology. Springer-Verlag, Berlin, pp. 167–190.

Sturkie, P.D., Poorvin, D.W., 1973. The avian neurotransmitter. Proc. Soc. Exp. Biol. Med. 143, 644–646.

Sturkie, P.D., Dirner, G., Gister, R., 1978. Role of renal portal valve in the shunting of blood flow in renal and hepatic circulations of chickens. Comp. Biochem. Physiol. C 59, 95–96.

Swart, J., Tate, K., Crossley II, D.A., 2014. Development of adrenergic and cholinergic receptor cardiovascular regulatory capacity in the Canada goose (*Branta Canadensis*) and domestic goose (*Anser anser domesticus*). Comp. Biochem. Physiol. A 167, 59–67.

Szabo, E., Viragh, S., Challice, C.E., 1986. The structure of the atrioventricular conducting system in the avian heart. Anat. Rec. 215, 1–9.

Szabuniewicz, M., McCrady, J.D., 1967. The electrocardiogram of the chicken. Southwest Vet. 20, 287–294.

Tagawa, T., Ando, K., Wasano, T., 1979. A histochemical study of the innervation of the cerebral blood vessels in the domestic fowl. Cell Tissue Res. 198, 43–51.

Takei, Y., 2000. Structural and functional evolution of the natriuretic peptide system in vertebrates. Int. Rev. Cytol. 194, 1–66.

Takei, Y., Hasegawa, Y., 1990. Vasopressor and depressor effects of native angiotensins and inhibition of these effects in the Japanese quail. Gen. Comp. Endocrinol. 79, 12–22.

Taylor, A.E., Townsley, M.I., 1987. Evaluation of the Starling fluid flux equation. NIPS 2, 48–52.

Taylor, M.G., 1964. Wave travel in arteries and the design of the cardiovascular system. In: Attinger, E.O. (Ed.), Pulsatile Arterial Blood Flow. McGraw–Hill, New York.

Tazawa, H., 1981. Effect of O_2 and CO_2 in N_2, He, and SF6 on chick embryo blood pressure and heart rate. J. Appl. Physiol. 51, 1017–1022.

Tazawa, H., Akiyama, R., Moriya, K., 2002. Development of cardiac rhythms in birds. Comp. Biochem. Physiol. A 132, 675–689.

Tazawa, H., Takenaka, H., 1985. Cardiovascular shunt and model analysis in the chick embryo. In: Johansen, K., Burggren, W.W. (Eds.), Cardiovascular Shunts. Munksgaard, Copenhagen, pp. 179–194.

Tazawa, H., Hashimoto, Y., Doi, K., 1992. Blood pressure and heart rate of chick embryo (*Gallus domesticus*). within the egg: responses to autonomic drugs. In: Hill, R.B. (Ed.), Phylogenetic Models in Functional Coupling of the CNS and the Cardiovascular System. Karger, Kingston, Rhode Island, pp. 86–96.

Tazawa, H., Lomholt, J.P., Johansen, K., 1985. Direct measurement of allantoic blood flow in the chicken, *Gallus domesticus*. Responses to alteration in ambient temperature and PO_2. Comp. Biochem. Physiol. A 81, 641–642.

Tcheng, K.T., Fu, S.K., 1962. The structure and innervation of the aortic body of the yellow-breasted bunting. Sci. Sin. 11, 221–232.

Tcheng, K.T., Fu, S.K., Chen, T.Y., 1963. Supracardial encapsulated receptors of the aorta and the pulmonary artery in birds. Sci. Sin. 12, 73–81.

Toop, T., Donald, J.A., 2004. Comparative aspects of natriuretic peptide physiology in non-mammalian vertebrates: a review. J. Comp. Physiol. B 174, 189–204.

Topouzis, S., Catravas, J., Ryan, J., Rosenquist, T., 1992. Influence of vascular smooth muscle heterogeneity on angiotensin converting enzyme activity in chicken embryonic aorta and in endothelial cells in culture. Circ. Res. 71, 923–931.

Torrella, J.R., Fouces, V., Viscor, G., 1999. Descriptive and functional morphometry of skeletal muscle fibers in wild birds. Can. J. Zool. 77, 724–736.

Trajanovska, S., Inoue, K., Takei, Y., Donald, J.A., 2007. Genomic analyses and cloning of novel chicken natriuretic peptide genes reveal new insights into natriuretic peptide evolution. Peptides 28, 2155–2163.

Tummons, J., Sturkie, P.D., 1968. Cardio-accelerator nerve stimulation in chickens. Life Sci. 7, 377–380.

Tummons, J.L., Sturkie, P.D., 1969. Nervous control of heart rate during excitement in the adult White Leghorn cock. Am. J. Physiol. 216, 1437–1440.

Tummons, J.L., Sturkie, P.D., 1970. Beta adrenergic and cholinergic stimulants from the cardioaccelerator nerve of the domestic fowl. Z. Vergl. Physiol. 68, 268–271.

Umans, J.G., Levi, R., 1995. Nitric oxide in the regulation of blood flow and arterial pressure. Ann. Rev. Physiol. 57, 771–790.

Ungureanu-longrois, D., Bezie, Y., Perret, C., Laurent, S., 1997. Effects of exogenous and endogenous nitric oxide on the contractile function of cultured chick embryo ventricular myocytes. J. Mol. Cell. Cardiol. 29, 677–687.

Vanhoutte, P.M., Leusen, I., 1969. The reactivity of isolated venous preparations to electrical stimulation. Pflügers Arch. 306, 341–353.

Vatner, S.F., Braunwald, E., 1975. Cardiovascular control mechanisms in the conscious state. New Engl. J. Med. 293, 970–976.

Verberne, M.E., Gittenberger-de Groot, A.C., Van Iperen, L., Poelmann, R.E., 1999. Contribution of the cervical sympathetic ganglia to the innervation of the pharyngeal arch arteries and the heart of in the chick embryo. Anat. Rec. 255, 407–419.

Villamor, E., Kessels, C.G.A., Ruijtenbeek, K., van Suylen, R.J., Belik, J., De Mey, J.G.R., Blanco, C.E., 2004. Chronic in ovo hypoxia decreases pulmonary arterial contractile reactivity and induces biventricular cardiac enlargement in the chicken embryo. Am. J. Physiol. 287, R642–R651.

Villamor, E., Ruijtenbeek, K., Pulgar, V., DeMey, J.G., Blanco, C.E., 2002. Vascular reactivity in intrapulmonary arteries of chicken embryos during transition to ex ovo life. Am. J. Physiol. 282, R917–R927.

Ward, S., Bishop, C.M., Woakes, A.J., Butler, P.J., 2002. Heart rate and the rate of oxygen consumption of flying and walking barnacle geese (*Branta leucopsis*) and Bar-headed geese (*Anser indicus*). J. Exp. Biol. 205, 3347–3356.

Webb, R.C., 2003. Smooth muscle contraction and relaxation. Adv. Physiol. Educ. 27, 201–206.

Weinstein, Y., Bernstein, M.H., Bickler, P.E., Gonzalez, D.V., Samaniego, F.C., Escobedo, M.A., 1985. Blood respiratory properties in pigeons at high altitudes: effects of acclimation. Am. J. Physiol. 249, R765–R775.

West, J.B., 2008. Respiratory Physiology—the Essentials, eighth ed. Lippincott, Williams and Wilkins, Baltimore, MD.

West, N.H., Langille, B.L., Jones, D.R., 1981. Cardiovascular system. In: King, A.S., McLelland, J. (Eds.), Form and Function in Birds, vol. 2. Academic Press, New York, pp. 235–339.

White, P.T., 1974. Experimental studies on the circulatory system of the late chick embryo. J. Exp. Biol. 61, 571–592.

Whittow, G.C., 2000. Avian Physiology, fifth ed. Springer-Verlag, Berlin.

Wideman, R.F., Gregg, C.M., 1988. Model for evaluating avian renal hemodynamics and glomerular filtration rate autoregulation. Am. J. Physiol. 254, R925–R932.

Wideman, R.F., Braun, E.J., Anderson, G.L., 1991. Microanatomy of the renal cortex in the domestic fowl. J. Morphol. 168, 249–267.

Wideman, R.F., Glahn, R.P., Bottje, W.G., Holmes, K.R., 1992. Use of a thermal pulse decay system to assess avian renal blood flow during reduced renal arterial perfusion pressure. Am. J. Physiol. 262, R90–R98.

Wilson, J.X., 1989. The renin-angiotensin system in birds. In: Hughes, M., Chadwick, A. (Eds.), Progress in Avian Osmoregulation. Leeds Philosophical and Literary Society, Leeds, UK, pp. 61–79.

Wilson, J.X., Butler, D.G., 1983. Catecholamine-mediated pressor responses to angiotensin II in the Pekin duck, *Anas platyrhynchos*. Gen. Comp. Endocrinol. 51, 477–489.

Wilson, J.X., West, N.H., 1986. Cardiovascular responses to neurohormones in conscious chickens and ducks. Gen. Comp. Endocrinol. 62, 268–280.

Wingard, C.J., Godt, R.E., 2002. Cardiac neural crest ablation alters aortic smooth muscle force and voltage-sensitive Ca^{2+} responses. J.Muscle Res. Cell. Motil. 23, 293–303.

Wolfenson, D., Frei, Y.F., Berman, A., 1982. Blood flow distribution during artificially induced respiratory alkalosis in the fowl. Resp. Physiol. 50, 87–92.

Wolfenson, D., Frei, Y.F., Snapir, N., Berman, A., 1978. Measurement of blood flow distribution by radioactive microspheres in the laying hen. Comp. Biochem. Physiol. A 61A, 549.

Womersley, J.R., 1957. The mathematical analysis of the arterial circulation in a state of oscillatory motion. Tech. Rep. Wright AFB Dev. Ctr. WADC-TR-56-614.

Wray, S., Smith, R.D., 2004. Mechanisms of action of pH-induced effects on vascular smooth muscle. Mol. Cell. Biochem. 263, 163–172.

Ying, L., James, T.N., Yamamoto, S., Terasaki, F., 1993. Cardiac conduction system in the chicken: gross anatomy plus light and electron microscopy. Anat. Rec. 236, 493–510.

Yoshigi, M., Ettel, J.M., Keller, B.B., 1997. Developmental changes in flow-wave propagation velocity in embryonic chick vascular system. Am. J. Physiol. 273, H1523–H1529.

Yousuf, N., 1965. The conducting system of the heart of the house sparrow, *Passer domesticus indicus*. Anat. Rec. 152, 235–250.

Zahka, K.G., Hu, N., Brin, K.P., Yin, F.C., Clark, E.B., 1989. Aortic impedance and hydraulic power in the chick embryo from stages 18 to 29. Circ. Res. 64, 1091–1095.

Zoer, B., Blanco, C.E., Villamor, E., 2010. Role of rho-kinase in mediating contraction of chicken embryo femoral arteries. J. Comp. Physiol. B 180, 427–435.

Chapter 12

Osmoregulatory Systems of Birds

Eldon J. Braun
Department of Physiology, College of Medicine, University of Arizona, AZ, USA

12.1 INTRODUCTION

To introduce the osmoregulatory strategies of birds, some aspects of osmoregulation that are common to all vertebrates are presented. Animals must have adequate water in the extracellular fluid compartment for normal cellular functions. However, this may be constrained because the environments in which animals live range from harsh deserts, where water is limited, to aquatic environments, where water is plentiful but may be brackish or marine. With respect to water and ion homeostasis, all animals can be categorized as either osmoconformers or osmoregulators of the internal environment.

Most vertebrates, including birds, are osmoregulators. The exceptions are those vertebrates that inhabit marine environments, such hagfish, skates, and sharks, which are osmoconformers (Somero, 1986). The composition of the cytosol of these osmoconformers is regulated; however, the extracellular fluids are isosmotic with the external marine aqueous media, but not isotonic with it with respect to individual ions (Somero, 1986). In some animals, the concentration of organic solutes in the extracellular fluid affects the movement of water into and out of cells. An example of animal in a marine environment is the shark, which is an osmocomforming animal (Somero, 1986). The extracellular fluid of these animals is isosmotic with seawater but not isotonic with respect to individual ions. The difference in total osmolality is made up of the organic solutes urea and trimethylamine (TMO) (Yancey and Somero, 1979). Urea cannot be the sole solute because it is toxic to proteins. TMO acts as a counteracting solute to prevent the potential negative effects of urea.

An example of a freshwater animal in which organic solutes prevent the negative effects of changes in the extracellular fluid is the tree frog in the north temperate climate zone (Zhang and Storey, 2012). In the late fall when ambient temperatures decline, the frogs burrow into the mud of freshwater ponds and freeze with their surrounding environment. The cells of the frogs do not suffer the typical damage incurred by ice crystals as tissues freeze. With onset of declining ambient temperatures, the liver of the frogs mobilizes large quantities of glucose, which functions to structure water to prevent the damaging effect of ice crystal formation (Dieni and Storey, 2010). The glucose in the cooled water forms a glass layer in a process referred to as vitrification, which prevents free flow of water across cell membranes (Brüggeller and Mayer, 1980). These are just two examples of the evolution of the effects of organic solutes on the movement of fluids across cell membranes.

12.1.1 Organs of Osmoregulation in Vertebrates

A number of avenues of osmoregulation have developed through the process of evolution to allow vertebrates to inhabit a wide variety of environments (Table 12.1). It is apparent that for all vertebrates except mammals, more than one organ and/or organ system contributes to the maintenance of osmoregulatory homeostasis. In the various groups of fish, four organs can contribute to osmoregulation. Within amphibians, individual species can be found to employ as many as five organs or organs systems in the maintenance of fluid and electrolyte balance. Reptiles are almost as diverse in the patterns of osmoregulation, as their kidneys, intestines, and bladder can function in this process. In addition, a large number of reptilian species have functional salt glands that contribute to osmoregulation. Mammals are the unusual group in terms of organs involved in osmoregulation in that only the kidney plays this role. Mammals can lose water and electrolytes though other avenues (skin, lungs, gastrointestinal tract), but not for the purpose of osmoregulation. The urinary bladder of mammals has a very limited capacity to transport ions and water; thus, it is not important in mammalian osmoregulation other than as a storage organ of urine until it can be conveniently voided (Lewis, 1977).

12.1.2 Osmoregulation by Birds

With respect to organs contributing to osmoregulation, birds fall between mammals and other nonmammalian groups because three organs can function for this purpose. Similar to reptiles, many avian species posses functional salt glands.

TABLE 12.1 Presence of Osmoregulatory Organs in Vertebrates

Organ	Fish	Amphibians	Reptiles	Birds	Mammals
Kidney	X	X	X	X	X
Intestine	X	X	X	X	
Bladder	X	X	X		
Gills	X	X			
Salt glands			X	X	
Skin		X			

Moreover, as in reptiles, the lower gastrointestinal tract is a very important component of osmoregulatory systems in birds. Although not thoroughly investigated for reptiles, the lower gastrointestinal tract of birds may be more important for osmoregulation than for other vertebrates.

In this chapter, osmoregulation by birds will be presented in terms of the three organs involved—the kidney, lower gastrointestinal tract, and the salt glands.

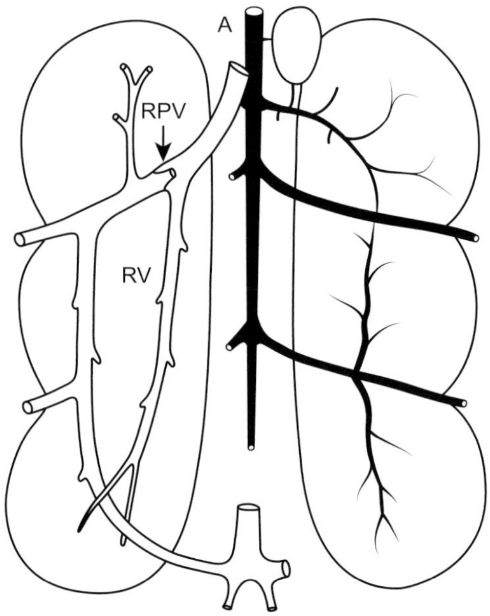

FIGURE 12.1 The surface anatomy of the avian kidney showing the lobes of the kidney, the arterial blood supply, venous drainage, and the renal portal blood flow. The arterial system is shown in black and venous system clear. RV; renal vein, RPV; renal portal valve. *Modified from Hodges (1974).*

12.2 THE AVIAN KIDNEY

From an embryologic viewpoint, the kidneys of vertebrates can be classified as being in one of four developmental categories, starting with the very primitive archinephric kidney to the most developed form found in amniotes—the metanephric kidney. Birds, being amniotes, have a metanephric kidney. Although the metanephric kidney is the functional excretory organ in posthatched birds, during embryonic development the excretory system passes through two primitive stages that recapitulate ontogeny. The first kidney forming at about 36 h of incubation is the pronephros. This kidney is present in the most primitive of the vertebrates but is wholly vestigial in birds. The mesonephros is the second excretory organ forming during development and is homologous with kidneys of adult fish and amphibians. This kidney begins to appear at about 55 h of incubation and is connected to the cloaca by the mesonephric duct, but it is not fully formed at 4 days. At about 5 days of incubation, circulation develops in the glomeruli of the mesonephros. The mesonephros' maximal period of function occurs during embryonic days 5–11. At about day 11, the metanephros becomes active and the mesonephros degenerates. During the peak functional period of the mesonephros urea is the main nitrogen excretory product. Once the metanephros becomes the active excretory organ, the metabolism switches to form uric acid as the major nitrogen excretory product (Patten, 1951).

The avian kidney does not take the classic bean shape of the unipapillate kidney found in many small mammals. It is compressed in the dorsoventral aspect of the body and typically has three divisions—anterior, middle, and posterior—although there are variations on this template (Figure 12.1; Hodges, 1974; Johnson, 1968; Braun and Dantzler, 1972). Moreover, internally the avian kidney is not divided into distinct cortical and medullary regions as is the unipapillate kidney of mammals. Typically, the cortical region of the avian kidney is composed of many cortical units that coalesce to form the medullary regions (Figure 12.2). This organization is reminiscent of the compound multirenculate kidneys of large mammals (Sperber, 1944). The cortical regions of the avian kidney are made up of a large number of small nephrons that do not have loops of Henle. These small loopless nephrons consist of a tubule that is folded upon itself four times and arranged radially around a central efferent vein. The nephrons connect to collecting ducts at right angles and therefore cannot function directly in the countercurrent multiplier (CCM) system. These nephrons essentially consist only of proximal and distal tubules. Deeper in the cortex are much larger nephrons that consist of a highly convoluted proximal tubule, loop of Henle, and a distal tubule (Figure 12.3). The transition between these morphological forms is not abrupt, but gradual. The number of loopless and looped nephrons varies from species to species, but in all species the loopless nephrons far outnumber the looped ones. At this time, there are data for very few species, but in general the total nephron

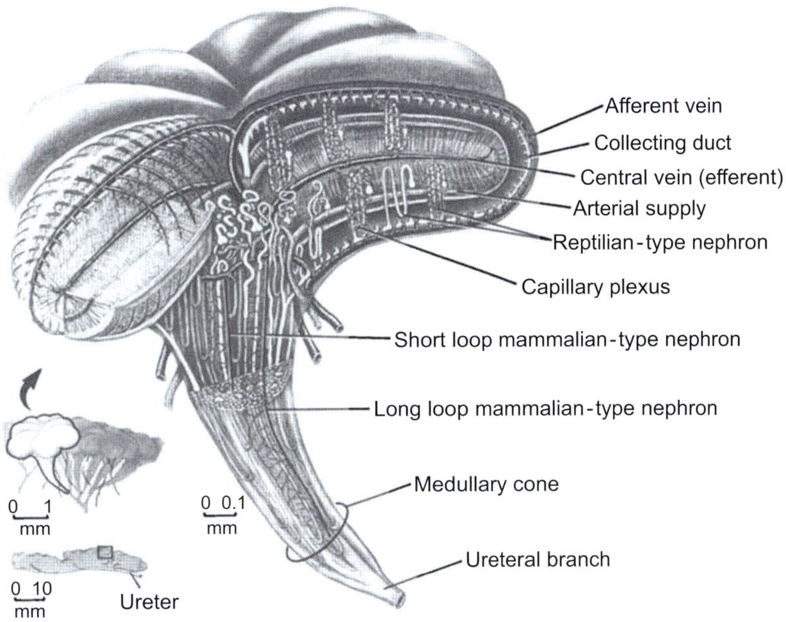

FIGURE 12.2 An illustration of the internal anatomy of the avian kidney based on dissection and observations of the desert quail kidney. The whole kidney is depicted at the lower left of the illustration. The cortical region of the kidney consists of loopless nephrons arranged around a central efferent vein. They enter collecting ducts at right angles and therefore do not function directly in countercurrent multiplication. The collecting of ducts from the loopless nephrons enter the medullary region of the kidney the output of which commingles with that of looped nephrons. Deep to the loopless nephrons are nephrons with loops of Henle of varying length. Note that there is a continuous duct from the shortest loopless nephron to the ureteral branch which eventually enters the ureter proper. Thus there is no renal calyx in the avian urinary system. *From Braun and Dantzler (1972).*

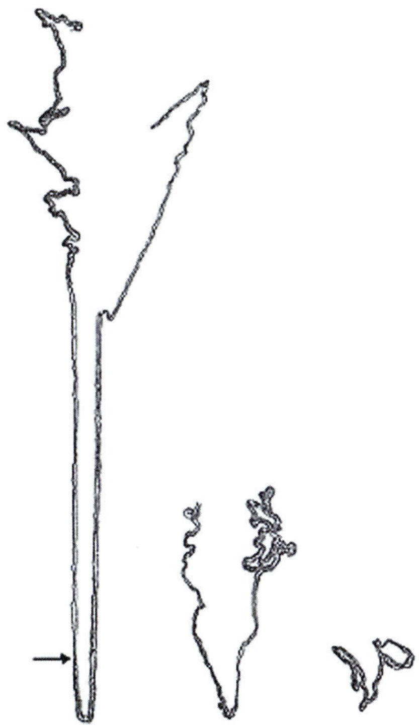

FIGURE 12.3 Nephrons isolated from the kidney of the desert quail. On the left is a large looped nephron. Arrow indicates transition from thin segment to thick descending segment. The image in the center is of shorter looped nephron. On the right is an image of a loopless nephron which appears to be a tubule folded upon itself four times. These nephrons enter collecting ducts at right angles and therefore do not function as part of countercurrent multipliers. *Modified from Braun and Dantzler (1972).*

population in avian kidneys consists of about 85% loopless and 15% looped nephrons (Braun and Dantzler, 1972).

The medullary region of the kidney is formed by small tapering structures called medullary cones. Within these cones are loops of Henle, collecting ducts, and vasa recta. This arrangement with counterflows facilitates the functioning of a CCM system, which permits the avian kidney to produce hyperosmotic urine.

12.2.1 Vascular Anatomy of the Kidney

The arterial blood supply to the avian kidney is derived from three branches of the inferior aorta (Figure 12.1). The smaller anterior and middle lobes of the kidney are supplied by separate branches and the larger posterior lobe of the kidney is supplied by a branch of the aorta that bifurcates on the surface of the kidney, such that there are two arteries that enter the parenchyma of this division. The arteries undergo a series of reductive branches until the afferent arterioles are formed, which enter the Bowman's capsule. After forming the glomerular capillaries, the vessel exits the capsule as the efferent arteriole, which forms a network around the renal tubule (peritubular capillary). The avian kidney, as with all other nonmammalian vertebrates, is also supplied with afferent venous blood by a functional renal portal system. The terminal vessels of the renal portal system join the same peritubular capillary network that is formed by the efferent arterioles (Figure 12.4). How flow in this network

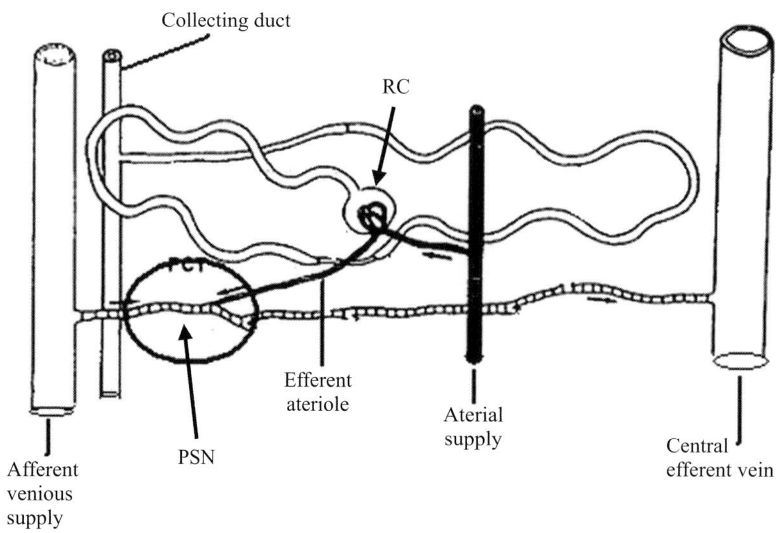

FIGURE 12.4 The pattern of blood flow within the renal cortex of the avian kidney. RC, renal corpuscle; PSN, peritubular capillary network. Note that the flow from the efferent arteriole and that of the renal portal, proximal convoluted tubule meet in the peritubular network as shown in added circle at left. *Modified from Hodges (1974).*

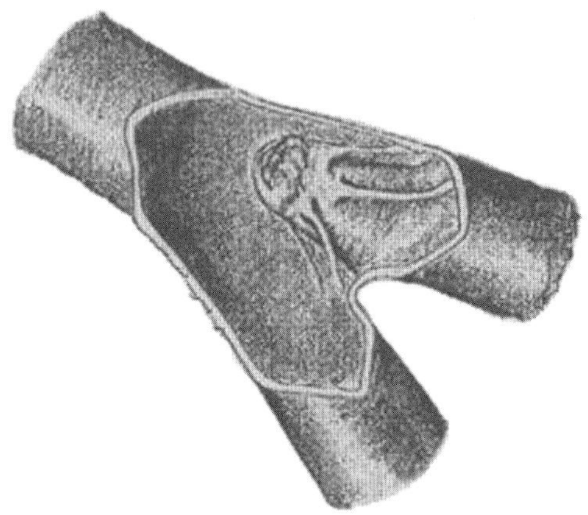

FIGURE 12.5 Artist's rendering of the valve within the renal portal system that directs blood flow into the parenchyma of the kidney or to bypass the secretory epithelia of the kidney. *Adapted from Burrows et al. (1983).*

is regulated is uncertain, but pressure measurements taken from these vessels have an oscillatory profile (Braun and Yokota, unpublished data). The renal portal system receives blood from the lower extremities (external iliac vein) and the lower gastrointestinal tract (caudal mesenteric and pudenal veins). Blood flow to the kidney from the renal portal vessels is controlled by a smooth muscle valve located in the common iliac vein (Figure 12.5). When the valve is open, flow directly enters the inferior vena cava and closure of the valve directs the flow into the renal parenchyma. Experimental data show that adrenergic stimulation closes the valve and cholinergic stimulation controls tone to open the valve (Burrows et al., 1983). One role the flow of blood in the portal system could play is to deliver substrates to the peritubular surface of nephrons to facilitate secretion by tubule cells. Portal blood flow does not enter the medullary cones (Wideman, 1988).

The venous outflow from the kidney originates from the peritubular capillary network, where the efferent arterioles and the terminal portions of the renal portal system unite. Flow from this network enters the central veins that form the core of the cortical lobules. The central veins coalesce to form interlobular veins, several of which unite before the surface of the kidney is reached, at which point they enter the renal vein. The renal veins enter the common iliac veins proximal to the point of the renal portal valve. The outflow of the vasa recta from within the medullary cones exits the base of the cones and enters the interlobular veins.

The capillaries forming the glomerular tuft within the Bowman's capsule of the avian kidney are much simpler and have a lesser degree of interconnecting branching than those that constitute glomerular tuft of most mammalian renal corpuscles. The capillary tufts of the small loopless nephrons in the cortical region of the kidney can consist of one loop—the afferent arteriole entering forming one loop and exiting as the efferent arteriole (Figure 12.6). The transition in size of the nephrons as the larger loop nephrons are reached is followed by an increase in complexity of the capillary tufts, although they are still not as complex as the mammalian capillary tufts (Figure 12.7). For many of these tufts, as the afferent arteriole enters the capsule, it forms loops at the periphery and exits as the efferent arteriole without interconnecting branches. Some of the tufts of the largest renal corpuscles do bifurcate once before the efferent arterioles are formed. As is true of all avian capillaries, those forming the glomerular tufts have larger inner diameter (7–8 μm) than those of mammals (3–4 μm). Avian red blood cells are fusiform in shape (8 × 15 μm), are nucleated, and therefore are much more ridged than those of mammals (which are 7 μm in diameter). These characteristics would make it difficult for the avian red blood cell to traverse the capillary tuft of a typical mammalian glomerulus.

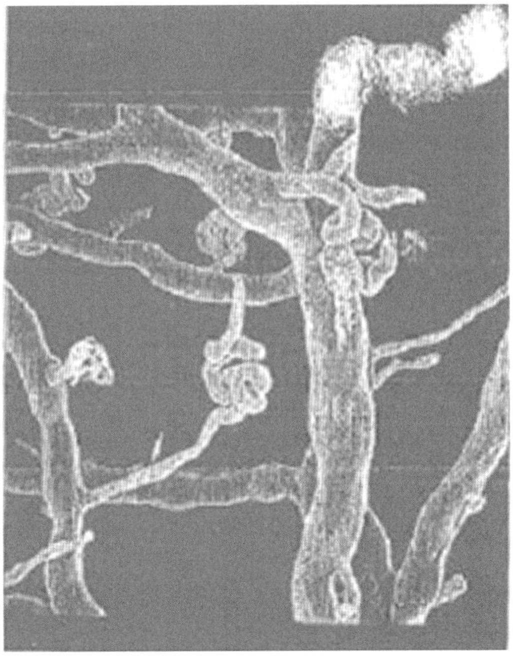

FIGURE 12.6 Methyl methacralyte cast of renal vasculature showing an afferent arteriole leading to the formation of a glomerular capillary of a loopless nephron. Note that there is no branching of the afferent arteriole forming the glomerular capillary. *Adapted from Braun (unpublished).*

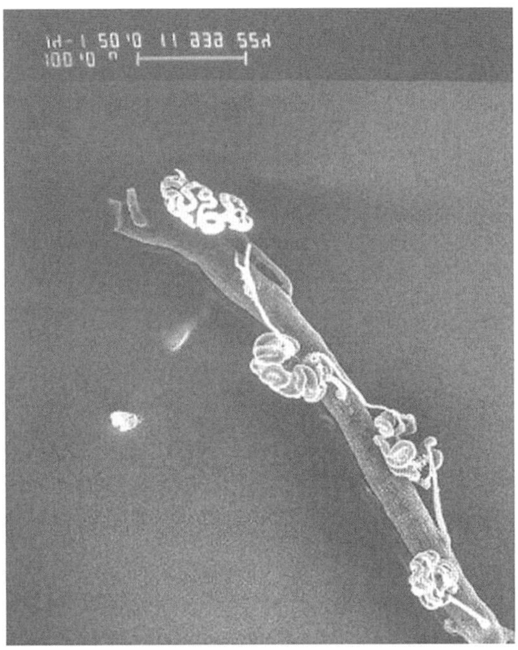

FIGURE 12.7 Scanning electron micrograph of renal vasculature showing a methylacrylate cast of a looped nephron glomerular capillary. Note simplicity of the network and/or the lack of anastomotic interconnecting branches. *Adapted from Braun (unpublished).*

Mammalian red blood cells are quite deformable and undergo "tank treading" as they pass along the small capillaries (Skotheim and Secomb, 2007).

Data on the structure of the filtration barrier formed by endothelial cells and the visceral layer of the Bowman's capsule of birds show that it is somewhat less restrictive than that of mammals (Casotti and Braun, 1996). The "pores" in the slit diaphragm of the chicken were found to be 40–80% larger than values reported for mammals. Moreover, the data indicated that there is less of a polyanionic charge on the filtration barrier of the chicken compared to data for mammals (rat) (Casotti and Braun, 1996). The combination of these two parameters suggests the filtration barrier of birds will allow somewhat larger molecules to pass and enter the filtrate. Indeed, avian ureteral urine contains on average 5 mg/mL protein, which is 100 times more than the urine of a typical mammal (Janes and Braun, 1997).

12.2.2 Glomerular Filtration

The movement of fluid and small solutes across the filtration barrier is a function of hydrostatic pressure produced by the heart. It favors movement across the barrier and colloid osmotic pressure exerted by plasma proteins, which opposes this movement. This is a passive process resulting only from the physical parameters mentioned. The rate of glomerular filtration by single nephrons (SNGFR) is rather low when birds and mammals of similar body mass are compared, but an allometric analysis shows that the total kidney glomerular filtration rates (GFR) are not significantly different between these two groups (bird $GFR = 1.24\,BM^{0.694}$, $n = 11$; mammal $GFR = 1.24\,BM^{0.765}$, $n = 4$; where BM is body mass in kg; Yokota et al., 1985). The lower SNGFR of avian kidneys is offset by a larger number of nephrons. For example, the desert quail (BM 140 g) kidney has approximately 48,000 nephrons and the laboratory rat (BM 300 g) has approximately 32,000.

SNGFRs have been quantified for only two avian species: the desert quail and the European starling (Braun and Dantzler, 1972; Laverty and Dantzler, 1982). The techniques used to make these measurements in the quail and starling were quite different. Using the sodium ferrocyanide technique, which theoretically allows for quantification of the SNGFRs of all the nephrons in the kidney at one point in time, the average SNGFR for the small loopless nephrons was 6.4 nL/min; for the larger looped nephrons, the average was 14.6 nL/min for the desert quail (Braun and Dantzler, 1972). Measurements for the starling were done by *in vivo* micropuncture, which allows the SNGFRs of only the nephrons at the very surface of the kidney to be quantified (Laverty and Dantzler, 1982). These nephrons at the surface are the smallest loopless nephrons. SNGFR values for these nephrons ranged from 0.25 to 0.5 nL/min. The data obtained by the sodium ferrocyanide technique represents an average value for the entire size range of loopless nephron. As mentioned, the transition from the smallest loopless nephrons to the largest loop nephrons is a gradual one. The distinction between the two types is made on the presence or absence of a loop of Henle and the nature of the connection to a collecting duct.

Whole-kidney GFRs of birds are more variable than those of mammals, with the state of hydration being the principle

TABLE 12.2 Orders of Birds Possessing Salt Glands

Order	Habitat	Order	Habitat
Tinamiformes	Tropical rainforests, to grasslands	Gruiformes	Aquatic, rainforests to grassland
Rheiformes	South American grasslands	Charadriiformes	Near water
Struthioniformes	Deserts to scrubland forests	Gaviiformes	Aquatic Fresh water ponds and lakes
Casuraiiformes	Tropical to temperate forests	Columbiformes	Grasslands and deserts
Aepyornithiformes	Madagascar	Psittaciformes	Tropical and subtropical
Dinornithiformes	Semi-arid and high rainfall ecological zones	Cuculiformes	Forests, savannas and wetlands
Podicipediformes	Freshwater ponds and lakes	Strigiformes	Cosmopolitan distribution
Sphenisciformes	Oceanic or coastal habitats	Caprimulgiformes	Grasslands, semi-arid deserts, woodlands
Procellariiformes	Islands, deserts	Apodiformes	Coniferous forests, desert, grassland
Pelecaniformes	Oceans, seashores	Coliiformes	Deciduous forests, grasslands
Anseriformes	Wetlands	Trogoniformes	Woodlands, grasslands, savannas
Phoenicopteriformes	Large lakes, shallow waters, inland or coastal	Coraciiformes	Tropical rainforests, woodlands near water, seacoasts
Ciconiiformes	Wetlands, swamps, forest streams,	Piciformes	Forests, woodlands
Falconiformes	Grasslands, open woodlands	Passeriformes	Grasslands, woodlands, scrublands, forests, deserts, mountains
Galliformes	Forests, deserts, scrub forests, cultivated lands		

Data on salt glands from Peaker and Linzell (1975).

determinate of this variation (Table 12.2). With water deprivation, the GFR can decrease by as much as 65% from the value at full hydration. These changes in GFR appear to be regulated in part by the avian antidiuretic hormone arginine vasotocin (AVT). Studies on desert quail, in which an osmotic load was administered by sodium chloride infusion, showed that GFR decreased with an increasing osmotic load and that this decrease could be attributed to a decreased number of filtering loopless nephrons (Braun and Dantzler, 1972). These data suggested that AVT exerts its effect through the vasculature, a vascular antidiuresis. In contrast, the infusion of increasing amount of AVT in conscious domestic fowl revealed the primary effect of AVT was on the renal tubules, presumably the collecting ducts (Figure 12.8, Stallone and Braun, 1985). Taken together, the data suggest that AVT functions to conserve water by affecting the renal vasculature and tubule epithelium. The GFR of some small birds (sun birds and probably hummingbirds) can change diurnally, driven primarily by the need to excrete large amounts of fluid containing metabolic substrates that was consumed to meet their energy requirements (del Rio et al., 2001).

The renal blood flow (RBF) and GFR of birds is autoregulated over a wide range of systemic blood pressures (Wideman, 1991). The regulation of renal blood flow has only been studied in the domestic fowl, where it was found that RBF and GFR were maintained at systemic blood pressures as low as 50 mm Hg.

12.2.3 Ion Transport by Renal Tubules

In general, the avian proximal renal tubule functions similarly to that of other vertebrate because it isosmotically reabsorbs a large fraction of the fluid that is filtered at the renal corpuscle. The fluid and electrolyte transport capabilities of the avian proximal renal tubule have been studied directly using *in vivo* micropuncture techniques, isolated tubules perfused *in vitro*, and primary cell cultures, as well as being inferred from whole animal studies (Brokl et al., 1994; Laverty and Alberici, 1987; Sutterlin and Laverty, 1998). Whole animal studies indicate that the proximal tubules as an aggregate reabsorb about 60% of the glomerular filtrate (Braun and Dantzler, 1972). To the contrary, data from *in vivo* micropuncture studies suggest that only 24% of the filtered load is reabsorbed by the proximal tubule (Laverty and Dantzler, 1982). This large discrepancy again may be due the sampling of fluid from the very small loopless nephrons at the surface,

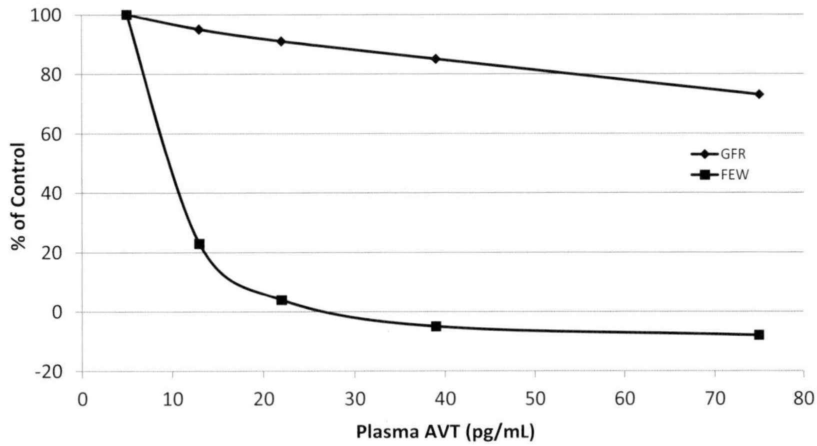

FIGURE 12.8 **The effect of arginine vasotocin (AVT), the avian antidiuretic hormone, on free water clearance and glomerular filtration rate in domestic fowl.** Note that free water clearance declines prior to a significant charge in the glomerular filtration rate suggesting that AVT has an effect on the renal tubules before it effects glomerular filtration. *Modified from Stallone and Braun (1985).*

which is a limitation of micropuncture techniques. Moreover, only the very early segment of the proximal tubule can be sampled because it is at the surface of the kidney. With this caveat, micropuncture data do indicate that proximal tubule reabsorption occurs in an isosmotic manner.

The avian kidney apparently has a very large capacity to reabsorb filtered glucose. Because birds have very high plasma glucose concentrations and GFRs that do not differ from that of mammals, the filtered load of glucose is very high (filtered load = plasma glucose concentration multiplied by GFR) (Morgan and Braun, 2001). Further, very little glucose appears in the ureteral urine. In mammals, glucose is reabsorbed by sodium-glucose cotransporters located in the proximal tubules (SGLTs), which have a finite capacity to reabsorb glucose (tubular maximum). Because the filtered load of glucose in birds is very high, the SGLTs must be upregulated in the avian kidney. The SGLTs of the avian kidney have not yet been studied.

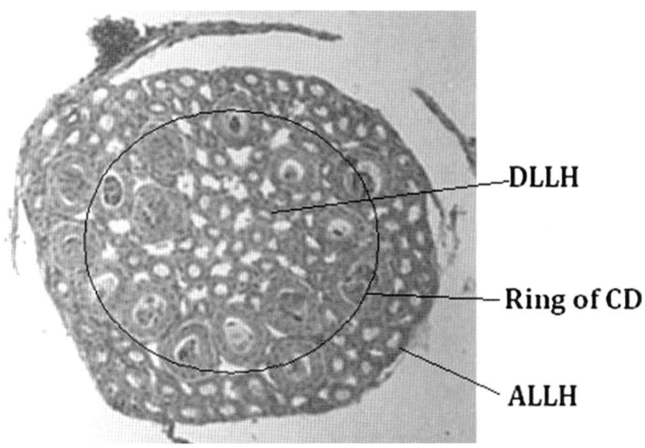

FIGURE 12.9 **Cross section of paraffin embedded tissue of a medullary cone of the desert quail kidney.** Note the ring-like pattern of collecting ducts with thick limbs of Henle's loop located peripheral to the collecting ducts and thin limbs tending to be located in the center of the collecting duct ring. *Adapted from Braun (unpublished).*

12.2.4 The Renal Medulla

The medullary region of the avian kidney is made up of a variable number of small tapering units called medullary cones (Figure 12.2). The number and size of the individual cones is somewhat species and habitat dependent (Johnson and Mugaas, 1970). Moreover, the size of the cones is not uniform throughout any given kidney, with the longer and larger cones generally found in the anterior division of the kidney (Johnson and Mugaas, 1970). Each cone terminates as a single, large collecting duct; several of these ducts join to form a primary ureteral branch, which enters the ureter proper on the ventral surface of the kidney. The medullary cones function as independent units because the output of one cone does not enter that of a second cone; that is, the cones function in a parallel manner and not in series. This ureteral anatomy is different from mammalian kidneys, where there is a discontinuity between the collecting ducts and the renal calyx and eventually the ureter. Therefore, in the avian kidney, there is a continuous duct from the smallest loopless nephron to the ureter. The functional significance of this anatomy is explained here.

There is definite pattern to the arrangement of the tubular and vasculature elements within the medullary cones. In the upper one-third of the cone when viewed in cross section, the collecting ducts form a ring with the thin limbs of the loop of Henle tending to be situated outside of the ring and the thick limbs of Henle inside the ring formed by the collecting ducts (Figure 12.9; Johnson and Mugaas, 1970). The vasa recta within the cones appear to form a lace-like pattern with few long, straight channels (Figure 12.10).

The pars recta of the proximal tubules enter the medullary cones to form the descending limb of Henle's loop. The transition from the pars recta to the thin descending limb of Henle's loop (DLLH) is very abrupt

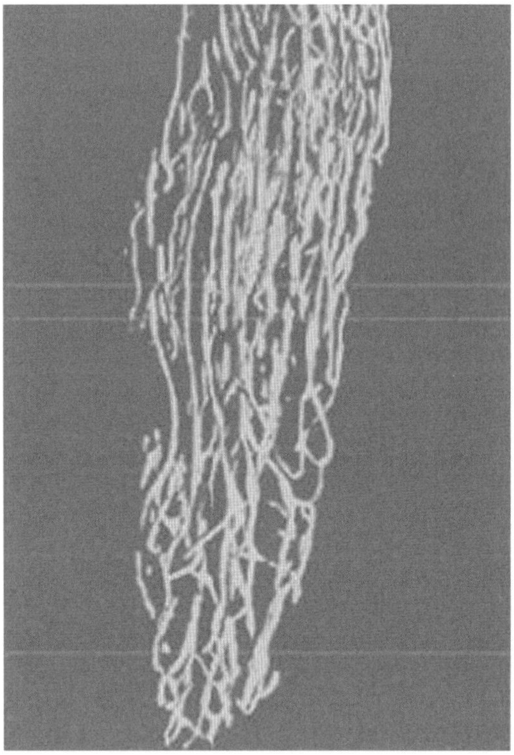

FIGURE 12.10 **The blood supply and venous drainage shown by a microfil cast of the vasa recta from within a medullary cone of the desert quail kidney.** The capillaries form a lace-like pattern within the medullary cone with few straight channels. *From Braun and published in previous edition of Sturkie.*

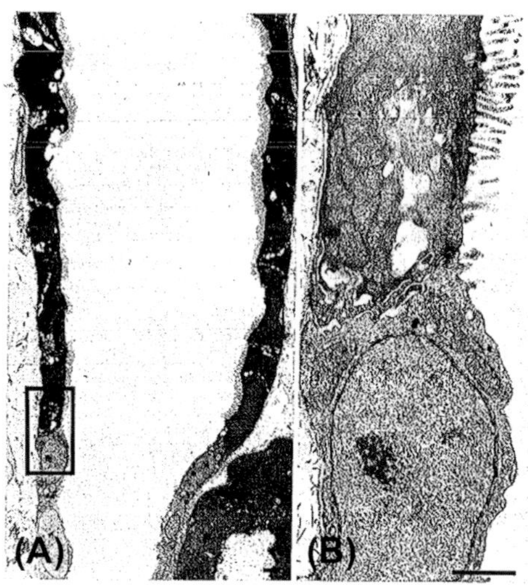

FIGURE 12.11 **An electron micrograph of a longitudinal section of a renal tubule showing the abrupt transition from the proximal straight tubule to the thin descending loop of Henle.** Inset shows transition at increased magnification. Note the presence and absence of microvilli from one cell to the next. *Adapted from Braun and Reimer (1988).*

(Figure 12.11; Braun and Reimer, 1988). The cuboidal cells of the pars recta give way to very low profile cells (type 3 epithelia), forming the DLLH (Braun and Reimer, 1988). The epithelium of the loop always thickens before the formation of the hairpin turn, with the thick descending limb being approximately 15% of the total DLLH length (Figure 12.3; Casotti et al., 2000). This morphology of the DLLH is markedly different from most nephrons of mammal kidneys, where the hairpin turn of the loop is formed by thin epithelium, with thick cells appearing at a variable point on the ALLH. The structure of the DLLH of the avian kidney is important to the CCM system, which functions within the medullary cones (Layton et al., 2000). NaCl transport from ascending limb thick segments contributes approximately 70% of urine concentrating capacity of the avian medullary cone (Layton et al., 2000). The ALLHs traverse up to the base of the cone, at which point they enter collecting ducts.

12.2.5 Concentration and Dilution of Urine

Along with mammals, birds have the capacity through their kidneys to conserve body water by eliminating solutes in excess of water; that is, the kidneys can produce urine with osmolalities that have greater osmotic potential than the plasma. Both groups of vertebrates have renal morphologies that permit the functioning of CCM systems. However, the details of the systems differ between birds and mammals. As described earlier, the avian renal medulla consists of many small units or medullary cones, within which there is a parallel arrangement of tubules. In these tubules, loops of Henle, (ascending and descending) collecting ducts, and vasa recta, there is countercurrent flow of fluid.

The primary excretory product of nitrogen in birds, uric acid, plays no role in the solute composition of the medullary cones because it is virtually insoluble in aqueous solutions. Moreover, birds excrete very little nitrogen as urea. On the other hand, urea, as the primary nitrogenous excretory product of mammals, makes up about 50% of the solutes in the renal medulla. In the avian kidney, there is a solute gradient from the base of the medullary cones to their tips, which is made up entirely of sodium chloride (Skadhauge and Schmidt-Nielsen, 1967). Sodium chloride is deposited in the cone interstitum by active transport from the thick segments of the loop of Henle, including the part before the hairpin turn is formed (Nishimura et al, 1986). From the interstitum, the sodium chloride passively enters the descending limb of the loop of Henle (DLLH) (Figure 12.12). Sodium chloride is thus recycled within the medullary cone. The avian DLLH has much higher diffusional permeability to sodium chloride than the DLLH of the hamster kidney (31.7×10^{-7} to $2.9 \times 10^{-7} cm^2/s$; Nishimura et al., 1986). The removal of sodium chloride from the ALLH which is impermeable to water, serves to dilute the tubule fluid as it moves to the distal convoluted tubule. In the cortical region of the kidney, the fluid in the distal convoluted tubule can equilibrate with the interstitum by water removal before the fluid enters collecting ducts to

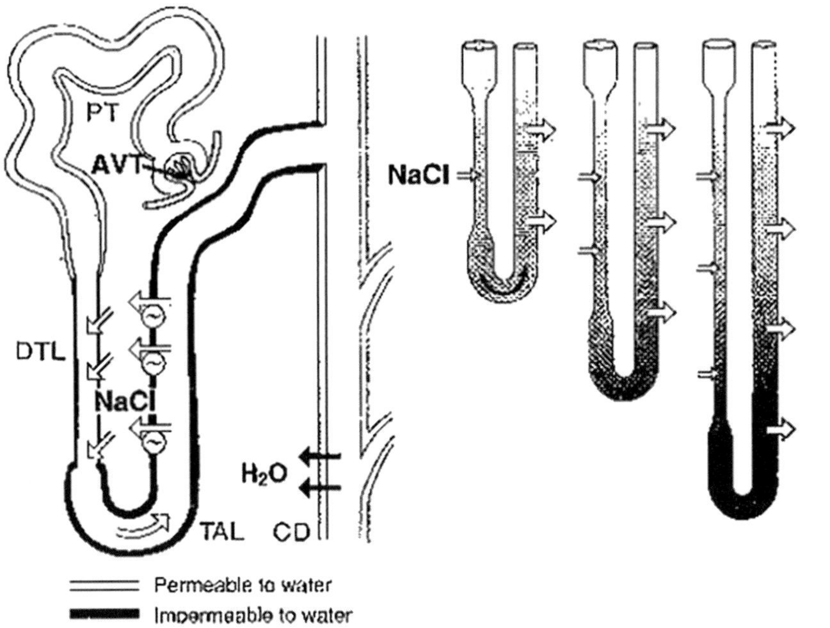

FIGURE 12.12 **Model of the avian countercurrent multiplier system.** Sodium chloride is actively extruded from the thick portions of Henle's loop and passively enters the thin descending limb of Henle from the high concentrations in the medullary interstitum. The process renders the fluid in the ascending limb dilute relative to the interstitum. In the presence of antidiuretic hormone (AVT), the fluid in collecting ducts equilibrates with the concentrated interstitum to become concentrated. *From Nishimura and published in previous edition of Sturkie.*

re-enter the medullary cone. In the collecting ducts within cones, the isosmotic tubule fluid is exposed to an increasing osmotic gradient and becomes hyperosmotic as it loses water to the cone interstitum. Water movement across the tubules in the cones is controlled by the antidiuretic hormone of birds' AVT.

Water flux across the tubules within the medullary cones is through water channels or aquaporins (AQPs). AQPs 3 and 4 have been localized to the basolateral membrane of collecting duct cells and the AVT-sensitive AQP 2 has been identified subapically in collecting duct cells (Nishimura, 2008). In contrast to mammalian renal medullary tissue, the avian collecting ducts are less sensitive to the antidiuretic hormone AVT in the promotion of water flux (Nishimura, 2008).

Compared with the kidneys of mammals, those of birds do not elaborate highly concentrated urine. The maximum urine to plasma osmolar ratios (U/Posms) of mammals range from 1 (mountain beaver) to 25 (desert rodents), whereas the maximum ratio for birds is approximately 2.5 (Braun, 1993). The lower capacity of birds is related to the architecture of their kidneys and the lack of urinary bladder, in which urine is stored until it can be conveniently excreted. The output from avian kidneys enters the cloaca, the terminal portion of the gastrointestinal tract. It does not remain in the cloaca but is moved by a retrograde peristalsis into the colon and cecae located at the junction between the colon and ileum. The urine is thus brought into contact with a second transporting epithelium.

As fluids move across the epithelium of the colon on concentration gradients, the lack of the ability to produce highly concentrated urine prevents the loss of body fluids by the movement of fluids in a serosal-to-mucosal direction. Data indicate that the colonic epithelium can move fluid in a mucosal-to-serosal direction against a concentration gradient of 200 mOsm/kg H_2O (Skadhauge and Kristensen, 1972). The bottom line is that the avian kidney and lower gastrointestinal tract function in concert to maintain fluid and electrolyte balance. The kidney should not produce a urine that is significantly more concentrated than the plasma as a mechanism to conserve body water.

As mentioned above, the anatomy of the avian kidney does not serve well the capacity to elaborate markedly concentrated urine. The urine of well-hydrated birds is isosmotic or only slightly hyperosmotic to the plasma. When water deprived, birds on average produce urine that is about 2.0 times more concentrated that the plasma (Table 12.3). Urine-to-plasma osmotic ratios for birds can be somewhat misleading because the plasma osmolality of birds is much more labile than that of mammals and tends to increase with water deprivation (Braun and Dantzler, 1972). There is one report in the literature of the salt marsh Savannah sparrow achieving 4–6 U/Posm when the birds were acclimated to drinking seawater (Poulson and Bartholomew, 1962). Inspection of these data suggests that the birds were not absorbing the saltwater across the gastrointestinal tract into the circulation to be excreted by the kidneys, but that the water was passing along the gastrointestinal tract and excreted. Water not being absorbed has also been observed in hummingbirds (del Rio et al., 2001).

TABLE 12.3 Glomerular Filtration Rates (GFR) of Selected Birds

Species	GFR (mL/kg/min)
Domestic fowl	2.5
Desert quail	1.8
Mallard duck	2.5
Canadian goose	1.2
Glaucous-winged gull	1.9
European starling	2.8
Budgerigar	4.4
Turkey	1.3

Data from Braun (1982).

TABLE 12.4 Urine-to-Plasma Osmolality Ratios for Selected Birds

	U/P Ratio	
Species	Ureteral Urine	Voided Fluid
Domestic fowl	2.0	
Ring-necked pheasant	1.5	
Kookaburra		2.7
Singing honey eater		2.4
Red wattlebird		2.4
Bobwhite quail		1.6
California quail		1.7
Gamble's quail		2.5
Senegal dove		1.7
English sparrow	1.7	
Song sparrow	2.2	
White crown sparrow	2.1	
White-winged dove	1.8	
Emu		1.4
Galah		2.5
Glaucous-winged gull	1.9	
Savannah sparrow	1.6	

Table modified from Braun (2009).

A second study of salt marsh Savannah sparrows carried out in the field suggested the U/Posm in these birds was 1.2 and that they did not take in seawater; rather, they stripped freshwater as dew from plant leaves (Goldstein et al., 1990).

12.2.6 Nitrogen Excretion

The major end product of nitrogen catabolism in birds is uric acid, accounting for 70–80% of the nitrogen excreted in the ureteral urine with other compounds such as urea, creatinine, and amino acids contributing minor amounts to total nitrogen excretion. Under some circumstances, such as high fluid intake required to maintain energy balance, nitrogen in the form of ammonia can account for more than 50% of the nitrogen excreted in the urine (Preest and Beuchat, 1997; Tsahar et al., 2005).

Uric acid as an end product of nitrogen catabolism is a very efficient form of excretion because it contains four nitrogen atoms and has a very low aqueous solubility compared to other nitrogen excretory products (Table 12.4). However, the metabolic cost of uric acid synthesis (nine adenosine triphosphates) is high compared with that of urea and ammonia. The low aqueous solubility has the potential of causing precipitation problems in the plasma, but using the Henderson–Hasselbach equation it can be calculated that 98% of uric acid in the plasma at pH 7.4 exists as the anionic form. The salts of uric acid have much higher aqueous solubility than the protonated form of the molecule (Table 12.4). The concentrations of urate in avian plasma vary somewhat but generally do not exceed 1.0 mmol/L. The average concentration of uric acid/urate of approximately 0.65 mmol/L is well within the solubility limit of sodium urate (8.32 mmol/L), which prevents crystal formation.

Urate, being a relatively small molecule (186 MW), freely passes the glomerular filtration barrier; however, five times more urate is excreted than is filtered. This suggests that urate is avidly secreted by the renal tubules. Indeed, data derived from different preparations substantiates this point (Brokl et al., 1994; Laverty and Dantzler, 1983; Dudas et al., 2005). *In vivo* micropuncture studies of the proximal tubules from superficial loopless nephrons demonstrated a marked secretion of urate. Moreover, preparations of isolated, perfused proximal tubules from deeper nephrons support the micropuncture data (Brokl et al., 1994). Data from the isolated, perfused proximal tubules also demonstrated that urate is taken up at the basolateral membrane against an electrochemical gradient and that this uptake could be competitively blocked by para-aminohippuric acid. The exit from the cell across the apical membrane is down an electrochemical gradient (Figure 12.13).

Once uric acid/urate enters the proximal tubule, there is the potential for crystal formation as the intraluminal concentration increases sufficiently to exceed the solubility limit of even the salt forms. The increased concentration is caused by the avid tubule secretion of urate and the normal removal of fluid from the tubule to recover water that was filtered at the glomerulus.

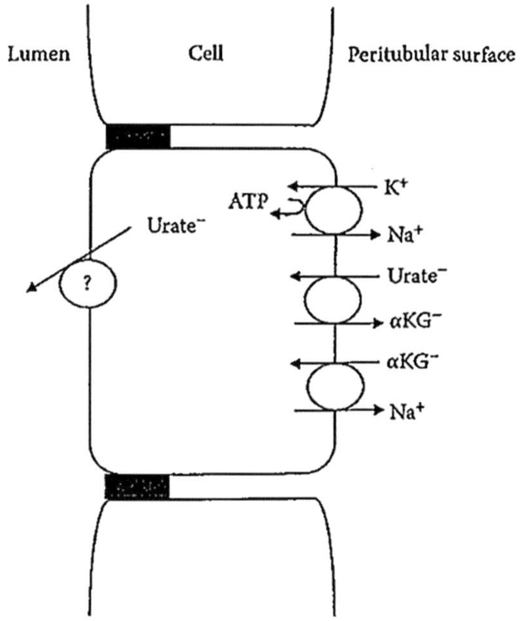

FIGURE 12.13 **The suggested mechanism of urate secretion by the proximal renal tubules.** Urate enters the cell on an organic acid transport protein (OAT1) in exchange for alpha-ketoglutaric acid. The movement of urate across the apical membrane into the lumen is facilitated by an unidentified carrier protein. *From Braun (2009).*

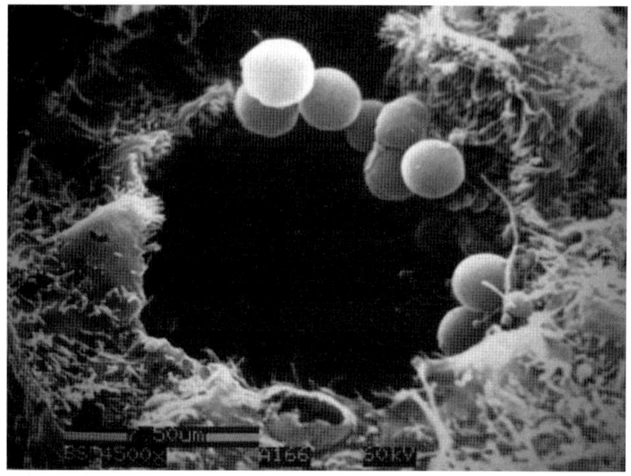

FIGURE 12.14 **Urate containing spheres as they form in the proximal tubule (as indicated by the presence of the brush border) of a nephron.** *Adapted from Braun (unpublished).*

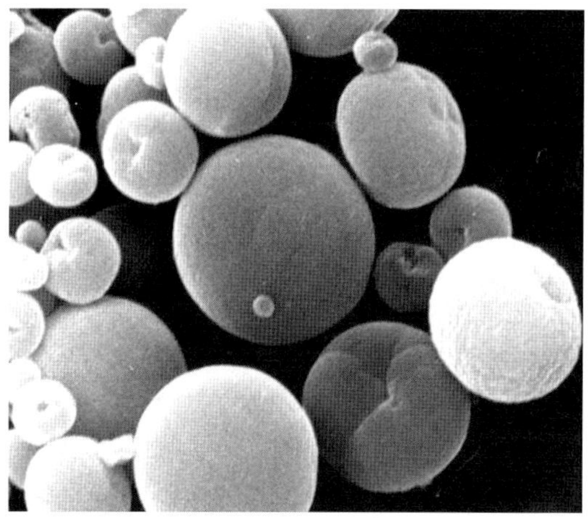

FIGURE 12.15 **Scanning electron micrograph (SEM) of ureteral urine.** The preparation was made by allowing ureteral urine to air dry on an electron microscope stub. Note the complete absence of urate crystals in the image. *Adapted from Braun (unpublished).*

12.2.7 Form of Uric Acid in Avian Urine

Because of the low aqueous solubility of uric acid and the salts of uric acid, the potential exists for them to precipitate from solution. As mentioned above, due to water abstraction and urate secretion, the concentration of the salts in the proximal tubule exceeds their solubility limits. Crystal formation is prevented by the urate anion binding to serum albumin in the tubule lumen. A small amount of albumin passes through the glomerular filtration barrier, as this barrier in birds appears to allow some macromolecules such as albumin to pass (Casotti and Braun, 1996). Indeed, avian ureteral urine contains 100 times more protein than the urine of mammals (5 mg/mL versus 0.05 mg/mL; Janes and Braun, 1997). Early in the proximal tubule, small, spherical structures begin to form (Figure 12.14). The final ureteral urine contains an array (1–14 μM diameter) of these spherical structures (Figure 12.15). Chemical analysis of these spheres shows that they consist of about 65% urate; no crystals of uric acid are observed when spheres are sectioned and viewed with an electron microscope (Figure 12.16). This suggests a chemical binding of urate to serum albumin, which takes urate out of solution and prevents crystal formation (Figure 12.17). Moreover, because urate is not in solution, it does not contribute to the osmolality of the final urine, as does urea in the urine of mammals. This highlights the point that birds and mammals cannot be directly compared with respect to urine-to-plasma osmolar ratios because their respective end products of nitrogen metabolism contribute unequally to the total osmolality of the urine.

Urine, along with the urate spheres, is conveyed to the urodeal compartment of the cloaca. All components of the urine do not remain in this part of the cloaca but are moved by retrograde peristaltic action into the colon and digestive ceca (Figure 12.18; Brummermann and Braun, 1995). The urine moves around a central fecal core and comes into contact with the epithelium of the colon and a large, diverse population of bacteria embedded in the brush border of the cells. This population of bacteria was studied and classified by Barnes and Impey (1970) and was found by Campbell and Braun (1986) to flourish on a culture containing only uric acid. Sodium dodecyl sulfate polyacrylamide gel electrophoresis gels of ureteral urine show large amounts of protein of varying molecular mass, with serum albumin being the densest mass (Figure 12.19; Janes and Braun, 1997). Similar gels of fluid excreted by birds show little or no protein (Figure 12.20; Janes and Braun, 1997). The protein is

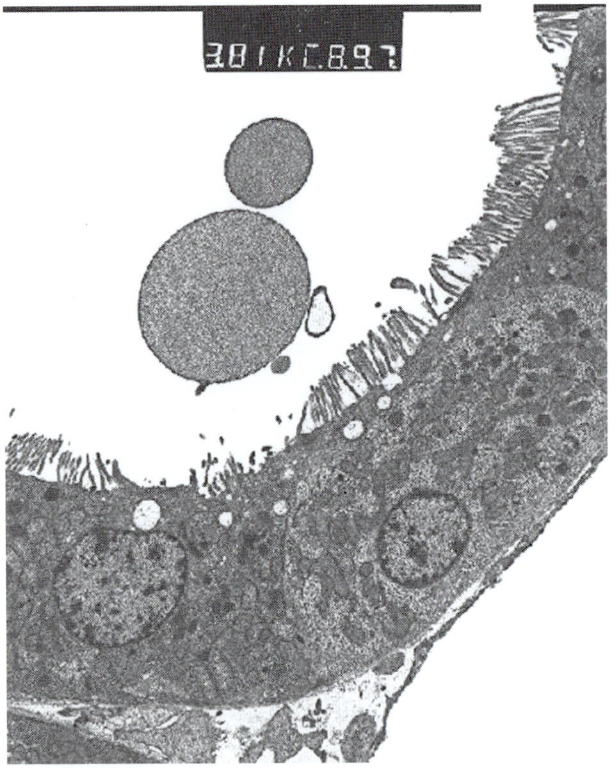

FIGURE 12.16 Transmission electron micrograph of a proximal renal tubule from the kidney of the desert quail showing a urate containing sphere sectioned. Note the absence of urate crystals. *Adapted from Braun (unpublished).*

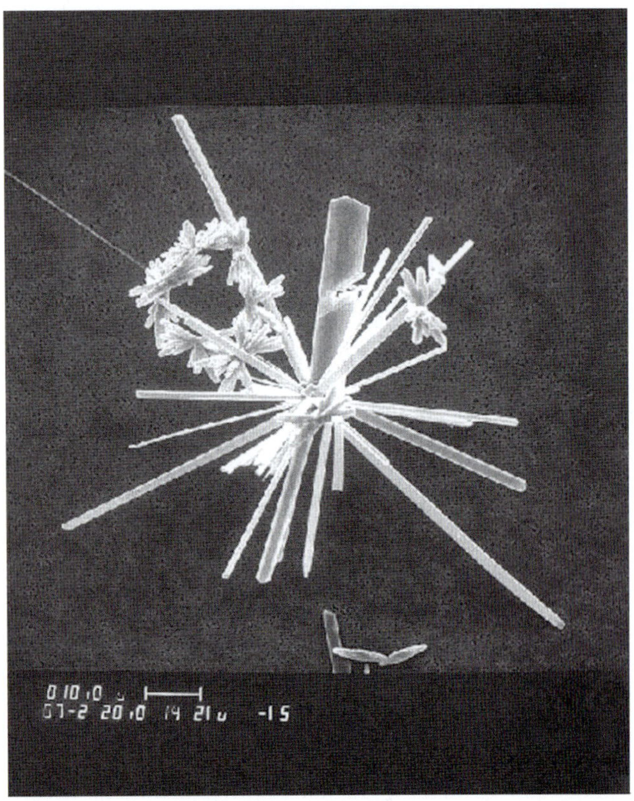

FIGURE 12.17 Scanning electron micrograph of urate crystals. Crystals of this type are not seen in the ureteral urine of birds in spite of there being high a concentration of urate in avian ureteral urine. (Refer back to Figure 12.16). *Adapted from Braun (unpublished).*

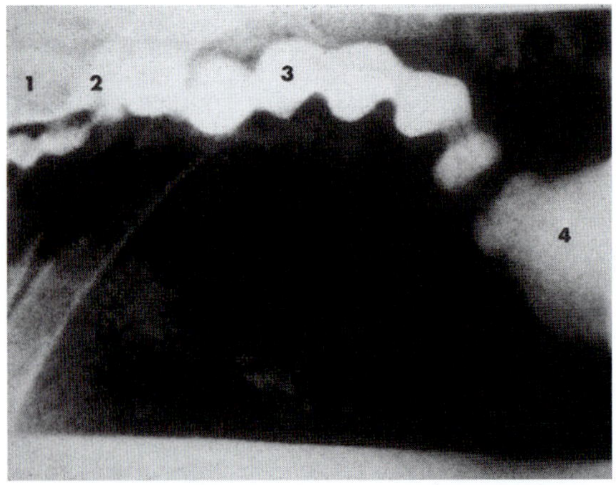

FIGURE 12.18 Radiograph of the lower intestine of a domestic fowl demonstrating the pattern of retrograde peristalsis. The cloaca is at the right (4) and the "rings" (1–3) leading from the cloaca show smooth muscle contractions producing the retrograde peristalsis. *Adapted from Brummermann and Braun (1995).*

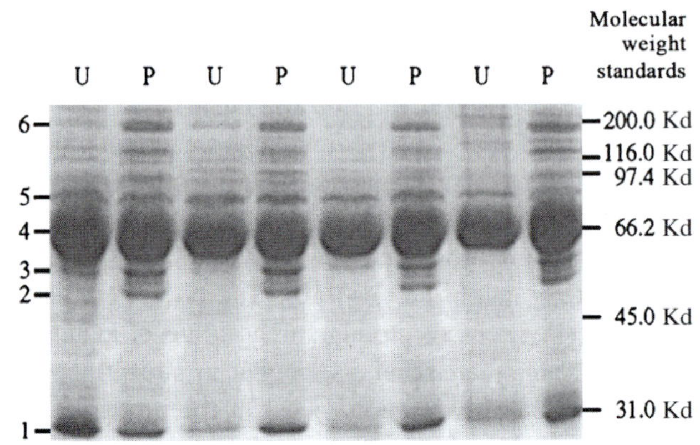

FIGURE 12.19 SDS-PAGE gel of ureteral urine and plasma samples from a domestic fowl. Note the similarity of the bands in the urine and plasma lanes. The band at 65 Kd is serum albumin. *Adapted from Janes and Braun (1997).*

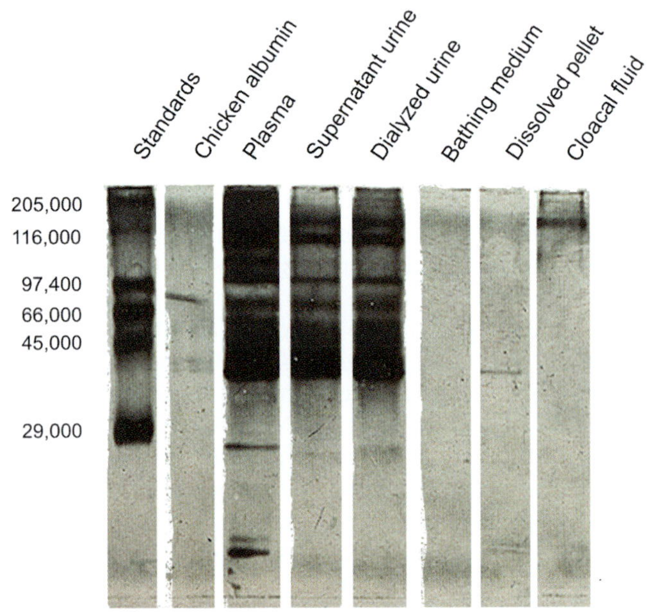

FIGURE 12.20 SDS-PAGE gel of plasma and excreted urine from a domestic fowl. Note the lack of bands in the lane containing the excreted urine. *Modified from Janes and Braun (1997).*

degraded to amino acids, di- and tri-peptides, and bacteria to produce short-chain volatile fatty acids that are absorbed by the colonic epithelium. The fatty acids serve as an energy source to enhance sodium transport and the other organics serve as substrates for the sodium/glucose cotransporter (Rice and Skadhauge, 1982). Thus, the large amount of protein (5 mg/mL) in ureteral urine is not excreted and lost by the animal. Not only is the free protein in the urine degraded, but also that binding urate in formation of the small spheres.

The uric acid molecule is also degraded by the bacteria within the lower gastrointestinal tract; among the products formed are ammonia and glutamic acid. Studies with radiolabeled glutamine have shown that it appears in the blood of the coccygeal mesenteric vein (Karasawa and Maeda, 1992). This vein enters the renal portal system, which conveys blood to the peritubular surfaces of the renal tubules. At this point, the glutamine can be transported by the basolateral membrane of cells, where the molecule can be de-ammoniated. The ammonia can be secreted into the tubule lumen to bind hydrogen ions to form neutral ammonia salts to be excreted. Thus, what was to be an excretory product can be recycled to function in acid/base balance. The recycling of the nitrogen of uric acid is very important in those birds, such as nectarivores and frugivores, which have diets low in nitrogen content.

12.3 THE AVIAN LOWER GASTROINTESTINAL TRACT

The lower gastrointestinal tract of birds (cloaca, colon, and digestive ceca) plays an important role in avian osmoregulation, although this role is variable depending on taxonomic group. As described earlier, urine produced by the kidneys enters the cloaca—specifically, the urodeum of the cloaca. The fluid is moved by a retrograde peristalsis into the coprodeum of the cloaca and farther into the colon and digestive ceca (Figure 12.18). The fluid is moved over and around a central fecal core and brought in close contact with the epithelium of the lower gastrointestinal tract. Brummermann and Braun (1995), using domestic fowl as experimental animals, demonstrated that the control of the retrograde peristalsis resides in the lower gastrointestinal tract and not with central osmoreceptors located in the hypothalamus. These experiments showed that the tonicity of the fluid entering the cloaca is monitored. If its tonicity is more than 200 mOsm/kg H_2O higher than that of the plasma, the retrograde peristalsis is slowed or stopped entirely. The *in vivo* perfusion of the colon of domestic fowl with hyperosmotic NaCl solutions decreased absorption by the colon, supporting the observation that motility was stopped by hyperosmotic solutions in the cloaca (Skadhauge and Kristensen, 1972).

Recent data generated by Vranish and Braun (2011) indicate that the element sensing the osmolality of the ureteral urine in the cloaca is a member of the transient receptor potential (TRP) family, a vanilloid type receptor; TRPV4 a moderately selective Ca^{2+}channel (Clapham et al., 2002). Vanilloid receptors are known to respond to noxious stimuli such as heat, pain, and extreme pH (Clapham et al., 2002). TRPV4 is unique in that it is the only vanilloid receptor shown to respond to deformation, such as what occurs when cells are exposed to changes in their osmotic environment (Clapham et al., 2002). TRPV4 has a basal level of inward Ca^{2+} current in isotonic conditions. When placed in hypotonic media, Ca^{2+} influx increases and a hypertonic environment leads to slowed or completely halted Ca^{2+} current (Plant and Strotmann, 2007). This protein is expressed in many mammalian tissues, including thick ascending limb and distal convoluted tubules of nephrons, alveolar epithelia, mechanosenstive neurons in the ear and eye, and the gastrointestinal tract (Brierley et al., 2008; Cohen, 2007; Holzer, 2007; Liedtke and Simon, 2004; Plant and Strotmann, 2007). Although the signaling mechanism of this channel is not fully understood, its transduction of osmolarity information—and in most cases, initiation of local responses—make it a good candidate for the sensing element in the cloaca of birds.

The transport activity of the avian colonic epithelium is modulated by several hormones. Data show that dietary salt intake affects sodium transport by electrogenic sodium channels (ENaC) in the lower intestine of the domestic fowl (Laverty and Skadhauge, 1999; Laverty et al., 2006). A low-sodium diet increases the plasma aldosterone levels, leading to a high rate of electrogenic sodium absorption in both coprodeum of the cloaca and colon. The effect in the coprodeum is marked because when aldosterone plasma concentrations are low there is little sodium transport. However, the effects on the colon are more complex: on a high-sodium diet, sodium absorption is still present but is mediated by sodium coupled co-transport with organic substrates, such as amino acids and glucose and not by EnaC channel activity. This transporter appears to be a member of the SGLT family of epithelial transporters (Laverty et al., 2001). The organic molecules are formed from the bacterial breakdown of protein that enters the lower gastrointestinal tract as a component of the urine. The continued transport of sodium facilitates the recovery of urinary water. Thus, the colonic co-transporters play an important physiological role not only in osmoregulation, but also in nutrition by assimilation of substrates derived from bacterial fermentation of protein and uric acid in the lower intestine.

A number of avian taxonomic groups (Pelecaniformes, Ciconiiformes, Falconiformes, Columbiformes, and Passeriformes) have small ceca that have been referred to as rudimentary or nonfunctional (Clench, 1999). However, a study of the house sparrow showed that these ceca have a very complex internal structure. There is a central channel with numerous side branches, giving them a bottle-brush appearance (Figure 12.21; Reyes and Braun, 2005). These channels are

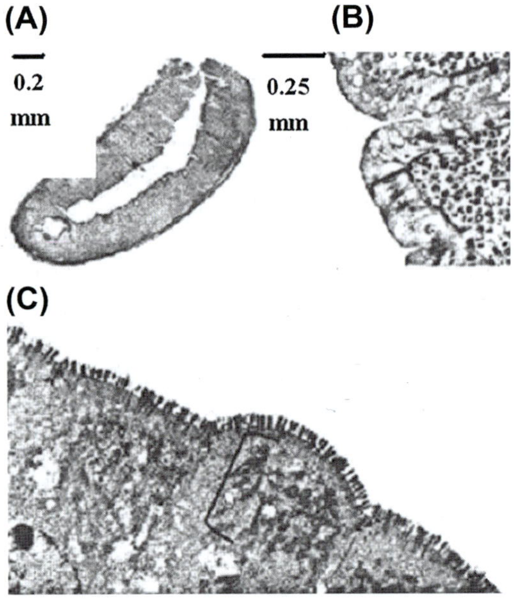

FIGURE 12.21 **Histological preparation of a House Sparrow cecum.** (A) At the upper left is a sagittal section of the whole structure showing a central channel. Cytological detail is depicted at two higher magnifications (B and C). Note the dense brush border (BB) and cells with a dense population of mitochondria highlighted in bracket. TJ, tight junctions. The cytological detail suggests that these small structures are not rudimentary. Moreover, the cells show high level ATPase activity. *Adapted from Reyes and Braun (2005).*

TABLE 12.5 Solubility of Nitrogen-Containing Compounds

Compound	Solubility (mmol/L)
Uric acid	0.381
Ammonium urate	3.21
Sodium urate	8.32
Potassium urate	14.75
Urea	16,650
Ammonia	∞

all lined with epithelia that have a very dense brush border and the cells show a very dense presence of mitochondria. This complex architecture strongly suggests that the small ceca are not rudimentary but may play an osmoregulatory role.

12.4 THE AVIAN SALT GLAND

As previously discussed, the avian kidney has a limited capacity to excrete solutes in excess of water—that is, to produce urine that is significantly hyperosmotic to plasma. Those birds that have a limited intake of freshwater do not rely entirely on the kidney to eliminate excess ions (primarily sodium) in the process of maintaining the homeostasis of the extracellular fluid. The alternate route for ion regulation is salt glands that can excrete very concentrated sodium chloride solutions. The location in the head of the birds can differ slightly: some of the glands are within the orbit of the eye, whereas others are located superior to the orbital and are commonly referred to as nasal salt glands (Peaker and Linzell, 1975). As presented in Table 12.5, salt glands have been identified in at least 15 orders of birds from a range of habitats where fresh drinking water is limited or the intake of high concentrations of electrolytes is part of the normal diet, such as carnivores or birds from desert regions of the world.

12.4.1 Structure of the Salt Glands

The anatomy of salt glands is complex, but it follows a pattern of parallel arrangement of epithelial tubules and blood vessels, which permits counter direction of fluid flow as is present within the renal medullae of birds and mammals (Figure 12.22; Gerstberger and Gray, 1993). The glands are formed from a series of secretory tubules that coalesce like the branches of a tree. The tubules begin as blind end ducts, the cells of which display the features of a strong secretory epithelium—numerous basolateral infoldings and large

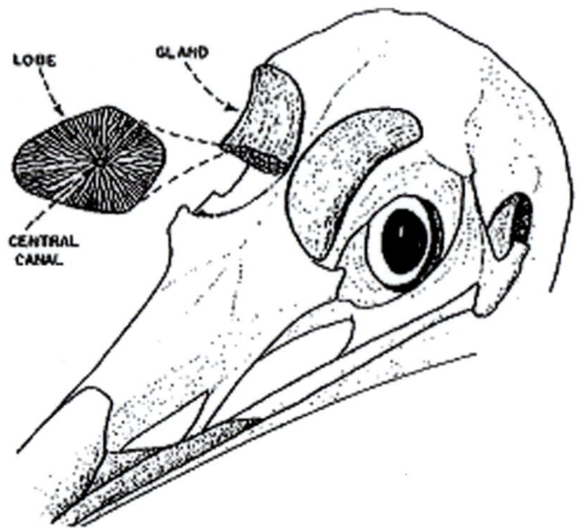

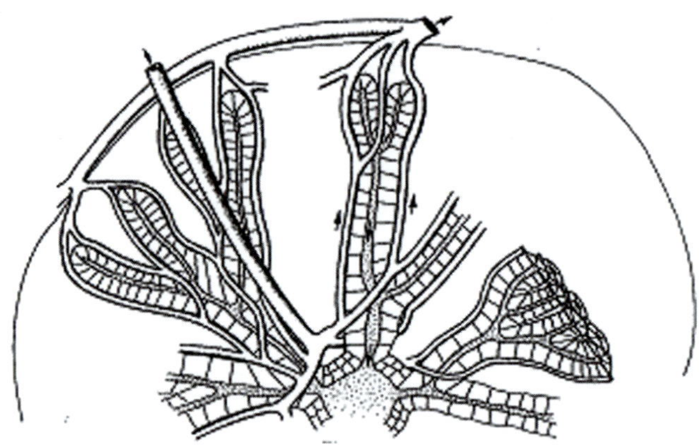

FIGURE 12.22 **Upper illustration shows the location of the salt secreting gland of some birds and the lower illustration shows the counter flow pattern of blood vessels and secretory tubules.** Arrows show direction of countercurrent flow in the secretory tubules and blood vessels. *Modified from Schmidt-Nielsen (1960).*

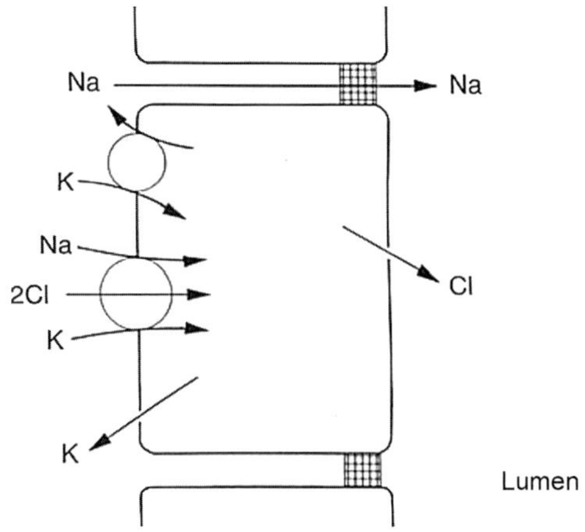

FIGURE 12.23 **The suggested mechanism by which salt glands produce a secretion of concentrated sodium chloride.** The process is driven by the Na/K 2Cl transporter. Chloride is extruded across the apical cell membrane with sodium moving passively through the paracellular channel. *From Shuttleworth (1995).*

numbers of mitochondria. The tubules coalesce to form two large ducts, which in most birds drain into the nasal cavity.

12.4.2 Product of Salt Gland Secretion

The salt glands produce a hyperosmotic fluid, such that the ions (primarily sodium chloride) are secreted with less water than that in which they were originally consumed. This leads to osmotically free water that is retained in the extracellular fluid. The proposed mechanism of ion secretion by the cells of the salt glands is illustrated in Figure 12.23 (Shuttleworth, 1995). The process is driven by gradients produced by the Na^+/K^+ exchange transporter on the basolateral side of the cells. The typical Na^+/K^+2Cl^- mechanism on the basolateral membrane moves Cl^- into the cell. The key process is the secondary active transport of Cl^- across the apical membrane of the cells, with Na moving passively through intercellular spaces. The net result of the process is the secretion of a fluid high in NaCl concentration into the lumen of the salt gland tubule.

12.4.3 Control of Salt Gland Secretion

Increases in the osmolality and volume of the extracellular fluid are stimuli that initiate or drive secretion by the cells of the salt gland ducts. These are due primarily to increases in sodium chloride consumption, which can in turn cause expansion of the extracellular fluid by drawing water from the intracellular compartment (Ash et al., 1969; Fänge et al., 1958). The change in extracellular fluid is reflected in changes in plasma composition, which can be detected by the central osmoreceptors. Cholinergic fibers running in the secretory nerve reach not only the tubules of the salt gland but also blood vessels running parallel to the tubules (Babonis and Brischoux, 2012). This leads to enhanced secretory activity and increased blood flow to facilitate secretion by delivering substrates to the tubules. Hormones appear not to initiate salt gland secretion, but they may play an important role in modulating the rate of secretion. For example, angiotensin II has been shown to inhibit secretion, whereas atrial natriuretic peptide has a positive effect on the gland (Gray et al., 1997).

With a kidney that does not produce large amounts free water by elaborating urine significantly more concentrated than the plasma, the capacity of salt glands to excrete excess osmotically active ions allows birds to inhabit a wide range of environments or habitats—some of which have limited sources of freshwater.

REFERENCES

Ash, R.W., Pearce, J.W., Silver, A., 1969. An investigation of the nerve supply to the salt gland of the duck. Quart. J. Exp. Physiol. 54, 281–295.
Babonis, L.S., Brischoux, F., 2012. Perspectives on the convergent evolution of tetrapod salt glands. Integr. Comp. Biol. 1–12.
Barnes, E.M., Impey, C.S., 1970. The isolation and properties of the predominant anaerobic bacteria in the caeca of chickens and turkeys. Br. Poult. Sci. 11, 467–481.
Braun, E.J., 2009. Osmotic and ionic regulation in birds. In: Evans, D.H. (Ed.), Osmotic and Ionic Regulation: Cells and Animals. CRC press, Boca Raton, FL.
Braun, E.J., 1993. Renal function in birds. In: Brown, J.A., Balment, R.J., Rankin, J.C. (Eds.), New Insights Vertebrate Kidney Function. Cambridge University Press, Cambridge, pp. 167–188.
Braun, E.J. 1982. Renal function. Comp. Biochem. Physiol. 71A, 511–517.
Braun, E.J., Dantzler, W.H., 1972. Function of mammalian-type and reptilian type nephrons in kidney of desert quail. Am. J. Physiol. 222 (3), 617–629.
Braun, E.J., Reimer, P.R., 1988. Structure of avian loop of Henle as related to countercurrent multiplier system. Am. J. Physiol. 255, F500–F512.
Brierley, S.M., Page, A.J., Hughes, P.A., Adam, B., Liebregts, T., Cooper, N.J., Holtmann, G., Liedtke, W., Blackshaw, L.A., 2008. A selective role for TRPV4 ion channels in visceral sensory pathways. Gastroenterology 134 (7), 2059–2069.
Brokl, O.H., Braun, E.J., Dantzler, W.H., 1994. Transport of PAH, urate, TEA, and fluid by isolated perfused and nonperfused avian renal proximal tubules. Am. J. Physiol. 266, R1085–R1094.
Brüggeller, P., Mayer, E., 1980. Complete vitrification in pure liquid water and dilute aqueous solutions. Nature 288, 569–571.
Brummermann, M., Braun, E.J., 1995. Effect of water deprivation on colonic motility of white leghorn roosters. Am. J. Physiol. 268, R690–R698.
Burrows, M.E., Braun, E.J., Duckles, S.P., 1983. Avian renal portal valve: a reexamination of its innervation. Am. J. Physiol. 245, H629–H634.
Campbell, C., Braun, E.J., 1986. Cecal degradation of uric acid in Gambel quail. Am. J. Physiol. 251, R59–R62.
Casotti, G., Braun, E.J., 1996. Functional morphology of the glomerular filtration barrier of *Gallus gallus*. J. Morphol. 228, 327–334.
Casotti, G., Lindberg, K.K., Braun, E.J., 2000. Functional morphology of the avian medullary cone. Am. J. Physiol. 279, R1722–R1730.
Clapham, D.E., Montell, C., Schultz, G., Julius, D., 2002. The TRP ion channel family. IUPHAR Compend. (TRP Channels).
Clench, M., 1999. The avian cecum: update and motility review. J. Exp. Zool. 283, 441–447.
Cohen, D.M., 2007. The role of TRPV4 in the kidney. Chapter 29. In: Liedtke, Heller (Eds.), TRP Ion Channel Function in Sensory Transduction and Cellular Signaling Cascades. CRC Press, Boca Raton, FL.

Del Rio, C.M., Schondube, J., Mcwhorter, T.J., Herrera, G., 2001. Intake responses in nectar feeding birds: digestive and metabolic causes, osmoregulatory consequences, and coevolutionary effects. Am. Zool. 41, 902–915.

Dieni, C.A., Storey, K.B., 2010. Regulation of glucose-6-phosphate dehydrogenase by reversible phosphorylation in liver of a freeze tolerant frog. J. Comp. Physiol. B 180, 1133–1142.

Dudas, P.L., Pelis, R.M., Braun, E.J., Renfro, J.L., 2005. Transepithelial urate transport by avian renal proximal tubule epithelium in primary culture. J. Exp. Biol. 208, 4305–4315.

Fänge, R., Schmidt-Nielsen, K., Robinson, M., 1958. Control of secretion from the avian salt gland. Am. J. Physiol. 195, 321–326.

Gerstberger, R., Gray, D.A., 1993. Fine structure, innervation, and functional control of avian salt glands. Int. Rev. Cytol. 144, 129–215.

Goldstein, D.L., Williams, J.B., Braun, E.J., 1990. Osmoregulation in the field by salt-marsh savannah sparrows (*Passerculus sandwichensis beldingi*). Physiol. Zool. 63, 669–682.

Gray, D., Downing, C., Sayed, N., 1997. Endogenous plasma atrial natriuretic peptide and the control of salt gland function in the Pekin duck. Am. J. Physiol. 273, R1080–R1085.

Hodges, R.D., 1974. The Histology of the Fowl. Academic Press, New York.

Holzer, P., 2007. Taste receptors in the gastrointestinal tract. Am. J. Physiol. Gastrointest. Liver Physiol. 292, 699–705.

Janes, D.N., Braun, E.J., 1997. Urinary protein excretion in red jungle fowl (*Gallus gallus*). Comp. Biochem. Physiol. A Physiol. 118 (4), 1273–1275.

Johnson, O.W., 1968. Some morphological features of avian kidneys. Auk 85, 216–228.

Johnson, O.W., Mugaas, J.N., 1970. Quantitative and organizational features of the avian renal medulla. Condor 72, 288–292.

Karasawa, Y., Maeda, M., 1992. Effect of colostomy on the utilisation of dietary nitrogen in the fowl fed on a low protein diet. Br. Poult. Sci. 33, 815–820.

Laverty, G., Elbrond, V.S., Arnason, S.S., Skadhauge, E., 2006. Endocrine regulation of ion transport in the avian lower intestine. Gen. Comp. Endocrinol. 147, 70–77.

Laverty, G., Bjarnadottir, S., Elbrond, V.S., Arnason, S.S., 2001. Aldosterone suppresses expression of an avian colonic sodium-glucose cotransporter. Am. J. Physiol. Regul. Integr. Comp. Physiol. 281, R1041–R1050.

Laverty, G., Skadhauge, E., 1999. Physiological roles and regulation of transport activities in the avian lower intestine. J. Exp. Zool. 283, 480–494.

Laverty, G., Alberici, M., 1987. Micropuncture study of proximal tubule pH in avian kidney. Am. J. Physiol. 253, R587–R591.

Laverty, G., Dantzler, W.H., 1982. Micropuncture of superficial nephrons in avian (*Sturnus vulgaris*) kidney. Am. J. Physiol. 243, F561.

Laverty, G., Dantzler, W.H., 1983. Micropuncture study of urate transport by superficial nephrons in avian (*Sturnus vulgaris*) kidney. Pflugers Arch. 397, 232–236.

Layton, H.E., Davis, J.M., Casotti, G., Braun, E.J., 2000. Mathematical model of an avian urine concentrating mechanism. Am. J. Physiol. Ren. Physiol. 279, F1139–F1160.

Lewis, S.A., 1977. A reinvestigation of the function of the mammalian urinary bladder. Am. J. Physiol. 232, F187–F195.

Liedtke, W., Simon, S.A., 2004. A possible role for TRPV4 receptors in asthma. Am. J. Physiol. Lung Cell Mol. Physiol. 287, L269–L271.

Morgan, C., Braun, E.J., 2001. Glucose handling by the kidney of the domestic fowl. 2001. FASEB J. 15, A854.

Nishimura, H., 2008. Urine concentration and avian aquaporin water channels. Pflugers Arch. 456, 755–768.

Nishimura, H., Imai, M., Ogawa, J., 1986. Diluting segment in avian kidney. I. Characterization of trans-epithelial voltages. Am. J. Physiol. 250, R333.

Patten, B.M., 1951. Early Embryology of the Chick. McGraw-Hill, New York.

Peaker, M., Linzell, J.L., 1975. Salt Glands in Birds and Reptiles. Cambridge University Press, Cambridge.

Plant, T.D., Strotmann, R., 2007. TRPV4: a multifunctional nonselective cation channel with complex regulation. Chapter 9. In: Liedtke, Heller (Eds.), TRP Ion Channel Function in Sensory Transduction and Cellular Signaling Cascades. CRC Press, Boca Raton, FL.

Poulson, T.L., Bartholomew, G.A., 1962. Salt balance in the savannah sparrow. Physiol. Zool. 35, 109–119.

Preest, M.R., Beuchat, C.A., 1997. Ammonia excretion by hummingbirds. Nature 386, 561–562.

Reyes, L., Braun, E.J., 2005. The functional morphology of the English sparrow cecum. Comp. Biochem. Physiol. 141, 292–297.

Rice, G.E., Skadhauge, E., 1982. Colonic and coprodeal transepithelial transport parameters in NaCl-loaded domestic fowl. J. Comp. Physiol. 147, 65–69.

Schmidt-Nielsen, K., 1960. The salt-secreting gland of marine birds. Circulation 21, 955–967.

Shuttleworth, T.J., 1995. Intracellular signals controlling ionic and acid-base regulation in avian nasal gland cells. Adv. Comp. Environ. Physiol. 22, 185–206.

Skadhauge, E., Kristensen, K., 1972. An analogue computer simulation of cloacal resorption of salt and water from ureteral urine in birds. J. Theor. Biol. 35, 473–487.

Skadhauge, E., Schmidt-Nielsen, B., 1967. Renal medullary electrolyte and urea gradient in chickens and turkeys. Am. J. Physiol. 212, 1313–1318.

Skotheim, J.M., Secomb, T.W., 2007. Red blood cells and other nonspherical capsules in shear flow: oscillatory dynamics and the tank-to-tumbling transition. Phys. Rev. Lett. 98, 1–4.

Somero, G.N., 1986. From dogfish to dog: trimethylamines protect proteins from urea. NIPS 1, 9–12.

Sperber, I., 1944. Studies on the Mammalian Kidney, 22, Zool. Bidrig, Uppsala. pp. 249–432.

Stallone, J.N., Braun, E.J., 1985. Contributions of glomerular and tubular mechanisms to antidiuresis in conscious domestic fowl. Am. J. Physiol. 249, F842.

Sutterlin, G.G., Laverty, G., 1998. Characterization of a primary cell culture model of the avian renal proximal tubule. Am. J. Physiol. 275, R220–R226.

Tsahar, E., Del Rio, C.M., Izhaki, I., Arad, Z., 2005. Can birds be ammonotelic? Nitrogen balance and excretion in two frugivores. J. Exp. Biol. 208, 1025–1034.

Vranish, J.R., Braun, E.J., March 2011. Isolation of a putative osmoreceptor from the avian GI tract. FASEB J. 25. 1047.3.

Wideman Jr, R.F., 1988. Avian kidney anatomy and physiology. In: CRC Critical Reviews in Poultry Biology, vol. 1, CRC Press, Boca Raton, FL, pp. 133–176.

Wideman Jr, R.F., 1991. Autoregulation of avian renal plasma flow: contribution of the renal portal system. J. Comp. Physiol. B 160, 663–669.

Yancey, P.H., Somero, G.N., 1979. Counteraction of urea destabilization of protein structure by methylamine osmoregulation compounds of elasmobranch fishes. Biochem. J. 183, 317–323.

Yokota, S.D., Benyajati, S., Dantzler, W.H., 1985. Comparative aspects of glomerular filtration in vertebrates. Ren. Physiol. 8, 193–221.

Zhang, J., Storey, K., 2012. Cell cycle regulation in the freeze tolerant wood frog (*Rana sylvatica*). Cell Cycle 11 (9), 1727–1742.

Chapter 13

Respiration

Frank L. Powell
Division of Physiology, Department of Medicine, University of California, San Diego, CA, USA

13.1 OVERVIEW

The primary function of the respiratory system is *gas exchange*—delivering oxygen from the environment to the tissues, and removing carbon dioxide from the tissues. Maintaining a constant body temperature in birds and mammals requires high levels of oxygen consumption and exercise in birds—namely flapping flight—creates the highest oxygen demand of any vertebrate. The structure of the avian respiratory system is very different from the mammalian lung and some of these differences support more efficient gas exchange, whereas others may be alternative evolutionary solutions to common problems in air breathing vertebrates.

Generally, the respiratory system acts as a servant to the rest of the organism by delivering enough oxygen and removing sufficient carbon dioxide for metabolic demands. As oxygen demand increases, a variety of respiratory responses ensure an adequate supply of oxygen, which involve the lungs, respiratory mechanics, the pulmonary circulation, transport of oxygen and carbon dioxide in blood, pulmonary and tissue gas exchange, and the coordination of all these mechanisms by the respiratory control system. Individual sections in this chapter focus on each of these physiological mechanisms. References are made to mammalian respiratory physiology, so the reader can consult the extensive literature available for more details on fundamentals concepts. In birds, the respiratory system is also critical for thermoregulation (by evaporative water loss) and nonrespiratory functions such as vocalization, but these are not covered in this chapter. Current research problems and important unanswered questions are highlighted in sections labeled *Frontiers*.

13.1.1 Oxygen Cascade

Figure 13.1 shows how these physiological transport steps function in series to transport oxygen from the environment to the cells. This is often referred to as the "oxygen cascade" because oxygen level (quantified as partial pressure, or P_{O_2}), decreases at each step in the model. Breathing movements bring fresh air into the lungs, and the heart pumps oxygen-poor blood to the lungs. Oxygen diffuses from gas to blood in the lungs, and this oxygen-rich blood returns to the heart via the pulmonary circulation. Arterialized blood is pumped to the various organs and tissues of the body via the systemic circulation. Finally, oxygen diffuses out of the systemic capillaries to metabolizing tissues and ultimately to the mitochondria inside cells. Carbon dioxide moves out of the cells to the environment through these same steps in the opposite direction from oxygen. Each of these steps is covered in the sections below, with an emphasis on respiratory structure–function relationships that are unique to birds, especially in comparison with mammals. Mammals are the only other vertebrates that achieve avian levels of oxygen demand during thermoregulation and activity.

13.1.2 Symbols and Units

Table 13.1 provides a list of abbreviations used in this chapter, which are based on a few simple conventions. Primary variables are symbolized with a capital letter, and a dot over the variable indicates the first derivative with respect to time (e.g., units for inspired ventilation, $\dot{V}_I$, are L/min). Modifiers are small capitals for the gas phase and lowercase letters for liquid or tissues. Finally, a specific gas species is indicated with a subscript. Respiratory gas volumes (e.g., ventilation) are reported for physiological conditions (body temperature and pressure, saturated) unless otherwise noted. Amounts of oxygen (O_2) or carbon dioxide (CO_2) are expressed in mmoles (e.g., O_2 concentration in mmol/L = mM). Pressure is expressed in Torr (7.5 Torr = 1 kPa).

13.2 ANATOMY OF THE AVIAN RESPIRATORY SYSTEM

The structure of the avian respiratory system is unique amongst the vertebrates, with small lungs that do not change volume during breathing, and nine large air sacs that act as bellows to ventilate the lung but do not participate directly in gas exchange (Figure 13.2). The total volume of the

Sturkie's Avian Physiology.

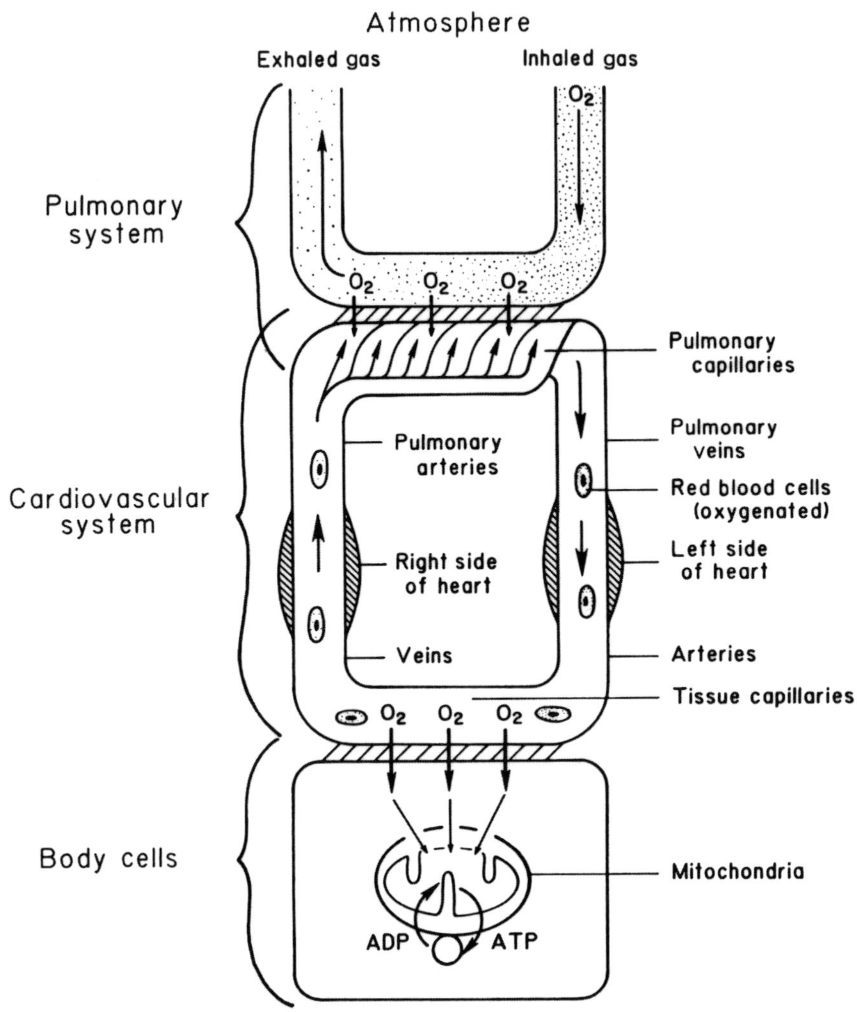

FIGURE 13.1 General model of the oxygen transport in birds. *Modified from Taylor and Weibel (1981).*

respiratory system in a bird (i.e., lungs and air sacs) is larger than that in a comparably sized mammal (~15% versus 7% of body volume), but the avian lung itself is smaller (~1–3% of body volume). Apparently, during evolution, birds segregated the functions of gas exchange and ventilation as the respiratory organ was subdivided into smaller functional units to increase gas exchange surface area. Such heterogeneous partitioning of the respiratory organ contrasts with the homogenous partitioning in mammals (Duncker, 1978). Alveoli in mammalian lungs perform both respiratory functions of ventilation and gas exchange.

Also in contrast to mammals, the avian thoracic cavity is essentially at atmospheric pressure (versus subatmospheric), and there is no diaphragm to functionally separate it from the abdominal cavity (see Section 13.3). This section covers the basic respiratory system anatomy necessary to understand respiratory function, but the reader is referred to several excellent monographs and reviews for more details (Duncker, 1971; King and Molony, 1971; Maina et al., 2010; McLelland, 1989b). Terminology according to the *Nomina Anatomica Avium* is used here (King, 1979). Details of the gas exchange surface and the anatomy of the pulmonary circulation are covered in later sections (Sections 13.4 and 13.6.3, respectively).

13.2.1 Upper Airways

Birds can breathe through the nares or mouth. Oronasal structures tend to heat and humidify inspired gas, and filter out large particles that could potentially damage the delicate respiratory surfaces. The oronasal cavity is separated from the trachea by the larynx, which opens into the trachea through the slit-like glottis. The laryngeal muscles contract with breathing, to open the glottis during inspiration and decrease the resistance to inspiratory airflow. This rhythmic opening of the glottis is useful when attempting to intubate a bird. The trachea has complete cartilaginous rings in most avian species and plentiful smooth muscle. Interesting exceptions include the "double trachea" of penguins, with a medial septum dividing the trachea into two

TABLE 13.1 Symbols in Respiratory Physiology

	Primary Variables (and Units)
C	Concentration or content (mM = mmol/L)
D	Diffusing capacity ($mmol_{O_2}/min\ Torr$)
P	Partial pressure or hydrostatic pressure (Torr or cm H_2O)
V	Gas volume (liters, L, or mL)
$\dot{V}$	Ventilation (L/min)
$\dot{Q}$	Blood flow or perfusion (L/min)
$\dot{M}$	Gas flux (mmol/min)
Modifying Symbols	
D	Dead space gas
E	Expired gas
$\bar{E}$	Mixed-expired gas
I	Inspired gas
T	Tidal gas
a	Arterial blood
c	Capillary blood
m	Membrane
t	Tissue
v	Venous blood
$\bar{v}$	Mixed-venous blood
Examples	
$P_{I_{O_2}}$	Partial pressure of O_2 in inspired gas
$P_{a_{O_2}}$	Partial pressure of O_2 in arterial blood
$P_{\bar{v}_{O_2}}$	Partial pressure of O_2 in mixed-venous blood
$\dot{M}_{O_2}$	O_2 consumption per unit time
$\dot{V}_P$	Ventilation of the parabronchi per unit time

tubes, and the slit-like opening on the ventral surface of the trachea in emus, which is responsible for their characteristic booming call (McLelland, 1989a). The anatomy and physiology of the larynx and trachea have been reviewed in detail (McLelland, 1989a).

Tracheal volume is an important determinant of "dead space" ventilation and therefore gas exchange. Hinds and Calder (1971) measured tracheal volume in 27 species of birds and found in situ volume (V in mL) was related to body mass (M_b in kg) as: $V = 3.7\,M_b^{1.09}$. This equation

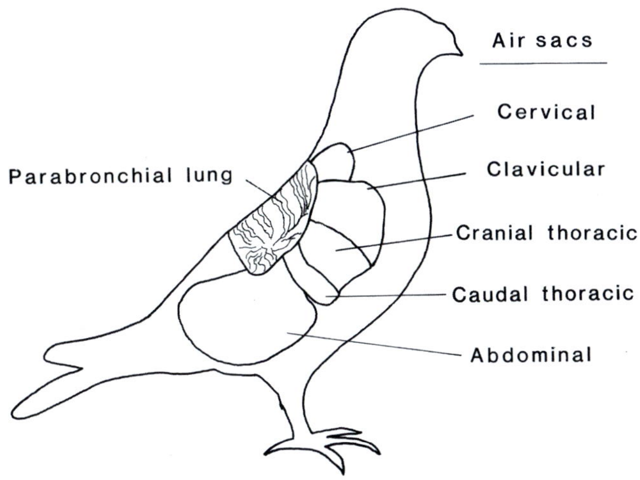

FIGURE 13.2 Respiratory system of a pigeon consisting of the parabronchial lung and air sacs.

underestimates tracheal volume in Pekin ducks (Bech et al., 1984; Hastings and Powell, 1986b) and pigeons (Powell, 1983b), and overestimates the value for male chickens (Kuhlmann and Fedde, 1976), but the error is less than 25%. Tracheal volume is 4.5 times larger in birds than in comparably sized mammals (Hinds and Calder, 1971) and birds generally compensate for this increased dead space with a deep and slow breathing pattern (Bouverot, 1978). Several species of birds possess tracheal elongations, which loop inside the neck, but their function is unknown (McLelland, 1989a).

The trachea bifurcates into two primary bronchi at the syrinx. In many species (e.g., chickens, ducks), but not all, this bifurcation occurs inside the thoracic cavity where the trachea runs through the clavicular air sac. The syrinx is responsible for vocalization in birds but relatively little is known about the precise mechanisms. Readers are referred to comprehensive reviews about this structure for details (King, 1989). Similar to the avian trachea, the syrinx shows considerable variation between species, and males in some species exhibit large bullae with unknown functions.

13.2.2 Lungs

The avian lung is located dorsally in the thoracoabdominal cavity of birds (Figure 13.2), with invaginations from spinal processes typically on the dorsal surface of the lung. Figure 13.3 shows the pulmonary bronchial branching pattern in a representative species.

13.2.2.1 Conducting Airways

In most cases, the extrapulmonary primary bronchi, between the syrinx and the lungs, are relatively short. The intrapulmonary primary bronchus travels through the entire length of the lung, entering on the medioventral aspect and

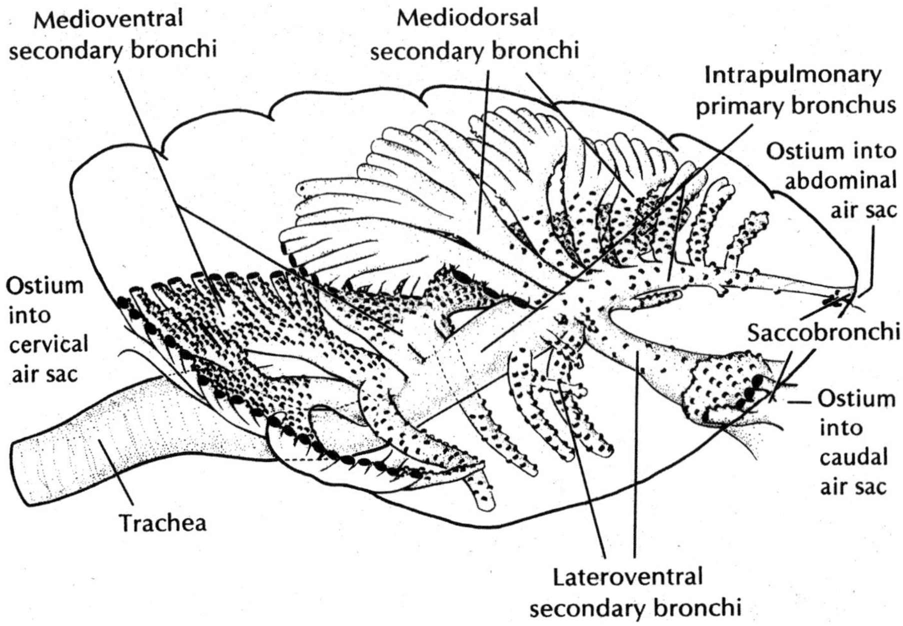

FIGURE 13.3 **Bronchial arrangement in the left lung of the mute swan (*Cygnus olor*).** *After Duncker (1971).*

exiting at the caudal border of the lung into the ostium of the abdominal air sac (see Section 13.2.3). The secondary bronchi can be considered in two functional groups based on their origin from the primary bronchus. The cranial group consists of four or five medioventral secondary bronchi, originating from the medioventral intrapulmonary primary bronchus. These cranial secondary bronchi branch further to form a fan covering the medioventral surface of the lung. The caudal group consists of 6–10 mediodorsal secondary bronchi, which also branch to form a fan over the mediodorsal surface of the lung.

A third group of secondary bronchi includes a variable number of lateroventral bronchi in most species, which also branch off caudal parts of the primary bronchus. The first or second laterobronchus forms a short connection to the posterior thoracic air sac (see below). Other lateroventral bronchi may penetrate lateroventral parts of the lung to variable degrees in different species, but they do not form a regular branching fan like the other secondary bronchi.

The primary and secondary bronchi are conducting airways because they do not participate in gas exchange. Cartilaginous semirings and smooth muscle support the primary bronchi, but the walls of the secondary bronchi are flaccid and require adhesion to the surrounding lung or pleura to remain open. The respiratory epithelium is ciliated with variable amounts of goblet cells in different species in the trachea, primary, and secondary bronchi (Duncker, 1974).

13.2.2.2 Parabronchi

Parabronchi are the functional unit of gas exchange in the avian lung. They are also called "tertiary bronchi" because they can originate from the secondary bronchi, but parabronchus is the preferred terminology because they also originate from further branches of secondary bronchi (Figure 13.3). Most of the parabronchi are organized as a parallel series of several hundred tubes connecting the medioventral and mediodorsal secondary bronchi (Figure 13.3). Such parabronchi are called *paleopulmonic* parabronchi. Together with the primary and cranial and caudal groups of secondary bronchi, they comprise the simplest scheme of bronchial branching in the avian lung (Duncker, 1972, 1974).

In all birds except some penguins, there are additional parabronchi called *neopulmonic* parabronchi (Duncker, 1972, 1974). These parabronchi are not organized as regular parallel stacks of tubes but may exhibit irregular branching patterns. Neopulmonic parabronchi may connect another set of caudal laterodorsal secondary bronchi to caudal air sacs (see Section 13.2.3), or other parabronchi. Neopulmonic parabronchi never comprise more than 25% of the parabronchi, and there are large species variations. However, it has not been possible to demonstrate any phylogenetic or evolutionary significance to neopulmonic versus paleopulmonic parabronchi, despite the implication of the terms (Maina, 1989). The functional significance of these different kinds of parabronchi is considered below (see Section 13.3.3.2).

Figure 13.4 shows the detailed structure of a parabronchus and how they are organized in parallel like a honeycomb in the lung. Gas exchange occurs in the walls of these tubes that are 0.5–2.0 mm in diameter, and they can be several millimeters long depending on the size of the bird (Duncker, 1971; Maina, 1989). The parabronchi are

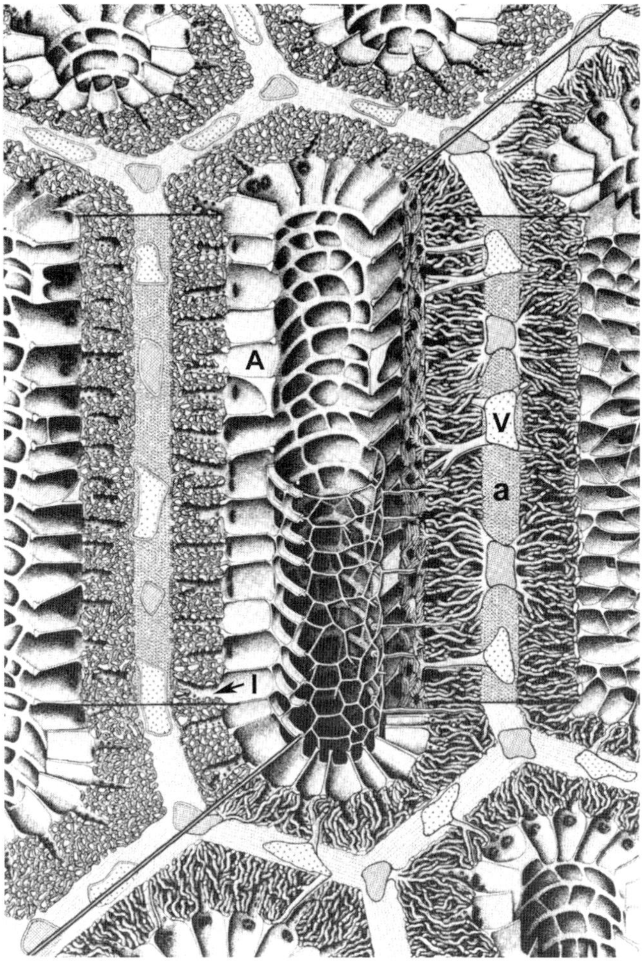

FIGURE 13.4 Cutaway drawing of a typical parabronchus (~1 mm diameter). Left side shows airflow from parabronchial lumen into the atria (A), infudibula (I), and air capillaries. Right side shows blood flow from interparabronchial arterioles (a) into blood capillaries that collect in venules near the base of the atria that flow to interparabronchial venules (v). *Modified from Duncker (1974).*

separated from each other along their length by a boundary of connective tissue and larger pulmonary blood vessels. The parabronchial lumen is lined by a meshwork of connective tissue and smooth muscle, which outlines the entrances to atria radiating from the parabronchial lumen. The atria lead to infundibulae and, ultimately, the air capillaries which are 2–10 μm in diameter and as long as about one-quarter of the total parabronchial diameter (Figure 13.4). The parabronchi are lined with cuboidal and squamous epithelial cells that become thinner moving into the atria and exclusively squamous in the infundibulae and air capillaries (Smith et al., 1986). The air capillaries intertwine with a similar network of pulmonary blood capillaries in the parabronchial mantle, where the air–blood capillary interface is the site of gas exchange.

13.2.2.3 Frontiers: Evolution of the Blood–Gas Barrier

The evolution of respiratory organs has been guided by at least two principles. First, the area of the blood gas barrier has to be very large and very thin to support oxygen uptake by diffusion (see Section 13.6.3). Subdividing the lungs into smaller compartments increases the surface-to-volume ratio, so the total surface area in a bird or mammal is more than 10 times that in a reptile with a simpler lung (Hsia et al., 2013). (Of course, the reptilian solution is adequate for the relatively lower levels of oxygen consumption in a reptile.) The second principle is keeping the blood–gas barrier extremely thin, also to facilitate diffusion, while maintaining structural integrity of the blood capillaries. This is particularly important in birds and mammals, which require high rates of blood flow to meet metabolic demands, and they have evolved separate pulmonary and systemic circulations allowing lower pressures in pulmonary than systemic capillaries. Even so, pressure can increase enough in exercise to cause stress-failure of pulmonary capillaries (West, 2011).

Birds have evolved a unique solution to these challenges, which is arguably more adaptive than the mammalian solution, with the total surface area for gas exchange being about 15% greater and 2.5 times thinner in a bird than in the same-sized mammal (Maina et al., 1989; West, 2009). The air capillaries in the avian lung are much smaller than the alveolar airspaces in mammalian lungs. This means the blood capillaries in the avian lung receive much more support than in an alveolar lung. Blood capillaries in the avian lung are extensively and uniformly supported by the epithelial bridges from the network of air capillaries (Figure 13.5). The exact biomechanics of this system remain to be determined (West et al., 2010), but such small airspaces are possible in the avian lung because they do not change volume with ventilation, in contrast to mammalian alveoli (see Section 13.3.3). Recent studies also show the air capillaries are less tubular and regular in shape than the blood capillaries (Figure 13.5), which may have consequences for gas transport (see Section 13.6.3.1).

13.2.3 Air Sacs

The air sacs are thin membranous structures connected to the primary or secondary bronchi via ostia and they comprise most of the volume of the respiratory system (Figure 13.2). Air sacs are poorly vascularized by the systemic circulation and do not directly participate in significant gas exchange but act as a bellows to ventilate the lungs. In most species, there are nine air sacs that can be considered in cranial and caudal functional groups (Duncker, 1971; Maina, 1989; McLelland, 1989b). Air sac diverticulae may also penetrate the skeleton, but there are large species differences and the

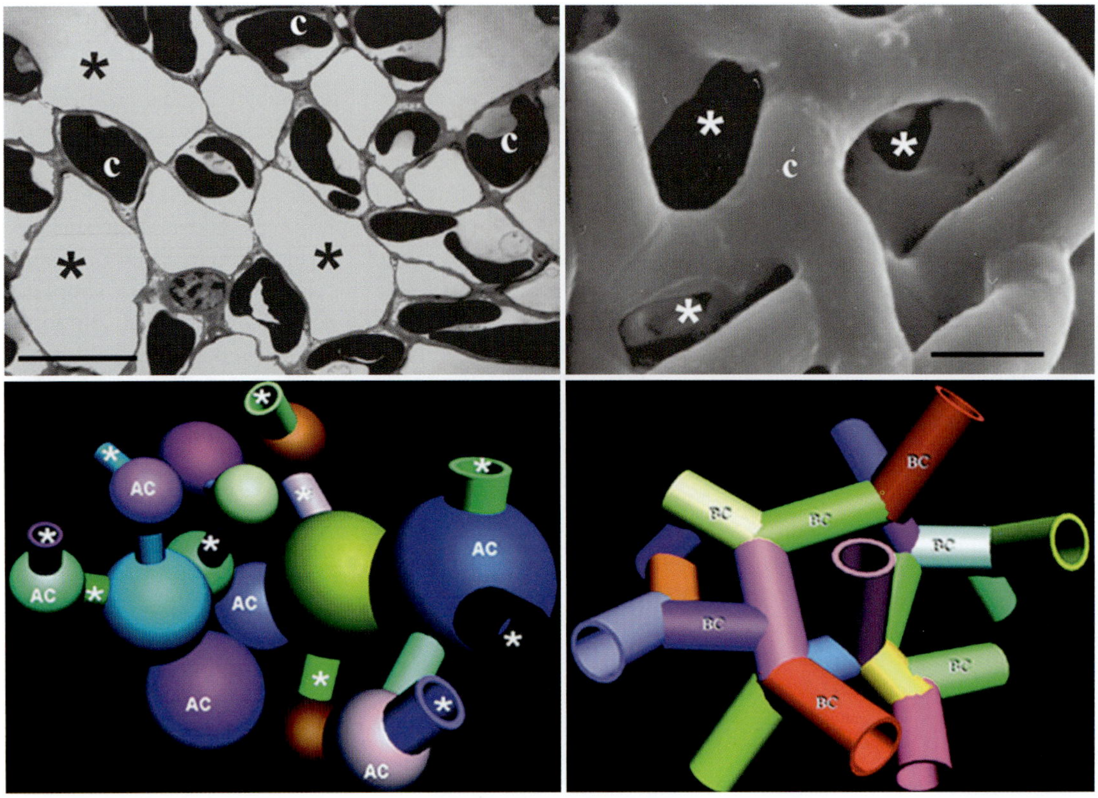

FIGURE 13.5 Top: Air (*)-blood (c) capillary network transmission electron micrograph (left) and scanning electron micrograph of case (right). Scale bar = 10 micron. Bottom: Computer-generated 3-D configuration of the air (left, AC) and blood capillaries (right, BC). AC consist of rotund spaces interconnected by short tubes (*) while BC are short interconnected tubes of similar length and diameter. *Modified from Maina (2007) and Maina et al. (2010).*

functional significance or such connections for respiration has not been established (Maina, 1989).

The cranial group consists of the paired cervical air sacs, the unpaired clavicular air sac, and the paired cranial thoracic air sacs. The cervical sacs directly connect to the first medioventral secondary bronchus. The clavicular air sac directly connects to the third medioventral secondary bronchus, and it may also have indirect connections via parabronchi to other cranial (medioventral) secondary bronchi in some species, such as chickens. The cranial thoracic air sacs generally connect to the third medioventral secondary bronchi, as well as to parabronchi originating from other cranial secondary bronchi in some species.

The caudal group consists of the paired caudal thoracic air sacs and paired abdominal air sacs. The caudal thoracic air sac is directly connected to the lateroventral secondary bronchus and may have indirect connections to other lateroventral, or even cranial (medioventral), secondary bronchi in species with large amounts of neopulmonic parabronchi, such as chickens. The abdominal air sacs connect to the caudal end of the intrapulmonary primary bronchus and may have more indirect connections to parabronchi from laterodorsal secondary bronchi and the last mediodorsal secondary bronchi. Air sac connections with parabronchi are frequently grouped into a funnel-like structure called the saccobronchus.

13.2.4 Respiratory System Volumes

The upper airways and bronchial branches proximal to the parabronchi comprise anatomical dead space, as described in Section 13.2.1. Such conducting airways are called "dead space" because they do not participate directly in gas exchange. The intrapulmonary conducting airways make a relatively small contribution to total dead space, but total dead space volume is generally larger than most mammals, consistent with the generally long neck of birds. Physiological measures of dead space are considered below (see Section 13.3.3.4).

The actual volume of air in the avian lung involved in gas exchange at any moment is in the air capillaries. This is considerably less than the volume in gas exchanging portions of alveolar lungs from comparably sized mammals (Powell and Mazzone, 1983). However, the unique pattern of airflow in the open-ended parabronchi renews this gas exchanging volume more frequently than does tidal ventilation in alveolar lungs (see Section 13.3.3.2). Therefore, birds do not need as large of a functional residual capacity (FRC) in the lungs as mammals to smooth out variations in gas exchange, O_2, and CO_2 levels that could occur during a breathing cycle.

Most of the respiratory system volume in birds is in the air sacs, and there is no comparable volume in the mammalian lung. Unlike mammalian alveoli, which change volume during ventilation, the air sacs are not important sites of gas exchange. There is tremendous variation in the volumes of air sacs reported in the literature, because the value is very sensitive to the method of measurement. For example, the volume of plastic casting material that can be instilled under a pressure head into the air sacs of a dead bird may be much greater than the gas volume in live birds with muscle tone in the thorax and abdominal wall. Also, air sac volume *in vivo* can vary with posture, digestive, and reproductive states, as different structures in the body (e.g., eggs) displace volume. Casting under controlled pressure conditions (Duncker, 1971) and gas dilution *in vivo* (Scheid et al., 1974) are probably the most accurate methods available for determining air sac volume.

13.3 VENTILATION AND RESPIRATORY MECHANICS

Respiratory muscles generate the forces (pressures) to move air in and out of the air sacs and through the parabronchial lung. The air sacs follow changes in body volume with respiratory muscle activity and act as a bellows to ventilate the parabronchial lung that is essentially constant in volume (Jones et al., 1985; Macklem et al., 1979). In contrast to mammals, the avian lung volume is maintained by attachments to the body wall, not by a subatmospheric pressure in an intrapleural space surrounding it. Also in contrast to mammals, birds have no diaphragm separating the body cavity into separate thoracic and abdominal compartments. Hence, pressures are relatively uniform in the avian thoracoabdominal cavity, which behaves mechanically as a single compartment (Scheid and Piiper, 1989).

Ventilation ($\dot{V}$) is the product of the volume per breath, or tidal volume (V_T) and the respiratory frequency (f_R), so $\dot{V}$ can be increased by breathing faster or deeper. The distribution of gas flow in the avian respiratory system depends upon the magnitude and pattern of respiratory muscle activity, as well as the mechanical properties of the body wall, lungs, and air sacs as described below.

13.3.1 Respiratory Muscles

Figure 13.6 shows the changes in thoracic skeleton between normal inspiration and expiration in a bird (King and Molony, 1971; Zimmer, 1935). During inspiration, the sternum rocks cranially and ventrally with the coracoids and furcula rotating at the shoulder. Simultaneously, the vertebral ribs move cranially to expand the sternal ribs and thoraco-abdominal cavity laterally.

In small birds (e.g., starlings) during flight, the furcula (wishbone) and sternum are mechanically coupled, such that the wing beat assists ventilation (Jenkins et al., 1988). However, wing movements in flight and ventilation are coordinated even in larger birds (e.g., geese), so this may involve coupling of neuromuscular circuits as much as respiratory mechanics (Funk et al., 1992a,b). During rest, both inspiration and expiration require active contraction of the respiratory muscles as listed in Table 13.2. The innervation for these muscles is summarized in de Wet et al. (1967). Increases in ventilatory volumes are achieved by recruiting more motor units in active muscles and additional respiratory muscles; during expiration, the opposite occurs (Fedde et al., 1963, 1964b, 1969; Kadono and Okada, 1962; Kadono et al., 1963). Therefore, the relaxed resting volume of the avian respiratory system is midway between inspiratory and expiratory volumes (Seifert, 1896), in contrast to mammals that relax to FRC at end-expiratory volume. The costoseptal muscles control the tension of the horizontal septum covering the ventral surface of the lung, but unlike the mammalian diaphragm, these are not effective at changing lung volume (Fedde et al., 1964a).

13.3.2 Mechanical Properties

13.3.2.1 Compliance

Compliance (C) defines the effectiveness of small pressure changes (ΔP) at inducing volume changes (ΔV):

$$C = \Delta V / \Delta P$$

Because the pressure changes with breathing are essentially uniform throughout the coelom in birds, they can be measured as the difference between pressure in an air sac (P_{AS}) and atmospheric pressure outside the bird. Changes in respiratory system volume (V_{RS}) can be measured with a plethysmograph, which measures changes in whole body volume during breathing, or a pneumotachograph, which quantifies the amount of air inhaled or exhaled at the mouth or trachea. Compliance, measured as the slope of the steepest part of a graph plotting V_{RS} versus P_{AS} in an artificially ventilated bird, ranges from 10 mL/cm H_2O in chickens (Scheid and Piiper, 1969) to 30 mL/cm H_2O in ducks (Gillespie et al., 1982b). These values are similar to compliance in mammals when correcting for body size.

Different results are obtained when compliance is measured by applying small oscillations in volume on spontaneously breathing birds. Compliance measured with this forced oscillation technique is only 7.7 mL/cm H_2O in ducks (Gillespie et al., 1982b), or much less than in mammals. In contrast, compliance measured with this technique in pigeons (2.8 mL/cm H_2O, Kampe and Crawford, 1973) is 3.7 times greater than the value predicted for a similar-sized mammal (Powell, 1983b). The reasons for these differences

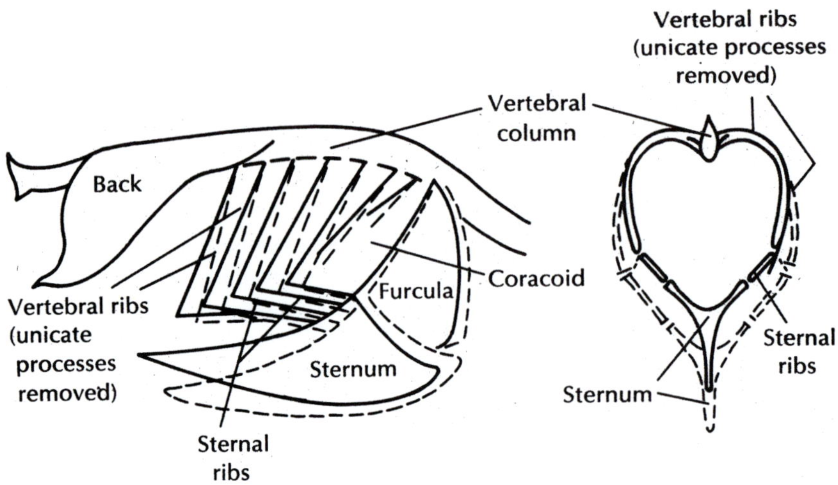

FIGURE 13.6 Changes in the position of the thoracic skeleton during breathing in a standing bird. Solid lines show thoracic position at the end of expiration and dotted lines show the end of inspiration. *After Zimmer (1935).*

TABLE 13.2 Respiratory Muscles of the Chicken

Inspiratory	Expiratory
M. scalenus	Mm. intercostales externi of fifth and sixth spaces
Mm. intercostales externi (except in fifth and sixth spaces)	Mm. intercostales interni of third to sixth spaces
Intercostalis interni in second space	M. costosternalis pars minor
M. costosternalis pars major	M. obiquus externus abdominis
Mm. levatores costarum	M. obliquus internus abdominis
M. serratus profundus	M. transversus abdominis
	M. rectus abdominis serratus superficialis, pars cranialis and caudalis
	M. costoseptalis

are not clear, but they suggest that compliance is exquisitely sensitivity to posture and muscular tone. Compliance primarily depends on the viscoelastic properties of the body wall and air sacs in birds, in contrast to the elastic properties of the lung in mammals (Macklem et al., 1979). Therefore, compliance in birds is high when volume changes occur by "unfolding" air sacs and stretching the abdominal wall, but it is low when volume changes are opposed by muscle tone in the body wall.

13.3.2.2 Resistance

Ohm's law defines the relationship between pressure, flow, and resistance (R) for the respiratory system as:

$$R = \Delta P / \dot{V}$$

where ΔP is the pressure gradient between the atmosphere and air sacs driving ventilatory airflow $(\dot{V})$. Expiration decreases air sac volume and creates a small positive pressure, which drives airflow out of the sac across small airway resistances; the opposite occurs during inspiration. Air sac pressure changes during breathing are small and similar in all of the air sacs (± 1 cm H_2O), so resistance can be measured by measuring pressure and volume changes during artificial or spontaneous breathing as described above for compliance measurements (Scheid and Piiper, 1989). Airway resistance ranges from 4.8 cm H_2O/(L/s) in ducks (Gillespie et al., 1982a) to 41 cm H_2O/(L/s) in pigeons at the resonant frequency of their respiratory system (Kampe and Crawford, 1973).

Airway geometry is an important determinant of resistance. Poiseuille's law predicts resistance to laminar airflow is directly proportional to the length of an airway, and inversely proportional to the fourth power of the airway radius. Therefore, resistance can vary between the different anatomical pathways possible for airflow in the avian respiratory system (see Section 13.3.2.2). Different pathways presumably explain why airway resistance is generally greater during inspiration than during expiration in birds (Brackenbury, 1971, 1972; Cohn and Shannon, 1968). However, measurements using the forced oscillation technique on unanesthestized ducks find similar resistance during inspiration and expiration (Gillespie et al., 1982a). Resistance measured with this technique does not include any contributions from the body wall (Scheid and Piiper, 1989), so some differences between inspiratory and expiratory resistance may reflect muscle tone.

Physiological factors also affect airway resistance. Decreased lung P_{CO_2} increases resistance by a local effect on the openings of the mediodorsal secondary bronchi into the primary bronchus (Molony et al., 1976). However, P_{CO_2} does not affect smooth muscle contraction in the parabronchi (Barnas and Mather, 1978). Turbulent flow, which may occur in large airways at high ventilation rates or at bronchial bifurcations, can increase resistance (Brackenbury, 1972; Molony et al., 1976). Finally, resistance can change with breathing frequency. The pathway for airflow may be different with ventilation at high frequencies and small volumes (Banzett and Lehr, 1982; Hastings and Powell, 1987) and therefore affect resistance as described above. Also, the respiratory system has a resonant frequency, at which the overall impedance to breathing is minimized (e.g., 9.4 breaths/s in pigeons, Kampe and Crawford, 1973).

13.3.2.3 Air Capillary Surface Forces

Surface tension at the gas–liquid interface of the respiratory exchange surface tends to collapse the air capillaries; this decreases interstitial pressure between the air and blood capillaries. A decrease in interstitial pressure increases capillary filtration and can lead to edema and a thickening of the blood–gas barrier (see Section 13.4.4). This tendency towards edema is probably a more important consequence of surface tension than airway collapse in the parabronchial lung because the air–blood capillary network tends to maintain constant volume (Maina, 2007). Similar to alveoli, air capillaries are lined by a surfactant, which lowers surface tension and counteracts these potentially deleterious effects (McLelland, 1989b). Lamellated osmophilic bodies in the parabronchial atria secrete the surfactant, which spreads as a trilaminar substance that is unique to birds over the air capillary surface (Bernhard et al., 2001; King and Molony, 1971; McLelland, 1989b; Pattle, 1978). Surface tension in avian surfactant is about threefold greater than in mammalian surfactant when subject to volume changes, but the maximum surface tension in avian surfactant is lower than in mammals (Bernhard et al., 2001). This is consistent with the chemical composition of avian versus mammalian surfactant and the fact that air capillaries do not change volume with ventilation, so there is no selective pressure to decrease surface tension for reducing the work of breathing.

13.3.3 Ventilatory Flow Patterns

13.3.3.1 Air Sac Ventilation

The air sacs are ventilated roughly in proportion to their volume, such that the cranial group (clavicular and cranial thoracic air sacs) and caudal group (caudal thoracic and abdominal air sacs) each receive about 50% of the inspired volume (Scheid et al., 1974). The ventilation–volume ratio affects the O_2 and CO_2 levels in the air sacs (see Section 13.3.3.3). Hence, the increase in air sac volume with changes in body wall muscle tone that may accompany anesthesia can alter air sac composition (Scheid and Piiper, 1969). There is no evidence for airflow between the sacs during normal breathing (Scheid and Piiper, 1989), although this has been postulated as a mechanism to enhance pulmonary gas exchange during breath-hold diving in birds (Boggs et al., 1996).

13.3.3.2 Pulmonary Ventilation

Figure 13.7 shows the general pattern of ventilatory airflow during inspiration and expiration in the avian lung. The unidirectional flow in a caudal-to-cranial direction through the paleopulmonic parabronchi during both phases of ventilation has been established by a variety of methods (Scheid and Piiper, 1989). Early researchers noted that soot deposited primarily in the caudal regions of the lungs of pigeons collected in train stations, suggesting that inspired gas entered the lungs from a caudal direction (Dotterweich, 1930). More recently, researchers have confirmed this pattern with direct measurements of airflow (Brackenbury, 1971; Bretz and Schmidt-Nielsen, 1971; Scheid et al., 1972) and respiratory gases (Powell et al., 1981) in different parts of the lung.

Upon inspiration, about half of the tidal volume goes into the caudal air sacs and half goes into the cranial air sacs. Figure 13.7 (filled arrows) shows how inspiratory flow bypasses the cranial secondary bronchial openings in the primary bronchus, flowing directly into the caudal air sacs and caudal secondary bronchi. When gas enters the caudal secondary bronchi, it continues through the paleopulmonic parabronchi in a caudal-to-cranial direction and enters the cranial air sacs via the cranial secondary bronchi. If tidal volume is large enough, some of the inspired gas may reach the cranial air sacs via the paleopulmonic parabronchi during the same breath.

Upon expiration, the air sacs expel gas that eventually leaves the bird via the primary bronchi and trachea. Figure 13.7 (filled arrows) shows how expiratory flow from the cranial air sacs leaves the lung via the cranial secondary bronchi emptying into the primary bronchi. Expiratory flow from the caudal air sacs is routed through the paleopulmonic parabronchi in a caudal-to-cranial direction via caudal secondary bronchi. If the expired volume is large

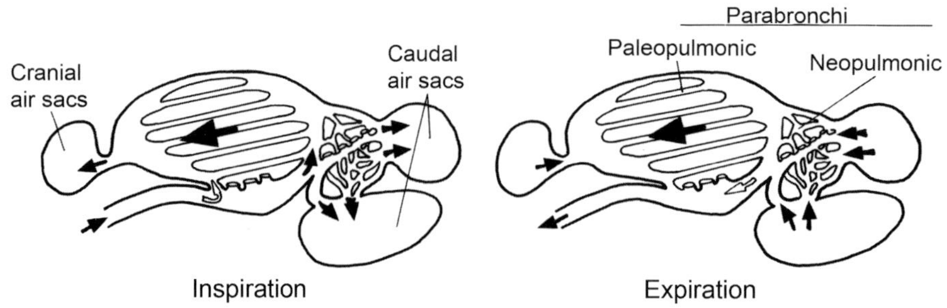

FIGURE 13.7 Pathway of airflow in the avian respiratory system during inspiration and expiration. Flow in paleopulmonic parabronchi is always caudal-to-cranial during both phases of breathing (large solid arrows) but neopulmonic flow is bidirectional. Open arrows show possible ventilatory shunts.

enough, gas from the caudal air sacs will also leave the lung through the cranial secondary bronchi and mix with the gas emptying from the cranial air sacs.

Hence, airflow in the paleopulmonic parabronchi is in a caudal-to-cranial direction during both inspiration and expiration. In contrast to the paleopulmonic parabronchi, flow is bidirectional in the neopulmonic parabronchi, which are functionally in series with the caudal air sacs (Figure 13.7). Inspiratory airflow is cranial-to-caudal through neopulmonic parabronchi and into caudal air sacs, and in the caudal-to-cranial direction during expiration (Scheid and Piiper, 1989). The implications of these flow patterns for gas exchange are discussed in Section 13.6.2.

The pattern of airflow in the avian lung is determined by "aerodynamic valving". Pressure measurements show that branch points, such as the openings of the cranial secondary bronchial into the primary bronchi, are more important than distal airway resistance for determining flow patterns (Kuethe, 1988; Molony et al., 1976). There is no evidence for anatomical valves, for example, closing the primary bronchial openings of the cranial secondary bronchi during inspiration (reviewed by Scheid and Piiper, 1989). As early as 1943, fluid dynamic models of the avian lung were used to show that these branches were critical for unidirectional caudal-to-cranial airflow in the paleopulmonic parabronchi (Hazelhoff, 1951). Modern theoretical models predict decreased effectiveness of aerodynamic valving with reduced gas density, and experiments with low-density gases show valve failure and inspiratory shunts in birds (Banzett et al., 1987, 1991; Butler and Turner, 1988; Wang et al., 1988, 1992). In contrast, expiratory aerodynamic valving is not sensitive to gas density (Brown et al., 1995). Future experiments are necessary to determine if decreased gas density at high altitude can affect aerodynamic valving.

13.3.3.3 Air Sac P_{O_2} and P_{CO_2}

This pattern of airflow is an important determinant of P_{O_2} and P_{CO_2} in the air sacs (Table 13.3). Cranial air sacs only receive gas from the parabronchi, so their P_{O_2} and P_{CO_2} levels are very near end-expired values. However, caudal air sacs contain a mixture of re-inhaled dead space gas (also end-expired P_{O_2} and P_{CO_2} levels) and fresh air, which raises their P_{O_2} and lowers their P_{CO_2}. Other factors that decrease P_{O_2} and increase P_{CO_2} in the air sacs include stratification, gas exchange across the air sac wall, and gas exchange in neopulmonic parabronchi in series with the air sacs (Geiser et al., 1984).

Gas exchange across the air sac walls is less than 5% of the total respiratory gas exchange and a minor factor in determining air sac P_{O_2} and P_{CO_2} (Magnussen et al., 1976). Stratification (i.e., incomplete mixing of freshly inspired gas and resident gas in the air sacs) has been observed in ducks (Torre-Bueno et al., 1980), but its effect on air sac gas concentrations is not clear (Powell and Hempleman, 1985). Gas exchange in neopulmonic parabronchi in series with caudal air sacs seems to be the most important factor causing differences in measured P_{O_2} and P_{CO_2} values and those predicted from re-inhaled dead space (Piiper, 1978). Bidirectional flow may also occur in a small fraction of parabronchi in the purely paleopulmonic lungs of penguins, as well as affect P_{O_2} and P_{CO_2} in the caudal lungs of these birds (Powell and Hempleman, 1985).

Considering the determinants of air sac P_{O_2} and P_{CO_2}, air sac sampling can offer valuable insights into respiratory function in birds under physiologically interesting conditions. For example, recent experiments sampling air sac gases in diving emperor penguins have been important for understanding how O_2 stores are used differently in diving birds and mammals; these differences result in similar O_2 levels in prolonged dives but the penguins maintain normal (breathing) levels of arterial saturation during the first 7 minutes of a deep dive in contrast to dramatic decreases in Weddell and elephant seals (Ponganis et al., 2011).

13.3.3.4 Effective Parabronchial Ventilation

A quantitative description of gas exchange requires a measure of the effective ventilation of the lung (see Section 13.3.3.4). By analogy with alveolar ventilation in mammals, this is defined as parabronchial ventilation $(\dot{V}_P)$ birds, and it differs from inspired ventilation $(\dot{V}_I)$ because of dead space

TABLE 13.3 Partial Pressure of O_2 and CO_2 in Air Sacs and in End-Expiratory Gas of Awake Birds

	Goose (Cohn and Shannon, 1968; Figure 13.6)	Goose (Scheid et al., 1991)	Chicken (Piiper et al., 1970)	Duck[1]	Pigeon[2]
Clavicular					
P_{CO_2} (Torr)	35	39	44.0	39.2	32
P_{O_2} (Torr)	100	92	83.9	99.4	109
Cranial Thoracic					
P_{CO_2} (Torr)	35	38	41.6	35.7	34
P_{O_2} (Torr)	100	95	99.1	104.3	105
Caudal Thoracic					
P_{CO_2} (Torr)	28	20	24.2	18.9	29
P_{O_2} (Torr)	115	124	120.3	123.9	111
Abdominal					
P_{CO_2} (Torr)	28	18	14.7	17.5	27
P_{O_2} (Torr)	115	128	130.0	126.7	110
End Expiratory					
P_{CO_2} (Torr)	35	39	36.7	35.7	–
P_{O_2} (Torr)	100	100	94.3	100.1	–

[1]*Vos (1935) calculated assuming 700 Torr dry pressure.*
[2]*Scharnke (1938) calculated assuming 700 Torr dry pressure.*

ventilation $(\dot{V}_D)$. $\dot{V}_D$ in birds includes not only anatomic dead space in the upper airways, but it may result from ventilatory shunts in which airflow bypasses the parabronchi. Figure 13.7 (open arrows) shows how gas might bypass the parabronchi during inspiration by directly entering the cranial secondary bronchi and air sacs or during expiration by flowing back out the primary bronchus (Powell, 1988).

P_{CO_2} measurements in the cranial secondary bronchi show that an inspiratory shunt, with inspired gas directly entering cranial air sacs from the primary bronchi, does not occur (Powell et al., 1981). Some inspiratory flow may enter the fourth medioventral (cranial secondary) bronchus, however, and flow in a cranial-to-caudal direction through some paleopulmonic parabronchi (Powell and Hempleman, 1985), but this would not be a shunt. In contrast, P_{CO_2} measurements indicate an expiratory shunt, with 10–25% of expired gas from the caudal air sacs gas flowing out through the primary bronchus (Powell, 1988; Powell et al., 1981). Other experiments indicate that the relative proportion of expiratory flow in a mesobronchial shunt may vary from 100 to 75% at the beginning of expiration to 0% near the midpoint of expiration (Hastings and Powell, 1986a). The magnitude of expiratory mesobronchial shunting depends on the pattern of ventilation, such as tidal volume, flow rate, and thermal panting (Bretz and Schmidt-Nielsen, 1971; Hastings and Powell, 1986a).

13.3.3.5 Artificial Ventilation

The flow-through design of the rigid parabronchial lung allows a unique form of artificial ventilation, called unidirectional ventilation (Burger and Lorenz, 1960). Fresh humidified gas can be insufflated through a cannula in the trachea or an air sac, so it flows through the parabronchi before leaving the body through another cannula. This technique can be used clinically to support gas exchange during surgery, which opens one of the air sacs (preventing effective spontaneous ventilation), to administer anesthetic gas or nebulized drugs (Fedde, 1978; Whittow and Ossorio., 1970), or for experimental studies (Burger et al., 1979; Fedde et al., 1974a). Artificial ventilation can also be performed manually by alternately compressing and lifting the sternum, such as in a bird that may be anesthetized too deeply. It is also important to note that the sternum should not be compressed when holding a bird because this may lead to suffocation.

13.3.3.6 Frontiers: Lung Structure-Function in Dinosaurs

There is a rich literature on the evolutionary relationship between birds and dinosaurs, and the modern avian respiratory system is an important part of this story. For example, the long trachea, voluminous air sacs, and pneumatized bones are hypothesized to enhance thermoregulation in extremely large dinosaurs (Perry et al., 2009). Also, models show that air sacs could facilitate the transition from quadrupedal to bipedal locomotion, which preceded flight by improving balance and agility (Farmer, 2006). Such theories are based on fossil skeletons, and answers to most important physiological questions are lost with the soft tissues of the dinosaurs. However, recent comparative physiology studies have shown the value of experimental versus descriptive and deductive approaches to the evolution of birds. For example, unidirectional airflow has been demonstrated in alligator lungs, which have a bronchial tree similar to modern birds but no air sacs (Farmer and Sanders, 2010). This means cross-current gas exchange is possible in animals using a hepatic piston mechanism of breathing, as found in alligators and presumably quardrupedal ancestors of birds; costal breathing or bipedal locomotion are not prerequisites for cross-current gas exchange. The efficiency of cross-current gas exchange could have given avian ancestors a competitive advantage over synapsids when the basal archosaurs were evolving in times of low environmental oxygen.

13.4 PULMONARY CIRCULATION

Chapter 11 covers the basic physiology of the circulation, but this section highlights details of the pulmonary circulation that are relevant to understanding pulmonary gas exchange. The pulmonary circulation is unique because the lung is the only organ to receive the entire cardiac output. The "series" arrangement of the systemic and pulmonary circulations in birds and mammals means that the lungs receive the same amount of blood flow as the whole rest of the body. However, the resistance to blood flow in the lungs is lower, and this allows lower perfusion pressure than in the systemic circulation, with the complete separation of the left and right ventricles. This section describes the structural and functional factors that determine pressures, volumes, and flows in the pulmonary capillaries, which are also important determinants of gas exchange.

13.4.1 Anatomy of the Pulmonary Circulation

The functional anatomy of the pulmonary circulation has been studied in detail for the domestic fowl (Abdalla and King, 1975). Interparabronchial arteries arise from main rami of the pulmonary arteries and run between the parabronchi and may perfuse more than one parabronchus. These vessels give rise to intraparabronchial arteries, which perfuse the parabronchial mantle at several points along a parabronchus. The intraparabronchial arteries branch into the pulmonary blood capillaries near the outside edge of the parabronchial mantle, which form a meshwork with air capillaries, as described in Section 13.2.2.2. Pulmonary capillary blood is collected in intraparabronchial veins near the parabronchial lumen. These veins deliver blood flow to interparabronchial veins located near the outside edges of the parabronchus. The interparabronchial veins run between the parabronchi and collect blood from several points along a parabronchus and from several parabronchi. There are no anastomoses between the arterioles and veins in the lung (Abdalla, 1989).

This anatomy has two important consequences for respiratory gas exchange. First, all of the parabronchi are perfused along their entire length by oxygen-poor mixed venous blood, and the oxygenated blood returning to the heart in the pulmonary vein (i.e., systemic arterial blood) is a mixture of blood draining the entire length of all the parabronchi. This allows cross-current gas exchange to occur, which is more efficient than alveolar gas exchange as described below (Section 13.6.2). Second, it means that blood flow within the parabronchial mantle is directed from the periphery towards the lumen, which can affect the efficiency of gas exchange in the air capillaries (see Section 13.6.3.1).

13.4.2 Pulmonary Capillary Volume

Pulmonary capillary blood volume is similar in avian parabronchial lungs and mammalian alveolar lungs. A notable exception is the domestic fowl, which has a pulmonary blood capillary volume about threefold less than predicted for other birds that size (Maina et al., 1989). However, pulmonary blood capillary volume in birds is essentially constant under all conditions, in contrast to mammals that can increase pulmonary capillary volume by recruitment and distention when flow or pressure increases. Recruitment or distention of blood capillaries in a parabronchial lung would collapse the adjacent air capillaries and reduce gas exchange efficiency by causing a shunt (see Sections 13.2.2.3 and 13.6.4.2).

13.4.3 Pulmonary Vascular Pressures, Resistance, and Flow

13.4.3.1 Pulmonary Vascular Resistance

Pulmonary vascular resistance (PVR), by analogy with Ohm's law, is as follows:

$$PVR = \dot{Q} / \Delta P$$

where $\dot{Q}$ is cardiac output and ΔP is the difference between mean pulmonary artery and left atrial pressure. Pulmonary vascular pressures, cardiac output, and therefore PVR are similar in resting birds and mammals. However, PVR increases more with increases in cardiac output in birds compared to mammals. For example, doubling the blood flow through one lung of a domestic duck almost doubles mean pulmonary artery pressure but causes no change in the resistance calculated for that lung, and no change in the capillary dimensions (Powell et al., 1985). Morphometric measurements of pulmonary blood capillaries in chickens with systematic increases in pulmonary artery pressure show no recruitment and very little distention (Watson et al., 2008). The lack of distention and recruitment of avian pulmonary capillaries also implies that the major effects of gravity and pulmonary vascular pressures on determining regional blood flow in mammalian lungs are not important in birds. The so-called zone 1 and zone 2 conditions of mammalian lungs, in which alveolar pressure can collapse blood capillaries and stop or determine flow, simply do not exist in birds (West et al., 2006). Hence, other mechanisms are responsible for the distribution of pulmonary blood flow in birds, as considered in the next section.

13.4.3.2 Distribution of Blood Flow

Local and regional changes in vascular resistance are more important than overall PVR for respiratory gas exchange. For example, regional control of blood flow between parabronchi has important effects on the efficiency of gas exchange (see Section 13.6.4.3). Hypoxia has been shown to decrease local parabronchial blood flow, and gradients in P_{O_2} can explain differences in perfusion along parabronchi (Holle et al., 1978; Parry and Yates, 1979). Hypoxia and hypercapnia can also redistribute blood flow between paleopulmonic and neopulmonic parabronchi (Weidner et al., 2012). The physiological mechanism of these responses is not known. It may involve hypoxic pulmonary vasoconstriction, which is a direct effect of alveolar hypoxia on pulmonary arterioles in mammals. Smooth muscle capable of controlling local blood flow has been described for interparabronchial arteries and veins in chickens, and this could be responsible for control of blood flow between and along the lengths of parabronchi (Abdalla, 1989). It may also explain pulmonary hypertension with hypoxia in birds (Black and Tenney, 1980a; Burton et al., 1968), but the effects of hypoxia on PVR independent of changes in blood flow remains to be determined in birds.

13.4.3.3 Frontiers: Pulmonary Vascular Pressures during Exercise in Birds

How the pulmonary circulation in birds copes with large increases in blood flow on exercise is a puzzle. The observation that there is no recruitment or significant distention of blood capillaries with increased flow and pressure in bird lungs contrasts sharply with mammals. Alveolar capillary recruitment and distention is important for facilitating gas exchange during exercise by increasing the surface area for diffusion and preserving transit time for red blood cells in the capillaries to absorb oxygen by diffusion (see Section 13.6.3). In mammals, if pressures increase too much during exercise (e.g., in elite human athletes or thoroughbred horses), then pulmonary capillaries can suffer "stress failure" and blood to leaks into the alveoli (West, 2011). The blood–gas barrier in the bird may be protected by the network of air capillaries (see Section 13.1.2.3), but it is also thinner and it is not clear how it avoids stress failure on exercise. Cardiac output and therefore pulmonary blood flow increases comparably in birds and mammals, so extremely high pulmonary artery pressures are predicted with limited compliance in avian pulmonary blood capillaries. To date, there are no measurements of pulmonary circulation pressure and flow in naturally exercising birds. Experiments simulating exercise, which used drug to increase metabolic rate, did not find the large increases in pulmonary artery pressure as predicted for the increase in pulmonary blood flow (West et al., 2010). The reasons for this unexpected result, and what happens to pulmonary artery pressure in a flying bird, are experimental challenges for the future.

13.4.4 Fluid Balance

Fluid balance in the lungs, as in all organs, depends on the balance of hydrostatic and colloid osmotic pressures across the capillaries and capillary permeability. These variables are similar in birds and mammals, although lymph drainage from avian lungs has not been measured directly (Weidner et al., 2006). Volume loading increases extravascular water in the lung interstitium, especially in the interparabronchial septum, and can lead to edema (Weidner et al., 1993; Weidner and Kinnison, 2002). The detailed anatomy of lymphatic vessels and rates pulmonary lymph drainage remains to be determined in birds.

A very important problem involving the pulmonary circulation and fluid balance in poultry husbandry is ascites in fast-growing broilers (chickens) bred for meat production (Julian, 1993). Ascites (i.e., fluid accumulation in the peritoneum) is associated with pulmonary arterial hypertension (PAH) and also occurs frequently in chickens raised at high altitude. Hence, early research on the problem focused on abnormal oxygen sensitivity, which could exaggerate increases in cardiac output with fast growth and cause hypoxic pulmonary vasoconstriction (Peacock et al., 1990). However, comparisons of PAH-susceptible and PAH-resistant broilers do not consistently reveal differences in cardiac output, although vascular capacity is limited in PAH susceptible broilers (Wideman et al., 2007). Factors that increase the pulmonary vascular resistance (e.g. hypoxia, thromboxane, endothelin-1 and serotonin) and reduce vasodilation (e.g., limited L-arginine, which is the substrate for nitric oxide) promote the onset of PAH and ascites, and appear to be under genetic control (Wideman et al., 2013).

13.5 GAS TRANSPORT BY BLOOD

Equilibrium curves, also called dissociation curves, quantify the amount of O_2 and CO_2 in blood as functions of partial pressure. It is necessary to consider *both* partial pressure *and* concentration because partial pressure gradients drive diffusive gas transport in lungs and tissues, but concentration differences determine convective gas transport rates in lungs and the circulation (see Section 13.6.1). The concentration of a physically dissolved gas in a liquid is directly, and linearly, proportional to its partial pressure according to Henry's law: $C=\alpha P$, where α=solubility in millimolar per Torr. This means that inert gases such as nitrogen, and even anesthetic gases, increase in blood in direct proportion to their partial pressure. However, O_2 and CO_2 also enter into chemical reactions with blood. These reactions result in more complex relationships between concentration and partial pressure, but they serve to (1) increase O_2 and CO_2 concentrations in blood; (2) allow physiological modulation of O_2 and CO_2 transport in blood; and (3) make respiratory CO_2 exchange an important mechanism of acid–base balance in the body.

13.5.1 Oxygen

Oxygen concentration in normal arterial blood $\left(C_{a_{O_2}}\right)$ of, for example, a pigeon is about 8.3 mM. However, the physical solubility of O_2 in blood (α_{O_2}) is only 0.00124 mM/Torr at 41°C, so only 0.117 mM of arterial O_2 content is dissolved gas with a normal arterial P_{O_2} of 95 Torr in pigeons (Powell, 1983b). Most of the O_2 in blood is chemically bound to hemoglobin.

13.5.1.1 Hemoglobin

Hemoglobin is a large molecule consisting of four individual polypeptide chains, each with a heme (iron-containing) protein that can bind O_2 when iron is in the *ferrous* (Fe^{2+}) form. Methemoglobin occurs when the iron is in the *ferric* form (Fe^{3+}), and it cannot bind O_2. Small amounts of methemoglobin, which occur under normal conditions, slightly reduce the amount of O_2 that can be bound to hemoglobin. One gram of pure mammalian hemoglobin can bind 0.060 mmol of O_2 when fully saturated, and this value appears similar in birds (Powell, 1983b). Hemoglobin is concentrated inside red blood cells or erythrocytes. This cellular packaging is important for the biophysics of the microcirculation, and it provides physiological control of O_2 binding through cellular changes in the hemoglobin microenvironment (see Section 13.4.1.3).

13.5.1.2 O_2-Blood Equilibrium Curves

Figure 13.8 shows the O_2 equilibrium curve for duck blood as saturation (S_{O_2}) versus P_{O_2}, where S_{O_2} is defined as the percentage of the total hemoglobin sites available for binding O_2 which are occupied by O_2. Therefore, the maximum S_{O_2} is 100% and independent of hemoglobin concentration in blood. In contrast, O_2-hemoglobin equilibrium curves plotting concentration versus P_{O_2} quantify the absolute amount of O_2 in blood at a given P_{O_2}, and the maximum O_2 concentration depends on the amount of hemoglobin available. O_2 capacity defines the maximum O_2 concentration in blood when hemoglobin is 100% saturated with O_2. Total O_2 concentration in blood (C_{O_2}), including chemically bound and dissolved O_2, can be calculated as:

$$C_{O_2} = [O_2 \text{ capacity } (S_{O_2}/100)] + (\alpha_{O_2} P_{O_2})$$

The sigmoidal (or "S") shape of the blood–O_2 equilibrium curve results from cooperative, *allosteric* interactions between the four subunits of hemoglobin, which determine the three-dimensional shape of the molecule. O_2 equilibrium curves for individual hemoglobin subunits are not sigmoidal but simple convex curves, such as the O_2 equilibrium curve for myoglobin (see Section 13.7). The shape of the blood–O_2 hemoglobin equilibrium curve facilitates O_2 loading on blood in the lungs and O_2 unloading from blood in the tissues.

The cooperativity between functional subunits of hemoglobin is quantified with Hill's coefficient, *n* (Powell and Scheid, 1989). High values of *n*, exceeding the theoretical

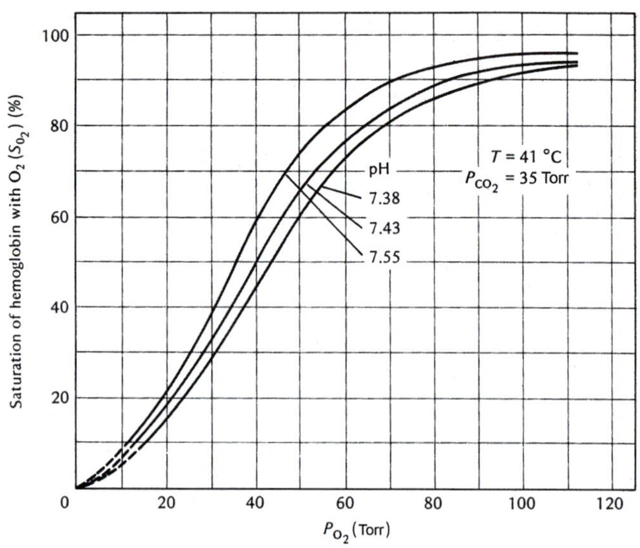

FIGURE 13.8 **O_2-blood equilibrium curves for duck.** Bohr effect is demonstrated by shifts of the curve as pH changes. *After Scheipers et al. (1975).*

limit of 4, have been observed in bird blood and may reflect increased cooperativity between aggregates of multiple hemoglobin molecules or interactions between different isoforms of hemoglobin within a blood sample (Black and Tenney, 1980b; Lapennas and Reeves, 1983).

The affinity of hemoglobin for O_2 is quantified with P_{50}, which is the P_{O_2} at 50% saturation. For example, a decrease in P_{50}, or a "left shift", indicates an increase in O_2 affinity because S_{O_2} is greater for a given P_{O_2}. In general, P_{50} in avian blood is greater than in mammalian blood. Table 13.4 shows P_{50} values for several common avian species in the 40 Torr range, while P_{50} in comparably sized mammals are nearer 30 Torr. Some studies have used erythrocyte suspensions (Lutz, 1980), and this might explain the low P_{50} values they have found. However, determinations of P_{50} in avian hemoglobin solutions agree well with other published values for whole blood, such as in pigeons (Powell, 1983b). Because the efficiency of O_2 uptake in the avian lung is greater than mammals, birds may have evolved blood with low O_2 affinity to maximize O_2 delivery in tissue.

Developmental changes in hemoglobin–O_2 affinity are explained by differences in the type of hemoglobin expressed in an individual. For example, the dramatic decrease of P_{50} in chickens, from 75 Torr at 8 days to 35 Torr at 14–16 days of development, is explained by the replacement fetal hemoglobin with adult hemoglobin expressed in the erythrocytes (Baumann and Baumann, 1978).

13.5.1.3 Physiological Control of O_2–Hemoglobin Affinity

The most important physiological factors that affect the P_{50} in a given species are (1) organic phosphate levels, (2) pH, and (3) temperature. These factors work by modifying the structure of the hemoglobin tetramer so it binds O_2 more or less efficiently.

Myinositol 1,3,4,5,6-pentophosphate (IPP) is the primary organic phosphate affecting P_{50} in birds (Weber and Wells, 1989). The effect of IPP binding with hemoglobin inside erythrocytes is considerable, for example, increasing P_{50} from less than 3 Torr in stripped hemoglobin from chickens, to over 40 Torr under *in vivo* conditions (Weingarten et al., 1978). Physiological changes in organic phosphates have not been studied extensively in birds (Maginniss and Kilgore, 1989; Weber and Wells, 1989). In mammals, acclimatization to conditions such as altitude can modulate P_{50} by altering organic phosphates (2,3 diphosphoglycerate, or 2,3-DPG). However, it is important to note that differences

TABLE 13.4 Respiratory Parameters in Avian Whole Blood

Reference	Burrowing Owl (*Athene cunicularia*)	Pigeon (Powell, 1983a)	Female Domestic Fowl (Bartels et al., 1966; Bauer et al., 1978; Baumann and Baumann, 1977; Henning et al., 1971; Hirsowitz et al., 1977; Holle et al., 1977; Lapennas and Reeves, 1983; Meyer et al., 1978; Wells, 1976)	Pekin Duck (Andersen and Lovo, 1967; Black and Tenney, 1980b)	Muscovy Duck (*Cairina moschata*) (Holle et al., 1977; Morgan and Chichester, 1935; Scheipers et al., 1975)	Bar-headed Goose (*Anser indicus*) (Meir and Milsom, 2013)
Hematocrit (%)	33.7 ± 2.1	48.7	26–30	45.4	37.3 ± 1.3	43.3 + 1.3
Hemoglobin (g%)	10.7 ± 0.4	14.3	8.6–9.3	15.5	–	17.1 + 1.24
O_2 capacity (mmol L^{-1})	–	8.6	8.6–12.3	≥8.9	7.3 ± 0.51	–
P_{50} (Torr)	42.3 ± 0.8	40.8 ± 1.4	47.7 ± 4.2	42.7–45.0	40.1 ± 3.7	31.2
Hill's *n* (–)	2.60–3.42	2.75	3.4 ± 0.1	4.3	2.9	–
Bohr coefficient (Δ log P_{50}/ΔpH)	0.42–0.46	0.42–0.53	0.50 ± 0.08	0.40–0.44	0.44–0.53	0.42–0.48
Temperature coefficient (Δ log P_{50}/ΔT)	–	0.015–0.026	0.014–0.015	–	–	0.024–0.032
Haldane effect ($\Delta C_{CO_2}/\Delta C_{O_2}$)	–	–	0.42	–	0.30	–

Where three or more measurements are available, the standard deviation is given; for less than three measurements, the range is indicated.

in P_{50} between birds adapted over generations to low or high altitude cannot be explained by different organic phosphate concentrations. Erythrocyte inorganic phosphate levels are similar in the greylag and Canada goose, natives to low altitude (with $P_{50}=39$ and 42 Torr, respectively) and the bar-headed goose, which is native to high altitude (with $P_{50}=29$ Torr) (Petschow et al., 1977). Differences in the binding of IPP to hemoglobin from different species, explains such differences in P_{50} (Rollema and Bauer, 1979).

Figure 13.8 shows the effect of pH on the O_2 affinity, which is known as the Bohr effect. Increases in pH cause decreases in P_{50} (i.e., increased O_2-hemoglobin affinity) and vice versa. H^+ binds to histidine residues in hemoglobin, which changes the molecular conformation and ability of heme sites to bind O_2. The physiological advantage of the Bohr effect is that it facilitates O_2 loading in the lungs, where CO_2 is low and pH is high (see Section 13.5.3). In muscles, the opposite occurs and decreased pH facilitates O_2 unloading to the tissues. The Bohr effect is independent of saturation in most birds (Lapennas and Reeves, 1983; Maginniss, 1985; Meyer et al., 1978) and similar to values reported for mammals (Table 13.4).

In most birds, and in contrast to mammals, there is no independent effect of CO_2 on P_{50} (Meyer et al., 1978). CO_2 forms carbamino compounds with hemoglobin in mammals, and these cause small increases in P_{50}. In some birds, such as sparrows and burrowing owls, the Bohr effect is greater when pH is changed with CO_2 compared with fixed acid (Maginniss, 1985; Maginniss and Kilgore, 1989). Therefore, carbamino formation does occur and can decrease O_2 affinity in stripped avian hemoglobin. However, strong binding of IPP to hemoglobin in most birds prevents an independent CO_2 Bohr effect (Lapennas and Reeves, 1983; Weingarten et al., 1978).

The *in vivo* physiological O_2 dissociation curve is steeper than the individual *in vitro* curves in Figure 13.8 because P_{CO_2} increases and pH decreases between arterial and venous blood. This is an advantage for gas exchange because it increases the change in O_2 concentration for a given change in P_{O_2}, thereby increasing O_2 uptake or delivery. The slope of the physiological O_2-blood equilibrium curve, in terms of O_2 concentration, is called βb_{O_2} (mM/Torr) and is used for quantitative descriptions of gas exchange (see Section 13.6.1).

Because the combination of O_2 with hemoglobin is a chemical reaction that releases heat, increased temperature reduces the affinity of hemoglobin for O_2. This facilitates O_2 off-loading at exercising muscle with relatively high temperatures. It has also been hypothesized to facilitate O_2 loading in the lungs of birds flying at high altitude, when high rates of ventilation with extremely cold air might cool the respiratory exchange surfaces (Faraci, 1986, 1991). However, experiments to date have not been able to demonstrate decreased temperature of blood in the lungs. The effect of temperature on P_{50} in birds is generally similar to the effect in mammals (Table 13.4) except in the bar-headed goose, described next.

13.5.1.4 Frontiers: Hemoglobin Adaptations to High Altitude

Hemoglobin–O_2 affinity is generally greater in birds native to high altitude compared to low-altitude species, presumably as an adaptation to hypoxia (Black and Tenney, 1980b; Jessen et al., 1991; Meir and Milsom, 2013; Petschow et al., 1977). For example, P_{50} in the bar-headed goose (discussed above) is 10 Torr less than P_{50} in the closely related greylag goose, which lives on the Indian planes year round (Petschow et al., 1977). Low P_{50} will enhance O_2 loading in alveolar lungs in extreme hypoxia (Bencowitz et al., 1982) and the trend for low P_{50} in high-altitude species is observed in both birds and mammals (Powell and Hopkins, 2010). Amino acid sequences from the bar-headed goose and greylag goose revealed a single substitution on one of the four hemoglobin subunits, where two α-chains and two β-chains comprise a hemoglobin molecule (Allen, 1983). Genetically engineering this substitution into human hemoglobin decreases P_{50} by a similar amount to the difference between the geese (Jessen et al., 1991). In general, it appears that a low P_{50} in high-altitude compared with low-altitude species requires only a few amino acid substitutions, and these occur at either (1) contact points between subunits that stabilize hemoglobin in either low or high O_2-affinity conformations, or (2) binding sites for physiological modulators of O_2-affinity (Weber, 2007). Additionally, bar-headed goose hemoglobin is more sensitive to temperature than other birds (Table 13.4) and this is hypothesized to confer additional advantages. With differences in temperature and CO_2 between lungs and exercising muscle during high-altitude flight, changes in P_{50} are predicted to double O_2 delivery, relative to a fixed P_{50} in bar-headed geese (Meir and Milsom, 2013).

13.5.1.5 Factors Affecting O_2 Capacity

Changes in hemoglobin concentration, [Hb], in blood will change the O_2 capacity and therefore the O_2 concentration at any P_{O_2} as described above. [Hb] depends on both the mean corpuscular hemoglobin concentration (MCHC) and the hematocrit. Typical hematocrit and [Hb] values are given in Table 13.4; [Hb] is expressed as g/100 mL of blood instead of millimolar because the molecular weight was not known for all hemoglobins when they were originally measured. Typical values for MCHC in birds are 30–40 g Hb/100 mL of erythrocytes (Palomeque et al., 1979), or similar to the MCHC in mammals.

If [Hb] decreases, such as with decreased hematocrit in anemia, then O_2 capacity and concentration decreases at any given P_{O_2}. The O_2 capacity increases when [Hb] increases, such as by the stimulation of red blood cell production in bone marrow by the hormone erythropoietin (EPO). EPO is released from cells in the kidneys in response to decreases in arterial O_2 levels (Sturkie, 1986). Significant gender differences in [Hb], for example in chickens, are explained by the effects of sex hormones on hematocrit (Sturkie, 1986).

13.5.2 Carbon Dioxide

CO_2–blood equilibrium (or dissociation) curves are nonlinear, but they have a different shape and position than O_2–blood equilibrium curves (Figure 13.9). CO_2 is carried in three forms by blood, so CO_2 concentrations in blood are generally much higher than O_2 concentrations. This results in a smaller range of P_{CO_2} values in the body, compared with the range P_{O_2} values, although the differences between arterial and venous *concentrations* are similar for CO_2 and O_2. The resulting physiological CO_2 dissociation curve between the arterial and venous points is much more linear than the physiological O_2 dissociation curve (Figure 13.9).

13.5.2.1 Forms of CO_2 in Blood

CO_2 solubility in water or plasma is 0.0278 mM/Torr, or about 20 times more soluble than O_2. Still, dissolved CO_2 only contributes about 5% of total CO_2 concentration in arterial blood. CO_2 can also combine with terminal amine groups in hemoglobin to form carbamino compounds (see Section 13.4.1.3). Bicarbonate ion (HCO_3^-) is the most important form of CO_2 carriage in blood. CO_2 combines with water to form carbonic acid, and this dissociates to HCO_3^- and H^+:

$$CO_2 + H_2O \leftrightarrow H_2CO_3 \leftrightarrow HCO_3^- + H^+$$

Carbonic anhydrase is the enzyme that catalyzes this reaction, and it occurs mainly in red blood cells (Maren, 1967). The reaction is almost instantaneous with carbonic anhydrase, but the uncatalyzed reaction will occur much more slowly in any aqueous medium (requiring over 4 minutes for equilibrium). The rapid conversion of CO_2 to bicarbonate results in about 90% of the CO_2 in arterial blood being carried in that form. The H^+ produced from CO_2 reacts with hemoglobin and affects both the O_2 dissociation curve (Bohr effect), and CO_2 dissociation curve as described next.

13.5.2.2 Factors Affecting Blood-CO_2 Equilibrium Curves

Hb–O_2 saturation is the major factor affecting the position of the CO_2 equilibrium curve. The Haldane effect increases CO_2 concentration when blood is deoxygenated, or decreases CO_2 concentration when blood is oxygenated, at any given P_{CO_2} (Figure 13.9). The Haldane effect is actually another view of the same molecular mechanism causing the Bohr effect on the O_2 equilibrium curve (see Section 13.4.1.3). H^+ ions and O_2 can be thought of as competing for hemoglobin binding, so increasing O_2 decreases the affinity of hemoglobin for H^+ (Haldane effect), and increased [H^+] decreases the affinity of hemoglobin for O_2 (Bohr effect).

The physiological advantages of the Haldane effect are to promote unloading of CO_2 in the lungs when blood is oxygenated, and CO_2 loading in the blood when O_2 is released to tissues. The Haldane effect also results in a steeper physiological CO_2-blood equilibrium curve (Figure 13.9), which has the physiological advantage of increasing CO_2 concentration differences for a given P_{CO_2} difference.

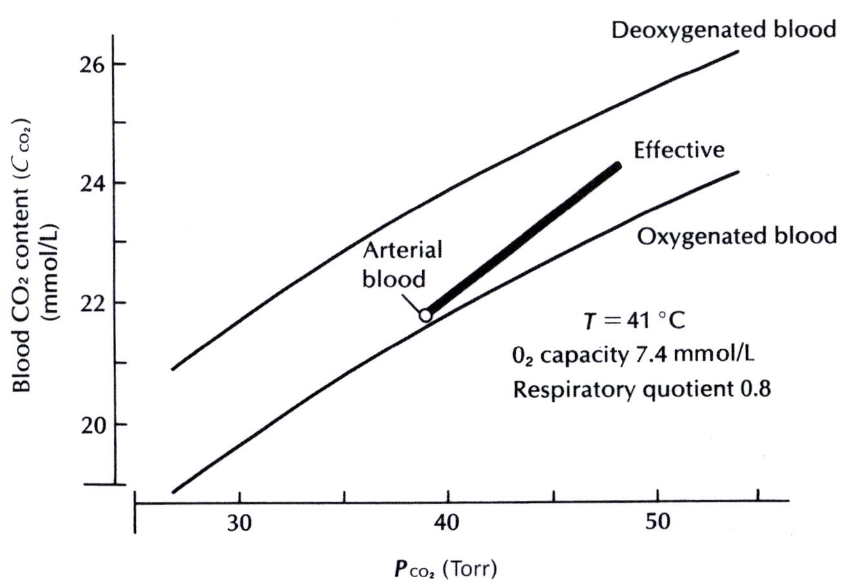

FIGURE 13.9 **CO_2-blood equilibrium curves from the duck.** Upper (deoxygenated blood) and lower (oxygenated blood) curves are derived from *in vitro* equilibration of blood samples. Heavy line is the physiological or *in vivo* dissociation curve from unanesthetized undisturbed birds. It illustrates the changes in CO_2 content (C_{CO_2}) in blood as it changes from arterial blood to venous blood in the tissue capillaries. *After Scheipers et al. (1975).*

Finally, the Haldane effect can cause apparent negative blood-gas CO_2 gradients when it is amplified by cross-current gas exchange in avian lungs (see Section 13.6.2.2).

13.5.3 Acid–Base

The chemical equilibrium between CO_2 and H^+/HCO_3^- ions has tremendous implications for acid–base physiology. Every mole of metabolic CO_2 produced results in 1 mol of acid, and over 95% of this acid is excreted by the lungs (Skadhauge, 1983). The ability to change blood P_{CO_2} levels rapidly by changing ventilation has a powerful effect on blood pH, so acid–base balance depends on the integrated function of respiratory and renal systems.

13.5.3.1 Henderson–Hasselbalch Equation

This equation describes the relationship between P_{CO_2}, pH, and $[HCO_3^-]$ in blood as:

$$\text{pH} = \text{p}K_a + \log\left([HCO_3^-]/\alpha P_{CO_2}\right)$$

where pK_a is $-\log_{10}$ of K_a, the dissociation constant for carbonic acid, $[HCO_3^-]$ = bicarbonate concentration in mEq/L or mM, and α is the physical solubility for CO_2 in water. A normal value for arterial pH (pH_a) in chickens is 7.52 (Table 13.5), which can be calculated from pK_a = 6.09 and $\alpha_{CO_2} = 0.03$ mM/Torr in chicken plasma at 41 °C (Helbacka et al., 1963), arterial $P_{CO_2} = 33$ Torr (Table 13.5), and arterial $[HCO_3^-]$ = 27.2 mM. The buffer value for plasma in ducks is similar to the value for humans (Scheipers et al., 1975) if corrections are made for differences in [Hb]. At pH = 7.5, the $[H^+]$ is only 30 nM, or significantly less than many other important ions in the body, such as Na^+, Cl_3^-, HCO_3^-, which occur in the mM range. Small changes in pH, corresponding to very small changes in $[H^+]$ (see Chapter 12), can lead to dramatic changes in physiological function.

The Henderson–Hasselbalch equation shows how the physiological control of pH depends on the ratio of $[HCO_3^-]$ to $[\alpha P_{CO_2}]$. Notice that a normal pH can occur with a variety of $[HCO_3^-]$ and P_{CO_2} values, so, for example, pH = 7.52 in a chicken does not necessarily indicate normal acid–base status. The primary cause of a chronic acid–base disturbance cannot be determined from P_{CO_2}, pH, and $[HCO_3^-]$ data alone. Other details of the disease history, pulmonary function, or blood chemistry must be obtained for a proper diagnosis.

The respiratory system controls pH primarily by changing arterial P_{CO_2} $\left(P_{aCO_2}\right)$. P_{aCO_2} is determined by parabronchial ventilation at any given metabolic rate (see Section 13.6.2). Increasing ventilation will decrease P_{aCO_2} and increase pHa, while decreasing ventilation will have the opposite effects. Therefore, ventilation is an extremely effective mechanism for changing pHa quickly, and ventilatory reflex responses to pH are the most important physiological mechanisms for rapid control of pH. The kidneys can also control pH by changing $[HCO_3^-]$ independent of CO_2 changes, as described in Chapter 12, but renal changes in pH generally take longer than respiratory changes in pH.

13.5.4 Blood Gas Measurements

Avian blood presents special challenges to the accurate measurement of P_{O_2}, P_{CO_2}, and pH with traditional equipment designed for humans. In contrast to mammals, avian erythrocytes are nucleated (as are most other vertebrates) and this may be the reason for high rates of O_2 consumption compared to mammals. Therefore, care must be taken to analyze arterial blood gases in birds as soon as possible after the sample is drawn, and to correct for any decreases in P_{O_2} with time if necessary. Storing the samples in ice water may help if immediate analysis is not possible, but the analyzed value can still differ from the *in vivo* value if any delay occurs, especially in normoxia where the blood–O_2 equilibrium curve is relatively flat and small O_2 content changes cause large changes in P_{O_2} (Scheid and Kawashiro, 1975). Sampling delays may also explain reports of 0 Torr P_{O_2} values in mixed venous blood from pigeons at simulated high altitude (Weinstein et al., 1985).

In addition to sampling delays, care must be taken to perform the analysis at body temperature, which is usually greater than the human value of 37 °C, or to apply temperature correction values established for avian blood (Kiley et al., 1979). Also, a blood–gas correction factor needs to be established with a tonometer to account for differences in P_{O_2} measured in the liquid phase after calibrating electrodes with a gas phase (Nightingale et al., 1968). These, and other factors which may affect O_2–hemoglobin saturation measurements, have been discussed in other reviews (Powell and Scheid, 1989).

Finally, the most important determinant of arterial blood gas values is the physiological state of the bird. Table 13.5 presents arterial blood gases for several species breathing room air, but these may not be "normal" if the bird was excited by the sampling procedure. Remote-controlled sampling devices have been used for resting ducks (Scheid and Slama, 1975), and this technology is being extended to other interesting conditions such as diving penguins (Ponganis et al., 2011) and flying geese (Meir et al., 2013).

13.6 PULMONARY GAS EXCHANGE

The unique anatomy of the avian respiratory system results, theoretically, in a model of gas exchange that is more efficient than the mammalian model (Piiper and Scheid, 1975). For a given level of ventilation to the gas exchange surfaces $(\dot{V}_P)$, cardiac output $(\dot{Q})$ and lung diffusing capacity $(D_{L_{O_2}})$, arterial O_2 loading, and CO_2 elimination are predicted to be better in a parabronchial lung, compared with an alveolar lung with the same inspired gases and

TABLE 13.5 Gas Exchange Variables in Awake Resting Birds

Reference	Pigeon (Bouverot et al., 1976)	Female Domestic Fowl (Piiper et al., 1970)	Pekin Duck (Bouverot et al., 1979)	Jones and Holeton (1972)	Muscovy Duck (*Cairina Moschata*) (Jones and Holeton, 1972)
M_b (kg)	0.38	1.6	2.37	2.4	2.16
$\dot{M}_{O_2}$ (mmol/min)	0.35	1.09	1.67	–	–
f_R (min^{-1})	27.3	23	15.6	8.2	10.5
V_T (mL)	7.5	33	58.5	98	69
$\dot{V}_E$ (L/min)	0.204	0.760	0.910	0.807	0.700
$\dot{Q}$ (L/min)	0.127	0.430	0.423	0.973	0.844
$P_{\bar{E}O_2}$ (Torr)	–	101.8	–	100.1	96.6
P_{aO_2} (Torr)	95	87	100	93.1	96.1
$P_{\bar{v}O_2}$ (Torr)	50	40.8	69.6	63.3	55.9
$P_{\bar{E}CO_2}$ (Torr)	–	33.0	–	34.2	34.2
P_{aCO_2} (Torr)	34	29.2	33.8	36.3	35.9
$P_{\bar{v}CO_2}$ (Torr)	–	39.3	–	37.3	42.6

Data collected from birds in body plethysmographs, except references Piiper et al., (1970) and Jones and Holeton (1972), which used endotracheal tubes in lightly restrained upright birds. $\dot{M}_{O_2}$ (mmol/min) is not given for reference d because mixed-expired gases were not measured.

metabolic demands (Powell and Scheid, 1989). This section describes the structural and functional basis for this model, and how it actually behaves in nature under physiological conditions.

13.6.1 Basic Principles of Oxygen Transport

13.6.1.1 Convection

Convection, or bulk flow of gas, is used to transport oxygen into the lungs by ventilation and to the tissues by blood flow. The Fick principle, which is simply conservation of mass applied to respiratory gas transport, can be used to quantify O_2 uptake as:

$$\dot{M}_{O_2} = \dot{V}\beta_{g_{O_2}}\left(P_{I_{O_2}} - P_{\dot{E}_{O_2}}\right)$$

where $\dot{M}_{O_2}$ is O_2 uptake, $\dot{V}$ is ventilation, $\beta_{g_{O_2}}$ is the capacitance coefficient for O_2 in the gas phase (0.512 mM/Torr at 41 °C) and $(P_{I_{O_2}} - P_{\bar{E}_{O_2}})$ is the difference between inspired and mixed-expired P_{O_2} (Piiper et al., 1971; Powell and Scheid, 1989). Inspired and expired ventilation are assumed equal in this formulation of the Fick principle, so the amount of O_2 consumed is the difference between the amount of O_2 inspired and the amount expired. Mixed-expired P_{O_2} is used when total ventilation is measured; end-expired P_{O_2} is used if parabronchial ventilation $(\dot{V}_P)$ is available (see Section 13.5.4.1). The same principles described for O_2, also apply to CO_2 exchange.

The Fick principle can be written for the cardiovascular transport of O_2 out of the lungs and to the tissues also:

$$\dot{M}_{O_2} = \dot{Q}\beta_{b_{O_2}}\left(P_{a_{O_2}} - P_{\dot{v}_{O_2}}\right)$$

where $\dot{Q}$ is cardiac output and $\beta_{b_{O_2}}$ is the physiological slope of the blood–O_2 equilibrium curve (see Section 13.4.1.3). Hence, the amount of O_2 taken up by blood in the lungs is the difference between the amount of O_2 leaving the lungs in arterial blood, and the amount that entered the lungs in mixed-venous blood. In a steady state, $\dot{M}_{O_2}$ is equal at each step of the O_2 cascade, so the Fick principle can be rearranged to calculate $\dot{M}_{O_2}$, $\dot{Q}$, $\dot{V}$ or P_{O_2} from measurements of other variables in the equation. Table 13.5 lists the important variables for quantifying gas exchange in several birds under resting conditions. Changes in these variables under different physiological conditions such as exercise, hypoxia, and thermal stress have been summarized in several reviews (Brackenbury, 1984; Butler, 1991; Faraci, 1991; Powell and Scheid, 1989).

13.6.1.2 Diffusion

O_2 movement over the very short distances across the blood-gas barrier occurs effectively by the "passive" mechanism of diffusion; active transport of O_2 does not occur in the body. Fick's law of diffusion describes O_2 transport from the air capillaries to the blood capillaries as:

$$\dot{M}_{O_2} = \Delta P_{O_2} \cdot D_{L_{O_2}}$$

where ΔP_{O_2} is the average P_{O_2} gradient between the air capillary and blood in the pulmonary capillary and $D_{L_{O_2}}$ is the diffusing capacity of the lung for O_2. The equation shows that a larger $D_{L_{O_2}}$ can transport more O_2 for a given P_{O_2} gradient. Determinants of $D_{L_{O_2}}$ are described in the section on lung diffusing capacity below (see Section 13.6.3).

13.6.2 Cross-Current Gas Exchange

13.6.2.1 O_2 Exchange

Figure 13.10 shows how airflow and blood flow in the parabronchus can be viewed as occurring perpendicular to one another, and this is the basis for describing avian gas exchange with a "cross-current" model (Piiper and Scheid, 1972, 1975). In this idealized model of cross-current exchange, airflow is assumed continuous through the parabronchus, which is uniformly perfused along its length by mixed-venous blood. At the inspiratory end of the parabronchus, there is a large P_{O_2} gradient driving diffusion of O_2 into the capillary blood, which raises capillary P_{O_2} and drops parabronchial P_{O_2}. As air flows along the parabronchus, the P_{O_2} gradient driving O_2 diffusion decreases with parabronchial P_{O_2}, while $P_{\bar{v}_{O_2}}$ is constant. At the expiratory end of the parabronchus, P_{O_2} has decreased to end-expired levels, and P_{O_2} in the capillary blood leaving this part of the parabronchus is correspondingly low. However, notice that arterialized blood returning to the heart is a mixture of capillary blood draining the entire length of the parabronchus. Therefore, arterial P_{O_2} is greater than end-expired P_{O_2} in ideal cross-current gas exchange.

A negative expired-arterial P_{O_2} difference, shown in Figure 13.10 as the overlap of P_{O_2} arrows, is not possible in alveolar gas exchange. The best situation that can be achieved in ideal alveolar gas lungs is equilibrium between expired and arterial P_{O_2}, so $(P_{E_{O_2}} - P_{a_{O_2}}) = 0$. However, negative expired-arterial P_{O_2} differences are not always observed in birds and arterial P_{O_2} is similar in resting, normoxic birds and mammals (Table 13.5). As explained in the following sections, cross-current O_2 exchange is theoretically more efficient than alveolar exchange, but similar magnitudes of gas exchange limitations have larger effects in birds than in mammas (see Section 13.6.4).

Notice that the efficiency of cross-current exchange should not depend on the direction of ventilatory flow in the parabronchus. This is supported by experiments which reversed the direction of parabronchial flow in ducks and chickens and found no difference in the expired-arterial P_{O_2} differences for O_2 or CO_2 (Powell, 1982; Scheid and Piiper, 1972). It is physiologically significant because airflow is presumably bidirectional in neopulmonic parabronchi (see Section 13.2.3).

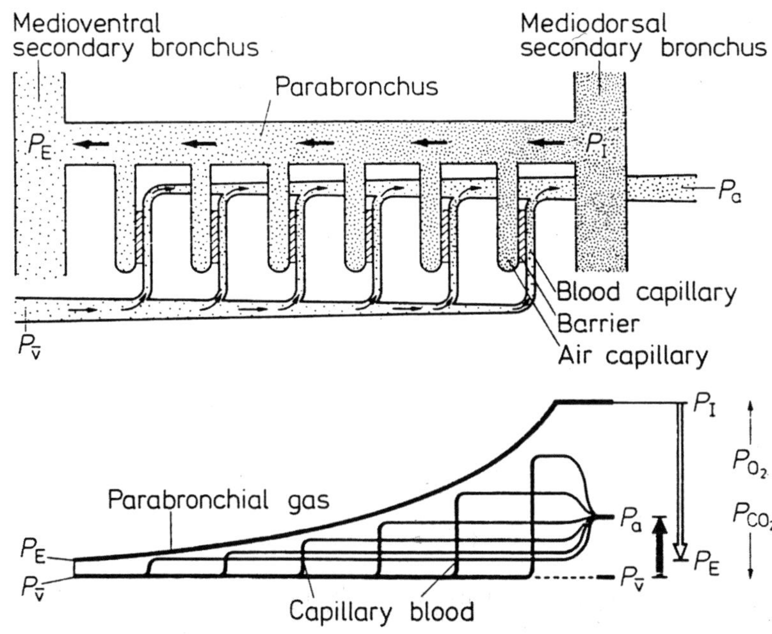

FIGURE 13.10 Cross-current model of gas exchange in the parabronchus as described in the text. Lower panel shows how $P_{a_{O_2}}$ results from a mixture of blood from capillaries all along the parabronchus, where P_{O_2} ranges from P_i to P_e. Overlap of P_{O_2} in gas (open arrows) and blood (filled arrows) shows how $P_{a_{O_2}}$ can exceed $P_{E_{O_2}}$ in birds. *From Scheid (1990).*

13.6.2.2 Cross-Current CO_2 Exchange

Figure 13.10 shows how expired P_{CO_2} can exceed arterial P_{CO_2} in a parabronchus, which is also impossible in alveolar gas exchange. This occurs similar to the "overlap" of arterial and expired P_{O_2} as explained above. However, experimental observations of expired P_{CO_2} exceeding *mixed-venous* P_{CO_2} in birds can only be explained by an interesting interaction of O_2 and CO_2 exchange in the parabronchus. Because of the shape of the O_2 and CO_2 blood equilibrium curves, the ratio of CO_2 elimination to O_2 uptake (R=respiratory exchange ratio) decreases near the expiratory ends of the parabronchus (Meyer et al., 1976). As originally postulated by Zeuthen (1942), this can lead to oxygenation of mixed-venous blood in expiratory ends of the parabronchi, and increase P_{CO_2} in capillaries by the Haldane effect. Because P_{CO_2} of oxygenated mixed-venous blood is greater than $P_{\dot{v}O_2}$ (Figure 13.9), P_{CO_2} in gas equilibrating with such blood at the expiratory end of the parabronchus (i.e., $P_{E_{CO_2}}$) can exceed true $P_{\dot{v}O_2}$.

Overlap between arterial and expired P_{CO_2} is more commonly observed than overlap for O_2 in birds (Table 13.5). This is because of differences in the O_2 and CO_2 blood equilibrium curves, and the fact that CO_2 is less sensitive than O_2 to some of the factors limiting the efficacy of cross-current gas exchange (Powell and Scheid, 1989).

CO_2 exchange can be affected by the same factors limiting O_2 exchange, which are considered in detail below. However, $P_{a_{CO_2}}$ is most sensitive to changes in $(\dot{V}_{D_{phys}})$ and effective $\dot{V}_P$. CO_2 should not be diffusion limited either because O_2 and CO_2 require similar times for diffusion equilibrium (Wagner, 1977). $\dot{V}/\dot{Q}$ mismatching and shunt increases $P_{a_{CO_2}}$ less than they decrease $P_{a_{O_2}}$ because of differences between the O_2 and CO_2–blood equilibrium curves (Powell and Scheid, 1989).

13.6.3 Lung Diffusing Capacity

The diffusing capacity of the lung for $O_2\,(D_{L_{O_2}})$ is a complex variable that depends on several physiological processes (Powell, 1982; Powell and Scheid, 1989). These include gas-phase diffusion in the air capillaries, diffusion across the blood–gas barrier, and the chemical reaction between O_2 and hemoglobin.

13.6.3.1 Gas Transport in Air Capillaries

Ventilatory flow in the parabronchi prevents significant P_{O_2} gradients from developing in the lumen of the parabronchus. However, the mechanism of gas transport in the air capillaries is not clear and P_{O_2} gradients could occur between the lumen and the depths of the parabronchial mantle. Originally, it was assumed that diffusion explained gas transport over the small dimensions of the air capillary, considering the small distances involved and effectiveness of diffusive gas transport in the functional gas exchange unit of alveolar lungs. Indeed, measurements show no significant diffusive resistance to gas exchange along the air capillaries (Burger et al., 1979; Crank and Gallagher, 1978; Scheid, 1978). If diffusion is the main gas transport mechanism within air capillaries, then air–blood capillary exchange could be enhanced by an arrangement similar to counter-current exchange (Scheid, 1978). O_2 would diffuse from the lumen towards the periphery in air capillaries, which is opposite to the direction of flow in the blood capillaries (Figures 13.4 and 13.10). Therefore, blood at the end of a capillary would equilibrate with the relatively high P_{O_2} levels in air capillaries near the parabronchial lumen. However, there is no experimental evidence for this; more recent morophometric studies of the air capillary network (Figure 13.5) show extensive connections and branching, in contrast to the dead end tubes in Figure 13.10. This could result in convective flow instead of only diffusion in air capillaries (Maina et al., 2010).

13.6.3.2 Blood–Gas Barrier Diffusion

The effect of the blood-gas barrier on diffusion in the lung can be evaluated by the membrane diffusing capacity $(D_{m_{O_2}})$, which can be estimated from morphometric measurements of the lung (Maina, 1989; Powell and Scheid, 1989). $D_{m_{O_2}}$ is directly proportional to the surface area, and inversely proportional to the thickness of the blood-gas barrier. These variables have been measured in several species of birds now, using perfusion fixation and rapid freezing to preserve tissue for electron microscopy and stereological analysis (Dubach, 1981; Maina et al., 1982, 1989; Powell and Mazzone, 1983; Powell and Mazzone, 1983). The values depend on body size (which determines metabolic levels; see Chapter 9), but in general they show that $D_{m_{O_2}}$ is larger in birds than in comparably sized terrestrial mammals (Maina, 1989). For example, $D_{m_{O_2}}$ in Canada geese is 1.7 times greater than a comparably sized mammal (Powell and Mazzone, 1983). Also, the thickness of the blood-gas barrier is 2.5 times thinner, and of much more uniform thickness, than in mammals (Maina et al., 1989; West, 2009). Interestingly, $D_{m_{O_2}}$ is similar in birds and bats of the same body size, indicating the importance of favorable diffusion for high O_2 consumption levels during flight (Maina, 1989).

13.6.3.3 O_2–Hemoglobin Reaction Rates

Finite reaction rates between O_2 and hemoglobin behave as an additional resistance to O_2 uptake across the blood–gas barrier. Consequently, $D_{L_{O_2}}$ increases if there is more hemoglobin available with increased capillary volume. Pulmonary capillary blood volume is similar in birds and terrestrial mammals of similar body sizes, except for chickens, which have a relatively low capillary volume (Maina, 1989). Also, estimates of the reaction rates between O_2 and avian hemoglobin in birds and mammals are similar (Phu et al., 1986). Therefore, O_2–hemoglobin reaction rates probably contribute a similar amount to diffusion resistances in avian and mammalian lungs. In mammals, this is estimated to comprise about half of the diffusive resistance to O_2 diffusion.

13.6.3.4 Physiological Estimates of $D_{L_{O_2}}$

$D_{L_{O_2}}$ can be estimated from experimental measurements on birds if certain conditions are met to satisfy the assumptions necessary for an ideal cross-current analysis of the data. For example, measurements should be made in hypoxia, where the O_2–blood equilibrium curve is linear, to satisfy the assumption of constant βb_{O_2}. Physiological measurements of $D_{L_{O_2}}$ that satisfy these assumptions have been made in ducks and chickens (Burger et al., 1979; Scheid and Piiper, 1970), and they generally agree with morphometric estimates (Maina and King, 1982). Physiological $D_{L_{O_2}}$ in ducks ranges from 38 to 68 μmol/(min Torr kg) (Hempleman and Powell, 1986). Other methods for estimating potential diffusion limitations indicate complete diffusion equilibrium for O_2 in resting birds (Powell and Scheid, 1989).

Exercise and pharmacological stimulation of metabolic rate increase $D_{L_{O_2}}$ in ducks (Geiser et al., 1984; Hempleman and Powell, 1986; Kiley et al., 1985), and the change is correlated closely with increased cardiac output (Hempleman and Powell, 1986). In alveolar lungs, $D_{L_{O_2}}$ increases with cardiac output by recruitment and distention of the pulmonary capillaries, which increases surface area and capillary volume. However, these enhancements can be offset by shorter transit times in the pulmonary capillaries allowing less time for diffusion equilibrium. Recruitment and distention do not occur in avian lungs (see Section 13.3.2), so mechanisms of increasing $D_{L_{O_2}}$ in birds during exercise are not known.

13.6.4 Heterogeneity in the Lung

The avian lung is a complex structure consisting of hundreds of parabronchi. Mismatching of ventilation, blood flow, or diffusing capacity between these functional units can reduce the efficacy of gas exchange. Temporal variations in flow rates and inspired gas composition can also impair gas exchange. Under normal resting conditions at sea level, such heterogeneity in lung function is the most important factor reducing gas exchange efficacy from ideal levels in birds and in mammals.

13.6.4.1 Physiological Dead Space

Physiological dead space is defined as the difference between total inspired or expired ventilation ($\dot{V}_E$ or $\dot{V}_I$) and effective parabronchial ventilation ($\dot{V}_P$):

$$\dot{V}_{D_{phys}} = \dot{V}_I - \dot{V}_P$$

$\dot{V}_{D_{phys}}$ includes anatomic dead space ventilation plus any heterogeneity such as inspiratory or expiratory mesobronchial shunts (see Section 13.3.3.4) or ventilation to regions of the lung with high $\dot{V}/\dot{Q}$ ratios (see Section 13.5.4.3). Therefore, $\dot{V}_{D_{phys}}$ considers ventilation "as if" total ventilation was partitioned between a single ideal parabronchus with $\dot{V}_P$ and anatomic dead space (Scheid and Piiper, 1989).

Many of the techniques used to estimate $\dot{V}_{D_{phys}}$ in alveolar lungs are not applicable to birds (Powell, 1988). However, with a computer model of O_2 and CO_2 dissociation curves and cross-current gas exchange, $\dot{V}_P$ can be calculated from measured ventilation, mixed-venous blood, and mixed-expired gas (Hastings and Powell, 1986b). $\dot{V}_{D_{phys}}$ in artificially ventilated ducks was almost 10 mL greater than anatomic plus instrument dead space, and two-thirds of the 15 mL anatomic dead space (Hastings and Powell, 1986b). This large amount of physiological dead space is consistent with relatively large amounts of total ventilation going to regions of the lung with high ventilation/perfusion ratios ($\dot{V}/\dot{Q}$; see Section 13.5.4.3), and indicates that ventilatory heterogeneity within the lung has a significant impact on $\dot{V}_P$ and gas exchange.

13.6.4.2 Shunt

Shunting of pulmonary blood flow past effective gas exchange areas is very small in birds. As described in Section 13.4.1, there are no arteriovenous anastomoses in the pulmonary circulation. Shunt ranges from less than 1% to 2.7% of cardiac output in anesthetized artificially ventilated geese and ducks, respectively, using inert gas methods to quantify true intrapulmonary shunt (Burger et al., 1979; Powell and Wagner, 1982). Oxygen can be used to quantify intrapulmonary plus extrapulmonary shunts, such as drainage of systemic venous blood from bronchial or Thebesian veins into pulmonary venous blood. Oxygen shunts average 6.3–8% of cardiac output in ducks, which is much greater than would be predicted given the magnitude of Thebesian and bronchial circulations in mammals (Bickler et al., 1986). One possible explanation for this large shunt is the connections between the vertebral venous and pulmonary circulations described for chickens (Burger and Estavillo, 1977), which also occur in ducks (Bickler et al., 1986). The sensitivity of extrapulmonary shunts to various physiological conditions has not been measured but $P_{a_{O_2}}$ is greater than predicted in hypoxia if such shunts persisted (see below).

13.6.4.3 $\dot{V}/\dot{Q}$ Mismatching

Differences in the ventilation/perfusion ratio ($\dot{V}/\dot{Q}$) between individual parabronchi is the main factor reducing arterial P_{O_2} from ideal cross-current levels in birds under resting conditions at sea level (Powell and Scheid, 1989). Ventilation and blood flow can differ between parabronchi depending on small differences in resistance or pressure gradients that can occur along the multiple parallel pathways through the lung. Physiological mechanisms, such as smooth muscle tone in the bronchi and parabronchi (see Section 13.2.2.2) or interparabronchial arteries (see Section 13.4.3.1) may act to reduce such heterogeneity, but matching is never perfect.

It is important to point out that the effect of such spatial $\dot{V}/\dot{Q}$ heterogeneity on gas exchange is different than the effect of changes in the *overall* $\dot{V}/\dot{Q}$ ratio. The overall $\dot{V}/\dot{Q}$ ratio can affect P_{aO_2}, so, for example, decreases in ventilation at constant cardiac output cause decreases in P_{aO_2}. Also, the overall $\dot{V}/\dot{Q}$ ratio affects the magnitude of the arterial-expired P_{O_2} difference in a perfectly homogeneous cross-current gas exchanger (Powell and Scheid, 1989). However, spatial mismatching of $\dot{V}/\dot{Q}$ ratios between parabronchi will decrease P_{aO_2} further and make the arterial-expired P_{O_2} difference more positive than predicted for a homogeneous cross-current exchanger with the same overall $\dot{V}/\dot{Q}$ ratio; this is analogous to increasing the ideal alveolar-arterial P_{O_2} difference in mammals.

Several techniques have been used to measure distributions of $\dot{V}/\dot{Q}$ ratios in the avian lung, but they are relatively complicated and have not been applied yet to awake birds under many physiological conditions lung (Burger et al., 1979; Hempleman and Powell, 1986; Powell, 1988; Powell and Wagner, 1982). Significant amounts of ventilation go to high $\dot{V}/\dot{Q}$ regions of the lung in some cases (Powell and Wagner, 1982), and this contributes to physiological dead space. Heterogeneity in $\dot{V}/\dot{Q}$ ratios near the overall *parabronchial* $\dot{V}/\dot{Q}$ ratio has a large impact on O_2 exchange in birds. Compared to mammals, $\dot{V}/\dot{Q}$ heterogeneity is slightly greater (Powell and Wagner, 1982) and cross-current gas exchange is more sensitive to $\dot{V}/\dot{Q}$ mismatching than alveolar gas exchange (Powell and Hempleman, 1988; Powell and Scheid, 1989). In normoxic artificially ventilated geese, P_{aO_2} is 25 Torr less than the ideal level predicted for homogeneous cross-current gas exchange, which is significantly greater than the ideal-measured (alveolar-arterial) P_{O_2} difference in typical mammals (Powell and Hempleman, 1988). Hence, most of this drop in P_{aO_2} from ideal levels (15 Torr) is explained by $\dot{V}/\dot{Q}$ heterogeneity (Powell, 1993). The remaining 10 Torr difference between measured and ideal P_{aO_2} can be explained by postpulmonary shunts, which are virtually nonexistent in mammals. Pulmonary shunt is small (Bickler et al., 1986; Powell and Wagner, 1982) and there is no evidence for a diffusion limitation in normoxia at rest (Powell and Scheid, 1989).

In addition to parallel $\dot{V}/\dot{Q}$ mismatching between parabronchi, serial $\dot{V}/\dot{Q}$ mismatch can occur along a single parabronchus if the longitudinal distribution of blood flow is not even. Several studies have shown that blood flow is greater at the inspiratory ends of the parabronchi (Holle et al., 1978; Jones, 1982; Parry and Yates, 1979). However, such serial heterogeneity does not affect gas exchange unless there is a diffusion limitation (Holle et al., 1978; Powell and Scheid, 1989), and this does not occur under most conditions (see Section 13.6.3.4).

13.6.4.4 Temporal Heterogeneity

Changes in instantaneous ventilation of the parabronchi during normal breathing could in temporal variations in the $\dot{V}/\dot{Q}$ ratio. In theory, this could significantly decrease P_{aO_2} relative to the ideal level predicted for continuous ventilation (Powell, 1988; Powell and Scheid, 1989). For example, a ventilatory pause could act like a breath-hold, and rapidly decrease P_{O_2} in the small gas volume of the avian lung (Powell and Scheid, 1989). However, experiments indicate that this does not occur during normal breathing because the effective parabronchial gas volume is increased by mixing with larger bronchi (Scheid et al., 1977). Temporal changes in P_{O_2} entering the parabronchi during breathing could also impact P_{aO_2}, but the effects are predicted to be relatively small (Powell, 1988). A time-averaged P_{IO_2} can be calculated for the parabronchi, and it is similar to caudal air sac P_{O_2} (Scheid et al., 1978).

13.6.5 Frontiers: Gas Exchange during High-Altitude Flight

How birds manage high levels of O_2 consumption for flight in extreme hypoxia at very high altitudes remains a fascinating and unanswered question. Birds are generally more tolerant of hypoxia than mammals (Tucker, 1972) and the first experiments actually measuring oxygen exchange during flight at high altitude used hummingbirds hovering in a hypobaric chamber (Berger, 1974). $\dot{M}_{O_2}$ during hovering at sea level was 32 mmol/(kg min), which is less than the maximal $\dot{M}_{O_2}$ for a hummingbird (Wells, 1993), However, this is greater than the maximal $\dot{M}_{O_2}$ for a comparably sized mammal. Furthermore, the hummingbird was able to maintain this high level of $\dot{M}_{O_2}$ at 6000 m simulated altitude, whereas maximal $\dot{M}_{O_2}$ is reduced to half the sea level value in mammals that have been studied at this altitude. Other experiments have studied $\dot{M}_{O_2}$ in hummingbirds with graded levels of exercise at different levels of O_2 (Chai and Dudley, 1996) but maximal $\dot{M}_{O_2}$ at different altitudes has not been determined in any bird to date.

Measuring the relevant physiological variables to understand how gas exchange changes in birds flying at high altitude is a huge experimental challenge. Consequently, most of the information available is from studies of resting or anesthetized birds exposed to hypoxia. Experiments on resting ducks and bar-headed geese show extremely high levels of ventilation at very high altitudes (Black and Tenney, 1980b; Powell et al., 2004), and these are predicted to eliminate the advantage of cross-current gas exchange, compared with alveolar exchange (Shams and Scheid, 1989). However, these predictions are extremely sensitive to cardiac output, which varies greatly in the same species studied during hypoxic rest in different laboratories (Black and Tenney, 1980b; Shams and Scheid, 1989). Depending on the cardiac output, the advantage of cross-current compared to alveolar gas exchange could increase P_{aO_2} a couple of Torr in birds at 11 km altitude. Such a small change is physiologically significant, however, when the maximum P_{O_2} gradient between inspired gas and mixed-venous blood is only 20 Torr (Powell, 1993).

Measurements of gas exchange limitations in anesthetized ducks during hypoxia show the difference between ideal and measured P_{aO_2} is only a few Torr (Powell, 1993), which contrasts with the relatively large effects of gas exchange limitations in normoxic birds (see above). This is because the effects of $\dot{V}/\dot{Q}$ mismatching on P_{aO_2} are decreased in hypoxia, when exchange operates on the steep portion of the O_2-hemoglobin equilibrium curve (see Section 13.6.3.4). Indeed, $\dot{V}/\dot{Q}$ distributions do not change with hypoxia versus normoxia in anesthetized ducks (Powell and Hastings, 1983). A small diffusion limitation may also contribute to the difference in measured P_{aO_2} from the ideal value predicted for cross-current exchange in hypoxia. However, postpulmonary shunts measured in normoxia (Bickler et al., 1986) must be absent in hypoxia or P_{aO_2} would be lower.

Other experiments measuring all of the variables necessary for a quantitative analysis of gas exchange during exercise in hypoxia have been done in running water-fowl (Kiley et al., 1985) and emus (Schmitt et al., 2002). Such studies can reveal unique features of the avian respiratory system that could provide an advantage over alveolar gas exchange but they are limited in only increasing $\dot{M}_{O_2}$ about threefold from resting levels. Studies on the emu showed no change in $\dot{V}/\dot{Q}$mismatching in hypoxic versus normoxia, similar to results in anesthetized ducks (Powell and Hastings, 1983). However, $\dot{V}/\dot{Q}$ mismatching did not change with exercise in the emus either, which contrasts with increased $\dot{V}/\dot{Q}$ mismatching in exercising mammals and may provide an avian advantage at altitude (Powell and Hopkins, 2010).

With advances in instrumentation, it is becoming possible to measure more of the relevant gas exchange variables in flying birds. Preliminary reports of metabolic rate in bar-headed geese flying in a wind tunnel show decreases in $\dot{M}_{O_2}$ when inspired O_2 is decreased to simulate altitudes of 5500 and 8500 m while flight speed is maintained (Meir et al., 2013). This suggests flight may become more efficient in hypoxia, similar to apparent increases in swimming efficiency during prolonged dives in penguins and marine mammals (Ponganis et al., 2011). More interesting for pulmonary gas exchange, arterial P_{O_2} in the flying bar-headed geese was equal to or greater than resting values. Emus running in hypoxia also increased P_{aO_2} and decreased P_{aCO_2} (Schmitt et al., 2002), suggesting part of the advantage birds have over mammals at high altitude is greater hyperventilation. This would involve differences in ventilatory control more than lung structure and function *per se* (see Section 13.7.3.4). Other factors that may allow birds to exercise at extreme altitudes are considered in later sections of this chapter and have been reviewed elsewhere (Faraci, 1991; Fedde, 1990; Scott and Milsom, 2006). However, determining how much of an advantage the unique structure and function of the avian respiratory system provides over alveolar lungs at altitude awaits more detailed physiological measurements from birds exercising maximally in hypoxia.

13.7 TISSUE GAS EXCHANGE

O_2 moves out of systemic capillaries to the mitochondria in cells by diffusion. Therefore, O_2 transport in tissues is described by Fick's first law of diffusion, similar to diffusion across the blood-gas barrier in the lung:

$$\dot{M}_{O_2} = \Delta P_{O_2} \cdot D_{t_{O_2}}$$

where ΔP_{O_2} is the *average* P_{O_2} gradient between capillary blood and the mitochondria, and $D_{t_{O_2}}$ is a tissue diffusing capacity for O_2, analogous to the lung diffusing capacity (see Section 13.6.1.2). The main difference between O_2 diffusion in tissue and in the lung is that diffusion pathways are much greater in tissue. Tissue capillaries may be 50 μm apart, so the distance from a capillary surface to mitochondria can be 50 times longer than the thickness of the blood-gas barrier (<0.5 μm).

13.7.1 Microcirculation

13.7.1.1 Skeletal Muscle

Long diffusion distances can lead to significant P_{O_2} gradients in muscle. Also, the P_{O_2} gradient varies along the length of a capillary as O_2 leaves the blood, and capillary P_{O_2} decreases from arterial to venous levels. However, birds have some unique structural features in the skeletal muscle microcirculation to minimize diffusion distances and enhance tissue gas exchange. For example, the number of capillaries per cross-sectional area of flight muscle fiber in hummingbirds is six times greater than the value for rat soleus muscle, and the value in pigeon flight muscle is three times greater (Mathieu-Costello, 1993). This clearly decreases radial diffusion distances for O_2 leaving capillaries in birds.

In addition, skeletal muscle capillaries in birds are very tortuous and have extensive manifolds connecting capillaries running along adjacent muscle fibers (Mathieu-Costello, 1991; Mathieu-Costello et al., 1992). This geometry increases the exchange surface area so a muscle fiber is functionally surrounded by a "sheet" of capillary blood. This provides better tissue oxygenation than the traditional mammalian model of straight capillaries running along a muscle fiber (i.e., the Krogh cylinder); Krogh's model predicts that P_{O_2} at the venous end of the capillary may be zero when O_2 supply decreases or demand increases (Mathieu-Costello, 1991).

13.7.1.2 Cerebral Circulation

Most evidence indicates cerebral blood flow increases with hypoxia but does not change with CO_2 in birds (Faraci, 1991). This results in significant improvements in tissue O_2 delivery during hypoxia (Faraci et al., 1984; Grubb et al., 1977). In mammals, hypoxia also increases cerebral blood flow, but this is partially offset by a vasoconstrictor effect of the decrease in P_{aco_2} that accompanies the reflex increase in ventilation during hypoxia (see Section 13.7.3.2). This difference in cerebral vascular control may help explain how some birds are able to tolerate severe hypoxia better than some mammals (Faraci, 1986).

13.7.2 Myoglobin

Myoglobin is an O_2-binding protein, similar to a single polypeptide chain of the hemoglobin molecule (see Section 13.4.1.1), which has an extremely high affinity for O_2. For example, the P_{50} for hummingbirds myoglobin is 2.5 Torr (Johansen et al., 1987), so myoglobin can readily accept O_2 from capillary blood. Consequently, myoglobin is thought to be important for facilitating O_2 diffusion in muscle, by shuttling O_2 to sites far away from capillaries, or towards the venous end of capillaries. High levels of myoglobin are present in the heart and skeletal muscles of diving birds (Giardina et al., 1985; Weber et al., 1974), birds native to high altitudes (Fedde, 1990) and in hummingbirds with extremely high metabolism (Johansen et al., 1987). Increases in myoglobin with physical training in birds provide further evidence for the physiological significance of myoglobin in tissue gas exchange (Butler and Turner, 1988). M_b is also a significant O_2 store in diving birds, carrying about one-third of the total O_2 stores while the respiratory system and blood carry similar amounts (Ponganis et al., 2011).

13.7.3 Effects of Hypoxia and Exercise

Generally, decreases in O_2 supply (e.g., hypoxia) or increases in O_2 demand (e.g., exercise) are satisfied at the tissue level by increased O_2 extraction from blood or increasing blood flow. This is illustrated by the Fick equation (see Section 13.5.1.1) applied to the cardiovascular system:

$$\dot{M}_{O_2} = \dot{Q}\left(C_{ao_2} - C_{\dot{v}o_2}\right)$$

Increased O_2 extraction from the blood decreases $C_{\dot{v}o_2}$ but $P_{\dot{v}o_2}$ does not decrease much because the slope of the blood–O_2 equilibrium curve is steep around the venous point (Figure 13.8). High $P_{\dot{v}o_2}$ levels are advantageous by keeping average capillary P_{O_2} levels high to drive O_2 diffusion into tissue. O_2 consumption is maintained in duck skeletal muscle during hypoxia without any change in blood flow (Grubb, 1981). During severe hypoxia in resting ducks and bar-headed geese, $C_{\dot{v}o_2}$ can be less than 0.5 mM (Black and Tenney, 1980b). This suggests nearly complete O_2 extraction, although such low venous O_2 values could also result from measurement error (see Section 13.4.5). O_2 extraction in the cerebral and coronary circulations is not known for birds.

In exercising birds, extraction and blood flow increase to satisfy metabolic demands (Faraci, 1986; Faraci et al., 1984; Fedde, 1990). Increasing blood flow helps maintain high average capillary P_{O_2} because it raises mixed-venous O_2 concentration for any given arterial concentration and O_2 consumption (see equation above). Increases in blood flow are also observed in the ventilatory muscles of resting birds during hypoxia, which presumably reflects increased work in these muscles during increased breathing (Faraci, 1986). Under some extreme conditions of hypoxic exercise, muscle blood flow or tissue O_2 diffusion may actually limit maximal $\dot{M}_{O_2}$ in birds (Fedde et al., 1989).

Capillaries in avian skeletal muscle also show physiological plasticity and apparent genetic adaptations to hypoxia. Chronic hypoxia increases the capillary-fiber surface area contact for aerobic flight muscles in pigeons (Mathieu-Costello and Agey, 1997) and bar-headed geese have more capillaries per muscle fiber than expected for their aerobic capacity, in comparison to other geese native to lower altitudes (Scott et al., 2009). Capillaries are also denser and more evenly spaced, and mitochondria are located closer to the capillaries in the bar-headed goose compared to other geese, but the respiratory capacities, O_2 kinetics, and phosphorylation efficiencies of isolated mitochondria were not different between species (Scott et al., 2009). Hence, acclimatization and adaptation to limited O_2 appear to involve changes in O_2 delivery versus utilization (i.e., mitochondrial respiration).

13.8 CONTROL OF BREATHING

Breathing originates as rhythmic motor output from the central nervous system. This basic respiratory rhythm is modulated by several reflexes, in response to changes in activity and the environment. These reflexes are examples of negative feedback control, and tend to maintain arterial blood gases and pH homeostasis (Table 13.6). For example, if dead space increases, then P_{aco_2} will increase if ventilation is constant. However, increased P_{aco_2} stimulates an increase in tidal volume, which compensates for the increased dead space and returns P_{aco_2} towards the control value. Like all reflexes, the ventilatory reflexes include (1) a sensory or afferent component, (2) an integrating component in the central nervous system (CNS), and (3) a motor or efferent component.

TABLE 13.6 Arterial Blood Gases and pH in Unanesthetized Birds Breathing Air

Bird	P_{O_2} (Torr)	P_{CO_2} (Torr)	pH
Female black bantam chicken (Calder and Schmidt-Nielsen, 1968)	–	29.9	7.48
Female white leghorn chicken (Kawashiro and Scheid, 1975)	82	33.0	7.52
Male white rock chicken (Calder and Schmidt-Nielsen, 1968)	–	29.2	7.53
Mallard duck (Butler and Taylor, 1983)	81	30.8	7.46
Muscovy duck (Kawashiro and Scheid, 1975)	82	38.0	7.49
Muscovy duck (Jones and Holeton, 1972)	96.1	35.9	7.46
Pekin duck (Black and Tenney, 1980b)	93.5	28.0	7.46
Pekin duck (Bouverot et al., 1979)	100	33.8	7.48
Emu (Jones et al., 1983)	99.7	33.5	7.45
Bar-headed goose (Black and Tenney, 1980)	92.5	31.6	7.47
Domestic goose (Scheid et al., 1991)	97	32	7.52
Herring gull (Calder and Schmidt-Nielsen, 1968)	–	27.2	7.56
Red-tailed hawk (Kollias and McLeish, 1978)	108	27.0	7.49
Burrowing owl (Kilgore et al., unpublished data)	97.6	32.6	7.46
White pelican (Calder and Schmidt-Nielsen, 1968)	–	28.5	7.50
Adelie penguin (Murrish, 1982)	83.8	36.9	7.51
Chinstrap penguin (Murrish, 1982)	89.1	37.1	7.52
Gentoo penguin (Murrish, 1982)	77.1	40.9	7.49
Pigeon (Powell, 1983a)	95	30	7.503
Roadrunner (Calder and Schmidt-Nielsen, 1968)	–	24.5	7.58
Abdim stork (Marder and Arad, 1975)	–	27.9	7.56
Mute swan (Bech and Johansen, 1980)	91.3	27.1	7.50
Turkey vulture (Calder and Schmidt-Nielsen, 1968)	–	27.5	7.51

This topic bridges respiratory physiology and neuroscience, but the emphasis here is on respiratory aspects and ventilatory reflex responses to changes in blood gases, often called the chemical control of ventilation. Several excellent reviews cover more details about the neuroscience of ventilatory control and control of breathing under different physiologic conditions (Bouverot, 1978; Davey and Seller, 1987; Gleeson and Molony, 1989; Jones and Milsom, 1982; Powell, 1983b; Scheid and Piiper, 1986; Taylor et al., 1999).

13.8.1 Respiratory Rhythm Generation

The basic respiratory rhythm is generated by a central pattern generator, composed of networks of neurons in the brainstem of the CNS. Respiratory rhythm can be measured in neural outputs from isolated hindbrain of chicks (Fortin et al., 1994), and transecting the brainstem between the XI and XII cranial nerve roots results in apnea and eventual death in pigeons (von Saalfeld, 1936). Reciprocal inhibition between medullary inspiratory and expiratory neurons are a common feature in birds (Peek et al., 1975), but experiments in mammals show that respiratory rhythm generation can occur without inhibitory synaptic interactions, implicating pacemaker processes (Smith et al., 1991). Other CNS structures important for sending motor outputs to ventilatory muscles in birds (Table 13.2) have been identified with stimulation experiments to study vocalization (Peek et al., 1975) and panting (Richards, 1971) in chickens or pigeons (Davey and Seller, 1987). The development of central pattern generators for bird song, which requires respiratory muscles, requires auditory feedback (Konishi, 2010) but such interactions have not been investigated for breathing.

13.8.2 Sensory Inputs

13.8.2.1 Central Chemoreceptors

In mammals, relatively discrete regions on the ventrolateral surface of the medulla in mammals show chemosensitivity to changes in P_{aCO_2} and local pH. These so-called central chemoreceptors can explain most of the reflex increase in ventilation when P_{aCO_2} increases in mammals (Bouverot, 1978; Guyenet et al., 2010). Central chemoreceptors have not been identified by neurophysiological or anatomic methods in birds. However, conscious ducks increase ventilation when P_{aCO_2} is increased in blood perfusing only the head, indicating an important physiological role for central chemoreceptors in birds (Milsom et al., 1981; Sèbert, 1979).

13.8.2.2 Arterial Chemoreceptors

Arterial chemoreceptors are sensitive to changes in P_{aO_2}, P_{aCO_2} and pH, and they explain all of the ventilatory response to hypoxia in birds and mammals (Bouverot, 1978). They are also important for the ventilatory response to CO_2 and pH. The carotid bodies are very small (<1 mm diameter) organs located bilaterally in the thorax between the carotid artery and the nodose ganglion of the vagus nerve (Adamson, 1958; Hempleman and Warburton, 2013; Kameda, 2002). These organs are richly perfused by a branch of the carotid artery, and they are innervated by a branch of the vagus. Carotid bodies are near the parathyroid and ultimobranchial glands in birds, and are enveloped within the parathyroid gland in some species (Kobayashi, 1969; Yamatsu and Kameda, 1995); the physiological significance of this association is unknown. Arterial chemoreceptors have also been identified in other regions of the neck, along the carotid artery and aorta by anatomical and physiological methods (reviewed by Gleeson and Molony, 1989).

Figure 13.11 shows the response of carotid body arterial chemoreceptors to changes in P_{aO_2} and P_{aCO_2} in a duck. Afferent information about hypoxia or hypercapnia is transmitted to the CNS via the vagus nerve as increased frequency of action potentials from the carotid body. The pattern of action potential firing can differ in single chemoreceptors depending on the stimulus modality (O_2 versus CO_2) but the physiological significance of this is not known (Nye and Powell, 1984; Powell and Hempleman, 1990). Avian arterial chemoreceptors can also respond to oscillations in P_{aO_2}, P_{aCO_2}, and pH that can occur during breathing (Hempleman et al., 1992). The cell types and ultrastructure of the avian and mammalian carotid bodies are similar (reviewed by Gleeson and Molony, 1989), and similar chemoreceptor mechanisms on glomus cells in the carotid body probably explain P_{O_2} and P_{CO_2}/H^+ sensitivity in both classes. However, cellular mechanisms of arterial chemosensitivity are not completely understood (Kumar and Prabhakar, 2012).

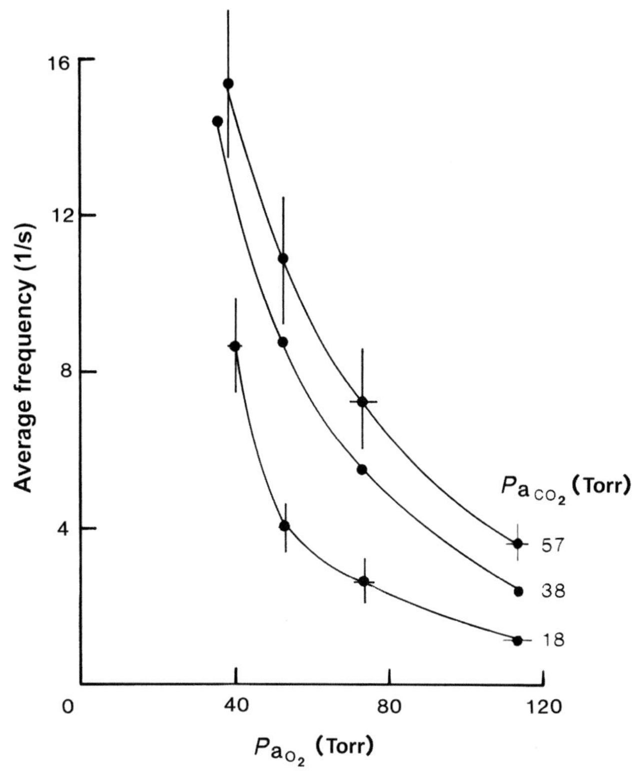

FIGURE 13.11 **Average frequency of action potentials (±SEM) for arterial chemoreceptors ($n=14$) in domestic ducks exposed to different combinations of P_{aO_2} and P_{aCO_2}.** *From Hempelman and Powell (unpublished data).*

13.8.2.3 Intrapulmonary Chemoreceptors

In contrast to mammals, the lungs of birds (and reptiles) contain intrapulmonary chemoreceptors (IPC), which respond to physiological changes in P_{CO_2} (Burger et al., 1974; Fedde et al., 1974a; Peterson and Fedde, 1968). The sensory endings of IPC have not been identified (reviewed by Gleeson and Molony, 1989), but physiological evidence indicates that these vagal afferents have multiple endings in the parabronchial mantle at several points along the length of one or more parabronchi (Hempleman and Burger, 1984). IPC can also respond to changes in CO_2 delivery to the lung by the pulmonary arteries (Banzett and Burger, 1977). Figure 13.12 shows how IPC are stimulated by *decreases* in P_{CO_2}, in contrast to arterial chemoreceptors that are inhibited by hypocapnia. However, increases in IPC activity cause ventilation to decrease, so the reflex response to CO_2 is similar in direction for central, arterial and intrapulmonary chemoreceptor reflexes (see Section 13.7.3). The mechanism of chemoreception in IPC depends on intracellular pH (Hempleman and Posner, 2004; Scheid et al., 1978), which is regulated by Na^+/H^+ exchangers (Hempleman et al., 2003), HCO_3^-/Cl^- exchangers (Shoemaker and Hempleman, 2001) and carbonic anhydrase (Hempleman et al., 2000). Calcium and potassium channels also modulate IPC sensitivity (Hempleman et al., 2006; Bina and Hempleman, 2007).

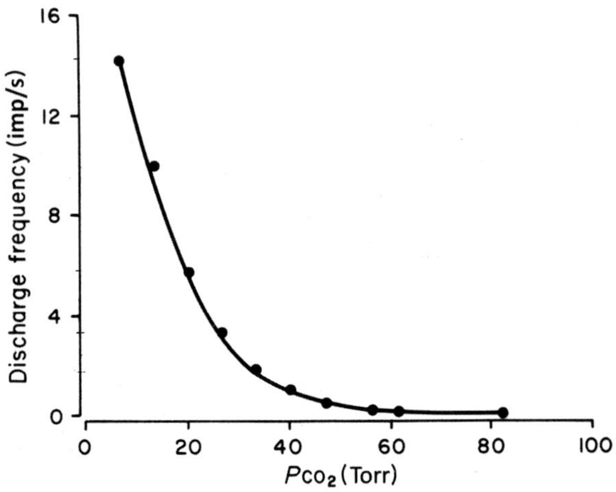

FIGURE 13.12 **Average frequency of action potentials for intrapulmonary chemoreceptors ($n=54$) exposed to CO_2 levels ranging from 7 to 82 Torr.** *From Nye and Burger (1978).*

IPC are extremely sensitive to changes in P_{CO_2}, and show large overshoots or undershoots in action potential frequency with the kinds of periodic changes in P_{CO_2} that occur in the lung during normal breathing (Fedde and Scheid, 1976; Scheid et al., 1978). This makes IPC well suited for fine-tuning the pattern of ventilation, similar to the role of vagal pulmonary stretch receptors in mammals (see Section 13.7.3). However, in contrast to mammalian or reptilian pulmonary stretch receptors that are sensitive to mechanical stimuli and P_{CO_2} (Powell et al., 1988), avian IPC are not sensitive mechanical stretching of the lung (Bouverot, 1978; Fedde et al., 1974b).

13.8.2.4 Other Receptors Affecting Breathing

Ventilation also responds to changes in activity from air sac mechanoreceptors, thermal receptors in the spinal cord, proprioceptors in the skin and may be skeletal muscle, upper airway receptors sensitive to irritants, cold and water, and, perhaps, arterial baroreceptors (reviewed by Gleeson and Molony, 1989).

13.8.3 Ventilatory Reflexes

13.8.3.1 CO_2 Response

Most birds are not normally exposed to increases in ambient CO_2 levels, except perhaps in specialized nests or burrows. However, ventilatory responses to common stimuli, such as exercise, hot or cold temperatures, and altitude, are influenced by CO_2, so it is important to understand the response to CO_2. Figure 13.13 shows how ventilation increases with increasing inspired CO_2 in conscious ducks. Most species show increased tidal volume but the frequency response can vary (Bouverot, 1978). At low levels of inspired CO_2, ventilation increases sufficiently to maintain P_{aco_2} at normal levels (Osborne and Mitchell, 1978).

Decreases in P_{aco_2} or intrapulmonary P_{CO_2} can decrease ventilation. Hence, when ventilation is increased by another stimulus, such as hypoxia or hyperthermia, the decrease in P_{aco_2} will act to inhibit ventilatory drive, and ventilation will be the net result of stimulation and inhibition. The hypercapnic ventilatory response scales with size in birds, so smaller birds have reduced CO_2 sensitivity (Williams et al., 1995). It is hypothesized that this may minimize inhibition of a strong hypoxic ventilatory response in small birds, considering the inverse scaling of the hypoxic sensitivity and size in birds (Kilgore et al., 2008 and see below).

Central chemoreceptors contribute to the ventilatory response to P_{aco_2} as described in Section 13.7.2.1. Arterial chemoreceptors play an important role in the response to dynamic changes in P_{aco_2} (Fedde et al., 1982; Jones and Purves, 1970; Seifert, 1896), such as P_{aco_2} oscillations that may occur during breathing, and perhaps in the response to static changes in P_{aco_2} also (Gleeson and Molony, 1989). IPC may contribute to the ventilatory response to changes in P_{aco_2} also, although their role in this response is controversial (Bouverot, 1978; Gleeson and Molony, 1989). Experimental evidence seems to favor a role for IPC in determining the pattern of breathing, but not the overall level of ventilation in conditions where P_{aco_2} increases (reviewed by Gleeson and Molony, 1989). IPC are well suited to sense breath-by-breath changes in ventilation as P_{CO_2} changes in the rigid avian lung. Therefore, IPC may play a similar role in the control of breathing to pulmonary stretch receptors in the alveolar lungs, which also sense instantaneous changes in ventilation as changes in lung volume.

13.8.3.2 Hypoxic Ventilatory Response

The hypoxic ventilatory response (Figure 13.14) is similar in birds and mammals (Black and Tenney, 1980b; Bouverot, 1978). Increasing P_{ao_2} above normal levels does not cause large decreases in ventilation, indicating that normoxic ventilatory drive from arterial chemoreceptors is relatively small (Gleeson and Molony, 1989). Also, ventilation does not change much until P_{ao_2} decreases below about 60 Torr in normal conditions, because small increases in ventilation will decrease P_{aco_2} and ventilatory drive. The hypoxic ventilatory response (Figure 13.14) has a similar shape to the arterial chemoreceptor response to P_{ao_2} (Figure 13.11) and arterial chemoreceptors are responsible for the hypoxic ventilatory response in birds (Jones and Purves, 1970; Seifert, 1896). However, the arterial chemoreceptor response in Figure 13.11 is measured under *isocapnic* conditions, in which P_{aco_2} was held constant by manipulating CO_2 levels in gas unidirectionally ventilating the lung of anesthetized birds. The hypoxic ventilatory response can be measured under isocapnic conditions also, by manipulating inspired CO_2 in awake birds. However, in contrast to mammals, such an isocapnic hypoxic ventilatory response does not measure

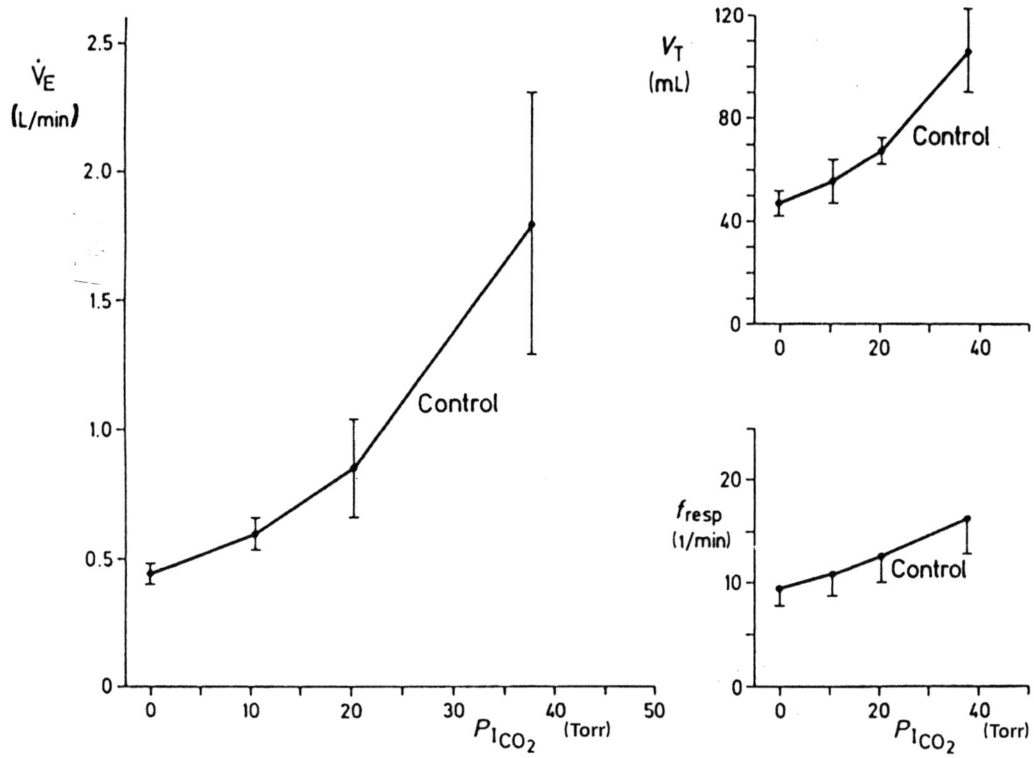

FIGURE 13.13 Ventilatory response to inhaled CO_2 in awake muscovy ducks (*Cairina moschata*). V_e=expired ventilation, V_T=tidal volume, fresp=frequency. *Modified from Powell et al. (1978).*

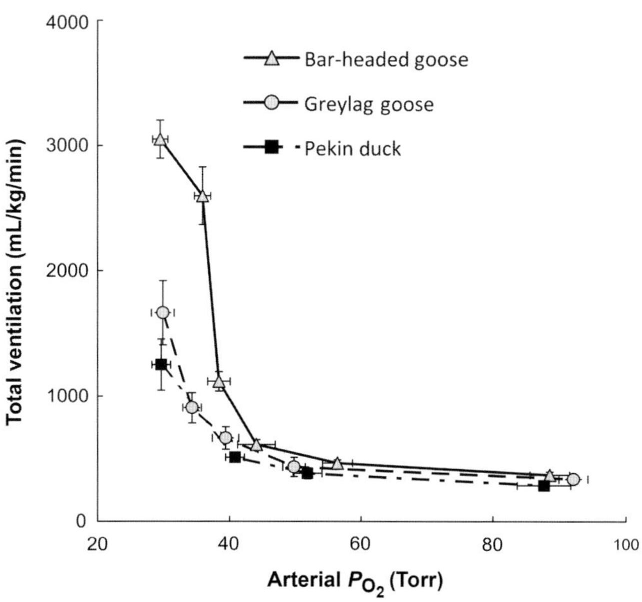

FIGURE 13.14 Hypoxic ventilatory response (isocapnic) for three species of waterfowl. *From Scott and Milsom (2007).*

of the pure effects of hypoxia on ventilation (independent of CO_2 effects). Changing inspired CO_2 can affect ventilation in birds through effects on IPC independently of $P_{a_{CO_2}}$ (Powell et al., 2000; Scott and Milsom, 2007).

Plasticity is observed in the hypoxic ventilatory response of birds with chronic changes in oxygen level. Chronic sustained hypoxia, as would be experienced during residence at high altitude, causes a time-dependent increase in ventilation above the acute response level (Black and Tenney, 1980b; Bouverot, 1985; Bouverot et al., 1976, 1979; Powell et al., 2004). This is called ventilatory acclimatization to hypoxia in mammals and involves changes in both O_2-sensitivity of arterial chemoreceptors and plasticity in respiratory centers in the CNS, which increases respiratory motor output for a given chemoreceptor afferent input (Powell et al., 1998). In mammals, ventilatory acclimatization to hypoxia causes time-dependent increases in $P_{a_{O_2}}$, but this is not observed in all avian studies. Acclimatizing Pekin ducks to 3800–5640 m altitude decreases $P_{a_{CO_2}} > 5$ Torr, indicating increased parabronchial ventilation, yet $P_{a_{O_2}}$ does not increase (Black and Tenney, 1980b; Powell et al., 2004). Apparently, other limitations to oxygen exchange increase during altitude acclimatization and offset the benefits of increased parabronchial ventilation for $P_{a_{O_2}}$.

Birds exhibit other forms of plasticity in the hypoxic ventilatory response too (Mitchell et al., 2001). Similar to mammals, ducks exhibit long-term facilitation of ventilation, in which breathing remains elevated above control levels following 3–10 brief episodes of hypoxia separated by normoxia. However, inhibitory mechanisms, such as hypoxic ventilatory decline or "rolloff" are not observed in birds, in contrast to mammals. Such a bias towards excitatory versus inhibitory effects may provide an advantage at high altitude (see Section 13.7.3.5).

13.8.4 Ventilatory Response to Exercise

Exercise is the most common cause of increased ventilation, but the exact physiological mechanism for this is still unknown. The best evidence to date indicates that ventilation increases during exercise through a combination of "feed forward" mechanisms and feedback from chemoreceptors (Dempsey et al., 1995). Feed-forward mechanisms, also called central command, are neural signals from higher centers in the CNS that may stimulate respiratory centers directly. For example, neural signals to locomotor muscles may also stimulate ventilatory muscles, and could contribute to some of the phase locking between wing beats and respiration (Funk et al., 1992a,b). Feedback from chemoreceptors prevents ventilation from increasing too much. P_{aCO_2} usually decreases in exercising birds (Kiley et al., 1979), and this hypocapnia would be even worst if ventilatory chemoreflexes did not inhibit ventilation. Other ventilatory stimuli such as body temperature and hypoxia at altitude can modify the response to exercise (reviewed by Gleeson and Molony, 1989).

13.8.5 Frontiers: Extreme Hyperventilation at High Altitude

Birds hyperventilate more at high altitude than mammals (Powell et al., 2004). This theme appears in evolutionary adaptations too as bar-headed geese, which are the archetypal high altitude bird, do not show a "blunted" hypoxic ventilatory response typical of some high altitude human populations (Scott and Milsom, 2007). Hyperventilation in hypoxia decreases P_{aCO_2}, which increases P_{aO_2} as well as exaggerating the Bohr effect to increase hemoglobin-O_2 affinity. Although P_{aO_2} does not increase further with ventilatory acclimatization to hypoxia in birds (see above), the time-dependent decrease in P_{aCO_2} will further increase arterial O_2 concentration by the Bohr effect. Hence, it appears that many of the advantages birds have at high altitude result from the extreme hyperventilation, hypocapnia and alkalosis in response to a strong hypoxic ventilatory response. Supporting this are observations of pigeons exposed to 9000 m simulated altitude showing only minimal stress while arterial pH is 7.85 (Lutz and Schmidt-Nielsen, 1977). Other studies show intracellular pH not changing while venous pH increases from 7.42 to 7.56 in pigeons exposed to 9000 m (Weinstein et al., 1985). This suggests that tighter regulation of the intracellular milieu may provide an avian advantage at altitude, although the hypothesis remains to be systematically tested. In any case, the control systems for ventilation and cerebral blood flow (see Section 13.7.1.2) that permit such extreme hypocapnia and alkalosis at altitude must have evolved with, or followed from, cellular adaptations preserving function with large pH changes.

REFERENCES

Abdalla, M.A., King, A.S., 1975. The functional anatomy of the pulmonary circulation of the domestic fowl. Respir. Physiol. 23, 267–290.

Abdalla, M.A., 1989. The blood supply to the lung. In: King, A.S., McLelland, J. (Eds.), Form and Function in Birds. Academic Press, London, pp. 281–306.

Adamson, T.P., 1958. The Comparative Morphology of the Carotid Body and Carotid Sinus. Chas C Thomas, Springfield.

Allen, R.L., 1983. Haemoglobins. In: Freeman, B.M. (Ed.), Physiology and Biochemistry of the Domestic Fowl. Academic Press, London, pp. 313–319.

Andersen, H.T., Lövö, A., 1967. Indirect estimation of partial pressure of oxygen in arterial blood of diving ducks. Respir. Physiol. 2, 163–167.

Banzett, R.B., Burger, R.E., 1977. Response of avian intrapulmonary chemoreceptors to venous CO_2 and ventilatory gas flow. Respir. Physiol. 29, 63–72.

Banzett, R.B., Lehr, J.L., 1982. Gas exchange during high-frequency ventilation of the chicken. J. Appl. Physiol. 53 (6), 1418–1422.

Banzett, R.B., Butler, J.P., Nations, C.S., Barnas, G.M., Lehr, J.L., Jones, J.H., 1987. Inspiratory aerodynamic valving in goose lungs depends on gas density and velocity. Respir. Physiol. 70, 287–300.

Banzett, R.B., Nations, C.S., Wang, N., Fredberg, J.J., Butler, J.P., 1991. Pressure profiles show features essential to aerodynamic valving in geese. Respir. Physiol. 84, 295–309.

Barnas, G.M., Mather, F.B., 1978. Response of avian intrapulmonary smooth muscle to changes in carbon dioxide concentration. Poult. Sci. 57, 1400–1407.

Bartels, H., Hiller, G., Reinhardt, W., 1966. Oxygen affinity of chicken blood before and after hatching. Respir. Physiol. 1, 345–356.

Bauer, C., Jelkmann, W., Rollema, H.S., 1978. Mechanisms controlling the oxygen affinity of bird and reptile blood: a comparison between the functional properties of chicken and crocodile haemoglobin. In: Piiper, J. (Ed.), Respiratory Function in Birds, Adult and Embryonic. Springer-Verlag, Berlin, pp. 61–66.

Baumann, F.H., Baumann, R., 1977. A comparative study of the respiratory properties of bird blood. Respir. Physiol. 31, 333–343.

Baumann, R., Baumann, F.H., 1978. Respiratory function of embryonic chicken hemoglobin. In: Piiper, J. (Ed.), Respiratory Functin in Birds, Adult and Embryonic. Springer-Verlag, Berlin, pp. 292–297.

Bech, C., Johansen, K., 1980. Ventilation and gas exchange in the mute swan, *Cygnus olor*. Respir. Physiol. 39, 285–295.

Bech, C., Johansen, K., Brent, R., Nicol, S., 1984. Ventilatory and circulatory changes during cold exposure in the Pekin duck *Anas platyrhynchos*. Respir. Physiol. 57, 103–112.

Bencowitz, H.Z., Wagner, P.D., West, J.B., 1982. Effect of change in P_{50} on exercise tolerance at high altitude: a theoretical study. J. Appl. Physiol. 53, 1487–1495.

Berger, M., 1974. Energiewechsel von Kolibris beim Schwirrflug unter Höhenbedingungen. J. Ornithol. 115, 273–288.

Bernhard, W., Gebert, A., Vieten, G., Rau, G.A., Hohlfeld, J.M., Postle, A.D., Freihorst, J., 2001. Pulmonary surfactant in birds: coping with surface tension in a tubular lung. Am. J. Physiol. Regul. Integr. Comp. Physiol. 281, R327–R337.

Bickler, P.E., Maginniss, L.A., Powell, F.L., 1986. Intrapulmonary and extrapulmonary shunt in ducks. Respir. Physiol. 63, 151–160.
Bina, R.W., Hempleman, S.C., 2007. Evidence for TREK-like tandem-pore domain channels in intrapulmonary chemoreceptor chemotransduction. Respir. Physiol. Neurobiol. 156, 120–131.
Black, C.P., Tenney, S.M., 1980a. Pulmonary hemodynamic responses to acute and chronic hypoxia in two waterfowl species. Comp. Biochem. Physiol. 67A, 291–293.
Black, C.P., Tenney, S.M., 1980b. Oxygen transport during progressive hypoxia in high-altitude and sea-level waterfowl. Respir. Physiol. 39, 217–239.
Boggs, D.F., Butler, P.J., Warner, M., 1996. Fluctuations in differential pressure between the anterior and posterior air sac of tufted ducks, *Aythya fuligula*, during breath-hold dives. Physiologist 39, A-27.
Bouverot, P., Hildwein, G., Oulhen, P., 1976. Ventilatory and circulatory O_2 convection at 4000 m in pigeon at neutral or cold temperature. Respir. Physiol. 28, 371–385.
Bouverot, P., 1978. Control of breathing in birds compared with mammals. Physiol. Rev. 58 (3), 604–655.
Bouverot, P., Douguest, D., Sèbert, P., 1979. Role of the arterial chemoreceptors in ventilatory and circulatory adjustments to hypoxia in awake Pekin ducks. J. Comp. Physiol. 133, 177–186.
Bouverot, P., 1985. Adaptation to Altitude-Hypoxia in Vertebrates. Springer-Verlag, Berlin.
Brackenbury, J., 1984. Physiological responses of birds to flight and running. Biol. Rev. 59, 559–575.
Brackenbury, J.H., 1971. Airflow dynamics in the avian lung as determined by direct and indirect methods. Respir. Physiol. 13, 319–329.
Brackenbury, J.H., 1972. Physical determinants of air flow pattern within the avian lung. Respir. Physiol. 15, 384.
Bretz, W.L., Schmidt-Nielsen, K., 1971. Bird respiration: flow patterns in the duck lung. J. Exp. Biol. 54, 103–118.
Brown, R.E., Kovacs, C.E., Butler, J.P., Wang, N., Lehr, J.L., Banzett, R.B., 1995. The avian lung: is there an aerodynamic expiratory valve? J. Exp. Biol. 198, 2349–2357.
Burger, R.E., Lorenz, F.W., 1960. Artificial respiration in birds by unidirectional air flow. Poult. Sci. 39 (1), 236–237.
Burger, R.E., Osborne, J.L., Banzett, R.B., 1974. Intrapulmonary chemoreceptors in *Gallus domesticus*: adequate stimulus and functional localization. Respir. Physiol. 22, 87–97.
Burger, R.E., Estavillo, J.A., 1977. Pulmonary circulation – vertebral venous interconnections in the chicken. Anat. Rec. 188 (1), 39–44.
Burger, R.E., Meyer, M., Graf, W., Scheid, P., 1979. Gas exchange in the parabronchial lung of birds: experiments in unidirectionally ventilated ducks. Respir. Physiol. 36, 19–37.
Burton, R.R., Besch, E.L., Smith, A.H., 1968. Effect of chronic hypoxia on the pulmonary arterial blood pressure of the chicken. Am. J. Physiol. 214 (6), 1438–1442.
Butler, P.J., Taylor, E.W., 1983. Factors affecting the respiratory and cardiovascular responses to hypercapnic hypoxia in mallard ducks. Respir. Physiol. 53, 109–127.
Butler, P.J., Turner, D.L., 1988. Effect of training on maximal oxygen uptake and aerobic capacity of locomotory muscles in tufted ducks, *Aythya fuligula*. J. Physiol. 401, 347–359.
Butler, P.J., 1991. Exercise in birds. J. Exp. Biol. 160, 233–262.
Calder, W.A., Schmidt-Nielsen, K., 1968. Panting and blood carbon dioxide in birds. Am. J. Physiol. 215, 477–482.
Chai, P., Dudley, R., 1996. Limits to flight energetics of hummingbirds hovering in hypodense and hypoxic gas mixtures. J. Exp. Biol. 199, 2285–2295.
Cohn, J.E., Shannon, R., 1968. Respiration in unanesthetized geese. Respir. Physiol. 5, 259–268.
Crank, W.D., Gallagher, R.R., 1978. Theory of gas exchange in the avian parabronchus. Respir. Physiol. 35, 9–25.
Davey, N.J., Seller, T.J., 1987. Brain mechanisms for respiratory control. In: Seller, T.J. (Ed.), Bird Respiration. CRC Press, Inc., Boca Raton, pp. 169–188.
Dempsey, J.A., Forster, H.V., Ainsworth, D.M., 1995. Regulation of hyperpnea, hyperventilation, and respiratory muscle recruitment during exercise. In: Dempsey, J.A., Pack, A.I. (Eds.), Regulation of Breathing. Marcel Dekker, Inc., New York, pp. 1065–1134.
Dotterweich, H., 1930. Versuche Über den weg der atemluft in der vogellunge. Z. Vergleich. Physiol. 11, 271–284.
Dubach, M., 1981. Quantitative analysis of the respiratory system of the house-sparrow, budgerigar and violet-eared hummingbird. Respir. Physiol. 46, 43–60.
Duncker, H.-R., 1978. General morphological principles of amniotic lungs. In: Piiper, J. (Ed.), Respiratory Function in Birds, Adult and Embryonic. Springer-Verlag, Berlin, pp. 1–18.
Duncker, H.R., 1971. The lung air sac system of birds. Adv. Anat. Embryol. Cell Biol. 45, 7–171.
Duncker, H.R., 1972. Structure of avian lungs. Respir. Physiol. 14, 44–63.
Duncker, H.R., 1974. Structure of the avian respiratory tract. Respir. Physiol. 22, 1–19.
Faraci, F.M., Kilgore, D.L., Fedde, M.R., 1984. Oxygen delivery to the heart and brain during hypoxia: Pekin duck vs. bar-headed goose. Am. J. Physiol. 16, R69–R75.
Faraci, F.M., 1986. Circulation during hypoxia in birds. Comp. Biochem. Physiol. 85A (4), 613–620.
Faraci, F.M., 1991. Adaptations to hypoxia in birds: how to fly high. Annu. Rev. Physiol. 53, 59–70.
Farmer, C.G., 2006. On the origin of avian air sacs. Respir. Physiol. Neurobiol. 154, 89–106.
Farmer, C.G., Sanders, K., 2010. Unidirectional airflow in the lungs of alligators. Science 327, 338–340.
Fedde, M.R., Burger, R.E., Kitchell, R.L., 1963. Electromyographic studies of the effects of bodily position and anesthesia on the activity of the respiratory muscles of the domestic cock. Poult. Sci. 43 (4), 839–846.
Fedde, M.R., Burger, R.E., Kitchell, R.L., 1964a. Anatomic and electromyographic studies of the costo-pulmonary muscles in the cock. Poult. Sci. 43 (5), 1177–1184.
Fedde, M.R., Burger, R.E., Kitchell, R.L., 1964b. Electromyographic studies of the effects of bilateral, cervical vagotomy on the action of the respiratory muscles of the domestic cock. Poult. Sci. 43 (5), 1119–1125.
Fedde, M.R., de Wet, P.D., Kitchell, R.L., 1969. Motor unit recruitment pattern and tonic activity in respiratory muscles of *Gallus domesticus*. J. Neurophysiol. 32, 995–1004.
Fedde, M.R., Gatz, R.N., Slama, H., Scheid, P., 1974a. Intrapulmonary CO_2 receptors in the duck: I. Stimulus specificity. Respir. Physiol. 22, 99–114.
Fedde, M.R., Gatz, R.N., Slama, H., Scheid, P., 1974b. Intrapulmonary CO_2 receptors in the duck: II. Comparison with mechanoreceptors. Respir. Physiol. 22, 115–121.
Fedde, M.R., Scheid, P., 1976. Intrapulmonary CO_2 receptors in the duck: IV. Discharge pattern of the population during a respiratory cycle. Respir. Physiol. 26, 223–227.
Fedde, M.R., 1978. Drugs used for avian anesthesia. Poult. Sci. 57, 1376–1399.

Fedde, M.R., Kiley, J.P., Powell, F.L., Scheid, P., 1982. Intrapulmonary CO_2 receptors and control of breathing in ducks: effects of prolonged circulation time to carotid bodies and brain. Respir. Physiol. 47, 121–140.

Fedde, M.R., Orr, J.A., Shams, H., Scheid, P., 1989. Cardiopulmonary function in exercising bar-headed geese during normoxia and hypoxia. Respir. Physiol. 77, 239–262.

Fedde, M.R., 1990. High-altitude bird flight: exercise in a hostile environment. NIPS 5, 191–193.

Fortin, G., Champagnat, J., Lumdsen, A., 1994. Onset and maturation of branchio-motor activites in the chick hindbrain. Neuroreport 5, 1149–1152.

Funk, G.D., Milsom, W.K., Steeves, J.D., 1992a. Coordination of wingbeat and respiration in the Canada goose. I. Passive wing flapping. J. Appl. Physiol. 73, 1014–1024.

Funk, G.D., Steeves, J.D., Milsom, W.K., 1992b. Coordination of wingbeat and respiration in birds. II. "Fictive" flight. J. Appl. Physiol. 73, 1025–1033.

Geiser, J., Gratz, R.K., Hiramoto, T., Scheid, P., 1984. Effects of increasing metabolism by 2,4-dinitrophenol on respiration and pulmonary gas exchange in the duck. Respir. Physiol. 57, 1–14.

Giardina, B., Corda, M., Pellegrini, M.G., Condo, S.G., Brunori, M., 1985. Functional properties of the hemoglobin system of two diving birds (*Podiceps nigricollis* and *Phalacrocorax carbo sinensis*). Molec. Physiol. 7, 281–292.

Gillespie, J.R., Gendner, J.P., Sagot, J.C., Bouverot, P., 1982a. Impedance of the lower respiratory system in ducks measured by forced oscillations during normal breathing. Respir. Physiol. 47, 51–68.

Gillespie, J.R., Sagot, J.C., Gendner, J.P., Bouverot, P., 1982b. Respiratory mechanics of Pekin ducks under four conditions: pressure breathing, anesthesia, paralysis or breathing CO_2-enriched gas. Respir. Physiol. 47, 177–191.

Gleeson, M., Molony, V., 1989. Control of breathing. In: King, A.S., McLelland, J. (Eds.), Form and Function in Birds. Academic Press, London, pp. 439–484.

Grubb, B., Mills, C.D., Colacino, J.M., Schmidt-Nielsen, K., 1977. Effect of arterial carbon dioxide on cerebral blood flow in ducks. Am. J. Physiol. 232 (6), H596–H601.

Grubb, B.R., 1981. Blood flow and oxygen consumption in avian skeletal muscle during hypoxia. J. Appl. Physiol. 50, 450–455.

Guyenet, P.G., Stornetta, R.L., Abbott, S.B., Depuy, S.D., Fortuna, M.G., Kanbar, R., 2010. Central CO_2 chemoreception and integrated neural mechanisms of cardiovascular and respiratory control. J. Appl. Physiol. 108, 995–1002.

Hastings, R.H., Powell, F.L., 1986a. Single breath CO_2 measurements of dead space in ducks. Respir. Physiol. 63, 139–149.

Hastings, R.H., Powell, F.L., 1986b. Physiological dead space and effective parabronchial ventilation in ducks. J. Appl. Physiol. 60 (1), 85–91.

Hastings, R.H., Powell, F.L., 1987. High-frequency ventilation of ducks and geese. J. Appl. Physiol. 63 (1), 413–417.

Hazelhoff, E.H., 1951. Structure and function of the lung of birds (reprinted from 1943). Poult. Sci. 30, 3–10.

Helbacka, N.V.L., Casterline Jr., J.L., Smith, C.J., Shaffner, C.S., 1963. Investigation of plasma carbonic acid pK' of the chicken. Poult. Sci. 43 (1), 138–144.

Hempleman, S.C., Burger, R.E., 1984. Receptive fields of intrapulmonary chemoreceptors in the Pekin duck. Respir. Physiol. 57, 317–330.

Hempleman, S.C., Posner, R.G., 2004. CO_2 transduction mechanisms in avian intrapulmonary chemoreceptors: experiments and models. Respir. Physiol. Neurobiol. 144, 203–214.

Hempleman, S.C., Powell, F.L., 1986. Influence of pulmonary blood flow and O_2 flux on D_{O2} in avian lungs. Respir. Physiol. 63, 285–292.

Hempleman, S.C., Warburton, S.J., 2013. Comparative embryology of the carotid body. Respir. Physiol. Neurobiol. 185, 3–8.

Hempleman, S.C., Adamson, T.P., Begay, R.S., Solomon, I.C., 2003. CO_2 transduction in avian intrapulmonary chemoreceptors is critically dependent on transmembrane Na^+/H^+ exchange. Am. J. Physiol. Regul. Integr. Comp. Physiol. 84, R1551–R1559.

Hempleman, S.C., Egan, S.X., Pilarski, J.Q., Adamson, T.P., Solomon, I.C., 2006. Calcium and avian intrapulmonary chemoreceptor response to CO_2. J. Appl. Physiol. 101, 1565–1575.

Hempleman, S.C., Powell, F.L., Prisk, G.K., 1992. Avian arterial chemoreceptor responses to steps of CO_2 and O_2. Respir. Physiol. 90, 325–340.

Hempleman, S.C., Rodriguez, T.A., Bhagat, Y.A., Begay, R.S., 2000. Benzolamide, acetazolamide, and signal transduction in avian intrapulmonary chemoreceptors. Am. J. Physiol. Regul. Integr. Comp. Physiol. 279, R1988–R1995.

Henning, B., Scheid, P., Piiper, J., 1971. Determination of the Haldane effect in chicken blood. Respir. Physiol. 11, 279–284.

Hinds, D.S., Calder, W.A., 1971. Tracheal dead space in the respiration of birds. Evolution 25, 429–440.

Hirsowitz, L.A., Fell, K., Torrance, J.D., 1977. Oxygen affinity of avian blood. Respir. Physiol. 31, 51–62.

Holle, J.P., Meyer, M., Scheid, P., 1977. Oxygen affinity of duck blood determined by *in vivo* and *in vitro* technique. Respir. Physiol. 29, 355–361.

Holle, J.P., Heisler, N., Scheid, P., 1978. Blood flow distribution in the duck lung and its control by respiratory gases. Am. J. Physiol. 234 (3), R146–R154.

Hsia, C.C.W., Schmitz, A., Lambertz, M., Perry, S.F., Maina, J.N., 2013. Evolution of air breathing: oxygen homeostasis and the transitions from water to land and sky. Compr. Physiol. 3, 849–915. doi: 10.1002/cphy.c120003.

Jenkins, F.A., Dial, K.P., Goslow, G.E., 1988. A cineradiographic analysis of bird flight: the wishbone in starlings is a spring. Science 241, 1495–1498.

Jessen, T.H., Weber, R.E., Fermi, G., Tame, J., Braunitzer, G., 1991. Adaptation of bird hemoglobins to high altitudes: demonstration of molecular mechanism by protein engineering. Proc. Natl. Acad. Sci. 88, 6519–6522.

Johansen, K., Berger, M., Bicudo, J.E.P., Ruschi, A., de Almeida, P.J., 1987. Respiratory properties of blood and myoglobin in hummingbirds. Physiol. Zool. 60 (2), 269–278.

Jones, D.R., Purves, M.J., 1970. The effect of carotid body denervation upon the respiratory response to hypoxia and hypercapnia in the duck. J. Physiol. 211 (2), 295–308.

Jones, D.R., Holeton, G.F., 1972. Cardiovascular and respiratory responses of ducks to progressive hypocapnic hypoxia. J. Exp. Biol. 56, 657–666.

Jones, D.R., Milsom, W.K., 1982. Peripheral receptors affecting breathing and cardiovascular function in non-mammalian vertebrates. J. Exp. Biol. 100, 59–91.

Jones, J.H., 1982. Pulmonary blood flow distribution in panting ostriches. J. Appl. Physiol. 53, 1411–1417.

Jones, J.H., Grubb, B., Schmidt-Nielsen, K., 1983. Panting in the emu causes arterial hypoxemia. Respir. Physiol. 54, 189–195.

Jones, J.H., Effmann, E.L., Schmidt-Nielsen, K., 1985. Lung volume changes during respiration in ducks. Respir. Physiol. 59, 15–25.

Julian, R.J., 1993. Ascites in poultry. Avian Pathol. 23, 419–454.
Kadono, H., Okada, T., 1962. Electromyographic studies on the respiratory muscles of the domestic fowl. Jpn. J. Vet. Sci. 24 (4), 215–223.
Kadono, H., Okada, T., Ono, K., 1963. Electromyographic studies on the respiratory muscles of the chicken. Poult. Sci. 42 (1), 121–128.
Kameda, Y., 2002. Carotid body and glomus cells distributed in the wall of the common carotid artery in the bird. Microsc. Res. Tech. 59, 196–206.
Kampe, G., Crawford, E.C., 1973. Oscillatory mechanics of the respiratory system of pigeons. Respir. Physiol. 18, 188–193.
Kawashiro, T., Scheid, P., 1975. Arterial blood gases in undisturbed resting birds: measurements in chicken and duck. Respir. Physiol. 23, 337–342.
Kiley, J.P., Kuhlmann, W.D., Fedde, M.R., 1979. Respiratory and cardiovascular responses to exercise in the duck. J. Appl. Physiol. 47, 827–833.
Kiley, J.P., Faraci, F.M., Fedde, M.R., 1985. Gas exchange during exercise in hypoxic ducks. Respir. Physiol. 59, 105–115.
Kilgore Jr, D.L., Boggs, D.F., Kilgore, T.J., Colby, C., Williams Jr, B.R., Bavis, R.W., 2008. Ventilatory and metabolic responses of burrowing owls, *Athene cunicularia*, tomoderate and extreme hypoxia: analysis of the hypoxic ventilatory threshold vs. hemoglobin oxygen affinity relationship in birds. Comp. Biochem. Physiol., A Mol. Integr. Physiol. 150, 247–257.
King, A.S., Molony, V., 1971. The anatomy of respiration. In: Bell, O.J., Freeman, B.M. (Eds.), Physiology and Biochemistry of the Domestic Fowl. Academic Press, New York, p. 227.
King, A.S., 1979. Systema respiratorium. In: Baumel, J.J., King, A.M., Lucas, A.M., Breazile, J.E., Evans, H.E. (Eds.), Nomina Anatomica Avium. Academic Press, London, pp. 227–265.
King, A.S., 1989. Functional anatomy of the syrinx. In: King, A.S., McLelland, J. (Eds.), Form and Function in Birds. Academic Press, London, pp. 105–192.
Kobayashi, S., 1969. Catecholamines in the avian carotid body. Specilia 25, 1075–1076.
Kollias, G.V., McLeish, I., 1978. Effects of ketamine hydrochloride in red-tailed hawks (*Buteo jamaicensis*) I. Arterial blood gas and acid base. Comp. Biochem. Physiol. 60, 57–59.
Konishi, M., 2010. From central pattern generator to sensory template in the evolution of birdsong. Brain Lang. 115, 18–20.
Kuethe, D.O., 1988. Fluid mechanical valving of air flow in bird lungs. J. Exp. Biol. 136, 1–12.
Kuhlmann, W.D., Fedde, M.R., 1976. Upper respiratory dead space in the chicken: its fraction of the tidal volume. Comp. Biochem. Physiol. 54A, 409–411.
Lapennas, G.N., Reeves, R.B., 1983. Oxygen affinity of blood of adult domestic chicken and red jungle fowl. Respir. Physiol. 52, 27–39.
Lutz, P.L., 1980. On the oxygen affinity of bird blood. Am. Zool. 20, 187–198.
Lutz, P.L., Schmidt-Nielsen, K., 1977. Effect of simulated altitude on blood gas transport in the pigeon. Respir. Physiol. 30, 383–388.
Macklem, P.T., Bouverot, P., Scheid, P., 1979. Measurement of the distensibility of the parabronchi in duck lungs. Respir. Physiol. 38, 23–35.
Maginniss, L.A., 1985. Red cell organic phosphates and Bohr effects in house sparrow blood. Respir. Physiol. 59, 93–103.
Maginniss, L.A., Kilgore, D.L., 1989. Blood oxygen binding properties for the burrowing owl, *Athene cunicularia*. Respir. Physiol. 76, 205–214.
Magnussen, H., Willmer, H., Scheid, P., 1976. Gas exchange in air sacs: contribution to respiratory gas exchange in ducks. Respir. Physiol. 26, 129–146.
Maina, J.N., Abdalla, A., King, A.S., 1982. Light microscopic morphometry of the lung of 19 avian species. Acta Anat. 112, 264–270.
Maina, J.N., King, A.S., 1982. Morphometrics of the avian lung. 2. The wild mallard (*Anas platyrhynchos*) and graying goose (*Anser anser*). Respir. Physiol. 50, 299–310.
Maina, J.N., 1989. The morphometry of the avian lung. In: King, A.S., McLelland, J. (Eds.), Form and Function in Birds. Academic Press, London, pp. 307–368.
Maina, J.N., King, A.S., Settle, G., 1989. An allometric study of pulmonary morphometric parameters in birds, with mammalian comparisons. Philos. Trans. R. Soc. Lond. B 326, 1–57.
Maina, J.N., 2007. Spectacularly robust! Tensegrity principle explains the mechanical strength of the avian lung. Respir. Physiol. Neurobiol. 155, 1–10.
Maina, J.N., West, J.B., Orgeig, S., Foot, N.J., Daniels, C.B., Kiama, S.G., Gehr, P., Mühlfeld, C., Blank, F., Müller, L., Lehmann, A., Brandenberger, C., Rothen-Rutishauser, B., 2010. Recent advances into understanding some aspects of the structure and function of mammalian and avian lungs. Physiol. Biochem. Zool. 83, 792–807.
Marder, J., Arad, Z., 1975. The acid bas balance of abdim's stork *(Sphenorhynchus abdimii)* during thermal panting. Comp. Biochem. Physiol. 51A, 887–889.
Maren, T.H., 1967. Carbonic anhydrase: chemistry, physiology and inhibition. Physiol. Rev. 47, 595–781.
Mathieu-Costello, O., 1991. Morphometric analysis of capillary geometry in pigeon pectoralis muscle. Am. J. Anat. 191, 74–84.
Mathieu-Costello, O., Suarez, R.K., Hochachka, P.W., 1992. Capillary-to-fiber geometry and mitochondrial density in hummingbird flight muscle. Respir. Physiol. 89, 113–132.
Mathieu-Costello, O., 1993. Comparative aspects of muscle capillary supply. Annu. Rev. Physiol. 55, 503–525.
Mathieu-Costello, O., Agey, P.J., 1997. Chronic hypoxia affects capillary density and geometry in pigeon pectoralis muscle. Respir. Physiol. 109, 39–52.
McLelland, J., 1989a. Larynx and trachea. In: King, A.S., McLelland, J. (Eds.), Form and Function in Birds. Academic Press, London, pp. 69–104.
McLelland, J., 1989b. Anatomy of the lungs and air sacs. In: King, A.S., McLelland, J. (Eds.), Form and Function in Birds. Academic Press, London, pp. 221–280.
Meir, J.U., Jardine, W., Yourk, J., Chua, B., Milsom, W.K., 2013. Heart rate and metabolic rate of bar-headed geese flying in hypoxia. FASEB J. 27, 1149.16 (abstract).
Meir, J.U., Milsom, W.K., 2013. High thermal sensitivity of blood enhances oxygen delivery in the high-flying bar-headed goose. J. Exp. Biol. 216, 2172–2175.
Meyer, M., Worth, H., Scheid, P., 1976. Gas-blood CO_2 equilibrium in parabronchial lungs of birds. J. Appl. Physiol. 41, 302–309.
Meyer, M., Holle, J.P., Scheid, P., 1978. Bohr effect induced by CO_2 and fixed acid at various levels of O_2 saturation in duck blood. Pflügers Arch. Ges. Physiol. 376, 237–240.
Milsom, W.K., Jones, D.R., Gabbott, G.R.J., 1981. On chemoreceptor control of ventilatory responses to CO_2 in unanesthetized ducks. J. Appl. Physiol. 50, 1121–1128.
Mitchell, G.S., Powell, F.L., Hopkins, S.R., Milsom, W.K., 2001. Time domains of the hypoxic ventilatory response in awake ducks: episodic and continuous hypoxia. Respir. Physiol. 124, 117–128.

Molony, V., Graf, W., Scheid, P., 1976. Effects of CO_2 on pulmonary air flow resistance in the duck. Respir. Physiol. 26, 333–349.

Morgan, V.E., Chichester, D.F., 1935. Properties of the blood of the domestic fowl. J. Biol. Chem. 110, 285–298.

Murrish, D.E., 1982. Acid-base balance in three species of antarctic penguins exposed to thermal stress. Physiol. Zool. 55 (2), 137–143.

Nightingale, T.E., Boster, R.A., Fedde, M.R., 1968. Use of the oxygen electrode in recording PO_2 in avian blood. J. Appl. Physiol. 25 (4), 371–375.

Nye, P.C.G., Burger, R.E., 1978. Chicken intrapulmonary chemoreceptors: discharge at static levels of intrapulmonary carbon dioxide and their location. Respir. Physiol. 33, 299–322.

Nye, P.C.G., Powell, F.L., 1984. Steady-state discharge and bursting of arterial chemoreceptors in the duck. Respir. Physiol. 56, 369–384.

Osborne, J.L., Mitchell, G.S., 1978. Intrapulmonary and systemic CO_2-chemoreceptor interaction in the control of avian respiration. Respir. Physiol. 33, 349–357.

Palomeque, J., Palacios, L., Planas, J., 1979. Comparative respiratory functions of blood in some passeriform birds. Comp. Biochem. Physiol. 66A, 619–624.

Parry, K., Yates, M.S., 1979. Observations on the avian pulmonary and bronchial circulation using labeled microspheres. Respir. Physiol. 38, 131–140.

Pattle, R.E., 1978. Lung surfactant and lung lining in birds. In: Piiper, J. (Ed.), Respiratory Function in Birds, Adult and Embryonic. Springer-Verlag, New York, p. 23.

Peacock, A.J., Pickett, C., Morris, K., Reeves, J.T., 1990. Spontaneous hypoxaemia and right ventricular hypertrophy in fast growing broiler checkens reared at sea level. Comp. Biochem. Physiol. 97A, 537–541.

Peek, F.W., Youngren, O.M., Phillips, R.E., 1975. Repetitive vocalizations evoked by electrical stimulation of avian brains. Brain Behav. Evol. 12, 1–41.

Perry, S.F., Christian, A., Breuer, T., Pajor, N., Codd, J.R., 2009. Implications of an avian-style respiratory system for gigantism in sauropod dinosaurs. J. Exp. Zool. A Ecol. Genet. Physiol. 311, 600–610.

Peterson, D.F., Fedde, M.R., 1968. Receptors sensitive to carbon dioxide in lungs of chicken. Science 162, 1499–1501.

Petschow, D., Wurdinger, I., Baumann, R., Duhm, J., Braunitzer, G., Bauer, C., 1977. Causes of high blood O_2 affinity of animals living at high altitude. J. Appl. Physiol. 42 (2), 139–143.

Phu, D.N., Yamaguchi, K., Scheid, P., Piiper, J., 1986. Kinetics of oxygen uptake and release by red blood cells of chicken and duck. J. Exp. Biol. 125, 15–27.

Piiper, J., Drees, F., Scheid, P., 1970. Gas exchange in the domestic fowl during spontaneous breathing and artificial ventilation. Respir. Physiol. 9, 234–245.

Piiper, J., Dejours, P., Haab, P., Rahn, H., 1971. Concepts and basic quantities in gas exchange physiology. Respir. Physiol. 13, 292–304.

Piiper, J., Scheid, P., 1972. Maximum gas transfer efficacy of models for fish gills, avian lungs and mammalian lungs. Respir. Physiol. 14, 115–124.

Piiper, J., Scheid, P., 1975. Gas transport efficacy of gills, lungs and skin: theory and experimental data. Respir. Physiol. 23, 209–221.

Piiper, J., 1978. Origin of carbon dioxide in caudal airsacs of birds. In: Piiper, J. (Ed.), Respiratory Function in Birds, Adult and Embryonic. Springer-Verlag, Berlin, pp. 148–153.

Ponganis, P.J., Meir, J.U., Williams, C.L., 2011. In pursuit of Irving and Scholander: a review of oxygen store management in seals and penguins. J. Exp. Biol. 214, 3325–3339.

Powell, F.L., Fedde, M.R., Gratz, R.K., Scheid, P., 1978. Ventilatory responses to CO_2 in birds. I. Measurements in the unanesthetized duck. Respir. Physiol. 35, 349–359.

Powell, F.L., Geiser, J., Gratz, R.K., Scheid, P., 1981. Airflow in the avian respiratory tract: variations of O_2 and CO_2 concentrations in the bronchi of the duck. Respir. Physiol. 44, 195–213.

Powell, F.L., 1982. Diffusion in avian lungs. Fed. Proc. 41, 2131–2133.

Powell, F.L., Wagner, P.D., 1982. Ventilation-perfusion inequality in avian lungs. Respir. Physiol. 48, 233–241.

Powell, F.L., 1983a. Circulation. In: Abs, M. (Ed.), Physiology and Behaviour of the Pigeon. Academic Press, New York, pp. 97–116.

Powell, F.L., 1983b. Respiration. In: Abs, M. (Ed.), Physiology and Behavior of the Pigeon. Academic Press, New York, pp. 73–95.

Powell, F.L., Hastings, R.H., 1983. Effects of hypoxia on ventilation-perfusion matching in birds. Physiologist 26, 50.

Powell, F.L., Mazzone, R.W., 1983. Morphometrics of rapidly frozen goose lungs. Respir. Physiol. 51, 319–332.

Powell, F.L., Hastings, R.H., Mazzone, R.W., 1985. Pulmonary vascular resistance during unilateral pulmonary arterial occlusion in ducks. Am. J. Physiol. 249 (18), R39–R43.

Powell, F.L., Hempleman, S.C., 1985. Sources of carbon dioxide in penguin air sacs. Am. J. Physiol. 248, R748–R752.

Powell, F.L., 1988. Lung structure and function. In: Wood, S.C. (Ed.), Comparative Pulmonary Physiology: Curr. Concepts. Marcel Dekker, Inc., New York, pp. 237–255.

Powell, F.L., Hempleman, S.C., 1988. Comparative physiology of oxygen transfer in lungs. In: Gonzalez, N.C., Fedde, M.R. (Eds.), Oxygen Transfer from Atmosphere to Tissues. Plenum Press, New York, pp. 53–65.

Powell, F.L., Milsom, W.K., Mitchell, G.S., 1988. Effects of intrapulmonary CO_2 and airway pressure on pulmonary vagal afferent activity in the alligator. Respir. Physiol. 74, 285–298.

Powell, F.L., Scheid, P., 1989. Physiology of gas exchange in the avian respiratory system. In: In: King, A.S., McLelland, J. (Eds.), Form and Function in Birds, vol. 4. Academic Press, London, pp. 393–437.

Powell, F.L., Hempleman, S.C., 1990. Information content of artreial chemoreceptro discharge pattern. In: Eyzaguirre, C., Fidone, S.J., Fitzgerald, R.S., Lahiri, S., McDonald, D.M. (Eds.), Arterial Chemoreception. Springer-Verlag, New York, pp. 247–253.

Powell, F.L., 1993. Birds at altitude. In: Scheid, P. (Ed.), Respiration in Health and Disease. G. Fisher, Stuttgart/New York, pp. 352–358.

Powell, F.L., Milsom, W.K., Mitchell, G.S., 1998. Time domains of the hypoxic ventilatory response. Respir. Physiol. 112, 123–134.

Powell, F.L., Dwinell, M.R., Aaron, E.A., 2000. Measuring ventilatory acclimatization to hypoxia: comparative aspects. Respir. Physiol. 122, 271–284.

Powell, F.L., Shams, H., Hempleman, S.C., Mitchell, G.S., 2004. Breathing in thin air: acclimatization to altitude in ducks. Respir. Physiol. Neurobiol. 144, 225–235.

Powell, F.L., Hopkins, S.R., 2010. Vertebrate life at high altitude. In: Nilsson, G.E. (Ed.), Respiratory Physiology of Vertebrates: Life with and without Oxygen. Cambridge University Press, Cambridge, pp. 265–299.

Kumar, P., Prabhakar, N.R., 2012. Peripheral chemoreceptors: function and plasticity of the carotid body. Compr. Physiol. 2, 141–219.

Richards, S.A., 1971. Brain stem control of polypnoea in the chicken and pigeon. Respir. Physiol. 11, 315–326.

Rollema, H.S., Bauer, C., 1979. The interaction of inositol pentaphosphate with the hemoglobin of the highland and the lowland geese. J. Biol. Chem. 254, 12038–12043.

von Saalfeld, E., 1936. Untersuchungen Über das hacheln bei tauben. Z. Vgl. Physiol. 23, 727–743.

Scharnke, H., 1938. Experimentelle Beiträge zur Kenntnis der Vogelatmung. Z. Vgl. Physiol. 25, 548–583.

Scheid, P., Piiper, J., 1969. Volume, ventilation and compliance of the respiratory system in the domestic fowl. Respir. Physiol. 6, 298–308.

Scheid, P., Piiper, J., 1970. Analysis of gas exchange in the avian lung: theory and experiments in the domestic fowl. Respir. Physiol. 9, 246–262.

Scheid, P., Piiper, J., 1972. Cross-current gas exchange in avian lungs: effects of reversed parabronchial air flow in ducks. Respir. Physiol. 16, 304–312.

Scheid, P., Slama, H., Piiper, J., 1972. Mechanisms of unidirectional flow in parabronchi of avian lungs: measurements in duck lung preparations. Respir. Physiol. 14, 83–95.

Scheid, P., Slama, H., Willmer, H., 1974. Volume and ventilation of air sacs in ducks studied by inert gas wash-out. Respir. Physiol. 21, 19–36.

Scheid, P., Kawashiro, T., 1975. Metabolic changes in avian blood and their effects on determination of blood gases and pH. Respir. Physiol. 23, 291–300.

Scheid, P., Slama, H., 1975. Remote-controlled device for sampling arterial blood in unrestrained animals. Pflügers Arch. 356, 373–376.

Scheid, P., Worth, H., Holle, J.P., Meyer, M., 1977. Effects of oscillating and intermittent ventilatory flow on efficacy of pulmonary O_2 transfer in the duck. Respir. Physiol. 31, 251–258.

Scheid, P., 1978. Analysis of gas exchange between air capillaries and blood capillaries in avian lungs. Respir. Physiol. 32, 27–49.

Scheid, P., Gratz, R.K., Powell, F.L., Fedde, M.R., 1978. Ventilatory response to CO_2 in birds. II. Contribution by intrapulmonary CO_2 receptors. Respir. Physiol. 35, 361–372.

Scheid, P., Piiper, J., 1986. Control of breathing in birds. In: Cherniack, N.S., Widdicombe, J.G. (Eds.), Handbood of Physiology: the Respiratory System – Control of Breathing. American Physiological Society, Bethesda, pp. 815–832.

Scheid, P., Piiper, J., 1989. Respiratory mechanics and air flow in birds. In: King, A.S., McLelland, J. (Eds.), Form and Function in Birds. Academic Press, London, pp. 369–391.

Scheid, P., 1990. Avian respiratory system and gas exchange. In: Sutton, J.R., Coates, G., Remmers, J.E. (Eds.), Hypoxia: the Adaptations. B.C.Decker, Inc., Toronto, pp. 4–7.

Scheid, P., Fedde, M.R., Piiper, J., 1991. Gas exchange and air-sac composition in the unanaesthetized, spontaneously breathing goose. J. Exp. Biol. 142, 373–385.

Scheipers, G., Kawashiro, T., Scheid, P., 1975. Oxygen and carbon dioxide dissociation of duck blood. Respir. Physiol. 24, 1–13.

Schmitt, P.M., Powell, F.L., Hopkins, S.R., 2002. Ventilation-perfusion inequality during normoxic and hypoxic exercise in the emu. J. Appl. Physiol. 93, 1980–1986.

Scott, G.R., Milsom, W.K., 2006. Flying high: a theoretical analysis of the factors limiting exercise performance in birds at altitude. Respir. Physiol. Neurobiol. 154, 284–301.

Scott, G.R., Milsom, W.K., 2007. Control of breathing and adaptation to high altitude in the bar-headed goose. Am. J. Physiol. Regul. Integr. Comp. Physiol. 293, R379–R391.

Scott, G.R., Egginton, S., Richards, J.G., Milsom, W.K., 2009. Evolution of muscle phenotype for extreme high altitude flight in the bar-headed goose. Proc. Biol. Sci. B 276, 3645–3653.

Sèbert, P., 1979. Mise en evidence de l'action centrale du stimulus $CO_2[H^+]$ de la ventilation chez le Canard Pekin. J. Physiol., Paris 75, 901–909.

Seifert, E., 1896. Über die Atmung der Reptilien und Vögel. Pflügers Arch. Ges. Physiol. 64, 321–506.

Shams, H., Scheid, P., 1989. Efficiency of parabronchial gas exchange in deep hypoxia: measurements in the resting duck. Respir. Physiol. 77, 135–146.

Shoemaker, J.M., Hempleman, S.C., 2001. Avian intrapulmonary chemoreceptor discharge rate is increased by anion exchange blocker 'DIDS'. Respir. Physiol. 128, 195–204.

Skadhauge, E., 1983. Formation and compostion of urine. In: Freeman, B.M. (Ed.), Physiology and Biochemistry of the Domestic Fowl. Academic Press, London, pp. 108–135.

Smith, J.C., Ellenberger, H.H., Ballanyi, K., Richter, D.W., Feldman, J.L., 1991. Pre-Bötzinger complex: a brainstem region that may generate respiratory rhythm in mammals. Science 254, 726–729.

Smith, J.H., Meier, J.L., Lamke, C., Neill, P.J., Box, E.D., 1986. Microscopic and submicroscopic anatomy of the parabronchi, air sacs, and respiratory space of the budgerigar (*Melopsittacus undulatus*). Am. J. Anat. 177, 221–242.

Sturkie, P.D., 1986. Body fluids: blood. In: Sturkie, P.D. (Ed.), Avian Physiology. Springer-Verlag, New York, pp. 102–139.

Taylor, E.W., Jordan, D., Coote, J.H., 1999. Central control of the cardiovascular and respiratory systems and their interactions in vertebrates. Physiol. Rev. 79, 855–916.

Taylor, C.R., Weibel, E.R., 1981. Design of the mammalian respiratory system. I. Problem and strategy. Respir. Physiol. 44, 1–10.

Torre-Bueno, J.R., Geiser, J., Scheid, P., 1980. Incomplete gas mixing in air sacs of the duck. Respir. Physiol. 42, 109–122.

Tucker, V., 1972. Respiration during flight in birds. Respir. Physiol. 14, 75–82.

Vos, H.F., 1935. Über den Weg der Atemluft in der Entenlunge. Z. Vgl. Physiol. 20, 552–578.

Wagner, P.D., 1977. Diffusion and chemical reaction in pulmonary gas exchange. Physiol. Rev. 57, 257–312.

Wang, N., Banzett, R.B., Butler, J.P., Fredberg, J.J., 1988. Bird lung models show that convective inertia effects inspiratory aerodynamic valving. Respir. Physiol. 73, 111–124.

Wang, N., Banzett, R.B., Nations, C.S., Jenkins, F.A., 1992. An aerodynamic valve in the avian primary bronchus. J. Exp. Biol. 262, 441–445.

Watson, R.R., Fu, Z., West, J.B., 2008. Minimal distensibility of pulmonary capillaries in avian lungs compared with mammalian lungs. Respir. Physiol. Neurobiol. 160, 208–214.

Weber, R.E., Hemmingsen, E.A., Johansen, K., 1974. Functional and biochemical studies of penguin myoglobin. Comp. Biochem. Physiol. 49B, 197–214.

Weber, R.E., Wells, R.M.G., 1989. Hemoglobin structure and function. In: Wood, S.C. (Ed.), Comparative Pulmonary Physiology: Current Concepts. Marcel Dekker, Inc., New York, pp. 279–310.

Weber, R.E., 2007. High-altitude adaptations in vertebrate hemoglobins. Respir. Physiol. Neurobiol. 158, 132–142.

Weidner, W.J., Selna, L.A., McClure, D.E., DeFouw, D.O., 1993. Effect of extracellular fluid volume expansion on avian lung fluid balance. Respir. Physiol. 91, 125–136.

Weidner, W.J., Kinnison, J.R., 2002. Effect of extracellular fluid volume expansion on the interparabronchial septum of the avian lung. J. Comp. Pathol. 127, 219–222.

Weidner, W.J., Waddell, D.S., Furlow, J.D., 2006. Measurement of the filtration coefficient (Kfc) in the lung of *Gallus domesticus* and the effects of increased microvascular permeability. J. Comp. Physiol. B 176, 567–574.

Weidner, W.J., Bradbury, C.A., Le, S.P., Wallace, S.R., 2012. Regional pulmonary blood flow in the lung of the chicken. Poult. Sci. 91, 1441–1443.

Weingarten, J.P., Rollema, H.S., Bauer, C., Scheid, P., 1978. Effects of inositol hexaphosphate on the Bohr effect induced by CO_2 and fixed acids in chicken hemoglobin. Pflügers Arch. 377, 135–141.

Weinstein, Y., Bernstein, M.H., Bickler, P.E., Gonzales, D.V., Samaniego, F.C., Escobedo, M.A., 1985. Blood respiratory properties in pigeons at high altitudes: effects of acclimation. Am. J. Physiol. 249, R765–R775.

Wells, D.J., 1993. Ecological correlates of hovering flight of hummingbirds. J. Exp. Biol. 178, 59–70.

Wells, R.M.G., 1976. The oxygen affinity of chicken hemoglobin in whole blood and erythrocyte suspensions. Respir. Physiol. 27, 24–31.

West, J.B., Watson, R.R., Fu, Z., 2006. Major differences in the pulmonary circulation between birds and mammals. Respir. Physiol. Neurobiol. 157, 382–390.

West, J.B., 2009. Comparative physiology of the pulmonary blood-gas barrier: the unique avian solution. Am. J. Physiol. Regul. Integr. Comp. Physiol. 297, R1625–R1634. doi: 10.1152/ajpregu.00459.2009.

West, J.B., Fu, Z., Deerinck, T.J., Mackey, M.R., Obayashi, J.T., Ellisman, M.H., 2010. Structure-function studies of blood and air capillaries in chicken lung using 3D electron microscopy. Respir. Physiol. Neurobiol. 170, 202–209. doi: 10.1016/j.resp.2009.12.010.

West, J.B., 2011. Comparative physiology of the pulmonary circulation. Compr. Physiol. 1, 1525–1539.

de Wet, P.D., Fedde, M.R., Kitchell, R.L., 1967. Innervation of the respiratory muscles of *Gallus domesticus*. J. Morphol. 123 (1), 17–34.

Whittow, G.C., Ossorio, N., 1970. A new technique for anesthetizing birds. Lab Anim. Care 20, 651–656.

Wideman, R.F., Chapman, M.E., Hamal, K.R., Bowen, O.T., Lorenzoni, A.G., Erf, G.F., Anthony, N.B., 2007. An inadequate pulmonary vascular capacity and susceptibility to pulmonary arterial hypertension in broilers. Poult. Sci. 86, 984–998.

Wideman, R.F., Rhoads, D.D., Erf, G.F., Anthony, N.B., 2013. Pulmonary arterial hypertension (ascites syndrome) in broilers: a review. Poult. Sci. 92, 64–83.

Williams Jr, B.R., Boggs, D.F., Kilgore Jr, D.L., 1995. Scaling of hypercapnic ventilatory responsiveness in birds and mammals. Respir. Physiol. 99, 313–319.

Yamatsu, Y., Kameda, Y., 1995. Accessory carotid body within the parathyroid gland III of the chicken. Histochemistry 103, 197–204.

Zeuthen, E., 1942. The ventilation of the respiratory tract in birds. K. Danske Vidensk. Selsk. Skr. 17, 1–50.

Zimmer, K., 1935. Beitrage zur Mechanik der Atmung be den Vogeln in Stand und Flug. Zoologica 33, 1–69.

Chapter 14

Gastrointestinal Anatomy and Physiology

D. Michael Denbow

Department of Animal and Poultry Sciences, Virginia Tech, Blacksburg, VA, USA

14.1 ANATOMY OF THE DIGESTIVE TRACT

The digestive tract is not only important for nutrient digestion and absorption, but it is the largest immunological organ in the body protecting against exogenous pathogens. The digestive system (Figure 14.1) has adaptations designed to facilitate flight. The length of the intestinal tract is shorter in birds relative to mammals (Table 14.1). Also, birds lack teeth and heavy jaw muscles, which have been replaced with a lightweight bill or beak. Food particles are swallowed whole and then reduced in size by the *ventriculus* or *gizzard* located within the body cavity. This chapter will not attempt to describe the many species variations in detail but will instead describe differences between birds and mammals. The reader is referred to the excellent reviews by McLelland (1975, 1979) for specific information on various species.

14.1.1 Beak, Mouth, and Pharynx

Birds have a bill or beak. This structure shows extensive anatomical variations among species, thus allowing for adaptations for different feeding styles. The upper bill is usually covered with a hard keratin. However, in some types of birds, the whole bill is relatively soft (i.e., Charadrii or shorebirds), whereas in others only the tip is hard (i.e., Anatidae or waterfowl). The culmen, the medial dorsal area of the upper beak, has a pointed protuberance in the embryo, the egg tooth, which drops off after hatching.

Birds, unlike mammals, have no sharp distinction between the pharynx and mouth. Birds lack a soft palate and a pharyngeal isthmus; the combined oral and pharyngeal cavities are referred to as the *oropharynx*. The palate contains a longitudinal fissure, the *choana*, which connects the oral and nasal cavities. Caudal to the choana is the *infundibular cleft*, which is medially located and is the common opening to the auditory tubes (Figure 14.2). The palate generally also has ridges that aid in opening the shell of seeds.

All vertebrates have a hyoid, which serves as an attachment site for muscles of the tongue and throat. Birds have a Y-shaped hyoid apparatus that extends to the tip of their tongue. Two long structures, the "horns" of the hyoid, grow backwards from the fork in the Y and provides insertion sites for protractor muscles originating on the lower jaw. The hyoid horns are modified in different species and can grow to the top of the head, around the eye socket and into the nasal cavity (Short, 1982; Wang et al., 2011). Contraction of the muscles attached to the hyoid apparatus pulls the apparatus forward and against the skull, thus thrusting the tongue outward. Lengthing the hyoid horns and attached muscles effectively gives the bird a longer tongue able to protrude further out the mouth.

The tongue shows adaptations for collecting, manipulating, and swallowing food (Harrison, 1964). In birds where the tongue is used for collecting food, it can be extended from the mouth during the collection process. Such tongues typically have lateral barbs and may be coated with mucous secreted by the mandibular salivary gland. The tongue may act as either a brush, a spear, or a dynamic tube (Rico-Guevara and Rubega, 2011), sometimes extending four times the length of the beak.

Tongues used to manipulate food, such as in piscivorous species, are nonprotruding and covered with stiff, sharp, caudally-directed papillae. In birds of prey, the tongue is a rasp-like structure, with the rostral portion frequently being very hard and rough (McLelland, 1979). On the tongue of birds that typically strain food particles (e.g., ducks), the rostral portion forms a scoop-like structure, with the lateral borders having a double row of overlapping bristles. The bristles work in conjunction with the lamellae of the bill to filter particles.

In birds where the tongue is utilized to aid in the swallowing of food, caudally-directed papillae tend to be located near the root of the tongue. These papillae function to propel food caudally. Tongues specialized for swallowing are nonprotruding.

Salivary glands also show considerable species variation. Although salivary glands are generally well developed in granivorous species, they are less developed in birds

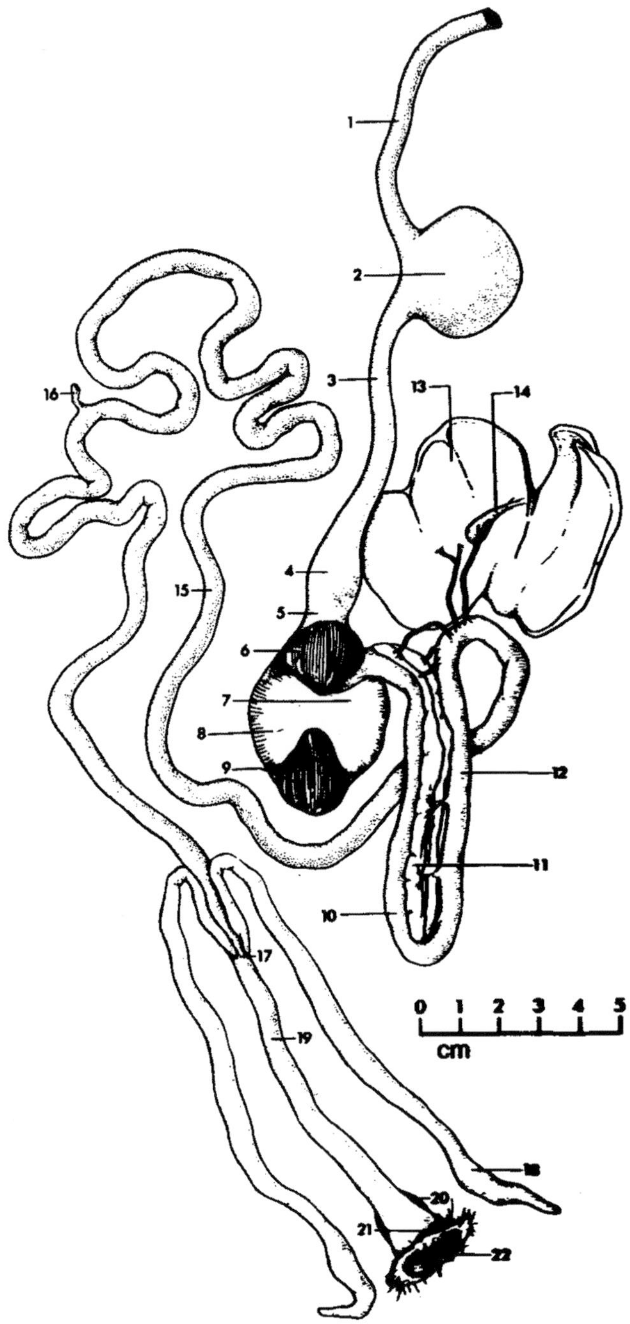

FIGURE 14.1 Digestive tract of a 12-week-old turkey. 1, precrop esophagus; 2, crop; 3, postcrop esophagus; 4, proventriculus; 5, isthmus; 6, thin craniodorsal muscle; 7, thick cranioventral muscle; 8, thick caudodorsal muscle; 9, thin caudoventral muscle (6–9, gizzard); 10, proximal duodenum; 11, pancreas; 12, distal duodenum; 13, liver; 14, gallbladder; 15, ileum; 16, Meckel's diverticulum; 17, ileocecal junction; 18, ceca; 19, rectum; 20, bursa of Fabricius; 21, cloaca; 22, vent. Scale is in centimeters. *Reprinted from Duke, G.E. Avian digestion. In: Swenson, M.J. (Ed.), Duke's Physiology of Domestic Animals, tenth ed. Copyright 1984 by Cornell University Press, Ithaca, NY, p. 360. Used by permission of the publisher.*

of prey, poorly developed in piscivores, and absent in the Anhinga and Great Cormorant (Antony, 1920). As a generalization, the *maxillary*, *palatine*, and *sphenopterygoid* glands are located in the roof of the mouth. The buccal gland is in the cheeks whereas the *mandibular*, *lingual*, and *cricoarytenoid* glands are in the floor of the mouth. Although the salivary glands of *Gallus* and *Meleagris* are reported to secrete little amylase, the house sparrow secretes considerable amounts (Jerrett and Goodge, 1973).

Taste buds are variably located on the upper beak epithelium, in the anterior mandible, and the mandibular epithelium posterior to the tongue. There are a small number of taste buds also located ventrolaterally on the anterior tongue. It is believed that chickens have as many as 300 taste buds, with the number in broilers being twice that of layer-type chickens (Ganchrow and Ganchrow, 1985; Kudo et al., 2008).

In some species of birds, the floor of the mouth contains sac-like diverticuli called *oral sacs*. These can act to either carry food or as a display apparatus during the breeding season.

Immediately behind the tongue is the *laryngeal mound*. It contains a narrow slit-like opening into the *glottis* of the larynx. The laryngeal mound generally contains rows of caudally-directed cornified papillae, which aid in moving food toward the esophagus during swallowing.

14.1.2 Esophagus and Crop

The esophagus is a thin-walled, distensible tube that transports food from the pharynx to the stomach, allowing birds to swallow their food whole. Thus, it contains a number of longitudinal folds that provide distensibility. The avian esophageal wall consists of four layers: mucosal, submucosal, muscle tunic, and the serosal layer; it generally contains only smooth muscle cells, with a circular muscle layer predominating (McLelland, 1979).

Unlike mammals, the avian esophagus is divided into a cervical and a thoracic region. In addition, the esophagus of birds lacks both upper and lower esophageal sphincters, which are present in mammals (see Mule, 1991).

In many, but not all (e.g., gulls, penguins, ostriches), species of birds, the cervical esophagus is expanded to form a *crop*. The crop functions to store food and may be spindle-shaped, bilobed, or unilobed. In the chicken, the crop is a ventral diverticulum of the esophagus and contains longitudinal folds on the inner surface, thus making it distensible. Beyond the crop, the esophagus continues as the thoracic esophagus to connect with the proventriculus.

A small number of species have a diverticulum or bilaterally symmetrical expansion of the cervical esophagus, the *esophageal sac*. In most species that have such a structure, it occurs only in the male and functions as a display during the breeding season and for the production of mating calls.

The esophagus and crop are lined with incompletely keratinized stratified squamous epithelia into which open numerous mucous glands. These glands are generally more numerous in the thoracic esophagus and may even be absent in the cervical region. Mucous glands are located in the crop only near the junction with the esophagus.

TABLE 14.1 Dimensions of the Digestive Tract of Various Species of Birds[1]

		Esophagus			Proventriculus and gizzard			Small intestine			Cecum		Rectum			Total	
Species	Body wt (kg)	Length (mm)	Total %	Wt (g)	Length (mm)	Total %	Wt (g)	Length (mm)	Total %	Wt (g)	Length (mm)	Wt (g)	Length (mm)	Total %	Wt (g)	Length (mm)	Length/ body wt
Chicken																	
Leghorn	1.2	136	9.9	8.2	86	6.3	26.7	1082	78.9	29.5	127	5.2	68	5.0	2.3	1372	1.14
Broiler	3.0	140	6.4	16.8	101	4.7	43.5	1796	82.7	73.6	188	10.7	134	6.2	5.1	2171	0.72
Turkey	3.0	123	5.7	8.5	110	5.1	52.9	1853	85.7	85.3	278	20.1	75	3.5	4.4	2161	0.72
Japanese quail[2]		75	11.5		38	5.8		510	78.1		100		30	4.6		653	
Domestic duck[3]	2.2	310	11.7		130	4.9		2110	79.9		140		90	3.4		2640	1.20
Emu[3]	53.0	790	12.1		260	4.0		5200	79.4		120		300	4.6		6550	0.12
Rhea[4]	25.0				310			1400			480		400				
Ostrich[4]	122.0				480			6400			940		8000				
Cedar waxwing[5]		51	16.2		36	11.4		171	54.3		0	0	57	18.1		315	

[1]*The length and weight (wt) of the gastrointestinal tract can change depending on the environment in which the birds are raised (see Deaton et al., 1985).*
[2]*From Fitzgerald (1969).*
[3]*From Herd (1985).*
[4]*From Fowler (1991).*
[5]*From Levey and Duke (1992).*

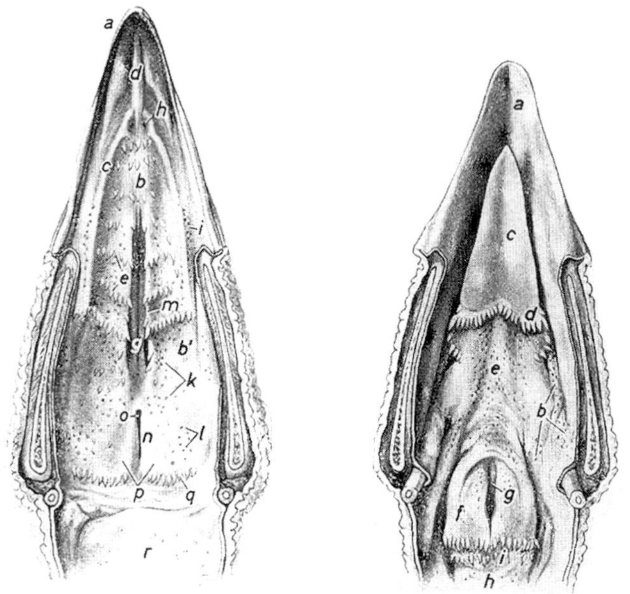

FIGURE 14.2 *Left*: Roof of the mouth cavity and pharynx of the fowl. a, upper beak; b, b', palate; c, lateral palatine ridges; d, median swelling and e, papillae of the palate; g, palatine cleft (choanal cleft); h, orifice of the gland. maxillaris; i, orifice of the glandula palantinae latteralis; k, orifice of the glandula palaninae meddialis; l, orifice of the glandula pterygoideae and glandula tubariae; m, orbital folds; n, pharyngeal folds; o, infundibular cleft; p, pharynx; q, rows of pharyngeal papillae; r, esophagus. *Right*: Floor of the mouth and pharyngeal cavities of the fowl. a, lower beak; b, orifice of the glandula mandibulares posterior; c, tongue; d, row of lingual papillae; e, orifice of the glandd. linguales posterior; f, larynx (larynx cran.); g, laryngeal cleft; h, esophagus; i, rows of pharyngeal papillae. *Reprinted from Nickel, R., Schummer, A., Seiferle, E., Siller, W.G., Wight, P.A.L. Anatomy of the Domestic Birds. Copyright 1977 by Springer-Verlag, New York, p. 43. Used by permission of the publisher.*

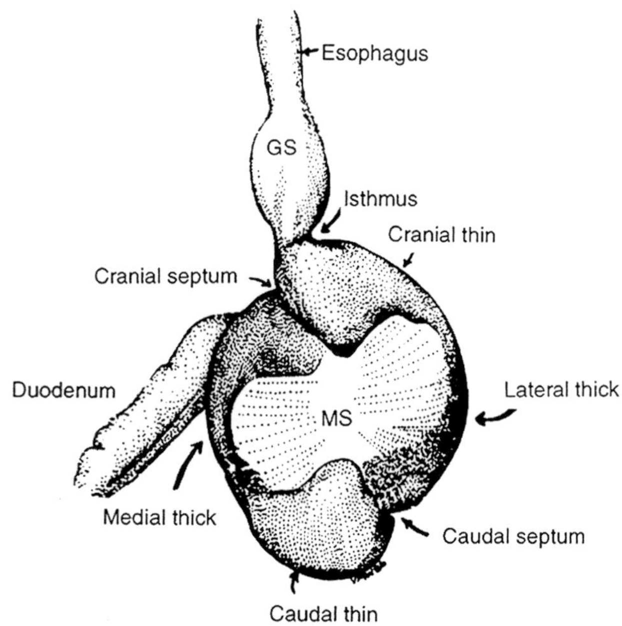

FIGURE 14.3 **Anatomical features of muscular stomach of domestic turkeys.** GS, glandular stomach; MS, muscular stomach. Note that cranial thin muscle of MS is continuous with lateral thick muscle and separated from caudal thin muscle at caudal septum. Similarly, the caudal thin muscle in continuous with the medial thick muscle and separated from cranial thin muscle by cranial septum. *From Chaplin and Duke (1990).*

The cervical esophagus is innervated by parasympathetic nerves. The thoracic esophagus is innervated by the vagus and the celiac plexus. The esophagus is innervated by a few adrenergic fibers, which synapse with the myenteric plexus rather than the muscles (McLelland, 1975; Mule, 1991).

14.1.3 Stomach

In mammals, the stomach consists of a single chamber. However, in birds, the stomach consists of two chambers, the proventriculus and gizzard, with the former being the mammalian counterpart. The proventriculus or glandular stomach is located orad to the gizzard or muscular stomach. The proventriculus is a fusiform organ varying in size and shape among species, being relatively large and distensible in aquatic carnivores while being relatively small in granivorous species. In ostriches, which lack a crop, the proventriculus is especially large. The proventriculus secretes mucus, hydrochloric acid, and pepsinogen, whereas the gizzard functions in mechanical digestion and is the site of gastric proteolysis. The pyloric region connects the gizzard to the duodenum.

There are two extremes in gastric anatomy (McLelland, 1979). The first type, characteristic of carnivorous and piscivorous species, is adapted for storage and the digestion of a relatively soft diet. The two stomach chambers contain little separation, although one chamber may be more developed than the other, depending on the species. The second stomach type, characteristic of omnivores, insectivores, herbivores, and granivores, is adapted for very hard diets. The proventriculus and gizzard are separated by an isthmus termed the *zona intermedia gastric*. The proventriculus is relatively small, while the gizzard is large and powerful. The gizzard consists of two pairs of opposing muscles, termed thick and thin pairs (Figure 14.3), which are composed of circular muscle. These pairs of muscles are lacking in some species (e.g., herons, hawks, owls).

The proventriculus, which generally lacks ridges except in fish- and meat-eating species, is lined with a mucous membrane. Projecting into the lumen are papillae on the surface, of which can be seen the openings of the compound glands that secrete gastric juices. These glands generally contain only oxynticopeptic cells and secrete hydrochloric acid, pepsin, and mucous. The proventriculus contains both a myenteric and submucosal plexus (Martinez et al., 2000).

The interior surface of the gizzard is lined with a cuticle, sometimes called *koilin*, which is produced by the mucosal glands. The cuticle protects the gizzard from acid and proteolytic enzymes secreted by the proventriculus and from injury during grinding of hard food items. The greenish or brownish color of the cuticle is due to the reflux of bile pigments from the duodenum. Although the cuticle is continuously worn away and replaced in most birds, it is shed in some species.

The pyloric portion of the stomach shows a considerable range in development. In *Gallus*, it is only 0.5 cm long. Its mucosal glands secrete mucous instead of a cuticle. In other species, the pyloric portion is enlarged and contains a cuticle. Although the function of this region is unknown, it is believed that it may slow the movement of large particles into the duodenum (Vergara et al., 1989).

The stomach is innervated by the vagus and perivascular nerve fibers from the celiac and mesenteric plexii. The muscle cells are innervated by cholinergic fibers, while noradrenergic fibers innervate mainly blood vessels. The myenteric nerve plexus lies just under the serosa, while the submucosal plexus is lacking. Because the longitudinal muscle layer is absent, the myenteric nerve plexus is normally visible through the transparent serosa.

14.1.4 Small Intestine

The small intestine is sometimes divided into the duodenum, jejunum, and ileum. Although there is a distinct duodenal loop, the yolk stalk (i.e., Meckel's diverticulum) is often used as a landmark to separate the jejunum and ileum. Intestinal length varies considerably between species, being relatively shorter in frugivores, carnivores, and insectivores and long in granivores, herbivores, and piscivores. However, the length appears to be relatively shorter than in mammals.

The intestinal wall can contain either folds or villi, depending on the species. The type of mucosal projections are not necessarily consistent between the small and large intestine. *Gallus* species have villi, which decrease in length from 1.5 mm in the duodenum to 0.4–0.6 mm in the ilium and rectum. The number of villi decreases from 1–10 days of age, but thereafter remains constant.

Genetic selection for growth has altered villi morphology (Yamauchi and Isshiki, 1991). Compared to white leghorns, the villi of broilers are larger and show more epithelial cell protrusions from the apical surface of the duodenal villi. Nevertheless, the villi from both types of chickens form a zigzag arrangement, which is thought to slow ingesta flow.

The intestinal wall contains the same four layers as the remainder of the tract including the mucosal, submucosal, muscle tunic, and the serosal layer. The mucosal layer consists of the muscularis mucosa, lamina propria, and epithelium. However, the muscularis mucosa and lamina propria are poorly developed in birds, possibly because the villi lack a central lacteal. Although Brunner's glands, common to mammals, are absent (Calhoun, 1954), *tubular* glands (possibly homologous to Brunner's glands) are present in some species (Ziswiler and Farner, 1972). The epithelium contains chief cells, goblet cells, and endocrine cells. The crypts of Lieberkuhn are the source of epithelial cells lining the villi. The crypts contain undifferentiated cells, goblet cells, endocrine cells, and lymphocytes. Globular leukocytes and Paneth cells appear near the base of the crypts.

The intestines contain extensive innervation from both the sympathetic and parasympathetic nervous system. As described by Bennett (1974), innervation is both cholinergic and adrenergic. The enteric nerve plexuses were described by Ali and McLelland (1978). Except in the rectum, the longitudinal muscle is sparsely innervated.

The *intestinal nerve* (nerve of Remak), which runs the length of the small and large intestine, is unique to birds. Although it has no mammalian homologue, it may be analogous to the prevertebral ganglion (Hodgkiss, 1984a). This nerve is thought to be a mixed nerve containing both sympathetic and parasympathetic autonomic fibers (Hodgkiss, 1986).

14.1.5 Ceca

Arising at the junction of the ileum and rectum are the ceca. In some species, the ceca may be absent (e.g., Psittaciformes, Apodiforms, and Piciforms) or rudimentary (e.g., Columbiformes and Piciformes). In other species, they are either paired (e.g., herbivores, most granivores, and owls), singular (Ardeidae), or consist of a double pair (secretary bird). McLelland (1979) has grouped ceca into four types based on morphology: (1) intestinal, which resemble the remainder of the intestine; (2) glandular, which are long and contain many actively secreting crypts; (3) lymphoid, which are reduced in size, containing many lymphocytes and occasional nonsecreting crypts; and (4) vestigial, which are small with a reduced lumen. A correlation between diet and cecal development or between the size of the ceca and the length and width of the rectum has not been discovered (McLelland, 1989).

In chickens, a cecum can be morphologically divided into three regions (Ferrer et al., 1991). Near the ileocecal junction is the *basis ceci*, where the villi are well developed. The medial cecal region (*corpus ceci*) has longitudinal folds with small villi, while the distal cecal region (*apex ceci*) similarly has small villi and contains both longitudinal and transverse folds. The combination of villi and musculature near the ileocecal junction effectively prevents even very small particles from entering the ceca (Ferrando et al., 1987), although fluid contents do enter.

There is growing understanding of the importance of the ceca. Cecectomy results in lower metabolizability of food, greater loss of amino acids, and lower digestibility of crude fiber (Chaplin, 1989). The role of the ceca in absorption is discussed later.

14.1.6 Colon (Rectum) and Cloaca

The colon, sometimes called the rectum, is relatively short, linking the ileum with the *coprodeal* compartment of the cloaca. Although the colon of mammals has no villi and many goblet cells, the colon of birds has numerous flat villi and relatively few goblet cells (Clauss et al., 1991). In addition, the avian rectum has relatively few crypts, and they are shorter than in mammals. The simple columnar epithelium lining the colon and ceca has a well-developed brush border. As discussed in Section 14.6, the cloaca and colon have an important role in water reabsorption.

The cloaca serves as a common pathway for excretory, reproductive, and digestive wastes. It contains three chambers: *coprodeum*, *urodeum*, and *proctodeum*. The coprodeum is the most cranial portion into which empties the colon. The coprodeum lacks villi, but has mucosal folds or rugae. The urodeum is the middle and smallest compartment of the cloaca, separated from the coprodeum and proctodeum by the coprourodeal fold and the uroproctodeal fold, respectively. The urinary and reproductive tracts empty into the urodeum. The final chamber, the proctodeum, opens externally through the anus. The bursa of Fabricius, involved in immune function, projects dorsally into the proctodeum. Also projecting into the dorsal portion of the proctodeum is the dorsal proctodeal gland, sometimes called the foam gland, which secretes a white, frothy fluid. Birds lack a urinary bladder, and it has been shown that urine enters the distal lower intestine and is forced back into the colon, ceca, and possibly the small intestine (Goldstein and Braun, 1986).

14.2 ANATOMY OF THE ACCESSORY ORGANS

14.2.1 Pancreas

The pancreas is a pale yellow or red organ located within the duodenal loop, although in some species, such as the budgerigar, part of it may be found outside the loop. The gland has both an endocrine function, which will be discussed later, and an exocrine function. It is relatively small in carnivores and granivores but large in piscivores and insectivores. The pancreas is generally divided into three lobes—dorsal, ventral, and splenic—but their function in unknown (Paik et al., 1974).

The exocrine pancreas consists of compound tubuloacinar glands, which are divided into lobules. The number of pancreatic ducts varies from one to three (three in the domestic fowl). The pancreatic ducts generally drain into the distal part of the ascending duodenum and rarely drain into the descending loop of the duodenum. In domestic birds, the pancreatic and bile ducts drain into the ascending loop of the duodenum by a common papilla (Figure 14.1). The pancreas secretes amylase, lipase, proteolytic enzymes, and sodium bicarbonate.

14.2.2 Liver

The liver also functions as both an endocrine and an exocrine gland. It is divided into right and left lobes, which are joined cranially at the midline. The right lobe is larger, and, in the domestic fowl and turkey, the left lobe is subdivided into the dorsal and ventral parts. The bile canaliculi drain into the interlobular ducts. The lobular ducts then combine to form the right and left hepatic ducts. In birds, unlike in mammals, bile is transported to the duodenum via two ducts. The right and left hepatic duct combine to form the common hepatoenteric duct, which then goes to the duodenum. However, a hepatocystic duct branches from the right hepatic duct and connects to the gall bladder which, in turn, is drained by the cysticoenteric duct into the duodenum. In species without a gall bladder (e.g., pigeons, some parrots and ostrich), the branch of the right hepatic duct, called the right hepatoenteric duct, drains directly into the duodenum. The bile ducts generally drain into the duodenum at a site very near the pancreatic ducts. This generally occurs on the ascending loop of the duodenum. However, in some species, including the ostrich and *Columba*, the ducts empty into the descending loop of the duodenum.

14.3 MOTILITY

As food is broken down, it not only needs to be transported along the length of the gastrointestinal tract (GIT) (i.e., peristalsis), but there is also local, nonpropagating motility that mixes food with gastric juices and brings particles in closer proximity to the enterocytes. Motility not only occurs while an animal is eating, but also while the animal is fasted (migrating motor complexes, MMCs). The latter type of motility probably helps keep the digestive tract free from indigestible particles and dead enterocytes, among others. The enteric nervous system, part of the autonomic nervous system, controls gastrointestinal tract function, including motility.

14.3.1 Esophagus

Deglutition has been studied in *Gallus* by White (1968, 1970) and Suzuki and Nomura (1975). For prehension, the head is first lowered, after which food is grasped with the beak and then moved towards the oropharynx with the tongue. The choana reflexively closes. The oral phase of swallowing involves the rapid rostrocaudal movements of the tongue for 1–3 s, which moves the food particles caudally (Suzuki and Nomura, 1975). This movement is assisted by the caudally directed papillae.

During the next phase, the pharyngeal phase, the infundibular mound and glottis close, the hyoid apparatus becomes concave, the tongue is moved backward, and the esophagus is moved forward, thus decreasing the distance between the oral cavity and the esophagus. The head is raised and further tongue movements, assisted by the rostrocaudal movements of the infundibular mound, propel the food particles from the tongue to the esophagus.

Primary peristalsis within the esophagus moves the bolus towards the stomach (esophageal phase). Contractions are more rapid in the cranial esophagus than in the thoracic esophagus. In a fasted bird, the longitudinal muscle layer obliterates the esophago-ingluvial fissura, thus preventing a bolus from entering the crop (Ashcraft, 1930). After partial gizzard filling, the esophago-ingluvial fissura is relaxed and food can either enter the crop or stomach depending on the contractile state of the gizzard. The crop acts as a temporary food storage site (Hill, 1971). The destination of food appears to be controlled by the contractile state of the gizzard, with food entering or bypassing the crop when the gizzard is contracting or relaxing, respectively. In 6- to 10-week-old turkeys fasted overnight, almost no ingested food enters the crop during the first 4–6 h after dawn. During each meal the gastrointestinal tract fills cephalic to the upper one-third of ileum. In late afternoon meals, the crop also fills (G. Duke, personal communication).

Food is evacuated from a crop as a result of contractions by the crop wall. Such contractions last about 6 s with a force of approximately 20 cm H_2O (Fileccia et al., 1984). During primary peristalsis, there is a cessation of spontaneous electrical activity, which is associated with relaxation of the muscular wall. This is followed by a propagated, long-lasting spike burst of high amplitude. As the peristaltic wave moves aborad, it is preceded by inhibition of the muscles directly in front of the wave. The rate of emptying of the crop is not influenced by particle size or whether the compounds are soluble or insoluble (Vergara et al., 1989).

Unlike mammals, spontaneous electrical activity and contractions have been recorded in the esophagi of birds (see Mule, 1991). This activity is myogenic in origin. It occurs independent of slow waves, which are absent in the esophagi of birds. Although the function of these spontaneous contractions is unknown, they may act to clear contents from the esophagus.

Primary peristalsis is mediated entirely by the extrinsic nervous system (Mule, 1991). The enteric nervous system of birds is not responsible for propagating peristalsis, at least beyond short distances. Sectioning the glossopharyngeal or vagus nerve can disrupt peristalsis in the cervical and thoracic esophagus, respectively. Denervation of the esophagus prevents the propagation of electrical wave activity, indicating that the muscle cells act similar to multiunit smooth muscle cells.

14.3.2 Gastrointestinal Cycle

Motility of the mammalian stomach is regulated by slow waves thought to arise from the interstitial cells of Cajal, which lie adjacent to the myenteric plexus (Sanders et al., 2006). These slow waves pace the "phasic muscles", increasing the probability that voltage-gated Ca^{2+} channels will open, thus exciting the muscles. Slow waves arise from distinct pacemaker regions within different parts of the GIT, thus producing regional differences in slow-wave frequency. In the stomach, the pacemaker cells reside in the greater curvature of the stomach. The slow waves do not propagate from the stomach to the small intestine, but instead the small intestine has its own pacemaker cells. The stomachs of most birds lack longitudinal smooth muscle and do not display slow waves. As a result, gastric motility is more complex in birds than in mammals (Dziuk and Duke, 1972).

During the gastrointestinal cycle of birds, the thin muscles of the muscular stomach contract and the isthmus closes, after which the pylorus opens and gastric contents flow into the duodenum (Figure 14.4). Next, the duodenum contracts, followed by relaxation of the isthmus and a contraction of the thick muscles of the muscular stomach. This precipitates a refluxing of the contents of the muscular stomach into the glandular stomach. The cycle is completed

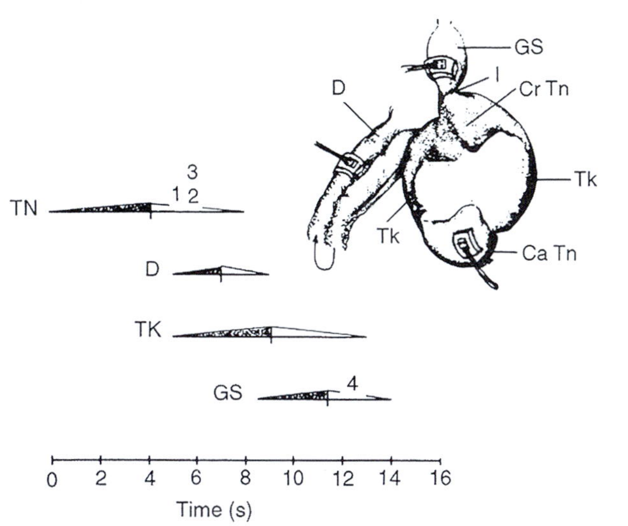

FIGURE 14.4 *Top right*: illustration of gastroduodenal apparatus of turkeys showing anatomical relationships of organs and placement of strain gauge transducers. Letters refer to organs as follows: GS, glandular stomach; I, isthmus; CrTN, cranial thin muscle of muscular stomach; TK, thick muscle pair (dorsal and ventral) of muscular stomach; CaTN, caudal thin muscle of muscular stomach; D, duodenum. *Bottom left*; triangles graphically represent sequence and duration of events in gastroduodenal cycle. Stippled areas indicate duration of contraction and clear areas indicate duration of relaxation of thin muscle pair (TN) and thick muscle pair (TK) of the muscular stomach, duodenum (D), and glandular stomach (GS). Numbers refer to noncontractile events in gastroduodenal cycle; 1, pylorus open; 2, isthmus open; 3, pylorus closed; 4, isthmus closed. *From Chaplin and Duke (1988).*

with contraction of the glandular stomach. A gastroduodenal cycle occurs with a frequency of 3.3 cycles/min in turkey (Duke, 1982). As might be expected, the largest change in intraluminal pressure is associated with contraction of the thick muscles (Figure 14.5).

As stated earlier, the pyloric region appears to control the movement of materials from the gizzard to the duodenum. Whereas soluble material is readily transported from the gizzard to the duodenum, larger particles are retained longer within the muscular stomach (Vergara et al., 1989).

Initiation of the gastrointestinal cycle is not dependent on extrinsic innervation in sated birds, suggesting that there is an intrinsic pacemaker controlling the cycle (Chaplin and Duke, 1988). Denervation of the stomach slows the rate of the gastrointestinal cycle in fasted birds and disrupts its normal synchronization (Figure 14.6), indicating that the gastrointestinal cycle is not independent of external neural input.

The pacemaker for the gastrointestinal cycle appears to reside in the isthmus (Chaplin and Duke, 1990). Destruction of the myenteric plexus in this area reduced contractions of the muscular stomach and duodenum by 50%, while simultaneously abolishing glandular stomach contractions.

The muscular stomach of raptors lack the two pairs of opposing muscles characteristic of other types of birds; as a result, the gastroduodenal cycle is simplified. It is described

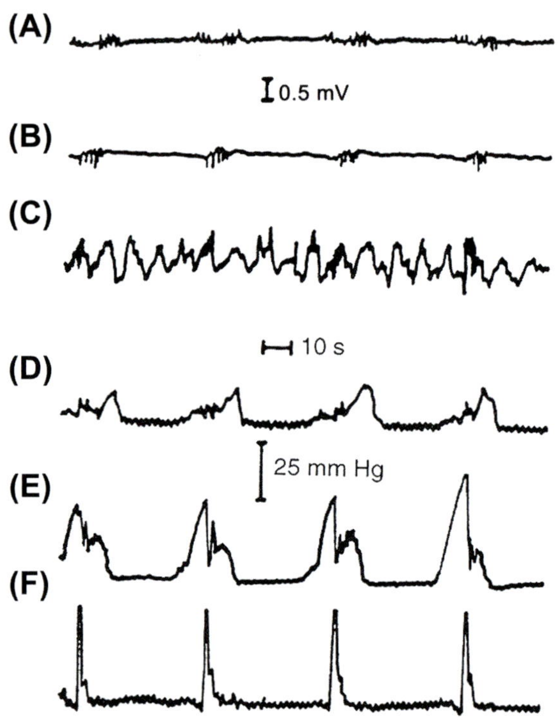

FIGURE 14.5 Tracings of typical records of electrical potential and intraluminal pressure changes from the stomach and duodenum of turkeys. (A), (B), and (C) are tracings of electrical potential changes recorded from the proventriculus, thick cranioventral muscle of the gizzard, and proximal duodenum, respectively. Slow waves with spikes are evident in the duodenum; only spikes associated with contractions are evident in the proventriculus and gizzard. (D), (E), and (F) are tracings of the corresponding intraluminal pressure changes recorded from these same organs, respectively. Time constant for electrical recordings was 1.1 s. *From Duke et al. (1975).*

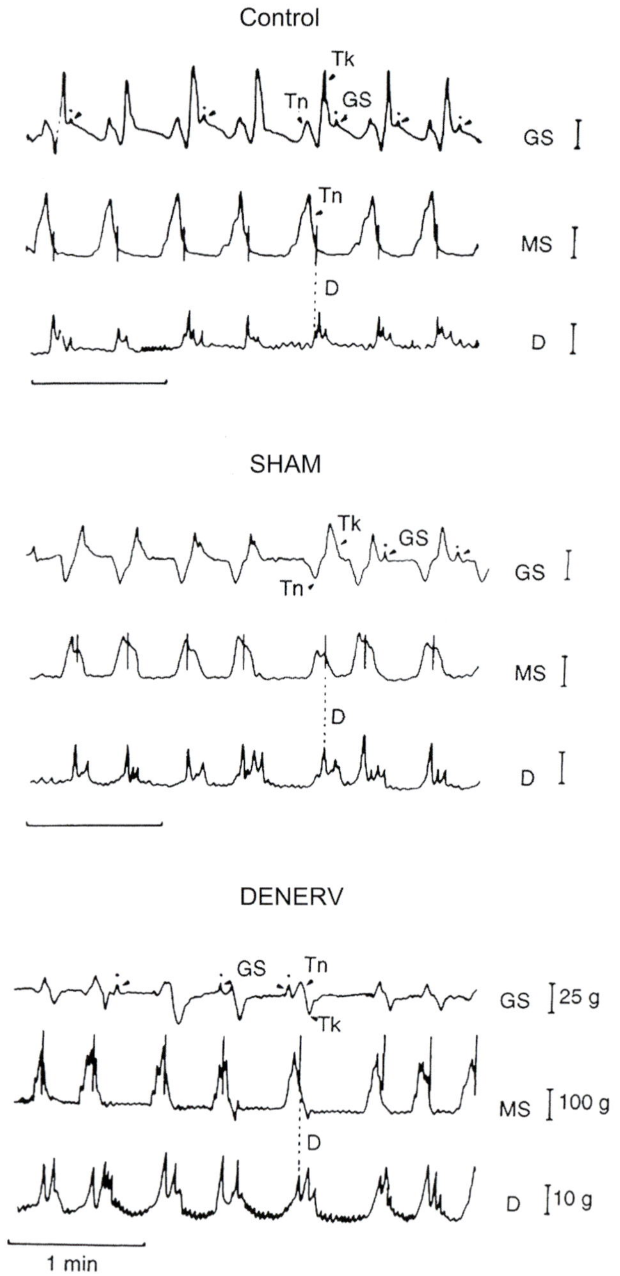

FIGURE 14.6 Tracings of recordings of glandular stomach (GS), muscular stomach (MS), and duodenal (D) contractions detected by implanted strain gauge transducers. Recordings were obtained on day 3 after surgery in three treatment groups used. Arrows and lettering identify contractions of each organ during one gastroduodenal cycle. Glandular stomach contractions are marked with an arrow and a closed circle; duodenal synchronization is identified with a dashed line. Note normal synchronization of contractions in the control tracings. *From Chaplin and Duke (1988).*

as having three phases, including mechanical, chemical, and pellet formation and egestion (Kostuch and Duke, 1975). A peristaltic wave originating in the proventriculus moves aborad through the gizzard and into the small intestine. Although the sequence of motility in the stomach of raptors resembles that in mammals, slow waves have not been recorded in these species.

Egestion is another gastrointestinal function unique to birds (Rea, 1973). This process, which involves the oral expulsion of nondigestible material, is more common in carnivores. Whenever bone, fur, or feathers are ingested, these materials are compacted and orally egested. This process is unlike regurgitation in ruminants or vomiting in mammals (Duke et al., 1976a,b). Beginning approximately 12 min prior to egestion, gizzard contractions increase in frequency and amplitude. This process both compacts the material into a pellet and moves it into the lower esophagus. Beginning 8–10 s prior to egestion, the pellet is moved orad by esophageal antiperistalsis. Neither the abdominal nor duodenal muscles are involved.

Enterogastric reflexes control gastric emptying. Simply increasing intraduodenal pressure, or administering intraduodenal injections of HCl, hypertonic NaCl, or amino acid solutions, inhibits gastric motility (Duke and Evanson, 1972; Duke et al., 1977, 1989). Inhibition was related to dose and volume; it generally occurred within 3–30 s, and it persisted for 2–35 min. Following an intraduodenal injection of a lipid solution, gastric motility decreased after 4–6 min and remained inhibited for 24–45 min. This latter response appears to involve hormonal regulation, presumably by an enterogastrone.

In addition, the presence of bile acids increases in the gizzard in proportion to the amount of wood shavings retained in the gizzard (Hetland et al., 2003). Because bile acids arrive in the intestine in the ascending loop of the duodenal duodenum, there is retrograde transport into the gizzard.

14.3.3 Small Intestine

The MMC is characterized by electrical potential changes known as slow waves, which travel aborad and are associated with periodic spike potentials and smooth muscle contraction (Figure 14.7). Although the MMC has not been extensively studied in birds, available data suggest that the MMC is similar in birds and mammals (Clench et al., 1989). The MMC has three phases: quiescence (phase 1); irregularly spaced spike activity superimposed on slow waves (phase 2); and high-amplitude, regular spike activity superimposed on slow waves (phase 3). In chickens, the MMC has a periodicity of 77–122 min, whereas in owls it is only 21 min. The duration of phase 3 lasts approximately 5–8 min (Figure 14.8). These values are similar to those in mammals. Nevertheless, the propagation speed in birds is relatively slow, ranging from 0.48 to 0.62 cm/min. Although the MMC has been observed in the duodenum of chickens, quail, and owls, it has only been observed aborad of the duodenum in turkeys (Mueller et al., 1990).

Intestinal refluxes appear unique to birds (Duke et al., 1973). Increases in intraluminal pressure of the duodenum normally follow increased pressure in the muscular stomach. However, approximately every 15–20 min in the turkey, large pressure changes were observed in the duodenum, which were associated with an inhibition of gastric motility and a reflux of intestinal contents (Figure 14.9). This has been observed in a number of other species as well, and appears more frequently as dietary fat levels increase (Duke et al., 1989).

14.3.4 Ceca

Motility in the ceca is not well understood. The ceca are filled as a result of the convergence of rectal antiperistaltic waves and ileal peristaltic waves. Due to the morphology of the ileo-cecal junction, only fluids or very small particles are allowed entrance to the ceca. In fact, 87–97% percent of cecal fluid originates from the urine (Bjornhag and Sperber, 1977). The importance of the movement of urine to the ceca is discussed below.

While MMC-like bursts of electrical activity have been observed within the cecum, this activity does not migrate and thus does not constitute an MMC (Clench et al., 1989). Two types of contractions have been recorded in the ceca of turkeys (Duke et al., 1980). One type has a low amplitude, occurring at a rate of 2.6/min, while the other has a higher amplitude and occurs at a rate of 1.2/min. Low-amplitude contractions are associated with mixing, whereas high amplitude contractions are propulsive. While aborad contractions were more common than orad contractions, the latter contractions had a greater amplitude and thus prevented the collection of digesta in the distal ceca. Peristalsis in a cecum appears to be myogenically mediated, with inhibitory neural input apparently

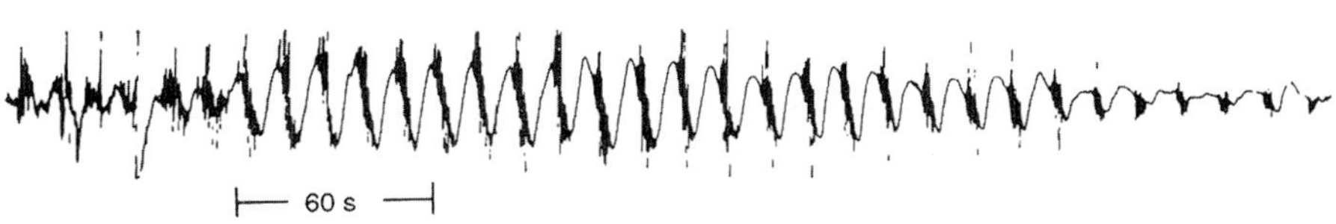

FIGURE 14.7 **Myoelectric recording of action potentials on slow waves (*Gallus*).** *From Clench et al. (1989).*

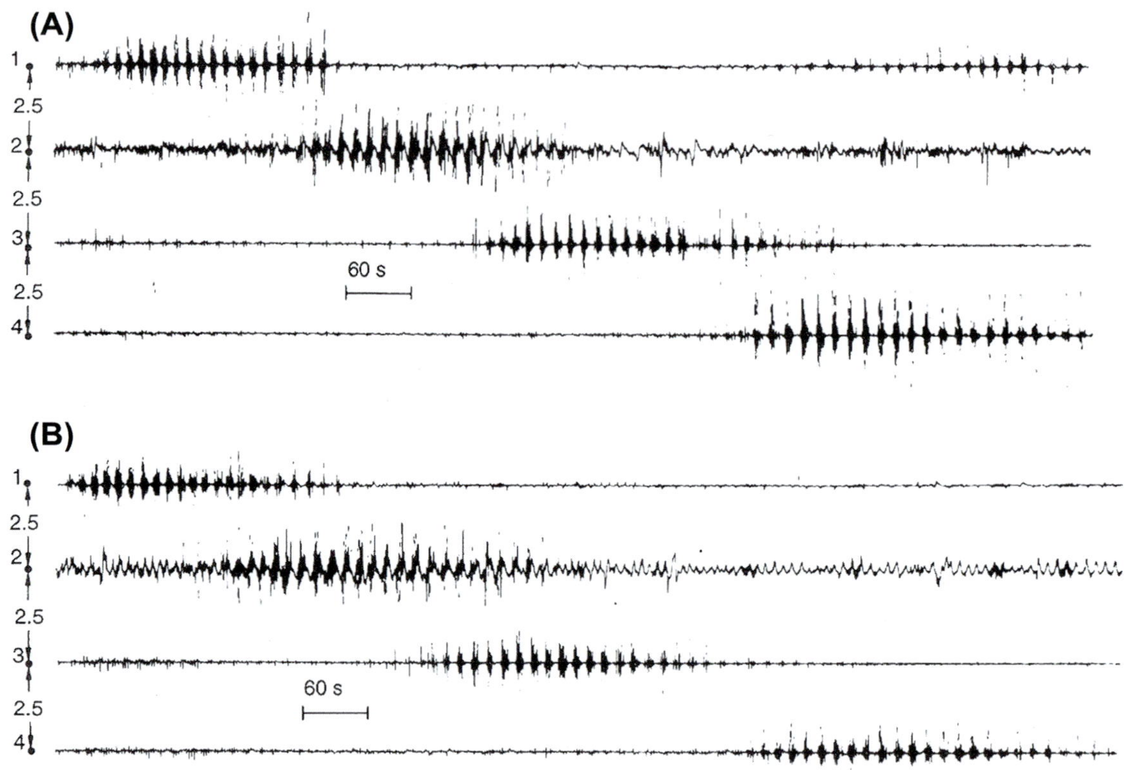

FIGURE 14.8 Representative myoelectric recording of a migrating myoelectric complex (MMC) in *Strix*. The electrodes were placed on the proximal ileum 2.5 cm apart. (A): bird in fed state. (B): bird in fasted state. High frequency of MMCs in this owl species in indicated in (A), with a second regular spike activity beginning on *lead 1* before previous complex has propagated through *lead 4*. *From Clench et al. (1989).*

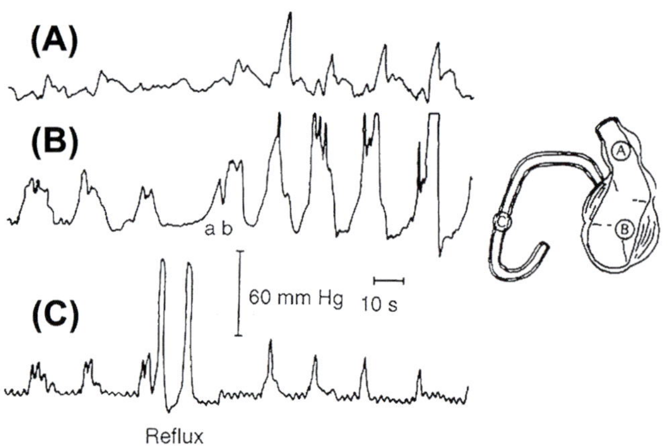

FIGURE 14.9 Tracings of typical records of pressure changes obtained from the proventriculus (A), gizzard (B), and upper proximal duodenum (C) of a turkey, showing pressure events occurring during a duodenal reflux. Positions of open-tipped tubes within GI tract are indicated by letters (A), (B), and (C) (circled) on the diagram of a sagittal section of the stomach. The biphasic patter (B) of the tracing representing the contraction of the gizzard is quite variable; two phases of one cycle are (*a*) pressure wave due to contraction of thin muscle pair; and (*b*) of thick muscle pair. *From Duke and Evanson (1972).*

able to suppress such contractions (Hodgkiss, 1984b). In mammals, distention causes an ascending stimulation and a descending inhibition, which is neurogenically mediated by the enteric nervous system and can be blocked by tetrodotoxin. By contrast, in birds, distention causes contraction of the circular muscle, which is unaffected by tetrodotoxin and thus apparently not controlled by the enteric nervous system.

The contents of the ceca are much different in consistency than that of the rectum and can therefore be easily distinguished from rectal feces. The ceca evacuate relatively infrequently compared to the rectum, with rectal evacuations preceding cecal evacuations (Duke et al., 1980). Cecal evacuations in young turkeys typically occur within 1–5 min of lights-on and again in late afternoon. There is an increase in the frequency of high-amplitude contractions, with four to seven such contractions occurring during the 2 min preceding cecal evacuation. These contractions are associated with high-amplitude electrical spiking. One high-amplitude contraction also occurs in the ileum and rectum at the time of cecal evacuation. The ratio of cecal to rectal evacuations is also influenced by diet, ranging from 1:7.3 when feeding barley to 1:11.5 when feeding corn (Roseler, 1929). As discussed in Section 14.6, the extended time that digesta spends in the ceca provides a unique role for this organ.

14.3.5 Colon

The rectum displays almost continuous antiperistalsis. Such motility is responsible for carrying urine from the urodeum

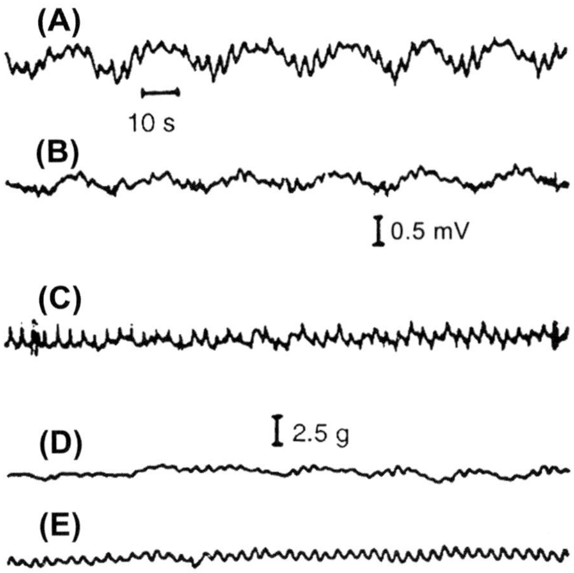

FIGURE 14.10 Electrical potential changes and contractile forces recorded from three bipolar electrodes ((A), (B), (C)) and two strain gauges ((D), (E)) in the colon of a turkey. Electrodes and strain gauges were implanted at 10, 6, and 1, and 8 and 3 cm from the cloaca, respectively. Both large and small slow waves are evident in tracings (A) and (B); only small slow waves are evident in tracing (C). Small contractions (antiperistalsis are evident in tracings (D) and (E)) but large contractions (peristalsis) can be seen only in tracings (D). *From Lai and Duke (1978).*

into the colon and ceca (Akester et al., 1967; Polin et al., 1967; Goldstein and Braun, 1986). Two types of colonic slow waves have been recorded (Lai and Duke, 1978). These include small, short duration (sSW) and large, long duration (lSW) slow waves (Figure 14.10). The sSW are associated with small contractions and, as shown with radiography, are antiperistaltic. The lSW are associated with large contractions but, radiographically, could not be associated with movements of rectal contents.

The frequencies of sSW and lSW are shown in Table 14.2. The frequency of sSW is highest in the distal colon. In contrast, the frequency of lSW is highest in the proximal colon and cannot be recorded from the distal colon. This pattern of slow waves, when compared to the motility pattern, suggests that sSW arise at the distal colon and are responsible for antiperistaltic movement, whereas the lSW begin in the proximal colon and are responsible for peristaltic movement of rectal contents.

The nearly continuous antiperistaltic activity of the colon is interrupted only near the time of defecation (Duke, 1982). Beginning approximately 10 min prior to defecation, the amplitude of the sSW begins to decrease while the frequency gradient of lSW increases. These conditions favor inhibition of antiperistalsis and stimulation of peristalsis. Defecation is associated with a strong contraction beginning at the proximal colon, moving aborally and carrying the contents the length of the colon in less than 4 s. Contraction of the rectum appears to be mediated by noncholinergic, nonadrenergic nerves (Bartlet, 1974; Takewaki et al., 1977).

14.3.6 Other Influences on Motility

Motility of the gastrointestinal tract shows diurnal variations. In turkeys, the frequency and amplitude of muscular stomach contractions increased during light periods relative to dark periods (Duke and Evanson, 1976). The increased and decreased gastric contractions coincided with just prior to "lights-on" and "lights-off", respectively. Although less pronounced, this diurnal rhythm of gastric activity continued when birds were fasted.

Birds also display cephalic, gastric, and intestinal phases of motility. In 24-hr fasted great horned owls, turkeys, and red-tailed hawks, the sight of food stimulated gastric contractions (Duke et al., 1976a). When allowed to eat, gastric activity increased further, indicative of the gastric phase. During the intestinal phase, entrance of food into the duodenum slows the frequency of gastric contractions to allow time for digestion (Duke et al., 1973). While the cephalic phase is mediated largely by the nervous system, there appears to also be an as yet unidentified endocrine component that increases motility.

Motility is influenced by many factors. For example, anesthetics, including nembutal and methoxyflurane, reduce gastroduodenal motility as well as decrease gastric secretion (Kessler et al., 1972; Duke et al., 1977). High environmental temperatures decrease gastrointestinal motility, whereas cold temperatures differentially affect motility in various parts of the digestive system but appear to have an overall effect of decreasing transit time (Tur and Rial, 1985).

14.4 NEURAL AND HORMONAL CONTROL OF MOTILITY

Extrinsic and intrinsic innervations of the avian gastrointestinal tract appears similar to mammals (Olsson and Holmgren, 2011). Extrinsic innervations is largely from the vagus nerve, nerve of Remak (ganglionic nerve running along the gut), and fibers from the splanchnic and pelvic nerves (Nilsson, 2011). Nitric oxide synthase is found in many neurons associated with the gut. Many of these neurons also contain vasoactive intestinal peptide (VIP) and pituitary adenylate cyclase-activating polypeptide. In pigeons, these nerves also contain galanin. In addition, most regions of the gut have nerves containing tachykinin-immunoreactivity, and the intestines have neurons containing gamma amino butyric acid. The myenteric plexus of the gizzard has aminergic neurons, although it has been reported that catecholamines are not present in the chicken intestine (Aisa et al., 1997).

TABLE 14.2 Mean Frequency and Amplitude of Small and Large Contractions and of Small and Large Slow Waves in the Rectum of Turkeys

	Contractions				Slow Waves			
	Small		Large		Small		Large	
	Frequency (cycles/min)	Amplitude (g)	Frequency (cycles/min)	Amplitude (g)	Frequency (per min)	Amplitude (mV)	Frequency (per min)	Amplitude (mV)
Proximal[1]	14.6±0.85	0.45±0.24	2.66±0.26	0.54±0.20	15.4±1.07	0.17±0.08	2.83±0.26	0.21±0.09
Middle	–	–	–	–	15.8±1.12	0.16±0.09	2.76±0.24	0.12±0.06
Distal	15.4±0.69	<0.70±0.33	–	–	16.4±2.16	0.25±0.12	–	–

[1]*Proximal, middle, and distal refer to electrode implant sites on the colon 10, 6, and 1 cm from the cloaca, respectively, or to strain gauge transducer implants at 8 (proximal) and 3.5 (distal) cm from the cloaca. (Large contractions were not recorded from the distal strain gauge nor were large slow waves recorded from the distal electrode.)*
Source: From Lai and Duke (1978).

Contractions of the esophagus are increased by acetylcholine (ACh) and vagal stimulation and unaffected by sympathetic nerve stimulation (Bowman and Everett, 1964; Ohashi and Ohgua, 1967; Taneike et al., 1988). However, there is evidence for nonadrenergic, noncholinergic (NANC) inhibition of esophageal smooth muscles (Sato et al., 1970; Postorino et al., 1985). Serotonin also causes contractions of the esophagus (Mule et al., 1987), and this effect appears to be mediated indirectly by activation of both cholinergic and NANC excitatory neurons (Fileccia et al., 1987). In addition, tachykinins cause contraction of the gut (Liu and Burcher, 2001). In contrast, epinephrine induces relaxation of the chicken rectum via alpha-adrenergic receptors (Ojewole, 1980).

The proventriculus has been shown to contain gastric inhibitory peptide (GIP), VIP, and nitric oxide synthase (nNOS), but not somatostatin, peptide histidine-isoleusine (PHI), peptide tyrosine-tyrosine (PYY), neuropeptide tyrosine (NPY), bombesin, met-enkephalin, serotonin, substance P, galanin, or calcitonin gene-related peptide (CGRP) (Martinez et al., 2000), VIP and nNOS. VIP and NO probably cause relaxation of smooth muscle. The proventriucular secretions are controlled by parasympathetic nerves and a diffuse endocrine system.

As shown in Table 14.3, many peptides have been identified within the gastrointestinal tract of birds. The function of these peptides remains to be elucidated. Although the identity of the NANC excitatory neurotransmitter was not identified, it has been shown that neurotensin can induce contraction of crop smooth muscle (Denac and Scharrer, 1987). This latter effect is a postjunctional response not mediated by acetylcholine, prostaglandins, or opioids.

Histamine interacts with cholinergic neurotransmission to control esophageal contractions (Taneike et al., 1988). Histamine dose-dependently induces esophageal contractions, but this effect is blocked by tetrodotoxin, suggesting that its effect is mediated by the release of ACh. Contractions induced by vagal stimulation are increased by the presence of histamine, whereas histamine-induced contractions were enhanced by acetylcholinesterase inhibitors. It appears, therefore, that histamine modulates, via H_1 receptors, the release of ACh to control the contraction of esophageal smooth muscle. As indicated in Figure 14.11, enkephalins also cause contraction of the esophagus, but their effect is mediated by serotonergic neurons.

The crop appears to be controlled similarly to the esophagus. Electrical stimulation of the crop causes contraction that is largely, but not completely, due to ACh release because it can be significantly blocked by atropine (Denac et al., 1990). While it is unclear as to which neurotransmitters are responsible for the contraction not blocked by atropine (i.e., NANC-induced), neurotensin, bombesin and substance P have been shown to cause atropine-resistant contractions following electrical stimulation of the crop (Denac and Scharrer, 1987, 1988). Norepinephrine caused crop muscle relaxation that was mediated by β-adrenoceptors. It appears that the crop and esophagus are controlled by three types of nerves: stimulatory cholinergic neurons, stimulatory NANC neurons (probably peptides), and inhibitory noradrenergic nerves.

Cholecystokinin octapeptide (CCK-8) is the most studied regulator of intestinal motility. Intravenous infusion of CCK-8 inhibited gastric and duodenal motility (Savory et al., 1981). CCK-8 and CCK-tetrapeptide (CCK-4) inhibited gastric electrical activity (Martinez et al., 1992). Whereas CCK-4 inhibited duodenal electrical activity, CCK-8 stimulated such activity (Figure 14.12). CCK-A and CCK-B receptor antagonists were unable to block the effects of CCK. It appears that the action of CCK on gastric and duodenal motility is similar in birds and mammals. CCK inhibits stomach motility, and the increase in duodenal activity is suggested to cause segmental contractions, and, thus, also delay gastric emptying (Martinez et al., 1992).

Vagotomy and hexamethonium blocked the response to CCK-8 in the chicken stomach (i.e., proventriculus and gizzard) but had no effect on the duodenal response (Martinez et al., 1993). Furthermore, the action of CCK-8 in the duodenum was not altered by atropine or methysergide. Phentolamine and propranolol had no effect on the gastric or duodenal response to CCK-8. It appears, therefore, that the action of CCK-8 on the stomach is mediated via the vagus, whereas the action in the duodenum probably involved a direct effect on smooth muscle cells. The action of CCK-8 on the stomach was blocked by N^G-nitro-L-arginine methyl ester (L-NAME), suggesting that the inhibitory effect of CCK involves the release of nitric oxide. Because L-NAME did not completely block the CCK effect on the caudodorsal thick muscle, it is likely that another neurotransmitter, possibly VIP, is involved in this latter response. Interestingly, CCK-8 caused an excitatory action in the stomach of vagotomized chickens, indicating that CCK-8 may, in addition to acting via the vagus, have a direct action on gastric muscles. The increase in electrical activity caused by CCK-8 in the duodenum was enhanced by L-NAME, indicating that nitric oxide may be a tonic inhibitor of duodenal electrical activity.

Chicken gastrin (cG), which belongs to the gastrin/CCK family, has been isolated from the chicken antrum (Dimaline et al., 1986). Intravenous infusions of cG caused effects similar to CCK-4 (Martinez et al., 1992). This suggests the existence of one receptor subtype in the stomach recognizing both CCK and cG, but possibly two receptors in the duodenum—one recognizing CCK-8 and the other CCK-4 and cG.

TABLE 14.3 Age of Chick Embryos (Days of Incubation) at First Appearance of Endocrine Cells Showing Immunoreactivity for Various Regulatory Peptides, Chromogranin and Serotonin in the Gut[1]

	Proventriculus	Gizzard	Pyloric Region	Duodenum	Upper Ileum	Lower Ileum	Ceca	Large Intestine
Chromogranin	9	12	12	10.5[2]	10.5[2]	15.5[2]	13.5[2]	10[2]
Enkephalin	21	–	–	–	–	–	–	–
Gastrin/CCK	12[3]	–	11	11	11	17	17	17
GRP	11	21	21	–	–	–	–	–
Glucagon	13	–	14	13	14	14	–	–
Motilin	–	21	21	21	13	17	17	17
Neurotensin	12	–	12	12	13	14	14	9
PP	14	–	14	13	13	14	–	21
Peptide YY	–	12	12	18	21	21	–	–
Serotonin	8	–	12	11	11	14	14	9
Somatostatin	12	–	12	11	13	14	17	13
Substance P	–	–	–	21	11	17	17	13
VIP	21	–	–	19	12[4]	19	21	21

[1]*–=no cells existed that were immunoreactive with the designated compound; CCK=cholecystokinin; GRP=gastrin releasing peptide; PP=pancreatic polypeptide; VIP=vasoactive intestinal peptide.*
[2]*Salvi et al. (1996).*
[3]*Aksoy and Cinar (2009).*
[4]*Parisi Salvi et al. (2004).*
Source: Data from Rawdon and Andrew (1999), except where noted.

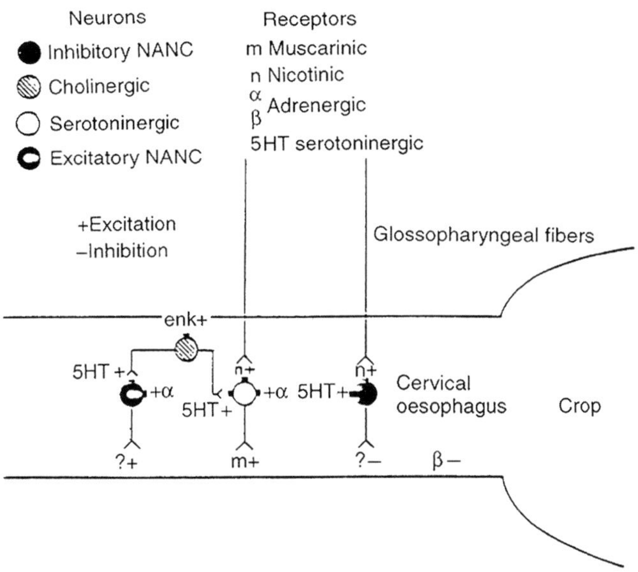

FIGURE 14.11 Schematic diagram illustrating the putative pattern of the intrinsic innervation of the pigeon cervical esophagus. Different neuron types are present in the intramural plexuses; excitatory cholinergic, excitatory NANC-, inhibitory NANC-, and serotoninergic neurons. *Reprinted from Mule (1991).*

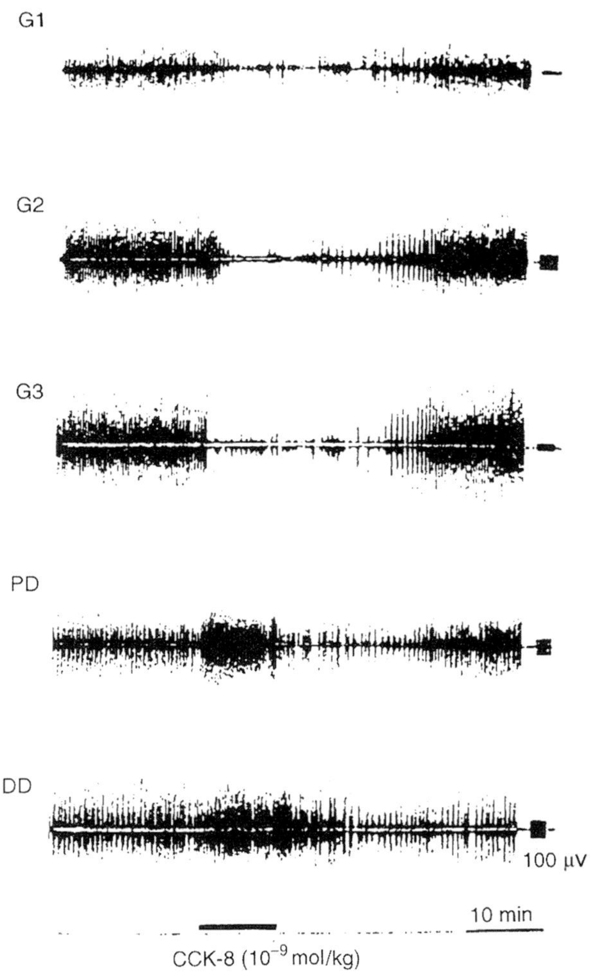

FIGURE 14.12 Recording of gastroduodenal electrical activity showing effect of cholecystokinin octapeptide (CCK-8; 10^{-9} mol kg^{-1}) infusion. Studied gastric areas are as follows: G1, proventriculus; G2, craniodorsal thin muscle; G3, caudodorsal thick muscle; PD, proximal duodenum; DD, distal duodenum. Similar responses were observed in all animals. *From Martinez et al. (1993).*

Opioid peptides appear to be involved in the MMC. Infusion of met-enkephalin, morphine, and β-casomorphin (5×10^{-7} mol/kg) induced intense electrical activity, similar to phase 3 of the MMC, in the distal duodenum which migrated through the small intestine (Jiménez et al., 1992). The effect of morphine was blocked by naloxone (5×10^{-7} mol/kg), while higher doses of naloxone reduced gastroduodenal motility. Simultaneous to increasing duodenal electrical activity, activity in the stomach is inhibited.

14.4.1 Rate of Passage

The rate of passage of material through the digestive tract has been measured in many ways. Because digesta consists of both solid and liquid components, different types of markers have been used. Insoluble markers such as chromium-mordanted rice, cerium-mordanted rice, Cr_2O_3, or radiopaque plastic pellets (Branch and Cummings, 1978; Uden et al., 1980; Ferrando et al., 1987) have been used as indicators of solid transit time, whereas a soluble marker such as Cr-EDTA (Vergara et al., 1989) or phenol red (Goñalons et al., 1982) has been used to measure liquid transit time. In general, it was found that larger particles are retained longer in the digestive tract.

In chickens, insoluble markers first appear in the excreta 1.6–2.6h after ingestion. However, mean retention time is a better indicator of transit time than is time to initial appearance of the marker. Mean retention time for insoluble markers can vary from 5–9h depending on the nature of the ingesta and its size.

Transit time of digesta is influenced by genetics. When comparing broiler and Leghorn-type chickens, the overall mean retention time is not different, but the time food spent in various parts of the digestive tract is different (Table 14.4.).

The rate of food passage is affected by many factors. Feed transit time through the small and large intestine increases with age (Shires et al., 1987). This may account for increases in metabolizable energy values of feedstuffs noted in older birds. Adding lipid (Sell et al., 1983) or protein (Sibbald, 1979) to the diet can increase passage time. Increases in environmental temperature also slows transit time.

14.5 SECRETIONS AND DIGESTION

At hatch, the chick has to make a transition from dependence on a lipid-rich endogenous diet coming from the

TABLE 14.4 Mean Retention Time (min) of Solid Phase Markers in Various Segments of the Digestive System of Broilers and Leghorns

Gastrointestinal Tract Segment	Broilers[1]	Leghorns[1]	Broilers[2]
Crop	31	48	41
Proventriculus + gizzard	39	71	33
Duodenum	10	7	5
Jejunum	84	85	71
Ileum	97	84	90
Ceca	119	112	–
Rectum	56	51	26

[1]*From Shires et al. (1987). The values have been adjusted for birds weighing 1800g.*
[2]*From Van Der Klis et al. (1990).*

yolk, to an exogenous diet rich in carbohydrate and protein (Sklan, 2001). Access to feed at hatch results in an approximately two-fold increase in small intestine length of chicks but only a 60% increase if birds remain fasted (Sklan and Noy, 2000). The increase in intestinal weight is correlated with increases in trypsin, amylase, and lipase activities, which remain unchanged in the absence of feed. Feeding a low-Na diet also decreased mucosal absorption because this is dependent on active transport of Na out of the enterocytes at the basolateral membrane.

14.5.1 Mouth

The salivary glands secrete mucous, and depending on the species, amylase. Although amylase is not present in the saliva of *Gallus* and *Meleagris*, it is found in the saliva of the house sparrow (Jerrett and Goodge, 1973) and other species (Bhattacharya and Ghose, 1971). The volume of daily salivary secretion in *Gallus* ranges from 7 to 25 ml (Leasure and Link, 1940).

Mucous functions to lubricate food and allow it to move down the esophagus. However, in some species, mucous also functions as an adhesive coating on the tongue to aid in capturing insects or as a material that cements components during the construction of nests.

14.5.2 Esophagus and Crop

The esophagus is not important in chemical digestion, as its major secretion is mucous. The secretion of esophageal mucous is nevertheless important because it is necessary to supplement the limited secretion of saliva. However, in some species, including the greater flamingo and male emperor penguin, a nutritive merocrine-type secretion is produced by the wall of the esophagus, which is fed to the young.

Some carbohydrate digestion may occur in the crop due to the presence of amylase activity (Philips and Fuller, 1983). Amylase activity at this site comes from either salivary secretions, intestinal reflux, or plant and/or bacterial sources. Bolton (1965) reported that starch is hydrolyzed within the crop where it can either be absorbed; converted to either alcohol, lactic acid, or other acids; or transported down the gastrointestinal tract. Pinchasov and Noy (1994) showed that substantial amylolysis occurs in the crop. Sucrose is also hydrolyzed within the crop. While absorption of sugars from the crop appears possible, it is probably minimal.

The crop is not essential for normal growth when access to food is sufficient. Cropectomy has no effect on growth rate of ad libitum fed chickens, but it does decrease growth rate when food intake is limited. This supports the view that the primary function of the crop is food storage, and it is not essential for digestion.

In pigeons and doves, "crop-milk" is produced during the breeding season under the influence of prolactin. Crop milk contains 12.4% protein, 8.6% lipids, 1.37% ash, and 74% water (Vandeputte-Poma, 1968). Therefore, while rich in protein and essential fatty acids (Desmeth, 1980), it is devoid of carbohydrates and calcium.

14.5.3 Stomach

The oxynticopeptic cells found in birds secrete both HCl and pepsinogen. Pepsinogen, under the influence of acid or pepsin that is already present, is converted to pepsin. While lipase has been found in gastric secretion, this is probably due to reflux from the duodenum. The basal gastric secretory rate is 15.4 mL/h and contains 93 mEq/L of acid and 247 Pu/ml of pepsin (Long, 1967) with a pH of 2.6 (Joyner and Kokas, 1971). The pH of gastric contents, however, is normally above 2.6 due to the presence of ingesta. The pH has been determined in several species immediately after sacrifice (Table 14.5.). Higher pH values have been reported when measurements were made on live birds. For example, Winget et al. (1962) reported the following values for chickens: mouth, 6.7; crop, 6.4; ileum, 6.7; rectum, 7.1. Age has no effect on pH of the digestive tract (Herpol, 1966). Acid secretion of chickens is high relative to mammals, possibly because of the rapid digestive transit time (Table 14.6).

While amylolysis occurs in the crop, it is not evident in the ventriculus. This is the result of the low pH of the stomach, which is unfavorable for amylase activity (Pinchasov and Noy, 1994).

TABLE 14.5 The pH of Contents of the Digestive Tract of Avian Species

	Chicken	Pigeon	Pheasant	Duck	Turkey
Crop	4.51	6.3[1]	5.8	4.9	6.0
			4.28		
Proventriculus	4.8	1.4[1]	4.7	3.4	4.7
		4.8			
Gizzard	4.74[2]	2.0	2.0	2.3	2.2
	2.50				
Duodenum	5.7–6.0	6.4[1]	5.6–6.0	6.0–6.2	5.8–6.5
	6.4[2]	5.2–5.4			
Jejunum	5.8–5.9	5.3–5.9	6.2–6.8	6.1–6.7	6.7–6.9
	6.6[2]				
Ileum	6.3–6.4	6.8[1]	6.8	6.9	6.8
		7.2[2]	5.6		
Rectum	6.3	5.4	6.6	6.7	6.5
		6.6[1]			
Ceca	5.7		5.4	5.9	5.9
	6.9[2]				
	5.5–7.0[3]				
Bile	7.7[4]		6.2	6.1	6.0
	6.6[2]				
	5.9				

[1]*From Herpol (1966).*
[2]*From Herpol and van Grembergen (1967).*
[3]*From Sudo and Duke (1980).*
[4]*From Lin et al. (1974).*
Source: From Sturkie (1976), based on work of Farner (1942).

There are three phases to gastric secretion: the cephalic phase, the gastric phase, and the intestinal phase. All three phases are present in birds (Burhol, 1982). The cephalic phase entails an increase in hydrogen ion (H^+) and pepsin secretion caused by the sight, smell, or expectation of food. This phase is under vagal control.

As summarized by Duke (1986), vagal stimulation increases both gastric secretion rate and pepsin secretion. Vagal stimulation causes a greater increase in gastric secretion than do cholinergic agents (Gibson et al., 1974). This suggests, as discussed below, that gastric secretion is stimulated by other neurotransmitters acting together with ACh.

In birds, vagal stimulation causes greater pepsin than H^+ accumulation (Burhol, 1982). In contrast, insulin injection inhibits gastric H^+ secretion without affecting pepsin secretion. Therefore, H^+ and pepsin secretion may be under different control.

TABLE 14.6 Basal Acid Secretion in Various Species

Species	Body Weight (kg)	Acid Output (mEq/kg/hr)	Pepsin Output (PU/kg/hr)
Man	70	0.03	862
Dog	15	0–0.004	0–62
Rat	0.35	0.25	2230
Monkey	2.5	0.12	730
Chicken	1.75	0.78	2430

Source: From Long (1967).

Pepsin from chicken and duck has been well characterized (see Pichová and Kostka, 1990). Duck pepsinogen and pepsin have 374 and 324 amino acids, respectively. Duck pepsin has a pH-optimum of 4, is stable up to pH 7.5, and is inactivated at pH 9.6.

Many hormones are involved in gastric secretion (Table 14.7). Gastrin plays a role in the gastric phase of secretion. Chicken gastrin has been isolated from the chicken pylorus, which is equivalent to the mammalian antrum (Dimaline et al., 1986). While structurally similar to cholecystokinin (CCK), cG has markedly different secretory functions (Dimaline and Lee, 1990). Infusion of cG increases both acid and pepsin secretion. Unlike CCK, gastrin has no effect on gall bladder contractions or pancreatic secretion. Another peptide, gastrin-releasing peptide, also induces acid secretion, but it is not known whether it acts via gastrin release (Campbell et al., 1991).

The intestinal phase of gastric secretion is controlled by several hormones including CCK, secretin, and avian pancreatic polypeptide (APP). APP, originally discovered in chickens (Kimmel et al., 1968; Larsson et al., 1974), is released from the pancreas postprandially in response to amino acids and HCl (Hazelwood et al., 1973; Duke et al., 1982). APP does not appear to act during the cephalic phase of gastric secretion because it is not released during sham feeding (Kimmel and Pollock, 1975). APP increases gastric acid and pepsin secretion, and this effect is independent of the vagus nerve (Hazelwood et al., 1973).

CCK (Dockray, 1977) and secretin (Nilsson, 1974) have been isolated from the duodenum and jejunum of birds. CCK stimulates gastric acid secretion while having no effect on pepsin secretion. Contrary to its effects in mammals, in which secretin inhibits acid and stimulates pepsin secretion, in chickens secretin stimulates both acid and pepsin secretion (Burhol, 1974).

TABLE 14.7 Gastrointestinal Hormones in the Domestic Fowl

Hormone	Site of Origin	Biological Actions
Gastrin	Proventriculus	Stimulates gastric acid and pepsin secretion
Cholecystokinin	Duodenum, jejunum	Stimulates gall-bladder contraction and pancreatic enzyme secretion and gastric acid secretion; inhibits gastric emptying; potentiates secretin-induced stimulation of pancreatic electrolyte secretion
Secretin	Duodenum, jejunum	Stimulates bicarbonate secretion by pancreas
Vasoactive intestinal peptide	Duodenum, jejunum	May be a more potent stimulator of pancreatic electrolyte secretion than secretin; inhibit smooth muscle contraction
Pancreatic polypeptide	Pancreas, proventriculus, duodenum	Stimulates gastric acid and pepsin secretion
Gastrin-releasing peptide (bombesin)	Proventriculus	Stimulates pancreatic enzyme secretion; stimulates crop contraction
Somatostatin	Pancreas, gizzard, proventriculus, duodenum, ileum	Inhibits secretion of other gut hormones

As in mammals, histamine is involved in gastric acid release. Injection of cimetidine, an H_2-receptor blocker, raised the pH of the proventriculus and duodenal contents (Ward et al., 1984). The increase in gastric acid secretion induced by the intravenous injection of 2-deoxy-D-glucose was also blocked by metiamide, a H_2-receptor blocker (Nakagawa et al., 1983).

14.5.4 Intestines

Intestinal digestion includes both luminal and brush-border digestion. The brush border contains sucrase-isomaltase, peptidases, and phosphatases. Chicks can hydrolyze disaccharides beginning prehatch, and sucrase-isomaltase expression increases posthatch. Glucose is transported across the enterocyte membrane via a sodium-glucose transporter (SGLT-1), which increases in expression beginning 2 days prehatch and involves a secondary active transport system driven by the active removal of sodium from the basal-lateral membrane of the enterocytes. Absorptive capacity of the intestines increases with proportionate to body weight, but the absorption rate for glucose is greatest during the first week and then declines, as does brush border SGLT-1 density (Barfull et al., 2002).

Amylase is produced by both the pancreas and intestine (Osman, 1982). While found in all parts of the small intestine, it is found in particularly high concentrations in the jejunum, with 80% of the activity found there. The high levels found in the jejunum are presumably due to the fact that the openings of the pancreatic ducts discharge near the anterior jejunum. Amylase is found in only trace amounts in the ceca. The optimum pH of the pancreatic and intestinal amylase is 7.5 and 6.9, respectively.

Intestinal enzymes provide the last step in digestion. These secretions are responsible for digesting starch, sucrose, fats, and protein (Table 14.8.). The small intestine of birds contains maltase, sucrase, palatinase, but does not contain trehalase (Siddons, 1969). Whether lactase is present appears to be debatable. However, it has been reported that lactase is not present in germ-free chicks, and that the rate of mortality of germ-free chicks fed lactose as the sole energy source is very high (Siddons and Coates, 1972). Enzyme activity is highest in the jejunum and decreases both proximally and distally. These enzymes are located in the epithelial cells of the villi. The maltase, sucrase, and palatinase activity found in the large intestine comes from the small intestine, whereas the lactase activity found in the large intestine is probably of a cecal bacterial origin.

TABLE 14.8 Enzymes Secreted by the Intestines

Enzyme	Substrate	Product or Function
Maltase	Maltose	Glucose
Isomaltase	Dextrins	Glucose
Sucrase	Sucrose	Glucose, fructose
Enterokinase	Trypsinogen	Trypsin
Lipase	Monoglycerides	Glycerol, fatty acids
Peptidases	Di- and tripeptides	Amino acids

Relatively little is known about the control of intestinal secretions in birds. Intestinal secretion is increased by duodenal distention, vagal stimulation, and secretin. Vagal stimulation has a greater effect on stimulating mucous secretion than enzyme secretion.

TABLE 14.9 Pancreatic Digestive Enzymes

Enzyme	Percent of Total
Trypsinogen	10
Chymotrypsinogen (A, B, and C)	20
Trypsin inhibitor	11.3
Amylase	28.9
Procarboxypeptidase (A and B)	29.8

Source: From Pubols (1991).

14.5.5 Colon

Chloride (Cl^-) ions are secreted in the rectum, ceca, and coprodeum. This is discussed in Section 14.6.

14.5.6 Pancreas

As mentioned, pancreatic and bile secretions enter the gastrointestinal tract near the anterior jejunum. Pancreatic secretions have a pH of 6.4–6.8 in chickens (Hulan et al., 1972) and 7.4–7.8 in turkeys (Duke, 1986). Secretions include an aqueous phase containing water and bicarbonate ions, as well as an enzymatic phase.

Digestive enzymes found in the pancreas of broiler type chickens are listed in Table 14.9. Although not shown, the pancreas is also reported to secrete ribonuclease and deoxyribonuclease (Dal Borgo et al., 1968). Amylase is found in the duodenum, jejunum, ileum, and colon. Both trypsin and amylase are found in highest concentrations in the jejunum (Bird, 1971; Osman, 1982), presumably because the pancreatic ducts enter near the end of the duodenum. Both pancreatic and intestinal amylase have a requirement for chloride ion. Characterization of these enzymes suggests that pancreatic amylase is similar to mammalian α-amylase, while intestinal amylase is similar to glucoamylase.

Pancreatic secretion is controlled by both nervous and hormonal mechanisms. The secretion rate is higher in birds than in mammals (Table 14.10). Secretion has both a cephalic and intestinal phase. When fasted birds are allowed to eat, pancreatic secretion increases immediately (Kokue and Hayama, 1972). This response is blocked by vagotomy or atropine and can be augmented by cholinergic agents (Hokin and Hokin, 1953).

Secretin-like activity is released in response to dilute HCl placed in the duodenum (Nilsson, 1974). Secretin, when injected intravenously, increases the aqueous component of pancreatic secretion. However, in contrast to mammals, VIP more potently stimulates the secretion of pancreatic juice (Vaillant et al., 1980). VIP is found in neurons both in the gastrointestinal tract and pancreas. It is believed that VIP, rather than secretin, is the primary regulator of pancreatic juice secretion and that this response may be either neuronally or hormonally mediated (Dockray, 1988). VIP does not stimulate pancreatic enzyme secretion.

TABLE 14.10 Pancreatic Secretory Rate and Influence of Fasting in the Chicken, Dog, Rat, and Sheep

Species	Starvation Time (hr)	Pancreatic Secretory Volume (mL/kg/hr)
Chicken	24	0.70
	48	0.68
	72	0.65
Dog	24	0.1–0.3
	48	Negligible
Rat	24	0.6–0.7
Sheep	24	0.13
	48	0.07

Source: Adapted from Kokue and Hayama (1972).

CCK is released in response to lipids and amino acids. The administration of CCK has been shown to increase pancreatic secretion in pigeons (Sahba et al., 1970) and to increase the flow rate and protein secretion rate in turkeys (Dockray, 1975).

Two gastrin-releasing peptides (GRP), which are structurally related to bombesin, have been isolated from the proventriculus (Campbell et al., 1991). These peptides are found in endocrine cells and contain either 27 or 6 amino acids. Distension of the proventriculus with peptone stimulates pancreatic juice and enzyme secretion. This effect appears to be mediated by GRP-27, with GRP-6 being ineffective. Distension with saline is less effective.

Diet can influence the secretory rate of pancreatic enzymes. Increasing the carbohydrate and fat content of the diet increases amylase and lipase activity in pancreatic secretions (Hulan and Bird, 1972).

14.5.7 Bile

Bile, produced and secreted by the liver, is essential for fat digestion. It acts to emulsify lipid so that it can be more efficiently digested by lipase. In addition, amylase begins to appear in chicken bile at 4–8 weeks of age (Farner, 1943). Therefore, bile is also involved in carbohydrate digestion.

Relatively little is known about biliary secretion in birds possibly because of the complex anatomy in which bile enters the small intestine via both the hepato-enteric duct and the cystico-enteric duct. The biliary secretion rate is 24.2 uL/min in fasted broilers (Lisbona et al., 1981). Chenodeoxycholyltaurine and cholyltaurine are the predominant bile acids in chickens and turkeys, while chenodeoxycholyltaurine and phocaecholyltaurine predominate in ducks (Elkin et al., 1990). These are secreted by an active transport system.

The bile salts glycocholate and taurocholate are readily absorbed through the intestinal wall. The absorption rate is higher near the distal end of the small intestine (Lindsay and March, 1967). This allows for recirculation of bile acids, thus allowing for their reuse in lipid digestion. It is estimated that 90% of bile salts are reabsorbed in the jejunum and ileum (Hurwitz et al., 1973).

Since chickens have low levels of liver glucuronyl transferase and little or no biliverdin reductase; the secretion rate of biliverdin is high relative to that of bilirubin (14.7 vs. 0.9 ug/kg/min). The green color observed in coprodeal droppings is likely due to biliverdin. The reason for the brown color associated with cecal droppings is unknown, but it may be due to bacterial reduction of biliverdin to bilirubin and a subsequent dehydrogenation (Hill, 1983).

14.6 ABSORPTION

14.6.1 Carbohydrates

Absorption of carbohydrates in birds occurs by mechanisms similar to those found in mammals. Absorption occurs more rapidly from the small intestine compared to the cecum with carbohydrates being absorbed by both active and passive mechanisms. Absorption of sugars from the distal regions of the intestinal tract may account for as much as 5% of the total capacity. Those sugars containing a six-membered ring with the hydroxyl group in the number three position oriented similar to glucose are actively transported. Active transport accounts for at least 80% of glucose absorption. D-Glucose and D-galactose are absorbed faster than D-xylose and D-fructose and D-arabinose. These sugars are absorbed faster than L-arabinose, L-xylose, D-ribose, D-mannose and D-cellobiose.

Glucose is absorbed via the apically located sodium-dependent SGLT-1 system found in both the small and large intestine. Fructose is transported by the apical GLUT5-type system (Garriga et al., 2004). Once inside the epithelial cells, these sugars are transported into the interstitial space by the basolateral GLUT2 transporter. Absorption rates appear greatest during the first week of life, and decline thereafter (Barfull et al., 2002). SGLT-1 expression is regulated by the rennin-antiotensin-aldosterone system (Garriga et al., 2004).

The absorption of sugars against a concentration gradient involves an apically located sodium-dependent, phloridzin-sensitive transport. This system is coupled to Na^+-K^+-ATPase located on the basolateral membrane. Contrary to what is believed to occur in mammals, the active transport of one molecule of carbohydrate in birds is coupled to the movement of two molecules of Na (Kimmich and Randles, 1984). Sugars leave the enterocytes on their way to the bloodstream crossing the basolateral membrane by either simple diffusion or facilitated by a Na-independent mechanism.

Within the small intestine, the greatest absorption of glucose occurs in the duodenum (Figure 14.13). Cumulatively, up to 65, 85, and 97% of ingested starch is digested through the duodenum, jejunum, and the terminal ileum, respectively (Riesenfeld et al., 1980). Virtually all of the glucose released from starch digestion is absorbed within the small intestine.

Significant absorption of glucose also occurs in the cecum (Savory and Mitchell, 1991). While the entire cecum is able of accumulate sugars at hatch, this ability is soon limited to the proximal region (Planas et al., 1986). The ability of the cecum to actively absorb sugars at low concentrations appears higher than that of the jejunum (Vinardell and Lopera, 1987).

There is a greater affinity for active transport of glucose in the ileum (Levin et al., 1983). The ileum appears well suited for transporting glucose which may not be absorbed within the jejunum.

The gastrointestinal tract also has a role in glucose homeostasis (Riesenfeld et al., 1982). Plasma glucose levels are maintained relatively constant in chickens when fed semipurified diets in which glucose is replaced with either fructose, soybean oil, or cellulose. While this is partially the result of varying glucose turnover rates, it is largely a result of varying rates of conversion of glucose to lactate by the intestine.

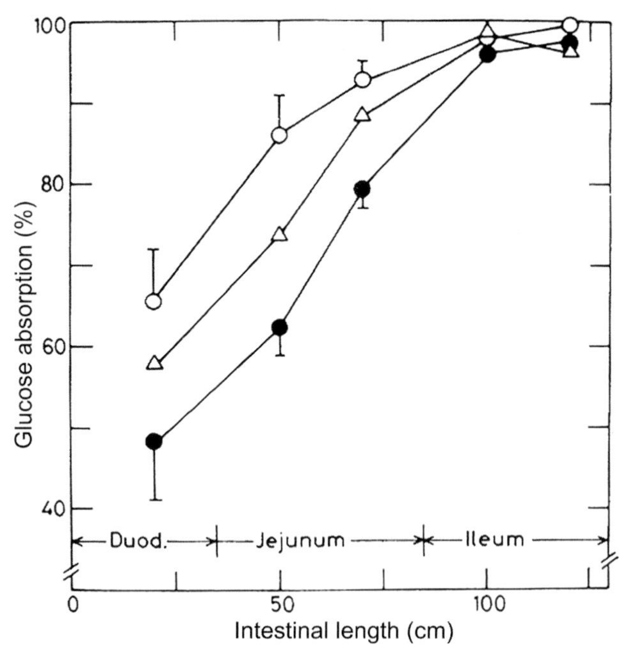

FIGURE 14.13 Cumulative percentage of glucose and starch disappearance from the intestine of 7-week-old chicks. Glucose absorption from glucose monohydrate as the sole dietary carbohydrate source (○), starch digestion (Δ), and absorption (●) in chicks fed starch as the sole dietary carbohydrate source. Values given in the figure are means of six chickens ± standard error. *From Riesenfeld et al. (1980).*

14.6.2 Amino Acids and Peptides

Amino acid and peptide absorption in birds (Gilbert et al., 2008) is similar to that in mammals. Amino acid transport occurs via a secondary active transport system as described for sugars. This process is saturable, coupled to Na^+ transport, and uses adenosine triphosphate for energy. The amino acid transport systems can be classified in four groups: (1) neutral amino acids; (2) proline, β-alanine, and related amino acids; (3) basic amino acids; and (4) acidic amino acids. Such a classification, however, is not rigid because many amino acids share transport with more than one group. For example, leucine, a neutral amino acid, can inhibit the uptake of proline and arginine, a basic amino acid. Glycine transport is partially inhibited by both proline and β-alanine.

TABLE 14.11 K_m (Mm) and J_{max} (Pmol $Cm^{-2}d^{-1}$) for Saturable Amino Acid Absorption in the Chicken Small Intestine

	Jejunum		Ileum	
Amino Acid	K_m	J_{max}	K_m	J_{max}
Amino-isobutyric acid	4.6±0.9	46±7	2.5±0.2	56±6
Glycine	4.2±0.4	37±5	2.7±0.2	55±6
Histidine	3.4±0.7	132±12	0.8±0.2	129±4
Methionine	4.9±0.6	147±14	1.9±0.6	148±5
Valine	3.2±0.7	38±7	1.5±0.2	82±13

Source: From Levin (1984).

The primary site for amino acid absorption is the small intestine, although absorption is also reported to occur in the crop, gizzard, and proventriculus. It is still unclear as to which section of the small intestine has the greatest capacity for absorption because there is a lack of agreement among studies. The rectum of hens is also capable of absorbing methionine via a saturable process. As with glucose transport, the K_m for various amino acids is lower in the ileum than in the duodenum (Table 14.11).

Peptide (i.e., di- and tripeptides) absorption can occur via paracellular movement and specific transport systems (Figure 14.14). Many amino acid transporters have been characterized. These include the neutral amino acid transporters with preference for leucine and other large hydrophobic neutral amino acids (system L), system A with preference for alanine and other small and polar neutral amino acids, and system ASC with preference for alanine, serine, and cysteine (Table 14.12). In addition, a separate nomenclature (x for anionic, y for cationic, z for neutral) has been applied to systems mediating transport of cationic amino acids (system y+) and anionic amino acids (system X^-_{AG}). With some exceptions (system L, system T), lower-case acronyms indicate Na^+-independent transporters, whereas uppercase acronyms are used for Na^+-dependent transporters.

For the absorption of di- and tripeptides, only one transport system, designated as PEPT1 (SLC15A1), is known. This is a low-affinity, high-capacity transport system and handles essentially all possible protein-derived di- and tripeptides as well as various peptidomimetics, such as

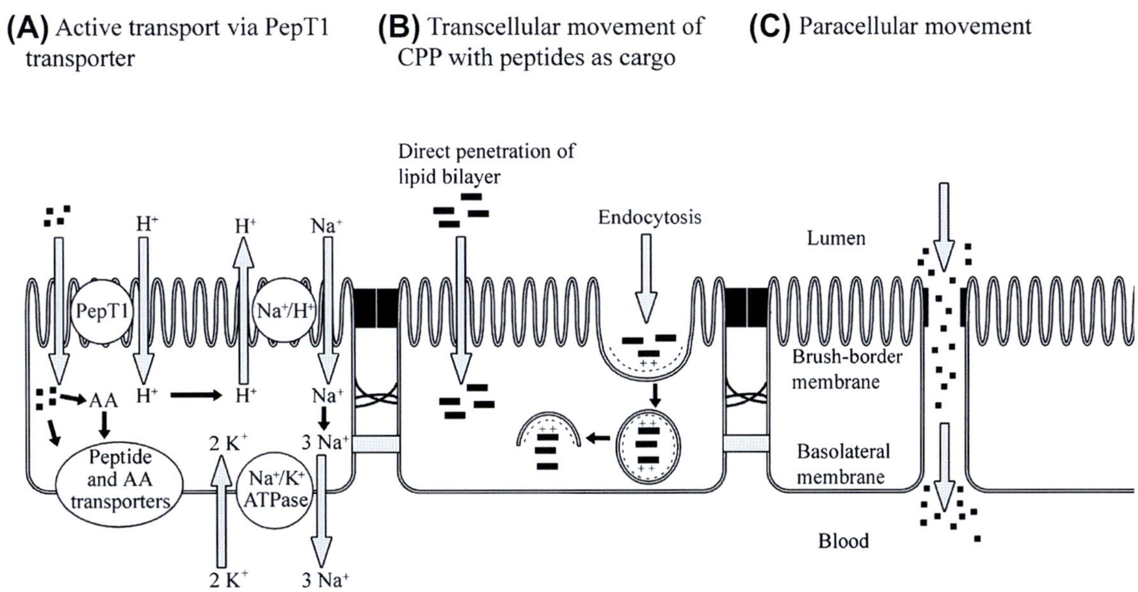

FIGURE 14.14 **Potential routes of peptide uptake in enterocytes.** (A) The primary route of di- and tripeptide absorption is through cotransport with H^+ by the peptide transporter, PepT1. (B) Cell-penetrating peptides (CPP) are capable of carrying cargo such as peptides to the inside of cells. (C) Increased permeability of tight junctions permits uptake of peptides via the paracellular route. *Gilbert et al. (2008).*

TABLE 14.12 Amino Acid Transporter Systems

System	cDNA*	SLC	Amino Acid Substrates
A	SNAT2	SLC38A2	G,P,A,S,C,Q,N,H,M
	SNAT4	SLC38A4	G,A,S,C,Q,N,M,AA^+
ASC	ASCT1	SLC1A4	A,S,C
	ASCT2	SLC1A5	A,S,C,T,Q
asc	4F2 hc/asc1	SLC3A2/SLC7A10	G,A,S,C,T
B^0	B^0AT1	SLC6A19	AA^0
	B^0AT2	SLC6A15	P,L,V,I,M
$B^{0,+}$	$ATB^{0,+}$	SLC6A14	AA^0, AA^+, β-Ala
$b^{0,+}$	rBAT/$b^{0,+}$AT	SLC3A1/SLC7A9	R,K,O,cysteine
β	TauT	SLC6A6	Tau, β-Ala
Gly	XT2	SLC6A18	G
Imino	Imino	SLC6A20	P, HO-P
L	4F2hc/LAT1	SLC3A2/SLC7A5	H,M,L,I,V,F,Y,W
	4F2hc/LAT2	SLC3A2/SLC7A8	AA^0 except P
	LAT3	SLC43A1	L,I,M,F
	LAT4	SLC43A2	L,I,M,F
N	SNAT3	SLC38A3	Q,N,H
	SNAT5	SLC38A5	Q,N,H,S,G
PAT (Imino acid)	PAT1	SLC36A1	P,G,A GABA, β-Ala
	PAT2	SLC36A2	P,G,A
T	TAT1	SLC16A10	F,Y,W
X^-_{AG}	EAAT2	SLC1A2	E,D
	EAAT3	SLC1A1	E,D
x^-_c	4F2 hc/xCT	SLC3A2/SLC7A11	E, $cystine^-$
y^+	CAT-1	SLC7A1	R,K,O,H
y^+L	4F2hc/y^+LAT1	SLC3A2/SLC7A7	K,R,Q,H,M,L
	4F2hc/y^+LAT2	SLC3A2/SLC7A6	K,R,Q,H,M,L,A,C

NR, not reported; A, antiport; AA^0, neutral amino acids; AA^+, cationic amino acids; U, uniport; S, symport; S-AA^0, symport together with neutral amino acids; K, kidney; I, intestine; AM, apical membrane; BM, basolateral membrane; Ub, ubiquitous. Amino acids are given in one-letter codes. O, ornithine; HO-P, hydroxyproline; GABA, gamma amino butyric acid. Affinity: high, <100 μM; medium, 100 μM to 1 mM; low, >1 mM.
*Expression in epithelial cells of kidney and intestine.
Source: Modified from Bröer (2008).

aminocephalosporins and various prodrugs. PepT1 involves the cotransport of protons. It has been well characterized in many species, including chickens and turkeys (Gilbert et al., 2008). Peptides appear to be absorbed more rapidly than amino acids. Absorption of peptides acts to eliminate competition between amino acids for uptake and thus helps increase speed of absorption.

PepT1 expression is upregulated during feed restriction. This may be a mechanism to compensate for decreased mucosal surface area caused by reduced villi height and crypt depth observed in broilers given delayed access to feed at hatch. Increased expression of PepT1 may be mediated by peroxisome-proliferator-activated receptor α (Shimakura et al., 2006).

The ceca are also important in amino acid absorption. Proline is absorbed in the cecum via a sodium-dependent carrier-mediated transport system, with transport rates being higher in the proximal than distal cecum (Obst and Diamond, 1989). In addition, leucine, phenylalanine, proline, and glycylsarcosine are absorbed against a concentration gradient (Gasaway, 1976; Calonge et al., 1990; Moretó et al., 1991). Proline and methionine have also been shown to be transported by a Na-independent system (Moretó et al., 1991).

The ceca has a greater ability to transport amino acids than sugars (Moretó et al., 1991). This may be functionally important when considering the following: (1) uric acid, which is retrogradely carried to the ceca from the coprodeum, may be microbially converted to amino acids; and (2) proteases are in high concentration within the ceca and may release amino acids from proteins. Therefore, the ceca may be important in amino acid absorption.

14.6.3 Fatty Acids and Bile Acids

Fatty acid absorption occurs in the distal half of the jejunum and, to a lesser extent, in the ileum. Because the bile duct enters the gastrointestinal tract of birds near the distal duodenum, emulsification of fats is delayed in birds relative to mammals.

In mammals, fatty acids enter the enterocytes where they are reesterified to triglycerides and packaged into chylomicrons which enter the lymphatic system. However, in birds, after being reesterified, lipids are packaged into portomicrons, which pass directly into the hepatic portal blood supply.

14.6.4 Volatile Fatty Acids

The concentration of volatile fatty acid (VFA), mainly acetate, but some propionate and butyrate, in the ceca is high. It can reach 125 mM in the chicken (Annison et al., 1968) and 70 mM in the goose (Clemens et al., 1975). VFAs are one of the byproducts of microbial decomposition of uric acid (see Braun and Campbell, 1989). Levels of VFAs are higher in the portal blood of conventionally raised compared to germ-free birds, suggesting that VFAs formed by bacteria are absorbed from the gastrointestinal tract. VFAs are absorbed from both the small intestine and the ceca by passive transport (Sudo and Duke, 1980). Because the concentration of VFAs is higher inside the lumen, absorption occurs by the passive movement of these compounds down their concentration gradient. While the rate of absorption of propionate and butyrate is equal between the small intestine and ceca, acetate is absorbed faster within the ceca. Microbial fermentation within the ceca of the ptarmigan can provide enough VFAs to meet 11–18% of the energy needs for basal metabolic rate (Gasaway, 1976).

14.6.5 Electrolytes

The major site of calcium and phosphate absorption, which are both actively absorbed, is the upper jejunum (Levin, 1984). When fed diets adequate in phosphorus and calcium, there is a net secretion of phosphate in the duodenum. However, when fed a phosphorous-deficient diet, there is a net absorption within the duodenum.

Calcium absorption occurs via an active transport system that is influenced by the vitamin D hormone 1,25-dihydroxyvitamin D_3 (1,25(0H)$_2$D$_3$). While calcium enters the mucosal membrane via a diffusion-type process, it is then secreted into the blood by an active transport system located on the basolateral membrane.

The hormonal byproduct of vitamin D, 1,25(0H)$_2$D$_3$, induces the synthesis of calbindin-D_{28k} in birds (Wasserman, 1992) and the calcium pump unit in the basolateral membrane of chicken enterocytes (Wasserman et al., 1992). Feeding chickens a calcium- or phosphorus-deficient diet, repleting vitamin D-deficient chickens with vitamin D, or injecting 1,25(0H)$_2$D$_3$ causes an increase in plasma membrane calcium pump mRNA in the duodenum, jejunum, ileum, and rectum (Cai et al., 1993).

The absorption and secretion of Na, potassium (K), calcium (Ca), and magnesium (Mg) in the gastrointestinal tract of broilers has recently been characterized (Van Der Klis et al., 1990). Na transport is discussed below. K, Ca, and Mg are secreted in the duodenum. In addition, there is some secretion of Mg in the ileum and rectum. The major site of absorption for these minerals is the proximal jejunum, with some absorption of Ca and Mg occurring in the gizzard. This coincides with the fact that dry matter spends approximately 25% of its time in the jejunum. There is little absorption of these minerals beyond the jejunum.

14.6.6 Water, Sodium, and Chloride

Water is absorbed throughout the small and large intestine and the ceca. Absorption of water occurs as a secondary response to the active absorption of other compounds such as glucose, Na, and amino acids.

Birds generally secrete a very dilute urine. Urine can travel into the coprodeum and, via antiperistalsis, continue orad into the ceca. Note that there are exceptions, such as the ostrich, which can concentrate urine and has no movement of urine back into the colon (Elbrond et al., 2009). The excreta of cecectomized, compared to intact chickens, is higher in water, indicating that the lower part of the intestinal tract has an important role in water and salt balance.

Sodium is secreted in the proximal portion of the intestinal tract (Van Der Klis et al., 1990). The major sites for Na absorption are the proximal jejunum followed by the colon. The ileum also has a small capacity for Na absorption.

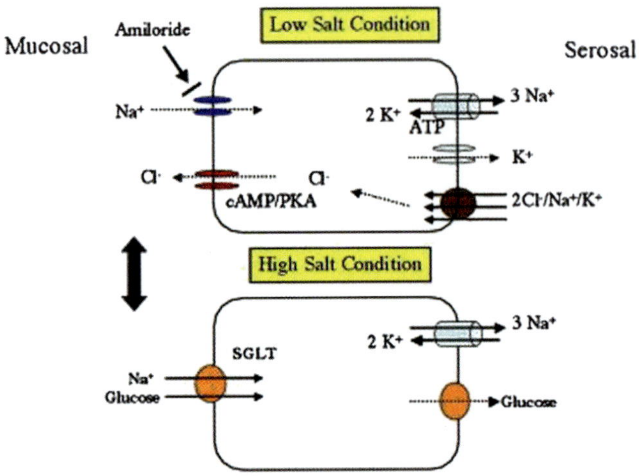

FIGURE 14.15 Working model for transport pathways in the hen lower intestine. The two panels show the different Na^+ absorptive pathways for tissues from low and high salt-acclimated hens. Example shown is for colon and would represent the extreme conditions of salt intake, in which there is a complete transition in transport pattern. SGLT-mediated glucose uptake and basolateral release via GLUT transporters is representative of several organic substrate cotransporter systems (amino acids and hexoses) expressed in the high salt-acclimated tissues. The diuretic amiloride blocks low salt ENaC channels at concentrations of 10^{-5} M. Potassium channels, shown on the basolateral or serosal side, may also be present on the apical surface and mediate K^+ secretion. For simplicity, the Cl^- secretory pathway is shown only in the low salt condition, even though it is also observed under high salt. *Laverty et al. (2006).*

The coprodeum, colon, and ceca serve an important role in Na, chloride (Cl) and water balance (Laverty et al., 2006). When fed a low-NaCl diet, there is a net absorption of Na from the coprodeum, colon, and ceca. Na^+ absorption occurs by active transport mediated by apical electrogenic and amiloride-sensitive Na^+ channels and basolateral Na^+ pumps (Na^+/K^+-ATPase) (Figure 14.15). The mucosal lining of the colon and coprodeum consist of simple, columnar epithelium with absorptive epithelial cells (AEC), goblet cells (GC), and mitochondria-rich cells (MR). On a low-salt diet, there is an extensive microvillous brush border. When fed a high-NaCl diet, absorption of Na^+ from the coprodeum nearly ceases, while that in the colon remains high provided a low concentration of hexoses and amino acids is maintained on the mucosal side. There is an increased expression of colon SGLT associated with increased dietary salt. This represents a dramatic change from electrogenic, channel-mediated Na^+ transport to organic substrate Na^+-cotransport consistent with that seen in the small intestine but not observed in the mammalian large intestine. On a high-salt diet, while the colon morphology remains the same, that of the coprodeum shows a decrease in density and length of microvilli. The adaptation to a low-salt diet is mediated by circulating aldosterone and maximizes Na^+ transport (absorption). Aldosterone suppresses SGLT expression (Laverty et al., 2001).

In addition, there is also a net secretion of Cl from the coprodeum and rectum, but the secretory ability of the coprodeum disappears in birds fed a high-NaCl diet (Skadhauge, 1989). The secretion of Cl is induced by serotonin (Hansen and Bindslev, 1989a). Receptor antagonists for 5-HT_{1A}, 5-HT_{1B}, 5-HT_{1C}, 5-HT_{1D}, 5-HT_2, 5-HT_3, adrenergic, cholinergic and histaminergic receptors were unable to block the response. However, the second messenger for serotonin-induced chloride secretion is cAMP (Hansen and Bindslev, 1989b).

The absorption of Na and secretion of Cl within the lower gut is regulated by plasma aldosterone (Clauss et al., 1991). Increases in plasma aldosterone stimulate amiloride-sensitive Na uptake in the coprodeum, rectum, and ceca. Feeding high-NaCl diets causes a decrease in plasma aldosterone, and a decrease in amiloride-dependent Na absorption. Furthermore, there is a switch to amiloride-independent hexose/amino acid stimulated Na-transport within the rectum and cecum, but a cessation of Na-transport in the coprodeum. The hexose/amino acid stimulated Na-transport in the rectum functions to counteract the osmotic water loss due to feeding a high-NaCl diet (Skadhauge et al., 1985), and is stimulated by increases in plasma arginine vasotocin and/or prolactin (Arnason and Skadhauge, 1991).

14.6.7 Vitamins

Absorption of vitamin B-6 occurs throughout the small intestine, but primarily in the duodenum (Heard and Annison, 1986). Absorption can also occur in the crop and cecum, but this is minimal. The mechanism of absorption is passive diffusion. Although microbial synthesis of vitamin B-6 does occur, this is not important in meeting a bird's vitamin requirement.

14.7 AGE-RELATED EFFECTS ON GASTROINTESTINAL FUNCTION

The small intestine grows rapidly during the later days of incubation, increasing from 1% of body weight at day 17 to 3.5% at hatch (day 21), followed by a two- to four-fold increase in intestine length through 12 days posthatch (Uni, 2006). Villi grow in length near the end of incubation and increase in size and number following hatch. Duodenal villus growth is nearly complete by day 7, whereas that in the jejunum and ileum continues beyond 14 days of age. Unlike in mammals, in which most enterocyte proliferation occurs within the crypts, in poultry proliferation occurs along the length of the villi. Enterocyte migration to the tip of the villi takes 3 days in 4-day-old chicks and 4 days in older birds.

It was hypothesized that postnatal growth rate is at least partially determined by the differential growth of various organs (Lilja, 1983). The intestinal tract grows rapidly following hatch, with relative growth being the greatest between 5 and 7 days-of-age. The intestinal length increases two- to four-fold by 12 days of age, with the small intestine increasing 7- to 10-fold during that time.

With regards to pancreatic enzyme activity within the intestines, trypsin, protease, and amylase activity increased rapidly during the first 21 days posthatch in turkey poults (Krogdahl and Sell, 1989). However, lipase activity did not begin to increase until after 21 days of age. Feeding a high-fat diet did not significantly increase lipase activity until after 21 days of age. At least during the first few weeks posthatch, it appeared that lipase activity may be a limiting factor in digestion.

Several notable changes have been found in the development of nutrient transport systems during ontogeny in broilers (Obst and Diamond, 1992). During the first week posthatch, proline uptake by the small intestine is high relative to glucose uptake. Because the relative growth rate of broilers is greatest during the first week, it appears that amino acid uptake may parallel this growth pattern. The uptake of glucose displays a transitory increase during week 2. This increase is hypothesized to be a result of a switch from lipid metabolism to carbohydrate metabolism due to depletion of the yolk reserves. Because of the allometric growth of the intestine, during the second week posthatch the intestine displays a decrease in size relative to body weight. This may be a second reason for the increase in uptake of glucose occurring at this time. There is a transitory increase in proline uptake during week 6. This increase parallels the first postjuvenile molt and a rise in absolute growth rate.

Interestingly, the intestinal uptake capacity of broilers closely matches a bird's nutrient needs. This contrasts with results in mammals, where uptake far exceeds the animal's needs. It remains to be determined whether this may indicate that nutrient uptake represents a potential constraint for increased growth in broilers, or whether it means that broilers are better allocating resources.

Feeding immediately posthatch stimulates intestinal development. Delaying feed for 24–48 h decreased villi length, enterocyte migration rate (Geyra et al., 2001), and enterocyte number but increased goblet cells producing acid and neutral mucine (Uni et al., 2003).

14.7.1 Microflora

The gut microflora is more complex than that of its host. In fact, it is now know that there are about 150 times more microbial genes (3.3 million versus 20,000–25,000) in the human gut than in the human genome. It is well documented that some gut microflora can cause disease (e.g., *Escherichia coli*, *Clostridium*, *Salmonella*). However, it is also recognized that many gut microbes are beneficial. For example, the enzymes that make vitamin B_{12} are produced by bacteria. Providing the GIT with beneficial microbes in order to reduce the presence of harmful bacteria is now an accepted strategy to improve health and productivity in poultry (Mead, 2000).

Evidence shows that microorganisms regulate an animal's genes and immune system (Lee and Mazmanian, 2010). Commensal microbiota may "program" aspects of T-cell differentiation, thus influencing the host genome and alterning the function of the adaptive immune system. It has been shown that mice that were raised in a sterile environment and therefore lack intestinal microbiota have defective mucosa-associated lymphoid tissues, fewer Peyer's patches, less lamina propria in the small intestine, and less mesenteric lymph nodes. The gut microflora has also been linked to obesity (Raoult, 2008) and type 2 diabetes. Obese patients have significantly more *Lactobacillus* sp., whereas diabetics have a lower proportion of *Firmicutes* and a higher proportion of *Bacteroidetes* and *Proteobactgetia* (Larsen et al., 2010).

Newly hatched chicks get bacteria from the surface of the eggshell. The GIT rapidly becomes colonized, with the maximum density occurring at 5 days of age (Apajalahti et al., 2004). The composition of the microflora then changes as the chicken grows (Lu et al., 2003). Subtherapeutic levels of antibiotics have been used in poultry diets for years in order to improve growth performance, and it is generally believed that their effects are at least partially mediated through an impact on the microflora (Coates et al., 1963; Knarreborg et al., 2002). Lactobacilli, streptococci, and *Clostridia perfringens* appear most commonly in the ileum, and lactobacilli and *C. perfringens* are most affected by dietary treatments.

Providing poultry with probiotics (i.e., competitive exclusion) has been used for poultry production to improve growth (Mead, 2000). Young chicks are very susceptible to *Salmonella typhimurium*, and this susceptibility can be greatly reduced if the chicks GIT is colonized with an adult-type microflora. Hence, competitive exclusion can be used to reduce the GIT colonization of *Salmonella* and *Campylobacter jejuni*—both potential causes of food poisoning. The growth promotion appears to be associated with increases in *Lactobacillus* sp. In broilers, chicks inoculated with *Lactobacillus* increase their weight gain significantly compared to control (Angelakis and Raoult, 2010).

REFERENCES

Aisa, J., Lahoz, M., Serrano, P.J., Junquera, C., Peg, M.T., Ver-Gil, A., 1997. Intrinsic innervations of the chicken lower digestive tract. Neurochem. Res. 22, 1425–1435.

Akester, A.R., Anderson, R.S., Hill, K.J., Osbaldiston, G.W., 1967. A radiographic study of urine flow in the domestic fowl. Br. Poult. Sci. 8, 209–212.

Aksoy, A., Cinar, K., 2009. Distribution and ontogeny of gastrin- and serotonin-immunoreactive cells in the proventriculus of developing chick, *Gallus gallus domestica*. J. Vet. Sci. 10, 9–13.

Ali, H.A., McLelland, J., 1978. Avian enteric nerve plexuses. A histochemical study. Cell Tissue Res. 189, 537–548.

Angelakis, E., Raoult, D., 2010. The increase in *Lactobacillus* species in the gut flora of newborn broiler chicks and ducks is associated with weight gain. PLoS One 5, 310463.

Annison, E.F., Hill, K.J., Kenworthy, R., 1968. Volatile fatty acids in the digestive tract of the fowl. Br. J. Nutr. 22, 207–216.

Antony, M., 1920. Uber die Speicheldrusen der Vogel. Zool. Jb. Abt. Anat. Ontog. Tiere 41, 547–660.

Apajalahti, J., Kettunen, A., Graham, H., 2004. Characteristics of the gastrointestinal microbial communities with special reference to the chicken. World's Poult. Sci. 60, 223–232.

Arnason, S.S., Skadhauge, E., 1991. Steady-state sodium absorption and chloride secretion of colon and coprodeum, and plasma levels of osmoregulatory hormones in hens in relation to sodium intake. J. Comp. Physiol. B 161, 1–14.

Ashcraft, D.W., 1930. The correlative activities of the alimentary canal of the fowl. Am. J. Physiol. 93, 105–110.

Barfull, A., Garriga, C., Tauler, A., Plana, J.M., 2002. Regulation of SGLT-1 expression in response to Na(+) intake. Am. J. Physiol. 282, R738–R743.

Bartlet, A.L., 1974. Actions of putative transmitters in the chicken vagus nerve/oesophagus and Remak nerve/rectum preparations. Br. J. Pharmacol. Chemother. 51, 549–558.

Bennett, T., 1974. Peripheral and autonomic nervous systems. In: Farner, D.S., King, J.R. (Eds.), Avian Biology. Academic Press, New York, pp. 1–77.

Bhattacharya, S., Ghose, K.C., 1971. Influence of food on amylase system in birds. Comp. Biochem. Physiol. B 40, 317–320.

Bird, F.H., 1971. Distribution of trypsin and amylase activities in the duodenum of the domestic fowl. Br. Poult. Sci. 12, 373–378.

Bjornhag, D., Sperber, I., 1977. Transport of various food components through the digestive tract of turkeys, geese and guinea fowl. Swedish J. Agric. Sci. 7, 57–66.

Bolton, W., 1965. Digestion in crop. Br. Poult. Sci. 6, 97–102.

Bowman, W.C., Everett, S.D., 1964. An isolated parasympathetically-innervated oesophagus preparation from the chick. Br. J. Pharmacol. 16, 72T–79T.

Branch, J., Cummings, J.H., 1978. Comparison of radio-opaque pellets and chromium sesquioxide as inert markers in studies requiring accurate fecal collections. Gut 19, 371–376.

Braun, E.J., Campbell, C.E., 1989. Uric acid decomposition in the lower gastrointestinal tract. J. Exp. Zool. Suppl. 3, 70–74.

Bröer, S., 2008. Amino acid transport across mammalian intestinal and renal epithelia. Physiol. Rev. 88, 249–286.

Burhol, P.G., 1974. Gastric stimulation by intravenous injection of cholecystokinin and secretin in fistula chickens. Scand. J. Gastroenterol. 9, 49–53.

Burhol, P.G., 1982. Regulation of gastric secretion in the chicken. Scan. J. Gastroenterol. 17, 321–323.

Cai, Q., Chandler, J.S., Wasserman, R.H., Kumar, R., Penniston, J.T., 1993. Vitamin D and adaptation to dietary calcium and phosphate deficiencies increase intestinal plasma membrane calcium pump gene expression. Proc. Natl. Acad. Sci. U.S.A 90, 1345–1349.

Calhoun, M., 1954. Microscopic Anatomy of the Digestive System. Iowa State College Press, Ames, Iowa.

Calonge, M.L., Ilundain, A., Bolufer, J., 1990. Glycylsarcosine transport by epithelial cells isolated from chicken proximal cecum and rectum. Am. J. Physiol. 258, G660–G664.

Campbell, B., Garner, A., Dimaline, R., Dockray, G.J., 1991. Hormonal control of avian pancreas. Am. J. Physiol. 261, G16–G21.

Chaplin, S.B., 1989. Effect of cecectomy on water and nutrient absorption of birds. J. Exp. Zool. Suppl. 3, 81–86.

Chaplin, S.B., Duke, G.E., 1988. Effect of denervation on initiation and coordination of gastroduodenal motility in turkeys. Am. J. Physiol. 255, G1–G6.

Chaplin, S.B., Duke, G.E., 1990. Effect of denervation of the myenteric plexus on gastroduodenal motility in turkeys. Am. J. Physiol. 259, G481–G489.

Clauss, W., Dantzer, V., Skadhauge, E., 1991. Aldosterone modulates Cl secretion in the colon of the hen (*Gallus domesticus*). Am. J. Physiol. 261, R1533–R1541.

Clemens, E.T., Stevens, C.E., Southworth, M., 1975. Sites of organic acid production and pattern of digesta movement in the gastrointestinal tract of geese. J. Nutr. 105, 1341–1350.

Clench, M.H., Pineiro-Carrero, V.M., Mathias, J.R., 1989. Migrating myoelectric complex demonstrated in four avian species. Am. J. Physiol. 256, G598–G603.

Coates, M.E., Fuller, R., Harrison, G.F., Lev, M., Suffolk, S.F., 1963. A comparison of the growth of chicks in the Gustafsson germ-free apparatus and in a conventional environment with the without dietary supplement of penicillin. Br. J. Nutr. 17, 141–150.

Dal Borgo, G.A., Salman, J., Pubols, M.H., McGinnis, J., 1968. Exocrine function of the chick pancreas as affected by dietary soybean meal and carbohydrate. Proc. Soc. Exp. Biol. Med. 129, 877–881.

Deaton, J.W., Branton, J.L., Lott, B.D., 1985. Noted difference in the digestive system in cages and floor-reared commercial egg-type pullets. Poult. Sci. 64, 1035–1037.

Denac, M., Scharrer, E., 1987. Effect of neurotensin on the smooth muscle of the chicken crop. Comp. Biochem. Physiol. 87C, 325–327.

Denac, M., Scharrer, E., 1988. Effect of bombesins and substance P on the smooth muscle of the chicken crop. Vet. Res. Comm. 12, 447–452.

Denac, M., Kumin, G., Scharrer, E., 1990. Effect of electrical field stimulation on muscle strips from chicken crop. Exp. Physiol. 75, 69–73.

Desmeth, M., 1980. Lipid composition of pigeon cropmilk – II. Fatty acids. Comp. Biochem. Physiol. 66B, 135–138.

Dimaline, R., Lee, C.M., 1990. Chicken gastrin: a member of the gastrin/CCK family with novel structure–activity relationships. Am. J. Physiol. 259, G882–G888.

Dimaline, R., Young, J., Gregory, H., 1986. Isolation from chicken antrum, and primary amino acid sequence of a novel 36-residue peptide of the gastrin/CCK family. FEBS Lett. 205, 318–322.

Dockray, G.J., 1975. Comparison of the actions of porcine secretin and extracts of chicken duodenum on pancreatic exocrine secretion in the cat and turkey. J. Physiol. 244, 625–637.

Dockray, G.J., 1977. Molecular evolution of gut hormones: application of comparative studies on the regulation of digestion. Gastroenterology 72, 344–358.

Dockray, G.J., 1988. Evolutionary aspects of gastrointestinal hormones. In: Advances in Metabolic Diseases, vol. 11, Academic Press, San Diego, p. 85.

Duke, G.E., 1982. Gastrointestinal motility and its regulation. Poult. Sci. 61, 1245–1256.

Duke, G.E., 1986. Alimentary canal: secretion and digestion, special digestive functions, and absorption. In: Sturkie, P.D. (Ed.), Avian Physiology, fourth ed. Springer-Verlag, New York, pp. 269–288.

Duke, G.E., Evanson, O.A., 1972. Inhibition of gastric motility by duodenal contents in turkeys. Poult. Sci. 51, 1625–1636.

Duke, G.E., Evanson, O.A., 1976. Diurnal cycles of gastric motility in normal and fasted turkeys. Poult. Sci. 55, 1802–1807.

Duke, G.E., Evanson, O.A., Ciganek, J.G., Miskowiec, J.F., Kostuch, T.E., 1973. Inhibition of gastric motility in turkeys by intraduodenal injections of amino acid solutions. Poult. Sci. 52, 1749–1756.

Duke, G.E., Kostuch, T.F., Evanson, O.A., 1975. Gastroduodenal electrical activity in turkeys. Am. J. Dig. Dis. 20, 1047–1058.

Duke, G.E., Evanson, O.A., Redig, P.T., 1976a. A cephalic influence on gastric motility upon seeing food in domestic turkeys, great horned owls (*Bubo virginianus*) and red-tailed hawks (*Buteo jamaicensis*). Poult. Sci. 55, 2155–2165.

Duke, G.E., Evanson, O.A., Redig, P.T., Rhoades, D.D., 1976b. Mechanism of pellet egestion in great horned owls (*Bubo virginianus*). Am. J. Physiol. 213, 1824–1829.

Duke, G.E., Dziuk, H.E., Evanson, O.A., Miller, J.E., 1977. Studies of methods for *in situ* observations of gastric motility in domestic turkeys. Poult. Sci. 56, 1575–1578.

Duke, G.E., Evanson, O.A., Huberty, B.J., 1980. Electrical potential changes and contractile activity of the distal cecum of turkeys. Poult. Sci. 59, 1925–1934.

Duke, G.E., Kimmel, J.R., Durham, K., Pollock, H.G., Bertoy, R., Rains-Epstein, D., 1982. Release of avian pancreatic polypeptide by various intraluminal contents in the stomach, duodenum or ileum or turkeys. Dig. Dis. Sci. 27, 782–786.

Duke, G.E., Place, A.R., Jones, B., 1989. Gastric emptying and gastrointestinal motility in Leach's Storm-Petrels chicks (*Oceanodroma leuchorhoa*). Auk 106, 80–85.

Dziuk, H.E., Duke, G.E., 1972. Cineradiographic studies of the gastric motility in turkeys. Am. J. Physiol. 222, 159–166.

Elbrond, V.S., Laverty, G., Dantzer, V., Grondahl, C., Skadhauge, E., 2009. Ultrastructure and electrolyte transport of the epithelium of coprodeum, colon and the proctodeal dicverticulum of *Rhea americana*. Comp. Biochem. Physiol. Part A 152, 357–365.

Elkin, R.G., Wood, K.V., Hagey, L.R., 1990. Biliary bile acid profiles of domestic fowl as determined by high performance liquid chromatography and fast atom bombardment mass spectrometry. Comp. Biochem. Physiol. 96B, 157–161.

Farner, D.S., 1942. The hydrogen ion concentration in avian digestive tracts. Poult. Sci. 21, 445–450.

Farner, D.S., 1943. Biliary amylase in the domestic fowl. Biol. Bull. 84, 240–243.

Ferrando, C., Vergara, P., Jiménez, M., Goñalons, E., 1987. Study of the rate of passage of food with chromium-mordanted plant cells in chickens (*Gallus gallus*). Quart. J. Exp. Physiol. 72, 251–259.

Ferrer, R., Planas, J.M., Durfort, M., Moretó, M., 1991. Morphological study of the caecal epithelium of the chicken (*Gallus Gallus domesticus* L.). Br. Poult. Sci. 32, 679–691.

Fileccia, R., Postorino, A., Serio, R., Mule, F., Abbadessa, U.S., 1984. Primary peristalsis in pigeon cervical oesophagus: two EMG patterns. Arch. Int. Physiol. Biochim. 92, 185–192.

Fileccia, R., Mule, F., Postorino, A., Serio, R., Abbadessa, U.S., 1987. 5-Hydroxytryptamine involvement in the intrinsic control of oesophageal EMG activity. Arch. Int. Physiol. Biochim. 95, 281–288.

Fitzgerald, T.C., 1969. The Coturnix Quail Anatomy and Histology. Iowa State University Press, Ames, Iowa. 207–260.

Fowler, M.E., 1991. Comparative clinical anatomy of ratites. J. Zoo. Wildl. Med. 22, 204.

Ganchrow, D., Ganchrow, J.R., 1985. Number and distribution of taste buds in the oral cavity of hatchling chicks. Physiol. Behav. 34, 889–894.

Garriga, C., Barfull, A., Plana, J.M., 2004. Kinetic characterization of apical D-fructose transport in chicken jejunum. J. Membr. Biol. 197, 71–76.

Gasaway, W.C., 1976. Seasonal variation in diet, volatile fatty acid production and size of the cecum of rock ptarmigan. Comp. Biochem. Physiol. 53A, 109–114.

Geyra, A., Uni, Z., Sklan, D., 2001. Enterocyte dynamics and mucosal development in the posthatch chick. Poult. Sci. 80, 776–782.

Gibson, R.G., Colvin Jr., H.W., Hirschowitz, B.I., 1974. Kinetics for gastric response in chickens to graded electrical vagal stimulation. Proc. Soc. Exp. Biol. Med. 145, 1058–1060.

Gilbert, E.R., Wong, E.A., Webb Jr., K.E., 2008. Peptide absorption and utilization: implications for animal nutrition and health. J. Anim. Sci. 86, 2135–2155.

Goldstein, D.L., Braun, E.J., 1986. Lower intestinal modification of ureteral urine in hydrated house sparrows. Am. J. Physiol. 250, R89–R95.

Goñalons, E., Rial, R., Turk, J.A., 1982. Phenol red as indicator of digestive tract motility in chickens. Poult. Sci. 61, 581–583.

Hansen, M.B., Bindslev, N., 1989a. Serotonin receptors for chloride secretion in hen colon. Comp. Biochem. Physiol. 94A, 189–197.

Hansen, M.B., Bindslev, N., 1989b. Serotonin-induced chloride secretion in hen colon, possible second messengers. Comp. Biochem. Physiol. 94A, 315–321.

Harrison, J.G., 1964. Tongue. In: Thomson, A.L. (Ed.), A New Dictionary of Birds. Nelson, London, pp. 825–827.

Hazelwood, R.L., Turner, S.D., Kimmel, J.R., Pollock, H.G., 1973. Spectrum effects of a new polypeptide (third hormone?) isolated from the chicken pancreas. Gen. Comp. Endocrinol. 21, 485–497.

Heard, G.S., Annison, E.F., 1986. Gastrointestinal absorption of vitamin B-6 in the chicken (*Gallus domesticus*). J. Nutr. 116, 107–120.

Herd, R.M., 1985. Anatomy and histology of the gut of the emu (*Dromaius novaehollandiae*). Emu 85, 43–46.

Herpol, C., 1966. Influence de l'ago sur le pH dans le tube digestif de *Gallus domesticus*. Ann. Biol. Anim. Biochim. Biophys. 4, 239–244.

Herpol, C., van Grembergen, G., 1967. La signification du pH dans le tube digestif de *Gallus domesticus*. Ann. Biol. Anim. Biochim. Biophys. 7, 33–38.

Hetland, H., Svihus, B., Krogdahl, A., 2003. Effects of oat hulls and wood shavings on digestion in broilers and layers fed diets based on whole or ground wheat. Brit. Poult. Sci. 44, 274–282.

Hill, K., 1971. Physiology of digestion. In: In: Bell, D.J., Freeman, B.M. (Eds.), Physiology and Biochemistry of the Domestic Fowl, vol. 1. Academic Press, New York, pp. 25–49.

Hill, K.J., 1983. Physiology of the digestive tract. In: In: Freeman, B.M. (Ed.), Physiology and Biochemistry of the Domestic Fowl, vol. 4. Academic Press, London, pp. 31–47.

Hodgkiss, J.P., 1984a. Evidence that enteric cholinergic neurones project orally in the intestinal nerve of the chicken. Quart. J. Exp. Physiol. 69, 797–807.

Hodgkiss, J.P., 1984b. Peristalsis and antiperistalsis in the chicken caecum are myogenic. Quart. J. Exp. Physiol. 69, 161–170.

Hodgkiss, J.P., 1986. The unmyelinated fibre spectrum of the main trunk and side branches of the intestinal nerve in the chicken (*Gallus gallus* var. *domesticus*). J. Anat. 148, 99–110.

Hokin, L.R., Hokin, M.R., 1953. Enzyme secretion and the incorporation of ^{32}P into phospholipids of pancreas slices. J. Biol. Chem. 203, 967–977.

Hulan, H.W., Bird, F.H., 1972. Effect of fat level in isonitrogenous diets on composition of avian pancreatic juice. J. Nutr. 102, 459–468.

Hurwitz, S., Bar, A., Katz, M., Sklan, D., Budowski, P., 1973. Absorption and secretion of fatty acids and bile acids in the intestine of the laying fowl. J. Nutr. 103, 543–547.

Jerrett, S.A., Goodge, W.R., 1973. Evidence for amylase in avian salivary glands. J. Morphol. 139, 27–46.

Jiménez, M., Martinez, V., Goñalons, E., Vergara, P., 1992. Opioid-induction of migrating motor activity in chickens. Life Sci. 50, 465–468.

Joyner, W.L., Kokas, E., 1971. Action of serotonin on gastric (proventriculus) secretion in chickens. Comp. Gen. Pharmacol. 2, 145–150.

Kessler, C.A., Hirschowitz, B.I., Burhol, P.G., Sachs, G., 1972. Methoxyflurane (penthrane) anesthesia effect on histamine stimulated gastric secretion in the chickens. Proc. Soc. Exp. Biol. Med. 139, 1340–1343.

Kimmel, J.R., Pollock, H.G., 1975. Factors affecting blood levels of avian pancreatic polypeptide (APP), a new pancreatic hormone. Fed. Proc. Fed. Am. Soc. Exp. Biol. 34, 454.

Kimmel, J.R., Pollock, H.G., Hazelwood, R.L., 1968. Isolation and characterization of chicken insulin. Endocrinology 83, 1323–1330.

Kimmich, G., Randles, J., 1984. Sodium-sugar coupling stoichiometry in chick intestinal cells. Am. J. Physiol. 247, C74–C82.

Knarreborg, A., Engberg, R.M., Jensen, S.K., Jensen, B.B., 2002. Quantitative determination of bile salt hydrolase activity in bacteria isolated from the small intestine of chickens. Appl. Environ. Microbiol. 68, 6425–6428.

Kokue, E., Hayama, T., 1972. Effects of starvation and feeding in the endocrine pancreas of chicken. Poult. Sci. 51, 1366–1370.

Kostuch, T.E., Duke, G.E., 1975. Gastric motility in great horned owls. Comp. Biochem. Physiol. A 51, 201–205.

Krogdahl, A., Sell, J.L., 1989. Influence of age on lipase, amylase, and protease activities in pancreatic tissue and intestinal contents of young turkeys. Poult. Sci. 68, 1561–1568.

Kudo, K-i., Nishimura, S., Tabata, L., 2008. Distribution of taste buds in layer-type chickens: scanning electron microscopic observations. Anim. Sci. J. 79, 680–685.

Lai, H.C., Duke, G.E., 1978. Colonic motility in domestic turkeys. Dig. Dis. 23, 673–681.

Larsen, N., Vogensen, F.K., van den Berg, F.W., Nielsen, D.S., Andreasen, A.S., Pedersen, B.K., Abu Al-Soud, W., Sørensen, S.J., Hansen, L.H., Jakobsen, M., 2010. Gut microbiota in human adults with type 2 diabetes differs from non-diabetic adults. PLoS One 5, e9085.

Larsson, L.I., Sundler, F., Hakanson, R., Pollock, H.G., Kimmel, J.R., 1974. Localization of APP, a postulated new hormone to a pancreatic cell type. Histochemistry 42, 377–382.

Laverty, G., Bjarmadpttor, S., Elbrond, V.S., Arnason, S.S., 2001. Alsosterone suppresses expression of an avian colonic sodium-glucose cotransporter. Am. J. Physiol. Regul. Integr. Comp. Physiol. 281, R1041–R1050.

Laverty, G., Elbrond, V.S., Arnason, S.S., Skadhauge, E., 2006. Endocrine regulation of ion transport in the avian lower intestine. Gen. Comp. Endocrinol. 147, 70–77.

Leasure, E.E., Link, R.P., 1940. Studies on the saliva of the hen. Poult. Sci. 19, 131–134.

Lee, Y.K., Mazmanian, S.K., 2010. Has the microbiota played a critical role in the evolution of the adaptive immune system? Science 330, 1768–1773.

Levey, D.J., Duke, G.E., 1992. How do frugivores process fruit? Gastrointestinal transit and glucose absorption in cedar waxwing (*Bambycilla cedrorum*). Auk 109, 722–730.

Levin, R.J., 1984. Absorption from the alimentary tract. In: In: Freeman, B.M. (Ed.), Physiology and Biochemistry of the Domestic Fowl, vol. 5. Academic Press, London, pp. 1–22.

Levin, R., Mitchell, M.A., Barber, D.C., 1983. Comparison of jejunal and ileal absorptive functions for glucose and valine *in vivo* – a technique for estimating real K_m and J_{max} in the domestic fowl. Comp. Biochem. Physiol. 74A, 961–966.

Lilja, C., 1983. A comparative study of postnatal growth and organ development in some species of birds. Growth 47, 317–339.

Lin, G.L., Himes, J.A., Cornelius, C.E., 1974. Bilirubin and biliverdin excretion by the chicken. Am. J. Physiol. 226, 881–885.

Lindsay, O.B., March, B.E., 1967. Intestinal absorption of bile salts in the cockerel. Poult. Sci. 46, 164–168.

Lisbona, F., Jiménez, M., Esteller, A., Lopez, M.A., 1981. Basal biliary secretion in conscious chicken and role of enterohepatic circulation. Comp. Biochem. Physiol. 69A, 341–344.

Liu, L., Burcher, E., 2001. Radioligand binding and functional characterization of tachykinin receptors in chicken small intestine. Naunyn Schmiedebers Arch. Pharmacol. 364, 305–313.

Long, J.F., 1967. Gastric secretion in unanesthetized chickens. Am. J. Physiol. 212, 1303–1307.

Lu, J., Idris, U., Harmon, B., Hofacre, C., Maurer, J.J., Lee, D., 2003. Diversity and succession of the intestinal bacterial community of the maturing broiler chicken. App. Environ. Microbiol. 69, 6816–6824.

Martinez, V., Jiménez, M., Goñalons, E., Vergara, P., 1992. Effects of cholecystokinin and gastrin on gastroduodenal motility and coordination in chickens. Life Sci. 52, 191–198.

Martinez, V., Jiménez, M., Goñalons, E., Vergara, P., 1993. Mechanism of action of CCK in avian gastroduodenal motility: evidence for nitric oxide involvement. Am. J. Physiol. 265, G842–G850.

Martinez, A., Lopez, J., Sesma, P., 2000. The nervous system of the chicken proventriculus: an immunocytochemical and ultrastructural study. Histochem. J. 32, 63–70.

McLelland, J., 1975. Aves digestive system. In: In: Getty, R. (Ed.), Sisson and Grossman's the Anatomy of the Domestic Animals, vol. 2. Saunders, Philadelphia, pp. 1857–1882.

McLelland, J., 1979. Digestive system. In: King, A.S., McLelland, J. (Eds.), Form and Function in Birds. Academic Press, London, pp. 69–181.

McLelland, J., 1989. Anatomy of the avian cecum. J. Exp. Zool. Suppl. 3, 2–9.

Mead, G.C., 2000. Prospects for competitive exclusion treatment to control salmonellas and other foodborne pathogens in poultry. Vet. J. 159, 111–123.

Moretó, M., Amat, C., Puchal, A., Buddington, R.K., Planas, J.M., 1991. Transport of L-proline and α-methyl-D-glucoside by chicken proximal cecum during development. Am. J. Physiol. 260, G457–G463.

Mueller, L.R., Duke, G.E., Evanson, O.A., 1990. Investigations of the migrating motor complex in domestic turkeys. Am. J. Physiol. 259, G329–G333.

Mule, F., 1991. The avian oesophageal motor function and its nervous control: some physiological, pharmacological and comparative aspects. Comp. Biochem. Physiol. 99A, 491–498.

Mule, F., Fileccia, R., Postorino, A., Serio, R., Abbadessa, U.S., La Grutta, G., 1987. Pigeon oesophageal EMG activity: analysis of intramural neural control. Arch. Int. Physiol. Biochim. 95, 269–280.

Nakagawa, H., Nishimura, M., Urakawa, N., 1983. 2-Deoxy-D-glucose stimulates acid secretion from chicken proventriculus. Jpn. J. Vet. Sci. 45, 721–726.

Nilsson, A., 1974. Isolation, amino acid composition and terminal amino acid residue of the vasoactive octacosapeptide from chicken intestine. Partial purification of chicken secretin. FEBS Lett. 47, 284–289.

Nilsson, S., 2011. Comparative anatomy of the autonomic nervous system. Auton. Neurosci. Basic Clin. 165, 3–9.

Obst, B.S., Diamond, J.M., 1989. Interspecific variation in sugar and amino acid transport by the avian cecum. J. Exp. Zool. Suppl. 3, 117–126.

Obst, B.S., Diamond, J., 1992. Ontogenesis of intestinal nutrient transport in domestic chickens (*Gallus gallus*) and its relation to growth. Auk 109, 451–464.

Ohashi, H., Ohgua, A., 1967. Transmission of excitation from the parasympathetic nerve to the smooth muscle. Nature 216, 291–292.

Ojewole, J.A., 1980. Studies on the responses of the isolated rectum of domestic chick (*Gallus domesticus*) to drugs. Biochem. Exp. Biol. 16, 413–420.

Olsson, C., Holmgren, S., 2011. Autonomic control of gut motility: a comparative view. Auton. Neurosci. Basic Clin. 165, 80–101.

Osman, A.M., 1982. Amylase in chicken intestine and pancreas. Comp. Biochem. Physiol. 73B, 571–574.

Paik, Y.K., Fujioka, T., Yasuda, M., 1974. Comparative and topographical anatomy of the fowl. LXXVIII. Division of pancreatic lobes and distribution of pancreatic ducts. Jpn. J. Vet. Sci. 36, 213–229.

Parisi Salvi, E., Vaccaro, R., Baglaj, S.M., Renda, T., 2004. Nervous system development in normal and atresic chick embryo intestine: an immunohistochemical study. Anat. Embryol. 209, 143–151.

Philips, S.M., Fuller, R., 1983. The activities of amylase and a trypsin-like protease in the gut contents of germ-free and conventional chickens. Br. Poult. Sci. 24, 115–121.

Pichová, I., Kostka, V., 1990. Molecular characteristics of pepsinogen and pepsin from duck glandular stomach. Comp. Biochem. Physiol. 97B, 89–94.

Pinchasov, Y., Noy, Y., 1994. Early postnatal amylolysis in the gastrointestinal tract of turkey poults *Meleagris gallopavo*. Comp. Biochem. Physiol. 107A, 221–226.

Planas, J.M., Villá, M.C., Ferrer, R., Moretó, M., 1986. Hexose transport by chicken cecum during development. Pfluegers Arch. 407, 216–220.

Polin, D., Wynosky, E.R., Loukides, M., Porter, C.C., 1967. A possible urinary back flow to ceca revealed by studies on chicks with artificial anus and fed amprolium-C^{14} or thiamine-C^{14}. Poult. Sci. 46, 89–93.

Postorino, A., Serio, R., Fileccia, R., Mule, F., Abbadessa, U.S., 1985. Electrical stimulation of glossopharyngeal nerve and oesophageal EMG response in the pigeon. Arch. Int. Physiol. Biochim. 93, 321–330.

Pubols, M.H., 1991. Ratio of digestive enzymes in the chick pancreas. Poult. Sci. 70, 337–342.

Raoult, D., Pedersen, B.K., Al-Soud, W.A., Sørensen, S.J., Hansen, L.H., Jakobsen, M., 2008. Human microbiome: take-home lesson on growth promoters? Nature 454, 690–691.

Rawdon, B.B., Andrew, A., 1999. Gut endocrine cells in birds: an overview, with particular reference to the chemistry of gut peptides and the distribution, ontogeny, embryonic origin and differentiation of the endocrine cells. Prog. Histochem. Cytochem. 34, 3–82.

Rea, A.M., 1973. Turkey vultures casting pellets. Auk 90, 209–210.

Rico-Guevara, A., Rubega, M.A., 2011. The hummingbird tongue is a fluid trap, not a capillary tube. PNAS 108, 9356–9360.

Riesenfeld, G., Sklan, D., Barr, A., Eisner, U., Hurwitz, S., 1980. Glucose absorption and starch digestion in the intestine of the chicken. J. Nutr. 110, 117–121.

Riesenfeld, G., Geva, A., Hurwitz, S., 1982. Glucose homeostasis in the chicken. J. Nutr. 112, 2261–2266.

Roseler, M., 1929. Die Bedeuntung der Blinddarme des Haushuhnes fur die Resorption der Nahrung und Verdauung der Rohfaser. Z. Tierz. Zuechtungsbid 13, 281–310.

Sahba, M.M., Morisset, J.A., Webster, P.D., 1970. Synthetic and secretory effects of cholecystokinin-pancreozymin on the pigeon pancreas. Proc. Soc. Exp. Biol. Med. 134, 728–732.

Salvi, E., Buffa, R., Renda, T.G., 1996. Ontogeny, distribution and amine/peptide content of chromogranin A- and B-immunoreactive endocrine cells in the small and large intestine of the chicken. Anat. Embryol. 194, 89–98.

Sanders, K.M., Koh, S.D., Ward, S.M., 2006. Interstitial cells of Cajal as pacemakers in the gastrointestinal tract. Ann. Rev. Physiol. 68, 307–343.

Sato, H., Ohga, A., Nakazato, Y., 1970. The excitatory and inhibitory innervation of the stomachs of the domestic fowl. Jpn. J. Pharmacol. 20, 382–397.

Savory, C.J., Mitchell, M.A., 1991. Absorption of hexose and pentose sugars *in vivo* in perfused intestinal segments in the fowl. Comp. Biochem. Physiol. 100A, 969–974.

Savory, C.J., Duke, G.E., Bertoy, R.W., 1981. Influence of intravenous injections of cholecystokinin on gastrointestinal motility in turkeys and domestic fowls. Comp. Biochem. Physiol. 70A, 179–189.

Sell, J.L., Eastwood, J.A., Mateos, G.G., 1983. Influence of supplemental fat on diet metabolizable energy and ingesta transit time in laying hens. Nutr. Rep. Intern. 28, 487–495.

Shimakura, J., Terada, T., Saito, H., Katsua, T., Inui, K.-I., 2006. The transcription factor Cdx2 regulates the intestine specific expression of human peptide transporter 1 through functional interaction with Sp1. Biochem. Pharamacol. 71, 1581–1588.

Shires, A., Thompson, J.R., Turner, B.V., Kennedy, P.M., Goh, Y.K., 1987. Rate of passage of corn-canola meal and corn-soybean meal diets through the gastrointestinal tract of broiler and White Leghorn chickens. Poult. Sci. 66, 289–298.

Short, L.L., 1982. Woodpeckers of the worlds. Del. Museum Nat. Hist.

Sibbald, I.R., 1979. Passage of feed through the adult rooster. Poult. Sci. 58, 446–459.

Siddons, R.C., 1969. Intestinal disaccharidase activities in the chick. Biochem. J. 112, 51–59.

Siddons, R.C., Coates, M.E., 1972. Effect of diet on disaccharidase activity in the chick. Br. J. Nutr. 27, 343–352.

Skadhauge, E., 1989. Hormonal regulation of sodium absorption and chloride secretion across the lower intestines of birds. Zool. Sci. 6, 437–444.

Skadhauge, E., Clauss, W., Arnason, S.S., Thomas, D.H., 1985. Mineralocorticoid regulation of lower intestinal ion transport. In: Gilles, R., Gilles-Baillien, M. (Eds.), Transport Processes, Iono- and Osmoregulation. Springer, Berlin, Heidelberg, New York, pp. 118–133.

Sklan, D., 2001. Development of the digestive tract of poultry. World's Poult. Sci. J. 57, 415–428.

Sklan, D., Noy, Y., 2000. Hydrolysis and absorption in the small intestines of posthatch chicks. Poult. Sci. 79, 1306–1310.

Sturkie, P.D., 1976. In: Stukie, P.D. (Ed.), Avian Physiology, third ed. Springer-Verlag, New York. p. 270, 280.

Sudo, S.Z., Duke, G.E., 1980. Kinetics of absorption of volatile fatty acids from the ceca of domestic turkeys. Comp. Biochem. Physiol. 67A, 231–237.

Suzuki, M., Nomura, S., 1975. Electromyographic studies on the deglutition movements in the fowl. Jpn. J. Vet. Sci. 37, 289–293.

Takewaki, T., Ohashi, H., Okada, T., 1977. Non-cholinergic and non-adrenergic mechanism in the contraction and relaxation of the chicken rectum. Jpn. J. Pharmacol. 27, 105–115.

Taneike, T., Miyazaki, H., Ohga, A., 1988. Histamine-induced potentiation of cholinergic transmission in chick oesophagus (*Gallus gallus*). Comp. Biochem. Physiol. 89C, 271–276.

Tur, J.A., Rial, R.V., 1985. The effect of temperature and relative humidity on the gastrointestinal motility of young broilers. Comp. Biochem. Physiol. 80A, 481–486.

Uden, P., Colucci, P.E., Van Soest, P.J., 1980. Investigation of chromium, cerium and cobalt as markers in digesta. Rate of passage studies. J. Sci. Food Agric. 31, 625–632.

Uni, Z., 2006. Early development of small intestinal function. In: Perry, G.C. (Ed.), Avian Gut Function in Health and Disease. Carfax Publishing Comp, Exfordshire, United Kingdom.

Uni, Z., Smirnov, A., Sklan, D., 2003. Pre- and posthatch development of goblet cells in the broiler small intestine: effect of delayed access to feed. Poult. Sci. 82, 320–327.

Vaillant, C., Dimaline, R., Dockray, G.J., 1980. The distribution and cellular origin of vasoactive intestinal polypeptide in the avian gastrointestinal tract and pancreas. Cell. Tissue Res. 211, 511–523.

Van Der Klis, J.G., Verstegen, M.W.A., De Wit, W., 1990. Absorption of minerals and retention time of dry matter in the gastrointestinal tract of broilers. Poult. Sci. 69, 2185–2194.

Vandeputte-Poma, J., 1968. Quelques donnees sur la composition du "Lait de Pigeon". Z. Verg. Physiol. 58, 356–363.

Vergara, P., Ferrando, C., Jiménez, M., Fernández, E., Goñalons, E., 1989. Factors determining gastrointestinal transit time of several markers in the domestic fowl. Quart. J. Exp. Physiol. 74, 867–874.

Vinardell, M.P., Lopera, M.T., 1987. Jejunal and cecal 3-oxy-methyl-D-glucose absorption in chicken using a perfusion system *in vivo*. Comp. Biochem. Physiol. 86A, 625–627.

Wang, L., Cheung, J.T.-M., Pu, F., Li, D., Zhang, M., Yubo, F., 2011. Why do woodpeckers resist head impact injury: a biomechanical investigation. PLoS One 6 (10), e26490–26498. doi: 10.1371.

Ward, N.E., Jones, J.E., Maurice, D.V., 1984. Increase in intestinal pH of chickens due to cimetidine injection. Fed. Proc. 43, 856.

Wasserman, R.H., 1992. In: Bronner, F., Peterlik, M. (Eds.), Extra- and Intracellular Calcium and Phosphate Regulation. CRC, Boca Raton.

Wasserman, R.J., Smith, C.A., Brindak, M.E., de Talamoni, N., Fullmer, C.S., Penniston, J.T., Kumar, R., 1992. Vitamin D and mineral deficiencies increases the plasma membrane calcium pump of chicken intestine. Gastroenterology 102, 886–894.

White, S.S., 1968. Mechanisms involved in deglutition in *Gallus domesticus*. J. Anat. 104, 177.

White, S.S., 1970. The Larynx of *Gallus domesticus* (Ph.D. Thesis). University of Liverpool, Liverpool.

Winget, C.M., Ashton, G.C., Cawley, A.J., 1962. Changes in gastrointestinal pH associated with fasting in laying hen. Poult. Sci. 41, 1115–1120.

Yamauchi, K.-E., Isshiki, Y., 1991. Scanning electron microscopic observations on the intestinal villi in growing White Leghorn and broiler chickens from 1 to 30 days of age. Br. Poult. Sci. 32, 67–78.

Ziswiler, V., Farner, D.S., 1972. Digestion and digestive system. In: Farner, D.S., King, James R. (Eds.), Avian Biology, vol. II. Academic Press, London, pp. 343–430.

FURTHER READING

Bird, F.H., Moreau, G.E., 1978. The effect of dietary protein levels in isocaloric diets on the composition of avian pancreatic juice. Poult. Sci. 57, 1622–1628.

Bray, G.A., 1991. Obestiy, a disorder of nutrient partitioning: the MONO LISA hypothesis. J. Nutr. 1212, 1146–1162.

Chaplin, S.B., Raven, J., Duke, G.E., 1992. The influence of the stomach on crop function and feeding behavior in domestic turkeys. Physiol. Behav. 52, 261–266.

Denbow, D., 1994a. Appetite and its control. Poult. Sci. Rev. 5, 209–229.

Denbow, D., 1994b. Peripheral regulation of food intake in poultry. J. Nutr. 124, 1349S–1354S.

Kitazawa, T., Ukai, H., Komori, S., Taneike, T., 2006. Pharmacological characterization of 5-hydroxytryptamine-induced contraction in the chicken gastrointestinal tract. Auton. Autocoid Pharmacol. 26, 157–168.

Kuenzel, W.J., 1989. Neuroanatomical substrates involved in the control of food intake. Poult. Sci. 68, 926–937.

Kuenzel, W.J., 1994. Central neuroanatomical systems involved in the regulation of food intake in birds and mammals. J. Nutr. 124, 1355S–1370S.

Lacy, M.P., Van Krey, H.P., Skewes, P.A., Denbow, D.M., 1985. Effect of intrahepatic glucose infusion on feeding in heavy and light breed chicks. Poult. Sci. 64, 751–756.

Lacy, M.P., Van Krey, H.P., Skewes, P.A., Denbow, D.M., 1986. Food intake in the domestic fowl: effect of intrahepatic lipid and amino acid infusion. Physiol. Behav. 36, 533–538.

Saito, I., 1966. Comparative anatomical studies of the oral organs of the poultry. V. Structures and distribution of taste buds of the fowl. Bull. Fac. Agric. Miyazaki Univ. 13, 95–102.

Savory, C.J., 1987a. How closely do circulating blood glucose levels reflect feeding state in fowls? Comp. Biochem. Physiol. 88A, 101–106.

Sturkie, P.D., 1965. In: Sturkie, P.D. (Ed.), Avian Physiology, second ed. Cornell University Press, Ithaca, NY, pp. 300–306.

Yamada, J., Kitamura, N., Yamashita, T., 1983. Avian gastrointestinal endocrine cells. In: Mikami, S-i., Homma, K., Wada, M. (Eds.), Avian Endocrinology: Environmental and Ecological Perspectives. Japan Scientific Societies Press, Tokyo, pp. 67–79.

Chapter 15

Poultry Bone Development and Bone Disorders

M. Pines
Institute of Animal Sciences, Volcani Center, Bet Dagan, Israel

R. Reshef
Department of Biology and Department of Evolutionary and Environmental Biology, University of Haifa, Haifa, Israel

15.1 INTRODUCTION

Chondrogenesis, the initial development of skeletal bones, starts during embryonic development in the early stages of somitogenesis and is considered to be the earliest phase of skeletal development. It involves progenitor cell specification, cell migration, epithelial-to-mesenchymal transition, and differentiation and maturation of chondrocytes. Mesenchymal cells that undergo chondrogenesis derive from neural crest cells, which originate from the neural crest ridge of the neural ectoderm; the presegmented paraxial mesoderm that differentiates into somites to elicit the sclerotome compartment, which leads to formation of the axis skeleton; and the somatopleure of the lateral plate mesoderm, which initiates limb buds for the appendage bones. Economically, the major skeletal disorders are associated with the rapid increase in growth rate achieved during the last few decades, and understanding of the relevant mechanisms will help in developing treatments for bone-related diseases. In this chapter, we portray the events leading chondrogenesis at the cellular and molecular levels and describe the various poultry bone disorders. The aim is to provide an introduction to the field, but by no means to cover all aspects of poultry bone development and disorders.

15.2 BONE DEVELOPMENT

15.2.1 Somitogenesis and Sclerotome Formation

During gastrulation, cells migrating through the various regions of the primitive streak establish the mesoderm. According to detailed fate map studies (Garcia-Martinez and Schoenwolf, 1992; Psychoyos and Stern, 1996) cells migrating via Hensen's node will contribute to formation of the notochord and the neural floor plate (Charrier et al., 2005). Cells that migrate through the anterior region of the primitive streak (posterior to the node) will establish the paraxial mesoderm that will bud off the somites. Cells that migrate through more posterior regions of the streak will give rise to the intermediate and lateral plate mesodermal tissues. In the presegmented region of the paraxial mesoderm, cells experience several waves of signaling that elicit the so-called segmentation clock, which ends in a precisely coordinated process of somite formation (Pourquié, 2011). This region, which contains the precursors of the sclerotome, flanks the notochord in the midline region and the intermediate mesoderm in the lateral aspect and underlies the neural plate. During neurulation and closure of the neural tube, the presegmented mesoderm buds and forms the somites, which now also neighbor the dorsal surface ectoderm. The somites mature from anterior to posterior; therefore, the anteriormost somite is the first to bud and the posteriormost one is the most recently formed (Christ and Ordahl, 1995). This anterior-to-posterior maturation gradient reflects changes in timing and location of an embryonic structure that is surrounded by several other developing tissues, which suggests a very dynamic process that involves external and internal events that lead to the compartmentalization and differentiation of somitic cells.

The somite buds into the form of an epithelial sphere surrounding the somitocoele. During its maturation, several compartments are formed sequentially, starting at the third new-formed somite. At this stage, two compartments are formed: the first is the sclerotome, which is undergoing an epithelial-to-mesenchymal transition at the ventromedial part of the somite; in parallel with this, the dorsal epithelial half forms the transitory structure of the dermomyotome (Ordahl and Le Douarin, 1992; Christ and Scaal, 2008). The mediodorsal part of the epithelial somite contains the muscle-progenitor cells that migrate anteriorly and dorsolaterally to form the scaffold of the future myotome (Kahane et al., 1998), whereas the central part of the dermomyotome will give rise to the back dermis (Ben-Yair et al., 2011). Also, Huang et al. (2000) found that thoracic dermomyotomal cells contain the precursors of the caudal part of the scapula, whereas the rostral region of this bone is of somatopleure origin. As development proceeds, the dermomyotome splits along its dorsoventral axis in a complicated sequence of events that form two waves of cell migration. The myotome forms in the ventral compartment, and the dermatome compartment is established at the center of the dorsolateral region of the somite (Ben-Yair et al., 2011). A fourth compartment—the syndetome—forms from sclerotomal cells and gives rise to the tendons; it is formed between the sclerotome and the myotome during later stages (Schweitzer et al., 2001; Brent et al., 2003).

Following the epithelial-to-mesenchymal transition, the sclerotome compartment consists of mesenchymal cells, which start to migrate along several pathways; they migrate ventromedially, dorsally, and laterally to populate various domains that will give rise to the various vertebra regions, the intervertebral discs and the ribs, before the cells start to differentiate into chondrocytes. The sclerotome is resegmented along the anterior-to-posterior axis and splits into anterior and posterior halves, each of which contributes cells to two sequential vertebrae. Thus, the body of a given vertebra consists of cells that originated from the posterior half of the rostrally adjacent sclerotome and the anterior half of the caudally adjacent sclerotome (Aoyama and Asamoto, 2000; Christ and Scaal, 2008). Subsequently, cells from the rostral half of the sclerotome populate the entire spinous process and the distal rib, whereas cells that originate in the caudal half of the sclerotome populate the rostral half of the spinous process, the vertebral arch, the transverse process, and the entire rib (Aoyama and Asamoto, 2000). This process of sclerotomal resegmentation is essential to enable the movement operated by the segmental muscles that now are attached by tendons to two resegmented vertebrae. The sclerotome is further subdivided into several subdomains, each of which becomes specified according to its location. The cells located at the medial part of the sclerotome will become the meninges and the blood vessels around the spinal cord, influenced by complex signals from the neural tube and notochord (Nimmagadda et al., 2007); this region is called the meningotome (Christ and Scaal, 2008). Another region, the arthrotome, is formed in a triangular shape from the somitocoele; it points toward the notochord and consists of mesenchymal cells that do not undergo the epithelial-to-mesenchymal transition. These cells will generate the vertebral joints, the intervertebral discs and the medial parts of the ribs (Mittapalli et al., 2005). For further reading on the various parts of the vertebrae, ribs, intervertebral discs, tendons, and blood vessels, see Christ and Scaal (2008).

15.2.2 Signaling Molecules Influence Sclerotome Formation

The somite is surrounded by several tissues that influence its maturation (reviewed in Christ and Scaal, 2008). These tissues secrete signaling molecules that diffuse over the somitic morphogenetic field and thus create gradients. At a specific stage, each cell in this somitic field experiences a different and inextricable combination of signaling factor concentrations and reacts according to its location and its competence to respond to these factors (Cairns et al., 2008; Piran et al., 2009). Four sources of external signaling factors affect the somitic field at the level of the newly formed somites. The neural tube that is polarized along the dorsal ventral axis and the notochord are the midline sources: the dorsal neural tube that, at various stages, contains the neural crest ridge and the roof plate is the dorsal medial signaling source; the notochord and the neural tube floor plate constitute the ventromedial signaling source. The surface ectoderm serves as the dorsal source of signaling molecules, and the intermediate and lateral plate mesodermal tissues constitute the fourth source of signaling molecules (Monsoro-Burq, 2005). All of these tissue sources secrete different sets of signaling molecules that affect the somitic field in various ways, causing it to differentiate into its compartments and to initiate the distinct molecular cascades that lead to various cell lineages.

Several Wnt family members are secreted from the dorsal part of the neural tube and the surface ectoderm: Sonic hedgehog (Shh) is secreted from the notochord and the neural tube floor plate; bone morphogenetic protein 4 (BMP4) is secreted from the lateral plate and intermediate mesodermal tissues at early stages of the paraxial mesoderm maturation, and from the neural roof plate from the level of the third newly formed somite at later stages (Pourquié et al., 1996; Reshef et al., 1998). The BMP antagonists, noggin and follistatin, exhibit their expression patterns within the paraxial mesoderm and affect the expression of BMP4 in the neighboring tissues (Amthor et al., 1996; Hirsinger et al., 1997; Marcelle et al., 1997; Reshef et al., 1998). These members of various families of paracrine signaling molecules create a very dynamic field within the somite; within it, they interact directly and indirectly with one another in various concentrations in diverse locations to elicit the compartmentalization of the somite (Cairns et al., 2008; Piran et al., 2009).

15.2.3 Tissue Interactions and Gene Activity in Limb Development

Limb chondrogenesis involves essentially the same molecular cassette as axial chondrogenesis, driving the same process but leading to bone formation in a different tissue—the lateral plate mesoderm. At particular locations along the anterior-posterior axis, as determined by specific combinations of *Hox* genes (Cohn and Tickle, 1999), the forelimb and hindlimb buds are formed from the flank ectoderm. These specific flank lateral plate regions express the mitogen factor FGF10 that induces the formation of a special tissue—the apical ectodermal ridge (AER)—in the competent corresponding flank ectoderm (Xu et al., 1998; Yonei-Tamura et al., 1999). This activity of FGF10 is mediated by the canonical Wnt3a, expressed in the ectoderm that induces expression of FGF8 in the AER (Niswander et al., 1993; Fallon et al., 1994; Fernandez-Teran and Ros, 2008). As the limb bud develops along the proximo–distal axis, the mesenchyme underlying the AER (a narrow ectodermal strip along the anteroposterior aspect of the distal tip of the limb bud)—the progress zone—proliferates extensively.

While cells close to the AER continue to proliferate, proximal cells gradually leave the cell cycle and start to differentiate into the various tissues of the limb, including tendons, cartilage, blood vessels, and muscle sheaths. Muscle progenitors invade the limb bud domain by migrating from the lateral aspect of the dermomyotome (Christ and Scaal, 2008). The core mesenchymal cells located in the center of the limb bud condense to form the cartilage anlagen. Condensing cells in the center of the limb bud are facilitated by TGF-β signaling factor secreted from neighboring cells and induce the expression of fibronectin and other cadherin molecules to stabilize the core of condensing cells. FGFs secreted from the surrounding ectoderm inhibit chondrogenesis via activation of the FGF receptor2, thereby keeping the cartilage anlagen (the condensed-cells core) in the center of the developing limb bud (Newman et al., 2008). Another important signaling center is the zone of polarizing activity (ZPA), which is a small piece of tissue located in the posterior region of the developing limb bud. Anterior implantation of ectopic ZPA results in a pattern of mirror-image digits, polydactyly, and can be mimicked by the signaling molecule Shh that normally is expressed in ZPA cells (Riddle et al., 1993). Complex positive and negative feedback loops of Shh, BMPs, FGFs, and their inhibitors permit the growth activity elicited by the AER. Gradient- and time-dependent exposure to Shh creates the anterior-posterior patterning of the limb in general and the digits pattern in particular. BMPs that are expressed throughout the distal mesenchyme are key factors in shaping the final form of the digits; they are expressed in the interdigital webbing mesenchyme and activate the apoptotic machinery. The strong BMP antagonists, noggin and gremlin, are expressed in the digits mesenchyme and protect these domains from apoptosis (Gilbert, 2010).

15.2.4 Molecular Mechanisms Controlling Chondrogenesis

The ventral midline tissues—the notochord and the neural tube floor plate—were shown in several studies to be the tissues that determine the ventral characteristics of the ventral somitic compartment—the sclerotome—by secreting a common signaling factor (Brand-Saberi et al., 1993; Dietrich et al., 1993, 1997; Goulding et al., 1993; Koseki et al., 1993; Pourquié et al., 1993; Ebensperger et al., 1995). The signaling properties of these tissues can be mimicked by the signaling factor Sonic Hedgehog (Shh) (Fan and Tessier-Lavigne, 1994; Johnson et al., 1994; Fan et al., 1995; Münsterberg et al., 1995; Borycki et al., 1998; Teillet et al., 1998). Mutant mice lacking Shh fail to form the sclerotome and therefore lack vertebrae (Chiang et al., 1996). Pax1 is a transcription factor that is expressed in early stages of sclerotome formation and was shown to be a Shh target (Ebensperger et al., 1995; Münsterberg et al., 1995; Šošić et al., 1997; Borycki et al., 1998; Teillet et al., 1998; Peters et al., 1999). As development proceeds, this transcription factor gradually disappears from migrating dorsal and lateral sclerotomal cells, and eventually from all cells undergoing chondrogenesis; this suggests that Shh is required only transiently in the early stages of chondrogenesis for the commitment of sclerotomal cells to become chondrocytes. This early molecular activity is regarded as the first phase of the endochondral ossification process.

The second phase is the condensation of mesenchymal cells into compact structures and their proliferation and differentiation into chondrocytes—the process that leads to cartilage cells. This process is dependent on BMP4 (Monsoro-Burq et al., 1996), which is an essential factor for limb chondrogenesis (Zou et al., 1997). However, in somite maturation and chondrocyte differentiation, its role is more complicated: several studies have shown that during early somite development BMP4 is a general inhibitor of the paraxial mesoderm, and can transform its development into the lateral plate mesoderm, as was demonstrated by the expression of lateral plate markers following overexpression of BMP4 in somite cultures and in *in vivo* experiments (Pourquié et al., 1996; Tonegawa et al., 1997; Reshef et al., 1998). Moreover, this factor is a myogenic-specific inhibitor, which suggests a role in determining the bone-to-muscle ratio that is reflected in the somitic compartment size of these two tissues (Piran et al., 2009). Several studies have shown that in the early stages of chondrogenesis there are two successive molecular phases: the first is Shh-dependent, whereas the second is Shh-independent and requires BMP4. It appears that Shh plays a role in determining the competence of sclerotomal cells to respond to a later BMP4 activity (Murtaugh et al., 1999) and that this process is mediated by the transcriptional repressor Nkx3.2 and the transcription factor Sox9, which suggests a role for these genes in maintaining the prochondrogenic response to BMP (Murtaugh et al., 2001; Zeng et al., 2002). Moreover, Sox9 and two other members of the Sox family—Sox5 and Sox6—create a complex of transcription factors that were shown to be expressed in all precartilaginous condensing cells and cartilage elements. This transcriptional complex is directly bind and activates specific genes of cartilage extracellular proteins (Lefebvre and Bhattaram, 2010). Sox9 upregulates the expression of Sox5/6 (Akiyama et al., 2002) and was proposed to upregulate its own expression (Kumar and Lassar, 2009). Sox9 is controlled positively by the activity of the signaling pathways of three factor families—Shh, BMP, and FGF—and negatively by the activity of Wnts/β-catenin, suggesting its expression absence in dorsal somitic regions (Lefebvre and Bhattaram, 2010). Heterozygous mutations in the human Sox9 gene resulted in severe chondrodysplasia, confirming its importance in chondrogenesis (Foster et al., 1994; Wagner et al., 1994).

The third and fourth phases of chondrogenesis comprise the proliferation, spatial rearrangement, and hypertrophy of chondrocytes. In these stages, proliferating chondrocyte cells are forming the cartilage model (the third phase) and later stop dividing, change morphology, and become hypertrophic chondrocytes (the fourth phase) (Lefebvre and Bhattaram, 2010) (Figure 15.1(A)). Further studies revealed a crucial role for the Runt transcription factor family member, Runx2, in these two stages. According to these studies, the hypertrophy of chondrocyte cells is Runx2-dependent (Yoshida et al., 2004), and the competence of prehypertrophic cells to respond to this transcription factor is mediated by Nkx3.2 and Sox9 (Kempf et al., 2007) Figure 15.1(B). Moreover, it was shown that the histone deacetylase4 (HDAC4) that is expressed in prehypertrophic cells plays a role in the regulation of Runx2 in these cells (Vega et al., 2004). Expression of Runx2 in prehypertrophic proliferating chondrocytes binds and transactivates the promoter of the paracrine factor Indian Hedgehog (Ihh) (Yoshida et al., 2004), which was found to be necessary and sufficient for the expression of the parathyroid-related peptide (PTHrP) (Vortkamp et al., 1996) which, in turn, inhibits Ihh. The tight negative feedback loop controls the proliferating phase of prehypertrophic chondrocyte cells. Once these cells mature and become hypertrophic chondrocytes, they secrete collagen X and other extracellular matrix proteins, thereby altering the matrix to become mineralized by calcium phosphate (Lefebvre and Bhattaram, 2010).

This concise description of the complex molecular networks and the tight regulation of the various phases of endochondral ossification (summarized in Figure 15.1) exemplifies the importance of the precise balance among the various stages of the process to achieving the right proportions, shape, and structure in the development of bones. Nevertheless, this review relates only to chondrogenesis per se and does not explore the potential influence of other tissues on this process. For example, recent studies have shown that during embryogenesis muscle force is able to direct the bone shape to bear the optimal load capacity, and this action is executed during chondrocyte convergent extension (Sharir et al., 2011; Shwartz et al., 2012). However, the molecular mechanisms of such effects remain to be elucidated.

15.3 BONE DISORDERS

Bone development and growth is controlled by precise cellular and molecular mechanisms. Any deviation from the normal process will result in bone disorders that present a major economic problem to the poultry industry. The regulation of bone development and growth appears to be complex, with various levels of interactions among the regulating agents. Several extensive reviews of various aspects of poultry bone disorders have been written during the last few years (Orth and Cook, 1994; Cook, 2000; Edwards, 2000; Rath et al., 2000; Pines et al., 2004).

15.3.1 Inherited and Rare Bone Disorders

Nanomelia is a recessively inherited connective tissue disorder that affects cartilage development and is characterized by chondrodystrophy and shortening of the long bones. It involves low aggrecan production and diminished aggrecan mRNA levels. A stop codon has been identified in the eighth repeat of the chondroitin sulfate 2 domain of the large tenth exon (Primorac et al., 1994). Ametapodia is a mutation associated with abnormal limb development that appeared in a strain of light brown leghorn chickens; the mutants are characterized by complete absence of the tarsometatarsals, and severely hypoplastic development of the metacarpals is also present. The disease is inherited as an autosomal recessive, and the affected chicks do not normally survive beyond 2–4 days of age (Smyth et al., 2000).

15.3.2 Bone Disorders of the Spine

Scoliosis is characterized by a lateral deviation of the spine, with rotation of the vertebrae. The effect of intense light on melatonin secretion by the pineal gland is probably involved in the incidence of the disease, which can be mimicked by pinealectomy (Machida et al., 2001). Microcomputed tomography (micro-CT) can be used to evaluate the concave and convex sides of vertebrae in pinealectomized scoliotic chickens (Fu et al., 2011). The incidence of expression of a scoliotic parent line implicated three major autosomal recessive genes (McCarrey et al., 1981).

Spondylolisthesis (kinky-back) is characterized by a ventral dislocation of the anterior end of the articulating vertebra—mainly but not only the fourth thoracic vertebra—with overriding of the posterior end by the fifth, which causes pinching of the spinal cord. Damage to the spinal cord causes leg weakness that is usually followed by partial posterior paralysis (McNamee et al., 1998). In some cases, the incidence was associated with *Enterococcus cecorum* (Makrai et al., 2011) and was greatest in chickens fed *ad libitum* and kept in batteries.

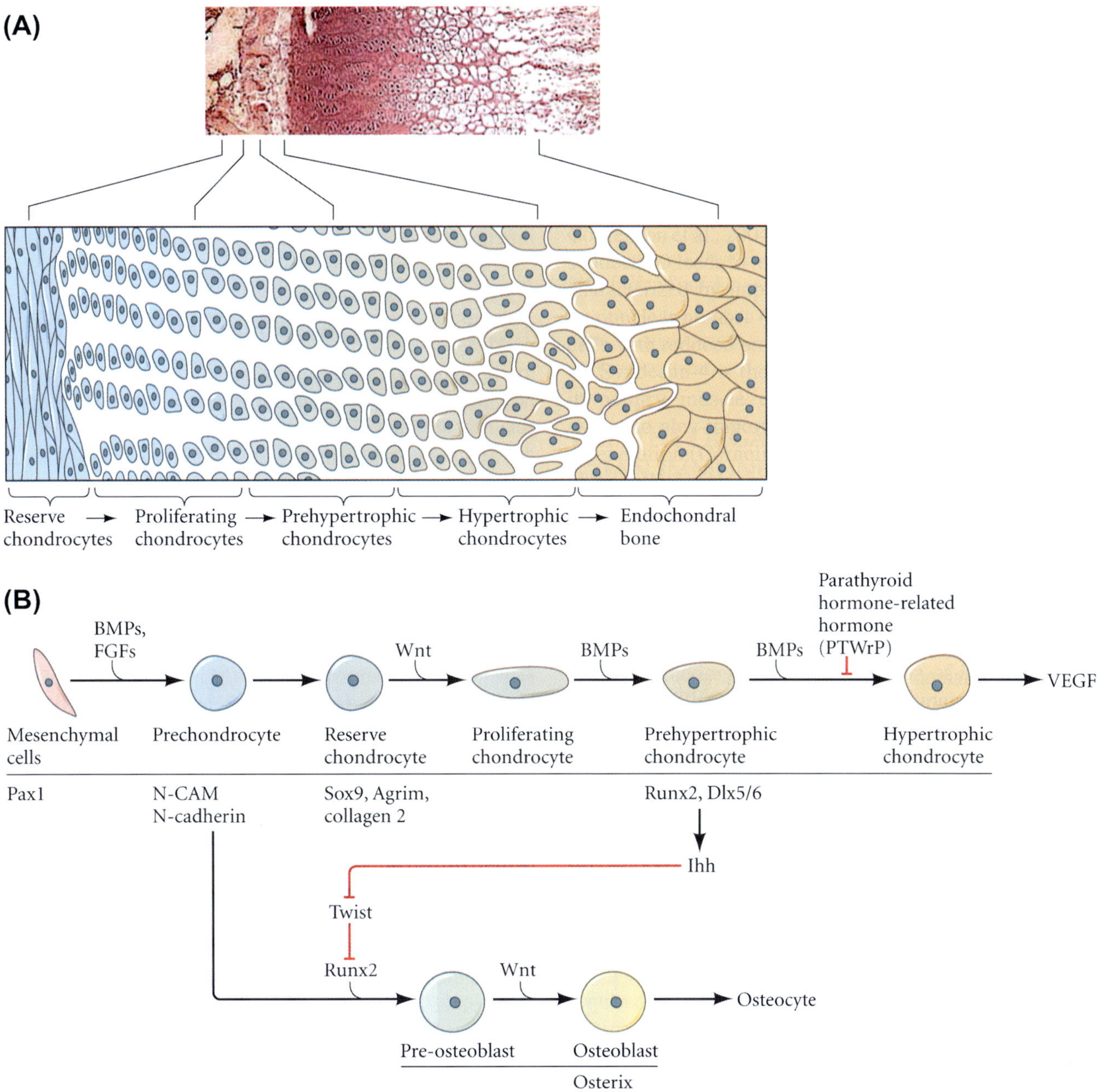

FIGURE 15.1 **Cell lineage and molecular mechanisms in endochondreal ossification.** (A) The cartilage is stained with alcian blue and the bone is stained with alizarin red. Below is a diagram of the transition zone wherein the chondrocytes (cartilage cells) divide, enlarge, die and are replaced by osteocytes (bone cells). (B) Paracrine factors and transcription factors active in the transition of cartilage to bone. The sclerotome cell can become a chondrocyte (characterized by the Sox9 transcription factor) or an osteocyte (characterized by the Osterix transcription factor) depending on the type of paracrine factors it experiences. The paracrine factor Indian hedgehog (Ihh), secreted by the growing chondrocytes, appears to repress Twist, an inhibitor of Runx2. Runx2 is critical for directing cell fate into the bone pathway, and activates Osterix, which in turn activates the bone-specific proteins. *With permission from Gilbert (2010), pp. 384–386.*

15.3.3 Bone Disorders due to Mycoplasma Infection

Leg weakness characterized by chondrodystrophy of the hock joints, clear fluid in hock joint spaces, valgus deformities and shortening of the tarsometatarsal bones, and curled toes were observed in turkey poults infected with *Mycoplasma* (Trampel and Goll, 1994). Joint lesions and curved toes were observed in turkey embryos inoculated with *Mycoplasma* and scanning electron microscopy of the tarso-metatarsal joints showed fissures in the cartilage (Lam et al., 2004). Vertebral chondrodystrophy was found in some cases to be associated with *Mycoplasma iowae* (Ley et al., 2010).

15.3.4 Bacterial Chondronecrosis with Ostemyelitis—Femoral Head Necrosis

Osteomyelitis was first reported as a cause of lameness in commercial broiler chickens in Australia, and *Staphylococcus aureus* was shown to be the cause. Bacterial chondronecrosis with ostemyelitis (BCO) is a naturally occurring bone disorder that also can be induced by *S. aureus* infection (Daum et al., 1990; McNamee et al., 1999). Although *S. aureus* is the bacterium most frequently reported in BCO, other bacteria also were isolated from lesions (McNamee et al., 1998; Dinev, 2009). BCO is one of the main reasons for lameness, and live affected birds usually have a characteristic limping gait (Wideman and Prisby, 2013). The most commonly affected sites in naturally occurring BCO are the proximal ends of the femur and tibiotarsus (McNamee et al., 1998, 1999). Lesions occur less frequently in the proximal tarsometatarsus, distal femur, distal tibiotarsus, proximal end of the humerus, and vertebrae (McNamee et al., 1998). Together with the cartilaginous lesions, osteomyelitis often occurs in the metaphysis of the femur and/or tibiotarsus, and occasionally in the secondary ossification center in the proximal end of the tibiotarsus. Macroscopically, BCO may appear as focal yellow areas of exudate or lytic areas. In *S. aureus*-induced lesions, there are usually clumps of basophilic bacteria in epiphyseal or physeal blood vessels, and these are surrounded by poorly stained cartilaginous matrix containing necrotic chondrocytes (McNamee and Smyth, 2000). Computed tomography revealed that the degenerative process involved loss of trabecular bone and lesion development in the mineralized matrix appeared to be coupled with increased bone resorption, associated with excessive proliferation of pathologically altered osteoclasts and lowered protein content of the bone organic matrix (Olkowski et al., 2011). In addition, immunosuppression in broilers has been implicated in the pathogenesis of BCO, and administration of steroids resulted in lameness associated with necrosis of the proximal tibial head and proximal femoral head (Wideman and Pevzner, 2012). Broilers kept on wire flooring developed significant levels of BCO without being intentionally exposed to known pathogens (Wideman and Pevzner, 2012).

15.3.5 Angular Limb Deformities

Angular or torsional limb deformities (twisted legs) are associated with abnormal diaphyseal curvature and physeal osteochondrosis. These deformities developed from a slight deviation or torsion of the distal tibiae. The progressive bowing of the tibiae and the diaphyseal cortical hypertrophy on the weight-bearing side are probably an adaptation to the disorder (Cruickshank and Sim, 1986). A structural abnormality in the distal tibiae may predispose the distal tibiae to a slight displacement of the gastrocnemius tendon, with consequently uneven stress on the distal condyles. The incidence of this abnormality was correlated with high growth rate and rapid weight gain (Fleming, 2008; Shim et al., 2012), early-age water restriction (Toghyani et al., 2011), cellulitis (Elfadil et al., 1996), organic trace minerals (Ferket et al., 2009), and vitamin D status (Newbrey et al., 1988). A model of hemicircumferential chick periosteal stripping of the left distal tibiotarsus that resulted in significant change in metaphyseal-diaphyseal angle with time can be used for study of this disorder (Rackard et al., 2002).

15.3.6 Vitamin and Mineral Deficiencies

There are numerous studies of the involvement of vitamins, minerals, and their interactions in skeletal disorders in poultry (Williams et al., 2000; Jin et al., 2001; Zhang et al., 2003). For example, broilers suffering from pyridoxine deficiency had tibiae of reduced dry weight and cortical thickness. Histomorphometry revealed a disproportionately large eroded surface, reductions in the amount of osteoid tissue, in the width of mineralized trabeculae and in the numbers of secondary ossification centers, along with coarse trabeculation. The metaphyseal cartilage showed irregular trabeculae and a markedly reduced amount of collagen and, probably, impaired collagen cross-linking (Masse et al., 1990, 1994). The effects of deficiencies of other minerals and vitamins such as manganese, biotin, and choline also have been studied (Stock and Latshaw, 1981; Liu et al., 1994).

The involvement of vitamin D and its metabolites is by far the most studied factor in relation to poultry bone disorder (Edwards, 2000; Dinev, 2012). Decreased extracellular phosphate, which is associated with vitamin D deficiency, probably plays a key role in rickets, and Ben-Bassat et al. (1999) studied the involvement of vitamin D in longitudinal bone growth and, especially, that of the growth plates at the ends of the long bones. In rickets, the width of the hypertrophic zone of the growth plate is increased and its mineralization is defective. The effects of vitamin D on the growth plate are mediated primarily through the vitamin D receptor that are expressed in intestinal epithelial cells, which exhibit increased calcium and phosphate absorption from the intestinal lumen. However, vitamin D metabolites also may act directly on the growth plate; for example, injection of 24,25-dihydroxyvitamin D_3 directly into rachitic chick growth plates resulted in healing (Lidor et al., 1987). *In vitro*, 24,25-dihydroxyvitamin D_3 stimulated differentiation but partly inhibited proliferation in resting-zone cells, whereas 1,25-dihydroxyvitamin D_3 decreased proliferation in both the resting and the proliferative zones (Boyan et al., 2002). The effects of vitamin D (Atencio et al., 2005), 1α-hydroxycholecalciferol (Driver et al., 2005), and 25-hydroxycholecalciferol (Bar et al., 2003) on bone development and disorders have been evaluated: the addition of phytase, in combination with 1α-hydroxycholecalciferol or 1,25-dihydroxyvitamin D_3, ameliorated leg problems or decreased the incidence of rickets, respectively (Mitchell and Edwards, 1996; Driver et al., 2005).

15.3.7 Tibial Dyschondroplasia

Tibial dyschondroplasia (TD) is one of the most common skeletal abnormalities that result in deformed bones and lameness (Leach and Gay, 1987; Orth and Cook, 1994; Pines et al., 2004); it is a disease of rapidly growing birds, especially of broilers (Leach and Lilburn, 1992) and turkeys (Wyers et al., 1991) growing at their maximal genetic potential. Comparison between fast-growing and slower-growing strains revealed less mineralization, more porous cortical bone, and increased Ca-to-P ratios in the former strain (Williams et al., 2000a). The porosity was a result of rapid primary osteon formation at the periosteal surface and incomplete infilling of the resulting canal with osteoblasts. These reductions in density and mineral content resulted in altered biomechanical properties, which caused high rates of bone breakage during catching, transport, and handling at the processing plant. Use of feed restriction led to the hypothesis that the rapid growth rate and not the genetic potential was responsible for the changes in the biochemical properties of the bones (Williams et al., 2004), which implies that genetic selection for growth rate has actually resulted in increased TD incidence (Figure 15.2).

Various factors have been found to be involved in the etiology of the disease, including dietary (Rennie et al., 1993), environmental (Riddel and Classen, 1992), and genetic (Leach and Lilburn, 1992; Kapell et al., 2012) factors. TD can occurs spontaneously in the field and can also be selected for (Ling et al., 2000) or induced by nutritional manipulations or by toxic agents (Orth and Cook, 1994; Rath et al., 2004). Thus, the various protocols that result in TD may initially act via distinct pathways, but downstream they probably share common pathway(s) that lead to the same phenotype (Pines et al., 2004).

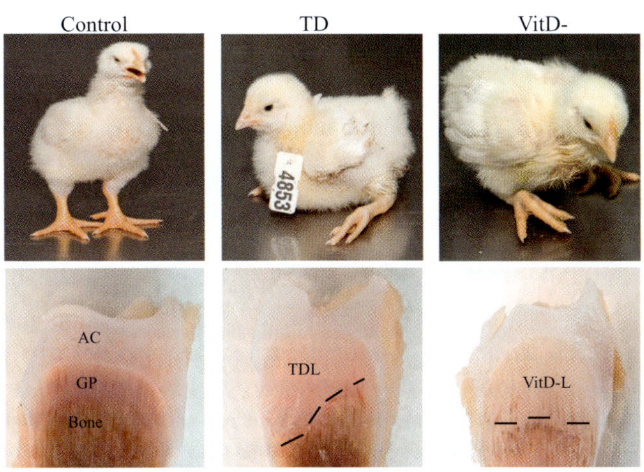

FIGURE 15.2 Chicks (upper panel) and their tibial growth plates (lower panel) of 12 day old broilers afflicted with thiram-dependent tibial dyschondroplasia (TD) and vitamin D-deficient rickets (VitD-). Note the lameness of the TD and VitD- chicks and the increase in the tibial growth plates (GP). AC, articular cartilage; TDL, tibial dyschondroplasia lesion; VitD-L, vitamin D-deficient lesion.

TD is a disease of the growth plates at the ends of the long bones: it is characterized by the appearance of a plug of unvascularized, unmineralized, opaque white cartilage, which dominates the proximal metaphysis of the tibiatarsus and, occasionally, the tarsometatarsus. In TD, the chondrocytes that populate the growth plates fail to undergo the complete differentiation that normally leads to cartilage vascularization and mineralization (Knopov et al., 1995; Pines et al., 1999). Various strategies were used in attempts to understand the causes of TD, and investigated aspects included mechanisms involved in chondrocyte differentiation (Knopov et al., 1997; Pines et al., 1999; Farquharson et al., 2001); chondrocyte apoptosis (Rath et al., 1998), effects of hormones and their receptors (Ben-Bassat et al., 1999; Webster et al., 2003), and fingerprint techniques to compare gene expression in normal and TD chondrocytes (Jefferies et al., 2000; Tian et al., 2013). TD was associated with copper deficiency (Lilburn and Leach, 1980) and could be induced by copper-chelating agents (Pines et al., 2004).

As in TD, the growth plate of rachitic chicks is enlarged, with an increased number of chondrocytes, and is unvascularized. Existence of an association between TD and rickets was demonstrated by the finding that supplementation of high doses of vitamin D or its analogs could ameliorate TD in selected lines (Edwards, 2000; Whitehead et al., 2004; Atencio et al., 2005). Divergent selection of broilers for low or high TD altered the physiological response to nutritionally inadequate levels of dietary D_3 (Shirley et al., 2003). Although TD and rickets have differing etiologies, inhibition of the heat shock protein 90 (Hsp90) resulted in a major improvement in the growth plate histopathology and ameliorated lameness of both TD and vitamin D-deficient chicks, suggesting similar pathways (Herzog et al., 2011). However, it is of the utmost importance to distinguish between the two lesions when studying and designing strategies to prevent TD or rickets; this is especially important because, in some cases, TD-afflicted birds have a background of rickets. It is possible to distinguish between the two conditions visually, but determination of molecular markers such as PTHR gene expression is much more accurate. In TD, normal expression of PTHR is observed, whereas in rickets the receptor is downregulated because of high parathyroid hormone levels, and no expression of the receptor is observed (Ben-Bassat et al., 1999).

15.4 CONCLUSIONS

A high degree of precision in the genetic control of developmental chondrogenesis is essential, and deviation beyond a certain threshold would cause abnormal bone growth. Multidisciplinary research at various levels, such as the genomic and proteomic approaches, cell culture methodology, genetic selection, nutritional manipulation, and environmental approaches, will provide us with a better understanding of the molecular mechanisms underlying bone disorders. Such understanding might enhance our knowledge and perhaps help in the design of rational strategies for the treatment of bone-related diseases. The enormous losses attributed to skeletal disorders in poultry are caused by increases in mortality and condemnations at the processing plant, and further downgrading caused by the trimming of breasts and legs. From the economic point of view, the major skeletal disorders are associated with the rapid increase in growth rate that has been achieved during the last few decades. Whether the disorders are direct effects of the rapid growth rate or indirect effects that result from increased body weight and inadequate development of bone, muscles and/or tendons is still to be determined. The economic dilemma facing the industry lies in the balance between continued selection for rapid growth, on the one hand, and the losses caused by bone and other metabolic disorders, on the other hand.

REFERENCES

Akiyama, H., Chaboissier, M.C., Martin, J.F., Schedl, A., de Crombrugghe, B., 2002. The transcription factor Sox9 has essential roles in successive steps of the chondrocyte differentiation pathway and is required for expression of Sox5 and Sox6. Genes Dev. 16, 2813–2828.

Amthor, H., Connolly, D., Patel, K., Brand-Saberi, B., Wilkinson, D.G., Cooke, J., Christ, B., 1996. The expression and regulation of follistatin and a follistatin-like gene during avian somite compartmentalization and myogenesis. Dev. Biol. 178, 343–362.

Aoyama, H., Asamoto, K., 2000. The developmental fate of the rostral/caudal half of a somite for vertebra and rib formation: experimental confirmation of the resegmentation theory using chick-quail chimeras. Mech. Dev. 99, 71–82.

Atencio, A., Edwards Jr., H.M., Pesti, G.M., 2005. Effect of the level of cholecalciferol supplementation of broiler breeder hen diets on the performance and bone abnormalities of the progeny fed diets containing various levels of calcium or 25-hydroxycholecalciferol. Poult. Sci. 84, 1593–1603.

Bar, A., Razaphkovsky, V., Vax, E., Plavnik, I., 2003. Performance and bone development in broiler chickens given 25-hydroxycholecalciferol. Br. Poult. Sci. 44, 224–233.

Ben-Bassat, S., Genina, O., Lavelin, I., Leach, R.M., Pines, M., 1999. Parathyroid receptor gene expression by epiphyseal growth plates in rickets and tibial dyschondroplasia. Mol. Cell Endocrinol. 149, 185–195.

Ben-Yair, R., Kahane, N., Kalcheim, C., 2011. LGN-dependent orientation of cell divisions in the dermomyotome controls lineage segregation into muscle and dermis. Development 138, 4155–4166.

Borycki, A.G., Mendham, L., Emerson Jr, C.P., 1998. Control of somite patterning by Sonic hedgehog and its downstream signal response genes. Development 125, 777–790.

Boyan, B.D., Sylvia, V.L., Dean, D.D., Del Toro, F., Schwartz, Z., 2002. Differential regulation of growth plate chondrocytes by 1alpha,25-(OH)2D3 and 24R,25-(OH)2D3 involves cell-maturation-specific membrane-receptor-activated phospholipid metabolism. Crit. Rev. Oral Biol. Med. 13, 143–154.

Brand-Saberi, B., Ebensperger, C., Willting, J., Balling, R., Christ, B., 1993. The ventralizing effect of the notochord on somite differentiation in chick embryos. Anat. Embryol. (Berl.) 188, 239–245.

Brent, A.E., Schweitzer, R., Tabin, C., 2003. A somitic compartment of tendon progenitors. Cell 113, 235–248.

Cairns, D.M., Sato, M.E., Lee, P.G., Lassar, A.B., Zeng, L., 2008. A gradient of Shh establishes mutually repressing somitic cell fates induced by Nkx3.2 and Pax3. Dev. Biol. 323, 152–165.

Charrier, J.-B., Catala, M., Le Douarin, N., Teillet, M.-A., 2005. Cellular dynamics and molecular control of the development of organizer-derived cells studied in quail-chick embryos. Int. J. Dev. Biol. 49, 181–191.

Chiang, C., Litingtung, Y., Lee, E., Young, K.E., Corden, J.L., Westphal, H., Beachy, P.A., 1996. Cyclopia and defective axial patterning in mice lacking Sonic hedgehog gene function. Nature 383, 407–413.

Christ, B., Ordahl, C.P., 1995. Early stages of chick somite development. Anat. Embryol. (Berl.) 191, 381–396.

Christ, B., Scaal, M., 2008. Formation and differentiation of avian somite derivatives. In: Maroto, M., Whittock, N.V. (Eds.), Somitogenesis. Landes Bioscience and Springer Science + Business Media.

Cohn, M.J., Tickle, C., 1999. Developmental basis of limblessness and axial patterning in snakes. Nature 399, 474–479.

Cook, M.E., 2000. Skeletal deformities and their causes: introduction. Poult. Sci. 79, 982–984.

Cruickshank, J.J., Sim, J.S., 1986. Morphometric and radiographic characteristics of tibial bone of broiler chickens with twisted leg disorders. Avian Dis. 30, 699–708.

Daum, R.S., Davis, W.H., Farris, K.B., Campeau, R.J., Mulvihill, D.M., Shane, S.M., 1990. A model of *Staphylococcus aureus* bacteremia, septic arthritis, and osteomyelitis in chickens. J. Orthop. Res. 8, 804–813.

Dietrich, S., Schubert, F.R., Gruss, P., 1993. Altered Pax gene expression in murine notochord mutants: the notochord is required to initiate and maintain ventral identity in the somite. Mech. Dev. 44, 189–207.

Dietrich, S., Schubert, F.R., Lumsden, A., 1997. Control of dorsoventral pattern in the chick paraxial mesoderm. Development 124, 3895–3908.

Dinev, I., 2009. Clinical and morphological investigations on the prevalence of lameness associated with femoral head necrosis in broilers. Br. Poult. Sci. 50, 284–290.

Dinev, I., 2012. Clinical and morphological investigations on the incidence of forms of rickets and their association with other pathological states in broiler chickens. Res. Vet. Sci. 92, 273–277.

Driver, J.P., Pesti, G.M., Bakalli, R.I., Edwards Jr, H.M., 2005. Phytase and 1alpha-hydroxycholecalciferol supplementation of broiler chickens during the starting and growing/finishing phases. Poult. Sci. 84, 1616–1628.

Ebensperger, C., Wilting, J., Brand-Saberi, B., Mizutani, Y., Christ, B., Balling, R., Koseki, H., 1995. Pax-1, a regulator of sclerotome development is induced by notochord and floor plate signals in avian embryos. Anat. Embryol. (Berl.) 191, 297–310.

Edwards Jr, H.M., 2000. Nutrition and skeletal problems in poultry. Poult. Sci. 79, 1018–1023.

Elfadil, A.A., Vaillancourt, J.P., Meek, A.H., Julian, R.J., Gyles, C.L., 1996. Description of cellulitis lesions and associations between cellulitis and other categories of condemnation. Avian Dis. 40, 690–698.

Fallon, J.F., Lopez, A., Ros, M.A., Savage, M.P., Olwin, B.B., Simandl, B.K., 1994. FGF-2: apical ectodermal ridge growth signal for chick limb development. Science 264, 104–107.

Fan, C.M., Tessier-Lavigne, M., 1994. Patterning of mammalian somites by surface ectoderm and notochord: evidence for sclerotome induction by a hedgehog homolog. Cell 79, 1175–1186.

Fan, C.M., Porter, J.A., Chiang, C., Chang, D.T., Beachy, P.A., Tessier-Lavigne, M., 1995. Long-range sclerotome induction by sonic hedgehog: direct role of the amino-terminal cleavage product and modulation by the cyclic amp signaling pathway. Cell 81, 457–465.

Farquharson, C., Jefferies, D., Seawright, E., Houston, B., 2001. Regulation of chondrocyte terminal differentiation in the postembryonic growth plate: the role of the PTHrP-Indian hedgehog axis. Endocrinology 142, 4131–4140.

Ferket, P.R., Oviedo-Rondón, E.O., Mente, P.L., Bohórquez, D.V., Santos Jr., A.A., Grimes, J.L., Richards, J.D., Dibner, J.J., Felts, V., 2009. Organic trace minerals and 25-hydroxycholecalciferol affect performance characteristics, leg abnormalities, and biomechanical properties of leg bones of turkeys. Poult. Sci. 88, 118–131.

Fernandez-Teran, M., Ros, M.A., 2008. The apical ectodermal ridge: morphological aspects and signaling pathways. Int. J. Dev. Biol. 52, 857–871.

Fleming, R.H., 2008. Nutritional factors affecting poultry bone health. Proc. Nutr. Soc. 67, 177–183.

Foster, J.W., Dominguez-Steglich, M.A., Guioli, S., Kwok, C., Weller, P.A., Stevanovi, M., Weissenbach, J., Mansour, S., Young, I.D., Goodfellow, P.N., Brook, J.D., Schafer, A.J., 1994. Campomelic dysplasia and autosomal sex reversal caused by mutations in an SRY-related gene. Nature 372, 525–530.

Fu, G., Yoshihara, H., Kawakami, N., Goto, M., Tsuji, T., Ohara, T., Imagama, S., 2011. Microcomputed tomographic evaluation of vertebral microarchitecture in pinealectomized scoliosis chickens. J. Pediatr. Orthop. B 20, 382–388.

Garcia-Martinez, V., Schoenwolf, G.C., 1992. Positional control of mesoderm movement and fate during avian gastrulation and neurulation. Dev. Dyn. 193, 249–256.

Gilbert, S.F., 2010. Developmental Biology, ninth ed. Sinauer Associates, Sunderland.

Goulding, M.D., Lumsden, A., Gruss, P., 1993. Signals from the notochord and floor plate regulate the region-specific expression of two pax genes in the developing spinal cord. Development 117, 1001–1016.

Herzog, A., Genin, O., Hasdai, A., Shinder, D., Pines, M., 2011. Hsp90 and angiogenesis in bone disorders-lessons from the avian growth plate. Am. J. Physiol. Regul. Integr. Comp. Physiol. 301, R140–R147.

Hirsinger, E., Duprez, D., Jouve, C., Malapert, P., Cooke, J., Pourquié, O., 1997. Noggin acts downstream of Wnt and Sonic Hedgehog to antagonize BMP4 in avian somite patterning. Development 124, 4605–4614.

Huang, R., Zhi, Q., Patel, K., Wilting, J., Christ, B., 2000. Dual origin and segmental organization of the avian scapula. Development 127, 3789–3794.

Jefferies, D., Houston, B., Lester, D., Whitehead, C.C., Thorp, B.H., Botman, M., Farquharson, C., 2000. Expression patterns of chondrocyte genes cloned by differential display in tibial dyschondroplasia. Biochim. Biophys. Acta 1501, 180–188.

Jin, S., Sell, J.L., Haynes, J.S., 2001. Effect of dietary vitamin K1 on selected plasma characteristics and bone ash in young turkeys fed diets adequate or deficient in vitamin D3. Poult. Sci. 80, 607–614.

Johnson, R.L., Laufer, E., Riddle, R.D., Tabin, C., 1994. Ectopic expression of sonic hedgehog alters dorsal-ventral patterning of somites. Cell 79, 1165–1173.

Kahane, N., Cinnamon, Y., Kalcheim, C., 1998. The origin and fate of pioneer myotomal cells in the avian embryo. Mech. Dev. 74, 59–73.

Kapell, D.N., Hill, W.G., Neeteson, A.M., McAdam, J., Koerhuis, A.N., Avendaño, S., 2012. Twenty-five years of selection for improved leg health in purebred broiler lines and underlying genetic parameters. Poult. Sci. 91, 3032–3043.

Kempf, H., Ionescu, A., Udager, A.M., Lassar, A.B., 2007. Prochondrogenic signals induce a competence for Runx2 to activate hypertrophic chondrocyte gene expression. Dev. Biol. 236, 1954–1962.

Knopov, V., Leach, R.M., Barak-Shalom, T., Hurwitz, S., Pines, M., 1995. Osteopontin gene expression and alkaline phosphatase activity in avian tibial dyschondroplasia. Bone 16, 329S–334S.

Knopov, V., Hadash, D., Hurwitz, S., Leach, R.M., Pines, M., 1997. Gene expression during cartilage differentiation in turkey tibial dyschondroplasia, evaluated by in situ hybridization. Avian Dis. 41, 62–72.

Koseki, H., Wallin, J., Wilting, J., Mizuanti, Y., Kispert, A., Ebensperger, C., Herrmann, B.G., Christ, B., Balling, R., 1993. A role for pax-1 as a mediator of notochordal signals during the dorsoventral specification of vertebrae. Development 119, 649–660.

Kumar, D., Lassar, A.B., 2009. The transcriptional activity of Sox9 in chondrocytes is regulated by RhoA signaling and actin polymerization. Mol. Cell Biol. 29, 4262–4273.

Lam, K.M., DaMassa, A.J., Ghazikhanian, G.Y., 2004. Mycoplasma meleagridis-induced lesions in the tarsometatarsal joints of turkey embryos. Avian Dis. 48, 505–511.

Leach, R.M., Gay, C.V., 1987. Role of epiphyseal cartilage in endochondral bone formation. J. Nutr. 117, 784–790.

Leach, R.M., Lilburn, M.S., 1992. Current knowledge on the etiology of tibial dyschondroplasia in the avian species. Poult. Sci. 4, 57–65.

Lefebvre, V., Bhattaram, P., 2010. Vertebrate skeletogenesis. Curr. Top. Dev. Biol. 90, 291–317.

Ley, D.H., Marusak, R.A., Vivas, E.J., Barnes, H.J., Fletcher, O.J., 2010. *Mycoplasma iowae* associated with chondrodystrophy in commercial turkeys. Avian Pathol. 39, 87–93.

Lidor, C., Atkin, I., Ornoy, A., Dekel, S., Edelstein, S., 1987. Healing of rachitic lesions in chicks by 24R,25-dihydroxycholecalciferol administered locally into bone. J. Bone Miner. Res. 2, 91–98.

Lilburn, M.S., Leach Jr, R.M., 1980. Metabolism of abnormal cartilage cells associated with tibial dyschondroplasia. Poult. Sci. 59, 1892–1896.

Ling, J., Kincaid, S.A., McDaniel, G.R., Waegell, W., 2000. Immunolocalization analysis of transforming growth factor-beta1 in the growth plates of broiler chickens with high and low incidences of tibial dyschondroplasia. Poult. Sci. 79, 1172–1178.

Liu, A.C., Heinrichs, B.S., Leach Jr, R.M., 1994. Influence of manganese deficiency on the characteristics of proteoglycans of avian epiphyseal growth plate cartilage. Poult. Sci. 73, 663–669.

Machida, M., Dubousset, J., Satoh, T., Murai, I., Wood, K.B., Yamada, T., Ryu, J., 2001. Pathologic mechanism of experimental scoliosis in pinealectomized chickens. Spine 26, E385–E391.

Makrai, L., Nemes, C., Simon, A., Ivanics, E., Dudás, Z., Fodor, L., Glávits, R., 2011. Association of *Enterococcus cecorum* with vertebral osteomyelitis and spondylolisthesis in broiler parent chicks. Acta Vet. Hung. 59, 11–21.

Marcelle, C., Stark, M.R., Bronner-Fraser, M., 1997. Coordinate actions of BMPs, Wnts, Shh and noggin mediate patterning of the dorsal somite. Development 124, 3955–3963.

Masse, P.G., Colombo, V.E., Gerber, F., Howell, D.S., Weiser, H., 1990. Morphological abnormalities in vitamin B6 deficient tarsometatarsal chick cartilage. Scanning. Microsc. 4, 667–673.

Masse, P.G., Pritzker, K.P., Mendes, M.G., Boskey, A.L., Weiser, H., 1994. Vitamin B6 deficiency experimentally-induced bone and joint disorder: microscopic, radiographic and biochemical evidence. Br. J. Nutr. 71, 919–932.

McCarrey, J.R., Abbott, U.K., Benson, D.R., Riggins, R.S., 1981. Genetics of scoliosis in chickens. J. Hered. 72, 6–10.

McNamee, P.T., Smyth, J.A., 2000. Bacterial chondronecrosis with osteomyelitis ('femoral head necrosis') of broiler chickens: a review. Avian Pathol. 29, 253–270.

McNamee, P.T., McCullagh, J.J., Thorp, B.H., Ball, H.J., Graham, D., McCullough, S.J., McConaghy, D., Smyth, J.A., 1998. Study of leg weakness in two commercial broiler flocks. Vet. Rec. 143, 131–135.

McNamee, P.T., McCullagh, J.J., Rodgers, J.D., Thorp, B.H., Ball, H.J., Connor, T.J., McConaghy, D., Smyth, J.A., 1999. Development of an experimental model of bacterial chondronecrosis with osteomyelitis in broilers following exposure to *Staphylococcus aureus* by aerosol, and inoculation with chicken anaemia and infectious bursal disease viruses. Avian Pathol. 28, 26–35.

Mitchell, R.D., Edwards Jr, H.M., 1996. Additive effects of 1,25-dihydroxycholecalciferol and phytase on phytate phosphorus utilization and related parameters in broiler chickens. Poult. Sci. 75, 111–119.

Mittapalli, V.R., Huang, R., Patel, K., Christ, B., Scaal, M., 2005. Arthrotome: a specific joint forming compartment in the avian somite. Dev. Dyn. 234, 48–53.

Monsoro-Burq, A.-H., 2005. Sclerotome development and morphogenesis: when experimental embryology meets genetics. Int. J. Dev. Biol. 49, 301–308.

Monsoro-Burq, A.-H., Duprez, D., Watanabe, Y., Bontoux, M., Vincent, C., Brickell, P., Le Douarin, N., 1996. The role of bone morphogenetic proteins in vertebral development. Development 122, 3607–3616.

Münsterberg, A.E., Kitajewski, J., Bumcrot, D.A., McMahon, A.P., Lassar, A.B., 1995. Combinatorial signaling by Sonic hedgehog and Wnt family members induces myogenic bHLH gene expression in the somite. Gen. Dev. 9, 2911–2922.

Murtaugh, L.C., Chyung, J.H., Lassar, A.B., 1999. Sonic hedgehog promotes somitic chondrogenesis by altering the cellular response to BMP signaling. Genes Dev. 13, 225–237.

Murtaugh, L.C., Zeng, L., Chyung, J.H., Lassar, A.B., 2001. The chick transcriptional repressor Nkx3.2 acts downstream of shh to promote BMP-dependent axial chondrogenesis. Dev. Cell 1, 411–422.

Newbrey, J.W., Baksi, S.N., Dhillon, A.S., Zimmerman, N.G., Truitt, S.G., Riedinger, R., 1988. Histomorphometry and vitamin D metabolism of valgus-varus deformity in broiler chickens. Avian Dis. 32, 704–712.

Newman, S.A., Christley, S., Glimm, T., Hentschel, H.G.E., Kazmierczak, B., Zhang, Y.-T., Zhu, J., Alber, M., 2008. Multiscale models for vertebrate limb development. Curr. Top. Dev. Biol. 81, 311–340.

Nimmagadda, S., Geetha-Loganathan, P., Scaal, M., Christ, B., Huang, R., 2007. FGFs, Wnts and BMPs mediate induction of VEGFR-2 (Quek-1) expression during avian somite development. Dev. Biol. 15, 421–429.

Niswander, L., Tickle, C., Vogel, A., Booth, I., Martin, G.R., 1993. FGF-4 replaces the apical ectodermal ridge and directs outgrowth and patterning of the limb. Cell 75, 579–587.

Olkowski, A.A., Laarveld, B., Wojnarowicz, C., Chirino-Trejo, M., Chapman, D., Wysokinski, T.W., Quaroni, L., 2011. Biochemical and physiological weaknesses associated with the pathogenesis of femoral bone degeneration in broiler chickens. Avian Pathol. 40, 639–650.

Ordahl, C.P., Le Douarin, N.M., 1992. Two myogenic lineages within the developing somite. Development 114, 339–353.

Orth, M.W., Cook, M.E., 1994. Avian tibial dyschondroplasia: a morphological and biochemical review of the growth plate lesion and its causes. Vet. Pathol. 31, 403–404.

Peters, H., Wilm, B., Sakai, N., Imai, K., Maas, R., Balling, R., 1999. Pax1 and Pax9 synergistically regulate vertebral column development. Development 126, 5399–5408.

Pines, M., Knopov, V., Genina, O., Hurwitz, S., Faerman, A., Gerstenfeld, L.C., Leach, R.M., 1999. Development of avian tibial dyschondroplasia: gene expression and protein synthesis. Calcif. Tissue. Int. 63, 521–527.

Pines, M., Hasdai, A., Monsonego-Ornan, E., 2004. Tibial dyschondroplasia – tools, new insights and future prospects. World's Poult. Sci. J. 61, 285–297.

Piran, R., Halperin, E., Guttmann-Raviv, N., Keinan, E., Reshef, R., 2009. Algorithm of myogenic differentiation in higher-order organisms. Development 136, 3831–3840.

Pourquié, O., 2011. Vertebrate segmentation: from cyclic gene networks to scoliosis. Cell 145, 650–663.

Pourquié, O., Coltey, M., Teillet, M.A., Ordahl, C., Le Douarin, N.M., 1993. Control of dorsoventral patterning of somitic derivatives by notochord and floor plate. Proc. Natl. Acad. Sci. U.S.A. 90, 5242–5246.

Pourquié, O., Fan, C.M., Coltey, M., Hirsinger, E., Watanabe, Y., Breant, C., Francis-West, P., Brickell, P., Tessier-Lavigne, M., Le Douarin, N.M., 1996. Lateral and axial signals involved in avian somite patterning: a role for BMP4. Cell 84, 461–471.

Primorac, D., Stover, M.L., Clark, S.H., Rowe, D.W., 1994. Molecular basis of nanomelia, a heritable chondrodystrophy of chicken. Matrix Biol. 14, 297–305.

Psychoyos, D., Stern, C.D., 1996. Fates and migratory routes of primitive streak cells in the chick embryo. Development 122, 1523–1534.

Rackard, S.M., Carr, A.J., Callanan, J.J., Bellenger, C.R., 2002. An avian model of limb deviation induced by periosteal surgery. Res. Vet. Sci. 73, 237–241.

Rath, N.C., Huff, W.E., Bayyari, G.R., Balog, J.M., 1998. Cell death in avian tibial dyschondroplasia. Avian Dis. 42, 72–79.

Rath, N.C., Huff, G.R., Huff, W.E., Balog, J.M., 2000. Factors regulating bone maturity and strength in poultry. Poult. Sci. 79, 1024–1032.

Rath, N.C., Huff, W.E., Balog, J.M., Huff, G.R., 2004. Comparative efficacy of different dithiocarbamates to induce tibial dyschondroplasia in poultry. Poult. Sci. 83, 266–274.

Rennie, S.J., Whitehead, C.C., Thorp, B.H., 1993. The effect of dietary 1,25-dihydroxycholecalciferol in preventing tibial dyschondroplasia in broilers fed on diets imbalanced in calcium and phosphorus. Br. J. Nutr. 69, 809–816.

Reshef, R., Maroto, M., Lassar, A.B., 1998. Regulation of dorsal somitic cell fates: BMPs and Noggin control the timing and pattern of myogenic regulator expression. Genes Dev. 12, 290–303.

Riddel, C., Classen, H.L., 1992. Effects of increasing photoperiod length and anticoccidials on performance and health of roaster chickens. Avian Dis. 36, 491–498.

Riddle, R.D., Johnson, R.L., Laufer, E., Tabin, C., 1993. Sonic hedgehog mediates the polarizing activity of the ZPA. Cell 75, 1401–1416.

Schweitzer, R., Chyung, J.H., Murtaugh, L.C., Brent, A.E., Rosen, V., Olson, E.N., Lassar, A., Tabin, C., 2001. Analysis of the tendon cell fate using Scleraxis, a specific marker for tendons and ligaments. Development 128, 3855–3866.

Sharir, A., Stern, T., Rot, C., Shahar, R., Zelzer, E., 2011. Muscle force regulates bone shaping for optimal load bearing capacity during embryogenesis. Development 138, 3247–3259.

Shim, M.Y., Karnuah, A.B., Anthony, N.B., Pesti, G.M., Aggrey, S.E., 2012. The effects of broiler chicken growth rate on valgus, varus, and tibial dyschondroplasia. Poult. Sci. 91, 62–65.

Shirley, R.B., Davis, A.J., Compton, M.M., Berry, W.D., 2003. The expression of calbindin in chicks that are divergently selected for low or high incidence of tibial dyschondroplasia. Poult. Sci. 82, 1965–1973.

Shwartz, Y., Farkas, Z., Stern, T., Aszo, A., Zelzer, E., 2012. Muscle contraction controls skeletal morphogenesis through regulation of chondrocyte convergent extension. Dev. Biol. 370, 154–163.

Smyth Jr., J.R., Sreekumar, G.P., Coyle, C.A., Bitgood, J.J., 2000. A new recessive ametapodia mutation in the chicken (*Gallus domesticus*). J. Hered. 91, 340–342.

Šošić, D., Brand-Saberi, B., Schmidt, C., Christ, B., Olson, E.N., 1997. Regulation of paraxis expression and somite formation by ectoderm- and neural tube-derived signals. Dev. Biol. 185, 229–243.

Stock, R.H., Latshaw, J.D., 1981. The effects of manganese, biotin, and choline on hexosamine and hydroxyproline content as related to leg weakness. Poult. Sci. 60, 1012–1016.

Teillet, M., Watanabe, Y., Jeffs, P., Duprez, D., Lapointe, F., Le Douarin, N.M., 1998. Sonic hedgehog is required for survival of both myogenic and chondrogenic somitic lineages. Development 125, 2019–2030.

Tian, W.X., Li, J.K., Qin, P., Wang, R., Ning, G.B., Qiao, J.G., Li, H.Q., Bi, D.R., Pan, S.Y., Guo, D.Z., 2013. Screening of differentially expressed genes in the growth plate of broiler chickens with tibial dyschondroplasia by microarray analysis. BMC Genomics 14, 276.

Toghyani, M., Toghyani, M., Shahryar, H.A., Zamanizad, M., 2011. Assessment of growth performance, immune responses, serum metabolites, and prevalence of leg weakness in broiler chicks submitted to early-age water restriction. Trop. Anim. Health Prod. 43, 1183–1189.

Tonegawa, A., Funayama, N., Ueno, N., Takahashi, Y., 1997. Mesodermal subdivision along the mediolateral axis in chicken controlled by different concentrations of BMP-4. Development 124, 1975–1984.

Trampel, D.W., Goll Jr, F., 1994. Outbreak of *Mycoplasma iowae* infection in commercial turkey poults. Avian Dis. 38, 905–909.

Vega, R.B., Matsuda, K., Oh, J., Barbosa, A.C., Yang, X., Meadows, E., McAnally, J., Pomajzl, C., Shelton, J.M., Richardson, J.A., Karsenty, G., Olson, E.N., 2004. Histone deacetylase 4 controls chondrocyte hypertrophy during skeletogenesis. Cell 119, 555–566.

Vortkamp, A., Lee, K., Lanske, B.L., Segre, G.V., Kronenberg, H.M., Tabin, C.J., 1996. Regulation of rate of cartilage differentiation by Indian hedgehog and PTH-related protein. Science 273, 613–622.

Wagner, T., Wirth, J., Meyer, J., Zabel, B., Held, M., Zimmer, J., Pasantes, J., Bricarelli, F.D., Keutel, J., Hustert, E., Wolf, U., Tommerup, N., Schempp, W., Scherer, G., 1994. Autosomal sex reversal and campomelic dysplasia are caused by mutations in and around the SRY-related gene. Cell 79, 1111–1120.

Webster, S.V., Farquharson, C., Jefferies, D., Kwan, A.P., 2003. Expression of type X collagen, Indian hedgehog and parathyroid hormone related-protein in normal and tibial dyschondroplastic chick growth plates. Avian Pathol. 32, 69–80.

Whitehead, C.C., McCormack, H.A., McTeir, L., Fleming, R.H., 2004. High vitamin D3 requirements in broilers for bone quality and prevention of tibial dyschondroplasia and interactions with dietary calcium, available phosphorus and vitamin A. Br. Poult. Sci. 45, 425–436.

Wideman Jr., R.F., Pevzner, I., 2012. Dexamethasone triggers lameness associated with necrosis of the proximal tibial head and proximal femoral head in broilers. Poult. Sci. 91, 2464–2474.

Wideman, R.F., Prisby, R.D., 2013. Bone circulatory disturbances in the development of spontaneous bacterial chondronecrosis with osteomyelitis: a translational model for the pathogenesis of femoral head necrosis. Front. Endocrinol. (Lausanne) 3, 183–195.

Williams, B., Waddington, D., Solomon, S., Farquharson, C., 2000. Dietary effects on bone quality and turnover, and Ca and P metabolism in chickens. Res. Vet. Sci. 69, 81–87.

Williams, B., Solomon, S., Waddington, D., Thorp, B., Farquharson, C., 2000a. Skeletal development in the meat-type chicken. Br. Poult. Sci. 41, 141–149.

Williams, B., Waddington, D., Murray, D.H., Farquharson, C., 2004. Bone strength during growth: influence of growth rate on cortical porosity and mineralization. Calcif. Tissue. Int. 74, 236–245.

Wyers, M., Cherel, Y., Plassiart, G., 1991. Late clinical expression of lameness related to associated osteomyelitis and tibial dyschondroplasia in male breeding turkeys. Avian Dis. 35, 408–414.

Xu, X., Weinstein, M., Li, C., Naski, M., Cohen, R.I., Ornitz, D.M., Leder, P., Deng, C., 1998. Fibroblast growth factor receptor 2 (FGFR2)-mediated reciprocal regulation loop between FGF8 and FGF10 is essential for limb induction. Development 125, 753–765.

Yonei-Tamura, S., Endo, T., Yajima, H., Ohuchi, H., Ide, H., Tamura, K., 1999. FGF7 and FGF10 directly induce the apical ectodermal ridge in chick embryos. Dev. Biol. 211, 133–143.

Yoshida, C.A., Yamamoto, H., Fujita, T., Furuichi, T., Ito, K., Inoue, K., Yamana, K., Zanma, A., Takada, K., Ito, Y., Komori, T., 2004. Runx2 and Runx3 are essential for chondrocyte maturation, and Runx2 regulates limb growth through induction of Indian hedgehog. Genes Dev. 18, 952–963.

Zeng, L., Kempf, H., Murtaugh, L.C., Sato, M.E., Lassar, A.B., 2002. Shh establishes an Nkx3.2/Sox9 autoregulatory loop that is maintained by BMP signals to induce somitic chondrogenesis. Genes Dev. 16, 1990–2005.

Zhang, C., Li, D., Wang, F., Dong, T., 2003. Effects of dietary vitamin K levels on bone quality in broilers. Arch. Tierernahr. 57, 197–206.

Zou, H., Choe, K.M., Lu, Y., Massagué, J., Niswander, L., 1997. BMP signaling and vertebrate limb development. Cold Spring Harb. Symp. Quant. Biol. 62, 269–272.

Chapter 16

Skeletal Muscle

Sandra G. Velleman and Douglas C. McFarland
The Ohio State University/OARDC, Wooster, OH, USA, South Dakota State University, Brookings, SD, USA

16.1 INTRODUCTION

This chapter covers the development and growth of avian skeletal muscle, beginning with a discussion of the structural diversity of skeletal muscle in different bird species. The chapter then proceeds to overview the embryonic origins of muscle, posthatch development, and growth, with a focus placed on the satellite cells. Satellite cells are adult myoblast stem cells that are responsible for all posthatch muscle growth. The next portion of the chapter deals with muscle fiber types, the mechanism of skeletal muscle contraction, and role of myogenic transcriptional regulatory factors. New emerging areas in avian skeletal muscle biology are also discussed, including satellite cell heterogeneity, extracellular matrix regulation of muscle growth and development, maternal inheritance of muscle morphological structure in turkeys, and the identification of novel genes involved in avian myogenesis.

16.2 DIVERSITY OF AVIAN SKELETAL MUSCLE

The ability to fly has allowed avian species the widespread utilization of many diverse environments and habitats. Perhaps due to this mobility, they may be the most successful terrestrial vertebrate in terms of numbers of species and body forms (Welty, 1982). There are approximately 8900 living species of birds compared to 3000 amphibians, 6000 reptiles, and 4100 mammals. The structure of the musculoskeletal system of birds varies tremendously among species depending on their flight characteristics or whether they are flightless. Most prominent on the sternum of flying birds is a large keel structure to which the powerful flight muscles are attached. Although the penguin is unable to fly, the keel serves as an anchor for muscles controlling the vestigial wings called flippers used for swimming. Flightless birds either lack a keel (ostrich) or the structure is minimal. Other skeletal structures peculiar to birds include a greatly lengthened pelvis fused with the synsacrum, covering approximately half the length of the body. Consequently, there is relatively little dorsal musculature in birds. The majority of the muscle tissue in birds is on the ventral side of the torso, but great differences exist among species. For instance, birds such as hummingbirds and swallows are highly accomplished fliers, but they are limited on the ground; approximately 25–35% of their body weight is devoted to flight musculature and only about 2% to the leg muscles. On the other hand, shore birds that rely more on running than flying have larger leg muscles than wing muscles. Additionally, raptors, which rely on strong grasping of prey, have very strong leg muscles.

The principal flight muscles are the large pectoralis and the supracoracoideus. In some birds, these two muscles may account for 25% of the total mass of the animal. In commercial poultry, genetic selection has largely focused on the size of these muscles because they are the most economically important part of the carcass. Both of these muscles arise from the keel with the supracoracoideus lying dorsal to and under the pectoralis muscle. The pectoralis attaches to the humerus and is responsible for depressing the wings. The supracoracoideus attaches to a tendon that passes dorsally through a foramen and attaches to the dorsal surface of the humerus. This "rope-and-pulley" arrangement allows for the upward movement of the wings. Additionally, there are many other smaller muscles that cause changes in the angle of the wings.

Throughout the skeletal muscular systems of birds, there are many variations in structure that allow this diverse class of vertebrates to adapt to environmental niches and feed sources. For instance, the jaw closing muscles of birds that feed on coarse seeds are stronger than those that feed on smaller grains or noncoarse materials. As will be discussed below, movement requirements of each muscle must be properly matched with the appropriate type of contractile fibers, metabolic characteristics, and neural signal.

Members of the class *Aves* are classified as either precocial or altricial in their developmental patterns. Precocial birds hatch with open eyes and a coat of down feathers; they leave the nest within 2 days. Altricial birds hatch with closed eyes, with little or no down feathers, and must be fed by parents. Included in the latter group are passerine birds, the largest order of birds. Additionally, there are birds that fall in between these two classification, which are referred to as semiprecocial or semialtricial. Among the precocial birds include aquatic species such as gulls, terns, ducks, geese, and rails. These species are able to run and swim in the first day following hatching. However, they cannot fly until fully grown (Dial and Carrier, 2012). The early maturation of the leg skeletal muscle system versus the wings allows these species to escape predators without relying on the ability to fly. Although not capable of walking following hatching, altricial birds are capable of performing alternating stepping movements following hatch, indicating that the locomotor program is in place early in development (Muir, 2000). Some of the differences between precocial and altricial species are explained by the immaturity of the neurological system in the latter species (Oppenheim, 1972). However, the sequence of movements leading to pipping and emergence from the egg remains the same in all avian species examined except for members of the family Megapodiidae (Oppenheim, 1972).

It has long been known that thyroid hormones play an important role in skeletal muscle development. Using the precocial Japanese quail and altricial Ring doves, McNabb et al. (1984) compared thyroid function between these two species. It was found that thyroid function and body growth were quite different between the quail and doves despite similar incubation periods, adult body weights, and adult serum thyroid hormone levels. The precocial quail thyroid activity was high in the perinatal period, declined shortly following hatching, and then gradually increased to adult levels. The altricial doves lacked a perinatal peak in thyroid activity; in fact, levels were low at hatching, increasing steadily to adult levels. It is likely that the differences in thyroid activity profiles and maturity of the neurological system play major roles in the marked differences in precocial and altricial skeletal muscle growth patterns.

16.3 EMBRYONIC ORIGINS OF SKELETAL MUSCLE

There are many excellent books on the embryology of vertebrates, including birds. Therefore, this section will focus only on those cells of the embryo that lead to the development of the skeletal musculature (for a review, see Gilbert, 2000). Following gastrulation, or the transformation of the spherical blastula into an elongated tubular structure, three primary germ layers are defined. The outer structure comprises the ectoderm and the inner surface comprises the endoderm. Between these layers is the mesoderm, which is the source of cells that lead to the formation of the skeletal muscle and other internal structures. As the embryo elongates due to lengthening of the structure called the notochord, the mesoderm grows as "slabs" on both sides of the notochord with the ectoderm above and the endoderm below. The mesodermal cells continue to proliferate during the elongation period. Eventually, the cells separate into blocks of cells called somites beginning at the anterior end and terminating at the posterior of the embryo. The cells of the somites become different structures depending on their location. Cells that are near the notochord are termed the sclerotome and eventually give rise to the vertebrae and ribs. Cells that are adjacent to the dorsal surface are the dermatome and give rise to the dermis of the trunk. The cells that lie between the sclerotome and dermatome are termed the myotome; they give rise to most of the skeletal muscles of the body.

At appropriate gestational ages, the primordial muscle cells (embryonic myoblasts) migrate out of the somites into future muscle beds and continue to proliferate. Some of the myoblasts fuse with one another and form primary muscle fibers. They have centrally located nuclei and form myofibrils. The remaining single-celled myoblasts continue to proliferate and eventually begin to fuse to form the secondary fibers. The secondary fibers form alongside the primary fibers, which serve to orient the developing muscles so that they will function properly. In most animals, fiber formation is essentially complete at the time of birth or hatching. Subsequent muscle growth is due to enlargement of the existing fibers and is called hypertrophy. The details as to how hypertrophy occurs remained a mystery until the 1960s; much of what is now known about posthatch or postnatal muscle growth was gained from studies with chickens. Smith (1963) demonstrated that fiber number increases occurred in chickens prior to hatching and subsequent growth was due to hypertrophy. The importance of initial fiber number following hatching was further demonstrated by studies with different lines of chickens. A growth-selected heavy strain of chickens had a greater number of fibers that were slightly smaller in diameter at hatch than a layer strain of chickens. Following 10 weeks of growth, fiber diameters of the growth selected line were much greater than those of the layer strain (Smith, 1963).

One of the major puzzles of muscle growth in the early years of this research was what the source of the nuclei is in growing and regenerating skeletal muscle. Were the myonuclei within the existing fibers dividing or were there other sources of nuclei? To answer this question, Stockdale and Holtzer (1961) isolated chick embryonic myoblasts and allowed them to fuse to form immature muscle fibers called myotubes in culture. Tritiated thymidine was administered to the cultures, but no mitotic figures or radiolabel were seen within the myotubes. However, if tritiated thymidine was administered to proliferating cells, labeled nuclei were detected. These findings demonstrated that myonuclei were incapable of mitotic division and that there was another cell responsible for DNA accretion. While it appeared clear that embryonic myoblasts were likely responsible for the development of muscle fibers prenatally, the source of cells leading to postnatal growth was not known.

16.4 POSTNATAL OR POSTHATCH SKELETAL MUSCLE DEVELOPMENT

In 1961, Alexander Mauro wrote a brief note about a cell that resided between the plasma membrane (or sarcolemma) and the basement membrane (or basal lamina) of frog muscle fibers (Mauro, 1961). Electron micrographic examination showed that the cell mass consisted mostly of nuclei, and there were little cytoplasm and few organelles such as mitochondria or golgi apparatus. Dr Mauro termed these cells "satellite cells" due to their peripheral location on muscle fibers. He speculated that satellite cells might be somehow activated and aid in the repair of damaged muscles. After conferring with other colleagues, he learned that these cells were also present in rat skeletal muscle, but not cardiac muscle. Were the satellite cells responsible for postnatal or posthatch growth of skeletal muscle? Evidence for the role of satellite cells in postnatal growth was reported by Moss and Leblond (1971). These researchers injected tritiated thymidine into young growing rats and followed the label in the muscle tissue. One hour following injection, the label was localized in satellite cells. For the next 24 h, the number of labeled satellite cells doubled, indicating that the satellite cells were dividing. Following this, label began to appear within myonuclei, indicating that the satellite cells had fused with the adjacent fibers. These observations established the key role of satellite cells in DNA accretion of muscle fibers.

The importance of DNA accretion in fiber hypertrophy was demonstrated by studies conducted by Moss (1968). In this work, it was shown that there was a constant ratio of muscle fiber diameter to the DNA content of the fiber. These findings demonstrated that DNA accretion via satellite cell incorporation was required for the growth of skeletal muscles. Although Mauro's initial observations reported inactive or dormant satellite cells, further work in other laboratories identified activated and dividing satellite cells, particularly in young animals (Snow, 1977). These cells had euchromatic nuclei and a prominent nucleolus. The cytoplasmic volume was larger and there were polyribosomes, more mitochondria, some rough endoplasmic reticulum, and Golgi apparatus. When examining muscles of older rats (8–30 months of age) Snow (1977) reported that the nuclei exhibited heterochromatin and a higher nuclear to cytoplasmic ratio compared to the young rats. The rough endoplasmic reticulum was fewer in number and the Golgi apparatus was poorly developed—all characteristics of dormant or inactive satellite cells. Schultz et al. (1978) injected tritiated thymidine into older mice and noted that there was a lack of radiolabel within the nuclei of satellite cells and the frequency of satellite cells was diminished compared with younger animals. In spite of the inactive state of satellite cells in mature animal muscles and their diminished numbers, satellite cells residing in older animals do have the capacity to become active and proliferate (Allen et al., 1980), but the proliferation potential is lower (Schultz and Lipton, 1982).

To determine if satellite cells were indeed involved in regeneration of damaged skeletal muscle as was proposed by Mauro (1961), Bischoff (1975) isolated individual muscle fibers from rat leg muscles and placed them in culture medium. During the first several hours, degeneration within the basement membrane occurred. Contraction clots formed and the myonuclei did not undergo mitosis. However, the satellite cells attached to the fibers began to enlarge, proliferate, and fuse and form myotubes within the sarcolemma. Eventually, the regenerated fibers exhibited contractions. These and other studies (Carlson and Faulkner, 1983; Ontell, 1986; Schultz, 1989) demonstrated the important role of satellite cells in skeletal muscle regeneration following damage. Furthermore, satellite cells have been observed to pass through the basement membrane, fuse with adjacent muscle fibers, and contribute to repair of fibers they were not associated with (Hughes and Blau, 1990).

Satellite cell activation is influenced by nutrition and exercise. The proportion of fiber nuclei residing in satellite cells versus myonuclei is diminished with malnutrition (Hansen-Smith et al., 1979). Furthermore, the satellite cells of malnourished individuals are largely quiescent. Following nutritional recovery, satellite cells enlarge and become mitotically active (Hansen-Smith et al., 1978). Darr and Schultz (1987) utilizing an untrained exercised rat model, demonstrated that exercise induced activation of satellite cells. Accompanying exercise was a small percentage of necrotic fibers (<3%). However, the extent of satellite cell activation was greater than that needed to support fiber regeneration. Microscopic examination of these muscles revealed fibers with centrally located myonuclei, which are characteristic of newly repaired muscle fibers.

16.5 SKELETAL MUSCLE GROWTH

Much of what has been learned about the factors regulating the proliferation and differentiation of satellite cells has been derived from cell culture studies and the development of serum-free media formulations for various species. Satellite cells were first isolated and cultured by Bischoff (1974) from the skeletal muscles of adult rats. The cells were capable of proliferating and differentiating into multinucleated myotubes. Since then, satellite cells have been isolated and cultured from many species including humans (Blau and Webster, 1981) and the major agriculturally important animals (for a review, see Dodson et al., 1996), including the chicken (Johnson et al., 1983; Matsuda et al., 1983; Yablonka-Reuveni et al., 1987) and turkey (McFarland et al., 1988).

As the serum component of typical cell culture media formulations contains variable levels of known and unknown hormones and growth factors that may influence cellular activity, they are of limited value in establishing cell requirements for proliferation, differentiation, and other activities. Serum-free medium formulations have been developed for chick embryonic myoblasts (Dollenmeier et al., 1981), turkey satellite cells (McFarland et al., 1991), and both chicken and turkey satellite cells (McFarland et al., 2011). Using these tools, researchers have been able to examine the effects of individual and combinations of growth factors on proliferation and differentiation. Similarities have emerged between findings with avian species and with mammals. The most studied growth factors are the insulin-like growth factors (IGF), fibroblast growth factors (FGF), platelet-derived growth factors (PDGF), hepatocyte growth factor (HGF), transforming growth factor-beta (TGF-β), and myostatin. At the present time, the role, if any, of epidermal growth factor (EGF) in avian skeletal muscle is unclear. McFarland et al. (1993) reported that EGF in combination with either IGF or FGF or both had no effect on turkey embryonic myoblast or satellite cell proliferation. However, EGF does stimulate proliferation of chick smooth muscle cells (Topouzis and Majesky, 1996) and stimulated chick cardiac cell development and function (Lau, 1993, 1994; Rabkin, 1996). A more comprehensive review of the influence of growth factors on avian satellite cells has been published previously (McFarland, 1999). The following is an overview of major findings and more recent material regarding growth factor effects on satellite cells.

Among the initial reports of IGF effects on avian skeletal muscle cells was a report by Schmid et al. (1983). This study reported that the two major forms of IGF (IGF-I and IGF-II) were equipotent in enhancing differentiation and fusion of chick embryonic myoblasts. In turkey embryonic myoblasts, however (McFarland et al., 1993), IGF-II was more potent than IGF-I in stimulating proliferation and differentiation. Exogenously added IGF-I or IGF-II were equipotent in stimulating proliferation of turkey satellite cells, but had no effect on differentiation. The presence of FGF was required for proliferation of either turkey muscle cell types to occur. FGF alone stimulated proliferation of both cell types, and there was a synergistic effect with both cells by the addition of FGF and IGF. Highest rates of proliferation of turkey satellite cells and embryonic myoblasts was seen with the combination of FGF, IGF, and PDGF. As the name implies, IGF have insulin-like activities, in both mammals and birds. For instance, IGF stimulates glucose uptake in the turkey (McFarland et al., 1994) and the chicken (Duclos et al., 1993a) satellite cells. Additionally, IGF stimulates protein synthesis and inhibits protein degradation in myotubes derived from chick embryonic myoblasts (Janeczko and Etlinger, 1984), chick satellite cells (Duclos et al., 1993b), and turkey embryonic myoblasts and satellite cells (McFarland et al., 1994). Recent evidence (Shinichi et al., 2012) demonstrated that polymorphism of the IGF-I gene is associated with pectoralis major muscle yields in chickens. This suggests that marker-assisted selection may be a useful tool to aid in improving breast muscle yields.

The FGFs are a large group of polypeptide growth factors with diverse actions on cells and tissues. Fibroblast growth factor is an important mitogen for avian skeletal muscle cells. In fact, omission of FGF from serum-free media results in no proliferation of turkey satellite cells or embryonic myoblasts, even if IGF, PDGF, or EGF is present (McFarland et al., 1993). Much of the muscle studies with FGF have focused on FGF-1 and FGF-2. Fibroblast growth factor 2 is a more potent mitogen for turkey (McFarland et al., 1993) and chicken (Wilkie et al., 1995) satellite cells than FGF-1. Fibroblast growth factor also acts as an inhibitor of differentiation. Addition of FGF-2 to cultures inhibits differentiation of chicken (McFarland et al., 1997a) and rodent (McFarland et al., 2000) satellite cells.

Platelet-derived growth factor is the major mitogen found in serum, where it is derived from the platelets following clotting of blood. Platelet-derived growth factor is also produced by many types of cells in the body, including skeletal muscle cells (Sejersen et al., 1986). In addition, PDGF was shown to be a potent mitogen for chick embryonic myoblasts (Yablonka-Reuveni and Seifert, 1993). It was shown that the BB isomer was the most potent, followed by the AB isomer, whereas the AA isomer was inactive. Similar findings were reported for the turkey embryonic myoblast and satellite cell (Ye et al., 1996). Platelet-derived growth factors also have a chemotactic effect on myoblasts. Using embryonic myoblasts isolated from Japanese quail (*Coturnix*) embryos, Venkatasubramanian and Solursh (1984) demonstrated that these cells migrated toward a gradient of PDGF, whereas nonmyogenic limb mesenchyme cells did not.

Early research by Bischoff and others described a mitogenic activity towards rodent satellite cells associated with saline extracts of crushed muscles. This so-called crushed muscle mitogen was capable of activating quiescent satellite cells residing on isolated muscle fibers (Bischoff, 1986). Using quiescent cultured satellite cells, Tatsumi et al. (2001) demonstrated that stretched cells entered the cell cycle earlier than quiescent cells cultured in static conditions. Furthermore, the conditioned media from the stretched cells activated quiescent cells. Stretch or injury released preexisting HGF from the extracellular matrix into the media. The administration of anti-HGF antibodies abolished the stretch activation of satellite cells and demonstrated that the crushed muscle mitogen was, indeed, HGF.

Hepatocyte growth factor exists in the extracellular matrix in a pro-HGF inactive form and is released in the active form by nitric oxide during the stretching process (Tatsumi and Allen, 2004). Gal-Levi et al. (1998) demonstrated that HGF affected chicken satellite cells as well. Hepatocyte growth factor increased DNA synthesis and decreased differentiation of cultured cells. Inhibition of differentiation was accomplished by inhibiting the activity of myogenic determination factors and subsequent muscle-specific protein expression. It was proposed that increasing proliferation and decreasing differentiation of satellite cells would increase the satellite cell pool size and eventually lead to more fiber formation. Hepatocyte growth factor has similar effects on the proliferation and differentiation of turkey satellite cells (Zeng et al., 2002). Addition of anti-HGF antibodies to turkey satellite cells in serum-free medium decreased proliferation, supporting the role of HGF as an autocrine or paracrine factor (i.e., locally-acting growth factor) in muscle. It was also reported that HGF has chemotactic activity toward satellite cells (Bischoff, 1997). Using chick embryos, Brand-Saberi et al. (1996) reported that HGF causes detachment of myogenic precursor cells from the somites and is therefore thought to be important in initiating the migration of these cells into muscle beds in the developing embryo.

An unknown factor secreted into media over rat liver cells was shown to inhibit the differentiation of rat L6 muscle cells (Evinger-Hodges et al., 1982). This "differentiation inhibitor" was studied in several laboratories and Florini et al. (1986) determined that this substance was identical to TGF-β. The mode of action appeared to be by blocking expression of myogenic differentiation genes, such as myogenin (Massague et al., 1991). The detection of TGF-β1 and TGF-β2 in chick embryos by day 1.5 of incubation (Jakowlew et al., 1994) and TGF-β4 expression by day 4 of incubation (Jakowlew et al., 1992) suggested that TGF-β is important in early development of avian species. In fact, these three forms of TGF-β were detected in all embryonic tissues examined and expression increased with embryonic age. Yun et al. (1997) reported that TGF-β1 inhibited both proliferation and differentiation of turkey satellite cells. Previously, Schofield and Wolpert (1990) demonstrated that, when added alone, TGF-β1 and TGF-β2 had no effect on chick embryonic myoblast differentiation. However, when added with FGF, TGF-β blocks the FGF inhibition of myoblast differentiation and the cells differentiate. It is not known whether the differences seen between the turkey and chicken cells represents species variation, media conditions used, or are due to the developmental ages of the cells.

Myostatin is a member of the TGF-β family of polypeptides. Defects in the myostatin gene result in dramatic increases in muscle mass seen in Belgian Blue and Piedmontese breeds of cattle (McPherron and Lee, 1997). Likewise, myostatin-null mice demonstrate enhanced musculature (McPherron et al., 1997). Srinivasan et al. (2004) demonstrated that levels of myostatin are elevated during muscle atrophy and Price et al. (2011) reported that photo-stimulated migrant sparrows had increased expression of myostatin. It is believed that elevated myostatin during muscle breakdown serves to prevent premature satellite cell activation while degeneration occurs. When muscle regeneration commences, myostatin levels diminish (Kirk et al., 2000). *In vitro* studies have shown that myostatin inhibits mouse C2C12 muscle cell proliferation, DNA synthesis, and protein synthesis but has no effect on protein degradation or apoptosis (Taylor et al., 2001). Administration of myostatin to muscle explants inhibits satellite cell activation and progression into the cell cycle (McCroskery et al., 2003).

Proliferation of both turkey embryonic myoblasts and satellite cells was inhibited between 26 and 45% in serum-free medium containing 20 ng/mL myostatin (McFarland et al., 2006). Myostatin also caused marked depression of differentiation in turkey satellite cells but had no effect on the differentiation of turkey embryonic myoblasts. Additionally, expression of decorin, an inhibitor of muscle growth, increased in the differentiating cells. Satellite cells isolated from the pectoralis major muscles of chickens were more responsive to the proliferation depressing effects of myostatin compared to cells from the biceps femoris (McFarland et al., 2007). Myostatin inhibited differentiation of satellite cells derived from the pectoralis major muscle but had no effect on the cells from the biceps femoris muscle. Administration of antimyostatin antibodies to proliferating cultures increased cell proliferation by 6–7% over 3 days, supporting a role of endogenous myostatin in autocrine regulation of muscle growth. Antimyostatin had no effect on differentiation of either pectoralis major or biceps femoris cells.

The above growth factors are among the most widely studied factors influencing satellite cell activity and muscle growth. Other factors influencing satellite cell or myoblast growth include granulocyte colony-stimulating factor (G-CSF). Receptors for G-CSF are expressed in developing somites and both the receptor and G-CSF are expressed in mouse embryonic myoblasts (Mie et al., 2011). Granulocyte colony-stimulating factor is also important in regeneration of damaged muscle because its receptor is transiently expressed in injured mouse muscle. Neutralizing antibodies to G-CSF impaired regeneration of damaged muscle. The G-CSF is important in muscle development and repair. Another factor, bone morphogenic protein (BMP), also appears to be an important influence on muscle growth (Ono et al., 2011). Bone morphogenic protein 4 stimulates satellite cell proliferation and inhibits differentiation. Addition of the BMP antagonist Noggin induced precocious differentiation of the cells. During satellite cell differentiation, Noggin expression increases and proliferation ceases and differentiation and cell fusion occurs. The authors conclude that BMP signaling occurs during expansion of the satellite cell pool and at the appropriate time Noggin is activated to antagonize BMP to initiate differentiation.

16.6 SKELETAL MUSCLE FIBER TYPES

Muscle fibers differ greatly in terms of contraction rate, metabolism, and function. There is generally accepted to be three fiber types: (1) slow-twitch, oxidative (SO), type I or red fiber; (2) fast-twitch glycolytic (FG), type IIB or white fiber; and (3) fast-twitch, oxidative/glycolytic (FOG), type IIA or intermediate fiber (Allen and Goll, 2003). The fiber types housed within muscle groups varies depending on the particular requirements for motion. Additionally, the same anatomical muscle may differ greatly in fiber composition between species. For instance, when comparing the pectoralis major muscle of a duck and a chicken or a pheasant, it is obvious that the duck pectoralis major muscle is primarily an oxidative muscle due to its high proportion of myoglobin, giving it a red color. This powerful flight muscle ideally suits the duck for long, sustained flights of several hundred miles as required for migration. In migratory wild birds, the muscles will have a higher proportion of type I fibers compared to type IIb (Ashmore et al., 1972). The type I fibers have a higher aerobic capacity and higher blood supply necessary for sustained activity. In contrast, the pectoralis major muscle of the domestic chicken is composed of type IIb fibers and has a lower blood supply because prolonged flight activity is not necessary and the muscle will fatigue more quickly.

Aerobic muscles oxidize glycogen and glucose through glycolysis and the tricarboxylic acid cycle and oxidize fatty acids via β-oxidation. Consequently, these muscles contain greater numbers of mitochondria, smaller fiber diameter to facilitate oxygen diffusion, greater triglyceride levels, and greater capillary density. The chicken or pheasant pectoralis major muscle is primarily composed of glycolytic fast twitch muscles and myoglobin levels are much lower, giving the muscle a white color. Glycolytic muscles oxidize glycogen and glucose to lactate and the lactate is returned to the liver for gluconeogenesis (Cori cycle). Therefore, much less energy is derived from oxidizing glucose in these muscles, so energy is expended quickly and serves to provide a rapid (but short) burst of activity.

In addition to differences in energy metabolism between fiber types, contraction properties are determined by the protein isoforms contained in the contractile apparatus of the fibers. Fast-twitch fibers contain myosin isoforms with fast ATPase activity (generating energy quickly for rapid movements) and slow-twitch fibers contain myosin isoforms with slow ATPase activity (generating energy more slowly for sustained movements). The functional unit of movement is the motor unit consisting of the neural cell body residing in the dorsal root of the spinal cord and its axon, which extends to the muscle being innervated. The single axon splits into numerous branches (a few to hundreds) and each branch innervates an individual muscle fiber. Motor units are interspersed throughout regions of the muscle group, so individual motor units influence the contractility of a wide area (Lieber, 1992). Patterns of action potentials generated from nerves innervating slow muscle fibers differ from those that innervate fast muscles. Nerve transplantation studies have revealed that these patterns influence myosin isoform expression (Cerny and Bandman, 1987). When fast muscles were innervated by nerves normally innervating slow muscles, fast myosin isoforms were replaced by slow myosin isoforms. However, neural stimulation is not the only factor affecting the contractile fate of muscles. Fiber type is also influenced by the myoblast lineage that makes up the muscle fiber (Dimario and Stockdale, 1997).

Cross-sectional area of muscle fibers is influenced by muscle fiber type. Type I muscles have a smaller cross-sectional area than type IIb fibers (Ashmore et al., 1972). The cross-sectional area of muscle fibers is not a static situation and will change based upon usage. The flight muscles of migratory birds change in fiber diameter in anticipation of migration and during the migratory flight (Price et al., 2011). For example, during a long migratory flight, the muscle fibers may decrease in diameter to coincide with a lighter bird weight as fat is oxidized (Lindström et al., 2000). During a stopover or rest period, the muscles may rebuild for the next flight and increase in diameter (Landys-Cianneli et al., 2003).

16.7 MUSCLE STRUCTURE AND CONTRACTION

Muscle is surrounded by a connective layer, the epimysium (Figure 16.1(A)), which separates individual muscles. Within the muscle, the muscle fibers form bundles or fasiculi. The bundles are separated from each other by a connective tissue layer called the perimysium. Muscle bundles can vary in size from 50 to up to 300 muscle fibers per bundle. Each muscle fiber is separated by the endomysial connective tissue layer. The connective tissue layers are all joined together at the myotendinous junction and thus are not independent from each other. The interconnection of the three connective tissue layers provides a strong structural framework for the muscle and contains capillary beds and water-holding molecules (Allen and Goll, 2003).

Mature muscle fibers can vary in their length and diameter. Skeletal muscle is multinucleated with myonuclei located at the periphery of the muscle fiber just underneath the sarcolemma. The sarcolemma is the plasma membrane of the muscle cell and contains invaginations into the muscle fiber called the transverse tubule system. Invaginations of the transverse tubule system into the muscle fiber occur at regular intervals along the sarcomere or contractile unit of the muscle. This system allows for calcium to be rapidly released upon sufficient membrane depolarization (Allen and Goll, 2003).

Muscle contraction is initiated by the myosin heads attaching to the actin thin filaments of the sarcomere. For a review of the muscle contraction process, see Alberts et al. (2008). Adenosine triphosphate is hydrolyzed to adenosine diphosphate, leading to the actin filaments being pushed toward the center of the sarcomere (Figure 16.1(B) and (C)). The distance from one Z band to another Z band constitutes an individual sarcomere. The thin filaments are composed primarily of actin, tropomyosin, troponins, tropomodulin, and nebulin. The filaments are anchored to the Z band through α-actinin and CapZ. The end of the thin filament closest to the M-line is capped with tropomodulin which, in part, regulates actin filament length (Fowler et al., 1993).

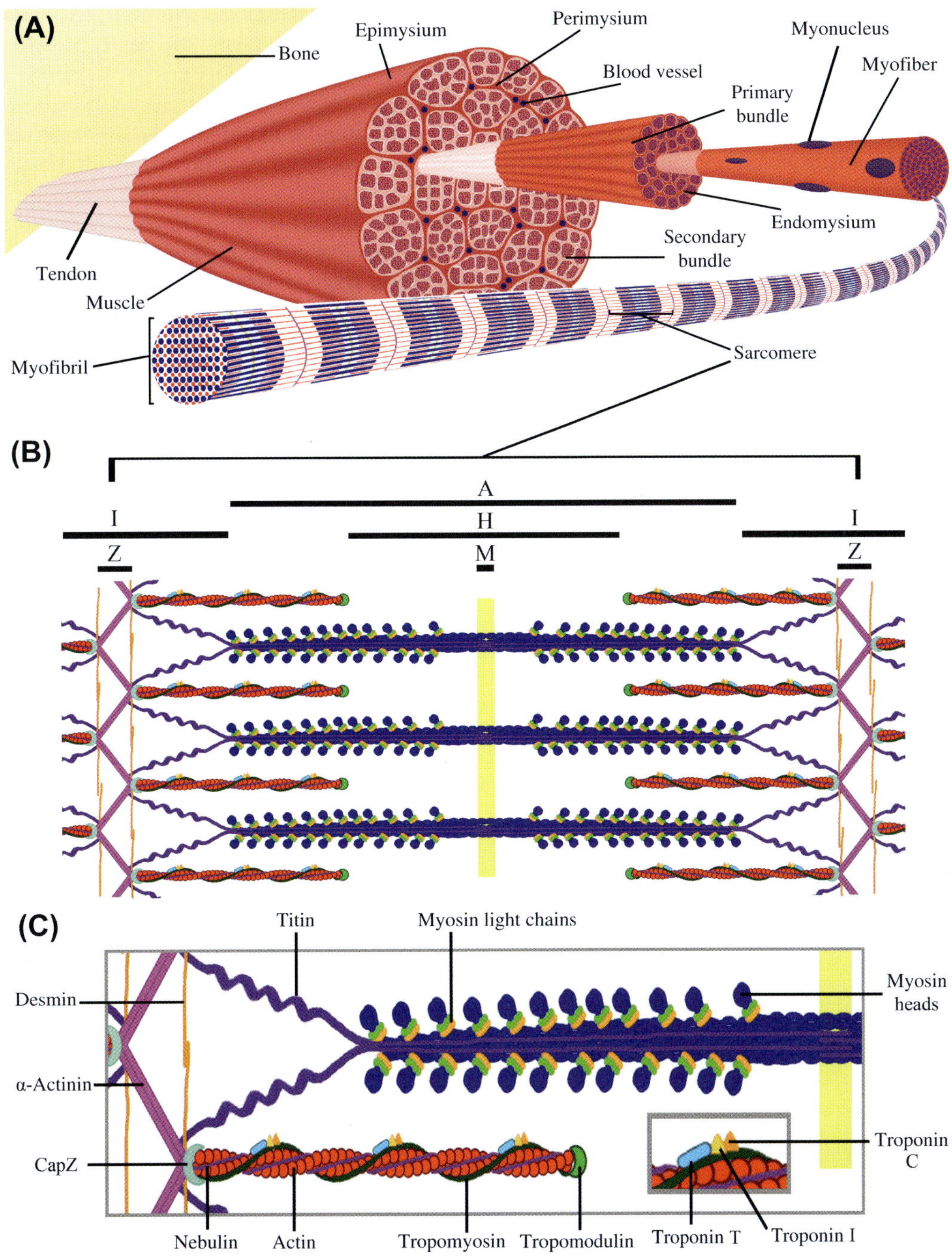

FIGURE 16.1 Schematic of muscle structure. (A) Overview of a cross-sectional area of muscle highlighting muscle fiber structure. (B) Sarcomere structure. (C) The myosin-actin overlap with the associated molecules.

Nebulin is an actin-binding protein. Nebulin knockout mice have reduced actin filament length and impaired contractile properties (Bang et al., 2006). In the I zone, the actin filaments do not overlap with the myosin filaments.

Myosin filaments in the sarcomere exists without any overlap with the actin filaments in the H zone, and in the A zone there is overlap of the myosin and actin filaments (Figure 16.1(B) and (C)). Myosin is a protein dimer composed of two heavy chains and two copies of two different light chains. The N-terminus globular domain is the region that binds to actin and is attached to an α-helical myosin tail region. Located near the globular myosin head is one copy of each of the two light chains.

Releasing the myosin heads from the actin filaments is necessary to prevent a constant state of contraction. The troponin complex containing troponin I, troponin T, and troponin C is involved in the regulation of the myosin head-actin filament interaction (Figure 16.1(C)). In the state of muscle relaxation, the troponin I–T complex causes tropomyosin to move from its natural position in the grooves of the actin helix to a position that prevents the myosin heads from binding to the actin. Following calcium being released from the sarcoplasmic reticulum, troponin C binds to the calcium, causing the troponin I–T to release from the actin. This allows tropomyosin to return to the actin groove and facilitate the binding of the myosin heads to the actin filament.

16.8 MUSCLE DEVELOPMENT: FUNCTION OF MYOGENIC REGULATORY FACTORS

When muscle precursor cells begin to differentiate into skeletal muscle, the activation of skeletal muscle-specific development during both embryonic and posthatch stages requires the precise expression of muscle regulatory factors (MRFs). The MRFs are a family of basic helix-loop-helix (bHLH) transcription factors that are expressed in a specific temporal order. They activate muscle-specific gene expression by binding to the E-box in the DNA (Olson, 1990). Myogenic determination factor 1 (MyoD) was the first identified MRF; its importance was immediately recognized because it could convert nonmyogenic cells to a skeletal muscle lineage (Davis et al., 1987; Weintraub, 1993). The expression of MyoD is necessary for myoblast and satellite cell proliferation. Myogenic factor 5 (Myf5) has functions that overlap with MyoD in that Myf5 also initiates proliferation. The absence of both MyoD and Myf5 will result in the absence of skeletal muscle formation (Rudnicki et al., 1993). However, MyoD and Myf5 are functionally redundant in that disruption of either MyoD or Myf5 will allow muscle development with just minor defects (Rudnicki et al., 1992, 1993). Myogenin is necessary to cause the expression of muscle-specific proteins needed for terminal differentiation with the fusion of myoblasts to form myotubes. Of the MRFs, myogenin is required for viability, as disruption of myogenin expression results in little or no muscle development even though myoblasts are present (Hasty et al., 1993). Myogenic regulatory factor 4 (MRF4, myogenic factor 6) is generally expressed after myogenin and is necessary for myofiber formation. However, MRF4 may play a role during cell proliferation in birds. Shin et al. (2012a) showed MRF4 expression to be high during turkey satellite cell proliferation. This biphasic expression pattern of MRF4 expression is supported by regeneration studies done in the mouse. Launay et al. (2001) reported that MRF4 expression was increased 2–3 days postinjury; its expression decreased at 8 days following the injury and subsequently upregulated at later stages of regeneration (30 days postinjury).

In addition to the MRFs, the myocyte enhancer factors (MEFs) are positive upregulators of muscle differentiation (Naya and Olson, 1999). The MEFs are not specific to skeletal muscle like the MRFs. They are also expressed in cardiac and smooth muscle. No skeletal muscle phenotype has been attributed directly to the MEFs, but they appear to function in a positive feedback manner, especially with myogenin (Ridgeway et al., 2000).

16.9 SATELLITE CELL AND MYOBLAST HETEROGENEITY

As would be expected, there is considerable heterogeneity in muscle fiber structure, metabolism, and response to stimuli, depending on the specific function of the muscle. Muscles for quick short burst movement differ from muscles for sustained movement. Postural muscles also have different properties than muscles not involved in posture.

Heterogeneity also exists among satellite cells, and subpopulations have been characterized by different criteria. Evidence for the existence of subpopulations expressing different myosin isoforms has been reported by several laboratories (Stockdale, 1990; Feldman and Stockdale, 1991; Hoh and Hughes, 1988). Schultz and Heckman-Jones (1990) demonstrated the presence of two populations of satellite cells *in vivo*. One population did not appear to divide during 7 days of extensive muscle growth. Clonal analysis of chicken satellite cells by Yablonka-Reuveni et al. (1987) identified differences in colony sizes. More than 90% of the clones gave rise to large colonies, whereas 8–9% gave rise to small colonies.

To test the hypothesis that at least some of the satellite cell heterogeneity was due to variation in response to growth factor responsiveness, McFarland et al. (1995a) isolated individual satellite cells from a suspension derived from one muscle (pectoralis major) of one 6 week old tom turkey. Using a robotic cell manipulator, 73 clones were developed, which varied greatly in the length of time taken to reach confluence in 25 cm^2 tissue culture flasks. Although there was a continuum of growth rates, the fastest growing clone reached confluence on day 17 following cloning and the slowest reached confluence on day 30. To further investigate differences between cells, a clone that reached confluence on day 19 and one that reached confluence on day 29 were chosen for further studies. The day 19 clone retained its characteristic of proliferating more rapidly than the day 29 clone. Furthermore, the day 19 clone was more responsive to the mitogenic effects of chicken serum at all levels examined (0.5% through 15%). Cells from the two clones exhibited similar morphology and absence of biochemically detectable DNA fragmentation characteristic of apoptotic cells. When near-confluent cultures of both clones were induced to differentiate and fuse in low serum-containing medium, there was no detectable difference in the rates of differentiation of the two clones. Further examination of the two clones was focused on comparing their responses to growth factors important in muscle growth and development (Yun et al., 1997). Using serum-free media, it was demonstrated that the day 17 clone was more responsive to the mitogenic effects of FGF-2, IGF-I, insulin, and PDGF-BB than the day 29 clone. Furthermore, the day 17 clone was also more responsive to the proliferation and differentiation depressing effects of administered TGF-β. Examination of the properties of the PDGF, FGF, and IGF receptors on these two clones revealed no differences in either dissociation constants or receptor numbers per cell. The results demonstrated that there is heterogeneity in satellite cell response to growth factors. To further examine differences between the clones, two additional fast-growing clones and two additional slow-growing clones were examined in conjunction with the original clones studied (McFarland et al., 2003). The fast growing clones used in these latter studies included clones that reached confluence in 25 cm^2 flasks at day 17, 18, or 19. The three slow-growing clones reached confluence on day 28 or day 29. Fast-growing clones were all more responsive to FGF-2 and expressed greater levels of FGF-2 and the FGF receptor-1 at the onset of proliferation than did the slow-growing clones. Fast-growing clones also expressed higher levels of heparan sulfate proteoglycans (HSPG), especially during differentiation, than did slower-growing clones. Heparan sulfate proteoglycans, which are important in FGF signaling, increased during proliferation of all clones tested and decreased in all but one of the clones during differentiation. Slow-growing clones increased their expression of FGF receptor-1 through proliferation and differentiation. However, expression of the receptor in fast growing clones decreased during differentiation. The FGF receptors-2 and -3 were not detected on turkey satellite cells or myotubes using reverse transcriptase-polymerase chain reaction methodology.

To determine if the differential responses to growth factor stimuli between satellite cell clones were due to variation in the levels of activated intracellular signaling proteins, the levels of phospho-MAPK (phospho-ERK 1/2) were determined in the six clonal populations described above (McFarland and Pesall, 2008). Western blotting of satellite cell extracts demonstrated a synergistic response to addition of IGF-I and FGF-2 in both satellite cells and turkey embryonic myoblasts. When administering IGF-I and FGF-2, two of the slow-growing clones exhibited lowest levels of phospho-MAPK. One of the slow-growing clones had levels of phospho-MAPK similar to the three fast-growing clones. The results suggest that variation in responsiveness to growth factor stimuli among satellite cell subpopulations within muscles may be due to several different reasons. Some differences in cell responsiveness appear to be due to variation in phospho-MAPK generation.

Variation in the characteristics of satellite cells has been noted between muscle types. A comparison of satellite cells derived from the turkey pectoralis major, which is composed primarily of glycolytic fibers and the biceps femoris, a muscle composed largely of oxidative fibers were compared in culture (McFarland et al., 1995b). Satellite cells derived from the pectoralis major muscle were more responsive to the mitogenic effects of serum than biceps femoris cells. When administered low serum-containing medium, near-confluent cultures of the biceps femoris cells differentiated and fused to form multinucleated myotubes more rapidly than pectoralis major cells. However, biceps femoris and pectoralis major cells did not respond differently to the mitogenic effects of IGF-I, insulin, or FGF-2. Examination of growth factor receptor characteristics revealed that there were no differences between the pectoralis major and bicep femoris cells in terms of FGF and IGF receptor affinities toward their ligands and there were no differences in receptor numbers per cell (McFarland et al., 1997b). However, both receptor affinities and receptor numbers were different for the PDGF receptors on these satellite cells. Although there were a greater number of PDGF receptors on bicep femoris cells, the receptor affinity was higher for pectoralis major cells. Rates of protein degradation and synthesis were higher in myotube cultures derived from bicep femoris cells than from pectoralis major cells. Several *in vivo* and isolated muscle studies have shown that protein synthesis rates are higher in oxidative muscles than in glycolytic muscles (reviewed in Beatty and Bocek, 1969). Additionally, there were greater rates of uptake of the glucose analog 2-deoxyglucose with bicep femoris cells compared with pectoralis major cells.

16.10 MATERNAL INHERITANCE AND GROWTH SELECTION ON BREAST MUSCLE MORPHOLOGY

Although muscle is structurally composed of muscle fibers and muscle fiber bundles, the size and morphology of the fibers and fiber bundles varies between lines. Breast muscle weight, as measured by breast width, is highly heritable. The unweighted average heritability of breast width using selected populations was 0.30 as summarized by Nestor et al. (1967). For breast width in a randomly bred population, McCartney (1961) reported a heritability of 0.42 and Nestor et al. (1967) obtained heritabilities of 0.77 (full-sib analysis) and 0.61 (mid-parent offspring regression). Studies were initiated by Velleman et al. (2003a) to determine the role of inheritance on breast muscle morphological structure. It was during these studies that the maternal inheritance of breast muscle morphological structure in turkeys was identified. Commercial turkeys are the result of a cross of a sire line (or sire line cross) and a dam line (or dam line cross). The sire lines are usually selected for increased growth rate and muscling, whereas selection within dam lines is on growth rate and reproduction traits.

In initial studies by Velleman et al. (2003a), two experiments were conducted with turkey lines having distinct breast muscle morphological structure (Figure 16.2). The lines used were the random bred control line 2 (RBC2), which is representative of the 1966 commercial turkey (Nestor et al., 1969); the F line, which was selected from the RBC2 line for only increased 16 week body weight (Nestor, 1977); and a commercial sire line B. The RBC2 line has well-defined extracellular matrix spacing (perimysial and endomysial space) and muscle fiber size was not as large as that of the commercial sire B line. The F line was characterized by greatly reduced extracellular matrix space and muscle fibers and poorly defined bundles. In the first experiment, the F line and B line were used. Offspring were obtained from the B and F lines and reciprocal crosses of the B and F lines as well as the pure F and B lines. In experiment 2, the F line and RBC2 line were used. Offspring from reciprocal crosses and the pure lines were obtained. From both the experiments, breast muscle samples were obtained and histologically evaluated. The results from both experiments consistently showed that the morphological structure of the breast muscle in males and females followed the female parent. Figure 16.3 contains representative breast muscle sections from the reciprocal crosses of the F and B lines. The maternal inheritance of breast muscle morphology in turkeys at 16 weeks of age was experimentally confirmed in subsequent experiments using F_1 and F_2 crosses of the F and RBC2 lines (Velleman and Nestor, 2004). The maternal inheritance of breast muscle morphological structure appears to be universal in turkeys as lines selected for egg production and not growth characteristics also exhibit maternal inheritance (Velleman and Nestor, 2006; Velleman et al., 2007a).

The biological cause of the maternal inheritance observed in the turkey breast muscle is not understood at this time. The observed maternal inheritance could result from mitochondrial inheritance, epigenetics or genetic imprinting, or factors in the egg. Mitochondrial DNA is predominantly maternal and gives rise to genetic asymmetry between males and females (for a review, see Wolff and Gemmell,

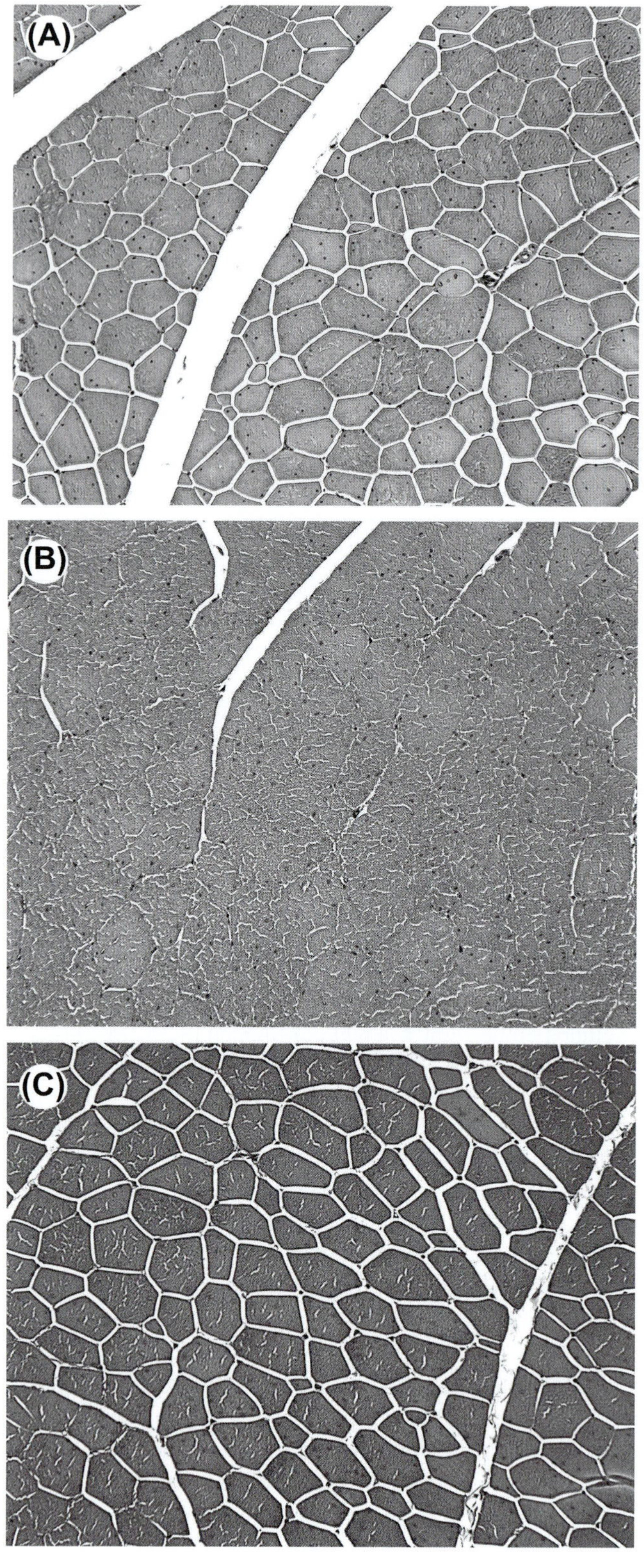

FIGURE 16.2 Representative samples at 16 weeks of age of the pectoralis major muscle from an experimental line (F) selected long term for increased body weight, a commercial line (B), and a random bred control line (RBC2) that was the base population of the F line. (A) = RBC2 line; (B) = F line; and (C) = B line. *Used by permission of Velleman et al. (2003a).*

2012). However, the mechanisms of mitochondrial inheritance and the relationship to muscle morphological structure are still not completely understood. Genetic imprinting has been shown to occur in angiosperms, mammals, and some protozoa (Pennisi, 2001). If genetic imprinting was of importance in the inheritance of turkey breast muscle morphology, some paternal inheritance should have been observed. Egg traits have been shown to influence the performance of turkeys posthatch. For example, egg weight has been shown to influence body weight up to 24 weeks of age (Scott and Phillips, 1936; Bray, 1965). It is quite likely that differences in egg traits, such as egg weight and nutritional composition of the egg, are not responsible for maternal inheritance.

Regardless of the cause of maternal inheritance of muscle morphological structure in turkeys, it is of importance to the commercial turkey industry that the type of mating used in the production of commercial crosses will have a major impact on performance. Lines with desirable characteristics in breast muscle morphology should be used as the female parent for commercial production. To use breast muscle morphological structure as a selection tool, Velleman et al. (2010) determined that the type of mating used to produce turkeys has a major effect on breast muscle morphology beginning at 12 weeks of age. At this time, it is not known if maternal inheritance of breast morphology occurs in other avian species or if other muscles exhibit maternal inheritance.

16.11 EFFECT OF SELECTION FOR INCREASED GROWTH RATE ON MUSCLE DAMAGE

The poultry breeding industry has placed major emphasis in selection for increased growth rate and breast muscling. Increased growth rates in poultry have helped the poultry industry keep up with consumer demand for economical chicken and turkey products, but these increases must be accompanied by consideration of the cellular mechanisms affecting muscle growth as both the morphological structure and biochemical properties of the muscle are affected. These changes in muscle, especially the breast muscle, have resulted in meat quality issues, which are the end consequence of muscle fiber defects. Typical muscle fiber defects found in poultry selected for increased growth rate include deep pectoral myopathy (Wilson et al., 1990; Sosnicki and Wilson, 1991), focal myopathy (Sosnicki, 1993), and pale, soft, and exudative (PSE) meat (Sosnicki, 1993; Pietrzak et al., 1997). In broiler chickens, a new meat quality issue of breast muscle white striping has recently

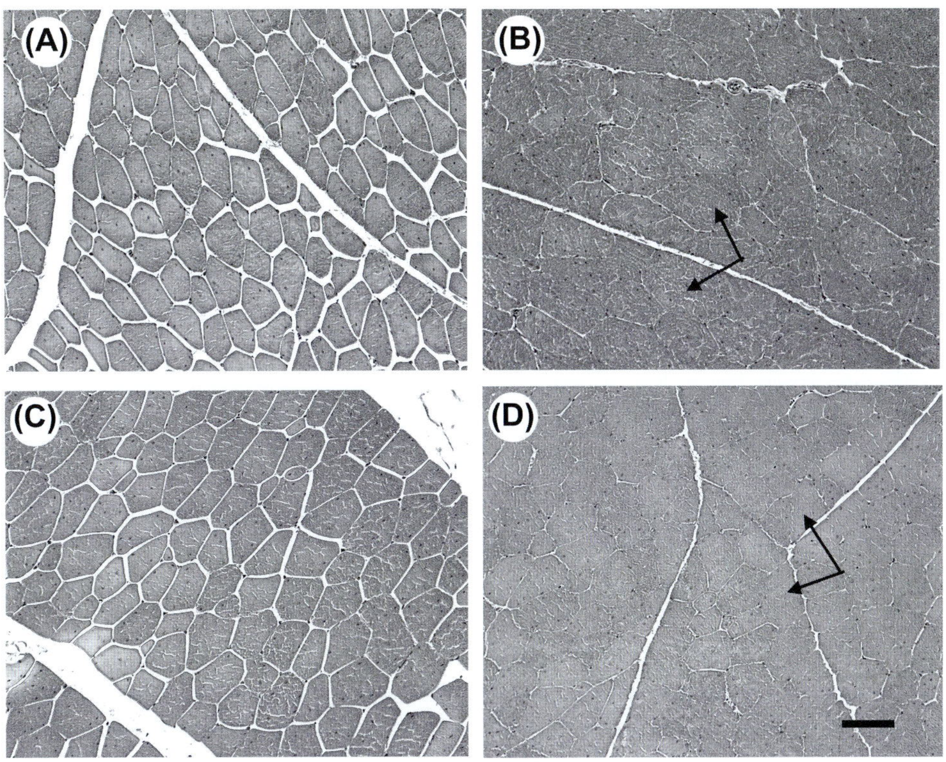

FIGURE 16.3 Representative pectoralis major muscle sections of the reciprocal crosses of a commercial sire line (B) and an experimental line (F) selected for long-term increased 16 week body weight. (A) = F sire × B dam; (B) = B sire × F dam; (C) = F sire × B dam; and (D) = B sire × F dam. Sections (A) and (B) are from males and sections (C) and (D) are from females. Scale bar = 50 μm. Arrows highlight muscle fiber fragmentation. *Used by permission of Velleman et al. (2003a).*

been reported, which is associated with heavier body weights or increased growth rates (Bauermeister et al., 2009; Kuttappan et al., 2012). The white striping defect results in white stripes or striations following the muscle fiber orientation. It does not appear to affect the quality of the cooked breast muscle but does affect the appearance. Because muscle fibers are surrounded by connective tissue, the white striping may arise from altered connective tissue deposition or structure. Kuttappan et al. (2013) reported that there is an increase in pathological changes in the breast muscle with white striping. The white striping can vary in severity and affect the morphology of the muscle. In breast muscle samples with severe or moderate levels of white striping, there is a loss of cross-striations, variable muscle fiber size, floccular or vacuolar degeneration and lysis of fibers, mild mineralization, occasional regeneration, mononuclear cell infiltration, lipidosis, and interstitial inflammation and fibrosis. In addition, the fat level is increased whereas protein is decreased in white striped muscle.

Also associated with many of these breast muscle fiber defects from chickens and turkeys selected for meat production are lighter or pale color and decreased water-holding capacity (Rémignon and Bihan-Duval, 2003). According to Owens et al. (2000), approximately 40% of the commercial turkey meat exhibits poor water-holding capacity. Muscling changes due to selection are not just limited to chickens and turkeys, but they have also been reported in ducks (Baeza et al., 1997).

Wilson et al. (1990) showed that in the turkey breast muscle, damage to the muscle fibers increased with age. Velleman et al. (2003b) traced the effect of growth selection on muscle fiber damage in turkeys from late embryonic development (day 25) through 20 weeks of posthatch age. In this study, muscle morphology comparisons were made between the RBC2, and F lines. Figures 16.4 and 16.5 show hematoxylin and eosin staining of the breast muscle from the RBC2 and F lines. Breast muscles from the unselected RBC2 line had well-structured muscle fiber and fiber bundles (Figure 16.4). Regardless of age, the muscle fiber and fiber bundles had sufficient spacing for endomysial and perimysial connective layers. In addition, the perimysial connective tissue layer contained large capillary beds, which are important in removing the byproducts of anaerobic respiration. However, at 20 weeks of age, some muscle fiber fragmentation was noted. In contrast, the F line had limited endomysial and perimysial spacing surrounding the fibers and fiber bundles by 8 weeks of age. At 16 weeks posthatch, hypercontracted muscle fibers were present in the F line and by 20 weeks significant fragmentation and degeneration of the muscle fiber bundles were present. Due to the limited connective tissue spacing, well-developed capillary beds are not present in the F line breast muscle.

The type of muscle morphological changes observed in the F line have led to some hypothesizing that poultry muscles have outgrown their muscle support systems, which results in the muscle damage. The muscle fibers undergoing fragmentation have reduced endomysial spacing. The breast muscle is a fast twitch anaerobic muscle composed predominantly of fast-twitch glycolytic (type IIb) muscle fibers. With increased breast muscle weight, there are more glycolytic fibers and anaerobic capacity is increased (Dransfield and Sosnicki, 1999; Yost et al., 2002). Lactic acid is produced by anaerobic respiration, which leads to higher acid concentrations in the muscle and decreased pH. Pale soft exudative meat is characterized by low postmortem pH. Muscles with reduced connective tissue spacing are likely to have limited circulation and lactic acid is largely removed from the muscle by the circulatory system, to be converted into glycogen by the liver (Bangsbo et al., 1991). Reduced capillaries have been found surrounding muscle in necrotic regions of turkey breast muscle fibers (Sosnicki and Wilson, 1991). The decreased ability to remove lactic acid from breast muscles with limited circulatory supply likely contributes to the reduced pH observed in PSE meat.

16.12 EXTRACELLULAR MATRIX REGULATION OF MUSCLE DEVELOPMENT AND GROWTH

As described above, muscle is surrounded by three distinct connective tissue layers: the epimysium, perimysium, and endomysium (Figure 16.1(A)). The epimysium surrounds the entire muscle and is continuous with the joining of various muscle group or the tendon attaching muscle to bone. Although the epimysium is very thick, it is not a factor in meat quality because it can easily be removed. The perimysium surrounds muscle fiber bundles and the endomysium surrounds individual muscle fibers. The connective tissue layers are responsible for many of the functional properties associated with muscle and factors affecting meat quality. Tissue structure, the elasticity of the tissue, vascular supply to the muscle tissue, and water-holding capacity are all properties linked to connective tissue, primarily of the perimysium, which can compose up to 90% of the intramuscular connective tissue. Connective tissue is made up of cells and the extracellular matrix.

The extracellular matrix is defined to include all secreted molecules that are immobilized outside cells. Thus, the cell makes its own extracellular environment. The major macromolecular components of the extracellular matrix include collagens, proteoglycans, and noncollagenous glycoproteins. The extracellular matrix was classically viewed as a structural substance that cells were embedded in and was thought to be passive in terms of affecting cell behaviors. Descriptions of the extracellular matrix referred to it as a ground substance filling space between cells. This view of the extracellular matrix

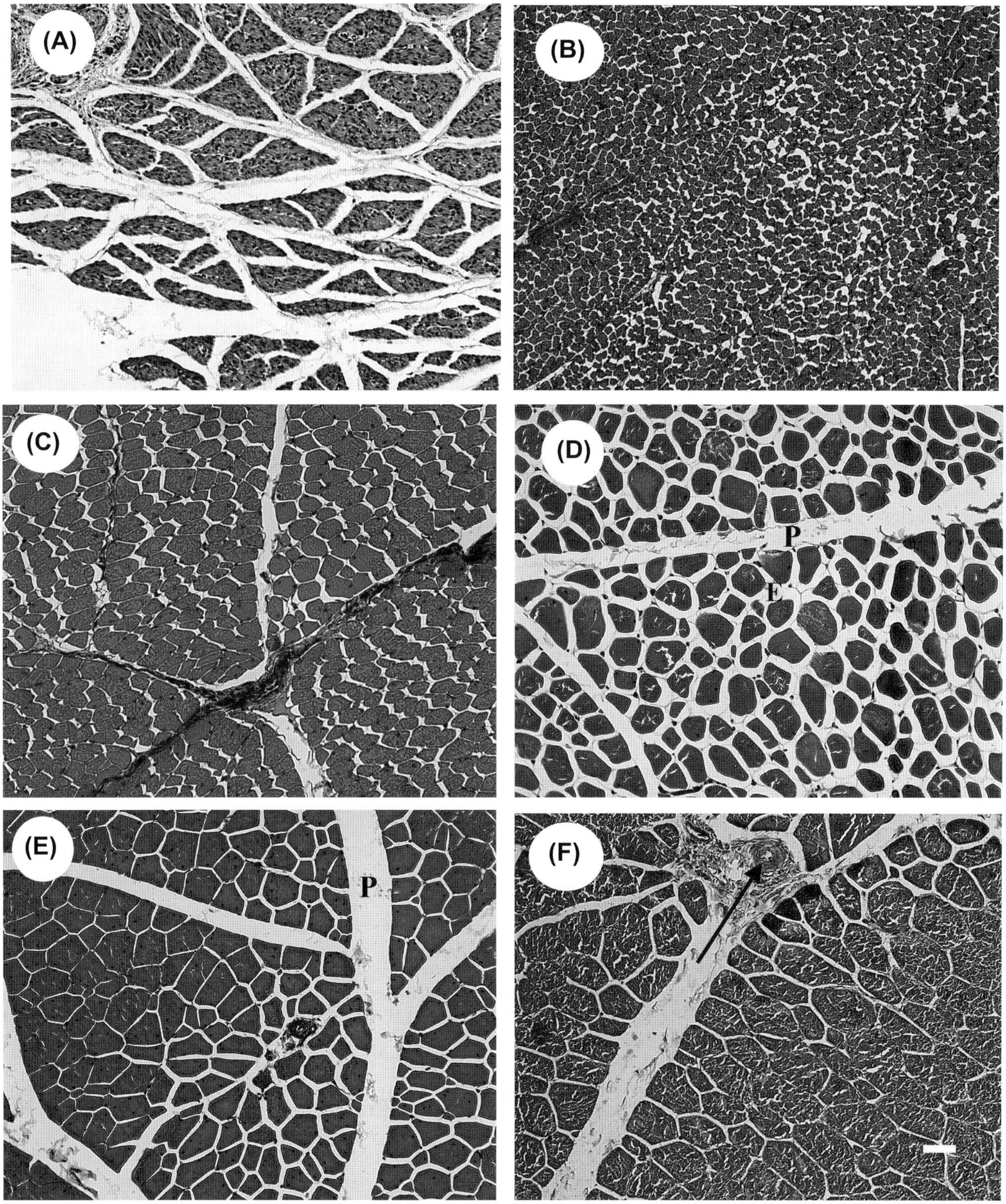

FIGURE 16.4 Hematoxylin and eosin staining of the pectoralis major muscle from the randombred control line at (A) 25 days of embryonic development; (B) 1 week posthatch; (C) 4 weeks posthatch; (D) 8 weeks posthatch; (E) 16 weeks posthatch; and (F) 20 weeks posthatch. P = perimysial connective tissue layer; E = indicates the endomysial connective tissue layer (in panel (D)); and the arrow shows capillary supply. Scale bar = 40 μm. *Used by permission of Velleman et al. (2003b).*

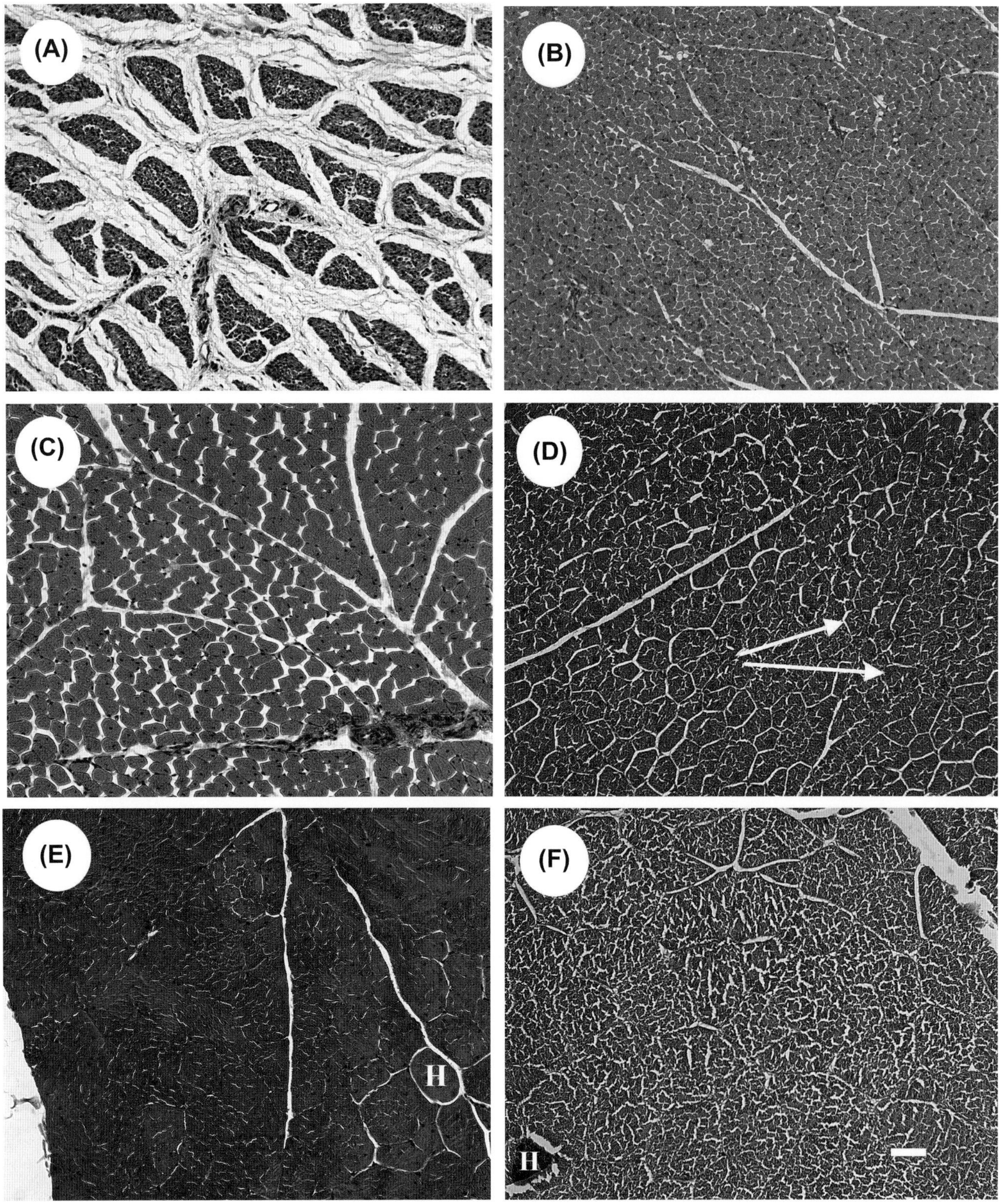

FIGURE 16.5 Hematoxylin and eosin staining of the pectoralis major muscle from the 16-week body weight selected F line at (A) 25 days of embryonic development; (B) 1 week posthatch; (C) 4 weeks posthatch; (D) 8 weeks posthatch; (E) 16 weeks posthatch; and (F) 20 weeks posthatch. The arrows highlight muscle fiber fragmentation; H = muscle fiber hypercontraction. Scale bar = 40 μm. *Used by permission of Velleman et al. (2003b).*

has been replaced by one showing a dynamic complex macromolecular network that regulates many mechanisms involved in tissue development, function, growth, and maintenance. The extracellular matrix regulates muscle function by being a substrate for cell migration; modulating growth factor activity; transmitting signals to the cell modulating proliferation and differentiation; and serving as a tissue structural framework necessary for normal tissue activity.

Muscle formation is dependent upon the migration of muscle cells to form multinucleated myotubes. The extracellular matrix functions as a substrate supporting cellular migration. The interaction of the cell with the extracellular matrix is largely governed by heterodimeric cell membrane integrin receptors. Integrins are widely expressed glycoproteins that are composed of α and β subunits that link the extracellular matrix to the cellular cytoskeleton. The integrins will frequently bind to the extracellular matrix glycoprotein fibronectin arginine–glycine–aspartate (RGD) cell attachment domain. This interaction of the integrins with fibronectin results in the formation of focal adhesions. Fibronectin is a multifunctional extracellular macromolecule secreted as a dimer held together by disulfide bonds at the C-terminus of the molecule. Each fibronectin subunit has a molecular weight of 235,000 to 270,000 Da. In addition to the cell binding to the RGD domain, fibronectin contains binding sites for heparin, fibrin, and collagen.

Cell migration involves a series of cellular changes, including cell extension at the leading edge, formation of stable contacts between the cell and extracellular matrix substrate, contraction of the cellular cytoskeleton, translocation of the cell, and release of the cell-extracellular matrix substrate attachment at the trail edge of the cell. Integrin expression is necessary for cellular migration and Boettiger et al. (1995) showed that inhibiting the expression of $\alpha5\beta1$ integrin blocked both muscle cell migration and myotube development. Furthermore, increased expression of $\alpha5\beta1$ integrin elevates the expression of antiapoptotic genes leading to cell survival and growth (Zhang et al., 1995). When muscle cells attach to the extracellular matrix substrate, they elongate and have more integrin-mediated focal adhesions, which will lead to more migration, formation of myotubes, and subsequent fiber development.

Collagen is the most abundant of the extracellular matrix molecules found in muscle, and it is localized in all three connective tissue layers (for a review, see McCormick, 1999). Types I and III fibrillar collagens are the predominant collagens found in skeletal muscle. These fibrillar collagens provide mechanical stability to the muscle and prevent overstretch of the muscle. The degree of stretch of the muscle is, in part, dependent upon the amount of crosslinking of the fibrillar collagens. The fibrillar collagens are composed of three α-polypeptide chains that form a right-handed triple helix. The helical domain of the triple helix contains repeats of the amino acids Gly–X–Y, where X and Y are frequently proline or lysine. In addition to the helical domain region, there are carboxy and amino terminal nonhelical regions that interact by interchain hydrogen bonds and the regions form intra- and interchain disulfide bonds.

Both intracellular and extracellular posttranslational modifications of the collagen triple helix take place. Intracellularly, precise lysine amino acids are glycosylated, and specific proline and lysine residues are hydroxylated. These posttranslational changes are required for collagen functionality. Once secreted into the extracellular matrix, the carboxy and amino termini are cleaved. At this point, the collagen molecules align in a quarter-stagger array pattern. The quarter-stagger array is the alignment of collagens overlapping the adjacent molecule by approximately one-quarter of its length. The array forms when the collagen fibril formation is initiated. Collagen fibers are the functional form of the collagen.

The initial collagen fibril is not a stable complex as the quarter-stagger array forms from hydrophobic and ionic interactions. Stabilization and functionality of the collagen fibril are derived from subsequent covalent bond formation called crosslinks. Both divalent and trivalent collagen crosslinking of the fibrillar collagens can form (for a review, see McCormick, 1999). The divalent or ketoamine crosslink is reversible in nature and can be transient. The ketoamine crosslink is replaced by a mature nonreversible trivalent hydroxylysl pyridinium crosslink (HP). The trivalent form of the HP crosslink increases with age and elevated levels of the HP crosslink are associated with less tender meat.

The formation of the collagen crosslinking is dependent upon the association of the proteoglycan decorin with the collagen fibril. Decorin is a member of the small leucine-rich proteoglycans, which consists of a core protein of approximately 45 kDa and a single covalently attached chondroitin or dermatan sulfate chain. Weber et al. (1996) determined that decorin is an arch shaped molecule and its concave inner surface wraps around a collagen triple helix approximately 54 nm from the amino terminal. A secondary binding site for decorin is 112 nm from the amino terminal end. Changes in the expression of decorin affect fibril formation and crosslinking (Danielson et al., 1997). Danielson et al. (1997) showed in a mouse decorin knockout model that collagen fibrils were irregular in size and diameter. Although the mice were viable, their skin was fragile and would tear with applied force. Thus decorin affects the maturation of collagen fibrils into larger fiber networks and affects the functional properties of a tissue. In the case of skeletal muscle, decorin expression levels would affect the elasticity of the muscle. An example of altered muscle elasticity is the chicken low score normal (LSN) genetic muscle weakness, which is characterized by subnormal muscle development and function.

The LSN condition was initially detected in F_2 progeny in an outcross between chickens with hereditary muscular dystrophy and a commercial white leghorn stock (Pierro and Haines, University of Connecticut, Storrs, CT, 06,268, unpublished data). The LSN nomenclature was used to distinguish birds that had an impaired ability to right themselves when placed on their backs (exhaustion score) but had a higher level of activity than the muscular dystrophy birds. In the pectoralis major muscle of the LSN birds at 20 days of embryonic development, there is a dramatic increase in decorin protein levels (Velleman et al., 1996). Following the increase in decorin levels, LSN pectoralis major collagen HP crosslink levels are elevated by close to 200% at 6 weeks of posthatch age (Velleman et al., 1996).

In addition to decorin-regulated collagen fibrillogenesis, decorin also influences both the proliferation and differentiation of muscle cells by binding to the growth factors TGF-β (Hildebrand et al., 1994) and myostatin (Miura et al., 2006). Decorin plays a key role as a regulator of the bioavailability of these two key growth factors by directing their availability to their cellular signal transduction receptors. Both TGF-β and myostatin are strong inhibitors of muscle cell proliferation and differentiation. Myostatin is a member of the TGF-β family of growth factors but is only found in skeletal muscle.

Transforming growth factor-β binds to the decorin core protein and is sequestered from binding to its receptors, type I and II, which will diminish muscle cell responsiveness to transforming growth factor-β (Drougett et al., 2006). The binding of TGF-β to its receptors results in the formation of a heterodimer, with the type II receptor phosphorylating and activating the type I receptor. The activated receptor then phosphorylates the receptor activated Smads, which will bind to a collaborating Smad, Smad-4, to form a Smad complex. Smads are intracellular proteins that transduce extracellular signals from TGF-β. The Smad complex is translocated to the nucleus where it binds to transcription factors and regulates the transcription of target genes (Mehra and Wrana, 2002). Circulating decorin levels may, in part, be controlled by its binding to low-density lipoprotein receptor (LRP-1) followed by decorin's subsequent catabolism following endocytosis (Brandan et al., 2006), which will affect TGF-β binding to its receptor.

Decorin expression is also regulated by TGF-β levels during muscle growth. Li et al. (2006) demonstrated that TGF-β can decrease decorin expression *in vitro*. The regulation of decorin expression by transforming growth factor-β also occurs *in vivo*. When exogenous TGF-β was injected into day 3 embryonic eggs decorin mRNA expression was decreased in the pectoralis major muscle by embryonic day 10 and protein levels were subsequently decreased at embryonic day 17 and 1 day posthatch compared with the untreated pectoralis major muscle (Li and Velleman, 2009).

Decorin also appears to be involved in the modulation of muscle fibrosis through its regulation of connective tissue growth factor. Connective tissue growth factor is produced by muscle cells in response to TGF-β and lysophosphatidic acid, and results in increased muscle fibrosis and the dedifferentiation of muscle (Vial et al., 2008). Decorin binds connective tissue growth factor at its core protein inhibiting the fibrotic effects of connective tissue growth factor (Vial et al., 2011).

Myostatin functions in an analogous manner to TGF-β and also binds to the decorin core protein (Miura et al., 2006). Similar to TGF-β, decorin functions as a presenter or sequester of myostatin to its Activin Type II receptor and similar to TGF-β, affects gene expression through Smad-mediated signal transduction. When myostatin is bound to the decorin core protein, myogenic proliferation and differentiation are enhanced due to the inability of myostatin to bind to its receptor (Kishioka et al., 2008). Mutations of the myostatin gene results in a phenotype of "double muscling" (Grobet et al., 1997; Kambadur et al., 1997; McPherron and Lee, 1997). Yang et al. (2003) showed that myostatin inhibits the proliferation and differentiation of chicken embryonic myoblasts, and McFarland et al. (2006) found that myostatin inhibits turkey satellite cell proliferation and differentiation. Regulation of myostatin expression may go beyond decorin and include the skeletal muscle basement membrane proteoglycan perlecan (Xu et al., 2010). In mice with a perlecan knockout, myostatin expression was decreased and muscle fiber hyptertrophy was increased especially in the type IIb fast muscle fibers.

Decorin is an extracellular matrix macromolecule that can function in the extracellular matrix domain imparting structure to the matrix and can work close to the cell surface or directly bind to cell surface receptors initiating signal transduction pathways regulating cell growth properties. Decorin can regulate cellular growth by directly binding to the insulin-like growth factor receptor (IGFR) and epidermal growth factor receptor (EGFR). Decorin binding to IGFR: phosphorylates IGFR, which creates docking or binding sites for the intracellular docking protein insulin receptor substrate-1 (IRS-1). IRS-1 is then phosphorylated; this can activate phosphatidylinositol-3-kinase, followed by phosphorylation of its cascade effector serine/tyrosine kinase (AKT). Activation of AKT will cause downregulation of an inhibitor of cyclin-dependent kinase activity, $p21^{WAF1/CIP1}$. This will allow G1-to-S phase transition of the cell cycle and increased abundance of cell cycle–related proteins and increased cell proliferation. As documented by Moscatello et al. (1998), when decorin binds EGFR, phosphorylation events ensue, mainly through the mitogen activated protein kinase pathway

(MEK1/2 and ERK1/2), leading to upregulation of p21. Upregulation of p21 results in suppression of cyclin and cyclin-dependent kinase activity.

16.13 REGULATION OF MUSCLE GROWTH PROPERTIES BY CELL-MEMBRANE ASSOCIATED EXTRACELLULAR MATRIX MACROMOLECULES

The cell membrane associated proteoglycans can have extracellular, transmembrane, and cytoplasmic domains. Thus, these proteoglycans can play roles in organizing extracellular matrix structure, cytoskeletal organization, cell-to-cell adhesions, and cell signal transduction. The syndecans and glypicans are two families of heparan sulfate proteoglycans that are integral components in growth factor signal transduction. Fibroblast growth factor 2 is a potent stimulator of muscle cell proliferation and a strong inhibitor of differentiation (Dollenmeier et al., 1981). In particular, the growth factor FGF2 binds with a high affinity to its tyrosine kinase receptor by binding to the heparan sulfate chains of the syndecans and glypicans (Steinfield et al., 1996).

The syndecans have four family members, syndecan-1 through -4; all four types are found in skeletal muscle (Larraín et al., 1997, 1998; Brandan and Larraín, 1998; Fuentealba et al., 1999; Liu et al., 2006). All the syndecans are type I membrane glycoproteins (Deepa et al., 2004) and usually contain at least three heparan sulfate chains as well as *N*-glycosylated chains. However, the core protein can also have covalently attached dermatan and chondroitin sulfate chains. All the syndecans have extracellular, transmembrane, and cytoplasmic domains. The cytoplasmic domain for syndecan-1 through -4 have two conserved regions (C1 and C2) separated by a variable region (V). The C terminus of the cytoplasmic domain has a conserved amino acid sequence, EFYA, which allows the syndecans to function as a diverse set of cell surface receptors (Choi et al., 2011). The following describes the known functions of each of the syndecans in skeletal muscle.

Syndecan-1 is a regulator of FGF2 signal transduction and syndecan-1 is expressed at higher levels during muscle cell proliferation than differentiation (Larraín et al., 1997; Liu et al., 2004). In growth-selected turkeys, syndecan-1 expression is higher than in turkeys without selection for growth (Liu et al., 2004). The higher expression of syndecan-1 may lead to a prolonged period of proliferation, leading to a larger pool of muscle cells available to differentiate into muscle fibers.

Syndecan-2, in a manner similar to syndecan-1, is expressed at higher levels during the proliferation of avian muscle cells and expression decreases with differentiation (Liu et al., 2006). Less is known about the function of syndecan-2 in muscle than the other syndecans, but there is evidence that the cytoplasmic domain may be involved in regulating cellular responsiveness to TGF-β (Chen et al., 2004).

Recent research on syndecan-3 suggests that it may play a role in the maintenance, proliferation, and differentiation of myogenic satellite cells (Fuentealba et al., 1999; Cornelison et al., 2004; Pisconti et al., 2010). The absence of syndecan-3 leads to muscular dystrophy types of conditions characterized by impaired locomotion, fibrosis, decreased myonuclear number (Cornelison et al., 2004), and satellite cells with reduced proliferation (Pisconti et al., 2010).

Syndecan-4 is the most well studied of the syndecans in poultry muscle development. Syndecan-4 has been shown to play important roles in muscle maintenance and regeneration (Cornelison et al., 2001, 2004). In turkey satellite cells, syndecan-4 is expressed at high levels during proliferation (Velleman et al., 2007b). The primary role for syndecan-4 in muscle is in its regulation of cytoskeletal organization and muscle cell migration (Shin et al., 2012b, 2013a; Song et al., 2012a,b). In the presence of phosphatidylinositol 4,5-bisphosphate (PIP_2) bound to the V region of the syndecan-4 cytoplasmic domain, syndecan-4 functions in the translocation of inactive protein kinase Cα (PKCα) to the muscle cell membrane resulting in its activation (Song et al., 2012c; Shin et al., 2012b). Once activated, PKCα mediates the activation of downstream RhoA signal transduction, which is involved in mediating cell migration (Dovas et al., 2006; Shin et al., 2013a). Deletion of the syndecan-4 cytoplasmic domain or a knockdown of RhoA inhibits cell migration in turkey satellite cells (Shin et al., 2013a). The binding of PIP_2 also functions in the stabilization of syndecan-4 leading to the formation of syndecan-4 oligomers in turkey satellite cells (Shin et al., 2012b). The oligomerization of syndecan-4 is required for the activation of syndecan-4 as syndecan-4 in the monomer form is not able to activate PKCα (Horowitz and Simons, 1998a,b).

Syndecan-4 may have varying functions with regard to FGF2 based on cell type. For example, in endothelial cells, both the syndecan-4 heparan sulfate chains and the cytoplasmic domain have been shown to be necessary for syndecan-4 mediated FGF2 cell signaling (Volk et al., 1999). However, in turkey satellite cells, the heparan sulfate chains are not required for syndecan-4 FGF2 signal transduction, but the cytoplasmic domain and cytoplasmic domain Ser^{178} in the conserved C1 region is required for modulating satellite cell responsiveness to FGF2 (Zhang et al., 2008; Song et al., 2012c,d). Horowitz and Simons (1998a,b) found that the Ser residue in the C1 domain can be phosphorylated by growth inhibitors and dephosphorylated by the addition of FGF2. The phosphorylation of the Ser in the C1 region will decrease PKCα activation. In turkey satellite cells, the

Ser[178] amino acid stimulates both PKCα activation and FGF2 induced proliferation (Song et al., 2012d).

16.14 REGULATION OF THE MYOGENIC REGULATORY FACTORS BY THE EXTRACELLULAR MATRIX

The heparan sulfate proteoglycans, syndecan-4 and glypican-1, have been shown to be associated with the expression of the MRFs during satellite cell proliferation and differentiation (Gutiérrez and Brandan, 2010; Shin et al., 2012a). In a study knocking down syndecan-4 and glypican-1 expression in turkey satellite cells, Shin et al. (2012a) measured the effect on the MRFs. Decreasing syndecan-4 levels primarily affected MyoD and MRF4 during proliferation by increasing the expression of these MRFs. These data suggest that syndecan-4 may affect the expression of MRFs regulating muscle number. In contrast to syndecan-4, decreasing glypican-1 expression in turkey satellite cells during proliferation and differentiation decreased the expression of all the MRFs. These data suggest that glypican-1 may be necessary for the expression of the MRFs.

16.15 NOVEL GENES INVOLVED IN AVIAN MYOGENESIS

Avian muscle growth may involve genes that have not been previously characterized to play a role in myogenesis. A complete transcriptome analysis was done in turkeys using a turkey skeletal muscle long oligonucleotide (TSKMLO) array that was constructed from the pectoralis major muscle at three developmental stages (Reed et al., 2008; Sporer et al., 2011a). The developmental stages of the TSKMLO array was the 18 day embryo and 1 day old and 16 week old turkeys. From the genes differentially expressed (Sporer et al., 2011b), three genes were studied further for their effects on the proliferation and differentiation of satellite cells. These genes were versican, matrix Gla protein, and death-associated protein.

Versican is a large chondroitin sulfate proteoglycan initially reported in cultured fibroblasts (Zimmerman and Ruoslahti, 1989). Zhang et al. (1998) showed that versican enhanced cell proliferation in NIH 3T3 fibroblasts. By knocking down versican expression in turkey satellite cells, satellite cell proliferation was increased, supporting a potential role for versican in the proliferation of satellite cells (Velleman et al., 2012). In addition, during the embryonic stages of muscle development when initial myofiber synthesis is taking place, large water-holding proteoglycans like versican may be involved. Versican has numerous chondroitin sulfate chains attached to its central core protein and ionically interacts with molecules. Because of its early expression during embryonic muscle formation, it has been hypothesized to play a role in the spacing of the developing muscle fibers (Fernandez et al., 1991).

Matrix Gla protein is highly expressed by vascular smooth muscle cells (Proudfoot and Shanahan, 2006), but prior to the report of Sporer et al. (2011b), matrix Gla protein had not been identified in skeletal muscle. Knocking down matrix Gla protein only decreased proliferation in satellite cells isolated from growth-selected turkeys (Velleman et al., 2012). At this time, it is not possible to hypothesize if matrix Gla protein has an overall function during myogenesis or is just expressed in growth selected lines.

Death-associated protein (DAP) is a highly conserved proline-rich phosphoprotein that was first identified in HeLa cells (Deiss et al., 1995). Recent studies have shown that DAP is a substrate of mTOR (Koren et al., 2010), which is a primary intracellular pathway controlling muscle hypertrophy (Bodine et al., 2001). In both turkey (Velleman et al., 2012) and chicken (Shin et al., 2013b) breast muscle satellite cells, knocking down DAP expression severely inhibits the formation of myotubes. Future studies on DAP will need to address the regulation of muscle DAP by mTOR because DAP could be a key gene in the regulation of avian myogenesis.

16.16 SUMMARY

This chapter on avian skeletal muscle biology provides a broad overview of muscle structure, the prehatch and posthatch development of muscle, effects of growth selection, extracellular matrix influences on muscle development and growth, and the identification of novel genes. Muscle development involves a wide range of cellular events that have distinct embryonic and posthatch phases. It is clear that many factors can influence the development and growth of avian skeletal muscle; as time passes, it is likely that more factors will be identified.

ACKNOWLEDGMENT

The authors thank Ms. Laura B. Harthan for technical graphic illustrations.

REFERENCES

Alberts, B., Johnson, A., Lewis, J., Raff, M., Roberts, K., Walter, P., 2008. Molecular Biology of the Cell. Garland Science, New York.

Allen, R.E., Goll, D.E., 2003. Cellular and developmental biology of skeletal muscle as related to muscle growth. In: Scanes, C.G. (Ed.), Biology of Growth of Domestic Animals. Iowa State Press, Ames, IA, pp. 148–169.

Allen, R.E., McAllister, P.K., Masak, K.C., 1980. Myogenic potential of satellite cells in skeletal muscle of old rats. A brief note. Mech. Ageing Dev. 13, 105–109.

Ashmore, C.R., Tompkins, G., Doerr, L., 1972. Postnatal development of muscle fiber type in domestic animals. J. Anim. Sci. 34, 37–41.

Baeza, E., de Carville, H., Salichon, M.R., Marche, G., Leclercq, B., 1997. Effects of selection, over three and four generations, on meat yield and fatness in Muscovy ducks. Br. Poult. Sci. 38, 359–365.

Bang, M.-L., Li, X., Littlefield, R., Bremner, S., Thor, A., Knowlton, K.U., Lieber, R.L., Chen, 2006. Nebulin-deficient mice exhibit shorter thin filament lengths and reduced contractile function in skeletal muscle. J. Cell Biol. 173, 905–916.

Bangsbo, J., Gollnick, P.D., Grahm, T.E., Saltin, B., 1991. Substrates for muscle glycogen synthesis in recovery from intense exercise in man. J. Physiol. 434, 423–440.

Bauermeister, L.J., Morey, A.U., Moran, E.T., Singh, M., Owens, C.M., McKee, S.R., 2009. Occurrence of white striping in chicken breast fillets in relation to broiler size. Poultry Sci. 88 (Suppl. 1), 33.

Beatty, C.H., Bocek, R.M., 1969. Biochemistry of the red and white muscle. In: Briskey, E.J., Cassens, R.G., Marsh, B.B. (Eds.), Physiology and Biochemistry of Muscle as a Food, vol. 2. University of Wisconsin Press, Madison, WI, pp. 155–191.

Bischoff, R., 1974. Enzymatic liberation of myogenic cells from adult rat muscle. Anat. Rec. 180, 645–661.

Bischoff, R., 1975. Regeneration of single muscle fibers in vitro. Anat. Rec. 182, 215–236.

Bischoff, R., 1986. A satellite cell mitogen from crushed adult muscle. Dev. Biol. 115, 140–147.

Bischoff, R., 1997. Chemotaxis of skeletal muscle satellite cells. Dev. Dyn. 208, 505–515.

Blau, H., Webster, C., 1981. Isolation and characterization of human muscle cells. Cell Biol. 78, 5623–5627.

Bodine, S.C., Stitt, T.N., Gonzalez, M., Kline, W.O., Stover, G.L., Bauerlein, R., Zlotchenko, E., Scrimgeour, A., Lawrence, J.C., Glass, D.J., Yancopoulos, G.D., 2001. Akt/mTOR pathway is a crucial regulator of skeletal muscle hypertrophy and can prevent muscle atrophy in vivo. Nat. Cell Biol. 3, 1014–1019.

Boettiger, D., Enomoto-Iwamot, M., Yoon, H.Y., Hofer, U., Menklo, A.S., Chiquet-Ehrismann, R., 1995. Regulation of integrin α5β1 affinity during myogenic differentiation. Dev. Biol. 169, 261–272.

Brandan, E., Larraín, J., 1998. Heparan sulfate proteoglycans during terminal skeletal muscle differentiation. Possible functions and regulation of their expression. Basic Appl. Myol. 8, 107–113.

Brandan, E., Retamal, C., Cabello-Verrugiom, C., Marzolo, M.P., 2006. The low density lipoprotein receptor-related protein, LRP, functions as an endocytic receptor for decorin. J. Biol. Chem. 281, 31562–31571.

Brand-Saberi, B., Muller, T.S., Wilting, J., Christ, B., Birchmeier, C., 1996. Scatter factor/hepatocyte growth factor (SF/HGF) induces emigration of myogenic cells at interlimb level in vivo. Dev. Biol. 179, 303–308.

Bray, D.F., 1965. Embryonic weight as a selection criterion in alteration of growth curves of domestic birds-Data from the domestic turkey. Can. J. Genet. Cytol. 7, 1–11.

Carlson, B.M., Faulkner, J.A., 1983. The regeneration of skeletal muscle fibers following injury: a review. Med. Sci. Sports Exerc. 15, 187–198.

Cerny, L.C., Bandman, E., 1987. Expression of myosin heavy chain isoforms in regenerating myotubes of innervated and denervated chicken pectoral muscle. Dev. Biol. 119, 350–362.

Chen, L., Klass, C., Woods, A., 2004. Syndecan-2 regulates transforming growth factor-beta signaling. J. Biol. Chem. 279, 15715–15718.

Choi, Y., Chung, H., Jung, H., Couchman, J.R., Oh, E.–S., 2011. Syndecans as cell surface receptors: unique structure equates with functional diversity. Matrix Biol. 30, 93–99.

Cornelison, D.D., Filla, M.S., Stanley, H.M., Rapraeger, A.C., Olwin, B.B., 2001. Syndecan-3 and syndecan-4 specifically mark skeletal muscle satellite cells and are implicated in satellite cell maintenance and muscle regeneration. Dev. Biol. 239, 79–94.

Cornelison, D.D., Wilcox-Adelman, S.A., Geotinck, P.F., Rauvala, H., Rapraeger, A.C., Olwin, B.B., 2004. Essential and separable roles for syndecan-3 and syndecan-4 in skeletal muscle development and regeneration. Genes Dev. 18, 32231–32236.

Danielson, K.G., Baribault, H., Holmes, D.F., Graham, H., Kadler, K.E., Iozzo, R.V., 1997. Targeted disruption of decorin leads to abnormal collagen fibril morphology and skin fragility. J. Cell Biol. 136, 729–743.

Darr, K.C., Schultz, E., 1987. Exercise-induced satellite cell activation in growing and mature skeletal muscle. J. Appl. Physiol. 63, 1816–1821.

Davis, R.L., Weintruab, H., Lassar, A.B., 1987. Expression of a single transfected cDNA converts fibroblasts to myoblasts. Cell 51, 987–1000.

Deepa, S.S., Yamada, S., Zako, M., Goldberg, O., Sughara, K., 2004. Chondroitin sulfate chains on syndecan-1 and syndecan-4 from normal murine mammary gland epithelial cells are structurally and functionally distinct and cooperate with heparan sulfate chains to bind growth factors. A novel function to control binding of midkine, pleiotrophin, and basic fibroblast growth factor. J. Biol. Chem. 279, 37368–37376.

Deiss, L.P., Feinstein, E., Berissi, H., Cohen, O., Kimchi, A., 1995. Identification of a novel serine/threonine kinase and a novel 15-kD protein as potential mediators of the gamma interferon-induced cell death. Genes Dev. 9, 15–30.

Dial, T.R., Carrier, D.R., 2012. Precocial hindlimbs and altricial forelimbs: partitioning ontogenetic strategies in mallards (*Anas platyrhynchos*). J. Exp. Biol. 215, 3703–3710.

Dimario, J.X., Stockdale, F.E., 1997. Both myoblast lineage and innervation determine fiber type and are required for expression of the slow myosin heavy chain 2 gene. Dev. Biol. 188, 167–180.

Dodson, M.V., McFarland, D.C., Grant, A.L., Doumit, M.E., Velleman, S.G., 1996. Extrinsic regulation of domestic animal-derived satellite cells. Domest. Anim. Endocrinol. 13, 107–126.

Dollenmeier, P., Turner, D.C., Eppenberger, H.M., 1981. Proliferation and differentiation of chick skeletal muscle cells cultured in a chemically defined medium. Exp. Cell Res. 135, 47–61.

Dovas, A., Yoneda, A., Couchman, J.R., 2006. PKC alpha-dependent activation by syndecan-4 during focal adhesion formation. J. Cell Sci. 119, 2837–2846.

Dransfield, E., Sosnicki, A.A., 1999. Relationship between muscle growth and poultry meat quality. Poultry Sci. 78, 743–746.

Drougett, R., Cabello-Verrugio, C., Riquelme, C., Brandan, E., 2006. Extracellular matrix proteoglycans modify TGF-beta bio-availability attenuating its signaling during skeletal muscle differentiation. Matrix Biol. 25, 332–341.

Duclos, M.J., Chevalier, B., Goddard, C., Simon, J., 1993b. Regulation of amino acid transport and protein metabolism in myotubes derived from chicken muscle satellite cells by insulin-like growth factor-I. J. Cell. Physiol. 157, 650–657.

Duclos, M.J., Chevalier, B., Le Marchand-Brustel, Y., Tanti, J.F., Goddard, C., Simon, J., 1993a. Insulin-like growth factor-I stimulated glucose transport in myotubes derived from chicken muscle satellite cells. J. Endocrinol. 137, 465–472.

Evinger-Hodges, M.J., Ewton, D.Z., Seifert, S.C., Florini, J.R., 1982. Inhibition of myoblast differentiation in vitro by a protein isolated from liver cell medium. J. Cell Biol. 93, 395–401.

Feldman, J.L., Stockdale, F.E., 1991. Skeletal muscle satellite cell diversity: satellite cells form fibers of different types in cell culture. Dev. Biol. 143, 320–334.

Fernandez, M.S., Dennis, J.E., Drushel, R.F., Carrino, D.A., Kimata, K., Yamagata, M., Caplan, A.I., 1991. The dynamics of compartmentalization of embryonic muscle by extracellular matrix molecules. Dev. Biol. 147, 46–61.

Florini, J.R., Roberts, A.B., Ewton, D.Z., Falen, S.L., Flanders, K.C., Sporn, M.B., 1986. Transforming growth factor-β. A very potent inhibitor of myoblast differentiation, identical to the differentiation inhibitor secreted by buffalo rat liver cells. J. Biol. Chem. 261, 16509–16513.

Fowler, V.M., Sussmann, M.A., Miller, P.G., Flucher, B.E., Daniels, M.P., 1993. Tropomodulin is associated with the free (pointed) ends of the thin filaments in rat skeletal muscle. J. Cell Biol. 120, 411–420.

Fuentealba, L., Carey, D.J., Brandan, E., 1999. Antisense inhibition of syndecan-3 expression during skeletal muscle differentiation accelerates myogenesis through a basic fibroblast growth factor-dependent mechanism. J. Biol. Chem. 274, 37876–37884.

Gal-Levi, R., Leshem, Y., Aoki, S., Nakamura, T., Halevy, O., 1998. Hepatocyte growth factor plays a dual role in regulating skeletal muscle satellite cell proliferation and differentiation. Biochim. Biophys. Acta 1402, 39–51.

Gilbert, S.F., 2000. Developmental Biology, sixth ed. Sinauer Associates, Inc.

Grobet, L., Martin, L.J., Pncelet, D., Pirottin, D., Brouwers, B., Riquet, J., Schoeberlein, A., Dunner, S., Ménisser, F., Massabanda, J., Fries, R., Hanset, R., Georges, M., 1997. A deletion in the bovine myostatin gene causes the double-muscled phenotype in cattle. Nat. Genet. 17, 71–74.

Gutiérrez, J., Brandan, E., 2010. A novel mechanism of sequestering fibroblast growth factor 2 by glypican in lipid rafts, allowing skeletal muscle differentiation. Mol. Cell. Biochem. 30, 1634–1649.

Hansen-Smith, F.M., Picou, D., Golden, M.N., 1978. Quantitative analysis of nuclear population in muscle from malnourished and recovered children. Pediatr. Res. 12, 167–170.

Hansen-Smith, F.M., Picou, D., Golden, M.N., 1979. Muscle satellite cells in malnourished and nutritionally rehabilitated children. J. Neurol. Sci. 41, 207–221.

Hasty, P., Bradley, A., Morris, J.H., Edmondson, D.G., Venuti, J.M., Olson, E.N., Klein, W.H., 1993. Muscle deficiency and neonatal death in mice with a targeted mutation in the myogenin gene. Nature 364, 501–506.

Hildebrand, A., Romaris, M., Rasmussen, L.M., Heinegård, D., Twadzik, D.R., Border, W.A., Ruoslahti, E., 1994. Interaction of the small interstitial proteoglycans biglycan, decorin and fibromodulin with transforming growth factor beta. Biochem. J. 302, 527–534.

Hoh, J.F.Y., Hughes, S., 1988. Myogenic and neurogenic regulation of myosin gene expression in cat jaw-closing muscles regenerating in fast and slow limb muscle beds. J. Muscle. Res. Cell Motil. 9, 59–72.

Horowitz, A., Simons, M., 1998a. Regulation of syndecan-4 phosphorylation in vivo. J. Biol. Chem. 273, 10914–10918.

Horowitz, A., Simons, M., 1998b. Phosphorylation of the cytoplasmic tail of syndecan-4 regulates activation of protein kinase C alpha. J. Biol. Chem. 273, 25548–25551.

Hughes, S.M., Blau, H.M., 1990. Migration of myoblasts across basal lamina during skeletal muscle development. Nature 345, 350–353.

Jakowlew, S.B., Ciment, G., Tuan, R.S., Sport, M.B., Roberts, A.B., 1992. Pattern of expression of transforming growth factor-beta 4 mRNA and protein in the developing chicken embryo. Dev. Dyn. 195, 276–289.

Jakowlew, S.B., Ciment, G., Tuan, R.S., Sport, M.B., Roberts, A.B., 1994. Expression of transforming growth factor-beta 2 and beta 3 mRNAs and proteins in the developing chicken embryo. Differentiation 55, 105–118.

Janeczko, R.A., Etlinger, J.D., 1984. Inhibition of intracellular proteolysis in muscle cultures by multiplication-stimulating activity and insulin on proteolysis, protein synthesis, amino acid uptake, and sugar transport. J. Biol. Chem. 259, 6292–6297.

Johnson, D.D., Wilcox, R., Wenger, B., 1983. Precocious in vitro development of satellite cells from dystrophic chicken muscle. In Vitro 19, 723–729.

Kambadur, R., Sharma, M., Smith, T.P., Bass, J.J., 1997. Mutations in myostatin (GDF8) in double-muscled Belgian Blue and Piedmontese cattle. Genome Res. 7, 910–916.

Kirk, S., Oldham, J., Kambadur, R., Sharma, M., Dobbie, Bass, J., 2000. Myostatin regulation during skeletal muscle regeneration. J. Cell. Physiol. 184, 356–363.

Kishioka, Y., Thomas, T., Wakamatsu, J., Hattori, A., Sharma, M., Kambadur, R., Mishimura, T., 2008. Decorin enhances the proliferation and differentiation of myogenic cells through suppressing myostatin activity. J. Cell Physiol. 215, 856–867.

Koren, I., Reem, E., Kimchi, A., 2010. DAP1, a novel substrate of mTOR, negatively regulates autophagy. Curr. Biol. 20, 1093–1098.

Kuttappan, V.A., Brewer, V.B., Waldroup, P.W., Owens, C.M., 2012. Influence of growth rate on the occurrence of white striping in broiler breast fillets. Poult. Sci. 91, 2677–2685.

Kuttappan, V.A., Shivaprasad, H.L., Shaw, D.P., Valentine, B.A., Hargis, B.M., Clark, F.D., McKee, S.R., Owens, C.M., 2013. Pathological changes associated with white striping in broiler breast muscles. Poultry Sci. 92, 331–338.

Landys-Cianneli, M.M., Piersma, T., Jukema, J., 2003. Strategic size changes of internal organs and muscle tissue in the bar-tailed godwit during fat storage on a spring stopover site. Funct. Ecol. 17, 151–159.

Larraín, J., Carey, D.J., Brandan, E., 1998. Syndecan-1 expression inhibits myoblast differentiation through basic fibroblast growth factor-dependent mechanisms. J. Biol. Chem. 273, 32288–32296.

Larraín, J., Cizmeci-Smith, G., Troncoso, V., Stahl, R.C., Carey, D.J., Brandan, E., 1997. Syndecan-1 expression is down-regulated during myoblast terminal differentiation. Modulation by growth factors and retinoic acid. J. Biol. Chem. 272, 18418–18424.

Lau, C.L., 1993. Behavior of embryonic chick heart cells in culture. 2. Cellular responses to epidermal growth factor and other growth signals. Tissue Cell 25, 681–693.

Lau, C.L., 1994. Behavior of embryonic chick heart cells in culture. 3. Spatial distribution of epidermal growth factor in heart muscle cells. Tissue Cell 26, 203–208.

Launay, T., Armand, A.S., Charbonnier, F., Mira, J.C., Donsez, E., Gallien, C.L., Chanoine, C., 2001. Expression and neural control of myogenic regulatory factor genes during regeneration of mouse soleus. J. Histochem. Cytochem. 49, 887–899.

Li, X., McFarland, D.C., Velleman, S.G., 2006. Effect of transforming growth factor-beta on decorin and beta 1 integrin expression during muscle development in chickens. Poultry Sci. 85, 326–332.

Li, X., Velleman, S.G., 2009. Effect of transforming growth factor-beta 1 on decorin expression and muscle morphology during chicken embryonic and posthatch growth and development. Poultry Sci. 88, 387–397.

Lieber, R.L., 1992. Skeletal muscle physiology. In: Butler, J.P. (Ed.), Skeletal Muscle Structure and Function: Implications for Rehabilitation and Sports Medicine. Williams & Wilkins, Baltimore, MD, pp. 49–110 (Chapter 2).

Lindström, Å., Kvist, A., Piersma, T., Dekinga, T., Dietz, M.W., 2000. Avian pectoral muscle size rapidly tracks body mass changes during flight, fasting and fueling. J. Exp. Biol. 203, 913–919.

Liu, C., McFarland, D.C., Nestor, K.E., Velleman, S.G., 2006. Differential expression of membrane-associated heparan sulfate proteoglycans in the skeletal muscle of turkeys with different growth rates. Poultry Sci. 85, 422–428.

Liu, X., McFarland, D.C., Nestor, K.E., Velleman, S.G., 2004. Developmental regulated expression of syndecan-1 and glypican in pectoralis major muscle in turkeys with different growth rates. Dev. Growth Differ. 46, 37–51.

Massague, J., Heino, J., Laiho, M., 1991. Mechanisms in TGF-beta action. Ciba Found. Symp. 157, 51–59.

Matsuda, R., Spector, D.H., Strohman, R.C., 1983. Regenerating adult chicken muscle and satellite cell cultures express embryonic patterns of myosin and tropomyosin isoforms. Dev. Biol. 100, 478–488.

Mauro, A., 1961. Satellite cell of skeletal muscle fibers. J. Biophys. Biochem. Cytol. 9, 493–495.

McCartney, M.G., 1961. Heritability and genetic correlations of body weight and conformation in a randombred population of turkeys. Poultry Sci. 40, 1694–1700.

McCormick, R.J., 1999. The flexibility of the collagen compartment of muscle. Poultry Sci. 78, 778–784.

McCroskery, S., Thomas, M., Maxwell, L., Sharma, M., Kambadur, R., 2003. Myostatin negatively regulates satellite cell activation and self-renewal. J. Cell Biol. 162, 1135–1147.

McFarland, D.C., 1999. Influence of growth factors on poultry myogenic satellite cells. Poultry Sci. 78, 747–758.

McFarland, D.C., Doumit, M.E., Minshall, R.D., 1988. The turkey myogenic satellite cell: optimization of in vitro proliferation and differentiation. Tissue Cell 20, 899–908.

McFarland, D.C., Gilkerson, K.K., Pesall, J.E., Ferrin, N.H., Wellenreiter, R.H., 1997a. In vitro characteristics of myogenic satellite cells derived from the pectoralis major and biceps femoris muscles of the chicken. Cytobios 91, 45–52.

McFarland, D.C., Gilkerson, K.K., Pesall, J.E., Wellenreiter, R.H., Ferrin, N.H., Ye, W.V., Yun, Y., Vander Wal, L.S., 1997b. Comparison of growth factor receptors and metabolic characteristics of satellite cells derived from the biceps femoris and pectoralis major muscles of the turkey. Gen. Comp. Endocrinol. 105, 114–120.

McFarland, D.C., Gilkerson, K.K., Pesall, J.E., Walker, J.S., Yun, Y., 1995a. Heterogeneity in growth characteristics of satellite cell populations. Cytobios 82, 21–27.

McFarland, D.C., Pesall, J.E., Gilkerson, K.K., Ye, W.V., Walker, J.S., Wellenreiter, R., 1995b. Comparison of in vitro properties of satellite cells derived from the pectoralis major and biceps muscles of growing turkeys. Basic Appl. Myol. 5, 27–31.

McFarland, D.C., Liu, X., Velleman, S.G., Zeng, C., Coy, C.S., Pesall, J.E., 2003. Variation in fibroblast growth factor response and heparan sulfate proteoglycan production in satellite cell populations. Comp. Biochem. Physiol. C 134, 341–351.

McFarland, D.C., Pesall, J.E., 2008. Phospho-MAPK as a marker of myogenic satellite cell responsiveness to growth factors. Comp. Biochem. Physiol. B 149, 463–467.

McFarland, D.C., Pesall, J.E., Gilkerson, K.K., 1993. The influence of growth factors on turkey embryonic myoblasts and satellite cells in vitro. Gen. Comp. Endocrinol. 89, 415–424.

McFarland, D.C., Pesall, J.E., Gilkerson, K.K., Ferrin, N.H., Ye, W.V., Swenning, T.A., 1994. Comparison of protein metabolism and glucose uptake in turkey (*Meleagris gallopavo*) satellite cells and embryonic myoblasts in vitro. Comp. Biochem. Physiol. A 107, 301–306.

McFarland, D.C., Pesall, J.E., Norberg, J.M., Dvoracek, M.A., 1991. Proliferation of the turkey myogenic satellite cell in a serum-free medium. Comp. Biochem. Physiol. A 99, 163–167.

McFarland, D.C., Singh, Y.N., Johnson, A.D., Pesall, J.E., Gilkerson, K.K., Vander Wal, L.S., 2000. Isolation and characterization of myogenic satellite cells from the muscular dystrophic hamster. Tissue Cell 32, 257–265.

McFarland, D.C., Velleman, S.G., Pesall, J.E., Coy, C.S., 2011. Effect of lipids on avian satellite cell proliferation, differentiation and heparin sulfate proteoglycan expression. Comp. Biochem. Physiol. A 159, 188–195.

McFarland, D.C., Velleman, S.G., Pesall, J.E., Liu, C., 2006. Effect of myostatin on turkey myogenic satellite cells and embryonic myoblasts. Comp. Biochem. Physiol. A 144, 501–508.

McFarland, D.C., Velleman, S.G., Pesall, J.E., Liu, C., 2007. The role of myostatin in chicken (*Gallus domesticus*) myogenic satellite cell proliferation and differentiation. Gen. Comp. Endocrinol. 151, 351–357.

McNabb, F.M.A., Stanton, F.W., Dicken, S.C., 1984. Post-hatching thyroid development and body growth in precocial vs altricial birds. Comp. Biochem. Physiol. A 78, 629–635.

McPherron, A.C., Lee, S.-J., 1997. Double muscling in cattle due to mutations in the myostatin gene. Proc. Natl. Acad. Sci. U.S.A. 94, 12457–12461.

McPherron, A.C., Lawler, A.M., Lee, S.-J., 1997. Regulation of skeletal muscle mass in mice by a new TGF-β superfamily member. Nature 387, 83–90.

Mehra, A., Wrana, J.L., 2002. TGF-beta and Smad signal transduction pathway. Biochem. Cell Biol. 80, 605–622.

Mie, H., Shinsuke, Y., Kenichiro, S., Takeshi, O., Nozomi, H., Yohei, O., Takahide, A., Fumiyuki, H., Ruri, K., Kensuke, K., Shinji, M., Motoaki, S., Keiichi, F., 2011. G-CSF influences mouse skeletal muscle development and regeneration by stimulating myoblast proliferation. J. Exp. Med. 208, 715–727.

Miura, T., Kishioka, Y., Wakamatsu, J., Hattori, A., Hennebry, A., Berry, C.J., Sharma, M., Kambadur, R., Nishimura, T., 2006. Decorin binds myostatin and modulates its activity to muscle cells. Biochem. Biophys. Res. Commun. 340, 675–680.

Moscatello, D.R., Santra, M., Mann, D.M., McQuillan, D.J., Wang, A.J., Iozzo, R.V., 1998. Decorin suppresses tumor cell growth by activating the epidermal growth factor receptor. J. Clin. Invest. 101, 406–412.

Moss, F.P., 1968. The relationship between the dimensions of the fibres and the number of nuclei during normal growth of skeletal muscle in the domestic fowl. Am. J. Anat. 122, 555–564.

Moss, F.P., Leblond, C.P., 1971. Satellite cells as the source of nuclei in muscles of growing rats. Anat. Rec. 170, 421–436.

Muir, G.D., 2000. Early ontogeny of locomotor behavior: a comparison between altricial and precocical animals. Brain. Res. Bull. 53, 719–726.

Naya, F.J., Olson, E., 1999. MEF2: a transcriptional target for signaling pathways controlling skeletal muscle growth and differentiation. Curr. Opin. Cell Biol. 6, 683–688.

Nestor, K.E., 1977. Genetics of growth and reproduction in the turkey. 5. Selection for increased body weight alone and in combination with increased egg production. Poultry Sci. 56, 337–347.

Nestor, K.E., McCartney, M.G., Bacheve, N., 1969. Relative contribution of genetics and environment to turkey improvement. Poultry Sci. 48, 1944–1949.

Nestor, K.E., McCartney, M.G., Harvey, W.R., 1967. Genetics of growth and reproduction in the turkey. 1. Genetic and non-genetic variation in body weight and body measurements. Poultry Sci. 46, 1374–1384.

Olson, E.N., 1990. MyoD family: a paradigm for development? Genes Dev. 4, 1454–1461.
Ono, Y., Calhabeu, F., Morgan, J.E., Katagiri, T., Amthor, H., Zammit, P.S., 2011. BMP signaling permits population expansion by preventing premature myogenic differentiation in muscle satellite cells. Cell Death. Differ. 18, 222–234.
Ontell, M., 1986. Muscular dystrophy and muscle regeneration. Hum. Pathol. 17, 673–682.
Oppenheim, R.W., 1972. Prehatching and hatching behavior in birds: a comparative study of altricial and precocial species. Anim. Behav. 20, 644–655.
Owens, C.M., Hirschler, E.M., McKee, S.R., Martinez-Dawson, R., Sams, A.R., 2000. The characterization of pale, soft, exudative turkey meat in a commercial plant. Poultry Sci. 79, 553–558.
Pennisi, E., 2001. Behind the scenes of gene expression. Science 10, 1064–1067.
Pietrzak, M., Greaser, M.L., Sosnicki, A.A., 1997. Effect of rapid rigor mortis processes on protein functionality in pectoralis major muscle of domestic turkeys. J. Anim. Sci. 75, 2106–2116.
Pisconti, A., Cornelison, D.D.W., Olguin, H.C., Antwine, T.L., Olwin, B.B., 2010. Syndecan-3 and Notch cooperate in regulating adult myogenesis. J. Cell Biol. 190, 427–441.
Price, E.R., Bauchinger, U., Zajac, D.M., Cerasale, D.J., McFarlan, J.T., Gerson, A.R., McWilliams, S.R., Guglielmo, C.G., 2011. Migration- and exercise-induced changes to flight muscle size in migratory birds and association with IGF1 and myostatin mRNA expression. J. Exp. Biol. 214, 2823–2831.
Proudfoot, D., Shanahan, C.M., 2006. Molecular mechanisms mediating vascular calcification: role of matrix Gla protein. Nephrology (Carlton) 11, 455–461.
Rabkin, S.W., 1996. Indapamide accentuates cardiac chronotropic responses to epidermal growth factor in chick cardiomyocytes. Tissue Cell 28, 469–472.
Reed, K.M., Mendoza, K.M., Juneja, B., Fahrenkrung, S.C., Velleman, S.G., Chiang, W., Strasburg, G.M., 2008. Characterization of expressed sequence tags from turkey skeletal muscle. Anim. Genet. 39, 635–644.
Rémignon, H., Bihan-Duval, E., 2003. Meat quality problems associated with selection for increased production. In: Muir, W.M., Aggrey, S.E. (Eds.), Poultry Genetics, Breeding and Biotechnology. CABI Publishing, Cambridge, MA.
Ridgeway, A.G., Wilton, S., Sherjanc, I.S., 2000. Myocyte enhance factor 2C and myogenin up-regulate each other's expression and induce the development of skeletal muscle in P19 cells. J. Biol. Chem. 275, 41–46.
Rudnicki, M.A., Braun, T., Hinuman, S., Jaenisch, R., 1992. Inactivation of *MyoD* in mice leads to upregulation of the myogenic HLH gene *Myf-5* and results in apparently normal muscle development. Cell 71, 383–390.
Rudnicki, M.A., Schnesgelsber, P.N., Stead, H., Braun, T., Arnold, H.H., Jaenisch, R., 1993. MyoD or Myof-5 is required for the formation of skeletal muscle. Cell 75, 1351–1359.
Schmid, C., Steiner, T., Froesch, E.R., 1983. Preferential enhancement of myoblast differentiation by insulin-like growth factors (IGF I and IGF II). FEBS Lett. 161, 117–121.
Schofield, J.N., Wolpert, L., 1990. Effect of TGF-beta 1 TGF-beta 2, and bFGF on chick cartilage and muscle cell differentiation. Exp. Cell Res. 191, 144–148.
Schultz, E., 1989. Satellite cell behavior during skeletal muscle growth and regeneration. Med. Sci. Sports Exerc. 21, S181–S186.
Schultz, E., Gibson, M.C., Champion, I., 1978. Satellite cells are mitotically quiescent in mature mouse muscle: an EM and radioautographic study. J. Exp. Zool. 206, 451–456.
Schultz, E., Heckman-Jones, L., 1990. Labeling characteristics of satellite cells in vivo. J. Cell Biol. 111, 34A (Abstr.).
Schultz, E., Lipton, B.H., 1982. Skeletal muscle satellite cells: changes in proliferation potential as a function of age. Mech. Ageing Dev. 20, 377–383.
Scott, H.M., Phillips, R.E., 1936. Egg size in relation to growth of Narragansett turkeys. Poultry Sci. 15, 435–438.
Sejersen, T., Betsholtz, C., Sjolund, M., Heldin, C.-H., Westermark, B., Thyberg, J., 1986. Rat skeletal myoblasts and arterial smooth muscle cells express the gene for the A chain but not the gene for B chain (c-sis) of platelet-derived growth factor (PDGF) and produce a PDGF-like protein. Proc. Natl. Acad. Sci. U.S.A. 83, 6844–6848.
Shin, J., McFarland, D.C., Velleman, S.G., 2012a. Heparan sulfate proteoglycans, syndecan-4 and glypican-1, differentially regulate myogenic regulatory transcription factors and paired box 7 expression during turkey satellite cell myogenesis: implication for muscle growth. Poultry Sci. 91, 201–207.
Shin, J., McFarland, D.C., Strasburg, G.M., Velleman, S.G., 2013b. The function of death-associated protein 1 in proliferation, differentiation, and apoptosis of chicken satellite cells. Muscle Nerve 48 (5), 777–790.
Shin, J., McFarland, D.C., Velleman, S.G., 2013a. Migration of turkey muscle satellite cells is enhanced by the syndecan-4 cytoplasmic domain through the activation of RhoA. Mol. Cell. Biochem. 375, 115–130.
Shin, J., Song, Y., McFarland, D.C., Velleman, S.G., 2012b. Function of the syndecan-4 cytoplasmic domain in oligomerization and association with α-actinin in turkey muscle satellite cells. Mol. Cell. Biochem. 363, 437–444.
Shinichi, S., Tsuyoshi, O., Yoshinobu, U., Yumi, O., Eiji, K., 2012. Polymorphism of insulin-like growth factor 1 gene is associated with breast muscle yields in chickens. Anim. Sci. J. 83, 1–6.
Smith, J.H., 1963. Relation of body size to muscle cell size and number in the chicken. Poult. Sci. 42, 283–290.
Snow, M.H., 1977. The effects of aging on satellite cells in skeletal muscles of mice and rats. Cell Tissue Res. 185, 399–408.
Song, Y., McFarland, D.C., Velleman, S.G., 2012a. Syndecan-4 cytoplasmic domain regulation of turkey satellite cell focal adhesions and apoptosis. Mol. Biol. Rep. 39, 8251–8264.
Song, Y., McFarland, D.C., Velleman, S.G., 2012b. Role of syndecan-4 side chains in turkey satellite cell apoptosis and focal adhesion formation. Cell Biol. Int. 36, 433–440.
Song, Y., McFarland, D.C., Velleman, S.G., 2012c. Fibroblast growth factor 2 and protein kinase C alpha are involved in syndecan-4 cytoplasmic domain modulation of turkey myogenic satellite cell proliferation. Comp. Biochem. Physiol. A 161, 44–52.
Song, Y., McFarland, D.C., Velleman, S.G., 2012d. Critical amino acids in syndecan-4 cytoplasmic domain modulation of turkey satellite cell growth and development. Comp. Biochem. Physiol. A 161, 271–278.
Sosnicki, A.A., 1993. Focal myonecrosis effects in turkey muscle tissue. Reciprocal Meat Conf. Proc. 46, 97–102.
Sosnicki, A.A., Wilson, B.W., 1991. Pathology of turkey skeletal muscle: implications for the poultry industry. Food Struct. 10, 317–326.

Sporer, K.R.B., Chiang, W., Tempelman, R.J., Ernst, C.W., Reed, K.M., Velleman, S.G., Strasburg, G.M., 2011a. Characterization of a 6K oligonucleotide turkey skeletal muscle microarray. Anim. Genet. 42, 75–82.

Sporer, K.R.B., Tempelman, R.J., Ernst, C.W., Reed, K.M., Velleman, S.G., Strasburg, G.M., 2011b. Transcriptional profiling identifies differentially expressed genes in developing turkey skeletal muscle. BMC Genomics 12, 143.

Srinivasan, D., Dodig, M., Muc Sean, M., Kalhan, S.C., McCullough, A.J., 2004. Skeletal muscle atrophy is associated with an increased expression of myostatin and impaired satellite cell function in the portacaval anastomosis rat. Am. J. Physiol. Gastrointest. Liver Physiol. 287, G1124–G1130.

Steinfield, R., Van Den Berghe, H., David, G., 1996. Stimulation of fibroblast growth factor receptor-1 occupancy and signaling by cell surface associated syndecans and glypican. J. Cell Biol. 133, 405–416.

Stockdale, F.E., 1990. The myogenic lineage: evidence for multiple precursors during avian limb development. Proc. Soc. Exp. Biol. Med. 194, 71–75.

Stockdale, F.E., Holtzer, H., 1961. DNA synthesis and myogenesis. Exp. Cell Res. 24, 508–520.

Tatsumi, R., Allen, R.E., 2004. Active hepatocyte growth factor is present in skeletal muscle extracellular matrix. Muscle Nerve 30, 654–658.

Tatsumi, R., Sheehan, S.M., Iwasaki, H., Hattori, A., Allen, R.E., 2001. Mechanical stretch induces activation of skeletal muscle satellite cells in vitro. Exp. Cell Res. 267, 107–114.

Taylor, W.E., Bhasin, S., Artaza, J., Byhower, F., Azam, M., Willard, D.H., Kull Jr., F.C., Gonzalez-Cadavid, N., 2001. Myostatin inhibits cell proliferation and protein synthesis in C2C12 muscle cells. Am. J. Physiol. Endocrinol. Metab. 280, E221–E228.

Topouzis, S., Majesky, M.W., 1996. Smooth muscle lineage diversity in the chick embryo. Two types of aortic smooth muscle cell growth and receptor-mediated transcriptional responses to transforming growth factor-beta. Dev. Biol. 178, 430–445.

Velleman, S.G., Anderson, J., Nestor, K.E., 2003a. Possible maternal inheritance of breast muscle morphology in turkeys at sixteen weeks of age. Poultry Sci. 82, 1479–1484.

Velleman, S.G., Anderson, J.W., Coy, C.S., Nestor, K.E., 2003b. Effect of selection for growth rate on muscle damage during turkey breast muscle development. Poultry Sci. 82, 1069–1074.

Velleman, S.G., Coy, C.S., Anderson, J.W., Nestor, K.E., 2007a. The effect of genetic increases in egg production and age and sex on breast muscle development of turkeys. Poultry Sci. 86, 2134–2138.

Velleman, S.G., Coy, C.S., McFarland, D.C., 2007b. Effect of syndecan-1, syndecan-4, and glypican-1 on turkey muscle satellite cell proliferation, differentiation, and responsiveness to fibroblast growth factor 2. Poultry Sci. 86, 1406–1413.

Velleman, S.G., Coy, C.S., Nestor, K.E., 2010. The influence of age on maternal inheritance of breast muscle morphology in turkeys. Poultry Sci. 89, 876–882.

Velleman, S.G., Nestor, K.E., 2004. Inheritance of breast muscle morphology in turkeys at sixteen weeks of age. Poultry Sci. 83, 1060–1066.

Velleman, S.G., Nestor, K.E., 2006. Inheritance of breast muscle morphology in a line of turkeys selected for increased egg production, its randombred control line, and reciprocal crosses between them. Poultry Sci. 85, 2130–2134.

Velleman, S.G., Sporer, K.R.B., Ernst, C.W., Reed, K.M., Strasburg, G.M., 2012. Versican, matrix Gla protein, death-associated protein expression affect muscle satellite cell proliferation and differentiation. Poultry Sci. 91, 1964–1973.

Velleman, S.G., Yeager, J.D., Krider, H., Carrino, D.A., Zimmerman, S.D., McCormick, R.J., 1996. The avian low score normal muscle weakness alters decorin expression and collagen crosslinking. Connect. Tissue Res. 34, 33–39.

Venkatasubramanian, K., Solursh, M., 1984. Chemotactic behavior of myoblasts. Dev. Biol. 104, 428–433.

Vial, C., Gutiérrez, J., Santander, C., Cabrera, D., Brandan, E., 2011. Decorin interacts with connective tissue growth factor (CTGF)/CCN2 by LRR12 inhibiting its biological activity. J. Biol. Chem. 286, 24242–24252.

Vial, C., Zúñiga, L.M., Cabello-Verrugio, C., Cañón, P., Fadic, R., Brandan, E., 2008. Skeletal muscle cells express the profibrotic cytokine connective tissue growth factor (CTGF/CCN2), which induces their differentiation. J. Cell Physiol. 215, 410–421.

Volk, R., Schwartz, J.J., Li, J., Rosenberg, R.D., Simons, M., 1999. The role of syndecan cytoplasmic domain in basic fibroblast growth factor-dependent signal transduction. J. Biol. Chem. 274, 24417–24424.

Weber, I.T., Harrison, R.W., Iozzo, R.V., 1996. Model structure of decorin and implications for collagen fibrillogenesis. J. Biol. Chem. 271, 31767–31770.

Weintraub, H., 1993. The *MyoD* family and myogenesis: redundancy, networks, and thresholds. Cell 75, 1241–1244.

Welty, J.C., 1982. Birds as flying machines; bones and muscles. In: The Life of Birds, third ed. Saunders College Publishing. pp. 1–8; 63-77.

Wilkie, R.S., O'Neill, I.E., Butterwith, S.C., Duclos, M.J., Goddard, C., 1995. Regulation of chick muscle satellite cells by fibroblast growth factors: interaction with insulin-like growth factor-I and heparin. Growth Regul. 5, 18–27.

Wilson, B.W., Nieberg, P.S., Buhr, R.J., 1990. Turkey muscle growth and focal myopathy. Poultry Sci. 69, 1553–1562.

Wolff, J.N., Gemmell, N.J., 2012. Mitochondria, maternal inheritance, and asymmetric fitness: why males die younger. Bioessays 35, 93–99.

Xu, Z., Ichikawa, N., Kosaki, K., Yamada, Y., Sasaki, T., Sakai, L.Y., Kurosawa, H., Hattori, N., Arikawa-Hirasawa, E., 2010. Perlecan deficiency causes muscle hypertrophy, a decrease in myostatin expression, and changes in muscle fiber composition. Matrix Biol. 29, 461–470.

Yablonka-Reuveni, Z., Quinn, L.S., Nameroff, M., 1987. Isolation and clonal analysis of satellite cells from chicken pectoralis muscle. Dev. Biol. 119, 252–259.

Yablonka-Reuveni, Z., Seifert, R.A., 1993. Proliferation of chicken myoblasts is regulated by specific isoforms of platelet-derived growth factor: evidence for differences between myoblasts and mid and late stages of embryogenesis. Dev. Biol. 156, 307–318.

Yang, W., Wang, K., Chen, Y., Zhang, Y., Huang, B., Zhu, D., 2003. Functional characterization of recombinant myostatin and its inhibitory role to chicken muscle development. Acta Biochem. Biophys. 35, 1016–1022.

Ye, W.V., McFarland, D.C., Gilkerson, K.K., Pesall, J.E., 1996. The role of platelet-derived growth factor in turkey skeletal muscle development. Cytobios 88, 53–62.

Yun, Y., McFarland, D.C., Pesall, J.E., Gilkerson, K.K., Vander Wal, L.S., Ferrin, N.H., 1997. Variation in response to growth factor stimuli in satellite cell populations. Comp. Biochem. Physiol. 117A, 463–470.

Yost, J.K., Kenney, P.B., Slider, S.D., Russell, R.W., Killefer, J., 2002. Influence of selection for breast muscle mass on myosin isoform composition and metabolism of deep pectoralis muscles of male and female turkeys. Poultry Sci. 81, 911–917.

Zeng, C.Y., Pesall, J.E., Gilkerson, K.K., McFarland, D.C., 2002. The effect of hepatocyte growth factor on turkey satellite cell proliferation and differentiation. Poultry Sci. 81, 1191–1198.

Zhang, Y., Cao, L., Yang, B.L., Yang, B.B., 1998. The G3 domain of versican enhances cell proliferation via epidermal growth factor-like motifs. J. Biol. Chem. 273, 21342–21351.

Zhang, X., Nestor, K.E., McFarland, D.C., Velleman, S.G., 2008. The role of syndecan-4 and attached glycosaminoglycan chains on myogenic satellite cell growth. Matrix Biol. 27, 619–630.

Zhang, Z., Vuori, K., Reed, J.C., Ruoslahti, E., 1995. The $\alpha5\beta1$ integrin supports survival of cells on fibronectin and up-regulates Bcl-2 expression. Proc. Natl. Acad. Sci. U.S.A. 92, 6161–6165.

Zimmerman, D.R., Ruoslahti, E., 1989. Multiple domains of the large fibroblast proteoglycan, versican. EMBO J. 8, 2975–2981.

Chapter 17

The Avian Immune System

Pete Kaiser and Adam Balic
The Roslin Institute & R(D)SVS, University of Edinburgh, Easter Bush, Midlothian, EH25 9RG, UK

17.1 INTRODUCTION

Although our knowledge of the avian immune system and the avian immune response to disease and vaccination still lags behind that of better studied biomedical model systems, such as the human and mouse, progress since the last edition of this book has been dramatic. Thanks to the chicken genome sequence, we now have far greater understanding of the genes and molecules available to the avian immune response and, therefore, access to the tools required to enable us to understand the biology of that response in far greater detail than previously. The focus of research in avian immunology, and therefore the focus of this chapter, remains the chicken because it is the main avian species of economic importance. As we apply our knowledge of the chicken's immune system to other avian species, it is obvious that, in broad terms, the same principles hold, although the details might vary. For example, different species have different numbers of members of immune multigene families (Huang et al., 2013).

In broad terms, the immune systems and responses of mammals and birds are similar. Both mount innate and adaptive immune responses, and the bird's adaptive immune responses include both cell-mediated and humoral immune responses, leading to immunological memory. However, when one looks at the organs, cells, and molecules of the immune response in birds, one begins to understand that mammals and birds achieve the same overall responses often in quite different ways and, in many respects (but not all), the avian immune response is different.

It would be very difficult to summarize all aspects of the avian immune system in this chapter. Instead, we will concentrate on the basic anatomy of the organs of the avian immune response, as well as a description of the major cell types and major areas where the cells and molecules of the immune response differ from those of mammals—in some cases, being unique to avians.

17.2 THE ORGANS AND CELLS OF THE AVIAN IMMUNE RESPONSE

We cannot understand the avian immune response without knowledge of its basic structure. Lymphoid tissues are either of epithelial (e.g., thymus and bursa of Fabricius) or mesenchymal (e.g., spleen and bone marrow) origin and are colonized by hematopoietic cells via the blood. The primary lymphoid organs, the thymus and the bursa of Fabricius, are colonized by stem cells of hematopoietic origin that develop in situ to become immunologically competent T and B cells, respectively. This development process involves educating the T and B cells to essentially ignore self-antigens but respond to foreign antigens. These immunologically mature cells then re-enter the circulation and colonize the peripheral lymphoid organs: these include the spleen, cecal tonsils, Peyer's patches, Meckel's diverticulum, the Harderian gland, and other gut-, bronchus-, skin-, nasal- and reproductive-associated lymphoid tissues. In these peripheral tissues, there are T- and B-dependent zones occupied primarily by T and B cells, respectively.

17.2.1 Primary Lymphoid Tissues

The two major arms of the adaptive immune response—T-dependent and B-dependent immune responses (more commonly called cell-mediated and humoral immune responses, respectively)—were first described in the chicken before they were described in any other vertebrate species. The bursa of Fabricius was first described in the seventeenth century by Hieronymous Fabricius. The existence of cell-mediated immune responses (involving macrophages and lymphocytes) and humoral immune responses (involving antibodies) was generally accepted by the 1960s, but it was not clear if the cells that produced antibodies and drove cell-mediated responses were the same or distinct populations. Through a series of bursectomy experiments carried out for other reasons, Glick et al. (1956) realized that lymphocytes from the bursa controlled antibody production; it was later confirmed that the bursa is the site of development of the B cell receptor repertoire in birds. Other workers then realized that cells from the thymus controlled cell-mediated immunity (Szenberg and Warner, 1962). Cooper et al. (1965) then proposed, because of the similarities between the immune systems and lymphoid tissues of birds and mammals, that there must be a mammalian equivalent of the bursa of Fabricius (later shown to be the bone marrow). Therefore, T cells are so-called because they develop in the thymus, and B cells because they develop in the bursa of Fabricius.

17.2.1.1 The Thymus, T Cells, and T Cell Receptors

The avian thymus lies in the neck, parallel to the vagus nerve and the internal jugular veins, and consists of 7–8 separate lobes on each side of the neck running from the third cervical vertebra to the upper thoracic vertebrae. The lobes reach a maximal size of 10–12 mm in diameter by 3–4 months of age and thereafter they begin to involute. Each lobe is surrounded by a fine fibrous connective tissue capsule and is embedded in adipose tissue. Each lobe is divided into a central medulla surrounded by a cortex. The thymus contains mostly T cells, fewer macrophages, which are scattered throughout both the cortex and medulla, and epithelial reticular cells. The subcapsular zone of the cortex is thought to be the major site of T cell proliferation. During T cell maturation, the cells migrate towards the cortico-medullary border, where thymic dendritic cells (DCs) and macrophages select the thymocytes, before the latter enter the medulla and then the circulation.

T cells recognize antigens presented in the context of major histocompatibility complex (MHC) molecules (see later) via the heterodimeric T cell receptor (TCR). Like mammals, chickens have both $\alpha\beta$ and $\gamma\delta$ TCRs (Chen et al., 1991). Each TCR is composed of two immunoglobulin (Ig) superfamily domains—a membrane-distal variable (V) region and a membrane-proximal constant (C) domain. The latter defines the class of the TCR. The Ig domains are anchored in the cell membrane by a connecting peptide and a transmembrane domain, followed by very short cytoplasmic regions that do not contain signaling motifs. Rather, signaling for all TCRs is mediated by the CD3 complex—a complex of five different proteins first identified in the chicken by Chen et al. (1986). There are two Vβ gene families in the chicken, Vβ1 and Vβ2, which are recognized by the antibodies TCR2 and TCR3, respectively (Chen et al., 1991). T cells expressing $\alpha\beta$ TCR are also subdivided by their expression of two other surface molecules, CD4 and CD8. CD4 is a co-receptor for MHC class II and CD8 for MHC class I. TCR-$\alpha\beta^+$CD4$^+$ T cells are described as T helper (Th) cells and TCR-$\alpha\beta^+$CD8$^+$ T cells as cytotoxic T (Tc) cells, and both shall be discussed in more detail later.

Precursor hematopoietic cells enter the thymus in three waves, and all three lineages are present in the first two waves (Chen et al., 1991). However, TCR-$\gamma\delta^+$ and TCR-$\alpha\beta_1^+$ mature T cells migrate to the spleen by 15 and 19 days of embryonic development (DE), respectively, whereas TCR-$\alpha\beta_2^+$ cells do not appear in the spleen until after hatching (Chen et al., 1991).

17.2.1.2 The Bursa of Fabricius

The bursa of Fabricius is essentially a diverticulum of the cloaca. In the chicken, it reaches its maximum size around 8 weeks of age and then, like the thymus, involutes. By sexual maturity, most bursae have largely disappeared. The bursa is surrounded by a thick coat of smooth muscle. Inside, the bursa is composed of 15–20 longitudinal folds, each containing many follicles, with total follicle numbers in a mature bursa being estimated at 8,000–12,000 (Olah and Glick, 1978a). The bursal anlage appears at 3–5 DE. Buds, which are the forerunner of the bursal follicles, begin to form thereafter and each goes on to be the medulla of a bursal follicle. The surface epithelium of each bud is made up of follicle-associated epithelium (FAE) and interfollicular epithelium (IFE). IFE produces a mucin-like substance that is released into the bursal lumen and lubricates the surface of the folds. FAE provides a direct connection between the follicular medulla and the bursal lumen. FAE can both take up antigen or particles from the lumen (e.g., Bockmann and Cooper, 1973; Olah and Glick, 1978a; Schaffner et al., 1974) and products of the bursal secretory DCs (BSDCs) in the opposite direction (Nagy et al., 2001). The bursal follicles themselves, like the thymic lobes, have both a cortex and a medulla (Figure 17.1).

17.2.1.3 B Cells

B cells colonize the bursa between 8 and 14 DE (Le Douarin et al., 1975; Houssaint et al., 1976), migrating there through the blood. Mature B cells expressing rearranged B cell receptors leave the bursa posthatch. There are at least three populations of B cells in the periphery posthatch. A short-lived (approximately 3 days) population expressing the LT2 antigen (Paramithiotis and Ratcliffe, 1996) comprise about 60% of peripheral B cells in the newly hatched bird but largely disappear by 3 weeks of age. A further 30–35% of peripheral B cells that do not express LT2 then disappear, having a lifespan of about 3 weeks (Paramithiotis and Ratcliffe, 1993, 1994, 1996). A third population (approximately 5% of peripheral B cells in the newly hatched bird) of short-lived B cells appear to be derived from rapidly dividing extrabursal precursors. These may also be the source of peripheral B cells in older birds after bursal involution.

The majority of the lymphoid cells in the bursa are B cells (98%), with only a scattering of T cells (2%). Macrophages are also present in both the cortex and medulla,

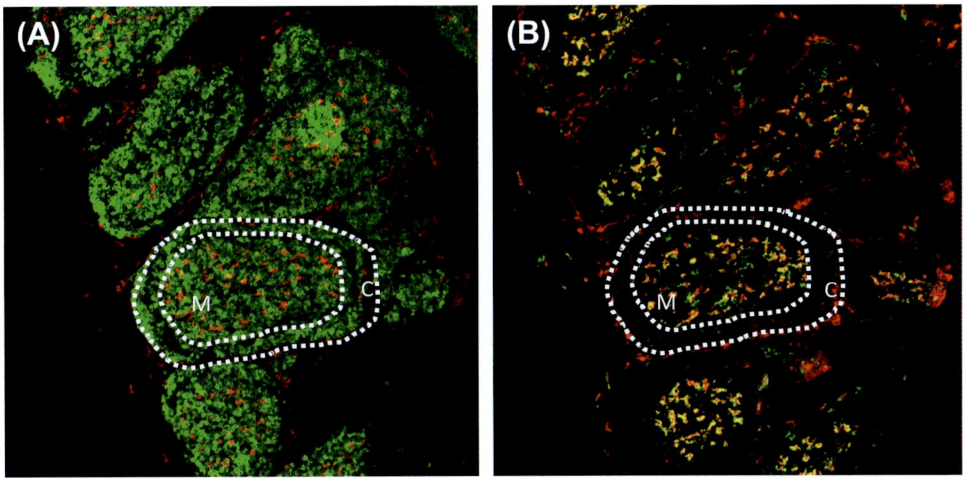

FIGURE 17.1 **Mononuclear phagocyte subsets in the bursa of Fabricius.** Immunofluorescence staining for Bu-1$^+$ B cells (A, green) or putative CD11 (B, green) and red fluorescent protein (RFP)$^+$ mononuclear phagocytes (red) in an 8 day old MacReporter chicken (a transgenic bird in which fluorescence is expressed in all cells of the macrophage lineage, driven by the CSF-1R promoter; Balic et al., unpublished). The structure of the B cell follicle in cross-section is shown (dotted lines). The B cell follicle medulla (M) contains a mononuclear phagocyte subset, the bursal secretory dendritic cell, that is positive for CD11. In contrast, the interfollicular mononuclear phagocyte subset adjacent to the B cell follicle cortex (C) expressed low levels of CD11.

as well as macrophage-like cells in the FAE. BSDCs, first identified by Olah and Glick (1978b), are found only in the medulla where they interact with lymphocytes; BSDCs are thought to play a role in promoting the education of B cells by controlling their interaction with antigens.

17.2.1.4 The B Cell Receptor/Antibody

Generation of the antibody repertoire (or more correctly, of the B cell receptor repertoire), of which the antibody represents a secreted form, in mammals takes place predominantly via gene rearrangement. For each immunoglobulin chain, there are multiple contiguous V (variable) segments, multiple D (diversity) segments for the heavy chains, and between 4 and 6 J (joining) segments; these are joined by recombination to produce functional VJ (light chain) or VDJ (heavy chain) Ig. This combinatorial diversity is augmented by junctional diversity at the joints between the different gene segments; the recombination mechanisms are not precise and nucleotides can be added or subtracted. Together, these mechanisms produce the B cell receptor and antibody repertoire, with as many as 10^8 different immunoglobulin specificities in an individual at any time.

In the chicken, the mechanism is somewhat different and is known as gene conversion. The chicken only has a single copy of functional V and J segments for both the light and heavy chains. Rearrangement of these would yield little or no diversity. However, both loci contain clusters of pseudogene V segments upstream of the functional V segment. VJ or VDJ rearrangement occurs in B cells in the bone marrow and spleen in chickens. B cells exhibiting this rearranged B cell receptor on their surface then migrate to the bursa of Fabricius. In the bursa, gene conversion involves random recombination between the pseudogene sequences and the canonical rearranged sequences, which leads to a B cell receptor/antibody repertoire similar in size to that of mammals.

Most mammals contain five classes of immunoglobulins: IgD, IgM, IgG (which undergoes subclass switching), IgA, and IgE. Three classes of avian Ig have been identified and represent the functional homologues of mammalian IgM, IgA, and IgG. IgD and IgE appear to be absent in all birds.

Chicken IgM is structurally and functionally homologous to its mammalian counterpart; during embryonic development, it is the first isotype expressed. Physical determinations of chicken IgM under various conditions have suggested a molecular weight (MW) of 823–954 kDa, with an average value of 890 kDa. The H chain has an MW of 70 kDa and the L chain 22 kDa, suggesting that chicken IgM is more likely to be tetrameric rather than pentameric; it could be a mixture of the two.

The avian homologue of mammalian IgG has similarities with both mammalian IgG and IgE and probably functionally shares the properties of both. It is the predominant form in sera, produced after IgM in the primary antibody response, and is the main isotype produced in the secondary response. It is described variously in the literature as IgY or avian IgG, but IgY in our opinion is preferable because the avian molecule appears to be the evolutionary predecessor of both IgG and IgE (Parvari et al., 1998; Bengten et al., 2000), sharing homology with each of the two mammalian isotypes. This nomenclature is also now well established for all nonmammalian vertebrates. The major difference between chicken IgY and mammalian IgG is that the H chain is longer in the chicken molecule. Avian IgY has five domains (V, C1–C4), as opposed to the four in mammalian

IgG, and the avian molecule does not possess a hinge. Instead, there are "switch" regions with limited flexibility at the C1-C2 and C3-C4 interfaces. Chicken IgY in serum is monomeric, with an MW of 165–206 kDa.

The predominant form of antibody activity in bodily secretions, such as tears, saliva, or bile, is IgA. In mammals, IgA is a dimer linked by a J chain that binds to its receptor on the surface of epithelial cells (Underdown and Schiff, 1986; Kerr, 1990). This receptor becomes integrated into the IgA molecule as a secretory component (SC), and the IgA complex is then transported through the epithelial cell and secreted into the lumen of the organ in question (Solari and Kraehenbuhl, 1985). SC promotes adhesion of IgA to the epithelial surface and protection from proteolytic degradation within cells. Chicken IgA, extracted from secretions such as bile, is usually larger than the IgA found in mammalian secretions, suggesting that the avian form is a trimer or a tetramer (Watanabe and Kobayashi, 1974). It also has a J chain and SC.

17.2.2 Secondary Lymphoid Tissues

17.2.2.1 The Spleen

In the chicken, the spleen is round/oval and lies behind and to the left of the proventriculus. It has an important role in embryonic lymphopoiesis; B cell progenitors undergo rearrangement of their Ig genes here before colonizing the bursa of Fabricius (Masteller and Thompson, 1994). At hatch, the spleen becomes a secondary lymphoid organ, providing an indispensable microenvironment for interactions between lymphoid and nonlymphoid cells. In fact, the contribution of the avian spleen to the immune system as a whole may be more important than it is in mammals because of the poorly developed lymphatic vessels and nodes in birds.

Mammalian and chicken spleens share the same basic structure, consisting of red and white pulp (Figure 17.2). The latter contains predominantly lymphocytes with few erythrocytes. Both lymphoid and nonlymphoid cells are found in the red pulp. $CD8^+TCR\gamma\delta^+$ T cells are plentiful in the sinuses, although $CD4^+TCR\alpha\beta1^+$ or $CD4^+TCR\alpha\beta2^+$ cells are also present.

In adult birds, most of the $TCR\gamma\delta^+$ cells are $CD8\alpha\beta^+$ (Tregaskes et al., 1995). Plasma cells expressing all three Ig isotypes are also present in the red pulp, especially near the large blood vessels. The red pulp also contains many macrophages, which are strongly acid-phosphatase positive and stain with the monoclonal antibodies 68.1 and 74.2 (Jeurissen et al., 1994) and KUL01 and MHC class II (Mast et al., 1998). Heterophils are also scattered throughout the red pulp sinuses.

The spleen is the largest lymphoid organ. It lacks afferent and efferent lymphatics, and it can therefore only obtain antigen from the blood. The white pulp surrounds the blood vessels in the spleen and has morphologically distinct areas. Periarteriolar lymphoid sheaths (PALS) surround the central arteries. Periellipsoid lymphoid sheaths (or PELS) analogous to the mammalian marginal zone, surround the penicillary capillaries. Germinal centres are found at the bifurcation of arteries, at the origin of the PALS.

The PALS contain predominantly T cells, mostly $CD4^+TCR\alpha\beta1^+$, but also some $CD4^+TCR\alpha\beta2^+$ and $CD8^+TCR\alpha\beta1^+$ cells. Chicken CD4 T cells are T helper cells and presumably play a role in B cell maturation as they surround the germinal centers. The PALS also contain KUL01+ MHC class II+ cells, which have been called

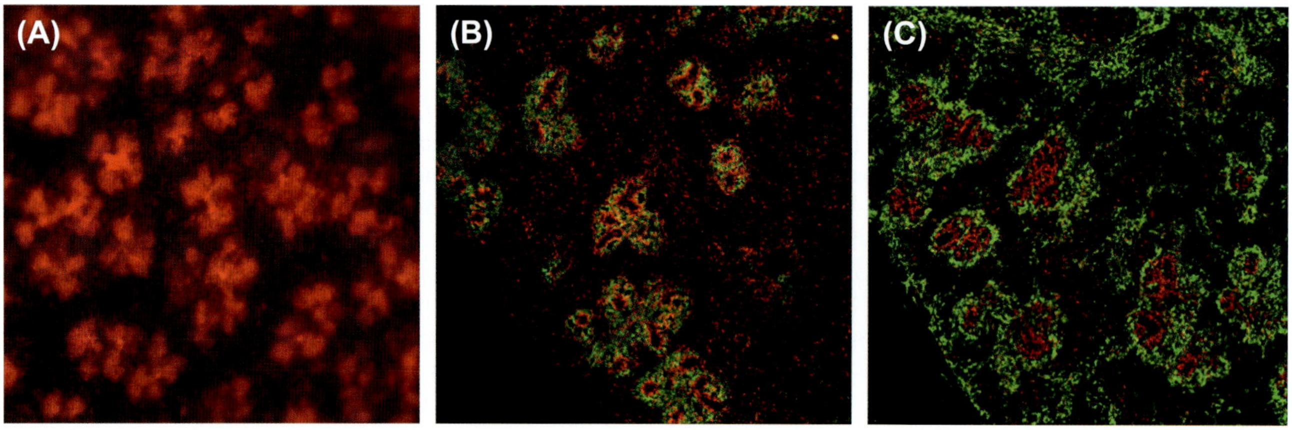

FIGURE 17.2 Whole-mount visualization (A) and immunofluorescence staining (B and C) of the spleen in a MacReporter chicken. (A) Whole mount visualization of the surface of the spleen, with mononuclear phagocytes expressing red fluorescent protein (RFP). RFP^+ cells are located throughout the spleen, but are particularly concentrated in the periellipsoid white pulp (PWP). The distinctive flower-like clustering of RFP^+ cells in the PWP reflects the structure of the penicillar capillaries as they branch off the central artery. (B) Immunofluorescence staining for $Bu\text{-}1^+$ B cells (green) and RFP^+ mononuclear phagocytes (red). RFP^+ mononuclear cells are concentrated in the B cell-rich periellipsoid lymphoid sheaths (PELS) and within the ellipsoid. (C) Immunofluorescence staining for CVI-ChNL-74.2 (green), which recognizes red pulp macrophages and a ring of macrophages surrounding the PELS and RFP^+ mononuclear phagocytes (red).

interdigitating dendritic cells (Igyarto et al., 2007). These cells are presumably the precursors of follicular dendritic cells (FDC), as they are either scattered or in aggregates approximately the size of germinal centers (Jeurissen et al., 1994; Igyarto et al., 2007; Olah and Glick, 1982; Gallego et al., 1993).

In the PELS, the penicillary capillaries are surrounded by ellipsoid-associated reticular cells. Ellipsoid-associated cells (EAC) are highly phagocytic and found at the surface of the ellipsoid (Olah and Glick, 1982). They are recognized by the monoclonal antibody 68.2 (Jeurissen, 1991) and are responsible for the clearance of antigens from the circulation. The ring of EAC is surrounded by Bu-1$^+$ B cells (Nagy et al., 2005; Mast and Goddeeris, 1998; Igyarto et al., 2008). B cells in the PELS primarily express IgM or IgA, but less IgY. Around the EAC, another ring is formed by 74.2$^+$ KUL01$^+$ macrophages (Nagy et al., 2005; Jeurissen et al., 1992; Mast et al., 1998).

Both innate and adaptive immune responses can be efficiently mounted in the spleen; it is therefore an important immune regulatory organ. Based on the localization of various lymphocyte and nonlymphoid cells in the spleen, the PALS seem most involved in adaptive immune responses, whereas the PELS are involved in both innate and adaptive immune responses.

17.2.2.2 The Harderian Gland

The Harderian gland (HG) is an immune-endocrine organ located in the orbit behind the eye. It produces secretions that lubricate and maintain the nictitating membrane via an excretory duct. The gland is divided into two areas based on differences in the surface epithelium and the organization of the underlying lymphoid tissues (Olah et al., 1996). The head of the gland has the structure of a typical secondary lymphoid organ with B cell-dependent germinal centers, FAE, and T cell-dependent interfollicular regions with scattered T cells and macrophages. The body of the gland contains many B and plasma cells and B and T cells cluster in separate areas in more mature glands.

The HG in young chicks contains CD45$^+$ cells, B cells, macrophages, interdigitating cells, and heterophils (Jeurissen et al., 1989; Bang and Bang, 1968). In older birds, the majority of T cells are CD4$^+$ and TCR$\alpha\beta$1$^+$, with a few scattered CD8$^+$TCR$\gamma\delta^+$ cells. The number of plasma cells increases dramatically with age, and uniquely they can proliferate in situ (Savage et al., 1992; Scott et al., 1993). IgM-, IgY- and IgA-producing plasma cells have all been described (Jeurissen et al., 1989; Olah et al., 1996; Jalkanen et al., 1984), reflecting the fact that all three Igs are found in tears.

17.2.2.3 The Cecal Tonsils

Cecal tonsils are located at head of the ceca near the ileocecal junction (Figure 17.3). The cecal tonsil primordium appears at about 10 DE, and lymphocytes are present by 18 DE (Payne, 1971). Posthatch lymphocyte numbers increase and germinal centers can be seen from the second week, increasing in number with age. The general structure of a cecal tonsil is comparable with that of a Peyer's patch, including a specialized lymphoepithelium, a subepithelial zone, germinal centers, and interfollicular areas. The subepithelial zone contains mainly IgM$^+$ B cells (with fewer IgY$^+$ B cells and occasional plasma cells) and a few CD4$^+$ and

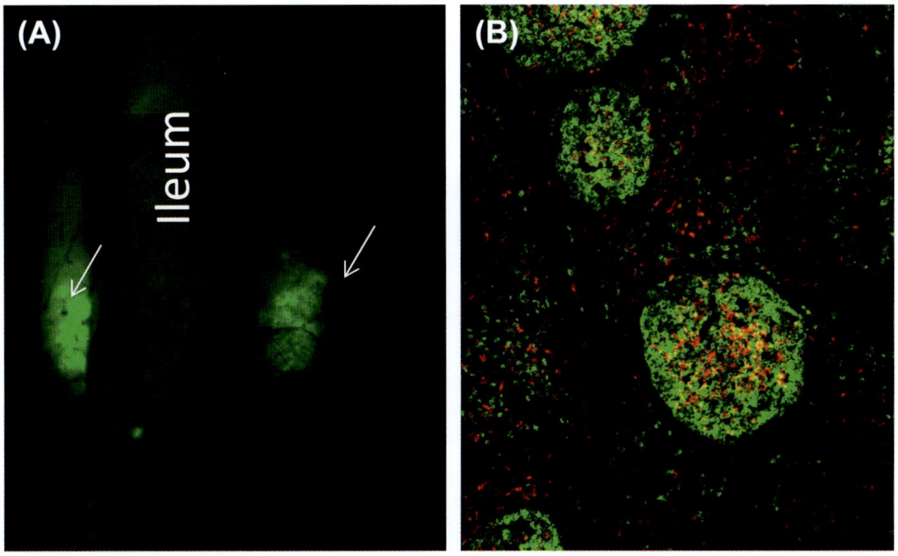

FIGURE 17.3 Whole-mount visualization (A) and immunofluorescence staining (B) of the cecal tonsil in a MacReporter chicken. (A) Whole-mount visualization of the surface of posthatch MacGreen cecal tonsils (arrows) with mononuclear phagocytes expressing green fluorescent protein (GFP). The cecal tonsil is composed of an extensive lymphoid tissue. (B) Immunofluorescence staining for Bu-1$^+$ B cells and mononuclear phagocytes (red) in the MacRed cecal tonsil identifies scattered B cells and mononuclear phagocytes, as well as germinal centers containing an extensive network of mononuclear phagocytes identified as avian follicular dendritic cells.

$CD8^+$ T cells expressing either TCRγδ or TCRαβ1. Monocytic cells are found throughout the tonsils but are mainly concentrated underneath the epithelium. The interfollicular areas are T cell-dependent and are mainly $CD4^+TCR\alpha\beta1^+$ cells (Olah and Glick, 1987; Bucy et al., 1988).

17.2.2.4 Peyer's Patches

Lymphoid aggregates with characteristics of mammalian Peyer's patches are present in the chicken intestine (Burns, 1982; Befus et al., 1980) (Figure 17.4). The exact number and distribution of chicken Peyer's patches is highly variable. The number of individual Peyer's patches peaks at 6–16 weeks of age, wherein they undergo involution, with the exception of one that is consistently found just before the ileocecal junction (Befus et al., 1980). This suggests that, as has been found in some mammals (Reynolds and Kirk, 1989), the chicken Peyer's patch may in fact include lymphoid structures of different ontogeny and function. Similarly to mammalian Peyer's patches, those in the chicken have widened villi, a follicular structure, a specialized epithelium containing M cells, active antigen uptake development in the embryo, and age-associated involution. Germinal centers, the subepithelial zone, and the interfollicular areas are similar to those in the cecal tonsils (see above). The subepithelial zone, where macrophages are more prevalent, is B cell dependent, and all B cells express the Bu-1/chB6 antigen (Jeurissen et al., 1989). The interfollicular zone is T cell dependent, and almost all T cells are $TCR\alpha\beta1^+$, with the majority being $CD4^+$ (Bucy et al., 1988). Only a few $TCR\gamma\delta^+$ cells (<5%) are present. The germinal centers and interfollicular areas contain mainly IgY^+ cells, with fewer IgA^+ and IgM^+ plasma cells.

17.2.2.5 Meckel's Diverticulum

The precise role of Meckel's diverticulum (MD) and its contribution to the bird's immune response remain to be determined. It is the remnant of the yolk stalk and is found halfway along the jejunum as an appendage of the small intestine. The wall of MD consists of four distinct layers. A serosa covers a thick layer of connective tissue. Next come bundles of smooth muscle. On the lumenal side of the muscle, the connective tissue contains large vessels and ganglions. This layer folds into the lumen of MD, and its surface is covered by columnar epithelium containing mucus-producing goblet cells. At hatch, MD contains no lymphoid cells. However, as the yolk sac regresses in the

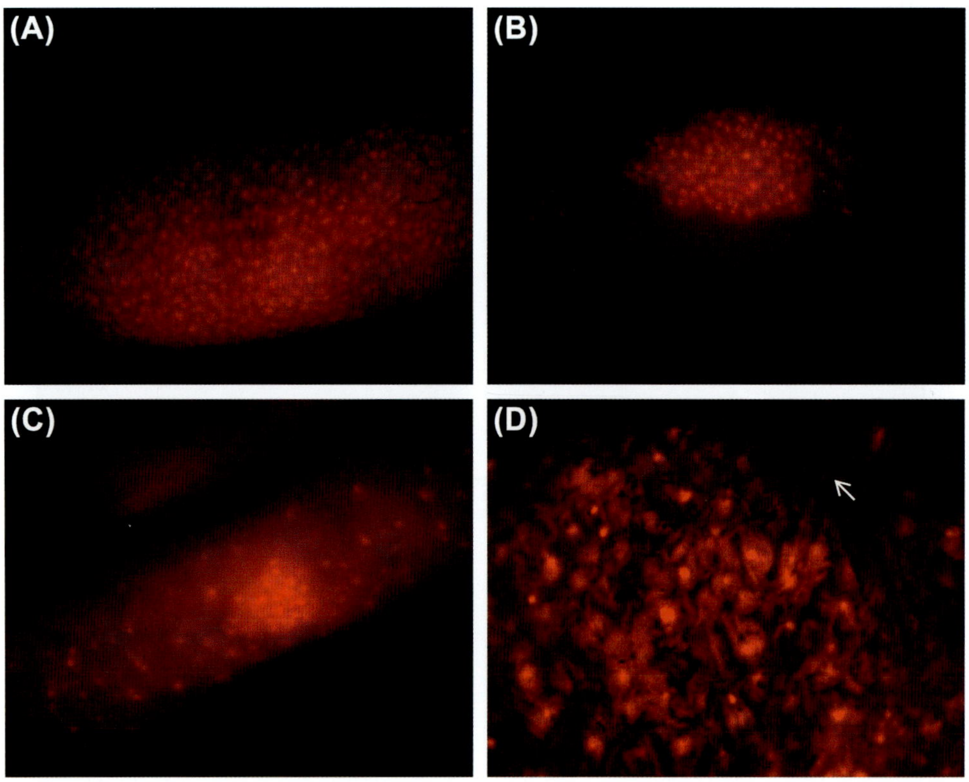

FIGURE 17.4 Whole-mount visualization (A–D) of Peyer's patches (PPs) in a 10-week-old MacReporter chicken. RFP expression in mononuclear phagocytes allows visualization of chicken gut-associated lymphoid tissues. Three different PP from the same individual shown at the same magnification (1.8×) demonstrate the wide range of sizes of chicken PPs. Individual PPs may contain tens to thousands of lymphoid follicles. Scattered non-PP lymphoid aggregates can also be seen in (C). (D) Ileum PP as seen from the mucosal surface. The normal chevron arrangement of intestinal villi (arrow) is disrupted by the lymphoid tissues of the PP.

first two weeks post-hatch myelopoietic and lymphopoietic tissues begin to appear. In the former tissue, there are three different zones. That closest to the yolk sac lumen contains monocytic cells, followed by a zone with large numbers of blast cells, and finally one containing immature granulocytes. Mature granulocytes are rare (Olah and Glick, 1984). In the lymphopoietic tissues, $CD45^+$ cells including IgM^+ B cells infiltrate the epithelium and connective tissue. Mononuclear phagocytes are found throughout the MD. The lymphoid tissue in MD gradually increases with age and fills the folds. As a result, the number of goblet cells is reduced, the epithelium is infiltrated by lymphocytes, and clusters of lymphoblats form around the antigen-presenting cells. Two to three months posthatch, germinal centers appear in large numbers located close to the muscle layer (Olah et al., 1984; Jeurissen et al., 1988). Dispersed throughout MD are IgA and IgY plasma cells. Later, B and T cell areas can be distinguished. The T cell areas are adjacent to germinal centers, whereas the B cell areas are generally beneath the epithelium (Jeurissen et al., 1988).

17.2.2.6 Other Mucosal Lymphoid Tissues and Cells

The combined mucosal surfaces of the gut, respiratory, and reproductive tracts represent by far the largest surface area in contact with the external milieu. They can also be considered the largest organ system in vertebrates, with each tract involved in diverse physiological functions, such as digestion and gas exchange. These functions come along with the continuous movement of external substances—nutrients and air, respectively—and the need to transport or exchange essential molecules across the mucosal surface. Hence, there is a continual challenge from new materials and microorganisms, including pathogenic microorganisms, which pass through the system. To prevent the entry of pathogens through the mucosal tissues, a wide variety of protective mechanisms has evolved. These range from barrier functions (e.g., keratinized skin, ciliated cells in the trachea, mucus secretion) to highly specialized immune cells (e.g., Langerhans cells in the skin) and organized lymphoid structures. In particular, the mucosal immune systems of the respiratory, gut, and reproductive tracts have highly developed lymphoid tissues, such as bronchus-associated lymphoid tissues and gut-associated lymphoid tissue (Figure 17.5), along with Peyer's patches, cecal tonsils, and lymphoid follicles.

In addition, the lamina propria harbors a wide range of immune cells, such as intraepithelial and lamina propria lymphocytes, macrophages, and dendritic cells. In mammalian species, the gut contains more lymphocytes than all secondary lymphoid tissues collectively. It is likely that this is also the case in avian species (reviewed in Schat and Meyers, 1991), particularly because birds lack lymph nodes. Numerous lymphoid follicles are found throughout the gut; typically these consist of one or more B cell follicles and a surrounding T cell zone. It is not clear if these structures represent secondary or tertiary lymphoid tissues; however, given the increasing number of different classes (Baptista et al., 2013) of similar structures that have been identified in mammals, it is likely that the lymphoid follicles of the chicken gut represent lymphoid tissues of diverse ontogenies and functions. Finally, the epithelial cells of the mucosal surfaces are able to sense pathogens and actively shape the response of the immune cells underneath.

For many years, research in mucosal immunology has focused on the question of how these cells become activated and interact to protect the host from invasion and dissemination of pathogenic microorganisms. Immunologists only recently began to appreciate that the gut harbors, and is in

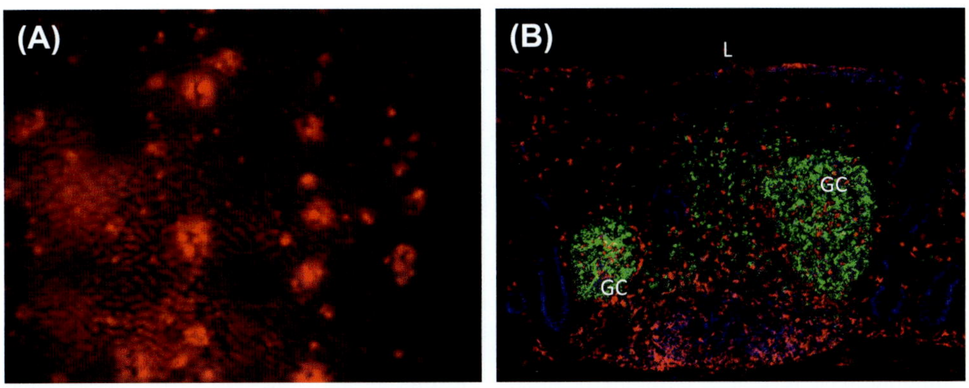

FIGURE 17.5 Whole-mount visualization (A) and immunofluorescence staining (B) of the rectal mucosa in a MacReporter chicken. (A) Whole-mount visualization of the rectum, as viewed from the mucosal surface, with mononuclear phagocytes expressing RFP. RFP^+ cells are located in the lamina propria of all mucosal villi, but are also concentrated in aggregates of variable size. (B) Immunofluorescence staining of a rectal mucosal lymphoid aggregate for $Bu\text{-}1^+$ B cells (green), RFP^+ mononuclear phagocytes (red) and cell nuclei (blue). RFP^+ mononuclear cells are found throughout the lamina propria, but are particularly associated with germinal centers (GC). L = lumen of the rectum.

constant contact with, trillions of commensal microorganisms that closely interact with the mucosal immune system (Molloy et al., 2012). This microbiome not only plays an important role in nutrient digestion; it is also a critical player in the development and function of the immune system and maintenance of intestinal integrity.

Under physiological conditions, the mucosal immune system has to remain tolerant of the microbiome to avoid chronic mucosal inflammation while at the same time being able to respond quickly and appropriately to pathogenic challenge. As a consequence, dysregulation of the cross-talk between the microbiome, the epithelium, and the mucosal immune system may result in local and systemic diseases. Although research on epithelial barrier biology has been a major topic in mammalian immunology in recent years, comparative studies in birds are only now being initiated (Mwangi et al., 2010).

17.3 REGULATION OF THE IMMUNE RESPONSE

17.3.1 Molecules and Cells of the Innate Immune Response

The chicken differs from mammals in its repertoire of lymphoid tissues, cells, and molecules, although many members of all three are the same. Arguably, the academically most interesting areas for study are where they differ, although it is vital to understand the similarities also.

As already discussed, the site of B cell development in the chicken is the bursa of Fabricius, which is not found in mammals. Chickens also lack lymph nodes, the primary site of antigen presentation in mammalian species. To date, the site of antigen presentation in the chicken has yet to be formally demonstrated, although it is thought to occur locally to the site of infection, probably in the many lymphoid aggregates found in the chicken's mucosal tissues. These aggregates include some unique structures, including Meckel's diverticulum and the cecal tonsils. Little has been published on immune responses in the former, but it is clear that the latter represent important sites at which immune responses are generated, particularly to pathogens such as *Campylobacter jejuni*, *Salmonella*, and *Eimeria tenella*.

Although one can detect cells from chicken blood that have the morphology of eosinophils, these have yet to be demonstrated to be functional. Eosinophils in mammals are key components of allergic and antihelminthic reactions (traditionally described as humoral immune responses)—in other words, of Th2 responses. There are many components of the mammalian Th2 response, particularly those that control the response and function of eosinophils, which are absent in the chicken. Th2 responses in mammals involve the cytokines IL-4, IL-13, and IL-5, all of which are upregulated following infection with an extracellular pathogen. Amongst other functions, they drive antibody isotype switching towards IgA and IgE, with the latter being crucial in eosinophil function because it binds the Fc-εR on eosinophils, triggering their degranulation. IgE production is driven by IL-5, after IL-4 induces isotype switching to IgE. In the chicken, IL-13 and IL-4 (the former up to 100-fold more than the latter) are both induced following infection with extracellular pathogens (Degen et al., 2005; Powell et al., 2009), but IL-5 expression is switched off (Powell et al., 2009). IgE is not produced in birds: chickens only have IgM, IgY (the functional equivalent of mammalian IgG, which does not have subtypes), and IgA. The eotaxins (chemokines) and the eotaxin receptor, which control migration of eosinophils in mammals, are absent in the chicken genome.

The other major immune cell type that is absent in the chicken is the neutrophil. These cells in mammals are essential effector cells of the induced innate response. In the chicken, they are replaced by heterophils; as yet, we lack the tools to determine if heterophils are a single set of cells with identical functions or if the term heterophil encompasses functionally different subsets of cells.

17.3.1.1 The CHIR Family

All pathogens (probably) have immune intervention/evasion mechanisms to manipulate the host's immune response to gain sufficient time to replicate. Viruses are particularly deft in this respect. To alert the adaptive immune system to viral infection, viral peptides are presented to CD8+ T cells in the context of MHC class I molecules. Many viruses use a variety of mechanisms to downregulate MHC class I expression in the cells they infect, to avoid triggering CD8+ T cells. In turn, the host has mechanisms to survey the levels of MHC class I expression on the surface of its cells, specifically by natural killer (NK) cells, which will target host cells with reduced surface MHC class I expression. In turn, some viruses have devised mechanisms to mimic MHC class I expression, either by encoding their own MHC class I mimics, selectively regulating host cell MHC class I expression, interfering with NK cell signaling, or modulating cytokine pathways that influence NK cells (reviewed in Orange et al., 2002).

The mechanism by which NK cells survey MHC class I expression levels and become activated is quite exquisite. In essence, a variety of receptors on the surface of NK cells recognize MHC class I and deliver a balance of activatory and inhibitory signals to the NK cell. Simply put, lack of detection of MHC class I on a host cell leads to a lack of inhibitory/suppressive signal for the NK cell, leading to NK killing of that cell.

In the human and mouse, these NK receptors are called KIR receptors (originally killer-cell immunoglobulin-like receptors for human, but now standing for killer-cell inhibitory receptors; the mouse uses a different family of receptors, dimeric lectins, to carry out the same function).

Immunoglobulin-like receptors form a superfamily and are involved in virtually all arms of the immune response. The human KIR receptors form part of a leukocyte receptor complex (LRC) on chromosome 19 that also include the leukocyte Ig-like receptor family (LILR). The equivalent LRC loci in the mouse are on chromosome 7.

The chicken LRC is located on microchromosome 31, and the genes have been given the name chicken immunoglobulin-like receptors, or CHIR (excellently reviewed in Viertlboeck and Gobel, 2011). In fact, the whole microchromosome seems only to encode the LRC, with no other genes yet assigned to it. Overall, the CHIR genes look equally like the mammalian KIR and LILR genes, and in terms of features look similar to both. CHIR genes are expressed on various leukocytes, like LILR genes, but are very diverse, like KIR genes. In humans, there are up to 17 KIR genes and 13 LILR genes. In the chicken, there are over 100 CHIR genes, a massive expansion compared with mammals, and the proteins encoded by the genes can be activatory, inhibitory, or bifunctional. The ligands for the vast majority of the CHIRs remain to be determined. The sole exception is CHIR-AB1, which binds to chicken IgY (Viertlboeck et al., 2007).

It is difficult to explain the reasons underlying this large expansion of genes in the chicken compared with mammals, and it remains the only major exception to the rule that chickens have the same immune gene families as mammals, but with fewer members. Further characterization of CHIR ligands will help to elucidate function, and it is tempting to speculate that this expansion of CHIR genes in some ways compensates for apparent contractions in the number of genes for other molecules, such as the MHC.

In comparison to mammals, chickens have different repertoires of Toll-like receptors (Boyd et al., 2007; Temperley et al., 2008; Cormican et al., 2009), defensins (Lynn et al., 2007), cytokines (Kaiser et al., 2005), chemokines (Kaiser et al., 2005; Hughes et al., 2007), antibodies (as discussed above), and other immune molecules (Kaiser, 2007, 2010). As can be seen from the references quoted, most of this knowledge postdates the sequencing of the chicken genome (ICGSC, 2004).

17.3.2 The Major Histocompatibility Complex

The adaptive immune response is only triggered when pathogen antigens are presented to T cells in the context of molecules called the major histocompatibility complex (MHC). The region of the genome that encodes the MHC is highly polymorphic and in mammals includes approximately 300 genes. The most relevant for antigen presentation are the class I genes, which present antigens from intracellular pathogens, mainly to CD8+ T cells, and the class II genes, which present antigens from extracellular pathogens, mainly to CD4+ T cells. MHC class I proteins are expressed on most cell types, whereas MHC class II proteins are mainly expressed on antigen-presenting cells such as dendritic cells (DC), macrophages, and B cells.

In contrast to the mammalian MHC, the chicken has a minimal essential MHC (Kaufman et al., 1999). In mammals, there are several class I and class II genes; precise numbers vary from species to species and in some cases between individuals of the same species. In contrast, in the chicken, the minimal essential MHC contains only two class I genes and two class IIβ genes, with only one of each dominantly expressed (reviewed in Kaufman, 2000). This has some striking consequences, particularly for the chicken's ability to mount an immune response to a pathogen, especially a virus.

Viral infections are cleared by cell-mediated immune responses, triggered by presentation of viral antigen via MHC class I. If a chicken expressed a single MHC class I gene, whose product will only fit certain peptides, then in order for an adaptive immune response to be triggered, the virus must encode a peptide that fits into that MHC class I molecule. Otherwise, the virus will not be seen by the adaptive immune response and will not be cleared. Most large viruses will encode so many peptides that the chances of none fitting the particular MHC class I gene in an individual is small. With smaller viruses, this chance becomes far greater. This has best been illustrated with Rous sarcoma virus (RSV) (Kaufman, 2000; Hofmann et al., 2003), which only encodes four genes. RSV is a transforming retrovirus that causes tumors, which progress and are fatal for some lines of chickens but are resolved in others. With this virus, the ability of an MHC class I gene product to bind a peptide from RSV (line B12) or not (line B4) absolutely defines the progression of the disease, with B12 birds regressing tumors and surviving and B4 birds succumbing to tumors.

17.3.3 Cytokines and Chemokines

Cytokines and chemokines are the molecules that control the precision of the immune response to disease or vaccination. Our understanding of the repertoire of cytokines and chemokines in the chicken was limited until recently, compared to those of mammalian species. A decade ago, only the type I interferon (Sekellick et al., 1994; Sick et al., 1996) and TGF-β family (Jakowlew et al., 1998, 1990; Burt and Jakowlew, 1992) had been characterized in the chicken. In general, chicken cytokines have only 25–35% amino acid identity with their mammalian orthologues. As a result, there are few, if any, cross-reactive monoclonal antibodies or bioassays. Moreover, cross-hybridization or degenerative (reverse-transcriptase) polymerase chain reaction approaches have been unsuccessful. Prior to the release of the chicken genome sequence (ICGCS, 2004), some progress had been made through a combination of expression cloning from expressed sequence tag (EST) libraries (Digby and Lowenthal, 1995; Sundick and Gill-Dixon,

1997; Weining et al., 1998), systematic sequencing of EST libraries (Lillehoj et al., 2001; Min and Lillehoj, 2002, 2004; Rothwell et al., 2004; Schneider et al., 2000, 2001) and genomics approaches based on the conservation of synteny (Avery et al., 2004; Balu and Kaiser, 2003). However, the availability of the chicken genome sequence has radically altered our ability to understand both the repertoire (Kaiser et al., 2005) and thereafter the biology of avian cytokines and chemokines.

Many of the cytokines and chemokines identified in mammals are also present in the chicken. The exceptions are in multigene families of cytokines and chemokines, where the chicken seems to have fewer members than do equivalent families in mammals. The absence of some of these multigene family members may explain some of the fundamental differences in the organs and cells of the avian immune system. Rather than give an exhaustive list of differences between the mammalian and chicken cytokine repertoires, we here focus on three multigene families—the chemokines, the tumor necrosis superfamily and the interleukin-1 family—where there are interesting differences that raise intriguing biological questions.

17.3.3.1 Chemokines

Chemokines are small chemical messengers that control the traffic of immune cells, both during inflammatory responses to infection and also to populate immune tissues and maintain homeostasis. As such, they are crucial to immune defense. They are broadly subdivided by structure into two main families (CC and CXC) and two much smaller families (XC and CX3C). A more useful division is arguably made by function; some chemokines have primarily a homeostatic role, and others are more involved in inflammatory responses. For the homeostatic chemokines, there is a single ligand-single receptor relationship. For the inflammatory chemokines, a single ligand can interact with several receptors, and a single receptor can have several ligands. The inflammatory chemokine genes and those of their receptors are grouped in multigene families at different loci in the genome. There is a single family of CXCL chemokines, and there are two families of CCL chemokines—the MCP and MIP families.

The chicken has slightly fewer homeostatic chemokines than mammals (Kaiser et al., 2005). For example, the eotaxins (CXCL9-11) and the eotaxin receptor (CXCR3) are absent, as discussed earlier, but for those present there are clear homologous relationships, which suggest similar functions. The one exception is CXCL13, which is a single copy gene in humans and mice, but at the same locus in the chicken genome there are three genes (CXCL13L1, CXCL13L2, and CXCL13L3). Of these, two interact with the cognate receptor, CXCR5, but the third (CXCL13L3) does not (Victoria Waters and Pete Kaiser, unpublished results).

The picture is less clear for the inflammatory chemokines and their receptors (Table 17.1). The mouse and human inflammatory CXCL families have five and ten

TABLE 17.1 Inflammatory Chemokine-Ligand Receptor Relationships in Humans and Chickens

Human		Chicken	
Receptor[1]	**Ligands[2]**	**Receptor[3]**	**Ligands[4]**
CXCR1	CXCL8, CXCL6	CXCR1	CXCLi1, CXCLi2 (Poh et al., 2008)
CXCR2	CXCL1–5, CXCL7	Unknown	CXCLi3
CCR1	*CCL4*, *CCL5*, *CCL6*, *CCL14*, *CCL15*, *CCL16*, *CCL23*	CCRa	CCLi1-10?
CCR2	**CCL2**, **CCL8**, *CCL16*	CCRb	CCLi1-10?
CCR3	**CCL11**, *CCL26*, **CCL7**, **CCL13**, CCL15, *CCL24*, *CCL5*, CCL28, *CCL18*		
CCR5	**CCL2**, *CCL3*, *CCL4*, *CCL5*, **CCL11**, **CCL13**, *CCL14*, *CCL16*		
CCR8	**CCL1**, *CCL16*	CCR8	Unknown
CCR4	*CCL3*, *CCL5*, CCL17, CCL20	CCRc	Unknown

[1]*The CCR genes are all found on human chromosome 3. CCR1, CCR2, CCR3, and CCR5 are at a single locus, CCR8 and CCR4 are at an increasing distance from this locus.*
[2]*Bold font indicates MCP family, whereas italic font indicates MIP family.*
[3]*At the locus in the chicken genome equivalent to that in the human genome where CCR1, CCR2, CCR3 and CCR5 are encoded, there are three chicken genes, CCRa, CCRb, and CCRc. In phylogenetic analysis, CCRa and CCRb cluster with human CCR1, CCR2, CCR3, and CCR5, whereas CCRc clusters with human CCR8.*
[4]*We hypothesize that CCLi1-10, the chicken MCP and MIP family chemokines, use CCRa and CCRb.*

members respectively, whereas the chicken family only has three members. Human IL-8 (CXCL8) is a member of this family; the mouse has no CXCL8 orthologue; in the chicken, there are two homologues, CXCLi1 and CXCLi2, which both share biological functions with human CXCL8 (Poh et al., 2008). The third chicken inflammatory CXCL, CXCLi3, has no direct mammalian orthologue. In mammals, the inflammatory CXCL chemokines signal through two receptors, CXCR1 and CXCR2, which are encoded at a single locus. The equivalent locus in the chicken genome only encodes a single receptor, CXCR1.

There are six members of the MCP family in humans, mice, and chickens. However, human and mouse MCP families share five members, and there are no direct orthologous relationships between the mammalian and chicken MCP families (Kaiser et al., 2005). The MIP families show greater differences across species, with eleven, five, and four members in the human, mouse, and chicken families, respectively. The human and mouse families share three members only. The chicken family appears to have an orthologue of mammalian CCL5, a potential orthologue of human CCL16, with no obvious orthologous relationship for the other two family members (Hughes et al., 2007). In mammals, the MCP and MIP family chemokines signal through four receptors (CCR1, CCR2, CCR3, and CCR5), which again are encoded at a single locus. At the equivalent locus in the chicken genome, there are only three genes. Two (CCRa and CCRb) look equally like all four mammalian receptors. The third (CCRc) clusters with the homeostatic CCR8 in phylogenetic analysis.

The general rule for chemokines is that the chicken has fewer members than mammals. Is this a "minimal essential chemokine repertoire"?

17.3.3.2 Tumor Necrosis Factors

The tumor necrosis factors form a superfamily (TNFSF) as do their receptors (TNFRSF). These molecules have crucial roles in all arms of immune responses, including induced innate responses and inflammation, apoptosis, and cell proliferation. The vast majority of the members of both families are membrane-bound, although certain of each can also be soluble proteins. Members of the TNFSF are primarily type II membrane proteins and function as homotrimers, whereas members of the TNFRSF are type I membrane proteins and function as monomers, although they often form higher level complexes with their ligands and other copies of themselves.

Members of the TNFSF in mammals and the chicken are present as small subfamilies in multigene loci on different chromosomes (see Table 17.2). As can be seen, the chicken has fewer TNFSF family members than mammals, and the rule seems to be that entire subfamilies are either present or absent in the chicken. Bearing in mind that the chicken genome is not complete, these absences are more likely to be real if the cognate TNFRSF members are also absent, and in general this is the case.

This has important implications for the chicken's immune response, particularly the apparent absence of the subfamily of genes encoding TNF-α and the two lymphotoxins (LT-α and LT-β). The LT genes are crucial in mammals for the development of secondary lymphoid organs including lymph nodes (reviewed in Fu and Chaplin, 1999). Their apparent absence in the chicken might explain the lack of lymph nodes. Ducks have primitive lymph nodes, but to date there is no evidence for the presence of the LT genes or TNF-α in the duck genome either. TWEAK and APRIL are involved in angiogenesis and immune regulation respectively in mammals, but both are absent in the chicken genome. 4-1BBL and LIGHT in mammals have costimulatory activity for activated T cells, enhancing T cell proliferation, secretion of IL-2, survival and cytolytic activity, as well as enhancing antitumor activity. CD27L (aka CD70) is expressed on activated T and B lymphocytes and mature DCs. Binding to its receptor, CD27, is important in priming, effector functions, differentiation and memory formation of T cells, and plasma and memory B cell generation. The absence of these latter five molecules in the chicken, but the presence of the functions which they are involved with in mammals, suggests again redundancy in the mammalian immune system. One could call the chicken TNFSF and TNFSRSF a "minimal essential TNF family."

17.3.3.3 The Interleukin-1 Family

Three members of the IL-1 family have been known for many years—IL-1α, IL-1β, and IL-1 receptor antagonist (IL-1RN). The three genes are encoded at a single locus in mammalian genomes. IL-1α and IL-1β are pro-inflammatory cytokines, whereas IL-1RN acts as a negative feedback mechanism to switch off the inflammation. A fourth member of the IL-1 family, IL-18, which drives Th1 adaptive immune responses, was discovered in 1995 at a different locus on a different chromosome. Genomic sequencing of the human and mouse identified seven more members of the family—six genes at the same locus as the original three family members and IL-33 at a third locus. The nomenclature of these genes in mammals was recently revised (Table 17.3). All members of the family are either pro- or anti-inflammatory.

To date, only four family members have been identified in the chicken. IL-1β (Weining et al., 1998) and IL-18 (Schneider et al., 2000) have been known for over a decade; they are still the only two IL-1 family members annotated in the chicken genome. We recently identified two new members of the chicken IL-1 family—IL-1RN (Gibson et al., 2012a) and IL-36RN (Gibson et al., 2012b)—from EST libraries. We have yet to identify the chromosomal

TABLE 17.2 Tumor Necrosis Factor Superfamily (TNFSF) and TNF Receptor Superfamily (TNFRSF) in Humans and Chickens Differ in Their Repertoires, with the Chicken Missing Three of the Mammalian Subfamilies

Human			Chicken		
TNFSF Member		Chromosome	TNFSF Member	Chromosome	TNFRSF Member
1	LT-α	6	No		No
2	TNF-α	6	No		Yes
3	LT-β	6	No		No
4	OX40L	1	Yes	21	Yes
18	AITRL	1	Yes	21	Yes
6	FASL	1	Yes	21	Yes
9	4-1BBL	19	No		Yes
7	CD27L	19	No		No
14	LIGHT	19	No		No
12	TWEAK	17	No		No
13	APRIL	17	No		No
15	VEGI	9	Yes	17	Yes
8	CD30L	9	Yes	17	Yes
5	CD40L	X	Yes	4	Yes
10	TRAIL	3	Yes	9	Yes
11	RANKL	13	Yes	1	Yes
13B	BAFF	13	Yes	1	Yes
–	–		TRAIL-L	4	?

location of the two novel chicken family members, but we have shown that they do not lie at the same locus as IL-1β. The large mammalian multigene locus, which encodes nine IL-1 family members, therefore seems to be absent in the chicken. This suggests that the family has undergone differential evolution since birds and mammals diverged from a common ancestor. It makes determination of the complete repertoire of chicken IL-1 family members challenging, especially as the two novel members are still not present in the genome sequence, suggesting they lie in regions of the genome that are refractory to sequencing.

The biological effects of the IL-1 family ligands are exerted through members of the IL-1 receptor (IL-1R) family, which are expressed on the surface of target cells or secreted as soluble receptors. In mammals, the family is comprised of 12 members (Figure 17.6) that are characterized by an IgG-like extracellular domain and a cytoplasmic Toll/IL-1R (TIR) domain. Despite the apparent paucity of IL-1 family members identified to date in the chicken, all of the IL-1R genes found in the human genome are present in the chicken at conserved loci (Gibson, Fife, and Kaiser, unpublished observations), suggesting that it is highly likely that more members of the chicken IL-1 family remain to be identified.

17.4 SUMMARY AND CONCLUSIONS

Birds live in the same geographical niches as mammals; share similar diets, ranges of body mass, and longevity; and are challenged with the same range of pathogens (e.g., viruses, bacteria, protozoan parasites, worms, ectoparasites). Birds and mammals share many general features of their immune responses, yet it is fair to say that birds' immune systems are somewhat simpler than those of mammals. It is difficult to hypothesize as to the selective pressures that have led to this increased complexity in mammals. It is hard to imagine that pressure being due to a particular pathogen challenge. Perhaps instead it has something to do with the nature of pregnancy in mammals and the requirement for nonrejection of the fetus by

TABLE 17.3 The Interleukin (IL)-1 Families in Humans and Chickens Differ Dramatically in Their Repertoires, with That of the Chicken Much Reduced

Human		Chicken	
Gene	Chromosome	Gene	Chromosome
IL-1α (IL-1F1)	2		
IL-1β (IL-1F2)	2	IL-1β	22
IL-36RN (IL-1F5)	2	IL-36RN	Unknown, but not in close proximity to IL-1β
IL-36α (IL-1F6)	2		
IL-37 (IL-1F7)	2		
IL-36β (IL-1F8)	2		
IL-36γ (IL-1F9)	2		
IL-38 (IL-1F10)	2		
IL-1rn (IL-1F3)	2	IL-1RN	Unknown, but not in close proximity to IL-1β
IL-18 (IL-1F4)	11	IL-18	24
IL-33 (IL-1F11)	9		

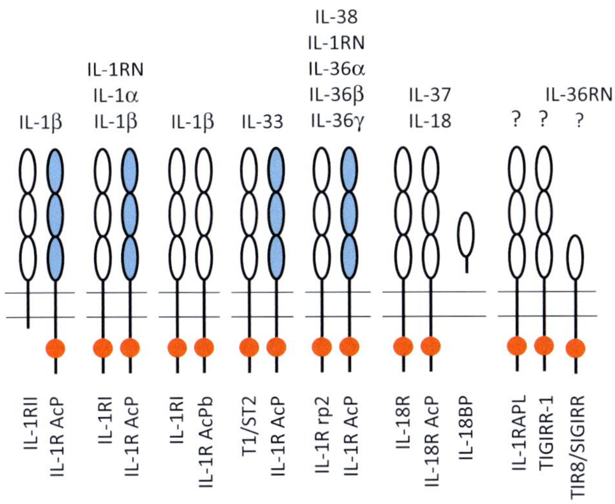

FIGURE 17.6 The IL-1 receptor family in humans, with their ligands shown above the molecules. The majority of the receptors contain cytoplasmic TIR domains (red circles), and several share the IL-1R accessory proteins (IL-1R AcP, shaded in blue). In the chicken, all 12 receptor family members are present, but only four of the IL-1 family members (IL-1β, IL-1RN, IL-36RN, and IL-18) have been identified.

the mother. However, that increased complexity seems to come with the price of increased prevalence of autoimmune disease in mammals—or have we simply failed to catalogue the same in birds?

It is an exciting time to be an avian immunologist. The chicken genome sequence and those of more than a hundred other avian species, either completed or in progress, and the rapid development of new tools and reagents to study host–pathogen interactions in birds, open the door to an improved understanding of avian immunophysiology.

REFERENCES

Avery, S., Rothwell, L., Degen, W.D.J., Schijns, V.E.C.J., Young, J., Kaufman, J., Kaiser, P., 2004. Characterization of the first non-mammalian T2 cytokine gene cluster: the cluster contains functional single-copy genes for IL-3, IL-4, IL-13 and GM-CSF, a gene for IL-5 which appears to be a pseudogene, and a gene encoding another cytokine-like transcript, KK34. J. Interferon. Cytokine Res. 24, 600–610.

Balu, S., Kaiser, P., 2003. Avian interleukin-12β (p40); cloning and characterisation of the cDNA and gene. J. Interferon. Cytokine Res. 23, 699–707.

Bang, B.G., Bang, F.B., 1968. Localized lymphoid tissues and plasma cells in paraocular and paranasal organ systems in chickens. Am. J. Pathol. 53, 735–751.

Baptista, A.P., Olivier, B.J., Goverse, G., Greuter, M., Knippenberg, M., Kusser, K., Domingues, R.G., Veiga-Fernandes, H., Luster, A.D., Lugering, A., Randall, T.D., Cupedo, T., Mebius, R.E., 2013. Colonic patch and colonic SILT development are independent and differentially regulated events. Mucosal Immunol. 6, 511–521.

Befus, A.D., Johnston, N., Leslie, G.A., Bienenstock, J., 1980. Gut-associated lymphoid tissue in the chicken. I. Morphology, ontogeny and some functional characteristics of Peyer's patches. J. Immunol. 125, 2626–2632.

Bengten, E., Wilson, M., Miller, N., Clem, L.W., Pilstrom, L., Warr, G.W., 2000. Immunoglobulin isotypes: structure, function, and genetics. Curr. Top. Microbiol. Immunol. 248, 189–219.

Bockmann, D.E., Cooper, M.D., 1973. Pinocytosis by epithelium associated with lymphoid follicles in the bursa of Fabricius, appendix and Peyer's patches. An electron microscopic study. Am. J. Anat. 136, 455–478.

Boyd, A., Philbin, V.J., Smith, A.L., 2007. Conserved and distinct aspects of the avian Toll-like receptor (TLR) system: implications for transmission and control of bird-borne zoonoses. Biochem. Soc. Trans. 35, 1504–1507.

Bucy, R.P., Chen, C.-L., Cihak, J., Loesch, U., Cooper, M., 1988. Avian T cells expressing γδ receptors localize in the splenic sinusoids and the intestinal epithelium. J. Immunol. 141, 2200–2205.

Burns, R.B., 1982. Histology and immunology of Peyer's patches in the domestic fowl (Gallus domesticus). Res. Vet. Sci. 32, 359–367.

Burt, D.W., Jakowlew, S.B., 1992. A new intrepretation of a chicken transforming growth factor-β4 complementary DNA. Mol. Endocrinol. 6, 989–992.

Chen, C.-L.H., Ager, L.L., Gartland, L.E., Cooper, M.D., 1986. Identification of a T3/T cell receptor complex in chickens. J. Exp. Med. 164, 375–380.

Chen, C.-L.H., Pickel, J.M., Lahti, J.M., Cooper, M.D., 1991. Surface markers on avian immune cells. In: Sharma, J.M. (Ed.), Avian Cellular Immunology. CRC Press, Boca Raton, FL, pp. 1–22.

Cooper, M.D., Peterson, R.D.A., Good, R.A., 1965. Delineation of the thymic and bursal lymphoid systems in the chicken. Nature 205, 143–147.

Cormican, P., Lloyd, A.T., Downing, T., Connell, S.J., Bradley, D., O'Farrelly, C., 2009. The avian Toll-like receptor pathway – subtle differences amidst general conformity. Dev. Comp. Immunol. 33, 967–973.

Degen, W.G.J., van Daal, N., Rothwell, L., Kaiser, P., Schijns, V.E.C.J., 2005. Th1/Th2 polarization by viral and helminth infection in birds. Vet. Microbiol. 105, 163–167.

Digby, M.R., Lowenthal, J.W., 1995. Cloning and expression of the chicken interferon-γ gene. J. Interferon. Cytokine Res. 15, 939–945.

Fu, Y.X., Chaplin, D.D., 1999. Development and maturation of secondary lymphoid tissues. Annu. Rev. Immunol. 17, 399–433.

Gallego, M., Olah, I., Del Cacho, E., Glick, B., 1993. Anti-S-100 antibody recognizes ellipsoid-associated cells and other dendritic cells in the chicken spleen. Dev. Comp. Immunol. 17, 77–83.

Gibson, M.S., Salmon, N., Bird, S., Fife, M., Kaiser, P., 2012a. Identification, cloning and functional characterisation of the IL-1 receptor antagonist in the chicken reveal important differences between the chicken and mammals. J. Immunol. 189, 539–550.

Gibson, M.S., Salmon, N., Bird, S., Kaiser, P., Fife, M., 2012b. Identification, cloning and characterization of interleukin-1F5 (IL-36RN) in the chicken. Dev. Comp. Immunol. 38, 136–147.

Glick, B., Chang, T.S., Jaap, R.G., 1956. The bursa of Fabricius and antibody production. Poult. Sci. 35, 224–225.

Hofmann, A., Plachy, J., Hunt, L., Kaufman, J., Hala, K., 2003. v-src oncogene-specific carboxy-terminal peptide is immunoprotective against Rous sarcoma growth in chickens with MHC class I allele B-F12. Vaccine 21, 4694–4699.

Houssaint, E., Belo, M., Le Douarin, N.M., 1976. Investigations on cell lineage and tissue interactions in the developing bursa of Fabricius through interspecific chimeras. Dev. Biol. 53, 250–264.

Huang, Y., Li, Y., Burt, D.W., Chen, H., Zhang, Y., Qian, W., Kim, H., Gan, S., Zhao, Y., Li, J., Yi, K., Feng, H., Zhu, P., Li, B., Liu, Q., Fairley, S., Magor, K.E., Du, Z., Hu, X., Goodman, L., Tafer, H., Vignal, A., Lee, T., Kim, K.W., Sheng, Z., An, Y., Searle, S., Herrero, J., Groenen, M.A., Crooijmans, R.P., Faraut, T., Cai, Q., Webster, R.G., Aldridge, J.R., Warren, W.C., Bartschat, S., Kehr, S., Marz, M., Stadler, P.F., Smith, J., Kraus, R.H., Zhao, Y., Ren, L., Fei, J., Morisson, M., Kaiser, P., Griffin, D.K., Rao, M., Pitel, F., Wang, J., Li, N., 2013. The duck genome and transcriptome provide insight into an avian influenza virus reservoir species. Nat. Genet. 45, 776–783.

Hughes, S., Poh, T.Y., Bumstead, N., Kaiser, P., 2007. Re-evaluation of the chicken MIP family of chemokines and their receptors suggests that CCL5 is the prototypic MIP family chemokine, and that different species have developed different repertoires of both the CC chemokines and their receptors. Dev. Comp. Immunol. 31, 72–86.

Igyarto, B.Z., Magyar, A., Olah, I., 2007. Origin of follicular dendritic cell in the chicken spleen. Cell Tissue Res. 327, 83–92.

Igyarto, B.Z., Nagy, N., Magyar, A., Olah, I., 2008. Identification of the avian B-cell-specific Bu-1 alloantigen by a novel monoclonal antibody. Poult. Sci. 87, 351–355.

International Chicken Genome Sequencing Consortium, 2004. Sequence and comparative analysis of the chicken genome provide unique perspectives on vertebrate evolution. Nature 432, 695–716.

Jakowlew, S.B., Dillard, P.J., Kondaiah, P., Sporn, M.B., Roberts, A.B., 1988. Complementary deoxyribonucleic acid cloning of a novel transforming growth factor-β messenger ribonucleic acid from chick embryo chondrocytes. Mol. Endocrinol. 2, 747–755.

Jakowlew, S.B., Dillard, P.J., Sporn, M.B., Roberts, A.B., 1990. Complementary deoxyribonucleic acid cloning of an mRNA encoding transforming growth factor-β2 from chicken embryo chondrocytes. Growth Factors 2, 123–133.

Jalkanen, S., Korpela, R., Granfors, K., Toivanen, P., 1984. Immune capacity of the chicken bursectomized at 60 hr of incubation: cytoplasmic immunoglobulins and histological findings. Clin. Immunol. Immunopathol. 30, 41–50.

Jeurissen, S.H.M., 1991. Structure and function of the chicken spleen. Res. Immunol. 142, 352–355.

Jeurissen, S.H.M., Claassen, E., Janse, E.M., 1992. Histological and functional differentiation of non-lymphoid cells in the chicken spleen. Immunology 77, 75–80.

Jeurissen, S.H.M., Janse, E.M., Koch, G., 1988. Meckel's diverticulum: a gut associated lymphoid organ in chickens. In: Fossum, S., Rolstad, B. (Eds.), Histophysiology of the Immune System. Plenum Publishing Corporation, New York, pp. 599–605.

Jeurissen, S.H.M., Janse, E.M., Koch, G., De Boer, G.F., 1989. Postnatal development of mucosa-associated lymphoid tissues in the chicken. Cell Tissue Res. 258, 119–124.

Jeurissen, S.H.M., Vervelde, L., Janse, E.M., 1994. Structure and function of lymphoid tissues of the chicken. Poult. Sci. Rev. 5, 183–207.

Kaiser, P., 2007. The avian immune genome – a glass half-full or half-empty? Cytogenet. Genome Res. 117, 221–230.

Kaiser, P., 2010. Advances in avian immunology – prospects for disease control: a review. Avian Pathol. 39, 309–324.

Kaiser, P., Poh, T.Y., Rothwell, L., Avery, S., Balu, S., Pathania, U.S., Hughes, S., Goodchild, M., Morrell, S., Watson, M., Bumstead, N., Kaufman, J., Young, J.R., 2005. A genomic analysis of chicken cytokines and chemokines. J. Interferon. Cytokine Res. 25, 467–484.

Kaufman, J., 2000. The simple chicken major histocompatibility complex: life and death in the face of pathogens and vaccines. Philos. Trans. R. Soc. Lond., B 355, 1077–1084.

Kaufman, J., Milne, S., Gobel, T.W.F., Walker, B.A., Jacob, J.P., Auffray, C., Zoorob, R., Beck, S., 1999. The chicken B locus is a minimal essential major histocompatibility complex. Nature 401, 923–925.

Kerr, M.A., 1990. The structure and function of human IgA. Biochem. J. 271, 285–296.

Le Douarin, N.M., Houssaint, E., Jotereau, F.V., Belo, M., 1975. Origin of hemopoietic stem cells in embryonic bursa of Fabricius and bone marrow studied through interspecific chimeras. Proc. Natl. Acad. Sci. USA 72, 2701–2705.

Lillehoj, H.S., Min, W., Choi, K.D., Babu, U.S., Burnside, J., Miyamoto, T., Rosenthal, B.M., Lillehoj, E.P., 2001. Molecular, cellular, and functional characterization of chicken cytokines homologous to mammalian IL-15 and IL-2. Vet. Immunol. Immunopathol. 82, 229–244.

Lynn, D.J., Higgs, R., Lloyd, A.T., O'Farrelly, C., Hervé-Grépinet, V., Nys, Y., Brinkman, F.S., Yu, P.L., Soulier, A., Kaiser, P., Zhang, G., Lehrer, R.I., 2007. Avian beta-defensin nomenclature: a community proposed update. Immunol. Lett. 110, 86–89.

Mast, J., Goddeeris, B.M., 1998. CD57, a marker for B-cell activation and splenic ellipsoid-associated reticular cells of the chicken. Cell Tissue Res. 291, 107–115.

Mast, J., Goddeeris, B.M., Peeters, K., Vandesande, F., Berghman, L.R., 1998. Characterisation of chicken monocytes, macrophages and interdigitating cells by monoclonal antibody KUL01. Vet. Immunol. Immunopathol. 61, 343–357.

Masteller, E.L., Thompson, C.B., 1994. B cell development in the chicken. Poult. Sci. 73, 998–1011.

Min, W., Lillehoj, H.S., 2002. Isolation and characterization of chicken interleukin-17 cDNA. J. Interferon. Cytokine Res. 22, 1123–1128.

Min, W., Lillehoj, H.S., 2004. Identification and characterization of chicken interleukin-16 cDNA. Dev. Comp. Immunol. 28, 153–162.

Molloy, M.J., Bouladoux, N., Belkaid, Y., 2012. Intestinal microbiota: shaping local and systemic immune responses. Semin. Immunol. 24, 58–66.

Mwangi, W.N., Beal, R.K., Powers, C., Wu, X., Humphrey, T., Watson, M., Bailey, M., Friedman, A., Smith, A.L., 2010. Regional and global changes in TCRalphabeta T cell repertoires in the gut are dependent upon the complexity of the enteric microflora. Dev. Comp. Immunol. 34, 406–417.

Nagy, N., Biro, E., Takacs, A., Polos, M., Magyar, A., Olah, I., 2005. Peripheral blood fibrocytes contribute to the formation of the avian spleen. Dev. Dyn. 232, 55–66.

Nagy, N., Magyar, A., David, C., Gumati, M.K., Olah, I., 2001. Development of the follicle-associated epithelium and the secretory dendritic cell in the bursa of Fabricius of the guinea fowl (*Numida meleagris*) studied by novel monoclonal antibodies. Anat. Rec. 262, 279–292.

Olah, I., Glick, B., 1978a. The number and size of the follicular epithelium (FE) and follicles in the bursa of Fabricius. Poult. Sci. 57, 1445–1450.

Olah, I., Glick, B., 1978b. Secretory cell in the medulla of the bursa of Fabricius. Experimentia 34, 1642–1643.

Olah, I., Glick, B., 1982. Splenic white pulp and associated vascular channels in chicken spleen. Am. J. Anat. 165, 445–480.

Olah, I., Glick, B., 1984. Meckel's diverticulum. I. Extramedullary myelopoiesis in the yolk sac of hatched chickens (Gallus domesticus). Anat. Rec. 208, 243–252.

Olah, I., Glick, B., 1987. Bursal secretory cells: an electron microscope study. Anat. Rec. 219, 268–274.

Olah, I., Glick, B., Taylor Jr, R.L., 1984. Meckel's diverticulum. II. A novel lymphoepithelial organ in the chicken. Anat. Rec. 208, 253–263.

Olah, I., Kupper, A., Kittner, Z., 1996. The lymphoid substance of the chicken's Harderian gland is organized in two histologically distinct compartments. Micros. Res. Tech. 34, 166–176.

Orange, J.S., Fassett, M.S., Koopman, L.A., Boyson, J.E., Strominger, J.L., 2002. Viral evasion of natural killer cells. Nat. Immunol. 3, 1006–1012.

Paramithiotis, E., Ratcliffe, M.J.H., 1993. Bursa dependent subpopulations of peripheral B lymphocytes in chicken blood. Eur. J. Immunol. 23, 96–102.

Paramithiotis, E., Ratcliffe, M.J.H., 1994. Survivors of bursal B cell production and emigration. Poult. Sci. 73, 991–997.

Paramithiotis, E., Ratcliffe, M.J.H., 1996. Evidence for phenotypic heterogeneity among B cells emigrating from the bursa of Fabricius. A reflection of functional diversity? Curr. Top. Microbiol. Immunol. 212, 27–34.

Parvari, R., Avivi, A., Lentner, F., Ziv, E., Tel-Or, S., Burstein, Y., Schechter, I., 1988. Chicken immunoglobulin gamma-heavy chains: limited VH gene repertoire, combinatorial diversification by D gene segments and evolution of the heavy chain locus. EMBO J. 7, 739–744.

Payne, L.N., 1971. The lymphoid system. In: Bell, D.J., Freeman, B.M. (Eds.), Physiology and Biochemistry of the Domestic Fowl. Academic Press, London, pp. 985–1037.

Poh, T.Y., Pease, J., Young, J., Bumstead, N., Kaiser, P., 2008. Re-evaluation of chicken CXCR1 determines the true gene structure; CXCLi1 (K60) and CXCLi2 (CAF/IL-8) are ligands for this receptor. J. Biol. Chem. 283, 16408–16415.

Powell, F., Rothwell, L., Clarkson, M., Kaiser, P., 2009. The turkey, compared to the chicken, fails to mount an effective early immune response to *Histomonas meleagridis* in the gut; towards an understanding of the mechanisms underlying the differential survival of poultry species. Parasite Immunol. 31, 312–327.

Reynolds, J.D., Kirk, D., 1989. Two types of sheep Peyer's patches: location along gut does not influence involution. Immunology 66, 308–311.

Rothwell, L., Young, J., Zoorob, R., Whittaker, C.A., Hesketh, P., Archer, A., Smith, A.L., Kaiser, P., 2004. Cloning and characterisation of chicken IL-10 and its role in the immune response to Eimeria maxima. J. Immunol. 173, 2675–2682.

Savage, M.L., Olah, I., Scott, T.R., 1992. Plasma cell proliferation in the chicken Harderian gland. Cell Prolif. 25, 337–344.

Schaffner, T., Mueller, J., Hess, M.W., Cottier, H., Sordat, B., Ropke, C., 1974. The bursa of Fabiricus: a central organ providing for contact between the lymphoid system and intestinal content. Cell. Immunol. 13, 304–312.

Schat, K.A., Meyers, T.J., 1991. Avian intestinal immunity. CRC Crit. Rev. Poult. Biol. 3, 19–34.

Schneider, K., Puehler, F., Baeuerle, D., Elvers, S., Staeheli, P., Kaspers, B., Weining, K.C., 2000. cDNA cloning of biologically active chicken interleukin-18. J. Interferon. Cytokine Res. 20, 879–883.

Schneider, K., Klaas, R., Kaspers, B., Staeheli, P., 2001. Chicken interleukin-6 cDNA structure and biological properties. Eur. J. Biochem. 268, 4200–4206.

Scott, T.R., Savage, M.L., Olah, I., 1993. Plasma cells of the chicken Harderian gland. Poult. Sci. 72, 1273–1279.

Sekellick, M.J., Ferrandino, A.F., Hopkins, D.A., Marcus, P.I., 1994. Chicken interferon gene: cloning, expression, and analysis. J. Interferon. Res. 14, 71–79.

Sick, C., Schultz, U., Staeheli, P., 1996. A family of genes coding for two serologically distinct chicken interferons. J. Biol. Chem. 271, 7635–7639.

Solari, R., Kraehenbuhl, J.P., 1985. The biosynthesis of secretory component and its role in the transepithelial transport of IgA dimer. Immunol. Today 6, 17–20.

Sundick, R.S., Gill-Dixon, C., 1997. A cloned chicken lymphokine homologous to both mammalian IL-2 and IL-15. J. Immunol. 159, 720–725.

Szenberg, A., Warner, N.L., 1962. Dissociation of immunological responsiveness in fowls with a hormonally arrested development of lymphoid tissue. Nature 194, 146.

Temperley, N.D., Berlin, S., Paton, I.R., Griffin, D.K., Burt, D.W., 2008. Evolution of the chicken Toll-like receptor gene family: a story of gene gain and gene loss. BMC Genomics 9, 62.

Tregaskes, C.A., Kong, F., Paramithiotis, E., Chen, C.-H., Ratcliffe, M.J.H., Davison, T.F., Young, J.R., 1995. Identification and analysis of the expression of CD8 and CD8 isoforms in chickens reveals a major TCRγδ-CD8 subset of intestinal intraepithelial lymphocytes. J. Immunol. 154, 4485–4494.

Underdown, B.J., Schiff, J.M., 1986. Immunoglobulin A: strategic defense initiative at the mucosal surface. Annu. Rev. Immunol. 4, 389–417.

Viertlboeck, B.C., Schweinsberg, S., Hanczaruk, M.A., Schmitt, R., Du Pasquier, L., Herberg, F.W., Gobel, T.W., 2007. The chicken leukocyte receptor complex encodes a primordial, activating, high-affinity IgY Fc receptor. Proc. Natl. Acad. Sci. USA 104, 11718–11723.

Viertlboeck, B.C., Gobel, T.W., 2011. The chicken leukocyte receptor cluster. Vet. Immunol. Immunopathol. 144, 1–10.

Watanabe, H., Kobayashi, K., 1974. Peculiar secretory IgA system identified in chickens. J. Immunol. 113, 1405–1409.

Weining, K.C., Sick, C., Kaspers, B., Staeheli, P., 1998. A chicken homolog of mammalian interleukin-1β: cDNA cloning and purification of active recombinant protein. Eur. J. Biochem. 258, 994–1000.

Part IV

Metabolism Theme

Chapter 18

Carbohydrate Metabolism

Colin G. Scanes
Department of Biological Sciences, University of Wisconsin, Milwaukee, WI, USA

18.1 OVERVIEW OF CARBOHYDRATE METABOLISM IN BIRDS

Carbohydrate metabolism in birds has close similarities to that in mammals. Glucose is absorbed in the small intestine. Glucose can be used for energy via glycolysis, the citric acid cycle, or the pentose pathway. Some tissues only use glucose, being incapable of using fatty acids. Glucose can be stored as the polysaccharide glycogen in the liver, muscles, and other tissues or used as a substrate to produce fatty acids (lipogenesis). Glucose can be synthesized from lactate, amino acids, and other glucogeneogenic precursors in the liver and kidneys. The differences between avian and mammalian carbohydrate metabolism include the following:

- The very high circulating concentrations of glucose in birds (discussed in more detail in Section 18.2)
- The site of lipogenesis being the liver

Several caveats need to be introduced. The energetic needs for extended flight are met using fatty acids, not glucose. Discussion of carbohydrate metabolism in birds is hampered by the preponderance of evidence coming from studies in domesticated birds, particularly from a single species, the chicken. The control of carbohydrate metabolism of (and hence circulating glucose concentrations by) the pancreatic hormones, glucagon, and insulin is addressed elsewhere in this volume.

The respiratory quotient (RQ) (carbon dioxide released divided by the oxygen consumed) informs as to the major substrate being used by an animal for energy. If the RQ is 1.0, carbohydrate (glucose) is the major substrate. If the RQ is 0.7, triglyceride (fatty acids) is the major substrate. If amino acids are being used for energy, the RQ is 0.8–0.9. Table 18.1 summarizes the RQ for a number of avian species. Birds seem to have an RQ of either 0.7 or 1.0, indicating that they are using either triglyceride/fatty acids or glucose. An RQ of 1.1 and greater, as reported in young chickens (Geelissen et al., 2006), reflects glucose utilization for energy production and fatty acid synthesis together with triglyceride deposition. When there is a shift from use of glucose to fatty acids predominantly, there is a decrease in RQ. For instance, the RQ decreases to 0.7 during fasting in the house sparrows (*Passer domesticus*), hummingbirds, and chickens during the nocturnal fast from ~0.9 to <0.7 (see Table 18.1).

18.2 CIRCULATING CONCENTRATIONS OF CARBOHYDRATES

18.2.1 Introduction: Circulating Concentrations of Glucose across Avian Species

The major circulating carbohydrate in birds is glucose. The mean circulating concentration of glucose in birds is 15.4 ± 0.32 (standard error of mean (SEM); number of species (n) = 139; Table 18.2). There are higher circulating concentrations of glucose in *Neognathae* than in the flightless rattites. There are marked differences between circulating concentrations of glucose between wild birds and mammals, with avian concentrations more than double those in mammals; for instance, plasma concentrations of glucose in fed and fasting rats were respectively 7.9 and 4.9 mM (Simon et al., 2011). The control of circulating concentrations of glucose by the pancreatic hormones, glucagon and insulin, is addressed elsewhere in this volume.

In mammals, there is a relationship between circulating concentration of glucose and body weight: the circulating concentration of glucose in mM = 7.6 − 0.44 log body mass in kilograms (Beuchat and Chong, 1998; Braun and Sweazea, 2008). In birds, a similar relationship was reported by Braun and Sweazea (2008): the circulating concentration of glucose in mM = 15.3 − 0.44 log body mass in kilograms. For birds, the y-intercept of the line at 1 kg is 15.3 mM (or 275 mg/dL) in wild birds (Braun and Sweazea, 2008). The basis for this relationship between circulating concentrations of glucose and body weight is not clearly established. No such relationship was observed by Beuchat and Chong (1998). There is frequently a dichotomy in the literature between research on wild and domesticated species. The analysis of circulating concentrations of glucose included only wild species.

TABLE 18.1 Examples of Respiratory Quotients (RQ) Reported in Birds

Species	RQ	Reference
Chicken (*Gallus gallus*) fed	~1.0 or 1.1 or greater	Brackenbury and El-Sayed, 1985; Geelissen et al., 2006
Chicken exercising	0.97	Brackenbury and El-Sayed, 1985
Chicken fasted	0.7	Boshouwers and Nicaise, 1981
House sparrow (*Passer domesticus*) fed	0.9 or ~1.0	Walsberg and Wolf, 1995
House sparrow fasted	0.7	Walsberg and Wolf, 1995; Khalilieh et al., 2012
Rufous hummingbirds (*Selasphorus rufus*) or Anna's hummingbirds (*Calypte anna*) upon feeding on nectar	1.0	Welch et al., 2007; Suarez et al., 2011
Rufous or Anna's hummingbirds fasted	0.7	Suarez et al., 1990, 2011; Welch et al., 2007
Carnivorous birds: Sparrow hawks (*Falco sparverius*), long-eared owls (*Asio otus*), saw whet owl (*Aegoliusa cadicus*)	0.74–0.78	Gatehouse and Markham, 1970

TABLE 18.2 Circulating Concentrations of Glucose Across the Class *Aves*

	Plasma/Serum Glucose mM (mmol/L) + SEM (Number of Species)
Order *Accipitriformes*	17.8 + 1.18 (11)
Order *Anseriformes*	12.2 + 1.34 (13)
Order *Apodiformes*	16.0 + 0.47 (3)
Order *Charadririiformes*	17.8 + 1.12 (8)
Order *Columbiformes*	16.4 + 0.94 (8)
Order *Ciconiformes*	14.5 + 1.01 (3)
Order *Falconiformes*	20.1 + 0.65 (8)
Order *Galliformes excluding poultry*	17.3 + 0.79 (14)
Order *Gruiformes*	13.1 + 0.57 (4)
Order *Passeriformes*	16.8 + 1.29 (12)
Order *Pelicaniformes*	14.6 + 0.90 (5)
Order *Phoenicopteriformes*	10.8 + 0.52 (4)
Order *Procellariiformes*	13.5 + 1.89 (5)
Order *Psittaciformes*	14.9 + 0.46 (20)
Order *Sphenisciformes*	14.1 + 0.83 (6)
Order *Suliformes*	10.6 + 1.32 (6)
Super-order *Neognathae*	15.5 + 0.32 (136)
Super-order *Palaeognathae*	11.1 + 1.18 (3)
Class *Aves*	15.4 + 0.32 (139)

18.2.2 Domestication and Circulating Concentrations of Glucose

There is conjecture that domestication and intensive selection have influenced glucose metabolism and circulating concentrations of glucose, although there is also some against this theory. The circulating concentrations of glucose in wild turkeys (*Meleagris gallopavo*) are more than 50% greater than those in domestic turkeys (Lisano and Kennamer, 1977; Anthony et al., 1990; see Table 18.3). In contrast, the circulating concentrations of glucose were higher in domestic than wild geese (Gee et al., 1981; Sitbon and Mialhe, 1979; see Table 18.3). A small study directly compared red jungle fowl (*Gallus gallus*, from which chickens are thought to have been domesticated), Asian village fowl, and commercial broiler chickens (Soleimani and Zulkifli, 2010). No differences were found between circulating concentrations of glucose between red jungle fowl and domesticated chickens (Table 18.3). Moreover, there are no discernible differences between the circulating concentrations of glucose in wild or domestic pigeons or ducks (Table 18.3). At present, there are no comprehensive studies comparing chickens with red jungle fowl or domestic turkeys, ducks and geese with wild turkeys, ducks and geese respectively. These should encompass multiple ages. Moreover, they should be performed in the same laboratory with circulating concentrations of glucose determined by the same methods to ensure consistency.

18.2.3 Fasting and Circulating Concentrations of Glucose

In some species of birds, circulating concentrations of glucose are depressed by fasting, such as in chickens (Dupont et al., 2008; Christensen et al., 2013), domestic ducks (*Anas platyrhynchos*; Zhang et al., 2005), domestic geese (*Anser anser*; Sitbon and Mialhe, 1979), house sparrows (*P. domesticus*; Khalilieh et al., 2012), hummingbirds (Anna's hummingbird (*Calypte anna*)), Costa's hummingbird (*Calypte costae*) and Ruby-throated hummingbirds

TABLE 18.3 Comparison of Circulating Concentrations of Glucose in Domesticated Birds and Wild Birds of the Same Species

Species	Plasma/Serum Concentration of Glucose (mM)	Reference
Red jungle fowl (*Gallus gallus*)	13.3	Soleimani and Zulkifli, 2010
Village chickens (*Gallus gallus domesticus*)	12.7	Soleimani and Zulkifli, 2010
Commercial broiler chickens (*Gallus gallus domesticus*)	15.0	Soleimani and Zulkifli, 2010
Fed chickens (15 studies)	13.0	Scanes, 2008
Fed chickens	13.2	Simon et al., 2011
Fasted chickens	12.6	Simon et al., 2011
Turkey (*Meleagris gallopavo*)		
Wild turkey	19.7	Lisano and Kennamer, 1977
Domestic turkey fasted	10.6	Anthony et al., 1990
Domestic turkey re-fed	11.5	Anthony et al., 1990
Duck (*Anas platyrhynchos*)		
Wild or mallard duck	11.5	Fairbrother et al., 1990
Domestic duck	10.0	Mean Applegate et al., 1999 & Zhang et al., 2005
Goose (*Anser anser*)		
Wild or Embden goose	12.3	Gee et al., 1981
Domestic goose	15.6	Sitbon and Mialhe, 1979
Pigeons (*Columba livia*)		
Wild or rock pigeon	15.5	Gayathri et al., 2004
Domestic or racing pigeon	16.1	Lumeij and de Brujne, 1985

(*Archilochus colubris*); Beuchat and Chong, 1998), and yellow-legged gulls (*Larus cachinnan*; Alonso-Alvarez and Ferrer, 2001). For instance, circulating concentrations of glucose are depressed by 25.3% in adult house sparrows fasted for 24 h (Khalilieh et al., 2012). Similarly, circulating concentrations of glucose decreased by fasting in late juvenile/adult domestic geese, being decreased by 25.4% in birds fasted for 24 h and by 32.5% in geese fasted for 4 days (Sitbon and Mialhe, 1979). In contrast, circulating concentrations of glucose were not depressed by fasting in other species/studies, such as adult ring doves (*Streptopelia risoria*; Lea et al., 1992), Japanese quail (*Coturnix japonica*; Sartori et al., 1996), Adélie penguins (*Pygoscelis adeliae*; Vleck and Vleck, 2002), Emperor penguins (*Aptenodytes forsteri*; Groscolas and Rodriguez, 1981), Garden warbler (*Silva borin*; Totzke et al., 1998), and juvenile herring gulls (*Larus argentatus*; Jeffrey et al., 1985); on the contrary, concentrations were increased in female ring doves (*Streptopelia risoria*; Lea et al., 1992).

18.2.4 Influence of Feeding

As might be expected, circulating concentrations of glucose are influenced by feeding. For instance, plasma concentrations of glucose are depressed at night in chickens when feeding is not occurring (Christensen et al., 2013). Moreover, concentrations of glucose are markedly increased following eating in hummingbirds, with circulating concentrations rising from already high concentrations to >26.5 mM—levels way into the diabetic range in mammals (Figure 18.1). This is not surprising with the consumption of nectar, which is high in both glucose and fructose. Plasma concentrations of glucose decline rapidly from 41 mM to basal (17 mM) within 1.5 h following feeding in, for instance, Anna's hummingbird (*Calypte anna*) (Beuchat and Chong, 1998). Similarly, oral loading with monosaccharides increases circulating concentrations of glucose in passerine birds (Figure 18.1) and chickens (Figure 18.2). Circulating concentrations of glucose return to basal within an hour (Sinsigalli et al., 1987; Figure 18.2).

18.2.5 Shifts in Circulating Concentration with Age, Reproductive State, and Migration

Circulating concentrations of glucose increase during the second half of embryonic development (Willemsen et al., 2011). There is a decrease in circulating concentrations of glucose immediately after hatching, followed by a gradual rise over 12 h (chickens: Rinaudo et al., 1982). Circulating concentrations of glucose decline during posthatch growth and development in chickens (*Gallus gallus*; Sinsigalli et al., 1987; Figure 18.2). The rate of clearance of glucose after a glucose challenge is slower in older but still juvenile birds, as shown in chickens by Sinsigalli et al. (1987; Figure 18.2).

There are some small sex differences in circulating concentrations of glucose in some birds. For instance, in adult

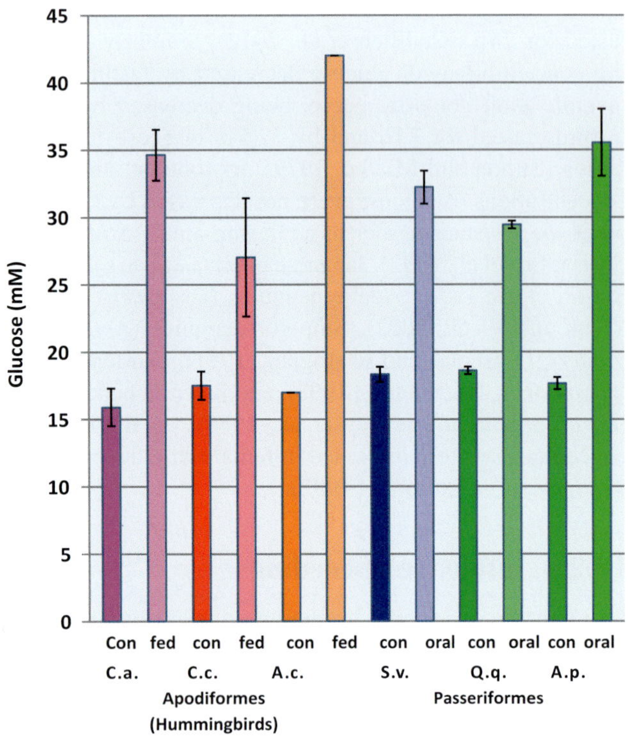

FIGURE 18.1 **Effects of nectar-feeding hummingbirds and administering an oral load of 1:1 glucose:fructose (intubated) to Passeriform birds (30 min later).** C.a., Anna's hummingbird (*Calypte anna*); C.c., Costa's hummingbird (*Calypte costae*); A.c., ruby-throated hummingbirds (*Archilochus colubris*); S.v., European starlings (*Sturnus vulgaris*); Q.q., common grackles (*Quiscalus quiscula*); A.p., red-winged blackbirds (*Agelaius phoeniceus*). Vertical bars: S.E.M. *Data calculated from Beuchat and Chong (1998) and Martinez del Rio et al. (1988).*

sexually mature quail, there is a small but consistent sex difference in circulating concentrations of glucose, being decreased 6.7% (Itoh et al., 1998) and 17.0% (Scholtz et al., 2009) in females in different reports. In contrast, circulating concentrations of glucose were lower in male than female mallards (Fairbrother et al., 1990; see Table 18.4).

Some changes have been reported in the circulating concentration of glucose during the breeding cycle. For instance, circulating concentrations of glucose are reduced by 53% in adult pigeons in the middle of period of incubating their eggs, but they recover late in incubation and then rise further after incubation is terminated (Gayathri et al., 2004). Similarly, in ring doves, circulating concentrations of glucose are increased after incubation and when the young squabs are receiving parental care (Lea et al., 1992). A similar reduction in circulating concentrations of glucose is reported in ducks (mallard duck (*A. platyrhynchos*); Fairbrother et al., 1990).

Prelaying circulating concentrations of glucose correlate well with later reproductive success in raising at least one chick in the greater sage grouse (Dunbar et al., 2005). What is not clear is the physiological basis of the circulating glucose concentrations having a predictive value for reproductive success.

There was little difference in the circulating concentrations of glucose in Canada geese (*Branta canadiensis interior*) irrespective of whether it was before or after the spring migration, during the breeding season, or in the period before the fall migration (Mori and George, 1978). An increase was observed after the fall migration (Mori and George, 1978). Circulating concentrations of glucose are markedly increased in Emperor penguins undergoing both fasting and

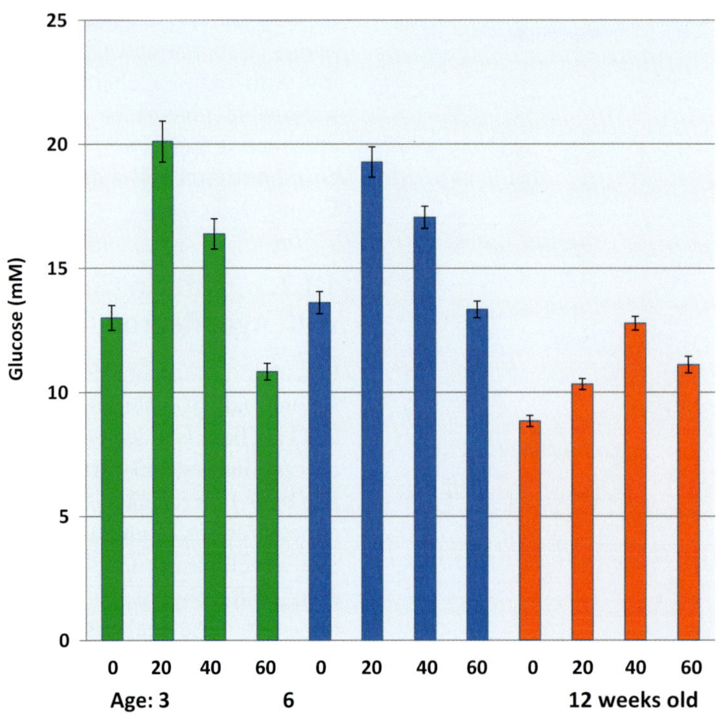

FIGURE 18.2 **Effect of glucose load (2 g glucose intubated per kg or 11.1 mmol/kg) on circulating concentrations of glucose at 3, 6 and 12 weeks of age in chickens prior to (0 min) and 20, 40, and 60 min after challenge.** *Data from Sinsigalli et al. (1987).*

TABLE 18.4 Effect of Reproductive Status on Serum Glucose Concentrations in Mallard Ducks (*Anas platyrhynchos*)

Reproductive Status	Serum Concentration of Glucose mM (mmol/L) ± SEM (*n*)
Pre-egg-laying females	13.2 ± 0.2 (34)[1]
Egg-laying females	14.3 ± 0.9 (10)[1]
Incubating females	11.7 ± 0.7 (20)[2]
Molting females	11.0 ± 0.5 (11)[2]
Nonreproductive females	11.9 ± 0.3 (34)[2]

[1,2]Different superscript numbers indicate significant difference $p < 0.01$.
Data from Fairbrother et al. (1990).

molting, increasing from a mean (SEM) of 19.7 ± 2.4 mM to 28.4 ± 6.5 mM (Groscolas and Rodriguez, 1981).

18.2.6 Shifts in Circulating Concentrations of Glucose due to Disease, Toxicants, and Husbandry Practices

There are pathological changes in the circulating concentration of glucose, with decreases with infectious disease and increases with exposure to toxicants. In the common pheasant, the circulating concentrations of glucose are depressed by severe infection with spironucleosis (Lloyd and Gibson, 2006). Similarly, in chickens with spiking mortality, there are extremely low circulating concentrations of glucose (Davis et al., 1995).

Environmental toxicants can elevate, depress, or have no effect on circulating concentrations of glucose. There is a marked increase in circulating concentrations of glucose in parakeets receiving the organophosphate insecticide quinalphos (roseringed parakeet, *Psittacula krameri*; Anan and Maitra, 1995). Circulating concentrations of glucose were depressed in mallard ducks exposed to the polychlorinated biphenol mixture Aroclor 1254 (Fowles et al., 1997). In contrast, there was no change in circulating concentrations of glucose in ducklings with toxicosis due to furazolidone (Webb et al., 1991).

Force feeding is used in some counties to produce *pate de fois gras* in ducks and geese. Circulating concentrations of glucose are increased by force feeding in geese (Sitbon and Mialhe, 1979) and ducks (Zhang et al., 2005).

18.2.7 Temperature and Circulating Concentrations of Glucose

There are effects of environmental temperature on circulating concentrations of glucose. Not only does fasting for 16 h depress circulating concentrations of glucose (Boussaid-Om Ezzine et al., 2010) but the effect is greater in birds exposed to elevated temperatures (32 °C for a week in broiler chickens; Boussaid-Om Ezzine et al., 2010). Moreover, glucose plus arginine administration to birds with free access to feed evoked a larger increase in circulating concentrations of glucose in the chickens at a higher environmental temperature (Boussaid-Om Ezzine et al., 2010).

18.3 GLUCOSE UTILIZATION

Glucose utilization can be very high in birds (see Table 18.5). For instance, in foraging hummingbirds, hovering flight is maintained by utilization of ingested sugars in rufous (*Selasphorus rufus*) and Anna's (*Calypte anna*) hummingbirds (Welch and Suarez, 2007). In contrast during migration, fats are employed as the energy source, with the Ruby-throated hummingbird (*Archilochus colubris*) losing most of the triglyceride stored in the adipose tissue during a 20 h flight over 960 km (600 miles; Hargrove, 2005). There are very high respiration rates of mitochondria from skeletal muscles of hummingbirds (7–10 mL O_2/cc/min); the rates are twice the maximum found in mammals (Suarez et al., 1991). This may be associated with increased mitochondrial surface area (Suarez et al., 1991). There is an increase in glucose turnover with exercise in chickens (Brackenbury and El-Sayed, 1985).

There are marked differences in glucose uptake by different organs (summarized in Table 18.6). The highest uptake is by the brain and heart. Insulin induced large increases in glucose uptake by the liver and some skeletal muscles. In contrast, glucose uptake is depressed in the brain of the chicken (Tokushima et al., 2005).

There are marked changes during the day in glucose uptake by tissues in chickens on a 12 h light/12 h dark cycle (Karaganis et al., 2009).

18.3.1 Developmental Changes

There is little glucose utilization for lipogenesis (fatty acid synthesis) in the liver of avian embryos and at hatching. They rise to a substantial plateau level within 8 days after hatching due to feeding, as shown in chickens (Goodridge, 1968a,b; Table 18.7).

18.3.2 Fasting and Glucose Utilization

There is reduced glucose utilization during fasting in birds. For instance, irreversible glucose turnover is depressed 32% by fasting (either associated with breeding or forced starvation) in Emperor penguins (*Aptenodytes forsteri*; Groscolas and Rodriguez, 1981):

- Fed 44.2 µmol/min/kg
- Fasted 30.1 µmol/min/kg

Similarly, there is a 57.6% decrease in glucose utilization in Japanese quail (*C. japonica*) fasted for 2 days (Sartori et al., 1996) and 64% in chickens fasted for 1 day (Belo et al.,

TABLE 18.5 Glucose Metabolism in Chickens

Parameter	Age	Method	Estimated Glucose Metabolism (μmol glucose/min/kg)	Reference
Total body glucose loss (utilization)	8 days	Disappearance of radioactive deoxyglucose from the blood	102	Calculated from Tokushima et al. (2005)
Total body glucose utilization	4 weeks	Heat produced	174	Calculated from Buyse et al. (1993)[1]
	5 weeks		112	
Total body glucose disappearance	5–6 months	Disappearance of radioactive glucose from the blood	79.4	Belo et al., 1976
Average			116	

[1]*This assumes that the respiratory quotient (RQ) of 1.0 with glucose being utilized for energy rather than either fats or amino-acids. Geelissen et al. (2006) reported RQ in similar-aged animals of 1.1 or greater. Data are calculated based on body weights of the same aged birds in the same laboratory (Buyse et al., 2004) and on catabolized energy from carbohydrate = 16.7 kJ/g substrate (3008.67 kJ per mole glucose) with RQ of 1.00 (Walsberg and Wolf, 1995).*

TABLE 18.6 Comparison of Glucose Uptake by Different Organs by Chickens

	Glucose Uptake (nmol/min/g)
Brain	315
Heart	223
Kidney	146
Small intestine	100
Skeletal muscle	38 (25–55)[1]
Adipose tissue	18.5
Liver	12.3

Data calculated From Tokushima et al. (2005) based on radioactive (^{3}H) 2-deoxyglucose uptake in 8-day-old chickens fasted for 12 h.
[1]*Range of muscles.*

1976). The decrease in glucose utilization is due to the following:

- Reduced glucose utilization by peripheral tissues such as skeletal muscles (Table 18.8)
- Reduced glucose utilization by the liver for:
  - lipogenesis, with as little as 2 h fasting (Figure 18.3). Rates of fatty acid synthesis are further decreased to 6.4% of control with 1 day fasting and to 0.2% of control with 3 days starvation (data from Yeh and Leveille (1971a,b)). Expression of genes for enzymes in lipogenesis is depressed by fasting in chickens (Désert et al., 2008; Sherlock et al., 2012).
  - Glycolysis (Goodridge, 1968a).

TABLE 18.7 Changes in Hepatic Utilization of Glucose for Glycolysis, Glycogenesis, and Lipogenesis during Early Posthatch Development in the Chicken

	Glucose Utilization[1] (As Percentage of Plateau Level)		
Age (days)	Glycolysis[2]	Glycogenesis[3]	Lipogenesis[4]
Late embryo and day 0	13.3 ± 0	<0.5	<0.5
2	51.1 ± 6.4	10.0 ± 4.1	4.0 ± 1.0
4	85.4 ± 5.4	8.3 ± 0	36.8 ± 9.8
8	**104.4 ± 4.6**	7.6 ± 1.1	**94.7 ± 14.7**
12	**84.5 ± 10.5**	34.8 ± 9.4	66.8 ± 10.4
16	**120.8 ± 12.5**	**105.1 ± 24.5**	**144.2 ± 29.0**

Bolded data are at plateau.
[1]*Determined by utilization of [U-^{14}C] glucose.*
[2]*Plateau conversion to CO_2: 1261 dpm/mg calculated from Goodridge (1968a).*
[3]*Plateau conversion to glycogen: 647 dpm/mg calculated from Goodridge (1968a).*
[4]*Plateau conversion to fatty acids: 582 dpm/mg calculated from Goodridge (1968a).*
Based on Goodridge (1968a).

18.4. GLUCOSE TRANSPORT

Glucose enters cells due to glucose transporters. There are five glucose transporter (GLUT) genes/proteins (chicken: Wagstaff et al., 1995; Kono et al., 2005) in birds, namely:

- *GLUT1*: Solute carrier family 2 facilitated glucose transporter member 1 (Wang et al., 1994) (chicken: Gene bank CGNC ID CGNC:49664)

TABLE 18.8 Glucose Generation/Utilization as Assessed from Venous-Arterial Differences in 8 Week Old Chickens either Fed or Starved for 6 Days

Venous-Arterial Difference (nmol blood/mL)		
	Fed	Starved (6 days)
Liver	+940[1]	+1450
Kidney	+44	+307
Hind quarter	−792	−199
Net glucose generation per unit organ weight (μmol/min/g)		
Liver[2]	1.8[1]	2.8 (Gluconeogenesis)
Kidney[3]	0.4	2.8 (Gluconeogenesis)
Net glucose generation (including gluconeogenesis) per unit body weight (μmol/min/kg)		
Liver	34	53
Kidney	2	17
Net glucose utilization per unit organ weight (nmol/min/g)		
Hind quarter/skeletal muscle[4]	40	10
Net glucose utilization per unit body weight (μmol/min/kg)		
Skeletal muscle[5]	22	5

[1]*Includes glucose not absorbed by liver from hepatic portal venous blood.*
[2]*Liver glucose generation calculated as the product of venous arterial difference and blood flow to the liver (1.92 mL/min; Tinker et al., 1986).*
[3]*Kidney glucose generation calculated as the product of venous arterial difference and arterial blood flow to the kidneys (9 mL/min/g; Merrill et al., 1981).*
[4]*Net glucose utilization by hindquarters calculated as the product of venous arterial difference and arterial blood flow to the skeletal muscle (50 μL/min/g; Merrill et al., 1981), using skeletal muscle as a surrogate for hindquarters.*
[5]*Assumes skeletal muscle represents 55% of body weight.*
Data from Tinker et al. (1986).

- *GLUT2*: Solute carrier family 2 facilitated glucose transporter member two GLUT3 (chicken: GenBank: Z22932.1)
- *GLUT3*: Solute carrier family 2 facilitated glucose transporter member 3 (SLC2A3) (chicken: Gene bank NM_205511.1)
- *GLUT5*: Solute carrier family 2 facilitated glucose fructose transporter member 5 (chicken: Gene bank CGNC ID1764)
- *GLUT8*: Solute carrier family 2 (facilitated glucose transporter member 8 (chicken: Gene bank CGNC ID378802))

There does not appear to be a homologue of the mammalian insulin dependent GLUT4 in birds (chickens: Seki et al., 2003; house sparrows *P. domesticus*: Sweazea and Braun, 2006). There is high expression of GLUT1 in adipose tissue and the brain (chicken: Kono et al., 2005) together with fibroblasts (Wagstaff et al., 1995). High GLUT2 expression is only found in the liver and kidneys (chicken: Kono et al., 2005; Lee et al., 2006), with low expression in the small intestine (chickens: Duarte et al., 2011). Epidermal growth factor (EGF) decreases GLUT2 protein levels in chicken hepatocytes (Lee et al., 2006). There is high GLUT3 expression in the brain (chicken: Kono et al., 2005) with chicken fibroblasts also expressing GLUT3 (Wagstaff et al., 1995) and all tissues expressing GLUT3 (house sparrows: Sweazea and Braun, 2006). Although GLUT1 is insulin independent in mammals, insulin stimulates glucose uptake along with both GLUT1 expression and protein content in avian myoblasts (chicken: Zhao et al., 2012). An insulin dependent GLUT8 is expressed in the chicken (high: adrenal, brain, kidney, lung, pancreas, spleen and testis; low: adipose tissue, liver, heart, skeletal muscle; chickens: Seki et al., 2003; Kono et al., 2005). GLUT expression decreases during the development of the avian erythrocyte (chicken: Johnstone et al., 1998) with only 200 copies of GLUT1 per mature erythrocyte (pigeon: Diamond and Carruthers, 1993). There is high expression of GLUT2 in the liver of the young chickens, together with some expression of both GLUT1 and GLUT3 (Humphrey et al., 2004). Skeletal muscle also exhibits high expression of GLUT3, with some expression of GLUT1 but no detectable expression of GLUT2 (Humphrey et al., 2004).

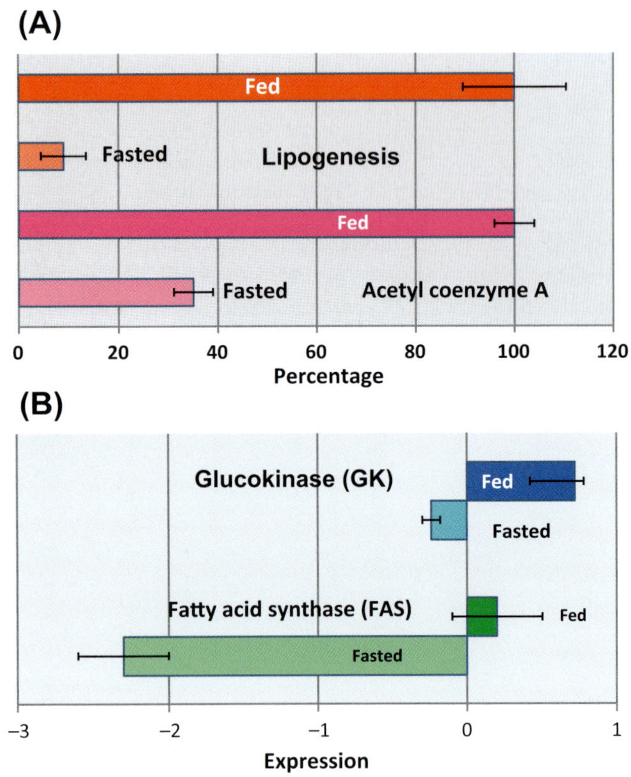

FIGURE 18.3 **Effects of fasting on carbohydrate metabolism in the liver.** (A) Effect of fasting (2h) on rate of lipogenesis in the chicken liver and on the liver concentration of acetyl coenzyme A. Data are expressed as a percentage of the control (fed) ± SEM. Calculated from Yeh and Leveille (1971a,b). (B) Effect of fasting for 5h on the hepatic expression of key enzymes in carbohydrate metabolism in the chicken. *Based on data from Dupont et al. (2008).*

18.5 INTERMEDIARY METABOLISM

18.5.1 Glucose Phosphorylation and Dephosphorylation

Glucose 6-phosphate plays a major role in the control of metabolism (Figure 18.4). There are rapid increases in hepatic concentrations of glucose 6-phosphate after hatching (turkeys: Rosebrough et al., 1979).

18.5.1.1 Glucose Phosphorylation to Glucose 6-Phosphate

Glucose is phosphorylated to glucose 6-phosphate catalyzed either by glucokinase or hexokinase:

Glucose + ATP → Glucose 6-phosphate

18.5.1.1.1 Enzyme: Glucokinase/Hexokinase

Glucokinase has been partially characterized both from chickens and mule ducks with 99% identity (Berradi et al., 2005, 2007). There is increased expression in the liver of mule ducks with overfeeding (Berradi et al., 2004) and feeding in chickens (Berradi et al., 2007). Fasting depresses glucokinase expression by 66% in chickens (Dupont et al., 2008). Low Michaelis constant (Km) hexokinase activity is present in the chicken liver and reduced with fasting (O'Neill and Langlow, 1978; Klandorf et al., 1986). High Km hexokinase activity is found in the mitochondria of the liver (chicken: Borrebaek et al., 2007). Hexokinase is present also in the muscles of hummingbirds (Suarez et al., 1990).

18.5.1.2 Glucose 6-Phosphatase

Glucose 6-phosphatase plays a critical role in glucose homeostasis:

Glucose 6-phosphate → Glucose + Phosphate

18.5.1.2.1 Glucose 6-Phosphatase

Glucose 6-phosphatase activity is present in the avian liver (chicken: O'Neill and Langlow, 1978). Thus, glucose 6-phosphate generated from glycogenolysis and glucogenesis is released from the liver into the circulation for peripheral use. There does not appear to be glucose 6-phosphatase in skeletal muscle; hence, muscle glycogen is not a source of circulating glucose. Along with elevated glycogenolysis and glucogenesis with fasting, there is increased glucose 6-phosphatase activity in the liver (chicken: O'Neill and Langlow, 1978).

18.5.2 Glycolysis

Glycolysis is summarized in Figure 18.4. The glycolytic pathway precedes carbohydrate metabolites entering the citric acid cycle and also allows the anaerobic metabolism of glucose.

18.5.2.1 Developmental Changes

There are low rates of hepatic glycolysis in the liver of avian embryos, rising quickly after hatching to a plateau level achieved after about a week (chicken: Goodridge, 1968a; Table 18.4).

18.5.2.2 Effects of Feeding and Fasting

The posthatch increase in glycosylation is induced by feeding (chicken: Goodridge, 1968b). Hepatics rates of glucose utilization for glycolysis and conversion to carbon dioxide are depressed by fasting. For instance, in young chickens, rates of glucose oxidation to carbon dioxide by liver slices are decreased by the following (Goodridge, 1968b):

- 1 day: 59.5% of control
- 3 days: 17.5%

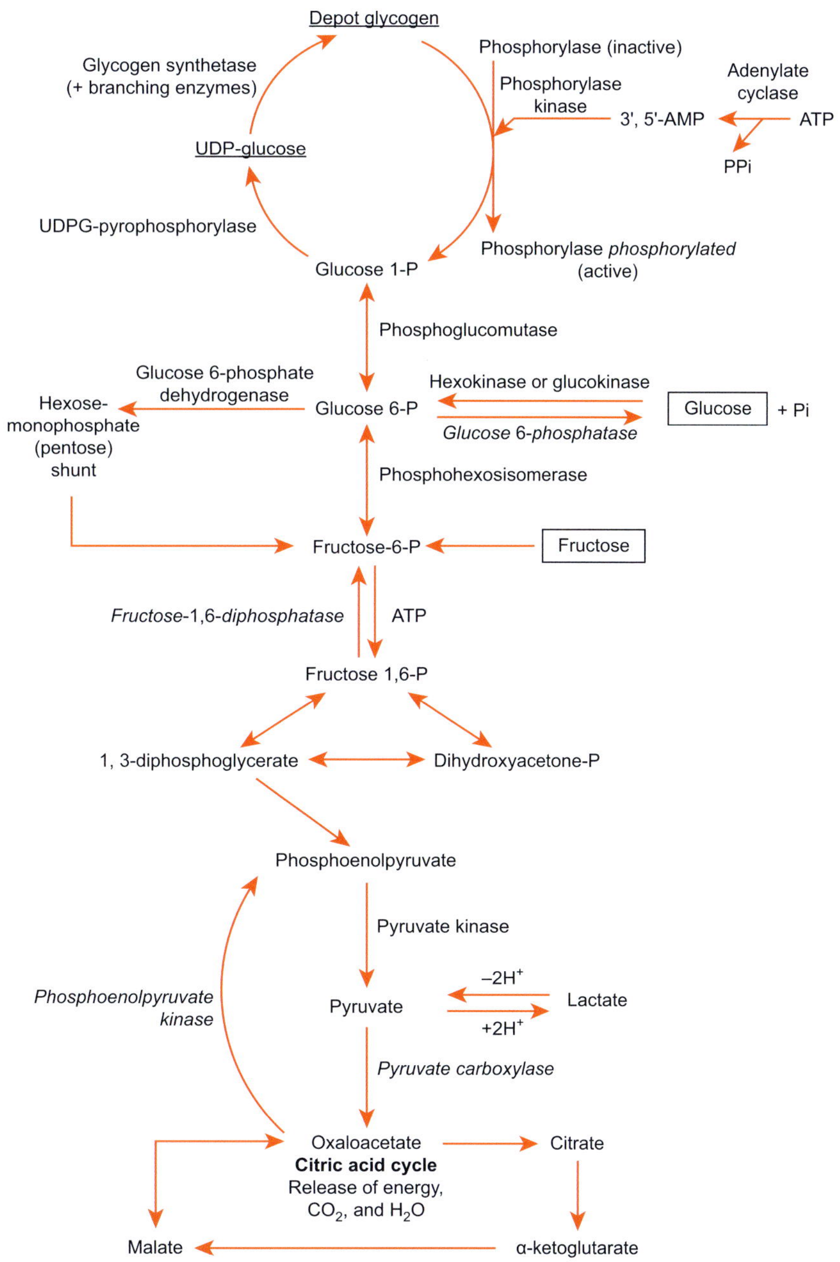

FIGURE 18.4 **Overview schematic diagram of carbohydrate metabolism (glucose phosphorylation/dephosphorylation, glycolysis, glycogenesis, glycogenolysis, and pentose phosphate shunt).**

Expression of genes for enzymes in the glycolytic pathway is depressed by fasting in chickens (Sherlock et al., 2012).

18.5.2.3 Erythrocytes

Glycolysis and pentose phosphate shunt are the pathways to obtain energy from glucose in the erythrocyte (pigeon: Kalomenopoulou et al., 1990). Phosphofructokinase activity, pyruvate kinase activity, and hexokinase activity are present in the erythrocyte (pigeon: Kalomenopoulou et al., 1990).

18.5.3 Citric Acid or Tricarboxylic Acid Cycle

18.5.3.1 Overview

Figure 18.5 summarizes the citric acid or tricarboxylic acid (TCA) cycle. There were increases in mitochondrial capacity (citrate synthase activity) at the end of embryonic and in early posthatch development (chicken: Walter et al., 2010). Citrate synthase (CS) and pyruvate kinase (PK) activities are found in skeletal muscle from chukar, pheasant, domestic turkey, and wild turkey (Shea et al., 2007). Citrate synthase activity in hummingbirds is similar to that

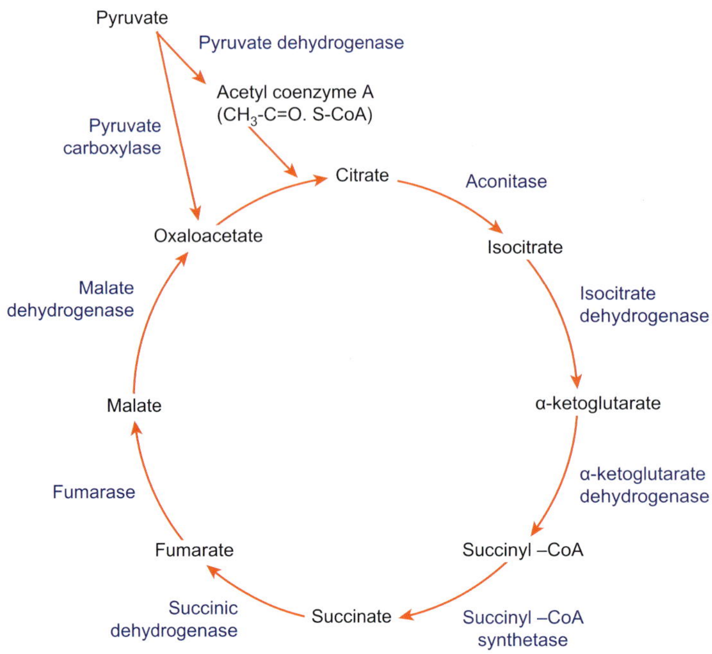

FIGURE 18.5 The citric acid or tricarboxylic acid (TCA) cycle.

TABLE 18.9 Comparison of Citrate Synthase Activity between Hummingbirds and Chickens

Species	Citrate Synthase Activity (nmol/min/g wet wt)	Reference
Hummingbird	448	Suarez et al., 2009
Chicken	21	Azad et al., 2010

in mammalian mitochondria (Suarez et al., 1991) but higher than that of the chicken (Table 18.9). PK has been identified by proteomics in pectoralis muscle with dietary protein deprivation (chicken: Corzo et al., 2006).

18.5.3.2 Erythrocytes

The citric acid cycle is absent in erythrocytes (pigeon: Kalomenopoulou et al., 1990).

18.6 GLUCONEOGENESIS

The gluconeogenic pathway is summarized in Figures 18.4 and 18.6. The major organs for gluconeogenesis in birds are the liver and kidneys (see below). There appears to be marked gluconeogenic potential for the yolk sac in the developing embryo, with high expression of fructose 1,6-bisphosphatase and phosphoenolpyruvate carboxykinase together with glucose 6-phosphatase (chicken: Yadgary et al., 2011). Muscle- and liver-type phosphofructokinase-1 (PFK-M and PFK-L) have been characterized in the chicken (Seki et al., 2006). Chicken PFK-L is expressed at a low level in the liver, skeletal muscle, and brain (Seki et al., 2006). Chicken PFK-M is expressed in skeletal muscle (Seki et al., 2006).

18.6.1 Gluconeogenesis and Fasting

Fasting increases gluconeogenesis in some but not all birds. There is greatly increased net release of glucose from both the liver and kidneys (Table 18.8), together with elevated hepatic glucose 6-phosphatase, phosphoenolpyruvate carboxykinase, alanine aminotransferase, and aspartate aminotransferase activities (chickens: Veiga et al., 1978). There are marked increases in gluconeogenesis during fasting in the Japanese quail (*C. japonica*), with increases in the rate of glucose formation from ^{14}C bicarbonate *in vivo* (Sartori et al., 1996, 2000). Moreover, with 2 days fasting (phase 2), there were increases in the activity of cytoplasmic phosphoenolpyruvate carboxykinase in the kidney (Sartori et al., 2000). Similarly, fasting increases gluconeogenesis in duckling liver-based perfusion studies in which the medium contains the gluconeogenic precursor, lactate (Bedu et al., 2001). Expression of genes for enzymes in the gluconeogenic pathway is increased by fasting (chicken: Désert et al., 2008; Sherlock et al., 2012). However, gluconeogenesis is depressed in fasted carnivorous birds (black vulture *Coragyps atratus*: Veiga et al., 1978).

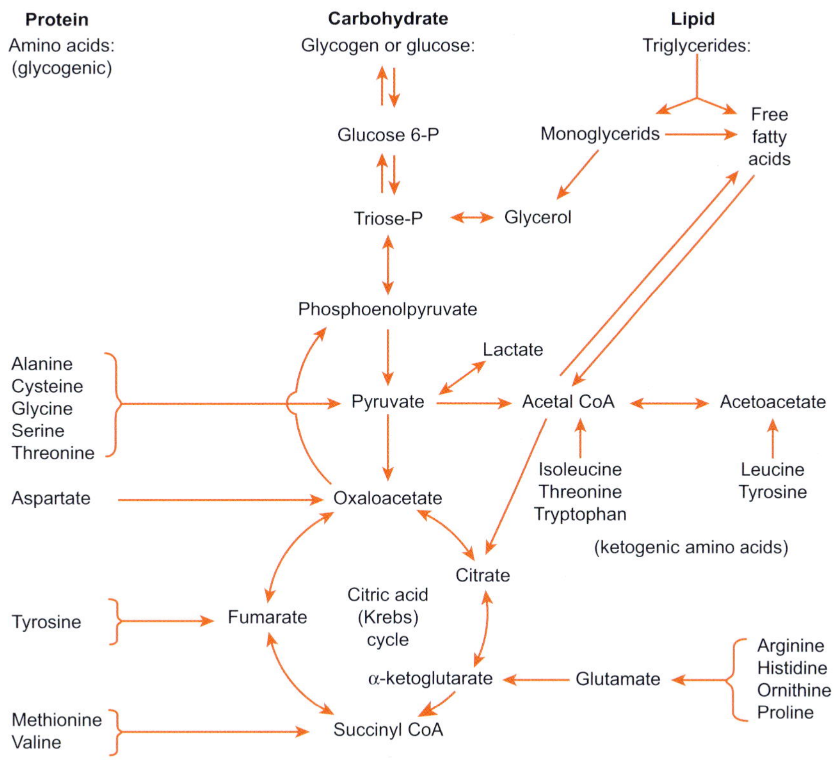

FIGURE 18.6 Overview schematic diagram of the relationships between lipid and protein metabolism with carbohydrate metabolism.

18.6.2 Relative Importance of the Liver and the Kidney

Gluconeogenesis occurs in both the liver and kidneys. The relative contribution is estimated from arteriovenous differences in fed and fasted birds, with the liver producing about 70% of the glucose and kidney about 30% in the fasted state (chicken: Tinker et al., 1986; Table 18.8). In fed carnivorous birds, there is high rates of gluconeogenesis (black vulture *C. atratus*: Veiga et al., 1978; also see Figure 18.9). There is likely to be some hepatic gluconeogenesis in graniferous birds. This is supported by studies of arterial-venous differences in glucose and lactate (Table 18.8). It is presumed that much of the lactate originates from gastrointestinal tract, with 37% of glucose converted to lactate in the chicken gastrointestinal tract (Riesenfeld et al., 1982).

18.7 GLYCOGEN

18.7.1 Overview

Glycogen is a branched polysaccharide composed of glucose monomers. It is the major storage form for carbohydrates, particularly in the liver and skeletal muscle. The synthesis and degradation of glycogen is summarized in Figure 18.4 (also see Figure 18.9). Glycogenolysis also plays an important role in astrocyte functioning (chicken: Gibbs et al., 2006).

This can be catalyzed by the enzyme glucokinase (expressed in the liver).

18.7.2 Synthesis and Breakdown

18.7.2.1 Glycogenesis (Synthesis)

Glycogenesis is synthesis of glycogen from glucose:

$$\text{Glucose 6-phosphate} \rightarrow \text{Glucose 1-phosphate}$$

18.7.2.1.1 Phosphoglucomutase

$$\text{Glucose 1-phosphate} + \text{Uridine triphosphate (UTP)} \rightarrow \text{Uridine diphosphoglucose}$$

18.7.2.1.2 UDP-Glucose Pyrophosphorylase

$$\text{Glycogen } (n) + \text{Uridine diphosphoglucose} \rightarrow \text{Glycogen } (n+1) + \text{UDP}$$

18.7.2.1.3 Glycogen Synthase

Glycogen synthase (GS) is expressed in pectoralis muscle (chicken: Sibut et al., 2008; Simon et al., 2012).

18.7.2.2 Glycogenolysis (Breakdown)

Glycogenolysis is the breakdown of glycogen to glucose 1-phosphate:

Glycogen (n)+ATP → Glycogen ($n-1$)+Glucose 1-phosphate

18.7.2.2.1 Glycogen Phosphorylase

Glycogen phosphorylase is expressed in pectoralis muscle (chicken: Sibut et al., 2008; Simon et al., 2012).

18.7.2.3 Glycogenesis (Breakdown)

There is little difference in hepatic phosphorylase activity between late avian embryos and newly hatched chicks (turkeys: Rosebrough et al., 1978b). As might be expected, cAMP increases phosphorylase activity in turkey liver tissue (Rosebrough and Von Vleck, 1990).

There are increases in AMP-activated protein kinase (AMPK) protein, the expression of regulatory and catalytic AMPK subunits, and phosphorylation of AMPK in the liver and skeletal muscle at the end of embryonic and in early posthatch development (chicken: Walter et al., 2010).

18.7.2.4 Developmental Changes

There are marked changes in the importance of glycogen predominantly during posthatch development of birds. Glycogen levels are low in the avian embryo with changes reported in some studies. For instance, liver glycogen is decreased between 15 and 19 days of embryonic development (chicken: Pulikanti et al., 2010) and between 19 days and prior to hatching (chicken: Zhai et al., 2011). These are shifts in the low levels of the storage form of glucose. In contrast, it is speculated that the increase in glycogen in the pipping muscle between 15 and 19 days of embryonic development is provide a ready source of easily mobilized glucose for the process of hatching (chicken: Pulikanti et al., 2010). At least in galliform birds, there are large increases in the amount of glycogen stored in the liver following hatching (turkey: Foye et al., 2006; see Table 18.10).

There are increases in liver glycogen during the second half of embryonic development to a maximum that occurs prior to the beginning of the process of hatching in the chicken (*Gallus gallus*):

- Glycogen concentrations increasing threefold between 11 days of embryonic development to a maximum at day 18 (Hamer and Dickson, 1987)
- Glycogen concentrations increasing 51% between 16 days of embryonic development to a maximum at day 18 (Willemsen et al., 2011)
- Glycogen content increasing between 18 days of embryonic development to a maximum at day 20 (Kornasio et al., 2011; see Table 18.11).
- A decrease (58%) decrease in the hepatic glycogen concentration was reported between internal pipping and hatching (chicken *Gallus gallus*: Willemsen et al., 2011) with glycogenolysis was calculated to yield 21 μmol glucose during the process of hatching.

After hatching, there are marked increased in liver glycogen in chicks receiving feed early (Kornasio et al., 2011; see Table 18.11). Before and after hatching, hepatic glycogen are increased by *in ovo* feeding (nutrient supplementation to the embryo; Kornasio et al., 2011).

Pectoralis muscle glycogen shows only small similar pattern to that of liver but with smaller changes, with a 33% increase between days 18 and 20 (Kornasio et al., 2011). Muscle glycogen is increased by *in ovo* feeding (Kornasio et al., 2011).

The rate of hepatic glycogenesis is very low in avian embryos, but it rises immediately after hatching to achieve a plateau at approximately 2 weeks old (chicken: Goodridge, 1968a; Table 18.4). The rise in glycogenesis does not occur in newly hatched chicks until after feeding, suggesting

TABLE 18.10 Changes in Liver Glycogen during Early Posthatch Embryonic Development of the Turkey

	Liver Glycogen	
Age (days)	Concentration (mg/g)	Content
0	16.4	20.2
7	84.7	591

Data from Foye et al. (2006).

TABLE 18.11 Changes in Glycogen in Liver and Muscles in the Chicken

	Glycogen Content (mg)	
Age	Liver	Muscle
ED 18	2	6
ED 19	6	8
ED 20	8	8
ED 21	3	7
Hatch	5	7
24 h	4	9
24 h, early feeding	26	12
36 h	5	6
36 h, early feeding	135	18

ED, embryonic day.
Data from Kornasio et al. (2011).

that nutrients induce expression of critical carbohydrate enzymes (chicken: Goodridge, 1968b). There are analogous increases in the activity of hepatic glycogen synthase in the first week of life after hatching (turkeys: Rosebrough et al., 1979).

18.7.2.5 Effect of Feeding and Fasting

There are marked effects of nutrition on liver glycogen concentrations. An example where hepatic glycogen storage is high is in chickens fed meals. Table 18.12 summarizes the hepatic glycogen concentrations and the daily estimated contribution of glycogenolysis in the postabsorptive period between meals to whole body metabolism. Moreover, the relatively high rates of glycogenolysis will be similar to those of glycogenesis in the fed state. Hepatic glycogen synthase activity is increased by glucose and, particularly, sucrose supplementation (turkeys: Rosebrough et al., 1979). There are also increases in skeletal muscle glycogen after hatching, reflecting increased glycogen synthase activity (turkeys: Rosebrough et al., 1979). Glycogen concentrations in the liver, kidneys, and skeletal muscle are depressed by fasting, as shown in chickens (Tinker et al., 1986; see Table 18.13) and turkeys (Rosebrough et al., 1978a).

18.7.3 *In ovo* Feeding

Liver glycogen concentrations rise in late embryonic development to a peak at day 20, with a subsequent decline before hatching (Kornasio et al., 2011). *In ovo* feeding increases glycogen concentrations in both the liver and skeletal muscle (Kornasio et al., 2011). There is a large increase in the concentrations of glycogen in both the liver and skeletal muscle after feeding (Kornasio et al., 2011).

18.7.4 Glycogen Body

The glycogen body is a gelatinous ovaloid organ found in dorsal area of the lumbosacral region of the spinal cord in birds (Watterson, 1949). It is composed of what was initially described as uniform glycogen body cells (Watterson, 1949). These are astroglial-like cells (Louis, 1959, 1993; De Gennaro, 1993). These cells have a high capacity to store glycogen and contain deposits of glycogen in the cytoplasm (Imagawa et al., 2006a). A second cell type has been identified based on co-incubation of glycogen body cells with cerebellar neurons. These have processes that extend and attach to neurons (Imagawa et al., 2006a). The glycogen body is derived from anterior section of the lumbosacral neural tube (De Gennaro, 1991). The lobes of Lachi have many metabolic similarities to glycogen bodies with glycogen synthase, glycogen phosphorylase, glucose 6-phosphate dehydrogenases, and 6-phosphogluconate dehydrogenase together with a lack of glucose 6-phosphatase (Benzo and DeGennaro, 1981).

Glycogen bodies contain multiple metabolic enzyme activities, including the following: glycogen synthase,

TABLE 18.13 Changes in Glycogen (μmol glucose equivalents/g) with Starvation in Chickens

	Glycogen (μmol glucose equivalents/g)	
Organ	**Fed**	**Starved 6 Days**
Liver	84.4	18.5
Muscle	54.4	25.8
Kidneys	2.8	1.3

Data from Tinker et al. (1986).

TABLE 18.12 Changes in Liver Glycogen with Meal Feeding in the Chicken, Allowing Determination of the Rate of Glycogenolysis

Time Relative to Feeding	Meal Fed	Liver Glycogen (mmol glucose equivalents per kg body weight)	Delta Liver Glycogen (mmol glucose equivalents per kg body weight)	Estimated Release of Glucose by Glycogenolysis (μmol glucose per min per kg body weight)
+12 h following meal	Once daily	7.7	4.2	5.8
Prior to feeding	Once daily	3.5		
+24 h following meal	Once on alternate days	15.9	9.0	6.2
Prior to feeding	Once on alternate days	6.9		

From De Beer et al. (2007).

glycogen phosphorylase, glucose 6-phosphate dehydrogenases, and 6-phosphogluconate dehydrogenase (Benzo and DeGennaro, 1974, 1981). There is lactate dehydrogenase (LDH) activity and LDH-B expression in the glycogen body (Imagawa et al., 2006b). It is suggested that the glycogen body consumes lactate body (Imagawa et al., 2006b). The glycogen level within the glycogen body cells is refractory to both nutritional status, being unchanged by starvation, and by hormones such as insulin, glucagon, and adrenocorticotropic hormone (Imagawa et al., 2006a). There is a small effect of norepinephrine on glycogenolysis with reduced glycogen levels in the glycogen body cells *in vitro* (Lee et al., 2001). It is suggested that the glycogen body and the lobes of Lachi function to provide precursors via the pentose phosphate pathway for myelin synthesis (Benzo and DeGennaro, 1981).

18.8 CARBOHYDRATE DIGESTION AND ABSORPTION

Carbohydrate digestion and absorption will be discussed in three parts: starch digestion, disaccharide digestion, and glucose absorption.

18.8.1 Starch Digestion

Starch is digested by amylase, which is produced by the small intestine (see Table 18.14) and exocrine pancreas (chicken: Osman, 1982). About 97% of corn starch in the diet is digested in the chicken (Riesenfeld et al., 1980). The sites of starch digestion in the chicken small intestine are the following (calculated from Riesenfeld et al., 1980):

- Duodenum 63%
- Jejunum 19%
- Ileum 12%

However, there is markedly higher proportion of digestion of other starches, for instance from legumes, in both the duodenum and ilium of chickens (Weurding et al., 2001). There is secretion of amylase in the upper gastro-intestinal tract and from the liver reported in some birds: from salivary glands *(e.g. spotted dove (*Streptopelia chinensis) and house crow (Corvus splendens*))*, the proventriculus (e.g. house sparrow (Passer domesticus)*) (*Bhattacharya and Ghose, 1971) and from the liver in bile (chicken: Farner, 1943).

TABLE 18.14 Enzymatic Activities in the Chicken Small Intestine

Enzyme	Duodenum	Jejunum
Amylase	13.4	19.0
Maltase	1.5	4.8
Sucrose	1.4	3.2

Based on Liu et al. (2008).

18.8.2 Disaccharide Digestion

Maltase cleaves the disaccharide maltose into its two constituent glucose moieties, whereas sucrose cleaves the disaccharide sucrose into its two constituent monosaccharide moieties, glucose and galactose. There is both maltase and sucrase activity in the duodenum and jejunum (chicken: Liu et al., 2008). Although maltase activity is higher than that of sucrose in a passerine bird, the rufous-tailed plantcutter (*Phytotoma rara*), there is a similar distribution along the small intestine (Meynard et al., 1999). Similarly, there is somewhat lower expression of isomaltase in the jejunum then duodenum (Liu et al., 2008). There are little changes in expression of isomaltase during embryonic development (Yadgary et al., 2011).

18.8.3 Glucose Absorption

There is active absorption of glucose in the small intestine, cecum and colon of birds (chickens: Riley et al., 1986; Vinardell and Lopera, 1987; Moreno et al., 1996; De La Hora et al., 1998, 2001). The major site of glucose absorption is the duodenum, accounting for more than 80% of glucose absorbed in young chickens fed a diet high in glucose (Riesenfeld et al., 1980). The very high glucose absorption from the intestines of hummingbirds is due to active, mediated transport (rufous (*Selasphorus rufus*) and Anna's (*Calypte anna*) hummingbirds) (McWhorter et al., 2006).

Glucose absorption from the lumen of the gastrointestinal tract into enterocytes is achieved by active transport, predominantly by sodium D-glucose and galactose co-transporter 1 (SGLT1). The SGLT1 protein has a high affinity for glucose (Gal-Garber et al., 2000) and is exclusively found in the brush border of enterocytes (chickens: De La Hora et al., 1998; Barfull et al., 2002). Expression of SGLT1 is high in the small intestine, being greatest in the jejunum as seen from the following values for the chicken (calculated from Gilbert et al., 2008, with expression shown as a percentage of that in the jejunum):

- Duodenum 77%
- Jejunum 100%
- Ileum 65%

Jejunal expression of sodium glucose transporter 1 is increased in fasted chickens (Gal-Garber et al., 2000; Duarte et al., 2011) and further elevated with re-feeding (Gal-Garber et al., 2000). During embryonic development expression of SGLT1 is increased (chicken: Yadgary et al., 2011; Speir et al. 2012; pigeon: Dong et al., 2012). During growth, the surface area increases but expression of SGLT1 declines (Barfull et al., 2002). Aldosterone influences expression of SGLT1 (chickens: De La Hora et al., 2001). There is also expression of sodium independent glucose (galactose and fructose) transporter 2 (GLUT2) in the small intestine. Jejunal expression of GLUT2 is unaffected by either fasting or re-feeding in young chickens (Duarte et al., 2011).

18.8.4 Gastrointestinal Storage of Ingesta

There is temporary storage of feed in the crop and gizzard of at least some graniferous birds (e.g., in the order *Galliformes*). For instance, substantial stores of feed are found in crop of in sexually immature meat-type chickens adapted to either daily or alternate day meal feeding (De Beer et al., 2008) (Figure 18.7). In chickens, feeding is restricted to the hours of light (day). There is a marked peak in consumption prior to lights off, with concomitant filling of both the crop and gizzard (Scanes et al., 1987; Buyse et al., 1993; Figure 18.7). Nocturnal energy needs are accommodated by reduced metabolic rate and the release of the stored ingesta (Buyse et al., 1993; Figure 18.8).

18.8.5 Intestinal Fermentation

Limited information is available on gastrointestinal fermentation in birds.

18.8.5.1 Cellulose

18.8.5.1.1 Foregut

A neotropic bird, the hoatzin (*Opisthocomus hoazin*), is said to be unique in having foregut fermentation (Grajal et al., 1989). The microbial communities in the muscular crop have been characterized and include methanogens as in the mammalian rumen (Wright et al., 2009; Godoy-Vitorino et al., 2012).

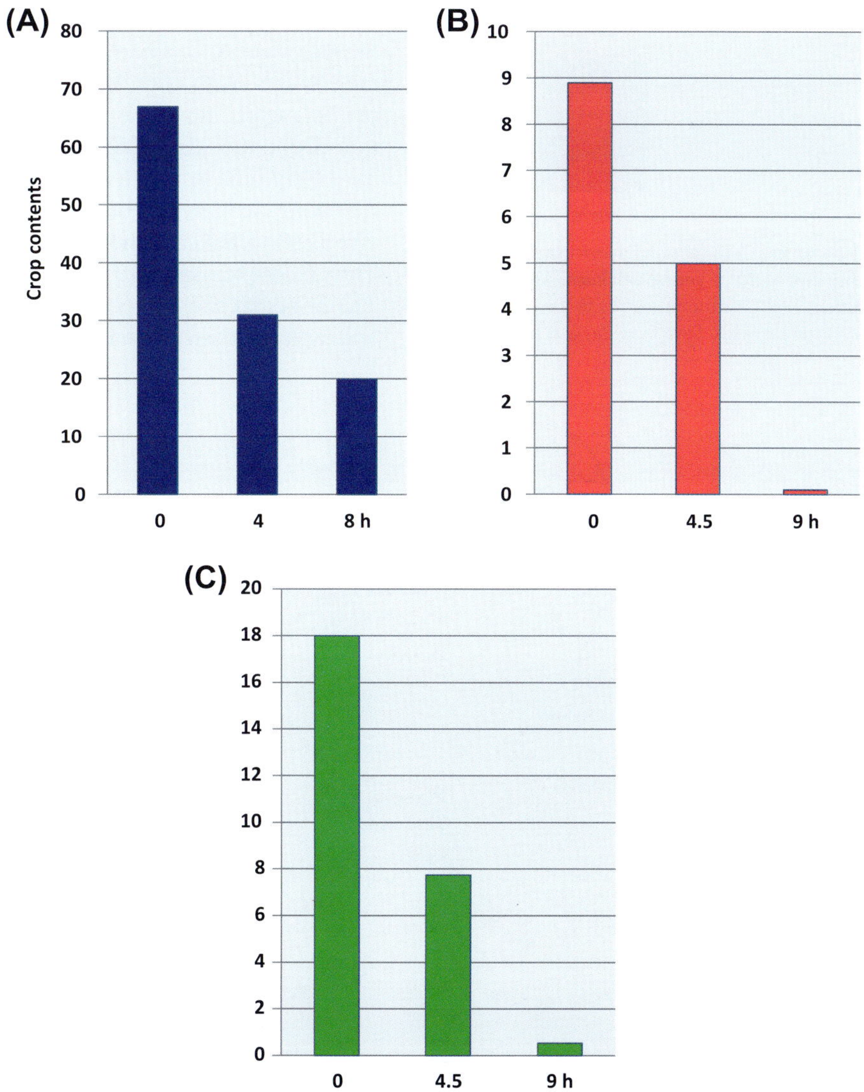

FIGURE 18.7 **Mechanical storage of ingesta in the crop (g) of chickens.** (A) Following meal feeding in broiler breeder pullets (sexually immature females) (De Beer et al., 2008) and (B) during the night following increased feeding prior to lights-off in young broiler chickens (Buyse et al., 1993) and during the night following increased feeding prior to lights-off in laying hens (Scanes et al., 1987).

18.8.5.1.2 Hindgut

Cellulose fermentation to volatile fatty acids in the ceca has been demonstrated in a galliform bird, rock ptarmigan (*Lagopus muta*) (Gasaway et al., 1976a,b,c).

18.8.5.2 Starch

There is evidence for fermentation of starch in the crop of domestic poultry. During the nocturnal storage of feed in the crop, there are marked increases in lactate acid concentrations but little increase in volatile simple fatty acids, except valeric and caproic acids (turkey: Johannsen et al., 2005). About 37% of glucose converted to lactate in the chicken gastrointestinal tract (Riesenfeld et al., 1982).

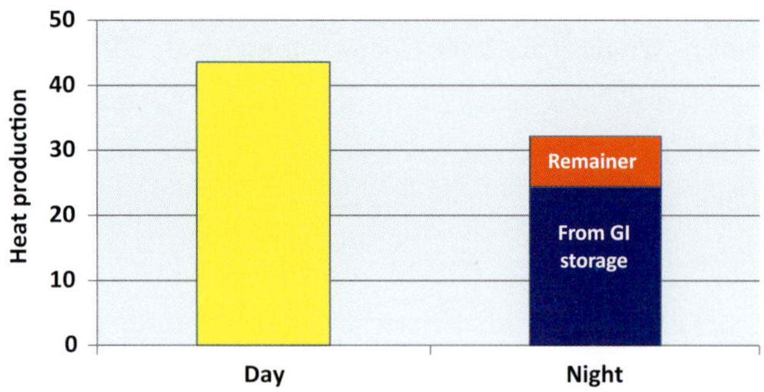

FIGURE 18.8 **Comparison of heat production/energy consumption (kJ kg$^{0.75}$ h) during the day and night in young chickens.** Also shown are the contributions of ingesta held (short-term stored) in the crop, gizzard, and proventriculus for nocturnal needs. *Data from Buyse et al. (1993).*

18.9 CONCLUSIONS

Figure 18.9 summarizes glucose flow in the metabolism in birds.

Although carbohydrate metabolism in birds is similar is that in mammals, there are distinct differences:

1. There are marked differences between circulating concentrations of glucose between wild birds and mammals, with avian concentrations double those in mammals.
2. Birds are very resistant to diabetes, with neither pancreatectomy nor insulin administration evoking large changes in circulating concentrations of glucose. However, chickens received passive immunization with antisera to insulin exhibit a diabetic-like shifts in metabolism, with circulating concentrations of glucose of 747 mg/dL reported 5 h after antisera administration (Dupont et al., 2008; Simon et al., 2012). Identical extremely high concentrations of glucose are observed following feeding in hummingbirds without adverse effects (Beuchat and Chong, 1998).
3. The glucose transporter GLUT4 is missing in birds.
4. The principal site of lipogenesis is the liver (compared with adipose tissue in mammals).

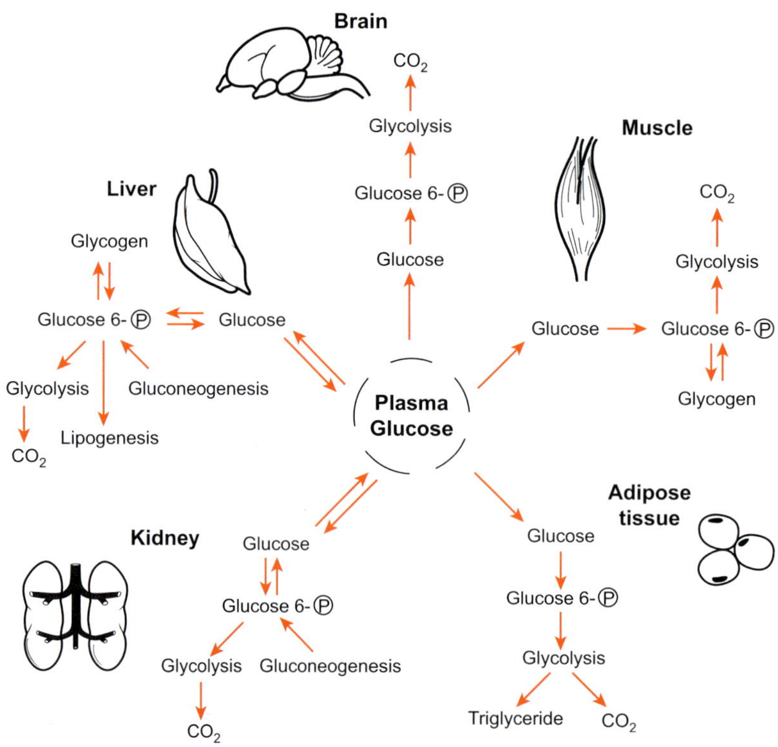

FIGURE 18.9 **Overview schematic diagram summarizing the flow of glucose to and from plasma to major organs.**

18.9.1 In starvation and metabolism

It is assumed that there are four metabolic phases:

- An absorptive phase following feeding, until ingesta carbohydrates in the gastrointestinal tract are completely depleted
- A postabsorptive phase or short-term fasting
- Phase 2 starvation: protein sparing and utilization of fatty acids
- Phase 3 starvation: fatty acids depleted and a shift to amino acids.

The absorptive phase is when glucose and other simple carbohydrates are being absorbed. It encompasses both the passage of ingesta with digestion, and any storage of ingesta in the foregut (crop, gizzard, and proventriculus) prior to digestion/absorption (Figure 18.7). In at least chickens and turkeys, there is nocturnal foregut storage of ingesta (laying hen: Scanes et al., 1987; young chicken: Buyse et al., 1993). This has been calculated to meet over 70% of energy requirements (Figure 18.8).

It has been generally assumed that, during the postabsorptive phase, glycogen is employed as the energy source. In contrast, during phase 2 starvation, there is protein sparing and fatty acids from the breakdown of adipose triglyceride used. During phase 3 starvation, there is a shift to muscle and other protein catabolism using amino acids for gluconeogenesis as triglyceride stores are exhausted (Goodman et al., 1980; Bernard et al., 2002).

During the postabsorptive phase, together with phase 2 and 3 starvation, there is a progressive decrease utilization of glucose for glycolysis, such as by muscle (see Table 18.8) and hepatic lipogenesis (Figure 18.3). The decrease precedes the end of absorptive phase as, for instance, the metabolic rate is decreased 26% in chickens during the nocturnal period (Figure 18.8), when feeding has essentially stopped. Hepatic glycogenolysis only meets a small proportion of body glucose during the postabsorptive phase. Table 18.12 shows hepatic glycogen contents in meal-fed chickens; it was assumed that the hepatic glycogen contents approach maximal in chickens fed meals either daily or on alternate days. Hepatic glycogenolysis in the postabsorptive phase between meals was calculated as follows:

- 6 μmol/min/kg compared with
- 55 μmol/min/kg total glucose utilization (Belo et al., 1976) or 116 μmol/min/kg total glucose utilization (Table 18.5).

Thus, only about 10% of whole body glucose utilization comes from hepatic glycogenolysis. This would indicate that hepatic gluconeogenesis is increased and hepatic glycolysis and lipogenesis are decreased during the postabsorptive phase and phase 2 starvation. Evidence for an immediate increase in gluconeogenesis and/or a decline in hepatic glycolysis comes from the observed reduction in the hepatic concentration of lactate with 2 h fasting (chicken: Yeh and Leveille, 1971b).

The glycogen content in both the liver and muscle decline during fasting, being depressed by respectively 78% and 53% with 6 days of starvation in the chicken (see Table 18.12; Tinker et al., 1986).

The conceptual framework for phases 2 and 3 starvation is supported in king penguins (*Aptenodytes patagonicus*), where body fat reverses are almost exhausted with markedly decreased circulating concentrations of fatty acids (Bernard et al., 2002). Evidence for increased gluconeogenesis comes from the increase in circulating concentrations of uric acid (Bernard et al., 2002). Moreover, in chickens, starvation for 6 days is accompanied by increased net release of glucose into hepatic and renal veins (Tinker et al., 1986); indicating elevated gluconeogenesis (see Table 18.8). This is calculated as gluconeogenesis providing 70 μmol glucose/min/kg.

Arterial venous differences demonstrate that there is a concomitant increase uptake of glucogenogenic precursors (amino acids and lactate) by the liver and kidney together with release from the hindquarter muscles (Tinker et al., 1986). Increased gluconeogenesis is indicated by the depressed circulating concentrations of glycine in chickens fasted for 24 h (Belo et al., 1976).

REFERENCES

Alonso-Alvarez, C., Ferrer, M., 2001. A biochemical study of fasting, subfeeding, and recovery processes in yellow-legged gulls. Physiol. Biochem. Zool. 74, 703–713.

Anan, K.K., Maitra, S.K., 1995. Impact of quinalphos on blood glucose and acetyl choline activity in the brain and pancreas in a roseringed parakeet (*Psittacula krameri borealis* Newmann). Arch. Environ. Contam. Toxicol. 29, 20–23.

Anthony, N.B., Vasilatos-Younken, R., Bacon, W.L., Lilburn, M.S., 1990. Secretory pattern of growth hormone, insulin, and related metabolites in growing male turkeys: effects of overnight fasting and refeeding. Poult. Sci. 69, 801–811.

Applegate, T.J., Ladwig, E., Weissert, L., Lilburn, M.S., 1999. Effect of hen age on intestinal development and glucose tolerance of the Pekin duckling. Poult. Sci. 78, 1485–1492.

Azad, M.A., Kikusato, M., Maekawa, T., Shirakawa, H., Toyomizu, M., 2010. Metabolic characteristics and oxidative damage to skeletal muscle in broiler chickens exposed to chronic heat stress. Comp. Biochem. Physiol. A Mol. Integr. Physiol. 155, 401–406.

Barfull, A., Garriga, C., Mitjans, M., Planas, J.M., 2002. Ontogenetic expression and regulation of Na(+)-D-glucose cotransporter in jejunum of domestic chicken. Am. J. Physiol. 282, G559–G564.

Bedu, E., Cohen-Adad, F., Dallevet, G., Garin, D., Barré, H., Duchamp, C., 2001. Effect of cold-acclimation on oxygen uptake and glucose production of perfused duckling liver. Comp. Biochem. Physiol. A Mol. Integr. Physiol. 128, 851–861.

Belo, P.S., Romsos, D.R., Leville, G.A., 1976. Blood metabolites and glucose metabolism in the fed and fasted chicken. J. Nutr. 106, 1135–1143.

Benzo, C.A., DeGennaro, L.D., 1974. Glycogen synthetase and phosphorylase in developing chick glycogen body. J. Exp. Zool. 188, 375–380.

Benzo, C.A., DeGennaro, L.D., 1981. Glycogen metabolism in the developing accessory lobes of Lachi in the nerve cord of the chick: metabolic correlations with the avian glycogen body. J. Exp. Zool. 215, 47–52.

Bernard, S.F., Fayolle, C., Robin, J.P., Groscolas, R., 2002. Glycerol and NEFA kinetics in long-term fasting king penguins: phase II versus phase III. J. Exp. Biol. 205, 2745–2754.

Berradi, H., Guy, G., Rideau, N., 2004. A glucokinase-like enzyme induced in Mule duck livers by overfeeding. Poult. Sci. 83, 161–168.

Berradi, H., Taouis, M., Cassy, S., Rideau, N., 2005. Glucokinase in chicken (*Gallus gallus*). Partial cDNA cloning, immunodetection and activity determination. Comp. Biochem. Physiol. B Biochem. Mol. Biol. 141, 129–139.

Berradi, H., Bernadet, M.D., Guy, G., Rideau, N., 2007. Expression of the glucokinase gene in mule duck liver and glucokinase activities in chicken and mule duck livers. Poult. Sci. 86, 2216–2220.

Beuchat, C.A., Chong, C.R., 1998. Hyperglycemia in hummingbirds and its consequences for hemoglobin glycation. Comp. Biochem. Physiol. A Mol. Integr. Physiol. 120, 409–416.

Bhattacharya, S., Ghose, K.C., 1971. Influence of food on the amylase system in birds. Comp. Biochem. Physiol. 40B, 317–320.

Boshouwers, F.M., Nicaise, E., 1981. Measurement of the respiratory metabolism of the fowl. Br. Poult. Sci. 22, 59–69.

Borrebaek, B., Christophersen, B., Tranulis, M.A., Aulie, A., 2007. Pre- and post-natal hepatic glucose phosphorylation in chicks (*Gallus domesticus*). Br. Poult. Sci. 48, 729–731.

Boussaid-Om Ezzine, S., Everaert, N., Métayer-Coustard, S., Rideau, N., Berri, C., Joubert, R., Temim, S., Collin, A., Tesseraud, S., 2010. Effects of heat exposure on Akt/S6K1 signaling and expression of genes related to protein and energy metabolism in chicken (*Gallus gallus*) pectoralis major muscle. Comp. Biochem. Physiol. B Biochem. Mol. Biol. 157, 281–287.

Brackenbury, J.H., El-Sayed, M.S., 1985. Respiratory metabolism of D[U-14C]glucose in hens and cocks during prolonged treadmill exercise. Comp. Biochem. Physiol. A Comp. Physiol. 82, 851–854.

Braun, E.J., Sweazea, K.L., 2008. Glucose regulation in birds. Comp. Biochem. Physiol. B Biochem. Mol. Biol. 151, 1–9.

Buyse, J., Adelsohn, D.S., Decuypere, E., Scanes, C.G., 1993. Diurnal-nocturnal changes in food intake, gut storage of ingesta food transit time and metabolism in growing broiler chickens: a model for temporal control of energy balance. Brit. Poult. Sci. 34, 699–709.

Buyse, J., Geypens, B., Malheiros, R.D., Moraes, V.M., Swennen, Q., Decuypere, E., 2004. Assessment of age-related glucose oxidation rates of broiler chickens by using stable isotopes. Life Sci. 75, 2245–2255.

Christensen, K., McMurtry, J.P., Thaxton, Y.V., Thaxton, J.P., Corzo, A., McDaniel, C., Scanes, C.G., 2013. Metabolic and hormonal responses of growing modern meat type chickens to fasting. Brit. Poult. Sci. 54, 199–205.

Corzo, A., Kidd, M.T., Dozier 3rd, W.A., Shack, L.A., Burgess, S.C., 2006. Protein expression of pectoralis major muscle in chickens in response to dietary methionine status. Br. J. Nutr. 95, 703–708.

Davis, J.F., Castro, A.E., de la Torre, J.C., Scanes, C.G., Radecki, S.V., Vasilatos-Younken, R., Doman, J.T., Teng, M., 1995. Hypoglycemia, enteritis, and spiking mortality in Georgia broiler chickens: experimental reproduction in broiler breeder chicks. Avian Dis. 39, 162–174.

De Beer, M., Rosebrough, R.W., Russell, B.A., Poch, S.M., Richards, M.P., Coon, C.N., 2007. An examination of the role of feeding regimens in regulating metabolism during the broiler breeder grower period. 1. Hepatic lipid metabolism. Poult. Sci. 86, 1726–1738.

De Beer, M., McMurtry, J.P., Brocht, D.M., Coon, C.N., 2008. An examination of the role of feeding regimens in regulating metabolism during the broiler breeder grower period. 2. Plasma hormones and metabolites. Poult. Sci. 87, 264–275.

De Gennaro, L.D., 1991. Origin of the avian glycogen body: I. Effects of tail bud removal in the chick embryo. Growth Dev. Aging 55, 19–26.

De Gennaro, L.D., 1993. Origin of the avian glycogen body. II. Observations in support of a glial nature in the chick embryo. Growth Dev. Aging 57, 275–281.

De La Horra, M.C., Calonge, M.L., Ilundáin, A.A., 1998. Effect of dehydration on apical Na^+-H^+ exchange activity and Na^+-dependent sugar transport in brush-border membrane vesicles isolated from chick intestine. Pflugers Arch. 436, 112–116.

De La Horra, M.C., Cano, M., Peral, M.J., Calonge, M.L., Ilundáin, A.A., 2001. Hormonal regulation of chicken intestinal NHE and SGLT-1 activities. Am. J. Physiol. 280, R655–R660.

Désert, C., Duclos, M.J., Blavy, P., Lecerf, F., Moreews, F., Klopp, C., Aubry, M., Herault, F., Le Roy, P., Berri, C., Douaire, M., Diot, C., Lagarrigue, S., 2008. Transcriptome profiling of the feeding-to-fasting transition in chicken liver. BMC Genomics 9, 611.

Diamond, D.L., Carruthers, A., 1993. Metabolic control of sugar transport by derepression of cell surface glucose transporters. An insulin-independent recruitment-independent mechanism of regulation. J. Biol. Chem. 268, 6437–6444.

Dong, X.Y., Wang, Y.M., Yuan, C., Zou, X.T., 2012. The ontogeny of nutrient transporter and digestive enzyme gene expression in domestic pigeon (*Columba livia*) intestine and yolk sac membrane during pre- and posthatch development. Poult. Sci. 91, 1974–1982.

Duarte, C.R., Vicentini-Paulino, M.L., Buratini Jr., J., Castilho, A.C., Pinheiro, D.F., 2011. Messenger ribonucleic acid abundance of intestinal enzymes and transporters in feed-restricted and refed chickens at different ages. Poult. Sci. 90, 863–868.

Dunbar, M.R., Gregg, M.A., Giordano, M.R., Davis, D.M., Byrne, M.W., Crawford, J.A., Tornquist, S.J., 2005. Normal hematologic and biochemical values for prelaying greater sage grouse (*Centrocercus urophasianus*) and their influence on chick survival. J. Zoo. Wildl. Med. 36, 422–429.

Dupont, J., Tesseraud, S., Derouet, M., Collin, A., Rideau, N., Crochet, S., Godet, E., Cailleau-Audouin, E., Métayer-Coustard, S., Duclos, M.J., Gespach, C., Porter, T.E., Cogburn, L.A., Simon, J., 2008. Insulin immuno-neutralization in chicken: effects on insulin signaling and gene expression in liver and muscle. J. Endocrinol. 197, 531–542.

Fairbrother, A., Craig, M.A., Walker, K., O'Loughlin, D., 1990. Changes in mallard (*Anas platyrhynchos*) serum chemistry due to age, sex, and reproductive condition. J. Wildl. Dis. 26, 67–77.

Farner, D.S., 1943. Biliary amylase in the domestic fowl. Biol. Bull. 84, 240–243.

Fowles, J.R., Fairbrother, A., Trust, K.A., Kerkvliet, N.I., 1997. Effects of Aroclor 1254 on the thyroid gland, immune function, and hepatic cytochrome P450 activity in mallards. Environ. Res. 75, 119–129.

Foye, O.T., Uni, Z., Ferket, P.R., 2006. Effect of in ovo feeding egg white protein, beta-hydroxy-beta-methylbutyrate, and carbohydrates on glycogen status and neonatal growth of turkeys. Poult. Sci. 85, 1185–1192.

Gal-Garber, O., Mabjeesh, S.J., Sklan, D., Uni, Z., 2000. Partial sequence and expression of the gene for and activity of the sodium glucose transporter in the small intestine of fed, starved and refed chickens. J. Nutr. 130, 2174–2179.

Gasaway, W.C., 1976a. Seasonal variation in diet, volatile fatty acid production and size of the cecum of rock ptarmigan. Comp. Biochem. Physiol. 53A, 109–114.

Gasaway, W.C., 1976b. Volatile fatty acids and metabolizable energy derived from cecal fermentation in the willow ptarmigan. Comp. Biochem. Physiol. 53A, 115–121.

Gasaway, W.C., 1976c. Cellulose digestion and metabolism by captive rock ptarmigan. Comp. Biochem. Physiol. 53A, 179–182.

Gatehouse, N., Markham, B.J., 1970. Respiratory metabolism in three species of raptors. Auk 87, 738–741.

Gayathri, K.L., Shenoy, K.B., Hegde, S.N., 2004. Blood profile of pigeons (*Columba livia*) during growth and breeding. Comp. Biochem. Physiol. A 138, 187–192.

Gee, C.F., Carpenter, J.W., Hensler, B.L., 1981. Species differences in hematological values of captive cranes, geese, raptors, and quail. J. Wildl. Manage 45, 463–483.

Geelissen, S.M., Swennen, Q., Geyten, S.V., Kühn, E.R., Kaiya, H., Kangawa, K., Decuypere, E., Buyse, J., Darras, V.M., 2006. Peripheral ghrelin reduces food intake and respiratory quotient in chicken. Domest. Anim. Endocrinol. 30, 108–116.

Gibbs, M.E., Anderson, D.G., Hertz, L., 2006. Inhibition of glycogenolysis in astrocytes interrupts memory consolidation in young chickens. Glia 54, 214–222.

Gilbert, E.R., Li, H., Emmerson, D.A., Webb Jr., K.E., Wong, E.A., 2008. Dietary protein quality and feed restriction influence abundance of nutrient transporter mRNA in the small intestine of broiler chicks. J. Nutr. 138, 262–271.

Godoy-Vitorino, F., Goldfarb, K.C., Karaoz, U., Leal, S., Garcia-Amado, M.A., Hugenholtz, P., Tringe, S.G., Brodie, E.L., Dominguez-Bello, M.G., 2012. Comparative analyses of foregut and hindgut bacterial communities in hoatzins and cows. ISME J. 6, 531–541.

Goodman, M.N., Larsen, P.R., Kaplan, M.M., Aoki, T.T., Young, V.R., Ruderman, N.B., 1980. Starvation in the rat. II. Effect of age and obesity on protein sparing and fuel metabolism. Am. J. Physiol. Endocrinol. Metab. 239, E277–E286.

Goodridge, A.G., 1968a. Conversion of [U-14C]glucose into carbon dioxide, glycogen, cholesterol and fatty acids in liver slices from embryonic and growing chicks. Biochem. J. 108, 655–661.

Goodridge, A.G., 1968b. The effect of starvation and starvation followed by feeding on enzyme activity and the metabolism of [U-14C] glucose in liver from growing chicks. Biochem. J. 108, 667–673.

Grajal, A., Strahl, S.D., Parra, R., Gloria Dominguez, M., Neher, A., 1989. Foregut fermentation in the hoatzin, a neotropical leaf-eating bird. Science 245, 1236–1238.

Groscolas, R., Rodriguez, A., 1981. Glucose metabolism in fed and fasted emperor penguins (*Aptenodytes Forsteri*). Comp. Biochem. Physiol. 70A, 191–198.

Hamer, M.J., Dickson, A.J., 1987. Developmental changes in hepatic fructose 2,6-bisphosphate content and phosphofructokinase-1 activity in the transition of chicks from embryonic to neonatal nutritional environment. Biochem. J. 245, 35–39.

Hargrove, J.L., 2005. Adipose energy stores, physical work, and the metabolic syndrome: lessons from hummingbirds. Nutr. J. 4, 36.

Humphrey, B.D., Stephensen, C.B., Calvert, C.C., Klasing, K.C., 2004. Glucose and cationic amino acid transporter expression in growing chickens (*Gallus gallus domesticus*). Comp. Biochem. Physiol. A 138, 515–525.

Imagawa, T., Shogaki, K., Uehara, M., 2006a. Interaction between glycogen body cell and neuron: examination in co-culture system. J. Vet. Med. Sci. 68, 1081–1087.

Imagawa, T., Yamamoto, E., Sawada, M., Okamoto, M., Uehara, M., 2006b. Expression of lactate dehydrogenase-A and -B messenger ribonucleic acids in chick glycogen body. Poult. Sci. 85, 1232–1238.

Itoh, N., Makita, T., Koiwa, M., 1998. Characteristics of blood chemical parameters in male and female quails. J. Vet. Med. Sci. 60, 1035–1037.

Jeffrey, D.A., Peakall, D.B., Miller, D.S., Herzberg, G.R., 1985. Blood chemistry changes in food-deprived herring gulls. Comp. Biochem. Physiol. A 81, 911–913.

Johannsen, S.A., Rasmussen, M.A., Hensley, M.J., Wilhelms, K., Griffith, R., Scanes, C.G., 2005. Effects of *Lactobacilli* and Lactose on *Salmonella typhimurium* Colonisation and Microbial Fermentation in the crop of the young turkey. Brit. Poult. Sci. 46, 708–716.

Johnstone, R.M., Mathew, A., Setchenska, M.S., Grdisa, M., White, M.K., 1998. Loss of glucose transport in developing avian red cells. Eur. J. Cell Biol. 75, 66–77.

Kalomenopoulou, M., Beis, I., 1990. Studies on the pigeon red blood cell metabolism. Comp. Biochem. Physiol. B 95, 677–684.

Karaganis, S.P., Bartell, P.A., Shende, V.R., Moore, A.F., Cassone, V.M., 2009. Modulation of metabolic and clock gene mRNA rhythms by pineal and retinal circadian oscillators. Gen. Comp. Endocrinol. 161, 179–192.

Khalilieh, A., McCue, M.D., Pinshow, B., 2012. Physiological responses to food deprivation in the house sparrow, a species not adapted to prolonged fasting. Am. J. Physiol. Regul. Integr. Comp. Physiol. 303, R551–R561.

Klandorf, H., Clarke, B.L., Scheck, A.C., Brown, J., 1986. Regulation of glucokinase activity in the domestic fowl. Biochem. Biophys. Res. Commun. 139, 1086–1093.

Kono, T., Nishida, M., Nishiki, Y., Seki, Y., Sato, K., Akiba, Y., 2005. Characterisation of glucose transporter (GLUT) gene expression in broiler chickens. Brit. Poult. Sci. 46, 510–515.

Kornasio, R., Halevy, O., Kedar, O., Uni, Z., 2011. Effect of in ovo feeding and its interaction with timing of first feed on glycogen reserves, muscle growth, and body weight. Poult. Sci. 90, 1467–1477.

Lea, R.W., Klandorf, H., Harvey, S., Hall, T.R., 1992. Thyroid and adrenal function in the ring dove (*Streptopelia risoria*) during food deprivation and a breeding cycle. Gen. Comp. Endocrinol. 86, 138–146.

Lee, K., Makino, S., Imagawa, T., Kim, M., Uehara, M., 2001. Effects of adrenergic agonists on glycogenolysis in primary cultures of glycogen body cells and telencephalon astrocytes of the chick. Poult. Sci. 80, 1736–1742.

Lee, M.Y., Park, S.H., Lee, Y.J., Heo, J.S., Lee, J.H., Han, H.J., 2006. EGF-induced inhibition of glucose transport is mediated by PKC and MAPK signal pathways in primary cultured chicken hepatocytes. Am. J. Physiol. 291, G744–G750.

Lisano, M.E., Kennamer, J.E., 1977. Values for several blood parameters in eastern wild turkeys. Poult. Sci. 56, 157–166.

Liu, N., Ru, Y.J., Li, F.D., Cowieson, A.J., 2008. Effect of diet containing phytate and phytase on the activity and messenger ribonucleic acid expression of carbohydrase and transporter in chickens. J. Anim. Sci. 86, 3432–3439.

Lloyd, S., Gibson, J.S., 2006. Haematology and biochemistry in healthy young pheasants and red-legged partridges and effects of spironucleosis on these parameters. Avian Pathol. 35, 335–340.

Louis, D.D., 1959. Differentiation of the glycogen body of the chick embryo under normal and experimental conditions. Growth 23, 235–249.

Louis, D.D., 1993. Origin of the avian glycogen body II. Observations in support of a glial nature in the chick embryo. Growth Dev. Aging 57, 275–281.

Lumeij, T.J., de Brujne, J.J., 1985. Blood chemistry reference values for racing pigeons. Avian Pathol. 14, 401–408.

Martínez del Rio, C., Stevens, B.R., Daneke, D.E. and Andreadis, P.T. 1988. Physiological correlates of preference and aversion for sugars in three species of birds. Physiol. Zool. 61, 222–229.

McWhorter, T.J., Bakken, B.H., Karasov, W.H., del Rio, C.M., 2006. Hummingbirds rely on both paracellular and carrier-mediated intestinal glucose absorption to fuel high metabolism. Biol. Lett. 2, 131–134.

Meynard, C., López-Calleja, V., Bozinovic, F., Sabat, P., 1999. Digestive enzymes of a small avian herbivore, the rufous-tailed plantcutter. Condor 101, 904–907.

Merrill, G.F., Russo, R.E., Halper, J.M., 1981. Cardiac output distribution before and after endotoxin challenge in the rooster. Am. J. Physiol. 241, R67–R71.

Moreno, M., Otero, M., Tur, J.A., Planas, J.M., Esteban, S., 1996. Kinetic constants of alpha-methyl-D-glucoside transport in the chick small intestine during perinatal development. Mech. Ageing Dev. 92, 11–20.

Mori, J.C., George, J.C., 1978. Seasonal changes in serum levels of certain metabolites, uric acid and calcium in the migratory Canada goose (*Branta canadiensis interior*). Comp. Biochem. Physiol. 59B, 263–269.

O'Neill, I.E., Langslow, D.R., 1978. Glucose phosphorylation and dephosphorylation in chicken liver. Comp. Biochem. Physiol. 59B, 317–325.

Osman, A.M., 1982. Amylase in chicken intestine and pancreas. Comp. Biochem. Physiol. 73B, 571–574.

Pulikanti, R., Peebles, E.D., Keirs, R.W., Bennett, L.W., Keralapurath, M.M., Gerard, P.D., 2010. Pipping muscle and liver metabolic profile changes and relationships in broiler embryos on days 15 and 19 of incubation. Poult. Sci. 89, 860–865.

Riesenfeld, G., Sklan, D., Bar, A., Eisner, U., Hurwitz, S., 1980. Glucose absorption and starch digestion in the intestine of the chicken. J. Nutr. 110, 117–121.

Riesenfeld, G., Geva, A., Hurwitz, S., 1982. Glucose homeostasis in the chicken. J. Nutr. 112, 2261–2266.

Riley Jr., W.W., Esteve-Garcia, E., Austic, R.E., 1986. Intestinal absorption of glucose and amino acids in chickens administered monensin. Poult. Sci. 65, 2292–2298.

Rinaudo, M.T., Curto, M., Bruno, R., 1982. Blood glucose and tissue glycogen concentrations in normal and deutectomised chickens during the first twelve hours after hatching. Brit. Poult. Sci. 23, 577–581.

Rosebrough, R.W., Von Vleck, M.F., 1990. Glucose production and glycogen cycle enzyme activities in avian liver explants: procedural optimization. Comp. Biochem. Physiol. 96B, 163–170.

Rosebrough, R.W., Geis, E., Henderson, K., Frobish, L.T., 1978a. Glycogen depletion and repletion in the chick. Poult. Sci. 57, 1460–1462.

Rosebrough, R.W., Geis, E., Henderson, K., Frobish, L.T., 1978b. Glycogen metabolism in the turkey embryo and poult. Poult. Sci. 57, 747–751.

Rosebrough, R.W., Geis, E., Henderson, K., Frobish, L.T., 1979. Control of glycogen metabolism in the developing turkey poult. Growth 43, 188–201.

Sartori, D.R., Kettelhut, I.C., Veiga, J.A., Migliorini, R.H., 1996. Gluconeogenesis and glucose replacement rate during long-term fasting of Japanese quails. Comp. Biochem. Physiol. A Physiol. 115, 121–125.

Sartori, D.R., Garofalo, M.A., Roselino, J.E., Kettelhut, I.C., Migliorini, R.H., 2000. Gluconeogenesis and P-enolpyruvate carboxykinase in liver and kidney of long-term fasted quails. J. Comp. Physiol. B. 170, 373–377.

Scanes, C.G., 2008. Perspectives on analytical techniques and standardization. Poult. Sci. 87, 2175–2177.

Scanes, C.G., Campbell, R., Grimminger, P., 1987. Control of energy balance during egg production in the laying hen. J. Nutr. 117, 605–611.

Scholtz, N., Halle, I., Flachowsky, G., Sauerwein, H., 2009. Serum chemistry reference values in adult Japanese quail (*Coturnix coturnix japonica*) including sex-related differences. Poult. Sci. 88, 1186–1190.

Seki, Y., Sato, K., Kono, T., Abe, H., Akiba, Y., 2003. Broiler chickens (Ross strain) lack insulin-responsive glucose transporter GLUT4 and have GLUT8 cDNA. Gen. Comp. Endocrinol. 133, 80–87.

Seki, Y., Sato, K., Kono, T., Akiba, Y., 2006. Two types of phosphofructokinase-1 differentially regulate the glycolytic pathway in insulin-stimulated chicken skeletal muscle. Comp. Biochem. Physiol. B Biochem. Mol. Biol. 143, 344–350.

Shea, R.E., Olson, J.M., Ricklefs, R.E., 2007. Growth rate, protein accumulation, and catabolic enzyme activity of skeletal muscles of galliform birds. Physiol. Biochem. Zool. 80, 306–316.

Sibut, V., Le Bihan-Duval, E., Tesseraud, S., Godet, E., Bordeau, T., Cailleau-Audouin, E., Chartrin, P., Duclos, M.J., Berri, C., 2008. Adenosine monophosphate-activated protein kinase involved in variations of muscle glycogen and breast meat quality between lean and fat chickens. J. Anim. Sci. 86, 2888–2896.

Simon, J., Rideau, N., Taouis, M., Dupont, J., 2011. Plasma insulin levels are rather similar in chicken and rat. Gen. Comp. Endocrinol. 171, 267–268.

Simon, J., Milenkovic, D., Godet, E., Cabau, C., Collin, A., Métayer-Coustard, S., Rideau, N., Tesseraud, S., Derouet, M., Crochet, S., Cailleau-Audouin, E., Hennequet-Antier, C., Gespach, C., Porter, T.E., Duclos, M.J., Dupont, J., Cogburn, L.A., 2012. Insulin immuno-neutralization in fed chickens: effects on liver and muscle transcriptome. Physiol. Genomics 44, 283–292.

Sinsigalli, N.A., McMurtry, J.P., Cherry, J.A., Siegel, P.B., 1987. Glucose tolerance, plasma insulin and immunoreactive glucagon in chickens selected for high and low body weight. J. Nutr. 117 (5).

Sitbon, G., Mialhe, P., 1979. Pancreatic hormones and plasma glucose: regulation mechanisms in the goose under physiological conditions. IV-effects of food ingestion and fasting on pancreatic hormones and gut GLI. Horm. Metab. Res. 11, 123–129.

Sherlock, L., Wathes, C.M., Cheng, Z., Wathes, D.C., 2012. Differential hepatic gene expression in the broiler chicken in response to the combined stressors of food withdrawal, catching and transport at the end of production. Stress 15, 293–305.

Soleimani, A.F., Zulkifli, I., 2010. Effects of high ambient temperature on blood parameters in red jungle fowl, village fowl and broiler chickens. J. Anim. Vet. Adv. 9, 1201–1207.

Suarez, R.K., Lighton, J.R., Moyes, C.D., Brown, G.S., Gass, C.L., Hochachka, P.W., 1990. Fuel selection in rufous hummingbirds: ecological implications of metabolic biochemistry. Proc. Natl. Acad. Sci. U. S. A. 87, 9207–9210.

Suarez, R.K., Lighton, J.R., Brown, G.S., Mathieu-Costello, O., 1991. Mitochondrial respiration in hummingbird flight muscles. Proc. Natl. Acad. Sci. U. S. A. , 4870–4873. http://www.ncbi.nlm.nih.gov/pubmed/205256888.

Suarez, R.K., Welch Jr., K.C., Hanna, S.K., Herrera, M.L.G., 2009. Flight muscle enzymes and metabolic flux rates during hovering flight of the nectar bat, *Glossophaga soricina*: further evidence of convergence with hummingbirds. Comp. Biochem. Physiol. 153A, 136–140.

Suarez, R.K., Herrera, M.L.G., Welch Jr, K.C., 2011. The sugar oxidation cascade: aerial refueling in hummingbirds and nectar bats. J. Exp. Biol. 214, 172–178.

Sweazea, K.L., Braun, E.J., 2006. Glucose transporter expression in English sparrows (*Passer domesticus*). Comp. Biochem. Physiol. 144B, 263–270.

Tinker, D.A., Brosnan, J.T., Herzberg, G.R., 1986. Interorgan metabolism of amino acids, glucose, lactate, glycerol and uric acid in the domestic fowl (*Gallus domesticus*). Biochem. J. 240, 829–836.

Tokushima, Y., Takahashi, K., Sato, K., Akiba, Y., 2005. Glucose uptake in vivo in skeletal muscles of insulin-injected chicks. Comp. Biochem. Physiol. 141B, 43–48.

Totzke, U., Hübinger, A., Bairlein, F., 1998. Glucose utilization rate and pancreatic hormone response to oral glucose loads are influenced by the migratory condition and fasting in the garden warbler (*Sylvia borin*). J. Endocrinol. 158, 191–196.

Veiga, J.A., Roselino, E.S., Migliorini, R.H., 1978. Fasting, adrenalectomy, and gluconeogenesis in the chicken and a carnivorous bird. Am. J. Physiol. 234, R115–R121.

Vinardell, M.P., Lopera, M.T., 1987. Jejunal and cecal 3-oxy-methyl-D-glucose absorption in chicken using a perfusion system in vivo. Comp. Biochem. Physiol. 86A, 625–657.

Vleck, C.M., Vleck, D., 2002. Physiological condition and reproductive consequences in adélie penguins. Integr. Comp. Biol. 42, 76–83.

Wagstaff, P., Kang, H.Y., Mylott, D., Robbins, P.J., White, M.K., 1995. Characterization of the avian GLUT1 glucose transporter: differential regulation of GLUT1 and GLUT3 in chicken embryo fibroblasts. Mol. Biol. Cell. 6, 1575–1589.

Walter, I., Hegarty, B., Seebacher, F., 2010. AMP-activated protein kinase controls metabolism and heat production during embryonic development in birds. J. Exp. Biol. 213, 3167–3176.

Walsberg, G.E., Wolf, B.O., 1995. Variation in the respiratory quotient of birds and implications for indirect calorimetry using measurements of carbon dioxide production. J. Exp. Biol. 198, 213–219.

Wang, M.Y., Tsai, M.Y., Wang, C., 1994. Identification of chicken liver glucose transporter. Arch. Biochem. Biophys. 310, 172–179.

Watterson, R.L., 1949. Development of the glycogen body of the chick spinal cord. J. Morphol. 85, 337–390.

Webb, D.M., DeNicola, D.B., Van Vleet, J.F., 1991. Serum chemistry alterations, including creatine kinase isoenzymes, in furazolidone toxicosis of ducklings: preliminary findings. Avian Dis. 35, 662–667.

Welch, K.C., Suarez, R.K., 2007. Oxidation rate and turnover of ingested sugar in hovering Anna's (*Calypte anna*) and rufous (*Selasphorus rufus*) hummingbirds. J. Exp. Biol. 210, 2154–2162.

Welch Jr., K.C., Altshuler, D.L., Suarez, R.K., 2007. Oxygen consumption rates in hovering hummingbirds reflect substrate-dependent differences in P/O ratios: carbohydrate as a 'premium fuel'. J. Exp. Biol. 210, 2146–2153.

Weurding, R.E., Veldman, A., Veen, W.A., van der Aar, P.J., Verstegen, M.W., 2001. Starch digestion rate in the small intestine of broiler chickens differs among feedstuffs. J. Nutr. 131, 2329–2335.

Wright, A.D., Northwood, K.S., Obispo, N.E., 2009. Rumen-like methanogens identified from the crop of the folivorous South American bird, the hoatzin (*Opisthocomus hoazin*). ISME J. 3, 1120–1126.

Willemsen, H., Li, Y., Willems, E., Franssens, L., Wang, Y., Decuypere, E., Everaert, N., 2011. Intermittent thermal manipulations of broiler embryos during late incubation and their immediate effect on the embryonic development and hatching process. Poult. Sci. 90, 1302–1312.

Yadgary, L., Yair, R., Uni, Z., 2011. The chick embryo yolk sac membrane expresses nutrient transporter and digestive enzyme genes. Poult. Sci. 90, 410–416.

Yeh, Y.Y., Leveille, G.A., 1971a. In vitro and in vivo restoration of hepatic lipogenesis in fasted chicks. J. Nutr. 101, 803–809.

Yeh, Y.Y., Leveille, G.A., 1971b. Studies on the relationship between lipogenesis and the level of coenzyme A derivatives, lactate and pyruvate in chick liver. J. Nutr. 101, 911–918.

Zhai, W., Bennett, L.W., Gerard, P.D., Pulikanti, R., Peebles, E.D., 2011. Effects of in ovo injection of carbohydrates on somatic characteristics and liver nutrient profiles of broiler embryos and hatchlings. Poult. Sci. 90, 2681–2688.

Zhang, C.L., Niu, Z.Y., Hou, S.S., Liu, F.Z., Huang, W., Xie, M., 2005. The effect of force-feeding, fasting and glucose saturated water intake on the contents of some biochemical parameters in plasma of Peking ducks. Int. J. Poult. Sci. 4, 202–205.

Zhao, J.P., Bao, J., Wang, X.J., Jiao, H.C., Song, Z.G., Lin, H., 2012. Altered gene and protein expression of glucose transporter1 underlies dexamethasone inhibition of insulin-stimulated glucose uptake in chicken muscles. J. Anim. Sci. 90, 4337–4345.

Chapter 19

Adipose Tissue and Lipid Metabolism

Johan Buyse and Eddy Decuypere
Laboratory of Livestock Physiology, Department of Biosystems, Faculty of Bioscience Engineering, KU Leuven, Leuven, Belgium

ABBREVIATIONS

ACC acetyl-CoA carboxylase
ATGL adipose triglyceride lipase
FAS fatty acid synthase
FTO fat mass and obesity-associated (FTO) gene
GH growth hormone
HDL high-density lipoproteins
IGF insulin-like growth factor
LDL low-density lipoproteins
LPL lipoprotein lipase
ME metabolizable energy
NEFA nonesterified fatty acids
PPARγ peroxisome proliferator-activated receptor γ
SREBP sterol regulatory element binding protein
TAG triacylglycerol
VLDL very-low-density lipoproteins
VLDLy yolk-targeted very-low-density lipoproteins

19.1 INTRODUCTION

The presumed predominant function of adipose tissue is to maintain energy homeostasis. In times of feed excess, energy is stored as hydrophobic triglycerides or triacylglycerols (TAGs) in expansive lipid droplets within the adipocytes. Triacylglycerols are very well suited for this as they contain a large amount of energy per unit of weight. In times of feed deprivation, however, adipocyte TAGs are mobilized (lipolysis) and the generated fatty acids are subsequently oxidized by various tissues to yield chemical energy. Adipose tissue is the most variable carcass component, and the amount that is deposited depends on genetic as well as on exogenous factors. However, it has become clear that adipose issue is more than only a storage buffer for energy as it also functions as an important endocrine and paracrine tissue involved in many bodily processes.

19.2 DEVELOPMENT OF ADIPOSE TISSUE

Adipose tissue cellular development consists of an increase in cell number (hyperplasia) and an increase in cell volume (hypertrophy). Both adipocyte number and volume increase with age and are positively correlated with fat pad weight and with body mass (Cartwright, 1991). Hence, both hyperplasia and hypertrophy of adipose cells contribute to the accumulation of fat in the chicken. However, the magnitude of the contribution of each of these two factors to the final size and weight of the distinct fat tissues is not equal.

At younger ages, increases in weight of fat depots would be mainly attributable to an increase in fat cell number, whereas at later ages, lipid accumulation in fat cells would be the major determinant. Literature data are not consistent about the age at which hypertrophy becomes the most important contributor to adipose tissue growth. It was first observed that in broiler chickens of a commercial strain, the number of abdominal adipocytes increased until about 12–14 weeks of age, whereas adipocyte volume only increased slowly. After this age, the filling of the existing adipose cells was then the predominant factor responsible for the increase in weight of the abdominal fat pad (Hood, 1982; March et al., 1984; Cartwright, 1991). In contrast, more recent studies revealed that at about 7 weeks of age, increases in fat deposition were primarily due to adipocyte hypertrophy in broiler chickens (Guo et al., 2011). The modern, growth-selected, meat-type chickens are proportionally fatter compared with the random-bred, unselected, or even growth-selected chickens of some decades ago. Therefore, it is reasonable to assume that the age at which hypertrophy becomes the major determinant for body fat accumulation is shifted to a younger age and even to a lower degree of maturity. Fast-growing chickens will probably also attain their genetically determined final number of fat cells earlier than slower growing chickens. Furthermore, sex (Hood, 1982), nutritional status (March et al., 1984; Cartwright, 1991), and surely direct and indirect divergent selection for fatness (Hood and Pym, 1982; Guo et al., 2011) all affect the age- or body weight-dependent developmental pattern of adipose cellularity. As a consequence of the increase in adipocyte number and size but with a different velocity, a bimodal frequency distribution (proportional number of cells per diameter class) is present in the abdominal fat pad of broilers by 7 weeks of age. With increasing age up to 22 weeks, the number of fat cells in the population of "large" cells increased slowly and the mean diameter shifted to larger values. In contrast, the number of cells in the population of "small" cells increased more rapidly, but the mean cell diameter remained unchanged (March et al., 1984). The same phenomenon was observed by Merkley and Cartwright (1989) for abdominal, thigh, back, and neck fat pads. It can be concluded that differences in fatness between populations, even when considered on a same-body-weight basis, are attributable to both the number and the size and volume of adipocytes, whereas reductions in body fat content by, for example, dietary manipulations are mainly caused by delipidation of adipocytes (hypotrophy).

19.3 ADIPOCYTE PROLIFERATION AND DIFFERENTIATION

The first step in adipogenesis is the differentiation of chicken embryonic stem cells into white preadipocytes, a process that is induced by insulin and dexamethasone (Li et al., 2011). The resulting adipocyte precursors are present in the stromal-vascular fraction of adipose tissue. These preadipocytes proliferate and then undergo differentiation to multilocular immature white adipocytes. When becoming filled with lipids, unilocular mature adipocytes appear that are characterized by distinct morphological and metabolic features (Butterwith, 1988; Evans, 1977). A mature avian fat cell (with an average diameter of 150 μm) is characterized by the presence of a large lipid droplet that occupies more than 80% of the total cell volume. As a consequence, all other cell organelles are displaced to the periphery, where the cytoplasm appears as a narrow rim. Another morphological feature of avian adipocytes is the presence of microfilaments at the interface between the cytoplasm and lipid droplet, which may provide structural support to the adipocyte (Evans, 1977; Hood, 1982).

Most of the studies concerning adipogenesis were done using cell culture systems with cell lines from mammalian and, to a much lesser extent, from avian origin and were mainly focused on its endocrine regulation. It was observed that the proliferation of chicken preadipocytes grown in culture was stimulated when insulin was added to the medium. When the medium was supplemented with 10% dialyzed chicken serum or when insulin was deleted from the medium, proliferation was retarded. Moreover, it was found that chicken preadipocyte proliferation *in vitro* was stimulated by insulin-like growth factor-1 and -2 (IGF1 and IGF2), fibroblast growth factor-1 and -2, epidermal growth factor, transforming growth factor-α and -β, and platelet-derived growth factor (reviewed by Butterwith (1997)). However, most of these growth factors also inhibit preadipocyte differentiation. These data suggest that these growth factors may be important autocrine, paracrine, and endocrine regulators of preadipocyte proliferation and differentiation. In addition, glucose (Qi et al., 2012), dexamethasone, and insulin (Ramsay and Rosebrough, 2003) also stimulate preadipocyte differentiation. Finally, insulin, heparin, chicken serum, dexamethasone, and the transcription factor peroxisome proliferator-activated receptor-γ (PPARγ) are involved in the maturation (lipid filling) of adipocytes.

19.4 DISTRIBUTION OF BODY FAT

The modern meat-type chickens contain between 150 and 220 g of lipids per kilogram of body weight at commercial slaughter age. Adipose tissue is distributed over several fat depots, of which the abdominal fat pad is the largest, followed by neck, thigh, back, and gizzard fat pads. These fat depots represent about 20% of the total body fat content. The skeleton and the skin contain, respectively, 15% and 19% of the total body lipid content, whereas several smaller fat depots represent about 8% of the total body fat content. The rest of the carcass, then, contains about one third of the total body lipid content. As avian striated muscle is characterized by only small amounts of inter- and intramuscular fat (1.5%), the remaining lipids must be located in the intestines, lungs, kidneys, glands, and so on (Nir et al., 1988).

The abdominal fat pad weight is highly positively correlated with total body fat content, and as it is rather easy to dissect it quantitatively, abdominal fat pad weight has been used extensively as an excellent estimator of total body fat content (Sonaiya, 1985). Indeed, selection against fatness in meat-type chickens by means of selection against abdominal fat content has been proven to be very successful (Leclercq, 1984; Baéza and Le Bihan-Duval, 2013).

The growth of higher vertebrates consists of not simply an increase in weight but also changes in body conformation and composition due to the differential growth rate of the constituent parts and of their functional abilities (Pálsson, 1955). The relationship between the weight of a part (Y) of the body in relation to total body weight (X) can be described by the Huxley equation:

$$Y = aX^b$$

Or, after logarithmic transformation:

$$\ln Y = \ln a + b \ln X$$

The regression coefficient b is also referred to as the allometric coefficient. When $b = 1$, the weight of the dependent variable Y increases proportionally with the weight of the independent variable X (isometric). When $b \neq 1$, the proportionality of absolute weights is not maintained, but the ratio between the relative growth rates of both *Y* and *X* remains constant during the growth trajectory.

As in mammals, the allometric coefficient for the growth of body fat in relation to body weight is also greater than unity in avian species. However, not all adipose tissues, including intramuscular and physiologically necessary fat, mature at the same rate as indicated by their different allometric coefficients. Merkley and Cartwright (1989) reported greater weight gains in abdominal and back fat than in neck and sartorius fat depots between 8 and 16 weeks of age in commercial female broilers. Nir et al. (1988) observed that in broiler lines selected for or against abdominal fat content, the allometric coefficients (pooled for both lines) of several fat depots increased in the following order: sartorial ≈ neck fat (0.83) < gizzard fat (1.030) < abdominal fat (1.17) < mesenteric fat (2.51). The allometric coefficients were higher for the high-fat-line chickens (average 1.24) compared to the low-fat-line chickens (average 1.02). From our studies on commercial broilers reared under continuous illumination, the following allometric coefficients were calculated: sartorial fat (1.137) ≈ scapular fat (1.171) < heart fat (1.284) < abdominal fat (1.615) ≈ stomach fat (1.623). It can also be concluded that the abdominal fat pad is relatively more affected by genetic and nongenetic factors than other fat depots and even the fat deposited in the rest of the carcass. However, the mechanisms involved in the differential allometric growth of adipose tissues and in the regulation of the discriminative effects of genetic and exogenous factors, according to their location in the body, remain to be elucidated further.

19.5 LIPID METABOLISM

19.5.1 Lipoprotein Metabolism

19.5.1.1 Portomicrons

After hydrolysis, simple sugars (hexoses and pentoses) and amino acids are transported by active or diffusion processes from the lumen of the small intestine into the blood circulation. Dietary lipids are first hydrolyzed by lipases, and the resulting free fatty acids are absorbed by the mucosa cells. Within the enterocytes, most fatty acids are re-esterified with glycerol to form new TAGs. These TAGs are then associated with phospholipids, cholesterol (esters), and specific apolipoproteins (mainly apoB100) to form lipoproteins. In mammals, these lipoproteins are transported via the lymphatic system and are therefore termed chylomicrons. In birds, however, the lymphatic system is poorly developed and these newly formed lipoproteins are released into the portal vein. It was therefore proposed to term these lipoproteins (average diameter of 150 nm) from dietary origin as portomicrons (Bensadoun and Rothfeld, 1972). The main characteristics as well as the lipid composition of these lipoproteins and other lipoprotein classes and remnants are given elsewhere (Sato et al., 2009; Alvarenga et al., 2011). It is worthy to note that lipoproteins of birds have no apoB48 and apoE (Walzem, 1996). However, the re-esterification of the dietary free fatty acids is not complete and the nonesterified fatty acids (NEFAs) are also released to the portal system, bound to albumin, and carried to peripheral tissues. The portomicron concentration in the blood is mainly dependent on the nutritional status of the bird and of the dietary fat content (Griffin and Hermier, 1988). After the portomicrons are released into the portal vein, they are transported to the liver. However, they are not metabolized by the liver (probably because they are too large to penetrate the cellular sieve of the hepatic capillary bed; Hermier (1997)) but pass through the liver to the extrahepatic tissues where they are partially hydrolyzed by lipoprotein lipase (LPL). The resulting portomicron remnants are taken up by the liver, probably by an apoB100-dependent lipoprotein remnant receptor as apoE is lacking in chickens (Sato et al., 2009). The endocytosed TAG can then be used to resynthesize new TAG-rich lipoproteins.

19.5.1.2 "De novo" Lipogenesis and Very-Low-Density Lipoprotein (VLDL) Synthesis

"De novo" lipogenesis refers to those metabolic pathways that are involved in the synthesis of TAGs from nonlipid precursors (e.g., carbohydrates from dietary origin, or from glycogenolysis or gluconeogenesis from glucogenic or ketogenic amino acids and metabolites such as lactate and glycerol). In contrast to most mammals, in which *de novo* lipogenesis predominantly takes place in the adipocytes, in avian species, the liver has been traditionally considered as the main site of *de novo* lipogenesis (Leveille et al., 1975). However, there are indications that the role of the liver in lipogenesis might be overestimated and that the contribution of other tissues (e.g., bones, skin, and intestines) must also be reconsidered (Griffin and Hermier, 1988; Nir and Lin, 1982).

Acetyl-CoA, formed from pyruvate by pyruvate dehydrogenase, plays a central role in *de novo* lipogenesis. Acetyl-CoA is then irreversibly converted to malonyl-CoA. This reaction is catalyzed by Acetyl-CoA carboxylase (ACC), and this step is considered to be rate limiting for fatty acid synthesis. Elongation of malonyl-CoA is catalyzed by the multienzyme complex fatty acid synthase (FAS). The newly formed fatty acids are then esterified with glycerol to form TAGs. These TAGs are then assembled with phospholipids, cholesterol (esters), and specific apolipoproteins (mainly apoB100) to form very low density lipoproteins (VLDLs). The newly synthesized VLDLs are then stored in vesicles and released into the bloodstream. As for mammals, many apolipoproteins have already been identified in birds, but the roles of the individual avian apolipoproteins are not yet fully elucidated (Griffin and Hermier, 1988; Walzem, 1996;

Sato et al., 2009). Insulin stimulates *de novo* hepatic lipogenesis and VLDL synthesis, whereas thyroxine, glucagon, and epinephrine have opposite effects. Plasma VLDL concentrations are highly positively correlated with total body fat content and can be used successfully as selection criteria for or against fat deposition in broiler lines (Whitehead, 1988; Guo et al., 2011).

19.5.1.3 LPL: A Key Enzyme

TAG-rich lipoproteins from the intestine (portomicrons) and from the liver (VLDL) secreted in the bloodstream are substrates for several lipase enzymes, of which LPL is by far the most important. LPL is synthesized by fat and by muscle cells and migrates to the luminal surface of the surrounding blood capillaries, where it becomes functional after being anchored to the capillary wall by heparin and must be activated by specific apolipoproteins on the surface coat of the lipoproteins (Butterwith, 1988; Griffin and Hermier, 1988). In mammals, apoCII is transferred from high-density lipoproteins (HDLs) to VLDLs and then activates LPL. In birds, however, the equivalent activator of LPL needs to be identified (Walzem, 1996) but is present on VLDLs from immature hens and portomicrons from mature and immature hens and HDLs (Griffin et al., 1982). LPL hydrolyzes TAGs from the core of the lipoprotein, after which the liberated free fatty acids are taken up by the cells. In muscle cells, fatty acids are then oxidized to yield energy, whereas in adipocytes, fatty acids are again incorporated into TAGs for storage. LPL activity is influenced by the age and, to some extent, nutritional state of the bird and is under the major stimulatory control of insulin (Griffin and Hermier, 1988).

19.5.1.4 Lipoprotein Catabolism

As the hydrolysis of the TAGs of the lipoproteins progresses, apolipoproteins are also lost and the lipoproteins decrease in size but increase in density. The partially metabolized VLDLs can be recovered from chicken plasma as intermediate-density lipoproteins (IDLs, or VLDL remnants). The IDL is remodeled in cholesterol-rich LDLs. The remaining VLDL remnants and portomicron remnants are substrate for hepatic lipase. LDLs are taken up by receptor-mediated endocytosis in the liver and other tissues. Premature HDLs probably originate in liver and intestines and receive cholesterol and phospholipids from VLDLs and after interaction with LDLs. As in mammals, HDLs are considered to be the main transporters of cholesterol and phospholipids in avian species.

19.5.1.5 Lipoproteins in Laying Hens

Plasma lipoprotein metabolism in laying hens differs substantially compared to that in immature hens or male chickens. As the ovary has very little capacity for lipid synthesis, all egg yolk lipids are synthesized mainly by the liver (stimulated by steroid hormones, i.e., estrogens and progesterone) and transported to the ovary in special yolk-targeted TAG-rich VLDLs (VLDLy) and in phospholipid-rich lipoproteins termed vitellogenins (Walzem, 1996). Compared with immature hen and normal VLDLs, laying hen VLDLy are 50% smaller in size (30 nm), contain relatively more TAGs and phospholipids but lower proportions of cholesterol (esters), have an unusual apoprotein composition (only apoB and apoVLDLII), and are sparely hydrolyzed by peripheral intravascular LPLs (Griffin et al., 1982; Griffin and Hermier, 1988). The sequence leading to the incorporation of lipoproteins into the oocytes starts with passing through the pores of the granulosa basal lamina (which acts as a sieve to exclude larger lipoproteins) and ends with the uptake of the lipoproteins into the developing oocytes by receptor-mediated endocytosis (the oocyte-specific apoB receptor, in contrast to the somatic apoB receptor). Apparently, apoB acts as the ligand and not apoVLDLII (Walzem, 1996). This oocyte-specific VLDL–vitellogenin receptor or LR8 receptor was then identified as a member of the LDL receptor superfamily (Bujo et al., 1994). In this way, only lipids of hepatic origin (except for polyunsaturated fatty acids) can be deposited in the egg yolk. Besides its role in mediating the diameter of VLDLy, it has been shown that the estrogen-dependent apoVLDLII acts as an inhibitor of LPL activity and in this way prevents TAGs from VLDLy to be hydrolyzed by LPLs (Schneider et al., 1990).

19.5.2 Hormonal Control of Lipid Metabolism

The earliest "endocrinological" observation that already had practical implications concerning growth of avian species was caponization. Since then, systematic studies of the hormonal system have revealed substantial effects of these endogenous molecules on protein, carbohydrate, and lipid metabolism. In addition to pancreatic hormones, two major axes can be identified: the somatotrophic axis and the thyrotrophic axis. These two axes seem to be very much interrelated in birds and especially in the determination of growth and body composition.

19.5.2.1 Somatotrophic Hormones

The pituitary hormone, growth hormone (GH), is a potent stimulator of growth in mammals. In the young, it promotes overall body growth, and in the adult, it stimulates the growth of the extremities. Therefore, it was obvious to suppose that this hormone had also a primary effect on growth in poultry (Scanes et al., 1984). The anabolic effects of GH in relation to its pulsatile secretory pattern are discussed elsewhere (Buyse and Decuypere, 1999).

Based on these observations, it became clear that GH is not the only determinant of growth and body composition in poultry. Also, other endogenous regulatory substances such as thyroid hormones and IGF must be considered. There is ample evidence for a positive relationship between plasma GH and IGF-I levels, which are known to stimulate several anabolic processes. Moreover, IGF-I stimulates adipogenesis while it inhibits the GH-induced lipolysis. Continuous administration of IGF-I reduces abdominal fat content (Buyse and Decuypere, 1999). Data on plasma IGF-II concentrations are scarce and its biological significance is unclear, except that IGF-II administration to broilers augmented fat deposition (Buyse and Decuypere, 1999). This also holds for the role of IGF-binding proteins in avian growth, development, and body composition.

Various sources of GH have been reported to affect the lipolytic activity of chicken adipose tissue explants *in vitro*. GH alone stimulates glycerol release but is also known to inhibit glucagon-induced lipolysis, although by different cellular mechanisms (Campbell and Scanes, 1988). The physiological importance of the observed effects of GH on lipolysis *in vitro* can be linked with a homeostatic role of GH in nutrient metabolism.

19.5.2.2 Thyroid Hormones

There is abundant evidence that thyroid hormones are important not only for embryogenesis but also for normal posthatch growth in birds. Thyroid hormones also affect body composition and fatness. A hyperthyroid status in chickens is associated with a decreased fat content, while in hypothyroid chickens, fat deposition is greatly enhanced (Decuypere and Buyse, 1988). Triiodothyronine has been shown to increase basal lipolytic activity and glucagon-induced lipolysis in cultured broiler adipocytes (Harden and Oscar, 1993). In addition, triiodothyronine stimulates hepatic malic enzyme activity, and hence lipogenesis.

19.5.2.3 Pancreatic Hormones

The pancreatic hormones (insulin, glucagon, avian pancreatic polypeptide, and somatostatin) are important in the regulation of carbohydrate, protein, and lipid metabolism and hence for normal growth and development of birds. The structure, secretion, and peripheral control of these pancreatic hormones are reviewed elsewhere (Simon, 1989). In brief, insulin acts to decrease circulating levels of glucose by increasing glucose uptake by several cell types and by inhibiting glycogenolysis and gluconeogenesis while glycogen synthesis is stimulated. Glucagon has the opposite effect on these processes. Therefore, it is suggested that the glucagon–insulin ratio would be more important in regulating nutrient metabolism. Insulin also stimulates hepatic lipogenesis mainly by stimulating the synthesis of lipogenic enzymes. In addition, insulin may increase the hydrolysis of VLDLs at the adipocyte level by stimulation of LPL activity (Simon, 1989). With respect to the role of insulin and glucagon on lipid mobilization, several profound differences between mammals and birds are apparent. First, in mammals, catecholamines are the main effectors of lipolysis, while in birds, glucagon is the most potent stimulator of lipolysis. Furthermore, in mammals insulin is clearly antilipolytic, while this is not the case in birds.

19.5.2.4 Other Hormones

There is ample evidence that adrenal glucocorticoids (mainly corticosterone in birds) depress body weight and increase fat deposition in birds as glucocorticoids exert a direct catabolic action and stimulate lipogenic activity in avian hepatocytes. The pituitary hormone prolactin is also known to be a lipogenic effector, and a possible temporal synergistic effect between corticosterone and prolactin on fat deposition has been demonstrated in broilers and wild-type birds (Decuypere and Buyse, 1988). The role of sex steroids in the hormonal control of growth in birds is not well established except that estrogens stimulate lipogenesis, whereas testosterone has the opposite effect.

19.5.2.5 Endocrine Control of Premigratory Hyperphagia and Fattening in Wild Birds

A peculiar feature of many bird species is the annual migration, which is a high-energy-demanding process. In preparation of vernal or autumnal migration over long distances, the occurrence of premigratory hyperphagia and fattening is well documented. The onset of premigratory morphological, behavioral, and body compositional changes is predominantly induced by alterations in day length and environmental temperature as principal Zeitgebers, and suggests a neuroendocrine involvement. It has been reported that besides gonadal hormones, both corticosterone and prolactin are involved in premigratory hyperphagia and fattening in wild-type birds (e.g., *Zonotrichia gambelii*: Meier and Farner (1964); and *Zonotrichia albicollis*: Meier and Martin (1971)). More in particular, the increased premigratory fat deposition is dependent on a temporal synergism between corticosterone and prolactin (Meier and Farner, 1964; Meier and Martin, 1971). The most potent lipolytic agent, glucagon, is also strongly implicated in premigratory and winter fattening (e.g., in red-winged blackbirds: Hintz, 2000). Indeed, significantly reduced glucagon levels were measured in the plasma of birds exposed to outdoor winter (from September to May, Ontario, Canada) conditions compared to birds maintained in summer conditions. It was concluded that the low glucagon levels are meant to preserve the deposited lipid stores. In addition, seasonal variations in adipocyte glucagon receptors and concomitant reduction in adipocyte sensitivity to glucagon may be involved in the preservation of body weight and fat.

19.6 FUNCTIONS OF ADIPOSE TISSUE

It has always been believed that the role of white adipose tissue is rather strictly limited to storing energy as TAGs and releasing these lipids for oxidation (primarily by skeletal muscle), and this has been known for many decades. Although their hydrophobic lipid content is variable between fat depots, it can amount to up to 90% of the fresh tissue weight (Evans, 1977; Nir et al., 1988). Taking into account the high energy content of lipids (±39 kJ/g) compared to protein (±23 kJ/g) and carbohydrates (±18 kJ/g), adipose tissue is indeed an appropriate reservoir of energy. In addition, fat tissue (e.g., subcutaneous fat) can play a role in insulation and can act as a protective shield for internal organs. More recent studies, however, have revealed that adipose tissue is also involved in numerous other processes and biological functions such as appetite regulation, reproduction, angiogenesis, coagulation, fibrinolysis, vascular tone control, and immunity.

Adipose tissue is not just equal to a sum of adipose cells. In adipose tissue, 90% of adipose mass may account for roughly 25% of the total cell population, the other cells being fibroblasts, blood cells, endothelial cells, pericytes, mesenchymal cells, and preadipocytes, all being targets for an extensive autocrine–paracrine cross-talk with each other and involved in adipogenesis (Frühbeck and Gómez-Ambrosi, 2005).

At least in mammals, and probably also in birds, new adipocyte formation is still possible even when a positive energy balance is maintained after hypertrophy; then hyperplasia resumes and a greater than normal number of adipose cells is developed. Moreover, adipocytes have been also reported to dedifferentiate: mature adipocytes lose their lipid-filled morphology, and they assume a fibroblast-like appearance and regain preadipocyte characteristics (Kokta et al., 2004). Especially the signaling interactions between muscle and fat cells are important to understand the relative fat and lean deposition and the efficiency of energy utilization in the complex processes of growth, development, and body composition. These signaling interactions are based on the notion of adipose tissue as a multifunctional organ as opposed to a passive organ for the storage of excess energy in the form of fat. This is linked with the ability of fat cells to secrete a large number of hormones, growth factors, cytokines, and matrix proteins collectively termed adipokines or adipocytokines (Table 19.1). At the same time, adipocytes have receptors for many of these adipokines as well as for several systemic hormonal factors (e.g., pituitary and pancreatic hormones).

Intra- and intermuscular fat is regulated by different factors than those regulating fat deposition in other tissues such as omental, mesenteric, and perirenal fat, which are not in close proximity to muscle and may therefore lack some specific paracrine signaling.

TABLE 19.1 Tentative List of Demonstrated Expression of Genes Involved in Lipid Metabolism of Avian Species[1] and Potential Avian Adipokines

Gene Expressions	Potential Adipokines
Stearoyl-CoA desaturase	Leptin
Hormone-sensitive lipase	Adiponectin
Adipose triglyceride lipase	Visfatin
AMP-activated protein kinase	Resistin
Adiponectin receptors R1 and R2	Adipsin
Insulin receptors (IRs)	Acylation-stimulating protein
IR-substrate protein	Adiponutrin
Peroxisome proliferator-activated receptor-γ	Interleukin-6
Sterol regulatory element binding protein-1 and -2	Tumor necrosis factor-α
Leptin receptor?	Transforming growth factor-β
Agouti-signaling peptide	Apelin
Melanocortin receptor	Angiotensinogen
Pro-opiomelanocortin	
Neuropeptide Y	
Fat mass and obesity-associated gene	
Lipoprotein lipase	
Fatty acid synthase	
Growth hormone receptor	
Interleukin-18	
Visfatin	
Hydroxy-3-methylglutaryl-CoA reductase	

[1]*A more comprehensive list of genes expressed in the adipose tissue of 7 week old fat- and lean-line broilers is provided by Wang et al. (2007).*
Source: After Richards et al., 2010.

The responsiveness of fat cells to neurohormonal signals may vary according to the adipose cell stage at the moment of exposure; this, together with local differences in the expression and presence of these numerous factors involved in the control of adipogenesis, may contribute to affecting the pattern of distribution of fat in birds, and this is in a species-specific way.

Fat tissue is central in energy substrate partitioning as it acts as a reservoir as a result of lipogenesis, on the one hand, and substrate mobilization, on the other hand. However, in energy balance equilibrium, lipogenesis and lipolysis are still going on continuously, making fat tissue a rather dynamic tissue with a turnover rate that is relatively high (e.g., the half-life of depot lipids in rodents is about 8 days; Frühbeck and Gómez-Ambrosi (2005)). NEFA mobilization (lipolysis) is followed by its uptake by diverse cell types for subsequent oxidation. In most cases, uptake of NEFAs is much faster than their oxidation except during heavy and/or prolonged exercise. Between 80% and 90% of the NEFAs used in muscle for oxidation comes from plasma NEFAs (coming from the diet and fat tissue), while only ±10% comes from muscle lipid itself. AMP-activated protein kinase plays a central role as energy sensor and positively affects cell processes that produce energy (NEFA oxidation and glucose uptake) while negatively affecting cell processes that consume energy (lipogenesis, protein synthesis, and gluconeogenesis). These processes are governed by endocrine factors as well as by adipokines.

The regulation of lipolysis in chicken adipose tissue is almost exclusively exerted by glucagon. No antilipolytic effect of insulin has been demonstrated in chicken adipocytes in culture, and only avian pancreatic polypeptide, somatostatin, and gut-glucagon-like immunoreactivity have antilipolytic effects (Dupont et al., 2012). Lipolysis in chicken adipose tissue is realized by a lipase (adipose triglyceride lipase (ATGL)) exclusively found in subcutaneous and abdominal fat (Lee et al., 2009), and this lipase is highly present after hatching and before feed is given. Hormone-sensitive lipase is also present in chicken (Anthonsen et al., 1997), and LPL, hepatic lipase, and endothelial and carboxylester lipase are all highly conserved in mammalian and bird (chicken) species (Sato et al., 2010). ATGL is stimulated by corticosteroids in broilers (Serr et al., 2011).

Adiponectin as well as its receptors (R_1 and R_2) are expressed in many tissues, including adipose tissue in the chicken (Hendricks et al., 2009; Ghazanfari et al., 2011). It stimulates NEFA oxidation in muscle, hence decreasing plasma NEFA as well as glucose levels (due to enhanced insulin sensitivity, at least in mammals). Fasting in chicken is known to decrease adiponectin receptor expression in adipose tissue (Ghazanfari et al., 2011), while plasma adiponectin is inversely related to abdominal fat pad weight in chicken (Hendricks et al., 2009), especially the plasma heavy-molecular-weight adiponectin isoform.

The insulin-signaling cascade in chicken adipose tissue, although similar in its downstream components (see Dupont et al. (2012)), is quantitatively markedly different from that operating in mammals. It has been shown that levels of IR (insulin receptor) and IRS1 (IR-substrate protein) in chicken are lower compared to those of mammals (rat), contributing to a certain insulin insensitivity or refractoriness.

PPARγ is a central regulator of adipogenesis, and it is expressed in fat tissue and even more in liver but not in muscle. Levels are affected by nutrition and are also higher at the onset of egg laying in layers in the liver and ovary, while it is then decreased in adipose tissue (Sato et al., 2004). Moreover, a single intraperitoneal injection of troglitazone, a synthetic PPARγ ligand, to newly hatched broiler chicks resulted in a reduction of the absolute and proportional abdominal fat pad weight at 48 days of age (Sato et al., 2008).

Sterol regulatory element binding protein-1 and -2 (SREBP1 and SREBP2) chicken genes, on two separate chromosomes and nearly perfectly conserved compared to their mammalian homologs (Assaf et al., 2003), also play a role in the regulation of lipogenesis in liver. SREBP1 is linked to FAS protein content or activity in adipose tissue and/or liver. Therefore, SREBP1/2 are key regulators in lipogenesis in both mammals and avian species, and their differential expression in different tissues is a major determinant of the site of fatty acid synthesis in the body (Gondret et al., 2001; Yen et al., 2005).

Visfatin is another adipokine hormone involved in immune system and glucose metabolism. Chicken visfatin is 92–94% homologous to mammalian visfatin and mainly expressed in skeletal muscle rather than in abdominal and subcutaneous fat in chicken; therefore it is rather a myokine than an adipokine (Krzysik-Walker et al., 2008; Li et al., 2012). Nevertheless, fat-line chickens selected for high abdominal fat have a higher expression of visfatin in adipose tissue compared to their lean-line counterparts (Cogburn et al., 2011).

There is controversy as to the leptin receptor expression in chicken adipose tissue. According to Wang et al. (2007), no leptin receptor expression in chicken adipose tissue was found, whereas Hausman et al. (2012) reported leptin receptor expression in broiler adipose tissue.

Also, local agouti-signaling peptide and melanocortin receptor expression were found in chicken adipose tissue, as well as pro-opiomelanocortin (Yabuuchi et al., 2010) and neuropeptide Y (Hausman et al., 2012), indicating that hypothalamic regulatory pathways for feed intake regulation, also in birds, may also play a role, locally, in adipose tissue metabolism.

In mammals, the fat mass and obesity-associated (FTO) gene is known to be implicated in the regulation of energy balance, more in particular in food intake and energy expenditure (Wang et al., 2012b). In chickens, FTO gene mRNA is present in many tissues, with high levels in the hypothalamus, cerebellum, liver, and visceral fat. However, the expression levels are dependent on breed (broiler versus layer), age, and nutritional status. More studies are needed to determine the functional role of FTO in avian species.

A specific role for other factors that were found as adipokines in mammalian species still have to be elucidated in birds and chicken in particular (e.g., resistin, adipsin, and plasminogen activator inhibitor-1; see Richards et al. (2010)).

19.7 FACTORS AFFECTING FAT METABOLISM AND DEPOSITION

Nutrition has a major impact on animal performance and body composition. Nutrition is a very broad term and includes quantitative (e.g., *ad libitum* versus feed restriction programs) and qualitative (e.g., metabolizable energy (ME), macro- and micronutrients, and toxins) aspects as well as feeding programs (e.g., sequential feeding or "skip-a-day" feeding).

It is common knowledge that feed deprivation–refeeding cycles are associated with fluctuations in intermediary metabolism induced by endocrine factors such as thyroid hormones, GH, and pancreatic hormones (Buyse et al., 2001). During feed deprivation, hepatic lipogenic activity is drastically reduced and the release of NEFA from adipose tissue is stimulated, whereas the opposite is true after refeeding. The reader is referred to recent elegant studies for detailed information from feed deprivation–refeeding schedules on the dynamics of hepatic mRNA levels of lipogenic genes and transcription factors (e.g., Wang et al., 2012a; Richards et al., 2003; Saneyasu et al., 2013). In addition, such feeding schedules also have a marked effect on hypothalamic orexigenic and anorexigenic neuropeptides and lipogenic genes such as FAS and ACC (Higgins et al., 2010; Song et al., 2013). A discussion of the relevance of these findings in the framework of the control and regulation of voluntary feed intake of avian species is, however, beyond the scope of this chapter.

With respect to diet quality, the impact of ME levels and macronutrient (lipids, carbohydrates, and proteins) ratios on performance and body composition has been widely investigated, mainly with broiler chickens. In general, diets with high ME levels (but normal protein levels) promote fat deposition in broiler chickens, but genetically determined body weight gain is rather maintained. This augmented fat deposition is the result of an excess in energy intake ("luxus" energy consumption) relative to the energy needed for maintenance and production. This luxus energy is then largely diverted to body fat deposition, although it is also dissipated as heat to some extent (Buyse et al., 1992). Widening of the ME–protein ratio due to a reduction in protein level results in not only a higher fat deposition but also a reduced growth rate. The higher fat deposition is caused by an enhanced hepatic lipogenesis (demonstrated *in vivo* and *in vitro*), as reflected in the increased expression and activity of lipogenic enzymes such as malic enzyme, ACC, and FAS (e.g., Donaldson, 1985; Adams and Davis, 2001; Rosebrough et al., 2011). The reader is referred to the work of Rosebrough, Richards, and coworkers with respect to short-term adaptations in hepatic lipogenesis when switching from high- to low-protein diets and vice versa (Rosebrough et al., 2011). The underlying causal mechanisms at the biochemical level of the hepatic intermediary metabolism are discussed in detail elsewhere (Swennen et al., 2007a). Such diets with a relative protein shortage also induce marked endocrine alterations (reviewed by Buyse et al. (2001), Rosebrough et al. (2011)). Indeed, protein-restricted chickens are characterized by an enhanced pulsatile GH secretion, lower plasma IGF-I and thyroxine levels, but higher 3,3′,5-triiodothyronine concentrations. Feeding broilers with diets with a too-high protein content (above requirements) will result in lean broilers (decreased *de novo* lipogenesis and expression and activity of lipogenic enzymes), but growth is likely to be stunted because of the energetic costs of nitrogen waste excretion. The effects of dietary fat level on body fat accretion and lipid metabolism are less unambiguous (Swennen et al., 2007a). The interpretation of studies focusing on dietary fat level is sometimes complicated by the fact that it is not clearly indicated whether the dietary fat is iso-energetically substituted with protein or carbohydrates or both. In general, increasing the dietary fat level will reduce hepatic lipogenic activity, probably due to the inhibitory effect of long-chain acyl-CoA on ACC activity (Leveille et al., 1975). Finally, high carbohydrate levels in broiler diets will enhance glycolysis and hence provide plenty of reducing equivalents for hepatic fatty acid production (Tanaka et al., 1983). However, these dietary fat and carbohydrate-induced changes in hepatic lipogenic activity do not necessarily result in changed body fat accretion (Swennen et al., 2007a).

Intensive selection of broiler chickens for high growth rate and feed efficiency has resulted in a tremendous progress in body weight gain and feed efficiency (direct selection response). However, several adverse indirect selection responses such as augmented fat deposition, leg problems, and metabolic diseases such as ascites and sudden death syndrome have also occurred. Havenstein and coworkers (Havenstein et al., 1994a,b) have elegantly assessed the differential impacts of selection for growth rate and of diet regimen by comparing a typical 1957 broiler strain (Athens-Canadian random-bred control strain) with a modern 1991 broiler strain (Arbor Acres) and feeding both on a typical 1957 or 1991 diet. A similar comparative study was done between the random-bred 1957 strain and a commercial 2001 (Ross 308) strain, with both fed on a typical 1957 or 2001 diet (Havenstein et al., 2003). These studies clearly revealed that genetics were the major contributing factor for the progress in performance and yield of carcass parts but also that carcass fat content increased due to selection for growth rate. In view of this augmented fat deposition and its associated negative consequences, divergent experimental fat and lean broiler lines were established in several poultry research institutes. Different selection strategies were used: divergent selection for or against abdominal fat content (France and Israel), selection for high body weight gain (fat line) or feed efficiency (lean line) (the Netherlands and Denmark), and divergent selection for high (fat line) or low (lean line) plasma VLDL concentration (United Kingdom). Detailed information on heritability estimates, genotypic and phenotypic correlations, as well as performance characteristics, energy, lipid and protein metabolism, endocrine profiles and intermediary metabolism, and breeder performance is provided elsewhere (Leclercq, 1988; Whitehead, 1988; Buyse et al., 1999; Baéza and Le Bihan-Duval, 2013). It is clear that either direct or indirect selection resulted in fat and lean broiler lines. With respect to lipid metabolism, all fat-line broilers were characterized by a higher hepatic lipogenic capacity compared to their lean-line counterparts, irrespective of the selection criteria. However, some indirect selection responses as well as the underlying endocrine mechanisms were different according to the selection strategy. Indeed, lean broilers produced by selection for feed efficiency had a higher basal as well as glucagon-stimulated lipolytic activity compared to their fat counterparts selected for high body weight gain. In contrast, direct selection for high or low abdominal fat content had apparently no effect on the lipolytic capacity of adipose tissue. These differences in lipolytic activity, amongst other parameters, according to the selection strategy can be related to the differential selection-induced alterations in endocrine profile. Indeed, direct selection rather triggered the thyrotrophic axis and pancreatic hormones, whereas indirect selection primarily affected the somatotrophic axis (Buyse et al., 1999). Another interesting genetic model to study the regulatory mechanisms of lipid metabolism is the selection model for the high (R+: fat line) or low (R−: lean line) residual feed intake of laying hens, which markedly differ in appetite; heat production, including diet-induced thermogenesis; body conformation; and composition (Bordas et al., 1992; Swennen et al., 2007b).

19.8 SUMMARY AND CONCLUSIONS

Adipose tissue is the most variable carcass component and is traditionally being considered as a rather inert storage tissue for energy under the form of lipids. Most of the fat in chickens is located in adipose depots, which are all late maturing. At the cellular level, preadipocyte differentiation and proliferation are under the control of multiple hormones and transcription factors. The growth of fat tissue is initially due to hyperplasia followed by hypertrophy of the mature adipocytes, and the amount of fat deposited depends on genetic and nutritional factors. Recent research has, however, revealed that adipose tissue must now be regarded as a dynamic tissue that secretes a considerable number of adipokines and hence plays a role in a multitude of bodily processes.

REFERENCES

Adams, K.A., Davis, J., 2001. Dietary protein concentration regulates the mRNA expression of chicken hepatic malic enzyme. J. Nutr. 131, 2269–2274.

Alvarenga, R.R., Zangeronimo, M.G., Pereira, L.J., Rodrigues, P.B., Gomide, E.M., 2011. Lipoprotein metabolism in poultry. World's Poult. Sci. J. 67, 431–440.

Anthonsen, M.W., Degerman, E., Holm, C., 1997. Partial purification and identification of hormone-sensitive lipase from chicken adipose tissue. Biochem. Biophys. Res. Commun. 236, 94–99.

Assaf, S., Hazard, D., Pitel, F., Morisson, M., Alizadeh, M., Gondret, F., Diot, C., Vignal, A., Douaire, M., Lagarrigue, S., 2003. Cloning of cDNA encoding the nuclear form of chicken sterol response element binding protein-2 (SREBP-2), chromosomal localization, and tissue expression of chicken SREBP-1 and -2 genes. Poult. Sci. 82, 54–61.

Baéza, E., Le Bihan-Duval, E., 2013. Chicken lines divergent for low or high abdominal fat deposition: a relevant model to study the regulation of energy metabolism. Animal 7, 965–973.

Bensadoun, A., Rothfeld, A., 1972. The form of absorption of lipids in the chicken, *Gallus domesticus*. Proc. Soc. Exp. Biol. Med. 141, 814–817.

Bordas, A., Tixier-Boichard, M., Mérat, P., 1992. Direct and correlated responses to divergent selection for residual food intake in Rhode Island Red laying hens. Br. Poult. Sci. 33, 741–754.

Bujo, H., Hermann, M., Kaderli, M.O., Jacobson, L., Sugawara, S., Nimpf, J., Yamamoto, T., Schneider, W.J., 1994. Chicken oocyte growth is mediated by an eight ligand binding repeat member of the LDL receptor family. EMBO J. 13, 5165–5175.

Butterwith, S.C., 1988. Avian adipose tissue: growth and metabolism. In: Leclercq, B., Whitehead, C.C. (Eds.), Leanness in Domestic Birds. Genetic, Metabolic and Hormonal Aspects. INRA-Butterworths, London, pp. 203–222.

Butterwith, S.C., 1997. Regulators of adipocyte precursor cells. Poult. Sci. 76, 118–123.

Buyse, J., Decuypere, E., Berghman, L., Kühn, E.R., Vandesande, F., 1992. The effect of dietary protein content on episodic growth hormone secretion and on heat production of male broilers. Br. Poult. Sci. 33, 1101–1109.

Buyse, J., Decuypere, E., 1999. The role of the somatotrophic axis in the metabolism of the chicken. Domest. Anim. Endocrinol. 17, 245–255.

Buyse, J., Leenstra, F.R., Zeman, M., Rahimi, G., Decuypere, E., 1999. A comparative study of different selection strategies to breed leaner meat-type poultry. Poult. Avian Biol. Rev. 10, 121–142.

Buyse, J., Darras, V.M., Kühn, E.R., Decuypere, E., 2001. Nutritional regulation of the somatotropic axis and intermediary metabolism in the chicken. In: Dawson, A., Chaturvedi, C.M. (Eds.), Avian Endocrinology. Narosa Publishing House, New Dehli, pp. 303–313.

Campbell, R.M., Scanes, C.G., 1988. Pharmacological investigations on the lipolytic and antilipolytic effects of growth hormone (GH) in chicken adipose tissue in vitro: evidence for involvement of calcium and polyamines. Proc. Soc. Exp. Biol. Med. 188, 177–184.

Cartwright, A.L., 1991. Adipose cellularity in *Gallus domesticus*: investigations to control body composition in growing chickens. J. Nutr. 121, 1486–1497.

Cogburn, L.A., Resnyk, C., Porter, T.E., Aggrey, S.E., Le Bihan-Duval, E., Duclos, M., Simon, J., 2011. Transcriptional analysis of abdominal fat accretion in genetically fat and lean chickens: a new polygenic model of visceral obesity. FASEB J. 25. 862.7.

Decuypere, E., Buyse, J., 1988. Thyroid hormones, corticosterone, growth hormone and somatomedins in avian species: general effects and possible implications in fattening. In: Leclercq, B., Whitehead, C.C. (Eds.), Leanness in Domestic Birds. Genetic, Metabolic and Hormonal Aspects. INRA-Butterworths, London, pp. 295–312.

Donaldson, W.E., 1985. Lipogenesis and body fat in chicks: effects of calorie-protein ratio and dietary fat. Poult. Sci. 64, 1199–1204.

Dupont, J., Métayer-Coustard, S., Ji, B., Ramé, C., Gespach, C., Voy, B., Simon, J., 2012. Characterization of major elements of insulin signaling cascade in chicken adipose tissue: apparent insulin refractoriness. Gen. Comp. Endocrinol. 176, 86–93.

Evans, A.J., 1977. The growth of fat. In: Boorman, K.N., Wilson, B.J. (Eds.), Growth and Poultry Meat Production. Br. Poult. Sci. Ltd, Edinburgh, pp. 29–64.

Frühbeck, G., Gómez-Ambrosi, J., 2005. Adipose tissue. In: Benjamin, C. (Ed.), Encyclopedia of Human Nutrition, second ed. Elsevier, Oxford, pp. 1–14.

Ghazanfari, S., Nobari, K., Yamauchi, T., 2011. Adiponectin: a novel hormone in birds. Asian J. Anim. Vet. Adv. 6, 429–439.

Gondret, F., Ferré, P., Dugail, I., 2001. ADD-1/SREBP-1 is a major determinant of tissue differential lipogenic capacity in mammalian and avian species. J. Lipid Res. 42, 106–113.

Griffin, H., Grant, G., Perry, M., 1982. Hydrolysis of plasma triacylglycerol-rich lipoproteins from immature and laying hens (*Gallus domesticus*) by lipoprotein lipase in vitro. Biochem. J. 206, 647–654.

Griffin, H., Hermier, D., 1988. Plasma lipoprotein metabolism and fattening in poultry. In: Leclercq, B., Whitehead, C.C. (Eds.), Leanness in Domestic Birds. Genetic, Metabolic and Hormonal Aspects. INRA-Butterworths, London, pp. 175–202.

Guo, L., Sun, B., Shang, Z., Leng, L., Wang, Y., Li, H., 2011. Comparison of adipose tissue cellularity in chicken lines divergently selected for fatness. Poult. Sci. 90, 2024–2034.

Harden, R.L., Oscar, T.P., 1993. Thyroid hormone and growth hormone regulation of broiler adipocyte lipolysis. Poult. Sci. 72, 669–676.

Hausman, G.J., Barb, C.R., Fairchild, B.D., Gamble, J., Lee-Rutherford, L., 2012. Expression of genes for interleukins, neuropeptides, growth hormone receptor, and leptin receptor in adipose tissue from growing broiler chickens. Domest. Anim. Endocrinol. 43, 260–263.

Havenstein, G.B., Ferket, P.R., Scheideler, S.E., Larson, B.T., 1994a. Growth, livability, and feed conversion of 1991 vs 1957 broilers when fed "typical" 1957 and 1991 diets. Poult. Sci. 73, 1785–1794.

Havenstein, G.B., Ferket, P.R., Scheideler, S.E., Larson, B.T., 1994b. Carcass composition and yield of 1991 vs 1957 broilers when fed "typical" 1957 and 1991 diets. Poult. Sci. 73, 1795–1804.

Havenstein, G.B., Ferket, P.R., Qureshi, M.A., 2003. Growth, livability, and feed conversion of 1957 versus 2001 broilers when fed representative 1957 and 2001 diets. Poult. Sci. 82, 1500–1508.

Hendricks, G.L., Hadley, J.A., Krzysik-Walker, S.M., Prabhu, K.S., Vasilatos-Younken, R., Ramachandran, R., 2009. Unique profile of chicken adiponectin, a predominantly heavy molecular weight multimer, and relationship to visceral adiposity. Endocrinology 150, 3092–3100.

Hermier, D., 1997. Lipoprotein metabolism and fattening in poultry. J. Nutr. 127, 805S–808S.

Higgins, S.E., Ellestad, L.E., Trakooljul, N., McCarthy, F., Saliba, J., Cogburn, L.A., Porter, T.E., 2010. Transcriptional and pathway analysis in the hypothalamus of newly hatched chicks during fasting and delayed feeding. BMC Genomics 11, 162.

Hintz, J.V., 2000. The hormonal regulation of premigratory fat deposition and winter fattening in red-winged blackbirds. Comp. Biochem. Physiol. B 125, 239–249.

Hood, R.L., 1982. The cellular basis for growth of the abdominal fat pad in broiler-type chickens. Poult. Sci. 61, 117–121.

Hood, R.L., Pym, R.A.E., 1982. Correlated responses for lipogenesis and adipose tissue cellularity in chickens selected for body weight gain, food consumption, and food conversion efficiency. Poult. Sci. 61, 122–127.

Kokta, T.A., Dodson, M.V., Gertler, A., Hill, R.A., 2004. Intercellular signaling between adipose tissue and muscle tissue. Domest. Anim. Endocrinol. 27, 303–331.

Krzysik-Walker, S.M., Ocón-Grove, O.M., Maddineni, S.R., Hendricks III, G.L., Ramachandran, R., 2008. Is visfatin an adipokine or myokine? Evidence for greater visfatin expression in skeletal muscle than visceral fat in chickens. Endocrinol 194, 1543–1550.

Leclercq, B., 1984. Adipose tissue metabolism and its control in birds. Poult. Sci. 63, 2044–2054.

Leclercq, B., 1988. Genetic selection of meat-type chickens for high or low abdominal fat content. In: Leclercq, B., Whitehead, C.C. (Eds.), Leanness in Domestic Birds. Genetic, Metabolic and Hormonal Aspects. INRA-Butterworths, London, pp. 25–40.

Lee, K., Shin, J., Latshaw, J.D., Suh, Y., Serr, J., 2009. Cloning of adipose triglyceride lipase complementary deoxyribonucleic acid in poultry and expression of adipose triglyceride lipase during development of adipose in chickens. Poult. Sci. 88, 620–630.

Leveille, G.A., Romsos, D.R., Yeh, Y.-Y., O'Hea, E.K., 1975. Lipid biosynthesis in the chick. A consideration of site of synthesis, influence of diet and possible regulatory mechanisms. Poult. Sci. 54, 1075–1093.

Li, B.C., Zhang, Y.I., Chen, X.N., Shi, Q.Q., Fu, D.Z., Yin, Y.H., Zhang, Z.T., Gao, B., Chen, G.H., 2011. Directional differentiation of chicken embryonic stem cells into osteoblasts, neuron-like cells and adipocytes. Afr. J. Biotechnol. 10, 7772–7779.

Li, J., Meng, F., Song, C., Wang, Y., Leung, F.C., 2012. Characterization of chicken visfatin gene: cDNA cloning, tissue distribution, and promoter analysis. Poult. Sci. 91, 2885–2894.

March, B.E., MacMillan, C., Chu, S., 1984. Characteristics of adipose tissue growth in broiler-type chickens to 22 weeks of age and the effects of dietary protein and lipid. Poult. Sci. 63, 2207–2216.

Meier, A.H., Farner, D.S., 1964. A possible endocrine basis for premigratory fatting in the white-crowned sparrow, *Zonotrichia leucophrys* gambelii (Nuttall). Gen. Comp. Endocrinol. 4, 584–595.

Meier, A.H., Martin, D.D., 1971. Temporal synergism of corticosterone and prolactin controlling fat storage in the white-throated sparrow, *Zonotrichia albicollis*. Gen. Comp. Endocrinol. 17, 311–318.

Merkley, J.W., Cartwright, A.L., 1989. Adipose tissue deposition and cellularity in cimaterol-treated female broilers. Poult. Sci. 68, 762–770.

Nir, I., Nitsan, Z., Keren-Zvi, S., 1988. Fat deposition in birds. In: Leclercq, B., Whitehead, C.C. (Eds.), Leanness in Domestic Birds. Genetic, Metabolic and Hormonal Aspects. INRA-Butterworths, London, pp. 141–174.

Nir, I., Lin, H., 1982. The skeleton, an important site of lipogenesis in the chick. Ann. Nutr. Metab. 26, 100–105.

Pálsson, H., 1955. Conformation and body composition. In: Hammond, J. (Ed.), Progress in the Physiology of Farm Animals, vol. 2. Butterworths, London, pp. 430–452.

Qi, R.L., Sun, C., Yan, J., Yang, H.L., Zhao, T.T., Zhao, X., 2012. Effects of glucose on differentiation and fat metabolism of chicken preadipocytes. J. Anim. Vet. Adv. 11, 1223–1229.

Ramsay, T.G., Rosebrough, R.W., 2003. Hormonal regulation of postnatal chicken preadipocyte differentiation in vitro. Comp. Biochem. Physiol. B 136, 245–253.

Richards, M.P., Poch, S.M., Coon, N.C., Rosebrough, R.W., Ashwell, C.M., McMurtry, J.P., 2003. Feed restriction significantly alters lipogenic gene expression in broiler breeder chickens. J. Nutr. 133, 707–715.

Richards, M.P., Rosebrough, R.W., Coon, C.N., McMurtry, J.P., 2010. Feed intake regulation for the female broiler breeder: in theory and in practice. J. Appl. Poult. Res. 19, 182–193.

Rosebrough, R.W., Russell, B.A., Richards, M.P., 2011. Further studies on short-term adaptations in the expression of lipogenic genes in broilers. Comp. Biochem. Physiol. A 159, 1–6.

Saneyasu, T., Shiragaki, M., Nakanishi, K., Kamisoyama, H., Honda, K., 2013. Effects of short term fasting on the expression of genes involved in lipid metabolism in chicks. Comp. Biochem. Physiol. B 165, 114–118.

Sato, K., Fukao, K., Seki, Y., Akiba, Y., 2004. Expression of the chicken peroxisome proliferator-activated receptor-γ gene is influenced by aging, nutrition, and agonist administration. Poult. Sci. 83, 1342–1347.

Sato, K., Matsushita, K., Matsubara, Y., Kamada, T., Akiba, Y., 2008. Adipose tissue fat accumulation is reduced by a single intraperitoneal injection of peroxisome proliferator-activated receptor gamma agonist when given to newly hatched chicks. Poult. Sci. 87, 2281–2286.

Sato, K., Suzuki, K., Akiba, Y., 2009. Characterization of chicken portomicron remnant and very low density lipoprotein remnant. J. Poult. Sci. 46, 35–39.

Sato, K., Seol, H.S., Kamata, T., 2010. Tissue distribution of lipase genes related to triglyceride metabolism in laying hens (*Gallus gallus*). Comp. Biochem. Physiol. B 155, 62–66.

Scanes, C.G., Harvey, S., Marsh, J.A., King, D.B., 1984. Hormones and growth in poultry. Poult. Sci. 63, 2062–2074.

Schneider, W.J., Carroll, R., Severson, D.L., Nimpf, J., 1990. Apolipoprotein VLDL-II inhibits lipolysis of triglyceride-rich lipoproteins in the laying hen. J. Lipid Res. 31, 507–513.

Serr, J., Suh, Y., Oh, S.A., Shin, S., Kim, M., Latshaw, J.D., Lee, K., 2011. Acute up-regulation of adipose triglyceride lipase and release of non-esterified fatty acids by dexamethasone in chicken adipose tissue. Lipids 46, 813–820.

Simon, J., 1989. Chicken as a useful species for the comprehension of insulin action. Crit. Rev. Poult. Biol. 2, 121–148.

Sonaiya, C.J., 1985. Abdominal fat weight and thickness as predictors of total body fat in broilers. Br. Poult. Sci. 26, 453–458.

Song, Z., Everaert, N., Wang, Y., Decuypere, E., Buyse, J., 2013. The endocrine control of energy homeostasis in chicken. Gen. Comp. Endocrinol. 190, 112–117.

Swennen, Q., Decuypere, E., Buyse, J., 2007a. Implications of dietary macronutrients for growth and metabolism in broiler chickens. World's Poult. Sci. J. 63, 541–556.

Swennen, Q., Verhulst, P.-J., Collin, A., Bordas, A., Decuypere, E., Verbeke, K., Vansant, G., Buyse, J., 2007b. Further investigations on the role of diet-induced thermogenesis in the regulation of feed intake in chickens: comparison of R+ and R− cockerels. Poult. Sci. 86, 1960–1971.

Tanaka, K., Ohtani, S., Shigeno, K., 1983. Effect of increasing dietary energy on hepatic lipogenesis in growing chicks. I. Increasing energy by carbohydrate supplementation. Poult. Sci. 62, 445–451.

Walzem, R.L., 1996. Lipoproteins and the laying hen: form follows function. Poult. Avian Biol. Rev. 7, 31–64.

Wang, H.-B., Li, H., Wang, Q.-G., Zhang, X.-Y., Wang, S.-Z., Wang, Y.-X., Wang, X.-P., 2007. Profiling of chicken adipose tissue gene expression by genome array. BMC Genomics 8, 193–206.

Wang, J.W., Chen, W., Kang, X.T., Huang, Y.Q., Tian, Y.D., Wang, Y.B., 2012a. Identification of differentially expressed genes induced by energy restriction using annealing control primer system from the liver and adipose tissues of broilers. Poult. Sci. 91, 972–978.

Wang, Y., Rao, K., Yuan, L., Everaert, N., Buyse, J., Grossmann, R., Zhao, R., 2012b. Chicken FTO gene: tissue-specific expression, brain distribution, breed difference and effect of fasting. Comp. Biochem. Physiol. A 163, 246–252.

Whitehead, C.C., 1988. Selection for leanness in broilers using plasma lipoprotein concentration as selection criterion. In: Leclercq, B., Whitehead, C.C. (Eds.), Leanness in Domestic Birds. Genetic, Metabolic and Hormonal Aspects. INRA-Butterworths, London, pp. 41–58.

Yabuuchi, M., Bando, K., Hiramatsu, M., Takahashi, S., Takeuchi, S., 2010. Local agouti signaling protein/melanocortin signaling system that possibly regulates lipid metabolism in adipose tissues of chickens. J. Poult. Sci. 47, 176–182.

Yen, C.F., Jiang, Y.N., Shen, T.F., Wong, I.M., Chen, C.C., Chen, K.C., Chang, W.C., Tsao, Y.K., Ding, S.T., 2005. Cloning and expression of the genes associated with lipid metabolism in Tsaiya ducks. Poult. Sci. 84, 67–74.

Chapter 20

Protein Metabolism

Colin G. Scanes
Department of Biological Sciences, University of Wisconsin, Milwaukee, WI, USA

20.1 INTRODUCTION

20.1.1 Protein Metabolism: Overview

Proteins play critical roles in birds, as in other animals, including the following: as structural elements, such as keratin in skin and feathers, collagens in connective tissue and cellular proteins (e.g., those of the cell membrane), cytoskeleton, protein synthetic machinery, and DNA-associated proteins (e.g., histones); as contractile elements in muscles (see Chapter 15); as oxygen-binding proteins, such as hemoglobin in erythrocytes and myoglobin in muscle; as receptors; as hormones, cytokines, and other messengers; as antibodies; as transporters; as nutrients for embryonic development in the yolk and egg albumin proteins (see Chapter 31); and as enzymes. Table 20.1 summarizes the protein composition of a young chicken as estimated by dissection and correction for muscles and associated connective tissue adhering to bone.

Proteins are synthesized from amino acids:

Plasma amino acid pool ↔ Intracellular amino acid pool ↔ Intracellular proteins

The essential amino acids in avian diets for protein production are as follows:

- Arginine, histidine, isoleucine, lysine, methionine (with a higher requirement if cysteine is not present), threonine, tryptophan, and valine.

Other important amino acids for protein production are the following:

- Alanine, aspartate, asparagine, cysteine, glutamate, glutamine, glycine, leucine, proline, phenylalanine, serine, and tyrosine.

Lysine and methionine are the first limiting amino acids for growth (Bornstein and Lipstein, 1975). Amino acids are synthesized in birds, predominantly in the liver. There is interconversion of some amino acids. For instance, glutamate can be converted to glutamine, catalyzed by glutamine synthase, while glutamine can be converted to glutamate by glutaminase. There are also extranutritional effects of amino acids (see Section 20.5), with amino acids converted to glucose in gluconeogenesis (see Chapter 17), to purines, to uric acid, to neurotransmitters, and to pigment (see Section 20.5).

Posttranslational modification of amino acids in proteins occurs, including:

- Methylation: Histidine residues in myofibrillar proteins (e.g., actin and myosin) are methylated (see Table 20.1).
- Phosphorylation: Phosphorylating serine, threonine, or tyrosine residues is essential for activating or deactivating enzymes.
- Glycosylation: This is to form glycoproteins.
- Acetylation: This forms lipoproteins.

20.1.2 Muscle

Over 20% of the wet weight of skeletal muscles is protein. The major proteins are myosin and actin of the thin and thick filaments, together with sarcoplasmic proteins.
Thin filaments:

- Actin
- Tropomyosin
- Troponin

Thick filaments:

- Myosin

Skeletal muscles are made of fibers. There are different types of fibers:

- Slow oxidative (SO) fibers; also called slow-twitch or type I fibers
- Fast oxidative glycolytic (FOG, or type IIA) fibers
- Fast glycolytic (FG, or type IIB) fibers
- Tonic (types IIIA and IIIB) fibers

These are characterized by their functional characteristics and biochemistry. Flight muscles, including pectoralis muscles, are FOG fibers capable of sustained rapid contraction in Anna's

Sturkie's Avian Physiology.

TABLE 20.1 Importance of Various Organs to Body Weight, Total Protein, and Myofibrillar Protein as Indicated by 3-Methyl-Histidine in Young Chickens

Organ	Weight (g) (% of body weight)	Protein as % of Total Body	3-Methyl-Histidine as % of Total Body
Skeletal muscle (and associated connective tissue)[1]	99.3 (37.7%)	48.6	84.0
Bone[1]	77.7 (29.2%)	22.9	1.6
Gastrointestinal tract	34.2 (13.0%)	11.9	9.8
Skin and feathers	33.3 (12.4%)	9.3	1.7
Liver	10.4 (3.9%)	5.3	1.0
Heart	3.2 (1.2%)	1.3	1.4
Kidneys	2.3 (0.8%)	0.9	0.0
Lung	2.2 (0.8%)	0.7	0.2
Brain	1.5 (0.6%)	0.4	0.1

[1]*Corrected for muscle and other soft tissues adhering to bones.*
Source: Calculated from data in Jones et al. (1986).

TABLE 20.2 Fiber Composition of Selected Chicken Muscles

Muscle	Fiber Types
Pectoralis	99% FG (type IIB)
Posterior latissimus dorsi	~90% FG (type IIB) and ~10% FOG (type IIA)
Sartorius white	~15% FOG (type IIA) and ~85% FG (type IIB)
Sartorius red	~40% SO (type I), ~40% FOG (type IIA), and ~20% FG (type IIB)
Tonic muscles such as anterior latissimus dorsi, plantaris, and adductor profundus	Almost entirely type III fibers (~70% type IIIA and ~30% type IIIB)

Source: Barnard et al. (1982).

hummingbirds (*Calypte anna*) and zebra finches (*Taeniopygia guttata*) (Welch and Altshuler, 2009). Table 20.2 summarizes the fiber types in various chicken muscles. There are multiple muscle fiber types in the gastrocnemius muscle, as seen in Anna's hummingbirds (*C. anna*: Welch and Altshuler, 2009), English sparrows (*Passer domesticus*: Marquez et al., 2006), ostrich (*Struthio camelus*: Velotto and Crasto, 2004), and zebra finches (*T. guttata*: Welch and Altshuler, 2009).

Multiple forms of myosin are expressed with differences between fiber types and during embryonic development and posthatch growth (Rosser et al., 1996, 1998). For instance, in adult chickens, pectoralis expresses fast myosin heavy chain (MyHC1), with slow myosin heavy chain (MyHC2) expressed in tonic muscles (Crew et al., 2010).

20.1.3 Feathers

Feathers make up 20–30% of body weight in wild birds (Griminger and Scanes, 1986). In contrast, they represent less than 12.4% of body weight in chickens (Table 20.1). Feathers are made of the protein β-keratin (Stettenheim, 2000). The proteins of the rachis of chicken feathers have a high content of four amino acids (Harrap and Woods, 1964):

- Serine: 13.7%
- Glycine: 13.3%
- Proline: 9.5%
- Cysteine: 7.2%

Chicken keratin genes A–D have been characterized, and these are expressed in feather tissue as early as day 14 of embryonic development (Presland et al., 1989).

Following molt (see Chapter 38), feathers are replaced. This replacement of feathers is critically important for flight. The rapidity of feather growth has been linked to feather abnormalities (Vágási et al., 2012). The time to replace feathers is longer in larger birds. Feathers scale with the body weight of birds by 0.32 power, while feather growth rate increases across species by much less, 0.17 power of body weight (Rohwer et al., 2009).

20.2 DIGESTION OF PROTEINS

Proteins are digested in the gastrointestinal tract to amino acids and dipeptides together with tripeptides. The predominant site for protein digestion is the small intestine. The amino acids and peptides are absorbed in the small intestine.

20.2.1 Protein Digestion in the Gizzard and Proventriculus

The digestion of proteins is initiated in the gizzard and proventriculus in birds by the following methods:

- Mechanical grinding by the gizzard
- Chemical denaturation by the low pH
- Enzymatic proteolysis by pepsin

Pepsinogen, a zymogen, has been purified in birds (chicken: Bohak, 1969). Avian pepsinogen is proteolytically activated to pepsin at low pH (Keilova et al., 1977). Pepsinogens A and C are expressed and secreted from the mucosal cells in the chicken proventriculus (Sakamoto et al., 1998).

20.2.2 Protein Digestion in the Small Intestine

The predominant site of protein digestion is the small intestine (Table 20.3). The digested protein (amino acids and dipeptides together with tripeptides) is absorbed in the small intestine (Table 20.3).

Proteins undergo enzymatic digestion in the small intestine by the enzymes trypsin and chymotrypsin. These are produced by the pancreas as inactive zymogens (respectively, trypsinogen and chymotrypsinogen) and transported to the small intestine in the pancreatic secretions. In the small intestine, both trypsinogen and chymotrypsinogen are activated by proteolytic cleavage:

- Trypsinogen → trypsin
- Chymotrypsinogen → chymotrypsin

In addition, there are peptidases that hydrolyze peptides generated by trypsin and chymotrypsin.

20.2.2.1 Trypsin

Avian trypsin is a 223 amino acid residue containing protein. Six members of the trypsinogen gene family have been characterized in the chicken (Wang et al., 1995), with one trypsinogen cDNA sequenced in the ostrich (Szenthe et al., 2005). These are expressed in the pancreas along with the liver, spleen, and thymus (Wang et al., 1995). Trypsinogen is proteolytically cleaved either by enterokinase or autocatalytically (ostrich: Szenthe et al., 2005). This proteolytic cleavage results in the formation of an active protease, trypsin.

20.2.2.2 Chymotrypsinogen

Chymotrypsinogen is activated in the small intestine by proteolytic cleavage, for instance by trypsin. Avian chymotrypsinogen has been isolated from pancreas (ostrich: van der Westhuizen et al., 1989; Japanese quail: Hou et al., 1990).

20.2.2.3 Aminopeptidases

Aminopeptidases are membrane-bound proteins of enterocytes. Most of the protein is extracellular, with a single transmembrane domain and a small intracellular domain. An avian aminopeptidase has been partially characterized (chicken: Gal-Garber and Uni, 2000). It is expressed throughout the small intestine with the highest expression in the ileum (chicken: Gal-Garber and Uni, 2000).

20.2.3 Amino Acid Absorption in the Small Intestine

Amino acids are absorbed in the small intestine (chicken jejunum and ileum: Tasaki and Takahashi, 1966). There is greater absorption of methionine in the ileum than in other regions of the small intestine (see Figure 20.1). Methionine is absorbed at the highest rate, followed by isoleucine, valine, and leucine, and glutamate is absorbed at the lowest rate, followed by aspartate, glycine, and arginine (Tasaki and Takahashi, 1966; Riley et al., 1986). There are series of amino acid transporters (10 in chickens) expressed in the enterocytes of the small intestine, with predominantly increases progressing down the small intestine from the duodenum, jejunum, and ileum (chickens: Gilbert et al., 2007). Moreover, there are increases in the expression of many amino acid transporters during the period of growth and development post hatching (chickens: Gilbert et al., 2007).

There is moderately high expression of the H^+-dependent peptide transporter-1 (PepT1) throughout the small intestine, with marked increases in the posthatch period (Gilbert et al., 2007). PepT1 functions to transport both dipeptides and tripeptides into the enterocyte.

Not only are proteins in the ingesta digested with the resultant amino acids and small peptides absorbed, but also mucosal intestinal cells are sloughed off and then digested. The turnover times for mucosal cells in the small intestine

TABLE 20.3 Protein Digestion and Absorption in the Regions of the Chicken Small Intestine

	Protein Digestion (%)	Protein Absorption (%)
End of duodenum	60	7
End of jejunum	85	70
End of ileum	93	83

Source: Based on Hurwitz et al. (1972).

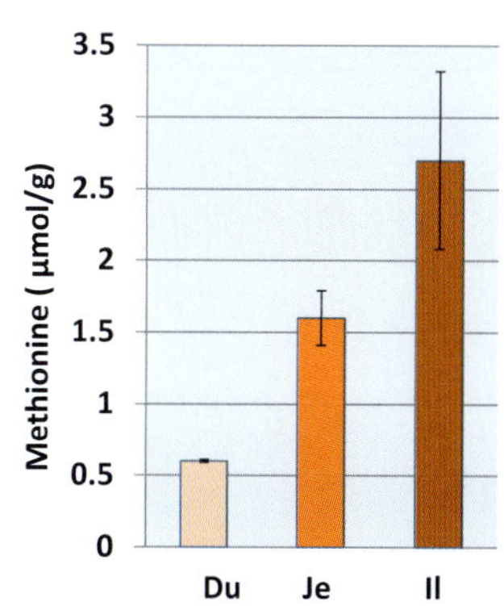

FIGURE 20.1 **Absorption of methionine by different regions of the chicken intestine *in vitro*.** Du, duodenum; Je, jejunum; Il, ileum; vertical lines, standard error of the mean. *Data from Knight and Dibner (1984).*

are estimated as the following (chicken: calculated from Imondi and Bird (1966)):

- Duodenum: 111 h
- Jejunum: 55 h
- Ileum: 89 h

20.2.4 Large Intestine and Protein Digestion

There is little information on the roles of the colon and ceca in protein and amino acid digestion and absorption. While most of the proteins in the ingesta are digested by the time the ingesta reaches the colon (Table 20.3), the fate of sloughed-off mucosa cells is not known. There are populations of proteolytic bacteria in the colon of birds (chicken: King et al., 2009). The impact of these proteolytic bacteria is not known. Interestingly, there is absorption of the methionine analog, 2-hydroxy-4-(methylthio) butanoic acid (Alimet), in the large intestine as well as the small intestine (chicken: Knight and Dibner, 1984; Dibner et al., 1988).

20.3 PROTEIN SYNTHESIS AND DEGRADATION

20.3.1 Whole-Body Synthesis and Degradation

Whole-body protein net synthesis (accretion)
= whole-body protein synthesis
− whole-body protein degradation/losses

Where whole-body protein losses includes in proteins in eggs or molted feathers.

If negative, whole-body protein net synthesis (accretion) is whole-body protein net degradation.

Whole-body net protein synthesis can be estimated by nitrogen balance. F nce, Kino and Okumura (1987) estimated nitrogen balance in 14–21 day old chickens as 0.29 g/day. It is also possible to experimentally determine whole-body protein net synthesis (accretion) or net degradation, and whole-body protein synthesis, using infusion of a radioactive form of a nonmetabolized amino acid such as phenyl alanine or tyrosine (see Table 20.4). As might be expected, net protein synthesis or accretion is negative in starved chicks or chicks on a diet deficient in specific amino acids (Table 20.5). Whole-body protein synthesis is markedly reduced by starvation and is depressed in birds on an amino acid–deficient diet (Kino and Okumura, 1987; Muramatsu et al., 1987b) (see Table 20.5). There is a concomitant increase in protein degradation in starved birds but not in those on amino acid–deficient diets (Kino and Okumura, 1987; Muramatsu et al., 1987b).

20.3.2 Muscle Protein Synthesis and Degradation

As with all proteins, muscle proteins are synthesized from the intramuscular amino acid pool:

TABLE 20.4 Effects of Age and Dietary Protein on Muscle Protein Synthesis and Degradation in Young Chickens

	Muscle			
	Breast (Pectoralis)		Leg	
	Fractional Protein Synthesis (% per day)	Fractional Protein Degradation (% per day)	Fractional Protein Synthesis (% per day)	Fractional Protein Degradation (% per day)
1 week old	38	26	25	18
4 weeks old	24	18	–	–
6 weeks old	17	14	–	–
Control	26	10	22	9
Lysine-deficient diet	24	22	20	19
Control	24	18	–	–
Refed protein-replete diet for 2 days	41	24	–	–
Control	19	10	–	–
Protein-deficient diet	13	5	–	–

Source: Based on Maruyama et al. (1978), MacDonald and Swick (1981), Urdaneta-Rincon and Leeson (2004).

Plasma amino acid pool ↔ Muscle amino acid pool ↔ Muscle protein

There are very similar rates of protein synthesis for sarcoplasmic and myofibrillar proteins (chicken: Laurent et al., 1978a). The fractional synthesis rates of noncollagen protein in muscles from adult chickens are the following (Laurent et al., 1978a):

- Anterior latissimus dorsi: 16% per day
- Posterior latissimus dorsi: 7% per day
- Heart: 14% per day
- Gizzard: 11% per day

Given the steady state situation in the adult (no net protein accretion or delta or net protein synthesis equals zero), the above data also describe the fractional rate of protein degradation. Tables 20.4 and 20.6 provide examples of fractional rates of protein synthesis, degradation and breakdown, and accretion in young chickens. The fractional rates of protein synthesis are usually over 10% per day, while those for protein degradation are also over 10% per day. Muscle protein is obviously turning over at a rapid rate, and this is occurring in both the growing and adult bird (Table 20.4).

Connective tissue makes up a significant proportion of muscle. This is exemplified by 23% of the proteins in the chicken anterior latissumus dorsi muscle being the connective tissue protein, collagen (Laurent et al., 1978b). Examples of the fractional synthesis rates for collagen are the following (Laurent et al., 1978b):

- Heart: 0.9% per day
- Anterior latissumus dorsi: 0.6% per day
- Posterior latissumus dorsi: 0.6% per day

These rates are much lower than for myofibrillar and sarcoplasmic proteins. The fractional rates of collagen synthesis are markedly increased in anterior latissimus dorsi muscles undergoing hypertrophy, as can be seen from the following (data from Laurent et al., 1978c):

- 0 days: 0.6% per day
- 1 day: 1.2% per day
- 3 days: 4.5% per day
- 7 days: 4.2% per day

There is a decline in the fractional rates of both protein synthesis and degradation during growth (Table 20.4). The rate of protein synthesis varies between muscles (e.g., between the anterior and posterior latissumus dorsi and between the pectoralis and gastrocnemius) (Tables 20.4 and 20.6).

20.3.2.1 Effects of Nutrition on Muscle Protein Synthesis and Degradation

The highest net rates of muscle protein synthesis (i.e., muscle protein synthesis and degradation) are observed with optimal nutrition. Dietary deficiencies reduce the net rate of muscle protein synthesis, with very low or negative protein accretion being reported with the most stringent deficiencies. This is observed, for instance, in young chickens on a lysine-deficient diet (Table 20.6). The effects of nutritional deficiency or restoration of adequate diet can be mediated via effects on muscle protein synthesis (refeeding with a protein-replete diet or supplementing a methionine diet with methionine) and/or degradation (as with lysine deficiency) (Tables 20.4 and 20.6).

TABLE 20.5 Effect of Nutritional Status on Whole-Body Protein Accretion (Net Synthesis), Synthesis, and Degradation in Young Broiler Chickens

	A	B	C
Study 1 (Muramatsu et al., 1987b)	Whole-Body Protein Synthesis[1] (g/day/kg body wt.)	Whole-Body Protein Degradation (g/day/kg body wt.)	Whole-Body Protein Net or Delta (Δ) synthesis (Accretion) or, If Negative, Degradation (g/day/kg body wt.)
Fed	21.3	11.2	10.1
Starved	14.7	19.5	−4.8
Study 2 (Kino and Okumura, 1987)	Fractional Rate of Protein Synthesis[1] (% per day)	Fractional Rate of Protein Degradation (% per day)	Whole-Body Rate of Protein Net or Delta (Δ) Synthesis (Accretion) or, If Negative, Degradation (g/day)
Control	23	18	+5
Histidine-free diet	17	18	−1
S-amino-acid-free diet	15	18	−3

[1]*Determined using ^{3}H -phenylalanine.*

TABLE 20.6 Effects of Methionine on Muscle Protein Synthesis in Broiler Chickens

Endpoint	Basal Methionine-Deficient Diet[1]	Basal Diet Plus Methionine (0.2% in diet)
Gain (g/day)	15.1±0.3	24.9±0.3[3]
Fractional Protein Synthesis Rate (% per day)		
Gastrocnemius	5.6±1.1	13.6±1.1[3]
Pectoralis	12.1±1.6	21.8±1.6[2]

[1]*Soy protein containing 0.5% sulfur amino acids.*
[2]*$p<0.05$.*
[3]*$p<0.01$.*
Source: Based on Barnes et al. (1995).

TABLE 20.7 Effects of Nervous Innervation, Immobilization, and Stretch on Protein Synthesis and on Degradation and Breakdown in the Anterior Latissimus Dorsi Muscle

	Denervated	Denervated, Immobilized, and Stretched	Denervated and Immobilized, But Not Stretched
Synthesis	+41%	+67%	−16%
Break-down	+82%	+111%	+52%

Source: Chicken, Goldspink (1978).

20.3.3.2 Effects of Stretch and Nervous Innervation on Muscle Protein Synthesis and Degradation

There are effects of both nervous innervation and stretch both on protein synthesis and on degradation and breakdown (chicken: Goldspink, 1978). For instance, denervating muscles results in shifts both in protein synthesis and in degradation and breakdown (chicken: Goldspink, 1978) (Table 20.7). Moreover, muscle hypertrophy can be induced by the application of a weight (and therefore muscle stretching). There are marked increases in muscle protein synthesis during this hypertrophy, resulting in accretion (Laurent et al., 1978c) (Figure 20.2). Surprisingly, there is a modest increase in muscle protein degradation during this hypertrophy (Laurent et al., 1978c) (Figure 20.2).

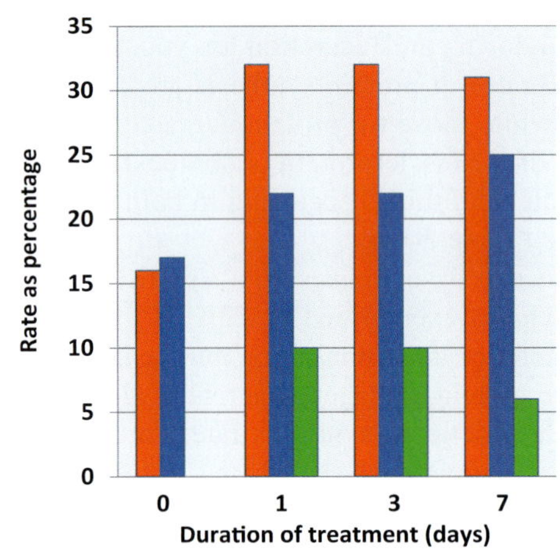

FIGURE 20.2 **Rates of protein synthesis, degradation, and accretion in anterior latissimus dorsi muscles undergoing hypertrophy due to long-term application of a weight around the humerus in adult chickens.** Red columns, protein synthesis; blue columns, protein degradation; green columns, protein accretion (net or delta synthesis). Note that at day 0, the rate of protein accretion (net or delta synthesis) is zero. *Data from Laurent et al. (1978c).*

20.3.2.3 Hormones and Muscle Protein Synthesis

Based on *in vitro* studies with myotubes derived from chicken pectoralis satellite cells, protein synthesis is stimulated by either insulin or insulin-like growth factor 1 (IGF1) (Duclos et al., 1993). Moreover, protein degradation is inhibited in the same system by both hormones (Duclos et al., 1993):

- IGF1 ↓ 32%
- Insulin ↓ 13%

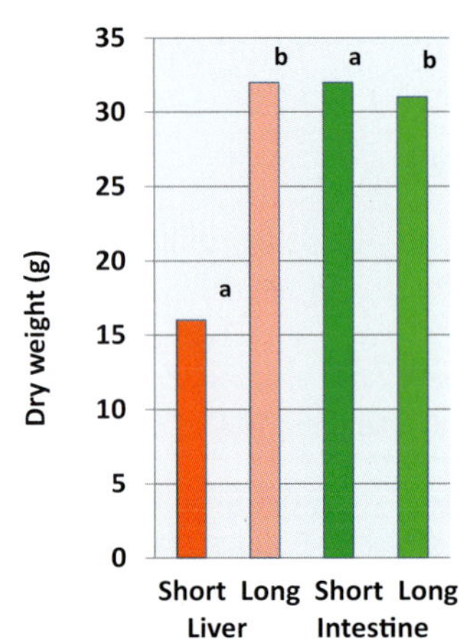

FIGURE 20.3 Effects of long versus short migration on the weight of the liver and intestine in barn swallows (*Hirundo rustica*). The letters a, b indicate significant difference $p<0.05$. *Adapted from Schwilch et al. (2002).*

20.3.2.4 Muscle Protein Degradation

Whole-body muscle protein degradation can be measured by the excretion of 3-methyl-histidine (Saunderson and Leslie, 1983). The histidine residues in the myofibrillar proteins of muscle are methylated (see Table 20.1) in the following manner:

Histidine residue + S-Adenosylmethionine ↔ 3-Methyl-histidine residue + S-Adenosylhomocysteine

When muscle proteins are broken down, 3-methyl-histidine is released but is not reutilized.

Measurement of whole-body muscle protein degradation has not been widely employed in birds. An exception is the study of molting. There is elevated excretion of 3-methyl-histidine in white-crowned sparrows (*Zonotrichia leucophrys*) during molting (Pearcy and Murphy, 1997):

- Nonmolting: 0.6 μmol/day
- Molting: 1.1 μmol/day

The increased degradation of muscle protein during molting is presumably to meet the needs of new feather formation. Increased excretion of 3-methyl-histidine has been reported in molting but not nonmolting white-crowned sparrows that were fed a diet deficient in sulfur amino acids when muscle protein degradation would be high (Pearcy and Murphy, 1997). In contrast, there was decreased excretion of 3-methyl-histidine in molting white-crowned sparrows fed a protein-deficient diet (Pearcy and Murphy, 1997).

20.3.3 Liver and Gastrointestinal Tract

Protein synthesis in the liver is high (see Table 20.8), representing 11% of all protein synthesis in the bird. Similarly, gastrointestinal protein synthesis is disproportionately high compared to in other tissue. For instance, in the young chick, 18.9% of protein synthesis occurs in the small intestine plus the pancreas and ceca (calculated from Muramatsu et al., 1987a) (see Table 20.9). This is in contrast to the gastrointestinal organs, which represent 5.2% of the protein in the bird or 5.9% of the body weight (see Table 20.8).

20.3.3.1 Physiological Effects on Protein Metabolism of the Liver and Intestine

There are physiological effects on the protein and weights of the liver and intestine. For instance, the weights of both the liver and intestine are decreased in pied flycatchers and in willow warblers following long migrations (Schwilch et al., 2002). This is due to amino acids from protein degradation supplying some of the energy needs of flight during migration.

Diet can influence the protein metabolism of the liver and intestine. For instance, feeding a diet that is high in fiber for 4 weeks is followed by marked hypertrophy (2.3-fold increase) of the gizzard in Japanese quail (Starck and Rahmaan, 2003). This is explicable for the following reason: there is a marked increase in the amount of feed consumed due to its lower nutritional density compared to the high-fiber diet, and this increase results in mechanical stretching of the gizzard by the volume of ingesta. Moreover, there are decreases in both the weights of the liver and small intestine and the rates of protein synthesis in these organs in chickens fed a germ-free diet

TABLE 20.8 Importance of Gastrointestinal Protein Metabolism in Young (21 Days) Chickens

			Protein Synthesis[1]		
	Weights at 23 Days (g/kg body wt.)	Protein (g/kg body wt.)	Fractional Synthesis (% per day)	Protein Synthesis (g/day/kg body wt.)	Protein Synthesis as a % of Whole-Body Protein Synthesis
Whole body	100[2]	190 ± 4	19 ± 3	35.7 ± 1.16	100[2]
Liver	2.9 ± 0.4	5.5 ± 0.1	73 ± 5	4.0 ± 0.2	11.0 ± 0.8
Pancreas	0.4 ± 0.02	0.8 ± 0.03	172 ± 12	1.5 ± 0.1	3.9 ± 0.3
Duodenum	1.5 ± 0.06	2.3 ± 0.01	77 ± 8	1.7 ± 0.2	4.5 ± 0.4
Jejunum and ileum	3.4 ± 0.11	5.9 ± 0.5	58 ± 2	3.5 ± 0.4	9.2 ± 0.9
Ceca	0.6 ± 0.01	0.8 ± 0.05	63 ± 5	0.5 ± 0.0	1.3 ± 0.1

[1]*Determined using ^{3}H-phenylalanine.*
[2]*By definition.*
Source: Data calculated from Muramatsu et al. (1987a).

TABLE 20.9 Effects of a Germ-Free Diet on Growth and Protein Synthesis[1] in Young Chickens

	Conventional Diet	Germ-Free Diet
Growth rate (g/day)	6.2 ± 0.1	6.9 ± 0.2[3]
Weights at 23 days (g/kg body wt.)		
Liver	2.9 ± 0.04	2.6 ± 0.06[3]
Duodenum	1.5 ± 0.06	1.2 ± 0.04[3]
Jejunum and ileum	3.4 ± 0.11	2.3 ± 0.08[3]
Nitrogen retention g/day	0.24 ± 0.05	0.27 ± 0.05
Protein synthesis (g/day/kg body wt.)		
Whole body	36 ± 1.2	34 ± 0.7
Liver	4.0 ± 0.2	3.2 ± 0.2
Duodenum	1.7 ± 0.2	1.3 ± 0.1
Jejunum and ileum	2.4 ± 0.2	3.5 ± 0.2[2]

[1]*Determined using ^{3}H-phenylalanine.*
[2]*$p < 0.05$.*
[3]*$p < 0.01$.*
Source: Data from Muramatsu et al. (1987a, 1988).

(Muramatsu et al., 1987a, 1988) (Table 20.8). This would suggest that microorganisms are stimulating gastrointestinal growth.

20.3.4 Protein Metabolism in Immune Tissues

As might be expected, there are high levels of protein synthesis in immune tissues. Protein synthesis in two immune tissues of the growing chicken is summarized in Table 20.10. Immune tissues' high rates of protein synthesis are disproportionate to their percentage of the body weight. The cationic amino acid transporter-1 through -3 (CAT1–3) is responsible for the transfer of amino acids across the cell membrane into cells. These are expressed in immune tissues—the bursa of Fabricius, the spleen, and the thymus—from 7 days old (chicken: Humphrey et al., 2004, 2006).

There are interactions between immune function, on one hand, and proteins and amino acids, on the other. Lysine increases the proliferation of chicken thymocytes (Humphrey and Klasing, 2005). Acute-phase reaction, as induced by challenge with *Salmonella typhimurium* lipopolysaccharide (LPS), decreased expression of CAT1 in the bursa of Fabricius and the thymus of 11 week old chickens (Humphrey and Klasing, 2005).

TABLE 20.10 Protein Metabolism in Immune Tissues of Young (21 Days) Chickens

	Weights at 23 Days (g/kg body wt.)	Fractional Synthesis (% per day)	Protein Synthesis (g/day/kg body wt.)	% of Whole-Body Protein Synthesis
Bursa of Fabricius	0.68 ± 0.04	67 ± 5	0.6 ± 0.1	1.6 ± 0.2
Spleen	0.15 ± 0.01	46 ± 2	0.09 ± 0.01	0.2 ± 0.03

[1]*Determined using ^{3}H-phenylalanine.*
Source: Calculated from data in Muramatsu et al. (1987a).

20.3.5 Proteins and Reproduction

20.3.5.1 Female Reproduction

There have not been definitive studies of the relative contributions of the reproductive organs to protein synthesis and degradation in birds. However, protein synthesis related to reproduction in the female is high. Birds' eggs are high in protein. For instance, the protein composition of a chicken hen's egg (assuming a 58 g egg) is the following (calculated from Belitz et al., 2009):

- Albumin: 3.6 g
- Yolk: 3.3 g
- Shell: 0.2 g

This is equivalent to 43% of the protein required in the diet of the chicken (see Table 20.11).

Virtually all of the yolk proteins are synthesized in the liver and are absorbed into the developing oocyte in the follicle (see Chapter 27). Yolk proteins consist of the following (Moran, 1987):

- Low density (very low-density lipoproteins LDF1 and LDF2): 66%
- Granules (phosvitins and lipovitellins): 23%
- Aqueous α livetin, β livetin, and γ livetin (IgG): 11%

Both phosvitin and lipovitellin are derived from vitellogenins (VTGs) produced in the liver and secreted into the plasma. The VTG is processed to lipovitellin and phosvitin in the ovary. There are multiple VTG genes in birds (e.g., chicken VTGI, VTGII, and VTGIII: Evans et al., 1988). These are only expressed in the liver in birds either producing estrogen or receiving exogenous estrogens (chicken: Evans et al., 1988; Japanese quail: Gupta and Kanungo, 1996). Progesterone can inhibit VTGII expression (Japanese quail: Gupta and Kanungo, 1996). Interestingly, possible endocrine disrupters of chemicals such as pesticides can be evaluated using either circulating concentrations of VTG or VTG gene expression in the Japanese quail (Shibuya et al., 2005).

TABLE 20.11 Characteristics of Sexually Mature Hens

Parameter	Metric
Body weight	1.5 kg
Carcass protein	23.9%
Carcass protein	358 g
Ovary	35.5 g
Oviduct	47.1 g
Oviduct protein	8.2 g
Liver	24 g
Egg Production	
Protein requirement in diet	16.5 g/day
Protein secreted in oviduct or taken up by developing ovum in the ovary	~7.1 g/day
Sexual Maturation	
Increase in body weight during sexual maturation	252 g
Increase in body protein during sexual maturation	2.2 g/day
Increase in ovarian and oviduct protein during sexual maturation	0.5 g/day

Source: Chickens; from or calculated from Reid (1976), Renema et al. (2001), Belitz et al. (2009).

What are not known are whole-body and specific-organ rates of protein synthesis and degradation in the reproductively mature female. This includes the turnover of structural and secretory proteins in the ovary and oviduct, and those related to the production of VTG and other yolk proteins in the liver.

In the photostimulated sexually immature female chicken (17 week old pullet), the development of the ovary and oviduct occurs over about 28 days (Renema et al., 2001). Oviductal growth is occurring under the influence of estrogen (estradiol) and progesterone with marked increases in protein synthesis (Muller et al., 1970).

20.3.5.2 Male Reproduction

The importance of the male reproductive organs to overall protein metabolism has received little attention in birds. The testes of birds are considerably larger than those of mammals, as can be readily observed from the following:

- Sexually mature Japanese quail testes: 2.5 g, or 2.5% of body weight (Vatsalya and Kashmiri, 2012)
- Sexually mature chicken testes: 33 g (González-Morán et al., 2008)

It is likely that there are high demands for protein synthesis in spermatogenesis. However, protein requirements for seminal fluid are likely to be low as semen volume is modest, for instance 0.4 ml, and seminal plasma protein concentrations are only 1.9% in turkeys (Kotłowska et al., 2005). There is no information on the protein requirements for epididymal, ductal, phallic, and secondary sexual characteristic development and maintenance.

20.4 AMINO ACIDS AND METABOLISM

20.4.1 Amino Acid Transfer into Muscle and Other Cells

There are transporters transferring amino acids into muscle cells and into the cells of other organs. Amino acid uptake by cells can be determined by uptake of [^{3}H]-aminoisobutyric acid (AIB), or ^{14}C-AIB. This approach prevents the confounding of data by measuring both the uptake of amino acids and their incorporation into proteins.

Hormones, including IGF1 and insulin, stimulate AIB uptake by muscle cells. The half-maximal effective doses are the following (Duclos et al., 1993):

- IGF1: 0.27 nM
- Insulin: 35 nM

These differences in the half-maximal effective doses are consistent with insulin acting via the IGF1 receptor. Earlier studies have also reported that insulin increases AIB uptake by chick embryo cardiac cells (Guidotti et al., 1968; Santora et al., 1979). It is not clear whether insulin is acting via the insulin receptor in the heart cells or as a surrogate for IGF1. Growth hormone does not influence AIB uptake by chicken myotubes (Duclos et al., 1993).

20.4.1.1 Cationic Amino Acid Transporters

The cationic amino acid transporters (CAT1–3) are expressed in the pectoralis muscle, heart, and liver (Humphrey et al., 2004, 2006). Expression of CAT1–3 is greatly decreased in birds on lysine-deficient diet (Humphrey et al., 2006). Acute-phase reaction as induced by challenge with *Salmonella typhimurium* LPS increased expression of CAT1–3 in the pectoralis of 11 week old chickens (Humphrey and Klasing, 2005).

20.4.4 Nitrogenous Waste

Most urinary nitrogenous waste in birds is in the form of uric acid. Estimates of the relative contributions of uric acid, urea, and ammonia to nitrogenous waste in birds are the following (Stevens, 1996):

- Uric acid: ~75%
- Urea: ~5%
- Ammonia: 12%

Ammonia is only a minor component and is little affected by dietary protein status. There is no effect of dietary protein on the concentration of ammonia in pigeon excreta (*Columbia livia*: McNabb et al., 1972). There are, however, some species differences. In the small hummingbird, the black-chinned hummingbird (*Archilochus alexandri*), over a quarter of nitrogenous waste is as ammonia (McWhorter et al., 2003).

20.4.2.1 Uric Acid

The major nitrogenous waste in birds is uric acid (Milroy, 1903). Infusion of ammonia or glutamine increases circulating concentrations of uric acid in chickens (Karasawa and Tasaki, 1973). The enzymes of the uric acid cycle are found in both the liver and kidney. For instance, phosphoribosylpyrophosphate amidotransferase and xanthine dehydrogenase activities are present in both avian liver and kidneys (chicken: McFarland and Coon, 1980, 1984). Glutamine is critically important for uric acid formation (see the Section 20.4.2.3). It is not surprising that glutaminase (catalyzing glutamine to glutamate) is either not present or at very low levels in the avian liver (chicken: Wu et al., 1998).

The predominant organ responsible for uric acid formation is the liver. Chicken hepatocytes have been demonstrated to synthesize uric acid *in vitro* (McFarland and Coon, 1984). The kidney synthesizes only about 17% of the uric acid found in the urine of birds (chicken: Chin and Quebbemann, 1978).

Uric acid formed from the amino acids glutamine, glycine, and aspartate by the following pathway (based on Stevens, 1996):

1. Ribose + ATP ↔ Ribose-5-phosphate + ADP
2. Ribose-5-phosphate + ATP ↔ 5-Phosphoribosyl-α –pyrophosphate (PRPP) + AMP
   Pyrophosphokinase
3. PRPP + Glutamine ↔ 5-β-Phosphoribosyl amine + Glutamate
   Phosphoribosylpyrophosphate amidotransferase (Aminophosphoribosyltransferase)
4. Phosphoribosylamine + Glycine ↔ Glycinamide ribotide (GAR) [Phosphoribosylglycineamine]
   GAR synthetase
5. GAR + N-formyl tetrahydrofolate ↔ Formylglyinamide ribotide (FGAR) [Phosphoribosyl-N-formylglycineamine] + Tetrahydrofolate
   GAR transformase
6. FGAR + Glutamine + ATP ↔ Formylglycineamidine ribotide (FGAM) + Glutamate + ADP
   FGAM Synthetase
7. FGAM + HCO_3^{-1} + ATP ↔ Phosphoribosyl-5-aminoimidazole-4 carboxylate + ADP
8. Phosphoribosyl-5-aminoimidazole-4 carboxylate + Asparate + ATP ↔ Inosine monophosphate + Fumarate + ADP
9. Inosine monophosphate ↔ Xanthine + PRPP
   IMP dehydrogenase

PRPP is cycled back to step 3 and hence uric acid cycle or used to generate purines.

10. Xanthine ↔ Uric acid
    Xanthine dehydrogenase

20.4.2.2 Urea

The existence of a urea cycle in birds has been questioned (Stevens, 1996). However, birds have both uric acid and urates as well as urea in the plasma (see Chapter 11).

The urea cycle is as follows:

1. Ornithine + Carbamyl phosphate ↔ Citrulline
2. Citrulline + Asparate + ATP ↔ Argininosuccinate + Fumarate + AMP
3. Argininosuccinate ↔ Arginine
4. Arginine ↔ Urea + Ornithine (and return to step 1)

20.4.2.3 Glutamine and Ammonia Detoxification

The primary ammonia-detoxifying enzyme in the avian liver is glutamine synthase. The gene has been characterized (Pu and Young, 1989). Glutamine synthase is expressed in the liver, brain, and retina (chicken: Satoh and Matsuno, 1983; Patejunas and Young, 1987). In contrast to the situation in mammals, there is little zonation of the enzyme in the avian liver (chicken and ducks: Smith and Campbell, 1988). Moreover, while glutamine synthase is cytosolic in mammals, the enzyme is located in the mitochondria in birds (chicken and ducks: Smith and Campbell, 1988).

20.4.3 Amino Acids as Energy Sources

20.4.3.1 Glutamine as an Energy Source

Glutamine is a significant energy source for at least some tissues in birds. For instance, it is the major energy source for avian erythrocytes (chicken: Mathew et al., 1993). Glutamine is also employed as a major energy source for enterocytes (chicken: Watford et al., 1979).

20.4.4 Amino Acid Derivatives

There are multiple compounds in the animals that are derived from amino acids via biosynthetic pathways. These include the following:

- Epinephrine (hormone) from tyrosine (and phenylalanine)
- Glutamate (neuromodulator)
- Histamine (neurotransmitter and inflammatory response) from histidine
- Melatonin (hormone) from tryptophan

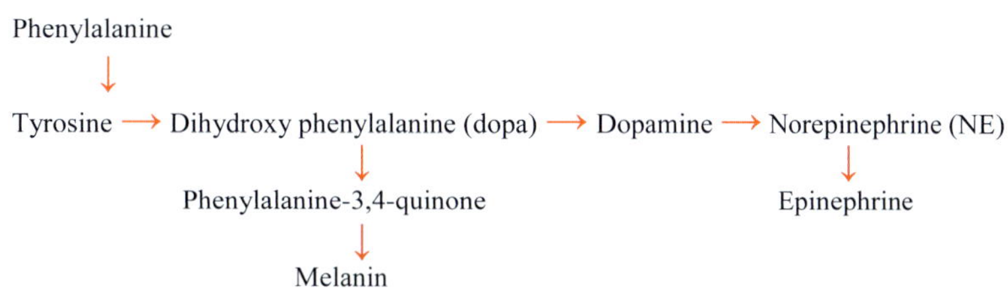

FIGURE 20.4 **Synthesis of melanin, dopamine, norepinephrine, and epinephrine from phenylalanine or tyrosine.**

- Melanin (pigment) from tyrosine
- Nitric oxide (signal molecule) from arginine catalyzed by nitric oxide synthase
- Norepinephrine (neurotransmitter and hormone) from tyrosine (and phenylalanine)
- Nucleotides (purines, e.g., adenine, and guanine together with uric acid) from glutamine, aspartate, and glycine
- Serotonin (neurotransmitter) from tryptophan
- Thyroxine and triiodothyronine (hormone) from thyroglobulin (tyrosine residues)
- Urea from arginine

An example of such biosynthetic pathways is shown in Figure 20.4.

20.5 EXTRANUTRITIONAL EFFECTS OF AMINO ACIDS

There is intriguing research supporting amino acids having regulatory roles.

20.5.1 Amino Acids in the Control of Metabolism

20.5.1.1 Glutamine and Muscle Growth

Glutamine has been proposed to be playing a role in the control of muscle protein synthesis (Wu et al., 1998). There are decreased levels of free glutamate and glutamine in leg but not pectoralis muscles from starved chickens along with the suppression of protein synthesis in leg but not pectoralis muscles (chicken: Watford and Wu, 2005) (see Table 20.12). Both glutamine synthase and glutaminase activities are reported in avian muscle. There is also much higher activity of glutamine synthase but not glutaminase in leg muscles than in pectoralis muscles (chicken and pigeon (*C. livia*): Watford and Wu, 2005).

TABLE 20.12 Effect of Starvation on Fractional Rate of Protein Synthesis, Together with Glutamine and Glutamate, in Different Muscles in Young Chickens

	Fed	Starved
Fractional Protein Synthesis (% per day)		
Pectoralis	11	10
Leg muscle	36	14
Glutamine (μmol/g)		
Pectoralis	1.4	1.4
Leg muscle	9.4	3.0
Glutamate (μmol/g)		
Pectoralis	0.9	0.9
Leg muscle	3.4	1.7

Source: Based on Watford and Wu (2005).

20.5.1.2 Glutamine and Intestinal Growth

Supplementing a corn–soybean diet with glutamine increased both duodenal and jejunal mucosal villi height in young chickens despite not impacting growth (Bartell and Batal, 2007). For instance, duodenal villi height was reported to be increased by 41% with glutamine supplementation in the following manner (Bartell and Batal, 2007):

- Controls on corn–soybean diet: 652 μm
- Supplemented with 4% glutamine: 921 μm

REFERENCES

Barnard, E.A., Lyles, J.M., Pizzey, J.A., 1982. Fibre types in chicken skeletal muscles and their changes in muscular dystrophy. J. Physiol. 331, 333–354.

Barnes, D.M., Calvert, C.C., Klasing, K.C., 1995. Methionine deficiency decreases protein accretion and synthesis but not tRNA acylation in muscles of chicks. J. Nutr. 125, 2623–2630.

Bartell, S.M., Batal, A.B., 2007. The effect of supplemental glutamine on growth performance, development of the gastrointestinal tract, and humoral immune response of broilers. Poult. Sci. 86, 1940–1947.

Belitz, H.-D., Grosch, W., Schieberle, P., 2009. Food Chemistry, fourth ed. Springer, Berlin, Germany.

Bohak, Z., 1969. Purification and characterization of chicken pepsinogen and chicken pepsin. J. Biol. Chem. 244, 4638–4648.

Bornstein, S., Lipstein, B., 1975. The replacement of some of the soybean meal by the first limiting amino acids in practical broiler diets. I. The value of special supplementation of chick diets with methionine and lysine. Br. Poult. Sci. 16, 177–188.

Chin, T.Y., Quebbemann, A.J., 1978. Quantitation of renal uric acid synthesis in the chicken. Am. J. Physiol. 234, F446–F451.

Crew, J.R., Falzari, K., DiMario, J.X., 2010. Muscle fiber type specific induction of slow myosin heavy chain 2 gene expression by electrical stimulation. Exp. Cell. Res. 316, 1039–1049.

Dibner, J.J., Knight, C.D., Swick, R.A., Ivey, F.J., 1988. Absorption of 14C-2-hydroxy-4-(methylthio)butanoic acid (Alimet) from the hindgut of the broiler chick. Poult. Sci. 67, 1314–1321.

Duclos, M.J., Chevalier, B., Goddard, C., Simon, J., 1993. Regulation of amino acid transport and protein metabolism in myotubes derived from chicken muscle satellite cells by insulin-like growth factor-I. J. Cell. Physiol. 157, 650–657.

Evans, M.I., Silva, R., Burch, J.B., 1988. Isolation of chicken vitellogenin I and III cDNAs and the developmental regulation of five estrogen-responsive genes in the embryonic liver. Genes Dev 2, 116–124.

Gal-Garber, O., Uni, Z., 2000. Chicken intestinal aminopeptidase: partial sequence of the gene, expression and activity. Poult. Sci. 79, 41–45.

Gilbert, E.R., Li, H., Emmerson, D.A., Webb Jr., K.E., Wong, E.A., 2007. Developmental regulation of nutrient transporter and enzyme mRNA abundance in the small intestine of broilers. Poult. Sci. 86, 1739–1753.

Goldspink, D.F., 1978. The influence of passive stretch on the growth and protein turnover of the denervated extensor digitorum longus muscle. Biochem. J. 174, 595–602.

González-Morán, M.G., Guerra-Araiza, C., Campos, M.G., Camacho-Arroyo, I., 2008. Histological and sex steroid hormone receptor changes in testes of immature, mature, and aged chickens. Domest. Anim. Endocrinol. 35, 371–379.

Griminger, P., Scanes, C.G., 1986. Protein metabolism. In: Sturkie, P.D. (Ed.), Avian Physiology, fourth ed. Springer Verlag, New York, pp. 326–344.

Guidotti, G.G., Borghetti, A.F., Gaja, G., Lo Reti, L., Ragnotti, G., Foà, P.P., 1968. Amino acid uptake in the developing chick embryo heart. The effect of insulin on α-aminoisobutyric acid accumulation. Biochem. J. 107, 565–574.

Gupta, S., Kanungo, M.S., 1996. Modulation of vitellogenin II gene by estradiol and progesterone in the Japanese quail. Biochem. Biophys. Res. Commun. 222, 181–185.

Harrap, B.S., Woods, E.F., 1964. Soluble derivatives of feather keratin. I. Isolation fractionation and amino acid composition. Biochem. J. 92, 8–18.

Hou, D.X., Maeda, Y., Okamoto, S., Hashiguchi, T., 1990. Purification and characterization of chymotrypsinogen from pancreas of Japanese quail (*Coturnix coturnix japonica*). Comp. Biochem. Physiol. B 97, 761–766.

Humphrey, B.D., Klasing, K.C., 2005. The acute phase response alters cationic amino acid transporter expression in growing chickens (*Gallus gallus domesticus*). Comp. Biochem. Physiol. A 142, 485–494.

Humphrey, B.D., Stephensen, C.B., Calvert, C.C., Klasing, K.C., 2004. Glucose and cationic amino acid transporter expression in growing chickens (*Gallus gallus domesticus*). Comp. Biochem. Physiol. A 138, 515–525.

Humphrey, B.D., Stephensen, C.B., Calvert, C.C., Klasing, K.C., 2006. Lysine deficiency and feed restriction independently alter cationic amino acid transporter expression in chickens (*Gallus gallus domesticus*). Comp. Biochem. Physiol. A 143, 218–227.

Hurwitz, S., Shamir, N., Bar, A., 1972. Protein digestion and absorption in the chick: effect of *Ascarida galli*. Am. J. Clin. Nutr. 25, 311–316.

Imondi, A.R., Bird, F.H., 1966. The turnover of intestinal epithelium in the chick. Poult. Sci. 45, 142–147.

Jones, S.J., Aberle, E.D., Judge, M.D., 1986. Estimation of the fractional breakdown rates of myofibrillar proteins in chickens from quantitation of 3-methylhistidine excretion. Poult. Sci. 65, 2142–2147.

Karasawa, Y., Tasaki, I., 1973. Effect on uric acid synthesis of the infusion of various levels of ammonium acetate or glutamine in chickens. J. Nutr. 103, 1727–1730.

Keilova, H., Kostka, V., Kay, J., 1977. The first step in the activation of chicken pepsinogen is similar to that of prochymosin. Biochem. J. 167, 855–858.

King, M.D., Guentzel, M.N., Arulanandam, B.P., Lupiani, B., Chambers, J.P., 2009. Proteolytic bacteria in the lower digestive tract of poultry may affect avian influenza virus pathogenicity. Poult. Sci. 88, 1388–1393.

Kino, K., Okumura, J., 1987. Whole-body protein turnover in chicks fed control, histidine, or methionine plus cystine-free diets. Poult. Sci. 66, 1392–1397.

Kotłowska, M., Glogowski, J., Dietrich, G.J., Kozłowski, K., Faruga, A., Jankowski, J., Ciereszko, A., 2005. Biochemical characteristics and sperm production of turkey semen in relation to strain and age of the males. Poult. Sci. 84, 1763–1768.

Knight, C.D., Dibner, J.J., 1984. Comparative absorption of 2-hydroxy-4-(methylthio)-butanoic acid and L-methionine in the broiler chick. J. Nutr. 114, 2179–2186.

Laurent, G.J., Sparrow, M.P., Bates, P.C., Millward, D.J., 1978a. Turnover of muscle protein in the fowl (*Gallus domesticus*). Rates of protein synthesis in fast and slow skeletal, cardiac and smooth muscle of the adult fowl. Biochem. J. 176, 393–401.

Laurent, G.J., Sparrow, M.P., Millward, D.J., 1978b. Turnover of muscle protein in the fowl. Changes in rates of protein synthesis and breakdown during hypertrophy of the anterior and posterior latissimus dorsi muscles. Biochem. J. 176, 407–417.

Laurent, G.J., Sparrow, M.P., Bates, P.C., Millward, D.J., 1978c. Turnover of muscle protein in the fowl. Collagen content and turnover in cardiac and skeletal muscles of the adult fowl and the changes during stretch-induced growth. Biochem. J. 176, 419–427.

MacDonald, M.L., Swick, R.W., 1981. The effect of protein depletion and repletion on muscle-protein turnover in the chick. Biochem. J. 194, 811–819.

Marquez, J., Sweazea, K.L., Braun, E.J., 2006. Skeletal muscle fiber composition of the English sparrow (*Passer domesticus*). Comp. Biochem. Physiol. B 143, 126–131.

Maruyama, K., Sunde, M.L., Swick, R.W., 1978. Growth and muscle protein turnover in the chick. Biochem. J. 176, 573–582.

Mathew, A., Grdisa, M., Johnstone, R.M., 1993. Nucleosides and glutamine are primary energy substrates for embryonic and adult chicken red cells. Biochem. Cell. Biol. 71, 288–295.

McFarland, D.C., Coon, C.N., 1980. Purine metabolism studies in the high and low uric acid containing lines of chickens: de novo uric acid synthesis and xanthine dehydrogenase activities. Poult. Sci. 59, 2250–2255.

McFarland, D.C., Coon, C.N., 1984. Purine metabolism in high and low uric acid lines of chickens: de novo uric acid synthesis in isolated hepatocytes and phosphoribosylpyrophosphate amidotransferase activities. Proc. Soc. Exp. Biol. Med. 177, 417–421.

McNabb, F.M.A., McNabb, R.A., Ward Jr, J.M., 1972. The effects of dietary protein content on water requirements and ammonia excretion in pigeons, *Columbia livia*. Comp. Biochem. Physiol. A 43, 181–185.

McWhorter, T.J., Powers, D.R., Martínez Del Rio, C., 2003. Are hummingbirds facultatively ammonotelic? Nitrogen excretion and requirements as a function of body size. Physiol. Biochem. Zool. 76, 731–743.

Milroy, T.H., 1903. The formation of uric acid in birds. J. Physiol. 30, 47–60.

Moran Jr, E.T., 1987. Protein requirement, egg formation and the hen's ovulatory cycle. J. Nutr. 117, 612–618.

Muller, K.R., Cox, R.F., Carey, N.H., 1970. Effects of progesterone on protein metabolism in chicken oviduct tissue pretreated with oestrogen. Biochem. J. 120, 337–344.

Muramatsu, T., Takasu, O., Furuse, M., Tasaki, I., Okumura, J., 1987a. Influence of the gut microflora on protein synthesis in tissues and in the whole body of chicks. Biochem. J. 246, 475–479.

Muramatsu, T., Aoyagi, Y., Okumura, J., Tasaki, I., 1987b. Contribution of whole-body protein synthesis to basal metabolism in layer and broiler chickens. Br. J. Nutr. 57, 269–277.

Muramatsu, T., Takasu, O., Furuse, M., Okumura, J., 1988. Effect of diet type on enhanced intestinal protein synthesis by the gut microflora in the chick. J. Nutr. 118, 1068–1074.

Patejunas, G., Young, A.P., 1987. Tissue-specific regulation of avian glutamine synthetase expression during development and in response to glucocorticoid hormones. Mol. Cell. Biol. 7, 1070–1077.

Pearcy, S.D., Murphy, M.E., 1997. 3-Methylhistidine excretion as an index of muscle protein breakdown in birds in different states of malnutrition. Comp. Biochem. Physiol. A 116, 267–272.

Presland, R.B., Gregg, K., Molloy, P.L., Morris, C.P., Crocker, L.A., Rogers, G.E., 1989. Avian keratin genes. I. A molecular analysis of the structure and expression of a group of feather keratin genes. J. Mol. Biol. 209, 549–559.

Pu, H.F., Young, A.P., 1989. The structure of the chicken glutamine synthetase-encoding gene. Gene 81, 169–175.

Reid, B.L., 1976. Estimated daily protein requirements of laying hens. Poult. Sci. 55, 1641–1645.

Renema, R.A., Robinson, F.E., Oosterhoff, H.H., Feddes, J.J., Wilson, J.L., 2001. Effects of photostimulatory light intensity on ovarian morphology and carcass traits at sexual maturity in modern and antique egg-type pullets. Poult. Sci. 80, 47–56.

Riley Jr., W.W., Esteve-Garcia, E., Austic, R.E., 1986. Intestinal absorption of glucose and amino acids in chickens administered monensin. Poult. Sci. 65, 2292–2298.

Rohwer, S., Ricklefs, R.E., Rohwer, V.G., Copple, M.M., 2009. Allometry of the duration of flight feather molt in birds. PLoS Biol. 7, e1000132.

Rosser, B.W., Wick, M., Waldbillig, D.M., Bandman, E., 1996. Heterogeneity of myosin heavy-chain expression in fast-twitch fiber types of mature avian pectoralis muscle. Biochem. Cell. Biol. 74, 715–728.

Rosser, B.W., Wick, M., Waldbillig, D.M., Wright, D.J., Farrar, C.M., Bandman, E., 1998. Expression of myosin heavy chain isoforms during development of domestic pigeon pectoralis muscle. Int. J. Dev. Biol. 42, 653–661.

Sakamoto, N., Saiga, H., Yasugi, S., 1998. Analysis of temporal expression pattern and cis-regulatory sequences of chicken pepsinogen A and C. Biochem. Biophys. Res. Commun. 250, 420–424.

Santora II, A.C., Wheeler, F.B., DeHaan, R.L., Elsas II, L.J., 1979. Relationship of insulin binding to amino acid transport by cultured 14-day embryonic chick heart cells. Endocrinology 104, 1059–1068.

Satoh, T., Matsuno, T., 1983. Purification and comparison of glutamine synthetase from chicken liver, brain and neural retina. Comp. Biochem. Physiol. B 75, 655–658.

Saunderson, C.L., Leslie, S., 1983. N tau-methyl histidine excretion by poultry: not all species excrete N tau-methyl histidine quantitatively. Br. J. Nutr. 50, 691–700.

Schwilch, R., Grattarola, A., Spina, F., Jenni, L., 2002. Protein loss during long-distance migratory flight in passerine birds: adaptation and constraint. J. Exp. Biol. 205, 687–695.

Shibuya, K., Wada, M., Mizutani, M., Sato, K., Itabashi, M., Sakamoto, T., 2005. Vitellogenin detection and chick pathology are useful endpoints to evaluate endocrine-disrupting effects in avian one-generation reproduction study. Environ. Toxicol. Chem. 24, 1654–1666.

Smith Jr., D.D., Campbell, J.W., 1988. Distribution of glutamine synthetase and carbamoyl-phosphate synthetase I in vertebrate liver. Proc. Natl. Acad. Sci. U. S. A. 85, 160–164.

Starck, J.M., Rahmaan, G.H., 2003. Phenotypic flexibility of structure and function of the digestive system of Japanese quail. J. Exp. Biol. 206, 1887–1897.

Stevens, L., 1996. Avian Biochemistry and Molecular Biology. Cambridge University Press, Cambridge, UK.

Stettenheim, P.R., 2000. The integumentary morphology of modern birds—an overview. Am. Zool. 40, 461–477.

Szenthe, B., Frost, C., Szilágyi, L., Patthy, A., Naudé, R., Gráf, L., 2005. Cloning and expression of ostrich trypsinogen: an avian trypsin with a highly sensitive autolysis site. Biochim. Biophys. Acta 1748, 35–42.

Tasaki, I., Takahashi, N., 1966. Absorption of amino acids from the small intestine of domestic fowl. J. Nutr. 88, 359–364.

Urdaneta-Rincon, M., Leeson, S., 2004. Muscle (pectoralis major) protein turnover in young broiler chickens fed graded levels of lysine and crude protein. Poult. Sci. 83, 1897–1903.

Vágási, C.I., Pap, P.L., Vincze, O., Benkő, Z., Marton, A., Barta, Z., 2012. Haste makes waste but condition matters: molt rate-feather quality trade-off in a sedentary songbird. PLoS Biol. 7, e40651.

van der Westhuizen, N., Naudé, R.J., Oelofsen, W., 1989. The isolation and partial characterization of chymotrypsinogen from the pancreas of the ostrich (*Struthio camelus*). Int. J. Biochem. 21, 91–97.

Vatsalya, V., Kashmiri, A.L., 2012. Allometric growth of testes in relation to age, body weight and selected blood parameters in male Japanese quail (*Coturnix japonica*). Int. J. Poult. Sci. 11, 251–258.

Velotto, S., Crasto, A., 2004. Histochemical and morphometrical characterization and distribution of fibre types in four muscles of ostrich (*Struthio camelus*). Anat. Histol. Embryol. 33, 251–256.

Wang, K., Gan, L., Lee, I., Hood, L., 1995. Isolation and characterization of the chicken trypsinogen gene family. Biochem. J. 307, 471–479.

Watford, M., Wu, G., 2005. Glutamine metabolism in uricotelic species: variation in skeletal muscle glutamine synthetase, glutaminase, glutamine levels and rates of protein synthesis. Comp. Biochem. Physiol. B 140, 607–614.

Watford, M., Lund, P., Krebs, H.A., 1979. Isolation and metabolic characteristics of rat and chicken enterocytes. Biochem. J. 178, 589–596.

Welch Jr., K.C., Altshuler, D.L., 2009. Fiber type homogeneity of the flight musculature in small birds. Comp. Biochem. Physiol. B 152, 324–331.

Wu, G., Chung-Bok, M.I., Vincent, N., Kowalski, T.J., Choi, Y.H., Watford, M., 1998. Distribution of phosphate-activated glutaminase isozymes in the chicken: absence from liver but presence of high activity in pectoralis muscle. Comp. Biochem. Physiol. B 120, 285–290.

Chapter 21

Food Intake Regulation

D. Michael Denbow and Mark A. Cline
Department of Animal and Poultry Sciences, Virginia Tech, Blacksburg, VA, USA

21.1 INTRODUCTION

Birds, like mammals, have complex mechanisms regulating food intake. Given a choice between more than one diet, turkeys, broilers, layers, and other avian species display the ability to self-select a diet adequate for growth or production (Denbow, 1999). Although compensation may not be complete, if exposed to severe feed restriction early in life, birds compensate by increasing their intake in order to increase weight gain following the restriction. In contrast, force feeding birds twice their ad libitum food intake causes Leghorns to stop eating for 7–10 days until their fat stores approach pre-force-feeding levels. Therefore, clearly, birds have the ability to regulate food intake.

Among the strongest evidence that food intake is regulated in birds is the temporal response to dietary amino acid imbalances. Broiler chicks decrease intake of a diet deficient in lysine, methionine, and tryptophan within 24 h posthatch (Picard et al., 1993). Preferences for certain diets are evident in as little as 7 h posthatch. The ability to discriminate among diets deficient in amino acids is dependent on the genetic background of the bird (Noble et al., 1993).

Intense genetic selection of poultry has resulted in strains optimized for either meat production or egg production. Genetic selection has resulted in broilers that are five times larger than layer-type chicks at 6 weeks of age (Zhao et al., 2004). Broilers now reach market weight 50 days earlier than 50 years ago. This increased growth rate has resulted largely from increased food intake.

Food intake regulation in birds has been reviewed (Kuenzel, 1994; Denbow, 2000; Furuse et al., 2007; Richards and Proszkowiec-Weglarz, 2007; Cline and Furuse, 2012). Food intake regulation involves sites both within and outside the central nervous system (CNS), and signals are integrated within the CNS. As in mammals, the CNS, and specifically the hypothalamus, has emerged as a key site in appetite control (Hussain and Bloom, 2013). Because the bulk of the research on avian food intake regulation has used poultry as a model, this chapter focuses mainly on domestic fowl species. However, research with other avian species is reported in several tables.

21.2 PERIPHERAL REGULATION OF FOOD INTAKE

Outside of the CNS, food intake is regulated by both the gastrointestinal tract and the liver (Denbow, 1994). The crop does not appear to have a direct role in food intake regulation in that it does not appear to have receptors that modulate intake. However, it may restrict intake physically when its capacity is reached. The gastrointestinal tract has osmoceptors that, when stimulated by hyperosmotic solutions, result in reduced food intake. Infusion of hyperosmotic solutions into the duodenum reduces food intake, suggesting the existence of osmoreceptors. Intraduodenal infusions of hyperosmotic solutions decrease food intake via a decrease in gastrointestinal motility (see Denbow, 1994).

After absorption from the gastrointestinal tract, nutrients generally travel directly to the liver. Whereas in mammals absorbed lipids are packaged into chylomicrons, enter the lymphatic system, and are transported to the subclavian vein, in birds lipids are packaged into portomicrons that enter the hepatic portal vein and travel directly to the liver. Therefore, the liver is strategically located to monitor nutrient intake and control food intake in birds.

Intrahepatic infusion of glucose, lysine, or lipids decreases food intake (Table 21.1), and these effects vary with the strain of bird. For example, intrahepatic infusion of glucose decreased food intake in Leghorns, but not broilers. Similarly, infusions of lipids decreased food intake in free-feeding Leghorns, but not broilers.

Although the liver appears able to alter food intake in response to changes in hepatic portal blood glucose concentration, there is little evidence that plasma or brain concentrations of glucose influence food intake. Savory (1987) reported no correlation between plasma glucose concentrations and food intake. Furthermore, injections of glucose, or its antimetabolites, into the lateral cerebroventricle have no effect on food intake (Denbow et al., 1982). Whereas insulin injections increase food intake in mammals, similar injections either have no effect or decrease food intake in birds (Cline and Furuse, 2012).

TABLE 21.1 Effects of Compounds on Food Intake when Infused into the Hepatic Portal System[1]

Compound	Broilers	Leghorns
Glucose	→	↓
Amino acid preparations[2] FreeAminelll®	ND	→
Lysine	ND	↓
Leucine	ND	Delayed ↓
Ammonium chloride	ND	→
Epinephrine	ND	↓
Fatty acids[3] Liposyn®	→	↓

[1]→, no effect; ↓, decrease in food intake; ND, not determined (Denbow, 1999).
[2]FreAmineIII® contains isoleucine, leucine, lysine, methionine, phenylalanine, threonine, tryptophan, valine, alanine, dl-, arginine, histidine, proline, serine, glycine, cysteine, sodium acetate, magnesium acetate, sodium chloride, potassium chloride, phosphoric acid, and potassium metabisulfite.
[3]Liposyn® is a commercial emulsified fat product.

While food intake is modified by direct injection of lipids into the liver, dietary lipids also alter food intake (Denbow, 1989). Feeding medium-chain triglycerides (glyceryl tricaprylate or glyceryl tricaprate) or corn oil containing long-chain triglycerides to Leghorns decreased food intake as soon as 30 min postfeeding. This agrees with studies showing that short- and medium-chain triacylglycerols were more effective in decreasing food intake than long-chain triacylglycerols when infused into the hepatic–portal system (see Denbow, 1994). However, medium- and long-chain triacylglycerols were equally effective in decreasing food intake when administered intragastrically. Ketone bodies, acting both peripherally and centrally, also decrease food intake in chicks (Sashihara et al., 2001).

The lipostatic theory of food intake proposed that the brain senses the state of fat depots and makes appropriate adjustments in food intake to maintain fat stores. In 1994, a protein coded by the Ob gene, named leptin, was cloned from the mouse (Zhang et al., 1994), and is believed to be part of a feedback system that indicates an animal's stores of white adipose tissue. As fat stores increase, circulating plasma leptin levels increase. This protein is lacking in ob/ob mice. Administering recombinant leptin to ob/ob mice reduces food intake, decreases body weight and lipid content, and increases thermogenesis. While many studies have shown the presence of leptin in birds, others have questioned these results (see Cerasale et al., 2011). The avian leptin gene has not been found, but its receptor has been reported to be present (Horev et al., 2000). Peripheral and central injections of leptin in rodents decrease food intake. Intracerebroventricular (ICV) injections of human leptin decreased food intake in both broiler and Leghorn chickens (Denbow et al., 2000).

Ghrelin, originally identified as a growth hormone-releasing peptide, is released from the stomach of mammals. ICV and intraperitoneal (IP) injections were shown to increase food intake in humans and rodents. In addition, its levels increase in plasma during fasting, and decrease upon re-feeding. Ghrelin is also found in the proventriculus of birds. Plasma ghrelin levels increase during fasting in Japanese quail and 6 day old layers, but not in 3 week old broilers (Richards and Proszkowiec-Weglarz, 2007; Richards and McMurtry, 2010). Ghrelin mRNA and peptide levels increase in the proventriculus during fasting and return to control levels after re-feeding. However, brain ghrelin mRNA does not change during fasting or re-feeding. While the changes in plasma ghrelin levels associated with fasting would suggest it may act as a peripheral orexigenic signal, in contrast to in rodents and humans, central administration of ghrelin decreases food intake in birds (Furuse et al., 2001). In Japanese quail, while IP injections of relatively low doses of ghrelin increase food intake, higher doses decrease food intake. Intravenous (IV) injections of ghrelin in chickens either weakly decreased food intake or had no effect.

21.3 CNS CONTROL OF FOOD INTAKE

Food intake regulation is highly conserved across animals, and thus neural and endocrine networks controlling this behavior are similarly conserved. The hypothalamus has emerged as the major site of food intake (Hussain and Bloom, 2013). The hypothalamus receives signals for the gut, pancreas, liver, and adipose tissue as well as other parts of the brain, and it integrates these inputs (Figure 21.1) (Richards and Proszkowiec-Weglarz, 2007).

As in mammals, lesioning the medial hypothalamus of avian species increases food intake, whereas lesioning the lateral hypothalamic area (LHA) decreases food intake (Kuenzel et al., 1999). While these sites were traditionally considered the satiety and feeding centers, respectively, it is currently believed that they are considered as parts of larger neural circuits involved in food intake regulation.

The neurochemical control of food intake is complex. Within the CNS, many neurochemicals have been shown to function in food intake control. As will be discussed in the "Classical Neurotransmitters" section, there are many classical neurotransmitters (Table 21.2) as well as peptides (Table 21.3) that have been shown to act in the CNS to alter food intake. In general, these compounds act similarly between mammals and avian species. However, there are some notable exceptions.

21.4 CLASSICAL NEUROTRANSMITTERS

Many classical neurotransmitters have been shown to have a role in the CNS in the control of food intake in birds, and this effect can vary with the type of bird (Table 21.2). For example, ICV injection of epinephrine (E) increased

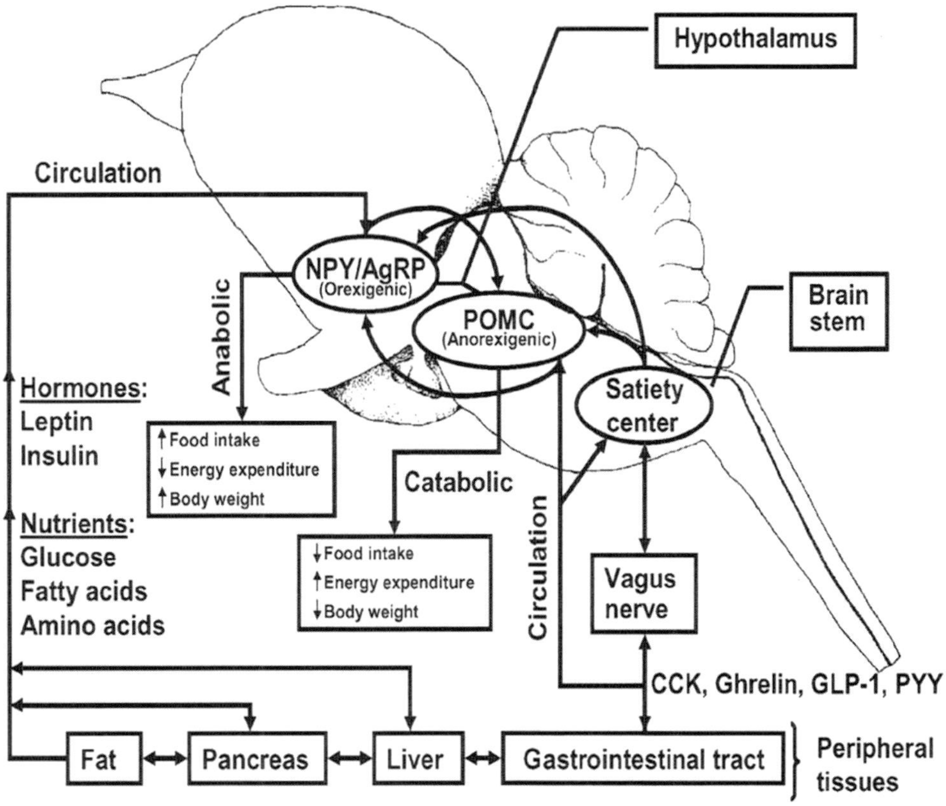

FIGURE 21.1 **A proposed model describing the long-term regulation of appetite and energy balance to achieve a stable body weight in poultry that integrates peripheral tissue and central nervous system circuits regulated by hormonal, neural, neuroendocrine, and nutrient signaling mechanisms.** NPY, neuropeptide Y; AgRP, agouti-related peptide; POMC, pro-opiomelanocortin; CCK, cholecystokinin; GLP-1, glucagon-like peptide-1; PYY, peptide YY.

food intake in broilers (Denbow et al., 1981), but was without effect in Leghorns (Denbow et al., 1983), and reduced food intake in turkeys (Denbow, 1983). More recently, Katayama et al. (2010) reported that high doses of norepinephrine (NE) injected ICV reduced food intake in broiler chicks, and this effect was not associated with changes in brain neuropeptide Y (NPY) or pro-opiomelanocortin (POMC) concentrations. However, Bungo et al. demonstrated a stimulatory effect of ICV NE on food intake in layer chicks. Dopamine, L-DOPA, and tyrosine have been reported to not affect feeding in layer chicks (Bungo et al., 2010). Serotonin (5-hydroxytryptamine; 5-HT) decreases food intake when injected into the brain of both Leghorns and broilers (Table 21.2). Furthermore, adding fenfluramine, a drug that increases the action of serotonin, to the diet of layers and broilers decreased food intake (Hocking and Bernard, 1993).

In mammals, NE, when injected into specific brain sites, both increases and decreases food intake depending on the site of injection. Injections into the ventromedial hypothalamus (VMH) or paraventricular nucleus (PVN) stimulate feeding, while injections into lateral sites including the perifornical region decrease feeding. In Leghorns, NE stimulated food intake when injected into the ventromedial nucleus, PVN, and medial septal sites (Denbow and Sheppard, 1993). NE inhibited food intake when injected near the lateral septal organ and the anterior portion of both the nucleus reticularis superior, pars dorsalis and the tractus occipitomesencephalicus.

Furthermore, food intake was increased in layer-type chicks by ICV injections of clonidine, an α_2-adrenergic agonist acting at presynaptic sites, and this effect was attenuated by yohimbine, an α_2-adrenergic antagonist (Tachibana et al., 2009b). Whereas ICV injections of 0.1 and 0.2 nmol of clonidine stimulated food intake in broilers (Bungo et al., 1999), 0.8 but not 0.4 nmol increased food intake in layer-type chicks. Therefore, it appears that selection for growth in broilers results in increased sensitivity to α_2-adrenergic stimulation. ICV injection of BRL37344, a β_3-adrenergic receptor agonist, suppressed food intake (Tachibana et al., 2003a).

The major inhibitory neurotransmitter in the CNS is γ-aminobutyric acid (GABA). GABA increases food intake in both broilers and turkeys (see Table 21.2). This effect appears mediated by $GABA_A$, but not $GABA_B$ (Jonaidi et al., 2002). Muscimol, a $GABA_A$ receptor agonist, was able to stimulate food intake in broilers, while baclofen, a $GABA_B$ receptors agonist, had no effect.

TABLE 21.2 Effects of Various Neurotransmitters or Their Agonists on Food Intake in Poultry when Injected Directly into the Central Nervous System[1]

Neurotransmitter	Broilers	Leghorns	Turkey
AMPA (α-amino-3-hydroxy-5-methyl-4-isoxazole propionic acid)[2]	→		
Bicuculline[3]	↓		
Dopamine	→	→	→
Epinephrine	↑	→	↓
Glutamate[3]	↓		
Isoproterenol[4]	↓		
Kainate[2]	→		
NMDA (*N*-methyl-D-aspartate)[2]	↑		
Norepinephrine	↑	↑	↓
5-HT (fed birds)	↓	↓	↓
5-HT (fasted birds)	→	↓	↓
Histamine[5]	↓	↓	ND
Carbachol (cholinergic agonist)	↓	ND	ND
Methacholine (muscarinic cholinergic agonist)	→	ND	ND
Muscimol[6] (GABA agonist)	↑	↑	↑
Propranolol[4]	↑		

[1]→, no effect; ↓, decrease in food intake; ↑, increase in food intake; ND, not determined. Data are from Denbow (1999a) unless otherwise stated.
[2]From Bungo et al. (2011). AMPA and NMDA reduced feed intake in pigeons (Zeni et al., 2000).
[3]From Zendehdel et al. (2009). Glutamate also reduced feed intake in pigeons (Zeni et al., 2000).
[4]From Baghbanzadeh et al. (2010).
[5]Results for broilers are from Cline and Furuse (2012); those for Leghorns are from Meade and Denbow (2001).
[6]Results for broilers are from Zendehdel et al. (2009), those for Leghorns are from Bungo et al. (2003).

Histamine was shown to inhibit food intake when injected ICV in both broiler and Leghorn chicks (Meade and Denbow, 2001). Histamine can work at three receptor subtypes: H_1, H_2, and H_3. The H_3 receptor is an autoreceptor that inhibits the release of endogenous histamine from histaminergic neurons. The anorexigenic effect of histamine was attenuated by the H_1 receptor antagonist chlorpheniramine and the H_2 receptor antagonist cimetidine. Taati et al. (2010) showed that injection of thioperaminde, an H_3 antagonist that enhanced histamine release, decreased food intake in 3-week-old broilers. In contrast, R-α-methylhistamine, an H_3 agonist, increased food intake. The H_1 receptor antagonist chlorpheniramine attenuated the effect of thioperaminde, whereas famotidine, an H_2 receptor antagonist, had no effect except at high doses, at which it decreased food intake. Therefore, while it appears that the H_1 receptor is involved in mediating the effect of a histamine-induced decrease in food intake, the role of the H_2 receptor is still unclear.

21.5 PEPTIDES

The list of peptides known to alter food intake is large and growing (Cline and Furuse, 2012). Many of these peptides have orexigenic effects (Table 21.3), and most have anorexigenic effects (Table 21.4).

21.5.1 Neuropeptide Y

First isolated from the pig brain in 1982 (Tatemoto et al., 1982), NPY is one of the most abundantly expressed peptides in the CNS. With a highly conserved sequence of 36 amino acids, the mammalian and avian amino acid sequences differ by a single residue (Blomqvist et al., 1992; Larhammar, 1996). In both mammals and avian species, NPY is one of the most potent orexigenic regulators of food intake. NPY stimulates food intake via NPY Y_2 and Y_5 receptors, but not the Y_1 receptor. Whereas in mammals peptide YY and pancreatic polypeptide suppress appetite, in broiler and Leghorn chicks peptide YY increases food intake (Kuenzel et al., 1987; Denbow et al., 1988). Peptide YY was much more potent than NPY in stimulating food intake, which may reflect a difference in the affinities of NPY and peptide YY to these NPY receptor subtypes.

In mammals, within the hypothalamus the arcuate nucleus (ARC) is a major site for food intake regulation (Hussain and Bloom, 2013). This nucleus has a semipermeable blood–brain barrier allowing peripheral signals to reach the brain. There are first-order orexigenic neurons in the ARC containing NPY and co-expressing agouti-related peptide (AGRP), and a second population of anorexigenic neurons containing POMC–CART (cocaine- and amphetamine-regulated transcript) neurons. These neurons then communicate with second-order neurons in other nuclei expressing such peptides as orexins and melanin-concentrating hormone. Upon fasting, NPY mRNA levels increase in the ARC and this coincides with an increase in NPY peptide content and release in the PVN.

The melanocortin 4 receptor (MC4R) is associated with inhibition of food intake. Release of AGRP, an endogenous melanocortin 4 receptor antagonist, from the NPY–AGRP neurons in the ARC blocks the MC4R, causing increased food intake. In addition, α-melanocyte-stimulating hormone (α-MSH), which is transcribed from the POMC gene, is an agonist at the MC4R and causes decreased food intake.

TABLE 21.3 Peptides Administered Centrally Having an Orexigenic Effect on Food Intake in Various Avian Species

Orexigenic Peptide	Broilers	Leghorns	Other birds
Avian pancreatic peptide		Denbow (1999a)	
Astressin	Emadi et al. (2011)		
CB65 (*N*-cyclohexyl-7-chloro-1-[2-(4-morph-olinyl)ethyl]quinolin-4(1*H*)-one-3-carboxamide)	Emadi et al. (2011)		
Clonidine	Bungo et al. (1999)	Tachibana et al. (2009b)	
CNQX (6-cyano-7-nitroquinoxaline-2, 3-dione)			Pigeons: Zeni et al. (2000)
B-Endorphin	Denbow (1999a)	Denbow (1999a)	
Galanin		Cline and Furuse (2012)	
GnIH (gonadotropin-inhibiting hormone)	Tachibana et al. (2008)	Cline and Furuse (2012)	
(Met^5)-enkephalin	McCormack and Denbow (1989)		
MK801			Pigeons: Zeni et al. (2000)
Neuropeptide Y	Denbow (1999a)	Cline and Furuse (2012)	Ring doves: Strader and Buntin (2001) White-crowned sparrows: Richardson et al. (1995)
Nociceptin	Cline and Furuse (2012)	Cline and Furuse (2012)	
Peptide YY	Cline and Furuse (2012)	Cline and Furuse (2012)	
Prolactin			Domestic turkey: Denbow (1999a)[1] Ring doves: Foreman et al. (1990) European quail: Boswell et al. (1995) Carneau pigeons: Miller and Riddle (1943)
Prolactin-releasing peptide		Cline and Furuse (2012)	
Somatostatin		Cline and Furuse (2012)	
Visfatin	Cline and Furuse (2012)		
26RFa (VGTALGSLAEELNGYNRKKGGFSFRF-NH 2)	Cline and Furuse (2012)	Cline and Furuse (2012)	
Δ^9-THA (Δ^9-tetrahydrocannabinol)	Abel et al. (1972)		

Authors and dates denote referenced papers, and blank spaces denote no reported data.
[1]*Had an anorexic effect.*

Therefore, fasting increases expression of AGRP mRNA and decreases expression of POMC mRNA.

In avian species, NPY-containing neurons are located in the infundibular nucleus, homologous to the ARC in mammals (Kuenzel, 1994; Wang et al., 2001), and these neurons co-express AGRP (Boswell et al., 2002). As in mammals, NPY gene expression increases in the hypothalamus in fasted and food-restricted avian species (Phillips-Singh et al., 2003; Song et al., 2012). Therefore, as in mammals, AGRP blocks the anorexigenic melanocortin pathway, while NPY mediates the orexigenic pathway.

ICV injections of NE do not alter brain NPY or POMC mRNA expression (Katayama et al., 2010). However, the increase in food intake induced by ICV NPY is

TABLE 21.4 Peptides Administered Centrally and Peripherally Having an Anorexigenic Effect (Causing Loss of Appetite) on Food Intake in Various Avian Species

	Central Administration			Peripheral Administration		
Anorexigenic Peptide	**Broilers**	**Leghorns**	**Other Avians**	**Broilers**	**Leghorns**	**Other Avians**
Alytesin	Cline and Furuse (2012)			Cline et al. (2008b)		
AM251 (1-(2,4-dichlorophenyl)-5-(4-iodophenyl)-4-methyl-*N*-(1-piperidyl) pyrazole-3-carboxamide)				Novoseletsky et al. (2011)		
Amylin	Cline et al. (2008d)			Cline et al. (2008d)		
Anserine	Cline and Furuse (2012)					
AVT (arginine vasotocin)		Cline and Furuse (2012)				
Bombesin	Denbow (1999a)	Meade and Denbow (2003)	Domestic turkey: Denbow (1999a)		Denbow (1999a)	Domestic turkey: Denbow (1999a)
Calcitonin	Cline and Furuse (2012)					
Carnosine	Cline and Furuse (2012)					
CCK (cholecystokinin)	Denbow (1999a)	Rodríguez-Sinovas et al. (1997)	Domestic turkey: Denbow (1999a) White-crowned sparrows: Richardson et al. (1993)	Denbow (1999a)	Denbow (1999a)	
CGRP (calcitonin gene-related peptide)	Cline and Furuse (2012)			Cline et al. (2009a)		
CRF (corticotropin-releasing factor)	Cline and Furuse (2012)	Cline and Furuse (2012)	White-crowned sparrows: Richardson et al. (2000)			
Gastrin	Cline and Furuse (2012)	Denbow (1999a)		Denbow (1994)		
Ghrelin	Cline and Furuse (2012)	Cline and Furuse (2012)	Japanese quail: Shousha et al. (2005a)	Geelissen et al. (2006)	Kaiya et al. (2007)[1]	Japanese quail: Shousha et al. (2005a)[2]
GLP (glucagon-like peptide)	Cline and Furuse (2012)	Cline and Furuse (2012)	Japanese quail: Shousha et al. (2007)			Japanese quail: Shousha et al. (2007)
Glucagon		Denbow (1999a)			Honda et al. (2007a)[1]	
GHRH (growth hormone-releasing hormone)	Cline and Furuse (2012)	Cline and Furuse (2012)				
Insulin	Cline and Furuse (2012)	Cline and Furuse (2012)			Bermudez et al. (1983)[1]	

TABLE 21.4 Peptides Administered Centrally and Peripherally Having an Anorexigenic Effect (Causing Loss of Appetite) on Food Intake in Various Avian Species—cont'd

	Central Administration			Peripheral Administration		
Anorexigenic Peptide	Broilers	Leghorns	Other Avians	Broilers	Leghorns	Other Avians
Leptin	Denbow et al. (2000)	Denbow et al. (2000)				
Litorin	Cline and Furuse (2012)	Cline and Furuse (2012)[1]				
LPLRF (Leu–Pro–Leu–Arg–Phe)	Cline and Furuse (2012)	Cline and Furuse (2012)[1]	Bobwhite quail: Cline et al. (2009c)			
MSH (α, β, and γ) (melanocyte-stimulating hormone)	Cline and Furuse (2012)	Cline and Furuse (2012) (α-MSH only)				
NAME (*N*-nitro-arginine methyl-ester)	Denbow (1999a)	Khan et al. (2007)[3]		Khan et al. (2007)	Khan et al. (2007)	
Neuromedin (B, C, and U)		Cline and Furuse (2012)	Japanese quail (NMU): Shousha et al. (2005b)			Japanese quail: (NMU) Shousha et al. (2005b)
Neuropeptide AF, FF, K, S, SF, and VF	Cline and Furuse (2012)					
Oxyntomodulin	Cline and Furuse (2012)					
PACAP (pituitary adenylate cyclase-activating polypeptide)		Cline and Furuse (2012)				
Stresscopin	Cline and Furuse (2012)	Cline et al. (2009d)				
Substance P		Cline and Furuse (2012)				
Urocortin	Cline and Furuse (2012)					
Urotensin	Cline and Furuse (2012)					
VIP (vasoactive intestinal polypeptide)		Cline and Furuse (2012)				
Xenin	Cline and Furuse (2012)			Cline et al. (2007b)		

Authors and dates denote referenced papers, and blank spaces denote no reported data.
[1]*Treatment did not affect food intake.*
[2]*At low doses, had an anorexigenic effect; at high doses, had an orexigenic effect.*
[3]*Treatment had an orexigenic effect.*

attenuated by yohimbine, an α2-adrenergic receptor antagonist (Tachibana et al., 2009b). This suggests that NPY neurons possibly work via communication with adrenergic neurons that act at α_2-adrenergic receptors. Furthermore, a $GABA_A$ antagonist is able to attenuate the response to NPY (Jonaidi and Noori, 2012).

21.5.2 Melanocortins

Melanocortins are derived from POMC, and in mammals most are released from neurons in the ARC. While these peptides can act at five different melanocortin receptor subtypes (MC1R through MC5R) distributed throughout the body, MC3R and MC4R are expressed in the mammalian brain. MC3R appears to regulate energy expenditure, while MC4R is involved with food intake. Agonists of the MC3R receptor decrease food intake.

ICV injection of α-melanocyte-stimulating hormone (α-MSH) significantly reduced food intake in fasted broiler chicks and attenuated NPY-induced food intake (Kawakami et al., 2000). Conversely, AGRP attenuated α-MSH-induced anorexia (Tachibana et al., 2001). CART, which is also found in the POMC neurons, also decreased food intake in broiler and layer chicks and attenuated NPY-induced feeding (Tachibana et al., 2003b).

In mammals, β-MSH binds MC4R with higher affinity than α-MSH (Harrold et al., 2003), suggesting that it may be the main endogenous melanocortin receptor agonist. However, in chickens, α-MSH binds to MC4R with greater affinity than does β-MSH (Ling et al., 2004). In broiler chicks, β-MSH decreased food and water intake while increasing plasma corticosterone concentrations (Smith et al., 2008). Gamma-MSH, a selective MC3R agonist, also reduced food intake in broiler chicks, but required a larger doses than α- or β-MSH (Smith et al., 2011). So it appears that the melanocortin system is involved in an anorexigenic effect in birds, but the role and location of the various receptors still need further elucidation.

21.5.3 Corticotrophin-Releasing Hormone

Corticotrophin-releasing hormone (CRF) acts within the brain to decrease food intake in both broilers and Leghorns (Furuse et al., 1997; Denbow et al., 1999). The ICV injection of the CRF homolog stresscopin also reduced food and water intake in broiler and White Leghorn chicks (Cline et al., 2009d). This was associated with an increased number of c-Fos immunoreactive cells in the VMH, the parvicellular and magnocellular divisions of the PVN, and the posterior hypothalamic nucleus (PHN), indicating increased neuronal activity in these areas. This anorexigenic effect was mediated via CRF receptors. Khan et al. (2008) showed that the anorexigenic affect of CRF was attenuated by blocking nitric oxide in broiler chicks.

As discussed above, the site of action of ghrelin is the CNS. In mammals, ghrelin potently stimulates food intake mainly by stimulating NPY neurons located in the ARC. In contrast, an ICV injection of ghrelin inhibits food intake in broiler chicks, an effect opposite that in mammals (see Furuse et al., 2007). In elucidating the mechanism of ghrelin in birds, Saito et al. (2005) showed that an ICV injection of ghrelin increased plasma corticosterone. Co-injection of the CRF receptor antagonist astressin attenuated the ghrelin-induced decrease in food intake and plasma corticosterone increase. Furthermore, co-injection of ghrelin with NPY inhibited the NPY-induced increase in food intake, and the ICV injection of ghrelin did not change NPY mRNA expression. In contrast to mammals, in which ghrelin induces the release of NPY from neurons in the ARC, in birds ghrelin in the CNS inhibits food intake through the release of endogenous CRF. Taati et al. (2010) demonstrated that blockade of histamine H_1 but not H_2 receptors attenuated ghrelin's anorexigenic effect. Therefore, in chicks, ghrelin, via the histaminergic system, likely causes CRF release, which is ultimately responsible for ghrelin-induced anorexia.

21.5.4 Obestatin

In addition to ghrelin, proghrelin also produces a 23 amino acid amidated peptide called obestatin, which was first isolated and purified from the rat stomach. This peptide is the endogenous ligand of the GPR39 receptor (Wolfgang et al., 2006). The role of obestatin is unclear since it has been reported to decrease food intake and body weight, but other studies failed to show an effect on food intake in rats (Seoane et al., 2006). Xu et al. (2011a) reported that ICV injection of obestatin increased food intake in a dose-dependent manner in a high-weight selected line of chickens. However, Song et al. (2013) found that IV injection of obestatin had no effect on food intake in broilers or Leghorns. Further studies are needed to elucidate the role of obestatin in food intake in birds.

21.5.5 AMP-Activated Protein Kinase

AMP-activated protein kinase (AMPK) is a heterotrimeric enzyme complex composed of one catalytic (alpha) subunit and two regulatory (beta and gamma) subunits. AMPK is activated by allosteric regulation of AMP and by phosphorylation of threonine (thr 172) on the alpha subunit by upstream kinases, including liver kinase B1 (LKB1), calcium/calmodulin-dependent protein kinase kinase (CaMKK), and TGF-beta-activated kinase-1 (TAK1). AMPK acts as an energy sensor. Glucose, leptin, and insulin inhibit hypothalamic AMPK activity and decrease food intake in mice, while ghrelin and adiponectin stimulate hypothalamic AMPK activity and increase food intake in rats (Xu et al., 2011a). ICV injection of ghrelin inhibited hypothalamic AMPK gene expression and phosphorylation of the alpha subunit of AMPK. Furthermore, ghrelin was more efficacious in decreasing food intake in low-weight selected birds than in high-weight selected birds, supporting the hypothesis that selection for lower body weight resulted in increased sensitivity to ghrelin and the AMPK system (Xu et al., 2011a).

21.5.6 Opioids

Opioids act as inhibitory neurotransmitters and are widely distributed throughout the CNS. Opioid receptors include μ, δ, or κ subtypes. The opioid antagonists naloxone and naltrexone were shown to decrease food intake in both broilers and layers (see Denbow, 1999). Opioids appear to alter food intake in birds acting at the δ- and κ-opioid receptors (Bungo et al., 2004). Dodo et al. (2005) reported that antagonists of the μ-opioid receptor, especially the μ_1-receptor, reduced NPY-induced feeding in neonatal chicks.

Nociceptin–orphanin FQ (N/OFQ) is structurally similar to opioids, and binds to the opioid-like G protein-coupled receptor 1, or nociceptin receptor (NOP). N/OFQ shows structural similarities to the classical opioid peptides, particularly dynorphin A, but N/OFQ does not interact with the dynorphin A–κ-opioid receptor system. N/OFQ has been shown to stimulate food intake in rats (Polidori et al., 2000). Similar to other opioids, ICV injection of N/OFQ stimulated food intake and increased pecking frequency in broiler chicks (Abbasnejad et al., 2005). Similarly, ICV N/OFQ stimulated food intake in White Leghorn chicks, and this was associated with increased agouti-related peptide (AGRP) and decreased CART mRNA expression in the diencephalon (Bungo et al., 2009). N/OFQ-induced increase in food intake was blocked by α-MSH. There was no change in NPY or POMC mRNA.

21.5.7 FMRFamides

Since molluscan neuropeptide Phe–Met–Arg–Phe–NH_2 (FMRFamide) was isolated from the ganglia of the clam, similar neuropeptides containing the RFamide sequence at their C-termini (RFamide peptides) have been characterized in various invertebrates. The chicken pentapeptide Leu–Pro–Leu–Arg–Phe–NH_2 (LPLRFamide) cross-reacts with the FMRFamide antibody and was the first RFamide isolated from a vertebrate (Dockray et al., 1983).

Several RFamides have been demonstrated to function as orexigenic factors in neonatal chicks, including prolactin-releasing peptide (PrRP) and gonadotropin-inhibitory hormone (GnIH). PrRP was originally proposed to be a stimulator of prolactin release in mammals (Hinuma et al., 1998). While central administration of PrRP decreased food intake in rats (Lawrence et al., 2000, 2002; Ellacott et al., 2002), ICV injection of PrRP significantly increased food intake in layer chicks (Tachibana et al., 2004b). PrRP has a significantly weaker orexigenic effect than NPY on an equimolar basis. The orexigenic effect of NPY was further enhanced with a co-injection of PrRP.

Tsutsui et al. (2000) isolated a novel decapeptide containing the C-terminal LPLRFamide motif, SIKPSAYL-PLRFamide, from the quail brain. This peptide inhibited gonadotropin-releasing factor reported in vertebrates, and it was termed GnIH. The GnIH precursor encodes one GnIH and two GnIH-related peptides (called GnIH-RP-1 and GnIH-RP-2). ICV injection of GnIH, GnIH-RP-1, and GnIH-RP-2 significantly stimulated food intake in layer chicks (Tachibana et al., 2005a). Tachibana et al. (2008) demonstrated that the effect of GnIH is likely mediated via the μ-, but not δ- or κ-, opioid receptor in broiler chicks. GnIH also did not appear to work via nitric oxide since a nitric oxide synthesis inhibitor had no effect on GnIH-induced feeding.

Ukena et al. (2010) recently demonstrated that another RFamide, 26RFa, was found in the anterior hypothalamus of Japanese quail. Its receptor, GPR103, was found in the cerebrum, diencephalon, mesencephalon, and cerebellum. Ukena et al. (2010) also demonstrated that central injection of 26RFa increased food intake of broiler but not layer-type chicks. Additionally, metastin, another RFamide first isolated from the placenta, causes increased food intake in broiler chicks, and its effects were mediated by μ-opioid receptors (Khan et al., 2009).

The RFamides, mentioned in this chapter, cause orexigenic effects in chicks when centrally administered. In contrast, there are other members of this family associated with anorexia. Chicks that received ICV LPLRFamide reduced their food intake at 30 min post injection, but this effect dissipated by 40 min post injection (Cline et al., 2009c). Since LPLRFamide-treated chicks spent more time in deep rest while other behaviors were unaffected, this may suggest that the short-term anorexigenic effect of LPLRFamide is secondary to this behavior. Further suggesting that LPLR-Famide may not have a specific effect on food intake is the report from Tachibana et al. (2005b) that it had no effect on food intake in layer chicks.

Several members of the neuropeptide FF (NPFF) subfamily of the RFamide family have also been shown to cause decreased food intake in chicks (see Cline and Furuse, 2012). For example, ICV injections of NPFF caused dose-dependent reduction in food intake in broilers (Cline et al., 2007a). The effect of NPFF is mediated through the μ and κ subtypes of opioid receptors, and NPFF attenuates NPY- and β-endorphin (END)-induced food intake stimulation (Cline and Mathews, 2008).

21.5.8 Galanin

Galanin, a peptide found in both the CNS and intestine, is conserved across species. Tachibana et al. (2008) demonstrated that ICV galanin also stimulated food but not water intake of both broiler and layer-type chicks. It appears to work via μ-opioid receptors, and α_2 adrenoceptors likely mediate this response.

21.5.9 Visfatin

Visfatin, also known as pre-B cell colony-enhancing factor, is a peptide that was originally isolated from visceral fat of humans and mice, and that mimics the actions of insulin (Fukuhara et al., 2005). Visfatin is associated with metabolic syndrome in humans (Filippatos et al., 2007). The effect of visfatin on food intake was first determined in broilers in which ICV injections nearly doubled food intake and pecking efficiency (Cline et al., 2008c). This was accompanied by an increase in the number of c-Fos immunoreactive cells in the LHA and a decrease in activity in the VMH. Brunetti et al. (2012) recently showed that in rats, visfatin injection into the ARC increased food intake and decreased CART and CRH mRNA levels, so visfatin may have a similar effect in birds and mammals.

21.5.10 Somatostatin

Somatostatin, known for the inhibition of growth hormone release, also affects food intake. While it has been shown to both increase and decrease food intake in rats depending on their feeding state, ICV injections of somatostatin increased food intake in broilers and layer-type chicks (Tachibana et al., 2009a). Its effects were mediated by μ-opioid and adrenergic α_2 receptors. By additional use of ICV injections of somatostatin analogs, it was shown that SSTR2, SSTR3, and SSTR5, but not SSTR4, receptors were associated with somatostatin-induced increased food intake (Tachibana et al., 2010).

21.5.11 Cannabinoids

The active ingredient in marijuana is Δ^9-tetrahydrocannabinol (Δ^9-THA), and it can bind to the cannabinoid receptors CB_1 and CB_2 found in the brain. The endogenous ligands for the cannabinoid receptors are 2-arachidonoylglycerol and anandamide, and endogenous antagonists include O-2050 and AM4113. The CB_1 receptor is present in the brain, as well as peripherally, in both birds and mammals, while the CB_2 receptor is found on immune and blood cells (Pagotto et al., 2006). While the effects of Δ^9-THA on food intake have been mixed, it is generally believed that it causes increased food intake and that decreased food intake is a secondary effect due to sedation (Pagotto et al., 2006). Abel et al. (1972) first showed an increase in food intake in chicks in response to Δ^9-THA. Giving an inverse agonist either via the feed or by IV injection attenuated food intake in broiler chicks (Novoseletsky et al., 2011), supporting an orexigenic role for cannabinoids in birds.

21.5.12 Glucagon-like Peptide

Glucagon-like peptide-1 (GLP1) is a member of the glucagon superfamily. Similar to results in mammals, ICV injection of GLP1 decreases food intake in broiler chicks (Bungo and Furuse, 2001). Growth hormone-releasing factor increases food intake in rats; however, it is a potent inhibitor of food intake in broiler chicks (Furuse et al., 2001). Similarly, other neuropeptides belonging to the glucagon superfamily, including vasoactive intestinal peptide (VIP) and pituitary adenylate cyclase-activating polypeptide (PACAP), reduced food intake in layer chicks (Tachibana et al., 2004a). ICV injection of PACAP and VIP increases plasma corticosterone concentrations, and these effects are attenuated by co-injection with the CRF receptor antagonist astressin, suggesting that PACAP and VIP function through activation of CRF neurons (Tachibana et al., 2004a). Similarly, glucagon (Honda et al., 2007a) and oxyntomodulin (Cline et al., 2008a) decreased food intake in chicks when injected ICV.

21.5.13 Cholecystokinin

Cholecystokinin (CCK), originally isolated from the porcine small intestine, decreases food intake when injected peripherally or centrally into mammals. CCK was also shown to decrease food intake when injected into the brain of broilers (Denbow and Myers, 1982). The structurally related peptide gastrin does not appear to decrease food intake in mammals. While the amino acid sequence of chicken CCK octapeptide (CCK8) is identical to that of mammalian species, differences in the amino acid sequence exist between chicken and mammalian gastrin. Furuse et al. (1999) demonstrated that ICV injection of chicken gastrin inhibits food intake and food passage in neonatal chicks. When further investigating the anorexigenic effects of CCK and gastrin, it was shown that the efficacy of these peptides depends on the length of their peptide sequence (Furuse et al., 2000).

21.5.14 Glucagon Superfamily

Like the glucagon superfamily, members of the calcitonin gene peptide superfamily, including calcitonin, calcitonin gene-related peptide (CGRP), and amylin, appear involved in food intake regulation. These peptides are derived via tissue-specific alternative splicing of the primary RNA transcripts of the calcitonin gene. Amylin has been shown to cause a dose-dependent decrease in food intake after ICV and IP injection in broilers (Cline et al., 2008d). This was associated with increased neuronal activity in the area postrema and the nucleus of the solitary tract. In addition, amylin caused increased alimentary canal transit time, increased plasma corticosterone concentration, and increased anxiety-related behaviors. Since amylin is co-secreted with insulin, it may be involved in terminating meals in chicks.

ICV injection of calcitonin in broiler chicks also reduced food and water intake (Layne et al., 2009). This was also associated with increased activation of the ARC, dorsomedial nucleus, and VMH. The CRF system also appears not to mediate this response. Similarly, broilers that received ICV and IP CGRP reduced both their food and water intake, but CGRP's effect on water intake appeared to be related to its effect on food intake (Cline et al., 2009a). CGRP-induced anorexia coincided with increased activation of the ARC, PVN, PHN, and VMH.

21.5.15 Insulin

Insulin decreases food intake when injected into the brain of mammals. Similarly, ICV injection of insulin decreased food intake in White Leghorn chicks (Honda et al., 2007b; Shiraishi et al., 2008a) while increasing expression of POMC, CART, and CRF, but not affecting NPY and AGRP mRNA hypothalamic expression (Honda et al., 2007b). Shiraishi et al. (2008a), however, reported a decrease in NPY expression after ICV insulin when including the brain stem. Blocking the melanocortin receptors prevented the anorexic effect of ICV insulin (Shiraishi et al., 2008a). Co-injection of β-endorphin (β-END) and insulin decreased central POMC expression more than just insulin injection, while blockade of μ-opioid receptors reduced the effect of insulin on food intake (Shiraishi et al., 2008b). Food intake is decreased in Single Comb Leghorns by chicken and pig insulin, but not by human or bovine insulin (Shiraishi et al., 2009). Kuenzel and McMurtry (1988) reported that ICV injection of insulin caused an increase in plasma insulin concentration. Insulin appears to act in the CNS to decrease food intake in chickens by both anorexigenic and orexigenic systems.

21.5.16 Bombesin

Bombesin, a tetradecapeptide originally from amphibian skin, inhibits food intake in mammalian species. Denbow (1989) reported that both ICV and IV injection of bombesin decreased food and water intake in turkeys. ICV injection of bombesin also inhibits food intake in broilers, and co-injection with NPY suppressed the orexigenic effect of NPY (Bungo et al., 2000). ICV and IP injections of bombesin analogs alytesin and litorin also decreased food intake in broiler chicks (Cline et al., 2008b, 2010a).

21.6 SELECTION FOR BODY WEIGHT ALTERS FOOD INTAKE CONTROL MECHANISMS

While broilers have been genetically selected for rapid body weight gain, Leghorns have been selected for egg production and, thus, a smaller body mass. As expected, selection for body weight alters the mechanisms controlling food intake. Several lines of evidence support this conclusion. Whereas ICV injections of NE and E increased food intake in broilers, they were without effect in Leghorns (see Denbow, 1985). While ICV injection of serotonin decreased food intake in fasted Leghorns, it had no effect in fasted broilers. So it appears that broilers are more sensitive to orexigenic compounds, whereas layer-type birds are more sensitive to anorexigenic compounds.

Much of the work exploring how selection for growth has altered the mechanisms controlling food intake has been conducted using a unique line of birds divergently selected for either low (LWS) or high (HWS) body weight at 8 weeks of age (Siegel, 1962; Rubin et al., 2010; Marquez et al., 2010). Selected for 55 generations, these lines at selection age (56 days) have a 10-fold difference in body weight (Figure 21.2) with correlated differences in food intake and body composition, and they are composed of anorexic and obese individuals.

The LWS line exhibits natural anorexia (Zelenka et al., 1988), and delayed (Dunnington and Siegel, 1996) or prevented sexual maturity (Dunnington et al., 1983, 1984). The HWS line contains compulsive eaters (Dunnington and Siegel, 1984).

Burkhart et al. (1983) reported that lesioning the VMH of low-weight-line chickens resulted in nonhyperphagic obesity, whereas similar lesions in high-weight-line chickens had no effect. These lines of birds originated from a common population selected divergently for 8 week juvenile body weight. The VMH is thought to be involved in satiety, and selection for body weight may have altered its function.

Injection of methoxamine, an α_1-adrenergic receptor agonist, stimulated food intake in the high-weight line while having no effect in the low-weight line (Denbow et al., 1986). In addition, the anorexigenic effect of 5-HT was longer lasting in the high-weight line compared to the low-weight line.

As was suggested by Bray (1991) for mammals, Denbow (1999a) suggested that differences in body weight in lines of chickens may have resulted from changes in the balance between the parasympathetic and sympathetic nervous system. Selection for increased body weight may have augmented the actions of the parasympathetic branch of the autonomic nervous system. Support for this hypothesis was provided when it was shown that LWS birds appear to have

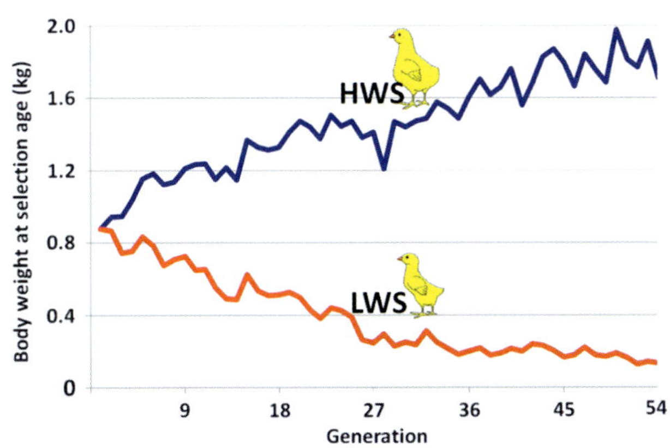

FIGURE 21.2 **Body weight at 56 days posthatch in male LWS and HWS chickens during the course of selection for low or high body weight.**

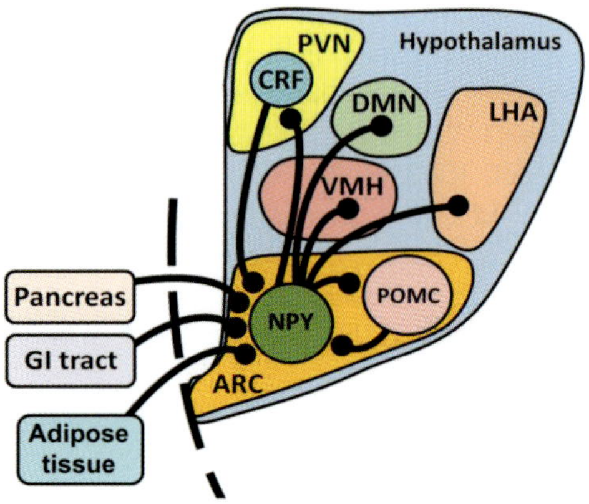

FIGURE 21.3 Based on the mammalian models, projections of NPY neurons from the ARC to other appetite-associated hypothalamic nuclei, reciprocal innervation between NPY and POMC neurons in the ARC, and reciprocal innervation between NPY and CRF in the ARC and PVN, receptively. Peripheral signals influence feed intake by affecting the ARC. *Adapted from Broberger et al. (1999); Tebbe et al. (2003); Mercer et al. (2011).*

greater sympathetic nervous system activity compared to the HWS line (Kuo et al., 2001).

For many of the anorexigenic neuropeptides, including α-MSH, CRF, insulin, amylin, and neuropeptide AF, the LWS birds respond to a lower dose than the HWS birds (Cline et al., 2010b) (Figure 21.3). For example, CRF decreases food intake in both the LWS and HWS lines, but the LWS line responded to a lower dose than did the HWS line (Cline et al., 2009b). Additionally, CRF-induced anorexia was attenuated in HWS line but not LWS line chicks by a CRF receptor antagonist. However, for neuropeptide S, calcitonin, and CGRP, the HWS respond to a lower dose than the LWS. The HWS and LWS respond similarly to ghrelin and galanin. Finally, the LWS does not respond to NPY or AGRP, while the HWS line does not reduce food intake when administered leptin.

21.7 DIFFERENCES BETWEEN BIRDS AND MAMMALS

In most cases, neurotransmitters and metabolites have similar effects in birds and mammals (Figure 21.4). However, there are several noteworthy differences. While peptide YY and pancreatic polypeptide decrease food intake in mammals, they stimulate food intake in birds. In contrast, while ghrelin is a potent orexigenic peptide in mammals, it is anorexigenic in birds. Other peptides shown to be orexigenic in mammals, including melanin-concentrating hormone, orexins (A and B), and motilin, have no effect on food intake in chickens. Finally, PrPP stimulates food intake in chickens but decreases food intake in mammals.

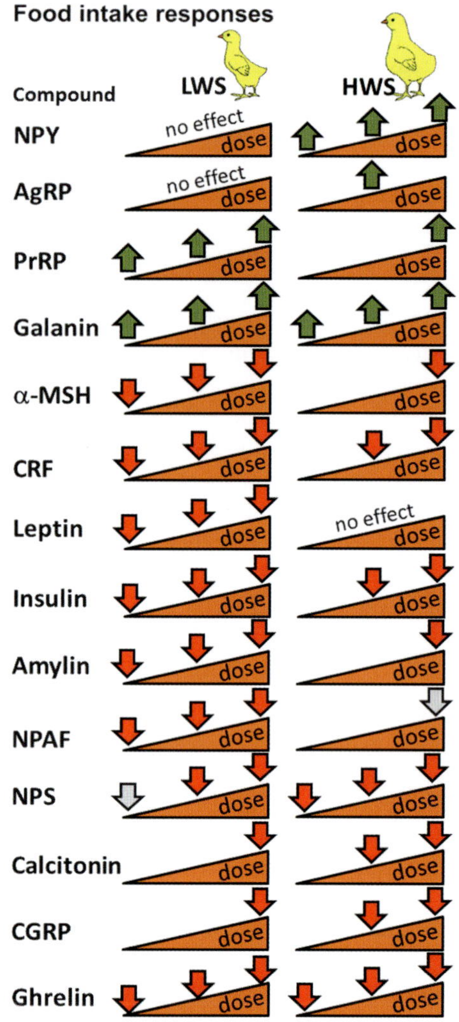

FIGURE 21.4 Summary of feed intake responses to exogenous neuropeptide injection in LWS and HWS chicks. The triangle represents an increasing dose of neuropeptide, whereas arrows indicate the relative threshold of either decreased (down red arrow) or increased (up green arrow) feed intake. Gray arrows indicate a lag in response time.

REFERENCES

Abbasnejad, M., Jonaidi, H., Denbow, D.M., Pour Rahimia, A.M., 2005. Feeding and locomotion responses to centrally injected nociceptin/orphanin FQ in chicks. Physiol. Behav. 85, 383–386.

Abel, E.L., McMillan, D.E., Harris, L.S., 1972. Tolerance to the behavioral and hypothermic effects of 1-9-tetrahydrocannabinol in neonatal chicks. Experientia 28, 1188–1189.

Baghbanzadeh, A., Hajinezhad, M.R., Shohreh, B., Maleklou, R., 2010. Intralateral hypothalamic area injection of isoproterenol and propranolol affects food and water intake in broilers. J. Comp. Physiol. A 196, 221–226.

Bermudez, F.F., Forbes, J.M., Injidi, M.H., 1983. Involvement of melatonin and thyroid hormones in the control of sleep, food intake and energy metabolism in the domestic fowl. J. Physiol. 337, 19–27.

Blomqvist, A.G., Soderberg, C., Lundell, I., Milner, R.J., Larhammar, D., 1992. Strong evolutionary conservation of neuropeptide Y: sequences of chicken, goldfish, and *Torpedo marmorata* DNA clones. Proc. Natl. Acad. Sci. 89, 2350–2354.

Boswell, T., Li, Q., Takeuchi, S., 2002. Neurons expressing neuropeptide Y mRNA in the infundibular hypothalamus of Japanese quail are activated by fasting and co-express agouti-related protein mRNA. Mol. Brain Res. 100, 31–42.

Boswell, T., Sharp, P.J., Hall, M.R., Goldsmith, A.R., 1995. Migratory fat deposition in European quail: a role for prolactin? J. Endocrinol. 146, 71–79.

Bray, G.A., 1991. Obesity, a disorder of nutrient partitioning: the MONA LISA hypothesis. J. Nutr. 121, 1146–1162.

Brunetti, L., Recinella, L., De Nisio, C., Chiavaroli, A., Leone, S., Ferrante, C., Orlando, G., Vacca, M., 2012. Effects of visfatin/PBEF/NAMPT on feeding behavior and hypothalamic neuromodulators in the rat. J. Biol. Reg. Homeost. Agents 26, 295–302.

Broberger, C., 1999. Hypothalamic cocaine-and amphetamine-regulated transcript (CART) neurons: histochemical relationship to thyrotropin-releasing hormone, melanin-concentrating hormone, orexin/hypocretin and neuropeptide. Y. Brain Res. 848, 101–113.

Bungo, T., Ando, R., Kawakami, S.-I., Ohgushi, A., Shimojo, M., Masuda, Y., Furuse, M., 2000. Central bombesin inhibits food intake and the orexigenic effect of neuropeptide Y in the neonatal chick. Physiol. Behav. 70, 573–576.

Bungo, T., Furuse, M., 2001. Glucagon-like peptide-1 (7–36) amide (GLP-1) is a potent satiety agent in chickens. In: Dawson, A., Chaturvedi, C.M. (Eds.), Avian Endocrinology. Narosa Publishing House, New Delhi, India, pp. 337–348.

Bungo, T., Izumi, T., Kawamura, K., Takagi, T., Ueda, H., Furuse, M., 2003. Intracerebroventricular injection of muscimol, baclofen or nipecotic acid stimulates food intake in layer-type, but not meat-type, chicks. Brain Res. 993, 235–238.

Bungo, T., Kawamura, K., Izumi, T., Dodo, K.-I., Ueda, H., 2004. Feeding responses to μ-, δ- and κ-opioid receptor agonists in the meat-type chick. Pharmacol. Biochem. Behav. 78, 707–710.

Bungo, T., Shimojo, M., Masuda, Y., Choi, Y.-H., Denbow, D.M., Furuse, M., 1999. Induction of food intake by a noradrenergic system using clonidine and fusaric acid in the neonatal chick. Brain Res. 826, 313–316.

Bungo, T., Yanagita, K., Shiraishi, J., 2010. Feed intake after infusion of noradrenalin, dopamine or its precursor into the lateral ventricles in neonatal chicks. J. Anim. Vet. Adv. 9, 760–763.

Bungo, T., Shiraishi, J.-I., Kawakami, S.-I., 2011. Feeding responses to central glutamatergic receptor agonist administration in meat-type chicks. J. Anim. Vet. Adv. 10, 955–958.

Bungo, T., Shiraishi, J.-I., Yanagita, Y., Ohta, Y., Fujita, M., 2009. Effect of nociceptin/orphanin FQ on feeding behavior and hypothalamic neuropeptide expression in layer-type chicks. Gen. Comp. Endocrinol. 163, 47–51.

Burkhart, C.A., Cherry, J.A., Van Krey, H.P., Siegel, P.B., 1983. Genetic selection for growth rate alters hypothalamic satiety mechanism in chickens. Behav. Genet. 13, 295–300.

Cerasale, D.J., Zajac, D.M., Guglielmo, C.G., 2011. Behavioral and physiological effects of photoperiod-induced migratory state and leptin on a migratory bird, *Zonotrichia albicollis*: I. Anorectic effects of leptin administration. Gen. Comp. Endocrinol. 174, 276–286.

Cline, M.A., Bowden, C.N., Nandar, W., Rogers, J.O., 2008a. Central oxyntomodulin causes anorexigenic effects associated with the hypothalamus and alimentary canal in chicks (*Gallus gallus*). Comp. Biochem. Physiol. A Mol. Integr. Physiol. 149, 405–410.

Cline, M.A., Fouse, D., Prall, B.C., 2008b. Central and peripheral alytesin cause short-term anorexigenic effects in neonatal chicks. Neuropeptides 42, 283–291.

Cline, M.A., Nandar, W., Prall, B.C., Bowden, C.N., Denbow, D.M., 2008c. Central visfatin causes orexigenic effects in chicks. Behav. Brain Res. 186, 293–297.

Cline, M.A., Nandar, W., Smith, M.L., Pittman, B.H., Kelly, M., Rogers, J.O., 2008d. Amylin causes anorexigenic effects via the hypothalamus and brain stem in chicks. Regul. Pept. 146, 140–146.

Cline, M.A., Calchary, W., Nandar, W., 2009a. Effect of calcitonin gene-related peptide (CGRP) on avian appetite-related processes. Behav. Brain Res. 196, 242–247.

Cline, M.A., Kuo, A., Smith, M., Nandar, W., Prall, B.C., Siegel, P.B., Denbow, D.M., 2009b. Differential food intake responses to central corticotrophin releasing factor in lines of chickens divergently selected for low or high body weight. Comp. Biochem. Physiol. A Mol. Integr. Physiol. 152, 130–134.

Cline, M.A., Layne, J.E., Calchary, W.A., Sheehy, R.S., Tachibana, T., Furuse, M., 2009c. LPLRFamide causes anorexigenic effects in broiler chicks and bobwhite quail. Gen. Comp. Endocrinol. 165, 315–320.

Cline, M.A., Prall, B., Rogers, J.O., Tachibana, T., 2009d. Satiety induced by central stresscopin is mediated by corticotrophin-releasing factor receptors and hypothalamic changes in chick. Pharmacol. Biochem. Behav. 92, 663–669.

Cline, M.A., Cofield, S.A., Tachibana, T., 2010a. Central litorin injection is associated with primary anorexigenic effects that coincide with activation of the magnocellular division of the paraventricular nucleus. Neuropeptides 44, 247–252.

Cline, M.A., Nandar, W., Bowden, C., Calchary, W., Smith, M., Prall, B., Newmyer, B., Rogers, J.O., Siegel, P.B., 2010b. The threshold of amylin-induced anorexia is lower in chicks selected for low compared to high juvenile body weight. Behav. Brain Res. 208, 650–654.

Cline, M.A., Furuse, M., 2012. Neuropeptide regulation of food intake in chicks. In: Morrison, J.L. (Ed.), Food Intake: Regulation, Assessing and Controlling. Nova Science Publishers, Inc., NY.

Cline, M.A., Mathews, D., 2008. Anoretic effects of neuropeptide FF are mediated via central mu and kappa subtypes of opioid receptors and receptor ligands. Gen. Comp. Endocrinol. 159, 125–129.

Cline, M.A., Nandar, W., Rogers, J.O., 2007a. Central neuropeptide FF reduces feed consumption and affects hypothalamic chemistry in chicks. Neuropeptides 41, 433–439.

Cline, M.A., Nandar, W., Rogers, J.O., 2007b. Xenin reduces feed intake by activating the ventromedial hypothalamus and influences gastrointestinal transit time in chicks. Behav. Brain Res. 179, 28–32.

Denbow, D.M., 1983. Food intake and temperature response to injections of catecholamines into the lateral ventricle of the turkey brain. Poult. Sci. 62, 1088–1092.

Denbow, D.M., 1985. Food intake control in birds. Neurosci. Biobehav. Rev. 9, 223–232.

Denbow, D.M., 1989. Centrally and peripherally administered bombesin decreases food intake in turkeys. Peptides 10, 275–279.

Denbow, D.M., 1994. Peripheral regulation of food intake. J. Nutr. 124, 1349S–1354S.

Denbow, D.M., 1999. Food intake regulation in birds. J. Exp. Zool. 283, 333–338.

Denbow, D.M., 2000. Gastrointestinal Anatomy and Physiology. In: Avian Physiology, 5th ed. (edited by G.C. Whittow). Academic Press: Orlando, Florida.

Denbow, D.M., Cherry, J.A., Siegel, P.B., Van Krey, H.P., 1981. Eating, drinking and temperature response of chicks to brain catecholamine injections. Physiol. Behav. 27, 265–269.

Denbow, D.M., Cherry, J.A., Van Krey, H.P., Siegel, P.B., 1982. Food and water intake following injection of glucose into the lateral ventricle of the brain of broiler-type chicks. Poult. Sci. 61, 1713–1719.

Denbow, D.M., Duke, G.E., Chaplin, S.B., 1988. Food intake, gastric secretion, and motility as affected by avian pancreatic polypeptide administered centrally in chickens. Peptides 9, 449–454.

Denbow, D.M., Meade, S., Robertson, A., McMurtry, J.P., Richards, M., Ashwell, C., 2000. Leptin induced decrease in food intake in chickens. Physiol. Behav. 69, 359–362.

Denbow, D.M., Myers, R.D., 1982. Eating, drinking, and temperature responses to intracerebroventricular cholecystokinin in the chick. Peptides 3, 739–743.

Denbow, D.M., Sheppard, B.J., 1993. Food and water intake responses of the domestic fowl to norepinephrine infusion at circumscribed neural sites. Brain Res. Bull. 31, 121–128.

Denbow, D.M., Snapir, N., Furuse, M., 1999. Inhibition of food intake by CRF in chickens. Physiol. Behav. 66, 645–649.

Denbow, D.M., Van Krey, H.P., Lacy, M.P., Dietrick, T.J., 1983. Feeding, drinking and body temperature of Leghorn chicks: effects of ICV injections of biogenic amines. Physiol. Behav. 31, 85–90.

Denbow, D.M., Van Krey, H.P., Siegel, P.B., 1986. Selection for growth alters the feeding response to brain injections of biogenic amines. Pharmacol. Biochem. Behav. 24, 39–42.

Dockray, G.J., Reeve Jr., J.R., Shively, J., Gayton, R.J., Barnard, C.S., 1983. A novel active pentapeptide from chicken brain identified by antibodies to FMRFamide. Nature 305, 328–330.

Dodo, K.-I., Izumi, T., Ueda, H., Bungo, T., 2005. Response of neuropeptide Y-induced feeding to μ-, δ- and κ-opioid receptor antagonists in the neonatal chick. Neurosci. Lett. 373, 85–88.

Dunnington, E.A., Siegel, P.B., 1984. Thermoregulation in newly hatched chicks. Poult. Sci 63, 1303–1313.

Dunnington, E.A., Siegel, P.B., 1996. Long-term divergent selection for eight-week body weight in white Plymouth Rock chickens. Poult. Sci. 75, 1168–1179.

Dunnington, E.A., Siegel, P.B., Cherry, J.A., Soller, M., 1983. Relationship of age and body weight at sexual maturity in selected lines of chickens. Arch. Geflugelk. 47, 85–89.

Dunnington, E.A., Siegel, P.B., Cherry, J.A., 1984. Delayed sexual maturity as a correlated response to selection for reduced 56-day body weight in White Plymouth Rock pullets. Arch. Geflugelk. 48, 111–113.

Ellacott, K.L., Lawrence, C.B., Rothwell, N.J., Luckman, S.M., 2002. PRL-releasing peptide interacts with leptin to reduce food intake and body weight. Endocrinology 143, 368–374.

Emadi, L., Jonaidi, H., Hosseini Amir Abad, E., 2011. The role of central CB2 cannabinoid receptors on food intake in neonatal chicks. J. Comp. Physiol. A 197, 1143–1147.

Filippatos, T.D., Derdemezis, C.S., Kiortsis, D.N., Tselepis, A.D., Elisaf, M.S., 2007. Increased plasma levels of visfatin/pre-B cell colony-enhancing factor in obese and overweight patients with metabolic syndrome. J. Endocrinol. Invest. 30, 323–326.

Foreman, R., Leaf, M., Buntin, J.P., 1990. Changes in feeding activity, plasma luteinizing hormone and testes weight in ring doves following hypothalamic injections of prolactin. J. Neuroendocrinol. 2, 667–673.

Fukuhara, A., Matsuda, M., Nishizawa, M., Segawa, K., Tanaka, M., Kishimoto, K., Matsuki, Y., Murakami, M., Ichisaka, T., Murakami, H., Watanabe, E., Takagi, T., Akiyoshi, M., Ohtsubo, T., Kihara, S., Yamashita, S., Makishima, M., Funahashi, T., Yamanaka, S., Hiramatsu, R., Matsuzawa, Y., Shimomura, I., 2005. Visfatin: a protein secreted by visceral fat that mimics the effects of insulin. Science 307, 426–430.

Furuse, M., Ao, R., Bungo, T., Ando, R., Shimojo, M., Masuda, Y., Saito, N., 1999. Central gastrin inhibits feeding behavior and food passage in neonatal chicks. Life Sci. 65, 305–311.

Furuse, M., Bungo, T., Ao, R., Ando, R., Shimojo, M., Masuda, Y., Denbow, D.M., 2000. Involvement of central gastrin and cholecystokinin in the regulation of food intake in the neonatal chick. J. Appl. Anim. Res. 18, 129–136.

Furuse, M., Matsumoto, M., Saito, N., Sugahara, K., Hasegawa, S., 1997. The central corticotrophin-releasing factor and glucagon-like peptide-1 in food intake of the neonatal chick. Eur. J. Pharmacol. 339, 211–213.

Furuse, M., Tachibana, T., Ohgushi, A., Ando, R., Yoshimatsu, T., Denbow, D.M., 2001. Intracerebroventricular injection of ghrelin and growth hormone releasing factor inhibits food intake in neonatal chicks. Neurosci. Lett. 301, 123–126.

Furuse, M., Yamane, H., Tomonaga, S., Tsuneyoshi, Y., Denbow, D.M., 2007. Neuropeptidergic regulation of food intake in the neonatal chick: a review. Poult. Sci. 44, 349–356.

Geelissen, S.M., Swennen, Q., Geyten, S.V., Kühn, E.R., Kaiya, H., Kangawa, K., Decuypere, E., Buyse, J., Darras, V.M., 2006. Peripheral ghrelin reduces food intake and respiratory quotient in chicken. Domest. Anim. Endocrinol. 30, 108–116.

Harrold, J.A., Widdowson, P.S., Williams, G., 2003. beta-MSH: a functional ligand that regulated energy homeostasis via hypothalamic MC4-R? Peptides 24, 397–405.

Hinuma, S., Habata, Y., Fujii, R., Kawamata, Y., Hosoya, M., Fukusumi, S., Kitada, C., Masuo, Y., Asano, T., Matsumoto, H., Sekiguchi, M., Kurokawa, T., Nishimura, O., Onda, H., Fujino, M.A., 1998. A prolactin-releasing peptide in the brain. Nature 393, 272–276.

Hocking, P.M., Bernard, R., 1993. Evaluation of putative appetite suppressants in the domestic fowl (*Gallus domesticus*). Brit. Poult. Sci. 34, 393–404.

Honda, K., Kamisoyama, H., Saito, N., Kurose, Y., Sugahara, K., Hasegawa, S., 2007a. Central administration of glucagon suppresses food intake in chicks. Neurosci. Lett. 416, 198–201.

Honda, K., Kamisoyama, H., Saneyasu, T., Sugahara, K., Hasegawa, S., 2007b. Central administration of insulin suppresses food intake in chicks. Neurosci. Lett. 423, 153–157.

Horev, G., Einat, P., Aharoni, T., Eshdat, Y., Friedman-Einat, M., 2000. Molecular cloning and properties of the chicken leptin-receptor (CLEPR) gene. Mol. Cell. Endocrinol. 162, 95–106.

Hussain, S.S., Bloom, S.R., 2013. The regulation of food intake by the gut-brain axis: implications for obesity. Int. J. Obes. 37, 625–633.

Jonaidi, H., Babapour, V., Denbow, D.M., 2002. GABAergic control of food intake in meat-type chickens. Physiol. Behav. 76, 465–468.

Jonaidi, H., Noori, Z., 2012. Neuropeptide Y-induced feeding is dependent on GABAA receptors in neonatal chicks. J. Comp. Physiol. A 198, 827–832.

Kaiya, H., Saito, E.S., Tachibana, T., Furuse, M., Kangawa, K., 2007. Changes in ghrelin levels of plasma and proventriculus and ghrelin mRNA of proventriculus in fasted and refed layer chicks. Domest. Anim. Endocrinol. 32, 247–259.

Katayama, S., Tomonaga, S., Sato, M., Yamane, H., Tsuneyoshi, Y., Denbow, D.M., Furuse, M., 2010. Norepinephrine does not alter NPY and POMC mRNA expression in neonatal chicks. Comp. Biochem. Physiol. A 156, 143–146.

Kawakami, S.-I., Bungo, T., Ando, R., Ohgushi, A., Shimojo, M., Masuda, Y., Furuse, M., 2000. Central administration of α-melanocyte stimulating hormone inhibits fasting- and neuropeptide Y-induced feeding in neonatal chicks. Eur. J. Pharmacol. 398, 361–364.

Khan, M.S.I., Nakano, Y., Tachibana, T., Ueda, H., 2008. Nitric oxide synthase inhibitor attenuates the anorexigenic effect of corticotropin-releasing hormone in neonatal chicks. Comp. Biochem. Physiol. A Mol. Integr. Physiol. 149, 325–329.

Khan, M.S.I., Ohkubo, T., Masuda, N., Tachibana, T., Ueda, H., 2009. Central administration of metastin increases food intake through opioid neurons in chicks. Comp. Biochem. Physiol. A Mol. Integr. Physiol. 153, 209–212.

Khan, M.S.I., Tachibana, T., Hasebe, Y., Masuda, N., Ueda, H., 2007. Peripheral or central administration of nitric oxide synthase inhibitor affects feeding behavior in chicks. Comp. Biochem. Physiol. A 148, 458–462.

Kuenzel, W.J., 1994. Central neuroanatomical systems involved in the regulation of food intake in birds and mammals. J. Nutr. 124, 1355S–1370S.

Kuenzel, W.J., Beck, M.M., Teruyama, R., 1999. Neural sites and pathways regulating food intake in birds: a comparative analysis to mammalian systems. J. Exp. Zool. 283, 348–364.

Kuenzel, W.J., Douglass, L.W., Davison, B.A., 1987. Robust feeding following central administration of neuropeptide Y or peptide YY in chicks, *Gallus domesticus*. Peptides 8, 823–828.

Kuenzel, M.J., McMurtry, J., 1988. Neuropeptide Y: Brain localization and central effects on plasma insulin levels in chicks. Physiol. Behav. 44, 669–678.

Kuo, A.Y., Lee, J.C., Siegel, P.B., Denbow, D.M., 2001. Differential cardiovascular effects of pharmacological agents in chickens selected for high and low body weight. Physiol. Behav. 74, 573–579.

Lawrence, C.B., Celsi, F., Brennand, J., Luckman, S.M., 2000. Alternative role for prolactin-releasing peptide in the regulation of food intake. Nat. Neurosci. 3, 645–646.

Larhammar, D., 1996. Evolution of neuropeptide Y, peptide YY and pancreatic polypeptide. Regul. Pept. 62, 1–11.

Lawrence, C.B., Ellacott, K.L., Luckman, S.M., 2002. PRL-releasing peptide reduces food intake and may mediate satiety signaling. Endocrinology 143, 360–367.

Layne, J.L., Hunt, K.E., True, T., Gill, R.S., Combos, R.E., Cline, M.A., 2009. Central calcitonin exerts anorectic effects via the hypothalamus in chicks. Pharmacol. Biochem. Behav. 92, 433–438.

Ling, M.K., Hotta, E., Kilianova, Z., Haitina, T., Ringholm, A., Johansson, L., Gallo-Payet, N., Takeuchi, S., Schiöth, H.B., 2004. The melanocortin receptor subtypes in chicken have high preference to ACTH-derived peptides. Br. J. Pharmacol. 143, 626–637.

Marquez, G.C., Siegel, P.B., Lewis, R.M., 2010. Genetic diversity and population structure in lines of chickens divergently selected for high and low 8-week body weight. Poult. Sci. 89, 2580–2588.

McCormack, J.F., Denbow, D.M., 1989. Ingestive responses to mu and delta opioid receptor agonists in the domestic fowl. Br. Poult. Sci. 30, 343–356.

Meade, S., Denbow, D.M., 2001. Feeding, drinking and temperature responses of chickens to intracerebroventricular histamine. Physiol. Behav. 73, 65–73.

Meade, S., Denbow, D.M., 2003. The interaction of bombesin and corticotropin-releasing hormone on ingestive behavior in the domestic fowl. Physiol. Behav. 78, 611–614.

Mercer, R.E., Chee, M.J., Colmers, W.F., 2011. The role of NPY in hypothalamic mediated food intake. Front. Neuroendocrinol. 32, 398–415.

Miller, R.A., Riddle, O., 1943. Effects of prolactin and corticohormones on body weight and food intake on adrenalectomized pigeons. Exp. Biol. Med. (Maywood) 52, 231–233.

Noble, D.O., Picard, M.L., Dunnington, E.A., Uzu, G., Larsen, A.S., Siegel, P.B., 1993. Food intake adjustments of chicks: short term reactions of genetic stocks to deficiencies in lysine, methionine or tryptophan. Br. Poult. Sci. 34, 725–735.

Novoseletsky, N., Nussinovitch, A., Friedman-Einat, M., 2011. Attenuation of food intake in chicks by an inverse agonist of cannabinoid receptor 1 administered by either injection or ingestion in hydrocolloid carriers. Gen. Comp. Endocrinol. 170, 522–527.

Pagotto, U., Marsicano, G., Cota, D., Lutz, B., Pasquali, R., 2006. The emerging role of the endocannabinoid system in endocrine regulation and energy balance. Endocr. Rev. 27, 73–100.

Phillips-Singh, D., Li, Q., Takeuchi, S., Ohkubo, T., Sharp, P.J., Boswell, T., 2003. Fasting differentially regulates expression of agouti-related peptide, pro-opiomelanocortin, prepro-orexin, and vasoactive intestinal polypeptide mRNAs in the hypothalamus of Japanese quail. Cell Tissue Res. 313, 217–225.

Picard, M.I., Uzu, G., Dunnington, E.A., Siegel, P.B., 1993. Food intake adjustments of chicks: short term reactions to deficiencies in lysine, methionine and tryptophan. Br. Poult. Sci. 34, 737–746.

Polidori, C., de Caro, G., Massi, M., 2000. The hyperphagic effect of nociceptin/orphanin FQ in rats. Peptides 21, 1051–1062.

Richards, M.P., McMurtry, J.P., 2010. The avian proghrelin system. Int. J. Pept. 2010, 1–14.

Richards, M.P., Proszkowiec-Weglarz, M., 2007. Mechanisms regulating feed intake, energy expenditure, and body weight in poultry. Poult. Sci. 86, 1478–1490.

Richardson, R.D., Boswell, T., Rafferty, B.D., Seely, R., Wingfield, J.C., Woods, C., 1995. NPY increases food intake in white-crowned sparrows: effect in short and long photoperiods. Am. J. Physiol. 268, R1418–R1422.

Richardson, R.D., Boswell, T., Weatherford, S.C., Wingfield, J.C., Woods, S.C., 1993. Cholecystokinin octapeptide decreases food intake in white-crowned sparrows. Am. J. Physiol. 264, R852–R856.

Richardson, R.D., Boswell, T., Woods, S.C., Wingfield, J.C., 2000. Intracerebroventricular corticotropin-releasing factor decreases food intake in white-crowned sparrows. Physiol. Behav. 71, 213–216.

Rodríguez-Sinovas, A., Fernández, E., Manteca, X., Fernández, A.G., Goñalons, E., 1997. CCK is involved in both peripheral and central mechanisms controlling food intake in chickens. Am. J. Physiol. 272, R334–R340.

Rubin, C.J., Zody, M.C., Eriksson, J., Meadows, J.R.S., Sherwood, E., Webster, M.T., Jiang, L., Ingman, L., Sharpe, T., Ka, S., Hallbook, F., Besnier, F., Carlborg, O., Bed'hom, B., Tixier-Boichard, B., Jensen, P., Siegel, P., Lindblad-Toh, K., Andersson, L., 2010. Whole-genome resequencing reveals loci under selection during chicken domestication. Nature 464, 587–591.

Saito, E.-S., Kaiya, H., Tachibana, T., Tomonaga, S., Denbow, D.M., Kangawa, K., Furuse, M., 2005. Inhibitory effect of ghrelin on food intake is mediated by the corticotropin-releasing factor system in neonatal chicks. Regul. Pept. 125, 201–208.

Sashihara, K., Miyamoto, M., Ohgushi, A., Denbow, D.M., Furuse, M., 2001. Influence of ketone body and the inhibition of fatty acid oxidation on food intake. Br. Poult. Sci. 42, 405–408.

Savory, C.J., 1987. An alternative explanation for apparent satiating properties of peripherally administered bombesin and cholecystokinin in domestic fowls. Physiol. Behav. 39, 191–202.

Seoane, L.M., Al-Massadi, O., Pazos, Y., Pagotto, U., Casanueva, F.F., 2006. Central obestatin administration does not modify either spontaneous or ghrelin-induced food intake in rats. J. Endocrinol. Invest. 29, RC13–RC15.

Shiraishi, J.I., Yanagita, K., Fujita, M., Bungo, T., 2008a. Central insulin suppresses feeding behavior via melanocortins in chicks. Domest. Anim. Endocrinol. 34, 223–228.

Shiraishi, J.I., Yanagita, K., Fujita, M., Bungo, T., 2008b. μ-Opioid receptor agonist diminishes POMC gene expression and anorexia by central insulin in neonatal chicks. Neurosci. Lett. 439, 227–229.

Shiraishi, J.I., Yanagita, K., Nishikawa, F., Tahara, Y., Fujita, M., McMurtry, J.P., Bungo, T., 2009. A comparison of the anorexic effects of chicken, porcine, human and bovine insulin on the central nervous system of chicks. Poult. Sci. 46, 144–148.

Shousha, S., Nakahara, K., Kojima, M., Miyazato, M., Hosoda, H., Kangawa, K., Murakami, N., 2005a. Different effects of peripheral and central ghrelin on regulation of food intake in the Japanese quail. Gen. Comp. Endocrinol. 141, 178–183.

Shousha, S., Nakahara, K., Miyazato, M., Kangawa, K., Murakami, N., 2005b. Endogenous neuromedin U has anorectic effects in the Japanese quail. Gen. Comp. Endocrinol. 140, 156–163.

Shousha, S., Nakahara, K., Nasub, T., Sakamoto, T., Murakami, N., 2007. Effect of glucagon-like peptide-1 and -2 on regulation of food intake, body temperature and locomotor activity in the Japanese quail. Neurosci. Lett. 415, 102–107.

Siegel, P.B., 1962. Selection for body weight at eight weeks of age. Poult. Sci. 41, 954–962.

Smith, M.L., Prall, B., Nandar, W., Cline, M.A., 2008. Beta-melanocyte-stimulating hormone potently reduces appetite via the hypothalamus in chicks. J. Neuroendocrinol. 20, 220–226.

Smith, M.L., Prall, B.P., Siegel, P.B., Cline, M.A., 2011. The threshold of insulin-induced hypophagia is lower in chicks selected for low rather than high juvenile body weight. Behav. Brain Res. 216, 719–722.

Song, Z., Everaert, N., Wang, Y., Decuypere, E., Buyse, J. 2013. The endocrine control of energy homeostasis in chickens. Gen. Comp. Gen. Comp. Endocrinol., 190, 112–117.

Song, Z., Liu, L., Yue, Y., Jiao, H., Lin, H., Sheikhahmadi, A., Everaert, N., Edcuypere, E., Buyse, J., 2012. Fasting alters protein expression of AMP-activated protein kinase in the hypothalamus of broiler chicks (*Gallus gallus domesticus*). Gen. Comp. Endocrinol. 178, 546–555.

Strader, A.D., Buntin, J.D., 2001. Neuropeptide-Y: a possible mediator of prolactin-induced feeding and regulator of energy balance in the ring dove (*Streptopelia risoria*). J. Neuroendocrinol. 13, 386–392.

Taati, M., Nayebzadeh, H., Khosravinia, H., Cheraghi, J., 2010. The role of the histaminergic system on the inhibitory effect of ghrelin on feed intake in broiler chickens. Iran. J. Vet. Res. 11, 38–45.

Tachibana, T., Sugahara, K., Ohgushi, A., Ando, R., Kawakami, S.-I., Yoshimatsu, T., Furuse, M., 2001. Intracerebroventricular injection of agouti-related protein attenuates the anorexigenic effect of alpha-melanocyte stimulating hormone in neonatal chicks. Neuro Let 305, 131–134.

Tachibana, T., Cline, M.A., Sakirul, M.D., Khan, I., Ueda, H., Hiramatsu, H., 2010. Feeding responses to central administration of several somatostatin analogs in chicks. Comp. Biochem. Physiol. A Mol. Integr. Physiol. 158, 47–51.

Tachibana, T., Cline, M.A., Sugahara, K., Ueda, H., Hiramatsu, K., 2009a. Central administration of somatostatin stimulates feeding behavior in chicks. Gen. Comp. Endocrinol. 161, 354–359.

Tachibana, T., Sugahara, K., Ueda, H., Cline, M.A., 2009b. Role of adrenergic alpha-2-receptors on feeding behavior in layer-type chicks. Gen. Comp. Endocrinol. 161, 407–411.

Tachibana, T., Masuda, N., Tsutsui, K., Ukena, K., Ueda, H., 2008. The orexigenic effect of GnIH is mediated by central opioid receptors in chicks. Comp. Biochem. Physiol. A Mol. Integr. Physiol. 150, 21–25.

Tachibana, T., Saito, E.-S., Takahashi, H., Saito, S., Takagi, T., Boswell, T., Furuse, M., 2004a. Anorexigenic effect of pituitary adenylate cyclase-activating polypeptide and vasoactive intestinal peptide in the chick brain are mediated by corticotrophin-releasing factor. Regul. Pept. 120, 99–105.

Tachibana, T., Saito, S., Tomonaga, S., Takagi, T., Saito, E.-S., Nakanishi, T., Koutoku, T., Tsukada, A., Ohkubo, T., Boswell, T., Furuse, M., 2004b. Effect of central administration of prolactin-releasing peptide on feeding in chicks. Physiol. Behav. 80, 713–719.

Tachibana, T., Sato, M., Takahashi, H., Ukena, K., Tsutsui, K., Furuse, M., 2005a. Gonadotropin-inhibiting hormone stimulates feeding behavior in chicks. Brain Res. 1050, 94–100.

Tachibana, T., Tsukada, A., Fujimoto, M., Takahashi, H., Ohkubo, T., Boswell, T., Furuse, M., 2005b. Comparison of mammalian prolactin-releasing peptide and Carassius RFamide for feeding behavior and prolactin secretion in chicks. Gen. Comp. Endocrinol. 144, 264–269.

Tachibana, T., Takagi, T., Saito, E.-S., Tomonaga, S., Zhang, R., Koga, Y., Kido, Y., Denbow, D.M., Furuse, M., 2003a. Beta 3-adrenergic receptor is involved in feeding regulation in chicks. Comp. Biochem. Physiol. A Mol. Integr. Physiol. 135A, 403–409.

Tachibana, T., Takagi, T., Tomonaga, S., Ohgushi, A., Ando, R., Denbow, D.M., Furuse, M., 2003b. Central administration of cocaine- and amphetamine-regulated transcript inhibits food intake in chicks. Neurosci. Lett. 337, 131–134.

Tatemoto, K., Carlquist, M., Mutt, V., 1982. Neuropeptide Y – a novel brain peptide with structural similarities to peptide YY and pancreatic polypeptide. Nature 296, 659–660.

Tebbe, J.J., Mronga, S., Schafer, M.K., Ruter, J., Kobelt, P., Monnikes, H., 2003. Stimulation of neurons in rat ARC inhibits gastric acid secretion via hypothalamic CRF1/2- and NPY-Y1 receptors. Am. J. Physiol. Gastr. Liver Physiol. 285, G1075–1083.

Tsutsui, K., Saigoh, E., Ukena, K., Teranishi, H., Fujisawa, Y., Kikuchi, M., Ishii, S., Sharp, P.J., 2000. A novel avian hypothalamic peptide inhibiting gonadotropin release. Biochem. Biophys. Res. Commun. 275, 661–667.

Ukena, K., Tachibana, T., Iwakoshi-Ukena, E., Saito, Y., Minakata, H., Kawaguchi, R., Osugi, T., Tobari, Y., Leprince, J., Vaudry, H., Tsutsui, K., 2010. Identification, localization, and function of a novel avian hypothalamic neuropeptide, 26RFa, and its cognate receptor, G protein-coupled receptor-103. Endocrinology 151, 2255–2264.

Wang, X., Day, J.R., Vasilatos-Younken, R., 2001. The distribution of neuropeptide Y gene expression in the chicken brain. Mol. Cell. Endocrinol. 174, 129–136.

Wolfgang, M.J., Jurama, T., Dai, Y., Suwa, A., Asaumi, M., Matsumoto, S.-I., Cha, S.H., Shimokawa, T., Lane, M.D., 2006. The brain carnitine palmitoyltranserase-1c regulates energy homeostasis. Proc. Natl. Acad. Sci. U.S.A. 103, 7282–7287.

Xu, P., Siegel, P.B., Denbow, D.M., 2011a. Genetic selection for body weight in chickens has altered responses of the brain's AMPK system to food intake regulation effect of ghrelin, but not obestatin. Behav. Brain Res. 221, 216–226.

Zhang, Y., Proenca, R., Maffei, M., Barone, M., Leopold, L., Friedman, J.M., 1994. Positional cloning on the mouse obese gene and its human homologue. Nature 374, 425–432.

Zhao, R., Muehlbauer, E., Decuypere, E., Grossmann, R., 2004. Effect of genotype-nutrition on growth and somatotripic gene expression in the chicken. Gen. Comp. Endocrinol. 136, 2–11.

Zelenka, D.J., Dunnington, E.A., Cherry, J.A., Siegel, P.B., 1988. Anorexia and sexual maturity in female white rock chickens. I. Increasing the feed intake. Behav. Genet. 18, 383–387.

Zendehdel, M., Baghbanzadeh, A., Babapour, V., Cheraghi, J., 2009. The effects of bicuculline and muscimol on glutamate-induced feeding behavior in broiler cockerels. Comp. Physiol. A 195, 715–720.

Zeni, L.A., Seidler, H.B., De Carvalho, N.A., Freitas, C.G., Marino-Neto, J., Paschoalini, M.A., 2000. Glutamatergic control of food intake in pigeons: effects of central injections of glutamate, NMDA, and AMPA receptor agonists and antagonists. Pharmacol. Biochem. Behav. 65, 67–74.

FURTHER READING

Xu, P., Siegel, P.B., Denbow, D.M., 2011b. AICAR and compound C regulate food intake independently of AMP-activated protein kinase in lines of chickens selected for high or low body weight. Comp. Biochem. Physiol. A 159, 401–412.

Part V

Endocrine Theme

Chapter 22

Avian Endocrine System

Colin G. Scanes
Department of Biological Sciences, University of Wisconsin, Milwaukee, WI, USA

22.1 INTRODUCTION

Hormones play a critical role in controlling many aspects of the physiology of birds, including reproduction, metabolism, and growth. Hormones are produced from ductless or endocrine glands. They are then released into the bloodstream. The hormones then act on specific tissues, cells, or organs (referred to as the target organ). Each hormone acts by interacting with receptors on the surface of the cells (proteins and polypeptides) or within the cytoplasm and nucleus (steroids and thyroid hormones). Birds have a complement of endocrine organs similar to those of mammals:

- Anterior pituitary gland (discussed elsewhere in this volume in Chapter on the pituitary gland)
- Hypothalamus posterior pituitary gland complex (also discussed in Chapter on the pituitary gland)
- Gonads (discussed in Chapters on male and female reproduction)
- Pancreatic islets (discussed in Chapter on the endocrine pancreas)
- Adrenal glands (discussed in Chapter on the adrenal glands)
- Thyroid glands (discussed in Chapter on the thyroid glands)
- Parathyroid glands (discussed in Chapter on the role of hormones in the regulation of bone turnover and eggshell calcification)
- Pineal body or gland (discussed in Chapter on the pineal gland)
- Ultimobranchial bodies or glands (C cells producing calcitonin). The ultimobranchial glands are found during development in mammals, with the C cells then becoming distributed in the thyroid and parathyroid glands. (This is also discussed in Chapter on the role of hormones in the regulation of bone turnover and eggshell calcification.)
- Endocrine cells of the gut (discussed in Chapter on the gastrointestinal anatomy and physiology).

The traditional view of a distinct endocrine system, with hormones only produced by specific endocrine glands that do little else, is overly restrictive. There is production of biologically active proteins and peptides by many, if not all, tissues. Organs that were thought of as simply target organs for hormones also produce hormones such as the following:

- Adipose tissue producing adiponectin and leptin (discussed in Chapter on adipose tissue)
- Heart cells producing natriuretic peptides
- Liver producing insulin-like growth factor-1 (IGF1)
- Kidney producing renin (and hence angiotensin-I, then angiotensin-II), 1,25-dihydroxy vitamin D3, and erythropoietin (discussed in Chapters on blood and osmoregulation)

22.2 AVIAN PHYLOGENY

Caution should be expressed about extrapolating studies for poultry species such as chickens to other birds due to the multiple generations of selection, with chickens domesticated about 10,000 years ago (Sawai et al., 2010). There are differences between the present poultry and the ancestral stock. For instance, the thyroid-stimulating hormone (TSH) receptor gene exhibits the same mutation throughout multiple lines of domesticated chickens but not in jungle fowl (Rubin et al., 2010). Moreover, most of the coding sequence of SH3RF2 is deleted (Rubin et al., 2010).

Even without domestication, it is argued that neither galliform (e.g., chickens, turkeys, and quail) nor anseriform birds (ducks and geese) are necessarily good models for other avian species. Within the class Aves (birds) and Neoornithes (modern birds, including all extant species), there are two branches:

1. *Palaeognathae* (including emus, ostriches, and rhea)
2. *Neognathae*, consisting of two distinct groups:
   a. *Galloanserae*, consisting of the orders *Anseriformes* and *Galliformes*
   b. *Neoaves* (most modern species of birds) (see the Tree of Life project, http://tolweb.org; and see Hackett et al., 2008; Jetz et al., 2012; McCormack et al., 2013)

Sturkie's Avian Physiology.

The two branches, Palaeognathae and Neognathae, diverged about 140 million years ago (Sanders and Lee, 2007), while the Galloanserae and Neoaves separated about 95 million years ago (Ericson et al., 2006). Within the *Neoaves*, there are three clades:

- Water-birds
- Land-birds
- A third clade containing bustards (family: Otididae), cuckoos (family: Cuculidae), flamingos (family: Phoenicopteridae), hummingbirds (family: Trochilidae), trumpeters (family: Psophiidae), and turacos (family: Musophagidae), exhibiting marked apparent diversity

Pigeons and doves (family: *Columbidae*) form a distinct side group within the *Neoaves* (Hackett et al., 2008; McCormack et al., 2013).

22.3 PEPTIDES AND OTHER CHEMICAL MESSENGERS CONTROLLING PHYSIOLOGY

The same chemical entity may act in one or more of the following manners:

- As a *hormone* traveling from endocrine cells to target cells via the bloodstream
- As a *paracrine factor* released from one cell type to affect another cell type passing through the extracellular fluid
- As an *autocrine factor* released from one cell type to affect another cell of the same type passing through the extracellular fluid
- As an *intracrine* factor acting with a cell type to affect another cell
- As a *neurotransmitter* or *neuromodulator* released from a neuron to act with another neuron
- As a *neuropeptide* released from a neuron to act with other cell types in close proximity
- As a *neuroendocrine factor* released from neurosecretory terminals of a neuron into the blood and then acting with other cell types

Many of these chemical messengers are neuropeptides. Table 22.1 summarizes a list of neuropeptides and protein–polypeptide hormones found in avian species. This is based on the genomics of both the chicken and zebra finch (*Taeniopygia guttata*) (Delfino et al. 2010; Xie et al., 2010). Table 22.2 lists neuropeptides found in other vertebrate groups but not in birds.

Throughout vertebrate evolution, there are multiple examples of gene duplication leading to new chemical signal peptides and proteins or to new receptors. There are also cases of gene loss. For instance, prolactin-like protein (PLP) is a gene found in both fish and birds but not mammals (Wang et al., 2010). Presumably, the gene was retained in the amphibian and reptilian ancestors of birds but lost in the ancestors of mammals.

Our ability to understand the physiology of birds is being facilitated with the tremendous progress in genomics (discussed in detail in Chapter on genomics). It should be noted that there can also be polymorphisms at the gene level, splicing differences of RNA, and posttranslational differences, including proteolytic cleavage (e.g., with the translated products of the POMC and glucagon genes) and glycosylation (e.g., with luteinizing hormone (LH), follicle-stimulating hormone (FSH), and TSH).

22.4 CHEMICAL MESSENGERS FOUND IN BIRDS BUT NOT MAMMALS

There are avian neuropeptides and their respective genes that are not present in mammals. These include corticotropin-releasing factor (CRF) amide, c-type natriuretic peptide 1 precursor, renal natriuretic peptide (Delfino et al., 2010), and PLP (Wang et al., 2010).

22.5 HORMONES PRODUCED BY NONTRADITIONAL ENDOCRINE ORGANS

22.5.1 Adiponectin

A nice example of a protein produced by organs not traditionally thought of as endocrine organs is the putative hormone adiponectin. This is expressed in multiple tissues of the chicken, including the adipose tissue (highest), liver (next highest), anterior pituitary, brain, kidney, and skeletal muscle (Maddineni et al., 2005). In addition, these tissues also express the adiponectin receptors, AdipoR1 and AdipoR2, with the highest expression of AdipoR1 being in skeletal muscle, adipose tissue, and brain and with some expression in anterior pituitary gland, kidney, liver, ovary, and spleen, and with the highest expression of AdipoR2 being in adipose tissue followed by skeletal muscle and some expression also in anterior pituitary gland, brain, kidney, liver, ovary, and spleen (Ramachandran et al., 2007). It can readily be envisioned that adiponectin may act as a hormone and as a paracrine factor.

22.6 UNIQUE ASPECTS OF BIRDS

22.6.1 Song

Passerine birds have a unique ability for song. This involves complex controls from the brain. There has been rapid molecular evolution of the song-related genes in passerine birds (Warren et al., 2010).

22.6.2 Salt Glands

Salt glands allow some birds to prosper in marine environments by secreting a hypertonic solution, predominantly sodium chloride (discussed in detail in Chapter on osmoregulation). The salt glands respond to hormones. Angiotensin

TABLE 22.1 Neuropeptides and Peptide–Protein Hormone in Birds

Pre-proneuropeptide or Neuropeptide	Official Gene Symbol	Present in Chicken Genome?[1]	Present in Zebra Finch Genome?[2]	Other, Avian-Specific Notes
Adenylate cyclase activating polypeptide 1 (pituitary)	Adcyap1	√	√	Expressed (e.g., in the brain and gastrointestinal (GI) tract)
Adrenomedullin	AMD	√	√	Expressed in lungs, adrenal glands, skeletal muscle, and GI tract (Zudaire et al., 2005) Expression increased in lungs of hypotensive chicken (Gomez et al., 2007)
Arginine vasopressin	AVP	√	√	Neuropeptide in birds arginine vasotocin (AVT)
C-RF amide peptide	LOC420716/ CRF	√	√	
C-type natriuretic peptide 1	CNP 1	√	√	Expressed in the chicken—discussed in narrative
C-type natriuretic peptide 3	CNP 2	√	√	
Calcitonin-related Polypeptide alpha [also called calcitonin gene-related peptide (CGRP)] Gene encodes calcitonin also	CALCA	√	√	
Cocaine- and amphetamine-regulated transcript protein	CART	X	√	CNS effect in chickens (Tachibana et al., 2003)
Corticotropin-releasing hormone (also referred to as corticoliberin or corticotropin-releasing factor (CRF))	CRH	√	√	Expressed in hypothalamus
Cholecystokinin	CCK	√	√	
Chromogranin A (parathyroid secretory protein 1)	CHGA	√	√	
Chromogranin B	CHGB	√	√	
Chromosome 12 open reading frame 39 (spexin)	C12orf39	√	√	
Chromosome 2 open reading frame 40	C2orf40	√	√	
Endothelin 1	EDN1	√	√	Expression increased in lungs of hypotensive chicken (Gomez et al., 2007)
Endothelin 2	EDN2	√	√	
Endothelin 3	EDN3	√	√	
Follicle stimulating hormone beta subunit	FSHB	√	√	Expressed in adenohypophyseal cells
Galanin pre-propeptide	GAL	√	√	Expressed in CNS and GI tract

Continued

TABLE 22.1 Neuropeptides and Peptide–Protein Hormone in Birds—cont'd

Pre-proneuropeptide or Neuropeptide	Official Gene Symbol	Present in Chicken Genome?[1]	Present in Zebra Finch Genome?[2]	Other, Avian-Specific Notes
Gastric-inhibitory polypeptide (GIP)	GIP	√	√	Expressed in CNS and GI tract
Gastrin-releasing peptide (GRP)	GRP	√	√	Expressed in GI tract and hypothalamus
Ghrelin or obestatin pre-propeptide	GHRL	√	√	Expressed in GI tract and hypothalamus
Glucagon pro-hormone also containing glucagon-like peptide 1 (GLP1) and GLP2	GCG	√	√	Expressed in pancreas, GI tract, and hypothalamus
Glycoprotein hormone, α polypeptide	CGA	√	√	Expressed in adenohypophyseal cells
Gonadotropin-releasing hormone (GnRHI) (formerly luteinizing hormone–releasing hormone I (LHRHI))	GnRH	√	√	Expressed in hypothalamus and second gene encoding GnRHII in birds
Growth hormone	GH	√	√	Expressed in adenohypophyseal cells and CNS Duplicate GH gene on chromosome 1 on passerine birds Expression of duplicate gene not well established
Growth hormone–releasing hormone (GHRH), also called somatoliberin	GHRH	√	√	Expressed in hypothalamus
Hypocretin (orexin) neuropeptide precursor	HCRT	√	√	Expressed in CNS
Insulin	INS	√	√	Expressed in pancreas
Insulin-like 5	INSL5	X	√	
Insulin like growth factor 1	IGF1	√	√	Expressed in liver and cartilage
Insulin like growth factor 2	IGF2	√	√	
Islet amyloid polypeptide	IAPP	√	√	Functionally similar to calcitonin-related polypeptide alpha; expressed in CNS, GI tract, and pancreas (Fan et al., 1994)
Kisspeptin	KISS1	X	X	Effects of kisspeptide 10 are reported in the chicken ovary (Xiao et al., 2011).
Leptin or leptin-like	OB	√?	√	See narrative in text for details
Luteinizing hormone beta subunit	LHB	√	√	Expressed in adenohypophyseal cells

TABLE 22.1 Neuropeptides and Peptide–Protein Hormone in Birds—cont'd

Pre-proneuropeptide or Neuropeptide	Official Gene Symbol	Present in Chicken Genome?[1]	Present in Zebra Finch Genome?[2]	Other, Avian-Specific Notes
Mesotocin–neurophysin 1	MST	√	√	Equivalent to oxytocin in mammals Expressed in CNS
Motilin	MLN	√	√	Expressed in CNS and GI tract
Natriuretic peptide precursor A	NPPA	√	√	
Neuromedin B	NMB	√	√	
Neuromedin U	NMU	√	√	
Neuropeptide S	NPS	√	√	
Neuropeptide VF (gonadotropin-inhibitory hormone (GnIH))	NPVF	√	√	
Neuropeptide W	NPW	X	√	
Neuropeptide Y	NPY	√	√	
Neurotensin	NTS	√	√	Expressed in CNS and GI tract
Osteocrin	OSTN	√	√	
Pancreatic polypeptide	PPY (or PP)	√	√	Expressed in pancreas and GI tract
Parathyroid hormone	PTH	√	√	Expressed in parathyroid gland
Parathyroid hormone–like hormone (also called PTH-related peptide (PTHrP))	PTHLH	√	√	Expressed in early embryonic development skeletal tissues For expression, also see Chapter on the pituitary gland
Pituitary adenylate cyclase–activating polypeptide 1	See "adenylate cyclase–activating polypeptide"			
Platelet-derived growth factor D	PDGFD			
Platelet-derived growth factor alpha polypeptide	PDGFA			
Platelet-derived growth factor beta polypeptide	PDGFB			
Pre-pronociceptin	PNOC			
Pro-melanin-concentrating hormone (MCH)	PMCH			
Prodynorphin	PDYN			
Proenkephalin	PENK			CNS
Prokineticin 2	PROK			
Prolactin	PRL	√	√	Adenohypophyseal cells
Prolactin B (also called prolactin-like protein (PLP))	PRLB	√	√	Reported in birds and fish but not mammals

Continued

TABLE 22.1 Neuropeptides and Peptide–Protein Hormone in Birds—cont'd

Pre-proneuropeptide or Neuropeptide	Official Gene Symbol	Present in Chicken Genome?[1]	Present in Zebra Finch Genome?[2]	Other, Avian-Specific Notes
Prolactin-releasing hormone	PRH	√	√	CNS
Pro-opiomelanocortin	POMC	√	√	CNS and adenohypophyseal cells
Pyroglutamylated RF amide				
Relaxin 3				
Secretin				CNS and GI tract
Secretogranin II (chromogranin C)				
Secretogranin V (7B2 protein)				
Somatostatin	SST1	√	√	CNS, GI tract, and pancreas
Somatostatin 2	SST2	√	√	
Tachykinin, precursor 1	TAC1	√	√	
Thyroid-stimulating hormone (or thyrotropin) beta subunit	TSHB	√	√	Adenohypophyseal cells
Thyrotropin-releasing hormone (also called thyroliberin)	TRH	√	√	CNS
Urocortin	UCN	√	√	
Urocortin 3	UCN3	√	√	
Urotensin 2	UTS2			
Urotensin 2 domain-containing	UTS2D			
Vascular endothelial growth factor C	VEGFC			
Vasoactive intestinal peptide	VIP			CNS and GI tract

[1]*Based on Delfino et al., 2010.*
[2]*Based on Xie et al., 2010.*

II directly decreases fluid flow from the duck salt gland (Butler, 2007). This effect is independent both of the adrenal gland and of catecholamines as angiotensin II is effective in adrenalectomized and chemically sympathectomized birds (Butler, 2007). Natriuretic peptides influence the functioning of the salt glands (Schütz and Gerstberger, 1990).

22.6.3 Unique Aspects of Metabolism

There are marked differences in circulating concentrations of glucose between wild birds and mammals, with avian concentrations double those in mammals and with physically smaller species of birds or mammals having higher circulating concentrations of glucose than larger ones (Braun and Sweazea, 2008) (discussed in detail in Chapters on carbohydrate metabolism and the pancreas). Moreover, birds do exhibit diabetes, with circulating concentrations of glucose held at a consistent level, except under one paradigm. Chickens that received passive immunization with antisera to insulin exhibit diabetic-like shifts in metabolism, with circulating concentrations of glucose of 747 mg/dL reported 5 h after antisera administration (Dupont et al., 2008; Simon et al., 2012). Interestingly, identical extremely high concentrations of glucose are observed following feeding in hummingbirds without adverse effects (Beuchat and Chong, 1998).

Birds also have high metabolic demands for both short-term flight and extended migratory flight (discussed in detail in Chapter on migration).

22.6.4 Reproduction

Birds exhibit marked differences from mammals in terms of reproduction (discussed elsewhere in this volume in Chapters on male and female reproduction). These differences include the internal testes, the production of large yolky eggs, the absence of pregnancy, the existence of only one ovary and oviduct (the Müllerian duct), and unique sexual, brooding, and parental behavior (discussed in Chapter on brooding).

TABLE 22.2 Neuropeptides Presently Not Demonstrated to Exist in Birds (Based on Delfino et al., 2010)

Apelin (APEL)	Proenkephalin-B (PDYN)
Cortistatin (CORT)[1]	Peptide YY (PYY)
Galanin-like peptide (GALP)	Putative peptide YY-2 (PYY2)
Hepcidin (HEPC)	Pro-relaxin 1 (REL1)
Insulin-like 3 (INSL3)	Pro-relaxin 2 (REL2)
Insulin-like 4 (INSL4)	Regulated endocrine-specific protein 18 (RES18)
Intermedin (ADM2)	Spexin (SPXN)
Metastasis-suppressor KiSS-1 (KISS1)[2]	Tachykinin, precursor 2 (TKN2)
Natriuretic factor B (ANFB)	Tachykinin, precursor 3 (TKN3)
Neuromedin-S (NMS)	Tachykinin, precursor 4 (TKN4)
Neuropeptide B (NPB)	Torsin family 2, member A (TOR2X)
Neuropeptide FF (NPFF)	
Neuropeptide S (NPS)	
Proprotein convertase subtilisin/kexin type 1 inhibitor (PCSK1N)	

[1]*Expression of cortistatin gene reported in chicken autonomic neurons (Nishi et al., 2010).*
[2]*Effects of kisspeptide 10 are reported on the chicken ovary (Xiao et al., 2011).*

22.6.5 Opportunities for Transgenic Poultry

Transgenic chickens have been produced that express human erythropoietin (hEPO) (Koo et al., 2010). Expression of hEPO is controlled by a tetracycline-inducible promoter (Koo et al., 2010), hence eliminating the potential of pathological effects of hEPO in the chickens had there been uncontrolled constitutive expression. Immunoreactive erythropoietin has been reported in birds (Wickramasinghe et al., 1994).

22.7 THE ENIGMA OF LEPTIN

An example of a hormone produced by adipose tissue in mammals is leptin. There is contradictory evidence for and against the existence of a chicken leptin (Taouis et al., 1998; Sharp et al., 2008; Simon et al., 2009). A leptin-like gene has been identified in the zebra finch (*T. guttata*) (leptin-like Gene ID: 101233729).

There is complete agreement in the literature that mammalian leptin exerts multiple effects in birds. The short form of the chicken leptin receptor has been characterized from the cDNA (Liu et al., 2007). The leptin receptor is present in multiple tissues, including the brain and liver in the chicken, based on expression (Liu et al., 2007) and the observed immunoreactive 180kDa protein in multiple tissues (Ohkubo et al., 2007). The leptin receptor is expressed in the anterior pituitary gland and basal hypothalamus of the chicken (Liu and Sharp, 2007). The leptin receptor has been characterized also in the turkey (Richards and Poch, 2003) and goose (Wang et al., 2011). The leptin receptor is expressed in the brain, lung, and spleen, with some expression also in the adipose tissue, duodenum, liver, pancreas, and skeletal muscle (Richards and Poch, 2003; Wang et al., 2011). Adipose expression of the leptin receptor is increased by oleic acid *in vitro* (Wang et al., 2011). *In ovo* administration of mammalian leptin into the albumin of chick embryos was reported to result in increased posthatching growth and circulating concentrations of triiodothyrone (T_3), with greater hepatic expression of growth hormone receptor and IGF1 mRNA (Li et al., 2011). *In vitro* mammalian leptin elevates expression and activities of deiodinase in chick embryo hepatocytes (Li et al., 2011). Administration of mammalian leptin decreases food intake and fat mass in wintering white-throated sparrows (Cerasale et al., 2011). The gene for the leptin receptor has been identified in the zebra finch (*T. guttata*) (leptin receptor (LEPR) Gene ID: 100229897).

REFERENCES

Beuchat, C.A., Chong, C.R., 1998. Hyperglycemia in hummingbirds and its consequences for hemoglobin glycation. Comp. Biochem. Physiol. A Mol. Integr. Physiol. 120, 409–416.

Braun, E.J., Sweazea, K.L., 2008. Glucose regulation in birds. Comp. Biochem. Physiol. B Biochem. Mol. Biol. 151, 1–9.

Butler, D.G., 2007. ANG II-induced attenuation of salt gland function in Pekin ducks is not catecholamine-dependent. J. Comp. Physiol. B. 177, 733–742.

Cerasale, D.J., Zajac, D.M., Guglielmo, C.G., 2011. Behavioral and physiological effects of photoperiod-induced migratory state and leptin on a migratory bird, *Zonotrichia albicollis*: I. Anorectic effects of leptin administration. Gen. Comp. Endocrinol. 174, 276–286.

Delfino, K.R., Southey, B.R., Sweedler, J.V., Rodriguez-Zas, S.L., 2010. Genome-wide census and expression profiling of chicken neuropeptide and prohormone convertase genes. Neuropeptides 44, 31–44.

Dupont, J., Tesseraud, S., Derouet, M., Collin, A., Rideau, N., Crochet, S., Godet, E., Cailleau-Audouin, E., Métayer-Coustard, S., Duclos, M.J., Gespach, C., Porter, T.E., Cogburn, L.A., Simon, J., 2008. Insulin immuno-neutralization in chicken: effects on insulin signaling and gene expression in liver and muscle. J. Endocrinol. 197, 531–542.

Ericson, P.G.P., Anderson, C.L., Britton, T., Elzanowski, A., Johansson, U.S., Källersjö, M., Ohlson, J.I., Parsons, T.J., Zuccon, D., Mayr, G., 2006. Diversification of Neoaves: integration of molecular sequence data and fossils. Biol. Lett. 4, 543–547.

Fan, L., Westermark, G., Chan, S.J., Steiner, D.F., 1994. Altered gene structure and tissue expression of islet amyloid polypeptide in the chicken. Mol. Endocrinol. 8, 713–721.

Gomez, A.P., Moreno, M.J., Iglesias, A., Coral, P.X., Hernández, A., 2007. Endothelin 1, its endothelin type A receptor, connective tissue growth factor, platelet-derived growth factor, and adrenomedullin expression in lungs of pulmonary hypertensive and nonhypertensive chickens. Poult. Sci. 86, 909–916.

Hackett, S.J., Kimball, R.T., Reddy, S., Bowie, R.C., Braun, E.L., Braun, M.J., Chojnowski, J.L., Cox, W.A., Han, K.L., Harshman, J., Huddleston, C.J., Marks, B.D., Miglia, K.J., Moore, W.S., Sheldon, F.H., Steadman, D.W., Witt, C.C., Yuri, T., 2008. A phylogenomic study of birds reveals their evolutionary history. Science 320, 1763–1768.

Jetz, W., Thomas, G.H., Joy, J.B., Hartmann, K., Mooers, A.O., 2012. The global diversity of birds in space and time. Nature 491, 444–448.

Koo, B.C., Kwon, M.S., Lee, H., Kim, M., Kim, D., Roh, J.Y., Park, Y.Y., Cui, X.S., Kim, N.H., Byun, S.J., Kim, T., 2010. Tetracycline-dependent expression of the human erythropoietin gene in transgenic chickens. Transgenic Res. 19, 437–447.

Li, R., Hu, Y., Ni, Y., Xia, D., Grossmann, R., Zhao, R., 2011. Leptin stimulates hepatic activation of thyroid hormones and promotes early posthatch growth in the chicken. Comp. Biochem. Physiol. A Mol. Integr. Physiol. 160, 200–206.

Liu, X., Sharp, P.J., 2007. Deletions in mRNA encoding the chicken leptin receptor gene binding domain. Comp. Biochem. Physiol. B Biochem. Mol. Biol. 146, 250–255.

Liu, X., Dunn, I.C., Sharp, P.J., Boswell, T., 2007. Molecular cloning and tissue distribution of a short form chicken leptin receptor mRNA. Domest. Anim. Endocrinol. 32, 155–166.

Maddineni, S., Metzger, S., Ocón, O., Hendricks 3rd, G., Ramachandran, R., 2005. Adiponectin gene is expressed in multiple tissues in the chicken: food deprivation influences adiponectin messenger ribonucleic acid expression. Endocrinology 146, 4250–4256.

McCormack, J.E., Harvey, M.G., Faircloth, B.C., Crawford, N.G., Glenn, T.C., Brumfield, R.T., 2013. A phylogeny of birds based on over 1500 loci collected by target enrichment and high-throughput sequencing. PLoS One 8, e54848.

Nishi, R., Stubbusch, J., Hulce, J.J., Hruska, M., Pappas, A., Bravo, M.C., Huber, L.P., Bakondi, B., Soltys, J., Rohrer, H., 2010. The cortistatin gene PSS2 rather than the somatostatin gene PSS1 is strongly expressed in developing avian autonomic neurons. J. Comp. Neurol. 518, 839–850.

Ohkubo, T., Nishio, M., Tsurudome, M., Ito, M., Ito, Y., 2007. Existence of leptin receptor protein in chicken tissues: isolation of a monoclonal antibody against chicken leptin receptor. Gen. Comp. Endocrinol. 151, 269–273.

Ramachandran, R., Ocón-Grove, O.M., Metzger, S.L., 2007. Molecular cloning and tissue expression of chicken AdipoR1 and AdipoR2 complementary deoxyribonucleic acids. Domest. Anim. Endocrinol. 33, 19–31.

Richards, M.P., Poch, P.M., 2003. Molecular cloning and expression of the turkey leptin receptor gene. Comp. Biochem. Physiol. B 136, 833–847.

Rubin, C.J., Zody, M.C., Eriksson, J., Meadows, J.R., Sherwood, E., Webster, M.T., Jiang, L., Ingman, M., Sharpe, T., Ka, S., Hallböök, F., Besnier, F., Carlborg, O., Bed'hom, B., Tixier-Boichard, M., Jensen, P., Siegel, P., Lindblad-Toh, K., Andersson, L., 2010. Whole-genome resequencing reveals loci under selection during chicken domestication. Nature 464, 587–591.

Sanders, K.L., Lee, M.S., 2007. Evaluating molecular clock calibrations using Bayesian analyses with soft and hard bounds. Biol. Lett. 3, 275–279.

Sawai, H., Kim, H.L., Kuno, K., Suzuki, S., Gotoh, H., Takada, M., Takahata, N., Satta, Y., Akishinonomiya, F., 2010. The origin and genetic variation of domestic chickens with special reference to junglefowls *Gallus g. gallus* and *G. varius*. PLoS One 5, e10639.

Schütz, H., Gerstberger, R., 1990. Atrial natriuretic factor controls salt gland secretion in the Pekin duck (*Anas platyrhynchos*) through interaction with high affinity receptors. Endocrinology 127, 1718–1726.

Sharp, P.J., Dunn, I.C., Waddington, D., Boswell, T., 2008. Chicken leptin. Gen. Comp. Endocrinol. 158, 2–4.

Simon, J., Rideau, N., Taouis, M., 2009. Reply to viewpoints by PJ Sharp, IC Dunn, D Waddington and T Boswell. Gen. Comp. Endocrinol. 161, 159.

Simon, J., Milenkovic, D., Godet, E., Cabau, C., Collin, A., Métayer-Coustard, S., Rideau, N., Tesseraud, S., Derouet, M., Crochet, S., Cailleau-Audouin, E., Hennequet-Antier, C., Gespach, C., Porter, T.E., Duclos, M.J., Dupont, J., Cogburn, L.A., 2012. Insulin immunoneutralization in fed chickens: effects on liver and muscle transcriptome. Physiol. Genomics 44, 283–292.

Tachibana, T., Takagi, T., Tomonaga, S., Ohgushi, A., Ando, R., Denbow, D.M., Furuse, M., 2003. Central administration of cocaine- and amphetamine-regulated transcript inhibits food intake in chicks. Neurosci. Lett. 337, 131–134.

Taouis, M., Chen, J.W., Daviaud, C., Dupont, J., Derouet, M., Simon, J., 1998. Cloning the chicken leptin gene. Gene 208, 239–242.

Wang, Y., Li, J., Yan Kwok, A.H., Ge, W., Leung, F.C., 2010. A novel prolactin-like protein (PRL-L) gene in chickens and zebrafish: cloning and characterization of its tissue expression. Gen. Comp. Endocrinol. 166, 200–210.

Wang, F., Lu, L., Yuan, H., Tian, Y., Li, J., Shen, J., Tao, Z., Fu, Y., 2011. Molecular cloning, expression, and regulation of goose leptin receptor gene in adipocytes. Mol. Cell. Biochem. 353, 267–274.

Warren, W.C., Clayton, D.F., Ellegren, H., Arnold, A.P., Hillier, L.W., Kunstner, A., Searle, S., White, S., Vilella, A.J., Fairley, S., Heger, A., Kong, L., Ponting, C.P., Jarvis, E.D., Mello, C.V., Minx, P., Lovell, P., Velho, T.A., Ferris, M., Balakrishnan, C.N., Sinha, S., Blatti, C., London, S.E., Li, Y., Lin, Y.C., George, J., Sweedler, J., Southey, B., Gunaratne, P., Watson, M., Nam, K., Backstrom, N., Smeds, L., Nabholz, B., Itoh, Y., Whitney, O., Pfenning, A.R., Howard, J., Volker, M., Skinner, B.M., Griffin, D.K., Ye, L., McLaren, W.M., Flicek, P., Quesada, V., Velasco, G., Lopez-Otin, C., Puente, X.S., Olender, T., Lancet, D., Smit, A.F., Hubley, R., Konkel, M.K., Walker, J.A., Batzer, M.A., Gu, W., Pollock, D.D., Chen, L., Cheng, Z., Eichler, E.E., Stapley, J., Slate, J., Ekblom, R., Birkhead, T., Burke, T., Burt, D., Scharff, C., Adam, I., Richard, H., Sultan, M., Soldatov, A., Lehrach, H., Edwards, S.V., Yang, S.P., Li, X., Graves, T., Fulton, L., Nelson, J., Chinwalla, A., Hou, S., Mardis, E.R., Wilson, R.K., 2010. The genome of a songbird. Nature 464, 757–762.

Wickramasinghe, S.N., Shiels, S., Wickramasinghe, P.S., 1994. Immunoreactive erythropoietin in teleosts, amphibians, reptiles, birds. Evidence that the teleost kidney is both an erythropoietic and erythropoietin-producing organ. Ann. N. Y. Acad. Sci. 718, 366–370.

Xiao, Y., Ni, Y., Huang, Y., Wu, J., Grossmann, R., Zhao, R., 2011. Effects of kisspeptin-10 on progesterone secretion in cultured chicken ovarian granulosa cells from preovulatory (F1–F3) follicles. Peptides 32, 2091–2097.

Xie, F., London, S.E., Southey, B.R., Annangudi, S.P., Amare, A., Rodriguez-Zas, S.L., Clayton, D.F., Sweedler, J.V., 2010. The zebra finch neuropeptidome: prediction, detection and expression. BMC Biol. 8, 28.

Zudaire, E., Cuesta, N., Martínez, A., Cuttitta, F., 2005. Characterization of adrenomedullin in birds. Gen. Comp. Endocrinol. 143, 10–20.

Chapter 23

Pituitary Gland

Colin G. Scanes
Department of Biological Sciences, University of Wisconsin, Milwaukee, WI, USA

23.1 INTRODUCTION

The pituitary gland (hypophysis) is intimately connected to the hypothalamus in both an anatomical and functional close relationship. The structure of the hypothalamic–hypophyseal complex is shown in Figure 23.1. The pituitary gland consists of two distinct structures:

- The anterior pituitary gland, which sits in a depression in the bone—the sella turcica
- The posterior pituitary gland, which is connected to the base of the brain by the infundibulum

The pituitary gland has a complex structure and an interesting embryonic development. Primary tissue can be classified as either adenohypophysis or neurohypophysis, each with a distinct embryonic origin. The adenohypophysis is derived from Rathke's pouch (probably ectoderm from the roof of the mouth), and the neurohypophysis is derived from the infundibulum (an outgrowth of the brain). The adenohypophysis, in mammals, goes on to form the pars distalis (anterior pituitary gland), the pars intermedia, and the pars tuberalis. However, there is not a pars intermedia in birds. The avian embryonic adenohypophysis goes on to form the pars distalis (or anterior pituitary gland) and the pars tuberalis. The neurohypophysis forms the pars nervosa (or posterior pituitary gland), the infundibular stalk, and the median eminence.

23.2 ANATOMY OF THE HYPOTHALAMIC–HYPOPHYSEAL COMPLEX

23.2.1 The Anterior Pituitary Gland or Pars Distalis

23.2.1.1 Introduction

The avian pars distalis consists of two distinct lobes: the cephalic and caudal lobes. The cephalic lobe develops from the upper and medial parts of the anterior wall of Rathke's pouch together with part of the posterior wall (Sasaki et al., 2003). The caudal lobe is derived from parts of the posterior Rathke's pouch together with the lower portion of the anterior wall of Rathke's pouch (Sasaki et al., 2003). The anterior pituitary gland is composed of secretory cells and folliculo-stellate cells together with extracellular colloid and fibrous materials (Mohanty et al., 2006).

23.2.1.2 Secretory Cells

The avian anterior pituitary gland has the full complement of secretory cells, including corticotrophs producing adrenocorticotropic hormone (ACTH), gonadotrophs producing either luteinizing hormone (LH) or follicle-stimulating hormone (FSH), lactotrophs producing prolactin, somatotrophs producing growth hormone (GH), and thyrotrophs producing thyrotropin (TSH) (Table 23.1). In addition, there are somatotolactotrophs producing both GH and prolactin (turkey: Ramesh *et al.*, 1998, 2001) and cells that co-express both the large protein, pro-opiomelanocortin (POMC), and the α subunit of LH, FSH, and TSH (chickens: Pals et al., 2006).

The pars distalis is well supplied with blood vessels, including the hypophyseal portal vessels. These latter provide a route from the neurosecretory nerve terminals in the median eminence to the anterior pituitary gland. Indeed, it is by way of the portal blood vessels that the anterior pituitary gland is controlled by *releasing hormones* (or hypothalamo-hypophysiotropic hormones) from the median eminence (Figure 23.1; also see Table 23.2).

23.2.1.3 Folliculo-Stellate Cells

The anterior pituitary gland also contains folliculo-stellate cells. These are of neuroectodermal origin, like the posterior pituitary gland. The presence of these agranular cells containing immunoreactive S-100 protein has been demonstrated in birds (Harrisson et al., 1982; Van Nassauw et al., 1987; Harrisson, 1989). Possible roles for the folliculo-stellate cells may include autocrine–paracrine regulation of hormone secretion from adjacent endocrine cells and an involvement in immune endocrine interactions. In birds, there is evidence of folliculo-stellate cells influencing thyrotrophs (Harrisson et al., 1982), with these cells expressing TSH receptor (Grommen et al., 2009), and also of their involvement in the pituitary response to an *Escherichia coli* challenge (Fernández et al., 1986). The putative influences of the folliculo-stellate cells may be mediated by gap junctions or their release of neuromodulators. The avian folliculo-stellate cells have received relatively little attention.

Follicular stellate cells are presumed to exert paracrine influences on secretory cells in the anterior pituitary gland. This may influence thyrotroph functioning in a short-loop feedback manner.

23.2.2 Pars Tuberalis

The pars tuberalis consists of adenohypophyseal cells located along the ventral surface of the median eminence. It is derived from Rathke's pouch, specifically the upper anterior portion (Sasaki et al., 2003). The pars tuberalis develops along the median eminence between days 8 and 14 of incubation in the chick embryo (Kameda et al., 2000). The pars tuberalis consists of the following cell types: (1) pars tuberalis secretory-specific cells; (2) pars distalis cell types present in the pars tuberalis, including gonadotrophs, thyrotrophs, corticotrophs, somatotrophs, and lactotrophs (Mohanty et al., 1997); (3) follicular cells (reviewed Yasuo and Korf, 2011); and (4) multiple macrophages (Sano and Murabe, 1980).

The pars tuberalis can communicate in a retrograde manner with the median eminence and in an anterograde with the pars distalis. There is high expression of the α subunit of LH, FSH, and TSH in the pars tuberalis as it develops (Kameda et al., 2000). Moreover, most cells in the pars tuberalis of the chicken co-localize LH and chromogranin A immunoreactivity (Kameda et al., 1998). There are progesterone receptors in the nuclei of cells in the pars tuberalis (Gasc and Baulieu, 1988).

23.2.3 Posterior Pituitary Gland or Pars Nervosa

The posterior pituitary gland consists of neurosecretory terminals that release either arginine vasotocin (AVT) or mesotocin (MT) (Table 23.1). These hormones are synthesized in cell bodies in nuclei in the hypothalamus and are transported to the pars nervosa through modified axons (Figure 23.1).

TABLE 23.1 Summary of the Chemistry of the Hormones of the Pituitary Gland

Pituitary Hormones	Alternate Names (IUPAC/IUB)	Chemistry
Anterior pituitary hormone		
Adrenocorticotropic hormone (ACTH)	Corticotropin	Synthesized as protein (pro-opiomelanocortin (POMC)), then proteolytically cleaved to polypeptide (ACTH)
Follicle-stimulating hormone (FSH)	Follitropin	Two glycoprotein subunits: α-subunit common to FSH, LH, and TSH β-specific subunit
Luteinizing hormone (LH)	Lutropin	Two glycoprotein subunits: α-subunit common to FSH, LH, and TSH β-specific subunit
Growth hormone (GH)	Somatotropin	Protein
Prolactin (PRL)	Lactotropin	Protein
Thyroid-stimulating hormone (TSH)	Thyrotropin	Two glycoprotein subunits: α-subunit common to FSH, LH, and TSH β specific subunit
Posterior pituitary hormone		
Arginine vasotocin (AVT)		Synthesized as protein, then proteolytically cleaved to peptide with 9 amino acid residues
Mesotocin (MT)		Synthesized as protein, then proteolytically cleaved to peptide with 9 amino acid residues

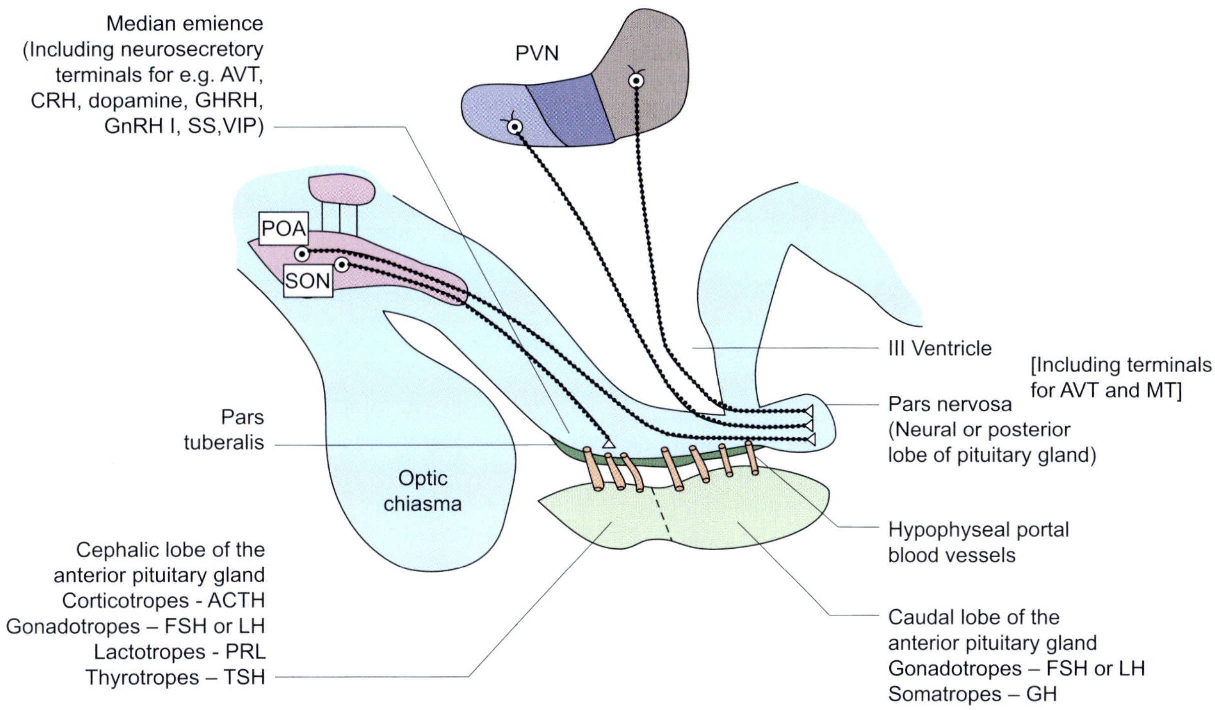

FIGURE 23.1 **Schematic diagram showing structure of the avian hypothalamus and pituitary gland.** Dark green: pars tuberalis; light blue: nervous tissue in the hypothalamus; light green: pars distalis or anterior pituitary gland (adenohypophyseal tissue); red: blood vessels of hypophyseal portal vessels bringing releasing hormones from the neurosecretory terminals in the median eminence to stimulate or inhibit hormone secretion from the cells of the anterior pituitary gland (neurosecretory cells have cell bodies in the nuclei in the hypothalamus and terminate either in the median eminence or the posterior pituitary gland); AVT, arginine vasotocin; MT, mesotocin, POA, pre-optic area, PVN, paraventricular nucleus; SON, supraoptic nucleus.

TABLE 23.2 Summary of Stimulatory and Inhibitory Releasing Hormones (Releasing Factors) in Birds, Including Neuropeptides Present in the Hypothalamus

Anterior Pituitary Hormones	Stimulatory Releasing Hormones	Inhibitory Releasing Hormones
Adrenocorticotropic hormone (ACTH)	Corticotropin-releasing hormone (CRH) Arginine vasotocin (AVT)	–
Follicle-stimulating hormone (FSH)	Gonadotropin-releasing hormone I? and II? (GnRH I and II)	Gonadotropin-inhibitory hormone (GnIH)
Luteinizing hormone (LH)	Gonadotropin-releasing hormone I and II (GnRH I and II)	Gonadotropin-releasing inhibitory hormone (GnIH)
Growth hormone (GH)	GH-releasing hormone (GHRH) Thyrotropin-releasing hormone (TRH) Ghrelin Pituitary adenylate cyclase–activating peptide (PACAP) Leptin? GnRH I?	Somatotostatin (SS or SRIF)
Prolactin (PRL)	Vasoactive intestinal peptide (VIP) Prolactin-releasing peptide (PrRP) AVT? TRH? Peptide histidine isoleucine (PHI)?	Dopamine?
Thyroid-stimulating hormone (TSH)	Thyrotropin-releasing hormone (TRH) CRH	SS?

23.3 GONADOTROPINS

23.3.1 Structure

Avian LH and FSH are glycoproteins with molecular weights of 30kDa (Burke et al., 1979a). Both LH and FSH consist of two glycoprotein subunits: an α-subunit that is common to LH, FSH, and TSH and a hormone-specific β-subunit (Table 23.1). Both subunits are required for biological activity (Burke et al., 1979b). The amino acid residue sequences have been deduced from the nucleotide sequences of the cDNA for α-subunit and both LH and FSH β-subunits (α-subunit: chicken: Foster et al., 1991; domestic and Muscovy duck: Hsieh et al., 2001; LH β-subunit: chicken: Noce et al., 1989; Japanese quail: Ando and Ishii, 1994; turkey: You et al., 1995; FSH β-subunit: Japanese quail: Kikuchi et al., 1998; chicken: Shen and Yu, 2002; Japanese crested ibis: Kawasaki et al., 2003).

23.3.2 Action of Gonadotropins

For more detailed discussion of the control of female and male reproduction, see Chapter 25

23.3.2.1 Actions of LH in the Female

23.3.2.1.1 Ovulation

A major role for LH is to induce ovulation. Premature ovulation is provoked following injection of LH (Imai, 1973). Administration of LH is followed by a series of events in the most mature follicle: breakdown of the germinal vesicle, dissociation of the junctions between the granulosa projections and the oocyte surface, development of a perivitelline space, and the formation of the first and then second maturation spindle (Yoshimura et al., 1993). During follicular development, there is remodeling involving proteases such as plasminogen activator. LH decreases both secreted and cell-associated plasminogen activator in granulosa cells from the largest follicle (F1) (Tilly and Johnson, 1987, 1988a,b).

23.3.2.1.2 Steroidogenesis

In vivo, mammalian LH increases progesterone and testosterone production by the hen ovary (Shahabi et al., 1975). LH stimulates the production of progesterone and androstenedione *in vitro* by granulosa cells from the largest follicle (F1) (Tilly and Johnson, 1987, 1988a,b). LH also stimulates production of androstenedione by theca cells from the next largest yellow follicle (F2) (Tilly and Johnson, 1989) and of progesterone, dehydroepiandrosterone (DHEA), androsterone, and estradiol from theca cells from small follicles (6–8mm) (Kowalski et al., 1991; Tilly et al., 1991a).

23.3.2.1.3 Ovarian Production of Peptide and Protein Growth Factors and Hormones

LH influences the production of growth factors and hormones within the ovary. For instance, LH reduced the expression of inhibin α-subunit in chicken granulosa cells from either F1 or F3 and F4 follicles (Davis et al., 1999). Expression of connective tissue growth factor (CTGF) is respectively decreased and increased by LH in granulosa cells from pre- and postovulatory follicles (Zhu et al., 2012).

23.3.2.1.4 Follicular Development

In embryonic development, LH increases the number of follicles together with the ratio of oocytes to oogonia while decreasing the mitosis of oogonia (González-Morán et al., 2007). Using chick embryo ovaries in culture, LH stimulates the entry of germ cells into meiosis as indicated by the increased expression of the premeiotic marker *Stra8* and meiotic marker *Scp3* (He et al., 2013).

23.3.2.2 Actions of FSH in the Female

23.3.2.2.1 Steroidogenesis

FSH influences the steroidogenesis of less mature large yolky follicles and small yellow follicles but not the large preovulatory (F1) follicles. FSH stimulates progesterone secretion by the granulosa of intermediate-stage follicles (e.g., F5, or the fifth-largest follicle) (Calvo and Bahr, 1983). In addition, FSH acutely increases progesterone, androstenedione, and estradiol production by theca cells from small follicles (6–8 and 9–12mm) (Kowalski et al., 1991). In addition, FSH stimulates maturation of the granulosa cells of small follicles such that they attain the ability to respond to LH, FSH, or cyclic adenosine monophosphate (cAMP) with increased progesterone production (Tilly et al., 1991b). This is due, at least in part, to the induction of the P450 side-chain cleavage enzyme mRNA (Li and Johnson, 1993).

23.3.2.2.2 Proliferation and Remodeling

FSH increases proliferation of both ovarian cells in hypophysectomized chicken embryos (Sánchez-Bringas et al., 2006) and epithelial cells on the surface of the ovary in intact embryos (Méndez et al., 2003). Moreover, FSH stimulates proliferation by granulosa cells from the F1 and smaller yellow F5–F6 follicles of the chicken (McElroy et al., 2004). In embryonic development, FSH suppresses the entry of germ cells into meiosis, as indicated by the decreased expression of the premeiotic marker *Stra8* and meiotic marker *Scp3* (He et al., 2013).

FSH increases expression of occludin, an important protein in tight junctions, in granulosa cells (chicken: Schuster et al., 2004). Thus FSH facilitates the rapid development of oocytes and yolk deposition. In addition, FSH in the presence of activin induces granulosa cell–specific differentiation markers, including zona pellucida protein C (chicken: Schierer et al., 2003).

23.3.2.2.3 Peptide and Protein Growth Factors and Hormones

FSH influences the production of growth factors and hormones within the ovary. For instance, FSH increases expression of inhibin α-subunit, together with inhibin protein levels, in granulosa cells from F1, F4–F5 follicles, small yellow follicles, and large white follicles (chicken: Davis et al., 1999, 2001). FSH depresses expression of connective tissue growth factor (CTGF) in granulosa cells from preovulatory follicles but elevates that from postovulatory follicles in the chicken (Zhu et al., 2012).

23.3.2.2.3.1 Mechanism The signal transduction mechanism for FSH involves activation of adenylate cyclase and increased production of cAMP (Calvo and Bahr, 1983; Kowalski et al., 1991).

23.3.2.3 Actions of LH in the Male

As in mammals, LH acts primarily to stimulate the Leydig cells to differentiate and produce testosterone. The administration of chicken LH to hypophysectomized Japanese quail greatly increases the number of mature Leydig cells and decreases the number of fibroblasts and transitional cell types in the interstitium (Brown et al., 1975). In Japanese quail, LH elevates plasma concentrations of testosterone (Maung and Follett, 1978) and increases testosterone synthesis *in vitro* (Maung and Follett, 1977).

23.3.2.4 Actions of FSH in the Male

The major role for FSH is stimulation of spermatogenesis. When FSH is injected into hypophysectomized quail, there are increases in testicular size (8.8-fold) and seminiferous tubule diameter (2.1-fold) with Sertoli differentiation promoted and spermatogenesis (Brown et al., 1975). FSH increases the proliferation of seminiferous tubule cells in the chick embryo (Méndez et al., 2003). FSH increases both the proliferation of chicken Sertoli cells and their secretion of inhibin *in vitro* (Guibert et al., 2011). FSH has little if any effect on testosterone production by bird testes (Maung and Follett, 1978). FSH upregulates its receptors (Tsuisui and Ishii, 1980).

23.3.2.5 Other Effects of FSH

An unexpected effect of FSH is on lipid metabolism. FSH increases adipose weight *in vivo* and influences lipid metabolism, as indicated by shifts in expression of enzymes *in vitro* (chicken: Cui et al., 2012). Evidence that the effect of FSH is specific comes from the observation that the FSH receptor (FSHR) is expressed in avian adipose tissue (chicken: Cui et al., 2012).

23.3.2.6 Chicken LH Receptor (LHR) and Signal Transduction

The chicken LHR has been characterized (Johnson et al., 1996). Expression of LHR is influenced by both peripheral and intraovarian hormones together with growth factors. Expression of LHR in granulosa cells from F1 follicles is increased by testosterone, LH, activin, and BMP6 (Johnson et al., 2006; Al-Musawi et al., 2007; Rangel et al., 2009). These effects may play a critical role in the preovulatory increase in both LH and progesterone. Activin also increases expression of LHR in granulosa cells from large (F1 and F3/F4) and small yellow follicles (Johnson et al., 2006). A soy isoflavone increases expression of LHR in chicken F1 follicles (Liu and Zhang, 2008).

The signal transduction mechanism by which LH exerts its effect on theca cells is via adenylate cyclase and cAMP (Tilly and Johnson, 1989). While in granulosa cells, LH acts by adenylate cyclase and cAMP (Johnson and Tilly, 1988) and subsequently calcium mobilization (Asem et al., 1987; Tilly and Johnson, 1988a,b).

23.3.2.7 FSHR and Signal Transduction

The cDNA for chicken FSHR has been characterized (Wakabayashi et al., 1997). Expression of the FSHR is influenced by FSH, intraovarian growth factors, and genomic variation. FSH upregulates FSHR expression by undifferentiated granulosa cells (Woods and Johnson, 2005). Activin increases expression of FSHR in granulosa cells from large (F1 and F3/F4) and small yellow follicles (Johnson et al., 2006). Bone morphogenetic protein-2 (BMP2) decreases FSHR expression by undifferentiated granulosa cells (Haugen and Johnson, 2010). Expression of FSHR in granulosa cells from large yellow follicles (F1, F2, and F3/F4) is increased by BMP6; theca cells produce BMPs, including BMP6 (Al-Musawi et al., 2007). There are polymorphisms of the FSHR gene that influence expression of FSHR (Li et al., 2011).

FSH activates the cAMP–protein kinase A, phosphatidylinositol 3-kinase–Akt (protein kinase B), and mitogen-activated protein kinase pathways in chicken Sertoli cells (Guibert et al., 2011). FSH effects on adipose tissue are mediated by peroxisome proliferator-activated receptor (PPAR) signaling (chicken: Cui et al., 2012).

23.3.3 Control of Gonadotropin Release

23.3.3.1 Introduction

Release of both LH and FSH is under predominantly stimulatory control by the hypothalamus based on de-afferentation and lesioning studies (Davies and Follett, 1980). LH is under stimulatory control by gonadotropin-releasing hormone-1 and -2 (GnRH1 and GnRH2) (Table 23.2). It is unclear whether the release of FSH is influenced by GnRH1 or GnRH2. There is also inhibitory control for both LH and FSH by gonadotropin-inhibitory hormone (GnIH).

23.3.3.2 Gonadotropin Hormone–Releasing Hormone

23.3.3.2.1 Chemistry

There are two GnRH peptides in the avian hypothalamus, each with separate genes, respectively, for GnRH1 and -2 (Dunn et al., 1993; Ikemoto and Park, 2006):

Chicken GnRH1 (cGnRH1)
pGlu–His–Trp–Ser–Tyr–Gly–Leu–Gln–Pro–Gly–NH_2
(King and Millar, 1982)

Chicken GnRH2 (cGnRH2)
pGlu–His–Trp–Ser–His–Gly–Trp–Tyr–Pro–Gly–NH_2
(Miyamoto et al., 1984)

An identical situation exists in other birds (ostrich: Powell et al., 1987). The cDNA for starling GnRH1 has been characterized (Ubaka et al., 2009). There is a 58% homology between the precursor for GnRH1 in starlings and chickens (Ubaka et al., 2009).

Chicken GnRH2 is considerably more potent than cGnRH1 in stimulating LH release *in vitro* (chicken: Chou et al., 1985; Millar et al., 1986) and *in vivo* (chicken: Wilson et al., 1989; Proudman et al., 2006). Administration of GnRH1 or -2 either evokes less release of FSH compared to that of LH in birds (Hattori et al., 1986) or is ineffective (Scanes et al., 1977; Krishnan et al., 1993; Proudman et al., 2006). The identity of the avian releasing hormone for FSH remains unclear. The actions of cGnRH1 and -2 are mediated by cAMP and inositol phosphate (Bonney and Cunningham, 1977; Joseph et al., 2009).

23.3.3.2.2 Control of cGnRH Release

Neurosecretory terminals release cGnRH1 or -2 in response to stimuli. Neurotransmitters and neuropeptide can either stimulate or inhibit release. Examples of stimulatory control include norepinephrine via α_1-adrenergic receptors (Knight et al., 1982; Millam et al., 1984) and neuropeptide Y (Fraley and Kuenzel, 1992; Contijoch et al., 1993a). Examples of inhibitory controls include dopamine (Knight et al., 1982; Contijoch et al., 1992), β endorphin (Contijoch et al., 1993b), and GnIH (discussed in Section 23.3.3.2.3).

23.3.3.2.3 Changes in Hypothalamic GnRH Content

The variation in hypothalamic GnRH1 immunoreactivity and/or expression is consistent with GnRH1 having a role in the control of reproduction in birds. There are changes with physiological state in the hypothalamic contents of cGnRH1 in turkeys (Rozenboim et al., 1993). The GnRH1 levels are higher in laying hens compared to nonphotostimulated birds (with photostimulated birds intermediate) (Rozenboim et al., 1993). Photostimulation of turkeys results in activation (c-fos expression) of GnRH1 in the nucleus commissura pallii and dopamine neurons in the nucleus premammillaris (Thayananuphat et al., 2007). There is decreased hypothalamic expression of GnRH1 in female chickens when incubating eggs (Dunn et al., 1996) and when young male chickens receive estradiol administration (Dunn and Sharp, 1999). GnRH1 expression and the GnRH1 peptide are co-localized in neurosecretory neurons in the pre-optic region of the hypothalamus in photostimulated starlings but are not detectible in photorefractory starlings (Ubaka et al., 2009). In contrast, photostimulation of chickens on short day lengths increased secretion of LH but did not influence expression of GnRH1 (Dunn and Sharp, 1999).

Changes in GnRH2 immunoreactivity and/or expression in the hypothalamus support it in having a major role in the control of avian reproduction. The levels of GnRH2 are increased by photostimulation; they are higher in laying hens, but reduced in incubating and photorefractory turkeys (Rozenboim et al., 1993). There are more and larger GnRH2 neurons in breeding than nonbreeding zebra finches (Perfito et al., 2011). During embryonic development of the chicken, there are changes in the brain concentrations of cGnRH1 and cGnRH2, but these are not in a parallel manner (Millam et al., 1993).

GnRH immunoreactivity is found not only in hypothalamic neurons but also in mast cells in the dove medial habenular regions. The number and distribution of these mast cells depend on reproductive behavior changing immediately after courtship behavior (Silver et al., 1992; Zhung et al., 1993).

23.3.3.2.4 GnRH Receptors

There are two GnRH receptor subtypes present in birds: GnRH-R1 and GnRH-R3. Chicken GnRH2 has a higher affinity than GnRH1 for both receptor types and is more potent in stimulating inositol phosphate production (Joseph et al., 2009). There is much higher expression of cGnRH-R3 than cGnRH-R1 in the chicken pituitary gland (Joseph et al., 2009), with higher expression in sexually mature birds (McFarlane et al., 2011). Some, albeit very low, expression of cGnRH-R3 is reported in the median eminence and small intestine (Joseph et al., 2009). There is expression of cGnRH-R1 in the pituitary gland, with some expression in the median eminence, small intestine, and testes (Joseph et al., 2009).

23.3.3.3 Gonadotropin Inhibitory Hormone

23.3.3.3.1 Chemistry

A novel neuropeptide was isolated from quail brain. This 12 amino acid residue-containing peptide inhibits LH release in quail and hence was named gonadotropin-inhibitory hormone (Tsutsui et al., 2000). The structure of GnIH and its precursor has been reported in birds (quail: Tsutsui et al., 2000; Satake et al., 2001; chicken: Ikemoto and Park, 2005; white-crowned sparrow: Osugi et al., 2004). Chicken LPLRF–NH_2 was the first RF-amide peptide found in any vertebrate (Dockray et al., 1983) and is identical to the C terminal 5-amino-acid residues of GnIH (discussed further in this chapter). The gene (neuropeptide VF precursor (NPVF) gene) also encodes GnIH-related peptide-1 (GnIH-RP1) and GnIH-RP2, each possessing an LPXRF-amide (X=L or Q) motif.

These are the structures of GnIH from three avian species:

Quail:	SIKPSAYLPLRF-NH_2
Chicken:	SIRPSAYLPLRF-NH_2
Sparrow:	SIKPFSNLPLRF-NH_2

23.3.3.3.2 Actions of GnIH

GnIH affects gonadotrophs producing either FSH or LH, with the GnIH receptor (GnIHR) co-localizing with both LH- and FSH-expressing cells (Maddineni et al., 2008). Release and synthesis of LH are depressed by GnIH. For instance, GnIH reduces circulating concentrations of LH following castration and GnRH1 stimulation in songbirds (white-crowned sparrow: Osugi et al., 2004). In addition, GnIH reduces pituitary expression of the common α- and LH β-subunits together with circulating concentrations of LH and testosterone in Japanese quail (Ubuka et al., 2006). GnIH decreases expression of α and FSH β *in vitro* (chicken: Ciccone et al., 2004; Maddineni et al., 2008).

23.3.3.3.3 Hypothalamic GnIH

Not only does GnIH act on the pituitary but also it inhibits gonadotropin release by influencing GnRH1 and GnRH2 neurons in the hypothalamus. GnRH1- and GnRH2-containing neurons in the hypothalamus have been reported to express GnIH receptors (Ubuka et al., 2008). There is also evidence suggesting the GnIH does not always have a critical role in the control of reproduction. For instance, administration of lipopolysaccharide mimicking acute infection decreases the number of hypothalamic neurons containing immunoreactive GnRH in zebra finches, but the number expressing GnIH is unchanged (Lopes et al., 2012). Similarly, no differences in GnRH1 and GnIH neurons were seen between breeding and nonbreeding zebra finches (Perfito et al., 2011).

Hypothalamic expression of GnIH is influenced by the reproductive state being greater in incubating than in laying chickens (Ciccone et al., 2004). Moreover, there is cross-talk between the pineal gland and the hypothalamus, with melatonin increasing GnIH release from hypothalamic explants (Chowdhury et al., 2010).

23.3.3.3.4 Receptor

The GnIHR has been characterized based on cDNA (chicken: Ikemoto and Park, 2005) and has been referred to as the RF-amide-related peptide (RFRP) receptor. Estrogen combined with progesterone decrease expression of GnIHR in the anterior pituitary gland (Maddineni et al., 2008).

23.3.3.4 Other Hormones and LH Release

Prolactin can act directly on the gonadotrophs that secrete LH, with prolactin depressing LH release from turkey anterior pituitary cells *in vitro* (turkey: You et al., 1995).

23.3.3.5 Negative and Positive Feedback

Sex steroids influence gonadotropin release and synthesis in birds, via direct effect on the anterior pituitary gland or shifts in the release of releasing hormones from the hypothalamus. This is predominantly an inhibitory control. Gonadectomy leads to large increases in the circulating concentrations of both LH and FSH (e.g., Japanese quail: Davies et al., 1980). The inhibitory effect of androgens on LH secretion is mediated via androgen and estrogen receptors (quail: Davies et al., 1980) or predominantly estrogen receptors (chickens: Fennell et al., 1990).

In females, estradiol exerts a *negative feedback* effect on LH secretion. Progesterone can either stimulate or inhibit LH release depending on the state of the bird, with the preovulatory LH surge in the hen induced by progesterone. In the ovariectomized domestic hen, plasma concentrations of LH are decreased by a single injection of either progesterone or estradiol (Wilson and Sharp, 1976a,b). Progesterone has a *positive feedback* effect on LH release in intact hens or in ovariectomized hens that have been primed with progesterone and estradiol (Wilson and Sharp, 1976a,b). Progesterone receptors are present in the chicken hypothalamus, with increases in immunoreactivity following estradiol administration (Gasc and Baulieu, 1988). Estrogens also induce the presence of progesterone receptors in gonadotrophs (LH-producing cells) (Gasc and Baulieu, 1988).

23.3.3.6 Cycles

23.3.3.6.1 Seasonal Breeding

Increasing day lengths increase LH and FSH secretion in temperate-zone birds (Japanese quail: Follett and Maung, 1978; Henare et al., 2011; starlings: Dawson et al., 1985). This facilitates reproduction occurring at the most advantageous season. As little as one long day stimulates LH and FSH release (Japanese quail: Follett et al., 1977). The role of the pars tuberalis in this is discussed in Section 23.9.1.

Temperature is an additional environmental cue for time reproduction, with low temperatures slowing the photoperiodic induction of reproduction (Wingfield et al., 2003).

Increasing temperature accelerates the timing of reproduction (great tits: Schaper et al., 2012). Availability of water increases the photoperiodic response in song sparrows (Wingfield et al., 2012). The importance of supplementary environmental cues becomes progressively more important at lower latitudes (Silverin et al., 2008).

23.3.3.6.2 Ovulation Cycle

The preovulatory surge in plasma LH concentrations occurs approximately 4–6h before ovulation (chicken: Furr et al., 1973) (Figure 23.2(A)). Preovulatory surges in LH are also reported in turkeys, but not where ovulation is arrested (Liu et al., 2001). The preovulatory LH surge is provoked by progesterone in the presence of estrogen priming (Wilson and Sharp, 1976a,b). There are increases in the expression of GnRHR in the chicken anterior pituitary gland during the ovulatory cycle (Lovell et al., 2005) (Figure 23.2(B)). Moreover, in immature hens treated with progesterone in the presence of estradiol, there are marked decreases in the expression of GnIH. There is little change in the circulatory concentrations of FSH during the ovulatory cycle of the hen (Scanes et al., 1977; Krishnan et al., 1993; Lovell et al., 2005).

23.3.4 Control of Expression of Gonadotropin Subunits

23.3.4.1 Control of Expression of the LH–FSH–TSH α-Subunit

Expression of the common LH–FSH–TSH α-subunit changes with physiology. It is not clear whether these changes are isolated to specific cell types. Starvation reduces the expression of the common α-subunit in Japanese quail (Kobayashi et al., 2002). Increasing food intake in previously feed-restricted hens is followed by increased α-subunit expression (Ciccone et al., 2007). There appears to be a negative feedback effect on expression of the α-subunit, as this is increased following ovariectomy in chickens, and this effect is reversed by estradiol replacement therapy (Terada et al., 1997). *In vivo* GnIH reduces pituitary expression of the α-subunit in Japanese quail (Ubuka et al., 2006).

23.3.4.2 Control of Expression of the LH β-Subunit

Expression of the LH β-subunit is influenced by developmental stage and by hypothalamic and ovarian hormones. It is expressed as early as day 11 of embryonic development (Grzegorzewska et al., 2009). GnRH1 increases its expression *in vitro* (turkey: You et al., 1995). Its expression is reduced by starvation (Japanese quail: Kobayashi et al., 2002), presumably due to decreased stimulation by GnRH1 and/or 2. GnIH reduces its expression of the LH β-subunit in anterior pituitary glands (Japanese quail: Ubuka et al., 2006). Expression of the LH β-subunit is also decreased, in a feedback manner, by ovarian hormones such as estradiol (Terada et al., 1997). Prolactin decreases expression of the LH β-subunit by turkey pituitary cells *in vitro* (You et al., 1995). LH β expression in hen anterior pituitary cells is elevated by prostaglandin D_2 (PGD_2) or PGJ_2, downstream metabolites of prostaglandin-D synthase via a PPAR signaling pathway (Chen et al., 2010).

23.3.4.3 Control of Expression of the FSH β-Subunit

FSH β-subunit expression changes with developmental stage by feedback from the gonadal hormones, and it seems to be affected by the hypothalamus. Chicken embryos as early as day 11 express the FSH β-subunit in their anterior pituitary glands (Grzegorzewska et al., 2009). Inhibin exerts a negative

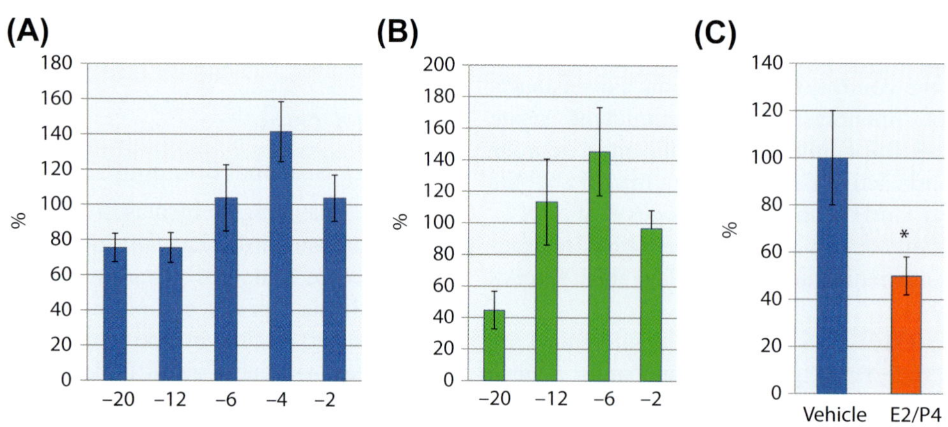

FIGURE 23.2 **Changes in LH during the ovulatory cycle of the chicken.** (A) Circulating concentrations of LH during the ovulatory cycle. (B) Pituitary GnRH receptor expression during the ovulatory cycle (*Data from Lovell et al. (2005)*). (C) Effect of *in vivo* treatment with estradiol (E_2) and progesterone (P_4) in immature hens on pituitary expression of GnIH receptors by chicken anterior pituitary tissue (Maddineni et al., 2008).

feedback effect on the expression of the FSH β-subunit in birds. Turkeys immunized against inhibin show increased pituitary FSH β-subunit expression and more small yellow follicles, presumably due to elevated FSH (Ahn et al., 2001).

There is stimulatory central nervous control of FSH β-subunit expression. The hypothalamic factors participating in birds have not been definitively identified. Evidence for the hypothalamic control comes from the following. Photostimulation is accompanied by increased FSH β-subunit expression (chickens: Li et al., 2009). Early puberty is induced by sulfamethazine with increased pituitary expression of the FSH β-subunit (chickens: Li et al., 2009). The expression of FSH β-subunits is reduced by starvation in Japanese quail (Kobayashi et al., 2002).

23.3.5 Pituitary Origin of Gonadotropins

Based on studies in the chicken, avian gonadotrophs seem to contain either LH or FSH (Proudman et al., 1999). The gonadotrophs containing LH are considerably more numerous than the gonadotrophs containing FSH (Proudman et al., 1999). Gonadotrophs containing either LH or FSH are found in both the cephalic and caudal lobes of the anterior pituitary gland, particularly in the periphery (Proudman et al., 1999; McFarlane et al., 2011). Most gonadotrophs producing LH also express the estrogen receptor α, while some also express the androgen receptor (Sun et al., 2012). FSH-producing gonadotrophs express betaglycan, the type 3 receptor for TGFβ and an accessory receptor for inhibin (chicken: Sweeney and Johnson, 2005).

23.3.6 Ontogeny of Gonadotropins

There is expression of the common α-subunit for LH, FSH, and TSH together with the presence of LH immunoreactivity in the basal–posterior region of Rathke's pouch on day 3.5 of development of the chick embryo prior to the formation of the pars distalis (Kameda et al., 2000). There is evidence of a role for corticosterone in the development of FSH-containing gonadotrophs. The glucocorticoid-induced leucine zipper (GILZ) is expressed in the chick embryo pituitary gland, with corticosterone increasing expression (Ellestad et al., 2009). Exogenous GILZ increases expression of the FSH β-subunit (Ellestad et al., 2009).

23.4 THYROTROPIN

23.4.1 Structure

Avian TSH is a heterodimer consisting of α- and β-subunits. Ostrich TSH has been purified and is a glycoprotein (Papkoff et al., 1982). The avian TSH β-subunit has been characterized based on the sequence of the cDNA (e.g., chicken: Gregory and Porter, 1997; Kato et al., 1998; Japanese quail: Catena et al., 2003; domestic and mule duck: Hsieh et al., 2007).

23.4.2 Actions of TSH

23.4.2.1 Role

TSH stimulates avian thyroid both to increase in size (Robinson et al., 1976) and to release thyroxine (T_4). Injection of TSH is followed by increases in the circulating concentrations of both T_3 and T_4 (chickens: MacKenzie, 1981; Williamson and Davison, 1985; doves and Japanese quail: McNichols and McNabb, 1988). It is likely that TSH does not directly increase T_3 release from the thyroid. This is supported by the extremely low level of T_3 in the avian thyroid gland (McNichols and McNabb, 1988) and the ready conversion of T_4 to T_3 (Decuypere et al., 1988; McNichols and McNabb, 1988). TSH stimulates T_4 release, which is rapidly converted peripherally to T_3.

23.4.2.2 Mechanism

TSH acts by binding to the TSH receptors (TSHRs), which are linked by G-proteins to adenylate cyclase. The TSHR has been characterized (chicken: Grommen et al., 2006). Activation of adenylate cyclase results in increased intracellular cAMP and activation of protein kinase A. TSH increases cAMP release from thyroid tissue *in vitro* (chicken: Tonoue and Kitch, 1978) and elevates the thyroid cAMP content *in vivo* (quail: McNichols and McNabb, 1988). TSH is expressed in the thyroid, as expected, along with the brain, anterior pituitary gland, retina, and pineal gland (chicken: Grommen et al., 2006). Five splice variants of TSHR are expressed in the thyroid gland (chicken: Grommen et al., 2006, 2008). Genomic sweeps indicate that the TSHR gene has been subject to genetic changes during the domestication of chickens (Rubin et al., 2010).

23.4.3 Control of Thyrotropin Release

The release of TSH from the hypothalamus is under predominantly stimulatory control with two stimulatory releasing hormones (thyrotropin-releasing hormone (TRH) and corticotropin-releasing hormone (CRH)) and also the inhibitory effects of somatostatin (SS or SRIF).

23.4.3.1 Thyrotropin-Releasing Hormone

The structure of TRH is the following:

Pyro-glutamyl-histidyl-proline (NH_2) (*Bøler et al.*, 1969)

Chicken pre-pro-TRH has been characterized, and it contains four TRH progenitor sequences (–K/RRQHPGK/RR) (Aoki et al., 2007). TRH stimulates the release of TSH (chicken: Scanes, 1974).

TRH is expressed in the chicken brain (nucleus preopticus periventricularis, nucleus preopticus medialis, regio lateralis hypothalami, paraventricular nucleus, nucleus periventricularis hypothalami, and nucleus ventromedialis hypothalamic) (Aoki et al., 2007). The chicken TRH receptor has been characterized (Sun et al., 1998).

23.4.3.2 Somatotostatin

There is inhibitory control of TSH release. Somatostatin depresses TSH release (Geris et al., 2003a). SS acts via the following somatostatin receptors (SSTRs): SSTR2 and SSTR5 (Geris et al., 2003a; De Groef et al., 2007).

23.4.3.3 Corticotropin-Releasing Hormone

CRH stimulates TSH release in birds (Geris et al., 2003b). This is mediated via the type 2 CRH receptors (CRH-R2s), these being expressed by chicken thyrotrophs (De Groef et al., 2003b).

23.4.3.4 Negative Feedback

As in mammals, T_3 exerts a *negative feedback* effect on TSH release in birds. This is supported by the ability of goitrogens to increase thyrotroph number and pituitary TSH content (Sharp et al., 1979).

23.4.3.5 Environmental Factors and TSH Release

Cold evokes a rapid increase in TSH release in birds. There are increased circulating concentrations of T_4 with cold exposure (Japanese quail: Herbute et al., 1984). This is likely to be mediated by TRH as cold stress increases hypothalamic expression of TRH (Wang and Xu, 2008). Circulating concentrations of TSH are decreased during fasting, with thyrotropes retaining sensitivity to either TRH or CRH (chicken: Geris et al., 1999).

23.4.4 Control of TSH β-Subunit Expression (in the Pars Distalis)

Studies in mule ducks (a hybrid of male Muscovy ducks and female domestic ducks) indicate preferential transcription of the TSH β-subunit from the maternal genome (Hsieh et al., 2007).

Expression of the TSH β-subunit is increased by TRH in duck pituitary tissue *in vitro* (Hsieh et al., 2007) (Figure 23.3). Thyroid hormones inhibit expression of the TSH β-subunit. Either T_3 or T_4 suppresses expression of the TSH β-subunit in pituitary tissue *in vitro* (duck: Hsieh et al., 2007) (Figure 23.3). Similarly, expression of the TSH β-subunit by chicken pituitary cells is depressed by T_3 (Gregory and Porter, 1997). There is increased expression of the TSH β-subunit by pituitary glands in goitrogen-treated quail (Catena et al., 2003). In addition, glucocorticoids such as corticosterone depress expression of the TSH β-subunit in duck pituitary tissue *in vitro* (Hsieh et al., 2007). There is no change in the expression of the TSH β-subunit with starvation in Japanese quail (Kobayashi et al., 2002; Kobayashi and Ishii, 2002).

23.4.5 Origin of Thyrotropin

23.4.5.1 Anterior Pituitary Gland

The thyrotrophs, the cells producing TSH, are located almost entirely in the cephalic (or rostral) lobe of the avian anterior pituitary gland (Figure 23.1). This is based on immunocytochemistry with antisera against mammalian the TSH β-subunit (Sharp et al., 1979). It is not surprising, based on the stimulatory control of TSH release by TRH, that TRH binding sites have been reported for membrane preparations from cephalic lobes of chicken pituitary gland (Harvey and Baidwan, 1989). The proportion of cells of the anterior pituitary gland that are thyrotrophs ranges from 2% to 6% (adult ring doves: Reichardt, 1993). Some thyrotrophs express the androgen receptor (Sun et al., 2012). The interaction between androgens and TSH secretion is not well established.

23.4.5.2 Extrapituitary TSH

The TSH β-subunit immunoreactivity is reported in the cells lining the spinal canal, the crop, the bronchial ducts, and the linings of the pleural and pericardial cavities of the chick embryo (Murphy and Harvey, 2001).

23.4.6 Ontogeny of Thyrotropin

Both TSH (pituitary) and TRH (hypothalamus) are detectible very early in the development of the chicken embryo (TSH at day 6.5: Thommes et al., 1983; the TSH β-subunit at day 12: Ellestad et al., 2011; and TRH at day 4.5: Thommes et al., 1985). Changes in cell number and TSH–TRH content occur during development. Self-organizing maps have been constructed on the genes expressed differentially in

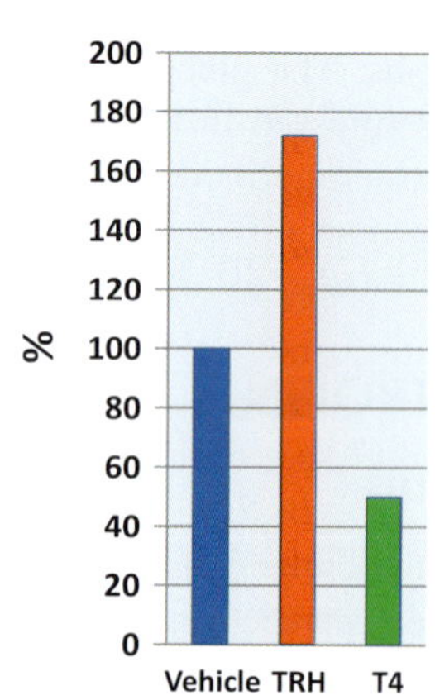

FIGURE 23.3 **Effect of TRH and thyroxine (T_4) on expression of TSH β subunit by duck anterior pituitary tissue *in vitro*.** *Data from Hsieh et al. (2007).*

the anterior pituitary during the development of thyrotrophs producing TSH β-subunits (Ellestad et al., 2006).

The hypothalamo–pituitary–thyroid axis appears to be functional by midembryonic development (Thommes et al., 1984). A second phase is the maturation of the axis occurs at about the time of hatching with deiodination pathways fully established (Decuypere et al., 1988; McNichols and McNabb, 1988).

23.5 GROWTH HORMONE

23.5.1 Chemistry

Chicken GH has been characterized based on cDNA (chickens: Lamb et al., 1988). An antagonist to chicken GH has been produced (Paczoska-Eliasiewicz et al., 2006).

23.5.2 Growth Hormone Variants

There are charge and size variants of avian GH. Unlike the situation with human GH, where there are multiple genes and where different splicing patterns of mRNA occur, GH variants in birds are due to posttranslation modification. There are reports of a glycosylated GH (Berghman et al., 1987), a phosphorylated GH (Arámburo et al., 1990, 1992), a cleaved 15 kDa GH (Arámburo et al., 2001), and dimeric and other oligomeric forms (Houston and Goddard, 1988) in the chicken.

Release of GH variants may be differentially controlled. There are shifts in the ratio of GH variants during growth and with GH-releasing hormone (GHRH) stimulation (Arámburo et al., 2000; Martínez-Coria et al., 2002). The ratio of glycosylated to nonglycosylated GH shifts during growth and development (chicken: Berumen et al., 2004). Glycosylated avian GH is present in the same secretory granules as nonglycosylated GH (chicken: Berumen et al., 2004).

23.5.3 Actions of Growth Hormone

Studies on the role of GH have been isolated predominantly to one domesticated species of bird—the chicken. There is a need for studies in multiple avian species.

23.5.3.1 GH and Growth

GH is likely to be required for normal posthatching growth. Hypophysectomy and the consequent lack of GH result in a decrease in the growth rate of over 50% in chickens (King and Scanes, 1986; also calculated from Kusnik et al., 2008) but either has no effect on growth rate in turkeys (Proudman and Opel, 1990a) or there is a modest depression of average daily gain (30%) (Proudman et al., 1994). The effect of pituitary ablation on growth rate is partially overcome by replacement therapy with GH in chicken (King and Scanes, 1986; Scanes et al., 1986), but no effect of GH is reported in turkeys (Proudman et al., 1994). In young intact birds, GH does not stimulate growth more than at most transitorily (chickens: Leung et al., 1986; Burke et al., 1987; Cogburn et al., 1989; turkeys: Proudman et al., 1994). Stimulation of growth in midgrowth-phase chickens is reported when GH is administered in a pulsatile manner (Vasilatos-Younken et al., 1988).

In mammals, GH exerts its effect on somatic growth by increasing production of insulin-like growth factor 1 (IGF1) by the liver. In birds, a similar situation seems to be the case. Plasma concentrations of IGF1 are reduced following hypophysectomy in young birds (chickens: Huybrechts et al., 1985; turkeys: Proudman et al., 1994) and restored somewhat by GH treatment (chickens: Scanes et al., 1986; turkeys: Proudman et al., 1994). Moreover, young female chickens that exhibited enhanced growth with pulsatile GH treatment also had elevated circulating concentrations of IGH1 (Vasilatos-Younken et al., 1988). *In vitro*, GH increases IGF1 release from chicken hepatocytes, particularly in the presence of insulin (Houston and O'Neill, 1991).

23.5.3.2 GH and Lipid Metabolism

GH exerts effects on lipid metabolism in birds. Administration of GH to chickens can influence the amount of adipose tissue, with the direction of the effect affected by the mode of administration. For instance, body fat is increased by daily injection (Cogburn et al., 1989) but decreased by pulsatile administration of GH (Vasilatos-Younken et al., 1988). GH receptors are expressed in adipose tissue (chicken: Hausman et al., 2012).

Acutely, GH influences the rate of lipolysis. GH stimulates lipolysis *in vitro* by adipose explants (chickens, turkeys, and pigeons: Harvey et al., 1977; Campbell and Scanes, 1985). This effect is blocked by a GH antagonist (chicken: Campbell et al., 1993). *In vivo*, GH increases circulating concentrations of nonesterified fatty acids (NEFAs) acutely (chickens: Hall et al., 1987; Scanes, 1992; hypophysectomized ducks: Foltzer and Mialhe, 1976) but was without an effect in intact or hypophysectomized young turkeys (Proudman et al., 1994). In addition, GH exerts an effect on avian adipose that is analogous to the insulin-like effect of GH in mammals. *In vitro*, GH reduces glucagon or cAMP-stimulated lipolysis (chicken: Campbell and Scanes, 1987).

GH inhibits lipogenesis in birds. GH suppressed insulin-stimulated synthesis of fatty acids by chick hepatocytes *in vitro* (Harvey et al., 1977). Moreover, pulsatile GH administration to young pullets *in vivo* reduced hepatic lipogenesis (Rosebrough et al., 1991).

23.5.3.3 GH and Thyroid Hormones

There are marked effects of GH on circulating concentrations of the thyroid hormone, triiodothyrone (T_3). It is not established whether GH influences thyroxine (T_4) release from the avian thyroid (chicken: MacKenzie, 1981). In young chicks, GH increases the circulating concentration of T_3 (Kuhn et al., 1985). This is due to GH reducing both the activity of hepatic type 3 monodeiodinase (the catabolic enzyme metabolizes T_3 to T_2; Darras et al., 1992, 1993) and the expression of this deiodinase (Van der Geyten et al., 1999). This is a physiological effect of GH as type 3 monodeiodinase is elevated following hypophysectomy and partially restored to normal by GH replacement therapy (Darras et al., 1993).

23.5.3.4 GH and Immune Functioning

GH exerts a stimulatory effect on the avian immune system. For instance, GH administration increases thymus weights in hypophysectomized chicks (King and Scanes, 1986; Scanes et al., 1986) and bursal weight if T_3 is also present in hypophysectomized chicks (Scanes et al., 1986), and it influences peripheral white blood cells (Johnson et al., 1993). GH reduces apoptosis of bursal cells (Rodríguez-Méndez et al., 2010). Moreover, GH is expressed by avian immune tissues (discussed further in this chapter).

23.5.3.5 GH and Reproduction

GH administration prior to sexual maturation increases ovarian weight, increasing cell proliferation in both the stroma and small follicles and depressing apoptosis *in vivo* (chickens: Hrabia et al., 2011). However, there is no effect on reproduction in dwarf male chickens with GH receptor (GHR) deficiency (Zheng et al., 2007). In addition, GH is expressed by gonadal tissues (discussed further in this chapter).

23.5.3.6 GH and Adrenocortical Hormones

GH can influence the avian adrenal cortex. Plasma concentrations of corticosterone are increased acutely following the administration of recombinant chicken GH (Cheung et al., 1988) or chronically following the continuous infusion of native GH (Rosebrough et al., 1991). There are effects of GH on corticosteroidogenesis *in vitro* (chicken: Carsia et al., 1985c).

23.5.3.7 GHR and Signal Transduction

Chicken GHR cDNA has been sequenced (Burnside et al., 1991). As with the mammalian GHR, the chicken GHR consists of three domains: an extracellular domain, a single transmembrane domain, and an intracellular domain. There are polymorphisms in the GHR associated with growth rate and adiposity (Ouyang et al., 2008). Sex-linked dwarf chickens have mutations such that the GHR is not expressed normally (Burnside et al., 1992).

GH downregulates the GHR. Hypophysectomy increases and GH administration decreases GH binding to chicken liver membranes (Vanderpooten et al., 1991a). Both hepatic GH binding (chicken: Vanderpooten et al., 1991b) and GHR mRNA expression (Burnside and Cogburn, 1992) are low in young birds when plasma concentrations of GH are high.

Following binding to the GHR, signal transduction is via Janus kinase 2 (JAK2). In the presence of GH, JAK2 is phosphorylated in liver and skeletal muscle (chicken: Wang et al., 2005). There is an association between polymorphisms in JAK2 and growth in chickens (Liu et al., 2010).

There is a circulating binding protein for avian GH (GHBP). A transcript for the GHBP derived from the GHR gene has been identified (chicken: Lau et al., 2007). Alternatively, GHBP is viewed as generated by proteolysis of the GHR (chicken: Vleurick et al., 1999).

23.5.4 Control of GH Secretion

The release of GH from the avian pituitary gland is under hypothalamic control, with multiple stimulatory releasing hormones (including GHRH, TRH, and ghrelin) and one inhibitory releasing hormone (SS) (Tables 23.2 and 23.3).

23.5.4.1 Growth Hormone–Releasing Hormone

The structure of chicken GHRH (shown here) and its precursor has been established based on the sequence of cDNA (Wang et al., 2006, 2007).

Chicken GHRH: HADAIFTDNYRKFLGQISARKFLQTIIGKRL-RNSESSPGEGVHKLLT

Two precursors for GHRH have been identified, one also containing the sequence for PACAP (Wang et al., 2007). GHRH is expressed in avian hypothalamus and also in the small intestine, liver, kidneys, lung, ovary, anterior pituitary gland, spleen, and testes (Wang et al., 2007). There is high expression of GHRH in the pituitary gland on embryonic day 8, but subsequently this declines to a low plateau level on days 12–20 (Wang et al., 2006). GHRH increases GH release in chickens *in vivo* and *in vitro* (chicken: Scanes et al., 1984). Avian somatotrophs respond to GHRH with increased intracellular calcium (chicken: Scanes et al., 2007) (Table 23.3).

TABLE 23.3 Effects of Neuropeptides on the Intracellular Concentrations in Responsive Chicken Somatotropes

Neuropeptide	$\Delta[Ca^{2+}]_i$
GHRH	120±6
TRH	222±16
Ghrelin	94±16
PACAP	90±14

Data from Scanes et al. (2007).

23.5.4.1.1 Receptor

The structure of the chicken GHRHR has been determined (Porter et al., 2006; Toogood et al., 2006). GHRH specifically binds to the GHRHR (Wang et al., 2007). GHRHR is specifically expressed in the avian anterior pituitary gland with expression increased by GHRH and decreased by either SS or corticosterone (Porter et al., 2006). Large increases in pituitary expression of GHRHR occur between days 8 and 12 and again between days 12 and 16 in embryonic development (Wang et al., 2006).

23.5.4.2 Thyrotropin-Releasing Hormone

A physiological stimulus for GH release in birds is the tripeptide TRH (Table 23.2). Furthermore, TRH is found in the avian hypothalamus (as discussed in this chapter), with TRH receptors present in the cells of the caudal lobe of the chicken pituitary gland (Harvey and Baidwan, 1989), where GH is produced. TRH stimulates GH secretion both *in vivo* and *in vitro* (chicken: Harvey et al., 1978a; turkey: Fehrer et al., 1985). Most somatotrophs respond to TRH with increased intracellular calcium (Scanes et al., 2007) (Table 23.3). TRH is a physiological regulator of GH secretion based on the ability of antiserum to TRH to suppress circulating concentrations of GH in young chicks (Klandorf et al., 1985).

23.5.4.3 Somatostatin

The structure of avian SS_{14} is identical to that of its mammalian homolog:

H–Ala–Gly–Cys–Lys–Asn–Phe–Phe–Trp–Lys–Thr–Phe–Thr–Ser–Cys–OH (pigeon: Spiess *et al.*, 1979; chicken: Hasegawa *et al.*, 1984).

Two somatostatin precursor variants have been identified in the avian brain: pro-somatostatin (PSS1) and the related peptide pro-cortistatin (PSS2) (Trabucchi et al., 2003). The only SS influencing anterior pituitary hormone secretion is most likely to be SS1.

Somatostatin inhibits GH release in birds. Not only are there SS neurons present in the hypothalamus (duck: Blahser et al., 1978), but also SS binding has been demonstrated in the chicken anterior pituitary gland (Harvey et al., 1990). Synthetic SS depresses both the *in vivo* and *in vitro* GH response to TRH (chicken: Harvey et al., 1978a) acting via SSTR2 (Bossis and Porter, 2001; Geris et al., 2003a).

23.5.4.4 Ghrelin

23.5.4.4.1 Actions and Chemistry

Ghrelin stimulates release of GH in the chicken (Ahmed and Harvey, 2002). Only a minority of chicken somatotrophs (about one-fifth) respond to ghrelin with increased intracellular calcium (Scanes et al., 2007). Ghrelin immunoreactivity is reported in the chicken hypothalamus (Ahmed and Harvey, 2002). The structure of avian ghrelin has been established (e.g., chicken: Kaiya et al., 2002):

Chicken:	GS<u>S</u>FLSPTYKNIQQQKDTRKPTARLH
Turkey:	GS<u>S</u>FLSPAYKNIQQQKDTRKPTARLH
Duck:	GS<u>S</u>FLSPEFKKIQQQNDPAKATAKIH
Goose:	GS<u>S</u>FLSPEEKKIQQQNDPTKTTAKIH
Emu:	GS<u>S</u>FLSPDYKKIQQRKDPRKPTTKLH

<u>**S**</u> octanyl [CH_3–$(CH2)_8C$=O.O–] acylation site (Based on Richards et al., 2006).

23.5.4.4.2 Receptor

Ghrelin acts via the GH secretagogue receptor (GHSR). This has been characterized in birds (e.g., Japanese quail: Kitazawa et al., 2009). There are at least two forms of GHSR in the avian anterior pituitary gland: GHS-R1a and GHS-R1c. Expression of these is depressed by ghrelin, GHRH, and corticosterone (chicken: Geelissen et al., 2003).

23.5.4.4.3 Extrahypothalamic Expression of Ghrelin

In addition to the hypothalamus, there is high expression of ghrelin in the proventriculus and proventriculus–gizzard junction, together with some in the gizzard and duodenum (Shao et al., 2010).

23.5.4.5 Other Neuropeptides and Hormones

Other neuropeptides influence the release of GH. GH release is also stimulated by PACAP, another peptide from the avian GHRH gene (Peeters et al., 1998). Subpopulations of somatotrophs respond to PACAP with increased intracellular calcium (Scanes et al., 2007). The structure of chicken PACAP is as follows:

Chicken PACAP 38: HIDGIFTDSYSRYRKQQMAVKKYLAAVL-GKRYKQRVKNK (Based on Wang et al., 2007).

PACAP binds specifically to the PACAPR (Wang et al., 2007). In addition, leptin increased intracellular calcium in chicken somatotrophs (Scanes et al., 2007). PRP31 depresses circulating concentrations of GH (Tachibana et al., 2011).

Growth hormone secretion in birds is inhibited by two peripheral hormones: IGF1 and T_3 in a negative feedback manner. IGF1 inhibits GH release *in vitro* and *in vivo* (chickens: Perez et al., 1985; Buonomo et al., 1987). Elevated levels of T_3 reduce GH release (chicken *in vivo*: Scanes and Harvey, 1989; *in vitro*: Donoghue and Scanes, 1991).

23.5.4.6 Nutrition and GH Secretion

Nutritional deprivation increases the plasma concentration of GH in birds. Plasma concentrations of GH are elevated by a short period (1–2 days) of starvation (chickens: Harvey et al., 1978b; turkeys: Anthony et al., 1990). Plasma concentrations of GH are also increased by calorie–protein deprivation in young chickens and turkeys (Engster et al., 1979). Similarly, long-term protein deficiency is accompanied by elevated plasma concentrations of GH in young chickens (Scanes et al., 1981).

23.5.5 GH Gene and Control of GH Expression

23.5.5.1 GH Gene

The structures of GH have been also reported deduced from the cDNA or genomic DNA from chicken (*Gallus gallus*) (NCBI Reference Sequence: NM_204359.2), chicken retinal GH (GenBank: AY373631.1), domestic duck (*Anas platyrhynchos*) GH (GenBank: X07079.1), domestic goose (*Anser anser*) GH (GenBank: AY149895.2), European pied flycatcher (*Ficedula hypoleuca*) (GenBank: DQ218278.1), Japanese quail (*Coturnix coturnix*) (GenBank: FJ458436.1), jungle crow (*Corvus macrorhynchos*) GH1A (GenBank: AB560855.1), ostrich (*Struthio camelus*) GH (GenBank: AB028191.1), turkey (*Meleagris gallopavo*) GH (GenBank: M33697.1), and zebra finch (*Taeniopygia guttata*) GH1 (NCBI Reference Sequence: XM_002196131.2).

In passerine birds, there is a duplicate GH gene with more rapid evolutionary changes (Yuri et al., 2008) and the structures of the second GH or GH-like proteins reported for the jungle crow (*C. macrorhynchos*) (Arai and Iigo, 2010) (GenBank: AB560856.1). Moreover, structures of GH-like proteins are reported in the zebra finch (*T. guttata*) (GH-like) (CBI Reference Sequence: XM_002187248.2). The reported structure of GH of the European pied flycatcher (*Ficedula hypoleuca*) (Buggiotti et al., 2006) appears to be more similar to GH-like or GH1B than to GH or GH1A. What are not known are the roles and control of GH-like and GH1B.

There are polymorphisms in the chicken GH gene associated with growth rate and the amount of adipose tissue (chicken: Nei et al., 2005; Zhang et al., 2007; Ouyang et al., 2008).

23.5.5.2 GH Expression

GH expression is under multiple stimulatory and inhibitory influences. For instance, GHRH increases GH mRNA *in vivo* (Vasilatos-Younken et al., 1992) and *in vitro* (Radecki et al., 1994) (Figure 23.4) via a cAMP-dependent protein kinase pathway (Kansaku et al., 1998). In addition, GH expression is increased by corticosterone in embryonic pituitary cells (Bossis and Porter, 2003; Heuck et al., 2009) and by TRH, PACAP, and ghrelin (Porter et al., 2006). In contrast, it is decreased by SRIF (Porter et al., 2006) and IGF1 (Scanes et al., 1999) (Figure 23.4). The changes in GH expression during growth are similar to those of plasma concentrations of GH (chicken: McCann-Levorse et al., 1993). The pattern of gene expression during the development of avian somatotrophs can be described by self-organizing maps (Ellestad et al., 2006).

23.5.6 Pituitary Origin of Growth Hormone

The somatotrophs are located in the caudal lobe of the anterior pituitary gland (Mikami, 1980; Reichardt et al., 1993) (Figure 23.1). During late embryological development and early posthatch life, the number of somatotrophs increases rapidly until a plateau level is achieved (dove: Reichardt et al., 1993; chicken: Malamed et al., 1997). There are ontogenic changes in the structure of the somatotroph. The number of secretory granule per unit volume of the cell increases during late embryonic development (chicken: Malamed et al., 1993), and there are also more in the adult than the young (chicken: Malamed et al., 1985, 1988). Somatotrophs are also reported in the pars tuberalis (see Section 23.2.1).

23.5.7 Extrapituitary Production of GH

In birds, GH and/or GH mRNA is present outside of the traditionally viewed site of production—the adenohypophyseal cells. GH has been found in gonads, retina, and immune tissues. An immunoreactive 17 KDa GH together with a 29 KDa GH have been reported in the testis of the adult chicken where an effect on spermatogenesis is implicated (Martínez-Moreno et al., 2011). There is also production of GH within the retina. *In vitro* chick embryo retinal ganglion cells contain immunoreactive 15 KDa GH and GHRs (reviewed Sanders et al., 2009). The GH exerts a neuroprotective role *in vitro* and *in ovo* (Sanders

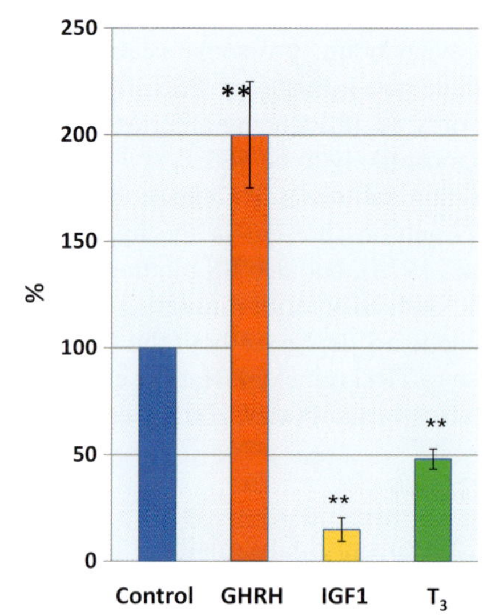

FIGURE 23.4 **Effect of GHRH *in vitro*, IGF1 infusion *in vivo*, and T_3 in the diet on expression of GH in the chicken anterior pituitary gland.** *Data from Radecki et al. (1994), Scanes et al. (1999).*

et al., 2010, 2011). There is evidence for GH exerting a paracrine effect on the ovary. Synthesis of progesterone by chicken granulosa cells is stimulated by GH (Ahumada-Solórzano et al., 2012). The presence of immunoreactive GH together with expression of GH have been reported in the chicken ovary, specifically the granulosa layer of the F5-2 follicles (Ahumada-Solórzano et al., 2012). Moreover, there is expression of GHR in both the F2 and F4 follicles (Ahumada-Solórzano et al., 2012). GH immunoreactivity is present in the bursa Fabricius, spleen, and thymus (chicken: Luna et al., 2005), with GH expression demonstrated in the bursa Fabricius (chicken: Luna et al., 2008). GH-containing neurons have been observed in dove and turkey brains (Ramesh et al., 2000).

23.5.8 Ontogeny of GH

23.5.8.1 Pou 1 (Pit 1)

Pou 1 (Pit 1) can transactivate a promoter for the GH gene (chicken: Murase et al., 2011). Avian Pou 1 (Pit 1) has been characterized (turkey: Wong et al., 1992; chicken: Tanaka et al., 1999) and is first expressed on embryonic day 5 (chicken: Van As et al., 2000). Two Pou 1 (Pit 1) transcripts, Pit1α and Pit1γ, are expressed in both the cephalic and caudal lobes of the chicken pituitary gland, with expression increasing in posthatching growth in a manner parallel to that of GH (chicken: Tanaka et al., 1999). Pou 1 (Pit 1) has been demonstrated to be present in somatotrophs (turkey: Weatherley et al., 2001). While no changes in Pou 1 (Pit 1)-expressing cells occur with corticosterone (chicken: Fu and Porter, 2004), TRH increases expression of Pou 1 (Pit 1) (Van As et al., 2004).

23.5.8.2 Somatotrophs and Their Differentiation

The embryonic increase in numbers of somatotrophs begins about day 14 of incubation (Porter et al., 1995). GHRH increases the number of somatotrophs (Porter et al., 1995). This increase is due to stimulatory effects of corticosterone and probably also of GHRH and thyroid hormones (Dean and Porter, 1999; Liu and Porter, 2004). The glucocorticoid induction of somatotroph differentiation involves both type 1 (mineralocorticoid) and type 2 (glucocorticoid) receptors (Bossis et al., 2004). Corticosterone induction of the GH gene is via a *cis*-acting glucocorticoid responsive element (Heuck-Knubel et al., 2012). It is likely that the corticosterone inducing differentiation of somatotrophs is of adrenal origin, but steroidogenic enzymes are expressed in the embryonic anterior pituitary gland (Zheng et al., 2008).

23.5.8.3 Secretion

A common pattern of circulating concentrations of GH has been observed in all species of birds examined. This consists of progressively rising circulating concentrations of GH in late embryonic development and early posthatch development, high plasma concentrations of GH during the period of rapid posthatching growth, and low GH concentrations in older and adult birds (Harvey et al., 1979a,b). The mechanistic basis for this ontogenic profile includes changes in GH synthesis, somatotroph numbers, and pituitary sensitivity to secretagogues. The decline in plasma concentrations of GH in the middevelopment and growth does not appear to be due to gonadal steroids as castration does not prevent the decrease (Scanes and Johnson, 1984). The ontogenic profile for GH mRNA is very similar to that of plasma concentrations of GH (McCann-Levorse et al., 1993).

23.6 PROLACTIN

23.6.1 Chemistry

Avian prolactin (PRL) has been isolated from pituitary tissue (Papkoff et al., 1982). In addition, recombinant chicken PRL has been purified following expression in *Escherichia coli* (Hanks et al., 1989a). The nucleotide sequence of avian PRL cDNA has been established, and the amino acid residue sequence deduced (chicken: Hank et al., 1989b; Watahiki et al., 1989).

23.6.2 Variants

As in mammals, posttranslationally modified variants of PRL exist in birds. In the turkey, two glycosylated PRL variants (one *O*- and one *N*-glycosylated) have been identified in addition to a nonglycosylated form (Corcoran and Proudman, 1991). These have reduced radioreceptor activity (Corcoran and Proudman, 1991). A glycosylated PRL is reported in the chicken (Berghman et al., 1992). Avian PRL can also be phosphorylated (Arámburo et al., 1992). Physiological state influences pituitary levels of PRL variants. For instance, there are shifts in the proportion of variants; between nonlaying, laying, and incubating hens (chicken: Hiyama et al., 2009); and during development (turkey: Bédécarrats et al., 1999a,b,c).

23.6.3 Actions of PRL

23.6.3.1 PRL and the Crop Sac Gland

PRL stimulates the production of "crop milk" from the crop sac gland in pigeons. There is considerable proliferation of the germinal cell layer of the crop, with expression of specific genes including keratin in nursing pigeons (Gillespie et al., 2013). Toward the end of when doves are incubating eggs, the crop increases in weight and there is a concomitant increase in the plasma concentrations of PRL (Goldsmith et al., 1981). This unique feature of the Columbidae (pigeons and doves) has been used as the basis of a biological assay for prolactin (Nicholl, 1967). The effects of prolactin on the crop sac gland are reported to be enhanced by synlactin, a putative PRL-induced liver factor (Nicoll et al., 1985). In an analogous manner, a serum factor potentiates the effect of prolactin on Nb2 rat lymphoma cells (McNeilly and Friesen, 1985). The identity of synlactin is not established but may be IGF1 or an IGF-binding protein. PRL also directly influences behavior as discrete PRL binding sites (receptors) have been reported in dove brains (Buntin et al., 1993).

23.6.3.2 PRL and Incubation Behavior and Broodiness

In birds, incubation behavior and broodiness are controlled by PRL. Administration of PRL induces broodiness (chickens: Riddle et al., 1935; turkey: Youngren et al., 1991). Incubation behavior is completely suppressed in turkey hens actively immunized against PRL (Crióstomo et al., 1998). In turkeys, PRL acts centrally to increase nesting activity and together with sex steroids to induce incubating behavior (El Halawani et al., 1986; Youngren et al., 1991). Similarly in bantam chickens, PRL increases incubation behavior, with increased nesting behavior in previously nest-deprived birds (Sharp et al., 1988). A positive feedback loop between PRL and incubation is envisaged.

Release of PRL and hence plasma concentrations of PRL are very high when birds are incubating their eggs (chickens: Sharp et al., 1988; turkeys: El Halawani et al., 1988; Wong et al., 1991). Circulating concentrations of PRL are also elevated during incubation in wild birds, for instance in both male and female Manx shearwater (Riou et al., 2010). Furthermore, the plasma concentrations of PRL fall rapidly if incubation behavior or nesting is interrupted (El Halawani et al., 1980; Tong et al., 1997).

23.6.3.3 PRL and Reproduction

Prolactin has both stimulatory and inhibitory effects on avian reproduction (Li et al., 2011), with effects at the hypothalamic levels inducing incubation behavior (considered in this chapter); the anterior pituitary with PRL decreases expression of the LH β-subunit (turkey: You et al., 1995) and possibly gonadal levels also.

23.6.3.4 PRL and Osmoregulation

PRL may have an osmoregulatory role in avian embryos. Allantoic concentrations of both sodium and chloride are depressed by ovine or bovine PRL, while metanephric kidney Na^+–K^+–ATPase activity is increased (Doneen and Smith, 1982; Murphy et al., 1986).

23.6.3.5 PRL Receptor (PRLR)

The amino acid sequence of the avian PRLR has been established based on the nucleotide sequence of PRLR cDNA (chicken: Tanaka et al., 1992; pigeon: Chen and Horseman, 1994; goose: Xing et al., 2011). The avian receptor, like its mammalian homolog, consists of extracellular, transmembrane, and intracellular domains. However, the PRLR has a unique feature with a double-antenna structure of two putative PRL binding sites in the extracellular domain.

The PRLR is expressed in the following tissues in the turkey: brain (and hypothalamus), brood patch, crop sac, duodenum, gizzard, heart, kidney, leukocytes, liver, lung, ovary, pancreas, pectoralis muscle, pituitary gland, shell gland, and spleen (Zhou et al., 1996). Similarly, PRLR is expressed in multiple tissues in the chicken (Tanaka et al., 1992; Kang et al., 2007). In the goose, PRLR is expressed in the testes, ovary, oviduct, kidney, and small and large intestine (Xing et al., 2011).

There are marked changes in PRLR expression with physiological state: more PRLR expression in the pituitary gland and shell gland but less in the hypothalamus in photostimulated laying hens, and further increases in incubating birds (turkey: Zhou et al., 1996). There are large perihatching increases in PRLR expression in the hypothalamus, pituitary gland, liver, pancreas, kidneys, and gonads (chickens and turkeys: Bole-Feysot et al., 2007; Leclerc et al., 2007).

23.6.4 Control of PRL Release

Prolactin secretion is predominantly under stimulatory hypothalamic control, although some inhibitory influence is possible.

23.6.4.1 Vasoactive Intestinal Peptide (VIP)

VIP is the major hypothalamo-hypophysiotropic factor stimulating PRL release (Table 23.2).

23.6.4.1.1 Chemistry

The amino acid sequence of chicken VIP is as follows:

HSDAVFTDNYSRFRKQMAVKKLYNSVLT (Nilsson, 1975)

23.6.4.1.2 Activity

Prolactin release is stimulated by VIP (chickens *in vivo*: Macnamee et al., 1986; turkeys *in vivo*: Opel and Proudman, 1988; turkeys *in vitro*: Proudman and Opel, 1988). Similarly, VIP evokes a 10-fold increase in circulating concentrations of PRL in white-crowned sparrow but is not effective in another passerine bird, the Florida scrub jay (Maney et al., 1999). VIP is found in the avian hypothalamus, particularly in the median eminence (MacNamee et al., 1986).

23.6.4.1.3 Physiological Role

The physiological role of VIP in the control of both the release and expression of avian PRL is supported by studies with antisera. Immunoneutralization of VIP decreases circulating concentrations of prolactin in chicken (Sharp et al., 1989; El Halawani et al., 2000), decreases PRL expression in chicken (Avital-Cohen et al., 2011), and abolishes all prolactin-releasing activity of turkey hypothalamic extracts (El Halawani et al., 1990).

23.6.4.1.4 Hypothalamic Content

The hypothalamic VIP content varies with physiological state in a manner that is consistent with VIP being the PRL-releasing hormone. For instance, hypothalamic content of VIP is elevated in photostimulated compared to nonphotostimulated turkeys and is increased further in incubating hens (Rozenboim et al., 1993). Dopamine increases VIP expression in the infundibular nuclear complex (INF) of the hypothalamus (turkey: Bhatt et al., 2003).

23.6.4.1.5 Release and Expression

Dopamine stimulates release of VIP from the hypothalamus via D1 dopamine receptors (turkey: Youngren et al., 1996; Chaiseha et al., 1997). In addition, serotonin increases PRL secretion via hypothalamic effects on dopaminergic neurons and then release of VIP (turkey: El Halawani et al., 1995; Chaiseha et al., 2010).

23.6.4.2 Arginine Vasotocin (AVT)

AVT has some PRL-releasing activity (Table 23.2). Posterior pituitary extracts can provoke PRL release from turkey pituitary cells due to both AVT and VIP. AVT stimulates prolactin release, while antiserum to AVT partially neutralizes the PRL-releasing effects of posterior pituitary extracts (El Halawani et al., 1992).

23.6.4.3 Other Stimulatory Factors

Other peptides stimulate PRL secretion in birds. PRL release is stimulated by TRH (turkey *in vivo*: Saeed and El Halawini, 1986) and peptide histidine isoleucine (PHI) (turkey: Proudman and Opel, 1990b).

23.6.4.4 Prolactin (PRL)-Releasing Peptide (PrRP)

23.6.4.4.1 Chemistry

PrRP is an RF-amide peptide. The precursor (pre-proprolactin-releasing peptide) for avian PrRP has been characterized in the chicken from its cDNA; there are PrRPs with 20- and 31-amino-acid residues (Tachibana et al., 2011):

PrRP31: SRPFKHQIDNRSPEIDPFWYVGRGVRPIGRF-NH_2
PrRP20: SPEIDPFWYVGRGVRPIGRF-NH_2

23.6.4.4.2 Actions

Avian PrRP is thought to influence PRL secretion in birds. There is a small increase in circulating concentration of PRL after peripheral administration of an intermediate dose of PrRP31 in chickens, presumably reflecting increased prolactin release (Tachibana et al., 2011) (Table 23.2). In contrast, central administration of PrRP31 into the brain consistently depresses circulating concentrations of PRL, indicating an ultrashort feedback effect (Tachibana et al., 2011).

23.6.4.4.3 Receptors

Three receptors have been characterized that bind PrRP in birds: PrRPR1, PrRPR2, and *Carassius* RF-amide peptide (C-RFa) receptor (C-RFaR). There is moderate to high expression of PrRP, C-RFa, PrRPR1, PrRPR2, and C-RFaR in specific regions of the chicken brain, including the hypothalamus, hindbrain, and telencephalon (Wang et al., 2012). There is high pituitary expression of PrRP, C-RFa, and PrRPR1 together with modest expression of C-RFaR (Wang et al., 2012).

23.6.4.5 Dopamine

Dopamine exerts both stimulatory and inhibitory effects on avian PRL secretion. Dopamine directly inhibits VIP-stimulated PRL release from the anterior pituitary gland via D2 dopamine receptors (turkey: Youngren et al., 1998) (Table 23.2). Moreover, there are reduced numbers of dopamine receptors in the cephalic lobe of the anterior pituitary gland in broody birds (bantam chickens: MacNamee and Sharp, 1989). The stimulatory effects are at the level of the hypothalamus.

23.6.4.6 Other Inhibitory Influences

Avian lactotrophs also contain calcitonin (chicken: Maddineni et al., 2007). Based on the inverse relationship between the concentrations of PRL and calcitonin, it has been proposed that calcitonin inhibits PRL release (chicken: Maddineni et al., 2007).

23.6.5 PRL Expression

Expression of PRL is under hypothalamic control. Photostimulation of young chickens is accompanied by increased PRL expression in the anterior pituitary gland (Li et al., 2009). Moreover, PRL expression is greatly increased during incubation (turkey: Wong et al., 1991). VIP increases expression of PRL (Kang et al., 2004), and the effect is mediated by cAMP signal transduction (Kansaku et al., 1998; Kang et al., 2002) and/or protein kinase C (Sun and El Halawani, 1995). Moreover, immunizing turkeys against VIP decreases pituitary expression of PRL (El-Halawani et al., 2000; Ahn et al., 2001).

23.6.6 Origin of PRL

The PRL-producing cells (lactotrophs) are largely restricted to the cephalic lobe of the anterior pituitary gland (Figure 23.1) (Mikami, 1980). The relative proportion of lactotrophs changes with physiological state. For instance, the proportion drops from 12% in newly hatched pigeons to less than 2% in adult pigeons (Reichardt, 1993). There are large numbers of lactotrophs in the anterior pituitary gland of incubating turkeys (Ramesh et al., 2001). A rapid decrease in the number of lactotrophs occurs if the birds are removed from their nests (Ramesh et al., 2001). The numbers of lactotrophs declines during molting (chickens: Sandhu et al., 2010).

Pituicytes containing both GH and PRL, which are somatolactotrophs, have been reported to be present in the caudal lobe both with the ventral portion of the cephalic lobe and at the cephalic–caudal junction of the anterior pituitary gland during incubation (turkeys: Ramesh et al., 1998). Somatolactotrophs rapidly disappear if the incubating birds are nest deprived (Ramesh et al., 2001). The somatolactotrophs contribute to the very high PRL secretion during incubation (Ramesh et al., 1998). There are also PRL-producing cells in the pars tuberalis (see Section 23.2.2).

23.6.7 Extrapituitary Production of Prolactin

Neurons containing immunoreactive PRL have been reported in the brain (e.g., in the hippocampus and hypothalamus; Ramesh et al., 2000). A prolactin-like protein (PLP) (also known as prolactin 22) has been characterized in the chicken and also in fish but not mammals (Wang et al., 2010). The precursor has 222 amino acids, is about 33% homologous to PRL, and can bind to the PRL receptor (Huang et al., 2009; Wang et al., 2010). It is not clear whether a different receptor exists for PLP. There is high expression of PLP in the chicken brain together with the lungs, ovary, and testes; moderate expression in the kidneys, skeletal muscle, and heart; but little expression in the pituitary gland (Wang et al., 2010).

23.6.8 Ontogeny of PRL

Corticosterone increases the number of lactotrophs in the embryonic anterior pituitary gland (Fu et al., 2004) via the zipper protein GILZ. Corticosterone increases GILZ expression in the chick embryo anterior pituitary gland (Ellestad et al., 2009). Exogenous GILZ increases expression of prolactin in the chick embryo anterior pituitary gland (Ellestad et al., 2009). The increased numbers of lactotroph occur by recruitment, with little effect of corticosterone on lactotroph proliferation (chicken: Fu and Porter, 2004).

While Pit1 can activate the promoter for prolactin, it is not present in lactotrophs (turkey: Weatherley et al., 2001). The pattern of gene expression during the development of lactotrophs has been described by self-organizing maps (Ellestad et al., 2006).

23.7 ADRENOCORTICOTROPIC HORMONE

23.7.1 Chemistry of ACTH and Other POMC-Derived Peptides

ACTH is synthesized as part of POMC (Table 23.1). This POMC also contains the sequences of β-endorphin (β-EP), which in itself is part of β-lipotropin (β-LPH), together with α- and β-melanocyte-stimulating hormone (MSH). The cDNA for chicken POMC has been characterized (Takeuchi et al., 1999). ACTH, β-EP, and β-LPH have been purified from several avian species (see Figure 23.2). Avian ACTH, like its mammalian homolog, is a simple polypeptide containing 39 amino acids (Li et al., 1978; Chang et al., 1980; Hayashi et al., 1991):

Chicken: SYSMEHFRWGKPVGRKRRPIKVYPNGVDEESAESYPMEF
Ostrich: SYSMEHFRWGKPVGRKRRPVKVYPNGVQEETSEGFPLEF

In the mammalian pars intermedia, proteolytic cleavage of POMC results in the production of α-MSH. There is not a pars intermedia in birds. Some α-MSH is found in the avian pars distalis (Hayashi et al., 1991). α-MSH may exert biological effects distinct from those evoked by ACTH. For instance, α-MSH but not ACTH stimulates sodium excretion from the nasal salt glands of ducks (Ituzziza et al., 1992). β-LPH and β-EP have been purified from ostrich and turkey adenohypophyseal tissue (Chang et al., 1980; Naude et al., 1980, 1981a,b). As in mammals, the 31 amino acid sequence of ostrich β-EP is identical to the C-terminal of ostrich β-LPH.

23.7.2 Actions of ACTH

23.7.2.1 Effects on Adrenal Cortical Cells

ACTH acts to stimulate avian adrenal cortical cells to produce corticosterone (the major glucocorticoid in birds), aldosterone (the major mineralocorticoid in birds), and deoxycorticosterone (chicken: Carsia et al., 1985a,b; ducks: Collie et al., 1992; greater rheas: Lèche et al., 2009). ACTH evokes marked changes in the morphology of avian adrenocortical cells due to changes in the cytoskeleton (Cronshaw et al., 1992).

The effect of ACTH is mediated by melanocortin receptors and cAMP (Carsia et al., 1985a). ACTH influences the expression of over 100 genes, including MC2R, cAMP response element modulator, and the steroidogenic acute regulatory protein (chicken: Bureau et al., 2009). The adrenal response to ACTH varies with physiological state (Carsia et al., 1985b). For instance, sensitivity to ACTH declines during growth.

23.7.2.2 Melanocortin Receptor (MCR)

ACTH acts by binding to MCRs. These are members of the G protein–coupled superfamily of receptors. There are five avian MCRs: MC1-R to MC5-R. These have been characterized in the chicken: MC1-R (Takeuchi et al., 1996), MC2-R (Takeuchi et al., 1998), MC3-R (Takeuchi and Takahashi, 1998a), and MC4-R and MC5-R (Takeuchi and Takahashi, 1998b).

MC2-R mediates the effects of ACTH in the adrenal and is expressed in the adrenal gland (chicken: Takeuchi et al., 1998). The expression of MC2-R is increased by ACTH (chicken: Bureau et al., 2009). There is also expression of MC3-R in avian adrenal glands (Takeuchi and Takahashi, 1998a). MC1-R, MC4-R, and MC5-R are expressed in the hypothalamus (chicken: Dridi et al., 2006), playing a role in the control of feeding.

There are differences in the specificity of the receptors to ACTH and α-MSH. ACTH is more potent than αMSH in inhibiting ligand binding to each of the MC receptor types (Ling et al., 2004). ACTH is more potent in stimulating cAMP formation with MC2-R, but α-MSH and ACTH have similar ability to stimulate cAMP formation with MC1-R, MC3-R, MC4-R, and MC5-R (Ling et al., 2004).

23.7.3 Control of ACTH Release

Stress induces the release of ACTH from the anterior pituitary gland via the releasing hormone, CRH. ACTH subsequently elevates corticosterone production.

23.7.3.1 Corticotropin-Releasing Hormone

CRH stimulates ACTH secretion (Table 23.2). The cDNA for the chicken CRH has been sequenced, and the deduced structure of chicken CRH is identical to that in mammals (Vandenborne et al., 2005a):

Chicken CRH: SEEPPISLDLTFHLLREVLEMARAEQLAQQA-HSNRKLMEIIGK

Immunoreactive CRH is present in the avian hypothalamus, particularly in the median eminence (Jozsa et al., 1984). CRH stimulates ACTH release from chicken and duck pituitary cells *in vitro* (Carsia et al., 1986; Castro et al., 1986). The chicken CRH receptor has been characterized, with both CRH and urotensin-1 being effective ligands (Yu et al., 1996). The effect of CRH is mediated by CRH-R1 (De Groef et al., 2003a). A binding protein for CRH is found in brain extracts (chicken: Seasholtz et al., 2002), but its role in birds is not well established.

23.7.3.2 Other Hormones

Both AVT and MT stimulate ACTH release (Castro et al., 1986). Central administration of ghrelin increases circulating concentrations of corticosterone in young chickens (Saito et al., 2005), presumably due to elevated CRH release. There is also intrapituitary control of ACTH release with parathyroid hormone–related peptide and calcitonin enhancing ACTH release (chicken: Nakayama et al., 2011a,b,c). Corticotrophs have calcitonin receptors, with calcitonin enhancing CRH-stimulated ACTH release (chicken: Nakayama et al., 2011b,c).

23.7.3.3 Feedback

Glucocorticoids inhibit the release of ACTH via effects at both the hypothalamic and pituitary levels. This has demonstrated *in vivo* (Herold et al., 1992) and *in vitro* (Carsia et al., 1986). Corticosterone also reduces expression of CRH in the hypothalamus (chicken: Vandenborne et al., 2005b). While corticosterone does not affect CRH-R expression in the anterior pituitary gland (chicken Vandenborne et al., 2005b), paradoxically pituitary expression of VT2R, presumably in corticotrophs, is upregulated by corticosterone (Sharma et al., 2009). Protein-deprived chicks that have elevated circulating concentrations of corticosterone also have reduced circulating concentrations of ACTH (Carsia et al., 1988), presumably due to increased glucocorticoid negative feedback.

23.7.4 Control of POMC Expression

Expression of POMC mRNA in the chicken anterior pituitary gland is decreased by either water deprivation (when there is elevated AVT release) or testosterone administration (Sharma and Chaturvedi, 2011). Corticosterone administration *in vivo* does not influence expression of POMC in the anterior pituitary gland, but corticosterone is effective in reducing anterior pituitary expression of POMC *in vitro* (chicken: Vandenborne et al., 2005b; Sharma et al., 2009).

23.7.5 Origin of ACTH

POMC is expressed and processed to ACTH in corticotrophs, these being acidophilic cells in the cephalic lobe of the anterior pituitary gland (Hayashi et al., 1991; Gerets et al., 2000) (Figure 23.1). The chicken anterior pituitary gland has a high content of ACTH (1.6 μg) but low levels of α-MSH (10 ng) (Hayashi et al., 1991), with ACTH and α-MSH coexisting in the same cells (Iturriza et al., 1980; Hayashi et al., 1991). The number of chicken anterior pituitary cells that express POMC increases when cultured in the presence of CRH (Pals et al., 2006). VT2R is found in avian corticotrophs (Jurkevich et al., 2005).

23.7.6 Extrapituitary Production

POMC is expressed in multiple organs in addition to the anterior pituitary gland. Expression of POMC has been reported in the brain, kidneys, adrenal glands, adipose tissue, gonads, and uropygial gland (chicken: Takeuchi et al., 1999). There are perikarya in the infundibular nucleus and median eminence together with fibers in the preoptic area and in the medial basal hypothalamus that contain POMC (Gerets et al., 2000). In addition, immunoreactive POMC products are present in immune tissues such as both the thymus and bursa Fabricius (Franchini et al., 1999). POMC and prohormone convertase 1 and 2 (PC1 and PC2) are expressed in feather follicles (Yoshihara et al., 2011).

The MCR is expressed in multiple tissues. There is expression of MC2-R in the spleen (Takeuchi et al., 1998). There is expression of both MC4-R and MC5-R in the brain, adrenal glands, gonads, and adipose tissues of the chicken, with expression of MC4-R also in the spleen and that of MC5-R also in the kidneys and the uropygial gland (Takeuchi and Takahashi, 1998b). The roles of these POMC products are not well established. Feather follicles also express MC1-R, which binds α-MSH or ACTH and thereby changes feather coloration (Yoshihara et al., 2011).

23.7.7 Ontogeny of ACTH

The corticotrophs are first observed on day 7 in embryonic development in the domestic fowl (Jozsa et al., 1979). Plasma concentrations of ACTH and corticosterone increase between embryonic days 11 and 17 (chicken: Jenkins et al., 2007). The rising circulating concentration of corticosterone subsequently increases both lactotroph and somatotroph cells in the anterior pituitary gland.

23.8 OTHER ANTERIOR PITUITARY PEPTIDES

In addition to the "classical" hormones of the anterior pituitary gland, other proteins are expressed. These include the following.

23.8.1 Adiponectin

The anterior pituitary expresses both adiponectin (chicken: Maddineni et al., 2005) and its receptor (AdipoR1 and AdipoR2) (chicken: Ramachandran et al., 2007). Pituitary expression of adiponectin is influenced by physiology, being increased by 48 h of feed withdrawal (Maddineni et al., 2005). There was a concomitant decrease in expression of both receptor subtypes (Ramachandran et al., 2007).

23.8.2 Calcitonin

Calcitonin is expressed in the chicken anterior pituitary gland, co-localizing with prolactin in many lactotrophs (Maddineni et al., 2007). Evidence of a physiological role for calcitonin comes from the report of an inverse relationship between pituitary expression of prolactin and calcitonin in different reproductive states (Maddineni et al., 2007). Corticotrophs have calcitonin receptors, with calcitonin enhancing CRH-stimulated ACTH release (chicken: Nakayama et al., 2011b).

23.8.3 Chromogranin A

Chromogranin A is an acidic peptide that can be secreted with protein hormones. A peptide with close (80%) homologies to mammalian chromogranin A has been purified and sequenced for ostrich adenohypophyseal tissue (Lazure et al., 1990). The function of chromogranin A or peptide fragments is not well established.

23.8.4 Glucagon Receptor

There is moderate expression of the glucagon receptor in the pituitary gland in the chicken (Wang et al., 2008). The cellular location and the function of the glucagon receptor are not yet established.

23.8.5 Ovoinhibitor

The avian egg white protein, ovoinhibitor, is co-localized in some adenohypophyseal cell types, such as somatotrophs, gonadotrophs that produce LH, and corticotrophs, but not in lactotrophs (Oubre et al., 2003).

23.8.6 Parathyroid-Related Peptide

There is parathyroid-related peptide (PTHrP) binding in the plasma membrane preparation from the anterior pituitary glands (chicken: Nakayama et al., 2011a). Administration of PTHrP into the third ventricle increases release of ACTH (Nakayama et al., 2011a). This may represent either an effect on the corticotrophs by leakage or PTHrP increasing CRH release. Evidence for a direct effect on the anterior pituitary gland comes from the reports of PTHrP expression together with the presence of a PTH-like peptide (PTH-L) in the pars distalis (Pinheiro et al., 2010).

23.8.7 Peptides

There is evidence that tetrapeptides in the anterior pituitary gland stimulate the thyroid and increase circulating thyroid hormone in hypophysectomized chickens (Kusnik et al., 2008).

23.8.8 Steroidogenic Enzymes

The embryonic anterior pituitary gland expresses steroidogenic enzymes, including P450 cholesterol side chain cleavage and 3β-hydroxysteroid dehydrogenase-1 (Zheng et al., 2008).

23.9 FUNCTIONING OF THE PARS TUBERALIS

Until relatively recently, the functioning of the avian pars tuberalis had received relatively little attention.

23.9.1 Pars Tuberalis and Photoperiodism

There is now abundant evidence that the pars tuberalis plays an important role in avian circadian rhythmicity and photoperiodism. In the Japanese quail, photoperiodic stimulation for 14 h increases expression of the TSH β-subunit in the pars tuberalis. This is followed within 4 h with elevated expression of type 2 monodeiodinase, and hence T_3, in the hypothalamus (Nakao et al., 2008). Photostimulation can be mimicked by TSH administered into the hypothalamus, indicating that TSH from the pars tuberalis is acting in a retrograde manner (Nakao et al., 2008).

23.9.2 Pineal Effects on the Pars Tuberalis

Communication with the pineal gland and the pars tuberalis is supported by the presence of melatonin-binding sites (Japanese quail: Cozzi et al., 1993) and melatonin receptor (Mel(1c)) expression in the pars tuberalis (Kameda et al., 2002). Moreover, pars tuberalis expression of the common LH–FSH–TSH glycoprotein α-subunit is increased following pinealectomy (chicken: Kameda et al., 2002). Moreover, the diurnal changes in the expression of the glycoprotein hormone α-subunit in the pars tuberalis are controlled by pineal melatonin (chicken: Arai and Kameda, 2004).

23.9.3 Circadian Rhythms and the Pars Tuberalis

There are endogenous circadian rhythms within the pars tuberalis. Circadian genes *Cry1* and *BMAL1* are expressed in the pars tuberalis of the Japanese quail (Yasuo et al., 2004; Ikegami et al., 2009).

There is communication between the pars tuberalis and both photoreceptors in other tissues. Neurons containing the photoreceptive protein opsin-5 extend to the outer region of the median eminence adjacent to the pars tuberalis (Halford et al., 2009; Nakane et al., 2010). Expression of the FSH–LH–TSH α-subunit in the pars tuberalis is decreased by continuous light regimens and increased by extended darkness irrespective of whether the birds are intact or pinealectomized (Kameda et al., 2002).

23.10 NEUROHYPOPHYSIS

23.10.1 Introduction

The anatomy of the hypothalamic neurosecretory tracts (neurons) leading to their secretory terminals in the pars nervosa is shown in Figure 23.1. The two hormones of the neurohypophysis are AVT and MT (Table 23.1). There are two genes. The first is the avian OXT encoding preprooxyphysin, the proteolytic products being MT and neurophysin-1. The second is the avian AVP gene encoding the precursor protein for AVT and neurophysin-2.

AVT and MT are produced by and secreted from predominantly separate neurosecretory neurons. The cell bodies of the AVT and MT neurons are located in both separate and overlapping areas within the hypothalamus (Goosens et al., 1977; Bons, 1980; Tennyson et al., 1985). The AVT- and MT- containing cell bodies and axons develop in the chicken embryo between days 6 and 17 (Tennyson et al., 1986).

MT and AVT are synthesized as part of the same precursor molecule as, respectively, neurophysin-1 and -2. The neurophysins act as carrier proteins in the transport of AVT and MT by axoplasmic transport. The hormones are then stored in the pars nervosa prior to release. Two avian neurophysins have been characterized and sequenced in birds (e.g., ostrich: Lazure et al., 1987, 1989; goose: Michel et al., 1990). The hormones of the avian neurohypophysis are AVT and MT (Acher et al., 1970):

AVT (8-arginine oxytocin): CyS–Tyr–Ile–Glu(NH2)–Asp(NH2)–CyS–Pro–Arg–Gly(NH2);

MT (8-isoleucine oxytocin): CyS–Tyr–Ile–Glu(NH2)–Asp(NH2)–CyS–Pro–Ile–Gly(NH2).

AVT differs from arginine vasopressin, the mammalian antidiuretic hormone, by a single amino acid residue substitution (isoleucine at position 3 for phenylalanine). MT differs from the mammalian homolog, oxytocin, by the substitution of isoleucine for leucine (position 8).

The posterior pituitary gland contains high levels of AVT and MT (chicken: 4.0 μg AVT and 0.9 μg MT) (Robinzon et al., 1988a). Low concentrations of these neuropeptides are also found throughout the brain (Robinzon et al., 1988a) and in the ovary (Saito et al., 1990). It is presumed that the major source of circulating concentrations of AVT and mesotocin is the posterior pituitary gland, although there is significant ovarian production (Saito et al., 1990).

23.10.2 Actions of AVT

The major roles of AVT are in renal functioning and reproduction.

23.10.2.1 Renal Functioning

AVT is the major antidiuretic hormone in birds (Goldstein, 2006). The absence of AVT, either following surgical removal of the pars nervosa or in AVT-deficient chicken, is accompanied by very large increases in the volume of urine produced (Shirley and Nalbandov, 1956; Dunson et al., 1972). The administration of AVT to birds has an antidiuretic effect (Ames et al., 1971). AVT exerts its principal effect on tubule function (Gerstberger et al., 1985; Stallone and Braun, 1985), with increased expression of aquaporin-2 (quail: Lau et al., 2009). Based on studies in the desert quail, AVT reduces GFR by decreasing both the number of reptilian-type nephrons filtering and the single-nephron GFR of mammalian-type nephrons (Braun and Dantzler, 1974). AVT also promotes conservation of water via a second effect on glomerular filtration together with renal vascular shifts (chickens: Stallone and Braun, 1985; gulls: Gray and Erasmus, 1988). AVT decreases urine flow rate and glomerular filtration rate in house sparrows, respectively, via V2R and VT1R (vascular, presumably renal) receptor mechanisms (house sparrow: Goecke and Goldstein, 1997). Chickens with hereditary diabetes insipidus show reduced sensitivity to AVT (Brummermann and Braun, 1995).

23.10.2.2 Cardiovascular Effects of AVT

AVT can have both vasodepressor and vasopressor activity in birds. Bolus injections of AVT to conscious adult chickens or adult or young ducks results in a marked drop in mean arterial pressure (Wilson and West, 1986; Robinzon et al., 1988b). In contrast, infusion of AVT can increase arterial blood pressure in conscious chickens (Robinzon et al., 1993). In an analogous manner, bolus injection of AVT elevates heart rate while infusion of AVT depresses cardiac frequency (Wilson and West, 1986; Robinzon et al., 1988b, 1993).

23.10.2.3 Oviposition

AVT causes oviposition, the physical process of laying eggs in birds. The hen uterus contracts in response to AVT but not to MT (Koike et al., 1988). The mechanism by which AVT provokes premature oviposition may involve AVT stimulation of local production of prostaglandins (probably E_1) by the uterus, which, in turn, causes uterine contractions (Rzasa, 1978, 1984). Circulating concentrations of AVT rise markedly at the time of oviposition (as discussed further in this chapter). There is also increased uterine sensitivity to AVT (Rzasa, 1978). A receptor for AVT, VT3R, is expressed in both the endometrium and myometrium of the chicken shell gland (Gubrij et al., 2005).

23.10.2.4 Other Roles

Neurohypophyseal peptides influence the secretion of other hormones in birds. For instance, AVT can stimulate the release of PRL (El Halawani et al., 1992) and ACTH (Castro et al., 1986).

23.10.2.5 Receptors

The receptors for AVT and MT are members of the vasotocin (VT) family of receptors within the superfamily of G protein–coupled receptors. There are four genes for the avian receptors for AVT and MT. These are the following: VT1R (expression widespread in brain), VT2R (V1b) (high expression widespread in pituitary gland), VT3R (oxytocin-like) (high expression widespread in brain), and VT4R (V1a) in birds (white-throated sparrow and zebra finch: Leung et al., 2011). The VT3R receptor has also been characterized in the chicken (Gubrij et al., 2005).

There is little expression of the VT2R (V1b) in the brain based on studies in songbirds (Leung et al., 2011). The expression of three VT receptors, VT1R, VT3R, and VT4R (V1a), has been reported in the brain of white-throated sparrows and zebra finches (Leung et al., 2011), and expression is high in regions associated with behavioral effects. The stimulatory effect of AVT on the release of avian ACTH is mediated by the VT2R, these being expressed on corticotrophs (Jurkevich et al., 2005). Surprisingly, in view of its negative feedback effect, administration of corticosterone to chickens increases both the expression of VT2R and VT2R immunoreactivity (Sharma et al., 2009). Expression of VT2R in the chicken anterior pituitary gland is increased by water deprivation or testosterone administration (Sharma and Chaturvedi, 2011). The VT3 receptor is expressed in the myometrium of the shell gland of reproductive Japanese quail (Srivastava et al., 2010).

23.10.3 Actions of MT

There is less information on the physiological role of MT in birds. While there are no changes in circulating concentrations of MT at the time of oviposition (Nouwen et al., 1984; Koike et al., 1988), MT potentiates the stimulatory effect of AVT on oviposition in laying hens by effects on AVT binding to the uterus (Takahashi and Kawashima, 2008a). Binding of mesotocin to receptors in the uterus is reported in the laying hen (Takahashi and Kawashima, 2008b). MT does not appear to influence the major cardiovascular indices of arterial blood pressure or heart rate. MT does influence blood flow to some organs (as indicated by reduced temperature of the shank and comb) (Robinzon et al., 1988b). In addition, MT infusion reduces circulating concentrations of aldosterone (Robinzon et al., 1988b).

23.10.4 Behavioral Effects of Mesotocin and AVT

Mesotocin and AVT, produced and acting within the brain, appear to influence behaviors such as bonding in birds. Administration of an oxytocin antagonist increased the latency to pair bonding and decreased pair bonding in zebra finches (Pedersen and Tomaszycki, 2012). The sociability of female zebra finches is reduced by infusions of an oxytocin antagonist into the septum (Goodson et al., 2009). There appears to be a link with flocking behavior, with a flocking- and winter-associated increase in mesotocin in the dorsal lateral septum (LS) and medial amygdala of field sparrows but not in territorial year-round song sparrows (Goodson et al., 2012). When turkey hens incubating eggs are exposed to young poults, there is c-fos mRNA expression of mesotocin-containing neurons in the nucleus paraventricularis magnocellularis (PVN) and nucleus supraopticus, pars ventralis (SOv) (Thayananuphat et al., 2011). Intracerebroventricular injection of an oxytocin agonist reduces brooding behavior when turkey hens incubating eggs are exposed to poults (Thayananuphat et al., 2011).

23.10.5 Control of AVT and MT Release

Table 23.4 summarizes the effects of factors that influence circulating concentrations of AVT. The release of AVT and MT is under independent control. For instance, heat stress increases plasma concentrations of AVT and depresses those of mesotocin (Wang et al., 1989).

23.10.5.1 Control of AVT Release Related to Renal Functioning

AVT is released when blood osmolality is high (see Table 23.4) and to prevent water loss. Frequently, conditions where blood osmolality is increased also reduce blood volume. The release of AVT is under osmotic rather than volemic control (Stallone and Braun, 1986). In the chicken, plasma concentrations of AVT are depressed by angiotensin-2, with the effects presumed to be direct on the neurohypophysis due to the presence of angiotensin receptors that, in turn, are reduced prior to oviposition (Takahashi et al., 2011).

23.10.5.2 Control of AVT Release Related to Oviposition

At the time of oviposition, circulating concentrations of AVT are greatly increased (Sturkie and Lin, 1966; Nouwen et al., 1984; Koike et al., 1988). In laying chickens, prostaglandin $F_2\alpha$ increases plasma concentrations of AVT (Shimada et al., 1987), with a decrease in prostaglandin $F_2\alpha$ receptors in the neurohypophysis (Takahashi and Kawashima, 2008c). The pars nervosa is the most likely source of AVT at the time of oviposition, but ovarian AVT may contribute. Follicular AVT levels decline immediately prior to oviposition (Saito et al., 1990).

There are also links with the pars nervosa and the ovary–oviduct. There is estrogen binding in the membrane fraction of the chicken neurohypophysis, with decreased binding prior to oviposition (Takahashi and Kawashima, 2009). Either estrogen or progesterone can decrease prostaglandin $F_2\alpha$ receptors in the neurohypophysis (Takahashi and Kawashima, 2008c, 2009).

Not surprisingly, with the large changes in calcium fluxes during egg shell deposition, there is cross-talk between calcium control and oviposition. There are calcitonin receptors in the hen neurohypophysis and preoptic area of the hypothalamus, with changes prior to oviposition (increased affinity and decreased capacity) (Nakayama et al., 2010, 2011b,c). Estrogen decreases CT receptor binding in the hen neurohypophysis (Nakayama et al., 2011d).

23.10.6 AVT and MT Expression

MT is expressed in cell bodies in the paraventricular nucleus, specifically the parvocellular, magnocellular, and periventricular subgroups (chicken: Barth et al., 1997). Some cell bodies express both MT and AVT (Barth et al., 1997). AVT is expressed in cell bodies in the supraoptic nucleus and specifically the ventral and external subgroups (chicken: Barth et al., 1997). Expression of the AVT gene is increased both at the time of oviposition (Japanese quail: Seth et al., 2004a; chicken: Seth et al., 2004b) and with water deprivation (Japanese quail: Seth et al., 2004a). Water deprivation increases the number of neurons containing AVT immunoreactivity (Japanese quail: Seth et al., 2004a).

TABLE 23.4 Factors Influencing Circulating Concentrations of AVT in Birds

Physiological Factor	Chicken	Duck	House Sparrow (*Passer domesticus*)
Oviposition	↑↑	–	–
Water deprivation	↑	↑	↑
Angiotensin	↑	→	–
Hemorrhage	↑	↑	–
Sodium chloride loading	↑	↑	–
Anesthesia	↓	–	–

Based on Scanes (2000).

REFERENCES

Acher, R., Chauvet, J., Chauvet, M.T., 1970. Phylogeny of the neurohypophyseal hormones: the avian active peptides. Eur. J. Biochem. 17, 509–513.

Ahmed, S., Harvey, S., 2002. Ghrelin: a hypothalamic GH-releasing factor in domestic fowl (*Gallus domesticus*). J. Endocrinol. 172, 117–125.

Ahn, J., You, S., Kim, H., Chaiseha, Y., El Halawani, M., 2001. Effects of active immunization with inhibin alpha subunit on reproductive characteristics of turkey hens. Biol. Reprod. 65, 1594–1600.

Ahumada-Solórzano, S.M., Carranza, M.E., Pedernera, E., Rodríguez-Méndez, A.J., Luna, M., Arámburo, C., 2012. Local expression and distribution of growth hormone and growth hormone receptor in the chicken ovary: effects of GH on steroidogenesis in cultured follicular granulosa cells. Gen. Comp. Endocrinol. 175, 297–310.

Al-Musawi, S.L., Gladwell, R.T., Knight, P.G., 2007. Bone morphogenetic protein-6 enhances gonadotrophin-dependent progesterone and inhibin secretion and expression of mRNA transcripts encoding gonadotrophin receptors and inhibin/activin subunits in chicken granulosa cells. Reproduction 134, 293–306.

Ames, E., Steven, K., Skadhauge, E., 1971. Effect of arginine vasotocin on renal excretion of Na^+, K^+, Cl^-, and urea in the hydrated chicken. Am. J. Physiol. 221, 1223–1228.

Ando, H., Ishii, S., 1994. Molecular cloning of complementary deoxyribonucleic acids for the pituitary glycoprotein hormone alpha-subunit and luteinizing hormone beta-subunit precursor molecules of Japanese quail (*Coturnix coturnix japonica*). Gen. Comp. Endocrinol. 93, 357–368.

Anthony, N.B., Vasilatos-Younken, R., Bacon, W.L., Lilburn, M.S., 1990. Secretory pattern of growth hormone, insulin and related metabolites in growing male turkeys: effect of overnight fasting and refeeding. Poult. Sci. 69, 801–811.

Aoki, Y., Ono, H., Yasuo, S., Masuda, T., Yoshimura, T., Ebihara, S., Iigo, M., Yanagisawa, T., 2007. Molecular evolution of prepro-thyrotropin-releasing hormone in the chicken (*Gallus gallus*) and its expression in the brain. Zoolog. Sci. 24, 686–692.

Arai, N., Iigo, M., 2010. Duplicated growth hormone genes in a passerine bird, the jungle crow (*Corvus macrorhynchos*). Biochem. Biophys. Res. Commun. 397, 553–558.

Arai, Y., Kameda, Y., 2004. Diurnal rhythms of common alpha-subunit mRNA expression in the pars tuberalis of hamsters and chickens. Cell Tissue Res. 317, 279–288.

Arámburo, C., Donoghue, D., Montiel, J.L., Berghman, L.R., Scanes, C.G., 1990. Phosphorylation of chicken growth hormone. Life Sci. 47, 947–952.

Arámburo, C., Montiel, J.L., Proudman, J.A., Berghman, L.R., Scanes, C.G., 1992. Phosphorylation of prolactin and growth hormone. J. Mol. Endocrinol. 8, 183–191.

Arámburo, C., Luna, M., Carranza, M., Reyes, M., Martínez-Coria, H., Scanes, C.G., 2000. Growth hormone size variants: changes in the pituitary during development of the chicken. Proc. Soc. Exp. Biol. Med. 223, 67–74.

Arámburo, C., Carranza, M., Reyes, M., Luna, M., Martinez-Coria, H., Berúmen, L., Scanes, C.G., 2001. Characterization of a bioactive 15 kDa fragment produced by proteolytic cleavage of chicken growth hormone. Endocrine 15, 231–240.

Asem, E.K., Molnar, M., Hertelendy, F., 1987. Luteinizing hormone-induced intracellular calcium mobilization in granulosa cells: comparison with forskolin and 8-bromo-adenosine 3′,5′ monophosphate. Endocrinology 120, 853–859.

Avital-Cohen, N., Heiblum, R., Argov, N., Rosenstrauch, A., Chaiseha, Y., Mobarkey, N., Rozenboim, I., 2011. The effect of active immunization against vasoactive intestinal peptide and inhibin on reproductive performance of young White Leghorn roosters. Poult. Sci. 90, 2321–2331.

Barth, S.W., Bathgate, R.A., Mess, A., Parry, L.J., Ivell, R., Grossmann, R., 1997. Mesotocin gene expression in the diencephalon of domestic fowl: cloning and sequencing of the MT cDNA and distribution of MT gene expressing neurons in the chicken hypothalamus. J. Neuroendocrinol. 9, 777–787.

Bédécarrats, G., Guémené, D., Morvan, C., Crisóstomo-Pinto, S., Kühnlein, U., Zadworny, D., 1999a. In vitro release of isoforms of prolactin from pituitary glands of turkey hens at different physiological stages. Gen. Comp. Endocrinol. 113, 105–111.

Bédécarrats, G., Guémené, D., Kühnlein, U., Zadworny, D., 1999b. Changes in levels of immunoreactive prolactin isoforms during a reproductive cycle in turkey hens. Gen. Comp. Endocrinol. 113, 96–104.

Bédécarrats, G., Guémené, D., Morvan, C., Kühnlein, U., Zadworny, D., 1999c. Quantification of prolactin messenger ribonucleic acid, pituitary content and plasma levels of prolactin, and detection of immunoreactive isoforms of prolactin in pituitaries from turkey embryos during ontogeny. Biol. Reprod. 61, 757–763.

Berghman, L.R., Grauwels, L., Vanhamme, L., Proudman, J.A., Foidart, A., Balthazart, J., Vandesande, F., 1992. Immunocytochemistry and immunoblotting of avian prolactins using polyclonal and monoclonal antibodies toward a synthetic fragment of chicken prolactin. Gen. Comp. Endocrinol. 85, 346–357.

Berghman, L.R., Lens, P., Decuypere, E., Kuhn, E.R., Vandesande, F., 1987. Glycosylated chicken growth hormone. Gen. Comp. Endocrinol. 68, 408–414.

Berumen, L.C., Luna, M., Carranza, M., Martínez-Coria, H., Reyes, M., Cárabez, A., Arámburo, C., 2004. Chicken growth hormone: further characterization and ontogenic changes of an N-glycosylated isoform in the anterior pituitary gland. Gen. Comp. Endocrinol. 139, 113–123.

Bhatt, R., Youngren, O., Kang, S., El Halawani, M., 2003. Dopamine infusion into the third ventricle increases gene expression of hypothalamic vasoactive intestinal peptide and pituitary prolactin and luteinizing hormone beta subunit in the turkey. Gen. Comp. Endocrinol. 130, 41–47.

Blahser, S., Fellman, D., Bugnon, C., 1978. Immunocytochemical demonstration of somatostatin-containing neurons in the hypothalamus of the domestic mallard. Cell Tissue Res. 195, 183–187.

Bole-Feysot, C., Goffin, V., Leclerc, B., Zadworny, D., Bédécarrats, G., Kühnlein, U., 2007. Development of a real-time (Q) PCR assay to measure variation in expression of prolactin receptor mRNA in the hypothalamus and pituitary gland during late embryogenesis in turkeys and chickens. Gen. Comp. Endocrinol. 150, 319–325.

Bøler, J., Enzmann, F., Bowers, C.Y., Schally, A.V., 1969. The identity of chemical and hormonal properties of the thyrotropin releasing hormone and pyroglutamyl-histidyl-proline amide. Biochem. Biophys. Res. Commun. 37, 705–710.

Bonney, R.C., Cunningham, F.J., 1977. A role for cyclic AMP as a mediator for the action of LH-RH on chicken anterior pituitary cells. Mol. Cell. Endocrinol. 7, 233–244.

Bons, N., 1980. The topography of mesotocin and vasotocin systems in the brain of the domestic mallard and Japanese quail: immunocytochemical identification. Cell Tissue Res. 213, 37–51.

Bossis, I., Porter, T.E., 2001. Identification of the somatostatin receptor subtypes involved in regulation of growth hormone secretion in chickens. Mol. Cell. Endocrinol. 182, 203–213.

Bossis, I., Porter, T.E., 2003. Evaluation of glucocorticoid-induced growth hormone gene expression in chicken embryonic pituitary cells using a novel in situ mRNA quantitation method. Mol. Cell. Endocrinol. 201, 13–23.

Bossis, I., Nishimura, S., Muchow, M., Porter, T.E., 2004. Pituitary expression of type I and type II glucocorticoid receptors during chicken embryonic development and their involvement in growth hormone cell differentiation. Endocrinology 145, 3523–3531.

Braun, E.J., Dantzler, W.D., 1974. Effects of ADH on single-nephron glomerular filtration rates in the avian kidney. Am. J. Physiol. 226, 1–8.

Brown, N.L., Bayle, J.D., Scanes, C.G., Follett, B.K., 1975. The actions of avian LH and FSH on the testes of hypophysectomized quail. Cell Tissue Res. 156, 499–520.

Brummermann, M., Braun, E.J., 1995. Renal response of roosters with diabetes insipidus to infusions of arginine vasotocin. Am. J. Physiol. 269, R57–R63.

Buggiotti, L., Primmer, C.R., 2006. Molecular evolution of the avian growth hormone gene and comparison with its mammalian counterpart. J. Evol. Biol. 19, 844–854.

Buntin, J.D., Ruzychi, E., Witebsky, J., 1993. Prolactin receptors in dove brain: autoradiographic analysis of binding characteristics in discrete brain regions and accessibility to blood-borne prolactin. Neuroendocrinology 57, 738–750.

Buonomo, F.C., Lauterio, T.J., Baile, C.A., Daughaday, W.H., 1987. Effects of insulin like growth factor I (IGF-I) on growth hormone-releasing hormone (GRF) and thyrotropin-releasing hormone (TRH) stimulation of growth hormone (GH) secretion in the domestic fowl (*Gallus domesticus*). Gen. Comp. Endocrinol. 66, 274–279.

Bureau, C., Hennequet-Antier, C., Couty, M., Guémené, D., 2009. Gene array analysis of adrenal glands in broiler chickens following ACTH treatment. BMC Genomics 10, 430.

Burke, W.H., Licht, P., Papkoff, H., Bona Gallo, A., 1979a. Isolation and characterization of luteinizing hormone and follicle-stimulating hormone for pituitary glands of the turkey (*Meleagris gallopavo*). Gen. Comp. Endocrinol. 37, 508–520.

Burke, W.H., Papkoff, H., Licht, P., Gallo, A.B., 1979b. Preparation and properties of luteinizing hormone (LH) subunits from the turkey (*Meleagris gallopavo*) and their recombination with subunits of ovine LH. Gen. Comp. Endocrinol. 37, 501–507.

Burke, W.H., Moore, J.A., Ogez, J.R., Builder, S.E., 1987. The properties of recombinant chicken growth hormone and its effects on growth, body composition, feed efficiency and other factors in broiler chickens. Endocrinology 120, 651–658.

Burnside, J., Cogburn, L.A., 1992. Developmental expression of hepatic growth hormone receptor and insulin-like growth factor-I mRNA in the chicken. Mol. Endocrinol. 89, 91–96.

Burnside, J., Liou, S.S., Zhong, C., Cogburn, L.A., 1992. Abnormal growth hormone receptor gene expression in the sex-linked dwarf chicken. Gen. Comp. Endocrinol. 88, 20–28.

Burnside, J., Liou, S.S., Cogburn, L.A., 1991. Molecular cloning of the chicken growth hormone receptor complimentary DNA: mutation of the gene in sex-linked dwarf chickens. Endocrinology 128, 3183–3192.

Calvo, F.O., Bahr, J.M., 1983. Adenylyl cyclase system of the small preovulatory follicles of the domestic hen: responsiveness to-stimulating hormone and luteinizing hormone. Biol. Reprod. 29, 542–547.

Campbell, R.M., Scanes, C.G., 1985. Lipolytic activity of purified pituitary and bacterially derived growth hormone on chicken adipose tissue in vitro. Proc. Soc. Exp. Biol. Med. 180, 513–517.

Campbell, R.M., Scanes, C.G., 1987. Growth hormone inhibition of glucagon- and cAMP-induced lipolysis by chicken adipose tissue in vitro. Proc. Soc. Exp. Biol. Med. 184, 456–460.

Campbell, R.M., Chen, W.Y., Wiehl, P., Kelder, B., Kopchick, J.J., Scanes, C.G., 1993. A growth hormone (GH) analog that antagonizes the lipolytic effect but retains full insulin-like (antilipolytic) activity of GH. Proc. Soc. Exp. Biol. Med. 203, 311–316.

Catena, M.L., Porter, T.E., McNabb, F.M., Ottinger, M.A., 2003. Cloning of a partial cDNA for Japanese quail thyroid-stimulating hormone and effects of methimazole on the thyroid and reproductive axes. Poult. Sci. 82, 381–387.

Carsia, R.V., Scanes, C.G., Malamed, S., 1985a. Isolated adrenocortical cells of the domestic fowl (*Gallus domesticus*): steroidogenic and ultrastructural properties. J. Steroid. Biochem. 22, 273–279.

Carsia, R.V., Scanes, C.G., Malamed, S., 1985b. Loss of sensitivity to ACTH of adrenocortical cells isolated from maturing domestic fowl. Proc. Soc. Exp. Biol. Med. 179, 279–282.

Carsia, R.V., Weber, H., King, D.B., Scanes, C.G., 1985c. Adrenocortical cell function in the hypophysectomized domestic fowl: effects of growth hormone and 3,5,3′-triiodothyronine. Endocrinology 117, 928–933.

Carsia, R.V., Weber, H., Perez, F.M., 1986. Corticotropin releasing factor stimulates the release of adrenocorticotropin from domestic fowl pituitary cells. Endocrinology 118, 143–148.

Carsia, R.V., Weber, H., Lauterio, T.J., 1988. Protein malnutrition in the domestic fowl induced alterations in adrenocortical function. Endocrinology 122, 673–680.

Castro, M.G., Estivariz, F.E., Iturriza, F.C., 1986. The regulation of the corticomelanotropic cell activity in aves. II. Effect of various peptides on the release of ACTH from dispensed, perfused duck pituitary cells. Comp. Biochem. Physiol. A 83, 71–75.

Chaiseha, Y., Youngren, O.M., El Halawani, M.E., 1997. Dopamine receptors influence vasoactive intestinal peptide release from turkey hypothalamic explants. Neuroendocrinology 65, 423–429.

Chaiseha, Y., Kang, S.W., Leclerc, B., Kosonsiriluk, S., Sartsoongnoen, N., El Halawani, M.E., 2010. Serotonin receptor subtypes influence prolactin secretion in the turkey. Gen. Comp. Endocrinol. 165, 170–175.

Chang, W.C., Chung, D., Li, C.H., 1980. Isolation and characterization of β lipotropin and adrenocorticotropin from turkey pituitary glands. Int. J. Pept. Protein Res. 15, 261–270.

Chen, X., Horseman, N.D., 1994. Cloning, expression, and mutational analysis of the pigeon prolactin receptor. Endocrinology 135, 269–276.

Chen, L.R., Lee, S.C., Lin, Y.P., Hsieh, Y.L., Chen, Y.L., Yang, J.R., Liou, J.F., Chen, C.F., Lee, Y.P., Shiue, Y.L., 2010. Prostaglandin-D synthetase induces transcription of the LH beta subunit in the primary culture of chicken anterior pituitary cells via the PPAR signaling pathway. Theriogenology 73, 367–382.

Cheung, A., Hall, T.R., Harvey, S., 1988. Stimulation of corticosterone release in the fowl by recombinant DNA-derived growth hormone. Gen. Comp. Endocrinol. 69, 128–132.

Chowdhury, V.S., Yamamoto, K., Ubuka, T., Bentley, G.E., Hattori, A., Tsutsui, K., 2010. Melatonin stimulates the release of gonadotropin-inhibitory hormone by the avian hypothalamus. Endocrinology 151, 271–280.

Chou, H.-F., Johnson, A.L., Williams, J.B., 1985. Luteinizing hormone releasing activity of [Gln8]-LHRH and [His5, Trp7, Tyr8]-LHRH in the cockerel, in vivo and in vitro. Life Sci. 37, 2459–2465.

Ciccone, N.A., Dunn, I.C., Boswell, T., Tsutsui, K., Ubuka, T., Ukena, K., Sharp, P.J., 2004. Gonadotrophin inhibitory hormone depresses gonadotrophin alpha and follicle-stimulating hormone beta subunit expression in the pituitary of the domestic chicken. J. Neuroendocrinol. 16, 999–1006.

Ciccone, N.A., Dunn, I.C., Sharp, P.J., 2007. Increased food intake stimulates GnRH-I, glycoprotein hormone alpha-subunit and follistatin mRNAs, and ovarian follicular numbers in laying broiler breeder hens. Domest. Anim. Endocrinol. 33, 62–76.

Cogburn, L.A., Liou, S.S., Rand, A.L., McMurtry, J.P., 1989. Growth, metabolic and endocrine responses of broiler cockerels given a daily subcutaneous injection of natural or biosynthetic chicken growth hormone. J. Nutr. 119, 1213–1222.

Collie, M.A., Holmes, W.N., Cronshaw, J., 1992. A comparison of the response of dispersed steroidogenic cells derived from embryonic adrenal tissue from the domestic chicken (*Gallus domesticus*), the domestic Pekin duck and the wild Mallard duck (*Anas platyrhynchos*), and the domestic Muscovy duck (*Cairina moschata*). Gen. Comp. Endocrinol. 88, 375–387.

Contijoch, A.M., Gonzolez, C., Singh, H.M., Malamed, S., Trancoso, S., Advis, J.P., 1992. Dopaminergic regulation of luteinizing releasing hormone release at the median eminence level: immunocytochemical and physiological evidence in hens. Neuroendocrinology 55, 290–300.

Contijoch, A.M., Malamed, S., McDonald, J.K., Advis, J.P., 1993a. Neuropeptide Y regulation of LHRH release in the median eminence: immunocytochemical and physiological evidence in hens. Neuroendocrinology 57, 135–145.

Contijoch, A.M., Malamed, S., Sarkar, D.K., Advis, J.P., 1993b. β-endorphin regulation of LHRH release at the median eminence level: immunocytochemical and physiological evidence in hens. Neuroendocrinology 57, 365–373. 83–90.

Corcaran, D.H., Proudman, J.A., 1991. Isoforms of turkey prolactin: evidence for differences in glycosylation and in tryptic-peptide mapping. Comp. Biochem. Physiol. B 99, 563–570.

Cozzi, B., Stankov, B., Viglietti-Panzica, C., Capsoni, S., Aste, N., Lucini, V., Fraschini, F., Panzica, G.C., 1993. Distribution and characterization of melatonin receptors in the brain of the Japanese quail, *Coturnix japonica*. Neurosci. Lett. 150, 149–152.

Crisóstomo, S., Guémené, D., Garreau-Mills, M., Morvan, C., Zadworny, D., 1998. Prevention of incubation behavior expression in turkey hens by active immunization against prolactin. Theriogenology 50, 675–690.

Cronshaw, J., Reese, B.K., Collie, M.A., Holmes, W.N., 1992. Cytoskeletal changes accompanying ACTH-induced steroidogenesis in cultured embryonic adrenal gland cells from the Pekin duck. Cell Tissue Res. 268, 157–165.

Cui, H., Zhao, G., Liu, R., Zheng, M., Chen, J., Wen, J., 2012. FSH stimulates lipid biosynthesis in chicken adipose tissue by upregulating the expression of its receptor FSHR. J. Lipid Res. 53, 909–917.

Darras, V.M., Berghman, L.R., Vanderpooten, A., Kuhn, E.R., 1992. Growth hormone acutely decreases type III iodothyronine deiodinase in chicken liver. FEBS Lett. 310, 5–8.

Darras, V.M., Rudas, P., Visser, T.J., Hall, T.R., Huybrechts, L.M., Vanderpoolen, A., Berghman, R., Decuypere, E., Kuhn, R., 1993. Endogenous growth hormone controls high plasma levels of 3,3′,5-triiodothyronine (T3) in growing chickens by increasing the T3-degrading type III deiodinase activity. Domest. Anim. Endocrinol. 10, 55–65.

Davies, D.T., Follett, B.K., 1980. Neuroendocrine regulation of gonadotrophin-releasing hormone secretion in the Japanese quail. Gen. Comp. Endocrinol. 40, 220–225.

Davies, D.T., Massa, R., James, R., 1980. Role of testosterone and of its metabolites in regulating gonadotrophin secretion in the Japanese quail. J. Endocrinol. 84, 211–222.

Davis, A.J., Brooks, C.F., Johnson, P.A., 1999. Gonadotropin regulation of inhibin alpha-subunit mRNA and immunoreactive protein in cultured chicken granulosa cells. Gen. Comp. Endocrinol. 116, 90–103.

Davis, A.J., Brooks, C.F., Johnson, P.A., 2001. Follicle-stimulating hormone regulation of inhibin alpha- and beta(B)-subunit and follistatin messenger ribonucleic acid in cultured avian granulosa cells. Biol. Reprod. 64, 100–106.

Dawson, A., Follett, B.K., Goldsmith, A.R., Nicholls, T.J., 1985. Hypothalamic gonadotrophin-releasing hormone and pituitary and plasma FSH and prolactin during photostimulation and photorefractoriness in intact and thyroidectomized starlings (*Sturnus vulgaris*). J. Endocrinol. 105, 71–77.

Dean, C.E., Porter, T.E., 1999. Regulation of somatotroph differentiation and growth hormone (GH) secretion by corticosterone and GH-releasing hormone during embryonic development. Endocrinology 140, 1104–1110.

Decuypere, E., Igbal, A., Michels, H., Kuhn, E.R., Scheider, R., ElAzeem, A.A., 1988. Thyroid hormone response to thyrotropin releasing factor after cold treatment during pre- and post-natal development in the domestic fowl. Horm. Metab. Res. 20, 484–489.

De Groef, B., Geris, K.L., Manzano, J., Bernal, J., Millar, R.P., Abou-Samra, A.B., Porter, T.E., Iwasawa, A., Kühn, E.R., Darras, V.M., 2003a. Involvement of thyrotropin-releasing hormone receptor, somatostatin receptor subtype 2 and corticotropin-releasing hormone receptor type 1 in the control of chicken thyrotropin secretion. Mol. Cell. Endocrinol. 203, 33–39.

De Groef, B., Goris, N., Arckens, L., Kuhn, E.R., Darras, V.M., 2003b. Corticotropin-releasing hormone (CRH)-induced thyrotropin release is directly mediated through CRH receptor type 2 on thyrotropes. Endocrinology 144, 5537–5544.

De Groef, B., Grommen, S.V., Darras, V.M., 2007. Feedback control of thyrotropin secretion in the chicken: thyroid hormones increase the expression of hypophyseal somatostatin receptor types 2 and 5. Gen. Comp. Endocrinol. 152, 178–182.

Dockray, G.J., Reeve Jr., J.R., Shively, J., Gayton, R.J., Barnard, C.S., 1983. A novel active pentapeptide from chicken brain identified by antibodies to FMRFamide. Nature 305, 328–330.

Doneen, B.A., Smith, T.E., 1982. Ontogeny of endocrine control of osmoregulation in chick embryo. II. Actions of prolactin, arginine vasopressin, and aldosterone. Gen. Comp. Endocrinol. 48, 310–318.

Donoghue, D.J., Scanes, C.G., 1991. Triiodothyronine (T3) inhibition of growth hormone secretion by chicken pituitary cells in vitro. Gen. Comp. Endocrinol. 84, 344–354.

Dridi, S., Ververken, C., Hillgartner, F.B., Arckens, L., Van der Gucht, E., Cnops, L., Decuypere, E., Buyse, J., 2006. FAS inhibitor cerulenin reduces food intake and melanocortin receptor gene expression without modulating the other (an)orexigenic neuropeptides in chickens. Am. J. Physiol. Regul. Integr. Comp. Physiol. 291, R138–R147.

Dunn, I.C., Chen, Y., Hook, C., Sharp, P.J., Sang, H.M., 1993. Characterization of the chicken preprogonadotrophin-releasing hormone-I gene. J. Mol. Endocrinol. 11, 19–29.

Dunn, I.C., Sharp, P.J., 1999. Photo-induction of hypothalamic gonadotrophin releasing hormone-I mRNA in the domestic chicken: a role for oestrogen? J. Neuroendocrinol. 11, 371–375.

Dunn, I.C., Beattie, K.K., Maney, D., Sang, H.M., Talbot, R.T., Wilson, P.W., Sharp, P.J., 1996. Regulation of chicken gonadotropin-releasing hormone-I mRNA in incubating, nest-deprived and laying bantam hens. Neuroendocrinology 63, 504–513.

Dunson, W.A., Buss, E.G., Sawyer, W.H., Sokol, H., 1972. Hereditary polydipsia and polyuria in chickens. Am. J. Physiol. 222, 1167–1176.

El Halawani, M.E., Burke, W.H., Dennison, P.T., 1980. Effect of nest deprivation on serum prolactin level in resting female turkeys. Biol. Reprod. 23, 118–123.

El Halawani, M.E., Silsby, J.L., Behnke, E.J., Fehrer, S.C., 1986. Hormonal induction of incubation behavior in ovariectomized female turkeys (*Meleagris gallopavo*). Biol. Reprod. 35, 59–67.

El Halawani, M.E., Silsby, J.L., Fehrer, S.C., 1988. Basal and hypothalamic extract-induced luteinizing hormone and prolactin secretion by cultured anterior pituitary cells from female turkeys in various stages of the reproductive cycle. Gen. Comp. Endocrinol. 71, 45–54.

El Halawani, M.E., Silsby, J.L., Mauro, L.J., 1990. Vasoactive intestinal peptide is a hypothalamic prolactin-releasing neuropeptide in the turkey (*Meleagris gallopavo*). Gen. Comp. Endocrinol. 78, 66–73.

El Halawani, M.E., Silsby, J.L., Koike, T.I., Robinzon, B., 1992. Evidence of a role for the turkey posterior pituitary in prolactin release. Gen. Comp. Endocrinol. 87, 436–442.

El Halawani, M.E., Youngren, O.M., Rozenboim, I., Pitts, G.R., Silsby, J.L., Phillips, R.E., 1995. Serotonergic stimulation of prolactin secretion is inhibited by vasoactive intestinal peptide immunoneutralization in the turkey. Gen. Comp. Endocrinol. 99, 69–74.

El-Halawani, M.E., Whiting, S.E., Silsby, J.L., Pitts, G.R., Chaiseha, Y., 2000. Active immunization with vasoactive intestinal peptide in turkey hens. Poult. Sci. 79, 349–354.

Ellestad, L.E., Carre, W., Muchow, M., Jenkins, S.A., Wang, X., Cogburn, L.A., Porter, T.E., 2006. Gene expression profiling during cellular differentiation in the embryonic pituitary gland using cDNA microarrays. Physiol. Genomics 25, 414–425.

Ellestad, L.E., Malkiewicz, S.A., Guthrie, H.D., Welch, G.R., Porter, T.E., 2009. Expression and regulation of glucocorticoid-induced leucine zipper in the developing anterior pituitary gland. J. Mol. Endocrinol. 42, 171–183.

Ellestad, L.E., Saliba, J., Porter, T.E., 2011. Ontogenic characterization of gene expression in the developing neuroendocrine system of the chick. Gen. Comp. Endocrinol. 171, 82–93.

Engster, H.M., Carew, L.B., Harvey, S., Scanes, C.G., 1979. Growth hormone metabolism in essential fatty acid-deficient and pair-fed non-deficient chicks. J. Nutr. 109, 330–338.

Fehrer, S.C., Silsby, J.L., Behnke, E.J., El Halawani, M.E., 1985. The influence of thyrotropin releasing hormone on in vivo prolactin release and in vitro prolactin, luteinizing hormone, and growth hormone release from dispersed pituitary cells of the young turkey (*Meleagris gallopavo*). Gen. Comp. Endocrinol. 59, 64–72.

Fennell, M.J., Johnson, A.L., Scanes, C.G., 1990. Influence of androgens on plasma concentrations of growth hormone in growing castrated and intact chickens. Gen. Comp. Endocrinol. 77, 466–475.

Fernández, A.J., Sierra, M.A., Méndez, A., Mozos, E., Moyano, M.C., Carrasco, L., 1986. Ultrastructural modifications of the cavities formed by folliculo-stellate cells in chicken adenohypophysis under septic shock conditions. Cell Struct. Funct. 11, 379–382.

Follett, B.K., Maung, S.L., 1978. Rate of testicular maturation, in relation to gonadotrophin and testosterone levels in quail exposed to various artificial photoperiods and to natural day lengths. J. Endocrinol. 78, 267–280.

Follett, B.K., Davies, D.T., Gledhill, B., 1977. Photoperiodic control of reproduction in Japanese quail: changes in gonadotrophin secretion on the first day of induction and their pharmacological blockade. J. Endocrinol. 74, 449–460.

Foltzer, C., Mialhe, P., 1976. Pituitary and adrenal control of pancreatic endocrine function in the duck. II. Plasma free fatty acids and insulin variations following hypophysectomy and replacement therapy with growth hormone and corticosterone. Diabet. Metab. 2, 101–105.

Foster, D.N., Galehouse, D., Giordano, T., Min, B., Lamb, I.C., Porter, D.A., Intehar, K.J., Bacon, W.L., 1991. Nucleotide sequence of the cDNA encoding the common α subunit of the chicken pituitary glycoprotein hormones. J. Mol. Endocrinol. 8, 21–27.

Fraley, G.S., Kuenzel, W.J., 1992. Precocious puberty in chicks (*Gallus domesticus*) induced by central injections of neuropeptide Y. Life Sci. 52, 1649–1956.

Franchini, A., Ottaviani, E., 1999. Immunoreactive POMC-derived peptides and cytokines in the chicken thymus and bursa of Fabricius microenvironments: age-related changes. J. Neuroendocrinol. 11, 685–692.

Fu, X., Porter, T.E., 2004. Glucocorticoid induction of lactotrophs and prolactin gene expression in chicken embryonic pituitary cells: a delayed response relative to stimulated growth hormone production. Endocrinology 145, 1322–1330.

Fu, X., Nishimura, S., Porter, T.E., 2004. Evidence that lactotrophs do not differentiate directly from somatotrophs during chick embryonic development. J. Endocrinol. 183, 417–425.

Furr, B.J.A., Bonney, R.C., England, R.J., Cunningham, F.J., 1973. Luteinizing hormone and progesterone in peripheral blood during the ovulatory cycle of the hen (*Gallus domesticus*). J. Endocrinol. 57, 159–169.

Gasc, J.M., Baulieu, E.E., 1988. Regulation by estradiol of the progesterone receptor in the hypothalamus and pituitary: an immunohistochemical study in the chicken. Endocrinology 122, 1357–1365.

Geelissen, S.M., Beck, I.M., Darras, V.M., Kühn, E.R., Van der Geyten, S., 2003. Distribution and regulation of chicken growth hormone secretagogue receptor isoforms. Gen. Comp. Endocrinol. 134, 167–174.

Gerets, H.H., Peeters, K., Arckens, L., Vandesande, F., Berghman, L.R., 2000. Sequence and distribution of pro-opiomelanocortin in the pituitary and the brain of the chicken (*Gallus gallus*). J. Comp. Neurol. 417, 250–262.

Geris, K.L., Berghman, L.R., Kühn, E.R., Darras, V.M., 1999. The drop in plasma thyrotropin concentrations in fasted chickens is caused by an action at the level of the hypothalamus: role of corticosterone. Domest. Anim. Endocrinol. 16, 231–237.

Geris, K.L., de Groef, B., Rohrer, S.P., Geelissen, S., Kühn, E.R., Darras, V.M., 2003a. Identification of somatostatin receptors controlling growth hormone and thyrotropin secretion in the chicken using receptor subtype-specific agonists. J. Endocrinol. 177, 279–286.

Geris, K.L., De Groef, B., Kühn, E.R., Darras, V.M., 2003b. In vitro study of corticotropin-releasing hormone-induced thyrotropin release: ontogeny and inhibition by somatostatin. Gen. Comp. Endocrinol. 132, 272–277.

Gerstberger, R., Kaul, R., Gray, D.A., Simon, E., 1985. Arginine vasotocin and glomerular filtration rate in saltwater-acclimated ducks. Am. J. Physiol. 248, F663–F667.

Gillespie, M.J., Crowley, T.M., Haring, V.R., Wilson, S.L., Harper, J.A., Payne, J.S., Green, D., Monaghan, P., Donald, J.A., Nicholas, K.R., Moore, R.J., 2013. Transcriptome analysis of pigeon milk production: role of cornification and triglyceride synthesis genes. BMC Genomics 14, 169.

Goecke, C.S., Goldstein, D.L., 1997. Renal glomerular and tubular effects of antidiuretic hormone and two antidiuretic hormone analogues in house sparrows (*Passer domesticus*). Physiol. Zool. 70, 283–291.

Goldsmith, A.R., Edwards, C., Koprucu, M., Silver, R., 1981. Concentrations of prolactin and luteinizing hormone in plasma of doves in relation to incubation and development of the crop. J. Endocrinol. 90, 437–443.

Goldstein, D.L., 2006. Regulation of the avian kidney by arginine vasotocin. Gen. Comp. Endocrinol. 147, 78–84.

González-Morán, M.G., 2007. Effects of luteinizing hormone treatment on oogenesis in ovarian germ cells of the chick (*Gallus domesticus*). Domest. Anim. Endocrinol. 33, 154–166.

Goodson, J.L., Schrock, S.E., Klatt, J.D., Kabelik, D., Kingsbury, M.A., 2009. Mesotocin and nonapeptide receptors promote estrildid flocking behavior. Science 325, 862–866.

Goodson, J.L., Wilson, L.C., Schrock, S.E., 2012. To flock or fight: neurochemical signatures of divergent life histories in sparrows. Proc. Natl. Acad. Sci. U. S. A. 109 (Suppl. 1), 10685–10692.

Goosens, N.S., Blahser, S., Oksche, A., Vandesande, F., Dierickx, K., 1977. Immunocytochemical investigation of the hypothalo-neurohypophyseal system in birds. Cell Tissue Res. 184, 1–13.

Gray, D.A., Erasmus, T., 1988. Glomerular filtration changes during vasotocin-induced antidiuresis in kelp gulls. Am. J. Physiol. 255, R936–R939.

Gregory, C.C., Porter, T.E., 1997. Cloning and sequence analysis of a cDNA for the beta subunit of chicken thyroid-stimulating hormone. Gen. Comp. Endocrinol. 107, 182–190.

Grommen, S.V., Taniuchi, S., Janssen, T., Schoofs, L., Takahashi, S., Takeuchi, S., Darras, V.M., De Groef, B., 2006. Molecular cloning, tissue distribution, and ontogenic thyroidal expression of the chicken thyrotropin receptor. Endocrinology 147, 3943–3951.

Grommen, S.V., Taniuchi, S., Darras, V.M., Takahashi, S., Takeuchi, S., De Groef, B., 2008. Identification of unique thyrotropin receptor (TSHR) splice variants in the chicken: the chicken TSHR gene revisited. Gen. Comp. Endocrinol. 156, 460–463.

Grommen, S.V., Geysens, S., Darras, V.M., De Groef, B., 2009. Chicken folliculo-stellate cells express thyrotropin receptor mRNA. Domest. Anim. Endocrinol. 37, 236–242.

Grzegorzewska, A.K., Sechman, A., Paczoska-Eliasiewicz, H.E., Rzasa, J., 2009. The expression of pituitary FSH beta and LH beta mRNA and gonadal FSH and LH receptor mRNA in the chicken embryo. Reprod. Biol. 9, 253–269.

Gubrij, K.I., Chaturvedi, C.M., Ali, N., Cornett, L.E., Kirby, J.D., Wilkerson, J., Mikhailova, M., Turner, M.L., Baeyens, D.A., 2005. Molecular cloning of an oxytocin-like receptor expressed in the chicken shell gland. Comp. Biochem. Physiol. B Biochem. Mol. Biol. 142, 37–45.

Guibert, E., Brière, S., Pelletier, R., Brillard, J.P., Froment, P., 2011. Characterization of chicken Sertoli cells in vitro. Poult. Sci. 90, 1276–1286.

Halford, S., Pires, S.S., Turton, M., Zheng, L., González-Menéndez, I., Davies, W.L., Peirson, S.N., García-Fernández, J.M., Hankins, M.W., Foster, R.G., 2009. VA opsin-based photoreceptors in the hypothalamus of birds. Curr. Biol. 19, 1396–1402.

Hall, T.R., Cheung, A., Harvey, S., 1987. Some biological activities of recombinant DNA-derived growth hormone on plasma metabolite concentrations in domestic fowl. Comp. Biochem. Physiol. A 86, 29–32.

Hanks, M.C., Talbot, R.T., Sang, H.M., 1989a. Expression biologically active recombinant-derived chicken prolactin in *Escherichia coli*. J. Mol. Endocrinol. 3, 15–21.

Hanks, M.C., Alonzi, J.A., Sharp, P.J., Sang, H.M., 1989b. Molecular cloning and sequence analysis of putative chicken prolactin cDNA. J. Mol. Endocrinol. 2, 21–30.

Harrisson, F., 1989. Primary cilia associated with striated rootlets in granulated and folliculo-stellate cells of the avian adenohypophysis. Anat. Embryol. (Berl). 180, 543–547.

Harrisson, F., Van Hoof, J., Vakaet, L., 1982. The relationship between the folliculo-stellate network and the thyrotropic cells of the avian adenohypophysis. Cell Tissue Res. 226, 97–111.

Harvey, S., Baidwan, J.S., 1989. Thyrotrophin-releasing hormone (TRH)-induced growth hormone secretion in fowl: binding of TRH to pituitary membranes. J. Mol. Endocrinol. 3, 23–32.

Harvey, S., Scanes, C.G., Howe, T., 1977. Growth hormone effects on in vitro metabolism of avian adipose and liver tissue. Gen. Comp. Endocrinol. 33, 322–328.

Harvey, S., Scanes, C.G., Chadwick, A., Bolton, N.J., 1978a. The effect of thyrotropin-releasing hormone (TRH) and somatostatin (GHRIH) on growth hormone and prolactin secretion in vitro and in vivo in the domestic fowl (*Gallus domesticus*). Neuroendocrinology 26, 249–260.

Harvey, S., Scanes, C.G., Chadwick, A., Bolton, N.J., 1978b. Influence of fasting, glucose and insulin on the levels of growth hormone and prolactin in the plasma of the domestic fowl (*Gallus domesticus*). J. Endocrinol. 78, 501–506.

Harvey, S., Davison, T.F., Chadwick, A., 1979a. Ontogeny of growth hormone and prolactin secretion in the domestic fowl (*Gallus domesticus*). Gen. Comp. Endocrinol. 39, 270–273.

Harvey, S., Scanes, C.G., Chadwick, A., Bolton, N.J., 1979b. Growth hormone and prolactin secretion growing domestic fowl; influence of sex and breed. Br. Poult. Sci. 20, 9–17.

Harvey, S., Baidwan, J.S., Attardo, D., 1990. Homologous and heterologous regulation of somatostatin-binding sites on chicken adenohypophysial membranes. J. Endocrinol. 127, 417–425.

Hasegawa, Y., Miyamoto, K., Nomura, M., Igarashi, M., Kangawa, K., Matsuo, H., 1984. Isolation and amino acid compositions of four somatostatin-like substances in chicken hypothalamic extract. Endocrinology 115, 433–435.

Hattori, A., Ishii, S., Wada, M., 1986. Effects of two kinds of chicken luteinizing hormone-releasing hormone (LHRH), mammalian LH-RH and its analogs on the release of LH and FSH in Japanese quail and chicken. Gen. Comp. Endocrinol. 64, 446–455.

Hausman, G.J., Barb, C.R., Fairchild, B.D., Gamble, J., Lee-Rutherford, L., 2012. Expression of genes for interleukins, neuropeptides, growth hormone receptor, and leptin receptor in adiposetissue from growing broiler chickens. Domest. Anim. Endocrinol. 43, 260–263.

Hayashi, H., Imai, K., Imai, K., 1991. Characterization of chicken ACTH and α-MSH: the primary sequence of chicken ACTH is more similar to xenopus ACTH than to other avian ACTH. Gen. Comp. Endocrinol. 82, 434–443.

Haugen, M.J., Johnson, A.L., 2010. Bone morphogenetic protein 2 inhibits FSH responsiveness in hen granulosa cells. Reproduction 140, 551–558.

He, B., Mi, Y., Zhang, C., 2013. Gonadotropins regulate ovarian germ cell mitosis/meiosis decision in the embryonic chicken. Mol. Cell. Endocrinol. 370, 32–41.

Henare, S.J., Kikuchi, M., Talbot, R.T., Cockrem, J.F., 2011. Changes in plasma gonadotrophins, testosterone, prolactin, thyroxine and triiodothyronine concentrations in male Japanese quail (*Coturnix coturnix japonica*) of a heavy body weight line during photo-induced testicular growth and regression. Br. Poult. Sci. 52, 782–791.

Herbute, S., Pintat, R., Ramade, F., Bayle, J.D., 1984. Effect of short exposure to cold on plasma thyroxine in *Coturnix* quail: role of the infundibular complex and its neural afferents. Gen. Comp. Endocrinol. 56, 1–8.

Herold, M., Brezinschek, H.P., Gruschwitz, M., Dietrich, H., Wick, G., 1992. Investigation of ACTH responses of chickens with autoimmune disease. Gen. Comp. Endocrinol. 88, 188–198.

Heuck, K.A., Ellestad, L.E., Proudman, J.A., Porter, T.E., 2009. Somatotropin response in vitro to corticosterone and triiodothyronine during chick embryonic development: involvement of type I and type II glucocorticoid receptors. Domest. Anim. Endocrinol. 36, 186–196.

Heuck-Knubel, K., Proszkowiec-Weglarz, M., Narayana, J., Ellestad, L.E., Prakobsaeng, N., Porter, T.E., 2012. Identification of *cis* elements necessary for glucocorticoid induction of growth hormone gene expression in chicken embryonic pituitary cells. Am. J. Physiol. Regul. Integr. Comp. Physiol.

Hiyama, G., Kansaku, N., Kinoshita, M., Sasanami, T., Nakamura, A., Noda, K., Tsukada, A., Shimada, K., Zadworny, D., 2009. Changes in post-translational modifications of prolactin during development and reproductive cycles in the chicken. Gen. Comp. Endocrinol. 161, 238–245.

Houston, B., Goddard, C., 1988. Molecular forms of growth hormone in the chicken pituitary gland. J. Endocrinol. 116, 35–41.

Houston, B., O'Neill, I.E., 1990. Insulin and growth hormone act synergistically to stimulate insulin-like growth factor-I production by cultured chicken hepatocytes. J. Endocrinol. 128, 389–393.

Hrabia, A., Sechman, A., Gertler, A., Rząsa, J., 2011. Effect of growth hormone on steroid content, proliferation and apoptosis in the chicken ovary during sexual maturation. Cell Tissue Res. 345, 191–202.

Hsieh, Y.L., Chatterjee, A., Chien, J.T., Yu, J.Y., 2001. Molecular cloning of the cDNAs for pituitary glycoprotein hormone alpha subunits of two species of duck and their gene regulation. J. Mol. Endocrinol. 27, 339–347.

Hsieh, Y.L., Chowdhury, I., Chien, J.T., Chatterjee, A., Yu, J.Y., 2007. Molecular cloning and sequence analysis of the cDNA encoding thyroid-stimulating hormone beta-subunit of common duck and mule duck pituitaries: in vitro regulation of steady-state TSH beta mRNA level. Comp. Biochem. Physiol. B Biochem. Mol. Biol. 146, 307–317.

Huang, X., Hui, M.N., Liu, Y., Yuen, D.S., Zhang, Y., Chan, W.Y., Lin, H.R., Cheng, S.H., Cheng, C.H., 2009. Discovery of a novel prolactin in non-mammalian vertebrates: evolutionary perspectives and its involvement in teleost retina development. PLoS 4, e6163.

Huybrechts, L.M., King, D.B., Lauterio, T.J., Marsh, J., Scanes, C.G., 1985. Plasma concentrations of somatomedin-C in hypophysectomized, dwarf and intact growing domestic fowl as determined by heterologous radioimmunoassay. J. Endocrinol. 94, 295–304.

Ikegami, K., Katou, Y., Higashi, K., Yoshimura, T., 2009. Localization of circadian clock protein BMAL1 in the photoperiodic signal transduction machinery in Japanese quail. J. Comp. Neurol. 517, 397–404.

Ikemoto, T., Park, M.K., 2005. Chicken RFamide-related peptide (GnIH) and two distinct receptor subtypes: identification, molecular characterization, and evolutionary considerations. J. Reprod. Dev. 51, 359–377.

Ikemoto, T., Park, M.K., 2006. Molecular and evolutionary characterization of the GnRH-II gene in the chicken: distinctive genomic organization, expression pattern, and precursor sequence. Gene 368, 28–36.

Imai, K., 1973. Effects of avian and mammalian pituitary preparations on induction of ovulation in the domestic fowl, *Gallus domesticus*. J. Reprod. Fertil. 33, 91–98.

Iturriza, F.C., Estivariz, F.E., Levitin, H.P., 1980. Coexistence of α-melanocyte stimulating hormone and adrenocorticotrophin in all cells containing either of the two hormones in the duck pituitary. Gen. Comp. Endocrinol. 42, 110–115.

Iturriza, F.C., Venosa, R.A., Pijol, M.G., Quintas, N.B., 1992. α-Melanocyte-stimulating hormone stimulates sodium excretion in the salt gland of the ducks. Gen. Comp. Endocrinol. 87, 369–374.

Jenkins, S.A., Muchow, M., Richards, M.P., McMurtry, J.P., Porter, T.E., 2007. Administration of adrenocorticotropic hormone during chicken embryonic development prematurely induces pituitary growth hormone cells. Endocrinology 148, 3914–3921.

Johnson, A.L., Bridgham, J.T., Wagner, B., 1996. Characterization of a chicken luteinizing hormone receptor (cLH-R) complementary deoxyribonucleic acid, and expression of cLH-R messenger ribonucleic acid in the ovary. Biol. Reprod. 55, 304–309.

Johnson, B.E., Scanes, C.G., King, D.B., Marsh, J.A., 1993. Effect of hypophysectomy and growth hormone on immune development in the domestic fowl. Dev. Comp. Immunol. 17, 331–339.

Johnson, A.L., Tilly, J.L., 1988. Effects of vasoactive intestinal peptide on steroid secretion and plasminogen activator activity in granulosa cells of the hen. Biol. Reprod. 38, 296–303.

Johnson, P.A., Woodcock, J.R., Kent, T.R., 2006. Effect of activin A and inhibin A on expression of the inhibin/activin beta-B-subunit and gonadotropin receptors in granulosa cells of the hen. Gen. Comp. Endocrinol. 147, 102–107.

Joseph, N.T., Morgan, K., Sellar, R., McBride, D., Millar, R.P., Dunn, I.C., 2009. The chicken type III GnRH receptor homologue is predominantly expressed in the pituitary, and exhibits similar ligand selectivity to the type I receptor. J. Endocrinol. 202, 179–190.

Jozsa, R., Scanes, C.G., Vigh, S., Mess, B., 1979. Functional differentiation of the embryonic chicken pituitary gland studied immunohistological approach. Gen. Comp. Endocrinol. 39, 158–163.

Jozsa, R., Vigh, S., Schally, A.V., Mess, B., 1984. Localization of corticotropin-releasing factor-containing neurons in the brain of the domestic fowl. Cell Tissue Res. 236, 245–248.

Jurkevich, A., Berghman, L.R., Cornett, L.E., Kuenzel, W.J., 2005. Characterization and immunohistochemical visualization of the vasotocin VT2 receptor in the pituitary gland of the chicken, *Gallus gallus*. Gen. Comp. Endocrinol. 143, 82–91.

Kaiya, H., Van Der Geyten, S., Kojima, M., Hosoda, H., Kitajima, Y., Matsumoto, M., Geelissen, S., Darras, V.M., Kangawa, K., 2002. Chicken ghrelin: purification, cDNA cloning, and biological activity. Endocrinology 143, 3454–3463.

Kameda, Y., Miura, M., Ohno, S., 1998. Localization and development of chromogranin A and luteinizing hormone immunoreactivities in the secretory-specific cells of the hypophyseal pars tuberalis of the chicken. Histochem. Cell. Biol. 109, 211–222.

Kameda, Y., Miura, M., Ohno, S., 2000. Expression of the common alpha-subunit mRNA of glycoprotein hormones during the chick pituitary organogenesis, with special reference to the pars tuberalis. Cell Tissue Res. 299, 71–80.

Kameda, Y., Miura, M., Maruyama, S., 2002. Effect of pinealectomy on the photoperiod-dependent changes of the specific secretory cells and alpha-subunit mRNA level in the chicken pars tuberalis. Cell Tissue Res. 308, 121–130.

Kang, S.W., Youngren, O.M., El Halawani, M.E., 2002. Influence of VIP on prolactinemia in turkey anterior pituitary cells: role of cAMP second messenger in VIP-induced prolactin gene expression. Regul. Pept. 109, 39–44.

Kang, S.W., Gazzillo, L.C., You, S., Wong, E.A., El Halawani, M.E., 2004. Turkey prolactin gene regulation by VIP through 35-bp *cis*-acting element in the proximal promoter. Gen. Comp. Endocrinol. 138, 157–165.

Kang, Z., Bédécarrats, G.Y., Zadworny, D., 2007. Expression patterns of the prolactin receptor gene in chicken lymphoid tissues during embryogenesis and posthatch period. Poult. Sci. 86, 2404–2412.

Kansaku, N., Shimada, K., Saito, N., Hidaka, H., 1998. Effects of protein kinase A inhibitor (H-89) on VIP- and GRF-induced release and mRNA expression of prolactin and growth hormone in the chicken pituitary gland. Comp. Biochem. Physiol. C Pharmacol. Toxicol. Endocrinol. 119, 89–95.

Kato, Y., Kato, T., Tomizawa, K., Kamiyoshi, M., Iwasawa, A., 1998. Complementary DNA sequence of chicken thyroid-stimulating hormone (TSH) beta subunit. Endocr. J. 45, 591–594.

Kawasaki, D., Aotsuka, T., Higashinakagawa, T., Ishii, S., 2003. Cloning of the genes for the pituitary glycoprotein hormone alpha and follicle-stimulating hormone beta subunits in the Japanese crested ibis. Nipponia Nippon. Zoolog. Sci. 20, 449–459.

Kikuchi, M., Kobayashi, M., Ito, T., Kato, Y., Ishii, S., 1998. Cloning of complementary deoxyribonucleic acid for the follicle-stimulating hormone-beta subunit in the Japanese quail. Gen. Comp. Endocrinol. 111, 376–385.

King, D.B., Scanes, C.G., 1986. Effect of mammalian growth hormone in the growth of hypophysectomized chickens. Proc. Exp. Biol. Med. 182, 201–207.

King, J.A., Millar, R.P., 1982. Structure of chicken hypothalamic luteinizing hormone-releasing hormone: II isolation and characterization. J. Biol. Chem. 257, 10729–10732.

Kitazawa, T., Maeda, Y., Kaiya, H., 2009. Molecular cloning of growth hormone secretagogue-receptor and effect of quail ghrelin on gastrointestinal motility in Japanese quail. Regul. Pept. 158, 132–142.

Klandorf, H., Harvey, S., Fraser, H.M., 1985. Physiological control of thyrotrophin-releasing hormone in the domestic fowl. J. Endocrinol. 105, 351–355.

Knight, P.G., Wilson, S.C., Gladwell, R.T., Cunningham, F.J., 1982. Evidence for the involvement of control catecholaminergic mechanisms in mediating the preovulatory surge of luteinizing, hormone in the domestic hen. J. Endocrinol. 94, 295–304.

Kobayashi, M., Ishii, S., 2002. Effects of starvation on gonadotropin and thyrotropin subunit mRNA levels and plasma hormone levels in the male Japanese quail (*Coturnix coturnix japonica*). Zoolog. Sci. 19, 331–342.

Kobayashi, M., Cockrem, J.F., Ishii, S., 2002. Effects of starvation and refeeding on gonadotropin and thyrotropin subunit mRNAs in male Japanese quail. Zoolog. Sci. 19, 449–461.

Koike, T.I., Shimada, K., Cornett, L.E., 1988. Plasma levels of immunoreactive mesotocin and vasotocin during oviposition in chickens: relationship to oxytocic action of the peptides in vitro and peptide interaction with myometrial membrane binding sites. Gen. Comp. Endocrinol. 70, 119–126.

Kowalski, K.I., Tilly, J.L., Johnson, A.L., 1991. Cytochrome P450 side-chain cleavage (P450scc) in the hen ovary. 1. Regulation of P450scc messenger RNA levels and steroidogenesis in theca cells of developing follicles. Biol. Reprod. 45, 955–966.

Krishnan, K.A., Proudman, J.A., Bolt, D.J., Bahr, J.M., 1993. Development of a homologous radioimmunoassay for chicken follicle-stimulating hormone and measurement of plasma FSH during the ovulatory cycle. Comp. Biochem. Physiol. A 105, 729–734.

Kuhn, E.R., Verheyer, G., Chiasson, R.B., Huts, C., Decuypere, E., 1985. Ovine growth hormone reverses the fasting induced decrease of plasma T3 in adult chickens. IRCS Med. Sci. 13, 451–452.

Kuznik, B.I., Pateyuk, A.V., Rusaeva, N.S., 2008. Effect of tetrapeptides Lys–Glu–Asp–Gly and Ala–Glu–Asp–Gly on the structure and function of the thyroid gland in neonatally hypophysectomized chickens. Bull. Exp. Biol. Med. 145, 104–107.

Lamb, I.C., Galehouse, D.M., Foster, D.N., 1988. Chicken growth hormone cDNA sequence. Nucl. Acids 16, 9339.

Lau, J.S., Yip, C.W., Law, K.M., Leung, F.C., 2007. Cloning and characterization of chicken growth hormone binding protein (cGHBP). Domest. Anim. Endocrinol. 33, 107–121.

Lau, K.K., Yang, Y., Cook, G.A., Wyatt, R.J., Nishimura, H., 2009. Control of aquaporin 2 expression in collecting ducts of quail kidneys. Gen. Comp. Endocrinol. 160, 288–294.

Lazure, C., Paquet, L., Litthauer, D., Naude, R.J., Oelofsen, W., Chretien, M., 1990. The ostrich pituitary contains a major peptide homologous to mammalian chromogranin A(1-76). Peptides 11, 79–89.

Lazure, C., Saayman, H.S., Naude, R.J., Oelofsen, W., Chretien, M., 1987. Complete amino acid sequence of a VLDV-type neurophysin from ostrich differs markedly from known mammalian neurophysins. Int. J. Pept. Prot. Res. 30, 634–645.

Lazure, C., Saayman, H.S., Naude, R.J., Oelofsen, W., Chretien, M., 1989. Ostrich MSEL-neurophysin belongs to the class of two-domain ''big'' neurophysin as indicated by complete amino acid sequence of the neurophysin/copeptin. Int. J. Pept. Prot. Res. 33, 46–58.

Lèche, A., Busso, J.M., Hansen, C., Navarro, J.L., Marín, R.H., Martella, M.B., 2009. Physiological stress in captive greater rheas (*Rhea americana*): highly sensitive plasma corticosterone response to an ACTH challenge. Gen. Comp. Endocrinol. 162, 188–191.

Leclerc, B., Zadworny, D., Bédécarrats, G., Kühnlein, U., 2007. Ontogenesis of the expression of prolactin receptor messenger ribonucleic acid during late embryogenesis in turkeys and chickens. Poult. Sci. 86, 1174–1179.

Leung, C.H., Abebe, D.F., Earp, S.E., Goode, C.T., Grozhik, A.V., Mididoddi, P., Maney, D.L., 2011. Neural distribution of vasotocin receptor mRNA in two species of songbird. Endocrinology 152, 4865–4881.

Leung, F.C., Taylor, J.E., Wien, S., Van Iderstine, A., 1986. Purified chicken growth hormone (GH) and a human pancreatic GH-releasing hormone increase body weight gain in chickens. Endocrinology 118, 1961–1965.

Li, C.H., Chung, D., Oelofsen, W., Naude, R.J., 1978. Adrenocorticotropin 53. The amino acid sequence of the hormone from the ostrich pituitary gland. Biochem. Biophys. Res. Commun. 84, 900–906.

Li, G., Sun, D.X., Yu, Y., Liu, W.J., Tang, S.Q., Zhang, Y., Wang, Y.C., Zhang, S.L., Zhang, Y., 2011. Genetic effect of the follicle-stimulating hormone receptor gene on reproductive traits in Beijing you chickens. Poult. Sci. 90, 2487–2492.

Li, H., Proudman, J., Kuenzel, W.J., 2009. Differential regulation of gene expression and release of FSH and prolactin by long day and sulfamethazine in chicks. Gen. Comp. Endocrinol. 161, 262–266.

Li, W.L., Liu, Y., Yu, Y.C., Huang, Y.M., Liang, S.D., Shi, Z.D., 2011. Prolactin plays a stimulatory role in ovarian follicular development and egg laying in chicken hens. Domest. Anim. Endocrinol. 41, 57–66.

Li, Z., Johnson, A.L., 1993. Regulation of P450 cholesterol side-chain cleavage messenger ribonucleic acid expression and progesterone production in hen granulosa cells. Biol. Reprod. 49, 463–469.

Ling, M.K., Hotta, E., Kilianova, Z., Haitina, T., Ringholm, A., Johansson, L., Gallo-Payet, N., Takeuchi, S., Schiöth, H.B., 2004. The melanocortin receptor subtypes in chicken have high preference to ACTH-derived peptides. Br. J. Pharmacol. 143, 626–637.

Liu, H.K., Long, D.W., Bacon, W.L., 2001. Concentration change patterns of luteinizing hormone and progesterone and distribution of

hierarchical follicles in normal and arrested laying turkey hens. Poult. Sci. 80, 1509–1518.
Liu, H.Y., Zhang, C.Q., 2008. Effects of daidzein on messenger ribonucleic acid expression of gonadotropin receptors in chicken ovarian follicles. Poult. Sci. 87, 541–545.
Liu, L., Porter, T.E., 2004. Endogenous thyroid hormones modulate pituitary somatotroph differentiation during chicken embryonic development. J. Endocrinol. 180, 45–53.
Liu, W.J., Sun, D.X., Yu, Y., Li, G., Tang, S.Q., Zhang, Y., Wang, Y.C., Zhang, Y., 2010. Association of Janus kinase 2 polymorphisms with growth and reproduction traits in chickens. Poult. Sci. 89, 2573–2579.
Lopes, P.C., Wingfield, J.C., Bentley, G.E., 2012. Lipopolysaccharide injection induces rapid decrease of hypothalamic GnRH mRNA and peptide, but does not affect GnIH in zebra finches. Horm. Behav. 62, 173–179.
Lovell, T.M., Knight, P.G., Gladwell, R.T., 2005. Variation in pituitary expression of mRNAs encoding the putative inhibin co-receptor (betaglycan) and type-I and type-II activin receptors during the chicken ovulatory cycle. J. Endocrinol. 186, 447–455.
Luna, M., Barraza, N., Berumen, L., Carranza, M., Pedernera, E., Harvey, S., Arámburo, C., 2005. Heterogeneity of growth hormone immunoreactivity in lymphoid tissues and changes during ontogeny in domestic fowl. Gen. Comp. Endocrinol. 144, 28–37.
Luna, M., Rodríguez-Méndez, A.J., Berumen, L., Carranza, M., Riesgo-Escovar, J., Baudet, M.L., Harvey, S., Arámburo, C., 2008. Immune growth hormone (GH): localization of GH and GH mRNA in the bursa of Fabricius. Dev. Comp. Immunol. 32, 1313–1325.
MacKenzie, D.S., 1981. In vivo thyroxine in day old cockerels in response to acute stimulation by mammalian and avian pituitary hormones. Poult. Sci. 60, 2136–2143.
MacNamee, M.C., Sharp, P.J., 1989. The functional activity of hypothalamic dopamine in broody bantam hens. J. Endocrinol. 121, 67–74.
MacNamee, M.C., Sharp, P.J., Lea, W., Sterling, R.J., Harvey, S., 1986. Evidence that vasoactive intestinal peptide is a physiological prolactin-releasing factor in the bantam hen. Gen. Comp. Endocrinol. 62, 470–478.
Maddineni, S., Metzger, S., Ocón, O., Hendricks, G., Ramachandran, R., 2005. Adiponectin gene is expressed in multiple tissues in the chicken: food deprivation influences adiponectin messenger ribonucleic acid expression. Endocrinology 146, 4250–4256.
Maddineni, S.R., Krzysik-Walker, S.M., Ocón-Grove, O.M., Motch, S.M., Hendricks, G.L., Ramachandran, R., 2007. Calcitonin is expressed in the chicken pituitary gland: influence of gonadal steroids and sexual maturation. Cell Tissue Res. 327, 521–528.
Maddineni, S., Ocón-Grove, O.M., Krzysik-Walker, S.M., Hendricks, G.L., Proudman, J.A., Ramachandran, R., 2008. Gonadotrophin-inhibitory hormone receptor expression in the chicken pituitary gland: potential influence of sexual maturation and ovarian steroids. J. Neuroendocrinol. 20, 1078–1088.
Malamed, S., Gibney, J.A., Loesser, K.E., Scanes, C.G., 1985. Age-related changes of the somatotrophs of the domestic fowl *Gallus domesticus*. Cell Tissue Res. 239, 87–91.
Malamed, S., Gibney, J.A., Scanes, C.G., 1988. Immunogold identification of the somatotrophs of domestic fowl of different ages. Cell Tissue Res. 251, 581–585.
Malamed, S., Gibney, J.A., Cain, L.D., Perez, F.M., Scanes, C.G., 1993. Immunocytochemical studies of chicken somatotroph granules before and after hatching. Cell Tissue Res. 272, 369–374.
Malamed, S., Deaver, D., Perez, F., Radecki, S., Gibney, J., Scanes, C.G., 1997. Quantitive studies of chicken somatotrophs during growth and development by morphometry, immunocytochemistry, and flow cytometry. Gen. Comp. Endocrinol. 108, 25–34.
Maney, D.L., Schoech, S.J., Sharp, P.J., Wingfield, J.C., 1999. Effects of vasoactive intestinal peptide on plasma prolactin in passerines. Gen. Comp. Endocrinol. 113, 323–330.
Martínez-Coria, H., López-Rosales, L.J., Carranza, M., Berumen, L., Luna, M., Arámburo, C., 2002. Differential secretion of chicken growth hormone variants after growth hormone-releasing hormone stimulation in vitro. Endocrine 17, 91–102.
Martínez-Moreno, C.G., Palma, L., Carranza, M., Harvey, S., Arámburo, C., Luna, M., 2011. Cellular and intracellular distribution of growth hormone in the adult chicken testis. Gen. Comp. Endocrinol. 172, 344–357.
Maung, Z.W., Follett, B.K., 1977. Effects of chicken and ovine luteinizing hormone on androgen release and cyclic AMP production by isolated cells from the quail testis. Gen. Comp. Endocrinol. 33, 242–253.
Maung, S.L., Follett, B.K., 1978. The endocrine control by luteinizing hormone of testosterone secretion from the testis of the Japanese quail. Gen. Comp. Endocrinol. 36, 79–89.
McCann-Levorse, L.M., Radecki, S.V., Donoghue, D.J., Malamed, S., Foster, D.M., Scanes, C.G., 1993. Ontogeny of pituitary hormone and growth hormone mRNA in the chicken. Proc. Soc. Exp. Biol. Med. 202, 109–113.
McElroy, A.P., Caldwell, D.J., Proudman, J.A., Hargis, B.M., 2004. Modulation of in vitro DNA synthesis in the chicken ovarian granulosa cell follicular hierarchy by follicle-stimulating hormone and luteinizing hormone. Poult. Sci. 83, 500–506.
McFarlane, H.O., Joseph, N.T., Maddineni, S.R., Ramachandran, R., Bédécarrats, G.Y., 2011. Development, validation, and utilization of a novel antibody specific to the type III chicken gonadotropin-releasing hormone receptor. Domest. Anim. Endocrinol. 40, 110–118.
McNeilly, A.S., Friesen, H.G., 1985. Presence of a nonlactogenic factor in human serum which synergistically enhances prolactin-stimulated growth of Nb2 rat lymphoma cells in vitro. J. Clin. Endocrinol. Metab. 61, 408–411.
McNichols, M.J., McNabb, F.M.A., 1988. Development of thyroid function and its pituitary control in embryonic and hatching precocial Japanese quail and altricial ring doves. Gen. Comp. Endocrinol. 69, 109–118.
Méndez, M.C., Ramírez, M., Varela, A.R., Chávez, B., Pedernera, E., 2003. Follicle-stimulating hormone increases cell proliferation in the ovary and the testis of the chick embryo. Gen. Comp. Endocrinol. 133, 181–188.
Michel, G., Lévy, B., Chauvet, M.T., Chauvet, J., Acher, R., 1990. Complete amino acid sequence of goose VLDV-neurophysin. Traces of a putative gene conversion between promesotocin and provasotocin genes. Int. J. Pept. Protein Res. 36, 457–464.
Mikami, S., 1980. Hypothalamic control of the avian adenohypophysis. In: Tanabe, Y., Tanaka, K., Ookawa, T. (Eds.), Biological Rhythms in Birds: Neural and Endocrine Aspects. Springer-Verlag, Tokyo, pp. 17–32.
Millam, J.R., Burke, W.H., El Halawani, M.E., 1984. Release of gonadotropin-releasing hormone from the Japanese quail hypothalamus in vitro. Gen. Comp. Endocrinol. 53, 293–301.
Millam, J.R., Craig-Veit, C.B., Petitte, J.N., 1993. Brain content of cGnRH I and II during embryonic development in chickens. Gen. Comp. Endocrinol. 92, 311.

Millar, R.P., Milton, R.C., Follett, B.K., King, J.A., 1986. Receptor binding and gonadotropin-releasing activity of a novel chicken gonadotropin-releasing hormone [His5, Trp7, Tyr8]-GnRH and a D-Arg6 and analog. Endocrinology 119, 224–231.

Miyamoto, K., Hasegawa, Y., Nomara, M., Igarashi, M., Kangawa, K., Matsuo, H., 1984. Identification of the second gonadotropin releasing hormone in chicken hypothalamus: evidence that gonadotropin secretion is probably controlled by two distinct gonadotropin releasing hormones in avian species. Proc. Natl. Acad. Sci. U. S. A. 81, 3874–3878.

Mohanty, B., 2006. Extracellular accumulations in the avian pituitary gland: histochemical analysis in two species of Indian wild birds. Cells Tissues Organs 183, 99–106.

Mohanty, B., Das, S., Naik, D.R., 1997. Immunocytochemistry of the pars tuberalis of the pituitary gland in some Indian wild birds: a comparative study. Gen. Comp. Endocrinol. 108, 109–118.

Murase, D., Taniuchi, S., Takeuchi, S., Adachi, H., Kansaku, N., Okazaki, K., Ohkubo, T., 2011. Role of chicken Pit-1 isoforms in activating growth hormone gene. Gen. Comp. Endocrinol. 173, 248–252.

Murphy, A.E., Harvey, S., 2001. Extrapituitary beta TSH and GH in early chick embryos. Mol. Cell. Endocrinol. 185, 161–171.

Murphy, M.J., Brown, P.S., Brown, S.C., 1986. Osmoregulatory effects of prolactin and growth hormone in embryonic chicks. Gen. Comp. Endocrinol. 62, 485–492.

Nakane, Y., Ikegami, K., Ono, H., Yamamoto, N., Yoshida, S., Hirunagi, K., Ebihara, S., Kubo, Y., Yoshimura, T., 2010. A mammalian neural tissue opsin (opsin 5) is a deep brain photoreceptor in birds. Proc. Natl. Acad. Sci. U. S. A. 107, 15264–15268.

Nakao, N., Ono, H., Yamamura, T., Anraku, T., Takagi, T., Higashi, K., Yasuo, S., Katou, Y., Kageyama, S., Uno, Y., Kasukawa, T., Iigo, M., Sharp, P.J., Iwasawa, A., Suzuki, Y., Sugano, S., Niimi, T., Mizutani, M., Namikawa, T., Ebihara, S., Ueda, H.R., Yoshimura, T., 2008. Thyrotrophin in the pars tuberalis triggers photoperiodic response. Nature 452, 317–322.

Nakayama, H., Nakagawa-Mizuyachi, K., Takahashi, T., Kawashima, M., 2010. Calcitonin receptor binding in the hen neurohypophysis before and after oviposition. Poult. Sci. 89, 1473–1480.

Nakayama, H., Takahashi, T., Oomatsu, Y., Nakagawa-Mizuyachi, K., Kawashima, M., 2011a. Parathyroid hormone-related peptide directly increases adrenocorticotropic hormone secretion from the anterior pituitary in hens. Poult. Sci. 90, 175–180.

Nakayama, H., Takahashi, T., Nakagawa-Mizuyachi, K., Kawashima, M., 2011b. Effect of calcitonin on adrenocorticotropic hormone secretion stimulated by corticotropin-releasing hormone in the hen anterior pituitary. Anim. Sci. J. 82, 475–480.

Nakayama, H., Takahashi, T., Funaki, W., Nakagawa-Mizuyachi, K., Kawashima, M., 2011c. Calcitonin receptor bindings in the hen hypothalamus before and after oviposition. Poult. Sci. 90, 642–647.

Nakayama, H., Takahashi, T., Nakagawa-Mizuyachi, K., Kawashima, M., 2011d. Effect of estradiol-17β on calcitonin receptor bindings in the hen neurohypophysis. Poult. Sci. 90, 191–194.

Naude, R.J., Chung, D., Li, C.H., Oelofsen, W., 1981a. β-lipotropin: primary structure of the hormone from the ostrich pituitary gland. Int. J. Pept. Protein Res. 18, 138–147.

Naude, R.J., Chung, D., Li, C.H., Oelofsen, W., 1981b. β-endorphin. Primary structure of the hormone from the ostrich pituitary gland. Biochem. Biophys. Res. Commun. 98, 108–114.

Naude, R.J., Oelofsen, W., Maske, R., 1980. Isolation, characterization and opiate activity of β–endorphin from the pituitary gland of the ostrich, *Struthio camelus*. Biochem. J. 187, 245–248.

Nie, Q., Sun, B., Zhang, D., Luo, C., Ishag, N.A., Lei, M., Yang, G., Zhang, X., 2005. High diversity of the chicken growth hormone gene and effects on growth and carcass traits. J. Hered. 96, 698–703.

Nicoll, C.S., 1967. Bio-assay of prolactin. Analysis of the pigeon crop-sac response to local prolactin injection by an objective and quantitative method. Endocrinology 80, 641–655.

Nicoll, S.C., Hebert, N.J., Russell, S.M., 1985. Lactogenic hormones stimulate the liver to secrete a factor that acts synergistically with prolactin to promote growth of the pigeon crop-sac mucosal epithelium in vivo. Endocrinology 116, 1449–1453.

Nilsson, A., 1975. Structure of the vasoactive intestinal octacosapeptide from chicken intestine: the amino acid sequence. FEBS Lett. 60, 322–326.

Noce, T., Ando, H., Ueda, T., Kubokawa, K., Higashinakagawa, T., Ishii, S., 1989. Molecular cloning and nucleotide sequence analysis of the putative cDNA for the precursor molecule of the chicken LH-β subunit. J. Mol. Endocrinol. 3, 129–137.

Nouwen, E.J., Decuypere, E., Kuhn, E.R., Michels, H., Hall, T.R., Chadwick, A., 1984. Effect of dehydration, hemorrhage and oviposition on serum concentrations of vasotocin, mesotocin, and prolactin in the chicken. J. Endocrinol. 102, 345–351.

Opel, H., Proudman, J., 1988. Stimulation of prolactin release in turkeys by vasoactive intestinal peptide. Proc. Soc. Exp. Biol. Med. 187, 455–460.

Osugi, T., Ukena, K., Bentley, G.E., O'Brien, S., Moore, I.T., Wingfield, J.C., Tsutsui, K., 2004. Gonadotropin-inhibitory hormone in Gambel's white-crowned sparrow (*Zonotrichia leucophrys gambelii*): cDNA identification, transcript localization and functional effects in laboratory and field experiments. J. Endocrinol. 182, 33–42.

Oubre, C.M., D'Hondt, E., Moore, R.W., Hargis, B.M., Berghman, L.R., 2003. The chicken pituitary expresses an ovoinhibitor-like protein in subpopulations of some, but not all, hormone-producing cell types. Domest. Anim. Endocrinol. 25, 389–397.

Ouyang, J.H., Xie, L., Nie, Q., Luo, C., Liang, Y., Zeng, H., Zhang, X., 2008. Single nucleotide polymorphism (SNP) at the GHR gene and its associations with chicken growth and fat deposition traits. Br. Poult. Sci. 49, 87–95.

Paczoska-Eliasiewicz, H.E., Salomon, G., Reicher, S., Gussakowsky, E.E., Hrabia, A., Gertler, A., 2006. Preparation and characterization of recombinant chicken growth hormone (chGH) and its putative antagonist chGH G119R mutein. Ann. NY Acad. Sci. 1091, 501–508.

Pals, K., Boussemaere, M., Swinnen, E., Vankelecom, H., Denef, C., 2006. A pituitary cell type coexpressing messenger ribonucleic acid of proopiomelanocortin and the glycoprotein hormone alpha-subunit in neonatal rat and chicken: rapid decline with age and reappearance in vitro under regulatory pressure of corticotropin-releasing hormone in the rat. Endocrinology 147, 4738–4752.

Papkoff, H., Licht, P., Bona-Gallo, A., MacKenzie, D.S., Oelofsen, W., Oosthuizen, M.M., 1982. Biochemical and immunological characterization of pituitary hormones from the ostrich (*Struthio camelus*). Gen. Comp. Endocrinol. 48, 181–195.

Peeters, K., Langouche, L., Vandesande, F., Darras, V.M., Berghman, L.R., 1998. Effects of pituitary adenylate cyclase-activating polypeptide (PACAP) on cAMP formation and growth hormone release from chicken anterior pituitary cells. Ann. NY Acad. Sci. 865, 471–474.

Perez, F.M., Malamed, S., Scanes, C.G., 1985. Biosynthetic human somatomedin C inhibits hpGRF (1-44) NH2-induced and TRH-induced GH release in a primary culture of chicken pituitary cells. IRCS Med. Sci. 13, 871–872.

Pedersen, A., Tomaszycki, M.L., May 24, 2012. Oxytocin antagonist treatments alter the formation of pair relationships in zebra finches of both sexes. Horm. Behav (Epub ahead of print). 62, 113–119.

Perfito, N., Zann, R., Ubuka, T., Bentley, G., Hau, M., 2011. Potential roles for GNIH and GNRH-II in reproductive axis regulation of an opportunistically breeding songbird. Gen. Comp. Endocrinol. 173, 20–26.

Pinheiro, P.L., Cardoso, J.C., Gomes, A.S., Fuentes, J., Power, D.M., Canário, A.V., 2010. Gene structure, transcripts and calciotropic effects of the PTH family of peptides in *Xenopus* and chicken. BMC Evol. Biol. 10, 373.

Porter, T.E., Couger, G.S., Dean, C.E., Hargis, B.M., 1995. Ontogeny of growth hormone (GH)-secreting cells during chicken embryonic development: initial somatotrophs are responsive to GH-releasing hormone. Endocrinology 136, 1850–1856.

Porter, T.E., Ellestad, L.E., Fay, A., Stewart, J.L., Bossis, I., 2006. Identification of the chicken growth hormone-releasing hormone receptor (GHRH-R) mRNA and gene: regulation of anterior pituitary GHRH-R mRNA levels by homologous and heterologous hormones. Endocrinology 147, 2535–2543.

Powell, R.C., Jach, H., Millar, R.P., King, J.A., 1987. Identification of Gln8-GnRH and His5, Trp7, Tyr8-GnRH in the hypothalamus and extrahypothalamic brain of the ostrich (*Struthio camelus*). Peptides 8, 185–190.

Proudman, J.A., Opel, H., 1988. Stimulation of prolactin secretion from turkey anterior pituitary cells in culture. Proc. Soc. Exp. Biol. Med. 187, 448–454.

Proudman, J.A., Opel, H., 1990a. Half-life and metabolic clearance rate of recombinant-derived chicken growth hormone and purified turkey growth hormone in intact and hypophysectomized turkeys. Poult. Sci. 69, 1569–1575.

Proudman, J.A., Opel, H., 1990b. Effect of peptide histidine isoleucine on in vitro and in vivo prolactin secretion in the turkey. Poult. Sci. 69, 1209–1214.

Proudman, J.A., McGuinness, M.C., Krishnan, K.A., Cogburn, L.A., 1994. Endocrine and metabolic responses of intact and hypophysectomized turkey poults given a daily injection of chicken growth hormone. Comp. Biochem. Physiol. C Pharmacol. Toxicol. Endocrinol. 109, 47–56.

Proudman, J.A., Vandesande, F., Berghman, L.R., 1999. Immunohistochemical evidence that follicle-stimulating hormone and luteinizing hormone reside in separate cells in the chicken pituitary. Biol. Reprod. 60, 1324–1328.

Proudman, J.A., Scanes, C.G., Johannsen, S.A., Berghman, L.R., Camp, M.J., 2006. Comparison of the ability of the three endogenous GnRHs to stimulate release of follicle-stimulating hormone and luteinizing hormone in chickens. Domest. Anim. Endocrinol. 31, 141–153.

Radecki, S.V., Deaver, D.R., Scanes, C.G., 1994. Triiodothyronine reduced growth hormone (GH) secretion and pituitary GH mRNA in the chicken in vivo and in vitro. Proc. Soc. Exp. Biol. Med. 205, 340–346.

Ramachandran, R., Ocón-Grove, O.M., Metzger, S.L., 2007. Molecular cloning and tissue expression of chicken AdipoR1 and AdipoR2 complementary deoxyribonucleic acids. Domest. Anim. Endocrinol. 33, 19–31.

Ramesh, R., Solow, R., Proudman, J.A., Kuenzel, W.J., 1998. Identification of mammosomatotrophs in the turkey hen pituitary: increased abundance during hyperprolactinemia. Endocrinology 139, 781–786.

Ramesh, R., Kuenzel, W.J., Buntin, J.D., Proudman, J.A., 2000. Identification of growth-hormone- and prolactin-containing neurons within the avian brain. Cell Tissue Res. 299, 371–383.

Ramesh, R., Kuenzel, W.J., Proudman, J.A., 2001. Increased proliferative activity and programmed cellular death in the turkey hen pituitary gland following interruption of incubation behavior. Biol. Reprod. 64, 611–618.

Rangel, P.L., Rodríguez, A., Rojas, S., Sharp, P.J., Gutierrez, C.G., 2009. Testosterone stimulates progesterone production and STAR, P450 cholesterol side-chain cleavage and LH receptor mRNAs expression in hen (*Gallus domesticus*) granulosa cells. Reproduction 138, 961–969.

Reichardt, A.K., 1993. Functional differentiation of the pituitary gland during development of the domestic ring dove (*Streptopelia roseogrisea*). Gen. Comp. Endocrinol. 92, 41–53.

Richards, M.P., Poch, S.M., McMurtry, J.P., 2006. Characterization of turkey and chicken ghrelin genes, and regulation of ghrelin and ghrelin receptor mRNA levels in broiler chickens. Gen. Comp. Endocrinol. 145, 298–310.

Riddle, O., Bates, R.W., Lahr, E.L., 1935. Prolactin induces broodiness in fowl. Am. J. Physiol. 111, 352–360.

Riou, S., Chastel, O., Lacroix, A., Hamer, K.C., 2010. Stress and parental care: prolactin responses to acute stress throughout the breeding cycle in a long-lived bird. Gen. Comp. Endocrinol. 168, 8–13.

Robinson, G.A., Wasnidge, D.C., Floto, F., Downie, S.E., 1976. Ovarian 125I transference in the laying Japanese quail: stimulation of FSH and lack of stimulation by TSH. Poult. Sci. 53, 398–401.

Robinzon, B., Koike, T.I., Marks, P.A., 1993. At low dose, arginine vasotocin has vasopressor rather than vasodepressor effect in chickens. Gen. Comp. Endocrinol. 91, 105–112.

Robinzon, B., Koike, T.I., Neldon, H.L., Kinzler, S.L., 1988a. Distribution of immunoreactive mesotocin and vasotocin in the brain and pituitary of chickens. Peptides 9, 829–833.

Robinzon, B., Koike, T.I., Neldon, H.L., Kinzler, S.L., Hendry, I.R., El Halawani, M.E., 1988b. Physiological effects of arginine vasotocin and mesotocin in cockerels. Br. Poult. Sci. 29, 639–652.

Rodríguez-Méndez, A.J., Luna-Acosta, J.L., Carranza, M., Harvey, S., Arámburo, C., Luna, M., 2010. Growth hormone expression in stromal and non-stromal cells in the bursa of Fabricius during bursal development and involution: causal relationships? Gen. Comp. Endocrinol. 167, 297–307.

Rosebrough, R.W., McMurtry, J.P., Vasilatos-Younken, J., 1991. Effect of pulsatile or continuous administration of pituitary-derived chicken growth hormone (p-cGH) on lipid metabolismin broiler pullets. Comp. Biochem. Physiol.A 99, 207–214.

Rozenboim, I., Silsby, J.L., Tabibzadeh, C., Pitts, G.R., Youngren, O.M., El Halawani, M.E., 1993. Hypothalamic and posteriorpituitary content of vasoactive intestinal peptide, gonadotropin releasing hormone I and II in the turkey hen. Biol. Reprod. 49, 622–627.

Rubin, C.J., Zody, M.C., Eriksson, J., Meadows, J.R., Sherwood, E., Webster, M.T., Jiang, L., Ingman, M., Sharpe, T., Ka, S., Hallböök, F., Besnier, F., Carlborg, O., Bed'hom, B., Tixier-Boichard, M., Jensen, P., Siegel, P., Lindblad-Toh, K., Andersson, L., 2010. Whole-genome resequencing reveals loci under selection during chicken domestication. Nature 464, 587–591.

Rzasa, J., 1978. Effects of arginine vasotocin and prostaglandin E1 on the hen uterus. Prostaglandins 16, 357–372.

Rzasa, J., 1984. The effect of arginine vasotocin on prostaglandin production of the hen uterus. Gen. Comp. Endocrinol. 53, 260–263.

Saeed, W., El Halawani, M.E., 1986. Modulation of prolactin response to thyrotropin releasing hormone by ovarian steroids inovariectomized turkeys (*Meleagris gallopavo*). Gen. Comp. Endocrinol. 62, 129–136.

Sandhu, M.A., Rahman, Z.U., Riaz, A., Rahman, S.U., Javed, I., Ullah, N., 2010. Somatotrophs and lactotrophs: an immunohistochemical study of *Gallus domesticus* pituitary gland at different stages of induced moult. Eur. J. Histochem. 54, e25.

Saito, N., Kinzler, S., Koike, T.I., 1990. Arginine vasotocin and mesotocin levels in theca and granulosa levels of the ovary during the oviposition cycle in hens (*Gallus domesticus*). Gen. Comp. Endocrinol. 79, 54–63.

Saito, E.-S., Kaiya, H., Tachibana, T., Tomonaga, S., Denbow, D.M., Kangawa, K., Furuse, M., 2005. Inhibitory effect of ghrelin on food intake is mediated by the corticotrophin-releasing factor system in neonatal chicks. Regul. Pept. 125, 201–208.

Sánchez-Bringas, G., Salazar, O., Pedernera, E., Méndez, C., 2006. Follicle-stimulating hormone treatment reverses the effect of hypophysectomy on cell proliferation in the chicken embryo ovary. Gen. Comp. Endocrinol. 149, 134–140.

Sanders, E.J., Baudet, M.L., Parker, E., Harvey, S., 2009. Signaling mechanisms mediating local GH action in the neural retina of the chick embryo. Gen. Comp. Endocrinol. 163, 63–69.

Sanders, E.J., Lin, W.Y., Parker, E., Harvey, S., 2010. Growth hormone expression and neuroprotective activity in a quail neural retina cell line. Gen. Comp. Endocrinol. 165, 111–119.

Sanders, E.J., Lin, W.Y., Parker, E., Harvey, S., 2011. Growth hormone promotes the survival of retinal cells in vivo. Gen. Comp. Endocrinol. 172, 140–150.

Sano, Y., Murabe, Y., 1980. Morphological and functional peculiarities of mesenchymal cells in the pars tuberalis of the pituitary gland. Cell Tissue Res. 206, 171–180.

Sasaki, F., Doshita, A., Matsumoto, Y., Kuwahara, S., Tsukamoto, Y., Ogawa, K., 2003. Embryonic development of the pituitary gland in the chick. Cells Tissues Organs 173, 65–74.

Satake, H., Hisada, M., Kawada, T., Minakata, H., Ukena, K., Tsutsui, K., 2001. Characterization of a cDNA encoding a novel avian hypothalamic neuropeptide exerting an inhibitory effect on gonadotropin release. Biochem. J. 354, 379–385.

Scanes, C.G., 1974. Some in vitro effects of synthetic thyrotrophin releasing factor on the secretion of thyroid stimulating hormone from the anterior pituitary gland of the domestic fowl. Neuroendocrinology 15, 1–9.

Scanes, C.G., 1992. Lipolytic and diabetogenic effects of native and biosynthetic growth hormone. Comp. Biochem. Physiol. 101A, 871–878.

Scanes, C.G., 2000. Introduction to endocrinology: pituitary gland. In: Whittow, G.C. (Ed.), Sturkie's Avian Physiology. Academic Press, San Diego, pp. 437–460.

Scanes, C.G., Harvey, S., 1989. Triiodothyronine inhibition of thyrotropin-releasing hormone- and growth hormone releasing factor induced growth hormone secretion in anesthetized chickens. Gen. Comp. Endocrinol. 73, 477–484.

Scanes, C.G., Johnson, A.L., 1984. Failure of castration to prevent the prepubescent decline in the circulating concentration of growth hormone in the domestic fowl. Gen. Comp. Endocrinol. 53, 398–401.

Scanes, C.G., Carsia, R.V., Lauterio, T.J., Huybrechts, L., Rivier, J., Vale, W., 1984. Synthetic human pancreatic growth hormone releasing factor (GRF) stimulates growth hormone secretion in the domestic fowl (*Gallus domesticus*). Life Sci. 34, 1127–1134.

Scanes, C.G., Duyka, D.R., Lauterio, T.T., Bowen, S.J., Huybrechts, L.M., Bacon, W.L., King, D.B., 1986. Effect of chicken growth hormone, triiodothyronine and hypophysectomy in growing domestic fowl. Growth 50, 12–31.

Scanes, C.G., Godden, P.M.M., Sharp, P.J., 1977. An homologous radioimmunoassay for chicken follicle-stimulating hormone: observations on the ovulatory cycle. J. Endocrinol. 73, 473–482.

Scanes, C.G., Griminger, P., Buonomo, F.C., 1981. Dietary protein restriction on circulating concentrations of growth hormone in growing domestic fowl (*Gallus domesticus*). Proc. Soc. Exp. Biol. Med. 168, 334–337.

Scanes, C.G., Proudman, J.A., Radecki, S.V., 1999. Influence of continuous growth hormone or insulin-like growth factor I administration in adult female chickens. Gen. Comp. Endocrinol. 114, 315–323.

Scanes, C.G., Glavaski-Joksimovic, A., Johannsen, S.A., Jeftinija, S., Anderson, L.L., 2007. Subpopulations of somatotropes with differing intracellular calcium concentration responses to secretagogues. Neuroendocrinology 85, 221–231.

Schaper, S.V., Dawson, A., Sharp, P.J., Gienapp, P., Caro, S.P., Visser, M.E., 2012. Increasing temperature, not mean temperature, is a cue for avian timing of reproduction. Am. Nat. 179, E55–E69.

Schmierer, B., Schuster, M.K., Shkumatava, A., Kuchler, K., 2003. Activin and follicle-stimulating hormone signaling are required for long-term culture of functionally differentiated primary granulosa cells from the chicken ovary. Biol. Reprod. 68, 620–627.

Schuster, M.K., Schmierer, B., Shkumatava, A., Kuchler, K., 2004. Activin A and follicle-stimulating hormone control tight junctions in avian granulosa cells by regulating occludin expression. Biol. Reprod. 70, 1493–1499.

Seasholtz, A.F., Valverde, R.A., Denver, R.J., 2002. Corticotropin-releasing hormone-binding protein: biochemistry and function from fishes to mammals. J. Endocrinol. 175, 89–97.

Seth, R., Köhler, A., Grossmann, R., Chaturvedi, C.M., 2004a. Expression of hypothalamic arginine vasotocin gene in response to water deprivation and sex steroid administration in female Japanese quail. J. Exp. Biol. 207, 3025–3033.

Seth, R., Xu, Y.X., Grossmann, R., Chaturvedi, C.M., 2004b. Changes in expression of AVT and AVT receptor (VT1) gene in hypothalamus and shell gland in relation to egg laying in white leghorn hen. Gen. Comp. Endocrinol. 137, 177–186.

Shao, Y., Liu, S., Tang, X., Gao, J., Wu, G., Li, Z., 2010. Ontogeny of ghrelin mRNA expression and identification of ghrelin-immunopositive cells in the gastrointestinal tract of the Peking duck, *Anas platyrhynchos*. Gen. Comp. Endocrinol. 166, 12–18.

Shahabi, N.A., Bahr, J.M., Nalbandov, A.V., 1975. Effect of LH injection on plasma and follicular steroids in the chicken. Endocrinology 96, 969–972.

Sharma, D., Cornett, L.E., Chaturvedi, C.M., 2009. Corticosterone- or metapyrone-induced alterations in adrenal function and expression of the arginine vasotocin receptor VT2 in the pituitary gland of domestic fowl, *Gallus gallus*. Gen. Comp. Endocrinol. 161, 208–215.

Sharma, D., Chaturvedi, C.M., 2011. Testosterone modulates pituitary vasotocin receptor expression and adrenal activity in osmotically stressed chicken. Comp. Biochem. Physiol. A Mol. Integr. Physiol. 158, 87–93.

Sharp, P.J., Chiasson, R.B., El Tounsy, M.M., Klandorf, H., Radke, W.J., 1979. Localization of cells producing thyroid- stimulating hormone in the pituitary gland of the domestic fowl. Cell Tissue Res. 198, 53–63.

Sharp, P.J., Macnamee, M.C., Sterling, R.J., Lea, R.W., Pedersen, H.C., 1988. Relationships between prolactin, LH and broody behavior in bantam hens. J. Endocrinol. 118, 279–286.

Sharp, P.J., Sterling, R.J., Talbot, R.T., Huskisson, N.S., 1989. The role of hypothalamic vasoactive intestinal polypeptide in the maintenance of prolactin secretion in incubating bantam hens: observations using passive immunization, radioimmunoassay and immunohistochemistry. J. Endocrinol. 122, 5–13.

Shen, S.T., Yu, J.Y., 2002. Cloning and gene expression of a cDNA for the chicken follicle-stimulating hormone (FSH)-beta-subunit. Gen. Comp. Endocrinol. 125, 375–386.

Shimada, K., Saito, N., Itogawa, K., Koike, T.I., 1987. Changes in plasma concentrations of arginine vasotocin after intrauterine injections of prostaglandin F 2α and acetylcholine at various times during the oviposition cycle of the domestic hen (*Gallus domesticus*). J. Reprod. Fertil. 80, 143–150.

Shirley, H.V., Nalbandov, A.V., 1956. Effects of neurohypophysectomy in domestic chickens. Endocrinology 58, 477–483.

Silver, R., Ramos, C.L., Silverman, A.-J., 1992. Sexual behavior triggers the appearance of non-neuronal cells containing gonadotropin-releasing hormone-like immunoreactivity. J. Neuroendocrinol. 4, 207–210.

Silverin, B., Wingfield, J., Stokkan, K.A., Massa, R., Järvinen, A., Andersson, N.A., Lambrechts, M., Sorace, A., Blomqvist, D., 2008. Ambient temperature effects on photo induced gonadal cycles and hormonal secretion patterns in great tits from three different breeding latitudes. Horm. Behav. 54, 60–68.

Spiess, J., Rivier, J.E., Rodkey, J.A., Bennett, C.D., Vale, W., 1979. Isolation and characterization of somatostatin from pigeon pancreas. Proc. Natl. Acad. Sci. U. S. A. 76, 2974–2978.

Srivastava, R., Cornett, L.E., Chaturvedi, C.M., 2010. Age-dependent expression of AVT and its oxytocic-like receptor VT3 in the shell gland of Japanese quail, *Coturnix coturnix japonica*. Gen. Comp. Endocrinol. 165, 47–52.

Stallone, J.N., Braun, E.J., 1985. Contributions of glomerular and tubular mechanisms to antidiuresis in conscious domestic fowl. Am. J. Physiol. 249, F842–F850.

Stallone, J.N., Braun, E.J., 1986. Osmotic and volemic regulation of plasma arginine vasotocin in conscious domestic fowl. Am. J. Physiol. 250, R644–R657.

Sturkie, P.D., Lin, Y.C., 1966. Release of vasotocin and oviposition in the hen. J. Endocrinol. 35, 325–326.

Sun, D., Cui, T., Luo, H., Li, R., Cui, S., Liu, J., 2012. Cell-specific distributions of estrogen receptor alpha (ERα) and androgen receptor (AR) in anterior pituitary glands from adult cockerels as revealed by immunohistochemistry. Cell Tissue Res. 348, 551–558.

Sun, S., El Halawani, M.E., 1995. Protein kinase-C mediates chicken vasoactive intestinal peptide stimulated prolactin secretion and gene expression in turkey primary pituitary cells. Gen. Comp. Endocrinol. 99, 289–297.

Sun, Y.M., Millar, R.P., Ho, H., Gershengorn, M.C., Illing, N., 1998. Cloning and characterization of the chicken thyrotropin-releasing hormone receptor. Endocrinology 139, 3390–3398.

Sweeney, S.A., Johnson, P.A., 2005. Messenger RNA and protein expression analysis of betaglycan in the pituitary and ovary of the domestic hen. Biol. Reprod. 72, 172–178.

Tachibana, T., Moriyama, S., Takahashi, A., Tsukada, A., Oda, A., Takeuchi, S., Sakamoto, T., 2011. Isolation and characterisation of prolactin-releasing peptide in chicks and its effect on prolactin release and feeding behaviour. J. Neuroendocrinol. 23, 74–81.

Takahashi, T., Kawashima, M., 2008a. Mesotocin increases the sensitivity of the hen oviduct uterus to arginine vasotocin. Poult. Sci. 87, 2107–2111.

Takahashi, T., Kawashima, M., 2008b. Mesotocin receptor binding in oviduct uterus of the hen before and after oviposition. Poult. Sci. 87, 546–550.

Takahashi, T., Kawashima, M., 2008c. Prostaglandin F(2alpha) receptor in the neurohypophysis of hens. Poult. Sci. 88, 1712–1718.

Takahashi, T., Kawashima, M., 2009. Properties of estrogen binding components in the plasma membrane of neurohypophysis in hens and changes in its binding before and after oviposition. Poult. Sci. 88, 2206–2211.

Takahashi, T., Nozaki, Y., Nakagawa-Mizuyachi, K., Nakayama, H., Kawashima, M., 2011. Changes in angiotensin II receptor bindings in the hen neurohypophysis before and after. Poult. Sci. 90, 2565–2572.

Takeuchi, S., Takahashi, S., 1998a. A possible involvement of melanocortin 3 receptor in the regulation of adrenal gland function in the chicken. Biochim. Biophys. Acta 1448, 512–518.

Takeuchi, S., Takahashi, S., 1998b. Melanocortin receptor genes in the chicken-tissue distributions. Gen. Comp. Endocrinol. 112, 220–231.

Takeuchi, S., Suzuki, S., Hirose, S., Yabuuchi, M., Sato, C., Yamamoto, H., Takahashi, S., 1996. Molecular cloning and sequence analysis of the chick melanocortin 1-receptor gene. Biochim. Biophys. Acta 1306, 122–126.

Takeuchi, S., Kudo, T., Takahashi, S., 1998. Molecular cloning of the chicken melanocortin 2 (ACTH)-receptor gene. Biochim. Biophys. Acta 1403, 102–108.

Takeuchi, S., Teshigawara, K., Takahashi, S., 1999. Molecular cloning and characterization of the chicken pro-opiomelanocortin (POMC) gene. Biochim. Biophys. Acta 1450, 452–459.

Tanaka, M., Maeda, K., Okubo, T., Natashima, K., 1992. Double antenna structure of chicken prolactin receptor deduced from the cDNA sequence. Biochem. Biophys. Res. Commun. 188, 490–496.

Tanaka, M., Yamamoto, I., Ohkubo, T., Wakita, M., Hoshino, S., Nakashima, K., 1999. cDNA cloning and developmental alterations in gene expression of the two Pit-1/GHF-1 transcription factors in the chicken pituitary. Gen. Comp. Endocrinol. 114, 441–448.

Tennyson, V.M., Hou-Yu, A., Nilaver, G., Zimmerman, E.A., 1985. Immunocytochemical studies of vasotocin and mesotocin in the hypothalamo-hypophyseal system of the chicken. Cell Tissue Res. 239, 279–291.

Tennyson, V.M., Nilaver, G., Hou-Yu, A., Valiquette, G., Zimmerman, E.A., 1986. Immunocytochemical study of the development of vasotocin/mesotocin in the hypothalamo-hypophysial system of the chick embryo. Cell Tissue Res. 243, 15–31.

Terada, O., Shimada, K., Saito, N., 1997. Effect of oestradiol replacement in ovariectomized chickens on pituitary LH concentrations and concentrations of mRNAs encoding LH beta and alpha subunits. J. Reprod. Fertil. 111, 59–64.

Thommes, R.C., Caliendo, J., Woods, J.E., 1985. Hypothalamo-adenohypophyseal-thyroid interrelationships in the developing chick embryo. VII. Immunocytochemical demonstration of thyrotrophin-releasing hormone. Gen. Comp. Endocrinol. 57, 1–9.

Thommes, R.C., Martens, J.B., Hopkins, W.E., Caliendo, J., Sorrentino, M.J., Woods, J.E., 1983. Hypothalamo adenohypophyseal-thyroid interrelationships in the chick embryo. IV. Immunocytochemical demonstration of TSH in the hypophyseal pars distalis. Gen. Comp. Endocrinol. 51, 434–443.

Thommes, R.C., Williams, D.J., Woods, J.E., 1984. Hypothalamo-adenohypophyseal-thyroid interrelationship in the chicken fowl. VI. Midgestational adenohypophyseal sensitivity to thyrotrophin-releasing hormone. Gen. Comp. Endocrinol. 55, 275–279.

Thayananuphat, A., Kang, S.W., Bakken, T., Millam, J.R., El Halawani, M.E., 2007. Rhythmic dependent light induction of gonadotrophin-releasing hormone-I expression and activation of dopaminergic neurones within the premammillary nucleus of the turkey hypothalamus. J. Neuroendocrinol. 19, 399–406.

Thayananuphat, A., Youngren, O.M., Kang, S.W., Bakken, T., Kosonsiriluk, S., Chaiseha, Y., El Halawani, M.E., 2011. Dopamine and mesotocin neurotransmission during the transition from incubation to brooding in the turkey. Horm. Behav. 60, 327–335.

Tilly, J.L., Johnson, A.L., 1987. Presence and hormonal control of plasminogen activator in granulosa cells of the domestic hen. Biol. Reprod. 37, 1156–1164.

Tilly, J.L., Johnson, A.L., 1988a. Effects of a phorbol ester, a calcium ionophore, and 3′,5′-adenosine monophosphate production on hen granulosa cell plasminogen activator activity. Endocrinology 123, 1433–1441.

Tilly, J.L., Johnson, A.L., 1988b. Attenuation of hen granulosa cell steroidogenesis by a phorbol ester and 1-oleoyl-2-acetylglycenol. Biol. Reprod. 39, 1–8.

Tilly, J.L., Johnson, A.L., 1989. Regulation of androstenedione production by adenosine 3′,5′-monophosphate and phorbol myristate acetate in ovarian thecal cells of the domestic fowl. Endocrinology 125, 1691–1699.

Tilly, J.L., Kowalski, K.I., Johnson, A.L., 1991a. Stage of ovarian follicular development associated with the initiation of steroidogenic competence in avian granulosa cells. Biol. Reprod. 44, 305–314.

Tilly, J.L., Kowalski, K.I., Johnson, A.L., 1991b. Cytochrome P450 side-chain cleavage (P450scc) in the hen ovary. II P450scc messenger RNA, immunoreactive protein, and enzyme activity in develop ing granulosa cells. Biol. Reprod. 45, 967–974.

Tonoue, T., Kitch, J., 1978. Release of cyclic AMP from the chicken thyroid stimulated with TSH in vitro. Endocrinol. Jpn. 25, 105–109.

Toogood, A.A., Harvey, S., Thorner, M.O., Gaylinn, B.D., 2006. Cloning of the chicken pituitary receptor for growth hormone-releasing hormone. Endocrinology 147, 1838–1846.

Tong, Z., Pitts, G.R., Foster, D.N., El Halawani, M., 1997. Transcriptional and post-transcriptional regulation of prolactin during the turkey reproductive cycle. J. Mol. Endocrinol. 18, 223–231.

Trabucchi, M., Tostivint, H., Lihrmann, I., Blähser, S., Vallarino, M., Vaudry, H., 2003. Characterization of the cDNA encoding a somatostatin variant in the chicken brain: comparison of the distribution of the two somatostatin precursor mRNAs. J. Comp. Neurol. 461, 441–451.

Tsutsui, K., Ishii, S., 1980. Hormonal regulations of follicle-stimulating hormone receptors in the tests of Japanese quail. J. Endocrinol. 85, 511–518.

Tsutsui, K., Saigoh, E., Ukena, K., Teranishi, H., Fujisawa, Y., Kikuchi, M., Ishii, S., Sharp, P.J., 2000. A novel avian hypothalamic peptide inhibiting gonadotropin release. Biochem. Biophys. Res. Commun. 275, 661–667.

Ubuka, T., Ukena, K., Sharp, P.J., Bentley, G.E., Tsutsui, K., 2006. Gonadotropin-inhibitory hormone inhibits gonadal development and maintenance by decreasing gonadotropin synthesis and release in male quail. Endocrinology 147, 1187–1194.

Ubuka, T., Kim, S., Huang, Y.C., Reid, J., Jiang, J., Osugi, T., Chowdhury, V.S., Tsutsui, K., Bentley, G.E., 2008. Gonadotropin-inhibitory hormone neurons interact directly with gonadotropin-releasing hormone-I and -II neurons in European starling brain. Endocrinology 149, 268–278.

Ubuka, T., Cadigan, P.A., Wang, A., Liu, J., Bentley, G.E., 2009. Identification of European starling GnRH-I precursor mRNA and its seasonal regulation. Gen. Comp. Endocrinol. 162, 301–306.

Van As, P., Buys, N., Onagbesan, O.M., Decuypere, E., 2000. Complementary DNA cloning and ontogenic expression of pituitary-specific transcription factor of chickens (*Gallus domesticus*) from the pituitary gland. Gen. Comp. Endocrinol. 120, 127–136.

Van As, P., Careghi, C., Bruggeman, V., Onagbesan, O.M., Van der Geyten, S., Darras, V.M., Decuypere, E., 2004. Regulation of growth hormone expression by thyrotropin-releasing hormone through the pituitary-specific transcription factor Pit-1 in chicken pituitary. Acta Vet. Hung. 52, 389–402.

Vandenborne, K., De Groef, B., Geelissen, S.M., Boorse, G.C., Denver, R.J., Kühn, E.R., Darras, V.M., Van der Geyten, S., 2005a. Molecular cloning and developmental expression of corticotropin-releasing factor in the chicken. Endocrinology 146, 301–308.

Vandenborne, K., De Groef, B., Geelissen, S.M., Kühn, E.R., Darras, V.M., Van der Geyten, S., 2005b. Corticosterone-induced negative feedback mechanisms within the hypothalamo-pituitary-adrenal axis of the chicken. J. Endocrinol. 185, 383–391.

Van der Geyten, S., Buys, N., Sanders, J.P., Decuypere, E., Visser, T.J., Kühn, E.R., Darras, V.M., 1999. Acute pretranslational regulation of type III iodothyronine deiodinase by growth hormone and dexamethasone in chicken embryos. Mol. Cell. Endocrinol. 147, 49–56.

Vanderpooten, A., Darras, V.M., Huybrechts, L.M., Rudas, P., Decuypere, E., Kuhn, E.R., 1991a. Effects of hypophysectomy andacute administration of growth hormone (GH) on GH- receptor binding in chick liver membranes. J. Endocrinol. 129, 275–281.

Vanderpooten, A., Huybrechts, L.M., Decuypere, E., Kuhn, E.R., 1991b. Differences in hepatic growth hormone receptor binding during development of normal and dwarf chickens. Reprod. Nutr. Dev. 31, 47–55.

Van Nassauw, L., Harrisson, F., Cras, P., Callebaut, M., 1987. Immunohistochemical localization of S-100 protein, glial fibrillary acidic protein, and neuron-specific enolase in the pars distalis of quail, rat, and human hypophyses. Histochemistry 86, 353–358.

Vasilatos-Younken, R., Gravener, T.L., Cogburn, L.A., Mast, M.G., Wellenreiter, R.H., 1988. Effect of pattern of administration on the response to exogenous pituitary-derived chicken growth hormone by broiler-strain pullets. Gen. Comp. Endocrinol. 71, 268–283.

Vasilatos-Younken, R., Tsao, P.H., Foster, D.N., Smiley, D.L., Bryant, H., Heiman, M.L., 1992. Restoration of juvenile baseline growth hormone secretion with preservation of the ultradian growth-hormone rhythm by continuous delivery of growth hormone-releasing factor. J. Endocrinol. 135, 371–382.

Vleurick, L., Kühn, E.R., Decuypere, E., Burnside, J., Pezet, A., Edery, M., 1999. Generation of chicken growth hormone-binding proteins by proteolysis. Gen. Comp. Endocrinol. 113, 283–289.

Wakabayashi, N., Suzuki, A., Hoshino, H., Nishimori, K., Mizuno, S., 1997. The cDNA cloning and transient expression of a chicken gene encoding a follicle-stimulating hormone receptor. Gene 197, 121–127.

Wang, J.T., Xu, S.W., 2008. Effects of cold stress on the messenger ribonucleic acid levels of corticotrophin-releasing hormone and thyrotropin-releasing hormone in hypothalami of broilers. Poult. Sci. 87, 973–978.

Wang, C.Y., Wang, Y., Li, J., Leung, F.C., 2006. Expression profiles of growth hormone-releasing hormone and growth hormone-releasing hormone receptor during chicken embryonic pituitary development. Poult. Sci. 85, 569–576.

Wang, S., Bottje, W.G., Kinzler, S., Neldon, H.L., Koike, T.I., 1989. Effect of heat stress on plasma levels of arginine vasotocin and mesotocin in domestic fowl (*Gallus domesticus*). Comp. Biochem. Physiol. A Comp. Physiol. 93, 721–724.

Wang, Y., Li, J., Wang, C.Y., Kwok, A.H., Leung, F.C., 2007. Identification of the endogenous ligands for chicken growth hormone-releasing hormone (GHRH) receptor: evidence for a separate gene encoding GHRH in submammalian vertebrates. Endocrinology 148, 2405–2416.

Wang, J., Wang, Y., Li, X., Li, J., Leung, F.C., 2008. Cloning, tissue distribution, and functional characterization of chicken glucagon receptor. Poult. Sci. 87, 2678–2688.

Wang, X., Hadley, J., Corey, S.J., Vasilatos-Younken, R., 2005. Regulation of JAK2 protein expression by chronic, pulsatile GH administration in vivo: a possible mechanism for ligand enhancement of signal transduction. Gen. Comp. Endocrinol. 144, 128–139.

Wang, Y., Li, J., Yan Kwok, A.H., Ge, W., Leung, F.C., 2010. A novel prolactin-like protein (PRL-L) gene in chickens and zebrafish: cloning and characterization of its tissue expression. Gen. Comp. Endocrinol. 166, 200–210.

Wang, Y., Wang, C.Y., Wu, Y., Huang, G., Li, J., Leung, F.C., 2012. Identification of the receptors for prolactin-releasing peptide (PrRP) and Carassius RFamide peptide (C-RFa) in chickens. Endocrinology 153, 1861–1874.

Watahiki, M., Tanaka, M., Masuda, N., Sugisaki, K., Yamamoto, M., Yamakawa, M., Nagai, J., Nakashima, K., 1989. Primary structure of chicken pituitary prolactin deduced from the cDNA sequence: conversed and specific amino acid resides in the domains of the prolactin. J. Biol. Chem. 264, 5535–5539.

Weatherly, K.L., Ramesh, R., Strange, H., Waite, K.L., Storrie, B., Proudman, J.A., Wong, E.A., 2001. The turkey transcription factor Pit-1/GHF-1 can activate the turkey prolactin and growth hormone gene promoters in vitro but is not detectable in lactotrophs in vivo. Gen. Comp. Endocrinol. 123, 244–253.

Williamson, R.A., Davison, T.F., 1985. The effects of a single injection of thyrotrophin on serum concentrations of thyroxine, triiodothyronine, and reverse triiodothyronine in the immature chicken (*Gallus domesticus*). Gen. Comp. Endocrinol. 58, 109–113.

Wilson, J.X., West, N.H., 1986. Cardiovascular responses to neurohormones in conscious chickens and ducks. Gen. Comp. Endocrinol. 62, 268–280.

Wilson, S.C., Sharp, P.J., 1976a. Induction of luteinizing hormone release by gonadal steroid in the ovariectomized domestic hen. J. Endocrinol. 71, 87–98.

Wilson, S.C., Sharp, P.J., 1976b. Effects of androgens, oestrogens and deoxycorticosterone acetate on plasma concentrations of luteinizing hormone in laying hens. J. Endocrinol. 69, 93–102.

Wilson, S.C., Cunningham, F.J., Chairil, R.A., Gladwell, R.T., 1989. Maturational changes in the LH response of domestic fowl to synthetic chicken LHRH-I and -II. J. Endocrinol. 123, 311–318.

Wingfield, J.C., Hahn, T.P., Maney, D.L., Schoech, S.J., Wada, M., Morton, M.L., 2003. Effects of temperature on photoperiodically induced reproductive development, circulating plasma luteinizing hormone and thyroid hormones, body mass, fat deposition and molt in mountain white-crowned sparrows, *Zonotrichia leucophrys oriantha*. Gen. Comp. Endocrinol. 131, 143–158.

Wingfield, J.C., Sullivan, K., Jaxion-Harm, J., Meddle, S.L., 2012. The presence of water influences reproductive function in the song sparrow (*Melospiza melodia morphna*). Gen. Comp. Endocrinol. 178, 485–493.

Wong, E.A., Silsby, J.L., El Halawani, M.E., 1992. Complimentary DNA cloning and expression of Pit-1/GHF-1 from the domestic turkey. DNA Cell Biol. 11, 651–660.

Woods, D.C., Johnson, A.L., 2005. Regulation of follicle-stimulating hormone-receptor messenger RNA in hen granulosa cells relative to follicle selection. Biol. Reprod. 72, 643–650.

Wong, E.A., Ferrin, N.H., Silsby, J.L., El Halawani, M.E., 1991. Cloning of a turkey prolactin mRNA throughout the reproductive cycle of the domestic turkey (*Meleagris gallopavo*). Gen. Comp. Endocrinol. 83, 18–26.

Xing, G., Zhao, Q., Mao, J., Liu, T., Wang, G., 2011. Identification and characterization of goose prolactin receptor. Poult. Sci. 90, 1050–1057.

Yasuo, S., Watanabe, M., Tsukada, A., Takagi, T., Iigo, M., Shimada, K., Ebihara, S., Yoshimura, T., 2004. Photoinducible phase-specific light induction of Cry1 gene in the pars tuberalis of Japanese quail. Endocrinology 145, 1612–1616.

Yasuo, S., Korf, H.W., 2011. The hypophysial pars tuberalis transduces photoperiodic signals via multiple pathways and messenger molecules. Gen. Comp. Endocrinol. 172, 15–22.

Youngren, O.M., Chaiseha, Y., El Halawani, M.E., 1998. Serotonergic stimulation of avian prolactin secretion requires an intact dopaminergic system. Gen. Comp. Endocrinol. 112, 63–68.

Yoshihara, C., Tashiro, Y., Taniuchi, S., Katayama, H., Takahashi, S., Takeuchi, S., 2011. Feather follicles express two classes of pro-opiomelanocortin (POMC) mRNA using alternative promoters in chickens. Gen. Comp. Endocrinol. 171, 46–51.

Yoshimura, Y., Okamoto, T., Tamura, T., 1993. Ultrastructural changes of oocyte and follicular wall during oocyte maturation in the Japanese quail (*Coturnix coturnix japonica*). J. Reprod. Fertil. 97, 189–196.

You, S., Foster, L.K., Silsby, J.L., el Halawani, M.E., Foster, D.N., 1995. Sequence analysis of the turkey LH beta subunit and its regulation by gonadotrophin-releasing hormone and prolactin in cultured pituitary cells. J. Mol. Endocrinol. 14, 117–129.

Youngen, O.M., Halawani, M.E., Silsby, J.L., Phillips, R.E., 1991. Intracranial prolactin perfusion induced incubation behavior in turkey hens. Biol. Reprod. 44, 425–431.

Youngren, O.M., Pitts, G.R., Phillips, R.E., El Halawani, M.E., 1996. Dopaminergic control of prolactin secretion in the turkey. Gen. Comp. Endocrinol. 104, 225–230.

Yu, J., Xie, L.Y., Abou-Samra, A.B., 1996. Molecular cloning of a type A chicken corticotropin-releasing factor receptor with high affinity for urotensin I. Endocrinology 137, 192–197.

Yuri, T., Kimball, R.T., Braun, E.L., Braun, M.J., 2008. Duplication of accelerated evolution and growth hormone gene in passerine birds. Mol. Biol. Evol. 25, 352–361.

Zhang, X.L., Jiang, X., Liu, Y.P., Du, H.R., Zhu, Q., 2007. Identification of Avai polymorphisms in the third intron of GH gene and their associations with abdominal fat in chickens. Poult. Sci. 86, 1079–1083.

Zheng, J., Takagi, H., Tsutsui, C., Adachi, A., Sakai, T., 2008. Hypophyseal corticosteroids stimulate somatotrope differentiation in the embryonic chicken pituitary gland. Histochem. Cell Biol. 129, 357–365.

Zheng, J.X., Liu, Z.Z., Yang, N., 2007. Deficiency of growth hormone receptor does not affect male reproduction in dwarf chickens. Poult. Sci. 86, 112–117.

Zhou, J.F., Zadworny, D., Guémené, D., Kühnlein, U., 1996. Molecular cloning, tissue distribution, and expression of the prolactin receptor during various reproductive states in *Meleagris gallopavo*. Biol. Reprod. 55, 1081–1090.

Zhu, G., Kang, L., Yang, C., Zhang, X., Wang, M., Jiang, Y., 2012. Differential expression of CTGF in pre- and post-ovulatory granulosa cells in the hen ovary is regulated by TGFβ1 and gonadotrophins. Gen. Comp. Endocrinol. 178, 314–322.

Zhung, X., Silverman, A.J., Silver, R., 1993. Reproductive behavior, endocrine state, and the distribution of GnRH-like immunoreactive mast cells on dove brain. Horm. Behav. 27, 283–295.

Chapter 24

Thyroids

F.M. Anne McNabb and Veerle M. Darras
Department of Biological Sciences, Virginia Tech, Blacksburg, VA, USA
Department of Biology, Katholieke Universiteit Leuven, Leuven, Belgium

24.1 ANATOMY, EMBRYOLOGY, AND HISTOLOGY OF THYROID GLANDS

Avian thyroid glands are paired, well-vascularized, oval glands located ventrolaterally to the trachea, and caudal to the junction of the subclavian and common carotid arteries (Figure 24.1). The histology and ultrastructure of avian thyroids (French and Hodges, 1977; Astier, 1980) are like those of other vertebrate classes. Avian thyroids contain follicles of epithelial cells that surround a colloid-filled lumen. The colloid contains thyroglobulin, the thyroid hormone–containing protein, which provides for large extracellular stores of hormone. Such extracellular hormone storage is unique to the thyroid and is considered an adaptation to the scarcity of the trace element iodine, a key component of thyroid hormones (for a review: see McNabb, 1992). Calcitonin cells, which are parafollicular in mammals, are lacking in the avian thyroid; they are contained in separate ultimobranchial organs (see Chapter 25 by Dacke et al.).

The thyroid gland, which arises from the ventral pharyngeal wall in vertebrates, appears very early (e.g., day 2 of the 21 day incubation in chicken embryos). This mass of epithelial cells is initially attached to the pharynx, but then detaches, and the thyroid glands assume their mature positions (by day 5). Follicle formation is initiated by midincubation and proceeds rapidly, along with functional maturation of the gland during the first half of embryonic life and establishment of hypothalamic–pituitary control in the latter half of embryonic life in precocial galliform birds. In contrast, in altricial ring doves, although some follicles appear early, little more follicular organization occurs during embryonic life, and most histological and functional development and control occur after hatch (reviews: Wentworth and Ringer, 1986; McNabb, 1992). In recent years, there have been significant advances in understanding both the intrinsic and extrinsic signals involved in the early steps of thyroid gland development. Although most of these studies have used mammalian models, investigations of a few specific aspects of this signaling in chicken embryos suggest general agreement with the more detailed information about thyroid developmental control in mammals (review: De Felice and Di Lauro, 2011). Study of the details of gene expression for most of the key components of thyroid gland function throughout embryonic life of chicken embryos has recently shown how thyroidal synthesis and release of thyroxine (tetraiodothyronine; T_4) are responsible for the increases in circulating T_4 during the latter half of incubation and during the perihatch period (Grommen et al., 2011).

24.2 THYROID HORMONES

24.2.1 Synthesis, Release, and Circulating Concentrations of Thyroid Hormones

In birds, as in other vertebrates, both T_4 and triiodothyronine (T_3) are considered thyroid hormones (Figure 24.2). Overall, investigations of avian thyroid function indicate that the mechanisms of hormone synthesis and release by avian thyroids are essentially equivalent to those in mammals (review: McNabb, 1992). Iodide is transported into the thyroid gland from the blood by a sodium-dependent iodide (Na-I) symporter, and another transporter, pendrin, is involved in transporting iodide from the follicle cells into the colloid (review on mammalian iodide transport: Bizhanova and Kopp, 2009). The effects of perchlorate on avian thyroids in several species suggest equivalency of Na-I symporter activity in birds to that in mammals and the few other vertebrates in which iodide transport has been studied (McNabb et al., 2006). The avian thyroid contains extremely high iodide concentrations and has prolonged iodide retention. Although avian thyroidal iodide can be influenced by extremes of iodide availability, circulating thyroid hormones and thyroid gland hormone content are well regulated over a wide range of iodide intakes in adult animals. Embryos and chicks also show a considerable amount of iodide regulation, although this is less than in adults (McNabb et al., 1985a,b; Stallard and McNabb, 1990).

Thyroid hormone synthesis begins with the iodination of tyrosine residues contained in thyroglobulin, the hormone storage protein of the colloid. Thyroglobulin is produced on ribosomes of the endoplasmic reticulum of follicle cells, carbohydrate components are added in the reticular lumen,

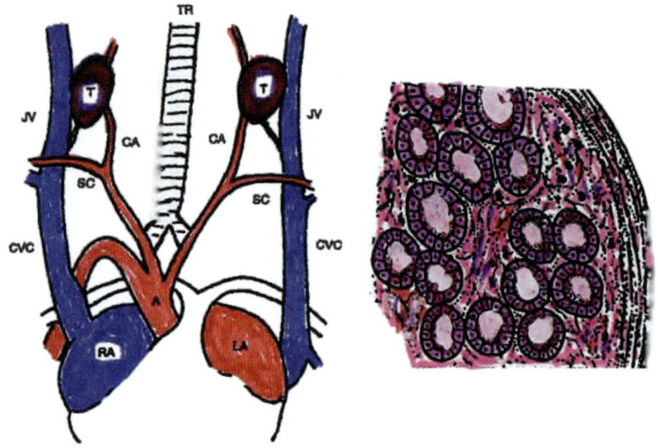

FIGURE 24.1 **Avian thyroid glands.** Left: ventral view of the general position of avian thyroids. In chickens, the right thyroid gland is more posterior than the left thyroid. A, aorta; CA, common carotid artery; CVC, cranial vena cava; JV, jugular vein; LA, left atrium; RA, right atrium; SC, subclavian artery; T, thyroid gland; TR, trachea. Right: diagram of part of a histological section through an avian thyroid gland, as seen with hematoxylin and eosin staining. Shown are follicles of various sizes, consisting of cuboidal cells (characteristic of a euthyroid state), surrounding a colloid-filled lumen. Part of the connective tissue capsule of the gland is shown on the right. The interstitial regions between the follicles contain blood vessels, nerves, and loose connective tissue.

further modifications and additions occur in the Golgi apparatus, and then the thyroglobulin is transported by vesicles to the apical cell border and extruded into the colloid by exocytosis. Iodination of the tyrosines in thyroglobulin requires the action of thyroid peroxidase and an oxidized form of iodide, and leads to the formation of monoiodotyrosines (MITs) and diiodotyrosines (DITs) within the thyroglobulin. These are coupled by thyroid peroxidase to yield the hormonal thyronines, T_4 (DIT+DIT) and T_3 (DIT+MIT). Hormone release from the thyroid involves endocytosis of droplets of colloid by the follicle cells, their fusion with lysosomes, digestion of thyroglobulin by lysosomal enzymes, and release of T_4 and T_3 into capillaries at the external surface of the follicle cells (review: McNabb, 1992).

In adult birds with adequate iodide intake, the thyroidal hormone content consists of primarily T_4, with lesser or undetectable amounts of T_3. When iodide is limiting, the T_3:T_4 ratio is increased and total hormone stores are decreased. Thyroid hormone secretion rates (TSRs), measured by a number of methods, are in the range of 1–3 µg T_4/100 g of body weight per day in chickens, quail, and pigeons. Cold temperatures increase the TSR, whereas iodide deficiency and aging tend to decrease the TSR (review: Wentworth and Ringer, 1986). As in other vertebrates, most control of thyroid function is by the hypothalamic–pituitary–thyroid (HPT) axis (see Section 24.3).

In embryos, the capability for hormone synthesis precedes organization of the thyroid gland into follicles. In precocial chicken embryos, thyroidal iodide concentration and some hormone production occur during the first one-fourth

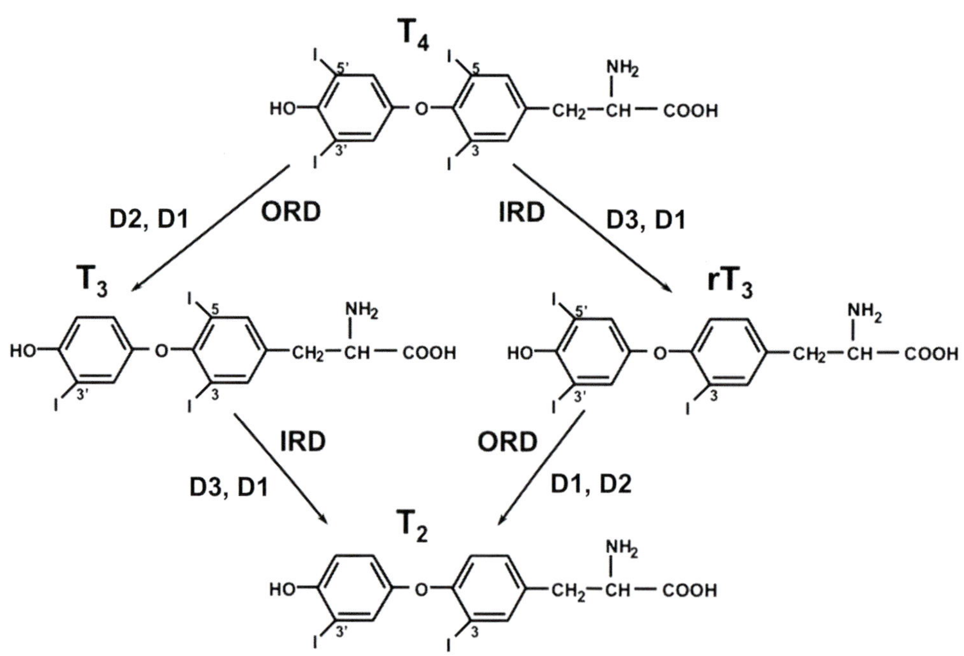

FIGURE 24.2 **Thyroid hormones and their deiodination pathways leading to thyroid hormone activation and inactivation.** T_4, 3,5,3′,5′-tetraiodothyronine or thyroxine; T_3, 3,5,3′-triiodothyronine; D1, type 1 deiodinase; D2, type 2 deiodinase; D3, type 3 deiodinase; IRD, inner ring deiodination; ORD, outer ring deiodination; rT_3, 3,3′,5′-triiodothyronine; T_2, 3,3′-diiodothyronine.

to one-third of incubation, but in altricial ring doves these events occur about midincubation (review: McNabb et al., 1998). Historically, circulating thyroid hormones were first measured as protein-bound iodine, then by competitive binding assay, then by radioimmunoassay (RIA) and enzyme-linked immunosorbent assay (ELISA), and most recently by liquid chromatography and mass spectrometry. Hormone measurements by RIAs and ELISAs account for the vast majority of the data in the literature since the late 1970s.

Concentrations of T_4 exceed those of T_3 in avian plasma by several-fold, although this ratio is much lower than in the thyroid gland. Adult birds of many species have plasma or serum T_4 concentrations in the range of 5–15 ng T_4/mL (6–19 pmol/mL) and T_3 concentrations in the range of 0.5–4 ng T_3/mL (0.7–1.5 pmol/mL). In comparison with mammals, avian plasma contains less T_4 (i.e., often 10-fold less) but similar concentrations of T_3 (McNabb, 2000). Circulating concentrations of both hormones have been measured in many avian species, but generalizations about species differences should be made with care because of technique differences and the many factors that influence thyroid status (e.g., food availability, seasonality, breed, age, and time of day).

Diurnal patterns of plasma thyroid hormones are readily demonstrated in birds. Plasma T_4 rises and peaks during the dark period, and plasma T_3 rises and is highest during the light period, reflecting the timing of highest extrathyroidal conversion of T_4 to T_3 during light. Patterns of food intake are a key factor influencing peripheral T_3 production and degradation and patterns of plasma thyroid hormones. Temperature also may be important; cold temperatures increase and warm temperatures depress plasma T_3 within these diurnal patterns. In general, temperature effects on plasma T_4 are opposite to those on T_3, although the T_4 effects are more complex (reviews: Cogburn and Freeman, 1987; Decuypere and Kühn, 1988; Sharp and Klandorf, 1985).

Circulating thyroid hormones in vertebrates are transported by binding proteins that maintain extrathyroidal hormone stores and help regulate hormone availability to the tissues. However, knowledge of free hormone concentrations alone is not sufficient to understand hormone entry into different tissues under different conditions (see Section 24.2.3 on hormone transport into tissues). The major thyroid hormone–binding proteins in birds are transthyretin, which is high affinity and low capacity, and albumin, which is low affinity and high capacity. Birds lack the very-high-affinity T_4-binding protein, thyroxine-binding globulin (TBG), which is present in the blood of large mammals. Several studies in birds suggest that binding proteins modulate the picture of thyroid hormone availability, for example during development and in the diurnal cycle (review: McNabb, 2000). Transthyretin, which is produced in the choroid plexus and liver of birds, has been characterized in several avian species and compared functionally with that in other vertebrates. Avian transthyretin has higher binding affinity for T_3 than T_4, the opposite of the binding pattern in mammals (Chang et al., 1999; Schreiber, 2002).

The ontogenic pattern of plasma thyroid hormones differs in birds with precocial versus altricial development. In precocial species, in which thyroid gland function and its control mature before hatching, plasma T_4 rises several-fold during the latter half of embryonic life but plasma T_3 remains low. During the perihatch period, both hormones rise dramatically to reach some of the highest concentrations that have been measured in avian plasma. This pattern is consistent among precocial galliform birds (McNabb et al., 1998) and in precocial mallard ducks (McNabb et al., 2006). The perihatch peaks in thyroid hormones are associated with the initiation of thermoregulatory responses to cooling and pulmonary respiration that occur in precocial species during this time. After the perihatch peak, plasma thyroid hormones decrease markedly, then gradually increase during posthatch life to reach adult concentrations. In altricial species from two avian orders (columbiform ring doves, and passeriform European starlings and redwing blackbirds), plasma concentrations of T_3 and T_4 are very low during embryonic life and the perihatch period, then gradually increase during the first 2–3 weeks posthatch to approach adult levels. Endothermic responses to cooling first appear several days to a week after hatching in these altricial birds, and the young are only homeothermic by the time of fledging (reviews: McNabb and Olson, 1996; McNabb et al., 1998).

24.2.2 Mechanism of Action of Thyroid Hormones

In birds, as in other vertebrate classes, many thyroid hormone actions are mediated through nuclear thyroid hormone receptors that are members of the nuclear receptor superfamily (a large group of hormone receptors that have similar domain structures). Thyroid hormone receptors, generally referred to as T_3 receptors because of their high affinity for T_3, are responsible for the transcriptional effects of thyroid hormones (i.e., the receptors are transcription factors that act on thyroid hormone–responsive genes; Figure 24.3). Thus, with respect to transcriptional effects, T_4, the predominant hormone released from the thyroid, is considered to be mainly a prohormone. The direct responses mediated by thyroid hormones are of two types, developmental and metabolic (see Sections 24.4.1 and 24.4.2), but there also are indirect or permissive effects of thyroid hormones as well as interactive effects with other hormones.

Thyroid receptors (TRs) in vertebrates are coded for by two genes, thyroid hormone receptor-alpha (*THRA*) and -beta (*THRB*), that generate a number of receptor isoforms with high affinity for T_3 (e.g., TRα1, TRβ1, TRβ2,

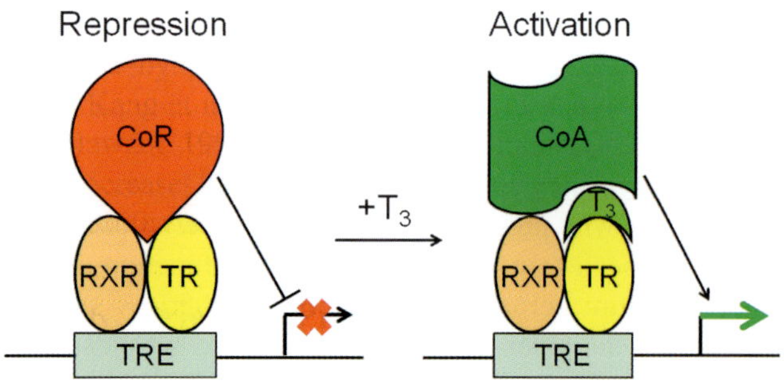

FIGURE 24.3 Thyroid hormone mechanism of action resulting in gene repression in the absence of T_3 and activation of gene transcription in the presence of T_3. CoA, co-activators; CoR, co-repressors; RXR; retinoic X receptor; TR, thyroid hormone receptor; TRE, thyroid hormone response element; T_3, 3,5,3′-triiodothyronine.

and TRβ3 in mammals, although the number and type of isoforms differ among vertebrate classes). The *THRA* gene also produces messenger RNAs (mRNAs) that can result in inhibitory effects on thyroid hormone action. In general, the TRα gene is expressed ubiquitously, although in avian brain it may vary during development. TRβ expression is restricted to fewer tissues and characteristically is regulated developmentally (reviews: mammals, Flamant et al., 2006; Cheng et al., 2010; birds, Decuypere et al., 2005; Grommen et al., 2008; Darras et al., 2011). In addition to T_3 (ligand) binding to the receptors, further regulation of transcription can result from differences in the hormone response elements located on the promoters of T_3 target genes, from differences in tissue receptor concentrations, from differences in developmental expression of receptor isoforms, and by effects of co-regulatory proteins in the nucleus. When TRs are not bound by T_3, co-regulatory proteins inhibit and repress the transcription of positively regulated thyroid-responsive genes; when TRs are bound to (i.e., occupied by) T_3, transcription is stimulated (Figure 24.3). The details of the molecular actions of thyroid hormones are best understood in mammals (review: Cheng et al., 2010). To the extent that receptor-mediated thyroid hormone actions have been investigated in birds, there is considerable similarity with mammals (Decuypere et al., 2005; Darras et al., 2011). The extent of receptor binding of T_3 (i.e., binding saturation) depends largely on deiodination reactions that supply T_3 to the circulation and play differential roles in intracellular T_3 (see Section 24.2.4) and on transport molecules that control hormone movement into specific tissues (see Section 24.2.3). In addition, other hormone degradation pathways play a role in intracellular T_3 availability.

Thyroid hormones also have effects through nongenomic actions at plasma membranes, in cytoplasm, and in mitochondria (i.e., these actions do not result from TR binding in the nucleus). These effects have only been investigated at the molecular level in mammals (Cheng et al., 2010; Scapin et al., 2010). However, one category of these nongenomic actions, thyroid hormone effects on thermogenesis, has received extensive attention at the organismal level in birds as well as mammals for decades (see Section 24.4.2).

Differences in the regulation of the different receptor isoforms in target tissues, as in mammals, have been demonstrated in birds by altering thyroid status with a goitrogen (methimazole) or by methimazole plus hormone supplementation. Hypothyroidism, sufficient to decrease growth and several developmental markers in ducklings, did not affect TRα mRNA but did result in a decrease in TRβ0 (presumed to be homologous to TRβ1) in some but not all muscles analyzed. Hyperthyroidism did not alter TRα mRNA but increased TRβ0 mRNA in cardiac, leg, and pectoralis muscles above that in euthyroid controls (Bishop et al., 2000). Developmental regulation of TR mRNA in the avian brain and pituitary has been demonstrated (review: Darras et al., 2011). TRβ2 is abundantly expressed in the pituitary, where it is thought to be important in feedback regulation (Grommen et al., 2008; and see Section 24.3).

24.2.3 Cellular Uptake of Thyroid Hormones

Since thyroid hormones are quite lipophilic, it was generally assumed that they cross the plasma membrane simply by passive diffusion. However, studies in rat hepatocytes already showed, in the late 1970s, that T_4 and T_3 enter the cell at least partly by active transport and via separate transport systems (Krenning et al., 1978, 1981). It took 30 more years before the first thyroid hormone transporter proteins were identified (Friesema et al., 2001, 2003), but we now know that several transmembrane proteins facilitate the transport of thyroid hormones in and out of cells. These proteins belong to the families of the monocarboxylate transporters (MCTs), the Na-taurocholate co-transporting polypeptides (NTCPs), the Na-independent organic anion-transporting polypeptides (OATPs), and the L-type amino acid transporters (LATs).

In humans and rodents, the transporters showing the highest specificity toward thyroid hormones are OATP1C1, which preferentially transports T_4 (and rT_3), and MCT8 and MCT10, which efficiently transport T_4 and T_3 (reviews: Friesema et al., 2005; Visser et al., 2008, 2011). So far, only one avian thyroid hormone transporter has been fully characterized, namely, chicken OATP1C1 (*SLCO1C1*). As in mammals, it is a highly specific transporter for T_4 (Nakao et al., 2006). Other transporters that are expressed in chicken are MCT8 (*SLC16A2*), MCT10 (*SLC16A10*), and LAT1 (*SLC7A5*), but they still need to be functionally characterized. These transporters are already present in the embryo, where they regulate thyroid hormone uptake in developing tissues, including brain (Geysens et al., 2012; Van Herck et al., 2012). In the adult quail hypothalamus, OATP1C1 is believed to be involved in the thyroid hormone–dependent stimulation of seasonal reproduction (Nakao et al., 2006; and see Chapter 34).

24.2.4 Thyroid Hormone Activation and Degradation

Once secreted by the thyroid gland, thyroid hormones can be metabolized in peripheral tissues by different pathways. Oxidative deamination and decarboxylation of the alanine side chain, as well as ether link cleavage, lead to irreversible degradation of the hormones. Sulfation and glucuronidation of the phenolic hydroxyl group are reversible processes that facilitate hormone solubility and subsequent excretion. The most important pathway, however, is deiodination consisting of removal of a single iodine from the phenolic outer ring (5′D or ORD) or from the tyrosyl or inner ring (5D or IRD) (Figure 24.2), processes catalyzed by enzymes called iodothyronine deiodinases. Since the avian thyroid gland secretes predominantly T_4, most of the T_3 present in the circulation and peripheral tissues is derived from extrathyroidal ORD of T_4.

The first direct proof for peripheral T_4 to T_3 conversion in birds was obtained in 1978 in Peking ducks (Astier and Newcomer, 1978). As for mammals, biochemical characterizations of deiodinating activities in birds first used liver homogenates from chickens (Borges et al., 1980; Rudas, 1986), quail (McNabb et al., 1986), and ring doves (Rieman and McNabb, 1991), and this was later extended to other tissues such as kidney, intestine, and brain. These studies suggested the presence of three types of deiodinases in birds (review: Darras et al., 2006) with activities very similar to those observed in mammalian tissues (review: Bianco et al., 2002). Cloning of the three chicken deiodinases D1 (*DIO1*), D2 (*DIO2*), and D3 (*DIO3*) finally confirmed their high homology with the mammalian homologs (Gereben et al., 1999; Van der Geyten et al., 1997). Partial information on deiodinase sequences also is available for some other avian species, including turkey, quail, tree sparrow, and zebra finch. It also is clear by now that the structures and functions of iodothyronine deiodinases have been strongly conserved throughout vertebrate evolution (reviews: Darras and Van Herck, 2012; Orozco et al., 2012).

All avian deiodinases are selenoproteins with a selenocysteine in their catalytic site. Therefore, they depend on the presence of a selenocysteine insertion sequence (SECIS) element in the 3′ untranslated region of their mRNA to allow selenocysteine incorporation (Bianco et al., 2002). The D1 enzyme is a nonselective enzyme catalyzing both ORD and IRD. In contrast, D2 only catalyzes ORD, while D3 only catalyzes IRD and is a purely inactivating enzyme (Figure 24.2). D1 is highly expressed in avian liver, kidney, and small intestine and contributes to peripheral T_3 production and rT_3 degradation. D2 is predominantly expressed in brain, where it is important in the local conversion of T_4 (that has entered the brain) to T_3. D3 is found in variable amounts in almost all chicken tissues and is believed to protect cells from overexposure to T_3 in specific situations such as early development, disease, and starvation (review: Darras et al., 2006).

Deiodinases play important roles in the pattern of circulating thyroid hormones during the perihatch period in precocial birds. D3 expression is particularly high in embryonic chicken liver. The sharp decline in hepatic D3 activity in the last days before hatching is the major factor contributing to the perihatch peak in circulating T_3 (Darras et al., 1992; Galton and Hiebert, 1987). At the level of the brain, the increase in T_3 content around hatching is facilitated by the enhanced local T_4 to T_3 conversion by D2 (Darras et al., 2009).

24.3 HYPOTHALAMIC–PITUITARY–THYROID AXIS

The avian thyroid gland is primarily under the control of the hypothalamic–pituitary axis. The avian hypothalamus produces two hormones (thyrotropin-releasing hormone (TRH) and corticotropin-releasing hormone (CRH)) that have stimulatory effects and somatostatin that has inhibitory effects on the thyrotrophs in the anterior pituitary. They produce thyroid-stimulating hormone (TSH, or thyrotropin), which is the major controller of the production and release of thyroid hormones by the thyroid gland. Thyroid hormones in turn exert a negative feedback on the pituitary and hypothalamus.

Historically, avian HPT axis control, based on investigations of chickens, appeared to be very similar to that in mammals (Decuypere and Kühn, 1988). The structure of TRH, a tripeptide, is identical in all vertebrates, and many studies have demonstrated that exogenous TRH could stimulate TSH release from the chicken pituitary. More recently, studies in birds, amphibians, reptiles, and fish have shown that CRH, rather than TRH, is often the most important hypothalamic stimulator of pituitary TSH in nonmammalian vertebrates. The most important role of TRH in influencing thyroid function in birds now appears to be through its growth hormone (GH)-releasing action, which results in circulating T_3 increases through growth hormone inhibition of T_3 degradation by type 3 deiodinase (review: De Groef et al., 2006; and see Sections 24.2.4 and 24.5).

The structure of TSH, which is a glycoprotein, differs in its β chain in different vertebrate classes, but several heterologous TSHs stimulate thyroid function in birds. The lack of antibodies specific to avian TSH meant that much of the background understanding of the development of TSH control was based on studies that used antibodies to heterologous TSHs. During development, the establishment of HPT axis control over the thyroid gland occurs about mid incubation (embryonic days 10.5–11.5) in precocial chickens. Pituitary removal by embryonic decapitation on day 9.5 does not alter plasma T_4 until day 11.5, but prevents the plasma T_4 increase that occurs after day 11.5. These findings suggest that sufficient TSH is produced to establish HPT control of the thyroid on day 11.5 but that prior thyroid gland development must be autonomous (review: Thommes et al., 1988). Identification of the presence of TSHβ peptide and mRNA at this same stage in embryonic development has provided highly specific confirmation of these findings (Ellestad et al., 2011). Maturation of the feedback response of pituitary thyrotrophs has been confirmed by embryonic day 19 by demonstrating that goitrogen suppression of circulating T_4 stimulates increased TSHβ mRNA (Muchow et al., 2005). Likewise, T_4 feedback on somatostatin inhibition of TSH production also matures between day 19 and hatch (De Groef et al., 2007). Studies on ring doves suggest that in altricial species, pituitary control of the thyroid appears to be established after hatch. In this species, the thyroid is insensitive to exogenous TSH until the second day posthatch, and circulating thyroid hormones remain very low, then gradually increase, during the first 3 weeks posthatch. Studies of the circulating thyroid hormone patterns in other altricial species (red-winged blackbirds and starlings) show patterns consistent with this developmental pattern (review: McNabb, 2007).

The effects of environmental and external factors (e.g., temperature and food consumption) on thyroid function are mediated largely through the HPT axis. Increased thyroid gland size and changes in thyroid secretion rate provide evidence of increased pituitary TSH release during cold exposure or winter conditions, and heat exposure has the opposite effect (review: Wentworth and Ringer, 1986). The picture of temperature effects on circulating thyroid hormones is complex, because in addition to the thyroid secretion rate, deiodination and other reactions involved in hormone turnover play a role (see Section 24.2.4). For the role of thyroid hormones in stimulating thermogenesis, see Section 24.4.2. The effects of food consumption, food restriction, and starvation have received extensive study in chickens and have been important to the understanding of GH involvement in HPT axis control and the metabolic effects of differences in food consumption (see Section 24.4.2).

24.4 EFFECTS OF THYROID HORMONES

24.4.1 Thyroid Hormone Effects on Development

Thyroid hormones influence both aspects of avian development (i.e., growth and differentiation–maturation). Growth (i.e., increase in mass) involves primarily cell proliferation (hyperplasia), but also may result from increases in cell size (hypertrophy). In general, thyroid hormones appear to act permissively or indirectly, in concert with other control substances, in their stimulation of growth in birds (review: McNabb and King, 1993). The primary direct hormonal stimulation of body growth results from circulating growth factors (such as insulin-like growth factor-1 (IGF1)) that are primarily under the control of GH from the pituitary (see Chapter 22).

The need for thyroid hormones for normal growth of birds has been demonstrated by the reduced growth, associated with hypothyroidism, that results from thyroidectomy or goitrogen administration. However, within the physiological range of circulating thyroid hormone concentrations, there is little consistent evidence of thyroid hormone stimulation of general body growth. At hyperthyroid extremes, growth is depressed because of high metabolic rates and a shift toward catabolism (reviews: McNabb and King, 1993; McNabb, 2000).

Thyroid hormones are important in triggering differentiation and maturation processes in many tissues. In mammals, their roles in development have been extensively studied in the gastrointestinal tract, heart, skeletal muscle, skin, bone, and neural tissue (reviews: McNabb, 1992; Pascual and Aranda, 2013). They are known to be crucial for development of the brain, eye, and ear (reviewed, respectively, in Bernal, 2007; Forrest and Swaroop, 2012; Rusch et al., 2001).

The information available on birds suggests similar roles for thyroid hormones in the development of avian tissues. In chicken TRα is expressed in neuroectoderm from the start of development, and receptor expression increases during neurulation (Flamant and Samarut, 1998). Thyroid hormones are present in the early embryonic brain, and transporters, deiodinases, and TRs show a dynamic and region-specific expression pattern throughout embryonic development (reviews: Darras et al., 2009; Forrest et al., 1991; Van Herck et al., 2012). Thyroid hormones are necessary for the development of normal brain architecture and neuronal connections critical to the function of signaling networks, as shown, for instance, in chicken cerebellum (Bouvet et al., 1987; Verhoelst et al., 2004). They stimulate the maturation of photoreceptors in the chicken retina (Fischer et al., 2011) and probably are involved in development of the inner ear (Geysens et al., 2012). Thyroid hormones also are essential for early learning, since they determine the start of the sensitive period for filial imprinting and may prime the brain for later learning (Yamaguchi et al., 2012).

In several tissues, the stimulation of differentiation–maturation by thyroid hormones involves interaction with other hormones. In chicken embryonic gut, thyroid hormones alone can stimulate cellular differentiation and induce digestive enzymes, but combination with glucocorticoids is necessary for the maturation of intestinal glucose transport (review: McNabb et al., 2006). Skeletal muscle differentiation and growth require both GH and thyroid hormones. GH alone seems to be sufficient to increase muscle weight in hypothyroid chickens, but both hormones are needed to reverse the effect on myosin. The appearance of specific myosin isoforms critical to functional maturation during late embryonic development is also modulated by T_3. Thyroid hormones also interact with IGF1 (e.g., in skeletal development). *In vitro* studies of pelvic cartilages from early chicken embryos have shown that thyroid hormones trigger cartilage differentiation–maturation by stimulating matrix production and ossification. In this case, T_3 stimulates chondrocyte hypertrophy but does not influence cell proliferation, which is only stimulated in the presence of IGF1 (reviews: McNabb and King, 1993; McNabb et al., 2006). *In ovo* exposure to increased levels of maternal T_4 also stimulates pelvic cartilage differentiation in Japanese quail embryos (Wilson and McNabb, 1997).

24.4.2 Thyroid Hormone Effects on Metabolism and Thermoregulation

Thyroid hormones control that part of metabolic, endothermic heat production that maintains the high and constant body temperature in homeothermic birds and mammals as well as additional adaptive changes in seasonal heat production (Danforth and Burger, 1984). The higher endothermic heat production in homeothermic birds and mammals, compared to ectothermic lower vertebrates, involves the uncoupling of aerobic metabolism from energy production as adenosine triphosphate (ATP). Historically, thyroid hormone–stimulated metabolic effects have been well documented in birds, with exogenous thyroid hormones stimulating and thyroidectomy or goitrogen administration depressing avian oxygen consumption (review: Wentworth and Ringer, 1986). Alterations in thyroid hormones also can influence the metabolic energy supply in other ways (e.g., increases in thyroid hormones facilitate liver glycogen storage, and hormone decreases lead to glycogen depletion and decreased plasma glucose).

There are fundamental differences in the regulation of endothermic heat production in mammals and birds. In mammals, brown adipose tissue (BAT), which is specialized for heat production, plays a key role in thermogenesis, but birds do not have BAT, so this thermogenesis must occur in other tissues. In all endotherms, there is uncoupling of the mitochondrial electron transport chain from ATP production, a phenomenon facilitated by anion carrier proteins in the inner mitochondrial membranes. Current evidence suggests that in mammals, uncoupling protein-1 (UCP1) is most important but that in birds the key uncoupling protein is adenine trinucleotide translocase (Walter and Seebacher, 2009), although avian uncoupling protein may play a role through interactions involving the β-adrenergic system (Joubert et al., 2010). Both T_4 and T_3 may play roles in endothermic heat production, and these responses may be nongenomic ones involving T_4 or T_3, or genomic ones resulting mainly from T_3 binding to nuclear TRs. Most of what is known about the nongenomic effects on thermogenesis is based on work in mammals, although there are a few studies in chicken embryos (review: Scapin et al., 2009).

The timing of thyroid development and HPT axis control appears to be critical to the development of thermoregulation in both precocial and altricial birds. Precocial birds, with most HPT axis maturation occurring during late embryonic life, exhibit endothermic responses to cooling during the perihatch period and gain the capability to maintain constant body temperatures (homeothermy) as chicks. Chicken embryos may exhibit some endothermic responses late in incubation. In contrast, altricial birds, with limited HPT axis maturation until several days or more posthatch, develop endothermic responses and homeothermy much more gradually in the weeks after hatch (Section 24.2.1; for a review, see McNabb et al., 1998).

Common to the perihatch period, in birds with both altricial and precocial development, are the initiation of pulmonary respiration, the maturation of those body parts involved in breaking out of the shell, yolk sac retraction, and the metabolic demands of hatching activities. Thyroid hormones are necessary for the maturation of lungs and surfactant production in the initiation of pulmonary respiration, and for a number of aspects of muscle functional maturation (i.e., genomic actions of thyroid hormones; see Sections 24.2.2 and 24.4.1). However, the circulating thyroid hormone concentrations during the perihatch period in altricial and precocial birds are very different, with only precocials showing dramatic increases in thyroid hormones at this time. This argues that the precocial peak of hormones in the circulation is primarily related to the initiation of endothermic responses to cooling (i.e., primarily nongenomic actions) that appear in the perihatch period in precocial but not in altricial species. It also suggests that the other maturational events stimulated by thyroid hormones during the perihatch period involve tissue- and organ-specific actions but do not require high circulating thyroid hormone concentrations. The exact roles of the HPT axis and its interactions with the hypothalamic–pituitary–adrenocortical axis remain to be clarified (Debonne et al., 2008; and see Sections 24.3 and 24.5).

24.4.3 Thyroid Hormone Effects on Reproduction and Maternal Influences on Developing Young

Thyroid hormones are among the factors important to avian reproductive processes, the photoperiodic initiation of reproductive cycles, and the timing of other energy-demanding processes that are timed to not compete with reproduction (e.g., molt). Egg laying is affected by thyroid status; for example, Japanese quail made hypothyroid by goitrogen administration stop laying, whereas at lower goitrogen doses, the hens may be able to maintain circulating thyroid hormones and continue to lay despite HPT axis activation. Because egg thyroid hormone content depends on circulating thyroid hormones in the hen, this indicates that hens may be able to "protect" maternal hormone supply to eggs for developing embryos to the extent that the hens are able to maintain euthyroid status. Japanese quail hens stopped laying when treated with a methimazole dose sufficient to decrease thyroid function, so they did not produce eggs deficient in thyroid hormones (Wilson and McNabb, 1997). In contrast, in a longer term study of methimazole-induced hypothyroidism in chickens, hens continued to lay despite decreases in their own circulating hormones and produced eggs deficient in maternal hormones and containing methimazole (Van Herck et al., 2013). It is not clear whether there are species differences in the effects of hypothyroidism on laying in chickens versus quail or whether these differences result from differences in the protocols used in these studies.

Both egg iodine and maternal thyroid hormones deposited in eggs influence embryonic thyroid status (review: McNabb and Wilson, 1997) as do many other components of the egg environment (Ho et al., 2011). When egg iodine availability was restricted by extremely low iodine in the diet of Japanese quail hens, egg iodine content was decreased, but adjustments in embryonic thyroid function maintained circulating T_3 concentrations (despite marked decreases in T_4 production), resulting in hatching and early posthatch growth that did not differ from that of controls (McNabb et al., 1985a,b; Stallard and NcNabb, 1990). Further adaptive potential of the thyroid system after hatch is shown by quail chicks that adapt their thyroid function to remain euthyroid despite sustained exposure to perchlorate, a thyroid inhibitor (McNabb et al., 2004; McNabb, 2006). It should, however, be mentioned that circulating thyroid hormone levels should not be assumed to reflect intracellular hormone availability in all target tissues because differences in hormone transport and deiodination capabilities in different tissues create individual local environments with respect to thyroid hormone availability. Few studies have measured tissue hormone content. In a recent study of chicken embryos from methimazole-treated hypothyroid hens, brain hormones were regulated independently of the low circulating hormone concentrations. However, because the hens deposited methimazole in the eggs, possible tissue-specific effects of the methimazole could have added to the complexity of this picture (Van Herck et al., 2013).

Most of the laboratory studies relating to thyroid effects on egg laying, egg components, and hatching have been done on precocial galliform birds (review: Decuypere et al., 1990). However, studies of neuroendocrine control of reproduction in wild birds (reviews: Yoshimura, 2010; and see Chapter 34 in this text) and studies of wild birds exposed to thyroid-disrupting environmental chemicals indicate the need for adequate thyroid function in altricial or semi-altricial species as well (review: McNabb, 2007). For example, historical studies of herring gulls exposed to a range of environmental polychlorinated biphenyl (PCB) concentrations suggest that exposure to this group of thyroid-disrupting chemicals contributes to developmental and hatching problems (reviews: Fox, 1993; McNabb and Fox, 2003; and see Chapter 43 in this text). Lab studies of specific PCBs (Roelens et al., 2005) and of dioxin (Bruggeman et al., 2003) have shown delayed hatching associated with thyroid disruption in chicken embryos.

Thyroid hormones have long been known to be important in postnuptial molt in both wild and domestic birds, and alterations in plasma thyroid hormones occur in conjunction with natural molts (review: Kuenzel, 2003; and see Chapter 39 in this text). Inhibition of reproductive activity (including cessation of egg laying) and molt occur concurrently in seasonally reproducing wild birds, and in commercial poultry practice, thyroid hormones and/or starvation have been used to induce molt in poultry. Prolactin also plays a role in molt, as do several other hormones (review: Kuenzel, 2003). Estrogen decreases appear to be important in the initiation of molt, whereas an increase in the thyroid hormone/estrogen ratio appears to be important in new feather formation (review: Decuypere and Verheyen, 1986).

24.5 THYROID INTERACTIONS WITH OTHER HORMONES

Thyroid hormones play important roles, both direct and indirect, in development, growth, and metabolism (see Section 24.4), and in each of these functions there are interactions with a number of other hormones and hormone systems. Thyroid hormones are required for body growth, but it appears that they act in a permissive or indirect way in conjunction with GH, which influences growth factors that directly stimulate cell proliferation (see Chapter 37). Thyroid hormones modulate GH production and release by the pituitary, by direct inhibition of pituitary somatotrophs, and by negative feedback effects on TRH, which stimulates somatotrophs. However, thyroid hormones also play an important role in stimulating the development of somatotrophs, probably in conjunction with glucocorticoids (Liu and Porter, 2004). This general picture differs from how thyroid function affects the growth axis in mammals, in which T_3 stimulates GH production and release. In birds, GH in turn stimulates thyroid function by increasing circulating T_3 through its inhibitory effects on T_3 degradation by IRD, a key factor in the increases in circulating T_3 concentrations during the perihatch period (see Sections 24.2.4 and 24.4.1).

Interactions between the HPT and other hypothalamic–pituitary control axes are important in several contexts. Hypothalamic TRH exerts some of its thyroid effects through stimulation of GH release and consequent effects on deiodination, while CRH stimulates pituitary thyrotrophs to release TSH, which exerts stimulatory control over the thyroid gland (see Section 24.3). During the perihatch period, glucocorticoids rise, and this may increase circulating T_3 levels by inhibiting T_3 degradation by type 3 deiodinase and by stimulating T_3 production by type 1 ORD (Darras et al., 1996). Thyroid hormones and glucocorticoids may both have important effects on organ differentiation in lung and gut (see Section 24.4.1). Although these hormonal interactions are conspicuous during the perihatch period, and thus have received considerable attention, they continue to be important in a variety of metabolic and developmental events at other ages. In adults, interactions with the reproductive axis are important, as thyroid hormones are necessary for the initiation and timing of reproductive cycles, and for the laying of viable eggs (see Section 24.4.3).

24.6 ENVIRONMENTAL INFLUENCES ON THYROID FUNCTION

Changes in temperature and food availability appear to be the natural environmental factors that have the greatest influence on thyroid function. As mentioned in this chapter, environmental cold can act through the HPT axis (see Section 24.3) to increase thyroid hormone release, can alter the peripheral thyroid hormone deiodination pathways (see Section 24.2.4), and can increase thyroid hormone turnover (Wentworth and Ringer, 1986). Overall, these changes provide increased thyroid hormone availability in association with the metabolic increases required for homeothermy at cold ambient temperatures. Plasma T_4 concentrations often change relatively little (i.e., proportionately less than T_3 changes) with sustained cold exposure, suggesting that the HPT axis response is less important than changes in deiodination pathways (Rudas and Pethes, 1984). In general, hot temperatures have the opposite effect (i.e., they depress thyroid function) (review: Sharp and Klandorf, 1985). However, as indicated in Section 24.4.2, the nuances of genomic and nongenomic effects on metabolism and the relative roles of T_4 and T_3 in metabolic stimulation have not been fully investigated in birds.

Like temperature change, food availability alters thyroid status. Either food restriction or starvation causes decreases in circulating T_3 concentrations, although the effects on T_4 are more variable (see review in Darras et al., 1995). There are some differences in the details of the thyroid effects and the relative causes of those effects (HPT axis role versus changes in deiodination) with starvation versus food restriction and in different vertebrate classes. In birds, both starvation and sustained partial food restriction lead to decreased circulating T_3 and increased T_4. The primary cause of the decrease in circulating T_3 is increased T_3 degradation by D3 IRD, without changes in T_4-to-T_3 conversion by D1 (Darras et al., 1995). Feeding, by increasing circulating T_3, appears to be one of the key factors that play a role in the diurnal patterns of plasma thyroid hormone concentrations (see Section 24.2.1). Likewise, it may be an important factor in some temperature-induced changes in thyroid hormone; for example, the depressed plasma T_3 concentrations in chickens exposed to 40 °C heat may result largely from the decreased food consumption at high temperatures (Williamson et al., 1985).

In addition to natural environmental events, endocrine-disrupting chemicals in the environment play a role in the thyroid function of birds exposed to such chemicals. A brief treatment of some examples is included in this chapter for the purpose of providing links to the more comprehensive coverage of endocrine disruptors in Chapter 43. Historically, large amounts of some persistent industrial chemicals, such as PCBs, that appeared to be thyroid disruptors were released into major bodies of water such as the Great Lakes. Major monitoring efforts in the Great Lakes, since the late 1960s when PCBs were last released, have provided circumstantial evidence that total PCB exposure in herring gulls, which are fish eating and experience bioaccumulation of such chemicals, results in population declines and thyroid disruption (Fox, 1993). Studies of thyroid function in gull embryos collected in 1998–2000 suggest that embryonic thyroid disruption, from the persistent environmental PCBs, continued at the most contaminated sites compared to the reference sites (McNabb and Fox, 2003). More recent studies have explored how commonly used PCB mixtures, such as Aroclor 1254, affect thyroid function through altering a key enzyme system (uridine diphosphate-glucuronosyltransferase (UDP-GT)) that increases T_4 clearance (Webb and McNabb, 2008) and how specific PCB isomers, such as the dioxin-like PCB77, induce changes in deiodinase patterns, resulting in decreases in plasma thyroid hormone concentrations (Beck et al., 2006). Thyroid hormone binding by transthyretin also is affected by PCBs (see reviews in Ishihara et al., 2003; Scanes and McNabb, 2003). Thus, PCBs affect thyroid function through a number of mechanisms, and understanding of their effects is further complicated by the large number of congeners and mixtures present in the environment.

Perchlorate compounds, which have been used as experimental goitrogens, also are environmental contaminants in many regions, so they affect thyroid function in animals exposed through drinking water or food sources. The primary effect of perchlorate ion is to competitively inhibit iodide uptake by the Na-I symporter in the thyroid, creating iodine deficiency and the consequent effects on thyroid function. There also is building evidence that perchlorate has additional effects within the thyroid and on HPT axis function. Most of what is known about perchlorate effects in birds comes from laboratory studies that include environmentally relevant perchlorate exposures (review: McNabb et al., 2006). Perchlorate-exposed quail hens transfer perchlorate to their eggs, where embryos may become hypothyroid and exhibit thyroid hypertrophy, decreased thyroidal hormone content, decreased body growth, delayed hatching, increased mortality, and some associated alterations in thyroid-responsive gene expression (Chen et al., 2008). Similar effects were observed in quail chicks exposed to perchlorate in drinking water (Chen et al., 2009). In both cases, the deiodinase responses appeared to "protect" T_3 supply to the brain but not the liver under hypothyroid conditions.

Other categories of environmental thyroid-disrupting chemicals that have received attention in recent years are those used as flame retardants (polybrominated diphenyl ethers (PBDEs)), those employed as nonstick coatings for cookware (perfluoroalkyl compounds (PFCs)), and bisphenol A, which is used in plastics and food can liners. These materials are widespread, persistent contaminants that are present in wild birds and have been shown to alter thyroid function. Flame retardants have been found in avian populations worldwide and have been shown to alter thyroid function in domestic galliform birds, model passerine species, and wild-caught kestrels and bald eagles (respectively, chickens: Farhat et al., 2013; zebra finches: Eng et al., 2013; kestrels: Fernie et al., 2005; eagles: Cesh et al., 2010). PFCs also have been shown to alter thyroid function in birds (Vongphachan et al., 2011; Cassone et al., 2012). Bisphenol A, which can act as a thyroid hormone analog for binding to TR (Zoeller, 2005), alters thyroid function in a number of animal species but does not appear to have been investigated in birds except with respect to reproductive disruption.

24.7 CONCLUSIONS AND SUMMARY

This chapter reviewed the development of the anatomy and function of the thyroid gland and the establishment of its control by the hypothalamic–pituitary–thyroid axis, with emphasis on differences in the timing of these events in precocial and altricial modes of avian development. The binding of thyroid hormones to proteins in the circulation, their transport in and out of cells via plasma membrane transport proteins, and deiodination by enzymes that activate and inactivate thyroid hormone are explained in the context of their functions in controlling intra-organ events. The mechanisms of thyroid hormone genomic actions, mediated primarily by nuclear receptors, as well as nongenomic actions are described in the context of their developmental and metabolic effects. Interactions of thyroid hormones with other hormones systems and environmental influences on thyroid function, including those by environmental contaminants, are discussed.

REFERENCES

Astier, H., 1980. Thyroid gland in birds: structure and function. In: Epple, A., Stetson, M.H. (Eds.), Avian Endocrinology. Academic Press, New York, pp. 167–189.

Astier, H.S., Newcomer, W.S., 1978. Extrathyroidal conversion of thyroxine to triiodothyronine in a bird: the Peking duck. Gen. Comp. Endocrinol. 35, 496–499.

Beck, V., Roelens, S.A., Darras, V.M., 2006. Exposure to PCB77 induces tissue-dependent changes in iodothyronine deiodinase activity patterns in the embryonic chicken. Gen. Comp. Endocrinol. 148, 327–335.

Bernal, J., 2007. Thyroid hormone receptors in brain development and function. Nat. Clin. Pract. Endocrinol. Metab. 3, 249–259.

Bianco, A.C., Salvatore, D., Gereben, B., Berry, M.J., Larsen, P.R., 2002. Biochemistry, cellular and molecular biology, and physiological roles of the iodothyronine selenodeiodinases. Endocr. Rev. 23, 38–89.

Bishop, C.M., McCabe, C.J., Gittoes, N.J.L., Butler, P.J., Franklyn, J.A., 2000. Tissue-specific regulation of thyroid hormone receptor mRNA isoforms and target gene proteins in domestic ducks. J. Endocrinol. 165, 607–615.

Bizhanova, A., Kopp, P., 2009. The sodium-iodide symporter NIS and pendrin in iodide homeostasis of the thyroid. Endocrinology 150, 1084–1090.

Borges, M., LaBourene, J., Ingbar, S.H., 1980. Changes in hepatic iodothyronine metabolism during ontogeny of the chick embryo. Endocrinology 107, 1751–1761.

Bouvet, J., Usson, Y., Legrand, J., 1987. Morphometric analysis of the cerebellar Purkinje cell in the developing normal and hypothyroid chick. Int. J. Dev. Neurosci. 5, 345–355.

Bruggeman, V., Swennen, Q., De Ketelaere, B., Onagbesan, O., Tona, K., Decuypere, E., 2003. Embryonic exposure to 2,3,7,8-tetracholorodibenzo-p-dioxin in chickens: effects of dose and embryonic stage on hatchability and growth. Comp. Biochem. Physiol. C Toxicol. Pharmacol. 136, 17–28.

Cassone, C.G., Vongphachan, V., Chiu, S., Williams, K.L., Letcher, R.J., Pelletier, E., Crump, D., Kennedy, S.W., 2012. *In ovo* effects of perfluorohexane sulfonate and perfluorohexanoate on pipping success, development, mRNA expression, and thyroid hormone levels in chicken embryos. Toxicol. Sci. 127, 216–224.

Cesh, L.S., Elliott, K.J., Quade, S., McKinney, M.A., Maisoneuve, F., Garcelon, D.K., Sandau, C.D., Letcher, R.J., Williams, T.D., Elliott, J.E., 2010. Polyhalogenated aromatic hydrocarbons and metabolites: relation to circulating thyroid hormone and retinol in nestling bald eagles (*Haliaeetus leucocephalus*). Environ. Toxicol. Chem. 29, 1301–1310.

Chang, L., Munro, S.L.A., Richardson, S.J., Schreiber, G., 1999. Evolution of thyroid hormone binding by transthyretins in birds and mammals. Eur. J. Biochem. 259, 534–542.

Chen, Y., Sible, J.C., McNabb, F.M.A., 2008. Effects of maternal exposure to ammonium perchlorate on thyroid function and the expression of thyroid-responsive genes in Japanese quail embryos. Gen. Comp. Endocrinol. 159, 196–207.

Chen, Y., Sible, J.C., McNabb, F.M.A., 2009. Perchlorate exposure induces hypothyroidism and affects thyroid-responsive genes in liver but not brain of quail chicks. Arch. Environ. Contam. Toxicol. 57, 598–607.

Cheng, S.-Y., Leonard, J.L., Davis, P.J., 2010. Molecular actions of thyroid hormone actions. Endocr. Rev. 31, 139–170.

Cogburn, L.A., Freeman, R.M., 1987. Response surface of daily thyroid hormone rhythms in young chickens exposed to constant ambient temperature. Gen. Comp. Endocrinol. 68, 113–123.

Danforth Jr., E., Burger, A., 1984. The role of thyroid hormones in the control of energy expenditure. Clin. Endocrinol. Metab. 13, 581–595.

Darras, V.M., Van Herck, S.L., 2012. Iodothyronine deiodinase structure and function: from ascidians to humans. J. Endocrinol. 215, 189–206.

Darras, V.M., Visser, T.J., Berghman, L.R., Kühn, E.R., 1992. Ontogeny of type I and type II deiodinase activities in embryonic and posthatch chicks: relationship with changes in plasma triiodothyronine and growth hormone levels. Comp. Biochem. Physiol. A 103, 131–136.

Darras, V.M., Cokelaere, M., Dewil, E., Arnouts, S., Decuypere, E., Kühn, 1995. Partial food restriction increases hepatic inner ring deiodinating activity in the chicken and the rat. Gen. Comp. Endocrinol. 100, 334–338.

Darras, V.M., Kotanen, S.P., Geris, K.L., Berghman, L.R., Kühn, E.R., 1996. Plasma thyroid hormone levels and iodothyronine deiodinase activity following an acute glucocorticoid challenge in embryonic compared with posthatch chickens. Gen. Comp. Endocrinol. 104, 203–212.

Darras, V.M., Verhoelst, C.H., Reyns, G.E., Kühn, E.R., Van der Geyten, S., 2006. Thyroid hormone deiodination in birds. Thyroid 16, 25–35.

Darras, V.M., Van Herck, S.L.J., Geysens, S., Reyns, G.E., 2009. Involvement of thyroid hormones in chicken embryonic brain development. Gen. Comp. Endocrinol. 163, 58–62.

Darras, V.M., Van Herck, S.L.J., Heijlen, M., De Groef, B., 2011. Thyroid hormone receptors in two model species for vertebrate embryonic development: chicken and zebrafish. J. Thyroid Res. 2011, 1–8.

De Felice, M., Di Lauro, R., 2011. Intrinsic and extrinsic factors in thyroid gland development: an update. Endocrinology 152, 2948–2956.

De Groef, B., Van der Geyten, S., Darras, V.M., Kühn, E.R., 2006. Role of corticotropin-releasing hormone as a thyrotropin-releasing factor in non-mammalian vertebrates. Gen. Comp. Endocrinol. 146, 62–68.

De Groef, B., Grommen, S.V.H., Darras, V.M., 2007. Feedback control of thyrotropin secretion in the chicken: thyroid hormones increase the expression of hypophyseal somatostatin receptor types 2 and 5. Gen. Comp. Endocrinol. 152, 178–182.

Debonne, M., Baarendse, P.J.J., Van Den Brand, H., Kemp, B., Bruggeman, V., Decuypere, E., 2008. Involvement of the hypothalamic-pituitary-thyroid axis and its interaction with the hypothalamic-pituitary-adrenal axis in the ontogeny of avian thermoregulation: a review. World's Poult. Sci. J. 64, 309–321.

Decuypere, E., Kühn, E.R., 1988. Thyroid hormone physiology in galliformes: age and strain related changes in physiological control. Am. Zool. 28, 401–415.

Decuypere, E., Verheyen, G., 1986. Physiological basis of induced moulting and tissue regeneration in fowls. World's Poult. Sci. J. 42, 56–66.

Decuypere, E., Dewil, E., Kühn, E.R., 1990. The hatching process and the role of hormones. In: Tullett, S.C. (Ed.), Avian Incubation. Butterworth & Co., London, UK, pp. 239–256.

Decuypere, E., Van As, P., Van der Geyten, S., Darras, V.M., 2005. Thyroid hormone availability and activity in avian species: a review. Dom. Anim. Endocrinol. 29, 63–77.

Ellestad, L.E., Saliba, J., Porter, T.E., 2011. Ontogenic characterization of gene expression in the developing neuroendocrine system of the chick. Gen. Comp. Endocrinol. 171, 82–93.

Eng, M.L., Williams, T.D., Elliott, J.E., 2013. Developmental exposure to a brominated flame retardant: an assessment of effects on physiology, growth, and reproduction in a songbird, the zebra finch. Environ. Pollut. 178, 343–349.

Farhat, A., Crump, D., Chiu, S., Williams, K.L., Letcher, R.J., Gauthier, L.T., Kennedy, S.W., 2013. *In ovo* effects of two organophosphate flame retardants – TCPP and TDCPP—on pipping success, development, mRNA expression, and thyroid hormone levels in chicken embryos. Toxicol. Sci. 134, 92–102.

Fernie, K.J., Shutt, J.L., Mayne, G., Hoffman, D., Letcher, R.J., Drouillard, K.G., Ritchie, I.J., 2005. Exposure to polybrominated diphenyl ethers (PBDEs): changes in thyroid, vitamin A, glutathione homeostasis, and oxidative stress in American kestrels (*Falco sparverius*). Toxicol. Sci. 88, 2375–2383.

Fischer, A.J., Bongini, R., Bastaki, N., Sherwood, P., 2011. The maturation of photoreceptors in the avian retina is stimulated by thyroid hormone. Neuroscience 178, 250–260.

Flamant, F., Samarut, J., 1998. Involvement of thyroid hormone and its alpha receptor in avian neurulation. Dev. Biol. 197, 1–11.

Flamant, F., Baxter, J.D., Forrest, D., Refetoff, F., Samuels, H., Scanlan, T.F., Vennstrom, B., Samarut, J., 2006. The pharmacology and classification of the nuclear receptor superfamily: thyroid hormone receptors. Pharmacol. Rev. 58, 705–711.

Forrest, D., Swaroop, A., 2012. Minireview: the role of nuclear receptors in photoreceptor differentiation and disease. Mol. Endocrinol. 26, 905–915.

Forrest, D., Hallbook, F., Persson, H., Vennstrom, B., 1991. Distinct functions for thyroid hormone receptors alpha and beta in brain development indicated by differential expression of receptor genes. EMBO J. 10, 269–275.

Fox, G.A., 1993. What have biomarkers told us about the effects of contaminants on the health of fish-eating birds in the Great Lakes? The theory and a literature review. J. Great Lakes Res. 19, 722–736.

French, E.I., Hodges, R.D., 1977. Fine structural studies on the thyroid gland of the normal domestic fowl. Cell Tissue Res. 178, 397–410.

Friesema, E.C., Docter, R., Moerings, E.P., Verrey, F., Krenning, E.P., Hennemann, G., Visser, T.J., 2001. Thyroid hormone transport by the heterodimeric human system L amino acid transporter. Endocrinology 142, 4339–4348.

Friesema, E.C., Ganguly, S., Abdalla, A., Fox, J.E.M., Halestrap, A.P., Visser, T.J., 2003. Identification of monocarboxylate transporter 8 as a specific thyroid hormone transporter. J. Biol. Chem. 278, 40128–40135.

Friesema, E.C., Jansen, J., Milici, C., Visser, T.J., 2005. Thyroid hormone transporters. Vitam. Horm. 70, 137–167.

Galton, V.A., Hiebert, A., 1987. The ontogeny of the enzyme systems for the 5'- and 5-deiodination of thyroid hormones in chick embryo liver. Endocrinology 120, 2604–2610.

Gereben, B., Bartha, T., Tu, H.M., Harney, J.W., Rudas, P., Larsen, P.R., 1999. Cloning and expression of the chicken type 2 iodothyronine 5'-deiodinase. J. Biol. Chem. 274, 13768–13776.

Geysens, S., Ferran, J.L., Van Herck, S.L., Tylzanowski, P., Puelles, L., Darras, V.M., 2012. Dynamic mRNA distribution pattern of thyroid hormone transporters and deiodinases during early embryonic chicken brain development. Neuroscience 221, 69–85.

Grommen, S.V.H., Arckens, L., Theuwissen, T., Darras, V.M., De Groef, B., 2008. Thyroid hormone receptor b2 is strongly up-regulated at all levels of the hypothalamo-pituitary-thyroidal axis during late embryogenesis in chicken. J. Endocrinol. 196, 519–528.

Grommen, S.V.H., Iwasawa, A., Beck, V., Darras, V.M., De Groef, B., 2011. Ontogenic expression profiles of thyroid-specific genes in embryonic and hatching chicks. Dom. Anim. Endocrinol. 40, 10–18.

Ho, D.H., Reed, W.L., Burggren, W.W., 2011. Egg yolk environment differentially influences physiological and morphological development of broiler and layer chicken embryos. J. Exp. Biol. 214, 619–628.

Ishihara, A., Nishiyama, N., Sugiyama, S.-I., Yamauchi, K., 2003. The effect of endocrine disrupting chemicals on thyroid hormone binding to Japanese quail transthyretin and thyroid hormone receptor. Gen. Comp. Endocrinol. 134, 36–43.

Joubert, R., Métayer Coustard, S., Swennen, Q., Sibut, V., Crochet, S., Cailleau-Audouin, E., Buyse, J., Decuypere, E., Wrutniak-Cabello, C., Cabello, G., Tesseraud, S., Collin, A., 2010. The beta-adrenergic system is involved in the regulation of the expression of avian uncoupling protein in the chicken. Dom. Anim. Endocrinol. 38, 115–125.

Krenning, E.P., Docter, R., Bernard, H.F., Visser, T.J., Hennemann, G., 1978. Active transport of triiodothyronine (T3) into isolated rat liver cells. FEBS Lett. 91, 113–116.

Krenning, E., Docter, R., Bernard, B., Visser, T., Hennemann, G., 1981. Characteristics of active transport of thyroid hormone into rat hepatocytes. Biochim. Biophys. Acta 676, 314–320.

Kuenzel, W.J., 2003. Neurobiology of molt in avian species. Poult. Sci. 82, 981–991.

Liu, L., Porter, T.E., 2004. Endogenous thyroid hormones modulate pituitary somatotroph differentiation during chicken embryonic development. J. Endocrinol. 180, 45–53.

McNabb, F.M.A., 1992. Thyroid Hormones. Prentice Hall, Englewood Cliffs. 283 pp.

McNabb, F.M.A., 2000. Thyroids. In: Whittow, G.C. (Ed.), Sturkie's Avian Physiology, fifth ed. Academic Press, San Diego, pp. 461–471 (Chapter 17).

McNabb, F.M.A., 2006. Avian thyroid development and adaptive plasticity. Gen. Comp. Endocrinol. 147, 93–101.

McNabb, F.M.A., 2007. The hypothalamic-pituitary-thyroid (HPT) axis in birds and its role in bird development and reproduction. Crit. Rev. Toxicol. 37, 163–193.

McNabb, F.M.A., Fox, G.A., 2003. Avian thyroid development in chemically contaminated environments: Is there evidence of alterations in thyroid function and development? Evol. Dev. 5, 76–82.

McNabb, F.M.A., King, D.B., 1993. Thyroid hormone effects on growth, development and metabolism. In: Schreibman, M.P., Scanes, C.G., Pang, P.K.T. (Eds.), The Endocrinology of Growth, Development, and Metabolism of Vertebrates. Academic Press, New York, pp. 393–417.

McNabb, F.M.A., Olson, J.M., 1996. Development of thermoregulation and its hormonal control in precocial and altricial birds. Poult. Avian Biol. Rev. 7, 111–125.

McNabb, F.M.A., Wilson, C.M., 1997. Thyroid hormone deposition in avian eggs and effects on embryonic development. Am. Zool. 37, 553–560.

McNabb, F.M.A., Blackman, J.R., Cherry, J.A., 1985a. The effects of different maternal dietary iodine concentrations on Japanese quail. I. Thyroid status of hens. Dom. Anim. Endocrinol. 2, 25–34.

McNabb, F.M.A., Dicken, S.G., Cherry, J.A., 1985b. The effects of different maternal dietary iodine concentrations on Japanese quail. II. Thyroid function in embryos and hatchlings. Dom. Anim. Endocrinol. 2, 35–42.

McNabb, F.M.A., Lyons, L.J., Hughes, T.E., 1986. Avian hepatic T3 generation by 5'-monodeiodination: characterization of two enzymatic pathways and the effects of goitrogens. Comp. Biochem. Physiol. A Comp. Physiol. 85, 249–255.

McNabb, F.M.A., Scanes, C.G., Zeman, M., 1998. Endocrine control of development. In: Starck, J.M., Ricklefs, R.E. (Eds.), Avian Growth and Development. Evolution within the Altricial-Precocial Spectrum. Oxford University Press, New York.

McNabb, F.M.A., Jang, D.A., Larsen, C.T., 2004. Does thyroid function in developing birds adapt to sustained ammonium perchlorate exposure? Toxicol. Sci. 82, 106–113.

McNabb, F.M.A., Hooper, M.J., Smith, E.E., McMurry, S., Gentles, A.B., 2006. Perchlorate effects in birds. In: Kendall, R., Smith, P.N. (Eds.), Perchlorate Ecotoxicology. Society for Environmental Toxicology and Chemistry Press, Pensacola, pp. 99–126 (Chapter 5).

Muchow, M., Bossis, I., Porter, T.E., 2005. Ontogeny of pituitary thyrotrophs and regulation by endogenous thyroid hormone feedback in the chick embryo. J. Endocrinol. 184, 407–416.

Nakao, N., Takagi, T., Iigo, M., Tsukamoto, T., Yasuo, S., Masuda, T., Yanagisawa, T., Ebihara, S., Yoshimura, T., 2006. Possible involvement of organic anion transporting polypeptide 1c1 in the photoperiodic response of gonads in birds. Endocrinology 147, 1067–1073.

Orozco, A., Valverde, R.C., Olvera, A., Garcia, G.C., 2012. Iodothyronine deiodinases: a functional and evolutionary perspective. J. Endocrinol. 215, 207–219.

Pascual, A., Aranda, A., 2013. Thyroid hormone receptors, cell growth and differentiation. Biochim. Biophys. Acta 1830, 3908–3916.

Rieman, J.D., McNabb, F.M., 1991. Assay validation and characterization of hepatic 5'-deiodinase activity in ring doves using reverse-T3 as substrate. Gen. Comp. Endocrinol. 82, 53–59.

Roelens, S.A., Beck, V., Maervoet, J., Aerts, G., Reyns, G.E., Schepens, P., Darras, V.M., 2005. The dioxin-like PCB 77 but not the *ortho*-substituted PCB 153 interferes with chicken embryo thyroid hormone homeostasis and delays hatching. Gen. Comp. Endocrinol. 143, 1–9.

Rudas, P., 1986. Comparison of type I 5'-deiodination of thyroxine and of reverse-triiodothyronine in rat and chicken liver homogenates. Gen. Comp. Endocrinol. 63, 400–407.

Rudas, P., Pethes, G., 1984. The importance of the peripheral thyroid hormone deiodination in adaptation to ambient temperature in the chicken (*Gallus domesticus*). Comp. Biochem. Physiol. 77, 567–571.

Rusch, A., Ng, L., Goodyear, R., Oliver, D., Lisoukov, I., Vennstrom, B., Richardson, G., Kelley, M.W., Forrest, D., 2001. Retardation of cochlear maturation and impaired hair cell function caused by deletion of all known thyroid hormone receptors. J. Neurosci. 21, 9792–9800.

Scanes, C.G., McNabb, F.M.A., 2003. Avian models for research in toxicology and endocrine disruption. Poult. Avian Biol. Rev. 14, 21–52.

Scapin, S., Leoni, S., Spagnuolo, S., Fiore, A.M., Incerpi, S., 2009. Short-term effects of thyroid hormones on Na+-K+-ATPase activity of chick embryo hepatocytes during development: focus on signal transduction. Am. J. Physiol. Cell Physiol. 296, C4–C12.

Scapin, S., Leoni, S., Spagnuolo, S., Gnocchi, D., De Vito, P., Luly, P., Pedersen, J.Z., Incerpi, S., 2010. Short-term effects of thyroid hormones during development: focus on signal transduction. Steroids 75, 576–584.

Schreiber, G., 2002. The evolutionary and integrative roles of transthyretin in thyroid hormone homeostasis. J. Endocrinol. 175, 61–73.

Sharp, P.J., Klandorf, H., 1985. Environmental and physiological factors controlling thyroid function in Galliformes. In: Follett, B.K., Ishii, S., Chandola, A. (Eds.), The Endocrine System and the Environment. Japan Scientific Societies Press, Tokyo and Springer-Verlag, Berlin, pp. 175–188.

Stallard, L.C., McNabb, F.M.A., 1990. The effects of different iodide availabilities on thyroid function during development in Japanese quail. Dom. Anim. Endocrinol. 7, 239–250.

Thommes, R.C., Hylka, V.W., Tonetta, S.A., Griesbach, D.A., Ropka, S.L., Woods, J.E., 1988. Hypothalamic regulation of the pituitary-thyroid unit in the developing chick embryo. Am. Zool. 28, 417–426.

Van der Geyten, S., Sanders, J.P., Kaptein, E., Darras, V.M., Kuhn, E.R., Leonard, J.L., Visser, T.J., 1997. Expression of chicken hepatic type I and type III iodothyronine deiodinases during embryonic development. Endocrinology 138, 5144–5152.

Van Herck, S.L., Geysens, S., Delbaere, J., Tylzanowski, P., Darras, V.M., 2012. Expression profile and thyroid hormone responsiveness of transporters and deiodinases in early embryonic chicken brain development. Mol. Cell. Endocrinol. 349, 289–297.

Van Herck, S.L.J., Geysens, S., Bald, E., Chwatko, G., Delezie, E., Dianati, E., Ahmed, R.G., Darras, V.M., 2013. Maternal transfer of methimazole and effects on thyroid hormone availability in embryonic tissues. J. Endocrinol. 218, 105–115.

Verhoelst, C.H., Darras, V.M., Doulabi, B.Z., Reyns, G., Kuhn, E.R., Van der Geyten, S., 2004. Type I iodothyronine deiodinase in euthyroid and hypothyroid chicken cerebellum. Mol. Cell. Endocrinol. 214, 97–105.

Visser, W.E., Friesema, E.C., Jansen, J., Visser, T.J., 2008. Thyroid hormone transport in and out of cells. Trends Endocrinol. Metab. 19, 50–56.

Visser, W.E., Friesema, E.C.H., Visser, T.J., 2011. Minireview: thyroid hormone transporters: the knowns and the unknowns. Mol. Endocrinol. 25, 1–14.

Vongphachan, V., Cassone, C.G., Wu, D., Chiu, S., Crump, D., Kennedy, S.W., 2011. Effects of perfluoroalkyl compounds on mRNA expression levels of thyroid hormone-responsive genes in primary cultures of avian neuronal cells. Toxicol. Sci. 120, 392–402.

Walter, I., Seebacher, F., 2009. Endothermy in birds: underlying molecular mechanisms. J. Exp. Biol. 212, 2328–2336.

Webb, C.M., McNabb, F.M.A., 2008. Polychlorinated biphenyl effects on avian hepatic enzyme induction and thyroid function. Gen. Comp. Endocrinol. 155, 650–657.

Wentworth, B.C., Ringer, R.K., 1986. Thyroids. In: Sturkie, P.D. (Ed.), Avian Physiology. Springer-Verlag, New York, pp. 452–465 (Chapter 20).

Williamson, R.A., Mission, B.H., Davison, T.F., 1985. The effect of exposure to 40 C on the heat production and the serum concentrations of triiodothyronine, thyroxine, and corticosterone in immature domestic fowl. Gen. Comp. Endocrinol. 60, 178–186.

Wilson, C.M., McNabb, F.M.A., 1997. Maternal thyroid hormones in Japanese quail eggs and their influence on embryonic development. Gen. Comp. Endocrinol. 107, 153–165.

Yamaguchi, S., Aoki, N., Kitajima, T., Iikubo, E., Katagiri, S., Matsushima, T., Homma, K.J., 2012. Thyroid hormone determines the start of the sensitive period of imprinting and primes later learning. Nat. Commun. 3, 1081.

Yoshimura, T., 2010. Neuroendocrine mechanism of seasonal reproduction in birds and mammals. Anim. Sci. J. 81, 403–410.

Zoeller, R.T., 2005. Environmental chemicals as thyroid hormone analogues: new studies indicate that thyroid hormone receptors are targets of industrial chemicals? Mol. Cell. Endocrinol. 242, 10–15.

Chapter 25

The Role of Hormones in the Regulation of Bone Turnover and Eggshell Calcification

Christopher G. Dacke*

Pharmacology Division, School of Pharmacy and Biomedical Science, University of Portsmouth, Portsmouth, UK

Toshie Sugiyama

Department of Agrobiology, Niigata University, Niigata, Japan

Carol V. Gay

Department of Biochemistry and Molecular Biology, Penn State University, University Park, PA, USA

ABBREVIATIONS

$1,25(OH)_2D_3$ or $25(OH)D_3$ Metabolites of vitamin D
Ca or Ca^{++} Calcium or ionic calcium
CA Carbonic anhydrase
cAMP Cyclic adenosine monophosphate
CGRP Calcitonin gene-related peptide
CT Calcitonin
DBP Vitamin D binding proteins
ESG Eggshell gland
OPG Osteoprotegerin
PGs Prostaglandins
PGE_2 Prostaglandin isoform E_2
Pi Inorganic phosphate
PKA, PKC Protein kinase A or C
PTGs Parathyroid glands
PTH Parathyroid hormone
PTHrP Parathyroid hormone-related peptide
PLP Parathyroid-like peptide
PTX Parathyroidectomized
RANK Receptor-activated nuclear factor κ-B
RANKL RANK ligand

25.1 INTRODUCTION

Calcium (Ca) is one of the most efficiently regulated plasma constituents in birds. Classical Ca regulating hormones, all recognized in birds, include parathyroid hormone (PTH), calcitonin (CT), and 1,25-dihydroxy vitamin D_3 ($1,25(OH)_2D_3$), although their actions and/or sensitivities can be quite different from those in mammals (Dacke, 1979) and it is clear that avian skeletal metabolism acts at an amplified rate compared with that in mammals (Gay, 1988). Other putative Ca and bone-regulating factors including prostaglandins (PGs) (Dacke, 1989), calcitonin gene-related peptides (CGRPs) (Dacke et al., 1993a), amylin (Guzel et al., 2009), and intracellular factors including the receptor-activated nuclear factor κ-B (RANK), RANK ligand (RANKL) and osteoprotegerin (OPG) system, also affect avian Ca metabolism in ways that are profoundly different from those in mammals. This may be partly related to enhanced requirements for Ca during eggshell formation when a dramatic change takes place in the bone biology of the hen at the onset of sexual maturity (Dacke, 2000; Whitehead, 2004).

More recent reviews of avian Ca and bone metabolism include those by Gay (1996), Sugiyama and Kusuhara (2001), Whitehead (2004), Stamford (2006), Bar (2008), de Matos (2008), and Kim et al. (2012). Avian Ca metabolism shares many features with other vertebrate classes, but is also typified by several characteristics related to the ability of birds to lay large megalecithal eggs with well-developed calcified eggshells (Romanoff and Romanoff, 1963). Other Archosaurian species, extant crocodilia and at least some extinct dinosaurs, the closest relatives of modern birds, also lay or laid eggs with calcified shells (Dacke, 1979; Schweitzer et al., 2005, 2007). The amount of Ca in each egg typically represents about 10% of the total body Ca stores in the bird (Kenny, 1986), an enormous amount. The Ca metabolism of an egg-laying hen in domesticated species

*The author has retired, and is no longer connected with the University of Portsmouth.

Sturkie's Avian Physiology.

such as chickens and Japanese quail on a daily basis compares with that of human females in 18 months of combined pregnancy and lactation. In order to provide a source of Ca for eggshell calcification to supplement the supply provided directly from the diet, egg-laying hens and some extinct therapod dinosaurs possess a highly labile reservoir in the form of medullary bone, developing within long bones of hens in response to gonadal steroids. It is the most overtly estrogen-sensitive form of vertebrate bone (Simkiss, 1967; Kusuhara and Schraer, 1982; Dacke et al., 1993b; Imamura et al., 2006; Hiyama et al., 2009).

A second feature influencing the evolution of avian Ca metabolism is the ability to fly. This probably led to development of light but robust skeletons in which long bones tend to be more hollowed out than in other vertebrates. Similar lightweight structures are found in fossilized bones of flying pterosaurs (Wellnhoser, 1991). This hollowing implies a high degree of remodeling during the growth phase of the skeleton, which probably influenced the evolution of certain hormonal activities and sensitivities in birds.

Mammals given either acute hyper- or hypocalcemic challenges respond within a few hours, whereas one-week-old chickens correct such challenges within minutes, recovery from the hypocalcemia being dependent upon the presence of PTH (Koch et al., 1984). An intravenous dose of ^{45}Ca in a week-old chick rapidly clears from the plasma pool, and by 15 min approximately 40% of the original dose is located within the skeleton (Shaw et al., 1989). By calculating the unidirectional plasma–bone clearance constant ($K_{pb}{}^{45}Ca$) and hence net Ca^{++} influx, and by estimating the total rate of Ca accretion into the skeleton, we find that total skeletal outflux of Ca^{++} is approximately 80% of influx, and that the plasma pool of Ca^{++} clears into the skeleton every few minutes. Typically, net Ca accumulation into the chick femur is about 0.28 µmol/min/g wet weight, although this will vary according to prevailing rates of dietary Ca absorption and other factors. Clearly any factor that modifies either rapid influx or outflux of Ca^{++} in this system will profoundly affect minute-to-minute plasma Ca modulation in the rapidly growing animal. Bronner and Stein (1992) calculated, using data from Shaw et al. (1989), that the t ½ for ^{45}Ca uptake by the chick femur is less than 10 min, compared with around 30 min in rabbit, dog, and rat.

The discovery of Ca- and bone-related disorders that affect both production and welfare has stimulated interest in the bone biology of the laying hen. Caged layer paralysis, identified soon after the introduction of battery cages (Couch, 1955), is associated with weakened skeletons characterized by osteoporosis (Stamford, 2006).

Avian osteoporosis is defined as a decrease in the amount of fully mineralized structural bone, leading to increased fragility and susceptibility to fracture (Whitehead and Fleming, 2000). Incidence and severity of avian osteoporosis are closely associated with egg-producing ability. High egg-producing hens are very susceptible to osteoporosis (Fleming et al., 2006); lines of egg-laying hens selected by their bone strength resistance to osteoporosis produce eggs with lower shell quality. The onset of osteoporosis is related to estrogen activity, which stimulates medullary bone formation for eggshell formation, and concomitantly starts to decrease the volume of cancellous bone and the thickness of cortical bone, finally contributing to weakened skeleton strength (Fleming et al., 1998; Turner et al., 1993; Wilson and Thorp, 1998). This osteoporosis is prevented by the antiestrogen tamoxifen, but egg laying also stops (Wilson and Thorp, 1998). There is a dramatic decrease in circulating estrogen in 70 week old hens compared with those in peak production (up to 29 week) with parallel changes in estrogen receptor-α populations in both kidney and duodenum. This results in decreasing $1,25(OH)_2D_3$ production and intestinal Ca absorption (Beck and Hansen, 2004; Hansen et al., 2003).

Much of this review is based around the physiology of Ca regulation in chickens, reflecting the abundant literature on this species, but other avian species are referred to where appropriate.

25.2 EVOLUTIONARY ASPECTS OF EGGLAY AND MEDULLARY BONE

25.2.1 Egglay and Evolution of Calcium Reservoirs

Reptiles and birds lay cleidoic eggs, which differ from those of lower vertebrates in that no primitive larval stage is produced at hatching. Instead, the hatchling is morphologically similar to the adult and has an ossified skeleton. This is true for Archosaurian (crocodilian, dinosaurian, and avian) eggs. This requires a supply of Ca provided by the egg yolk. Reptilian eggs tend to have leathery shells, although turtle shells are thick and impregnated with deposits of calcium carbonate ($CaCO_3$). Apart from its protective function, the avian and presumably other archosaurian eggshells are a major source of Ca for skeletal calcification (Johnston and Comar, 1955; Simkiss, 1967; Tuan and Ono, 1986; Dacke, 1979). It has been suggested that slow-developing (precocial) avian species such as the Japanese quail (*Coturnix japonica*) move more Ca from the eggshell and therefore have more ossified skeletons compared with fast-developing (altricial) species such as the starling (*Sturnus vulgaris*) (Blom and Lilja, 2004; Karlsson and Lilja, 2008).

The large throughput of Ca in these species means that the reproducing female requires a highly labile Ca reservoir in addition to the more stable cortical bone. These range from the turtle plastron (Dacke, 1979) to the heavily calcified scutes (scales) in crocodilia (Elsey et al., unpublished data), while in some archosaurians the unique presence of medullary bone within the cavities of long bones has evolved (Schweitzer et al., 2005).

25.2.2 Evolution of Medullary Bone

Medullary bone is described as a secondary bone tissue that develops within marrow cavities of long bones in reproducing female birds, provides Ca for the eggshell, and has no biomechanical function. Avian medullary bone is a very brittle and fragile type of bone; the spicules of the medullary bone are easily separated from the originating layer. It was hitherto considered to be a unique avian phenomenon (Bloom et al., 1941; Simkiss, 1967; Dacke, 1979). Anecdotal suggestions that this bone could be found in crocodilians (Whitehead, 2004) are now largely discounted following a comprehensive study in reproducing alligators where no evidence of medullary bone was found during the reproductive cycle (Schweitzer et al., 2007).

A groundbreaking paper (Schweitzer et al., 2005) provided unambiguous evidence for the presence of endosteally derived bone tissues lining the interior marrow cavities of portions of a fossilized therapod dinosaur (*Tyrannosaurus rex*) hindlimb specimen, and hypothesized that these tissues are homologous to avian medullary bone. In the same paper, they showed similarities of *T. rex* medullary bone to that in ratites, the most primitive extant birds. Because medullary bone was hitherto considered unique to female birds, its discovery in dinosaurs solidified the link between dinosaurs and birds, and suggests similar reproductive strategies within this archosaur clade. Subsequently, Lee and Werning (2008) found medullary bone in another saurischian dinosaur, *Allosaurus*, and the ornithischian dinosaur *Tenontosaurus*.

25.2.3 Bone Formation and Resorption during the Egg-Laying Cycle

Medullary bone is a labile source of Ca for eggshell formation. It lines the endosteal surface of structural bone and also occurs as spicules within the marrow cavity (Simkiss, 1967; Dacke, 1979; Sugiyama and Kusuhara, 2001). It has little inherent strength but can contribute to fracture resistance.

There are bone-forming osteoblasts, bone-resorbing osteoclasts on medullary bone surfaces, as well as osteocytes embedded in the matrix (Gay, 1988; Gay et al., 2000; Gay and Weber, 2000; Miller, 1992; Turner et al., 1994). The space between the spicules is filled with hematopoietic cells, erythrocytes, and marrow stromal cells (Bloom et al., 1941; Simkiss, 1967).

Osteoclasts resorb both medullary and structural bone so that during the period while hens remain in reproductive condition, there is a progressive loss of skeletal structural bone, which is characteristic of osteoporosis. Increasing fragility makes the bones more susceptible to fractures. The dynamics of bone loss can be affected by a number of nutritional, environmental, and genetic factors (Whitehead, 2004). About 30% of battery hens experience fractures, either during the production period or during depopulation (Gregory and Wilkins, 1989).

Similar to cortical bone, medullary bone matrix consists mainly of inorganic and organic phases. The organic matrix is divided into collagenous and noncollagenous proteins. There is a much greater amount of protein in medullary bone than in the cortical bone. The principle component of the protein is type I collagen, the content of which is one-half or one-quarter of that in cortical bone, and spicule density and direction are coarse and irregular (Ascenzi et al., 1963; Bonucci and Gherardi, 1975; Hiyama et al., 1998; Knott and Bailey, 1998). Most of the noncollagenous protein consists of proteoglycans, and medullary bone contains 2–3 times as much proteoglycan as in cortical bone (Candlish and Holt, 1971). Keratan sulfate is the major proteoglycan in medullary bone (Candlish and Holt, 1971; Hunter and Schraer, 1981), which is also co-localized with the calcified matrix, suggesting close involvement in the regulation of medullary bone calcification (Yamamoto et al., 2001, 2005). Interestingly, tartrate-resistant acid phosphatase (TRAP), a bone-degrading enzyme, accumulates in the medullary bone matrix during bone formation (Yamamoto and Nagai, 1994). Most of the inorganic phase of medullary bone is hydroxyapatite, similar to cortical and trabecular bones. However, it also contains significant amounts of $CaCO_3$ (calcite) (Lorcher and Newesely, 1969; Pellegrino and Biltz, 1970), which may be closely correlated with rapid formation and resorption.

Calcification of medullary bone initially occurs in matrix vesicles found between the collagen fibrils of the osteoid tissue, and spreads into the surrounding interfibrillar matrix (Bonucci and Gherardi, 1975). However, Ca and apatite contents are similar to those in cortical bone (Ascenzi et al., 1963). The characteristics of medullary bone matrix are clear, but their detailed functions are not fully understood. Thus, the matrix of medullary bone is substantially different from that of cortical bone. This is probably an important factor in rapid bone turnover during eggshell formation.

In nonseasonal layers such as domestic hens, medullary bone formation begins 12–14 days preceding the first ovulation, whereas seasonal layers such as pigeons develop medullary bone before the first ovulation, and it disappears after the last egg in the clutch has been laid (Bloom et al., 1941; Miller, 1992; Turner et al., 1994). Simultaneously, the ovarian interstitial cells together with the growing ovarian follicles secrete increasing amounts of estrogen during the onset of sexual development. The final rapid growth of the follicles occurs within a 7 to 11 day period prior to ovulation, during which time estrogen secretion by the thecal cells reaches a peak (Phillips et al., 1985; Johnson, 2000). Thus medullary bone formation in the female birds coincides with the initiation of ovarian follicle maturation and estrogen secretion. The cycle of medullary bone formation and resorption during the egg-laying cycle is shown in Figure 25.1.

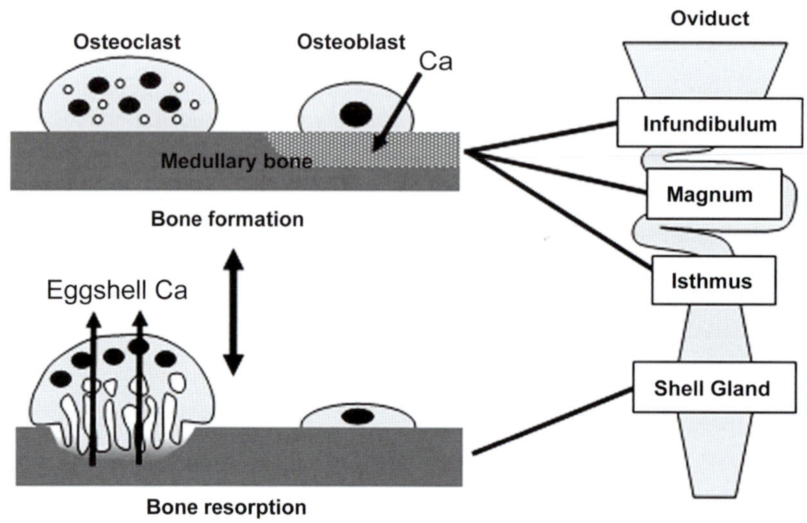

FIGURE 25.1 **Medullary bone formation and resorption during the egg-laying cycle.** *Modified after Sugiyama and Kusuhara (2001).*

The important role of gonadal hormones in regulating medullary bone formation has been recognized for many years, this process requiring the combined influence of estrogen and androgen (Bloom et al., 1942). Adult male pigeons, quails, ducks, and chickens form medullary bone if estrogen is administered to them, as will immature female birds if both hormones are provided. Medullary bone will not form if only one of these hormones is present (Ascenzi et al., 1963; Bloom et al., 1942; Miller and Bowman, 1981). Injection of estrogen into male quails promotes proliferation of the endosteal bone lining cells and then differentiation into osteoblasts (Kusuhara and Schraer, 1982). In addition, estrogen receptors are present in medullary bone osteogenic cells, including osteoblasts, preosteoblasts, lining cells, and marrow stromal cells, of egg-laying and estrogen-treated chickens or quails (Ohashi and Kusuhara, 1993; Ohashi et al., 1991; Turner et al., 1993). This suggests that new medullary bone at the onset of egg laying is formed by osteoblasts differentiated from endosteal bone-lining cells, and later the osteoblasts involved in the cyclic bone formation during the egg-laying cycle are supplied from bone marrow stromal cells (Sugiyama and Kusuhara, 2001). The structure of medullary bone is shown in Figure 25.2.

It was demonstrated, using reverse transcriptase polymerase chain reaction (RT-PCR) and *in situ* hybridization, that estrogen receptor-α, a typical receptor, is highly expressed in hen medullary bone osteoblasts (Imamura et al., 2006; Hiyama et al., 2009), while the estrogen receptor antagonists tamoxifen and trioxifene (LY133314) inhibit medullary bone formation in estrogen-treated male birds (Williams et al., 1991). Estrogen is synthesized by conversion of testosterone to estradiol by the action of aromatase in follicles, and letrozole, an aromatase inhibitor, inhibits medullary bone formation in prelay pullets (Deng et al., 2010). If the hen goes out of reproductive condition, estrogen levels fall, osteoblasts resume structural bone formation, and skeletal regeneration occurs.

Periods of medullary bone formation and resorption relate to the position of an egg in the oviduct during the egg-laying cycle. Following ovulation, the ovum enters the infundibulum, staying for 15–20 min before passing to the magnum, where it remains for 3–3.5 h during which time albumen is secreted. Thereafter, both inner and outer shell membranes are formed in the isthmus, taking 1.25–2 h, and the outer membranes are deposited with $CaCO_3$ in the shell gland forming the eggshell, over a period of 18 h. The calcified egg then passes to the vagina and is expelled within a few minutes. In domestic hens, the egg-laying cycle takes about 24 h, and osteoblasts and osteoclasts periodically modify their morphology and functions during such a cycle (Figure 25.1).

While an egg is in the magnum, cuboidal osteoblasts actively form medullary bone (matrix formation and calcification). When the egg enters the shell gland, spindle-shaped osteoblasts no longer form bone (Bloom et al., 1942; Ohashi et al., 1990a). Nevertheless, van de Velde et al. (1985) and Wilson and Duff (1990) reported little significant change in medullary bone volume, regardless of the degree of mineralization during the egg-laying cycle, suggesting that osteoblasts form the nonmineralized bone matrix (osteoid) and later mineralize the bone. It is known that the medullary bone formation in domestic hens is regulated by estrogen fluctuations during the egg-laying cycle, plasma estrogen concentration being highest immediately before oviposition and very low in the latter period of eggshell formation (Shahabi et al., 1975; Shodono et al., 1975). Treatment of egg-laying hens with the estrogen receptor modulator tamoxifen inhibits osteoblastic bone formation during the inactive period of shell formation (Ohashi et al., 1990a).

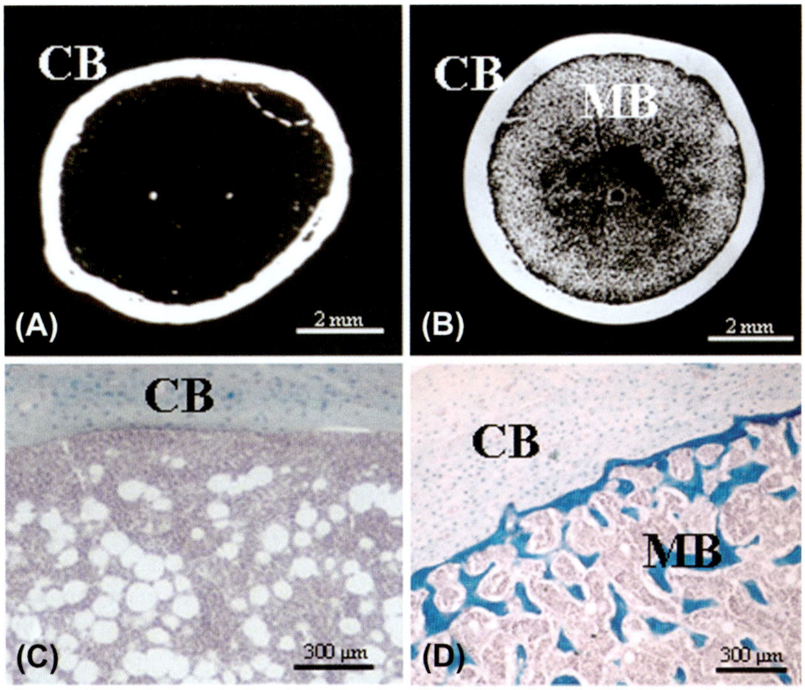

FIGURE 25.2 Micrographs (A and B) and light microscopy (C and D) of cross-sections of femurs from male chickens (A and C) and egg-laying hens (B and D). CB, Cortical bone; MB, medullary bone. Reticular medullary bone is developed within the marrow cavities of long bones in the egg-laying hen. The calcified density of the medullary bone is lower than that of the cortical bone. Medullary bone is intensely stained with alcian blue staining, suggesting that it has an abundance of acid mucopolysaccharide Medullary bone is not present in the male chicken.

Medullary bone osteoclasts alternatively cease and accelerate resorption during the egg-laying cycle, thus moderating supplies of Ca for eggshell formation. While osteoclast populations do not change during the egg-laying cycle in chickens and quails, their morphology and ultrastructure dramatically modify, concomitant with Ca requirements (Miller, 1977, 1981; Sugiyama and Kusuhara, 1993, 1994b; van de Velde et al., 1984b). Thus, when an egg is in the infundibulum, isthmus, or magnum, osteoclasts lose their ruffled borders (bone resorption sites) and cease to resorb bone. However, when an egg is calcifying in the shell gland, osteoclasts develop ruffled borders and resorb the bone for eggshell formation.

In general, the modification of osteoclast activities is regulated by both PTH and CT. PTH is secreted by parathyroid glands in the response to low plasma Ca^{++} levels. CT blocks osteoclastic resorption of medullary bone when an egg is in the magnum of the oviduct, and PTH stimulates osteoclastic bone resorption for eggshell formation during the period when an egg is in the shell gland. Injection of PTH into hens when an eggshell is not being calcified causes osteoclast ruffled border development, thus increasing the plasma Ca levels (Miller, 1978; Kusuhara, 1982). PTH, in chicken medullary bone organ culture systems, stimulates the development of ruffled borders and the acid phosphatase activity of osteoclasts, while CT inhibits them (Sugiyama et al., 1993; Sugiyama and Kusuhara, 1994a, 1996a, 1996b).

25.2.4 Specific Calcium Metabolism for Eggshell Formation

In laying hens, approximately 2.4 g Ca is required over 18 h to produce a shelled egg of 60 g. Only 60–75% of eggshell Ca can be provided directly by feed, the remainder coming from body stores (Whitehead, 2004). A hen produces up to one egg per day for a 52 week period. Eggshell formation requires large amounts of the ionic precursors for $CaCO_3$ deposition. Both ionic species ($Ca^{++}$and HCO_3^-) are supplied by the blood via transepithelial transport through the shell gland mucosa. This is one of the most rapid of biomineralization processes. The process by which Ca is secreted by the oviduct into the eggshell has attracted considerable attention because of its commercial implications (Whitehead, 2004).

Avian eggshells consist largely of $CaCO_3$. Carbonate ions are formed from metabolic CO_2 (i.e., $CO_2+H_2O \leftrightarrow H^+ + HCO_3^-$) catalyzed by carbonic anhydrase (CA), an enzyme family consisting of at least 14 different isozymes (reviewed in Chegwidden and Carter, 2000). Disruption of this process results in profound effects on eggshell thickness and quality (Berg et al., 2004). CA is localized in epithelial columnar cells and the tubular gland cells of the eggshell gland (ESG) (Gay et al., 1974; Arai et al., 1996). Ca^{++}-ATPase, however, is distributed in the cytoplasm of the tubular glands (Yamamoto et al., 1985; Arai et al., 1996). These results suggest that HCO_3^- and Ca^{++} are supplied independently from different parts of the shell gland.

$CaCO_3$ secretion by the ESG is a thermodynamically active process (Schraer and Schraer, 1965). The egg-laying quail oviduct contains a Ca^{++}–Mg^{++}-activated ATPase within the microsomal fraction of the ESG mucosa. The activity of this enzyme is between one and a half and three times greater than that in precalcifying or immature birds (Pike and Alvarado, 1975). The role of uterine (i.e., ESG) ion transporters for mineralization precursors of the avian eggshell, according to Jonchere et al. (2012), is summarized in Figure 25.3.

25.3 CHEMISTRY AND SECRETION OF CALCIUM-REGULATING HORMONES

25.3.1 Parathyroid Hormone and Related Peptides

25.3.1.1 The Parathyroid Glands

Two to four parathyroid glands (PTGs) are present in birds; two pairs in chickens slightly caudal to the thyroid and one pair in Japanese quail are less closely associated with the thyroids than the four or more found in mammals (Dacke, 1979; Clark and Sasayama, 1981; Kenny, 1986). PTGs derived from the third and fourth branchial pouches in the embryo are encapsulated by connective tissue and composed mainly of chief cells, a situation similar to that in rats (Roth and Schiller, 1976). Oxyphil cells, present in mammalian PTGs, are absent in birds. The low granular content of the avian chief cell is consistent with a low level of PTH secretion (Kenny, 1986).

The importance of PTH in maintaining avian Ca levels has been recognized for years. Parathyroidectomy (PTX) in birds leads to hypocalcemia, tetany, and death depending on factors such as dietary Ca intake, the presence of accessory parathyroid tissue, and reproductive status (Kenny, 1986). PTH injections into Japanese quail or chickens increase plasma Ca levels (Kenny and Dacke, 1974). The hypercalcemic effects of PTH are greater in egg-laying hens than in cockerels, due to either Ca^{++} binding by yolk proteins in the plasma or additional PTH receptors present in medullary bone and oviduct (Dacke, 1979). Immature birds show rapid and sensitive responses to intravenous PTH, this being formerly used as a bioassay (Dacke and Kenny, 1973; Parsons et al., 1973). Primary targets for PTH in birds are bone and kidney. The major physiological stimulus for PTH secretion from the chief cells is a fall in plasma Ca^{++} concentration, while increasing plasma Ca^{++} suppresses it (Brown, 1991).

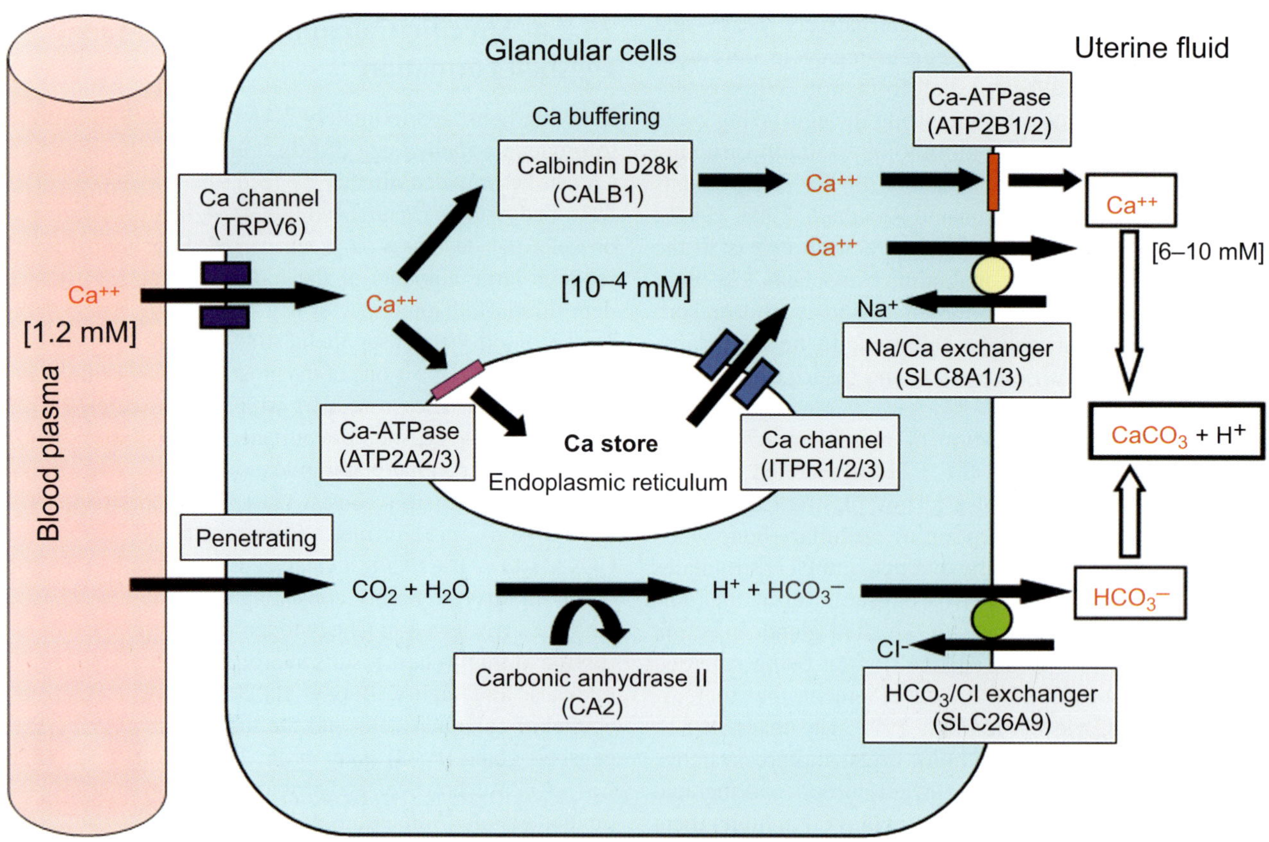

FIGURE 25.3 General model describing ion transporters in the uterine (shell) gland during eggshell calcification. This model is modified from Jonchere et al. (2012). Parentheses represent gene symbols. Brackets indicate the concentration of Ca^{++}.

25.3.1.2 Chemistry of Parathyroid Hormone and Related Peptides

Mammalian PTH is an 84 amino acid polypeptide, while chicken PTH is 88 amino acids (Khosla et al., 1988; Limm et al., 1991). However, only the first 32–34 amino acids from the N terminus are necessary for biological activity (Tregear et al., 1973). The nucleotide sequence of chicken pre-pro-PTH mRNA was determined from a 2.3-kilobase fragment of complementary PTH DNA cloned in *Escherichia coli*. The mRNA (2.3 kilobases) for chicken hormone precursor was approximately three times the size of that for mammalian pre-pro-PTH. The amino acid sequence deduced from the DNA showed that chicken pre-pro-PTH mRNA encoded a 119 amino acid precursor peptide and an 88 amino acid hormone which is four residues longer than all known mammalian homologs and includes gene deletions and insertions. There is significant homology of sequence in the biologically active 1–34 region with mammalian hormones, but much less in the middle and carboxyl terminal regions (Russell and Sherwood, 1989).

PTH and PTH-related peptide (PTHrP) belong to a family of endocrine factors sharing a highly conserved N-terminal region (amino acids 1–34) and play key roles in Ca homeostasis, bone formation, and skeletal development. Recently, a third PTH-like peptide (PTH-L) was identified in teleost fish, amphibians (*Xenopus*), and birds (chicken), raising questions about the evolution of these proteins (Guerreiro et al., 2007; Pinheiro et al., 2010). The function of PTHrP in nonmammalian vertebrates is poorly documented. Amphibians and birds occupy phylogenetic positions, the former at the transition from aquatic to terrestrial life and the latter at the transition to homeothermy. The PTH-L gene is present throughout vertebrates with the exception of placental mammals. Gene structure of PTH and PTH-L seems to be conserved in vertebrates, while PTHrP gene structure is divergent and has acquired new exons and alternative promoters. Splice variants of PTHrP and PTH-L are common in *Xenopus* and chicken and transcripts of the former have a widespread tissue distribution, although PTH-L is more restricted. PTH is widely expressed in fish tissue, but from *Xenopus* to mammals it becomes largely restricted to the PTG. The N-terminal (1–34) regions of PTH, PTHrP, and PTH-L in *Xenopus* and chicken share high sequence conservation and the capacity to modify Ca fluxes across epithelia, suggesting a conserved role in Ca metabolism possibly via similar receptors (Guerreiro et al., 2007).

In teleosts, five genes encode PTHrP (2), PTH (2), and PTH-L (1); in tetrapods, including birds, there are three (PTHrP, PTH, and PTH-L); while placental mammals lack PTH-L. It is hypothesized that genes of the PTH family appeared at approximately the same time during the vertebrate radiation and evolved via gene duplication and deletion events. PTH, which has a paracrine distribution in lower vertebrates, became the product of a specific endocrine tissue, the PTG first evolving in amphibians. The PTHrP and PTH-L gene organization diverged and retained its widespread tissue distribution complex in vertebrates, which is congruent with its paracrine nature (Pinheiro et al., 2010).

At the time of writing, there appear to be few reports of calcemic or other actions of the third member of the PTH-peptide family, PTH-L, in either birds or mammals. Guerreiro et al. (2007) suggest that PTH-L is the ancestral from which PTH and PTHrP have evolved, while Pinheiro et al. (2012) suggest that PTH-L can influence Ca fluxes across membranes in *Xenopus* and chickens as well as in fish.

25.3.1.3 Parathyroid Hormone Release

Similar to mammals, a Ca^{++}-sensing receptor is found in the avian PTG. The Ca-receptor (CaR) gene is expressed by PTG chief cells, which also store and secrete PTH. Increases in plasma Ca^{++} increase expression of the CaR gene. Low plasma Ca^{++}, before or after eggshell formation, is associated with decreased CaR gene expression in the PTG. The level of CaR gene expression is inversely correlated with PTH content of the PTG. Thus, in contrast to mammals, CaR gene expression in the PTG of the chicken is inversely associated with changes in plasma Ca^{++} (Yarden et al., 2000).

25.3.1.4 Circulating Parathyroid Hormone

Relatively few reports exist of circulating PTH levels in birds, reflecting the fact that they are normally low relative to those in mammals and that pure avian PTH was hitherto unavailable for antibody production. Van de Velde et al. (1984a) measured plasma PTH-like bioactivity during the egg-laying cycle of the chicken by cytochemical bioassay. This is elevated during eggshell calcification, after which it falls to a low level but is slightly raised again 2h after ovulation. Singh et al. (1986) similarly measured levels of "PTH" during the egg cycle of chickens using *in vitro* bioassay. PTH was higher in hens fed Ca-deficient diets compared with those on a high Ca diet. The levels were highest during shell calcification, then shortly after ovulation, in both groups of hens; furthermore, they showed an inverse relationship with plasma ionized Ca levels (see Dacke, 2000), suggesting that changes in bioactive PTH play an important role in the Ca metabolism of the chicken under physiological Ca stress. At least part of the PTH-like activity measured in these studies was probably PTHrP, but at the time of writing there are no reports distinguishing between the two peptides to ascertain their respective roles in the avian egg cycle or in avian Ca metabolism in general.

25.3.2 Calcitonin and Related Peptides

CT contains 32 amino acids and a seven-membered N-terminal ring, all of which are necessary for biological activity. Variations in amino acid sequence of CT from different species give rise to differences in bioactivity, with CTs of fish origin being particularly potent (Dacke, 1979). Structures of different CTs, including that from chicken, are reviewed by Zaidi et al. (1990a), who also give detailed consideration to their structure–activity relationships.

CT secretion is regulated primarily by rising plasma Ca levels leading to an increased secretion from ultimobranchial glands (Dacke, 1979). In egg-laying quails, plasma CT concentrations are at their highest after oviposition and fall as eggshell calcification proceeds, rising at the end of calcification (Dacke, 2000; Dacke et al., 1972).

Eliam-Cisse et al. (1993) studied influences of Ca and vitamin D deficient diet on CT gene expression in the CT-secreting ultimobranchial cells of the developing chicken. They exhibited a striking reduction of CT biosynthesis by decreasing the number of secretory cells and not by triggering modifications of the biosynthetic activity of the ultimobranchial endocrine cells.

25.3.2.1 Circulating Calcitonin Levels

High levels of circulating bioreactive CT are present in submammalian species, including birds (Dacke, 1979). They are higher in adult males than in egg-laying females except for a brief period immediately before commence of lay. In quail hens, they are at their highest shortly after ovulation and fall as eggshell calcification proceeds, rising at the end of this period. In this species at least, gonadal steroids, especially androgens, appear to have a major influence on circulating CT levels (see Taylor and Dacke, 1984; Kenny, 1986). In chickens, plasma CT levels positively correlated with dietary Ca intake and thus to circulating Ca levels (Taylor and Dacke, 1984), while in chick embryos they are very low until just prior to hatching (Abbas et al., 1985).

Figure 25.4 shows the relationships between PTH, CT, $1,25(OH)_2D_3$, estrogen, and Ca levels during the hen's egglay cycle.

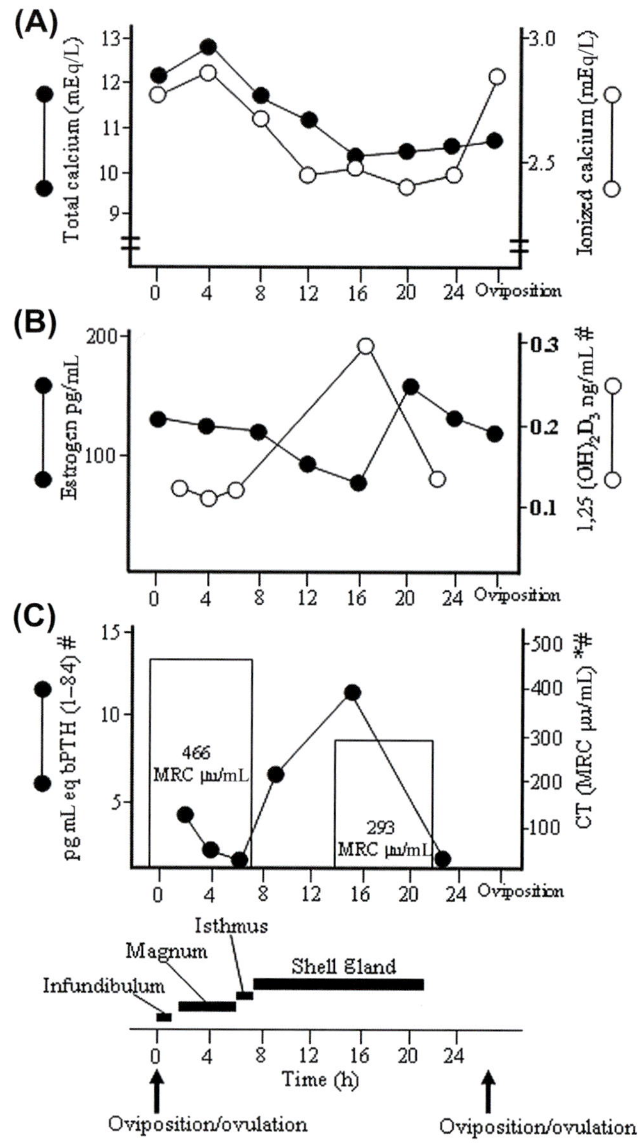

FIGURE 25.4 Concentrations of plasma calcium and hormone levels in the domestic hen during the egg-laying cycle. (A) Plasma total Ca and Ca^{++} (Parsons and Combs, 1981); (B) estrogen (Shodono et al., 1975) and $1,25\ (OH)_2D_3$ (Abe et al., 1979); and (C) PTH (van de Velde 1984b) and CT (Dacke et al., 1972). *: Data from egg-laying quails; the others are from egg-laying chickens. #: Data after ovulation; the others are after oviposition. $1\ mEq/L = 0.5\ mmol^{-1}$. A: ●—● Total Ca (mEq/L), ○—○Ca^{++} (mEq/L). B: ●—● Estrogen pg/mL plasma, ○—○ $1,25(OH)_2D_3$ ng/mL plasma#. C: ●—● bPTH (1–84) pg mL eq#; square bar represents plasma CT (MRC µu/mL) #*. *Originally from Sugiyama and Kusuhara (2001).*

25.3.3 The Vitamin D System

Hou (1931) first reported the importance of the preen gland in vitamin D metabolism and the prevention of rickets in growing chickens.

Vitamin D_3 regulation in birds was reviewed by Taylor and Dacke (1984), Norman (1987), Hurwitz (1989a), Nys (1993), and Bar (2008). It is well established that chickens metabolize vitamin D_3 to $25(OH)D_3$ and $1,25(OH)_2D_3$ in their liver and kidneys, respectively (Holick, 1989). Birds, unlike mammals, discriminate between vitamin D_2 and D_3, and chickens are unable to utilize vitamin D_2 as efficiently as D_3 (Taylor and Dacke, 1984). This appears to be due to the fact that in birds, plasma vitamin D–binding protein has relatively low affinity for vitamin D_2, which is thus more rapidly broken down (Holick, 1989). The avian kidney, as in other vertebrates, synthesizes and secretes $1,25(OH)_2D_3$. Numerous factors stimulate production

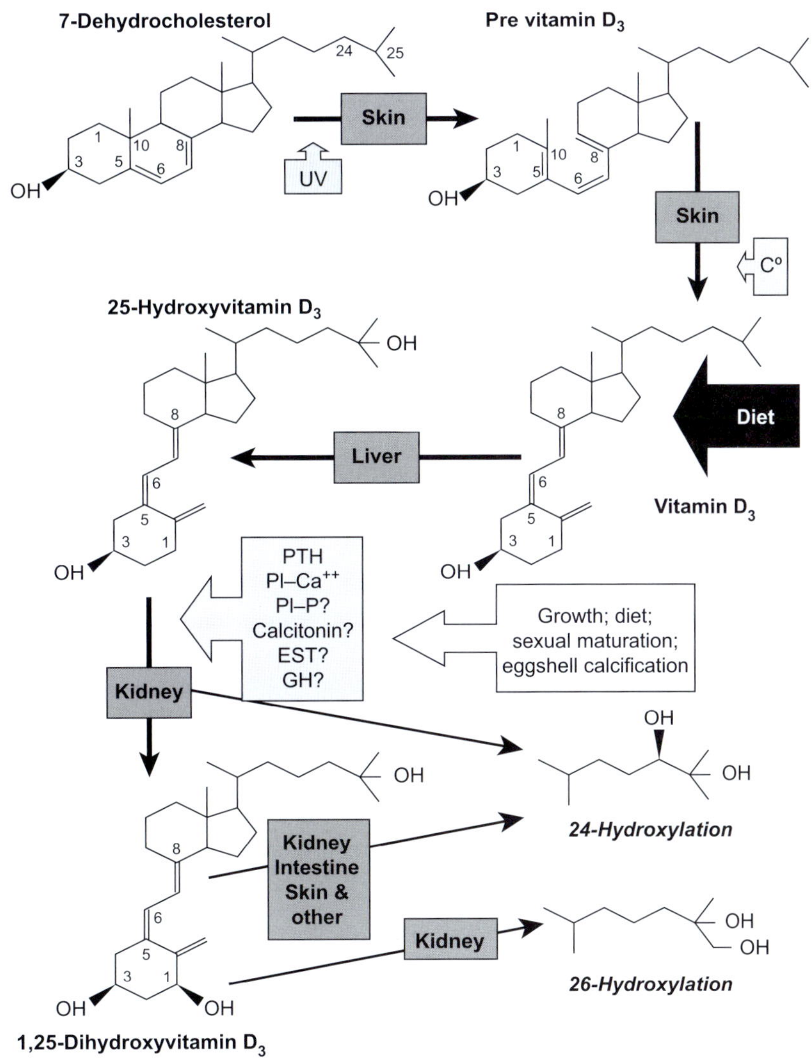

FIGURE 25.5 **Vitamin D metabolism and regulation in the laying bird.** EST: estrogens; GH: growth hormone; Pl–Ca^{++}: plasma Ca^{++}; Pl–P: plasma phosphate. *Originally from Bar (2008).*

of $1,25(OH)_2D_3$, including PTH, $1,25(OH)_2D_3$ itself, and prolactin (Kenny, 1981; Henry and Norman, 1984). There are contradictory reports as to whether CT influences $1,25(OH)_2D_3$ production by avian kidney, but Kenny (1986) concluded that CT is not a major regulator of vitamin D_3 in birds. The metabolism of vitamin D to its active forms in birds is summarized in Figure 25.5. Egg yolk is a rich source of vitamin D, of which 90% is in the form of vitamin D_3 and 5% in the form of $25(OH)D_3$. This vitamin D is probably derived from the hen rather than the embryo (Fraser and Emtage, 1976).

The extra demands on Ca homeostasis, especially in domestic birds characterized by their long clutch lengths, stimulate vitamin D metabolism and expression. Vitamin D metabolism is also influenced by other processes associated with increased demands for Ca, including growth and bone formation. A series of intestinal, renal, or bone proteins are consequently expressed in the target organs via mechanisms involving a vitamin D receptor. Some of these proteins (CA, calbindin, and Ca-ATPase) are found in the ESG and are believed to be involved in Ca transport as they are in the intestine or kidney. These processes are reviewed by Bar (2008).

Bar et al. (1999) investigated relationships between age, eggshell thickness, and vitamin D metabolism and its expression in laying hens. Hens forming thin shells synthesize less $1,25(OH)_2D_3$ and less duodenal and eggshell gland (ESG) calbindin than normal laying hens. Hens forming thin shells had lower intestinal and ESG calbindin and its mRNA. Reducing ESG Ca^{++} transport by the CA inhibitor acetazolamide, but not by dietary Ca restriction, reduced ESG calbindin and its mRNA. It was suggested that the mechanism responsible for Ca^{++} transport to the eggshell consists of a vitamin D-dependent absorption of Ca^{++} and a multifactor-dependent transfer of

Ca^{++} to the shell. Both steps are most likely calbindin mediated; however, the induction of calbindin gene expression in the ESG is predominantly Ca^{++} dependent, and the apparent defect in vitamin D metabolism or its expression seen in older hens is typical of, or even exclusive to, thin-shell-forming hens.

25.3.3.1 Circulating Levels of Vitamin D Metabolites

Vitamin D_3 and its metabolites are transported in association with vitamin D binding–protein (DBP), with plasma albumins, and in some avian species with α- and β-globulins as well. The DBP exists as two 4S forms that differ in the number of neuraminic acid residues present. DBP probably has a single binding site for all vitamin D metabolites, and its affinities for 25- and $24,25(OH)_2D_3$ are similar, whereas $1,25(OH)_2D_3$ has a 10-fold lower affinity. The plasma concentration of DBP is more than twice as high in laying hens than in immature birds or adult males (reviewed by Taylor and Dacke, 1984; Hurwitz, 1989a).

Levels of circulating vitamin D_3 metabolites in Japanese quail and egg-laying chickens have been determined. Increased intestinal Ca absorption occurring after sexual maturity and during eggshell formation is related to renal 25-hydroxycholecalciferol-1-hydroxylase activity (Kenny, 1976), which in turn enhances circulating $1,25(OH)_2D_3$ levels (Castillo et al., 1979) and the accumulation of this metabolite in the intestinal mucosa (Bar et al., 1978). Increases in $25(OH)D_3$-1-hydroxylase activity are induced by injecting estrogen into immature birds (Baksi and Kenny, 1977; Sedrani et al., 1981). Abe et al. (1979) reported that plasma concentrations of $25(OH)D_3$ and $1,25(OH)_2D_3$ but not $24,25(OH)_2D_3$ in egg-laying hens fluctuate during the eggshell calcification cycle. Nys et al. (1986) confirmed these results and also demonstrated that hens laying soft-shelled eggs do not show these fluctuations. Circulating levels of $1,25(OH)_2D_3$ increase in hens during the prelaying period and again at the onset of egg production (Nys, 1993). Bar and Hurwitz (1979) demonstrated that the stimulatory effect of estrogen on renal $25(OH)D_3$-hydroxylase in Ca-deficient birds is eliminated, suggesting that increased $1,25(OH)_2D_3$ production results from Ca deficiency induced by estrogens.

Plasma $1,25(OH)_2D_3$ concentrations fluctuate during the egg-laying cycle of domestic hens, reaching a peak during eggshell formation (Abdulrahim et al., 1979; Castillo et al., 1979). $1,25(OH)_2D_3$ is necessary for osteoblastic bone calcification, and osteoblasts possess its receptors (Boivin et al., 1987; Hurwitz, 1992). Studies on the effects of $1,25(OH)_2D_3$ deficiency on hen Ca metabolism suggest that feeding vitamin D deficient diets results in increased osteoid (Takahashi et al., 1983; Wilson and Duff, 1991). Therefore, $1,25(OH)_2D_3$ is important for the calcification of medullary bone during the egg-laying cycle.

25.4 ACTIONS OF PARATHYROID HORMONE, CALCITONIN, AND VITAMIN D ON TARGET ORGANS

25.4.1 Actions on Skeleton (Bone and Cartilage)

25.4.1.1 Actions of PTH

PTH has multiple catabolic and anabolic effects on the skeleton. The slow hypercalcemic response in mammals involves recruitment and activation of osteoclasts mediated via receptors located in osteoblasts (Hurwitz, 1989b). Birds are exquisitely sensitive to PTH, with hypercalcemic responses occurring in egg-laying hens as early as 8 min after administration (Candlish and Taylor, 1970), too brief a time for significant osteoclastic resorption (Hurwitz, 1989a) and unlikely to be due to intestinal or renal transport mechanisms. Using acute ^{45}Ca labelling, Kenny and Dacke (1974) demonstrated that the initial (0–30 min) response to PTH involved inhibition of plasma Ca^{++} clearance. Using a method of temporal microwave fixation of injected radioisotopes in skeletal and other tissues, it was demonstrated that decreased plasma ^{45}Ca clearance in chicks is accounted for by an inhibition of net Ca uptake into the skeleton (Shaw and Dacke, 1985; Shaw et al., 1989). These dose-dependent responses were elicited using the active bPTH(1–34) fragment, but could also be seen following intravenous injection of a stable PGE_2 analog, 16,16-dimethyl PGE_2. They are rapid and most apparent in the femur and, to some extent, calvaria (Dacke and Shaw, 1987). Phosphodiesterase inhibitors, caffeine or 3-isobutyl-1-methylxanthine, mimic effects of PTH on skeletal ^{45}Ca uptake in chicks, indicating a role for cAMP in this response (Shaw and Dacke, 1989).

PTH receptors are located on osteoblast surfaces but have been considered absent in osteoclasts (Hurwitz, 1989b), although this is disputed (Pandala and Gay, 1990; Teti et al., 1991; May et al., 1993). Considerable evidence exists to suggest that PTH can induce rapid changes in Ca transfer by osteoblasts and osteocytes. Thus, PTH-stimulated increases in Ca uptake by these cells have been observed, while others have reported either no response in intracellular Ca^{++} concentration or a net Ca efflux from bone cells, at least in embryonic chick bone *in vitro* (Hurwitz, 1989b; Malgaroli et al., 1989). At least two types of voltage-controlled ionic channels were described using patch clamp techniques in cultured embryonic chick osteoblasts (Ypey et al., 1988), which predicted a role for these channels in the response to PTH.

PTH also affects cell spread area in avian osteoclasts (Miller, 1977, 1978) in studies that used electron microscopy to determine responses of medullary bone osteoclasts in egg-laying Japanese quail *in vivo* during the inactive phase of their eggshell calcification cycle. PTH induced the osteoclasts to form ruffled borders bounded by filamentous-rich clear zones within 20 min of hormone treatment, these changes being characteristic of "active" bone-resorbing cells found during shell calcification. Similarly, Sugiyama and Kusuhara (1994a, 1996a) demonstrated that PTH induces ruffled border formation in osteoclasts located within hen medullary bone maintained in culture. Zambonin-Zallone et al. (1982) adopted a procedure whereby domestic hens are prefed a Ca-deficient diet in order to increase cell yield, and then a prolonged isolation technique is used consisting of unit sedimentation and filtration. Osteoclasts from tibiae of Ca-deficient chicks *in situ* were shown to increase their cell spread area by 40% within 2–4 min of PTH challenge (Pandala and Gay, 1990). Whether this response represents direct or indirect effects of the hormone on these cells, it is remarkably fast compared with classical responses to PTH (Dacke, 1979; Hurwitz, 1989a). Bronner (1996) speculated that an important mechanism underlying minute-to-minute regulation of blood Ca levels in both birds and mammals is the ability of bone-lining cells, osteoclasts and osteoblasts, to alter their size and shape and migrate to and from areas of the bone surface where high- or low-affinity binding sites for Ca^{++} are located; this is an interesting hypothesis that remains to be tested experimentally.

Biochemical mechanisms underlying avian osteoclast function are similar to those of mammalian osteoclasts, although in some respects they are functionally distinct from the latter variety (Gay, 1988). Thus the ruffled border contains a proton pump-ATPase and an Na^+, K^+-ATPase. It also contains carbonic anhydrase, which is closely associated with the cytoplasmic side of the membrane. Ca^{++}-ATPase is present on the plasma membrane of the marrow side of osteoclasts but absent in the ruffled borders; its role is presumably to direct the outward flow of transmembrane Ca^{++} flux. May et al. (1993) demonstrated a direct effect of PTH resulting in increased acid production by chick osteoclasts, the mechanism involving activation of adenylate cyclase via a G_s-type protein. Stimulation of acidification by PTH and cAMP is blocked by estradiol. Estradiol was inhibitory to the same extent as CT; these effects were not additive. Estradiol-17β in micromolar, but not in nanomolar, amounts blocked proton pumping in isolated plasma membrane vesicles (Gay et al., 1993). The actions of PTH and other calciotrophic factors on bone cell function are summarized in Figure 25.6.

25.4.1.2 *Action of PTH on Cartilage*

Both PTH and PTHrP enhance cAMP and inhibit collagen synthesis in avian epiphysial cartilage cells, an effect that is blockadeable by the PTH antagonist PTH (3–34) (Pines et al., 1990). Two reports show that PTH and PTHrP can modulate growth of avian tibial and sternal cartilage via chondrocytic maturation and proliferation. Regulation of phenotype in chick tibial growth plate chondrocytes by PTHrP is facilitated via signaling through three pathways: protein kinase A (PKA), protein kinase C (PKC), and inositol-1,4,5-trisphosphate-induced Ca^{++} transients

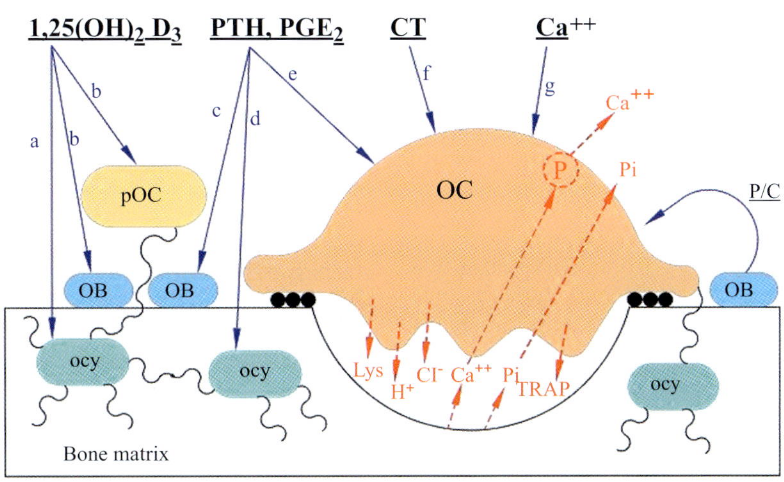

FIGURE 25.6 Summary diagram showing key features of avian osteoclast regulation and activities during remodeling of the skeleton. Osteoblasts control osteoclasts during the early phases of bone growth through interaction of RANKL and RANK (not shown, see text). RANK-expressing osteocytes become the command cells during bone remodeling, as recent *in vivo* studies show. Vitamin D and PTH often work synergistically and can be either anabolic or catabolic. PGE2 effects on osteoclasts are powerful but are likely to be indirect. Cells: pOC, pre-osteoclast; OC, osteoclast; OB, osteoblast; ocy, osteocyte; P, Ca-ATPase pump. Secretions through the ruffled border into the resorption lacuna include the following: Lys, lysosomal contents; TRAP, tartrate-resistant acid phosphatase; and various ions. OC effectors include $1,25(OH)_2D_3$, PTH, CT, Ca^{++}, and paracrine and cytocrine factors (P/C). Podosomes (●) contain αvβ3 integrin, which binds to osteopontin, secreted by osteoclasts onto the sealing zone. Solid lines: controlling factors; dashed lines: secretions and ion fluxes. (a) Nakashima et al. (2011); (b) Norman and Hurwitz (1993); (c) Hurwitz (1989b) and Dacke (1989); (d) Bonewald (2011); (e) Dacke (1989), Pandala and Gay (1990), Teti et al. (1991), and May et al. (1993); (f) Eliam et al. (1988), Pandala and Gay (1990), Hall et al. (1994), and Sugiyama and Kusuhara (1996b); and (g) see section 25.4.1.4.

(Zuscik et al., 2002). PTH (10^{-7}–10^{-15} M) decreased terminal chondrogenesis in the avian sterna. During the first half of an 8 day culture, 100 nM PTH (1–34) significantly increased sternal length and downregulated the deposition of type X collagen and its mRNA expression. There was less type II and X collagen in PTH-treated sterna with concomitant decreases in mRNA production, suggesting that proliferation is the major contributor to cartilage growth in the presence of PTH and PTH-related peptide receptor activation. These experiments demonstrated that PTH increased cartilage growth by upregulating cell proliferation and the expression and secretion of extracellular matrix components (Harrington et al., 2007).

25.4.1.3 Actions of Calcitonin

Despite the discovery of CT more than five decades ago, its role in bone and Ca metabolism remains an enigma. Only in mammalian species has CT been shown to regulate plasma Ca levels, the basis of its hypocalcemic action lying in an ability to inhibit osteoclastic bone resorption (Copp and Kline, 1989). Plasma Ca levels in submammalian vertebrates are refractory to dosage with CT (Dacke, 1979). Whether or not this is due to the high circulating levels of biologically active hormone causing receptor downregulation is not clear. Initial studies in avian osteoclasts showed no postreceptor responses to CT (Miyaura et al., 1981; Ito et al., 1985; Nicholson et al., 1986; Dempster et al., 1987); however, cAMP responses were demonstrated in osteoclasts from chicks maintained on low-Ca or rachitogenic diets (Eliam et al., 1988; Rifkin et al., 1988). Moreover, osteoclasts from Ca-deficient chicks respond to CT *in vitro* within 4 min by a 58% reduction in cell spread area (Pandala and Gay, 1990) and also by an inhibition of their bone-resorptive activity (de Vernejoul et al., 1988). Hall et al. (1994) reported specific surface binding and rapid clearance (within 10 min) of CT applied to isolated chick osteoclasts. CT also causes the disappearance of ruffled borders in cultured medullary bone osteoclasts (Sugiyama et al., 1993; Sugiyama and Kusuhara, 1996a), these findings being consistent with the idea that CT receptors in chicks are downregulated under normal physiological conditions. Ancill et al. (1991) found that dosing heavily (22 h) fasted chicks with salmon CT *in vivo* caused a rapid (10 min) but variable inhibition of net ^{45}Ca uptake into the skeleton. This effect is similar to that of PTH and PGE_2, but its physiological significance is obscure.

Salmon CT acutely increased embryonic chick calvarial cell proliferation *in vitro*, [^{3}H]-thymidine incorporation into DNA, [^{3}H]-proline incorporation into bone matrix collagen, and [^{3}H]-hydroxyproline in intact calvaria and tibiae (Farley et al., 1988). The increased [^{3}H]-hydroxyproline incorporation was associated with proportional increases in bone alkaline phosphatase activity. [^{3}H]-proline incorporation in embryonic chick calvariae also increased during 3 days of limited exposure (i.e., 4 h/day) to CT. These actions of CT also occur in cultures of neonatal mouse calvaria.

Zambonin-Zallone et al. (1982) isolated chicken medullary bone osteoclasts that lacked CT receptors and showed no increase in cAMP levels in response to CT. Similar results were reported by Ito et al. (1985) and Nicholson et al. (1986, 1987). However, isolated osteoclasts from chicks fed low-Ca diets do possess CT receptors; hence, osteoclastic bone resorption is inhibited by the CT treatment (Rifkin et al., 1988; Eliam et al., 1988). In these reports, interestingly, plasma Ca levels influence the induction of CT receptors in osteoclasts, suggesting that CT receptors in chicken osteoclasts are remarkably downregulated with high plasma Ca levels. Furthermore, Gay (1988) and Gay and Weber (2000) reported that isolated chick osteoclasts possess CT-binding sites that mediate the signal without the increase of cAMP levels, and CT inhibits the osteoclastic bone resorption within a few minutes. PTH-binding sites were also detected in chick and hen medullary bone osteoclasts (Teti et al., 1991; Agarwala and Gay, 1992), and the PTH induction of the osteoclastic bone resorption is directly mediated by PTH receptors on osteoclasts, contrary to the mammalian observations where PTH indirectly stimulates the osteoclastic bone resorption via PTH receptors on osteogenic cells (May et al., 1993). Estrogen receptors are detected in isolated medullary bone osteoclasts (Oursler et al., 1993), and receptor occupancy inhibits the lysosome-related genes and adhesion activity of the medullary bone osteoclasts (Oursler et al., 1993; Sugiyama et al., 1999). However, estrogen receptors were not detected in these osteoclasts by histochemical techniques (Ohashi et al., 1990a, 1990b; Imamura et al., 2006), Brubaker and Gay (1999) suggested that bone resorption of chick osteoclasts is directly inhibited by a membrane-mediated nongenomic action of estrogen.

Yasuoka et al. (2001) studied developmental changes of PTH, PTHrP, and CT receptor binding properties in chicken calvaria and kidneys. The equilibrium dissociation constant (Kd) and the maximum binding capacity (Bmax) of receptors for PTH, PTHrP, and CT in the membrane fraction of calvariae and kidney were examined from one day after the hatch up to 24 weeks of age using radioligand binding assays. Kd values of the PTH and PTHrP receptor in both tissues were decreased at 10 and 24 weeks in female birds, whereas values were increased at 24 weeks in males. Bmax values of the PTH and PTHrP receptor in both tissues decreased at 10 weeks and returned to baseline at 24 weeks in females but then increased at 24 weeks in males. The Kd and Bmax values of CT receptors in both tissues were constant during these periods in both sexes. This suggested that binding properties of PTH–PTHrP receptors but not CT receptors may be influenced by gonadal hormones relating to sexual maturation.

25.4.1.4 Role of Extracellular Calcium

Freshly isolated quail medullary bone osteoclasts, unlike those from neonatal rats, do not respond to elevated extracellular Ca $[Ca^{++}]_e$ by a rise in intracellular Ca $[Ca^{++}]_i$ (Bascal et al., 1992). Unlike rat cells, cultured chicken osteoclasts are sensitive to changes in membrane voltage and to dihydropyridine-type Ca^{++} channel blockers (Miyauchi et al., 1990). When medullary bone osteoclasts are cultured away from bone substrata for several days, they recover an ability to respond to elevated $[Ca^{++}]_e$, but neither the fresh nor cultured cells exhibit any response to CT in terms of raised $[Ca^{++}]_i$ (Arkle et al., 1994). Freshly isolated quail medullary osteoclasts are also refractory to $[Ca^{++}]_e$ in terms of cell spread area; however, unlike neonatal rat cells, they respond to ionomycin (a Ca^{++} ionophore) with modest reductions in cell spread area. This suggests that the quail cells retain intracellular mechanisms necessary for elaboration of the aforementioned responses, but lack receptor mechanisms for detecting changes in $[Ca^{++}]_e$. When medullary bone osteoclasts are cultured for several days, the stimulus causes a reduction in cell spread area similar to that in fresh rat cells (Bascal et al., 1994). These results indicate that putative Ca^{++} "receptors" on freshly isolated quail medullary bone osteoclasts are normally downregulated but that this disappears upon culturing the cells for several days. This suggests that during the eggshell calcification cycle, when resorption of medullary bone prevails and raised local Ca^{++} levels are generated by intense osteoclastic activity, osteoclasts become insensitive to inhibitory factors such as elevated $[Ca^{++}]_e$ (Figure 25.6).

Ca-sensing receptors (i.e., CaRs) are present in many cell types. In skeletal tissues, CaRs are present in osteoblasts, osteocytes, and chondrocytes (Rodriguez et al., 2005). Discovering the signaling pathways utilized by CaRs has been fraught with difficulty. This is perhaps not surprising since Ca ions are utilized by cells in a plethora of ways, including maintaining membrane potential and serving as an intracellular signal. Ca^{++} moves across the plasma membrane through a variety of channels, including stretch-activated channels. Ca^{++}-ATPase pumps are also employed. There is general agreement that raised extracellular Ca^{++} stimulates proliferation of osteoblasts and inhibits osteoclastogenesis (Theman and Collins, 2009). While the precise roles of CaR in bone cells are unknown, one possibility worthy of investigation is that early-stage osteoblasts may be able to follow a Ca^{++} gradient and migrate to sites of bone resorption. Also, inhibition of osteoclasts by elevated local $[Ca^{++}]$ closes a negative feedback loop.

25.4.1.5 Actions of Vitamin D

Feeding laying hens vitamin D deficient diets results in medullary bone resorption, while in nonlaying birds osteodystrophy results (Wilson and Duff, 1991). The vitamin D metabolite, $1,25(OH)_2D_3$, appears to facilitate bone formation by inducing biosynthesis of osteocalcin (bone γ-carboxy-glutamic acid protein). The biological function of this noncollagenous vitamin D binding protein in skeletal mineralization is obscure, although it appears to be a specific product of osteoblasts during bone formation. This small MW bone protein has been purified and sequenced for several species, including chicken. Its precise function is unknown, but it binds Ca^{++} and shows affinity for hydroxyapatite, suggesting an involvement in mineral dynamics (Hauschka et al., 1989). Osteocalcin is released into the circulation, where it provides a convenient index of bone turnover, reflecting new osteoblast formation rather than release of matrix protein during bone resorption (Nys, 1993). Osteocalcin synthesis is stimulated by binding $1,25(OH)_2D_3$ to promoter elements and enhancing osteocalcin gene transcription. However, unlike intestinal calbindin, substantial osteocalcin synthesis occurs in vitamin D-deficient chicks (Lian et al., 1982).

Approximately 30–40% of eggshell Ca derives from medullary bone, the remainder being provided directly by dietary sources (Simkiss, 1967; Dacke, 1979; Hurwitz, 1989a). Formation of medullary bone matrix is induced by gonadal steroids regardless of vitamin D status, although it only becomes fully mineralized when both vitamin D_3 and the steroids are present (Takahashi et al., 1983). Nys (1993) reports that changes in blood osteocalcin levels parallel those of $1,25(OH)_2D_3$ in laying hens so that concentrations of blood osteocalcin rise in hens fed a low-Ca diet and decrease in hens laying thin-shelled eggs. It is possible that increased osteocalcin levels in response to estrogen are a reflection of increased vitamin D receptor expression by osteoblasts (Liel et al., 1992). However, osteoclasts from medullary bone as well as from rat bone appear to be devoid of $1,25(OH)_2D_3$ receptors, and the effects of the metabolite are considered to be mediated via the osteoblasts (Merke et al., 1986). Harrison and Clark (1986) cultured medullary bone obtained from egg-laying hens and demonstrated that these cultures could respond to $1,25(OH)_2D_3$ by a dose-dependent inhibition of $[^3H]$-proline uptake.

The vitamin D receptor (VDR) shares a conserved structural and functional organization with other nuclear receptor (NR) superfamily members. For many NRs, N-terminal variant isoforms that display distinct cell-, stage- and promoter-specific actions have been identified. The novel VDR isoform VDRB1, with a 50 amino acid N-terminal extension, is produced from low-abundance transcripts that contain exon 1d of the human VDR locus. There is evidence for the conservation of this exon in other mammalian and avian species. The transactivation differences between VDRB1 and the original VDR provide insights into mechanisms that may contribute to functional differences and potentially distinct physiological roles for these two VDR isoforms (Gardiner et al., 2004). A detailed review of the molecular mechanisms of vitamin D actions in mammals is provided by Haussler et al. (2013).

25.4.1.6 The RANK–RANKL–OPG System

Elucidation of how bone cells interact through the RANK–RANKL–OPG system is currently a very exciting area of skeletal research. A review by Nakashima and colleagues (2012) provides an historical and up-to-date account of the discoveries so far. This system is essential for osteoclast differentiation and activation. RANK is a transmembrane receptor found on osteoclasts and osteoclast progenitor cells. RANKL, the ligand for RANK, occurs in both soluble and membrane-bound forms and is expressed in a number of cell types, including osteoblasts and osteocytes. Osteoprotegerin (OPG), so named because of its protective role for bone, is a soluble decoy receptor that binds to RANKL and thereby blocks the signaling pathways that lead to osteoclast formation and activation. The RANKL–OPG ratio is useful for estimating the extent of osteoclastogenesis. Genes for all three proteins are known, making it possible to overexpress or ablate expression in specific cell types both *in vitro* and *in vivo*. Much information comes from gene manipulation in mice, but a number of studies with avian systems reveal the necessity of RANK, RANKL, and OPG in birds as well. The expression of the components of this system is regulated by a number of osteotropic hormones.

OPG was the first of the triad to be discovered by two independent groups (Simonet et al., 1997; Yasuda et al., 1998a) using a variety of molecular approaches described in a review by Khosla (2001). OPG is widespread in occurrence, being expressed in lung, heart, kidney, liver, stomach, intestine, brain, spinal cord, mesenchymal cells in bone marrow, osteoblasts, and osteocytes. Identification of OPG quickly led to the realization that OPG likely bound to an unidentified component that stimulated osteoclastogenesis. This unidentified component turned out to be the elusive osteoclast-differentiating factor (ODF). Using expression cloning with OPG as a probe, ODF was found to be identical to RANKL (Lacey et al., 1998; Yasuda et al., 1998b). These two groups of investigators developed the concept and proved that direct contact between RANKL-expressing osteoblasts and RANK-possessing osteoclasts, and their progenitors, is essential for osteoclastogenesis during bone development and growth to proceed. This was an important, major advance in understanding osteoclastogenesis.

RANKL, the second member of the triad to be discovered in bone, occurs in a number of cell types, including osteoblasts, osteocytes, bone marrow stromal cells, and several lymphoid cells. Since 1998, a vast literature has emerged that shows how RANKL expressed by osteoblasts controls osteoclasts, as a review by Suda et al. (2012) describes. Meanwhile, osteocytes have been found to express RANKL in great abundance, at a level several-fold higher than in osteoblasts, and to control RANK-expressing osteoclasts during bone remodeling (Bonewald, 2011; Xiong et al., 2011; Nakashima et al., 2012). This conclusion is based on gene deletion studies. In both osteoblasts and osteocytes, RANKL synthesis is increased by $1,25(OH)_2D_3$, PTH, PGE_2, glucocorticoids, and a number of proresorptive cytokines (e.g., interleukins) and is downregulated by estrogen (Khosla, 2001; Nakashima et al., 2012).

RANK, the third triad member, is the receptor for RANKL and is found on osteoclasts and their progenitor cells. RANK–RANKL interaction is essential for the development of osteoclasts from progenitor cells to fully functional, multinucleate resorptive cells. M-CSF (macrophage colony-stimulating factor) is also required. Both RANK and its ligand, RANKL, had been identified previously (and named) by immunologists, so the search for RANK in osteoclasts fell readily into place (Hsu et al., 1999). Many of the molecular details of the signaling pathway that follow RANK–RANKL interaction are coming to light through use of the tools of gene expression in mice, as described in recent reviews (Ross, 2008; Nakashima et al., 2012; Bonewald et al., 2013).

While elucidation of the RANK–RANKL–OPG story has been carried out largely in mice, studies with avian skeletal cells make it clear that avian osteoblasts, osteocytes, and chondrocytes control avian osteoclasts through RANK–RANKL interaction. Human recombinant (hr) RANKL has been shown to stimulate the resorptive activity of osteoclasts isolated from embryonic chick tibia (Boissy et al., 2001) and from bone marrow of Muscovy ducks (Gu et al., 2009). A model for examining podosome development utilized the transition of avian monocytes to osteoclasts by treating monocyte cultures with hrRANKL (Pfaff and Jurdic, 2001). Primary sternal chondrocytes isolated from chickens were found to express cell surface RANKL and stimulate osteoclast formation and resorptive activity at the eroding face of the growth plate (Masuyama et al., 2006). Chicken RANKL has been shown to induce osteoclast formation from chicken bone marrow and to activate primary chicken osteoclasts (Wang et al., 2008). Bone morphogenetic protein-2 (BMP2) induced RANKL in chicken and mouse chondrocytes and stimulated osteoclastic resorption (Usui et al., 2008). Embryonic chick calvaria are widely used as a source of osteocytes for RANK–RANKL investigations. These studies indicate that there is considerable conservation of RANKL and its receptor, RANK, among avian and mammalian species. OPG has been similarly conserved. Chicken OPG has been shown to inhibit Ca^{++} release, to reduce the amount of TRAP expressed, and to increase apoptosis of chick embryo-derived osteoclasts (Hou et al., 2011) and of primary chicken osteoclasts

(Wang et al., 2008). The signaling pathways of the controlling factors of osteoclastic bone resorption (RANKL, vitamin D, PTH, etc.) have been and are being studied in great detail in mammalian systems; it is likely that these pathways will be similar or the same in avian species, given the close parallels found so far.

With the advent of techniques to isolate highly pure, viable osteocytes, it has become apparent that RANKL is expressed by osteocytes in great abundance and that direct contact of osteocyte processes with osteoclasts has profound effects on osteoclast function. How do matrix-embedded osteocytes gain control of surface-bound osteoclasts and their progenitors in bone marrow? Anatomically, osteocyte processes have been shown to reach through and beyond the osteoblast layer of chick calvarial bone surfaces (Kamioka et al., 2001), and the processes can be as long as 1 mm, a length ~100 times the diameter of an osteocyte (Sugawara et al., 2005). This makes it possible for osteocytes to physically connect with osteoclasts and cells in the bone marrow compartment. Additionally, the soluble form of RANKL has a molecular mass of ~31 kDa (Lacey et al., 1998), small enough to diffuse through canaliculi and have effects quite distant from sites of synthesis. Osteocytes also secrete the small, diffusible OPG.

As the osteocyte transforms from an osteoblast on a bone surface, it must change shape, develop long processes (about 50 per cell), secrete unmineralized matrix (osteoid), regulate the mineralization of osteoid, and develop and maintain contact with other bone cells (i.e., osteoblasts, bone-lining cells, osteoclasts, and each other). The molecular composition of the osteocyte changes profoundly during this transition (Bonewald, 2011). Osteocytes, rather than being simple, passive cells that serve only to maintain bone, are substantially different from osteoblasts and are, in fact, the regulators of bone remodeling. An entire issue of the journal *Bone*, edited by Bonewald et al. (2013), provides considerable information on osteocytes as a network of communicating cells that respond to hormones, vitamins, cytokines, and mechanical forces. The roles of RANK, RANKL, and OPG in medullary bone, the formation of which requires estrogen, await investigation. Are medullary bone osteoclasts regulated by osteoblasts or by osteocyte processes reaching into the bone marrow compartment? Or will a different mechanism be found that controls the daily demands supplying Ca^{++} eggshell formation?

25.4.2 Renal Actions

PTX leads to increased urinary Ca excretion and decreased inorganic phosphate (Pi) excretion in starlings (Clark and Wideman, 1977) and Japanese quail (Clark and Sasayama 1981). Mechanisms of renal Ca and Pi regulation were reviewed by Laverty and Clark (1989). In immature males and egg-laying females, intravenous injections of PTH transiently reduce plasma Ca and Pi levels, followed by an increase peaking 20–30 min after the injection. In contrast, adult cockerels seldom show any variation in plasma Ca or Pi levels. In all birds, PTH increased the glomerular filtration rate, the urine flow rate, and Pi and Ca clearance. Infusion of CT decreases plasma Ca only in PTX chickens, but in all birds it increased GFR, urine volume, and Ca excretion, and in PTG-intact birds renal Pi clearance also increased.

Clark (1991) measured serum and renal clearance of Pi and Ca and compared these in vitamin D–deficient and vitamin D–replete chickens. She observed that most renal functions studied after Ca loading, PTH administration, or PTX are unaltered by vitamin D deficiency in the chicken. The major significant finding was that vitamin D-deficient chickens do not excrete increased amounts of Pi in response to PTH stimulus.

Renal Ca and Pi homeostasis in birds, as in mammals, depends on a balanced release of PTH and CT. However, in the chicken, response times are faster, thus minimizing fluctuations in plasma Ca and Pi levels. This rapid homeostatic response is less effective when Ca demands are high in growing or laying birds (Sommerville and Fox, 1987).

Goldstein et al. (1999) studied second messenger production in avian medullary nephron segments in response to peptide hormones. Activity of adenylyl cyclase in the thin descending limb was stimulated approximately twofold by PTH but not by other hormones tested: arginine vasotocin (AVT), glucagon, atrial natriuretic peptide (ANP), or isoproterenol, each at 10^{-6} M. Activities in the thick ascending limb were stimulated two- to threefold by both AVT and PTH. These data support a tubular action of AVT and PTH in the avian renal medulla.

Dudas et al. (2002) studied transepithelial Pi transport regulation by PTH in chicken proximal tubule epithelium. The effect of PTH and activation of PKC and PKA on transepithelial Pi transport was examined in monolayers of chick proximal tubule cells (PTCs) in primary culture. Acute exposure of the PTCs to PTH (10^{-9} M, basolateral side) significantly decreased the net reabsorption of Pi by approximately 66%. There was no effect after the addition of PTH to the luminal side. PTH inhibition was blocked by either Bisindolylmaleimide I (a highly selective PKC inhibitor) or H-89 (a potent inhibitor of PKA). Tissue electrophysiology remained stable after all treatments. PTH treatment decreased immunoreactivity of the luminal Na^+–Pi cotransporter NaPi-IIa. These data indicate that PTH inhibition of Pi reabsorption in this *in vitro* system is mediated by PKC and PKA.

Electrophysiological responses to PTH were studied by Laverty et al. (2003) in primary cell confluent monolayer cultures of chick proximal tubules. These exhibited vectorial transport, including glucose-stimulated current. Under short-circuit conditions, PTH induced dose-positive current (short-circuit current (I(sc))) responses, with a 2 min peak response over baseline followed by slow decay. Responses were dose dependent, and half-maximal at 5×10^{-9} M. They were nearly abolished by apical addition of EIPA, an inhibitor of Na^+–H^+ exchangers, and partially blocked by Cl^- channel blockers. Bilateral reduction of Cl^- in the buffer from 137 to 2.6 mM abolished the response. The authors concluded that in the chick proximal tubule, PTH activates both a Na^+–H^+ exchanger and a Cl^- channel that may be functionally linked. Yasuoka et al. (2001) documented developmental changes of PTH, PTHrP, and CT receptor binding in chicken kidneys (and calvaria).

Bouizar et al. (1989) studied distributions of renal CT-binding sites in vertebrates. No renal ^{125}I-sCT binding sites were detected in fish, amphibians, or reptiles. In rat, chicken, and quail binding, CT binding sites observed in the medulla and in the cortex. The pattern followed the distribution of the glomeruli and/or the collecting tubules, suggesting that CT renal receptors appeared late in evolution and regulation of renal function by CT is effective in only birds and mammals. As mentioned in this chapter, CT has effects on renal Ca and Pi excretion, being most pronounced in PTX chickens.

25.4.3 Actions on Intestine and Oviduct

Ogawa et al. (2000) and Ieda et al. (2000) described PTH receptor binding in the endometrium of the shell gland (uterus) of the oviduct of, respectively, the guinea fowl and domestic hens during an oviposition cycle. Receptor binding in the membrane fraction was analyzed by the use of [^{125}I] PTH-related peptide (PTHrP) binding assays. Specificity, reversibility, and saturation of binding were demonstrated, while in both studies, Scatchard plots revealed a single class of binding sites. The equilibrium dissociation constant (Kd) was 0.50–1.15 nM in laying birds and 1.07–1.16 nM in nonlaying birds. Maximum binding capacity (Bmax) per milligram of membrane protein was 65.2–110.9 fmol in laying birds and 105.8–120.6 fmol in nonlaying birds. Both Kd and Bmax values changed within the above range during an oviposition cycle in laying birds, showing a decrease during the period of eggshell formation. No change was found in nonlaying birds. These results collectively suggest that PTH and PTHrP act on the endometrium of the ESG during eggshell calcification (discussed further in this chapter).

Intestinal absorption of Ca is regulated by $1,25(OH)_2D_3$ by inducing RNA transcription and synthesis of proteins such as calbindin D28k. The physiological function of calbindin D_{28k} is not well established, but its concentration in the intestine is reflected by the ability to absorb Ca. Three forms of calbindin D28k with varying size have been identified in the chicken intestine, with the smallest being most abundant (Nys, 1993). In vitamin D deficient chicks, intestinal calbindin mRNA is barely detectable but increases dramatically following $1,25(OH)_2D_3$ injection (Mayel-Afshar et al., 1988). The onset of ovulation in hens is associated with increased intestinal Ca absorption and elevated concentrations of calbindin D28k (Nys, 1993) coinciding with increased plasma $1,25(OH)_2D_3$ concentrations (Castillo et al., 1979; Nys et al., 1992). A similar or identical protein is present in the ESG (Fullmer et al., 1976). The uterus of the laying hen contains receptors for $1,25(OH)_2D_3$, although a large part of the calbindin appears to be independent of this metabolite (Coty, 1981). Uterine calbindin and its mRNA concentration increase in immature pullets treated with estrogen and are higher in hens laying thin-shelled eggs (Navickis et al., 1979). However, the effect of the sex steroid on uterine calbindin appears to be an indirect one, probably related to the general development and maturation of the oviduct (Nys, 1993). Although concentrations of uterine calbindin and its mRNA rise during formation of the first and subsequent eggs in chickens and quail, they are not directly related to the activity of $1,25(OH)_2D_3$ (Bar et al., 1992; Nys, 1993). The function of uterine calbindin is not precisely established, but its presence is indicated wherever there is a physiological requirement for Ca translocation across the uterine wall (Hurwitz, 1989a).

Sugiyama et al. (2007) investigated the expression and localization of calbindin D28k in all intestinal segments of laying hens. Western blotting analysis showed that the entire intestine expressed CaBP-D28k to the following degree: duodenum>jejunum>cecum>ileum>colon; while immunohistochemistry showed strong CaBP-D28k localization in enterocytes along the villus tip-crypt axis in the duodenum and in villus tips in the cecum and colon. The jejunum and ileum had moderate localization with respect to the number of immunoreactive cells and staining intensity. This suggests that laying hens actively absorb Ca in both the large and small intestines.

Ogawa et al. (2003), using radioligand binding assays of membrane fractions of the endometrium, identified CT receptors in the shell gland of the guinea fowl and changes in binding properties during an oviposition cycle. The equilibrium Kd obtained by Scatchard analyses was 0.50–1.25 nM in laying birds and 1.12–1.19 nM in nonlaying birds, while the Bmax per milligram of protein was 33.1–107.5 fmol in laying birds and 101.4–114.9 fmol in nonlaying birds. Both Kd and Bmax values changed during eggshell formation in laying birds, suggesting that CT receptor binding may be related to eggshell formation.

Krzysik-Walker et al. (2007) studied CT expression in the chicken ovary and identified influences of follicular maturation and ovarian steroids. CT has been linked to reproduction specifically as a marker for embryo implantation in the mammalian uterus. Although CT expression is documented in several tissues, there are no previous reports of CT production by the ovary of any vertebrate species. Using RT-PCR, they detected CT mRNA and the CT receptor (CTR) mRNA in the granulosa and thecal layers of preovulatory and prehierarchical follicles. Both CT and CT mRNA were localized in granulosa and thecal cells. Using quantitative PCR analysis, the F1 follicle granulosa layer was found to contain significantly greater CT mRNA and CTR mRNA levels compared with those of any other preovulatory or prehierarchical follicles. Progesterone treatment of sexually immature chickens resulted in a significantly greater abundance of ovarian CT mRNA, whereas estradiol or prostaglandin combined with estradiol treatment significantly reduced ovarian CT mRNA. These results suggested that follicular maturation and gonadal steroids influence CT and CTR gene expression in the chicken ovary.

25.4.4 Actions on Smooth Muscle

PTH has potent hypotensive activity in several vertebrate classes, including birds. In the domestic hen, a PTH bolus reduces bone blood flow before 3 min, which is associated with transient hypocalcemia, followed by hyperemia within 30 min associated with an increased venous–arterial Ca gradient and hypercalcemia (Boelkins et al., 1976).

More recently, dose-dependent relaxant effects of PTH and PTHrP were demonstrated on oviduct motility in bird and mouse oviductal tissues *in vitro* (Francis et al., 2003). Effects of N-terminal PTH and PTHrP on the spontaneous *in vitro* contractility of oviductal smooth muscle using tissues from egg-laying Japanese quail (10–15 h post ovulation) and 4–9 days pregnant mouse uterus were investigated. Myometrial tissues from both species contracted vigorously for several hours when incubated in organ baths in De Jalon's solution. Contractions were enhanced in high (1.2–2.5 mM) compared with low (0.1–0.5 mM) Ca-containing media. Bovine PTH (1–34 amide) (bPTH (1–34)), human PTH (1–34) (hPTHrP (1–34) amide), and hPTHrP (1–40) caused similar concentration-related inhibition of contractions in media containing 1.2 mM Ca over dose ranges of 10^{-9}–10^{-7} M, whereas C-terminal hPTHrP (107–139) was devoid of activity. Responses to bPTH (1–34) were inhibited by both nonselective and selective neuronal nitric oxide synthase (NOS) inhibitors, which were more effective in inhibiting bPTH (1–34)-induced relaxation in the absence of L-arginine compared with in the presence of 1 mM L-arginine (an NOS substrate) in the incubation media. This suggests that relaxant responses to N-terminal PTH and PTHrP peptides are well conserved in oviductal and uterine tissues from avian and mammalian species. The results also suggest that NO may be responsible for mediating relaxant activities of these peptides in a variety of uterine tissues.

25.5 PARATHYROID HORMONE RELATED PEPTIDES

PTHrP is found in the circulation of certain patients with malignancy-associated hypercalcemia and also as the predominant peptide in fetal mammals. PTH and PTHrP genes are located on different chromosomes but share common organizational features, and the peptides show much structural homology, suggesting a common ancestral gene. PTHrP exists in three known forms, ranging in size from 139 to 173 amino acids, as the result of alternative gene splicing. It has a wide spectrum of actions in mammals, many in common with PTH, ranging from stimulation of osteoclastic bone resorption to enhancement of placental mineral transport (Mallette, 1991). PTHrP is expressed in a variety of tissues in chick embryos, this molecule having a highly conservative structural homology with the human sequence, the first 21 residues being identical (Schermer et al., 1991). It is also expressed in the isthmus and shell gland of the hen's oviduct, where it has a potential role as a local modulator of vascular smooth muscle tension and shell gland motility during the egg-laying cycle (Thiede et al., 1991). They followed the expression of PTHrP in the shell gland at different times in the laying cycle and found levels of PTHrP to transiently increase as the egg moves through the oviduct, gradually returning to basal levels during the calcification period. PTHrP mRNA and immunoreactive PTHrP were localized to the shell gland serosal and smooth muscle layer, suggesting that the peptide may modulate vascular smooth muscle activity. In support of this hypothesis, synthetic chicken PTHrP (1–34) NH_2 was found to relax the resting tension of isolated shell gland blood vessels in a dose-dependent manner (Thiede et al., 1991). Together, these data indicate that expression of the PTHrP gene in the avian oviduct is both temporally and spatially regulated during the egg-laying cycle and that PTHrP functions as an autocrine–paracrine modulator of shell gland smooth muscle activity. The vasorelaxant properties of N-terminal fragments of PTHrP support a role for this molecule in the temporal increase in blood flow to the shell gland during egg calcification.

PTHrP (1–34) was tested in the chick hypercalcemic assay and showed only slight PTH agonist activity with respect to either plasma Ca levels or ^{45}Ca clearance. In femur, the peptide caused a substantial decrease in ^{45}Ca uptake, while in calvarium the opposite effect apparently occurred (Dacke et al., 1993a). Fenton et al. (1994) studied carboxyl-terminal peptides from PTHrP for their effect on bone resorption by embryonic chick osteoclasts. Basal bone resorption was directly inhibited by chicken and human PTHrP (107–139) and by the pentapeptide PTHrP (107–111). The number of resorption pits and total area resorbed per bone slice were reduced by PTHrP (107–139), while resorption stimulated by hPTH (1–34) in co-cultured chicken osteoclasts and osteoblasts was also inhibited by cPTHrP (107–139). These results suggest that c-terminal PTHrP may be a paracrine regulator of bone cell activity.

Schermer et al. (1994) studied functional properties of synthetic chicken PTHrP fragments in chicken renal plasma membrane, 19-day chick embryonic bone cells, canine renal plasma membranes, and rat osteosarcoma cells (UMR-106-H5). The biologic activities of human PTHrP (1–34) and bovine PTH (1–34) are remarkably similar despite marked sequence divergence in their primary binding domain, residues 25–34. In both avian and mammalian systems, the binding affinity of [36Tyr]cPTHrP(1–36)NH_2 is half that of hPTHrP(1–34)NH_2. Potencies of [36Tyr] cPTHrP(1–36)NH_2 and hPTHrP(1–34)NH_2 for activation of adenylate cyclase were similar in canine renal membranes and chick bone cells. In UMR-106 cells and chicken renal membranes, the potency of [36Tyr]cPTHrP(1–36)NH_2 for activation of adenylate cyclase was half that of [36Tyr] hPTHrP(1–36)NH_2. Binding of ^{125}I-[36Tyr]cPTHrP(1–36) NH_2 to chick bone cells and chicken renal membranes was completely displaced by bPTH(1–34) and hPTHrP(1–34) NH_2, suggesting no evidence for a distinct chicken PTHrP N-terminal receptor.

Medill et al. (2001) studied PTHrP expression in the epiphyseal growth plate of chick bones and concluded that most of the PTHrP present originates in the growth plate itself. Furthermore, the presence of large amounts of PTHrP protein in the hypertrophic zone supports the concept that PTHrP has other paracrine functions in addition to regulating chondrocyte differentiation.

25.6 CALCITONIN GENE-RELATED PEPTIDE AND AMYLIN

CGRP is a 37 amino acid neuropeptide derived from the same gene as CT, both belonging to the amylin superfamily (Rosenfield et al., 1983). Unlike CT, which is expressed mainly within the avian ultimobranchial body, CGRP is found within the central and peripheral nervous system, including the chicken retina (Kiyama et al., 1985), carotid body (Kameda, 1989), spinal cord motor neurons of developing embryos and posthatch chicks (Villar et al., 1988), and bone neurones (Bjurholm et al., 1985), where its distribution corresponds with that of substance P (Mallette, 1991). It is also expressed within the dense distributions of peptidergic neurones found within the chick ultimobranchial gland (Kameda, 1991). CGRP molecules from human, rat, cow, salmon, and chicken have been sequenced; the structure is well conserved with around 90% structural homology between chicken and human CGRPs, compared with only 50% between respective CT molecules (Zaidi et al., 1990a). CGRP presence in bone neurones, coupled with the interaction of this peptide with osteoclastic CT receptors (Goltzman and Mitchell, 1985; Zaidi et al., 1987), suggests a paracrine role, modulating bone turnover and hence Ca homeostasis. A further member of the CT–CGRP family, amylin, a peptide from pancreatic islet cells, is the most potent non-CT peptide thus far discovered, at least in mammalian assay systems (Zaidi et al., 1990a). CGRP has several putative physiological functions. It is a potent vasodilator and is implicated in central and peripheral neurotransmission and modulation (Zaidi et al., 1990a). Its role in bone and Ca metabolism is also recognized. CGRP shares the acute hypocalcemic effects of CT, albeit at around 1000-fold less potency in rodents, in inhibiting bone resorption, stimulating cAMP production in mouse calvaria, and inhibiting neonatal rat osteoclastic spreading (Zaidi et al., 1990a). In the rabbit, *in vivo* CGRP causes transient hypocalcemia followed by a more sustained hypercalcemia (Tippins et al., 1984), and the same paper reported *in vivo* hypercalcemic effects of the peptide in chicks. These findings in chicks were repeated and extended (Bevis et al., 1990; Ancill et al., 1991). Bevis et al. (1990) gave details of comparative dose response curves for CGRP and PTH in chicks, the two peptides being approximately equipotent on a molar basis. In Ancill et al. (1991), we investigated the effects of the peptide on a simultaneously injected ^{45}Ca label in the plasma. Intravenous injection of rat CGRP gave a rapid hypercalcemic response that was sustained for at least 1 h. This was most evident in nonfasted chicks, while fasted chicks gave a hypophosphatemic response and increased plasma ^{45}Ca clearance. The effect of CGRP on ^{45}Ca uptake into the chick skeleton was subsequently investigated (Ancill et al., 1991). Both rat and chicken CGRP sequences caused transient (10 min) increases in ^{45}Ca uptake into a variety of bone types, these responses being well developed in fasted chicks but absent in fed ones; the greatest were in calvariae and vertebrae. Furthermore, with low doses of CGRP, reversal of the response was noticed in the calvaria but not in other bone types, while in fed animals this was the only response seen, again in the calvaria. These findings indicate that CGRP may have a variety of effects on bone and Ca metabolism in the chick that involve acute effects on net movement of Ca into and out of the skeleton. However, while consistent with changes in plasma ^{45}Ca clearance, they seem too transient to account for them and would not account for the hypercalcemic responses found, although some alternative target site such as the kidney (Zaidi et al., 1990b) may be responsible for these.

Mixed bone cell cultures obtained by sequential collagenase-trypsin digestion of neonate chick and rodent calvariae respond to CGRP with increases in cAMP formation (Michelangeli et al., 1989). This effect was not the result of an action as a weak CT agonist, since in most instances no CT effect was observed. They concluded that chick, rat, and mouse bones contain osteoblast-rich populations that respond specifically to CGRP by a rise in cAMP.

Another member of the CT family, amylin, is co-secreted with insulin from the pancreatic beta cells and can act as an osteoblast mitogen and as an inhibitor of bone resorption in mice and humans. No effect of amylin on Ca metabolism was found in chicks *in vivo* (Dacke et al., 1993a). However, Guzel et al. (2009) studied effects of amylin on bone development and egg production in hens. Hens injected subcutaneously with amylin at a 75 μg/kg dose every other day and along with control hens were slaughtered at 14, 16, 18, and 20 week of age, and serum and bone parameters were compared. In amylin-treated hens, bone Ca levels increased, whereas serum Ca decreased versus controls. Amylin also increased the cortical width of tibiotarsae in hens. Eggshell thickness was thicker in the treated groups than in controls. These results suggest that amylin may stimulate the bone and eggshell quality by increasing Ca uptake from the bloodstream and may influence the sustainability of yield in hens.

25.7 PROSTAGLANDINS AND OTHER FACTORS

Prostaglandins (PGs) are lipids that act in an autocrine–paracrine manner and mainly utilize G protein–coupled receptors that span the plasma membrane for signaling. PGs are synthesized by most cell types in the body by the action of the inducible enzyme cyclooxygenase-2 (COX2). COX2 is induced by many systemic and local factors, including cytokines (IL1, IL6, and TNFα), growth factors (TGFα, TGFβ, fibroblast growth factor-2 (FGF2), and BMP2), systemic hormones (PTH, 1,25 $(OH)_2D_3$), calcium, fluid shear and mechanical loading (Pilbeam et al., 2008). PGs are not stored but are synthesized locally as needed. They are rapidly metabolized by passage through the lungs. PGs can have both anabolic and catabolic effects. Reviews of the roles of PGs and other eicosanoids in Ca metabolism in vertebrates (including avians) include Dacke (1989) and Pilbeam et al. (2008). PG effects on mammalian bone cells are similar to those of PTH in that they stimulate cAMP production, cause transient increases in Ca^{++} influx, stimulate carbonic anhydrase synthesis, release lysosomal enzymes, and may inhibit collagen synthesis. They also elicit morphological responses in osteoclasts and osteoblasts similar to those with other osteolytic agents (Dacke, 1989). It was demonstrated that a stable methylated PGE_2 analog, 16,16-dimethyl PGE_2, is hypercalcemic when injected into chicks, the response being more profound than that in mammals (Kirby and Dacke, 1983), while indomethacin, a drug that interferes with PG synthesis, produces hypocalcemia in egg-laying chickens (Hammond and Ringer, 1978) and quail (Dacke and Kenny, 1982). In chickens, this is accompanied by a considerable delay in oviposition and thicker eggshells. PGE_2 and other eicosanoids are powerful stimulators of bone resorption (Dacke, 1989). The effects of CT, PTH, and PGE_2 on cAMP production were studied in osteoclast-rich cultures derived from avian medullary bone and the long bones of newborn rats. PGE_2 increased cAMP production in both types of osteoclasts, suggesting essentially similar mechanisms (Nicholson et al., 1986; Arnett and Dempster, 1987). In addition to PTH, CGRP, and possibly CT, PGs can acutely influence Ca exchange between avian blood and bone compartments. We previously published a simple model (Shaw et al., 1989) in which the rapid effects of PTH and PGE_2 on skeletal ^{45}Ca uptake could be explained in terms of cAMP-mediated inhibitions of outwardly directed Ca^{++} pumps located in the membranes of bone-lining cells. Such a model has its disadvantages; for example, it would be metabolically wasteful since the intracellular free Ca level would have to remain undisturbed in the face of a high rate of bidirectional transcellular Ca^{++} flux. Because of the transient nature of PGs, *in vitro* studies require careful interpretation since PGs added to cultured cells will be present over much longer periods of time than in the *in vivo* situation. However, clear-cut results from *in vivo* studies are also difficult to obtain due to the presence of several types of cells in a given locale. Adding to the complexity is that regulation by PGs occurs at multiple levels and utilizes multiple receptors. PGE_2 is the most widely utilized prostaglandin in bone. It is synthesized by all bone cell types and by hematopoietic cells. How PGE_2 is released from cells is largely unknown, with the exception that avian osteocytes have been shown to release PGE_2 through hemi-channels (Cherian et al., 2005). There are four known PGE_2 receptors in bone. In osteoblasts, PGE_2 signals through the cAMP–PKA pathway. There is evidence for a strong direct effect of PGE_2 on osteoclast activity with signaling through cAMP–PKA (Dacke, 1989; Mano et al., 2000; Okamoto et al., 2004). In addition, indirect regulation of osteoclasts can occur by increased expression of RANKL–OPG in osteoblasts and osteocytes that have been stimulated by PGE_2.

Osteopontin (OPN) is a noncollagenous protein found in many tissues, and it has a number of functions. In bone, it is secreted into matrix by osteoblasts, where it appears to have a number of roles. For example, it appears ahead of the mineralization front (Roach, 1994) and may be involved in mineralization as well as in hemopoietic stem cell regulation (Haylock and Nilsson, 2006). OPN is secreted by osteoclasts and has been postulated to be an autocrine factor that stimulates osteoclast motility (Chelliah and Hruska, 2003). It binds to αvβ3 integrin found in podosomes of osteoclasts and thereby forms a seal around the resorption pit (Pfaff and Jurdic, 2001). OPN is speculated to mediate disuse-induced osteoclastic resorption. In a study by Gross et al. (2005), OPN was found to be upregulated in immobilized long bones of turkeys.

BMPs are expressed by a number of cells, including mesenchymal cells, chondrocytes, osteoblasts, preosteoclasts,

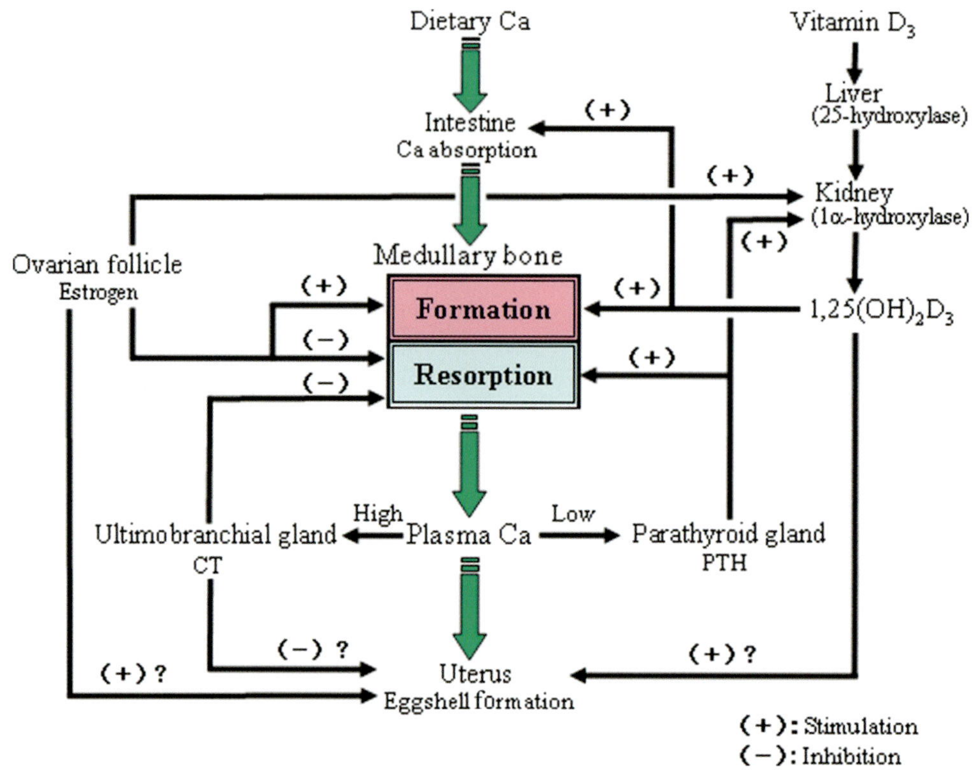

FIGURE 25.7 **Summary of Ca metabolism in egg-laying birds.** Transfer of Ca between diet, plasma, and medullary bone and across the oviduct may be influenced by a variety of recognized and putative hormonal factors (see text). *Modified after Sugiyama and Kusuhara (2001).*

osteoclasts, and endothelial cells. BMPs are produced locally, signaling through transmembrane receptors. The signal is relayed by Smads (Sma- and Mad-related proteins) in the cytoplasm and, upon binding to Smad4, enters the nucleus to stimulate cell proliferation. BMPs can have both anabolic and catabolic effects, depending on cross-communication among other signaling pathways and various other factors. The osteogenic function of BMPs, notably BMP2 and BMP7, has been studied extensively, as a review by Kaymia and Mishina (2011) describes. BMP2 has been shown to stimulate osteoclastic resorption through binding to BMP receptors on preosteoclasts and osteoclasts (Kaneko et al., 2000; Itoh et al., 2001) as well as hypertrophic chondrocytes in chicks and mice (Usui et al., 2008). Signaling in these latter cases is through the RANK–RANKL pathway.

25.8 CONCLUSIONS

Figure 25.7 summarizes our present understanding of the regulation of avian Ca and bone metabolism.

Within the last two decades, it has become apparent that a variety of nonclassical factors can influence bone and Ca metabolism in birds as well as in mammals. Avian responses are often more explicit than equivalent ones in mammals, for example the hypercalcemic responses to PGs. Other factors such as PTHrP and CGRP are represented in birds, while novel factors such as the RANK–RANKL–OPG system are also now recognized in birds and the next few years can be expected to provide fertile ground for new research on their role in bone and Ca metabolism in general, to which avian models are likely to make an important contribution. The interactions of these novel agents with more classical hormones in avian bone remain to be elucidated. Their interactions with gonadal steroids in forming and maintaining avian medullary bone may prove a particularly rewarding area for future studies. Medullary bone represents the most overtly estrogen sensitive of all vertebrate bone types. It forms in male birds dosed with estrogens within a matter of days, and this process can be blocked by the simultaneous administration of anti-estrogenic compounds such as tamoxifen (Ohashi et al., 1987; Williams et al., 1999). Upon cessation of estrogen treatment, medullary bone resorbs just as rapidly. Medullary bone represents an excellent rapidly responding model for studies of the effects of anti-osteoporotic drugs such as bisphosphonates. Preliminary studies indicate that the bisphosphonate alendronate can protect structural bone and inhibit medullary bone formation if given to hens before the commencement of egglay. When given during egglay, the drug reduces medullary bone volume and, at higher doses, eggshell quality (Thorp et al., 1993). It is anticipated that this model will become exploited to a much greater extent than hitherto.

REFERENCES

Abbas, S.K., Fox, J., Care, A.D., 1985. Calcium homeostasis in the chick embryo. Comp. Biochem. Physiol. 81B, 975–980.

Abdulrahim, S.M., Patel, M.B., Mcginnis, J., 1979. Effects of vitamin-D3 and D3-metabolites on production parameters and hatchability of eggs. Poult. Sci. 58, 858–863.

Abe, E., Tanabe, R., Suda, T., Yoshiki, S., 1979. Circadian rhythm of 1 alpha, 25-dihydroxyvitamin D_3 production in egg-laying hens. Biochem. Biophys. Res. Commun. 88, 500–507.

Agarwala, N., Gay, C.V., 1992. Specific binding of parathyroid hormone to living osteoclasts. J. Bone Miner. Res. 7, 531–539.

Ancill, A.K., Bascal, Z.A., Whitaker, G., Dacke, C.G., 1991. Calcitonin gene-related peptide promotes transient radiocalcium uptake into chick bone *in vivo*. Exp. Physiol. 76, 143–146.

Arai, T., Sugiyama, T., Kusuhara, S., 1996. Immunohistochemical studies of Ca^{2+}-ATPase and carbonic anhydrase II in the shell gland of egg-laying hens. Jpn. Poult. Sci. 33, 371–376.

Arkle, S., Wormstone, I.M., Bascal, Z.A., Dacke, C.G., 1994. Estimation of intracellular calcium activity in confluent monolayers of primary cultures of quail medullary bone osteoclasts. Exp. Physiol. 79, 975–982.

Arnett, T.R., Dempster, D.W., 1987. A comparative study of disaggregated chick and rat osteoclasts *in vitro:* effects of calcitonin and prostaglandins. Endocrinology 120, 602–608.

Ascenzi, A., Francois, C., Bocciarellids, S., 1963. On the bone induced by estrogens in birds. J. Ultrastruct. Res. 8, 491–505.

Baksi, S.N., Kenny, A.D., 1977. Vitamin D_3 metabolism in immature Japanese quail: effects of ovarian hormones. Endocrinology 101, 1216–1220.

Bar, A., 2008. Calcium homeostasis and vitamin D metabolism and expression in strongly calcifying laying birds. Comp. Biochem. Physiol. A. Mol. Integr. 151, 477–490.

Bar, A., Hurwitz, S., 1979. The interaction between dietary calcium and gonadal hormones in their effect on plasma calcium, bone, 25-hydroxycholecalciferol-1-hydroxylase and duodenal calcium-binding protein, measured by a radio immunoassay in chicks. Endocrinology 104, 1455–1460.

Bar, A., Cohen, A., Edelstein, S., Shemesh, M., Montecuccoli, G., Hurwitz, S., 1978. Involvement of cholecalciferol metabolism in birds in the adaptation of calcium absorption to the needs during reproduction. Comp. Biochem. Physiol. 59B, 245–249.

Bar, A., Striem, S., Vax, E., Talpaz, H., Hurwitz, S., 1992. Regulation of calbindin turnover in intestine and shell gland of the chicken. Am. J. Physiol. 262, R800–R805.

Bar, A., Vax, E., Striem, S., 1999. Relationships among age, eggshell thickness and vitamin D metabolism and its expression in the laying hen. Comp. Biochem. Physiol. A. Mol. Integr. Physiol. 123, 147–154.

Bascal, Z.A., Alam, A.S.M.T., Zaidi, M., Dacke, C.G., 1994. Effect of raised extracellular calcium on cell spread area in quail medullary bone osteoclasts. Exp. Physiol. 79, 15–24.

Bascal, Z.A., Moonga, B.S., Zaidi, M., Dacke, C.G., 1992. Osteoclasts from medullary bone of egg-laying Japanese quail do not possess the putative calcium receptor. Exp. Physiol. 77, 501–504.

Beck, M.M., Hansen, K.K., 2004. Role of estrogen in avian osteoporosis. Poult. Sci. 83, 200–206.

Berg, C., Blomqvist, A., Holm, L., Brandt, I., Brunström, B., Ridderstråle, Y., 2004. Embryonic exposure to oestrogen causes eggshell thinning and altered shell gland carbonic anhydrase expression in the domestic hen. Reproduction 128, 455–461.

Bevis, P.J.R., Zaidi, M., MacIntyre, I., 1990. A dual effect of calcitonin gene-related peptide on plasma calcium levels in the chick. Biochem. Biophys. Res. Commun. 169, 846–850.

Bjurholm, A., Kreicbergs, A., Brodin, E., Schultzberg, M., 1985. Substance P- and CGRP-immunoreactive nerves in bone. Peptides 9, 165–171.

Blom, J., Lilja, C., 2004. A comparative study of growth, skeletal development and eggshell composition in some species of birds. J. Zool. 262, 361–369.

Bloom, M., Bloom, M.A., McLean, F.C., 1941. Calcification and ossification. Medullary bone changes in the reproductive cycle of female pigeons. Anat. Rec. 81, 433–475.

Bloom, M.A., McLean, F.C., Bloom, W., 1942. Calcification and ossification. The formation of medullary bone in male and castrate pigeons under influence of sex hormones. Anat. Rec. 83, 99–120.

Boelkins, J.N., Mazurkiewicz, M., Mazur, P.E.P.E., Mueller, W.J., 1976. Changes in blood flow to bone during hypocalcemic and hypercalcemic phases of the response to parathyroid hormone. Endocrinology 98, 403–412.

Boissy, P., Destaing, O., Jurdic, P., 2001. RANKL induces formation of avian osteoclasts from macrophages but not from macrophage polykaryons. Biochem. Biophys. Res. Commun. 288, 340–346.

Bonewald, L.F., 2011. The amazing osteocyte. J. Bone Miner. Res. 26, 229–238 (Review).

Bonewald, L.F., Kneissel, M., Johnson, M. (Eds.), 2013. The osteocyte. Bone, vol. 54. pp. 181–306.

Bonucci, E., Gherardi, G., 1975. Histochemical and electron microscopy investigations on medullary bone. Cell Tissue Res. 163, 81–97.

Boivin, G., Mesguich, P., Pike, J.W., Bouillon, R., Meunier, P.J., Haussler, M.R., Dubois, P.M., Morel, G., 1987. Ultrastructural immunocytochemical localization of endogenous 1,25-dihydroxyvitamin D_3 and its receptors in osteoblasts and osteocytes from neonatal mouse and rat calvaria. Bone Miner. 3, 125–136.

Bouizar, Z., Khattab, M., Taboulet, J., Rostene, W., Milhaud, G., Treilhou-Lahille, F., Moukhtar, M.S., 1989. Distribution of renal calcitonin binding sites in mammalian and non mammalian vertebrates. Gen. Comp. Endocrinol. 76, 364–370.

Bronner, F., 1996. Calcium metabolism in birds and mammals. In: Dacke, C.G., Danks, J., Flik, G., Caple, I. (Eds.), Comparative Endocrinology of Calcium Regulating Hormones. J. Endocrinol. Ltd, Bristol, pp. 131–135.

Bronner, F., Stein, W.D., 1992. Modulation of bone-calcium binding sites regulates plasma calcium: an hypothesis. Calcif. Tissue Int. 50, 483–489.

Brown, E.M., 1991. Extracellular Ca^{2+} sensing, regulation of parathyroid cell function and role of Ca^{2+} and other ions as extracellular (first) messengers. Physiol. Rev. 71, 371–411.

Brubaker, K.D., Gay, C.V., 1999. Depolarization of osteoclast plasma membrane potential by 17β-estradiol. J. Bone Miner. Res. 14, 1861–1866.

Candlish, J.K., Taylor, T.G., 1970. The response-time to the parathyroid hormone in the laying fowl. J. Endocrinol. 48, 143–144.

Candlish, J.K., Holt, F.J., 1971. The proteoglycans of fowl cortical and medullary bone. Comp. Biochem. Physiol. B 40, 283–293.

Castillo, L., Tanaka, Y., Wineland, M.J., Jowsey, J.O., Deluca, H.F., 1979. Production of 1,25-dihydroxyvitamin D_3 and formation of medullary bone in the egg laying hen. Endocrinology 104, 1598–1606.

Chegwidden, W.R., Carter, N.D., 2000. Introduction to the carbonic anhydrases. In: Chegwidden, W.R., Carter, N.D., Edwards, Y.H. (Eds.), The Carbonic Anhydrases: New Horizons. Birkhäuser Verlag, Berlin, pp. 13–28.

Chellaiah, M.A., Hruska, K.A., 2003. The integrin alpha(v)beta(3) and CD44 regulate the actions of osteopontin on osteoclast motility. Calcif. Tissue Int. 72, 197–205.

Cherian, P.P., Siller-Jackson, A.J., Gu, S., Wang, X., Bonewald, L.F., Sprague, E., Jiang, J.X., 2005. Mechanical strain opens connexin 43 hemichannels in osteocytes: a novel mechanism for the release of prostaglandin. Mol. Biol. Cell. 16, 3100–3106.

Clark, N.B., 1991. Renal clearance of phosphate and calcium in vitamin D-deficient chicks: effect of calcium loading, parathyroidectomy, and parathyroid hormone administration. J. Exp. Zool. 259, 188–195.

Clark, N.B., Sasayama, Y., 1981. The role of parathyroid hormone on renal excretion of calcium and phosphate in the Japanese quail. Gen. Comp. Endocrinol. 43, 234–241.

Clark, N.B., Wideman, R.F., 1977. Renal excretion of phosphate and calcium in parathyroidectomised starlings. Am. J. Physiol. 233, F138–F144.

Copp, D.H., Kline, L.W., 1989. Calcitonin. In: In: Pang, P.K.T., Schreibman, M.P. (Eds.), Vertebrate Endocrinology: Fundamentals and Biomedical Implications, vol. 3. Academic Press, New York, pp. 79–103.

Cory, W.A., 1981. A specific, high affinity binding protein for 1,25-dihydroxyvitamin D in the chick oviduct shell gland. Biochem. Biophys. Res. Commun. 93, 285–292.

Couch, J.R., 1955. Cage layer fatigue. Feed Age 5, 55–57.

Dacke, C.G., 1989. Eicosanoids, steroids and miscellaneous hormones. In: In: Pang, P.K.T., Schreibman, M.P. (Eds.), Vertebrate Endocrinology: Fundamentals and Biomedical Implications, vol. 3. Academic Press, New York, pp. 171–210.

Dacke, C.G., 1979. Calcium Regulation in Sub-mammalian Vertebrates. Academic Press, London.

Dacke, C.G., Kenny, A.D., 1973. Avian bioassay for parathyroid hormone. Endocrinology 92, 463–470.

Dacke, C.G., Kenny, A.D., 1982. Prostaglandins: are they involved in avian calcium homeostasis? In: Scanes., C.G., Ottinger., M.A., Kenny., A.D., Balthazart., J., Cronshaw, J., Chester-Jones, I. (Eds.), Aspects of Avian Endocrinology: Practical and Theoretical Implications. Texas Tech Univ. Press, Lubbock, pp. 255–262.

Dacke, C.G., Ancill, A.K., Whitaker, G., Bascal, Z.A., 1993a. Calcitrophic peptides and rapid calcium fluxes into chicken bone *in vivo*. In: Sharp, P.J. (Ed.), Avian Endocrinology. J. Endocrinol. Ltd, Bristol, pp. 239–248.

Dacke, C.G., Arkle, S.A., Cook, D.J., Wormstone, I.M., Jones, S., Zaidi, M., Bascal, Z.A., 1993b. Medullary bone and avian calcium metabolism. J. Exp. Biol. 184, 63–88.

Dacke, C.G., 2000. The parathyroids, calcitonin, and vitamin D. In: Whittow, G.C. (Ed.), Avian Physiology. Academic Press, London, pp. 473–488.

Dacke, C.G., Boelkins, J.N., Smith, W.K., Kenny, A.D., 1972. Plasma calcitonin levels in birds during the ovulation cycle. J. Endocrinol. 54, 369–370.

Dacke, C.G., Shaw, A.J., 1987. Studies of the rapid effects of parathyroid hormone and prostaglandins on ^{45}Ca uptake into chick and rat bone *in vivo*. J. Endocrinol. 115, 369–377.

de Matos, R., 2008. Calcium metabolism in birds. Vet. Clin. North Am. Exot. Anim. Pract. 11, 59–82.

de Vernejoul, M.C., Horowitz, M., Demignon, J., Neff, L., Baron, R., 1988. Bone resorption by isolated chick osteoclasts in culture is stimulated by murine spleen cell supernatant fluids (osteoclast activating factor) and inhibited by calcitonin and prostaglandin E_2. J. Bone Miner. Res. 3, 69–80.

Dempster, D.W., Murrils, F.J., Horbert, W.R., Arnett, T.R., 1987. Biological activity of chicken calcitonin: effects on neonatal rat and embryonic chick osteoclasts. J. Bone Miner. Res. 2, 443–448.

Deng, Y.F., Chen, X.X., Zhou, Z.L., Hou, J.F., 2010. Letrozole inhibits the osteogenesis of medullary bone in prelay pullets. Poult. Sci. 89, 917–923.

Dudas, P.L., Villalobos, A.R., Gocek-Sutterlin, G., Laverty, G., Renfro, J.L., 2002. Regulation of transepithelial phosphate transport by PTH in chicken proximal tubule epithelium. Am. J. Physiol. Regul. Integr. Comp. Physiol. 282, R139–R146.

Eliam, M.C., Baslé, M., Bouziar, B., Bielakoff, J., Moukhtar, M., de Vernejoul, M.C., 1988. Influence of blood calcium on calcitonin receptors in isolated chick osteoclasts. J. Endocrinol. 119, 243–248.

Eliam-Cisse, M.C., Taboulet, J., Bielakoff, J., Lasmoles, F., de-Vernejoul, M.C., Treilhou-Lahille, F., 1993. Influence of calcium and vitamin D deficient diet on calcitonin gene expression in the ultimobranchial cells of the developing chicken. Gen. Comp. Endocrinol. 89, 195–205.

Farley, J.R., Tarbaux, N.M., Hall, S.L., Linkhart, T.A., Baylink, D.J., 1988. The anti-bone-resorptive agent calcitonin also acts *in vitro* to directly increase bone formation and bone cell proliferation. Endocrinology 123, 159–167.

Fenton, A.J., Martin, T.J., Nicholson, G.C., 1994. Carboxyl-terminal parathyroid hormone-related protein inhibits bone resorption by isolated chicken osteoclasts. J. Bone Miner. Res. 9, 515–519.

Fleming, R.H., McCormack, H.A., McTeir, L., Whitehead, C.C., 1998. Medullary bone and humeral breaking strength in laying hens. Res. Vet. Sci. 64, 63–67.

Fleming, R.H., McCormack, H.A., McTeir, L., Whitehead, C.C., 2006. Relationships between genetic, environmental and nutritional factors influencing osteoporosis in laying hens. Br. Poult. Sci. 47, 742–755.

Francis, M., Arkle, M., Martin, L., Butler, T.M., Cruz, M.C., Opare-Aryee, G., Dacke, C.G., Brown, J.F., 2003. Relaxant effects of parathyroid hormone and parathyroid hormone-related peptides on oviduct motility in birds and mammals: possible role of nitric oxide. Gen. Comp. Endrocrinol. 133, 243–251.

Fraser, D.R., Emtage, J.S., 1976. Vitamin D in the avian egg: Its molecular identity and mechanism of incorporation into the yolk. Biochem. J. 160, 671–682.

Fullmer, C.S., Bridak, M.E., Bar, A., Wasserman, R.H., 1976. The purification of calcium binding protein from the uterus of laying hens. Proc. Roy. Soc. Exptl. Biol. Med. 152, 237–241.

Gardiner, E.M., Esteban, L.M., Fong, C., Allison, S.J., Flanagan, J.L., Kouzmenko, A.P., Eisman, J.A., 2004. Vitamin D receptor B1 and exon 1d: functional and evolutionary analysis. J. Steroid. Biochem. Mol. Biol. 89–90, 233–238.

Gay, C.V., 1988. Avian bone resorption at the cellular level. CRC Crit. Rev. Poult. Biol. 1, 197–210.

Gay, C.V., 1996. Avian bone turnover and the role of bone cells. In: Dacke, C.G., Danks, J., Flik, G., Caple, I. (Eds.), Comparative Endocrinology of Calcium Regulating Hormones. J. Endocrinol. Ltd, Bristol, pp. 113–121.

Gay, C.V., Weber, J.A., 2000. Regulation of differentiated osteoclasts. Crit. Rev. Eukaryot. Gene Expr. 10, 213–230.

Gay, C.V., Faleski, E.J., Schraer, H., Schraer, R., 1974. Localization of carbonic anhydrase in avian gastric mucosa, shell gland and bone by immunohistochemistry. J. Histochem. Cytochem. 22, 819–825.

Gay, C.V., Gilman, V.R., Sugiyama, T., 2000. Perspectives on osteoblast and osteoclast function. Poult. Sci. 79, 1005–1008.

Gay, C.V., Kief, N.L., Bekker, P.J., 1993. Effect of estrogen on acidification in osteoclasts. Biochem. Biophys. Res. Commun. 192, 1251–1259.

Goldstein, D.L., Reddy, V., Plaga, K., 1999. Second messenger production in avian medullary nephron segments in response to peptide hormones. Am. J. Physiol. 276, R847–R854.

Goltzman, D., Mitchell, J., 1985. Interaction of calcitonin and calcitonin gene-related peptide at receptor sites in target tissues. Science 227, 1343–1345.

Gregory, N.G., Wilkins, L.J., 1989. Broken bones in domestic fowl: handling and processing damage in end-of-lay battery hens. Br. Poult. Sci. 30, 555–562.

Gross, T.S., King, K.A., Rabaia, N.A., Pathare, P., Srinivasan, S., 2005. Upregulation of osteopontin by osteocytes deprived of mechanical loading or oxygen. J. Bone Miner. Res. 20, 250–256.

Gu, J.H., Liu, J.D., Shen, Y., Liu, Z.P., 2009. Effects of RANKL, osteoprotegerin, calcium and phosphorous on survival and activation of Muscovy duck osteoclasts *in vitro*. Vet. J. 181, 321–325.

Guerreiro, P.M., Renfro, J.L., Power, D.M., Canario, A.V., 2007. The parathyroid hormone family of peptides: structure, tissue distribution, regulation, and potential functional roles in calcium and phosphate balance in fish. Am. J. Physiol. Regul. Integr. Comp. Physiol. 292, R679–R696.

Guzel, S., Gunes, N., Yildiz, H., Yilmaz, B., 2009. Effects of amylin on bone development and egg production in hens. Poult. Sci. 88, 1719–1724.

Hall, M.R., Kief, N.L., Gilman, V.R., Gay, C.V., 1994. Surface binding and clearance of calcitonin by avian osteoclasts. Comp. Biochem. Physiol. Comp. Physiol. 108, 59–63.

Hammond, R.W., Ringer, R.K., 1978. Effect of indomethacin on the laying cycle, plasma calcium and shell thickness in the laying hen. Poult. Sci. 57, 1141.

Hansen, K.K., Kittok, R.J., Sarath, G., Toombs, C.F., Caceres, N., Beck, M.M., 2003. Estrogen receptor-alpha populations change with age in commercial laying hens. Poult. Sci. 82, 1624–1629.

Harrington, E.K., Roddy, G.W., West, R., Svoboda, K.K., 2007. Parathyroid hormone/parathyroid hormone-related peptide modulates growth of avian sternal cartilage via chondrocytic proliferation. Anat. Rec. (Hoboken) 290, 155–167.

Harrison, J.R., Clark, N.B., 1986. Avian medullary bone in organ culture: effects of vitamin D metabolites on collagen synthesis. Calcif. Tissue Int. 39, 35–43.

Hauschka, P.V., Lian, J.B., Cole, D.E., Gundberg, C.M., 1989. Osteocalcin and matrix Gla protein; vitamin K-dependent protein in bone. Physiol. Rev. 69, 990–1034.

Haussler, M.R., Whitfield, G.K., Kaneko, I., Haussler, C.A., Hsieh, D., Hsieh, J.C., Jurutka, P.W., 2013. Molecular mechanisms of vitamin D action. Calcif. Tissue Int. 92, 77–98.

Haylock, D.N., Nilsson, S.K., 2006. Osteopontin: a bridge between bone and blood. Br. J. Haematol. 134, 467–474.

Henry, H.L., Norman, A.W., 1984. Vitamin D: metabolism and biological actions. Ann. Rev. Nutr. 4, 493–520.

Hiyama, S., Sugiyama, T., Kusuhara, S., 1998. Characteristics of the bone matrix formed by osteogenic cells isolated from embryonic chick calvaria and hen medullary bones. Jpn. Poult. Sci. 35, 228–233.

Hiyama, S., Sugiyama, T., Kusuhara, S., Uchida, T., 2009. Evidence for the expression of estrogen receptors in osteogenic cells isolated from hen medullary bone. Acta Histochem. 111, 501–507.

Holick, M.F., 1989. Phylogenetic and evolutionary aspects of vitamin D from phytoplankton to humans. In: In: Pang, P.K.T., Schreibman, M.P. (Eds.), Vertebrate Endocrinology, vol. 3. Academic Press, New York, pp. 7–43.

Hou, H.C., 1931. Relation of preen gland of birds to rickets. III. Site of activation during irradiation. Chin. J. Physiol. 5, 11–18.

Hou, L., Hou, J., Yao, J., Zhou, Z., 2011. Effects of osteoprotegerin from transfection of pcDNA3.1(+)/chOPG on bioactivity of chicken osteoclasts. Acta Vet. Scand. 53. article 21.

Hsu, H., Lacey, D.L., Dunstan, C.R., Solovyev, I., Colombero, A., Timms, E., Tan, H.L., Elliott, G., Kelley, M.J., Sarosi, I., Wang, L., Xia, X.Z., Elliott, R., Chiu, L., Black, T., Scully, S., Capparelli, C., Morony, S., Shimamoto, G., Bass, M.B., Boyle, W.J., 1999. Tumor necrosis factor receptor family member RANK mediates osteoclast differentiation and activation induced by osteogrotogerin ligand. Proc. Natl. Acad. Sci. U. S. A. 96, 3540–3545.

Hunter, S.J., Schraer, H., 1981. *In vitro* synthesis of the glycosaminoglycans in estrogen-induced medullary bone in Japanese quail. Arch. Biochem. Biophys. 210, 647–652.

Hurwitz, S., 1989a. Calcium homeostasis in birds. Vitamins Horm. 45, 173–221.

Hurwitz, S., 1989b. Parathyroid hormone. In: In: Pang, P.K.T., Schreibman, M.P. (Eds.), Vertebrate Endocrinology: Fundamentals and Biomedical Implications, vol. 3. Academic Press, New York, pp. 45–77.

Hurwitz, S., 1992. The role of vitamin D in poultry bone biology. In: In: Whithead, C.C. (Ed.), Bone Biology and Skeletal Disorders in Poultry, vol. 23. Carfax Publishing Co., Abington, pp. 87–102.

Ieda, T., Takahashi, T., Saito, N., Yasuoka, T., Kawashima, M., Shimada, K., 2000. Changes in parathyroid hormone-related peptide receptor binding in the shell gland of laying hens (*Gallus domesticus*) during the oviposition cycle. Gen. Comp. Endocrinol. 117, 182–188.

Imamura, T., Sugiyama, T., Kusuhara, S., 2006. Expression and localization of estrogen receptors α and β mRNA in medullary bone of laying hens. Anim. Sci. J. 77, 223–229.

Ito, M.B., Schraer, H., Gay, C.V., 1985. The effects of calcitonin, parathyroid hormone and prostaglandin E_2 on cyclic AMP levels of isolated osteoclasts. Comp. Biochem. Physiol. 81A, 653–657.

Itoh, K., Udagawa, N., Katagiri, T., Iemura, S., Ueno, N., Yasuda, H., Higashio, K., Quinn, J.M., Gillespie, M.T., Martin, T.J., Suda, T., Takahashi, N., 2001. Bone morphogenetic protein 2 stimulates osteoclast differentiation and survival supported by receptor activator of nuclear factor-kappaB ligand. Endocrinology 142, 3656–3662.

Johnson, A.L., 2000. Reproduction in the female. In: Whittow, G.C. (Ed.), Avian Physiology, fifth ed. Academic Press, New York, pp. 569–596.

Johnston, P.M., Comar, C.L., 1955. Distribution and contribution of calcium from the albumen, yolk and shell to the developing chick embryo. Am. J. Physiol. 183, 365–370.

Jonchere, V., Brionne, A., Gautron, J., Nys, Y., 2012. Identification of uterine ion transporters for mineralisation precursors of the avian eggshell. BMC Physiol. 12. 10.

Kameda, Y., 1989. Distribution of CGRP-, somatostatin-, galanin-, VIP-, and substance P-immunoreactive nerve fibers in the chicken carotid body. Cell. Tissue Res. 257, 623–629.

Kameda, Y., 1991. Immunocytochemical localisation and development of multiple kinds of neuropeptides and neuroendocrine proteins in the chick ultimobranchial gland. J. Comp. Neurol. 304, 373–386.

Kamioka, H., Honjo, T., Takano-Yamamoto, T., 2001. A three-dimensional distribution of osteocyte processes revealed by the combination of confocal laser scanning microscopy and differential interference contrast microscopy. Bone 28, 145–149.

Kamiya, N., Mishina, Y., 2011. New insights on the roles of BMP signaling in bone: a review of recent mouse genetic studies. Biofactors 37, 75–82.

Kaneko, H., Arakawa, T., Mano, H., Kaneda, T., Ogasawara, A., Nakagawa, M., Toyama, Y., Yabe, Y., Kumegawa, M., Hakeda, Y.,

2000. Direct stimulation of osteoclastic bone resorption by bone morphogenetic protein (BMP)-2 and expression of BMP receptors in mature osteoclasts. Bone 27, 479–486.

Karlsson, O., Lilja, C., 2008. Eggshell structure, mode of development and growth rate in birds. Zoology 111, 494–502.

Kenny, A.D., 1976. Vitamin D metabolism: physiological regulation in egg laying Japanese quail. Am. J. Physiol. 230, 1609–1615.

Kenny, A.D., 1981. Intestinal Absorption of Calcium and Its Regulation. CRS Press, Boca Raton.

Kenny, A.D., 1986. Parathyroid and ultimobranchial glands. In: Sturkie, P.D. (Ed.), Avian Physiology, fourth ed. Springer-Verlag, New York, pp. 466–478.

Kenny, A.D., Dacke, C.G., 1974. The hypercalcaemic response to parathyroid hormone in Japanese quail. J. Endocrinol. 62, 15–23.

Khosla, S., 2001. Minireview: the OPG/RANKL/RANK system. Endocrinology 142, 5050–5055 (Review).

Khosla, S., Demay, M., Pines, M., Hurwitz, S., Potts Jr., J.T., Kronenberg, H.M., 1988. Nucleotide sequence of cloned cDNAs encoding chicken preproparathyroid hormone. J. Bone Miner. Res. 3, 689–698.

Kim, W.K., Bloomfield, S.A., Sugiyama, T., Ricke, S.C., 2012. Concepts and methods for understanding bone metabolism in laying hens. World's Poult. Sci. J. 68, 71–82.

Kirby, G.C., Dacke, C.G., 1983. Hypercalcaemic responses to 16,16-dimethyl prostaglandin E_2, a stable prostaglandin E_2 analogue, in chicks. J. Endocrinol. 99, 115–122.

Kiyama, H., Katayama, Y., Hillyard, C.J., Girgis, S., MacIntyre, I., Emson, P.C., Tohyama, M., 1985. Occurrence of calcitonin gene-related peptide in the chicken amacrine cells. Brain Res. 327, 367–369.

Knott, L., Bailey, A.J., 1998. Collagen biochemistry of avian bone: comparison of bone type and skeletal site. Br. Poult. Sci. 40, 371–379.

Koch, J., Wideman, R.F., Buss, E.G., 1984. Blood ionic calcium response to hypocalcemia in the chicken induced by ethylene glycol-bis-(B-aminoethylether)-N,N'-tetraacetic acid: role of the parathyroids. Poult. Sci. 63, 167–171.

Krzysik-Walker, S.M., Ocón-Grove, O.M., Maddineni, S.B., Hendricks 3rd, G.L., Ramachandran, R., 2007. Identification of calcitonin expression in the chicken ovary: influence of follicular maturation and ovarian steroids. Biol. Reprod. 77, 626–635.

Kusuhara, S., 1982. Influence of parathyroid hormone on the resorption of chicken medullary bone developed by the administration of sexual hormones. Jpn. J. Zootech. Sci. 53, 332–337.

Kusuhara, S., Schraer, H., 1982. Cytology and autoradiography of estrogen-induced differentiation of avian endosteal cells. Calcif. Tissue Int. 34, 352–358.

Lacey, D.L., Timms, E., Tan, H.-L., Kelley, M.J., Dunstan, C.R., Burgess, T., Elliott, R., Colombero, A., Elliott, G., Scully, S., Hsu, H., Sullivan, J., Hawkins, N., Davy, E., Capparelli, C., Eli, A., Qian, Y.X., Kaufman, S., Sarosi, I., Shalhoub, V., Senaldi, G., Guo, J., Delaney, J., Boyle, W.J., 1998. Osteoprotegerin ligand is a cytokine that regulates osteoclast differentiation and activation. Cell 93, 165–176.

Laverty, G., Clark, N.B., 1989. The kidney. In: In: Pang, P.K.T., Schreibman, M.P. (Eds.), Vertebrate Endocrinology: Fundamentals and Biomedical Implications, vol. 3. Academic Press, New York, pp. 277–317.

Laverty, G., McWilliams, C., Sheldon, A., Arnason, S.S., 2003. PTH stimulates a Cl(-)-dependent and EIPA-sensitive current in chick proximal tubule cells in culture. Am. J. Physiol. Ren. Physiol. 284, F987–F995.

Lee, A.H., Werning, S., 2008. Sexual maturity in growing dinosaurs does not fit reptilian growth models. Proc. Natl. Acad. Sci. U. S. A. 105, 582–587.

Lian, J.B., Glimcher, M.J., Roufosse, A.H., Hauscha, P.V., Gallop, P.M., Cohen-Solal, L., Reit, B., 1982. Alterations in the gamma-carboxyglutamic acid and osteocalcin concentrations in vitamin D deficient chick bone. J. Biol. Chem. 257, 4999–5003.

Liel, Y., Kraus, S., Levy, J., Shany, S., 1992. Evidence that estrogens modulate activity and increase the number of 1,25-dihydroxyvitamin D receptors in osteoblast-like cells. Endocrinology 130, 2597–2601.

Limm, S.K., Gardella, T., Thompson, A., Rosenberg, J., Keutmann, H., Potts, J., Kronenberg, H., Nussbaum, S., 1991. Full-length chicken parathyroid hormone. Biosynthesis in *Escherichia coli* and analysis of biologic activity. J. Biol. Chem. 266, 3709–3714.

Lorcher, K., Newesely, H., 1969. Calcium carbonate (calcite) as a separate phase besides calcium phosphate apatite in medullary bone of laying hens. Calcif. Tissue Res. 3, 358–362.

Malgaroli, A., Meldolesi, J., Zambonin-Zallone, A., Teti, A., 1989. Control of cytosolic free calcium in rat and chicken osteoclasts. The role of extracellular calcium and calcitonin. J. Biol. Chem. 264, 14342–14347.

Mallette, L.E., 1991. The parathyroid polyhormones: new concepts in the spectrum of peptide hormone action. Endocr. Rev. 12, 110–117.

Mano, M., Arakawa, T., Mano, H., Nakagawa, M., Kaneda, T., Kaneko, H., Yamada, T., Miyata, K., Kiyomura, H., Kumegawa, M., Hakeda, Y., 2000. Prostaglandin E2 directly inhibits bone-resorbing activity of isolated mature osteoclasts mainly through the EP4 receptor. Calcif. Tissue Int. 67, 85–92.

Masuyama, R., Stockmans, I., Torrekens, S., Van Looveren, R., Maes, C., Carmeliet, P., Bouillon, R., Carmeliet, G., 2006. Vitamin D receptor in chondrocytes promotes osteoclastogenesis and regulates FGF23 production in osteoblasts. J. Clin. Invest. 116, 3150–3159.

May, L.G., Gilman, V.R., Gay, C.V., 1993. Parathyroid regulation of avian osteoclasts. In: Sharp, P.J. (Ed.), Avian Endocrinology. J. Endocrinol. Ltd, Bristol, pp. 227–237.

Mayel-Afshar, S., Lane, S.M., Lawson, D.E.M., 1988. Relationship between the levels of calbindin synthesis and calbindin mRNA. J. Biol. Chem. 263, 4355–4361.

Medill, N.J., Praul, C.A., Ford, B.C., Leach, R.M., 2001. Parathyroid hormone-related peptide expression in the epiphyseal growth plate of the juvenile chicken: evidence for the origin of the parathyroid hormone-related peptide found in the epiphyseal growth plate. J. Cell Biochem. 80, 504–511.

Merke, J., Klaus, G., Hugel, U., Waldherr, R., Ritz, E., 1986. No 1,25-dihydroxyvitamin D_3 receptors on osteoclasts of calcium-deficient chickens despite demonstratable receptors on circulating monocytes. J. Clin. Invest. 77, 312–314.

Michelangeli, V.P., Fletcher, A.E., Allan, E.H., Nicholson, G.C., Martin, T.J., 1989. Effects of calcitonin gene-related peptide on cyclic AMP formation in chicken, rat, and mouse bone cells. J. Bone Miner. Res. 4, 269–272.

Miller, S.C., 1977. Osteoclast cell-surface changes during the egg-laying cycle in Japanese quail. J. Cell Biol. 75, 104–118.

Miller, S.C., 1978. Rapid activation of the medullary bone osteoclast cell surface by parathyroid hormone. J. Cell Biol. 76, 615–618.

Miller, S.C., 1981. Osteoclast cell-surface specializations and nuclear kinetics during egg-laying in Japanese quail. Am. J. Anat. 162, 35–43.

Miller, S.C., 1992. Calcium homeostasis and mineral turnover in the laying hen. In: Whitehead, C.C. (Ed.), Bone Biology and Skeletal Disorders in Poultry. Carfax Publishing Co., Oxfordshire, pp. 103–116.

Miller, S.C., Bowman, B.M., 1981. Medullary bone osteogenesis following estrogen administration to mature male Japanese quail. Dev. Biol. 87, 52–63.

Miyauchi, A., Hruska, K.A., Greenfield, E.M., Duncan, R., Alverez, J., Barattolo, R., Colucci, S., Zambonin-Zallone, A., Teitlelbaum, S.L., Teti, A., 1990. Osteoclast cytosolic calcium regulated by voltage-gated calcium channels and extracellular calcium, controls podosome assembly and bone resorption. J. Cell Biol. 111, 2543–2552.

Miyaura, C., Nagata, N., Suda, T., 1981. Failure to demonstrate the stimulatory effect of calcitonin on cyclic AMP accumulation in avian bone *in vitro*. Endocrinol. Jpn. 28, 403–408.

Nakashima, T., Hayashi, M., Takayanagi, H., 2012. New insights into osteoclastogenic mechanisms. Trends Endocrin. Met. 23, 582–590.

Nakashima, T., Hayashi, M., Fukunaga, T., Kurata, K., Oh-Hora, M., Feng, J.Q., Bonewald, L.F., Kodama, T., Wutz, A., Wagner, E.F., Penninger, J.M., Takayanagi, H., 2011. Evidence for osteocyte regulation of bone homeostasis through RANKL expression. Nat. Med. 17, 1231–1234.

Navickis, R.J., Katzenellenbogen, B.S., Nalbandov, A.V., 1979. Effects of the sex steroid and vitamin D_3 on calcium-binding protein in the chick shell gland. Biol. Reprod. 21, 1153–1162.

Nicholson, G.C., Livesey, S.A., Moseley, J.M., Martin, T.J., 1986. Actions of calcitonin, parathyroid hormone, and prostaglandin E_2 on cyclic AMP formation in chicken and rat osteoclasts. J. Cell. Biochem. 31, 229–241.

Nicholson, G.C., Moseley, J.M., Sexton, P.M., Martin, T.J., 1987. Chicken osteoclasts do not possess calcitonin receptors. J. Bone Miner. Res. 2, 53–59.

Norman, A.W., 1987. Studies on the vitamin D endocrine system in the avian. J. Nutr. 117, 797–807.

Norman, A.W., Hurwitz, S., 1993. The role of the vitamin D endocrine system in avian bone biology. J. Nutr. 123 (Suppl. 2), 310–316 (Review).

Nys, Y., 1993. Regulation of plasma 1,25-$(OH)_2D_3$, of osteocalcin and of intestinal and uterine calbindin in hens. In: Sharp, P.J. (Ed.), Avian Endocrinology. J. Endocrinol. Ltd, Bristol, pp. 345–357.

Nys, Y., N'Guyen, T.M., Williams, J., Etches, R.J., 1986. Blood levels of ionised calcium, inorganic phosphorus, 1,25-dihydroxycholecalciferol and gonadal hormones in hens laying hard shelled or shell-less eggs. J. Endocrinol. 111, 151–157.

Nys, Y., Van Baelen, H., Bouillon, R., 1992. Plasma 1,25-dihydroxycholecalciferol and its free index are potentiated by the ovulation dependent factors and shell formation induced hypocalcemia in the laying hen. Domest. Anim. Endocrin. 9, 37–47.

Ogawa, H., Takahashi, T., Kuwayama, T., Kawashima, M., 2003. Presence of calcitonin receptors in shell gland of the guineafowl and changes in binding property during an oviposition cycle. Poult. Sci. 82, 1302–1306.

Ogawa, H., Takahashi, T., Yasuoka, T., Kuwayama, T., Tanaka, K., Kawashima, M., 2000. Parathyroid hormone receptor binding property in the shell gland of oviduct of the guineafowl during an oviposition cycle. Poult. Sci. 79, 575–579.

Ohashi, T., Kusuhara, S., 1993. Immunoelectron microscopic detection of estrogen target cells in the bone marrow of estrogen-treated male Japanese quail. Bone Miner. 20, 31–39.

Ohashi, T., Kusuhara, S., Ishida, K., 1987. Effects of oestrogen and anti-oestrogen on the cells of the endosteal surface of male Japanese quail. Br. Poult. Sci. 28, 727–732.

Ohashi, T., Kusuhara, S., Ishida, K., 1990a. Electron microscopic observations of osteoblasts and osteoclasts on the medullary bone of tamoxifen-treated hens. Jpn. Poult. Sci. 27, 122–127.

Ohashi, T., Kusuhara, S., Ishida, K., 1990b. Immunohistochemical demonstration of estrogen receptors in the medullary bone of Japanese quail. Jpn. J. Zootech. Sci. 61, 919–923.

Ohashi, T., Kusuhara, S., Ishida, K., 1991. Estrogen target cells during the early stage of medullary bone osteogenesis: immunohistochemical detection of estrogen receptors in osteogenic cells of estrogen-treated male Japanese quail. Calcif. Tissue Int. 49, 124–127.

Okamoto, F., Kajiya, H., Fukushima, H., Jimi, E., Okabe, K., 2004. Prostaglandin E2 activates outwardly rectifying Cl(−) channels via a cAMP-dependent pathway and reduces cell motility in rat osteoclasts. Am. J. Physiol. Cell. Physiol. 287, C114–C124.

Oursler, M.J., Landers, J.P., Riggs, B.L., Spelsberg, T.C., 1993. Oestrogen effects on osteoblasts and osteoclasts. Ann. Med. 25, 361–371.

Pandala, S., Gay, C.V., 1990. Effects of parathyroid hormone, calcitonin, and dibutyryl–cyclic AMP on osteoclast area in cultured chick tibiae. J. Bone Miner. Res. 5, 701–705.

Parsons, A.H.., Combs Jr., G.F., 1981. Blood ionized calcium cycles in the chicken. Poult. Sci. 60, 1520–1524.

Parsons, J.A., Reit, B., Robinson, C.J., 1973. Bioassay for parathyroid hormone using chicks. Endocrinology 92, 454–462.

Pellegrino, E.D., Biltz, R.M., 1970. Calcium carbonate in medullary bone. Calcif. Tissue Res. 6, 168–171.

Pfaff, M., Jurdic, P., 2001. Podosomes in osteoclast-like cells: structural analysis and cooperative roles of paxillin, proline-rich tyrosine kinase 2 (Pyk2) and integrin alphaVbeta3. J. Cell Sci. 114, 2775–2786.

Pike, J.W., Alvarado, R.H., 1975. Ca^{2+}-Mg^{+}-activated ATPase in the shell gland of Japanese quail (*Coturnix coturnix japonica*). Comp. Biochem. Physiol. B 51, 119–125.

Pilbeam, C.C., Choudhary, S., Blackwell, K., Raisz, L.G., 2008. Prostaglandins and bone metabolism. In: Bilezikian, J.P., Raisz, L.G., Martin, T.J. (Eds.), Principles of Bone Biology. Elsevier Inc., San Diego, pp. 1235–1271.

Pines, M., Granot, I., Hurwitz, S., 1990. Cyclic AMP-dependent inhibition of collagen synthesis in avian epiphysial cartilage cells: effect of chicken and human parathyroid hormone and parathyroid hormone-related peptide. Bone Miner. 9, 23–34.

Pinheiro, P.L., Cardoso, J.C., Gomes, A.S., Fuentes, J., Power, D.M., Canário, A.V., 2010. Gene structure, transcripts and calciotropic effects of the PTH family of peptides in *Xenopus* and chicken. BMC Evol. Biol. 10, 373.

Pinheiro, P.L., Cardoso, J.C., Power, D.M., Canário, A.V., 2012. Functional characterization and evolution of PTH/PTHrP receptors: insights from the chicken. BMC Evol. Biol. 12, 110.

Phillps, J.G., Butler, J., Sharp, P.J., 1985. Physiological Strategies in Avian Biology. Blackie, Glasgow. pp. 112–139.

Rifkin, B.R., Auszmann, J.M., Kleckner, A.P., Vernillo, A.T., Fine, A.S., 1988. Calcitonin stimulates cAMP accumulation in chick osteoclasts. Life Sci. 42, 799–804.

Roach, H.I., 1994. Why does bone matrix contain non-collagenous proteins? The possible roles of osteocalcin, osteonectin, osteopontin and bone sialoprotein in bone mineralisation and resorption. Cell Biol. Int. 18, 617–628.

Rodriguez, L., Tu, C., Cheng, Z., Chen, T.H., Bikle, D., Shoback, D., Chang, W., 2005. Expression and functional assessment of an alternatively spliced extracellular Ca^{2+}-sensing receptor in growth plate chondrocytes. Endocrinology 146, 5294–5303.

Romanoff, A.L., Romanoff, A.J., 1963. The Avian Egg, second ed. Wiley, New York.

Rosenfield, M.G., Mermod, J.J., Amara, S.G., Swanson, L.W., Sawchenko, P.E., Rivier, J., Vale, W.W., Evans, R.M., 1983. Production of a novel peptide encoded by the calcitonin gene via tissue specific RNA processing. Nature 304, 129–135.

Ross, F.P., 2008. Osteoclast biology and bone resorption. In: Rosen, C.J. (Ed.), Primer on the Metabolic Bone Diseases and Disorders of Mineral Metabolism. Am. Soc. for Bone and Mineral Research, Washington, DC, pp. 16–22.

Roth, S.I., Schiller A.L., 1976. Comparative anatomy of the parathyroid glands. In: Aurbach, G.D. (Ed.), Parathyroid Gland, In: Greep, R.O., Astwood, E.B. (Eds.), Handbook of Physiology Section 7, American Physiological Society, Washington, DC, pp. 281–311.

Russell, J., Sherwood, L.M., 1989. Nucleotide sequence of the DNA complementary to avian (chicken) preproparathyroid hormone mRNA and the deduced sequence of the hormone precursor. Mol. Endocrinol. 3, 325–331.

Schermer, D.T., Chan, S.D.H., Bruce, R., Nissenson, R.A., Wood, W.I., Strewler, G.J., 1991. Chicken parathyroid hormone-related protein and its expression during embryologic development. J. Bone Miner. Res. 6, 149–155.

Schermer, D.T., Bradley, M.S., Bambino, T.H., Nissenson, R.A., Strewler, G.J., 1994. Functional properties of a synthetic chicken parathyroid hormone-related protein 1-36 fragment. J. Bone Miner. Res. 9, 1041–1046.

Schraer, R., Schraer, H., 1965. Changes in metal distribution of the avian oviduct during the ovulation cycle. Proc. Soc. Exp. Biol. Med. 119, 937–942.

Schweitzer, M.H., Elsey, R.M., Dacke, C.G., Horner, J.R., Lamm, E.T., 2007. Do egg-laying crocodilian (*Alligator mississippiensis*) archosaurs form medullary bone? Bone 40, 1152–1158.

Schweitzer, M.H., Wittmeyer, J.L., Horner, J.R., 2005. Gender-specific reproductive tissue in ratites and *Tyrannosaurus rex*. Science 308, 1456–1460.

Sedrani, S.H., Taylor, T.G., Akbtar, M., 1981. The regulation of 25-hydroxylase-calciferol metabolism in the kidney of the Japanese quail by sex hormones and by parathyroid extract. Gen. Comp. Endocrinol. 44, 514–523.

Shahabi, N.A., Norton, H.W., Nalbandov, A.V., 1975. Steroid levels in follicles and the plasma of hens during the ovulatory cycle. Endocrinology 96, 962–968.

Shaw, A.J., Dacke, C.G., 1989. Cyclic nucleotides and the rapid inhibitions of bone ^{45}Ca uptake in response to bovine parathyroid hormone and 16,16-dimethyl prostaglandin E2 in chicks. Calcif. Tissue Int. 44, 209–213.

Shaw, A.J., Dacke, C.G., 1985. Evidence for a novel inhibition of calcium uptake into chick bone in response to bovine parathyroid hormone (1-34) or 16,16-dimethyl prostaglandin E_2 *in vivo*. J. Endocrinol. 105, R5–R8.

Shaw, A.J., Whitaker, G., Dacke, C.G., 1989. Kinetics of rapid ^{45}Ca uptake into chick skeleton *in vivo*: effects of microwave fixation. Q. J. Exp. Physiol. 74, 907–915.

Shodono, M., Nakamura, T., Tanabe, Y., Wakabayashi, K., 1975. Simultaneous determinations of oestradiol-17 beta, progesterone and luteinizing hormone in the plasma during the ovulatory cycle of the hen. Acta Endocrinol. Copenh. 78, 565–573.

Simkiss, K., 1967. Calcium in Reproductive Physiology. Chapman and Hall Ltd, London. pp. 155–197.

Simonet, W.S., Lacey, D.L., Dunstan, C.R., Kelley, M., Chang, M.S., Lüthy, R., Nguyen, H.Q., Wooden, S., Bennett, L., Boone, T., Shimamoto, G., DeRose, M., Elliott, R., Colombero, A., Tan, H.L., Trail, G., Sullivan, J., Davy, E., Bucay, N., Renshaw-Gegg, L., Hughes, T.M., Hill, D., Pattison, W., Campbell, P., Sander, S., Van, G., Tarpley, J., Derby, P., Lee, R., Boyle, W.J., 1997. Osteoprotegerin: a novel secreted protein involved in the regulation of bone density. Cell 89, 309–319.

Singh, R., Joyner, C.J., Peddie, M.J., Taylor, T.G., 1986. Changes in the concentrations of parathyroid hormone and ionic calcium in the plasma of laying hens during the egg cycle in relation to dietary deficiencies of calcium and vitamin D. Gen. Comp. Endocrinol. 61, 20–28.

Sommerville, B.A., Fox, J., 1987. Changes in renal function of the chicken associated with calcitonin and parathyroid hormone. Gen. Comp. Endocrinol. 66, 381–386.

Stamford, M., 2006. Calcium metabolism. In: In: Harrison, G.J., Lightfoot, T.L. (Eds.), Clinical Avian Medicine, vol. 1. Spix Publishing, Inc., Palm Beach, FL, pp. 141–151.

Suda, T., Takahashi, F., Takahashi, N., 2012. Bone effects of vitamin D – discrepancies between *in vivo* and *in vitro* studies. Arch. Biochem. Biophys. 523, 22–29 (Review).

Sugawara, Y., Kamioka, H., Honjo, T., Tezuka, K., Takano-Yamamoto, T., 2005. Three-dimensional reconstruction of chick calvarial osteocytes and their cell processes using confocal microscopy. Bone 36, 877–883.

Sugiyama, T., Kusuhara, S., 1993. Ultrastructural changes of osteoclasts on hen medullary bone during the egg-laying cycle. Br. Poult. Sci. 34, 471–477.

Sugiyama, T., Kusuhara, S., 1994a. Effect of parathyroid hormone on osteoclasts in organ-cultured medullary bone. Jpn. Poult. Sci. 31, 392–399.

Sugiyama, T., Kusuhara, S., 1994b. The kinetics of actin filaments in osteoclasts on chicken medullary bone during the egg-laying cycle. Bone 15, 351–353.

Sugiyama, T., Kusuhara, S., 1996a. Morphological changes of osteoclasts on hen medullary bone during the egg-laying cycle and their regulation. In: Dacke, C.G., Danks, J., Flik, G., Caple, I. (Eds.), Comparative Endocrinology of Calcium Regulating Hormones. J. Endocrinol. Ltd, Bristol, pp. 149–150.

Sugiyama, T., Kusuhara, S., 1996b. Effects of parathyroid hormone and calcitonin on kinetics of actin filaments in chicken osteoclast of cultured medullary bone. Anim. Sci. J. 67, 526–532.

Sugiyama, T., Kusuhara, S., 2001. Avian calcium metabolism and bone function. Asian Australas. J. Anim. Sci. 14, 82–90.

Sugiyama, T., Kusuhara, S., Gay, C.V., 1999. Parathyroid hormone and estrogen effects on adhesion of chicken medullary bone osteoclasts. In: Danks, J., Dacke, C., Flik, G., Gay, C. (Eds.), Calcium Metabolism: Comparative Endocrinology. BioScientifica Ltd, Bristol, pp. 107–111.

Sugiyama, T., Ohashi, T., Kusuhara, S., 1993. Inhibition of osteoclastic bone resorption by calcitonin in the cultured medullary bone of laying hens. Jpn. Poult. Sci. 30, 16–23.

Sugiyama, T., Kikuchi, H., Hiyama, S., Nishizawa, K., Kusuhara, S., 2007. Expression and localisation of calbindin D28k in all intestinal segments of the laying hen. Br. Poult. Sci. 48, 233–238.

Takahashi, N., Shinki, T., Abe, E., Morinchi, N., Yamaguchi, A., Yoshiki, S., Suda, T., 1983. The role of vitamin D in the medullary bone formation in egg-laying Japanese quails and in immature chicks treated with sex hormone. Calcif. Tissue Int. 35, 465–471.

Taylor, T.G., Dacke, C.G., 1984. Calcium metabolism and its regulation. In: Freeman, B.M. (Ed.), Physiology and Biochemistry of the Domestic Fowl, vol. 5. Academic Press, London, pp. 125–170.

Teti, A., Rizzoi, R., Zambonin-Zallone, A., 1991. Parathyroid hormone binding to cultured avian osteoclasts. Biochem. Biophys. Res. Commun. 174, 1217–1222.

Theman, T.A., Collins, M.T., 2009. The role of the calcium-sensing receptor in bone biology and pathophysiology. Curr. Pharm. Biotechnol. 10, 289–301 (Review).

Thiede, M.A., Harm, S.C., McKee, R.L., Grasser, W.A., Duong, L.T., Leach, R.M., 1991. Expression of the parathyroid hormone-related protein gene in the avian oviduct: potential role as a local modulator of vascular smooth muscle tension and shell gland motility during the egg-laying cycle. Endocrinology 129, 1958–1966.

Thorp, B.H., Wilson, S., Rennie, S., Solomons, S., 1993. The effect of a bisphosphonate on bone volume and eggshell structure in the hen. Avian Pathol. 22, 671–682.

Tippins, J.R., Morris, H.R., Panico, M., Etienne, T., Bevis, P., Girgis, S., MacIntyre, I., Azria, M., Attinger, M., 1984. The myotropic and plasma-calcium modulating effects of calcitonin gene-related peptide (CGRP). Neuropeptides 4, 425–434.

Tregear, G.W., van Rietschoten, J., Greene, E., Keutmann, H.T., Niall, H.D., Reit, B., Parsons, J.A., Potts, J.T., 1973. Bovine parathyroid hormone: minimum chain length of synthetic peptide required for biological activity. Endocrinology 93, 1349–1353.

Tuan, R.S., Ono, T., 1986. Regulation of extraembryonic calcium mobilization by the developing chick embryo. J. Embryol. Exp. Morphol. 97, 63–74.

Turner, R.T., Bell, N.H., Gay, C.V., 1993. Evidence that estrogen binding-sites are present in bone-cells and mediate medullary bone-formation in Japanese-quail. Poult. Sci. 72, 728–740.

Turner, R.T., Riggs, B.L., Spelsberg, T.C., 1994. Skeletal effects of estrogen. Endocr. Rev. 15, 275–300.

Usui, M., Xing, L., Drissi, H., Zuscik, M., O'Keefe, R., Chen, D., Boyce, B.F., 2008. Murine and chicken chondrocytes regulate osteoclastogenesis by producing RANKL in response to BMP2. J. Bone Miner. Res. 23, 314–325.

van de Velde, J.P., Loveridge, N., Vermeiden, J.P., 1984a. Parathyroid hormone responses to calcium stress during eggshell calcification. Endocrinology 115, 1901–1904.

van de Velde, J.P., Vermeiden, J.P., Bloot, A.M., 1985. Medullary bone matrix formation, mineralization, and remodeling related to the daily egg-laying cycle of Japanese quail: a histological and radiological study. Bone 6, 321–327.

van de Velde, J.P., Vermeiden, J.P., Touw, J.J., Veldhuijzen, J.P., 1984b. Changes in activity of chicken medullary bone cell populations in relation to the egg-laying cycle. Metab. Bone Dis. Relat. Res. 5, 191–193.

Villar, M.J., Huchet, M., Hokfelt, T., Changeux, J.P., Fahrenkrug, J., Brown, J.C., 1988. Existence and coexistence of calcitonin gene-related peptide, vasoactive intestinal polypeptide- and somatostatin-like immunoreactivities in spinal cord motoneurons of developing embryos and post-hatch chicks. Neurosci. Lett. 86, 114–118.

Wang, Y., Hou, J.-F., Zhou, Z.-L., 2008. Chicken receptor activator of nuclear factor activator of nuclear factor-kappa B ligand induces formation of chicken osteoclasts from bone marrow cells and also directly activates mature osteoclasts. Poult. Sci. 87, 2344–2349.

Wellnhoser, P., 1991. The Illustrated Encyclopaedia of Pterosaurs. Salamander Books Ltd, London.

Whitehead, C.C., 2004. Overview of bone biology in the egg-laying hen. Poult. Sci. 83, 193–199.

Whitehead, C.C., Fleming, R.H., 2000. Osteoporosis in cage layers. Poult. Sci. 79, 1033–1041.

Williams, D.C., Paul, D.C., Herring, J.R., 1991. Effects of antiestrogenic compounds on avian medullary bone formation. J. Bone Miner. Res. 6, 1249–1256.

Wilson, S., Duff, S.R., 1990. Morphology of medullary bone during the egg formation cycle. Res. Vet. Sci. 48, 216–220.

Wilson, S., Duff, S.R., 1991. Effects of vitamin or mineral deficiency on the morphology of medullary bone in laying hens. Res. Vet. Sci. 50, 216–221.

Wilson, S., Thorp, B.H., 1998. Estrogen and cancellous bone loss in the fowl. Calcif. Tissue Int. 62, 506–511.

Xiong, J., Onal, M., Jilka, R.L., Weinstein, R.S., Manolagas, S.C., O'Brien, C.A., 2011. Matrix-embedded cells control osteoclast formation. Nat. Med. 17, 1235–1241.

Yamamoto, T., Nagai, H., 1994. Tartrate-resistant acid phosphatase accumulated in the matrix of developing medullary bone induced by estrogen treatment of male Japanese quail. J. Bone Miner. Res. 9, 1153–1157.

Yamamoto, T., Nagaoka, N., Hirata, A., Nakamura, H., Inoue, M., Kawai, M., Ikegame, M., 2005. Ultrastructural and immunohistochemical studies of medullary bone calcification, with special reference to sulphated glycosaminoglycans. J. Electron Microsc. (Tokyo) 54, 29–34.

Yamamoto, T., Nakamura, H., Tsuji, T., Hirata, A., 2001. Ultracytochemical study of medullary bone calcification in estrogen injected male Japanese quail. Anat. Rec. 264, 25–31.

Yamamoto, T., Ozawa, H., Nagai, H., 1985. Histochemical studies of Ca-ATPase, succinate and NAD^+-dependent isocitrate dehydrogenases in the shell gland of laying Japanese quails: with special reference to calcium-transporting cells. Histochemistry 83, 221–226.

Yarden, N., Lavelin, I., Genina, O., Hurwitz, S., Diaz, R., Brown, E.M., Pines, M., 2000. Expression of calcium-sensing receptor gene by avian parathyroid gland *in vivo*: relationship to plasma calcium. Gen. Comp. Endocrinol. 117, 173–181.

Yasuda, H., Shima, N., Nakagawa, N., Mochizuki, S.I., Yano, K., Fujise, N., Sato, Y., Goto, M., Yamaguchi, K., Kuriyama, M., Kanno, T., Murakami, A., Tsuda, E., Morinaga, T., Higashio, K., 1998a. Identity of osteoclastogenesis inhibitory factor (OCIF) and osteoprotegerin (OPG): a mechanism by which OPG/OCIF inhibits osteoclastogenesis *in vitro*. Endocrinology 39, 1329–1337.

Yasuda, H., Shima, N., Nakagawa, N., Yamaguchi, K., Kinosaki, M., Mochizuki, S., Tomoyasu, A., Yano, K., Goto, M., Murakami, A., Tsuda, E., Morinaga, T., Higashio, K., Udagawa, N., Takahashi, N., Suda, T., 1998b. Osteoclast differentiation factor is a ligand for osteoprotegerin/osteoclastogenesis-inhibitory factor and is identical to TRANCE/RANKL. Proc. Natl. Acad. Sci. U. S. A. 95, 3597–3602.

Yasuoka, T., Takahashi, T., Tanaka, K., Kawashima, M., 2001. Developmental changes of parathyroid hormone/parathyroid hormone-related peptide and calcitonin receptor binding properties in the chicken calvaria and kidney. Poult. Sci. 80, 1231–1235.

Ypes, D.L., Ravesloot, J.H., Buisman, H.P., Nijweide, P.J., 1988. Voltage activated ionic channels and conductance's in embryonic chick osteoblast cultures. J. Membr. Biol. 101, 141–150.

Zaidi, M., Chambers, T.J., Gaines-Das, R.E., Morris, H.R., MacIntyre, I., 1987. A direct action of human calcitonin gene-related peptide on isolated human osteoclasts. J. Endocrinol. 155, 511–518.

Zaidi, M., Datta, H.K., Bevis, P.J.R., 1990b. Kidney: a target organ for calcitonin gene-related peptide. Exp. Physiol. 75, 27–32.

Zaidi, M., Moonga, B., Bevis, J.R., Bascal, Z.A., Breimer, M.D., 1990a. The calcitonin gene peptides: biology and clinical relevance. Crit. Rev. Clin. Lab. Sci. 28, 109–174.

Zambonin-Zallone, A., Teti, A., Primavera, M.V., 1982. Isolated osteoclasts in primary culture: first observations on structure and survival time in culture media. Anat. Embryol. 165, 405–413.

Zuscik, M.J., O'Keefe, R.J., Gunter, T.E., Puzas, J.E., Schwarz, E.M., Rosier, R.N., 2002. Parathyroid hormone-related peptide regulation of chick tibial growth plate chondrocyte maturation requires protein kinase A. J. Orthop. Res. 20, 1079–1090.

Chapter 26

Adrenals

Rocco V. Carsia
Department of Cell Biology, Rowan University School of Osteopathic Medicine, Stratford, NJ, USA

26.1 ANATOMY

26.1.1 Gross Anatomy, Blood Supply, and Innervation

The paired adrenal glands are located anterior and medial to the cephalic lobes of the kidneys (Figure 26.1). Although their intraclass shape is highly variable, they are generally flat and lie close together and are closely applied to the dorsal aorta and posterior (caudal) vena cava. The gland is enclosed in a highly vascularized but extremely friable connective tissue capsule (Assenmacher, 1973; Holmes and Phillips, 1976; Chester Jones and Phillips, 1986). In the domestic chicken (*Gallus gallus domesticus*), the glands differ in shape and dimensions with the left gland being consistently heavier (Kober et al., 2012). The glands receive direct arterial blood supply from the cranial renal arteries and occasionally from the aorta. Each gland is drained by a single adrenal vein that enters the posterior (caudal) vena cava or its caudal bifurcation. In the chicken, each gland also receives one or two lymphatic vessels.

Each gland generally has two postganglionic sympathetic ganglia (cranial and caudal) embedded within the pericapsular sheath or more deeply within the outer gland substance (e.g., see Figure 26.2(A)). In the duck (*Anas platyrhynchos*), these ganglia are composed of 30–60 cell bodies (Squillacioti et al., 2008). The caudal ganglion mainly supplies surrounding organs, including the gonads. Smaller clusters of ganglion cells are found within the gland (e.g., see Figure 26.2(D)). Preganglionic sympathetic fibers in thoracic and synsacral splanchnic nerves converge on the ganglia. The preponderance of the fibers (cholinergic) course through the ganglia without synapsing and penetrate the gland substance to innervate clusters of adrenal chromaffin cells consisting of 12 or more chromaffin cells. Indeed, gland denervation leads to atrophy of the chromaffin tissue (Holmes and Phillips, 1976). In addition, postganglionic sympathetic fibers emanating from the ganglia mainly innervate the adrenal blood vessel (vasomotor), and the adrenergic type innervates the chromaffin cells. Current thinking is that the outer cranial ganglion and intraglandular ganglion cells form the adrenal intrinsic innervation; however, studies supporting this postulate are limited to chromaffin cells. In any study, endings on adrenocortical cells are rarely observed. Overall, these studies indicate that the postganglionic adrenergic innervation is, in part, involved in the regulation of catecholamine synthesis and secretion, although a nonneurogenic mechanism may operate as well in some species (Ghosh et al., 2001). Finally, there is an extrinsic innervation of additional postganglionic sympathetic fibers and possibly preganglionic parasympathetic fibers from the vagus nerve, but the extent of this innervation and its function are poorly understood. Evidence indicates that the vagus innervation affects the structural and functional integrity of both adrenocortical and chromaffin tissues (Ghosh et al., 2001).

26.1.2 Microanatomy

26.1.2.1 The Chromaffin Tissue

The adrenal gland of birds consists of an intermingling of steroid-secreting cells (adrenocortical cells) and chromaffin (adrenomedullary) cells (Figures 26.2 and 26.3). The adrenocortical cells are arranged in parallel cords of columnar cells (Figures 26.2 and 26.3) and are described later in this chapter. The chromaffin cells are larger, polygonal cells arranged in nests or clusters surrounded by a thin basal lamina. A single nerve bundle innervates the cluster such that one neuronal terminal synapses with up to three chromaffin cells of the same type (i.e., norepinephrine (NE) or epinephrine (E) chromaffin cells) (Unsicker, 1973). The consensus of studies suggests that the ratio of adrenocortical to chromaffin cells is about 1.5:1 (Holmes and Phillips, 1976). Indeed, a similar ratio has been demonstrated in dispersed chicken adrenal cells (Carsia et al., 1985a). However, there is a tendency for this ratio to increase in the peripheral region of the gland and to decrease approaching more central regions of the gland (Kober et al., 2010). Indeed, in the chicken, there is a subcapsular layer of essentially pure chromaffin cells (Kober et al., 2012).

Sturkie's Avian Physiology.

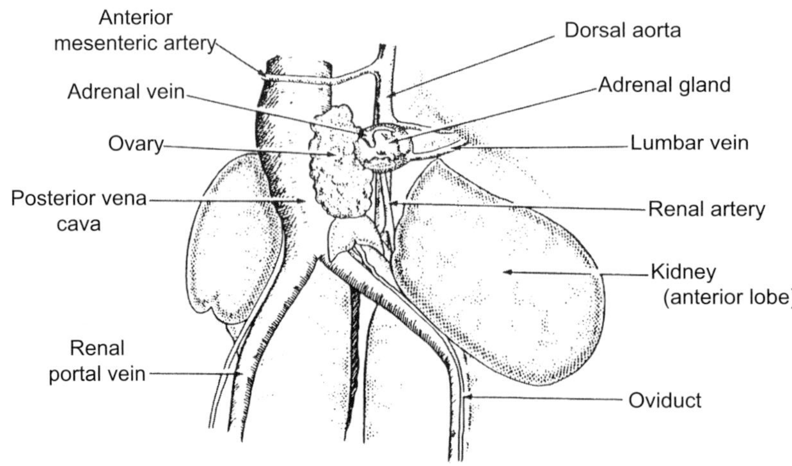

FIGURE 26.1 **The position of the left adrenal gland and its vascular supply in the female gull (*Larus argentatus*).** *Preparation and drawing by I. Carthy and J. G. Phillips. Taken from Holmes and Phillips (1976).*

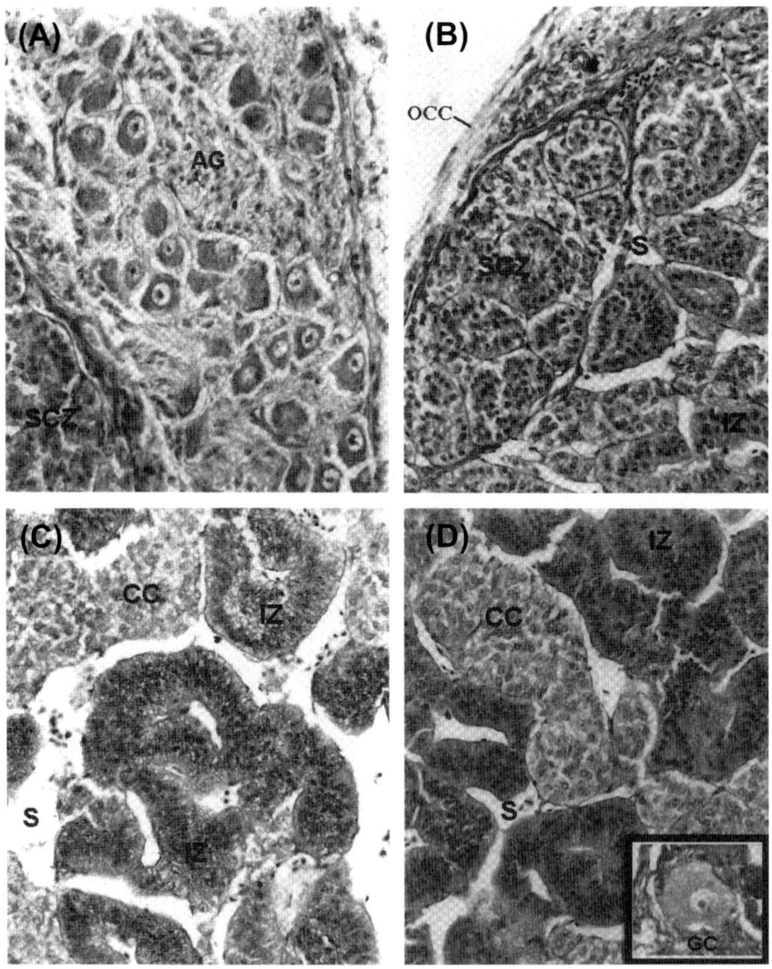

FIGURE 26.2 **Light micrographs of portions of sections through the adrenal gland of the immature (4 weeks old) domestic turkey (*Meleagris gallopavo*).** (A) Pericapsular adrenal ganglion (AG) and adjacent subcapsular zonal (SZ) cell cords. Note the large ganglion cell bodies with conspicuous nuclei. (B) Outer connective tissue capsule (OCC) with SZ cell cords, subjacent inner zonal (IZ) cell cords, and sinusoid (S). (C) IZ cell cords with S and chromaffin cell (CC) islet. (D) CC islet with intraglandular ganglion cell (GC; inset), IZ cell cords, and S. (X250). *Micrographs courtesy of R.G. Nagele, PhD, New Jersey Institute for Successful Aging and Department of Cell Biology, Rowan University School of Osteopathic Medicine.*

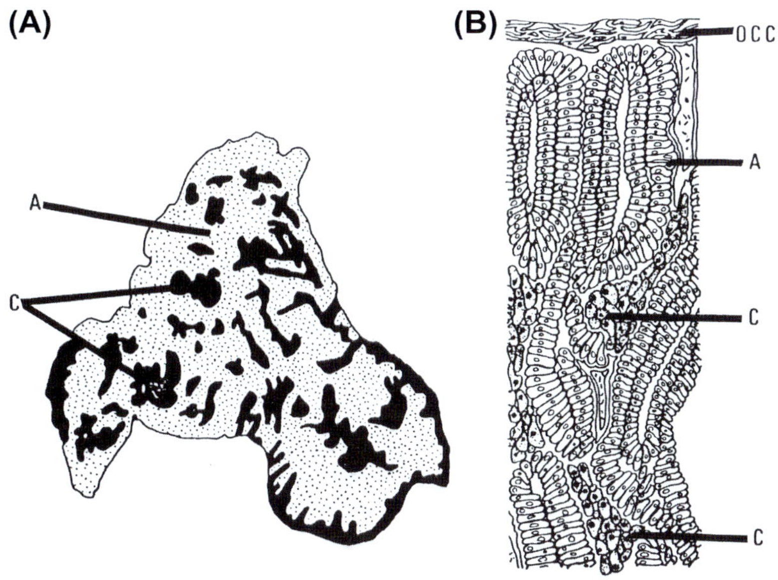

FIGURE 26.3 **Microanatomy of a typical avian adrenal gland (Florida quail, *Colinus virginiatus floridanus*).** (A) Distribution of chromaffin (C) and adrenocortical (A) tissue. (B) Structure of the looped cords of adrenocortical cells with intermingled chromaffin cell islets. The outer connective tissue (OCC; adrenal capsule) is also indicated. *Taken from Chester Jones and Phillips (1986).*

The ratio of E- to NE-secreting chromaffin cells varies with species. There is some suggestion that the ratio is smaller (i.e., a greater proportion of NE cells) in orders with more primitive ancestry and greater in more recently evolved birds (Ghosh et al., 2001; Kober et al., 2010). In the domestic chicken, 70% of the chromaffin cells are E-secreting cells (Ohmori, 1998). Complicating this picture is that this ratio also varies with age within the same species. Also variable are the sizes of the secretory vesicles within chromaffin cells. In galliform and anseriform species, NE vesicles tend to be larger than E vesicles. Across avian species, vesicles can range from 80 to 500 nM (Ghosh et al., 2001). In general, in NE cells, the granules are compact, electron opaque, and eccentrically placed, not completely filling the vesicles, whereas in E cells, the granules are more rarefied, are less electron opaque, and fill the vesicles (Holmes et al., 1991).

The avian adrenal gland is endowed with an intrinsic innervation. A proposed scheme based on comparative work on the chromaffin and neural components of the intrinsic adrenal regulation is shown in Figure 26.4.

The preponderance of ganglion cells are catecholaminergic; however, there appears to be noncatecholaminergic cells as well. Ganglion cells and the chromaffin cells elaborate a number of biogenic amines and neuropeptides (García-Arrarás et al., 1992; Wolfensberger et al., 1995; Ehrhart-Borstein et al., 1998; Ohmori, 1998; De Falco et al., 2008; Squillacioti et al., 2008), and avian adrenal tissues exhibit a robust expression of cognate receptors. The ganglion cells and the chromaffin cells are primarily innervated by preganglionic sympathetic fibers containing acetylcholine and neuronal nitric oxide synthase (nNOS) (Squillacioti et al., 2008). Nitric oxide appears to be mainly inhibitory on corticosteroid secretion, but the presence of nNOS in multiple postganglionic cell types, chromaffin cells, and subpopulations of ganglion cells, and within pre- and postganglionic terminals, indicates its importance. For example, it appears to be important in maintaining the normal ratio of adrenocortical tissue to chromaffin tissue (Chaturvedi and Kumar, 2007).

Another prominent peptide is vasoactive intestinal peptide (VIP) and a related peptide, pituitary adenylyl cyclase-activating peptide (PACAP). VIP is mainly present in intrinsic and extrinsic postganglionic sympathetic terminals. In addition, VIP and PACAP are synthesized in the chromaffin cells. Furthermore, receptors for these peptides exist on both adrenocortical cells and chromaffin cells, providing a robust synaptic and paracrine interaction. VIP and PACAP directly stimulate corticosteroid release from adrenocortical cells. However, VIP and PACAP also stimulate catecholamine secretion, and this, in turn, can activate β-adrenergic receptors on adrenocortical cells (Mazzocchi et al., 1997b). Here again is another example of direct synaptic and paracrine interactions controlling corticosteroid secretion and also epinephrine secretion since corticosterone stimulates epinephrine synthesis.

Concerning adrenocortical function, the avian adrenal chromaffin cells have an intrinsic corticotropin-releasing hormone (CRH)–melanocortin system and the requisite cognate receptors (Takeuchi and Takahashi, 1998, 1999; Takeuchi et al., 1998, 1999, 2000; De Groef et al., 2004; Mirabella et al., 2004). Intraadrenal ACTH can activate melanocortin-2 (ACTH) receptors on adrenocortical cells. Chromaffin cells also synthesize atrial natriuretic peptide

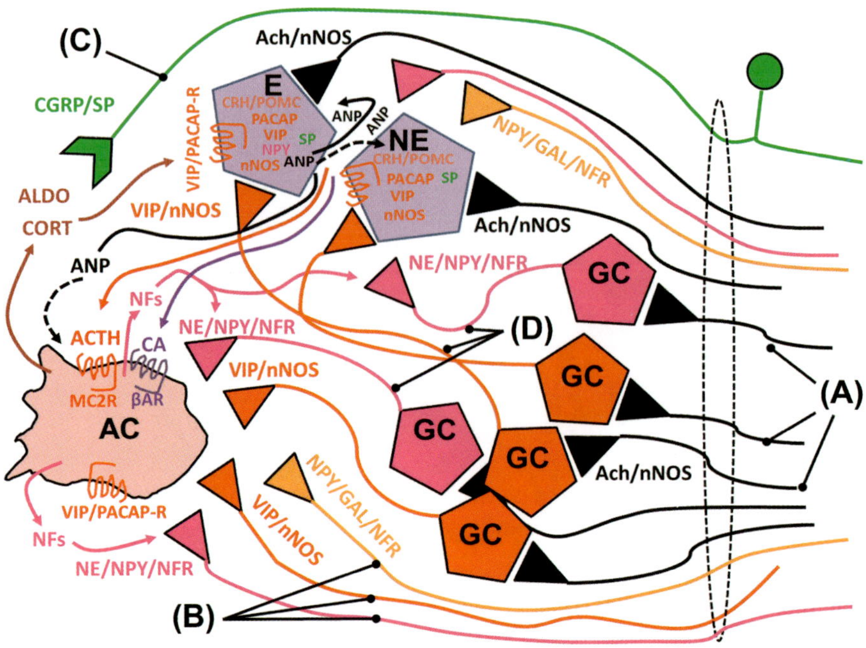

FIGURE 26.4 **A proposed model for the intraadrenal neuroendocrine regulation based on comparative studies of reptiles, birds, and mammals.** The neural regulation has both extrinsic and intrinsic components of the sympathetic nervous system. Its axonic terminals are represented by *black and colored triangles*. The extrinsic innervation is composed of (A) preganglionic sympathetic efferent axons containing acetylcholine (Ach) and neuronal nitric oxide synthetase (nNOS) (*black*); (B) postganglionic sympathetic efferent axons containing vasoactive intestinal peptide (VIP) and nNOS (*red*), neuropeptide tyrosine (NPY) and galanine (GAL) (*orange*), or NPY and norepinephrine (NE) (*pink*); and (C) sympathetic afferent (sensory) peripheral processes (dendrites) containing substance P (SP) and calcitonin gene-related peptide (CGRP) (*green*). *Dotted oval* represents the splanchnic nerve bundle that contains these extrinsic fibers. The intrinsic innervation (D) is composed of the postganglionic sympathetic ganglion cells (GCs; *smaller colored pentagons*) of the cranial and caudal ganglia and the scattered ganglion cells throughout the gland. A GC contains VIP and nNOS (*red*) or NPY and NE (*pink*). There is evidence that in some species, GCs also have phenylethanolamine *N*-methyltransferase for the synthesis of epinephrine, but it is not clear if this enzyme is active or if epinephrine is released from axonic terminals. Postganglionic axons containing NPY probably also express neurotrophic factor receptors (NFRs) for axonal growth. Growth of NPY axons is stimulated by neurotrophic factors (NFs) released by adrenocortical cells. This neurotrophic loop probably remodels the adrenocortical tissue during life-history cycle or stage transitions and periods of prolonged stress. The intrinsic endocrine component consists of the chromaffin cells (*larger purple pentagons*) and the adrenocortical cells (ACs). The chromaffin cells are also modifications of postganglionic sympathetic neurons and are of two functional types: epinephrine-secreting cells (E) and NE-secreting cells. These cells also have the enzymatic machinery for an intrinsic corticotropin-releasing hormone (CRH)–adrenocorticotropin (ACTH) system in that they contain and process pro-opiomelanocortin (POMC). They also contain many neurotransmitters: pituitary adenylyl cyclase-activating peptide (PACAP), VIP, SP, and the nitric oxide–producing enzyme, nNOS. In addition, the E cells contain atrial natriuretic peptide (ANP). The intraadrenal ACTH interacts with melanocortin 2 (ACTH) receptors (MC2R) on adrenocortical cells to stimulate corticosteroid (corticosterone (CORT) and aldosterone (ALDO)) synthesis and secretion (*brown arrows*). The intraadrenal CORT, in turn, stimulates E production. Whereas the intraadrenal ANP dampens corticosteroid production in adrenocortical cells and inhibits NE cell function (*dotted black arrows*), it may stimulate the conversion of NE cells to E cells (*solid black curved arrow*). Both adrenocortical and chromaffin cells have VIP–PACAP receptors (VIP–PACAP-R). In general, activation of these receptors stimulates corticosteroid and catecholamine production. Furthermore, intraadrenal catecholamines interact with β-adrenergic receptors (βARs) on adrenocortical cells to augment corticosteroid production. SP from sensory endings stimulates adrenocortical cells to produce corticosterone and aldosterone. In addition, it stimulates the release of catecholamines. CGRP and GAL also appear to be stimulatory for adrenocortical cells. In this scheme, the parasympathetic components are omitted because of the absence of definitive work. See the text for additional details. *Scheme drawn by R. V. Carsia.*

(ANP) (Wolfensberger et al., 1995). Intraadrenal ANP not only dampens corticosteroid secretion but also influences the chromaffin cell composition by inhibiting norepinephrine cells and stimulating the conversion of norepinephrine cells to epinephrine cells.

Another ubiquitous peptide is neuropeptide tyrosine (NPY). It is present in many extrinsic postganglionic terminals and in subpopulations of ganglion cells. It is also synthesized in chromaffin cells, predominantly the epinephrine cells (Squillacioti et al., 2008). Its widespread, intraadrenal distribution is matched by the number of receptor subtypes expressed in the avian adrenal gland (Bromée et al., 2006). In many cells and in axonic terminals, NPY is accompanied by galanine. NPY has the broad effect of stimulating catecholamine secretion and, indirectly, corticosteroid secretion through β-adrenergic receptor stimulation by catecholamines. However, its main effect may be to regulate the adrenocortical cell composition, that is, the rudimentary zonation, to meet physiological needs (see Tran et al., 2010). In this scenario, physiological demands or stressors stimulate the release of neurotrophic factors from adrenocortical cells, which, in turn, activate cognate receptors on postganglionic sympathetic NPY terminals to stimulate axonal growth. The enhanced innervation induces hyperplasia of the appropriate adrenocortical cell subpopulation to address the physiological demand or stressor. When the stressor subsides, the axons regress and the zonation gradually returns to the prestressor state.

In addition to the pre- and postganglionic sympathetic efferent components, there are extrinsic afferent components, the cell bodies of which reside in the dorsal root ganglia (Figure 26.4). The peripheral processes or dendrites of type B sensory neurons release substance P and calcitonin gene-related peptide (CGRP). Substance P is also present in the chromaffin and ganglion cells. Substance P stimulates adrenocortical cells to produce corticosterone and aldosterone. In addition, it stimulates the release of catecholamines. CGRP also appears to be stimulatory for adrenocortical cells. However, in studies with the rat adrenal gland, it does not appear to function in the reorganization of the adrenocortical cell subtypes with various physiological demands and stressors (see Tran et al., 2010).

Thus, as in mammals, in addition to the direct intraadrenal innervation, there is supporting evidence for a rich environment of autocrine, juxtacrine, and paracrine interactions among chromaffin and adrenocortical cells. These interactions are further complicated when one considers the intraglandular cytokines, growth factors, and endothelial-derived substances. The impact of this extrinsic and intraadrenal regulation is poorly understood, but it may support an intrinsic clock for corticosteroid secretion rhythmicity and adrenocortical sensitivity to ACTH (Dickmeis, 2009). In short, the immunocytochemical and molecular biological observations on this regulation far exceed frank *in vivo* and *in vitro* studies in determining its importance in avian adrenal function over life history cycle and life history stage.[1] Although the emerging story of the avian intraadrenal regulation may describe additional levels of control of adrenocortical and chromaffin secretion, the "traditional" concepts of the regulation of avian adrenal function (i.e. the splanchnic innervation and the hypothalamus–pituitary–adrenal (HPA) axis) remain valid.

26.1.2.2 The Adrenocortical Tissue

The adrenocortical cells are arranged in cords that radiate from the deep part of the gland, branching and anastomosing freely (Figures 26.2 and 26.3(B)). At the periphery, the cords form tighter loops of cells that in some species resemble the glomerular structure seen in mammals (Figure 26.2). The cords consist of a double row of adrenocortical cells oriented with their columnar axes perpendicular to the cord (Holmes and Phillips, 1976). This pattern is highly variable among species, and there appears to be no phylogenetic trend (Bhattacharyya et al., 1972).

The preponderance of histological, ultrastructural, and functional evidence indicates that birds have a modest segregation of adrenocortical cell subpopulations that resembles the zonation of the mammalian adrenal cortex (Bhattacharyya et al., 1972; Pierce et al., 1978; Holmes and Cronshaw, 1984, 1993). Histological preparations reveal at least two different zones of cells: a subcapsular zone (SZ) and an inner zone (IZ). More recently, based on studies of the chicken adrenal gland and the amount of intermingling of adrenocortical and chromaffin tissues (Kober et al., 2012), a three-zone nomenclature is proposed: a subcapsular zone of essential pure, small chromaffin cells immediately deep to the capsule; a peripheral zone consisting of mainly adrenocortical cells with scattered, small islets of chromaffin cells; and a central zone in which the two tissues are evenly intermingled. However, this alternative zonal description is not apparent in other species, nor is there functional evidence to support it. Thus, for consistency in further discussions in this chapter, we shall adhere to the SZ–IZ pattern exemplified in the duck (*A. platyrhynchos*) adrenal gland.

Adrenocortical cells of the subcapsular zone are large, often binucleated, and replete with lipid droplets. Like most steroid-secreting cells, they have the troika of organelles: stores of lipid, mainly in lipid droplets; varying amounts of smooth endoplasmic reticulum; and mitochondria. The mitochondria bear regularly arranged tubular cristae (incorrectly described as platelike), similar to mammalian zona glomerulosa cells. Functionally, the tissue secretes predominantly aldosterone in response to ACTH and exclusively aldosterone in response to angiotensin II, and it is insensitive to adrenocorticotropin (ACTH) withdrawal (adenohypophysectomy) (Klingbeil, 1985; Pierce et al., 1978; Holmes and Cronshaw, 1984, 1993).

Adrenocortical cells of the IZ are smaller, are more elongated or columnar, contain fewer lipid droplets, have more abundant smooth endoplasmic reticulum, and have mitochondria bearing tubulovesicular cristae, similar to mammalian zona fasciculata cells. Functionally, the tissue secretes predominantly corticosterone in response to ACTH, is refractory to angiotensin II, and undergoes atrophy with ACTH withdrawal (adenohypophysectomy), with some cells converting to the SZ type (Klingbeil, 1985; Pierce et al., 1978; Holmes and Cronshaw, 1984, 1993).

Additional evidence for a segregation or zonation of avian adrenocortical tissue comes from work with dispersed adrenocortical cells from the domestic turkey (*Meleagris gallopavo*) subjected to dietary Na^+ restriction and in which functionally distinct cell subpopulations were segregated

[1] Life-history cycle and life-history stage are sometimes used interchangeably in the literature. Here we define life-history cycle as the avian developmental and maturational transitions: the embryonic period, hatch, posthatch (postnatal), and sexual maturation. We define life-history stage as the predicable periods or phases of sexually mature birds that are coordinated with habitat, such as the seasonal, prelaying, breeding, late-breeding, and molting periods.

based on density (Kocsis et al., 1995a). A population of high-density cells, the proportion of which was enhanced by the Na^+ restriction, had functional and ultrastructural features similar to those of mammalian zona glomerulosa cells (Figure 26.5). In addition, these cells have a disproportionately enhanced aldosterone response to angiotensin II and K^+ (Kocsis et al., 1995a).

Unfortunately, much of what is known about the functional zonation of the avian adrenal gland comes from work on a limited number of precocial species (see Holmes and Cronshaw, 1993) using tissue slices from the SCZ and IZ or dispersed adrenocortical cells. Clearly, similar work needs to be expanded to semiprecocial and altricial species. In addition, these *in vitro* studies, although facile and informative, have the inherent problem of organ disruption and at best are a declension of the *in vivo* situation. Tissue slices are conceivably better than dispersed cells. For example, whereas PACAP (Figure 26.4) induces release of corticosterone and aldosterone from chicken adrenal quarters, it has no effect on dispersed chicken adrenal cells (Mazzocchi et al., 1997a), presumably because its effect in this species is indirect through the release of catecholamines from nearby chromaffin cell islets. Nevertheless, it can be generally concluded that extrapolation of findings from any *in vitro* study to the *in vivo* condition must be viewed with caution (see Vinson et al., 1985).

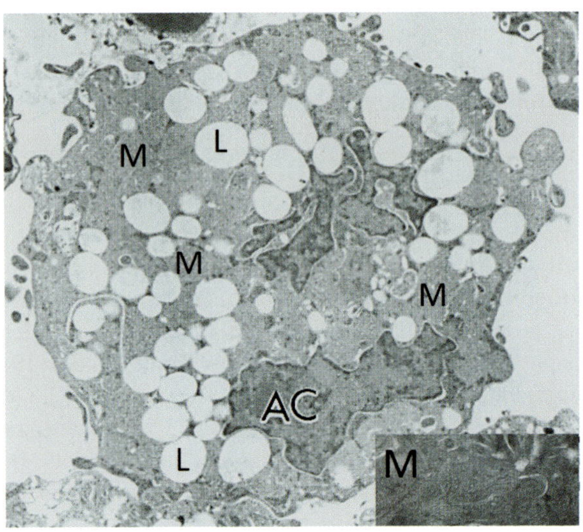

FIGURE 26.5 **Section through a high-density adrenocortical cell (AC) of the growing (6 weeks old) domestic turkey (*Meleagris gallopavo*) subjected to dietary Na^+ restriction (0.04%) for 8 days (×8700).** Inset: higher magnification of a section through a mitochondrion (×41,700). Features are similar to subcapsular zonal cells of the duck (*Anas platyrhynchos*) (Pierce et al., 1978). Note the numerous lipid droplets (L) and mitochondria (M) and scant endoplasmic reticulum. Inset: mitochondria have regularly arranged tubular cristae. These cells have a disproportionately enhanced aldosterone response to angiotensin II and K^+ compared with other density-separated subpopulations (Kocsis et al., 1995a). *Electron micrographs courtesy of Jean Gibney, Department of Neuroscience and Cell Biology, Rutgers University–Robert Wood Johnson Medical School.*

26.2 ADRENOCORTICAL HORMONES

26.2.1 Corticosteroid Secretory Products

The glucocorticoid corticosterone is the principal corticosteroid released by the avian adrenal gland. The principle mineralocorticoid, aldosterone, is produced in considerably less quantity in the posthatch bird. *In vitro* studies with anseriform and galliform adrenal tissue and dispersed adrenal cells indicate that the ratio of aldosterone to corticosterone production rates is about 1:19. However, based on studies in which both corticosteroids were measured, the average ratio of the baseline circulating concentrations of aldosterone to corticosterone is about 1:175 (Table 26.1). Younger birds tend to have higher circulating concentrations of each compared to those of adults. This ratio discrepancy is probably due to varying sample stress and corticosteroid half-life in the avian circulation. In a very limited number of earlier studies, the average half-lives of corticosterone and aldosterone in the circulation are similar, about 15 and 13 min, respectively. However, more recent work shows these half-lives to be considerably less, perhaps 8–9 min.

The avian adrenal gland secretes additional steroids that presumably have important ontogenetic function and play a role in life history cycle transitions (Landys et al., 2006; Wada, 2008; Wingfield, 2013). Cortisol is a significant secretory product during the embryonic and early posthatch periods, and then declines thereafter. Indeed, earlier work with chick embryos reports equivalent concentrations of

TABLE 26.1 Baseline Peripheral Plasma Concentrations of Corticosterone and Aldosterone in Some Galliform and Anseriform Birds in Which both Corticosteroids were Measured in the Same Samples by Radioimmunoassay (Averages from 26 Studies)

Bird Order	Life-History Cycle	Corticosterone (ng/ml)	Aldosterone (pg/ml)
Galliform birds[1]	Hatch	11	33
	Immature	11	46
	Adult	2	19
Anseriform birds[2]	Hatch	16	345
	Immature	8	308
	Adult	18	60

[1]*Values derived from studies using the domestic chicken (Gallus gallus domesticus), the domestic turkey (Meleagris gallopavo), and the Japanese quail (Coturnix japonica).*
[2]*Values derived from studies using the wild mallard duck and the Pekin duck (Anas platyrhynchos) and the Muscovy duck (Cairina moschata).*

adrenal corticosteroids (progesterone, corticosterone, cortisol, and cortisone) in the circulation that reach a peak near hatch and then decline after hatching (Kalliecharan and Hall, 1974). Presumably, cortisol has a role in the development and maturation of several organ systems (see Carsia et al., 1987a). During posthatch life, circulating cortisol may serve as a depot for modulation of the active hormone at the tissue level. For example, in some psittacine species, circulating cortisol is insensitive to exogenous ACTH (Lothrop et al., 1985; Walsh et al., 1985). Thus, the availability and action of cortisol may be less dependent on the HPA axis and is more dependent on regulation at the tissue level (as discussed further in this chapter).

The efficacy of corticosterone and cortisol during development and during posthatch and adult life history cycle and stage transitions is influenced by a number of factors. For example, there appear to be isoforms of corticosteroid-binding globulin that preferentially bind cortisol and, thus, influence its half-life for delivery to sensitive tissues (Schmidt et al., 2008, 2010), and there appear to be isoforms of glucocorticoid receptors (GRs) in various target tissues that selectively bind cortisol (Schmidt et al., 2010). In addition, there is interplay between systemic delivery of corticosteroids to target tissues and the intrinsic local production within tissues (Schmidt and Soma, 2008; Schmidt et al., 2008). For example, corticosterone may be synthesized or regenerated in the brain (Newman et al., 2008), and locally may mediate the acute effect of stress on brain aromatase activity in a sex-dependent manner (Dickens et al., 2011a). The picture is further complicated by the local expression of 11β- and 20-hydroxysteroid dehydrogenases that can degrade and regenerate active glucocorticoids within cells. Overall, despite its relatively low circulating concentration relative to corticosterone, cortisol produced locally throughout the posthatch life of birds (e.g., within immune tissues and possibly the brain) suggests it has unique and important functions in both an organ-specific and age-specific manner.

In addition to cortisol, the adrenal is an important source of sex steroids (Tanabe et al., 1986; Ottinger et al., 2001) during the embryonic and perinatal periods, and it complements the gonads for establishing adequate levels of androgens and estradiol. This "adrenal–gonadal unit" is thought to be important for the sexual differentiation of the hypothalamus.

Here again, as seen with cortisol, the adrenal continues to complement the gonads in the regulation of circulating sex steroids through maturity and during life history stage transitions (e.g., breeding and nonbreeding periods) (Bhujle and Nadkarni, 1976; Soma and Wingfield, 2001). Furthermore, a similar theme between systemic and locally produced sex steroids is evident, and its expression is dependent on life history cycle, life history stage, and species-specific adaptations (Schmidt et al., 2008). A growing body of evidence supports the notion that the adrenal gland assists the gonads in establishing the balance between systemic delivery of sex steroids to the brain and their local production within the brain. This dynamic balance between systemic and locally produced sex steroids establishes important central nervous signals that support breeding, territorial, and courtship behaviors (Soma and Wingfield, 2001; Schmidt et al., 2008; Pintér et al., 2011).

26.2.2 Other Secretory Products

The chicken adrenocortical tissue is a major source of inhibin during embryonic development (Rombauts et al., 1994; Decuypere et al., 1997). Dexamethasone depresses inhibin release *in ovo*, whereas ACTH elicits a rise in inhibin release that parallels corticosterone release in primary cultures of embryonic chick adrenal cells. Furthermore, the adrenal gland continues to be a complementary extragonadal source of inhibin in the adult (see Vanmontfort et al. (1997), Bruggeman et al. (2003)), and molecular studies suggest activin as well (Chen and Johnson, 1996). The roles of locally produced activin and inhibin in adrenal gland function *per se* remain to be investigated. Their release from cultured adrenal cells are directly inhibited by dexamethasone (Vanmontfort et al., 1997). Based on comparative studies, the adrenal activin–inhibin system may have a negative autocrine–paracrine effect on intraglandular steroid synthesis (Vänttinen et al., 2003).

26.2.3 Synthesis and Intraadrenal Degradation of Corticosteroids

The framework for avian adrenocortical steroidogenesis is composed of the organelles, mitochondria, smooth endoplasmic reticulum and lipid droplets, and cytoskeleton (reviewed in Feuilloley and Vaudry (1996)) (Figure 26.5). Under the action of ACTH, the cytoskeleton and a variety of molecular motors and shuttles bring about a clustered organelle deployment, primarily the tethering of the smooth endoplasmic reticulum to mitochondria, for steroidogenesis (Poderoso et al., 2013; Sewer and Li, 2013). In addition, corticosteroidogenesis involves intracellular depots of cholesterol in both membranes and lipid droplets. Cholesterol stores for steroidogenesis are apparently initially derived from plasma lipoproteins (Hertelendy et al., 1992; Latour et al., 1995). Avian adrenocortical cells are laden with cholesterol–ester-rich lipid droplets (Holmes and Cronshaw, 1980; Carsia et al., 1985a; Holmes et al., 1991), and thus, not surprisingly, their intrinsic ability to synthesize cholesterol is normally low (Lehoux et al., 1977). Cholesterol is accessed from esterified stores through the action of hormone-sensitive lipase. Once free, cholesterol is intracellularly transported by steroidogenic acute regulatory (StAR)-related lipid transfer domain (StarD) proteins.

StarD proteins are responsible for loading free cholesterol to the outer mitochondrial membrane. StAR is a hormone-sensitive protein that delivers the free cholesterol from the outer mitochondrial membrane to the inner mitochondrial membrane, where it can access the cholesterol side-chain cleavage enzyme, P450scc, and be converted to pregnenolone (Miller and Auchus, 2011). In response to ACTH, StAR is acutely synthesized as a 37 kDa protein that has a mitochondrial localization sequence, and then is cleaved into its active, 30 kDa fragment upon mitochondrial entry. StAR is recycled such that each molecule moves hundreds of molecules of cholesterol before being degraded. Not surprisingly, StAR is highly conserved in vertebrate steroidogenic tissue (Bauer et al., 2000). StAR activity requires the peripheral benzodiazepine receptor (PBR), also known as translocator protein (TSPO). PBR may have additional significance in birds in that certain associated proteins may underlie changes in adrenocortical sensitivity to ACTH (Bureau et al., 2009).

The synthetic pathways of the major steroids secreted by the avian adrenal gland during the life cycle are shown in Figure 26.6. The pattern of enzymes carrying out these synthetic steps is highly conserved among the vertebrates and homologous to that characterized in the mammal (Miller and Auchus, 2011). Although there are substantial differences in the primary structure of steroidogenic enzymes studied thus far between mammalian and avian species, conserved domains of enzymatically active sites are apparent (see Nomura et al. (1999)). This conservation has aided in assessing the expression of steroidogenic enzymes in developing and adult birds (e.g., Nomura et al. (1999), Freking et al. (2000), Kanda et al. (2000), Kamata et al. (2004)).

Although steroidogenic enzymes in homogenates can be bidirectional, they are almost uniformly unidirectional in intact cells and *in vivo,* and accumulation of products does not drive flux back to the immediate precursor. In this steroidogenic scheme, mitochondrial cytochrome P450SCC (or CYP11A1) (Nomura et al., 1997) mediates 20α-/22-hydroxylation and scission of the C20-22 carbon bond, producing pregnenolone and isocaproaldelhyde, a set of reactions traditionally called "20, 22-desmolase". Pregnenolone is converted to progesterone by 3βHSD2 (HSD3B2) (Nakabayashi et al., 1995), a non-P450 microsomal–mitochondrial dehydrogenase–isomerase complex that mediates 3β-hydroxysteroid dehydrogenase and isomerase (Δ^5 double bond to Δ^4 double bond) activities. Microsomal P450c21 (CYP21A2) then mediates 21-hydroxylation steps converting progesterone to 11-deoxycorticosterone. Corticosterone is the predominant corticosteroid product of the avian adrenal gland, although 11-deoxycorticosterone is also a significant secretory product. Corticosterone is formed from 11-deoxycorticosterone by an 11β-hydroxylation via a mitochondrial enzyme, P450c11β. To date, this enzyme has not been cloned from reptiles and birds.

Aldosterone is largely formed from 11-deoxycorticosterone and not from corticosterone because corticosterone is a poor substrate. Thus, 11-deoxycorticosterone undergoes an 11β-hydroxylation, converting it to enzyme-bound corticosterone that serves as an intermediate. This is followed by an 18-hydroxylation and lastly an 18-methyl oxidase step, producing the potent mineralocorticoid aldosterone. In birds, it is not clear whether the same enzyme catalyzes the conversion of 11-deoxycorticosterone to aldosterone (CYP11B0), as it does in amphibians (Nonaka et al., 1995) and cattle (Morohashi et al., 1987), or whether in addition to P450c11β (CYP11B1), a separate enzyme, as in humans (CYP11B2 (P450c11AS), or aldosterone synthase) (Mornet et al., 1989), or two additional enzymes, as in rats (CYP11B2 and CYP11B3) (Mellon et al., 1995), catalyze the final steps to aldosterone. In addition, gene array analysis of chicken adrenals from birds stimulated with ACTH does not reveal a differential expression of these key enzymes (Bureau et al., 2009). However, if perhaps it is just one enzyme, there must be additional controls for the predominant or exclusive secretion of aldosterone: first, SCZ slices from the duck adrenal gland secrete predominantly aldosterone in response to ACTH and exclusively aldosterone in response to angiotensin II, whereas the IZ tissue secretes predominantly corticosterone in response to ACTH and is refractory to angiotensin II (Klingbeil, 1985; Pierce et al., 1978; Holmes and Cronshaw, 1984, 1993); and, second, dietary Na^+ restriction in the domestic turkey induces a subpopulation of adrenocortical cells that have a disproportionately enhanced aldosterone response to angiotensin II and K^+ (Kocsis et al., 1995a).

Other enzymatic activities are also apparent mainly during the embryonic and neonatal periods and at modest levels in the adult. Formation of cortisol is predominantly via 17α-hydroxylation of progesterone by a microsomal P450c17 (CYP17A1) (Ono et al., 1988) to form 17α-hydroxyprogesterone, then a 21-hydroxylation by microsomal P450c21 (CYP21A1) to form 11-deoxycortisol, and finally an 11β-hydroxylation step by a mitochondrial P450c11β to form cortisol. A minor pathway is pregnenolone to 17α-hydroxypregnenolone via P450c17, and then to 17α-hydroxyprogesterone via 3βHSD2. There is some evidence that 11β-hydroxyprogesterone serves as a precursor for corticosterone (via a 21-hydroxylation step) and cortisol (via 17α- and 21-hydroxylation steps), although these pathways appear to be of little significance.

The pathway for adrenal sex steroids uses a microsomal P450c17 (CYP17A1) (Ono et al., 1988), which has both 17α-hydroxylase and 17,20-lyase activities, and efficiently converts pregnenolone to 17α-hydroxypregnenolone and then to dehydroepiandrosterone (DHEA). In addition, the 3β-hydroxyl group of DHEA can be sulfated by an adrenal sulfotransferase. As mentioned in this chapter, DHEA and DHEA-sulfate are in the avian circulation

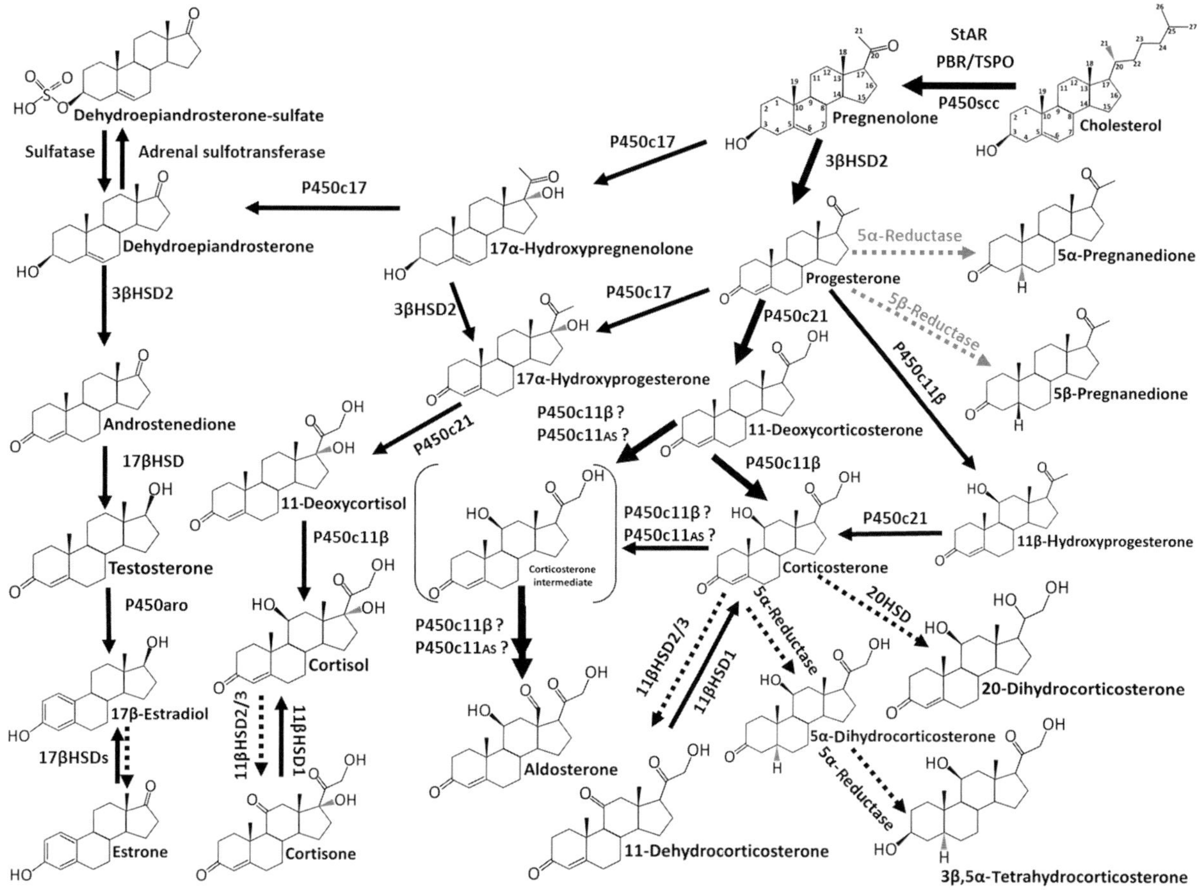

FIGURE 26.6 Steroidogenesis in the avian adrenal gland. *Thick arrows* indicate the predominant pathway of posthatch (postnatal) birds leading to corticosterone and aldosterone. *Thin arrows* indicate additional pathways that are more prominent in the embryonic and early-posthatch (postnatal) periods, and during life-history stage transitions; these are the pathways to cortisol and to the sex steroids. *Dotted arrows* indicate steroid-inactivating or degrading pathways. Dotted *gray arrows* indicated degradative pathways of progesterone during the embryonic period. Solid black and gray bonds on steroids represent key alpha (down) and beta (up) positions; however, not all hydrogen bond positions are shown. StAR, steroidogenic acute regulatory protein; PBR–TSPO, peripheral benzodiazepine receptor–translocator protein; P450scc, mitochondrial cytochrome P450 cholesterol side-chain cleavage enzyme; 3βHSD2, microsomal–mitochondrial 3B-hydroxysteroid dehydrogenase–Δ5→Δ4 isomerase, type 2; P450c21, microsomal cytochrome P450 21-hydroxylase; P450C11β, mitochondrial cytochrome P450 11β-hydroxylase; P450C11AS?, mitochondrial aldosterone synthase (it is not clear if the avian adrenal gland has just one enzyme, P450C11β, that catalyzes the 11β-hydroxylation of 11-deoxycorticosterone to corticosterone and of 11-deoxycortisol to cortisol and also, under certain conditions, catalyzes (1) the 11β-hydroxylation of 11-deoxycorticosterone to a corticosterone intermediate, then (2) an 18-hydroxylation step taking the corticosterone intermediate to an 18-hydroxycorticosterone intermediate, and then (3) an 18-methyl oxidase step (which is the key function of aldosterone synthase) to aldosterone. *Double superimposed arrows* indicates steps 2 and 3.); P450c17, microsomal cytochrome P450 17α-hydroxylase/17,20-lyase; 17βHSD, microsomal 17-ketosteroid reductase; P450aro, microsomal cytochrome P450 aromatase; 17βHSDs, not fully characterized adrenal microsomal enzymes in birds; 11βHSD2/3, microsomal 11β-hydroxysteroid dehydrogenase, types 2 and 3; 11βHSD1, 11β-hydroxysteroid dehydrogenase type 1 (this is largely microsomal, although there is evidence for a significant cytosolic type-1-like enzyme); 20HSD, microsomal 20-hydroxysteroid dehydrogenase. See text for additional details. *Pathway drawn by R. V. Carsia.*

(Soma and Wingfield, 2001) and in the adrenal gland. Although P450c17 can also convert small amounts of progesterone to 17α-hydroxyprogesterone and then to androstenedione, it does so poorly.

Adrenal androstendione can be converted to testosterone by a17βHSD (Nomura et al., 1999), a microsomal, non-P450 enzyme. Indeed, androgens are detected in the avian adrenal gland (Ottinger et al., 2001; Soma and Wingfield, 2001). Furthermore, there is evidence that testosterone is aromatized to estradiol by a microsomal aromatase, P450aro (CYP19A1) (Shen et al., 1994), in that estradiol has been detected in the embryonic adrenal gland (Tanabe et al., 1979, 1986; Ottinger et al., 2001), and that it persists into the posthatch period (Ottinger et al., 2001). But this sex steroid pattern may not be uniform among the avian species. For example, in the developing zebra finch, the adrenal expresses P450scc and 3βHSD2 but not P450c17 and P450aro, and the converse is true in the gonadal tissue (Freking et al., 2000). However, in the adult zebra finch, P450c17 activity is only detected in the male adrenal gland, and this level of activity is about 1/10th to 1/100th that of the ovary and testis (Schlinger et al., 1999). Thus, in some species, there may be an "adrenal–gonadal unit" in which the adrenals contribute precursors for gonadal steroidogenesis. This may have significance in influencing primary sex bias in offspring. In the domestic hen, raising plasma testosterone at the critical period of meiotic segregation produces a significant bias toward male embryos (Pinson et al., 2011).

Degradation pathways exist in the avian adrenal gland during development, in the posthatch period, and in the adult. Metabolites of progesterone, corticosterone, cortisol, estradiol, and testosterone have been detected. Thus, the adrenal gland has the sets of enzymes to carry out both degradative and regenerative reactions. For example, during the embryonic period, progesterone can undergo a 5β- and, to a lesser extent, a 5α-reduction to 5β- and 5α-pregnanedione (5β- and 5α-pregnan-3,20-dione) (Gonzalez et al., 1983).

In the adult gland, a more prominent enzyme is 5α-reductase (Miller and Auchus, 2011; Langlois et al., 2010). The activity of this enzyme has been shown to degrade corticosterone to 5α-dihydrocorticosterone (5α-pregnan-11β,21-diol-3,20-dione) and 5α-tetrahydrocorticosterone (5α-pregnan-3α,11β,21-triol-20-one) (Carsia et al., 1984). In dispersed chicken adrenocortical cells stimulated with ACTH, 5α-reductase activity is acutely enhanced by exogenous corticosteroids, and is part of a short-looped negative feedback on the avian adrenocortical cell (Carsia et al., 1984, 1987b). The importance of this negative feedback is demonstrated by the fact that may be overridden with certain stressors, such as protein restriction stress, resulting in enhanced corticosterone secretion (McIlroy et al., 1999). It is also acutely inhibited by prolactin (Carsia et al., 1984, 1987b). This acute action of prolactin is Ca^{2+} dependent and occurs with a half-effective concentration (EC_{50}) of 55 ng/ml, a concentration that is well within the physiological range of circulating prolactin. Furthermore, this *in vitro* effect of prolactin on the avian adrenocortical cells to facilitate corticosterone secretion by limiting its degradation may have broader physiological significance. Corticosteroid production by dispersed chicken adrenocortical cells increases abruptly at hatch and is maximal at one day posthatch (Carsia et al., 1987a). Similarly, prolactin and the expression of its receptor increase abruptly approaching the perihatch period (Leclerc et al., 2007), and in the adult adrenal gland, the receptor is expressed at a level as high as the gonad (Ohkubo et al., 1998). In addition, some studies suggest that prolactin levels can promote corticosterone secretion (Koch et al., 2004; Miller et al., 2009), and injections into chick embryos dramatically raise plasma corticosterone within 2 h (Kühn et al., 1996). Furthermore, the plasma levels of the two hormones appear linked with migratory condition (Holberton et al., 2008). Finally, depending on species and life history stage context, corticosterone response and prolactin response to certain stressors may be complementary (Árnason et al., 1986; Angelier and Chastel, 2009); however, the consistency of this relationship is controversial (Angelier et al., 2012).

The adrenal glands of birds, like that of mammals (Cole, 1995), probably have 11β-hydroxysteroid dehydrogenase activities that metabolize (11βHSD2) and regenerate (11βHSD1) corticosterone and cortisol and 17βHSD activity to degrade estradiol to estrone (see Figure 26.6). Although *in vitro* studies support the presence of these metabolites in homogenates of avian adrenal glands (DERoos, 1961; Tanabe et al., 1986), the regulation of these pathways awaits future studies.

26.2.4 Transport of Corticosteroids

In birds, most circulating corticosterone and cortisol are transported bound in dynamic equilibrium to plasma proteins. More than 90% of circulating corticosteroids are bound to the α-2 globulin, corticosteroid-binding globulin (CBG) (that has a high affinity and low binding capacity for corticosteroids). Minor amounts are bound to low-affinity, high-capacity, nonspecific binding proteins (mostly albumin) (Wingfield et al., 1984). Complementing a previous study (Malisch and Breuner, 2010), a random survey containing more recent studies from different species suggests a fairly wide range of affinities (K_d), 1.2–8.3 nM, and capacities, 38–225 nM, for corticosterone (white-crowned sparrow, *Zonotrichia leucophrys gambelii* (Lynn et al., 2003; Charlier et al., 2009); barn owl, *Tyto alba* (Almasi et al., 2009); zebra finch, *Taeniopygia guttata* (Schmidt et al., 2010); Japanese quail, *Coturnix japonica* (Malisch et al., 2010); American kestrel, *Falco sparverius* (Whitman et al., 2011); house sparrow, *Passer domesticus*; northern mockingbird, *Mimus polyglottos*; curved-bill thrasher, *Toxostoma curvirostre*; Albert's towhee, *Pipilo alberti*; and canyon towhee, *Pipilo fuscus* (Fokidis et al., 2009)). In chickens, corticosterone binds to CBG with an average K_d of about 2 nM and a capacity of about 50 nM (Fässler et al., 1986). In some species, cortisol binds to avian CBG with a similar affinity and capacity (Schmidt et al., 2010); however, in many species, it binds with affinities similar to those of the sex steroids (see Malisch and Breuner (2010)). CBG also binds progesterone with high affinity, and testosterone and dihydrotestosterone with a lower affinity, but within the physiological range of these sex steroids. Binding of estrogen is negligible.

A number of factors affect the circulating concentration of CBG, and these responses vary with species and life history stage (Malisch and Breuner, 2010). Various perturbations or stressors have been shown to increase plasma CBG capacity (Malisch and Breuner, 2010), such as poor environmental conditions (Almasi et al., 2009), territorial intrusion (Charlier et al., 2009), and the composite stress of urbanization (Fokidis et al., 2009). However, decreases in CBG capacity are seen with fasting stress (Lynn et al., 2003), handling stress (Malisch et al., 2010), and neonatal handling (Whitman et al., 2011). In addition, changes in CBG capacity in response to stress can be acute (within 30 min) (Breuner et al., 2006; Charlier et al., 2009; Malisch et al., 2010). Thus, as in rats (Qian et al., 2011), the CBG response in birds to stressors can be rapid, but its direction is context and species specific. Furthermore, CBG capacity is influenced by ontogenic processes, hormonal manipulations, and other physiological states (see Landys et al. (2006), Wingfield (2013)).

The binding of corticosteroids to CBG and other plasma proteins influences the bioavailability of the corticosteroid for target cells, but this can be altered by changes in competing binding steroids, such as progesterone (Malisch and Breuner, 2010). It is thought that only unbound glucocorticoids enter cells (Breuner et al., 2003; Landys et al., 2006; Malisch and Breuner, 2010). However, there may be differential tissue attraction or targeting of CBG-bound glucocorticoids that may finely regulate their local concentrations and hence their bioavailability and action. In addition, there appear to be isoforms of CBG that preferentially bind cortisol and, thus, influence its half-life for delivery to sensitive tissues (Schmidt et al., 2008, 2010). Furthermore, it is possible that the CBG–glucocorticoid complex may initiate unique cellular processes mediated through unexplored receptors (Hryb et al. (1986), Hammond (1990), Maitra et al. (1993)). Thus, the interaction between glucocorticoids, their cognate binding proteins, and their target tissues is complex, and the precise role played by CBGs in mediating the actions of glucocorticoids awaits further investigation.

26.2.5 Circulating Concentrations of Corticosterone and Aldosterone

Investigations in which the circulating concentrations of corticosterone and aldosterone were measured in the same study are restricted to galliform (chicken, turkey, and Japanese quail) and anseriform (duck) species. The circulating concentration of corticosterone in unstressed birds is in the nanogrampermililiter (ng/mL) range, whereas that for aldosterone is in the picogram-permililiter (pg/mL) range (Table 26.1). There is a tendency for the circulating values to fall from hatch to maturity. Part of this is explained by the decrease in relative adrenal weight with age (Holmes and Phillips, 1976; Carsia and Weber, 1986). However, it also may be due to the decrease in adrenal sensitivity and maximal steroidogenic capacity with posthatch age (Carsia et al., 1985c, 1987a; Carsia and Weber, 1986).

In recent years, carefully controlled studies have demonstrated a good correlation between plasma corticosterone and the concentration of corticosterone in feathers (Koren et al., 2011; Crossin et al., 2013). This relatively noninvasive approach may be useful in predicting survival and reproductive success in various avian species. A more detailed discussion of the use and value of feather corticosterone can be found in the next chapter.

26.2.6 Secretion, Clearance, and Metabolism of Corticosterone and Aldosterone

Not surprisingly, corticosteroid concentrations that are initially high in the adrenal gland and adrenal venous effluent (in micrograms per milliliter), and then are secreted at about a microgram per minute per kilogram (Assenmacher, 1973), are then rapidly diluted in the circulation and distributed to the extracellular fluid.

Circulating concentrations of corticosteroids are maintained by the dynamic balance between metabolic clearance and adrenal secretion. Transmembrane passage of corticosteroids is rapid for all cells (~10^{-4} cm/s) (Giorgi and Stein, 1981), thus facilitating rapid clearance and metabolism in competent cells such as the kidney and liver. Indeed, in a very limited number of studies, the average half-lives of corticosterone and aldosterone in the circulation are similar, about 15 and 13 min, respectively. These averages, however, do not reflect the species differences in half-lives. For example, in the duck, the half-lives for corticosterone and aldosterone are considerably less—about 8 and 6 min, respectively—and are increased by adenohypophysectomy threefold to eightfold, indicating a dramatic decrease in the metabolic clearance rate (Holmes et al., 1972, 1974). Indeed, these altered circulatory parameters and the altered profile of metabolites in cloacal fluid in response to adenohypophysectomy are restored to normal with corticosterone replacement (Holmes and Slikker, 1976).

The metabolic clearance rate for corticosterone varies with physiological status; it is decreased with age (Holmes and Kelly, 1976) and by thyroidectomy (Kovács and Péczely, 1983). In addition, it appears to be influenced by nutritional status. For example, in chickens subjected to dietary protein restriction, it is increased 85%, which translates to a fourfold increase in secretion rate (Carsia et al., 1988a). Clearance from the circulation is facilitated by intracellular binding and sequestration and ultimately metabolism, primarily by the liver. The metabolizing enzymes include hepatic 5α-reductase, which converts corticosterone and aldosterone to inactive metabolites. The activity of this 5α-reductase appears to be augmented by acute stress (Daniel and Assenmacher, 1971). The metabolites in urine exist as several polar and nonpolar forms, including intact corticosteroids, some free and some conjugated to sulfate and glucuronide (Helton and Holmes, 1973; Holmes et al., 1974; Holmes and Slikker, 1976; Rettenbacher et al., 2004).

Glucocorticoids are the end product signals of a system largely concerned with regulating energy flow—a system of which the HPA axis is a component. By contrast, mineralocorticoids are the end product signals of a system largely concerned with maintaining salt and water balance—a system of which the renin–angiotensin–aldosterone system is a component. These steroid hormones work on specific effector tissues of each system. However, glucocorticoids are also essential to permit the full expression of almost all metabolic processes.

The effector molecules in the effector tissues are the glucocorticoid and mineralocorticoid receptors (MRs), which are ligand-inducible transcription factors. However, the MR binds corticosterone with an affinity that is more than 30 times that of the GR. Since the circulating concentration of corticosterone is about 20–300 times that of aldosterone, almost all of the MRs are fully occupied by corticosterone. Thus, in order for aldosterone to represent the system regulating salt and water balance, corticosterone (and cortisol) in the effector tissues needs to be either sequestered from the MR or rapidly degraded so that its intracellular concentrations exposed to the MR are extremely low.

The main enzymes that carry out the local degradation of corticosterone and cortisol are 11β-hydroxysteroid dehydrogenase 2 (11βHSD2) and 20-hydroxysteroid dehydrogenase (20HSD) (Figure 26.7). 11βHSD2 catalyzes the oxidation of corticosterone and cortisol to the inactive metabolites, 11-dehydrocorticosterone and cortisone, respectively, and 20HSD converts corticosterone to 20-dihydrocorticosterone (see Figure 26.7). 11βHSD2 and 20HSD are found in the chicken intestine, liver and kidney, testis, oviduct, and brain (Vylitová et al., 1998; Ahmed et al., 2013). There also appears to be a second corticosterone (cortisol)-degrading enzyme, 11βHSD3, expressed in the chicken colon and, to a greater extent, in the kidney (Katz et al., 2008). These enzymes are obviously expressed in other avian groups. A recent study shows robust expression of 11βHSD2 and 20HSD in the zebra finch brain and several peripheral tissues analyzed, and, to a lesser extent, in the gonads, moreso in the ovary than in the testis (Katz et al., 2010).

Also expressed in chickens is the enzyme catalyzing the reverse reactions (reductive), 11βHSD1, regenerating corticosterone (and cortisol). There is also evidence for a significant cytosolic form of an 11βHSD1-like enzyme (Katz et al., 2007). Thus, a balance in inactivation and reactivation of glucocorticoids provides a prereceptor modulation of corticosteroid signals in avian effector tissues (Kučka et al., 2006).

Corticosterone, along with sex steroids, does accumulate in the egg (Hahn, 2011) and in a species-specific manner (Quillfeldt et al., 2011), although the role of follicle HSDs in this process is unclear. It is suggested that the bulk of corticosterone that accumulates in the egg is derived from the adrenal glands and is delivered to the egg from the circulation (Hahn, 2011). Indeed, elevating circulating corticosterone concentrations within the physiological range in the barn owl (*T. alba*) result in the deposition of corticosterone in the egg (Almasi et al., 2012). In addition, restraint stress in the Japanese quail (*C. japonica*) increases corticosterone and decreases testosterone in the egg, thereby possibly serving as a signal about adverse environmental conditions from

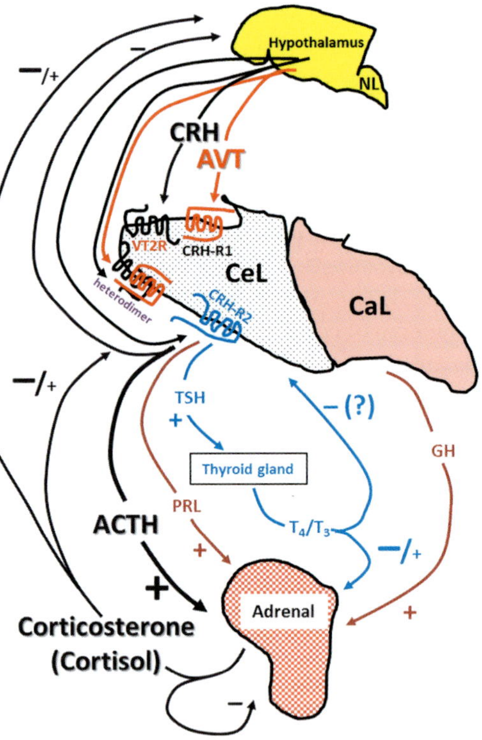

FIGURE 26.7 **The hypothalamus–pituitary–adrenal (HPA) axis in birds.** The hypothalamus is depicted in a sagittal view and with yellow fill. NL is the neural lobe. The pars distalis of the pituitary gland is depicted in a sagittal view with a dot pattern and light brown fill. CeL is the cephalic lobe of the pars distalis, and CaL is the caudal lobe of the pars distalis. The adrenal gland is in a brown pattern to depict the intermingling of chromaffin and adrenocortical tissues. Cartoons of structures are not drawn to actual relative size, with the pituitary disproportionately enlarged. Negative (−) and positive characters (+) and their relative sizes when paired indicate the general consensus of the effect of a particular hormone. CRH, corticotropin-releasing hormone; AVT, arginine vasotocin; CRH-R1, corticotropin-releasing hormone receptor type 1; CRH-R2, corticotropin-releasing hormone receptor type 2; VT2R, vasotocin receptor type 2; ACTH, adrenocorticotropic hormone (corticotropin); TSH, thyroid-stimulating hormone (thyrotropin); PRL, prolactin; GH, growth hormone; T4, thyroid hormone 3,5,3′,5′-tetraiodothyronine; T3, thyroid hormone 3,5,3′-triiodothyronine. CRH and AVT (and sometimes mesotocin) interact with their respective receptors, CRH-R1 and VT2R, to stimulate the release of ACTH in a synergistic manner. ACTH most likely exerts a negative short-loop feedback on CRH release. Part of the synergism of CRH-R1 and VT2R is due to their separate signaling pathways, and it is suggested that part is also due to their receptors undergoing heterodimerization. CRH also interacts with CRH-R2 to stimulate the release of TSH. TSH acts on the thyrocytes of the thyroid gland to release T4 and T3. ACTH stimulates the secretion of glucocorticoids from adrenocortical cells of the adrenal gland. Its action is assisted by the action of PRL, which is also released from the CeL, and GH, which is secreted from the CaL. Sufficient evidence indicates that in addition to ACTH, PRL and probably GH are released during stress. Glucocorticoids exert a negative long-loop feedback on both the pituitary gland and the hypothalamus to curtail ACTH release. However, some brain CRH neurons and pituitary VT2Rs are upregulated by glucocorticoids. Further illustrating the complexity of HPA axis regulation is the role of thyroid hormones. Thyroid hormones appear to decrease VT2R expression and also inhibit the expression of prohormone convertase 1, an enzyme that processes POMC to produce ACTH and other melanocortins. In addition, most studies suggest that thyroid hormones have a negative impact on corticosteroid secretion. Intraadrenal and circulating glucocorticoids also exert a negative short-loop feedback on the adrenocortical cells to reduce the release glucocorticoids. See text for details of additional modulators of the HPA axis. *Cartoon drawn by R. V. Carsia.*

the mother to progeny (Okuliarová et al., 2010). However, it is unclear whether egg corticosterone has a direct effect on embryonic and posthatch development or an indirect effect on egg quality and subsequent progeny development (Henriksen et al., 2013). Perhaps the influence of egg corticosterone on lowering egg androgen may be more significant. For example, in domestic chicks, yolk androgen, in particular androstenedione, which is deposited into egg yolk in substantially higher concentrations than testosterone, is linked with sex-dependent alterations in growth and behavior (Benowitz-Fredericks and Hodge, 2013).

26.2.7 General Regulation of Adrenocortical Function

26.2.7.1 Adrenocorticotropin

The most efficacious stimulator of avian corticosteroid secretion is ACTH, a 39aminoacid peptide derived from pro-opiomelanocortin (POMC) (Takeuchi et al., 1999). Numerous studies *in vivo* and *in vitro* with isolated adrenocortical cells (Carsia, 1990) from several species demonstrate a dose-dependent and rapid (less than 5 min) corticosteroid response to ACTH. Not surprisingly, this rapid response needs to be considered when assessing corticosterone rhythms in birds (Breuner et al., 1999; Romero and Reed, 2005; de Jong et al., 2001).

Of the avian species studied, there are differences between the primary sequences of ACTH in that the chicken ACTH sequence has greater identity with an amphibian ACTH (*Xenopus laevis*) than with that of another galliform (turkey, *M. gallopavo*) and that of a ratite (ostrich, *Struthio camelus*) (Hayashi et al., 1991). Interestingly, despite this difference in primary sequences, both chicken and turkey adrenocortical cells exhibit similar steroidogenic responses to mammalian and avian ACTH peptides (Carsia et al., 1985a, 1988a; Kocsis and Carsia, 1989; Carsia, 1990).

Studies with isolated chicken adrenocortical cells suggest that ACTH operates through specific receptors that are present at a concentration of about 2000–3000 sites per cell (Carsia and Weber, 1988). In birds, ACTH-induced corticosteroid production operates through a G protein–coupled receptor (GPCR), the melanocortin-2 receptor (MC2R) (Takeuchi et al., 1998; Boswell and Takeuchi, 2005), which is activated exclusively by ACTH (Veo et al., 2011). In addition, ACTH is a potent stimulator of the other four avian melanocortin receptor subtypes (Ling et al., 2004). Yet, despite the high identity of the avian MC2R with mammalian orthologs, there are comparative differences in ACTH receptor signaling. For example, unlike rat adrenocortical cells, chicken adrenocortical cells fail to distinguish between mammalian αACTH-(1–24) and its 9-tryptophan-(*O*-nitrophenylsulfenyl) derivative (Carsia et al., 1985a). However, like rat adrenocortical cells, human ACTH-(7–38) is a complete antagonist of ACTH in chicken adrenocortical cells (Carsia and Weber, 1988).

Comparative molecular studies (Veo et al., 2011) and numerous *in vitro* studies with isolated adrenocortical cells from diverse avian species confirm that dissemination of the ACTH signal in avian adrenocortical cells is largely through the cyclic adenosine monophosphate (cAMP)-dependent pathway involving a stimulatory guanine nucleotide binding protein, adenylyl cyclase, cAMP, and cAMP-dependent protein kinase (see Carsia (1990), McIlroy et al. (1999)). In addition, physiological concentrations of Ca^{2+} (Kocsis et al., 1994a) and a background level of calmodulin activity (Kocsis et al., 1995b) appear to be critical for optimal cellular response to ACTH.

Cell surface expression and function of the MC2R require melanocortin receptor accessory proteins (MRAPs) (Hinkle and Sebag, 2009; Webb and Clark, 2010; Cooray and Clark, 2011; Liang et al., 2011; Veo et al., 2011; Cerdá-Reverter et al., 2013). MRAPs are small linear proteins (~14 kDa) of about 100–170 amino acids, having a single hydrophobic transmembrane domain. Cytoplasmic MRAPs form antiparallel homodimers that interact with the MC2R and probably assist folding of the MC2R into a functional conformation for ACTH binding and signal transduction. Formation of the heterotrimeric complex is essential for trafficking to the cell surface. MRAP1 facilitates full receptor function, that is, high-affinity ACTH binding and optimal signal transduction (cAMP production). By contrast, MRAP2 competes with MRAP1, and although it still assists with trafficking of MC2R to the cell surface, it serves as an endogenous inhibitor of MC2R activation by ACTH. MRAPs, StAR, and CYP11A are key steroidogenic proteins comprising the rapid transcriptional response to ACTH (Liu et al., 2013). Future investigations of the avian MRAPs may be important in understanding the changes in adrenal sensitivity and response to ACTH during development, life history cycle and stage transitions, and stress (Lattin et al., 2011a). Furthermore, MRAP2 proteins appear to regulate energy expenditure and energy intake in vertebrates by influencing melanocortin receptors in various brain regions both postnatally and during the transition from embryonic to postnatal life (Asai et al. (2013), Sebag et al. (2013)). Thus, these accessory proteins and their cognate receptors may play similar roles in regulating food intake and energy expenditure in birds.

26.2.7.2 Angiotensins

There is now a strong body of evidence supporting the role of angiotensins in the regulation of avian adrenocortical function (reviewed in Holmes and Cronshaw (1993)). *In vivo* studies and *in vitro* work with adrenal tissue and cells suggest that angiotensins, notably angiotensin II, play a direct role in adrenocortical regulation. This topic is discussed in detail under the "Regulation of Aldosterone Secretion" section.

26.2.7.3 Other Putative Regulators

Studies indicate that avian adrenocortical function is modulated by secretory products of the immune system and other endocrine organs. Interpretations of results derived from these studies should be viewed with caution since they are restricted to anseriform and galliform species and in many instances are based on *in vitro* systems that have not been substantiated *in vivo*. Nevertheless, these studies deserve attention. Such peripheral agents may directly stimulate or inhibit adrenocortical function or may act as positive or negative modulators of tropic-hormone-induced responses.

26.2.7.3.1 Stimulators and Positive Modulators

26.2.7.3.1.1 Prolactin and Growth Hormone The pituitary secretory products, prolactin and growth hormone, appear to have a positive influence on avian adrenocortical function (Figure 26.7). The effect of prolactin has been mentioned here. The actions of growth hormone are less clear. Nevertheless, the importance of growth hormone in avian adrenocortical maintenance is indicated by its hyperstimulation of cellular steroidogenic capacity after hypophysectomy in the absence of ACTH and other hormonal replacement (Carsia et al., 1985c). Additionally, like prolactin, growth hormone is increased with various stressors. Furthermore, growth hormone assists in the overall maintenance of the avian adrenal gland (Harvey et al., 1995).

26.2.7.3.1.2 Calciotropic Hormones In chicken adrenocortical cell preparations, chicken parathyroid hormone (PTH) is a potent stimulator of cAMP, corticosterone secretion, and especially aldosterone secretion (Rosenberg et al., 1987, 1988a,b, 1989a). It is important to point out that the latter studies used purified extracts derived from chicken parathyroid glands. By contrast, a more recent study with recombinant chicken PTH did not demonstrate this novel steroidogenic action (Lim et al., 1991). Nevertheless, avian forms of the PTH and parathyroid hormone-related peptide (PTH–PTHrP) receptor have been characterized (Pinheiro et al., 2012) and are expressed in many tissues. In addition, a PTH–PTHrP receptor in chicken adrenocortical cells, which is absent in chromaffin cells, has been demonstrated. Furthermore, PTHrP enhances ACTH-induced corticosterone production by these dispersed cells (Kawashima et al., 2005).

This corticosteroidogenic effect is not limited to PTH–PTHrP peptides. A similar potentiation of ACTH action is seen with chicken calcitonin, and here again, it appears to act through specific receptors on avian adrenocortical cells and not chromaffin cells (Nakagawa-Mizuyachi et al., 2009). Thus, calciotropic hormones appear to influence the action of ACTH on avian adrenocortical cells.

26.2.7.3.1.3 Humoral Immune System There is evidence suggesting that the humoral immune system may exert trophic and tropic influences on the avian adrenal gland. Although a large part of the ACTH is from the pituitary, it appears that significant quantities of ACTH are released from activated lymphocytes. For example, an antigen challenge or CRH induces the release of ACTH from chicken lymphocytes that is sufficient to elicit a corticosterone response from co-incubated chicken adrenocortical cells (Hendricks et al., 1991, 1995a), and this direct effect of CRH on chicken leukocyte ACTH release is inhibited by corticosterone (Hendricks et al., 1995b). In addition, the bursa of Fabricius may elaborate substances necessary for adequate corticosteroidogenic capacity during development and maturation (Pedernera et al., 1980) and for normal corticosterone response of adrenocortical cells to ACTH (El-Far et al., 1994). However, a bursal suppressive factor has received more attention (*vide infra*).

26.2.7.3.2 Inhibitors and Negative Modulators

26.2.7.3.2.1 Humoral Immune System A bidirectional link between the immune system and the adrenal gland, similar to that which exists in mammals (Besedovsky and del Rey, 1996), is present in birds. There is evidence that the immune system elaborates adrenal suppressor substances. Highly purified extracts of prepubescent chicken bursae suppress inducible steroid synthesis from chicken granulosa and adrenocortical cells and from mammalian adrenocortical cells. The active principle has been termed bursal antisteroidogenic peptide (BASP) (Byrd et al., 1994). BASP appears to be a late-pathway inhibitor in that although it stimulates cAMP production, it inhibits cAMP analog-stimulated steroidogenesis. This peptide awaits full characterization, but recent work suggests that BASP is an ovoinhibitor or is a variant of an ovoinhibitor (Moore et al., 2004).

26.2.7.3.2.2 Thyroid Hormones Evidence indicates that the thyroid hormone, 3,5,3′-triiodothyronine (T_3), is an important modulator of avian adrenocortical function (Figure 26.7). In many cases, there is an inverse relationship between circulating T_3 concentrations and adrenocortical function (Carsia and Weber, 1986; Carsia et al., 1988a). For example, thyroidectomy in chickens results in a marked enhancement in adrenocortical function, and the enhanced function is normalized with T_3 replacement. At least part of this effect of T_3 appears to be direct at the adrenocortical cell level (Carsia et al., 1997). Furthermore, in hypophysectomized cockerels, T_3 replacement alone reduces cellular corticosterone production even more than that due to hypophysectomy in the absence of hormonal replacement (Carsia et al., 1985c). Also, there is some suggestion that thyroid hormones negatively modulate POMC processing in that work with the rat demonstrates that T_3 suppresses pituitary prohormone convertase-1 expression (Li et al., 2001). It should be pointed out, however, that other studies show that thyroid hormones stimulate avian adrenocortical activity (see Sharma and Chaturvedi (2009)), and therefore the thyroid hormones may fall into the biphasic category of avian adrenocortical regulation (*vide infra*).

26.2.7.3.2.3 ***Androgens*** The effect of androgens on adrenocortical secretion appears to be variable depending on the species and type of endocrine manipulation. Exogenous testosterone in wild species seems to enhance circulating corticosterone, suggesting an increase in adrenocortical function (Ketterson et al., 1991; Schoech et al., 1999). Early work suggests that testosterone suppresses adrenocortical function in orchiectomized chickens (Nagra et al., 1965). Nevertheless, more recent work with orchiectomized cockerels suggests that although androgen (testosterone and 5α-dihydrotestosterone) replacement suppresses adrenocortical cell response to ACTH, it maintains relative adrenal weight (Carsia et al., 1987c). Interestingly, orchiectomy is required to demonstrate the effects of androgens on adrenocortical function and domestic fowl (Nagra et al., 1965; Carsia et al., 1987c). Thus, it is quite possible that the testis elaborates both inhibitory and trophic substances that influence avian adrenocortical function.

In this connection, there are reported sex differences in adrenocortical activity, and some of these are correlated to sexual maturity (see Madison et al. (2008)). In addition, sex-dependent alterations in adrenocortical cellular sensitivity and maximal corticosterone response to ACTH have been reported (Carsia et al., 1987a, 1988b; Kocsis and Carsia, 1989). Sex differences are also apparent at more proximal steps of the HPA axis, such as the axis response to corticotropin-releasing hormone and arginine vasotocin (Madison et al., 2008). However, the fact that these alterations at the cellular level are apparent during the embryonic and early posthatch periods and prior to sexual maturity (Carsia et al., 1987a; Kocsis and Carsia, 1989) argues that they are in part sex linked and independent of the higher levels of sex steroids accompanying sexual maturity.

26.2.7.3.3 Biphasic Modulators

26.2.7.3.3.1 ***Thyroid Hormones*** Whereas the use of isolated adrenocortical cells has a caveat when extrapolating findings to *in vivo* physiology, it is amenable to gaining insights into the fine regulation of adrenal steroidogenic function that would be difficult or impossible to obtain from *in vivo* studies. An example is the understanding that a hormone may elicit biphasic influences on steroidogenic function. Thyroid hormones may fall into this category since T_3 suppresses corticosteroidogenic capacity but preserves cellular sensitivity to ACTH in the hypophysectomized chicken (Carsia et al., 1985c). However, in the thyroidectomized chicken, T_3 replacement influences cellular steroidogenic capacity but does not alter cellular sensitivity to ACTH (Carsia et al., 1997). Thus, the prevailing overall hormonal status in which T_3 acts will influence its effect on adrenocortical function. Indeed, as mentioned here, there are studies showing that thyroid hormones stimulate adrenal activity (Sharma and Chaturvedi, 2009).

26.2.7.3.3.2 ***Prostaglandins*** Prostaglandins also seem to fall into this category. Prostaglandins appear to be potent and efficacious stimulators of avian adrenocortical function. In the Japanese quail (*C. japonica*), prostaglandin E_2 is more than six times greater than immobilization stress in raising circulating corticosterone. In addition, it synergizes with immobilization stress to rapidly increase plasma corticosterone (Satterlee et al., 1989). A large part of the effect of prostaglandins appears to be due to their interaction with prevailing concentrations of ACTH. In both chicken and turkey adrenocortical cells, prostaglandins exert a biphasic influence on corticosterone production, depending on their concentration. At low concentrations (10^{-8}–10^{-5} M), they synergize with submaximal concentrations of ACTH, whereas at high concentrations (10^{-5}–10^{-3} M), they are inhibitory. The most potent and efficacious prostaglandin for stimulation is prostaglandin E_2, whereas prostaglandin A_2 is the most potent for inhibition. Prostaglandins have a similar but considerably weaker biphasic influence on basal corticosterone secretion (Kocsis et al., 1999).

26.2.8 Regulation of Aldosterone Secretion

26.2.8.1 Role of Angiotensin II (Ang II)

As in other vertebrates (Wilson, 1984; Holmes and Cronshaw, 1993; Kempf and Corvol, 2001; Nishimura, 2001), the avian kidney contains juxtaglomerular cells capable of producing renin. Renin cleaves fowl angiotensinogen ([Asp^1, Val^5, Ser^9] tetradecapeptide renin substrate) to form the decapeptide angiotensin I ([Asp^1, Val^5, Ser^9] Ang I). The Ang II octapeptide produced from this decapeptide ([Val^5] Ang II; $M_r = 1031.53$) is that of most vertebrates, including ovine and bovine species, whereas other mammalian species produce [Ile^5] Ang II. In several species studied thus far (Japanese quail, duck, and turkey), plasma Ang II concentrations range from 30 to 70 pg/ml (i.e., from 2.9×10^{-11} to 6.8×10^{-11} M) but can increase to 100–400 pg/ml (from 9.7×10^{-11}–3.9×10^{-10} M) with perturbations to electrolyte and hemodynamic balance and with adaptation to saltwater (see Kocsis et al. (1995a)).

Earlier *in vivo* studies demonstrated a direct regulation of aldosterone production by Ang II. An intraperitoneal injection of Ang II into Japanese quail (Kobayashi and Takei, 1982) and infusions of Ang II into ducks (Gray et al., 1989) increase circulating aldosterone, with the latter study showing a dose-dependent aldosterone response exclusive of alterations in circulating corticosterone. In addition, an exclusive, dose-dependent aldosterone response to Ang II has been demonstrated with duck adrenal tissue, specifically from the subcapsular region (Klingbeil, 1985; Holmes and Cronshaw, 1993). This is also true for isolated turkey adrenocortical cells (Kocsis et al., 1994b, 1995a), largely because Ang II does not stimulate cholesterol side chain cleavage in these cells (Kocsis et al., 1995a). Presumably, the prevailing basal level of precursors within the cells is sufficient to support aldosterone synthesis. However, the distinction of functional zones exhibited by the duck adrenal gland may be less apparent or absent in other species. For example, in studies with isolated adrenocortical cells from normal turkeys, there are few functional differences between density-separated subpopulations. However, after dietary Na^+ restriction, one subpopulation, which comprises about 5% of the total adrenocortical cell population, exhibits disproportionately greater aldosterone responses to Ang II and potassium (K^+) but not to ACTH (Kocsis et al., 1995a). It may be that in other avian species, perturbations in electrolyte and hemodynamic homeostasis or other stressors are required to induce zona glomerulosa–like function in a subpopulation of adrenal steroidogenic cells. Furthermore, conceivably other cell subpopulations with distinct function may be induced by different stressors. Indeed, this is the case in the domestic turkey subjected to protein restriction (Carsia and McIlroy, 1998; Carsia and Weber, 2000) and in which Ang II may play a specific role. In this stress model, Ang II may induce cell death in some cell subpopulations, permitting other subpopulations with different steroidogenic function to predominate.

Finally, although it is fairly clear what the releasing factors for aldosterone are in birds with decreased plasma Na^+ and no salt glands, that is, gallinaceous birds, it is not clear what the aldosterone-releasing factors are in birds with salt glands and excessive Na^+ intake or when birds with salt glands undergo dehydration. It does not appear to be Ang II or changes in plasma K^+. Furthermore, the interaction of hormones regulating aldosterone secretion has not been studied in birds that live exclusively in marine and desert environments (see Hughes (2003)).

26.2.8.2 Mechanism of Ang II Action in Avian Adrenocortical Cells

Studies with duck adrenal membrane preparations (Gray et al., 1989) and isolated turkey (Kocsis et al., 1994a,b, 1995a,b,c; Carsia and McIlroy, 1998) and chicken (Holmes and Cronshaw, 1993) adrenocortical cells indicate that the avian adrenal gland possesses high-affinity Ang II receptors, having a K_d of 0.9–2 nM and a concentration of about 30,000–150,000 sites per cell. It should be pointed out that chickens are odd in that mature birds exhibit a complete aldosterone tachyphylaxis to Ang II both *in vivo* and *in vitro*, even though their adrenocortical cells express high-affinity Ang II receptors (Holmes and Cronshaw, 1993). In addition, the receptor concentration may vary among cell subpopulations, as demonstrated in the turkey adrenal gland, and the receptor concentration among cell subpopulations may be differentially modulated by physiological stressors, such as dietary Na+ restriction (Kocsis et al., 1995a).

Turkey (Murphy et al., 1993) and chicken (Kempf and Corvol, 2001) Ang II receptors have also been cloned. They are both 359-amino-acid proteins (sharing 99.7% identity) and exhibit a 75% sequence identity with the mammalian type 1 receptor, although there are frank differences in the pharmacological and physicochemical properties between mammalian and avian forms. Both avian forms have been shown to couple to the phosphoinositide–lipid–protein kinase C pathway (Murphy et al., 1993; Kempf and Corvol, 2001). However, differences in tissue expression exist between turkeys and chickens. For example, the cloned turkey receptor is exclusively expressed in the adrenal gland (Carsia et al., 1993), whereas the chicken receptor has a broader tissue expression (Kempf and Corvol, 2001). In support of ligand binding isotherms with turkey adrenal cells, the galliform adrenal Ang II receptor is not expressed in adrenal chromaffin cells (Kocsis et al., 1995b; Kempf and Corvol, 2001) but is expressed in the postganglionic catecholaminergic cells of the ganglia adjacent to the gland (Kempf and Corvol, 2001; Nishimura, 2001). Furthermore, it is expressed in chicken vascular endothelial cells but not in vascular smooth muscle cells (Kempf and Corvol, 2001; Nishimura, 2001). These observations support the model to explain the biphasic action of Ang II on chicken vascular tonus (Nishimura, 2001): the rapid vasorelaxation after infusion is due to the response of vascular smooth muscle cells to cyclic guanosine monophosphate (cGMP) subsequent to nitric oxide generation by Ang II–stimulated endothelial cells; the later response of α-adrenergic-dependent vasoconstriction is due to the more delayed catecholamine release from Ang II–stimulated postganglionic neuronal endings.

In turkey adrenocortical cells, Ang II induces a rapid and sustained increase in cytosolic Ca^{2+} (Kocsis et al., 1994b, 1995b). However, demonstrating a link between increases in cytosolic Ca^{2+} and aldosterone production has been difficult. Avian corticosteroidogenesis is fairly refractory to agonists and antagonists of the Ca^{2+}–phosphoinositide–lipid–protein kinase C pathway (Rosenberg et al., 1988b; Kocsis et al., 1995b). However, extracellular Ca^{2+} is required for optimal Ang II–induced aldosterone secretion (Kocsis et al., 1994a), and calmodulin activity appears to be essential for aldosterone synthesis in turkey adrenocortical cells (Kocsis et al., 1995b). Even more puzzling is the action of K^+. Extracellular K^+ does nothing to aldosterone secretion in chicken adrenocortical cells (Rosenberg et al., 1988b). By contrast, it is an absolute requirement for Ang II–induced aldosterone production by turkey adrenocortical cells (Kocsis et al., 1994a, 1995b), having a maximal effect at a physiological concentration (about 5 mM; Rosenberg and Hurwitz, 1987). Here again, this effect of K^+ in turkey adrenocortical cells is not linked to changes in cytosolic Ca^{2+} (Kocsis et al., 1995b). In addition, in turkey adrenocortical cells, Ang II–induced increases in cytosolic Ca^{2+} and aldosterone production appear to be dissociated (Kocsis et al., 1995b). Thus, in agreement with other studies of nonmammalian vertebrates, there appear to be unidentified Ang II receptors awaiting molecular characterization (Kempf and Corvol, 2001; Nishimura, 2001).

26.2.8.3 Action of ACTH on Aldosterone Secretion

Whereas there is an ample number of studies demonstrating the selective action of Ang II on avian aldosterone secretion, the action of ACTH is less specific, stimulating both aldosterone and corticosterone secretion. Overall, most studies *in vivo* and *in vitro* indicate that ACTH is a competent stimulator of aldosterone secretion. Indeed, although less potent, ACTH is 2- to 10-fold more efficacious than Ang II in stimulating aldosterone secretion in studies with avian adrenal tissue and avian adrenocortical cell preparations. These observations raise further questions concerning the importance of Ang II. However, the specificity and importance of Ang II reside not in its ability to greatly stimulate aldosterone secretion, but in its ability to link perturbations in electrolyte and hemodynamic balance to aldosterone secretion—a role not readily fulfilled by ACTH. Furthermore, with turkey adrenocortical cells, threshold and suboptimal concentrations of ACTH greatly augment the efficacy of Ang II in stimulating aldosterone secretion (Kocsis et al., 1994a). Thus, against an ACTH background *in vivo,* Ang II probably stimulates aldosterone secretion to a magnitude that exceeds what is revealed when tested alone *in vitro.*

26.2.8.4 Action of Atrial Natriuretic Peptide (ANP) on Aldosterone Secretion

In general, there are few studies on the regulation of aldosterone secretion in avian species, but even less is known about interacting factors in its regulation. Somatostatin has been shown to inhibit Ang II–stimulated, but not ACTH-stimulated, aldosterone production by turkey adrenocortical cells (Mazzocchi et al., 1997c). However, more attention has been paid to the role of ANP in avian aldosterone secretion.

Avian ANP was originally demonstrated as a diuretic agent in extracts from chicken hearts (Gregg and Wideman, 1986). It is a 29-amino-acid peptide (M_r=3158.45) that has significant structural homology to mammalian ANPs (Miyata et al., 1988). Avian ANP is stored and released from both atrial and ventricular cardiocytes (Toshimori et al., 1990), but it is also produced by adrenal chromaffin cells (Wolfensberger et al., 1995). In ducks and chickens, the plasma concentration of ANP is about 70–80 pg/ml (~2.5×10^{-11} M) and is inversely related to changes in blood volume in response to hemodynamic and electrolyte perturbations (Gray et al., 1991a; Gray, 1993). ANP receptors are found in both renal and adrenal tissue of the duck (Gray et al., 1991b). In adrenocortical cells, it is present at a concentration of ~90,000 sites per cell (Kocsis et al., 1995c). The receptors have an apparent K_d of 1–3 nM (Gray et al., 1991b; Kocsis et al., 1995c). In the duck, infusions of avian ANP inhibit plasma aldosterone responses to Ang II without affecting circulating corticosterone (Gray et al., 1991b). In addition, in chicken adrenocortical cells, ANP is a potent (EC_{50}~1 nM) and efficacious inhibitor (>90%) of maximal ACTH- and PTH-induced aldosterone secretion without affecting corticosterone secretion (Rosenberg et al., 1988b, 1989b). ANP appears to act at several intracellular loci (Rosenberg et al., 1989b), and cGMP appears to be its second messenger (Rosenberg et al., 1988b, 1989b). Interestingly, in turkey adrenocortical cells, ANPs are as efficacious as Ang II in stimulating aldosterone production and actually augment maximal aldosterone production induced by Ang II, K^+ and ACTH (Kocsis et al., 1995c). Obviously, further work with diverse avian species is required to resolve this inconsistency and to understand the physiological relevance of ANP.

26.2.9 Overview of the HPA Axis

The HPA axis (Figure 26.7) is composed of hypothalamic neurons that contain the releasing hormones CRH, arginine vasotocin (AVT), and mesotocin (MT); the corticotropes of the cephalic lobe of the pituitary gland, which contain enzymes that synthesize and process (prohormone convertases) POMC to produce ACTH (corticotropin); and the steroid-secreting cells of the adrenal gland, which express MC2R, which exclusively binds ACTH.

It is well established that glucocorticoids exert a negative feedback on both the hypothalamus and pituitary gland to decrease CRH and POMC in a classical negative feedback (Denver, 2009). However, this negative feedback is finely modulated by changes in the pituitary AVT receptor (vide infra). Furthermore, ACTH itself appears to exerts a short-loop feedback on the hypothalamic CRH neurons (Jankowski et al., 2009), and, as mentioned in this chapter, glucocorticoids induced a short-loop feedback on the adrenocortical cell, probably by reducing MC2R expression and signal transduction (McIlroy et al., 1999).

The components of the HPA axis interact to produce corticosterone rhythms that are thought to regulate homeostatic mechanisms (Breuner et al., 1999; de Jong et al., 2001). However, the different components can have a level of independent but complex regulation suited for life history stage and its interaction with season, suited for a changing and unpredictable environment (Romero et al., 1998; Romero and Rich, 2007; Cyr and Romero, 2009; Lattin et al., 2011a), or as a result of genetic selection (Carsia and Weber, 1986; Carsia et al., 1988b; Hazard et al., 2007). In addition, the HPA axis operates within a larger regulatory system for energy flow and partitioning, especially for the central nervous system.

CRH and AVT are the main stimulators of ACTH secretion from corticotropes, the ACTH-secreting cells of the cephalic lobe (Berghman et al., 1998). CRH is a 41-amino-acid peptide that binds to the type 1 CRH receptor (CRH-R1) on corticotropes and activates adenylyl cyclase to elevate intracellular cAMP, resulting in the release of ACTH (Carsia et al., 1986; Kuenzel et al., 2013). It also binds to the type 2 CRH receptor (CRH-R2) on thyrotropes and stimulates the secretion of thyroid-stimulating hormone (thyrotropin) (TSH) (De Groef et al., 2005, 2006). This is not surprising since thyroid hormones regulate intermediary metabolism, and the HPA and hypothalamic–pituitary–thyroid (HPT) axes interact tightly to regulate certain developmental processes and hatching in birds. Furthermore, the hypothalamus–pituitary–gonadal axis may interact with the HPA axis as well since there are sex differences in the plasma corticosterone response to CRH in AVT (Madison et al., 2008) and in AVT release and hypothalamic AVT receptor expression in response to osmotic stress (Chaturvedi et al., 2000). Finally, there are periodic resurgences of interest in a possible interaction between the pineal gland and the HPA axis, since both melatonin and glucocorticoids exhibit circadian rhythms (see Barriga et al. (2002)). However, a definitive link remains elusive.

AVT is a nonapeptide that binds to the type 2 VT receptor (VT2R) (and the type 4 VT receptor) on corticotropes. VTRs activate phospholipase C, leading to the generation of inositol-1,4,5-triphosphate and diacylglycerol and mobilization of intracellular Ca^{2+} (see Kuenzel et al. (2013)). There is evidence that a low thyroid hormone status (Sharma and Chaturvedi, 2009) and high circulating corticosterone (Sharma et al., 2009a) increase VT2R expression and that chemical adrenalectomy decreases it (Sharma et al., 2009a), but POMC expression is unaltered. The effect of glucocorticoids on VT2R expression may be the basis for the observed glucocorticoid programming of the HPA axis for enhanced stress response in zebra finches (Spencer et al., 2009). Also, here again, these studies show the intersection of both the HPA and HPT axes. Furthermore, AVT may recruit the HPA axis in response to osmotic stress (Sharma et al., 2009b). Thus, the intersection of CRH and AVT in regulating stress response in a number of avian species is clearly indicated. It should be pointed out, however, that these molecular studies using semiquantitative methods to measure messenger RNA levels are complex and are better understood when coupled with adrenal function studies. It is not unusual for receptor and POMC expression to be at variance with the adrenocortical response because of the less understood posttranscriptional regulation (Sharma et al., 2009b).

Because CRH and AVT operate through different signal transduction cascades, their combined action leads to synergistic release of ACTH (Mikhailova et al., 2007). In addition, there is evidence that this synergistic effect is also due to heterodimerization of CRH-R1 and VT2R (Mikhailova et al., 2007; Cornett et al., 2013).

Taken collectively, studies indicate that there are multiple pathways to elicit ACTH release from the cephalic lobe for normal rhythms of corticosteroid secretion and for greater corticosteroid responses to address a variety of environmental and homeostatic stressors. The adrenal gland, in turn, has intrinsic mechanisms to permit the episodic release of corticosteroids of varying magnitude in response to varying ACTH levels. This involves not only the rapid transcriptional regulation of such important proteins (MC2R, MRAP, StAR, and CYP11A) but also the rapid induction of inactivation pathways (Liu et al., 2013). Finally, the resultant circadian and ultradian rhythms of free glucocorticoids are highly synchronized between the circulation, the subcutaneous tissue, and the brain (Qian et al., 2012). This underscores the precision of how glucocorticoids are distributed to different body compartments for target tissue-specific functions.

26.2.10 Adrenocortical Function in Development, Maturation, and Senescence

The domestic fowl (*G. gallus domesticus*) is the most studied model to describe the development of the avian adrenal gland. Thus, the pattern of development of this precocial species described herewith may be somewhat different from those of semiprecocial, semialtricial, and altricial species.

The chick adrenal begins as an adrenogenital primordium that is apparent at about 3 days of development. Development of steroidogenic cells requires the nuclear receptor, steroidogenic factor-1 (SF1, or NR5A1), which is expressed in the adrenogenital primordium (Smith et al., 1999). SF1 promotes cell growth, limits cell death (apoptosis), and regulates the expression of genes involved in steroidogenesis, the MC2R, and StAR. However, work with knockout mice suggests that additional transcription factors are necessary for adrenocortical cell differentiation from steroidogenic progenitor cells and for the final cytoarchitecture of gland (Pihlajoki et al., 2013). At this time, the expression of key steroidogenic enzymes is detected in the chick embryo (Nomura et al., 1999) and in the developing zebra finch adrenal gland (Freking et al., 2000). However, *in situ* hybridization detects these key enzymes at about 5 days of development (Kanda et al., 2000), and this is in agreement with a more recent gene expression study (Kamata et al., 2004). Also, at about 5 days of development, the segregation of the adrenogenital primordium into the adrenal gland and the gonad is apparent (Smith et al., 1999) as well as the histological identification of steroid-secreting cells arranged in cords (Chimenti and Accordi, 2008). From days 7 to 15 of development, chromaffin cells penetrate gland, and complete morphogenesis of the adrenal gland is complete by day 15.

It is thought that the adrenal gland is capable of steroidogenic function as early as day 5 of development, well before morphogenesis is complete. Embryonic circulating corticosterone levels slowly rise to day 15, but are independent of pituitary ACTH, even though immunoreactive ACTH is present in corticotropes at day 7 and embryonic adrenocortical cells can respond to ACTH. However, negative feedback is probably established by day 11. After day 15 and until hatch, the gland is under ACTH control, and CRH neurons become detectable in the hypothalamus a day earlier (see Jenkins and Porter (2004), De Groef et al. (2008), Ellestad et al. (2011)). So, there is general agreement that the adrenal gland of precocial species is under the control of pituitary ACTH at day 15 of development. From that day to hatch, circulating corticosteroids reach their peak. This rise is matched by an abrupt increase in maximal cellular corticosteroid response and maximal cellular sensitivity to ACTH (Carsia et al., 1987a). However, 1–2 days posthatch, chicks have a stress hyporesponsive period (see Wada (2008)). This is due to a proximal HPA axial block since adrenocortical cells are fully responsive to ACTH (Carsia et al., 1987a). Interestingly, there is evidence for an age-dependent window for "imprinting" the avian adrenal gland for a long-term, enhanced setpoint of response to ACTH (Avrutina et al., 1985). In chickens, this appears to be about 2–4 weeks posthatch. For example, subjecting cockerels to transient protein restriction over this period induces long-term potentiation of adrenocortical cell function (Weber et al., 1990).

In altricial birds, the time frame for HPA axis development is shifted later in the posthatch period. It is hypothesized that the peak corticosteroid response to a fully developed axis occurs at fledging (Wada, 2008). Here again, birds of an earlier age can respond to an ACTH challenge, indicating that the reduced corticosteroid response prior to fledging is proximal to the adrenal gland. It is thought that in altricial species, this uncoupling of the hypothalamus–pituitary unit from the adrenal prevents increases in circulating corticosteroids in response to stress, which could have an adverse effect on growth and immune function. In addition, the body condition of nestlings, such as fat stores, impact the HPA axis response to stress and circulating CBG, which modulates the target tissue effects of the prevailing circulating levels of corticosterone (Müller et al., 2010).

Overall, a limited number of studies indicate a gradual decline in adrenocortical function with age (Schmeling and Nockels, 1978; Avrutina et al., 1985; Davis and Siopes, 1987) and during senescence. In chickens, this decline, in part, may be due to a decrease in adrenocortical cell sensitivity and response to ACTH (Carsia et al., 1985b, 1987a). However, there is some suggestion from work with wild house sparrows (*P. domesticus*) that there is an increase in glucocorticoids that heralds death in birds susceptible to the stresses accompanying the winter season (Koren et al., 2011).

26.3 PHYSIOLOGY OF ADRENOCORTICAL HORMONES

Corticosteroids are the end products of the integration of environmental stimuli that impinges on the HPA axis. They are the transducers of these environmental cues and therefore regulate physiological and behavior processes appropriate to the environmental situation. The action of corticosteroids in birds is highly complex, and sometimes quixotic and paradoxical, depending on species, gender, life history stage, habitat, and a changing environment within a particular habitat. This corticosteroid hormonal pleiotropy is largely due to the expression and interaction of their cognate receptors and the intracellular coactivators and transcription factors. Many studies are associative, linking endogenous or exogenous glucocorticoids to an effect in birds. Given the fact that the MR, GR, and mGR have decreasing affinities to glucocorticoids (~0.2 nM (~69 pg/ml), ~6 nM (~2 ng/ml), and ~20 nM (~6.9 ng/ml), respectively), it is presumed that the effective concentration or dose of free glucocorticoid operates through the corresponding receptor. These studies are relatively underpowered to determine a causal association. Other studies use available mammalian receptor antagonists; however, complete antagonists are lacking. Finally, little is known about the repertoire of intracellular coactivators and other transcription factors that mediate the activated forms of the corticosteroid receptors. Nevertheless, an overview of the corticosteroid receptors and the complex functions that they are thought to regulate are discussed in the remainder of this section.

26.3.1 Corticosteroid Receptors and Their Action in Target Cells

The structure of avian forms of the GR (Kwok et al., 2007) and MR (Proszkowiec-Weglarz and Porter, 2010) has been determined. In traditional ligand-binding and competitive-binding studies, these manifest as the low-affinity site and high-affinity site for corticosterone binding. Both share high homology with mammalian corticosteroid receptors and, for the most part, function as homodimers after corticosteroid binding. However, as seen in mammalian studies, they probably can function as activated monomers in some instances (see Adams et al. (2003)).

Localization of the unstimulated GR and MR is mainly to the cytoplasm, and there is active translocation to the nucleus with corticosteroid treatment (Proszkowiec-Weglarz and Porter, 2010). Transfection of receptors shows that aldosterone is 2–3 times more effective than corticosterone for receptor activation. Interestingly, like aldosterone, the glucocorticoid corticosterone is 10 times more effective in activating the MR than the GR (Proszkowiec-Weglarz and Porter, 2010). However, the mammalian GR antagonist, onapristone, only partially blocked the corticosterone-activated chicken GR, and the mammalian MR antagonist, spironolactone, was a very weak antagonist of the activated chicken MR and actually had agonistic activity. Thus, the mammalian corticosteroid receptor antagonists (Figure 26.8) do not always work with the same fidelity in avian studies. This functional observation in transfection studies supports and explains the receptor-binding affinities observed in ligand-binding and competitive-binding studies, with cytosol and membrane preparations from the tissues of diverse avian species.

Like mammalian GRs, the chicken GR has a putative palmitoylation site (cysteine 660) providing the opportunity for membrane localization. However, like mammalian MRs, the chicken MR has an isoleucine in this site (isoleucine 773), although it is recognized that other portions of the hormone-binding domains of each are sufficient for membrane localization (Grossmann et al., 2008; Samarasinghe et al., 2011). Roughly 5–10% of classical corticosteroid receptors are associated with the plasma membrane. Localization to the plasma membrane involves association with caveolae, specialized plasma membrane invaginations within sphingolipid-rich lipid raft domains. Within the caveolae, the corticosteroid receptors are associated with caveolin-1 protein and other scaffolding proteins. Caveolae are important "signalsomes" in which activated steroid receptors are brought into close contact with various tyrosine kinases, other adaptor proteins, and also GPCRs. It is thought the membrane localization orientates the ligand-binding domain to face the extracellular environment where free corticosteroids can bind. Like the membrane-associated estrogen receptor, it is also thought that homodimerization occurs after binding of corticosteroids to membrane-associated corticosteroid receptors. In general, membrane localization most often leads to extracellular signal-regulated kinases (ERKs) for relatively rapid cellular responses. This type of signaling contrasts with the relatively slower, classical genomic signaling. Because the ERKs phosphorylate proteins that play a role in genomic signaling, it is thought that plasma membrane signaling modulates corticosteroid nuclear signaling—an "outside–inside" regulation of steroid hormonal action (Hammes and Levin, 2011; Hammes and Mendelson, 2012). In birds, this regulation may permit the uncoupling of receptor occupancy from transcriptional processes controlling protein and glucose metabolism in a tissue-specific manner (Orchinik, 1998). It should be pointed out that there is also support for nonclassical membrane receptors for corticosteroids, that is, heptahelical-GPCRs, especially for aldosterone (Wendler et al., 2012), although this novel mode of corticosteroid signaling awaits validation and characterization.

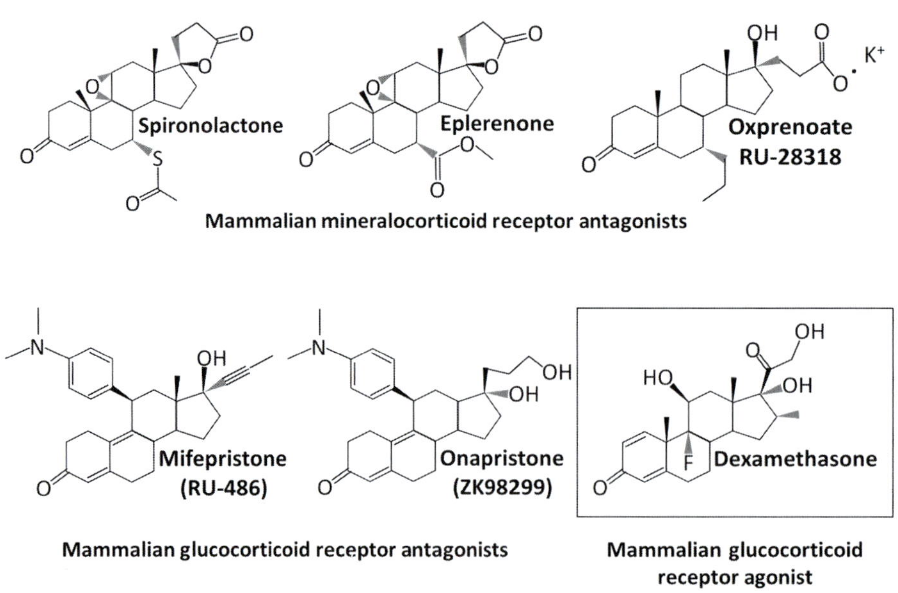

FIGURE 26.8 **Chemical structure of mammalian mineralocorticoid and glucocorticoid receptor antagonists, and the potent glucocorticoid receptor agonist, dexamethasone.** The use of these compounds in avian corticosteroid receptor studies has yielded mixed results, since they sometimes are only partially active or have agonist or antagonist activity opposite their usually complete action in mammalian receptor studies. *Molecules drawn by R. V. Carsia.*

Corticosteroid nuclear signaling involves the activation of monomeric receptors bound to heat shock protein 90 (HSP90). After binding, there is a conformational change freed by the release of HSP90. The receptor undergoes homodimerization and recruits a coactivator complex that contains histone acetyltransferase activity. This favors an opening of the chromatin structure and the recruitment of proteins for transcriptional activation. The complement of genes transcribed is tissue specific. Activated GRs can also interact directly with transcription factors instead of a particular DNA response element. Thus, the effect of corticosterone-activated GR on the immune system is in part by direct interaction with transcription factor nuclear factor κB. This leads to repression of transcription of genes encoding cytokines and cyclooxygenase-2 and, thus, a reduction in inflammatory response.

Corticosteroid receptors have been described through standard ligand- and competitive-binding studies on cytosolic and nuclear preparations of many tissues from diverse species, such as the house sparrow (*P. domesticus*) (Breuner and Orchinik, 2001, 2009; Lattin et al., 2011b) and the zebra finch (*T. guttata*) (Schmidt et al., 2010). The binding sites are sufficiently different from mammalian types in that known mammalian MR antagonists fail to displace corticosterone from the high-affinity site (MR), and, similarly, the mammalian GR antagonist milfepristone (RU-486) is a variable and often weak antagonist of the low-affinity site (GR). The affinities of these mineralocorticoid and GRs for corticosterone are fairly consistent: K_ds of about 0.2 nM and 6 nM, respectively, with a membrane-binding site of about 14–30 nM. Since the affinity of avian CBG is about 2 nM, the high-affinity site (MR) can easily draw corticosterone from CBG compared to the low-affinity site (GR). Thus, corticosterone most likely activates the avian MR and only in certain tissues with high 11βHSD2 and 20HSD activity is aldosterone selectively more active. Interestingly, cytosolic and membrane-associated corticosteroid receptors differentially bind corticosterone and cortisol in a tissue-specific manner (Schmidt et al., 2010). This differential binding of glucocorticoids is most likely due to posttranslational modification of the GR and MR. It is plausible that the membrane and cytosolic GR and MR and related isomorphs from posttranslational modifications and perhaps even novel corticosteroid receptors interact to provide a complex regulation of effector tissues.

The cloned avian GR and MR also serve as molecular tools to assess receptor expression in the avian brain (Hodgson et al., 2007; Dickens et al., 2009, 2011b; Banerjee et al., 2012). For example, various stressors such as maternal care deprivation (Banerjee et al., 2012); chronic, combined stress protocols (Dickens et al., 2009); and genetic selection for heightened stress response (Hodgson et al., 2007) all show differential changes in MR and GR expression in various regions of the brain, whereas translocation stress (Dickens et al., 2011b) is not associated with changes in expression. However, studies of corticosteroid receptor expression based on messenger RNA should be viewed with caution because of poor correlation with actual acceptor protein (i.e., corticosteroid receptor) levels in neuronal tissue (Medina et al., 2013).

26.3.2 Corticosteroids and Intermediary Metabolism

Since the HPA axis is within a larger system that regulates energy flow and partitioning, the overall action of glucocorticoids is to maintain adequate circulating concentrations of glucose. Studies in birds indicate that glucocorticoids exert profound effects on protein, carbohydrate, and lipid metabolism for this purpose (see Landys et al. (2006), Scanes (2009)). As a result, there is a tendency for prolonged elevated glucocorticoids to decrease body mass, but this is not a consistent effect among wild and domestic species because the increase in fat depots sometimes compensates for protein loss. However, baseline glucocorticoids are permissive for processes that maintain adequate food intake and optimal body mass. Similarly, some studies indicate that high glucocorticoids increase food intake and, in defense of protein partitioning, stimulate a preference for diets with higher protein value and negatively impact feed efficiency. Indeed, glucocorticoids operating through an interaction between the MR and GR seem to influence food choice. Furthermore, blocking GR activation (e.g., RU-486) suppresses the compensatory increase in food intake during periods of high energy demand, but not other types of feeding behavior during periods of low energy demand. Moreover, the variations in expression of brain corticosteroid receptors may underlie the changes in feeding behavior in different life history stages, as for example migration (discussed further in this chapter). Glucocorticoids also appear to work on the alimentary system by increasing transit time and the uptake of glucose, calcium, and phosphate. Further complicating the picture is that high glucocorticoids may differentially increase oxidative stress in tissues, impacting the interpretations of changes in key metabolic indicators (Costantini et al., 2008).

Since most studies use some form of glucocorticoid depot or glucocorticoid feeding regimen, the normal rhythms of glucocorticoids and, in turn, the complementary rhythms of delivery to target tissues, as well as the normal responses of interacting endocrines to those rhythms, are largely disrupted. Thus, it is not clear whether the responses in birds to exogenous glucocorticoids are physiological or pharmacological. Nevertheless, some conclusions as to their action can be drawn from the large number of studies conducted in diverse avian species.

26.3.2.1 Protein Metabolism

The action of glucocorticoids largely results in protein catabolism (see Landys et al. (2006), Scanes (2009)). The amino acids enter the citric acid cycle as an energy source and also shunt to hepatic, renal, and possibly intestinal sites (Watford, 2005) for gluconeogenesis to maintain adequate glucose levels. In domestic fowl, the important enzyme for gluconeogenesis is cytosolic phosphoenolpyruvate carboxykinase. In young chickens, the activity of this enzyme is localized in both the kidney and liver, whereas in the adult, it is predominantly in the kidney.

Frank changes in circulating glucose are not always apparent because of facilitated entry into cells through the action of insulin. Indeed, in domestic fowl, corticosterone induces hyperinsulinemia. However, a number of molecular markers also indicate insulin resistance in the liver and muscle and an increase in muscle glycogen (Lin et al., 2007). This paradox may be due to species, dose of exogenous glucocorticoid, and route of delivery and duration of treatment.

Since one of the largest sources of protein is muscle, it is not surprising that there is muscle breakdown. Strong indicators of generalized protein catabolism and more specific muscle protein breakdown are increases in circulating uric acid and the concentration of 3-methyhistidine in muscle, respectively. In domestic fowl, both sarcoplasmic and myofibrillar protein breakdown is apparent, and protein synthesis is depressed (Dong et al., 2007). However, in wild species, there is evidence for a differential effect in which sarcoplasmic reticular proteins are preferentially degraded, whereas myofibrillar proteins are initially spared, presumably to preserve flight muscle. The effect of glucocorticoids on protein metabolism is clearly concentration dependent, suggesting that these steroid hormones finely modulate protein metabolism.

26.3.2.2 Lipid Metabolism

Exogenous glucocorticoids, at concentrations reflecting ranges just exceeding circulating baseline levels and probably into stress levels, do have an impact on lipid metabolism (see Landys et al. (2006), Scanes (2009)). Again, as with protein catabolism, lipid stores are accessed to preserve circulating glucose and to provide an alternate energy source. Evidence suggests that the action of glucocorticoids on lipid metabolism is mediated through the low-affinity GR in that some actions are blocked by RU-486. Indeed, thus far, only the low-affinity, GR-type receptor is detected in avian adipose tissue (Lattin et al., 2011b).

Glucocorticoids restrict glucose utilization in many tissues, including adipose tissue. This drives some lipolysis but mostly depresses free fatty acid re-esterification, some of which is used as an energy source via the citric acid cycle within adipose tissue and some of which is released into the circulation. In addition, glucocorticoids regulate amino acid availability and utilization within adipose tissue to provide enzyme intermediates within the citric acid cycle.

In birds with positive energy balance, glucocorticoids also stimulate lipogenesis and fat deposition in the liver, and increase abdominal fat pad size. This lipogenic effect may be driven largely by insulin. In adequately fed, growing poultry, all glucocorticoid-driven effects on lipid metabolism are seen: an increase in adiposity and hepatic lipogenesis, an increase in circulating free fatty acids, and a decrease in glucose utilization.

In mammals, leptin is the circulating hormone derived from the adipose secretome that integrates the HPA axis, feeding behavior, and fat stores (e.g., see Glasgow et al. (1998)). Many wild avian species have extreme seasonal and life history stage–specific differences in appetite and fat accumulation. Some undergo fat depot changes that comprise up to 30% of their body mass. Yet, a factor like leptin of the avian adipose secretome remains elusive, and it is suggested that the cloned avian (chicken) leptin receptor is an orphan receptor (see Hen et al. (2008), Kordonowy et al. (2010), Pitel et al. (2010), Yosefi et al. (2010), Gogga et al. (2013)).

26.3.3 Corticosteroids and Electrolyte Balance

Birds regulate both plasma electrolytes and plasma volume within fairly narrow limits. Corticosteroids are critical for maintaining electrolyte and water balance. Most of the action of corticosteroids is mediated by the high-affinity MR. Thus, prevailing glucocorticoids are sufficient to maintain this balance. However, specific detection systems of perturbations in this balance activate the MRs in the main target tissues—the kidney, intestine, and hindgut—by deploying the potent mineralocorticoid aldosterone. Since some extreme perturbations also call in the HPA axis, it is especially important that glucocorticoid-inactivating enzymes (11βHSD2 and 20HSD) are upregulated and/or activated in these target tissues to permit aldosterone action (see Hughes (2003), Landys et al. (2006), Laverty et al. (2006)). This is especially important in birds with life history stages between freshwater and marine habitats, or between water-replete environments and arid environments.

Interestingly, the mechanism of aldosterone action on these target tissues is similar between birds without and with salt glands. Birds without salt glands cannot tolerate high Na^+ intake, but they do respond to low Na^+ intake. In this situation, there is an expected increase in circulating aldosterone in order to maintain adequate plasma Na^+ concentrations. Aldosterone acts on the kidney to reabsorb Na^+ from the glomerular filtrate. In addition, it increases the Na^+ channels and their activity in the colon and coprodeum to reabsorb lost Na^+ from the ureteral fluid. Furthermore, it induces ureteral fluid reflux from the coprodeum to more orad (proximal) segments for additional cycles of Na^+ reabsorption.

In estuarian and marine birds with salt glands, aldosterone levels are maintained and elevated with high Na^+ intake through an unknown mechanism. The actions of aldosterone are for the most part the same as in birds without salt glands. Birds with salt glands tend to have larger kidneys and greater rates of glomerular filtration compared to birds without salt glands. Thus, there is an increase in Na^+ and water reabsorption at the kidney and hindgut with refluxing to augment reabsorption. However, in birds with salt glands, the resultant excess circulating Na^+ is secreted from the salt glands, which are stimulated by high plasma osmolality and high extracellular fluid volume but are not under the control of aldosterone. In this way, water is added to the extracellular fluid without the high concentrations of Na^+, thus maintaining a plasma volume and osmolality compatible with life.

26.3.4 Corticosteroids and Immune Function

Some aspects of the immune system influences on avian adrenocortical function have been mentioned in previous sections of this chapter. The focus here is on the action of corticosteroids on immune tissues. The obvious importance of corticosteroids in immune regulation is evinced by a growing body of literature indicating that the immune tissues of birds have an intrinsic system for either the local production of corticosteroids (Lechner et al., 2001) and/or the enzymes capable of reactivating circulating corticosteroid metabolites (Schmidt et al., 2008). Furthermore, it appears that avian immune tissues possess both MRs and GRs, with the suggestion that the MRs are activated preferentially by corticosterone (and aldosterone?) and the GRs by cortisol (Schmidt et al., 2010; Lattin et al., 2011b).

In general, birds have a different repertoire of immune organs, immune cells, and elaborated signaling molecules and immune-regulating genes compared to those of mammals. For example, birds lack lymph nodes for antigen presentation and rely predominantly on tissue dendritic cells. Nevertheless, the observed effects of glucocorticoids on the avian immune system appear consistent with those seen in the mammalian immune system, that is, the effects are largely immunosuppressive (see Mumma et al. (2006), Kaiser et al. (2009), Shini et al. (2010)). Exogenous corticosterone reduces the relative weight of the spleen and the bursa of Fabricius. As mentioned here, an antigen challenge or CRH induces the release of ACTH from chicken lymphocytes that is sufficient to elicit a corticosterone response from co-incubated chicken adrenocortical cells (Hendricks et al., 1991, 1995a), and this direct effect of CRH on chicken leukocytes is inhibited by corticosterone (Hendricks et al., 1995b). Conversely, immune stimulation upregulates high-affinity ACTH and GRs in lymphocytes (Mumma et al., 2006). However, there is evidence showing an initial phase of stimulating antibody response and a general positive influence on the adaptive immune response. Exogenous corticosterone stimulates heterophils and reduces leukocytes, thereby increasing the heterophil–leukocyte ratio. This response is different than that stimulated by lipopolysaccharide treatment, suggesting that different cytokines are stimulated by corticosterone. Indeed, corticosterone treatment initially upregulates the expression of proinflammatory chemokines. However, prolonged corticosterone treatment, thought to mimic prolonged stress, downregulates proinflammatory cytokines and chemokines (Shini et al., 2010). Thus, corticosterone may finely modulate immune response, and its group of corticosteroid receptors may serve as an important avian immune rheostat to ensure optimal immune response in birds.

26.3.5 Corticosteroids and Behavior

Behavior strategies used by birds, especially breeding behavior, are energy costly. Glucocorticoids and other hormones function to partition energy flow in order to balance breeding-related activity and self-maintenance and survival-oriented activities (Landys et al., 2006; Cornelius et al., 2013a,b; Wingfield, 2013). This balance is strongly influenced by the investment in breeding for the perpetuation of the species (Cornelius et al., 2013b). Those species with a high investment in a breeding cycle because of limited opportunities for breeding or because of short life spans employ mechanisms to dampen the HPA axis, allowing for just basic maintenance levels of corticosterone for permissive functions that support other hormones involved in energy partitioning. For those species with greater life spans that permit numerous opportunities for breeding, greater emphasis is on maintaining an active HPA axis in order to exploit those opportunities. Since many activities are associated with preparation for breeding, migration, territorial behavior, nest building, and so on, it is not surprising that glucocorticoids regulate these activities.

In general, glucocorticoids affect activity patterns, cognitive and learning processes, reproductive parameters such as courtship, copulation, and parental behavior such as interaction with offspring and procurement of food for offspring (Mostl and Palme, 2002; Rubolini et al., 2005). Since glucocorticoids regulate a number of locomotory patterns of birds, it is not surprising that locomotory activity and circulating glucocorticoid levels are tightly linked (Landys et al., 2006). The effect of glucocorticoids is rapid, is concentration dependent, and is not blocked by the GR antagonist RU-486. This suggests that the receptor mode employed to regulate locomotory activity is nongenomic and perhaps even of the nonclassical membrane type.

Glucocorticoids are essential for processes related to migration. They regulate the replenishment of energy stores at stopover points and at final breeding grounds by increasing brain neurotransmitters controlling the increase in food intake, by decreasing thermogenesis, and by promoting fat deposition (Cornelius et al., 2013a). They also regulate the transition from migration to breeding in that high circulating levels inhibit the progression into breeding if environmental conditions are not conducive for nesting.

Of recent interest and of great importance is the role of glucocorticoids in the behavioral responses of birds to anthropogenic stressors and the worldwide encroachment of urbanization (Bonier, 2012; Wingfield, 2013). Evidence indicates that these persistent abiotic stressors can shape endocrine traits, including glucocorticoid responses. As mentioned here, glucocorticoids can be deleterious to survival or to species propagation by disrupting breeding. The emerging picture is that due to the variations in HPA axis plasticity among avian species, urban stressors may differentiate species into urban avoiders, urban adaptors, and urban exploiters (Bonier, 2012).

26.4 ADRENAL CHROMAFFIN TISSUE HORMONES

26.4.1 Catecholamine Synthesis and Secretion

Catecholamines are widely synthesized throughout the postganglionic sympathetic nervous system, which includes the adrenal chromaffin tissue. The biosynthetic pathway for the adrenal catecholamines (Ghosh et al., 2001; Mahata et al., 2002; Trifaró, 2002) is outlined in Figure 26.9. The rate-limiting step in catecholamine biosynthesis is tyrosine hydroxylase. Catecholamines exert a negative feedback on tyrosine hydroxylase. Acetycholine released from preganglionic sympathetic (splanchnic) nerve terminals interacts with both nicotinic and muscarinic receptors to stimulate phosphorylation of tyrosine hydroxylase, for rapid catecholamine synthesis and release, and it upregulates tyrosine hydroxylase synthesis with more long-term stimulation. Phosphorylation of tyrosine hydroxylase lowers its K_m for the cofactor tetrahydrobiopterin. VIP and PACAP also stimulate the phosphorylation of tyrosine hydroxylase.

Tyrosine hydroxylase catalyzes the conversion of tyrosine to dihyrophenylalanine (L-DOPA). Dihydrophenylalanine is converted to dopamine by aromatic L-amino acid decarboxylase (DOPA decarboxylase). Dopamine either is secreted or undergoes further conversion to norepinephrine (NE) by dopamine β-hydroxylase with ascorbic acid as a cofactor.

The final step to epinephrine (E) is carried out by phenylethanolamine *N*-methyltransferase (PNMT) in which the *S*-methyl group of *S*-adenosyl methionine is transferred to the primary nitrogen group of NE. All of these enzymes have some requirement for innervation and are affected by numerous intraadrenal peptides. However, PNMT is strictly GR dependent.

The release of catecholamines into the circulation occurs by the well-studied stimulus secretion coupling mechanism. In this mechanism, preganglionic nerve impulses release acetylcholine, which in turn activates its cognate receptors to depolarize the chromaffin cells. This results in an influx in Ca^{2+}, which induces exocytosis of the secretory granules. In addition to catecholamines, there are a number of neuropeptides that are co-released (see Section 26.1.2.1). Furthermore, chromagranins and secretogranins are also released. The peptides derived from chromagranins in particular have autocrine and paracrine effects on catecholamine release (Trifaró, 2002).

There is no preferential release of E or NE from isolated chicken chromaffin cells in response to acetylcholine analogs (Knight and Baker, 1986). *In vivo,* however, the secretion of E and NE from the avian adrenal gland is thought to be finely regulated by a number of neural-derived and blood-borne factors and hormones resulting in the differential release of catecholamines (Ghosh et al., 2001). Yet, it is not known precisely how the differential adrenal secretion of E and NE occurs in response to disparate physiological conditions and stressors. What is fairly clear is that the adrenal gland is nearly the sole contributor to circulating E and most of the NE in response to acute stressors (Butler and Wilson, 1985; Lacombe and Jones, 1990).

26.4.2 Circulating Catecholamines and Stress Response

Studies on the circulating concentrations of catecholamines in avian species have yielded disparate and conflicting values. The most recent and consistent values are obtained from studies with the domestic chicken (*G. gallus domesticus*) and Muscovy duck (*Cairina moschata*) using high-pressure liquid chromatography with electrochemical detection. The plasma concentrations of NE and E in growing and mature chickens average 0.52 and 0.96 ng/ml, respectively, whereas those of the duck average 0.45 and 2.10 ng/ml, respectively. Thus, the circulating concentrations of NE tend to be greater than those of E. Dopamine also exists at variable but comparable concentrations (Butler and Wilson, 1985; Cheng et al., 2001; Pohle and Cheng, 2009). In the chicken, NE, but not E, shows a diurnal rhythm (de Jong et al., 2001). Catecholamines are degraded mainly in the liver by catecholamine-*O*-methyltransferase and monoamine oxidase, and the inactive products are removed by the kidney. Monoamine oxidases are also present in the adrenal gland and thus can modulate intraadrenal levels of catecholamines (Ghosh et al., 2001).

Adrenal catecholamines are part of the first wave of the acute stress response (i.e., the "fight-or-flight response"). The epinephrine response to stress stimulates immediate

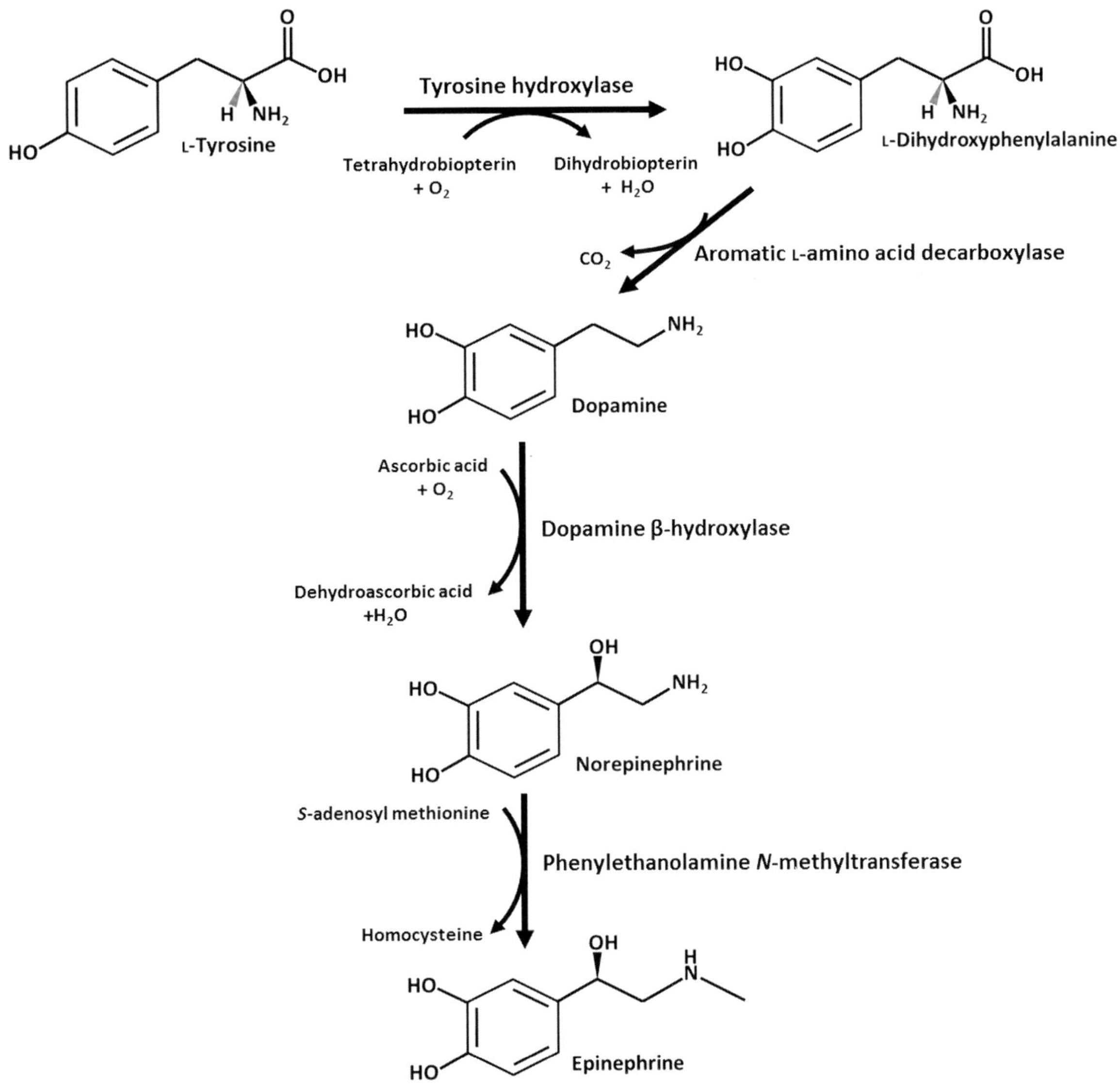

FIGURE 26.9 **Biosynthesis of norepinephrine and epinephrine.** Solid black and gray bonds on molecules represent key alpha (down) and beta (up) positions; however, not all hydrogen bond positions are shown. See text for additional details. *Pathway drawn by R. V. Carsia.*

hepatic glycogenolysis to provide glucose for critical tissues during energetic stress. In addition, catecholamines appear to interact with glucagon to regulate thermogenesis in cold stress (Abdelmelek et al., 2001; Filali-Zegzouti et al., 2005) Furthermore, catecholamines enhance the activity of the HPA axis at all levels ultimately driving more glucocorticoid release. In turn, the action of catecholamines is dependent on glucocorticoids (Sapolsky et al., 2000; Cyr and Romero, 2009). Glucocorticoids enhance cardiovascular sensitivity to catecholamines. The effect of glucocorticoids occurs at multiple levels of catecholamine action. They inhibit re-uptake of catecholamines by sympathetic terminals, reduce catecholamine-degrading enzymes, catechol-*O*-methyltransferase and monoamine oxidase, and increase the binding parameters and transmembranous signaling of β-adrenergic receptors. Indeed, a large part of the modulation of catecholamine action occurs through the regulation of adrenergic receptor function.

26.4.3 Some Physiological Actions of Norepinephrine and Epinephrine

Catecholamines have diverse effects on immune cells. For example, continuous infusion of physiological concentrations of NE and E enhance chicken leukocyte migration and differentially affect the phytohemagglutinin wattle response (McCorkle and Taylor, 1993). In addition, NE and E, *in vivo* and *in vitro,* have contrasting effects on immunoglobulin plaque-forming cells from splenic lymphocytes, and these contrasting effects are probably mediated by α- and β-adrenergic receptors (Denno et al., 1994). Catecholamines also differentially affect macrophage effector functions, *in vitro* (Ali et al., 1994).

The metabolic effects of catecholamines are widespread in that they interact with other regulating hormones such as insulin, prolactin, thyroid hormones, growth hormone, and glucagon. Catecholamines influence carbohydrate and lipid metabolism (also see Chapter X). E stimulates glycogenolysis in chicken hepatocytes via β-adrenergic receptor activation and cAMP production, which, in turn, activates glycogen phosphorylase. Thus, the effect of catecholamines leads to a rapid increase in blood glucose levels (Thurston et al., 1993). Catecholamines (NE and presumably E) also stimulate gluconeogenesis via α-adrenergic receptor activation of intracellular calcium mobilization (Cramb et al., 1982). It is probable that the effects of E on glycogenolysis, gluconeogenesis, and also lipogenesis (discussed further in this section) are physiological roles for adrenal E, as these metabolic parameters are influenced by concentrations of E that are found in the circulation of a stressed bird.

In chickens, during periods of behavioral inactivity, there is a positive relationship between circulating catecholamines and free fatty acids. Thus, catecholamines, at least in the liver, may be physiological regulators of lipid metabolism. The synthesis of fatty acid (lipogenesis) by liver tissue and cells is inhibited by E and, to a lesser extent, by NE. This effect is mediated via both α- and β-adrenergic receptors and, at least partially, by cAMP acting as the intracellular messenger (Campbell and Scanes, 1985). Lipolysis in fat cells is stimulated by E in several species of birds and appears to be mediated via β-adrenergic receptors and cAMP (Campbell and Scanes, 1985).

26.4.4 Changes in Development, Maturation, and Senescence

Experiments with precocial species indicate that the adrenal chromaffin cells are derived from caudal thoracic (the region of somites 18–24) populations of neural crest cells (sympathoadrenal progenitor cells) (see Shtukmaster et al. (2013)). Recent evidence indicates that the adrenal chromaffin cells and postganglionic sympathetic neurons are derived from a common precursor cell population. In the chick embryo at about day 2, this population delaminates from the neural tube and migrates ventrally to the area of the dorsal aorta. Under the induction of bone morphogenic proteins elaborated from the aorta, the cells acquire catecholaminergic features, such as the expression of tyrosine hydroxylase. Cells that reach the region of the adrenal gland at around embryonic day 6 gradually lose neuronal features such as the expression of neurofilament proteins and gain chromaffin cell features. This transformation from neuronal type to chromaffin type does not require the GR or even adrenocortical tissue (Gut et al., 2005; Huber, 2006). However, the differentiation to E cells and the expression of PNMT is absolutely GR dependent and is apparent in chromaffin cells at day 5, although E cell differentiation does not require penetration into the developing adrenocortical tissue (Chimenti and Accordi, 2008). Indeed, chromaffin cell migration and penetration into the developing adrenocortical tissue do not occur until around day 8 or 9 and are not complete until day 15. In the chick, in addition to catecholamines, a variety of peptides are elaborated by the adrenal chromaffin cells postnatally. Some of these peptides were described in this chapter.

E is detected in the chick embryonic circulation somewhere between embryonic days 10 and 13 (see von Blumröder and Tönhardt (2002)). The importance of the adrenal gland in providing catecholamines for embryonic homeostasis is unclear since catecholamines can come from both adrenal and neuronal sources. However, a consistent observation is that circulating levels of NE predominate over those of dopamine and E. In the chick embryo, plasma catecholamines increase approaching hatch (von Blumröder and Tönhardt, 2002). NE appears to be an important catecholamine that supports the embryo for pipping and hatching, and the metabolic changes associate with the perihatch transition.

There appears to be an extreme range of variation in the adrenal NE and E content of birds of different phylogenetic groups (see Ghosh et al. (2001)). Although there has been recent interest in birds as comparative models to study the mechanisms of aging (Holmes et al., 2001; Holmes and Ottinger, 2005; Ricklefs, 2010), there is little information concerning the endocrine changes associated with aging. The chromaffin catecholamine content and content response to endocrine alterations appear to vary among different species with age. However, a consistent emerging picture that is shared with the limited number of vertebrate species studied thus far is that chromaffin activity increases with age in disparate avian species (Ghosh et al., 2001). Since there is a suggestion that glucocorticoids increase with age-related failure in adaptation to environmental and seasonal stressors (Koren et al., 2011), presumably similar changes occur with adrenal catecholamines. Clearly, more definitive studies are needed to understand the specific age-related changes in adrenocortical and adrenal chromaffin function in birds.

ACKNOWLEDGMENTS

The author thanks Walter L. Miller, MD, for helpful discussions on adrenal steroidogenic enzymes and Stephen R. Hammes, MD, PhD, for clarification on the types of glucocorticoid receptors. The author is especially indebted to Colin G. Scanes, PhD, ScD, for his invitation to write this chapter and for his invaluable guidance throughout its development.

REFERENCES

Abdelmelek, H., Fechtali, T., Filali-Zegzouti, Y., Rouanet, J.L., Cottet-Emard, J.M., Pequignot, J.M., Barré, H., 2001. J. Neural Transm. 108, 793–801.

Adams, M., Meijer, O.C., Wang, J., Bhargava, A., Pearce, D., 2003. Homodimerization of the glucocorticoid receptor is not essential for response element binding: activation of the phenylethanolamine *N*-methyltransferase gene by dimerization-defective mutants. Mol. Endocrinol. 17, 2583–2592.

Ahmed, A.A., Ma, W., Guo, F., Ni, Y., Grossman, R., Zhao, R., 2013. Differences in egg deposition of corticosterone and embryonic expression of corticosterone metabolic enzymes between slow and fast growing broiler chickens. Comp. Biochem. Physiol. A 164, 200–206.

Ali, R.A., Qureshi, M.A., McCorkle, F.M., 1994. Profile of chicken macrophage functions after exposure to catecholamines in vitro. Immunopharmacol. Immunotoxicol. 16, 611–625.

Almasi, B., Rettenbacher, S., Müller, C., Brill, S., Wagner, H., Jenni, L., 2012. Maternal corticosterone is transferred into the egg yolk. Gen. Comp. Endocrinol. 178, 139–144.

Almasi, B., Roulin, A., Jenni-Eiermann, S., Breuner, C.W., Jenni, L., 2009. Regulation of free corticosterone and CBG capacity under different environmental conditions in altricial nestlings. Gen. Comp. Endocrinol. 164, 117–124.

Angelier, F., Chastel, O., 2009. Stress, prolactin and parental investment in birds: a review. Gen. Comp. Endocrinol. 163, 142–148.

Angelier, F., Wingfield, J.C., Trouvé, C., de Grissac, S., Chastel, O., 2012. Modulation of the prolactin and the corticosterone response: do they tell the same story in a long-lived bird, the Cape petrel? Gen. Comp. Endocrinol. 182, 7–15.

Árnason, S.S., Rice, G.E., Chadwick, A., Skadhauge, E., 1986. Plasma levels of arginine vasotocin, prolactin, aldosterone and corticosterone during prolonged dehydration in the domestic fowl: effect of dietary NaCl. J. Comp. Physiol. B 156, 383–397.

Asai, M., Ramachandrappa, S., Joachim, M., Shen, Y., Zhang, R., Nuthalapati, N., Ramanathan, V., Strochlic, D.E., Ferket, P., Linhart, K., Ho, C., Novoselova, T.V., Garg, S., Ridderstråle, M., Marcus, C., Hirschhorn, J.N., Keogh, J.M., O'Rahilly, S., Chan, L.F., Clark, A.J., Farooqi, I.S., Majzoub, J.A., 2013. Loss of function of the melanocortin 2 receptor accessory protein 2 is associated with mammalian obesity. Science 341, 275–278.

Avrutina, A.J., Galpern, I.L., Kisljuk, S.M., 1985. Stimulation of adrenals during the critical periods of development and production in fowls. World's Poult. Sci. J. 41, 108–114.

Assenmacher, I., 1973. The peripheral endocrine glands. In: Farner, D.S., King, J.R., Parkes, K.C. (Eds.), Avian Biology, vol. III. Academic Press, New York, pp. 183–286.

Banerjee, S.B., Arterbery, A.S., Fergus, D.J., Adkins-Regan, E., 2012. Deprivation of maternal care has long-lasting consequences for the hypothalamic-pituitary-adrenal axis of zebra finches. Proc. R. Soc. B 279, 759–766.

Barriga, C., Marchena, J.M., Lea, R.W., Harvey, S., Rodríguez, A.B., 2002. Effect of stress and dexamethasone treatment on circadian rhythms of melatonin and corticosterone in ring dove (*Streptopelia risoria*). Mol. Cell. Biochem. 232, 27–31.

Bauer, M.P., Bridgham, J.T., Langenau, D.M., Johnson, A.L., Goetz, F.W., 2000. Conservation of steroidogenic acute regulatory (StAR) protein structure and expression in vertebrates. Mol. Cell. Endocrinol. 168, 119–125.

Benowitz-Fredericks, Z.M., Hodge, M., 2013. Yolk androstenedione in domestic chicks (*Gallus gallus domesticus*): uptake and sex-dependent alterations of growth and behavior. Gen. Comp. Endocrinol. 193, 48–55.

Berghman, L.R., Devreese, B., Verhaert, P., Gerets, H., Arckens, L., Vanden Broeck, J., Van Beeumen, J., Vaudry, H., Vandesande, F., 1998. The molecular characterization of chicken N-terminal pro-opiomelanocortin (POMC). Mol. Cell. Endocrinol. 142, 119–130.

Besedovsky, H.O., del Rey, A., 1996. Immune-neuro-endocrine interactions: facts and hypotheses. Endocr. Rev. 17, 64–102.

Bhattacharyya, T.K., Sinha, D., Ghosh, A., 1972. A comparative histological survey of the avian adrenocortical homologue. Arch. Histol. Jpn. 34, 419–432.

Bhujle, B.V., Nadkarni, V.B., 1976. Steroid dehydrogenases in the adrenal gland of four species of birds: a histochemical study. Histochem. J. 8, 591–596.

Bonier, F., 2012. Hormones in the city: endocrine ecology of urban birds. Horm. Behav. 61, 763–772.

Boswell, T., Takeuchi, S., 2005. Recent developments in our understanding of the avian melanocortin system: its involvement in the regulation of pigmentation and energy homeostasis. Peptides 26, 1733–1743.

Bromée, T., Sjödin, P., Fredriksson, R., Boswell, T., Larsson, T.A., Salaneck, E., Soorob, R., Mohell, N., Larhammar, D., 2006. Neuropeptide Y-family receptors Y_6 and Y_7 in chicken: cloning, pharmacological characterization, tissue distribution and conserved synteny with human chromosome region. FEBS J. 273, 2048–2063.

Breuner, C.W., Lynn, S.E., Julian, G.E., Cornelius, J.M., Heidinger, B.J., Love, O.P., Sprague, R.S., Wada, H., Whitman, B.A., 2006. Plasma binding globulins and acute stress response. Horm. Metab. Res. 38, 260–268.

Breuner, C.W., Orchinik, M., 2001. Seasonal regulation of membrane and intracellular corticosteroid receptors in the house sparrow brain. J. Neuroendocrinol. 13, 412–420.

Breuner, C.W., Orchinik, M., 2009. Pharmacological characterization of intracellular, membrane, and plasma binding sites for corticosterone in house sparrows. Gen. Comp. Endocrinol. 163, 214–224.

Breuner, C.W., Orchinik, M., Hahn, T.P., Meddle, S.L., Moore, I.T., Owen-Ashley, N.T., Sperry, T.S., Wingfield, J.C., 2003. Differential mechanisms for plasticity of the stress response across latitudinal gradients. Am. J. Physiol. Regul. Integr. Comp. Physiol. 285, R594–R600.

Breuner, C.W., Wingfield, J.C., Romero, L.M., 1999. Diel rhythms of basal and stress-induced corticosterone in a wild, seasonal vertebrate, Gambel's white-crowned sparrow. J. Exp. Zool. 284, 334–342.

Bruggeman, V., Room, G., Vanmontfort, D., Verhoeven, G., Decuypere, E., 2003. Effect of embryonic 19-nortestosterone treatment and surgical bursectomy on plasma concentrations of reproductive hormones, on inhibin content in adrenals and gonads and on the histological appearance of the gonads in the young chicken. Gen. Comp. Endocrinol. 131, 106–116.

Bureau, C., Hennequet-Antier, C., Couty, M., Guémené, D., 2009. Gene array analysis of adrenal glands in broiler chickens following ACTH treatment. BMC Genomics 10, 430.

Butler, D.G., Wilson, J.X., 1985. Cardiovascular function in adrenalectomized Pekin ducks (*Anas platyrhynchos*). Comp. Biochem. Physiol. A 81, 353–358.

Byrd, J.A., Dean, C.E., Hargis, B.M., 1994. The effect of the humoral immune system-derived bursal anti-steroidogenic peptide (BASP) on corticosteroid biosynthesis in avian, porcine and canine adrenal cortical cells. Comp. Biochem. Physiol. C 108, 221–227.

Campbell, R.M., Scanes, C.G., 1985. Adrenergic control of lipogenesis and lipolysis in the chicken in vitro. Comp. Biochem. Physiol. C 82, 137–142.

Carsia, R.V., 1990. Hormonal control of avian adrenocortical function: cellular and molecular aspects. In: Epple, A., Scanes, C.G., Stetson, M.H. (Eds.), Progress in Comparative Endocrinology. Wiley-Liss, Inc., Delaware, pp. 439–444.

Carsia, R.V., Lamm, E.-T., Marsh, J.A., Scanes, C.G., King, D.B., 1997. The thyroid hormone, 3,5,3′-triiodothyronine, is a negative modulator of domestic fowl (*Gallus gallus domesticus*) adrenal steroidogenic function. Gen. Comp. Endocrinol. 170, 251–261.

Carsia, R.V., McIlroy, P.J., 1998. Dietary protein restriction stress in the domestic turkey (*Meleagris gallopavo*) induces hypofunction and remodeling of adrenal steroidogenic tissue. Gen. Comp. Endocrinol. 109, 140–153.

Carsia, R.V., McIlroy, P.J., Kowalski, K.I., Tilly, J.L., 1993. Isolation of turkey adrenocortical cell angiotensin II (AII) receptor partial cDNA: evidence for a single-copy gene expressed predominantly in the adrenal gland. Biochem. Biophys. Res. Commun. 191, 1073–1080.

Carsia, R.V., Morin, M.E., Rosen, H.D., Weber, H., 1987a. Ontogenic corticosteroidogenesis of the domestic fowl: response of isolated adrenocortical cells. Proc. Soc. Exp. Biol. Med. 184, 436–445.

Carsia, R.V., Scanes, C.G., Malamed, S., 1987b. Polyhormonal regulation of avian and mammalian corticosteroidogenesis in vitro. Comp. Biochem. Physiol. 88A, 131–140.

Carsia, R.V., Reisch, N.M., Fennell, M.J., Weber, H., 1987c. Adrenocortical function of the domestic fowl: effects of orchiectomy and androgen replacement. Proc. Soc. Exp. Biol. Med. 185, 223–232.

Carsia, R.V., Scanes, C.G., Malamed, S., 1984. Self-suppression of corticosteroidogenesis: evidence for the role of adrenal 5α-reductase. Endocrinology 115, 2464–2472.

Carsia, R.V., Scanes, C.G., Malamed, S., 1985a. Isolated adrenocortical cells of the domestic fowl (*Gallus domesticus*): steroidogenic and ultrastructural properties. J. Steroid. Biochem. 22, 273–279.

Carsia, R.V., Scanes, C.G., Malamed, S., 1985b. Loss of sensitivity to ACTH of adrenocortical cells isolated from maturing domestic fowl. Proc. Soc. Exp. Biol. Med. 179, 279–282.

Carsia, R.V., Weber, H., King, D.B., Scanes, C.G., 1985c. Adrenocortical cell function in the hypophysectomized domestic fowl: effects of growth hormone and 3,5,3′-triiodothyronine replacement. Endocrinology 117, 928–933.

Carsia, R.V., Weber, H., 1986. Genetic-dependent alterations in adrenal stress response and adrenocortical cell function of the domestic fowl (*Gallus domesticus*). Proc. Soc. Exp. Biol. Med. 183, 99–105.

Carsia, R.V., Weber, H., 1988. Protein malnutrition in the domestic fowl induces alterations in adrenocortical cell adrenocorticotropin receptors. Endocrinology 122, 681–688.

Carsia, R.V., Weber, H., 2000. Remodeling of turkey steroidogenic tissue induced by dietary protein restriction: the potential role of cell death. Gen. Comp. Endocrinol. 118, 471–479.

Carsia, R.V., Weber, H., Lauterio, T.J., 1988a. Protein malnutrition in the domestic fowl induces alterations in adrenocortical function. Endocrinology 122, 673–680.

Carsia, R.V., Weber, H., Satterlee, D.G., 1988b. Steroidogenic properties of isolated adrenocortical cells from Japanese quail selected for high serum corticosterone response to immobilization. Domest. Anim. Endocrinol. 5, 231–240.

Carsia, R.V., Weber, H., Perez Jr., F.M., 1986. Corticotropin-releasing factor stimulates the release of adrenocorticotropin from domestic fowl pituitary cells. Endocrinology 118, 143–148.

Cerdá-Reverter, J.M., Agulleiro, M.J., Cortés, R., Sánchez, E., Guillot, R., Leal, E., Fernández-Durán, B., Puchol, S., Eley, M., 2013. Involvement of melanocortin receptor proteins (MRAPs) in the function of melanocortin receptors. Gen. Comp. Endocrinol. 188, 133–136.

Charlier, T.D., Underhill, C., Hammond, G.L., Soma, K.K., 2009. Effects of aggressive encounters on plasma corticosteroid-binding globulin and its ligands in white-crowned sparrows. Horm. Behav. 56, 339–347.

Chaturvedi, C.M., Chowdhary, A., Wall, P.T., Koike, T.I., Cornett, L.E., 2000. A sexual dimorphism in hypothalamic arginine vasotocin (AVT) gene expression and AVT plasma levels in the Japanese quail (*Coturnix coturnix japonica*) in response to water deprivation. Gen. Comp. Endocrinol. 117, 129–137.

Chaturvedi, C.M., Kumar, P., 2007. Nitric oxide modulates gonadal and adrenal function in Japanese quail (*Coturnix coturnix japonica*). Gen. Comp. Endocrinol. 151, 285–299.

Chen, C.-C., Johnson, P.A., 1996. Molecular cloning of inhibin/activin β_A-subunit complementary deoxyribonucleic acid and expression of inhibin/activin α- and β_A-subunits in the domestic hen. Biol. Reprod. 54, 429–435.

Cheng, H.-W., Dillworth, G., Singleton, P., Chen, Y., Muir, W.M., 2001. Poult. Sci. 80, 1278–1285.

Chester Jones, I., Phillips, J.G., 1986. The adrenal and interrenal glands. In: Pang, P.K.T., Schreibman, M.P. (Eds.), Vertebrate Endocrinology: Fundamental and Biomedical Implications, vol.1. Academic Press, New York, pp. 319–350.

Chimenti, C., Accordi, F., 2008. Development and evolution of the adrenal gland and its homolog in teleosts, anurans, chelonians and birds. In: Capaldo, A. (Ed.), Recent Advances in Non-mammalian Adrenal Gland Research. Research Signpost, Kerala, India, pp. 1–29.

Cole, T.J., 1995. Cloning of the mouse 11β-hydroxysteroid dehydrogenase type 2 gene: tissue specific expression and localization in distal convoluted tubules and collecting ducts of kidney. Endocrinology 136, 4693–4696.

Cooray, S.N., Clark, A.J.L., 2011. Melanocortin receptors and their accessory proteins. Mol. Cell. Endocrinol. 331, 215–221.

Cornelius, J.M., Boswell, T., Jenni-Eiermann, S., Breuner, C.W., Ramenofsky, M., 2013a. Contributions of endocrinology to the migration life history of birds. Gen. Comp. Endocrinol. 190, 47–60.

Cornelius, J.M., Watts, H.E., Dingle, H., Hahn, T.P., 2013b. Obligate versus rich patch opportunism: evolution and endocrine mechanisms. Gen. Comp. Endocrinol. 190, 76–80.

Cornett, L.E., Kang, S.W., Kuenzel, W.J., 2013. A possible mechanism contributing to the synergistic action of vasotocin (VT) and corticotropin-releasing hormone (CRH) receptors on corticosterone release in birds. Gen. Comp. Endocrinol. 188, 46–53.

Costantini, D., Fanfani, A., Dell'Omo, G., 2008. Effects of corticosteroids on oxidative damage and circulating carotenoids in captive adult kestrels (*Falco tinnunculus*). J. Comp. Physiol. B 178, 829–835.

Cramb, G., Langslow, D.R., Phillips, J.H., 1982. Hormonal effects of cyclic nucleotides and carbohydrate and lipid metabolism in isolated chicken hepatocytes. Gen. Comp. Endocrinol. 46, 310–321.

Crossin, G.T., Phillips, R.A., Lattin, C.R., Romero, L.M., Williams, T.D., 2013. Corticosterone mediated costs of reproduction link current to future breeding. Gen. Comp. Endocrinol. 193, 112–120.

Cyr, N.E., Romero, L.M., 2009. Identifying hormonal habituation in field studies of stress. Gen. Comp. Endocrinol. 161, 295–303.

Daniel, J.Y., Assenmacher, I., 1971. Early appearance of metabolites after single i.v. injection of 3H-corticosterone in rabbit and duck. Steroids 18, 325–340.

Davis, G.S., Siopes, T.D., 1987. Plasma corticosterone response of turkeys to adrenocorticotropic hormone: age, dose and route of administration effects. Poult. Sci. 66, 1727–1732.

Decuypere, E., Rombauts, L., Vanmontfort, D., Verhoeven, G., 1997. Inhibin from the embryo to the adult hen. In: Harvey, S., Etches, R.J. (Eds.), Perspectives in Avian Endocrinology. Journal of Endocrinology Ltd, Bristol, pp. 71–80.

De Groef, B., Geris, K.L., Vandeborne, K., Darras, V.M., Kühn, E.R., 2005. CRH control of thyroid function in the chicken. In: Dawson, A., Sharp, P.J. (Eds.), Functional Avian Endocrinology. Narosa Publishing House, New Delhi, India, pp. 415–426.

De Groef, B., Grommen, S.V.H., Darras, V.M., 2008. The chicken embryo as a model for developmental endocrinology: development of the thyrotropic, corticotropic and somatotropic axes. Mol. Cell. Endocrinol. 293, 17–24.

De Groef, B., Grommen, S.V.H., Mertens, I., Schoofs, L., Kühn, E., Darras, V.M., 2004. Cloning and tissue distribution of the chicken type 2 corticotropin-releasing hormone receptor. Gen. Comp. Endocrinol. 138, 89–95.

De Groef, B., Van der Geyten, S., Darras, V.M., Kühn, E.R., 2006. Role of corticotropin-releasing hormone as a thyrotropin-releasing factor in non-mammalian vertebrates. Gen. Comp. Endocrinol. 146, 62–68.

De Falco, M., Capaldo, A., Sciarrillo, R., Valiante, S., Laforgia, V., 2008. Neuropeptide regulation of the adrenal gland in reptiles. In: Capaldo, A. (Ed.), Recent Advances in Non-mammalian Adrenal Gland Research. Research Signpost, Kerala, India, pp. 243–255.

de Jong, I.C., van Voorst, A.S., Erkens, J.H.F., Ehlhardt, D.A., Blokhuis, H.J., 2001. Determination of the circadian rhythm in plasma corticosterone and catecholamine concentrations in growing broiler breeders using intravenous cannulation. Physiol. Behav. 74, 200–304.

Denno, K.M., McCorkle, F.M., Taylor Jr., R.L., 1994. Catecholamines modulate chicken immunoglobulin M and immunoglobulin G plaque-forming cells. Poult. Sci. 73, 1858–1866.

Denver, R.J., 2009. Structural and functional evolution of vertebrate neuroendocrine stress systems. In "Trends in Comparative Endocrinology and Neurobiology". Ann. NY Acad. Sci. 1163, 1–16.

DERoos, R., 1961. The corticoids of the avian adrenal gland. Gen. Comp. Endocrinol. 1, 494–512.

Dickens, M.J., Cornil, C.A., Balthazart, J., 2011a. Acute stress differentially affects aromatase activity in specific brain nuclei of adult male and female quail. Neuroendocrinology 152, 4242–4251.

Dickens, M.J., Meddle, S.L., Romero, L.M., 2011b. Mineralocorticoid and glucocorticoid receptor mRNA expression in the brain of translocated chukar (*Alectoris chukar*). Gen. Comp. Endocrinol. 170, 569–574.

Dickens, M., Romero, L.M., Cyr, N.E., Dunn, I.C., Meddle, S.L., 2009. Chronic stress alters glucocorticoid receptor and mineralocorticoid receptor mRNA expression in the European starling (*Sturnus vulgaris*). J. Neuroendocrinol. 21, 832–840.

Dickmeis, T., 2009. Glucocorticoids and the circadian clock. J. Endocrinol. 200, 3–22.

Dong, H., Lin, H., Jiao, H.C., Song, Z.G., Zhao, J.P., Jiang, K.J., 2007. Altered development and protein metabolism in skeletal muscles of broiler chickens (*Gallus gallus domesticus*) by corticosterone. Comp. Biochem. Physiol. A, 189–195.

Ehrhart-Borstein, M., Hinson, J.P., Bornstein, S.R., Scherbaum, W.A., Vinson, G.P., 1998. Intraadrenal interactions in the regulation of adrenocortical steroidogenesis. Endocr. Rev. 19, 101–143.

El-Far, A.A., Mashaly, M.M., Kamar, G.A., 1994. Bursectomy and in vitro response of adrenal gland to adrenocorticotropic hormone and testis to human chorionic gonadotropin in immature male chickens. Poult. Sci. 73, 113–117.

Ellestad, L.E., Saliba, J., Porter, T.E., 2011. Ontogenic characterization of gene expression in the developing neuroendocrine system of the chick. Gen. Comp. Endocrinol. 171, 82–93.

Fässler, R., Schauenstein, K., Krömer, G., Schwarz, S., Wick, G., 1986. Elevation of corticosteroid-binding globulin in obese strain (OS) chickens: possible implications for the disturbed immunoregulation and the development of spontaneous autoimmune thyroiditis. J. Immunol. 136, 3657–3661.

Feuilloley, M., Vaudry, H., 1996. Role of the cytoskeleton in adrenocortical cells. Endocr. Rev. 17, 269–288.

Filali-Zegzouti, Y., Abdelmelek, H., Rouanet, J.L., Cottet-Emard, J.M., Pequignot, J.M., Barré, H., 2005. Role of catecholamines in glucagon-induced thermogenesis. J. Neural Transm. 112, 481–489.

Fokidis, H.B., Orchinik, M., Deviche, P., 2009. Corticosterone and corticosteroid binding globulin in birds: relation to urbanization in a desert city. Gen. Comp. Endocrinol. 160, 259–270.

Freking, F., Nazairians, T., Schlinger, B.A., 2000. The expression of the sex steroid-synthesizing enzymes CYP11A1, 3B-HSD, CYP17, and CYP19 in gonads and adrenals of adult and developing zebra finches. Gen. Comp. Endocrinol. 119, 140–151.

García-Arrarás, J., Lugo-Chinchilla, A., Chévere-Colón, I., 1992. The expression of neuropeptide Y immunoreactivity in the avian sympathoadrenal system conforms with two models of coexpression development for neurons and chromaffin cells. Development 115, 617–627.

Ghosh, A., Carmichael, S.W., Mukherjee, M., 2001. Avian adrenal medulla: cytomorphology and function. Acta Biol. Szeged. 45, 1–11.

Giorgi, E.P., Stein, W.D., 1981. The transport of steroids into animal cells in culture. Endocrinology 108, 688–697.

Glasgow, A., Haidan, A., Hilbers, U., Breidert, M., Gillespie, J., Scherbaum, W.A., Chrousos, G.P., Bornstein, S.R., 1998. Expression of Ob receptor in normal human adrenals: differential regulation of adrenocortical and adrenomedullary function by leptin. J. Clin. Endocrinol. Metab. 83, 4459–4466.

Gogga, P., Karbowska, J., Kochan, Z., Meissner, W., 2013. Circulating leptin levels do not reflect the amount of body fat in the dunlin *Calidris alpina* during migration. Gen. Comp. Endocrinol. 187, 74–78.

Gonzalez, C.B., Cozza, E.N., De Bedners, M.E.O., Lantos, C.P., Aragones, A., 1983. Progesterone and its reductive metabolism in steroidogenic tissues of the developing hen embryo. Gen. Comp. Endocrinol. 51, 384–393.

Gray, D.A., Gerstberger, R., Simon, E., 1989. Role of angiotensin II in aldosterone regulation in the Pekin duck. J. Endocrinol. 123, 445–452.

Gray, D.A., 1993. Plasma atrial natriuretic factor concentrations and renal actions in the domestic fowl. J. Comp. Physiol. B 163, 519–523.

Gray, D.A., Schülz, H., Gerstberger, R., 1991a. Interaction of atrial natriuretic factor and osmoregulatory hormones in the Pekin duck. Gen. Comp. Endocrinol. 81, 246–255.

Gray, D.A., Schülz, H., Gerstberger, R., 1991b. Plasma atrial natriuretic factor responses to blood volume changes in the Pekin duck. Endocrinology 128, 1655–1660.

Gregg, C.M., Wideman Jr., R.F., 1986. Effect of atriopeptin and chicken heart extract in *Gallus domesticus*. Am. J. Physiol. 251, R453–R551.

Grossmann, C., Freudinger, R., Mildenberger, S., Husse, B., Gekle, M., 2008. EF domains are sufficient for nongenomic mineralocorticoid receptor action. J. Biol. Chem. 283, 7109–7116.

Gut, P., Huber, K., Lohr, J., Brühl, B., Oberle, S., Treier, M., Ernsberger, U., Kalcheim, C., Unsicker, K., 2005. Lack of an adrenal cortex in *Sf1* mutant mice is compatible with the generation and differentiation of chromaffin cells. Development 132, 4611–4619.

Hahn, D.C., 2011. Patterns of maternal yolk hormones in eastern screech owl eggs (*Megascops asio*). Gen. Comp. Endocrinol. 172, 423–429.

Hammes, S.R., Levin, E.R., 2011. Minireview: recent advances in extranuclear steroid receptor actions. Endocrinology 152, 4489–4495.

Hammes, S.R., Mendelson, C.R., 2012. Mechanism of hormone action. In: Kovacs, W.J., Ojeda, S.R. (Eds.), Textbook of Endocrinology, sixth ed. Oxford University Press, Inc., New York, pp. 58–98.

Hammond, G.L., 1990. Molecular properties of corticosteroid binding globulin and the sex-steroid binding proteins. Endocr. Rev. 11, 65–79.

Harvey, S., Scanes, C.G., Daughaday, W.H., 1995. IV. Adrenocortical hormones. In: Growth Hormone. CRC Press, Inc., Boca Raton, Florida, p. 426.

Hayashi, H., Imai, K., Imai, K., 1991. Characterization of chicken ACTH and α-MSH: the primary sequence of chicken ACTH is more similar to *Xenopus* ACTH than to other avian ACTH. Gen. Comp. Endocrinol. 82, 434–443.

Hazard, D., County, M., Guémené, D., 2007. Characterization of CRF, AVT and ACTH cDNA and pituitary-adrenal axis function in Japanese quail divergently selected for tonic mobility. Am. J. Physiol. Regul. Integr. Comp. Physiol. 293, R1421–R1429.

Helton, E.D., Holmes, W.N., 1973. The distribution and metabolism of labeled corticosteroids in the duck (*Anas platyrhynchos*). J. Endocrinol. 56, 361–385.

Hen, G., Yosefi, S., Ronin, A., Einat, P., Rosenblum, C.I., Denver, R.J., Friedman-Einat, M., 2008. Monitoring leptin activity using the chicken leptin receptor. J. Endocrinol. 197, 325–333.

Hendricks III, G.L., Siegel, H.S., Mashaly, M.M., 1991. Ovine corticotropin-releasing factor increases endocrine and immunological activity of avian leukocytes. Proc. Soc. Exp. Biol. Med. 196, 390–395.

Hendricks III, G.L., Mashaly, M.M., Siegel, H.S., 1995a. Validation of an assay to measure adrenocorticotropin in plasma and from chicken leukocytes. Poult. Sci. 74, 337–342.

Hendricks III, G.L., Mashaly, M.M., Siegel, H.S., 1995b. Effect of corticosterone in vivo and in vitro on adrenocorticotropic hormone production by corticotropin-releasing factor leukocytes. Proc. Soc. Exp. Biol. Med. 209, 382–386.

Henriksen, R., Rettenbacher, S., Groothuis, T.G.G., 2013. Maternal corticosterone elevation during egg formation in chickens (*Gallus gallus domesticus*) influences offspring traits, partly via prenatal undernutrition. Gen. Comp. Endocrinol. 191, 83–91.

Hertelendy, F., Todd, H., Molnár, M., 1992. Influence of chicken and human lipoproteins on steroidogenesis in granulosa cells of the domestic fowl (*Gallus domesticus*). Gen. Comp. Endocrinol. 85, 335–340.

Hinkle, P.M., Sebag, J.A., 2009. Structure and function of the melanocortin 2 receptor accessory protein (MRAP). Mol. Cell. Endocrinol. 300, 25–31.

Hodgson, Z.G., Meddle, S.L., Roberts, M.L., Buchanan, K.L., Evans, M.R., Metzdorf, R., Gahr, M., Healy, S.D., 2007. Spatial ability is impaired and hippocampal mineralocorticoid receptor mRNA expression reduced in zebra finches (*Taeniopygia guttata*) selected for acute high corticosterone response to stress. Proc. R. Soc. B 274, 239–245.

Holberton, R.L., Boswell, T., Hunter, M.J., 2008. Circulating prolactin and corticosterone concentrations during the development of migratory condition in the dark-eyed Junco, *Junco hyemalis*. Gen. Comp. Endocrinol. 155, 641–649.

Holmes, D.J., Flückiger, R., Austad, S.N., 2001. Comparative biology of aging in birds: an update. Exp. Gerontol. 36, 869–883.

Holmes, D.J., Ottinger, M.A., 2005. Aging in birds. Age 27, iii–v.

Holmes, W.N., Al-Ghawas, S.C., Cronshaw, J., Rohde, K.E., 1991. The structural organization and the steroidogenic responsiveness in vitro of adrenal gland tissue from the neonatal mallard duck (*Anas platyrhynchos*). Cell Tissue Res. 263, 557–566.

Holmes, W.N., Bradley, E.L., Helton, E.D., Chan, M.Y., 1972. The distribution and metabolism of corticosterone in birds. Gen. Comp. Endocrinol. 3 (Suppl.), 266–278.

Holmes, W.N., Broock, R.L., Devlin, J.M., 1974. Tritiated corticosteroid metabolism in intact and adenohypophysectomized ducks (*Anas platyrhynchos*). Gen. Comp. Endocrinol. 22, 417–427.

Holmes, W.N., Cronshaw, J., 1980. Adrenal cortex: structure and function. In: Epple, A., Stetson, M. (Eds.), Avian Endocrinology. Academic Press, New York, pp. 271–299.

Holmes, W.N., Cronshaw, J., 1984. Adrenal gland: some evidence for the structural and functional zonation of the steroidogenic tissues. J. Exp. Zool. 232, 627–631.

Holmes, W.N., Cronshaw, J., 1993. Some actions of angiotensin II in gallinaceous and anseriform birds. In: Sharp, P.J. (Ed.), Avian Endocrinology. Journal of Endocrinology Ltd, Bristol, pp. 201–216.

Holmes, W.N., Kelly, M.E., 1976. The turnover and distribution of labelled corticosterone during post-natal development of the duckling (*Anas platyrhynchos*). Pflügers Arch. 365, 145–150.

Holmes, W.N., Phillips, J.G., 1976. The adrenal cortex of birds. In: Chester Jones, I., Henderson, I.W. (Eds.), General, Comparative and Clinical Endocrinology of the Adrenal Cortex, vol. 1. Academic Press, New York, pp. 293–413.

Holmes, W.N., Slikker, W., 1976. Some properties of the labelled material excreted by intact and adenohypophysectomized ducks (*Anas platyrhynchos*) given single doses of labeled corticosterone. Gen. Comp. Endocrinol. 29, 128–140.

Hryb, D.J., Khan, M.S., Romas, N.A., Rosner, W., 1986. Specific binding of human corticosteroid-binding globulin to cell membranes. Proc. Soc. Exp. Biol. Med. U.S.A. 83, 3253–3256.

Huber, K., 2006. The sympathoadrenal cell lineage: specification, diversification, and new perspectives. Dev. Biol. 298, 335–343.

Hughes, M.R., 2003. Regulation of salt gland, gut and kidney interactions. Comp. Biochem. Physiol. A 136, 507–524.

Jankowski, M.D., Wittwer, D.J., Heisey, D.M., Franson, J.C., Hofmeister, E.K., 2009. The adrenocortical response of greater sage grouse (*Centrocercus urophasianus*) to capture, ACTH injection, and confinement, as measured in fecal samples. Physiol. Biochem. Zool. 82, 190–201.

Jenkins, S.A., Porter, T.E., 2004. Ontogeny of the hypothalamo-pituitary-adrenocortical axis in the chicken embryo: a review. Domest. Anim. Endocrinol. 26, 267–275.

Kaiser, P., Wu, Z., Rothwell, L., Fife, M., Gibson, M., Poh, T.-Y., Shini, A., Bryden, W., Shini, S., 2009. Prospects for understanding immune-endocrine interactions in the chicken. Gen. Comp. Endocrinol. 163, 83–91.

Kalliecharan, R., Hall, B.K., 1974. A developmental study of the levels of progesterone, corticosterone, cortisol, and cortisone circulating in the plasma of chick embryos. Gen. Comp. Endocrinol. 24, 364–372.

Kamata, R., Takahashi, S., Morita, M., 2004. Gene expression of sex-determining factors and steroidogenic enzymes in the chicken embryo: influence of xenoestrogens. Gen. Comp. Endocrinol. 138, 148–156.

Kanda, I., Akazome, Y., Ogasawara, O., Mori, T., 2000. Expression of cytochrome P450 cholesterol side chain cleavage and 3β-hydroxysteroid dehydrogenase during embryogenesis in chicken adrenal glands and gonads. Gen. Comp. Endocrinol. 118, 96–104.

Katz, A., Heiblum, R., Meidan, R., Robinzon, B., 2007. Distinct features of dehydrocorticosterone reduction into corticosterone in the liver and duodenum of the domestic fowl (*Gallus gallus domesticus*). Gen. Comp. Endocrinol. 154, 67–74.

Katz, A., Heiblum, R., Meidan, R., Robinzon, B., 2008. Corticosterone oxidative neutralization by 11-β hydroxysteroid dehydrogenases in kidney and colon of the domestic fowl. Gen. Comp. Endocrinol. 155, 814–820.

Katz, A., Oyama, R.K., Feng, N., Chen, X., Schlinger, B.A., 2010. 11-β hydroxysteroid dehydrogenase type 2 in zebra finch brain and peripheral tissues. Gen. Comp. Endocrinol. 166, 600–605.

Kawashima, M., Takahashi, T., Yanai, H., Ogawa, H., Yasuoka, T., 2005. Direct action of parathyroid hormone-related peptide to enhance corticosterone production stimulated by adrenocorticotropic hormone in adrenocortical cells. Poult. Sci. 84, 1463–1469.

Kempf, H., Corvol, P., 2001. Angiotensin receptor(s) in fowl. Comp. Biochem. Physiol. A 128, 77–88.

Ketterson, E.D., Nolan Jr., V., Wolf, L., Ziegenfus, C., Dufty Jr., A., Ball, G.F., Johnsen, T.S., 1991. Testosterone and avian life histories: the effect of experimentally elevated testosterone on corticosterone and body mass in dark-eyed juncos. Horm. Behav. 25, 489–503.

Klingbeil, C., 1985. Corticosterone and aldosterone dose-dependent responses to ACTH and angiotensin II in the duck (*Anas platyrhynchos*). Gen. Comp. Endocrinol. 59, 382–390.

Knight, D.E., Baker, P.F., 1986. Observations on the muscarinic activation of catecholamine secretion in the chicken adrenal gland. Neuroscience 19, 357–366.

Kobayashi, H., Takei, Y., 1982. Mechanisms for induction of drinking with special reference to angiotensin II. Comp. Biochem. Physiol. A 71, 485–594.

Kober, A.K.M.H., Aoyama, M., Sugita, S., 2010. Immunohistochemical localization of catecholamine biosynthetic enzymes in the adrenal gland of the domestic fowl (*Gallus domesticus*). Poult. Sci. 89, 1709–1715.

Kober, A.K.M.H., Aoyama, M., Sugita, S., 2012. Morphological and histological studies on the adrenal gland of the chicken (*Gallus domesticus*). J. Poult. Sci. 49, 39–45.

Koch, K.A., Wingfield, J.C., Buntin, J.D., 2004. Prolactin-induced parental hyperphagia in ring doves: are glucocorticoids involved? Horm. Behav. 46, 498–505.

Kocsis, J.F., Boyette, M.H., McIlroy, P.J., Carsia, R.V., 1994a. Regulation of aldosteronogenesis in domestic turkey (*Meleagris gallopavo*) adrenal steroidogenic cells. Gen. Comp. Endocrinol. 96, 108–121.

Kocsis, J.F., McIlroy, P.J., Chiu, A.T., Schimmel, R.J., Carsia, R.V., 1994b. Properties of angiotensin II receptors of domestic turkey (*Meleagris gallopavo*) adrenal steroidogenic cells. Gen. Comp. Endocrinol. 96, 92–107.

Kocsis, J.F., Carsia, R.V., 1989. Steroidogenic properties of isolated turkey adrenocortical cells. Domest. Anim. Endocrinol. 6, 121–131.

Kocsis, J.F., Lamm, E.-T., McIlroy, P.J., Scanes, C.G., Carsia, R.V., 1995a. Evidence for functionally distinct subpopulations of steroidogenic cells in the domestic turkey (*Meleagris gallopavo*) adrenal gland. Gen. Comp. Endocrinol. 98, 57–72.

Kocsis, J.F., Schimmel, R.J., McIlroy, P.J., Carsia, R.V., 1995b. Dissociation of increases in intracellular calcium and aldosterone production induced by angiotensin II (AII): evidence for regulation by distinct AII receptor subtypes or isomorphs. Endocrinology 136, 1626–1634.

Kocsis, J.F., McIlroy, P.J., Carsia, R.V., 1995c. Atrial natriuretic peptide stimulates aldosterone production by turkey (*Meleagris gallopavo*) adrenal steroidogenic cells. Gen. Comp. Endocrinol. 99, 364–372.

Kocsis, J.F., Rinkardt, N.E., Satterlee, D.G., Weber, H., Carsia, R.V., 1999. Concentration-dependent, biphasic effect of prostaglandins on avian corticosteroidogenesis in vitro. Gen. Comp. Endocrinol. 115, 132–142.

Kordonowy, L.L., McMurtry, J.P., Williams, T.D., 2010. Variation in plasma leptin-like immunoreactivity in free-living European starlings (*Sturnus vulgaris*). Gen. Comp. Endocrinol. 166, 47–53.

Koren, L., Nakagawa, S., Burke, T., Soma, K.K., Wynne-Edwards, K.E., Geffen, E., 2011. Non-breeding feather concentrations of testosterone, corticosterone, and cortisol are associated with subsequent survival in wild house sparrows. Proc. R. Soc. B 279, 1560–1566.

Kovács, K., Péczely, P., 1983. Phase shifts in circadian rhythmicity of total, free corticosterone and transcortine plasma levels in hypothyroid male Japanese quail. Gen. Comp. Endocrinol. 50, 483–489.

Kučka, M., Vagnerová, K., Klusoňová, P., Mikšík, I., Pácha, J., 2006. Corticosterone metabolism in chicken tissues: evidence for tissue-specific distribution of steroid dehydrogenases. Gen. Comp. Endocrinol. 147, 377–383.

Kuenzel, W.J., Kang, S.W., Jurkevich, A., 2013. Neuroendocrine regulation of stress in birds with an emphasis on vasotocin receptors (VTRs). Gen. Comp. Endocrinol. 190, 18–23.

Kühn, E.R., Shimada, K., Ohkubo, T., Vleurick, L.M., Berghman, L.R., Darras, V.M., 1996. Influence of recombinant chicken prolactin on thyroid hormone metabolism in the chick embryo. Gen. Comp. Endocrinol. 103, 349–358.

Kwok, A.H.Y., Wang, Y., Wang, C.Y., Leung, F.C., 2007. Cloning of chicken glucocorticoid receptor (GR) and characterization of its expression in pituitary and extrapituitary tissues. Poult. Sci. 86, 423–430.

Lacombe, A.M.A., Jones, D.R., 1990. The source of circulating catecholamines in forced dived ducks. Gen. Comp. Endocrinol. 80, 41–47.

Landys, M.M., Ramenofsky, M., Wingfield, J.C., 2006. Actions of glucocorticoids at a seasonal baseline as compared to stress-related levels in the regulation of periodic life processes. Gen. Comp. Endocrinol. 148, 132–149.

Langlois, V.S., Zhang, D., Cooke, G.M., Trudeau, V.L., 2010. Evolution of steroid-5α-reductases and comparison of their function with 5β-reductase. Gen. Comp. Endocrinol. 166, 489–497.

Latour, M.A., Peebles, E.D., Boyle, C.R., Brake, J.D., Kellogg, T.F., 1995. Changes in serum lipid, lipoprotein and corticosterone concentrations during neonatal chick development. Biol. Neonate 67, 381–386.

Lattin, C.R., Bauer, C.M., de Bruijn, R., Romero, L.M., 2011a. Hypothalamus-pituitary-adrenal axis activity and the subsequent response to chronic stress differ depending upon life history stage. Gen. Comp. Endocrinol. 178, 494–501.

Lattin, C.R., Waldron-Francis, K., Richardson, J.W., de Bruijn, R., Bauer, C.M., Breuner, C.W., Romero, L.M., 2011b. Pharmacological characterization of intracellular glucocorticoid receptors in nine tissues from house sparrow (*Passer domesticus*). Gen. Comp. Endocrinol. 179, 214–220.

Laverty, G., Elbrønd, V.S., Árnason, S.S., Skadhauge, E., 2006. Endocrine regulation of ion transport in the avian lower intestine. Gen. Comp. Endocrinol. 147, 70–77.

Lechner, O., Dietrich, H., Wiegers, G.J., Vacchio, M., Wick, G., 2001. Glucocorticoid production in the chicken bursa and thymus. Int. Immunol. 13, 769–776.

Leclerc, B., Zadworny, D., Bédécarrats, G., Kühnlein, U., 2007. Ontogenesis of the expression of prolactin receptor messenger ribonucleic acid during late embryogenesis in turkeys and chickens. Poult. Sci. 86, 1174–1179.

Lehoux, J.-G., Bellabarba, D., Beaudry, C., 1977. A comparative study of 3-hydroxy-3-methyglutaryl coenzyme A reductase activity in vertebrate adrenocortical cells. Gen. Comp. Endocrinol. 53, 116–125.

Li, Q.-L., Jansen, E., Brent, G.A., Friedman, T.C., 2001. Regulation of prohormone convertase 1 (PC1) by thyroid hormone. Am. J. Physiol. Endocrinol. Metab. 280, E160–E170.

Liang, L., Sebag, J.A., Eagelston, L., Serasinghe, M.N., Veo, K., Reinick, C., Angleson, J., Hinkle, P.M., Dores, R.M., 2011. Functional expression of frog and rainbow trout melanocortin 2 receptors using heterologous MRAP1s. Gen. Comp. Endocrinol. 174, 5–14.

Lim, S.-K., Gardella, T., Thompson, A., Rosenberg, J., Keutmann, H., Potts Jr., J., Kronenberg, H., Nussbaum, S., 1991. Full-length chicken parathyroid hormone. J. Biol. Chem. 266, 3709–3714.

Lin, H., Sui, S.J., Jiao, H.C., Jiang, K.J., Zhao, J.P., Dong, H., 2007. Effects of diet and stress mimicked by corticosterone administration on early postmortem muscle metabolism of broiler chicks. Poult. Sci. 86, 545–554.

Ling, M.K., Hotta, E., Kilianova, Z., Haitaina, T., Ringholm, A., Johansson, L., Gallo-Payet, N., Takeuchi, S., Schiöth, H.B., 2004. The melanocortin receptor subtypes in chicken have high preference to ACTH-derived peptides. Br. J. Pharmacol. 143, 626–637.

Liu, Y., Smith, L.I., Huang, V., Poon, V., Coello, A., Olah, M., Spiga, F., Lightman, S.L., Aguilera, G., 2013. Transcriptional regulation of episodic glucocorticoid secretion. Mol. Cell. Endocrinol. 371, 62–70.

Lothrop, C.D., Olsen, J.H., Loomis, M.R., Jensen, J.M., Lenhard, A., 1985. Evaluation of adrenal function in psittacine birds, using the ACTH stimulation test. J. Am. Vet. Med. Assoc. 187, 1113–1115.

Lynn, S.E., Breuner, C.W., Wingfield, J.C., 2003. Short-term fasting affects locomotor activity, corticosterone, and corticosterone binding globulin in a migratory songbird. Horm. Behav. 43, 150–157.

Madison, F.N., Jurkevich, A., Kuenzel, W.J., 2008. Sex differences in plasma corticosterone release in undisturbed chickens (*Gallus gallus*) in response to arginine vasotocin and corticotropin releasing hormone. Gen. Comp. Endocrinol. 155, 566–573.

Mahata, S.K., Mahapatra, N.R., Mahata, M., O'Connor, D.T., 2002. Neuroendocrine cell type-specific and inducible expression of chromogranin/secretogranin genes. Crucial promotor motifs. Ann. NY Acad. Sci. 971, 27–38.

Maitra, U.S., Khan, M.S., Rosner, W., 1993. Corticosteroid-binding globulin receptor of the rat hepatic membrane: solubilization, partial characterization, and the effect of steroids on binding. Endocrinology 133, 1817–1822.

Malisch, J.L., Breuner, C.W., 2010. Steroid-binding proteins and free steroids in birds. Mol. Cell. Endocrinol. 316, 42–52.

Malisch, J.L., Satterlee, D.G., Cockrem, J.F., Wada, H., Breuner, C.W., 2010. How acute is the acute stress response? Baseline corticosterone and corticosteroid-binding globulin levels change 24 h after an acute stressor in Japanese quail. Gen. Comp. Endocrinol. 165, 345–350.

Mazzocchi, G., Gottardo, G., Nussdorfer, G.G., 1997a. Pituitary adenylate cyclase-activating peptide enhances steroid production by interrenal glands in fowls: evidence for an indirect mechanism of action. Horm. Metab. Res. 29, 86–87.

Mazzocchi, G., Gottardo, G., Nussdorfer, G.G., 1997b. Catecholamines stimulates steroid secretion of dispersed fowl adrenocortical cells, acting through the β-adrenergic subtype. Horm. Metab. Res. 29, 190–192.

Mazzocchi, G., Gottardo, G., Nussdorfer, G.G., 1997c. Effects of somatostatin on steroid production by adrenocortical cells of the domestic turkey and fowl. Zool. Sci. 14, 359–361.

McCorkle, F.M., Taylor Jr., R.L., 1993. Biogenic amines regulate avian immunity. Poult. Sci. 72, 1285–1288.

McIlroy, P.J., Kocsis, J.F., Weber, H., Carsia, R.V., 1999. Dietary protein restriction stress in the domestic fowl (*Gallus gallus domesticus*) alters adrenocorticotropin-transmembranous signaling and corticosterone negative feedback in adrenal steroidogenic cells. Gen. Comp. Endocrinol. 113, 255–266.

Medina, C.O., Lattin, C.R., McVey, M., Romero, L.M., 2013. There is no correlation between glucocorticoid receptor mRNA expression and protein binding in the brains of house sparrows (*Passer domesticus*). Gen. Comp. Endocrinol. 193, 27–36.

Mellon, S.H., Bair, S.R., Monis, H., 1995. P45011B3 mRNA, transcribed from a third P450c11 gene, is expressed in a tissue-specific, developmentally and hormonally regulated fashion in the rodent adrenal, and encodes a protein with both 11-hydroxylase and 18-hydroxylase activities. J. Biol. Chem. 270, 1643–1649.

Mikhailova, M.V., Mayeux, P.R., Jurkevich, A., Kuenzel, W.J., Madison, F., Periasamy, A., Chen, Y., Cornett, L.E., 2007. Heterooligomerization between vasotocin and corticotropin-releasing hormone (CRH) receptors augments CRH-stimulated 3′,5′-cyclic adenosine monophosphate production. Mol. Endocrinol. 21, 2178–2188.

Miller, C.M., Vleck, C.M., Otis, D.L., 2009. Individual variation in baseline and stress-induced corticosterone and prolactin levels predicts parental effort by nesting mourning doves. Horm. Behav. 56, 457–464.

Miller, W.L., Auchus, R.J., 2011. The molecular biology, biochemistry and physiology of human steroidogenesis and its disorders. Endocr. Rev. 32, 81–151.

Mirabella, N., Esposito, V., Squillacioti, C., De Luca, A., Paino, G., 2004. Expression of agouti-related protein (AgRP) in the hypothalamus and adrenal gland of the duck (*Anas platyrhynchos*). Anat. Embryol. 209, 137–141.

Miyata, A., Minamino, N., Kangawa, K., Matsuo, H., 1988. Identification of a 29-amino acid natriuretic peptide in chicken heart. Biochem. Biophys. Res. Commun. 155, 1330–1337.

Moore, R.W., Hargis, B.M., Porter, T.E., Caldwell, D.Y., Oubre, C.M., Vandesande, F., Berghman, L.R., 2004. Ovoinhibitor in the chicken bursa of Fabricius: identification, isolation, and localization. Cell Tissue Res. 317, 247–251.

Mornet, E., Dupont, J., Vitek, A., White, P.C., 1989. Characterization of two genes encoding human steroid 11β-hydroxylase (P45011β). J. Biol. Chem. 264, 20961–20967.

Morohashi, K., Yoshioka, H., Gotoh, O., Okada, Y., Yamamoto, K., Miyata, T., Sogawa, K., Fujii-Kuriyama, Y., Omura, T., 1987. Molecular cloning and nucleotide sequence of DNA of mitochondrial P-450(11) of bovine adrenal cortex. J. Biochem. 102, 559–568.

Mostl, E., Palme, R., 2002. Hormones as indicators of stress. Domest. Anim. Endocrinol. 23, 67–74.

Müller, C., Jenni-Eirmann, S., Jenni, L., 2010. Development of the adrenocortical response to stress in Eurasian kestrel nestlings: defence ability, age, brood hierarchy and condition. Gen. Comp. Endocrinol. 168, 474–483.

Mumma, J.O., Thaxton, J.P., Vizzier-Thaxton, Y., Dodson, W.L., 2006. Physiological stress in laying hens. Poult. Sci. 85, 761–769.

Murphy, T.J., Nakamura, Y., Takeuchi, K., Alexander, R.W., 1993. A coned angiotensin receptor isoform from the turkey adrenal gland is pharmacologically distinct from mammalian angiotensin II receptors. Mol. Pharmacol. 44, 1–7.

Nagra, C.L., Sauers, A.K., Wittmaier, H.N., 1965. Effect of testosterone, progetagens, and metopirone on adrenal activity in cockerels. Gen. Comp. Endocrinol. 5, 69–73.

Nakagawa-Mizuyachi, K., Takahashi, T., Kawashima, M., 2009. Calcitonin directly increases adrenocorticotropic hormone-stimulated corticosterone production in the hen adrenal gland. Poult. Sci. 88, 2199–2205.

Nakabayashi, O., Nomura, O., Nishimori, K., Yasue, H., Mizuno, S., 1995. The cDNA cloning and transient expression of a chicken gene encoding a 3 beta-hydroxysteroid dehydrogenase/delta 5 → 4 isomerase unique to major steroidogenic tissues. Gene 162, 261–265.

Newman, A.E., Chin, E.H., Schmidt, K.L., Bond, L., Wynne-Edwards, K.E., Soma, K.K., 2008. Analysis of steroids in songbird plasma and brain by coupling solid phase extraction to radioimmunoassay. Gen. Comp. Endocrinol. 155, 503–510.

Nishimura, H., 2001. Angiotensin receptors—evolutionary overview and perspectives. Comp. Biochem. Physiol. A 128, 11–30.

Nomura, O., Nakabayashi, O., Nishimori, K., Yasue, H., Mizuno, S., 1997. The cDNA cloning and transient expression of a chicken gene encoding cytochrome P-450scc. Gene 185, 217–222.

Nomura, O., Nakabayashi, O., Nishimori, K., Yasue, H., Mizuno, S., 1999. Expression of five steroidogenic genes including aromatase gene at early developmental stages of chicken male and female embryos. J. Steroid. Biochem. Mol. Biol. 71, 103–109.

Nonaka, Y., Takemori, H., Halder, S.K., Sun, T., Ohta, M., Hatano, O., Takakusu, A., Okamoto, M., 1995. Frog cytochrome *P*-450 (11β,aldo), a single enzyme involved in the final steps of glucocorticoid and mineralocorticoid biosynthesis. Eur. J. Biochem. 229, 249–256.

Ohkubo, T., Tanaka, M., Nakashima, K., Talbot, R.T., Sharp, P.J., 1998. Prolactin receptor gene expression in the brain and peripheral tissues in broody and nonbroody breeds of domestic hen. Gen. Comp. Endocrinol. 109, 60–68.

Ohmori, Y., 1998. Localization of biogenic amines and neuropeptides in adrenal medullary cells of birds. Horm. Metab. Res. 30, 384–388.

Okuliarová, M., Šárniková, B., Rettenbacher, S., Škrobánek, P., Zeman, M., 2010. Yolk testosterone and corticosterone in hierarchical follicles and laid eggs of Japanese quail exposed to long-term restraint stress. Gen. Comp. Endocrinol. 165, 91–96.

Ono, H., Iwasaki, M., Sakamoto, N., Mizuno, S., 1988. cDNA cloning and sequence analysis of a chicken gene expressed during the gonadal development and homologous to mammalian cytochrome P-450c17. Gene 66, 77–85.

Orchinik, M., 1998. Glucocorticoids, stress, and behavior: shifting the timeframe. Horm. Behav. 34, 320–327.

Ottinger, M.A., Pitts, S., Abdelnabi, M.A., 2001. Steroid hormones during development in Japanese quail: plasma, gonadal and adrenal levels. Poult. Sci. 80, 795–799.

Pedernera, E.A., Romano, M., Besedovsky, H.O., Aguilar, M.D.C., 1980. The bursa of Fabricius is required for normal endocrine development in chicken. Gen. Comp. Endocrinol. 42, 413–419.

Pierce, R.B., Cronshaw, J., Holmes, W.N., 1978. Evidence for the zonation of interrenal tissue in the adrenal gland of the duck (*Anas platyrhynchos*). Cell Tissue Res. 192, 363–379.

Pihlajoki, M., Gretzinger, E., Cochran, R., Kyrönlahti, A., Schrade, A., Hiller, T., Sullivan, L., Shoykhet, M., Schoeller, E.L., Brooks, M.D., Heikinheimo, M., Wilson, D.B., 2013. Conditional mutagenesis of Gata6 in SF1-positive cells causes gonadal-like differentiation in the adrenal cortex of mice. Endocrinology 154, 1754–1767.

Pinheiro, P.L.C., Cardosa, J.C.R., Power, D.M., Canário, A.V.M., 2012. Functional characterization and evolution of PTH/PTHrP receptors: insights from chicken. BMC Evol. Biol. 12, 110.

Pinson, S.E., Wilson, J.L., Navara, K.J., 2011. Elevated testosterone during meiotic segregation stimulates laying hens to produce more sons than daughters. Gen. Comp. Endocrinol. 174, 195–201.

Pintér, O., Péczely, P., Zsebők, S., Zelena, D., 2011. Seasonal changes in courtship behavior, plasma androgen levels and in hypothalamic aromatase immunoreactivity in male free-living European starlings (*Sturnus vulgaris*). Gen. Comp. Endocrinol. 172, 151–157.

Pitel, F., Faraut, T., Bruneau, G., Monget, P., 2010. Is there a leptin gene in the chicken genome? Lessons from phylogenetics, bioinformatics and genomics. Gen. Comp. Endocrinol. 167, 1–5.

Poderoso, C., Duarte, A., Cooke, M., Orlando, U., Gottifredi, V., Solano, A.R., Lemos, J.R., Podestá, E.J., 2013. The spatial and temporal regulation of the hormonal signal. Role of mitochondria in the formation of a protein complex required for the activation of cholesterol transport and steroid synthesis. Mol. Cell. Endocrinol. 371, 26–33.

Pohle, K., Cheng, H.-W., 2009. Comparative effects of furnished and battery cages on egg production and physiological parameters in white leghorn hens. Poult. Sci. 88, 2042–2051.

Proszkowiec-Weglarz, M., Porter, T.E., 2010. Functional characterization of chicken glucocorticoid and mineralocorticoid receptors. Am. J. Physiol. Regul. Integr. Comp. Physiol. 298, R1257–R1268.

Qian, X., Droste, S.K., Gutièrrez-Mecinas, M., Collins, A., Kersanté, F., Reul, J.M.H.M., Linthorst, A.C.E., 2011. A rapid release of corticosteroid-binding globulin from the liver restrains the glucocorticoid hormone response to acute stress. Endocrinology 152, 3738–3748.

Qian, X., Droste, S.K., Lightman, S.L., Reul, J.M.H.M., Linthorst, C.E., 2012. Circadian and ultradian rhythms of free glucocorticoid hormone are highly synchronized between blood, the subcutaneous tissue, and the brain. Endocrinology 153, 4346–4353.

Quillfeldt, P., Poisbleau, M., Parenteau, C., Trouvé, C., Demongin, L., van Noordwijk, H.J., Möstl, E., 2011. Measuring corticosterone in seabird egg yolk and the presence of high yolk gestagen concentrations. Gen. Comp. Endocrinol. 173, 11–14.

Rettenbacher, S.E., Möstl, E., Hackl, R., Ghareeb, K., Palme, R., 2004. Measurement of corticosterone metabolites in chicken droppings. Br. Poult. Sci. 45, 704–711.

Ricklefs, R.E., 2010. Insights from comparative analyses of aging in birds and mammals. Aging Cell 9, 273–284.

Rombauts, L., Vanmontfort, D., Berghman, L.R., Decuypere, E., Verhoeven, G., 1994. Contribution of the fetal adrenal to circulating immunoactive inhibin in the chicken embryo. Biol. Reprod. 51, 926–933.

Romero, L.M., Reed, J.M., 2005. Collecting baseline corticosterone samples in the field: is under 3 min good enough? Comp. Biochem. Physiol. A 140, 73–79.

Romero, L.M., Rich, E.L., 2007. Photoperiodically-induced changes in hypothalamic-pituitary-adrenal axis sensitivity in captive house sparrow (*Passer domesticus*). Comp. Biochem. Physiol. A 147, 562–568.

Romero, L.M., Soma, K.K., Wingfield, J.C., 1998. Hypothalamic-pituitary-adrenal axis changes allow seasonal modulation of corticosterone in a bird. Am. J. Physiol. Regul. Integr. Comp. Physiol. 274, R1338–R1344.

Rosenberg, J., Hurwitz, S., 1987. Concentration of adrenocortical hormones in relation to cation homeostasis in birds. Am. J. Physiol. 253, R20–R24.

Rosenberg, J., Pines, M., Hurwitz, S., 1987. Response of adrenal cells to parathyroid hormone stimulation. J. Endocrinol. 112, 431–437.

Rosenberg, J., Pines, M., Hurwitz, S., 1988a. Stimulation of chick adrenal steroidogenesis by avian parathyroid hormone. J. Endocrinol. 116, 91–95.

Rosenberg, J., Pines, M., Hurwitz, S., 1988b. Regulation of aldosterone secretion by avian adrenocortical cells. J. Endocrinol. 118, 447–453.

Rosenberg, J., Pines, M., Levy, J.J., Nutt, R.F., Caulfield, M.P., Russell, J., Sherwood, L.M., Hurwitz, S., 1989a. Renal and adrenal adenosine 3′,5′-monophosphate production and corticosteroid secretion in response to synthetic chicken parathyroid hormone-(1–34). Endocrinology 125, 1082–1089.

Rosenberg, J., Pines, M., Hurwitz, S., 1989b. Inhibition of aldosterone secretion by atrial natriuretic peptide in chicken adrenocortical cells. Biochem. Biophys. Acta 1014, 189–194.

Rubolini, D., Romano, M., Boncoraglio, G., Ferrari, R.P., Martinelli, R., Galeotti, P., Fasola, M., Saino, N., 2005. Effects of elevated egg corticosterone levels on behavior, growth, and immunity of yellow-legged gull (*Larus michahellis*) chicks. Horm. Behav. 47, 592–605.

Samarasinghe, R.A., Di Maio, R., Volonte, D., Galbiati, F., Lewis, M., Romero, G., DeFranco, D.B., 2011. Nongenomic glucocorticoid receptor action regulates gap junction intercellular communication and neural progenitor cell proliferation. Proc. Natl. Acad. Sci. 108, 16657–16662.

Sapolsky, R.M., Romero, L.M., Munck, A.U., 2000. How do glucocorticoids influence stress responses? Integrative, permissive, suppressive, stimulatory and preparative actions. Endocr. Rev. 21, 55–89.

Satterlee, D.G., Comeaux, A.P., Johnson, W.A., Munn, B.J., 1989. Influence of exogenous prostaglandin E_2 and stress on circulating corticosterone concentrations in *Coturnix*. Poult. Sci. 68, 1289–1293.

Scanes, C.G., 2009. Perspectives on the endocrinology of poultry growth and metabolism. Gen. Comp. Endocrinol. 163, 24–32.

Schlinger, B.A., Lane, N.I., Grisham, W., Thompson, L., 1999. Androgen synthesis in a songbird: a study of Cyp17 (17α-hydroxylase/C17,20-Lyase) activity in zebra finch. Gen. Comp. Endocrinol. 113, 46–58.

Schmeling, S.K., Nockels, C.F., 1978. Effects of age, sex, and ascorbic acid ingestion on chicken plasma corticosterone levels. Poult. Sci. 57, 527–533.

Schmidt, K.L., Malisch, J.L., Breuner, C.W., Soma, K.K., 2010. Corticosterone and cortisol binding sites in plasma, immune organs and brain of developing zebra finches: intracellular and membrane-associated receptors. Brain Behav. Immun. 24, 908–918.

Schmidt, K.L., Pradhan, D.S., Shah, A.H., Charlier, T.D., Chin, E.H., Soma, K.K., 2008. Neurosteroids, immunosteroids, and the balkanization of endocrinology. Gen. Comp. Endocrinol. 157, 266–274.

Schmidt, K.L., Soma, K.K., 2008. Cortisol and corticosterone in the songbird immune and nervous systems: local vs. systemic levels during development. Am. J. Physiol. Regul. Integr. Comp. Physiol. 295, R103–R110.

Schoech, S.J., Ketterson, E.D., Nolan Jr., V., 1999. Exogenous testosterone and the adrenocortical response to dark-eyed juncos. Auk 116, 64–72.

Sebag, J.A., Zhang, C., Hinkle, P.M., Bradshaw, A.M., Cone, R.D., 2013. Developmental control of the melanocortin-4 receptor by MRAP2 proteins in zebrafish. Science 341, 278–281.

Sewer, M.B., Li, D., 2013. Regulation of adrenocortical steroid hormone production by RhoA-diaphanous 1 signaling and the cytoskeleton. Mol. Cell. Endocrinol. 371, 79–86.

Sharma, D., Chaturvedi, C.M., 2009. Effects of thyroid status on arginine vasotocin receptor VT2R expression and adrenal function in osmotically stimulated domestic fowl. J. Comp. Physiol. B 179, 811–819.

Sharma, D., Cornett, L.E., Chaturvedi, C.M., 2009a. Corticosterone- or metapyrone-induced alterations in adrenal function and expression of the arginine vasotocin receptor VT2 in the pituitary gland of domestic fowl, *Gallus gallus*. Gen. Comp. Endocrinol. 161, 208–215.

Sharma, D., Cornett, L.E., Chaturvedi, C.M., 2009b. Osmotic stress induced alterations in the expression of arginine vasotocin receptor VT2 in the pituitary gland and adrenal function of domestic fowl. Gen. Comp. Endocrinol. 160, 216–222.

Shen, P., Campagnoni, C.W., Kampf, K., Schlinger, B.A., Arnold, A.P., Campagnoni, A.T., 1994. Isolation and characterization of a zebra finch aromatase cDNA: *in situ* hybridization reveals high aromatase expression in the brain. Mol. Brain Res. 24, 227–237.

Shini, S., Huff, G.R., Shini, A., Kaiser, P., 2010. Understanding stress-induced immunosuppression: exploration of cytokine and chemokine gene profiles in chicken peripheral leukocytes. Poult. Sci. 89, 841–851.

Shtukmaster, S., Schier, M.C., Huber, K., Krispin, S., Kalcheim, C., Unsicker, K., 2013. Sympathetic neurons and chromaffin cells share a common progenitor in the neural crest in vivo. Neural Dev. 8, 12.

Smith, C.A., Smith, M.J., Sinclair, A.H., 1999. Expression of chicken steroidogenic factor-1 during gonadal development. Gen. Comp. Endocrinol. 113, 187–196.

Soma, K.K., Wingfield, J.C., 2001. Dehydroepiandrosterone in songbird plasma: seasonal regulation and relationship to territorial aggression. Gen. Comp. Endocrinol. 123, 144–155.

Spencer, K.A., Evans, N.P., Monaghan, P., 2009. Postnatal stress in birds: a novel model of glucocorticoid programming of the hypothalamic-pituitary-adrenal axis. Endocrinology 150, 1931–1934.

Squillacioti, C., De Luca, A., Paino, S., Nicola, M., 2008. Innervation and neurochemical characteristics of medullary cells in the duck adrenal gland. In: Capaldo, A. (Ed.), Recent Advances in Non-mammalian Adrenal Gland Research. Research Signpost, Kerala, India, pp. 189–202.

Takeuchi, S., Kudo, T., Takahashi, S., 1998. Molecular cloning of the chicken melanocortin 2 (ACTH)-receptor gene. Biochim. Biophys. Acta 1403, 102–108.

Takeuchi, S., Takahashi, S., 1998. Melanocortin receptor genes in the chicken—tissue distribution. Gen. Comp. Endocrinol. 112, 220–231.

Takeuchi, S., Takahashi, S., 1999. A possible involvement of melanocortin 3 receptor in the regulation of adrenal gland function in the chicken. Biochim. Biophys. Acta 148, 512–518.

Takeuchi, S., Teshigawara, K., Takahashi, S., 1999. Molecular cloning and characterization of the chicken pro-opiomelanocortin (POMC) gene. Biochim. Biophys. Acta 1450, 452–459.

Takeuchi, S., Teshigawara, K., Takahashi, S., 2000. Widespread expression of agouti-related protein (AGRP) in the chicken: a possible involvement of AGRP in regulating peripheral melanocortin systems in the chicken. Biochim. Biophys. Acta 1496, 261–269.

Tanabe, Y., Nakamura, T., Fujioka, K., Doi, O., 1979. Production and secretion of sex steroid hormones by the testes, the ovary, and the adrenal glands of embryonic and young chickens (*Gallus domesticus*). Gen. Comp. Endocrinol. 39, 26–33.

Tanabe, Y., Saito, N., Nakamura, T., 1986. Ontogenetic steroidogenesis by testes, ovary, and adrenals of embryonic and postembryonic chickens (*Gallus domesticus*). Gen. Comp. Endocrinol. 63, 456–463.

Thurston, R.J., Bryant, C.C., Korn, N., 1993. The effects of corticosterone and catecholamine infusion on plasma glucose levels in chicken (*Gallus domesticus*) and turkey (*Meleagris gallopavo*). Comp. Biochem. Physiol. C 106, 59–62.

Toshimori, H., Toshimori, K., Minamino, N., Kangawa, K., Oura, C., Matsukura, S., Matsuo, H., 1990. Chicken atrial natriuretic peptide (chANP) and its secretion. Cell Tissue Res. 259, 293–298.

Tran, P.V., Georgieff, M.K., Engeland, W.C., 2010. Sodium depletion increases sympathetic neurite outgrowth and expression of a novel TMEM35 gene-derived protein (TUF1) in the rat adrenal zona glomerulosa. Endocrinology 151, 4852–4860.

Trifaró, J.-M., 2002. Molecular biology of the chromaffin cell. Ann. NY Acad. Sci. 971, 11–18.

Unsicker, K., 1973. Fine structure and innervation of the avian adrenal gland. II. Cholinergic innervation of adrenal chromaffin cells. Z. Zellforsch. 145, 417–442.

Vänttinen, T., Liu, J., Kuulasmaa, T., Kivinen, P., Voutilainen, R., 2003. Expression of activin/inhibin signaling components in the human adrenal gland and the effects of activins and inhibins on adrenocortical steroidogenesis and apoptosis. J. Endocrinol. 178, 479–489.

Vanmontfort, D., Room, G., Bruggeman, V., Rombauts, L., Berghman, L.R., Verhoeven, G., Decuypere, E., 1997. Ovarian and extraovarian sources of immunoreactive inhibin in the chicken: effects of dexamethasone. Gen. Comp. Endocrinol. 105, 333–343.

Veo, K., Reinick, C., Liang, L., Moser, E., Angleson, J.K., Dores, R.M., 2011. Observations on the ligand selectivity of the melanocortin 2 receptor. Gen. Comp. Endocrinol. 172, 3–9.

Vinson, G.P., Hinson, J.P., Raven, P.W., 1985. The relationship between tissue preparation and function; methods for the study of control of aldosterone secretion: a review. Cell Biochem. Funct. 3, 235–253.

von Blumröder, D., Tönhardt, H., 2002. Influence of long-term changes in incubation temperature on catecholamine levels in plasma of chicken embryos (*Gallus gallus f. domestica*). Comp. Biochem. Physiol. A 131, 701–711.

Vylitová, M., Mikšík, I., Pácha, J., 1998. Metabolism of corticosterone in mammalian and avian intestine. Gen. Comp. Endocrinol. 109, 315–324.

Wada, H., 2008. Glucocorticoids: mediators of vertebrate ontogenetic transitions. Gen. Comp. Endocrinol. 156, 441–453.

Walsh, M.T., Beldegreen, R.A., Clubb, S.L., Chen, C.L., 1985. Effect of exogenous ACTH on serum corticosterone and cortisol concentrations in the Moluccan cockatoo (*Cacatua moluccensis*). Am. J. Vet. Res. 46, 1584–1588.

Watford, M., 2005. Is the small intestine a gluconeogenic organ? Nutr. Rev. 63, 356–360.

Webb, T.R., Clark, A.J.L., 2010. Minireview: the melanocortin 2 receptor accessory proteins. Mol. Endocrinol. 24, 475–484.

Weber, H., Kocsis, J.F., Lauterio, T.J., Carsia, R.V., 1990. Dietary protein restriction stress and adrenocortical function: evidence for transient and long-term induction of enhanced cellular function. Endocrinology 127, 3138–3150.

Wendler, A., Albrecht, C., Wehling, M., 2012. Nongenomic actions of aldosterone and progesterone revisited. Steroids 77, 1002–1006.

Whitman, B.A., Breuner, C.W., Dufty Jr., A.M., 2011. The effects of neonatal handling on adrenocortical responsiveness, morphological development and corticosterone binding globulin in nestling American kestrels (*Falco sparverius*). Gen. Comp. Endocrinol. 172, 260–267.

Wilson, J.X., 1984. The renin-angiotensin system in nonmammalian vertebrates. Endocr. Rev. 5, 45–61.

Wingfield, J.C., 2013. Ecological processes and the ecology of stress: the impacts of abiotic environmental factors. Funct. Ecol. 27, 37–44.

Wingfield, J.C., Matt, K.S., Farner, D.S., 1984. Physiological properties of steroid hormone-binding proteins in avian blood. Gen. Comp. Endocrinol. 53, 281–292.

Wolfensberger, M., Forssmann, W.G., Reinecke, M., 1995. Localization and coexistence of atrial natriuretic peptide (ANP) and neuropeptide Y (NPY) in vertebrate adrenal chromaffin cells immunoreactive to TH, DBH and PNMT. Cell Tissue Res. 267–276.

Yosefi, S., Hen, G., Rosenblum, C.I., Cerasale, D.J., Beaulieu, M., Criscuolo, F., Friedman-Einat, M., 2010. Lack of leptin activity in blood samples of Adélie penguin and bar-tailed godwit. J. Endocrinol. 207, 113–122.

Chapter 27

Endocrine Pancreas

Joëlle Dupont
Unité de Physiologie de la Reproduction et des Comportements, Institut National de la Recherche Agronomique, 37380 Nouzilly, France

Nicole Rideau and Jean Simon
Unité de Recherches Avicoles, Institut National de la Recherche Agronomique, 37380 Nouzilly, France

27.1 INTRODUCTION

The endocrine pancreas of the chicken (and of birds in general) displays several peculiarities (Hazelwood, 2000; Simon, 1989). Before discussing the current knowledge, we will recall the basics from these previous reviews. Most of these are certainly connected and interacting.

- Birds have high plasma uric acid and glucose levels, despite normal plasma insulin levels (see Simon et al., 2011, for chicken). Such high plasma glucose levels (2–2.15 g/L in the fasting state) would be harmful for humans.
- In chickens and ducks, high doses of exogenous insulin are required to induce hypoglycemia, and huge doses are not lethal. The chicken features high blood glucose and low insulin sensitivity and, as will become apparent in this chapter, the weak insulin release elicited by glucose in the isolated chicken pancreas is characteristic of type 2 diabetes in humans.
- Experimental diabetes models are not available for chickens. Pancreatectomies are rarely complete; plasma glucose regulation is, however, impaired during re-feeding. Diabetogenic drugs (alloxan or streptozotocin) are inefficient, even when glycemia is decreased before streptozotocin injection. Some isolated cases of diabetes have been described in pet birds, although not cited here.
- Glucagon exerts a potent hyperglycemic effect; it is the lipolytic hormone in birds. The glucagon-induced cyclic adenosine monophosphate (cAMP) increase is comparatively small but long-lived; the feedback decrease of cAMP by fatty acids appears inefficient in chicken adipocytes. Insulin lacks an antilipolytic effect in chicken adipocytes. Several peptides have been shown to have an antilipolytic effect *in vitro*: APP (avian pancreatic polypeptide), somatostatin, gastrin, and GLI (gut glucagon-like immunoactivity); however, their function in chicken physiology has not been fully quantified *in vivo*.
- Chicken insulin is "hyperactive", whereas duck insulin is "hypoactive". Chicken insulin has accordingly been suggested as more potent at inducing hypoglycemia in chickens than sheep or bovine insulin (in Hazelwood, 2000). The extent of the difference may have been overestimated: at high concentrations, a similar nadir (maximal hypoglycemic effect) should be achieved irrespective of insulin type (this is the case for insulin binding and phosphorylation of artificial substrate in solubilized chicken liver or brain insulin receptors).
- The importance of the insulin–glucagon ratio in the control of metabolism has been demonstrated in chickens and ducks. *In vivo*, this ratio is governed by opposite insulin–glucagon variations in response to nutritional status (fed versus fasting). Glucagon–insulin interactions are, however, more complex. Insulin immuno-neutralization in fed chickens (i.e., a nutritional status for high insulin demand, a protocol previously developed in ducks; Mirsky et al., 1964 in Dupont et al., 2008) evokes a rapid (less than 30 min) and large rise in plasma glucose levels. Surprisingly, injection of a glucagon antagonist (des-His^{1} [Glu^{9}] glucagon) has no effect on plasma glucose levels in fed chickens (Dupont et al., 2008); a hypoglycemic effect might be expected. This strongly suggests that in normally fed chickens, the control of plasma glucose level relies more on insulin than on glucagon (in Dupont et al., 2008, 2012). The presence of glucagon is, however, essential: in the pathological model of "hypoglycemia-spiking mortality syndrome", where pancreatic glucagon content is extensively depleted following a virus infection, chickens do not eat and ultimately die from starvation in a profound hypoglycemic status (Davis and Vasilatos-Younken, 1995). To inhibit pancreatic glucagon release, glucose requires the presence of insulin (Dupont et al., 2008), which confirms previous results obtained in ducks and later on in mammals (see Miahle's group references in Simon, 1989).

- Chicken (and bird) metabolism is also peculiar in several ways (as discussed throughout this book). In short, gluconeogenesis is active in liver and kidneys but appears able to be regulated only in kidneys. Liver is the organ for *de novo* lipogenesis.
- Insulin exerts pleiotropic effects in chickens during embryonic and posthatch development. The major typical effects on carbohydrate, lipid, and protein metabolism have been demonstrated. Insulin stimulates glucose and amino acid transport in various cells. The insulin effect on glucose transport is, however, limited in muscle and still doubtful in chicken adipocytes. Insulin stimulation of amino acid transport requires protein synthesis. Insulin greatly stimulates liver lipogenesis and expression of lipogenic enzymes. In contrast, insulin stimulation of lipogenesis remains marginal, at best, in chicken adipocytes (lipogenesis is also intrinsically low in human adipocytes). Glucagon counteracts insulin stimulation of lipogenesis in chicken liver.
- Finally, birds have a high body temperature (about 42 °C in chickens) and a high metabolic rate and anabolism. They are able to adapt to long catabolic states and to survive long periods without food.

This chapter offers a synthetic view of current knowledge, focusing on major aspects of glucagon and insulin control for the sake of brevity. References have been kept as few as possible; early major original contributions (including ours) for chickens and birds are acknowledged and listed only by the first author's name and date, but not referenced, when quoted in recent publications. For evident reasons, knowledge of the role of the endocrine pancreas is far more advanced in mammals; on several occasions, it was necessary to refer to it. Again, to save space, these data are listed in the text by the first author's name and date. Readers will need to connect to PubMed to have access to full references.

27.2 PANCREAS EMBRYOGENESIS AND DEVELOPMENT

27.2.1 Morphology of Avian Pancreas (Figure 27.1)

In most birds, the pancreas is situated in the duodenal loop and consists of three lobes: the dorsal, ventral, and splenic lobes. The splenic lobe is contiguous with the rest of the pancreas, but in some birds it can be completely separated. A fourth lobe, sometimes confusingly called the "third lobe", is a part of the ventral lobe and is distinguishable in galliform birds (which include chickens and quail).

27.2.2 Distribution of Avian Pancreatic Endocrine Cells between the Different Pancreatic Lobes

The endocrine pancreas contains clusters of cells called the islets of Langerhans, which are distributed throughout the exocrine parenchyma. These represent only 1–2% of the whole gland. In contrast to mammals, in which only one type of islets occurs, three types of islets are found in birds: the light, the dark, and the mixed islets. They differ in shape, size, and staining characteristics. The light islets (or B-islets) remain clear after classical staining procedures (Heidenhain's iron hematoxylin); they are essentially composed of β- and δ-cells secreting insulin and somatostatin, respectively, and occasionally very few α-cells secreting glucagon. The dark islets (or A-islets) are composed essentially of α- and δ-cells secreting glucagon and somatostatin, respectively. They are larger in size than the B-islets. The mixed islets are composed of numerous β-cells and a few α- and δ-cells. Some wild bird species have mixed islets (randomly distributed) rather than light and dark islets (Steiner et al., 2010). A fourth type of endocrine cell is found in the pancreas: the F-cells. These cells synthesize and release APP; they occur as single cells or in groups disseminated in the exocrine parenchyma, mainly in the dorsal and ventral lobes. Some F-cells can be found within the islets. Isolated δ-cells (as numerous as F-cells) are also found as dispersed extra-insular endocrine elements. The splenic lobe and, to a lesser extent, the third lobe (when present) notably contain more islet tissue per unit volume than do the dorsal and ventral lobes (Rideau, 1988).

The feature unique to the avian pancreas is the predominance of the glucagon-producing cells over the insulin-producing cells in the ratio of approximately 2:1 (see Andrew, 1984 in Manakova and Titlbach, 2007). In addition, somatostatin-producing δ-cells are also more numerous in the avian pancreas than in that of mammals (Rideau, 1988). The existence, distribution, and relative frequency of various (neuro-)endocrine cells (biotin, serotonin, neuropeptide

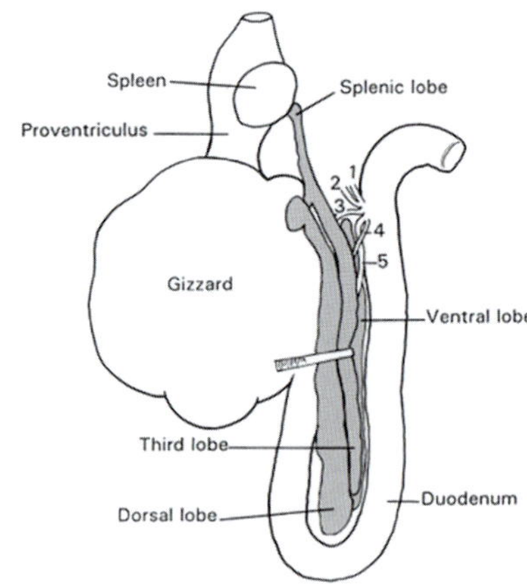

FIGURE 27.1 **Morphology of the avian pancreas.** The anatomical relationship in the chicken of the four lobes of the pancreas (shaded) and neighboring organs (blood supply and liver not shown) viewed from the dorsal aspect. (1) cystic duct; (2) hepatic duct; (3) dorsal pancreatic duct; (4) duct of the third lobe; and (5) ventral pancreatic duct. *After Mikami, S.I., Ono, K., 1962. Glucagon deficiency induced by extirpation of alpha islets of the fowl pancreas. Endocrinology 71, 464–473; The Endocrine Society.*

Y (NPY), and chromogranin) have been demonstrated in the pancreas of avian species (Ku et al., 2000; Lucini et al., 2000).

The islet tissue is richly vascularized, irrigated by the pancreaticoduodenal artery and drained by the pancreaticoduodenal vein. Innervation of the pancreatic islets of birds is not as apparent as it is in mammals. Electron microscopic studies have clearly shown that the main endocrine cellular types of the pancreas of the domestic fowl are associated with nerve terminals. The nerve supply to the islets is, however, less abundant than that to the acinar cells; it is also less marked to the A-islets than to the B-islets. In contrast, δ-cells are richly and directly innervated in chickens (Rideau, 1988).

27.2.3 Development of Avian Pancreas

The pancreas is of endodermal origin; it is formed from the embryonic foregut. In birds it arises from three buds: a dorsal bud and right and left ventro-lateral buds (Rawdon, 1998) that appear at the stage of 22–31 somites (ss) (56–67 h) and begin to fuse at day 7 of embryonic development (Hamburger and Hamilton stage 30 (HH30)) (Matsuura et al., 2009).

The determination of endoderm to form dorsal and ventral bud derivatives occurs before formation of the buds. In mammals, the signaling pathways underlying the pancreatic development process include the Hedgehog system, the homeobox gene Pdx1, and Notch signaling. The vast majority of neurogenin-3 (ngn3)-expressing cells are committed to the endocrine lineage (Rosenberg et al., 2010). Two transcription factors, Pax4 and Arx, have competing roles in the commitment of the first of the ngn3-positive cells to generate either the α/pancreatic polypeptide (PP)-cell lineage or the β/δ-cell lineage. Subsequently, within cells that are committed to the β/δ-cell lineage, the persistence of pax4 expression seems to select for the β-cell lineage, whereas suppression of pax4 expression selects for the δ-cell lineage (Gittes, 2009).

Early *in vitro* studies by Rawdon using the dorsal pancreatic bud of 5 day chick embryo established that pancreatic mesenchyme may play an instructive role with respect to the early development of the chick pancreas. Results also suggested that the notochord plays an early role in induction of the pancreas, although experimental evidence was lacking (Rawdon, 1998). Studies of the past decade confirmed and extended Rawdon's observations (Rawdon, 1998). Prepancreatic endoderm first receives pancreas-inducing signals from the underlying somatic mesoderm, inducing Pdx1 expression as early as 4 ss (i.e., in 1.5 day old embryos) (Katsumoto et al., 2009). The regulation of pancreatic endocrine development by Notch is also found in chicken: inhibition of Notch signaling results in increased endocrine differentiation, whereas activation of the Notch1 receptor blocks endocrine development and maintains the proliferation of pancreatic progenitor cells in the embryonic chicken pancreas (Ahnfelt-Ronne et al., 2007). Neurogenin 3, which is transiently expressed in pancreatic endocrine progenitors, is responsible for the activation of a transcription factor cascade that ultimately leads to the maturation of endocrine cells in chicken endoderm (Rosenberg et al., 2010). The origin of the ventral pancreatic precursor cells has been studied less extensively; they are located at sites in which Pdx1 expression is detected by *in situ* hybridization (Matsuura et al., 2009).

Organization, volume, and ratio among the different endocrine cell types vary throughout embryonic development and postnatal life. The dorsal bud is the major source of insulin, glucagon, and somatostatin cells, whereas the ventral buds generate PP cells (Rawdon, 1998). In the dorsal bud, glucagon immunoreactive cells appear first at 2.25–2.5 days of incubation, then insulin cells at 3–5 days and somatostatin cells at 3.25 days (Manakova and Titlbach, 2007). PP cells appear only later, at 7 days in the splenic lobe (Cowap, 1985). In contrast to the dorsal bud, insulin cells have not been identified in prospective ventral bud derivatives until 13 days of incubation (Manakova and Titlbach, 2007). Development of exocrine tissue is delayed in comparison to that of endocrine tissue (Manakova and Titlbach, 2007).

27.3 INSULIN AND GLUCAGON PEPTIDES

27.3.1 Insulin and Pre-pro-insulin, Pro-insulin, and C-Peptides

The chicken insulin gene was cloned and sequenced as early as 1980 (Perler et al. in Simon et al., 2004); it is localized on GGA5 and exhibits the "ancestral" structure containing two introns: intron-1 interrupts the 5′ untranslated region, and intron-2 (about 3.5 kb, versus 141–786 bp in other animal species; Steiner et al, 1985 in Simon et al., 2004) interrupts the coding sequence. It encodes a pre-prohormone (i.e., pre-proinsulin) that contains the signal peptide (rich in hydrophobic residues), the insulin B-chain, the connecting peptide (C-peptide), and the insulin A-chain. To date, there is no evidence for the existence of a second insulin gene in birds (in Simon et al., 2004).

Insulin has been purified and sequenced in chickens, turkeys, ostriches, geese, and Pekin and Muscovy ducks (in Chevalier et al., 1996). The insulin gene has been cloned and sequenced in other bird species, including hummingbird and 28 other species (Fan et al., 1993 in Simon et al., 2004). From deduced amino acid sequences, bird insulins appear highly conserved. As compared to human insulin, minor neutral changes on the B-chain concern few residues at both ends: B1, B2, B3, B20, B27, or B30 (out of 30 positions). On the A-chain, changes concern positions A8–10 (out of 21 positions). In our study (Simon et al., 2004), some uncertainty remains for the last six amino acids of the A-chain because the reverse primer was targeting the corresponding codons and the stop codon. Changes on the A-chain (at A8–10 positions) between bird and human insulins are not neutral. This region is the most variable and is considered as the most immunogenic part. Accordingly, most antibodies prepared against mammalian insulins do not cross-react with chicken or duck insulin.

At the A8 position, a histidine residue is found in most bird species, as opposed to a glutamic acid residue in six species of the Anseriform order and a threonine in human insulin. The A8 position is now thought to be involved in the insulin-binding site to the insulin receptor (site 1; Sajid et al., 2009); the synthesis of human insulin analogs wearing different residues at the A8 position confirmed an early hypothesis, namely, the presence of histidine at A8 confers the properties of chicken insulin to such analogs (high binding affinity and low dissociation rate; see Weiss et al., 2001, 2002; Wan et al., 2004 and references quoted in Chevalier et al., 1996). Chicken insulin is the only natural super analog thus far described (at least twice as potent as human or pig insulin in bioassays or binding assays in many species; see Kemmler et al., 1978; Schauder and Buck, 1969; Simon et al., 1974, 1977 in Chevalier et al., 1996).

A8 glutamic acid, as found in the Anseriform order, increases the thermodynamic stability of the molecule, which should make "duck-type" insulins even more potent than "chicken-type" insulins (Weiss, 2001). In fact, the presence of negatively charged residues such as glutamic or aspartic acid at this position decreases the affinity of analogs for the receptor (Weiss, 2001). Duck insulin accordingly exhibits low affinity in rat liver membranes (Chevalier et al., 1996) and low bioactivity in rat fat cells (Moody, personal communication). The fact that chicken or Pekin and Muscovy duck insulin receptors (IRs) do not discriminate so extensively, compared to potencies observed with rat IR, may be specific to bird IRs (see Section 27.5.2). So, thus far, only two types of insulin have been found in 30 bird species: the hyperactive "chicken-type" insulin and the hypoactive "duck-type" insulin. From alignments of nonmammalian insulins (see references in Simon et al., 2004), A8 histidine was present early in the ancestor insulin molecule and disappeared in mammals (which possibly decreases the risk of hypoglycemic events and tumorigenicity). The appearance of A8 glutamic acid in anseriform birds and some snakes may have occurred independently in these species.

As in other species, bird C-peptides have been much less conserved; 14 changes out of 28 amino acid residues are found in 29 bird species (Simon et al., 2004). In other animal species, C-peptides are longer (31–38 amino acids; see Gross et al., 1989; Wahren et al., 2000 in Simon et al., 2004). Whether the shortness of bird C-peptide modifies the rate and yield of insulin synthesis is unknown. C-peptide is linked to B- and A-insulin chains through two dibasic residues at both ends (RR or KR, as in other species). In mammals, proteolytic cleavage and removal of these two pairs are performed by specific endopeptidases: PC3/PC1 and PC2 prohormone convertases and carboxypeptidase E. Such endoproteases have been detected in chicken embryos using heterologous antibodies (see Hernandez-Sanchez et al., 2002; Teshigawara et al., 2001 in Simon et al., 2004). Though predicted to be 28 residues long in all bird species, purified duck (*Anas platyrhyncos*) C-peptide is only 26 amino acids long (Marksen and Sundby, 1973 in Simon et al., 2004). Such extensive cleavages are also suggested in all the birds studied (with a possible exception in *Passer* in Simon et al., 2004). In mammalian species, at least two acidic amino acids surrounding a hydrophobic residue are present at the NH2-terminal end of C-peptide (e.g., EAED, EVE, and EAE); such a structure has been found to be critical for a normal rate of conversion of proinsulin to insulin. A very similar structure is observed in bird species studied (DVE, DIE, or DAE; Simon et al., 2004). C-peptide has long been considered as devoid of physiological functions on its own (other than structuring the precursor during insulin biosynthesis). C-peptide is now considered to control e-NOS, Na^+, K^+-ATPase, alpha-enolase, and several transcription factors (Wahren et al., 2012; Ishii et al., 2012). The central segment (ELGGGPGAG) or the COOH-terminal pentapeptide fragment (EGSLQ) of human C-peptide mimics the action of the entire peptide in several assays. The central part is very different in bird species, and C-terminal fragments are acidic (QEEYE, HEEYQ, QEEFE, or PEEYE). The C-peptide may therefore have acquired specific biological functions only in mammals, unless the still putative receptor co-evolved in bird species.

The length of available pre-proinsulin gene reading frames is similar in all bird species (187 nucleotides at the 5′ end and 134 nucleotides at the 3′ end; Simon et al., 2004). In contrast, the size of the PCR products varies greatly (from 2.7 kb in Passeriformes (*Passer*, *Turdus*, and *Pica*) to about 4.5 kb in Paleognathae species (*Struthio*, *Rea*, and *Dromiceius*)). Phylogenetic analyses supported the monophyletic origin of birds and identified the Paleognathae group as an isolated group at 100%, strongly suggesting this group is basal. After the branching of Paleognathae, a gallo-anserae was identified (at 63%), whereas all other Neognathae clustered at 99%. If we accept that Paleognathae, rather than Passeriformes, are basal to Neornithes, intron-2 must have evolved from a very long ancient form (likely close to that found in present day Paleognathae species) toward the shorter form found in Passeriformes.

In humans, some insulin gene polymorphisms are associated with diabetes. Some polymorphisms have been described in chicken (Bennett et al., 2006; Qiu et al., 2006) and in duck (one paper in Chinese: Kong et al., 2008) insulin genes. In the chicken, polymorphisms (mostly single nucleotide polymorphisms), thus far identified, are located in 5′- and 3′-UTR or intron. In these pioneer studies, significant associations were found between body weight at hatching, 28 days, and 55 weeks and small intestine length (but not with abdominal fat weight) at 13 weeks of age. Further studies conducted at other ages and involving other physiological parameters may reveal other associations.

27.3.2 Glucagon and Glucagon-like Peptides

27.3.2.1 Glucagon Structure and Physiological Effects

Glucagon (a member of the vasoactive intestinal polypeptide (VIP)–secretin family of peptide hormones) is a 29 amino acid peptide that is very similar across species. In mammals, glucagon is encoded by a single proglucagon gene, and there is a single mRNA that generates a single protein precursor that contains glucagon and two structurally related peptides at its carboxyl terminus: glucagon-like peptide 1 (GLP-1) and glucagon-like peptide 2 (GLP-2). The posttranslational proteolytic processing of this unique precursor is tissue specific. In the pancreas, glucagon is one of the major peptides released by the α cells of A-islets in response to low glucose levels, whereas GLP-1 and GLP-2, but not glucagon, are produced in intestinal L cells and the central nervous system.

Chicken glucagon is also a 29 amino acid peptide that is derived from a single proglucagon gene, which expresses multiple mRNA transcripts with different coding potentials (Richards and McMurtry, 2008). In the case of the domestic fowl, the pre-proglucagon precursor is 151 amino residues long. Chicken glucagon differs from rat and human glucagon by only one amino acid (ser versus asn at position 28) and from duck glucagon by two amino acids (ser versus asn at position 28 and thr versus ser at position 16). In contrast to mammals, the chicken proglucagon gene expresses two classes of proglucagon mRNA transcripts (PGA and PGB): class A mRNA (PGA) codes for glucagon and GLP-1 and resembles lower vertebrate transcripts such as those found in fish, while class B mRNA (PGB) codes for glucagon, GLP-1, and GLP-2 and is more like mammalian transcripts. Pancreas and proventriculus displayed the highest expression levels of class A and class B mRNA. Two prohormone convertases (PCs: PC2 and PC1/3) cleave the precursor into active peptides: PC2 mRNA is predominantly expressed in the pancreas and proventriculus, whereas PC1/3 mRNA is more highly expressed in the duodenum and brain. Fasting and re-feeding have no effects on proglucagon mRNA expression in various chicken tissues: pancreas, proventriculus, duodenum, or whole brain. This indicates that transcriptional regulation does not account for acute changes in plasma glucagon resulting from short-term changes in energy status in birds (Richards and McMurtry, 2008, 2009). The proglucagon gene is also expressed in various neurons of the avian retina (Fischer et al., 2006) and in chicken adipose tissue (Cogburn, personal communication), whereas, in contrast to in other tissues, PGB mRNA levels appear insulin dependent (Ji et al., 2012).

In birds, glucagon has several physiological roles and peculiarities that already have been described in Hazelwood, 2000, 5th Edition. Plasma glucagon levels are much higher in birds than in mammals. Chicken, ostrich, and pigeon glucagons were initially found to be less potent than mammalian glucagon, but this was not confirmed in a subsequent study (see the following: McCumbee and Hazelwood, 1978; Ferreira et al., 1991; Tung et al., 1977; Huang et al., 1987 in Simon, 1995). In turkeys, pig and chicken glucagon were found to be equally effective in promoting hyperglycemia, an increase in plasma nonesterified free fatty acid levels, and a decline in circulating levels of T_3 and T_4 (McMurtry et al., 1996). Long-term injection of glucagon has been shown to induce uncoupling protein (UCP) expression in the skeletal muscle of ducks, consistent with the thermogenic action of this hormone in birds (Raimbault et al., 2001). Glucagon is also important in the chicken eye for regulating ocular growth (Vessey et al., 2005). Honda et al. found that central, but not peripheral, administration of glucagon suppressed food intake and induced hyperglycemia in chickens and that these effects were mediated by the downstream actions of corticotrophin-releasing factor (CRF; Honda et al., 2007 in Honda et al., 2012). Thus, glucagon may serve as a neurotransmitter in the avian central nervous system, as has been suggested for mammals (Mayo et al., 2003).

27.3.2.2 GLP-1

GLP-1 slows down nutrient assimilation by inhibiting gastric emptying and acid secretion. Central but not peripheral administration of GLP-1 inhibits food intake and crop emptying in chickens and shifts fuel utilization from carbohydrates to lipids without affecting overall energy expenditure (Tachibana et al., 2007). The hypothalamus is involved in the anorexic effect of GLP-1 in chicks (Tachibana et al., 2004 in Tachibana et al., 2007). Moreover, the anorexigenic effect of central GLP-1 administration is mediated by the GLP-1 receptor because co-administration of a specific GLP-1 receptor antagonist (exendin 5–39) or N-terminal fragments attenuates the anorexia, while injection of the antagonist alone increases food intake in layer-type but not broiler chicks (Furuse et al., 1998). These results indicate the potential for variable responses in food intake to endogenous GLP-1 among different strains of poultry. Expression of the GLP-1 receptor mRNA in chicken pancreas is consistent with a possible incretin role for GLP-1 in birds (although it is not yet demonstrated in chicken).

27.3.2.3 GLP-2

In mammals, GLP-2 plays a role in intestinal growth and nutrient absorption by maintaining the integrity of mucosal epithelial cells. Secreted by enteroendocrine L-cells in response to the presence of nutrients, GLP-2 promotes crypt cell proliferation and suppresses apoptosis in mucosal epithelial cells. Elevated expression of GLP-2 receptor mRNA in chicken duodenum is consistent with the proposed role for GLP-2 in intestinal growth and function. Shousha et al. reported no effect of intracerebroventricular or intraperitoneal administration of GLP-2 on food intake, body temperature, or locomotor activity in Japanese quail (Shousha et al., 2007). However, these studies involved the use of rat GLP-2, which shares only 52% amino acid identity with chicken GLP-2 and may lack activity in avian species.

27.4 INSULIN AND GLUCAGON RELEASE

27.4.1 Pancreatic Hormone Levels During Development and After Hatching

Circulating concentrations of insulin and glucagon rise during development of the chicken embryo. In Cobb 500 chick embryos and hatched chicks, plasma insulin increased from 10 days of embryonic life (E10) (30 pg/mL), reaching 1000 pg/mL at 17 days after hatching (Lu et al., 2007). Plasma insulin concentrations, at 3–4 ng/mL in fed growing ducks and chickens, are similar to those of mammals (Simon et al., 2011). They are low at one day post hatch and increase with age up to 28 days, with higher insulin sensitivity in one-day-old chicks than in 21 day old broiler chickens, as observed in the latest strain of broiler chickens (Shiraishi et al., 2011b) and in previous reports on egg-laying or meat-type chickens. Plasma insulin concentration decreased by 45–50% after 6 h of fasting in 19 to 47 day old male and female modern meat-type chickens (Christensen et al., 2013). A low plasma glucagon concentration (59 pg/mL) was found in embryos at E10, remaining low until E17 and then increasing approximately threefold by the time of hatching (Lu et al., 2007). Values between 152 pg/mL and 450 pg/mL were measured in fed and 6 h fasted modern meat-type chickens, respectively (Christensen et al., 2013). It should be noted that in some of these experiments, the feeding pattern of chickens was not always synchronized with light–dark cycles.

27.4.2 Insulin Release

Whereas in mammals glucose is the primary physiological regulator of insulin secretion, the insulinotropic effect of glucose is less obvious in birds *in vivo* and *in vitro*. *In vivo*, raising blood glucose levels in ducks and chicken only transiently increases insulin levels (Rideau, 1988). The high basal plasma glucose concentrations found in chickens may themselves contribute to the relative insensitivity of the chicken β-cell to glucose: impairment of glucose-induced insulin release has been reported in hyperglycemic mammalian models. The mechanisms are, however, presently unknown in chickens. Somatostatin, which is present at high concentration in the chicken pancreas (Weir, 1976 in Rideau, 1988), may also exert inhibitory paracrine effects on the β-cell and partly explain the insensitivity of the β-cell to fuel nutrients. However, somatostatin immunoneutralization in isolated-perfused chicken pancreas has only transitory effects on insulin release at high glucose concentration (42 mM) and no effect at low levels (14 mM) (Honey, 1981 in Rideau, 1988). *In vitro* studies provided further insight into the mechanisms responsible for the relative glucose insensitivity of chicken β-cell. Using isolated and perfused duodenum-pancreas from adult hens (King et al., 1976 in Rideau, 1988, 1998) first showed that the pancreatic β-cell is relatively insensitive to glucose as very high and nonphysiological glucose concentrations (30–40 mM) are required to induce a modest insulin release. This insensitivity was confirmed and extended to other fuel nutrients, which are potent insulinotropic agents in mammals, using isolated duodenum-pancreas from young chickens. D-glucose and D-glyceraldehyde require high concentrations to evoke either very weak or delayed insulin release; leucine and its ketoacid (α-KIC) are even totally ineffective. The chicken β-cell, however, remains sensitive to depolarizing agents (K^+), tolbutamide (which closes K^+ATP-dependent channels), and potentiators of insulin secretion (acetylcholine or cAMP; Rideau, 1988). Therefore, the initiation of insulin secretion by fuel nutrients, the critical and specific step of the β-cell, appears different in chickens and mammals, whereas potentiating mechanisms (arginine, acetylcholine, and cAMP in the presence of low glucose concentration) and membrane depolarization events (K^+ and arginine) are present in chickens as in mammals.

The key role of glucokinase (GK) in the regulation of insulin release and hepatic glucose utilization has been well described in mammals. Although the GK gene is not referenced in the chicken genome at Ensembl, we isolated a chicken GK cDNA and showed that the GK messenger is expressed in chicken liver and pancreas (Berradi, 2005 in Rideau et al., 2010). A human GK antibody identified a GK protein in chicken liver, the amount and activity of which were insulin dependent (Dupont et al., 2008). Furthermore, RO0281675 (a specific GK activator in mammals) caused severe hypoglycemia in chickens but, surprisingly, did so without significantly increasing insulin levels, contrary to results obtained in mammals (Rideau et al., 2010). Thus, liver glucose utilization is substantially enhanced following liver GK activation. Although GK is present in the chicken pancreas, pancreatic GK activation may not be sufficient to cause insulin secretion, which confirms the *in vitro* studies reported here and supports the conclusion that the coupling between intra-B-islet metabolism and insulin release is different in chicken B-islets. The lack of insulinotropic effect of leucine or α-KIC (metabolized in the mitochondria) suggests that the inefficiency is located beyond the pyruvate step in the chicken β-cell.

Isolated chicken B-islets are necessary to further understand the insensitivity of the chicken β-cell to "insulinotropic" fuel nutrients. Although functional islets were isolated in various mammalian species more than 40 years ago, isolation of avian islets of Langerhans has proven to be especially difficult due to the fact that the morphology of the isolated A- and B-islets had not been characterized. Ruffier et al. described isolation of functional A- (glucagon) chicken islets (Ruffier et al., 1998). Glucose-stimulated insulin release from B-islets isolated from the dorsal lobe of the chick pancreas has been reported (Datar et al., 2006). In our opinion, however, the specificity of these preparations remains to be established as the dithiocarbazone staining method stains both B- and A-islets in chickens (Ruffier et al., 1998).

Regarding resistance to diabetogenic drugs, the extreme resistance of birds to the diabetogenic drugs alloxan and streptozotocin (STZ) was described very early. In mammals, the β-cell specificity of these drugs is mainly due to the fact that they are selectively transported by the β-cell glucose transporter, GLUT2 (Lenzen, 2008). GLUT2 expression was reported in the liver and kidney in chickens; unfortunately, the pancreas was not examined (Kono et al., 2005). Within the mammalian β-cell, alloxan selectively inhibits glucose-induced insulin secretion through specific inhibition of glucokinase; in addition, it induces ROS formation, resulting in the selective necrosis of β-cells. STZ's effects are attributed to its specific alkylating potency, which modifies biological macromolecules and DNA fragments and ultimately destroys the β-cells (Lenzen, 2008). Compared with mouse islets, chicken islets generate smaller amounts of ROS under STZ. Consequently, less damage to proteins, lipids, and DNA can be expected in chicken than in mouse islets. These results may partly explain why STZ does not induce diabetes in chicken, even at high doses (Modak et al., 2007).

Despite the peculiarities present in chickens in the coupling between metabolism and insulin release, insulin release is under the control of multiple components *in vivo* and *in vitro*. Although glucose or a mixture of amino acids separately increases plasma insulin to a small extent, they act synergistically when combined *in vivo*. Re-feeding with gradually increasing amounts of food results in gradual rises in plasma insulin levels, adjusting glucose to a similar level. Changes in the glucose–insulin balance are observed in several experimental conditions or models (Table 27.1), including genetically fat and lean lines of chickens, dwarf chickens (*dw/dw*), hypothyroidism, corticosterone treatment, and chronic exposure to high temperature (Rideau, 1997). The incretin effect is not documented in chickens, but a variety of other stimuli, including nutrients (glucose and arginine), hormones (cholecystokinin, glucagon, corticosterone, and GH) (Rideau, 1988), and neuronal stimuli (Karmann et al., 1992), increase plasma insulin levels. The effects of ghrelin–obestatin and GLP-2 on insulin release are not known in birds. Conversely, insulin secretion is inhibited by epinephrine and somatostatin (Rideau, 1988). Although the existence of leptin in the chicken is still debated, a recombinant "chicken" leptin (or leptin analog) potently inhibits insulin secretion from the perfused chicken pancreas (Benomar et al., 2003).

27.4.3 Glucagon Release

Fasting markedly increases circulating concentrations of glucagon in chickens (Edwards, 1999 in Scanes, 2009). In modern meat-type chickens, plasma glucagon levels increased 3.5- to 3.7-fold after 6h of fasting while plasma glucose was depressed by 11% (Christensen et al., 2013). Glucagon secretion is inhibited by glucose *in vivo* in domestic birds (Honey, 1980 in Rideau, 1988) and in wild species (Tzotze, 1998 in Scanes and Braun, 2012). Exogenous glucose (42 mM) inhibited glucagon release by 27% from isolated A-islets of Langerhans (Ruffier et al., 1998). Conversely, glucagon release is stimulated by amino acids, free fatty acids, cholecystokinin, somatostatin, and insulin *in vivo* (Scanes and Braun, 2012). A tonic suppression effect on glucagon by insulin is suggested in a model of insulin immune-neutralization in chickens (Dupont et al., 2008). As for insulin, the effect of gastrointestinal peptides and adipose tissue hormones on glucagon secretion is unknown in birds.

The mechanisms underlying the secretion of glucagon from α-cells are largely unknown in chickens. In mammals, α-cells contain a comparable secretory "machinery" as in β-cells: a glucose transporter (in this case, the ubiquitous GLUT-1), glucokinase, K^+ATP channels, L-type voltage-gated calcium channels, and secretory granules (Le Marchand and Piston, 2010). However, α-cell secretory activity is quite different from that of β-cells. The mechanisms underlying the glucose inhibition of α-cell secretory activity are poorly understood. Two nonexclusive models have been proposed: a direct inhibition by glucose, and an intra-islet paracrine inhibition arising from non-α-cells. The paracrine control might involve somatostatin released from δ-cells and insulin, zinc, or gamma-aminobutyric acid released from β-cells. Recently, Le Marchand and Piston suggested that in the absence of cell–cell contacts, isolated α-cells function similarly to β-cells (i.e., glucagon secretion increases as a function of glucose concentration). However, in the islet, glucose inhibits glucagon release downstream the calcium step, presumably at the level of vesicle trafficking or exocytotic machinery (Le Marchand and Piston, 2010).

27.4.4 Somatostatin

Somatostatin (also known as growth hormone–inhibiting hormone (GHIH) or somatotropin release-inhibiting factor

TABLE 27.1 Experimental or Genetical Models Exhibiting Changes in the Plasma Glucose–Insulin Relationship in Chickens[1]

	Fasting Plasma Levels		Glucose Tolerance			
Models	Glucose	Insulin	Glucose Disposal	Insulin Levels	Insulin Sensitivity	Fattening
Intermittent (IT) versus *ad libitum* feeding (C)	IT≥C	IT=C	IT>C	IT<C	ND	IT≥C
Corticosterone (cort) treatment.	Cort>C	Cort>C	Cort<C	Cort>C	Cort<C	Cort>C
Fat–lean lines (FL-LL)	FL<LL	FL≥LL	FL>LL	FL>LL	FL≥LL	FL>LL
High- and low-glucose lines (LGl–HGl)	LGl<HGl	LGl=HGl	LGl=HGl	LGl=HGl	LGl=HGl	LGlL>HGlL
High- and low-growth lines (HG–LH)	HG≤LG	HG>LG	HG=LG	HG>LG	ND	HG>LG
Dwarf (*dw*)–Normal (N) chickens	*dw*=N	*dw*=N	*dw*=N	*dw*<N	*dw*>N	*dw*>N

[1]*ND: not determined. Refeeding experiments have been performed following a fast in several of these models; results, not summarized in the table, can be found in Simon (1988) and in Simon et al. (2000) (LGl–HGl). Corticosterone was injected in doses of 5–6mg/kg. Insulin sensitivity was evaluated through the hypoglycemic effect of exogenous insulin. Results of glucose tolerance in HG–LG chickens are unpublished data (Simon et al.).*

(SRIF)) is a 14 amino acid peptide that regulates the endocrine system. Chicken somatostatin is present in two forms, somatostatin-14 (SST-14) and somatostatin-28 (SST-28), which are encoded by the same gene (PSS1) as in mammals. Chicken pancreatic A-islets, B-islets, and D-cells contain only the SST-14 peptide form (Takayagani et al., 1996 in Trabucchi et al., 2003). Another somatostatin gene variant (PSS2) exists in many species, including chickens. PSS2 messenger is expressed in chicken pancreatic islets or pancreas and in a few specific brain structures, fewer than the structures expressing PSS1 messenger (Trabucchi et al., 2003).

27.4.5 Avian Pancreatic Polypeptide

APP was first isolated and characterized from chicken pancreas (Kimmel et al., 1968); it is a member of the neuropeptide Y family (PP, peptide YY (PYY), and NPY). PYY and PP are gut endocrine peptides almost exclusively expressed in the digestive system, whereas NPY is expressed in the central and peripheral nervous systems at all levels of the gut–brain axis. All three peptides are 36 amino acids long and act on G protein–coupled receptors; five receptor subtypes have been described. In mammals, PP is released from the endocrine pancreas following a meal and acts preferentially via Y4 receptors. It inhibits gastric emptying, intestinal electrolyte, water secretion intestinal motor activity, and peristalsis. PP also reduces appetite (Holzer et al., 2012). In chickens, APP is released in response to gut peptides and amino acids (Colca, 1982 in Rideau, 1988). Y4 receptors are expressed in a large number of peripheral tissues and in the brain (Lundell et al., 2002). APP primarily acts by inhibiting gastrointestinal tract motility and gall bladder and exocrine pancreas secretion in chickens (Hazelwood, 1993 in Hazelwood, 2000). It also exerts metabolic effects: liver glycogenolysis, hypoglycerolemia, and lowering of the plasma free fatty acid level. One study, however, suggests that APP increases food intake in white leghorn chickens (Denbow et al., 1988).

27.5 GLUCAGON AND INSULIN RECEPTORS

27.5.1 Glucagon Receptors

In mammals, the biological actions of glucagon are mediated by a specific glucagon receptor (GCGR; Authier and Desbuquois, 2008), a receptor belonging to G protein–coupled receptor subfamily B, which also includes the receptors for GLP-1, GLP-2, VIP, growth hormone-releasing hormone, and growth hormone–releasing hormone-like peptide. Stimulation of the glucagon receptor modulates adenylate cyclase, initiates the production of cAMP, and thereby activates extracellular signal-regulated protein kinase-1 and -2 via cAMP-dependent protein kinase A (cAMP–PKA; Figure 27.2). In addition, it also activates signaling pathways via cAMP-independent interactions, leading to stimulation of phospholipase C and release of Ca^{2+} from IP_3-sensitive intracellular Ca^{2+} stores.

The chicken glucagon receptor gene was cloned from adult brain tissue (Wang et al., 2008a); it encodes for a short glucagon receptor of 496 AA (GCGR-s) or a long receptor variant of 554 AA, designated glucagon receptor variant 1 (GCGR-v1). The chicken GCGR-s shares high AA sequence identity with that of human (70%), rat (69%), mouse (69%),

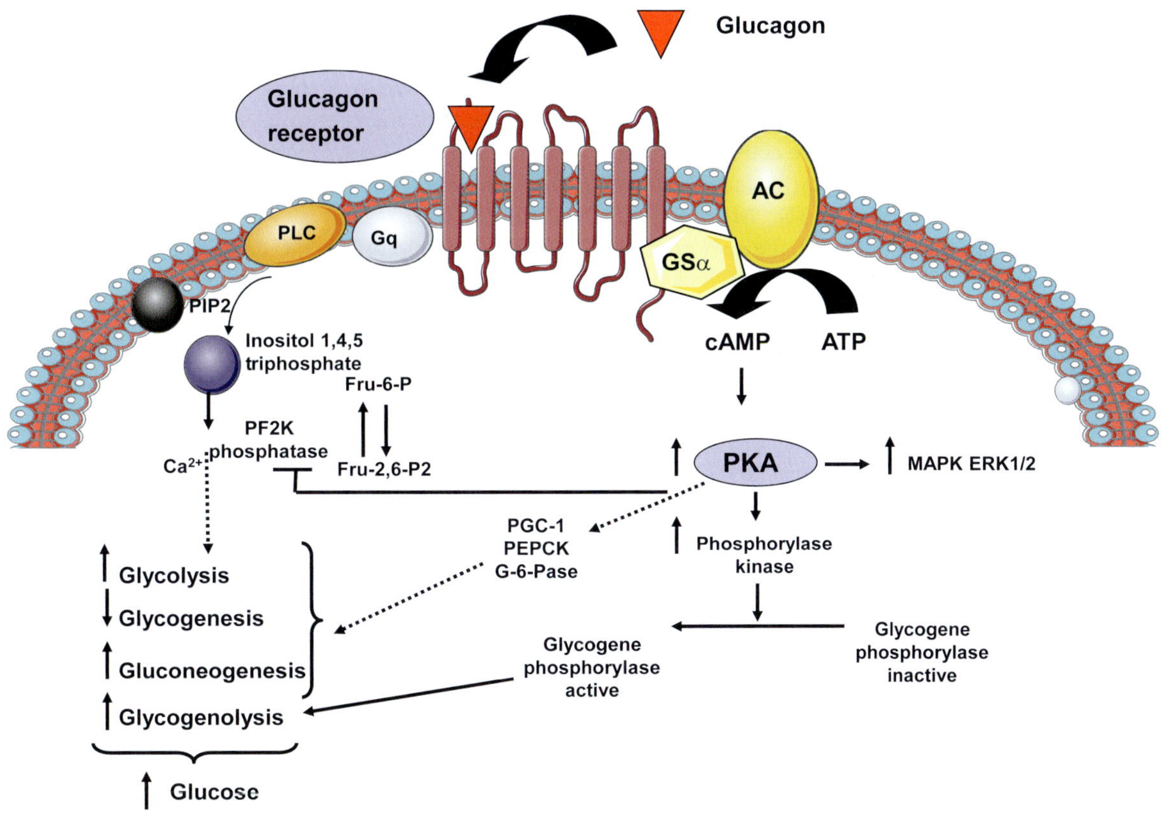

FIGURE 27.2 Glucagon receptor signaling.

and *Xenopus* (64%). Functional analysis confirmed that both GCGR-s and GCGR-v1 can be activated by glucagon and are functionally coupled to the cAMP-PKA signaling pathway. The glucagon receptor is widely expressed in all tissues, including the hypothalamus, with the greatest expression found in the liver, suggesting that glucagon plays a wide range of roles in both hepatic and nonhepatic chicken tissues through the two receptors, GCGR-s and GCGR-v1 (Richards and McMurtry, 2008). Recently, Wang et al. identified a novel ligand–receptor pair, named GCGL and GCGL receptor (GCGLR), from chicken brain (Wang et al., 2012). GCGL is a novel GCG-like peptide of 29 amino acids that shares high amino acid sequence similarity with mammalian and chicken GCG (62–66%). GCGLR is a GCGL-specific receptor, which shares relatively high amino acid sequence identity with chicken GCGR and structurally related receptors (GLP1R, GIPR, and GLP2R). GCGL mRNA expression is mainly detected in the central nervous system, spinal cord, and testis, whereas GCGLR is more widely expressed in chicken tissues (Wang et al., 2012).

27.5.2 Insulin Receptor

Early studies on ontogeny, structure, binding affinity, binding characteristics, number, internalization, and up- and downregulation of insulin receptors (IR) in chicken or turkey tissues have already been reviewed (Simon, 1989). Since then, changes in IR number have been described in the hypothalamus, at least at the level of IR messenger, in layer and broiler-type chicks under ad libitum conditions (Shiraishi et al., 2011a); changes in hypothalamic IR mRNA in response to photoperiodism are dependent upon gonad hormones (Anraku et al., 2007). In chicken liver plasma membranes, hepatocytes, and thymocytes, insulin-binding capacity (IR number) is lower than in rat or veal (in Simon, 1989), which may contribute to the low insulin sensitivity of chickens. However, in crude liver membranes (i.e., plasma and intracellular organite membranes), the IR number is about the same in chickens and rats (Dupont et al., 2004). A high proportion of IR may therefore be internalized in chicken liver (Krup and Lane, 1981, 1982 in Simon, 1989). Chicken IR shows a heterotetrameric structure (two α and two β subunits linked by disulfide bonds, as described in mammals and other species; reviewed in Simon, 1989). The α subunit (about 135 kDa) is entirely within the extracellular compartment and binds insulin; the β subunit (about 95 kDa) contains a large hydrophobic fragment (inserted into the membrane), crosses the cell membrane, and extends into the intracellular compartment. In chickens, as in other species, the α subunit of brain or muscle receptors is smaller than that of liver receptor α subunit, in part as a result of differences in glycosylation (reviewed in Simon, 1989).

In mammals, both IR α and β chains are synthesized from a single mRNA, which is encoded by 22 exons. In mammals, exon 11 is alternatively spliced, which results in two IR isoforms (IR-A and IR-B) that differ by the absence or presence of 12 amino acids at the C-terminus of the α subunit, respectively. IR-A and IR-B isoforms display differences in ligand affinity binding, kinase activity, receptor internalization, and recycling as well as intracellular signaling capacity and tissue distribution (De Meyts, 2012). According to Hernández-Sanchez et al., the region corresponding to exon-11 is missing in chicken IR; chicken tissues would only express the IR-A type (i.e., the IR type showing the highest insulin affinity) (Hernandez-Sanchez et al., 2008). This may account for the fact that chicken IRs exhibit much higher affinity for IGFs than mammalian IRs. It may also explain why chicken and duck IRs do not discriminate among chicken, pork, and duck insulins very well (Chevalier et al., 1996).

In its intracellular part, the IR β subunit is a tyrosine kinase activated by autophosphorylation following insulin binding to the α subunit. The chicken IR tyrosine kinase activity shows the same features as those from other species (reviewed in Kato et al., 2000; Simon, 1989). Chicken liver or kidney IR tyrosine kinase activities (basal and maximal activities) are regulated, for instance, by nutritional conditions (e.g., fasting or re-feeding), as in mammals. In contrast, muscle and adipose tissue IR kinase activity, as well as β subunit tyrosine phosphorylation, do not decrease in the fasting state or increase in re-fed animals, despite alterations in plasma insulin (Adamo et al., 1987; Dupont et al., 2008, 2012). Thus, for still unknown reasons, the regulation of the tyrosine phosphorylation of IR β subunit and kinase activity differs between species and tissues. As depicted in Figure 27.3 (steps highlighted in blue for chicken tissues), once activated, the IR tyrosine kinase phosphorylates several intracellular substrates on specific tyrosine residues, which represents the proximal and early events initiating and propagating the insulin signal through a cascade of reactions leading to the multiple actions of insulin. Phosphorylated IR substrates activate directly or indirectly various protein (serine–threonine) or lipid kinases, including phosphatidylinositol 3′-kinase (PI3K), protein kinase B (Akt), P70S6K, and mitogen-activated protein kinase (MAPK)–extracellular signal-regulated kinase (ERK)-1/2. Several of these postreceptor steps also regulate insulin cell sensitivity, enabling alterations of insulin sensitivity without changes in insulin receptor number or kinase activity.

27.5.2.1 Proximal IR Substrates

In mammals, at least 11 intracellular IR substrates have been identified (reviewed in Taniguchi et al., 2006; Siddle, 2012). These include the family of IRS (insulin receptor substrate) proteins (IRS-1–6) and Shc (Src homology collagen) protein. Four different IRS proteins (IRS-1–4) have been reported in various mammals, and two additional IRS proteins have been detected in humans (IRS5/DOK4 and IRS6/DOK5). These last two additional IRS proteins are involved in insulin signaling but are not involved in PI3K activation (the major insulin-signaling step). Other IR substrates such as Gab1, Dok, APS, SH2B, Cbl, and Crk have been also described in mammalian tissues or cells (cited by Siddle, 2012).

The coding region of chicken IRS-1 gene has been cloned and sequenced (Taouis et al., 1996). The deduced chicken IRS-1 protein (cIRS-1) presents high sequence identity with its human, rat, and mouse homologs. cIRS-1 is at least present in the liver, muscle, and adipose tissue and interacts with the insulin receptor (Dupont et al., 1998 in Dupont et al., 2009, 2012). Interestingly or surprisingly enough, the tyrosine phosphorylation of cIRS-1 and PI3K activity (as discussed further in this chapter) is accordingly dependent upon nutritional conditions or insulin status in the liver but not in muscle or adipose tissue (Dupont et al., 1998 in Dupont et al., 2009, 2008, 2012), suggesting a tissue-specific regulation and a relative insulin refractoriness in chicken muscle and adipose tissue. Coding sequences have recently been described for IRS-2, IRS-4, Crk, Cbl, and APS chicken homologs (Ensemble release 71, April 2013, at http://www.ensembl.org/Gallus_gallus/index.html); they are located on chromosomes 1, 4, 19, 1, and 19, respectively. However, the presence of these substrates as proteins and their insulin sensitivity remain to be established in chicken tissues. Cbl and Crk, which in mammals mediate insulin signals and are associated with lipid rafts, are upregulated, at the messenger level, in chicken adipose tissue by fasting (Ji et al., 2012). In mammals, this pathway stimulates glucose uptake, in addition to and independently of PI3K activation. Whether these potential substrates contribute to the apparent refractoriness of PI3K activity to insulin that is observed in chicken skeletal muscle and adipose tissue remain unknown. It should be noticed, however, that insulin stimulation of glucose transport in chicken adipose tissue, if any, is very poor (as discussed further in this chapter).

In a chicken hepatoma cell line (LMH cells), cIRS-1 repression substantially increases Shc expression (Taouis et al., 1998). Shc is phosphorylated on tyrosine residues in response to insulin and therefore serves as an additional IR substrate in the absence of IRS-1. Chicken liver, muscle, and adipose tissue express homologs of the three mammalian Shc isoforms (Dupont et al., 2008, 2012). Interestingly, tyrosine phosphorylation of Shc (52 kDa) is dependent on insulin status: it increases after re-feeding or insulin injection and is inhibited after insulin immuno-neutralization in chicken liver but also, surprisingly, in leg muscles, in contrast to IRS-1. Also surprisingly, neither insulin privation nor fasting alters Shc tyrosine phosphorylation in adipose tissue (Dupont et al., 2009, 2012). In chicken liver, Shc, mainly the 52 kDa isoform, interacts with IR, whereas the

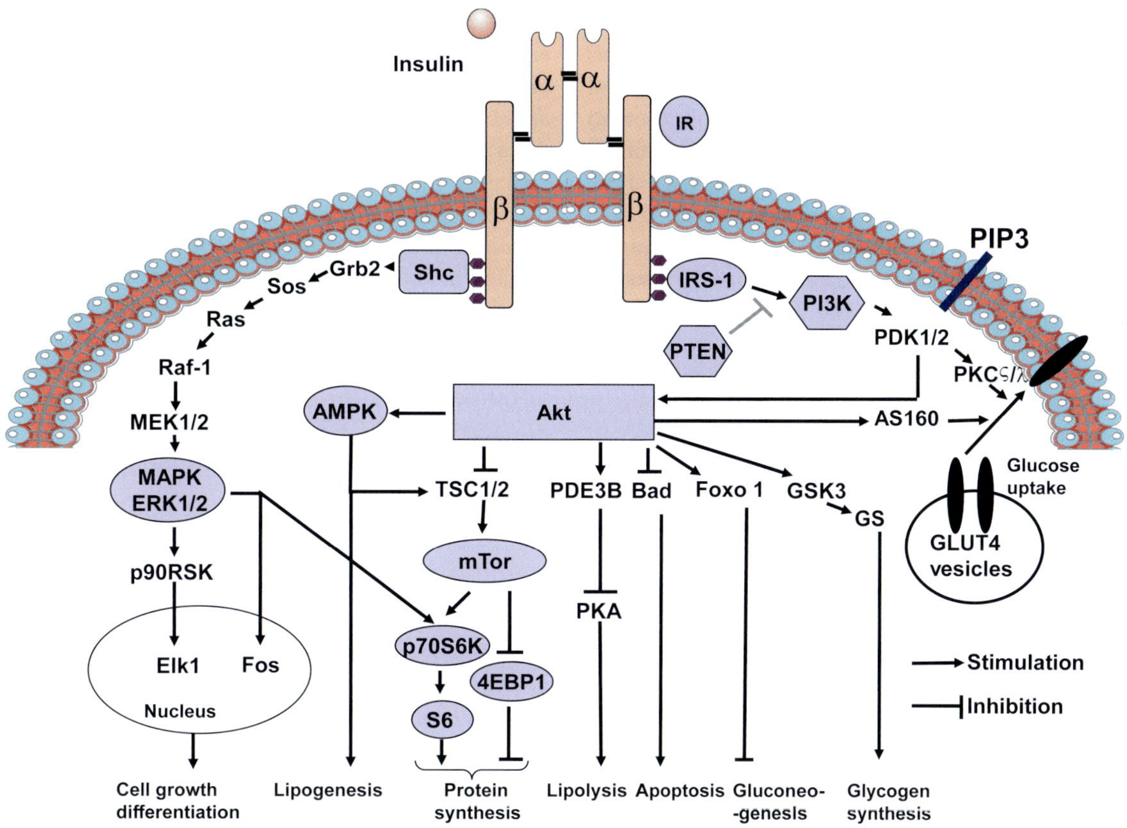

FIGURE 27.3 Insulin receptor signaling. (Adapted from mammals; only steps highlighted in blue have been characterized in chicken tissues).

46 kDa isoform associates with IRS-1 (Dupont et al., 1998 in Dupont et al., 2009). The 52 kDa Shc isoform also interacts with the p85 regulatory subunit of PI3K. Shc (52 kDa) is thus a key substrate for the insulin receptor in the chicken because, first, it is able to associate with PI3K and, second, in contrast to IRS-1, the extent of its tyrosine phosphorylation is dependent on insulin levels. Furthermore, in liver, Shc (52 kDa) phosphorylation is inhibited after insulin immuno-neutralization (Dupont et al., 2008). The possible existence of different subtypes of IR complexes signaling through IRS-1 or Shc has been investigated in the liver. The presence of a large complex involving IR, IRS-1, Shc (mainly the 52 kDa isoform), and the regulatory subunits of PI3K has been demonstrated; and a third complex, including IRS-1 and the 46 kDa Shc isoform, is also present in chicken liver (Dupont et al., 1998 in Dupont et al., 2009). No such interactions have been described in mammalian tissues, and the functions of these complexes in chicken liver and their presence in other chicken tissues remain to be established.

27.5.2.2 PI3K–Akt

PI3K has been characterized in chicken liver, leg muscle, and adipose tissue (Dupont et al., 1998 in Dupont et al., 2009, 2012) and in LMH cells (Taouis et al., 1998 in Dupont et al., 2009). Liver PI3K activity depends upon the nutritional state (prolonged fasting versus refeeding) in several experimental models (Dupont et al., 1998, 1999 in Dupont et al., 2009). Furthermore, PI3K activity increases in response to insulin injection and is inhibited following insulin immuno-neutralization (Dupont et al., 1998 in Dupont et al., 2009; see also Dupont et al., 2008). This is in agreement with the changes observed for early steps of the insulin-signaling cascade (tyrosine phosphorylation of IR β subunit, IRS-1, and Shc). Conversely, in muscle, the PI3K activity and, as previously mentioned, tyrosine phosphorylation of IR and IRS-1 are totally unresponsive to insulin levels in experimental models that alter insulin sensitivity, such as long-term corticosterone treatment and genetically selected fat and lean lines of chickens (Dupont et al., 1999 in Dupont et al., 2009). As mentioned in this chapter, only the phosphorylation of the 52 kDa Shc isoform is reduced in muscle by fasting and increased by re-feeding. Comparison of the responses to exogenous insulin in chickens and rats suggests that this phenomenon is specific to chicken. Furthermore, PI3K activity in leg muscle is about 30 times greater in the chicken than in the rat (Dupont et al., 2004). This "constitutive" hyperactivity of PI3K in chicken muscle may overstimulate the feedback inhibitory pathway described in mammals (as discussed in this chapter), thereby desensitizing chicken muscle to insulin. The p85 regulatory subunit of PI3K itself exerts strong inhibitory control of IRS signaling in mammals (Taniguchi et al., 2006). However, lipid phosphatases such as PTEN (**p**hosphatase and **ten**sin homolog deleted on chromosome 10) may also contribute to this relative insulin insensitivity of the early steps of insulin signaling in chicken muscle. We have detected PTEN in chicken tissues, including muscle (Vaudin et al., 2006); however, the role of PTEN in the control of insulin signaling in chicken tissues remains to be determined. In adipose tissue, insulin signaling is even more peculiar; insulin privation did not alter early steps and PI3K activity, but also other downstream elements such Akt (see next paragraph; Dupont et al., 2012).

Akt has been detected in several chicken tissues including liver, muscle (Bigot et al., 2003; Dupont et al., 2008), adipose tissue, and granulosa cells (Dupont et al., 2012; Tosca et al., 2006). The phosphorylation of Akt on Ser 473 is increased following feeding of posthatched chicks (Bigot et al., 2003); this is prevented when feeding is delayed (48h), then rapidly restored following food intake. In fed chickens, insulin immuno-neutralization significantly decreases Akt phosphorylation in both liver and muscle, but not in adipose tissue (Dupont et al., 2008, 2012). Thus, insulin is presumably involved in Akt activation in chicken muscle, despite the fact that chickens exhibit particular features in the early steps of the insulin-signaling cascade (IR, IRS-1, and PI3K). This issue remains unresolved. Potential mechanisms involved in the atypical feature found for insulin signaling in chicken muscle and adipose tissue have been discussed in Dupont et al. (2008, 2012).

27.5.2.3 P70S6K

p70S6 kinase, a further downstream step, has been characterized in a quail muscle (QM7) cell line and in chicken skeletal muscle (Tesseraud et al., 2003, http://www.sciencedirect.com/science/article/pii/S001664800800378X; Duchêne et al., 2008 in Dupont et al., 2009). p70S6K activity increases significantly in both pectoralis and gastrocnemius muscles after re-feeding for 30min following 16h starvation, and also in fasting chickens after a single insulin injection (Bigot et al., 2003). Similarly, p70S6K activity in the muscle of neonatal chicks is correlated with the plasma insulin concentration, strongly suggesting an insulin-dependent p70S6K activation (Bigot et al., 2003). In the liver, stimulation of phosphorylation of P70S6K1 and activation of S6K (using specific artificial substrate) was not detectable (Dupont et al., 2008), although insulin stimulation of phosphorylation of S6 ribosomal protein was observed in the same conditions (Duchene et al., 2008b). Furthermore, p70S6K phosphorylation decreases in leg muscles but not in adipose tissue in response to insulin immuno-neutralization (Dupont et al., 2008, 2012). These findings and those for Akt are again surprising in view of the relative insulin refractoriness of chicken muscle to exogenous insulin (Dupont et al., 2004). However, p70S6K has been found to be involved in another feedback loop, which attenuates the ability of insulin to activate PI3K. Indeed, there is evidence in mammals that p70S6K inhibits insulin signaling by increasing the serine–threonine phosphorylation of IRS-1. In chicken muscle, re-feeding and insulin injection increase the phosphorylation of both p70S6K on key residues (T389, T229, and T421/S424) and its downstream target, the ribosomal protein S6 (downstream insulin-signaling steps). However, in these conditions, the phosphorylation of IRS1 on serine 632 and serine 635 residues is also induced (Duchene et al., 2008b). In mammals, the phosphorylation of these sites inhibits the insulin signal. Thus, the AKT/p70S6K may inhibit IRS1 tyrosine phosphorylation in chicken muscle.

27.5.2.4 MAPK–ERK1/2 Pathway

MAPK–ERK has been identified in various chicken tissues (Duchene et al., 2008a). In contrast to mammals, ERK1 is not present in bird tissues; ERK2 is the only phosphorylated isoform detected (Duchene et al., 2008a). It is activated in chicken myoblasts and LMH cells *in vitro* by insulin (Duchene et al., 2008a). U0126, an inhibitor of the MAPK–ERK pathway, completely abolishes insulin-induced S6K1 phosphorylation and activity in chicken myoblasts *in vitro*, implicating MAPK–ERK2 in the control of S6K1 by insulin in avian cells (Duchene et al., 2008a). MAPK–ERK2 phosphorylation in chicken liver muscle and adipose tissue *in vivo* is strongly reduced by insulin immuno-neutralization (Dupont et al., 2012, 2008). However, the role of MAPK–ERK2 in insulin action *in vivo* in chicken has not been investigated. This signaling could be crucial to chicken growth, and this pathway may participate in the insulin-dependent regulation of p70S6K, because Shc, which is upstream, seems to be a key IR substrate in chicken muscle.

From the elements of insulin signaling thus far studied in chickens, insulin signaling appears to work in chicken liver as expected from the knowledge gained from mammals. In contrast, in muscle and adipose tissue, the insulin cascade appears refractory, suggesting the presence of inhibitory counterregulatory mechanisms in these tissues.

27.6 GENERAL EFFECTS OF GLUCAGON AND INSULIN

27.6.1 Insulin and Embryonic or Posthatch Development

The insulin gene is expressed early during embryonic development in extra-pancreatic tissues prior to development of the pancreas, showing specific regulations (de la Rosa and de Pablo, 2011). Insulin accelerates the growth of embryos explanted at Hamburger Hamilton stage 4, without increasing mitoses; some specific developmental genes are upregulated (Patwardhan et al., 2004). A fine tuning of insulin gene expression is required to sustain adequate morphogenesis. In chicken neurulating embryos and chicken embryo retina, proinsulin is expressed but not cleaved to insulin. Whereas this precursor is considered to exert marginal (if any) biological activity in peripheral tissues, it exerts anti-apoptotic effects in those tissues at this developmental stage, probably through a hybrid insulin–IGF receptor (in de la Rosa and de Pablo, 2011). Furthermore, two transcription-induced chimerisms have been identified with the adjacent tyrosine hydroxylase gene (*TH*-insulin; in de la Rosa and de Pablo, 2011). (In humans, a similar mechanism has been described between the insulin gene and *IGF2*, the other adjacent gene.) The insulin gene is also expressed in the neural tube and somites of chicken embryos. The addition of insulin (or IGFs) to other specific factors in the incubation medium cocktail stimulates somite cell proliferation and myogenesis (Pirskanen et al., 2000). This effect is blocked by an insulin antibody and insulin receptor antibody, showing that insulin acts as a myogenic signal early in embryogenesis. Insulin also supports *in vitro* multiplication of 1 day old broiler muscle satellite cells and delays their differentiation (Sato et al., 2012). Surprisingly, an opposite effect is observed between insulin and IGF1 for MyoD and myogenin messengers. Insulin or tolbutamide administration to day-old broiler chicken increases Pax7 messenger (a striated cell marker) in breast muscle at 3 days of age and their body weight at 50 days without changing their body composition (Sato et al., 2012). These interesting effects deserve further characterization. In satellite cell–derived myotubes, insulin (at high doses, though, compared with IGF1) stimulates glucose (as discussed in this chapter) and amino acid transport and protein synthesis and inhibits proteolysis (Duclos et al., 1993a,b).

27.6.2 Insulin–Glucagon and Food Intake

Pioneer studies suggested that insulin contributes to the regulation of food intake (through inhibitory or stimulatory mechanisms; in Simon, 1989). Intracerebroventricular injection of insulin (through the melanocortin system) or glucagon inhibits food intake in young leghorn chickens (Honda et al., 2007; Shiraishi et al., 2011b and quoted references). Surprisingly, insulin is inefficient in young broiler chickens in fasted or fed conditions (Shiraishi et al., 2011b and quoted references), suggesting a difference in their set point or sensitivity; in broiler chickens, plasma insulin levels are higher, but they appear to decrease hypothalamic IR mRNA expression (Shiraishi et al., 2011b). Insulin immuno-neutralization in fed growing chickens decreased food intake, most likely by inducing very high plasma glucose levels (Dupont et al., 2008).

27.6.3 Insulin and the Endocrine System

Insulin interacts with glucagon (see Section 27.1) and with other endocrine systems. As early as 1 h after insulin immuno-neutralization (Dupont et al., 2008), plasma T_3 levels decreased (no reliable measurements were obtained for T_4). After 5 h of insulin deprivation, liver D2- and D3-deiodinase messengers decreased or increased, respectively. Thus, insulin is another partner involved in bird liver T_4 metabolism (Darras et al., 2006 in Dupont et al., 2008). The IGF system also appears partly insulin dependent. Liver *IGFBP1* messenger (microarray and qRTPCR measurements) and plasma IGFBP1 protein levels increased after insulin immuno-neutralization (plasma IGF1 was not measured). Microarrays also revealed that liver *IGFBP2* and muscle *IGF1R* messengers increased or decreased, respectively (similar impairments are found in diabetic patients; Paalsgard et al., 2009 in Simon et al., 2012). In another study (Nagao et al., 2001), 2 day fasting decreased plasma IGF1 and increased *IGFBP2* messenger in the liver and gizzard of white leghorn chickens (but not in brain or kidney); exogenous insulin further decreased plasma IGF1 levels and largely inhibited liver and gizzard *IGFBP2* messenger (6 h re-feeding also inhibited liver and gizzard *IGFBP2* messenger but accordingly increased plasma IGF1 levels).

27.6.4 Insulin and Glucose Metabolism

The severe and rapid hyperglycemia induced by insulin deprivation in fed chickens certainly depends upon multiple changes in glucose metabolism in several tissues. In liver, where glucose transport is not insulin dependent, glucose utilization was probably greatly depressed as suggested by the drop in liver GK messenger and protein contents (see the earlier discussion about the existence and the regulation of liver GK activity). In mammals, other insulin-dependent mechanisms are inhibition of liver gluconeogenesis and glucose production. Gluconeogenesis is considered to escape insulin control in chicken liver (Section 27.1). In our transcriptome study, 27 messengers from the glycolysis–gluconeogenesis or pyruvate metabolism pathways were differentially expressed in liver following insulin deprivation, including *G6PC2* (glucose 6-phosphatase catalytic, 2) messenger, which increased, suggesting that liver glucose production may also be insulin inhibitable in chicken liver. Unfortunately, this was not assessed in euglycemic–hyperinsulinemic clamp experiments performed on chickens (Chou and Scanes, 1988; Hamano, 2006). Indirect evidence (prolonged fasting and corticosterone experiments) suggests that insulin also inhibits kidney gluconeogenesis by inhibiting cytosolic phosphoenolpyruvate carboxykinase activity (Bisbis et al., 1994a,b and quoted references).

Insulin greatly stimulates glucose uptake in muscles and adipose tissues of mammals by recruiting insulin-sensitive

Glut-4 transporters at the cell surface from an intracellular pool (Figure 27.3). So far, no Glut-4 glucose transporter has been identified in chickens or house sparrows (*Passer domesticus*) as protein or messenger. This may be the consequence of the lack of cross-reactivity of antibodies or the absence of Glut4 in chicken or birds since no Glut-4 homolog sequence is found in the chicken Ensembl database (an ortholog gene to the human Glut-4 (*SLC2A4*) gene is, however, described in the anole lizard and several fish species). Functional and immunoreactive Glut-4 transporters have been observed in duck leg muscle (Thomas-Delloye et al., 1999 in Dupont et al., 2009). Only chicken Glut-1, -2, -3, and -8 have been characterized as messengers and/or proteins (Carver et al., 2001; Seki et al., 2003; Kono et al., 2005; Swaezea and Braun, 2006 in Zhao et al., 2012; Humphrey et al., 2004). It has been hypothesized that Glut-8 (encoded by the *SLC2A8* gene) may act in chickens as Glut-4 does in mammals (Kono et al., 2005; Zhao et al., 2012). *In vivo*, exogenous insulin increases glucose uptake moderately in some muscles but not in adipose tissue (Tokushima et al., 2005). (Surprisingly, insulin stimulated or inhibited ^{3}H-2DG uptake in liver and brain, respectively.) Thus, an insulin-sensitive glucose transport is present in chicken; however, it requires extensive characterization.

Currently, an increasing number of studies report quantitative trait loci (QTLs) as controlling growth and/or body composition and some physiological parameters in various avian species. In several cases, these QTL regions harbor glucose- and/or insulin-dependent genes. This aspect cannot be reviewed here, but some references can be found in Nadaf et al. (2009).

27.6.5 Insulin–Glucagon and Lipid Metabolism

Several transcription factors controlling mRNA expression of lipogenic enzymes also appear insulin-sensitive in chicken liver *in vivo* (SREBP1, PPAR-gamma and THRSP alpha (spot-14), and FASN messengers; Dupont et al., 2008). Carbohydrate-responsive element-binding protein (ChREBP, encoded by the *MLXIPL* gene) is another transcriptional factor involved in glucose metabolism, glucotoxicity, and/or lipogenesis in liver, adipose tissue, and pancreas in mammals (Leclerc et al., 2012). ChREBP is also activated in parallel with lipogenesis in chicken liver (Proszkowiec-Weglarz et al., 2009). In our insulin neutralization experiment, several other enzyme messengers involved in lipogenesis or fatty acid oxidation decreased or increased, respectively, following insulin deprivation (Simon et al., 2012), extending the role of insulin in chicken liver lipid metabolism. In abdominal fat tissue, 5 h of fasting altered the level of 2016 messengers. Plasma glucagon increase certainly accounts for a major part of these changes (Ji et al., 2012). Insulin neutralization in fed chickens altered the expression of only 92 genes, 72 of which were also differentially expressed by 5 h of fasting. All genes that were affected by both treatments changed in the same direction (up- or downregulated in both groups). Most messengers that were specifically altered by insulin neutralization (but not by fasting) were either poorly or not described; however, glucagon precursor (class B-preproglucagon) and glucagon receptor precursor were in this category and were up- and downregulated, respectively. A local insulin-dependent paracrine glucagon regulation, which needs full characterization, is suggested in chicken adipose tissue. In general, a contribution of insulin to adipose tissue development and metabolism (in addition to its control of liver lipogenesis) can be expected, despite the very atypical feature exhibited by chicken adipocytes.

Differentiation of chicken adipocyte precursors (obtained from the stromal vascular fraction of fat tissue) or transdifferentiation of fibroblasts, epithelial oviduct, or primordial germ cells into adipocytes has been successfully performed using media rather similar to those used for mammalian adipocyte differentiation (Khuong and Jeong, 2011; Li et al., 2010; Liu et al., 2010; Ramsay and Rosebrough, 2003). Insulin is required but unable to support differentiation alone; IGF1 or T3 are unable to replace insulin (Ramsay and Rosebrough, 2003). Chicken serum and dexamethasone can be deleted, provided oleic acid is introduced into the differentiation medium. Unless oleic acid exerts a specific effect, this supports the conclusion that to overcome their intrinsically low lipogenic activity and achieve a full adipocyte phenotype, chicken adipocyte precursors need exogenous fatty acids (Liu et al., 2009; Matsubara et al., 2005, 2008). Several transcription factors are involved in adipogenesis: C/EBPalpha, PPARgamma, and SREBP1 in chicken (Liu et al., 2009, 2010; Sato et al., 2009; Wang et al., 2008b) and duck (Xiong et al., 2010).

In adipocytes of mammals, lipolysis is under the control of several lipases, including the adipose triglyceride lipase, encoded by the *PNPLA2* gene, and the hormone-sensitive lipase (HSL), encoded by the *LIPE* gene (Zechner et al., 2009). HSL (*LIPE*) disruption in normal or *ob/ob* mice leads to several changes in the endocrine pancreas and adipocytes that resemble the chicken feature (see Osuga et al., 2000; Wang et al., 2001; Haemmerle et al., 2002 quoted in Zechner et al., 2009; Mulder et al., 2003; Harada et al., 2003; Sekia et al., 2009). Although not identified in chicken, the *LIPE* gene is present in the anole lizard and several fish. If the *LIPE* gene is really missing in chicken, some of the chicken physiological peculiarities may be due to this.

27.6.6 Insulin and Protein Metabolism

In addition to its general effects on protein metabolism (see Section 27.1; Tesseraud et al., 2007a), insulin acts in the ubiquitin–proteasome proteolytic pathway. It inhibits the expression of atrogin-1 messenger in quail muscle (QT6) fibroblasts (Tesseraud et al., 2007b) and *in vivo* in chicken muscle (Dupont et al., 2008). Following insulin immunoneutralization, several messengers coding for enzymes of the oxidative pathway decreased. This pathway is also altered in muscle of human diabetics (in Simon et al., 2012).

27.6.7 Insulin and Gene Expression

In mammals, insulin stimulates the translation of preexisting mRNA and coordinates (stimulates or inhibits) the expression of multiple genes in various tissues. Multiple insulin responsive elements (IRE) have been identified on target genes (other IREs are likely to be identified; Mounier and Posner, 2006). To our knowledge, the characterization and the functionality of IREs, as such, are still missing in chickens. The fact that several transcription factors or enzymes involved in lipid metabolism are also insulin sensitive in chickens (as discussed in this chapter) suggests that major IREs have been conserved or have co-evolved in chickens. A total of 1573 and 1225 signals were altered by insulin deprivation in liver and muscle, respectively, including several transcription factors. Several messengers have been associated with energy expenditure, insulin resistance, metabolism sensing, obesity, and/or diabetes in mammals (see Table 3 in Simon et al., 2012). The chicken elongation factor-2 (EF-2) gene, a major factor in protein synthesis in mammals, is also under the control of insulin and 8-bromo-cAMP or phorbol ester (Lim and Kim, 2007). Insulin deprivation in fed (but not fasting) chickens decreased *EGR1* (early growth response 1) messenger in liver and muscle but not in adipose tissue. However, *TOE1* (target of egr 1, member 1) was also specifically decreased by insulin deprivation and not by fasting in adipose tissue (Ji et al., 2012). *EGR1* target genes are largely unknown, particularly in the field of metabolism.

Finally, microRNAs (miRNAs) represent another mechanism in the regulation of gene translation. At least miR-196 and miR-200b showed significant and opposite changes in response to insulin neutralization (an increase in liver versus a decrease in muscle; Bouhallier, Rouault, and Simon, unpublished data). Gene targets of miR-196 and miR-200b need to be identified; so far, miR-196 has been involved in chicken embryo development (McGlinn et al., 2009) and miR-200b in chicken liver lipid metabolism (Hu et al., 2012).

27.7 EXPERIMENTAL OR GENETICAL MODELS

Table 27.1 summarizes results about fasting plasma glucose–insulin relationships, glucose tolerance, insulin sensitivity, and fattening in various experimental or genetic chicken models. Most of these results have been discussed in previous publications (the effect of prolonged fasting has also been investigated, as discussed in this chapter). The table is shown to support some general considerations. It should be noted that in all these experiments, the feeding pattern of chickens has been consistently synchronized with light–dark cycles. In these conditions, as in mammals, the fasting plasma glucose level appears as an indicator or a central node, which reflects or integrates general metabolism. Although the primary mechanism is unknown, changes in glucose–insulin balance and/or insulin sensitivity are associated with changes in fattening. In contrast to obese mammals, fat chickens have low plasma glucose and do not develop insulin resistance. Conversely, chickens from a low-glucose line are fatter than their counterparts. Only the corticosterone model behaves like obese mammals (showing higher plasma glucose and impaired insulin sensitivity). As discussed here, changes in insulin signaling are present in the models thus far investigated. Changes in glucose tolerance are not consistently associated with changes in body composition: in a commercial broiler cross, chickens exhibiting fast or slow plasma glucose regulation at 2 and 4 weeks of age following an oral glucose load did exhibit fast or slow glucose disposal at 8 weeks of age but did not differ in fasting or glucose-induced plasma insulin levels and body composition, which remains unexplained (Simon, 1980). One can hypothesize that, as in FL chickens, enhanced glucose-induced insulin levels are required to increase fattening without necessarily leading to insulin resistance. *In vitro*, perfused pancreas from fat chickens released less insulin during the first phase in response to glucose, or to arginine or acetylcholine in the presence of glucose, than pancreas from lean chickens, which remains unexplained in Rideau (1988). In general, chickens appear to be controlled within a much narrower range for their body composition than mammals. Whether the endocrine pancreas of birds plays a part in this remains an open question and a challenge for the future.

27.8 SUMMARY AND CONCLUSION

Previous reviews, summarized in the Section 27.1, revealed several features of the chicken endocrine pancreatic physiology. The present chapter is focused on glucagon and insulin; it highlights new knowledge about chicken embryo development, glucagon (and glucagon-related peptides), bird insulins, glucagon and insulin release, glucagon and insulin receptors, and major general effects of these hormones. Many results highlighted in this review point out further chicken peculiarities at the level of endocrine pancreatic physiology. Chickens (and birds in general) developed several species specificities during evolution. The most original ones concern the refractoriness of glucose-induced insulin release and the insulin signaling in muscle and adipose tissue. One unidentified inefficient metabolic step makes glucose a poor insulinotropic nutrient in chickens. Although several elements of insulin signaling pathways have not been investigated yet, several steps of the IR cascade appear insensitive in muscle; the feature is even worse in adipose tissue. In contrast, IR signaling appears "normal" in chicken liver. Revealing mechanisms or steps accounting for these chicken peculiarities may provide important information, including for human diabetes pathology. Glucagon physiology appears more complex in chicken than in mammals: the glucagon gene encodes several transcripts, it is expressed in adipose tissue, two types of glucagon receptors are present, and an insulin-dependent paracrine glucagon regulation is suggested in adipose tissue. In the fed status, insulin is coordinating the endocrine balance to permit the development or the maintenance of an active anabolism. From a practical point of view, finding the way to favor pleiotropic anabolic actions of insulin in muscle and minimize liver lipogenic and adipogenic actions of insulin would be highly profitable for human nutrition and poultry production. This chapter also pointed out many shortcomings or new areas to explore. In short, the secretory "machinery" of A- and B-islets needs to be fully characterized. Chicken is the unique model to provide such possibilities. In addition, nothing is known about the physiological actions of GLP-2, oxyntomodulin, miniglucagon, glicentin, or GRPP (gastrin-releasing pancreatic peptide) in chicken. Full characterization of IR signaling and the subsequent steps up to the control of gene expression certainly also offer large areas for future research. Finally, chicken glucose transporters (and their presence and functionality) need to be characterized. Although chicken is inconvenient because it requires the development of technical tools for each area to explore, it remains an almost unexplored experimental animal model of large interest for comparative endocrinology and physiology.

REFERENCES

Adamo, M., Simon, J., Rosebrough, R.W., McMurtry, J.P., Steele, N.C., LeRoith, D., 1987. Characterization of the chicken muscle insulin receptor. Gen. Comp. Endocrinol. 68, 456–465.

Ahnfelt-Ronne, J., Hald, J., Bodker, A., Yassin, H., Serup, P., Hecksher-Sorensen, J.J., 2007. Preservation of proliferating pancreatic progenitor cells by Delta-Notch signaling in the embryonic chicken pancreas. BMC Dev. Biol. 7, 63.

Anraku, T., Takagi, T., Nakao, N., Watanabe, M., Yasuo, S., Katou, Y., Ueda, Y., Murai, A., Iigo, M., Ebihara, S., Yoshimura, T., 2007. Photoperiodic changes in hypothalamic insulin receptor gene expression are regulated by gonadal testosterone. Brain Res. 1163, 86–90.

Bennett, A.K., Hester, P.Y., Spurlock, D.E., 2006. Polymorphisms in vitamin D receptor, osteopontin, insulin-like growth factor 1 and insulin, and their associations with bone, egg and growth traits in a layer–broiler cross in chickens. Anim. Genet. 37, 283–286.

Benomar, Y., Rideau, N., Crochet, S., Derouet, M., Taouis, M., 2003. Leptin fully suppresses acetylcholine-induced insulin secretion and is reversed by tolbutamide in isolated perfused chicken pancreas. Horm. Metab. Res. 35, 81–85.

Bigot, K., Taouis, M., Tesseraud, S., 2003. Refeeding and insulin regulate S6K1 activity in chicken skeletal muscles. J. Nutr. 133, 369–373.

Bisbis, S., Derouet, M., Simon, J., 1994a. Characterization of insulin receptors in chicken kidneys: effect of nutritional status. Gen. Comp. Endocrinol. 96, 37–49.

Bisbis, S., Taouis, M., Derouet, M., Chevalier, B., Simon, J., 1994b. Corticosterone-induced insulin resistance is not associated with alterations of insulin receptor number and kinase activity in chicken kidney. Gen. Comp. Endocrinol. 96, 370–377.

Chevalier, B., Anglade, P., Derouet, M., Mollé, D., Simon, J., 1996. Isolation and characterization of Muscovy (*Cairina moschata*) duck insulin. Comp. Biochem. Physiol. B Biochem. Mol. Biol. 114, 19–26.

Chou, H.F., Scanes, C.G., 1988. Influence of age, strain, and beta-adrenergic agonist on insulin sensitivity in chicks as determined by an adaptation of the euglycemic clamp technique. Poult. Sci. 67, 470–475.

Christensen, K., McMurtry, J.P., Thaxton, Y.V., Corzo, A., McDaniel, C., Scanes, C.G., 2013. Metabolic and hormonal responses of growing modern meat-type chickens to fasting. Br. Poult. Sci. 54, 199–205.

Cowap, J., 1985. The first appearance of endocrine cells in the splenic lobe of the embryonic chick pancreas. Gen. Comp. Endocrinol. 60, 131–137.

Datar, S.P., Suryavanshi, D.S., Bhonde, R.R., 2006. Chick pancreatic B islets as an alternative in vitro model for screening insulin secretagogues. Poult. Sci. 85, 2260–2264.

Davis, R., Vasilatos-Younken, R., 1995. Markedly reduced pancreatic glucagon levels in broiler chickens with spiking mortality syndrome. Avian Dis. 39, 417–419.

Denbow, D.M., Duke, G.E., Chaplin, S.B., 1988. Food intake, gastric secretion, and motility as affected by avian pancreatic polypeptide administered centrally in chickens. Peptides 9, 449–454.

Duchene, S., Audouin, E., Crochet, S., Duclos, M.J., Dupont, J., Tesseraud, S., 2008a. Involvement of the ERK1/2 MAPK pathway in insulin-induced S6K1 activation in avian cells. Domest. Anim. Endocrinol. 34, 63–73.

Duchene, S., Audouin, E., Berri, C., Dupont, J., Tesseraud, S., 2008b. Refeeding and insulin activate the AKT/p70S6 kinase pathway without affecting IRS1 tyrosine phosphorylation in chicken muscle. Domest. Anim. Endocrinol. 34, 1–13.

Duclos, M.J., Chevalier, B., Goddard, C., Simon, J., 1993a. Regulation of amino acid transport and protein metabolism in myotubes derived from chicken muscle satellite cells by insulin-like growth factor-I. J. Cell Physiol. 157, 650–657.

Duclos, M.J., Chevalier, B., Le Marchand-Brustel, Y., Tanti, J.F., Goddard, C., Simon, J., 1993b. Insulin-like growth factor-I-stimulated glucose transport in myotubes derived from chicken muscle satellite cells. J. Endocrinol. 137, 465–472.

Dupont, J., Dagou, C., Derouet, M., Simon, J., Taouis, M., 2004. Early steps of insulin receptor signaling in chicken and rat: apparent refractoriness in chicken muscle. Domest. Anim. Endocrinol. 26, 127–142.

Dupont, J., Tesseraud, S., Derouet, M., Collin, A., Rideau, N., Crochet, S., Godet, E., Cailleau-Audouin, E., Metayer-Coustard, S., Duclos, M.J., Gespach, C., Porter, T.E., Cogburn, L.A., Simon, J., 2008. Insulin immuno-neutralization in chicken: effects on insulin signaling and gene expression in liver and muscle. J. Endocrinol. 197, 531–542.

Dupont, J., Tesseraud, S., Simon, J., 2009. Insulin signaling in chicken liver and muscle. Gen. Comp. Endocrinol. 163, 52–57.

Dupont, J., Metayer-Coustard, S., Ji, B., Rame, C., Gespach, C., Voy, B., Simon, J., 2012. Characterization of major elements of insulin signaling cascade in chicken adipose tissue: apparent insulin refractoriness. Gen. Comp. Endocrinol. 176, 86–93.

Fischer, A.J., Skorupa, D., Schonberg, D.L., Walton, N.A., 2006. Characterization of glucagon-expressing neurons in the chicken retina. J. Comp. Neurol. 496, 479–494.

Furuse, M., Bungo, T., Shimojo, M., Masuda, Y., Saito, N., Hasegawa, S., Sugahara, K., 1998. Effects of various N-terminal fragments of glucagon-like peptide-1(7-36) on food intake in the neonatal chick. Brain Res. 807, 214–217.

Gittes, G.K., 2009. Developmental biology of the pancreas: a comprehensive review. Dev. Biol. 326, 4–35.

Hamano, Y., 2006. Effects of dietary lipoic acid on plasma lipid, in vivo insulin sensitivity, metabolic response to corticosterone and in vitro lipolysis in broiler chickens. Br. J. Nutr. 95, 1094–1101.

Hazelwood, R.L., 2000. Pancreas. In: Whittow, G. (Ed.), Sturkie's Avian Physiology, fifth ed. Academic Press, New York, NY, pp. 539–555.

Hernandez-Sanchez, C., Mansilla, A., de Pablo, F., Zardoya, R., 2008. Evolution of the insulin receptor family and receptor isoform expression in vertebrates. Mol. Biol. Evol. 25, 1043–1053.

Holzer, P., Reichmann, F., Farzi, A., 2012. Neuropeptide Y, peptide YY and pancreatic polypeptide in the gut-brain axis. Neuropeptides 46, 261–274.

Honda, K., Kamisoyama, H., Saito, N., Kurose, Y., Sugahara, K., Hasegawa, S., 2007. Central administration of glucagon suppresses food intake in chicks. Neurosci. Lett. 416, 198–201.

Honda, K., Kamisoyama, H., Uemura, T., Yanagi, T., Saito, N., Kurose, Y., Sugahara, K., Katoh, K., Hasegawa, S., 2012. The mechanism underlying the central glucagon-induced hyperglycemia and anorexia in chicks. Comp. Biochem. Physiol. A Mol. Integr. Physiol. 163, 260–264.

Hu, Y., Zhang, R., Zhang, Y., Li, J., Grossmann, R., Zhao, R., 2012. In ovo leptin administration affects hepatic lipid metabolism and microRNA expression in newly hatched broiler chickens. J. Anim. Sci. Biotechnol. 3, 16.

Humphrey, B.D., Stephensen, C.B., Calvert, C.C., Klasing, K.C., 2004. Glucose and cationic amino acid transporter expression in growing chickens (*Gallus gallus domesticus*). Comp. Biochem. Physiol. A Mol. Integr. Physiol. 138, 515–525.

Ji, B., Ernest, B., Gooding, J.R., Das, S., Saxton, A.M., Simon, J., Dupont, J., Metayer-Coustard, S., Campagna, S.R., Voy, B.H., 2012. Transcriptomic and metabolomic profiling of chicken adipose tissue in response to insulin neutralization and fasting. BMC Genomics 13, 441.

Karmann, H., Rideau, N., Zorn, T., Malan, A., Le Maho, Y., 1992. Early insulin response after food intake in geese. Am. J. Physiol. 263, R782–R784.

Kato, H., Okubo, Y., Matsumura, Y., Roberts Jr., C.T., Sugahara, K., LeRoith, D., 2000. The tyrosine kinase activity of the chicken insulin receptor is similar to that of the human insulin receptor. Biosci. Biotechnol. Biochem. 64, 903–906.

Katsumoto, K., Fukuda, K., Kimura, W., Shimamura, K., Yasugi, S., Kume, S., 2009. Origin of pancreatic precursors in the chick embryo and the mechanism of endoderm regionalization. Mech. Dev. 126, 539–551.

Khuong, T.T., Jeong, D.K., 2011. Adipogenic differentiation of chicken epithelial oviduct cells using only chicken serum. In Vitro Cell. Dev. Biol. Anim. 47, 609–614.

Kimmel, J.R., Pollock, H.G., Hazelwood, R.L., 1968. Isolation and characterization of chicken insulin. Endocrinology 83, 1323–1330.

Kono, T., Nishida, M., Nishiki, Y., Seki, Y., Sato, K., Akiba, Y., 2005. Characterisation of glucose transporter (GLUT) gene expression in broiler chickens. Br. Poult. Sci. 46, 510–515.

Ku, S.K., Lee, J.H., Lee, H.S., 2000. An immunohistochemical study of the insulin-, glucagon- and somatostatin-immunoreactive cells in the developing pancreas of the chicken embryo. Tissue Cell 32, 58–65.

Lenzen, S., 2008. The mechanisms of alloxan- and streptozotocin-induced diabetes. Diabetologia 51, 216–226.

Li, B.C., Tian, Z.Q., Sun, M., Xu, Q., Wang, X.Y., Qin, Y.R., Xu, F., Gao, B., Wang, K.H., Sun, H.C., Chen, G.H., 2010. Directional differentiation of chicken primordial germ cells into adipocytes, neuron-like cells, and osteoblasts. Mol. Reprod. Dev. 77, 795–801.

Lim, E.J., Kim, C.W., 2007. Functional characterization of the promoter region of the chicken elongation factor-2 gene. Gene 386, 183–190.

Liu, S., Wang, L., Wang, N., Wang, Y., Shi, H., Li, H., 2009. Oleate induces transdifferentiation of chicken fibroblasts into adipocyte-like cells. Comp. Biochem. Physiol. A Mol. Integr. Physiol. 154, 135–141.

Liu, S., Wang, Y., Wang, L., Wang, N., Li, Y., Li, H., 2010. Transdifferentiation of fibroblasts into adipocyte-like cells by chicken adipogenic transcription factors. Comp. Biochem. Physiol. A Mol. Integr. Physiol. 156, 502–508.

Lu, J.W., McMurtry, J.P., Coon, C.N., 2007. Developmental changes of plasma insulin, glucagon, insulin-like growth factors, thyroid hormones, and glucose concentrations in chick embryos and hatched chicks. Poult. Sci. 86, 673–683.

Lucini, C., Romano, A., Castaldo, L., 2000. NPY immunoreactivity in endocrine cells of duck pancreas: an ontogenetic study. Anat. Rec. 259, 35–40.

Lundell, I., Boswell, T., Larhammar, D., 2002. Chicken neuropeptide Y-family receptor Y4: a receptor with equal affinity for pancreatic polypeptide, neuropeptide Y and peptide YY. J. Mol. Endocrinol. 28, 225–235.

Manakova, E., Titlbach, M., 2007. Development of the chick pancreas with regard to estimation of the relative occurrence and growth of endocrine tissue. Anat. Histol. Embryol. 36, 127–134.

Matsubara, Y., Sato, K., Ishii, H., Akiba, Y., 2005. Changes in mRNA expression of regulatory factors involved in adipocyte differentiation during fatty acid induced adipogenesis in chicken. Comp. Biochem. Physiol. A Mol. Integr. Physiol. 141, 108–115.

Matsubara, Y., Endo, T., Kano, K., 2008. Fatty acids but not dexamethasone are essential inducers for chick adipocyte differentiation in vitro. Comp. Biochem. Physiol. A Mol. Integr. Physiol. 151, 511–518.

Matsuura, K., Katsumoto, K., Fukuda, K., Kume, K., Kume, S., 2009. Conserved origin of the ventral pancreas in chicken. Mech. Dev. 126, 817–827.

Le Marchand, S.J., Piston, D.W., 2010. Glucose suppression of glucagon secretion: metabolic and calcium responses from alpha-cells in intact mouse pancreatic islets. J. Biol. Chem. 285, 14389–14398.

Mayo, K.E., Miller, L.J., Bataille, D., Dalle, S., Goke, B., Thorens, B., Drucker, D.J., 2003. International Union of Pharmacology. XXXV. The glucagon receptor family. Pharmacol. Rev. 55, 167–194.

De Meyts, P., 2012. The insulin receptor isoform A: a mitogenic proinsulin receptor? Endocrinology 153, 2054–2056.

McGlinn, E., Yekta, S., Mansfield, J.H., Soutschek, J., Bartel, D.P., Tabin, C.J., 2009. In ovo application of antagomiRs indicates a role for miR-196 in patterning the chick axial skeleton through Hox gene regulation. Proc. Natl. Acad. Sci. U. S. A. 106, 18610–18615.

McMurtry, J.P., Tsark, W., Cogburn, L., Rosebrough, R., Brocht, D., 1996. Metabolic responses of the turkey hen (*Meleagris gallopavo*) to an intravenous injection of chicken or porcine glucagon. Comp. Biochem. Physiol. C Pharmacol. Toxicol. Endocrinol. 114, 159–163.

Modak, M.A., Datar, S.P., Bhonde, R.R., Ghaskadbi, S.S., 2007. Differential susceptibility of chick and mouse islets to streptozotocin and its co-relation with islet antioxidant status. J. Comp. Physiol. B. 177, 247–257.

Nadaf, J., Pitel, F., Gilbert, H., Duclos, M.J., Vignoles, F., Beaumont, C., Vignal, A., Porter, T.E., Cogburn, L.A., Aggrey, S.E., Simon, J., Le Bihan-Duval, E., 2009. QTL for several metabolic traits map to loci controlling growth and body composition in an F2 intercross between high- and low-growth chicken lines. Physiol. Genomics 38, 241–249.
Nagao, K., Aman Yaman, M., Murai, A., Sasaki, T., Saito, N., Okumura, J., Kita, K., 2001. Insulin administration suppresses an increase in insulin-like growth factor binding protein-2 gene expression stimulated by fasting in the chicken. Br. Poult. Sci. 42, 501–504.
Patwardhan, V., Gokhale, M., Ghaskadbi, S., 2004. Acceleration of early chick embryo morphogenesis by insulin is associated with altered expression of embryonic genes. Int. J. Dev. Biol. 48, 319–326.
Pirskanen, A., Kiefer, J.C., Hauschka, S.D., 2000. IGFs, insulin, Shh, bFGF, and TGF-beta1 interact synergistically to promote somite myogenesis in vitro. Dev. Biol. 224, 189–203.
Proszkowiec-Weglarz, M., Richards, M.P., Humphrey, B.D., Rosebrough, R.W., McMurtry, J.P., 2009. AMP-activated protein kinase and carbohydrate response element binding protein: a study of two potential regulatory factors in the hepatic lipogenic program of broiler chickens. Comp. Biochem. Physiol. B Biochem. Mol. Biol. 154, 68–79.
Qiu, F.F., Nie, Q.H., Luo, C.L., Zhang, D.X., Lin, S.M., Zhang, X.Q., 2006. Association of single nucleotide polymorphisms of the insulin gene with chicken early growth and fat deposition. Poult. Sci. 85, 980–985.
Raimbault, S., Dridi, S., Denjean, F., Lachuer, J., Couplan, E., Bouillaud, F., Bordas, A., Duchamp, C., Taouis, M., Ricquier, D., 2001. An uncoupling protein homologue putatively involved in facultative muscle thermogenesis in birds. Biochem. J. 353, 441–444.
Ramsay, T.G., Rosebrough, R.W., 2003. Hormonal regulation of postnatal chicken preadipocyte differentiation in vitro. Comp. Biochem. Physiol. B Biochem. Mol. Biol. 136, 245–253.
Rawdon, B.B., 1998. Morphogenesis and differentiation of the avian endocrine pancreas, with particular reference to experimental studies on the chick embryo. Microsc. Res. Tech. 43, 292–305.
Richards, M.P., McMurtry, J.P., 2008. Expression of proglucagon and proglucagon-derived peptide hormone receptor genes in the chicken. Gen. Comp. Endocrinol. 156, 323–338.
Richards, M.P., McMurtry, J.P., 2009. The avian proglucagon system. Gen. Comp. Endocrinol. 163, 39–46.
Rideau, N., 1988. Insulin secretion in birds. In: Leclercq, B., Whitehead, C.C. (Eds.), Leanness in Domestic Birds. Genetic, Metabolic and Hormonal Aspects. Butterworths and Co. (Publishers) Ltd-INRA, pp. 269–294.
Rideau, N., 1997. Insulin secretion. In: Harvey, S., Etches, R.J. (Eds.), Perspectives in Avian Endocrinology. Journal of Endocrinology Ltd, Bristol, pp. 329–334.
Rideau, N., 1998. Peculiarities of insulin secretion in chickens. Ann. N. Y Acad. Sci. 839, 162–165.
Rideau, N., Derouet, M., Grimsby, J., Simon, J., 2010. Glucokinase activation induces potent hypoglycemia without recruiting insulin and inhibits food intake in chicken. Gen. Comp. Endocrinol. 169, 276–283.
de la Rosa, E.J., de Pablo, F., 2011. Proinsulin: from hormonal precursor to neuroprotective factor. Front. Mol. Neurosci. 4, 20.
Rosenberg, L.C., Lafon, M.L., Pedersen, J.K., Yassin, H., Jensen, J.N., Serup, P., Hecksher-Sorensen, J., 2010. The transcriptional activity of Neurog3 affects migration and differentiation of ectopic endocrine cells in chicken endoderm. Dev. Dyn. 239, 1950–1966.
Ruffier, L., Simon, J., Rideau, N., 1998. Isolation of functional glucagon islets of Langerhans from the chicken pancreas. Gen. Comp. Endocrinol. 112, 153–162.
Sato, K., Yonemura, T., Ishii, H., Toyomizu, M., Kamada, T., Akiba, Y., 2009. Role of peroxisome proliferator-activated receptor beta/delta in chicken adipogenesis. Comp. Biochem. Physiol. A Mol. Integr. Physiol. 154, 370–375.
Sato, K., Aoki, M., Kondo, R., Matsushita, K., Akiba, Y., Kamada, T., 2012. Administration of insulin to newly hatched chicks improves growth performance via impairment of MyoD gene expression and enhancement of cell proliferation in chicken myoblasts. Gen. Comp. Endocrinol. 175, 457–463.
Scanes, C.G., 2009. Perspectives on the endocrinology of poultry growth and metabolism. Gen. Comp. Endocrinol. 163, 24–32.
Scanes, C.G., Braun, E., 2012. Avian metabolism: its control and evolution. Front. Biol., 1–26.
Shiraishi, J., Tanizawa, H., Fujita, M., Kawakami, S., Bungo, T., 2011a. Localization of hypothalamic insulin receptor in neonatal chicks: evidence for insulinergic system control of feeding behavior. Neurosci. Lett. 491, 177–180.
Shiraishi, J., Yanagita, K., Fukumori, R., Sugino, T., Fujita, M., Kawakami, S., McMurtry, J.P., Bungo, T., 2011b. Comparisons of insulin related parameters in commercial-type chicks: evidence for insulin resistance in broiler chicks. Physiol. Behav. 103, 233–239.
Shousha, S., Nakahara, K., Nasu, T., Sakamoto, T., Murakami, N., 2007. Effect of glucagon-like peptide-1 and -2 on regulation of food intake, body temperature and locomotor activity in the Japanese quail. Neurosci. Lett. 415, 102–107.
Simon, J., 1980. Apparent independence of glucose tolerance and body fat content in 8-week-old broilers. Br. Poult. Sci. 21, 309–313.
Simon, J., 1988. Insulin in birds: Metabolic effects and possible implications in genetically fat and lean chickens. In: Leclercq, B., Whitehead, C.C. (Eds.), Leanness in Domestic Birds. Genetic, Metabolic and Hormonal Aspects. Butterworth and Co (Publishers) Ltd-INRA, pp. 253–268.
Simon, J., 1989. Chicken as a useful species for the comprehension of insulin action. Crit. Rev. Poult. Biol. 2, 121–148.
Simon, J., 1995. Insulin-glucagon and growth in broilers. Arch. Gefluegelkunde, 14–17. Special issue on OECD Workshop, June 7–9, 1994, Celle, Germany.
Simon, J., Guillaumin, S., Chevalier, B., Derouet, M., Guy, G., Marche, G., Ricard, F.H., Leclercq, B., 2000. Plasma glucose-insulin relationship in chicken lines selected for high or low fasting glycaemia. Br. Poult. Sci. 41, 424–429.
Simon, J., Laurent, S., Grolleau, G., Thoraval, P., Soubieux, D., Rasschaert, D., 2004. Evolution of preproinsulin gene in birds. Mol. Phylogenet. Evol. 30, 755–766.
Simon, J., Rideau, N., Taouis, M., Dupont, J., 2011. Plasma insulin levels are rather similar in chicken and rat. Gen. Comp. Endocrinol. 171, 267–268.
Simon, J., Milenkovic, D., Godet, E., Cabau, C., Collin, A., Metayer-Coustard, S., Rideau, N., Tesseraud, S., Derouet, M., Crochet, S., Cailleau-Audouin, E., Hennequet-Antier, C., Gespach, C., Porter, T.E., Duclos, M.J., Dupont, J., Cogburn, L.A., 2012. Insulin immunoneutralization in fed chickens: effects on liver and muscle transcriptome. Physiol. Genomics 44, 283–292.
Steiner, D.J., Kim, A., Miller, K., Hara, M., 2010. Pancreatic islet plasticity: interspecies comparison of islet architecture and composition. Islets 2, 135–145.

Tachibana, T., Oikawa, D., Adachi, N., Boswell, T., Furuse, M., 2007. Central administration of alpha-melanocyte-stimulating hormone changes lipid metabolism in chicks. Comp. Biochem. Physiol. A Mol. Integr. Physiol. 148, 408–412.

Taouis, M., Taylor, S.I., Reitman, M., 1996. Cloning of the chicken insulin receptor substrate 1 gene. Gene 178, 51–55.

Taouis, M., Dupont, J., Gillet, A., Derouet, M., Simon, J., 1998. Insulin receptor substrate 1 antisense expression in an hepatoma cell line reduces cell proliferation and induces overexpression of the Src homology 2 domain and collagen protein (SHC). Mol. Cell. Endocrinol. 137, 177–186.

Tesseraud, S., Metayer, S., Duchene, S., Bigot, K., Grizard, J., Dupont, J., 2007a. Regulation of protein metabolism by insulin: value of different approaches and animal models. Domest. Anim. Endocrinol. 33, 123–142.

Tesseraud, S., Metayer-Coustard, S., Boussaid, S., Crochet, S., Audouin, E., Derouet, M., Seiliez, I., 2007b. Insulin and amino acid availability regulate atrogin-1 in avian QT6 cells. Biochem. Biophys. Res. Commun. 357, 181–186.

Tokushima, Y., Takahashi, K., Sato, K., Akiba, Y., 2005. Glucose uptake in vivo in skeletal muscles of insulin-injected chicks. Comp. Biochem. Physiol. B Biochem. Mol. Biol. 141, 43–48.

Tosca, L., Crochet, S., Ferre, P., Foufelle, F., Tesseraud, S., Dupont, J., 2006. AMP-activated protein kinase activation modulates progesterone secretion in granulosa cells from hen preovulatory follicles. J. Endocrinol. 190, 85–97.

Trabucchi, M., Tostivint, H., Lihrmann, I., Blahser, S., Vallarino, M., Vaudry, H., 2003. Characterization of the cDNA encoding a somatostatin variant in the chicken brain: comparison of the distribution of the two somatostatin precursor mRNAs. J. Comp. Neurol. 461, 441–451.

Vaudin, P., Dupont, J., Duchene, S., Audouin, E., Crochet, S., Berri, C., Tesseraud, S., 2006. Phosphatase PTEN in chicken muscle is regulated during ontogenesis. Domest. Anim. Endocrinol. 31, 123–140.

Vessey, K.A., Rushforth, D.A., Stell, W.K., 2005. Glucagon- and secretin-related peptides differentially alter ocular growth and the development of form-deprivation myopia in chicks. Invest. Ophthalmol. Vis. Sci. 46, 3932–3942.

Wang, J., Wang, Y., Li, X., Li, J., Leung, F.C., 2008a. Cloning, tissue distribution, and functional characterization of chicken glucagon receptor. Poult. Sci. 87, 2678–2688.

Wang, Y., Mu, Y., Li, H., Ding, N., Wang, Q., Wang, S., Wang, N., 2008b. Peroxisome proliferator-activated receptor-gamma gene: a key regulator of adipocyte differentiation in chickens. Poult. Sci. 87, 226–232.

Wang, Y., Meng, F., Zhong, Y., Huang, G., Li, J., 2012. Discovery of a novel glucagon-like peptide (GCGL) and its receptor (GCGLR) in chickens: evidence for the existence of GCGL and GCGLR genes in nonmammalian vertebrates. Endocrinology 153, 5247–5260.

Xiong, M., Li, S., Peng, X., Feng, Y., Yu, G., Xin, Q., Gong, Y., 2010. Adipogenesis in ducks interfered by small interfering ribonucleic acids of peroxisome proliferator-activated receptor gamma gene. Poult. Sci. 89, 88–95.

Zhao, J.P., Bao, J., Wang, X.J., Jiao, H.C., Song, Z.G., Lin, H., 2012. Altered gene and protein expression of glucose transporter1 underlies dexamethasone inhibition of insulin-stimulated glucose uptake in chicken muscles. J. Anim. Sci. 90, 4337–4345.

Part VI

Reproductive Theme

Chapter 28

Reproduction in the Female

Alan L. Johnson
Center for Reproductive Biology and Health, The Pennsylvania State University, University Park, PA, USA

28.1 INTRODUCTION

The avian lineage evolved from ancestral, oviparous amniotes that include reptiles and the archosaurs (theropod dinosaurs and crocodilians). Oviparity represents the ancestral form of reproduction in all archosaurs, and it is the only form of reproduction utilized by birds. Whereas all modern crocodilians have two functional ovaries, only the left ovary is functional in the majority of avian species (Figure 28.1). Significantly, avian fossils from two species of an extinct lineage (enantiornithine birds of the Early Cretaceous period) have revealed that these ancestors possessed a single, functional left ovary (Zheng et al., 2013). This is consistent with the hypothesis that the loss of one ovary to reduce body weight occurred early in avian evolution, perhaps even preceding the capacity for true flight. Interestingly, viviparity has evolved nearly 100 times within the reptilian lineage that includes lizards and snakes; thus, it is unclear why there is no incidence of viviparity within the avian lineage. One line of reasoning is that because extant birds are endothermic and can precisely control the process of egg incubation and embryo development, there may be no thermoregulatory advantage to viviparity in birds.

28.2 DEVELOPMENT AND FUNCTION OF THE FEMALE REPRODUCTIVE SYSTEM

The female gonads develop from a thickening of the ventromedial surface of the embryonic mesonephros and are initially composed of an outer epithelial cortex and an underlying medulla. Cortical and medullary cords are formed and colonized by the primordial germ cells (PGCs) that originate within the epiblast, move to the germinal crescent, then migrate to the ovary via the vascular system. By day 2.5 of incubation, PGCs, initially numbering about 450, begin to exit the vasculature at the germinal ridge by extravasation in response to chemotactic signals, including stromal cell-derived factor 1 (SDF1) and its receptor, C–X–C chemokine receptor type 4 (CXCR4) (Stebler et al., 2004). By day 4 of incubation, the distribution of PGCs within the left ovary of the chicken exceeds that in the right ovary, and this increase may be facilitated by factors that inhibit meiosis and promote germ cell proliferation; these include stem cell factor, ciliary neurotrophic factor, fibroblast growth factor, and follicle-stimulating hormone (FSH) (Gonzalez-Moran, 2007; He et al. 2012, 2013; Karagenç and Petitte, 2000), plus several microRNAs (miRNAs, which are endogenous noncoding small regulatory RNAs) (Lee et al., 2011). Proliferating PGCs form aggregates in which cells are joined via intercellular bridges and their cell cycles become synchronized. The isolation and culture of avian PGCs *in vitro* have been reported (van de Lavoir et al., 2006), and such methods may provide an opportunity to genetically modify avian PGCs and to produce transgenic birds.

Genetic sex in all birds is determined at fertilization, with the female being the heterogametic sex (ZW) and the male homogametic (ZZ). The evolutionary origin of genetic sex determination in Aves is obscure since all crocodilians

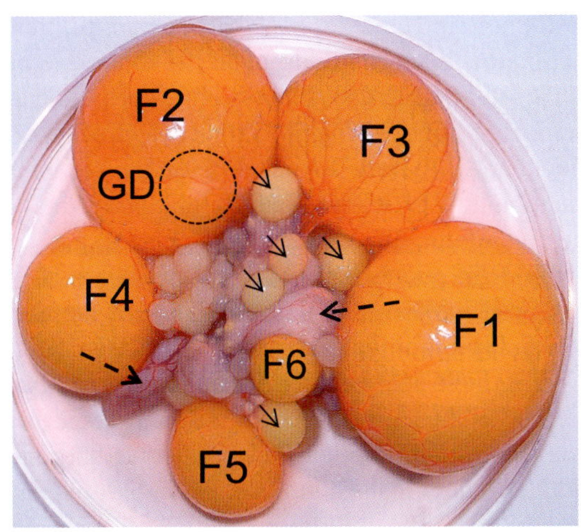

FIGURE 28.1 **Functional left ovary of the laying hen.** F1–F5 represent hierarchal, preovulatory follicles; and F6 represents the preovulatory follicle most recently selected from the cohort of prehierarchal follicles (arrow heads). Dashed arrows, postovulatory follicles. Dotted line surrounds the opaque germinal disc (GD) region. *Photograph by Dr O. Ocón-Grove.*

display temperature-dependent sex determination. While the double sex and mab-3 related transcription factor 1 (*DMRT1*) gene represents a likely candidate for male sex determination (much like the sex-determining region Y gene in mammals), the mechanisms by which the DMRT1 protein directs sex determination remain unclear. *In ovo* knockdown of *DMRT1* mRNA during early development using RNA interference selectively reduces DMRT1 protein expression and results in the feminization of genetically male gonads (Smith et al., 2009). By day 10 of incubation, anatomical asymmetry between the left and right ovary becomes evident, and the degeneration of the right ovary is initiated via programmed cell death (apoptosis). In genetic females, R-spondin 1 (*RSPO1*) gene expression (one of the earliest known genes controlling the female genetic developmental program in mammals) becomes elevated within the left ovary at the time of differentiation. This coincides with female-specific activation of the Forkhead box L2 (*FOXL2*) gene, increased expression of aromatase, and estrogen synthesis (Hudson et al., 2005). R-spondin 1, the protein product of *RSPO1*, activates the canonical Wnt/β-β-catenin signaling pathway required for female somatic cell differentiation and germ cell commitment into meiosis. In the chicken, the R-spondin 1 protein is localized to the outer cortical zone of the developing ovary, the site of primordial follicle formation and germ cell differentiation. Finally, there is evidence for a number of miRNAs that are differentially expressed in the gonad as early as day 3 of incubation. Several of these miRNAs have been implicated in gonadal sexual differentiation (Cutting et al., 2012).

The initial development of the left ovary is directed via the homeobox gene, *PITX2*, which promotes somatic development of the gonad cortex and supports the proliferation and differentiation of PGCs (Guioli and Lovell-Badge, 2007; Rodríguez-León et al., 2008). With increased expression of R-spondin 1, the left gonad develops a differentiated surface epithelium where eventual follicle development occurs (Smith et al., 2008). Expression of mRNAs encoding luteinizing hormone (LH) and FSH receptors occurs by day 4 of incubation (before sexual differentiation), and FSH treatment on day 14 stimulates proliferation of the ovarian surface epithelium, oogonia within the cortex, and somatic cells within the medullary cords (Akazome et al., 2002; Méndez et al., 2003). Germ cell meiosis in the domestic chicken embryo is initiated by incubation day 15.5 by the actions of retinoic acid and LH (Smith et al., 2008; He et al., 2013). This sequence of events is not unlike that in mice, where PGC migration occurs from about embryonic day 8.5 to 12.5, and meiosis is initiated by embryonic day 13.5 (Saitou et al., 2012).

In the domestic hen, the appearance of a meiotic nucleus and Balbiani body (a concentration of cell organelles shifted to one pole of the cell) is characteristic of germ cells located in the left ovarian cortex, but neither characteristic is found in the left medulla or the right ovary. Aromatase activity and estrogen synthesis within medullary tissue play key roles in the differentiation of the ovary, and differences between left versus right ovarian development depend upon the selective expression of the estrogen receptor within developing cortical tissue of the left ovary (Smith and Sinclair, 2004; Ishimaru et al., 2008). Unlike mammals, the differentiation of avian embryonic gonads is not as strictly regulated by genetic sex since the administration of an estrogen synthesis inhibitor prior to gonad differentiation will induce testis development in a genetic female (reviewed in Ayers et al., 2013).

The female embryo develops undifferentiated Müllerian (paramesonephric) ducts early in embryonic development. Both embryonic ovaries express anti-Müllerian hormone (AMH) and produce estrogen. AMH expression is regulated via the transcription factor, steroidogenic factor-1 (SF1), and can be initiated, *in vitro*, by several factors (including vitamin D and bone morphogenetic proteins-4 and -6 (BMP4 and BMP6, respectively)) (Johnson et al., 2008; Ocón-Grove et al., 2012). The right Müllerian duct eventually regresses (beginning by day 8 in the duck) in response to AMH. By contrast, estrogen derived from the left ovary protects the left Müllerian duct from the apoptosis-inducing effects of AMH, thus enabling development of the left oviduct (Takada et al., 2006).

28.2.1 Late-Embryonic and Posthatch Ovary

While the reproductive system of virtually all females within the avian order Galliformes consists of a single left ovary and its oviduct, on rare occasions a functional right ovary and oviduct may be present. By comparison, among the Falconiformes and in the brown kiwi (a ratite; order Struthioniformes), both left and right gonads and associated oviducts may be functional, although the ovaries often remain asymmetrical in size. In sparrows (Passeriformes) and pigeons (Columbiformes), about 5% of specimens are reported to have two developed ovaries, but again, some size asymmetry and difference in ovulation frequency often exist between the two ovaries (Kinsky, 1971).

The left ovary develops within the abdominal region ventral to the caudal vena cava and adjacent to the left kidney and adrenal gland. The developing ovary is eventually suspended from the body wall primarily by the mesovarium. The hilus serves as a conduit for vasculature, nervous innervation, and smooth muscle that integrates ovarian function with environmental, endocrine, and central nervous system input (reviewed in Gilbert, 1979). The number of oocytes in the chick embryo increases from approximately 28,000 on the ninth day of development to an approximate 680,000 on the 17th day, and subsequently decreases to some 480,000 by the time of hatching, when oogenesis is believed to terminate (Hughes, 1963). This is in contrast to essentially all amphibians, many fishes, and some reptiles where oogenesis continues within the adult female. The ovary of immature birds consists of a mass of small ova, of which as many as 2000 are visible to the naked eye. At the time of hatch, avian oocytes are mostly in meiotic prophase I. Only a relatively few of these (250–500) will reach maturity and ovulate within the life span of most domesticated species, and considerably fewer mature in wild species.

The organization of oocytes into primordial follicles occurs around the time of hatch in most birds. Following hatch and prior to sexual maturation, the left ovary remains small and undeveloped. PGCs become surrounded by a single epithelial granulosa cell layer, and these primordial follicles are loosely arranged in clusters (or nests). The oogonia are mostly in the diplotene stage of meiotic prophase I, and chromosomes display the typical lampbrush configuration from which a high rate of RNA synthesis occurs. An inner perivitelline membrane layer (homologous to the mammalian zona pellucida) is deposited between the granulosa cells and the oocyte cell membrane (Figure 28.2). Glycoproteins of the perivitelline membrane form a matrix that subsequently serves functions critical to the fertilization process, including the interaction of spermatozoa with the oocyte and inducing acrosomal exocytosis (Rodler et al., 2012).

Granulosa cells surrounding the perivitelline layer eventually form cytoplasmic projections that extend into this layer and into the oocyte. Initially, a single layer of theca cells surrounds the granulosa layer separated by the basal lamina (Gilbert, 1979). The basal lamina of the domestic hen follicle is approximately 1 μm thick and consists of a chondroitin sulfate–modified collagen and the glycoprotein, fibronectin, and other components deposited by both the granulosa and theca layers.

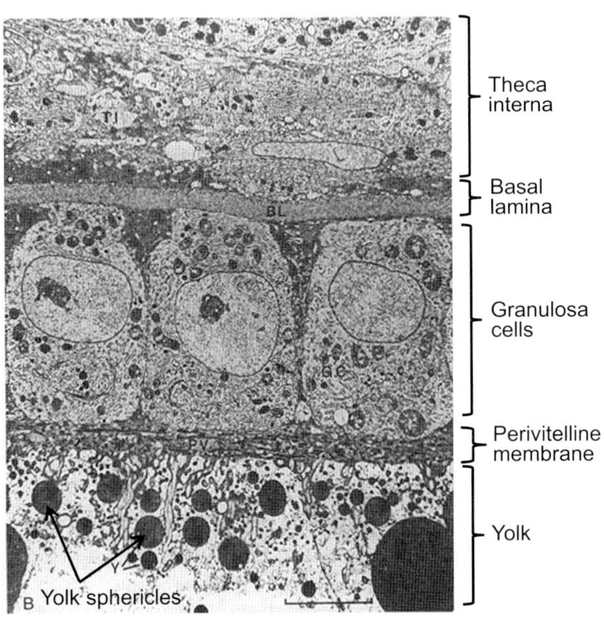

FIGURE 28.2 Layers of tissue from slow-growing follicles. Note that the granulosa layer is cuboidal in shape at this early stage of development, and that the outer theca layer (externa) is not shown. *From Perry and Gilbert (1979).*

28.2.2 Sexually Mature Ovary

The length of time required to reach sexual maturation depends upon the species, and it may be as short as several weeks after hatch (Japanese quail) to as long as 2 to 3 or more years (e.g., kiwis, penguins, and condors). Sexual maturation in the domestic hen will occur at about 5 months of age when hens are raised under a nonstimulatory photoperiod (less than 12h of light per day), but it can be advanced significantly by photostimulation beginning at about 2 months of age. The functionally mature ovary of most birds is arranged with an obvious hierarchy of follicles (Figure 28.1). As in mammals, both sexual maturation and recrudescence of the gonads at the onset of seasonal breeding are initiated by changes within the neuroendocrine axis and involve, among other factors, thyroid hormones (Nakao et al., 2008) and gonadotropin-inhibitory hormone (GnIH) (see Chapter 30 on the hypothalamus). These and additional hormones and growth factors indirectly (e.g. 3,3′,5-triiodothyronine; T_3) or directly (gonadotropins) impact gene expression at the level of the ovary. Gene expression can also be modified by miRNAs acting at a posttranscriptional level. For instance, there is evidence that a large number of miRNAs are differentially expressed in the ovary from sexually immature versus mature chickens, and in developing ovarian follicles relative to the stage of maturation (Kang et al., 2013). The growth of follicles at puberty or with seasonal breeding is an energetically expensive process and increases resting metabolic rates by 27% in the great tit (*Parus major*; Nilsson and Råberg, 2001) and 22% in the European starling (*Sturnis vulgaris*; Vézina and Williams, 2002).

The activation and growth of primordial follicles to the primary follicle stage are associated with the formation of the theca interna layer (from mesenchymal cells). The developing theca interna layer is separated from the cuboidal-shaped granulosa layer by the basal lamina (Figure 28.2). Growth to the prehierarchal follicle stage (6–8 mm) entails the accumulation of lipoprotein-rich white yolk, plus the differentiation of the theca into interna and externa layers. Vasculature and nervous innervation reaches the follicle through the pedicle and radiates through the theca layer. Immediately following selection into the preovulatory hierarchy, preovulatory follicles begin to rapidly grow from approximately 9 mm in diameter to 40+ mm over the course of days (Figure 28.3). Granulosa cells from preovulatory follicles become squamous in shape to facilitate the uptake of large amounts of vitellogenin and very-low-density lipoprotein (VLDL), except within the germinal disc region. Ovulation of the largest preovulatory follicle eventually

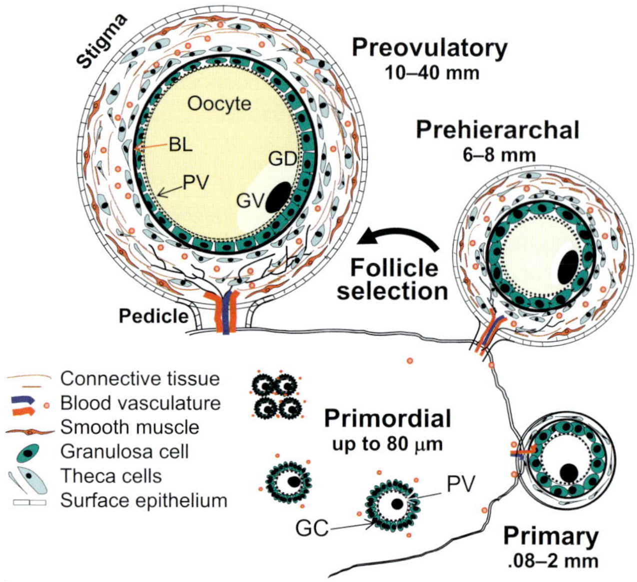

FIGURE 28.3 Follicles within the mature hen ovary. Primordial follicles consist of the germinal vesicle (GV) contained within a perivitelline membrane (PV) and surrounded by granulosa cells (GC). Primary follicles become further organized by a basal lamina (BL) separating the GC layer from a single layer of theca cells in primary follicles and eventually theca interna and externa layers in prehierarchal and preovulatory follicles. Increasing amounts of yellow yolk (vitellogenin and very-low-density lipoprotein) are accumulated beginning at follicle selection. BL, basal lamina; GD, germinal disc; GV, germinal vesicle; PV, perivitelline membrane. See the text for further details. *Adapted from Johnson and Woods (2007).*

occurs at the region of the comparatively avascular stigma region. The mature ovary is supported by a blood supply from the ovarian artery, which usually arises from the left renolumbar artery but may branch directly from the dorsal aorta (Hodges, 1965). Within the fully developed ovary, the percentage of blood flow is greatest to the five largest preovulatory follicles and significantly increases with increased follicle size and at the time of ovulation (Scanes et al., 1982). The ovarian artery divides into many branches and eventually leads to each follicle stalk (pedicle). All veins from the ovary unite into the two main anterior and posterior veins, which subsequently drain into the posterior vena cava.

Both adaptive and innate immune responses have been described within the domestic hen ovary, and these systems function to protect against colonization and infection by microbial pathogens, as well as to maintain normal functions of the ovary. Monocytes, macrophages, and activated B and T cells expressing major histocompatibility complex class II (MHC-II$^+$) molecules have been localized to the theca layer of healthy cortical follicles plus growing prehierarchal and preovulatory follicles, and more prominently in both theca and granulosa layers of postovulatory and atretic follicles. Cell-mediated immunity within ovarian stromal and theca tissues is mediated via T cells (CD4$^+$ and CD8$^+$ cells) and represents an effective measure against intracellular pathogens such as *Salmonella* (Barua et al., 2001; Barua and Yoshimura, 2004). Moreover, there is considerable evidence for a functional innate immune system consisting of β-defensins (small cationic antimicrobial peptides that exhibit a direct lytic action on microorganisms; Abdelsalam et al., 2012) and Toll-like receptors (pattern recognition receptors that activate cytokines, antimicrobial peptides, and immunocompetent cells; Woods et al., 2009). A total of 14 avian β-defensin genes (AvβD1–14) have been identified and these attack a wide range of microorganisms, including Gram-positive and Gram-negative bacteria, fungi, and yeast.

Innervation to the ovary is provided by both sympathetic (primarily adrenergic) and parasympathetic fibers (Gilbert, 1969; Dahl, 1970; Unsicker et al., 1983). This innervation consists of both extrinsic and intrinsic neural systems (Madekurozwa, 2008), and provides a number of neurochemicals (e.g., catecholamines, serotonin, and acetylcholine) and neuropeptides (e.g., neurotropins, vasoactive intestinal peptide (VIP), and galanin) to the ovary and its follicles. Such factors function in diverse roles as in the initial organization of primordial follicles in the embryonic and early posthatch ovary, the growth and differentiation of follicles in the adult ovary, the regulation of blood flow, and steroidogenesis.

As puberty approaches and the slow growth of follicles resumes, the theca layer of primary follicles differentiates into discrete theca interna and externa layers with distinct steroidogenic profiles. As the follicle grows, the theca layer becomes increasingly more vascularized except for the stigma (the point of rupture during ovulation), which contains fewer underlying small veins and arteries. The largest arteries from the follicle stalk are directed toward the fastest growing follicles, branch into arterioles, and pass through the theca to the basal lamina to form arterial capillaries. There are also an increasingly greater number of neurons present within the theca layer as the follicle progressively matures. While granulosa cells from early-developing follicles (<2 mm) are closely packed and cuboidal, they eventually form a single cell layer and become squamous in shape as the follicle grows (Gilbert, 1979).

The germinal vesicle is localized within the germinal disc region and includes the oocyte nucleus and the overlying granulosa layer (Figure 28.3). The disc appears as a white plaque due to a localized, restricted passage of yolk precursors. This region contains maternal RNA used for protein synthesis prior to the onset of the transcription from the zygotic genome. Additional extra-embryonic RNA is localized within the cytoplasmic layer directly under the perivitelline membrane, and is proposed to be required for oocyte development (Olszanska and Stepinska, 2008). The total amount of maternal RNA deposited in the avian oocyte is considerably greater than in mammals, but comparable to that in *Xenopus*. In preovulatory follicles, granulosa cells within the germinal disc region remain relatively undifferentiated (steroidogenically inactive) and mitotically active, while the outer granulosa layer becomes steroidogenically active but mitotically inactive (Yao and Bahr, 2001). Oocytes remain arrested at the first meiotic prophase stage during follicle development, and, in the domestic hen, resume meiosis (undergo germinal vesicle breakdown) 4–6h prior to ovulation. One microarray study identified no less than five genes differentially expressed within the germinal disc region that correlate with enhanced fertility; these genes may eventually be useful as markers of fertility in the domestic hen (Elis et al., 2009).

The ovary of the sexually mature chicken maintains numerous primordial follicles smaller than 1 mm diameter that remain embedded within the outer ovarian cortex. These follicles can exist in a quiescent state for months to years until activated (possibly by the removal from inhibitory AMH signaling) to begin an extended period of growth and differentiation. As follicles initiate development ("initial recruitment"), they project from the surface of the cortex and become suspended by a stalk (an extension of the ovarian cortex) through which blood vasculature and several nerve bundles reach each follicle. Typically, there are numerous slow-growing follicles measuring approximately 2–6 mm in diameter that contain small amounts of a protein-rich white yolk. The selection of a single follicle per day into the preovulatory hierarchy occurs from a cohort of 6–8 mm prehierarchal follicles (numbering approximately 8–12; Figure 28.1). The most recently selected follicle (9–12 mm) represents that which has initiated differentiation and rapid growth. In the domestic hen, there are typically four to six large, yolk-filled, preovulatory follicles (ranging in size from >12 to 45^{+} mm in diameter). This final growth phase prior to ovulation in domestic fowl, ducks, and pigeons generally lasts for 4–6 days (up to 25 days or more in the brown kiwi; Jensen and Durant, 2006). The number of preovulatory follicles can range from one or perhaps two for species laying a single egg (e.g., kiwi and emperor penguin) to eight or more in species that lay 12 to 18 eggs in a clutch (e.g., wild turkey and ring-neck pheasant). Importantly, the maintenance of both prehierarchal follicles in an undifferentiated state together with rapidly differentiating preovulatory follicles represents a fundamental distinction between the avian and mammalian reproductive strategies. Specifically, in eutherian mammals the selection of a dominant follicle or follicles in each estrous or menstrual cycle is rapidly followed by atresia in all growing but subdominant follicles. By contrast, the ability to produce a full clutch of eggs requires that a viable, prehierarchal cohort of follicles be maintained such that a single follicle can be selected for final growth and differentiation on a daily or near-daily basis.

Interestingly, the avian perivitelline layer does not prevent polyspermy. Accordingly, at insemination, more than one male pronuclei may be found in the egg cytoplasm, but only the one located in the center of the germinal disc forms the zygote nucleus. It is not yet known how a specific sperm is selected or at which stage between insemination and fertilization sperm selection takes place. Nevertheless, the remaining sperm degenerate and disappear, presumably due to the actions of DNases I and II (Stepinska and Bakst, 2006). Fusion of the male and female pronuclei (syngamy) initiates embryo development almost immediately, and a number of cell divisions occur by the time the ovum has become incorporated into the egg and the egg is laid.

28.2.3 Follicle Selection and Establishment of the Preovulatory Hierarchy

The process of follicular selection (also referred to as "cyclic recruitment") represents the rate-limiting stage for the development of mature oocytes. This process ultimately determines the reproductive capacity for vertebrate species whose strategy emphasizes a limited number of offspring combined with significant parental investment (as compared to most bony fish and amphibians). A fundamental question yet to be defined pertains to the mechanisms by which a single follicle is selected from the prehierarchal cohort each day. To date, the most proximal marker of follicle selection in the hen (chicken and pheasant) ovary is the initial ability of the FSH receptor to initiate cell signaling via cyclic adenosine monophosphate (cAMP) specifically within the granulosa layer from a single prehierarchal follicle per day (Tilly et al., 1991a; Kim et al., 2013). One hypothesis currently under investigation is that this selection process entails the removal of inhibitory cell signaling that prevents FSH receptor signaling (e.g., promotes receptor desensitization) until the time of selection; this inhibitory process also precludes the selection of more than one prehierarchal follicle per day (Johnson, 2011). At the time of selection, the initial activation of the cAMP signaling pathway rapidly (1) induces the transcription and translation of key steroidogenic enzymes (e.g., CYP11A and CYP17) plus steroidogenic acute regulatory (STAR) protein and, consequently, initiates the process of steroidogenesis; (2) increases the expression of antiapoptotic proteins, thus enhancing cell survival; (3) promotes a transition from FSH receptor dominance to LH receptor dominance; and (4) initiates lipid-rich yolk accumulation within the selected follicle. During this final phase of development, the follicle of the domestic hen will, on average, transport up to 2g of yolk per day and increase in volume by a factor of 3500–8000.

With the exception of immunoglobulins, the major yolk proteins are formed in the liver. Yolk synthesis is regulated primarily by gonadotropins and steroid hormones (predominantly estrogen, and to a lesser extent androgens). Uptake of VLDLs (which function mainly to transport triglycerides, phospholipids, and cholesterol) and vitellogenins (VTG1, VTG2, and VTG3; these are large, multidomain phosphoglycolipoproteins) (Finn, 2007) from blood capillaries into the follicle occurs through spaces within the theca layer. From here, they pass through the basal lamina via a selective process that depends on size and molecular charge. These particles then pass between granulosa cells to the perivitelline membrane of the oocyte. The size of these intercellular spaces is dependent upon the stage of follicular growth. For instance, the relatively narrow intercellular space of the primordial follicle allows ferritin and other small molecules to cross. At the time follicles reach the prehierarchal stage, follicles begin to accumulate lipoprotein-rich, white yolk, but transport remains limited by tight junctions between granulosa cells (Schuster et al., 2004). By comparison, during the rapid growth of preovulatory follicles, the shape and organization of granulosa cells change to accommodate an increasing flow of egg yolk precursors that accumulate at the basal lamina of the follicle. These substances finally diffuse through the perivitelline membrane to reach the major yolk precursor receptor, LR8, on the oocyte (Schneider, 2009). This receptor is related by sequence homology to mammalian low-density lipoprotein (LDL) receptors. Significantly, a point mutation within the extracellular domain of the chicken LR8 results in a conformational change in the translated protein that significantly limits oocyte uptake of serum-derived yolk precursors; the site of this point mutation is identical to that in the human LDL receptor that causes familial hypercholesterolemia (Elkin et al., 2012). Finally, yolk proteins are further processed within the oocyte to phosvitin, lipovitellin, triglycerides, cholesterol, and phospholipids by one or more of the proteases and protease inhibitors found in egg yolk or the perivitelline membrane (Bourin et al., 2012). It is estimated that deposition of yolk into the maturing follicle is terminated approximately 24 h before ovulation.

In the highly productive laying hen, the decrease in egg production at the end of the first-year laying cycle results from a reduction in the number of follicles selected into the preovulatory hierarchy and to an increase in the incidence of follicle atresia. As a result, fewer follicles receive a proportionately greater quantity of yolk, resulting in a larger egg size. Eventually, yolk proteins are taken up by the developing embryo primarily for tissue growth, while the yolk lipids represent the main source of energy. The ultrastructure of follicles at different stages of development has been well described in Wyburn et al. (1965), Rothwell and Solomon (1977), Perry et al. (1978a,b), and Gilbert et al. (1980).

28.2.4 Follicle Atresia

Ovarian follicles that initially begin to grow but fail to reach the fully differentiated stage at which they are ovulated become atretic; the loss of viability and resorption of such follicles occur via apoptosis (Tilly et al., 1991b). Atresia can be passively initiated by the loss of gonadotropin support (e.g., with the initiation of incubation or onset of molt, and following hypophysectomy) or by destruction of the germinal disc region (Yoshimura et al., 1994), or actively initiated via one or more death domain-containing receptors from the tumor necrosis factor (TNF) receptor superfamily (Bridgham et al., 2003). Each of these mechanisms indirectly initiates a cascade of enzymatic activity within the granulosa cell layer consisting of initiator and effector caspases (Johnson and Bridgham, 2002) and endonucleases, with subsequent cleavage of chromatin DNA into internucleosomal fragments. This is followed shortly thereafter by apoptotic cell death within the theca layer (Madekurozwa and Kimaro, 2008). Both 3β-HSD and 17β-HSD enzyme activities have been identified in follicles during the early stage of follicle atresia; however, this activity and consequently all steroid production are rapidly lost as atresia progresses. Atresia occurs with a comparatively high incidence among cortical and primordial follicles and slow-growing (<1–5 mm) follicles, occurs occasionally within the prehierarchal cohort of follicles (6–8 mm), and under normal physiological conditions is rarely found among preovulatory (9–45 mm) hierarchal follicles (the latter sometimes referred to as "bursting atresia").

The virtual absence of atresia in preovulatory follicles is associated with increased resistance to apoptosis within the granulosa layer (Tilly et al., 1991b). This increased resistance results in large part from increased expression of antiapoptotic proteins induced by cAMP signaling. Factors that promote cell viability include B cell lymphoma-1 (Bcl2) members (e.g., BCL2, BCLx, and MCL1), inhibitor of apoptosis proteins (IAPs), Flice-like inhibitory protein (FLIP), and survivin. Included among the factors implicated in mediating apoptosis are members of the BCL2-related family of proteins (BAX and BAD), and FADD-like IL1 beta-converting enzyme (FLICE), the interleukin-converting enzyme (ICE)-related family of enzymes, as well as a select group of proto-oncogenes (e.g., c-MYC) and tumor suppressor genes (e.g., p53). It is significant to note that (1) genes from each of these groups are expressed in developing chicken ovarian follicles, (2) the level of constitutive expression for each gene (apoptosis-inducing or apoptosis-suppressing) is correlated to inherent granulosa cell susceptibility or resistance to cell death, and (3) the level of mRNA expression for such genes is generally regulated by the same physiological factors (e.g., growth factors and gonadotropins) that promote hen follicle viability (Johnson, 2003).

28.2.5 Ovulation and Postovulatory Follicles

The preovulatory surge of LH is a primary stimulus for both germinal vesicle breakdown (e.g., condensation of chromatin, extrusion of the first polar body, and formation of the second maturation spindle) in the F1 follicle, followed by the physical process of ovulation. In avian species studied to date, peak circulating levels of LH typically precede ovulation by 4–6 h. The preovulatory LH surge occurs coincident with increased secretion of progesterone, and studies conducted *in vivo* have demonstrated a stimulatory interaction between LH and progesterone such that in the absence of progesterone secretion, a fully potentiated LH surge fails to occur. In the domestic hen, the largest preovulatory (F1) follicle produces a significantly greater amount of progesterone during the LH surge compared to less mature follicles. This enhanced capacity for progesterone production is related to particularly high levels of STAR protein (which regulates the transport of cholesterol from the outer to inner mitochondrial membrane) expression specifically within the F1 follicle granulosa layer (Johnson and Bridgham, 2001). Unlike mammals, there is no apparent increase in circulating FSH coincident with the preovulatory LH surge in the domestic hen.

Rupture of the F1 follicle and the release of the oocyte at ovulation are facilitated by coordinated events localized along the stigma. These include enzyme activation, physical forces brought about by smooth muscle contraction, and localized apoptosis. Postovulatory follicles contain the granulosa and theca layers that remain after ovulation (Figure 28.1). Immediately following ovulation, steroid production decreases in the postovulatory follicle (within approximately 24 h) due in large part to the loss of gonadotropin-induced signaling via cAMP and the consequent decline in STAR protein expression plus the steroidogenic enzymes, 3β-HSD and 17β-HSD (Dick et al., 1978; Johnson et al., 2002). Structural regression of the postovulatory follicle is associated with an infiltration of immune cells plus the production of cytokines, and it occurs via the process of apoptosis (Sundaresan et al., 2008; Tilly et al., 1991b). Almost complete resorption of the structure occurs over a period of 6–10 days in the chicken, but may take several weeks or even months in the mallard, pheasant, and grouse. In these species, the residual structures provide a means for studying ovulation rates within wild populations. Removal of the most recent postovulatory follicle is reported to abolish nesting behavior and, in particular, the granulosa layer has been demonstrated to be functional in timing oviposition (Gilbert et al., 1978). There is no structure in birds functionally analogous to the corpus luteum of mammals.

28.2.6 Domesticated Hen Ovary as a Model for Human Ovarian Cancer

Recent studies have provided evidence for the domestic hen ovary as an important, nonprimate model for human ovarian cancer. The high rate of ovarian cancers in the laying hen (up to 35% develop the disease by 3.5 years of age) is attributed to the near-daily incidence of ovulation and the prolonged duration of ovulation (up to 365 eggs produced in one year). This results in the development of an epithelial-origin cancer with a number of common pathological features and histological subtypes with the human cancer. In particular, ovarian cancers in both women and the domestic hen produce, with high frequency, copious amounts of ascites fluid and are highly metastatic (Johnson and Giles, 2013).

28.2.7 Reproductive Tract and Sperm Storage Glands

The left oviduct of the domestic hen develops rapidly after 16 weeks of age and becomes fully functional just prior to the onset of egg production (at approximately 20 weeks). The oviduct is suspended within the peritoneal cavity by dorsal and ventral ligaments, and consists of five distinguishable regions: the infundibulum, magnum, isthmus, shell gland, and vagina (Figure 28.4). (For details on general oviduct histology, see Aitken, 1971; Bakst, 1998; Parizzi et al., 2008; on the infundibulum and magnum, see Wyburn et al., 1970; for the isthmus, see Hoffer, 1971; Solomon (1975); and for the shell gland, see Breen and de Bruyn, 1969; Nevalainen (1969); Wyburn et al. (1973).)

Subsequent to ovulation, the ovum is engulfed by the infundibulum (a tissue not directly connected to the ovary), where it resides for approximately 18 min (range: 15–30 min). Occasionally, the ovum fails to be picked up by the infundibulum (an "internal ovulation") and is reabsorbed, usually within 48–72 h. Infundibular activity appears not to be controlled by ovulation *per se*, as foreign objects placed into the abdominal cavity prior to ovulation will also be taken up, and there are reports that entire unovulated follicles may be engulfed and later laid as fully developed eggs. Fertilization of the ovum occurs in the infundibulum, and it is here that the first layer of albumen is produced in the chicken.

The ovum next passes to the largest portion of the oviduct, the magnum (a length of 33 cm in the chicken), where in the mature female the majority of albumen is formed. Estrogen stimulates epithelial stem cells to develop into three

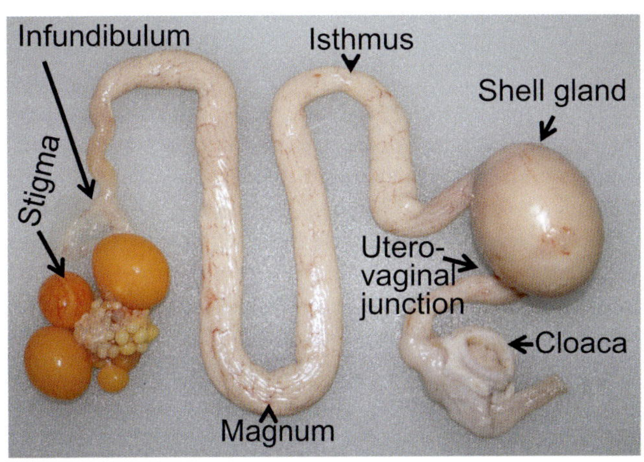

FIGURE 28.4 **Ovary and reproductive tract from the domestic hen.** Note the prominent stigma (site of ovulation) visible on the third largest preovulatory follicle. *Photograph by Dr O. Ocón-Grove.*

morphologically different cell types: tubular gland cells, ciliated cells, and goblet cells. Tubular glands are responsible for the production of ovalbumin, lysozyme, and conalbumin under the stimulatory control of estrogen, while goblet cells synthesize avidin following exposure to progesterone and estrogen (Tuohimaa et al., 1989). A measurable amount of calcium secretion occurs within the magnum, but at no time during the laying cycle is calcium secretion greater than the basal rate of secretion found in the shell gland (Eastin and Spaziani, 1978a). The ovum remains in the magnum approximately 2–3 h.

The isthmus is clearly distinguishable from the magnum. It has a thick circular muscle layer muscle, and glandular tissue is less developed compared to in the magnum. This tissue is characterized by a layer of epithelial cells with underlying tubular gland cells. The production of mucin by glandular tissues within the lower reproductive tract (isthmus, shell gland, and vagina) provides a barrier to prevent invasion by pathogenic agents. Both inner and outer shell membranes are formed during the 1–2 h (mean time: 1 h and 14 min) passage through the isthmus. There is some evidence to suggest that shell formation, particularly the mammillary cores, is initiated in the distal portion of the isthmus.

The shell gland (uterus) is characterized by a prominent longitudinal muscle layer lined medially with both tubular gland and unicellular goblet cells. Maturation of these characteristics, together with the induction of the arginine vasotocin receptor (VT3), is promoted by estrogen. Prior to calcification, the egg takes up salts and approximately 15 g fluid into the albumin from the tubular glands, a process termed "plumping"; this fluid contains carbonic anhydrase, acid phosphatase, and esterase activity, plus bicarbonate and a variety of additional ions (Salevsky and Leach, 1980). The ovum remains in the shell gland for 18–26 h, depending on the cycle length.

Calcification within the shell gland is associated with stimuli initiated by ovulation or by neuroendocrine factors that control and coordinate both ovulation and calcium secretion. Osteopontin (a phosphoprotein) is secreted from epithelial cells of the shell gland and is incorporated within the developing eggshell; its function has been linked to both the structural integrity of the egg and the termination of shell calcification. In addition, expression of the calcium-binding and calcium transport protein, calbindin, occurs at the onset of egg production and is lost with the termination of egg laying. Evidence suggests that distension of the shell gland by the egg is not a stimulus for initiating a high rate of calcium secretion, which is characteristic of calcification, nor is autonomic innervation involved. Calcification of the egg at first occurs slowly, increases to a rate of up to 300 mg/h over a duration of 15 h, then again slows during the last 2 h before oviposition; shell pigments (discussed further in this chapter) are deposited via ciliated cells of the shell gland epithelium during the period from 3 to 0.5 h prior to oviposition.

The vagina is separated from the shell gland by the uterovaginal sphincter muscle and terminates at the urodeum of the cloaca. There are numerous folds of mucosa, which are lined by ciliated and nonciliated cells, and a few tubular mucosal glands that may have a secretory function. The vagina has no role in the formation of the egg but, in coordination with the shell gland, participates in the expulsion of the egg. The cloaca is divided into three compartments, the copradeum (most anterior), urodeum, and proctodeum. During oviposition, a urodeal fold can form a thin membrane across the rectum to prevent feces contamination of the egg. Prolapse of the oviduct (eversion of the vagina and protrusion through the ventral opening) can occur in commercial chickens and turkey hens, and may result in death due to shock. Prolapse has been associated with premature photostimulation, inadequate body size due to early lay, large (often double-yolked) eggs, and obesity.

In birds, spermatozoa are stored in specialized sperm storage tubules (SSTs) located at the uterovaginal junction (the primary site of storage) and infundibulum (a secondary site of storage). It is estimated that only 1–2% of inseminated sperm (only morphologically "normal") enter the SSTs, indicating that some process of selection occurs. The SSTs are lined by a single layer of nonciliated cells and provide a mucosal surface where spermatozoa can remain viable for a period of 7–14 days in the chicken and for up to 10 weeks in the turkey hen; the longer storage time in the turkey is attributed to a greater total number of SSTs. This storage capacity enables a precise synchrony between ovulation and fertilization on a daily basis. Following oviposition of each egg, spermatozoa are released from these tubules and migrate to the infundibulum for fertilization. Uterovaginal junction glands are apparently devoid of innervation and contractile tissue, but possess a well-developed vascular system (Tingari and Lake, 1973; Das et al., 2008; Bakst, 2011). Evidence suggests that spermatozoa fill the uterovaginal glands in a sequential fashion without mixing so that with successive inseminations, sperm from the latest insemination is most likely to fertilize an ovum. Tubular glands are also found in the turkey urodeum and, to a lesser extent, the ventral proctodeum. Bakst and Akuffo (2008) have proposed that sperm that initially accumulated within these glands at insemination may eventually be carried by tubular secretions to vaginal SST.

Both the innate and acquired immune systems are well developed within the oviduct (Ozoe et al., 2009), yet SSTs represent an immuno-privileged site. Overall immunocompetence, including increased expression of defensins, is enhanced at sexual maturation and is supported by adequate levels of estrogen (Anastasiadou et al., 2013). The blood supply to the oviduct and shell gland of the domestic hen has been described by Hodges (1965). For a review of the vasculature to the oviduct of additional avian species, see Gilbert (1979). Blood flow to the shell gland of the hen is increased during the presence of a calcifying egg compared to times when no egg is present. For additional details on blood flow, see Chapter 30.

The oviduct is innervated by both sympathetic and parasympathetic nerves. Innervation of the infundibulum is via the aortic plexus, and the magnum by the aortic and renal plexuses (Hodges, 1974). Sympathetic innervation of the shell gland is via the hypogastric nerve, which is the direct continuation of the aortic plexus. Parasympathetic pelvic nerves, which constitute the left pelvic plexus, arise from the pelvic visceral rami of spinal nerves 30 to 33. Cilia are found along the entire length of the oviduct, and a likely function of these cilia is that of sperm transport. Egg transport is primarily accomplished by contractions of the oviduct; oviduct musculature functions as a stretch receptor, and the mechanical stimulus is produced by the ovum itself (Ariamaa and Talo, 1983). Changes in electrical activity and oviduct motility have been recorded in the magnum, isthmus, and shell gland during the ovulatory cycle, with the greatest frequency of electrical activity and contractions occurring in the shell gland at the time of oviposition (Shimada and Asai, 1978). Both α- and β-adrenergic receptors are present throughout the length of the oviduct and have been shown to affect oviduct motility (Crossley et al., 1980).

28.3 OVARIAN HORMONES

28.3.1 Embryo and Posthatch Ovary

The developing chick embryo begins to express steroidogenic enzymes and synthesize measurable amounts of steroids within the first week of incubation. These steroidogenic enzymes include CYP11A1, 3β-HSD, CYP17, 17β-HSD, and CYP19A. Genetic females selectively express CYP19 and synthesize estradiol-17β. Asymmetric expression of the CYP19 gene in the ovary occurs by day 6.5, when gonadal differentiation is first recognized. There is clear evidence for the local action of several steroids as indicated by ovarian expression of intracellular estrogen receptor α, androgen receptor, and progesterone receptors A and B, plus the membrane progesterone receptor γ, beginning early in embryonic development (Gonzalez-Moran et al., 2013). Expression of both LH and FSH receptors occurs by mid- to late embryonic development. Following hatch, the production of ovarian steroids within the ovarian stroma remains relatively low and nonresponsive to gonadotropins up to the beginning of sexual maturation.

Between the time of hatch and the onset of lay in the domestic hen, there is an early LH peak at about 1 week of age, and a prepubertal LH rise beginning approximately 15 weeks post hatch that is highest about 3 weeks prior to sexual maturity. The hen ovary is unresponsive to stimulation by either mammalian gonadotropins or avian pituitary extracts until 16–18 weeks of age. In contrast, the pituitary becomes less responsive to exogenous gonadotropin-releasing hormone (GnRH) as the onset of puberty approaches (Wilson and Sharp, 1975); this change is probably the result of negative feedback by increasing steroid secretion by the ovary.

Ovarian steroidogenesis can be demonstrated as early as day 3.5 of incubation. Circulating concentrations of estradiol-17β increase 2–3 weeks before the onset of lay, whereas levels of progesterone begin to increase about one week prior to lay, and both steroids initiate the development of the reproductive tract (Williams and Sharp, 1977). Moreover, the surge in estrogen production alters the function of osteoblasts to begin forming medullary bone that serves as a labile source of calcium for impending egg production.

Several protein hormones and growth factors have been described in the immature ovary in addition to AMH (discussed in this chapter). Bone morphogenetic protein 7 (BMP7) is preferentially expressed in the left ovary during early embryo development (Hoshino et al., 2005; Oréal et al., 2002), but it is not clear whether BMP7 represents part of the differentiation mechanism or is simply a result of the process. Other factors reportedly expressed in the immature ovary without any specific function yet ascribed include BMP2, BMP3, BMP4, transforming growth factor beta (TGFβ), growth hormone (and its receptor), GnRH-I, inhibins, and activins (Onagbesan et al., 2004; Johnson et al., 2006; Hrabia et al., 2008).

28.3.2 Mature Ovary

It is now well established that ovarian function is regulated not only by hypothalamic GnRH and pituitary gonadotropins (FSH and LH), but also by a diverse group of endocrines, neuropeptides, growth factors, and cytokines acting in a paracrine and/or autocrine fashion within the mature ovary. These factors have variably been associated with regulating granulosa and theca cell survival and apoptosis, cell proliferation, gonadotropin receptor expression, and steroidogenesis plus the maintenance of the oocyte.

28.3.2.1 Ovarian Stromal Tissue and Slow-Growing Follicles

Prior to sexual maturation, whole, small white (<1–3 mm diameter) follicles produce dehydroepiandrosterone, androstenedione, and estradiol in response to LH *in vitro*. The inability of these follicles to produce progesterone at this stage of development is indicative of steroidogenesis via the Δ5 pathway (Robinson and Etches, 1986). The theca layer from prehierarchal (6–8 mm) follicles, as well as ovarian stromal tissue, is also steroidogenically active primarily via the Δ5 steroidogenic pathway (Kowalski et al., 1991; Levorse and Johnson, 1994). Both LH and FSH, acting via their respective receptors, can initiate cAMP signaling and stimulate steroid production from theca tissue.

By contrast, granulosa cells from prehierarchal (6–8 mm) follicles are incapable of synthesizing progesterone due to the absence of both CYP11A1 and STAR expression. While granulosa cells from prehierarchal follicles express 3β-HSD activity and can actively convert exogenously added pregnenolone to progesterone *in vitro*, these cells fail to produce any significant amount of progesterone *in vivo*. Despite the inability of granulosa to produce detectable levels of sex steroids at this stage of development, these cells express both FSH receptor mRNA and VIP receptor (*VPAC1* and *VPAC2*) mRNAs, but undetectable levels of LH receptor mRNA. FSH receptor mRNA expression within the granulosa layer is maintained at elevated levels by several autocrine and paracrine factors, including TGFβ1, activin A, bone morphogenetic protein-6 (BMP6), and BMP4 (Woods and Johnson, 2005; Kim et al., 2013). Nevertheless, at this stage of development, both the FSH receptor and the VIP receptors (all of which represent G protein-coupled receptors) remain in a desensitized state (e.g., fail to generate cAMP in response to their respective agonist). Significantly, this inhibition of cell signaling has been linked to inhibitory mitogen-activated protein kinase activity (Johnson and Woods, 2009). Several epidermal growth factor receptor ligands (EGFRLs) together with their receptors, ErbB1, ErB2, and ErbB4, are expressed within these follicles. The EGFRLs include epidermal growth factor (EGF), TGFα, betacellulin, heparin-binding EGF-like growth factor, amphiregulin, epiregulin, and neuregulins (Woods et al., 2005). Collectively, these findings are significant in that they help to explain how prehierarchal follicles are maintained in a relatively undifferentiated state *in vivo*, despite a continued exposure to circulating concentrations of FSH and VIP, the latter of which is provided, at least in part, via theca innervation (Johnson et al., 1994).

Kit ligand mRNA levels are highest in granulosa cells from follicles <5 mm, and its expression is increased when cultured with oocyte-conditioned medium; this cytokine is proposed to play a role in regulating the slow growth of small, undifferentiated follicles (Kundu et al., 2012). Additional autocrine and paracrine factors found expressed within slow-growing follicles prior to selection include brain-derived neurotropic factor, growth hormone, growth and differentiation factor-9 (GDF9), inhibins, activins, follistatin, and TNFα (Ahumada-Solórzano et al., 2012; Onagbesan et al., 2009).

28.3.2.2 Hierarchal Follicles

At follicle selection, the granulosa layer acquires FSH and VIP responsiveness (e.g., initiates receptor-mediated cAMP production) and begins the transition from being predominantly FSH dependent (in granulosa cells from follicles ~12 mm in diameter) to becoming primarily LH dependent (in hierarchal follicles >12 mm). Subsequent to follicle selection, active receptor-mediated cAMP signaling within the granulosa layer is directly responsible for the production of CYP11a enzyme and STAR protein to initiate steroidogenesis and LH receptor expression (Johnson and Bridgham, 2001). LH-induced adenylyl cyclase responsiveness and progesterone production continues to increase within the granulosa layer as a follicle approaches the time of ovulation. Progesterone is the primary steroid involved in potentiating the preovulatory surge of LH that precedes ovulation by 4–6 h. By contrast, FSH-induced cAMP formation and steroidogenesis are comparatively low in hierarchal follicles. Fully potentiated steroid production by preovulatory follicles is best described by a three-cell model (Porter et al., 1989) and occurs predominantly via the Δ4 steroidogenic pathway. The granulosa layer produces predominantly progesterone that serves as a precursor for androstenedione and testosterone synthesis within the theca layer, and to a lesser extent by granulosa cells. The granulosa layer does not express CYP19 enzyme, and thus does not synthesize estrogen.

The theca layer also expresses CYP11A activity; however, the predominant steroid product of the theca interna layer is androstenedione, while cells in the theca externa synthesize estrogen. Steroid production by the theca layer is predominantly under the regulatory control of LH acting via the adenylyl cyclase–cAMP second messenger pathway. LH receptor mRNA is expressed in theca tissue of all hierarchal follicles (Johnson et al., 1996). The theca adenylyl cyclase system is highly sensitive to LH, and much less responsive to FSH. There is also evidence for LH-stimulated steroid production occurring within each of the follicle layers via the phosphoinositide 3-kinase–inositol 1,4,5-trisphosphate (IP3) pathway and calcium mobilization. In addition, activation of the protein kinase C second messenger system attenuates LH-induced steroid production by granulosa and theca cells. Included among the potential physiological factors that may act via protein kinase C are growth factors (e.g., TGFα and EGF) and prostaglandins (Tilly and Johnson, 1991).

A primary function of the somatic cell layers (granulosa and theca) is to support the viability of the oocyte and maintain meiotic arrest until ovulation. In recent years, an ever-increasing number of factors have been implicated in directly or indirectly modulating the function of the granulosa and theca layers (see Table 28.1). For instance, inhibin A is produced at the highest levels by the granulosa layer of the four largest preovulatory follicles and influences ovarian function by inhibiting FSH (but not LH) secretion (Johnson PA et al., 1993). By comparison, activin A is produced by the theca layer and increases both LH receptor and FSH receptor expression in cultured granulosa cells (Johnson et al., 2006). BMP15 and GDF9 are preferentially expressed by granulosa cells within the germinal disc, and have been implicated in regulating cell proliferation and steroid production (Elis et al., 2007; Johnson et al., 2005).

TABLE 28.1 Partial List of Factors Reported to Be Expressed and/or to Act within the Avian Ovary

Factor	Proposed Function(s)	Reference(s)
Activins, follistatin, and inhibins	Follicle-stimulating hormone (FSH) secretion; FSH and luteinizing hormone receptor expression	Onagbesan et al. (2004)
Adiponectin	Steroidogenesis (?)	Chabrolle et al. (2007)
Anti-Müllerian hormone	Follicle recruitment; FSH responsiveness	Wojtusik and Johnson (2012)
Arginine vasotocin	Oviposition	Baeyens and Cornett (2006)
Bone morphogenetic proteins (BMPs)		
BMP2, -3, -4, -5, -6, and -7	FSH receptor expression; granulosa differentiation	Onagbesan et al. (2003)
BMP15	Steroidogenesis; cell proliferation	Elis et al. (2007)
Growth and differentiation factor-9	Granulosa cell proliferation	Johnson et al. (2005)
Calcitonin	Follicle maturation (?)	Krzysik-Walker et al. (2007)
Cytokines		
Tumor necrosis factor α	Regulates steroidogenesis; apoptosis	Witty et al. (1996) and Onagbesan et al. (2000)
Interleukins	Postovulatory follicle regression; innate immunity	Sundaresan et al. (2008) and Abdelsalam et al. (2012)
Fpidermal growth factor (EGF) receptor ligands		
EGF, transforming growth factor α, betacellulin, and amphiregulin	Cell differentiation, proliferation, and apoptosis	Woods et al. (2005) and Woods and Johnson (2007)
Fibroblast growth factors	Granulosa cell proliferation	Lin et al. (2012)
Follicle-stimulating hormone	Granulosa cell differentiation	Johnson and Woods (2009) and Bruggeman et al. (2002)
Gonadotropin-inhibitory protein	Follicle differentiation (?)	Bédécarrats et al. (2009)
Ghrelin	Steroidogenesis, cell proliferation, and apoptosis	Sirotkin et al. (2006)
Growth hormone	Modulates steroidogenesis	Ahumada-Solorzano et al. (2012)
Insulin-like growth factors (IGF1 and 2)	Follicle growth and differentiation	Onagbesan et al. (1999)
Luteinizing hormone	Steroidogenesis, ovulation, and cell viability	Johnson and Bridgham (2001)
Melatonin	Gonadotropin responsiveness (?)	Sundaresan et al. (2009)
Neurotropins (nerve growth factor and brain-derived neurotrophic factor)	Steroidogenesis	Jensen and Johnson (2001)
Prolactin	Steroidogenesis	Tabibzadeh et al. (1995)
Prostaglandins (PGE and PGF2α)	Apoptosis; cell proliferation	Li et al. (1997) and Manchanda et al. (2001)
Transforming growth factors (TGFβ1)	FSH receptor expression	Woods and Johnson (2005)
Thyroxine and triodothyronine	Modulates steroidogenesis	Sechman (2013)
Vasoactive intestinal peptide	Granulosa cell differentiation; steroidogenesis	Johnson et al. (1994)
1,25-dihydroxyvitamin D_3	Regulates anti-Müllerian hormone expression	Wojtusik and Johnson (2012)

Sturkie's Avian Physiology, G. Causey Whittow, Ed., Academic Press.

28.4 ENDOCRINE AND PHYSIOLOGIC FACTORS AFFECTING OVULATION AND OVIPOSITION

Ovulation in the domestic hen generally follows an oviposition within 15–75 min except for the first ovulation of a sequence, which is not associated with oviposition. Neither the premature expulsion of the egg (which can be induced with prostaglandins or other agents) nor a delayed oviposition (affected by epinephrine or progesterone) influences the time of ovulation. Ovulation occurs via a rupture along the follicle stigma, and this process likely involves multiple, often redundant factors that include proteolytic enzymes (e.g., collagenase and plasminogen activator), vasoactive substances, and the process of apoptosis.

28.4.1 Ovulation

28.4.1.1 Gonadotropin-Releasing Hormone (LH-Releasing Hormone)

Avian species express two different forms of gonadotropin-releasing hormone (GnRH-I or GnRH-II; see Chapter 30); however, only GnRH-I appears to be directly involved in regulating LH secretion. Injection of GnRH-I induces ovarian steroid production and premature ovulation, whereas *in vivo* administration of GnRH antiserum has been shown to block ovulation. These effects are mediated via the pituitary and LH secretion, and there is no evidence of a direct ovulation-inducing effect of either GnRH moiety within the ovary.

28.4.1.2 Luteinizing Hormone

Plasma concentrations of LH in the domestic hen (and most other birds studied to date) peak 4–6 h prior to ovulation (Johnson and van Tienhoven, 1980). This preovulatory surge provides a direct stimulus for germinal vesicle breakdown and for subsequent ovulation. Some workers have reported an additional peak of LH at 11–14 h prior to ovulation; however, the significance of this second peak has not been established. In addition to these ovulatory–oviposition cycle-related peaks, there occurs a crepuscular (occurring at the onset of darkness) peak of LH, which has a periodicity of 24 h and has been proposed to act as a timing cue for the subsequent preovulatory LH surge (Wilson et al., 1983).

Injection of LH into laying hens almost always increases plasma concentrations of progesterone, estrogens, and androgens; however, the ovulatory response is dependent on the stage within the sequence. For instance, LH treatment 11–14 h prior to the first ovulation of a sequence results in premature ovulation, whereas the same treatment before a midsequence ovulation commonly results in follicle atresia and blocked ovulation (Gilbert et al., 1981). Passive immunization of laying hens with antiserum generated against partially purified chicken LH results in the cessation of ovulation for approximately 5 days and extensive atresia of existing follicles (Sharp et al., 1978).

28.4.1.3 Follicle-Stimulating Hormone

A rise in plasma FSH has been observed 15 h prior to ovulation in the domestic hen (Scanes et al., 1977), and this occurs coincident with an increase in FSH binding to ovarian tissue (Etches and Cheng, 1981). Other researchers find less pronounced cycle-related fluctuations in circulating FSH (Krishnan et al., 1993). A primary role for FSH is related to granulosa cell differentiation and the induction of steroidogenesis in prehierarchal follicle granulosa cells (as discussed in this chapter), but there is little evidence for circulating levels of FSH serving to initiate ovulation. In fact, FSH receptor mRNA expression decreases as preovulatory follicles mature (You et al., 1996). Although injection of recombinant FSH is capable of inducing ovulation in mammals (Tapanainen et al., 1993), this apparently has not been tested in avian species.

28.4.1.4 Ovarian Steroids

Highest concentrations of plasma progesterone are found 4–6 h before ovulation and coincide with the preovulatory LH peak. This increase is predominantly the result of progesterone secretion by the granulosa layer from the largest preovulatory (F1) follicle (Etches, 1990). Systemic and intraventricular injection of progesterone can induce both a preovulatory surge of LH and premature ovulation, while administration of progesterone antiserum prior to the preovulatory surge of progesterone can block ovulation. Evidence suggests that the preovulatory LH surge is preceded by an increase in progesterone in view of the findings that (1) in the absence of the normal preovulatory rise in progesterone (blocked by the steroidogenesis inhibitor, aminoglutethimide), there is no preovulatory increase in plasma LH; and (2) intramuscular injection of progesterone in aminoglutethimide-treated laying hens induces a normal preovulatory LH surge in the absence of any increase in plasma testosterone or estrogen (Johnson and van Tienhoven, 1984). Moreover, results from a study of hypophysectomized hens suggest that progesterone, in the absence of preovulatory gonadotropin, can induce ovulation (Nakada et al., 1994).

Progesterone circulates in the blood bound to a high-affinity corticoid-binding globulin or to albumin and other γ-globulins. Progesterone-specific receptors have been demonstrated in the hypothalamus and pituitary of the domestic hen, and the number of receptors is influenced by the reproductive state. In the oviduct, progesterone receptors are present in surface epithelial cells, tubular gland cells, stromal fibroblasts, and smooth muscle cells in the arterial walls and myometrium (Yoshimura and Bahr, 1991a). These findings provide evidence for a role of progesterone in the production of avidin, contraction of the myometrium, and shell formation. Moreover, progesterone receptors are expressed within granulosa, theca, and germinal epithelial cells of preovulatory follicles (Isola et al., 1987; Yoshimura and Bahr, 1991b). The localization of such receptors associated with the stigma region provides a physiological mechanism by which progesterone may mediate a direct effect on ovulation.

Peak preovulatory concentrations of testosterone occur 6–10 h prior to ovulation, whereas highest levels of 5α-dihydrotestosterone (DHT) occur approximately 6 h before ovulation. The increase in testosterone at this time is the result of secretion from at least the four largest follicles. The role of androgens in ovulation is, as yet, unclear. Intraventricular injections of testosterone fail to release LH or induce premature ovulation, and peripheral injections of testosterone that generally result in unphysiologically high circulating concentrations of testosterone are required to stimulate LH secretion and induce ovulation. Moreover, ovulation can occur in the absence of any preovulatory increase in plasma testosterone. Both granulosa and theca cells of prehierarchal and hierarchal follicles express an androgen receptor, and androgens regulate steroidogenesis via a paracrine and/or autocrine action (Lee and Bahr, 1990; Yoshimura et al., 1993). As in the male, androgens are responsible for the growth and coloring of the comb and wattles at the time of sexual maturation, and they synergize with estrogen to induce medullary ossification.

Both estradiol-17β and estrone show highest plasma concentrations 4–6 h before ovulation with a smaller, more inconsistent rise in estrogens 18–23 h before ovulation. Preovulatory follicle secretion of estradiol increases in each of the four largest follicles 3–6 h prior to ovulation and is greatest in the third and fourth largest follicles. Overall, however, the majority of estrogen produced by the ovary originates from small (1–2 mm) follicles (Robinson and Etches, 1986). Like testosterone, estrogens are unlikely to be involved in the direct induction of LH secretion or ovulation, as injection of estradiol in laying hens either inhibits or has no effect on ovulation or LH secretion. Moreover, ovulation can occur in the absence of a preovulatory increase in plasma estrogens.

Ovarian estrogens have a variety of additional functions related to reproduction, including the regulation of calcium metabolism for shell formation, induction of its own receptor in the oviduct, and induction of progesterone receptor expression in the ovary and reproductive tract. Treatment of immature Japanese quail and young female chickens with estradiol enhances growth of the oviduct and promotes the formation of tubular secretory glands and epithelial differentiation. Moreover, estradiol induces the synthesis of ovalbumin, conalbumin, ovomucoid, and lysozyme in the oviduct and vitellogenin in the liver. Secondary sex characteristics, such as the color and shape of plumage, and sexual behavior, are also under the control of estrogens. Sexual differentiation of the female brain is under the influence of estrogen, as indicated by thc finding that treatment of female quail embryo with an antiestrogen results in behavioral masculinization.

28.4.1.5 Corticosterone

Plasma concentrations of corticosterone display both a daily (photoperiod-related) rhythm and a peak coincident with oviposition. The role of corticoids on the ovary and in the ovulatory process is unclear, as both facilitatory and inhibitory actions have been described. While there is some evidence to suggest that the adrenal gland, via corticosterone secretion, regulates the timing of the preovulatory LH surge (Wilson and Cunningham, 1980), there is no ovulation-related increase in circulating corticosterone. Injection of corticosterone, deoxycorticosterone, or ACTH induces premature ovulation; however, it is unlikely to act at the level of the hypothalamus and/or pituitary to directly induce LH release. In several wild species, seasonal increases of plasma corticosterone are associated with egg laying (e.g., white-crowned sparrow, European starling, and Western meadowlark) or with brooding behavior (e.g., pied flycatcher), whereas in others (e.g., Canada goose) there is apparently little change or a decrease in corticosterone during the breeding season. Absolute plasma levels of corticosterone can be misleading, however, as the capacity of the corticosteroid-binding globulin may also fluctuate with season (Breuner et al., 2003).

28.4.1.6 Prolactin

It is well documented that prolactin is associated with parental behavior and may also play a role in osmoregulation, especially in marine birds; there no clear relationship between circulating prolactin and the LH surge. Prolactin secretion from the anterior pituitary is primarily under the stimulatory control of VIP (see Chapter 30). Circulating levels of prolactin increase at the onset of egg laying through incubation in many species of birds, and prolonged elevated prolactin levels have an antisteroidogenic effect on the ovary, in part via inhibition of steroidogenic enzyme gene expression (Tabibzadeh et al., 1995), and act on the neuroendocrine system to reduce hypothalamic GnRH levels and inhibit LH stimulation of the ovary (Rozenboim et al., 1993). During the ovulation–oviposition cycle of the domestic hen, circulating levels of prolactin are highest approximately 10 h and lowest 6 h prior to ovulation, though it is unlikely that prolactin plays a direct role in the ovulatory process. Prolactin levels are also elevated during the summer months in response to increasing photoperiod. Expression of prolactin receptor mRNA and prolactin binding has been detected in crop sac mucosa, liver, brain, ovary, and oviduct (Tanaka et al., 1992).

28.4.1.7 Other Factors

Preovulatory follicles produce prostaglandins, and the secretion of prostaglandin (PG) F2α by the F1 follicle is greatest around the time of ovulation. PGF2α, PGE2, acetylcholine, and oxytocin increase the contraction of smooth muscles in the connective tissue wall of the follicle *in vitro* and may, in combination with proteolytic enzymes, play a role in the rupture of the stigma at ovulation (Yoshimura et al., 1983). Nevertheless, it is clear that PGs are not obligatory for ovulation to occur as neither the administration of prostaglandins nor the injection of indomethacin into the follicle affects the timing of ovulation.

Gonadotropin-inhibitory hormone (GnIH) was originally described as a factor that suppresses pituitary LH secretion, but both GnIH and GnIH receptor have subsequently been identified in the chicken ovary (Bédécarrats et al., 2009). Ligand binding to the GnIH receptor inhibits cAMP signaling, and it is speculated that GnIH may serve to modulate the growth and differentiation of ovarian follicles.

28.4.2 Oviposition

Expulsion of the egg (oviposition) involves the relaxation of abdominal muscles and sphincter between the shell gland and vagina and the contraction of smooth muscles of the shell gland. The majority of research to elucidate the mechanisms controlling oviposition has been performed in the quail and domestic hen, and has implicated neurohypophyseal hormones, prostaglandins, acetylcholine, galanin, and hormones of the preovulatory and postovulatory follicles.

Administration of oxytocin and arginine vasotocin (both of neurohypophyseal origin) can induce premature oviposition in laying hens. Arginine vasotocin activity in the blood of the laying hen is reported to be highest at the time of oviposition, and its actions are mediated via its receptor, VT3. Release of arginine vasotocin from the neurohypophysis is stimulated by various factors, such as prostaglandins (PGF2α, PGE1, and PGE2), estradiol-17β and progesterone, angiotensin-II, and acetylcholine. Although removal of the neurohypophysis fails to affect the pattern of timing of oviposition, it is likely that an additional source of arginine vasotocin is the preovulatory follicle granulosa and/or theca layer. The oviposition-inducing actions of oxytocin may be mediated via prostaglandins, as administration of indomethacin, a prostaglandin synthesis inhibitor, blocks oxytocin-induced premature oviposition.

There is much evidence suggesting that prostaglandins directly induce uterine (shell gland) contractions, and that this is mediated via extracellular Ca^{++} entry and myosin light chain kinase phosphorylation. Exogenous administration of PGF2α stimulates shell gland contractility, relaxes the vagina, and induces premature oviposition (Shimada and Asai, 1979; Takahashi et al., 2011). In contrast, treatment with indomethacin or aspirin depresses the peak of prostaglandins in the plasma and, in pre- and postovulatory follicles, suppresses uterine contractility and delays oviposition. Injection of PGE1 induces premature oviposition in chickens and Japanese quail, while passive immunization with PGE1 antiserum delays oviposition in the domestic hen. Further evidence that prostaglandins mediate the oviposition inducing activity of arginine vasotocin is that plasma prostaglandins are significantly elevated at the time of arginine vasotocin-induced premature oviposition, and that arginine vasotocin stimulates the biosynthesis and release of prostaglandin from the shell gland.

Endogenous concentrations of prostaglandins E and F increase within ovarian follicles beginning 4–6 h prior to the expected time of oviposition. The highest and next highest concentrations were observed at oviposition in the largest preovulatory follicle and most recent postovulatory follicle (Saito et al., 1987). Increased prostaglandin production and secretion from one or more of these sources is reflected by significantly elevated plasma concentrations of the prostaglandin metabolite 13,14-dihydro-15-keto prostaglandin F2α, coincident with the time of oviposition. Cyclooxygenase, the rate-limiting enzyme of prostaglandin synthesis, is selectively expressed within granulosa cell layer, the cortical interstitium, the ovarian surface epithelium, and within the most recent postovulatory follicle (Hales et al., 2008). Among factors that have been shown to promote prostaglandin production within the granulosa layer are TGFα and TGFβ (Li et al., 1994). Removal of either the most recent postovulatory follicle or the largest preovulatory follicle results in a 1–7 day delay in oviposition (Rothchild and Fraps, 1944); the factor(s) may be a peptide, possibly analogous or identical to the neurohypophyseal hormones or prostaglandins. Despite the fact that the shell gland contractility is influenced by epinephrine and norepinephrine, there is no evidence that these catecholamines influence the timing of oviposition.

28.5 REPRODUCTIVE SEASONALITY, BREEDING, AND OVULATION–OVIPOSITION CYCLES

Seasonal reproduction is regulated according to a circannual rhythm, and this rhythm is phased by an increasing photoperiod that is detected by photoreceptors within the brain. A recent model suggests that photoperiod-induced thyrotropin hormone (TSH) from the pars tuberalis initiates triiodothyronine (T3) production within the mediobasal hypothalamus. In turn, T3 initiates seasonal gonadotropin-releasing hormone production and pituitary gonadotropin secretion (Yoshimura, 2013; for additional details, see Chapter 30 (Hypothalamus/Pituitary)).

Breeding cycles among species of birds can be classified according to the length of the cycle and the time of the year at which each species becomes reproductively active. Continuous breeders, such as the domestic hen or Khaki Campbell duck, are under optimal conditions reproductively active throughout the year, but these are not common. Most wild species that breed in temperate, subarctic, and arctic zones display yearly cycles, while birds adapted to tropical or desert climates may breed with cycles less than a year, at 6-month intervals, or when favorable conditions exist (e.g., rain and food availability). One example of an opportunistic breeder is the zebra finch (*Taeniopygia guttata*), which has adapted to the semiarid and arid regions of Australia. Ovarian growth in the female begins shortly after hatch and this continues throughout the first 3 months posthatch, when it enters an indefinite resting stage. Following an appropriate ultimate cue (e.g., rainfall), follicles are selected into the rapid growth phase and are ovulated within 1 to 2 weeks (Sossinka, 1980). This raises two important questions as to how the ovary maintains an appropriate milieu to support long-term prehierarchal follicle viability (e.g., prevent cellular apoptosis and consequent atresia over a period from weeks to months), and how the presence of ultimate factors becomes translated into appropriate cellular signals to initiate follicle selection into the preovulatory hierarchy (for additional information on opportunistic breeders, see Hau, 2001). Given that the process of reproduction in the majority of birds is closely synchronized with the availability of food and an appropriate climate, it is possible that any significant climate change could negatively impact the timing of migration and the optimal time for reproduction.

Seasonal growth of the ovary is stimulated in almost all species by increasing photoperiod. Such species initially respond to increasing day length with increased hypothalamic secretion of gonadotropin releasing hormone (GnRH), which initiates pituitary gonadotropin (FSH and LH) production and secretion. In some birds (e.g., Japanese quail), an increase in GnRH secretion occurs within the first day of photostimulation. Subsequently, additional stimuli (proximal factors such as quality and abundance of food or nesting sites, temperature, or interactions with a male) are often required to initiate follicular growth and maturation. Nevertheless, there are a few species known to initiate reproductive activity during a season of decreasing or short (<12 hours light) photoperiod (e.g., emperor penguin, *Aptenodytes forsteri*; crossbills, *Loxia* spp.); this is likely a reflection of seasonally available food resources required for raising offspring.

Wild birds generally produce one or more eggs in a clutch and then terminate lay to incubate the eggs. The number of eggs per clutch and total number of clutches vary with species, age of the female and season. For example, some birds, such as the sooty tern (*Sterna fuscata*), lay a single egg to incubate. The king penguin (*Aptenodytes patogonica*) will lay but one egg, and breeding will not necessarily occur on a yearly basis. In contrast, the European partridge will lay a single clutch per year consisting of as many as 12 to 20 eggs. The pigeon usually lays two eggs per clutch and averages eight clutches per year; the interval between clutches is approximately 45 days in the fall and winter and from 30 to 32 days in the spring and early summer. Within a species, the size and number of eggs laid tend to increase with age, and the largest improvement occurs between the first and second breeding seasons. In starlings, this increase in performance subsequent to the first year is associated with a greater amount of vitellogenin production in females that have experienced a prior period of photostimulation (Sockman et al., 2004). In bobwhite quail, clutch size declines with advancing season, from a mean high of 19.2 eggs in early May to 11.3 eggs in late July.

Some species of birds produce a replacement clutch when eggs are destroyed or removed from the nest early enough in the breeding season (examples are the Mallard duck, *Anas platyrhynchos*; Northern Lapwings, *Vanellus vanellus*; and Gentoo Penguins, *Pygoscelis papua*). These species are considered indeterminate layers because they produce a number of developing follicles that is greater than the number of eggs normally laid. If eggs are removed as they become laid, such species will continue to produce eggs well beyond the normal clutch size. The stimulus (or cue) to continue laying due to egg loss may be visual or tactile. Determinate layers fail to lay additional eggs on the removal or destruction of eggs in the nest. In these species, the number of preovulatory follicles generated is equal to the number of eggs ovulated. Examples of determinant layers include Yellow Warblers (*Dendroica petechia*), Red-winged Blackbirds (*Agelaius phoeniceus*), and Pied Flycatchers (*Ficedula hypolueca*). In seasonal (or noncontinuous) breeders, the ovary undergoes periods of growth and regression. The weight of the European starling ovary may vary from 8 mg during the regression phase to 1400 mg at the height of the breeding season.

28.5.1 Ovulation–Oviposition Cycle and Rate of Lay

The ovulation–oviposition cycle (the time from ovulation of an ovum to the oviposition of the calcified egg) of the domestic hen generally ranges from somewhat longer than 24–28 h in length, and ovulations proceed uninterrupted for several days, or as long as 1 year or more in extreme instances. The number of eggs laid on successive days is called a sequence, and each sequence is separated by one or more pause days on which no egg is laid. The term "clutch" is sometimes used synonymously with sequence, although the former term is generally considered more appropriate to describe the group of eggs laid by nondomesticated species prior to incubation behavior. The 24+h ovulation–oviposition cycle is characteristic of the chicken, turkey, and bobwhite quail. By contrast, the interval between successively laid eggs in the pigeon is reported to be 40–44 h and in the Khaki Campbell duck is 23.5–24.5 h.

In the domestic hen, the longer the sequence, the shorter the duration of the ovulation–oviposition cycle. The delay, or "lag", in hours between the oviposition of successive eggs in a sequence is not constant (Table 28.2). The differences in lag time within a sequence represent mainly variation in the amount of time between oviposition and the subsequent ovulation; this interval typically ranges from 15 to 75 min. It is clear that even when the lag between the oviposition of successive eggs approaches 24 h, there remains a progressive shift toward laying the egg later and later in the day. As the sequence becomes shorter, the lag becomes greater and the length of the ovulation oviposition cycle deviates more from 24 h. The total lag time between the first ovulation of a sequence (C1 ovulation) and last ovulation of a sequence (Ct ovulation) is greater in chickens (4–8 h depending on the sequence length) than in Japanese quail (1.5–5 h). The normal release of LH in the domestic hen is restricted to a 4–11 h period (the "open period") beginning at the onset of the scotophase (dark phase). The timing and regulation of the "open period" are as yet incompletely understood. For further details concerning the ovulation–oviposition cycle and the sequence, see Fraps (1955) and Etches (1990).

A chicken typically ovulates the first egg of a sequence early in the photophase, and the timing of ovulation is synchronized by the onset of the scotophase. In comparison, Japanese quail ovulate the first egg of the clutch 8–9 h after the onset of the photophase, and this appears to be synchronized by the timing of the light phase. The "rate of lay" refers to the number of eggs laid in a given period of time, irrespective of the regularity or pattern of laying. For example, a chicken laying at the rate of 50% for 60 days produces 30 eggs, the same number of eggs as another that lays at a rate of 75% for 40 days. On reaching sexual maturity (at ~20–22 weeks of age), the domestic hen lays with an erratic pattern (sequences with two or more pause days, or occasionally more than one egg per day) and with a high incidence of abnormal (soft-shelled and double-yolked) eggs for the first 2 weeks. Six to ten weeks after the onset of egg laying, production peaks (frequently at a rate of 90% or more) and gradually decreases over a period of 40–50 weeks, depending on whether the chicken is a laying or broiler breed. Subsequently, egg production progressively declines and the hen begins a period of molt. A typical laying strain of chicken will produce 280 or more eggs in a 50-week production period.

The physiological mechanisms responsible for regulating ovulation cycles in birds have been the topic of study for decades (Bastian and Zarrow, 1955; Etches and Schoch, 1984; Silver, 1986). Conventional models clearly predict that the primary signals involve multiple oscillatory sites at a central level (likely including the retina, pineal, and medial basal hypothalamus) and a local ovarian clock. It has been demonstrated that clock-related genes (e.g., Per2, Per3, Clock, Cry1, and Bmal1) are expressed in ovarian follicles from Japanese quail (Nakao et al., 2007) and the chicken (Tischkau et al., 2011); levels of clock proteins vary with photoperiod and stage of ovulatory cycle within both the granulosa and thecal layers. Such data provide evidence not only for a functional clock within the avian ovary, but also for the role of this local clock in regulating follicle growth, steroidogenesis, the rate of lay, and the timing of ovulation.

28.5.2 Parthenogenesis

Parthenogenesis (embryonic development from an unfertilized oocyte) has been documented in turkeys; 32–49% of infertile eggs may initiate development, but most embryos

TABLE 28.2 Composition of the Hen's Egg

		Albumen				
	Yolk	Outer	Middle	Inner	Chalaziferous	Shell
Weight (g)	18.7	7.6	18.9	5.5	0.9	6.2
Water (%)	48.7	88.8	87.6	86.4	84.3	1.6
Solids (%)	51.3	11.2	12.4	13.6	15.7	98.4

	All Layers		
	Yolk	Albumen	Shell
Proteins (%)	16.0	10.6	3.3
Carbohydrates (%)	1.0	0.9	–
Fats (%)	32.6	Trace	0.03
Minerals (%)	1.1	0.6	95.10

From Romanoff and Romanoff (1949).

die at an early stage (Olsen, 1975; Cassar et al., 1998). Genetic selection can increase the incidence of parthenogenesis in turkeys, and viable poults are homozygous diploid males that are often sexually competent. Cytologic studies indicate that parthenogenesis in turkeys is initiated from a haploid oocyte and proceeds to the diploid state after a meiosis that is unaccompanied by cytokinesis (mitosis). Parthenogenesis has also been reported to occur in the domestic hen (<5% of eggs, with only a single parthenogenetic adult reported), in Chinese painted quail (*Coturnix chinensis*), and in the zebra finch (*T. guttata*) (Sarvella, 1970; Parker et al., 2012; Schut et al., 2008).

28.5.3 Maternal and Environmental Effects on the Embryo

Almost all birds provide some form of parental care that can be described as female-only care, male-only care, biparental care, cooperative breeding, or brood parasitism (conspecific or heterospecific) (Cockburn, 2006). Perhaps the exception is represented by the Australian brush turkey (*Alectura lathami*; one species from a lineage of galliformes, the megapodites), which uses geothermal heat provided by mounded soil and vegetation for egg incubation. At hatch, the chicks are fully developed and independent from any parental care. Moreover, there is evidence for temperature modification of the sex ratio at hatch in these birds in that a comparatively low incubation temperature (~31 °C) results in a 75:25% male–female ratio, whereas incubation at 34 °C results in approximately 28% male offspring. Many bird species can directly influence embryonic development not only via genetics but also by nongenetic input (e.g., the type and amount of yolk deposition, various hormones, and RNA transcripts). There is additional evidence that the embryo itself can perceive and actively adjust their own development in response to their environment. For instance, some embryos sense and respond to photoperiodic, olfactory, and auditory cues, and some are known to communicate via vocalizations prior to hatch (reviewed in Reed and Clark, 2011).

Several studies provide evidence for maternal modifications of offspring sex ratio based upon the predicted cost and survival of producing one sex or the other. This possibility is afforded largely because the avian female is heterogametic and because the sex-determining division in avian meiosis occurs prior to ovulation and fertilization. Among the mechanisms proposed to influence the sex ratio in birds are epigenetic modifications to bias the segregation of sex chromosomes, sex-specific fertilization, sex-specific incubation, and effects of maternal sex steroids (Rutkowska and Badyaev, 2008; Pike and Petrie, 2003). With respect to the latter, it is suggested that maternal plasma steroid levels may affect preovulatory mechanisms to influence the primary sex ratio (the ratio at fertilization), whereas steroids deposited during egg formation would influence the secondary sex ratio (ratio at hatch) (Goerlich-Jansson et al., 2013). Finally, the amount of maternally deposited progesterone (and/or androgen) in the eggs of the rockhopper penguin (*Eudyptes chrysocome*) is inversely related to body mass in the embryos, chicks, and adults; in turn, body mass is directly related to survivability. Thus, it is proposed that the selective regulation of yolk progesterone deposition may represent a strategy to regulate brood numbers according to food availability and expected survival (Poisbleau et al., 2011).

28.5.4 Photorefractoriness, Broodiness, Molt, and Reproductive Senescence

The final stage of an annual reproductive cycle consists of photorefractoriness, a state where the reproductive system no longer responds to changes in photoperiodic or supplemental cues. Absolute photorefractoriness is analogous to the prepubertal state, and both states are associated with a very low synthesis and content of hypothalamic GnRH and the absence of ovarian support by circulating gonadotropins. In passerine birds, photorefractoriness is programmed early in the reproductive season by the presence of thyroid hormones and a long photoperiod. High circulating levels of prolactin are thought to hasten the onset of photorefractoriness. In starlings, a transition toward photorefractoriness begins when the photoperiod exceeds 12L:12D and corresponds to a time when circulating concentrations of prolactin (PRL) are increasing. Peak concentrations of circulating PRL in starlings coincide with the onset of ovarian regression. This photorefractory state is terminated by a short photoperiod (less than 12L:12D) for some critical, species-dependent interval of time, after which GnRH production and secretion can be renewed. By comparison, relative photorefractoriness, such as occurs in quail, entails the regression of the ovary, but there is no marked decline in hypothalamic GnRH. Thus, recrudescence can occur during continued exposure to this photoperiod (Dawson, 2008). Nonphotoperiodic cues such as temperature, rainfall, social cues, and food availability can affect the timing of the onset of photorefractoriness, but there is no evidence that such cues influence the termination of photorefractoriness (Dawson and Sharp, 2007).

Often occurring coincident with photorefractoriness and regression of the ovary is the process of molt. This represents the replacement of old feathers with new, and is an important but one of the most energetically costly and inefficient endeavors experienced by birds. Accordingly, the timing of molt usually does not overlap with migration or reproduction; nevertheless, there are several examples when molt is initiated during egg laying, incubation, or chick rearing (Williams, 2012). When overlap does occur, molt proceeds over a longer period of time, the growth rate of feathers is significantly reduced, and/or birds devote more time to feeding and less time to feather care. While little is yet known regarding the proximal environmental or physiological control of molt, in starlings there appears to be a clear association with prolactin levels and, specifically, decreasing levels of prolactin after a seasonal peak (Dawson, 2006). Immunization against VIP, the prolactin-releasing hormone in birds, inhibits photoperiodically induced prolactin secretion and molt. In commercial egg-laying flocks, forced molt represents a method to synchronize the termination of lay to maximize the production rate plus size and quality of eggs during a second year of lay. Forced-molting programs generally involve some form of stress such as withdrawing feed for the short term, providing a low-density diet, or creating an imbalance of nutrients.

Broody behavior is induced by elevated prolactin and consists of terminating egg production, initiating the incubation of eggs, and caring for the young. Broody females exhibit increased body temperature, reduced food and water consumption, plus aggressive and defensive behaviors. The onset of incubation occurs coincident with regression of the ovary and accessory reproductive tissues such as the oviduct and comb. In present-day commercial egg-producing hens, broodiness is virtually nonexistent (especially in Mediterranean breeds such as the white leghorn), but still exists in the bantam hen and broiler breeder, and is common in turkeys and geese. Broody behavior may be accompanied by the development of an incubation (or brood) patch, and the increased vascularity, edema, and thickening of the epidermis occur under the regulation of estrogen and prolactin. Incubation activity stimulates the development of the crop sac in the pigeon; proliferation of this gland is under the direct control of prolactin. In the ring dove, endogenous estrogen production, stimulated by FSH secretion, induces the female to nest build. In turn, the initiation of nest-building activity in the male is dependent on the hormonal environment and behavior of the female.

As occurs with photorefractoriness, the regulation of prolactin gene expression and secretion with broodiness occurs primarily via hypothalamic VIP. Induction of incubation behavior in turkey hens can be initiated by administration of prolactin, and active immunization against prolactin reduced the incidence or delayed the onset of broodiness in turkeys and bantam hens (March et al., 1994; El Halawani et al., 1995). VIP secretion is promoted by hypothalamic dopamine via a stimulatory dopamine receptor (DRD1). Dopamine can also inhibit the secretion of prolactin via an inhibitory receptor (DRD2) at the level of the pituitary (Youngren et al., 2002). Treatment of hens with a dopamine receptor antagonist resulted in the termination of broodiness by inhibiting secretion of prolactin. Behavioral broodiness in the female is preceded by decreases in circulating concentrations of LH, progesterone, testosterone, and estradiol, while growth hormone tends to be lower only while hens are caring for the young. Beginning with and during the period of lay, a basal production of prolactin is maintained by ovarian estrogens. With the initiation of broodiness, the circulating concentrations of prolactin are increased dramatically and remain elevated throughout incubation. This elevation in prolactin is accomplished in part by a transdifferentiation of pituitary somatotropes (growth hormone secreting cells) into cells that produce both prolactin and growth hormone (mammosomatotropes). Among species in which the male assumes the predominant role of incubation (e.g., Wilson's phalarope, *Phalaropus tricolor*), levels of prolactin are higher in the male than in the female (Oring et al., 1988). For further information, see Sharp (2009).

Evolutionary theories predict that any decline in reproductive performance with age (e.g., reproductive senescence) is related not only to environmental conditions to which the species or individuals within a species are exposed, but also to the rate of lay early in life. Thus, both species and individuals within a species that invest the most in early reproduction are most likely to undergo early senescence and also die sooner (Reed et al., 2008). Reproductive aging can be defined as any decline in reproductive performance, and includes a decrease in fitness of the offspring. By comparison, ovarian senescence constitutes a significant decline or loss of the primary follicle reserve within the ovary, and/or the significant loss of oocyte quality and capacity for fertilization. Reproductive senescence and the gradual decline in egg production in the domestic hen are associated with an increase in atresia of slow-growing follicles, plus a decrease in the rate of follicle selection into the preovulatory hierarchy (Waddington et al., 1985).

Holmes et al. (2003) have described female reproductive aging in birds according to three paradigms. The first is characterized by a relatively short life span with moderately rapid declines in fertility (e.g., many Galliformes, including chicken, quail, turkey, and pheasant). A second model consists of moderately slow-aging birds with a somewhat longer life span where the slow decline in reproductive success correlates closely with increased mortality (e.g., passerine songbirds, including the great tit, *P. major*, and small raptors like the European sparrow hawk, *Accipiter nisus*). The third model is characterized by long life spans, slow aging, and negligible reproductive declines in either sex. For instance, neither the common tern (*Sterna hirundo*; Nisbet et al., 1999) nor the Nazca boobie (*Sula granti*; Anderson and Apanius, 2003) appears to show an appreciable decline in reproductive effort even after 20 years of reproductive activity. Some terns and gulls have been reported to actually increase the number of offspring they successfully fledge with advancing age, although this may be more related to posthatch survival and parental behaviors than to enhanced ovarian fecundity. On the other hand, some species within the order Psittaciformes (parrots and cockatoos, which represent some of the longest living birds) may breed for up to two decades, but they also live for many years past the termination of breeding (Young et al., 2011). These proposed models, although not yet rigorously tested, raise important questions regarding the role of the ovarian environment in promoting long-term germ cell survival and viability.

28.6 COMPOSITION AND FORMATION OF THE YOLK, ALBUMEN, ORGANIC MATRIX, AND SHELL

The avian amniotic egg is unique among vertebrates and, not unexpectedly, expresses several genes that have no apparent orthologs within nonavian lineages. These include the eggshell proteins, ovocalyxin-36 (a proposed pattern recognition molecule that contributes to innate immunity), ovocleidin-17, and ovocleidin-116 (a novel eggshell matrix protein), plus two unique ovalbumin proteins (ovalbumin X and Y) resulting from gene duplication (Tian et al., 2010). Components of the avian egg include the yolk, the albumen, the organic matrix, and the crystalline shell. Each component has been described in reviews by a number of workers, and the reader is referred to the following literature for more details: Gilbert (1971a,b), Simkiss and Taylor (1971), Hincke et al. (2008), and Solomon (2010). There appears to be some uniformity among many species of birds as to the density of yolk and albumen and the proportion of egg yolk relative to egg weight. However, variations do occur in the lipid, amino acid, and carbohydrate content of albumen and yolk, with precocial species (e.g., from the orders Anseriformes and Galliformes) having a higher energy density than altricial species (e.g., the pigeon, *Columba livia*, and pelican, *Pelecanus occidentalis*) (Roca et al., 1984; Bucher, 1987). Many of the egg membrane and shell matrix–related proteins (e.g., collagen, carbonic anhydrase, calbindin, vitamin D receptor, and osteopontin) are formed specifically in the eggshell gland or are present there in high concentrations (Arias and Fernandez, 2001; Soledad Fernandez et al., 2001).

The calcified eggshell is a unique mineralized structure that protects the growing embryo from dehydration, physical trauma, and infection by microorganisms, and it allows gaseous exchange while buffering against temperature fluctuations to maintain appropriate developmental conditions for the embryo. In addition to these functions, the eggshell contributes to embryo growth and development by acting as the major source of calcium; by late incubation, approximately 80% of the chick's calcium is derived from the eggshell. The two most likely routes of egg contamination by bacteria originate before shell formation from an infection of reproductive organs, or before and/or after oviposition via the penetration of pores in the eggshell; however, there is no apparent correlation between pore number and incidence of contamination.

28.6.1 Yolk

Egg yolk serves to provide the lipids and many, varied proteins required for embryonic growth. It is composed of a clear, yellow fluid (plasma) and suspended particles (granules containing lipovitellin and phosvitin, both derived from vitellogenin). The final composition of yolk in the hen's egg consists of approximately 36% lipid (by wet weight) compared to 17% protein. The greatest proportion of yolk is water (~50%) with lesser amounts of free carbohydrates (~1%) and inorganic elements (1–2%) (Anton, 2007). Of the major classes of lipids, approximately 70% are triacylglycerols, 25% are phospholipids, and 5% are cholesterol and cholesterol esters. Several antioxidants (e.g., carotenoids and vitamins) are also abundant in the egg yolk. Yolk is usually deposited in concentric bands that result from a nonuniform pattern of daily feed intake. One component responsible for conveying passive immunity to the developing embryo is the avian-specific antibody, IgY. Recently, a chicken yolk sac IgY receptor was characterized (FcRY, a member of the mannose receptor family); this receptor is responsible for selective IgY transport from yolk to the embryonic circulation (Tesar et al., 2008).

28.6.2 Albumen

There are four distinct layers of albumen in the fully formed egg: (1) the chalaziferous (inner thick) layer attached to the yolk, (2) the inner thin (liquid) layer, (3) the outer thick layer, and (4) the outer thin (fluid) layer. Approximately one-fourth of the total albumen by weight is found in the outer thin layer, and slightly greater than one-half in the outer thick layer. The inner layer represents ~17% of the total albumin, while the chalaziferous layer plus chalazae provides 2.7% of the total. One function of the egg albumen layer is to prevent invasion of microorganisms into the yolk (as discussed here), as well as to serve as a source of water, protein, and minerals to the embryo during development. The chalazae are two fiber-like structures at each end of the egg that permit a limited amount of rotation but little lateral displacement of the yolk.

In the domestic hen, the initial albumen layer is deposited by the caudal region of the infundibulum. The majority of albumen proteins are deposited by tubular gland cells of the magnum, while avidin is synthesized by the goblet cells under the influence of progesterone and estrogen. The discrete layers of albumen are the result of consecutive deposition either by different regions of the magnum or by changes that occur within the shell gland and with movement of the egg through the oviduct.

The major proteins found in albumen include (1) ovalbumen, 54% of total the albumen; (2) ovotransferrin (binds iron), 13%; (3) ovomucoid, 11%; (4) ovoglobulins (type G2 and G3), 8%; (5) lysozyme, 3.5%; and (6) α and β ovomucin, 1.5–3.0%. There are also several characteristic proteins present in lesser concentrations, including avidin (approximately 0.05% of total protein), flavoprotein- and thiamine-binding proteins, ovomacroglobulin, and cystatin. Ovalbumen, synthesized in the laying hen oviduct, serves as a source of amino acids for the embryo during development, and, based on its homology to the serpin family of protease inhibitors, it may also suppress enzyme activity in the egg. Ovotransferrin acts primarily as an iron chelator, thereby helping to prevent bacterial growth within the egg. Ovomucin is a serine protease inhibitor and represents the major inhibitor of protease (primarily trypsin) activity in the egg; targeted enzymes may be produced by invading microorganisms. Alternatively, ovomucoid may serve to modulate the enzymatic degradation of albumin during embryo development.

Alpha-ovomucin (a glycoprotein) and β ovomucin (60% carbohydrate by weight) are insoluble, fibrous proteins that are responsible for the gel-like qualities of egg white, particularly in the thick layers. Ovomucins block the invasion of microorganisms, and they may express antiviral properties. The main biologic function of egg albumen lysozyme is lytic activity against gram-negative bacterial cell walls.

Several egg proteins are known for their ability to bind specific substrates (e.g., ovotransferrin binds iron, copper, and zinc; flavoprotein binds riboflavin; and avidin binds biotin and iron), and others act as protease inhibitors (ovomucoid, ovoinhibitor, cystatin, and ovomacroglobulin). Avidin is a progesterone-dependent secretory protein; its biologic role in the egg is likely related to its ability to sequester essential nutrients from bacteria by chelating minerals and inhibiting bacterial growth. Cystatin has antimicrobial and antiviral activities via its ability to inactivate proteases necessary for microbial metabolism or invasion of host tissues. While hens carrying a load of pathogenic bacteria will actively increase the export IgY to the egg yolk, it appears that antimicrobial proteins and peptides (e.g., defensins) are constitutively transferred from the oviduct to the albumen plus eggshell membranes and the eggshell; such proteins are active against a wide range of microbes (Bedrani et al., 2013).

28.6.3 Organic Matrix

The organic fraction of the eggshell consists of shell membranes, the mammillary cores, the shell matrix, and the cuticle. Although these components constitute only a small fraction of the entire eggshell, their integrity is critical to its formation and strength.

28.6.3.1 Shell Membranes

Shell membranes are organized into an inner membrane (~50–70 μm thick; remains uncalcified) and outer membrane (~15–25 μm; penetrates the adjacent mammillary cones of the calcified shell), and they are produced by the proximal isthmus region of the oviduct over a 1–2 h period. These membranes interface with the chorioallantoic membrane during embryonic development (Figure 28.5). The proteinaceous, disulfide-rich membranes lie adjacent to one another except at one end, where they part to form the air cell. There is some question as to whether epithelial cells or tubular gland cells of the isthmus are responsible for the secretion of the membranes. The membranes consist of a meshwork of protein fibers, cross-linked by disulfide and lysine-derived bonds, with small fibrous protuberances of uncertain function. The membranes are composed of type I, V, and X collagens (10%). The remainder of the fibrous component is composed of protein (70–75%) and glycoprotein (Leach, 1982). The membranes are semipermeable and permit the passage of gases, water, and crystalloids but not albumen. There is no relationship between thickness of the membranes and thickness of the shell, but membrane thickness does decrease with the age of the hen. Shell membrane thickness also decreases during incubation, which allows for increased water and oxygen permeability.

28.6.3.2 Mammillary Cores

The mammillary cores, which are projections from the outer membrane surface (Figure 28.5), represent the initial sites of calcification. They are composed largely of protein, but they also contain carbohydrate and mucopolysaccharides, and are formed by the epithelial cells of the isthmus. The mammillary cores represent the greatest proportion of organic material in the eggshell. Different species of birds have evolved eggs with shells containing different mammillary densities that meet calcium requirements imposed by different growth rates and modes of development. Precocial bird species grow slowly but have high mammillary density, while altricial bird species grow rapidly but have low mammillary density (Osterström and Lilja, 2012).

28.6.3.3 Shell Matrix

The organic shell matrix is a series of layers of proteins (glycoproteins and proteoglycans) plus acid mucopolysaccharide on which calcification takes place. The nonmineral components represent approximately 2% of the total organic composition of the eggshell. The matrix plus calcified crystals make up the outer palisade layer of the shell. The innermost region of the shell has a greater density of matrix compared to the outer regions. Both calcium-binding protein and carbonic anhydrase have been identified within the matrix. Deposition of the matrix occurs soon after the egg reaches the shell gland.

28.6.3.4 Cuticle

The outermost surface of the egg is often (but not always) covered by a thin waxy cuticle composed of polysaccharide,

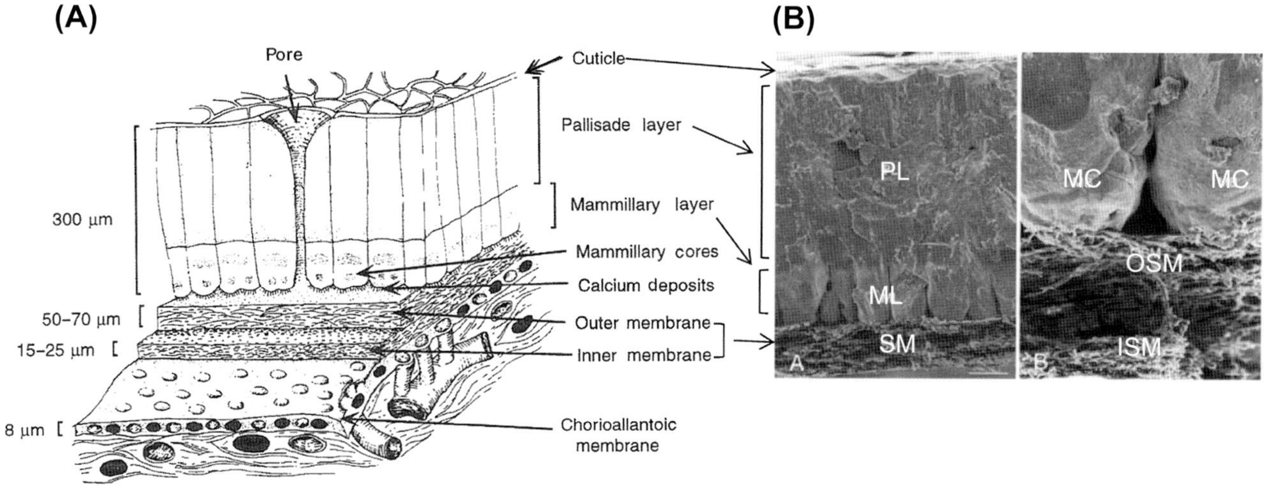

FIGURE 28.5 **(A) Diagram of a section through the shell and membranes of a hen's egg, showing the outer crystalline structure and the inner organic material that remains after calcium carbonate has been dissolved.** *(Adapted from Rahn et al. (1979). Copyright 1979 by Scientific American, Inc. All rights reserved.)* **(B) Scanning electron micrographs of a shell cross-section. ISM, inner shell membrane; OSM, outer shell membrane; MC, mammillary core; ML, mammillary layer; PL, palisade layer.** *From Hincke et al. (2000).*

lipid, and protein. Among the 47 different proteins identified are those with known antimicrobial activity (e.g., lysozyme C, ovotransferrin, ovocalyxin-32, cystatin, and ovoinhibitor) (Rose-Martel et al., 2012). The cuticle may be unevenly distributed over the entire surface, ranging in depth from 0.5 to 12.8 μm; the average dry weight of the cuticle in a 60 g chicken egg is 12 mg. Its function is primarily to protect the egg from water evaporation and microbial invasion; it is unlikely to add to the structural integrity of the shell. The cuticle, when present, is formed during the last 30 min prior to oviposition.

28.6.4 Layers of Crystallization

The eggshell is composed of three calcified layers; the mammillary, palisade, and vertical layers. The amount of calcium deposited in each egg typically represents about 10% of the body stores of calcium of the bird. Endogenous stores of calcium in females from smaller bird species may not be sufficient to meet calcium demand during egg laying, and thus they may be forced to forage specifically for calcium-rich food items (e.g., snails) during egg production and when providing nourishment for the nestlings (Wilkin et al., 2009).

The calcified portion of the shell can be arbitrarily divided into the mammillary knob layer, the palisade layer, and the outer surface, the vertical crystal layer (Figure 28.5). Collectively, these layers represent the major portion of the avian eggshell, are largely responsible for its mechanical strength, and consist of approximately 97% inorganic material. The degree of elasticity within the shell reflects interactions between the inorganic and protein components. While calcium is the predominant cation, the shell also sequesters a significant amount of magnesium in the form of magnesium carbonate; manganese is likely required for the development of the mammillae because of its role in the synthesis of mucopolysaccharides (Leach and Gross, 1983; reviewed by Hincke et al., 2012).

28.6.4.1 Mammillary Knob Layer

Outward crystallization of the mammillary cores results in the formation of the mammillary knob layer; some crystals grow inward to fuse with the outer egg membrane. In addition, calcium deposits radiate from the bottom of the mammillary cores to penetrate the outer shell membrane. Formation of the mammillary knobs occurs in the shell gland during the first 5 h of calcification. The passage of water during plumping stretches the shell membranes and increases the distance between the tips of the mammillae. Crystals that form laterally grow and eventually abut with crystals from other mammillae, whereas those that grow outward may extend to the shell surface. At some points, the crystals do not grow completely together, leaving pores with a diameter of 0.3–0.9 μm.

28.6.4.2 Palisade Layer

The crystallized palisade (or spongy) layer (approximately 200 μm thick) is composed mainly of crystalline calcium carbonate in the form of calcite and represents the greatest portion of the shell (Solomon, 2010). Calcification of this layer is initiated 5–6 h after the entrance of the egg into the shell gland, during the process of plumping. The palisade layer is arranged as columns situated directly over mammillary knobs that are perpendicular to the surface. As calcification proceeds, adjacent columns fuse to enhance the structural integrity.

28.6.4.3 Surface Crystal Layer

The outermost layer of calcification is designated the surface crystal layer. The crystalline structure is denser than that of the palisade region and lies perpendicular to the shell surface. The overall thickness of this layer ranges from 3 to 8 μm. The slow phase of shell formation corresponds to the formation of the mammillary layer, while the fast phase represents growth of the palisade and surface crystal layers.

28.6.4.4 Respiration via the Eggshell

Shell pores within the egg are simple funnel-shaped openings that arise at the shell surface and protrude through to the mammillary knob layer; they occupy approximately 0.02% of the eggshell surface. In most species, pores traverse the shell radially and branch longitudinally along the axis of the egg. The pores are the result of areas of incomplete crystallization. The number of pores (ranging from 7000 to 17,000) is generally related to metabolic demand prior to the initiation of lung function within the embryo; the number of pores per unit area decreases with increasing egg weight. The function of eggshell pores is to serve as a mechanism of chemical communication between the air cell of the egg and the external environment during embryo development (Tullett, 1984; Rahn and Paganelli, 1990). The exchange of oxygen, carbon dioxide, and water occurs predominantly via passive diffusion, and a 79 g turkey egg is estimated to exchange the equivalent of 271 of gas (oxygen, carbon dioxide, and water vapor) during the 28-day incubation period (Rahn et al., 1987). The average diffusive water loss during incubation for 117 species of birds is 15%; the loss is determined by the difference in water vapor pressure between the egg and nest environment. Inner and outer shell membranes play a minor role in limiting the flow of oxygen, carbon dioxide, and water vapor, while the major resistance to passage of these substances comes from the inner end of the pore. Therefore, air cell gas tensions are determined predominantly by the number and size of the pores and thickness of the shell relative to the metabolic rate of the embryo. The total area of pores on a turkey egg with a surface area of $90\,cm^2$ is calculated to be $2.2\,mm^2$.

Almost all avian eggs require turning during the process of incubation. Failure to do so prevents an even reduction of shell membrane thickness that reduces oxygen uptake and water permeability; in turn, this results in both decreased hatchability and delayed hatching. The elliptical shape of the avian egg provides for an air chamber at the blunt end that increases in volume as incubation progresses. There is a progressive increase in gas exchange and in pore density from the sharp end to the blunt end of the shell. The asymmetrical shape assures that the sharp end faces down during incubation to promote maximal gas exchange at the blunt end (Mao et al., 2007).

28.6.4.5 Eggshell Pigmentation

Egg coloration has been proposed to function as a mechanism to camouflage, mimic host eggs, filter solar radiation, or even indicate a female's genetic fitness and egg quality. Eggshell pigments include protoporphyrin (brown eggs; a pro-oxidant synthesized by shell gland epithelial cells), bilirubin (yellowish), and biliverdin (blue–green). In particular, the latter two are blood-derived bile pigments or can be synthesized in the shell gland, and both possess strong antioxidant activities. It has been suggested that eggshell color intensity (positively related to the concentration and amount of carotenoids) may serve as an indication of egg viability and subsequent fitness of the offspring. The synthesis and accumulation of pigment in the shell gland appear to be stimulated by progesterone.

28.6.5 Calcium Availability

Calcium requirements are increased dramatically both at the initiation of ovarian activity during puberty to promote medullary bone formation, and subsequently during each ovulatory cycle. The shell gland transports 2.0–2.5 g of calcium within a period of 11–15 h for the calcification of a single egg. A commercial hen that lays 280 eggs in a production year will use a quantity of calcium for the purpose of shell formation corresponding to 30 times the calcium content of the entire body. In general, calcium transport and homeostasis related to eggshell calcification are regulated by parathyroid hormone, calcitonin, and 1,25 dihydroxyvitamin D_3, and are mediated via at least five groups of proteins (calbindins, Na^+–Ca^{2+} exchangers, membrane localized calcium-ATPases, epithelial calcium channels, and tight junction proteins; this latter group is required for paracellular transport of calcium). A detailed description concerning mechanisms of absorption and changes in uptake by the intestine and shell gland relative to age and egg production are beyond the scope of the present discussion, and the reader is referred to Chapter 30 (on digestive tract) and Bar (2009).

28.6.5.1 Sources of Eggshell Calcium

Calcium for eggshell formation is provided via the blood following absorption from the intestine (duodenum and upper jejunum) or resorption from bone (principally medullary bone, but also cortical bone under conditions of calcium deficiency). Significantly, Ca^{2+} absorption is noticeably increased specifically during the period of shell calcification, and the mechanisms mediating this rhythmicity appear to be vitamin D related. Resorption of calcium from bone is regulated by both parathyroid hormone and 1,25-dihydroxyvitamin D_3, while absorption through the gut is facilitated by 1,25-dihydroxyvitamin D_3 only. The relative importance of the two organs as sources of calcium depends on the concentration of dietary calcium. Laying hens consume approximately 25% more feed on days when eggshell formation occurs than on days when it does not. When the recommended concentrations of calcium in the food are 3.6–4.0%, a majority of the eggshell calcium can be derived directly from the intestine. Yet increased intestinal absorption during shell calcification is not sufficient to satisfy the high Ca^{2+} requirement for shell calcification; thus, as much as 20–40% of the shell calcium is derived from bone. The bone reservoir is subsequently replenished with Ca^{2+} from the intestinal source during the periods of eggshell gland inactivity. It is likely that these relationships vary depending on the time of day. When hens are provided constant free access to feed, much of the daily intake occurs early in the photoperiod; the remainder is consumed at the end of the photoperiod. However, much of the shell is formed during the night, when generally no calcium is consumed and when the calcium content of the digestive tract is gradually decreasing. Therefore, medullary bone is the primary source of shell calcium during the latter hours of darkness.

Blood calcium circulates in two forms, as nondiffusible protein bound calcium and as diffusible ionized calcium. Nondiffusible calcium is bound by plasma calcium–binding proteins, vitellogenin and albumin. Estrogen treatment increases total plasma calcium, in part by stimulating the production of blood calcium–binding proteins. Similarly, total plasma calcium increases several weeks before laying, and the increase is attributable to an increase in the protein-bound, but not in the diffusible, fraction. During the ovulation–oviposition cycle, concentrations of ionized calcium peak (0.057 mg/mL) 4 h after oviposition, then decrease significantly during the period of shell calcification (minimum of 0.049 mg/mL). On the other hand, plasma concentrations of total calcium (ionized plus inionized) fluctuate only minimally (between 0.2 and 0.26 mg/mL) during the ovulation–oviposition cycle.

Feeding a calcium-deficient diet to laying hens causes a significant decrease in plasma ionized-calcium concentrations, a significant decrease or complete cessation of lay, and regression of the ovary within 6–9 days. However, birds on the same diet injected with crude chicken pituitary extracts continued to lay for a more prolonged period, despite a comparable decrease in ionized calcium (Luck and Scanes, 1979a). Basal concentrations of LH in birds fed a calcium-deficient diet (0.2% Ca) are significantly lower compared to birds on a calcium-rich diet (3.0% Ca), but between the two groups the LH-releasing activity of exogenously administered GNRH is similar (Luck and Scanes, 1979b). These results suggest that pituitary LH synthesis and release, and the ability of the ovary to respond to gonadotropins, are resistant to a calcium deficiency. Mongin and Sauveurl (1974) determined that voluntary calcium consumption by laying hens increases at the time the egg enters the shell gland, and given a choice between a calcium-deficient diet (1.0% Ca) and a diet supplemented with calcium-rich oyster shell, hens will preferentially consume the calcium-rich diet during the period of eggshell calcification. (For a recent review, see Hincke et al., 2012.)

28.6.5.2 Vitamin D

Vitamin D plays an important role in the regulation of calcium metabolism via the metabolically active metabolite, 1,25-dihydroxyvitamin D_3. Conversion to this form occurs by the 1-hydroxylation of 25-hydroxyvitamin D_3 in the kidney under the hormonal control of estradiol and parathyroid hormone. For instance, renal 1-hydroxylase activity increases just prior to the initiation of egg laying, at a time corresponding with an increase in circulating estrogen and total plasma calcium. In addition, 1-hydroxylase activity during the ovulation–oviposition cycle increases at the time of ovulation. The increase in activity is followed by an increase in circulating concentrations of 1,25-dihydroxyvitamin D_3 4 h after ovulation, and elevated levels persist until 10 h after ovulation or after the initiation of eggshell formation. Among the proteins induced by 1,25-dihydroxyvitamin D_3 is osteopontin, a multifunctional, highly phosphorylated protein expressed by epithelial cells of the isthmus and shell gland specifically during the period of egg calcification. Once secreted, osteopontin is found in nonmineralized shell membrane fibers, the mamillary cores, and the outermost part of the palisade (Fernandez et al., 2003). For additional information regarding vitamin D and calcium metabolism, see Bar (2008).

28.6.5.3 Calcium Mobilization from Medullary Bone

In the long bones (e.g., the humerus, femur, and tibia), medullary bone lines the endosteal surface and grows with a system of interconnecting spicules that may completely fill the narrow spaces. Medullary bone forms in female birds during the final 10 days before egg laying under the influence of primarily estrogen, which stimulates osteoblast function. This type of bone is unique to birds and crocodilians, and it can be readily induced in intact male birds by administration of estrogen or in castrated males by estrogen and testosterone. Gonadal steroids apparently act directly on osteoblasts in the medullary cavity, and independently of calcium intake. In addition, it is unlikely that 1,25-dihydroxyvitamin D_3 mediates the transfer of calcium to medullary bone at this time, as the appearance of this bone occurs 1–2 weeks prior to the increase in renal 1 hydroxylase activity and the elevation of plasma total calcium.

During the ovulation–oviposition cycle, periods of intense medullary bone formation alternate with periods of severe bone depletion. Commercial hens fed a high-calcium diet are generally able to replenish the calcium lost from medullary bone during shell calcification when shell formation is not taking place, but on a low-calcium diet the cortical bone of the femur is eroded, while medullary bone is maintained in a fairly constant amount. Under these conditions, the new medullary bone that forms is only partially calcified, and an increase in the number of osteoblasts is indicative of a more rapid turnover rate. Accordingly, continuous lay over a prolonged period of time is associated with increased susceptibility to osteoporosis. At the termination of lay, medullary bone is gradually lost and the formation of new structural bone is initiated. For additional information, see Whitehead (2004).

28.6.5.4 Calcium Absorption and Secretion by the Shell Gland

Our understanding of mechanisms mediating calcium transfer across the shell gland, as well as the hormonal regulation of the process, is still incomplete. The presence of Ca^{2+}-ATPase and calbindin, two critical components of the transcellular transport pathway, provide support in which Ca^{2+} is transported via both a transcellular pathway and an alternative or complementary paracellular passive transport pathway (Bar, 2009). As discussed in this chapter, the initiation of egg production is associated with increased 1,25-dihydroxyvitamin D_3 synthesis and accumulation in the intestine and shell gland, and calcium transport as well as concentrations of the calcium-binding protein, calbindin-D28K, increase in both organ systems during egg production. Calbindin is expressed in tubular gland cells of the shell gland and in the distal portion of the isthmus, but apparently not in the proximal isthmus or magnum. Calbindin synthesis in the intestine is regulated primarily by 1,25-dihydroxyvitamin D_3, yet its production in the shell gland apparently requires ovarian steroids and other, as yet unidentified, factors in addition to 1,25-dihydroxyvitamin D_3 (Bar et al., 1990). While levels of calbindin mRNA increase during the ovulatory cycle at a time coincident with shell calcification, there is little to no change in the tissue levels of calbindin protein; such results indicate posttranslational control of calbindin levels (Nys et al., 1989).

Calcium secretion from the shell gland increases approximately 7 h after ovulation, reaches a maximum as the shell is being formed, and decreases to the basal secretion rate after shell formation is complete but before the expulsion of the egg. The presence of an egg within the shell gland is not likely to be the major stimulus for the initiation of calcium secretion but appears to be more closely related to ovulation. The hormonal signals that affect changes in the rate of calcium secretion are unknown, although estrogen involvement has been suggested (Eastin and Spaziani, 1978a). Similarly, termination of a high rate of calcium secretion is not related to egg removal, as calcium deposition decreases fully 2 h before eggs are laid. Movement of calcium across the shell gland, which may occur by both diffusion and active transport (Eastin and Spaziani, 1978b), involves expenditure of metabolic energy. In the aged hen, calcium deposition and eggshell weight and density are related to shell gland levels of calbindin (Bar et al., 1992).

28.6.5.5 Carbonate Formation and Deposition

The mineral content of the eggshell consists of about 97–98% calcium carbonate, while the remainder consists of magnesium carbonate and tricalcium phosphate. Carbonate in the shell is derived from blood bicarbonate or is synthesized from CO_2 by carbonic anhydrase located in oviduct tissue and the shell. In addition, luminal carbonate ion content is supplemented by bicarbonate in fluid flowing from the magnum to the shell gland. The secretion from the shell gland of HCO_3, like that of calcium, is accomplished by active transport and may be influenced by luminal HCO_3 concentrations. The net amount of calcium secretion is functionally linked to HCO_3 production and to luminal HCO_3 concentrations. For further information on carbonate–bicarbonate production and calcium bicarbonate interactions and secretion, see Eastin and Spaziani (1978b).

REFERENCES

Abdelsalam, M., Isobe, N., Yoshimura, Y., 2012. Effects of lipopolysaccharide and interleukins on the expression of avian β-defensins in hen ovarian follicular tissue. Poult. Sci. 91 (11), 2877–2884.

Ahumada-Solórzano, S.M., Carranza, M.E., Pedernera, E., Rodríguez-Méndez, A.J., Luna, M., Arámburo, C., 2012. Local expression and distribution of growth hormone and growth hormone receptor in the chicken ovary: effects of GH on steroidogenesis in cultured follicular granulosa cells. Gen. Comp. Endocrinol. 175 (2), 297–310.

Aitken, R.N.C., 1971. The oviduct. In: In: Bell, J., Freeman, B.M. (Eds.), Physiology and Biochemistry of the Domestic Fowl, vol. 3. Academic Press, London and New York (Chapter 53).

Akazome, Y., Abe, T., Mori, T., 2002. Differentiation of chicken gonad as an endocrine organ: expression of LH receptor, FSH receptor, cytochrome P450c17 and aromatase genes. Reproduction 123 (5), 721–728.

Anastasiadou, M., Avdi, M., Theodoridis, A., Michailidis, G., 2013. Temporal changes in the expression of avian β-defensins in the chicken vagina during sexual maturation and *Salmonella* infection. Vet. Res. Commun..

Anderson, D.J., Apanius, V., 2003. Actuarial and reproductive senescence in a long-lived seabird: preliminary evidence. Exp. Gerontol. 38, 757–760.

Anton, M., 2007. Composition and structure of hen egg yolk. In: Huopalahti, R., López-Fandino, R., Anton, M., Schade, R. (Eds.), Bioactive Egg Compounds. Springer-Verlag, Berlin Heidelberg, pp. 1–6.

Ariamaa, O., Talo, A., 1983. The membrane potential of the Japanese quail's oviductal smooth muscle during ovum transport. Acta Physiol. Scand. 118, 349–353.

Arias, J.L., Fernandez, M.S., 2001. Role of extracellular matrix molecules in shell formation and structure. Worlds Poult. Sci. J. 57, 349–357.

Ayers, K.L., Smith, C.A., Lambeth, L.S., 2013. The molecular genetics of avian sex determination and its manipulation. Genesis 51 (5), 325–336.

Baeyens, D.A., Cornett, L.E., 2006. The cloned avian neurohypophysial hormone receptors. Comp. Biochem. Physiol. B, Biochem. Mol. Biol. 143 (1), 12–19.

Bakst, M.R., 1998. Structure of the avian oviduct with emphasis on sperm storage in poultry. J. Exp. Zool. 282 (4–5), 618–626.

Bakst, M.R., Akuffo, V., April 2008. Turkey sperm reside in the tubular glands in the urodeum following artificial insemination. Poult. Sci. 87 (4), 790–792.

Bakst, M.R., May 2011. Physiology and endocrinology symposium: role of the oviduct in maintaining sustained fertility in hens. J. Anim. Sci. 89 (5), 1323–1329.

Bar, A., Striem, S., Mayel-Afshar, S., Lawson, D.E.M., 1990. Differential regulation of calbindin-D_{28K} mRNA in the intestine and eggshell gland of the laying hen. J. Mol. Endocrinol. 4, 93–99.

Bar, A., Vax, E., Striem, S., 1992. Relationships between calbindin (M_r 28,000) and calcium transport by the eggshell gland. Comp. Biochem. Physiol. 101A, 845–848.

Bar, A., 2008. Calcium homeostasis and vitamin D metabolism and expression in strongly calcifying laying birds. Comp. Biochem. Physiol. A, Mol. Integr. Physiol. 151 (4), 477–490.

Bar, A., 2009. Calcium transport in strongly calcifying laying birds: mechanisms and regulation. Comp. Biochem. Physiol. A, Mol. Integr. Physiol. 152 (4), 447–469.

Barua, A., Michiue, H., Yoshimura, Y., 2001. Changes in the localization of MHC class II positive cells in hen ovarian follicles during the processes of follicular growth, postovulatory regression and atresia. Reproduction 121, 953–957.

Barua, A., Yoshimura, Y., June 2004. Ovarian cell-mediated immune response to *Salmonella enteritidis* infection in laying hens (*Gallus domesticus*). Poult. Sci. 83 (6), 997–1002.

Bastian, J.W., Zarrow, M.X., 1955. A new hypothesis for the asynchronous ovulatory cycle of the domestic hen (*Gallus domesticus*). Poult. Sci. 34, 776–788.

Bédécarrats, G.Y., McFarlane, H., Maddineni, S.R., Ramachandran, R., 2009. Gonadotropin-inhibitory hormone receptor signaling and its impact on reproduction in chickens. Gen. Comp. Endocrinol. 163 (1–2), 7–11.

Bedrani, L., Helloin, E., Guyot, N., Réhault-Godbert, S., Nys, Y., 2013. Passive maternal exposure to environmental microbes selectively modulates the innate defences of chicken egg white by increasing some of its antibacterial activities. BMC Microbiol. 13, 128.

Bourin, M., Gautron, J., Berges, M., Nys, Y., Réhault-Godbert, S., 2012. Sex- and tissue-specific expression of "similar to nothepsin" and cathepsin D in relation to egg yolk formation in *Gallus gallus*. Poult. Sci. 91 (9), 2288–2293.

Breen, P.C., de Bruyn, P.P.H., 1969. The fine structure of the secretory cells of the uterus (shell gland) of the chicken. J. Morphol. 128, 35–66.

Breuner, C.W., Orchinik, M., Hahn, T.P., Meddle, S.L., Moore, I.T., Owen-Ashley, N.T., Sperry, T.S., Wingfield, J.C., 2003. Differential mechanisms for regulation of the stress response across latitudinal gradients. Am. J. Physiol. Regul. Integr. Comp. Physiol. 285 (3), R594–R600.

Bridgham, J.T., Wilder, J.A., Hollocher, H., Johnson, A.L., 2003. All in the family: evolutionary and functional relationships among death receptors. Cell Death Differ. 10 (1), 19–25.

Bruggeman, V., van As, P., Decuypere, E., 2002. Developmental endocrinology of the reproductive axis in the chicken embryo. Comp. Biochem. Physiol. A, Mol. Integr. Physiol. 131 (4), 839–846.

Bucher, T.L., 1987. Patterns in the mass-independent energetics of avian development. J. Exp. Zool. Suppl. 1, 139–150.

Cassar, G., Mohammed, M., John, T.M., Gazdzinski, P., Etches, R.J., 1998. Differentiating between parthenogenetic and "positive development" embryos in turkeys by molecular sexing. Poult. Sci. 77 (10), 1463–1468.

Chabrolle, C., Tosca, L., Crochet, S., Tesseraud, S., Dupont, J., 2007. Expression of adiponectin and its receptors (AdipoR1 and AdipoR2) in chicken ovary: potential role in ovarian steroidogenesis. Domest. Anim. Endocrinol. 33 (4), 480–487.

Cockburn, A., 2006. Prevalence of different modes of parental care in birds. Proc. Biol. Sci. 273 (1592), 1375–1383.

Crossley, J., Ferrando, G., Eiler, H., 1980. Distribution of adrenergic receptors in the domestic fowl oviduct. Poult. Sci. 59, 2331–2335.

Cutting, A.D., Bannister, S.C., Doran, T.J., Sinclair, A.H., Tizard, M.V., Smith, C.A., 2012. The potential role of microRNAs in regulating gonadal sex differentiation in the chicken embryo. Chromosome Res. 20 (1), 201–213.

Dahl, E., 1970. Studies of the fine structure of ovarian interstitial tissue. 3. The innervation of the thecal gland of the domestic fowl. Z. Zellforsch. Mikrosk. Anat. 109, 212–226.

Das, S.C., Isobe, N., Yoshimura, Y., December 2008. Mechanism of prolonged sperm storage and sperm survivability in hen oviduct: a review. Am. J. Reprod. Immunol. 60 (6), 477–481.

Dawson, A., 2006. Control of molt in birds: association with prolactin and gonadal regression in starlings. Gen. Comp. Endocrinol. 147, 314–322.

Dawson, A., 2008. Control of the annual cycle in birds: endocrine constraints and plasticity in response to ecological variability. Philos. Trans. R. Soc. B 363, 1621–1633.

Dawson, A., Sharp, P.J., 2007. Photorefractoriness in birds: photoperiodic and non-photoperiodic control. Gen. Comp. Endocrinol. 153 (1–3), 378–384.

Dick, H.R., Culbert, J., Wells, J.W., Gilbert, A.B., Davidson, M.F., 1978. Steroid hormones in the postovulatory follicle of the domestic fowl (*Gallus domesticus*). J. Reprod. Fert. 53, 103–107.

Eastin, W.C., Spaziani, E., 1978a. On the control of calcium secretion in the avian shell gland (uterus). Biol. Reprod. 19, 493–504.

Eastin, W.C., Spaziani, E., 1978b. On the mechanism of calcium secretion in the avian shell gland (uterus). Biol. Reprod. 19, 505–518.

El Halawani, M.E., Silsby, J.L., Rozenboim, I., Pitt, G.R., 1995. Increase egg production by active immunization against vasoactive intestinal peptide in the turkey (*Meleagris gallipavo*). Biol. Reprod. 52, 179–183.

Elis, S., Dupont, J., Couty, I., Persani, L., Govoroun, M., Blesbois, E., Batellier, F., Monget, P., 2007. Expression and biological effects of bone morphogenetic protein-15 in the hen ovary. J. Endocrinol. 194, 485–497.

Elis, S., Blesbois, E., Couty, I., Balzergue, S., Martin-Magniette, M.L., Batellier, F., Govoroun, M.S., 2009. Identification of germinal disk region derived genes potentially involved in hen fertility. Mol. Reprod. Dev. 76 (11), 1043–1055.

Elkin, R.G., Bauer, R., Schneider, W.J., 2012. The restricted ovulator chicken strain: an oviparous vertebrate model of reproductive dysfunction caused by a gene defect affecting an oocyte-specific receptor. Anim. Reprod. Sci. 136 (1–2), 1–13.

Etches, R.J., Schoch, J.P., 1984. A mathematical representation of the ovulatory cycle of the domestic hen. Br. Poult. Sci. 25, 65–76.

Etches, R.J., 1990. The ovulatory cycle of the hen. CRC Crit. Rev. Poult. Biol. 2, 293–318.

Etches, R.J., Cheng, K.W., 1981. Changes in the plasma concentrations of luteinizing hormone, progesterone, oestradiol and testosterone and in the binding of follicle-stimulating hormone to the theca of follicles during the ovulation cycle of the hen (*Gallus domesticus*). J. Endocrinol. 91 (1), 11–22.

Fraps, R.M., 1955. Egg production and fertility in poultry. In: In: Hammond, J. (Ed.), Progress in the Physiology of Farm Animals, vol. II. Butterworth, London.

Fernandez, M.S., Escobar, C., Lavelin, I., Pines, M., Arias, J.L., 2003. Localization of osteopontin in oviduct tissue and eggshell during different stages of the avian egg laying cycle. J. Struct. Biol. 143, 171–180.

Finn, R.N., 2007. Vertebrate yolk complexes and the functional implications of phosvitins and other subdomains in vitellogenins. Biol. Reprod. 76 (6), 926–935.

Gilbert, A.B., 1969. Innervation of the ovary of the domestic hen. Q. J. Exp. Physiol. 54, 404–411.

Gilbert, A.B., 1971a. Egg albumin and its formation. In: In: Bell, D.J., Freeman, B.M. (Eds.), Physiology and Biochemistry of the Domestic Fowl, vol. 3. Academic Press, London and New York, pp. 1291–1329 (Chapter 54).

Gilbert, A.B., 1971b. The egg: its physical and chemical aspects. In: In: Bell, D.J., Freeman, B.M. (Eds.), Physiology and Biochemistry of the Domestic Fowl, vol. 3. Academic Press, London and New York, pp. 1379–1399 (Chapter 58).

Gilbert, A.B., 1979. Female genital organs. In: In: King, A.S., McLelland, J. (Eds.), Form and Function in Birds, vol. I. Academic Press, London and New York, pp. 237–360 (Chapter 5).

Gilbert, A.B., Hardie, M.A., Perry, M.M., Dick, H.R., Wells, J.W., 1980. Cellular changes in the granulosa layer of the maturing ovarian follicle of the domestic fowl. Br. Poult. Sci. 21, 257–263.

Gilbert, A.B., Davidson, M.F., Wells, J.W., 1978. Role of the granulosa cells of the postovulatory follicle of the domestic fowl in oviposition. J. Reprod. Fertil. 52 (2), 227–229.

Gilbert, A.B., Davidson, M.F., Hardie, M.A., Wells, J.W., 1981. The induction of atresia in the domestic fowl (*Gallus domesticus*) by ovine LH. Gen. Comp. Endocrinol. 44, 344–349.

Goerlich-Jansson, V.C., Müller, M.S., Groothuis, T.G., May 28, 2013. Manipulation of primary sex ratio in birds: lessons from the homing pigeon (*Columba livia domestica*). Integr. Comp. Biol. 53 (6), 902–912 (Epub ahead of print).

González-Morán, M.G., 2007. Effects of luteinizing hormone treatment on oogenesis in ovarian germ cells of the chick (*Gallus domesticus*). Domest. Anim. Endocrinol. 33 (2), 154–166.

González-Morán, M.G., González-Arenas, A., Germán-Castelán, L., Camacho-Arroyo, I., 2013. Changes in the content of sex steroid hormone receptors in the growing and regressing ovaries of *Gallus domesticus* during development. Gen. Comp. Endocrinol. 189C, 51–58.

Guioli, S., Lovell-Badge, R., 2007. PITX2 controls asymmetric gonadal development in both sexes of the chick and can rescue the degeneration of the right ovary. Development 134, 4199–4208.

Hales, D.B., Zhuge, Y., Lagman, J.A., Ansenberger, K., Mahon, C., Barua, A., Luborsky, J.L., Bahr, J.M., 2008. Cyclooxygenases expression and distribution in the normal ovary and their role in ovarian cancer in the domestic hen (*Gallus domesticus*). Endocrine 33 (3), 235–244.

Hau, M., 2001. Timing of breeding in variable environments: tropical birds as model systems. Horm. Behav. 40 (2), 281–290.

He, B., Lin, J., Li, J., Mi, Y., Zeng, W., Zhang, C., 2012. Basic fibroblast growth factor suppresses meiosis and promotes mitosis of ovarian germ cells in embryonic chickens. Gen. Comp. Endocrinol. 176 (2), 173–181.

He, B., Mi, Y., Zhang, C., 2013. Gonadotropins regulate ovarian germ cell mitosis/meiosis decision in the embryonic chicken. Mol. Cell Endocrinol. 370 (1–2), 32–41.

Hincke, M.T., Gautron, J., Panheleux, M., Garcia-Ruiz, J., McKee, M.D., Nys, Y., 2000. Identification and localization of lysozyme as a component of eggshell membranes and eggshell matrix. Matrix Biol. 19 (5), 443–453.

Hincke, M.T., Nys, Y., Gautron, J., Mann, K., Rodriguez-Navarro, A.B., McKee, M.D., 2012. The eggshell: structure, composition and mineralization. Front. Biosci. 17, 1266–1280.

Hincke, M.T., Wellman-Labadie, O., McKee, M.D., Gautron, J., Nys, Y., Mann, K., 2008. Biosynthesis and structural assembly of eggshell components. In: Mine, Y. (Ed.), Egg Bioscience and Biotechnology. Wiley-Interscience, Hoboken, N.J, pp. 97–128.

Hodges, R.D., 1965. The blood supply to the avian oviduct, with special reference to the shell gland. J. Anat. 99, 485–506.

Hodges, R.D., 1974. The Histology of the Fowl. Academic Press, New York and London.

Hoffer, A.P., 1971. The ultrastructure and cytochemistry of the shell membrane-secreting region of the Japanese quail oviduct. Am. J. Anat. 131, 253–288.

Holmes, D.J., Thompson, S.L., Wu, J., Ottinger, M.A., 2003. Reproductive aging in female birds. Exp. Gerontol. 38, 751–756.

Hoshino, A., Koide, M., Ono, T., Yasugi, S., 2005. Sex-specific and left-right asymmetric expression pattern of Bmp7 in the gonad of normal and sex-reversed chicken embryos. Dev. Growth Differ. 47 (2), 65–74.

Hrabia, A., Paczoska-Eliasiewicz, H.E., Berghman, L.R., Harvey, S., Rzasa, J., 2008. Expression and localization of growth hormone and its receptors in the chicken ovary during sexual maturation. Cell Tissue Res. 332 (2), 317–328.

Hudson, Q.J., Smith, C.A., Sinclair, A.H., July 2005. Aromatase inhibition reduces expression of FOXL2 in the embryonic chicken ovary. Dev. Dyn. 233 (3), 1052–1055.

Hughes, G.C., 1963. The population of germ cells in the developing female chick. J. Embryol. Exp. Morph 11 (3), 513–536.

Ishimaru, Y., Komatsu, T., Kasahara, M., Katoh-Fukui, Y., Ogawa, H., Toyama, Y., Maekawa, M., Toshimori, K., Chandraratna, R.A.S., Morohashi, K., Yoshioka, H., 2008. Mechanism of asymmetric ovarian development in chick embryos. Development 135, 677–685.

Isola, J., Korte, J.M., Tuohimaa, P., 1987. Immunocytochemical localization of progesterone receptor in the chick ovary. Endocrinology 121, 1034–1040.

Jensen, T., Durant, B., 2006. Assessment of reproductive status and ovulation in female brown kiwi (*Apteryx mantelli*) using fecal steroids and ovarian follicle size. Zoo Biol. 25, 25–34.

Jensen, T., Johnson, A.L., 2001. Expression and function of brain-derived neurotrophin factor and its receptor, TrkB, in ovarian follicles from the domestic hen (*Gallus gallus domesticus*). J. Exp. Biol. 204 (Pt 12), 2087–2095.

Johnson, A.L., 2003. Intracellular mechanisms regulating cell survival in ovarian follicles. Anim. Reprod. Sci. 78 (3–4), 185–201.

Johnson, A.L., 2011. Chapter 3. Organization and functional dynamics of the avian ovary. In: In: Norris, D.O., Lopez, K.H. (Eds.), Hormones and Reproduction of Vertebrates, vol. 4. Academic Press, pp. 71–90.

Johnson, A.L., van Tienhoven, A., 1980. Plasma concentrations of six steroids and LH during the ovulatory cycle of the hen, *Gallus domesticus*. Biol. Reprod. 23, 386–393.

Johnson, A.L., van Tienhoven, A., 1984. Effects of aminoglutethimide on luteinizing hormone and steroid secretion, and ovulation in the hen, *Gallus domesticus*. Endocrinology 114, 2276–2283.

Johnson, A.L., Li, Z., Gibney, J.A., Malamed, S., 1994. Vasoactive intestinal peptide-induced expression of cytochrome P450 cholesterol side-chain cleavage and 17α-hydroxylase enzyme activity in hen granulosa cells. Biol. Reprod. 51, 327–333.

Johnson, A.L., Bridgham, J.T., Wagner, B., 1996. Characterization of a chicken luteinizing hormone receptor (cLH-R) cDNA, and expression of cLH-R mRNA in the ovary. Biol. Reprod. 55, 304–309.

Johnson, A.L., Bridgham, J.T., 2001. Regulation of steroidogenic acute regulatory (StAR) protein messenger RNA in hen granulosa cells. Endocrinology 142 (7), 3116–3124.

Johnson, A.L., Bridgham, J.T., 2002. Caspase mediated apoptosis in the vertebrate ovary. Reproduction 124 (1), 19–27.

Johnson, A.L., Solovieva, E.V., Bridgham, J.T., 2002. Relationship between StAR expression and progesterone production in hen granulosa cells during follicle development. Biol. Reprod. 67, 1313–1320.

Johnson, A.L., Woods, D.C., 2009. Dynamics of avian follicle development: cellular mechanisms of granulosa cell development. Gen. Comp. Endocrinol. 163, 12–17.

Johnson, A.L., Woods, D.C., 2007. Chapter 6. Ovarian dynamics and follicle development. In: Jamieson, B.G.M. (Ed.), Reproductive Biology and Phylogeny of Aves. Science Publishers, Inc., pp. 243–277.

Johnson, P.A., Brooks, C., Wang, S.-Y., Chen, C.C., 1993. Plasma concentrations of immunoreactive inhibin and gonadotropins following removal of ovarian follicles in the domestic hen. Biol. Reprod. 49, 1026–1031.

Johnson, P.A., Dickens, M.J., Kent, T.R., Giles, J.R., 2005. Expression and function of growth differentiation factor-9 in an oviparous species, *Gallus domesticus*. Biol. Reprod. 72 (5), 1095–1100.

Johnson, P.A., Woodcock, J.R., Kent, T.R., 2006. Effect of activin A and inhibin A on expression of the inhibin/activin beta-B-subunit and gonadotropin receptors in granulosa cells of the hen. Gen. Comp. Endocrinol. 147 (2), 102–107.

Johnson, P.A., Kent, T.R., Urick, M.E., Giles, J.R., 2008. Expression and regulation of anti-Mullerian hormone in an oviparous species, the hen. Biol. Reprod. 78 (1), 13–19.

Johnson, P.A., Giles, J.R., 2013. The hen as a model of ovarian cancer. Nat. Rev. Cancer 13 (6), 432–436.

Kang, L., Cui, X., Zhang, Y., Yang, C., Jiang, Y., May 26, 2013. Identification of miRNAs associated with sexual maturity in chicken ovary by illumina small RNA deep sequencing. BMC Genomics 14 (1), 352.

Karagenç, L., Petitte, J.N., 2000. Soluble factors and the emergence of chick primordial germ cells in vitro. Poult. Sci. 79 (1), 80–85.

Krzysik-Walker, S.M., Ocón-Grove, O.M., Maddineni, S.B., Hendricks 3rd, G.L., Ramachandran, R., 2007. Identification of calcitonin expression in the chicken ovary: influence of follicular maturation and ovarian steroids. Biol. Reprod. 77 (4), 626–635.

Kim, D., Ocón-Grove, O., Johnson, A.L., 2013. Bone morphogenetic protein 4 (BMP4) supports the initial differentiation of hen (*Gallus gallus*) granulosa cells. Biol. Reprod. 88 (6), 161.

Kinsky, F.C., 1971. The consistent presence of paired ovaries in the kiwi (*Aptryx*) with some discussion of this condition in other birds. J. Ornithol. 112, 334–357.

Kowalski, K.I., Tilly, J.L., Johnson, A.L., 1991. Cytochrome P450 side-chain cleavage (P450scc) in the hen ovary. I. Regulation of P450scc messenger RNA levels and steroidogenesis in theca cells of developing follicles. Biol. Reprod. 45, 955–966.

Krishnan, K.A., Proudman, J.A., Bolt, D.J., Bahr, J.M., 1993. Development of an homologous radioimmunoassay for chicken follicle-stimulating hormone and measurement of plasma FSH during the ovulatory cycle. Comp. Biochem. Physiol. 105A, 729–734.

Kundu, M.C., Wojtusik, J., Johnson, P.A., 2012. Expression and regulation of kit ligand in the ovary of the hen. Gen. Comp. Endocrinol. 179 (1), 47–52.

van de Lavoir, M.C., Diamond, J.H., Leighton, P.A., Mather-Love, C., Heyer, B.S., Bradshaw, R., Kerchner, A., Hooi, L.T., Gessaro, T.M., Swanberg, S.E., Delany, M.E., Etches, R.J., 2006. Germline transmission of genetically modified primordial germ cells. Nature 441 (7094), 766–769.

Leach, R.M., 1982. Biochemistry of the organic matrix of the eggshell. Poult. Sci 61, 2040–2047.

Leach, R.M., Gross, J.R., 1983. The effect of manganese deficiency upon the ultrastructure of the eggshell. Poult. Sci. 62, 499–504.

Lee, H.T., Bahr, J.M., 1990. Inhibition of the activities of P450 cholesterol side-chain cleavage and 3β-hydroxysteroid dehydrogenase and the amount of P450 cholesterol side-chain cleavage by testosterone and estradiol-17β in hen granulosa cells. Endocrinology 126, 779–786.

Lee, S.I., Lee, B.R., Hwang, Y.S., Lee, H.C., Rengaraj, D., Song, G., Park, T.S., Han, J.Y., 2011. MicroRNA-mediated posttranscriptional regulation is required for maintaining undifferentiated properties of blastoderm and primordial germ cells in chickens. Proc. Natl. Acad. Sci. U. S. A. 108 (26), 10426–10431.

Levorse, J.M., Johnson, A.L., 1994. Regulation of steroid production in ovarian stromal tissue from 5- to 8-week-old pullets and laying hens. J. Reprod. Fertil. 100, 195–202.

Li, J., Li, M., Lafrance, M., Tsang, B.K., 1994. Avian granulosa cell prostaglandin secretion is regulated by transforming growth factor α and β and does not control plasminogen activator activity during follicular development. Biol. Reprod. 51, 787–794.

Li, J., Li, M., Lafrance, M., Simmons, D.L., Tsang, B.K., 1997. Role and regulation of prostaglandin synthesis in the mitogenic response of ovarian granulosa cells to transforming growth factor alpha. Adv. Exp. Med. Biol. 407, 509–514.

Lin, J., Jia, Y., Zeng, W., Mi, Y., Zhang, C., 2012. Basic FGF promotes proliferation of ovarian granulosa cells in the laying chickens via FGFR1 and PKC pathway. Reprod. Domest. Anim. 47 (1), 135–142.

Luck, M.R., Scanes, C.G., 1979a. Plasma levels of ionized calcium in the laying hen (*Gallus domestics*). Comp. Biochem. Physiol. 63A, 177–181.

Luck, M.R., Scanes, C.G., 1979b. The relationship between reproductive activity and blood calcium in the calcium-deficient hen. Br. Poult. Sci. 20, 559–564.

Madekurozwa, M.C., 2008. An immunohistochemical study of ovarian innervation in the emu (*Dromaius novaehollandiae*). Onderstepoort J. Vet. Res. 75, 59–65.

Madekurozwa, M.C., Kimaro, W.H., 2008. An ultrastructural characterization of the ooplasm in ovarian follicles of the immature ostrich (*Struthio camelus*). Anat. Histol. Embryol. 37 (3), 214–218.

Manchanda, R., Kim, J.M., Tsang, B.K., 2001. Role of prostaglandins in the suppression of apoptosis in hen granulosa cells by transforming growth factor alpha. Reproduction 122 (1), 91–101.

Mao, K.M., Murakami, A., Iwasawa, A., Yoshizaki, N., 2007. The asymmetry of avian egg-shape: an adaptation for reproduction on dry land. J. Anat. 210 (6), 741–748.

March, J.B., Sharp, P.J., Wilson, P.W., Sang, H.M., 1994. Effect of active immunization against recombinant-derived chicken prolactin fusion protein on the onset of broodiness and photoinduced egg laying in bantam hens. J. Reprod. Fertil. 101, 227–233.

Méndez, M.C., Ramírez, M., Varela, A.R., Chávez, B., Pedernera, E., 2003. Follicle-stimulating hormone increases cell proliferation in the ovary and the testis of the chick embryo. Gen. Comp. Endocrinol. 133 (2), 181–188.

Mongin, P., Sauveurl, B., 1974. Voluntary food and calcium intake by the laying hen. Br. Poult. Sci. 15, 349–359.

Nakada, T., Koja, Z., Tanaka, K., 1994. Effect of progesterone on ovulation in the hypophysectomized hen. Br. Poult. Sci. 35, 153–156.

Nakao, N., Yasuo, S., Nishimura, A., Yamamura, T., Watanabe, T., Anraku, T., Okano, T., Fukada, Y., Sharp, P.J., Ebihara, S., Yoshimura, T., 2007. Circadian clock gene regulation of steroidogenic acute regulatory protein gene expression in preovulatory ovarian follicles. Endocrinology 148, 3031–3038.

Nakao, N., Ono, H., Yoshimura, T., 2008. Thyroid hormones and seasonal reproductive neuroendocrine interactions. Reproduction 136 (1), 1–8.

Nevalainen, T.J., 1969. Electron microscope observations on the shell gland mucosa of calcium deficient hens (*Gallus domesticus*). Anat. Rec. 164, 127–140.

Nilsson, J.–Å., Råberg, L., 2001. The resting metabolic cost of egg laying and nestling feeding in great tits. Oecologia 128, 187–192.

Nisbet, I.C., Finch, C.E., Thompson, N., Russek-Cohen, E., Proudman, J.A., Ottinger, M.A., 1999. Endocrine patterns during aging in the common tern (*Sterna hirundo*). Gen. Comp. Endocrinol. 114, 279–286.

Nys, Y., Mayel-Afshar, S., Bouillon, R., Van Baelen, H., Lawson, D.E.M., 1989. Increases in calbindin D 28K mRNA in the uterus of the domestic fowl induced by sexual maturity and shell formation. Gen. Comp. Endocrinol. 76, 322–329.

Ocón-Grove, O.M., Poole, D.H., Johnson, A.L., 2012. Bone morphogenetic protein 6 promotes FSH receptor and anti-Müllerian hormone mRNA expression in granulosa cells from hen prehierarchal follicles. Reproduction 143 (6), 825–833.

Olsen, M.W., 1975. Avian Parthenogenesis. Agricultural Research Service, USDA, ARS-NE. 65. 1–82.

Olszanska, B., Stepinska, U., 2008. Molecular aspects of avian oogenesis and fertilisation. Int. J. Dev. Biol. 52 (2–3), 187–194.

Onagbesan, O., Bruggeman, V., Decuypere, E., 2009. Intra-ovarian growth factors regulating ovarian function in avian species: a review. Anim. Reprod. Sci. 111 (2–4), 121–140.

Onagbesan, O.M., Bruggeman, V., Van As, P., Tona, K., Williams, J., Decuypere, E., 2003. BMPs and BMPRs in chicken ovary and effects of BMP-4 and -7 on granulosa cell proliferation and progesterone production in vitro. Am. J. Physiol. Endocrinol. Metab. 285 (5), E973–E983.

Onagbesan, O.M., Mast, J., Goddeeris, B., Decuypere, E., 2000. Effect of TNF-alpha on LH and IGF-I modulated chicken granulosa cell proliferation and progesterone production during follicular development. J. Reprod. Fertil. 120 (2), 433–442.

Onagbesan, O.M., Safi, M., Decuypere, E., Bruggeman, V., 2004. Developmental changes in inhibin alpha and inhibin/activin betaA and betaB mRNA levels in the gonads during post-hatch prepubertal development of male and female chickens. Mol. Reprod. Dev. 68 (3), 319–326.

Onagbesan, O.M., Vleugels, B., Buys, N., Bruggeman, V., Safi, M., Decuypere, E., 1999. Insulin-like growth factors in the regulation of avian ovarian functions. Domest. Anim. Endocrinol. 17 (2–3), 299–313.

Oréal, E., Mazaud, S., Picard, J.Y., Magre, S., Carré-Eusèbe, D., 2002. Different patterns of anti-Müllerian hormone expression, as related to DMRT1, SF-1, WT1, GATA-4, Wnt-4, and Lhx9 expression, in the chick differentiating gonads. Dev. Dyn. 225 (3), 221–232.

Oring, L.W., Fivizzani, A.J., Colwell, M.A., El Halawani, M.E., 1988. Hormonal changes associated with natural and manipulated incubation in the sex-role reversed Wilson's phalarope. Gen. Comp. Endocrinol. 72, 247–256.

Osterström, O., Lilja, C., 2012. Evolution of avian eggshell structure. J. Morphol. 273 (3), 241–247.

Ozoe, A., Isobe, N., Yoshimura, Y., February 15, 2009. Expression of Toll-like receptors (TLRs) and TLR4 response to lipopolysaccharide in hen oviduct. Vet. Immunol. Immunopathol. 127 (3–4), 259–268.

Parizzi, R.C., Santos, J.M., Oliveira, M.F., Maia, M.O., Sousa, J.A., Miglino, M.A., Santos, T.C., 2008. Macroscopic and microscopic anatomy of the oviduct in the sexually mature rhea (*Rhea americana*). Anat. Histol. Embryol. 37 (3), 169–176.

Parker, H.M., Kiess, A.S., Robertson, M.L., Wells, J.B., McDaniel, C.D., 2012. The relationship of parthenogenesis in virgin Chinese painted quail (*Coturnix chinensis*) hens with embryonic mortality and hatchability following mating. Poult. Sci. 91 (6), 1425–1431.

Perry, M.M., Gilbert, A.B., Evans, A.J., 1978a. Electron microscope observations on the ovarian follicle of the domestic fowl during the rapid growth phase. J. Anat. 125, 481–497.

Perry, M.M., Gilbert, A.B., Evans, A.J., 1978b. The structure of the germinal disc region of the hen's ovarian follicle during the rapid growth phase. J. Anat. 127, 379–392.

Perry, M.M., Gilbert, A.B., 1979. Yolk transport in the ovarian follicle of the hen (*Gallus domesticus*): lipoprotein-like particles at the periphery of the oocyte in the rapid growth phase. J. Cell Sci. 39, 257–272.

Pike, T.W., Petrie, M., 2003. Potential mechanisms of avian sex manipulation. Biol. Rev. Camb. Philos. Soc. 78 (4), 553–574.

Poisbleau, M., Demongin, L., Parenteau, C., Eens, M., 2011. Intra-clutch ratio of yolk progesterone level changes with laying date in rockhopper penguins: a strategy to influence brood reduction? PLoS One 6 (11), e27765.

Porter, T.E., Hargis, B.M., Silsby, J.L., El Halawani, M.E., 1989. Differential steroid production between theca interna and theca externa cells: a three cell model for follicular steroidogenesis in avian species. Endocrinology 125, 109–116.

Rahn, H., Paganelli, C.V., Ar, A., 1987. Pores and gas exchange of avian eggs: a review. J. Exp. Zool. Suppl. 1, 165–172.

Rahn, H., Ar, A., Paganelli, C.V., 1979. How eggs breathe. Sci. Am. 240, 46–55.

Rahn, H., Paganelli, C.V., 1990. Gas fluxes in avian eggs: driving forces and the pathway for exchange. Comp. Biochem. Physiol. 95A, 1–15.

Reed, W.L., Clark, M.E., 2011. Beyond maternal effects in birds: responses of the embryo to the environment. Integr. Comp. Biol. 51 (1), 73–80.

Reed, T.E., Kruuk, L.E., Wanless, S., Frederiksen, M., Cunningham, E.J., Harris, M.P., 2008. Reproductive senescence in a long-lived seabird: rates of decline in late-life performance are associated with varying costs of early reproduction. Am. Nat. 171 (2), E89–E101.

Robinson, F.E., Etches, R.J., 1986. Ovarian steroidogenesis during follicular maturation in the domestic fowl (*Gallus domesticus*). Biol. Reprod. 35 (5), 1096–1105.

Roca, P., Sainz, F., Gonzalez, M., Alemany, M., 1984. Structure and composition of the eggs from several avian species. Comp. Biochem. Physiol. 77A, 307–310.

Rodríguez-León, J., Rodríguez Esteban, C., Martí, M., Santiago-Josefat, B., Dubova, I., Rubiralta, X., Izpisúa Belmonte, J.C., 2008. Pitx2 regulates gonad morphogenesis. Proc. Natl. Acad. Sci. U. S. A. 105 (32), 11242–11247.

Rodler, D., Sasanami, T., Sinowatz, F., 2012. Assembly of the inner perivitelline layer, a homolog of the mammalian zona pellucida: an immunohistochemical and ultrastructural study. Cells Tissues Organs. 195 (4), 330–339.

Romanoff, A.L., Romanoff, A.J., 1949. The Avian Egg. Wiley, New York.

Rose-Martel, M., Du, J., Hincke, M.T., May 17, 2012. Proteomic analysis provides new insight into the chicken eggshell cuticle. J. Proteomics 75 (9), 2697–2706.

Rothchild, I., Fraps, R.M., 1944. On the function of the ruptured ovarian follicle of the domestic fowl. Proc. Soc. Exp. Biol. Med. 56, 79–82.

Rothwell, B., Solomon, S.E., 1977. The ultrastructure of the follicle wall of the domestic fowl during the phase of rapid growth. Br. Poult. Sci. 18, 605–610.

Rozenboim, I., Tabibzadeh, C., Silsby, J.L., El Halawani, M.E., 1993. Effect of ovine prolactin administration on hypothalamic vasoactive intestinal peptide (VIP), gonadotropin releasing hormone I and II content, and anterior pituitary VIP receptors in laying turkey hens. Biol. Reprod. 48, 1246–1250.

Rutkowska, J., Badyaev, A.V., 2008. Review. Meiotic drive and sex determination: molecular and cytological mechanisms of sex ratio adjustment in birds. Philos. Trans. R. Soc. Lond. B, Biol. Sci. 363 (1497), 1675–1686.

Saitou, M., Kagiwada, S., Kurimoto, K., 2012. Epigenetic reprogramming in mouse pre-implantation development and primordial germ cells. Development 139 (1), 15–31.

Saito, N., Sato, K., Shimada, K., 1987. Prostaglandin levels in peripheral and follicular plasma, the isolated theca and granulosa layers of pre- and postovulatory follicles, and the myometrium and mucosa of the shell gland (uterus) during a midsequence-oviposition of the hen (*Gallus domesticus*). Biol. Reprod. 36 (1), 89–96.

Salevsky, E., Leach, R.M., 1980. Studies on the organic components of shell gland fluid and the hen's egg shell. Poult. Sci. 59, 438–443.

Sarvella, P., 1970. Sporadic occurrence of parthenogenesis in poultry. J. Hered. 61 (5), 215–219.

Scanes, C.G., Godden, P.M.M., Sharp, P.J., 1977. An homologous radioimmunoassay for chicken follicle-stimulating hormone: observations on the ovulatory cycle. J. Endocrinol. 73, 473–481.

Scanes, C.G., Mozelic, H., Kavanagh, E., Merrill, G., Rabii, J., 1982. Distribution of blood flow in the ovary of domestic fowl (*Gallus domesticus*) and changes after prostaglandin F-2alpha treatment. J. Reprod. Fertil. 64, 227–231.

Schneider, W.J., 2009. Receptor-mediated mechanisms in ovarian follicle and oocyte development. Gen. Comp. Endocrinol. 163 (1–2), 18–23.

Schuster, M.K., Schmierer, B., Shkumatava, A., Kuchler, K., 2004. Activin A and follicle-stimulating hormone control tight junctions in avian granulosa cells by regulating occluding expression. Biol. Reprod. 70, 1493–1499.

Schut, E., Hemmings, N., Birkhead, T.R., 2008. Parthenogenesis in a passerine bird, the zebra finch, *Taeniopygia guttata*. Ibis 150, 197–199.

Sechman, A., 2013. The role of thyroid hormones in regulation of chicken ovarian steroidogenesis. Gen. Comp. Endocrinol..

Sharp, P.J., Scanes, C.G., Gilbert, A.B., 1978. *In vivo* effects of an antiserum to partially purified chicken luteinizing hormone (CM2) in laying hens. Gen. Comp. Endocrinol. 34, 296–299.

Sharp, P.J., 2009. Broodiness and broody control. In: Hocking, P.M. (Ed.), Biology of Breeding Poultry.

Shimada, K., Asai, I., 1978. Uterine contraction during the ovulatory cycle of the hen. Biol. Reprod. 19, 1057–1062.

Shimada, K., Asai, I., 1979. Effects of prostaglandin F2a and indomethacin on uterine contraction in hens. Biol. Reprod. 21, 523–527.

Silver, R., 1986. Circadian and interval timing mechanisms in the ovulatory cycle of the hen. Poult. Sci. 65, 2355–2362.

Simkiss, K., Taylor, T.G., 1971. Shell formation. In: In: Bell, D.J., Freeman, B.M. (Eds.), Physiology and Biochemistry of the Domestic Fowl, vol. 3. Academic Press, New York, pp. 1331–1343 (Chapter 55).

Sirotkin, A.V., Grossmann, R., María-Peon, M.T., Roa, J., Tena-Sempere, M., Klein, S., 2006. Novel expression and functional role of ghrelin in chicken ovary. Mol. Cell Endocrinol. 257–258, 15–25.

Smith, C.A., Sinclair, A.H., 2004. Sex determination: insights from the chicken. Bioessays 26, 120–132.

Smith, C.A., Shoemaker, C.M., Roeszler, K.N., Queen, J., Crews, D., Sinclair, A.H., 2008. Cloning and expression of R-spondin1 in different vertebrates suggests a conserved role in ovarian development. BMC Dev. Biol. 8, 72.

Smith, C.A., Roeszler, K.N., Ohnesorg, T., Cummins, D.M., Farlie, P.G., Doran, T.J., Sinclair, A.H., 2009. The avian Z-linked gene DMRT1 is required for male sex determination in the chicken. Nature 461, 267–271.

Sockman, K.W., Williams, T.D., Daeson, A., Ball, G.F., 2004. Prior experience with photostimulation enhances photo-induced reproductive development in female European starlings: a possible basis for the age-related increase in avian reproductive performance. Biol. Reprod. 71, 979–986.

Soledad Fernandez, M., Moya, A., Lopez, L., Arias, J.L., 2001. Secretion pattern, ultrastructural localization and function of extracellular matrix molecules involved in eggshell formation. Matrix Biol. 19, 793–803.

Solomon, S.E., 1975. Studies on the isthmus region of the domestic fowl. Br. Poult. Sci. 16, 255–258.

Solomon, S.E., 2010. The eggshell: strength, structure and function. Br. Poult. Sci. 51 (Suppl. 1), 52–59.

Sossinka, R., 1980. Ovarian development in an opportunistic breeder, the zebra finch, *Poephila guttata castanotis*. J. Exp. Zool. 211, 225–230.

Stebler, J., Spieler, D., Slanchev, K., Molyneaux, K.A., Richter, U., Cojocaru, V., Tarabykin, V., Wylie, C., Kessel, M., Raz, E., 2004. Primordial germ cell migration in the chick and mouse embryo: the role of the chemokine SDF-1/CXCL12. Dev. Biol. 272, 351–361.

Stepinska, U., Bakst, M.R., 2006. Fertilization. In: Jamieson, B.G.M. (Ed.), Reproductive Biology and Phylogeny of Aves. Science Publishers, Enfield, NH; Plymouth, UK, pp. 553–587.

Sundaresan, N.R., Marcus Leo, M.D., Subramani, J., Anish, D., Sudhagar, M., Ahmed, K.A., Saxena, M., Tyagi, J.S., Sastry, K.V., Saxena, V.K., 2009. Expression analysis of melatonin receptor subtypes in the ovary of domestic chicken. Vet. Res. Commun. 33 (1), 49–56.

Sundaresan, N.R., Saxena, V.K., Sastry, K.V., Nagarajan, K., Jain, P., Singh, R., Anish, D., Ravindra, P.V., Saxena, M., Ahmed, K.A., 2008. Cytokines and chemokines in postovulatory follicle regression of domestic chicken (*Gallus gallus domesticus*). Dev. Comp. Immunol. 32 (3), 253–264.

Tabibzadeh, C., Rozenboim, I., Silsby, J.L., Pitts, G.R., Foster, D.N., El Halawani, M.E., 1995. Modulation of ovarian cytochrome P450 17α-hydroxylase and cytochrome aromatase messenger ribonucleic acid by prolactin in the domestic turkey. Biol. Reprod. 52, 600–608.

Takada, S., Wada, T., Kaneda, R., Choi, Y.L., Yamashita, Y., Mano, H., 2006. Evidence for activation of Amh gene expression by steroidogenic factor 1. Mech. Dev. 123 (6), 472–480.

Takahashi, T., Tajima, H., Nakagawa-Mizuyachi, K., Nakayama, H., Kawashima, M., 2011. Changes in prostaglandin F2α receptor bindings in the hen oviduct uterus before and after oviposition. Poult. Sci. 90 (8), 1767–1773.

Tanaka, M., Maeda, K., Okubo, T., Nakashima, K., 1992. Double antenna structure of chicken prolactin receptor deduced from the cDNA sequence. Biochem. Biophys. Res. Comm. 188, 490–496.

Tapanainen, J.S., Lapolt, P.S., Perlas, E., Hsueh, A.J., 1993. Induction of ovarian follicle luteinization by recombinant follicle-stimulating hormone. Endocrinology 133 (6), 2875–2880.

Tesar, D.B., Cheung, E.J., Bjorkman, P.J., 2008. The chicken yolk sac IgY receptor, a mammalian mannose receptor family member, transcytoses IgY across polarized epithelial cells. Mol. Biol. Cell 19, 1587–1593.

Tian, X., Gautron, J., Monget, P., Pascal, G., 2010. What makes an egg unique? Clues from evolutionary scenarios of egg-specific genes. Biol. Reprod. 83 (6), 893–900.

Tilly, J.L., Johnson, A.L., 1991. Protein kinase C in preovulatory follicles from the hen ovary. Domest. Anim. Endocrinol. 8, 1–13.

Tilly, J.L., Kowalski, K.I., Johnson, A.L., 1991a. Stage of follicular development associated with the initiation of steroidogenic competence in avian granulosa cells. Biol. Reprod. 44, 305–314.

Tilly, J.L., Kowalski, K.I., Johnson, A.L., Hsueh, A.J.W., 1991b. Involvement of apoptosis in ovarian follicular atresia and postovulatory regression. Endocrinology 129, 2799–2801.

Tingari, M.D., Lake, P.E., 1973. Ultrastructural studies on the uterovaginal sperm host glands of the domestic hen, *Gallus domesticus*. J. Reprod. Fert. 34, 423–431.

Tischkau, S.A., Howell, R.E., Hickok, J.R., Krager, S.L., Bahr, J.M., 2011. The luteinizing hormone surge regulates circadian clock gene expression in the chicken ovary. Chronobiol. Int. 28 (1), 10–20.

Tullett, S.G., 1984. The porosity of avian eggshells. Comp. Biochem. Physiol. 78A, 5–13.

Tuohimaa, P., Joensuu, T., Isola, J., Keinanen, R., Kunnas, T., Niemala, A., Pekki, A., Wallen, M., Ylikomi, T., Kulomaa, M., 1989. Development of progestin-specific response in the chicken oviduct. Int. J. Dev. Biol. 33, 125–134.

Unsicker, K., Seidel, F., Hofmann, H.D., Muller, T.H., Schmidt, R., Wilson, A., 1983. Catecholaminergic innervation of the chicken ovary. Cell Tissue Res. 230, 431–450.

Vézina, F., Williams, T.D., 2002. Metabolic costs of egg production in the European starling (*Sturnus vulgaris*). Physiol. Biochem. Zool. 75, 377–385.

Waddington, D., Perry, M.M., Gilbert, A.B., Hardie, M.A., 1985. Follicular growth and atresia in the ovaries of hens (*Gallus domesticus*) with diminished egg production rates. J. Reprod. Fertil. 74, 399–405.

Whitehead, C.C., 2004. Overview of bone biology in the egg-laying hen. Poult. Sci. 83 (2), 193–199.

Wilkin, T.A., Gosler, A.G., Garant, D., Reynolds, S.J., Sheldon, B.C., 2009. Calcium effects on life-history traits in a wild population of the great tit (*Parus major*): analysis of long-term data at several spatial scales. Oecologia 159 (2), 463–472.

Williams, J.B., Sharp, P.J., 1977. A comparison of plasma progesterone and luteinizing hormone in growing hens from eight weeks of age to sexual maturity. J. Endocrinol. 75, 447–448.

Williams, T.D., 2012. Hormones, life-history, and phenotypic variation: opportunities in evolutionary avian endocrinology. Gen. Comp. Endocrinol. 176 (3), 286–295.

Wilson, S.C., Sharp, P.J., 1975. Effects of progesterone and synthetic luteinizing hormone releasing hormone on the release of luteinizing hormone during sexual maturation in the hen (*Gallus domesticus*). J. Endocrinol. 67, 359–369.

Wilson, S.C., Cunningham, F.J., 1980. Modification by metyrapone of the "open period" for pre ovulatory LH release in the hen. Br. Poult. Sci. 21, 351–361.

Wilson, S.C., Jennings, R.C., Cunningham, F.J., 1983. An investigation of diurnal and cyclic changes in the secretion of luteinizing hormone in the domestic hen. J. Endocrinol. 98, 137–145.

Witty, J.P., Bridgham, J.T., Johnson, A.L., 1996. Induction of apoptotic cell death in hen granulosa cells by ceramide. Endocrinology 137 (12), 5269–5277.

Wojtusik, J., Johnson, P.A., 2012. Vitamin D regulates anti-Mullerian hormone expression in granulosa cells of the hen. Biol. Reprod. 86 (3), 1–7.

Woods, D.C., Johnson, A.L., 2005. Regulation of follicle-stimulating hormone receptor mRNA in hen granulosa cells relative to follicle selection. Biol. Reprod. 72, 643–650.

Woods, D.C., Haugen, M.J., Johnson, A.L., 2005. Opposing actions of TGFbeta and MAP kinase signaling in undifferentiated hen granulosa cells. Biochem. Biophys. Res. Commun. 336 (2), 450–457.

Woods, D.C., Johnson, A.L., 2007. Protein kinase C activity mediates LH-induced ErbB/Erk signaling in differentiated hen granulosa cells. Reproduction 133, 733–741.

Woods, D.C., Schorey, J.S., Johnson, A.L., 2009. Toll-like receptor signaling in hen ovarian granulosa cells is dependent on stage of follicle maturation. Reproduction 137 (6), 987–996.

Wyburn, G.M., Aitken, R.N.C., Johnsron, H.S., 1965. The ultrastructure of the zona radiata of the ovarian follicle of the domestic hen. J. Anat. 99, 469–484.

Wyburn, G.M., Johnston, H.S., Draper, M.H., Davidson, M.F., 1970. The fine structure of the infundibulum and magnum of the oviduct of *Gallus domesticus*. Q. J. Exp. Physiol. 55, 213–232.

Wyburn, G.M., Johnston, H.S., Draper, M.H., Davidson, M.F., 1973. The ultrastructure of the shell forming region of the oviduct and the development of the shell of *Gallus domesticus*. Q. J. Exp. Physiol. 58, 143–151.

Yao, H.H., Bahr, J.M., 2001. Chicken granulosa cells show differential expression of epidermal growth factor (EGF) and luteinizing hormone (LH) receptor messenger RNA and differential responsiveness to EGF and LH dependent upon location of granulosa cells to the germinal disc. Biol. Reprod. 64, 1790–1796.

Yoshimura, Y., Tanaka, K., Koga, O., 1983. Studies on the contractility of follicular wall with special reference to the mechanism of ovulation in hens. Br. Poult. Sci. 24, 213–218.

Yoshimura, Y., Bahr, J.M., 1991a. Localization of progesterone receptors in the shell gland of laying and nonlaying chickens. Poult. Sci. 70, 1246–1251.

Yoshimura, Y., Bahr, J.M., 1991b. Localization of progesterone receptors in pre- and postovulatory follicles of the domestic hen. Endocrinology 128, 323–330.

Yoshimura, Y., Chang, C., Okamoto, T., Tamura, T., 1993. Immunolocalization of androgen receptor in the small, preovulatory, and postovulatory follicles of laying hens. Gen. Comp. Endocrinol. 91, 81–89.

Yoshimura, Y., Tischkau, S.A., Bahr, J.M., 1994. Destruction of the germinal disc region of an immature preovulatory follicle suppresses follicular maturation and ovulation. Biol. Reprod. 51, 229–233.

Yoshimura, T., 2013. Thyroid hormone and seasonal regulation of reproduction. Front. Neuroendocrinol..

You, S., Bridgham, J.T., Foster, D.N., Johnson, A.L., 1996. Characterization of a chicken follicle-stimulating hormone receptor (cFSH-R) cDNA, and expression of cFSH-R mRNA in the ovary. Biol. Reprod. 55, 1055–1062.

Youngren, O., Chaiseha, Y., Al-Zailaie, K., Whiting, S., Kang, S.W., El Halawani, M., 2002. Regulation of prolactin secretion by dopamine at the level of the hypothalamus in the turkey. Neuroendocrinology 75, 185–192.

Young, A.M., Hobson, E.A., Bingaman Lackey, L., Wright, T.F., 2011. Survival on the ark: life-history trends in captive parrots. Anim. Conserv. 15, 28–43.

Zheng, X., O'Connor, J., Huchzermeyer, F., Wang, X., Wang, Y., Wang, M., Zhou, Z., 2013. Preservation of ovarian follicles reveals early evolution of avian reproductive behaviour. Nature 495 (7442), 507–511.

FURTHER READING

Smith, C.A., Roeszler, K.N., Bowles, J., Koopman, P., Sinclair, A.H., 2004. Onset of meiosis in the chicken embryo; evidence of a role for retinoic acid. BMC Dev. Biol. 8, 85.

Chapter 29

Reproduction in Male Birds

Jorge Vizcarra
Department of Food and Animal Sciences, Alabama A&M University, Huntsville, AL, USA

Rebecca Alan and John Kirby
College of the Environment and Life Sciences, University of Rhode Island, Kingston, RI, USA

29.1 INTRODUCTION

Reproduction is a process that can be organized into distinct developmental and functional phases. In the case of the male, these include fertilization, formation of a patent reproductive tract, production of sperm, manifestation of male-specific behavioral patterns, and expulsion of sperm from the body. This perspective can provide insight that may be missed if reproduction is viewed primarily as an isolated act. For example, although the reproductive tract is fully functional only in the adult, it is formed, for the most part, prior to hatching. Furthermore, although spermatogenesis is associated with puberty, spermatogenesis is not constrained by chronological age but rather by the extent to which testicular cells proliferate and differentiate which, in turn, is coupled to the developmental limitations of gonadotropin secretion. Finally, androgens essential to the function of the reproductive tract, the appearance of secondary sexual attributes, and male behavior may have detrimental effects upon the development of immune and connective tissue if the hormonal signal appears during the period of rapid prepubertal growth and differentiation. Thus, this chapter will discuss the process of reproductive system development and function in the male bird.

29.2 REPRODUCTIVE TRACT ANATOMY

29.2.1 Testis

The gross morphology and relative position of the male reproductive organs are shown in Figure 29.1. For detailed reviews, see King (1979); Lake (1981); Nickel et al. (1977). Paired reproductive tracts lie along the dorsal body wall. Each tract consists of a testis, an epididymis, and a highly convoluted deferent duct running alongside the ureter. The testes are connected to the body wall by a mesorchium. This peritoneal fold not only serves as an attachment for the testis but also as a conduit for nerves and blood vessels as well. Each testis is an aggregate of anastomosing seminiferous tubules, with associated interstitial tissue enveloped by a connective tissue capsule. The testicular capsule is an important component of the contractile mechanism of the testes and is composed of three layers, including thc tunica serosa, tunica albuginea, and tunica vasculosa. Of these three layers, the tunica albuginea represents the main tissue layer and comprises cellular elements that alternate with thick bundles of collagen fibers. The testicular capsule is, in general, thinner in birds than in mammals; however, studies in ratite birds have revealed a capsule that is considerably thicker than that of other bird species (Aire and Ozegbe, 2007; Ozegbe et al., 2008).

The testis contains two types of parenchymal tissue: the interstitial tissue and the seminiferous epithelium. The interstitial tissue contains blood and lymphatic vessels, nerves, peritubular epithelial cells, and Leydig cells, whereas thin concentric layers of myoepithelial cells, fibroblasts, and connective tissue fibrils overlie the basal lamina of the seminiferous tubule (Rothwell and Tingari, 1973). The seminiferous epithelium within the seminiferous tubules of sexually mature males is compartmentalized into basal and adluminal regions via tight junctions between adjacent Sertoli cells (Bergmann and Schindelmeiser, 1987; Osman, 1980). The seminiferous epithelium contains developing germ cells in distinct associations referred to as *stages*. The stages are arranged sequentially in a helix that extends along the length of the seminiferous tubule (Lin and Jones, 1990).

In birds, the left testis is commonly larger than the right, which may either be a byproduct of evolutionary selection for asymmetry in the female reproductive tract or an adaptation to facilitate flight (Birkhead et al., 1998; Denk and Kempenaers, 2005). The overall size of the testis scales

Sturkie's Avian Physiology.

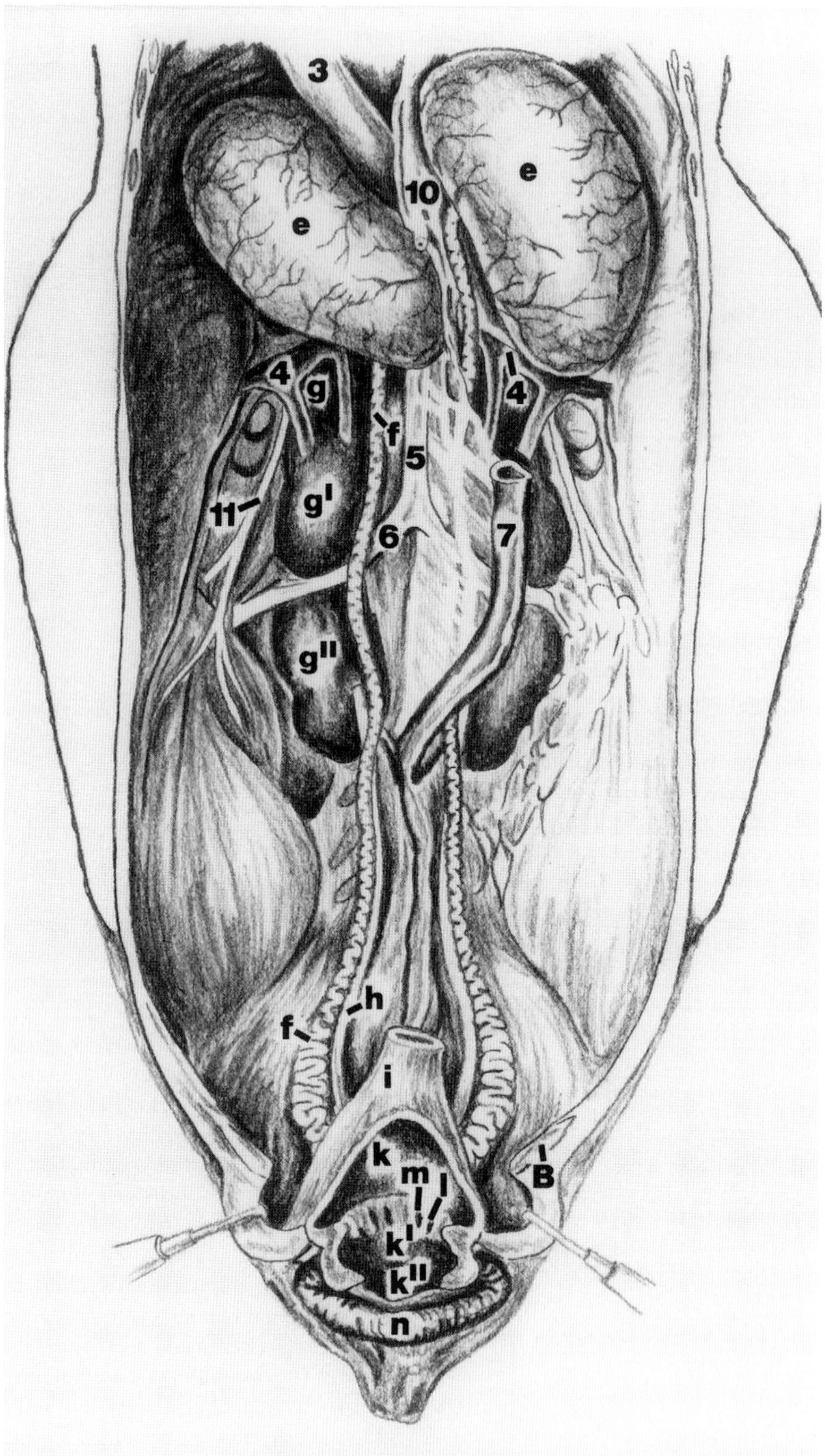

FIGURE 29.1 Topography of the dorsal body wall of the rooster. Paired testes (e) are located anterior to the cranial lobe of the kidneys (g). The deferent ducts (f) run alongside the ureters (h) towards the cloaca (k–k″), ultimately opening (l) into the urodeum (k′). Abbreviations: b, pubis; e, testis; f, ductus deferens; g, cranial; g′, middle, and g″, caudal lobes of the kidney; h, ureter; I, colon; k, coprodeum; k′, urodeum, k″, proctodeum; l, opening of left ductus deferens; m, opening of left ureter; n, anus. *Modified from Nickel et al. (1977).*

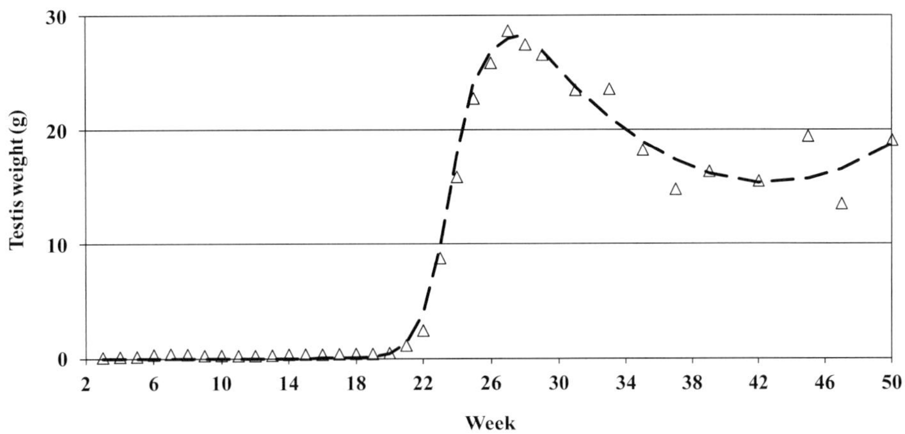

FIGURE 29.2 **Least squares regression (lines) and means (symbols) for testis weight in male broiler breeders.** Males on a "pedigree" breeder program were reared on a 23L:1D photoperiod and unrestricted food and water intake for 6 weeks. At 7 weeks, males were placed on a restricted diet and the photoperiod was reduced to 8L:16D. At week 18, birds were photostimulated (16L:8D) until the end of the experiment. Each symbol represents an average of 30 testes. *Modified from Vizcarra et al. (2010).*

allometrically to body mass; however, once controlled for body mass, testes size is often larger in bird species that experience a higher degree of sperm competition (Moller, 1991; Moller and Briskie, 1995; Pitcher et al., 2005). Sperm competition occurs when the sperm of two or more males is present in the female reproductive tract and must compete for egg fertilization. Although many bird species are socially monogamous, extrapair copulations often occur. Species that participate in extrapair copulations more frequently experience a more intense degree of sperm competition and have developed larger testis than species that do not frequently participate in extrapair copulations (Moller, 1991; Moller and Briskie, 1995; Pitcher et al., 2005).

In male broiler breeders, no significant differences were observed in the weight of the right and left testis from 2 to 50 weeks of age ($n = 2700$; Vizcarra and Kirby, unpublished results). Testes of male broiler breeders reared under a "pedigree" breeding program had an exponential increase in weight after photostimulation (18 weeks of age; Figure 29.2). Within 10 weeks (i.e., week 18–28), testis weight increased from 0.4 to 28 g. However, after 28 weeks of age, testis weight was significantly decreased, reaching a minimum weight of 15 g by week 42 (Figure 29.2). The reason for the testicular weight regression after 28 weeks of age, is not completely clear at this time. Decreased follicle-stimulating hormone (FSH) concentrations might be responsible for testicular weight regression (Vizcarra et al., 2010).

29.2.2 Excurrent Ducts

The excurrent ducts or excurrent canals are ducts associated with the male reproductive tract that derive from the mesonephros and encompass the efferent ducts, epididymis, and deferens ducts. The epididymis is located on the dorsomedial aspect of the testis, which is referred to as the hilus

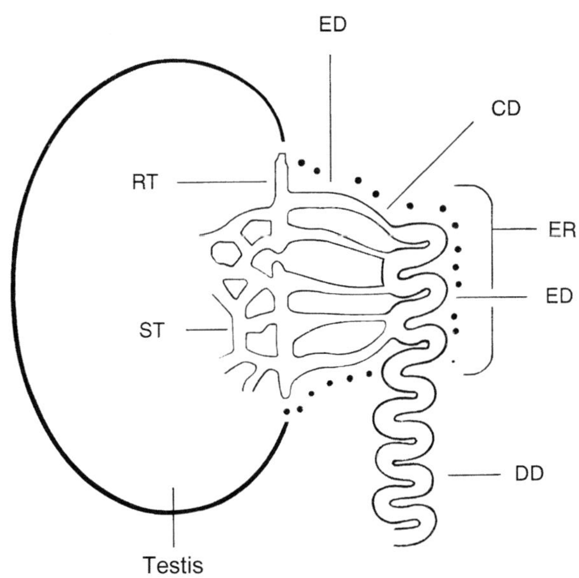

FIGURE 29.3 **Schematic of the excurrent ducts of the testis.** ST, seminiferous tubules; RT, rete testis; ED, efferent duct; CD, connecting duct; ER, epididymal region; ED, epididymal duct; DD, deferent duct. *Reproduced with permission from Academic Press, London and redrawn from Lake (1981).*

(Figure 29.3). The epididymis is actually a series of ducts that ultimately empty into the deferent duct. The ducts within the epididymis include the rete testis, efferent ducts, connecting ducts, and the epididymal duct. The epididymal ducts and the deferent duct are referred to collectively as the excurrent ducts of the testis. Seminiferous tubules connect with the rete testis at discrete sites along the testis–epididymal interface (Tingari, 1971). Osman (1980) identified three distinct types of junctures between the seminiferous tubules and rete testis of the rooster: (1) a germ cell free epithelium of modified Sertoli cells that abruptly gives way to a rete-like epithelium, referred to as a terminal segment and tubulus rectus, respectively; (2) a terminal segment that connects directly to the rete testis; and (3) anastomosis of a seminiferous tubule with the rete testis. Tubuli recti have also been observed in quail testes (Aire, 1979a).

The rete testis has intratesticular, intracapsular, and extratesticular regions (Aire, 1982). As depicted in Figure 29.3, the rete testis exists as lacunae. In some avian species, including domestic fowl, quail, guinea fowl, and ducks, the rete testis is lined with simple cuboidal and simple squamous epithelial cells (Aire, 1982), whereas in ratite birds, such as the ostrich, the epithelium contains cuboidal and columnar cell types (Aire and Soley, 2000, 2003). In gallinaceous birds, the rete testis exits the hilus after coursing through the testicular capsule before entering the epididymis, which is in close proximity. In contrast, the rete testis of ratite birds makes up a much lower volumetric proportion of the epididymal tissue, and the rete testis travels a relatively larger distance from the testicular capsule before entering the epididymis (Ozegbe et al., 2010). Unlike the rete testis, the efferent ducts are wide at their proximal ends and narrow at their distal ends in many bird species (Aire and Soley, 2003; Simões et al., 2004); however, in ratite birds, both ducts are of similar height and diameter, which makes them less easily distinguished from one another (Aire and Soley, 2000; Ozegbe et al., 2006). On a volumetric basis (Table 29.1), the efferent ducts are the principal excurrent duct within the epididymis (Aire, 1979b; Aire and Josling, 2000). In most bird species, the efferent duct mucosa is highly folded, especially in the proximal portion (Aire, 1979a; Aire and Josling, 2000; Bakst, 1980; Simões et al., 2004; Tingari, 1971). In ratite birds, however, neither the proximal nor distal efferent ducts possess mucosal folds and they are regular in outline both internally and externally, which differs from other species (Ozegbe et al., 2006). In general, the efferent duct mucosa is characterized by a pseudostratified columnar epithelium that contains ciliated and nonciliated epithelial cells (Aire, 1980; Budras and Sauer, 1975; Hess and Thurston, 1977; Simões et al., 2004; Tingari, 1972) as well as sparsely distributed intraepithelial lymphocytes (Aire and Malmqvist, 1979a). The epithelium of the connecting ducts is also composed of pseudostratified columnar cells (Aire, 1979a; Aire and Josling, 2000; Bakst, 1980; Hess and Thurston, 1977; Tingari, 1971, 1972). The distinguishing attributes of the connecting tubules are a narrow luminal diameter relative to that of adjacent excurrent ducts and the smooth contour of the mucosal surface (Aire, 1979b; Aire and Josling, 2000; Bakst, 1980; Tingari, 1971).

TABLE 29.1 Species Variation in Volumetric Proportions (%) of Epididymal Structures[a]

	Species		
Structures	**Chicken**	**Japanese Quail**	**Guinea Fowl**
Rete testis	13.3	9.9	10.7
Proximal efferent ductules	27.6	40.8	45.7
Distal efferent ductules	7.7	15.2	16.2
Connecting ducts	2.3	1.7	0.7
Epididymal duct	7.6	2.4	1.8
Connective tissue	38.7	27.3	22.6
Blood vessels	2.5	2.7	2.3
Aberrant ducts	0.3	—	—

[a]*Adapted from Aire (1979a).*

The tortuous epididymal and deferent ducts are characterized by low mucosal folds covered with nonciliated pseudostratified columnar epithelial cells (Aire, 1979b; Aire and Josling, 2000; Bakst, 1980; Simões et al., 2004; Tingari, 1971). In gallinaceous birds, the epididymal duct constitutes between 2.5 and 10% of the total epididymal volume (Aire and Josling, 2000), whereas in ratite birds the volumetric proportion of the epididymal duct is as much as tenfold higher (Ozegbe et al., 2010). Based upon histological and morphological evidence, the deferent duct is a continuation of the epididymal duct. Luminal diameter gradually increases by threefold between the cranial epididymal duct and the distal deferent duct (Tingari, 1971). One notable difference between the epididymal duct and the deferent duct are the layers of dense connective tissue and smooth muscle surrounding the mucosa of the latter (Aire and Josling, 2000; Tingari, 1971). A dense capillary network envelopes the excurrent ducts from the level of the proximal efferent ducts to the deferent duct (Nakai et al., 1988). The distal deferent duct straightens and then abruptly widens at its juncture with the cloaca. This structure, known as the receptacle of the deferent duct, has a bean-shaped appearance when it is engorged with semen. Each deferent duct terminates in the cloacal urodeum as a papilla immediately below the ostium of a ureter (Figure 29.4).

29.2.3 Accessory Organs

Accessory reproductive organs include the paracloacal vascular bodies, dorsal proctodeal gland, and lymphatic folds (Fujihara, 1992). The accessory reproductive organs are either in proximity to or are an integral part of the cloaca. As shown in Figure 29.4, the paracloacal vascular body is found alongside the receptacle of the deferent duct, and lymphatic folds exist within the wall of the proctodeum.

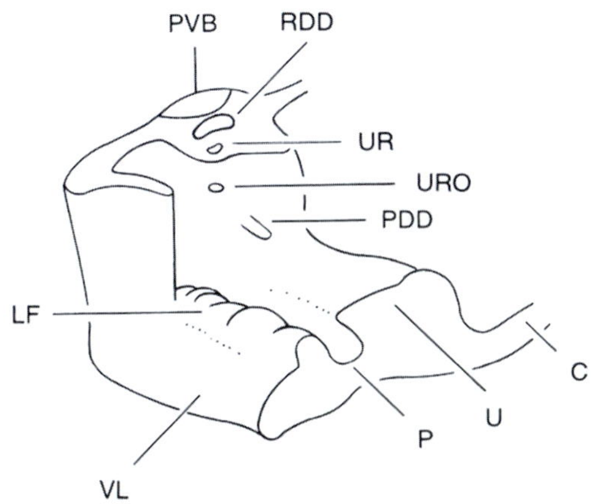

FIGURE 29.4 **Schematic of the lower left quadrant of the cloaca.** Before entering the cloacal wall, the deferent duct straightens then widens to form the receptacle of the deferent duct. The paracloacal body, also seen sectioned but shaded, is attached to the lateral aspect of the receptacle. The papilla of the deferent duct is within the urodeum of the cloaca. *PVB, paracloacal vascular body; RDD, receptacle of the deferent duct; C, ventral wall of coprodeum; U, ventral wall of urodeum; P, ventral wall of proctodeum; VL, ventral lip of vent; LF, lymphatic folds. Reproduced from King (1981), with permission from Academic Press.*

An intromittent phallus exists in only 3% of bird species (Herrera et al., 2013), including ratites and the waterfowl. In chickens, the phallus is nonintromittent and sperm is passed to the female via the "cloacal kiss" during mating (Briskie and Montgomerie, 2001). The cloacal kiss entails the union of the male and female cloacae for a few seconds. During copulation or in response to massage, a nonintromittent phallus forms a tumescent lymphatic tissue, which is everted through the vent immediately before ejaculation. As reviewed by Fujihara (1992), the paracloacal vascular bodies are essential for lymphatic tissue tumescence, as they are the sites where lymph is formed by ultrafiltration of blood. Thus, phallus erection in birds is lymphatic. In mammals, penile erection is blood-vascular and, during ejaculation, semen is transported via the urethra. However, in birds, semen is transported by the sulcus spermaticus, an external groove present in the phallus (Brennan et al., 2010).

Courtship behavior in male birds consists of searching and approaching a potential female (for an extensive review, see Ball and Balthazart, 2004). For instance, quails display a series of sexual behaviors that include mounting and aggressive pursuit for mating (Hutchison, 1978). Most of the evidence suggests that the medial portion of the preoptic area (POA) of the brain is responsible for the control of male behavior (Adkins and Adler, 1972; Konkle and Balthazart, 2011). The POA is significantly larger in males than females, and male castration causes a reduction of the volume of this area (Panzica et al., 1987). In addition, castration reduces the normal sexual behavior in males and exogenous testosterone restores the normal behavior (Ball and Balthazart, 2004). In castrated males, estrogen can also restore normal behavior, but co-injection of testosterone and an aromatase inhibitor blocks this effect (Balthazart et al., 1997). It is now widely recognized that aromatization of testosterone in the POA is responsible for the normal courtship behavior observed in male birds (Cornil et al., 2011; Wade and Arnold, 1996).

29.3 ONTOGENY OF THE REPRODUCTIVE TRACT

29.3.1 Overview

The adult male reproductive tract is derived from two embryonic organs: a gonad and its associated mesonephros. However, the formation of the mesonephros precedes that of the gonad (Figure 29.5). In fact, the gonad arises from the ventromesial surface of the mesonephros (Figure 29.6). As reviewed by Romanoff (1960), three distinct pairs of excretory organs appear sequentially during embryonic development: the pronephros, mesonephros, and metanephros. The most anterior organ, the pronephros, disappears by day 4 of incubation in the chicken. Nonetheless, the pronephric duct, also known as the Wolffian duct, persists through time to (1) induce the formation of the mesonephric tubules, (2) induce the formation of the Müllerian duct, (3) temporarily link the mesonephros with the cloaca, and (4) ultimately serve as the deferent duct in males. The critical point in the ontogeny of the reproductive tract is gonadal differentiation; prior to this point in time, the embryo has bipotential gonads and the rudiments of oviducts and deferent ducts in the form of the Müllerian and Wolffian ducts, respectively.

Sex determination in mammalian species is associated with the expression of the *SRY* gene located in the Y chromosome of the heterogametic XY male. The major function of the *SRY* gene is to facilitate Sertoli cell differentiation (via Sox9) that, in turn, promotes testis development as previously reviewed (Knower et al., 2003). In contrast to mammalian species, male birds are homogametic and females are heterogametic (ZZ and ZW gonosomes, respectively). Thus, the molecular mechanism of sex determination in birds is different from mammalian species. Because females are heterogametic, one possibility is that the W chromosome is responsible for female development. In this scenario, a factor (similar to the *SRY* gene in male mammals) will be coupled to female differentiation. However, very few W-linked genes have been reported in the literature (Hori et al., 2000; O'Neill et al., 2000), suggesting that other mechanisms are responsible for male development. A second possibility is that a gene in the Z chromosome triggers male development when two copies of the gene are present. The Z chromosome "dosage" theory states that higher doses of a gene can initiate male differentiation if the additional copy is transcribed and

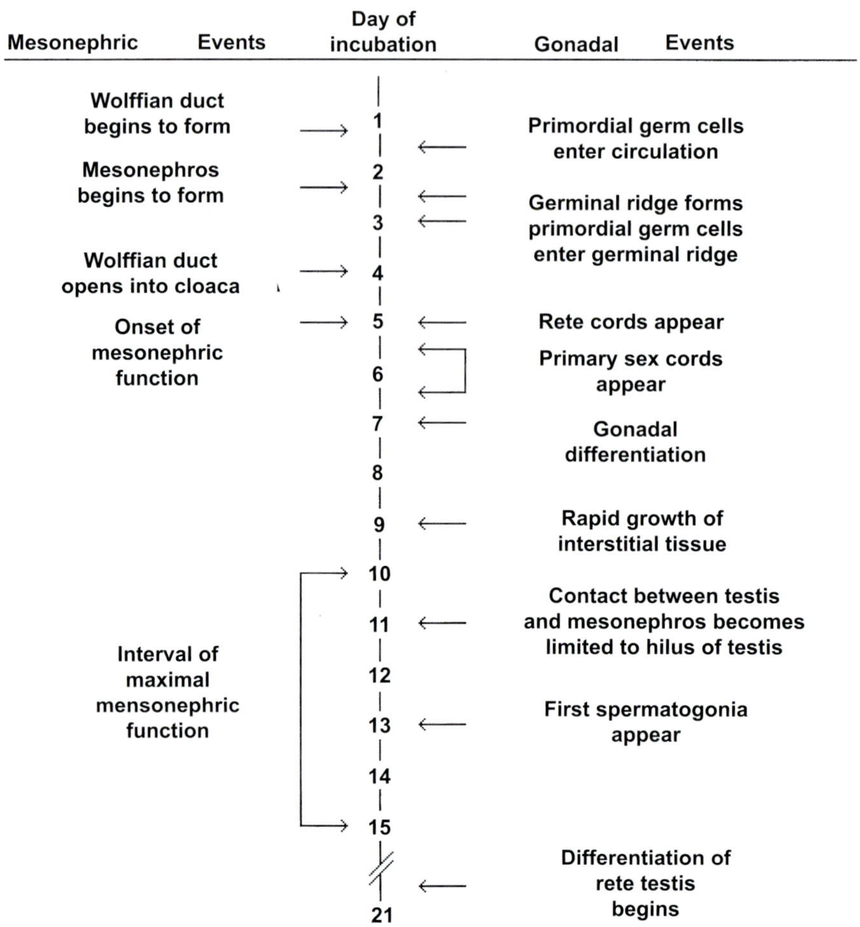

FIGURE 29.5 **Chronology of developmental events *in ovo* associated with the formation of the male reproductive tract in *Gallus domesticus*.** See text for details.

translated in higher doses of the resulting protein. It has been shown that the Doublesex and Mab-3 Related Transcription factor 1, encoded by the *DMRT1* gene, is a male sex-determining factor in chickens (Smith et al., 2009). In all birds studied to date, the *DMRT1* gene is present in the Z chromosome and absent in the W chromosome (Chue and Smith, 2011). Reduction of *DMART1* protein expression, using RNA interference (RNAi) technology in early male embryos, resulted in feminization of the gonads. These data and other support the Z chromosome "dosage" theory in birds (Chue and Smith, 2011; Nanda et al., 2008). Similar to mammalian species, there is evidence that activation of the male determining factor in birds facilitates Sertoli cell differentiation via Sox9 (Bagheri-Fam et al., 2010; Morais da Silva et al., 1996). Taken together, mammals and birds use a different molecular mechanism for sex differentiation (*SRY* and *DMRT1*, respectively). After genes are expressed, a common pathway that involves a Sertoli cell differentiation factor (Sox9) promotes testicular differentiation and development (Figure 29.7).

29.3.2 Formation of the Undifferentiated Gonad

Even though the gonad forms upon the ventromesial surface of the mesonephros, a fully differentiated mesonephros is not required for gonadal development (Merchant-Larios et al., 1984). The gonad is formed by the invasion of primordial germ cells into coelomic epithelium overlying a portion of the mesonephros (Figure 29.5), known as the germinal ridge (Clinton, 1998; Romanoff, 1960; Shimada, 2002). Primordial germ cells arise from endoderm along the anterior interface of the blastoderm's area opaca and area pellucida. After entering embryonic circulation, primordial germ cells are randomly distributed throughout the vasculature of the embryo (Meyer, 1964; Shimada, 2002). Subsequently, primordial germ cells colonize the germinal ridge (Fujimoto et al., 1976; Shimada, 2002). The undifferentiated gonad contains rete cords and primary sex cords in the anterior and middle regions, respectively (Romanoff, 1960).

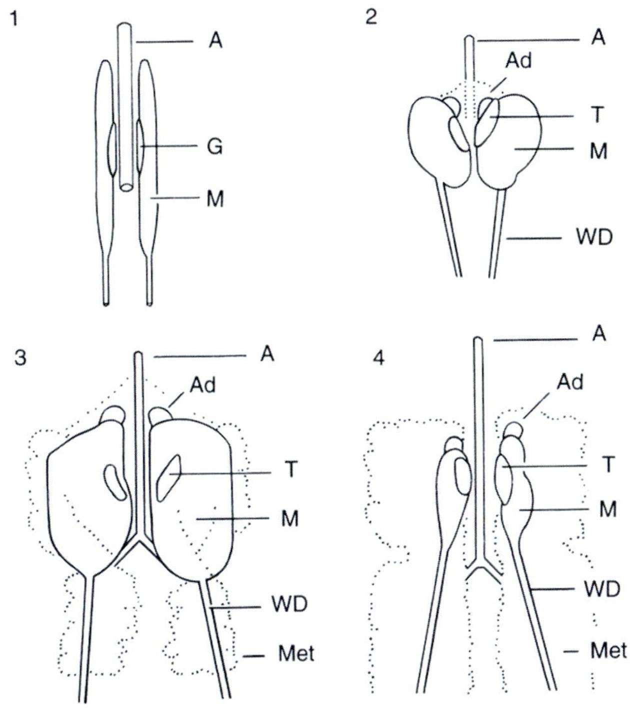

FIGURE 29.6 Stages in the development of the male urogential system. Numbers denote the first, second, third, or fourth quarter of incubation. The mesonephros forms before the differentiation of the gonad. After the metanephros becomes functional midway through incubation, mesonephric function declines and the mesonephros recedes apart from where it contacts the testis. A, aorta; G, undifferentiated gonad; M, mesonephros; Ad, adrenal gland; T, testis; WD, Wolffian duct; Met, metanephros. *Reproduced from Romanoff (1960), with permission from the MacMillan.*

29.3.3 Gonadal Differentiation and Müllerian Duct Regression

Gonadal differentiation in chick embryos occurs at approximately 6.5–7 days of incubation (Figure 29.5). In comparison to the ovary, the embryonic testis is characterized by a germinal epithelium that recedes with time, a thicker capsule, the absence of secondary or cortical sex cords, as well as the presence of primary sex cords surrounded by stroma (Clinton and Haines, 1999; Romanoff, 1960; Shimada, 2002). The primary sex cords contain numerous primordial germ cells and are anlage of the seminiferous tubules. Biochemically, gonadal differentiation is evident in terms of increased cyclic nucleotide concentrations (Teng, 1982), increased protein synthesis (Samsel et al., 1986), and the pattern of sex steroid synthesis (Guichard et al., 1977; Imataka et al., 1988; Mizuno et al., 1993). It is noteworthy that these traits become evident prior to the development of the pituitary–gonadal axis, which appears to become functional at 13.5 days of incubation in the chick (Woods and Weeks, 1969). It must also be noted that the interval of sexual bipotentiality differs among species.

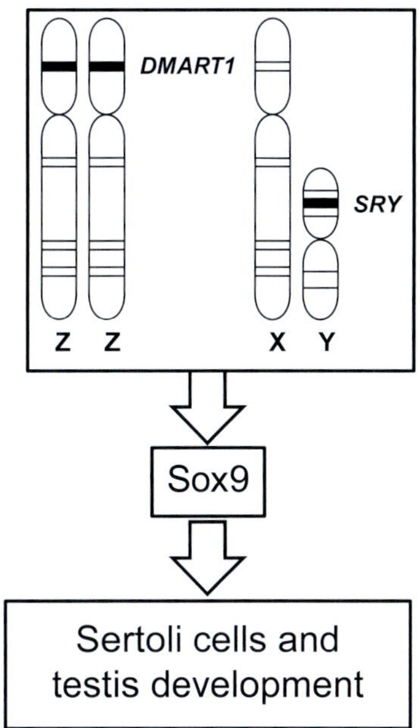

FIGURE 29.7 Differential mechanisms for sex differentiation in mammalian and avian species. In mammals, the XY male uses the *SRY* gene, while in birds the ZZ male uses the *DMRT1* gene to facilitate Sertoli cell differentiation (via *Sox9*) that, in turn, promotes testis development.

Once the testes have formed, the Müllerian ducts cease development and undergo regression in a caudocranial direction (Hutson et al., 1985; Romanoff, 1960). Müllerian duct regression is attributable to anti-Müllerian hormone (AMH), also known as Müllerian inhibiting substance (Bruggeman et al., 2002; Smith and Sinclair, 2004). This hormone is a 7.4kDa gonad-specific glycoprotein (Teng et al., 1987), which, in the male, is derived from the progenitors of Sertoli cells found within the primary sex cords (Oreal et al., 2002; Stoll and Maraud, 1973). The gene encoding AMH is expressed in both males and females, but is consistently higher in males during sexual differentiation (Smith et al., 1999). Regression of the Müllerian ducts in males occurs because they contain insufficient circulating estrogen levels to counteract the effect of AMH, which may be a direct result of aromatase suppression by AMH itself (Bruggeman et al., 2002; Hutson et al., 1985; MacLaughlin et al., 1983).

29.3.4 Formation of the Excurrent Ducts

The excurrent ducts are derived from the mesonephros. As shown in Figures 29.5 and 29.6, the mesonephros is active throughout the second and third quarters of incubation. The functional unit of the mesonephros includes a malphigian corpuscle, proximal, intermediate, and distal tubules, as well as a connecting tubule. The latter connects the series of mesonephric tubules with the Wolffian duct. The malpighian corpuscle is a capillary tuft or glomerulus within a double-walled epithelial enclosure known as the Bowman's capsule.

The visceral epithelial layer adheres to the glomerulus. At the pole of the corpuscle where the afferent and efferent arterioles are attached to the glomerulus within, the visceral epithelium folds back upon itself to form the parietal epithelium. As the parietal epithelium extends away from the vascular pole, it is separated from the visceral epithelium by a lumen. While the mesonephros is functional, glomerular filtrate enters the lumen of the proximal tubule. After the metanephros becomes functional during midincubation (Romanoff, 1960), the mesonephros begins to degenerate except where there is contact with the embryonic testis. Consequently, the mesonephros undergoes a profound change in size during embryonic development (Figure 29.6).

During the last third of incubation, a subset of mesonephric tubules are converted into the excurrent ducts of the testis (Budras and Sauer, 1975) and the malphighian corpuscles in proximity to the rete cords fuse. The glomeruli undergo progressive degeneration while the simple squamous epithelial cells of the parietal epithelium differentiate into simple columnar cells. This tissue becomes the epithelium of the proximal efferent ducts. Proximal, intermediate, and distal mesonephric tubules are transformed into distal efferent ducts, connecting ductules become the connecting ducts, the Wolffian duct associated with the mesonephros becomes the epididymal duct, and the distal Wolffian duct becomes the deferent duct.

The transformation of the mesonephric ductules into the excurrent ducts of the testis is dependent on androgen production by the Leydig cells (Bruggeman et al., 2002; Maraud and Stoll, 1955; Stoll and Maraud, 1974), and is initiated shortly after the concentration of blood plasma testosterone has peaked *in ovo* (Woods et al., 1975). However, the entire process of ductule conversion in the chicken requires an interval of 8–10 weeks (Budras and Sauer, 1975; Marvan, 1969). During this time, mean plasma testosterone levels stay constant at about 12–13% of those observed in sexually mature males (Driot et al., 1979; Tanabe et al., 1979). In contrast to the mesonephric tubules, differentiation of the rete cords begins at about the time of hatching and is complete by 5 weeks of age (Budras and Sauer, 1975). The convolutions of the newly formed epididymal and deferent ducts increase until sexual maturity (Budras and Sauer, 1975). The final length of the deferent duct has been estimated to be 4 times greater than the linear distance between the epididymis and cloaca due to these convolutions (Marvan, 1969).

29.4 DEVELOPMENT AND GROWTH OF THE TESTIS

29.4.1 Proliferation of Somatic and Stem Cells in the Testis

The testis of the mature bird is organized into discrete, easily discernible cellular associations and functional compartments (Figure 29.8(D)). However, during embryonic and early posthatch development this organization is less apparent. The posthatch development of the fowl's testis can be divided into three distinct phases: (1) proliferation of spermatogonia and the somatic cells that support spermatogenesis (Sertoli, peritubular myoid, and interstitial cells); (2) differentiation and the acquisition of functional competence by somatic support cells; and (3) spermatogonial differentiation resulting in the initiation of meiosis. Although the boundaries of these phases are not clearly defined, this three-step process results in functional seminiferous tubules that can sustain spermatogenesis when the appropriate hormonal cues are present. There is some evidence that FSH may play an important role in embryonic gonad growth by stimulating cell division. Treatment with FSH increased the mitotic index in both the seminiferous tubules and peritubular cells of chicken embryo testis at around 14 days of development (Mi et al., 2004).

The testis of the young cockerel (approximately 0–6 weeks of age) is characterized by an abundance of interstitial tissue and seminiferous tubules with only a single layer of cells within the basal lamina (Figure 29.8(A)). The majority of cells located within the seminiferous tubules at this time are Sertoli cells and spermatogonia, with macrophages and mast cells observed as well (de Reviers, 1971a). The seminiferous tubules at this time are narrow (approximately 40 μm in diameter) with either a poorly formed lumen or no apparent lumen at all. This is a period of rapid cellular proliferation: although the tubules are only slowly growing in diameter, they are growing rapidly in length (de Reviers, 1971b). Even though the somatic and stem cells are proliferating, the absolute weight and volume of the testis is increasing slowly as the seminiferous tubules displace interstitial cells (de Reviers, 1971a; Marvan, 1969).

29.4.2 Differentiation of Somatic Cells within the Testis

Following a posthatching period of Sertoli cell proliferation, these cells become mitotically quiescent and differentiate (for a review see Russell and Griswold, 1993). In a study by de Reviers and Williams (1984), hemicastration was shown to result in compensatory hypertrophy of the remaining testis of males up to 8 weeks of age. As hemicastration typically results in significant hypertrophy only during the period of Sertoli cell proliferation in mammals, it is possible to postulate that Sertoli cells are proliferating in the fowl to about 8 weeks of age. A more recent study by Bozkurt et al. (2007) provides support for this hypothesis. In this study, the mitotic proliferation of Sertoli cells was observed in the testis of fowl from one to 10 weeks of age. Initially, the density of germ cell nuclei was less than that of Sertoli cell nuclei; however, by week 6, the volume density of the two cell types had converged and remained

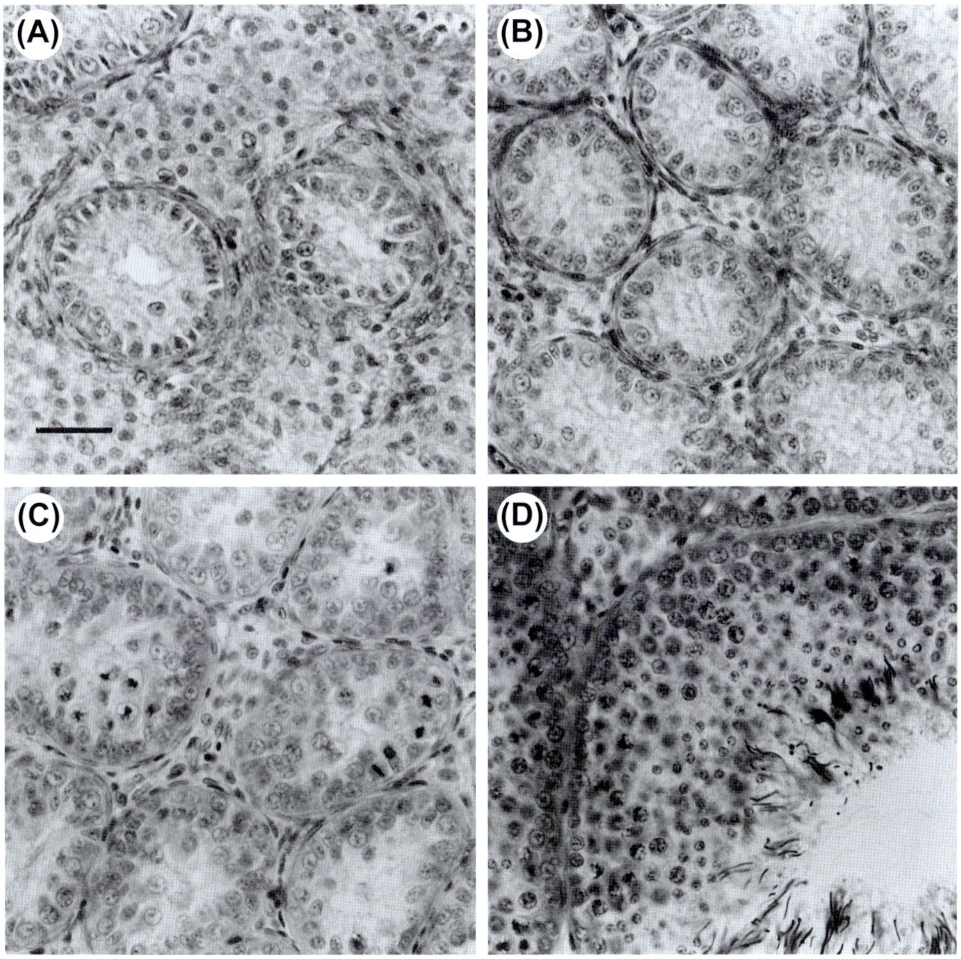

FIGURE 29.8 **Cross-sections of testes of cockerels reared under 8h of light (A–C) and of an adult male on 14h of light (D).** (A) Cross-section of a testis from a 14 day old male. Notice the simple single layer of spermatogonia and Sertoli cells in the tubules and the relative abundance of interstitial space. (B) Cross-section of a testis from a 56 day old cockerel. The single layer of cells within the seminiferous tubule has become taller and more complex. (C) Cross-section of a testis from a 140 day old cockerel. Notice the appearance of spermatocytes (cells with darkly stained chromatin) that have moved away from the basement membrane. (D) Cross-section of a testis from an adult male, which clearly demonstrates active spermatogenesis and the dramatic reduction in the relative area of the interstitium. The bar in (A) equals 20 μm; all of the micrographs are of equivalent magnification. *Photographs by the author.*

constant until the commencement of meiosis. In addition, clear labeling of both Sertoli and germ cells was observed from weeks one through seven but, by week eight, the only cells still labeled were germ cells. These data confirm that the cessation of Sertoli cell proliferation occurs around 8 weeks in domestic fowl.

Sertoli cells differentiate into highly polarized cells that extend from the basal lamina to the lumen of the seminiferous tubule. The mature Sertoli cell is a complex, columnar cell that contains numerous crypts into which developing sperm cells are embedded (Nagano, 1962; Sertoli, 1865, 1878). Sertoli cell functions are thought to be regulated by a myriad of endocrine and paracrine factors, although FSH and testosterone have been most thoroughly studied (Brown et al., 1975; Brown and Follett, 1977). A primary function of Sertoli cells is to provide a carefully regulated environment, including the sequestration of postmeiotic germ cells into the adluminal compartment is accomplished by the formation of tight junctions between adjacent Sertoli cells at a point basal to the maturing, meiotic, germ cell (Bergmann and Schindelmeiser, 1987; Pelletier, 1990). While tight junction formation appears to be androgen dependent in the fowl, tight junctions are maintained during periods of spermatogenic quiescence in the mallard (*Anas platyrhynchos*) (Pelletier, 1990). Another indicator of mature Sertoli cell function is the complex complement of proteins synthesized and secreted in response to androgen and/or FSH stimulation. While studied only superficially in male birds, spermatogenic stage-specific protein synthesis and secretion is well documented in other groups (see Russell and Griswold, 1993). As observed in other species, the close association of Sertoli and developing germ cells in the fowl suggests that they are intimately involved in the regulation of germ cell development (Cooksey and Rothwell, 1973).

The growth of the seminiferous tubules is associated with a reorganization and reduction of the testicular interstitium (Figure 29.8(A)–(D)). Although the seminiferous tubule is the compartment where spermatogenesis will occur in the adult, Leydig cells of the interstitium are responsible for testicular androgen production. In the interstitium, there is the presence of mesenchymal cell-like Leydig cell precursors, which have vesicles of smooth endoplasmic reticulum and lipid droplets (Connell, 1972). The differentiation of these cells into their mature form is under the influence of luteinizing hormone (LH) (Brown et al., 1975; Driot et al., 1979; Narbaitz and Adler, 1966). Mature Leydig cells are formed following a continuum of cellular differentiation that occurs during the period of testicular growth leading to the prepubertal increases of circulating LH levels (Rothwell, 1973). Although limited testosterone production is observed during embryogenesis and early development, the capacity to produce testosterone at sexually mature levels requires the presence of a fully differentiated Leydig cell population (for a review of testicular steroidogenesis see Johnson, 1986). The mature Leydig cell of the rooster has extensive smooth endoplasmic reticulum, a prominent Golgi complex, mitochondria with tubular cristae, abundant lipid droplets, and lysosomal elements (Rothwell, 1973).

29.4.3 Initiation of Meiosis

During the period of Sertoli and Leydig cell differentiation, the population of spermatogonia increases in size and complexity. As described later in the chapter, spermatogonia undergo a series of transformations prior to committing to meiosis. However, once a spermatogonium is transformed into a preleptotene spermatocyte, it separates from the basement membrane and shifts toward the adluminal compartment of the seminiferous tubule. The onset of active meiosis is marked by the presence of pachytene and zygotene spermatocytes which are easily discernible due to their highly condensed and darkly stained chromatin (Figure 29.8(C)). The initiation of meiosis appears to occur only after the completion of Sertoli cell proliferation and an increase in circulating androgens is encountered (de Reviers, 1971a,b). The onset of meiosis can be significantly altered by manipulation of the photoperiod (Ingkasuwan and Ogasawara, 1966). When broiler breeders were reared on a 23L:1D photoperiod and *ad libitum* food and water intake for 6 weeks males showed precocious testicular maturation with zygotene or pachytene, or both, spermatocytes visible in all seminiferous tubules at 7 weeks of age. In contrast, when broiler breeders were reared on a 15L:9D photoperiod and *ad libitum* food and water intake for only 2 weeks, the structures previously described were not present at 7 weeks of age (Vizcarra et al., 2010).

29.4.4 Altering the Pattern of Testis Growth and Maturation

The temporal component of testicular maturation can be altered by manipulation of the endocrine milieu. Specifically, onset of meiosis and sustained spermatogenesis can be altered dramatically in the fowl by manipulating the photoperiod (Ingkasuwan and Ogasawara, 1966). Male fowl reared on a long photoperiod (14 h light or more per day) will typically have reached sexual maturity by 16–20 weeks of age (Ingkasuwan and Ogasawara, 1966; Sharp and Gow, 1983). However, as shown in Figure 29.8(C), when males are reared on short days (in this case 8 h of light per day) the onset of spermatogenesis is delayed. Eventually, male fowl became refractory to short days, FSH and LH levels increase, and spermatogenesis commences. Conversely, precocious puberty and the onset of spermatogenesis can be accelerated in the fowl by chronic treatment with tamoxifen, an estrogen receptor antagonist (Rozenboim et al., 1986). Males treated with tamoxifen exhibited adult behaviors and produced viable spermatozoa by 9 weeks of age, months ahead of their nontreated siblings (Rozenboim et al., 1986).

Testicular maturation can also be altered by exposure to estrogen and estrogen-like compounds. Rooster embryos exposed to ethinylestradiol, a synthetic form of estrogen, exhibited significant reduction in the average area of their seminiferous tubules. In addition, exposed birds had a decreased proportion of seminiferous tubules and an increase in the proportion of interstitial tissue (Blomqvist et al., 2006). In Japanese quail, male embryos exposed to ethinylestradiol became feminized and had ovary-like tissue within the testis along with persistent Müllerian ducts (Berg et al., 1999).

29.5 HORMONAL CONTROL OF TESTICULAR FUNCTION

29.5.1 Central Control of Testicular Function

Spermatogenesis is the process in which the division of spermatogonial stem cells ultimately yields sperm cells while a population of stems cells is maintained. This complex phenomenon occurs within the seminiferous epithelium and depends upon the availability of testosterone and FSH, the activity of Sertoli cells, as well as interactions between germ cells and Sertoli cells (Sharpe, 1994). Although spermatogenesis is controlled, in part, by cells within the testis, the process is ultimately controlled by neurons within the central nervous system (CNS; see review by Sharp and Gow, 1983). Neurosecretory activity, in turn, is affected by somatic and environmental stimuli. Wingfield et al. (1992), have proposed a model accounting for the central integration of environmental information that affects gonadotropin secretion on a seasonal basis. The model is based upon the interplay of endogenous "predictive" rhythms and supplementary environmental information. The duration of the period of light in a subjective day (24h period) is the principal environmental factor that stimulates spermatogenesis in galliform birds. As reviewed by Kuenzel (1993), photoreceptors that affect spermatogenesis are neither found within the retina nor the pineal gland. Most of the evidence suggests that the most plausible site associated with the photoperiodic response is the medial basal hypothalamus (MBH) via direct innervations of GnRH neurons by encephalic photoreceptors (Saldanha et al., 2001), or via the pars tuberalis of the adenohypophysis (Yasuo and Yoshimura, 2009; Yoshimura, 2006, 2010). It has been reported that the pars tuberalis synthesize and secretes β-TSH (thyroid stimulating hormone) when Japanese quails are exposed to long-day stimulation (Yoshimura et al., 2003). In turn, TSH is retrogradely transported to the MBH via non-classical mechanisms (Krieger and Liotta, 1979; Mezey et al., 1979). In the MBH, TSH regulates the expression of genes encoding type 2 (*Dio2*) and type 3 (*Dio3*) deiodinases. Thyroxine (T4) is converted (within the MBH) to the bioactive T3 (triiodothyronine) form that, subsequently, stimulates the photoperiodic response of the gonadal axis (Yoshimura et al., 2003).

29.5.2 Control of Adenohypophyseal Function in Males

Gonadotropin-releasing hormone (GnRH) is a decapeptide that plays a fundamental role in the release of gonadotropins from the pituitary gland, and is the primary hormone that regulates reproduction. Andrew Schally and his team isolated and synthesized for the first time the decapeptide GnRH, after the extraction of more than 250,000 pig hypothalami (Matsuo et al., 1971; Wade, 1978). In addition to GnRH, the occurrence of other two GnRH isoforms was first reported in chickens more than 10years after of Dr Schally's initial report. The nomenclature used to distinguish different GnRH isoforms between mammalian and non-mammalian species have been described using a variety of phylogenetic and genomic synteny analyses (Kim et al., 2011; Millar et al., 2004; Roch et al., 2011; Tostivint, 2011). For the purpose of this chapter, we adopted the nomenclature based on the species in which they were first discovered, depicted in Table 29.2, and described elsewhere (Millar et al., 2004).

Chicken GnRH-I (King and Millar, 1982; Miyamoto et al., 1982) differs in only one amino acid compared with the mammalian form (cGnRH-I; [Gln8]-GnRH); whereas chicken GnRH-II (Miyamoto et al., 1984) differs in three amino acids (cGnRH-II; [His5-Trp7-Tyr8]-GnRH). The coordination of gonadotropin secretion via cGnRH is also modulated by the interaction of the GnRH peptide with its receptors. The Gonadotropin-releasing-hormone receptors (GnRHR) have the characteristic feature of a classical seven-transmembrane, G-protein-coupled receptor (Millar, 2003, 2005; Neill, 2002). Four vertebrate GnRHR lineages have been proposed using genome synteny and phylogenetic analyses; nonmammalian type I, nonmammalian type II, nonmammalian type III/mammalian type II, and mammalian type I (Kim et al., 2011).

Two GnRHRs have been identified in chickens (McFarlane et al., 2011; Shimizu and Bedecarrats, 2006). The non-mammalian type I receptor is predominantly express in the chicken pituitary gland (Joseph et al., 2009; Shimizu and Bedecarrats, 2006), whereas the nonmammalian type II receptor is not only expressed in the in the pituitary and the brain, but also in the in the gonads and other tissues (Sun et al., 2001a). The nonmammalian type I receptor

TABLE 29.2 Amino Acid (AA) Sequence of GnRH Isoforms

	1 2 3 4 5 6 7 8 9 10
Mammal (mGnRH)	**pGlu-His-Trp-Ser**-Tyr-Gly-Leu-Arg-**Pro-Gly**
Chicken-I (cGnRH-I)	**pGlu-His-Trp-Ser**-Tyr-Gly-Leu-Gln-**Pro-Gly**
Chicken-II (cGnRH-II)	**pGlu-His-Trp-Ser**-His-Gly-Trp-Tyr-**Pro-Gly**

The bolded regions represent the conserved NH_3- and COOH- terminal residues. The numbers represent the relative position of each AA in the GnRH peptide (1 represents the N-terminal AA).

in chickens (NCBI accession number: NP_001012627) has been previously referred as cGnRH-R-III (McFarlane et al., 2011), GnRHR1/III (Kim et al., 2011), or GnRHR2 (Shimizu and Bedecarrats, 2006), whereas the nonmammalian type II receptor in chickens (NCBI accession number: NP_989984) has been previously referred as cGnRH-R (Kim et al., 2011; Sun et al., 2001b) or cGnRHR 1 (Shimizu and Bedecarrats, 2006).

The affinity of the cGnRH-I peptide to the nonmammalian type II receptor is higher than the affinity of cGnRH-II peptide to the same receptor (Sun et al., 2001b). The nonmammalian type I receptor in the chicken pituitary is differentially expressed with respect to the reproductive status, and is associated with the control of gonadotropin secretion (McFarlane et al., 2011).

In the chicken, GnRHergic neurons arise within the olfactory epithelium on day 4.5 of embryonic development; reach the CNS by migrating along the olfactory nerve, and stop migrating by day 12 of embryonic development (Sullivan and Silverman, 1993). GnRHergic neurons are found within extrahypothalamic and hypothalamic sites (Kuenzel and Blahser, 1991). However, only those neurons whose axons terminate within the median eminence are believed to induce gonadotropin secretion *in vivo* (Mikami et al., 1988). Based upon immunocytochemical analysis, GNRH-positive axons are found within the median eminence by day 14 of embryonic development (Sullivan and Silverman, 1993). The significance of GnRHergic neurons was demonstrated well in advance of the purification of chicken GnRH with experiments that induced functional castration by either electrolytic lesions or deafferentation within the hypothalamus (Davies and Follett, 1980; Ravona et al., 1973). Axons of GnRHergic neurons terminate in proximity to capillaries within the median eminence of the hypothalamus. GnRH secreted from these axons reaches target cells within the adenohypophysis via the hypothalamo-hypophyseal portal vessels (Gilbert, 1979).

In avian species, only indirect measurements of the GnRH pulse generator is available by measuring plasma LH concentrations in frequent samples or in pituitary extracts (Chou and Johnson, 1987; Sharp and Gow, 1983; Wilson and Sharp, 1975). In addition, the episodic nature of LH and FSH was evaluated in unrestrained male broiler breeders using serial blood sampling (Vizcarra et al., 2004). Gonadotropin secretion in males is characterized by a pulsatile pattern with LH pulses being more frequent and having greater amplitude than FSH pulses (Figure 29.9). In chickens, LH- and FSH-containing gonadotrophs reside in separate cells within the pituitary gland (Proudman et al., 1999). We observed that there was a lack of synchrony between the episodic release of LH and FSH. Only 23% of the LH pulses were associated with FSH episodes, suggesting that in the adult male fowl LH and FSH secretion are regulated independently (Vizcarra et al., 2004).

Both cGnRH-I and -II stimulates gonadotropin release *in vivo* and *in vitro* in the chicken. However, most of the evidence indicates that cGnRH-I is the prime regulator of gonadotropin release in birds (Dawson and Sharp, 2007; Katz et al., 1990; Sharp et al., 1990; Stevenson et al., 2012; Ubuka and Bentley, 2009).

Nevertheless, the relative contribution and the potency of each cGnRH isoforms on the hypothalamic–pituitary–gonadal (HPG) axis of male birds is controversial. For instance, cGnRH-II was not found in the median eminence of the male white-crowned sparrow (*Zonotrichia leucophrys gambelii*), suggesting that in these species cGnRH-II does not regulate pituitary gonadotropin secretion (Meddle et al., 2006). On the other hand, concentrations of FSH in small cockerels were not affected by cGnRH-I challenges (Krishnan et al., 1993).

In immature cockerels (Chou et al., 1985) and male turkeys (Guemene and Williams, 1992), GnRH-II was more potent than GnRH-I *in vitro* and *in vivo* to release LH, suggesting that very small doses of GnRH-II at the pituitary level might be sufficient to induce gonadotropin release. However, no significant differences were observed between cGnRH-I and cGnRH-II, in LH releasing activity from young cockerels' pituitary samples (Hattori et al., 1986). When the LH response to cGnRH-I and -II was evaluated in cockerels, both peptides were equally potent (Sharp et al., 1987). However, in a similar experiment, the response to cGnRH-II, in terms of LH release was consistently greater than the response to GnRH-I in male chickens (Wilson et al., 1989). These data suggests that there may be age differences in the LH response to both decapeptides. Whether this differential response is due to different steroid inhibitory feedback mechanism is still unknown. The negative feedback of testicular steroids on GnRH-I have been reported in starlings (*Sturnus vulgaris*). Castrated male starlings significantly increased immunoreactive GnRH-I in the rostral POA of the hypothalamus, suggesting that this may represent the neuroanatomical location that mediates steroid negative feedback (Stevenson and Ball, 2009).

Active immunization against cGnRH-I and cGnRH-II in adult broiler breeder males was associated with a differential response on the ability to produce an immune reaction (Vizcarra et al., 2000). Titers were increased in cGnRH-I but not in cGnRH-II treated birds compared with BSA immunized males (Figure 29.10). Concentrations of LH and FSH in frequent samples were not affected by treatment; however, testis weight was significantly decreased in cGnRH-I birds compared with the other treatments (Figure 29.11). Taken together, these data support the idea that cGnRH-I is the prime regulator of gonadotropin release in male birds.

Behavioral, visual/auditory cues, and the breeding season also regulates GnRH-I and -II in birds. In male ring doves (*Streptopelia risoria*) and male starlings (*S. vulgaris*), courtship interaction increases cGnRH-I synthesis (Mantei

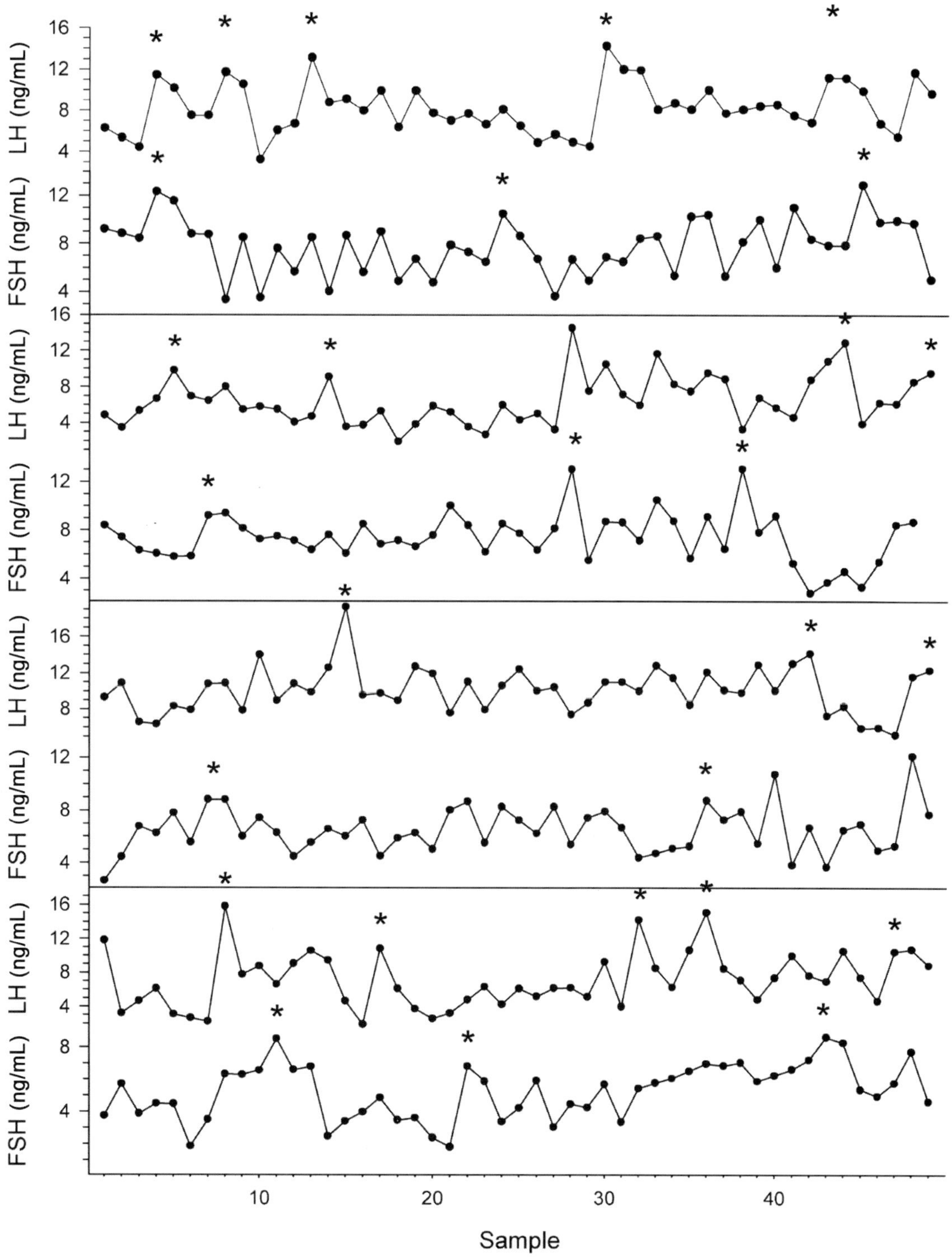

FIGURE 29.9 Pulsatile secretion of luteinizing hormone (LH) and follicle-stimulating hormone (FSH) in plasma of four birds. Blood samples were obtained every 10 min for 8 h. Asterisks indicate the presence of a pulse of LH or FSH, as determined by Pulsar. *Adapted from Vizcarra et al. (2004).*

et al., 2008; Stevenson and Ball, 2009), suggesting that sexual behavior can have a significant effect on the HPA axis. In most seasonal breeder birds, the stimulation of the gonadotropic axis is influenced by photoperiod (for a review see Dawson and Sharp, 2007). Hypothalamic photoreceptors transduce the energy from light into a biological signal that regulates the secretion of GnRH. Encephalic photoreceptors are in close proximity with GnRH-I neurons, indicating that brain photoreceptors communicate directly with GnRH-neurons and terminals (Saldanha et al., 2001). During the breeding season, GnRH-I increases in males birds including the house finches (*Carpodacus mexicanus*), the dark-eyed juncos

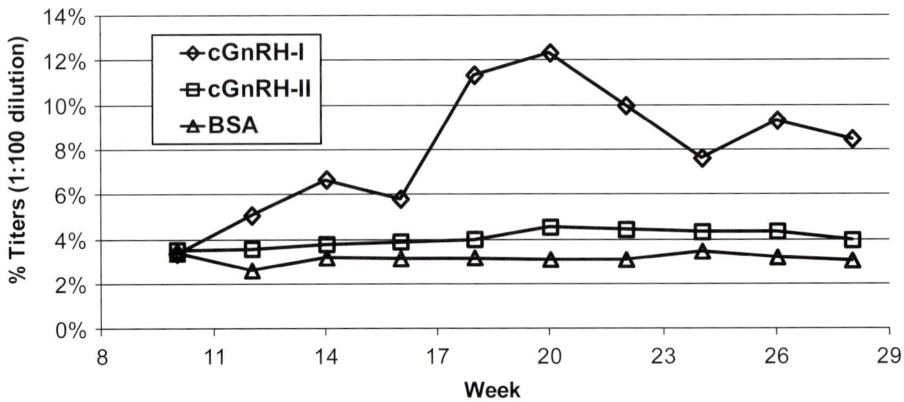

FIGURE 29.10 Antibody titers (^{125}I-cGnRH bound, %) of male broiler breeders immunized against cGnRH-I, cGnRH-II, and BSA. Titers were increased ($p<0.05$) in cGnRH-I but not in cGnRH-II treated birds compared with BSA-immunized males. *Adapted from Vizcarra et al. (2000).*

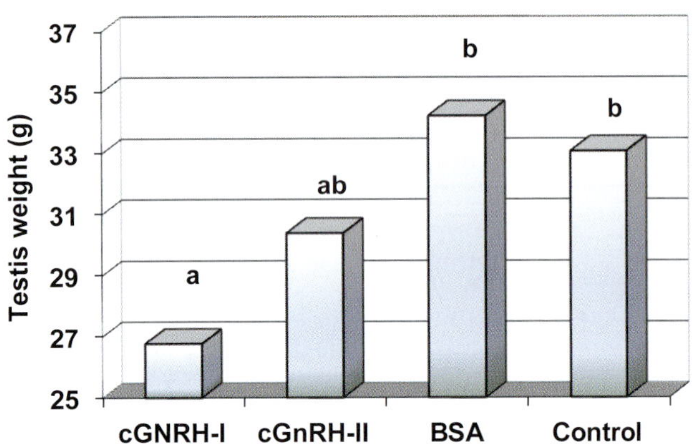

FIGURE 29.11 Testis weight of male broiler breeders immunized against cGnRH-I, cGnRH-II, BSA, and not immunized (control) birds. Different letters indicate significant differences ($p<0.05$). *Adapted from Vizcarra et al. (2000).*

(*Junco hyemalis*), and the European starlings (*S. vulgaris*) (Cho et al., 1998; Deviche et al., 2000; Foster et al., 1987). However, the contribution of GnRH-II during the breeding season (if any) is far from being fully elucidated. In nonbreeding mature male Zebra finch (*Taeniopygia guttata*) the number of cGnRH-II neurons is significantly reduced during the nonbreeding compared with the breeding season, but the area and optical density of the stained cells did not differ (Perfito et al., 2011).

Until recently, there was no known neuropeptide that directly inhibited gonadotropin secretion; however, Tsutsui et al. (2000) isolated gonadotropin-inhibitory hormone (GnIH), a hypothalamic dodecapeptide from the Japanese quail. This dodecapeptide consists of 12 amino acids in the sequence Ser(62)-Ile(252)-Lys(233)-Pro(226)-Ser(38)-Ala(194)-Tyr(173)-Leu(148)-Pro(104)-Leu(108)-Arg(45)-Phe(52) with RFamide at the C-terminus. GnIH significantly inhibits LH and FSH release in cultured quail cells (Tsutsui et al., 2000) and also inhibits circulating LH *in vivo* when administered continuously via osmotic pump (Ubuka et al., 2006). Interestingly, chronic GnIH treatment inhibits photoinduced testicular development, causes decreased plasma testosterone and testicular apoptosis, and results in decreased spermatogenic activity in the testis of adult quail (Ubuka et al., 2006). Similarly, testicular growth and plasma testosterone are suppressed in immature quail (Ubuka et al., 2006). Osugi et al. (2004) found that GnIH injected simultaneously with GnRH inhibited plasma LH surge in song sparrows and decreased LH concentrations in breeding free-living Gambel's white-crowned sparrows. GnIH reactive neurons were found to be localized within the paraventricular nucleus and mesencephalic regions of the hypothalamus in quail, white-crowned sparrows, and European starlings (Bentley et al., 2003; Tsutsui et al., 2000; Ubuka et al., 2003; Ukena et al., 2003). In addition, the GnIH receptor was found to be expressed in the thecal cells, Leydig cells, germ cells, and epididymis in male quail, which suggests roles for the hormone in regulating: gonadal steroid synthesis and release, differentiation of germ cells, and sperm maturation (Bentley et al., 2003, 2008).

Melatonin may be a key factor involved in GnIH neural functions because the neurons of male quail have been found to contain a receptor for melatonin (Ubuka et al., 2005). Administration of melatonin via injection stimulated the expression of GnIH in a dose-dependent way in birds that had undergone pinealectomy and orbital enucleation (Ubuka et al., 2005). In addition, hypothalamic explants from quail exposed to long-day photoperiods released greater amounts of GnIH during dark periods than during light periods *in vitro*. Conversely, LH concentrations decreased during dark periods. GnIH expression also increased under short-day periods when the duration of nocturnal melatonin secretion typically increases (Chowdhury et al., 2010). These results suggest a role for melatonin in stimulating the release of GnIH in birds and subsequently inhibiting gonadal activity and sex steroid levels by deceasing gonadotropin synthesis and release.

29.5.3 Effects of Gonadotropins on Testicular Function

Gonadotropins exert their effects on the testis by binding to specific cell-surface receptors on two distinct types of testicular parenchymal cells: Leydig and Sertoli cells. These cells have been described in detail by Rothwell (1973) and by Cooksey and Rothwell (1973), respectively. The principal role played by each type of cell has been determined by experiments in which one type of gonadotropin has been administered to hypophysectomized males, sexually immature males, or, in the case of Leydig cells, cells in culture (Brown et al., 1975; Ishii and Furuya, 1975; Ishii and Yamamoto, 1976). Such experiments have demonstrated that the principal effects of LH and FSH are exerted upon Leydig cells and Sertoli cells, respectively.

Leydig cells contain the steroidogenic enzymes necessary for the production of androgens (reviewed in Johnson, 1986) and respond rapidly to LH through rapid increases in the second messenger cAMP (Maung and Follett, 1977). The principal steroids secreted by Leydig cells include testosterone and androstenedione, a precursor of testosterone (Galli et al., 1973; Nakamura and Tanabe, 1972; Sharp et al., 1977). The concentration of testosterone in the general circulation is in the range of 5–15 nM, secreted as discrete pulses (Figure 29.12), which closely follow LH pulses (Bacon et al., 1991; Driot et al., 1979; Vizcarra et al., 2004), and are several times less than that in the testicular vein (Ottinger and Brinkley, 1979). The association between LH and testosterone pulses averaged 83% in unrestrained male broiler breeders fitted with jugular cannulas and free access to feed and water (Vizcarra et al., 2004). Testosterone in the adult male is essential for spermatogenesis, maintenance of the excurrent ducts and secondary sexual attributes, the expression of specific behaviors, and, as mentioned above, altering the pattern of GnRH secretion. These actions, as well as the inactivation of the hormone, depend upon transformation of testosterone by enzymes such as aromatase, 5α-reductase, and 5β-reductase (for reviews see Balthazart, 1989; Ottinger, 1983).

Although FSH is known to affect Sertoli cells, the means by which FSH works is only poorly understood in birds. As reviewed by Walker and Cheng (2005), studies in other vertebrates have shown that the FSH receptor is a seven transmembrane G-protein, and FSH binding is known to activate at least five different signaling pathways in Sertoli cells. These pathways include the cAMP-PKA, MAP kinase, phosphatidylinositol 3-kinase (pI3-K), calcium, and phospholipase A_2 (PLA_2) pathways. It has been proposed that oligomerization of FSH receptors may provide a mechanism for bringing together the components of these multiple signaling pathways in order to refine and control intracellular signaling. It is

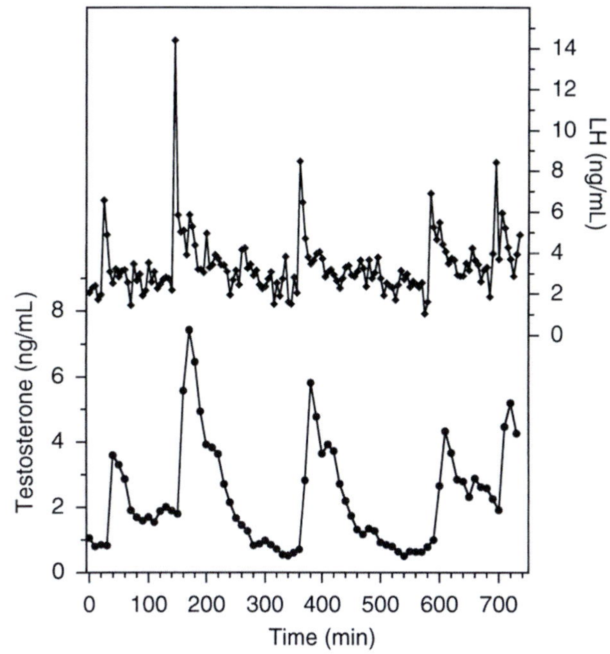

FIGURE 29.12 **The pattern of pulsatile changes in circulating luteinizing hormone (LH; upper panel) and testosterone (lower panel) in an adult male domestic turkey (*Meleagris gallopavo*).** The peaks of the discrete LH pulses occur prior to those of testosterone, demonstrating the close relationship between LH secretion and that of testosterone. *Figure and data kindly provided by Dr. Wayne Bacon, Ohio State University.*

well established that testosterone is absolutely essential for maintaining spermatogenesis and, unlike LH, the effect of FSH is potentiated by testosterone (Tsutsui and Ishii, 1978). As evidenced by the inability of exogenous testosterone to maintain spermatogenesis in hypophysectomized quail (Brown and Follett, 1977), LH and FSH appear to be essential for spermatogenesis in galliform birds. However, in male broiler breeders, more than 90% of the variation on testis weight was explained by changes in FSH, while only 35% of the variation in testis weight was explained by changes in LH concentrations (Vizcarra et al., 2010). These results indicate that FSH is the most important regulator of Sertoli cell proliferation and differentiation. Although sperm production depends on the number of Sertoli cells (de Reviers and Williams, 1984; Sharpe, 1994), the process is ultimately controlled by the hypothalamus-pituitary-gonadal axis.

29.6 SPERMATOGENESIS AND EXTRAGONADAL SPERM MATURATION

29.6.1 Spermatogenesis

To date, spermatogenesis in galliform birds has been most fully characterized in the Japanese quail (*Coturnix coturnix japonica*). Four types of spermatogonial cells have been described in *Coturnix* (Lin and Jones, 1992; Lin et al., 1990).

These have been designated as spermatogonia A_d, A_{p1}, A_{p2}, and B. The letter subscripts indicate intensity of staining and denote "dark" and "pale," respectively. The numerical subscripts denote ultrastructural differences (Lin and Jones, 1992). Spermatogonia A_d are stem cells. Based upon the model of stem cell renewal and spermatogonial proliferation proposed by Lin and Jones (1992), each division of an A_d spermatogonium yields A_d and A_{p1} daughter cells. Thus, spermatogenesis in *Coturnix* commences with the mitosis of an A_d spermatogonium. Cell types derived from each A_{p1} spermatogonium include A_{p2} spermatogonia ($n=2$), B spermatogonia ($n=4$), primary spermatocytes ($n=8$), secondary spermatocytes ($n=16$), and spermatids ($n=32$). Secondary spermatocytes and spermatids are formed by the first and second meiotic divisions, respectively.

Spermiogenesis is the transformation of spermatids into sperm cells without further cell division. Spermiogenesis in *Coturnix* entails 12 distinct morphological *steps* (Lin and Jones, 1993; Lin et al., 1990). In comparison, 8–10 steps have been reported for *Gallus* (de Reviers, 1971b; Gunawardana, 1977; Tiba et al., 1993), 10 steps for the guinea fowl (Aire et al., 1980), 12 steps for the turkey (Aire, 2003), and 6 steps for the house sparrow (Goes and Dolder, 2002). Spermiogenesis entails formation of an acrosome and an axoneme, loss of cytoplasm, and replacement of nucleohistones with nucleoprotamine, which accompanies nuclear condensation (Gunawardana, 1977; Gunawardana and Scott, 1977; Nagano, 1962; Okamura and Nishiyama, 1976; Oliva and Mezquita, 1986; Sprando and Russell, 1988; Tingari, 1973). The reductions in cytoplasmic and nuclear volumes are striking; mature rooster sperm embody only 3% of the initial spermatid cell volume (Sprando and Russell, 1988). In summary, the seminiferous tubules contain Sertoli cells and a broad array of differentiating germinal cells including the various types of spermatogonia, spermatocytes, and spermatids.

However, germinal cells are not found in a physical continuum of differentiation within the seminiferous epithelium. Rather, they are found in distinct cellular associations. The cellular associations in *Coturnix* occupy an average of 17,902 μm^2 and contain an average of 13.5 Sertoli cells per association (Lin and Jones, 1990). There are 10 cellular associations in *Coturnix*, and each is referred to as a *stage* of the seminiferous epithelium (Figure 29.13). The seminiferous epithelium passes through each successive stage as a function of time at any fixed point. Once a complete series has been finished, a new series is initiated. Consequently, the series of stages is referred to as the cycle of the seminiferous epithelium. The duration of the cycle in *Coturnix* is 2.69 days (Lin et al., 1990). In comparison, the duration of the cycle of the seminiferous epithelium in *Gallus* has been estimated to be 3–4 days (de Reviers, 1968; Tiba et al., 1993). The time between the division of an A_d spermatogonium and spermiation, known as the duration of spermatogenesis is 12.8 days in *Coturnix* (Lin and Jones, 1992). Thus, 4.75 cycles of the seminiferous epithelium are required to produce 32 sperm cells from a single A_d spermatogonium at any fixed point within the seminiferous epithelium.

Types of cells										
	11	11	11	12	12					
	1	2	3	4	5	6	7	8	9	10
	P	P	P	P	P	P	Dp	Dk	An M	II
	B	B	B	L	Z	P	P	P	P	P
	Ap1	Ap1	Ap2	Ap2	Ap2	Ap2	B	B	B	B
	Ad	Ad	Ad	Ad	Ad	Ad	Ad	Ad	Ad	Ap1 Ad
Stage	I	II	III	IV	V	VI	VII	VIII	IX	X
Relative frequency	.12	.15	.24	.10	.08	.06	.09	.06	.04	.05

FIGURE 29.13 The cycle of the seminiferous epithelium of *Coturnix coturnix japonica*. The relative frequency of each cellular association, or stage, is shown below the stage's number. Ad, dark type A spermatogonia (stem cell); Ap1 and 2, pale type A spermatogonia; B, type B spermatogonia; L, leptotene primary spermatocytes; Z, zygotene primary spermatocytes; P, pachytene primary spermatocytes; Dp, diplotene primary spermatocytes; DK, diakinesis of primart spermatocytes; M, metaphase primary spermatocytes; An, anaphase primary spermatocytes; II, secondary spermatocytes; 1–12, step 1 through step 12 spermatids. *Adapted from Lin et al. (1990).*

A complete series of stages in the context of a seminiferous tubule is referred to as a *spermatogenic wave*. As shown in Figure 29.14, these waves in *Coturnix* are arranged helically along the length of seminiferous tubules (Lin and Jones, 1990). It must be noted that while sequential stages may be contiguous in space (Figure 29.14), they are not observed at a common frequency (Figure 29.13) because duration of each of the stages ranges from 2.5 to 15.5 h (Lin et al., 1990). Consequently, the prevalence of any given stage is directly proportional to the stage's duration (Lin et al., 1990).

Daily sperm production may be defined as the number of sperm produced per gram of testis per day. Daily sperm production in *Coturnix* has been estimated to be 92.5×10^6 sperm per gram of testis per day (Clulow and Jones, 1982). This estimate is equivalent to the DSP of $80–120 \times 10^6$ sperm per gram of testis reported for *Gallus* (de Reviers and Williams, 1984), though the estimate was much higher for barred Plymouth Rock and Nigerian domestic fowl (*Gallus domesticus*) ($2.41 \pm 1.17 \times 10^9$ and $0.076 \pm 0.71 \times 10^9$, respectively) (Orlu and Egbunike, 2009). These values denote the number of fully formed sperm released per day from the seminiferous

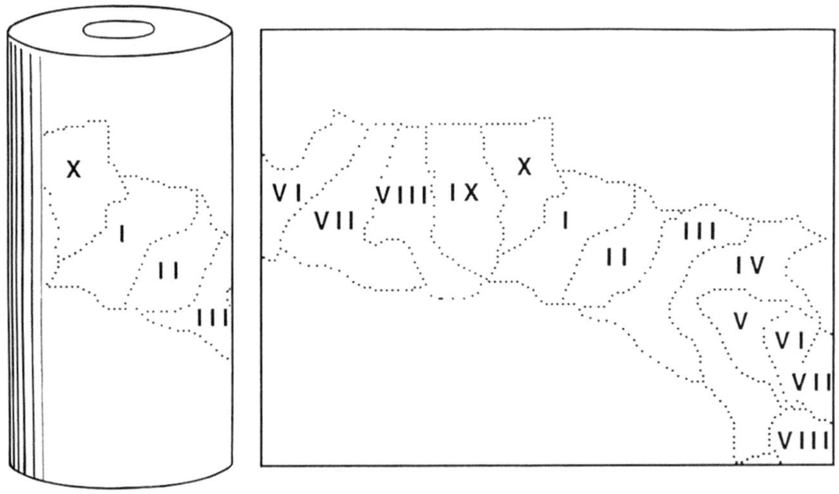

FIGURE 29.14 Spatial arrangement of the wave of spermatogenesis within a seminiferous tubule of *Coturnix coturnic japonica*. The cylinder to the right represents a section of a seminiferous tubule. The rectangle to its left denotes the two-dimensional representation of two contiguous cycles of the seminiferous epithelium. The cycle is arranged helically along the length of the tubule, Roman numerals denote stages. *Adapted from Lin and Jones (1990).*

epithelium into the lumen of the seminiferous tubules. This phenomenon is defined as spermiation, which in *Coturnix* would be limited to the seminiferous epithelium in stage V (Figure 29.13). At the time of spermiation, superfluous cytoplasm found alongside the sperm cell's head is jettisoned as a residual body (Lin and Jones, 1993; Sprando and Russell, 1988). Sperm cells released from the seminiferous epithelium are immotile (Ashizawa and Sano, 1990). Galliform sperm are vermiform cells (Figure 29.15) with a maximum width of 0.5–0.7 μm and a length of 75–90 μm (Thurston and Hess, 1987). They contain a conical acrosome, a slightly bent cylindrical nucleus, and a helix of 25–30 mitochondria surrounding the proximal portion of a long flagellum, which accounts for approximately 84% of the cell's length (for review, see Thurston and Hess, 1987). The sperm of passerine birds tends to have a pointed, helical acrosome and ranges in length from 46.8 to 287.6 μm with the flagellum accounting for 79–95% of the total length (Helfenstein et al., 2009; Lupold et al., 2009). One exception to this general trend is the Eurasian bullfinch, in which the sperm acrosome is rounded, not helical, and the midpiece size is extremely short (Birkhead et al., 2006). In these birds, extra-pair copulation frequency is rare in comparison to other passerines. It may be that a lack of sperm competition in this species could explain the marked difference in sperm morphology (Birkhead et al., 2006).

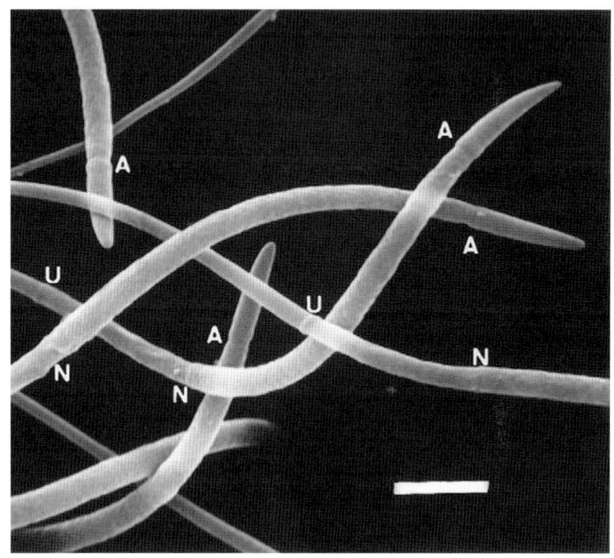

FIGURE 29.15 Scanning electron micrograph of chicken sperm cells. The constriction at the anterior end of the sperm cell (A) marks posterior boundary of the acrosome. The nucleus extends posteriorly from (A) to the neck region (N), which marks the anterior end of the midpiece. The midpiece, site of the anterior portion of the axoneme and the highly modified mitochondria, extends back to the raised annulus (U). The tail of the sperm extends from the annulus to the cell's termination. At the nucleus, chicken sperm are about 0.5 μm in diameter, with the overall length of the cell approximately 90 μm. Bar = 2 μm. *Adapted from Bahr, J.M., Bakst, M.R., 1987.* Reproduction in Farm Animals, *fifth ed. Philadelphia, PA: © Lea & Febiger.*

29.6.2 Extragonadal Sperm Transport and Maturation

Following spermiation, sperm cells are suspended within fluid secreted by Sertoli cells. Sperm passage through the labyrinth of seminiferous tubules most likely depends upon the hydrostatic pressure of seminiferous tubule fluid and the contractility of the myoepithelial cells overlying the outer surface of the seminiferous tubule (Rothwell and Tingari, 1973). Sperm passage through the distal excurrent ducts, in particular the deferent duct, presumably depends upon peristalsis (de Reviers, 1975). Like spermatogenesis, sperm transport through the excurrent ducts has been characterized most fully in the Japanese quail. Thus, an analysis of the phenomena and time-course of sperm transport through the excurrent ducts of *Gallus* has been estimated to take several days (de Reviers, 1975; Munro, 1938). Perhaps the most evident phenomenon that occurs during sperm transport through the excurrent ducts; as evidenced by a change in sperm concentration, is the absorption of seminiferous tubule fluid (Table 29.3). The concentration of sperm in the seminiferous tubule fluid of *Coturnix* is 3.8×10^4 per µl whereas the concentration of sperm within the distal deferent duct is 2.3×10^6 per µl (Clulow and Jones, 1988). This 60-fold increase in sperm concentration is largely due to seminiferous tubule fluid absorption at the level of efferent ducts. Nakai et al. (1989), demonstrated that the nonciliated epithelial cells in the proximal efferent ducts of *Gallus* incorporate fluid by pinocytosis.

The efferent ducts, which represent the principal excurrent duct within the epididymis (Figure 29.3), may be a critical site for sperm maturation. Due to extensive mucosal folding, apocrine secretion, the presence of ciliated cells, and epithelial cells with abundant microvilli, the efferent ducts appear to be a site where sperm are mixed with secretions as they are concentrated. Additional evidence of the importance of the efferent duct comes from studies of reproductive anomalies in turkeys and chickens. For instance, turkeys may ejaculate yellow rather than white semen (Thurston et al., 1982a). Epithelial cells within the efferent ducts of such turkeys have been shown to be hypertrophied, engorged with lipid droplets, and characterized by heightened phagocytosis of sperm (Hess et al., 1982). The latter phenomenon appears to be mediated by macrophages within the rete testis under normal conditions (Aire and Malmqvist, 1979b; Nakai et al., 1989), but can be mediated by epithelial cells in the excurrent ducts when the deferent duct is not patent (Tingari and Lake, 1972). Subfertile roosters with malformed proximal efferent ducts (Figure 29.16) provide a second example of the relationship between efferent duct function and reproductive performance (Kirby et al., 1990). The biochemical imbalances observed in semen from such roosters have been attributed to excurrent duct dysfunction (al-Aghbari et al., 1992).

TABLE 29.3 Estimates of Plasma Flux (Reabsorption) across the Epithelium Lining the Excurrent Ducts of Japanese Quail[a]

	Testicular Plasma	
Region	**Output (%)**	**$\mu l/cm^2/h$**
Rete testis to proximal efferent ducts	6.3	8.0
Proximal to distal efferent ducts	85.8	100.4
Distal efferent ducts to connecting ducts	6.5	21.6
Connecting ducts to epididymal duct	0.4	2.1
Epididymal duct to proximal deferent duct	0.2	0.1
Proximal to distal deferent duct	0.2	0.2

[a]*Adapted from Clulow and Jones (1988).*

In review, sperm are suspended in seminiferous tubule fluid prior to their passage through the excurrent ducts. The volume of this suspensory fluid and its composition (Esponda and Bedford, 1985) are altered prior to sperm entry into the epididymal duct. The resultant medium is seminal plasma. The chemical composition of seminal plasma is distinct from blood plasma as evidenced by differences in electrolyte, free amino acid, and protein composition (Freeman, 1984; Lake, 1966; Lake and Hatton, 1968; Stratil, 1970; Thurston et al., 1982b). The maintenance of these differences along the length of the excurrent ducts has been attributed to tight junctions between adjacent epithelial cells lining the ducts (Nakai and Nasu, 1991).

Apart from proteins specific to the reproductive tract (Esponda and Bedford, 1985; Stratil, 1970), the principal differences between seminal and blood plasma involve glucose, glutamate, K^+, Cl^-, and Ca^{2+} (Table 29.4). At present, it is not understood how the composition of excurrent duct fluid affects sperm motility. While Esponda and Bedford (1985) have demonstrated that sperm maturation proteins do exist in *Gallus* and while Ashizawa et al. (1988) have reported that the Na^+-to-K^+ ratio associated with the midpiece of rooster sperm increases as a function of sperm passage through the deferent duct, such changes do not induce motility; for sperm within the deferent duct are immotile (Ashizawa and Sano, 1990). Nonetheless, sperm acquire the potential for motility as they pass through the excurrent ducts (Ashizawa and Sano, 1990; Clulow and Jones, 1982; Howarth, 1983; Munro, 1938). This potential is distinct from fertilizing ability in that testicular sperm have the capacity to fertilize oocytes, providing they are placed in the oviduct above the vaginal sphincter (Howarth, 1983). As shown by Allen and Grigg (1957), sperm need not be motile to ascend the oviduct above the vaginal sphincter. Thus, the fertilizing ability of testicular sperm may not depend upon motility as much as the ability of sperm cells to undergo an acrosome reaction in response to contact with the oocyte's inner perivitelline layer (Okamura and Nishiyama, 1978).

The deferent duct, particularly the distal portion, contains the bulk of the extragonadal sperm reserve (ESR).

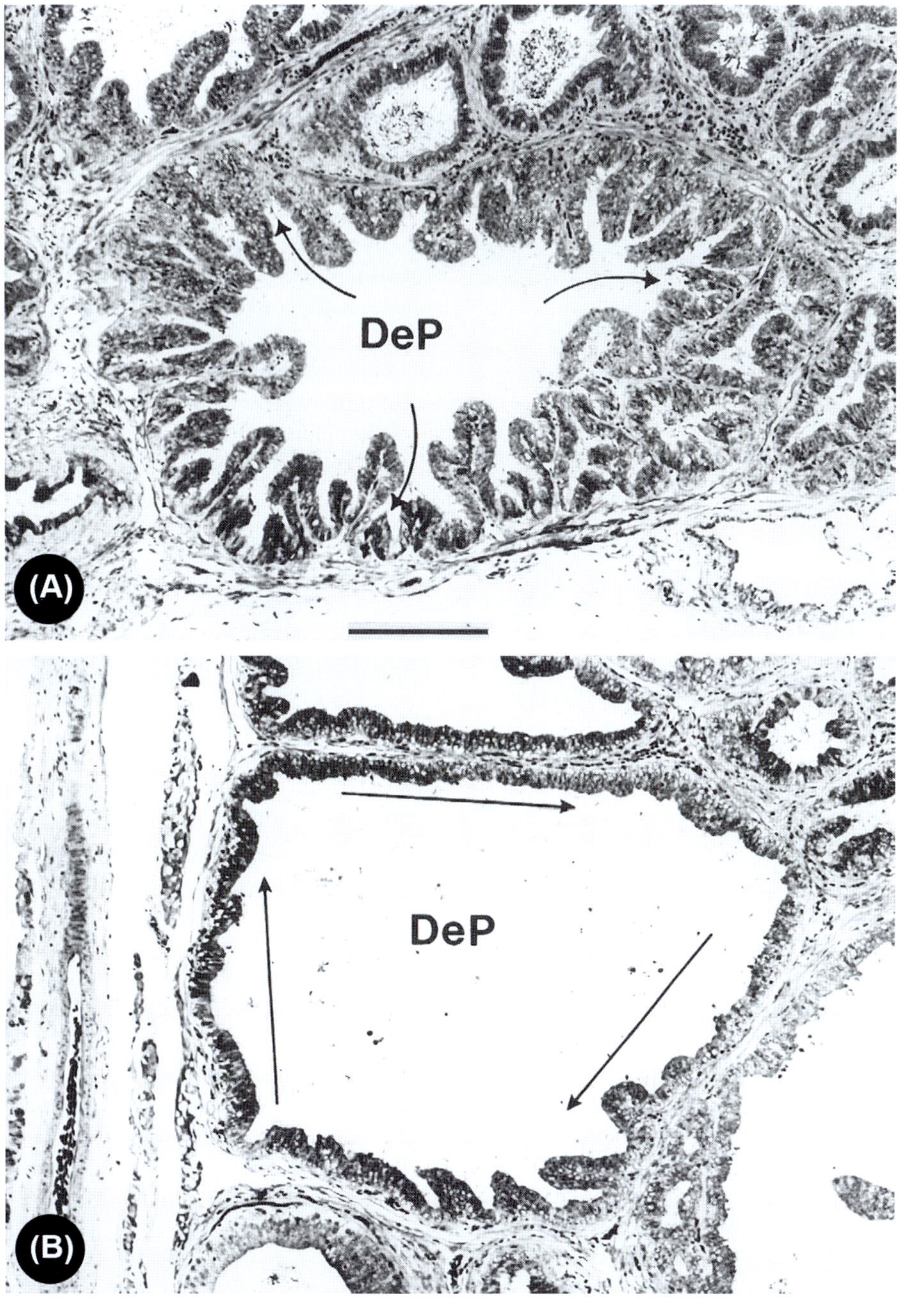

FIGURE 29.16 **Cross sections through the proximal deferent ducts (DeP) of fertile (A) and subfertile (B) roosters.** The presence of the *sd* allele in crossbred roosters (B) resulted in a twofold increase in DeP cross-sectional area and a 75% decrease in luminal surface area as compared to the wild-type (A) male. Males exhibiting this genetic defect of the DeP are characterized by poor semen quality, frequently exhibiting 30–90% dead and degenerate sperm in an ejaculate. *Reproduced with permission from the Society for the Study of Reproduction. From Kirby et al. (1990).*

In *Coturnix*, the number of sperm in the ESR is equivalent to the number produced daily by the testes and 92% of these are found within the different duct (Clulow and Jones, 1982). In contrast, the ESR in *Gallus* is equivalent to ≤3.5 times the daily sperm production, but 95% of the ESR is found in the deferent duct (de Reviers, 1975). In either case, the duration of sperm storage within the deferent duct is relatively brief. As stated above, these sperm are immotile prior to ejaculation. While Ashizawa and Sano (1990) have proposed that a temperature change accompanying ejaculation may initiate sperm motility in *Gallus*, it is also likely that accessory reproductive fluids play a role in the initiation of sperm motility (for reviews, see Ashizawa et al., 2000; Fujihara, 1992). In this regard, Ca^{2+} and HCO_3^- have been shown to be motility agonists *in vitro* (Ashizawa and Sano, 1990; Ashizawa and Wishart, 1987).

TABLE 29.4 Comparison of Selected Substances in Rooster Blood Plasma and Deferent Duct Fluid[a]

Substance	Blood Plasma	Deferent Duct Fluid
Glucose (nM/L)	15	—
Glutamate (mM/L)	< 1	100
Sodium (mEq/L)	160	142
Potassium (mEq/L)	5	34
Chloride (mEq/L)	115	40
Calcium (mEq/L)	5	1

[a]*Adapted from appendices in "Physiology and Biochemistry of the Domestic Fowl", Vol. 5, by B. M. Freeman.*

29.7 SEASONAL GONADAL RECRUDESCENCE AND REGRESSION

29.7.1 Photoperiodic Control of Gonadal Regression and Recrudescence

Many bird species, especially long-distance migrants, face the challenge of a narrow reproductive window because of an intense selective pressure to restrict breeding to the time of year when food is most abundant (reviewed in Dawson, 2008). Typically, male birds have regressed gonads for most of the year with gonadal recrudescence occurring only immediately prior to reproduction. Birds are generally viewed as being long-day breeders (with some exceptions) such that the HPG axis responds to increasing day lengths with a marked increase in gonadotropin secretion and gonadal growth (reviewed in Ball and Ketterson, 2008). The exact time and rate of gonadal maturation varies between bird species; however, it has been suggested that, as photoperiod increases during the spring, a critical threshold point is reached at which gonadal recrudescence is triggered. This threshold has been termed the critical photoperiod. A study in Japanese quail provides support for this hypothesis and found that the rate of gonadal maturation was low when the photoperiod was less than 11.5 h of light per day and maturation became rapid light exposure was increased to 12 h per day (Follett and Maung, 1978).

While gonadal maturation in the spring is in response to an increase in photoperiod, longer photoperiods at the end of the breeding season induce a condition termed photorefractoriness, which causes the gonads to regress (reviewed in Dawson, 2008; Robinson and Follett, 1982). Two forms of photorefractoriness, absolute and relative, have been identified in various bird species. In the absolute form, continued exposure to long photoperiods leads to spontaneous gonadal regression after which a further increase in photoperiod does not induce renewed gonadal maturation until the next breeding season. This state has been successfully demonstrated in species such as the European starling, tree sparrow, and white-crowned sparrow and is only dissipated when birds experience short days in autumn. Such exposure to short photoperiods is thought to induce reacquisition of photosensitivity (reviewed in Ball and Ketterson, 2008). In contrast, relative photorefractoriness is evident when a decrease in photoperiod is necessary to induce gonadal regression, but this regression occurs under a photoperiod that is still longer than the one which induced maturation earlier in the year. In addition, exposure to long photoperiods can inhibit regression and even induce recrudescence (reviewed by Dawson, 2008). In general, it is thought that birds with early, predictable breeding seasons tend to become absolutely photorefractory whereas those with later or less predictable breeding seasons are more likely to exhibit relative photorefractoriness.

29.7.2 Other Factors Affecting Gonadal Maturation and Regression

Due to the vast variety between and within species, birds exhibit plasticity in both gonadal maturation and regression. Though photoperiod plays an important role in regulating these processes, non-photoperiodic cues such as breeding location, temperature, rainfall, and food availability are also important in determining the rate and timing of recrudescence and regression. For example, Caro et al. (2006) observed two separate populations of blue tits and noticed that, not only did one population breed earlier than the other, but earlier breeding birds exhibited a faster rate of testicular development than did their later breeding counterparts. White-crowned sparrows breeding at low latitudes experienced inhibition of gonadal maturation and delayed onset of photorefractoriness at low temperatures, but the same were not true for birds of the same species breeding at mid latitudes (Wingfield et al., 2003). Retarded rates of testicular recrudescence were also observed in juncos subjected to low temperatures between 4 and 8 °C (Engels and Jenner, 1956). In willow tits, high temperatures were shown to accelerate testicular maturation, but the same effect was not evident in great tits (Silverin and Viebke, 1994). These data may suggest that temperature is more likely to affect species with longer and more flexible breeding seasons.

Interestingly, rainfall has been shown to affect gonadal maturation in several tropical bird species. For example, two populations of rufous-collared sparrows separated by only 25 km were observed to have highly asynchronous reproductive physiology based on local rainfall patterns (Moore et al., 2006). In the rufous-winged sparrow, though seasonality is known to be photoperiodically controlled, the exact time of breeding occurs in response to rainfall (Deviche et al., 2006). Darwin's finches undergo gonadal maturation at drastically different times in different years in direct response to rainfall, yet the gonads have been shown to fully regress between breeding seasons (Hau et al., 2004). Another factor which may influence seasonal recrudescence is the availability of food resources. Food availability at stopover has been directly linked to gonadal maturation during spring migration. For example, Garden Warblers experiencing limited food conditions at stopover had significantly slower testicular recrudescence and decreased testosterone in comparison to birds allowed to eat *ad libitum* (Bauchinger et al., 2009). It is clear that many factors, both intrinsic and extrinsic, have the potential to influence patterns of seasonal gonadal recrudescence and regression. The mechanisms driving such patterns are complex and likely overlap with one another. As research in this area progresses, it will be important to investigate how these mechanisms interact in order to better understand the various factors that drive reproductive success in the male bird.

REFERENCES

Adkins, E.K., Adler, N.T., 1972. Hormonal control of behavior in the Japanese quail. J. Comp. Physiol. Psychol. 81, 27–36.

Aire, T.A., 1979a. The epididymal region of the Japanese quail (*Coturnix coturnix japonica*). Acta Anat. 103, 305–312.

Aire, T.A., 1979b. Micro-stereological study of the avian epididymal region. J. Anat. 129, 703–706.

Aire, T.A., Malmqvist, M., 1979a. Intraepithelial lymphocytes in the excurrent ducts of the testis of the domestic fowl (*Gallus domesticus*). Acta Anat. 103, 142–149.

Aire, T.A., Malmqvist, M., 1979b. Macrophages in the excurrent ducts of the testes of normal domestic fowl (*Gallus domesticus*). Anat. Histol. Embryol. 8, 172–176.

Aire, T.A., 1980. The ductuli efferentes of the epididymal region of birds. J. Anat. 130, 707–723.

Aire, T.A., Olowo-okorun, M.O., Ayeni, J.S., 1980. The seminiferous epithelium in the guinea fowl (*Numida meleagris*). Cell Tissue Res. 205, 319–325.

Aire, T.A., 1982. The rete testis of birds. J. Anat. 135, 97–110.

Aire, T.A., Josling, D., 2000. Ultrastructural study of the luminal surface of the ducts of the epididymis of gallinaceous birds. Onderstepoort J. Vet. Res. 67, 191–199.

Aire, T.A., Soley, J.T., 2000. The surface features of the epithelial lining of the ducts of the epididymis of the ostrich (*Struthio camelus*). Anat. Histol. Embryol. 29, 119–126.

Aire, T.A., 2003. Ultrastructural study of spermiogenesis in the turkey, *Meleagris gallopavo*. Br. Poult. Sci. 44, 674–682.

Aire, T.A., Soley, J.T., 2003. The morphological features of the rete testis of the ostrich (*Struthio camelus*). Anat. Embryol. 207, 355–361.

Aire, T.A., Ozegbe, P.C., 2007. The testicular capsule and peritubular tissue of birds: morphometry, histology, ultrastructure and immunohistochemistry. J. Anat. 210, 731–740.

al-Aghbari, A., Engel Jr., H.N., Froman, D.P., 1992. Analysis of seminal plasma from roosters carrying the Sd (sperm degeneration) allele. Biol. Reprod. 47, 1059–1063.

Allen, T., Grigg, G., 1957. Sperm transport in the fowl. Aust. J. Agric. Res. 8, 788–789.

Ashizawa, K., Wishart, G.J., 1987. Resolution of the sperm motility-stimulating principle of fowl seminal plasma into Ca2+ and an unidentified low molecular weight factor. J. Reprod. Fertil. 81, 495–499.

Ashizawa, K., Ozawa, Y., Okauchi, K., 1988. Changes of elemental concentrations around and on the surface of fowl sperm membrane during maturation in the male reproductive tract and after in vitro storage. Gamete Res. 21, 23–28.

Ashizawa, K., Sano, R., 1990. Effects of temperature on the immobilization and the initiation of motility of spermatozoa in the male reproductive tract of the domestic fowl, *Gallus domesticus*. Comp. Biochem. Physiol. A Comp. Physiol. 96, 297–301.

Ashizawa, K., Wishart, G.J., Tsuzuki, Y., 2000. Avian sperm motility: environmental and intracellular regulation. Avian Poult. Biol. Rev. 11, 161–172.

Bacon, W.L., Proudman, J.A., Foster, D.N., Renner, P.A., 1991. Pattern of secretion of luteinizing hormone and testosterone in the sexually mature male turkey. Gen. Comp. Endocrinol. 84, 447–460.

Bagheri-Fam, S., Sinclair, A.H., Koopman, P., Harley, V.R., 2010. Conserved regulatory modules in the Sox9 testis-specific enhancer predict roles for SOX, TCF/LEF, Forkhead, DMRT, and GATA proteins in vertebrate sex determination. Int. J. Biochem. Cell Biol. 42, 472–477.

Bakst, M.R., 1980. Luminal topography of the male chicken and turkey excurrent duct system. Scanning Electron Microsc. 3, 419–425.

Ball, G.F., Balthazart, J., 2004. Hormonal regulation of brain circuits mediating male sexual behavior in birds. Physiol. Behav. 83, 329–346.

Ball, G.F., Ketterson, E.D., 2008. Sex differences in the response to environmental cues regulating seasonal reproduction in birds. Philos. Trans. R. Soc. Lond. B Biol. Sci. 363, 231–246.

Balthazart, J., 1989. Steroid metabolism and the activation of social behavior. In: Advances in Comparitive and Environmental Physiology, 3, pp. 103–159.

Balthazart, J., Castagna, C., Ball, G.F., 1997. Aromatase inhibition blocks the activation and sexual differentiation of appetitive male sexual behavior in Japanese quail. Behav. Neurosci. 111, 381–397.

Bauchinger, U., Van't Hof, T., Biebach, H., 2009. Food availability during migratory stopover affects testis growth and reproductive behaviour in a migratory passerine. Horm. Behav. 55, 425–433.

Bentley, G.E., Perfito, N., Ukena, K., Tsutsui, K., Wingfield, J.C., 2003. Gonadotropin-inhibitory peptide in song sparrows (*Melospiza melodia*) in different reproductive conditions, and in house sparrows (*Passer domesticus*) relative to chicken-gonadotropin-releasing hormone. J. Neuroendocrinol. 15, 794–802.

Bentley, G.E., Ubuka, T., McGuire, N.L., Chowdhury, V.S., Morita, Y., Yano, T., Hasunuma, I., Binns, M., Wingfield, J.C., Tsutsui, K., 2008. Gonadotropin-inhibitory hormone and its receptor in the avian reproductive system. Gen. Comp. Endocrinol. 156, 34–43.

Berg, C., Halldin, K., Fridolfsson, A.K., Brandt, I., Brunstrom, B., 1999. The avian egg as a test system for endocrine disrupters: effects of diethylstilbestrol and ethynylestradiol on sex organ development. Sci. Total Environ. 233, 57–66.

Bergmann, M., Schindelmeiser, J., 1987. Development of the blood-testis barrier in the domestic fowl (*Gallus domesticus*). Int. J. Androl. 10, 481–488.

Birkhead, T.R., Fletcher, F., Pellatt, E.J., 1998. Testes asymmetry, condition and sexual selection in birds: an experimental test. Proc. R. Soc. B Biol. Sci. 265, 1185–1189.

Birkhead, T.R., Immler, S., Pellatt, E.J., Freckleton, R., 2006. Unusual sperm morphology in the Eurasian bullfinch (*Pyrrhula Pyrrhula*). Auk 123, 383.

Blomqvist, A., Berg, C., Holm, L., Brandt, I., Ridderstrale, Y., Brunstrom, B., 2006. Defective reproductive organ morphology and function in domestic rooster embryonically exposed to o,p'-DDT or ethynylestradiol. Biol. Reprod. 74, 481–486.

Bozkurt, H.H., Aktas, A., Ulkay, M.B., Firat, U.B., 2007. Sertoli cell proliferation during the post hatching period in domestic fowl. J. Vet. Sci. 8, 219–222.

Brennan, P.L., Clark, C.J., Prum, R.O., 2010. Explosive eversion and functional morphology of the duck penis supports sexual conflict in waterfowl genitalia. Proc. Biol. Sci. 277, 1309–1314.

Briskie, J.V., Montgomerie, R., 2001. Efficient copulation and the evolutionary loss of the avian intromittent organ. J. Avian Biol. 32, 184–187.

Brown, N.L., Baylé, J.D., Scanes, C.G., Follett, B.K., 1975. Chicken gonadotrophins: their effects on the testes of immature and hypophysectomized Japanese quail. Cell Tissue Res. 156, 499–520.

Brown, N.L., Follett, B.K., 1977. Effects of androgens on the testes of intact and hypophysectomized Japanese quail. Gen. Comp. Endocrinol. 33, 267–277.

Bruggeman, V., Van As, P., Decuypere, E., 2002. Developmental endocrinology of the reproductive axis in the chicken embryo. Comparative biochemistry and physiology. Part A, Mol. Integr. Physiol. 131, 839–846.

Budras, K.D., Sauer, T., 1975. Morphology of the epididymis of the cock (*Gallus domesticus*) and its effect upon the steroid sex hormone synthesis. Anat. Embryol. 148, 175–196.

Caro, S.P., Lambrechts, M.M., Chastel, O., Sharp, P.J., Thomas, D.W., Balthazart, J., 2006. Simultaneous pituitary-gonadal recrudescence in two Corsican populations of male blue tits with asynchronous breeding dates. Horm. Behav. 50, 347–360.

Cho, R.N., Hahn, T.P., MacDougall-Shackleton, S., Ball, G.F., 1998. Seasonal variation in brain GnRH in free-living breeding and photorefractory house finches (*Carpodacus mexicanus*). Gen. Comp. Endocrinol. 109, 244–250.

Chou, H.F., Johnson, A.L., Williams, J.B., 1985. Luteinizing hormone releasing activity of [Gln8]-LHRH and [His5, Trp7, Tyr8]-LHRH in the cockerel, in vivo and in vitro. Life Sci. 37, 2459–2465.

Chou, H.F., Johnson, A.L., 1987. Luteinizing hormone secretion from anterior pituitary cells of the cockerel: evidence for an ultradian rhythm. Poult. Sci. 66, 732–740.

Chowdhury, V.S., Yamamoto, K., Ubuka, T., Bentley, G.E., Hattori, A., Tsutsui, K., 2010. Melatonin stimulates the release of gonadotropin-inhibitory hormone by the avian hypothalamus. Endocrinology 151, 271–280.

Chue, J., Smith, C.A., 2011. Sex determination and sexual differentiation in the avian model. FEBS J. 278, 1027–1034.

Clinton, M., 1998. Sex determination and gonadal development: a bird's eye view. J. Exp. Zool. 281, 457–465.

Clinton, M., Haines, L.C., 1999. An overview of factors influencing sex determination and gonadal development in birds. Cell Mol. Life Sci. 55, 876.

Clulow, J., Jones, R.C., 1982. Production, transport, maturation, storage and survival of spermatozoa in the male Japanese quail, *Coturnix coturnix*. J. Reprod. Fertil. 64, 259–266.

Clulow, J., Jones, R.C., 1988. Studies of fluid and spermatozoal transport in the extratesticular genital ducts of the Japanese quail. J. Anat. 157, 1–11.

Connell, C., 1972. The effect of luteinizing hormone on the ultrastructure of the leydig cell of the chick. Z. Zellforsch 128, 139–151.

Cooksey, E.J., Rothwell, B., 1973. The ultrastructure of the Sertoli cell and its differentiation in the domestic fowl (*Gallus domesticus*). J. Anat. 114, 329–345.

Cornil, C.A., Ball, G.F., Balthazart, J., Charlier, T.D., 2011. Organizing effects of sex steroids on brain aromatase activity in quail. PLoS One 6, e19196.

Davies, D.T., Follett, B.K., 1980. Neuroendocrine regulation of gonadotrophin-releasing hormone secretion in the Japanese quail. Gen. Comp. Endocrinol. 40, 220–225.

Dawson, A., Sharp, P.J., 2007. Photorefractoriness in birds–photoperiodic and non-photoperiodic control. Gen. Comp. Endocrinol. 153, 378–384.

Dawson, A., 2008. Control of the annual cycle in birds: endocrine constraints and plasticity in response to ecological variability. Philos. Trans. R. Soc. Lond. B Biol. Sci. 363, 1621–1633.

de Reviers, M., 1968. Determination de la duree des processus spermatogenetiques chez le coq a l'aide de thymidine tritice. 6th Int. Congr. Anim. Reprod. Paris 1, 183–185.

de Reviers, M., 1971a. Le developpement testiculaire chez le coq. II. Morphologie de l'epithelium seminifere et etablissement de la spermatogenese. Ann. Biol. Anim. Biochim. Biophys. 11, 531–546.

de Reviers, M., 1971b. Le developpement testiculaire chez le coq. I. Criissance Ponderale des testicules et devloppement des tubes seminiferes. Ann. Biol. Anim. Biochim. Biophys. 11, 519–530.

de Reviers, M., 1975. Sperm transport and survival in male birds. In: Hafez, E.S.E., Thibault, C.G. (Eds.), The Biology of Spermatozoa, pp. 10–16.

de Reviers, M., Williams, J.B., 1984. Testis development and production of spermatozoa in the cockerel (*Gallus domesticus*). In: Cunningham, F.J., Lake, P.E., Hewitt, D. (Eds.), Reproductive Biology of Poultry, pp. 183–202.

Denk, A.G., Kempenaers, B., 2005. Testosterone and testes size in mallards (*Anas platyrhynchos*). J. Ornithol. 147, 436–440.

Deviche, P., Saldanha, C.J., Silver, R., 2000. Changes in brain gonadotropin-releasing hormone- and vasoactive intestinal polypeptide-like immunoreactivity accompanying reestablishment of photosensitivity in male dark-eyed juncos (*Junco hyemalis*). Gen. Comp. Endocrinol. 117, 8–19.

Deviche, P., Small, T., Sharp, P., Tsutsui, K., 2006. Control of luteinizing hormone and testosterone secretion in a flexibly breeding male passerine, the Rufous-winged Sparrow, *Aimophila carpalis*. Gen. Comp. Endocrinol. 149, 226–235.

Driot, F.J., de Reviers, M., Williams, J., 1979. Plasma testosterone levels in intact and hemicastrated growing cockerels. J. Endocrinol. 81, 169–174.

Engels, W.L., Jenner, C.E., 1956. The effect of temperature on testicular recrudescense in juncos at different photoperiods. Biol. Bull. 110, 129–137.

Esponda, P., Bedford, J.M., 1985. Surface of the rooster spermatozoon changes in passing through the Wolffian duct. J. Exp. Zool. 234, 441–449.

Follett, B.K., Maung, S.L., 1978. Rate of testicular maturation, in relation to gonadotrophin and testosterone levels, in quail exposed to various artificial photoperiods and to natural daylengths. J. Endocrinol. 78, 267–280.

Foster, R.G., Plowman, G., Goldsmith, A.R., Follett, B.K., 1987. Immunohistochemical demonstration of marked changes in the LHRH system of photosensitive and photorefractory European starlings (*Sturnus vulgaris*). J. Endocrinol. 115, 211–220.

Freeman, B.M., 1984. Appendix X. Reproduction: semen. In: In: Freeman, B.M. (Ed.), Physiology and Biochemistry of the Domestic Fowl, 5, pp. 422–423.

Fujihara, N., 1992. Accessory reproductive fludis and organs in male domestic birds. Worlds Poult. Sci. J. 48, 39–56.

Fujimoto, T., Ukeshima, A., Kiyofuji, R., 1976. The origin, migration and morphology of the primordial germ cells in the chick embryo. Anat. Rec. 185, 139–145.

Galli, F.E., Irusta, O., Wassermann, G.F., 1973. Androgen production by testes of *Gallus domesticus* during postembryonic development. Gen. Comp. Endocrinol. 21, 262–266.

Gilbert, A.B., 1979. Glandulae endocrinae. In: Baumel, J.J. (Ed.), Nomina Anatomica Avium, pp. 337–342.

Goes, R.M., Dolder, H., 2002. Cytological steps during spermiogenesis in the house sparrow (*Passer domesticus, Linnaeus*). Tissue Cell 34, 273–282.

Guemene, D., Williams, J.B., 1992. In-vitro and in-vivo responses to chicken LHRH-I and chicken LHRH-II in male turkeys (*Meleagris gallopavo*). J. Endocrinol. 132, 387–393.

Guichard, A., Cedard, L., Mignot, T.M., Scheib, D., Haffen, K., 1977. Radioimmunoassay of steroids produced by cultured chick embryonic gonads: differences according to age, sex, and side. Gen. Comp. Endocrinol. 32, 255–265.

Gunawardana, V.E., 1977. Stages of spermatids in the domestic fowl: a light microscope study using Araldite sections. J. Anat. 123, 351–360.

Gunawardana, V.K., Scott, M.G., 1977. Ultrastructural studies on the differentiation of spermatids in the domestic fowl. J. Anat. 124, 741–755.

Hattori, A., Ishii, S., Wada, M., 1986. Effects of two kinds of chicken luteinizing hormone-releasing hormone (LH-RH), mammalian LH-RH and its analogs on the release of LH and FSH in Japanese quail and chicken. Gen. Comp. Endocrinol. 64, 446–455.

Hau, M., Wikelski, M., Gwinner, H., Gwinner, E., 2004. Timing of reproduction in a Darwin's finch: temporal opportunism under spatial constraints. Oikos 106, 489–500.

Helfenstein, F., Podevin, M., Richner, H., 2009. Sperm morphology, swimming velocity, and longevity in the house sparrow *Passer domesticus*. Behav. Ecol. Sociobiol. 64, 557–565.

Herrera, Ana M., Shuster, Simone G., Perriton, Claire L., Cohn, Martin J., 2013. Developmental basis of phallus reduction during bird evolution. Curr. Biol. 23, 1065–1074.

Hess, R.A., Thurston, R.J., 1977. Ultrastructure of epithelial cells in the epididymal region of the turkey (*Meleagris gallopavo*). J. Anat. 124, 765–778.

Hess, R.A., Thurston, R.J., Biellier, H.V., 1982. Morphology of the epididymal region of turkeys producing abnormal yellow semen. Poult. Sci. 61, 531–539.

Hori, T., Asakawa, S., Itoh, Y., Shimizu, N., Mizuno, S., 2000. Wpkci, encoding an altered form of PKCI, is conserved widely on the avian W chromosome and expressed in early female embryos: implication of its role in female sex determination. Mol. Biol. Cell 11, 3645–3660.

Howarth Jr., B., 1983. Fertilizing ability of cock spermatozoa from the testis epididymis and vas deferens following intramagnal insemination. Biol. Reprod. 28, 586–590.

Hutchison, R.E., 1978. Hormonal differentiation of sexual behavior in Japanese quail. Horm. Behav. 11, 363–387.

Hutson, J.M., Donahoe, P.K., MacLaughlin, D.T., 1985. Steroid modulation of Mullerian duct regression in the chick embryo. Gen. Comp. Endocrinol. 57, 88–102.

Imataka, H., Suzuki, K., Inano, H., Kohmoto, K., Tamaoki, B., 1988. Sexual differences of steroidogenic enzymes in embryonic gonads of the chicken (*Gallus domesticus*). Gen. Comp. Endocrinol. 69, 153–162.

Ingkasuwan, P., Ogasawara, F.X., 1966. The effect of light and temperature and their interaction on the semen production of white Leghorn males. Poult. Sci. 45, 1199–1206.

Ishii, S., Furuya, T., 1975. Effects of purified chicken gonadotropins on the chick testis. Gen. Comp. Endocrinol. 25, 1–8.

Ishii, S., Yamamoto, K., 1976. Demonstration of follicle stimulating hormone (FSH) activity in hypophyseal extracts of various vertebrates by the response of the Sertoli cells of the chick. Gen. Comp. Endocrinol. 29, 506–510.

Johnson, A.L., 1986. Reproduction in the male. In: Sturkie, P.D. (Ed.), Avian Physiology, pp. 432–451.

Joseph, N.T., Morgan, K., Sellar, R., McBride, D., Millar, R.P., Dunn, I.C., 2009. The chicken type III GnRH receptor homologue is predominantly expressed in the pituitary, and exhibits similar ligand selectivity to the type I receptor. J. Endocrinol. 202, 179–190.

Katz, I.A., Millar, R.P., King, J.A., 1990. Differential regional distribution and release of two forms of gonadotropin-releasing hormone in the chicken brain. Peptides 11, 443–450.

Kim, D.K., Cho, E.B., Moon, M.J., Park, S., Hwang, J.I., Kah, O., Sower, S.A., Vaudry, H., Seong, J.Y., 2011. Revisiting the evolution of gonadotropin-releasing hormones and their receptors in vertebrates: secrets hidden in genomes. Gen. Comp. Endocrinol. 170, 68–78.

King, A.S., 1979. Systema urogenitale. In: Baumel, J.J. (Ed.), Nomina Anatomica Avium, pp. 289–335.

King, A.S., 1981. Phallus. In: King, A.S., and McLelland, J. (Eds.), Form and Function in Birds, vol. 2, Academic Press, London, pp. 10–147.

King, J.A., Millar, R.P., 1982. Structure of chicken hypothalamic luteinizing hormone-releasing hormone. II. Isolation and characterization. J. Biol. Chem. 257, 10729–10732.

Kirby, J.D., Froman, D.P., Engel Jr., H.N., Bernier, P.E., Hess, R.A., 1990. Decreased spermatozoal survivability associated with aberrant morphology of the ductuli efferentes proximales of the chicken (*Gallus domesticus*). Biol. Reprod. 42, 383–389.

Knower, K.C., Kelly, S., Harley, V.R., 2003. Turning on the male–SRY, SOX9 and sex determination in mammals. Cytogenet. Genome Res. 101, 185–198.

Konkle, A.T., Balthazart, J., 2011. Sex differences in the rapid control of aromatase activity in the quail preoptic area. J. Neuroendocrinol. 23, 424–434.

Krieger, D.T., Liotta, A.S., 1979. Pituitary hormones in brain: where, how, and why? Science (New York, NY) 205, 366–372.

Krishnan, K.A., Proudman, J.A., Bolt, D.J., Bahr, J.M., 1993. Development of an homologous radioimmunoassay for chicken follicle-stimulating hormone and measurement of plasma FSH during the ovulatory cycle. Comp. Biochem. Physiol. Comp. Physiol. 105, 729–734.

Kuenzel, W.J., Blahser, S., 1991. The distribution of gonadotropin-releasing hormone (GnRH) neurons and fibers throughout the chick brain (*Gallus domesticus*). Cell Tissue Res. 264, 481–495.

Kuenzel, W.J., 1993. The search for deep encephalic photoreceptors within the avian brain, using gonadal development as a primary indicator. Poult. Sci. 72, 959–967.

Lake, P.E., 1966. Physiology and biochemistry of poultry semen. In: McLaren, A. (Ed.), Advances in Reproductive Physiology, 1, pp. 93–123.

Lake, P.E., Hatton, M., 1968. Free amino acids in the vas deferens, semen, transparent fluid and blood plasma of the domestic rooster, *Gallus domesticus*. J. Reprod. Fertil. 15, 139–143.

Lake, P.E., 1981. Male genital organs. In: King, A.S., McLelland, J. (Eds.), Form and Function in Birds, 2, pp. 2–61.

Lin, M., Jones, R.C., 1990. Spatial arrangement of the stages of the cycle of the seminiferous epithelium in the Japanese quail, *Coturnix coturnix japonica*. J. Reprod. Fertil. 90, 361–367.

Lin, M., Jones, R.C., Blackshaw, A.W., 1990. The cycle of the seminiferous epithelium in the Japanese quail (*Coturnix coturnix japonica*) and estimation of its duration. J. Reprod. Fertil. 88, 481–490.

Lin, M., Jones, R.C., 1992. Renewal and proliferation of spermatogonia during spermatogenesis in the Japanese quail, *Coturnix coturnix japonica*. Cell Tissue Res. 267, 591–601.

Lin, M., Jones, R.C., 1993. Spermiogenesis and spermiation in the Japanese quail (*Coturnix coturnix japonica*). J. Anat. 183 (Pt 3), 525–535.

Lupold, S., Calhim, S., Immler, S., Birkhead, T.R., 2009. Sperm morphology and sperm velocity in passerine birds. Proc. Biol. Sci. 276, 1175–1181.

MacLaughlin, D.T., Hutson, J.M., Donahoe, P.K., 1983. Specific estradiol binding in embryonic Mullerian ducts: a potential modulator of regression in the male and female chick. Endocrinology 113, 141–145.

Mantei, K.E., Ramakrishnan, S., Sharp, P.J., Buntin, J.D., 2008. Courtship interactions stimulate rapid changes in GnRH synthesis in male ring doves. Horm. Behav. 54, 669–675.

Maraud, R., Stoll, R., 1955. Action de la testosterone sur la constitution de l' epididme du. Coq. C. R. Soc. Biol. 149, 704–707.

Marvan, F., 1969. Postnatal development of the male genital tract of the *Gallus domesticus*. Anat. Anz. 124, 443–462.

Matsuo, H., Baba, Y., Nair, R.M., Arimura, A., Schally, A.V., 1971. Structure of the porcine LH- and FSH-releasing hormone. I. The proposed amino acid sequence. Biochem. Biophys. Res. Commun. 43, 1334–1339.

Maung, Z.W., Follett, B.K., 1977. Effects of chicken and ovine luteinizing hormone on androgen release and cyclic AMP production by isolated cells from the quail testis. Gen. Comp. Endocrinol. 33, 242–253.

McFarlane, H.O., Joseph, N.T., Maddineni, S.R., Ramachandran, R., Bedecarrats, G.Y., 2011. Development, validation, and utilization of a novel antibody specific to the type III chicken gonadotropin-releasing hormone receptor. Domest. Anim. Endocrinol. 40, 110–118.

Meddle, S.L., Bush, S., Sharp, P.J., Millar, R.P., Wingfield, J.C., 2006. Hypothalamic pro-GnRH-GAP, GnRH-I and GnRH-II during the onset of photorefractoriness in the white-crowned sparrow (*Zonotrichia leucophrys gambelii*). J. Neuroendocrinol. 18, 217–226.

Merchant-Larios, H., Popova, L., Reyss-Brion, M., 1984. Early morphogenesis of chick gonad in the absence of mesonephros (morphogenesis/gonad/mesonephric agenesis). Dev. Growth Differ. 26, 403–417.

Meyer, D.B., 1964. The migration of primordial germ cells in the chick embryo. Dev. Biol. 10, 154–190.

Mezey, É., Kivovics, P., Palkovits, M., 1979. Pituitary-brain retrograde transport. Trends Neurosci. 2, 57–60.

Mi, Y., Zhang, C., Xie, M., Zeng, W., 2004. Effects of follicle-stimulating hormone and androgen on proliferation of cultured testicular germ cells of embryonic chickens. Gen. Comp. Endocrinol. 138, 237–246.

Mikami, S., Yamada, S., Hasegawa, Y., Miyamoto, K., 1988. Localization of avian LHRH-immunoreactive neurons in the hypothalamus of the domestic fowl, *Gallus domesticus*, and the Japanese quail, *Coturnix coturnix*. Cell Tissue Res. 251, 51–58.

Millar, R.P., 2003. GnRH II and type II GnRH receptors. Trends Endocrinol. Metab. 14, 35–43.

Millar, R.P., Lu, Z.L., Pawson, A.J., Flanagan, C.A., Morgan, K., Maudsley, S.R., 2004. Gonadotropin-releasing hormone receptors. Endocr. Rev. 25, 235–275.

Millar, R.P., 2005. GnRHs and GnRH receptors. Anim. Reprod. Sci. 88, 5–28.

Miyamoto, K., Hasegawa, Y., Minegishi, T., Nomura, M., Takahashi, Y., Igarashi, M., Kangawa, K., Matsuo, H., 1982. Isolation and characterization of chicken hypothalamic luteinizing hormone-releasing hormone. Biochem. Biophys. Res. Commun. 107, 820–827.

Miyamoto, K., Hasegawa, Y., Nomura, M., Igarashi, M., Kangawa, K., Matsuo, H., 1984. Identification of the second gonadotropin-releasing hormone in chicken hypothalamus: evidence that gonadotropin secretion is probably controlled by two distinct gonadotropin-releasing hormones in avian species. Proc. Natl. Acad. Sci. U.S.A. 81, 3874–3878.

Mizuno, S., Saitoh, Y., Nomura, O., Kunita, R., Ohtomo, K., Nishimori, K., Ono, H., Saitoh, H., 1993. Sex-specific DNA sequences in galliformes and their application to the study of sex differentiation. In: Etches, R.J., Gibbins, A.M.V. (Eds.), Manipulation of the Avian Genome, pp. 257–274.

Moller, A.P., 1991. Sperm competition, sperm depletion, paternal care, and relative testis size in birds. Am. Nat. 137, 882–906.

Moller, A.P., Briskie, J.V., 1995. Extra-pair paternity, sperm competition and the evolution of testis size in birds. Behav. Ecol. Sociobiol. 36, 357–365.

Moore, I.T., Bentley, G.E., Wotus, C., Wingfield, J.C., 2006. Photoperiod-independent changes in immunoreactive brain gonadotropin-releasing hormone (GnRH) in a free-living, tropical bird. Brain Behav. Evol. 68, 37–44.

Morais da Silva, S., Hacker, A., Harley, V., Goodfellow, P., Swain, A., Lovell-Badge, R., 1996. Sox9 expression during gonadal development implies a conserved role for the gene in testis differentiation in mammals and birds. Nat. Genet. 14, 62–68.

Munro, S.S., 1938. Functional changes in fowl sperm during their passage through the excurrent ducts of the male. J. Exp. Zool. 79, 71–92.

Nagano, T., 1962. Observations on the fine structure of the developing spermatid in the domestic chicken. J. Cell Biol. 14, 193–205.

Nakai, M., Hashimoto, Y., Kitagawa, H., Kon, Y., Kudo, N., 1988. Microvasculature of the epididymis and ductus deferens of domestic fowls. Nihon Juigaku Zasshi 50, 371–381.

Nakai, M., Hashimoto, Y., Kitagawa, H., Kon, Y., Kudo, N., 1989. Histological study on seminal plasma absorption and spermiophagy in the epididymal region of domestic fowl. Poult. Sci. 68, 582–589.

Nakai, M., Nasu, T., 1991. Ultrastructural study on junctional complexes of the excurrent duct epithelia in the epididymal region in the fowl. J. Vet. Med. Sci. 53, 677–681.

Nakamura, T., Tanabe, Y., 1972. In vitro steroidogenesis by testes of the chicken (*Gallus domesticus*). Gen. Comp. Endocrinol. 19, 432–440.

Nanda, I., Schlegelmilch, K., Haaf, T., Schartl, M., Schmid, M., 2008. Synteny conservation of the Z chromosome in 14 avian species (11 families) supports a role for Z dosage in avian sex determination. Cytogenet. Genome Res. 122, 150–156.

Narbaitz, R., Adler, R., 1966. Submicroscopic observations on the differentiation of the chick gonads. J. Embryol. Exp. Morphol. 15, 41–47.

Neill, J.D., 2002. GnRH and GnRH receptor genes in the human genome. Endocrinology 143, 737–743.

Nickel, R., Schummer, A., Seiferle, E., Siller, W.G., Wright, P.A.L., 1977. Urogenital system. In: Anatomy of the Domestic Birds, pp. 70–84.

O'Neill, M., Binder, M., Smith, C., Andrews, J., Reed, K., Smith, M., Millar, C., Lambert, D., Sinclair, A., 2000. ASW: a gene with conserved avian W-linkage and female specific expression in chick embryonic gonad. Dev. Genes Evol. 210, 243–249.

Okamura, F., Nishiyama, H., 1976. The early development of the tail and the transformation of the shape of the nucleus of the spermatid of the domestic fowl, *Gallus gallus*. Cell Tissue Res. 169, 345–359.

Okamura, F., Nishiyama, H., 1978. The passage of spermatozoa through the vitelline membrane in the domestic fowl, *Gallus gallus*. Cell Tissue Res. 188, 497–508.

Oliva, R., Mezquita, C., 1986. Marked differences in the ability of distinct protamines to disassemble nucleosomal core particles in vitro. Biochemistry 25, 6508–6511.

Oreal, E., Mazaud, S., Picard, J.Y., Magre, S., Carre-Eusebe, D., 2002. Different patterns of anti-Mullerian hormone expression, as related to DMRT1, SF-1, WT1, GATA-4, Wnt-4, and Lhx9 expression, in the chick differentiating gonads. Dev. Dyn. 225, 221–232.

Orlu, E.E., Egbunike, G.N., 2009. Daily sperm production of the domestic fowl (*Gallus domesticus*) as determined by quantitative testicular histology and homogenate methods. Pak. J. Biol. Sci. 12, 1359–1364.

Osman, D.I., 1980. The connection between the seminiferous tubules and the rete testis in the domestic fowl (*Gallus domesticus*). Morphological study. Int. J. Androl. 3, 177–187.

Osugi, T., Ukena, K., Bentley, G.E., O'Brien, S., Moore, I.T., Wingfield, J.C., Tsutsui, K., 2004. Gonadotropin-inhibitory hormone in Gambel's white-crowned sparrow (*Zonotrichia leucophrys gambelii)*: cDNA identification, transcript localization and functional effects in laboratory and field experiments. J. Endocrinol. 182, 33–42.

Ottinger, M.A., Brinkley, H.J., 1979. Testosterone and sex related physical characteristics during the maturation of the male Japanese quail (*coturnix coturnix japonica*). Biol. Reprod. 20, 905 909.

Ottinger, M.A., 1983. Hormonal control of reproductive behavior in the avian male. Poult. Sci. 62, 1690–1699.

Ozegbe, P.C., Aire, T.A., Soley, J.T., 2006. The morphology of the efferent ducts of the testis of the ostrich, a primitive bird. Anat. Embryol. 211, 559–565.

Ozegbe, P.C., Aire, T.A., Madekurozwa, M.C., Soley, J.T., 2008. Morphological and immunohistochemical study of testicular capsule and peritubular tissue of emu (*Dromaius novaehollandiae*) and ostrich (*Struthio camelus*). Cell Tissue Res. 332, 151–158.

Ozegbe, P.C., Kimaro, W., Madekurozwa, M.C., Soley, J.T., Aire, T.A., 2010. The excurrent ducts of the testis of the emu (*Dromaius novaehollandiae*) and ostrich (*Struthio camelus*): Microstereology of the epididymis and immunohistochemistry of its cytoskeletal systems. Anat. Histol. Embryol. 39, 7–16.

Panzica, G.C., Viglietti-Panzica, C., Calacagni, M., Anselmetti, G.C., Schumacher, M., Balthazart, J., 1987. Sexual differentiation and hormonal control of the sexually dimorphic medial preoptic nucleus in the quail. Brain Res. 416, 59–68.

Pelletier, R.M., 1990. A novel perspective: the occluding zonule encircles the apex of the Sertoli cell as observed in birds. Am. J. Anat. 188, 87–108.

Perfito, N., Zann, R., Ubuka, T., Bentley, G., Hau, M., 2011. Potential roles for GNIH and GNRH-II in reproductive axis regulation of an opportunistically breeding songbird. Gen. Comp. Endocrinol. 173, 20–26.

Pitcher, T.E., Dunn, P.O., Whittingham, L.A., 2005. Sperm competition and the evolution of testes size in birds. J. Evol. Biol. 18, 557–567.

Proudman, J.A., Vandesande, F., Berghman, L.R., 1999. Immunohistochemical evidence that follicle-stimulating hormone and luteinizing hormone reside in separate cells in the chicken pituitary. Biol. Reprod. 60, 1324–1328.

Ravona, H., Snapir, N., Perek, M., 1973. The effect on the gonadal axis in cockerels of electrolytic lesions in various regions of the basal hypothalamus. Gen. Comp. Endocrinol. 20, 112–124.

Robinson, J.E., Follett, B.K., 1982. Photoperiodism in Japanese quail: the termination of seasonal breeding by photorefractoriness. Containing papers of a Biological character. Royal Society Proc. R. Soc. Lond. B Biol. Sci. 215, 95–116.

Roch, G.J., Busby, E.R., Sherwood, N.M., 2011. Evolution of GnRH: diving deeper. Gen. Comp. Endocrinol. 171, 1–16.

Romanoff, A.L., 1960. The urogenital system. In: The Avian Embryo. pp. 783–862.

Rothwell, B., 1973. The ultrastructure of Leydig cells in the testis of the domestic fowl. J. Anat. 116, 245–253.

Rothwell, B., Tingari, M.D., 1973. The ultrastructure of the boundary tissue of the seminiferous tubule in the testis of the domestic fowl (*Gallus domesticus*). J. Anat. 114, 321–328.

Rozenboim, I., Gvaryahu, G., Robinzon, B., Sayag, N., Snapir, N., 1986. Induction of precocious development of reproductive function in cockerels by tamoxifen administration. Poult. Sci. 65, 1980–1983.

Russell, L.D., Griswold, M.D., 1993. The Sertoli Cell. Cache River Press.

Saldanha, C.J., Silverman, A.J., Silver, R., 2001. Direct innervation of GnRH neurons by encephalic photoreceptors in birds. J. Biol. Rhythms 16, 39–49.

Samsel, J., Lorber, B., Petit, A., Weniger, J.P., 1986. Analysis of the cytosolic proteins of chick embryo gonads by two-dimensional gel electrophoresis. J. Embryol. Exp. Morphol. 94, 221–230.

Sertoli, E., 1865. De l'esistenze di particulari cellule ramifacate nei canalicoli seminiferi dell'testicolo umano. Morgagni 7, 31–40.

Sertoli, E., 1878. Sulla sturttura dei canalicoli seminiferi dei testicolo. Arch. Sci. Med. 2, 107–146. 267–295.

Sharp, P.J., Culbert, J., Wells, J.W., 1977. Variations in stored and plasma concentrations of androgens and luteinizing hormone during sexual development in the cockerel. J. Endocrinol. 74, 467–476.

Sharp, P.J., Gow, C.B., 1983. Neuroendocrine control of reproduction in the cockerel. Poult. Sci. 62, 1671–1675.

Sharp, P.J., Dunn, I.C., Talbot, R.T., 1987. Sex differences in the LH responses to chicken LHRH-I and -II in the domestic fowl. J. Endocrinol. 115, 323–331.

Sharp, P.J., Talbot, R.T., Main, G.M., Dunn, I.C., Fraser, H.M., Huskisson, N.S., 1990. Physiological roles of chicken LHRH-I and -II in the control of gonadotrophin release in the domestic chicken. J. Endocrinol. 124, 291–299.

Sharpe, R.M., 1994. Regulation of spermatogenesis. In: second ed. In: Knobil, E., Neill, J.D. (Eds.), The Physiology of Reproduction, 1, pp. 1363–1434.

Shimada, K., 2002. Sex determination and sex differentiation. Avian Poult. Biol. Rev. 13, 1–14.

Shimizu, M., Bedecarrats, G.Y., 2006. Identification of a novel pituitary-specific chicken gonadotropin-releasing hormone receptor and its splice variants. Biol. Reprod. 75, 800–808.

Silverin, B., Viebke, P.A., 1994. Low temperatures affect the photoperiodically induced LH and testicular cycles differently in closely related species of tits (*Parus spp.*). Horm. Behav. 28, 199–206.

Simões, K., Orsi, A.M., Artoni, S.M.B., Cruz, C. da., Schimming, B.C., Pinheiro, P.F.F., 2004. Structural features of the epididymal region of the domestic duck (*Anas plathyrynchos*). Braz. J. Vet. Res. Anim. Sci. 41, 92–97.

Smith, C.A., Smith, M.J., Sinclair, A.H., 1999. Gene expression during gonadogenesis in the chicken embryo. Gene 234, 395–402.

Smith, C.A., Sinclair, A.H., 2004. Sex determination: insights from the chicken. BioEssays: News Rev. Mol. Cell. Dev. Biol. 26, 120–132.

Smith, C.A., Roeszler, K.N., Ohnesorg, T., Cummins, D.M., Farlie, P.G., Doran, T.J., Sinclair, A.H., 2009. The avian Z-linked gene DMRT1 is required for male sex determination in the chicken. Nature 461, 267–271.

Sprando, R.L., Russell, L.D., 1988. Spermiogenesis in the red-ear turtle (*Pseudemys scripta*) and the domestic fowl (*Gallus domesticus*): a study of cytoplasmic events including cell volume changes and cytoplasmic elimination. J. Morphol. 198, 95–118.

Stevenson, T.J., Ball, G.F., 2009. Anatomical localization of the effects of reproductive state, castration, and social milieu on cells immunoreactive for gonadotropin-releasing hormone-I in male European starlings (*Sturnus vulgaris*). J. Comp. Neurol. 517, 146–155.

Stevenson, T.J., Hahn, T.P., Macdougall-Shackleton, S.A., Ball, G.F., 2012. Gonadotropin-releasing hormone plasticity: a comparative perspective. Front. Neuroendocrinol. 33, 287–300.

Stoll, R., Maraud, R., 1974. Le role du testicule dans la differenciation sexuell des gonoductes chez l' embryon des vertebres amniotes. Bull. Assoc. Anat. 58. 699–764.

Stoll, R.L.L., Maraud, R., 1973. Sur l' origine de l'hormone testiculaire responsible de la regression des canaux de Muller de l' embryon de Poulet. C.R. Soc. Biol. 167, 1092–1096.

Stratil, A., 1970. Studies on proteins of seminal fluid from the vasa deferentia of the cock, *Gallus gallus L*. Int. J. Biochem. 1, 728–734.

Sullivan, K.A., Silverman, A.J., 1993. The ontogeny of gonadotropin-releasing hormone neurons in the chick. Neuroendocrinology 58, 597–608.

Sun, Y.M., Dunn, I.C., Baines, E., Talbot, R.T., Illing, N., Millar, R.P., Sharp, P.J., 2001a. Distribution and regulation by oestrogen of fully processed and variant transcripts of gonadotropin releasing hormone I and gonadotropin releasing hormone receptor mRNAs in the male chicken. J. Neuroendocrinol. 13, 37–49.

Sun, Y.M., Flanagan, C.A., Illing, N., Ott, T.R., Sellar, R., Fromme, B.J., Hapgood, J., Sharp, P., Sealfon, S.C., Millar, R.P., 2001b. A chicken gonadotropin-releasing hormone receptor that confers agonist activity to mammalian antagonists. Identification of D-Lys(6) in the ligand and extracellular loop two of the receptor as determinants. J. Biol. Chem. 276, 7754–7761.

Tanabe, Y., Nakamura, T., Fujioka, K., Doi, O., 1979. Production and secretion of sex steroid hormones by the testes, the ovary, and the adrenal glands of embryonic and young chickens (*Gallus domesticus*). Gen. Comp. Endocrinol. 39, 26–33.

Teng, C., 1982. Ontogeny of cyclic nucleotides in embryonic chick gonads. Biol. Neonate 41, 123–131.

Teng, C.S., Wang, J.J., Teng, J.I., 1987. Purification of chicken testicular mullerian inhibiting substance by ion exchange and high-performance liquid chromatography. Dev. Biol. 123, 245–254.

Thurston, R.J., Hess, R.A., Froman, D.P., Biellier, H.V., 1982a. Elevated seminal plasma protein: a characteristic of Yellow turkey semen. Poult. Sci. 61, 1905–1911.

Thurston, R.J., Hess, R.A., Hughes, B.L., Froman, D.P., 1982b. Seminal plasma free amino acids and seminal and blood plasma proteins of the guinea fowl (*Numidia meleagris*). Poult. Sci. 61, 1744–1747.

Thurston, R.J., Hess, R.A., 1987. Ultrastructure of spermatozoa from domesticated birds: comparative study of turkey, chicken and guinea fowl. Scanning Microsc. 1, 1829–1838.

Tiba, T., Yoshida, K., Miyake, M., Tsuchiya, K., Kita, I., Tsubota, T., 1993. Regularities and irregularities in the structure of the seminiferous epithelium in the domestic fowl (*Gallus domesticus*). I. Suggestion of the presence of the seminiferous epithelial cycle. Anat. Histol. Embryol. 22, 241–253.

Tingari, M.D., 1971. On the structure of the epididymal region and ductus deferens of the domestic fowl (*Gallus domesticus*). J. Anat. 109, 423–435.

Tingari, M.D., 1972. The fine structure of the epithelial lining of the ex-current duct system of the testis of the domestic fowl (*Gallus domesticus*). Q. J. Exp. Physiol. Cogn. Med. Sci. 57, 271–295.

Tingari, M.D., Lake, P.E., 1972. Ultrastructural evidence for resorption of spermatozoa and testicular fluid in the excurrent ducts of the testis of the domestic fowl, *Gallus domesticus*. J. Reprod. Fertil. 31, 373–381.

Tingari, M.D., 1973. Observations on the fine structure of spermatozoa in the testis and excurrent ducts of the male fowl, *Gallus domesticus*. J. Reprod. Fertil. 34, 255–265.

Tostivint, H., 2011. Evolution of the gonadotropin-releasing hormone (GnRH) gene family in relation to vertebrate tetraploidizations. Gen. Comp. Endocrinol. 170, 575–581.

Tsutsui, K., Ishii, S., 1978. Effects of follicle-stimulating hormone and testosterone on receptors of follicle-stimulating hormone in the testis of the immature Japanese quail. Gen. Comp. Endocrinol. 36, 297–305.

Tsutsui, K., Saigoh, E., Ukena, K., Teranishi, H., Fujisawa, Y., Kikuchi, M., Ishii, S., Sharp, P.J., 2000. A novel avian hypothalamic peptide inhibiting gonadotropin release. Biochem. Biophys. Res. Commun. 275, 661–667.

Ubuka, T., Ueno, M., Ukena, K., Tsutsui, K., 2003. Developmental changes in gonadotropin-inhibitory hormone in the Japanese quail (*Coturnix japonica*) hypothalamo-hypophysial system. J. Endocrinol. 178, 311–318.

Ubuka, T., Bentley, G.E., Ukena, K., Wingfield, J.C., Tsutsui, K., 2005. Melatonin induces the expression of gonadotropin-inhibitory hormone in the avian brain. Proc. Natl. Acad. Sci. U.S.A. 102, 3052–3057.

Ubuka, T., Ukena, K., Sharp, P.J., Bentley, G.E., Tsutsui, K., 2006. Gonadotropin-inhibitory hormone inhibits gonadal development and maintenance by decreasing gonadotropin synthesis and release in male quail. Endocrinology 147, 1187–1194.

Ubuka, T., Bentley, G.E., 2009. Identification, localization, and regulation of passerine GnRH-I messenger RNA. J. Endocrinol. 201, 81–87.

Ukena, K., Ubuka, T., Tsutsui, K., 2003. Distribution of a novel avian gonadotropin-inhibitory hormone in the quail brain. Cell Tissue Res. 312, 73–79.

Vizcarra, J., Rhoads, M., Hsu, C., Washington, J., Morgan, J., Yang, J., Tang, H., Warren, J., Kirby, J., 2000. Effect of immunization against chicken gonadotropin releasing hormone-I (cGnRH-I) and cGnRH-II on reproductive function in adult broiler breeder males. Poult. Sci 79. Abs 343.

Vizcarra, J.A., Kreider, D.L., Kirby, J.D., 2004. Episodic gonadotropin secretion in the mature fowl: serial blood sampling from unrestrained male broiler breeders (*Gallus domesticus*). Biol. Reprod. 70, 1798–1805.

Vizcarra, J.A., Kirby, J.D., Kreider, D.L., 2010. Testis development and gonadotropin secretion in broiler breeder males. Poult. Sci. 89, 328–334.

Wade, J., Arnold, A.P., 1996. Functional testicular tissue does not masculinize development of the zebra finch song system. Proc. Natl. Acad. Sci. U.S.A. 93, 5264–5268.

Wade, N., 1978. Guillemin and schally: the years in the wilderness. Science 200, 279–282.

Walker, W.H., Cheng, J., 2005. FSH and testosterone signaling in Sertoli cells. Reproduction 130, 15–28.

Wilson, S.C., Sharp, P.J., 1975. Episodic release of luteinizing hormone in the domestic fowl. J. Endocrinol. 64, 77–86.

Wilson, S.C., Cunningham, F.J., Chairil, R.A., Gladwell, R.T., 1989. Maturational changes in the LH response of domestic fowl to synthetic chicken LHRH-I and -II. J. Endocrinol. 123, 311–318.

Wingfield, J.C., Hahn, T.P., Levin, R., Honey, P., 1992. Environmental predictability and control of gonadal cycles in birds. J. Exp. Zool. 261, 214–231.

Wingfield, J.C., Hahn, T.P., Maney, D.L., Schoech, S.J., Wada, M., Morton, M.L., 2003. Effects of temperature on photoperiodically induced reproductive development, circulating plasma luteinizing hormone and thyroid hormones, body mass, fat deposition and molt in mountain white-crowned sparrows, *Zonotrichia leucophrys oriantha*. Gen. Comp. Endocrinol. 131, 143–158.

Woods, J.E., Weeks, R.L., 1969. Ontogenesis of the pituitary-gonadal axis in the chick embryo. Gen. Comp. Endocrinol. 13, 242–254.

Woods, J.E., Simpson, R.M., Moore, P.L., 1975. Plasma testosterone levels in the chick embryo. Gen. Comp. Endocrinol. 27, 543–547.

Yasuo, S., Yoshimura, T., 2009. Comparative analysis of the molecular basis of photoperiodic signal transduction in vertebrates. Integr. Comp. Biol. 49, 507–518.

Yoshimura, T., Yasuo, S., Watanabe, M., Iigo, M., Yamamura, T., Hirunagi, K., Ebihara, S., 2003. Light-induced hormone conversion of T4 to T3 regulates photoperiodic response of gonads in birds. Nature 426, 178–181.

Yoshimura, T., 2006. Molecular mechanism of the photoperiodic response of gonads in birds and mammals. Comp. Biochem. Physiol. A Mol. Integr. Physiol. 144, 345–350.

Yoshimura, T., 2010. Neuroendocrine mechanism of seasonal reproduction in birds and mammals. Anim. Sci. J. 81, 403–410.

Chapter 30

Reproductive Behavior

Pierre Deviche
School of Life Sciences, Arizona State University, Tempe, AZ, USA

ABBREVIATIONS

ACTH Adrenocorticotropic hormone
AVT Arginine vasotocin
cGnRH-I Chicken gonadotropin-releasing hormone-I
CORT Corticosterone
DHT 5α-dihydrotestosterone
FSH Follicle-stimulating hormone
GnIH Gonadotropin-inhibitory hormone
GnRH Gonadotropin-releasing hormone
HPA Hypothalamo–pituitary–adrenal axis
HPG Hypothalamo–pituitary–gonadal axis
LH Luteinizing hormone
PRL Prolactin
POA Preoptic area
T Testosterone

30.1 INTRODUCTION

For a species to persist, reproduction is arguably the most important life history stage. In most species, reproduction is associated with the expression of some type of reproductive behavior, an orderly sequence of behavioral events that can include visual and auditory displays, copulation, nest building, oviposition, incubation, and care (including food provisioning) to young. Reproductive behavior is a crucial target of selection, and understanding its proximate bases can, therefore, help us address issues related to the persistence and evolution of organisms. Within vertebrates, reproductive behavior is widespread and exhibited by representatives of all classes, and it is thought to have evolved well before the emergence of birds. Indeed, similar to modern birds, nonavian maniraptoran dinosaurs probably exhibited a daily oviposition pattern, produced eggs with a complex shell microstructure, and protected their clutch (Zheng et al., 2013). As well, based on correlative analysis of clutch size and adult bone morphology, Varricchio et al. (2008) conclude that some Cretaceous dinosaurs exhibited paternal care and possibly had a polygamous mating system (Varricchio et al., 2008). More recently, ancestral phoenicopterids from the early Miocene are thought to have built floating nests, as do some of their extant relatives (grebes: Grellet-Tinner et al., 2012).

Given its distant origin, it is not surprising that within each vertebrate class, reproductive behavior varies enormously with respect to complexity, seasonal timing, duration, and control mechanisms. Birds have long been considered to be choice models to investigate the mechanisms that control the expression of reproductive behavior. Being endotherms, they thrive in diverse environments ranging from cold and hot deserts across a wide range of latitudes and altitudes to tropical and equatorial forests, as well as oceanic islands. Many species are, therefore, readily accessible to researchers. In addition, all aspects of avian reproductive behavior, as well as avian social systems, which range from social monogamy to polygyny and polyandry (dunnock, *Prunella modularis*: Langmore et al., 2002; Owens, 2002; spotted sandpiper, *Actitis macularius*: Rissman and Wingfield, 1984), are extremely diverse, and this diversity presents boundless opportunities for comparative work. Furthermore, many species are diurnal and colorful, and they sport a sexually dimorphic adult plumage. Their reproductive behavior often includes conspicuous visual (e.g., social displays, copulation, and parental feeding) and auditory (e.g., song) signals that are relatively easy to observe and quantify. Also of note, the expression of avian reproductive behavior is in many species seasonal and generally restricted to the time of year when environmental resources, in particular food, that are necessary for the optimal development of the offspring are most abundant and accessible. This expression is often regulated by hormones whose secretion likewise fluctuates seasonally. Not unexpectedly, therefore, birds have long been used as primary model organisms for correlative studies on hormones and reproductive behavior, and for mechanistic research aimed at identifying the hormonal and neural bases of this behavior during ontogeny and in adulthood. Finally, several species (chicken, *Gallus domesticus*; Japanese quail, *Coturnix coturnix japonica*; domesticated

Sturkie's Avian Physiology.

turkey, *Meleagris gallopavo*; ring dove, *Streptopelia risoria*; domesticated canary, *Serinus canaria*; and zebra finch, *Taeniopygia guttata*) have long been domesticated, and many aspects of their reproductive behavior and physiology have been studied extensively (see Sections 30.4.1, 30.6.2, and 30.6.5). This Knowledge gained from these studies, along with ready access, presents unsurpassed opportunities for in-depth mechanistic investigations (Ball and Balthazart, 2004; Balthazart et al., 2003).

30.2 REGULATION OF REPRODUCTIVE BEHAVIOR

The expression of reproductive behavior at maturity is regulated by a multitude of external factors—environmental and social—as well as internal factors, the nature and relative importance of which often vary interspecifically, between sexes, during ontogeny, as a function of adult age and experience, and according to the reproductive stage. A comprehensive review of the vast literature related to this regulation well exceeds the scope of the present chapter. Thus, we focus on a limited number of representative topics based on their potential interest to readers, progress made in the recent past, and the generality of conclusions that can be reached based on the available information. We begin with an analysis of environmental factors—in particular, light and food resources—and then social factors with well-documented effects on the reproductive system and behavior. Age and breeding experience can profoundly impact the adult reproductive system and the expression of reproductive behavior, and so we address the respective roles of these factors. Many studies have investigated the mechanisms by which peripheral hormones and neurohormones regulate reproductive behavior. We follow those sections with an analysis of this regulation, with particular attention paid to the role of gonadotropin-releasing hormone (GnRH), gonadotropin-inhibitory hormone (GnIH), arginine vasotocin (AVT), prolactin (PRL), gonadal steroids, and neurosteroids.

Three underlying themes run through the present chapter. First, we emphasize that gonadal hormones play an overwhelming role in the control of avian reproductive behavior. Any analysis of the mechanisms that regulate this behavior must, therefore, necessarily encompass a discussion of the mechanisms that control the activity of the reproductive system itself. For this reason, we provide examples drawn not only from the behavioral literature but also from the endocrine literature. Second, when addressing the (neuro)endocrine mechanisms that regulate reproductive behavior, one needs to take into account that this regulation is frequently reciprocal. For example, gonadal steroids profoundly influence the expression of reproductive behavior, but social and perhaps chemical signals from conspecifics can markedly affect the secretion of gonadal hormones. In some cases, even an individual's own behavior causes marked endocrine changes within this individual. Third, at the individual level, most endocrine systems do not operate independently but are functionally interrelated. Illustrating this point, gonadal steroids can markedly influence the production and secretion of behaviorally active neuropeptides such as adrenocorticotropic hormone and AVT. Furthermore, during stress, elevated secretion of hypothalamo-pituitary-adrenal (HPA) axis hormones can negatively affect aspects of reproductive physiology and behavior.

30.3 ENVIRONMENTAL FACTORS

Many environmental factors have the potential to influence the reproductive physiology and behavior of birds. These factors include weather-related events such as droughts (Reichert et al., 2012) and storms (Astheimer et al., 1995; Bolger et al., 2005), temperature (Ardia et al., 2009), and humidity (Cynx, 2001). Understanding the effects of these factors and how they affect reproduction is of considerable importance given the potential outcome of global climate change on biological systems (Beale et al., 2006; Senapathi et al., 2011; Shine and Brown, 2008; Moller, 2013), but this understanding is complicated by several factors. For example, weather-related events such as storms are generally perceived as stressful and often have negative reproductive consequences (e.g., they delay the onset of breeding and incubation behavior), but whether this is the case can be reproductive stage dependent (Wingfield, 1984). In addition, the effects of weather- or climate-related factors on the reproductive system can be direct (temperature: Ardia et al., 2009; Visser et al., 2009) but, alternatively or in addition, be indirect and result from changes, such as in food resources (drought: Bolger et al., 2005). Here we consider the role of two factors that have garnered considerable attention and have clear effects on the reproductive physiology and behavior of birds: light and food resources.

30.3.1 Light

In most avian species studied to date, exposure to sufficiently long days (i.e., above a minimum, species-specific threshold) stimulates a cascade of physiological events, including synthesis and secretion of chicken gonadotropin-releasing hormone-I (cGnRH-I). This stimulation ultimately causes gonadal development and increases the secretion of gonadal steroids (Dawson, 1999; Dawson et al., 2001; Stevenson et al., 2013), which play a primary role in the control of reproductive behavior (see Section 30.6.4). It is notable that even species such as the opportunistic zebra finch, which has traditionally been considered to be nonphotoperiodic, respond to very long days by activating their reproductive system (Bentley et al., 2000b). Due to its effects on the hypothalamo–pituitary–gonadal (HPG) axis, photostimulation in many cases exert effects that are indirect and mediated by elevated circulating concentrations of gonadal hormones. Limited evidence, however, suggests in some conditions that photoperiod influences avian reproductive behavior independent of its effects of the HPG axis. For example, castrated, androgen-treated ring doves exposed to long photoperiods exhibit higher levels of nest-building activity than short-photoperiod-exposed birds (McDonald and Liley, 1978). The mechanism mediating this effect remains unclear and may be related to changes in body condition. Indeed, birds exposed to long days have more time to feed and gain energy daily than short-day-exposed birds, and they may consequently be able to devote more resources to energy-consuming activities such as those associated with breeding (Perfito et al., 2008). Stimulatory effects of artificial urban lights on the timing of the dawn chorus of songbirds, as well as on their laying dates and male pairing success (Kempenaers et al., 2010; Longcore, 2010), may likewise result from extension of the daily feeding time and access to more food resources. In free-ranging birds, separating behavioral effects of photoperiod that are mediated by changes in energy intake from those that are secondary to elevated HPG hormone secretion has proved challenging, and additional work on this subject is clearly warranted.

30.3.2 Food Resources

Resource availability has long been thought to exert a critical influence on the timing of reproduction, including the expression of reproductive behavior, with parents generally raising young when food is seasonally most abundant. This view is supported by correlative and manipulative studies. For example, crossbills (*Loxia* spp.) feed primarily on conifer seeds, the abundance of which varies greatly and often unpredictably in time and space (Benkman, 1992). In these birds, food availability (along with day length and social factors) plays a determining role in the timing of reproduction (Benkman, 1990; Hahn, 1995; Deviche and Sharp, 2001). Another illustration is provided by research on blue tits (*Cyanistes caeruleus*). Blue tits preferentially feed their young noctuid lepidopteran caterpillars, but will also feed them less preferred tortricid lepidopteran caterpillars, which are smaller but easier to obtain and have a different phenology than noctuids. In this species, an adjustment of the timing of reproduction that is associated with an increased proportion of noctuids fed to young results in heavier nestlings (Garcia-Navas and Sanz, 2011).

Food resource manipulations—supplementation or restriction—are commonly used to investigate the role of these resources. With some exceptions (Harrison et al., 2010), food supplementation generally promotes the activity of the reproductive system as assessed by gonadal development, plasma levels of reproductive hormones, laying date, and expression of reproductive behavior (Hahn, 1995; Brommer et al., 2004; Schoech et al., 2004; Watts and Hahn, 2012). It is noted that these effects are, however, susceptible to modulation by various factors, including photoperiod and social interactions (Hahn, 1995) and predation pressure (Zanette et al., 2003). photoperiod and social interactions (Hahn, 1995), and predation pressure (Zanette et al., 2003). In addition, the interpretation of the findings from food manipulation studies needs to consider that short-term reproductive effects of food supplementation may differ from long-term effects. For example, food supplementation to song sparrow (*Melospiza melodia*) parents has transgenerational behavioral effects, causing these parents to produce more offspring than control parents. However, this apparent advantage is mitigated by the fact that food-supplemented parents produce smaller eggs and nestlings than control birds and that their male offspring have a smaller song repertoire—and are thus presumably less fit—than the offspring of control parents (Zanette et al., 2009; Figure 30.1).

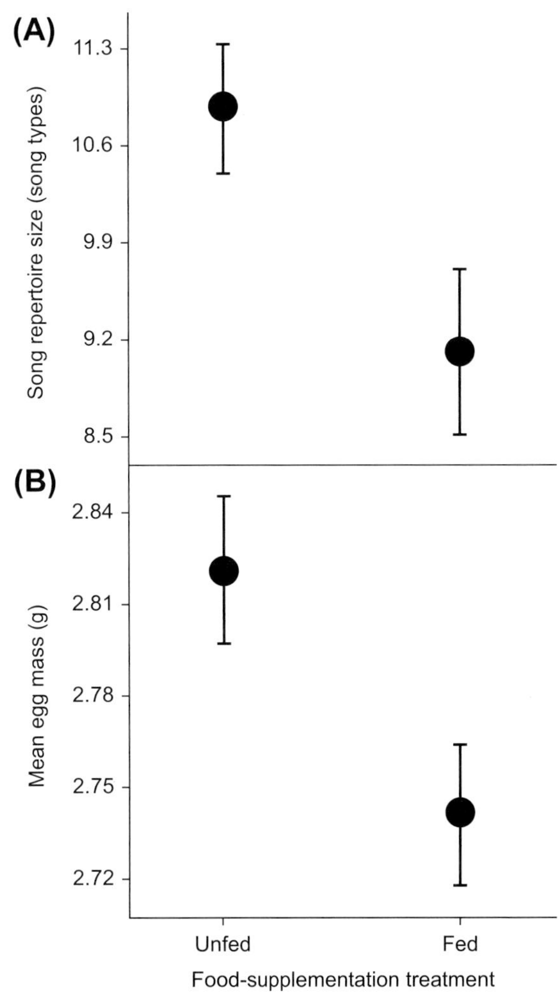

FIGURE 30.1 Food supplementation to song sparrow parents decreases (A) the song repertoire size of their sons ($p=0.030$), and (B) the mean egg mass ($p=0.017$). *Reprinted from Zanette et al. (2009) with permission.*

Food restriction generally results in effects in the opposite direction compared to those of food supplementation (Bauchinger et al., 2009). In the hen, this manipulation leads to lower plasma luteinizing hormone (LH) and gonadal steroids, as well as decreased mass of reproductive organs, and these changes are reversed in response to re-feeding (Richard-Yris et al., 1987). Endocrine effects of food restriction in hens are reproductive stage specific, however, with plasma follicle-stimulating hormone being lower in food-restricted than *ad libitum*–fed birds prior to sexual maturity, but not in egg-laying birds (Bruggeman et al., 1998). Similar to the situation in females, food restriction in males leads to reproductive deficits such as reduced sperm quality and semen volume (Cerolini et al., 1995). In the male zebra finch, no testicular development occurs following transfer from short to long day length in food-restricted males, but this transfer stimulates testicular recrudescence in *ad libitum*–fed birds (Perfito et al., 2008). In this opportunistic breeder, food availability, therefore, apparently plays a more important role than day length in the control of HPG axis activity.

30.3.3 Case Study: Urbanization

The diversity of environmental factors to which birds can respond and of the effects of these factors (e.g., migration: Tryjanowski et al., 2013; reproduction: Partecke and Gwinner, 2007; Davies et al., 2013; community structure: Gagne and Fahrig, 2011; Schlesinger et al., 2008; parasite infection: Sitko and Zalesny, 2012) is well illustrated by research on the effects of urbanization. Urbanization is associated with massive and in most cases largely irreversible environmental changes, including an increase in ambient noise, temperature (heat island effect: Zhang et al., 2010b), and nocturnal illumination, and differences in plant community composition and phenology (Mimet et al., 2009) that can in turn drive alterations in the food resources that are available to herbivores and carnivores. These changes, acting separately and/or cumulatively, and directly and/or indirectly, have the potential to profoundly affect the reproductive physiology and behavior of urban birds. The net consequences of urbanization on fitness remain, however, debated and poorly understood. On the one hand, urban birds frequently have an advanced reproductive phenology compared to corresponding nonurban birds (Davies et al., 2013; Dominoni et al., 2013), and this advancement has the potential to increase their reproductive output and fitness by increasing the duration of the annual breeding period. On the other hand, urban birds are exposed to physical factors that can be perceived as stressful and that adversely affect physiological and behavioral processes. One such factor is anthropogenic noise (noise pollution). Correlative evidence strongly suggests that noise pollution has a detrimental influence on breeding success and individual fitness (eastern bluebird, *Sialia sialis*: Kight et al., 2012; house sparrow, *Passer domesticus*: Schroeder et al., 2012). This conclusion is supported by experimental studies. For example, greater sage grouse (*Centrocercus urophasianus*) that are experimentally exposed to chronic noise show elevated excretion of glucocorticoid metabolites, suggesting enhanced stress levels, and decreased lek attendance (Blickley et al., 2012). The reproductive effects of noise pollution are, however, often species specific and can be complex. Illustrating this, exposure to elevated ambient noise level does not substantially affect plasma corticosterone (CORT) or reproductive behavior in the Australian black swan (*Cygnus atratus*) (Payne et al., 2012). In addition, by disrupting predator–prey interactions, ambient noise can in some situations in fact *increase* breeding success (Francis et al., 2009). A few studies have examined the relationships in birds between urbanization and the acute stress response (increased plasma glucocorticoids in response to capture and restraint: Fokidis et al., 2009; Partecke et al., 2006). Urban European blackbirds (*Turdus merula*) show a lower response to acute stress than forest conspecifics (Partecke et al., 2006). For the most part, however, the physiological (including endocrine) bases of differences between urban and corresponding nonurban birds remain poorly understood, and this research topic is ripe for new correlative and manipulative investigations.

30.4 SOCIAL FACTORS

Studies over many years, using a variety of experimental models and carried out in field and laboratory settings, consistently demonstrate that social interactions between conspecifics can influence all aspects of their reproductive behavior. Research on this subject also shows that an individual's own behavior can influence its physiology, including the production and secretion of behaviorally active hormones and neurochemicals, and this influence can in turn lead to behavioral modifications. The modulation of reproductive behavior by social factors is often complex, and with a few exceptions (e.g., the ring dove; see Section 30.4.1) the proximate underlying mechanisms—including the sensory modalities involved and the pathways that underlie sensory information processing—remain poorly understood. This topic provides rich opportunities for new and exciting research.

30.4.1 Effects of Males on Conspecific Females

Intraspecific social interactions can profoundly affect the behavior and physiology of individuals of the opposite sex. In the female Japanese quail, plasma CORT increases as a result of mating (Rutkowska et al., 2011). This increase is not seen in females that interact with a male without opportunity for mating, but it is observed in females that interact socially with other females. Thus, direct social interactions rather than the act of mating *per se* apparently elevate plasma CORT in these birds. The role of social interactions is shown also by work on Gouldian finches (*Erythrura gouldiae*), in which females that are experimentally paired with a non-preferred, poor-quality male have chronically higher plasma CORT than females paired with a preferred male (Griffith et al., 2011; Figure 30.2). Across species, lower stress levels in females that are paired with attractive mates may contribute to these females investing more in reproduction, as measured by egg size and food-provisioning behavior (Horvathova et al., 2012). In another type of investigation, exposure of females to conspecific song was found to induce behavioral and neural effects. For example, in several species, this exposure enhances female sexual responsiveness (song sparrow: Searcy and Marler, 1981; Pasteau et al., 2012; brown-headed cowbird, *Molothrus ater*: Freed-Brown and White, 2009) and activity level (European starling, *Sturnus vulgaris*: Riters and Teague, 2003). Exposing females to male songs also has physiological consequences that may influence fitness, as shown in canaries by the fact that females exposed to attractive male song repertoires deposit more testosterone (T) in their eggs than females exposed to unattractive male song repertoires (Gil et al., 2004).

The role of social interactions on the female reproductive system is further exemplified by work on captive willow tits (*Poecile montanus*), in which ovarian development is advanced in females paired with a male compared to females held in isolation (Silverin and Westin, 1995). This study did not isolate the factors that promote ovarian development, but other studies have addressed this

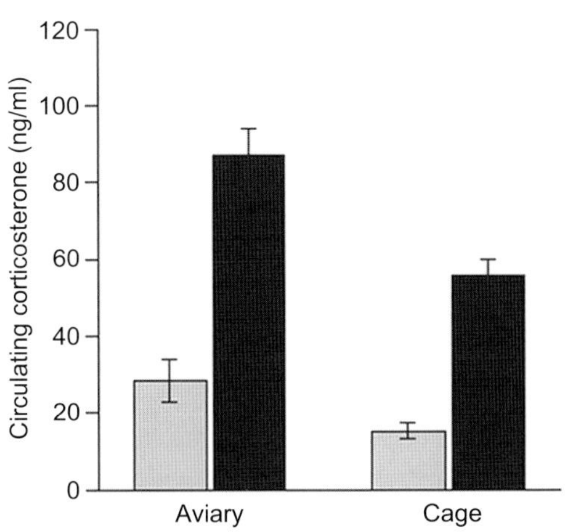

FIGURE 30.2 Baseline plasma corticosterone (CORT) during the egg-laying period in female Gouldian finches paired either with a genetically compatible (gray bars) or with a genetically incompatible (black bars) conspecific male. During breeding, birds were held either in a group aviary (aviary) or in cages containing one pair of birds (cage). Irrespective of the setting, females paired with an incompatible male have higher plasma CORT than females paired with a compatible male. *Reprinted from Griffith et al. (2011) with permission.*

question and demonstrate an important role for acoustic signals. For example, in song sparrows and domesticated canaries, exposure to conspecific song stimulates follicular development (Bentley et al., 2000a). The neuroendocrine mechanism mediating this stimulation is explored in the white-throated sparrow (*Zonotrichia albicollis*). In this species, male song rapidly (within 1 h) increases the secretion of LH and the expression of the immediate early gene, early growth response-1, in the mediobasal hypothalamus, but not in GnRH-producing neurons, suggesting behaviorally induced secretion of the peptide (Maney et al., 2007).

The most detailed investigations on the neural pathways that mediate reproductive effects of conspecific vocalizations in females have been conducted in the ring dove, a domesticated species that has long been used as an excellent model to unravel interactions between hormones and reproductive behavior (Lehrman, 1964; Lehrman and Friedman, 1969; McDonald and Liley, 1978; White, 1975). In the female ring dove, visual and auditory (but not physical) contact with a male stimulates the expression of courtship vocalizations (nest-coos) and ovarian development, and this effect is more pronounced after exposure to intact than castrated males (e.g., Lott et al., 1967). Females whose hypoglossal nerves (which innervate the vocal organ or syrinx) are sectioned show a reduction in their expression of nest-coos and, concurrently, in male courtship–induced ovarian development (Cohen and Cheng, 1979), leading to the remarkable hypothesis that a female's vocal behavior promotes her own ovarian development. The selectivity of this activation is demonstrated by the observation that plasma LH increases more in female doves that are exposed to female than male nest-coos (Cheng, 2008; Cheng et al., 1998). As neuronal track-tracing studies demonstrate, the self-stimulation of ovarian development involves the relay of auditory information to the midbrain nucleus intercollicularis (ICo). This brain region in turn communicates with the anterior hypothalamus, where some neurons themselves respond to auditory stimulation, through enkephalinergic projections (Cheng and Zuo, 1994). When stimulated, some hypothalamic neurons increase their secretion of GnRH, and this increase is in turn responsible for elevated gonadotropin secretion and ovarian development.

30.4.2 Effects of Females on Conspecific Males

The above studies conclusively show effects of males on the reproductive system of conspecific females, but there is also strong experimental evidence that females can in some cases influence the reproductive system of conspecific males. In the domesticated Japanese quail, the reproductive maturation rate of photostimulated males that are held with females is faster than in isolated males; paired males also have higher plasma LH and T than isolated males (Delville et al., 1984). Parallel findings are reported in free-ranging birds: treatment of free-ranging female song sparrows with estradiol at the beginning of the breeding season increases their sexual behavior, and this increase correlates with elevated plasma T and territorial behavior in males residing in the same area (Wingfield and Monk, 1994). Similarly, estradiol administration to free-ranging female pied flycatchers (*Ficedula hypoleuca*) in breeding condition stimulates the aggressive behavior of their mates (Silverin, 1991). It must, however, be pointed out that the presence of females does not necessarily have stimulatory effects on the male reproductive system, and so great care must be exercised when trying to generalize findings across species. Illustrating this, in captive male willow tits exposed to long photoperiods, testicular maturation is *delayed* and birds regress their testes *earlier* when kept in the presence of a female than when in isolation (Silverin and Westin, 1995). It also needs to be stressed that experiments using estrogen-treated versus control females or comparing paired versus isolated males to study the role of social factors do not, by themselves, inform about the specific factors (e.g., visual and auditory) to which males respond behaviorally and physiologically. Even though most birds have a relatively poor sense of olfaction relative to many other vertebrates, olfaction may play a role to regulate the behavioral response of males to conspecific females. Supporting this conclusion, the courtship behavior of male chickens is apparently influenced by olfactory signals emanating from the female uropygial gland, a structure that has long been considered to be a potential source of pheromones (Zhang et al., 2010a; Hirao et al., 2009). Consistent with these findings, spotless starlings (*Sturnus unicolor*) can recognize the sex of conspecific individuals based on chemical olfactory cues that may be released by the uropygial gland (Amo et al., 2012).

30.4.3 Effects of Males on Conspecific Males

A great deal of research has investigated the effect of male behavior on the reproductive physiology and behavior of conspecific males, with particular attention given to the role of visual and auditory stimuli. Long-term (several weeks) studies find stimulating effects of hearing conspecific song on the HPG axis. For example, in the male rufous-winged sparrow (*Peucaea carpalis*), plasma LH is higher and testes develop faster in response to photostimulation for several weeks when birds are exposed daily to conspecific song than in control males not hearing this song (Small et al., 2008; Figure 30.3). Social stimulation can also induce short-term effects on the reproductive endocrine system. These effects are often assessed experimentally by presenting males with prerecorded conspecific vocalizations combined with a decoy bird (Balthazart et al., 2009; Wingfield, 1994b; Silverin et al., 2004). This situation, generally referred to as simulated territorial intrusion (STI), often provokes a strong aggressive response on the part of the challenged male (Wingfield, 1994a), and this response is in some cases associated with a rapid (within minutes) increase in plasma T (Wingfield and Hahn, 1994). Testosterone can exert rapid (within minutes to hours) physiological (Sachs and Leipheimer, 1988) and behavioral (Wright et al., 2009) effects, and it can enhance attention to novel conspecifics and other relevant stimuli (Archer, 1977). According to the Challenge Hypothesis (Wingfield et al., 1987; Wingfield and Goldsmith, 1990), elevated plasma T in response to social challenge may, therefore, function to enhance the persistence of a challenged individual's behavioral response even after the end of the challenge (Oyegbile and Marler, 2005; Wingfield, 1994b; Lynn et al., 2005). It should be pointed out that a social challenge of the above-described type often either fails to detectably increase (Addis et al., 2010; Deviche et al., 2012) or even decreases plasma androgens (Landys et al., 2007; for a review, see Goymann, 2009). This variability suggests a modulation of the response to STI by other factors. Indeed, it has been proposed that this response is contingent upon the mating system (Wingfield and Goldsmith, 1990; single versus multiple broodedness: Landys et al., 2007), the contribution of males to incubation (Hirschenhauser et al., 2003), and the duration of the breeding season and the breeding state (Goymann, 2009).

30.5 AGE AND EXPERIENCE

Ample evidence exists in birds that breeding performance is age related (Forslund and Part, 1995). For example, the reproductive output of stitchbirds (*Notiomystis cincta*) increases during the first years of life and then declines as birds become senescent (Low et al., 2007). As discussed by Forslund and Part (1995), age-related differences in breeding performance may result from many nonexclusive factors that include age *per se* but also breeding experience, foraging success, and reproductive effort. Disentangling the respective contribution of these factors has been a major impetus of numerous studies. Research on free-ranging birds clearly indicates in some species that breeding experience gained during previous breeding attempts influences the reproductive output positively and independently of age. Using sophisticated statistical approaches, these studies show that breeding experience *per se* increases the probability of breeding and/or the reproductive output in birds belonging to various families (greater flamingo, *Phoenicopterus roseus*: Pradel et al., 2012; black-legged kittiwake, *Rissa tridactyla*: Desprez et al., 2011; snail kite, *Rostrhamus sociabilis*: Reichert et al., 2012). The interplay of age and experience is illustrated by research on free-ranging

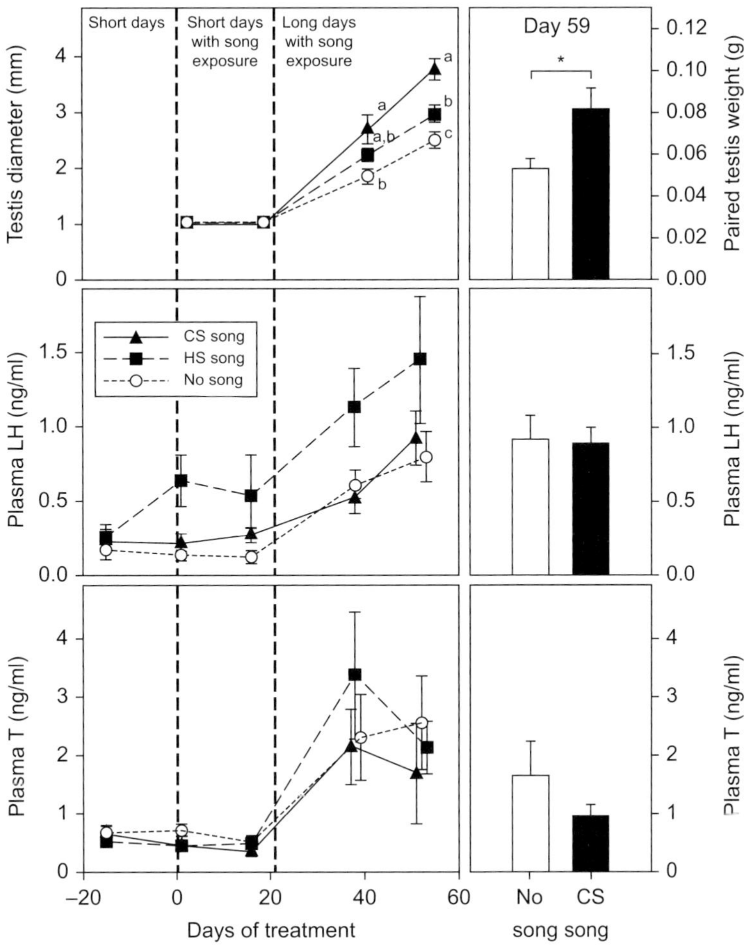

FIGURE 30.3 Left panels: Testis diameter (top left) and plasma luteinizing hormone (LH) and testosterone (T) in captive adult male rufous-winged sparrows exposed daily to conspecific song (CS Song), to heterospecific song (HS Song), or to no song (No Song). Birds were exposed to short days followed with exposure to a stimulating long photoperiod. On the same sampling day, different letters indicate significant differences between corresponding groups. Right panels: Testis mass and plasma LH and T measured at the time of euthanasia in the same birds as shown on the left panels. *Reprinted from Small et al. (2008) with permission.*

lance-tailed manakins (*Chiroxiphia lanceolata*; DuVal, 2012). In this species, males begin to breed when they reach high (alpha) social status and at an age that varies individually. The reproductive success of these males increases as a function of age and breeding experience. However, as birds get older, senescence negatively affects their breeding success, but this effect is compensated by the positive influence of social experience. Manipulative studies likewise support the idea of age-independent effects of breeding experience. For example, inexperienced male ring doves display less courtship behavior than age-matched experienced males (Cheng et al., 1986). This difference is specific to this behavior, as indicated by the observation that the two groups of males show similar levels of agonistic behavior. In another study, Cichon (2003) compared the reproductive output of inexperienced and experienced female collared flycatchers (*Ficedula albicollis*) (Figure 30.4). These two groups of birds had similar breeding success, but broods from experienced females were heavier and larger than those of inexperienced females, suggesting a positive effect of experience on fitness.

An insight into the mechanisms by which experience proximately influences reproductive behavior is provided by research on Japanese quail (Cornil and Ball, 2010). In this species, the probability of copulation is higher in sexually experienced than naïve males. However, visual (without physical) contact with a female (versus a male or no bird) also enhances the occurrence of subsequent precopulatory behavior (latency to mate), indicating that male sexual performance is stimulated by sensory cues from females without actual sexual encounter taking place. Importantly, this study used castrated, T-treated birds, and behavioral effects of social experience were, therefore, not related to plasma T. Similarly, behavioral effects of social experience in the male ring dove (Cheng et al., 1986) do not result from differences in plasma T between naïve and experienced males. That the influence of a bird's social experience on its behavior can be modulated by sensory

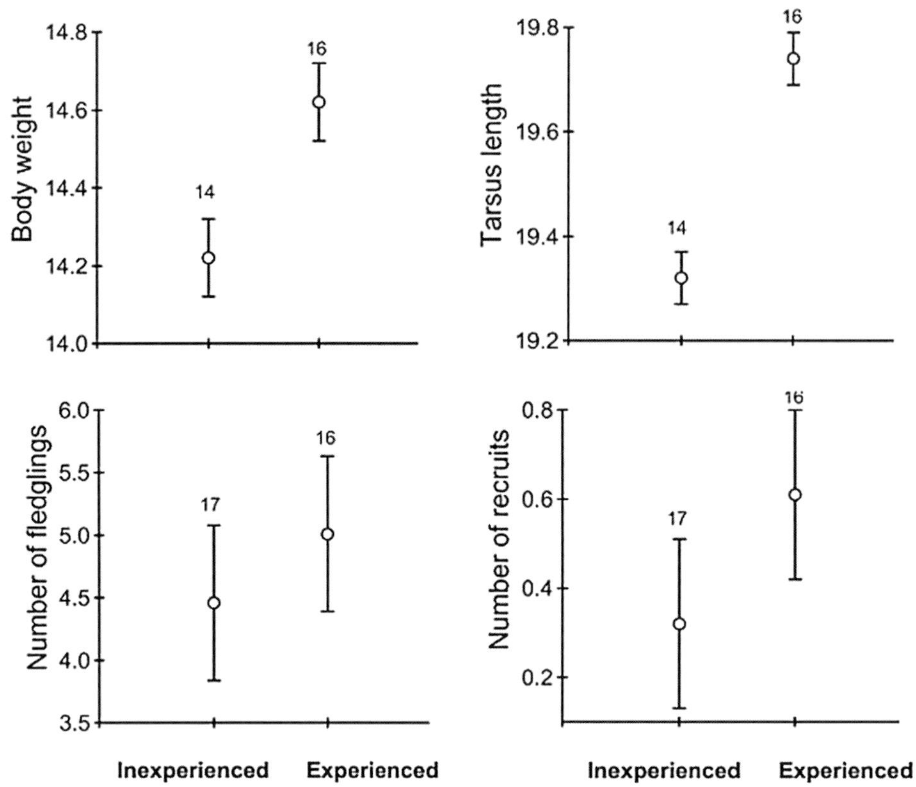

FIGURE 30.4 Breeding performance of 2 year old female collared flycatchers with (experienced) or without (inexperienced) previous breeding experience. Nestlings produced by inexperienced females were lighter (body mass on day 12 after hatching: $p=0.006$) and smaller (tarsus length: $p<0.001$) than those produced by experienced females, but there was no difference in the number of fledglings produced or subsequently recruited. Numbers above symbols refer to the number of nests studied. *Reprinted from Cichon (2003) with permission.*

cues provided by another bird is demonstrated by other type of work (Michel, 1976). In the ring dove, birds with previous nest-building breeding experience are more likely to initiate incubation in response to progesterone treatment than inexperienced birds. However, the propensity of a bird to enter the nest area and stand on the nest also depends on the breeding experience of its mate.

30.6 ENDOCRINE AND NEUROENDOCRINE REGULATION OF REPRODUCTIVE BEHAVIOR

Based on the observation of changes in aggressive and mating behavior resulting from castration or castration followed with testis reinsertion into the abdominal cavity of roosters, Berthold was among the first to postulate that the reproductive behavior of birds is influenced by internally produced substances (Quiring, 1944). Since then, this behavior has been shown to be regulated by a great variety of hormones and neuropeptides. This section presents an overview of the influence of the best-studied endocrine and neuroendocrine factors on specific aspects of reproductive behavior. However, the reader should be aware that other factors that are not discussed here due to space limitations play an influential role as well. These factors include opioid peptides and corticotropin-releasing hormone, which can inhibit copulation solicitation in females (Maney and Wingfield, 1998) and sexual behavior (opioids: Kotegawa et al., 1997), and catecholamines (Barclay et al., 1992; Rauceo et al., 2008).

30.6.1 Gonadal Steroids

Gonadal steroids (androgens, estrogens, and progestagens) have been the subject of most attention by far in terms of endocrine regulation of avian reproductive behavior. Their action mechanisms and behavioral effects are summarized in several recent reviews (Ball and Balthazart, 2010; Balthazart et al., 2010; Fusani, 2008), and the reader is referred to these publications for detailed information on this subject. Many studies on this topic are correlative and take advantage of the fact that most birds are photoperiodic (see Section 30.3.1). In these species, day length plays a critical role in the regulation of seasonal cycles, including gonadal development and the secretion of gonadal hormones, this secretion being dramatically higher during (compared to outside of) the breeding season. The seasonality of the reproductive system activity offers excellent opportunities to relate cyclic (neuro)endocrine changes (e.g., in plasma steroid concentrations) (Osorno et al., 2010; Landys et al., 2010) and the brain expression of steroid-metabolizing enzymes (Soma et al., 2003) or steroid receptors (Fraley et al., 2010), with the occurrence of reproductive behavior. Numerous studies on the behavioral role of gonadal steroids have also been manipulative. Commonly used approaches in these studies include gonadectomy (El Halawani et al., 1986; Hagelin, 2001), peripheral or central administration of hormones (Seredynski et al., 2013; Hunt and Wingfield, 2004; Komisaruk, 1967; also see Sections 30.6.4 and 30.6.6), and treatment with pharmacological agents. Some such agents serve as specific agonists or antagonists for steroid hormone receptors (Delville and Balthazart, 1987), whereas others affect the production or metabolism of steroid hormones (Belle et al., 2005; Schlinger and Callard, 1990). Investigations on the centrally mediated effects of gonadal steroids also take advantage of the coupling of hormonal manipulations with localized brain lesions (Bailhache et al., 1993; Del Negro et al., 1998) or electrophysiological approaches (Meitzen et al., 2007).

Gonadal steroids influence aspects of reproductive behavior at all life stages, including during sexual differentiation (Adkins-Regan, 2009; Banerjee et al., 2012) and in adulthood, when they impact courtship (Fusani, 2008), singing (see Section 30.6.6), sexual receptivity, copulation, incubation, and parental behavior (O'Neal et al., 2008; Van Roo, 2004). Gonadal steroids also play important roles in the regulation of life history stages outside of the reproductive period, such as during preparation for migration (Tonra et al., 2011). They additionally influence metabolic processes and body condition (Deviche, 1992; Buchanan et al., 2001; Jaccoby et al., 1995) and immunity (Deviche and Cortez, 2005; Selvaraj and Pitchappan, 1985), although the pathways (direct versus indirect and mediated, e.g., by glucocorticoids) involved in this influence remain poorly understood (Buttemer et al., 2008; Owen-Ashley et al., 2004). We discussed relationships between gonadal steroids and aspects of social behavior in Section 30.4. The remainder of this section further exemplifies the importance of gonadal steroids in the control of reproductive behavior, in particular their relationships with the AVT system (see Section 30.6.4) and their critical influence on brain regions that are responsible for song production in oscine passerines (see Section 30.6.6).

30.6.2 Neurosteroids

It has become clear in the past 15 years that the vertebrate (including avian) brain is capable of neurosteroidogenesis (i.e., the synthesis of steroids *de novo* from cholesterol) (Schlinger and Remage-Healey, 2012; Ubuka et al., 2009). This conclusion in birds is based on the discovery, among others, that the brain expresses enzymes that are involved in the synthesis of steroids, including pregnenolone, progesterone, T, and estradiol, and this expression is region and developmental stage specific (London et al., 2006). The roles of these enzymes and their products in songbirds are well illustrated by research on brain aromatase, the enzyme that converts androgens to estrogens. Aromatase in the songbird brain is widely distributed and present, among others, in areas that are involved in the control of song production (e.g., the medial border of the HVC and robust nucleus of the arcopallium (RA); see Section 30.6.6) and auditory information processing (e.g., the caudiomedial mediopallium (NCM); Schlinger and Balthazart, 2012). Physiological and behavioral studies reveal that rapid changes in the brain production of estrogen result in behavioral changes. For example, local pharmacological inhibition of estrogen production within the left NCM of male zebra finches rapidly disrupts their ability to preferentially respond behaviorally to their own song versus conspecific song (Remage-Healey et al., 2010; Figure 30.5). Reciprocally, a bird's own behavior, as well as relevant external stimuli, can rapidly modify the brain production of estrogen. Indeed, in the male zebra finch forebrain, the expression of aromatase rapidly increases in response to the act of singing (but not to hearing song; Remage-Healey et al., 2009), and the forebrain concentration of estradiol rapidly increases during social interactions with females (Remage Healey et al., 2008). In the male white-crowned sparrow (*Zonotrichia leucophrys*), a social challenge (simulated territorial intrusion) rapidly decreases the local concentration of estrogen in several brain regions (Charlier et al., 2011). Rapid, steroid hormone–mediated, and centrally mediated changes in behavior suggest that these hormones act via nonclassic (i.e. nongenomic) mechanisms (discussed further in this section). This proposition is supported by the fact that steroids can bind to cell membrane receptors and

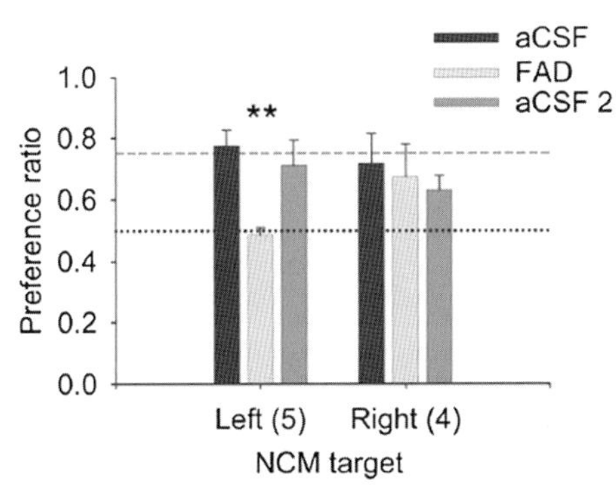

FIGURE 30.5 Influence of the aromatase inhibitor fadrozole (FAD) administered by *in vivo* retrodialysis into the auditory region caudal nidopallium (NCM) of the male zebra finch brain on behavioral preference for the bird's own song versus conspecific song. Control vehicle injections consisted of artificial cerebrospinal fluid (aCSF). Song preference was abolished by FAD administration into the left but not right NCM. *Reprinted from Remage-Healey et al. (2010) with permission.*

induce rapid nongenomic cellular effects (Heimovics et al., 2012).

The Japanese quail has proved to be a particularly profitable experimental model in which to investigate the rapid behavioral effects of neurosteroids and in particularly neuroestrogens. As anatomical and functional studies demonstrate, the quail preoptic area (POA) plays a critical role in the regulation of male sexual behavior (Balthazart et al., 1992; for reviews, see Balthazart et al., 2006; Panzica et al., 1996). This brain region is sexually dimorphic (it is larger in males than females, which do not show masculine sexual behavior) and is a target region for sex steroids: it contains androgen and estrogen receptors, and its volume decreases after castration and is restored following T treatment (Seredynski et al., 2011). The POA contains aromatase, and estrogen in the quail plays a critical role in the control of sexual behavior (Balthazart et al., 2009). Brain aromatase and, therefore, the local production of estrogens in the POA are regulated not only by slow-acting steroid-mediated genomic mechanisms that involve changes in transcription (Voigt et al., 2011) but also by nongenomic mechanisms that can rapidly (within minutes) enhance the local availability of estrogen (Balthazart et al., 2004). This stimulation can in turn stimulate a short-term increase in expression of some aspects of sexual behavior (Cornil et al., 2006a,b; Seredynski et al., 2011). Collectively, research on Japanese quail and the studies described here on songbirds firmly establish neurosteroids as important modulators of reproductive behavior (Schlinger and Remage-Healey, 2012).

30.6.3 Gonadotropin-Releasing and Gonadotropin-Inhibitory Hormones

GnRH in vertebrates plays an essential stimulatory role in the regulation of anterior pituitary gland gonadotropins. In mammals, GnRH also acts as a neurotransmitter and/or in a paracrine fashion to modulate sexual behavior, an effect that is presumably mediated by central GnRH receptors (Hsueh and Schaeffer, 1985). Limited evidence supports the view that GnRH has centrally mediated behavioral actions also in birds. Some of the original work on this subject was conducted in ring doves, in which synthetic GnRH administration stimulates courtship behavior, and this effect is apparently not mediated by LH or progesterone (Cheng, 1977). The brain of many vertebrates expresses at least two forms of GnRH: GnRH-I (cGnRH-I), which is released into the median eminence and stimulates the pituitary gland, and GnRH-II, which is normally not released into the median eminence but into extrahypothalamic brain regions. Central injection of GnRH-II to female white-crowned sparrows rapidly increases their expression of courtship behavior in response to hearing conspecific male song, and this effect is specific to GnRH-II, not being induced by cGnRH-I administration (Maney et al., 1997b). Thus, consistent with the situation in mammals (Hsueh and Schaeffer, 1985), GnRH in birds may act centrally to influence reproductive behavior. It should also be pointed out that, at least in some avian species, the brain contains a nonneuronal source of GnRH. This conclusion is drawn from studies on ring doves, in which the medial habenula contains immune (mast) cells that synthesize GnRH as well as other neurochemicals (Wilhelm, 2011). The activity of these cells is regulated by the endocrine (gonadal steroid) environment, and their number increases within hours in response to courtship (Zhuang et al., 1993), suggesting that their secretory products and in particular GnRH are involved in the control of reproductive behavior.

The discovery of GnIH, a neuropeptide with inhibitory effects on gonadotropin secretion and with fibers extending to a large number of brain areas (Tsutsui et al., 2000; Ukena et al., 2003), has prompted considerable research into its endocrine and behavioral functions. GnIH in birds is produced in the paraventricular nucleus (PVN). Centrally mediated roles for GnIH are suggested by the widespread brain distribution of immunostained GnIH fibers (Ubuka et al., 2008). In the female white-crowned sparrow, central treatment with GnIH inhibits courtship behavior (Bentley et al., 2006), thereby exerting effects that are opposite of those induced by GnRH-II treatment (Maney et al., 1997b; and discussed in this chapter). This behavioral inhibition may result from negative effects on GnRH function because midbrain regions where GnRH-II is produced contain GnIH fibers, and in the European starling, cGnRH-I- and GnRH-II-producing neurons express GnIH receptor mRNA (Bentley et al., 2008). Recent work using RNA interference (RNAi) of the GnIH gene in the Japanese quail provides further support for a centrally mediated behavioral role of this neuropeptide. Here, brain injection of small interfering RNA (siRNA) against GnIH precursor mRNA (GnIH siRNA) stimulated sexual behavior, and this stimulation was blocked by GnIH treatment (Ubuka et al., 2013). An emergent conclusion from studies on the behavioral role of GnIH is that it may function as an important inhibitory modulator of social interactions, including reproductive behavior (Calisi et al., 2011).

30.6.4 Arginine Vasotocin

A peptide whose behavioral role in birds has received considerable attention is AVT, the avian homolog of vasopressin. Given peripherally to zebra finches, AVT decreases courtship behavior, and this decrease is attenuated by T administration (Harding and Rowe, 2003). Zebra finches are opportunistic breeders whose reproduction is induced by water availability. AVT plays a major role in the control of water balance (Sharma et al., 2009). Thus, it is postulated in these finches that drought conditions stimulate

AVT secretion and the peptide in turn inhibits the reproductive system, including T secretion, thereby preventing reproduction during periods of unfavorable environmental conditions. Contrasting with its suppressive action after peripheral administration, AVT exerts centrally mediated stimulatory behavioral effects, as shown by studies reporting that brain injection of the neuropeptide stimulates singing (Voorhuis et al., 1991a; de Kloet et al., 1993; Maney et al., 1997a). Arginine vasotocin is synthesized in several brain regions, including the paraventricular and supraoptic nuclei and the nucleus of the stria terminalis (Panzica et al., 1999b), and AVT-immunostained fibers extend to a large number of brain areas, some of which are involved in the control of vocal behavior (the RA and mesencephalic nucleus intercollicularis: Kiss et al., 1987; Panzica et al., 1999b; Voorhuis and de Kloet, 1992). Binding sites for AVT in the avian brain are similarly widespread (Leung et al., 2009, 2011), and regions containing these sites are innervated by AVT fibers and terminals (Voorhuis et al., 1988b). Collectively, the neuroanatomical distribution of AVT and its receptors suggests that the peptide is involved in the control of multiple functions, including sexual behavior and GnRH secretion (Panzica et al., 2001).

As shown by a number of investigations, the central AVT system and the behavioral effects of the neuropeptide are steroid hormone sensitive. This conclusion is supported by work on canaries showing a sex difference in brain AVT immunostaining (Voorhuis et al., 1988a) and showing, in males of this species, that seasonal changes in AVT immunostaining parallel changes in plasma T (Voorhuis et al., 1991b). Manipulative studies, including gonadectomy and hormone replacement experiments, confirm an important, region-specific stimulatory role for gonadal steroids in the brain expression of AVT (Kimura et al., 1999; Plumari et al., 2004; Viglietti-Panzica et al., 1994) and AVT mRNA (Panzica et al., 1999a). In the castrated male Japanese quail, the effects of T administration on the brain expression of AVT are mimicked by treatment with estradiol but not the nonaromatizable T metabolite 5α-dihydrotestosterone (5α-DHT) (Viglietti-Panzica et al., 2001). Effects of T on brain AVT in this species may, therefore, require aromatization of this steroid. Similarly, when given to female quail, estradiol but not 5α-DHT mimics the stimulatory influence of T on the brain expression of AVT mRNA (Aste et al., 2013; Fig. 30.6).

It should be noted that while estradiol influences the central AVT system, there is also evidence that AVT may modulate the brain production of estrogen. Indeed, in the Japanese quail, aromatase-containing cells in brain regions that control reproductive behavior (e.g., the medial POA; see Section 30.6.2) apparently receive AVT innervation (Balthazart et al., 1997). Acute stress in quail rapidly (within minutes) upregulates their medial POA aromatase activity (Dickens et al., 2011), and there are indications that AVT contributes to mediating this effect (Dickens et al., 2013). Arginine vasotocin may accordingly be an important component of the neural system that controls rapid stress-induced changes in reproductive behavior. New research is clearly warranted to confirm this hypothesis and elucidate the precise nature of this regulation.

30.6.5 Prolactin

The secretion of PRL in birds is under complex regulation by environmental and physical factors. The main environmental factor that controls PRL secretion in many species is day length, with long days stimulating this secretion (Dawson et al., 2001; Sharp et al., 1998). Plasma PRL in these species is accordingly elevated in spring and summer (i.e. during the breeding period) (Gahali et al., 2001; Sharp et al., 1998), and then declines. Elevated plasma PRL toward the end of the breeding period is believed to contribute to, but not be responsible for, the seasonal gonadal regression that signals the end of the reproductive season (Dawson et al., 2001). On a shorter term basis, the secretion of PRL is influenced by factors associated with incubation. For example, the introduction of an artificial egg into the nest of yellow-eyed penguins (*Megadyptes antipodes*), before birds normally start laying, stimulates PRL secretion in females (but has the opposite effect in males; Massaro et al., 2007). Inversely, plasma PRL rapidly decreases in captive female common eiders (*Somateria mollissima*) that are deprived of their eggs (Criscuolo et al., 2002). And in the chicken hen, a precocial species, physical contact with chicks stimulates brooding behavior and precipitates a decrease in plasma PRL, suggesting a role for the hormone in the transition from incubation to brooding (Richard-Yris et al., 1998). Consistent with this conclusion, plasma PRL in this species is high during incubation and then rapidly decreases after hatching (Kuwayama et al., 1992). In the altricial canary, however, plasma PRL remains elevated in females after hatching and decreases gradually during rearing of the young (Goldsmith et al., 1981). Furthermore, plasma PRL in male canaries, which do not incubate or feed young, increases only slightly during the parental phase of the reproductive cycle (Goldsmith et al., 1981). Taken together, these studies support the hypothesis that PRL is involved in the control of parental behavior. It should be noted in some species (Adélie penguin, *Pygoscelis adeliae*; and emperor penguin, *Aptenodytes forsteri*) that the increase in plasma PRL that occurs during reproduction is apparently controlled endogenously rather than by day length or tactile stimulation (Lormee et al., 1999; Vleck et al., 2000). Sustained plasma PRL in the absence of tactile input in these species is thought to be necessary for breeding birds to maintain parental behavior despite the fact that they are absent from their nest for prolonged periods while foraging at sea.

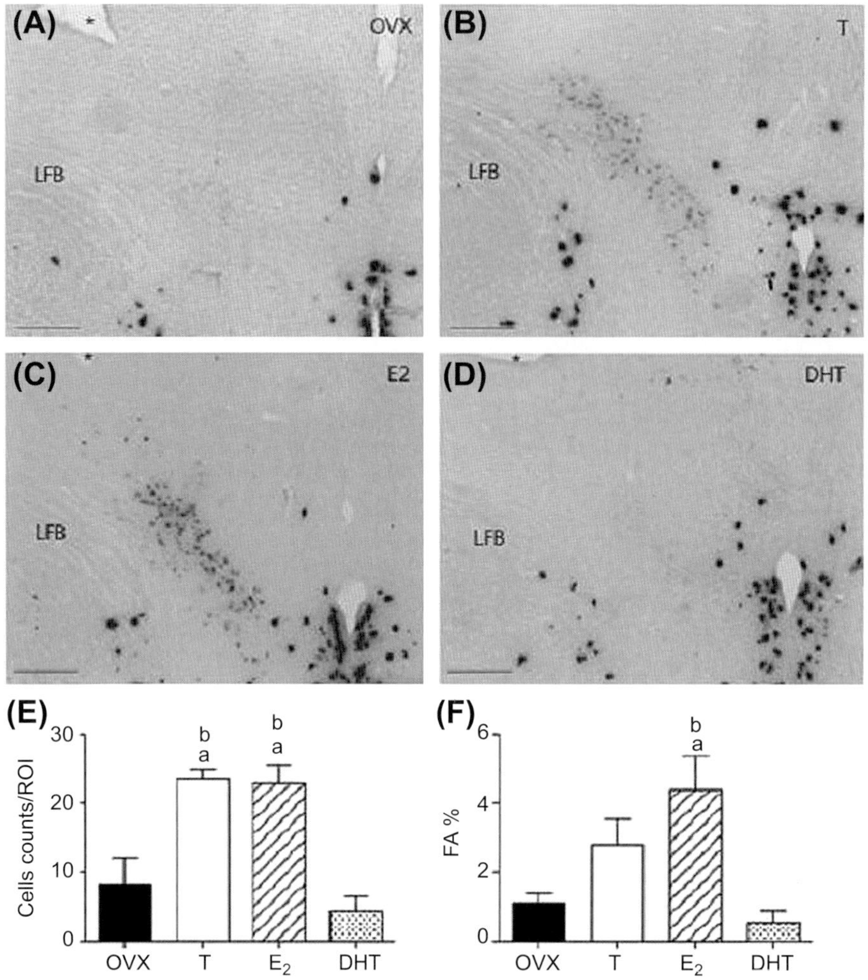

FIGURE 30.6 Regulation of arginine vasotocin (AVT) mRNA expression by gonadal steroids in the female Japanese quail brain. (A–D): Photomicrographs showing the distribution of AVT mRNA-expressing neurons in the caudomedial part of the bed nucleus of the stria terminalis after ovariectomy and one systemic injection of vehicle solution (OVX), testosterone (T), estradiol (E2), or 5α-dihydrotestosterone (5α-DHT). (E): Number of AVT mRNA-containing cells in the region of interest (ROI). This number is increased by T or E2, but not 5α-DHT, treatment. (F): Fractional area (FA) covered by mRNA hybridization signal in the same region. The FA is increased only after E2 administration. *Reprinted from Aste et al. (2013) with permission.*

The ring dove has proved to be a particularly informative model to study the involvement of PRL in the control of parental behavior. In contrast to the situation in many species, PRL secretion in doves does not change seasonally and is not photoperiod regulated, and these birds can breed year-round (Lea et al., 1986a). A role for nest-associated stimuli in the control of PRL secretion in this species is indicated by the observation that plasma PRL increases during incubation, but nest deprivation causes a rapid decline in hormone level (Lea et al., 1986b). Furthermore, injections of PRL enhance the persistence of incubation behavior (Janik and Buntin, 1985) and stimulate food provisioning to squabs (Buntin et al., 1991). Stimulatory effects on parental behavior (including feeding squabs) are also observed after intracerebroventricular administration of the hormone at doses that do not stimulate PRL-sensitive peripheral tissues such as the crop sac, indicating that these effects are mediated centrally (Buntin et al., 1991; Figure 30.7). Consistent with this proposition, the ring dove brain contains PRL-binding sites (Buntin and Ruzycki, 1987) and chemical lesion of the POA produces a deficit in parental behavior in response to PRL administration (Slawski and Buntin, 1995). Centrally acting PRL may be of peripheral (pituitary gland) origin. However, the dove brain contains PRL-immunopositive cells and fibers that project to several hypothalamic regions (Ramesh et al., 2000). Thus, in this species, centrally produced PRL binding to local receptors, perhaps in addition to PRL reaching the brain through systemic circulation, may activate parental behavior.

Stress is often associated with elevated secretion of CORT and can negatively influence parental behavior. Exposure to stressful conditions also often depresses plasma PRL (Angelier et al., 2013). These observations have led researchers to ask whether CORT and PRL normally exert opposite actions on parental behavior, and whether stress-related changes in

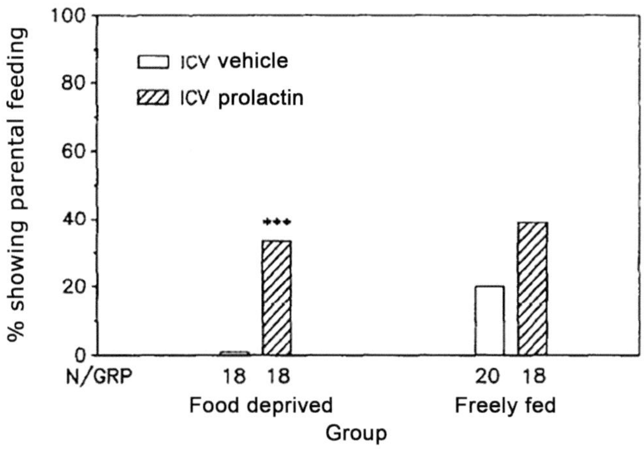

FIGURE 30.7 Stimulation of parental feeding behavior in nonbreeding ring doves by daily intracerebroventricular (ICV) injection of prolactin or vehicle solution. Birds were either fed *ad libitum* or food deprived for 16h before behavioral testing. N/GRP refers to the number of birds in each experimental group. *** Denotes a significant difference with the vehicle-treated group. *Reprinted from Buntin et al. (1991) with permission.*

plasma CORT and PRL are mechanistically related. These questions have been studied extensively in some seabirds. Correlative evidence is provided by work on black-legged kittiwakes (Chastel et al., 2005). In this species, acute stress decreases PRL more in failed breeders than in breeding birds, suggesting that maintaining sufficiently high plasma PRL during stress may result in birds continuing to breed rather than abandoning their nest. Supporting this proposition, plasma PRL in Adélie penguins is low in birds that abandon their nest compared to birds that do not do so (Spee et al., 2010). However, pharmacological blockade of PRL secretion in this species does not cause nest desertion (Thierry et al., (2013), suggesting the involvement of additional factors. One such factor may be CORT because male penguins that abandon their nest have high plasma CORT (Spee et al., 2010), and CORT administration leads to nest desertion (Kleven et al., 2009). Thus, nest abandonment in this species may require high plasma CORT combined with low plasma PRL. In a more general sense, the above studies suggest in breeding birds that during periods of stress the parental decision to continue to breed can be at the expense of their own condition and survival, and this decision is critically influenced by baseline (prestress) and stress-induced plasma CORT and PRL (Angelier et al., 2009). The extent to which the PRL (decrease) and CORT (increase) stress responses are mechanistically related, however, remains largely conjectural (Angelier et al., 2013).

30.6.6 Case Study: The Oscine Vocal Control System

Acoustic signaling in some species is mediated by nonvocal signals such as those produced by wings (e.g., manakins: Schlinger et al., 2008) and tail feathers (e.g., some hummingbirds: Clark and Feo, 2008), but in many species this signaling is mediated primarily by vocalizations. The neural circuit that controls the perception and expression of vocal behavior in passerines has been the object of enormous attention. This circuit provides an exquisite model in which to investigate how hormones regulate reproductive behavior, the central mechanisms that mediate this regulation, and the reciprocal interactions between behavior and the (neuro)endocrine system. The control of vocal perception and output has been particularly well studied in oscine passerines. Many oscines produce complex, learned vocalizations (songs), the learning and production of which are controlled by a network of discrete and interconnected brain regions called the vocal control system (Nottebohm et al., 1976, 1982; Figure 30.7). Many suboscine passerines also produce complex songs, but in contrast to oscines, vocalizations in these species are generally thought to be innate rather than learned (but see Saranathan et al., 2007), and suboscine song control regions are correspondingly absent or rudimentary (Liu et al., 2013). In oscines, song control regions are typically larger in males than females in species in which males do most of the singing (Tobari et al., 2005; Tramontin et al., 1998), and this morphological difference is associated with cytoarchitectural (Del Negro and Edeline, 2002; Nealen, 2005), biochemical (Charlier et al., 2003; Riters and Ball, 2002; Sakaguchi et al., 2000), and physiological (Adret and Margoliash, 2002; Del Negro and Edeline, 2001) differences between sexes. The exact role of song control regions in females that do not normally sing remains incompletely understood. Some of these regions serve an auditory as well as motor function (Figure 30.8), and in females they may, therefore, regulate vocal behavior perception rather than output (Hamilton et al., 1997).

A remarkable feature of the adult oscine vocal control system is its neuroplasticity. In seasonal breeders, the size of song control regions such as the HVC and RA changes seasonally. These regions are larger during the breeding season, when birds are exposed to long photoperiods, gonads are developed and release large amounts of steroid hormones, and birds sing at a high frequency, than outside this season (Meitzen and Thompson, 2008; Nottebohm, 1981; Smith et al., 1997). The basis of these changes has been researched in particular detail for the HVC, a telencephalic region that plays a major role in song production. Seasonal changes in HVC volume are associated with cytoarchitectural changes, including neuron number, density, and life span (Bowers et al., 2011; Meitzen and Thompson, 2008; Nottebohm et al., 1994). Some vocal control regions including the HVC contain sex steroid receptors (Fraley et al., 2010; Gahr et al., 1987, 1993; Nordeen et al., 1987) whose density is regulated seasonally (Fusani et al., 2000; Soma et al., 1999), suggesting locally mediated roles for these hormones. Consistent with this view, considerable evidence indicates that gonadal steroids play many-faceted roles in the control of the vocal control system's plasticity in adulthood (Bottjer et al., 1986; Ball et al., 2003). These roles include regulation

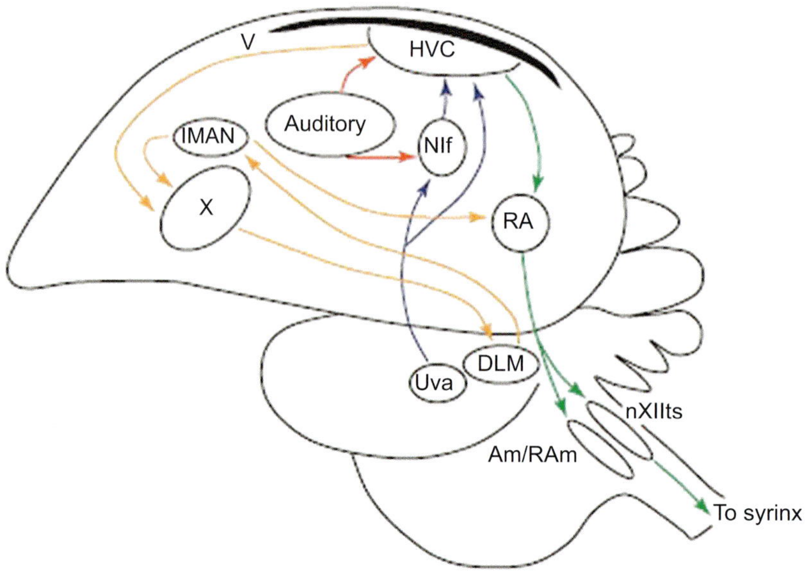

FIGURE 30.8 Projections of the major regions of the oscine brain vocal control system. The system consists of a motor pathway (green) that controls vocal output and includes projections from the HVC (correct full name; not an abbreviation) to the robust nucleus of the arcopallium (RA). A second pathway (orange) is essential for song learning and perception, and includes the medial portion of dorsolateral thalamus, the lateral magnocellular nucleus of the anterior neostriatum, area X, HVC, and RA. *Reprinted from Brenowitz and Beecher (2005) with permission.*

of the size of song control regions (Thompson et al., 2007; Tramontin et al., 2003), newborn neuron migration into the HVC, and the density and survival of existing neurons (MacDougall-Shackleton, 2012). Considerable attention has been devoted to the cellular and molecular mechanisms of action of steroids in song control regions (Thompson et al., 2012), and these studies have identified an important role for brain-derived neurotrophic factor (BDNF). In concert with T, BDNF is thought to promote the proliferation and survival of HVC neurons (Brenowitz, 2013; Alvarez-Borda et al., 2004; Rasika et al., 1999) and also to contribute to seasonal plasticity in the RA (Wissman and Brenowitz, 2009; Li et al., 2000). Furthermore, in the vocal control system the expression of numerous genes that are involved with diverse functions such as neurogenesis, apoptosis, and axonal growth changes seasonally (Thompson et al., 2012). A major challenge in the future is to identify the role of these genes' products and to determine whether and how steroids regulate the activity of these genes. Adding to this complexity, factors such as photoperiod (Dloniak and Deviche, 2001; Gulledge and Deviche, 1998) and social factors (Boseret et al., 2006) influence the size and cytoarchitectural characteristics of song control regions, and this influence is in some situations independent of gonadal steroids. Finally, while this chapter's discussion makes it clear that song control regions play an important role in the expression of vocal behavior, some of these regions are also influenced by sensory (acoustic) stimulation. For example, T administration to female canaries increases their HVC volume, but this increase is attenuated after deafening (i.e., in the absence of auditory stimulation) (Bottjer et al., 1986). The mechanism mediating this difference is not fully understood, and it may involve a role for the auditory stimulus-regulated production of neurosteroids (see Section 30.6.2), in particular estradiol (MacDougall-Shackleton, 2012; Remage-Healey et al., 2012; Soma et al., 2004).

ACKNOWLEDGMENTS

The writing of this chapter was supported by National Science Foundation Award IOB 1026620 to the author, who thanks Scott Davies for comments on an early version of the manuscript.

REFERENCES

Addis, E.A., Busch, D.S., Clark, A.D., Wingfield, J.C., 2010. Seasonal and social modulation of testosterone in Costa Rican rufous-collared sparrows (*Zonotrichia capensis costaricensis*). Gen. Comp. Endocrinol. 166, 581–589.

Adkins-Regan, E., 2009. Hormones and sexual differentiation of avian social behavior. Dev. Neurosci. 31, 342–350.

Adret, P., Margoliash, D., 2002. Metabolic and neural activity in the song system nucleus robustus archistriatalis: effect of age and gender. J. Comp. Neurol. 454, 409–423.

Alvarez-Borda, B., Haripal, B., Nottebohm, F., 2004. Timing of brain-derived neurotrophic factor exposure affects life expectancy of new neurons. Proc. Natl. Acad. Sci. U.S.A 101, 3957–3961.

Amo, L., Aviles, J.M., Parejo, D., Pena, A., Rodriguez, J., Tomas, G., 2012. Sex recognition by odour and variation in the uropygial gland secretion in starlings. J. Anim. Ecol. 81, 605–613.

Angelier, F., Moe, B., Blanc, S., Chastel, O., 2009. What factors drive prolactin and corticosterone responses to stress in a long-lived bird species (snow petrel *Pagodroma nivea*)? Physiol. Biochem. Zool. 82, 590–602.

Angelier, F., Wingfield, J.C., Trouve, C., de Grissac, G.S., Chastel, O., 2013. Modulation of the prolactin and the corticosterone stress responses: do they tell the same story in a long-lived bird, the cape petrel? Gen. Comp. Endocrinol. 182, 7–15.

Archer, J., 1977. Testosterone and persistence in mice. Anim. Behav. 25, 479–488.

Ardia, D.R., Perez, J.H., Chad, E.K., Voss, M.A., Clotfelter, E.D., 2009. Temperature and life history: experimental heating leads female tree swallows to modulate egg temperature and incubation behaviour. J. Anim. Ecol. 78, 4–13.

Aste, N., Sakamoto, E., Kagami, M., Saito, N., 2013. Vasotocin mRNA expression is sensitive to testosterone and oestradiol in the bed nucleus of the stria terminalis in female Japanese quail. J. Neuroendocrinol. 25, 811–825.

Astheimer, L.B., Buttemer, W.A., Wingfield, J.C., 1995. Seasonal and acute changes in adrenocortical responsiveness in an arctic-breeding bird. Horm. Behav. 29, 442–457.

Bailhache, T., Surlemont, C., Balthazart, J., 1993. Effects of neurochemical lesions of the preoptic area on male sexual behavior in the Japanese quail. Brain Res. Bull. 32, 273–283.

Ball, G.F., Balthazart, J., 2004. Hormonal regulation of brain circuits mediating male sexual behavior in birds. Physiol. Behav. 83, 329–346.

Ball, G.F., Balthazart, J., 2010. Japanese quail as a model system for studying the neuroendocrine control of reproductive and social behaviors. ILAR. J. 51, 310–325.

Ball, G.F., Castelino, C.B., Maney, D.L., Appeltants, D., Balthazart, J., 2003. The activation of birdsong by testosterone: multiple sites of action and role of ascending catecholamine projections. Ann. N.Y Acad. Sci. 1007, 211–231.

Balthazart, J., Surlemont, C., Harada, N., 1992. Aromatase as a cellular marker of testosterone action in the preoptic area. Physiol. Behav. 51, 395–409.

Balthazart, J., Absil, P., Viglietti-Panzica, C., Panzica, G.C., 1997. Vasotocinergic innervation of areas containing aromatase-immunoreactive cells in the quail forebrain. J. Neurobiol. 33, 45–60.

Balthazart, J., Baillien, M., Charlier, T.D., Cornil, C.A., Ball, G.F., 2003. The neuroendocrinology of reproductive behavior in Japanese quail. Domest. Anim. Endocrinol. 25, 69–82.

Balthazart, J., Baillien, M., Cornil, C.A., Ball, G.F., 2004. Preoptic aromatase modulates male sexual behavior: slow and fast mechanisms of action. Physiol. Behav. 83, 247–270.

Balthazart, J., Cornil, C.A., Taziaux, M., Charlier, T.D., Baillien, M., Ball, G.F., 2006. Rapid changes in production and behavioral action of estrogens. Neuroscience 138, 783–791.

Balthazart, J., Cornil, C.A., Charlier, T.D., Taziaux, M., Ball, G.F., 2009. Estradiol, a key endocrine signal in the sexual differentiation and activation of reproductive behavior in quail. J. Exp. Zool. A Ecol. Genet. Physiol. 311, 323–345.

Balthazart, J., Charlier, T.D., Barker, J.M., Yamamura, T., Ball, G.F., 2010. Sex steroid-induced neuroplasticity and behavioral activation in birds. Eur. J. Neurosci. 32, 2116–2132.

Banerjee, S.B., Arterbery, A.S., Fergus, D.J., Adkins-Regan, E., 2012. Deprivation of maternal care has long-lasting consequences for the hypothalamic-pituitary-adrenal axis of zebra finches. Proc. Biol. Sci. 279, 759–766.

Barclay, S.R., Harding, C.F., Waterman, S.A., 1992. Correlations between catecholamine levels and sexual behavior in male zebra finches. Pharmacol. Biochem. Behav. 41, 195–201.

Bauchinger, U., Van't Hof, T., Biebach, H., 2009. Food availability during migratory stopover affects testis growth and reproductive behaviour in a migratory passerine. Horm. Behav. 55, 425–433.

Beale, C.M., Burfield, I.J., Sim, I.M., Rebecca, G.W., Pearce-Higgins, J.W., Grant, M.C., 2006. Climate change may account for the decline in British ring ouzels *Turdus torquatus*. J. Anim. Ecol. 75, 826–835.

Belle, M.D., Sharp, P.J., Lea, R.W., 2005. Aromatase inhibition abolishes courtship behaviours in the ring dove (*Streptopelia risoria*) and reduces androgen and progesterone receptors in the hypothalamus and anterior pituitary gland. Mol. Cell. Biochem. 276, 193–204.

Benkman, C.W., 1990. Intake rates and the timing of crossbill reproduction. Auk 107, 376–386.

Benkman, C.W., 1992. White-winged crossbill. No. 27. In: Poole, A., Gill, F. (Eds.), The Birds of North America. Academy of Natural Sciences, Philadelphia. American Ornithologists' Union, Washington, DC, 18 pp.

Bentley, G.E., Wingfield, J.C., Morton, M.L., Ball, G.F., 2000a. Stimulatory effects on the reproductive axis in female songbirds by conspecific and heterospecific male song. Horm. Behav. 37, 179–189.

Bentley, G.E., Spar, B.D., Macdougall-Shackleton, S.A., Hahn, T.P., Ball, G.F., 2000b. Photoperiodic regulation of the reproductive axis in male zebra finches, *Taeniopygia guttata*. Gen. Comp. Endocrinol. 117, 449–455.

Bentley, G.E., Jensen, J.P., Kaur, G.J., Wacker, D.W., Tsutsui, K., Wingfield, J.C., 2006. Rapid inhibition of female sexual behavior by gonadotropin-inhibitory hormone (GnIH). Horm. Behav. 49, 550–555.

Bentley, G.E., Ubuka, T., McGuire, N.L., Chowdhury, V.S., Morita, Y., Yano, T., Hasunuma, I., Binns, M., Wingfield, J.C., Tsutsui, K., 2008. Gonadotropin-inhibitory hormone and its receptor in the avian reproductive system. Gen. Comp. Endocrinol. 156, 34–43.

Blickley, J.L., Word, K.R., Krakauer, A.H., Phillips, J.L., Sells, S.N., Taff, C.C., Wingfield, J.C., Patricelli, G.L., 2012. Experimental chronic noise is related to elevated fecal corticosteroid metabolites in lekking male greater sage grouse (*Centrocercus urophasianus*). PLoS One 7, e50462.

Bolger, D.T., Patten, M.A., Bostock, D.C., 2005. Avian reproductive failure in response to an extreme climatic event. Oecologia 142, 398–406.

Boseret, G., Carere, C., Ball, G.F., Balthazart, J., 2006. Social context affects testosterone-induced singing and the volume of song control nuclei in male canaries (*Serinus canaria*). J. Neurobiol. 66, 1044–1060.

Bottjer, S.W., Schoonmaker, J.N., Arnold, A.P., 1986. Auditory and hormonal stimulation interact to produce neural growth in adult canaries. J. Neurobiol. 17, 605–612.

Bowers, E.K., Sakaluk, S.K., Thompson, C.F., 2011. Adaptive sex allocation in relation to hatching synchrony and offspring quality in house wrens. Am. Nat. 177, 617–629.

Brenowitz, E.A., 2013. Testosterone and brain-derived neurotrophic factor interactions in the avian song control system. Neuroscience 239, 115–123.

Brenowitz, E.A., Beecher, M.D., 2005. Song learning in birds: diversity and plasticity, opportunities and challenges. Trends Neurosci. 28, 127–132.

Brommer, J.E., Karell, P., Pietiainen, H., 2004. Supplementary fed Ural owls increase their reproductive output with a one year time lag. Oecologia 139, 354–358.

Bruggeman, V., Onagbesan, O., Vanmontfort, D., Berghman, L., Verhoeven, G., Decuypere, E., 1998. Effect of long-term food restriction on pituitary sensitivity to cLHRH-I in broiler breeder females. J. Reprod. Fertil. 114, 267–276.

Buchanan, K.L., Evans, M.R., Goldsmith, A.R., Bryant, D.M., Rowe, L.V., 2001. Testosterone influences basal metabolic rate in male house sparrows: a new cost of dominance signalling? Proc. Biol. Sci. 268, 1337–1344.

Buntin, J.D., Ruzycki, E., 1987. Characteristics of prolactin binding sites in the brain of the ring dove (*Streptopelia risoria*). Gen. Comp. Endocrinol. 65, 243–253.

Buntin, J.D., Becker, G.M., Ruzycki, E., 1991. Facilitation of parental behavior in ring doves by systemic or intracranial injections of prolactin. Horm. Behav. 25, 424–444.

Buttemer, W.A., Warne, S., Bech, C., Astheimer, L.B., 2008. Testosterone effects on avian basal metabolic rate and aerobic performance: facts and artefacts. Comp. Biochem. Physiol. A Mol. Integr. Physiol. 150, 204–210.

Calisi, R.M., Diaz-Munoz, S.L., Wingfield, J.C., Bentley, G.E., 2011. Social and breeding status are associated with the expression of GnIH. Genes Brain Behav. 10, 557–564.

Cerolini, S., Mantovani, C., Bellagamba, F., Mangiagalli, M.G., Cavalchini, L.G., Reniero, R., 1995. Effect of restricted and *ad libitum* feeding on semen production and fertility in broiler breeder males. Br. Poult. Sci. 36, 677–682.

Charlier, T.D., Balthazart, J., Ball, G.F., 2003. Sex differences in the distribution of the steroid receptor coactivator SRC-1 in the song control nuclei of male and female canaries. Brain Res. 959, 263–274.

Charlier, T.D., Newman, A.E., Heimovics, S.A., Po, K.W., Saldanha, C.J., Soma, K.K., 2011. Rapid effects of aggressive interactions on aromatase activity and oestradiol in discrete brain regions of wild male white-crowned sparrows. J. Neuroendocrinol. 23, 742–753.

Chastel, O., Lacroix, A., Weimerskirch, H., Gabrielsen, G.W., 2005. Modulation of prolactin but not corticosterone responses to stress in relation to parental effort in a long-lived bird. Horm. Behav. 47, 459–466.

Cheng, M.F., 1977. Role of gonadotrophin releasing hormones in the reproductive behaviour of female ring doves (*Streptopelia risoria*). J. Endocrinol. 74, 37–45.

Cheng, M.F., 2008. The role of vocal self-stimulation in female responses to males: implications for state-reading. Horm. Behav. 53, 1–10.

Cheng, M.F., Zuo, M., 1994. Proposed pathways for vocal self-stimulation: met-enkephalinergic projections linking the midbrain vocal nucleus, auditory-responsive thalamic regions and neurosecretory hypothalamus. J. Neurobiol. 25, 361–379.

Cheng, M.F., Klint, T., Johnson, A.L., 1986. Breeding experience modulating androgen dependent courtship behavior in male ring doves (*Streptopelia risoria*). Physiol. Behav. 36, 625–630.

Cheng, M.F., Peng, J.P., Johnson, P., 1998. Hypothalamic neurons preferentially respond to female nest coo stimulation: demonstration of direct acoustic stimulation of luteinizing hormone release. J. Neurosci. 18, 5477–5489.

Cichon, M., 2003. Does prior breeding experience improve reproductive success in collared flycatcher females? Oecologia 134, 78–81.

Clark, C.J., Feo, T.J., 2008. The Anna's hummingbird chirps with its tail: a new mechanism of sonation in birds. Proc. Biol. Sci. 275, 955–962.

Cohen, J., Cheng, M.F., 1979. Role of vocalizations in the reproductive cycle of ring doves (*Streptopelia risoria*): effects of hypoglossal nerve section on the reproductive behavior and physiology of the female. Horm. Behav. 13, 113–127.

Cornil, C.A., Ball, G.F., 2010. Effects of social experience on subsequent sexual performance in naive male Japanese quail (*Coturnix japonica*). Horm. Behav. 57, 515–522.

Cornil, C.A., Dalla, C., Papadopoulou-Daifoti, Z., Baillien, M., Balthazart, J., 2006a. Estradiol rapidly activates male sexual behavior and affects brain monoamine levels in the quail brain. Behav. Brain Res. 166, 110–123.

Cornil, C.A., Taziaux, M., Baillien, M., Ball, G.F., Balthazart, J., 2006b. Rapid effects of aromatase inhibition on male reproductive behaviors in Japanese quail. Horm. Behav. 49, 45–67.

Criscuolo, F., Chastel, O., Gabrielsen, G.W., Lacroix, A., Le, M.Y., 2002. Factors affecting plasma concentrations of prolactin in the common eider *Somateria mollissima*. Gen. Comp. Endocrinol. 125, 399–409.

Cynx, J., 2001. Effects of humidity on reproductive behavior in male and female zebra finches (*Taeniopygia guttata*). J. Comp. Psychol. 115, 196–200.

Davies, S., Rodriguez, N.S., Sweazea, K.L., Deviche, P., 2013. The effect of acute stress and long-term corticosteroid administration on plasma metabolites in an urban and desert songbird. Physiol. Biochem. Zool. 86, 47–60.

Dawson, A., 1999. Photoperiodic control of gonadotropin-releasing hormone secretion in seasonally breeding birds. In: Rao, P., Kluwer, P. (Eds.), Neural Regulation in the Vertebrate Endocrine System. Academic/Plenum Pub., New York, pp. 141–159.

Dawson, A., King, V.M., Bentley, G.E., Ball, G.F., 2001. Photoperiodic control of seasonality in birds. J. Biol. Rhythms 16, 365–380.

de Kloet, E.R., Elands, J., Voorhuis, D.A., 1993. Implication of central neurohypophyseal hormone receptor-mediated action in timing of reproductive events: evidence from novel observations on the effect of a vasotocin analogue on singing behaviour of the canary. Regul. Pept. 45, 85–89.

Del Negro, N.C., Edeline, J.M., 2001. Differences in auditory and physiological properties of HVc neurons between reproductively active male and female canaries (*Serinus canaria*). Eur. J. Neurosci. 14, 1377–1389.

Del Negro, N.C., Edeline, J.M., 2002. Sex and season influence the proportion of thin spike cells in the canary HVc. Neuroreport 13, 2005–2009.

Del Negro, C., Gahr, M., Leboucher, G., Kreutzer, M., 1998. The selectivity of sexual responses to song displays: effects of partial chemical lesion of the HVC in female canaries. Behav. Brain Res. 96, 151–159.

Delville, Y., Balthazart, J., 1987. Hormonal control of female sexual behavior in the Japanese quail. Horm. Behav. 21, 288–309.

Delville, Y., Sulon, J., Hendrick, J.C., Balthazart, J., 1984. Effect of the presence of females on the pituitary-testicular activity in male Japanese quail (*Coturnix coturnix japonica*). Gen. Comp. Endocrinol. 55, 295–305.

Desprez, M., Pradel, R., Cam, E., Monnat, J.Y., Gimenez, O., 2011. Now you see him, now you don't: experience, not age, is related to reproduction in kittiwakes. Proc. Biol. Sci. 278, 3060–3066.

Deviche, P., 1992. Testosterone and opioids interact to regulate feeding in a male migratory songbird. Horm. Behav. 26, 394–405.

Deviche, P., Sharp, P.J., 2001. Reproductive endocrinology of a free-living, opportunistically breeding passerine (white-winged crossbill, *Loxia leucoptera*). Gen. Comp. Endocrinol. 123, 268–279.

Deviche, P., Cortez, L., 2005. Androgen control of immunocompetence in the male house finch, *Carpodacus mexicanus* Muller. J. Exp. Biol. 208, 1287–1295.

Deviche, P., Sharp, P.J., Dawson, A., Sabo, J., Fokidis, B., Davies, S., Hurley, L., 2012. Up to the challenge? Hormonal and behavioral responses of free-ranging male Cassin's sparrows, *Peucaea cassinii*, to conspecific song playback. Horm. Behav. 61, 741–749.

Dickens, M.J., Cornil, C.A., Balthazart, J., 2011. Acute stress differentially affects aromatase activity in specific brain nuclei of adult male and female quail. Endocrinology 152, 4242–4251.

Dickens, M.J., Cornil, C.A., Balthazart, J., 2013. Neurochemical control of rapid stress-induced changes in brain aromatase activity. J. Neuroendocrinol. 25, 329–339.

Dloniak, S.M., Deviche, P., 2001. Effects of testosterone and photoperiodic condition on song production and vocal control region volumes in adult male dark-eyed juncos (*Junco hyemalis*). Horm. Behav. 39, 95–105.

Dominoni, D., Quetting, M., Partecke, J., 2013. Artificial light at night advances avian reproductive physiology. Proc. Biol. Sci. 280, 20123017.

DuVal, E.H., 2012. Variation in annual and lifetime reproductive success of lance-tailed manakins: alpha experience mitigates effects of senescence on siring success. Proc. Biol. Sci. 279, 1551–1559.

El Halawani, M.E., Silsby, J.L., Behnke, E.J., Fehrer, S.C., 1986. Hormonal induction of incubation behavior in ovariectomized female turkeys (*Meleagris gallopavo*). Biol. Reprod. 35, 59–67.

Fokidis, H.B., Orchinik, M., Deviche, P., 2009. Corticosterone and corticosteroid binding globulin in birds: relation to urbanization in a desert city. Gen. Comp. Endocrinol. 160, 259–270.

Forslund, P., Part, T., 1995. Age and reproduction in birds—hypotheses and tests. Trends Ecol. Evol. 10, 374–378.

Fraley, G.S., Steiner, R.A., Lent, K.L., Brenowitz, E.A., 2010. Seasonal changes in androgen receptor mRNA in the brain of the white-crowned sparrow. Gen. Comp. Endocrinol. 166, 66–71.

Francis, C.D., Ortega, C.P., Cruz, A., 2009. Noise pollution changes avian communities and species interactions. Curr. Biol. 19, 1415–1419.

Freed-Brown, G., White, D.J., 2009. Acoustic mate copying: female cowbirds attend to other females' vocalizations to modify their song preferences. Proc. Biol. Sci. 276, 3319–3325.

Fusani, L., 2008. Testosterone control of male courtship in birds. Horm. Behav. 54, 227–233.

Fusani, L., Van't Hof, T., Hutchison, J.B., Gahr, M., 2000. Seasonal expression of androgen receptors, estrogen receptors, and aromatase in the canary brain in relation to circulating androgens and estrogens. J. Neurobiol. 43, 254–268.

Gagne, S.A., Fahrig, L., 2011. Do birds and beetles show similar responses to urbanization? Ecol. Appl. 21, 2297–2312.

Gahali, K., El Halawani, M.E., Rozenboim, I., 2001. Photostimulated prolactin release in the turkey hen: effect of ovariectomy and environmental temperature. Gen. Comp. Endocrinol. 124, 166–172.

Gahr, M., Flugge, G., Guttinger, H.R., 1987. Immunocytochemical localization of estrogen-binding neurons in the songbird brain. Brain Res. 402, 173–177.

Gahr, M., Guttinger, H.R., Kroodsma, D.E., 1993. Estrogen receptors in the avian brain: survey reveals general distribution and forebrain areas unique to songbirds. J. Comp. Neurol. 327, 112–122.

Garcia-Navas, V., Sanz, J.J., 2011. The importance of a main dish: nestling diet and foraging behaviour in Mediterranean blue tits in relation to prey phenology. Oecologia 165, 639–649.

Gil, D., Leboucher, G., Lacroix, A., Cue, R., Kreutzer, M., 2004. Female canaries produce eggs with greater amounts of testosterone when exposed to preferred male song. Horm. Behav. 45, 64–70.

Goldsmith, A.R., Edwards, C., Koprucu, M., Silver, R., 1981. Concentrations of prolactin and luteinizing hormone in plasma of doves in relation to incubation and development of the crop gland. J. Endocrinol. 90, 437–443.

Goymann, W., 2009. Social modulation of androgens in male birds. Gen. Comp. Endocrinol. 163, 149–157.

Grellet-Tinner, G., Murelaga, X., Larrasoana, J.C., Silveira, L.F., Olivares, M., Ortega, L.A., Trimby, P.W., Pascual, A., 2012. The first occurrence in the fossil record of an aquatic avian twig-nest with phoenicopteriformes eggs: evolutionary implications. PLoS One 7, e46972.

Griffith, S.C., Pryke, S.R., Buttemer, W.A., 2011. Constrained mate choice in social monogamy and the stress of having an unattractive partner. Proc. Biol. Sci. 278, 2798–2805.

Gulledge, C.C., Deviche, P., 1998. Photoperiod and testosterone independently affect vocal control region volumes in adolescent male songbirds. J. Neurobiol. 36, 550–558.

Hagelin, J.C., 2001. Castration in Gambel's and Scaled quail: ornate plumage and dominance persist, but courtship and threat behaviors do not. Horm. Behav. 39, 1–10.

Hall, Z.J., MacDougall-Shackleton, S.A., 2012. Influence of testosterone metabolites on song-control system neuroplasticity during photostimulation in adult European starlings (*Sturnus vulgaris*). PLoS One 7, e40060.

Hahn, T.P., 1995. Integration of photoperiodic and food cues to time changes in reproductive physiology by an opportunistic breeder, the red crossbill, *Loxia curvirostra* (Aves: Carduelinae). J. Exper. Zool. 272, 213–226.

Hamilton, K.S., King, A.P., Sengelaub, D.R., West, M.J., 1997. A brain of her own: a neural correlate of song assessment in a female songbird. Neurobiol. Learn. Mem. 68, 325–332.

Harding, C.F., Rowe, S.A., 2003. Vasotocin treatment inhibits courtship in male zebra finches; concomitant androgen treatment inhibits this effect. Horm. Behav. 44, 413–418.

Harrison, T.J., Smith, J.A., Martin, G.R., Chamberlain, D.E., Bearhop, S., Robb, G.N., Reynolds, S.J., 2010. Does food supplementation really enhance productivity of breeding birds? Oecologia 164, 311–320.

Heimovics, S.A., Prior, N.H., Maddison, C.J., Soma, K.K., 2012. Rapid and widespread effects of 17beta-estradiol on intracellular signaling in the male songbird brain: a seasonal comparison. Endocrinology 153, 1364–1376.

Hirao, A., Aoyama, M., Sugita, S., 2009. The role of uropygial gland on sexual behavior in domestic chicken *Gallus gallus domesticus*. Behav. Processes 80, 115–120.

Hirschenhauser, K., Winkler, H., Oliveira, R.F., 2003. Comparative analysis of male androgen responsiveness to social environment in birds: the effects of mating system and paternal incubation. Horm. Behav. 43, 508–519.

Horvathova, T., Nakagawa, S., Uller, T., 2012. Strategic female reproductive investment in response to male attractiveness in birds. Proc. Biol. Sci. 279, 163–170.

Hsueh, A.J., Schaeffer, J.M., 1985. Gonadotropin-releasing hormone as a paracrine hormone and neurotransmitter in extra-pituitary sites. J. Steroid. Biochem. 23, 757–764.

Hunt, K.E., Wingfield, J.C., 2004. Effect of estradiol implants on reproductive behavior of female Lapland longspurs (*Calcarius lapponicus*). Gen. Comp. Endocrinol. 137, 248–262.

Jaccoby, S., Arnon, E., Snapir, N., Robinzon, B., 1995. Effects of estradiol and tamoxifen on feeding, fattiness, and some endocrine criteria in hypothalamic obese hens. Pharmacol. Biochem. Behav. 50, 55–63.

Janik, D.S., Buntin, J.D., 1985. Behavioural and physiological effects of prolactin in incubating ring doves. J. Endocrinol. 105, 201–209.

Kempenaers, B., Borgstrom, P., Loes, P., Schlicht, E., Valcu, M., 2010. Artificial night lighting affects dawn song, extra-pair siring success, and lay date in songbirds. Curr. Biol. 20, 1735–1739.

Kight, C.R., Saha, M.S., Swaddle, J.P., 2012. Anthropogenic noise is associated with reductions in the productivity of breeding Eastern bluebirds (*Sialia sialis*). Ecol. Appl. 22, 1989–1996.

Kimura, T., Okanoya, K., Wada, M., 1999. Effect of testosterone on the distribution of vasotocin immunoreactivity in the brain of the zebra finch, *Taeniopygia guttata castanotis*. Life Sci. 65, 1663–1670.

Kiss, J.Z., Voorhuis, T.A., van Eekelen, J.A., de Kloet, E.R., De, W.D., 1987. Organization of vasotocin-immunoreactive cells and fibers in the canary brain. J. Comp. Neurol. 263, 347–364.

Kleven, O., Fossoy, F., Laskemoen, T., Robertson, R.J., Rudolfsen, G., Lifjeld, J.T., 2009. Comparative evidence for the evolution of sperm swimming speed by sperm competition and female sperm storage duration in passerine birds. Evolution 63, 2466–2473.

Komisaruk, B.R., 1967. Effects of local brain implants of progesterone on reproductive behavior in ring doves. J. Comp. Physiol. Psychol. 64, 219–224.

Kotegawa, T., Abe, T., Tsutsui, K., 1997. Inhibitory role of opioid peptides in the regulation of aggressive and sexual behaviors in male Japanese quails. J. Exp. Zool. 277, 146–154.

Kuwayama, T., Shimada, K., Saito, N., Ohkubo, T., Sato, K., Wada, M., Ichinoe, K., 1992. Effects of removal of chicks from hens on concentrations of prolactin, luteinizing hormone and oestradiol in plasma of brooding Gifujidori hens. J. Reprod. Fertil. 95, 617–622.

Landys, M.M., Goymann, W., Raess, M., Slagsvold, T., 2007. Hormonal responses to male-male social challenge in the blue tit *Cyanistes caeruleus*: single-broodedness as an explanatory variable. Physiol. Biochem. Zool. 80, 228–240.

Landys, M.M., Goymann, W., Schwabl, I., Trapschuh, M., Slagsvold, T., 2010. Impact of season and social challenge on testosterone and corticosterone levels in a year-round territorial bird. Horm. Behav. 58, 317–325.

Langmore, N.E., Cockrem, J.F., Candy, E.J., 2002. Competition for male reproductive investment elevates testosterone levels in female dunnocks, *Prunella modularis*. Proc. Biol. Sci. 269, 2473–2478.

Lea, R.W., Vowles, D.M., Dick, H.R., 1986a. Factors affecting prolactin secretion during the breeding cycle of the ring dove (*Streptopelia risoria*) and its possible role in incubation. J. Endocrinol. 110, 447–458.

Lea, R.W., Sharp, P.J., Klandorf, H., Harvey, S., Dunn, I.C., Vowles, D.M., 1986b. Seasonal changes in concentrations of plasma hormones in the male ring dove (*Streptopelia risoria*). J. Endocrinol. 108, 385–391.

Lehrman, D.S., 1964. The reproductive behavior of ring doves. Sci. Am. 211, 48–54.

Lehrman, D.S., Friedman, M., 1969. Auditory stimulation of ovarian activity in the ring dove (*Streptopelia risoria*). Anim. Behav. 17, 494–497.

Leung, C.H., Goode, C.T., Young, L.J., Maney, D.L., 2009. Neural distribution of nonapeptide binding sites in two species of songbird. J. Comp. Neurol. 513, 197–208.

Leung, C.H., Abebe, D.F., Earp, S.E., Goode, C.T., Grozhik, A.V., Mididoddi, P., Maney, D.L., 2011. Neural distribution of vasotocin receptor mRNA in two species of songbird. Endocrinology 152, 4865–4881.

Li, X.C., Jarvis, E.D., Alvarez-Borda, B., Lim, D.A., Nottebohm, F., 2000. A relationship between behavior, neurotrophin expression, and new neuron survival. Proc. Natl. Acad. Sci. U.S.A 97, 8584–8589.

Liu, W.C., Wada, K., Jarvis, E., Nottebohm, F., 2013. Rudimentary substrates for vocal learning in a suboscine. Nat. Commun. 4, 2082.

London, S.E., Monks, D.A., Wade, J., Schlinger, B.A., 2006. Widespread capacity for steroid synthesis in the avian brain and song system. Endocrinology 147, 5975–5987.

Longcore, T., 2010. Sensory ecology: night lights alter reproductive behavior of blue tits. Curr. Biol. 20, R893–R895.

Lormee, H., Jouventin, P., Chastel, O., Mauget, R., 1999. Endocrine correlates of parental care in an Antarctic winter breeding seabird, the emperor penguin, *Aptenodytes Forsteri*. Horm. Behav. 35, 9–17.

Lott, D., Scholz, S.D., Lehrman, D.S., 1967. Exteroceptive stimulation of the reproductive system of the female ring dove (*Streptopella risoria*) by the mate and by the colony milieu. Anim. Behav. 15, 433–437.

Low, M., Part, T., Forslund, P., 2007. Age-specific variation in reproduction is largely explained by the timing of territory establishment in the New Zealand stitchbird *Notiomystis cincta*. J. Anim. Ecol. 76, 459–470.

Lynn, S.E., Walker, B.G., Wingfield, J.C., 2005. A phylogenetically controlled test of hypotheses for behavioral insensitivity to testosterone in birds. Horm. Behav. 47, 170–177.

Maney, D.L., Wingfield, J.C., 1998. Neuroendocrine suppression of female courtship in a wild passerine: corticotropin-releasing factor and endogenous opioids. J. Neuroendocrinol. 10, 593–599.

Maney, D.L., Goode, C.T., Wingfield, J.C., 1997a. Intraventricular infusion of arginine vasotocin induces singing in a female songbird. J. Neuroendocrinol. 9, 487–491.

Maney, D.L., Richardson, R.D., Wingfield, J.C., 1997b. Central administration of chicken gonadotropin-releasing hormone-II enhances courtship behavior in a female sparrow. Horm. Behav. 32, 11–18.

Maney, D.L., Goode, C.T., Lake, J.I., Lange, H.S., O'Brien, S., 2007. Rapid neuroendocrine responses to auditory courtship signals. Endocrinology 148, 5614–5623.

Massaro, M., Setiawan, A.N., Davis, L.S., 2007. Effects of artificial eggs on prolactin secretion, steroid levels, brood patch development, incubation onset and clutch size in the yellow-eyed penguin (*Megadyptes antipodes*). Gen. Comp. Endocrinol. 151, 220–229.

McDonald, P.A., Liley, N.R., 1978. The effects of photoperiod on androgen-induced reproductive behavior in male ring doves, *Streptopelia risoria*. Horm. Behav. 10, 85–96.

Meitzen, J., Moore, I.T., Lent, K., Brenowitz, E.A., Perkel, D.J., 2007. Steroid hormones act transsynaptically within the forebrain to regulate neuronal phenotype and song stereotypy. J. Neurosci. 27, 12045–12057.

Meitzen, J., Thompson, C.K., 2008. Seasonal-like growth and regression of the avian song control system: neural and behavioral plasticity in adult male Gambel's white-crowned sparrows. Gen. Comp. Endocrinol. 157, 259–265.

Michel, G.F., 1976. Role of mate's previous experience in ring dove hormone-induced incubation. J. Comp. Physiol. Psychol. 90, 468–472.

Mimet, A., Pellissier, V., Quenol, H., Aguejdad, R., Dubreuil, V., Roze, F., 2009. Urbanisation induces early flowering: evidence from *Platanus acerifolia* and *Prunus cerasus*. Int. J. Biometeorol. 53, 287–298.

Moller, A.P., 2013. Biological consequences of global change for birds. Integr. Zool. 8, 136–144.

Nealen, P.M., 2005. An interspecific comparison using immunofluorescence reveals that synapse density in the avian song system is related to sex but not to male song repertoire size. Brain Res. 1032, 50–62.

Nordeen, K.W., Nordeen, E.J., Arnold, A.P., 1987. Estrogen accumulation in zebra finch song control nuclei: implications for sexual differentiation and adult activation of song behavior. J. Neurobiol. 18, 569–582.

Nottebohm, F., 1981. A brain for all seasons: cyclical anatomical changes in song control nuclei of the canary brain. Science 214, 1368–1370.

Nottebohm, F., Stokes, T.M., Leonard, C.M., 1976. Central control of song in the canary, *Serinus canarius*. J. Comp. Neurol. 165, 457–486.

Nottebohm, F., Kelley, D.B., Paton, J.A., 1982. Connections of vocal control nuclei in the canary telencephalon. J. Comp. Neurol. 207, 344–357.

Nottebohm, F., O'Loughlin, B., Gould, K., Yohay, K., Alvarez-Buylla, A., 1994. The life span of new neurons in a song control nucleus of the adult canary brain depends on time of year when these cells are born. Proc. Natl. Acad. Sci. U.S.A 91, 7849–7853.

O'Neal, D.M., Reichard, D.G., Pavilis, K., Ketterson, E.D., 2008. Experimentally-elevated testosterone, female parental care, and reproductive success in a songbird, the dark-eyed junco (*Junco hyemalis*). Horm. Behav. 54, 571–578.

Osorno, J.L., Nunez-de la-Mora, A., D'Alba, L., Wingfield, J.C., 2010. Hormonal correlates of breeding behavior and pouch color in the magnificent frigatebird, *Fregata magnificens*. Gen. Comp. Endocrinol. 169, 18–22.

Owen-Ashley, N.T., Hasselquist, D., Wingfield, J.C., 2004. Androgens and the immunocompetence handicap hypothesis: unraveling direct and indirect pathways of immunosuppression in song sparrows. Am. Nat. 164, 490–505.

Owens, I.P., 2002. Male-only care and classical polyandry in birds: phylogeny, ecology and sex differences in remating opportunities. Philos. Trans. R. Soc. Lond. B Biol. Sci. 357, 283–293.

Oyegbile, T.O., Marler, C.A., 2005. Winning fights elevates testosterone levels in California mice and enhances future ability to win fights. Horm. Behav. 48, 259–267.

Panzica, G.C., Viglietti-Panzica, C., Balthazart, J., 1996. The sexually dimorphic medial preoptic nucleus of quail: a key brain area mediating steroid action on male sexual behavior. Front. Neuroendocrinol 17, 51–125.

Panzica, G., Pessatti, M., Viglietti-Panzica, C., Grossmann, R., Balthazart, J., 1999a. Effects of testosterone on sexually dimorphic parvocellular neurons expressing vasotocin mRNA in the male quail brain. Brain Res. 850, 55–62.

Panzica, G.C., Plumari, L., Garcia-Ojeda, E., Deviche, P., 1999b. Central vasotocin-immunoreactive system in a male passerine bird (*Junco hyemalis*). J. Comp. Neurol. 409, 105–117.

Panzica, G.C., Aste, N., Castagna, C., Viglietti-Panzica, C., Balthazart, J., 2001. Steroid-induced plasticity in the sexually dimorphic vasotocinergic innervation of the avian brain: behavioral implications. Brain Res. Brain Res. Rev. 37, 178–200.

Partecke, J., Gwinner, E., 2007. Increased sedentariness in European blackbirds following urbanization: a consequence of local adaptation? Ecology 88, 882–890.

Partecke, J., Schwabl, I., Gwinner, E., 2006. Stress and the city: urbanization and its effects on the stress physiology in European blackbirds. Ecology 87, 1945–1952.

Pasteau, M., Ung, D., Kreutzer, M., Aubin, T., 2012. Amplitude modulation of sexy phrases is salient for song attractiveness in female canaries (*Serinus canaria*). Anim. Cogn. 15, 639–645.

Payne, C.J., Jessop, T.S., Guay, P.J., Johnstone, M., Feore, M., Mulder, R.A., 2012. Population, behavioural and physiological responses of an urban population of black swans to an intense annual noise event. PLoS One 7, e45014.

Perfito, N., Kwong, J.M., Bentley, G.E., Hau, M., 2008. Cue hierarchies and testicular development: is food a more potent stimulus than day length in an opportunistic breeder (*Taeniopygia g. guttata*)? Horm. Behav. 53, 567–572.

Plumari, L., Plateroti, S., Deviche, P., Panzica, G.C., 2004. Region-specific testosterone modulation of the vasotocin-immunoreactive system in male dark-eyed junco, Junco Hyemalis. Brain Res. 999, 1–8.

Pradel, R., Choquet, R., Bechet, A., 2012. Breeding experience might be a major determinant of breeding probability in long-lived species: the case of the greater flamingo. PLoS One 7, e51016.

Quiring, D.P., 1944. Transplantation of testes (by A.A. Berthold). Bull. Histol. Med. 16, 399–401.

Ramesh, R., Kuenzel, W.J., Buntin, J.D., Proudman, J.A., 2000. Identification of growth-hormone- and prolactin-containing neurons within the avian brain. Cell Tissue Res. 299, 371–383.

Rasika, S., Alvarez-Buylla, A., Nottebohm, F., 1999. BDNF mediates the effects of testosterone on the survival of new neurons in an adult brain. Neuron 22, 53–62.

Rauceo, S., Harding, C.F., Maldonado, A., Gaysinkaya, L., Tulloch, I., Rodriguez, E., 2008. Dopaminergic modulation of reproductive behavior and activity in male zebra finches. Behav. Brain Res. 187, 133–139.

Reichert, B.E., Cattau, C.E., Fletcher Jr., R.J., Kendall, W.L., Kitchens, W.M., 2012. Extreme weather and experience influence reproduction in an endangered bird. Ecology 93, 2580–2589.

Remage-Healey, L., Maidment, N.T., Schlinger, B.A., 2008. Forebrain steroid levels fluctuate rapidly during social interactions. Nat. Neurosci. 11, 1327–1334.

Remage-Healey, L., Oyama, R.K., Schlinger, B.A., 2009. Elevated aromatase activity in forebrain synaptic terminals during song. J. Neuroendocrinol. 21, 191–199.

Remage-Healey, L., Coleman, M.J., Oyama, R.K., Schlinger, B.A., 2010. Brain estrogens rapidly strengthen auditory encoding and guide song preference in a songbird. Proc. Natl. Acad. Sci. U.S.A. 107, 3852–3857.

Remage-Healey, L., Dong, S.M., Chao, A., Schlinger, B.A., 2012. Sex-specific, rapid neuroestrogen fluctuations and neurophysiological actions in the songbird auditory forebrain. J. Neurophysiol. 107, 1621–1631.

Richard-Yris, M.A., Leboucher, G., Williams, J., Garnier, D.H., 1987. Influence of food restriction and of the presence of chicks on the reproductive system of the domestic hen. Br. Poult. Sci. 28, 251–260.

Richard-Yris, M.A., Sharp, P.J., Wauters, A.M., Guemene, D., Richard, J.P., Foraste, M., 1998. Influence of stimuli from chicks on behavior and concentrations of plasma prolactin and luteinizing hormone in incubating hens. Horm. Behav. 33, 139–148.

Rissman, E.F., Wingfield, J.C., 1984. Hormonal correlates of polyandry in the spotted sandpiper, Actitis Macularia. Gen. Comp. Endocrinol. 56, 401–405.

Riters, L.V., Ball, G.F., 2002. Sex differences in the densities of alpha 2-adrenergic receptors in the song control system, but not the medial preoptic nucleus in zebra finches. J. Chem. Neuroanat. 23, 269–277.

Riters, L.V., Teague, D.P., 2003. The volumes of song control nuclei, HVC and lMAN, relate to differential behavioral responses of female European starlings to male songs produced within and outside of the breeding season. Brain Res. 978, 91–98.

Rutkowska, J., Place, N.J., Vincent, S., Adkins-Regan, E., 2011. Adrenocortical response to mating, social interaction and restraint in the female Japanese quail. Physiol. Behav. 104, 1037–1040.

Sachs, B.D., Leipheimer, R.E., 1988. Rapid effect of testosterone on striated muscle activity in rats. Neuroendocrinology 48, 453–458.

Sakaguchi, H., Li, R., Taniguchi, I., 2000. Sex differences in the ventral paleostriatum of the zebra finch: origin of the cholinergic innervation of the song control nuclei. Neuroreport 11, 2727–2731.

Saranathan, V., Hamilton, D., Powell, G.V., Kroodsma, D.E., Prum, R.O., 2007. Genetic evidence supports song learning in the three-wattled bellbird *Procnias tricarunculata* (Cotingidae). Mol. Ecol. 16, 3689–3702.

Schlesinger, M.D., Manley, P.N., Holyoak, M., 2008. Distinguishing stressors acting on land bird communities in an urbanizing environment. Ecology 89, 2302–2314.

Schlinger, B.A., Callard, G.V., 1990. Aromatization mediates aggressive behavior in quail. Gen. Comp. Endocrinol. 79, 39–53.

Schlinger, B.A., and Balthazart, J. 2012. Aromatase and Behavior: Concepts Gained from Studies of Aromatase in the Avian Brain. In Brain Aromatase, Estrogens, and Behavior (Balthazart J. and Ball G., Eds). Oxford Scholarship Online.

Schlinger, B.A., Remage-Healey, L., 2012. Neurosteroidogenesis: insights from studies of songbirds. J. Neuroendocrinol. 24, 16–21.

Schlinger, B.A., Day, L.B., Fusani, L., 2008. Behavior, natural history and neuroendocrinology of a tropical bird. Gen. Comp. Endocrinol. 157, 254–258.

Schoech, S.J., Bowman, R., Reynolds, S.J., 2004. Food supplementation and possible mechanisms underlying early breeding in the Florida Scrub-Jay (*Aphelocoma coerulescens*). Horm. Behav. 46, 565–573.

Schroeder, J., Nakagawa, S., Cleasby, I.R., Burke, T., 2012. Passerine birds breeding under chronic noise experience reduced fitness. PLoS One 7, e39200.

Searcy, W.A., Marler, P., 1981. A test for responsiveness to song structure and programming in female sparrows. Science 213, 926–928.

Selvaraj, P., Pitchappan, R.M., 1985. Effect of oestradiol dipropionate on the immune system of the pigeon, *Columba livia*. Dev. Comp. Immunol 9, 669–677.

Senapathi, D., Nicoll, M.A., Teplitsky, C., Jones, C.G., Norris, K., 2011. Climate change and the risks associated with delayed breeding in a tropical wild bird population. Proc. Biol. Sci. 278, 3184–3190.

Seredynski, A.L., Ball, G.F., Balthazart, J., Charlier, T.D., 2011. Specific activation of estrogen receptor alpha and beta enhances male sexual behavior and neuroplasticity in male Japanese quail. PLoS One 6, e18627.

Seredynski, A.L., Balthazart, J., Christophe, V.J., Ball, G.F., Cornil, C.A., 2013. Neuroestrogens rapidly regulate sexual motivation but not performance. J. Neurosci. 33, 164–174.

Sharma, D., Cornett, L.E., Chaturvedi, C.M., 2009. Osmotic stress induced alteration in the expression of arginine vasotocin receptor VT2 in the pituitary gland and adrenal function of domestic fowl. Gen. Comp. Endocrinol. 160, 216–222.

Sharp, P.J., Dawson, A., Lea, R.W., 1998. Control of luteinizing hormone and prolactin secretion in birds. Comp. Biochem. Physiol. C Pharmacol. Toxicol. Endocrinol. 119, 275–282.

Shine, R., Brown, G.P., 2008. Adapting to the unpredictable: reproductive biology of vertebrates in the Australian wet-dry tropics. Philos. Trans. R. Soc. Lond. B Biol. Sci. 363, 363–373.

Silverin, B., 1991. Behavioral, hormonal, and morphological responses of free-living male pied flycatchers to estradiol treatment of their mates. Horm. Behav. 25, 38–56.

Silverin, B., Baillien, M., Balthazart, J., 2004. Territorial aggression, circulating levels of testosterone, and brain aromatase activity in free-living pied flycatchers. Horm. Behav. 45, 225–234.

Silverin, B., Westin, J., 1995. Influence of the opposite sex on photoperiodically induced LH and gonadal cycles in willow tit (*Parus montanus*). Horm. Behav. 29, 207–215.

Sitko, J., Zalesny, G., 2012. The effect of urbanization on helminth communities in the Eurasian blackbird (*Turdus merula* L.) from the eastern part of the Czech Republic. J. Helminthol., 1–8.

Slawski, B.A., Buntin, J.D., 1995. Preoptic area lesions disrupt prolactin-induced parental feeding behavior in ring doves. Horm. Behav. 29, 248–266.

Small, T.W., Sharp, P.J., Bentley, G.E., Millar, R.P., Tsutsui, K., Strand, C., Deviche, P., 2008. Auditory stimulation of reproductive function in male rufous-winged sparrows, *Aimophila carpalis*. Horm. Behav. 53, 28–39.

Smith, G.T., Brenowitz, E.A., Wingfield, J.C., 1997. Seasonal changes in the size of the avian song control nucleus HVC defined by multiple histological markers. J. Comp. Neurol. 381, 253–261.

Soma, K.K., Hartman, V.N., Wingfield, J.C., Brenowitz, E.A., 1999. Seasonal changes in androgen receptor immunoreactivity in the song nucleus HVc of a wild bird. J. Comp. Neurol. 409, 224–236.

Soma, K.K., Schlinger, B.A., Wingfield, J.C., Saldanha, C.J., 2003. Brain aromatase, 5 alpha-reductase, and 5 beta-reductase change seasonally in wild male song sparrows: relationship to aggressive and sexual behavior. J. Neurobiol. 56, 209–221.

Soma, K.K., Tramontin, A.D., Featherstone, J., Brenowitz, E.A., 2004. Estrogen contributes to seasonal plasticity of the adult avian song control system. J. Neurobiol. 58, 413–422.

Spee, M., Beaulieu, M., Dervaux, A., Chastel, O., Le, M.Y., Raclot, T., 2010. Should I stay or should I go? Hormonal control of nest abandonment in a long-lived bird, the Adélie penguin. Horm. Behav. 58, 762–768.

Stevenson, T.J., Bernard, D.J., McCarthy, M.M., Ball, G.F., 2013. Photoperiod-dependent regulation of gonadotropin-releasing hormone 1 messenger ribonucleic acid levels in the songbird brain. Gen. Comp. Endocrinol. 190, 81–87.

Thierry, A.M., Brajon, S., Massemin, S., Handrich, Y., Chastel, O., and Raclot, T., 2013. Decreased prolactin levels reduce parental commitment, egg temperatures, and breeding success of incubating male Adelie penguins. Horm. Behav. 64, 737–747.

Thompson, C.K., Bentley, G.E., Brenowitz, E.A., 2007. Rapid seasonal-like regression of the adult avian song control system. Proc. Natl. Acad. Sci. U.S.A. 104, 15520–15525.

Thompson, C.K., Meitzen, J., Replogle, K., Drnevich, J., Lent, K.L., Wissman, A.M., Farin, F.M., Bammler, T.K., Beyer, R.P., Clayton, D.F., Perkel, D.J., Brenowitz, E.A., 2012. Seasonal changes in patterns of gene expression in avian song control brain regions. PLoS One 7, e35119.

Tobari, Y., Nakamura, K.Z., Okanoya, K., 2005. Sex differences in the telencephalic song control circuitry in Bengalese finches (*Lonchura striata var. domestica*). Zoolog. Sci. 22, 1089–1094.

Tonra, C.M., Marra, P.P., Holberton, R.L., 2011. Early elevation of testosterone advances migratory preparation in a songbird. J. Exp. Biol. 214, 2761–2767.

Tramontin, A.D., Smith, G.T., Breuner, C.W., Brenowitz, E.A., 1998. Seasonal plasticity and sexual dimorphism in the avian song control system: stereological measurement of neuron density and number. J. Comp. Neurol. 396, 186–192.

Tramontin, A.D., Wingfield, J.C., Brenowitz, E.A., 2003. Androgens and estrogens induce seasonal-like growth of song nuclei in the adult songbird brain. J. Neurobiol. 57, 130–140.

Tryjanowski, P., Sparks, T.H., Kuzniak, S., Czechowski, P., Jerzak, L., 2013. Bird migration advances more strongly in urban environments. PLoS One 8, e63482.

Tsutsui, K., Saigoh, E., Ukena, K., Teranishi, H., Fujisawa, Y., Kikuchi, M., Ishii, S., Sharp, P.J., 2000. A novel avian hypothalamic peptide inhibiting gonadotropin release. Biochem. Biophys. Res. Commun. 275, 661–667.

Ubuka, T., McGuire, N.L., Calisi, R.M., Perfito, N., Bentley, G.E., 2008. The control of reproductive physiology and behavior by gonadotropin-inhibitory hormone. Integr. Comp. Biol. 48, 560–569.

Ubuka, T., Lai, H., Kitani, M., Suzuuchi, A., Pham, V., Cadigan, P.A., Wang, A., Chowdhury, V.S., Tsutsui, K., Bentley, G.E., 2009. Gonadotropin-inhibitory hormone identification, cDNA cloning, and distribution in rhesus macaque brain. J. Comp. Neurol. 517, 841–855.

Ubuka, T., Mizuno, T., Fukuda, Y., Bentley, G.E., Wingfield, J.C., Tsutsui, K., 2013. RNA interference of gonadotropin-inhibitory hormone gene induces aggressive and sexual behaviors in birds. Gen. Comp. Endocrinol. 181, 179–186.

Ukena, K., Ubuka, T., Tsutsui, K., 2003. Distribution of a novel avian gonadotropin-inhibitory hormone in the quail brain. Cell Tissue Res. 312, 73–79.

Van Roo, B.L., 2004. Exogenous testosterone inhibits several forms of male parental behavior and stimulates song in a monogamous songbird: the blue-headed vireo (*Vireo solitarius*). Horm. Behav. 46, 678–683.

Varricchio, D.J., Moore, J.R., Erickson, G.M., Norell, M.A., Jackson, F.D., Borkowski, J.J., 2008. Avian paternal care had dinosaur origin. Science 322, 1826–1828.

Viglietti-Panzica, C., Aste, N., Balthazart, J., Panzica, G.C., 1994. Vasotocinergic innervation of sexually dimorphic medial preoptic nucleus of the male Japanese quail: influence of testosterone. Brain Res. 657, 171–184.

Viglietti-Panzica, C., Balthazart, J., Plumari, L., Fratesi, S., Absil, P., Panzica, G.C., 2001. Estradiol mediates effects of testosterone on vasotocin immunoreactivity in the adult quail brain. Horm. Behav. 40, 445–461.

Visser, M.E., Holleman, L.J., Caro, S.P., 2009. Temperature has a causal effect on avian timing of reproduction. Proc. Biol. Sci. 276, 2323–2331.

Vleck, C.M., Ross, L.L., Vleck, D., Bucher, T.L., 2000. Prolactin and parental behavior in Adélie penguins: effects of absence from nest, incubation length, and nest failure. Horm. Behav. 38, 149–158.

Voigt, C., Ball, G.F., Balthazart, J., 2011. Effects of sex steroids on aromatase mRNA expression in the male and female quail brain. Gen. Comp. Endocrinol. 170, 180–188.

Voorhuis, T.A., de Kloet, E.R., 1992. Immunoreactive vasotocin in the zebra finch brain (*Taeniopygia guttata*). Brain Res. Dev. Brain Res. 69, 1–10.

Voorhuis, T.A., Kiss, J.Z., de Kloet, E.R., De Wied, W.D., 1988a. Testosterone-sensitive vasotocin-immunoreactive cells and fibers in the canary brain. Brain Res. 442, 139–146.

Voorhuis, T.A., de Kloet, E.R., De Wied, W.D., 1988b. The distribution and plasticity of [3H]vasopressin-labelled specific binding sites in the canary brain. Brain Res. 457, 148–153.

Voorhuis, T.A., de Kloet, E.R., De Wied, W.D., 1991a. Effect of a vasotocin analog on singing behavior in the canary. Horm. Behav. 25, 549–559.

Voorhuis, T.A., de Kloet, E.R., De Wied, W.D., 1991b. Ontogenetic and seasonal changes in immunoreactive vasotocin in the canary brain. Brain Res. Dev. Brain Res. 61, 23–31.

Watts, H.E., Hahn, T.P., 2012. Non-photoperiodic regulation of reproductive physiology in the flexibly breeding pine siskin (*Spinus pinus*). Gen. Comp. Endocrinol. 178, 259–264.

White, S.J., 1975. Effects of stimuli emanating from the nest on the reproductive cycle in the ring dove. I: pre-laying behaviour. Anim. Behav. 23, 854–868.

Wilhelm, M., 2011. Neuro-immune interactions in the dove brain. Gen. Comp. Endocrinol. 172, 173–180.

Wingfield, J.C., 1984. Influence of weather on reproduction. J. Exp. Zool. 232, 589–594.

Wingfield, J.C., 1994a. Control of territorial aggression in a changing environment. Psychoneuroendocrinology 19, 709–721.

Wingfield, J.C., 1994b. Regulation of territorial behavior in the sedentary song sparrow, *Melospiza melodia morphna*. Horm. Behav. 28, 1–15.

Wingfield, J.C., Goldsmith, A.R., 1990. Plasma levels of prolactin and gonadal steroids in relation to multiple-brooding and renesting in free-living populations of the song sparrow, *Melospiza melodia*. Horm. Behav. 24, 89–103.

Wingfield, J.C., Hahn, T.P., 1994. Testosterone and territorial behaviour in sedentary and migratory sparrows. Anim. Behav. 47, 77–89.

Wingfield, J.C., Monk, D., 1994. Behavioral and hormonal responses of male song sparrows to estradiol-treated females during the non-breeding season. Horm. Behav. 28, 146–154.

Wingfield, J., Ball, G.F., Dufty, A.M., Hegner, R.E., Ramenofsky, M., 1987. Testosterone and aggression in birds. Am. Sci. 75, 602–608.

Wissman, A.M., Brenowitz, E.A., 2009. The role of neurotrophins in the seasonal-like growth of the avian song control system. J. Neurosci. 29, 6461–6471.

Wright, L.J., Hoblyn, R.A., Green, R.E., Bowden, C.G., Mallord, J.W., Sutherland, W.J., Dolman, P.M., 2009. Importance of climatic and environmental change in the demography of a multi-brooded passerine, the woodlark *Lullula arborea*. J. Anim. Ecol. 78, 1191–1202.

Zanette, L., Smith, J.N., van, O.H., Clinchy, M., 2003. Synergistic effects of food and predators on annual reproductive success in song sparrows. Proc. Biol. Sci. 270, 799–803.

Zanette, L., Clinchy, M., Sung, H.C., 2009. Food-supplementing parents reduces their sons' song repertoire size. Proc. Biol. Sci. 276, 2855–2860.

Zhang, J.X., Wei, W., Zhang, J.H., Yang, W.H., 2010a. Uropygial gland-secreted alkanols contribute to olfactory sex signals in budgerigars. Chem. Senses 35, 375–382.

Zhang, K., Wang, R., Shen, C., Da, L., 2010b. Temporal and spatial characteristics of the urban heat island during rapid urbanization in Shanghai, China. Environ. Monit. Assess. 169, 101–112.

Zheng, X., O'Connor, J., Huchzermeyer, F., Wang, X., Wang, Y., Wang, M., Zhou, Z., 2013. Preservation of ovarian follicles reveals early evolution of avian reproductive behaviour. Nature 495, 507–511.

Zhuang, X., Silverman, A.J., Silver, R., 1993. Reproductive behavior, endocrine state, and the distribution of GnRH-like immunoreactive mast cells in dove brain. Horm. Behav. 27, 283–295.

Chapter 31

Brooding

Yupaporn Chaiseha
School of Biology, Institute of Science, Suranaree University of Technology, Thailand

Mohamed E. El Halawani
Department of Animal Science, University of Minnesota, St. Paul, MN, USA

31.1 INTRODUCTION

Parental care promotes the survival and well-being of the offspring, and species, at a cost of resources to the parents as well as the apparently sacrificial acts of individual parents (Gross, 2005). This important vertebrate behavior is the most obvious and pervasive example of altruism in the kingdom Animalia representing the original form of prosocial behavior from which all other animal behaviors are presumed to be derived (Rilling, 2013). The period between birth/hatching and recruitment into a breeding population is a critical phase and may be considered one of the least understood components of animal life histories (Clutton-Brock, 1991). The survival of the young until reproduction has marked effects on population growth and is more sensitive to environmental changes than adult survival (Stearns, 1992; Clark and Martin, 2007).

Among the major vertebrate taxa, behaviors in species in the class Aves are the most extensively studied, as birds exhibit an incomparable balance of tractability, diversity, and cognitive complexity (Rosenblatt, 2003; Kentner et al., 2010; Goodson et al., 2012). Reproductive efforts in birds are extended past fertilization with a variety of parental behaviors. Patterns of parental care range from brood parasitism, in which eggs are laid in the nests of a host species and no display of parental behavior, to intensively prolonged egg incubation and care of offspring by parents. More than 99% of the 9000 species of birds exhibit parental behaviors (Silver et al., 1985). Unlike other vertebrate species, more than 90% avian species exhibit biparental care (Kendeigh, 1952; Lack, 1968) in various degrees between species (Schradin and Anzenberger, 1999; Vleck and Vleck, 2011). More than 70% rear altricial young (Silver et al., 1985). These are poorly developed and helpless at hatch. They depend on their parents to provide food, heat, and protection for an extended period (Buntin, 1996). The remaining 30% species of birds rear precocial young (Buntin, 2010). These are well developed and able to leave the nest, follow their parents, and forage for food on their own after hatching. In precocial species, parental care after hatching consists of brooding the chicks to keep them warm, defense of the chicks, and leading them to food sources and shelter (Buntin, 2010).

Parental behavior is defined as the behavior of the parents that contributes to the survival of their offspring (Numan and Insel, 2003). Maternal behavior is defined as the collection of behaviors by the mother that can increase offspring survival (Krasnegor and Bridges, 1990). Nurturing behaviors analogous to maternal behaviors are called paternal behavior by fathers/male mating partners and alloparental behavior by older conspecifics (Kuroda et al., 2011). In birds, parental behavior includes nest preparation, egg laying into a preferred site such as a nest, egg incubation, and posthatch care of the young to independence (Rosenblatt, 2003; Ruscio and Adkins-Regan, 2004). Maternal care is limited to incubation and brooding or rearing behaviors (El Halawani et al., 1988a). Paternal care refers to behaviors performed by the mature male, which have a positive influence on development, growth, well-being, and survival of the offspring (Fernandez-Duque et al., 2009).

In the vast majority of birds, it is essential that one or both parents incubate their eggs until hatching and then provide posthatching care. The extent of parental care for the eggs and chicks is depended on the developmental maturity of the hatchling such as precocial and altricial chicks. Care of the young ranges from guarding and guiding in the most precocial species such as those within the orders Anseriformes and Galliformes to provisioning of all food and intensive brooding for thermoregulation such as within species within the orders Passeriformes and Psittaciformes (Vleck, 1998). However, there are some species with the

Sturkie's Avian Physiology.

most precocial chicks that require no posthatching care. For instance, in the family *Megapodiidae* (order *Galliformes*), the Australian brush-turkey (*Alectura lathami*) shows no parental care. They lay eggs in underground nests, and these eggs are subsequently incubated by external heat sources. The chicks hatch, then dig out of their nests by themselves, and develop independently of their parents and their siblings (Goth, 2002; Goth and Vogel, 2003). Common cuckoos (*Cuculus canorus*) and brown-headed cowbirds (*Molothrus ater*) are parasitic species. They lay their eggs in the nests of other (host) species and let them raise their young (Winfree, 1999; Kruger, 2007). Likewise, Goldeneye ducks (*Bucephala clangula*) show intraspecific brood parasitism, with some females laying their eggs in the nest of other females of the same species (Andersson and Eriksson, 1982).

31.2 BROODING (BROODINESS)

The term "brooding" (broodiness) in gallinaceous birds refers both to hens incubating their eggs (incubation behavior) or those caring for their young after hatching (brooding behavior) (Ramsay, 1953; El Halawani et al., 1988a). Thus, broodiness has two components: incubation and brooding behaviors. The former can only be induced in laying hens while the latter can be induced in both laying and nonlaying hens together with males. These two behaviors enable the development of fertilized eggs to hatching and the subsequent care of young until independence. Also, they are regulated by different neuronal circuits and associated with different patterns of hormone secretion (Sharp, 2009). Incubation behavior in birds is defined by sitting continually on their eggs until they hatch, whereas brooding behavior is related to the care of newly hatched chicks (Opel and Proudman, 1988; Sharp, 2009). Brooding behavior is characterized by adults of either or both sexes allowing chicks to approach and remain underneath the adults' wings. The adults assume a distinctive crouching posture exhibiting preening and feather fluffing, leading the chicks to food or away from danger, and in some species, calling to the young. Generally, hens develop maternal behavior gradually in four stages: brooding, titbitting (food calling), clucking, and showing normal broody behavior (Ramsay, 1953; Sherry, 1981). The incidence of maternal behaviors concurs with a pause in laying and a significant long-term fall in the plasma levels of reproductive hormones including pituitary hormones and gonadal steroids (Richard-Yris et al., 1983, 1988).

31.2.1 Physiology and Behavior Characteristics Marked by Cessation of Egg Laying and Readiness to Incubate

The physiology and behavior associated with incubation behavior are multifaceted. The behavioral patterns in the domestic turkey (*Meleagris gallopavo*) include the following: increased nesting activity, nest protection, and anorexia (El Halawani et al., 1988a). In chickens, incubation behavior is usually associated with increased body temperature, reduced feed and water intake, frequent nest occupancy, turning and retrieval of eggs, aggressive or defensive behaviors, characteristic clucking, and cessation of egg laying (Romanov et al., 2002). In bantam hens (*Gallus gallus domesticus*), the onset of incubation is related to an increase in nesting frequency and egg laying until the first day of incubation, when hens terminate their egg laying (Lea et al., 1981). A similar increase in nesting frequency is observed in turkeys before the onset of incubation (Haller and Cherms, 1961). Hens sit on their clutches and persistently turn their eggs, rearranging them to guarantee that they are all well covered. In addition to a dramatic increase in nesting time, the incubating bird generally undergoes gonadal regression, shows aggressive nest protection activity, issues warning vocalization, and develops brood patches (El Halawani et al., 1984).

Most incubating avian species develop a so-called "brood patch". This is a defeathered, edematous, and hyperemic (elevated blood flow) area of the skin and includes most of the caudal ventral thoracic and a portion of the cranial ventral abdominal regions. The brood patch develops prior to the initiation of incubation. It functions to facilitate heat transfer from the hen to the eggs and facilitates transmission of tactile stimuli from the embryos to the hen (Jones, 1971; El Halawani et al., 1988a). Birds eat and drink very little and lose body weights during incubation. Weight loss during the incubation period has been recognized in turkeys (Zadworny et al., 1985), bantams (Savory, 1979), Canada geese (*Branta canadensis*; Akesson and Raveling, 1981), mallard ducks (*Anas platyrhynchos*; Gatti, 1983), and native Thai chickens (*G. gallus domesticus*; Kosonsiriluk et al., 2008).

Termination of egg laying and the initiation of incubation behavior normally start after the hens accumulate a full clutch of eggs. Bantam hens accumulate about 10–20 eggs per clutch (Lea and Sharp, 1982). In some birds, such as red grouse (*Lagopus lagopus scoticus*), the same number of eggs are laid whether or not eggs are removed from the nests while the birds are still laying (Moss and Watson, 1982). Incubation behavior is terminated when the chicks are hatched, but it may persist for a prolonged period if the nest contains unhatched eggs. Many wild bird species will continue incubating infertile eggs for about 50% longer than the time normally required for hatching (Skutch, 1962).

31.2.1.1 Neuroendocrine Regulation of Incubation Behavior

The control of avian reproduction involves the interaction of external stimuli with neuroendocrine mechanisms. These critical environmental stimuli include sensory information such as photoperiod, ambient temperature, and the presence of eggs or offspring (Curlewis, 1992). The primary components of the integrated reproductive neuroendocrine system in birds are the brain, especially the hypothalamus, the pituitary gland, and the gonad. This system is referred to as the hypothalamic–pituitary–gonadal (HPG) axis and includes neurotransmitters, neurohormones, neuromodulators, and hormones that play a pivotal role in the regulation of the avian reproductive cycle.

The neuroendocrine components of the reproductive system include the following:

- Gonadotropin releasing hormone-I (GnRH-I) stimulating the release of luteinizing hormone (LH) and follicle stimulating hormone (FSH).
- Vasoactive intestinal peptide (VIP), which increases prolactin (PRL) release.

Both systems are influenced by dopaminergic (DAergic) neurotransmission. The GnRH-I/FSH/LH system regulates the period of egg laying. GnRH-I stimulates the secretion of FSH and LH, which in turn are responsible for ovarian follicular growth and ovulation. On the other hand, the VIP/PRL system initiates and maintains maternal behaviors and may influence the onset of gonadal regression (Bhatt et al., 2003; Chaiseha and El Halawani, 2005).

31.2.1.1.1 Inhibition of the GnRH-I/FSH/LH System and the Termination of Egg Laying

The commencement of sexual maturity and the activation of the reproductive neuroendocrine system include the stimulation of both the GnRH-I/FSH/LH and VIP/PRL systems. Activation of the GnRH-I/FSH-LH system stimulates the development of the ovary and the formation of the hierarchy of follicles; the follicular production of estrogen, androgens, and progesterone; and the initiation of egg laying (sexual maturity) (Wineland and Wentworth, 1975; El Halawani et al., 1986). The combined action of estrogen, progesterone, and nesting activity further stimulates the VIP/PRL system (El Halawani et al., 1983; Porter et al., 1991; Chaiseha and El Halawani, 2005).

The onset of incubation behavior is marked by an increase in PRL and a decrease in circulating concentrations of LH and FSH; resulting in ovarian regression and decline in the plasma concentrations of the ovarian steroids levels, estrogen, and progesterone. Thereafter, circulating concentrations of LH continue to decline during the incubating period while those of PRL increase dramatically (Lea et al., 1981; El Halawani et al., 1984; Myers et al., 1989; Sharp et al., 1989; Porter et al., 1991) (Figure 31.1). These increasing PRL levels suppress the activity of the GnRH-I/FSH/LH system causing reduced ovarian steroid secretion, a cessation of ovulation, and ovarian regression (Youngren et al., 1991; Rozenboim et al., 1993b; Tabibzadeh et al., 1995). This signals the transition from egg laying activity to incubation behavior.

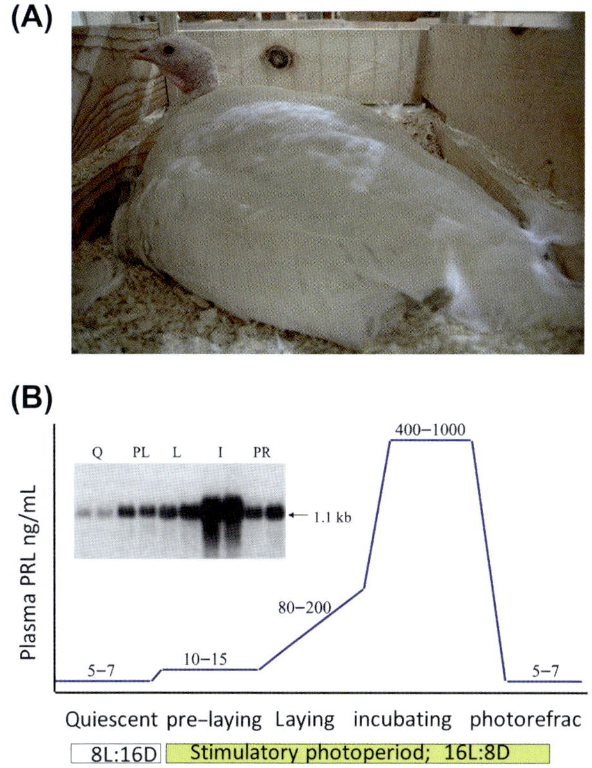

FIGURE 31.1 (A) Incubating turkey hen and (B) schematic profile of circulating prolactin (PRL) and PRL gene expression in turkey hens during various stages of their reproductive life cycles. *Based on Wong et al. (1991).*

It is well established that PRL is a causative factor for the reduced FSH and LH levels and subsequent ovarian regression, when birds make the transition from egg laying to egg incubation in bantam hens (*G. gallus domesticus*; Lea et al., 1981), canaries (*Serinus canaria*; Goldsmith et al., 1984), cockatiels (*Nymphicus hollandicus*; Myers et al., 1989), cowbirds (*Molothrus ater*; Hohn, 1959), domestic chickens (*G. gallus domesticus*; Sharp et al., 1977, 1979; Bedrak et al., 1981), Japanese quails (*Coturnix japonica*; Goldsmith and Hall, 1980), mallard ducks (*A. platyrhynchos*; Bluhm et al., 1983), native Thai chickens (*G. gallus domesticus*; Kosonsiriluk et al., 2008), pheasants (*Phasianus colchicus*; Breitenbach and Meyer, 1959), pied flycatchers (*Ficedula hypoleuca*; Silverin and Goldsmith, 1983), pigeons (*Columba livia*; Riddle et al., 1935), ring doves (*Streptopelia risoria*; Goldsmith et al., 1981), snow geese (*Anser caerulescens caerulescens*; Campbell et al., 1978), spotted sandpipers (*Actitis macularia*; Oring et al., 1986), turkeys (*M. gallopavo*; Cogger et al., 1979; Burke and Dennison, 1980; El Halawani et al., 1997), white-crowned sparrows (*Zonotrichia leucophrys pugetensis*; Wingfield and Farner, 1978; Hiatt et al., 1987), European wide starlings (*Sturnus vulgaris*; Dawson and Goldsmith, 1982), and zebra finches (*Poephila guttata*; Vleck and Priedkalns, 1985) (Figure 31.1).

High concentrations of PRL act both on the neuroendocrine system to inhibit LH secretion by reducing GnRH-I levels in the hypothalamus (Adkins-Regan et al., 2013) and directly on the ovary, inducing ovarian regression and inhibiting LH-induced ovarian steroid production (Rozenboim et al., 1993b). An inverse relationship exists between expression of LH and PRL genes in incubating turkey hens (Wong et al., 1992). Administration of PRL to laying domestic turkey hens has been demonstrated to suppress ovariectomy-induced increases in LH release while delaying the onset of egg laying and inducing incubation behavior (El Halawani et al., 1991). When incubation behavior is terminated, there is a decrease in circulating concentrations of PRL and a concomitant increase in circulating concentrations of LH levels (domestic turkey: El Halawani et al., 1988b; Knapp et al., 1988). In seasonally breeding birds, reproductive activity is terminated by increasing circulating concentrations of PRL, which in turn suppresses LH release, inhibits follicular development, and finally ends egg laying activity (Magang goose: Huang et al., 2008).

In the Thai hen (*G. gallus domesticus*), nest deprivation and the disruption of incubation behavior is accompanied by decreased circulating concentrations of PRL (Prakobsaeng et al., 2011) (Figure 31.2). Moreover, there is increased expression of GnRH-I mRNA in nest-deprived bantam chickens (Dunn et al., 1996) together with elevated circulating concentrations of LH and ovarian steroids (domestic turkeys: El Halawani et al., 1980; bantam chickens: Richard-Yris et al., 1998a). These neuroendocrine changes were reversed when hens renested (Sharp et al., 1988).

Circulating FSH, LH, and gonadal steroids levels are regulated by hypothalamic GnRH-I (El Halawani et al., 1988b; Sharp et al., 1990). GnRH-I is the primary hypophysiotropic factor stimulating the release of LH and FSH (Sharp et al., 1990). An association between hypothalamic GnRH-I mRNA, GnRH-I peptide content, and pituitary gonadotropins secretion during the avian reproductive cycle has been reported (Millam et al., 1989; Rozenboim et al., 1993a; Dunn and Sharp, 1999; Dawson et al., 2002). In seasonal breeders, GnRH-I neuronal activity is regulated by photoperiod (Sharp and Blache, 2003). Photostimulation increases the sensitivity of pituitary cells to GnRH-I (Davies and Follett, 1975) and increases GnRH-I gene transcription and its secretion (Dunn and Sharp, 1999). In temperate-zone birds, GnRH-I mRNA is abundant in the nucleus commissuraepallii (nCPa) of the hypothalamus with expression greater in laying hens than either those of the nonphotostimulated and incubating hens or photorefractory hens (domestic turkey: Kang et al., 2006).

In continuously breeding domesticated bird such as native Thai chickens, the number of GnRH-I neurons in the nCPa is highest in laying hens compared with those at other reproductive stages (Sartsoongnoen et al., 2012). Nest deprivation of incubating hens increases the number of GnRH-I immunoreactive neurons in the nCPa compared with those of the incubating hens. Taken together, these data indicate an association of the GnRH-I system in the nCPa with maternal behavior in the nonphotoperiodic native Thai chicken (Sartsoongnoen et al., 2012) (Figures 31.3 and 31.4).

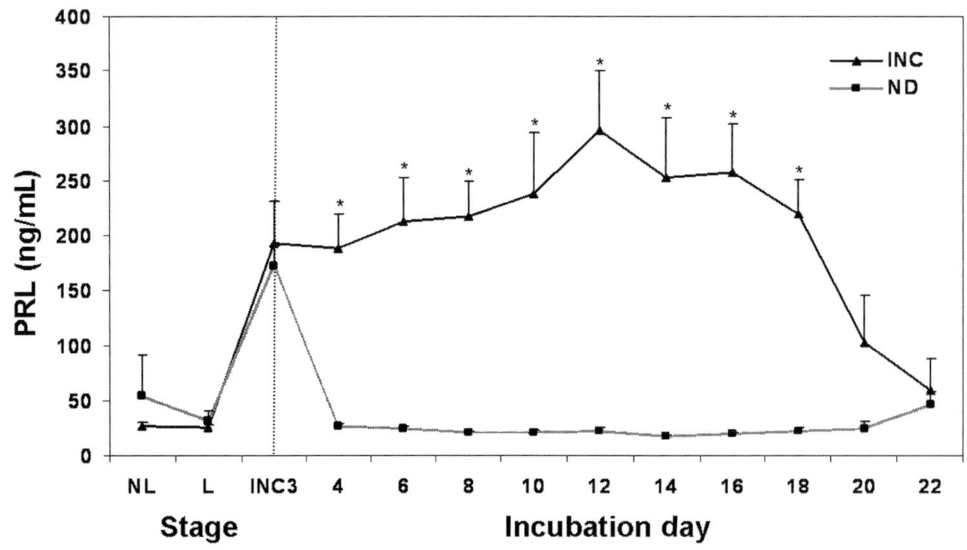

FIGURE 31.2 Changes in prolactin (PRL) level before and after initiation of incubation and nest deprivation of native Thai chickens. Hens were divided into two groups after day 3 of incubation (INC3); one group continued to incubate their eggs (INC) and birds in the second group were nest deprived (ND). Blood samples were collected prior to egg laying (NL), during egg laying (L), and following incubation and nest deprivation for determination of PRL levels. *From Prakobsaeng et al. (2011).*

31.2.1.1.2 Enhanced Prolactin Gene Expression and Secretion

PRL was discovered in 1932 (Riddle et al., 1932); its name is based on the findings that it both causes growth of the crop sac and the production of crop milk in pigeons together with promoting lactation in rabbits (Bern and Nicoll, 1968). The protein hormone, PRL, is synthesized in and secreted from the lactotrophs of the anterior pituitary gland (Freeman et al., 2000). There are more than 300 different physiological functions ascribed to PRL such as in growth, development, reproduction, behavior, metabolism, osmoregulation, immunoregulation, migration, and nurturing of the young in multiple vertebrates (Sinha, 1995; Bole-Feysot et al., 1998; Harris et al., 2004).

Prolactin is associated with reproductive cycles in multiple avian species (El Halawani et al., 1997). In both temperate zone and equatorial chickens, hyperprolactinemia is associated with incubation behavior and ovarian regression (Kosonsiriluk et al., 2008; Prakobsaeng et al., 2011). Moreover, circulating concentrations of PRL are very low in reproductively quiescent chickens and turkeys, increase during egg laying with further increases at the initiation of incubation, leading to the cessation of ovulation, egg laying, ovarian regression, and induction and maintenance of incubation behavior. Thus, it is well established that PRL plays a significant role in the onset and maintenance of incubation in birds (Sharp, 2009; Buntin, 2010). Circulating concentrations of PRL increase gradually at the onset of incubation behavior and are maintained at high levels during the incubation period and then decrease to the levels of reproductively quiescent birds when incubation behavior is terminated (El Halawani et al., 1980; Wentworth et al., 1983). Changes in PRL gene expression correlate well with circulating concentrations of PRL in the avian reproductive cycle, with PRL mRNA reaching a maximal during incubation (Talbot et al., 1991; Wong et al., 1991; Tong et al., 1997; Karatzas et al., 1997; Liang et al., 2006; Huang et al., 2008) (Figure 31.1).

Prolactin is involved in many aspects of reproductive physiology and behaviors in birds in the orders Galliformes and Columbiformes. It plays an important role in a series of parental behaviors by mediating increases in incubation

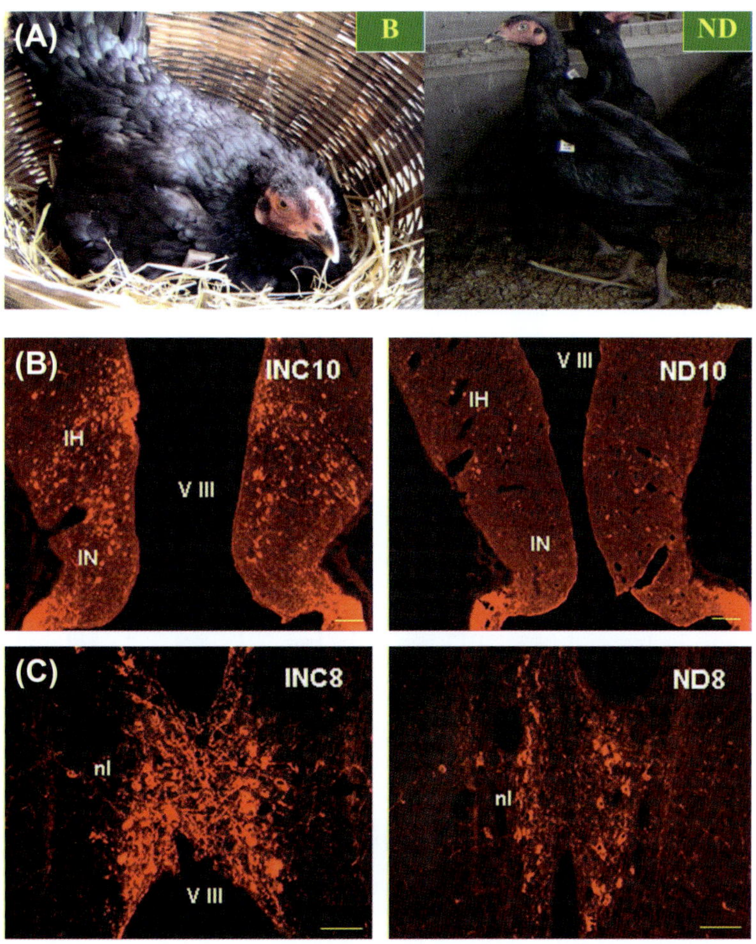

FIGURE 31.3 The distribution of (B) vasoactive intestinal peptide-immunoreactive (VIP-ir) neurons in the nucleus inferioris hypothalami-nucleus infundibula hypothalami (IH-IN) at day 10 and (C) tyrosine hydroxylase-immunoreactive (TH-ir) neurons in the nucleus intramedialis (nI) at day 8 following the initiation of incubation or nest deprivation of (A; B) incubating and (A; ND) nest-deprived native Thai hens.

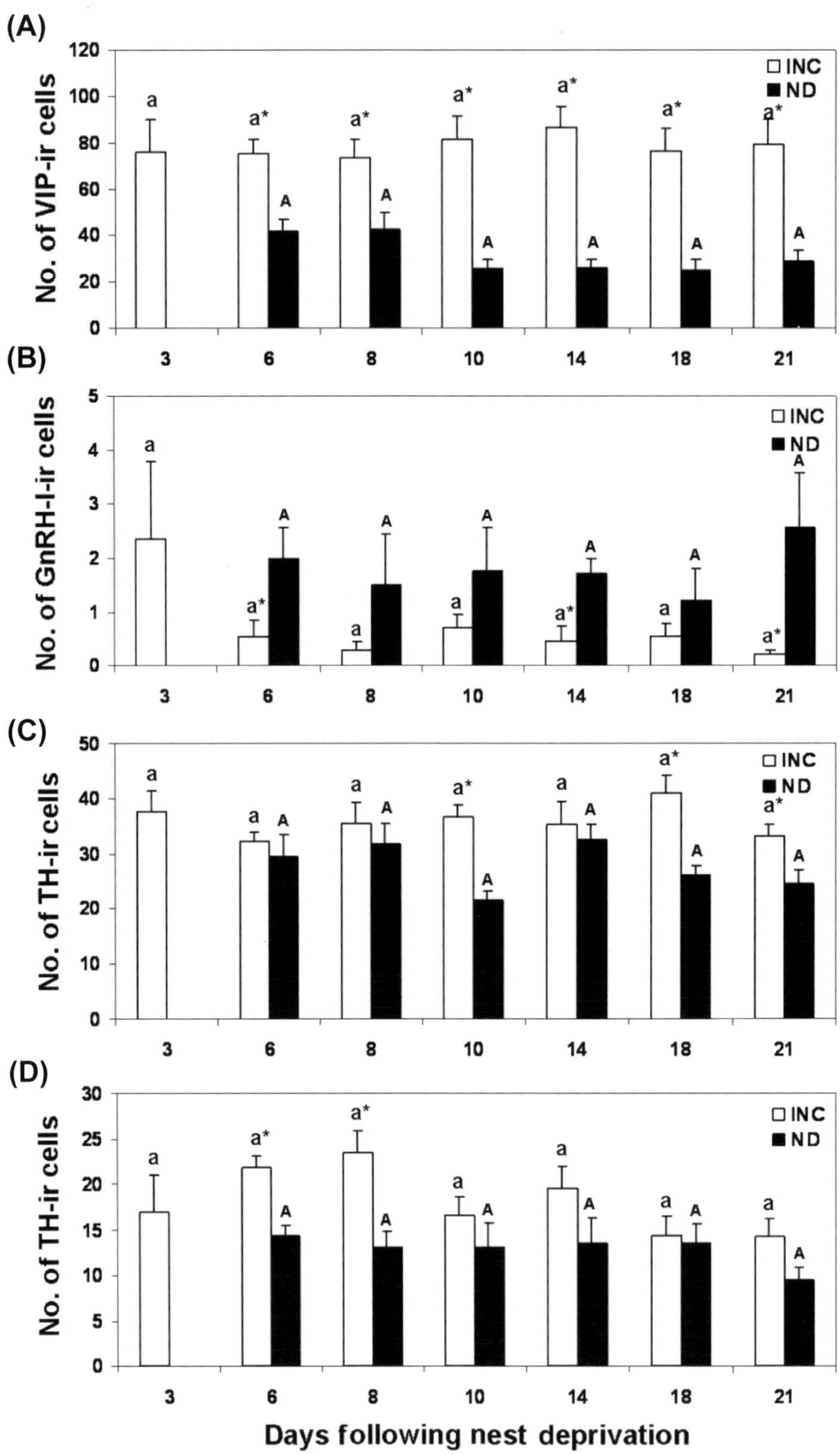

FIGURE 31.4 Changes in the numbers of: (A) vasoactive intestinal peptide-immunoreactive(VIP-ir) neurons in the nucleus inferioris hypothalami-nucleus infundibula hypothalami (IH-IN); (B) gonadotropin releasing hormone-I-immunoreactive neurons in the nucleus commissuraepallii (nCPa); (C) tyrosine hydroxylase-immunoreactive (TH-ir) neurons in the nucleus intramedialis (nI); (D) TH-ir neurons in the mamillaris of incubating (INC) and nest-deprived (ND) native Thai hens. *From Prakobsaeng et al. (2011).*

behavior, crop milk secretion, feeding of young, and nest defense (Silver, 1984; Buntin, 2010). Administration of exogenous PRL results in increases in parental behaviors (Pedersen, 1989; Buntin et al., 1991; Youngren et al., 1991), and active immune-neutralization against PRL reduces the incidence, delays the development, or completely prevents the occurrence of incubation behavior (March et al., 1994; Crisostomo et al., 1998).

Prolactin plays a role in terminating egg laying; thus it regulates the clutch size in species that lay more than two eggs per clutch, with the cessation of egg laying being associated with an increase in circulating concentrations of PRL (Bluhm et al., 1983; Hall and Goldsmith, 1983; Sockman et al., 2000). During the incubating period, there are indications that the suppression of FSH and LH secretion is independent of increased PRL secretion (Lea et al., 1996). In contrast, PRL may directly inhibit ovarian steroidogenesis; resulting in involution of the ovary with reduced ovarian steroidogenesis and regression of the oviduct (Porter et al., 1991; Rozenboim et al., 1993b).

31.2.1.2 Neuronal Regulation of Incubation Behavior

A wealth of information has confirmed the involvement of hypothalamic VIP, dopamine (DA), and GnRH-I in the regulation of the avian reproductive cycle (El Halawani et al., 1997, 2001; Sharp, 2009). The final common pathway governing the secretion of PRL, FSH, and LH is a system of peptidergic neurons whose axons terminate around portal capillaries in the external layer of the median eminence (ME). Among the best characterized hypophysiotropic releasing/inhibiting factors in birds are VIP, DA, and GnRH-I (Chaiseha and El Halawani, 2005).

31.2.1.2.1 Enhanced Dopaminergic Neurotransmission Regulating Incubation Behavior

Avian PRL secretion is regulated by both stimulatory and inhibitory hypothalamic factors. It is predominantly under tonic inhibitory control in mammals (Ben-Jonathan et al., 1989), with DA being the major PRL-inhibiting factor (PIF) (Ben-Jonathan and Hnasko, 2001). This is not the case in birds, with PRL being under tonic stimulatory control by the hypothalamus (Kragt and Meites, 1965; Bern and Nicoll, 1968). Removal of hypothalamic inputs results in the cessation of PRL secretion (Tixier-Vidal et al., 1966; Hall et al., 1986). It is now established that VIP is the avian hypothalamic PRL-releasing factor (PRF) with VIP increasing both PRL secretion and gene expression (El Halawani et al., 1997).

Dopaminergic neurotransmission is involved in both stimulation and inhibition of PRL secretion in birds and depends on multiple DA receptor subtypes (Youngren et al., 1995; Chaiseha et al., 1997) with an intact VIPergic system required for the stimulatory effects (Youngren et al., 1996b). There is co-expression of D_2 DA receptor mRNA in VIP neurons within the lateral hypothalamus (LHy) and infundibular nuclear complex (INF) (Chaiseha et al., 2003). It has been suggested that DA stimulates PRL secretion at the hypothalamic level via D_1 DA receptors residing in the INF, where the VIP neurons are located. Dopamine inhibits PRL secretion directly at the pituitary level via D_2 DA receptors by blocking the action of VIP (Chaiseha et al., 2003). Neurotransmitters and peptides including dynorphin, serotonin (5-HT), DA, and VIP stimulate PRL secretion along a common pathway. It has been proposed that synapses expressing κ opioid, 5-HTergic, DAergic, and VIPergic receptors are arranged serially with the VIPergic system as the final mediator (El Halawani et al., 2001) (Figure 31.5).

There are changes in DAergic activity and DA receptor subtype mRNA expression with reproductive/behavioral stage of the bird. Dopaminergic activity in the anterior hypothalamus is markedly increased in incubating hens compared to laying or nest-deprived bantam hens (Macnamee and Sharp, 1989). Moreover, DAergic activity is elevated at this time in the periventricular regions of ring doves (Lea et al., 2001). Hypothalamic expression of the stimulatory D_1 DA receptors is increased in hyperprolactinemic incubating laying hens. In contrast, the pituitary expression of the inhibitory D_2 DA receptors is increased in hypoprolactinemic photorefractory hens (Schnell et al., 1999; Chaiseha et al., 2003).

The hypothalamic distribution of DA neurons has been established in multiple avian species including domestic chickens (Knigge and Piekut, 1985; Moons et al., 1995), Japanese quails (Absil et al., 2001), pigeons (Kiss and Peczely, 1987; Durstewitz et al., 1998), zebra finches (Mello et al., 1998), budgerigars (*Melopsittacus undulatus*; Roberts et al., 2001), collared doves (*Streptopelia decaocto*; den Boer-Visser and Dubbeldam, 2002), turkeys (Al-Zailaie and El Halawani, 2000), canaries (Appeltants et al., 2001), and native Thai chickens (Sartsoongnoen et al., 2008). Dopaminergic neurons are dispersed in multiple hypothalamic nuclei (Reiner et al., 1994) being intermingled with VIP neurons in the INF, GnRH neurons in the preoptic area (POA), and with both VIP and GnRH terminals in the external layer of the ME (Contijoch et al., 1992; Fraley and Kuenzel, 1993). There appears to be a regional specificity in those DA neurons controlling either the VIP/PRL or the GnRH-I/FSH/LH systems. The ability of exogenous DA to increase VIP expression is limited to neurons in the INF, with this being correlated with pituitary expression of PRL and LH-β (Bhatt et al., 2003). Activation of the DA neurons in the nucleus mamillaris (ML) is associated with the activation of GnRH-I and VIP neurons and the subsequent release of LH and PRL (Al-Zailaie et al., 2006).

The distribution of tyrosine hydroxylase immunoreactivity (TH-ir), the rate-limiting enzyme for DA synthesis and

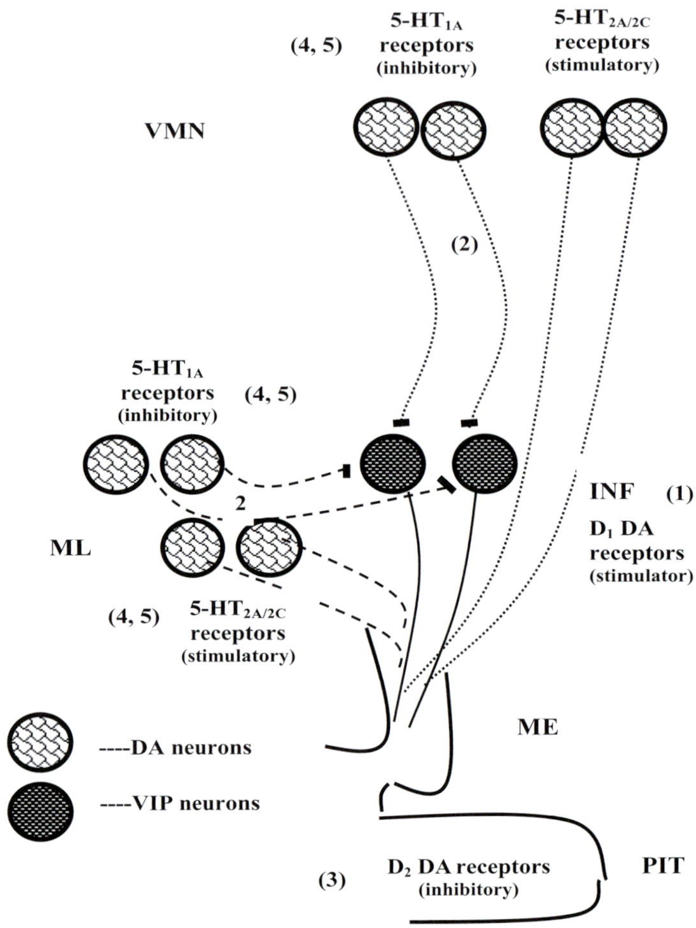

FIGURE 31.5 **Schematic diagram showing the neuronal interactions between serotonin (5-HT) and dopamine (DA) and the control of vasoactive intestinal peptide/prolactin secretion.** *From Chaiseha and El Halawani (2005).*

hence a marker for DAergic activity, has been reported in turkeys and other avian species. It is predominantly located within the diencephalon and mesencephalon (Kiss and Peczely, 1987; Bailhache and Balthazart, 1993; Moons et al., 1994; Reiner et al., 1994; den Boer-Visser and Dubbeldam, 2002). In the native Thai chickens, the number of TH-ir neurons in the nucleus intramedialis (nI) parallel the PRL levels; increasing during incubation and decreasing when incubation behavior is disrupted (Sartsoongnoen et al., 2008; Prakobsaeng et al., 2011) (Figures 31.4 and 31.6).

31.2.1.2.2 VIP Neurotransmission

VIP was first isolated from porcine duodenum (Said and Mutt, 1970; Mutt and Said, 1974). It is well established that VIP is the avian PRF, with PRL secretion and expression regulated by VIP secreted from neurons located in the INF of the caudomedial hypothalamus (El Halawani et al., 2001). Immunoneutralization against VIP prevents increased PRL secretion and expression and induction of incubation behavior while upregulating LH and FSH gene expression and extending the duration of egg laying. It does not, however, prevent gonadal regression and molting (Sharp et al., 1989; El Halawani et al., 1995, 1996; Talbot et al., 1991; Ahn et al., 2001).

Prolactin secretion parallels hypothalamic VIP immunoreactivity, VIP content in both the INF and ME, and VIP expression in the INF throughout the avian reproductive cycle (Mauro et al., 1989; Chaiseha et al., 1998; Chaiseha and El Halawani, 1999; Kosonsiriluk et al., 2008). VIP acts directly on the lactotrophs to stimulate PRL secretion (El Halawani et al., 1997). There are increases in the number and cell size of VIP neurons reported in the pigeons and ring doves when circulating concentrations of PRL are elevated (Peczely and Kiss, 1988; Cloues et al., 1990). Increased pituitary VIP receptor expression is also found in the laying or incubating turkey hens supporting PRL secretion being principally regulated by VIP at the pituitary level (Chaiseha et al., 2004). Tactile stimuli from the nests and eggs maintain the elevated circulating concentrations of PRL and upregulate VIP gene expression in the incubating hens (Silver et al., 1988; Buntin et al., 1991; Massaro et al., 2007).

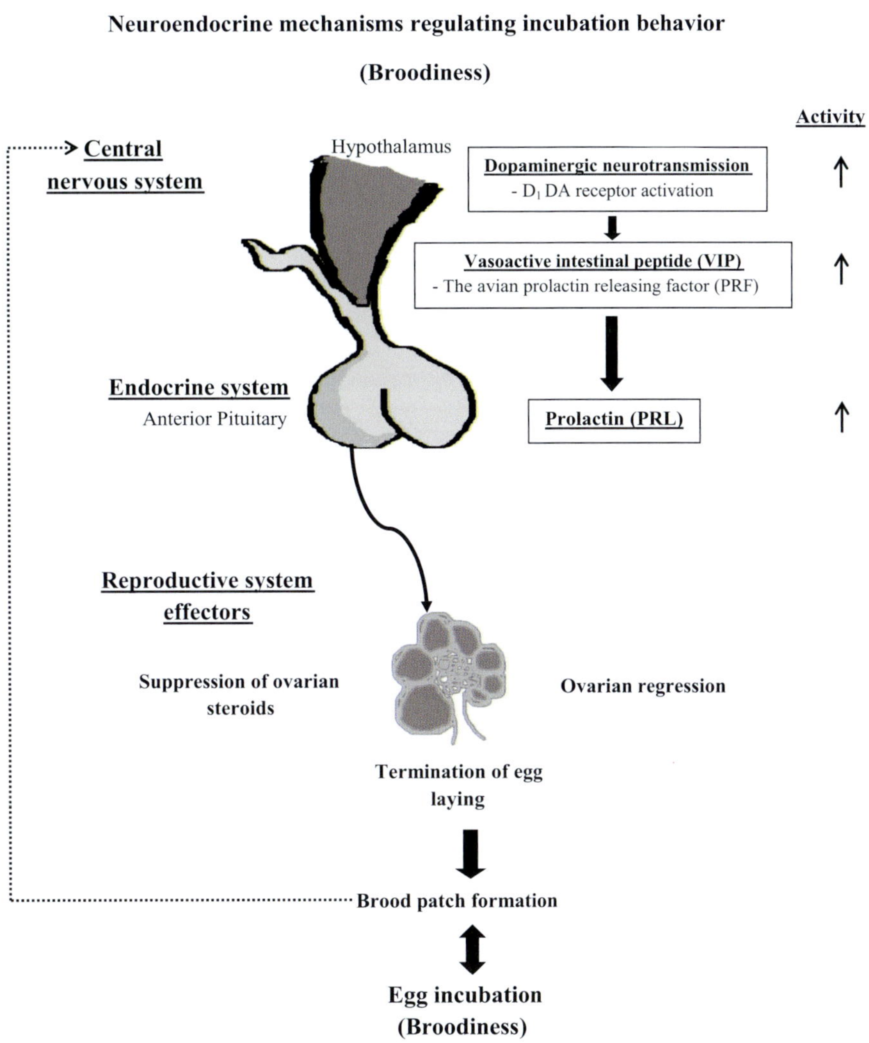

FIGURE 31.6 Schematic diagram showing the neuroendocrine mechanisms regulating incubation behavior.

VIP-containing neurons have been mapped in many birds such as Peking ducks (*A. platyrhynchos*; Korf and Fahrenkrug, 1984), Japanese quails (Teruyama and Beck, 2001), turkeys (Chaiseha and El Halawani, 1999), pigeons (Cloues et al., 1990), ring doves (Norgren and Silver, 1990), chickens (Kuenzel et al., 1997), dark-eyed juncos (*Junco hyemalis*; Saldanha et al., 1994), zebra finches (Bottjer and Alexander, 1995), and native Thai chickens (Kosonsiriluk et al., 2008). VIP neurons are extensively distributed throughout the hypothalamus (Macnamee et al., 1986; Hof et al., 1991; Chaiseha and El Halawani, 1999; den Boer-Visser and Dubbeldam, 2002; Kosonsiriluk et al., 2008). The number, distributed area, and density of VIP neurons are all higher in incubating than in laying hens (Sharp et al., 1989; Chaiseha and El Halawani, 1999; Kosonsiriluk et al., 2008). In native Thai chickens, VIP neurons are concentrated within the nucleus inferioris hypothalami (IH) and nucleus infundibula hypothalami (IN) with changes in the number of VIP neurons correlating with PRL levels in the reproductive cycle. The number of VIP neurons decreases concurrently with PRL levels in nest-deprived incubating birds, suggesting that VIP in the IH-IN plays a regulatory role in year-round reproductive activity in these continuous breeding birds (Kosonsiriluk et al., 2008; Sartsoongnoen et al., 2008; Prakobsaeng et al., 2011) (Figures 31.2, 31.4 and 31.6).

31.3 REARING BEHAVIOR

For reproduction to be successful, not only is sexual activity critical, but also care of the young is essential. The offspring need one or both parents to provide food, heat, and protection. This behavior, providing care and defense for the young, must be performed immediately after hatching of the offspring (Brunton and Russell, 2008). The onset of responsiveness is a prerequisite for all mother–offspring interactions. When the mother is exposed to the newly hatched chick, she receives a unique constellation of tactile, visual, auditory, and olfactory external stimuli as well as changes in neuro/hormonal state (Numan and Woodside, 2010). These induce other behaviors promoting care and survival of the chicks (Swain et al., 2007).

Maternal experiences, neurotransmitters, neurohormones, neuromodulators, hormones, and stimuli from the young interact in a complex manner to promote maternal responsiveness in birds. The presence of chicks results in the expression of rearing behavior including such specific maternal behaviors as clucking and food-calling vocalizations. Hens display physical contact with the chicks by brooding them for longer durations after hatching, while clucking and food calling are regular behaviors exhibited in hens rearing chicks (Richard-Yris et al., 1998b). Galliforms are precocial with newly hatched chicks able to walk, feed, see, and hear. However, the chicks cannot effectively thermoregulate during the first two weeks of posthatch life. Therefore, brooding by the hens helps them to survive (Mills et al., 1997).

Rearing behavior consists of the hens allowing the chicks to nestle underneath their slightly raised wings while assuming a distinct crouching posture (Hess et al., 1976). Stimuli from the chicks are clearly involved in the establishment, appearance, and maintenance of this behavior. It can be induced in chickens, turkeys, and Japanese quail by introducing newly hatched chicks to them. The hens respond almost immediately with maternal care behaviors (Richard-Yris and Leboucher, 1987; Opel and Proudman, 1989). These behaviors are induced by physical contact between the hen and chicks, alone or in combination with visual and/or auditory stimuli from the chicks (Opel and Proudman, 1988; Richard-Yris et al., 1998b). A mother–offspring bond is then formed, with the chicks learning to respond to the food calling, distress call, and purring sound of the hens. The mother–offspring bond is strengthened by repeated exposure of the chicks to the hen (Wauters and Richard-Yris, 2001, 2002; Edgar et al., 2011), with this repeated exposure important for the development of post-hatch species-specific maternal call recognition (Gottlieb, 1976; Jain et al., 2004).

31.3.1 Neuroendocrine Regulation of Rearing Behavior

Dopaminergic neurotransmission plays a pivotal role in both GnRH-I/FSH/LH and VIP/PRL systems in avian reproduction. A role for PRL in inducing avian parental behavior was first seen when injections of PRL induced brooding behavior in domestic chickens (Riddle et al., 1935). Neuronal interactions between GnRH-Iergic, VIPergic, DAergic, and mesotocinergic (MTergic) systems have been reported to be involved in rearing behavior (Thayananuphat et al., 2011; Chaiyachet et al., 2013a,b; Chokchaloemwong et al., 2012, 2013).

31.3.1.1 VIP/PRL-DA Neurotransmission and Suppression of the GnRH-I/FSH/LH System

PRL has a well-established role as an incubation-promoting hormone. However, it has also been implicated to be involved as a crucial factor to the onset and maintenance of the rearing behavior (Vleck, 1998; Sharp, 2009; Buntin, 2010). High circulating concentrations of PRL are associated with rearing behavior in chickens (Sharp et al., 1988; Hoshino and Wakita, 1989), turkeys (Proudman and Opel, 1981), mallard ducks (Goldsmith and Williams, 1980), Australian black swans (*Cygnus atratus*; Goldsmith, 1982), ring doves (Buntin, 1996), and native Thai chickens (Chaiyachet et al., 2013b).

Prolactin secretion is stimulated by exposure of hens to tactile and visual stimuli from the chicks with PRL facilitating the expression of maternal behaviors such as incubating, rearing and feeding (Angelier and Chastel, 2009). The transition from sexual to parental activity involved elevated PRL secretion (Sharp et al., 1998) with the circulating maternal care responses (Buntin, 1996). Either sharp declines (bantams; Sharp et al., 1979, 1988; Lea et al., 1981; barheaded geese; *Anser indicus*; Dittami, 1981; common eiders; *Somateria mollissima* Criscuolo et al., 2002; domestic ducks; Hall and Goldsmith, 1983; Hall, 1987; Japanese bantams; Zadworny et al., 1988; mallard ducks; Goldsmith and Williams, 1980; swans; Goldsmith, 1982; native Thai chickens; Chaiyachet et al., 2013b; turkeys; Wentworth et al., 1983) or slow decreases (spotted sandpipers, Oring et al., 1986; Wilson's phalarope, *Phalaropus tricolor*; Oring et al., 1988; red-necked phalarope, *Phalaropus lobatus*; Gratto-Trevor et al., 1990) of circulating concentrations of PRL in hens after hatching of the chicks have been reported in different birds (Schradin and Anzenberger, 1999). In precocial species, PRL secretion declines shortly after hatching, is moderately elevated during the rearing period, and thereafter decreases. In altricial species, PRL secretion increases after hatching while the chicks are being intensively fed and guarded (Buntin, 2010). Among species in which parental care is performed by the females only, there is a markedly larger increase in PRL secretion in females than in males (pied flycatcher; Silverin and Goldsmith, 1984; Angelier and Chastel, 2009). In contrast, in species in which parental care is performed only by the male, PRL secretion is higher in males than in females (Wilson's phalarope and red-necked phalarope; *P. lobatus*; Buntin, 1996; Buntin et al., 1998). A positive correlation between PRL secretion and the intensity/quality of parental care has been observed in many parental birds and cooperative breeders. In cooperatively breeding species, Florida Scrub-Jays (*Aphelocoma coerulescens*; Schoech et al., 1996) and Harris' hawks (*Parabuteo unicinctus*; Dawson and Mannan, 1991), PRL secretion in helpers are positively correlated with the rate of nestling provisioning (Angelier and Chastel, 2009).

In galliforms, PRL is released throughout the rearing period, with its secretion facilitated by the presence of chicks and the high PRL secretion maintaining the rearing behavior (Richard-Yris et al., 1995). In the presence of chicks, circulating concentrations of PRL begin to drop immediately after hatching and continue to decline for about one week after hatching (Zadworny et al., 1988; Leboucher et al., 1990). The presence of chicks modifies the rate of decline in PRL secretion (Opel and Proudman, 1989; Chaiyachet et al., 2013b). In native Thai chickens, circulating concentrations of PRL were higher in rearing than nonrearing hens (Chaiyachet et al., 2013b). The presence of chicks induces the emergence of specific maternal behaviors in many species (Richard-Yris et al., 1983; Richard-Yris and Leboucher, 1987; Opel and Proudman, 1988; Leboucher et al., 1990, 1993; Ruscio and Adkins-Regan, 2004). In the incubating hens, the substitution of chicks for eggs, or the appearance of chicks at hatching, is associated with an increase in circulating concentrations of LH and a marked decrease in circulating concentrations of PRL from the high levels presented during incubation period (Zadworny et al., 1988; Leboucher et al., 1993; Richard-Yris et al., 1998b). Furthermore, exposure to chicks can induce maternal behavior in the incubating, nonincubating, and ovariectomized hens, which exhibit marked differences in circulating concentrations of gonadal steroids and patterns of PRL secretion (Richard-Yris et al., 1987, 1988; Leboucher et al., 1993; Lea et al., 1996; Wang and Buntin, 1999).

In altricial species in the order Columbiformes, the parents jointly rear the young. Pigeons and doves feed their newly hatched chicks by regurgitating crop milk produced by epithelial mucosa cells of the crop sac gland. The epithelial cells proliferate in response to PRL and ultimately slough from the wall of crop sac (Wang and Buntin, 1999; Buntin, 2010). This rise in PRL secretion is accompanied with the onset and maintenance of incubation and rearing behaviors (Columbiformes) and also in free-living passerine species (Goldsmith, 1991; Buntin, 1996). The elevated PRL secretion during the early posthatching period may also promote the display of parental behaviors that are essential for transferring the crop milk to the young squabs in Columbiformes. After chicks achieve thermal independence, PRL secretion declines (Goldsmith, 1991; Buntin, 2010) (Figure 31.7).

The onset of PRL secretion, and its timing of elevation in relation to parental behaviors, varies considerably among biparental species. In some species, rearing behavior and

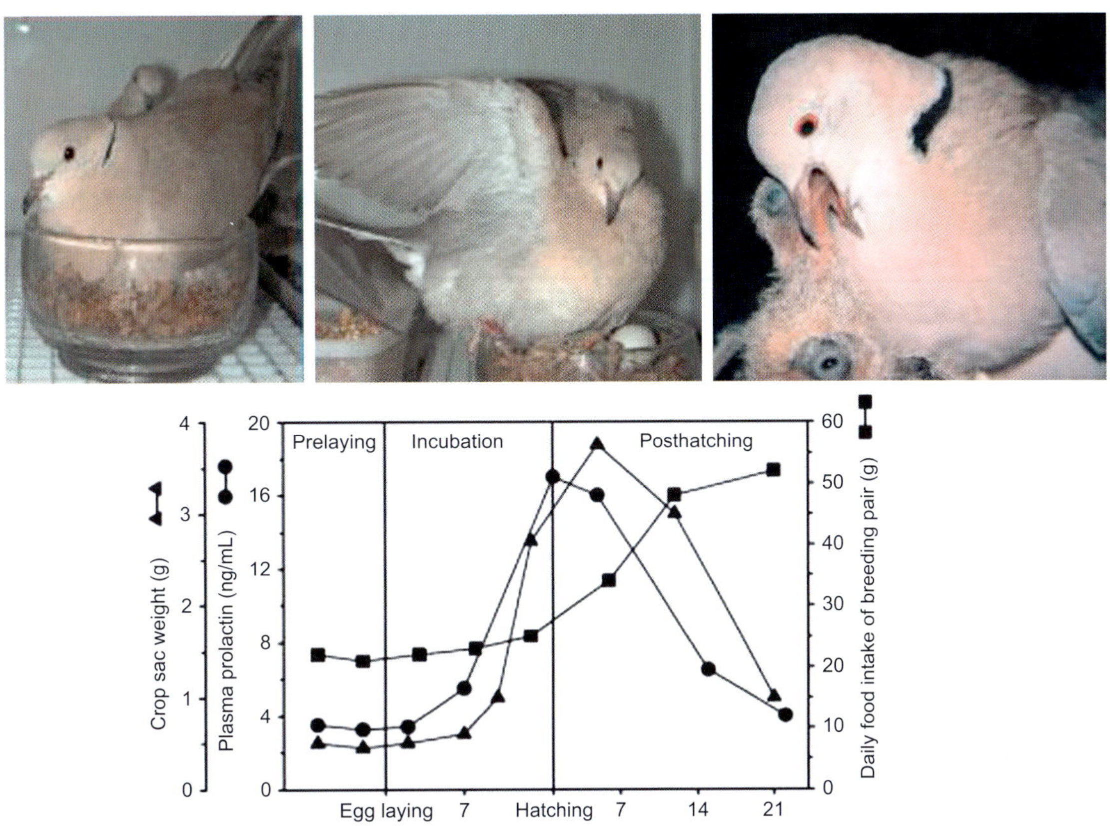

FIGURE 31.7 Prolactin (PRL) and parental behavior in ring doves. Rising PRL levels during the incubation and post-hatching periods of the breeding cycle maintain incubation behavior, facilitate nest defense, stimulate crop sac development and crop milk formation, increase parental foraging behavior, and promote parental regurgitation behavior to transfer crop milk and seeds to the young by both parents. *From Buntin (2010).*

elevated PRL secretion are responsive to visual and tactile stimulation from the nest and eggs; the onset of the increase of PRL secretion beginning near the time of hatching (bantams; Sharp et al., 1988; ring doves, Buntin, 2010; turkeys; Opel and Proudman, 1988). In other species, the onset occurs prior to incubation and chick rearing (king penguin; *Aptenodytes patagonicus*; Garcia et al., 1996; Adélie penguins; *Pygoscelis adeliae*; Vleck and Vleck, 2011). Male and female king penguins and ring doves show a steady increase in PRL levels as the incubation of eggs advances and the chick rearing period begins (Garcia et al., 1996). Male scrub-jay (*Aphelocoma c. coerulescens*; Schoech et al., 1996), canaries (Goldsmith, 1982), white-crowned sparrows (Hiatt et al., 1987), song sparrows (*Melospiza melodia*; Wingfield and Goldsmith, 1990) and European starlings (Dawson and Goldsmith, 1985) all exhibit elevated PRL secretion prior to hatching regardless of whether the males participate in incubation. This suggests that elevated PRL secretion may mediate the feeding of incubating females or of young nestlings in cooperatively breeding species. Thus, elevated PRL secretion in these birds is not solely a response to the stimulation from the eggs or chicks (Ziegler, 2000). There appears to be an inverse relationship between circulating concentrations of PRL and those of testosterone during the avian reproductive cycle. The circulating concentrations of testosterone are elevated for a longer time in polygynous than in monogamous species where there is an abbreviated peak of testosterone (Wingfield and Goldsmith, 1990).

Circulating concentrations of PRL decrease when the incubation period ends and the adults are allowed to hatch and rear their chicks in chickens, ducks, and Australian black swans (Sharp et al., 1979; Goldsmith and Williams, 1980; Goldsmith, 1982). In turkeys, this drop in circulating concentrations of PRL might be related to the pipping and hatching eggs, and the consequent transition to maternal behaviors (Wentworth et al., 1983; Opel and Proudman, 1989).

Rearing behavior is associated with low circulating concentrations of LH and ovarian steroids. The onset of maternal behaviors is accompanied by a prolonged decrease in circulating concentrations of LH, which in turn may be inhibited by increased PRL secretion (Opel and Proudman, 1980; Zadworny et al., 1989). The presence of chicks or the introduction of chicks induces a fall in PRL secretion and a moderate increase in circulating concentrations of LH and ovarian steroids in the incubating chickens (Richard-Yris et al., 1998b; Leboucher et al., 1993; Lea et al., 1996) and turkeys (Opel and Proudman, 1989). This indicates that maternal care, and in particular physical contact with the young chicks, plays a key role in producing these hormonal differences (Richard-Yris et al., 1995). Native Thai hens rearing chicks return to lay later than those not rearing, indicating that the physical stimuli from the chicks slows down the decrease of PRL secretion and inhibits secretion of gonadotropins and ovarian steroids. Disruption of rearing behavior results in an immediate increase in circulating concentrations of LH and estradiol together with a concurrent marked decrease in PRL levels (Chaiyachet et al., 2013b). Both circulating concentrations of LH and estradiol gradually increase after hatching or the presence of chicks to reach a maximum immediately after removal of the chicks. This increase is associated with increased GnRH-I mRNA and is thought to be regulated by the removal of the inhibitory effects of PRL (Kuwayama et al., 1992; Dunn et al., 1996; Richard-Yris et al., 1998b; Dunn and Sharp, 1999). Ovarian mass is positively correlated with GnRH-I gene expression in sexually mature passerine songbirds (Ubuka and Bentley, 2009).

Although PRL is not released at an increased rate in hens that are caring for their young, it appears to be involved in the initiation and/or maintenance of rearing (Sharp et al., 1979). This corresponds to the time when the levels of LH increase to the levels found during laying (Sharp et al., 1979). As the chicks grow older and become fledged, there is a progressive decline in rearing behavior in the adult, leading to a drop in PRL secretion (Boos et al., 2007; Riou et al., 2010). Shorebirds and red-necked phalaropes are facultatively polyandrous and only males care for the eggs and chicks. In these species, PRL secretion in broody males decreases gradually with increasing age of the brood (Gratto-Trevor et al., 1990). In some species (domestic chickens; Leboucher et al., 1990; pied flycatchers; ring doves; Buntin, 1996), stimuli from the young or from the parent–young interactions may promote or sustain the elevated in PRL levels. A definite threshold in PRL levels appears to be essential to promote and/or maintain posthatching care behavior. Although PRL levels are lower during the rearing period than in the incubation period, they are higher than in nonrearing birds, inferring that PRL is involved in parental care after hatching (Criscuolo et al., 2002; Boos et al., 2007).

Although there are high numbers of GnRH-I neurons in the nCPa of nonrearing native Thai hens, there are considerably fewer neurons when the birds are rearing ones. The number of GnRH-I neurons increases after removal of the chicks, suggesting that changes in GnRH-I neurons in the nCPa are key regulators of the avian reproduction irrespective of whether the birds are photoperiodic or equatorial breeders (Chaiyachet et al., 2013b). The number of VIP neurons, together with immunoreactivity and cell size, markedly increases during the incubating period and then decreases sharply during the rearing period (Cloues et al., 1990). At the onset of rearing behavior, the number of VIP neurons of hens caring for the young is lower than during incubation but higher in hens deprived of their chicks (Chaiyachet et al., 2013b). The increase in the number of VIP neurons correlates with an upregulation of VIP peptide and the amount of VIP in the hypophysial portal vasculature

(Mauro et al., 1989; Sharp et al., 1989; Cloues et al., 1990; Youngren et al., 1996a). It is concluded that VIP neurotransmission is playing a role in rearing behavior and, further, that the activity of the VIP/PRL system coupled to its contribution to rearing behavior are related to the extent to which maternal care is required for the hatched young.

The VIP/PRL system is not only a key regulator of incubation behavior but also has an involvement in the regulation of rearing behavior. There appears to be, however, a lower VIPergic threshold required for rearing behavior. Disruption of rearing behavior markedly decreases PRL secretion with a parallel decline in the number of VIP neurons and an accompanying increase in the number of GnRH-I neurons (Chaiyachet et al., 2013a,b) (Figure 31.8).

An association between the DAergic and GnRH/FSH/LH systems has been demonstrated in the turkeys. There are parallel changes in the number of activated DA neurons in the nucleus premamillaris (PMM), in the number of GnRH-I neurons in the nCPa, and an upregulation of GnRH-I gene expression in the nCPa (Kang et al., 2006; Thayananuphat et al., 2007). As might be expected, GnRH axons terminate in the ME and in close proximity to the terminals of the tuberoinfundibular DAergic neurons (Ugrumov et al., 1989). Dopaminergic axons and fibers are found intermingled with VIP neurons in the INF, GnRH neurons in the POA, and with both VIP and GnRH terminals in the external layer of the ME (Contijoch et al., 1992; Fraley and Kuenzel, 1993). The neuronal interactions between the VIP and DAergic neurons are also reported (Teruyama and Beck, 2001). Activation of DAergic neurons in the nucleus mamillaris lateralis is associated with the activation of GnRH-I and VIP neurons and secretion of LH and PRL (Al-Zailaie et al., 2006). Rearing native Thai hens had enhanced DAergic activity and elevated PRL levels when compared to that of hens deprived of their young. This suggests a role for the DAergic system in both incubation and rearing behaviors. Furthermore, the decline in DAergic activity and PRL levels during the disruption of rearing behavior supports the involvement of DA in rearing behavior in this equatorial precocial bird. These findings provide strong evidence for the association of the neuronal interactions between the GnRH-Iergic, DAergic, and VIPergic systems in the neuroendocrine regulation of reproductive activity and care of the chicks (Prakobsaeng et al., 2011; Chaiyachet et al., 2013a,b; Chokchaloemwong et al., 2012).

31.3.1.2 Enhanced Hypothalamic Mesotocin Expression

The hormonal basis of parental care behavior has been extensively studied in multiple vertebrates and collectively these

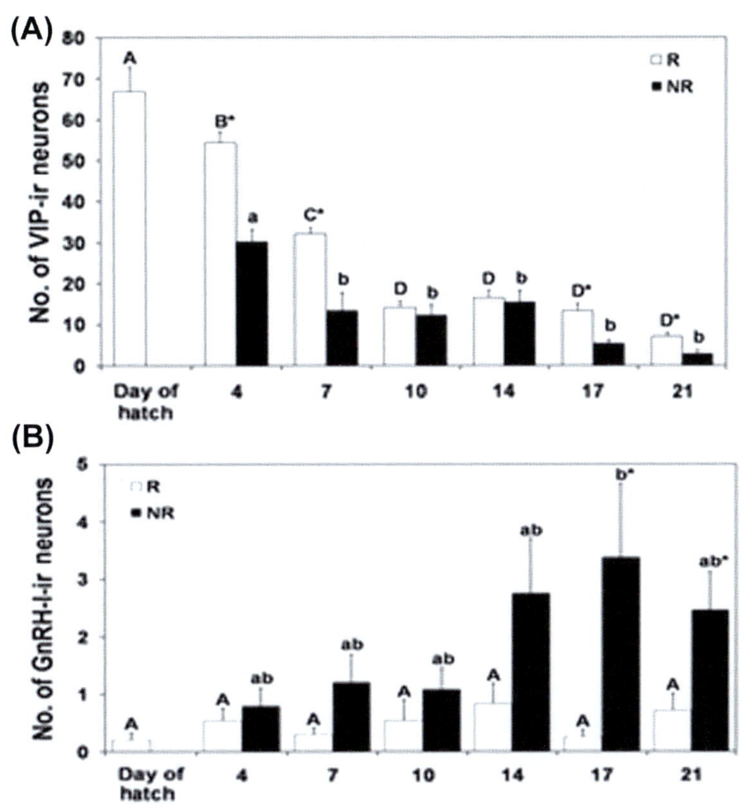

FIGURE 31.8 Changes in the numbers: (A) vasoactive intestinal peptide-immunoreactive neurons in the nucleus inferioris hypothalami-nucleus infundibula hypothalami (IH-IN); (B) gonadotropin releasing hormone-I-immunoreactive neurons in the nucleus commissuraepallii (nCPa) of rearing (R) and non-rearing (NR) native Thai hens. *Modified from Chaiyachet et al. (2013a,b).*

indicate a role of neurohypophysial nonapeptides (Acher, 1972; Goodson et al., 2012). These nonapeptides evolved more than 600 million years ago (Donaldson and Young, 2008). These neurohypophysial hormones are synthesized in magnocellular neurons of the hypothalamus and released from nerve terminals of the posterior pituitary gland or into the hypophysial portal blood via the ME. The best known neurohypophysial nonapeptides are arginine vasopressin (AVP) and oxytocin; these being found in most mammalian species. There are at least 14 additional neurohypophysial hormones that are found in non-mammalian vertebrates. In birds, the two hormones are arginine vasotocin (AVT) and mesotocin (MT), the avian homologs of AVP and oxytocin, respectively (Acher et al., 1970; Hoyle, 1998). They are synthesized in magnocellular neurons of the supraoptic and paraventricular nuclei (SON and PVN). MT neurons are found in several brain areas such as the nucleus supraopticus, pars ventralis (SOv), PVN, cerebellum, lateral septum (LS), optic lobe, pons, and medulla oblongata (Goossens et al., 1977; Bons, 1980; Robinzon et al., 1988).

Little is known of the physiological function(s) of MT in birds. The first evidence for a role of MT in rearing behavior was reported in turkeys (Thayananuphat et al., 2011). The number of MT neurons in the PVN and SOv increases in incubating hens compared with laying hens. In the incubating hens, there is high expression of the immediate early gene, c-fos, observed in MT neurons within the PVN and SOv. The majority of c-fos expression is observed in the DAergic neurons in the ventral part of the nucleus preopticus medialis (POM). The induction of c-fos mRNA expression in the MT neurons within the PVN and SOv in the incubating hens is stimulated by the presence of poults. Blocking MT or D_2 DA receptors prevents poults from brooding. This suggests that MT is involved in the onset of maternal activities (Thayananuphat et al., 2011). In birds, the brain areas that have been implicated in the regulation of parental behaviors are the POA, PVN, and ventromedial nucleus (Slawski and Buntin, 1995; Schoech et al., 1998; Lea et al., 2001). The expression of fos-ir in the brain of brooding ring doves and Japanese quail is higher in the POA, LHy, LS, MPOA, and bed nucleus of the stria terminalis than that of the parents not allowed to rear their young (Ruscio and Adkins-Regan, 2004; Buntin et al., 2006). Similarly, the distribution of MT neurons and fibers of native Thai chickens has been reported. The greatest densities are found predominantly within the SOv, POM, and PVN. The number of MT neurons is maximal after the hens have shifted from egg laying to rearing (Chokchaloemwong et al., 2013). This supports the view that the MTergic system plays an important role in neuroendocrine reorganization to establish and maintain maternal behavior. This is further supported by the analogous role of oxytocin neurons in similar nuclei in mammals (Chokchaloemwong et al., 2013) (Figures 31.9 and 31.10).

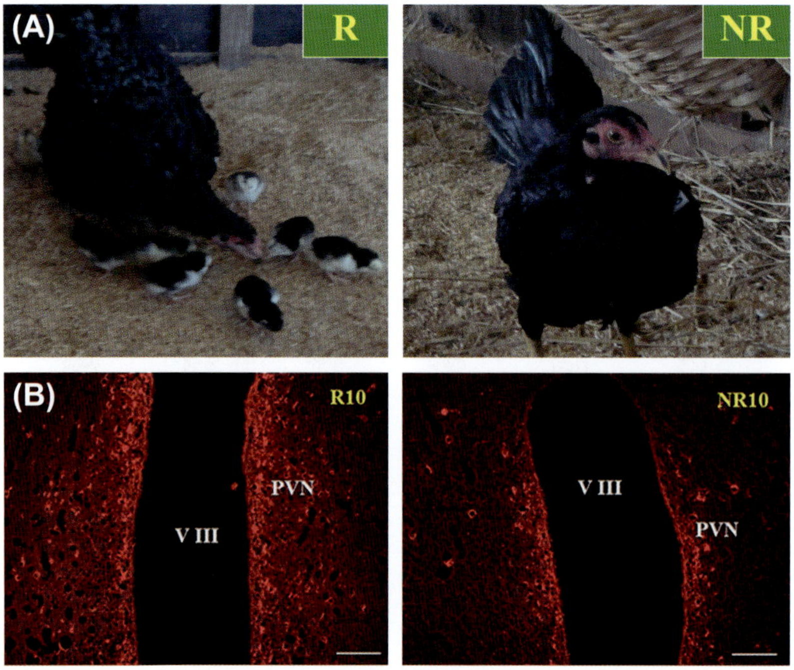

FIGURE 31.9 The distribution of mesotocin-immunoreactive (MT-ir) neurons in the paraventricular nucleus (PVN) at day 10 (B) following the initiation of rearing (A; R) or non-rearing chicks (A; NR) native Thai hen. *Modified from Chokchaloemwong et al. (2013).*

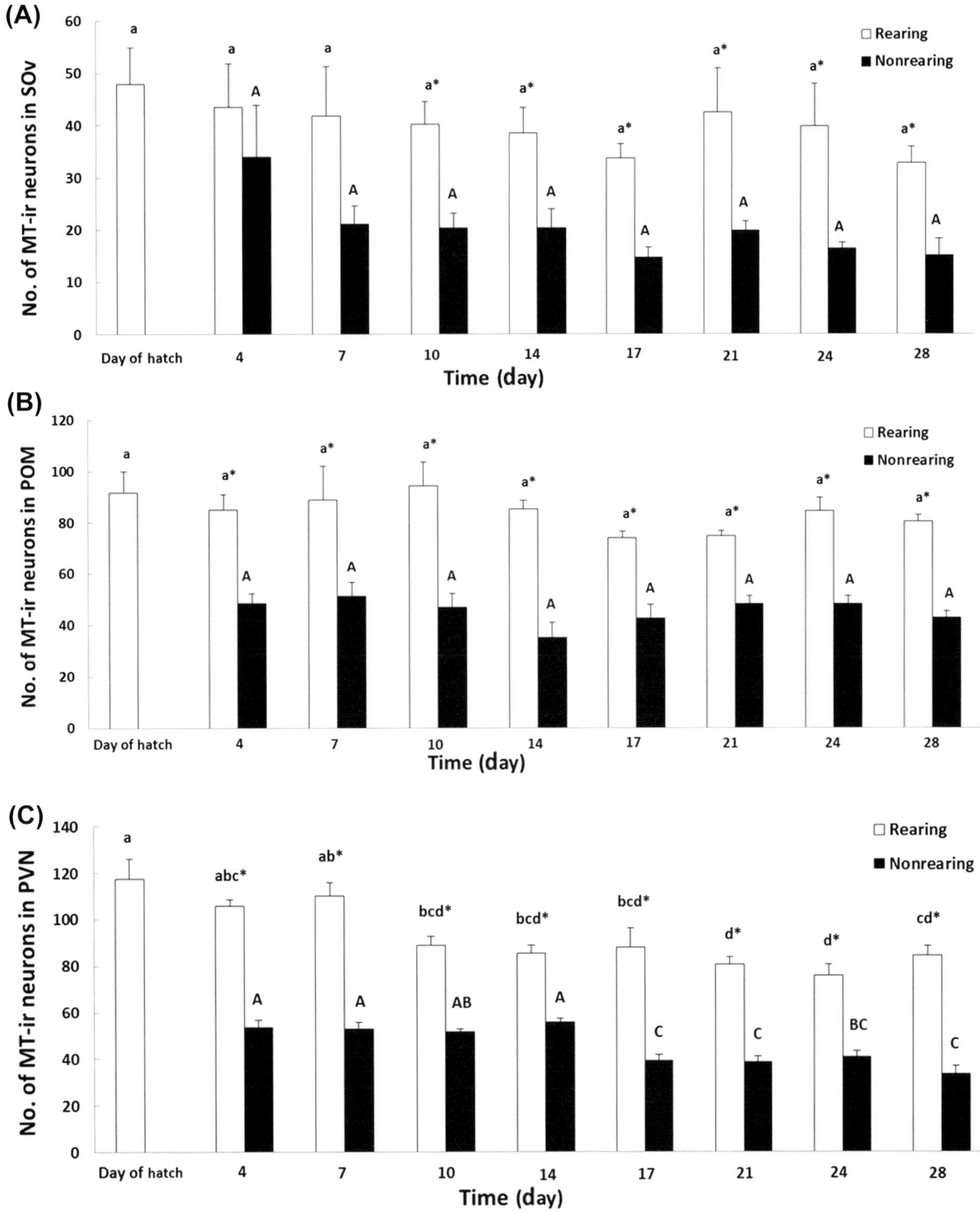

FIGURE 31.10 Changes in the number of mesotocin-immunoreactive (MT-ir) neurons in the (A) supraopticus, pars ventralis (SOv), (B) preopticus medialis (POM), and (C) paraventricular nuclei (PVN) of rearing and non-rearing native Thai hens. *From Chokchaloemwong et al. (2013).*

REFERENCES

Absil, P., Foidart, A., Hemmings Jr., H.C., Steinbusch, H.W., Ball, G.F., Balthazart, J., 2001. Distribution of DARPP-32 immunoreactive structures in the quail brain: anatomical relationship with dopamine and aromatase. J. Chem. Neuroanat. 21, 23–39.

Acher, R., 1972. Chemistry of the Neurohypophysial Hormones: An Example of Molecular Evolution, Endocrinology. American Physiological Society, Washington DC, USA. pp. 119–130.

Acher, R., Chauvet, J., Chauvet, M.T., 1970. Phylogeny of the neurohypophysial hormones. The avian active peptides. Eur. J. Biochem. 17, 509–513.

Adkins-Regan, E., Banerjee, S.B., Correa, S.M., Schweitzer, C., 2013. Maternal effects in quail and zebra finches: Behavior and hormones. Gen. Comp. Endocrinol. 190, 34–41.

Ahn, J., You, S.K., Kim, H., Chaiseha, Y., El Halawani, M.E., 2001. Effects of active immunization with inhibin α subunit on reproductive characteristics of turkey hens. Biol. Reprod. 65, 1594–1600.

Akesson, T.R., Raveling, D.G., 1981. Endocrine and body weight changes of non-nesting Canada geese. Biol. Reprod. 25, 792–804.

Al-Zailaie, K.A., El Halawani, M.E., 2000. Neuroanatomical relationship between immunoreactive dopamine and vasoactive intestinal peptide neurons in the turkey hypothalamus. Poult. Sci. 79 (Suppl. 1), 50.

Al-Zailaie, K.A., Kang, S.W., Youngren, O.M., Thayananuphat, A., Bakken, T., Chaiseha, Y., Millam, J.R., Proudman, J.A., El Halawani, M.E., 2006. Identification of dopamine, gonadotrophin-releasing hormone-I, and vasoactive intestinal peptide neurones activated by electrical stimulation to the medial preoptic area of the turkey hypothalamus: a potential reproductive neuroendocrine circuit. J. Neuroendocrinol. 18, 514–525.

Andersson, M., Eriksson, M.O.G., 1982. Nest parasitism in goldeneyes *Bucephala clangula*: some evolutionary aspects. Am. Nat. 120, 1–16.

Angelier, F., Chastel, O., 2009. Stress, prolactin and parental investment in birds: a review. Gen. Comp. Endocrinol. 163, 142–148.

Appeltants, D., Ball, G.F., Balthazart, J., 2001. The distribution of tyrosine hydroxylase in the canary brain: demonstration of a specific and sexually dimorphic catecholaminergic innervation of the telencephalic song control nuclei. Cell Tissue Res. 304, 237–259.

Bailhache, T., Balthazart, J., 1993. The catecholaminergic system of the quail brain: immunocytochemical studies of dopamine β-hydroxylase and tyrosine hydroxylase. J. Comp. Neurol. 329, 230–256.

Bedrak, E., Harvey, S., Chadwick, A., 1981. Concentrations of pituitary, gonadal and adrenal hormones in serum of laying and broody white rock hens (*Gallus domesticus*). J. Endocrinol. 89, 197–204.

Ben-Jonathan, N., Hnasko, R., 2001. Dopamine as a prolactin (PRL) inhibitor. Endocr. Rev. 22, 724–763.

Ben-Jonathan, N., Arbogast, L.A., Hyde, J.F., 1989. Neuroendocrine regulation of prolactin release. Prog. Neurobiol. 33, 399–447.

Bern, H.A., Nicoll, C.S., 1968. The comparative endocrinology of prolactin. Recent Prog. Horm. Res. 24, 681–720.

Bhatt, R., Youngren, O.M., Kang, S.W., El Halawani, M.E., 2003. Dopamine infusion in the third ventricle increases gene expression of hypothalamic vasoactive intestinal peptide and pituitary prolactin and luteinizing hormone beta subunit in the turkey. Gen. Comp. Endocrinol. 130, 41–47.

Bluhm, C.K., Phillips, R.E., Burke, W.H., 1983. Serum levels of luteinizing hormone, prolactin, estradiol and progesterone in laying and nonlaying mallards (*Anas platyrhynchos*). Biol. Reprod. 28, 295–305.

Bole-Feysot, C., Goffin, V., Edery, M., Binart, N., Kelly, P.A., 1998. Prolactin (PRL) and its receptor: actions signal transduction pathways and phenotypes observed in PRL receptor knockout mice. Endocr. Rev. 19, 225–268.

Bons, N., 1980. The topography of mesotocin and vasotocin systems in the brain of the domestic mallard and Japanese quail: immunocytochemical identification. Cell Tissue Res. 213, 37–51.

Boos, M., Zimmer, C., Carriere, A., Robin, J.P., Petit, O., 2007. Post-hatching parental care behaviour and hormonal status in a precocial bird. Behav. Process. 76, 206–214.

Bottjer, S.W., Alexander, G., 1995. Localization of met-enkephalin and vasoactive intestinal polypeptide in the brains of male zebra finches. Brain Behav. Evol. 45, 153–177.

Breitenbach, R.P., Meyer, R.K., 1959. Pituitary prolactin levels in laying, incubating and brooding pheasants (*Phasianus colchicus*). Proc. Soc. Exp. Biol. Med. 101, 16–19.

Brunton, P.J., Russell, J.A., 2008. The expectant brain: adapting for motherhood. Nat. Rev. Neurosci. 9, 11–25.

Buntin, J.D., 1996. Neural and hormonal control of parental behavior in birds. Adv. Stud. Behav. 25, 161–213.

Buntin, J.D., 2010. Parental behavior and hormones in non-mammalian vertebrates. In: In: Breed, M.D., Moore, J. (Eds.), Encyclopedia of Animal Behavior, vol. 2. Academic Press, Oxford, pp. 664–671.

Buntin, J.D., Becker, G.M., Ruzycky, E., 1991. Facilitation of parental behavior in ring doves by systemic or intracranial injections of prolactin. Horm. Behav. 25, 424–444.

Buntin, L., Berghman, L.R., Buntin, J.D., 2006. Patterns of fos-like immunoreactivity in the brains of parent ring doves (*Streptopelia risoria*) given tactile and nontactile exposure to their young. Behav. Neurosci. 120, 651–664.

Buntin, J.D., EL Halawani, M.E., Ottinger, M.A., Fan, Y., Fivizzani, A.J., 1998. An analysis of sex and breeding stage differences in prolactin binding activity in brain and hypothalamic GnRH concentration in Wilson's phalarope, a sex role-reversed species. Gen. Comp. Endocrinol. 109, 119–132.

Burke, W.H., Dennison, P.T., 1980. Prolactin and luteinizing hormone levels in female turkeys (*Meleagris gallopavo*) during a photoinduced reproductive cycle and broodiness. Gen. Comp. Endocrinol. 41, 92–100.

Campbell, R.R., Ashton, S.A., Follet, B.K., Leatherland, J.F., 1978. Seasonal changes in plasma concentrations of LH in the lesser snow goose (*Anser caerulescens caerulescens*). Biol. Reprod. 18, 663–668.

Chaiseha, Y., El Halawani, M.E., 1999. Expression of vasoactive intestinal peptide/peptide histidine isoleucine in several hypothalamic areas during the turkey reproductive cycle: relationship to prolactin secretion. Neuroendocrinology 70, 402–412.

Chaiseha, Y., El Halawani, M.E., 2005. Neuroendocrinology of the female turkey reproductive cycle. J. Poult. Sci. 42, 87–100.

Chaiseha, Y., Tong, Z., Youngren, O.M., El Halawani, M.E., 1998. Transcriptional changes in hypothalamic vasoactive intestinal peptide during a photo-induced reproductive cycle in the turkey. J. Mol. Endocrinol. 21, 267–275.

Chaiseha, Y., Youngren, O.M., Al-Zailaie, K.A., El Halawani, M.E., 2003. Expression of D1 and D2 dopamine receptors in the hypothalamus and pituitary during the turkey reproductive cycle: colocalization with vasoactive intestinal peptide. Neuroendocrinology 77, 105–118.

Chaiseha, Y., Youngren, O.M., El Halawani, M.E., 1997. Dopamine receptors influence vasoactive intestinal peptide release from turkey hypothalamic explants. Neuroendocrinology 65, 423–429.

Chaiseha, Y., Youngren, O.M., El Halawani, M.E., 2004. Expression of vasoactive intestinal peptide receptor messenger RNA in the hypothalamus and pituitary throughout the turkey reproductive cycle. Biol. Reprod. 70, 593–599.

Chaiyachet, O.A., Chokchaloemwong, D., Prakobsaeng, N., Sartsoongnoen, N., Kosonsiriluk, S., Chaiseha, Y., 2013a. Gonadotropin releasing hormone and brooding behavior in the native Thai hen. Acta Histochem. 115, 626–636.

Chaiyachet, O.A., Chokchaloemwong, D., Prakobsaeng, N., Sartsoongnoen, N., Kosonsiriluk, S., Rozenboim, I., El Halawani, M.E., Porter, T.E., Chaiseha, Y., 2013b. Neuroendocrine regulation of rearing behavior in the native Thai hen. Acta Histochem. 115, 209–218.

Chokchaloemwong, D., Chaiyachet, O., Prakobsaeng, N., Sartsoongnoen, N., Kosonsiriluk, S., El Halawani, M.E., Chaiseha, Y., June 2012. The dopaminergic-PRL system involvement in rearing behavior of Gallinacious birds. In: The 10th International Symposium on Avian Endocrinology, Gifu, Japan.

Chokchaloemwong, D., Prakobsaeng, N., Sartsoongnoen, N., Kosonsiriluk, S., El Halawani, M.E., Chaiseha, Y., 2013. Mesotocin and maternal care of chicks in native Thai hens (*Gallus domesticus*). Horm. Behav. 64, 53–69.

Clark, M.E., Martin, T.E., 2007. Modeling tradeoffs in avian life history traits and consequences for population growth. Ecol. Model. 209, 110–120.

Cloues, R., Ramos, C., Silver, R., 1990. Vasoactive intestinal polypeptide-like immunoreactivity during reproduction in doves: influence of experience and number of offspring. Horm. Behav. 24, 215–231.

Clutton-Brock, T.H., 1991. The Evolution of Parental Care. Princeton University Press, Princeton, NJ.

Cogger, E.A., Burke, W.H., Ogren, L.A., 1979. Serum luteinizing hormone, progesterone, and estradiol levels in relation to broodiness in the turkey (*Meleagris gallopavo*). Poult. Sci. 58, 1355–1360.

Contijoch, A.M., Gonzalez, C., Singh, H.N., Malamed, S., Troncoso, S., Advis, J.P., 1992. Dopaminergic regulation of luteinizing hormone-releasing hormone release at the median eminence level: immunocytochemical and physiological evidence in hens. Neuroendocrinology 55, 290–300.

Criscuolo, F., Chastel, O., Gabrielsen, G.W., Lacroix, A., Le Maho, Y., 2002. Factors affecting plasma concentrations of prolactin in the common eider *Somateria mollissima*. Gen. Comp. Endocrinol. 125, 399–409.

Crisostomo, S., Guemene, D., Garreau-Mills, M., Morvan, C., Zadworny, D., 1998. Prevention of incubation behavior expression in Turkey hens by active immunization against prolactin. Theriogenology 50, 675–690.

Curlewis, J.D., 1992. Seasonal prolactin secretion and its role in seasonal reproduction: a review. Reprod. Fertil. Dev. 4, 1–23.

Davies, D.T., Follett, B.K., 1975. The neuroendocrine control of gonadotrophin release in the Japanese quail. I. The role of the tuberal hypothalamus. Proc. R. Soc. Lond. B, Biol. Sci. 191, 285–301.

Dawson, A., Goldsmith, A.R., 1982. Prolactin and gonadotrophin secretion in wild starlings (*Sturnus vulgaris*) during the annual cycle and in relation to nesting, incubation, and rearing young. Gen. Comp. Endocrinol. 48, 213–221.

Dawson, A., Goldsmith, A.R., 1985. Modulation of gonadotropin and prolactin secretion by daylength and breeding behaviour in free-living starlings, *Sturnus vulgaris*. J. Zool. Lond. 206, 241–252.

Dawson, A., Talbot, R.T., Dunn, I.C., Sharp, P.J., 2002. Changes in basal hypothalamic chicken gonadotropin-releasing hormone-I and vasoactive intestinal polypeptide associated with a photo-induced cycle in gonadal maturation and prolactin secretion in intact and thyroidectomized starlings (*Sturnus vulgaris*). J. Neuroendocrinol. 14, 533–539.

Dawson, J.W., Mannan, R.W., 1991. Dominance hierarchies and helper contributions in Harris' hawks. Auk 108, 649–660.

den Boer-Visser, A.M., Dubbeldam, J.L., 2002. The distribution of dopamine, substance P, vasoactive intestinal polypeptide and neuropeptide Y immunoreactivity in the brain of the collared dove, Streptopelia decaocto. J. Chem. Neuroanat. 23, 1–27.

Dittami, J.P., 1981. Seasonal changes in the behavior and plasma titers of various hormones in barheaded geese, *Anser indicus*. Z. Tierpsychol. 55, 289–324.

Donaldson, Z.R., Young, L.J., 2008. Oxytocin, vasopressin, and the neurogenetics of sociality. Science 322, 900–904.

Dunn, I.C., Sharp, P.J., 1999. Photo-induction of hypothalamic gonadotrophin releasing hormone-I mRNA in the domestic chicken: a role for oestrogen? J. Neuroendocrinol. 11, 371–375.

Dunn, I.C., Beattie, K.K., Maney, D., Sang, H.M., Talbot, R.T., Wilson, P.W., Sharp, P.J., 1996. Regulation of chicken gonadotropin-releasing hormone-I mRNA in incubating, nest-deprived and laying bantam hens. Neuroendocrinology 63, 504–513.

Durstewitz, D., Kroner, S., Hemmings Jr., H.C., Gunturkun, O., 1998. The dopaminergic innervation of the pigeon telencephalon: distribution of DARPP-32 and co-occurrence with glutamate decarboxylase and tyrosine hydroxylase. Neuroscience 83, 763–779.

Edgar, J.L., Lowe, J.C., Paul, E.S., Nicol, C.J., 2011. Avian maternal response to chick distress. Proc. Biol. Sci. 278, 3129–3134.

El Halawani, M.E., Burke, W.H., Dennison, P.T., 1980. Effect of nest-deprivation on serum prolactin level in nesting female turkeys. Biol. Reprod. 23, 118–123.

El Halawani, M.E., Burke, W.H., Millam, J.R., Fehrer, S.C., Hargis, B.M., 1984. Regulation of prolactin and its role in gallinaceous bird reproduction. J. Exp. Zool. 232, 521–529.

El Halawani, M.E., Fehrer, S.C., Hargis, B.M., Porter, T.E., 1988a. Incubation behavior in the domestic turkey: physiological correlates. CRC Crit. Rev. Poult. Biol. 1, 285–314.

El Halawani, M.E., Silsby, J.L., Fehrer, S.C., 1988b. Basal and hypothalamic extract-induced luteinizing hormone and prolactin secretion by cultured anterior pituitary cells from female turkeys in various stages of the reproductive cycle. Gen. Comp. Endocrinol. 71, 45–54.

El Halawani, M.E., Pitts, G.R., Sun, S., Silsby, J.L., Sivanandan, V., 1996. Active immunization against vasoactive intestinal peptide prevents photo-induced prolactin secretion in turkeys. Gen. Comp. Endocrinol. 104, 76–83.

El Halawani, M.E., Silsby, J.L., Behnke, E.J., Fehrer, S.C., 1986. Hormonal induction of incubation behavior in ovariectomized female turkeys (*Meleagris gallopavo*). Biol. Reprod. 35, 59–67.

El Halawani, M.E., Silsby, J.L., Fehrer, S.C., Behnke, E.J., 1983. Effects of estrogen and progesterone on serum prolactin and luteinizing hormone levels in ovariectomized turkeys (*Meleagris gallopavo*). Gen. Comp. Endocrinol. 52, 67–78.

El Halawani, M.E., Silsby, J.L., Rozenboim, I., Pitts, G.R., 1995. Increased egg production by active immunization against vasoactive intestinal peptide in the turkey (*Meleagris gallopavo*). Biol. Reprod. 52, 179–183.

El Halawani, M.E., Silsby, J.L., Youngren, O.M., Phillips, R.E., 1991. Exogenous prolactin delays photo-induced sexual maturity and suppresses ovariectomy-induced luteinizing hormone secretion in the turkey (*Meleagris gallopavo*). Biol. Reprod. 44, 420–424.

El Halawani, M.E., Youngren, O.M., Chaiseha, Y., 2001. Neuroendocrinology of PRL regulation in the domestic turkey. In: Dawson, A., Chaturvedi, C.M. (Eds.), Avian Endocrinology. Narosa Publishing House, New Delhi, India, pp. 233–244.

El Halawani, M.E., Youngren, O.M., Pitts, G.R., 1997. Vasoactive intestinal peptide as the avian prolactin-releasing factor. In: Harvey, S., Etches, R.J. (Eds.), Perspectives in Avian Endocrinology. Journal of Endocrinology, Bristol, UK, pp. 403–416.

Fernandez-Duque, E., Valeggia, C.R., Mendoza, S.P., 2009. The biology of paternal care in human and nonhuman primates. Annu. Rev. Anthropol. 38, 115–130.

Fraley, G.S., Kuenzel, W.J., 1993. Immunocytochemical and histochemical analyses of gonadotrophin releasing hormone, tyrosine hydroxylase, and cytochrome oxidase reactivity within the hypothalamus of chicks showing early sexual maturation. Histochemistry 99, 221–229.

Freeman, M.E., Kanyicska, B., Lerant, A., Nagy, G., 2000. Prolactin: structure, function, and regulation of secretion. Physiol. Rev. 80, 1523–1631.

Garcia, V., Jouventin, P., Mauget, R., 1996. Parental care and the prolactin secretion pattern in the king penguin: an endogenously timed mechanism? Horm. Behav. 30, 259–265.

Gatti, K.C., 1983. Incubation weight loss in the mallard. Can. J. Zool. 61, 565–569.

Goldsmith, A.R., 1982. The Australian black swan (*Cygnus atratus*): prolactin and gonadotrophin secretion during breeding including incubation. Gen. Comp. Endocrinol. 46, 458–462.

Goldsmith, A.R., 1991. Prolactin and avian reproductive strategies. Acta Congr. Int. Ornithol. 20, 2063–2071.

Goldsmith, A.R., Hall, M., 1980. Prolactin concentrations in the pituitary gland and plasma of Japanese quail in relation to photoperiodically induced sexual maturation and egg laying. Gen. Comp. Endocrinol. 42, 449–454.

Goldsmith, A.R., Williams, D.M., 1980. Incubation in mallards (*Anas platyrhynchos*): changes in plasma levels of prolactin and luteinizing hormone. J. Endocrinol. 86, 371–379.

Goldsmith, A.R., Burke, S., Prosser, J.M., 1984. Inverse changes in plasma prolactin and UI concentrations in female canaries after disruption and reinitiation of incubation. J. Endocrinol. 103, 251–256.

Goldsmith, A.R., Edwards, C., Koprucu, M., Silver, R., 1981. Concentrations of prolactin and luteinizing hormone in plasma of doves in relation to incubation and development of the crop gland. J. Endocrinol. 90, 437–443.

Goodson, J.L., Kelly, A.M., Kingsbury, M.A., 2012. Evolving nonapeptide mechanisms of gregariousness and social diversity in birds. Horm. Behav. 61, 239–250.

Goossens, N., Blahser, S., Oksche, A., Vandesande, F., Dierickx, K., 1977. Immunocytochemical investigation of the hypothalamo-neurohypophysial system in birds. Cell Tissue Res. 184, 1–13.

Goth, A., 2002. Behaviour of Australian Brush-turkey (*Alectura lathami*, Galliformes: Megapodiidae) chicks following underground hatching. J. Ornithol. 143, 477–488.

Goth, A., Vogel, U., 2003. Juvenile dispersal and habitat selectivity in the megapode *Alectura lathami* (Australian Brush-turkey). Wildl. Res. 30, 69–74.

Gottleib, G., 1976. Consequences of prenatal development: behavioural embryology. Psychol. Rev. 83, 215–234.

Gratto-Trevor, C.L., Fivizzani, A.J., Oring, L.W., Cooke, F., 1990. Seasonal changes in gonadal steroids of a monogamous versus a polyandrous shorebird. Gen. Comp. Endocrinol. 80, 407–418.

Gross, M.R., 2005. The evolution of parental care. Q. Rev. Biol. 80, 37–46.

Hall, M.R., Goldsmith, A.R., 1983. Factors affecting prolactin secretion during breeding and incubation in the domestic duck (*Anas platyrhynchos*). Gen. Comp. Endocrinol. 49, 270–276.

Hall, M.R., 1987. External stimuli affecting incubation behavior and prolactin secretion in the duck (*Anas platyrhynchos*). Horm. Behav. 21, 269–287.

Hall, T.R., Harvey, S., Chadwick, A., 1986. Control of prolactin secretion in birds: a review. Gen. Comp. Endocrinol. 62, 171–184.

Haller, R.W., Cherms, F.L., 1961. A comparison of several treatments on terminating broodiness in broad breasted bronze turkeys. Poult. Sci. 40, 15–163.

Harris, J., Stanford, P.M., Oakes, S.R., Ormandy, C.J., 2004. Prolactin and the prolactin receptor: new targets of an old hormone. Ann. Med. 36, 414–425.

Hess, E.H., Petrovich, S.B., Goodwin, E.B., 1976. Induction of parental behavior in Japanese quail (*Coturnix japonica*). J. Comp. Physiol. Psychol. 90, 244–251.

Hiatt, E.S., Goldsmith, A.R., Farner, D.S., 1987. Plasma levels of prolactin and gonadotropins during the reproductive cycle of white-crowned sparrows (*Zonotrichia leucophrys*). Auk 104, 208–217.

Hof, P.R., Dietl, M.M., Charnay, Y., Martin, J.L., Bouras, C., Palacios, J.M., Magistretti, P.J., 1991. Vasoactive intestinal peptide binding sites and fibers in the brain of the pigeon (*Columba livia*): an autoradiographic and immunohistochemical study. J. Comp. Neurol. 305, 393–411.

Hohn, E.O., 1959. Prolactin in the cowbird's pituitary in relation to avian brood parasitism. Nature 184, 2030.

Hoshino, S., Wakita, M., 1989. Increased synthesis of prolactin and growth hormone during incubation in the pituitary of broody Nagoya hens. Horm. Metab. Res. 21, 480–482.

Hoyle, C.H., 1998. Neuropeptide families: evolutionary perspectives. Regul. Pept. 73, 1–33.

Huang, Y.M., Shi, Z.D., Liu, Z., Liu, Y., Li, X.W., 2008. Endocrine regulations of reproductive seasonality, follicular development and incubation in Magang geese. Anim. Reprod. Sci. 104, 344–358.

Jain, S., Sharma, R., Wadhwa, S., 2004. Effect of prenatal species-specific and music stimulation on the postnatal auditory preference of domestic chick. Ind. J. Physiol. Pharmacol. 48, 174–183.

Jones, R.E., 1971. The incubation patch of birds. Biol. Rev. 46, 315–339.

Kang, S.W., Thayananuphat, A., Rozenboim, I., Millam, J.R., Proudman, J.A., El Halawani, M.E., 2006. Expression of hypothalamic GnRH-I mRNA in the female turkey at different reproductive states and following photostimulation. Gen. Comp. Endocrinol. 146, 86–94.

Karatzas, C.N., Guemene, D., Zadworny, D., Kuhnlein, U., 1997. Changes in expression of the prolactin and growth hormone gene during different reproductive stages in the pituitary gland of turkeys. Reprod. Nutr. Dev. 37, 69–79.

Kendeigh, S.C., 1952. Parental care and its evolution in birds. Ill. Biol. Mon. 22, 1–358.

Kentner, A.C., Abizaid, A., Bielajew, C., 2010. Modeling dad: animal models of paternal behavior. Neurosci. Biobehav. Rev. 34, 438–451.

Kiss, J.Z., Peczely, P., 1987. Distribution of tyrosine-hydroxylase (TH) immunoreactive neurons in the diencephalon of the pigeon (*Columba livia domestica*). J. Comp. Neurol. 257, 333–346.

Knapp, T.R., Fehrer, S.C., Silsby, J.L., Porter, T.E., Behnke, E.J., El Halawani, M.E., 1988. Gonodal steroid modulation of basal and vasoactive intestinal polypeptide-stimulated prolactin release by turkey anterior pituitary cells. Gen. Comp. Endocrinol. 72, 226–236.

Knigge, K.M., Piekut, D.T., 1985. Distribution of CRF- and tyrosine hydroxylase-immunoreactive neurons in the brainstem of the domestic fowl (*Gallus domesticus*). Peptides 6, 97–101.

Korf, H.W., Fahrenkrug, J., 1984. Ependymal and neuronal specializations in the lateral ventricle of the Pekin duck, *Anas platyrhynchos*. Cell Tissue Res. 236, 217–227.

Kosonsiriluk, S., Sartsoongnoen, N., Chaiyachet, O.A., Prakobsaeng, N., Songserm, T., Rozenboim, I., El Halawani, M.E., Chaiseha, Y., 2008. Vasoactive intestinal peptide and its role in continuous and seasonal reproduction in birds. Gen. Comp. Endocrinol. 159, 88–97.

Kragt, C.L., Meites, J., 1965. Stimulation of pigeon pituitary prolactin release by pigeon hypothalamic extracts in vitro. Endocrinology 76, 1169–1176.

Krasnegor, N.A., Bridges, R.S., 1990. Mammalian Parenting: Biochemical, Neurobiological, and Behavioral Determinants. Oxford University Press, New York.

Kruger, O., 2007. Cuckoos, cowbirds and hosts: adaptations, trade-offs and constraints. Philos. Trans. R. Soc. Lond., B, Biol. Sci. 362, 1873–1886.

Kuenzel, W.J., Mccune, S.K., Talbot, R.T., Sharp, P.J., Hill, J.M., 1997. Sites of gene expression for vasoactive intestinal polypeptide throughout the brain of the chick (*Gallus domesticus*). J. Comp. Neurol. 381, 101–118.

Kuroda, K.O., Tachikawa, K., Yoshida, S., Tsuneoka, Y., Numan, M., 2011. Prog. Neuropsychopharmacol. Biol. Psychiatry 35, 1205–1231.

Kuwayama, T., Shimada, K., Saito, N., Ohkubo, T., Sato, K., Wada, M., Ichinoe, K., 1992. Effects of removal of chicks from hens on concentrations of prolactin, luteinizing hormone and oestradiol in plasma of brooding Gifujidori hens. J. Reprod. Fertil. 95, 617–622.

Lack, D., 1968. Ecological Adaptations for Breeding in Birds. Methuen, London.

Lea, R.W., Sharp, P.J., 1982. Plasma prolactin concentrations in broody turkeys: lack of agreement between homologous chicken and turkey prolactin radioimmunoassays. Br. Poult. Sci. 23, 451–459.

Lea, R.W., Clark, J.A., Tsutsui, K., 2001. Changes in central steroid receptor expression, steroid synthesis, and dopaminergic activity related to the reproductive cycle of the ring dove. Microsc. Res. Tech. 55, 12–26.

Lea, R.W., Dods, A.S., Sharp, P.J., Chadwick, A., 1981. The possible role of prolactin in the regulation of nesting behavior and the secretion of luteinizing hormone in broody bantams. J. Endocrinol. 91, 89–97.

Lea, R.W., Richard-Yris, M.A., Sharp, P.J., 1996. The effect of ovariectomy on concentrations of plasma prolactin and LH and parental behavior in the domestic fowl. Gen. Comp. Endocrinol. 101, 115–121.

Leboucher, G., Richard-Yris, M.A., Guemene, D., Chadwick, A., 1993. Respective effects of chicks and nest on behavior and hormonal concentrations of incubating domestic hens. Physiol. Behav. 54, 135–140.

Leboucher, G., Richard-Yris, M.A., Williams, J., Chadwick, A., 1990. Incubation and maternal behaviour in domestic hens: influence of the presence of chicks on circulating luteinising hormone, prolactin and oestradiol and on behaviour. Br. Poult. 31, 851–862.

Liang, Y., Cui, J., Yang, G., Leung, F.C., Zhang, X., 2006. Polymorphisms of 5' flanking region of chicken prolactin gene. Domest. Anim. Endocrinol. 30, 1–16.

Macnamee, M.C., Sharp, P.J., 1989. The functional activity of hypothalamic dopamine in broody bantam hens. J. Endocrinol. 121, 67–74.

Macnamee, M.C., Sharp, P.J., Lea, R.W., Sterling, R.J., Harvey, S., 1986. Evidence that vasoactive intestinal polypeptide is a physiological prolactin-releasing factor in the bantam hen. Gen. Comp. Endocrinol. 62, 470–478.

March, J.B., Sharp, P.J., Wilson, P.W., Sang, H.M., 1994. Effect of active immunization against recombinant-derived chicken prolactin fusion protein on the onset of broodiness and photoinduced egg laying in bantam hens. J. Reprod. Fertil. 101, 227–233.

Massaro, M., Setiawan, A.N., Davis, L.S., 2007. Effects of artificial eggs on prolactin secretion, steroid levels, brood patch development, incubation onset and clutch size in the yellow-eyed penguin (*Megadyptes antipodes*). Gen. Comp. Endocrinol. 151, 220–229.

Mauro, L.J., Elde, R.P., Youngren, O.M., Phillips, R.E., El Halawani, M.E., 1989. Alterations in hypothalamic vasoactive intestinal peptide-like immunoreactivity are associated with reproduction and prolactin release in the female turkey (*Meleagris gallopavo*). Endocrinology 125, 1795–1804.

Mello, C.V., Pinaud, R., Ribeiro, S., 1998. Noradrenergic system of the zebra finch brain: immunocytochemical study of dopamine-β-hydroxylase. J. Comp. Neurol. 400, 207–228.

Millam, J.R., Craig-Veit, C.B., Adams, T.E., Adams, B.M., 1989. Avian gonadotrophin-releasing hormones I and II in the brain and other tissues in turkey hens. Comp. Biochem. Physiol. 94A, 771–776.

Mills, A.D., Crawford, L.L., Domjan, M., Faure, J.M., 1997. The behavior of the Japanese or domestic quail *Coturnix japonica*. Neurosci. Biobehav. Rev. 21, 261–281.

Moons, L., D'Hondt, E., Pijcke, K., Vandesande, F., 1995. Noradrenergic system in the chicken brain: immunocytochemical study with antibodies to noradrenaline and dopamine-β-hydroxylase. J. Comp. Neurol. 360, 331–348.

Moons, L., van Gils, J., Ghijsels, E., Vandesande, F., 1994. Immunocytochemical localization of L-DOPA and dopamine in the brain of the chicken (*Gallus domesticus*). J. Comp. Neurol. 346, 97–118.

Moss, R., Watson, A., 1982. Heritability of egg size, hatch weight, body weight, and viability in red grouse (*Lagopus lagopus scoticus*). Auk 99, 638–686.

Mutt, V., Said, S.I., 1974. Structure of porcine vasoactive intestinal octacosapeptide. The amino-acid sequence. Use of kallikrein in its determination. Eur. J. Biochem. 42, 581–589.

Myers, S.A., Millam, J.R., El Halawani, M.E., 1989. Plasma LH and prolactin levels during the reproductive cycle of the cockatiel (*Nymphicus hollandicus*). Gen. Comp. Endocrinol. 73, 85–91.

Norgren, R.B., Silver, R., 1990. Distribution of vasoactive intestinal peptide-like and neurophysin-like immunoreactive neurons and acetylcholinesterase staining in the ring dove hypothalamus with emphasis on the question of an avian suprachiasmatic nucleus. Cell Tissue Res. 259, 331–339.

Numan, M., Insel, T.R., 2003. The Neurobiology of Parental Behavior. Springer-Verlag, New York.

Numan, M., Woodside, B., 2010. Maternity: neural mechanisms, motivational processes, and physiological adaptations. Behav. Neurosci. 124, 715–741.

Opel, H., Proudman, J.A., 1980. Failure of mammalian prolactin to induce incubation behavior in chickens and turkeys. Poult. Sci. 59, 2550–2558.

Opel, H., Proudman, J.A., 1988. Effects of poults on plasma concentrations of prolactin in turkey hens incubating with or without eggs or nest. Br. Poult. Sci. 31, 791–800.

Opel, H., Proudman, J.A., 1989. Plasma prolactin levels in incubating turkey hens during pipping of the eggs and after introduction of poults into the nest. Biol. Reprod. 40, 981–987.

Oring, L.W., Fivizzani, A.J., Colwell, M.A., El Halawani, M.E., 1988. Hormonal changes associated with natural and manipulated incubation in the sex-role reversed Wilson's phalarope. Gen. Comp. Endocrinol. 72, 247–256.

Oring, L.W., Fivizzani, A.J., El Halawani, M.E., Goldsmith, A.R., 1986. Seasonal changes in prolactin and luteinizing hormone in the polyandrous spotted sandpiper, *Acistis macularia*. Gen. Comp. Endocrinol. 62, 394–403.

Peczely, P., Kiss, J.Z., 1988. Immunoreactivity to vasoactive intestinal polypeptide (VIP) and thyreotropin-releasing hormone (TRH) in hypothalamic neurons of the domesticated pigeon (*Columba livia*). Alterations following lactation and exposure to cold. Cell Tissue Res. 251, 485–494.

Pedersen, H.C., 1989. Effects of exogenous prolactin on parental behavior in free-living female willow ptarmigan, *Lagopus l. lagopus*. Anim. Behav. 38, 926–934.

Porter, T.E., Silsby, J.L., Behnke, E.J., Knapp, T.R., El Halawani, M.E., 1991. Ovarian steroid production in vitro during gonadal regression in the turkey. I. Changes associated with incubation behaviour. Biol. Reprod. 45, 581–586.

Prakobsaeng, N., Sartsoongnoen, N., Kosonsiriluk, S., Chaiyachet, O.A., Chokchaloemwong, D., Rozenboim, I., El Halawani, M.E., Porter, T.E., Chaiseha, Y., 2011. Changes in vasoactive intestinal peptide and tyrosine hydroxylase immunoreactivity in the brain of nest-deprived native Thai hen. Gen. Comp. Endocrinol. 171, 189–196.

Proudman, J.A., Opel, H., 1981. Turkey prolactin: validation of a radioimmunoassay and measurement of changes associated with broodiness. Biol. Reprod. 25, 573–580.

Ramsay, A.O., 1953. Variations in the development of broodiness in fowl. Behaviour 5, 51–57.

Reiner, A., Karle, E.J., Anderson, K.D., Medina, L., 1994. Catecholaminergic perikarya and fibers in the avian nervous system. In: Smeets, W.J.A.J., Reiner, A. (Eds.), Phylogeny and Development of Catecholamine Systems in CNS of Vertebrates. Cambridge University Press, Cambridge, UK, pp. 135–181.

Richard-Yris, M.A., Leboucher, G., 1987. Responses to successive test of induction of maternal behavior in hens. Behav. Process. 15, 17–26.

Richard-Yris, M.A., Chadwick, A., Guemene, D., Grillou-Schuelke, H., Leboucher, G., 1995. Influence of the presence of chicks on the ability to resume incubation behavior in domestic hens (*Gallus domesticus*). Horm. Behav. 29, 425–441.

Richard-Yris, M.A., Garnier, D.H., Leboucher, G., 1983. Induction of maternal behavior and some hormonal and physiological correlates in the domestic hen. Horm. Behav. 17, 345–355.

Richard-Yris, M.A., Leboucher, G., Chadwick, A., Garnier, D.H., 1987. Induction of maternal behavior in incubating and non-incubating hens: influence of hormones. Physiol. Behav. 40, 193–199.

Richard-Yris, M.A., Leboucher, G., Williams, J., Chadwick, A., 1988. Tendency to display spontaneous incubation does not affect maternal responsiveness in the domestic hen. Behav. Neural Biol. 49, 165–173.

Richard-Yris, M.A., Guemene, D., Lea, R.W., Sharp, P.J., Bedecarrats, G., Foraste, M., Wauters, A.M., 1998a. Behaviour and hormone concentrations in nest deprived and renesting hens. Br. Poult. Sci. 39, 309–317.

Richard-Yris, M.A., Sharp, P.J., Wauters, A.M., Guemene, D., Richard, J.P., Foraste, M., 1998b. Influence of stimuli from chicks on behavior and concentrations of plasma prolactin and luteinizing hormone in incubating hens. Horm. Behav. 33, 139–148.

Riddle, O., Bates, R.W., Dykshorn, S.W., 1932. A new hormone of the anterior pituitary. Proc. Soc. Exp. Biol. Med. 29, 1211–1212.

Riddle, O., Bates, R.W., Lahr, E.L., 1935. Prolactin induces broodiness in fowl. Am. J. Physiol. 111, 352–360.

Rilling, J.K., 2013. The neural and hormonal bases of human parental care. Neuropsychologia 51, 731–747.

Riou, S., Chastel, O., Lacroix, A., Hamer, K.C., 2010. Stress and parental care: prolactin responses to acute stress throughout the breeding cycle in a long-lived bird. Gen. Comp. Endocrinol. 168, 8–13.

Roberts, T.F., Cookson, K.K., Heaton, K.J., Hall, W.S., Brauth, S.E., 2001. Distribution of tyrosine hydroxylase-containing neurons and fibers in the brain of the budgerigar (*Melopsittacus undulatus*): general patterns and labeling in vocal control nuclei. J. Comp. Neurol. 429, 436–454.

Robinzon, B., Koike, T.I., Neldon, H.L., Kinzler, S.L., 1988. Distribution of immunoreactive mesotocin and vasotocin in the brain and pituitary of chickens. Peptides 9, 829–833.

Romanov, M.N., Talbot, R.T., Wilson, P.W., Sharp, P.J., 2002. Genetic control of incubation behavior in the domestic hen. Poult. Sci. 81, 928–931.

Rosenblatt, J.S., 2003. Outline of the evolution of behavioral and non-behavioral patterns of parental care among the vertebrates: critical characteristics of mammalian and avian parental behavior. Scand. J. Psychol. 44, 265–271.

Rozenboim, I., Silsby, J.L., Tabibzadeh, C., Pitts, G.R., Youngren, O.M., El Halawani, M.E., 1993a. Hypothalamic and posterior pituitary content of vasoactive intestinal peptide and gonadotrophin releasing hormones I and II in the turkey hen. Biol. Reprod. 49, 622–626.

Rozenboim, I., Tabibzadeh, C., Silsby, J.L., El Halawani, M.E., 1993b. Effect of ovine prolactin administration on hypothalamic vasoactive intestinal peptide (VIP), gonadotropin releasing hormone I and II content, and anterior pituitary VIP receptors in laying turkey hens. Biol. Reprod. 48, 1246–1250.

Ruscio, M.G., Adkins-Regan, E., 2004. Immediate early gene expression associated with induction of brooding behavior in Japanese quail. Horm. Behav. 46, 19–29.

Said, S.I., Mutt, V., 1970. Polypeptide with broad biological activity: isolation from small intestine. Science 169, 1217–1218.

Saldanha, C.J., Deviche, P.J., Silver, R., 1994. Increased VIP and decreased GnRH expression in photorefractory dark-eyed juncos (*Junco hyemalis*). Gen. Comp. Endocrinol. 93, 128–136.

Sartsoongnoen, N., Kosonsiriluk, S., Prakobsaeng, N., Songserm, T., Rozenboim, I., El Halawani, M.E., Chaiseha, Y., 2008. The dopaminergic system in the brain of the native Thai chicken, *Gallus domesticus*: localization and differential expression across the reproductive cycle. Gen. Comp. Endocrinol. 159, 107–115.

Sartsoongnoen, N., Prakobsaeng, N., Kosonsiriluk, S., Chaiyachet, O.A., Chokchaloemwong, D., EL Halawani, M.E., Chaiseha, Y., 2012. Distribution and variation in gonadotropin releasing hormone-I (GnRH-I) immunoreactive neurons in the brain of the native Thai chicken during the reproductive cycle. Acta Histochem. 114, 409–420.

Savory, C.J., 1979. Changes in food intake and body weight of bantam hens during breeding. Appl. Anim. Ethol. 5, 283–288.

Schnell, S.A., You, S.K., El Halawani, M.E., 1999. D1 and D2 dopamine receptor messenger ribonucleic acid in brain and pituitary during the reproductive cycle of the turkey hen. Biol. Reprod. 60, 1378–1383.

Schoech, S.J., Ketterson, E.D., Nolan Jr., V., Sharp, P.J., Buntin, J.D., 1998. The effect of exogenous testosterone on parental behavior, plasma prolactin, and prolactin binding sites in dark-eyed juncos. Horm. Behav. 34, 1–10.

Schoech, S.J., Mumme, R.L., Wingfield, J.C., 1996. Prolactin and helping behaviour in the cooperatively breeding Florida scrub-jay, *Aphelocoma c. coerulescens*. Anim. Behav. 52, 445–456.

Schradin, C., Anzenberger, G., 1999. Prolactin, the hormone of paternity. News Physiol. Sci. 14, 223–231.

Sharp, P.J., 2009. Broodiness and broody control. In: Hocking, P.M. (Ed.), Biology of Breeding Poultry. CAB International, Wallingford, UK, pp. 181–205.

Sharp, P.J., Blache, D., 2003. A neuroendocrine model for prolactin as the key mediator of seasonal breeding in birds under long- and short-day photoperiods. Can. J. Physiol. Pharmacol. 81, 350–358.

Sharp, P.J., Culbert, J., Wells, J.W., 1977. Variations in stored and plasma concentrations of androgens and luteinizing hormone during sexual development in the cockerel. J. Endocrinol. 74, 467–476.

Sharp, P.J., Dawson, A., Lea, R.W., 1998. Control of luteinizing hormone and prolactin secretion in birds. Comp. Biochem. Physiol. C, Pharmacol. Toxicol. Endocrinol. 119, 275–282.

Sharp, P.J., Macnamee, M.C., Sterling, R.J., Lea, R.W., Pedersen, H.C., 1988. Relationships between prolactin, LH and broody behavior in bantam hens. J. Endocrinol. 118, 279–286.

Sharp, P.J., Scanes, C.G., Williams, J.B., Harvey, S., Chadwick, A., 1979. Variations in concentrations of prolactin, luteinizing hormone, growth hormone and progesterone in the plasma of broody bantams (*Gallus domesticus*). J. Endocrinol. 80, 51–57.

Sharp, P.J., Sterling, R.J., Talbot, R.T., Huskisson, N.S., 1989. The role of hypothalamic vasoactive intestinal polypeptide in the maintenance of prolactin secretion in incubating bantam hens: observations using passive immunization, radioimmunoassay and immunohistochemistry. J. Endocrinol. 122, 5–13.

Sharp, P.J., Talbot, R.T., Main, G.M., Dunn, I.C., Fraser, H.M., Huskisson, N.S., 1990. Physiological roles of chicken LHRH-I and -II in the control of gonadotrophin release in the domestic chicken. J. Endocrinol. 124, 291–299.

Sherry, D.F., 1981. Parental care and the development of thermoregulation in red jungle fowl. Behaviour 76, 250–279.

Silver, R., 1984. Prolactin and parenting in the pigeon family. J. Exp. Zool. 232, 617–625.

Silver, R., Andrews, H., Ball, G.F., 1985. Parental care in an ecological perspective: a quantitative analysis of avian subfamilies. Am. Zool. 25, 823–840.

Silver, R., Witkovsky, P., Horvath, P., Alones, V., Barnstable, C.J., Lehman, M.N., 1988. Coexpression of opsin- and VIP-like immunoreactivity in CSF-contacting neurons of the avian brain. Cell Tissue Res. 253, 189–198.

Silverin, B., Goldsmith, A.R., 1983. Reproductive endocrinology of free living pied flycatchers (*Ficedula hypoleuca*): prolactin and FSH secretion in relation to incubation and clutch size. J. Zool. 200, 119–130.

Silverin, B., Goldsmith, A.R., 1984. The effects of modifying incubation on prolactin secretion in free-living pied flycatchers. Gen. Comp. Endocrinol. 55, 239–244.

Sinha, Y.N., 1995. Structural variants of prolactin: occurrence and physiological significance. Endocr. Rev. 16, 354–369.

Skutch, A.F., 1962. The constancy of incubation. Wilson Bull. 74, 115–151.

Slawski, B.A., Buntin, J.D., 1995. Preoptic area lesions disrupt prolactin-induced parental feeding behavior in ring doves. Horm. Behav. 29, 248–266.

Sockman, K.W., Schwabl, H., Sharp, P.J., 2000. The role of prolactin in the regulation of clutch size and onset of incubation behavior in the American kestrel. Horm. Behav. 38, 168–176.

Stearns, S.C., 1992. The Evolution of Life Histories. Oxford University Press, Oxford.

Swain, J.E., Lorberbaum, J.P., Kose, S., Strathearn, L., 2007. Brain basis of early parent-infant interactions: psychology, physiology, and in vivo functional neuroimaging studies. J. Child Psychol. Psychiatry 48, 262–287.

Tabibzadeh, C., Rozenboim, I., Silsby, J.L., Pitts, G.R., Foster, D.N., El Halawani, M.E., 1995. Modulation of ovarian cytochrome P450 17 alpha-hydroxylase and cytochrome aromatase messenger ribonucleic acid by prolactin in the domestic turkey. Biol. Reprod. 52, 600–608.

Talbot, R.T., Hanks, M.C., Sterling, R.J., Sang, H.M., Sharp, P.J., 1991. Pituitary prolactin messenger ribonucleic acid levels in incubating and laying hens: effects of manipulating plasma levels of vasoactive intestinal polypeptide. Endocrinology 129, 496–502.

Teruyama, R., Beck, M.M., 2001. Double immunocytochemistry of vasoactive intestinal peptide and cGnRH-I in male quail: photoperiodic effects. Cell Tissue Res. 303, 403–414.

Thayananuphat, A., Kang, S.W., Bakken, T., Millam, J.R., El Halawani, M.E., 2007. Rhythmic dependent light induction of gonadotrophin-releasing hormone-I expression and activation of dopaminergic neurones within the premammillary nucleus of the turkey hypothalamus. J. Neuroendocrinol. 19, 399–406.

Thayananuphat, A., Youngren, O.M., Kang, S.W., Bakken, T., Kosonsiriluk, S., Chaiseha, Y., El Halawani, M.E., 2011. Dopamine and mesotocin neurotransmission during the transition from incubation to brooding in the turkey. Horm. Behav. 60, 327–335.

Tixier-Vidal, A., Bayle, J.D., Assenmacher, I., 1966. Ultrastructural cytologic study of the pituitary of the pigeon after ectopic autograft. Absence of stimulation of the cells by prolactin. C. R. Acad. Sci. Hebd. Seances Acad. Sci. D 262, 675–678.

Tong, Z., Pitts, G.R., Foster, D.N., El Halawani, M.E., 1997. Transcriptional and post-transcriptional regulation of prolactin during the turkey reproductive cycle. J. Mol. Endocrinol. 18, 223–231.

Ubuka, T., Bentley, G.E., 2009. Identification, localization, and regulation of passerine GnRH-I messenger RNA. J. Endocrinol. 201, 81–87.

Ugrumov, M., Hisano, S., Daikoku, S., 1989. Topographic relation between tyrosine hydroxylase- and luteinizing hormone-releasing hormone-immunoreactive fibers in the median eminence. Neurosci. Lett. 102, 159–164.

Vleck, C.M., 1998. Hormonal control of incubation/brooding behavior: lessons from wild birds. In: Proceeding of the WPSA 10th European Poultry Conference, Jerusalem, Israel, pp. 163–169.

Vleck, C.M., Priedkalns, J., 1985. Reproduction in zebra finches: hormone levels and effects of dehydration. Condor 87, 37–46.

Vleck, C.M., Vleck, D., 2011. Hormones and regulation of parental behavior in birds. In: Norris, D.O., Lopez, K.H. (Eds.), Hormones and Reproduction of Vertebrates. Birds, vol. 4. Academic Press, San Diego, pp. 181–203.

Wang, Q., Buntin, J.D., 1999. The roles of stimuli from young, previous breeding experience, and prolactin in regulating parental behavior in ring doves (*Streptopelia risoria*). Horm. Behav. 35, 241–253.

Wauters, A.M., Richard-Yris, M.A., 2001. Experience modulates emission of food calls in broody hens. C. R. Acad. Sci. III 324, 1021–1027.

Wauters, A.M., Richard-Yris, M.A., 2002. Mutual influence of the maternal hen's food calling and feeding behavior on the behavior of her chicks. Dev. Psychobiol. 41, 25–36.

Wentworth, B.C., Proudman, J.A., Opel, H., Wineland, M.J., Zimmerman, N.G., Lapp, A., 1983. Endocrine changes in the incubating and brooding turkey hen. Biol. Reprod. 29, 87–92.

Wineland, M.J., Wentworth, B.C., 1975. Peripheral serum levels of 17beta estradiol in growing turkey hens. Poult. Sci. 54, 381–387.

Winfree, R., 1999. Cuckoos, cowbirds and the persistence of brood parasitism. Trends Ecol. Evol. 14, 338–343.

Wingfield, J.C., Farner, D.S., 1978. The annual cycle in plasma irLH and steroid hormones in feral populations of the white-crowned sparrow, *Zonotrichia leucophrys gambelii*. Biol. Reprod. 19, 1046–1056.

Wingfield, J.C., Goldsmith, A.R., 1990. Plasma levels of prolactin and gonadal steroids in relation to multiple brooding and renesting in free-living populations of the song sparrow, *Melospiza melodia*. Horm. Behav. 24, 89–103.

Wong, E.A., Ferrin, N.H., Silsby, J.L., El Halawani, M.E., 1991. Cloning of a turkey prolactin cDNA: expression of prolactin mRNA throughout the reproductive cycle of the domestic turkey (*Meleagris gallopavo*). Gen. Comp. Endocrinol. 83, 18–26.

Wong, E.A., Silsby, J.L., Ishii, S., El Halawani, M.E., 1992. Pituitary luteinizing hormone and prolactin messenger ribonucleic acid levels are inversely related in laying and incubating turkey hens. Biol. Reprod. 47, 598–602.

Youngren, O.M., Chaiseha, Y., Phillips, R.E., El Halawani, M.E., 1996a. Vasoactive intestinal peptide concentrations in turkey hypophysial portal blood differ across the reproductive cycle. Gen. Comp. Endocrinol. 103, 323–330.

Youngren, O.M., El Halawani, M.E., Silsby, J.L., Phillips, R.E., 1991. Intracranial prolactin perfusion induces incubation behavior in turkey hens. Biol. Reprod. 44, 425–431.

Youngren, O.M., Pitts, G.R., Phillips, R.E., El Halawani, M.E., 1995. The stimulatory and inhibitory effects of dopamine on prolactin secretion in the turkey. Gen. Comp. Endocrinol. 98, 111–117.

Youngren, O.M., Pitts, G.R., Phillips, R.E., El Halawani, M.E., 1996b. Dopaminergic control of prolactin secretion in the turkey. Gen. Comp. Endocrinol. 104, 225–230.

Zadworny, D., Shimada, K., Ishida, H., Sato, K., 1989. Gonadotropin-stimulated estradiol production in small ovarian follicles of the hen is suppressed by physiological concentrations of prolactin in vitro. Gen. Comp. Endocrinol. 74, 468–473.

Zadworny, D., Shimada, K., Ishida, H., Sumi, C., Sato, K., 1988. Changes in plasma levels of prolactin and estradiol, nutrient intake, and time spent nesting during the incubation phase of broodiness in the Chabo hen (*Japanese bantam*). Gen. Comp. Endocrinol. 71, 406–412.

Zadworny, D., Walton, J.S., Etches, R.J., 1985. The relationship between plasma concentrations of prolactin and consumption of feed and water during the reproductive cycle of the domestic turkey. Poult. Sci. 64, 401–410.

Ziegler, T.E., 2000. Hormones associated with non-maternal infant care: a review of mammalian and avian studies. Folia. Primatol. (Basel) 71, 6–21.

Chapter 32

The Physiology of the Avian Embryo

Casey A. Mueller, Warren W. Burggren and Hiroshi Tazawa
Developmental and Integrative Biology, Department of Biological Science, University of North Texas, Denton, TX, USA

ABBREVIATIONS

A_p effective pore area
CAM chorioallantoic membrane
CO cardiac output
CO_2 carbon dioxide
dCO_2 carbon dioxide diffusion coefficient
dH_2O water vapor diffusion coefficient
dO_2 oxygen diffusion coefficient
DO_2 oxygen diffusing capacity
EP external pipping
$F\bar{c}_{OX}$ mean corpuscular oxygenation velocity
GCO_2 carbon dioxide conductance
GH_2O water vapor conductance
GO_2 oxygen conductance
Hb hemoglobin
Hct hematocrit
HR heart rate
I incubation period
IHR instantaneous heart rate
IP internal pipping
IRR instantaneous respiratory rate
L shell thickness
$\dot{M}CO_2$ carbon dioxide elimination rate
MHR mean heart rate
$\dot{M}H_2O$ rate of water loss
$\dot{M}O_2$ oxygen consumption rate
O_2 oxygen
P_aCO_2 arterialized blood carbon dioxide partial pressure
P_aO_2 arterialized blood oxygen partial pressure
$PACO_2$ airspace carbon dioxide partial pressure
PAO_2 airspace oxygen partial pressure
P_B atmospheric pressure
$P\bar{c}_{O_2}$ mean capillary oxygen partial pressure
PH_2O water vapor pressure
$PICO_2$ effective environmental carbon dioxide partial pressure
PIO_2 effective environmental oxygen partial pressure
P_a arterial blood pressure
P_{sys} systolic blood pressure
$\dot{Q}_a$ allantoic blood flow
Q_{10} temperature coefficient
R gas constant
T_a ambient temperature
T_{egg} egg temperature
t_c contact time of erythrocytes in chorioallantoic capillary with O_2
V_c capillary volume
$[HCO_3^-]$ bicarbonate concentration
$[La^-]$ lactate concentration

32.1 INTRODUCTION

The freshly laid avian egg contains most of the materials needed for embryonic growth and development, but lacks the oxygen and heat needed for successful development. Microscopic pores in the eggshell allow O_2 to diffuse into the egg from the environment and water vapor and CO_2 produced by the embryo to diffuse out. The adult bird has a key role in incubation, providing not only the heat necessary for embryonic development but also controlling the microclimate of the egg. In the poultry industry and for research purposes, the adult bird can be conveniently replaced by an incubator. The majority of research on avian incubation is undertaken using artificially incubated chicken (*Gallus gallus domesticus*) eggs. Thus, in this chapter, the chicken embryo is used to elucidate the development of physiological function during avian incubation, supplemented by additional species when data are available. Developmental physiology of the gas exchange, acid–base, cardiovascular, osmoregulatory and thermoregulatory systems are examined. The optimal conditions for artificial incubation are outlined and embryonic responses to incubation extremes are described.

32.2 THE FRESHLY LAID EGG

The mass of the freshly laid bird egg ranges from ~0.8 g in the bee hummingbird (*Mellisuga helenae*) to ~2 kg in the ostrich (*Struthio camelus*). The egg is composed of the eggshell and outer and inner shell membranes that encompass the albumen, which serves as a source of water and protein; and yolk, a source of necessary nutrients. The composition of the freshly laid egg is related to the maturity of the hatchling, which differs considerably between species.

Sturkie's Avian Physiology.

Hatchlings are divided into four major categories based on criteria such as mobility, amount of down, ability to feed, and locomotion. Precocial species are the most mature, and can walk, swim, or dive soon after hatching, while altricial species are the least mature and hatch naked, with eyes closed and are incapable of coordinated locomotion. Two intermediate categories—semi-precocial and semi-altricial—describe species that are within these extremes. The amount of yolk in the freshly laid egg is much greater in precocial species than in altricial species (Sotherland and Rahn, 1987). The yolk is only 16% of egg mass in the altricial red-footed booby (*Sula sula*), while in the highly precocial kiwi (*Apteryx australis*) it is 69% of egg mass. As most of the energy contained in the egg is in the lipid fraction, most of which is in the yolk, the energy content of precocial eggs is higher than that of altricial eggs. Furthermore, most of the water in the egg is in the albumen, and as the yolk/albumen ratio is lower in altricial than in precocial eggs, altricial eggs have relatively higher water compared with precocial eggs (Sotherland and Rahn, 1987).

32.3 INCUBATION

32.3.1 Incubation Period

Incubation encompasses the prenatal period, prior to "internal pipping" (IP), when the embryo penetrates its beak through the chorioallantoic and inner shell membranes into the air cell; and the perinatal or paranatal period from IP to when the embryo fractures the shell ("external pipping," EP) and hatches. The incubation period (*I*, days) increases with egg size and can be represented as follows:

$$I = 12 \cdot (\text{Egg mass})^{0.22} \tag{32.1}$$

where egg mass is in g (Rahn and Ar, 1974). In addition, as a first approximation, the product of incubation period and O_2 consumption ($\dot{M}O_2$) at the stage where $\dot{M}O_2$ remains unchanged (i.e., plateau) is proportional to egg mass (Rahn et al., 1974):

$$I \cdot \dot{M}O_2 = c \cdot (\text{Egg mass}) \tag{32.2}$$

where *c* is a constant. Thus, for a given egg mass an egg that consumes less O_2 at the plateau stage needs a longer incubation period. Furthermore, because the $\dot{M}O_2$ at the plateau stage is matched to the shell gas conductance, for a given egg mass an egg with lower shell conductance requires a longer incubation period.

Apart from these general trends, eggs that are left unattended by the incubating parent for a period of time will exhibit cooling, and this results in slower embryonic growth and a longer incubation period. Some birds, notably tropical seabirds and petrels (Procellariiformes), have much longer incubation periods than that suggested by the general relationship (Whittow, 1980), even though they are incubated continuously.

32.3.2 Egg Water Content and Shell Conductance

As the embryo grows within the egg, water is lost and both the yolk and albumen diminish in mass. However, metabolic water is produced when fat in the yolk is oxidized. Part of the water is inevitably lost by diffusion through the pores in the shell to the microclimate of the egg. The rate of water loss ($\dot{M}H_2O$, mg/day) is determined by two factors: the water vapor conductance of the shell and shell membranes (GH_2O, mg/day/kPa) and the difference in water vapor pressure between the contents of the egg and the environment (ΔPH_2O, kPa) (Rahn and Ar, 1974):

$$\dot{M}H_2O = GH_2O \cdot \Delta PH_2O \tag{32.3}$$

GH_2O is a function of multiple factors, including shell geometry (effective pore area, A_p, thickness of the shell combined with shell membranes, *L*); water vapor diffusion coefficient (dH_2O); and the inverse product of the gas constant (*R*) and absolute temperature (*T*):

$$GH_2O = [(A_p/L) \cdot dH_2O]/RT \tag{32.4}$$

Both pore area and shell thickness increase in larger eggs (Ar et al., 1974). Although an increase in pore area increases GH_2O, an increase in shell thickness has the opposite effect, because it lengthens the diffusion pathway for water vapor. The water lost from the egg is replaced by air, enlarging the air cell at the blunt pole of the egg during development. Upon IP and EP, the rate of water loss from the egg increases, as water vapor can diffuse through the cracks (pip-hole) in the shell. The water loss from the egg over the entire incubation period amounts to ~18% of the mass of the freshly laid egg.

32.3.3 Heat Transfer

Many adult birds (typically but not always the female) play an important role in incubation of their eggs, by using their body and nesting material to alter the environment of the eggs. Most birds develop a seasonal bare patch of skin, the brood patch, on part of the thorax and abdomen. This directly contacts with the eggs, permitting a greater rate of heat transfer than if the patch were covered with plumage. Accompanying the loss of feathers is an increase in the size and number of blood vessels in the bare skin. The adult can adjust the rate of heat transfer to the egg by standing over or leaving the vicinity of the egg, but also by the closeness with which the bird applies its patch to the egg. The bird responds physiologically to variations in egg temperature (T_{egg}), increasing its metabolic heat production in response to cooling of the egg (Tøien et al., 1986; Rahn, 1991).

The amount of heat transferred to the eggs is directly proportional to the increase in heat production (Tøien et al., 1986). The efficiency of heat transfer to the eggs diminishes with decreasing ambient temperature and clutch size. Heat stored in the incubating bird while flying and foraging (i.e., while not incubating the eggs) can be transferred to the eggs on return to the nest (Biebach, 1986). If the egg is very cold, "cold vasodilation" occurs in the brood patch, increasing the patch blood flow and temperature and therefore heat transfer to the egg (Mitgard et al., 1985). The brood patch temperature varies from 34.9 °C in the bonin petrel (*Pterodroma hypoleuca*) to 42.4 °C in the dusky flycatcher (*Empidonax oberholseri*). The brood patch temperature is 1.1–5.5 °C higher than the T_{egg} in numerous species (Rahn, 1991).

At the beginning of incubation, heat is transferred through the egg by conduction, and the surface of the egg on the opposite side to the brood patch may be 4 °C or more below the brood patch temperature (Rahn, 1991). As incubation proceeds, this temperature difference diminishes as the embryo's developing circulation assists in the distribution of heat and its increasing metabolism provides additional heat (Turner, 1987; Rahn, 1991). This effect of blood flow on heat flow is more important in large eggs than in small ones (Tazawa et al., 1988a). The main barrier to heat loss from the egg is a thin layer of air immediately adjacent to the shell, the boundary layer (Sotherland et al., 1987). If the egg is in a nest, the nest itself imposes an additional resistance to heat loss.

As the embryo grows, incubation likely becomes more a matter of regulating the balance between heat gain from the mother and heat loss through radiation from the egg surfaces, because the near-term embryo is producing considerable heat itself. The rate of heat loss from the egg is going to depend on many factors, though one of the more important ones may be the surface area-to-volume ratio of the egg. Thus, larger eggs (e.g., emu, ostrich) may have lower capacity for heat loss because of their smaller surface area-to-volume ratio. However, larger embryos may also show conventional metabolic scaling, with lower per-gram body mass rates of temperature production than smaller embryos. A study that integrates egg size, embryo metabolic rate, and incubation behavior changes during embryonic development is highly warranted.

32.3.4 Energy Use

As chicken embryos grow, their mass increases geometrically until growth rate slows during the last stages of development (Romanoff, 1967; Van Mierop and Bertuch, 1967; Tazawa et al., 1971a; Lemez, 1972; Clark et al., 1986; Haque et al., 1996). From the mean values of the above references, the geometrical increase in embryo wet body mass to day 16 of incubation is expressed by:

$$\text{Body mass (mg)} = 0.24 \cdot I^4 \qquad (32.5)$$

Of the energy measured as oxygen uptake $(\dot{M}O_2)$, the majority is utilized by the embryo to synthesize new tissues for growth and to meet the physiological demands for maintenance during development. In addition, toward the end of incubation, the embryo uses energy for active control of body temperature and hatching. Prior to this, it is assumed that prenatal $\dot{M}O_2$ is used for maintenance, which is proportional to body mass, and for growth, which is proportional to growth rate ($GR = \Delta Body\ mass/time$) as follows (Vleck et al., 1980; Mortola and Cooney, 2008):

$$\dot{M}O_2 = a \cdot (\text{Body mass}) + b \cdot \text{GR} \qquad (32.6)$$

where coefficients a and b express the daily average cost of maintaining 1 g of tissue (mL O_2/g/day) and the cost of growing it (mL O_2/g), respectively. During days 9 to 18 of the ~21 day incubation period of chickens, and excluding early development when $\dot{M}O_2$ of small embryos does not reflect total egg $\dot{M}O_2$ because of uptake by extra-embryonic tissues, a and b are estimated as ~15 mL O_2/g/day and ~41 mL O_2/g, respectively (Mortola and Cooney, 2008). Therefore, the cost of growing 1 g of tissue averages ~3 times the cost of maintaining it. In addition, cooling due to low temperature incubation (35 °C) only decreases GR with a decrease in the cost of growth, but hypoxic incubation decreases not only GR but also the cost of maintenance. That is, cold incubation decreases GR without altering the partitioning of energy expenditure while hypoxia decreases GR and alters the partitioning of the energy.

Daily energy use increases as the embryo grows and is reflected in an increase in $\dot{M}O_2$. However, precocial and altricial birds show a different developmental pattern in embryonic $\dot{M}O_2$ (e.g., Vleck and Vleck, 1987). In altricials, $\dot{M}O_2$ increases at an increasing rate throughout incubation, lacking a plateau. In precocials, the rate of increase in $\dot{M}O_2$ plateaus before the egg is pipped and then increases again after pipping. For a given egg size and incubation period, altricials use less energy, and so have a lower incubation energy cost to hatching than precocials (Vleck et al., 1980). The high cost of incubation in precocials is due to their rapid growth early in development, which results in higher maintenance costs—that is, there is more tissue to maintain for a longer time (Vleck et al., 1980).

Based on the assumption that both precocials and altricials have an identical basic pattern of increase in $\dot{M}O_2$, including a plateau phase attributable to the shift of the gas exchange from the chorioallantoic membrane (CAM) to the lungs (Rahn and Ar, 1974), Prinzinger et al. (1995) defines the plateau phase as a clear interruption of the continuous exponential increase in $\dot{M}O_2$. This demonstrates no fundamental difference between both modes with respect to the presence of a plateau (Figure 32.1) (Prinzinger et al., 1995; Prinzinger and Dietz, 1995). They suggest that in small altricials the plateau lasts only a few hours and can be overlooked when $\dot{M}O_2$ measurement is not continuous or when many individual measurements are averaged. The extremely precocial mound-building birds (Megapodiidae) are one exception having no plateau phase as they lack the CAM-pulmonary transition within the egg (Vleck et al., 1984).

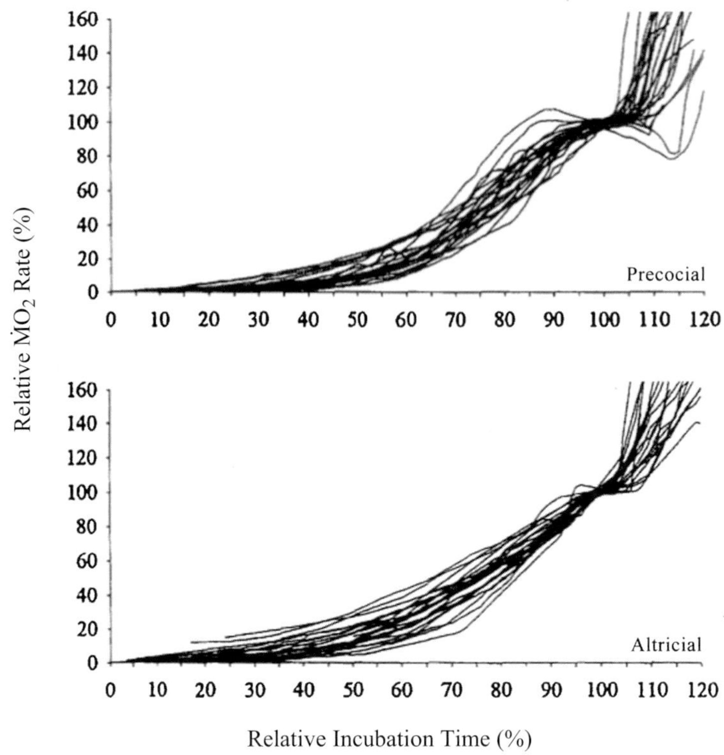

FIGURE 32.1 Developmental patterns of relative oxygen uptake ($\dot{M}O_2$, %) in precocial and semiprecocial species (N=27) (top) and altricial and semi-altricial species (N=24) (bottom) plotted against relative incubation time (%). The O_2 uptake and incubation time at the middle of plateau stage are both set at 100%. *Modified from Prinzinger and Dietz (1995), with permission from Elsevier.*

32.4 DEVELOPMENT OF PHYSIOLOGICAL SYSTEMS

32.4.1 Gas Exchange

The egg with its hard shell does not enable embryonic ventilatory movements, and thus there is no convective gas exchange until the embryo's lungs begin to function. Early in incubation, prior to a formation of the heart and even initially after the forming heart beats, O_2 can be adequately supplied from the environment to the embryo through diffusion. $\dot{M}O_2$ is normally maintained without blood convection, even in embryos whose main vessel from the heart is ligated or hemoglobin (Hb) is rendered functionless due to carbon monoxide exposure (Burggren et al., 2000, 2004; Mortola et al., 2010; Burggren, 2013). When gas transport by diffusion alone becomes inadequate, blood convection begins to meet O_2 demand during development, and three different gas exchangers sequentially function in the egg; the area vasculosa, the chorioallantois, and the lungs. Figure 32.2 shows the growth rate of the functional surface area of the area vasculosa and the chorioallantois (Ackerman and Rahn, 1981) and the associated $\dot{M}O_2$ of the chicken embryo. The area vasculosa is a well-vascularized region of the yolk sac that fans out from the embryo and rapidly grows around the yolk during days 3 to 5 of incubation. The blood vessels of the yolk sac connect with the dorsal aorta of the embryo by day 2 and blood begins to circulate through the embryo and the area vasculosa. The fine reticulation of the vitelline circulatory system plays the role of the main gas exchanger until the chorioallantois makes contact with the inner shell membrane around day 6 (Ackerman and Rahn, 1981). Subsequently, respiratory function transitions from the area vasculosa to the chorioallantois.

Beginning on day 5 of incubation, the mesenchyme covering the fundus of the allantoic sac comes into contact with the mesenchyme lining the chorion. The two membranes begin to fuse, and the growing allantoic sac flattens out beneath the chorion, which lies close to the eggshell. The outer limb of the flattened allantois, composed of the cohesive chorion and allantois, is the chorioallantois. The chorioallantois grows rapidly (Figure 32.2). It reaches almost the same size as the embryo by the time it makes contact with the shell membranes on day 6. By day 12, the chorioallantois extends to envelop the contents of the whole egg, lining the entire surface of the inner shell membrane.

The outer surface of the chorioallantois, the chorioallantoic membrane (CAM), is well vascularized (Wangensteen et al., 1970/71; Tazawa and Ono, 1974; Wangensteen and Weibel, 1982). Early in incubation, the capillaries lie on the mesenchymal surface of the CAM. They begin to migrate through the ectoderm on day 10 and lie on its thin layer late in incubation. In addition, relocation of the nuclei of the

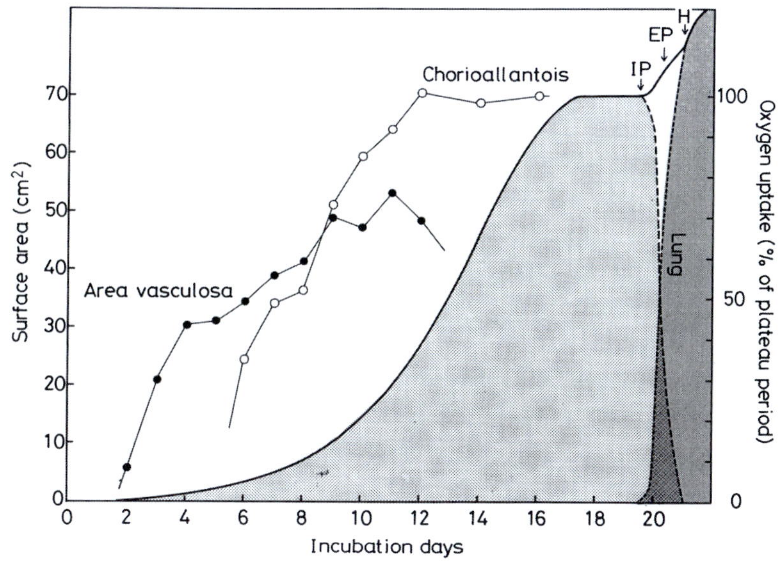

FIGURE 32.2 Daily changes in functional surface area of the area vasculosa and the chorioallantois (left ordinate), and developmental pattern of oxygen uptake ($\dot{M}O_2$) during the prenatal period (until internal pipping, IP) and perinatal period (from IP to hatching, H) (right ordinate). External pipping (EP) occurs during the perinatal period. $\dot{M}O_2$ is drawn diagrammatically with the plateau set at 100%. The lightly shaded area indicates $\dot{M}O_2$ by diffusion through the area vasculosa/chorioallantois and the heavily shaded area $\dot{M}O_2$ by the lungs. *From Ackerman and Rahn (1981), with permission from Elsevier.*

CAM capillaries occurs (Mayer et al., 1995). The endothelial nuclei randomly distribute around the capillary lumen early in incubation and are located progressively on the portion of the capillaries away from the shell membrane after the chorioallantois envelops the whole contents of the egg. Together with capillary migration, the relocation of endothelial nuclei results in progressive thinning of the gas diffusion pathway between the interstices of the inner shell membrane (airspace) and the capillary blood.

The CAM functions as a gas exchanger during prenatal development (prior to IP) until it degenerates when the embryo pips it and the inner shell membrane with the beak. The increase in $\dot{M}O_2$ of the chicken embryo during the first ~16 days of incubation is enabled by the extension of the CAM (Figure 32.2). The gas exchange of embryos before IP takes place by diffusive transport across the porous shell, two shell membranes, and capillary blood of the CAM (Wangensteen and Rahn, 1970/71; Wangensteen et al., 1970/71, Wangensteen and Weibel, 1982). Then, the chick breaks the shell and breathes environmental air by lung ventilation (EP). As the near-term embryo begins ventilating the lungs, they replace the gas exchange function of the CAM.

When the CAM envelopes all contents of the egg, the gas exchange by diffusive transport is expressed by:

$$\dot{M}O_2 = GO_2 \cdot (PIO_2 - PAO_2) \quad (32.7)$$

$$= DO_2 \cdot (PAO_2 - P\bar{c}_{O_2}) \quad (32.8)$$

where $\dot{M}O_2$ is oxygen uptake (mL/min), GO_2 is oxygen conductance of the shell and shell membrane (mL O_2/min/kPa), DO_2 is the diffusing capacity of the CAM and capillary blood (mL O_2/min/kPa), PIO_2 is the effective environmental oxygen partial pressure (kPa); PAO_2 is the airspace oxygen partial pressure (kPa), and $P\bar{c}_{O_2}$ is mean capillary oxygen partial pressure (kPa).

GO_2 is related to GH_2O by the diffusion coefficient ratio (dO_2/dH_2O =0.23/0.27). GO_2 (in mL/min/kPa) is derived from GH_2O (in mg/min/kPa) by multiplying by a factor of 1.06. Because GO_2 in air is nearly constant during the prenatal incubation, PAO_2 decreases as the embryo grows and consumes more O_2 (Table 32.1). The increased difference of O_2 partial pressure between the environmental air and airspace is the driving force that meets the increased O_2 demand of the developing embryo. Concurrently with the decrease in PAO_2, $P\bar{c}_{O_2}$ also decreases. The O_2 flux from the airspace to the Hb of capillary blood (inner diffusion barrier) is facilitated by an increase in DO_2, which is expressed by:

$$DO_2 = 60 \cdot V_c \cdot F\bar{c}_{OX} \cdot Hct \quad (32.9)$$

$$= \dot{Q}_a \cdot t_c \cdot F\bar{c}_{OX} \cdot Hct \quad (32.10)$$

where V_c is capillary volume of the CAM (μL), $F\bar{c}_{OX}$ is mean corpuscular oxygenation velocity during the contact time per s per kPa, Hct is hematocrit, $\dot{Q}_a$ is blood flow through the CAM (mL/min), and t_c is the contact time of erythrocytes with O_2 when they pass through the CAM capillaries (s) (Tazawa et al., 1976b; Tazawa and Mochizuki, 1976) (Table 32.1).

DO_2 is increased by an increase in V_c, which depends on $\dot{Q}_a$ and t_c; $V_c = (\dot{Q}_a/60) \times t_c$ (Table 32.1). While $\dot{Q}_a$ increases about five-fold from days 10 to 18, the t_c halves,

TABLE 32.1 Gas Exchange Variables Used to Determine the Diffusing Capacity of the Inner Diffusion Barrier (Chorioallantoic Membrane and Capillary Blood) in Developing Chicken Embryos

Age (days)	10	12	14	16	18
Body mass	2.4	5.2	9.3	16.0	22.9
PAO_2	18.5	17.2	16.3	14.7	14.4
$PACO_2$	1.5	2.1	3.3	3.8	4.2
P_aO_2	10.9	10.7	9.9	8.6	7.8
pH_a	7.64	7.62	7.54	7.50	7.48
S_aO_2	87.2	88.2	87.0	88.0	84.5
PvO_2	4.5	3.7	3.4	2.5	2.4
$PvCO_2$	2.5	3.3	4.9	5.4	5.5
S_vO_2	10.4	11.2	17.7	17.3	25.2
O_2 capacity	9.4	10.1	10.5	11.4	12.3
$\dot{M}O_2$	0.08	0.15	0.25	0.36	0.42
$\dot{Q}_a$	1.11	1.93	3.42	4.48	5.78
t_c	0.87	0.74	0.57	0.49	0.36
V_c	16.1	23.8	32.5	36.6	34.7
F_{COX}	41.9	51.0	60.8	62.3	65.3
Hct	20.5	22.6	27.6	32.8	36.5
DO_2	0.83	1.65	3.27	4.49	4.96

Body mass (g); from Tazawa and Mochizuki (1976); PAO_2 (airspace PO_2, kPa), $PACO_2$ (airspace PCO_2, kPa), PvO_2 (mixed venous blood PO_2, kPa), $PvCO_2$ (mixed venous blood PCO_2, kPa); calculated from data in Tazawa (1973) and Wangensteen and Rahn (1970/71). Note: Unit of partial pressure is converted from 'mmHg' in original sources to 'kPa' in current text; P_aO_2 (arterialized PO_2 in the allantoic vein, kPa); from Tazawa (1973); pH_a (arterialized blood pH corresponding to the allantoic vein); from Tazawa et al. (1971a); S_aO_2 (oxygen saturation of arterialized blood, %); calculated by substituting P_aO_2 and pH_a into the modified Hill's equation of the O_2 dissociation curve (Tazawa et al., 1976a); S_vO_2 (oxygen saturation of mixed venous blood, %); determined by micro-photometer (Tazawa and Mochizuki, 1976); O_2 capacity (vol%); determined from Tazawa (1971) and Tazawa and Mochizuki (1977); $\dot{M}O_2$ (oxygen uptake, mL/min); from Tazawa (1973); $\dot{Q}_a$ (allantoic blood flow, mL/min); calculated by $\dot{M}O_2/[O_2 \text{ capacity } (S_aO_2-S_vO_2)]$; t_c (contact time, sec); determined by micro-photometer (Tazawa and Mochizuki, 1976); V_c (capillary volume, 10^{-3} mL), calculated by $\dot{Q}_a/60 \times t_c$; F_{COX} (oxygenation velocity factor per s per kPa); from Tazawa et al. (1976b); Hct (hematocrit, %); from Tazawa et al. (1971a); DO_2 (diffusing capacity of inner diffusion barrier, mL/min/kPa); calculated by $60 \cdot V_c \cdot F_{COX} \cdot Hct$ or $\dot{Q}_a \cdot t_c \cdot F_{COX} \cdot Hct$.

probably because of shortening of the blood circulation time, as cardiac output increases more rapidly than total blood volume. Consequently, V_c increases even after the CAM spreads over the whole surface of the inner shell membrane on day 12 and reaches a maximum volume on days 14 and 15. Nevertheless, DO_2 increases further after day 14. This is largely due to the increase in $\dot{Q}_a$ and Hct, as both variables increase after the CAM spreads over the inner shell membrane. Towards the end of prenatal development, the increase in DO_2 slows down and $\dot{M}O_2$ reaches a plateau. This suggests an important contribution of the inner diffusion barrier to gas exchange, which contributes, along with the outer diffusion barrier (GO_2), to the plateau status of $\dot{M}O_2$. This results in the developmental pattern of $\dot{M}O_2$ paralleling the daily changes in DO_2 (Table 32.1).

In chicken eggs, the variability of shell GO_2 is large, higher than that of egg mass. Mass-specific $\dot{M}O_2$, measured on days 16 to 19 of incubation, is maximal at medium GO_2, decreasing at both lower and higher GO_2 (Visschedijk et al., 1985). The maximum $\dot{M}O_2$ at medium GO_2 values is considered to be optimal for embryo development, and the decrease in $\dot{M}O_2$ at both higher and lower GO_2 is a sign of compromised development. When GO_2 is decreased by partially covering the shell with impermeable material or increased by partially removing the shell over the air cell at the beginning of incubation, $\dot{M}O_2$ of day 16 embryos increases hyperbolically with increasing GO_2, reaching a maximum at the control GO_2 of intact eggs and decreasing with further increase in GO_2 (Okuda and Tazawa, 1988). When part of the shell is covered and the other is exposed to hyperoxia, $\dot{M}O_2$ and growth rate do not change, indicating uniform GO_2, and therefore uniform chorioallantoic perfusion is not required to maintain $\dot{M}O_2$ (Wagner-Amos and Seymour, 2002).

Because the oxygen diffusion coefficient (dO_2) affects shell conductance, GO_2 can be changed by replacing N_2 in air with an inert background gas whose density is different from that of N_2 (e.g., He or SF_6), thus changing dO_2 (Erasmus and Rahn, 1976; Ar et al., 1980; Tazawa, 1981a; Tazawa et al., 1981). Accordingly, the gas exchange of the egg can be manipulated by changing GO_2 with He or SF_6. The dO_2 is also inversely related to atmospheric pressure (P_B), and thus gas exchange of the egg is increased at high altitude because the shell GO_2 increases inversely with P_B. The reduction of GO_2 in eggs laid by birds incubating at high altitude occurs as a natural adaptation to altitude (Rahn et al., 1977).

As incubation proceeds, diffusive gas exchange governed by GO_2 and DO_2 is gradually replaced by convective gas exchange through the lungs during the perinatal period, beginning at IP (Figure 32.2). Breathing movements of perinatal embryos can be recorded as pressure changes using an optical system or pressure transducer (Romijn, 1948; Vince and Salter, 1967; Dawes, 1976). Conveniently, prenatal cardiac rhythms at the onset of IP and EP and the subsequent development of respiratory rhythms until hatching can be demonstrated by continuous measurements of cardiogenic, ventilatory and hatching activities in chickens by means of a condenser-microphone measuring

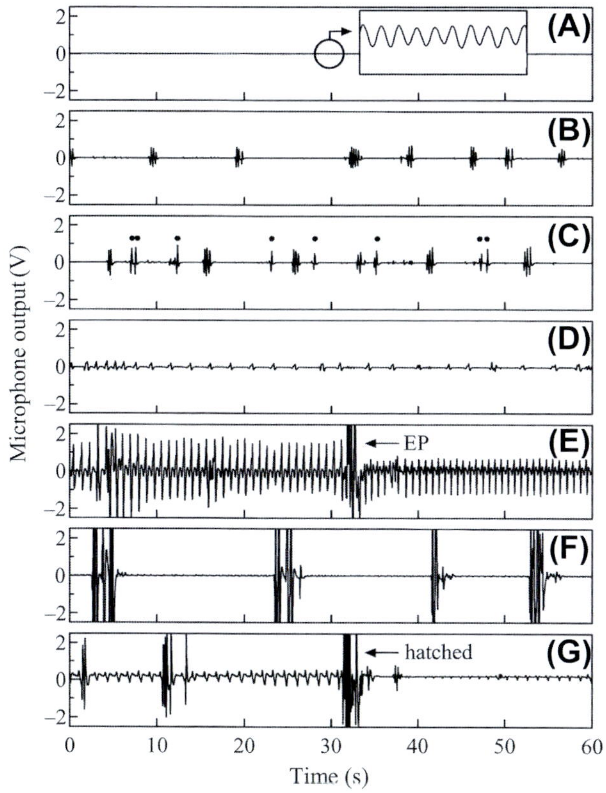

FIGURE 32.3 The cardiogenic and breathing activities of a chicken embryo from the prenatal through perinatal period to hatching, recorded by a condenser microphone attached hermetically on the shell. A 60 s recording is shown in each panel. (A) Cardiogenic signal (acoustocardiogram, ACG) during the final stages of the prenatal period; ACG is evident in the enlarged inset. (B) Beak-clapping signals just prior to IP (C) Onset of breathing signals occurring as single, irregular deflections (approximately once per 5 s) among beak clapping after IP (acoustorespirogram, ARG). (D) Rhythms become fast and somewhat irregular. (E) Breathing signals are more regular before and after EP. The onset of EP is evident by a sudden decrease in magnitude of ARG. (F) Approximately 1 h prior to hatching, large deflections due to hatching activities (e.g. climax) begin to occur and the breathing becomes intermittent. (G) The hatching activities continue before breathing regulates and finally the embryo hatches. Breathing signals are recorded after hatching, because the microphone is located close to the beak of hatchling. *From Chiba et al. (2002), with permission from Elsevier.*

system (Figure 32.3) (Chiba et al., 2002). Using the breathing patterns of chickens from the onset of IP through EP to hatching, the length of the IP, EP, and whole perinatal periods are determined as 14.1 ± 2.0 (SEM) h, 13.6 ± 1.3 h, and 27.6 ± 1.7 h long, respectively. The duration of climax (hatching) takes 48 ± 6 min. The instantaneous respiratory rate (IRR in breaths per min) is approximately 10 to 15 breaths/min at the beginning of IP; IRR baseline increases up to 80–100 breaths/min with large fluctuations of IRR of 60 to 180 breaths/min after the onset of EP.

Pulmonary ventilation and gas exchange during the perinatal period has been quantified using a barometric plethysmograph or pneumotachograph (Pettit and Whittow, 1982a; Menna and Mortola, 2003; Sbong and Dzialowski, 2007; Szdzuy and Mortola, 2007) and by measurement of gases in air cell and allantoic blood (Pettit and Whittow, 1982b; Tazawa et al., 1983b). The embryonic development of pulmonary ventilation and its regulation has been reviewed by Mortola (2009).

32.4.2 Acid–Base Regulation

While the embryo consumes O_2, it produces CO_2, which is partially dissolved and stored in the blood and body fluids, but mostly eliminated through the eggshell. As with O_2, CO_2 elimination ($\dot{M}CO_2$) depends upon the eggshell CO_2 conductance (GCO_2) and CO_2 partial pressure difference between air cell ($PACO_2$) and atmosphere ($PICO_2$):

$$\dot{M}CO_2 = GCO_2 \cdot (PACO_2 - PICO_2) \qquad (32.11)$$

GCO_2 is a function of eggshell geometry (A_p and L), CO_2 diffusion coefficient (dCO_2), and the inverse product of R and T. $PICO_2$ is very close to zero when the egg is in air.

As embryos develop and their body mass increases, they produce more CO_2, which accumulates in the egg. $PACO_2$ increases, thus increasing arterialized blood PCO_2, P_aCO_2. Consequently, arterialized blood pH is lowered (Figure 32.4) or stays relatively constant in the allantoic vein during the last half of prenatal development (Dawes and Simkiss, 1969; Girard, 1971; Erasmus et al., 1970/71; Boutilier et al., 1977; Everaert et al., 2011). The rate of decrease in pH during embryonic development slows, however, late in incubation. Although the plasma bicarbonate concentration ($[HCO_3^-]$) increases with development, the increase is more than would be expected from changes in pH and the buffer value (−16 mmol / L, Burggren et al., 2012). The pH change due to CO_2 accumulation is mitigated by an increase in nonrespiratory HCO_3^- (Figure 32.4) (Tazawa, 1986, 1987). Hemoglobin (Hb), which serves as the noncarbonate buffer in blood, increases during the last half of incubation and thus is partly responsible for the mitigated change in pH. Reflecting the developmental increase in Hct and [Hb], the buffer value increases from ~−8 to −10 mmol/L on days 9 to 10 to ~−17 mmol/L on days 15 to 18 (Erasmus et al., 1970/1971; Tazawa and Piiper, 1984). In day 16 embryos, values ranging from −12.7 to −15 mmol/L are reported (Tazawa et al., 1981; Tazawa, 1980a, 1981b, 1982, 1986). Other studies have not found an increase in buffer value during development (Tazawa et al., 1983a; Andrewartha et al., 2011b; Burggren et al., 2012). The determination of the buffer value is a controversial issue, and clearly not all sources of variation have been identified. However, the value of −16 mmol/L reported recently (Burggren et al., 2012) is depicted in a pH-$[HCO_3^-]$ diagram (Davenport diagram, Figure 32.4) to show and estimate respiratory changes in acid–base balance.

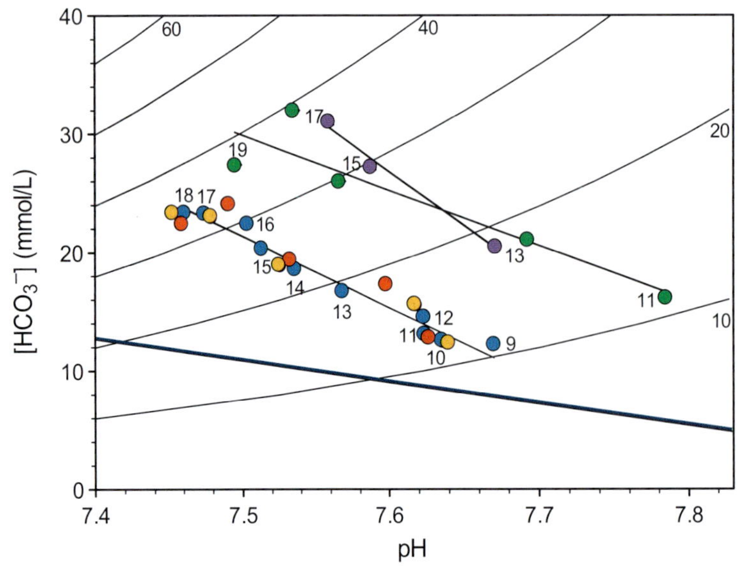

FIGURE 32.4 Davenport diagram illustrating the daily changes in control acid–base balance of arterialized blood in chicken embryos developing in air. The dark blue line is an arbitrary buffer line with the determined slope of −16 mmol/L pH/unit. Numbered curves are PCO_2 isopleths in mm Hg. The numbers near the data points represent the ages of the embryos. *Data are collected from the following studies; blue circles: Tazawa et al. (1971a), red circles: Tazawa et al. (1971b), yellow circles: Tazawa (1973), purple circles: Burggren et al. (2012), green circles: Tazawa et al. (2012). The solid line drawn through the data points corresponds to the following regression equation; (1) $[HCO_3^-] = -60.2\ pH + 472.6$ for combined data from Tazawa et al. (1971a, 1971b) and Tazawa (1973), (2) $[HCO_3^-] = -90.9\ pH + 717.4$ from Burggren et al. (2012), and (3) $[HCO_3^-] = -46.6\ pH + 379.1$ from Tazawa et al. (2012).*

Responses in acid–base balance to 1 day exposure to altered environmental gas mixtures differ depending on the gas mixture and age of chicken embryos (Figure 32.5) (Burggren et al., 2012). One day of hypercapnic exposure (5% CO_2, 20% O_2) increases P_aCO_2 and decreases pH_a, producing respiratory acidosis that is partially (~50%) compensated by metabolic alkalosis at all embryonic stages examined (days 13, 15, and 17) (Figure 32.5). Similar patterns of partially compensated respiratory acidosis have been reported in embryos exposed to 9% CO_2 in air for >3 days (Dawes and Simkiss, 1969). One day of exposure to hypercapnic hypoxia (5% CO_2, 15% O_2) abolishes compensatory metabolic alkalosis in day 15 and day 17 embryos, but a metabolic compensation of ~37% still occurs in day 13 embryos (Figure 32.5). This suggests that a relatively high O_2 level is required for metabolic compensation to occur. The lower $\dot{M}O_2$ and overall higher allantoic P_aO_2 in air (Tazawa, 1971, 1980a; Tazawa et al., 1971a,b) in day 13 embryos compared with more advanced embryos preserves the metabolic compensation during hypoxia. One day of hypoxic exposure (15% O_2) creates a metabolic acidosis in day 15 and day 17 embryos, but not in day 13 embryos (Figure 32.5). Hypoxia produces hypometabolism without anaerobic energy compensation (Bjønnes et al., 1987; Mortola and Besterman, 2007). However, once a lower threshold $\dot{M}O_2$ is reached, embryos turn to anaerobic glycolysis and blood lactate concentration ($[La^-]$) increases (Grabowski, 1961, 1966; Bjønnes et al., 1987). Therefore, embryos exposed to severe hypoxia (e.g., 10% O_2) encounter metabolic acidosis caused by glycolysis (Tazawa et al., 2012). However, metabolic acidosis occurring in moderate hypoxia is attributed only slightly to glycolysis and other unverified mechanisms, such as O_2 level influencing HCO_3^- transfer across the CAM. One day exposure to hyperoxia (40% O_2) causes respiratory acidosis that varies with embryonic age (Figure 32.5). Hyperoxia causes hypermetabolism (and a consequent increase in $\dot{M}CO_2$) (Visschedijk et al., 1980; Høiby et al., 1983; Stock et al., 1985; Tazawa et al., 1992b). Increased CO_2 is accumulated in the blood and hydrated to create H^+ and HCO_3^-. Coupled with a nearly fixed eggshell gas GCO_2, this results in respiratory acidosis. Age-specific differences in the magnitude of respiratory acidosis are due to hyperoxia causing a greater hypermetabolism in advanced embryos (Stock et al., 1985; Tazawa et al., 1992b).

Because eggshell GCO_2 is governed by dCO_2, respiratory acidosis also occurs in embryos exposed to a SF_6/O_2 gas mixture, which reduces GCO_2. Inversely, respiratory alkalosis occurs in embryos exposed to a He/O_2 atmosphere, which increases GCO_2 (Tazawa et al., 1981). Partially covering the eggshell with a gas-impermeable material or opening the eggshell over the air cell also produce respiratory acidosis and respiratory alkalosis, respectively. These respiratory disturbances to the acid–base balance occur rapidly and blood P_aCO_2 reaches a plateau ~10 min after changing GCO_2 (Tazawa, 1981a). Concurrently, blood pH changes to the value predicted by buffer capacity during the following 30–60 min (i.e., noncompensated respiratory

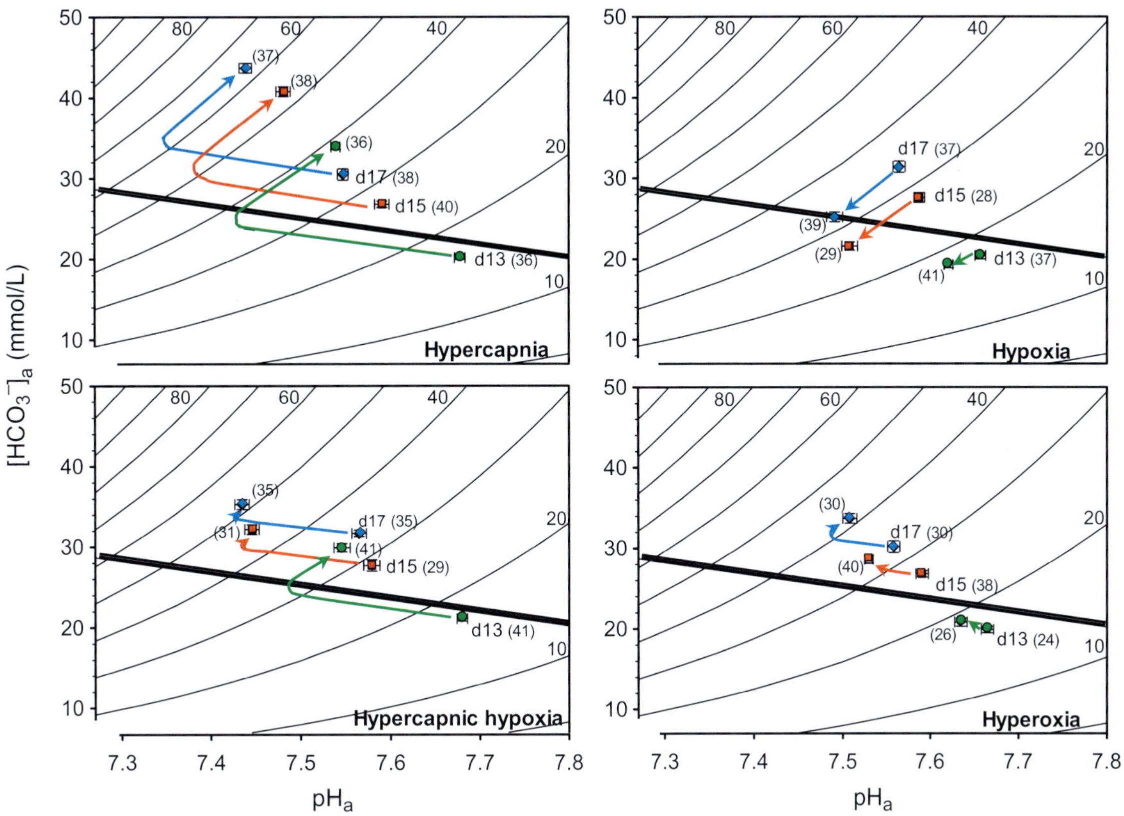

FIGURE 32.5 Regulation of arterialized blood acid–base balance of chicken embryos exposed to air (control) or altered [CO_2] and/or [O_2] for the last day of 13 (green), 15 (red), and 17 (blue) days of incubation. The colored arrows indicate changes in acid–base status from control. The dark blue line is the same as in Figure 32.4. *N* values are presented in parentheses. Mean values ±1 S.E.M. are shown. *Modified from Burggren et al. (2012), with permission from Elsevier.*

acidosis/alkalosis) with a subsequent (2–6h), but incomplete, change toward the control level, while P_aCO_2 is maintained at a constant value. Once partial metabolic compensation is achieved at ~4–6h, the state of acid–base disturbance remains constant over the next 24h in normoxia or hyperoxia (e.g., 5% CO_2/20% O_2 or 5% CO_2/40% O_2), but metabolic compensation is not preserved after 24h in hypoxia (5% CO_2/15% O_2) (Mueller et al., 2013b). Therefore, preservation of metabolic compensation requires an O_2 concentration above 20%.

Although partial metabolic compensation in response to hypercapnia or moderate hypoxia with hypercapnia (e.g., 15% O_2/5% CO_2) proceeds over the course of several hours, responses to severe hypoxia (10% O_2), with or without CO_2 (e.g., 5%), progress more quickly (Figure 32.6) (Tazawa et al., 2012). This is due to anaerobic glycolysis and the attendant rapid increase in [La^-] that creates severe metabolic acidosis. If hypoxic embryos can preserve [HCO_3^-] above ~10mmol/L (in the case of day 15 embryos), which is generally reached beyond 2h, embryos can survive and recover to the control state of acid–base balance in ~2h after being returned in air (Figure 32.6) (Tazawa et al., 2012). Accordingly, hypoxia-induced acid–base disturbances are only transient and may not affect long-term survival.

In nature, the natural variations in eggshell conductance cause large differences in P_aCO_2 among eggs, but blood pH variations are kept to a minimum (Tazawa et al., 1983a). In eggs with low GCO_2, the Hct (and thus Hb) increases. In part, increased [Hb] may be responsible for the minimum change in pH.

The acid–base balance of chicken embryos also reacts promptly to metabolic disturbances, as shown by the time course changes in metabolic acid–base alterations created by infusion of electrolyte solution ($NaHCO_3$ and NH_4Cl) (see Figure 32.5 in Tazawa, 1982). For instance, the infusion of 15μL 1M $NaHCO_3$ increases plasma [HCO_3^-] and blood PCO_2. Because the embryo has no convective ventilation, the metabolic disturbance is not subjected to respiratory compensation. The increase in PCO_2 1h after infusion indicates that the infused HCO_3^- is partly eliminated as dissolved CO_2. The acid–base status returns to control values 6h after infusion. Besides elimination of CO_2 from the CAM, the increased fluid volume and hypertonicity of the infused $NaHCO_3$ solution may be partially responsible for the decrease in [HCO_3^-]. Additionally, penetration of HCO_3^- into the intracellular space and excretion of HCO_3^- into allantoic fluid through the CAM may contribute to the regulation (Tazawa, 1982, 1986).

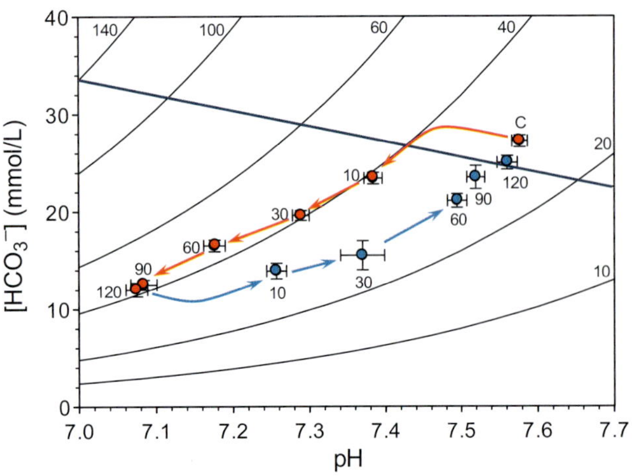

FIGURE 32.6 **Time-course of acid–base disturbance in day 15 chicken embryos exposed to hypercapnic hypoxia (5% CO_2, 10% O_2) for 120 min (red circles) with recovery in air for 120 min (blue circles).** Other details as in Figures 32.4 and 32.5. *Modified from Tazawa et al. (2012), with permission from Elsevier.*

In summary, acid–base regulation develops in concert with the increased CO_2 production as avian embryos grow. The ability of chicken embryos to tolerate and respond to acid–base disturbance, induced by environmental gas challenges, increases or decreases in GCO_2, or infusion of electrolytes, increases with development. Comparative approaches are required to examine species differences in acid–base regulation, and how tolerance to perturbations may be related to the incubation environment.

32.4.3 Cardiovascular System

32.4.3.1 Basic Cardiovascular Parameters

The primordial bird embryo's heart is a paired tubular structure that soon becomes a single tube. The heart begins to elongate more rapidly than the pericardial cavity containing it, and this space limitation forces the tubular heart to bend. Only the ventricle and bulbus are present on days 1.5 to 2 of chicken incubation. The impact of bloodstreams upon the inner surface of the contorted tube forms the external configuration and internal structure of the heart (Taber, 2001; Alford and Taber, 2003; Tobita et al., 2005) as well as contributing to blood vessel formation (le Noble et al., 2008; Burggren, 2013), although the specific role of pressure and flow fluctuations to the contribution of developing blood vessels remains somewhat enigmatic (Branum et al., 2013).

The structural alterations that separate atrium from ventricle, ventricle from aorta, and the left from the right chambers take place during days 3 to 8 of incubation, resulting in a four-chambered heart by days 8 to 9 (Pattern, 1951; see also multiple papers in Burggren and Keller, 1997).

The mass of the heart increases as a power function of incubation day (*I*) (Romanoff, 1967; Clark et al., 1986):

$$\text{Heart mass } (\mu g) = 12.62 \cdot I^{3.26} \qquad (32.12)$$

The growth of the heart relative to the whole body is greatest early in development and declines during incubation. The ratio of the heart mass to the whole body mass falls from 1.8% on day 4 to 0.7% on day 18 of incubation.

The heart begins to beat at about 30 h of incubation and blood begins to circulate after about 40 h, when the connections between the dorsal aorta and the vessels of the yolk sac complete the circulation (Pattern, 1951). Despite the early generation of heartbeat and blood flow, heart rate is not initially required for convective blood flow to the tissues, but likely plays a role in angiogenesis (Burggren et al., 2000, 2004; Burggren, 2004, 2013; Branum et al., 2013).

Blood volume increases with incubation day (Figure 32.7(B)) (Yosphe-Purer et al., 1953; Barnes and Jensen, 1959; Lemez, 1972; Kind, 1975), and can be approximated by the following function (Tazawa and Hou, 1997):

$$\text{Blood volume } (\mu L) = 1.85 \cdot I^{2.64} \qquad (32.13)$$

The increase in blood volume during development is not as rapid as growth rate (Figure 32.7(A)), and therefore embryo mass-specific blood volume decreases during development from about 1 mL/g on day 4 to 0.15 mL/g on day 18.

As both the heart mass and blood volume increase, the stroke volume of the heart increases (Hughes, 1949; Faber et al., 1974; Hu and Clark, 1989). During early development, prior to the left and right separation of the heart, stroke volume depends on blood volume and increases in parallel with embryonic growth (Faber et al., 1974). The relationship between stroke volume and incubation day (*I*, days 2–6) (Hu and Clark, 1989) can be described as:

$$\text{Stroke volume } (\mu L) = 0.002 \cdot I^{3.46} \qquad (32.14)$$

Even after the second week of incubation, stroke volume seems to increase as a power function of incubation day (Figure 32.7(C)) (Hughes, 1949).

For early embryos, the dorsal aortic blood flow increases as a power function of incubation day during days 2 to 6 (Hu and Clark, 1989). The mass (embryo)-specific value is 0.5–1 mL/min/g. The mass-specific cardiac output of day 3 to 5 embryos, determined from the stroke volume, is similar at 1 mL/min/g (Faber et al., 1974). Cardiac output calculated from the stroke volume of young embryos and that of day 16 chicken embryos estimated by model analysis and blood O_2 measurement (White, 1974; Rahn et al., 1985; Tazawa and Johansen, 1987) indicate that the mass-specific cardiac output of young/late embryos is in the narrow range of 0.5–1.5 mL/min/g, and thus the cardiac output seems to increase as a power function of incubation day (Figure 32.7(D)). Eventually, cardiac output may increase almost in parallel with embryonic growth. If it is assumed that the

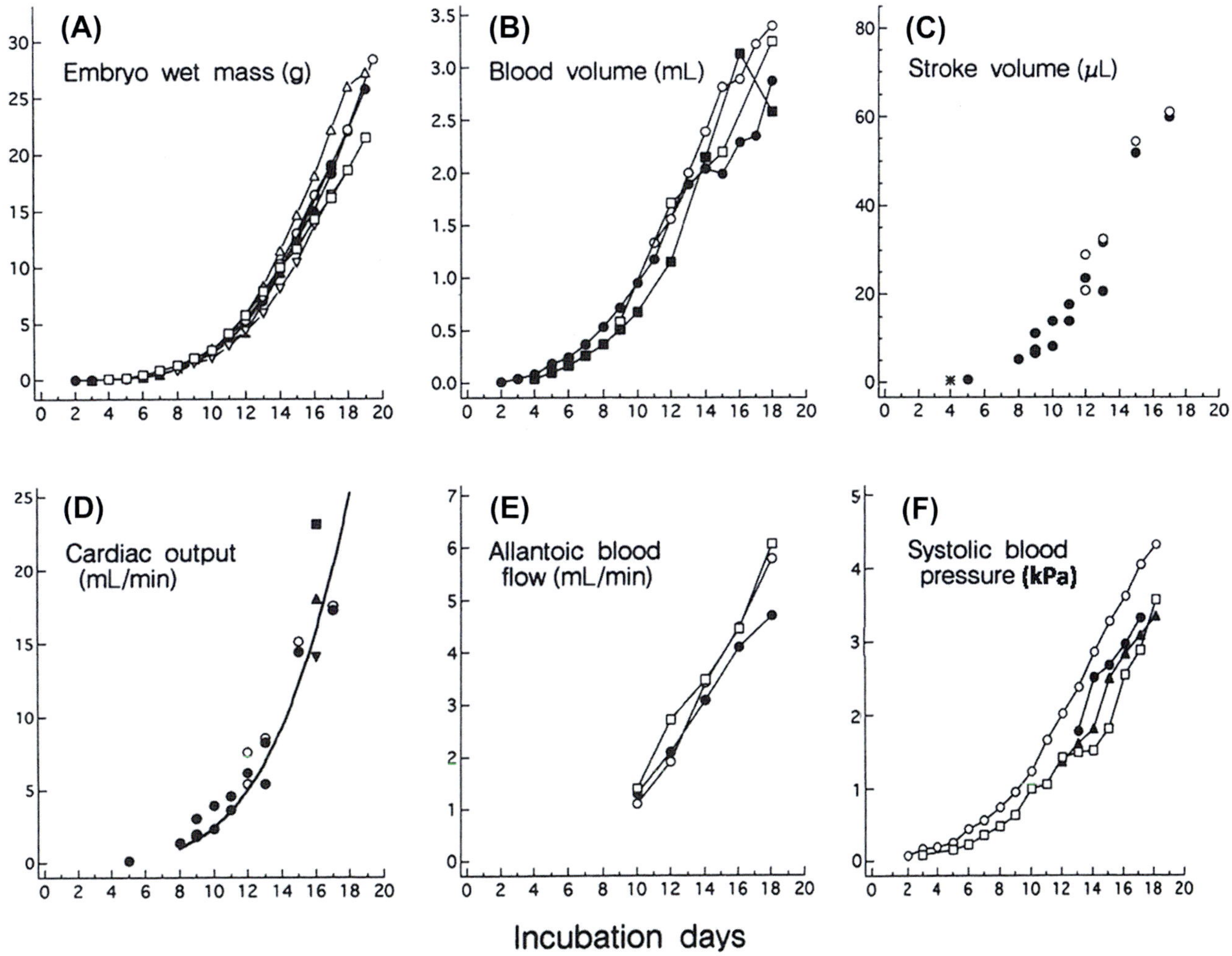

FIGURE 32.7 Developmental patterns of embryo wet mass (A), total blood volume (B), cardiac stroke volume (C), cardiac output (D), allantoic blood flow (E), and arterial systolic blood pressure (F) during the prenatal stage. Symbols connected by solid lines in (A), (B), (E), and (F) indicate the developmental patterns from the papers cited in the text.

mass-specific cardiac output during the last 2 weeks of prenatal development is 1 mL/min/g, the cardiac output of young embryos is related to incubation day (I) as follows,

$$\text{Cardiac output }(\mu\text{L/min}) = 0.24 \cdot I^4 \qquad (32.15)$$

which is the same expression as that relating body mass of developing embryos to incubation day (Eqn (32.5), Figure 32.7(A), (D)).

Blood flow through the CAM, determined from $\dot{M}O_2$, blood gas analysis, or via flow probe (Tazawa and Mochizuki, 1976, 1977; Bissonnette and Metcalfe, 1978; Van Golde et al., 1997), increases with embryonic growth (Figure 32.7(E)), but the mass-specific value decreases from ~0.5 mL/min/g on day 10 to ~0.25 mL/min/g on day 18. Although cardiac output increases in parallel with embryonic growth, its partition to the CAM decreases as embryos grow. On days 10 to 13, about half of the cardiac output goes to the CAM and this decreases to ~40% on days 17 to 19 (Mulder et al., 1997; Dzialowski et al., 2011 for review) or to ~17–20% on days 16 to 18 (Tazawa and Johansen, 1987). Consequently, the percentage of cardiac output to the organs and tissues increases as development proceeds.

Arterial pressure (P_a) is measured via a vitelline vessel in early embryonic chick development and an allantoic artery thereafter. Arterial systolic pressure (P_{sys}), measured by micropipette (Van Mierop and Bertuch, 1967; Girard, 1973) or needle-catheter techniques that maintain adequate gas exchange through the eggshell (Tazawa, 1981c; Tazawa and Nakagawa, 1985), increases with incubation day (I) in the following manner (Figure 32.7(F)):

$$P_{sys}\text{ (kPa)} = 0.015 \cdot I^{1.86} \qquad (32.16)$$

P_a of the vitelline artery is pulsatile by day 2 of chicken incubation (Van Mierop and Bertuch, 1967; Hu and Clark, 1989; Taber, 2001). While the presence of a dicrotic notch is reported in early embryos (Hu and Clark, 1989; Yoshigi et al., 1996), a clear dicrotic notch is absent from allantoic P_a waves determined by others (Van Mierop and Bertuch,

1967; Tazawa and Nakagawa, 1985). However, the configuration of the P_a tracings and the ventricular pressure waves suggest that an arterial valve mechanism is present in early embryos, which lack true cardiac valves (Van Mierop and Vertuch, 1967; Faber, 1968).

32.4.3.2 Mean Heart Rate

According to the equations expressing the relationship between incubation day and body mass (Eqn (32.5)) or cardiac output (Eqn (32.15)), both body mass and cardiac output in day 16 chicken embryos are ~16 g and ~16 mL/min. Assuming a mean heart rate (MHR) of 280 beats per min (bpm), the stroke volume and mass-specific stroke volume are estimated as ~60 μL and ~4 μL/g. The cardiac contraction generating this amount of stroke volume physically moves the body of the embryo by an almost imperceptible amount, and these movements are transferred to the whole egg, producing cardiogenic ballistic movement of the egg, referred to as a ballistocardiogram (BCG) of the egg (Tazawa et al., 1989b). The egg movement determined by a laser displacement meter is ~1 μm (Sakamoto et al., 1995). The BCG measured by various means offers a convenient way to noninvasively determine MHR of avian embryos (Tazawa et al., 1999; Tazawa, 2005). Additionally, the heartbeat of the embryo inside the eggshell produces not only the ballistic movements of the egg, but also acoustic pressure changes outside the eggshell. A conventional condenser microphone hermetically fixed on the eggshell detects the cardiogenic acoustic pressure changes, designated as an acoustocardiogram (ACG) (Rahn et al., 1990), which can conveniently determine MHR of avian embryos.

In contrast to developmental patterns of the circulatory variables that increase steadily with embryonic growth, the embryonic MHR of chickens increases rapidly after the heart commences beating and then becomes asymptotic until early in the second week of incubation (Cain et al., 1967; Van Mierop and Bertuch, 1967; Girard, 1973; Hu and Clark, 1989; Tazawa et al., 1991a; Burggren and Warburton, 1994; Howe et al., 1995). During the last half of incubation, the daily change in MHR is small, with a temporal increase at 60–70% of incubation followed by a decrease toward IP and increase during hatching (Tazawa et al., 1991a). The change in MHR near the end of development in other precocial birds is shown in Table 32.2. Precocial species, from the smallest (king quail, *Coturnix chinensis*) to the largest (ostrich), show either a plateau or decrease in MHR prior to IP (Tazawa et al., 1991a, 1998a; 1998b, 2000; Pearson et al., 1998; Kato et al., 2002).

In comparison, HR increases rapidly toward hatching in small altricial embryos (Table 32.3) (Tazawa et al., 1994; Pearson et al., 1999). However, the increase in MHR during pre-pipping development slows in larger altricial eggs, such as those of crow (*Corvus corone*) and cattle egret (*Bubulcus ibis*), tending to remain unchanged during the last half of the prenatal period (Pearson and Tazawa, 1999; Tazawa et al., 2001b). The trend for MHR to remain unchanged during the end of prenatal development also occurs in semi-precocial seabirds such as the brown noddy (*Anous stolidus*) and laysan albatross (*Phoebastria immutabilis*) (Table 32.2) (Tazawa et al., 1991b; Tazawa and Whittow, 1994; Pearson et al., 2000). Hence, the pattern of change in MHR during development, particularly at the end of incubation, can be somewhat predicted by the degree of development at hatching. Precocial species tend to show a plateau or decrease in MHR near hatching, while MHR tends to increase in altricial species.

These determinations of embryonic MHR along with egg mass of various species (Tables 32.2 and 32.3) reveal a significant relationship between MHR at 80% of *I* and egg mass (Tazawa et al., 2001b). The allometric relationship derived for 20 species of altricial and semi-altricial (ASA) birds with egg mass ranging from 0.96 g of the zebra finch to ~41 g of the lanner falcon (*Falco biarmicus*) is:

$$\text{MHR at 80\% of } I = 371 \cdot (\text{Egg mass})^{-0.121} \quad (r = -0.846,\ P < 0.001) \tag{32.17}$$

The relationship for 13 species of precocial and semi-precocial (PSP) birds with egg mass ranging from 6 g of the king quail to ~1400 g of the ostrich is:

$$\text{MHR at 80\% of } I = 433 \cdot (\text{Egg mass})^{-0.121} \quad (r = -0.963,\ P < 0.001) \tag{32.18}$$

The slopes for MHR are parallel, but the MHR of ASA embryos is low compared with PSP embryos from the same egg mass. In ASA embryos, HR becomes maximal during the pipping period, and the maximum HR is significantly related to egg mass. Consequently, the allometric relationship of maximum HR and egg mass in ASA and that of 80% of *I* in PSP is statistically identical, expressed by a single allometric equation:

$$\text{MHR} = 437 \cdot (\text{Egg mass})^{-0.123} \quad (r = -0.948,\ P < 0.001,\ N = 33) \tag{32.19}$$

In addition, determination of MHR in embryos from the same clutch (siblings) in pigeons and bank swallows demonstrates that developmental patterns of MHR in siblings are more alike than those of embryos from other clutches of the same species (Burggren et al., 1994). These findings may be termed the "sibling effect," in which siblings are predisposed to show particular physiological patterns. Whether these effects are genetically based (i.e., resulting from the common genetic heritage of the F_1 offspring) or involve epigenetic influences is uncertain. Certainly, physiological process can be transferred across generations (see Ho and Burggren, 2010, 2012; Burggren, in press) through

TABLE 32.2 Egg Mass (g), Incubation Period (*I*, days), Heart Rate (HR, beats/min) at 80% of Incubation and Maximum Heart Rate (Max HR) during Prenatal Development in Precocial Birds and Heart Rate Prior to Internal Pipping (Pre-IP) and at the First Star Fracture of the Eggshell in Semi-Precocial Birds

	Species	Egg Mass (g)	Incubation Period (*I*, days)	HR at 80% of *I* (beats/min)	Max HR (beats/min)
Precocial	King quail (Pearson et al., 1998) *Coturnix chinensis*	6.0±0.4	16	341±8 (81%)	341±8 (81%)
	Japanese quail (Tazawa et al., 1991a) *Coturnix coturnix japonica*	10.7±0.7	17	319±8 (76%)	326±7 (76%)
	Chicken (Tazawa et al., 1991a) *Gallus gallus domesticus*	64.9±2.5	21	287±9 (81%)	287±9 (81%)
	Duck (Tazawa et al., 1991a) *Anas platyrhynchos*	79.0±2.5	28	247±15 (82%)	258±11 (61%)
	Turkey (Tazawa et al., 1991a) *Meleagris gallopavo*	82.9±2.6	28	246±10 (79%)	248±10 (75%)
	Peafowl (Tazawa et al., 1991a) *Pavo cristatus*	111.3±9.3	28	262±12 (79%)	267±9 (86%)
	Goose (Tazawa et al., 1991a) *Anser cygnoides*	158.3±11.3	30	224±8 (80%)	248±10 (60%)
	Emu (Tazawa et al., 2000) *Dromaius novaehollandiae*	634±9	50	192±7 (80%)	199±11 (72%)
	Ostrich (Tazawa et al., 1998a) *Struthio camelus*	1395±199	42	185±12 (81%)	208±9 (55%)
				Pre-IP HR (beats min^{-1})	**HR at First Star Fracture of Shell (beats min^{-1})**
Semi-precocial	Brown noddy (Tazawa et al., 1991a) *Anous stolidus*	37.9±2.2	35	298±7	303±13
	Wedge-tailed shearwater (Tazawa and Whittow, 1994) *Puffinus pacificus*	57.2±2.3	52	244±10	252±11
	Laysan albatross (Tazawa and Whittow, 1994) *Diomedea immutabilis*	288±18	65	232±15	233±15

Values were measured at 38 °C or converted to that at 38 °C using $HR(38\,°C) = HR(T\,°C)e^{[0.0639(38-T)]}$; For precocial species: if HR was not measured at 80% of *I*, the value closest to 80% of *I* was used with the % of *I* at which HR was measured shown in parentheses. The % of *I* of max HR is shown in parentheses; Data are mean ±S.D.
Modified From Tables 32.1 and 32.2 in Tazawa et al. (2001b).

TABLE 32.3 Egg Mass (g), Incubation Period (*I*, days), and Heart Rate (HR, beats/min) at 80% of Incubation and during Internal Pipping (IP) and External Pipping (EP) in Altricial and Semi-altricial Birds

Species	Egg Mass (g)	Incubation Period (*I*, days)	HR at 80% of *I* (beats/min)	HR during IP (beats/min)	HR during EP (beats/min)
Zebra finch (Pearson et al., 1999) *Taeniopygia guttata*	0.96±0.13	14	335±10	376±20	405±12
Bengalese finch (Pearson et al., 1999) *Lonchura striata* Var. *domestica*	1.10±0.12	15	404±36	409±25	448±35
Marsh tit (Pearson et al., 1999) *Parus palustris*	1.39±0.04	14	363±17	409±19	–
Bank swallow (Tazawa et al., 1994) *Riparia riparia*	1.42±0.10	14	298±12	–	352±16
Great tit (Pearson et al., 1999) *Parus varius*	1.59±0.14	14	348±11	432±13	495±14
Varied tit (Pearson et al., 1999) *Parus varius*	1.69±0.01	14	356±7	434±11	–
Tree sparrow (Pearson et al., 1999) *Passer montanus*	2.09±0.07	12	335±13	411±32	–
Budgerigar (Pearson et al., 1999) *Melopsittacus undulates*	2.19±0.19	18	314±14	339±15	364±12
House martin (Pearson et al., 1999) *Delichon urbica*	2.25±0.04	15	357±7	369±8	367±11
Japanese bunting (Pearson et al., 1999) *Emberiza spodocephala*	2.56±0.09	13	370±5	426±1	–
Red-cheeked starling (Pearson et al., 1999) *Sturnus philippensis*	4.14±0.01	14	358±1	409±5	–
Cockatiel (Pearson et al., 1999) *Nymphicus hollandicus*	5.08±0.18	20	300±8	318±25	344±19
Brown-eared bulbul (Pearson et al., 1999) *Hypsipetes amaurotis*	6.4±0.5	16	333±7	402±8	–
Domestic pigeon (Tazawa et al., 1994) *Columba domestica*	17.1±1.0	18	247±17	–	276±13
Fantail pigeon (Tazawa et al., 1994) *Columba domestica*	19.7±2.4	18	267±10	–	293±6
Homing pigeon (Tazawa et al., 1994) *Columba domestica*	19.8±1.2	18	230±16	–	273±4
Crow (Pearson and Tazawa, 1999) *Corvus corone*	20.5±2.2	20	297±11	348±35	366±22
Barn owl (Tazawa et al., 2001b) *Tyto alba*	20.1±0.6	30	219±11	–	276±13
Cattle egret (Tazawa et al., 2001b) *Bubulcus ibis*	27.5±3.3	23	251±8	–	283±12
Lanner falcon (Tazawa et al., 2001b) *Falco biarmicus*	41.2±0.4	33	242±9	–	276±6

Values were measured at 38 °C or converted to that at 38 °C using $HR(38\,°C) = HR(T\,°C)e^{[0.0639(38-T)]}$; Data are mean ±S.D.
Modified From Tazawa et al. (2001b).

epigenetic mechanisms. However, environmental effects on the mother may also cause siblings to undergo similar patterns of physiological development through a "maternal effect" or even direct effects on the gamete cells of the mother (Burggren et al., 1994; Dzialowski and Sotherland, 2004; Ho et al., 2011; Burggren, in press).

32.4.3.3 Instantaneous Heart Rate

In addition to daily changes in MHR during development, HR varies from beat to beat (i.e., instantaneous HR; IHR). At rest, the baseline IHR in chicken embryos is more or less constant before ~day 11, indicating that neither IHR accelerations of adrenergic origin nor IHR decelerations of cholinergic origin occurs during early development. The flat baseline of IHR begins to fluctuate with the appearance of rapid, transient decelerations on ~days 12 to 13, with a subsequent augmented frequency of appearance (Figure 32.8) (Höchel et al., 1998; Tazawa et al., 1999; Chiba et al., 2004). Thereafter, IHR becomes arrhythmic with the appearance of transient accelerations on ~days 15 to 16 and more arrhythmic with augmented decelerations and accelerations toward hatching (Höchel et al., 1998; Tazawa et al., 1999, 2002a; Moriya et al., 2000; Khandoker et al., 2003). These IHR decelerations are eliminated, and baseline HR is elevated, by intravenous administration of atropine (Figure 32.8). In

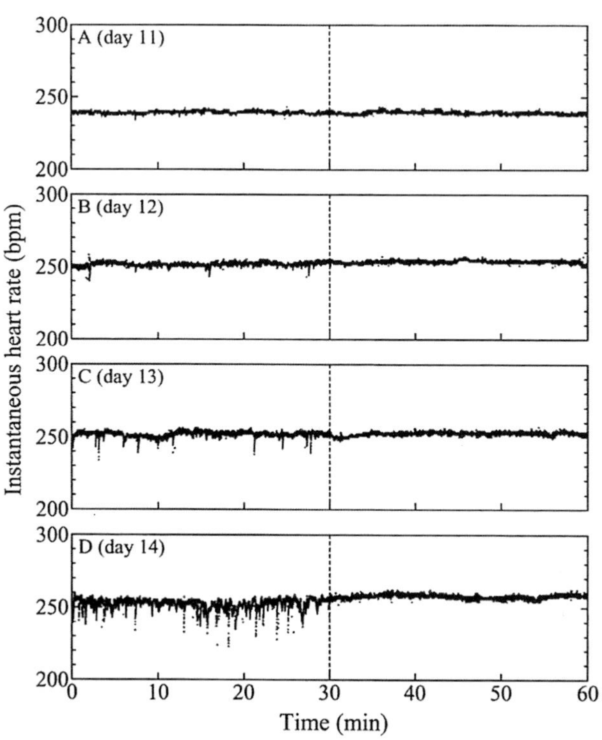

FIGURE 32.8 Instantaneous heart rate fluctuations before and after intravenous infusion of 20 µg atropine at 30 min in day 11 to day 14 chicken embryos. *From Chiba et al. (2004), with permission from Elsevier.*

late embryos whose IHR fluctuations consist of complicated decelerations and accelerations with augmented magnitude and frequency, atropine diminishes decelerations only but preserves accelerations, while the HR baseline is markedly elevated (e.g., Figure 32.3 of Chiba et al., 2004), indicating that IHR decelerations are mediated by the vagus nerve. Accordingly, vagal tone begins to appear on around days 12–13 of incubation in chickens, maturing with development, correcting previous reports on no vagal control of HR determined from blood pressure signals (Tazawa et al., 1992a). Furthermore, cholinergic chronotropic control of HR occurs on days 12 to 13 of incubation in both broiler and White Leghorn embryos (Yoneta et al., 2006c). These events coincide with the presence of a complete cholinergic effector pathway from 60% of chicken incubation (Pappano and Löffelholz, 1974).

Some controversy exists as to whether cholinergic tone is present at this developmental timepoint; blood pressure studies indicate no vagal tone on the heart during embryonic development (Crossley and Altimiras, 2000). Further, blood pressure studies using adrenoreceptor stimulation via pharmacological manipulation indicate that a positive β-adrenergic chronotropic tone can be induced from 60% of incubation (Crossley and Altimiras, 2000). The effect is enacted by circulating catecholamines, as ganglionic blockade with hexamethonium has no impact on HR. In comparison, pharmacological manipulation of blood pressure indicates that the emu heart is under both β-adrenergic and cholinergic control by 70% of incubation (Crossley et al., 2003).

Long-term measurement of IHR reveals the occurrence and development of various cardiac rhythms and irregularities. The developmental pattern of MHR (constructed from IHR measurement) of an embryo before and after hatching demonstrates the occurrence of infradian rhythm before and during IP, a gradual increase in HR baseline during IP, a sudden drop with a subsequent sudden increase in HR baseline during EP, and the first occurrence of circadian rhythm of HR after hatching (Figure 32.9) (Moriya et al., 2000). Recordings of IHR reveal respiratory arrhythmia associated with EP, together with three unique patterns of IHR during the final stage of EP (i.e., relatively long-lasting cyclic small accelerations, irregular intermittent large accelerations, and short-term repeated large accelerations) (Tazawa et al., 1999; Moriya et al., 2000). Short-term repeated large accelerations also occur in EP emu embryos, indicating imminent hatching (Figure 32.10) (Kato et al., 2002).

After hatching, three types of IHR fluctuations (Types I, II and III) are demonstrated in chickens (Moriya et al., 1999, 2000; Tazawa et al., 2002a). Type I is characterized as a widespread baseline HR (20–50 bpm) due to respiratory arrhythmia with a mean oscillatory frequency of ~0.7 Hz. Type II is evident by low-frequency oscillations of baseline HR at ~0.07 Hz, occurring at low T_a or when hatchlings are

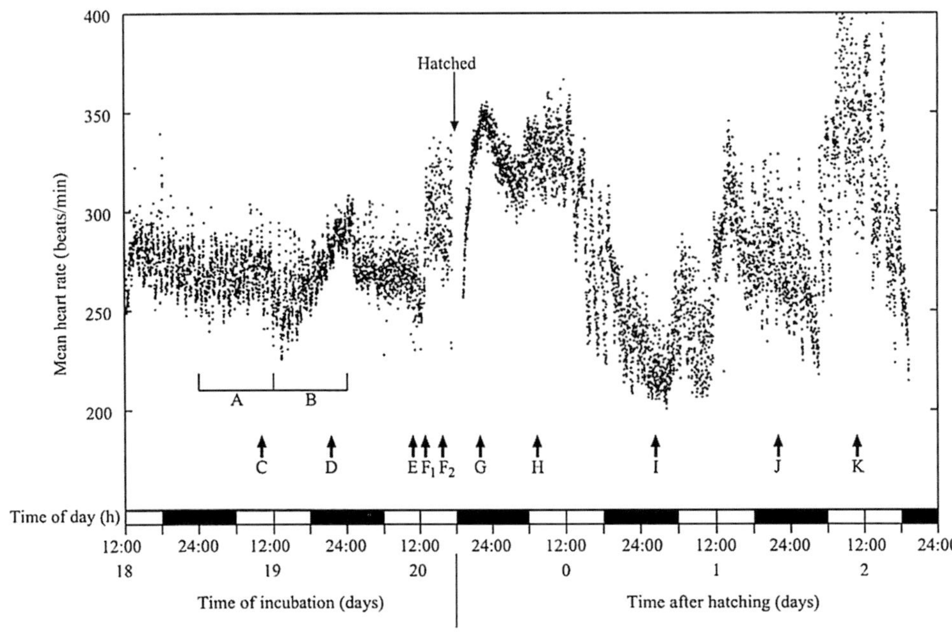

FIGURE 32.9 Typical developmental pattern of mean heart rate (MHR) of a chick, recorded from day 18 of incubation through hatching on day 20 to day 2 of post-hatching. Each point indicates the MHR over a 1 min period determined from the continuous recording of instantaneous heart rate (IHR). IP begins at around 12:00 on day 19 when MHR temporarily decreases with subsequent gradual increase during IP. The MHR oscillates with a period of about 42 min from the beginning of recording to middle of IP, i.e. infradian rhythm. EP begins at the start of day 20 when MHR suddenly drops. Following the drop, which lasts for the first half of day 20, MHR increases sharply at 12:00 and maintains high baseline until hatching. During the last period of EP, characteristic patterns of IHR occur, e.g., a sharp increase in MHR (pointed by F_1) consisting of relatively long-lasting cyclic small accelerations. The wide baseline of MHR during the last 1% of incubation is composed of repeated alternate occurrence of irregular intermittent large accelerations and short-term repeated large accelerations indicating imminent hatching (see Figure 32.10). After hatching, circadian rhythms occur on days one and 2. *Reproduced from Moriya et al. (2000), with permission from The Company of Biologists.*

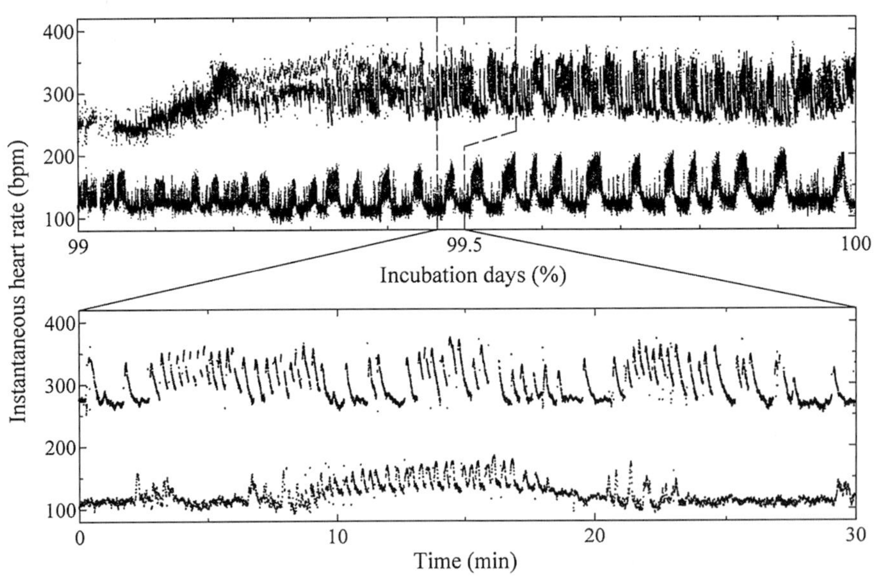

FIGURE 32.10 Instantaneous heart rate (IHR) during the last 1% of incubation of chicken (upper recording) and emu (lower recording) embryos (top panel) and the 60-min recordings taken from the middle of the last 1% of incubation indicated by broken lines (bottom panel). Short-term repeated large IHR accelerations with intermittent large accelerations continuously appear towards hatching in both embryos. HR of small chicken embryos is higher than that of large emu embryos (allometry). *From Kato et al. (2002), with permission from Elsevier.*

exposed to lowered T_a, and is thus related to thermoregulation. Type III is characterized as noncyclic irregularities, dominated by frequent transient accelerations. The image-processing system developed to capture the movements of wing or whole body of the hatchling reveals that Type I and Type II HR oscillations are related to the periodic movements of the wing and Type III HR irregularities occur simultaneously with spontaneous whole body movements (Yoneta et al., 2006a). Consequently, it is likely that the chick moves the body at the same frequencies as Type I and Type II oscillations and simultaneously with Type III HR irregularities, and these HR fluctuations and body movements are likely to be attributed to the same origins.

32.4.3.4 Blood Pressure Regulation

The sympathetic nervous system is heavily involved in regulation of avian embryonic blood pressure (Altimiras et al., 2009). Injection with the α-adrenereceptor antagonist phentolamine causes hypotension from 60% of chicken incubation, indicating α-adrenergic tone on the vasculature (Girard, 1973; Saint Petery and Van Mierop, 1974; Tazawa et al., 1992a; Crossley and Altimiras, 2000). Phentolamine also causes bradycardia, a likely indirect effect of vasodilation, and reduced venous return. β-adrenergic tone on the vasculature is also present from 60% of incubation, as evident by a P_a increase in response to the β-adrenereceptor antagonist propranolol (Girard, 1973; Saint Petery and Van Mierop, 1974; Tazawa et al., 1992a; Crossley and Altimiras, 2000). The magnitude of both α- and β-adrenergic tone is maximal at 90% of incubation, just prior to IP (Crossley and Altimiras, 2000). The increase in adrenergic tone is matched to the maximal plasma catecholamine release on day 19. Likewise, emu embryos show signs of increasing α- and β-adrenergic tone on the vasculature matched to increasing levels of catecholamines as incubation proceeds (Crossley et al., 2003).

In addition to regulation via catecholamines, other hormones may also influence baseline cardiovascular function of avian embryos. The most extensively studied in chicken embryos is angiotensin II (ANG II), a strong vasoconstrictor and the active peptide of the renin-angiotensin system (RAS). Components of the RAS, including renin, angiotensin-converting enzyme, ANG II and its receptors, are present early in chicken ontogeny (Nishimura et al., 2003; Savary et al., 2005; Crossley et al., 2010). ANG II levels are elevated in embryos compared to adults; the peptide produces hypertension from at least 60% of incubation (Crossley et al., 2010), and contributes to baseline P_a at 90% of incubation (Mueller et al., unpublished). Despite an increase in P_a, ANG II does not alter MHR as it instead attenuates the embryonic cardiac baroreflex. The baroreflex is an important compensatory mechanism that buffers short-term changes in P_a and is composed of a peripheral limb adjusting vascular resistance and a cardiac limb that changes HR. A functioning baroreflex is present in chicken embryos from 80% of incubation (Altimiras and Crossley, 2000; Elfwing et al., 2011). ANG II decreases the sensitivity of the cardiac baroreflex at 90% of incubation so that reflex changes in HR in response to P_a changes are blunted (Mueller et al., 2013a). ANG II also raises the operating P_a of embryos, and it is via these short-term changes that the hormone partly enacts some of its influence on long-term P_a.

Other potentially important moderators of cardiovascular function include endothelin-1 (ET-1), a potent vasoconstrictor produced primarily in the endothelium, and natriuretic peptides (NP), strong vasodilators excreted by heart muscle cells. Components necessary for functional ET-1, including mRNA and converting enzymes, are found in chicken embryos from 15% of incubation (Hall et al., 2004; Groenendijk et al., 2008) and ET-1 alters chicken embryo hemodynamics (Groenendijk et al., 2008; Moonen and Villamor, 2011). Likewise, NP is present in the chicken heart and most likely contributes to hemodynamics from at least day 14 (Maksimov and Korostyshevskaya, 2013). Further research from the molecular to whole-organism level is required to understand the contribution of these hormones to P_a and HR control, and therefore embryonic cardiovascular regulation.

32.4.4 Osmoregulation

During development, various avian embryos face one of two osmoregulatory challenges: water loss through the pores of the eggshell, in desiccating arid environments; or excess water gain from the metabolic production of water as part of metabolizing the yolk stores (Ar and Rahn, 1980). Irrespective of the species-specific challenge, the developing kidney and extra-embryonic structures, including the yolk, CAM, and allantoic and amniotic fluids work in concert to regulate ion and water balance. The embryonic kidney of chickens has three stages, actually comprising separate structures: the pronephros, mesonephros, and metanephros. The pronephros appears first and functions until day 5 to 6 of incubation (Abdel-Malek, 1950; Himura and Nakamura, 2003). Mesonephros function takes over from day 5, is maximal on days 10 to 15 (Romanoff, 1960), and then degrades between days 18 and 19 (Atwell and Hanan, 1926). The mesonephros functions simultaneously with the metanephros, which begins developing from day 4 (Abdel-Malek, 1950) and continues to develop post-hatching. The metanephros is the most complex of the three embryonic kidney structures, and comprises the functioning structure in adults. The allantoic sac first appears at about days 3 to 4 of incubation, and acts as a repository for kidney secretions, as evident by the increase in uric acid content throughout development (Romanoff, 1967). The allantoic epithelium actively transports sodium across its membrane from days 12 to 19 (Stewart and Terepka, 1969; Graves et al., 1986; Gabrielli and Accili, 2010), thus maintaining the fluid hypo-osmotic to the blood and permitting reabsorption of water by the embryo (Hoyt, 1979).

Osmoregulation is strongly tied to cardiovascular function through the synergistic maintenance of blood pressure and osmotic homeostasis. Regulators of blood pressure, such as ANG II and NP (see above), may also contribute to ion and water balance in embryos. Chronic removal of ANG II from chicken eggs not only decreases P_a but also lowers osmolality, decreases Na^+, and increases K^+ concentration of the blood at 90% of incubation. These actions remove the osmolality gradient between the blood and allantoic fluid (Mueller et al., unpublished). Therefore, ANG II appears to influence osmotic balance and may do so by direct vasoconstrictive action or by stimulating the release of aldosterone and arginine vasotocin, which promote proximal tubular sodium reabsorption. Aldosterone is present in chicken embryo adrenal glands from day 15 (Pedernera and Lantos, 1973) and arginine vasotocin is present in the brain from day 6 and the plasma from days 14 to 16 (Klempt et al., 1992; Mühlbauer et al., 1993). Both aldosterone and arginine vasotocin alter allantoic fluid volume, allantoic salt content, and renal enzyme activity in chicken embryos (Doneen and Smith, 1982). Prolactin and growth hormone also have osmoregulatory effects in chicken embryos (Doneen and Smith, 1982; Murphy et al., 1986). The function of these hormones in osmoregulation requires further study, including how they may contribute to interactions between the developing cardiovascular and osmoregulatory systems and the maintenance of homeostasis.

32.4.5 Thermoregulation

Embryos produce metabolic heat, which increases as development progresses. Metabolic heat production of the chicken embryo increases from about 35 mW on day 12 to 130–140 mW on days 17 to 18 of incubation, and reaches 160–170 mW at EP (Tazawa et al., 1988b). Over the same developmental range, the egg's thermal conductance is ~70 mW/°C (Tazawa et al., 2001a). This means that metabolic heat production at 38 °C elevates egg temperature (T_{egg}) above ambient temperature (T_a) by ~0.5 °C on day 12 and ~2.5 °C in EP eggs. When eggs are cooled to 28 °C, a new quasi-equilibrium state is reached in 5 h (defined as a change in T_{egg} of less than 0.2 °C/h) (Tazawa and Rahn, 1987). The new difference between T_{egg} and T_a is lower, so that even in EP embryos T_{egg} is only 1.2 °C above T_a. In addition, during the quasi-equilibrium state at lowered T_a, embryos consume the amount of O_2 predicted by a temperature coefficient (Q_{10}) of 2. On the other hand, hatchlings, which are subjected to the same test, have body temperatures ~6 °C above T_a, even right after hatching, and consume much more O_2 than that predicted by Q_{10} of 2 (Tazawa and Rahn, 1987). These observations suggest that chickens are essentially poikilothermic during embryonic stages, even at EP, and rapidly develop the capacity to maintain body temperature upon cold exposure soon after hatching.

When eggs incubated at 38 °C are exposed to a T_a only 2 °C lower, the heat loss from the egg is greater than the heat produced by young embryos and comparable with the heat production in late embryos. If the eggs are exposed to air 10 °C cooler than the egg, the embryos would have to generate heat of about 800 mW to keep T_{egg} steady (Turner, 1986). Yet, even EP embryos can generate at most 170 mW. Consequently, heat loss during cooling exceeds the embryo's maximum rate of heat production. A feeble compensatory capacity, if any, may be overwhelmed by the much larger losses of heat. The egg cools and its metabolic rate decreases as a result of the van't Hoff-Arrhenius effect. The detection of homeothermic capacity thus requires a procedure in which the heat loss from the egg does not overwhelm the heat production of the embryo. The gradual cooling test fulfills this requirement, and it shows that prenatal chicken embryos near term respond to lowered T_a by maintaining $\dot{M}O_2$ until T_a falls below 35 °C (Tazawa et al., 1988b). This response in late embryos is evidently different from that in young embryos. In the prolonged cooling test, late chicken embryos are exposed to a slightly lowered T_a for a prolonged period, the $\dot{M}O_2$ is maintained at a level above that predicted by a Q_{10} of 2. This response also differs from that shown by young embryos (Tazawa et al., 1989a). Consequently, precocial chickens exhibit a feeble, incipient metabolic response to cooling, indicating endothermic homeothermy before hatching. This coincides with enhanced activity of the thyroid gland and increasing concentrations of peripheral thyroid hormones during the last stages of incubation (Decuypere et al., 1979; McNabb, 1987). In fact, while late prenatal embryos in eggs injected with saline show a feeble homeothermic metabolic response to gradual cooling, this response is absent in eggs treated with thiourea, which antagonizes the metabolic effect of thyroid hormones (Tazawa et al., 1989c). Additionally, while the compensatory metabolic response disappears in embryos exposed to hypoxia, it is augmented in perinatal embryos in hyperoxia or following improved oxygenation by opening the shell over the air cell (Tazawa et al., 1989c; Dzialowski et al., 2007; Szdzuy et al., 2008). These results indicate that the homeothermic metabolic response in late embryos is "O_2 conductance limited" in precocial chickens (Tazawa et al., 1988b).

The development of homeothermy in precocial and altricial birds has been modeled (Figure 32.11) (Tazawa et al., 1988b; Whittow and Tazawa, 1991; Tzschentke and Rumpf, 2011 for review). The transition for a precocial bird takes place in four stages; (1) an Arrhenius-limited stage in which $\dot{M}O_2$ is directly related to the temperature with a Q_{10} value

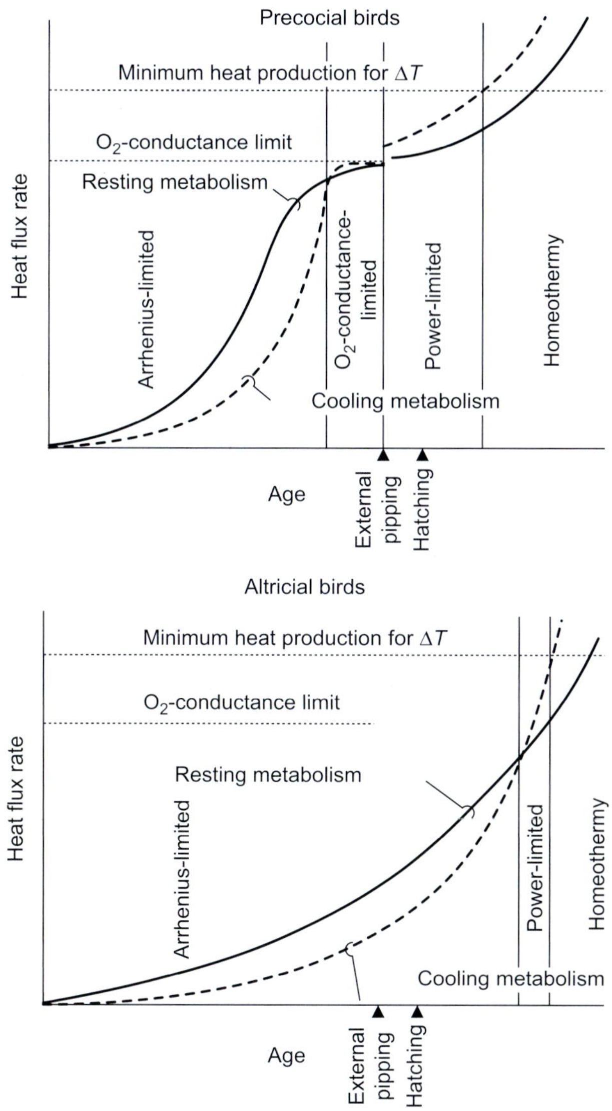

FIGURE 32.11 Models of the development of homeothermy in precocial and altricial birds. Minimum Heat Production for ΔT equals the amount of heat needed to keep the egg temperature warmer than ambient temperature by ΔT and homeothermy occurs when the embryo heat production reaches this level. *From Tazawa et al. (1988b), with permission from Elsevier.*

of approximately 2; (2) an O_2-conductance-limited stage in which $\dot{M}O_2$ is limited by the rate of diffusion of O_2 through the shell and CAM; (3) a power-limited stage in which the embryo has a limited capacity to generate heat in response to cooling, which is a function of the maturity of the tissues and development of thyroid activity; and (4) "full-blown" homeothermy. An altricial species never passes through the O_2-conductance-limited stage and is subject to the Arrhenius limitation until after it hatches.

The comparative metabolic responses to prolonged cooling have been examined in the precocial duck, semi-precocial brown noddy, and altricial pigeon (*Columba livia domestica*) (Matsunaga et al., 1989; Kuroda et al., 1990; Tazawa et al., 2001a). Incipient homeothermic ability appears in the duck during prenatal development, but it is not evident in the pigeon even after emergence from the shell. The precocial chicken and semi-precocial noddy are intermediate in their metabolic response between the duck and the pigeon.

The development and maturation of homeothermy is expected to occur and progress even during the prenatal period in the highly precocial emu. In addition to the precocial nature of the emu, its large egg with a smaller surface area-to-volume ratio and greater contribution of blood circulation compared to smaller eggs most likely enhances thermoregulatory capacity. For emus, a more apparent metabolic response test is used in which embryos and hatchlings are exposed for a prolonged period (e.g., 1.5h) to T_a altered sequentially by 10°C, for example, 35-25-35°C for embryos or 25-35-25°C for hatchlings (sequential cooling-warming test, with ΔT_a of ±10°C) (Dzialowski et al., 2007). Hatchlings and EP embryos respond to ΔT_a with an endothermic change in $\dot{M}O_2$, showing an inverse metabolic response with marked increase and decrease in $\dot{M}O_2$ in response to sequential cooling and warming bouts. Late prenatal (day 45) and IP (day 49) embryos do not change $\dot{M}O_2$ in response to ΔT_a in air, but demonstrate partial (day 45) or apparent (IP) endothermic change in $\dot{M}O_2$ when the test is run in 40% O_2. This suggests that the late prenatal emu embryo already possesses homeothermic ability, but it is limited by the eggshell gas GO_2.

Changes in IHR, with apparent increases or decreases in HR baseline with oscillating pattern, in response to ΔT_a, are also effective in investigating endothermic capacity in precocial embryos and hatchlings (Tazawa et al., 2001a, 2004; Tamura et al., 2003; Khandoker et al., 2004). In the sequential cooling-warming test with ΔT_a of ±10°C, the sequence of exposure (cooling-warming and vice versa) does not affect the endothermic HR response (Yoneta et al., 2006b). In chick hatchlings the endothermic HR response is demonstrated to be advanced by ~1 day in broiler compared with White Leghorn during 2 days of post-hatch life (Yoneta et al., 2007).

In the chicken embryo, the baseline of IHR shows a thermo-conformity pattern in response to a decrease in T_a during EP (day 20), but it changes little on day 21. IHR rises accompanying HR oscillation on day 22, when the embryo stays inside the egg (Tazawa et al., 2001a; Andrewartha et al., 2011b). Although the embryo fails to hatch after EP, it matures equivalently to a hatchling on day 21 and day 22. In newly hatched chicks, HR oscillations with a period of 10–25s occur frequently in air, reported as Type II HR fluctuation (Moriya et al., 1999, 2000). In addition, IHR oscillates with lowering T_a, resulting in Type II low-frequency HR oscillation, which disappears when hatchlings are transferred to high T_a (Tazawa et al., 2002b; Khandoker et al., 2004). Consequently, chicken hatchlings possess low-frequency HR oscillation (Type II HR fluctuation) in relation to their thermoregulation (Tazawa, 2005 for review).

In duck embryos (Figure 32.12 (A); day 24), the HR response during pre-IP stage indicates thermoconformity (Andrewartha et al., 2011b). However, the recovery of HR (and T_{egg}) at T_a of 38 °C is faster than chicken embryos (cf. Figure 32.1 of Tazawa et al., 2001a). By EP (Figure 32.12(C), (D); day 27), duck embryos demonstrate an endothermic HR response (increase during cold exposure) larger than chicken EP embryos. Immediately after hatching, the wet duckling cannot maintain the increased HR, succumbs to the cold, and HR decreases (E). However, the HR of ducklings blotted dry within 2h is maintained at similar values at T_a of 35 °C (F), demonstrating incomplete endothermic capacity. By 2 and 13h post-hatching (Figure 32.12(G), (H)), the plumage of the ducklings has naturally dried and their HR shows an apparent endothermic response with a small decrease in body temperature, T_b ($\Delta T_b = -1.9$ and -2.0 °C, respectively), close to full-blown homeothermy. Ducklings must attain thermoregulatory competence early (relative to the domestic chicken) during perinatal development to live in an aquatic environment soon after hatching.

In the more highly precocial emu, pre-IP and IP embryos exhibit thermoconformity and incomplete endothermic response of HR, respectively, but the apparent endothermic HR response occurs in EP embryos just like the endothermic metabolic response (Fukuoka et al., 2006; Dzialowski et al., 2007; Andrewartha et al., 2011b).

Assessment of $\dot{M}O_2$ and HR responses to various tests that alter incubation temperature have been successfully utilized to assess thermoregulatory ability of avian embryos. The timing of endothermic responses differs along the precocial to altricial continuum, with precocial species generally showing thermoregulatory ability earlier than altricial species. Examination of mitochondrial, cellular, and tissue-level mechanisms are now required to further our understanding of how thermoregulatory ability develops in embryos and hatchlings.

32.5 ARTIFICIAL INCUBATION

32.5.1 Preincubation Egg Storage

Many precocial birds laying multiple eggs in a clutch start their incubation with the penultimate or ultimate eggs. Consequently, first-laid eggs are stored in the nest until incubation starts, sometimes for many days. Egg storage is not just a phenomenon of nature, of course. Egg storage commonly occurs in commercial conditions of artificial incubation. If the storage temperature for freshly laid chicken eggs is kept below physiological zero (25–27 °C), dormancy of the embryo is maintained and fertile eggs can be stored for several days without a major loss of hatchability (Butler, 1991; Wilson, 1991). The optimal temperature for 3 to 7 days storage of chicken eggs is 16–17 °C and it drops to 10–12 °C for eggs stored for more than 7 days (Butler, 1991; Wilson, 1991). However, prolonged preincubation egg storage results in malformations and retarded growth of the embryos, decreased hatchability, and increased incubation period, and it even influences hatchling growth (Arora and Kosin, 1966; Mather and Laughlin, 1979). The hatchability of the northern bobwhite quail (*Colinus virginianus*) remains above 70% after 14 days of preincubation storage at 20–22 °C, but decreases to below 30% after 21 days of storage (Reyna, 2010). These deleterious effects are related to not only the length of storage but also the environmental and physical conditions such as temperature, relative humidity, atmospheric gas composition, orientation, and positional changes during storage (Brake et al., 1997).

Prolonged preincubation egg storage also affects the physiological function of developing chicken embryos (Haque et al., 1996; Fasenko, 2007). Albumen quality, an indicator of overall egg quality, decreases with storage (Scott and Silversides, 2000; Reyna, 2010). Furthermore, while the developmental patterns of $\dot{M}O_2$ are consistent among unstored (control) eggs, those stored for 20 and 30 days at 10–11 °C are varied and depressed among eggs; the depression of the $\dot{M}O_2$ increases in severity as the storage duration increases. The developmental trajectories of HR in stored eggs are flattened compared with those of control eggs. As a result, the O_2 pulse (O_2 uptake every heartbeat) is markedly lowered by preincubation storage, decreasing blood O_2 transport and retarding embryo growth in stored eggs, resulting in death during the last days of incubation (Haque et al., 1996).

32.5.2 Egg Turning

In many avian species, incubating adults actively move their eggs in the nest (egg-turning). The critical period for lack of egg-turning in artificial incubation of the chicken egg ranges from days 3 to 7 of artificial incubation (New, 1957; Deeming, 1989a). Chicken eggs should be turned minimally 3 times a day. Turning more than 24 times a day does not further enhance hatchability. Lack of egg-turning is detrimental not only to hatchability, but can slow incubation, the development of the CAM, and embryo growth (New, 1957; Tazawa, 1980b; Deeming et al., 1987; Tullet and Deeming, 1987; Deeming, 1989a,b).

Failure to turn eggs during incubation also produces adverse effects on gas exchange through the CAM (Tazawa, 1980b). The movement of albumen into the amnion is retarded, and the unabsorbed albumen, which becomes more viscid and heavy as it loses water early in incubation, sinks towards the lower end of the egg, where it remains. The chorioallantois fails to fold around the unabsorbed

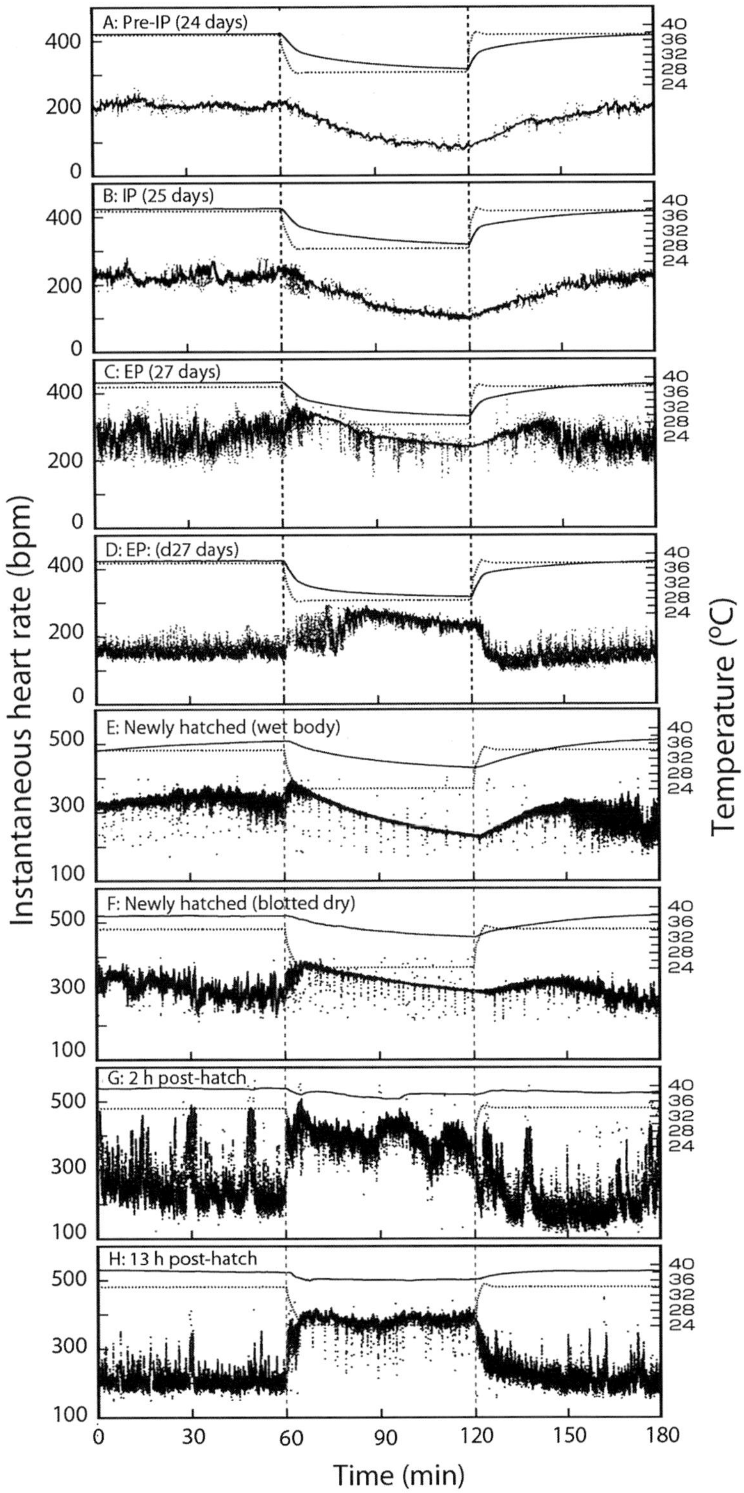

FIGURE 32.12 Responses of instantaneous heart rate (IHR) in duck embryos and hatchlings exposed sequentially to high ambient temperature (T_a = 38 °C for embryos and 35 °C for hatchlings; dotted line), low ambient temperature (T_a = 28 °C for embryos and 25 °C for ducklings), and high temperature again for 60 min, respectively. The solid line represents egg temperature or body temperature. *From Andrewartha et al. (2011a), with permission from Elsevier.*

albumen, and the interposition of the albumen between the CAM and inner shell membrane reduces gas exchange. These effects cause a pronounced fall in the arterialized blood PO_2 of late embryos, which is accompanied by an increase in Hct. The inhibition of gas exchange is reflected in the decreased $\dot{M}O_2$ of unturned eggs compared to turned eggs (Pearson et al., 1996).

32.5.3 Ambient Temperature and Incubation

Freshly laid eggs can be stored at a low temperature to maintain dormancy of embryos. However, once incubation starts, T_a must be kept within a certain range so that the embryo temperature is maintained and cell proliferation can proceed. In artificial incubation of chicken eggs at constant T_a, hatchability decreases when T_a is lowered to 35 °C or elevated to 40 °C from the optimal temperature of 37.5 °C. At 35 °C, incubation period and embryonic development of chickens are ~3 days longer than at 38 °C (Tazawa, 1973; Black and Burggren, 2004a). Survival during total incubation period is reduced at 35 °C (Black and Burggren, 2004a) and temperatures below this threshold are lethal. Hypothermal incubation decreases $\dot{M}O_2$ in proportion to retarded development of embryos and preserves the relative timing of IP and EP (Tazawa, 1973; Black and Burggren, 2004a). However, at the final stages of hypothermal incubation, hematological development and blood O_2-carrying capacity are retarded (Black and Burggren, 2004b), and eventually hypothermal incubation causes a significant delay of the relative timing of the onset of thermoregulatory ability (Tzschentke et al., 2001; Nichelmann, 2004; Black and Burggren, 2004b; Mortola, 2006).

The tolerance limits of developing embryos to acutely lowered and elevated T_a have been investigated in chickens in reference to their HR (Tazawa and Rahn, 1986; Ono et al., 1994). When day 10 embryos are exposed to a T_a of 28 °C or 18 °C, HR decreases in an exponential fashion to reach plateau values during 2–3 h of exposure. The plateau values are ~100 bpm at 28 °C and ~30 bpm at 18 °C, which are maintained until irreversible cardiac arrest occurs at ~100 and ~60 h after exposure to 28 °C and 18 °C, respectively. Accordingly, day 10 embryos survive exposure to 28 °C for no less than 4 days and to 18 °C for about 2.5 days. Reduction of T_a to 8 °C forces the heartbeat of day 10 embryos to cease after ~3 h. However, the cardiac arrest at this low T_a (8 °C) does not mean the death of embryos. Ten-day embryos survive the low T_a, without a heartbeat, for 18 h more, and the heart begins to beat again after rewarming at 38 °C. The survival time at 8 °C is reduced as the embryos grow. While the heart of day 6 embryos begins to beat even after 1 day exposure to 8 °C, the heart of day 20 embryos fails to beat after an 8 h exposure (Tazawa and Rahn, 1986).

Although chicken embryos can withstand a lowered T_a for a prolonged period without a heartbeat, at an increased T_a the cardiac arrest following arrhythmia is irreversible and the embryos cannot withstand exposure to an increased T_a for a prolonged period (Ono et al., 1994). The HR of embryos increases in an exponential fashion at increased T_a. When T_{egg} reaches 46–47 °C, regardless of the developmental stage of embryos, the HR becomes arrhythmic and irreversible cardiac arrest follows. The lethal T_a and tolerance time of chicken embryos depend upon the time T_{egg} takes to reach a lethal critical value of 46–47 °C. At a T_a of 48 °C, the tolerance time of day 12 embryos is ~100 min and that of day 20 embryos shortens by about half. The tolerance time to 48 °C exposure is shortened as the embryos grow. This may be in part due to the higher metabolic heat produced by late embryos compared to earlier embryos. The HR reaches ~450 bpm at the critical T_{egg} (46–47 °C) (Ono et al., 1994).

The embryos of ground-nesting birds in semiarid and arid areas, in particular, experience environmental temperatures well in excess of typical avian incubation temperatures. For example, the northern bobwhite quail can experience nest temperatures as high as 45 °C in the wild. Not surprisingly, this species shows remarkable tolerance to brief preincubation hyperthermia. Bobwhite quail embryos can survive 6 h exposure to 46 °C, 3 h exposure to 49 °C, and, remarkably, 1 h exposure to 50 °C (Reyna and Burggren, 2012). While these temperatures certainly increase mortality and decrease hatching rates, the high temperature tolerance of the bobwhite quail highlights the importance of a comparative approach to examining temperature tolerance. Extreme temperatures in the natural environment may result in increased tolerance in avian species, which warrants further investigation.

32.5.4 Humidity

Successful artificial incubation of chicken eggs occurs over a relative humidity (RH) range of 40–70% (Robertson, 1961). Within the successful incubation range, 53% RH is optimal for embryo survival and hatchability (Bruzual et al., 2000).

High (>85%) and low (<30%) humidity increases mortality in chicken embryos (Ar and Rahn, 1980; Bolin, 2009). Under low humidity, inadequate water content can result from increased water loss across the porous eggshell due to a high vapor pressure gradient between the inside of the egg and surrounding air. Under high humidity, water loss is reduced and this can exert its osmotic stress within the embryo. Allantoic fluid volume, which is a balance between renal filtrate production and water reabsorption into the embryo, is reduced under high water loss conditions, and conversely increased when water loss is low (Davis et al., 1988). Under high water loss, water is reabsorbed from the allantois, uric acid levels in the allantois increase, and sodium is actively transported to maintain the allantoic fluid hypotonic to the blood and aid water reabsorption (Hoyt, 1979; Davis et al., 1988). Plasma calcium, sodium, and potassium levels are elevated in late-stage embryos, a sign of osmotic stress (Davis et al., 1988). After exposure to <30% RH, the kidneys of day 18 chicken embryos have additional glomeruli, more functioning glomeruli, and high cloacal fluid osmolality, illustrating an increased filtering capacity (Bolin, 2009).

Extremes in humidity or water loss reduce hatching success (Davis et al., 1988). Water balance appears to have the greatest effects on hatchability early in incubation, with hatchability related to water loss during the first half of incubation, rather than total water loss (Snyder and Birchard, 1982). Humidity extremes also affect late-stage embryo wet mass; that is, embryos from low-humidity or high water loss environments have smaller wet masses while those from high-humidity environments have higher wet masses (Bruzual et al., 2000). In low water loss environments, excess water is not always incorporated into wet mass but is instead left behind in the albumen (Davis et al., 1988). Growth and oxygen consumption can be retarded after exposure to extreme humidity (Bolin, 2009), but embryos that do hatch under extreme humidity reach normal weight within 7 to 10 days post hatch (Davis et al., 1988).

32.6 CONCLUSIONS AND FUTURE DIRECTIONS

Incubation is the first important process experienced by the embryonic life stages of any avian species. The characteristics of the egg itself and the surrounding environment determine the success and timing of embryonic development. As this chapter has demonstrated, the study of avian embryonic physiology has provided extensive information on the complex processes that occur during avian development. We examined the physical processes of incubation, including the transfer of water and heat between the egg and environment. We provided an overview of the development of the form and function of the major physiological systems, including the gas exchange, acid–base, cardiovascular, osmoregulatory, and thermoregulatory systems. The conditions for successful artificial incubation were discussed, as well as the effects of incubation extremes. However, numerous possibilities still exist to expand our knowledge of avian development by utilizing new physiological techniques and undertaking studies using a cross-discipline approach. Molecular and cellular studies will be important in revealing the mechanistic underpinnings of system- and organism-level observations. Furthermore, many fruitful areas of research at all levels, from genes to the whole organism, remain. These include the timing and control of important physiological events during development, how and when systems are perturbed during development and if recovery is possible, and how systems work together during development. Comparative approaches are also needed to place physiological processes in an ecological and evolutionary context. Addressing these research areas will allow us to further understand avian embryos for the complex, multisystem organisms that they are, while also contributing to the broader field of vertebrate development.

ACKNOWLEDGMENTS

The authors thank the U.S. National Science Foundation (grant # IOS-1025823) for financial support.

REFERENCES

Abdel-Malek, E.T., 1950. Early development of the urinogenital system in the chick. J. Morphol. 86, 599–626.

Ackerman, R.A., Rahn, H., 1981. *In vivo* O_2 and water vapor permeability of the hen's eggshell during early development. Respir. Physiol. 45, 1–8.

Alford, P.W., Taber, L.A., 2003. Regional epicardial strain in the embryonic chick heart during the early looping stages. J. Biomech. 36, 1135–1141.

Altimiras, J., Crossley II, D.A., 2000. Control of blood pressure mediated by baroreflex changes of heart rate in the chicken embryo (*Gallus gallus*). Am. J. Physiol. Reg. Int. Comp. Physiol. 278, R980–R986.

Altimiras, J., Crossley II, D.A., Villamor, E., 2009. Prenatal development of cardiovascular regulation in avian species. In: Glass, M.L., Wood, S.C. (Eds.), Cardio-respiratory control in vertebrates. Springer Verlag, Berlin, pp. 397–427.

Andrewartha, S.J., Tazawa, H., Burggren, W.W., 2011a. Hematocrit and blood osmolality in developing chicken embryos (*Gallus gallus*): *in vivo* and *in vitro* regulation. Respir. Physiol. Neurobiol. 179, 142–150.

Andrewartha, S.J., Tazawa, H., Burggren, W.W., 2011b. Embryonic control of heart rate: examining developmental patterns and temperature and oxygenation influences using embryonic avian models. Respir. Physiol. Neurobiol. 178, 84–96.

Ar, A., Rahn, H., 1980. Water in the avian egg overall budget of incubation. Am. Zool. 20, 373–384.

Ar, A., Paganelli, C.V., Reeves, R.B., Greene, D.G., Rahn, H., 1974. The avian egg: water vapor conductance, shell thickness, and functional pore area. Condor 76, 153–158.

Ar, A., Visschedijk, A.H.J., Rahn, H., Piiper, J., 1980. Carbon dioxide in the chick embryo towards end of development: effects of He and SF6 in breathing mixtures. Respir. Physiol. 40, 293–307.

Arora, K.L., Kosin, I.L., 1966. Changes in the gross morphological appearance of chicken and turkey blastoderms during preincubation. Poult. Sci. 45, 819–825.

Atwell, W.J., Hanan, E.B., 1926. The time during which the mesonephros and the metanephros of the developing chick are able to store trypan blue. Anat. Rec. 32, 228.

Barnes, A.E., Jensen, W.N., 1959. Blood volume and red cell concentration in the normal chick embryo. Am. J. Physiol. 197, 403–405.

Biebach, H., 1986. Energetics of rewarming a clutch in starlings (*Sturnus vulgaris*). Physiol. Zool. 59, 69–75.

Bissonnette, J.M., Metcalfe, J., 1978. Gas exchange of the fertile hen's egg: components of resistance. Respir. Physiol. 34, 209–218.

Bjønnes, P.O., Aulie, A., Høiby, M., 1987. Effects of hypoxia on the metabolism of embryos and chicks of domestic fowl. J. Exp. Zool. 1 (Suppl.), 209–212.

Black, J.L., Burggren, W.W., 2004a. Acclimation to hypothermic incubation in developing chicken embryos (*Gallus domesticus*) I. Developmental effects and chronic and acute metabolic adjustments. J. Exp. Biol. 207, 1543–1552.

Black, J.L., Burggren, W.W., 2004b. Acclimation to hypothermic incubation in developing chicken embryos (*Gallus domesticus*) II. Hematology and blood O_2 transport. J. Exp. Biol. 207, 1553–1561.

Bolin, G.M., 2009. Incubation humidity as a biological stressor on the osmoregulatory developmental program of the chicken, *Gallus gallus domesticus*. Doctor of Philosophy, University of North Texas, Denton.

Boutilier, R.G., Gibson, M.A., Toews, D.P., Anderson, N.G., 1977. Gas exchange and acid-base regulation in the blood and extraembryonic fluids of the developing chicken embryo. Respir. Physiol. 31, 81–89.

Brake, J., Walsh, T.J., Benton Jr., C.E., Petitte, J.N., Meijerhof, R., Penalva, G., 1997. Egg handling and storage. Poult. Sci. 76, 144–151.

Branum, S.B., Yamada-Fisher, M., Burggren, W., 2013. Reduced heart rate and cardiac output differentially affect angiogenesis, growth, and development in early chick embryo (*Gallus domesticus*). Physiol. Biochem. Zool. 86, 370–382.

Bruzual, J., Peak, S., Brake, J., Peebles, E., 2000. Effects of relative humidity during incubation on hatchability and body weight of broiler chicks from young breeder flocks. Poult. Sci. 79, 827–830.

Burggren, W.W., 2004. What is the purpose of the embryonic heart beat? Or how facts can ultimately prevail over physiological dogma. Physiol. Biochem. Zool. 77, 333–345.

Burggren, W.W., 2013. Cardiovascular development and angiogenesis in the early vertebrate embryo. Cardiovasc. Eng. Tech., 4, 234–245.

Burggren, W.W. Epigenetics as a source of variation in comparative animal physiology – or Lamarck is lookin' pretty good these days. J. Exp. Biol., in press (It may be printed in March).

Burggren, W.W., Andrewartha, S.J., Tazawa, H., 2012. Interactions of acid-base balance and hematocrit regulation during environmental respiratory gas challenges in developing chicken embryos (*Gallus gallus*). Respir. Physiol. Neurobiol. 183, 135–148.

Burggren, W.W., Keller, B. (Eds.), 1997. Development of cardiovascular systems: molecules to organisms. Cambridge Univ. Press, New York.

Burggren, W.W., Khorrami, S., Pinder, A., Sun, T., 2004. Body, eye, and chorioallantoic vessel growth are not dependent on cardiac output level in day 3–4 chicken embryos. Am. J. Physiol. Reg. Int. Comp. Physiol. 287, R1399–R1406.

Burggren, W.W., Tazawa, H., Thompson, D., 1994. Genetic and maternal environmental influences on embryonic physiology: intraspecific variability in avian embryonic heart rates. Isr. J. Zool. 40, 351–362.

Burggren, W.W., Warburton, S., 1994. Patterns of form and function in developing hearts: contributions from non-mammalian vertebrates. Cardioscience 5, 183–191.

Burggren, W.W., Warburton, S.J., Slivkoff, M.D., 2000. Interruption of cardiac output does not affect short-term growth and metabolic rate in day 3 and 4 chick embryos. J. Exp. Biol. 203, 3831–3838.

Butler, D.E., 1991. Egg handling and storage at the farm and hatchery. In: Tullett, S.G. (Ed.), Avian incubation. Butterworth, London, pp. 195–203.

Cain, J.R., Abbott, U.K., Rogallo, V.L., 1967. Heart rate of the developing chick embryo. Proc. Soc. Exp. Biol. Med. 126, 507–510.

Chiba, Y., Fukuoka, S., Niiya, A., Akiyama, R., Tazawa, H., 2004. Development of cholinergic chronotropic control in chick (*Gallus gallus domesticus*) embryos. Comp. Biochem. Physiol. A 137, 65–73.

Chiba, Y., Khandoker, A.H., Nobuta, M., Moriya, K., Akiyama, R., Tazawa, H., 2002. Development of respiratory rhythms in perinatal chick embryos. Comp. Biochem. Physiol. A 131, 817–824.

Clark, E.B., Hu, N., Dummett, J.L., Vandekieft, G.K., Olson, C., Tomanek, R., 1986. Ventricular function and morphology in chick embryo from stages 18 and 29. Am. J. Physiol. 250, H407–H413.

Crossley II, D., Altimiras, J., 2000. Ontogeny of cholinergic and adrenergic cardiovascular regulation in the domestic chicken (*Gallus gallus*). Am. J. Physiol. Regul. Integr. Comp. Physiol. 279, R1091–R1098.

Crossley II, D.A., Bagatto, B.P., Dzialowski, E.M., Burggren, W.W., 2003. Maturation of cardiovascular control mechanisms in the embryonic emu (*Dromiceius novaehollandiae*). J. Exp. Biol. 206, 2703–2710.

Crossley, D., Jonker, S., Hicks, J., Thornburg, K., 2010. Maturation of the angiotensin II cardiovascular response in the embryonic White Leghorn chicken (*Gallus gallus*). J. Comp. Physiol. B. 180, 1057–1065.

Davis, T.A., Shen, S.S., Ackerman, R.A., 1988. Embryonic osmoregulation: consequences of high and low water loss during incubation of the chicken egg. J. Exp. Zool. 245, 144–156.

Dawes, C.M., 1976. A method for recording the respiratory and hatching movements of the chick embryo. J. Exp. Biol. 64, 379–383.

Dawes, C.M., Simkiss, K., 1969. The acid-base status of the blood of the developing chick embryo. J. Exp. Biol. 50, 79–86.

Decuypere, E., Nouwen, E.J., Kühn, E.R., Geers, R., Michels, H., 1979. Differences in serum iodohormone concentration between chick embryos with and without the bill in the air chamber at different incubation temperatures. Gen. Comp. Endocrinol. 37, 264–267.

Deeming, D.C., 1989a. Characteristics of unturned eggs: critical period, retarded embryonic growth and poor albumen utilisation. Br. Poult. Sci. 30, 239–249.

Deeming, D.C., 1989b. Importance of sub-embryonic fluid and albumen in the embryo's response to turning of the egg during incubation. Br. Poult. Sci. 30, 591–606.

Deeming, D.C., Rowlett, K., Simkiss, K., 1987. Physical influences on embryo development. J. Exp. Zool. 1 (Suppl.), 341–345.

Dzialowski, E.M., Burggren, W.W., Komoro, T., Tazawa, H., 2007. Development of endothermic metabolic response in embryos and hatchlings of the emu (*Dromaius novaehollandiae*). Respir. Physiol. Neurobiol. 155, 286–292.

Dzialowski, E.M., Sirsat, T., van der Sterren, S., Villamor, E., 2011. Prenatal cardiovascular shunts in amniotic vertebrates. Respir. Physiol. Neurobiol. 178, 66–74.

Dzialowski, E.M., Sotherland, P.R., 2004. Maternal effects of egg size on emu *Dromaius novaehollandiae* egg composition and hatchling phenotype. J. Exp. Biol. 207, 597–606.

Doneen, B.A., Smith, T.E., 1982. Ontogeny of endocrine control of osmoregulation in chick embryo: II. Actions of prolactin, arginine vasopressin, and aldosterone. Gen. Comp. Endocrinol. 48, 310–318.

Elfwing, M., Lundengård, K., Altimiras, J., 2011. Fetal development of baroreflex sensitivity: the chicken embryo as a case model. Respir. Physiol. Neurobiol. 178, 75–83.

Erasmus, B., De, W., Howell, B.J., Rahn, H., 1970/71. Ontogeny of acid-base balance in the bullfrog and chicken. Respir. Physiol. 11, 46–53.

Erasmus, B., De, W., Rahn, H., 1976. Effects of ambient pressures, He and SF_6 on O_2 and CO_2 transport in the avian egg. Respir. Physiol. 27, 53–64.

Everaert, N., Willemsen, H., Willems, E., Franssens, L., Decuypere, E., 2011. Acid-base regulation during embryonic development in amniotes, with particular reference to birds. Respir. Physiol. Neurobiol. 178, 118–128.

Faber, J.J., 1968. Mechanical function of the septating embryonic heart. Am. J. Physiol. 214, 475–481.

Faber, J.J., Green, T.J., Thornburg, K.L., 1974. Embryonic stroke volume and cardiac output in the chick. Dev. Biol. 41, 14–21.

Fasenko, G.M., 2007. Egg storage and the embryo. Poult. Sci. 86, 1020–1204.

Fukuoka, S., Khandoker, A.H., Dzialowski, E.M., Burggren, W.W., Tazawa, H., 2006. Development of endothermic heart rate response in emu (*Dromaius novaehollandiae*) embryos. In: Yahav, S., Tzschentke, B. (Eds.), New insights into fundamental physiology and peri-natal adaptation of domestic fowl. Nottingham Univ. Press, pp. 29–42.

Gabrielli, M.G., Accili, D., 2010. The chick chorioallantoic membrane: a model of molecular, structural, and functional adaptation to transepithelial ion transport and barrier function during embryonic development. J. Biomed. Biotechnol. 2010, 12. http://dx.doi.org/10.1155/2010/940741. ID940741.

Girard, H., 1971. Respiratory acidosis with partial metabolic compensation in chick embryo blood during normal development. Respir. Physiol. 13, 343–351.

Girard, H., 1973. Arterial pressure in the chick embryo. Am. J. Physiol. 224, 454–460.

Grabowski, C.T., 1961. Lactic acid accumulation as a cause of hypoxia-induced malformations in the chick embryo. Science 134, 1359–1360.

Grabowski, C.T., 1966. Physiological changes in the bloodstream of chick embryos exposed to teratogenic doses of hypoxia. Dev. Biol. 13, 199–213.

Graves, J.S., Dunn, B.E., Brown, S.C., 1986. Embryonic chick allantois: functional isolation and development of sodium transport. Am. J. Physiol. Cell Physiol. 251, C787–C794.

Groenendijk, B.C.W., Stekelenburg-de Vos, S., Vennemann, P., Wladimiroff, J.W., Nieuwstadt, F.T.M., Lindken, R., Westerweel, J., Hierck, B.P., Ursem, N.T.C., Poelmann, R.E., 2008. The Endothelin-1 pathway and the development of cardiovascular defects in the haemodynamically challenged chicken embryo. J. Vasc. Res. 45, 54–68.

Hall, C.E., Hurtado, R., Hewett, K.W., Shulimovich, M., Poma, C.P., Reckova, M., Justus, C., Pennisi, D.J., Tobita, K., Sedmera, D., Gourdie, R.G., Mikawa, T., 2004. Hemodynamic-dependent patterning of endothelin converting enzyme 1 expression and differentiation of impulse-conducting Purkinje fibers in the embryonic heart. Development 131, 581–592.

Haque, M.A., Pearson, J.T., Hou, P.C.L., Tazawa, H., 1996. Effects of pre-incubation egg storage on embryonic functions and growth. Respir. Physiol. 103, 89–98.

Hiruma, T., Nakamura, H., 2003. Origin and development of the pronephros in the chick embryo. J. Anat. 203, 539–552.

Ho, D., Burggren, W.W., 2010. Epigenetics and transgenerational transfer: a physiological perspective. J. Exp. Biol. 213, 3–16.

Ho, D., Burggren, W.W., 2012. Parental hypoxic exposure confers offspring hypoxia resistance in zebrafish (*Danio rerio*). J. Exp. Biol. 215, 4208–4216.

Ho, D., Reed, W.L., Burggren, W.W., 2011. Egg yolk environment differentially influence physiological and morphological development of broiler and layer chicken embryos. J. Exp. Biol. 214, 619–628.

Höchel, J., Akiyama, R., Masuko, T., Pearson, J., Nickelmann, M., Tazawa, H., 1998. Development of heart rate irregularities in chick embryos. Am. J. Physiol. Heart Cir. Physiol. 275, H527–H533.

Høiby, M., Aulie, A., Reite, O.B., 1983. Oxygen uptake in fowl eggs incubated in air and pure oxygen. Comp. Biochem. Physiol. A 74, 315–318.

Hoyt, D.F., 1979. Osmoregulation by avian embryos: the allantois functions like a toad's bladder. Physiol. Zool. 52, 354–362.

Howe, R.S., Burggren, W.W., Warburton, S.J., 1995. Fixed patterns of bradycardia during late embryonic development in domestic fowl with C locus mutation. Am. J. Physiol. 268, H56–H60.

Hu, N., Clark, E.B., 1989. Hemodynamics of the stage 12 to stage 29 chick embryo. Cir. Res. 65, 1665–1670.

Hughes, A.F.W., 1949. The heart output of the chick embryo. J. Roy. Microsc. Soc. 69, 145–152.

Kato, K., Moriya, K., Dzialowski, E.M., Burggren, W.W., Tazawa, H., 2002. Cardiac rhythms in prenatal and perinatal emu embryos. Comp. Biochem. Physiol. A 131, 775–785.

Khandoker, A.H., Dzialowski, E.M., Burggren, W.W., Tazawa, H., 2003. Cardiac rhythms of late pre-pipped and pipped chick embryos exposed to altered oxygen environments. Comp. Biochem. Physiol. A 136, 289–299.

Khandoker, A.H., Fukazawa, K., Dzialowski, E.M., Burggren, W.W., Tazawa, H., 2004. Maturation of homeothermic response of heart rate to altered ambient temperature in developing chick hatchlings. Am. J. Physiol. Reg. Int. Comp. Physiol. 286, R129–R137.

Kind, C., 1975. The development of the circulating blood volume of the chick embryo. Anat. Embryol. 147, 127–132.

Klempt, M., Ellendorff, F., Grossmann, R., 1992. Functional maturation of arginine vasotocin secretory responses to osmotic stimulation in the chick embryo and the newborn chicken. J. Endocrinol. 133, 269–274.

Kuroda, O., Matsunaga, C., Whittow, G.C., Tazawa, H., 1990. Comparative metabolic responses to prolonged cooling in precocial duck (*Anas domestica*) and altricial pigeon (*Columba domestica*) embryos. Comp. Biochem. Physiol. A 95, 407–410.

le Noble, F., Klein, C., Tintu, A., Pries, A., Buschmann, I., 2008. Neural guidance molecules, tip cells, and mechanical factors in vascular development. Cardiovasc. Res. 78, 232–241.

Lemez, L., 1972. Thrombocytes of chick embryos from the 2nd day of incubation till the 1st postembryonic day. Acta Univ. Carol. Ser. Med. Mono. 53-54, 365–371.

Maksimov, V.F., Korostyshevskaya, I.M., 2013. Morphogenesis and reaction to hypoxia of atrial myoendocrine cells in chick embryos (*Gallus gallus*). J. Evol. Biochem. Physiol. 49, 251–258.

Mather, C.M., Laughlin, K.F., 1979. Storage of hatching eggs: the interaction between parental age and early embryonic development. Br. Poult. Sci. 20, 595–604.

Matsunaga, C., Mathiu, P.M., Whittow, G.C., Tazawa, H., 1989. Oxygen consumption of Brown Noddy (*Anous stolidus*) embryos in a quasi-equilibrium state at lowered ambient temperatures. Comp. Biochem. Physiol. A 93, 707–710.

Mayer, A.A., Metcalfe, J., Stock, M.K., 1995. Relocation during incubation of endothelial nuclei in the chick chorioallantois. Respir. Physiol. 100, 171–176.

Menna, T.M., Mortola, J.P., 2003. Ventilatory chemosensitivity in the chick embryo. Respir. Physiol. Neurobiol. 137, 69–79.

McNabb, F.M.A., 1987. Comparative thyroid development in precocial Japanese quail and altricial ring doves. J. Exp. Zool. 1 (Suppl.), 281–290.

Mitgard, V., Sejrsen, P., Johansen, K., 1985. Blood flow in the brood patch of Bantam hens: evidence of cold vasodilation. J. Comp. Physiol. B 155, 703–709.

Moonen, R.M., Villamor, E., 2011. Developmental changes in mesenteric artery reactivity in embryonic and newly hatched chicks. J. Comp. Physiol. B 181, 1063–1073.

Moriya, K., Höchel, J., Pearson, J., Tazawa, H., 1999. Cardiac rhythms in developing chicks. Comp. Biochem. Physiol. A 124, 461–468.

Moriya, K., Pearson, J.T., Burggren, W.W., Ar, A., Tazawa, H., 2000. Continuous measurements of instantaneous heart rate and its fluctuations before and after hatching in chickens. J. Exp. Biol. 203, 895–903.

Mortola, J.P., 2006. Metabolic response to cooling temperatures in chicken embryos and hatchlings after cold incubation. Comp. Biochem. Physiol. A 145, 441–448.

Mortola, J.P., 2009. Gas exchange in avian embryos and hatchlings. Comp. Biochem. Physiol. A 153, 359–377.

Mortola, J.P., Besterman, A.D., 2007. Gaseous metabolism of the chicken embryo and hatchling during post-hypoxic recovery. Respir. Physiol. Neurobiol. 156, 212–219.

Mortola, J.P., Cooney, E., 2008. Cost of growth and maintenance in chicken embryos during normoxic or hypoxic conditions. Respir. Physiol. Neurobiol. 162, 223–229.

Mortola, J.P., Wills, K., Trippenbach, T., Al Awam, K., 2010. Interactive effects of temperature and hypoxia on heart rate and oxygen consumption of the 3-day old chicken embryo. Comp. Biochem. Physiol. A 155, 301–308.

Mueller, C.A., Burggren, W.W., Crossley II, D.A., 2013a. Angiotensin II and baroreflex control of heart rate in embryonic chickens (*Gallus gallus domesticus*). Am. J. Physiol. Reg. Int. Comp. Physiol 305, R855–R863.

Mueller, C.A., Tazawa, H., Burggren, W.W., 2013b. Dynamics of metabolic compensation and hematological changes in chicken (*Gallus gallus*) embryos exposed to hypercapnia with varying oxygen. Respir. Physiol. Neurobiol. 185, 272–280.

Mühlbauer, E., Hamannt, D., Xu, B., Ivell, R., Udovic, B., Ellendorff, F., Grossmann, R., 1993. Arginine vasotocin gene expression and hormone synthesis during ontogeny of the chicken embryo and the newborn chick. J. Neuroendocrinol. 5, 281–288.

Mulder, T.L.M.C., van Golde, J., Prinzen, F.W., Blanco, C.E., 1997. Cardiac output distribution in the chick embryo from stage 36 to 45. Cardiovasc. Res. 34, 525–528.

Murphy, M.J., Brown, P.S., Brown, S.C., 1986. Osmoregulatory effects of prolactin and growth hormone in embryonic chicks. Gen. Comp. Endocrinol. 62, 485–492.

New, D.A.T., 1957. A critical period for the turning of hen's eggs. J. Embryol. Exp. Morphol. 5, 293–299.

Nichelmann, M., 2004. Perinatal epigenetic temperature adaptation in avian species: comparison of turkey and Muscovy duck. J. Therm. Biol. 29, 613–619.

Nishimura, H., Yang, Y., Hubert, C., Gasc, J.-M., Ruijtenbeek, K., De Mey, J., Boudier, H.A.J.S., Corvol, P., 2003. Maturation-dependent changes of angiotensin receptor expression in fowl. Am. J. Physiol. Reg. Int. Comp. Physiol. 285, R231–R242.

Okuda, A., Tazawa, H., 1988. Gas exchange and development of chicken embryos with widely altered shell conductance from the beginning of incubation. Respir. Physiol. 74, 187–198.

Ono, H., Hou, P.-C., Tazawa, H., 1994. Responses of developing chicken embryos to acute changes in ambient temperature: noninvasive study of heart rate. Isr. J. Zool. 40, 467–479.

Pappano, A.J., Löffelholz, K., 1974. Ontogenesis of adrenergic and cholinergic neuroeffector tranmission in chick embryo heart. J. Pharmacol. Exp. Ther. 191, 468–478.

Pattern, B.M., 1951. Early Embryology of the Chick. McGraw-Hill Book Company, New York.

Pearson, J.T., Haque, M.A., Hou, P.-C.L., Tazawa, H., 1996. Developmental patterns of O_2 consumption, heart rate and O_2 pulse in unturned eggs. Respir. Physiol. 103, 83–87.

Pearson, J.T., Moriya, K., Yanone, M., Tazawa, H., 2000. Development and regulation of heart rate in embryos and hatchlings of gulls (*Larus schistisagus* and *Larus crassirostris*) in relation to growth. J. Comp. Physiol. B 170, 429–438.

Pearson, J.T., Noma, Y., Tazawa, H., 1999. Developmental patterns of heart rate in altricial avian embryos and hatchlings. J. Exp. Biol. 202, 1545–1550.

Pearson, J.T., Tazawa, H., 1999. Ontogeny of heart rate in embryonic and nestling crows (*Corvus corone* and *Corvus macrorhynchos*). J. Comp. Physiol. B 169, 256–262.

Pearson, J.T., Tsuzuki, M., Nakane, Y., Akiyama, R., Tazawa, H., 1998. Development of heart rate in the precocial king quail, *Coturnix chinensis*. J. Exp. Biol. 201, 931–941.

Pedernera, E.A., Lantos, C.P., 1973. The biosynthesis of adrenal steroids by the 15-day-old chick embryo. Gen. Comp. Endocrinol. 20, 331–341.

Pettit, T.N., Whittow, G.C., 1982a. The initiation of pulmonary respiration in a bird embryo: blood and air cell gas tension. Respir. Physiol. 48, 199–208.

Pettit, T.N., Whittow, G.C., 1982b. The initiation of pulmonary respiration in a bird embryo: tidal volume and frequency. Respir. Physiol. 48, 209–218.

Prinzinger, R., Dietz, V., 1995. Qualitative course of embryonic O_2 consumption in altricial and precocial birds. Respir. Physiol. 100, 289–294.

Prinzinger, R., Schmidt, M., Dietz, V., 1995. Embryogeny of oxygen consumption in 13 altricial and precocial birds. Respir. Physiol. 100, 283–287.

Rahn, H., 1991. Why birds lay eggs. In: Deeming, D.C., Ferguson, M.W.J. (Eds.), Egg Incubation: Its Effects on Embryonic Development in Birds and Reptiles. Cambridge Univ. Press, Cambridge, pp. 345–360.

Rahn, H., Ar, A., 1974. The avian egg: incubation time and water loss. Condor 76, 147–152.

Rahn, H., Carey, C., Balmas, K., Bhatia, B., Paganelli, C., 1977. Reduction of pore area of the avian eggshell as an adaptation to altitude. Proc. Natl. Acad. Sci. U.S.A. 74, 3095–3098.

Rahn, H., Matalon, S., Sotherland, P.R., 1985. Circulatory changes and oxygen delivery in the chick embryo prior to hatching. In: Johansen, K., Burggren, W. (Eds.), Cardiovascular Shunts: Phylogeneic, Ontogenetic and Clinical Aspects. Munksggard, Copenhagen, pp. 179–198.

Rahn, H., Paganelli, C.V., Ar, A., 1974. The avian egg: air-cell gas tension, metabolism and incubation time. Respir. Physiol. 22, 297–309.

Rahn, H., Poturalski, S.A., Paganelli, C.V., 1990. The acoustocardiogram: a non-invasive method for measuring heart rate of avian embryos in ovo. J. Appl. Physiol. 69, 1546–1548.

Reyna, K.S. (2010). Thermal Stress during Pre-incubation Induces Subsequent Developmental Plasticity in Northern Bobwhites. Doctor of Philosophy, University of North Texas.

Reyna, K.S., Burggren, W.W., 2012. Upper lethal temperatures of Northern Bobwhite embryos and the thermal properties of their eggs. Poult. Sci. 91, 41–46.

Robertson, I.S., 1961. Studies on the effect of humidity on the hatchability of hen's eggs I. The determination of optimum humidity for incubation. J. Agric. Sci. 57, 185–194.

Romanoff, A.L., 1960. The Avian Embryo. Macmillan, New York.

Romanoff, A.L., 1967. Biochemistry of the Avian Embryo. Wiley, New York.

Romijn, C., 1948. Respiratory movements of the chicken during the parafoetal period. Physiol. Comp. Oecol. 1, 24–28.

Sakamoto, Y., Haque, M.A., Ono, H., Pearson, J., Tazawa, 1995. Two-dimensional cardiogenic ballistic movements of avian eggs. Med. Biol. Eng. Comput. 33, 611–614.

Savary, K., Michaud, A., Favier, J., Larger, E., Corvol, P., Gasc, J., 2005. Role of the renin-angiotensin system in primitive erythropoiesis in the chick embryo. Blood 105, 103–110.

Saint Petery, L.B., Van Mierop, L.H.S., 1974. Evidence for presence of adrenergic receptors in the 6 day-old chick embryo. Am. J. Physiol. 227, 1406–1410.

Sbong, S., Dzialowski, E.M., 2007. Respiratory and cardiovascular responses to acute hypoxia and hyperoxia in internally pipped chicken embryos. Comp. Biochem. Physiol. A 148, 761–768.

Scott, T.A., Silversides, F.G., 2000. The effect of storage and strain of hen on egg quality. Poult. Sci. 79, 1725–1729.

Snyder, G.K., Birchard, G.F., 1982. Water loss and survival in embryos of the domestic chicken. J. Exp. Zool. 219, 115–117.

Sotherland, P.R., Rahn, H., 1987. On the composition of bird eggs. Condor 89, 48–65.

Sotherland, P.R., Spotila, J.R., Paganelli, C.V., 1987. Avian eggs: barriers to the exchange of heat and mass. J. Exp. Zool. 1 (Suppl.), 81–86.

Stewart, M.E., Terepka, A., 1969. Transport functions of the chick chorio-allantoic membrane: I. Normal histology and evidence for active electrolyte transport from the allantoic fluid, in vivo. Exp. Cell. Res. 58, 93–106.

Stock, M.K., Asson-Batres, M.A., Metcalfe, J., 1985. Stimulatory and persistent effect of acute hyperoxia on respiratory gas exchange of the chick embryo. Respir. Physiol. 62, 217–230.

Szdzuy, K., Fong, L.M., Mortola, J.P., 2008. Oxygenation and establishment of thermogenesis in the avian embryo. Life Sci. 82, 50–58.

Szdzuy, K., Mortola, J.P., 2007. Monitoring breathing in avian embryos and hatchlings by the barometric technique. Respir. Physiol. Neurobiol. 159, 241–244.

Taber, L.A., 2001. Biomechanics of cardiovascular development. Annu. Rev. Biomed. Eng. 3, 1–25.

Tamura, A., Akiyama, R., Chiba, Y., Moriya, K., Dzialowski, E.M., Burggren, W.W., Tazawa, H., 2003. Heart rate responses to cooling in emu hatchlings. Comp. Biochem. Physiol. A 134, 829–838.

Tazawa, H., 1971. Measurement of respiratory parameters in blood of chicken embryo. J. Appl. Physiol. 30, 17–20.

Tazawa, H., 1973. Hypothermal effect on the gas exchange in chicken embryo. Respir. Physiol. 17, 21–31.

Tazawa, H., 1980a. Oxygen and CO_2 exchange and acid-base regulation in the avian embryo. Am. Zool. 20, 395–4004.

Tazawa, H., 1980b. Adverse effect of failure to turn the avian egg on embryo oxygen exchange. Respir. Physiol. 41, 137–142.

Tazawa, H., 1981a. Compensation of diffusive respiratory disturbances of the acid-base balance in the chick embryo. Comp. Biochem. Physiol. A 69, 333–336.

Tazawa, H., 1981b. Effect of O_2 and CO_2 in N_2, He and SF_6 on chick embryo blood pressure and heart rate. J. Appl. Physiol. 51, 1017–1022.

Tazawa, H., 1981c. Measurement of blood pressure of chick embryo with an implanted needle catheter. J. Appl. Physiol. 51, 1023–1026.

Tazawa, H., 1982. Regulatory process of metabolic and respiratory acid-base disturbances in embryos. J. Appl. Physiol. 53, 1449–1454.

Tazawa, H., 1986. Acid-base equilibrium in birds and eggs. In: Heisler, N. (Ed.), Acid-Base Regulation in Animals. Elsevier, Amsterdam, pp. 203–233.

Tazawa, H., 1987. Embryonic respiration. In: In: Seller, T.J. (Ed.), Bird Respiration., vol. II. CRC Press, Boca Raton, FL, pp. 3–41.

Tazawa, H., 2005. Cardiac rhythms in avian embryos and hatchlings. Avian Poult. Biol. Rev. 16, 123–150.

Tazawa, H., Akiyama, R., Moriya, K., 2002a. Development of cadiac rhythms in birds. Comp. Biochem. Physiol. A 132, 675–689.

Tazawa, H., Andrewartha, S.J., Burggren, W.W., 2012. Acute regulation of hematocrit and blood acid-base balance during severe hypoxic challenges in late chicken embryos (*Gallus gallus*). Respir. Physiol. Neurobiol. 184, 86–96.

Tazawa, H., Ar, A., Gefen, E., Moriya, K., Pearson, J.T., 1998b. Effects of incubator humidity on embryonic heart rate in the ostrich. In: Proc. 10th European Poult. Conf., pp. 843–847.

Tazawa, H., Ar, A., Moriya, K., Gefen, E., Pearson, J.T., 2000. Embryonic heart rate measurements during artificial incubation of emu eggs. Br. Poult. Sci. 41, 89–93.

Tazawa, H., Ar, A., Pearson, J.T., Moriya, K., Gefen, E., 1998a. Heart rate in developing ostrich embryos. Br. Poult. Sci. 39, 161–166.

Tazawa, H., Chiba, Y., Khandoker, A.H., Dzialowski, E.M., Burggren, W.W., 2004. Early development of thermoregulatory competence in chickens: responses of heart rate and oxygen uptake to altered ambient temperatures. Avian Poult. Biol. Rev. 15, 166–176.

Tazawa, H., Hashimoto, Y., Doi, K., 1992a. Blood pressure and heart rate of chick embryo (*Gallus domesticus*) within the egg: responses to autonomic drugs. In: Hill, R.B., Kuwasawa, K. (Eds.), Phylogenetic Models in Functional Coupling of the CNS and the Cardiovascular System. Karger, Amsterdam, pp. 86–96.

Tazawa, H., Hashimoto, Y., Nakazawa, S., Whittow, G.C., 1992b. Metabolic responses of chicken embryos and hatchlings to altered O_2 environments. Respir. Physiol. 88, 37–50.

Tazawa, H., Hiraguchi, T., Kuroda, O., Tullett, S.G., Deeming, O.C., 1991a. Embryonic heart rate during development of domesticated birds. Physiol. Zool. 64, 1002–1022.

Tazawa, H., Hou, P.-C.L., 1997. Avian cardiovascular development. In: Burggren, W.W., Keller, B. (Eds.), Development of Cardiovascular Systems: Molecules to Organisms. Cambridge Univ. Press, New York, pp. 193–210.

Tazawa, H., Johansen, K., 1987. Comparative model analysis of central shunts in vertebrate cardiovascular systems. Comp. Biochem. Physiol. A 86, 595–607.

Tazawa, H., Kuroda, O., Whittow, G.C., 1991b. Noninvasive determination of the embryonic heart rate during hatching in the Brown Noddy (*Anous stolidus*). Auk 108, 594–601.

Tazawa, H., Mikami, T., Yoshimoto, C., 1971a. Respiratory properties of chicken embryonic blood during development. Respir. Physiol. 13, 160–170.

Tazawa, H., Mikami, T., Yoshimoto, C., 1971b. Effect of reducing the shell area on the respiratory properties of chicken embryonic blood. Respir. Physiol. 13, 352–360.

Tazawa, H., Mitsubayashi, H., Hirata, M., Höchel, J., Pearson, J.T., 1999. Cardiac rhythms in chick embryos during hatching. Comp. Biochem. Physiol. A 124, 511–521.

Tazawa, H., Mochizuki, M., 1976. Estimation of contact time and diffusing capacity for oxygen in the chorioallantoic vascular plexus. Respir. Physiol. 28, 119–128.

Tazawa, H., Mochizuki, M., 1977. Oxygen analysis of chicken embryo blood. Respir. Physiol. 31, 203–215.

Tazawa, H., Moriya, K., Tamura, A., Akiyama, R., 2002b. Low frequency oscillation of instantaneous heart rate in newly hatched chicks. Comp. Biochem. Physiol. A 131, 797–803.

Tazawa, H., Moriya, K., Tamura, A., Komoro, T., Akiyama, R., 2001a. Ontogenetic study of thermoregulation in birds. J. Therm. Biol. 26, 281–286.

Tazawa, H., Nakagawa, S., 1985. Response of egg temperature, heart rate and blood pressure in the chick embryo to hypothermal stress. J. Comp. Physiol. B 155, 195–200.

Tazawa, H., Okuda, A., Nakazawa, S., Whittow, G.C., 1989a. Metabolic responses of chicken embryos to graded, prolonged alterations in ambient temperature. Comp. Biochem. Physiol. A 92, 613–617.

Tazawa, H., Ono, T., 1974. Microscopic observation of the chorioallantoic capillary bed of chicken embryo. Respir. Physiol. 20, 81–90.

Tazawa, H., Ono, T., Mochizuki, M., 1976a. Oxygen dissociation curve for chorioallantoic capillary blood of chicken embryo. J. Appl. Physiol. 40, 393–398.

Tazawa, H., Ono, T., Mochizuki, M., 1976b. Oxygenation and deoxygenation velocity factors of chorioallantoic capillary blood. J. Appl. Physiol. 40, 399–403.

Tazawa, H., Pearson, J.T., Komoro, T., Ar, A., 2001b. Allometric relationships between embryonic heart rate and fresh egg mass in birds. J. Exp. Biol. 204, 165–174.

Tazawa, H., Piiper, J., 1984. Carbon dioxide dissociation and buffering in chicken blood during development. Respir. Physiol. 57, 123–134.

Tazawa, H., Piiper, J., Ar, A., Rahn, H., 1981. Changes in acid-base balance of chick embryos exposed to a He and SF_6 atmosphere. J. Appl. Physiol. 50, 819–823.

Tazawa, H., Rahn, H., 1986. Tolerance of chick embryos to low temperatures in reference to the heart rate. Comp. Biochem. Physiol. A 85, 531–534.

Tazawa, H., Rahn, H., 1987. Temperature and metabolism of chick embryos and hatchlings after prolonged cooling. J. Exp. Zool. 1 (Suppl.), 105–109.

Tazawa, H., Suzuki, Y., Musashi, H., 1989b. Simultaneous acquisition of ECG, BCG and blood pressure from chick embryos in the egg. J. Appl. Physiol. 67, 478–483.

Tazawa, H., Turner, J.S., Paganelli, C.V., 1988a. Cooling rates of living and killed chicken and quail eggs in air and in helium-oxygen gas mixture. Comp. Biochem. Physiol. A 90, 99–102.

Tazawa, H., Visschedijk, A.H.J., Piiper, J., 1983a. Blood gases and acid-base status in chicken embryos with naturally varying egg shell conductance. Respir. Physiol. 54, 137–144.

Tazawa, H., Visschedijk, A.H.J., Wittmann, J., Piiper, J., 1983b. Gas exchange, blood gases and acid-base status in the chick before, during and after hatching. Respir. Physiol. 53, 173–185.

Tazawa, H., Wakayama, H., Turner, J.S., Paganelli, C.V., 1988b. Metabolic compensation for gradual cooling in developing chick embryos. Comp. Biochem. Physiol. A 89, 125–129.

Tazawa, H., Watanabe, W., Burggren, W.W., 1994. Embryonic heart rate in altricial birds, the pigeon (*Columba domestica*) and the bank swallow (*Riparia riparia*). Physiol. Zool. 40, 1448–1460.

Tazawa, H., Whittow, G.C., 1994. Embryonic heart rate and oxygen pulse in two procellariiform seabirds, *Diomedea immutabilis* and *Puffinus pacificus*. J. Comp. Physiol. B 163, 642–648.

Tazawa, H., Whittow, G.C., Turner, J.S., Paganelli, C.V., 1989c. Metabolic responses to gradual cooling in chicken eggs treated with thiourea and oxygen. Comp. Biochem. Physiol. A 92, 619–622.

Tobita, K., Garrison, J.B., Liu, L.J., Tinney, J.P., Keller, B.B., 2005. Three-dimensional myofiber architecture of the embryonic left ventricle during normal development and altered mechanical loads. Anat. Rec. A Discov. Mol. Cell. Evol. Biol. 283A, 193–201.

Tøien, O., Aulie, A., Steen, J.B., 1986. Thermoregulatory responses to egg cooling in incubating bantam hens. J. Comp. Physiol. B. 156, 303–307.

Tullett, S.G., Deeming, D.C., 1987. Failure to turn eggs during incubation: effects on embryo weight, development of the chorioallantois and absorption of albumen. Br. Poult. Sci. 28, 239–243.

Turner, J.S., 1986. Cooling rate and size of bird's eggs – a natural isomorphic body. J. Therm. Biol. 10, 101–104.

Turner, J.S., 1987. Blood circulation and the flows of heat in an incubated egg. J. Exp. Zool. 1 (Suppl.), 99–104.

Tzschentke, B., Basta, D., Nichelmann, M., 2001. Epigenetic temperature adaptation in birds: peculiarities and similarities in comparison to acclimation. News Biomed. Sci. 1, 26–31.

Tzschentke, B., Rumpf, M., 2011. Embryonic development of endothermy. Respir. Physiol. Neurobiol. 178, 97–107.

Van Golde, J., Mulder, T., Blanco, C.E., 1997. Changes in mean chorioallantoic artery blood flow and heart rate produced by hypoxia in the developing chick embryo. Pediatr. Res. 42, 293–298.

Van Mierop, L.H.S., Bertuch Jr, C.J., 1967. Development of arterial blood pressure in the chick embryo. Am. J. Physiol. 212, 43–48.

Vince, M., Salter, S.H., 1967. Respiration and clicking in quail embryos. Nature 216, 582–583.

Visschedijk, A.H.J., Ar, A., Rahn, H., Piiper, J., 1980. The independent effects of atmospheric pressure and oxygen partial pressure on gas exchange of the chicken embryo. Respir. Physiol. 39, 33–44.

Visschedijk, A.H.J., Tazawa, H., Piiper, J., 1985. Variability of shell conductance and gas exchange of chicken eggs. Respir. Physiol. 59, 339–345.

Vleck, C.M., Vleck, D., 1987. Metabolism and energetic of avian embryos. J. Exp. Zool. 1 (Suppl.), 111–125.

Vleck, C.M., Vleck, D., Hoyt, D.F., 1980. Patterns of metabolism and growth in avian embryos. Am. Zool. 20, 405–416.

Vleck, D., Vleck, C.M., Seymour, R.S., 1984. Energetics of embryonic development in the megapode birds, mallee fowl *Leipoa ocellata* and brush turkey *Alectura lathami*. Physiol. Zool. 57, 444–456.

Wagner-Amos, K., Seymour, R.S., 2002. Effect of regional changes to shell conductance on oxygen consumption and growth of chicken embryos. Respir. Physiol. 129, 385–395.

Wangensteen, O.D., Rahn, H., 1970/71. Respiratory gas exchange by the avian embryo. Respir. Physiol. 11, 31–45.

Wangensteen, O.D., Weibel, E.R., 1982. Morphometric evaluation of chorioallantoic oxygen transport in the chick embryo. Respir. Physiol. 47, 1–20.

Wangensteen, O.D., Wilson, D., Rahn, H., 1970/71. Diffusion of gases across the shell of the hen's egg. Respir. Physiol. 11, 16–30.

White, P.T., 1974. Experimental studies on the circulatory system of the late chick embryo. J. Exp. Biol. 61, 571–592.

Whittow, G.C., 1980. Physiological and ecological correlates of prolonged incubation in sea birds. Am. Zool. 20, 427–436.

Whittow, G.C., Tazawa, H., 1991. The early development of thermoregulation in birds. Physiol. Zool. 64, 1371–1390.

Wilson, H.R., 1991. Physiological requirements of the developing embryo: temperature and turning. In: Tullett, S.G. (Ed.), Avian Incubation. Butterworth, London, pp. 145–156.

Yoneta, H., Akiyama, R., Nakata, W., Moriya, K., Tazawa, H., 2006a. Video analysis of body movements and their relation to the heart rate fluctuations in chicken hatchlings. In: Yahav, S., Tzschentke, B. (Eds.), New Insights into Fundamental Physiology and Peri-natal Adaptation of Domestic Fowl. Nottingham Univ. Press, pp. 57–68.

Yoneta, H., Dzialowski, E.M., Burggren, W.W., Tazawa, H., 2007. Endothermic heart rate response in broiler and White Leghorn chicks (*Gallus gallus domesticus*) during the first two days of post hatch life. Comp. Biochem. Physiol. A 147, 529–535.

Yoneta, H., Fukazawa, K., Dzialowski, E.M., Burggren, W.W., Tazawa, H., 2006b. Does sequence of exposure to altered ambient temperature affect the endothermic heart rate response of newly hatched chicks? In: Yahav, S., Tzschentke, B. (Eds.), New Insights into Fundamental Physiology and Peri-natal Adaptation of Domestic Fowl. Nottingham Univ. Press, pp. 15–28.

Yoneta, H., Fukuoka, S., Akiyama, R., Tazawa, H., 2006c. Early development of cholinergic heart rate control in embryos of broiler and White Leghorn chickens. In: Yahav, S., Tzschentke, B. (Eds.), New Insights into Fundamental Physiology and Peri-natal Adaptation of Domestic Fowl. Nottingham Univ. Press, pp. 1–14.

Yoshigi, M., Hu, N., Keller, B.B., 1996. Dorsal aortic impedance in stage 24 chick embryo following acute changes in circulating blood volume. Am. J. Physiol. Heart Circ. Physiol. 270, H1597–H1606.

Yosphe-Purer, Y., Fendrich, J., Davies, A.M., 1953. Estimation of blood volumes of embryonated hen eggs at different ages. Am. J. Physiol. 175, 178–180.

Part VII

Cross cutting Themes

Chapter 33

Stress in Birds

Julio Blas
Estación Biológica de Doñana, Consejo Superior de Investigaciones Científicas (CSIC), Seville, Spain

33.1 INTRODUCTION

Birds undergo profound changes in physiology, morphology, and behavior across their lifecycles, and the endocrine system plays a fundamental role integrating external and internal signals and orchestrating adequate responses aimed at maximizing individual fitness. Some of these changes occur in response to *predictable fluctuations* in the environment like night–day and seasons, which allow year-round anticipatory organization of major life history stages in cyclic patterns. Superimposed on predictable components, *unpredictable perturbations* like severe weather, loss of social rank, habitat destruction, or human disturbance require emergency adjustments in physiology and behavior. The *adrenocortical response to "stress"*, which results in a rapid elevation of circulating *glucocorticoid* levels (Figure 33.1; Chapter 26), provides a major physiological mechanism allowing birds to cope with environmental perturbations. *Glucocorticoids,* the *stress hormones,* allow mobilization of body energy, increase cardiovascular tone, regulate the immune system, and inhibit a variety of costly anabolic processes including digestion, energy storage, growth, and reproduction in response to "stress" (Sapolsky et al., 2000; Sapolsky, 2002). But these hormones are also fundamental for normal body functions and normal regulation of energy balance, physiology, morphology, and behavior under nonstress conditions (Landys et al., 2006). Unfortunately, it is relatively easy to find ornithology texts where the "stress of reproduction" or the "stress of migration" are invoked: authors are indeed referring to physiological demands of *normal, predictable life history stages* that *are not truly related to stress.* Even the common use of the word "stress" among comparative biologists and physiological ecologists often says *"virtually nothing about the underlying physiological or behavioral mechanisms"* (McEwen and Wingfield, 2003b). Therefore, before continuing this chapter dedicated to stress in birds, we should clarify (1) what we mean when using the term "stress", and (2) what alternative, unequivocal terms should be used to avoid confusion. Because the specific terminology is sometimes redundant and can be initially confusing, a final section (Section 33.6) provides concise definitions for many of the terms highlited in italics throughout this chapter.

33.2 UNDERSTANDING STRESS: FROM ENERGY TO GLUCOCORTICOIDS

"Stress" is a polysemous word that, depending on the context, may mean: (1) the *stimuli* that challenge homeostasis (i.e., *stressors, perturbations*), (2) the *emergency responses* to perturbations *(stress response),* or (3) the *chronic state of imbalance* that follows over-activation of the adrenocortical axis (*pathology, chronic stress*). This lack of specificity often implies that *"stress means whatever the author wants it to mean"* (Romero, 2012). In the biomedical literature this word remains ambiguous since it was first brought from the field of engineering by Cannon (1932) and Selye (1946). *Modern field endocrinologists* are making (ongoing) efforts to define and consolidate alternative and unequivocal terms, leading to the recent introduction of *"allostasis*" and an accompanying set of modern vocables to precisely define physiological responses to challenging and normal life history conditions, which are key aspects for this chapter.

Allostasis is the process of maintaining homeostasis through change. This concept was introduced decades ago to redefine stress (Sterling and Eyer, 1988), but it only started to reach ecological studies after McEwen and Wingfield proposed the *Allostasis Model* (McEwen and Wingfield, 2003a). Through the key concepts of *allostatic state, allostasis load,* and *allostatic overload,* the model incorporates classical *homeostasis* in the context of an organism's lifecycle and in relation to individual experience and how the individual responds to the ever-changing physical and social environments (Figure 33.2; McEwen and Wingfield, 2010). This model uses the balance between energy input and energy requirements to predict when an

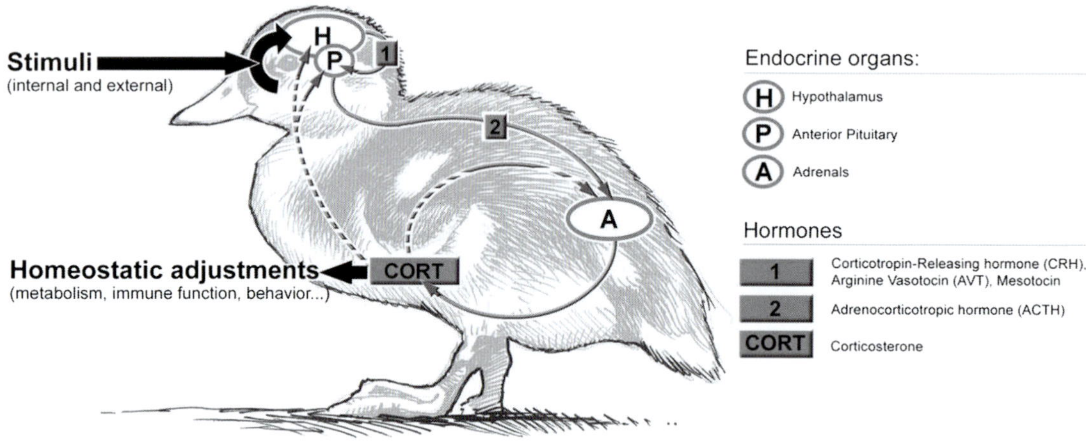

FIGURE 33.1 **The adrenocortical response to stress in birds.** Following exposure to perturbations of exogenous or endogenous origin (e.g., predation attempts, energy imbalances; see upper black arrows), the hypothalamus (H) releases a number of hormones including corticotropin-releasing hormone CRH. These, in turn, stimulate the anterior pituitary (P) to secrete adrenocorticotropic hormone ACTH into circulation. In birds, the adrenals (A) respond to increased ACTH levels secreting corticosterone (CORT). Within minutes to hours following exposure to stress, the resulting corticosterone elevations promote multiple changes in physiology and behavior (lower black arrow) including increased gluconeogenesis, suppression of reproductive behaviors, regulation of immune function, irruptive migration, and increased night restfulness. These adjustments promote the maintenance of homeostasis through change (i.e., allostasis). Corticosterone secretion is subjected to negative feedback mechanisms, as indicated with dashed arrows. In addition to stress-related fluctuations, baseline corticosterone levels show circadian and circannual rhythms in birds, allowing endogenous regulation of numerous physiological processes including body energy balance. *Figure from Baos and Blas, Adrenal toxicology, first edition, copyright © 2009, Informa Healthcare. Reproduced with permission of Informa Healthcare.*

animal should move away from the normal lifecycle into an emergency survival mode, and when a pathological imbalance would occur.

The *Reactive Scope Model* is an offshoot of the *Allostasis Model*, but presents a different terminology for contextualizing the impact of stress in the body and a new classification for the levels of the *mediators* (e.g., *corticosterone* in birds; see Figure 33.3) that allows the incorporation of a process known as *"wear and tear"*. Combining traditional notions of stress and homeostasis with the more recent terminology of *allostasis* and *allostatic load,* the *Reactive Scope Model* addresses the mechanisms of response to perturbations and the roles of different mediators, including glucocorticoids. It is critical to understand these two models and their associated terminology before we can understand the patterns of adrenocortical responses to stress and the underlying physiological mechanisms in birds, within a life history context.

- "Stress" is a misleading and highly unspecific term, which, depending on the context, may mean: (1) the *stimuli* that challenge homeostasis (i.e., stressors, perturbations), (2) the *emergency responses* to perturbations (stress response), or (3) the chronic *state of imbalance* that follows overactivation of the adrenocortical axis (pathology).
- Field endocrinologists are making ongoing efforts to define and consolidate alternative and unequivocal terms, leading to the recent introduction of the *Allostasis Model* and the *Reactive Scope Model.*

33.2.1 Allostasis

Homeostasis (as used in Cannon, 1932) *is the stability of physiological systems that are essential to life.* As discussed in McEwen and Wingfield (2010), the concept of homeostasis is *restricted to a few truly essential systems* (pH, body temperature, glucose levels, and oxygen tension): *those aspects of physiology that "keep us alive"* (McEwen and Wingfield, 2003a). *Allostasis is the process of achieving stability through change:* it maintains homeostasis even though the set points and other boundaries of control may change with environmental conditions. *Allostasis* thus refers to *those aspects of physiology that "help us adapt"* (McEwen and Wingfield, 2010). *Circulating glucocorticoids are major, but not the only allostasis mediators* (additional mediators include cathecolamines, cytokines, behavior, hearth rate, blood pressure, and antibody titers; see table 1 in Romero et al., 2009). Glucocorticoid levels can markedly *change to help the body maintain homeostasis* within narrow limits. To allow for this, glucocorticoids *levels fluctuate in daily and seasonal patterns* (i.e. not only in response to perturbations, despite being equivocally called "stress" hormones), *and also rise rapidly when a perturbation occurs.* Free-living birds respond to potentially noxious stimuli (like storms, social challenge, or human disturbance that reduce access to food and shelter) by increasing glucocorticoid secretion, which facilitates foraging and gluconeogenesis. In addition, elevated glucocorticoids inhibit

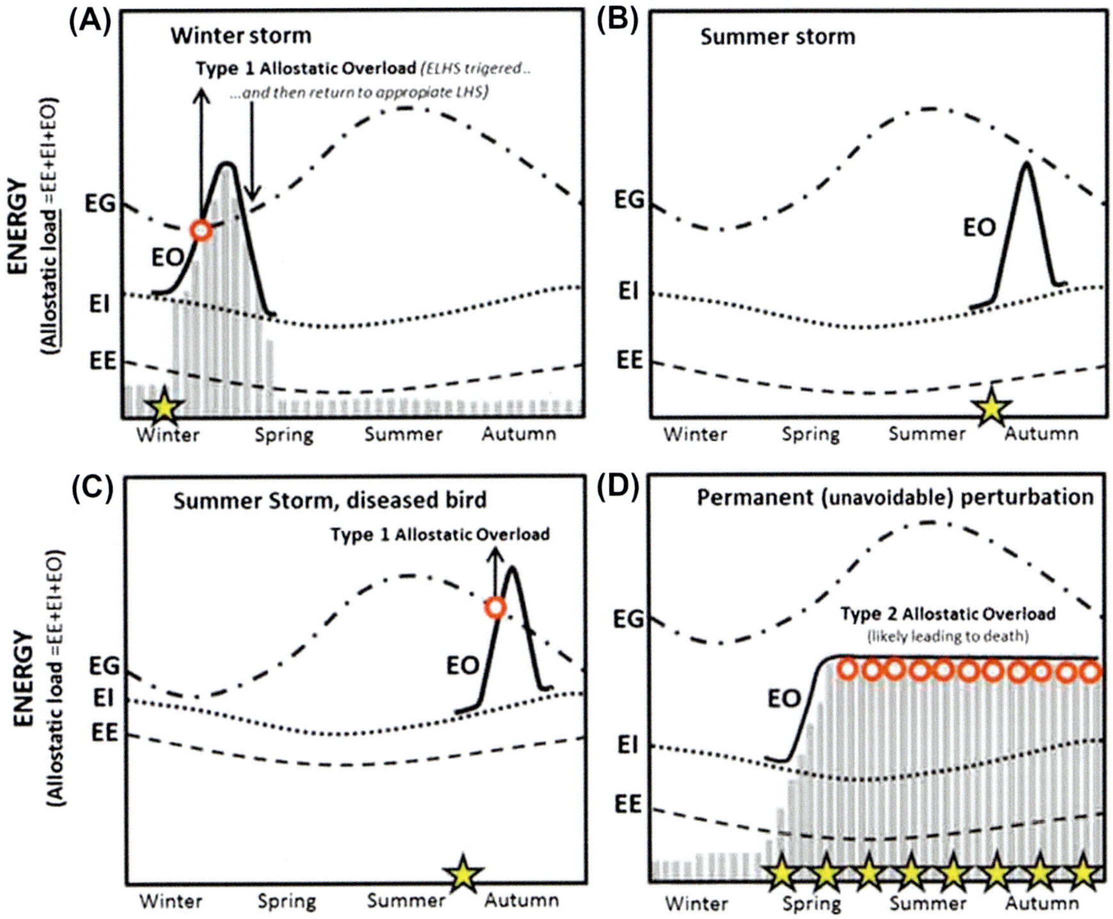

FIGURE 33.2 **The *Allostasis Model*.** The *Allostasis Model* provides a framework to understand the energetic requirements of organisms during their life, and predictions for when allostatic overload (red circles) occurs, under a range of environmental, social, and individual-based scenarios. The *y*-axis represents potential nutritional requirements (including energy *per se* and nutrients) during one year (seasons in *x*-axis). The lines represent several energy components: EG (energy to be gained) represents the amount of energy in food available in the environment. In temperate areas it is expected to rise dramatically in spring and summer and then decline through autumn and winter when primary productivity is low. EE (existence energy) is the minimum existence energy (resting metabolism) required for basic homeostasis in any life history stage. It may decline in spring and summer with ambient temperature. EI (ideal energy) is the extra energy required to go out, find food, process and assimilate it under ideal conditions (when there are no perturbations) while performing seasonal routines (breeding, molting, migrating). It likely changes seasonally and for clarity we assume that it varies in parallel with EE. EO-(energy following perturbation) is the additional energy required to go out, find food, process and assimilate it under non-ideal conditions (following a perturbation such as a storm). Allostatic load is the energy resulting from adding EE+EI+EO. In the figures, the star symbol defines when a perturbation (for example, severe weather) starts. Type 1 allostatic overload (panels A, and C) occurs when allostatic load surpasses the energy available in the environment EG, promoting corticosterone elevations (gray bars, only shown in panels A and D for clarity) that trigger an Emergency Life History Stage (ELHS). One result is suppressing the expression of another life history stages (LHS) like migration or breeding, and bring allostatic load below the level of EG (e.g., the bird may abandon spring migration to obtain more food and only continue migrating when the perturbation ends). The bird can now survive the perturbation in positive energy balance and glucocorticoid elevations subside, avoiding pathologies associated with chronic high levels. Note that the same perturbation (for example, a storm) can result in Type 1 overload during winter but not in summer (compare panels A and B) because there is more food available in the latter season (and EG>EE+EI+EO). However, an individual heavily infected by parasites or with a permanent injury or chronic disease (panel C) would enter Type 1 allostatic overload facing the same storm even in summer because its intrinsic (disease-related) EE and EI levels are higher compared to a healthy bird (compare panels B and C). A second type of overload (Type 2 overload; panel D) can take place in situations where energy from the environment EG is not a limiting factor but a long-term perturbation (for example climate change, captivity, permanent exposure to predators, permanent social challenge) triggers an allostatic state. The behavioral or physiological adjustments promoted by allostasis mediators (e.g. corticosterone) do not allow evading/copying with the noxious stimuli (the individual must be unable to escape the perturbation in order for Type 2 allostasis overload to occur). Corticosterone levels (gray bars) become chronically elevated, generating pathologies, and individuals likely die early. In the figures, the levels of allostasis mediators (for example, corticosterone) are only indicated as gray bars in the first and last panels (A and D) for simplicity, noting that their associations with energy levels (the metrics for *y*-axis in all the figures) are not necessarily linear. *Figures modified and expanded from McEwen and Wingfield, (2003a) with permission.*

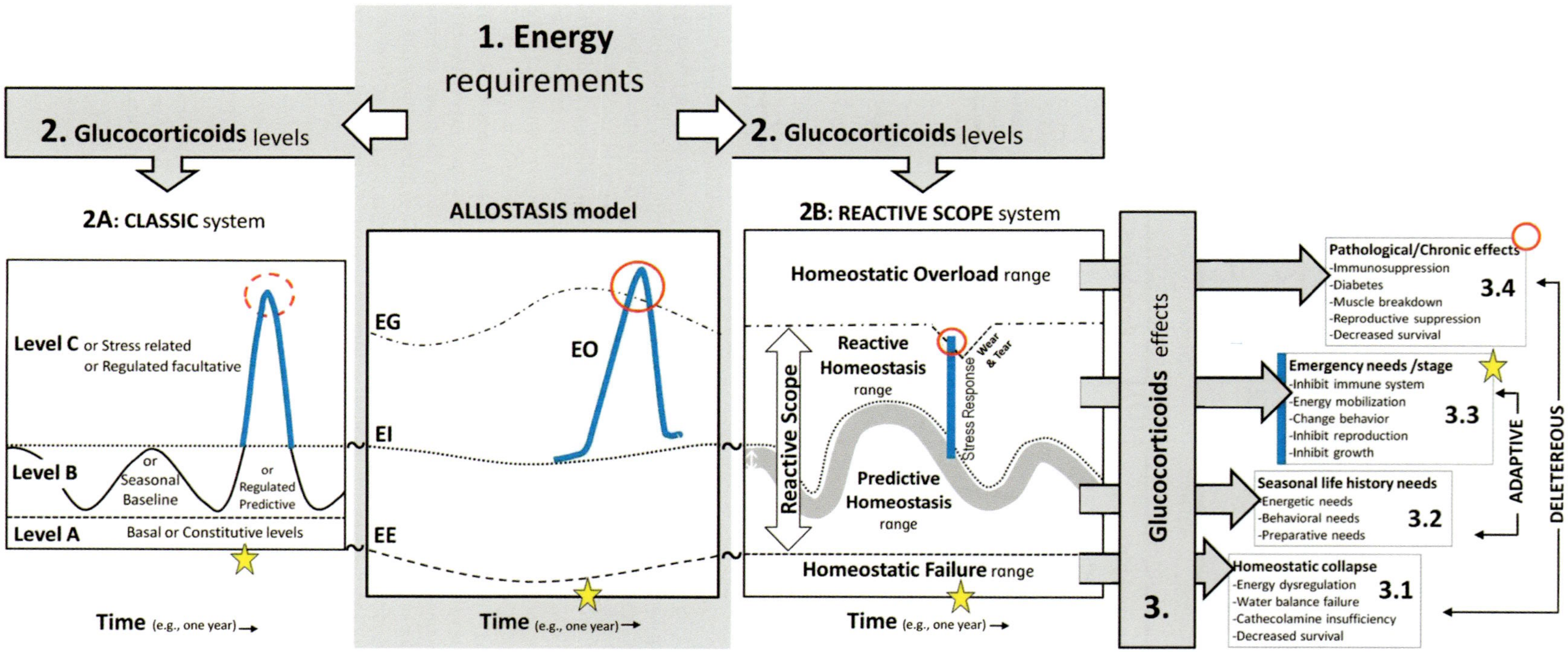

FIGURE 33.3 **Classification of corticosterone levels.** Classification of functional GC ranges according to the "Classic system" (**panel 2A**) and the "Reactive Scope system" (**panel 2B**). Both GC classification systems are presented in parallel to highlight the degree of overlapping and equivalence among system-specific terms (see main text and Section 33.6 for definitions), and in relation to the individual energy balance introduced by the *Allostasis Model* (energy requirements precede changes in GC levels, and thus the *Allostasis Model* is here presented as a central **panel 1**). Examples of GC effects on physiology and behavior are also presented in line with the corresponding GC ranges (**panel 3**). Within any given chart the x-axis represents the timeline (for example one year) and the star symbol indicates exposure to a perturbation, which first results in increased energy demands (depicted as a blue line in panel 1, implying increased allostatic load) and secondly in concomitant increases of GC levels reaching a range known as Level C or reactive homeostasis (blue lines in panels 2a and 2b respectively). The perturbation may eventually bring the individual into negative energy balance (depicted as a red circle in panel 1, implying allostatic overload type I), and GC elevations will reach an upper range (red circles in panels 2A and 2B) potentially causing deleterious effects on body systems (panel 3, box 3.4). To avoid the damage associated to both the energy imbalance and the high GC levels, a facultative emergency life history stage is typically triggered (panel 3, box 3.3). If the negative energy balance subsides (because the perturbation extinguishes, or the individual relocates) GC levels return to a normal seasonal baseline known as level B or predictive homeostasis (in panels 2A and 2B respectively) allowing resuming normal life history activities (panel 3, box 3.2). Note that the deleterious effects of GCs can occur both at very high and very low concentrations (panel 3, boxes 3.1 vs. 3.4). *Composite figure integrates graphs modified and expanded from several sources, including McEwen and Wingfield, 2003a; Landys et al., 2006 and Romero et al., 2009, with permission.*

processes that are not essential to immediate survival (e.g., reproduction, migration) and increase physical activity allowing an individual to move away from the perturbation or to find shelter as well as saving energy by increasing night restfulness (Wingfield, 1994; Wingfield and Ramenofsky, 1999; Wingfield et al., 1998). When the body experiences altered and sustained activity levels of *allostasis mediators* (e.g., *corticosterone,* the main avian glucocorticoid), it enters an *allostatic state,* which can only be maintained for limited periods before damaging the body.

The *Allostasis Model* (Figure 33.2) provides a framework to understand when and how emergency responses are activated (reaching an *allostatic state*) through modeling the energetic requirements of organisms during their life, and in relation to the energy available in the environment. In order to understand the model, the following energy components need to be considered:

EG (energy to be gained) represents the amount of energy in food available in the environment. In temperate areas it is expected to rise dramatically in spring and summer and then decline through autumn and winter when primary productivity is low.

EE (existence energy) is the minimum existence energy (resting metabolism) required for basic homeostasis in any life history stage. It may decline in spring and summer with increasing ambient temperature (if it becomes easier maintaining body functions).

EI (ideal energy) is the extra energy required to go out, find food, process and assimilate it under ideal conditions (when there are no perturbations) while performing seasonal routines (breeding, molting, migrating). It changes seasonally: it can follow EE and ambient temperature (for example, in a nonbreeding bird) or can increase during reproduction (spring-summer) and decrease during wintering. For clarity, in the Figure 33.2 (A) it varies in parallel with EE.

EO (energy following perturbation) is the additional energy required to go out, find food, and process and assimilate it under non-ideal conditions (following a perturbation such as a storm). It is always higher than EE+EI. In the Figure 33.2, the star symbol defines when the *perturbation* starts, setting the beginning point for elevations in EO energy demands.

Allostatic load can be now defined as the cumulative energetic requirements of the organism at a given time, and equals EE+EI+EO (although it is sometimes noted simply as EO). This concept relates to *all the energy requirements in a broad sense (encompassing nutrients and all other potentially limiting resources)*, and it can be interpreted as a cost: the cumulative cost of being there, responding to the current situations whether they are predictable or not; or the "workload" of a particular moment (at certain times, an animal has to work harder—expend more energy to maintain homeostasis). A detailed definition of *allostatic load* in McEwen and Wingfield (2010) is "*all of the energy and nutrients an organism needs to go about its daily and seasonal routines as the life cycle progresses, and to deal with unpredictable events from the physical and social environment that have the potential to be stressful.*" In the graphical representation, *allostatic load* fluctuates along a continuum with important transitional points (see red circles in Figure 33.2) that determine when the individual can cope with daily activities (when is the load affordable?) and when it needs to trigger facultative physiological and behavioral responses designed to reduce costs (when is the load unaffordable, becoming an overload?). In other words, transitional points determine when the individual enters *overload* and triggers an *allostatic state*.

There are two different outcomes. First, if energy demands exceed energy income and what can be mobilized from stores (i.e., if *allostatic load* surpasses EG), then *Type 1 allostatic overload* occurs. For example, breeding birds take advantage of the increasing food abundance in spring to raise their broods. If inclement weather then increases the cost of maintaining homeostasis in addition to the demands of breeding, and at the same time reduces the food available to fuel this *allostatic load,* then negative energy balance (*overload*) results in a loss of body mass and suppression of reproduction (Wingfield et al., 1983). This negative energy balance is *Type 1 allostatic overload*, which leads to an *allostatic state* (increased levels of glucocorticoids), which then triggers physiological changes (use of lipid stores resulting in body mass loss) and behavioral changes (brood desertion). This response has been classically referred to as the *emergency life-history stage (ELHS),* and will be discussed more in detail in Section 33.3.3. Note for now, however, that short-term elevations of glucocorticoids may promote rapid physiological and behavioral changes, allowing a decrease of *allostatic load* below the EG level, before it is required to abandon the present life history stage. Only above a certain threshold (likely variable among individuals) would corticosterone trigger *ELHS* (Wingfield and Romero, 2001). The final outcome of an *allostatic state* is normally survival because a positive energy balance can be reestablished: *allostatic load* will return to affordable levels when the individual relocates (or seeks refuge until the perturbations ends) and restores depleted condition through foraging. All of these actions are facilitated by the *allostasis mediators* (e.g., *corticosterone* levels, but also changes in behavior, hearth rate, blood pressure, cytokines, and antibody titers; see Romero et al., 2009), which are highly adaptive when elevated for a short period and then returned to normal (pre-perturbation) levels. Only in those circumstances when individuals are unable to avoid the perturbation even after entering the *ELHS*, elevated corticosterone can be maintained for long periods, resulting in serious pathological effects and eventually death.

The heuristic value of this *Allostasis Model* becomes clear when we incorporate individual variation due to habitat quality, body condition, disease and parasites, reproductive state, experience, genetics, social status, and many other additional parameters. For example, because the energy available in the environment (EG) fluctuates (e.g., food is more abundant in spring than in winter), a given perturbation will result in *Type 1 allostatic overload* faster at specific times of the year. This aspect is illustrated in Figure 33.2: the same storm (star symbol) promoting *Type 1 allostatic overload* in winter (panel (A)) fails to do so in summer (panel (B)) because there is sufficient energy in the environment to afford the increased *allostatic load*. However, an individual heavily affected by parasites or disease, or with a permanent injury (panel (C)), would enter *Type 1 allostatic overload* even in summer because its intrinsic (disease-related) energy requirement levels EE and EI are higher than average (in other words, its *allostatic load* was already elevated before the storm, and any additional EO load surpasses the energy available in the environment EG, rapidly resulting in *overload*). As an extreme example, habitats degraded by human transformation (not shown in graphics) often reduce food availability in the environment below the requirements for EE and EI. In the figures, the levels of *allostasis mediators* (for example, circulating corticosterone levels) are only indicated as gray bars in panels (A) and (D) for simplicity, but note that the metrics in the *Allostasis Model* refer to energy levels rather that the levels of any mediator. The relationships between energy requirement at a given moment and the level of a specific mediator are not necessarily linear because different *allostasis mediators* (e.g., glucocorticoids, cathecolamines, cytokines) act in concert and interact in complex physiological ways to meet allostatic demands (McEwen and Wingfield, 2010).

A second type of *allostatic overload* (qualitatively different from the Type 1 overload patterns described above) can take place in situations where energy from the environment EG is not limiting: then it is called *Type 2 allostatic overload* (see panel (D) in Figure 33.2). Here a long-term, rather than a short-term, perturbation (for example, climate change instead of weather inclemency, captivity or permanent exposure to predators instead of a single predation attempt, permanent social subordination instead of temporary social instability; see Section 33.3.3 for details) triggers an *allostatic state,* but concomitant elevations of glucocorticoids are unable to promote behavioral or physiological adjustments to cope with or to evade the noxious stimuli. For *Type 2 allostatic overload* to occur, the individual must be unable to evade the perturbation due to physical or social constraints. Corticosterone levels then become chronically elevated, generating pathologies despite adequate food resources (*Type 2 overload* occurs despite $EE+EI+EO<EG$; see Figure 33.2(D)). The effects on the body are similar to experimental exposure to exogenous corticosterone and metabolic pathologies arise (e.g., hyperinsulenemia, hyperphagia, obesity). This situation is most likely to occur in vertebrates living in complex hierarchical systems, where subordinate (or depending on the specific system, sometimes dominant) status imposes heavy *allostatic loads.* However, it can also affect birds under certain conditions, such as (1) natural populations exposed to permanent habitat disturbances or climate change, (2) avian social systems where dominance structure predominates over food and shelter as a source of *allostatic load* (see review in Goymann and Wingfield, 2004), and (3) captive settings. The hormonal imbalances and metabolic costs imposed by *Type 2 allostatic overload* are high and likely result in premature death and chronic disease (e.g., even birds with extreme fat stores are more likely predated due to decreased maneuverability; Lind et al., 1999; Gosler et al., 2002), providing strong selection for the mechanisms limiting *allostatic load.* But in captive animals (and human societies), the selection pressure is not there, and pathologies can persist long-term.

In summary, the concept of *allostasis* (maintaining *homeostasis through change*) combines the energetic costs or demands associated with daily rhythms and life history stages (predictable) with those costs accompanying environmental *perturbations* and social challenges (unpredictable and either short- or long-term, see Section 33.3.1) into a continuum named *allostatic load* (see figure 3 in Landys et al., 2006). Then *allostatic overload* is defined as either a short-term state in which the costs of the life history stage and accompanying challenges (=*allostatic load*) exceeds the food resources available to provide sufficient energy (*Type 1 allostatic overload*), or a long-term state in which deleterious challenges are chronic and lead to a sustained *allostatic state* independent of seasonal changes in the environment (*Type 2 allostatic overload*). The terminology introduced by the *Allostasis Model* allows physiologists to replace the overused word "stress" by a set of distinct, unambiguous, and precisely defined terms that are being gradually incorporated into modern ecological studies.

However, an acknowledged weakness of the *Allostasis Model* is the use of energy as both the underlying mechanism and the universal metric (for a discussion of this topic see Walsberg, 2003; McEwen and Wingfield, 2003b; Romero et al., 2009; McEwen and Wingfield, 2010). This becomes a problem that is mainly a consequence of our current inability to measure energy demand (*allostatic load*) and availability (EG) in free-living animals. Until we develop ways of measuring direct components of *allostatic load* (see Porter et al., 2002; Ricklefs and Wikelski, 2002 for research challenges and anticipating progress for the coming years), we will have to rely on the levels of *allostasis mediators* (e.g., circulating glucocorticoids) as best proxies to understand *allostasis*. The expected changes in *corticosterone* levels as a function of energy dynamics were only partially addressed in the original presentation of the *Allostasis Model*, and this will be the subject of the following subsection.

- *Homeostasis* is the stability of physiological systems that are essential to life, including pH, body temperature, glucose levels, and oxygen tension (those aspects of physiology that "keep us alive"). In contrast, *allostasis* is the process of achieving homeostasis through change (those aspects of physiology that "help us adapt").
- The *Allostasis Model* provides a framework to understand the energetic requirements of organisms during their life, and in relation to the energy available. The model incorporates three concepts (*allostasis, allostatic load,* and *allostatic overload*) that allow replacing the overused and unambiguous word "stress".
  - *Allostatic load* is the cumulative energetic requirements of the organism in a broad sense (the "workload" in a particular moment), including predictable and unpredictable demands. *Allostatic load* fluctuates along a continuum, and transitional points determine when the individual can cope with daily activities and when it needs to trigger emergency responses aimed at reducing costs.
  - *Type 1 allostatic overload* is a short-term state of energy imbalance in which *allostatic load* exceeds the energy available from the environment plus the internal reserves.
  - *Type 2 allostatic overload* is a long-term state in which environmental or socially demanding challenges become chronic and exert deleterious effects on the individual, regardless of the energy available.
- The metric or "currency" for the *Allostasis Model* is energy. This represents a problem due to the difficulty of estimating energy demands and availability outside controlled conditions. However, energy management is achieved through changing levels of *allostasis mediators* (e.g., glucocorticoids, cathecolamines, cytokines, behavior, hearth rate, blood pressure, antibody titers) that can be measured accurately. The relationships between energy requirement at a given moment and the level of a specific mediator are not necessarily linear because different mediators act in concert and interact in complex ways.
- *Corticosterone* is the predominant *glucocorticoid* in birds and a major *allostasis mediator*. Circulating corticosterone levels *change in response to internal and external demands to manage energy balance*. When the body experiences altered and sustained corticosterone elevations, it enters an *allostatic state*, which can only be sustained for limited periods before damaging the body.
- The *Allostasis Model* is a largely theoretical paradigm requiring empirical tests. However, it provides a framework to understand when and how emergency responses are activated, and how individuals respond to changing physical and social environments.

33.2.2 Classification of Glucocorticoid Levels

Glucocorticoids (GCs) are major *allostatic mediators:* they allow lifetime modulation of avian energy balance throughout changing internal and external conditions. Their role in gluconeogenesis explains much of their relevance as mediators, but they also act facilitating morphological and behavioral changes and balancing the immune system in ways that are crucial for individual fitness (see details in Chapter 26). The year-round dynamics of avian energy requirements and constraints introduced with the *Allostasis Model* now provide a reference to describe expected changes in GC levels (Figure 33.3). It is critical to assimilate and properly use the range-descriptor terms provided below, because GCs can exert either housekeeping actions or devastating effects on the organism depending on what levels are involved: this requires specific terminology. The paradox affects most *allostasis mediators* (they are *"good and bad", "stress and antistress"* depending on the range), and an adequate term usage constitutes the only way to avoid confusion and to understand how opposing results regarding GC actions among studies are not necessarily a contradiction.

The nomenclature describing GC ranges and functional thresholds differ, however, between what we will here call the *Classic System* and the *Reactive Scope System.* The two classifications are graphically presented in Figure 33.3 as two parallel panels (2A and 2B) and in relation to the *Allostasis Model* (but note that the latter refers to energy balance and not GC levels) for ease of comparison. The *Classic System* combines three hormonal thresholds (levels A, B, and C) and a set of *corticosterone terms* traditionally used (*basal* versus *baseline*, *constitutive* versus *facultative,* seasonal versus stress related) as proposed by Wingfield et al. (1997) and further refined in Landys et al. (2006). The *Reactive Scope System* corresponds to a more recent classification given in the original proposal of a model (Romero et al., 2009) which will be explained in the following section (Section 33.2.3). Despite the possibility that parallel terminologies may initially add some confusion in understanding the complexity of GC actions (additional terms are not exempt from controversy, McEwen and Wingfield, 2010), it is relevant to become familiar with both systems. These ultimately agree that GC levels are presumed to exist in five ranges of increasing magnitude (Figure 33.3):

1. ***A pathological and physiologically unsustainable range where GC levels are too low*** for minimum homeostatic processes. This pathological range can be simulated by removing GCs through adrenalectomy (in mammals, Darlington et al., 1990) or chemical treatment (in birds, where adrenocortical cells are also located in the kidney tissue; see Chapter 26), and death rapidly occurs. In the *Allostasis Model* this would be equivalent to an

individual unable to meet the minimum energy requirements for existence (i.e., below EE level).

  a. In the Classic System, this range corresponds to GC concentrations *below level A* (this classification system does not have a specific place or level to represent a deadly physiological level or stage).
  b. In the Reactive Scope System, this is the GC range of *homeostatic failure*.

2. ***A minimum GC level allowing basic homeostatic processes*** (allowing meeting the existence energy demands EE in the *Allostasis Model*).
  a. In the Classic System, this range corresponds to GC concentrations *within level (range) A,* and corticosterone levels are termed *basal or constitutive*, (allowing "a physiological *state A* representative of undisturbed animals at rest, in which hormone action maintains internal systems at a basal operating level to support the most fundamental requirements of life"; Landys et al., 2006).
  b. In the Reactive Scope System this will be the lower end of (but already *inside) the predictive homeostasis range.*
3. ***A range where GC levels fluctuate according to predictable environmental changes,*** making it possible to cope with daily and seasonal routines of the life cycle (i.e., meeting ideal energy needs EI, in the *Allostasis Model*).
  a. In the Classic System, this range corresponds to GC concentrations *within levels (range) B* and corticosterone levels are termed *seasonal baseline* levels or *regulated predictive* level (allowing a physiological *state B* where "hormone action maintains systems within a heightened operated range to support increasing demands—*allostatic load,* especially in association with predictable changes in the environment or life history"; Landys et al., 2006).
  b. In the Reactive Scope System, this will be the *predictive homeostasis range*, where GCs encompass normal circadian, daily, and seasonal changes. In the graphical representation of the *Reactive Scope Model* (Figure 33.3, panel 2B), the width of the gray area represents circadian variation (driven by predictable daily changes in light and dark; see, e.g., Dallman et al., 1987), while the convolutions of this gray line along one year reflect seasonal variation (predictable changes normally driven by photoperiod), which allow anticipation and progression of life history stages such as breeding (see, e.g., Romero, 2002). The thin line above the gray line represents corticosterone fluctuations in response to normal, and still predictable daily activities like foraging (using other terminology, "the circadian peak within the gray line corresponds to resting metabolic rate, while the thin line marks the active metabolic rate for that time of the year"; Romero et al., 2009).
4. ***An upper, facultative range of GC levels*** (superimposed above the normal circadian, daily, and seasonal fluctuations), ***where short-term elevations allow counteracting unpredictable changes*** in the environment, and either maintain or return the body to homeostasis (i.e., meeting *allostatic load* EO: the increased energy demands that follow perturbations).
  a. In the Classic System, this range corresponds to GC concentrations *within level (range) C* and corticosterone levels are termed *stress-regulated* levels or *regulated facultative* levels (entering a physiological *state C* that "allow animals to survive threatening perturbations when *allostatic load* exceeds the immediate ability to cope, e.g. during encounters with a predator, infection, severe weather, energy storage and/or social instability"; Landys et al., 2006).
  b. In the Reactive Scope System, this will be the *reactive homeostasis range*.

Both classification systems agree to term these facultative GC elevations as the "***stress response***", and link the physiological, morphological, and behavioral changes to a distinct ***emergency life-history stage (ELHS)*** (Wingfield et al., 1998), which will be presented in detail in Section 33.3.3. In panels 2A and 2B of Figure 33.3, the occurrence of a *perturbation* is depicted by a star symbol in the time line (x-axis), and concomitant GC elevations (the GC stress response) are represented with a thicker blue line (the same perturbation is also depicted in the central panel dedicated to the *Allostasis Model* with the predicted changes in energy demand EO, *allostatic load,* represented with a thicker blue line). Note in the *Reactive Scope Model* (panel 2B), that the combination of *predictive* and *reactive homeostasis* ranges establishes the *normal reactive scope* for the individual, a concept that defines the upper and lower physiological constraints of a healthy animal and provides a name for the model, as we will explain shortly.

5. ***Finally, GC levels may reach a higher range where pathologies also develop.*** In the *Allostasis Model* this would be equivalent to reaching *Type 1 allostatic overload* (the demands exceed the energy available).
  a. The Classic System (panel 2A) does not have a specific graphical notation for this range, likely because the development of pathology also involves a prolonged duration (time) of the GC elevations in a range that could still be indicative of typical regulated facultative levels. Prolonged GC elevations in this range are termed *chronic GC levels*. For ease of interpretation, the red circle in panel 2A depicts the highest GC concentrations that would lead to pathology if maintained for days.
  b. In the Reactive Scope System, the upper GC levels would move into a *homeostasis overload* range (when corticosterone levels overpass the *normal*

reactive scope above the upper end of the *reactive homeostasis* range). A key aspect in the *Reactive Scope Model* (a conceptual novelty that was missing in the Classic System and in the *Allostasis Model*) is that, unlike the threshold for *homeostatic failure*, the actual threshold for *homeostatic overload* can change within and between individuals in response to certain stimuli through a mechanism known as *wear and tear*. Therefore the amplitude of the *reactive scope* is NOT a fixed trait in the individual. The effect of *wear and tear* is graphically represented in panel 2B, and its implications for *reactive scope* will be specifically discussed in the next section.

Because the absolute GC concentration defining each of the ranges described above varies between species (and even among populations within a given species, Wingfield and Romero, 2001), and GC actions on physiology and behavior are very different depending on the functional range (compare boxes 3.3 and 3.4 in Figure 33.3), a particular value or concentration of GC is generally meaningless *per se*, and unable to define the state of an individual. Unless it is close to zero or extremely high compared to the published records in other taxa (in both cases it would be indicative of serious body malfunction), a GC level or range presented in isolation says nothing about the physiological condition of that particular individual or population. As a consequence, the starting point for any field or laboratory study on avian adrenocortical function requires an initial assessment of the normal reactive scope (*predictive* and *reactive homeostasis* GC ranges) for the particular species or populations under research, and a simultaneous assessment of the conditions characterizing the same study subjects (e.g., health conditions, behavior, environmental conditions) that may affect *allostasis* and thus GC levels at a given time.

It is also important to remember that, for simplicity, we have been assuming a direct association between circulating GC levels and GC actions (panel 2 versus panel 3 in Figure 33.3). However, GC effects can be regulated at several levels (see the Chapter 26 and Wingfield, 2013b), including at least (1) regulation of secretion and clearance rates, (2) regulation of transport (over 90% of circulating GCs are bound to carrier proteins, and only the remaining 10% constitute free and biological active hormone; Malisch and Breuner, 2010), (3) regulation of receptors (there are three types of receptors differing in affinities and exerting different effects on the organism), and (4) regulation of steroidogenic enzymes.

- GCs can exert housekeeping actions or devastating effects on the organism depending on what levels are involved (they can be "*good and bad*", "*stress and antistress*"). The absolute GC concentrations defining these ranges are species-specific.
- Two nomenclature systems (the *Classic System* and the *Reactive Scope System*) propose *alternative terms* to name GC ranges within adaptive and nonadaptive (pathological) levels.
- Healthy individuals show circulating GC levels within two narrow ranges bellow and above which serious pathologies occur:
  - *Seasonal baseline* levels or *predictive homeostasis* range: GCs fluctuate according to *predictable* environmental changes, allowing the individual to cope with daily and seasonal routines.
  - *Regulated facultative* levels or *reactive homeostasis* range: short-term GC elevations (above *seasonal baseline*) allow counteracting *unpredictable* changes in the environment. Facultative elevations are also termed the "*stress response*", and allow rapid physiological and behavioral changes characteristic of a distinct *emergency life history stage* (ELHS).

33.2.3 "Wear and Tear" and the Reactive Scope

Allostatic load (the "workload" to maintain homeostasis) is recognized as one the most relevant conceptual advances of the *Allostasis Model* (McEwen and Wingfield, 2003a). However, this concept needed to be further refined and expanded to incorporate additional body costs that were not explicitly acknowledged in the original formulation of the model. In particular, *the elevation and maintenance of allostasis mediators incur a cost themselves*. For example, the long-term behavioral and cardiovascular responses that characterize *allostasis* result in cardiovascular disease (Sapolsky, 2001), and GC elevations are known to exert a negative impact on the immune system, leading to a greater susceptibility to infections (Spencer et al., 2001). These costs are directly related to the *allostasis mediators* themselves and not to *allostatic load per se*, leading to a recent proposal of the term "*wear and tear*": the cost of maintaining and using the physiological systems that mediate *allostasis* (Romero et al., 2009). The concept of *wear and tear* was originally introduced as a distinct attribute within a wider graphical interface, the *Reactive Scope Model* (Romero et al., 2009), which was proposed as a tool to characterize the physiological state of individuals across time, in response to *perturbations*, and according to the levels of *mediators* (not according to energy, which is the specific metric system of the *Allostasis Model* and generated ample criticism; Walsberg, 2003; Romero et al., 2009; McEwen and Wingfield, 2010). The *Reactive Scope Model* is therefore an extension (not a rebuttal) of the *Allostasis Model* and the seminal article leading to its formulation was supported by a thorough review of the potential *allostasis mediators* and their effects at different ranges, including immune factors, cardiovascular responses, behavior, the central nervous system, and circulating GCs (Romero et al., 2009). For simplicity, only the later mediators (GCs; *corticosterone)* will be used here to explain *wear and tear* and the dynamics of the *Reactive Scope Model* across time and in response to *perturbations.*

As we detailed in the previous section, the *Reactive Scope Model* also proposed a specific nomenclature to characterize the functional GC ranges (Figure 33.3, panel 2B). This new classification was justified to avoid confusion with other terms previously defined (what we called the *Classic System* in Figure 33.3, panel 2A), and circulating GC levels were presumed to exist in four ranges of increasing magnitude: *homeostatic failure, predictive homeostasis, reactive homeostasis*, and *homeostatic overload*. The combination of *predictive* and *reactive homeostasis ranges* establishes the *normal reactive scope* for the individual (which lies in between the upper and lower physiological constraints of a healthy animal, *above and below which serious pathologies arise*) and provides a name for the model (Figure 33.4)

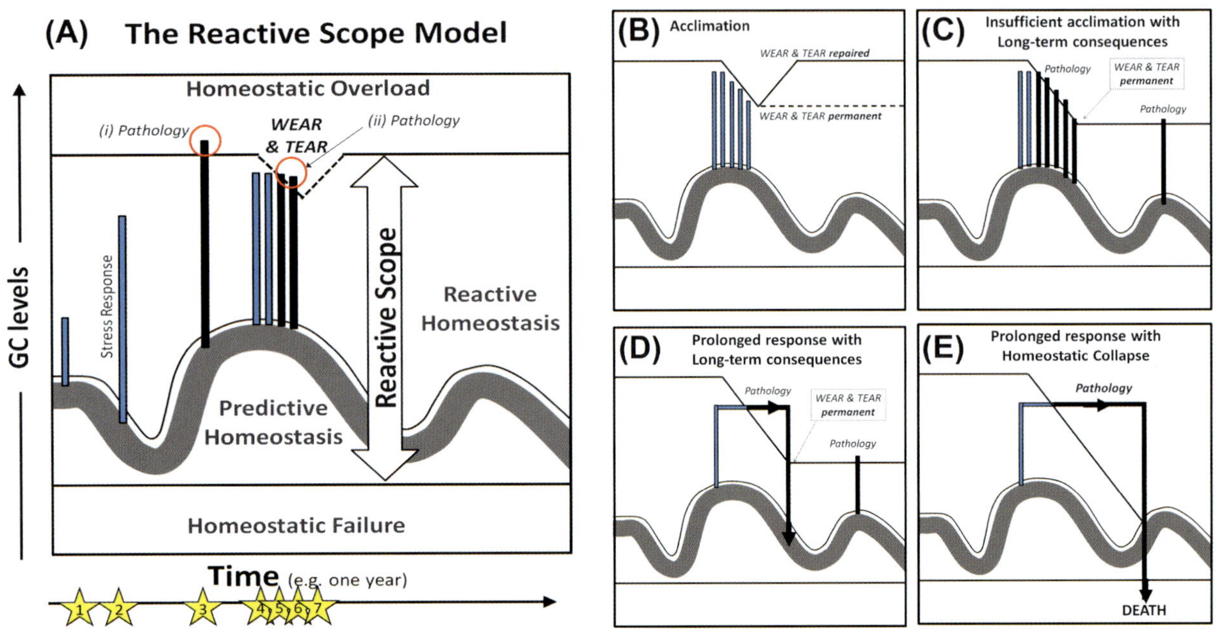

FIGURE 33.4 **The Reactive Scope Model.** Changes in circulating GC levels (*y*-axis) across time (*x*-axis) according to the Reactive Scope Model. ***Panel A:*** GC levels exist in four nominal ranges: the predictive homeostasis range varies according to seasons and life history stages, and the reactive homeostasis range represents the facultative range of GC elevations in response to unpredictable perturbations. Both ranges combined define the normal reactive scope of the individual. Below and above this range, GC levels result in pathology (homeostatic failure and homeostatic overload ranges, respectively). In response to perturbations (numbered stars in *x*-axis), each vertical line represents both a rapid GC spike in the reactive homeostasis range and a rapid decrease when the perturbation ends. In panel A, perturbation #2 is stronger than #1 and thus requires a stronger GC elevation to maintain homeostasis. Perturbations #2 and #3 are of similar strength, but occur at different times of year and only the latter results in GC levels extending inside the homeostatic *overload* range (depicted as black rather than blue bars) and cause pathology. Perturbations #4–5 do not cause pathology, but repeated GC elevations exert "wear and tear" as depicted by a progressive decrease in the threshold between reactive homeostasis and homeostasis overload (decreasing the reactive scope of the individual). Wear and tear implies that GC response to subsequent perturbations #6–7 will generate pathology. In this example, wear and tear is rapidly repaired and there are no long-term consequences on the ability to respond to future perturbations (the reactive scope becomes normal after the last perturbation #7). ***Panel B*** represents acclimation. Repeated responses to stressors exert wear and tear but the animal habituates or acclimates when GC elevations decrease over time. Since GC levels never cross the threshold of homeostatic overload, pathology does not develop (note that all the bars are blue). The accumulated wear and tear can be repaired once the stressor ends (upper continuum line), or alternatively remain permanent (lower dashed line). ***Panel C*** represents an insufficient acclimation with long-term consequences. Here the animal acclimates to a repeated perturbation but despite GC elevations progressively decrease, they enter the GC homeostatic overload range and pathology develops (denoted with black bars). The threshold of reactive homeostasis could be repaired, but in this example wear and tear exert permanent effects, increasing susceptibility to future perturbations. As a consequence, even a moderate GC elevation elicited to face the last perturbation in this panel (C), enters the homeostasis overload range and generates pathology. ***Panel D*** presents an example of long-term exposure to a perturbation (e.g., chronic social stress, habitat degradation). Long-term exposure elicits prolonged GC elevations (note that the light bars now indicate extended GC secretion), which also exert wear and tear. When the sustained increase in GC levels extends above the adjusted threshold (see blue color in bars turning to black color), pathology resulting from the elevated GCs themselves appears. Once the perturbation ends, GCs return to the predictive homeostasis range, but in this example wear and tear becomes permanent as it likely occurs under social stress. Consequently, a later response to perturbation pushes GCs immediately into the homeostatic overload range. ***Panel E*** presents an extreme example of long-term exposure to perturbations. It implies that GC elevations remain in the homeostatic overload range for long periods, generating long-term pathology and accumulating a continuous wear and tear. Once the threshold between GC reactive homeostasis and homeostatic overload ranges intersect with the predictive homeostasis range, GC levels collapse into the homeostatic failure range, turning pathology into death. *Figure modified and expanded from Romero et al. (2000) with permission.*

In response to unpredictable *perturbations,* GC levels rapidly elevate outside the *predictive* range into the *reactive homeostasis range*. These elevations are graphically represented as rapid GC spikes (blue vertical bars in Figure 33.4), which counteract the negative effects of *perturbations* (depicted as numbered star symbols along the time line) though physiological and behavioral changes and then quickly return to the *predictive homeostasis range*. This is the adrenocortical "*stress response*" aimed at maintaining *homeostasis* in the face of a *perturbation*. The sequence of perturbations experienced by a model avian individual in Figure 33.4 (A) indicate first that the magnitude of GC elevations depends of the type of perturbations. For example, the GC increase in response to perturbation 1 (e.g., a mild storm in winter) is significantly smaller than the elevations in response to perturbation 2 (a severe storm in winter). In these two examples, GC elevations remain within the *reactive homeostasis range*. In contrast, the same type of perturbation occurring at a different time of the year and even eliciting the exact same release of GCs can exert dramatically different effects on the individual. In Figure 33.4 (A), perturbations number 2 and 3 are two storms of similar quality (e.g., a severe storm occurring at the end of winter, and then again in early spring) eliciting similar GC elevations. However, only the latter causes *pathology* (red circle) as GC elevations reach the *homeostatic overload range* (GC spikes are represented now with a black, rather than a blue bar to indicate pathological effects). A similar feature was already suggested in the *Allostasis Model* with regards to an individual's likelihood of reaching states of negative energy balance at different times of the year, but some clarifications are needed. The *Allostasis Model* postulates that birds would be more resistant in winter because their energy requirements are lowest (winter perturbations should be stronger and demand more energy in order to reach *allostatic overload* Type 1; McEwen and Wingfield, 2003a), while the *Reactive Scope Model* postulates that birds are more resistant in winter because their GC *reactive homeostasis range* is wider (and thus reaching *homeostatic overload* would require a higher GC stress response, and likely a stronger *perturbation;* Romero et al., 2009). Energy demand (the *Allostasis Model* currency) and GC levels (the *Reactive Scope Model* currency) coincide in this example, as they both change in the same direction and result in deleterious effects, but this may not always be the case.

As we pointed out, the task of maintaining GC levels in the *reactive homeostasis* range during a *"stress response"* incurs a cost in itself known as *wear and tear. Wear and tear can be defined as the cumulative cost of maintaining GC levels within the reactive homeostasis range,* and this cost has a consequence: a decrease in the threshold level between the *reactive homeostasis* range and *homeostatic overload* (see decreased threshold after perturbations 4 and 5 in Figure 33.4 (A)). Note that the concept of *wear and tear* is different from the concept of *pathology*: in the latter, GCs themselves are causing damage, whereas in the former the likelihood of GCs causing damage increases. To illustrate this conceptual difference, compare the effect of perturbation 3 with the effects of perturbations number 4 and 5 in Figure 33.4 (A): only the former perturbation 3 positively generates pathology. *Wear and tear* can therefore be also interpreted as a *gradual decrease in the ability to cope with perturbations.* As the individual continues to respond to sequential or prolonged perturbations, GC elevations repeatedly enter the *reactive homeostasis* range, and the ability to counteract further stressors diminishes. At some point, the elevated GCs will cross a threshold and will start to create problems themselves. It is very important to note that, through *wear and tear,* GC levels can enter the *homeostatic overload* range even though their concentrations may not have changed. In other words the *Reactive Scope Model* proposes two ways of reaching *homeostatic overload*: either GC levels extend beyond the normal *reactive scope*, or GC levels remain in the *reactive homeostasis* range for an extended period, "shrinking" the *reactive scope* of the individual. Both scenarios are graphically depicted in Figure 33.4 (A). In response to perturbation 3, GC levels elevate above the *normal reactive scope* of the individual, entering the *homeostatic overload* range and generating *pathology.* Later on that year, the same individual is exposed to perturbations number 4 and 5, which trigger GC elevations within the *reactive homeostasis* range. These *perturbations* do not cause pathology *per se* because concomitant GC elevations do not reach the *homeostatic overload* range. However, they exerted *wear and tear* on the individual, increasing susceptibility to enter *homeostatic overload* in response to subsequent perturbations (in fact, perturbations number 6 and 7 trigger GC elevations similar in magnitude to perturbations 4 and 5, but only the former cause pathology as the *reactive scope* range "shrunk"; see Figure 33.4 (A)). In healthy individuals, *wear and tear* can be repaired after the perturbations end, but it could also remain for life, imposing a permanent decrease of the *reactive scope* of the individual and a permanent susceptibility to pathologies (the two options are represented within Figure 33.4 (B)). *Wear and tear* and its graphical representation is an alternative way to express the concepts encompassed by *allostatic load.* For examples and detailed explanations of the physiological mechanisms underlying *wear and tear,* see Romero et al. (2009).

Modeling *wear and tear* the way explained above illustrates why *equivalent GC responses might be adaptive early, but cause problems later*. It also illustrates why animals will be more resistant to stress-related pathologies during some life history stages compared to others. But in addition, the model suggests relevant evolutionary implications. For example, the authors propose that the slope of *wear and tear* in a specific individual could be determined empirically, because the threshold separating *reactive homeostasis* from *homeostatic overload* is equivalent to the maximal GC secretion of the *normal reactive scope* (Romero et al., 2009). Maximum GC secretion in a young, healthy, naïve bird could be determined experimentally (for example, through exogenous adrenocorticotropin ACTH treatment; see Section 33.5.2), providing the actual limit of *reactive homeostasis*. Then, if a moderate *perturbation* is experimentally applied repeatedly to the animal for a short period, acclimation may occur (e.g., Walker et al., 2006) and the magnitude of the response will decrease (a visual example is provided as panel (B) in (Figure 33.4) As long as pathological symptoms do not appear, thereby revealing that the individual has entered the *homeostatic overload* range, the acclimation over time will occur and the slope of *wear and tear* will be manifested as a time-related decrease in GC secretion. As we already mentioned, there are situations where repeated GC elevations cause permanent changes to the individual's physiology, and this can be illustrated by a permanent decrease in the *reactive homeostasis range* (e.g., in Figure 33.4 (C), the decreased *reactive scope* implies that this individual will be lifelong susceptible to pathologies). Also, the examples above only depict scenarios of repeated, but short-term perturbations followed by episodic spikes of GC into the *reactive homeostasis* range. These scenarios would be equivalent to *Type 1 overload* in the *Allostasis Model* (McEwen and Wingfield, 2003a,b). However, a single but long-term *perturbation* may similarly cause prolonged GC elevations, a situation that was named *Type 2 overload* in the *Allostasis Model* and required separate conceptual explanations (McEwen and Wingfield, 2003a,b). With the graphical proposal of the *Reactive Scope Model* and the associated concept of *wear and tear*, the upper end of the *reactive homeostasis* range is simply expected to show a long-term (permanent) decrease in response to long-term perturbations, as depicted in panel (D). The graphical models for the response to both (1) repeated short-term perturbations (panels (A)–(C) in Figure 33.4) and (2) prolonged perturbations like subordinate social status or habitat degradation (panels (D) and (E)) are very similar under the *Reactive Scope Model* and the dichotomy between *Type 1* and *Type 2 overloads* is no longer necessary according to Romero et al. (2009). For example, low social status may impose a long-term cost to subordinate individuals compared with dominants (the reverse situation may also occur depending on the dynamics of the specific social system; Goyman and Wingfield, 2004), leading to what has been classically termed "*chronic social stress*". In the *Reactive Scope Model* this is depicted in Figure 33.4 (D) (social conflict elicits a prolonged *adrenocortical response*, likely "shrinking" the *reactive scope* of individuals).

In all of the examples above we represented scenarios where the repeated/prolonged GC responses eventually end. However, the model predicts that as long as a prolonged stress response remains in the *homeostatic overload* range, *wear and tear* should continue to occur (Figure 33.4 (E)). The threshold between *predictive homeostasis* range and *homeostatic overload* will decrease, until actually becoming lower than the *predictive homeostasis* range itself. In this situation, very low GC levels will fall into the *homeostatic failure* range, where normal *homeostasis* cannot be maintained and death should follow. *The Reactive Scope Model* further provides predictions for how long GC levels can remain in the *homeostatic overload* range until collapse will occur. Examples of chronic malnutrition, changes in social status, and lifelong allostatic changes resulting from early life experiences are provided in the original formulation of the *Reactive Scope Model* (Romero et al., 2009), establishing comparisons to the *Allostasis Model* and highlighting the benefits of an alternative interpretation for scenarios that generate enormous interest for both biomedical researchers and ecologists. The *Reactive Scope Model* provides, for example, a new framework to understand the concept of eustress (Selye, 1976), and those changes that occur during development and can reset an animal's *reactive scope* resulting in lower (rather than higher) vulnerability to enter *homeostatic overload* later in life, a concept that challenges traditional models (Saino et al., 2005).

Finally, note that despite the fact that considerable empirical grounds led to the formulation of both the *Allostasis Model* and the *Reactive Scope Model* these are still today largely theoretical paradigms. Experimental demonstration of the models and their predictions in free-living birds under diverse ecological scenarios are among the major current challenges for avian physiologists.

- The *Reactive Scope Model* is an extension of the *Allostasis Model* presented as a graphical tool for characterizing the physiological state of individuals according to the levels of mediators (e.g., circulating *corticosterone* titers) rather than energy levels.
- The combination of the two adaptive corticosterone ranges of a healthy animal (i.e., *predictive* and *reactive homeostasis* ranges) form the normal *reactive scope* of the individual. *Corticosterone levels above and below the reactive scope cause serious pathologies*. Therefore, the reactive scope defines the upper and lower physiological constraints of a healthy individual.
- In response to unpredictable perturbations, GC levels rapidly elevate into the *reactive homeostasis range* during a "*stress response*". Even when GC elevations do not reach *homeostatic overload* to cause pathology, as the individual continues to respond to sequential or prolonged perturbations GC levels will exert *wear and tear.*
- *Wear and tear* is defined as the cost of maintaining and using the physiological systems that mediate *allostasis,* and can be graphically depicted as a decreased threshold level between the *reactive homeostasis* range and *homeostatic overload* (and therefore a decreased *reactive scope*).

- *Wear and tear* implies a gradual decrease in the ability to cope with perturbations. This concept is different from pathology: GCs themselves are not causing damage, but increase the likelihood of damage due to a progressively decreased *reactive scope*. *Wear and tear* illustrates why equivalent GC responses might be adaptive early, but cause problems later.
- The *Reactive Scope Model* proposes two ways of reaching *homeostatic overload*: (1) either GC levels extend beyond the normal *reactive scope* and cause pathology; or (2) GC levels remain in the *reactive homeostasis range* for an extended period, "shrinking" the *reactive scope* of the individual and increasing the likelihood of pathology.
- The *Reactive Scope Model* is still a largely theoretical graphical tool requiring empirical demonstrations.

33.3 ADRENOCORTICAL RESPONSE TO ENVIRONMENTAL CHANGE

A cornerstone question for evolutionary biologists and field endocrinologists is: How do birds respond and adapt to an ever changing environment? There is large amount of evidence that *the mechanisms by which birds perceive potential challenges and transduce this information into neuronal and endocrine responses are fundamentally different when facing predictable versus unpredictable environmental events* (Wingfield, 2013a). In this section we will characterize the unpredictable events of the environment (the *perturbations*), especially those that disrupt the normal life cycle, and will summarize the adrenocortical responses that allow adaptive orchestration of facultative physiological and behavioral mechanisms aimed at maximizing fitness. This subject is critically relevant for basic and conservation-oriented research in the light of global change, because human-induced *perturbations* can generate similar adrenocortical responses to natural stressors and the individuals´ inability to cope inevitably precede all extinction processes.

33.3.1 Predictable versus Unpredictable Environmental Change

As we have previously pointed out, because some seasonality in climatic, biological, and social conditions exists to different degrees in all avian habitats, birds have evolved mechanisms to organize their life cycles in synchrony with other individuals and the predictable environment. Internal and external (temperature, photoperiod, rainfall, social interactions) environmental cues allow anticipatory adjustments of major life history events (Wingfield, 2008). Individuals can thus prepare for energy demanding times of year such as a rainy winter or a hot and dry season, and GC levels will fluctuate remaining within the *baseline, predictive homeostasis range*. Regardless of how stressful local conditions may appear to the human eye, GCs do not elevate into the *reactive homeostasis range* in response to the *predictable environment* (these are *not perturbations*). For example, American goldfinches (*Carduelis tristis*) do not elevate GC levels in winter despite exposure to severe weather and temperatures below −20 °C (Dawson et al., 1991), and adult king penguins (*Aptenodytes patagonica*) fast for weeks, relying on their fat stores during the austral winters without neccesarily increasing circulating GCs (Cherel et al., 1988). Extreme cold and lack of food constitute, within certain limits, the *normal environmental conditions* under which these two species evolved their life cycles. However, there is no habitat subjected to perfectly foreseeable changes. Sudden severe storms outside the expected season, attacks from predators or dominant individuals, human disturbance, and parasites/disease are generally not predictable and have the potential to disrupt the life cycle of the individual, potentially even causing death. *Unpredictable events* in the environment are generally termed *perturbations* or *stressors*. These could be characterized in the light of the *Allostasis Model* described earlier (Section 33.3.1), across a gradient of increased energy demands *(allostatic load),* and their effects *(allostasis)* in terms of GC secretion and subsequent adjustments of life history stages, which should depend on the energy available to each particular individual at each particular moment (see figure 3 in Landys et al., 2006). The fact than some types of *perturbations* eventually bring the individual into *allostatic overload* makes a qualitative difference for the role of circulating GCs in behavior and physiology, as will be discussed below. A widely accepted *classification for the types of perturbations* is provided by Wingfield (2013a,) (see also Wingfield and Romero, 2001), who proposes discriminating between *labile* (short-term, transient) and *permanent* (long-term, modifying factors). *Labile perturbation factors (LPFs)* are in turn roughly divided into two major groups: *indirect* LPF (lasting seconds to a few minutes) and *direct* LPF (lasting minutes to hours). The formulation for this accepted classification takes into account the duration of the unpredictable stimuli, the actual duration of the adrenocortical response, and its resulting effects on the normal progression of the individual's life cycle (all of these factors combined). Figure 33.5 synthetizes information about these parameters and provides examples for ease of interpretation of the text below.

- The mechanisms by which birds perceive potential environmental challenges and transduce this information into endocrine responses are fundamentally different when facing *predictable versus unpredictable* events.
- Regardless of how stressful local conditions may appear to the human eye, circulating corticosterone levels do not elevate into the *reactive homeostasis range* in response to birds' *predictable* environment.

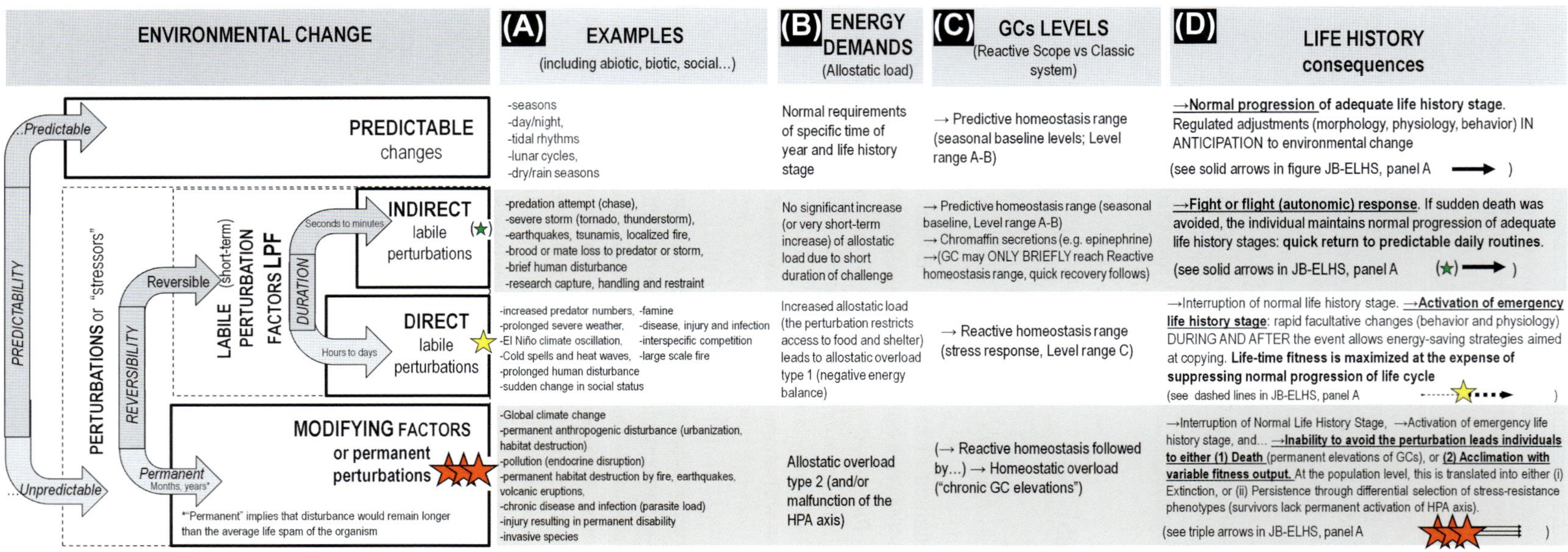

FIGURE 33.5 **Environmental change and adrenocortical responses.** Characterization of environmental change and its associated adrenocortical responses. Environmental changes can be predictable, or otherwise are termed "perturbations" or "stressors" (***column A*** lists potential examples). These can be classified as either labile (short-term, transient) or permanent (long-term, modifying factors). Labile perturbation factors (LPFs) are in turn roughly divided into two major groups: indirect LPF (lasting seconds to a few minutes) and direct LPF (lasting minutes to hours). The formulation for this accepted classification does not only take into account the duration of the unpredictable noxious stimuli, but also the energy demands (allostatic load) imposed on the individual ***(column B),*** the actual elevation of GC titers ***(column C)*** and its resulting effects on the normal progression of the individual's life cycle ***(column D).*** The fact than some types of perturbations may bring the individual into allostatic overload makes a qualitative difference for the actions of circulating GC, which may trigger an emergency life history stage (adaptive response) aimed at avoiding the perturbation, or create permanent damage (maladaptive response) if the individual cannot habituate to permanent stressors. *Figure created after Wingfield (2013a) and expanded to incorporate biotic and social examples from other sources and to implement the allostatic and reactive scope concepts.*

- *Unpredictable* and potentially noxious events in the environment are termed *perturbations* (or *stressors*), and elicit corticosterone elevations into the *reactive homeostasis range.*
- *Perturbations* can be classified as either (1) *labile perturbation factors* or *LPFs* (short-term, transient) or (2) *permanent perturbations* (long-term, also called *modifying factors*). *LPFs* can in turn be classified in two major groups: *indirect* LPFs (lasting seconds to a few minutes) and *direct* LPFs (lasting minutes to hours).
- This classification takes into account the duration of the unpredictable stimuli, the duration of the associated adrenocortical response, and the resulting effects on the normal progression of the individual's life cycle.

33.3.2 Indirect, Labile (Short-Term) Perturbations

Indirect LPFs are very short-lived perturbations, lasting seconds to a few minutes and have little or no long-term effects on an individual (as long as it manages to rapidly survive the sudden exposure to noxious stimuli). Examples may include a predation attempt/chase, a conspecific agonistic interaction, the loss of nest/offspring to a short severe storm, or a brief human disturbance. Here a "*flight or fight*" response is typically triggered with little or no elevation of GCs into the *reactive homeostasis range*, and resumption of normal activities (e.g,. social behavior, renesting) takes place within minutes to hours (Figure 33.5). The *flight or fight* response implies stimulation of the cholinergic sympathetic fibers that innervate the chromaffin tissue of the adrenal, causing the local release of acetylcholine, which triggers the release of predominately epinephrine in the blood. This rapid allostatic response promotes increased heart rate, vasodilation of arterioles in skeletal muscle, general venoconstriction, relaxation of bronchiolar muscles, pupilar dilation, piloerection, and (very importantly) mobilization of liver glycogen and free fatty acids (Axelrod and Reisine, 1984; Romero et al., 2009). Only if the perturbation persists beyond the first few minutes will the adrenocortical response be activated. Then an elevation of corticotropin (ACTH) secretion from the adenohypophysis will trigger the increase of circulating GC levels into the *reactive homeostasis range* (Figure 33.1; see details for the complete cascade of hormonal secretions in Chapter 26). Sustained elevation of GCs in the *reactive homeotasis range* mark the short-term response. At this point, the individual will require additional energy resources to cope with the noxious stimuli. Even though the perturbation is still *short-term or labile*, after a few minutes it will only be considered *indirect* if the individual readily resumes its normal life cycle when the perturbation is over. For example, the experimental *capture, restraint, and serial blood sampling protocol* that is routinely applied to quantify adrenocortical responses in wild birds (Section 33.5.1) typically lasts 30 to 60 min. Birds during this protocol elevate circulating GCs within the *reactive homeostasis range,* but they typically resume normal activities (e.g., nest attendance) within minutes following release. Capture and blood sampling would not be considered an *indirect LPF* according to its duration and associated GC elevations (allostatic responses need to go beyond the *flight or fight*, autonomic response), but this event does not qualify for a *direct perturbation* (see below) provided a fast resumption of normal life cycle activities and no apparent energy imbalance.

- *Indirect LPFs* are short-lived, lasting seconds to a few minutes and have little or no long-term effects on an individual. Examples include short predator chases, agonistic interactions, the loss of nest/offspring to a short severe storm, and brief human disturbance.
- *Indirect LPFs* elicit a *flight or fight* response largely mediated by the sympathetic nervous system, with little or no elevation of GCs into the *reactive homeostasis range*. Resumption of normal activities takes place within minutes to hours.

33.3.3 Direct, Labile (Short-Term) Perturbations and the "Emergency Life History Stage"

Direct LPFs are perturbations lasting hours or even days, when they impose high energetic demands upon the individual and result in a temporary disruption of the normal life cycle (for example breeding, resulting in brood desertion). Potential examples include exposure to prolonged severe weather, intra and interspecific competition for food or shelter, increased number of predators, reduced food resources, pollution, and prolonged human disturbance (Figure 33.5 (A)). The cumulative *allostatic load* imposed on the individual brings a negative energy balance (*allostatic overload Type 1*), and the individual redirects physiology and behavior into a facultative mode known as the *emergency life history stage (ELHS)* (Wingfield et al., 1998). Figure 33.6 illustrates a sequence of life history events of a model avian species (panel (A)) and empirical data from white-crowned sparrows *Zonotrichia leucophrys pugetensis* (panel (B)). Under normal conditions (in absence of perturbations), *the life cycle of a bird is composed of a finite number of life history stages LHS:* winter or nonbreeding, breeding and moult (depending on the species these could be more, and include for example, migration). The temporal sequence of LHS varies from species to species and each stage is expressed at a particular time of year for which it has evolved to maximize fitness (Jacobs and Wingfield, 2000). Each LHS has a varying number of unique substages (for example, the breeding stage involves gonadal maturation, courtship, ovulation, etc.) following a specific "one-way"

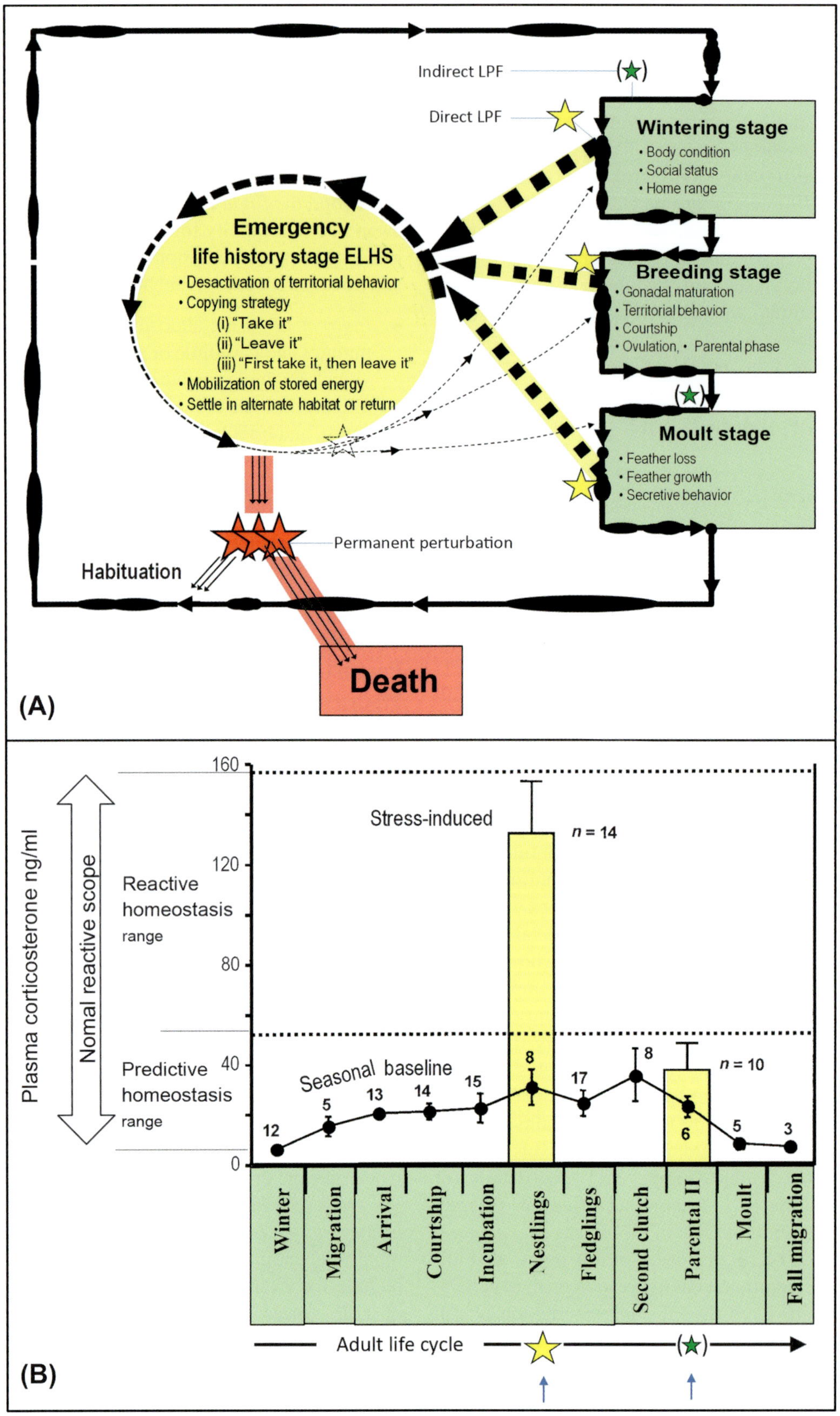
Indirect LPF
Direct LPF
Wintering stage
• Body condition
• Social status
• Home range
Emergency
life history stage ELHS
• Desactivation of territorial behavior
• Copying strategy
(i) "Take it"
(ii) "Leave it"
(iii) "First take it, then leave it"
• Mobilization of stored energy
• Settle in alternate habitat or return
Breeding stage
• Gonadal maturation
• Territorial behavior
• Courtship
• Ovulation, • Parental phase
Moult stage
• Feather loss
• Feather growth
• Secretive behavior
Permanent perturbation
Habituation
Death
(A)
Plasma corticosterone ng/ml
Nomal reactive scope
Reactive homeostasis range
Predictive homeostasis range
160
120
80
40
0
Stress-induced
n = 14
n = 10
Seasonal baseline
12
5
13
14
15
8
17
8
6
5
3
Winter
Migration
Arrival
Courtship
Incubation
Nestlings
Fledglings
Second clutch
Parental II
Moult
Fall migration
Adult life cycle
(B)

sequence as indicated with a solid thick arrow in Figure 33.6(A). The development of each stage and the sequential transitions between stages are regulated by endocrine secretions, which follow *predictable environmental cues* such as the annual photocycle and allow adequate stage-specific changes in physiology, morphology, and behavior in anticipation to energy demands (i.e., as dictated by the predictable portion of environmental change). In contrast, exposure to a *direct LPF* (represented with a yellow star symbol in the figure) can occur at any time and it will trigger a facultative "*emergency stage*" (the *ELHS*) at any point of the life cycle, as illustrated by dashed arrows in Figure 33.6(A). The substages within an ELHS are remarkably constant in all bird species studied to date, and serve first to redirect the individual into a survival mode and then to allow it returning to the normal LHS once the perturbation passes. The behavioral and physiological components that make up the substages of an ELHS (Wingfield and Kitaysky, 2002; Wingfield and Romero, 2001) are:

1. Disactivation of territorial behavior or social hierarchies in the home range of a group.
2. Adoption of one of the following alternative strategies: (1) "*leave it*" strategy: movement away from the direct LPFs; (2) "*take it*" strategy: switch to an alternate set of energy-conserving behavioral and physiological traits; or (3) "*take it at first, and then leave it*" strategy: switch to energy-conserving mode first, and then move away if conditions do not improve.
3. Once a strategy has been adopted, *mobilization of stored energy* (fat and protein) to fuel movement away, or finding refuge and endure while sheltering.
4. Finally, once the *direct LPF* passes (or the individual evades), the individual must *settle in an alternative habitat if an appropriate site is located or return* to the original site and resume the normal sequence of LHSs (returning to the most appropriate for that time of year, whether it is the same as before the LPF or any other; see thin dashed arrows in the figure).

FIGURE 33.6 **The emergency life history stage (ELHS).** ***Panel A*** presents a schematic representation of the adult life cycle in a model avian species (and concomitant hypothetic changes in circulating GC levels, depicted with different arrow line patterns), including a facultative emergency life history stage ELHS. The cycle is composed of a finite number of life history stages LHS (square boxes: winter or non-breeding, reproduction, moulting) occurring in a specific sequence. This sequence has evolved to maximize individual fitness, and each LHS occur at specific times of the year, when and where the environmental and social conditions are likely most adequate to use the energy available (e.g. food, fat stores) and invest it in performing specific life tasks. Each LHS is composed of a number of substages that also follow a pre-set sequence (for example, the breeding stage involves gonadal maturation, territory acquisition, courtship, laying etc). The external black arrow indicates the direction of the sequence and its width represent an hypothetical magnitude of the seasonal baseline GC levels, which fluctuate within a predictive homeostasis range, and allow regulation of the energy balance (allostasis) following predictable environmental cues (e.g. seasons, day/night rhythms). Superimposed on predictable environmental changes are unpredictable perturbations (star symbols), which may occur at any time of the year. Perturbations can be short-term and then they are termed labile perturbation factors LPF. When they require little energy investment to survive (for example, a short predator chase), they are termed *indirect labile perturbation factors* (green star symbol, within brackets), which trigger the autonomous "flight or fight" response, result in little or no elevations of circulating GC above seasonal baseline and allow the individual to continue its life cycle. However, labile perturbations may last hours and even days and require a larger energy investment to cope (for example, increased predator numbers). They are then termed *direct labile perturbations* (yellow star symbol), and trigger a facultative and considerable elevation of GC levels above the normal baseline, into the GC reactive homeostasis range or stress-induced levels (dashed arrows in the figure). These perturbations redirect individual´s life cycle into a facultative emergency life history stage (ELHS, central box), aimed at copying with the imposed extra energy demands (allostatic load) and survive at the expense of temporary suppression of normal LHS. Here reactive GC elevations (above seasonal baseline as depicted by the ticker width of the dashed arrows) promote dramatic changes in physiology and behavior, including the use of lipid and protein stores as energy to survive while sheltering or moving away (for example, during irruptive avian migrations). Once the direct LPF is over or the individual relocates, circulating GC levels decrease back to the predictive homeostasis range and the individual returns to the most appropriate LHS, which may or may not be the one when the perturbation occurred. Perturbations, however, may not always subside and eventually they may become long-term or permanent (for example, habitat destruction or climate change). These perturbations are then termed *modifying factors* (triple red star symbol). If the individual cannot avoid them, long-term or chronic GC elevations (triple arrow lines) can cause deleterious effects on the organism and eventually result in death. Alternatively, individuals may habituate to the perturbation and remain alive but fitness and normal progression of LHSs would only be achieved if GC function returns to normal levels. ***Panel B*** presents an empirical example of year-round changes in circulating GC levels (y-axis, average plasma corticosterone ± SE) across the life history stages and substages (x-axis) of adult male white-crowned sparrows *Zonotrichia leucophrys pugetensis*. The numbers indicates blood sample sizes. The solid black line connecting circles shows corticosterone levels in males breeding normally (i.e. GC levels within predictive homeostasis range or seasonal baseline levels). In contrast, the two vertical bars depict corticosterone levels following two separate perturbation events of severe weather, which occurred in a different year. A first severe storm resulted in birds abandoning their nests and territories and can be therefore considered a direct LPF (yellow star symbol). Note that corticosterone levels elevated well above the seasonal baseline, into a higher reactive homeostasis range allowing activation of an emergency life history stage ELHS (birds temporarily suspended breeding activities, ranging for days over a large area in loose flocks). After this perturbation subsided corticosterone levels returned to the predictive homeostasis range, the birds returned to their territories and renested. Then a second severe weather event occurred after renesting (during the parental phase II). However, because the second storm failed to elevate corticosterone levels above the predictive homeostasis range and no ELHS was activated, this labile perturbation is considered indirect (depicted with a green start symbol within brackets). *Panel A modified and expanded after Jacobs and Wingfield, (2000), with permission; panel B expanded from Wingfield and Kitaysky, (2002) by permission of Oxford University Press.*

These dramatic changes in behavior and physiology can occur within minutes to hours of exposure to *direct LPF,* and a combination of laboratory experiments and field studies indicates that the elevation and maintenance of GC levels *above seasonal baseline (predictive homeostasis) into the reactive homeostasis range (or stress-induced GC levels;* see Figure 33.3) underlies this facultative ELHS (Figure 33.6). A few examples follow, and additional examples can be found in other reviews (Wingfield and Romero, 2001; Wingfield and Kitaysky, 2002; Wingfield et al., 2011). Young but not adult European blackbirds (*Turdus merula*) migrated southwards when exposed to a severe weather event (*direct LPF*) in Germany that resulted in snow and ice covering their winter feeding grounds. The energy demanded by digging for food likely exceeded the energy to be gained, but only among unexperienced, first year individuals. As predicted, the group of young, unexperienced birds that left their regular winter areas showed concomitant elevations of circulating corticosterone levels compared with nonmigrating adults (Schwabl et al., 1985). There are similar studies linking unusually severe winter weather (a *direct LPF*) to elevated corticosterone levels and irruptive migrations (a "*leave it*" ELHS strategy) in ground-feeding birds because the snow covering food resources may impose an (unpredictable and) considerable increase to *allostatic load.* In North America, dark-eyed juncos (*Junco hyemalis*) showed significantly higher plasma corticosterone levels during a severe snowstorm compared with before or after it; circulating GC levels were highest when home ranges were abandoned and lower when the birds found refuge after the storm (Rogers et al., 1993). In contrast, Harris sparrows became inactive to ride out a severe winter storm (a "*take it*" strategy), but also showed concomitant elevations of circulating corticosterone (Rohwer and Wingfield, 1981). Note that the influences of inclement weather on the ELHS depend upon the severity of the storm and when it occurs.

If we now focus on the reproductive season, white-crowned sparrows (*Zonotrichia leucophris pugetensis*) delay the onset of breeding without elevating corticosterone levels if severe weather occurs before egg laying. This suggests that storms early in the season do not qualify as LPFs, and that they rather act as supplementary information (predictive environmental change) to fine-tune gonadal development. Alternatively, it is possible that the energy demands for that time of the year were low compared to the energy available in the environment plus the body reserves, allowing individuals to cope with a perturbation without triggering GC elevations above the *baseline, predictive homeostasis range*. However, during incubation and chick brooding, the additional energetic costs associated with these normal LHSs increases vulnerability to perturbations, and equivalent severe storms result in elevated corticosterone levels and the abandonment of nest and territory, which are typical examples of a "*leave it*" strategy within the ELHS (Wingfield, 1984).

Because sex roles during reproduction impose different *allostatic loads* for males and females in many bird species, we may also expect that a given LPF should differentially trigger corticosterone elevations and an ELHS depending on the energy invested by each sex in each substage. For example, territory acquisition is largely a male role in raptors. Male black kites *Milvus migrans* likely face higher *allostatic loads* compared to females at the start of the breeding season, and they showed higher corticosterone elevations in response to the experimental LPF of capture and restraint (Blas et al., 2011). *Allostatic load* should be even higher among young males because they rely on the competitive displacement of an older male to initiate breeding; and as expected, corticosterone levels following capture and restraint were higher in young compared to older males. This pattern was reversed among females, because young, floating individuals incur neither the *allostatic loads* of competition for territories nor the energy demands for egg formation: young females showed the lowest levels of corticosterone after exposure to capture, handling, and restraint and likely the lowest susceptibility to trigger an ELHS in response to perturbations (Blas et al., 2011). The effect of sex roles on the susceptibility to LPF has been also suggested in white-crowned sparrows. Severe weather may affect females more than males during the incubation period because males do not incubate, but a storm during the fledging phase would affect males more, as their parental effort is at its highest point (Wingfield, 1984; Wingfield et al., 1983). Here it is worth reminding readers that although some LHSs and substages (territory acquisition, incubation, and brooding) are energetically demanding, they are totally predictable and do not generate GC elevations into the *reactive homeostasis range per se*. However, because these life cycle activities increase *allostatic load,* they can generate a greater susceptibility to secrete GCs into the *reactive homeostasis range* after exposure to unpredictable perturbations (and therefore a greater susceptibility to enter ELHS).

The examples above strongly suggest that stress-induced GC elevations in the *reactive homeostasis range* underlie activation on the ELHS, but additional field and laboratory experiments (many of which include experimental manipulations of systemic GC titers) have allowed a more precise characterization of GC actions on the particular behavioral and physiological traits that make up the substages of an ELHS. We will briefly comment on some of these GC actions occurring within the *stress-induced* or *reactive homeostasis range* (i.e., those that accompany life-threatening perturbations). However, it is important to note that GC actions at the lower range that characterizes daily and seasonal life processes (that is, within the *predictive homeostasis range* or *baseline levels*) can modify the same traits in different and even opposite ways. These aspects are thoroughly reviewed elsewhere (Landys et al., 2006; Wingfield and Romero, 2001). Major GC actions at *stress-induced* or *reactive* levels include:

1. ***Reactive GC levels increase protein catabolism and gluconeogenesis.*** Administration of exogenous GCs to mimic *reactive homeostasis levels* induces protein loss and muscle atrophy in a variety of avian species. For example, domestic fowls *Gallus gallus* respond to corticosterone treatment showing a rapid elevation of plasma uric acid, a byproduct of amino acid breakdown (Saadoun et al., 1987). In house sparrows (*Passer domesticus*), corticosterone implants decrease the mass of *pectoralis* muscles, but differentially promoting catabolism of the soluble (sarcoplasmatic) protein fraction rather than contractile (myofibrilar) components, which allows maintenance of flight capabilities (Honey, 1990). By making available amino acids, reactive GC levels support increased *allostatic load* through indirect provisioning of intermediates for the citric acid cycle and substrates for hepatic gluconeogenesis.
2. ***Reactive GC levels promote availability of lipid energy from adipose tissue stores*** (Dallman et al., 1993). Stored triglycerides may be released as free fatty acids, through either an increase of trygliceride breakdown or a decrease of in the rate of re-esterification. In addition, elevated GCs seem to support enzymatic mechanisms that allow delivery of fatty acids to working tissues (Mantha and Deshaies, 2000; Landys et al., 2006 and citations therein). Note that artificial elevation of GC levels may also result in increased fat depots when the experimental animals are in positive energy balance (Wingfield and Silverin, 1986).
3. ***Reactive GC levels decrease body mass.*** For example, exogenous corticosterone elevations inhibit weight gain in house sparrows and rufous hummingbirds *Selasphorus rufus* (Honey, 1990; Hiebert et al., 2000). Mass loss might be a consequence of the two metabolic effects described above (increased energy mobilization) plus a decreased maintenance of structural tissues.
4. ***Reactive GC levels modulate locomotor activity.*** For example, white-crowned sparrows *Z. leucophris* show a rapid increase in activity after corticosterone ingestion (Breuner et al., 1998). This effect likely depends of GC concentration: while moderate elevations promote intense activity (a "*leave it*" ELHS strategy), very high doses may actually stimulate inactivity (a "*take it*" strategy; Wingfield and Ramenofsky, 1999). In addition, energy availability seems to modulate this response: while corticosterone implants increase activity in fasting birds, they decrease activity when food is *ad libitum* (Astheimer et al., 1992). Both the dose- and the context-dependent effects of GC elevations match the expected behavioral responses under an ELHS: if the LPF does not reduce food resources it may be adaptive to "*take it*", but if the perturbation persists then a reduction of the available energy (food) would trigger higher GC elevations and a "*leave it*" strategy aimed at searching for energy somewhere else. GC effects on locomotor activity may be particularly relevant in avian species adapted to severe habitats such as deserts or high latitudes, which show irruptive movements in response to LPF (Wingfield and Ramenofsky, 1997). Note, however, that seasonal GC levels are typically elevated during those life history stages that demand high physical activity such as migration and dispersal (Landys-Ciannelli et al., 2002; Belthoff and Dufty, 1998), but these effects occur within the *predictive homeostasis range*.
5. ***Reactive GC levels may increase food intake.*** Although some studies propose that GCs regulate feeding rate in a permissive capacity (Dallman et al., 1993), recent research suggest that this effect is mediated through a threshold action of glucocorticoid receptors rather than by the extent of GC elevations above the *seasonal baseline* (see details in Landys et al., 2006). There is ample correlational evidence that foraging activities are normally highest during the times of day when baseline GC levels are maximal, and that feeding intensity can be shifted by *corticosterone* manipulations (Dallman et al., 1993), but again, these effects occur within the predictive homeostasis range.
6. ***Reactive GC levels suppress territorial and reproductive behavior.*** For example, male song-sparows (*Melospiza melodia*) reduce territorial aggression when treated with exogenous corticosterone (Wingfield and Silverin, 1986). Similarly, free-living pied flycatchers (*Ficedula hypoleuca*) decrease nestlings feeding rates when treated with exogenous corticosterone (bringing a concomitant decrease in nestlings' body condition and survival), and even cease all reproductive activities (abandoning the nest and the acquired territories) when given high-dose GC implants (Silverin, 1986).
7. ***Reactive GC levels increase night restfulness.*** Exogenous corticosterone elevations in captive white-crowned sparrows (*Z. leucophris*) and pine siskins (*Spinus spinus*) result in a uniformly low oxygen consumption pattern, with an estimated 20% energy savings overnight compared with controls (Buttemer et al., 1991). However, similar treatment applied to other species results in contradictory results (Palokangas and Hissa, 1971), likely because GC-induced gluconeogenesis also affects metabolic rate and thus oxygen consumption. Corticosterone administration to rufous hummingbirds (*S. rufus*) results in nocturnal torpor, a form of seeking *"refuge/take it"* strategy that these birds acquire through decreasing body temperature (Hiebert et al., 2000).

Summarizing, GC elevations into the *reactive homeostasis range*, which are typically several-fold higher than the *seasonal baseline or predictive range* (Figure 33.6(B)), allow critical changes in physiology and behavior aimed at coping with LPFs and thus have a great adaptive value in maximizing survival and lifetime reproductive success, even at the cost of a temporary suppression of the life cycle. Once the direct LPF is over or the individual relocates, circulating GC levels normally decrease back to the *predictive homeostasis range* and the individual returns to the most appropriate LHS, which may or may not be the one when the perturbation occurred. Perturbations, however, may not always subside and eventually they may become *long-term or permanent* (for example, habitat destruction, climate change). Then, they are termed *modifying factors*.

- *Direct LPFs* last hours or days, impose high energetic demands *(allostatic load),* and result in a temporary disruption of the normal life cycle. Examples include exposure to prolonged severe weather, intraspecific competition, increased predators pressure, reduced food resources, and prolonged human disturbance.
- The cumulative *allostatic load* imposed on the individual brings a negative energy balance (*allostatic overload Type 1*), and the individual redirects physiology and behavior into a facultative life stage known as the *emergency life-history stage (ELHS).* Such an emergency stage is largely driven by the elevation and maintenance of GC levels *above seasonal baseline (predictive homeostasis)* and into the *reactive homeostasis range* (or *stress-induced* levels).
- Corticosterone elevations in the *reactive homeostasis range* during an ELHS promote (1) increased protein metabolism and gluconeogenesis, (2) availability of lipid energy from adipose tissue stores, (3) decreased body mass, (4) changes in locomotor activity and foraging, (5) suppression of reproductive behavior, and (6) increased night restfulness. These corticosterone-mediated changes in physiology and behavior have a great adaptive value, maximizing individual fitness at the cost of a temporary suppression of the life cycle.
- Once the *direct LPF* is over or the individual relocates, circulating GC levels normally decrease back to the *predictive homeostasis range* and the individual returns to normal life history stages.

33.3.4 Permanent (Long-Term) Perturbations or "Modifying Factors"

Modifying factors are perturbations lasting months or years (they are termed "*permanent*" because the disturbance may remain longer than the average lifespan of the organism), when they impose high energetic demands upon the individual and result in a permanent disruption, even disintegration of the normal life cycle (Wingfield and Romero, 2001; Wingfield, 2013a). Potential examples include chronic disease and infection (e.g., high parasite load), injury resulting in permanent disability, global climate change, and sustained anthropogenic disturbance (Figure 33.5 (A)). The cumulative *allostatic load* imposed on the individual may bring a negative energy balance, leading to death if a prior activation of the ELHS did not allow evading the perturbation (for example, when relocation was not possible). Here the sustained long-term GC elevations exert *wear and tear* (see Section 33.2.3), reaching the *homeostatic overload range* and resulting in deleterious effects on critical body systems including inhibition of gonadotropin secretion and cessation of reproductive function (e.g., delay of puberty onset, gonadal involution); suppressed secretion of growth hormone (e.g., stress-induced dwarfism); complete breakdown of the skeletal muscle (e.g., suppressed flight and movement capabilities); suppression of T-lymphocyte response to infections (increasing susceptibility to disease); severe neuron loss in the hippocampus through impaired glutamate and calcium regulation; and decreased generation of arachidonic acid (for a review of effects, see Sapolsky, 1996; Wingfield and Romero, 2001; Romero et al., 2009). In contrast to the highly adaptive GC responses that characterize the ELHS, chronic GC elevations lack adaptive value. Note also that at a given point, chronic GC exposure will bring a disruption of the *hypothalamus-pituitary-adrenal (HPA)* axis, and the deleterious effects on the organism may occur even if the long-term perturbation subsides and the energy available in the environment becomes above the energy needs of a healthy individual (this will be equivalent to *allostatic overload Type 2,* to a critically reduced *reactive scope* generated by *wear and tear,* or to the ecological scenarios of social stress that we have already commented upon). Pollution is also considered a type of *modifying factor*, but its impact on avian adrenocortical function has been subjected to scarce research and there is a great need for field investigations (reviewed in Baos and Blas, 2009).

A critical point regarding to the effects of *modifying factors* is that not all the individuals within a population perceive and respond to perturbations in the same way. As a consequence, under particular scenarios (which should depend on the species and the type of perturbation) it is possible that even a drastic initial response to *modifying factors* (e.g. disappearance of a fraction of the population through death or emigration) is followed by a recovery through differential selection of stress-resistance phenotypes (Blas et al., 2007; Cockrem, 2007). In other words, perturbations may become selective pressures, potentially allowing adaptation, and population persistence could occur provided that (1) there exists some genetic variability in the response to stress, whereby specific genotypes do not activate adrenocortical responses to that specific perturbation; (2) individuals acclimatize to that particular perturbation, progressively reducing their adrenocortical responses and surviving without permanent activation of the HPA axis; or (3) epigenetic effects (e.g., *maternal programming*) reduce the impact of the perturbation through modifying the offspring's adrenocortical responses. Examples for these potential responses will be provided in the following section.

- *Permanent perturbations* (also called *modifying factors)* last months or years, imposing high energetic demands upon the individual and resulting in a permanent disruption of life cycles. Potential examples include chronic disease and infection, global climate change, and sustained anthropogenic disturbance (e.g., urbanization, habitat destruction)
- After the initial activation of the ELHS, inability to evade the perturbation implies that the cumulated *allostatic load* will bring the individuals into a negative energy balance, and death may follow.
- The sustained long-term GC elevations exert *wear and tear,* reaching the *homeostatic overload range* and resulting in deleterious effects on critical body systems, including (1) inhibition of gonadotropin secretion and cessation of reproductive function, (2) suppressed secretion of growth hormone, (3) complete breakdown of the skeletal muscle, (4) suppression of immune system, and (5) neuronal death.
- *Modifying factors* can lead to population extinctions. However, population persistence through local adaptations could also potentially occur if genetic variability and phenotypic plasticity in the response to stress allow a differential survival of stress-resistant individuals.

33.4 PHENOTYPIC PLASTICITY AND SELECTION ON THE STRESS RESPONSE

Avian speciation has taken place under contrasting environmental scenarios (e.g., from deserts to poles), and *adrenocortical responses* have evolved in parallel, shaped by the prevailing *perturbation* factors and likely explaining the strong interspecies variability. Within a given species, different subspecies, races, morphs, and even populations have been reported to differ in the patterns of adrenocortical secretion (Wingfield and Romero, 2001). On the one hand, such variability reveals a genetic basis that controls HPA function and, for example, has allowed humans to domesticate wild animals and generate stress-resistant phenotypes through artificial selection (Keith et al., 1973; Satterlee et al., 2000). In addition to such genetically determined patterns, the adrenocortical response to stress shows a strong phenotypic plasticity. For example, the environmental conditions experienced early in life can prime lifelong differences in the patterns of adrenocortical responses, revealing adaptive epigenetic patterns (Love et al., 2013). Furthermore, individuals often show the ability to modulate their adrenocortical response according to the specific requirements characterizing different life history stages, thereby maximizing the cost-benefit balance associated with the actions of corticosterone on behavior and physiology (Lendvai et al., 2007). Phenotypic plasticity is also manifested when repeated exposure to the same perturbations provoke gradual changes in adrenocortical function, revealing that experience-related patterns occur through acclimation, and highlighting the close connection between cognition (i.e., an individual's perception of a potential threat) and the adrenal response (Cyr and Romero, 2009). Overall, the HPA axis shows all the features of a trait subjected to natural selection (Love et al., 2013; Cockrem, 2013): large individual variation (Williams, 2008; Cockrem, 2013), repeatability under consistent conditions (Ouyang et al., 2011), heritability (Bartels et al., 2003; Evans et al., 2006; Solberg et al., 2006), and actual response to artificial selection (Satterlee and Johnson, 1988; Evans et al., 2006).

- The HPA axis shows all the features of a trait subjected to natural selection: large individual variation, repeatability under consistent conditions, heritability and response to artificial selection.
- Superimposed on genetically determined patterns, individuals show phenotypic plasticity in their responses to stress

33.4.1 The Stress Response during Development

Young birds display highly variable HPA responses during development (reviewed in Blas and Baos, 2008). On one hand, interspecies differences reflect the wealth of variability in developmental strategies within the altricial-precocial spectrum (Figure 33.7(A); Starck and Ricklefs, 1998). Although precocial species such as mallards (*Anas platyrhynchos,* which hatch with sight, covered with down, and can thermoregulate, locomote, and feed independently from parents) can elevate circulating corticosterone in response to human handling one day after hatching (Holmes et al., 1990), altricial species such as northern mockinbirds (*Mimus polyglottos*, hatching blind, almost naked, unable to thermoregulate and locomote) show little HPA response to the same stimuli during the first few days post-hatch (Sims and Holberton, 2000), indicating a *"stress hyporesponsive period"*. On the other hand, within a given developmental mode and species, HPA responses gradually increase as the individual grows and develops. For example, white storks (*Ciconia ciconia*) a semi-altricial species, show very moderate GC responses to capture and handling until 20 days post-hatching (Figure 33.7(B)), but the same type of experimental stimuli gradually elicit more robust corticosterone secretions as birds grow older, reaching maximal corticosterone secretion (resembling typical adult-like responses) near fledging (Blas et al., 2006a).

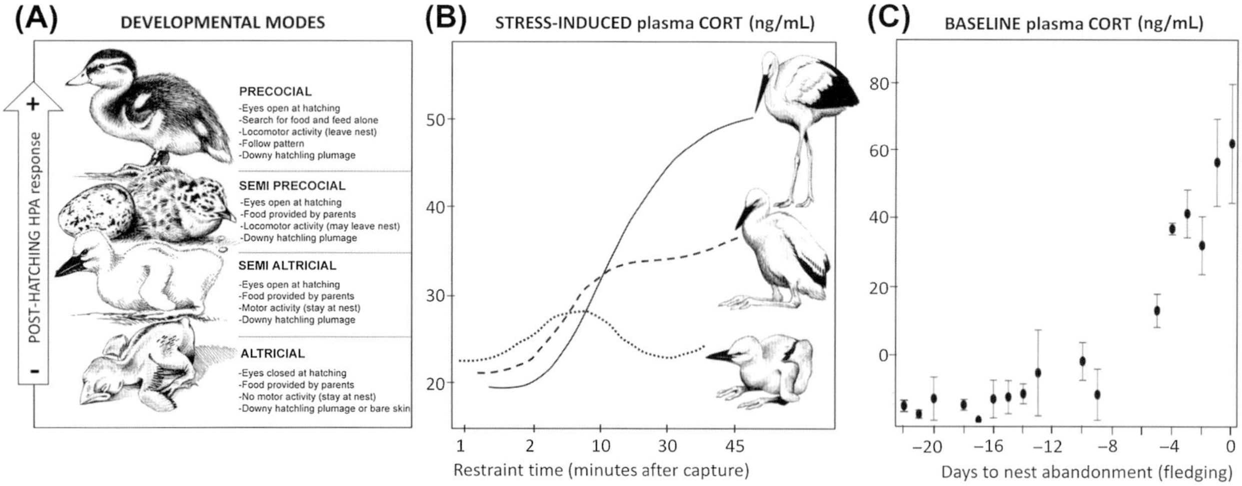

FIGURE 33.7 The stress response during development. ***Panel A.*** *Avian developmental modes* range across a gradient defined by two extremes: precocial and altricial species. Altricial species (e.g., Passeriforms, a house sparrow *Passer domesticus* in the drawing) hatch with closed eyes, lack of feather coat, have reduced motor capabilities, and are unable to thermoregulate or forage without parental care. Precocial species (e.g., Galliforms, a mallard *Anas platyrhynchos* in the drawing) hatch in an advanced developmental stage, with open eyes, feather coat, and independing foraging and locomotor capabilities. Intermediate developmental modes include semi-altricial (e.g., a white stork *Ciconia ciconia*) and semi-precocial species (e.g., a yellow-legged gull, *Larus michahellis*). The ability of responding to environmental and social perturbations through HPA function (e.g., ability to elevate corticosterone in response to capture and handling) increases across the developmental gradient of hatchlings: altricial species show a *stress hyporesponsive period* characterized by inability to show Glucocorticoid elevations until days following hatching, while altricial species show robust HPA responses within one day post-hatch. ***Panel B.*** *Developmental changes in HPA function.* Within a given species (e.g., white storks, *Ciconia ciconia*) HPA function progressively increases during ontogenic development. The figure shows circulating corticosterone levels (*y*-axis) during the 45 min following capture and restraint (reactive homeostasis range) for three nestlings aged 24, 40, and 60 days old. ***Panel C.*** *Transition from nestling to fledging stages.* Circulating baseline corticosterone levels typically elevate shortly before hatching, fledging, and during dispersal, allowing birds to meet the increasing energy demands associated to ontogenic transitions. The figure shows baseline circulating corticosterone (predictive homeostasis range) in thin-billed prion *Pachyptila belcheri* chicks, relative to the observed time of fledging. Note maximum corticosterone levels during the day prior to departure. *Panel A modified from Blas and Baos (2008), with permission; panel B redrawn from Blas et al. (2006a), with permission; panel C from Quillfeldt et al. (2007) with kind permission from Springer Science + Business Media.*

The *Developmental Hypothesis* proposes that the patterns of maturation of HPA responses within and across species evolved to balance the costs and benefits of GC actions with the individual's progressive ability to overcome environmental perturbations (Blas and Baos, 2008). Inability to show robust responses during initial developmental stages avoids exposure to high and prolonged GC levels, which reduces growth and impairs thyroid function and cognitive and competitive abilities (Schwabl, 1999; Kitaysky et al., 2003; Hayward and Wingfield, 2004). However, as the individual grows, its ability to show *stress-induced* responses (i.e., GC elevations within the *reactive homeostasis* range) acquires a critical adaptive value. For example, corticosterone elevations in nestling black-legged kittiwakes *Rissa tridactyla* at mid-developmental stages (15–20 days old) facilitate begging and aggression (Kitaysky et al., 2001, 2003), two behavioral responses highly advantageous to cope with food shortages early in life despite being still dependent on parental care. In addition to a progressive, developmental increase in the ability to trigger robust HPA responses, circulating GCs in the *predictive homeostasis* range (i.e., *baseline* levels) typically show age-related changes that serve to regulate ontogenic transitions across vertebrate taxa (reviewed in Wada, 2008). *Baseline* (*predictive*) GC elevations promote maturation of critical organs and possibly control both the ability to hatch and the timing of fledging. For example, *baseline* corticosterone levels peaks around hatchling in domestic fowls *Gallus domesticus* (Carsia et al., 1987) and graylag geese *Anser anser* (Frigerio et al., 2001), and exogenous corticosterone administration in turkey *Meleagris gallopavo* embryos 2 days before hatching significantly increases hatching success (Wentworth and Hussein, 1985). In white storks *C. ciconia*, *baseline* corticosterone levels increase prior to fledging (Corbel and Groscolas, 2008), and similar patterns have been found in American kestrels *Falco sparverius* (Heath, 1997), pied flycatchers *F. hypoleuca* (Kern et al., 2001), canaries *Serinus canaria* (Schwabl, 1999), thin-billed prions *Pachyptila belcheri* (Quillfeldt et al., 2007), and Laysan albatross *Phoebastria immutabilis* (Seabury and Breuner, 2005). Interestingly, *baseline* corticosterone elevations at the time of nest departure may be related to the species requirements to perform the first flights, and thus nest location. Avian species nesting on the ground (e.g., snowy owls, *Nyctea scandiaca*) or in low brush (e.g., northen mockingbirds) do not perform their first flights until a few days following nest departure, and the later transition is not associated to *baseline* corticosterone elevations (Romero et al., 2006; Sims and Holberton, 2000). In contrast, avian species nesting in cavities or higher substrates (high trees, cliffs) perform their first flights as they abandon the nest, and typically show increased *baseline* corticosterone levels (e.g., cavity-nesting American kestrels, Heath, 1997; screech-owls *Otus* sp, Belthoff and Dufty, 1998; or burrow-nestling thin-billed prions, Quillfeldt et al., 2007; see Figure 33.7(C)). Furthermore, corticosterone levels elevate before and during the activity periods in fledging screech-owls (Belthoff and Dufty, 1998), suggesting a role of this hormone in natal dispersal. Corticosterone elevations during avian development may thus serve to meet with the increasing energy demands *(allostatic load)* associated with ontogenic transitions including hatching, fledging, and dispersal (Wada, 2008).

- Avian developmental modes range across a gradient defined by two extremes: altricial and precocial.
  - The *stress response* of hatchlings increases across the developmental gradient: while altricial species show a "*stress hyporesponsive period*" characterized by an early inability to show GC elevations, altricial species show robust HPA responses soon after hatch.
  - Inability to show robust responses early during development allows avoiding exposure to high and prolonged GC levels, which reduce growth and impair thyroid function, cognitive and competitive abilities.
  - The *Developmental Hypothesis* proposes that the patterns of maturation of HPA responses evolved to match the costs and benefits of GC actions with the individual's progressive ability to cope with *perturbations* (e.g., increased behavioral performance and parental independence).
  - *Baseline* corticosterone levels (*predictive homeostasis* range) increase prior to hatching, fledging, and during dispersal, making it possible to meet the increased energy demands *(allostatic load)* associated with ontogenic transitions.

33.4.2 Maternal Effects

Studies in humans and laboratory models indicate a profound effect of parental care early in life on offspring's HPA function and associated behavioral responses (McGowan and Szyf, 2010; McGowan et al., 2011). GC actions during development exert permanent *organizational effects* on the individual, with a stronger impact on performance (fitness: reproduction and survival), compared to reversible *activational effects* of the same hormones when the endocrine systems are already developed (Williams, 2008). In birds, developmental exposure to high GC levels can occur prenatally (i.e., maternally mediated changes or *in ovo* composition, e.g., Almasi et al., 2012) and also postnatally or *ex ovo* (e.g., according to the amount of offpring care: food provisioning, shelter, social competition; see Blas and Boas, 2008 for a review), and has the potential to translate ecological and environmental conditions into permanent offspring responses through a process termed *maternal programming* (Figure 33.8(A–B), reviewed in Love et al., 2013). As a consequence, mothers can "*prime*" lifelong responses in offspring through exposure to high or low levels of GCs, resulting in a permanent modification of HPA function (Hayward and Wingfield, 2004; Love et al., 2005; Saino et al., 2005; Love and Williams, 2008) that will likely affect the ability to cope with ecological perturbations. Such priming can be highly adaptive and allow animals to adjust to changing environments, because exposure to a low level of perturbations early in life often result in individuals that are better able to cope with exposure to higher levels of perturbations on subsequent occasions (a process known as *hormesis*; Constantini et al., 2010). Note that these adaptations occur *epigenetically*: the genetic composition in the offspring can be the same, but the expression of different genes in response to perturbations and GCs during development will result in different phenotypes. For example, cross-fostering the biological offspring between high and low caring mothers can permanently reverse offspring's HPA response in laboratory animals (Francis et al., 1999). The fitness of the modified phenotypes through *maternal programming* is expected to be context-specific, and determined by the environmental conditions encountered after development, and how much they resemble natal conditions (*Maternal-Match Hypothesis*; Love et al., 2013). In those situations when the early priming was a reliable predictor of the offspring's future environment, *maternal programming* is expected to increase offspring and even maternal fitness (Figure 33.8(B)). However, if maternal signaling was a poor predictor of offspring's future environment, the programming would exert negative (maladaptive) effects on the offspring ("*thrifty phenotypes*"; Hales and Baker, 1992). The *Maternal-Match Hypothesis* (Love et al., 2013) is a recently proposed paradigm refining the broader *Environmental-Match Hypothesis* (Sheriff et al., 2009, 2010, 2011a). Although these hypotheses require testing in wild birds, there is evidence for a number of environment-maternal-offspring connections mediated through corticosterone, which constitute the pillars for *maternal programming* summarized in Figure 33.8:

1. ***Environmental and social conditions affect circulating GC levels in adult birds, including reproducing females.***

   For example, increased risk of predation (Hawlena and Schmitz, 2010; see Clinchy et al., 2013 for a review on the "*ecology of fear*"), unpredictable changes in food availability (Kitaysky et al., 1999a,b, 2007; Shultz and

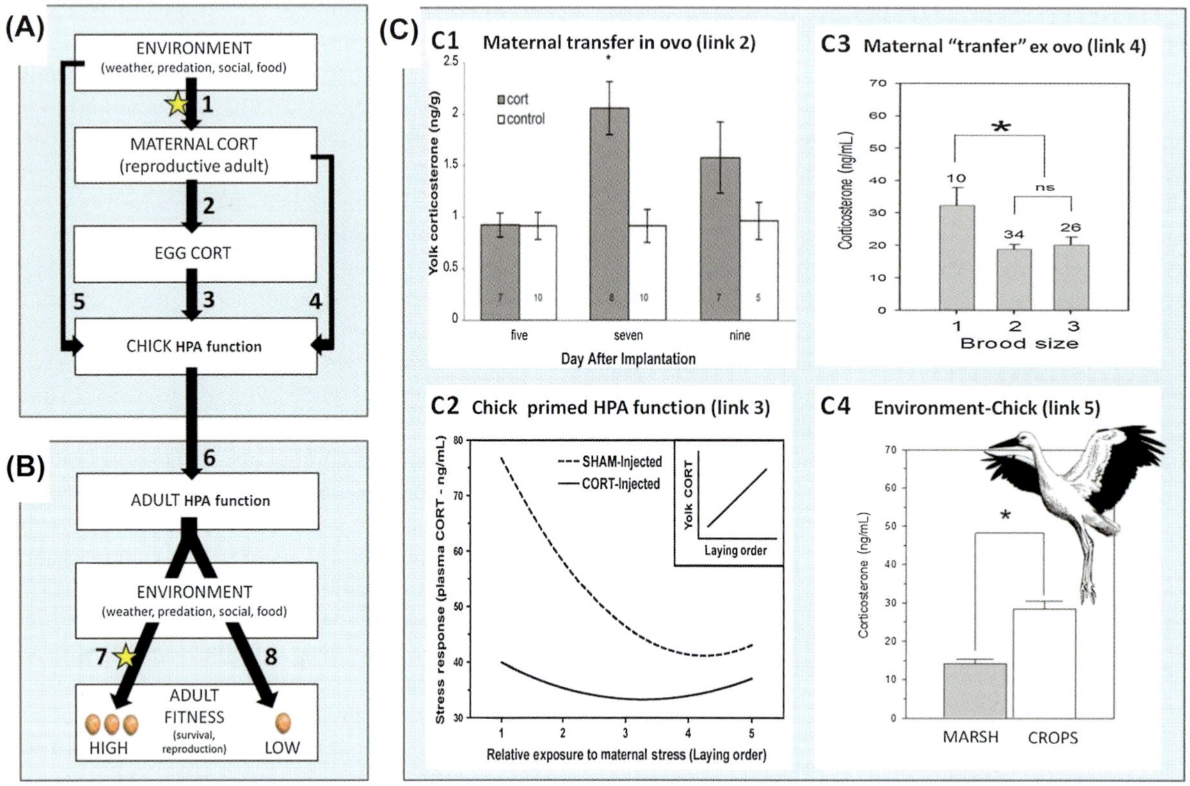

FIGURE 33.8 **Maternal effects.** ***Panel A.*** Summary of the proposed mechanisms whereby environmental perturbations transduce into permanent changes of the HPA-axis function in developing birds, through maternal effects (maternal programming). Environmental perturbations (star symbols) activate the HPA axis of reproducing, adult female birds (#1). High circulating corticosterone levels in pre-laying and laying females result in high corticosterone levels being deposited in the eggs (#2), and these will hatch nestlings developing a "primed", permanent modification of their HPA response (#3). In addition, variability in maternal care such as reduced nestlings´ attendance and food provisioning or increased within-brood social conflicts, can elicit further corticosterone elevations in the nestlings (#4). Because parental care cannot completely buffer environmental perturbations, these may also exert a direct effect on nestlings´ corticosterone levels (#5) and contribute to permanent modifications of the HPA-axis. The "priming" of HPA responses experienced by growing birds during development can be long-lasting, and the resulting adult phenotypes will show a permanent epi-genetic modification of their response (#6). ***Panel B.*** *The Maternal-Match Hypothesis.* Depending on the GC exposure that birds received during development (a function of maternal care and natal environmental conditions) different phenotypes regarding HPA function will result. For example, high- vs. low-HPA responders could be the offspring of the same couple of birds in different years. Phenotypes may differ in the amount of corticosterone secreted in response to standardized perturbations, and therefore in the ability to cope with environmental challenge later in life. The maternal-match hypothesis proposes that the fitness of each phenotype will be determined by the environmental conditions encountered after development. In those situations when the early "priming" was a reliable predictor of the offspring´s future environment (#7; note the same star as #1 in Panel A) offspring fitness will increase. However, if maternal signaling was a poor predictor of offspring´s future environment, offspring fitness will be reduced (#8). ***Panel C.*** *Examples illustrating maternal effects.* Corticosterone implants in female Japanese quail (*Coturnix coturnix japonica*) significantly increased the concentrations of corticosterone in the yolk of eggs laid seven and nine days after implantation. This figure illustrates arrow #2 in panel A. *C2*: Relative responsiveness of the stress axis (reactive corticosterone levels in plasma) in fledging, free-living European starling (*Sturnus vulgaris*) in relation to natural variation in exposure to maternal stress (note that yolk corticosterone naturally increases across laying order; see insert and Love et al., 2008). The stress response of nestlings was further decreased when the eggs had been previously treated with exogenous corticosterone (continuous line) compared to controls (discontinuous line). This figure illustrates arrow #3 in panel A. *C3*: Nestling white storks (*Ciconia ciconia*) showed higher baseline corticosterone (predictive plasma levels) when they were singletons, reared in nests that suffered a strong brood reduction (insufficient parental care), compared to nestlings from two- and three-chick broods. This figure illustrates arrow #4 in panel A. *C4*: Baseline corticosterone (predictive plasma levels) in nestling white storks (*Ciconia ciconia*) were higher when reared in the less productive environment (crop fields) compared to nestlings reared beside wetlands (marsh), where food availability is higher. This figure illustrates arrow #5 in panel A. *Panel C1 from Hayward and Wingfield 2004; panel C2: from Love et al. 2013; panels C3–C4 modified from Blas et al. 2005; and Blas and Baos, 2008, all with permission.*

Kitaysky, 2008), human disturbance (Thiel et al., 2008; Zhang et al., 2011), and social competition consistently promote plasma GC elevations in females that may be forming eggs, laying, or brooding offspring. Therefore, the potential for GC transfer into eggs and developing young exists.

2. ***Environmental and social conditions affect offspring GC levels (in ovo and developing birds).***

Although the precise mechanism whereby circulating GCs are transferred from the mother to the eggs requires additional research (Groothuis et al., 2005), positive correlations between maternal (plasma) and egg (yolk) levels have been shown in European starlings *Sturnus vulgaris* (Love et al., 2005) and barn owls *Tyto alba* (Almasi et al., 2012). Additional studies have reported a concomitant increase in egg GC levels in response to maternal exposure to environmental challenge (barn swallows *Hirundo rustica,* Saino et al., 2005; European starlings *S. vulgaris,* Love et al., 2008; song sparrows *M. melodia,* Travers et al., 2010), including predation pressure and social conditions. For example, solitary breeding females of the semicolonial European starling deposit higher egg GC levels (Love and Williams, 2008), and reduced female condition often results in decreased food provisioning rates to nestlings, elevating post-hatching GC levels (Love et al., 2004; see Angelier et al., 2007b, 2009 for seabirds). Single chicks in white storks nests are the result of partial hatching failure and brood reduction (indicating insufficient parental care), and show higher circulating GC levels (*baseline* or *predictive homeostasis* range) compared to nestlings from two- and three-chick broods (Figure 33.8, panel C3; Blas et al., 2005).

3. ***Environmental challenge and elevated GC levels in developing offspring (in ovo, and post-hatch) generate long-term differences in HPA function, behavior, and performance.***

   For example, in nestling black-legged kittiwakes *R. tridactyla,* experimental food restriction during development results in sustained elevations in 30 day old chicks (Kitaysky et al., 1999a,b), and experimentally elevated corticosterone levels promote begging and aggression (Kitaysky et al., 2001) but result in long-term impairment of cognitive abilities persisting at least 8 months later (Kitaysky et al., 2003). In European starlings *S. vulgaris,* reduced maternal provisioning rates increase fledglings' adrenocortical responsiveness (Love and Williams, 2008). Within a given clutch, maternal GC levels (deposited in egg yolks) sequentially increase (Love et al., 2008) and the subsequent HPA responses of nestlings are negatively correlated with *in ovo* GC exposure (Figure 33.8, panel C2). The latter association can be further enhanced by exogenous administration of GC in the eggs, which results in fledgings showing the lowest HPA responses (Love and Williams, unpublished data in Love et al., 2013). Somehow opposing results have been reported in Japanese quail *Coturnix coturnix japonica,* where experimentally elevated plasma corticosterone in laying females is transferred to egg yolk (Figure 33.8, panel C1), but the hatched chicks (which grow at a slower rate) display higher, rather than lower responsiveness of the HPA axis as adults (Hayward and Wingfield, 2004). Unfortunately there is a lack of studies in wild birds monitoring whether such changes in HPA function, behavior, and performance persist in the phenotypes after fledging, but a negative relationship between stress-induced corticosterone levels in nestlings and survival into adulthood was found in wild, unmanipulated white storks *C. ciconia* (Blas et al., 2007).

Maternal programming has been proved to be an adaptive (GC-mediated, epigenetic) mechanism bridging maternal and offspring environment to explain, for example, the cyclic population dynamics of wild mammals (see Sheriff et al., 2009, 2010, 2011a). Whether this mechanism also applies to wild birds urgently requires additional field research under contrasting environmental scenarios and in association with long-term monitoring of the individual's performance and HPA function.

- Glucocorticoids actions during development exert *permanent organizational effects* on the individuals.
- Developmental exposure to glucocorticoids occur pre-hatch (*in ovo*; during egg formation), and post-hatch (*ex ovo*: according to the amount of maternal and paternal care).
- Glucocorticoid transfer from mothers to offspring has the potential to translate ecological and environmental conditions into permanent offspring responses through a process termed "*maternal programming*".
- Such "*priming*" can be highly adaptive and allow animals to adjust to changing environments.
- The fitness of the modified phenotypes is likely context-specific, and determined by the environmental conditions encountered after development (*Maternal-Match Hypothesis*).

33.4.3 Modulation of the Stress Response

The *emergency life history stage ELHS* (Figure 33.6 and Section 33.3.3) redirects energy investment away from normal activities into a survival mode. Such energy shift (mediated through corticosterone elevations into the *reactive homeostasis* range) can be highly adaptive in most circumstances, but the individual also incurs important costs. In fact, life history theory postulates that organisms are continually facing trade-offs in the allocation of a limited quantity of resources/energy between competing functions (Stearns, 1992). For example, breeding birds have to trade off energy and resources between current reproductive investment and survival (which determines future reproduction). Whether an individual should allocate more resources to one or the other life history trait depends on the relative importance in maximizing lifetime fitness, and natural selection has likely filtered optimal decisions. This is a core argument for widely accepted models on the cost of reproduction (Williams, 1966), leading to a recently proposed "*Brood Value Hypothesis* (Heidinger et al., 2006; Lendvai et al., 2007; Lendvai and Chastel, 2008). The *Brood Value Hypothesis* proposes that the stress response should be modulated as a function of the relative importance of the current reproductive attempt: when the value of the current reproduction is relatively high, the stress response should be mitigated to ensure that the current breeding attempt is not compromised. This hypothesis has found empirical support in numerous field studies, within and across species. For example, maximum corticosterone in response to capture and restraint (*reactive homeostasis*, plasma levels) in red-legged partridges *Alectoris rufa* declines from the pre-laying to the laying stage, suggesting an adaptive down-modulation of adrenocortical responses as the value of reproduction increases (Figure 33.9(A); Blas and Marchant, unpublished data). The first experimental demonstration that individuals can flexibly modulate their stress response with respect to the reproductive value of their brood was performed in free-living house sparrows *P. domesticus*. By means of an experimental increase or reduction of clutch size during the nestling period, Lendvai et al. (2007) found that parents tending enlarged clutches responded less strongly to a stressor than those tending reduced clutches (Figure 33.9(B)). In addition, they examined whether individuals responded less strongly to a stressor as the breeding season progressed and future reproductive opportunities declined. The stress response decreased with breeding date during the birds' first breeding attempt, but it remained constant during their second breeding attempt. The same line of hypothesis applies to interspecies comparisons: avian species breeding in northern latitudes have limited opportunities to renest due to a short breeding season, and as predicted by the *Brood Value Hypothesis*, they show attenuated stress responses compared to southern populations/species (having a wider reproductive season, and more possibilities to breed within the same year; Silverin et al., 1997; O'Reilly and Wingfield, 2001; Breuner et al., 2003; but see Wingfield et al., 1995a). Similarly, older individuals have progressively reduced breeding opportunities, making it adaptive to increase reproductive effort as they age ("*Terminal Investment Hypothesis*": Clutton-Brock, 1984; Velando et al., 2006). Consistent with the *Brood Value Hypothesis*, older birds show attenuated secretion of GCs in response to challenges (Heidinger et al., 2006), although this pattern may not always occur (Angelier et al., 2007a). The response to stress have been shown to be reduced in the sex providing the most parental care (Wingfield et al., 1995b; O'Reilly and Wingfield, 2001; Holberton and Wingfield, 2003), during the specific breeding substages requiring higher parental investment (Meddle et al., 2003; see Figure 33.9(A)), and even within a given substage (e.g., brooding) as offspring quality or quantity increases (Lendvai et al., 2007; Lendvai and Chastel, 2008). A recent phylogenetic comparative analysis on 34 wild avian species found that those with a higher value of the current brood relative to future breeding mounted weaker corticosterone responses during acute stress (Figure 33.9(C)), and that females in the species with more female-biased parental care had weaker corticosterone responses (Bókony et al., 2009). Collectively, correlational, experimental, and comparative evidence suggest that modulation of the stress response functions as a physiological mediator for the adaptive allocation of energy and resources between current and future reproduction. Also very important, these studies point to a particular hormonal mechanism, the HPA axis function, as the physiological bases for life history trade-offs, indicating that HPA

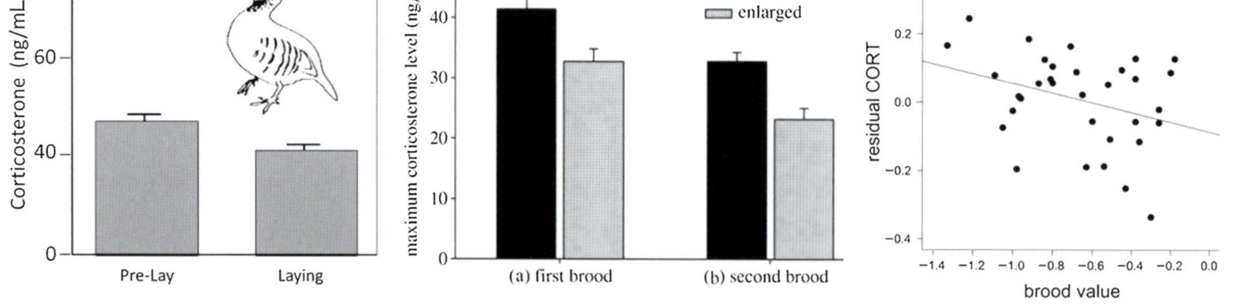

FIGURE 33.9 **Modulation of the stress response during reproduction.** ***Panel A.*** Maximum corticosterone (reactive, plasma levels) in response to capture and restraint in captive red-legged partridges (*Alectoris rufa*), across breeding sub-stages. Stress-induced corticosterone declined as birds started laying eggs, suggesting an adaptive down-modulation of HPA responses. ***Panel B.*** Maximum corticosterone titers (reactive, plasma levels) in free-living house sparrows (*Passer domesticus*) as a function of a manipulation of brood size (number of nestlings) performed in the (a) first and (b) second breeding attempt of a year. Experimentally increased broods elevated the brood value to adult birds, which down-modulated their HPA responses. ***Panel C.*** Residual, maximum corticosterone titers (peak plasma level controlled for baseline and breeding latitude) in relation to brood value in 34 wild avian species. Brood value expresses the putative importance of current reproduction as LOG(clutch size/[clutch size × broods per year × average reproductive life span]). *Panel A from J. Blas and T.A. Marchant, unpublished data; panel B reproduced from Lendvai et al. 2007 with permission from the Royal Society; panel C reproduced from Bókony et al. 2009 with permission from the University of Chicago Press.*

function and GC secretion may be a target of selection just like other traits (Bókony et al., 2009).

- Individuals can flexibly modulate their stress response with respect to the reproductive value of their brood (*Brood Value Hypothesis*).
- The intra-individual patterns of modulation are consistent with interspecies differences in stress responses, suggesting similar underlying ecological constraints. Avian species breeding in northern latitudes have limited opportunities to re-nest due to short breeding seasons, and show attenuated stress responses. Females in avian species showing female-biased parental care have weaker corticosterone responses.
- Modulation of corticosterone secretions makes it possible to trade off energy and resources between current reproductive investment and survival, providing a proximate mechanism for life-history evolution.

33.5 FIELD METHODS TO STUDY ADRENOCORTICAL FUNCTION

Our current understanding of avian adrenocortical function is to a large extent based on experimental studies performed in laboratory conditions using captive birds. This allows researchers to control a considerable number of variables known to affect *hypothalamus-pituitary-adrenal (HPA)* function (e.g., photoperiod, temperature, food and water, social interactions, status, age, sex, and life history stage). Captive birds are easily accessible for repeated sampling and trait manipulations, and most lab facilities have readily available infrastructure to obtain, process, and store the collected samples in repeatable conditions.

However, laboratory settings impose a considerable bias when extrapolating results to the real world (Fusani et al., 2005). To start with, only a few avian species can be kept in captivity for experimentation, limiting research conclusions to specific species or phylogenetic groups (e.g., poultry species), and even to specific breeds whose HPA system has been voluntarily or involuntarily modified through artificial selection (Keith et al., 1973; Satterlee et al., 2000). Furthermore, when a wild avian species is brought to captivity for experimentation, the artificial settings are unable to replicate the richness of natural stimuli and their complex interactions, and the recorded physiological and behavioral responses likely represent a minimum diversity compared to what occurs in natural habitats, when birds are exposed to real ecological scenarios (Fusani et al., 2005; Fusani, 2008). For example, wild *Passerine* species show reduced hippocampal volume, lower levels of circulating testosterone, and contrasting responses to pharmacological treatment when brought to captivity (Wingfield et al., 1990; Smulders et al., 2000; Canoine and Gwinner, 2002), strongly indicating that deprivation of environmental and social cues can dramatically modify neuroendocrinological and behavioral responses.

The last four decades have witnessed an increasing number of behavioral endocrinologists focusing their research on wild species under natural settings, largely following the pioneering work of John Wingfield (Wingfield and Farner, 1975, 1993) who implemented methods to quantify hormones in small blood samples, avoiding the need for killing the study subjects and establishing the discipline of *Field Endocrinology*. Working in the field imposes at least five important challenges compared to laboratory settings (Fusani et al., 2005; Fusani, 2008) and has led to the use of alternative, noninvasive tools to monitor endocrine function (Bortolotti et al., 2008; Sheriff et al., 2011b). First, wild specimens inevitably differ in nutritional condition, social status, life history stage, and many other factors whose variability should be taken into account during the design of a field experiment, the collection of samples, and the subsequent statistical treatment of data. Second, catching wild birds may take a considerable amount of effort, and the trapping method should be carefully chosen to avoid additional interferences on adrenocortical function. For example, the use of visual decoys and playbacks may facilitate capture, but using social stimulation as "bait" also stimulates rapid hormone secretions (e.g., elevating testosterone levels, Wingfield and Wada, 1989), potentially effecting GC levels (Charlier et al., 2009). Even the use of passive trapping methods like mist nets should be carefully supervised because the capture event itself is stressful, and the time elapsed until sampling need to be standardized, controlled for, or at least meticulously recorded (see Section 33.5.1). Third, sample sizes are often reduced compared to laboratory studies ($N \leq 10$ are not uncommon) due to the difficulty of capturing wild birds and also to minimize the impact on wild populations. Whenever possible, repeated-measures designs are advisable to increase the statistical power, although recapturing the same individuals may not be always possible as individuals learn avoidance behaviors. Fourth, because conditions can be primitive when working in field or remote locations, specific logistics may be required for sample processing and transportation. For example, the separation of plasma from blood samples for subsequent determination of corticosterone levels may require using a battery- or manually-operated centrifuge, while plasma preservation, storage, and transport in frozen conditions may require handling dry ice or liquid nitrogen in specific containers adapted for field and travel (details are in Fusani et al., 2005). Finally, special care should be taken regarding the ethical and legal considerations associated with the manipulation of wild birds (Fair et al., 2010), aimed and minimizing potential impacts on the well-being of study subjects and the viability of natural populations.

All the constraints and the special considerations above imply that a first step for any field study of avian adrenocortical function requires a carefully planned experimental design. Environmental stochasticity (unpredictable events) occur frequently during field studies, and this may actually result in "natural experiments" that provide unique opportunities to understand the role of adrenocortical function in birds, for example, during inclement weather (Wingfield et al., 1983), social conflict (Goymann and Wingfield, 2004), famine conditions (Kitaysky et al., 1999a,b) and pollution accidents (Baos et al., 2006; Baos and Blas, 2009). Below we will summarize common field techniques that are frequently used for assessing adrenocortical function in wild birds, some of which have been developed or largely implemented in recent years (laboratory techniques aimed at quantifying GC levels have been reviewed elsewhere; see e.g., Sheriff et al., 2011b).

- Deprivation of environmental and social cues can modify avian neuroendocrinological and behavioral responses, preventing the extrapolation of results from laboratory studies to the real world.
- Field endocrinologists focus their research on wild species under natural settings, but face important challenges during sample collection, processing, and transportation. These have led to the development of alternative and noninvasive tools to monitor endocrine function.
- Field research requires carefully planned experimental designs and special care to minimize impacts on the well-being of individuals and the viability of natural populations.
- Environmental stochasticity often results in "natural experiments" that provide unique opportunities to understand the role of the stress response in avian ecology and evolution.

33.5.1 Obtaining Adequate Blood Samples: Capture and Restraint Protocols and the "Stress Series"

Field endocrinologists face an obvious methodological difficulty when collecting samples aimed at assessing HPA function in wild animals: an adequate "*baseline*" blood sample containing GC levels truly indicative of the individual's *predictive homeostasis* range should not be affected by the investigator disturbance. However, the trapping and capturing activities that necessarily precede blood sampling are perceived by wild birds as real predation attempts (Scheuerlein et al., 2001; Canoine et al., 2002). Although many behavioral and neuroendocrine components of the "*fight or flight*" response occur within seconds (e.g., epinephrine/norepinephrine elevations accompanied by escape behaviors and alarm calls; see details in Chapter 26) the elevation of circulating corticosterone levels typically shows a 2–3 min delay (Romero and Reed, 2005). This short timeframe constitutes, in many field situations, a feasible period for obtaining a first blood sample *in situ*. This can be done through venipuncture of the wing (brachial) vein using a sterilized needle and collecting blood drops in heparinized microhematocrit tubes (e.g., one or several 50 μL tubes) or using heparinized syringes that also allow accessing the leg or jugular veins. These procedures are generally harmless provided adequate training and observing the recommended blood sampling volume limitations according to body size (Fair et al., 2010; though see Brown and Brown, 2009). Following capture and initial collection of a baseline blood sample, birds should be kept in cloth bags or immobilized with cloth corsets, allowing adequate ventilation but preventing injury if the bird struggles. Birds should also be placed in a safe place, sheltered from direct negative effects of weather, and their vision deprived (large birds may require using falconry hoods, Blas et al., 2010). It is very important to follow a strict capture, handling, and restraint protocol, because subsequent collection of additional blood samples at predefined intervals of time (for example, 10, 30, and 45 min postcapture, Blas et al., 2006a) will allow the characterization of GC elevations at the individual level during a "*stress series*" (Figure 33.10(A)). Researchers can therefore transform/adapt their normal capture and sampling activities into standardized experimental protocols that will allow establishing formal group comparisons in a repeatable manner within and across studies (for example, in relation to age, sex, and body condition). In some field situations it may not be possible to obtain rapid samples from all the individuals in unstressed/*baseline* conditions (for example, when the trapping method involves simultaneous capture of many birds; Blas et al., 2011) or repeated samples from the same individuals (for example, in small species or during early developmental stages, when blood volume is highly limited; Fairhurst et al., 2013b). In all the cases it is still critical to keep accurate records of the times elapsed from capture to blood sampling each individual, and GC responses can be subjected to group comparisons through regression analyses (Figure 33.10(B)).

- Capturing wild birds stimulates their stress response. As a consequence, *baseline* corticosterone levels (representative of the individuals' *predictive homeostasis range*) can only be accurately assessed in blood samples obtained within 2–3 min postcapture. Additional blood samples will allow characterizing the *adrenocortical response to stress* at the individual level ("*stress series*").
- Blood sampling requires keeping meticulous records of the times elapsed from capture to blood sampling for each individual. Samples can be collected at predefined intervals of time or along a continuum, but a standardized *capture, handling, and restraint protocol* is required to establish group comparisons.

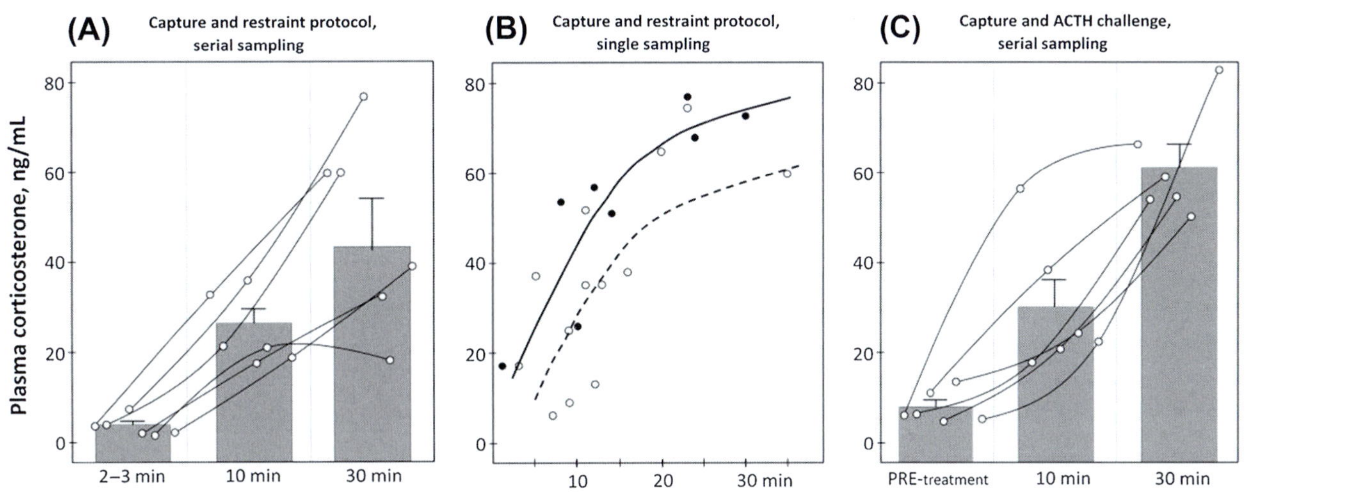

FIGURE 33.10 **Protocols for assessing the "stress response".** ***Panel A.*** *Capture and restraint with repeated blood sampling.* Plasma corticosterone levels in nestlings black kites (*Milvus migrans*) after standardized capture, restraint and serial bleeding of the same individuals (one line per individual, N=6 birds) at three fixed times (X-axis). The bars in the background represent mean and S.E. per sampling episode. ***Panel B.*** *Capture and restraint with single blood sampling.* Plasma corticosterone levels in adult male black kites (*Milvus migrans*) after standardized capture, restraint and collection of one single sample per individual (single dots are independent individuals). The solid and dashed lines indicate logistic regression estimates for non-breeding (black dots) and adult breeding birds (white dots). ***Panel C.*** *Capture and restraint with ACTH challenge and serial blood sampling.* Plasma corticosterone levels in nestlings black kites (*Milvus migrans*) after standardized capture, restraint and intravenous injection of ACTH. Serial bleeding was performed before and after (10 and 30 minutes) treatment (one line per individual, N=6 birds). ACTH injections were performed within 2 minutes from capture for ease of comparison with the corticosterone responses shown in Panel A. The bars in the background represent mean and S.E. per sampling episode. *Panel A and C from L. López et al., unpublished data; panel B redrawn from Blas et al. (2011).*

33.5.2 Quantifying Adrenocortical Sensitivity and Robustness

Although the *handling and restraint protocol* accompanied by serial collection of blood samples is a widely used method for assessing adrenocortical function in wild birds, the resulting GC elevations may be affected by perception differences among and within individuals. For example, wild penguins exposed to human presence show moderate GC elevations following the same capture and restraint protocol that triggers strong adrenocortical responses in unexposed conspecifics (Walker et al., 2006), suggesting habituation to human presence (Cyr and Romero, 2009). The reverse outcome may also occur when predator-naïve species learn that humans can be a potential danger, resulting in gradually stronger GC responses (Rödl et al., 2007). For these reasons, field studies may benefit by incorporating additional research methods to characterize adrenocortical function regardless of perceptional biases (see Canoine et al., 2002). For example, the sensitivity and the robustness of an individual's adrenocortical response can be estimated by means of administration of two commercially available drugs followed by the collection of blood samples to monitor concomitant, short-term (e.g., within 30–90 min) changes in endogenous GC levels (Figure 33.10(C)). The adrenals' *maximal ability or capacity* to secrete GC can be experimentally assessed through intravenous, intramuscular, or intraperitoneal administration of exogenous adrenocorticotropin (ATCH). Porcine ACTH is commercially available for research purposes, and it is known to elicit corticosterone release in the avian adrenals (Astheimer et al., 1994; Wasser et al., 1997; Rich and Romero, 2005). The commercial product is normally provided lyophilized, requiring suspension in an injectable isotonic solution (Ringer's solution). Dosage choice and injection volumes can be assessed according to previous references, and a pilot study using a small sample size (e.g., four to six birds) would allow selecting the minimal dose eliciting maximal response (*adrenal saturation*) and the optimal postinjection time for the subsequent collection of blood samples. Unfortunately, ACTH degrades at room temperature or following thaw–freeze cycles, requiring the preparation of aliquots maintained frozen until a few minutes before injection. The *robustness* of an individual's adrenocortical response can be characterized through exogenous (intravenous, intramuscular, or intraperitoneal) administration of *dexamethasone* (DEX). DEX is a synthetic GC that competes with endogenous corticosterone for binding receptors in the brain, and artificially stimulates negative feedback (Dallman et al., 1992). Because DEX does not bind to the antibodies commonly used in radioimmunoassays, a decrease of endogenous GC levels can be monitored by performing serial collection of blood samples within a relatively short period (e.g., 30, 60, and 90 min) after chemical challenge, allowing standardized comparisons among experimental groups. Negative feedback is a critical component of HPA function, as a failure in this system can lead to persistent elevated corticosterone levels that lengthen and strengthen the overall stress response (Romero, 2004), potentially resulting in stress-related disease (Sapolsky, 1992; Romero and Wikelski, 2010). Contrary to ACTH, commercial DEX solutions are generally stable at room temperature, and because it is a common antiinflammatory drug the product is easily accessible in veterinary clinics.

- The individual's *maximal capacity* to secrete GCs can be experimentally assessed through ACTH injections, followed by serial collection of blood samples and quantification of corticosterone elevations.
- The *robustness* of an individual's adrenocortical response can be characterized through exogenous administration of dexamethasone (DEX), a synthetic GC that competes with endogenous corticosterone and stimulates negative feedback.

33.5.3 Phenotypic Engineering

One of the best ways to explore whether and how traits of organisms are currently adaptive is to manipulate them experimentally and compare the relative fitness of altered and unaltered individuals. This method has been termed *phenotypic engineering*, and it is a powerful tool in *field endocrinology* studies (Ketterson et al., 1996). For example, one could manipulate body condition and monitor the effects on circulating hormones (Figure 33.11(A)). More often, field endocrinologists are interested in a continuous manipulation of the individual's hormone levels (over the course of a few days or weeks) in order to assess the effects on physiology, morphology, behavior, and performance (Figure 33.11(B–C)). Repeated administration of a substance (e.g., daily injections) is impossible in most field conditions because focal individuals avoid being regularly recaptured. As a consequence, field researchers rely on subcutaneous implant devices to deliver the target substance (e.g., corticosterone, dexamethasone) more or less continuously. Depending on the chemical properties of the target substance and other considerations (including the researcher's budget), this can be achieved through three alternative subcutaneous implant devices (described below; see also Fusani, 2008). With the required training, the associated surgical procedures are simple, fast, and minimally invasive: after applying local anesthesia (e.g., external or subcutaneous lidocaine solution between the wings, the lower neck, or the upper leg area where the implant will be placed) and superficially cleaning the area with a cotton swab impregnated in povidone-iodine solution (alcohol should be avoided), a small incision using a sterile scalpel allows subsequent insertion of the implant device with forceps. The skin is subsequently closed using a small dab of tissue adhesive (Fairhurst et al., 2013b) or surgical stitches (Blas et al., 2006b).

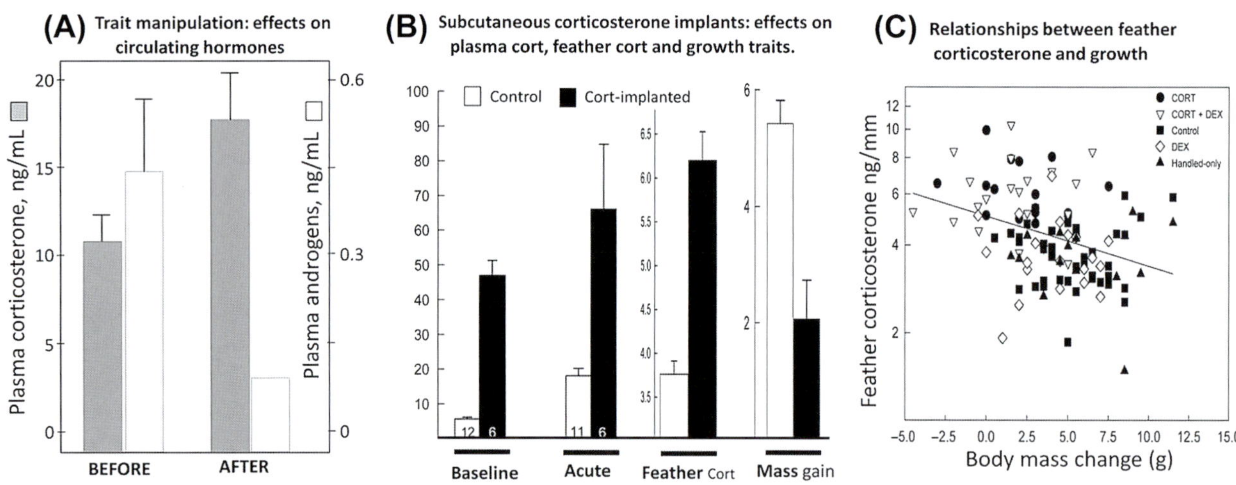

FIGURE 33.11 **Phenotypic engineering and feather corticosterone.** ***Panel A.*** *Effects of trait manipulations on circulating hormones.* Plasma levels of baseline corticosterone (left axis, gray bars) and androgens (right axis, white bars) before and after an experimental food restriction that caused a decrease in the body condition of adult male red-legged partridges (*Alectoris rufa*). Bars indicate means and S.E. All plasma samples were collected 3 minutes post-capture. ***Panel B.*** *Effects of sustained corticosterone manipulations on plasma cort, feather cort and body mass.* Plasma corticosterone levels (ng/mL), feather corticosterone levels (pg/mm) and body mass gain (g) in nestling tree swallows (*Tachycineta bicolor*) subcutaneously implanted time-release corticosterone pellets or drug-free pellets (black and white bars respectively) on day 7th post-hatch. Baseline blood samples were collected within 2–3 minutes following capture, and acute (stress-induced) blood samples were collected after 30 minutes of restraint (all samples collected 3 days after implanting). The feathers were growing during the experimental period, and were collected on day 15 post-hatch. Body mass change was estimated between days 7 and 11 post-hatch. All bars indicate means and S.E. ***Panel C.*** *Relationships between feather corticosterone and growth.* Negative association between feather corticosterone and body mass gain during development in nestling tree swallows (*Tachycineta bicolor*) subjected to systemic manipulations of hormone levels through time-release pellets. *Panel A redrawn from Pérez-Rodriguez et al. (2006) with permission; panels B and C adapted from Fairhurst et al., (2013b), with permission.*

1. ***Silicone implants.*** Silastic Medical Grade silicone tubing (Dow Corning, Inc.) can be filled with crystalline hormone and the two ends closed with silicon glue. After subcutaneous insertion, the substance diffuses thought the semipermeable walls for a variable period determined by the length of the tube, the thickness of the walls, the amount of drug, and its lipophilic nature. Although this inexpensive technique can be useful for highly lipophilic hormones such as testosterone (Blas et al., 2006b), silastic implants have low permeability for hydrophilic substances such as corticosterone (Kinel et al., 1968). In addition, *continuous* release does not mean *consistent* release (see Fusani, 2008). The choice of perforating the tube or leaving one end open causes a fast release of the hormone and thus inability to maintain a sustained elevation within or between individuals (Newman et al., 2010), which may confound experimental results.
2. ***Osmotic pumps.*** Osmotic minipumps (Alzet, Durect Corporation, Cupertino, CA, USA) allow for up to four weeks of *continuous and consistent* hormone release independent of its solubility (e.g., Soma et al., 2000; Fusani et al., 2001). This method is relatively expensive but provides high repeatability (Fusani, 2008).
3. ***Time-release pellets.*** Innovative Research of America (Sarasota, Florida, USA) manufactures pellets for the most common hormones and antihormone drugs as well as placebo (vehicle only) pellets (Figure 33.11(B–C)). The hormone is included in a cholesterol matrix that is slowly reabsorbed when placed subcutaneously. Although commercial pellets are relatively expensive, it is not necessary to recapture the animal to remove any experimental device after the study, providing a considerable advantage compared to silicone implants and osmotic pumps. Caution should be taken regarding dosages and expected release times despite manufacturer specifications. These estimations result from studies performed in laboratory rodents, and a number of studies have pointed out that systemic elevation of corticosterone and dexamethasone from subcutaneous pellets lasts for shorter times in a range of avian species (Muller et al., 2009; Fairhurst et al., 2013b), likely due to the higher metabolic rates of birds.

Note that continuing hormonal treatment for longer than a few days or one week can bring endogenous readjustment of the individual's endocrine system (evolved to maintain homeostasis). Therefore, long-term administration of an exogenous hormone may result in reduced endogenous production through negative feedback, and eventually in a regression of endocrine glands. These effects may last longer and be opposite to the intended initial manipulation (Fusani, 2008), potentially confounding the interpretation of results.

- One method to explore how endocrine traits are adaptive is to alter them experimentally and compare the fitness of altered and unaltered individuals ("*phenotypic engineering*").
- Field endocrinologists rely on subcutaneous implant devices to deliver the target substances (e.g., corticosterone) more or less continuously. Depending on the chemical properties of the target substance this can be achieved through silicone implants, osmotic pumps, or time-release pellets.
- Hormonal treatment for longer than a few days may result in a reduced endogenous production and caution should be taken to avoid potentially confounding results.

33.5.4 Corticosterone in Feathers

Quantifying circulating (plasma) corticosterone levels is the best way to characterize individual adrenocortical status at a particular moment and to understand how the HPA system works mechanistically and in relation to the environment. However, blood levels are instantaneous measurements, and the circulating GC titers within a given individual show dramatic fluctuations according to the immediate internal and external circumstances (e.g., social, environmental, daily rhythms). Field avian researchers face a challenge when trying to establish associations between long-term metrics (for example, fitness) and individual adrenocortical function, because GCs at a particular instant in time (a blood sample) hardly represent the long-term allostatic responses of an individual. Establishing robust eco-physiological associations would require repeatedly sampling the same individuals across time and/or large samples sizes. Such tasks are difficult to accomplish in field scenarios, explaining, for example, why GC-fitness associations are rarely reported and remain generally elusive to ecological studies (Breuner et al., 2008; Bonier et al., 2009; Crespi et al., 2013). In addition, obtaining blood samples requires capturing wild birds, with a potential artifact of the investigator disturbance on the results (Table 33.1) and concerns related to the well-being of the individuals and the conservation of wild populations (Sheriff et al., 2011b). For this and other methodological and logistical reasons summarized in Table 33.1, field researchers may chose alternative biological samples to study adrenocortical function in wild birds.

Recently it has been discovered (Bortolotti et al., 2008, 2009a) and independently confirmed (Lattin et al., 2011; Koren et al., 2012; Fairhurst et al., 2013b) that corticosterone is also present in feathers, and adequate laboratory protocols for its quantitative determination have been provided (Bortolotti et al., 2008, 2009a). Although the precise

TABLE 33.1 Monitoring Adrenocortical Function in Blood, Feathers, and Excreta

	Blood Samples (Corticosterone, CORT)	Feather Samples (Corticosterone, CORT)	Excreta Samples (Corticosterone metabolites, CM)
Trapping and capture required	**Yes**	**No** Capturing is optional. Feathers can be collected after natural molt (e.g., near nest site) or from dead/museum specimens.	**No** Capturing is optional. Direct field observation may be required to collect fresh samples and assign bird identity.
Potential researcher's interference on results	**High** Blood GC levels increase 2–3 min postcapture. Variability in handling and restraint affects circulating CORT in the collected samples.	**Neglegible** Even if capture is applied for sampling, the potential effect would only affect currently growing feathers (and even there it will be negligible on the overall feather CORT levels). Only longer-term perturbations (such as captivity) would affect CORT levels in growing feathers.	**Low** If capture is applied, time until defecation may elevate excreta CM levels (the extent of interference depends on the frequency of defecations and the species' metabolism).
Training required to obtain samples	**Yes** Proper blood sampling techniques require initial training	**No** Training not required. Fresh feathers can be collected during natural molt in the field, or be plucked/clipped from research specimens.	**No** Training not required. Fresh excreta can be collected in the field, or after restraining captured birds in cloth bags.
Time scale (of adrenocortical activity) reflected in samples	**Short** Minutes (possibility of accurate, short-term characterization of adrenocortical responses).	**Medium-long** Days to weeks or months (depending on feather growth duration). Adult birds may have feathers grown at different times and in different locations. Feather morphology allows discrimination of different portions grown across consecutive days.	**Medium-short** Minutes to hours (timeframes depend on the actual delay between feeding and excretion in any given sample).
Logistics for sample collection and initial processing	**High** Venipuncture and blood sampling require using syringes or sterilized needles and microhematocrit tubes (preferably heparinized). Plasma should be separated from whole blood through centrifugation (field centrifuge, pipettes, tubes, etc. required).	**Low** Plastic or paper bags.	**Medium-low** Excreta samples can be collected in vials or plastic bags.
Logistics for sample preservation, storage, and transport	**High** Coolers are required to preserve blood samples fresh until centrifugation within the same day. Storage and transport requires freezing (freezer, dry ice, or liquid nitrogen and specific containers).	**Low** Feathers can be stored in paper envelopes or plastic bags at room temperature.	**Medium-high** Considerable microbial degradation requires freezing samples immediately. Alternatively, samples can be preserved in alcohol, dried in an oven or lyophilized.
Possibility of subsampling to characterize timeframes	**No** Repeated blood sampling is required to analyze time-related patterns (e.g., capture and restraint series).	**Yes** Feathers can be clipped and subsampled to select portions grown across different days.	**No** Subsampling should be avoided. CM are not uniformly distributed within a sample. Undigested materials can cause interference with final results.

TABLE 33.1 Monitoring Adrenocortical Function in Blood, Feathers, and Excreta—cont'd

	Blood Samples (Corticosterone, CORT)	Feather Samples (Corticosterone, CORT)	Excreta Samples (Corticosterone metabolites, CM)
Retrospective potential	**No** Circulating CORT reflects current adrenal activity and therefore has very little retrospective potential.	**High** GCs are incorporated in the feather structure during growth, but birds maintain the same feathers for months or years (depending on the feather tract and the species' molt patterns), allowing retrospective analyses (for example, feathers collected this year can reflect circulating CORT levels of a previous year when they were grown).	**Low** Excreted CM reflects secretion, metabolism, and excretion during a variable period (minutes to hours) prior to deposition, allowing for short-term retrospective analyses (for example, researchers may decide to collect samples minutes to hours after a storm or a bout of aggressive interactions, and focal individuals can be chosen *posthoc*).
Discrimination between predictive and reactive homeostasis GC levels (i.e., baseline *versus* stress-induced levels)	**High** Capture and restraint (and serial blood sampling) allows assessing baseline and stress-induced CORT levels in the individual. An evaluation of the internal (e.g., nutritional), environmental, and social conditions is required to determine whether baseline blood samples represent predictive versus reactive homeostasis CORT levels.	**Low** Feather CORT integrates the amplitude and duration of circulating CORT levels during a growth period. Interpretation of results should take into account the potential variation in both predictive and reactive homeostasis ranges within and between feather samples.	**Low** Excreta CMs integrate the amplitude and duration of circulating CORT levels between food ingestion and excreta deposition. Interpretation of results should take into account the potential variation in both predictive and predictive homeostasis ranges within and between excreta samples.

Summary of methodological advantages and disadvantages associated with the use of blood, feathers, and excreta samples for assessing adrenocortical function in wild birds. Differences in laboratory assay methods required for the quantification of corticosterone or corticosterone metabolites levels in each type of sample should also be taken into account.

proximate mechanism remains largely unknown, there is solid evidence that circulating (systemic) corticosterone is gradually deposited in the feather structure as it grows from dermal follicles (likely through diffusion), and remains sequestered within the feather keratin structure rather than being externally deposited (for example, through preen oils; Bortolotti et al., 2008; Lattin et al., 2011). In a recent study manipulating systemic corticoterone levels in wild tree swallows (*Tachycineta bicolor*), Fairhurst and colleagues provide robust experimental evidence that feather corticosterone can correlate with both baseline and stress-induced measurements of plasma GC levels, but they emphasize that such a relationship likely occurs only when the latter reflect sustained HPA activity (Figure 33.11(C); Fairhurst et al., 2013b).

The slow growth rate of feathers implies that the period over which researchers can assess an individual's GC levels is days or weeks, rather than minutes or hours. This represents a considerable methodological advantage for field studies, as feather corticosterone levels likely reflect a more representative and ecological meaningful time scale of individual allostatic responses, including exposure and response to environmental perturbations. Despite the novelty of the discovery, feather corticosterone levels have already been shown to correlate with individual fitness and performance (Figure 33.11(B–C); Bortolotti et al., 2008; Koren et al., 2012; Fairhurst et al., 2012b, 2013a; Kouwenberg et al., 2013), quality-dependent traits (Bortolotti et al., 2009b; Mougeot et al., 2010; Kennedy et al., 2013), and environmental conditions (Fairhurst et al., 2011, 2012a, 2013a; Carrete et al., 2013).

Feather corticosterone levels are best expressed as pg/mm because length reflects time during feather growth (Bortolotti et al., 2008, 2009a; Bortolotti, 2010). Depending on the researcher interest, feathers can be used to assess changes in HPA activity across short- or long-term scales. For example, because daily feather growth can be determined through visual examination of growth bands (one pair of dark/light bands is typically grown every 24 h; Jovani et al., 2010), by clipping and subsampling different feather portions it is possible to assess changes in GC levels across the few days or weeks required to complete feather formation. Alternatively, a complete feather could be used to assess GC activity over a longer period of days or weeks, depending on the total amount of time required to grow each feather type. Because birds regularly replace all body feathers in a species-specific molting sequence, different feathers from the same individuals allow assessing HPA activity at different times of the year and at different locations (Carrete et al., 2013). The analysis of GC levels can be combined with stable isotopes in the same samples to establish the detailed spatial-temporal dynamics of adrenocortical function (Fairhurst et al., 2013a). Sample storage, transport, and preservation does not require freezing, and feathers can be stored in plastic bags or paper envelopes at room temperature. The analysis of feather hormones has the unique advantages of allowing for experimentation and sampling at any time of the year with minimal investigator-induced impacts and offering a large retrospective potential (Table 33.1). For example, molted feathers can be collected without disturbing wild birds, and also from museum specimens. Considering these benefits, feather corticosterone may well provide the ultimate noninvasive physiological measure of adrenocortical function in birds (Bortolotti et al., 2009a).

- Circulating corticosterone is deposited in growing feathers, remaining sequestered within the keratin structure and allowing a retrospective assessment of adrenocortical activity.
- Feather corticosterone levels are best expressed as pg/mm because length reflects time during feather growth. The slow growth rate allows inferring an individual's exposure and response to environmental perturbations across days or weeks (rather than minutes or hours).
- Feather plucking induces regrowth, allowing for experimentation with minimal investigator-induced artifacts. In addition, feathers are naturally replaced in a specific molt sequence, allowing noninvasive, long-term monitoring of adrenocortical function.
- Different feathers from a captured individual or a museum specimen allow inferring adrenocortical activity at different times of the year and at different locations.
- Feather corticosterone provides the ultimate noninvasive physiological measure of adrenocortical function.

33.5.5 Corticosterone Metabolites in Droppings (Excreta)

Circulating corticosterone levels can be also traced using bird "droppings", the excretory products including urine and feces (both are normally mixed in the cloaca prior to excretion; Klasing, 2005). This can be done using either a homogenate of the entre dropping (Washburn et al., 2003) or collecting only the fecal portion (Hirschenhauser et al., 2005). However, it is very important to point out that there are no native GCs in excreta samples (Touma and Palme, 2005): GCs are extensively metabolized in the liver and excreted into urine (via the kidneys), or into the guts (via the bile ducts), undergoing further chemical modifications by the intestinal flora (Palme, 2005; Palme et al., 2005). This process results in complex and varied excreted ***glucocorticoid metabolites*** (GCMs) whose proportion and structure depends on the species and its metabolism at the time of dropping production (Goymann et al., 2006). Importantly, this requires a proper validation study to ensure that the final levels of the analyzed byproduct (GCMs) reliably reflects the original GC levels (e.g., corticosterone) in the circulatory system of the research model (Millspaugh and Washburn, 2004; Goymann, 2005; Palme et al., 2005). Because excreta samples are also subjected to considerable microbial degradation after defecation (Möstl et al., 1999, 2005), samples should be collected fresh and preserved frozen immediately. Alternatively, they can be preserved in alcohol or dried in an oven (Wasser et al., 1997; Khan et al., 2002; Terio et al., 2002): the choice of method will determine whether GCMs are expressed as per wet or dry mass (these two values may correlate; Wasser et al., 2000). It is also important to avoid excreta subsampling (GCM may not be uniformly distributed with a sample) and remove undigested materials if differences in diet among samples are observed, because both factors may introduce considerable interference with final results.

Excreta sample collection in the field is noninvasive (see Table 33.1), but obtaining fresh samples implies the need to perform observations of marked/controlled individuals to monitor defecation events either visually (Wasser et al., 1997) or through remote spatial tracking (Thiel et al., 2008). Alternatively, birds can be trapped and restrained until defecation occurs (Garamszegi et al., 2012). Whether the stress caused by capture and manipulation affects excreta GCM levels depends on the frequency of defecations and the species' metabolism (Palme et al., 2005), making it advisable to keep detailed records of the times from capture to sample collection and preservation, and restrict the analyses to those samples collected within a short time after capture (for example, within 5 min: Garamszegi et al., 2012). GCM levels in droppings are assumed to reflect an integrated average of three processes: GC secretion, metabolism, and excretion. Although it is debated whether GCMs best represent baseline or stress-induced GCs, it is more likely that they reflect overall GC levels in circulation within a variable period of time (Sheriff et al., 2011b). The noninvasive nature of excreta sampling procedures make it a very suitable substrate for the study of avian adrenocortical function in relation to behavior (Lucas et al., 2006; Carere et al., 2003), habitat quality (Wasser et al., 1997), and to monitor the effects of human-related perturbations in natural environments (Thiel et al., 2008), having a wide array of applications in animal welfare and conservation biology (for reviews, see Busch and Hayward, 2009; Millspaugh and Washburn, 2004).

- Adrenocortical activity can be monitored analyzing avian excretory products (bird "dropings"). However, ***native corticosterone does not exist in excreta samples***.

- Before excretion, circulating glucocorticoids are chemically transformed into ***glucocorticoid metabolites*** (GCMs), whose proportion and structure depends on the avian species. As a consequence, proper validation studies are required before assuming a direct association between circulating corticosterone titers and excreted GCM levels.
- Excreta sample collection in the field is noninvasive and provides a suitable substrate for the study of avian adrenocortical activity. Because bird "droppings" are subjected to microbial degradation, it is important to control the times from defecation to sample collection and to preserve samples adequately.

33.6 GLOSARY OF TERMS AND ABBREVIATIONS

ACTH: *corticotropin; adrenocorticotropic hormone.*

Acute levels *(of corticosterone)*: [also termed *stress-regulated* levels or *regulated facultative* levels]. Corticosterone levels within the *reactive homeostasis* range.

Adrenocortical response: physiological response to stress involving the hypothalamic–pituitary–adrenal (HPA) axis and culminating with the release of glucocorticoids (e.g. *corticosterone*) from the adrenal glands [33.5.1]; [Fig. 33.1].

Allostasis: process of maintaining stability (i.e. *homeostasis*) through change. It refers to those aspects of physiology that "*help us adapt*". This concept is gradually replacing one of the meanings of the ambiguous term "*stress*" [33.2.1]; [Fig. 33.2]; [Fig. 33.3].

Allostasis mediators: processes and body systems allowing allostasis (e.g. *glucocorticoids*, corticosterone, cytokines, cathecolamines, hearth rate, blood pressure, antibody titers) [33.2.3].

Allostasis Model: theoretical paradigm and framework to understand when and how animals activate emergency ("*stress* related") responses through modeling their energetic requirements across life and in relation to the energy available. Using energy as study metrics, the model introduces the key concepts of *allostasis*, *allostatic state*, *allostatic load* and *allostatic overload* [33.2.1]; [Fig. 33.2]; [Fig. 33.3].

Allostatic overload type 1: short-term, deleterious state of energy imbalance in which *allostatic load* exceeds the energy available from the environment plus the internal energy reserves. As a response, individuals typically acquire an *allostatic state* (e.g. sustained *corticosterone* elevations) leading to *the emergency life history stage ELHS* [33.2.1]; [33.3.3]; [Fig. 33.2]; [Fig. 33.3].

Allostatic overload type 2: long-term, deleterious state in which environmental or socially demanding challenges become *chronic* (e.g. social stress, captivity) and the sustained *corticosterone* elevations generate pathologies, regardless of the energy available [33.2.1]; [33.3.4]; [Fig. 33.2].

Allostatic load: cumulative energetic requirements of the organism in a broad sense (the "*workload*" to maintain *homeostasis* in a particular moment), including predictable and unpredictable demands [33.2.1]; [Fig. 33.2]; [Fig. 33.3].

Allostatic state: altered and sustained activity levels of allostasis mediators (e.g., elevated *corticosterone*), which can only be maintained for limited periods before damaging the body [33.2.1]; [Fig. 33.2].

Basal levels *(of corticosterone)*: [also termed *constitutive levels*]. Minimum levels allowing basic homeostatic processes. Minimum levels allowing a *physiological "state A"* representative of undisturbed animals at rest [33.2.2]; [Fig. 33.3].

Baseline levels *(of corticosterone)*: [also termed *seasonal baseline* levels, *regulated predictive* levels or *predictive homeostasis* range]. A range where *corticosterone* levels fluctuate according to *predictable environmental changes*, allowing copying with daily and seasonal routines of the life cycle but being insufficient to cope with *unpredictable perturbations*. *Baseline levels* allow a *physiological "state B"* where hormone actions maintain systems within a heightened operated range to support *allostatic load* [33.2.2]; [Fig. 33.3].

Brood Value Hypothesis: hypothesis proposing that birds can flexibly modulate their adrenocortical response according to the value of their brood (as the brood value increases, the adrenocortical response decreases) [33.4.3]; [Fig. 33.9].

Capture, handling and restraint protocol: standardized method for capturing, handling and restraining wild birds in repeatable conditions, required to establish adequate comparisons of their *adrenocortical response*. This protocol is frequently accompanied by the collection of a *stress series* [33.5.1]; [Fig. 33.10].

Chronic stress: pathological state characterized by a malfunctioning of critical body systems as a consequence of prolonged or repeated exposure to *perturbations*. *Allostatic overload type 2* [Fig. 33.3].

Classic System: traditional nomenclature describing corticosterone ranges and functional thresholds. This classification system combines three hormonal thresholds (*levels A, B, and C*) and a set of contrasting terms including *basal* and *baseline*, *constitutive* and *facultative*, *seasonal* and *stress-related* [33.2.2]; [Fig. 33.3].

Constitutive levels *(of glucocorticoids)*: *basal levels.*

Corticosterone (CORT): predominant *glucocorticoid* in birds, and a major *allostasis mediator*. Circulating corticosterone levels change in response to internal and external demands (both predictable and unpredictable) to manage energy balance [Fig. 33.1]; [Fig. 33.3].

Corticotropin (ACTH): *adrenocorticotropic hormone*. Hormone produced by the anterior pituitary (in response to hypothalamic signals) that stimulates the release *glucocorticoids* in the adrenal glands [33.5.2]; [Fig. 33.1]; [Fig. 33.10].

Developmental Hypothesis: hypothesis proposing that the patterns of maturation of *adrenocortical responses* within and between avian species evolved to match the costs and benefits of *glucocorticoid* actions with the individual's progressive ability to cope with *perturbations* (e.g., increased behavioral performance and parental independence) [33.4.1]; [Fig. 33.7].

DEX: dexamethasone.

Dexamethasone (DEX): synthetic glucocorticoid that competes with corticosterone for binding sites and stimulates *negative feedback* [33.5.2]; [Fig. 33.11].

Direct LPF *(direct labile perturbation factors)*: short-lived *perturbations* lasting hours or days, imposing high energetic demands and resulting in activation of the *emergency life history stage (ELHS)* temporary disrupting the normal lifecycle [33.3.3]; [Fig. 33.5]; [Fig. 33.6].

ELHS: *emergency life history stage.*

Emergency life history stage (ELHS): facultative life stage activated in response to *direct labile perturbations*, and driven by the elevation and maintenance of *glucocorticoid* levels above *baseline* titers (i.e. into the *reactive homeostasis range* or *stress regulated levels*). During an *ELHS*, birds change their behavior and physiology to maximize survival at the cost of temporary suppressing their normal lifecycle [33.3.3]; [Fig. 33.2]; [Fig. 33.6].

Field Endocrinology: emerging scientific discipline aimed at studying the interplay between the endocrine system and the environment, using wild animals in their natural habitats as study models [33.5].

Flight or fight *response*: physiological and behavioral responses to short-term perturbations, mediated by the *sympathetic nervous system* with little or no elevation of *glucocorticoids* [33.3.2]; [Fig. 33.6].

GCM: glucocorticoid metabolites [33.5.5]; [Table 33.1].

GCs: *glucocorticoids* [Fig. 33.1].

Glucocorticoids (GCs): [also termed *stress hormones*] steroid hormones produced in the adrenals and released in response to ACTH. GCs are major *allostasis mediators* in vertebrates, control carbohydrate, fat and protein metabolism and their circulating levels change in response to internal and external demands to manage energy balance. The predominant GC in birds is *corticosterone* [Fig. 33.1]; [Fig. 33.3].

Homeostasis: stability of the physiological systems essential to life. This concept is often restricted to a few truly essential systems such as *pH, body temperature, glucose levels,* and *oxygen tension.* Homeostasis refers to those aspects of physiology that "*keep us alive*" [33.2.1]; [Fig. 33.1]; [Fig. 33.4].

Homeostatic failure: pathological state characterized by unsustainably low levels of *corticosterone*, insufficient to maintain *homeostasis* and resulting in death. Term associated to the *Reactive Scope Model* [33.2.2]; [33.2.3]; [Fig. 33.3]; [Fig. 33.4].

Homeostatic overload: pathological state characterized by abnormally elevated levels of *corticosterone* that disrupt normal body functions. The actual threshold for *homeostatic overload* can change within and between individuals in response to certain stimuli through a mechanism known as *wear and tear* [33.2.2]; [33.2.3]; [Fig. 33.2]; [Fig. 33.4].

HPA axis: hypothalamic–pituitary–adrenal axis [Fig. 33.1].

Indirect LPF (*indirect labile perturbation factors*): short-lived perturbations lasting seconds to a few minutes and inducing a *flight or fight* response with little or no elevation of glucocorticoids into the *reactive homeostasis range* [33.3.2]; [Fig. 33.5]; [Fig. 33.6].

Labile perturbation factor (LPF): short-term *perturbation*. According to their duration, *labile perturbations* can be classified as either direct (direct LPF) or indirect (indirect LPF) [33.3.2]; [Fig. 33.5].

LHS: life history stage (e.g. migration, moult, breeding) [Fig. 33.6].

LPF: *labile perturbation factor.*

Maternal Programming Hypothesis: hypothesis proposing that the transfer of *glucocorticoids* from mothers to offspring is an adaptation, evolved to translate ecological and environmental conditions into permanent offspring responses [33.4.2]; [Fig. 33.8].

Maternal-Match Hypothesis: offset from the Maternal Programming Hypothesis, proposing that the fitness of the modified offspring phenotypes (through *maternal programming*) is context specific, and highest when the environmental conditions experienced during adulthood are similar to those experienced during development [33.4.2]; [Fig. 33.8].

Modifying factor: [also termed *long-term* or *permanent perturbation*] perturbation lasting months or years, imposing high energetic demands and resulting in a permanent disruption of lifecycles, potentially leading to populations extinction [33.3.4]; [Fig. 33.5]; [Fig. 33.6].

Permanent perturbation: *modifying factor.*

Perturbation: [also termed *stressor*] unpredictable and potentially noxious environmental stimulus. Perturbations can be classified as either *labile* (*LPF*; and these can be direct or indirect) or *permanent* (also known as *modifying factors*) [33.3]; [Fig. 33.2]; [Fig. 33.4]; [Fig. 33.5].

Phenotypic engineering: experimental method consisting in altering a trait (e.g. elevating systemic hormone levels through subcutaneous implants) to explore whether and how such trait is adaptive. This method is a powerful tool in field endocrinology studies [33.5.3]; [Fig. 33.11].

Predictive homeostasis range *(corticosterone levels)*: intermediate levels of *corticosterone* allowing circadian and seasonal adjustments in response to *predictable environmental changes.* [Equivalent terms: *baseline* levels, *seasonal baseline* levels and *regulated predictive* levels] [33.2.2]; [Fig. 33.3]; [Fig. 33.4].

Reactive homeostasis range *(corticosterone levels)*: elevated levels of *corticosterone* needed to reestablish *homeostasis* after a *perturbation*. Upper, facultative range of *corticosterone* levels required to cope with unpredictable challenges. The elevations of *corticosterone* levels within this range are known as the *stress response*, and can trigger an *emergency life history stage ELHS.* [Equivalent terms: *acute* levels, *stress-regulated* levels, *regulated facultative* levels] [33.2.2]; [33.2.3]; [Fig. 33.3]; [Fig. 33.4].

Reactive scope: range of *corticosterone* levels defining the upper and lower physiological constraints of a healthy animal, according to the *Reactive Scope Model*. Combination of the *predictive* and the *reactive homeostasis ranges* above and below which serious pathologies arise. The amplitude of the *reactive scope* can be reduced through *wear and tear* [33.2.3]; [Fig. 33.3]; [Fig. 33.4].

Reactive Scope Model: graphical tool for characterizing the physiological state of animals across time, in response to *perturbations*, and according to the levels of allostasis mediators such as circulating *corticosterone*. Offshoot of the *Allostasis Model*, presenting a different terminology for contextualizing the impact of stress in the body, a new classification of corticosterone levels (see *Reactive Scope System*) and introducing the concept of *wear and tear* [33.2.3]; [Fig. 33.3]; [Fig. 33.4].

Reactive Scope System: recently proposed nomenclature describing *corticosterone* ranges and functional thresholds within the framework of the *Reactive Scope Model. Corticosterone* levels are presumed to exist in four ranges of increasing magnitude: *homeostatic failure, predictive homeostasis, reactive homeostasis,* and *homeostatic overload* [33.2.2]; [33.2.3]; [Fig. 33.3]; [Fig. 33.4].

Regulated facultative levels *(of corticosterone)*: *acute* levels, *stress-regulated* levels, corticosterone levels within the *reactive homeostasis range.*

Regulated predictive levels *(of corticosterone)*: *baseline* levels, *seasonal baseline* levels, corticosterone levels within the *predictive homeostasis range.*

Seasonal baseline levels *(of corticosterone)*: *baseline* levels, *regulated predictive baseline* levels, corticosterone levels within the *predictive homeostasis range.*

Stress: polysemous term that depending on the context may mean: (1) the stimuli that challenges homeostasis (i.e., *stressor, perturbation*), (2) the emergency responses to perturbations (*stress response*), or (3) the chronic state of imbalance that follows over-activation of the *adrenocortical axis* (*pathology; chronic stress*). Modern field endocrinologists are making efforts to define and consolidate alternative and unequivocal terms, leading to the recent introduction of the word *allostasis* [33.2].

Stress hormones: *glucocorticoids* [Fig. 33.1].

Stress hyporesponsive period: temporary inability to elevate circulating *glucocorticoids* in response to *perturbations* [33.4.1]; [Fig. 33.7].

Stress regulated levels *(of corticosterone)*: *acute* levels, *regulated facultative* levels, corticosterone levels within the *reactive homeostasis* range.

Stress response: facultative and short-term elevations of *corticosterone* levels in response to *perturbations*, allowing rapid physiological and behavioral changes aimed at maximizing survival [33.2.2]; [33.2.3]; [Fig. 33.3]; [Fig. 33.4].

Stress series: serial collection of blood samples at predefined intervals of time following capture, aimed at characterizing the adrenocortical response of the individual through subsequent determination of *glucocorticoids* levels [33.5.1]; [Fig. 33.10].

Stressor: perturbation.

Wear and tear: the cost of maintaining *corticosterone* elevations during a *stress response*, as proposed in the *Reactive Scope Model*. Cost of maintaining and using the physiological systems that mediate *allostasis*. *Wear and tear* can be graphically represented as a decrease in the threshold level between the *reactive homeostasis range* and *homeostasis overload* which results in a decreased ability to cope with *perturbations* and increased likelihood of pathology [33.2.3]; [Fig. 33.3]; [Fig. 33.4].

ACKNOWLEDGMENTS

I am grateful to R. Carsia for proposing me for this contribution. R. Carsia C. Scannes and P. González provided valuable assistance during the preparation of this chapter. G. Fairhurst, T.A. Marchant, and S. Cabezas reviewed a previous manuscript. J.A. Sencianes made the bird drawings for Figures 33.1, 33.7, and 33.8. S. Palacios provided invaluable logistic assistance. L. López, F. Sergio, A. Tanferna, S. Cabezas, and T. Marchant kindly provided unpublished data for Figures 33.9 and 33.10. J. Blas was supported by a Ramon y Cajal contract from the CSIC, a research project CGL2012-32544 (Spanish Ministry of Economy and Competitiveness and FEDER funds), and grant 511/2012 (Organismo Autónomo de Parques Nacionales) from the Spanish Ministry of Agriculture, Food and the Environment.

REFERENCES

Almasi, B., Rettenbacher, S., Müeller, C., Brill, S., Wagner, H., Jenni, L., 2012. Maternal corticosterone is transferred into the egg yolk. Gen. Comp. Endocrinol. 178, 139–144.

Angelier, F., Clément-Chastel, C., Welcker, J., Gabrielsen, G.W., Chastel, O., 2009. How does corticosterone affect parental behavior and reproductive success? A study of prolactin in black-legged kittiwakes. Funct. Ecol. 23, 784–793.

Angelier, F., Moe, B., Weimerskirch, H., Chastel, O., 2007a. Age-specific reproductive success in a long-lived bird: do older parents resist stress better? J. Anim. Ecol. 76, 1181–1191.

Angelier, F., Shaffer, S.A., Weimerskirch, H., Trouvé, C., Chastel, O., 2007b. Corticosterone and foraging behavior in a pelagic seabird. Physiol. Biochem. Zool. 80, 283–292.

Astheimer, L.B., Buttemer, W.A., Wingfield, J.C., 1992. Interactions of corticosterone with feeding, activity and metabolism in passerine birds. Ornis Scand. 23, 355–365.

Astheimer, L.B., Buttemer, W.A., Wingfield, J.C., 1994. Gender and seasonal differences in the adrenocortical response to ACTH challenge in an arctic passerine, *Zonotrichia leucophrys gambelii*. Gen. Comp. Endocrinol. 94, 33–43.

Axelrod, J., Reisine, T.D., 1984. Stress hormones: their interaction and regulation. Science 224, 452–459.

Baos, R., Blas, J., Bortolotti, G.R., Marchant, T.A., Hiraldo, F., 2006. Adrenocortical response to stress and thyroid hormone status in free-living nestling white storks (*Ciconia ciconia*) exposed to heavy metal and arsenic contamination. Environ. Health Perspect. 114, 1497–1501.

Baos, R., Blas, J., 2009. Adrenocortical toxicology in birds: environmental contaminants and the avian response to stress. In: Harvey, P.W., Everett, D., Springall, C. (Eds.), Adrenal Toxicology, Part IV Adrenal Dysfunction in Environmental Sentinel Species. Informa Healthcare USA, Inc., New York, pp. 257–293 (Chapter 11).

Bartels, M., Van den Berg, M., Sluyter, F., Boomsma, D.I., de Geus, E.J.C., 2003. Heritability of cortisol levels: review and simultaneous analysis of twin studies. Psychoneuroendocrinology 28, 121–137.

Belthoff, J.R., Dufty, J.A.M., 1998. Corticosterone, body condition and locomotor activity: a model for dispersal in screech-owls. Anim. Behav. 55, 405–415.

Blas, J., Baos, R., 2008. Stress in the nest: causes and consequences of adrenocortical secretion in developing birds. In: Capaldo, A. (Ed.), Recent Advances in Non-mammalian Adrenal Gland Research. Research Signpost, Kerala, pp. 89–128 (Chapter 4).

Blas, J., Baos, R., Bortolotti, G.R., Marchant, T.A., Hiraldo, F., 2005. A multi-tier approach to identifying environmental stress in altricial nestling birds. Funct. Ecol. 19, 315–322.

Blas, J., Baos, R., Bortolotti, G.R., Marchant, T.A., Hiraldo, F., 2006a. Age-related variation in the adrenocortical response to stress in nestling white storks (*Ciconia ciconia*) supports the developmental hypothesis. Gen. Comp. Endocrinol. 148, 172–180.

Blas, J., Pérez-Rodríguez, L., Bortolotti, G., Viñuela, J., Marchant, T., 2006b. Testosterone increases bioavailability of carotenoids: insights into the honesty of sexual signalling. Proc. Natl. Acad. Sci. U.S.A. 103, 18633–18637.

Blas, J., Bortolotti, G., Tella, J.L., Baos, R., Marchant, T., 2007. Stress response during development predicts fitness in a wild, long lived vertebrate. Proc. Natl. Acad. Sci. U.S.A. 104, 8080–8084.

Blas, J., López, L., Tanferna, A., Sergio, F., Hiraldo, F., 2010. Reproductive endocrinology of wild, long-lived raptors. Gen. Comp. Endocrinol. 168, 22–28.

Blas, J., Sergio, F., Wingfield, J.C., Hiraldo, F., 2011. Experimental tests of endocrine function in breeding and non-breeding raptors. Physiol. Biochem. Zool. 84, 406–416.

Bókony, V., Lendvai, A.Z., Liker, A., Angelier, F., Wingfield, J.C., Chastel, O., 2009. Stress response and the value of reproduction: are birds prudent parents? Am. Nat. 173, 589–598.

Bonier, F., Martin, P.R., Moore, I.T., Wingfield, J.C., 2009. Do baseline glucocorticoids predict fitness? Trends Ecol. Evol. 24, 634–642.

Bortolotti, G.R., 2010. Flaws and pitfalls in the chemical analysis of feathers: bad news good news for avian chemoecology and toxicology. Ecol. Appl. 20, 1766–1774.

Bortolotti, G.R., Marchant, T.A., Blas, J., German, T., 2008. Corticosterone in feathers is a long-term, integrated measure of avian stress physiology. Funct. Ecol. 22, 494–500.

Bortolotti, G.R., Marchant, T., Blas, J., Cabezas, S., 2009a. Tracking stress: localisation, deposition and stability of corticosterone in feathers. J. Exp. Biol. 212, 1477–1482.

Bortolotti, G.R., Mougeot, F., Martinez-Padilla, J., Webster, L.M.I., Piertney, S.B., 2009b. Physiological stress mediates the honesty of social signals. PLoS One 4, e4983.

Breuner, C.W., Greenberg, A.L., Wingfield, J.C., 1998. Non-invasive corticosterone treatment rapidly increases activity in Gambel's white-crowned sparrows (*Zonotrichia leucophrys gambelii*). Gen. Comp. Endocrinol. 111, 386–394.

Breuner, C.W., Orchinik, M., Hahn, T.P., Meddle, S.L., Moore, I.T., Owen-Ashley, N.-T., Sperry, T.S., Wingfield, J.C., 2003. Differential mechanisms for regulation of the stress response across latitudinal gradients. Am. J. Physiol. 285, R594–R600.

Breuner, C.W., Patterson, S.H., Hahn, T.P., 2008. In search of relationships between the acute adrenocortical response and fitness. Gen. Comp. Endocrinol. 157, 288–295.

Brown, M.B., Brown, C.R., 2009. Blood sampling reduces annual survival in cliff swallows (*Petrochelidon pyrrhonota*). Auk 126, 853–861.

Busch, D.S., Hayward, L.S., 2009. Stress in a conservation context: a discussion of glucocorticoid actions and how levels change with conservation-relevant variables. Biol. Conserv. 142, 2844–2853.

Buttemer, W.A., Astheimer, L.B., Wingfield, J.C., 1991. The effect of corticosterone on standard metabolic rates of small passerine birds. J. Comp. Physiol. 161B, 427–431.

Cannon, W.B., 1932. The Wisdom of the Body. W.W. Norton & Company Inc., New York.

Canoine, V., Gwinner, E., 2002. Seasonality in androgenic control of aggressive behavior in captive European stonechats. Horm. Behav. 41, 446.

Canoine, V., Hayden, T.J., Rowe, K., Goymann, W., 2002. The stress response of European stonechats depends on the type of stressor. Behaviour 139, 1303–1311.

Carsia, R.V., Morin, M.E., Rosen, H.D., Weber, H., 1987. Ontogenic corticosteroidogenesis of the domestic fowl response of isolated adrenocortical cells. Proc. Soc. Exp. Biol. Med. 184, 436–445.

Carere, C., Groothuis, T.G.G., Möstl, E., Daan, S., Koolhaas, J.M., 2003. Fecal corticosteroids in a territorial bird selected for different personalities: daily rhythm and the response to social stress. Horm. Behav. 43, 540–548.

Carrete, M., Bortolotti, G.R., Sánchez-Zapata, J.A., Delgado, A., Cortés-Avizanda, A., Grande, J.M., Donázar, J.A., 2013. Stressful conditions experienced by endangered Egyptian vultures on African wintering areas. Anim. Conserv. 16, 353–358.

Charlier, T.D., Underhill, C., Hammond, G.L., Soma, K., 2009. Effects of aggressive encounters on plasma corticosteroid-binding globulin and its ligands in white-crowned sparrows. Horm. Behav. 56, 339–347.

Cherel, Y., Robin, J.P., Walch, O., Karmann, H., Netchitailo, P., Le Maho, Y., 1988. Fasting in king penguin. I. Hormonal and metabolic changes during breeding. Am. J. Physiol. 254, R170–R177.

Clinchy, M., Sheriff, M.J., Zanette, L., 2013. Predator-induced stress and the ecology of fear. Funct. Ecol. 27, 56–65.

Clutton-Brock, T.H., 1984. Reproductive effort and terminal investment in iteroparous animals. Am. Nat. 123, 212–229.

Cockrem, J.F., 2007. Stress, corticosterone responses and avian personalities. J. Ornithol. 148, 169–S178.

Cockrem, J.F., 2013. Individual variation in glucocorticoid stress responses in animals. Gen. Comp. Endocrinol. 181, 45–58.

Constantini, D., Metcalfe, N.B., Monagham, P., 2010. Ecological processess in a hermetic framework. Ecol. Lett. 13, 1435–1447.

Corbel, H., Groscolas, R., 2008. A role for corticosterone and food restriction in the fledging of nestling White storks. Horm. Behav. 53, 557–566.

Crespi, E.J., Williams, T.D., Jessop, T.S., Delehanty, B., 2013. Life history and the ecology of stress: how do glucocorticoid hormones influence life-history variation in animals? Funct. Ecol. 27, 93–106.

Cyr, N.E., Romero, L.M., 2009. Identifying hormonal habituation in field studies of stress. Gen. Comp. Endocrinol. 161, 295–303.

Dallman, M.F., Akana, S.F., Cascio, C.S., Darlington, D.N., Jacobson, L., Levin, N., 1987. Regulation of ACTH secretion: variations on a theme of B. Recent Prog. Horm. Res. 43, 113–173.

Dallman, M.F., Strack, A.M., Akana, S.F., Bradbury, M.J., Hanson, E.S., Scribner, K.A., Smith, M., 1993. Feast and famine: critical role of glucocorticoids with insulin in daily energy flow. Front. Neuroendocrinol. 14, 303–347.

Darlington, D.N., Chew, G., Ha, T., Keil, L.C., Dallman, M.F., 1990. Corticosterone, but not glucose, treatment enables fasted adrenalectomized rats to survive moderate hemorrhage. Endocrinology 127, 766–772.

Dawson, W.R., Carey, C., Van't Hof, T.J., 1991. Metabolic aspects of shivering thermogenesis in passerines during winter. Proceedings of the Symposium Hormones, Physiology and Non-reproductive Behaviour in Birds Ornis Scand. 23, 21–23.

Dallman, M.F., Akana, S.F., Scribner, K.A., Bradbury, M.J., Walker, C.-D., Strack, A.M., Cascio, C.S., 1992. Stress, feedback and facilitation in the hypothalamo–pituitary–adrenal axis. J. Neuroendocrinol. 4, 517–526.

Evans, M.R., Roberts, M.L., Buchanan, K.L., Goldsmith, A.R., 2006. Heritability of corticosterone response and changes in life history traits during selection in the zebra finch. J. Evol. Biol. 19, 343–352.

Fair, J., Paul, E., Jones, J. (Eds.), 2010. Guidelines to the Use of Wild Birds in Research. third ed. The Ornithological Council, Washington, D.C. Available online at: http://www.nmnh.si.edu/BIRDNET/guide/guidelines.html?Operation=ENTER+HERE+%7E+English.

Fairhurst, G.D., Frey, M.D., Reichert, J.F., Szelest, I., Kelly, D.M., Bortolotti, G.R., 2011. Does environmental enrichment reduce stress? an integrated measure of corticosterone from feathers provides a novel perspective. PLoS One 6, e17663.

Fairhurst, G.D., Treen, G.D., Clark, R.G., Bortolotti, G.R., 2012a. Nestling corticosterone response to microclimate in an altricial bird. Can. J. Zool. 90, 1422–1430.

Fairhurst, G.D., Navarro, J., González-Solís, J., Marchant, T.A., Bortolotti, G.R., 2012b. Feather corticosterone of a nestling seabird reveals consequences of sex-specific parental investment. Proc. R. Soc. B 279, 177–184.

Fairhurst, G.D., Vögeli, M., Serrano, D., Delgado, A., Tella, J.L., Bortolotti, G.R., 2013. Can synchronizing feather-based measures of corticosterone and stable isotopes help us better understand habitat-physiology relationships? Oecologia 173, 731–743.

Fairhurst, G.D., Marchant, T.A., Soos, C., Machin, K.L., Clark, R.G., 2013b. Experimental relationships between levels of corticosterone in plasma and feathers in a free-living bird. J. Exp. Biol. 216, 4071–4081. http://dx.doi.org/10.1242/jeb.091280.

Francis, D., Diorio, J., Liu, D., Meaney, M.J., 1999. Nongenomic transmission across generations of maternal behavior and stress responses in the rat. Science 286, 1155–1158.

Frigerio, D., Moestl, E., Kotrschal, K., 2001. Excreted metabolites of gonadal steroid hormones and corticosterone in greylag geese (*Anser anser*) from hatching to fledging. Gen. Comp. Endocrinol. 124, 246–255.

Fusani, L., 2008. Endocrinology in field studies: problems and solutions for the experimental design. Gen. Comp. Endocrinol. 157, 249–253.

Fusani, L., Canoine, V., Goymann, W., Wikelski, M., Hau, M., 2005. Difficulties and special issues associated with field research in behavioral neuroendocrinology. Horm. Behav. 48, 484–491.

Fusani, L., Gahr, M., Hutchison, J.B., 2001. Aromatase inhibition reduces specifically one display of the ring dove courtship behavior. Gen. Comp. Endocrinol. 122, 23–30.

Garamszegi, L.Z., Rosivall, B., Rettenbacher, S., Markó, G., Zsebők, S., Szöllősi, E., Eens, M., Potti, J., Török, J., 2012. Corticosterone, avoidance of novelty, risk-taking and aggression in a wild bird: no evidence for pleiotropic effects. Ethology 118, 621–635.

Goymann, W., Wingfield, J.C., 2004. Allostatic load, social status, and stress hormones – the costs of social status matter. Anim. Behav. 67, 591–602.

Goymann, W., 2005. Noninvasive monitoring of hormones in bird droppings: physiological validation, sampling, extraction, sex differences, and the influence of diet on hormone metabolite levels. Ann. N.Y. Acad. Sci. 1046, 35–53.

Goymann, W., Trappschuh, M., Jensen, W., Schwabl, I., 2006. Low ambient temperature increases food intake and dropping production, leading to incorrect estimates of hormone metabolite concentrations in European stonechats. Horm. Behav. 49, 644–653.

Gosler, A.G., Greenwood, J.J.D., Perrins, C., 2002. Predation risk and the cost of being fat. Nature 377, 621–623.

Groothuis, T.G.G., Müller, W., von Engelhardt, N., Carere, C., Eising, C., 2005. Maternal hormones as a tool to adjust offspring phenotype in avian species. Neurosci. Biobehav. Rev. 29, 329–352.

Hales, C.N., Barker, D.J., 1992. Type 2 (non-insulin-dependent) diabetes mellitus: the thrifty phenotype hypothesis. Diabetologia 35, 595–601.

Hayward, L.S., Wingfield, J.C., 2004. Maternal corticosterone is transferred to avian yolk and may alter offspring growth and adult phenotype. Gen. Comp. Endocrinol. 135, 365–371.

Hawlena, D., Schmitz, O.J., 2010. Physiological stress as a fundamental mechanism linking predation to ecosystem functioning. Am. Nat. 175, 537–556.

Heath, J., 1997. Corticosterone levels during nest departure of juvenile American kestrels. Condor 99, 806–811.

Heidinger, B.J., Nisbet, I.C.T., Ketterson, E.D., 2006. Older parents are less responsive to a stressor in a long-lived seabird: a mechanism for increased reproductive performance with age? Proc. R. Soc. Lond. B 273, 2227–2231.

Hiebert, S.M., Salvante, K.G., Ramonofsky, M., Wingfield, J.C., 2000. Corticosterone and nocturnal torpor in the rufous hummingfbird (*Selasphorus rufus*). Gen. Comp. Endocrinol. 120, 220–234.

Hirschenhauser, K., Kotrschal, K., Möstl, E., 2005. Synthesis of measuring steroid metabolites in goose feces. Ann. N.Y. Acad. Sci. 1046, 138–153.

Holberton, R.L., Wingfield, J.C., 2003. Modulating the corticosterone stress response: a mechanism for balancing individual risk and reproductive success in arctic-breeding sparrows? Auk 120, 1140–1150.

Holmes, W.N., Crinshow, J., Redonde, J.L., 1990. The ontogeny of adrenal steroidogenic function in the mallard duck (*Anas platyrhynchos*). In: Wada, M., Ishii, S., Scanes, C.G. (Eds.), Endocrinology of Birds: Molecular to Behavioral. Jpn Sci Soc Press, Tokyo/Springer, Berlin, Heidelberg, New York, pp. 143–158.

Honey, P.K., 1990. Avian Flight Muscle *Pectoralis Major* as a Reserve of Proteins and Amino Acids (M.S. Thesis). University of Washington.

Jacobs, J.D., Wingfield, J.C., 2000. Endocrine control of lifecycle stages: a constraint on response to the environment? Condor 102, 35–51.

Jovani, R., Blas, J., Navarro, C., Mougeot, F., 2010. Feather growth bands and photoperiod. J. Avian Biol. 41, 1–5.

Keith, I., Brown, K.I., Nestor, K.E., 1973. Some physiological responses of turkeys selected for high and low adrenal response to cold stress. Poult. Sci. 52, 1948–1954.

Kennedy, E.A., Lattin, C.R., Romero, L.M., Dearborn, D.C., 2013. Feather coloration in museum specimens is related to feather corticosterone. Behav. Ecol. Sociobiol. 67, 341–348.

Kern, M., Bacon, W., Long, D., Cowie, R.J., 2001. Possible roles for corticosterone and critical size in the fledging of nestling pied flycatchers. Physiol. Biochem. Zool. 74, 651.

Ketterson, E.D., Nolan Jr, V., Cawthorn, M.J., Parker, P.G., Ziegenfus, C., 1996. Phenotypic engineering: using hormones to explore the mechanistic and functional bases of phenotypic variation in nature. Ibis 138, 70–86.

Khan, M.Z., Altmann, J., Isani, S.S., Yu, J., 2002. A matter of time: evaluating the storage of fecal samples for steroid analysis. Gen. Comp. Endocrinol. 128, 64–67.

Kinel, F.A., Benagian.G, Angee, I., 1968. Sustained release hormonal preparations I. Diffusion of various steroids through polymer membranes. Steroids 11, 673–680.

Kitaysky, A.S., Kitaiskaia, E.V., Piatt, J.F., Wingfield, J.C., 2003. Benefits and costs of increased levels of corticosterone in seabird chicks. Horm. Behav. 43, 140–149.

Kitaysky, A.S., Piatt, J.F., Wingfield, J.C., 2007. Stress hormones link food availability and population processes in seabirds. Mar. Ecol. Prog. Ser. 352, 245–258.

Kitaysky, A.S., Piatt, J.F., Wingfield, J.C., Romano, M., 1999a. The adrenocortical stress-response of black-legged kittiwake chicks in relation to dietary restrictions. J. Com. Physiol. B 169, 303–310.

Kitaysky, A.S., Wingfield, J.C., Piatt, J.F., 1999b. Food availability, body condition and physiological stress response in breeding black-legged kittiwakes. Funct. Ecol. 13, 577–584.

Kitaysky, A.S., Wingfield, J.C., Piatt, J.F., 2001. Corticosterone facilitates begging and affects resource allocation in the black-legged kittiwake. Behav. Ecol. 12, 619–625.

Klasing, K.C., 2005. Potential impact of nutritional strategy on noninvasive measurements of hormones in birds. Ann. N.Y. Acad. Sci. 1046, 5–16.

Koren, L., Nakagawa, S., Burke, T., Soma, K.K., Wynne-Edwards, K.E., Geffen, E., 2012. Non-breeding feather concentrations of testosterone, corticosterone and cortisol are associated with subsequent survival in wild house sparrows. Proc. R. Soc. B 279, 1560–1566.

Kouwenberg, A.L., Hipfner, J.M., McKay, D.W., Storey, A.E., 2013. Corticosterone and stable isotopes in feathers predict egg size in Atlantic puffins *Fratercula Arctica*. Ibis 155, 413–418.

Landys, M., Ramenofsky, M., Wingfield, J.C., 2006. Actions of glucocorticoids at a seasonal baseline as compared to stress-related levels in the regulation of periodic life processes. Gen. Comp. Endocrinol. 148, 132–149.

Landys-Ciannelli, M.M., Ramenofsky, M., Piersma, T., Jukema, J., Wingfield, J.C., 2002. Baseline and stress-induced plasma corticosterone during long-distance migration in the bar-tailed godwit, *Limosa lapponica*. Physiol. Biochem. Zool. 75, 101–110.

Lattin, C.R., Reed, J.M., DesRochers, D.W., Romero, L.M., 2011. Elevated corticosterone in feathers correlates with corticosterone-induced decreased feather quality: a validation study. J. Avian Biol. 42, 247–252.

Lendvai, Á.Z., Chastel, O., 2008. Experimental mate-removal increases the stress response of female house sparrows: the effects of offspring value? Horm. Behav. 52, 395–401.

Lendvai, Á.Z., Giraudeau, M., Chastel, O., 2007. Reproduction and modulation of the stress response: an experimental test in the house sparrow. Proc. R. Soc. Lond. B 274, 391–397.

Lind, J., Fransson, T., Jakobson, S., Kullberg, C., 1999. Reduced take-off ability in robins (Erithacus rubecula) due to migratory fuel load. Behav. Ecol. Sociobiol. 46, 65–70.

Love, O.P., Breuner, C.W., Vezina, F., Williams, T.D., 2004. Mediation of a corticosterone-induced reproductive conflict. Horm. Behav. 46, 59–65.

Love, O.P., Chin, E.H., Wynne-Edwards, K.E., Williams, T.D., 2005. Stress hormones: a link between maternal condition and sex-biased reproductive investment. Am. Nat. 166, 751–766.

Love, O.P., McGowan, P.O., Sheriff, M.J., 2013. Maternal adversity and ecological stressors in natural populations: the role of stress axis programming in individuals, with implications for populations and communities. Funct. Ecol. 27, 81–92.

Love, O.P., Williams, T.D., 2008. Plasticity in the adrenocortical response of a free-living vertebrate: the role of pre- and post-natal developmental stress. Horm. Behav. 54, 496–505.

Love, O.P., Wynne-Edwards, K.E., Bond, L., Williams, T.D., 2008. Determinants of within- and among-clutch variation of yolk corticosterone in the European starling. Horm. Behav. 53, 104–111.

Lucas, J.R., Freeberg, T.M., Egbert, J., Schwabl, H., 2006. Fecal corticosterone, body mass, and caching rates of Carolina chickadees (*Poecile carolinensis*) from disturbed and undisturbed sites. Horm. Behav. 49, 634–643.

Malisch, J., Breuner, C., 2010. Steroid-binding proteins and free steroids in birds. Mol. Cell. Endocrinol. 316, 42–52.

Mantha, L., Deshaies, Y., 2000. Energy intake-independent modulation of triglyceride metabolism by glucocorticoids in the rat. Am. J. Physiol. Regul. Integr. Comp. Physiol. 278, R1424–R1432.

McEwen, B.S., Wingfield, J.C., 2003a. The concept of allostasis in biology and biomedicine. Horm. Behav. 43, 2–15.

McEwen, B.S., Wingfield, J.C., 2003b. Response to commentaries on the concept of allostasis. Horm. Behav. 43, 28–30.

McEwen, B.S., Wingfield, J.C., 2010. What's in a name? Integrating homeostasis, allostasis and stress. Horm. Behav. 57, 105.

McGowan, P.O., Szyf, M., 2010. The epigenetics of social adversity in early life: implications for mental health outcomes. Neurobiol. Dis. 39, 66–72.

McGowan, P.O., Suderman, M., Sasaki, A., Huang, T.C., Hallett, M., Meaney, M.J., Szyf, M., 2011. Broad epigenetic signature of maternal care in the brain of adult rats. PLoS One 6, e14739.

Meddle, S.L., Owen-Ashley, N.T., Richardson, M.I., Wingfield, J.C., 2003. Modulation of the hypothalamic-pituitary-adrenal axis of an arctic-breeding polygynandrous songbird, the Smith's longspur, *Calcarius pictus*. Proc. R. Soc. Lond. B 70, 1849–1856.

Millspaugh, J.J., Washburn, B.E., 2004. Use of fecal glucocorticoid. metabolite measures in conservation biology research: considerations for application and interpretation. Gen. Comp. Endocrinol. 138, 189–199.

Möstl, E., Messman, S., Bagu, E., Robia, C., Palme, R., 1999. Measurement of glucocorticoid metabolite concentrations in faeces of domestic livestock. J. Vet. Med. A 46, 621–632.

Möstl, E., Rettenbacher, S., Palme, R., 2005. Measurement of corticosterone metabolites in birds' droppings: an analytical approach. Ann. N.Y. Acad. Sci. 1046, 17–34.

Mougeot, F., Martinez-Padilla, J., Bortolotti, G.R., Webster, L.M.I., Piertney, S.B., 2010. Physiological stress links parasites to carotenoid-based colour signals. J. Evol. Biol. 23, 643–650.

Muller, C., Almasi, B., Roulin, A., Breuner, C.W., Jenni-Eiermann, S., Jenni, L., 2009. Effects of corticosterone pellets on baseline and stress-induced corticosterone and corticosteroid-binding-globulin. Gen. Comp. Endocrinol. 160, 59–66.

Newman, A.E.M., MacDougall-Shackleton, S.A., An, Y.S., Kriengwatana, B., Soma, K.K., 2010. Corticosterone and dehydroepiandrosterone have opposing effects on adult neuroplasticity in the avian song control system. J. Comp. Neurol. 518, 3662–3678.

O'Reilly, K.M., Wingfield, J.C., 2001. Ecological factors underlying the adrenocortical response to capture stress in arctic-breeding shorebirds. Gen. Comp. Endocrinol. 124, 1–11.

Ouyang, J.Q., Hau, M., Bonier, F., 2011. Within seasons and among years: when are corticosterone levels repeatable? Horm. Behav. 60, 559–564.

Palme, R., 2005. Measuring fecal steroids: guidelines for practical application. Ann. N.Y. Acad. Sci. 1046, 75–80.

Palme, R., Rettenbacher, S., Touma, C., El-Bahr, S.M., Möstl, E., 2005. Stress hormones in mammals and birds: comparative aspects regarding metabolism, excretion, and noninvasive measurement in fecal samples. Ann. N.Y. Acad. Sci. 1040, 162–171.

Palokangas, R., Hissa, R., 1971. Thermorregulation in young black-headed young (Larus ribibundus L.). Comp. Biochem. Physiol. 38A, 743–750.

Pérez-Rodríguez, L., Blas, J., Viñuela, J., Marchant, T.A., Bortolotti, G.R., 2006. Condition and androgen levels: are condition-dependent and testosterone-mediated traits two sides of the same coin? Anim. Behav. 72, 97–103.

Porter, W.P., Sabo, J.L., Tracy, C.R., Reichman, O.J., Ramankutty, N., 2002. Physiology on a landscape scale: plant animal interactions. Integr. Comp. Biol. 42, 431–453.

Quillfeldt, P., Poisbleau, M., Chastel, O., Masello, J., 2007. Corticosterone in thin-billed prion *Pachyptila belcheri* chicks: diel rhythm, timing of fledging and nutritional stress. Naturwissenschaften 94, 919–925.

Rich, E.L., Romero, L.M., 2005. Exposure to chronic stress downregulates corticosterone responses to acute stressors. Am. J. Physiol. Regul. Integ Comp. Physiol. 288, R1628–R1636.

Ricklefs, R.E., Wikelski, M., 2002. The physiology/life-history nexus. Trends Ecol. Evol. 17, 462–468.

Rödl, T., Berger, S., Romero, L.M., Wikelski, M., 2007. Tameness and stress physiology in a predator-naïve island species confronted with novel predation threat. Proc. R. Soc. B 274, 577–582.

Rogers, C.M., Ramenofsky, M., Ketterson, E.D., Nolan, V., Wingfield, J.C., 1993. Plasma corticosterone, adrenal mass, winter weather, and season in nonbreeding populations of dark-eyed juncos (Junco hyemalis hyemalis). Auk 110, 279–285.

Romero, L.M., 2002. Seasonal changes in plasma glucocorticoid concentration in free living vertebrates. Gen. Comp. Endocrinol. 128, 1–24.

Romero, L.M., 2004. Physiological stress in ecology: lessons from biomedical research. Trends Ecol. Evol. 19, 249–255.

Romero, L.M., 2012. Using the reactive scope model to understand why stress physiology predicts survival during starvation in Galápagos marine iguanas. Gen. Comp. Endocrinol. 176, 296–299.

Romero, L.M., Reed, J.M., 2005. Collecting baseline corticosterone samples in the field: is under 3 min good enough? Comp. Biochem. Physiol. A Mol. Integr. Physiol. 40, 73–79.

Romero, L.M., Wikelski, M., 2010. Stress physiology as a predictor of survival in Galapagos marine iguanas. Proc. R. Soc. Lond. B 277, 3157–3162.

Romero, L.M., Dickens, M.J., Cyr, N.E., 2009. The reactive scope model – a new model integrating homeostasis, allostasis and stress. Horm. Behav. 55, 375–389.

Romero, L.M., Holt, D.W., Maples, M., Wingfield, J.C., 2006. Corticosterone is not correlated with nest departure in snowy owl chicks (*Nyctea scandiaca*). Gen. Comp. Endocrinol. 149, 119–123.

Rohwer, S., Wingfield, J.C., 1981. A field study of social dominance, plasma levels of luteinizing hormone and steroid hormones in wintering Harris' sparrows. Z. Tierpsychol. 57, 173–183.

Saadoun, A., Simon, J., Leclercq, B., 1987. Effects of exogenous corticosterone in genetically fat and lean chickens. Br. Poult. Sci. 28, 519–528.

Saino, N., Romano, M., Ferrari, R.P., Martinelli, R., Møller, A.P., 2005. Stressed mothers lay eggs with high corticosterone levels which produce low-quality offspring. J. Exp. Zool. 303A, 998–1006.

Sapolsky, R.M., 1992. Cortisol concentrations and the social significance of rank instability among wild baboons. Psychoneuroendocrinology 17, 701–709.

Sapolsky, R.M., 1996. Why stress is bad for your brain. Science 273, 749–750.

Sapolsky, R.M., 2001. Physiological and pathophysiological implications of social stress in mammals. In: McEwen, B.S., Goodman, H.M. (Eds.), Handbook of Physiology, Section 7: The Endocrine System. Coping with the Environment: Neural and Endocrine Mechanisms, Vol. IV. Oxford University Press, New York, pp. 517–532.

Sapolsky, R.M., 2002. Endocrinology of the stress response. In: Becker, J.B., Breedlove, S.M., Crews, D., McCarthy, M. (Eds.), Behavioral Endocrinology, second ed. MIT Press, Cambridge, Massachusetts, pp. 409–450.

Sapolsky, R.M., Romero, L.M., Munck, A.U., 2000. How do glucocorticosteroids influence stress responses? Integrating permissive, suppressive, stimulatory, and preparative actions. Endocr. Rev. 21, 55–89.

Satterlee, D.G., Cadd, G.G., Jones, R.B., 2000. Developmental instability in Japanese quail genetically selected for contrasting adrenocortical responsiveness. Poult. Sci. 79, 1710–1714.

Satterlee, D.G., Johnson, W.A., 1988. Selection of Japanese quail for contrasting blood corticosterone response to immobilization. Poult. Sci. 67, 25–32.

Scheuerlein, A., Van't Hof, T.J., Gwinner, E., 2001. Predators as stressors? Physiological and reproductive consequences of predation risk in tropical stonechats (*Saxicola torquata axillaris*). Proc. R. Soc. Lond. B 268, 1575–1582.

Schwabl, H., 1999. Developmental changes and among-sibling variation of corticosterone levels in an altricial avian species. Gen. Comp. Endocrinol. 116, 403–408.

Schwabl, H., Wingfield, J.C., Farner, D.S., 1985. Influence of winter on endocrine state and behavior in European blackbirds (*Turdus merula*). Z. Tierpsychol. 68, 244–252.

Seabury, S.R., Breuner, C.W., 2005. Timing of fledging, body condition, and corticosteroid binding globulin in Laysan Albatross. Integr. Comp. Biol. 45, 1070.

Selye, H., 1946. The general adaptation syndrome and the diseases of adaptation. J. Clin. Endocrinol. 6, 117–230.

Selye, H., 1976. The Stress of Life. McGraw-Hill, New York.

Sheriff, M.J., Dantzer, B., Delehanty, B., Palme, R., Boonstra, R., 2011b. Measuring stress in wildlife: techniques for quantifying glucocorticoids. Oecologia 166, 869–887.

Sheriff, M.J., Krebs, C.J., Boonstra, R., 2009. The sensitive hare: sublethal effects of predator stress on reproduction in snowshoe hares. J. Anim. Ecol. 78, 1249–1258.

Sheriff, M.J., Krebs, C.J., Boonstra, R., 2010. The ghosts of predators past: population cycles and the role of maternal effects under fluctuating predation risk. Ecology 91, 2983–2994.

Sheriff, M.J., Krebs, C.J., Boonstra, R., 2011a. From process to pattern: how fluctuating predation risk impacts the stress axis of snowshoe hares during the 10-year cycle. Oecologia 166, 593–605.

Shultz, M.T., Kitaysky, A.S., 2008. Spatial and temporal dynamics of corticosterone and corticosterone binding globulin are driven by environmental heterogeneity. Gen. Comp. Endocrinol. 155, 717–728.

Silverin, B., 1986. Corticosterone binding proteins and behavioral effects of high plasma levels of corticosterone during the breeding period in the pied flycatcher. Gen. Comp. Endocrinol. 64, 67–74.

Silverin, B., Arvidsson, B., Wingfield, J.C., 1997. The adrenocortical response to stress in breeding willow warblers *Phylloscopus trochilus* in Sweden: effects of latitude and gender. Funct. Ecol. 11, 376–384.

Sims, C.G., Holberton, R.L., 2000. Development of the corticosterone stress response in young northern mockingbirds (*Mimus polyglottos*). Gen. Comp. Endocrinol. 119, 193–201.

Smulders, T.V., Casto, J.M., Nolan, V., Ketterson, E.D., DeVoogd, T.J., 2000. Effects of captivity and testosterone on the volumes of four brain regions in the dark-eyed junco (*Junco hyemalis*). J. Neurobiol. 43, 244–253.

Solberg, L.C., Baum, A.E., Ahmadiyeh, N., Shimomura, K., Li, R., Turek, F.W., Takahashi, J.S., Churchill, G.A., Redei, E.E., 2006. Genetic analysis of the stress-responsive adrenocortical axis. Physiol. Genomics 27, 362–369.

Soma, K.K., Tramontin, A.D., Wingfield, J.C., 2000. Oestrogen regulates male aggression in the non-breeding season. Proc. R. Soc. Lond. B 267, 1089–1096.

Spencer, R.L., Kalman, B.A., Dhabhar, F.S., 2001. Role of endogenous glucocorticoids in immune system function: regulation and counter-regulation. In: McEwen, B., Goodman, H.M. (Eds.), Handbook of Physiology, Section 7: The Endocrine System. Coping with the Environment: Neural and Endocrine Mechanisms, vol. IV. Oxford University Press, New York, pp. 381–423.

Starck, J.M., Ricklefs, R.E., 1998. Avian Growth and Development: Evolution within the Altricial-Precocial Spectrum. Oxfod University, New York.

Stearns, S.C., 1992. The Evolution of Life History. Academic Press, London.

Sterling, P., Eyer, J., 1988. Allostasis a new paradigm to explain arousal pathology. In: Fisher, S., Reason, J. (Eds.), Handbook of Life Stress Cognition and Health. John Wiley and Sons, Inc., New York, pp. 629–650.

Terio, K.A., Brown, J.L., Moreland, R., Munson, L., 2002. Comparison of different drying and storage methods on quantifiable concentrations of fecal steroids in the cheetah. Zoo Biol. 21, 119–134.

Thiel, D., Jenni-Eiermann, S., Braunisch, V., Palme, R., Jenni, L., 2008. Ski tourism affects habitat use and evokes a physiological stress response in capercaillie *Tetrao urogallus*: a new methodological approach. J. Appl. Ecol. 45, 845–853.

Touma, C., Palme, R., 2005. Measuring fecal glucocorticoid metabolites in mammals and birds: the importance of validation. Ann. N.Y. Acad. Sci. 1046, 54–74.

Travers, M., Clinchy, M., Zanette, L., Boonstra, R., Williams, T., 2010. Indirect predator effects on clutch size and the cost of egg production. Ecol. Lett. 13, 980–988.

Velando, A., Drummond, H., Torres, R., 2006. Senescent birds redouble reproductive effort when ill: confirmation of the terminal investment hypothesis. Proc. Biol. Sci. 273, 1443–1448.

Wada, H., 2008. Glucocorticoids: mediators of vertebrate ontogenetic transitions. Gen. Comp. Endocrinol. 156, 441–453.

Walker, B.G., Boersma, P.D., Wingfield, J.C., 2006. Habituation of adult Magellanic penguins to human visitation as expressed through behavior and corticosterone secretion. Conserv. Biol. 20, 146–154.

Walsberg, G.E., 2003. How useful is energy balance as an overall index of stress in animals? Horm. Behav. 43, 16–17.

Washburn, B.E., Millspaugh, J.J., Schulz, J.H., Jones, S.B., Mong, T., 2003. Using fecal glucocorticoids for stress assessment in mourning doves. Condor 105, 696–706.

Wasser, S.K., Bevis, K., King, G., Hanson, E., 1997. Noninvasive physiological measures of disturbance in the northern spotted owl. Conserv. Biol. 11, 1019–1022.

Wasser, S.K., Hunt, K.E., Brown, J.L., Cooper, K., Crockett, C.M., Bechert, U., Millspaugh, J.J., Larson, S., Monfort, S.L., 2000. A generalized fecal glucocorticoid assay for use in a diverse assay of non-domestic mammalian and avian species. Gen. Comp. Endocrinol. 120, 260–275.

Williams, G.C., 1966. Natural selection, the costs of reproduction and a refinement of Lack's principle. Am. Nat. 100, 687–690.

Williams, T.D., 2008. Individual variation in endocrine systems: moving beyond the 'tyranny of the Golden Mean'. Philos. Trans. R. Soc. Lond. B 363, 1687–1698.

Wentworth, B.C., Hussein, M.O., 1985. Serum corticosterone levels in embryos newly hatched and young turkey poults. Poult. Sci. 64, 2195–2201.

Wingfield, J.C., 1984. Influence of weather on reproduction. J. Exp. Biol. 232, 589–594.

Wingfield, J.C., 1994. Modulation of the adrenocortical response to stress in birds. In: Davey, K.G., Peter, R.E., Tobe, S.S. (Eds.), Perspectives in Comparative Endocrinology. National Research Council, Canada, Ottawa, pp. 520–528.

Wingfield, J.C., 2008. Organization of vertebrate annual cycles: implications for control mechanisms. Philos. Trans. R. Soc. Lond. B 363, 425–441.

Wingfield, J.C., 2013a. Ecological processess and the ecology of stress: the impacts of abiotic environmental factors. Funct. Ecol. 27, 37–44.

Wingfield, J.C., 2013b. The comparative biology of environmental stress: behavioural endocrinology and variation in ability to cope with novel, changing environments. Anim. Behav. 85, 1127–1133.

Wingfield, J.C., Farner, D.S., 1975. The determination of five steroids in avian plasma by radioimmunoassay and competitive protein-binding. Steroids 26, 311–327.

Wingfield, J.C., Farner, D.S., 1993. Endocrinology of reproduction in wild species. In: Farner, D.S., et al. (Ed.), Avian Biology. Academic Press, San Diego, pp. 163–327.

Wingfield, J.C., Wada, M., 1989. Changes in plasma levels of testosterone during male–male interactions in the song sparrow, *Melospiza melodia*: time course and specificity of response. J. Comp. Physiol. A 166, 189–194.

Wingfield, J.C., Breuner, C.W., Jacobs, J., 1997. Corticosterone and behavioral responses to unpredictable events. In: Harvey, S., Etches, R.J. (Eds.), Perspectives in Avian Endocrinology. J. Endocrinology Press, Bristol, pp. 267–278.

Wingfield, J.C., Breuner, C., Jacobs, J., Lynn, S., Maney, D., Ramenofsky, M., Richardson, R., 1998. Ecological bases of hormone-behavior interactions: the ''emergency life history stage.'' Am. Zool. 38, 191–206.

Wingfield, J.C., Hegner, R.E., Dufty, A.M., Ball, G.F., 1990. The''challenge hypothesis'': theoretical implications for patterns of testosterone secretion, mating systems, and breeding strategies. Am. Nat. 136, 829–846.

Wingfield, J.C., Kelley, J.P., Angelier, F., Chastel, O., Lei, F., Lynn, S.E., Miner, B., Davis, J.S., Li, D., Wang, G., 2011. Organism-environment interactions in a changing world: a mechanistic approach. J. Ornithol. 152 (Suppl. 1), S279–S288.

Wingfield, J.C., Kitaysky, A.S., 2002. Endocrine responses to unpredictable environmental events: stress or anti-stress hormones. Integ. Comp. Biol. 42, 600–609.

Wingfield, J.C., Kubokawa, K., Ishida, K., Ishii, S., Wada, M., 1995a. The adrenocortical response to stress in male bush warblers, *Cettia diphone*: a comparison of breeding populations in Honshu and Hokkaido, Japan. Zool. Sci. 12, 615–621.

Wingfield, J.C., Moore, M.C., Farner, D.S., 1983. Endocrine responses to inclement weather in naturally breeding populations of white-crowned sparrows (Zonotrichia leucophrys pugetensis). Auk 100, 56–62.

Wingfield, J.C., O'Reilly, K.M., Astheimer, L.B., 1995b. Modulation of the adrenocortical responses to acute stress in arctic birds: a possible ecological basis. Am. Zool. 35, 285–294.

Wingfield, J.C., Ramenofsky, R., 1999. Hormones and the behavioral ecology of stress. In: Balm, P.H.M. (Ed.), Stress Physiology in Animals. CRC Press, Boca Raton, pp. 1–51.

Wingfield, J.C., Ramenofsky, R., 1997. Corticosterone and facultative dispersal in response to unpredictable events. Ardea 85, 155–166.

Wingfield, J.C., Romero, L.M., 2001. Adrenocortical responses to stress and their modulation in free-living vertebrates. In: McEwen, B.S. (Ed.), Handbook of Physiology, Section 7: The Endocrine System. Coping with the Environment: Neural and Endocrine Mechanisms, vol. IV. Oxford University Press, Oxford, pp. 211–236.

Wingfield, J.C., Silverin, B., 1986. Effects of corticosterone on territorial behavior of free-living male song sparrows *Melospiza melodia*. Horm. Behav. 20, 405–417.

Zhang, S., Lei, F., Liu, S., Li, D., Chen, C., Wang, P., 2011. Variation in baseline corticosterone levels of tree sparrow (*Passer montanus*) populations along an urban gradient in Beijing, China. J. Ornithol. 152, 801–806.

Chapter 34

Circadian Rhythms

Vincent M. Cassone
Department of Biology, University of Kentucky, Lexington, KY, USA

Vinod Kumar
Department of Zoology, University of Delhi, Delhi, India

Biological circadian rhythms and the clocks that control them are fundamental properties of all freely living organisms studied in any detail. These are expressed by organisms ranging from cyanobacteria to humans with highly conserved formal properties. Among animals, the molecular details of biological clocks are highly conserved as well. The conserved nature of these rhythms and clocks likely derive from the fact that they are all adaptations to a single selective pressure—the daily change from night to day and the abiotic and biotic rhythms that accompany it. Biological clock function is particularly apparent in birds, partly due to the fact that they share with us a primarily diurnal lifestyle, but also because time of day and time of year pervades all aspects of avian physiology.

34.1 ENVIRONMENTAL CYCLES

34.1.1 Light Cycles

Although we all know that the cycle of day to night derives from the rotation of the Earth on its axis approximately every 24 h, we experience this process through the apparent rise of the sun on the eastern horizon (dawn), its traverse across the sky, its setting in the west (dusk), and subsequent night. The most salient feature of the *solar day* is the dramatic change in visible and invisible light intensity, polarization, and wavelength spectra. Daylight intensity can be measured in a variety of radiometric and photometric ways, extending beyond the scope of this chapter. One can gather the scale of daily change in lux (lx), the International System of Units (SI) measure of illuminance, or the intensity of white light as perceived by the human eye. These values range from 120,000 lx in the brightest sunlight to 400 lx at dusk or dawn on a clear day to 0.0001 lx under overcast, moonless night sky. This vast range in intensity may go unperceived consciously, because our and birds' visual systems adapt to changes in intensity. However, as discussed in this chapter, changes in absolute intensity and timing of illumination are detected and processed by the circadian clock. In addition to intensity, the wavelength of ambient light changes due to diffraction of shorter wavelengths at dawn and dusk, resulting in the perception of reddish hues at sunrise and sunset.

34.1.2 Temperature

Due to the daily changes in solar illumination, the atmosphere and surface of the Earth undergo daily changes in temperature that generally rise as the day progresses but vary dramatically due to latitude, season, and local conditions such as urban growth, proximity to bodies of water, desertification, and variations in land use (Geerts, 2002). The daily temperature range (DTR) is highly variable at different latitudes. For example, at the equator, the DTR is 8 °C in the summer and 9 °C in the winter. In contrast, at 40° latitude in North America, the DTR in summer is 18 °C in the summer and only 10 °C in the winter. Proximity of oceans also ameliorates daily temperature ranges. For example, in July in Australia, the DTR increases from 6 °C on the coast to 17 °C 150 km inland. These changes are clearly important for natural ecosystems as well as agricultural activity—perhaps being as important as the mean temperature itself. There is increasing evidence that, in addition to the gradual increase in mean global temperatures, the daily temperature ranges have changed as well, decreasing the daily range 0.4 °C from 1950 to 1993.

34.1.3 Other Physical Cycles

Other physical characteristics vary over the day as a derivative of the diurnal solar cycle. These include daily changes

Sturkie's Avian Physiology.

in barometric pressure, characteristics of the magnetic field, and, in many locations, precipitation. Pressure variations derive from the thermal energy arising from the heating of the upper atmosphere that moves in a westward wave the speed of the Earth's rotation (Hardy et al., 1998). These variations depend upon latitude, altitude, and season. Aspects of the Earth's magnetic field also vary in intensity and inclination depending on the time of day, as the solar wind emanating from the sun interacts with the magnetosphere on a regular 24 h cycle.

34.1.4 Rhythms in the Biotic Environment

Because daily cycles in many aspects of the physical, abiotic environment are so pervasive, it should not be surprising that nearly all free-living organisms have evolved adaptations to the changing environment. Thus, in addition to daily changes in abiotic factors, birds' biotic environment is also changing on a daily basis. The presence of food, competitors, predators, parasites and potential mates may be present at one time of day and absent another (Bradshaw and Holzapfel, 2010). Hence, birds, like nearly all organisms on Earth, have evolved an internal biological clock that synchronizes patterns of many aspects of behavior, physiology, biochemistry and molecular biology to external time (Pittendrigh, 1993) and coordinate internal processes to temporally maximize efficient processes within (Navara and Nelson, 2007).

34.2 CIRCADIAN RHYTHMS

34.2.1 Formal Properties

The internal biological clock shares fundamental formal properties among most living organisms through the expression of endogenously generated circadian (*circa* = approximately; *dian* = a day) oscillations that entrain to local time through the process of entrainment. Rhythmic processes cannot be identified as circadian unless they are experimentally observed to persist when the organism in question is placed in constant environmental conditions of either constant darkness (DD) or constant dim light (dimLL). (Constant high light (LL) may have other effects, frequently abolishing circadian rhythms altogether and/or damaging photoreceptive elements in the system (Aschoff, 1979).) In this scenario, organisms will repeatedly express patterns of behavior, physiology, or biochemical processes with a period, τ, of close to but rarely exactly 24 h (Figure 34.1). These endogenously driven rhythms must then be entrained to the relevant environmental cycle, typically the light:dark cycle (LD) of day and night, such that the internal phase, φ_i, of the organism's clock corresponds appropriately to the external phase, φ_e, of the LD cycle. Thus, a diurnal bird's locomotor activity pattern entrains to the LD cycle so that activity onset φ_i corresponds approximately to dawn φ_e, maintaining a stable phase relationship, ψ_{ie}. Conversely, a nocturnal bird's ψ_{ie} would be approximately 12 h later than that of a diurnal bird.

34.2.2 Stability and Lability of Circadian Rhythms

In DD, the circadian τ of diurnal birds is generally longer than 24 h, while τ in owls and nightjars tends to be shorter than 24 h (Aschoff, 1979). However, there is quite a bit of inter- and intraspecific variability in τ and even consistent changes within an individual. For example, many house sparrows, *Passer domesticus*, will express a τ of less than 24 h upon transfer from an LD cycle to DD (see Figure 34.1). However, birds' τ gradually lengthens over several days to weeks, depending on the individual, eventually stabilizing with a τ of approximately 25 h. This is a phenomenon known as Eskin's knee, because the actogram, described by Dr. Arnold Eskin in late 1970s (Menaker et al., 1978), appears bent as a knee. Once stabilized, the τ remains stable until environmental factors change.

The major environmental factor affecting τ in birds is ambient light. If diurnal birds are experimentally maintained in dimLL rather than DD, birds' τ will shorten, and as illuminance is increased, τ decreases until at constant illuminances of approximately 50–100 lx, circadian patterns of locomotor activity are lost (Aschoff, 1979); birds are arrhythmic. Nocturnal birds such as owls and nightjars are more highly variable than are diurnal birds. On average, however, their τs lengthen (increase) with increasing illuminance. Other environmental factors have not been studied in as much detail as has light, although birds' circadian patterns appear to be relatively insensitive to altered constant temperatures, as circadian patterns in all organisms tend to be temperature compensated with a Q_{10} of 0.8–1.1 (Pittendrigh, 1993).

34.2.3 Entrainment

A biological clock with a τ that is different from 24 h would not be a particularly useful clock if it could not confer a sense of internal time relative to external, sidereal time, or the time it takes the Earth to rotate relative to the vernal equinox, the solar day. Thus, organisms sense environmental cues called Zeitgebers (German for "time givers"), interpret them, and then adjust their internal time to external time through the process of entrainment. The formal properties of entrainment vary, depending on the Zeitgeber and on the process being measured.

Photic entrainment. The major Zeitgeber for nearly all organisms is the daily change in light intensity, and light affects circadian rhythms in birds profoundly. Photic entrainment in birds is of two types: parametric entrainment

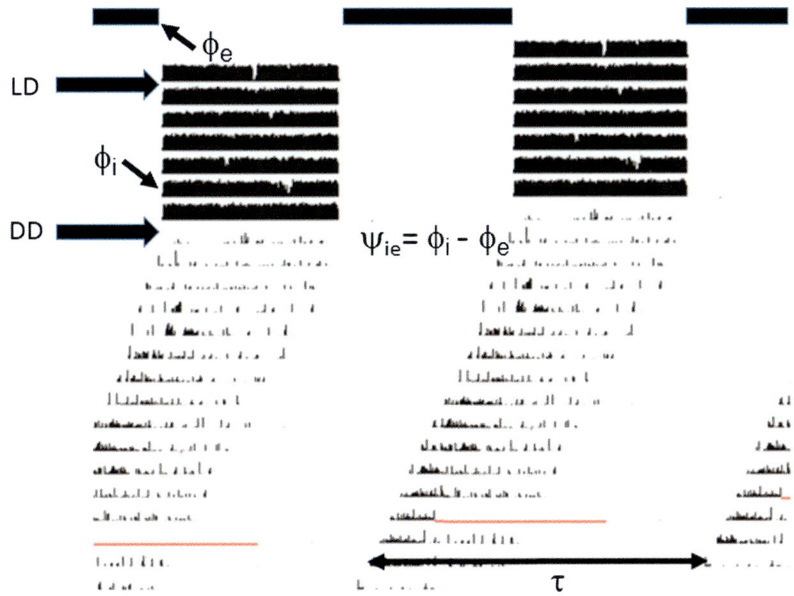

FIGURE 34.1 **Actogram of locomotor activity from a single zebra finch, *Taeniopygia guttata*.** The top bars indicate the times during which lights are off (black) versus on (white). These are plotted in a 48 h timespan in order to "double-plot" the data. The time off lights on, φ_e, indicated by the arrow, is a phase reference determined by the investigator. The internal phase, φ_i, is determined to be the activity onset. The relationship between φ_i and φ_e is called ψ_{ie}. This relationship may change depending on the time of year and physiological condition of the bird. The internal period, τ, is indicated here as the average interval between activity onsets.

and nonparametric entrainment. Nonparametric entrainment entails the process by which the internal clock is differentially sensitive to the Zeitgeber depending on the time of day rather than the amount (the parameter) of the Zeitgeber present (Aschoff and Pohl, 1978). This process is similar to other organisms in that their circadian system is relatively insensitive to light during the subjective day. If a house sparrow is free-running in DD and she is illuminated for 15–30 min during her subjectively diurnal activity (subjective day), no or little effect on her φ_i will occur. However, if she is illuminated in the beginning of her subjective night, when she is inactive, light will change her φ_i ($\Delta\varphi$) by delaying her activity onset ($-\Delta\varphi$). If she is illuminated during the late subjective night, her subsequent activity onset will be advanced ($+\Delta\varphi$). Through this process, birds' φ_i is maintained in a stable phase relationship (ψ_{ie}) with external time (φ_e). In diurnal birds, however, because they are active during times of great illumination, parametric entrainment can also contribute to the maintenance of an advantageous ψ_{ie}. Because diurnal birds tend to express a τ greater than 24 h and illumination shortens τ, illumination of increasing intensity (a parameter) during the day under natural conditions decreases the amplitude of $\Delta\varphi$ required to maintain a stable ψ_{ie}. This is an underappreciated process that may even be relevant to human circadian organization (Roenneberg et al., 2010).

Nonphotic entrainment. Several nonphotic factors entrain birds' circadian clocks. Homoeothermic animals, such as the vast majority of birds, do not synchronize to daily patterns of ambient temperature, although tissue cultures from avian tissues entrain to cycles of high and low temperature (Barrett and Takahashi, 1995; Csernus et al., 2005). Further, ambient temperature can affect the rate by which birds entrain to novel light cycles (Rensing and Ruoff, 2002).

The presentation of food at a particular time of day can synchronize locomotor rhythms in many species of birds, and the data have suggested that birds, as in mammals, possess both light-entrainable oscillators and experimentally separable food-entrainable oscillators underlying food-seeking behavior. House sparrows maintained in dimLL entrain both locomotor and feeding patterns to the period of the feeding cycle, such that locomotor behavior occurs prior to the time of day at which food is presented (Hau and Gwinner, 1992). In domestic pigeons, *Columba livia*, presentation of food every 23.5 h to birds on a 24 h LD cycle results in an uncoupling of food-associated anticipatory changes in body temperature (Tb) and O_2 consumption with activity associated with LD (Rashotte and Stephan, 1996). Similarly, for Svalbard ptarmigan, *Lagopus mutus hyperboreus*, and Indian weaver birds, *Ploceus phillipinus*, food entrainment is coupled to birds' entrainment to the light cycle, suggesting in both species the food-entrainable oscillator is coupled to the light-entrainable pacemaker (Reierth and Stokkan, 1998; Rani et al., 2009).

There is a very little evidence for social cues influencing circadian patterns in birds. Daily presentation of taped sounds recorded from an aviary entrains circadian locomotor rhythms of house sparrows maintained in dimLL (Menaker and Eskin, 1966). These data suggest that birdsong and/or calls may influence the circadian clock.

34.3 PHOTORECEPTORS

Because light is the major Zeitgeber involved in circadian entrainment to LD and an environmental factor that strongly influences τ under constant environmental conditions, the pathways by which light influences clocks provide profound clues to the mechanisms underlying biological clocks. Intuitively, we might think that these photoreceptive processes are the exclusive domain of retinal photoreceptors, but this would be the furthest from the truth. Although the retinae participate in photic entrainment and/or parametric responses to light, they are minor participants.

34.3.1 Encephalic Photoreceptors

In addition to photoreceptors in the lateral eyes shared by all vertebrate classes, it has been known for some time that nonmammalian vertebrates express functional photopigments within the brain that are critical for entrainment of both circadian rhythms and circannual cycles (Okano and Fukada, 2000). Early studies by Benoit in the 1930s showed that domestic ducks, *Anas platyrhynchos*, that had been blinded (enucleated) continued to exhibit reproductive responses to changing photoperiod (Benoit and Assenmacher, 1954). Work by Menaker and colleagues in the 1960s and 1970s in passerine birds clearly showed that the eyes are not necessary for circadian entrainment (Menaker, 1968; Menaker and Underwood, 1976). In a classic series of experiments, Menaker's group demonstrated that enucleated house sparrows could entrain to a series of LD cycles of dimmer and dimmer illuminances. Once birds were no longer capable of entrainment, they showed that the responsible photoreceptor resided inside the head by simply plucking feathers, and entrainment was reinstated. They then blocked entrainment by injecting India ink beneath the scalp (Menaker and Underwood, 1976).

Subsequent research has now identified at least four distinct structures within the brain that are functionally photoreceptive, containing several opsin-based photopigments and photoisomerases (Nakane and Yoshimura, 2010; Peirson et al., 2009; Bailey and Cassone, 2004) (Figure 34.2). These include the pineal gland, which expresses a pineal-specific opsin, pinopsin (Bailey et al., 2003; Okano et al., 1994; Max et al., 1995), as well as melanopsin (OPN4) (Bailey and Cassone, 2005; Bailey et al., 2003; Chaurasia et al., 2005) and iodopsin (OPN1) (Masuda et al., 1994; Natesan et al., 2002) and whose photoreceptive function will be discussed further below. In addition, neurons within the preoptic area express vertebrate ancient (VA) opsin (Halford et al., 2009; Davies et al., 2010, 2012), and project to the

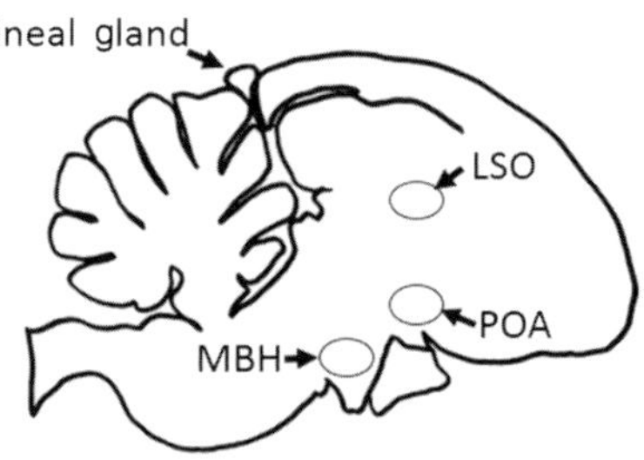

FIGURE 34.2 Schematic of a sagittal view of an avian brain. The general locations of extraocular photoreceptors are delineated in the pineal gland, the lateral septal organ (LSO), preoptic area (POA), and mediobasal hypothalamus (MBH).

tuberal hypothalamus, while the tuberal hypothalamus itself expresses a plethora of photoreceptive cells that appear to be divergent among avian species. In Japanese quail, *Coturnix coturnix*, cerebrospinal fluid-contacting neurons in the mediobasal hypothalamus (MBH) express both OPN4 and neuropsin (OPN5) (Nakane et al., 2010). In house sparrows, neurons within the arcuate nucleus express rhodopsin (OPN2) itself, in addition to OPN4 and OPN5 (Wang and Wingfield, 2011). Finally, the lateral septal organ expresses rhodopsin-like immunoreactivity (Wada et al., 1998). It is not clear whether each of these photoreceptive organs and/or their photopigments subserve mutually exclusive physiological processes or whether these overlap in their functions. In addition to opsin-based photopigments, birds express flavin-based cryptochromes (Bailey and Cassone, 2005; Chaurasia et al., 2005; Kubo et al., 2006). Although cryptochrome is the major photopigment responsible for photoentrainment in *Drosophila* (Emery et al., 1998), the multiple cryptochromes expressed by vertebrates have not been established as photoresponsive molecules.

34.3.2 Pineal Gland

The avian pineal gland emerges from the epithalamus of the diencephalon and extends to the dorsal surface of the brain, where it resides nestled between the two cerebral hemispheres and the cerebellum (Figure 34.2). There, the gland's narrow stalk extends to the skull itself, enlarges in its distal aspects, and attaches to the dura mater. There is considerable variability in pineal structure among birds (Quay, 1965; Menaker and Oksche, 1974). For example, nocturnal owls, shearwaters, and petrels exhibit only a vestigial or rudimentary pineal complex (Quay, 1965), whereas diurnal birds express well-developed pineal glands of differing overt anatomy. According to Menaker and Oksche (1974), avian pineal glands can be categorized as saccular (passerine birds), tubulofollicular (columbiform and anseriform birds), and lobular (galliform birds). At the cellular level, pineal glands comprise at least four classes of cell-types. The photoreceptor-like pinealocytes express outer segments reminiscent of retinal photoreceptors of a sensory ciliated type with a characteristic 7:0 microtubule configuration, although these are reduced in size and frequently coiled (Oksche et al., 1972; Menaker and Oksche, 1974). These cells, as stated above, express several opsin-based photopigments as well as flavin-based cryptochromes, as well as phototransduction signaling molecules (Masuda et al., 1994; Bailey et al., 2003), clearly indicating that pinealocytes are directly photosensitive. This will be discussed further below. In addition, pineal glands contain interstitial cells that include ependymal and astrocytic neuroglia as well as leukocytes and leukoblasts (Oksche et al., 1972), as well as neurons that project to the epithalamic habenular complex (Sato and Ebisawa, 1988). Finally, the pineal gland is vascularized and contains endothelial capillary cells (Oksche et al., 1972).

34.3.3 Retina

It is axiomatic that the retinae are photoreceptive, but their role as photoreceptors in avian circadian entrainment is less clear. In mammals, which do not possess extraocular photoreceptors (Doyle and Menaker, 2007), circadian entrainment is predominantly mediated by intrinsically photoreceptive retinal ganglion cells, expressing melanopsin (OPN4; Panda, 2007). These cells project directly to circadian pacemakers in the hypothalamic suprachiasmatic nuclei (SCN). In birds, as stated above, the eyes are not necessary for circadian entrainment to LD cycles, since enucleated house sparrows and several other species entrain to LD cycles. However, that is not to say that the eyes are not involved. Multiple layers of the retinae of domestic chicks, *Gallus gallus domesticus*, express OPN4, including retinal ganglion cells (Bailey and Cassone, 2005; Chaurasia et al., 2005), similar to the situation in mammals.

Employing a retinal degenerate chick model developed by Semple-Rowland (Semple-Rowland and Cheng, 1999; Guido et al., 2010), Guido and colleagues have shown that the chicks possessing only inner retinal photoreceptors are capable of entraining feeding rhythms to an LD cycle (Valdez et al., 2009, 2013). When encephalic photoreceptors are covered with a blackened hood, the chicks expressed a free-running rhythm with a 24.5 h τ. When the birds were enucleated, hooded birds free-ran, but illumination of the skull reestablished entrainment. The picture that emerges is that multiple photoreceptors in the brain, the pineal gland and the retina redundantly contribute to photic entrainment in birds (Doyle and Menaker, 2007; Guido et al., 2010).

34.4 PACEMAKERS

As we will see below, the capacity to express circadian rhythms is widely distributed among tissues and cells in multicellular organisms (Bell-Pedersen et al., 2005). However, certain tissues have been designated "pacemakers" because of their capacity for sustained circadian rhythmicity and their importance for circadian organization. In birds, these structures are the pineal gland, the retinae, and the hypothalamic SCN, whose mutual interactions are critical for overt circadian organization.

34.4.1 Pineal Gland and Melatonin

Searching for the location of the intracranial, extraretinal photoreceptors, Gaston and Menaker (1968) surgically removed the pineal gland (PINX) from house sparrows (Figure 34.3). Although the birds retained their ability to entrain to LD, they became arrhythmic when placed in DD, demonstrating that the pineal gland is necessary for self-sustained circadian rhythmicity. However, the data also showed that the pineal gland is part of a system of circadian clock components, because PINX sparrows could anticipate the time of lights on in an LD cycle and because birds only gradually became arrhythmic over 5–15 days following transfer from LD to DD (Gaston and Menaker, 1968; Binkley et al., 1971; Ebihara and Kawamura, 1981; Lu and Cassone, 1993a; Wang et al., 2012). Further, the effect of PINX is not universal among avian species. PINX of European starlings, *Sturnus vulgaris*, results in a range of behavioral changes ranging from arrhythmicity akin to those seen in house sparrows to slight disruption of behavioral locomotor rhythmicity (Gwinner, 1978; Gwinner et al., 1987). Circadian rhythms of locomotor behavior in columbiform and galliform birds are little or not affected at all by PINX (Ebihara et al., 1984; Underwood and Siopes, 1984).

Even so, the pineal gland represents both the capacity for rhythmicity and time of day. In an elegant experiment, Zimmerman and Menaker (1979) transplanted pineal glands from two groups of house sparrows into the anterior chambers of the eye of PINX, arrhythmic sparrows maintained in DD. The first group of donor birds were entrained to an early LD cycle, with lights on at midnight, while the second set of donors were entrained to a late LD cycle, with lights on at 11 AM. Transplantation restored circadian rhythms to both groups of recipients within one day. Moreover, birds that received pineal glands from early donors exhibited an early φ_i , while the recipients of late donor pineal glands exhibited a late φ_i. Thus, the pineal gland is not only necessary for circadian rhythmicity in these birds, but it contains a correlate that confers time of day to recipient birds.

That hormone was known even then to be the indoleamine melatonin from earlier work of Lerner and later of Axelrod, Klein, and their co-workers (Klein et al., 1997), who explored the biochemical basis for melatonin biosynthesis in the pineal gland of the chick, *Gallus domesticus* (Figure 34.4). Research from a large number of investigators have shown that pinealocytes—the photoreceptive, secretory cells of the avian pineal gland—take up the amino acid tryptophan, which is converted to 5-hydroxytryptophan

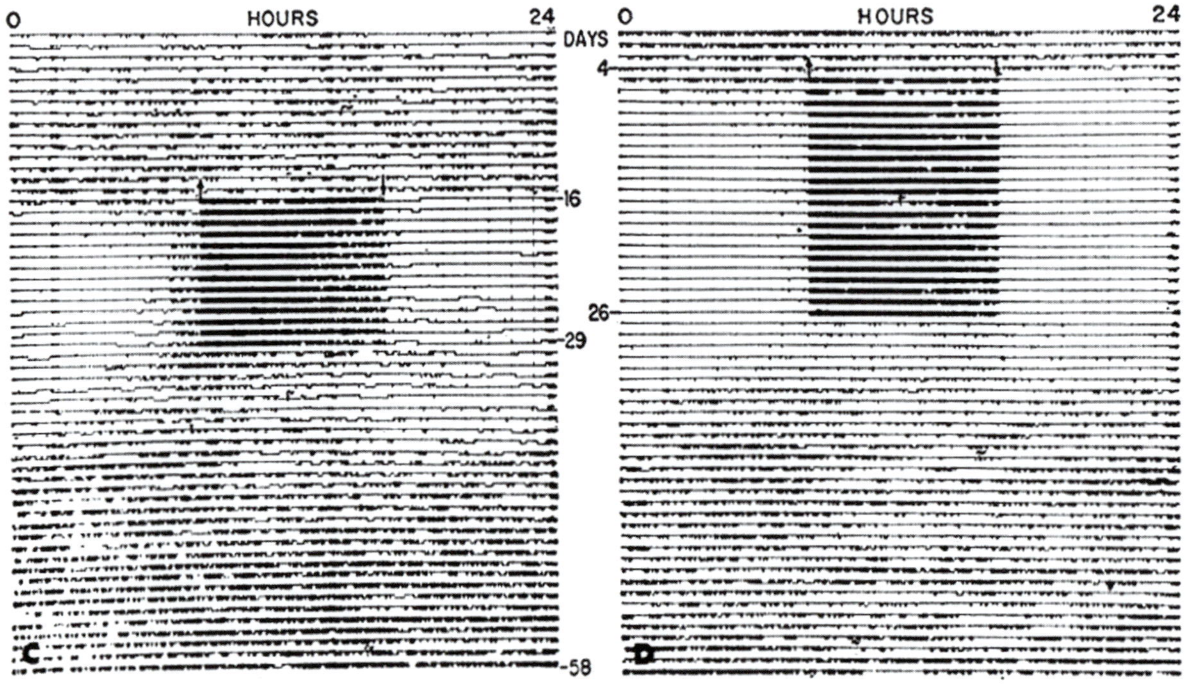

FIGURE 34.3 **Effects of pinealectomy on locomotor activity rhythms in house sparrows.** Birds are arrhythmic in constant darkness (DD) but are able to entrain to light:dark (LD) cycles (small arrows up and down, respectively). When birds are released into DD again, they take 5–15 days to become arrhythmic again. *From Gaston and Menaker (1968).*

by tryptophan hydroxylase (TrH; EC 1.14.16.4; Chong et al., 1998) and then decarboxylated to produce serotonin (5HT) by aromatic L-amino acid decarboxylase (AAADC; EC 4.1.1.28). During the night in LD and subjective night in DD, 5HT is converted to *N*-acetylserotonin (NAS) by arylalkylamine (or serotonin)-*N*-acetyltransferase (AANAT; EC 2.3.1.87; Bernard et al., 1997). NAS is then converted to melatonin by hydroxyindole-*O*-methyltransferase (HIOMT; EC 2.1.1.4; Bernard et al., 1991). The genes encoding each of these enzymes have been isolated, cloned, and sequenced in several avian species. In chicks, at least, TrH, AANAT, and HIOMT are regulated by both the molecular clockworks within the pinealocytes and directly by light at the transcriptional, translational, and posttranslational levels, so that the enzymatic regulation of pineal melatonin is a dynamic, rhythmic process (Klein et al., 1997).

Rhythmic administration of melatonin to PINX house sparrows, European starlings, and zebra finches, *Taeniopygia guttata*, or to EX/PINX pigeons (Chabot and Menaker, 1992; Lu and Cassone, 1993b; Gwinner et al., 1997; Heigl and Gwinner, 1995; Cassone et al., 1992, 2008; Wang et al., 2012) restores a daily pattern of locomotor behavior (Figure 34.5). Activity typically decreases following melatonin administration and rebounds following the daily dosage, suggesting that melatonin induces a soporific state and/or sleep, which it may well do. In addition, however, the synchronization of locomotor behavior by rhythmic melatonin administration represents entrainment of circadian clockworks in the PINX bird because melatonin administration in a T-cycle different from 24 h results in systematic changes in the phase relationship (ψ) of melatonin to the onset of locomotor activity (Figure 34.5; Chabot and Menaker, 1992; Gwinner et al., 1997). Further, administration of three different behaviors—locomotor behavior, call, and song—by a single melatonin regime results in entrainment at differential rates (Wang et al., 2012).

Further, constant administration of melatonin has profound effects on overt rhythmicity (Gwinner and Brandstatter, 2001). Subcutaneous implants of beeswax (Turek et al., 1976) or Silastic (Hau and Gwinner, 1995; Abraham et al., 2000) containing crystalline melatonin change τ at low concentrations and abolishes rhythmicity altogether at higher dosages. Interestingly, constant melatonin decreases the times taken to reentrain to food cycle (Hau and Gwinner, 1995) or LD cycle (Abraham et al., 2000). This is similar to the effect of PINX itself (Kumar and Gwinner, 2005), punctuating the view that it is the rhythm of endogenous melatonin that preeminently affects circadian organization and that this effect acts through the entrainment of downstream oscillators (Cassone and Westneat, 2012).

Avian pineal glands contain the circadian clockworks and photoreceptors to generate circadian patterns of melatonin biosynthesis *in vitro* as well as *in vivo*, which can be entrained to LD cycles directly (Binkley et al., 1978; Brandstatter et al., 2000; Natesan et al., 2002). Pineal tissue and pinealocyte cultures express circadian patterns of AANAT activity (Binkley et al., 1977; Deguchi, 1979),

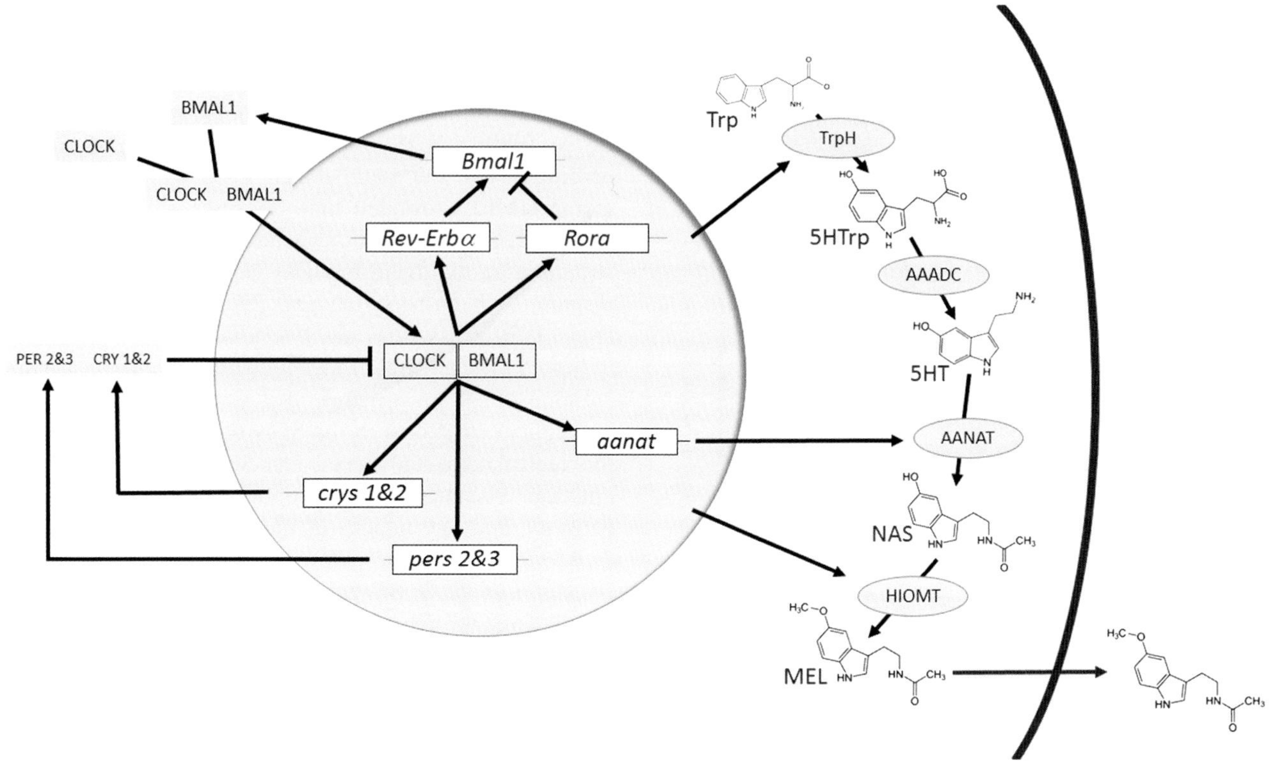

FIGURE 34.4 Schematic of the molecular clockworks regulating circadian patterns of melatonin biosynthesis in a pinealocytes or retinal photoreceptor. Positive elements CLOCK and BMAL1 enter the nucleus and activate expression of genes whose promoters contain an E-Box. Among these are the negative elements *periods 2* and *3* (*per2* and *per3*) and *cryptochromes 1* and *2* (*crys1* and *crys2*), *Rev-Erbα* and *Rora*, which form a secondary loop regulating *Bmal1* transcription, and output, clock-controlled genes such as arylalkylamine-*N*-actyltransferase (*aanat*). The *pers* and *crys* are translated, form heterodimers with other components, such as the casein kinases, and reenter the nucleus to interfere with CLOCK/BMAL1 activation. Melatonin biosynthesis pathways are indicated on the right. Amino acid tryptophan is converted to 5-hydroxytryptophan (Trp) by tryptophan hydroxylase (TrpH). Aromatic amino acid decarboxylase (AAADC) then converts 5-hydroxytryptophan (5HTrp) to 5-hydroxytryptamine (5HT; serotonin). Then, during the night, AANAT converts 5HT to *N*-acetylserotonin (NAS), a substrate for hydroxyindole-*O*-methyltransferase (HIOMT), which produces melatonin (MEL) itself. Presumably, melatonin diffuses out of the cell at this time, although a release mechanism may exist.

gene expression (Karaganis et al., 2008), and melatonin efflux (Figure 34.6; Takahashi et al., 1980), such that levels are high during the night and low during the day in LD. These rhythms persist for 4–10 days in DD before damping to arrhythmicity. Exposure to light has three effects on cultured pineal rhythms: light inhibits melatonin biosynthesis, light increases amplitude and decreases damping, and light phase-shifts the clock within pineal cells (Zatz et al., 1988).

34.4.2 Retinae

Interestingly, the photoreceptors in the retinae of the lateral eyes also synthesize and release melatonin in many vertebrate species (Underwood et al., 1984). In fact, in Japanese quail and domestic pigeon, *C. livia*, the retinae release almost as much melatonin into the systemic circulation as does the pineal gland and removal of this source by enucleation or retinectomy in addition to PINX results in arrhythmic circadian locomotor behavior, similar to the effects of PINX alone in passerine birds (Ebihara et al., 1984, 1997; Underwood and Siopes, 1984). Thus, the variability of the effects of PINX among birds may in part be due to this retinal component in some species, and that it is not the pineal per se but rhythmic melatonin that is important for circadian locomotor behavior.

34.4.3 Suprachiasmatic Nuclei

In birds, two sets of structures have been associated with SCN function: the medial suprachiasmatic nuclei (mSCN) and the visual suprachiasmatic nuclei (vSCN) (Cassone and Moore, 1987; Cantwell and Cassone, 2006a,b). These structures are connected via neuronal projections and are contiguous in terms of their cellular populations, especially in the distribution of astrocytes. The vSCN, but not the mSCN, expresses metabolic rhythmicity and electrical activity such that levels are high during the day and low during the night (Cassone, 1988; Lu and Cassone, 1993a,b; Juss et al., 1994; Cantwell and Cassone, 2002). Further, the vSCN, but not the mSCN, receives retinohypothalamic (RHT) input (Cassone and Moore, 1987; Cassone, 1988; Cantwell and Cassone, 2002, 2006a,b), and the vSCN, but not the mSCN, contains melatonin receptor binding (Rivkees et al., 1989;

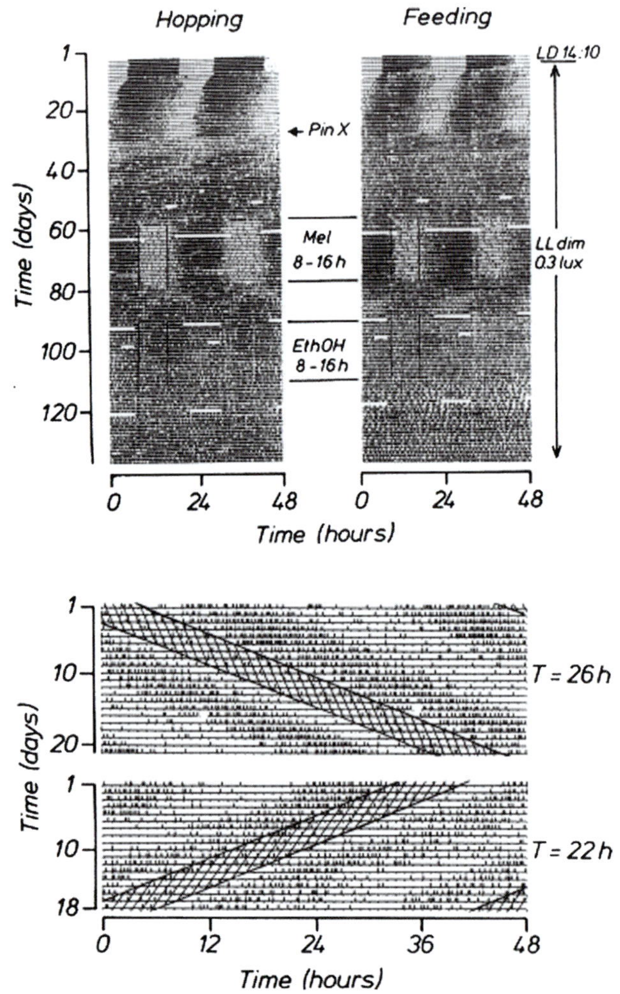

FIGURE 34.5 Effects of rhythmic administration of melatonin to house sparrows. Top: Actograms of perch-hopping and feeding behavior of house sparrows maintained in DD. Birds become arrhythmic in both behaviors following pinealectomy (PINX) and daily rhythms of behavior are reestablished with melatonin administration. Bottom: This synchronization is entrainment rather masking because locomotor activity maintains a systematic phase relationship with the administration of the hormone. When melatonin is presented every 26h, activity occurs after the administration, whereas a melatonin cycle every 22h results in behavior that anticipates the melatonin. *From Gwinner et al. (1997).*

Cassone et al., 1995). Administration of melatonin to chicks and house sparrows decreases glucose utilization within the vSCN (Lu and Cassone, 1993b; Cantwell and Cassone, 2002). Finally, light activates *c*-fos expression in the vSCN, but not in the mSCN (King and Follett, 1997). In quail, only the mSCN expresses clock gene rhythmicity (Yoshimura et al., 2001; Yasuo et al., 2002), while in the house sparrow, both structures rhythmically express the expression of the "clock gene" *per2* (Abraham et al., 2002, 2003). Importantly, lesions directed at the mSCN in Java sparrows, *Padda oryzivora*, and house sparrows result in arrhythmicity similar to that observed following PINX (Ebihara and Kawamura, 1981; Takahashi and Menaker, 1982). However, it is not clear whether these lesions also affected vSCN integrity.

34.5 SITES OF MELATONIN ACTION

34.5.1 Melatonin Receptors

In the 1980s and 1990s, high affinity melatonin receptor binding using the radiolabeled agonist 2[^{125}I]-iodomelatonin (IMEL) revealed high densities of IMEL binding in retinal, retinorecipient structures and visual integrative structures in the avian brain as well as peripheral tissues (Figure 34.7; Dubocovich and Takahashi, 1987; Rivkees et al., 1989; Cassone et al., 1995). Binding affinity studies indicated kDs in the pM range with high specificity for melatonin itself. Brain structures that bind IMEL included retinorecipient structures in the SCN of the circadian system, the ventrolateral and dorsal geniculate nuclei of the thalamofugal visual pathway, the optic tectum of the tectofugal pathway, and the nucleus of the basal optic root (nBOR) or the accessory optic pathway. In all species, integrative structures of the tectofugal pathway such as nucleus rotundus (Rt) and the ectopallium (Ep) also bind IMEL (Rivkees et al., 1989; Cassone et al., 1995). In some but not all species, hyperpallial structures, including the visual Wulst, are sites of IMEL binding. In male passerine birds but not females, structures associated with bird song learning and control also revealed high affinity IMEL binding (Gahr and Kosar, 1996; Whitfield-Rucker and Cassone, 1996).

Reppert and colleagues were able to isolate and clone genes encoding two high affinity melatonin receptors; these were designated the Mel_{1A} and Mel_{1C} receptors (Reppert et al., 1995). Independent work isolated partial sequences encoding an ortholog of the Mel_{1B} receptor in the same year (Liu et al., 1995). Subsequent work has confirmed in birds that there are at least three melatonin receptors, the Mel_{1A}, Mel_{1B}, and the Mel_{1C} receptors (Reppert, 1997). All three melatonin receptor subtypes represent 7-transmembrane domain, GTP-binding protein structures, and all three are in the G_i GTP-binding protein category, although some cross-talk with G_q has been documented (Reppert, 1997). The distributions of these three receptor subtypes are not uniform in chicks, zebra finches, and house sparrows. The Mel_{1A} receptor predominates in central nervous neurons and peripheral tissues (Natesan and Cassone, 2002; Karaganis et al., 2009), whereas the Mel_{1B} receptor is expressed in inner retinal neurons and photoreceptors as well as other central nervous neurons (Natesan and Cassone, 2002). In passerines, the Mel_{1B} receptor is the major receptor subtype in song control nuclei, but the other two are expressed as well (Jansen et al., 2005; Bentley et al., 2012). The Mel_{1C} receptor, on the other hand, predominates in nonneuronal elements of the central nervous system (Reppert et al., 1995; Adachi et al., 2002). Culture studies with chick astrocytes (Adachi et al., 2002) show 95–100% diencephalic astrocytes express the Mel_{1C} receptor, while an overlapping 5–10% expresses Mel_{1A}. Astrocytes do not appear to express Mel_{1B}.

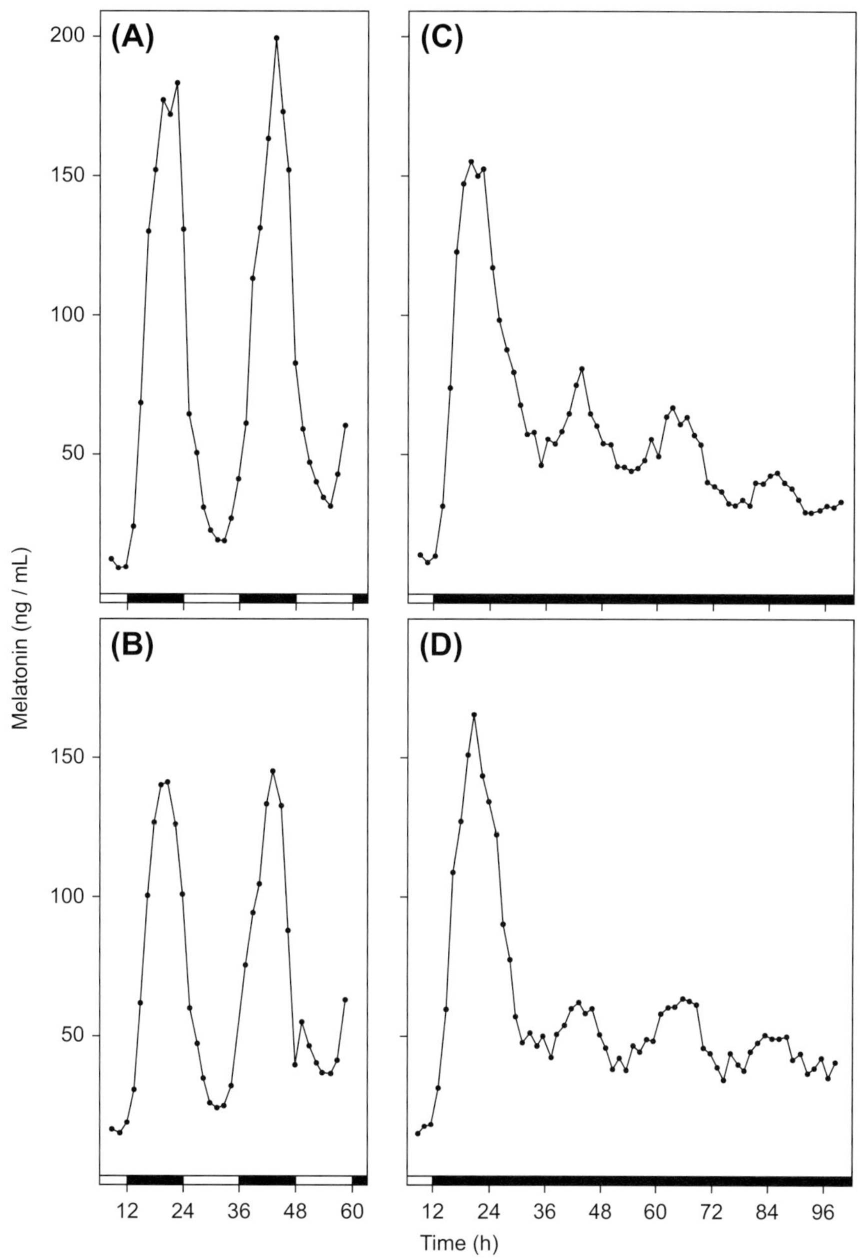

FIGURE 34.6 Melatonin rhythms from cultured chick pineal glands. Melatonin is released during the dark phase in glands maintained in LD *in vitro* (left). These rhythms continue in DD for several days at a lower amplitude but eventually damp out. *From Takahashi et al. (1980).*

34.5.2 Mechanisms of Action

As 7-transmembrane domain receptor proteins in the G_i GTP-binding protein group (Reppert, 1997), the melatonin receptors collectively act via inhibition of adenylyl cyclase activity in cells that express the receptors. There is some evidence for downstream cross-talk with other signaling pathways, however. For example, calcium signaling among chick astrocytes is mediated by a Mel_{1C} melatonin receptor through IP3-depending processes (Peters et al., 2005). In addition, there is evidence that melatonin may act via a receptor-independent and/or by receptor molecules as yet undiscovered. However, these have not been described in birds.

34.6 AVIAN CIRCADIAN ORGANIZATION

Overt circadian rhythms in birds are controlled by multiple circadian pacemakers that are entrained by multiple photoreceptive elements in the central nervous system. The relative importance of these pacemakers varies among the few species studied, but in each, the core of the circadian system can be identified as the pineal gland, the retinae, and the SCN (mSCN and vSCN). Early analyses of the interactions of circadian pacemakers and photoreceptors suggested a hierarchical relationship in which the pineal gland imposes and/or entrains rhythmicity and phase on downstream processes or oscillators respectively. The discovery of SCN and

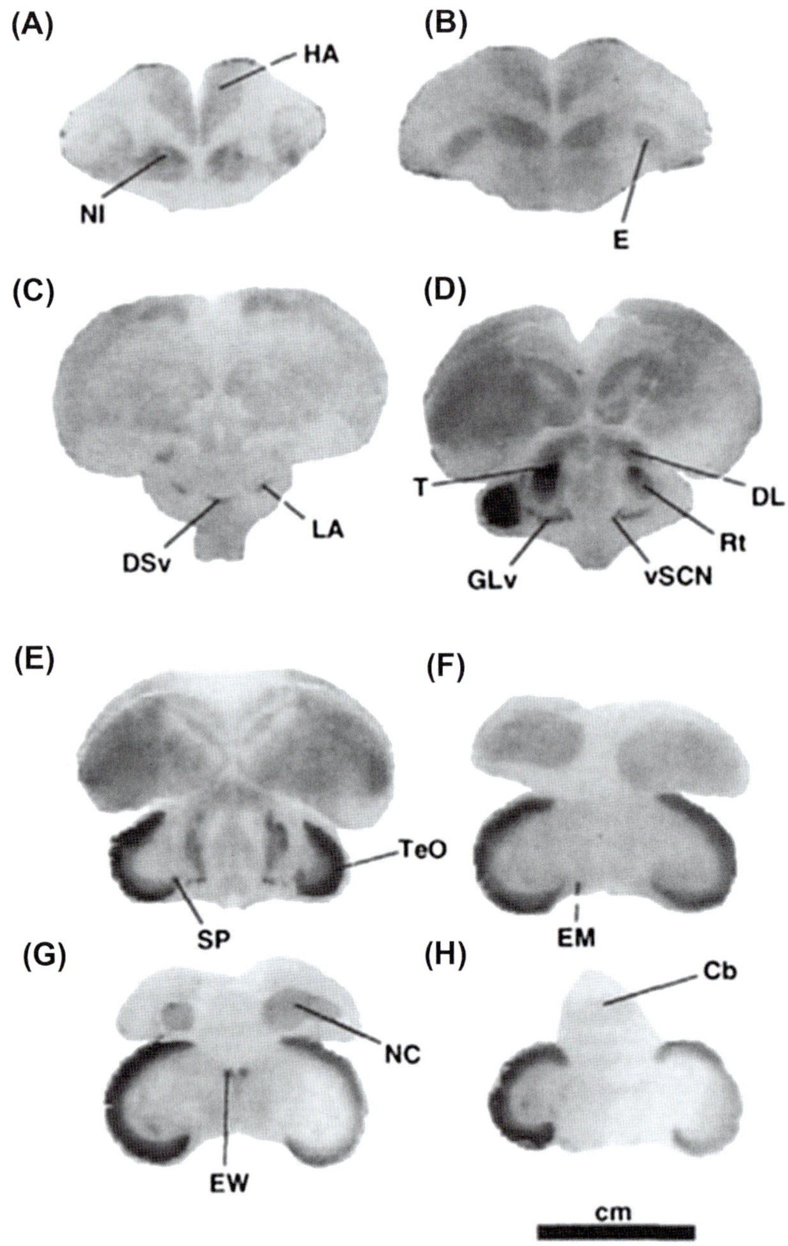

FIGURE 34.7 **Melatonin receptor binding in the brain of the ring-necked pheasant, *Phasianus colchicus*.** Video digitized images of autoradiographs depicting 2[^{125}I]iodomelatonin binding (IMEL) at 50 pM from the rostral (A) to caudal (H) extents. Densest binding is present in retinorecipient and integrative structures of the circadian (vSCN), tectofugal (optic tectum (TeO), *nucleus rotundus* (Rt), and ectopallium (E)), thalamofugal (lateral anterior nucleus (LA) and dorsolateral nucleus (DL) in the thalamus and the hyperpallium anterior (HA)), and accessory optic (ectomammillary (EM; also called nucleus of the basal optic root) and nucleus of Edinger-Westphal (EW)) visual pathways. *From Cassone et al. (1995).*

retinal pacemaker activity suggested a more complex relationship (Figure 34.8).

There is little doubt that the pineal gland and/or rhythmic melatonin secretion is important for circadian organization, but it is clear it is part of a more complex system of structures and process we shall call avian circadian organization or the avian circadian system. Although PINX indeed does abolish circadian locomotor, Tb, brain metabolism and singing behavior rhythms in oscine passeriform birds, there is compelling evidence for other pacemakers in the system delineated above. First, PINX birds entrain to LD cycles and typically commence locomotor and singing behavior before dawn (Gaston and Menaker, 1968; Menaker et al., 1978; Gwinner et al., 1994; Wang et al., 2012), indicating some capacity for time-keeping. Secondly, when PINX house sparrows, zebra finches and other fringillid passerine birds (sparrows and finches) are transferred from LD to DD or dimLL, they express 5–10 days of damped activity that

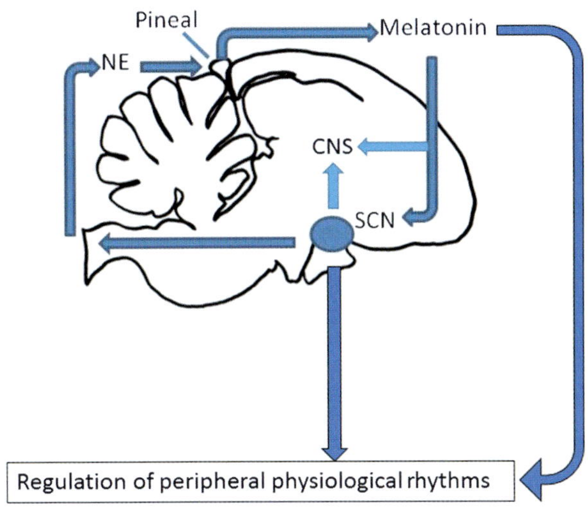

FIGURE 34.8 Neuroendocrine loop model for avian circadian organization. Pacemakers in th pineal gland and SCN interact via mutual inhibition to maintain a stable antiphase relationship. Each pacemaker can affect downstream oscillators and processes independently or in concert in the CNS and periphery. *Modified from Cassone and Westneat (2012).*

gradually descend into arrhythmicity (Figure 34.3), suggesting a remnant pacemaker underlying rhythmic behavior. Thirdly, sturnid passerines (e.g., starlings) and columbiform birds (e.g., pigeons) express a variable response to PINX, with some birds becoming arrhythmic, others disrupted, and still others unaffected (Gwinner, 1978; Ebihara et al., 1984; Gwinner et al., 1997). Finally, as stated above, timed melatonin administration differentially entrains different behaviors in zebra finches (Wang et al., 2012).

At least one of the sites of melatonin action in synchronizing circadian locomotor rhythms is the vSCN. The vSCN express melatonin receptors (Rivkees et al., 1989; Reppert et al., 1995), and melatonin affects the vSCN physiologically. When intact house sparrows are placed in DD and injected with 2DG at different times during the subjective day to map metabolic activity, the vSCN expresses a rhythm in 2DG uptake for at least 10 days in DD (Lu and Cassone, 1993a). However, when PINX birds are similarly placed in DD, 2DG uptake rhythms in the vSCN decline in amplitude and damp to arrhythmicity at the same rate as does locomotor activity (Lu and Cassone, 1993a). When PINX birds are administered melatonin on a daily basis, locomotor activity and vSCN 2DG uptake rhythms are both reestablished (Lu and Cassone, 1993b). Further, injection of melatonin during the late subjective day acutely decreases vSCN 2DG uptake (Cantwell and Cassone, 2002).

Conversely, although it is certain the avian pineal gland contains patent circadian oscillators and photoreceptors to establish overt circadian rhythms in melatonin secretion *in vivo* and *in vitro* (Binkley et al., 1977; Deguchi, 1979; Takahashi et al., 1980; Kumar and Follett, 1993; Csernus et al., 2005; Karaganis et al., 2008), these rhythms are not self-sustained *in vitro* in DD (Figure 34.6). Cultured pineal glands express rhythms of melatonin biosynthesis and secretion such that melatonin levels are high during the night and low during the day in LD, but the amplitude of this rhythm declines in DD, such that the rhythm damps to arrhythmicity in 2–8 days (Cassone and Menaker, 1984; Csernus et al., 2005).

In vivo, the avian pineal gland is innervated by postganglionic sympathetic nerves (Ueck, 1979), and receives daily and circadian input through release of norepinephrine (NE) during the day and subjective day (Cassone et al., 1986). This rhythm of NE turnover is dependent on the vSCN because surgical destruction of the vSCN, but not the mSCN, abolishes the circadian NE turnover rhythms (Cassone et al., 1990). When the pineal gland of domestic hens is denervated of its sympathetic innervation *in vivo*, plasma melatonin rhythms dampen within 4 days (Cassone and Menaker, 1983), similar to the situation of pineal glands *in vitro*. Administration of NE to chick pineal glands *in vivo* and *in vitro* has two effects on pineal melatonin rhythms: (1) NE inhibits melatonin biosynthesis, and (2) NE increases amplitude and decreases damping, but does *not* phase-shift the pineal circadian clock (Cassone and Menaker, 1983; Zatz and Mullen, 1988).

To account for these observations in house sparrows, at least, Cassone and Menaker (1984), Cassone (1990) proposed the neuroendocrine loop model for avian circadian organization (Figure 34.8). In this scenario, pineal and SCN pacemakers are damped circadian oscillators. We do not know if this damping is due to individual cells within these structures themselves damping or whether these cells are self-sustained oscillators that drift out of phase from one another. Light inhibits the output of the pineal gland oscillators through photoreceptors within the gland themselves, and light activates SCN output as well as possessing the capacity to phase-shift the clocks within each. During the night and subjective night in DD, pineal oscillators secrete melatonin and influence a wide array of downstream processes and structures. Among these are the vSCN, which are inhibited by melatonin. However, because the pineal is an endogenous oscillator, its output declines as dawn and subjective dawn in DD approach, disinhibiting SCN (vSCN and mSCN) output. The SCN in turn are active during the day and subjective day in DD, affecting a wide array of downstream processes and structures. Among these are the circadian rhythm in sympathetic outflow in NE turnover in the pineal gland at least, which are high during the day and low during the night. There, NE inhibits the biosynthesis and release of melatonin until, because the SCN are circadian oscillators, their output diminishes, and pineal oscillators are disinhibited. This mutually inhibitory relationship of SCN and pineal oscillators maintains stable phase relationships and enable each pacemaker to affect downstream processes singularly or in concert. Gwinner (1989) suggested that the relationship between pineal and SCN oscillators conveyed an "internal resonance" in which each oscillator increased the amplitude of its partner. This idea is not mutually exclusive from the neuroendocrine loop model. Indeed, it is clear that application of NE to chick pineal glands increases the amplitude of melatonin output *in vivo* (Cassone and Menaker, 1983) and *in vitro* (Zatz and Mullen, 1988), even as it inhibits melatonin output. The pineal gland releases rebounding high levels of melatonin release following cessation of the NE administration.

Where do the retinae fit into this scheme? For columbiform and galliform species whose retinae secrete melatonin rhythmically, at least part of their role is to serve as a "second (and third) pineal gland" (Menaker, 1985), secreting melatonin during the night, presumably influencing SCN and other downstream oscillators in parallel with pineal output. In addition, inner retinal photoreceptors and circadian oscillators may directly influence circadian patterns of visual perception and activity through retinohypothalamic and other visual pathways (Guido et al., 2010). It is clear, however, that the retinae are the predominant pacemakers in galliform birds, but the nature of this function is not apparent at this stage (Steele et al., 2006).

The mechanism by which this system influences downstream processes is not at all clear at this stage. However, new and exciting data pointing to distributed capacities for rhythmicity are becoming available, and these are described below. First, however, we must explore the molecular aspects of avian circadian clocks.

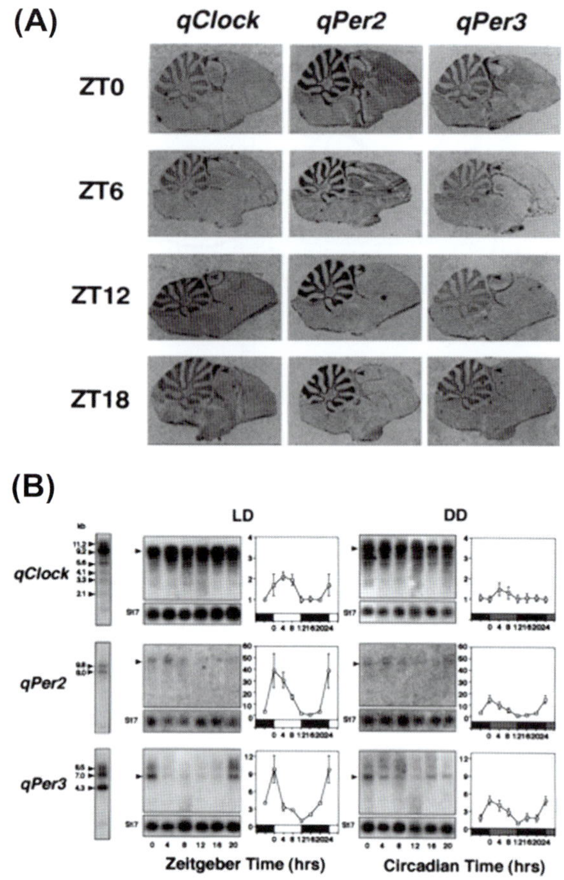

FIGURE 34.9 Circadian expression of clock genes in the (A) pineal gland by *in situ* hybridization and (B) retinae by northern blot analysis in the Japanese quail. In both tissues, *clock* is expressed during the day, whereas *per2* and *per3* are expressed in the late night. *From Yoshimura et al. (2000).*

34.7 MOLECULAR BIOLOGY

The idea that behavior and other complex processes derive from genes and their expression arises from early mutational analyses in search of mutants and the identification of the genes underlying the process. Among the early entrants in this study was the analysis of mutations affecting circadian rhythms in *Drosophila* (Konopka and Benzer, 1971; Rosbash et al., 2007).

34.7.1 Identification, Characterization, and Localization of Molecular Clockworks in Birds

Circadian rhythms are regulated by a highly conserved set of genes, collectively called "clock genes," whose products are believed to dynamically interact to elicit rhythmic patterns of transcription, translation, biochemical and physiological processes, and behavior (Reppert and Weaver, 2002; Bell-Pedersen et al., 2005; Rosbash et al., 2007). In animals ranging from *Drosophila* to humans, the central core of this gene network can be broadly characterized as "positive elements" *clock* and *bmal1* and "negative elements" *period 1* (*Per1*), *period 2* (*per2*), *period 3* (*per3*), and the cryptochromes *cryptochrome 1* (*cry1*) and *cryptochrome 2* (*cry2*).

In contrast to mammals, birds do not express *per1* and have been shown to only express only *per2* and *per3* (Figure 34.9; Yoshimura et al., 2000; Yasuo et al., 2002; Bailey et al., 2003, 2004). *Clock* and *bmal1* are transcribed and then translated in the cytoplasm, where they dimerize and reenter the nucleus and activate transcription of the negative elements through the activation of E-box promoter elements (Figure 34.4; Haque et al., 2010). The *pers* and *crys* in turn are transcribed and translated in the cytoplasm, where the PER proteins are targeted for proteosomal proteolysis by a series of protein kinases, most notably casein kinase 1ε (CK1ε) and CK1δ. This process slows the accumulation of the cytoplasmic PER and thereby increases the period of the molecular cycle. In the cytoplasm, PER and CRY proteins form oligomers that reenter the nucleus and interfere with the CLOCK/BMAL1-mediated activation. A secondary cycle involving two genes containing E-box promoters, *Reverbα* and *rorA*, amplify the cycle by activating and inhibiting *bmal1* transcription respectively. Disruption and/or knock-out of these genes' action has profound effects on the expression of circadian rhythms in animals in which these technologies are possible (i.e., mice and *Drosophila*) ranging from changes in period to arrhythmicity.

34.7.2 Peripheral Oscillators in Avian Circadian Clocks

Although clock genes are expressed by cells within pacemaker tissues (Yoshimura et al., 2000; Yasuo et al., 2002; Bailey et al., 2002, 2003, 2004), it was surprising to find rhythmic expression of clock genes in other parts of the brain as well as in peripheral tissues, such as heart, liver, lungs, and gonads (Chong et al., 2003; Helfer et al., 2006; Karaganis et al., 2009; Zeman et al., 2009). To determine whether pacemakers in the pineal and eyes were responsible for peripheral clock gene rhythms, Karaganis et al. (2009) showed that PINX or EX of chicks decreased the amplitudes of clock gene expression and changed the φ_i of rhythms in *cry1*, *per3*, and *bmal1* rhythms but did not abolish them. It is not clear whether combined PINX and EX would abolish these rhythms, but at this stage, one must conclude the peripheral oscillations persist and are only modulated by the pineal and retinae.

34.7.3 Prospects for Transgenesis and Molecular Manipulation of Avian Clocks

At this stage, no stably transformed birds are available for circadian biology. However, there are now many reports of development of transgenic reporter and knock-out lines of chickens (Poynter and Lansford, 2008; Nishijima and Iijima, 2013). Clearly, this is the line of future research in this field. Even so, some progress has been made with viral transductions in avian retina models (Semple-Rowland et al., 2010). Further, application of microRNA vectors associated with CLOCK and BMAL1 expression affect AANAT expression in chick retina (Haque et al., 2010). Thus, it is possible to manipulate the avian circadian clock at the molecular level. The proof now will be in the pudding.

34.8 CONCLUSION AND PERSPECTIVE

Biological clocks in birds are critical components of their physiology and behavior (Cassone, 1990; Gwinner and Brandstatter, 2001). Early research studying avian circadian organization was critical in the understanding of structure–function relationships in the central nervous system. Indeed, the identification of the pineal gland as a "master pacemaker" conveyed a sense that the system was hierarchically organized. However, as the properties of the pineal gland's function became clearer and the identification of new pacemakers in the hypothalamus and retinae made the system appear more complex, it became clear that the avian circadian clock is a system of multiple circadian pacemakers in the SCN, the pineal gland, and in at least some species the retinae. Each of these were entrained to environmental light cycles by photoreceptors in the retinae, the pineal gland, and the brain. These were in turn thought to interact to accomplish a self-sustained oscillation that drove rhythmic processes downstream. The advent of molecular biological techniques revealed that even this distributed system of pacemakers act via entrainment of downstream oscillators rather than via direct action. It remains to be seen how molecular oscillations in the brain and peripheral tissues are coupled to physiological output, but this is the physiology of circadian clocks in the future.

REFERENCES

Abraham, U., Gwinner, E., Van't Hof, T.J., 2000. Exogenous melatonin reduces the resynchronization time after phase shifts of a nonphotic zeitgeber in the house sparrow (*Passer domesticus*). J. Biol. Rhythms 15 (1), 48–56.

Abraham, U., Albrecht, U., Gwinner, E., Brandstätter, R., 2002. Spatial and temporal variation of passer Per2 gene expression in two distinct cell groups of the suprachiasmatic hypothalamus in the house sparrow (*Passer domesticus*). Eur. J. Neurosci. 16 (3), 429–436.

Abraham, U., Albrecht, U., Brandstätter, R., 2003. Hypothalamic circadian organization in birds. II. Clock gene expression. Chronobiol. Int. 20 (4), 657–669.

Adachi, A., Natesan, A.K., Whitfield-Rucker, M.G., Weigum, S.E., Cassone, V.M., September 2002. Functional melatonin receptors and metabolic coupling in cultured chick astrocytes. Glia 39 (3), 268–278.

Aschoff, J., 1979. Circadian rhythms: influences of internal and external factors on the period measured in constant conditions. Z. Tierpsychol. 49 (3), 225–249.

Aschoff, J., Pohl, H., 1978. Phase relations between a circadian rhythm and its zeitgeber within the range of entrainment. Naturwissenschaften 65 (2), 80–84.

Bailey, M.J., Cassone, V.M., 2004. Opsin photoisomerases in the chick retina and pineal gland: characterization, localization, and circadian regulation. Invest. Ophthalmol. Vis. Sci. 45 (3), 769–775.

Bailey, M.J., Cassone, V.M., 2005. Melanopsin expression in the chick retina and pineal gland. Brain Res. Mol. Brain Res. 134 (2), 345–348.

Bailey, M.J., Chong, N.W., Xiong, J., Cassone, V.M., 2002. Chickens' Cry2: molecular analysis of an avian cryptochrome in retinal and pineal photoreceptors. FEBS Lett. 513 (2–3), 169–174.

Bailey, M.J., Beremand, P.D., Hammer, R., Bell-Pedersen, D., Thomas, T.L., Cassone, V.M., 2003. Transcriptional profiling of the chick pineal gland, a photoreceptive circadian oscillator and pacemaker. Mol. Endocrinol. 17 (10), 2084–2095.

Bailey, M.J., Beremand, P.D., Hammer, R., Reidel, E., Thomas, T.L., Cassone, V.M., 2004. Transcriptional profiling of circadian patterns of mRNA expression in the chick retina. J. Biol. Chem. 279 (50), 52247–52254.

Barrett, R.K., Takahashi, J.S., 1995. Temperature compensation and temperature entrainment of the chick pineal cell circadian clock. J. Neurosci. 15 (8), 5681–5692.

Bell-Pedersen, D., Cassone, V.M., Earnest, D.J., Golden, S.S., Hardin, P.E., Thomas, T.L., Zoran, M.J., 2005. Circadian rhythms from multiple oscillators: lessons from diverse organisms. Nat. Rev. Genet. 6 (7), 544–556.

Benoit, J., Assenmacher, I., 1954. Comparative sensitivity of superficial and deep receptors in photosexual reflex in duck. C. R. Hebd Seances Acad. Sci. 239 (1), 105–107.

Bentley, G.E., Perfito, N., Calisi, R.M., 2012. Season- and context-dependent sex differences in melatonin receptor activity in a forebrain song control nucleus. Horm. Behav. (Epub ahead of print).

Bernard, M., Voisin, P., Guerlotté, J., Collin, J.P., 1991. Molecular and cellular aspects of hydroxyindole-O-methyltransferase expression in the developing chick pineal gland. Brain Res. Dev. Brain Res. 59 (1), 75–81.

Bernard, M., Iuvone, P.M., Cassone, V.M., Roseboom, P.H., Coon, S.L., Klein, D.C., 1997. Avian melatonin synthesis: photic and circadian regulation of serotonin N-acetyltransferase mRNA in the chicken pineal gland and retina. J. Neurochem. 68 (1), 213–224.

Binkley, S., Kluth, E., Menaker, M., 1971. Pineal function in sparrows: circadian rhythms and body temperature. Science 174 (4006), 311–314.

Binkley, S., Stephens, J.L., Riebman, J.B., Reilly, K.B., 1977. Regulation of pineal rhythms in chickens: photoperiod and dark-time sensitivity. Gen. Comp. Endocrinol. 32 (4), 411–416.

Binkley, S.A., Riebman, J.B., Reilly, K.B., 1978. The pineal gland: a biological clock in vitro. Science 202 (4373), 1198–1220.

Bradshaw, W.E., Holzapfel, C.M., 2010. What season is it anyway? Circadian tracking vs. photoperiodic anticipation in insects. J. Biol. Rhythms 25 (3), 155–165.

Brandstätter, R., Kumar, V., Abraham, U., Gwinner, E., 2000. Photoperiodic information acquired and stored in vivo is retained in vitro by a circadian oscillator, the avian pineal gland. Proc. Natl. Acad. Sci. U. S. A. 97 (22), 12324–12328.

Cantwell, E.L., Cassone, V.M., 2002. Daily and circadian fluctuation in 2-deoxy[(14)C]-glucose uptake in circadian and visual system structures of the chick brain: effects of exogenous melatonin. Brain Res. Bull. 57 (5), 603–611.

Cantwell, E.L., Cassone, V.M., 2006a. Chicken suprachiasmatic nuclei: I. Efferent and afferent connections. J. Comp. Neurol. 496 (1), 97–120.

Cantwell, E.L., Cassone, V.M., 2006b. Chicken suprachiasmatic nuclei: II. Autoradiographic and immunohistochemical analysis. J. Comp. Neurol. 499 (3), 442–457.

Cassone, V.M., 1988. Circadian variation of [14C]2-deoxyglucose uptake within the suprachiasmatic nucleus of the house sparrow, *Passer domesticus*. Brain Res. 459 (1), 178–182.

Cassone, V.M., 1990. Melatonin: time in a bottle. Oxford Rev. Reprod. Biol. 12, 319–367.

Cassone, V.M., Menaker, M., 1983. Sympathetic regulation of chicken pineal rhythms. Brain Res. 272 (2), 311–317.

Cassone, V.M., Menaker, M., 1984. Is the avian circadian system a neuroendocrine loop? J. Exp. Zool. 232 (3), 539–549.

Cassone, V.M., Moore, R.Y., 1987. Retinohypothalamic projection and suprachiasmatic nucleus of the house sparrow, *Passer domesticus*. J. Comp. Neurol. 266 (2), 171–182.

Cassone, V.M., Westneat, D.F., 2012. The bird of time: cognition and the avian biological clock. Front. Mol. Neurosci. 5, 32.

Cassone, V.M., Takahashi, J.S., Blaha, C.D., Lane, R.F., Menaker, M., 1986. Dynamics of noradrenergic circadian input to the chicken pineal gland. Brain Res. 384 (2), 334–341.

Cassone, V.M., Forsyth, A.M., Woodlee, G.L., 1990. Hypothalamic regulation of circadian noradrenergic input to the chick pineal gland. J. Comp. Physiol. A 167 (2), 187–192.

Cassone, V.M., Brooks, D.S., Hodges, D.B., Kelm, T.S., Lu, J., Warren, W.S., 1992. Integration of circadian and visual function in mammals and birds: brain imaging and the role of melatonin in biological clock regulation. In: Gonzalez-Lima, F., Finkenstaedt, T., Scheich, H. (Eds.), Advances in Metabolic Mapping Techniques for Brain Imaging of Behavioral and Learning Functions. Kluwer Academic Publishers, Dordrecht/Boston/London, pp. 299–318.

Cassone, V.M., Brooks, D.S., Kelm, T.A., 1995. Comparative distribution of 2[^{125}I]iodomelatonin binding in the brains of diurnal birds: outgroup analysis with turtles. Brain Behav. Evol. 45 (5), 241–256.

Cassone, V.M., Bartell, P.A., Earnest, B.J., Kumar, V., 2008. Duration of melatonin regulates seasonal changes in song control nuclei of the house sparrow, *Passer domesticus*: independence from gonads and circadian entrainment. J. Biol. Rhythms 23 (1), 49–58.

Hau, M., Gwinner, E., 1995. Continuous melatonin administration accelerates resynchronization following phase shifts of a light-dark cycle. Physiol. Behav. 58 (1), 89–95.

Heigl, S., Gwinner, E., September 1995. Synchronization of circadian rhythms of house sparrows by oral melatonin: effects of changing period. J. Biol. Rhythms 10 (3), 225–233.

Chabot, C.C., Menaker, M., 1992. Effects of physiological cycles of infused melatonin on circadian rhythmicity in pigeons. J. Comp. Physiol. A 170 (5), 615–622.

Chaurasia, S.S., Rollag, M.D., Jiang, G., Hayes, W.P., Haque, R., Natesan, A., Zatz, M., Tosini, G., Liu, C., Korf, H.W., Iuvone, P.M., Provencio, I., 2005. Molecular cloning, localization and circadian expression of chicken melanopsin (Opn4): differential regulation of expression in pineal and retinal cell types. J. Neurochem. 92 (1), 158–170.

Chong, N.W., Cassone, V.M., Bernard, M., Klein, D.C., Iuvone, P.M., 1998. Circadian expression of tryptophan hydroxylase mRNA in the chicken retina. Brain Res. Mol. Brain Res. 61 (1–2), 243–250.

Chong, N.W., Chaurasia, S.S., Haque, R., Klein, D.C., Iuvone, P.M., 2003. Temporal-spatial characterization of chicken clock genes: circadian expression in retina, pineal gland, and peripheral tissues. J. Neurochem. 85 (4), 851–860.

Csernus, V., Faluhelyi, N., Nagy, A.D., 2005. Features of the circadian clock in the avian pineal gland. Ann. N. Y. Acad. Sci. 1040, 281–287.

Davies, W.L., Hankins, M.W., Foster, R.G., 2010. Vertebrate ancient opsin and melanopsin: divergent irradiance detectors. Photochem. Photobiol. Sci. 9 (11), 1444–1457.

Davies, W.I., Turton, M., Peirson, S.N., Follett, B.K., Halford, S., Garcia-Fernandez, J.M., Sharp, P.J., Hankins, M.W., Foster, R.G., 2012. Vertebrate ancient opsin photopigment spectra and the avian photoperiodic response. Biol. Lett. 8 (2), 291–294.

Deguchi, T., 1979. Circadian rhythm of serotonin N-acetyltransferase activity in organ culture of chicken pineal gland. Science 203 (4386), 1245–1247.

Doyle, S., Menaker, M., 2007. Circadian photoreception in vertebrates. Cold Spring Harb. Symp. Quant. Biol. 72, 499–508.

Dubocovich, M.L., Takahashi, J.S., 1987. Use of 2-[125I]iodomelatonin to characterize melatonin binding sites in chicken retina. Proc. Natl. Acad. Sci. U. S. A. 84 (11), 3916–3920.

Ebihara, S., Kawamura, H., 1981. The role of the pineal organ and the suprachiasmatic nucleus in the control of circadian locomotor rhythms in the Java sparrow, *Padda oryzivora*. J. Comp. Physiol. 141, 207–214.

Ebihara, S., Uchiyama, K., Oshima, I., 1984. Circadian organization in the pigeon, *Columba livia*: the role of the pineal organ and the eye. J. Comp. Physiol. A 154, 59–69.

Ebihara, S., Adachi, A., Hasegawa, M., Nogi, T., Yoshimura, T., Hirunagi, K., 1997. In vivo microdialysis studies of pineal and ocular melatonin rhythms in birds. Biol. Signals 6 (4–6), 233–240.

Emery, P., So, W.V., Kaneko, M., Hall, J.C., Rosbash, M., 1998. CRY, a Drosophila clock and light-regulated cryptochrome, is a major contributor to circadian rhythm resetting and photosensitivity. Cell 95 (5), 669–679.

Gahr, M., Kosar, E., 1996. Identification, distribution, and developmental changes of a melatonin binding site in the song control system of the zebra finch. J. Comp. Neurol. 367 (2), 308–318.

Gaston, S., Menaker, M., June 7, 1968. Pineal function: the biological clock in the sparrow? Science 160 (3832), 1125–1127.

Geerts, B., 2002. Empirical estimation of the monthly-mean daily temperature range. Theor. Appl. Climatol., 1–20.

Guido, M.E., Garbarino-Pico, E., Contin, M.A., Valdez, D.J., Nieto, P.S., Verra, D.M., Acosta-Rodriguez, V.A., de Zavalía, N., Rosenstein, R.E., 2010. Inner retinal circadian clocks and non-visual photoreceptors: novel players in the circadian system. Prog. Neurobiol. 92 (4), 484–504.

Gwinner, E., 1978. Effects of pinealectomy on circadian locomotor activity rhythms in European starlings, *Sturnus vulgaris*. J. Comp. Physiol. 126, 123–129.

Gwinner, E., 1989. Melatonin and the circadian system of birds: model of internal resonance. In: Hiroshige, T., Honma, K. (Eds.), Circadian Clocks and Ecology. Hokkado Univ Press, Sapporo, pp. 127–145.

Gwinner, E., Brandstätter, R., 2001. Complex bird clocks. Philos. Trans. R. Soc. Lond., B, Biol. Sci. 356 (1415), 1801–1810.

Gwinner, E., Subbaraj, R., Bluhm, C.K., Gerkema, M., 1987. Differential effects of pinealectomy on circadian rhythms of feeding and perch hopping in the European starling. J. Biol. Rhythms 2 (2), 109–120.

Gwinner, E., Hau, M., Heigl, S., 1997. Melatonin: generation and modulation of avian circadian rhythms. Brain Res. Bull. 44 (4), 439–444.

Halford, S., Pires, S.S., Turton, M., Zheng, L., González-Menéndez, I., Davies, W.L., Peirson, S.N., García-Fernández, J.M., Hankins, M.W., Foster, R.G., 2009. VA opsin-based photoreceptors in the hypothalamus of birds. Curr. Biol. 19 (16), 1396–1402.

Haque, R., Ali, F.G., Biscoglia, R., Abey, J., Weller, J., Klein, D., Iuvone, P.M., June 2010. CLOCK and NPAS2 have overlapping roles in the circadian oscillation of arylalkylamine N-acetyltransferase mRNA in chicken cone photoreceptors. J. Neurochem. 113 (5), 1296–1306.

Hardy, D.R., Vuille, M., Braun, C., Kemig, F., Bradley, R.S., 1998. Annual and daily meteorological cycles at high altitude on a tropical mountain. Bull. Am. Meteor. Soc. 79, 1899–1913.

Hau, M., Gwinner, E., 1992. Circadian entrainment by feeding cycles in house sparrows, *Passer domesticus*. J. Comp. Physiol. A 170 (4), 403–409.

Helfer, G., Fidler, A.E., Vallone, D., Foulkes, N.S., Brandstaetter, R., 2006. Molecular analysis of clock gene expression in the avian brain. Chronobiol. Int. 23 (1–2), 113–127.

Jansen, R., Metzdorf, R., van der Roest, M., Fusani, L., ter Maat, A., Gahr, M., 2005. Melatonin affects the temporal organization of the song of the zebra finch. FASEB J. 19 (7), 848–850.

Juss, T.S., Davies, I.R., Follett, B.K., Mason, R., 1994. Circadian rhythm in neuronal discharge activity in the quail lateral hypothalamic retinorecipient nucleus (LHRN) recorded in vitro. J. Physiol. 475, 132.

Karaganis, S.P., Kumar, V., Beremand, P.D., Bailey, M.J., Thomas, T.L., Cassone, V.M., 2008. Circadian genomics of the chick pineal gland in vitro. BMC Genomics 9, 206. http://dx.doi.org/10.1186/1471-2164-9-206.

Karaganis, S.P., Bartell, P.A., Shende, V.R., Moore, A.F., Cassone, V.M., 2009. Modulation of metabolic and clock gene mRNA rhythms by pineal and retinal circadian oscillators. Gen. Comp. Endocrinol. 161 (2), 179–192.

King, V.M., Follett, B.K., 1997. c-fos expression in the putative avian suprachiasmatic nucleus. J. Comp. Physiol. A 180 (5), 541–551.

Klein, D.C., Coon, S.L., Roseboom, P.H., Weller, J.L., Bernard, M., Gastel, J.A., Zatz, M., Iuvone, P.M., Rodriguez, I.R., Bégay, V., Falcón, J., Cahill, G.M., Cassone, V.M., Baler, R., 1997. The melatonin rhythm-generating enzyme: molecular regulation of serotonin N-acetyltransferase in the pineal gland. Recent Prog. Horm. Res. 52, 307–357.

Konopka, R.J., Benzer, S., 1971. Clock mutants of Drosophila melanogaster. Proc. Natl. Acad. Sci. U. S. A. 68 (9), 2112–2116.

Kubo, Y., Akiyama, M., Fukada, Y., Okano, T., 2006. Molecular cloning, mRNA expression, and immunocytochemical localization of a putative blue-light photoreceptor CRY4 in the chicken pineal gland. J. Neurochem. 97 (4), 1155–1165.

Kumar, V., Follett, B.K., 1993. The circadian nature of melatonin secretion in Japanese quail (*Coturnix coturnix japonica*). J. Pineal Res. 14 (4), 192–200.

Kumar, V., Gwinner, E., 2005. Pinealectomy shortens resynchronisation times of house sparrow (*Passer domesticus*) circadian rhythms. Naturwissenschaften 92 (9), 419–422.

Liu, F., Yuan, H., Sugamori, K.S., Hamadanizadeh, A., Lee, F.J., Pang, S.F., Brown, G.M., Pristupa, Z.B., Niznik, H.B., 1995. Molecular and functional characterization of a partial cDNA encoding a novel chicken brain melatonin receptor. FEBS Lett. 374 (2), 273–278.

Lu, J., Cassone, V.M., 1993a. Pineal regulation of circadian rhythms of 2-deoxy[^{14}C]glucose uptake and 2[^{125}I]iodomelatonin binding in the visual system of the house sparrow, *Passer domesticus*. J. Comp. Physiol. A 173, 765–774.

Lu, J., Cassone, V.M., 1993b. Daily melatonin administration synchronizes circadian patterns of brain metabolism and behavior in pinealectomized house sparrows, *Passer domesticus*. J. Comp. Physiol. A 173, 775–782.

Masuda, H., Oishi, T., Ohtani, M., Michinomae, M., Fukada, Y., Shichida, Y., Yoshizawa, T., 1994. Visual pigments in the pineal complex of the Japanese quail, Japanese grass lizard and bullfrog: immunocytochemistry and HPLC analysis. Tissue Cell 26 (1), 101–113.

Max, M., McKinnon, P.J., Seidenman, K.J., Barrett, R.K., Applebury, M.L., Takahashi, J.S., Margolskee, R.F., 1995. Pineal opsin: a nonvisual opsin expressed in chick pineal. Science 267 (5203), 1502–1506.

Menaker, M., 1968. Extraretinal light perception in the sparrow. I. Entrainment of the biological clock. Proc. Natl. Acad. Sci. U. S. A. 59 (2), 414–421.

Menaker, M., 1985. Eyes – the second (and third) pineal glands? Ciba Found Symp. 117, 78–92.

Menaker, M., Eskin, A., 1966. Entrainment of circadian rhythms by sound in *Passer domesticus*. Science 154 (3756), 1579–1581.

Menaker, M., Oksche, A., 1974. The avian pineal. In: Avian Biology, vol. IV, Academic Press, NY, pp. 80–114.

Menaker, M., Underwood, H., 1976. Extraretinal photoreception in birds. Photophysiology 23 (4), 299–306.

Menaker, M., Takahashi, J.S., Eskin, A., 1978. The physiology of circadian pacemakers. Annu. Rev. Physiol. 40, 501–526.

Nakane, Y., Yoshimura, T., 2010. Deep brain photoreceptors and a seasonal signal transduction cascade in birds. Cell Tissue Res. 342 (3), 341–344.

Nakane, Y., Ikegami, K., Ono, H., Yamamoto, N., Yoshida, S., Hirunagi, K., Ebihara, S., Kubo, Y., Yoshimura, T., August 24, 2010. A mammalian neural tissue opsin (Opsin 5) is a deep in birds. Proc. Natl. Acad. Sci. U. S. A. 107 (34), 15264–15268.

Natesan, A.K., Cassone, V.M., 2002. Melatonin receptor mRNA localization and rhythmicity in the retina of the domestic chick, *Gallus domesticus*. Vis. Neurosci. 19 (3), 265–274.

Natesan, A., Geetha, L., Zatz, M., 2002. Rhythm and soul in the avian pineal. Cell Tissue Res. 309 (1), 35–45.

Navara, K.J., Nelson, R.J., 2007. The dark side of light at night: physiological, epidemiological, and ecological consequences. J. Pineal Res. 43 (3), 215–224.

Nishijima, K., Iijima, S., January 2013. Transgenic chickens. Dev. Growth Differ. 55 (1), 207–216. http://dx.doi.org/10.1111/dgd.12032 (Epub December 26, 2012).

Okano, T., Fukada, Y., 2000. Photoreceptors in pineal gland and brain: cloning, localization, and overexpression. Methods Enzymol. 316, 278–291.

Okano, T., Yoshizawa, T., Fukada, Y., 1994. Pinopsin is a chicken pineal photoreceptive molecule. Nature 372 (6501), 94–97.

Oksche, A., Kirschstein, H., Kobayashi, H., Farner, D.S., 1972. Electron microscopic and experimental studies of the pineal organ in the white-crowned sparrow, *Zonotrichia leucophrys gambelii*. Z. Zellforsch Mikrosk Anat. 124 (2), 247–274.

Panda, S., 2007. Multiple photopigments entrain the Mammalian circadian oscillator. Neuron 53 (5), 619–621.

Peirson, S.N., Halford, S., Foster, R.G., 2009. The evolution of irradiance detection: melanopsin and the non-visual opsins. Philos. Trans. R. Soc. Lond., B, Biol. Sci. 364 (1531), 2849–2865.

Peters, J.L., Cassone, V.M., Zoran, M.J., 2005. Melatonin modulates intercellular communication among cultured chick astrocytes. Brain Res. 1031 (1), 10–19.

Pittendrigh, C.S., 1993. Temporal organization: reflections of a Darwinian clock-watcher. Annu. Rev. Physiol. 55, 16–54.

Poynter, G., Lansford, R., 2008. Generating transgenic quail using lentiviruses. Methods Cell Biol. 87, 281–293.

Quay, W.B., 1965. Histological structure and cytology of the pineal gland in birds and mammals. Prog. Brain Res. 10, 49–86.

Rani, S., Singh, S., Malik, S., Singh, J., Kumar, V., 2009. Synchronization of Indian weaver bird circadian rhythms to food and light zeitgebers: role of pineal. Chronobiol. Int. 26 (4), 653–665.

Rashotte, M.E., Stephan, F.K., 1996. Coupling between light- and food-entrainable circadian oscillators in pigeons. Physiol. Behav. 59 (4–5), 1005–1010.

Reierth, E., Stokkan, K.A., 1998. Dual entrainment by light and food in the Svalbard ptarmigan (*Lagopus mutus hyperboreus*). J. Biol. Rhythms 13 (5), 393–402.

Rensing, L., Ruoff, P., 2002. Temperature effect on entrainment, phase shifting, and amplitude of circadian clocks and its molecular bases. Chronobiol. Int. 19 (5), 807–864.

Reppert, S.M., 1997. Melatonin receptors: molecular biology of a new family of G protein-coupled receptors. J. Biol. Rhythms 12 (6), 528–531.

Reppert, S.M., Weaver, D.R., 2002. Coordination of circadian timing in mammals. Nature 418 (6901), 935–941.

Reppert, S.M., Weaver, D.R., Cassone, V.M., Godson, C., Kolakowski Jr, L.F., 1995. Melatonin receptors are for the birds: molecular analysis of two receptor subtypes differentially expressed in chick brain. Neuron 15 (5), 1003–1015.

Rivkees, S.A., Cassone, V.M., Weaver, D.R., Reppert, S.M., 1989. Melatonin receptors in chick brain: characterization and localization. Endocrinology 125 (1), 363–368.

Roenneberg, T., Hut, R., Daan, S., Merrow, M., 2010. Entrainment concepts revisited. J. Biol. Rhythms 25 (5), 329–339.

Rosbash, M., Bradley, S., Kadener, S., Li, Y., Luo, W., Menet, J.S., Nagoshi, E., Palm, K., Schoer, R., Shang, Y., Tang, C.H., 2007. Transcriptional feedback and definition of the circadian pacemaker in Drosophila and animals. Cold Spring Harb. Symp. Quant. Biol. 72, 75–83.

Sato, T., Ebisawa, S., 1988. A pineal ganglion associated with the pineal tract in the domestic fowl. Cell Tissue Res. 252 (2), 287–292.

Semple-Rowland, S.L., Cheng, K.M., 1999. rd and rc chickens carry the same GC1 null allele (GUCY1*). Exp. Eye Res. 69 (5), 579–581.

Semple-Rowland, S.L., Coggin, W.E., Geesey, M., Eccles, K.S., Abraham, L., Pachigar, K., Ludlow, R., Khani, S.C., Smith, W.C., 2010. Expression characteristics of dual-promoter lentiviral vectors targeting retinal photoreceptors and Müller cells. Mol. Vis. 16, 916–934.

Steele, C.T., Tosini, G., Siopes, T., Underwood, H., 2006. Time keeping by the quail's eye: circadian regulation of melatonin production. Gen. Comp. Endocrinol. 145 (3), 232–236.

Takahashi, J.S., Menaker, M., 1982. Role of the suprachiasmatic nuclei in the circadian system of the house sparrow, *Passer domesticus*. J. Neurosci. 2 (6), 815–828.

Takahashi, J.S., Hamm, H., Menaker, M., 1980. Circadian rhythms of melatonin release from individual superfused chicken pineal glands in vitro. Proc. Natl. Acad. Sci. U. S. A. 77 (4), 2319–2322.

Turek, F.W., McMillan, J.P., Menaker, M., 1976. Melatonin: effects on the circadian locomotor rhythm of sparrows. Science 194 (4272), 1441–1443.

Ueck, M., 1979. Innervation of the vertebrate pineal. Prog. Brain Res. 52, 45–88.

Underwood, H., Siopes, T., 1984. Circadian organization in Japanese quail. J. Exp. Zool. 232, 557–566.

Underwood, H., Binkley, S., Siopes, T., Mosher, K., 1984. Melatonin rhythms in the eyes, pineal bodies, and blood of Japanese quail (*Coturnix coturnix japonica*). Gen. Comp. Endocrinol. 56 (1), 70–81.

Valdez, D.J., Nieto, P.S., Garbarino-Pico, E., Avalle, L.B., Díaz-Fajreldines, H., Schurrer, C., Cheng, K.M., Guido, M.E., 2009. A nonmammalian vertebrate model of blindness reveals functional photoreceptors in the inner retina. FASEB J. 23 (4), 1186–1195.

Valdez, D.J., Nieto, P.S., Díaz, N.M., Garbarino-Pico, E., Guido, M.E., 2013. Differential regulation of feeding rhythms through a multiple-photoreceptor system in an avian model of blindness. FASEB J. (Epub ahead of print).

Wada, Y., Okano, T., Adachi, A., Ebihara, S., Fukada, Y., March 6, 1998. Identification of rhodopsin in the pigeon deep brain. FEBS Lett. 424 (1–2), 53–56.

Wang, G., Wingfield, J.C., 2011. Immunocytochemical study of rhodopsin-containing putative encephalic photoreceptors in house sparrow, *Passer domesticus*. Gen. Comp. Endocrinol. 170 (3), 589–596.

Wang, G., Harpole, C.E., Trivedi, A.K., Cassone, V.M., 2012. Circadian regulation of bird song call, and locomotor behavior by pineal melatonin in the zebra finch. J. Biol. Rhythms 27 (2), 145–155.

Whitfield-Rucker, M.G., Cassone, V.M., 1996. Melatonin binding in the house sparrow song control system: sexual dimorphism and the effect of photoperiod. Horm. Behav. 30 (4), 528–537.

Yasuo, S., Yoshimura, T., Bartell, P.A., Iigo, M., Makino, E., Okabayashi, N., Ebihara, S., 2002. Effect of melatonin administration on qPer2, qPer3, and qClock gene expression in the suprachiasmatic nucleus of Japanese quail. Eur. J. Neurosci. 16 (8), 1541–1546.

Yoshimura, T., Suzuki, Y., Makino, E., Suzuki, T., Kuroiwa, A., Matsuda, Y., Namikawa, T., Ebihara, S., 2000. Molecular analysis of avian circadian clock genes. Brain Res. Mol. Brain Res. 78 (1–2), 207–215.
Yoshimura, T., Yasuo, S., Suzuki, Y., Makino, E., Yokota, Y., Ebihara, S., 2001. Identification of the suprachiasmatic nucleus in birds. Am. J. Physiol. Regul. Integr. Comp. Physiol. 280 (4), R1185–R1189.
Zatz, M., Mullen, D.A., 1988. Norepinephrine, acting via adenylate cyclase, inhibits melatonin output but does not phase-shift the pacemaker in cultured chick pineal cells. Brain Res. 450 (1–2), 137–143.
Zatz, M., Mullen, D.A., Moskal, J.R., 1988. Photoendocrine transduction in cultured chick pineal cells: effects of light, dark, and potassium on the melatonin rhythm. Brain Res. 438 (1–2), 199–215.
Zeman, M., Szántóová, K., Herichová, I., 2009. Ontogeny of circadian oscillations in the heart and liver in chicken. Comp. Biochem. Physiol. A Mol. Integr. Physiol. 154 (1), 78–83.
Zimmerman, N.H., Menaker, M., 1979. The pineal gland: a pacemaker within the circadian system of the house sparrow. Proc. Natl. Acad. Sci. U. S. A. 76 (2), 999–1003.

Chapter 35

Circannual Cycles and Photoperiodism

Vincent M. Cassone
Department of Biology, University of Kentucky, Lexington, KY, USA

Takashi Yoshimura
Laboratory of Animal Physiology, Graduate School of Bioagricultural Sciences, Nagoya University, Furo-cho, Chikusa-ku, Nagoya, Japan, Avian Bioscience Research Center, Graduate School of Bioagricultural Sciences, Nagoya University, Furo-cho, Chikusa-ku, Nagoya, Japan

In addition to the effects of daily patterns on physiology and behavior, birds are also very sensitive to annual patterns in the abiotic and biotic environments. This is particularly true for species of birds in circumpolar and temperate zones, but annual cycles of avian physiology are also evident in subtropical and tropical zones of the Earth. In all cases, these annual cycles derive primarily from annual changes in solar irradiance due to the axial tilt of the Earth's rotation and the elliptical orbit of the Earth around the sun, resulting in seasons (Pianka, 1978). In addition, long-term trends in climate, such as global climate change, whether human-derived or astronomical, may have significant effects on avian physiology (Lockwood, 2010).

35.1 ANNUAL CYCLES

35.1.1 Abiotic

35.1.1.1 Photoperiod

Seasons result from the annual revolution of the Earth around the sun and its axial tilt (Figure 35.1). When the Earth's axial tilt is parallel with the trajectory of the Earth's movement relative to the sun, all surfaces receive equal amounts of solar irradiance—12 h of light and 12 h of darkness per 24 h day (LD 12:12); these days are the equinoxes, occurring on March 20 and September 22. The duration of the day phase is the photoperiod, whereas the duration of the night phase is the scotoperiod. In the northern hemisphere, this is the vernal equinox (the beginning of spring), while in the southern hemisphere, it is the autumnal equinox (the beginning of autumn).

As the Earth progresses in its orbit from March to June, the axial tilt exposes the northern hemisphere to more direct and longer durations (photoperiod) of solar illumination. Conversely, the southern hemisphere undergoes less direct and shorter photoperiods until June 21 of each year, when the Earth's axial tilt points the northern hemisphere toward the sun, perpendicular to its orbit. This is the solstice. In the northern hemisphere, it is the longest day, while in the southern, it is the shortest, corresponding to the beginning of summer and winter, respectively. As the Earth then progresses from June to September, the seasons progress to the autumnal equinox in the north and the vernal equinox in the south on September 22. As the Earth progresses from September to December, the Earth's axial tilt once again becomes perpendicular to the Earth's orbit. However, this time, the axial tilt exposes the southern hemisphere to direct light and longer photoperiods, creating the southern summer solstice and conversely plunging the northern hemisphere into the winter solstice.

On Earth, a major result of these astronomical progressions is the annual change in photoperiod, which is in turn dependent on latitude (Figure 35.2). In contrast to the situation in the laboratory in which lights are turned on and off in a timed square-wave, described below, natural photoperiod depends on the definition of global illuminance one defines as twilight, in which the zenith angle of the sun ranges from 85° as in civil twilight (the sun is just below the horizon) to 105° as in astronomical twilight (the sun is 12° below the horizon). At the equator, photoperiod at civil twilight remains static at LD 12:12 throughout the year, but as latitude increases toward the poles, the annual change in photoperiod increases in amplitude. At the Tropics of Cancer and Capricorn (23.4378°), the photoperiod at summer solstice is approximately LD 14:10, while the winter solstice is LD 10:14. At more temperate New York, NY, USA (43.5°), the photoperiod at summer solstice is approximately LD 15:9 with a winter solstice of LD 9:15, while the range of photoperiods above the Arctic Circle (66.6°) is from LD 24:0 to LD

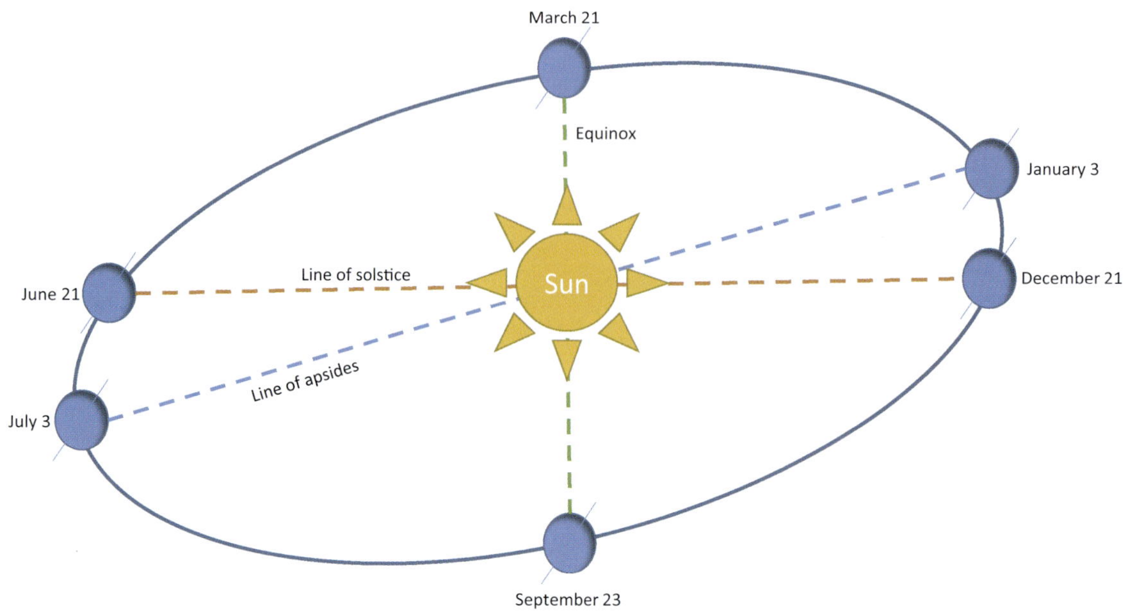

FIGURE 35.1 **The astronomical basis for the seasons on Earth.** Earth's rotation on its axis orients the poles asymmetrically toward or away from the sun depending on the time of year.

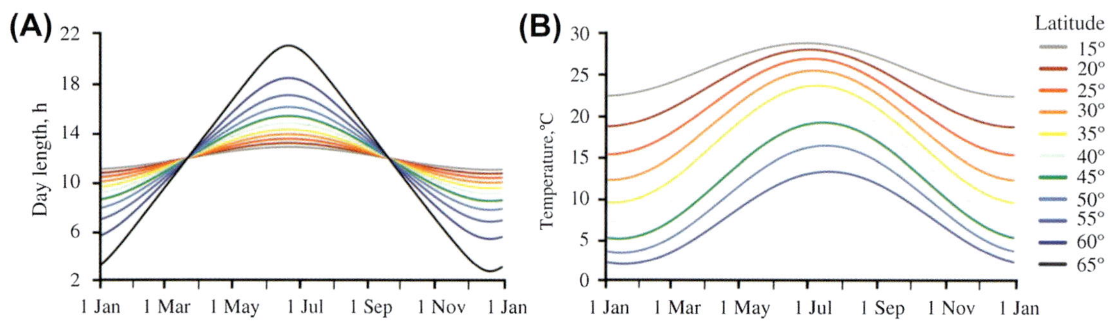

FIGURE 35.2 (A) Annual changes in photoperiod depend upon the latitude in which the amplitude of photoperiodic change is greatest at higher latitudes. (B) Similarly, the seasonal changes in ambient temperature are also dependent on latitude. However, the average temperature at lower latitudes is higher than at higher latitudes. *From Wilczek et al. (2010).*

0:24. Thus, as latitude increases from the equator to the poles, the range of photoperiod dramatically increases over the year.

35.1.1.2 Temperature

One of the major consequences of changing photoperiod as seasons progress is a parallel change in ambient temperature (T_A). Average annual ranges in T_A (ΔT_A), the differences between the coldest month from the hottest month of the year (Glickman, 2000), increase with increases in latitude such that ΔT_A is 0 at the equator to 40 °C at 90° latitude in the northern hemisphere and 31 °C in the south. The principal cause of the asymmetry in ΔT_A between north and south is that a higher percentage of Earth's surface is covered by ocean in the south, absorbing much of the heat. Earth's elliptical orbit in part moderates this difference, because the Earth is closest to the sun in January (Figure 35.1) during the southern summer and northern winter.

In addition to latitude, altitude can dramatically affect T_A as well as ΔT_A. T_A decreases on average 6.4 °C/km altitude, a phenomenon defined as the lapse rate (Glickman, 2000). The lapse rate depends on several variables, however, especially humidity, which decreases lapse rate. For example, an average dry lapse rate is 10 °C/km altitude, while wet lapse rates average 5.5 °C/km altitude. Other abiotic factors that affect T_A and ΔT_A include urbanization, proximity to coasts (continentality), and precipitation.

35.1.1.3 Precipitation and Wind

In addition to annual cycles of photoperiod and temperature, precipitation patterns can also provide temporal cues to birds and other organisms (Pianka, 1978; Glickman, 2000). This is especially the case in subtropical and tropical zones, which experience predictable wet and dry seasons. The intertropical convergence zone results from the Coriolis effect from the

Earth's rotation as well as the evaporation of oceanic water in equatorial areas. Variation in the intertropical convergence zone affects rainfall in many equatorial regions, resulting in predictable wet seasons and dry seasons in subtropical and tropical zones as opposed to the cold and warm seasons at higher latitudes. For example, in tropical Queensland, Australia, precipitation ranges from 65% humidity and 400 mm rain/month the wet season (December–March) to only 17% humidity and 25 mm rain/month in the dry season (June–October). In West Africa, the rainy season extends from April to July, while the dry season falls between October and March. The absolute times of wet and dry seasons are not as predictable globally as is photoperiod, because other factors such as wind direction, mountains, and other physical barriers affect them. However, they tend to be locally predictable, such that many organisms' reproductive processes are synchronized to these abiotic factors.

35.1.2 Biotic

Due to the fact that many abiotic factors vary depending on time of year, it is not surprising that aspects of birds' biotic environment are equally variable. These factors include, but are not limited to, presence and absence of food, competitors, predators and mates. As a consequence, selective pressures for systems to tune physiological processes accurately to the timing of these biotic factors are very high (Pianka, 1978).

35.2 ANNUAL CYCLES OF BIRDS

Birds' life history strategies must adapt to and predict these changing environmental factors. These life history strategies can be complex and species-specific, and the timing of the processes will determine a bird's survival and fitness, perhaps more than any other vertebrate taxon. For an individual and species to survive, time must be allocated to each essential process such as reproduction, migration, molt, and other processes that may interfere with each other (King, 1974). These processes are coordinated such that life history stages are timed to an appropriate time of year relative both to the external environmental conditions described above and among the diverse processes that are coordinated. For example, annual patterns of breeding, migration, body mass, and multiple molt patterns in long-distance migratory birds such as red knots (*Calidrus canutus*) and ruffs (*Philomachus pugnax*) are orchestrated such that each stage of the species' annual cycle occurs at the appropriate time (Helm et al., 2012, Figure 35.3). Thus, birds exhibit hyperphagia and fatten up in anticipation of the vernal migration from southern hemisphere wintering grounds to northern hemisphere breeding grounds, where courtship, sexual behavior, and reproduction occur. Multiple molts occur to replace flight feathers following migratory stints, while breast molt occurs as incubation pads are prepared for incubation of young. Phenological analyses (the study of the temporal aspects of recurrent natural phenomena and their relation to weather and climate of annual cycles; Lincoln et al., 1998) of multiple species of birds indicate similarly complex life history cycles. Obviously, transequatorial migrants, such as ruffs and red knots, exhibit more complex cycles, but these temporal programs are common among avian species.

35.3 CIRCANNUAL RHYTHMS

35.3.1 Circannual Rhythms in the Laboratory

How do birds synchronize their internal biological processes to prevailing environmental cycles? As with circadian rhythms, many physiological and behavioral rhythms in birds exhibit circannual rhythms (Gwinner, 2003). Circannual rhythms can only be demonstrated in the laboratory. Typically, birds are maintained continuously in equatorial photoperiods (LD 12:12) for several years. Under these conditions, many birds will express a free-running rhythm of approximately 1 year (hence, circannual), with periods of slightly less than or more than 365 days. As an example, circannual rhythms of testicular activity and molt in a single African stonechat, *Saxicola torquata axillaris*, maintained in captivity for up to 12 years in LD 12:12 persisted with a period, τ, of approximately 9 months until the bird died, presumably of old age (Figure 35.4; Gwinner, 2003). The expression of circannual rhythms is a widespread phenomenon among avian species, especially among species that experience transequatorial migration (Rani and Kumar, 2013), but circannual patterns are highly variable among species and among individuals within species (Hazlerigg and Loudon, 2008).

The mechanisms by which circannual rhythms are generated, however, are not known. System-level and molecular aspects of circadian clocks appear to interact with the expression of circannual rhythms, but they do not appear to be necessary. For example, pinealectomy (PINX) or constant light (LL), which abolishes circadian patterns of perch-hopping in the spotted munia, *Lonchura punctulata,* as well as other species of oscine passerine birds does not affect circannual patterns of testicular growth and regression or premigratory fattening (Pant and Chandola-Saklani, 1992). Conversely, PINX of migratory white-crowned sparrows, *Zonotrichia leuchophrys*, abolishes both circadian patterns of perch-hopping behavior and circannual patterns of nocturnal migratory restlessness, or *Zugunruhe* (see below; McMillan, 1970). At the molecular level, studies focusing on allelic differences in C-terminal polyglutamine repeats in the CLOCK protein in breeding and migratory patterns of barn swallows, *Hirundo rustica*, suggest a negative selection on "deviant" genotypes (Caprioli et al., 2012; Saino et al., 2013). Even so, other studies in several different swallow species in the genus *Tachycineta* failed to show similar associations (Dor et al., 2012).

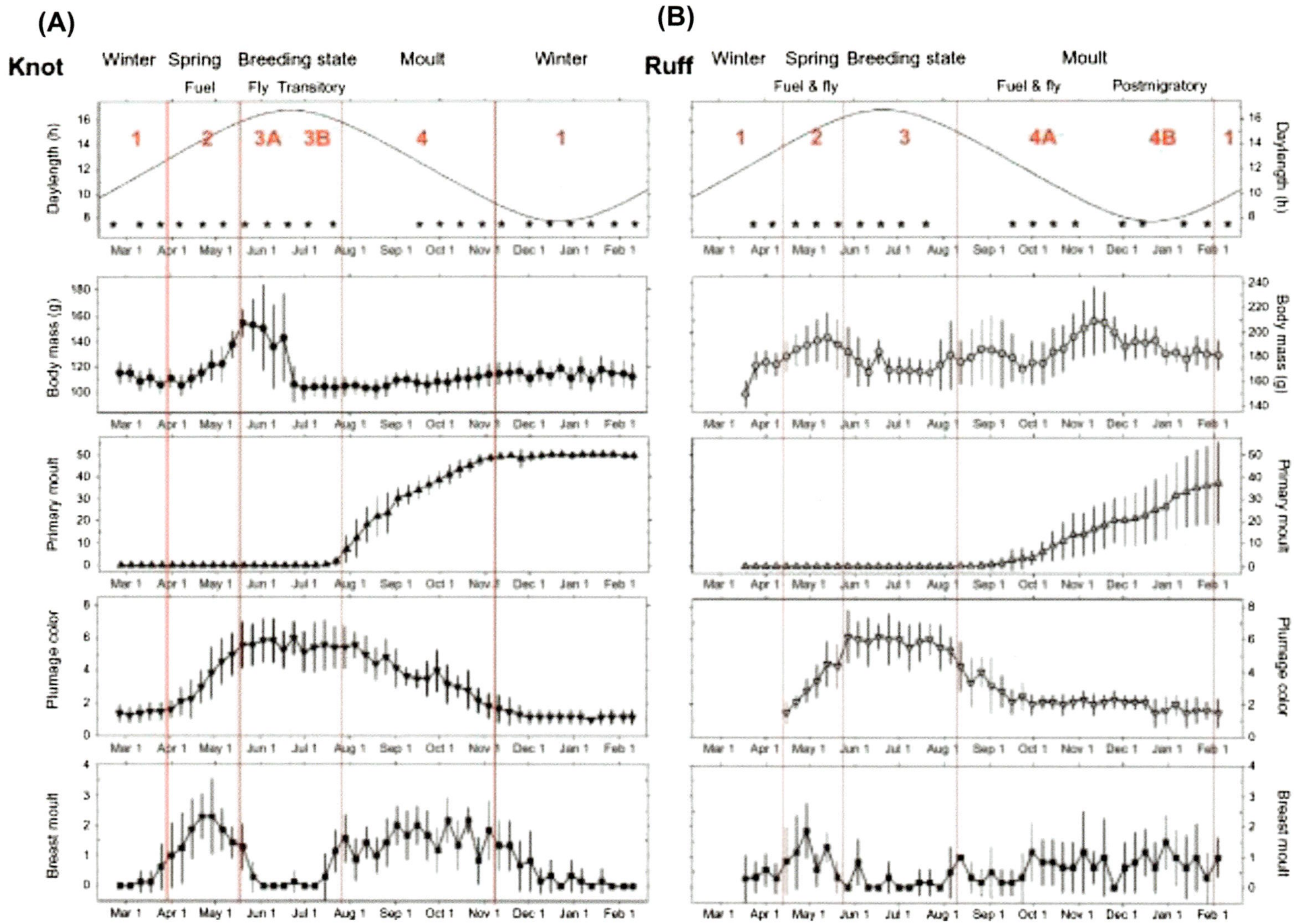

FIGURE 35.3 **The complex annual life histories of two species of wading birds.** (A) The knot, *Calidrus canutus*; and (B) the ruff, *Philomachus pugnax*. Annual changes in body mass, vernal and autumnal molt, and plumage coloration occur in coordinated fashion in species-specific fashion. *From Helm et al. (2012).*

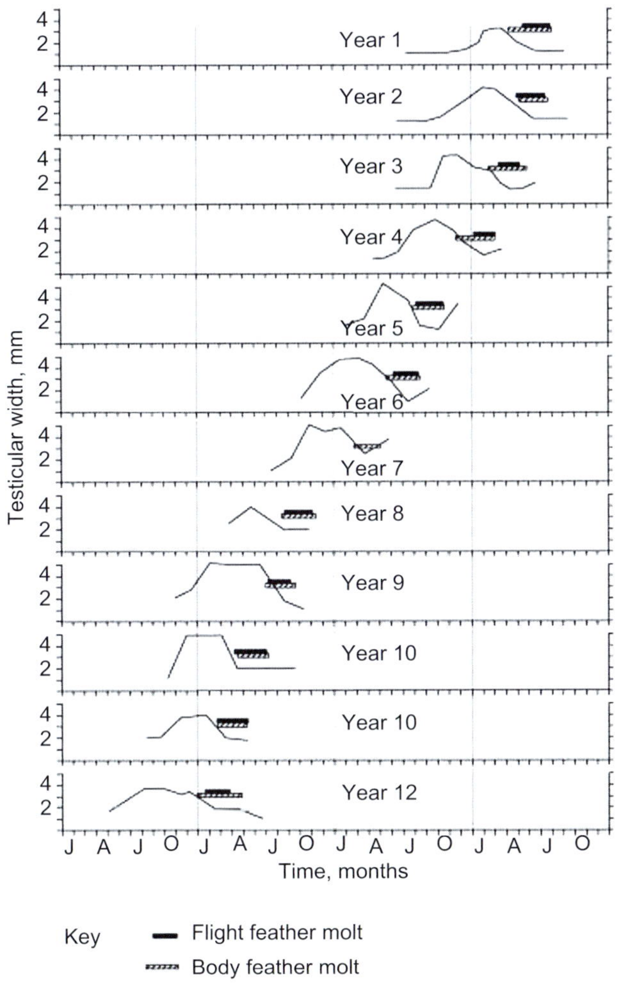

FIGURE 35.4 **Circannual cycles of a single African stonechat, *Saxicola torquata axillaris*, of testicular size and molt through 12 consecutive years maintained in constant LD 12:12.** *From Gwinner (2003).*

35.3.2 Synchronization of Circannual Rhythms to Environmental Cues

The expression of circannual patterns whose τ are different from 365 days in laboratory conditions but exactly 365 days under natural conditions suggests that, like circadian rhythms, circannual rhythms must be synchronized or entrained to prevailing environmental cues (Gwinner, 2003; Rani and Kumar 2013). In most cases, alternating cycles of increasing and decreasing photoperiod synchronizes or entrains circannual rhythms of birds within a limit of entrainment. For example, exposure of European stonechats, *Saxicola torquata rubicola*, to alternating photoperiods of high amplitude (LD 16:6 to LD 10:14) entrain circannual patterns of testicular width and molt to 12 and 6 months in duration, respectively, but lower amplitude cycles of LD 13:11 to LD 11:13 were unable to entrain these processes to the 6 month cycle. There is little evidence of other Zeitgebers entraining circannual rhythms in birds, although periodic presentation of novel food has transitory effects on circannual cycles in African stonechats (Gwinner and Scheirlein, 1998; Scheuerlein and Gwinner, 2002).

35.4 PHOTOPERIODISM

35.4.1 Effects of Photoperiod on Avian Physiological Function

Aside from the entraining effects of photoperiod on circannual cycles, changes in photoperiod have direct effects on many annual cycles in birds, although it is not clear that these are necessarily mutually exclusive. In fact, it has been known for some time that seasonally breeding, temperate-zone birds time reproduction and migration in a clock-like fashion in response to changing photoperiod (Rowan, 1926). Birds living in temperate-zone latitudes generally restrict breeding to the spring and summer, maximizing the likelihood that young will be hatched during times at which food is plentiful (Gwinner, 1989, 2003). The vast majority of seasonally breeding birds are long-day breeders; that is, they begin sexual maturation, courtship, and mating as the photoperiod transitions from the winter solstice to longer and longer days. There is one well-documented short-day breeding species, the emu, *Dromaius novaehollandiae*, which breeds as the photoperiod shortens (Blache et al., 2001).

The exact photoperiod that induces or suppresses reproductive function in birds is species-, latitude-, and function-specific, called the *critical photoperiod*. This is the minimum photoperiod that induces or suppresses a seasonally regulated process (Figure 35.5). For example, annual testicular growth is induced earlier and by shorter photoperiods in the temperate zone European starling than in the Nearctic-breeding white-crowned sparrow (Dawson, 2008). This corresponds to the likelihood of food availability occurring later in the season at higher versus lower latitudes at a species level. Similarly, within species (especially those that are widely distributed), critical photoperiod tends to be longer in individuals of the same species residing at higher latitudes than at lower latitudes (Dawson, 2013). Thirdly, the critical photoperiod that induces or suppresses different functions, such as reproduction, molt, metabolism, and migration, usually differ among these processes. Obviously, molt while migration would be disadvantageous for any bird.

Subtropical and tropical birds are not immune to these changes, as the breeding seasons of these species are commonly timed as well (Scheuerlein and Gwinner, 2002; Rani and Kumar, 2013). Although the amplitude of photoperiodic change is much reduced at these latitudes (Figure 35.2), there are nonetheless many seasonal signals to which birds can synchronize reproductive and other processes. For example, the spotted antbird, *Hylophylax naevioides*, lives in Panamanian rainforests around 9°N, experiencing a very low-amplitude photoperiodic change. These birds breed in the rainy season (May–December), when insect prey are abundant (Wikelski et al., 2000). Although the rainy season occurs regularly from year to year, there is significant variation in its inception and duration (Hau et al., 2000). When male antbirds are experimentally transferred from LD 12:12 to LD 13:11, testes volume increases dramatically, while birds maintained in LD 12:12 do not, demonstrating these birds do respond to photoperiodic changes. Conversely, birds maintained in the equinoctial photoperiod stimulated with an additional dietary supplement of live crickets exhibit larger testes than birds maintained on the standard insectivorous diet of mealworms only (Hau et al., 2000). The scenario suggests that slight changes in photoperiod prime the birds' reproductive system to respond to the increase in diet necessary to raise young (Hau, 2001). This is just one example of what is likely a wide array of adaptations to temporal changes in the subtropics and tropics. Because avian biodiversity is 5 times the level in the tropics than in Nearctic regions, the variability in these adaptations is likely just as diverse.

In seasonally reproducing, long-day birds, gonadal activity and gonad size regress and become inactive as the photoperiod decreases, whereas gonads recrudesce, becoming more active in response to increased photoperiod (Figure 35.5(A)). If birds are maintained in long photoperiods for long periods of time, their reproductive systems become insensitive to the photostimulatory effects of long photoperiods and spontaneous regress (Figure 35.5(A)). This process is called photorefractoriness, and birds remain photorefractory until they are placed in short days for some time to make them photosensitive again. Similarly, other photoperiodically regulated processes wax and wane depending on the photoperiod, although the absolute time at which they occur may differ from gonadal function (Figure 35.5(B) and (C)).

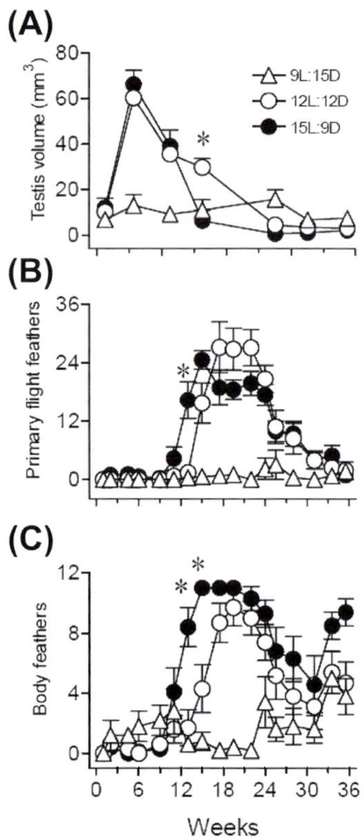

FIGURE 35.5 The effect of differing photoperiods on (A) testes volume, (B) primary, and (C) body molt in captive spotted munia, *Lonchura punctulata. From Rani and Kumar (2013).*

35.4.2 Role of Photoreceptors

Multiple photoreceptors are involved in circadian entrainment, and these photoreceptors reside in the retinae, pineal gland, and at least three encephalic sites: lateral septal organ (LSO), preoptic area (POA), and tuberal hypothalamus/mediobasal hypothalamus (MBH) (Foster and Hankins, 2002). As with circadian entrainment, the retinae are not necessary for photoperiodic responses (Benoit and Assenmacher, 1954; Menaker, 1968; Menaker and Keatts, 1968; Menaker and Underwood, 1976), although there is some species variability. In the house sparrow, *Passer domesticus*, and American tree sparrow, *Spizella arborea*, removal of the eyes, the pineal gland, or both has no effect on gonadal responses to increases or decreases in photoperiod (Menaker et al., 1970; Wilson, 1990). However, enucleated Japanese quail, *Coturnix coturnix japonica*, exhibit normal photoperiodic responses to increases in photoperiod in egg-laying in females and enlargement of the cloacal gland in males, similar to the situation in sparrows. Still, termination of reproductive activity by exposure to short photoperiods requires that birds experience long days at or before the time of blinding (Homma et al., 1972). Because Japanese quail and house sparrows also differ in the role played by retinal melatonin in circadian organization (Cassone and Menaker, 1984; Underwood and Siopes, 1984), it is not clear if this difference is due to differences in photoreceptors *per se* or due to differences in circadian organization (see below).

Local illumination of the MBH or LSO leads to gonadal growth (Benoit, 1935; Benoit and Assenmacher, 1954; Homma et al., 1979). Nomographic analyses of the action spectra for photoperiodic responses suggest a rhodopsin-like photopigment, with a peak sensitivity of 480 nm to ultraviolet wavelengths (Foster et al., 1985). More recently, several genes encoding rhodopsin superfamily photopigments have been identified in the brains of several avian species, corroborating this view. These include melanopsin (OPN4), vertebrate ancient opsin (VA opsin), neuropsin (OPN5), and rhodopsin itself (OPN2) (Silver et al., 1988; Bailey and Cassone, 2005; Chaurasia et al., 2005; Halford et al., 2009; Nakane et al., 2010; Kang et al., 2010; Wang and Wingfield, 2011). OPN5 is expressed in CSF-contacting neurons within the periventricular organ (PVO) of the Japanese quail (Figure 35.6; Nakane et al., 2010). OPN4-immunoreactivity and mRNA is widely expressed in the brain of the turkey, *Meleagris gallopavo*, including the preoptic area, suprachiasmatic region (mSCN and vSCN) of the hypothalamus, lateral septal region, *pars distalis* of the *adenohypophysis* and the premammillary nucleus (PMM) just dorsal of the PVO (Kang et al., 2010).

35.4.3 Role of Circadian Clocks in Photoperiodic Time Measurement

The role of the circadian clock in annual cycles has been known for some time (Bünning, 1969; Follett et al., 1992; Menaker and Eskin, 1967). Although there are differences among species of birds and between birds and other taxa, neither the absolute length of the photoperiod, length of the scotoperiod (the duration of the dark phase) or their ratio is the proximal causes of gonadal induction. Rather, it is the circadian phase (φ) at which light impinges on photoreceptive elements that causes reproductive changes. For example, male Japanese quail, *Coturnix japonica*, and white-crowned sparrows, *Zonotrichia leucophys*, which are maintained in LD 6:18 will exhibit regressed testes. However, if the last hour of the 6h photoperiod is extended each long night to a specific "photoinducible phase", φ_{pi}, usually 11–12h following the onset of the short photoperiod,

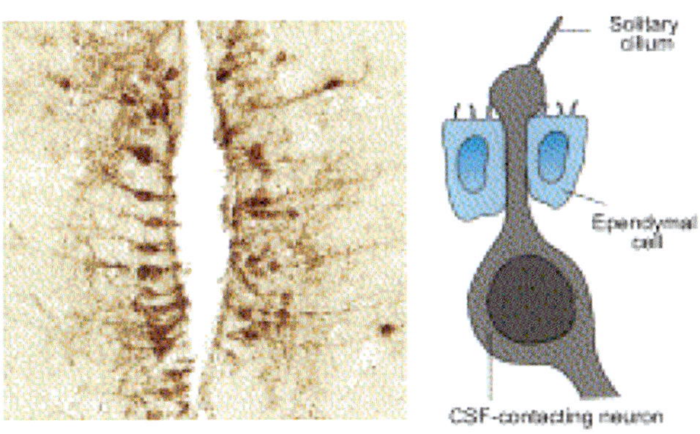

FIGURE 35.6 Expression of neuropsin (OPN5) in the periventricular organ of the mediobasal hypothalamus in the Japanese quail, *Coturnix coturnix*. *From Yoshimura (2013).*

reproductive activity is commenced. Thus, birds exposed to LD 6:18 or L5: D1: L1: D17 (a single 1h light pulse interrupting the night) will retain regressed gonads, but if the 1hr pulse occurs 5h later (L5:D6:L1:D12), gonads will recrudesce (Menaker and Eskin, 1967; Follett et al., 1974; Sharp, 2005). The same total amount of light (and dark) is present for each 24h, but the effect is dramatically different. It is the *timing* of light coinciding with an internal process that induces or prevents reproductive activity, suggesting a circadian clock underlies photoperiodic time measurement. Nanda and Hamner (1958) had shown this to be the case in plants through an elegant series of experiments in which soy bean plants exposed to a 6h photoperiod coupled with scotoperiods of varying lengths that were multiples of a 24h cycle (e.g. LD 6:18; LD 6:42 or LD 6:66) did not flower. However, when plants were exposed to 6 h photoperiods followed by scotoperiods of lengths that did not resonate with 24h (e.g. LD 6:6; LD 6:30 or LD 6:54) flowering was observed. Similar studies in several species of birds showed that seasonal changes in both gonadal recrudescence and regression are regulated by a circadian clock (Follett et al., 1974, Follett and Pearce-Kelly, 1991; Kumar et al., 1996).

These and other observations have led to two competing models for a role of circadian clocks in photoperiodic time measurement (Figure 35.7). In one, the "external coincidence model" proposed by Bünning (1969) suggests that light has two complementary roles. First, light entrains the circadian clock with a stable ψ_{ie} such that the photoinducible phase, φ_{pi}, is also maintained with a stable ψ approximately 11.5h following the beginning of the photoperiod (lights on). When light coincides with the φ_{pi} either because the photoperiod is long under natural conditions or through exposure to an experimental light pulse, the reproductive axis is induced.

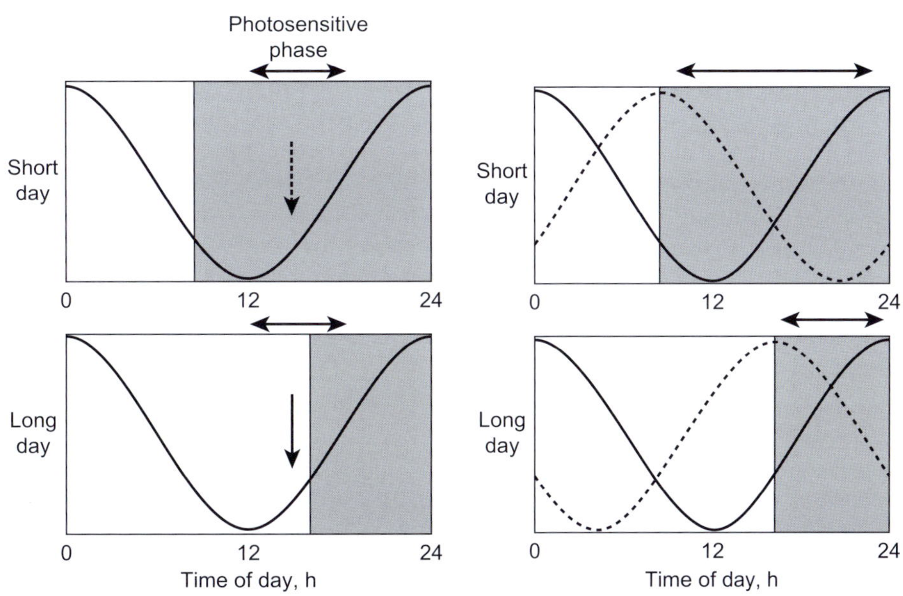

FIGURE 35.7 The difference between the external coincidence (left) and the internal coincidence (right) models for photoperiodic induction. In the case of external coincidence, light (1) entrains the circadian system to a stable phase relationship to the photoperiod and (2) stimulates or inhibits a process such as gonadal induction at a particular photoinducible phase. In the internal coincidence model, at least two circadian oscillators entrain to the photoperiod, and the phase relationship between these two oscillators induces or suppresses the seasonally controlled process. *From Yoshimura (2013).*

The competing notion, the "internal coincidence model" proposed by Pittendrigh (1993) stems from the observation that many circadian systems behave as if they are composed of at least two oscillators, one entrained to dawn and the other to dusk. In the internal coincidence model the phase relationship between the dawn oscillator and the dusk oscillator, ψ_{dawndusk}, induces seasonal changes in reproduction. As we will see below, each of these models has support from different experimental systems in birds. However, until specific structures and/or molecules become associated with these models, they are essentially untestable at the physiological level.

35.4.4 Role of Circadian System Structures (or Lack Thereof)

As stated in Chapter 31, the avian circadian system is comprises the pineal gland, the suprachiasmatic nuclei, the retinae and photoreceptors in the retinae, pineal and brain (Cassone and Menaker, 1984). These structures mutually interact to elicit circadian patterns of a broad array of processes and to entrain them to LD cycles. In view of the fact that photoperiodic time measurement relies on circadian rhythmicity, one might expect that these structures are critical for photoperiodism as well. Intriguingly, in spite of the fact that the rhythmic production of melatonin is critical for the expression of circadian locomotor rhythms in birds, melatonin does not affect seasonal changes in primary reproductive function in these species. As in mammals, pineal melatonin levels faithfully reflect the length of the scotoperiod both in vivo and in vitro (Binkley et al., 1977; Brandstätter et al., 2000). However, PINX and/or EX of several species of birds has little effect on seasonal changes in gonad size or activity (Siopes, 1983; Bentley, 2001; Kumar et al., 2002). Moreover, administration of exogenous melatonin of different durations has little effect on primary reproductive function (Cassone et al., 2008). This corresponds to the relative absence of melatonin receptor activity in the tuberal hypothalamus and hypophysis (Cassone et al., 1995), in stark contrast to the situation in seasonally reproducing mammals, where IMEL binding and melatonin receptor expression in pars tuberalis is a major site of melatonin action (Goldman, 2001).

Circadian clock function has been localized within the mediobasal hypothalamus (MBH) itself of Japanese quail controlling photoperiodic time measurement for reproductive function (Yoshimura, 2010, 2013). Earlier studies had shown that lesion of the MBH blocked testicular recrudescence in response to lengthening photoperiods (Sharp, 2005) and illumination of this area had resulted in excitation of the tuberal hypothalamus and testicular growth (Foster et al., 1985; Meddle and Follett, 1997). As stated above, the MBH of quail and the PMM of turkeys have been shown to express both OPN4 and OPN5 in cerebrospinal fluid (CSF)-contacting neurons (Kang et al., 2010; Nakane and Yoshimura, 2010; Nakane et al., 2010). Noting that PINX or EX or even SCN lesion had little effect on photoperiodic regulation of gonadal function, rhythmic expression of the clock genes *per2*, *per3*, *clock*, and *Bmal1* were identified in the MBH, suggesting that this structure contains the circadian pacemaker associated with photoperiodic time measurement (Yoshimura, 2010, 2013).

35.5 NEUROENDOCRINE REGULATION OF PHOTOPERIODIC TIME MEASUREMENT

35.5.1 Photoperiodic Control of Gonadotropins and Prolactin in Seasonal Reproduction

The circadian pacemaker presumably residing in the MBH controls seasonal changes in the release of gonadotropin releasing hormone (GnRH) and vasoactive intestinal polypeptide (VIP) via the hypothalamohypophysial portal system (HHPS), which in turn regulate the synthesis and release the gonadotropin luteinizing hormone (LH) and prolactin (PRL), respectively (Sharp et al., 1998). GnRH neurons predominate in the preoptic area and anterior hypothalamus and project to the median eminence (Saldanha et al., 2001; Teruyama and Beck, 2001), whereas the VIP neurons are found throughout the basal hypothalamus. The GnRH cells, at least, are in direct contact with terminals arising from rhodopsin-immunoreactive cells, presumably extraocular photoreceptors in the lateral septal organ (Saldanha et al., 2001). The VIP cells project throughout the hypothalamus but also innervate the capillaries in the median eminence (Teruyama and Beck, 2001). GnRH- and VIP-immunoreactive terminals are more dense and in closer proximity to median eminence during long photostimulatory photoperiods than in short photoinhibitory photoperiods.

In long-day breeding birds, lengthening photoperiods induce an increase in the release of GnRH from terminals in the median eminence into the HHPS. *Ex vivo* hypothalamic explants from male Japanese quail release GnRH in culture in response to long photoperiods (Perera and Follett, 1992). GnRH in turn induces the synthesis and release of LH, which remains high during the breeding season. However, when photoperiod decreases, LH levels also decrease. In addition, in long-day breeders, continued exposure to long days induces photorefractoriness, in which birds fail to respond to photostimulatory photoperiods, and LH levels decline (Sharp, 2005).

Lengthening photoperiod also presumably increases the release of VIP from median eminence terminals into the HHPS, although this has not been demonstrated directly. Even so, administration of VIP potently induces the synthesis and release of PRL by the adenohypophysis in several species of passerine, galliform, and columbiform species (Maney et al., 1999; Kosonsiriluk et al., 2008). The rise in PRL is typically delayed relative to LH, declining only after LH declines. Immunization of European starlings, *Sturnus vulgaris*, against VIP delays and attenuates the decline in LH levels during photorefractoriness (Maney et al., 1999).

35.5.2 Role of the Thyroid in Avian Photoperiodism

Thyroidectomy induces gonadal recrudescence in starlings and Japanese quail (Follett and Nicholls, 1985; Dawson et al., 1986), and injection of T3 into the MBH induces quail gonadal growth (Follett and Nicholls, 1985; Yoshimura et al., 2003). Using differential subtractive hybridization, Yoshimura's group found type 2 iodothyronine deiodinase (*Dio2*) to be induced in the MBH by a light pulse associated with long-day induction. *Dio2* encodes an enzyme that catalyzes the conversion of inactive thyroxine (T4) to active triiodothyroine (T3) (Nakao et al., 2008a,b). Later, they showed that type 3 iodothyronine deiodinase (*Dio3*), which inactivates T3, was induced in the MBH by exposure to short days. The scenario they envision is that photoperiod is perceived by photopigments in the MBH, which entrain a circadian oscillator within the MBH. In long days, the circadian clock induces *Dio2*, while short days induce *Dio3*. These two gene switches fine tune local thyroid hormone concentration within the MBH. Although precise mechanisms of thyroid hormone action remain unclear, it is proposed that locally activated thyroid hormone causes morphological changes in neuro–glial interaction between GnRH nerve terminals and the endfeet of the glial processes in the median eminence that regulates or modulates the seasonal GnRH secretion (Figure 35.8; Yamamura et al., 2004).

35.5.3 Role of Gonadal and Neural Steroids

As stated previously, the reproductive systems of long-day breeders recrudesce as the photoperiod lengthens in response to gonadotropin activation. The exact timing of these is species and gender specific. For example, in European starlings, seasonal testes growth begins more than a month earlier than does female follicle formation (Williams, 2012). As testes recrudesce, Sertoli cells synthesize and secrete estrogens, such as estradiol, while Leydig cells synthesize and secrete androgens such as testosterone, androstenedione and dehydroepiandrosterone. Ovarian granulosa cells begin to synthesize and secrete estrogens from androgens generated by adjacent thecal cells. These gonadal steroids induce gender-appropriate seasonal growth and maturation of primary and secondary sexual characteristics (Adkins-Regan, 2012). In males, many of the genomic effects of testosterone are mediated by estrogens generated by local conversion by aromatase. When photoperiod decreases, LH also decreases, and the steroidogenic capacity of the gonads decline in parallel.

Among structures that annually recrudesce and regress are brain structures associated with reproductive behavior (Figure 35.9; Balthazart et al., 2009, 2010). The medial preoptic nucleus (POM) within the POA of Japanese quail has been well established as a structure involved in male sexual behavior (Ball and Balthazart, 2004; Balthazart and Ball, 2007), in which the size

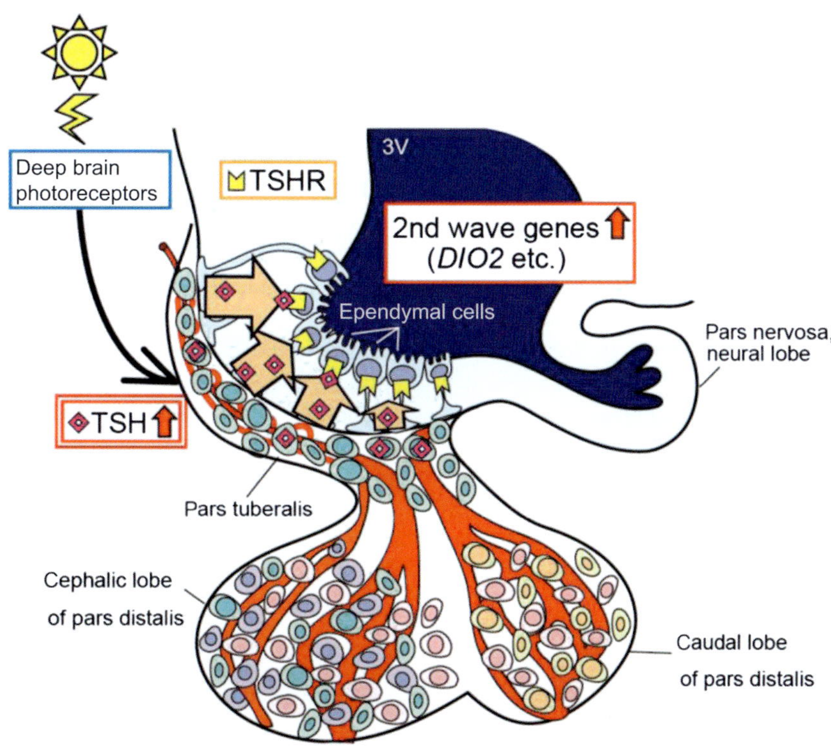

FIGURE 35.8 **Schematic of the quail tuberal hypothalamus.** Extraocular photoreceptors induce an increase in TSH signaling in the pars tuberalis, which in turn induce changes in ependymal cells or tanycytes. These cells affect the ability of neuroendocrine signals entering the hypothalamohypophysial portal system. *From Yoshimura (2013).*

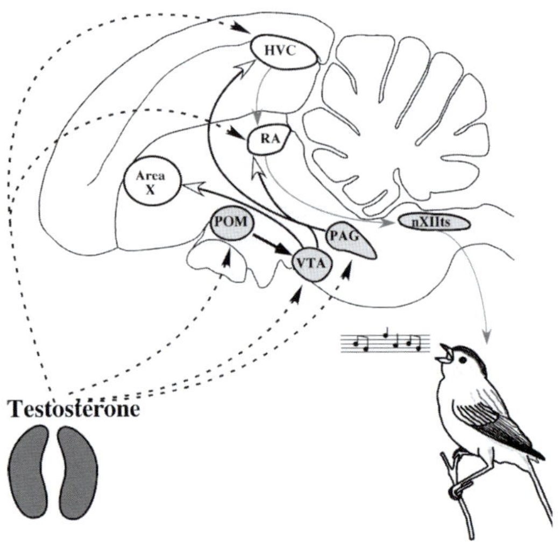

FIGURE 35.9 **Schematic of the effects of testosterone on brain structures associated with reproduction and song.** Steroid receptors residing in preoptic, brainstem, and forebrain structures mediate primary reproductive behavior as well as bird song. *From Ball and Balthazart (2004).*

of POM is greater in males than in females (Viglietti-Panzica et al., 1986; Ball and Ketterson, 2008). In male quail, the POM increases in volume during the spring and decreases in the fall (Thompson and Adkins-Regan, 1994). Castration of these birds abolishes this pattern, and administration of testosterone, mediated by aromatase-induced estrogens, increases POM volume through increases in cell size and dendritic branching within 2 weeks of administration. In contrast to the situation involving song control (see below), there is no evidence that POM growth involves *de novo* neurogenesis.

However, the gonads and adrenal glands are not the only sources of biologically active steroid hormone (Tsutsui et al., 1999; Schlinger and Remage-Healey, 2012). The brains of many species of birds express the enzymes capable of neurosteroidogenesis, biosynthesis of biologically active steroid hormones by the brain (Tsutsui et al., 1999). In the Lapland longspur (*Calcarus lapponicus*) and song sparrow (*Melospiza melodia*), for example, the hypothalamus, hippocampus, and ventral telencephalon express aromatase, which converts androgens to estrogens, and 5-β-reductase, which deactivates testosterone (Soma et al., 1999, 2003). This suggests that the avian brain can convert circulating gonadal steroids into biologically active estrogens. However, there is growing evidence that the brain can also synthesize steroids from circulating cholesterol (Schlinger et al., 2008; Schlinger and Remage-Healey, 2012).

35.5.4 Mechanisms of Photoperiodic Regulation of Bird Song

The song control system of oscine passeriform birds is a specialized network of brain nuclei involved in singing and song learning (Nottebohm, 1981; Mooney, 2009). This system receives auditory input from ascending, primary auditory pathways beginning in the cochlear nuclei (Co), which project to the lateral dorsal mesencephalic nuclei (MLd). MLd in turn projects to the thalamic *nucleus ovoidalis* (Ov), which in turn projects to Field L in the forebrain. Song processing begins in secondary auditory areas in the caudal mesopallium (cM) and caudomedial nidopallium (NCM), which interact with the anterior forebrain pathway for song plasticity and learning. This pathway includes the HVC in the dorsal forebrain, which projects to Area X, whose projections form a loop between the dorsolateral thalamus (DLM) and the lateral magnocellular nucleus of the anterior nidopallium (LMAN). Then, both HVC and LMAN project to the robust nucleus of the archipallium (RA), which forms the song motor output pathway.

This system enables birds to process a complex species-specific identification of both self and con-specifics as well as other dynamics in birds' acoustic environments (e.g., competitors, prey, predators; Mooney, 2009). These acoustic environments as well as their reproductive and survival relevance are not constant. Territories change hands and the available range of mates fluctuates. In order to effectively interpret the acoustic features in song, the auditory system must structure its representation in such a way that allows for dynamic behavioral goals of the organism. At the same time, the structure and behavior of song itself must be tuned to reproductively appropriate situations.

Both the probability of a male bird to sing in response to a given stimulus as well as the size and complexity of the song control nuclei within its brain vary depending upon the time of year (Nottebohm, 1981). Photosensitive birds in the short days of winter possess small HVC, RA, and (depending on the species) other structures in the system. When birds are photostimulated as photoperiod increases, the song control nuclei grow in parallel with the growth of the testes. As they become photorefractory, these structures regress in both size and complexity. Some of the growth in these structures, especially in HVC, is due to de novo neurogenesis from neural stem cells residing in periventricular areas of the lateral ventricle (Alvarez-Buyla et al., 1988). These cells are born predominantly during the fall and migrate into the HVC over the course of the winter, where they differentiate and extend axonal projections to RA and other structures. In addition, increases in neuropil and glia contribute to this growth. Regression of these structures is characterized by programmed cell death as well as retraction of neuropil.

Seasonal fluctuations in androgens and estrogens appear to be critical for changes in song control. Song control nuclei contain both androgen receptors (AR) and both estrogen receptor subtypes (ERα and ERβ), as well as aromatase, which are capable of converting androgens into biologically active estrogens (Ball and Balthazart, 2010). Further, the rate of male song in several songbird species increases when testosterone levels are high; castration decreases this rate, and hormone replacement reestablishes vernal song patterns (Balthazart et al., 2010).

Even so, a few studies have indicated gonad-independent regulation of song control nuclei under different photoperiods.

In both American tree sparrows (Bernard et al., 1997) and house sparrows (Whitfield-Rucker and Cassone, 2000), HVC and RA increase in size in response to a change in photoperiod from a short day to long day (Figure 35.9). Castrated birds in these studies also exhibited photostimulated song control nuclei, although the level of induction was not as great as in the sham-operated birds. Thus, while the song system certainly responds to the seasonal changes in gonadal steroids, regulation of song control nuclei comprises a gonad-independent as well as a gonad-dependent aspect. As stated above, recent studies have implicated neurosteroids in this supplemental signal to the song system (Soma et al., 1999, 2003).

Alternatively, studies employing *in vitro* receptor binding and 2[^{125}I]iodomelatonin (IMEL) in the 1980s and 1990s demonstrated high affinity melatonin receptor binding and the identity of three receptor subtypes, Mel_{1A}, Mel_{1B}, and Mel_{1C} melatonin receptors, in a variety of vertebrates (Reppert et al., 1995, 1996). In birds and reptiles, IMEL binding predominates in retinorecipient and integrative structures within the visual system (Cassone et al., 1995), which has led to the view that visual sensitivity and contrast detection are under circadian control. In male house sparrows (Whitfield-Rucker and Cassone, 1996) and European starlings (Bentley and Ball, 2000), but not female birds, high affinity IMEL binding was present in song control nuclei HVC, RA, LMAN, and, to a lesser extent, Area X. Both HVC and RA express the Mel_{1B} melatonin receptor (Jansen et al., 2005).

These observations point to a role for melatonin in song behavior and in the growth and regression of song control nuclei. Bentley et al. (1999) have shown that continuous administration exogenous melatonin attenuated the long-day-induced volumetric increase in HVC and also decreased the volume of another song-control nucleus, area X, in European starlings. This effect was independent of the birds' reproductive state. House sparrows exposed to constant light (LL), which abolishes circadian patterns of behavior, and subsequent rhythmic administration of summer-like (short duration) or winter-like (long duration) melatonin, which entrains their behavior, also affects the volumes of song control nuclei HVC and RA (Cassone et al., 2008). Birds that received no melatonin or the short duration melatonin cycle (who entrained to the regime) exhibited large HVC and RA, but birds that received the long duration melatonin entrained to the melatonin regime but showed regressed HVC and RA. This showed that melatonin affects song control nuclei independently of gonadal state and whether they entrained to the melatonin cycle (Figure 35.10).

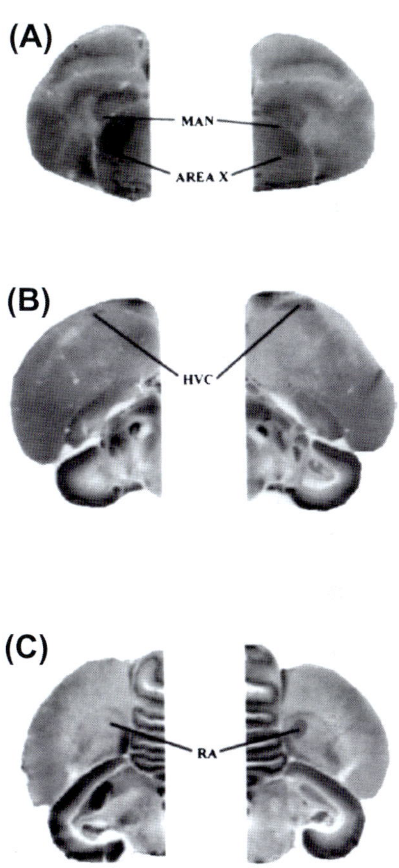

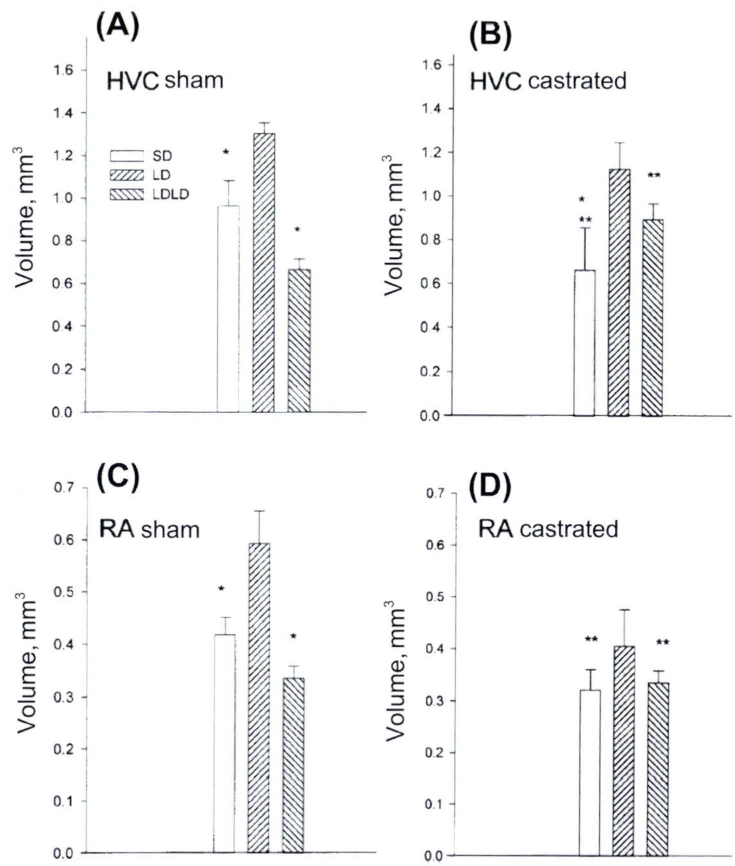

FIGURE 35.10 Bird song structures, such as HVC and the robust nucleus of the archipallium (RA) change in size and complexity depending on the photoperiod. HVC and RA are small in short days (left) and large in long days (right). Castration attenuates but does not abolish these seasonal cycles. *From Whitfield-Rucker and Cassone (2000).*

35.6 MOLECULAR MECHANISMS OF PHOTOPERIODISM

Yoshimura and coworkers envisioned an external coincidence model in which circadian oscillators within the MBH are entrained by photoperiod co-localized in that structure. When the length of the photoperiod coincides with a φ_{pi}, *Dio2* is induced, enabling a metabolic cascade in response to T3 hormone, and gonadal induction occurs. It is not clear, at this stage, what molecular components link the circadian clock to *Dio2* or *Dio3*. The MBH of quail and the PMM of turkeys rhythmically express clock genes (Ikegami et al., 2009; Ikegami and Yoshimura, 2012; Leclerc et al., 2010). The availability of chicken genome sequences has permitted the expansion of studies from single-gene to genome-wide transcriptome analyses in avian species. Microarray analysis of photoperiodism have identified waves of genes that were induced or reduced by long day stimulus. Long day-induced thyrotropin (thyroid-stimulating hormone, TSH) in the pars tuberalis of the pituitary gland induces *DIO2* expression and reduces *DIO3* expression through the TSH receptor (TSHR)-Gαs-cAMP signaling pathway (Nakao et al., 2008b). Because chronic administration of TSH into the hypothalamus causes full testicular development, it is suggested that the thyrotropin secreted from the pars tuberalis is the master factor regulating seasonal reproduction. This result was unexpected because of the well-known function of thyrotropin in stimulation of thyroid gland.

35.7 COMPARISON TO OTHER VERTEBRATE TAXA

In marked contrast to nonmammalian vertebrates, the eyes are the only photoreceptor organs in mammals. Light information received by the eye is transmitted to the pineal gland through the suprachiasmatic nucleus. Melatonin is secreted during the night as in the case of birds, but the mode of melatonin action on seasonal reproduction had been unclear until recently. It has been demonstrated that melatonin receptors are expressed in the pars tuberalis of mammals and that melatonin suppresses TSH expression in the pars tuberalis via the MT1 melatonin receptor (Ono et al., 2008; Hanon et al., 2008; Yasuo et al., 2009) (Figure 35.11).

Most fish living in temperate zones also exhibit a photoperiodic response and the involvement of thyroid hormone in seasonal reproduction has been reported extensively. However, fish do not possess anatomically distinct pars tuberalis. Recent work has demonstrated that the saccus vasculosus (SV) of fish acts as a sensor of seasonal reproduction (Nakane et al., 2013). All the photoperiodic signaling pathways from photoreceptors (opsins) to neuroendocrine output (thyrotropin and DIO2) are localized in the coronet cells of the SV, and isolated SV responds to photoperiodic changes *in vitro*. Thus, the coronet cells of the SV may be the ancestral seasonal sensor in vertebrates (Figure 35.12).

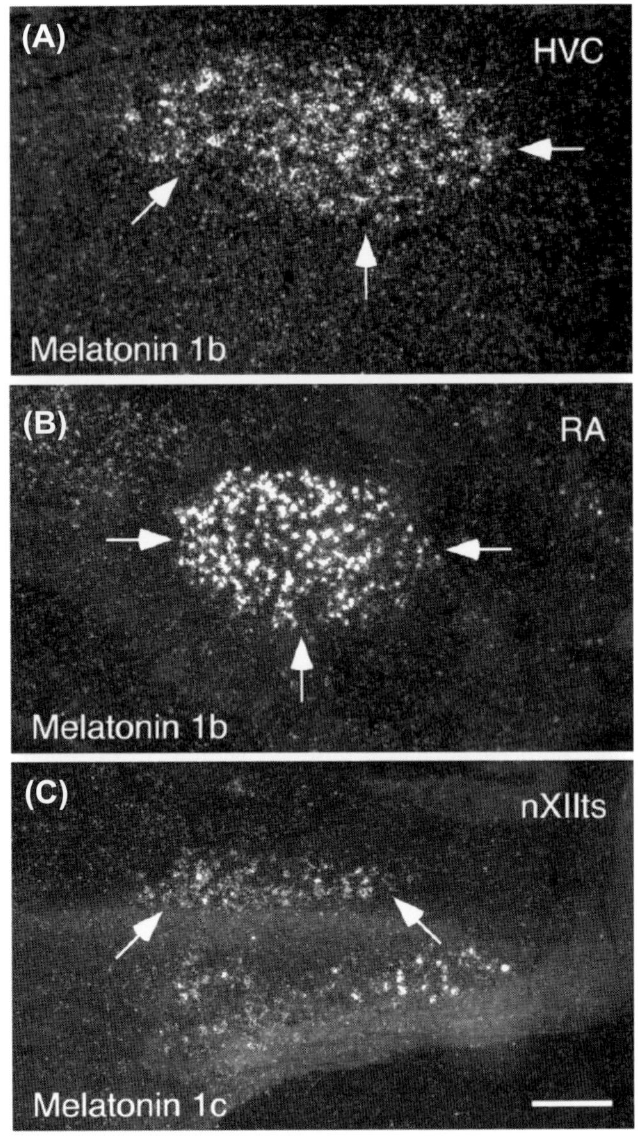

FIGURE 35.11 In situ hybridization of melatonin receptors in (A) HVC, (B) robust nucleus of the archipallium (RA), and (C) the nucleus tractus solitaries (nXIIts) of the zebra finch. *From Jansen et al. (2005).*

35.8 CONCLUSION

The selective pressures to breed and reproduce at the most advantageous times of year, when food is most likely to be present for the provision of young, are particularly strong in the many species of birds that express seasonal rhythms. This is largely due to the relatively large efforts birds expend raising their young. In addition, birds' complex life cycles require a concerted orchestration of seasonal events, ranging from migration to molt to courtship to reproduction and so on. This complex phenology shares a variety of mechanisms,

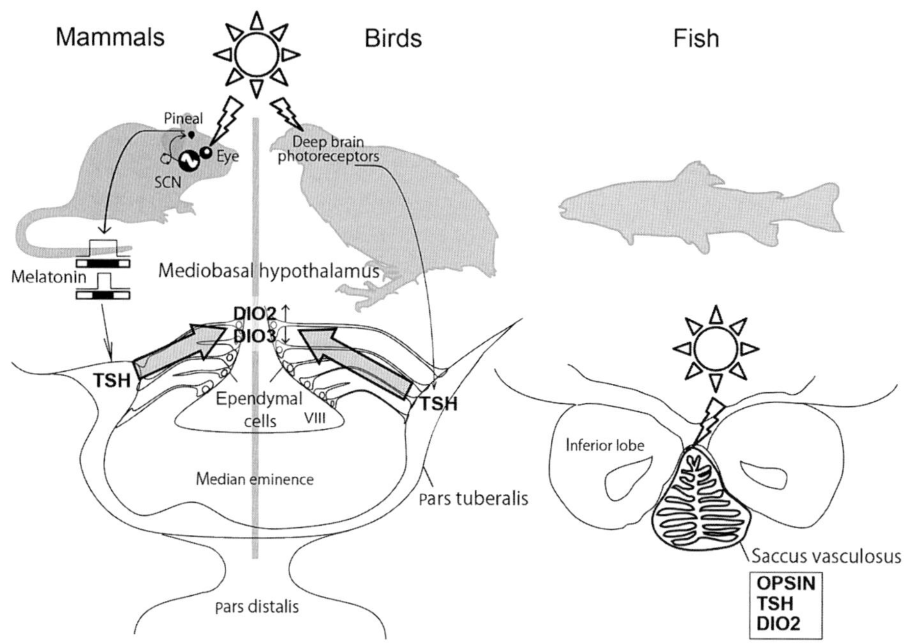

FIGURE 35.12 Vertebrates share common pathways for reproductive seasonality but vary in several aspects. In mammals, light information arrives in the tuberal hypothalamus via transduction of the pineal hormone melatonin within the pars tuberalis, while in birds and fish, tuberal hypothamic structures are directly light sensitive. The signal converge, however, through a common regulation of *DIO2* and *DIO3*.

but these are not completely known. In oscine passeriforms, for example, the mechanisms regulating primary reproductive function (which appear to be oscillators within the tuberal hypothalamus) seem to be functionally separated from those associated with courtship and song, which in part involve pineal melatonin. Future research focused on the coordination of other processes, such as migration, molt, and female reproduction, will go a long way toward fully understanding the complex seasonality of birds.

REFERENCES

Adkins-Regan, E., 2012. Hormonal organization and activation: evolutionary implications and questions. Gen. Comp. Endocrinol. 176 (3), 279–285.

Alvarez-Buyla, A., Theelen, M., Nottebohm, F., 1988. Birth of projection neurons in the higher vocal center of the carary forebrain before, during and after song learning. Proc. Natl. Acad. Sci. U. S. A. 85, 8722–8726.

Bailey, M.J., Cassone, V.M., 2005. Melanopsin expression in the chick retina and pineal gland. Brain Res. Mol. Brain Res. 134 (2), 345–348.

Ball, G.F., Balthazart, J., 2004. Hormonal regulation of brain circuits mediating male sexual behavior in birds. Physiol. Behav. 83 (2), 329–346.

Ball, G.F., Balthazart, J., 2010. Seasonal and hormonal modulation of neurotransmitter systems in the song control circuit. J. Chem. Neuroanat. 39 (2), 82–95.

Ball, G.F., Ketterson, E.D., 2008. Sex differences in the response to environmental cues regulating seasonal reproduction in birds. Philos. Trans. R. Soc. Lond., B, Biol. Sci. 363 (1490), 231–246.

Balthazart, J., Ball, G.F., 2007. Topography in the preoptic region: differential regulation of appetitive and consummatory male sexual behaviors. Front. Neuroendocrinol. 28 (4), 161–178.

Balthazart, J., Cornil, C.A., Charlier, T.D., Taziaux, M., Ball, G.F., 2009. Estradiol, a key endocrine signal in the sexual differentiation and activation of reproductive behavior in quail. J. Exp. Zool. A Ecol. Genet. Physiol. 311 (5), 323–345.

Balthazart, J., Charlier, T.D., Barker, J.M., Yamamura, T., Ball, G.F., 2010. Sex steroid-induced neuroplasticity and behavioral activation in birds. Eur. J. Neurosci. 32 (12), 2116–2132.

Benoit, J., 1935. Le role des yeux dans l'action stimulante de la lumiere sure le development testiculaire chez le canard. C. R. Soc. Biol. (Paris) 118, 669–671.

Benoit, J., Assenmacher, I., 1954. Comparative sensitivity of superficial and deep receptors in photosexual reflex in duck. C. R. Hebd. Seances Acad. Sci. 239 (1), 105–107.

Bentley, G.E., 2001. Unraveling the enigma: the role of melatonin in seasonal processes in birds. Microsc. Res. Tech. 53 (1), 63–71.

Bentley, G.E., Ball, G.F., 2000. Photoperiod-dependent and independent regulation of melatonin receptors in the forebrain of songbirds. J. Neuroendocrinol. 12 (8), 745–752.

Bentley, G.E., Van't Hof, T.J., Ball, G.F., April 13, 1999. Seasonal neuroplasticity in the songbird telencephalon: a role for melatonin. Proc. Natl. Acad. Sci. U. S. A. 96 (8), 4674–4679.

Bernard, D.J., Wilson, F.E., Ball, G.F., 1997. Testis-dependent and independent effects of photoperiod on volumes of song control nuclei in American tree sparrows (*Spizella arborea*). Brain Res. 760 (1–2), 163–169.

Binkley, S., Stephens, J.L., Riebman, J.B., Reilly, K.B., 1977. Regulation of pineal rhythms in chickens: photoperiod and dark-time sensitivity. Gen. Comp. Endocrinol. 32 (4), 411–416.

Blache, D., Talbot, R.T., Blackberry, M.A., Williams, K.M., Martin, G.B., Sharp, P.J., 2001. Photoperiodic control of the concentration of luteinizing hormone, prolactin and testosterone in the male emu (*Dromaius novaehollandiae*), a bird that breeds on short days. J. Neuroendocrinol. 13 (11), 998–1006.

Brandstätter, R., Kumar, V., Abraham, U., Gwinner, E., 2000. Photoperiodic information acquired and stored in vivo is retained in vitro by a circadian oscillator, the avian pineal gland. Proc. Natl. Acad. Sci. U. S. A. 97 (22), 12324–12328.

Bünning, E., 1969. Common features of photoperiodism in plants and animals. Photochem. Photobiol. 9 (3), 219–228.

Caprioli, M., Ambrosini, R., Boncoraglio, G., Gatti, E., Romano, A., Romano, M., Rubolini, D., Gianfranceschi, L., Saino, N., 2012. Clock gene variation is associated with breeding phenology and maybe under directional selection in the migratory barn swallow. PLoS One 7 (4), e35140. http://dx.doi.org/10.1371/journal.pone.0035140.

Cassone, V.M., Menaker, M., 1984. Is the avian circadian system a neuroendocrine loop? J. Exp. Zool. 232 (3), 539–549.

Cassone, V.M., Brooks, D.S., Kelm, T.A., 1995. Comparative distribution of 2[^{125}I]iodomelatonin binding in the brains of diurnal birds: outgroup analysis with turtles. Brain Behav. Evol. 45 (5), 241–256.

Cassone, V.M., Bartell, P.A., Earnest, B.J., Kumar, V., 2008. Duration of melatonin regulates seasonal changes in song control nuclei of the house sparrow, *Passer domesticus*: independence from gonads and circadian entrainment. J. Biol. Rhythms 23 (1), 49–58.

Chaurasia, S.S., Rollag, M.D., Jiang, G., Hayes, W.P., Haque, R., Natesan, A., Zatz, M., Tosini, G., Liu, C., Korf, H.W., Iuvone, P.M., Provencio, I., 2005. Molecular cloning, localization and circadian expression of chicken melanopsin (Opn4): differential regulation of expression in pineal and retinal cell types. J. Neurochem. 92 (1), 158–170.

Dawson, A., 2008. Control of the annual cycle in birds: endocrine constraints and plasticity in response to ecological variability. Philos. Trans. R Soc. Lond., B, Biol. Sci. 363 (1497), 1621–1633.

Dawson, A., 2013. The effect of latitude on photoperiodic control of gonadal maturation, regression and molt in birds. Gen. Comp. Endocrinol. 190, 129–133.

Dawson, A., Goldsmith, A.R., Nicholls, T.J., Follett, B.K., 1986. Endocrine changes associated with the termination of photorefractoriness by short daylengths and thyroidectomy in starlings (*Sturnus vulgaris*). J. Endocrinol. 110 (1), 73–79.

Dor, R., Cooper, C.B., Lovette, I.J., Massoni, V., Bulit, F., Liljesthrom, M., Winkler, D.W., 2012. Clock gene variation in Tachycineta swallows. Ecol. Evol. 2 (1), 95–105.

Follett, B.K., Mattocks Jr., P.W., Farner, D.S., May 1974. Circadian function in the photoperiodic induction of gonadotropin secretion in the white-crowned sparrow, *Zonotrichia leucophrys gambelii*. Proc. Natl. Acad. Sci. U. S. A. 71 (5), 1666–1669.

Follett, B.K., Nicholls, T.J., 1985. Influences of thyroidectomy and thyroxine replacement on photoperiodically controlled reproduction in quail. J. Endocrinol. 107 (2), 211–221.

Follett, B.K., Pearce-Kelly, A.S., 1991. Photoperiodic induction in quail as a function of the period of the light-dark cycle: implications for models of time measurement. J. Biol. Rhythms 6 (4), 331–341.

Follett, B.K., Kumar, V., Juss, T.S., 1992. Circadian nature of the photoperiodic clock in Japanese quail. J. Comp. Physiol. A 171 (4), 533–540.

Foster, R.G., Hankins, M.W., 2002. Non-rod, non-cone photoreception in the vertebrates. Prog. Retin. Eye Res. 21 (6), 507–527.

Foster, R.G., Follett, B.K., Lythgoe, J.N., 1985. Rhodopsin-like sensitivity of extra-retinal photoreceptors mediating the photoperiodic response in quail. Nature 313 (5997), 50–52.

Glickman, T., 2000. Glossary of Meteorology, second ed. American Meteorological Society, Boston.

Goldman, B.D., 2001. Mammalian photoperiodic system: formal properties and neuroendocrine mechanisms of photoperiodic time measurement. J. Biol. Rhythms 16 (4), 283–301.

Gwinner, E., 1989. Photoperiod as a modifying and limiting factor in the expression of avian circannual rhythms. J. Biol. Rhythms 4 (2), 237–250.

Gwinner, E., 2003. Circannual rhythms in birds. Curr. Opin. Neurobiol. 13 (6), 770–778.

Gwinner, E., Scheirlein, A., 1998. Seasonal changes in day-light intensity as a potential zeitgeber of circannual rhythms in equatorial Stonechats. J. Ornithol. 139, 407–412.

Halford, S., Pires, S.S., Turton, M., Zheng, L., González-Menéndez, I., Davies, W.L., Peirson, S.N., García-Fernández, J.M., Hankins, M.W., Foster, R.G., 2009. VA opsin-based photoreceptors in the hypothalamus of birds. Curr. Biol. 19 (16), 1396–1402.

Hanon, E.A., Lincoln, G.A., Fustin, J.M., Dardante, H., Masson-Pevet, M., Morgan, P.J., Hazlerigg, D.G., 2008. Ancestral TSH mechanism signals summer in a photoperiodic mammal. Curr. Biol. 18, 1147–1152.

Hau, M., 2001. Timing of breeding in variable environments: tropical birds as model systems. Horm. Behav. 40 (2), 281–290.

Hau, M., Wikelski, M., Wingfield, J.C., 2000. Visual and nutritional food cues fine-tune timing of reproduction in a neotropical rainforest bird. J. Exp. Zool. 286 (5), 494–504.

Hazlerigg, D., Loudon, A., 2008. New insights into ancient seasonal life timers. Curr. Biol. 18 (17), R795–R804.

Helm, B., Gwinner, E., Koolhaas, A., Battley, P., Schwabl, I., Dekinga, A., Piersma, T., 2012. Avian migration: temporal multitasking and a case study of melatonin cycles in waders. Prog. Brain Res. 199, 457–479.

Homma, K., Wilson, W.O., Siopes, T.D., October 27, 1972. Eyes have a role in photoperiodic control of sexual activity of coturnix. Science 178 (4059), 421–423.

Homma, K., Ohta, M., Sakakibara, Y., 1979. Photoinducible phase of the Japanese quail detected by direct stimulation of the brain. In: Suda, M., Hayaishi, O., Nakagawa, H. (Eds.), Biological Rhythms and Their Central Mechanism. Elsevier, Amsterdam, pp. 85–94.

Ikegami, K., Yoshimura, T., 2012. Circadian clocks and the measurement of daylength in seasonal reproduction. Mol. Cell. Endocrinol. 349, 76–81.

Ikegami, K., Katou, Y., Higashi, K., Yoshimura, T., 2009. Localization of circadian clock protein BMAL1 in the photoperiodic signal transduction machinery in Japanese quail. J. Comp. Neurol. 517 (3), 397–404.

Jansen, R., Metzdorf, R., van der Roest, M., Fusani, L., ter Maat, A., Gahr, M., 2005. Melatonin affects the temporal organization of the song of the zebra finch. FASEB J. 19 (7), 848–850.

Kang, S.W., Leclerc, B., Kosonsiriluk, S., Mauro, L.J., Iwasawa, A., El Halawani, M.E., 2010. Melanopsin expression in dopamine-melatonin neurons of the premammillary nucleus of the hypothalamus and seasonal reproduction in birds. Neuroscience 170 (1), 200–213.

King, J.R., 1974. Seasonal allocation of time and energy resources. In: Paynter, R.A. (Ed.), Avian Energetics. Nuttall Ornithological Club, Cambridge, pp. 4–70.

Kosonsiriluk, S., Sartsoongnoen, N., Chaiyachet, O.A., Prakobsaeng, N., Songserm, T., Rozenboim, I., El Halawani, M., Chaiseha, Y., 2008. Vasoactive intestinal peptide and its role in continuous and seasonal reproduction in birds. Gen. Comp. Endocrinol. 159 (1), 88–97.

Kumar, V., Jain, N., Follett, B.K., 1996. The photoperiodic clock in blackheaded buntings (*Emberiza melanocephala*) is mediated by a self-sustaining circadian system. J. Comp. Physiol. A 179 (1), 59–64.

Kumar, V., Singh, S., Misra, M., Malik, S., Rani, S., 2002. Role of melatonin in photoperiodic time measurement in the migratory redheaded bunting (*Emberiza bruniceps*) and the nonmigratory Indian weaver bird (*Ploceus philippinus*). J. Exp. Zool. 292 (3), 277–286.

Leclerc, B., Kang, S.W., Mauro, L.J., Kosonsiriluk, S., Chaiseha, Y., El Halawani, M.E., 2010. Photoperiodic modulation of clock gene expression in the avian premammillary nucleus. J. Neuroendocrinol. 22 (2), 119–128.

Lincoln, R.J., Boxshall, G.A., Clark, P.F., 1998. A Dictionary of Ecology, Evolution and Systematics. .

Lockwood, M., 2010. Solar change and climate: an update in the light of the current exceptional minimum. Proc. R. Soc. A 466, 303–329.

Maney, D.L., Schoech, S.J., Sharp, P.J., Wingfield, J.C., 1999. Effects of vasoactive intestinal peptide on plasma prolactin in passerines. Gen. Comp. Endocrinol. 113 (3), 323–330.

McMillan, J.P., 1970. Pinealectomy abolishes circadian rhythm of migratory restlessness. J. Comp. Physiol. 79, 105–110.

Meddle, S.L., Follett, B.K., 1997. Photoperiodically driven changes in Fos expression within the basal tuberal hypothalamus and median eminence of Japanese quail. J. Neurosci. 17 (22), 8909–8918.

Menaker, M., 1968. Extraretinal light perception in the sparrow. I. Entrainment of the biological clock. Proc. Natl. Acad. Sci. U. S. A. 59 (2), 414–421.

Menaker, M., Eskin, A., 1967. Circadian clock in photoperiodic time measurement: a test of the Bünning hypothesis. Science 157 (3793), 1182–1185.

Menaker, M., Keatts, H., 1968. Extraretinal light perception in the sparrow. II. Photoperiodic stimulation of testis growth. Proc. Natl. Acad. Sci. U. S. A. 60 (1), 146–151.

Menaker, M., Underwood, H., 1976. Extraretinal photoreception in birds. Photophysiology 23 (4), 299–306.

Menaker, M., Roberts, R., Elliott, J., Underwood, H., 1970. Extraretinal light perception in the sparrow. 3. The eyes do not participate in photoperiodic photoreception. Proc. Natl. Acad. Sci. U. S. A. 67 (1), 320–325.

Mooney, R., 2009. Neurobiology of song learning. Curr. Opin. Neuobiol. 19, 654–660.

Nakane, Y., Yoshimura, T., 2010. Deep brain photoreceptors and a seasonal signal transduction cascade in birds. Cell Tissue Res. 342 (3), 341–344.

Nakane, Y., Ikegami, K., Ono, H., Yamamoto, N., Yoshida, S., Hirunagi, K., Ebihara, S., Kubo, Y., Yoshimura, T., 2010. A mammalian neural tissue opsin (Opsin 5) is a deep brain photoreceptor in birds. Proc. Natl. Acad. Sci. U. S. A. 107 (34), 15264–15268.

Nakane, Y., Ikegami, K., Iigo, M., Ono, H., Takeda, K., Takahashi, D., Uesaka, M., Kimijima, M., Hashimoto, R., Arai, N., Suga, T., Kosuge, K., Abe, T., Maeda, R., Senga, T., Amiya, N., Azuma, T., Amano, M., Abe, H., Yamamoto, N., Yoshimura, T., 2013. The saccus vasculosus of fish is a sensor of seasonal changes in day length. Nat. Commun. 4, 2108.

Nakao, N., Ono, H., Yoshimura, T., 2008a. Thyroid hormones and seasonal reproductive neuroendocrine interactions. Reproduction 136 (1), 1–8.

Nakao, N., Ono, H., Yamamura, T., Anraku, T., Takagi, T., Higashi, K., Yasuo, S., Katou, Y., Kageyama, S., Uno, Y., Kasukawa, T., Iigo, M., Sharp, P.J., Iwasawa, A., Suzuki, Y., Sugano, S., Niimi, T., Mizutani, M., Namikawa, T., Ebihara, S., Ueda, H.R., Yoshimura, T., 2008b. Thyrotrophin in the pars tuberalis triggers photoperiodic response. Nature 452 (7185), 317–322.

Nanda, K.K., Hamner, K.C., 1958. Studies on the nature of the endogenous rhythm affecting photoperiodic response in the biloxi soy bean. Bot. Gaz. 120, 14–25.

Nottebohm, F., 1981. A brain for all seasons: cyclical anatomical changes in song control nuclei of the canary brain. Science 214, 1368–1370.

Ono, H., Hoshino, Y., Yasuo, S., Watanabe, M., Nakane, Y., Murai, A., Ebihara, S., Korf, H.W., Yoshimura, T., 2008. Involvement of thyrotropin in photoperiodic signal transduction in mice. Proc. Natl. Acad. Sci. U. S. A. 105, 18238–18242.

Pant, K., Chandola-Saklani, A., 1992. Pinealectomy and LL abolished circadian perching rhythms but did not alter circannual reproductive or fattening rhythms in finches. Chronobiol. Int. 9 (6), 413–420.

Perera, A.D., Follett, B., 1992. Photoperiodic induction in vitro: dynamics of gonadotrophin releasing hormone release from hypothalamic explants of the Japanese quail. Endocrinology 131, 2898–2908.

Pianka, E.R., 1978. Evolutionary Ecology, second ed. Harper and Row, New York. 397 pp.

Pittendrigh, C.S., 1993. Temporal organization: reflections of a Darwinian clock-watcher. Annu. Rev. Physiol. 55, 16–54.

Rani, S., Kumar, V., 2013. Avian circannual systems: persistence and sex differences. Gen. Comp. Endocrinol. 190, 61–67.

Reppert, S.M., Weaver, D.R., Cassone, V.M., Godson, C., Kolakowski Jr, L.F., 1995. Melatonin receptors are for the birds: molecular analysis of two receptor subtypes differentially expressed in chick brain. Neuron 15 (5), 1003–1015.

Reppert, S.M., Weaver, D.R., Godson, C., 1996. Melatonin receptors step into the light: cloning and classification of subtypes. Trends Pharmacol. Sci. 17 (3), 100–102.

Rowan, W., 1926. On photoperiodism, reproductive periodicity, and the annual migrations of certain birds and fishes. Proc. Boston Soc. Nat. Hist. 38, 147–189.

Saino, N., Romano, M., Caprioli, M., Fasola, M., Lardelli, R., Micheloni, P., Scandolara, C., Rubolini, D., Gianfranceschi, L., 2013. Timing of molt of barn swallows is delayed in a rare clock genotype. PeerJ 1, e17. http://dx.doi.org/10.7717/peerj.17.

Saldanha, C.J., Silverman, A.J., Rilver, R., 2001. Direct innervation of GnRH neurons by encephalic photoreceptors in birds. J. Biol. Rhythms 16, 39–49.

Scheuerlein, A., Gwinner, E., 2002. Is food availability a circannual zeitgeber in tropical birds? A field experiment on stonechats in tropical Africa. J. Biol. Rhythms 17 (2), 171–180.

Schlinger, B.A., Remage-Healey, L., 2012. Neurosteroidogenesis: insights from studies of songbirds. J. Neuroendocrinol. 24 (1), 16–21.

Schlinger, B.A., Pradhan, D.S., Soma, K.K., 2008. 3beta-HSD activates DHEA in the songbird brain. Neurochem. Int. 52 (4–5), 611–620.

Sharp, P.J., 2005. Photoperiodic regulation of seasonal breeding in birds. Ann. N. Y. Acad. Sci. 1040, 189–199.

Sharp, P.J., Dawson, A., Lea, R.W., 1998. Control of luteinizing hormone and prolactin secretion in birds. Comp. Biochem. Physiol. 119, 275–282.

Silver, R., Witkovsky, P., Horvath, P., Alones, V., Barnstable, C.J., Lehman, M.N., 1988. Coexpression of opsin- and VIP-like-immunoreactivity in CSF-contacting neurons of the avian brain. Cell Tissue Res. 253 (1), 189–198.

Siopes, T.D., 1983. Effect of pinealectomy on broodiness of turkey hens. Poult. Sci. 62 (11), 2245–2248.

Siopes, T.D., 1994. Critical day lengths for egg production and photorefractoriness in the domestic turkey. Poult. Sci. 73 (12), 1906–1913.

Soma, K.K., Bindra, R.K., Gee, J., Wingfield, J.C., Schlinger, B.A., 1999. Androgen-metabolizing enzymes show region-specific changes across the breeding season in the brain of a wild songbird. J. Neurobiol. 41 (2), 176–188.

Soma, K.K., Schlinger, B.A., Wingfield, J.C., Saldanha, C.J., 2003. Brain aromatase, 5alpha-reductase, and 5 beta-reductase change seasonally in wild male song sparrows: relationship to aggressive and sexual behavior. J. Neurobiol. 56 (3), 209–221.

Teruyama, R., Beck, M.M., 2001. Double immunocytochemistry of vasoactive intestinal peptide and GnRH-I in male quail: photoperiodic effects. Cell Tissue Res. 303, 403–414.

Thompson, R.R., Adkins-Regan, E., 1994. Photoperiod affects the morphology of a sexually dimorphic nucleus within the preoptic area of male Japanese quail. Brain Res. 667, 201–208.

Tsutsui, K., Ukena, K., Takase, M., Kohchi, C., Lea, R.W., 1999. Neurosteroid biosynthesis in vertebrate brains. Comp. Biochem. Physiol. C Pharmacol. Toxicol. Endocrinol. 124 (2), 121–129.

Underwood, H., Siopes, T., 1984. Circadian organization in Japanese quail. J. Exp. Zool. 232, 557–566.

Viglietti-Panzica, C., Panzica, G.C., Fiori, M.G., Calcagni, M., Anselmetti, G.C., Balthazart, J., 1986. A sexually dimorphic nucleus in the quail preoptic area. Neurosci. Lett. 64, 129–134.

Wang, G., Wingfield, J.C., 2011. Immunocytochemical study of rhodopsin-containing putative encephalic photoreceptors in house sparrow, *Passer domesticus*. Gen. Comp. Endocrinol. 170 (3), 589–596.

Whitfield-Rucker, M.G., Cassone, V.M., 1996. Melatonin binding in the house sparrow song control system: sexual dimorphism and the effect of photoperiod. Horm. Behav. 30 (4), 528–537.

Whitfield-Rucker, M.G., Cassone, V.M., 2000. Photoperiodic regulation of the male house sparrow song control system: gonadal dependent and independent mechanisms. Gen. Comp. Endocrinol. 118, 173–183.

Wikelski, M., Hau, M., Wingfield, J.C., 2000. Seasonality of reproduction in a neotropical rainforest bird. Ecology 81, 2458–2472.

Wilczek, A.M., Burghardt, L.T., Cobb, A.R., Cooper, M.D., Welch, S.M., Schmitt, J., 2010. Genetic and physiological bases for phonological responses to current and predicted climates. Philos. Trans. Soc. R. Soc. B 365, 3129–3147.

Williams, T.D., 2012. Hormones, life-history, and phenotypic variation: opportunities in evolutionary avian endocrinology. Gen. Comp. Endocrinol. 176 (3), 286–295.

Wilson, F.E., March 1990. Extraocular control of seasonal reproduction in female tree sparrows (*Spizella arborea*). Gen. Comp. Endocrinol. 77 (3), 397–402.

Yamamura, T., Hirunagi, K., Ebihara, S., Yoshimura, T., 2004. Seasonal morphological changes in the neuro-glial interaction between gonadotropin-releasing hormone nerve terminals and glial endfeet in Japanese quail. Endocrinology 145, 4264–4267.

Yasuo, S., Yoshimura, T., Ebihara, S., Korf, H.W., 2009. Melatonin transmits photoperiodic signals through the MT1 melatonin receptor. J. Neurosci. 29, 2885–2889.

Yoshimura, T., 2010. Neuroendocrine mechanism of seasonal reproduction in birds and mammals. Anim. Sci. J. 81 (4), 403–410.

Yoshimura, T., 2013. Thyroid hormone and seasonal regulation of reproduction. Front. Neuroendocrinol. 34 (3), 157–166.

Yoshimura, T., Yasuo, S., Watanabe, M., Iigo, M., Yamamura, T., Hirunagi, K., Ebihara, S., 2003. Light-induced hormone conversion of T4 to T3 regulates photoperiodic response of gonads in birds. Nature 426 (6963), 178–181.

FURTHER READING

Adkins-Regan, E., 2009. Hormones and sexual differentiation of avian social behavior. Dev. Neurosci. 31 (4), 342–350.

Ball, G.F., Silver, R., 1983. Timing of incubation bouts by ring doves (*Streptopelia risoria*). J. Comp. Psychol. 97 (3), 213–225.

Ball, G.F., Riters, L.V., Balthazart, J., 1997. Neuroendocrinology of song behavior and avian brain plasticity: multiple sites of action of sex steroid hormones. Front. Neuroendocrinol. 23 (2), 137–178.

Brandstätter, R., 2003. Encoding time of day and time of year by the avian circadian system. J. Neuroendocrinol. 15 (4), 398–404.

Chue, J., Smith, C.A., 2011. Sex determination and sexual differentiation in the avian model. FEBS J. 278 (7), 1027–1034.

Davies, W.I., Turton, M., Peirson, S.N., Follett, B.K., Halford, S., Garcia-Fernandez, J.M., Sharp, P.J., Hankins, M.W., Foster, R.G., 2012. Vertebrate ancient opsin photopigment spectra and the avian photoperiodic response. Biol. Lett. 8 (2), 291–294.

Dawson, A., Sharp, P.J., 1998. The role of prolactin in the development of photorefractoriness and postnuptial molt in the European starling (*Sturnus vulgaris*). J. Endocrinol. 139, 485–490.

Dawson, A., King, V.M., Bentley, G.E., Ball, G.F., 2001. Photoperiodic control of seasonality in birds. J. Biol. Rhythms 16 (4), 365–380.

El Halawani, M.E., Kang, S.W., Leclerc, B., Kosonsiriluk, S., Chaiseha, Y., 2009. Dopamine-melatonin neurons in the avian hypothalamus and their role as photoperiodic clocks. Gen. Comp. Endocrinol. 163 (1–2), 123–127.

Fusani, L., 2008. Testosterone control of male courtship in birds. Horm. Behav. 54 (2), 227–233.

Gahr, M., 2007. Sexual differentiation of the vocal control system of birds. Adv. Genet. 59, 67–105.

Gahr, M., Kosar, E., 1996. Identification, distribution, and developmental changes of a melatonin binding site in the song control system of the zebra finch. J. Comp. Neurol. 367 (2), 308–318.

Gwinner, E., Brandstätter, R., 2001. Complex bird clocks. Philos. Trans. R. Soc. Lond., B, Biol. Sci. 356 (1415), 1801–1810.

Hut, R.A., Paolucci, S., Dor, R., Kyriacou, C.P., Daan, S., July 3, 2013. Latitudinal clines: an evolutionary view on biological rhythms. Proc. Biol. Sci. 280 (1765), 20130433. http://dx.doi.org/10.1098/rspb.2013.0433.

Jurkevich, A., Grossmann, R., 2003. Vasotocin and reproductive functions of the domestic chicken. Domest. Anim. Endocrinol. 25 (1), 93–99.

Kang, S.W., Thayananuphat, A., Bakken, T., El Halawani, M.E., 2007. Dopamine-melatonin neurons in the avian hypothalamus controlling seasonal reproduction. Neuroscience 150 (1), 223–233.

Konkle, A.T., Balthazart, J., 2011. Sex differences in the rapid control of aromatase activity in the quail preoptic area. J. Neuroendocrinol. 23 (5), 424–434.

Kosonsiriluk, S., Mauro, L.J., Chaiworakul, V., Chaiseha, Y., El Halawani, M.E., 2013. Photoreceptive oscillators within neurons of the premammillary nucleus (PMM) and seasonal reproduction in temperate zone birds. Gen. Comp. Endocrinol. 190, 149–155.

Mewaldt, L.R., King, J.R., 1977. The annual cycle of white-crowned sparrows (*Zonotrichia leuchophrys nuttalli*) in coastal California. Condor 79, 445–455.

Nakao, N., Yasuo, S., Nishimura, A., Yamamura, T., Watanabe, T., Anraku, T., Okano, T., Fukada, Y., Sharp, P.J., Ebihara, S., Yoshimura, T., 2007. Circadian clock gene regulation of steroidogenic acute regulatory protein gene expression in preovulatory ovarian follicles. Endocrinology 148 (7), 3031–3038.

O'Connell, M.E., Reboulleau, C., Feder, H.H., Silver, R., 1981. Social interactions and androgen levels in birds. I. Female characteristics associated with increased plasma androgen levels in the male ring dove (*Streptopelia risoria*). Gen. Comp. Endocrinol. 44 (4), 454–463.

Price, T.D., 2002. Domesticated birds as a model for the genetics of speciation by sexual selection. Genetica 116 (2–3), 311–327.

Proudman, J.A., Siopes, T.D., 2002. Relative and absolute photorefractoriness in turkey hens: profiles of prolactin, thyroxine, and triiodothyronine early in the reproductive cycle. Poult. Sci. 81 (8), 1218–1223.

Sharp, P.J., Blache, D., 2003. A neuroendocrine model for prolactin as the key mediator of seasonal breeding in birds under long- and short-day photoperiods. Can. J. Physiol. 81, 350–358.

Steinman, M.Q., Valenzuela, A.E., Siopes, T.D., Millam, J.R., 2013. Tuberal hypothalamic expression of the glial intermediate filaments, glial fibrillary acidic protein and vimentin across the turkey hen (*Meleagris gallopavo*) reproductive cycle: further evidence for a role of glial structural plasticity in seasonal reproduction. Gen. Comp. Endocrinol. 193C, 141–148.

Tramontin, A.D., Brenowitz, E.A., 2000. Seasonal plasticity in the adult brain. Trends Neurosci. 23 (6), 251–258.

Trivedi, A.K., Rani, S., Kumar, V., 2006. Control of annual reproductive cycle in the subtropical house sparrow (*Passer domesticus*): evidence for conservation of photoperiodic control mechanisms in birds. Front. Zool. 3, 12.

de Vries, G.J., Panzica, G.C., 2006. Sexual differentiation of central vasopressin and vasotocin systems in vertebrates: different mechanisms, similar endpoints. Neuroscience 138 (3), 947–955.

Wang, G., Harpole, C.E., Trivedi, A.K., Cassone, V.M., 2012. Circadian regulation of bird song call, and locomotor behavior by pineal melatonin in the zebra finch. J. Biol. Rhythms 27 (2), 145–155.

Wilson, F.E., 1986. A testosterone-independent reduction in net photoperiodic drive triggers photorefractoriness in male tree sparrows (*Spizella arborea*). J. Endocrinol. 109 (1), 133–137.

Wilson, F.E., Reinert, B.D., 1993. The thyroid and photoperiodic control of seasonal reproduction in American tree sparrows (*Spizella arborea*). J. Comp. Physiol. B 163 (7), 563–573.

Wingfield, J.C., Jacobs, J., Hillgarth, N., 1997. Ecological constraints and the evolution of hormone-behavior interrelationships. Ann. N. Y. Acad. Sci. 807, 22–41.

Chapter 36

Annual Schedules

Thomas P. Hahn, Kathleen R. Brazeal, Elizabeth M. Schultz, Helen E. Chmura and Jamie M. Cornelius
Department of Neurobiology, Physiology and Behavior, University of California, Davis, CA, USA

Heather E. Watts
Department of Biology, Loyola Marymount University, Los Angeles, CA, USA

Scott A. MacDougall-Shackleton
Departments of Psychology and Biology, University of Western Ontario, Canada

36.1 INTRODUCTION

Research into the mechanisms underlying the annual cycles of birds began in the 1920s with the pioneering work of Rowan (1925, 1926) exploring effects of changing photoperiod on migratory and reproductive physiology and behavior. Since that time, there has been major progress toward a deep understanding of these mechanisms at the organismal, cellular, and molecular levels. There also has been increasing investigation of the orchestration of transitions among stages of the entire annual schedule—not only the timing of breeding and migration—and exploration of seasonal changes in demanding processes, such as immune function, that must be maintained at all times. Current interest is high in the implications of timing mechanisms for responses to human-induced environmental changes, such as climate change. This chapter focuses on the current state of research on avian annual schedules, highlighting the areas of recent progress and interest. Birds continue to be exciting and productive models for understanding the basic mechanisms by which organisms respond to the environment, from molecular to whole-organism levels, and for revealing the importance of these mechanisms in evolution.

36.2 BACKGROUND: PATTERNS OF ENVIRONMENTAL VARIATION AND AVIAN ANNUAL SCHEDULES

Long-term changes of physiology, behavior, and morphology, orchestrated by the neuroendocrine and endocrine systems, are prominent in most species of birds. In many cases, these represent consistent annual cycles, particularly for species that spend all or part of the year at mid to high latitudes, where relevant changes in environmental conditions (the "ultimate factors" of Baker, 1938) track the seasons, but also in tropical areas with regular seasonal patterns of rainfall. Less regular long-term changes in physiology, behavior, and morphology prevail in species occupying environments where conditions change more erratically, such as certain deserts where rainfall is unpredictable (Perfito et al., 2007), or even in otherwise seasonal environments where variation in the food supply is not regularly seasonal for some reason (Hahn et al., 1997, 2008). In at least some of these species (e.g., crossbills), temporally flexible changes in physiology, behavior, and morphology are superimposed on underlying seasonal cycles (Hahn, 1995, 1998; Hahn et al., 1997, 2008; Cornelius and Hahn, 2012). In others (e.g., Darwin's finches), it remains unclear to what extent changes in reproductive physiology are entirely opportunistic versus based in part on an underlying seasonal program (Hau et al., 2004; Hahn et al., 2008).

Whether regularly seasonal or more flexible or erratic, annual schedules all must at least coordinate reproduction and plumage molt, and often also migrations and a nonbreeding "overwintering" stage (Wingfield, 2008). Timing and coordination of these processes are crucial to fitness, and mechanisms permitting this have responded to selection (Bradshaw and Holzapfel, 2007, 2010). The timing mechanisms that evolve are based on cue response systems that link neuroendocrine and endocrine regulatory mechanisms to "proximate factors" (Baker, 1938)—cues such as day length, temperature, food supply, and social interactions—that provide predictive information either in the short or long term regarding changing ultimate factors. These cue response systems generally include a long-term component that sets the time windows when particular cycle stages can occur and that dictates general features of physiological preparation for and termination of the stages, based on some combination of response to changing photoperiod and an endogenous program (the so-called initial predictive cues of Wingfield, 1980, 1985; Gwinner, 1986; Wingfield and Farner, 1993; Wingfield and Kenagy, 1991; Dawson, 2002; Bradshaw and Holzapfel, 2007; Wikelski et al., 2008; Helm et al., 2013). This long-term component is then combined with fine-tuning of precise timing by responses to more immediate cues that provide accurate short-term information about local conditions, such as changing temperature, weather, food supply itself, and behavioral cues from other individuals (Wingfield, 1980, 1985).

Annual schedules of different species vary in their complexity and therefore potentially in the mechanisms required to orchestrate them. The different components (e.g., reproduction, migration, plumage molt, overwintering) can theoretically overlap temporally or be separated into very distinct "finite states" of the annual schedule (Jacobs and Wingfield, 2000; Wingfield, 2008). Certain activities simply cannot overlap (e.g., egg laying, incubation, and care of nestlings are incompatible with migration), but many components of "breeding," such as the gonadal maturation process (Wingfield et al., 1992; Bauchinger et al., 2007, 2008, 2009), pair formation (Ganter et al., 2005), and extended parental care of fully mobile youngsters can be compatible with migration, and plumage molt need not be incompatible with any stage of reproduction (Bond et al., 2013) or migration (Yuri and Rohwer, 1997; Voelker and Rohwer, 1998). Coordination of reproduction, migration, and molt can be viewed as an overlap continuum, from near-complete temporal separation at one extreme to near-complete temporal overlap (within constraints noted above) at the other. A critical and fundamental trade-off is associated with this continuum. Specifically, greater overlap of components confers greater temporal flexibility but at the expense of scope for tolerance of variation in environmental conditions. Conversely, greater separation of components into distinct stages ("finite states") confers improved tolerance scope for environmental variation but at a cost to temporal flexibility.

Two relatively extreme examples illustrate this concept (Figure 36.1; Wingfield, 2008). A regular seasonal migrant, the Gambel's white-crowned sparrow (*Zonotrichia leucophrys gambelii*), displays a six-part annual schedule consisting of separate and minimally overlapping stages of overwintering, partial prealternate molt, vernal migration, breeding, complete prebasic molt, and autumnal migration. The phenotypic traits expressed during each of these stages broaden the scope of environmental variation (e.g., temperatures that can be endured, foods that are acceptable) the birds can tolerate across the entire year as compared with tolerances within any one stage. This benefit comes at a cost of temporal flexibility, however; breaking the year into multiple stages, each with unique phenotypic characteristics requiring time and energy to gear-up and terminate, reduces how nimbly the animal can adjust to any environmental changes that deviate from the normal routine it has evolved to track. In contrast, certain desert-dwelling reproductive opportunists, such as zebra finches (*Taeniopygia guttata*) in parts of the arid interior of Australia, display such extensive overlap between stages that no truly separate stages exist. Although the birds do not breed continuously, they can maintain near-readiness to breed for long periods of time, they apparently can move in search of better conditions at any time, and they display a low-intensity but essentially continuous plumage molt that may overlap with both breeding and any migratory movements that do occur (Cornelius et al., 2011). Expression of this single chimaera stage confers high temporal flexibility because any activity is possible at any time, but at the cost of scope for tolerating environmental extremes. This strategy is an example of the jack-of-all-trades being master of none (Huey and Hertz, 1984).

The capacity to express these different patterns depends on coordination mechanisms. Specifically, separation of the annual schedule into distinct stages requires mechanisms synchronizing the cycle with the environment, as well as mechanisms of coordination that limit overlap between stages, maintain the stages in an adaptive sequence, and orchestrate transitions from stage to stage so that processes involved in termination of one do not interfere with processes involved in development and onset of the next (Wingfield, 2008). In contrast, if components are to overlap partially or completely, mechanisms that facilitate this overlap must be present. The next section will address recent advances in understanding the mechanisms by which diverse annual schedules, from relatively fixed and seasonal to extremely flexible or opportunistic, are generated, with particular focus on the rapidly advancing field of avian photoperiodism. We return later to the question of whether these diverse patterns reflect adaptively specialized mechanisms or plastic outputs of common mechanistic underpinnings.

36.3 EFFECTS OF AND MECHANISMS OF RESPONSE TO PHOTOPERIOD AND OTHER ENVIRONMENTAL CUES

36.3.1 Photoperiodic Response

Annual schedules of reproduction, migration, plumage molt, and overwintering that track consistent seasonal changes in the environment have long been known to depend on proximate responses to changing photoperiod (Rowan, 1925, 1926; Dawson et al., 2001; Dawson, 2002; Sharp,

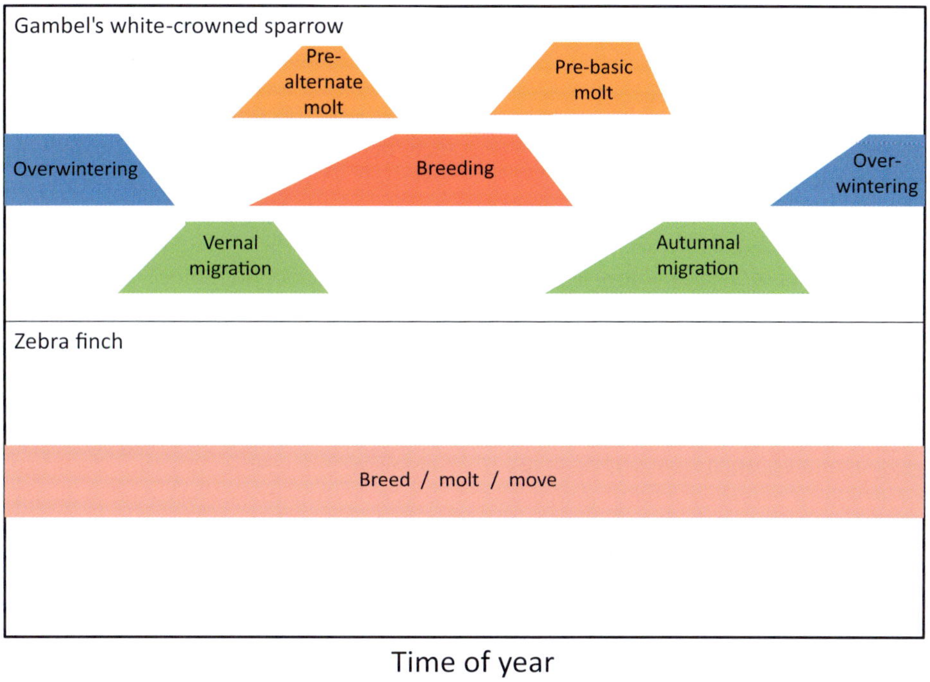

FIGURE 36.1 Diagrammatic representation of the multiple life cycle stages of a long-distance migrant seasonal breeder, the Gambel's white-crowned sparrow (*Zonotrichia leucophrys gambelii*, top panel), and of the single chimaera life stage of a nomadic reproductive opportunist, the zebra finch (*Taeniopygia guttata*, bottom panel). *Adapted from Wingfield (2008).*

2005; Dawson and Sharp, 2007; Bradshaw and Holzapfel, 2007). A system that uses photoperiod as a proximate cue for regulation of annual schedules, either as a driver or as an entrainer of an endogenous program, requires a combination of a mechanism for detecting light (photoreceptors), a mechanism for determining the length of the day (known in birds to depend on a circadian clock), and a mechanism for transducing the photoperiod information into neuroendocrine and endocrine signals that regulate the different components of the schedule, such as the septo-infundibular GnRH-I system that regulates reproduction (see Follett, 1984).

Recent advances in our understanding of how photoperiod may be measured and transduced into regulatory signals have been particularly impressive regarding photoperiodic regulation of reproductive cycles, although much remains to be done to flesh out details and to determine which mechanistic features are universal versus unique to particular species or populations. Most birds are long day breeders, and the annual reproductive cycle of seasonal breeders breaks naturally into three stages: sensitivity, stimulation, and refractoriness (see Goodson et al., 2005; Ball, 1993; Dawson et al., 2001). The transition from being *sensitive* to long days but reproductively inactive to *stimulated* by long days depends on interpretation of increasing day length as an activational cue. Birds possess a circadian cycle of photoinducibility, whereby light occurring during a window of circadian times is interpreted as a long day, and activation of the hypothalamo-pituitary-gonad axis results

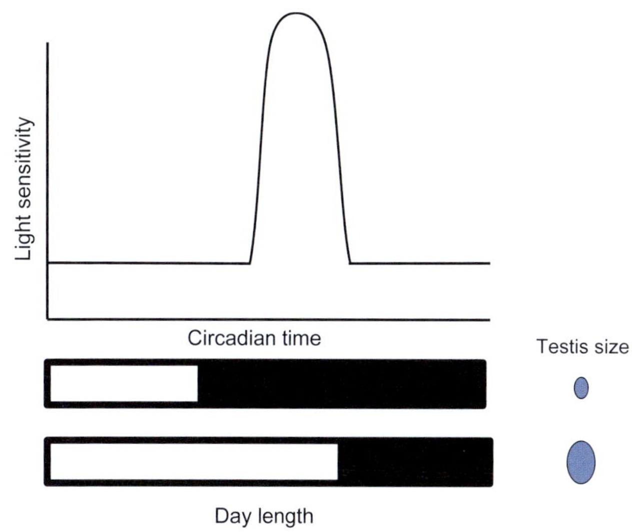

FIGURE 36.2 **Illustration of the circadian cycle of photoinducibility displayed by photoperiodic birds.** The reproductive system is responsive to the stimulatory effects of light during a particular phase of the circadian cycle, represented by the peak between about circadian hours 12 and 20 in the top panel. On short days, the light period always ends before this inducible phase begins; hypothalamo–pituitary activity remains minimal and gonads remain small (middle panel). On long days, the light period extends beyond the beginning of the inducible phase; hypothalamo–pituitary activity increases, and the gonads grow. Experiments manipulating the circadian time when short (8h or less) pulses of light occur demonstrate that it is not how long it is light, but what circadian time it is light that determines whether the bird activates the reproductive axis (Hamner, 1963, 1964; Follett et al., 1974).

(Hamner, 1963, 1964; Follett et al., 1974; Figure 36.2). Recent discoveries have revealed new ideas regarding the process of light detection, evaluation of the day length, and transduction into neuroendocrine signals regulating changes in physiology, behavior, and morphology.

Detection of the light cue regulating seasonal cycles of birds has long been known not to be retinal (Benoit, 1935), and has been confirmed not to require either the retina or the pineal; all key components of avian photoperiodism—photostimulation and photorefractoriness in response to long days, and dissipation of the refractory state in response to short days—are expressed normally in pinealectomized/enucleated American tree sparrows (*Spizella arborea*; Wilson, 1991). Deep brain photoreceptors are responsible for photodetection regulating seasonality in birds and other nonmammalian vertebrates (Silver et al., 1988; see reviews in Foster et al., 1994; Foster and Soni, 1998). These cerebrospinal fluid contacting neurons are thought to reside in the mediobasal hypothalamus (MBH) and/or lateral septum (LS); recent studies increasingly point to the cells in the LS as the key photodetectors in the avian photoneuroendocrine system (see Li et al., 2004). Details of how the long day cue is transduced are beginning to emerge, although virtually all of the work to date is restricted to Japanese quail (Figure 36.3). In quail, these photoreceptors (Opsin 5 positive CSF-contacting neurons) detect light, and when light occurs during the inducible phase of the circadian cycle of photoinducibility (see Figure 36.2), this long day signal is transmitted to thyroid-stimulating hormone (TSH)-producing cells in pars tuberalis (PT). TSH from these cells putatively acts on the nearby ependymal cells (ECs) lining the third ventricle (VIII), inducing transcription of the Dio2 gene that codes for the type 2 deiodinase enzyme responsible for converting T4 (thyroxine) to the physiologically active form T3 (3,5,3′-triiodothyronine). This TSH also induces downregulation of expression of the Dio3 gene in these cells, thus reducing the deactivation paths from T4 to reverse T3 (rT3), and from T3 to T2 (3,3′-diiodothyronine), both inactive forms. Inside these ependymal cells, short days reduce T3 production by enhancing Dio3 expression, leading to increased conversion of T4 to rT3 and T3 to T2. When T3 is low, glial endfeet appear to ensheath axon terminals of gonadotropin-releasing hormone (GnRH-I) cells, separating them physically from the basal lamina in the median eminence (ME), thereby presumably obstructing GnRH-I secretion into the hypothalamo-hypophysial portal blood. As a result, gonadotropin release from the anterior pituitary remains low under short days. In contrast, under long days, Dio2 expression in the ECs is enhanced so T4 conversion to the biologically active T3 is emphasized. T3 release from the ECs is therefore high under long days, and this may cause retraction of the glial endfeet sheathes so that

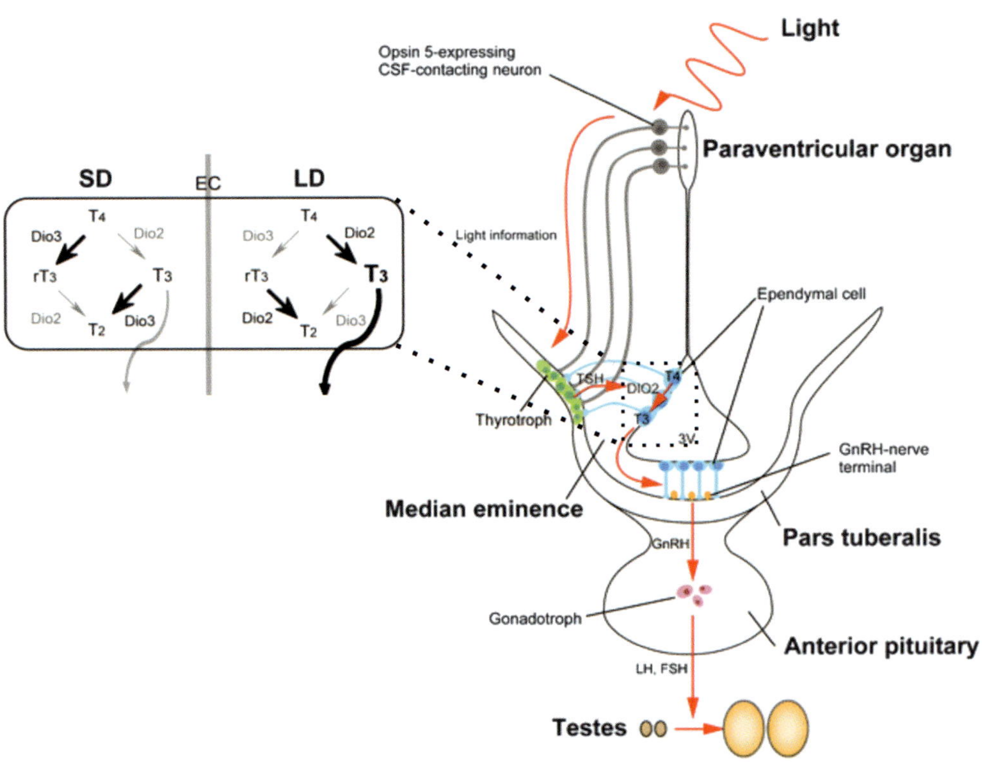

FIGURE 36.3 Illustration of the key components of the photoneuroendocrine system of the Japanese quail (*Coturnix japonica*). Inset at left shows the enzymatic pathways emphasized under short (SD, left) and long (LD, right) days in the ependymal cells (EC) lining the third ventricle. Full explanation in text. *After Ikegami and Yoshimura (2012).*

GnRH-I axon terminals are closely apposed to basal lamina and GnRH-I release is high, leading to high gonadotropin secretion and gonadal growth (see Ikegami and Yoshimura, 2012 for review).

Whether the scenario outlined above applies generally to photoperiodic regulation of avian seasonality or only to Japanese quail exposed to abrupt changes in photoperiod remains unclear at present. Seasonal changes in Dio2 and GnRH-I expression in European starlings (*Sturnus vulgaris*) allowed to breed in outdoor aviaries on naturally changing photoperiods are not consistent with the above quail-based model (Bentley et al., 2013). Specifically, breeding males with maximal testis volume and GnRH-I gene expression did not have significantly higher Dio2 expression in the MBH than did refractory individuals on declining photoperiods. Further, photosensitive individuals with small gonads on still-declining photoperiods in autumn had higher Dio2 expression than did photostimulated individuals in spring. It is not known whether this deviation from the pattern described for Japanese quail relates to phylogenetic differences, to differences in photoperiodic characteristics (starlings becoming absolutely refractory, whereas quail become only relatively refractory), or to differences resulting from the experimental conditions used (naturally changing photocycles for starlings, stepwise changes in photoperiod for quail). Regardless of the explanation, it is clear that more work testing the effects of different photoperiod conditions on a wider range of species is needed to identify robust general features of the neural mechanisms underlying photoperiodic regulation of annual schedules in birds (Bentley et al., 2013).

The mechanistic details of the connection between a circadian cycle of photoinducibility and the decision of the photoneuroendocrine system to increase TSH within the PT remain obscure at present. However, it is noteworthy that cells within the MBH maintain steady expression profiles of key clock genes under different photo-regimes, consistent with the possibility that they are responsible fundamentally for maintaining the temporal phasing of photoinducibility (Yasuo et al., 2003; see also Ikegami and Yoshimura, 2012).

The above details apply to the process of photostimulation, which leads to development of the reproductive axis as well as preparations for spring migration. Mechanisms underlying photorefractoriness, regulation of the timing of molt, autumn migration, and the dissipation of the refractory state are much less well-studied, but the basic structure of the entire annual schedule likely depends on the seasonallychanging responses to photoperiod. The transition out of the reproductive state generally depends on development of photorefractoriness (Nicholls et al., 1988; Hahn et al., 1997; Hahn and MacDougall-Shackleton, 2008). Expression of absolute photorefractoriness is a two-stage process, first involving cessation of release of GnRH-I at the median eminence, followed by a dramatic decline in the presence of the GnRH-I peptide and reduction in transcription of the GnRH-I gene (Ubuka et al., 2009; Stevenson et al., 2009, 2011; see Stevenson et al., 2012 for review). Return of photosensitivity is associated with reappearance of the GnRH-I peptide first in the preoptic area, followed by the median eminence in European starlings (Dawson and Goldsmith, 1997); as refractoriness dissipates, birds become progressively more responsive to long days (Hamner, 1968). Species vary in the extent to which steroid feedback during fall and winter helps to keep gonadotropin secretion of photosensitive individuals in check until days increase sufficiently for photostimulation (Cockrem, 1995).

36.3.2 Processing of Nonphotic Cues

There is much more information on the phenomenology and physiological consequences of the effects of nonphotic cues on annual schedules than there is on the mechanisms by which these cues are processed. This section will discuss effects of a variety of nonphotic cues on annual schedules.

36.3.2.1 Effects of Temperature

Ambient temperature is correlated with local variation in vegetation and food phenology and consequently can be a valuable supplementary proximate cue fine-tuning timing of transitions between life cycle stages. Experimental studies reveal effects of temperature on reproductive development under photostimulation (Wingfield et al., 1997, 2003; Perfito et al., 2005; Silverin et al., 2008; Caro and Visser, 2009; Visser et al., 2009, 2011; Schaper et al., 2012), on the transition from reproduction to photorefractoriness and molt (Wingfield et al., 1996, 1997, 2003; Wada, 1993; Wada et al., 1990; Dawson, 2005; Silverin et al., 2008; Visser et al., 2011), and on preparations for spring and/or fall migration (white-crowned sparrows: Wingfield et al., 1996, 1997, 2003). The effects on reproductive development appear to be locally adapted as functions of breeding latitude (Wingfield et al., 1996, 1997, 2003; Silverin et al., 2008, Figure 36.4), elevation (Perfito et al., 2005), and status as a migrant or resident (Meijer et al., 1999; Dawson, 2005). These findings support theoretical predictions that populations inhabiting more predictable environments will be less responsive to supplementary environmental cues than those inhabiting less predictable environments (Wingfield et al., 1992, 1993). Birds appear to be more responsive to temperature increases across the season than to average daily temperatures or daily temperature fluctuations (Schaper et al., 2012). It is important to note that different results have been obtained when using different artificial fixed photoperiods (Silverin and Viebke, 1994; Silverin et al., 2008) or using naturally changing photoperiods versus artificially long or short days (Dawson, 2005).

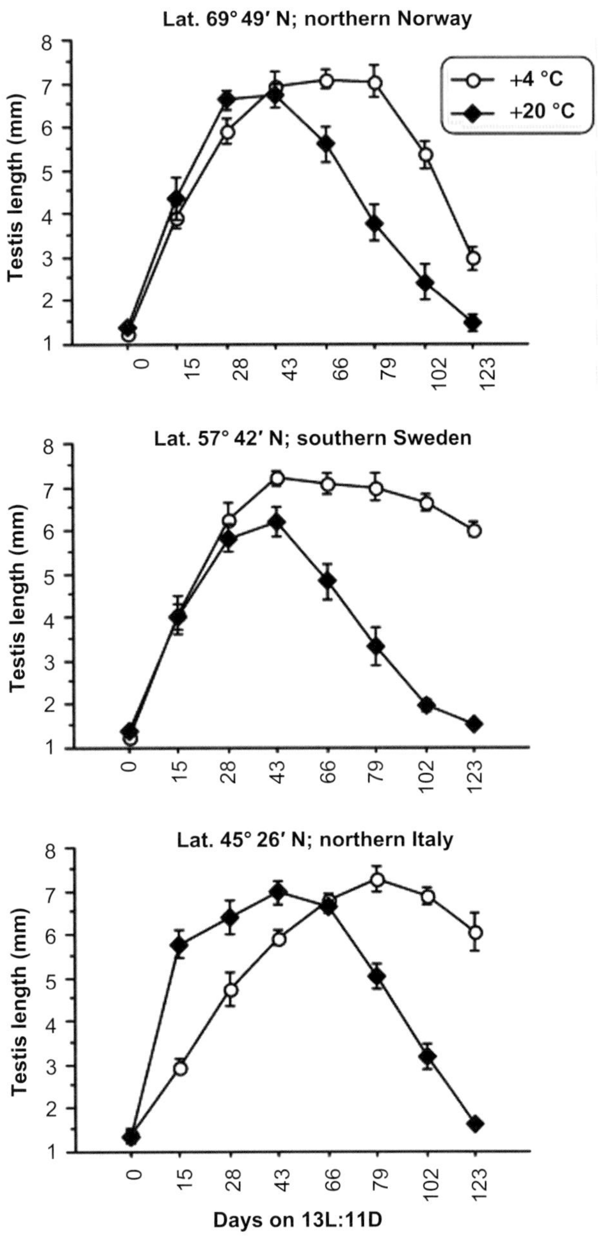

FIGURE 36.4 Effects of different temperature treatments on photo-induced gonadal development in male great tits (*Parus major*) from populations living at three different latitudes. Low temperatures delayed termination of reproductive competence in all three populations but only affected rate of gonadal development in the lowest latitude population. *From Silverin et al. (2008).*

The influence of temperature on the onset of refractoriness and the transition to molt and fall migration has been less well studied. Persistent low temperatures under photostimulation retard the transition to gonadal regression and progression of molt while persistent high temperatures advance these life history stages (Wingfield et al., 1997, 2003). High temperatures under photostimulation also cause earlier gonadal regression in European starlings (Dawson, 2005) and great tits (Silverin et al., 2008; Visser et al., 2011). Some of the most detailed investigations on the role of temperature during the transition from reproduction to photorefractoriness have been conducted in Japanese quail. Some strains of Japanese quail express relative photorefractoriness simply when exposed to a decline in photoperiod after a period of reproductive maturity (Robinson and Follett, 1982). Others do not express refractoriness without a concomitant decline in ambient temperature (Wada, 1993; Wada et al., 1990).

A few studies suggest that the annual schedule of males and females responds differently to temperature cues. High temperatures advanced brood patch and follicular development in female Puget Sound white-crowned sparrows, while testis development in males was unaffected (Wingfield et al., 1997).

Several pathways may be involved in the integration of temperature information with other physiological systems. Early studies found no difference in luteinizing hormone (LH) secretion associated with temperature treatment despite differences in gonadal development (Wingfield et al., 1996, 1997, 2003). However, three populations of great tits did display differences in timing of LH secretion with temperature treatment (Silverin et al., 2008). Thyroid hormones and prolactin are also likely candidates for mediating temperature effects (Caro et al., 2013). In one strain of domesticated Japanese quail, the transition to refractoriness appears to require temperature-dependent conversion of T4 to T3 (Wada, 1993; Wada et al., 1990). It would be particularly interesting to explore the temperature dependence of local production of T3 in the MBH of this strain of quail in light of recent discoveries regarding this process in quail photoperiodism (see above). Additional studies propose that thyroid hormones, as mediators of metabolism, may play a role in reproductive timing through changes in energy balance (see below.)

Results of studies investigating the role of prolactin in transducing temperature signals to gonadal development and regression have been inconclusive (Maney et al., 1999; Perfito et al., 2005; Dawson and Sharp, 2010). Differences in prolactin were associated with temperature treatments and gonadal development in female *Z. l. pugetensis* and male and female *Z. l. oriantha* but not in *Z. l. gambelii*, consistent with observed patterns of temperature sensitivity in these taxa (Maney et al., 1999). However, prolactin levels did not differ in high and low temperature treatments in two populations of song sparrows (Perfito et al., 2005); work in starlings provides no support for changes in prolactin levels leading to the advancement of photorefractoriness under high temperatures (Dawson and Sharp, 2010).

Temperature could influence seasonal timing by acting directly on effectors or indirectly through energy balance (Caro et al., 2013; Meijer et al., 1999). Because temperature affects thermoregulatory costs, the amount of energy available for various processes will be altered by ambient temperature, and this may affect timing of transitions between stages (Meijer et al., 1999). Although this is an intriguing possibility, this hypothesis has not been rigorously tested and the mechanisms by which physiological decisions about energy allocation are made remain largely unknown. Great tits exposed to high and low temperature treatments did differ in oxygen consumption, but without coincident differences in testis growth rate (Caro and Visser, 2009). This implies no energetic constraint of temperature on gonadal growth in males; however, other trade-offs, such as a decrease in immune function, were not investigated. In house sparrows, basal metabolic rate and thyroid hormone levels were positively associated, and early breeding birds had an earlier peak in thyroid hormone than late breeding birds (Chastel et al., 2003). Perhaps late breeders face a metabolic constraint that prevents them from breeding early, and thyroid hormones play a role in mediating this relationship (Chastel et al., 2003). Further studies, especially investigations including females who face greater energy demands during follicle maturation, are needed.

The mechanisms by which environmental temperatures influence behavioral timing remain a mystery. It is still unclear how the temperature cues that influence seasonal timing are detected or how are those cues are integrated with systems that control changes in morphology, physiology and behavior. Typically, ambient temperature is detected by thermoreceptors in the periphery. It is possible that these receptors may detect seasonal changes in temperature, although this has not been explored (Caro et al., 2013).

36.3.2.2 Effects of Food

Birds living in seasonally fluctuating environments time reproduction to coincide with periods of high food availability as well as availability of specific foods required by the young (Lack, 1968; Perrins, 1970); individuals that match rearing of young with peak food supply generally enjoy relatively high reproductive success (Perrins, 1991; Nager and van Noordwijk, 1995; van Noordwijk et al., 1995). Indeed, the evolution of the entire photoneuroendocrine timing system in birds has presumably evolved as a consequence of selection for tracking seasonal changes in food supply. However, food itself is also a potent supplementary proximate factor (Hahn et al., 2005) that can assist in fine tuning transitions of the annual schedule to local conditions. Studies of the proximate effects of food on annual schedules include correlative field studies, experimental field studies, and experimental studies with captive birds.

It can be challenging to relate food supply to annual scheduling when birds consume a wide variety of foods, but specialist taxa, such as crossbills (*Loxia* spp.), provide unique opportunities in this regard. Field studies of crossbills show close correspondence between breeding and intake rates of conifer seeds in these nomadic opportunistic breeders (Benkman, 1990). Crossbills also make the transition from breeding to molt earlier in years of low cone crops (Cornelius, Schultz, and Hahn, unpublished data), but this could be simply because the birds forego breeding or terminate it earlier in years with low food supply, not because low food "stimulates" molt. However, reducing food is a well-established means of stimulating the transition from egg-laying to molt in domestic poultry (see Yousaf and Chaudhry, 2008 for review), so it remains plausible that this transition responds directly to food supply as a proximate cue in free-living populations as well. There is some evidence that natural changes in food supply affect migration schedule. For instance, free-living American redstarts alter spring departure date interannually as a function of variation in rainfall and arthropod abundance (Studds and Marra, 2011). Likewise, redstarts wintering in higher quality (wetter, better food supply) habitats arrive on the breeding grounds earlier than those from lower quality wintering habitat (Tonra et al., 2011; Marra et al., 1998). Drought in stopover areas has been implicated in dramatic delays in arrival on breeding grounds by trans-Saharan migrants, probably because of reduced refueling rates owing to poor food supplies (Tøttrup et al., 2012). Some experimental evidence also supports the interpretation that food supply can affect timing of migration. Free-living American redstarts that are experimentally "upgraded" from low-quality habitat by creating vacancies for them in nearby high-quality habitat show earlier spring migration departure (Studds and Marra, 2005). Likewise, dark-eyed juncos brought temporarily into captivity and subjected to experimental diets (restricted versus copious) varied in migration departure date (inferred from resighting surveys after release), with individuals displaying higher body condition indices departing earlier (Bridge et al., 2010). Together these data are consistent with the idea that food supply can act as a supplementary cue, influencing details of migration timing under major control by photoperiod and/or an endogenous program. Some irruptive species, such as crossbills, appear to be able to initiate long-distance movements at most times of year, and these movements probably are in direct response to declines in food supply. Experimental studies with captive crossbills indicate that these birds are very sensitive to changes in food supply, modifying activity, fat deposition, and circulating corticosterone levels in ways consistent with the hypothesis that they use food availability as a cue to regulate timing of migratory movements (Cornelius et al., 2010).

Numerous studies have been performed in the field and a few in captivity manipulating aspects of food supply to determine how responsive the reproductive axis is to food as a proximate cue. Food supplementation of free-living birds generally advances timing of egg-laying (Arcese and Smith, 1988; Meijer et al., 1990; Schoech, 1996; Nager et al., 1997). Availability of calories is not all that matters.

For instance, in calcium-poor environments, calcium supplementation can significantly advance egg laying date (Mand et al., 2000). Some of the most elegant field experiments have provided groups of free-living Florida scrub jays with isocaloric food supplementation that differs in protein content. Although food supplementation with either high protein or low protein pellets advanced egg-laying date compared with unsupplemented controls, the high protein supplement advanced lay date significantly more than did the low protein supplement (Schoech et al., 2004). Meta-analyses of the results of field experiments reveal a latitude effect on responsiveness to food supplementation; high latitude populations are less responsive to the advancing effects of food supplementation than are low latitude populations (Schoech and Hahn, 2007). However the representation of low latitude populations in this analysis is very small. Further, the apparent latitude effect is probably more directly the consequence of whether populations are single or multiple brooded (Dhondt, 2010). In any case, these analyses are consistent with the interpretation that populations are locally adapted in terms of responsiveness to supplementary cues fine tuning reproductive timing.

Food manipulations with captive birds generally reveal at most very subtle effects of food alone on reproductive development (in the absence of concomitant photostimulation), even in flexible breeders such as crossbills (Hahn et al., 2005). However effects of food manipulations in photostimulated birds can be quite dramatic. For instance, pine siskins (*Spinus pinus*) on modestly stimulatory photoperiod provided seeds in addition to commercial pellets show dramatically advanced gonadal development compared with those lacking seeds, and this effect occurs irrespective of social environment (Watts and Hahn, 2012, Figure 36.5).

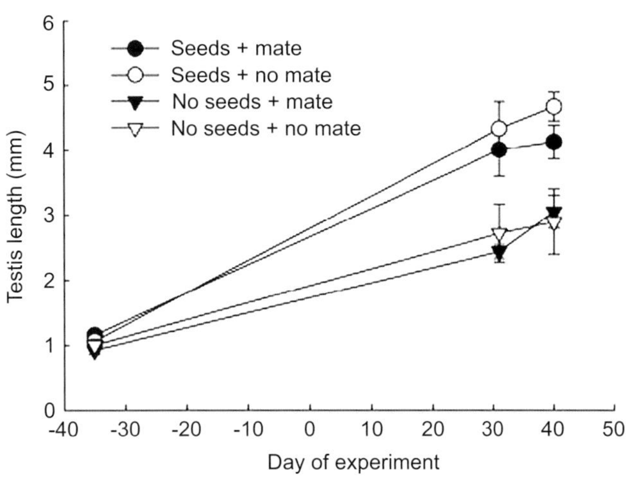

FIGURE 36.5 Effect of provisioning with sunflower seeds in addition to commercial pellet diet as compared with pellet diet alone in pine siskins (*Spinus pinus*). Access to seeds accelerated photoinduced gonadal development comparably whether or not the males were housed with a conspecific female. *From Watts and Hahn (2012).*

In tropical spotted antbirds (*Hylophylax naevioides*), food cues can have dramatic effects on timing of reproductive development (Hau et al., 2000; O'Brien and Hau, 2005), and at least part of this effect appears to be via nonnutritional pathways; exposure to live insect food, even if it cannot be eaten, has a pronounced positive effect on singing behavior (Hau et al., 2000). This is similar to some flexibly breeding temperate zone species, such as pinyon jays, which can show late summer reproductive development if allowed to handle as well as consume seeds from fresh developing pinyon pine cones (Ligon, 1974). It is important to note that effects of long days on reproductive development could potentially be explained in part by increases in food intake under longer days, however experimental restriction of time available to feed or amount of food provided to short-day levels reveal only modest effects of food as compared with photoperiod, even in the very flexible crossbills (see Dawson, 1986; Hahn, 1995; Hahn et al., 2005).

It remains important to consider the extent to which effects of food are direct ("income" effects) or operate through effects on body condition and reserves ("capital" effects; Meijer and Drent, 1999). At least part of the effect of elevated protein on laying date in scrub-jays is likely to be through improved condition and protein reserves in females (i.e., may be a capital effect; see Schoech et al., 2004). Capital effects on reproductive output are well-known in large birds (e.g., snow geese; Ankney and MacInnes, 1978), but some of the effects in food supplementation experiments are consistent with nutrient reserves (capital effects) influencing reproductive timing of small passerines as well (see Arcese and Smith, 1988).

36.3.2.3 Effects of Behavioral Factors

Behavioral interactions between conspecifics can provide valuable information for fine-tuning timing of transitions of physiology, behavior, and morphology, and are essential for coordination of pair, family, and group activities (Wingfield and Marler, 1988).

Two excellent examples of the influence of social factors on annual scheduling come from experimental studies of the transition from reproduction to molt in songbirds. Photostimulated captive male European starlings delay gonadal collapse and onset of molt when housed with females (Schwab and Lott, 1969). Co-habiting with photostimulated females is now known to affect GnRH-I gene transcription in the septo-infundibular GnRH-I system of male starlings (Stevenson and Ball, 2009), supporting a centrally-mediated signalized mechanism for the delaying effects of females on males' transition out of the breeding state to refractoriness and molt. Likewise, free-living male song sparrows (*Melospiza melodia melodia*) delay gonadal collapse and molt when paired to females whose reproductive competence is experimentally prolonged with exogenous estradiol. This effect is unique to males: female song sparrows paired with testosterone-implanted males show no such socially-mediated delay of the breeding-molt transition, despite their mates remaining reproductively active, territorial, and delaying molt themselves.

Although there are potentially multiple sensory modalities through which behavioral interactions might influence individual schedules, acoustic stimuli from conspecifics have received more attention than any others. Male song has potent effects on female reproductive development (see Bentley et al., 2000 for review), and these effects can be very rapid. For instance, female white-throated sparrows (*Zonotrichia albicollis*) exposed to male song show rapid (within 1 h) elevation of circulating LH and enhanced Egr-1 (ZENK) immediate early gene expression in the mediobasal hypothalamus (Maney et al., 2007a,b). These studies reveal at least parts of the picture regarding how the annual schedule may be affected by these types of cues, and again highlight the role of the MBH in processing environmental cues that regulate annual schedules.

36.3.3 Integration of Multiple Cue Types: Parallel or Serial Processing?

One fundamental question about how different kinds of cues are processed and integrated relates to whether there is a hierarchy of cues, one or more of which must be stimulatory or at least permissive before others can have their effects (i.e., cues acting in series), or whether different cues can substitute for one another (i.e., cues acting in parallel). As noted above, at least for many species, the general role of endogenous mechanisms combined with seasonal changes in photoperiod take a primary role in a hierarchy of cue processing; days must first be above a certain threshold in length and/or the bird must be at the correct phase of an annual program before other cues can have an influence. This fits well with the general paradigm envisioned by Wingfield (1980, 1985), in which supplementary cues and synchronizing and integrating cues serve to fine-tune the development of a life cycle stage that has been initiated by initial predictive cues. Few studies permit evaluation of whether this cue hierarchy exists, but female white-crowned sparrows (*Z. l. gambelii*) only show enhanced reproductive development with exposure to male song if they are also photostimulated (Morton et al., 1985). It is possible, however, for cues to operate in parallel. That is, the same outcome of initiation of a life cycle stage might be achieved either through stimulation with one type of cue or with another, but both are not necessary, and neither takes hierarchical priority (i.e., is necessary before the other can have its effect). Red crossbills (*Loxia curvirostra*) provide a possible example of this type of processing. These birds can be in full reproductive condition and breed in the wild on both the shortest and longest days of the year (Berthold and Gwinner, 1978; Hahn, 1998), and in outdoor aviaries they can come into full breeding condition in January, when days have barely begun to increase in length and when temperatures are low, as long as they have unlimited food and a mate (Hahn et al., 1995). They also come into near-full breeding condition under long-day stimulation even if they do not have mates (Hahn, 1995). These findings are consistent with the interpretation that long days and some combination of abundant food and mates can essentially substitute for one another in terms of stimulating the reproductive system in some species.

In some cases, cue hierarchies may deviate from those described above. An experimental study of zebra finches showed that enhanced food alone or in combination with long days could accelerate reproductive development, whereas long days were not stimulatory if food supply was mildly restricted (Perfito et al., 2008). These data are consistent with the interpretation that changes in food may actually act as the initial predictive cues for certain opportunists (see Hahn et al., 1997), and any effects of photoperiod may either be permissive or mediated through time available to eat, unlike in many temperature zone photoperiodic species.

It is unclear whether it is more productive to think of endogenous programs and photoperiod—both considered to be forms of initial predictive cues by Wingfield (1980, 1985)—as interacting in a hierarchical fashion, or simply as complementary components of the annual scheduling process. The phase of the annual cycle (e.g., whether refractory or not) definitely determines the nature of the response to photoperiod, but photoperiodic history also has dramatic effects on whether and when transitions between stages of the annual schedule occur (see Gwinner, 1986). However, the fact that annual schedules of many species persist in the absence of clear timing cues from the environment seems consistent with the idea that it may be most appropriate to consider the endogenous ("innate") component as primary, and the modulatory effects of the environment as secondary, at least for some species.

36.4 ADAPTIVE VARIATION IN CUE PROCESSING MECHANISMS AS IT RELATES TO LIFE IN DIFFERENT ENVIRONMENTS

There is abundant evidence of adaptive variation in life cycle scheduling in different environments, and some evidence of adaptive variation in underlying mechanisms. There has been substantial focus on variation in photoperiodic regulation of reproductive development (Lofts and Murton, 1968; Lambrechts and Perret, 2000; Lambrechts et al., 1996, 1997; Hahn et al., 2009; Helm, 2009; Liedvogel et al., 2009; Caprioli et al., 2012), photorefractoriness (Lofts and Murton, 1968; Nicholls et al., 1988; Hahn and MacDougall-Shackleton, 2008; Hahn et al., 2009), as well as timing of molt and migratory restlessness (see Gwinner, 1986 for review).

One of the key distinctions that must be established is whether interpopulation variation in annual schedules is the consequence of evolved differences in endogenous timing mechanisms and responsiveness to environmental cues or simply a result of populations experiencing different environmental conditions. This problem can be addressed in several different ways. Examination of the phylogenetic distribution of reproductive photorefractoriness is consistent with the interpretation that failure to demonstrate refractoriness is a derived trait associated specifically with the most flexible or opportunistic annual schedules (Hahn and MacDougall-Shackleton, 2008). Thus, birds such as crossbills, zebra finches, and some tropical taxa with particularly flexible annual schedules fail to develop absolute photorefractoriness when exposed to extended periods of constant long days. Such taxa wait for input from the environment before making the transition from reproduction to molt and reproductive quiescence, in contrast to most temperate zone seasonal breeders, which spontaneously enter molt and collapse the reproductive system after some period of reproductive activity, even if the environment provides no direct information about impending deterioration of environmental conditions (Nicholls et al., 1988). Evolutionary loss of absolute photorefractoriness is, then, an adaptive specialization in these highly flexible taxa (Hahn and MacDougall-Shackleton, 2008).

Another way of investigating whether intertaxon variation in schedules represents adaptation to different environments is to look for evidence of a genetic basis for the observed variation. This has been reviewed copiously in the context of avian circannual rhythms by Gwinner (1986); a number of species of migratory songbirds display good evidence of robust differences in timing of diverse elements of their annual schedules (e.g., molt, migratory restlessness, body mass, reproductive condition) that persist in the absence of changing environmental conditions (see Gwinner, 1986). In addition, many studies demonstrate persistent interpopulation differences in scheduling under common garden constant conditions and intermediate scheduling phenotypes in F1 hybrids between different populations (see Gwinner, 1986 for review). Some more recent studies look specifically at heritability (a prerequisite for adaptive change in response to selection) as well as population-level response to artificially applied selection. For instance, variation in timing of onset of autumn migratory activity of blackcaps (*Sylvia atricapilla*) is heritable, and two generations of directional selection for later migration onset led to a 10-day shift in mean date of onset of autumn migratory restlessness (Pulido et al., 2001). This result supports the interpretation that at least some populations contain heritable variation in scheduling traits that is responsive to selection.

An elegant demonstration of adaptive variation in annual scheduling combined common garden manipulations of photoperiod with studies of hybrids between two subspecies of stonechats (*Saxicola torquata*) that differ in natural annual schedules (Helm et al., 2009, Figure 36.6). European stonechats undergo earlier reproductive development, remain in reproductive condition longer, and show a more protracted postbreeding plumage molt than do Siberian

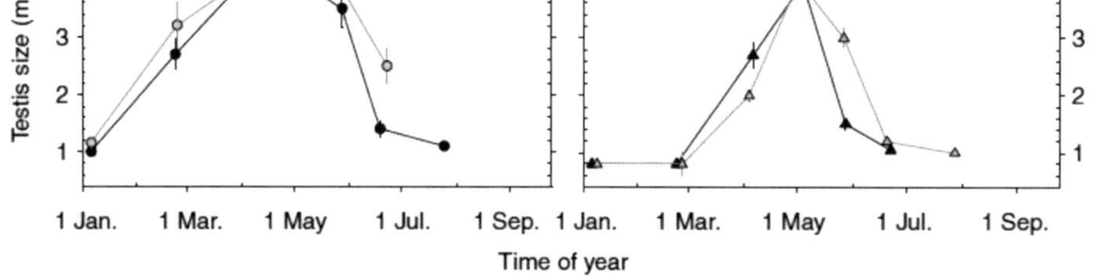

FIGURE 36.6 Comparison of patterns of gonadal development and plumage molt in European and Siberian stonechats (*Saxicola torquata*) exposed either to the photocycle typical for their own population or that typical of the other subspecies. In the bottom left panel, response of European stonechats exposed to photocycle typical of free-living members of their population is illustrated by black line and filled dots, and by darker molt bar; response of those exposed to photocycle typical of the Siberian population is illustrated by finer line and open circles, and by the paler molt bar. In the bottom right panel, response of Siberian stonechats exposed to photocycle typical of free-living members of their population is illustrated by finer line and open triangles, and by paler molt bar; response of those exposed to photocycle typical of the European population is illustrated by the darker line and black triangles, and by the darker molt bar. *From Helm et al. (2009).*

stonechats. These differences could in theory result from environmental condition-dependent plasticity. However, birds from the two different populations show persistence of the key differences in schedules when held under identical constant photoperiod, and F1 hybrids between the two display intermediate scheduling characteristics. Further, exposure to the subtly different photocycles that the different populations would experience on their different migration and wintering areas (summer photoperiods for the two are similar) produce different outcomes in the two populations. European stonechats exposed to seasonal photoregimes typical of the wintering and migration experience of Siberian stonechats shift onset of reproductive development even earlier and retain their prolonged period of reproductive competence and protracted molt. In contrast, Siberian stonechats exposed to the European seasonal photoregime only slightly advance onset of reproductive development, and retain their brief periods of reproduction and molt (Helm et al., 2009). This study thus shows that the specific annual schedules of these two subspecies differ because of an interaction between differences in both endogenous programming and in the phase-dependent response of the system to environmental inputs, not simply because the environments they experience differ.

36.5 INTEGRATED COORDINATION OF STAGES AND CARRYOVER EFFECTS

As noted above, stages of the annual cycle do not exist as separate entities but are intricately linked with the stages preceding and following them. The transition from one stage to the next involves termination of one stage and development of the next (Wingfield, 2008), each of which requires mechanistic control. This may involve one mechanism controlling both termination of one and development of the next or separate mechanisms. The first scenario could ensure that stages follow in the correct sequence throughout the year. For instance, in the transition from breeding to prebasic molt, the inactivation of the hypothalamus-pituitary-gonadal axis (HPG) axis as photorefractoriness develops leads to a precipitous decline in circulating sex steroids. This decline results in regression of secondary sex characteristics (e.g., cloacal protuberance, beak color, brood patch) and declines in song and courtship as well as facilitating the final stages of parental care in males (Nicholls et al., 1988; Hahn et al., 1997; Hahn and MacDougall-Shackleton, 2008). Because sex steroids inhibit molt, the decline in their levels allows molt to proceed (Schleussner et al., 1985; Nolan et al., 1992; Dawson, 1994, 2004).

However, stages occurring in the correct sequence could also result from two separate mechanisms controlling termination and development phases of adjacent stages. This is possible if one mechanism provides a trigger for the next. As with the breeding-molt transition example above, although photorefractoriness and the resulting decline in sex steroids is important for terminating breeding and the onset of molt, other physiological changes are occurring. Changes in prolactin and thyroid hormones have long been thought to be necessary for molt, but whether these are simply correlated events or play causal roles remains unresolved (Dawson, 2006, 2008; Mishra et al., 2004).

How transitions between stages are coordinated (e.g., whether stages overlap or remain distinct) has important consequences. One major source of these consequences is a phenomenon called carry-over effects, which occur when processes or conditions in one stage affect timing or performance in subsequent stages (Norris and Marra, 2007, Harrison et al., 2011). Most studies of carry-over effects examine individuals, but population-level consequences also may occur (Norris and Marra, 2007). At the individual level, carry-over effects often result from the constraint that transitional stage overlap can place on energy or resource allocation. An example of stage overlap and resulting carry-over effects occurs when the development of the breeding stage overlaps with the final phases of the preceding vernal migration stage. Many physiological changes take place during the transition between migration and breeding, including changes in musculature, nutrient and energy reserves, fat deposition, organ size and functioning associated with gonadal development and migration termination, as well as hormone and yolk precursor level changes (Arizmendi-Mejía et al., 2013). For example, female Macaroni penguins (*Eudyptes chrysolophus*) initiate vitellogenesis (yolk production) while they are still at sea migrating to the breeding territory (Crossin et al., 2010). With greater overlap of breeding preparation during migration, Macaroni penguins are able to start laying eggs sooner upon arrival. However, simultaneous expression of these functions constrains the rate of egg formation, as evidenced by lower vitellogenin levels, and therefore lower reproductive readiness, upon arrival in late-arriving females. This effect carries over to influence the size difference between the two eggs in the clutch (Crossin et al., 2010, Figure 36.7).

Similarly, the transition from migration to breeding stages in female black-browed albatrosses (*Thalassarche melanophrys*) involves a carry-over effect on the breeding stage as a result of processes occurring during migration. Postmigratory condition (weight and hematocrit) and reproductive readiness (progesterone (P4), testosterone (T), and vitellogenin levels (VTG)) were lower in females that ended up deferring breeding compared with those who did lay eggs. This suggests a resource- or condition-based allocation trade-off underlying this carry-over effect in which returning females in better condition (those that cope more effectively with the metabolic demands of migration) are more likely to initiate breeding. Furthermore, of the birds that attempted breeding, success was associated with low T and high VTG levels, compared with failed breeders whose

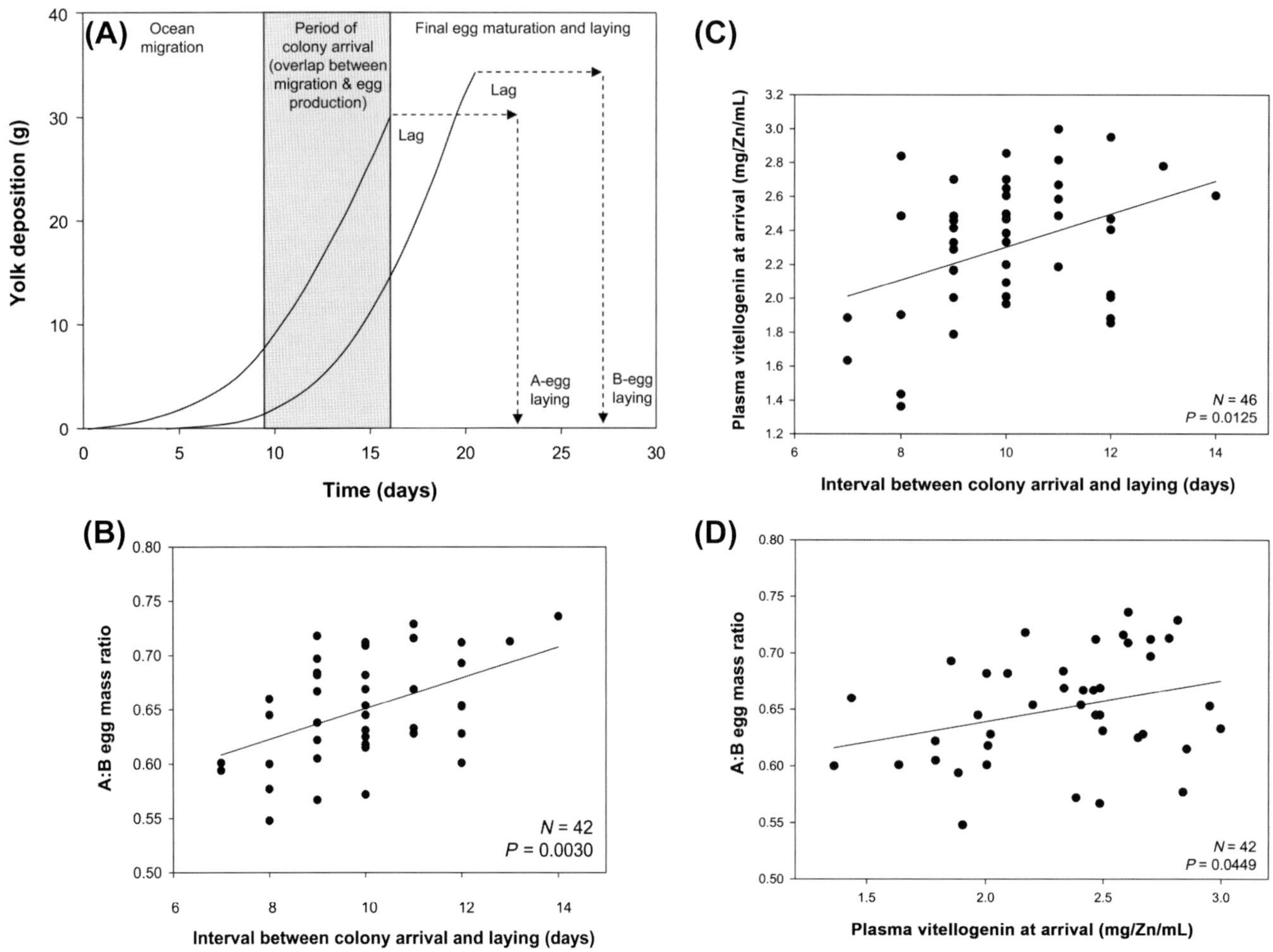

FIGURE 36.7 Egg formation process and timing during migration and colony arrival in female macaroni penguins (graph A). Degree of overlap between migration and yolk deposition influences the interval between colony arrival and laying (lag time). This interval affects egg size dimorphism (graph B) and plasma vitellogenin levels at arrival (graph C). Vitellogenin at arrival also predicts egg size dimorphism (graph D). Lines are best linear fit. *From Crossin et al. (2010).*

eggs did not hatch. This finding suggests another possible mechanism underlying carry-over effects: conflicting physiological processes. The differences in T and VTG levels between successful and failed breeders indicates a potential difference in the levels of aromatization from T to estradiol (E2), a necessary step in the process of vitellogensis (E2 stimulates vitellogenesis in the liver). The inhibition of aromatization, possibly by a stress-related mechanism, may be influencing the migration-breeding transition and its resulting carry-over effects.

Reproductive decisions and outcomes are influenced by processes occurring earlier, during the migration stage; however, processes even earlier, during the wintering stage, can have carry-over effects on migration and breeding (Marra et al., 1998; Studds & Marra, 2005; Norris et al., 2004). For example, American redstarts wintering in low-quality, drier scrub habitats are in poorer body condition and have higher corticosterone levels than those in wetter habitats (Marra et al., 1998). Both males and females wintering in drier areas, as well as those in poorer physical condition in general, tend to depart for migration later (Marra et al., 1998). This effect was shown experimentally, when birds that were upgraded from low- to high-quality winter habitats had higher body mass and departed earlier than controls (Studds and Marra, 2005). High-quality wintering grounds translate into earlier arrival on breeding grounds and higher reproductive success Norris et al., 2004). Males arriving earlier from better wintering grounds sired more offspring, both within-pair and extra-pair, and had greater fledging success (Norris et al., 2004). Early-arriving females started laying and fledging young earlier; high-quality winter habitat predicted production of at least two more young than for females from poor-quality wintering grounds (Norris et al., 2004). Similarly, as already described above, food supplementation during the overwintering stage can lead to advancement of onset of egg laying and improved reproductive success, processes occurring weeks to months postsupplementation, in song sparrows, kestrels, and scrub jays (Arcese and Smith, 1988; Meijer et al., 1988; Schoech, 1996). Further, female snow geese arriving on breeding grounds with higher nutrient reserves had more ovarian follicles, suggesting a larger potential clutch (Ankney and MacInnes, 1978). All of these examples illustrate potential carryover effects of the ability to obtain food during one stage on transition to, and performance in, following stages.

Carry-over effects such as these often are presumed to result from resource allocation trade-offs. However, studies documenting such tradeoffs often lack direct evidence to support this underlying mechanism (Harrison et al., 2011). In Cassin's auklets (*Ptychoramphus aleuticus*), a resource allocation trade-off based on quality of pre-breeding diet, as measured by stable isotopes in body feathers grown during prealternate molt, directly influenced laying date. Female auklets that consumed more energy-rich copepods during the period prior to breeding began laying earlier and laid larger eggs than those that consumed more energy-poor juvenile rockfish. Although studies like these help to illuminate the potential mechanisms responsible for carry-over effects (and also provide additional examples of responses of annual schedule timing to food cues), much more research is needed (Norris and Marra, 2007; Harrison et al., 2011). Most studies involve observational field work, but experimental manipulations are necessary to learn more about the physiological mechanisms (Harrison et al., 2011).

In addition, more research examining carry-over effects related to other aspects of the annual cycle is needed. Much attention has been paid to carry-over effects between migration and breeding (Norris and Marra, 2007), but fewer studies have looked at carry-over effects involving molt timing. Late breeding can lead to delayed molt or increased overlap of breeding with prebasic molt (Morton & Morton, 1990). Delayed molt can carry-over to result in later migratory departure in Wood thrushes, and can also require a faster molt that produces inferior feathers (Dawson et al., 2000). Future research should also attempt to track longer-term carry-over effects. Most studies examine effects between adjacent stages, but carry-over effects may trickle across the entire annual cycle and may even exert effects interannually.

As evidence accumulates about how carry-over effects lead to changes in the timing of all aspects of the annual cycle, it becomes even more apparent that changes in the environment can have long-term and far-reaching consequences. Indeed, processes during the wintering stage can influence timing of breeding, a stage that often occurs thousands of miles away (Norris et al., 2004). Therefore, understanding how rapid human-induced environmental change can affect seasonal timing is important. This is the topic of the next section.

36.6 VARIATION IN SCHEDULING MECHANISMS AND RESPONSES TO HUMAN-INDUCED RAPID ENVIRONMENTAL CHANGE

Climate change, as well as other human-induced changes to the environment such as urbanization, can have rapid and substantial effects on the timing of seasonal activities (Helm et al., 2013). Photoperiod (usually a primary proximate driver, as noted above) is unchanging with respect to climate change; however, temperatures are changing rapidly and varying stochastically between years and regions where birds spend different parts of their annual cycle (Visser et al., 2004). Although birds can adjust flexibly to changing temperatures (see above), climate change is causing temperatures to deviate from historic ranges and is affecting different parts of the year differently. For these reasons, and because reproductive decisions must be made well in advance of when selection pressures are acting (i.e., the timing of food peaks), climate change poses a significant challenge to birds' ability to time processes appropriately (Visser et al., 2004). Conversely, phenology in species at other trophic levels, particularly of insect prey, directly respond to temperature changes because they have very different response mechanisms than birds (Bradshaw and Holzapfel, 2010). This can contribute further to mistimed breeding in birds (Visser et al., 2004; Bradshaw and Holzapfel, 2010). A temporal mismatch between peak abundance of insects and the peak of need for feeding nestlings has occurred in a population of great tits. Over a span of 23 years in which spring temperatures continued to increase, the tit egg-laying dates did not advance enough to match the much earlier peak in caterpillar abundance (Visser et al., 1998, Figure 36.8). Similar reproductive mistiming has been documented in many species and has been shown to lead to population declines in some (Visser et al., 2004; Both et al., 2006); it is specifically populations that have not advanced breeding times that are in decline. In at least some cases it is through adaptive individual plasticity, rather than evolved changes in cue responsiveness, that some populations have been able to closely track environmental changes (Charmantier et al., 2008).

Species differences like these emphasize the importance of how species variation in seasonal timing mechanisms can lead to important differences in responses to climate change. As described in detail throughout this chapter, the relative reliance on photic versus nonphotic cues varies among species. For example, rigidly seasonal species tend to rely more heavily on photoperiod, whereas opportunistic species show stronger responses to nonphotic cues in addition to their photoperiodic response (Hahn et al., 1997, 2008). Therefore, species differences in responsiveness to nonphotic cues could correlate with ability to cope with climate change.

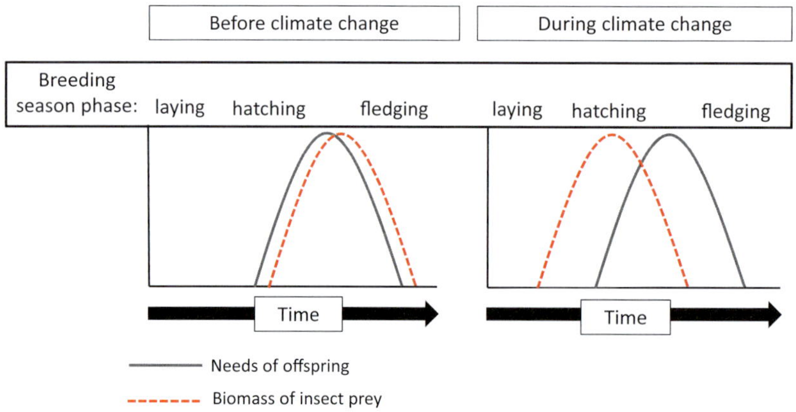

FIGURE 36.8 **Schematic showing reproductive mistiming in great tits.** The need for food for offspring (gray solid line) is highest between the hatching and fledging stages. Laying dates have remained similar over the years, yet the peak of caterpillar biomass (dotted red line) has shifted much earlier due to climate change, which results in a mismatch between the time nestlings are fed and the peak food abundance. *Adapted from Visser et al. (2004).*

In contrast to the hypothesis that mismatched timing is a maladaptive result of the fact that photoperiodic cues no longer accurately predict the best time to initiate breeding (cues hypothesis), an interesting and counterintuitive alternative is that mismatched breeding may actually be adaptive (Visser et al., 2012). The constraint hypothesis states that reproductive mistiming could be adaptive if the costs resulting from trying to breed early enough to match insect peak abundance outweighs the benefit that would be gained if they did match (Visser et al., 2012).

36.7 EFFECTS OF SEASONALITY ON IMMUNE FUNCTION

Although annual schedules often consist of several distinct life cycle stages, birds also adjust investment in processes that span multiple stages. These processes may be maintained at all times, during every stage, yet the appropriate level of investment may vary as a function of resource availability, environmental conditions, individual age or condition, and so forth. In this section, we will discuss one particularly interesting example of such processes that has been receiving increasing attention in the context of annual scheduling: seasonal adjustments of investment in immune function.

36.8 SEASONAL MODULATION OF IMMUNE FUNCTION

36.8.1 Overview

Across the annual cycle, dynamic environments expose organisms to extensive seasonal variation in weather, resource availability, and disease potential (King, 1974). Natural selection dictates that longer-lived organisms, such as birds, should time costly energetic processes to maximize fitness by balancing seasonal allocation to both reproduction and survival-related processes such as immune function (see Box 36.1 for summary of key immune system components). Immune function contributes to survival by detecting pathogens and limiting infection, but because maintenance of immunity can be costly (Schmid-Hempel and Ebert, 2003; Nelson, 2004; Sheldon and Verhulst, 1996; Klasing, 2004), especially if immune activation is overused (Lochmiller and Deerenberg, 2000) or not sufficient to clear an infection (Nelson et al., 2002), investment in immunity is often variable (Martin et al., 2008). Many environmental and physiological variables can contribute to these changes in investment patterns in immunity (Buehler et al. 2008a, 2008b; Martin et al., 2008; Nelson et al., 2002). Examples of these contributing environmental factors include endogenous and/or photoperiodically regulated annual cycles, changes in disease threat or parasite pressure, food availability and weather and temperature conditions (Nelson and Demas, 1996; Nelson et al., 2002; Martin et al., 2008; Adelman et al., 2013; Hasselquist, 2007). Other factors that contribute to changes in immune function include competing physiological processes such as reproduction (Sheldon and Verhulst, 1996; Lochmiller and Deerenberg, 2000) and other survival-related processes such as plumage molt (Moreno-Rueda, 2010; Buehler et al., 2008b), growth (Prendergast et al., 2004), or migration (Nebel et al., 2011, 2013).

36.8.2 Seasonality of Immune Function

Much of what is known about the seasonality of immune function has focused on mammals, specifically small rodents, with many studies documenting the effects of temperature, photoperiod, and food availability (Demas and Nelson, 1998; Zysling et al., 2009; Bilbo and Nelson, 2003). To summarize, a small mammal's investment in immune function tends to decrease during the summer months, while peaking during the winter months (Nelson and Demas, 1996; Sinclair and Lochmiller, 2000), mainly

Box 36.1 Specific Components of the Immune System and Their Costs

In general, all components of the immune system are not created equal in terms of energetic cost of maintenance and use, in that some components (described below) are substantially more costly to maintain than others. Broadly, the immune system can be categorized along an innate (nonspecific)-acquired (specific) axis and a constitutive (noninduced)-induced axis (Schmid-Hempel and Ebert, 2003). Further, a constitutive response is a baseline response or a response that will deploy without previous exposure, whereas an acquired response requires previous exposure to that pathogen or antigen (Schmid-Hempel and Ebert, 2003). Innate immune function provides an immediate and nonspecific response to a pathogen, whereas acquired immunity is activated by the innate response to produce specific antibodies against the pathogen (reviewed in Martin et al., 2008; Lee, 2006) and is typically slower in response time. Because of the differing costs of the different components of immunity, seasonal variation in immune function is expected to differ for different immune axes.

Constitutive Innate Immunity. Examples of innate defenses include anatomical barriers (e.g., skin, mucus membranes), resident microflora such as nonpathogenic bacteria, humoral factors (complement and acute phase proteins), and cellular components (white blood cells responsible for phagocytosis and an inflammatory response) and other antimicrobial proteins such as lysozyme, defensins (reviewed in Demas et al., 2011; Lee, 2006). The costs of constitutive innate immunity are not well studied, but overall are thought to be low due to the lower cell turnover rate that occurs when an immune response is not required and overall small combined tissue mass of the innate cells and proteins listed above (reviewed in Lee, 2006; Klasing and Leshchinsky, 1999).

Induced Innate Defense. If necessary, cells and proteins of the constitutive innate response will induce local inflammation via inflammatory cytokines, and if the infection is large enough, a systemic inflammatory response will be further induced. This response is also called the acute phase response (APR) and is characterized by increased production of acute phase proteins by the liver, changes in energy and nutrient metabolism, activation of the hypothalamo-pituitary-adrenal axis (HPA), inhibition of the HPG axis, decreased locomotor and social activities, anorexia, and fever (Klasing and Leshchinsky, 1999). This portion of the immune system is effective at quickly eliminating invading pathogens but can be extremely energetically expensive and potentially self-damaging (Klasing, 2004).

Constitutive Acquired (Adaptive) Defense. Natural, nonspecific antibodies such as IgM provide a baseline defense against pathogens and tend to circulate in the blood without requiring previous exposure to a pathogen (unlike most other antibodies) (reviewed in Lee, 2006). Overall, this response is not considered energetically costly to maintain.

Induced Acquired (Adaptive) Defense. This component of the adaptive response includes two types: cell-mediated and humoral. The cell-mediated component is responsible for killing infector host cells, and develops immunological memory for intracellular pathogens such as viruses via T-helper cells (Th-1) and B-cells. This response can often induce systemic inflammation, and the cellular turnover rate required for this response and developmental costs of specific antibodies makes this response substantially energetically demanding when induced (reviewed in Lee, 2006; Janeway et al., 2004). The humoral response is responsible for killing extracellular host cells and develops immunological memory for extra-cellular pathogens and parasites via T-helper cells (Th-2) and B cells. Overall, the energetic cost of this response is much lower than the cell-mediated component (Janeway et al., 2004).

to counteract the environmentally induced immunosuppressive effects (such as stress caused by inclement weather and low food availability) that occur during winter (Demas and Nelson, 1998). Overall, it appears that these mammals use photoperiod as their primary cue to modulate investment in immunity (with some effects of food levels and ambient temperature), so as to be best prepared for the upcoming and potentially inclement season (Nelson et al., 2002).

Far less information is available on overall seasonal patterns of avian immunity, with most published studies focusing on single components of the annual cycle, such as breeding compared to nonbreeding or breeding compared to molting (reviewed in Martin et al., 2008). For example, experimental manipulations of brood size in collared flycatchers (*Ficedula albicollis*) found that brood size and immunity were inversely related (Gustafsson et al., 1994). A meta-analysis of experiments increasing and decreasing clutch size found that immune function and brood size were inversely related (Knowles et al., 2009). Additionally, when incubation, mating effort, flight cost, or thermoregulatory costs were increased, immunity decreased (reviewed in Hasselquist and Nilsson, 2012). Another meta-analysis by Møller and colleagues (2003) found that T-cell cellular-based immunity increased during the summer months in 13 passerine species. Finally, a common garden experiment by Martin and colleagues (2004) demonstrated that house sparrows exhibited lower unspecific T-cell mediated response (to PHA, a nonpathogenic, stimulatory novel protein) during the early breeding season when compared to late breeding or fall.

Mechanistically, these effects have been attributed to changes in seasonal hormone profiles, such as glucocorticoids, androgens, and estrogens (which all can suppress immunity in certain contexts) (reviewed in Martin et al., 2008). Some of the same hormones that affect timing of territory establishment, reproduction, parental care, and molt affect immune function (reviewed in Martin et al., 2008; Koustos and Klasing, 2013). For example, both glucocorticoids and androgens can depress immunity, but their effect on the immune system is highly dependent on the overall level and duration of increase, as well as what immune component is measured (reviewed in Koustos and Klasing, 2013). In general, acute stress or quick elevations of glucocorticoids increases levels of immunity, while chronic or persisting levels of stress dampen immune responses. For testosterone and other androgens, it is difficult to disentangle the direct effects from the indirect effects of glucocorticoids; that is, elevated levels of testosterone may depress immunity by elevating levels of glucocorticoids (Koustos and Klasing, 2013).

36.8.2.1 Captive Studies

Of the few studies of annual cycles of immunity, most have occurred in captivity. These studies have shown that both cellular and humoral immunity have either been higher during the winter months or show no change when compared to summer months (Martin et al., 2008; Lee, 2006). In a study of red knot immunity sampled over the annual cycle (Buehler et al., 2008b), both higher energetic cost phagocyte-based (cellular) immunity and lower-cost lymphocyte-based (antibody) immunity were higher during periods of mass change (i.e., during fattening and spring migration in wild birds), but there was a clear shift toward lower-cost lymphocyte-based immunity during peak molt. Due to the high costs associated with synthesizing feathers (Haake and Sawyer, 1986), it is preferable for birds to use a lower cost, less phagocytic-based immune strategy, as was demonstrated in the above study. In a subsequent experiment using low, high, and variable temperatures, there was little effect on annual variation in immune function (Buehler et al., 2009), perhaps due to the presence of *adlibitum* food. However, when food levels were restricted alone in a follow-up experiment, no change in constitutive immunity (natural antibody and complement levels) was observed, although food restriction did diminish investment in the acute-phase or induced immune response that is substantially more energetically demanding (Buehler et al., 2009). Conversely, a captive study of white-crowned sparrows did not find robust seasonal changes in immunity (specifically the acute-phase response) (Owen-Ashley and Wingfield, 2007); however, the acute phase response was more pronounced in sparrows during the winter months but was highly correlated to body condition and depended on which population was sampled (Owen-Ashley and Wingfield, 2007).

36.8.2.2 Studies of Free-Living Birds

Because captivity can reduce both inflammatory and phagocytic (cellular)-based immunity (either by reducing the pathogen pressures or due to the increased stress associated with being held in captivity) (Buehler et al., 2008a), it is important to study immunity both in the wild and captivity however, few studies do so (Adelman et al., 2013). In free-living birds, very few studies have looked at multiple immune indices over several components of the entire annual cycle (as opposed to just comparing two stages such as breeding to nonbreeding as the studies mentioned above). One study on nonmigratory great tits (*Parus major*; Pap et al., 2010) found significant seasonal changes in immunity. Specifically, total immunoglobulin and heterophil levels were highest during the summer breeding months and continued to increase until September when the birds began molting and then decreased until the following spring. Lymphocyte levels, however, were higher in the winter than in the summer. Another study, in skylarks (Hegemann et al., 2012), is the only published study to date to look at multiple immune indices over several years. Metrics of constitutive innate immunity (complement, natural antibody, and haptoglobin levels) varied interannually but were consistently lower during fall migration than during the breeding season.

Migration is an important and energetically demanding component of a bird's life history (Wikelski et al., 2003). Thus, energetic tradeoffs with immune function often occur, as has been documented in several studies. For example, a study in thrushes (Owen and Moore, 2008) and Bewick's swans (*Cygnus bewickii*) (Van Gils et al., 2007) found that migratory activity can be immunosuppressive. Two recent studies using a flight chamber found, similar to the aforementioned studies, that simulated migration depresses several aspects of immunity. The first study on European Starlings (*Sturnus vulgaris*) found that three of the four immune measures (haptoglobin, natural antibody and complement levels) decreased after the birds were placed in the wind tunnel for 1–4 h of continuous flight (Nebel et al., 2013). Another related study on long-distance migrant western sandpipers (*Calidris mauri*) found that flight performance (flight duration of 3 h) was not diminished by a previous immune challenge of LPS (lipopolysaccharide, which induces a "sickness response"), but flight while mounting a sickness response negatively affected other aspects of immune function including bacterial killing ability and natural antibody levels (Nebel et al., 2013).

REFERENCES

Adelman, J.S., Ardia, D.R., Schat, K.A., 2013. Chapter 22. Ecoimmunology. Avian Immunology. Elsevier Ltd. pp. 1–23.

Ankney, C.D., MacInnes, C.D., 1978. Nutrient reserves and reproductive performance of female lesser snow geese. Auk 95, 459–471.

Arcese, P., Smith, J.N.M., 1988. Effects of population density and food on reproduction in song sparrows. J. Anim. Ecol. 57, 119–136.

Arizmendi-Mejía, R., Militão, T., Viscor, G., González-Solís, J., 2013. Pre-breeding ecophysiology of a long-distance migratory seabird. J. Exp. Mar. Biol. Ecol. 443, 162–168.

Baker, J.R., 1938. The evolution of breeding seasons. In: DeBeer, G.B. (Ed.), Evolution: Essays on Aspects of Evolutionary Biology. Clarendon Press, Oxford, UK, pp. 161–177.

Ball, G.F., 1993. The neural integration of environmental information by seasonally breeding birds. Am. Zool. 33, 185–199.

Bauchinger, U., Van't Hof, T., Biebach, H., 2007. Testicular development during long-distance migration. Horm. Behav. 51, 295–305.

Bauchinger, U., Van't Hof, T., Biebach, H., 2008. Migratory stopover conditions affect the developmental state of male gonads in garden warblers (*Sylvia borin*). Horm. Behav. 54, 312–318.

Bauchinger, U., Van't Hof, T., Biebach, H., 2009. Food availability during migratory stopover affects testis growth and reproductive behaviour in a migratory passerine. Horm. Behav. 55, 425–433.

Benkman, C.W., 1990. Foraging rates and the timing of crossbill reproduction. Auk 107, 376–386.

Benoit, J., 1935. Stimulation par la lumiere artificiell du developpement testiculaire chez des canards aveugles par enucleation des globes oculaires. C R. Soc. Biol. (Paris) 120, 136–139.

Bentley, G.E., Wingfield, J.C., Morton, M.L., Ball, G.F., 2000. Stimulatory effects on the reproductive axis in female songbirds by conspecific and heterospecific male song. Horm. Behav. 37, 179–189.

Bentley, G.E., Tucker, S., Chou, H., Hau, M., Perfito, N., 2013. Testicular growth and regression are not correlated with *Dio2* expression in a wild male songbird, *Sturnus vulgaris*, exposed to natural changes in photoperiod. Endocrinology 154, 1813–1819.

Berthold, P., Gwinner, E., 1978. Jahresperiodik der Gonadengrosse beim Fichtenkreuzschnabel (*Loxia curvirostra*). J. Ornithol. 119, 338–339.

Bilbo, S.D., Nelson, R.J., 2003. Photoperiod influences the effects of Exercise and food restriction on an antigen-specific immune response in Siberian hamsters. Endocrinology 145, 556–564.

Bond, A.L., Konyukhov, N.B., Jones, I.L., 2013. Variation in primary molt in the least auklet. Condor 115, 348–355.

Both, C., Bouwhuis, S., Lessells, C.M., Visser, M.E., 2006. Climate change and population declines in a long-distance migratory bird. Nat. Lett. 441, 81–83.

Bradshaw, W.E., Holzapfel, C.M., 2007. Evolution of animal photoperiodism. Annual Review of Ecology. Evol. Syst. 38, 1–25.

Bradshaw, W.E., Holzapfel, C.M., 2010. Light, time, and the physiology of biotic response to rapid climate change in animals. Ann. Rev. Ecol. Evol. Syst. 72, 147–166.

Bridge, E.S., Kelly, J.F., Bjornen, P.E., Curry, C.M., Crawford, P.H.C., Paritte, J.M., 2010. Effects of nutritional condition on spring migration: do migrants use resource availability to keep pace with a changing world? J. Exp. Biol. 213, 2424–2429.

Buehler, D.M., Encinas Viso, F., Petit, M., Vézina, F., Tieleman, B.I., Piersma, T., 2009. Limited access to food and physiological trade–offs in a long–distance migrant shorebird. II. Constitutive immune function and the acute–phase response. Physiol. Biochem. Zool. 82 (5), 561–571.

Buehler, D.M., Piersma, T., Matson, K., Tieleman, B.I., 2008b. Seasonal redistribution of immune function in a migrant shorebird: annual–cycle effects override adjustments to thermal regime. Am. Nat. 172, 783–796.

Buehler, D., Piersma, T., Tieleman, B.I., 2008a. Captive and free-living red knots *Calidris canutus* exhibit differences in non-induced immunity that suggest different immune strategies in different environments. J. Avian Biol. 39, 560–566.

Caro, S.P., Visser, M.E., 2009. Temperature-induced elevation of basal metabolic rate does not affect testis growth in great tits. J. Exp. Biol. 212, 1994–1998.

Caro, S.P., Schaper, S.V., Hut, R.A., Ball, G.F., Visser, M.E., 2013. The case of the missing mechanism: how does temperature influence seasonal timing in endotherms? PLoS Biol. 11, e1001517.

Caprioli, M., Ambrosini, R., Boncoraglio, G., Gatti, E., Romano, A., Romano, M., Rubolini, D., Gianfranceschi, L., Saino, N., 2012. *Clock* gene variation is associated with breeding phenology and may be under directional selection in the migratory barn swallow. PLoS One 7, e35140.

Charmantier, A., McCleery, R.H., Cole, L.R., Perrins, C., Kruuk, L.E.B., Sheldon, B.C., 2008. Adaptive phenotypic plasticity in response to climate change in a wild bird population. Science 320, 800–803.

Chastel, O., Lacroix, A., Kersten, M., 2003. Pre-breeding energy requirements: thyroid hormone, metabolism and the timing of reproduction in house sparrows *Passer domesticus*. J. Avian Biol. 34, 298–306.

Cockrem, J.F., 1995. Timing of seasonal breeding in birds, with particular reference to New Zealand birds. Reprod. Fertil. Dev. 7, 1–19.

Cornelius, J.M., Hahn, T.P., 2012. Seasonal pre-migratory fattening and increased activity in a nomadic and irruptive migrant, the Red Crossbill *Loxia curvirostra*. Ibis 154, 693–702.

Cornelius, J.M., Breuner, C.W., Hahn, T.P., 2010. Under a neighbour's influence: public information affects stress hormones and behaviour of a songbird. Proc. R. Soc. B 277, 2399–2404.

Cornelius, J.M., Perfito, N., Zann, R., Breuner, C.W., Hahn, T.P., 2011. Physiological trade-offs in self-maintenance: plumage molt and stress physiology in birds. J. Exp. Biol. 214, 2768–2777.

Crossin, G., Trathan, P., Phillips, A., Alistair Dawson, A., Le Bouard, F., Williams, T., 2010. A Carryover Effect of Migration Underlies Individual Variationin Reproductive Readiness and Extreme Egg Size Dimorphismin Macaroni Penguins. Am. Nat. 176, 357–366.

Dawson, A., 1986. The effect of restricting the daily period of food availability on testicular growth of starlings *Sturnus vulgaris*. Ibis 128, 572–575.

Dawson, A., 1994. The effects of daylength and testosterone on the initiation and progress of moult in starlings, *Sturnus vulgaris*. Ibis 136, 335–340.

Dawson, A., 2002. Photoperiodic control of the annual cycle in birds and comparison with mammals. In: Both, C., Piersma, T. (Eds.), The Avian Calendar: Exploring the Biological Hurdles in the Annual CycleProc. 3rd Conf. European Orn. Union, Groningen, August 2001. Ardea, vol. 90, pp. 355–367(special issue).

Dawson, A., 2004. The effects of delaying the start of moult on the duration of moult, primary feather growth rates and feather mass in common starlings, *Sturnus vulgaris*. Ibis 196, 493–500.

Dawson, A., 2005. The effect of temperature on photoperiodically regulated gonadal maturation, regression and moult in starlings – potential consequences of climate change. Funct. Ecol. 19, 995–1000.

Dawson, A., 2006. Control of molt in birds: association with prolactin and gonadal regression in starlings. Gen. Comp. Endocrinol. 147, 314–322.

Dawson, A., 2008. Control of the annual cycle in birds: endocrine constraints and plasticity in response to ecological variability. Philos. Trans. R. Soc. B 363, 1621–1633.

Dawson, A., Goldsmith, A.R., 1997. Changes in gonadotropin-releasing hormone in the preoptic area and median eminence of starlings during the recovery of photosensitivity and during photostimulation. J. Reprod. Fertil. 111, 1–6.

Dawson, A., Sharp, P., 2007. Photorefractoriness in birds – photoperiodic and non-photoperiodic control. Gen. Comp. Endocrinol. 153, 378–384.

Dawson, A., Sharp, P., 2010. Seasonal changes in concentrations of plasma LH and prolactin associated with the advancement in the development of photo-refractoriness and molt by high temperature in the startling. Gen. Comp. Endocrinol. 167, 122–127.

Dawson, A., Hinsley, S.A., Ferns, P.N., Bonser, R.H.C., Eccleston, L., 2000. Rate of moult affects feather quality: a mechanism linking current reproductive effort to future survival. Proc. R. Soc. B 267, 2093–2098.

Dawson, A., King, V.M., Bentley, G.E., Ball, G.F., 2001. Photoperiodic control of seasonality in birds. J. Biol. Rhythms 16, 366–381.

Demas, G.E., Nelson, R.J., 1998. Photoperiod, Ambient Temperature, and Food Availability Interact to Affect Reproductive and Immune Function in Adult Male Deer Mice (*Peromyscus maniculatus*). J. Biol. Rhythms 13, 253–262.

Demas, G.E., Zysling, D.A., Beechler, B.R., Muehlenbein, M.P., French, S.S., 2011. Beyond phytohaemagglutinin: assessing vertebrate immune function across ecological contexts. J. Anim. Ecol. 80, 710–730.

Dhont, A.A., 2010. Broodedness, not latitude, affects the response of reproductive timing of birds to food supplementation. J. Ornithol. 151, 955–957.

Follett, B.K., 1984. Birds. In: In: Lamming, G.E. (Ed.), Marshall's Physiology of Reproduction, vol. 1. Longman Green, Edingburgh, pp. 283–350.

Follett, B.K., Mattocks Jr., P.W., Farner, D.S., 1974. Circadian function in the photoperiodic induction of gonadotropin secretion in the white-crowned sparrow, *Zonotrichia leucophrys gambelii*. Proc. Natl. Acad. Sci. U.S.A. 71, 1666–1669.

Foster, R.G., Soni, B.G., 1998. Extraretinal photoreceptors and their regulation of temporal physiology. Rev. Reprod. 3, 145–150.

Foster, R.G., Grace, M.S., Provencio, I., DeGrip, W.J., Garcia-Fernandez, J.M., 1994. Identification of vertebrate deep brain photoreceptors. Neurosci. Biobehav. Rev. 18, 541–546.

Ganter, B., Boyd, W.S., Baranyuk, V.V., Cooke, F., 2005. First pairing in snow geese *Anser caerulescens*: at what age and at what time of year does it occur? Ibis 147, 57–66.

Goodson, J.L., Saldanha, C.J., Hahn, T.P., Soma, K.K., 2005. Recent advances in behavioral neuroendocrinology: insights from studies on birds. Horm. Behav. 48, 461–473.

Gustafsson, L., Nordling, D., Andersson, M.S., Sheldon, B.C., Qvarnstrom, A., 1994. Infectious-diseases, reproductive effort and the cost of reproduction in birds. Philos. Trans. R. Soc. B 346, 323–331.

Gwinner, E., 1986. Circannual Rhythms. Springer-Verlag, Berlin, Heidelberg, New York.

Hasselquist, D., Nilsson, J.A., 2012. Physiological mechanisms mediating costs of immune responses: what can we learn from studies of birds? Anim. Behav. 83, 1303–1312.

Haake, A.R., Sawyer, R.H., 1986. Differences in the histogenesis and keratin expression of avian extraembryonic ectoderm and endoderm recombined with dermis. Dev. Biol. 113, 295–304.

Hahn, T.P., 1995. Integration of photoperiodic and food cues to time changes in reproductive physiology by an opportunistic breeder, the red crossbill, *Loxia curvirostra* (Aves: Carduelinae). J. Exp. Zool. 272, 213–226.

Hahn, T.P., 1998. Reproductive seasonality in an opportunistic breeder, the red crossbill, *Loxia curvirostra*. Ecology 79, 2365–2375.

Hahn, T.P., Boswell, T., Wingfield, J.C., Ball, G.F., 1997. Temporal flexibility in avian reproduction: patterns and mechanisms. Current Ornithology 14, 39–80.

Hahn, T.P., Cornelius, J.M., Sewall, K.B., Kelsey, T.R., Hau., M., Perfito, N., 2008. Environmental regulation of annual schedules in opportunistically-breeding songbirds: adaptive specializations or variations on a theme of white-crowned sparrow? Gen. Comp. Endocrinol. 157, 217–226.

Hahn, T.P., Pereyra, M.E., Katti, M., Ward, G.M., MacDougall-Shackleton, S.A., 2005. Effects of food availability on the reproductive system. In: Dawson, A., Sharp, P.J. (Eds.), Functional Avian Endocrinology. Narosa Publishing House, New Delhi, India.

Hahn, T.P., Watts, H.E., Cornelius, J.M., Brazeal, K.R., MacDougall-Shackleton, S.A., 2009. Evolution of environmental cue response mechanisms: adaptive variation in photorefractoriness. Gen. Comp. Endocrinol. 163, 193–200.

Hahn, T.P., MacDougall-Shackleton, S.A., 2008. Adaptive specialization, conditional plasticity and phylogenetic history in the reproductive cue response systems of birds. Phil. Trans. R. Soc. B 363, 267–286.

Hahn, T.P., Wingfield, J.C., Mullen, R., Deviche, P.J., 1995. Endocrine bases of spatial and temporal opportunism in arctic breeding birds. Am. Zool. 35, 259–273.

Hamner, W.M., 1963. Diurnal rhythm and photoperiodism in testicular recrudescence of the house finch. Science 142, 1294–1295.

Hamner, W.M., 1964. Circadian control of photoperiodism in the house finch demonstrated by interrupted-night experiments. Nature 203, 1400–1401.

Hamner, W.M., 1968. The photorefractory period of the house finch. Ecology 49, 211–227.

Harrison, X.A., Blount, J.D., Inger, R., Norris, D.R., Bearhop, S., 2011. Carry-over effects as drivers of fitness differences in animals. J. Anim. Ecol. 80, 4–18.

Hasselquist, D., 2007. Comparative immunoecology in birds: hypotheses and tests. J. Ornithol. 148, 571–582.

Hau, M., Wikelski, M., Gwinner, H., Gwinner, E., 2004. Timing of reproduction in a Darwin's finch: temporal opportunism under spatial constraints. Oikos 106, 489–500.

Hau, M., Wikelski, M., Wingfield, J.C., 2000. Visual and nutritional food cues fine-tune timing of reproduction in a neotropical rainforest bird. J. Exp. Zool. 286, 494–504.

Hegemann, A., Matson, K.D., Both, C., Tieleman, B.I., 2012. Immune function in a free-living bird varies over the annual cycle, but seasonal patterns differ between years. Oecologia 170, 605–618.

Helm, B., Ben-Shlomo, R., Sheriff, M.J., Hut, R.A., Foster, R., Barnes, B.M., Dominoni, D., 2013. Annual rhythms that underly phenology: biological time-keeping meets environmental change. Proc. R. Soc. B 280, 20130016.

Helm, B., 2009. Geographically distinct reproductive schedules in a changing world: costly implications in captive stonechats. Integr. Comp. Biol. 49, 563–579.

Helm, B., Schwabl, I., Gwinner, E., 2009. Circannual basis of geographcailly distinct bird schedules. J. Exp. Biol. 212, 1259–1269.

Huey, R.B., Hertz, P.E., 1984. Is a jack-of-all-temperatures a master of none? Evolution 38, 441–444.

Ikegami, K., Yoshimura, T., 2012. Circadian clocks and the measurement of daylength in seasonal reproduction. Mol. Cell. Endocrinol. 349, 76–81.

Jacobs, J.D., Wingfield, J.C., 2000. Endocrine control of life-cycle stages: a constraint on response to the environment? Condor 102, 35–51.

Janeway, C.A., Travers, P., Walport, M., Shlomchik, M., 2004. Immunobiology: The Immune System in Health and Disease. Garland, New York.

King, J.R., 1974. Seasonal allocation of time and energy resources in birds. In: Paynter, R.A.J. (Ed.), Avian Energetics. Nuttall Ornithological Club, Cambridge, MA, pp. 4–85.

Klasing, K.C., 2004. The costs of immunity. Acta Zool. Sinica 50, 961–969.

Klasing, K.C., Leshchinsky, T.V., 1999. Functions, costs, and benefits of the immune system during development and growth. Ostrich 69, 2817–2832.

Knowles, S.C.L., Nakagawa, S., Sheldon, B.C., 2009. Elevated reproductive effort increases blood parasitaemia and decreases immune function in birds: a meta-regression approach. Funct. Ecol. 23, 405–415.

Koutsos, E.A., Klasing, K.C., 2013. Factors Modulating the Avian Immune System. Avian Immunology. Elsevier Ltd. pp. 1–17.

Lack, D., 1968. Ecological Adaptations for Breeding in Birds. Methuen, London.

Lambrechts, M.M., Perret, P., 2000. A long photoperiod overrides non-photoperiodic factors in blue tits' timing of reproduction. Proc. R. Soc. B 267, 585–588.

Lambrechts, M.M., Perret, P., Blondel, J., 1996. Adaptive differences in the timing of egg laying between different populations of birds result from variation in photoresponsiveness. Proc. R. Soc. B 263, 19–22.

Lambrechts, M.M., Blondel, J., Maistre, M., Perret, P., 1997. A single response mechanism is responsible for evolutionary adaptive variation in a bird's laying date. Proc. Natl. Acad. Sci. U.S.A. 94, 5153–5155.

Lee, K., 2006. Linking immune defenses and life history at the levels of the individual and the species. Integr. Comp. Biol. 46, 1000.

Li, H., Ferrari, M.B., Kuenzel, W.J., 2004. Light-induced reduction of cytoplasmic free calcium in neurons proposed to be encephalic photoreceptors in chick brain. Dev. Brain Res. 153, 153–161.

Liedvogel, M., Szulkin, M., Knowles, S., Wood, M.J., Sheldon, B.C., 2009. Phenotypic correlates of *clock* gene variation in a wild blue tit population: evidence for a role in seasonal timing of reproduction. Mol. Ecol. 18, 2444–2456.

Ligon, J.D., 1974. Green cones of the pinon pine stimulate late summer breeding in the pinon jay. Nature 250, 80–82.

Lochmiller, R., Deerenberg, C., 2000. Trade-offs in evolutionary immunology: just what is the cost of immunity? Oikos 88, 87–98.

Lofts, B., Murton, R.K., 1968. Photoperiodic and physiological adaptations regulating avian breweding cycles and their ecological significance. J. Zool. 155, 327–394.

Mand, R., Tilgar, V., Leivits, A., 2000. Reproductive response of great tits, *Parus major*, in a naturally base-poor forest habitat to calcium supplementation. Can. J. Zool. 78, 689–695.

Maney, D.L., Hahn, T.P., Schoesch, S.J., Sharp, P.J., Morton, M.L., Wingfield, J.C., 1999. Effects of ambient temperature on photo-induced prolactin secretion in three subspecies of white-crowned sparrow, Zonotrichia leucophrys. Gen. Comp. Endocrinol. 113, 445–456.

Maney, D.L., Goode, C.T., Lake, J.I., Lange, H.S., O'Brien, S., 2007a. Rapid neuroendocrine responses to auditory courtship signals. Endocrinology 148, 5614–5623.

Maney, D.L., Goode, C.T., Ball, G.F., 2007b. Transduction of a non-photic cue: from auditory system to a neuroendocrine response? J. Ornithol. 148 (Suppl. 2), S527–S538.

Marra, P.P., Hobson, K.A., Holmes, R.T., 1998. Linking winter and summer events in a migratory bird by using stable-carbon isotopes. Science 282, 1884–1886.

Martin II, L., Pless, M., Svoboda, J., Wikelski, M., 2004. Immune activity in temperate and tropical house sparrows: a common-garden experiment. Ecology 85, 2323–2331.

Martin, L., Weil, Z., Nelson, R., 2008. Seasonal changes in vertebrate immune activity: mediation by physiological trade-offs. Philos. Trans. R. Soc. B 363, 321–339.

Meijer, T., Daan, S., Hall, M., 1990. Family planning in the kestrel (*Falco tinnunculus*): the proximate control of covariation of laying date and clutch size. Behaviour 114, 117–136.

Meijer, T., Daan, S., Dijkstra, C., 1988. Female condition and reproduction—effects of food manipulation in free-living and captive kestrels. Ardea 76, 141–154.

Meijer, T., Drent, R., 1999. Re-examination of the capital and income dichotomy in breeding birds. Ibis 141, 399–414.

Meijer, T., Nienaber, U., Langer, U., Trillmich, F., 1999. Temperature and timing of egg laying of European starlings. Condor 101, 124–132.

Mishra, M.K., Wilson, F.E., Scanlan, T.S., Chiellini, G., 2004. Thyroid hormone-dependent seasonality in American tree sparrows (*Spizella arborea*): effects of GC-1, a thyroid receptor β-selective agonist, and of iopanoic acid, a deiodinase inhibitor. J. Comp. Physiol. B Biochem. Syst. Environ. Physiol. 174, 471–479.

Møller, A.P., Erritzøe, J., Saino, N., 2003. Seasonal changes in immune response and parasite impact on hosts. Am. Nat. 161, 657–671.

Moreno-Rueda, G., 2010. Experimental test of a trade-off between moult and immune response in house sparrows *Passer domesticus*. J. Evol. Biol. 23, 2229–2237.

Morton, M.L., Pereyra, M.E., Baptista, L.F., 1985. Photoperiodically induced ovarian growth in the white-crowned sparrow (*Zonotrichia leucophrys gambelii*) and its augmentation by song. Comp. Biochem. Physiol. 80A, 93–97.

Morton, G., Morton, M.L., 1990. Dynamics of postnuptial molt in free-living Mountain White-Crowned Sparrows. Condor 92, 813–828.

Nager, R.G., Ruegger, C., van Noordwijc, A.J., 1997. Nutrient or energy limitation on egg formation: a feeding experiment in great tits. J. Anim. Ecol. 66, 495–507.

Nager, R.G., van Noordwijk, A.J., 1995. Proximate and ultimate aspects of phenotypic plasticity in timing of great tit breeding in a heterogeneous environment. Am. Nat. 146, 454–474.

Nebel, S., Bauchinger, U., Buehler, D.M., Langlois, L.A., Boyles, M., Gerson, A.R., 2011. Constitutive immune function in European starlings, *Sturnus vulgaris*, is decreased immediately after an endurance flight in a wind tunnel. J. Exp. Biol. 215, 272–278.

Nebel, S., Buehler, D.M., MacMillan, A., Guglielmo, C.G., 2013. Flight performance of western sandpipers, *Calidris mauri*, remains uncompromised when mounting an acute phase immune response. J. Exp. Biol. 216, 2752–2759.

Nelson, R., Demas, G., 1996. Seasonal changes in immune function. Q. Rev Biol. 71, 511–548.

Nelson, R.J., Demas, G.E., Klein, S.L., Kriegsfeld, L.J., 2002. Seasonal Patterns of Stress, Immune Function and Disease. Cambridge University Press, Cambridge.

Nelson, R., 2004. Seasonal immune function and sickness responses. Trends Immunol. 25, 187–192.

Nicholls, T.J., Goldsmith, A.R., Dawson, A., 1988. Photorefractoriness in birds and comparison with mammals. Physiol. Rev. 68, 133–176.

Nolan, V., Ketterson, E.D., Ziegenfus, C., Cullen, D.P., Chandler, C.R., 1992. Testosterone and avian life histories—effects of experimentally elevated testosterone on prebasic molt and survival in male dark-eyed juncos. Condor 94, 364–370.

Norris, D.R., Marra, P.P., 2007. Seasonal interactions, habitat quality, and population dynamics in migratory birds. Condor 109, 535–547.

Norris, D.R., Marra, P.P., Kyser, T.K., Sherry, T.W., Ratcliffe, L.M., 2004. Tropical winter habitat limits reproductive success on the temperate breeding grounds in a migratory bird. Proc. R. Soc. B 271, 59–64.

O'Brien, S., Hau, M., 2005. Food cues and gonadal development in neotropical spotted antbirds (*Hylophylax naevioides*). J. Ornithol. 146, 332–337.

Owen, J.C., Moore, F.R., 2008. Swainson's thrushes in migratory disposition exhibit reduced immune function. J. Ethol. 26, 383–388.

Owen-Ashley, N.T., Wingfield, J.C., 2007. Acute phase responses of passerine birds: characterization and seasonal variation. J. Ornithol. 148, S583–S591.

Pap, P.L., Vagasi, C.I., Tokolyi, J., Czirjak, G.A., Barta, Z., 2010. Variation in haematological indices and immune function during the annual cycle in the Great Tit *Parus major*. Ardea 98, 105–112.

Perfito, N., Kwong, J.M.Y., Bentley, G.E., Hau, M., 2008. Cue hierarchies and testicular development: is food a more potent stimulus than day length in an opportunistic breeder (*Taeniopygia g. guttata*)? Horm. Behav. 53, 567–572.

Perfito, N., Meddle, S.L., Tramontin, A.D., Sharp, P.J., Wingfield, J.C., 2005. Seasonal gonadal recrudescence in song sparrows: response to temperature cues. Gen. Comp. Endocrinol. 143, 121–128.

Perfito, N., Zann, R.A., Bentley, G.E., Hau, M., 2007. Opportunism at work: habitat predictability affects reproductive readiness in free-living zebra finches. Funct. Ecol. 21, 291–301.

Perrins, C.M., 1970. The timing of birds' breeding seasons. Ibis 112, 242–255.

Perrins, C.M., 1991. Tits and their caterpillar food supply. Ibis 133 (Suppl.), 49–54.

Pulido, F., Berthold, P., Mohr, G., Querner, U., 2001. Heritability of the timing of autumn migration in a natural bird population. Proc. R. Soc. B 268, 953–959.

Prendergast, B.J., Hotchkiss, A.K., Bilbo, S.D., Nelson, R.J., 2004. Peripubertal immune challenges attenuate reproductive development in male Siberian hamsters (*Phodopus sungorus*). Biol. Reprod. 70, 813–820.

Robinson, J.E., Follett, B.K., 1982. Photoperiodism in Japanese quail: the termination of seasonal breeding by photorefractoriness. Proc. R. Soc. Lond. B 215, 95–116.

Rowan, W., 1925. Relation of light to bird migration and developmental changes. Nature 115, 494–495.

Rowan, W., 1926. On photoperiodism, reproductive periodicity, and the annual migrations of birds and certain fishes. Proc. Boston Soc. Nat. Hist. 38, 147–189.

Schaper, S.V., Dawson, A.V., Sharp, P.J., Gienapp, P., Caro, S.P., Visser, M.E., 2012. Increasing temperature, not mean temperature, is a cue for avian timing of reproduction. Am. Nat. 179, E55–E69.

Schleussner, G., Dittami, J.P., Gwinner, E., 1985. Testosterone implants affect molt in male European starlings, *Sturnus vulgaris*. Physiol. Zool. 58, 597–604.

Schmid-Hempel, P., Ebert, D., 2003. On the evolutionary ecology of specific immune defense. Trends Ecol. Evol. 18, 27–32.

Schoech, S.J., 1996. The effect of supplemental food on body condition and the timing of reproduction in a cooperative breeder, the Florida scrub-jay. Condor 98, 234–244.

Schoech, S.J., Bowman, R., Reynolds, S.J., 2004. Food supplementation and possible mechanisms underlying early breeding in the Florida scrub-jay (*Aphelocoma coerulescens*). Horm. Behav. 46, 565–573.

Schoech, S.J., Hahn, T.P., 2007. Food supplementation and timing of reproduction: does the responsiveness to supplementary information vary with latitude? J. Ornithol. 148 (Suppl.), S625–S632.

Schwab, R.G., Lott, D.F., 1969. Testis growth and regression in starlings (*Sturnus vulgaris*) as a function of the presence of females. J. Exp. Zool. 171, 39–42.

Sharp, P.J., 2005. Photoperiodic regulation of seasonal breeding in birds. Ann. NY Acad. Sci. 1040, 189–199.

Sheldon, B.C., Verhulst, S., 1996. Ecological immunology: costly parasite defences and trade-offs in evolutionary ecology. Trends Ecol. Evol. 11, 317–321.

Silver, R., Witkovsky, P., Horvath, P., Alones, V., Barnstable, C.J., Lehman, M.N., 1988. Co-expression of opsin- and VIP-like immunoreactivity in CSF-contacting neurons of the avian brain. Cell Tissue Res. 253, 189–198.

Silverin, B., Viebke, P.A., 1994. Low temperatures affect the photoperiodically induced LH and testicular cycles differently in closely related species of tits (*Parus* spp.). Horm. Behav. 28, 199–206.

Silverin, B., Wingfield, J., Stokkan, K.A., Massa, R., Jarvinen, A., Andersson, N.A., Lambrechts, M., Sorace, A., Blomqvist, D., 2008. Ambient temperature effects on photoinduced gonadal cycles and hormonal secretion patterns in Great Tits from three different breeding latitudes. Horm. Behav. 54, 60–68.

Sinclair, J., Lochmiller, R., 2000. The winter immunoenhancement hypothesis: associations among immunity, density, and survival in prairie vole (*Microtus ochrogaster*) populations. Can. J. Zool. 78, 254–264.

Stevenson, T.J., Ball, G.F., 2009. Anatomical localization of the effects of reproductive state, castration, and social milieu on cells immunoreactive for gonadotropin-releasing hormone-I in male European starlings (*Sturnus vulgaris*). J. Comp. Neurol. 517, 146–155.

Stevenson, T.J., Bernard, D.J., Ball, G.F., 2009. Photoperiodic condition is associated with region-specific expression of GNRH1 mRNA in the preoptic area of the male starling (*Sturnus vulgaris*). Biol. Reprod. 81, 674–680.

Stevenson, T.J., Hahn, T.P., Ball, G.F., 2011. Variation in gonadotrophin-releasing hormone-1 gene expression in the preoptic area predicts transitions in seasonal state. J. Neuroendocrinol. 24, 267–274.

Stevenson, T.J., Hahn, T.P., MacDougall-Shackleton, S.A., Ball, G.F., 2012. Gonadotropin-releasing hormone plasticity: a comparative perspective. Front. Neuroendocrinol. 33, 287–300.

Studds, C.E., Marra, P.P., 2005. Nonbreeding habitat occupancy and population processes: an upgrade experiment with a migratory bird. Ecology 86, 2380–2385.

Studds, C.E., Marra, P.P., 2011. Rainfall-induced changes in food availability modify the spring departure programme of a migratory bird. Proc. R. Soc. B 278, 3437–3443.

Tonra, C.M., Marra, P.P., Holberton, R.L., 2011. Migration phenology and winter habitat quality are related to circulating androgen in a long-distance migratory bird. J. Avian Biol. 42, 397–404.

Tøttrup, A.P., Klaassen, R.H.G., Kristensen, M.W., Strandberg, R., Vardanis, Y., Lindstrom, A., Rahbek, C., Alerstam, T., Thorup, K., 2012. Drought in Africa caused delayed arrival of European songbirds. Science 338, 1307.

Ubuka, T., Cadigan, P.A., Wang, A., Liu, J., Bentley, G.E., 2009. Identification of European starling GnRH-I precursor mRNA and its seasonal regulation. Gen. Comp. Endocrinol. 162, 301–306.

van Gils, J.A., Munster, V.J., Radersma, R., Liefhebber, D., Fouchier, R.A., Klaassen, M., 2007. Hampered foraging and migratory performance in swans infected with low-pathogenic avian influenza A virus. PLoS One 2, e184.

van Noordwijc, A.J., McCleery, R.H., Perrins, C.M., 1995. Selection for the synchronization of great tit (*Parus major*) breeding with caterpillar growth, using temperature as a predictor. J. Anim. Ecol. 64, 451–458.

Visser, M.E., Both, C., Lambrechts, M.M., 2004. Global climate change leads to mistimed avian reproduction. Adv. Ecol. Res. 35, 89–110.

Visser, M.E., van Noordwijk, A.J., Tinbergen, J.M., Lessells, C.M., 1998. Warmer springs lead to mistimed reproduction in great tits (*Parus major*). Proc. R. Soc. B Biol. Sci. 265, 1867–1870.

Visser, M.E., Holleman, L.J.M., Caro, S.P., 2009. Temperature has a causal effect on avian timing of reproduction. Proc. R. Soc. Biol. Sci. 276, 2323–2331.

Visser, M.E., te Marvelde, L., Lof, M.E., 2012. Adaptive phenological mismatches of birds and their food in a warming world. J. Ornithol. 153, S75–S84.

Visser, M.E., Schaper, S.V., Holleman, L.J.M., Dawson, A., Sharp, P., Gienapp, P., Caro, S.P., 2011. Genetic variation in cue sensitivity involved in avian timing of reproduction. Funct. Ecol. 25, 868–877.

Voelker, G., Rohwer, S., 1998. Contrasts in scheduling of molt and migration in eastern and western warbling vireos. Auk 115, 142–155.

Wada, M., 1993. Low temperature and short days together induce thyroid activation and suppression of LH release in Japanese quail. Gen. Comp. Endocrinol. 90, 355–363.

Wada, M., Hatanaka, F., Tsuyoshi, H., Sonoda, Y., 1990. Temperature modulation of photoperiodically induced LH secretion and its termination in Japanese quail (*Coturnix coturnix japonica*). Gen. Comp. Endocrinol. 80, 465–472.

Watts, H.E., Hahn, T.P., 2012. Non-photoperiodic regulation of reproductive physiology in the flexibly breeding pine siskins (*Spinus pinus*). Gen. Comp. Endocrinol. 178, 259–264.

Wingfield, J.C., 1980. Fine temporal adjustment of reproductive function. In: Epple, A., Stetson, M.H. (Eds.), Avian Endocrinology. Academic Press, New York, pp. 367–389.

Wingfield, J.C., 1985. Environmental factors influencing the termination of reproduction in finches. In: Ilyichev, V.D., Gavrilov, V.M. (Eds.), Acta XVIII Congressus Internationalis Ornithologici. Nauka, Moscow, pp. 478–487.

Wikelski, M., Tarlow, E.M., Raim, A., Diehl, R.H., Larkin, R.P., Visser, G.H., 2003. Costs of migration in free-flying songbirds. Nature 423, 704–705.

Wikelski, M., Martin, L.B., Scheuerlein, A., Robinson, M.T., Robinson, N.D., Helm, B., Hau, M., Gwinner, E., 2008. Avian circannual clocks: adaptive significance and possible involvement of energy turnover in their proximate control. Philos. Trans. R. Soc. B 363, 411–423.

Wilson, F.E., 1991. Neither retinal nor pineal photoreceptors mediate photoperiodic control of seasonal reproduction in American tree sparrows (*Spizella arborea*). J. Exp. Zool. 259, 117–127.

Wingfield, J.C., 2008. Organization of vertebrate annual cycles: implications for control mechanisms. Philos. Trans. R. Soc. B 363, 425–441.

Wingfield, J.C., Farner, D.S., 1993. Endocrinology of reproduction in wild species. Avian Biol. 9, 163–327.

Wingfield, J.C., Marler, P., 1988. Endocrine basis of communication in reproduction and aggression. In: Knobil, E., Neill, J. (Eds.), The Physiology of Reproduction. Raven Press, Ltd, New York.

Wingfield, J.C., Hahn, T.P., Doak, D., 1993. Integration of environmental factors regulating transition of physiological state, morphology and behaviour. In: Sharp, P.J. (Ed.), Avian Endocrinology. Society for Endocrinology, Bristol, pp. 111–122.

Wingfield, J.C., Hahn, T.P., Levin, R., Honey, P., 1992. Environmental predictability and control of gonadal cycles in birds. J. Exp. Zool. 261, 214–231.

Wingfield, J.C., Hahn, T.P., Maney, D.L., Schoesch, S.J., Wada, M., Morton, M.L., 2003. Effects of temperature on photoperiodically induced reproductive development, circulating plasma luteinizing hormone and thyroid hormones, body mass, fat deposition and molt in mountain white-crowned sparrows, *Zonotrichia leucophrys oriantha*. Gen. Comp. Endocrinol. 131, 143–158.

Wingfield, J.C., Hahn, T.P., Wada, M., Astheimer, L.B., Schoech, S., 1996. Interrelationship of day length and temperature on the control of gonadal development, body mass, and fat score in white-crowned sparrows, *Zonotrichia leucophrys gambelii*. Gen. Comp. Endocrinol. 101, 242–255.

Wingfield, J.C., Hahn, T.P., Wada, M., Schoech, S.J., 1997. Effects of day length and temperature on gonadal development, body mass, and fat depots in white-crowned sparrows, *Zonotrichia leucophrys pugetensis*. Gen. Comp. Endocrinol. 107, 44–62.

Wingfield, J.D., Kenagy, G.J., 1991. Natural control of reproduction. In: In: Pang, P.K.T., Schreibman, M.P. (Eds.), Handbook of Comparative Endocrinology, vol 4. Academic Press, New York, pp. 181–241.

Yasuo, S., Watanabe, M., Okabayashi, N., Ebihara, S., Yoshimura, T., 2003. Circadian clock genes and photoperiodism: comprehensive analysis of clock gene expression in the mediobasal hypothalamus, the suprachiasmatic nucleus, and the pineal gland of Japanese quail under various light schedules. Endocrinology 144, 3742–3748.

Yuri, T., Rohwer, S., 1997. Molt and migration in the northern rough-winged swallow. Auk 114, 249–262.

Yousaf, M., Chaudhry, A.S., 2008. History, changing scenarios and future strategies to induce moulting in laying hens. Worlds Poult. Sci. J. 64, 65–75.

Zysling, D., Garst, A., Demas, G., 2009. Photoperiod and food restriction differentially affect reproductive and immune responses in Siberian hamsters *Phodopus sungorus*. Funct. Ecol. 23, 979–988.

Chapter 37

Regulation of Body Temperature: Strategies and Mechanisms

Shlomo Yahav

Department of Poultry and Aquaculture Sciences, Institute of Animal Sciences, ARO, The Volcani Center, Bet-Dagan, Israel

LIST OF ABBREVIATIONS

AVT arginine vasotocin
BDNF brain-derived neurotrophic factor
BMR basal metabolic rate
CAM chorioallantoic membrane
CNS central nervous system
CVS cardiovascular system
CWL cutaneous water loss
E evaporative heat loss
E embryonic day
EST eggshell temperature
EWL evaporative water loss
HR heart rate
HSP heat shock protein
IR infrared
LCT lower critical temperature
M metabolic energy
NGF nerve growth factor
NST non-shivering thermogenesis
PO/AH preoptic anterior hypothalamus
Q_c convective heat loss
Q_r radiative heat loss
RH relative humidity
RMR resting metabolic rate
ROI Region of Interest, as defined by the camera software
RTSR rapid thermal stress response
RWL respiratory water loss
SC stratum corneum
SHL sensible heat loss
ST shivering thermogenesis
T_3 triiodothyronine
T_4 thyroxine
T_a ambient temperature
T_{af} allantoic fluid temperature
T_b body temperature
T_{fs} facial temperature
T_s skin temperature
T_{set} set point temperature
TM thermal manipulation
TSH thyroid stimulating hormone
UCP uncoupling protein
UCT upper critical temperature
VO_2 oxygen consumption
avUCP avian uncoupling protein

DEFINITIONS—ACCORDING TO IUPS THERMAL COMMISSION (2001)

Endothermy The pattern of thermoregulation of animals in which the body temperature depends on a high and controlled rate of heat production.

Ectothermy The pattern of temperature regulation of animals in which body temperature depends mainly on the behaviorally controlled exchange of heat with the environment.

Poikilothermy Wide variability of an organism's body temperature as a function of ambient temperature, without effective autonomic temperature regulation.

Homeothermy The pattern of temperature regulation in a tachymetabolic species in which cyclic variation in core temperature—either diurnally or seasonally—is maintained within arbitrarily defined limits, despite much wider variations in ambient temperature, i.e., the body temperature is held within a narrow range.

Heterothermy A pattern of temperature regulation in a tachymetabolic species, in which the variation in core temperature, either diurnal or seasonal, exceeds that which characterizes homeothermy.

Basal Metabolic rate (BMR) Metabolic energy transformation calculated from measurements of heat production or oxygen consumption in a rested and awake organism that has fasted sufficiently long to be in a postabsorptive state and a thermoneutral zone.

Resting Metabolic Rate (RMR) The metabolic rate of an animal that is resting in the thermoneutral environment but not in the postabsorptive state.

Body temperature (T_b) regulation, which is an exclusive characteristic of endothermic birds and mammals, allows for the adaptation of processes and survival under varied environmental conditions. This chapter focuses on the evolution of endothermy, which enables maintenance of T_b within a relatively small range; the mechanisms of heat production and heat loss; various strategies to maintain T_b; the development of endothermy during embryogenesis; and the cost of endothermy in various ecological niches.

37.1 INTRODUCTION

Anaximander—a Greek philosopher of about 570 B.C.E asserted: "Living things had been produced in the course of natural processes that were under the necessity of either adapting themselves to their environment or perishing".

Body temperature (T_b) control is a unique feature of endothermic organisms. Endothermy was adopted by birds and mammals in the course of their evolution, in order to better adapt to the varying environment.

Endothermy in general is a collective outcome of physiological adjustments coupled with morphological modifications; it profoundly affects the biology of birds, in conferring various physiological, ecological, and behavioral benefits that enable the wide aerial, aquatic, and terrestrial distribution of birds. Endothermic species maintain T_b within a relatively narrow range, such that T_b is the most physiologically guarded parameter of the body. Therefore, the thermoregulatory system in these animals operates at a very high gain, in order to control T_b within a relatively narrow range despite moderate to extreme changes in the environment.

The ability to maintain a relatively stable T_b springs from the mechanisms that control heat production and heat loss—mechanisms that changed in the course of evolution to enable endothermy to replace ectothermy. Both processes, especially heat production, are probably older than endothermy (Silva, 2006; Morrison et al., 2008; Richards and Proszkowiec-Weglarz, 2007). The mechanism of heat production involves increases in lung ventilation rates, cardiovascular system capacity, oxygen consumption (VO_2), and internal heat production that is based on aerobic metabolism. The enhanced aerobic mechanism provides avians with the ability to sustain high levels of activity which, in turn, enable birds to migrate over long distances and to inhabit varied environmental niches (Ruben, 1995). The evolutionary changes from ectothermy to endothermy were achieved because the developmental regulatory mechanisms maintained a balance between heat production and heat loss. It is not clear when birds attained endothermic status; however, the close phylogenetic affinity of birds to the extinct but successful vertebrates—the dinosaurs, which dominated the Mesozoic Era—received considerable attention as the phylogenetic source (Horner et al., 2001). Understanding the processes leading to the changes in metabolic rate and heat loss that result in the transition from ectothermy to endothermy is of great importance for understanding the mechanisms that enable control of T_b.

Birds are divided into altricial and precocial species. Altricial species are hatched at an undeveloped stage and are poikilothermic, being able to develop endothermic thermoregulation when they grow to one-third to one-half of adult size (Cooper and Geiser, 2008). In contrast, precocial birds hatch with a developed endothermic thermoregulatory system that continues its maturation process, mainly the feedback mechanism, up to 10 days posthatch (Arad and Itsaki-Glucklish, 1991).

This chapter will focus on the evolution of endothermy, which enables maintenance of T_b within a relatively narrow range; the mechanisms of heat production and heat loss; different strategies to maintain T_b; the development of endothermy during embryogenesis; and the cost of endothermy in various ecological niches.

37.2 THE EVOLUTION OF ENDOTHERMY

The evolution of endothermy attracts great attention as one of the most significant transitions in vertebrate evolution. Endothermy in birds is a major physiological specialization that profoundly affects their biology. It imparts a variety of physiological, ecological, and behavioral advantages that enable the distribution of birds among a wide variety of environments.

Study of the evolutionary development of endothermy must take into account the fact that endotherms are assumed to be typically homeothermic (i.e., characterized by stable and high T_b). However, some birds are characterized by heterothermy rather than homeothermy (i.e., they exhibit daily and seasonal variations in T_b; McKechnie and Lovegrove, 2002) or torpor, which involves controlled reductions of metabolic rate and T_b (Cooper and Geizer, 2008).

During the course of evolution, the transition from ectothermic to endothermic ancestors is theoretically attributed to either direct selection for high T_b and high resting metabolic rate (RMR) or selection for increased endurance and sustainable activity (the aerobic-capacity model). The first of these theoretical models of endothermic selection was based on various models implied that body warmth enabled higher RMR, which had enhanced selective value. This selective value enabled endothermic birds to support RMRs 5–15 times as great as those of ectotherms of similar body size (Bennett et al., 2000), and field metabolic rates that in general exceeded the reptilian rate by 20- to 30-fold (Nagy, 1987). The second model excluded warmth as a sufficient causative factor for higher RMR, but rather considered the high metabolic rate as a consequence of selection for postural changes that led to increases in physical activity and brain size and allowed enhanced aerobic capacity. In turn, selection for increased aerobic capacity supported higher levels of sustained activity that led to higher RMR and higher heat expenditure, thus enabling the transition from ecto- to endothermy (Grigg, 2004).

The transition from ecto- to endothermy during the course of evolution necessitated changes related to heat production and controlled lowering of heat conductance. Changes in heat production generally resulted from locomotor activity, shivering thermogenesis, high resting metabolic rate, nonshivering thermogenesis. All four were sufficiently enabled by the ability of birds to maintain gradients across plasma membranes, generated by sodium/potassium ATPase pumps—a process that apparently requires generation of significantly increased levels of metabolic energy (Else et al., 2004). On the other hand, control by means of lowered heat conductance is achieved by insulation based on plumage and feathers, body-fat distribution, and changes in peripheral blood flow (i.e., vasoconstriction and vasodilatation).

These physiological changes could not take place without anatomical changes, among the most important of which, in birds, involve the respiratory system; they enable the higher metabolic rate and T_b than those of mammals. Collectively, the respiratory organs of birds are different from those of mammals; the small, compact lungs communicate with voluminous, thin-walled air sacs. Such a respiratory system has been considered as an adaptation to flight, in light of its volume and the co-current air flow, which is a product of the anatomical modifications. Comparison between the respiratory volume of a 1kg bird with that of a mammal of identical body weight shows total respiratory volumes of 160.8 and 54.4ml for birds and mammals, respectively (Shmidt-Nielsen, 1975). The anatomical modification in birds permits air to flow through the lung and continuously past the exchange surface, whereas in mammals air must flow in and out. The continuous air flow through the bird lung exchange surface significantly improved the efficacy of O_2/CO_2 exchange in the lungs; it is facilitated by the positions of the air sacs (anterior and posterior) in the body, which enable co-current air flow.

Collectively, the evolutionary processes combined anatomical and physiological modifications/adaptations that enabled exploitation of the ecological opening for development of endothermic birds from developed ectothermic reptiles. The evolutionary changes from ectothermy to endothermy were achieved as a result of the developmental regulatory mechanisms that maintained a balance between heat production and heat loss; the mechanisms of both are permanently activated and regulated by both neuronal and hormonal signals (Silva, 2006; Richards and Proszkowiec-Weglarz, 2007; Morrison et al., 2008).

37.3 DIFFERENT STRATEGIES TO MAINTAIN ENDOTHERMY

Two vital resources are crucial for any organism to exist—energy and water. Therefore, maintaining energy and water balances is the goal of life. For an endothermic organism, the "fire of life" means an energy balance that facilitates functioning of the body systems and enables maintenance of the dynamic steady state of the internal systems. Endothermy facilitates homeothermy and/or heterothermy, as expressed in T_b. Bligh (1966) formulated the two-tier theory of thermoregulation, according to which there are two distinct levels of thermal control: the fine or narrow-band control that acts to hold T_b within its normal variation range (i.e., the condition of normothermia) and the wide-band control, which is activated when T_b deviates markedly from the normal range (i.e., a condition of hyperthermia or hypothermia).

During their life span, birds of each and every species experience extreme environmental conditions that may cause difficulties in holding T_b within the normothermic range; therefore, they must be able to thermally tolerate the environmental changes.

To sustain thermal tolerance and avoid the deleterious consequences of thermal stresses, three direct responses are elicited in birds: the rapid thermal stress response (RTSR) characterized by response times ranging from minutes to hours (Parsell and Lindquist, 1994; Yahav, 2009); acclimation/acclimatization, characterized by response times of days to weeks (Yahav et al., 1997; Horowitz, 2002); and epigenetic adaptation during the perinatal period, which is based on timing (Nichelmann et al., 1999; Tzschentke et al., 2001; Piestun et al., 2008). The RTSR can be defined as an immediate response to temperature stress; it initiates the recruitment of body resources to maintain the organism's dynamic steady state. This response cannot be maintained for a long period; therefore, under continuous severe temperature stress, a cascade of events may lead to temperature stroke. Acclimation is defined as physiological or behavioral changes that occur within the lifetime of an organism and which reduce or enhance tolerance of the strain caused by experimentally induced stressful changes—specifically, climatic factors (IUPS Thermal Commission, 2001). Whereas acclimation refers to experimental procedures, acclimatization refers to natural processes. Epigenetic adaptation has been defined as a lifelong adaptation that occurs during prenatal (embryogenesis) or early posthatching ontogeny; it takes place within critical developmental phases that affect gene expression. Its presumed occurrence is based on the assumption that during critical developmental phases, environmental factors, especially ambient temperature (T_a), have a strong influence on determination of the set point for physiological control systems. Dörner (1974) first described this as the determination rule.

37.4 REGULATORY MECHANISM OF ENDOTHERMY

37.4.1 Neuronal Signals to and from the Preoptic Anterior Hypothalamus

The body temperature of endothermic birds is regulated through a variety of thermoregulatory responses, such as shivering thermogenesis (ST), nonshivering thermogenesis (NST), panting, cutaneous vasomotion, and behavior (Figure 37.1). All these responses are controlled and

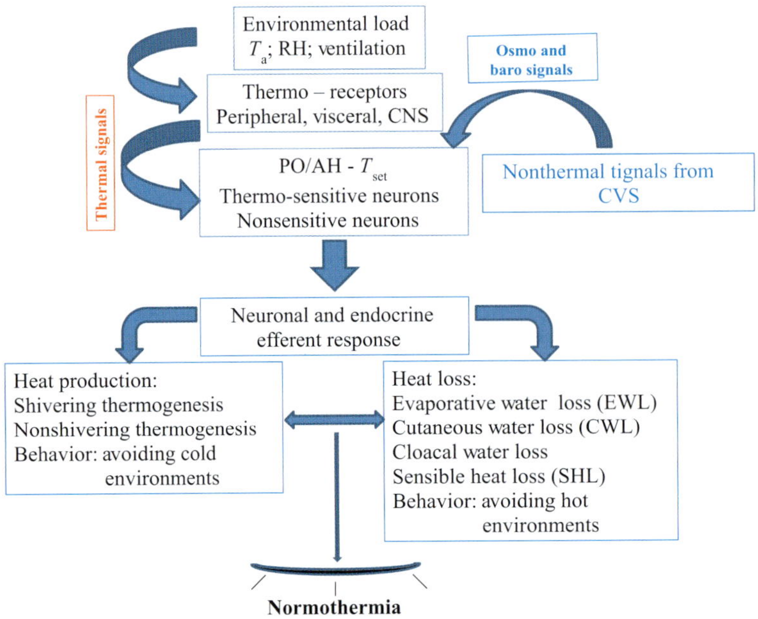

FIGURE 37.1 Schematic flow chart describing how environmental heat load is transformed to thermal information by thermoreceptors, and transferred to the preoptic anterior hypothalamus (PO/AH) to elicit thermoregulatory responses in the animal (heat production and heat loss efferent response) mediated by the neuronal and endocrine systems. Nonthermal signals, mainly from the blood system, can modify the response. ST, shivering thermogenesis; NST, nonshivering thermogenesis; CNS, central nervous system; CVS, cardiovascular system; EWL, evaporative water loss; CWL, cutaneous water loss; SHL, Sensible heat loss; T_{set}, set-point temperature.

orchestrated by the brain to optimize the internal T_b in order to facilitate appropriate molecular activities and reactions (involving enzymes, ion channels, etc.; Nakamura, 2011). The thermoregulatory system that enables maintenance of a relatively constant T_b consists of a sensory afferent part (involving thermo-, osmo-, and baro-receptors), an integrating part (the thermoregulatory center), and the command efferent part (involving neurological and endocrine signals; Figure 37.1).

The *sensory afferent part* comprises thermal information transferred by thermoreceptors from the bird's surface (skin), all internal organs, the spinal cord, and the brain itself. Whereas the thermoreceptors in the internal organs, spinal cord, and brain sense a relatively constant and protected internal environment, those of the skin sense the variable environment outside the body. Changes in the external environment encompass its temperature, relative humidity (RH), and ventilation, all of which affect the animal's surface temperature, and form the data base transferred by the peripheral thermoreceptors to the integrating center. Skin thermoreceptors are also continuously affected by the body core temperature, as heat is transferred from body core to the periphery by blood flow (i.e., the vasomotor response).

One of the nonthermal information items transferred to the integrating center (Figure 37.1) refers to another homeostatic parameter, the osmolality, which includes afferent signals coming from peripheral and central receptors that are sensitive to osmotic pressure, volume, and sodium concentration of the extracellular fluid and that can be closely represented by the blood system (Keil et al., 1994). The homeostasis systems that control thermoregulation and osmolality, respectively, are linked via the evaporative heat-loss mechanism: thermogenesis is balanced by energy loss, which depends mainly on evaporative water loss (EWL) which, in turn, depends on the body water and salt equilibria. It is well established that thermal homeostasis is modified under conditions that shift the osmotic balance (Simon and Notel, 1990; Keil et al., 1994) and vice versa (Landgraf et al., 1994), and this interaction was demonstrated in many endothermic animals. In general, at high T_a, the frequency of thermal polypnea decreased because of the reduction of extracellular fluid volume that resulted from the high evaporation rate; this leads to dehydration, hypovolemia, and hyperosmolarity (Yahav et al., 2005). In such cases, hyperthermia continues to develop.

The *integrating part* is neuroanatomically located in the preoptic area, which is located in the rostral pole of the hypothalamus (i.e., the preoptic anterior hypothalamus, PO/AH; Boulant, 1996; DiMicco and Zaretsky, 2007). The tight correlation between thermoregulation and osmoregulation focuses the maintenance of T_b and water/salt equilibrium in the PO/AH, so that lesions of this site elicit both severe thermoregulatory disturbances and deficiencies in drinking behavior and osmoregulation (Bligh, 1973; Hatton, 1990; Kuznetsov and Kazakov, 2000).

This area plays a dual role: it monitors local temperature changes and integrates temperature information received from the periphery (Hellon and Taylor, 1982; Patapoutain et al., 2003; Wechselberger et al., 2006). In mammals and birds, temperature is monitored by temperature-sensitive neurons, which change their firing rate in accordance with the hypothalamic temperature (Griffin et al., 2001) and can be described as the hypothalamic setting point (Boulant, 2006). These temperature-sensitive neurons form 40% of the PO/AH neurons; the rest are temperature-insensitive neurons. About 75% of the temperature-sensitive neurons are warmth sensitive (their firing rate increases strongly when the hypothalamic temperature increases), whereas about 25% of them are cold sensitive with a firing rate that changes in the opposite sense to the hypothalamus temperature (see Tzschentke et al., 2004 for birds; Boulant, 2006 for mammals). However, in endothermic mammals (in which most of the research was done) the evidence confirms that the PO/AH comprises 70% temperature-insensitive neurons and 20% temperature-sensitive ones (Nakayama et al., 1961).

In various models (Hammel, 1965; Boulant, 2006), the PO/AH serves to define a temperature set point (T_{set}) in order to promote heat production and/or heat loss. When the PO/AH site is cooled below its T_{set}, efferent neuronal signals will promote heat production by means that include ST and NST, peripheral vasoconstriction (in the skin), and behavioral responses that include huddling and seeking a warm environment. When PO/AH is warmed above T_{set}, heat loss mechanisms will promote panting, skin vasodilatation, and behavior that facilitates cooling, such as seeking shade and/or a ventilated environment or exposing nonfeathered skin areas. Each of these responses has its own T_{set}, and the magnitude of the response depends on the hypothalamic T_{set}.

However, additional neurons are encompassed by the concept of hypothalamic T_{set} based on temperature-sensitive neurons. These are the temperature-guardian neurons (Basta et al., 1997), which were detected in the PO/AH of Muscovy ducks. These neurons reacted exclusively to critical brain temperature signals and possibly stimulated a second type of thermoregulatory effector activity. This type of neuron falls within Bligh's (1966) two-tier concept which, in this case, operates when T_b shifts outside the normothermic range, in cases of hyperthermy or hypothermy.

The *efferent neurons* can be characterized according to the Hammel (1965) model, as updated by Boulant (2006). Basically, the warmth- and cold-sensitive and temperature-insensitive neurons, together with the efferent neurons, control the thermoregulatory responses. The efferent neurons are located within the PO/AH or in sites such as the posterior hypothalamus and brain stem; they are either spontaneously firing neurons or silent ones that are synaptically excited by warmth- and cold-sensitive neurons, or are inhibited by temperature-insensitive neurons.

The *endocrine efferent response* is mediated mainly by the hypothalamus–pituitary–thyroid axis. In birds, the metabolic rate is associated with secretion of the hormone thyroxine (T_4), which is secreted by the thyroid gland and is deiodinated to triiodothyronine (T_3) in all peripheral tissues. T_3 is the main metabolism-stimulating hormone (McNabb and King, 1993; Gabarrou et al., 1997) that was found to be associated with T_b regulation (McNabb and King, 1993; Yahav, 2000).

In summary, the mechanisms of both heat production and heat loss are permanently activated and are regulated by both neuronal and hormonal signals.

37.5 PHYSIOLOGICAL PROCESSES THAT ENABLE ENDOTHERMY

37.5.1 The Energy Balance Equation

In endotherms, the initial exposure to the thermal environment starts at the skin surface and extends outward into surroundings that are characterized by varied air temperature and humidity, radiation, and barometric pressure that varies with altitude. The last of these is most important for birds that spend migratory periods at high altitudes, exposed to low barometric pressure, coupled with high radiation and low ambient temperatures. In addition, the environment, including all the aforementioned factors, plays a crucial role in modifying heat transfer from the bird's surface via four principal processes: radiation, convection, conduction, and evaporation.

The thermogenic response of endotherms to changes in the environment can be divided into three zones (Figure 37.2): a zone of body cooling, a zone of evaporative regulation, and a zone of comfort. In the first zone, the regulation of T_b is controlled by heat retention (through vasoconstriction), followed by ST and NST. In the zone of evaporative regulation, T_b is regulated by vasodilation, followed by evaporation and sensible heat loss (SHL). In the comfort zone—the "thermoneutral zone"—T_b is controlled mainly by conduction. In both the cooling and the heating zones, the main concern is how tolerant the bird will be of extreme conditions and how long it will take to acclimate to such environments by changing the response thresholds of the physiological reactions. Thus, the thermogenic response will affect the development of hypothermy (in a cold environment) or hyperthermy (in a hot environment).

Collectively, heat transfer modeling has been used to understand thermoregulation in endotherms in general. The models assume a constant T_b for an animal not performing external work and are represented by the following heat-balance equation, based on the first law of thermodynamics:

$$S = M - E \pm R \pm C \pm K \tag{37.1}$$

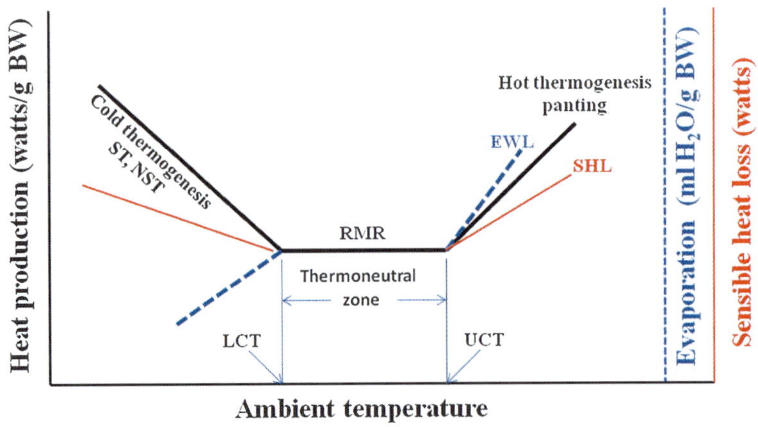

FIGURE 37.2 A general thermogenetic curve of endothermic organisms. The thermoneutral zone ranged between the lower and upper critical temperatures (LCT and UCT, respectively); it represents the organism's resting metabolic rate (RMR). Below the LCT heat generation by shivering thermogenesis (ST) and nonshivering thermogenesis (NST) increases in response to cold exposure. Above the UCT, thermogenesis increases to permit evaporative water loss (EWL), mainly via panting. Sensible heat loss (SHL) increases above the UCT as a result of increased ventilation, whereas below the LCT it increases as a result of the increased differences between the organism surface temperature and the ambient temperature.

in which S is the bodily heat gain or loss that must be balanced by M, metabolic heat production; E, evaporative heat loss; R, radiative heat gain or loss; C, convective heat loss or gain; and K, conductive heat loss or gain. Body temperature will remain unchanged when S is zero—that is, heat gain matches heat loss. If more heat is produced and/or gained than lost, then S will be positive and T_b will rise and vice versa.

Within Eqn (37.1), thermogenesis (heat production, M) can be divided into obligatory and facultative thermogenesis (Silva, 2006). Obligatory thermogenesis refers to the energy required to maintain T_b as long as the ambient temperature (T_a) lies in the thermoneutral zone—the range within which the body is in thermal equilibrium with the environment and produces energy at a level termed the resting metabolic rate (RMR) (Gordon, 1993). Facultative thermogenesis refers to stimulated energy production, which is required when T_a deviates below or, to some extent, above the thermoneutral zone.

37.5.1.1 Metabolic Energy

The metabolic rate represents the free energy produced by the transformation of chemical energy during aerobic and anaerobic metabolic activities within the organism. Energy is obtained through oxidation of feedstuffs; therefore, VO_2 can be used as measure of energy metabolism. However, it must be noted that under hypoxic conditions, when anaerobic mechanisms play a role, VO_2 will not be an adequate measure of energy metabolism.

Metabolic energy as measured per unit of time is termed the metabolic rate; it can be determined from the differences between energy intake (i.e., the energy value of all feed consumed) and that of all excreta (mainly feces and urine, which can be determined with a bomb calorimeter). This method requires measurements over a relatively long period, during which the bird has to have a relatively constant body weight, which is not the case with poultry. A second means of determining the metabolic rate is based on providing accurate information about the substance (feed stuff) that is oxidized and the amounts of energy produced for 1 g of carbohydrate, fat, and protein—4.2, 9.4, and 4.25 kcal, respectively. A third method uses direct measurement of VO_2, either in a calorimetric cage or an open air flow system. The calorimetric cage involves long-term measurements that encompass feed and water intake, feces and urine excretion, evaporation, etc. An open air flow system is based on short-term measurements (Buffenstein and Yahav, 1991), applying to a resting animal that is in a relatively steady-state condition (i.e., constant body weight, relatively empty gastrointestinal tract, and stress-free). A fourth method is based on the double-labeled water technique that uses $D_2{}^{18}O$ and $^3H_2{}^{18}O$, which is considered the gold standard for measuring energy expenditure under free-living conditions (Shaffer, 2011). The reason that oxygen can be used as a practical index of metabolic rate is based on the relatively constant amounts of heat produced for each liter of oxygen used in metabolism—5.0, 4.7, and 4.4 kcal for carbohydrate, fat, and protein, respectively.

37.5.1.2 Metabolic Rates of Passerine and Nonpasserine Birds

The total VO_2 of birds cannot be used to provide an accurate comparison among different species. However, calculation of VO_2 per units of weight and time, as well as conversion of the results to values appropriate to conditions of 0 °C, barometric pressure at sea level (760 mm Hg), and a dry environment, will provide a sufficient basis to compare bird species.

A comparative study (Lasiewski and Dawson, 1967) of metabolic rates of 58 non-passerine birds of body weights ranging from 3 g (hummingbird) up to 100 kg (ostrich) established an equation with a regression line corresponding to the following:

$$V\mathrm{o}_2 = 4.6 \times \mathrm{M}_{\mathrm{b}}^{0.723} \quad (37.2)$$

in which $V\mathrm{o}_2$ is the entire nonpasserine VO_2, M_b is body mass, and 0.723 is the slope of the regression line.

A comparative study of 36 passerine species of body weights ranging from 6.1 to 866 g yielded the following equation:

$$V\mathrm{o}_2 = 7.54 \times M_{\mathrm{b}}^{0.724} \quad (37.3)$$

The regression slopes for passerine and nonpasserine birds were similar, but the metabolic rate of a passerine bird can be expected to be higher by about 65% than that of a nonpasserine bird of similar body weight (Lasiewski and Dawson, 1967). It must be noted that birds that deviated from the regression line were adapted to environments such as deserts, high altitudes, or polar regions, which differed from a so-called regular environment.

37.5.1.3 *Metabolic Rates of Domesticated Birds*

There has been massive genetic selection processes for over 50 years in domestic poultry; for example, broiler chickens and turkeys are selected for high growth of meat (muscle) and laying hens for high egg production. The development in the genetic selection of meat-type fowl—broilers (Havenstein et al., 1994, 2003a) and turkeys (Havenstein et al., 2007)—led to rapid growth, with increased feed efficiency and metabolic rate (Janke et al., 2004), which has led to heavy poultry with relatively short growth periods. Such development logically necessitates parallel increases in the size and efficacy of visceral systems, such as the cardiovascular and the respiratory systems, to support body energy balance. However, inferior development of such major systems (Havenstein et al., 2003b) has led to a relatively low capability to balance the significantly increased energy expenditure and, consequently, to a relatively low capability to control T_b efficiently under extreme environmental conditions (Yahav, 2009). These difficulties, in turn, lead to severe difficulties in maintaining a dynamic steady state—difficulties that can impair productivity. The dynamic changes in energy expenditure of poultry prevented acquisition of information that could lead to formulation of a general equation for comparing $V\mathrm{o}_2$ of domestic birds of differing body weights, as was accomplished for passerine and nonpasserine birds (Eqns (37.2) and (37.3)).

37.5.1.4 *Nonshivering Thermogenesis*

Although birds have no distinct brown adipose tissue or related thermogenic tissue, there is strong evidence that the NST mechanism in birds plays an important role during exposure to cold; the NST process has been implicated in the regulation of T_b and metabolic rate. For many decades, birds were known to meet their increased burden of thermogenic requirements in response to cold stress by adopting ST (West, 1965), and only recently was NST recognized in birds too, despite their lack of brown adipose tissue. This recognition sprang from the finding that NST occurred in skeletal muscle tissue too, as previously was recorded in mammals (Vidal-Puig et al., 1997), and also in studies on European finches that, under severe cold conditions, exhibited an independent rate of VO_2, distinct from their integrated pectoral electrographic activity; this suggested the involvement of NST (Saarela et al., 1995). However, it must be noted that the general assumption is that under severe cold exposure, birds will activate both ST and NST simultaneously to ensure sufficient control of T_b.

In both ST and NST, mitochondrial functioning is of prime importance because it combines the production of energy through adenosine triphosphate (ATP) synthesis coupled with a chain of redox reactions and production of reactive oxygen species (ROS). Mitochondrial respiration may be uncoupled from ATP synthesis by a proton leak induced by a thermogenic uncoupling protein (UCP). Uncoupling proteins are transporters present in the mitochondrial inner membrane; some of them (mainly UCP1, UCP2, and UCP3) were described in mammals, whereas in avian species (broilers, duckling) they were designated as avUCP (Collin et al., 2005). The avUCP is predominantly muscular UCP, which was reported to be an ortholog of mammalian UCP3 and that could be involved in glucose metabolism, cold-induced thermogenesis, lipid metabolism, and limitation of ROS production. The expression of avUCP is modulated by genotype; hormones such as thyroid and glucagon, which are the most lipolytic hormones in birds; low ambient temperatures; and nutritional status. Furthermore, several studies in birds had yielded some evidence that a β-adrenergic system controls mRNA expression of avUCP in chicken muscle (Joubert et al., 2010).

In summary, it can be concluded that both ST and NST are the main mechanisms that control T_b below the thermoneutral zone.

37.5.1.5 *Evaporative Heat Loss*

The evolution and success of endothermy enabled the control of T_b to be independent of the external thermal environment when ambient temperatures were low. However, it required the development of appropriate strategies for enhancing heat loss in hot environments. Loss of heat in the form of latent heat of evaporation of water is one of the physiological mechanisms available to birds for controlling T_b in the face of variable environments and excess metabolic heat. In birds, EWL occurs across the skin as cutaneous water loss (CWL), through the respiratory tract as respiratory water loss (RWL), and via the cloaca. The rate of heat dissipation and the proportional contribution of each pathway to the total evaporative water loss (TEWL) are strongly influenced by acclimatization and natural selection, and they vary among species (Williams and Tieleman, 2002).

Heat transfer by evaporation is unique in its coupling of heat loss with resource loss (i.e., of body water content). This underlines an important biological conflict: the birds with least access to water, such as desert birds, are the ones with the greatest need to lose water to maintain T_b. This conflict between competing needs—for water retention and for water evaporation—leads to a tradeoff between avoiding overheating and avoiding dehydration.

In many species of birds, evaporative water demand is relatively high because most birds are diurnal and, therefore, are exposed to higher solar radiation, higher T_a and higher ventilation than nocturnal animals (Maclean, 1996). Furthermore, their high metabolic rate which, in turn, results in high oxygen demand, positively influences RWL (Tieleman et al., 1999), which, for many years was regarded as the main mechanism for EWL. All other routes were neglected, on the assumption that their role in EWL was negligible (Hillman et al., 1985). However, it was well established during recent decades that CWL is an efficient and advantageous means to maintain and control T_b, and that its contribution to overall evaporative loss varies with the environmental conditions. Only recently was another route—the cloacal one—recognized as a path for EWL (Hoffman et al., 2007).

Phenotypic flexibility/plasticity is one of the factors involved in adjustment. Adjustment of evaporation necessitates changes in the evaporative conductance and/or the surface area of the exposed epithelia. Dissipation of latent heat of vaporization of water by panting and or cutaneous routes demonstrates phenotypic flexibility, especially when adaptation to diverse environments is considered. All three evaporation routes—respiratory, cutaneous, and cloacal—may be subjected to selective pressures as a result of birds' exposure to high T_a, or scarcity of drinking water and food in the desert or during migration (McKechnie, 2008; Versteegh et al., 2008).

37.5.1.6 Respiratory Evaporation: Panting

The respiratory system plays a determinant role in the regulation of T_b and water balance. Respiratory temperature control is exercised through evaporative and convective cooling. During inspiration cool air passes through the system, where it becomes heated and saturated with water vapor, and during expiration energy taken up by the air is transported out of the respiratory system. Because respiratory gas exchange requires the humidification of inspired air, an increase in respiratory ventilation will also elevate evaporation. However, respiratory ventilation must take into account the exchange of oxygen and carbon dioxide that is needed to preserve the requirements of gas exchange and dynamic steady-state pH level (i.e., homeostasis). The increase in ventilation to drive EWL should be confined to the dead space in which humidification takes place and should not comprise gas exchange. Thus, increasing respiratory frequency while proportionately decreasing tidal volume will achieve the E required to control T_b, and thermal panting occurs in most birds during exposure to high T_a (Marder and Arad, 1989). Heat stress markedly increased respiratory frequency while the tidal volume decreased, and the net effect was found to be a sixfold to sevenfold increase in minute ventilation (Ludders, 2004). An increase in ventilation subsequently decreases arterial P_aCO_2 and increase pH, causing blood respiratory alkalosis (Houpt, 2004). Birds, unlike mammals, have peripheral receptors named intrapulmonary chemoreceptors located in the parabronchi of the lungs; they are sensitive to the partial pressure of CO_2 in the lung, but not to that of oxygen (Burger et al., 1974). These receptors act as an afferent control for PCO_2; they become stimulated, i.e., increase their rate of discharge when the CO_2 partial pressure decreases, and thus the ventilation rate declines. This chemoreceptor site enables control of P_aCO_2 from breath to breath in spite of possible variations in conditions that depend on the ecological niches occupied by different species (e.g., desert compared with mesic).

Total evaporative heat loss (E) can be calculated from the following equation:

$$E = \dot{m}_e \cdot \lambda \tag{37.4}$$

in which ($\dot{m}_e$) is total water loss and λ is the latent heat of water vaporization, which is approximately 40.8 W h/g or 2.4 J/g. Total water loss involves respiratory loss E_{res} and cutaneous loss E_{cuta} as in Eqn (37.5):

$$E = E_{res} + E_{cuta} \tag{37.5}$$

A simple way to measure EWL is based on changes in body mass of a bird over a defined period, with changes in body mass through defecations taken into account. Then Eqn (37.5) becomes:

$$E = \dot{m} \cdot \lambda / A_D \tag{37.6}$$

in which $\dot{m}$ is obtained from continuous measurements of rate of change of body mass (g/min) and A_D is the skin total surface area.

To distinguish respiratory from CWL, the following equation has to be used:

$$E_{res} = V\left(\text{pex} - \phi_a p_{in}\left[10^{-2}\right]\right)\lambda\,(\text{W}) \qquad (37.7)$$

in which V is respiratory volume per minute (l/min), pex is water content (g/l) of expired saturated air at the expiration temperature, ϕ_a is the RH (%) of inspired air, p_{in} is the water content (g/l) of the inspired saturated air at ambient air temperature, and λ is the latent heat of water vaporization in the expired air (J/g).

37.5.1.7 Cutaneous Evaporation

Cutaneous water loss (CWL) is one of the mechanisms that control T_b under varying environmental conditions. Until the early 1980s, no mechanisms for control of CWL were known, and the respiratory avenue was considered as the main evaporation route in birds (Dawson, 1982). This perception was challenged by several studies that revealed cutaneous evaporation to be the predominant avenue of water loss while resting and at high ambient temperatures, and that its regulation involved, to some extent, the outer skin layer, the stratum corneum (SC), which exhibits phenotypic plasticity. Phenotypic plasticity can be expressed within individuals; it can be the reversible phenotypic plasticity of adults or the developmental plasticity of young, and it can be influenced by natural selection (Piersma and Drent, 2003).

The difference between RWL and CWL lies in the disorders associated with respiratory alkalosis caused by RWL, which do not occur when CWL predominates. Indeed in many studies CWL was found to contribute at least half of total water loss in birds (Wolf and Walsberg, 1996), thereby reducing the deleterious effect of respiratory alkalosis.

A putative model of the mechanism involved in CWL can be summarized in the following equation:

$$\text{CWL} = (\rho_s - \rho_a)/r_v \qquad (37.8)$$

in which CWL is expressed in g/cm/s; ϱ_s is absolute humidity (i.e., water content, g/m^3) below the surface of the skin, assumed to be saturated at skin temperature; ϱ_a is absolute humidity (g/m^3) above the skin; and r_v is the total resistance to vapor diffusion.

The total resistance to vapor diffusion comprises the resistances of the skin, the boundary layer, and the feathers. However, the last two of these parameters contribute less than 10%, which focuses attention onto the skin as presenting the primary resistance to water loss, and onto the SC, as the outermost layer of the epidermis in birds (Boouwstra, 1997). One of the main questions regarding CWL is whether water permeation through the skin is a passive diffusion process or whether it involves an active physiological control mechanism; some studies suggested the former, whereas others suggested an active mechanism, driven by changes in ion gradients (Falkenberg and Georgiadis, 2008) or affected by the vascular blood supply to the dermis (Ophir et al., 2004). If the CWL process is under active physiological control, there should be differences between live and dead birds, in light of the facts that ion gradients are maintained by metabolic activity and, therefore, would not persist after death, so that there would be no blood flow to the periphery in dead birds. Ro and Williams (2010) conducted experiments on 12 bird species that inhabit various ecological areas; in dead birds, they found a reduction of about 16% in CWL, suggesting that at least part of CWL was controlled by active physiological mechanisms. However, different adaptations may be adopted under differing, possibly harsh environmental conditions. A study on the heat-acclimated rock pigeon (*Columba livia*) had shown a complex regulatory pathway for CWL; it consisted of α_2 and β_2 adrenergic receptors.

Whether the CWL is passive or active depends mainly on the intracellular lipids in the SC, which forms the physical barrier to water vapor diffusion. The intracellular lipids in the SC of birds had been identified as a mixture of cholesterol, free fatty acids, ceramides, and cerebrosides. The rate of water loss through the skin did not appear to be determined by the total quantity of lipids in the SC, but rather by the specific mixture of lipid classes, such as free fatty acids, ceramides, and cerebrosides, which were found to be tightly associated with differences in CWL (Muñoz-Garcia and Williams, 2005). In general, desert birds have lower rates of CWL, which suggests that natural selection has the potential to modify water loss through the SC.

It can be concluded that CWL represents an important component of water evaporation; it is higher in mesic-zone-inhabiting birds than in arid-zone ones. The most important boundary layer, which to some extent controls evaporation, is the SC. The fact that studies found only limited physiological control of CWL may explain why, when facing shortage of water, desert-inhabiting birds use mainly the respiratory water loss (RWL) pathway—the physiologically controlled one.

37.5.1.8 Cloacal Evaporation

Only one study on birds (Hoffman et al., 2007) addressed cloacal EWL; it described its importance for thermoregulation in adult Inca doves. The idea of cloacal evaporation evolved from a study on a desert reptile, the Gila monster; this study showed for the first time in any animal the importance of cloacal evaporation for thermoregulation (DeNardo et al., 2004).

The study by Hoffman et al. (2007) on the Inca dove—a bird that inhabits desert areas and can tolerate high ambient temperatures—showed that at 42 °C these doves can evaporate as much water from the cloaca as from the respiratory tract: 21.2 and 25.4%, respectively, of total EWL. These results suggest that Inca doves can employ all three routes to regulate their thermal status; or, in another words, that all three routes contribute to the control of T_b, at least under harsh environmental conditions. Under low T_a, CWL is minimized and cloacal evaporation is eliminated by constricting the cloacal sphincter.

In desert inhabitants, the use of respiratory evaporation is limited by respiratory alkalosis and cutaneous evaporation is maximized to its limit, which is determined by the integument structure. In such cases, the cloacal epithelium is exposed in order to provide further evaporation, which enables these birds to survive the harsh environmental conditions.

37.5.1.9 Sensible Heat Loss

Heat transfer by evaporation is unique in that it couples heat loss with loss of a resource—body water content. This highlights an important biological conflict: that the birds with least access to water, such as desert birds, are the ones with the greatest need to lose water to maintain T_b. This conflict between the competing needs for water retention and water evaporation leads to a tradeoff between avoiding overheating and avoiding dehydration. Furthermore, E, which uses panting as the main mechanism for evaporation, causes respiratory alkalosis, which, in turn, affects the steady state of the bird. The two evaporation-limiting effects can be reduced by efficient sensible-heat loss through radiation, convection, and conduction.

It has been assumed that SHL does not play an important part when T_a is above the upper boundary of the thermoneutral zone (Hillman et al., 1985). This assumption was based on the small difference between the body surface temperature (T_s) and T_a, as well as the fact that in fully feathered birds only limited areas (i.e., legs, head, wattle, and comb) are unfeathered. However, use of infrared (IR) thermal imaging enables noninvasive measurement of birds' surface temperature to identify the various routes of nonevaporative heat loss, according to different models.

Infrared thermal imaging technology has undergone major developments in the early twenty-first century; it has been developed rapidly, through use of automated systems. Infrared thermal imaging cameras measure the amount of near-IR radiation (characterized by the wavelength range of 8–12 μm) that is emitted from a surface, and convert it to a radiative temperature reading, according to the Stefan–Boltzmann Equation:

$$R = \varepsilon \sigma \mathrm{T}^4 \tag{37.9}$$

in which ε is the emissivity of the surface, defined as its ability to emit and absorb radiation, and which, for biological surface tissues, ranges between 0.94 and 1.0 (Monteith and Unsworth, 1990); σ is the Stefan–Boltzmann Constant ($5.67 \times 10^{-8}\,\mathrm{W/m^2/K^4}$); and T is the surface absolute temperature in kelvins (K).

The major advantage of IR is that it provides a noninvasive, contact-free method of measuring surface temperature, at either short or relatively long distances, depending on the goal of the thermal study. For any thermal study, a thermal resolution of 0.1 °C is recommended, and spatial resolution must be sufficient and appropriate for the size of the studied object and its distance from the IR camera. In general, a spatial resolution of (320 × 240) pixels, as determined by the characteristics of the detector array, is used and is found sufficient. Use of a lens with a 24-degree field of view, with the camera placed 60–80 cm from the object, resulted in images in which each pixel corresponded to a rectangular detail of 0.9–1.5 mm² (Figure 37.3; Yahav and Giloh, 2012). Infrared thermal imaging provides accurate measurements of the surface temperature of the whole object—measurements that are undoubtedly more precise and accurate than those provided by thermocouples (Mohler and Heath, 1988).

The IR technique plays a major role in biological research applications, including thermal physiology of mammals (Klir and Heath, 1992) and birds (Yahav et al., 2004; Stewart et al., 2005); fever diagnosis in homeothermic birds and mammals, including humans (Teunissen and Daanen, 2011); cancer diagnosis (Kontos et al., 2011); and animal population counts. For birds, the actual temperature measured by IR thermal imaging represents the temperature beneath the outer physical surface of the insulating layer. The matrix nature of plumage (which in general lasts for up to 3 weeks on poultry) and of the subsequent feathers causes radiation to be exchanged at varied depths within the outer layer, with the result that the radiative temperature is usually higher than the temperature at the outermost physical surface (McCafferty et al., 2011). Therefore, the IR technology measures the temperature of the feathers/plumage several millimeters below the surface. However, several regions, such as the face, wattle, comb, legs, and beak, and unfeathered areas below the wings radiate directly; their measured temperatures will be those at the surface (Figure 37.3).

Surface temperature is significantly correlated to the vasomotor responses of the bird. It is well documented that the vasomotor response—vasodilatation or vasoconstriction—is affected by the bird's T_b, whether it is hyperthermic or hypothermic, respectively. Heat is transferred from the body core via various internal tissues, and also through external insulating layers, by conduction, and the amount conducted (q_{cond}) is determined by various factors (McCafferty et al., 2011) (Eqn (37.10)):

$$q_{cond} = Ak/d\,(T_{bpart} - T_g) \tag{37.10}$$

in which: d (m) is the depth; k (W/m/C) is the thermal conductivity; T_{bpart} is the body part temperature (°C); T_g is the

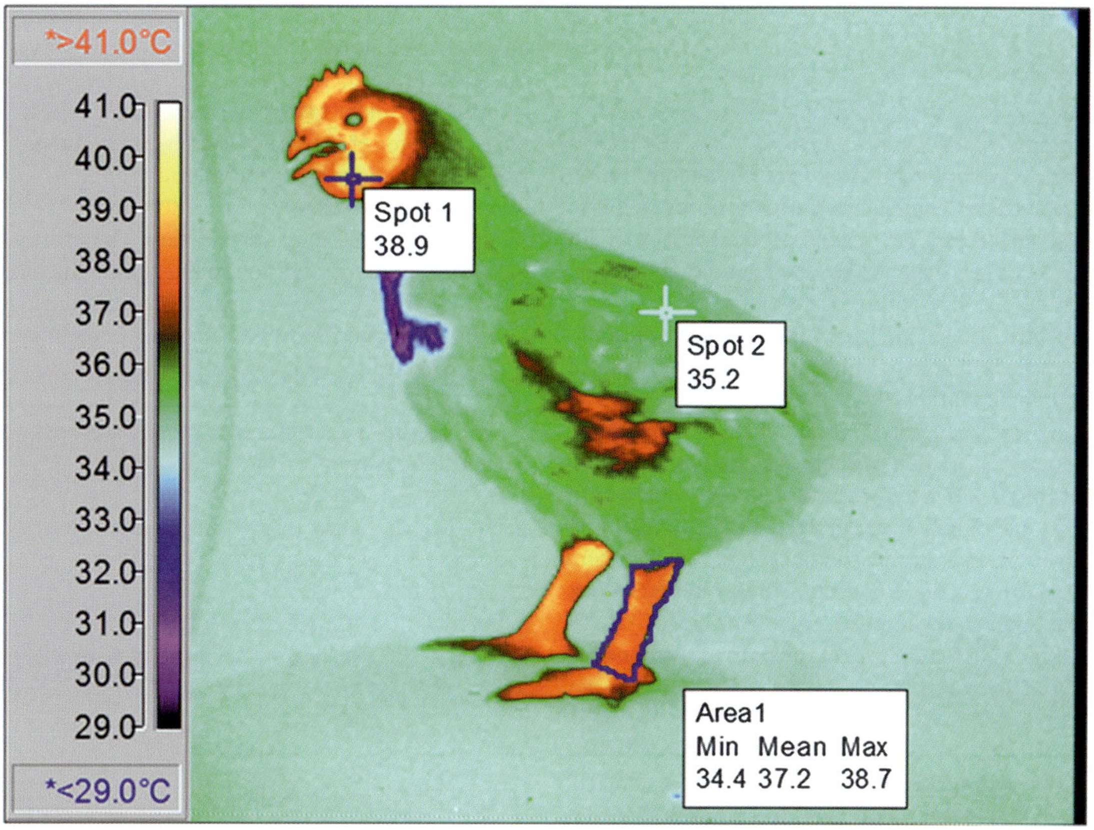

FIGURE 37.3 **Thermal image of a broiler chicken exposed to 35 °C.** The blue outline on the leg defines a region of interest (ROI, as defined by the camera software) that comprises a section of surface area for which minimum, average and maximum surface temperatures were determined. Spots 1 and 2 exhibit the surface temperature in each spot. Thermal images and their analyses were obtained with the Thermacam P 2.8 SR-1 program. *From Yahav and Giloh (2012).*

ground temperature (°C); and A (m^2) is the contacting surface area.

Most of the models are based on steady-state conditions of the animal, in which heat production is balanced by heat loss. The following calculations for poultry follow that of Yahav et al. (2005). To calculate heat transfer from the fowl, each part of the surface is represented by a corresponding geometrical shape. For each part, radiative and convective heat transfers are estimated by means of available and especially developed heat transfer relationships.

37.5.1.10 Convective Heat Transfer

Heat is transferred by convection when a body at a given temperature is in contact with air at another temperature. The convective heat flux, q_c, depends on the temperature difference ΔT, between the body and the air, the area of contact A, and the heat-transfer coefficient h:

$$q_c = hA\,\Delta T \quad (37.11)$$

The average heat transfer coefficient, h, depends on the geometry of the body, the physical properties of the air, and the flow regime. The major difficulty in calculating q_c stems from the strong dependence of h on the flow regime; it is expressed through the nondimensional group known as the Nusselt number:

$$\mathrm{Nu} = \frac{hD}{k} \quad (37.12)$$

in which D is the body-length scale (i.e., diameter in the case of a sphere or a cylinder) and k is the thermal conductivity of the air. Heat-transfer relationships given in the literature relate the Nusselt number to two other nondimensional groups. The first is the Reynolds number:

$$Re = \frac{UD}{\nu} \quad (37.13)$$

in which U is the air velocity and ν is the kinematic viscosity. The second is the Prandtl number:

$$Pr = \frac{\nu}{\kappa} \quad (37.14)$$

in which κ is the thermal diffusivity of the air. The three groups are related, in general, by:

$$\mathrm{Nu} = f(Pr, Re) \quad (37.15)$$

in which the function f is specific to each geometry and flow regime.

37.5.1.11 Radiative Heat Transfer

Heat transfer by radiation involves electromagnetic radiation from one surface to another because of the temperature difference between them. The rate of radiative heat transfer between two surfaces depends on their temperatures, the view area, and the respective surface emissivities.

Radiative heat transfer between the fowl and its environment and among adjacent organs (e.g., the two legs) of the fowl itself occurs if their temperatures are different. The view area changes frequently because of movement of the bird. In the model that leads to Table 37.1 values, radiation among the parts of the bird's surface was neglected and it was assumed that radiative heat transfer takes place only between the fowl and its environment. The environment was treated as a large surface at uniform temperature that surrounds the relatively small bird.

Radiative heat flux from (or to) the animal is estimated as:

$$q_r = \varepsilon_1 \sigma A_1 \left(T_1^4 - T_2^4\right) \qquad (37.16)$$

in which subscript r stands for radiation; indices 1 and 2 represent, respectively, the body surface and the environment; ε (=0.96) is the emissivity of a biological tissue; σ is the Stefan–Boltzmann constant ($5.669 \times 10^{-8}\,\mathrm{W/m^2/K^4}$); A is the surface area; and T is the absolute temperature (K).

The model presented above was used to determine the importance of SHL to thermoregulation above the critical point of the thermoneutral zone, and its association with the water balance of poultry. This was studied in detail (Yahav et al., 2004, 2005) in broiler chickens exposed to 35 °C and a range of ventilation rates from 0.8 to 3.0 m/s. It was obvious that ventilation rate did not affect Q_r over this wide range of air velocities (Table 37.1). However, a significant ($r^2 = 0.998$), linear increase of convective heat loss, Q_c, with increasing ventilation rate was observed. It was well established that exposure of endothermic birds to high ambient temperatures will cause a cascade of thermal events, which include a decline in heat production. Therefore, to establish the contribution of SHL to total heat loss, the overall energy expended for maintenance had to be evaluated. This can be calculated by means of the following equation:

$$M = EI - (EXE + RETE) \qquad (37.17)$$

in which M is the energy required for maintenance; EI is energy intake; EXE is excreted energy; and RETE is retained energy, as evaluated from daily weight gain, on the assumption that 85% of the intake is retained as fat and 15% as protein (Boekholt et al., 1994). Increasing the ventilation rate during exposure to a high T_a (35 °C) had a dramatic effect on the ratio of SHL to maintenance energy. Whereas the ratio was only 29% in broilers exposed to air flows of 0.8 and 1.5 m/s, it exceeded 44% of the energy expended for maintenance in broilers subjected to 3.0 m/s (Table 37.1). This demonstrates that, depending on ventilation rate, SHL (i.e., convection plus radiation) may make a major contribution to heat loss, despite the small temperature differences between the bird surface and the environment.

37.5.1.12 The Association between Energy and Water Balance

Scientifically, it is difficult to devise a process that distinguishes between the energy and water balances maintained by the body in response to exposure to extreme (mainly hot)

TABLE 37.1 The Effect of Ventilation Rate at High Ambient Temperature on Osmoregulation and Sensible Heat Loss by Radiation (Q_r), Convection (Q_c) and (Radiation Plus Convection) (Q_t) as a Percentage of Energy Expended for Maintenance

	Ventilation Rate (m/s)			
Variables	0.8	1.5	2.0	3.0
Q_r (watt)	25.2 ± 4.7	25.5 ± 3.6	31.9 ± 4.5	31.6 ± 4.7
Q_c (watt)	55.7 ± 7.0^d	78.3 ± 7.4^c	104.9 ± 7.1^b	132.6 ± 7.0^a
Q_t (watt)	80.9 ± 10.2^b	104.9 ± 9.4^b	136.8 ± 10.4^a	164.2 ± 10.1^a
Q_t (% of energy expended for maintenance)	24.1 ± 3.7^b	29.1 ± 4.4^b	36.8 ± 2.2^{ab}	44.7 ± 4.7^a
T_b (°C)	43.9 ± 0.08^a	42.9 ± 0.08^c	42.8 ± 0.09^c	43.2 ± 0.10^b
Arginine vasotocin (pg/ml)	25.3 ± 1.7^{ab}	22.4 ± 2.4^{bc}	19.2 ± 1.5^c	28.0 ± 1.8^a
Osmolality (mOsm/l)	324 ± 3.0^a	317 ± 2.1^{bc}	314 ± 2.2^c	323 ± 1.4^{ab}

Within rows, values designated by different superscript letters (a, b, c, d) differ significantly ($P < 0.05$; $n = 8$).

environmental conditions. The SHL and RH studies provide the opportunity to understand the dynamic relationship between energy balance and water balance.

In the study of broiler SHL (Yahav et al., 2004, 2005), heat loss by radiation and convection increased linearly with increasing ventilation rate (Table 37.1). Consequently, at the highest ventilation rate (3.0 m/s), T_b of the birds would be expected to be similar to or even less than that monitored at 2.0 m/s. However, increasing the ventilation rate to 3 m/s negatively affected thermoregulation by eliciting negative effects on T_b, body weight, feed intake, and feed efficiency. It was suggested that the effect on body water balance was the main reason for the deterioration in the ability to maintain energy balance and performance. Examination of two osmoregulatory parameters—plasma osmolality and plasma arginine vasotocin (AVT) concentration—supports this hypothesis. Changes in plasma osmolality activate osmo-receptors in the blood system, thereby sending signals to the PO/AH (Figure 37.1) which initiate the "debate" between balancing energy and balancing water. Furthermore, AVT is an important hormone in regulating water balance; it was well documented that water deprivation resulted in increased plasma osmolality and plasma AVT concentration, and the two parameters were positively correlated (Arad et al., 1985; Saito and Grossmann, 1998). The results of the study of broilers (Yahav et al., 2004, 2005) (Table 37.1) reveal a significant ($R^2=0.84$) and positive correlation between plasma osmolality and plasma AVT concentration. The significant increases of these two parameters in broilers exposed to high ventilation rates indicate difficulties in balancing body water as a consequence of increased RWL and CWL and/or inability of broilers to drink sufficient water under extreme environmental conditions. Thus, a high degree of water loss from the respiratory tract and the cutaneous surface, together with insufficient drinking capacity lead to some degree of dehydration, associated with increases in plasma osmolality and AVT concentration, leading to increase in T_b and to the debate between energy balance and water balance.

The RH study (Yahav et al., 1995) was based on exposing broilers to high T_a, coupled with several different RHs; in this study, the association between T_b and panting rate was evaluated. Yahav et al. (1995) demonstrated a decline in broiler panting rate despite increases in T_b; it was speculated that mild dehydration depressed panting rate and reduced water loss at low RH of 40%, leading to increased T_b. In both studies, it was speculated that hyperthermia developed as a result of body water deficit. However, whereas the deficit at high ventilation rate resulted, most likely from CWL, the deficit observed at low RH most likely resulted from a high panting rate.

The effort to control T_b within the normothermic range is energetically far more expensive in domesticated birds than in wild ones as a result of their high productivity. The broiler study shed new light on the difficulties of maintaining energy and water balances at high ambient temperatures. The efforts to control T_b and total body water directed a large amount of energy toward maintenance in broilers exposed to an optimal ventilation rate and thereby prevented the hyperthermia that can develop at high temperatures. In such cases, it would have been expected that unbalanced domesticated birds would expend an enhanced amount of energy for maintenance, especially when energy consumption was reduced. However, the results obtained in the broiler study (Yahav et al., 2005) suggested an opposite response, namely a decline in energy invested for maintenance, which was manifested in a development of severe hyperthermia (T_b of 43.9 °C; Table 37.1). It can only be speculated that the genetic selection for growth is so dominant in these fowls that even hyperthermia and dehydration do not significantly alter the amount of energy invested in maintenance.

37.5.2 Body Temperature

Balance or imbalance of the parameters presented in Eqn (37.1) will cause the development of hyper- or hypothermy in birds. Body temperature is the main parameter that represents the thermal status of an organism; therefore, endotherms aim to maintain a relatively constant temperature that will permit efficient functioning of the cells, tissues, and organs. However, differences are maintained among the temperatures of different organs as a result of their locations (i.e., visceral or peripheral), activities, and diverse biological functions (e.g., smooth or striated muscle). With no exception, endotherms will maintain a temperature gradient between the core (visceral) and the surface (peripheral) in order to control heat exchange with the environment.

For many years, endothermic birds and mammals were characterized as homeothermic. However, use of modern techniques for measuring T_b established endothermic organisms as heterothermic. Heterothermy involves changes in T_b in response to changes in the environment, such as alterations in T_a or in food and water availability (Table 37.2). Changing T_b from normothermic values in response to environmental alterations saves energy and is a key component of thermoregulation in birds of different sizes exposed to different environments: tropical, subtropical, arid, and mesic (McKechnie and Mzilikazi, 2011).

It is known, from studies that were conducted during many decades and were summarized by Prinzinger et al. (1991), that in birds T_b cycles diurnally and differs among resting, active, and high-activity phases. In resting birds, T_b may range from 35.0 to 40.8 °C; in active ones, between 35.6 and 44.6 °C; and high activity increases the range to 40.4–47.7 °C. Despite the fact that endotherms are considered to be homeothermic, this is not the case, and their T_b fluctuates diurnally by 2.48–2.84 °C, depending on

TABLE 37.2 Deep Body Temperature of Selected Birds at Rest under Thermoneutral Conditions[1]

Species	Body Mass (kg)	Deep Body Temperature[2] (°C)	Reference[3]
Ostrich (*Struthio camelus*)	100.0	38.3	*
Emu ♂♂ (*Dromaius novaehollandiae*)	40.7	37.7 S	1
Emu ♀♀	45.4	38.3 S	1
Emu ♂♂	39.7	37.7	1
Emu ♀♀	37.0	38.2	1
Rhea (*Rhea americana*)	21.7	39.7	*
Mute swan (*Cygnus olor*)	8.3	39.5	*
Domestic goose (*Anser anser*)	5.0	41.0	*
Gentoo penguin (*Pygoscelis papua*)	4.9	38.3	*
Giant petrel (*Macronectes giganteus*)	3.9	39.2	2
Peruvian penguin (*Spheniscus humboldti*)	3.9	39.0	*
Domestic turkey (*Meleagris gallopavo*)	3.7	41.2	*
Adélie penguin (*Pygoscelis adeliae*)	3.5	38.5	*
Chinstrap penguin (*Pygoscelis antarctica*)	3.1	39.4	*
Great spotted kiwi (*Apteryx haastii*)	2.5	38.4	3
Domestic fowl (*Gallus gallus*)	2.4	41.5	*
Domestic duck (*Anas platyrhynchos*)	1.9	42.1	*
Double-crested cormorant (*Phalacrocorax auritus*)	1.33	41.2 D, 40.2 N	*
South polar skua (*Catharcta maccormicki*)	1.250	40.9	2
Black grouse (*Lyrurus tetrix*)	1.079	41.3 S	*
	0.931	40.2 W	
Kelp gull (*Larus dominicanus*)	0.98	41.0	2
Anhinga (*Anhinga anhinga*)	1.33	39.9 D	*
		39.1 N	*
Great horned owl (*Bubo virginianus*)	1.00	39.9	4
Little penguin (*Eudyptula minor*)	0.9	38.4	*
Brünnich's guillemot (*Uria lomvia*)	0.819	39.6	5
Fulmar (*Fulmarus glacialis*)	0.651	38.7	5
Brown-necked raven (*Corvus corax ruficollis*)	0.610	39.9	*
Willow ptarmigan (*Lagopus lagopus*)	0.573	39.9	*
Mexican spotted owl (*Strix occidentalis lucida*)	0.571	39.1	4
Tawny frogmouth (*Podargus striatus*)	0.420	38.6[4]	6
European coot (*Fulica atra*)	0.387	39.6[4]	7
Black-legged kittiwake (*Rissa tridactyla*)	0.365	40.2	5
Black guillemot (*Cepphus grylle*)	0.342	39.9	5
Papuan frogmouth (*Podargus papuensis*)	0.315	38.8[4]	6
Rock pigeon (*Columbia livia*)	0.3	42.2	*

TABLE 37.2 Deep Body Temperature of Selected Birds at Rest under Thermoneutral Conditions[1]—cont'd

Species	Body Mass (kg)	Deep Body Temperature[2] (°C)	Reference[3]
Bobwhite (*Colinus virginianus*)	0.210	38.9 DS[4]	8
		37.0 NS[4]	8
	0.228	37.7 DW[4]	8
		37.4 NW[4]	8
Brown noddy (*Anous stolidus*)	0.142	40.3	9
California quail (*Callipepla californica*)	0.139	41.3	*
Tengmalm's owl (*Aegolius funereus*)	0.127	39.4	10
American kestrel (*Falco sparverius*)	0.119	39.3	*
Acorn woodpecker (*Melanerpes formicivorus*)	0.082	42.4[4]	11
Green woodhoopoe (*Phoeniculus purpureus*)			
♂♂	0.080	39.6 N	12
♀♀	0.072	39.7 N	12
Evening grosbeak (*Coccothraustes vespertinus*)	0.060	41.0	*
Barred button quail (*Turnix suscitator*)	0.058	39.5	13
Blue-breasted quail (*Coturnix chinensis*)	0.053	39.0	13
Speckled mousebird (*Colius striatus*)	0.053	39.0	*
Wilson's storm petrel (*Oceanites oceanicus*)	0.034	39.2	2
Common redpoll (*Carduelis flammea*)	0.015	40.1	*
Sunbirds[5]	0.007–0.017	42.5 D, 38.9 N	14
Zebra finch (*Poephila guttata*)	0.012	40.3	*
Anna hummingbird (*Calypte anna*)	0.005	42.0 (Median T_b)	15

[1]Modified and expanded from Whittow (1976, 1986), which should be consulted for references marked by an asterisk. Thermoneutral temperatures are those requiring neither regulatory thermogenesis nor active evaporative cooling. The vast majority of the values for T_b were obtained during the daytime, the active phase of the daily cycle for most species with the exception of such birds as owls and frogmouths. In cases where measurements were made during both day and night, the particular period is specified.
[2]D, N, S, and W refer to daytime, nighttime, summer, and winter measurements, respectively.
[3]References: 1, Maloney and Dawson (1993); 2, Morgan et al. (1992); 3, McNab (1996); 4, Ganey et al. (1993); 5, Gabrielsen et al. (1988); 6, McNab and Bonaccorso (1995); 7, Brent et al. (1985); 8, Swanson and Weinacht (1997); 9, Ellis et al. (1995); 10, Hohtola et al. (1994); 11, Weathers et al. (1990); 12, Williams et al. (1991); 13, Prinzinger et al. (1993); 14, Prinzinger et al. (1989); 15, Powers (1992).
[4]T_b independent of T_a over a range of temperatures that includes or closely approaches zone of thermal neutrality.
[5]Values for T_b at 26.5 °C T_a calculated from equations relating T_b to T_a based on data for five species of sunbirds: Aethopyga siparaja, Anthreptes collaris, Nectarinia cuprea, Nectarinia tacazze, and Nectarinia klimensis.
Scanned from Avian Physiology 5th edition.

the bird's prevailing metabolic phase. The highest T_b ever recorded in a wild bird was 47.7 °C in the white crowned sparrow *Zonotrichia leucophrys* (Southwick, 1973). Such extreme development of hyperthermia is in most cases irreversible.

Body temperature varies among species also as a result of body mass, ecological niche (e.g., aquatic), and varied availability of diet during the year (Table 37.2). In genetically selected poultry, T_b in the resting and active phases is significantly higher than those recorded for undomesticated birds; it is related to their higher metabolic rates. The values presented in Table 37.3 show various environmental factors that have considerable effects on domesticated birds' T_b.

In general, both invasive and noninvasive techniques are used to measure T_b in birds. The commonly used ways to record T_b in domesticated birds involve invasive techniques, which are time-consuming and may be traumatic as a result of handling, and may affect T_b within a few minutes by the induction of fever that increases T_b by up to 0.5 °C (Cabanac and Aizawa, 2000). In both wild and domesticated birds, a telemetry technique is applied, which involves implantation of temperature-sensitive transmitters, usually into

TABLE 37.3 The Effects of Ventilation Rate at Various Ambient Temperatures on Body Temperature of 6 Week Old Broilers and Turkeys, and 8 to 9 Month Old Laying Hens

	Ventilation Rate (m/s)			
	0.8	1.5	2.0	2.5
	Broiler body temperature (°C)[2]			
T_a=35 °C	43.9±0.10[a]	42.9±0.08[c]	42.8±0.09[c]	[1]43.2±0.10[b]
T_a=30 °C	40.5±0.10[b]	40.4±0.06[b]	40.9±0.09[a]	41.0±0.09[a]
T_a=25 °C	40.7±0.06[b]	41.1±0.07[a]	41.1±0.06[a]	41.0±0.05[a]
	Turkey body temperature (°C)[2]			
T_a=35 °C	41.5±0.06[a]	41.2±0.01[a,b]	40.6±0.09[a]	41.2±0.07[a,b]
T_a=30 °C	40.9±0.12[b]	41.1±0.08[a,b]	41.4±0.07[a]	41.3±0.08[a]
T_a=25 °C	40.6±0.06	40.8±0.04	40.7±0.09	40.7±0.13
T_a=20 °C	41.4±0.06	41.0±0.09	41.3±0.10	41.2±0.09
	Laying hens body temperature (°C)[3]			
T_a=35 °C	41.6±0.06[a]	41.5±0.09[a,b]	41.6±0.07[a]	[1]41.3±0.14[b]
T_a=30 °C	41.5±0.11	41.6±0.05	41.6±0.10	[1]41.6±0.14
T_a=25 °C	41.5±0.06[a]	41.4±0.03[a]	41.4±0.11[a]	[1]41.5±0.08[a]
T_a=20 °C	41.5±0.04	41.5±0.10	41.7±0.08	[1]41.6±0.07

Within rows, values designated by different superscript letters (a, b, c) differ significantly ($P \le 0.05$).
[1]*This value was recorded at ventilation rate of 3.0 m/s.*
[2]*Yahav et al. (2009).*
[3]*Yahav (unpublished data).*

the abdominal cavity. This enables long-term recording of undisturbed birds (Bligh and Heal, 1974).

Whereas the telemetry technique involves handling and implantation, the IR thermography technique is an entirely noninvasive and contact-free method of measuring surface temperature, at either short or relatively long distances (Figure 37.3; Giloh et al., 2012). Thermographic imaging for T_b determination can be relevant only if a satisfactory correlation exists between the body and surface temperatures; in the case of birds, the surface must be unfeathered. Use of the temperature of an unfeathered surface is based on its significant correlation with the vasomotor responses of the bird; it is well documented that the vasomotor response—vasodilatation or vasoconstriction—is affected by the bird's T_b. Studies conducted on broiler chickens demonstrated that such a correlation could indeed be found; they also demonstrated a high correlation between facial surface temperature (T_{fs}) and T_b. In laboratory studies, where the climatic conditions were relatively similar for all the chickens, individual body temperatures correlated well with individual facial surface temperatures for chickens that were exposed to acute environmental heat conditions, both with and without ventilation, and to persistent heat without ventilation (Figure 37.4; Yahav and Giloh, 2012).

It can be concluded that, regarding T_b, birds are mainly endothermic, heterothermic organisms.

37.6 THE DEVELOPMENT OF ENDOTHERMY DURING EMBRYOGENESIS

The bird model used to investigate the metabolic transition from ectothermy to endothermy during embryogenesis is based on the increase of aerobic capacity which occurs at two functional levels that are regulated independently of one another: upregulation of gene expression and a significant increase in the catalytic activity of the main oxidative control enzymes. This switch occurs because of the development of the physiological regulatory mechanisms that govern metabolism. These regulatory pathways involve changes in gene and protein expression, coupled with changes in cellular physiology. It is of great interest to understand the process leading to the high metabolic rate that enables endothermy, and to know whether these changes are fixed in time or can be modified by environmental changes.

The development of thermoregulatory competence in birds has been studied in both altricial and precocial species during the perinatal period. Although the development of

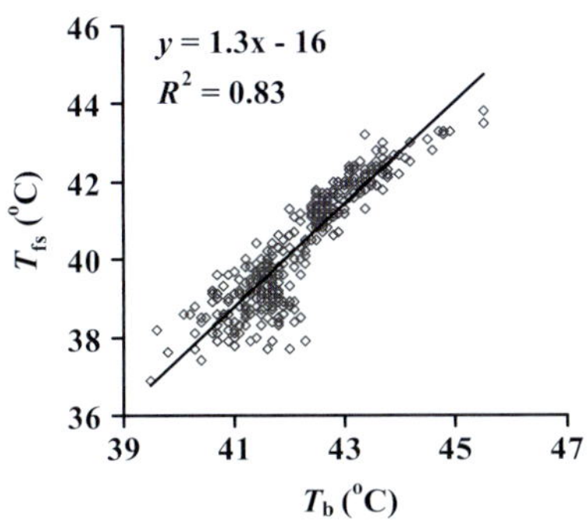

FIGURE 37.4 **Correlation of individual facial temperature (T_{fs}) with body temperature (T_b).** Chickens were reared under controlled conditions. On each of days 8, 15, 22, 29, and 36, a group was randomly chosen for exposure to three different heat treatments: acute heat only, acute heat together with ventilation, and persistent heat. Each treatment included nine chickens for monitoring the responses of body and facial temperatures. The data presented in the graph represent all measurements of individual chickens of all ages in all treatments. *From Yahav and Giloh (2012).*

thermoregulation in altricial birds is an event that mostly takes place after hatch, precocial species exhibit incipient thermogenesis during the later stages of incubation and during the first 10 days posthatch (Tazawa et al., 1989; Arad and Itsaki-Glucklish, 1991). Therefore, precocial birds were used as a model for elucidating the cellular and molecular physiological mechanisms involved in the transition towards endothermy in embryos.

For precocial birds, the transition has been suggested to comprise four stages during embryogenesis (Tazawa et al., 2001): the Arrhenius-limited stage, in which the metabolic rate is directly related to environmental temperature; the O_2-conductance-limited stage, in which VO_2 is limited by oxygen diffusion through the shell and the chorioallantoic membrane (CAM) in response to the embryo's oxygen demands; the power-limited stage, in which the embryo has limited capacity to generate heat in response to cold exposure; and the full-blown stage, in which the organism is homeothermic. These four stages are typical of precocial birds, whereas the altricial birds seem to be confined to the Arrhenius-limited stage until after hatching.

37.6.1 The Transition from Ectothermic to Endothermic Embryo: Cellular and Molecular Aspects

The BMR of endothermic birds is 5–10 times greater than that of ectothermic vertebrates (Bennett et al., 2000), despite the fact that metabolic pathways are highly conserved among vertebrates (Smith and Morowitz, 2004). The transition from ecto- to endothermy is based on modifications to the systems and mechanisms that regulate metabolic enzyme activities. Such a transition necessitates increased capacity of the respiratory system, changes in the cardiovascular system (CVS) such as possession of four heart chambers, increased efficiency of the gastrointestinal tract, and modifications to the neural and endocrine systems (Chiba et al., 2004). Mechanisms that may regulate metabolic enzyme activities include changes in protein structures, modifications to membranes, and up- or down-regulation of genes (Hulbert and Else, 2005).

The egg temperature is maintained (in nature by the incubating parents or in domesticated birds by industrial incubators) at a level conducive to the development of the embryonic tissues. During early stages of embryogenesis, the survival of embryos confronted by alterations of the incubation temperature is entirely dependent on their tolerance, and not on thermoregulatory mechanisms. At this stage, the embryos' thermotolerance protects them, to some extent, against incubation temperature fluctuations. However, there is no doubt that, close to hatch, thermoregulatory mechanisms become necessary. The questions of interest concern when the ability to effectively maintain T_b appears, and what is the molecular and physiological evidence that confirms that the embryo reached the endothermic stage. The time-point at which to expect incipient thermoregulation is close to hatch, when tissues, organs, and the control axis achieve the maximal degree of development (Whittow and Tazawa, 1991).

During the development of precocial birds, VO_2 increases until shortly before hatching, at which point (i.e., the oxygen plateau) it stabilizes as embryos reach their hatching mass (Prinzinger et al., 1995). The increase in oxidative metabolic rate during embryogenesis is facilitated by mitochondrial density enhancement, mitochondrial uncoupling, proton leakage across the inner mitochondrial membrane, and the increase in enzyme activity that coincides with development of oxidative pathways. Reaching this stage of metabolic capacity requires the activity of enzymes of the oxidative phosphorylation and citric acid cycles (Walter and Seebacher, 2009). Metabolic regulation of enzymes enables cells to use the existing pool of mitochondria to match energy production to energy demands. Cells can regulate energy by altering the structure of mitochondria, but they can do so without changes in structure or enzymology (Harper and Himms-Hagen, 2001), by alteration of the magnitude of the proton leakage across the inner mitochondrial membrane, which determines the proportion of energy that is released as heat (Porter et al., 1996). Another important characteristic of endothermy is the permeability of the plasma membrane to Na^+ and K^+ ions, which leads to enhanced activity of Na^+/K^+-ATPase, in order to maintain the ion gradient across the membrane. The increased enzyme activity increases energy demand, which requires increased proton flux to increase ATP synthesis in the mitochondria. The capacity to maintain endothermy is controlled by several nuclear hormone receptors, such the thyroid receptor, retinoic acid receptor, peroxisome proliferator-activated receptors (PPARα, γ and δ) and, most importantly, their coactivators PGC-1α and β (Puigserver and Spiegelman, 2003). Thyroid hormones, mainly triiodothyronine (T_3), can modulate aerobic metabolic capacity by regulating mitochondrial biogenesis and PGC-1α gene expression, which control transcription of genes that encode enzymes involved in metabolic pathways (Lin et al., 2005).

Endothermy, therefore, can be defined as interactions among uncoupling and proton leakage in mitochondria, cellular energy demand, and oxidative capacity. It is usually accepted that activities of these mechanisms are completed close to hatch, but at no specific time-point.

37.6.2 The Transition from Ectothermy to Endothermy: Physiological Parameters

Three physiological parameters of the embryo—T_b, heart rate (HR), and VO_2—evidently can shed light on this transition. Research into the physiological transition is based mainly on precocial models, in which the incubation temperature is the most important vital parameter during embryogenesis; and for several decades the practice in breeding poultry was, and still is to keep this parameter stable. Optimizing the incubation temperature is essential for achieving the best hatchability (Swann and Brake, 1990; French, 1997) and chick quality (Wilson, 1991; Decuypere and Michels, 1992), while avoiding possible deleterious effects on the development of the chicken embryo (Krausova and Peterka, 2007). The temperature of the embryo does, indeed, depend on its incubation temperature, but also on other environmental parameters such as RH, air speed, and air flow, which are basically the parameters of the physical microenvironment around the egg (Meir and Ar, 1987; Swann and Brake, 1990). These factors, coupled with the heat exchange between the egg and its microenvironment, and the capacity of the embryo to produce heat, will determine the embryo's T_b (Van Brecht et al., 2005).

37.6.2.1 Embryonic Body Temperature

The thermoregulatory capability of the embryo may not be effective in the sense that of enabling the embryo to maintain T_b during environmental cooling or warming but, in fact, the embryo responds to changes in incubation temperature with an appropriate change in T_b. The ability to respond is pronounced in the poultry that has a precocial mode of development, which results in a mature state at hatch, whereas the altricial mode of development results in a very immature hatchling. Therefore, the precocial mode can provide a better model for observing the development of endothermy.

The confirmation of complete metabolic responses to altered environmental incubation temperatures that would provide proof of the development of endothermy prior to hatch will depend on the measurement of T_b. Previous studies measured embryonic temperature within the egg adjacent to the embryo (Tazawa and Rahn, 1987; Holland et al., 1998). One of the techniques used involved measurement of the allantoic fluid temperature (T_{af}), and recently (French, 1997; Joseph et al., 2006; Piestun et al., 2008; Zimerman et al., 2013) eggshell temperature (EST) was also found to be a reliable proxy for T_b. However, measurement of T_{af} or EST rather than embryonic T_b is a compromise based on the following facts. The allantoic fluid and the egg shell both lie within the temperature gradient (i.e., the path of heat conduction between the embryo and the environment), and the advantage of EST is that it is a noninvasive IR temperature measurement. The disadvantage of EST stems from the effect of convection on the measurement. However, in all the studies that used EST measurement to determine the embryonic T_b the convective effects were minimized.

Although there is no doubt that both T_{af} and EST are lower than the embryonic T_b, they significantly reflect embryonic T_b and its pattern of response to changes in incubation temperature, as was demonstrated by Piestun et al. (2008) and Shinder et al. (2009). Figure 37.5 shows that EST remained relatively constant during the periods of organ development (E9 to E11 in broilers and E13 to E15 in turkeys). Subsequently, energy was directed mainly towards embryo growth—a process that coincided with elevation of heat production by the embryo, which led to increases in EST through E16 and E22 in broiler and turkey embryos, respectively. This was followed by 3 days of relatively constant EST, which parallels the VO_2 plateau and results from the limited ability of the CAM to deliver an adequate amount of oxygen and which, in turn, initiates the preparation for internal and external piping, during which EST increases dramatically.

Tazawa et al. (1988) reported evidence of a weak metabolic response to lowered T_a in chicken embryos at E18 of incubation. Nichelmann et al. (1998) decreased the incubation temperature of Muscovy ducks to about 31 °C for 4 h during the last third of incubation; the study found that although they responded with declines in T_{af} and heat production, T_{af} remained always higher than the incubation temperature and there was only a moderate drop in heat production. This demonstrated a thermoregulatory response, although it was weak.

Yahav and Piestun (unpublished data) exposed embryos of broilers and laying hens from embryonic day (E) E10 through E19 to 39.5 and 34.5 °C, for 10 h/d (Figure 37.6) and turkey embryos to the same temperatures from E16 through E25 (Figure 37.6), and measured EST to determine the birds' responses to the alterations in incubation temperature. The results are presented as the difference (Δ°C) between EST prior to the exposure and its average values during the exposure. In all three species—broilers, laying hens, and turkeys—exposure to 39.5 °C induced a decline in Δ°C of EST with days of exposure. However, whereas the slope for broilers and turkeys was −0.061 and −0.04, respectively, in laying hens it was −0.11.

The ability of laying hens to maintain a lower decline in Δ°C of EST with time than broilers and turkeys reflects, most probably, their embryonic heat production. It was well documented that genetic selection of these species during recent decades significantly increased their heat production during embryogenesis, and that broilers and turkeys exhibited the most dramatic alterations of heat production (Hulet, 2007). Thus, maintaining EST during exposure to an increased incubation temperature turns out to be a difficult task for broiler and turkey embryos. However, the decline in Δ°C EST with time may shed light on the development of thermoregulatory capacity.

Exposing the embryos to 34.5 °C induced progressive increase in EST with accumulating days of embryogenesis. However, in this case, turkey embryos were able to maintain much better EST with increasing age, despite the lower T_a; their EST slope increased by 0.10, compared with 0.049 and 0.06 for broilers and laying hens, respectively. It is not clear what caused these differences. One of the reasons why turkey eggs maintained much better EST could be the size of the eggs; turkey eggs weighed 80–100 g, whereas those of broilers and laying hens weighed between 60 and 70 g; therefore the ratio of surface area to volume favored turkey eggs. In all three species, although coping with changes in incubation temperature was observed to improve with age, Δ°C EST at E19 indicated weak thermoregulatory capacity. For all three species maintenance of an optimal incubation temperature resulted in Δ°C EST close to zero, emphasizing that the industry's incubation temperature is within the comfort zone for embryogenesis.

In a study (Shinder et al., 2009) in which broiler embryos were exposed to 15 °C for 30 or 60 min on E18 and E19, the capacity to withstand such extreme environmental conditions was observed. On E18, during the first cold exposure, EST dropped from 38.3 °C to 26.9 and 21.1 °C after 30 and 60 min of exposure, respectively. On E19, EST dropped to 27.5 and 22.0 °C after 30 and 60 min of exposure, respectively. The ability to maintain an EST significantly higher than the incubation temperature despite the extreme drop in

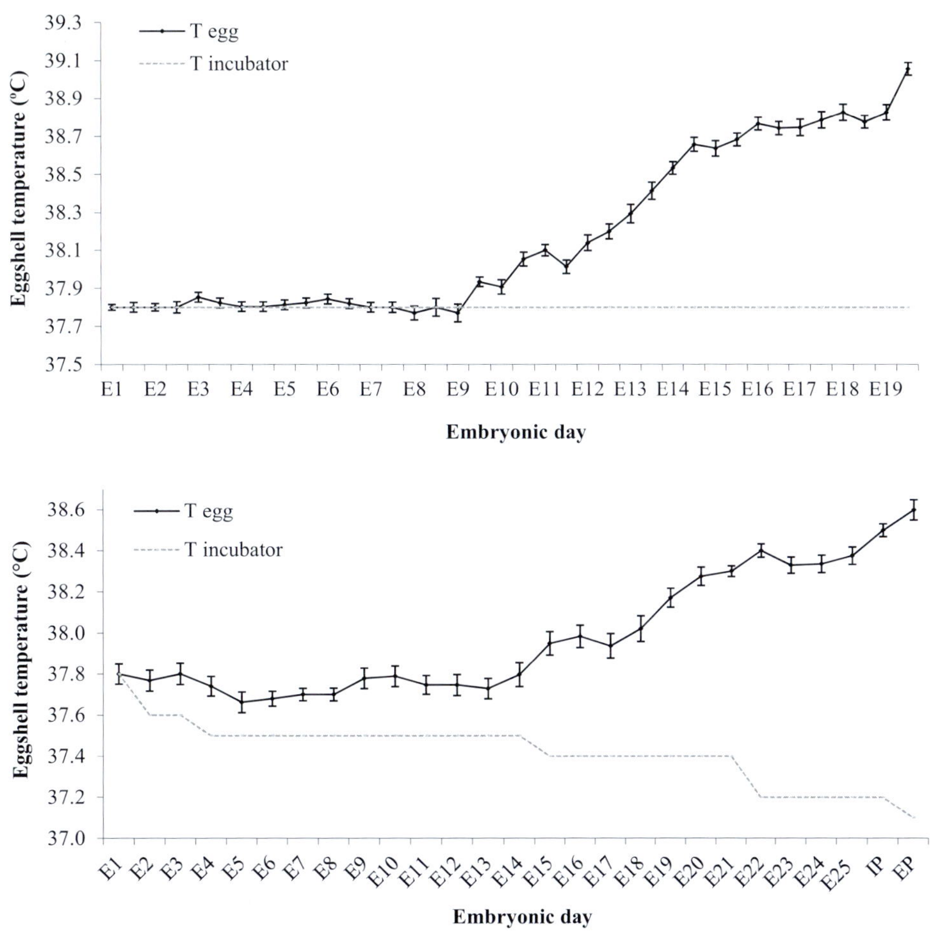

FIGURE 37.5 Eggshell temperature (°C) measured by infrared thermometer during 21 days of incubation of broilers (upper graph) and 28 days of incubation of turkeys (lower graph). The gray line indicates the incubation temperature.

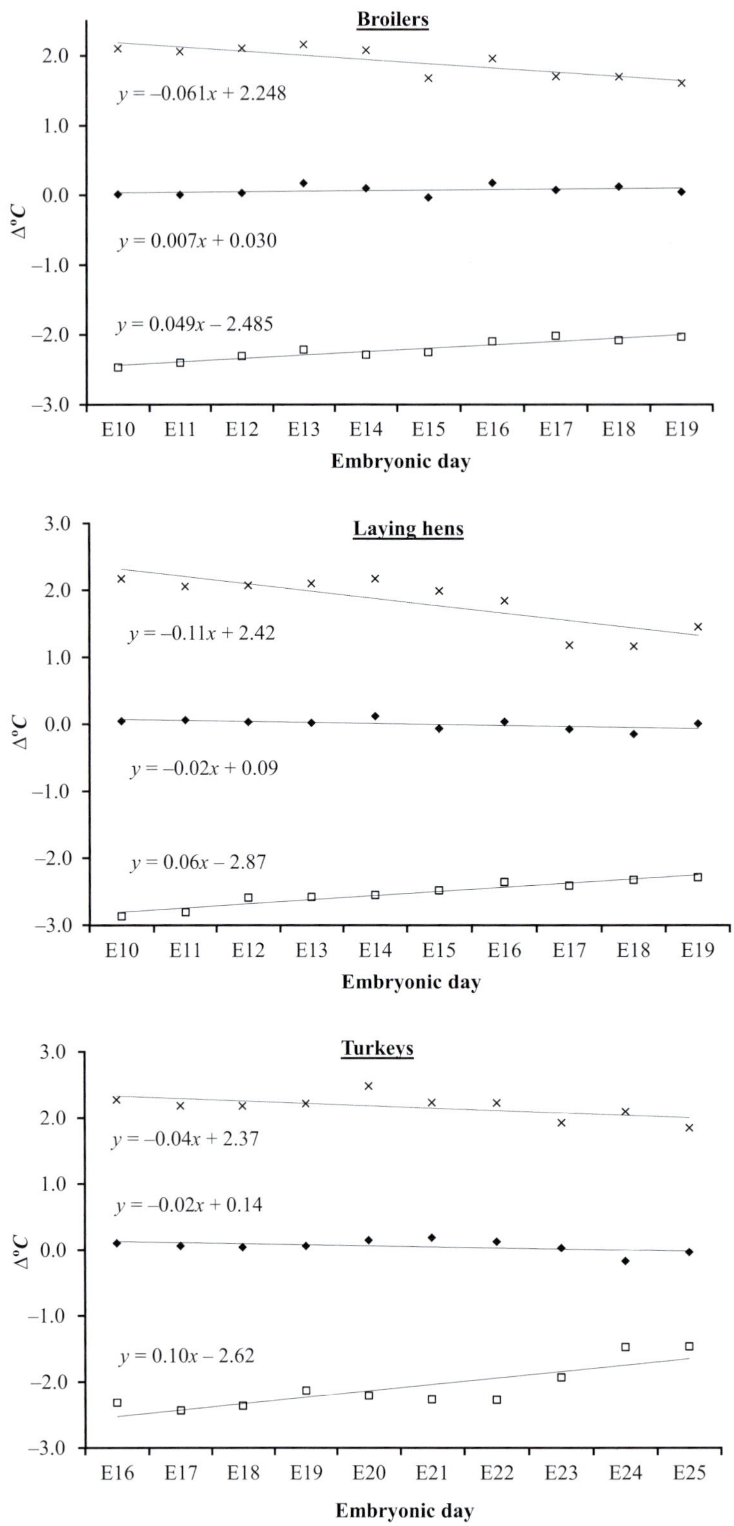

FIGURE 37.6 The effects of increasing incubation temperature from 37.5 °C (♦-line) to 39.5 °C (×-line) or decreasing it to 34.5 °C (□-line) for 10 h/d from embryonic day (E) E10 through E19 (broiler and laying hen embryos) and from E16 through E25 (turkey embryos) on the difference (Δ°C) between eggshell temperature prior to the exposure and its average values during the exposure, in embryos of broilers (upper graph), laying hens (middle graph) and turkey (lower graph).

the latter on E18 and E19 can be attributed to the thermoregulatory ability of broilers close to hatch. This evidently was the case with Muscovy ducks at E34 (Nichelmann et al., 2001).

37.6.2.2 Embryonic Heart Rate

An appropriate HR is essential to produce blood flow in order to transport respiratory gases, nutrients, hormones, and metabolic wastes. In the early stages of birds' embryogenesis, after the heart is formed and has started to beat, a high HR is not required for transportation of gases from the environment to the embryo tissues. In these stages, HR and blood flow are relatively insensitive to changes in environmental oxygen partial pressure, but they play a role in the angiogenesis processes of the embryo (Burggren, 2004).

However, as the embryo develops (from embryonic day ~E11 during 21 days of embryogenesis in chickens and laying hens, and from ~E15 during 28 days of incubation in turkeys and ducks), the tissues' demand for oxygen increases dramatically and the heart provides the mechanical thrust to supply oxygen to the embryo's body. Although HR alone provides little insight into the developmental mechanisms because it is relatively stable during embryogenesis, with the exception of the internal and external period (Figure 37.7), it can be used to provide information about thermoregulatory development (Andrewartha et al., 2011).

By exposing embryos to lower and higher incubation temperature of 34.5 and 39.5 °C, respectively, Yahav and Piestun demonstrated a dramatic effect on HR. Broiler and laying hen embryos exhibited a more pronounced HR response than turkey embryos (Figure 37.8) during exposure to 39.5 °C. They both reduced HR with increasing embryonic age; by E19, their HR was close to that monitored prior to heat exposure. However, this did not occur in turkeys, which on E19 still maintained an HR higher than the control level. Opposite responses were observed when the three species were exposed to 34.5 °C: there were relatively slow increases in HR as embryonic development of broilers and laying hens progressed, whereas the turkey embryos exhibited the pronounced response. It can be concluded from the HR responses that embryos cope better with exposure to high incubation temperatures than to low ones. This may be related to the 1 °C difference between the temperature elevation (2 °C) and the reduction (3 °C). Nevertheless, it still could be related to a better physiological capacity to cope with higher incubation temperature, because of the wide range of abnormalities that high incubation temperature might cause.

In all three species, a significant correlation was found between EST and HR at the high and low incubation temperatures, whereas a very low correlation between these parameters was exhibited at the regular incubation temperature (37.5 °C), which highlights the designation of this temperature as the physiological comfort temperature.

37.6.2.3 Embryonic Oxygen Consumption

The general perception of the developing embryo is that precocial birds pass through four stages during embryogenesis (Tazawa et al., 2001), whereas altricial ones seem to remain in the Arrhenius-limited stage until post hatch. Thus, differences between the qualitative patterns of O_2 consumption of altricial and precocial embryos were suggested by different authors.

In precocial birds O_2 consumption is described as exhibiting a plateau phase of metabolic rate in late incubation, whereas in precocial ones the metabolic rate continued to increase exponentially till hatch. Several recent studies demonstrated that in both altricial and precocial birds the developmental mode exhibited a plateau phase (Prinzinger and Dietz, 1995; Prinzinger et al., 1995). From these studies, it appears that the plateau was present, irrespective of egg mass and length of incubation period; it basically represents the shift of respiration from the CAM to the lungs. However, the two modes of hatching differ in the level and duration of their plateau phase. In the study by Prinzinger and Dietz (1995), the altricial birds exhibited an VO_2 plateau that was lower by 35% than that of precocial ones, whereas the durations of the plateau, as percentages of the total incubation duration, were 8.9 and 17.6%, respectively. The two modes differed also in total energy consumption during incubation, being lower by 15% in the altricial than in the precocial birds.

In general, heat production of endothermic birds at temperatures below and above the thermoneutral zone is a result of two processes—thermoregulatory heat production and energy metabolism, of which the latter follows the van't Hoff rule. In avian embryos, a drop or increase in T_b as a result of changing incubation temperature will cause a decrease or increase, respectively, in net heat production, but these changes maybe moderate compared with that expected from the van't Hoff rule. The available information on heat production during embryonic development is contradictory. Most of the studies were based on cooling the environment during incubation. Romijn and Lockhorst (1955) did not find any changes in heat production in poultry embryos following exposure to cooling until internal piping occurred, whereas Freeman (1964) found a transient metabolic response. Piestun et al. (2009) demonstrated a significant decline in VO_2 from E17 in broiler embryos that had been continuously exposed to an elevated

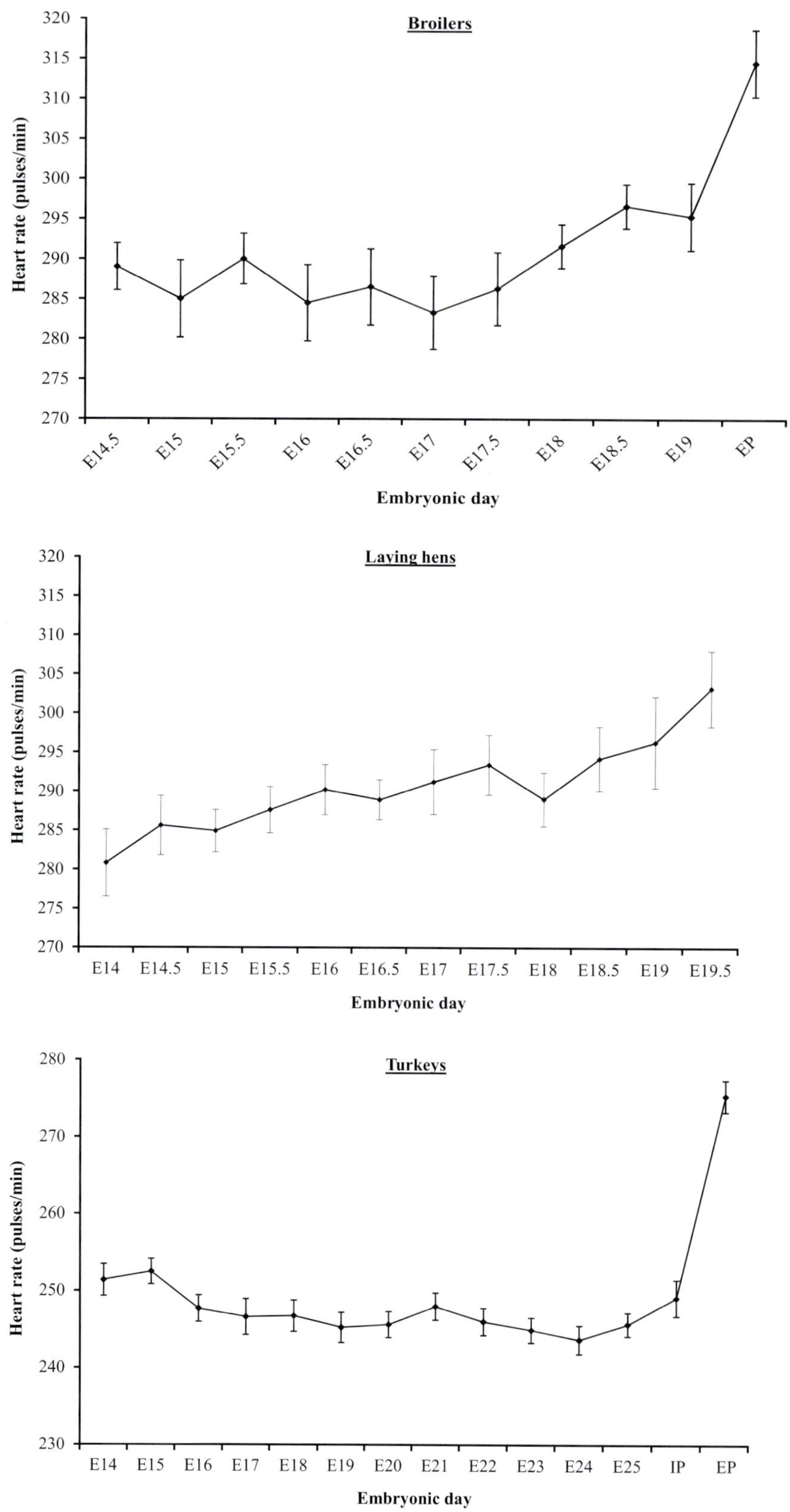

FIGURE 37.7 **Embryonic heart rate (HR, pulses/min), as measured by Buddy digital embryo HR monitor during incubation of broilers (upper graph), laying hens (middle graph), and turkeys (lower graph).**

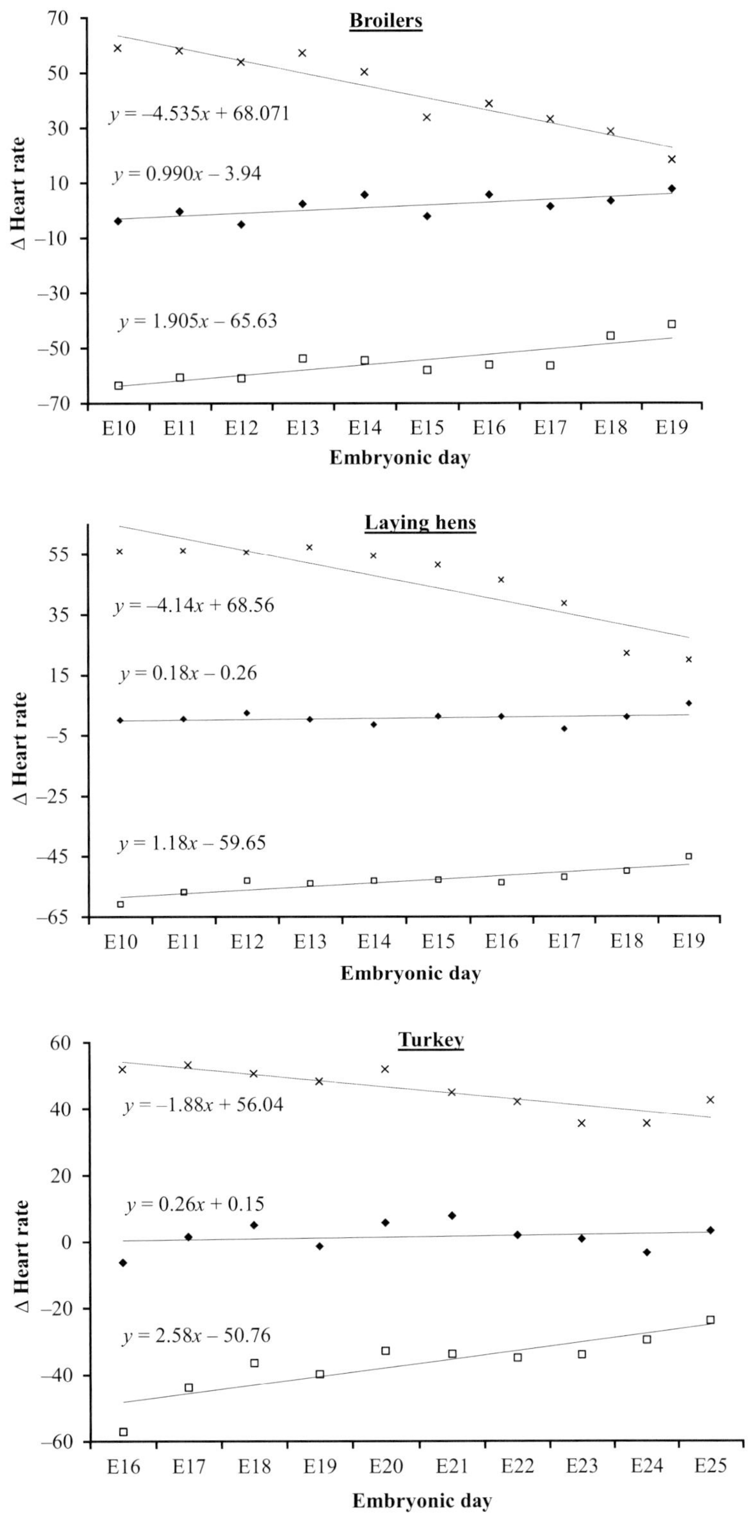

FIGURE 37.8 The effect of increasing incubation temperature from 37.5 °C (♦-line) to 39.5 °C (×-line) or decreasing it to 34.5 °C (□-line) for 10 h/d from embryonic day (E) E10 through E19 (broiler and laying hen embryos) and from E16 through E25 (turkey embryos) on the difference between heart rate (Δ HR) prior to the exposure and its average values during the exposure, in embryos of broilers (upper graph), laying hens (middle graph), and turkeys (lower graph).

incubation temperature of 39.5 °C from E7 through E16 (Figure 37.9). The ability to maintain lower VO_2 than that of the control from E17 onwards, despite being exposed to a similar incubation temperature, may indicate thermoregulatory capacity.

In Eqn (37.18), Q_{10}, whose value indicates endo- or ectothermy, is based on O_2 consumption; it is defined as the increase in VO_2 that results from an increase of 10 °C in T_a:

$$Q_{10} = [VO_2(2)/VO_2(1)]^{10/T_2 - T_1} \quad (37.18)$$

in which $VO_2(2)$ and $VO_2(1)$ represent the VO_2 when T_a is T_2 and T_1, respectively.

A Q_{10} value higher or lower than 2 indicates ectothermy or endothermy, respectively (Whittow and Tazawa, 1991); these authors had shown that in chicken embryos at a very early stage of development in which no endothermic reactions occurred, Q_{10} was between 2.0 and 2.4. A study by Nichelmann et al. (1998) found a value of Q_{10} lower than 2.0 in chicken embryos from day E14 onward, suggesting endothermic capability from this E onward.

In summary, it can be concluded that the transition from ectothermy to endothermy occurs in precocial birds during embryogenesis, as early as the oxygen plateau phase, which is associated with the development mode of becoming independent immediately post hatch. In contrast, altricial birds seem to undergo a transition towards endothermy via a poikilothermic stage that lasts until the end of the nestling stage.

37.6.3 Reducing Body Temperature by Using the Epigenetic Temperature Adaptation Approach during Embryogenesis

37.6.3.1 The Epigenetic Approach

Epigenetic temperature adaptation is based on the assumption that, during critical developmental phases, environmental factors, especially T_a, play a strong role in determination of the set point for physiological control systems. Dörner (1974) first referred to this as the determination rule. Epigenetic adaptation, which has been defined as a lifelong adaptation that occurs during prenatal (embryogenesis) or early post-hatching ontogeny, takes place within critical developmental phases that affect gene expression (Tzschentke et al., 2004), and seems to offer a suitable opportunity to reach the goal of improved acquisition of thermotolerance in poultry. During early development, most functional systems evolve, according to the transformation rule (Dörner, 1974), from an open-loop type without feedback to a closed-loop control type with feedback. Application of thermal manipulation (TM) during the critical phases of this development process may induce alterations in the thermoregulatory control system.

37.6.3.2 The Epigenetic Temperature Adaptation Approach: The Embryogenesis Model

In nature, incubation conditions are not uniform (Webb, 1987) because of searching for food, escaping from

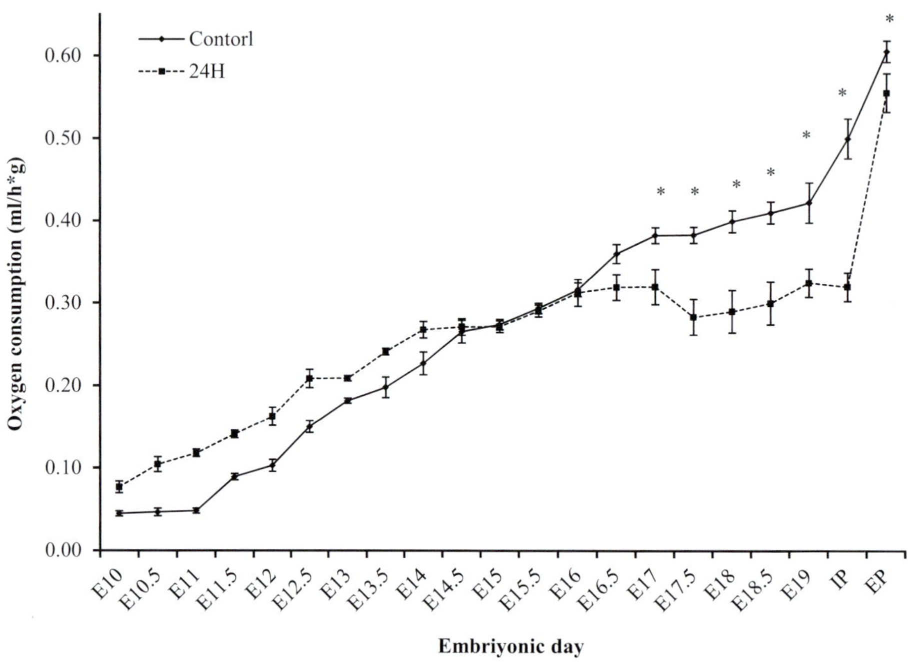

FIGURE 37.9 Oxygen consumption of broiler embryos incubated under control (37.8 °C, 56% RH) or thermally manipulated continuously (TM—24H treatment; 39.5 °C, 65% RH) from E7 through E16 (180 through 408 h of incubation). For each day of incubation, an asterisk indicates significant differences ($p \leq 0.05$) between treatments.

predators, and nonuniform nest insulation. This may be one of the reasons why birds in the wild are quite able to cope with extreme environmental temperatures, while maintaining T_b as close as possible to the normothermic range through efficient use of thermoregulatory mechanisms. In contrast to the nonuniform conditions in nature, commercial incubation conditions are strictly uniform. The uniform incubation conditions, coupled with the consequences of genetic selection for high production, confront the mature poultry with a great challenge to cope with extreme environmental conditions, especially with regard to T_a. The uniformity of incubation conditions makes possible the use of precocial domesticated embryos as a model for epigenetic temperature adaption by using altered incubation temperature—the TM model.

Three critical parameters have to be considered in using the TM approach during chick embryogenesis: the timing of the critical phase, the temperature level to which the embryo is exposed, and the duration of exposure. Determination of the critical phase of embryogenesis that would enable successful application of TM to improve acquisition of thermotolerance is based on the hypothesis that the set point or response threshold of controlling systems related to metabolic rate could be altered most efficiently during the development/maturation of the hypothalamus-hypophyseal-thyroid axis, which is associated with thermoregulation, and the hypothalamus-hypophyseal-adrenal axis, which is associated with stress but which plays a major role in activation of the thyroid axis during incubation.

Until mid-incubation, the thyroid gland exhibits only limited ability to synthesize hormones. This period is characterized by the synthesis of monoiodotyrosine on E8, of diiodotyrosine on E9, and of T_4 and thyroid-stimulating hormone (TSH) by the hypophysis on E10. The linkage of the hypothalamic-pituitary-thyroid axis is formed between E10.5 and E11.5. Levels of T_3 start to increase on E12, and increase significantly prior to hatching, in preparation for their roles in the final maturation of many tissues and in the physiological integration of hatching. Therefore, application of TM during the sensitive period of development of this axis may affect the set point of the heat-production threshold. Although the major function of the adrenal axis relates to stress, during incubation corticotrophin-releasing hormone stimulates the secretion of TSH and corticosterone prevents T_3 from being deiodinated to T_2. In this situation, both axes are involved in determining the embryo metabolic rate, and determination of the sensitive period for application of TM must consider both. Furthermore, it was suggested that the embryo is susceptible to stress. Therefore, increasing the incubation T_a during the development and maturation of the hypothalamic-hypophyseal-adrenal axis may also affect the stress response of the posthatch chick and, in turn, its metabolic rate.

The other two factors that play significant roles with respect to planning TM use are the temperature level and the exposure duration. Neither must affect the performance of the genetically selected organism, but they should significantly improve its thermotolerance.

Studies that addressed all three parameters were conducted by Piestun et al. (2008, 2009). In these studies, TM of 39.5 °C and 65% RH was applied between E7 and E16 for 12 h per day (12H treatment). The TM alterations with respect to the regular incubation conditions were 1.7 °C and 9% RH. Application of TM caused a significant decline in EST, HR, and VO_2 from E18 onward (Figure 37.10), and significantly affected the metabolic rate of the hatched chicks. This was manifested in a significantly reduced T_b, accompanied by significantly reduced plasma T_4 and T_3 concentrations posthatch. Although T_3 is the potent hormone with regard to the chick's metabolic rate, the fact that plasma T_4 concentration was also significantly reduced in the hatched TM chicks suggests that there was reduction in the activity of the thyroid gland upon hatch, which emphasizes the significant effect of TM in reducing the metabolic rate. However, the plasma corticosterone concentration was significantly greater in the TM chicks than in the control ones, which may have been due to the hatching difficulties exhibited by the treated chicks. Despite the changes in the hatched chicks' metabolic rate, there was no significant effect on their hatchability or body weight at hatch.

A similar approach was applied to turkey embryos, except that the sensitive period was changed to match the development and maturation of thyroid and adrenal axis in turkeys (i.e., E10 through E22; Zimerman et al., 2013). This application caused a reduction in the embryos' metabolic rate, as indicated by changes in EST, HR, and VO_2 (Figure 37.11) from E23 onwards (i.e., after the TM treatment was terminated).

To study whether TM has a long-lasting effect, posthatch thermal challenges must be applied. It was previously reported that exposing embryos to high or low temperatures during incubation improved their capacity to adapt to hot or cold environments, respectively, in the posthatch phase (for review, see Yahav, 2009). However, most studies challenged the broilers only at a young age or tested their potential with a mild heat stress. A study that evaluated the potential of broilers to cope with severe heat stress at marketing age (Collin et al., 2007), following TM at E8 through E10, E16, through E18 or a combination of the two treatments during embryogenesis, could not demonstrate that TM conferred any advantage in coping with deleterious environmental temperatures, mainly because TM was not applied during the most sensitive period. To be able to evaluate thermotolerance acquisition of chickens that were exposed to TM from E7 through E16 (Piestun et al., 2008), hatched chicks were raised up to marketing age (35–42 days of age) under environmental comfort conditions and were exposed to heat challenge (high T_a and low RH) for

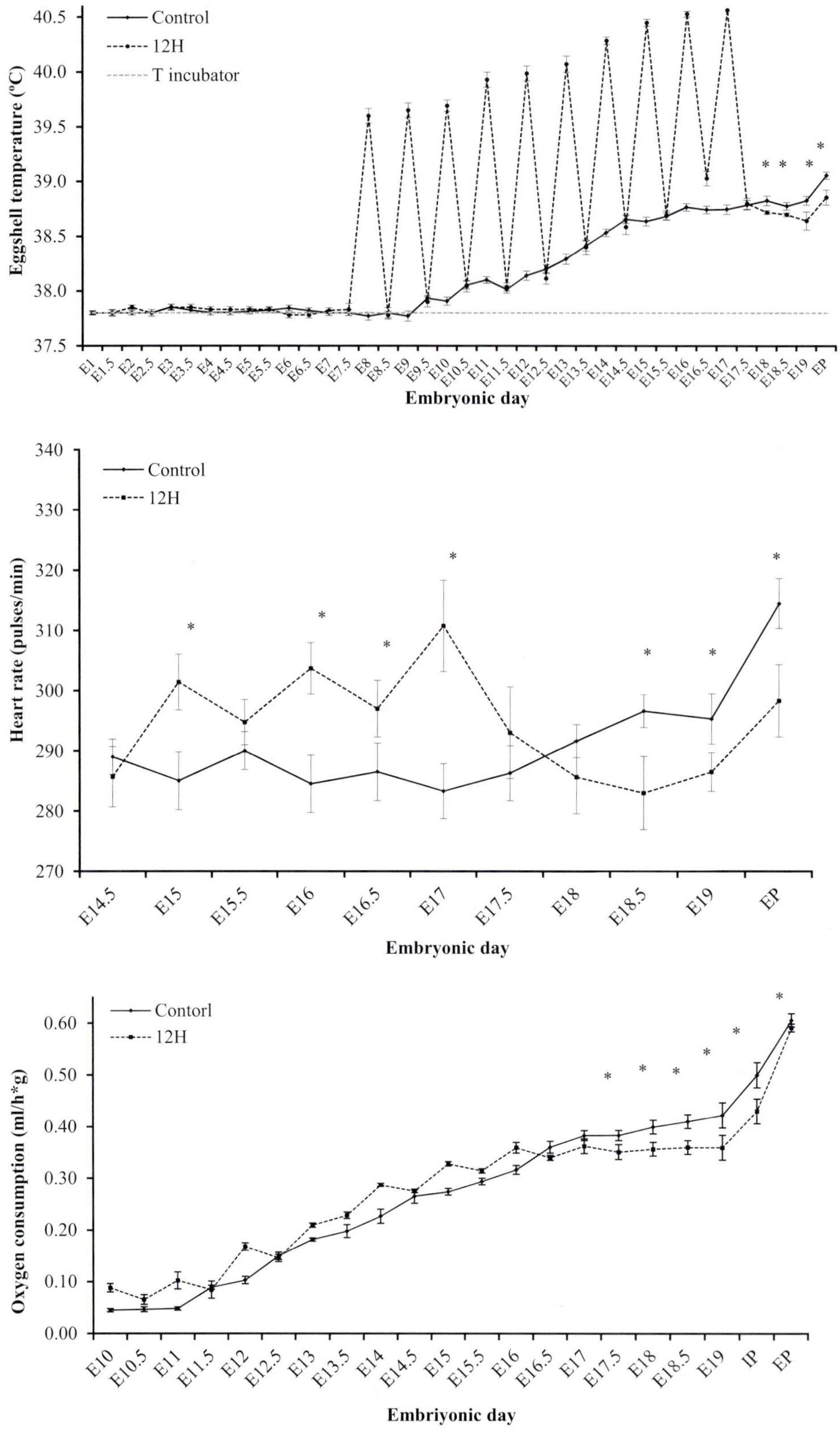

FIGURE 37.10 Eggshell temperature, heart rate, and oxygen consumption of broiler embryos incubated under control (37.8 °C, 56% RH) or intermittently thermally manipulated (TM—12H treatment; 39.5 °C, 65% RH) from E7 through E16 (180 through 408 h of incubation). For each day of incubation, an asterisk indicates significant differences ($p \leq 0.05$) between treatments.

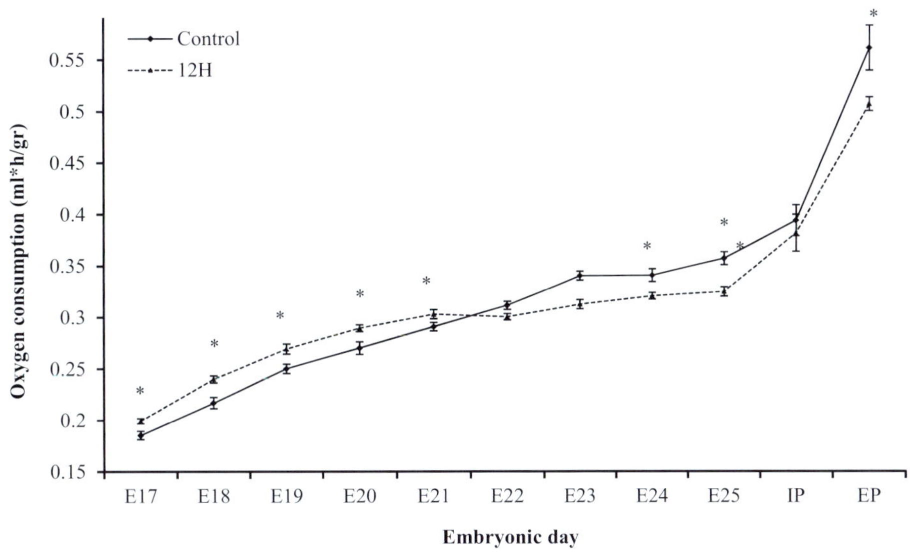

FIGURE 37.11 **Oxygen consumption of turkey embryos incubated under control (37.5 °C, 56% RH) or intermittently thermally manipulated (TM—12H treatment; 39.2 °C, 65% RH) from E10 through E22 (240 through 528 h of incubation).** For each day of incubation, an asterisk indicates significant differences ($p \leq 0.05$) between treatments.

several hours. Thermotolerance acquisition can be defined by reference to several parameters, such as T_b, thyroid and corticosterone hormones, and vasomotor responses. Thermal challenge induced a severe hyperthermia in all birds but, when the distributions of T_b were examined, the advantage of 12H TM was pronounced (Piestun et al., 2008): the number of birds that developed T_b in the range between 43 and 44 °C was much higher among the TM-treated ones, whereas the number of those that developed T_b of 44–45 °C was much higher among the nontreated ones, and the ability to recover from a hyperthermic condition with T_b of 44–45 °C was very low (Yahav, 2000). The ability of the TM-challenged chickens to maintain lower T_b springs from a sufficient reduction in the activity of the thyroid gland; a reduction in the peripheral deiodination of T_4 to T_3, coupled, most probably, with enhanced activity of 5-monodeiodinase (which would transform T_4 to the nonactive form of T_3; that is, the rT_3); and an adequate vasomotor response, as measured by radiative and convective heat loss. Figure 37.12 demonstrates heat loss by radiation and convection of broilers exposed to 23 °C (the regular T_a) and then to the challenge temperature of 35 °C. At 23 °C, heat loss was significantly lower in the 12H treatment than in the controls, whereas exposure to 35 °C caused a complete change in the response: exposing birds with the lower metabolic rate (12H treatment) to a lower T_a (23 °C) elicited vasoconstriction to retain heat, which was expressed in lower SHL, whereas exposure to a high T_a (35 °C) elicited vasodilatation, which was expressed in high SHL. Inducing different thermoregulatory pathways in order to reduce the development of hyperthermia caused only a mild increase in plasma corticosterone concentration, which indicated mild stress of the TM-treated chickens.

In sum, the accumulating evidence precludes a regular physiological adaptation, in light of the time elapsed between the TM treatment and the posthatch challenge. However, the evidence shows that the epigenetic adaptation approach and its association with changes in the incubation environment—with emphasis on fine-tuning the level and duration of stress to coincide with the critical phase—can elicit efficient epigenetic temperature adaptation in broiler chickens.

37.6.3.3 The Epigenetic Temperature Adaptation Using the Posthatch Model

Broiler chicks complete development of their brain and T_b regulation at the age of 10 days posthatch (Arad and Itsaki-Glucklish, 1991). Until this time point, body and brain temperatures are maintained at lower levels than in adult chickens. Subsequently, as age increases, the difference between body and brain temperatures increases linearly and significantly. The epigenetic response has been successfully modulated by early-age TM of posthatching chicks, by exploiting the incomplete maturation of the thermoregulatory system: TM involving exposure of a 3 day old broiler to 37–38 °C at 60–80% RH for 24 h was found to improve acquisition of thermotolerance. The improvement achieved was manifested in the ability of the TM chicks to reduce heat production efficiently during exposure to acute thermal challenge at marketing age. This was accompanied by the following (Yahav, 2009): an alteration in SHL through convection and radiation; a significant reduction in the stress level of the TM chickens, as indicated by their plasma corticosterone concentration; and pronounced decrease in the

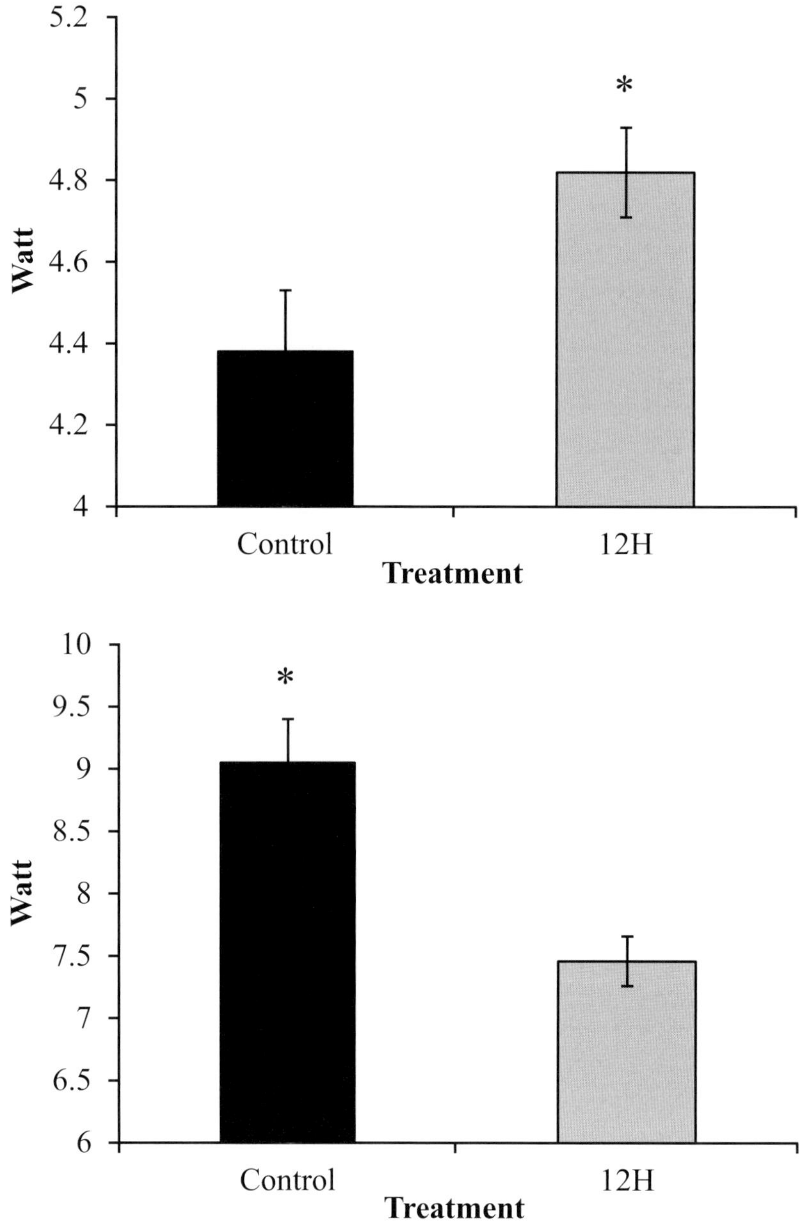

FIGURE 37.12 **Sensible heat loss (via convection and radiation) from 35 day old broilers exposed to control conditions (23 °C, lower graph) or to thermal challenge of 35 °C for 5 h (upper graph).** The broilers were incubated under control (37.8 °C, 56% RH) or thermally manipulated (TM; 39.5 °C, 65% RH) intermittently for 12 h (12H treatment) from E7 through E16 (180 through 408 h of incubation). Asterisk indicates significant difference between treatments ($p \leq 0.05$).

27, 70, and 90 kDa heat-shock proteins (HSPs) in the heart muscle and lung tissue of TM chickens during thermal challenge. As a result of this study, it has been suggested that the induction of HSPs is correlated with T_b, and that the HSP response is not part of the long-term mechanism elicited by the early-age TM. The reduction in heat production, coupled with increased SHL, enabled relatively slow development of hyperthermia and thus dramatically reduced mortality. The TM application at 3 days of age also induced compensatory growth, leading to improvement of performance and muscle growth, because of proliferation of skeletal muscle satellite cells during TM (Halevy et al., 2001). It was further found to positively affect the post-TM development of the gastrointestinal tract (Uni et al., 2001).

37.6.3.4 The Plasticity of the PO/AH and Its Association with Posthatch TM

Although the effect of TM on long-term storage of information was not studied in embryos, there was great interest in elucidating the mechanisms by which sensory information affects neural circuits during other critical developmental periods. Like other sensory mechanisms, the thermal response set point is probably elicited by TM causing alterations in cellular properties in the hypothalamus (Boulant, 2006).

Posthatch TM, which induces long-term changes in the ability of chicks to withstand heat stress later in life, recently stimulated research that aims to understand the alteration in the PO/AH thermoregulatory control site. Recently, the signal-transduction pathways underlying postnatal hypothalamic-neuronal-network reorganization were characterized, revealing that in chicks there are three known neurotrophic factors: nerve growth factor (NGF), brain-derived neurotrophic factor (BDNF), and neurotrophin-3 (NT-3). Of these three factors, BDNF plays a pivotal role in developmental plasticity. Katz and Meiri (2006) found that during TM of 3-day-old chicks, expression of BDNF mRNA, but not that of NGF or NT3, was induced in the PO/AH; a threefold increase in mRNA expression occurred 6h after the initiation of TM. In order to test the hypothesis that BDNF activation was critical for the association of PO/AH plasticity with TM, BDNF was "knocked down" by using antisense, thereby providing the opportunity to induce transient changes in gene expression (downregulation) during significant time windows (Kramer and Cohen, 2004). Indeed, use of BDNF antisense inhibited BDNF expression in the PO/AH by 80% during the TM experiment (Katz and Meiri, 2006) and thereby impaired the TM chick's thermal tolerance. This confirmed the critical involvement of BDNF in thermotolerance acquisition as a result of TM, in that BDNF acted not only as an effector during TM but also as a mediator whose expression was correlated with long-term thermal set-point determination. Furthermore, it was found that the acquisition of thermotolerance was mediated in the PO/AH by R-Ras3, which is known to participate in growth-related signal transduction in the brain (Labunskay and Meiri, 2005), and 14-3-3ε, which is known as a chaperon of phosphorylation and of determination of cellular localization in establishing thermal control. In its role as a chaperon, 14-3-3ε activates transcription via Jun (Meiri, 2007). The process has been shown to be regulated at the translational level by induction of eIF2B expression.

In sum, induction of growth signaling during TM probably can account for the changes in cellular properties in the PO/AH that may, in turn, lead to changes in the thermal set point that induce thermotolerance acquisition.

37.7 THE COST OF MAINTAINING BODY TEMPERATURE IN POULTRY COMPARED WITH THAT IN OTHER BIRD SPECIES

The costs of maintaining T_b in various species with differing backgrounds (i.e., domesticated versus undomesticated) and diverse environmental situations (e.g., exposure to extreme environmental conditions, migration) necessitate a variety of physiological responses in order to adjust to the new challenges. These responses may differ according to any specific challenge. Irrespective of the differences among the challenges, the adjustment will always aim to maintain T_b within the limits of the two-tier theory of thermoregulation.

37.7.1 Domesticated Fowl and Thermotolerance: Acute Heat Exposure

Recent decades have seen significant developments in the genetic selection of meat-type fowl—that is, broilers (Havenstein et al., 1994, 2003a) and turkeys (Havenstein et al., 2007). These developments led to rapid growth, accompanied by increased feed efficiency and metabolic rate (Janke et al., 2004) but relatively poor enhancements in functional efficiency of several visceral systems, such as the cardiovascular and respiratory systems. Thus, acute exposure of chickens to extreme conditions, such as hot spells, confronts broilers, and, to a lesser extent, turkeys with difficulties in maintaining T_b. This is an outcome of the conflict between high production and thermotolerance, which leads to the relatively diminished ability of domesticated fowl to maintain a sufficiently dynamic steady state in face of severe changes in the environment (Yahav, 2009).

The level of hyperthermia from which a sufficient recovery can occur was found to be between 44.0 and 44.5 °C for broilers and turkeys (Yahav et al., unpublished data). It can be speculated that above this level, heat stroke is developed; it is characterized by detrimental changes at the cellular and molecular levels—changes that lead to the fatal cascade of events comprising decreased blood pressure, brain hypoxia, neuronal dysfunction, cell fatigue, etc. (Hales et al., 1996). Thus, heat stroke may be the reason for insufficient recovery that is characterized by neuronal disorders, or for complete inability to recover. However, to avoid reaching this stage of hyperthermia, the body will strongly activate two main mechanisms—EHL and vasodilatation—although one important mechanism, facultative heat production, will decline.

The vasodilatation response increases blood flow to the skin, especially to the nonfeathered areas of the fowl, as well as to the upper respiratory passageways, in order to transport heat from the viscera to the periphery. Vasodilatation will follow the development of hyperthermia. Development of hyperthermia, followed by increase in surface temperature, was monitored in broiler chickens by Giloh et al. (2012), who found that, irrespective of the broiler's age, hyperthermia developed in a short time. T_b reached values of 42.9–44.9, 42.6–43.7, 42.6–43.3, and 43.1–44.6 °C at the ages of 9, 16, 22, and 36 days, respectively; the increased facial surface temperatures (e.g., vasodilatation) that developed in parallel reached 40.9–43.0, 40.9–42.3, 41.2–41.7, and 42.5–44.0 °C, respectively. The increase in surface temperature enhances heat loss by radiation convection and conductance (i.e., SHL). However, whereas before the acute exposure to heat stress a difference of approximately 2.0 °C between T_b and facial surface temperature was measured, this difference was significantly reduced during the development of hyperthermia, suggesting a limited capacity for vasodilatation despite the progressive development of hyperthermia.

In parallel to vasodilatation, also EWL is activated. However, evaporative cooling through panting may lead to dehydration, which may dramatically affect the functioning of the blood system, leading to heat stroke. During the first stages of acute heat stress, before the development of dehydration, although the blood supply to the skin and the upper respiratory passageways is enhanced, there is no impact on the blood supply to other important tissues. This is mainly because of the increase in cardiac output that results from the increases in HR and, to a much lesser extent, in stroke volume, and from the redistribution of blood flow among tissues, by which blood flow is diverted from nonvital tissues to vital ones. The resulting increases in HR and venous return support an overall increase in cardiac output that refills the arterial pressure reservoir more rapidly and prevents further reduction in arterial pressure (Whittow et al., 1964; Sturkie, 1967; Zhou, 2000). However, as dehydration develops, depletion of the blood volume will cause a reduction in venous pressure which, in turn, will diminish blood flow to the skin and the upper respiratory tract, thereby impairing the efficacy of the E routes. This will lead to lethal levels of hyperthermia, with T_b that could reach 46–47 °C.

A further combined mechanism involves modulation of heat production, which declines with increasing duration and level of heat stress, to prevent excessive accumulation of heat in the body. Under acute heat stress, the main and immediate mechanism to reduce T_3 and, thereby, heat production acts by reducing the deiodination of peripheral T_4 to T_3 and increasing the transformation of T_3 to rT_3.

It can be concluded that activation of the physiological mechanisms that lead to enhancement of heat loss via every route, coupled with reduction of heat production will minimize the deleterious effect of heat stress. However, long-term (hours) incidence of extreme heat stress may eventually lead to the cascade of heat stroke events that eventually will end in death.

37.7.2 Physiological Adjustments of Birds to Arid, Cold, and Aquatic Environments

Living in a desert environment exposes endothermic birds to shortage of energy and water while exposing them to high T_a and solar radiation during the day. Such environmental conditions force birds to exhibit physiological and behavioral adaptations to enable them to cope with such deleterious environmental conditions. Most desert species of birds are diurnal, and they may be exposed to shade temperatures (T_a) around 50 °C. Exposure to high T_a not only necessitates efficient thermotolerance, but it also puts an additional strain on water and energy balance in an environment that lacks these elements. On the other hand, low T_a increases the energetic costs required for maintenance of homeothermy. For some small birds or those living in extreme environments, the level of heat production that would be required to maintain T_b will reach levels that necessitate the adoption of thermoregulatory adjustments, such as cyclic torpor or hypothermia, which enable energy reserves to last through the stressful period (Körtner et al., 2000; McKechnie and Lovegrove, 2002).

37.7.2.1 Desert Environment

In desert mammals, BMR and total evaporative water loss are significantly lower than in nondesert species. Across-species comparison between desert and nondesert birds demonstrated that, on average, arid zone birds exhibited lower RMR (Tieleman and Williams, 2000) and lower total evaporative water loss (Williams, 1996) than those of other zones. Other strategies are shifting the thermoneutral zone towards higher environmental temperatures and minimizing the dry heat gain from the environment. *Dry heat gain* is calculated as:

$$\text{Dry Heat Gain} = h\,(T_b - T_a) \qquad (37.19)$$

in which h is the dry heat transfer coefficient, which is influenced by skin vasodilatation, feather insulation, the bird's volume-to-surface ratio, and subcutaneous fat tissue; and $(T_b - T_a)$ is the difference between T_b and the environmental temperature. When T_a exceeds T_b, heat will flow into the body, and the physiological adaptation requires h to decrease to a minimal level.

However, it must be pointed out that a comparison between desert and nondesert species may be misleading. Species differ not only in habitats, but also in their phylogenetic backgrounds, behavior, and diets.

A study by Tieleman et al. (2002), which tried to avoid differences in these parameters addressed various species of larks (*Alaudidae*), compared birds from the arid zone (e.g., Hoope larks and Dunn's larks with those from the mesic zone (e.g., skylarks and woodlarks)). The study found that in desert larks BMR and total water turnover were lower by 43 and 27%, respectively than those observed in the mesic species, and the T_b monitored in the desert species was lower by 1.1 °C. However, the value of h in the desert species was not lower when T_b was equal to T_a, and a similar pattern of h at high environmental temperatures was found in other small-bird species. Minimizing the heat transfer coefficient should minimize heat gain from the environment, in order to reduce the loss of water through evaporation, which is the only mechanism that can cool the body when T_b is equal to or below T_a.

An important determinant for BMR is the size of internal organs, such as the heart, kidney, liver and intestine (Chappell et al., 1999). The selection for low energy requirement that the environment imposes on desert birds may lead to minimal BMR and ratio of metabolic organ size to body mass. On the other hand, reduction in total evaporative water loss may be associated with the structure of the SC of the skin that influences CWL (Peltonen et al., 1998).

37.7.2.2 Cold and Aquatic Environments

As mentioned above, the main concern raised by exposure to cold environments involves the cost of the energy required for maintaining T_b. This cost is even higher for aquatic species because the thermal conductance of water is 25 times as high as that of air, and because of the drastic reduction of the insulation provided by plumage or feathers in water. Thus, several physiological adjustments are required in order to reduce or maintain the activity of the organism.

Birds, and especially, small species, are able to employ torpor (i.e., controlled reductions of metabolic rate and T_b) to substantially reduce energy expenditure during certain parts of the day or year, and thereby to minimize the impact of energy challenges. This response is most pronounced during long and severe winters, when low T_a is combined with reduced availability of feed. However, temporary variations in food resources also may trigger heterothermy/torpor; a study of torpor in Malachite Sunbirds (*Nectarinia famosa*) found that T_b fell to 26.8 °C when they were exposed to T_a of 5 °C (Downs and Brown, 2002). Although torpor was associated mainly with small birds, McKechnie and Lovegrove (2002) found $T_b < 15$ °C in avian species with body mass larger than 70 g, indicating the importance of food availability. Although measurements of T_b in free-ranging birds are quite common, birds may be torpid and save energy and water by utilizing metabolic depression, which would not be manifested in their T_b (McKechnie and Mzilikazi, 2011).

Whereas low T_a coupled with food scarcity will cause torpor, in diving birds searching for food low T_a necessitates different physiological adjustments. In such cases prevention of body cooling should be the strategy employed by diving birds such as penguins. To achieve this, a dramatic decrease in the peripheral T_b is used, to reduce peripheral heat loss. Such reduction is an outcome of compression of the air in the feathers at depth and speed, combined with peripheral vasoconstriction. The peripheral vasoconstriction causes the warm blood to bypass the cold-exposed tissues and enhances their cooling, while the internal tissues are isolated and the core temperature is maintained (Schmidt et al., 2006).

37.7.3 Physiological Adjustments of Birds to Migration

Each year in autumn, billions of birds fly from their breeding grounds to wintering grounds, from which they return in spring. This is a highly energetic process that enables the birds to escape adverse local conditions during winter and to return in spring to search for breeding sites. Thus, the process is influenced by the thermoregulatory difficulties caused by the energy cost of transport. Modern tracking of migration records revealed round trips of more than 64,000 km/year by some marine birds, and another physical achievement was demonstrated in a single nonstop flight of 11,000 km from Alaska to New Zealand by the bar-tailed godwit (*Limosa lapponica* L.) (Weber, 2009). Such extreme physical accomplishments must involve the evolution of specific adaptations that make these records possible.

37.7.3.1 Variations in BMR

Mobility of any organism has been identified as a major determinant of selection that affects maintenance energy demands. Therefore, it would be expected that migrating birds would exhibit variations in BMR, because migration is associated with wide changes in a bird's body weight in the course of long-distance movements, sometimes with no breaks. Migration may affect BMR in three different ways: long-distance migration may necessitate high rates of energy acquisition in preparation for migration, thus leading to higher BMR than that of nonmigrating birds (Kvist and Lindstrom, 2001); seasonal presence in the breeding region may remove some of the selection pressure on BMR that affects nonmigrating birds, such as the need to survive through the cold winter; and alternating occupation of differing environments may cause differences in BMR when migrating and nonmigrating birds are compared. Among migrants, BMR may vary strongly both within and between individuals during preparations for migration, whereas in nonmigrants little variation in BMR was observed previously (McKechnie, 2008).

37.7.3.2 Minimizing the Cost of Migration

Migrating birds evolved various strategies to minimize migration costs. For instance, the use of locomotory muscles can generate force with very high efficiency (Kvist et al., 2001). They minimize body mass by atrophying the digestive and reproductive organs before taking off, in parallel with escaping the winter shortage of food, and decrease the maintenance costs of other tissues that are not essential for migration (McWilliams and Karasov, 2004). They use selection of lipids to store energy for muscle activity, because lipids provide the highest ATP yield per gram of fuel, and generate more metabolic water upon oxidation than proteins or carbohydrates; the premigration lipid reserve can reach 50% of body mass.

The amount of energy produced during migration must cause an elevation of T_b, but there is almost no information in the literature on migrants' T_b. The most up-to-date report focuses on T_b fluctuations in migrating birds during stopovers. Stopovers to refuel may be efficient if the bird is able to reduce energy costs by becoming hypothermic or even entering torpor while at rest; facultative reduction of a bird's T_b is known to be a common procedure. A study on the Blackcaps passerine (*Sylvia atricapilla* L.) by Wojciechowski and Pinshow (2009) shed some light on the thermal physiology of this species during stopover. This species breeds in Europe, migrates to northern Africa in autumn, and returns in spring. During stopover, the birds reduced their T_b by up to 10 °C below normothermic, which ranges between 39 and 44 °C (Prinzinger et al., 1991). The decline in T_b reduces the difference between T_b and T_a and thereby significantly reduces the energy expenditure needed to maintain T_b, which, in turn, accelerates the rate of fuel accumulation during stopover. Small migrating birds may also reduce energy expenditure and maintain mild hypothermy by huddling (Gilbert et al., 2010).

The hypothermic response, whether during stopover or at the end of the migration, facilitates efficient recovery from the harmful consequences of migration.

37.8 SUMMARY AND CONCLUSIONS

Body temperature control exclusively characterizes endothermic birds and mammals. It results from physiological adjustments coupled with morphological modifications. In birds, its profoundly varied physiological, ecological, and behavioral benefits enable wide aerial, aquatic and terrestrial distributions. Controlling T_b necessitates a very-high-gain thermoregulatory system that prevents deleterious environmental effects; it operates via control of heat production, such as by shivering and NST, and heat loss, such as via evaporation and SHL. In the course of evolution, to enable endothermy to replace ectothermy, these control mechanisms changed through modifications to the neuronal and endocrine regulatory systems that maintain a balance between heat production and heat loss. Understanding the processes that lead to changes in metabolic rate and heat loss, and the transition from ectothermy to endothermy, is essential for understanding the mechanisms that enable control of T_b.

It can be concluded that body temperature control is a unique characteristic of endothermic birds and mammals. Endothermy was selected during evolution in order to better adapt to a varying environment. T_b is the most physiologically guarded parameter of the body; its maintenance operates at a very high gain, achieved by the developmental regulatory mechanisms that maintain a balance between heat production and heat loss, and that are activated and regulated by both neuronal and hormonal signals. T_b of endothermic birds is regulated through a variety of thermoregulatory responses, controlled and orchestrated by the brain, that optimize the internal T_b in order to enable appropriate molecular activities and reactions.

Furthermore, to sustain thermal tolerance and avoid the deleterious consequences of thermal stresses, three direct responses are elicited by birds: the RTSR, characterized by response times ranging from minutes to hours; acclimation/acclimatization, characterized by response times ranging from days to weeks; and epigenetic adaptation during the perinatal period, which is based on timing. Endothermy heterothermy involves changing T_b in response to changes in the environment, such as alterations in T_a, and food and water availability. Furthermore, changing T_b from normothermic values in response to environmental alterations saves energy and is a key component of thermoregulation in birds of diverse sizes exposed to differing environments: tropical, subtropical, arid, and mesic. The transition from ectothermy to endothermy necessitates increased capacity of the respiratory system, changes in the CVS, increased efficiency of the gastrointestinal tract, and modifications to the neural and endocrine systems that regulate metabolic enzyme activities. These modifications include changes in protein structure, modifications to membranes and up- or down-regulation of genes. This transition occurs in precocial birds during embryogenesis, in which it takes place as early as the plateau phase, whereas in altricial birds the transition takes place during poikilothermic stage, which lasts until the end of nestling.

Accumulating evidence on TM precludes a regular physiological adaptation but emphasizes the occurrence of an efficient epigenetic temperature adaptation process. TM induces growth signaling, which probably can account for the changes in cellular properties in the PO/AH that may lead to the changes in the thermal set point that induce thermotolerance acquisition. Finally, living in extremely low T_a environments necessitates the adoption of thermoregulatory adjustments, such as cyclic torpor or hypothermia, whereas extreme high-T_a environments necessitate adjustments to RMR and changes in the CVS to permit efficient SHL.

REFERENCES

Andrewartha, S.J., Tazawa, H., Burggren, W.W., 2011. Embryonic control of heart rate: examining developmental patterns and temperature and oxygenation influences using embryonic avian model. Res. Physiol. Neurobiol. 178, 84–96.

Arad, Z., Itsaki-Glucklish, S., 1991. Ontogeny of brain temperature in quail chicks (*Coturnix coturnix japonica*). Physiol. Zool. 64, 1356–1370.

Arad, Z., Sighvatur, S., Arnason, S., Chadwick, A., Skadhauge, E., 1985. Osmotic and hormonal responses to heat and dehydration in the fowl. J. Comp. Physiol. B 155, 227–234.

Basta, D., Tzschentke, B., Nichelmann, M., 1997. Temperature guardian neurons in the preoptic area of the hypothalamus. Brain Res. 767, 361–362.

Bennett, A.F., Hicks, J.W., Cullum, A.J., 2000. An experimental test of thermoregulatory hypothesis for the evolution of endothermy. Evolution 54, 1768–1773.

Bligh, J., 1966. The thermosensitivity of the hypothalamus and thermoregulation in mammals. Biol. Evol. 41, 125–130.

Bligh, J. (Ed.), 1973. Temperature Regulation in Mammals and Other Vertebrates. North-Holland Publishing Company, Amsterdam, London, p. 260.

Bligh, J., Heal, J.W., 1974. The use of radio-telemetry in the study of animal physiology. Proc. Nutr. Soc. 33, 173–181.

Boekholt, H.A., van der Grinten, P., Schreurs, V.V., Los, M.J., Leffering, C.P., 1994. Effect of dietary energy restriction on retention of protein, fat and energy in broiler chickens. Br. Poult. Sci. 35, 603–614.

Boouwstra, J.A., 1997. The skin barrier, a well organized membrane. Colloids Surf. A Physicochem. Eng. Asp. 123, 403–413.

Boulant, J.A., 1996. Environmental physiology: hypothalamic neurons regulating body temperature. In: Fregly, M.J., Blatteis, C.M. (Eds.), Handbook of physiology, vol. 4. APS, Oxford University Press, New York, pp. 105–126.

Boulant, J.A., 2006. Neuronal basis of Hammel's model for set-point thermoregulation. J. Appl. Physiol. 100, 1347–1354.

Brent, R., Pedersen, P.F., Bech, C., Johansen, K., 1985. Thermal balance in the European coot *Fulica atra* exposed to temperatures from −28°C to 40°C. Ornis. Scand. 16, 145–150.

Buffenstein, R., Yahav, S., 1991. Is the naked mole-rat *Heterocephalus glaber* an endothermic yet poikilothermic mammal? J. Therm. Biol. 16, 227–232.

Burger, R.E., Osborne, J.L., Banzett, R.B., 1974. Intra-pulmonary chemoreceptors in *Gallus domesticus*: adequate stimulus and functional localization. Respir. Physiol. 22, 87–97.

Burggren, W.W., 2004. What is the purpose of the embryonic heart rate? or how facts can ultimately prevail over physiological dogma. Physiol. Biochem. Zool. 77, 333–345.

Cabanac, M., Aizawa, S., 2000. Fever and tachycardia in a bird (*Gallus domesticus*) after simple handling. Physiol. Behav. 69, 541–545.

Chappell, M.A., Bech, C., Buttmer, W.A., 1999. The relationship of central and peripheral organ masses to aerobic performance variation in house sparrow. J. Exp. Biol. 202, 2269–2279.

Chiba, Y., Fukuoka, S., Niiya, A., Akiyama, R., Tazawa, H., 2004. Development of cholinergic chronotropic control in chick (*Gallus gallus domesticus*) embryos. Comp. Biochem. Physiol. A 137, 65–73.

Collin, A., Cassy, S., Buyse, J., Decuypere, E., Damon, M., 2005. Potential involvement of mammalian and avian uncoupling proteins in the thermogenic effect of thyroid hormones. Domes. Anim. Endocrinol. 29, 78–87.

Collin, A., Berri, C., Tesseraud, S., Rodón, F.E., Skiba-Cassy, S., Crochet, S., Duclos, M.J., Rideau, N., Tona, K., Buyse, J., Bruggeman, V., Decuypere, E., Picard, M., Yahav, S., 2007. Effects of thermal manipulation during early and late embryogenesis on thermotolerance and breast muscle characteristics in broiler chickens. Poult. Sci. 86, 795–800.

Cooper, C.H., Geiser, F., 2008. The "minimal boundary curve for endothermy" as a predictor of heterothermy in mammals and birds: a review. J. Comp. Physiol. B 178, 1–8.

Dawson, W.R., 1982. Evaporative losses of water by birds. Comp. Biochem. Physiol. 71A, 495–509.

Decuypere, E., Michels, H., 1992. Incubation temperature as a management tool: a review. World's Poult. Sci. J. 48, 28–38.

DeNardo, D.F., Zubal, T.F., Hoffman, T.C.M., 2004. Cloacal evaporative cooling: a previously undescribed means of increasing evaporative water loss at high temperatures in a desert ectothermy, the Gila monster *Heloderma suspectum*. J. Exp. Biol. 207, 945–953.

DiMicco, J.A., Zaretsky, D.V., 2007. The dorsomedial hypothalamus: a new player in thermoregulation. Am. J. Physiol. Regul. Integr. Comp. Physiol. 292, R47–R63.

Dörner, G., 1974. Environment-dependent brain differentiation and fundamental processes of life. Acta Biol. Med. Germ 33, 129–148.

Downs, C.T., Brown, M., 2002. Nocturnal heterothermy and torpor in the Malachite sunbird (*Nectarinia famosa*). Auk 119, 251–260.

Ellis, H.I., Maskrey, M., Pettite, T.N., Whittow, G.C., 1995. Thermoregulation in the brown noddy (*Anous stolidus*). J. Therm. Biol. 20, 307–313.

Else, P.L., Turner, N., Hulbert, A.J., 2004. The evolution of endothermy: role for membranes and molecular activity. Physiol. Biochem. Zool. 77, 950–958.

Falkenberg, C.V., Georgiadis, J.G., 2008. Water and solute active transport through human epidermis: contribution of electromigration. Int. J. Heat Mass Transf. 51, 5623–5632.

Freeman, B.M., 1964. The emergence of homeothermic metabolic response in the fowl (*Gallus domesticus*). Comp. Biochem. Physiol. 13, 413–422.

French, N.A., 1997. Modeling incubation temperature: the effect of incubator design, embryonic development, and egg size. Poult. Sci. 76, 124–133.

Gabarrou, J.F., Duchamp, C., Williams, J., Géraert, P.A., 1997. A role for thyroid hormones in the regulation of diet-induced thermogenesis in birds. Br. J. Nutr. 78, 963–973.

Gabrielsen, G.W., Mehlum, F., Karlsen, H.E., 1988. Thermoregulation in four species of arctic seabirds. J. Comp. Physiol. B 157, 703–708.

Ganey, J.L., Balda, R.P., King, R.M., 1993. Metabolic rate and evaporative water loss of Mexican spotted and great horned owls. Wilson Bull. 105, 645–656.

Gilbert, C., McCafferty, D., Le Maho, Y., Martrette, J.M., Giroud, S., Blanc, S., Ancel, A., 2010. One for all and all for one: the energetic benefits of huddling in endotherms. Biol. Rev. 85, 545–569.

Giloh, M., Shinder, D., Yahav, S., 2012. Skin surface temperature of broiler chickens is correlated to body core temperature and indicative of the chicken's thermoregulatory status. Poult. Sci. 91, 175–188.

Gordon, C.J., 1993. Temperature Regulation in Laboratory Rodents. Cambridge University Press, New York.

Griffin, J.D., Saper, C.B., Boulant, J.A., 2001. Synaptic and morphological characteristic of temperature-sensitive and -insensitive rat hypothalamic neurons. J. Physiol. 537, 521–535.

Grigg, G.C., 2004. An evolution framework for studies of hibernation and short term torpor. In: Barnes, M., Carey, H.V. (Eds.), Life in the Cold: Evolution, Adaptation, Mechanisms, and Applications. University of Alaska, Fairbanks, pp. 1–11. 12th International Hibernation Symposium. Biological Papers of the University of Alaska.

Hales, J.R.S., Hubbard, R.W., Gaffin, S.L., 1996. Limitation of heat tolerance. In: Fregly, M.J., Blattris, C.M. (Eds.), Handbook of Physiology, Section 4, Environmental Physiology, vol. 1. Oxford University Press, Oxford, pp. 285–359.

Halevy, O., Krispin, A., Leshem, Y., McMurtry, J.F., Yahav, S., 2001. Early age heat stress accelerates skeletal muscle satellite cell proliferation and differentiation in chicks. Am. J. Physiol. 281, R302–R317.

Hammel, H.T., 1965. Neurons and temperature regulation. In: Yamamoto, W.S., Brobeck, J.R. (Eds.), Physiological controls and regulations. WB Saunders USA, Philadelphia, pp. 71–79.

Harper, M.E., Himms-Hagen, J., 2001. Mitochondrial efficiency: lessons learned from transgenic mice. Biochem. Biophys. Acta 1504, 159–172.

Hatton, G.I., 1990. Emerging concept of structure-function dynamics in adult brain: the hypothalamo-neurohypophysial system. Prog. Neurobiol. 34, 437–504.

Havenstein, G.B., Ferket, P.R., Scheideler, S.E., Larson, B.T., 1994. Growth, livability, and feed conversion of 1991 versus 1957 broilers when fed "typical" 1957 and 1991 broiler diets. Poult. Sci. 73, 1785–1794.

Havenstein, G.B., Ferket, P.R., Qureshi, M.A., 2003a. Growth, livability, and feed conversion of 1957 versus 2001 broilers when fed representative 1957 and 2001 broiler diets. Poult. Sci. 82, 1500–1508.

Havenstein, G.B., Ferket, P.R., Qureshi, M.A., 2003b. Carcass composition and yield of 1957 versus 2001 broilers when fed representative 1957 and 2001 broiler diets. Poult. Sci. 82, 1509–1518.

Havenstein, G.B., Ferket, P.R., Grimes, J.L., Qureshi, M.A., Nestor, K.E., 2007. Comparison of the performance of 1966- versus 2003-type turkeys when fed representative 1966 and 2003 turkey diets: growth rate, livability, and feed conversion. Poult. Sci. 86, 232–240.

Hellon, R.F., Taylor, D.C.M., 1982. An analysis of a thermal afferent pathway in the rat. J. Physiol. 326, 319–328.

Hillman, P.E., Scott, N.R., Van Tienhoven, A., 1985. Physiological responses and adaptations to hot and cold environments. In: Yousef, M.K. (Ed.), Stress Physiology in Livestock. CRC Press, Boca Raton, pp. 27–71.

Hoffman, T.C.M., Welsberg, G.E., Denardo, D.F., 2007. Cloacal evaporation: an important and previously undescribed mechanism for avian thermoregulation. J. Exp. Biol. 210, 741–749.

Hohtola, E., Pyörnilä, A., Rintamäki, H., 1994. Fasting endurance and cold resistance without hypothermia in a small predatory bird: the metabolic strategy of Tengmalam's owl, *Aegolius funereus*. J. Comp. Physiol. B 164, 430–437.

Holland, S., Höchel, J., Burmeister, A., Janke, O., Nichelmann, M., 1998. A method for measuring deep body temperature in avian embryos. J. Therm. Biol. 23, 123–129.

Horner, J.R., Padian, K., de Ricqlès, A., 2001. Comparative osteohistology of some embryonic and perinatal archosaurs: developmental and behavioral implications for dinosaurs. Paleobiology 27, 39–58.

Horowitz, M., 2002. From molecular and cellular to integrative heat defense during exposure to chronic heat. Comp. Biochem. Physiol. 131, 475–483.

Houpt, T.R., 2004. Acid base balance. In: Reece, W.O. (Ed.), Dukes' Physiology of Domestic Animals. Cornell University Press, Ithaca, pp. 162–177.

Hulbert, A.J., Else, P., 2005. Membranes and the setting of energy demands. J. Exp. Biol. 208, 1593–1599.

Hulet, R.M., 2007. Managing incubation: where are we and why? Poult. Sci. 86, 1017–1019.

IUPS Thermal Commission, 2001. Glossary of terms for thermal physiology. Jpn. J. Physiol. 51, 245–280.

Janke, O., Tzschentke, B., Boerjan, B., 2004. Comparative investigations of heat production and body temperature in modern chicken breeds. Avian Poult. Biol. Rev. 15, 191–196.

Joseph, N.S., Lourens, A., Moran Jr, E.T., 2006. The effects of suboptimal EST during incubation on broiler chick quality, live performance, and further processing yield. Poult. Sci. 85, 932–938.

Joubert, R., Métayer Coustard, S., Swennen, Q., Sibut, V., Crochet, S., Cailleau-Audouin, E., Buyse, J., Decuypere, E., Wrutniak-Cabello, C., Cabello, G., Tesseraud, S., Collin, A., 2010. The beta-adrenergic system is involved in the regulation of the expression of avian uncoupling protein in the chicken. Domes. Anim. Endocrinol. 38, 115–125.

Katz, A., Meiri, N., 2006. Brain-derived neurotrophic factor is critically involved in thermal-experience-dependent developmental plasticity. J. Neurosci. 12, 3899–3907.

Keil, R., Gerstberger, R., Simon, E., 1994. Hypothalamic integration of osmo- and thermoregulatory signals and the release of osmoregulatory hormones. In: Pleschka, K., Gerstberger, R. (Eds.), Integrative and Cellular Aspects of Autonomic Function: Temperature and Osmoregulation. John Libbey Eurotext, Paris.

Klir, J.J., Heath, J.E., 1992. An infrared thermographic study of surface temperature in relation to thermal stress in three species of foxes: the red fox (*Vulpes vulpes*), arctic fox (*Alopex lagopus*), and kit fox (*Vulpes macrotis*). Physiol. Zool. 65, 1011–1021.

Kontos, M., Wilson, R., Fentiman, I., 2011. Digital infrared thermal imaging (DITI) of breast lesions: sensitivity and specificity of detection of primary breast cancer. Clin. Radiol. 66, 536–539.

Körtner, G., Brigham, R.M., Geiser, F., 2000. Winter torpor in large birds. Nature 407, 318.

Kramer, R., Cohen, D., 2004. Functional genomics to new drug targets. Nat. Rev. Drug. Discov. 3, 965–972.

Krausova, T., Peterka, M., 2007. Teratogenic and lethal effects of 2–24 h hyperthermia episodes on chick embryos. J. Therm. Biol. 32, 193–203.

Kuznetsov, I.E., Kazakov, V.N., 2000. Integration of thermal and osmotic signals in the preoptic/anterior hypothalamic neurons. Neuroscience 99, 363–371.

Kvist, A., Lindstrom, A., 2001. Basal metabolic rate in migratory waders: intra-individual, intra-specific, interspecific, and seasonal variation. Funct. Ecol. 15, 465–473.

Kvist, A., Lindstrom, A., Green, M., Piersma, T., Visser, G.H., 2001. Carrying large fuel loads during sustained bird flight is cheaper than expected. Nature 413, 730–732.

Labunskay, G., Meiri, N., 2005. R-Ras3/(M-Ras) is involved in thermal acquisition in the critical period of thermal control establishment. J. Neurobiol. 66, 56–70.

Landgraf, R., Horn, T., Neumann, I., 1994. Effects of osmotic stimulation of the supraoptic nucleus on central and peripheral release of vasopressin and on baro- and thermoregulation. In: Pleschka, K., Gerstberger, R. (Eds.), Integrative and Cellular Aspects of Autonomic Function: Temperature and Osmoregulation. John Libbey Eurotext, Paris.

Lasiewski, R.C., Dawson, W.R., 1967. A re-examination of the relation between standard metabolic rate and body weight in birds. Condor 69, 13–23.

Lin, J., Handschin, C., Spiegelman, B.M., 2005. Metabolic control through the PGC-1 family of transcription coactivators. Cell Metab. 1, 361–370.

Ludders, J.W., 2004. Respiration in birds. In: Reece, W.O. (Ed.), Dukes' Physiology of Domestic Animals. Cornell University Press, Ithaca, pp. 149–161.

Maclean, G.L., 1996. Adaptation of desert organism. In: Maclean, G.L. (Ed.), The Ecophysiology of Desert Birds. Springer Verlag, Berlin, Heidelberg.

Maloney, S.K., Dawson, T.J., 1993. Sexual dimorphism in basal metabolism and body temperature of a large bird, the emu. Condor 95, 1034–1037.

Marder, J., Arad, Z., 1989. Panting and acid-base regulation in heat-stressed birds. Comp. Biochem. Physiol. 94A, 395–400.

McCafferty, D.J., Gilbert, C., Paterson, W., Pomeroy, P.P., Tompson, C., Currie, J.I., Ancel, A., 2011. Estimating metabolic heat loss in birds and mammals by combining infrared thermography with biophysical models. Comp. Biochem. Physiol. A 158, 337–345.

McKechnie, A.E., 2008. Phenotypic flexibility in basal metabolic rate and the changing view of avian physiological diversity: a review. J. Comp. Physiol. B 178, 235–247.

McKechnie, A.E., Lovegrove, B.G., 2002. Avian facultative hypothermic responses: a review. Condor 104, 705–724.

McKechnie, A.E., Mzilikazi, N., 2011. Heterothermy in afrotropical mammals and birds: a review. Integ. Comp. Biol. 51, 349–363.

McNab, B.K., 1996. Metabolism and temperature regulation of kiwis (Apterygidae). Auk 113, 687–692.

McNab, B.K., Bonaccorso, F.J., 1995. The energetic of Australian swifts, frogmouths, and nightjars. Physiol. Zool. 68, 245–261.

McNabb, F., King, D.B., 1993. Thyroid hormones effects on growth development and metabolism. In: Schreibman, M.P., Scanes, C.G., Pang, P.K.T. (Eds.), The Endocrinology of Growth Development and Metabolism in Vertebrates. Academic Press, New York, pp. 393–417.

McWilliams, S.R., Karasov, W.H., 2004. Migration takes guts: digestive physiology of migratory birds and its ecological significance. In: Marra, P., Greenberg, R. (Eds.), Birds of Two Worlds. Johns Hopkins University Press, Baltimore.

Meir, M., Ar, A., 1987. Improving turkey poult quality by correcting incubator humidity to match eggshell conductance. Br. Poult. Sci. 28, 337–342.

Meiri, N., 2007. 14-3-3ε Expression is induced during the critical period of thermal control establishment. Develop. Neurobiol. 68, 62–72.

Mohler, F.S., Heath, J.E., 1988. Comparison of IR thermography and thermocouple measurement of heat loss from rabbit pinna. Am. J. Physiol. 254, 389–395.

Monteith, J.L., Unsworth, M.H., 1990. Radiation environment. In: Principals of Environmental Physics. Edward Arnold, New York, pp. 53–54.

Morgan, K.R., Chappell, M.A., Bucher, T.L., 1992. Ventilatory oxygen extraction in relation to ambient temperature in four Antarctic seabirds. Physiol. Zool. 65, 1092–1113.

Morrison, S.F., Nakamura, K., Madden, C.J., 2008. Central control of thermogenesis in mammals. Exp. Physiol. 93, 773–797.

Muñoz-Garcia, A., Williams, J.B., 2005. Cutaneous water loss and lipids of the stratum corneum in house sparrows *Passer domesticus* from arid and mesic environments. J. Exp. Biol. 208, 3689–3700.

Nagy, K.A., 1987. Field metabolic rate and food requirement scaling in mammals and birds. Ecol. Monogr. 57, 111–128.

Nakamura, K., 2011. Central circuitries for body temperature regulation and fever. Am. J. Physiol. Regul. Integr. Comp. Physiol. 301, R1207–R1228.

Nakayama, T., Eisenman, J.S., Hardy, J.D., 1961. Single unit activity of anterior hypothalamus during local heating. Science 134, 560–561.

Nichelmann, M., Burmeister, A., Janke, O., Höchel, J., Tzschentke, B., 1998. Avian embryonic thermoregulation: role of Q_{10} in interpretation of endothermic reactions. J. Therm. Biol. 23, 369–376.

Nichelmann, M., Höchel, J., Tzschentke, B., 1999. Biological rhythms in birds—development, insights and perspectives. Comp. Biochem. Physiol. 124A, 437–439.

Nichelmann, M., Janke, O., Tzschentke, B., 2001. Efficiency of thermoregulation in precocial avian species during the pre-natal period. J. Therm. Biol. 26, 273–280.

Ophir, E., Arieli, Y., Marder, J., 2004. The effect of α_2-adrenergic receptors on cutaneous water evaporation in the rock pigeon (*Columba livia*). Comp. Biochem. Physiol. A 139, 411–415.

Parsell, D.A., Lindquist, S., 1994. Heat shock proteins and stress tolerance. In: Morimoto, R.I., Tissieres, A., Georgopoulos, C. (Eds.), Biology of Heat Shock Proteins and Molecular Chaperones. Cold Spring Harbor Laboratory Press, Cold Spring Harbor, pp. 457–494.

Patapoutain, A., Peier, A.M., Story, G.M., Viswanath, V., 2003. ThermoTRP channels and beyond: mechanisms of temperature sensation. Nat. Rev. Neurosci. 4, 529–539.

Peltonen, L., Arieli, Y., Marder, J., 1998. Adaptive changes in the epidermal structure of the heat acclimated rock pigeon (*Columba livia*): a comparative electron microscopy study. J. Morphol. 235, 17–29.

Piersma, T., Drent, J., 2003. Phenotypic flexibility and the evolution of the organismal design. Trends Ecol. Evol. 18, 228–233.

Piestun, Y., Shinder, D., Ruzal, M., Halevy, O., Yahav, S., 2008. The effect of thermal manipulations during the development of the thyroid and adrenal axes on in-hatch and post-hatch thermoregulation. J. Therm. Biol. 33, 413–418.

Piestun, Y., Halevy, O., Yahav, S., 2009. Thermal manipulations of broiler embryos—the effect on thermoregulation and development during embryogenesis. Poult. Sci. 88, 2677–2688.

Porter, R.K., Hulbert, A.J., Brand, M.D., 1996. Algometry of mitochondrial proton leak: influence of membrane surface area and fatty acid composition. Am. J. Physiol. Regul. Integr. Comp. Physiol. 40, R1550–R1560.

Powers, D.R., 1992. Effect of temperature and humidity on evaporative water loss in Anna's hummingbirds (*Calypte annai*). J. Comp. Physiol. B 162, 74–84.

Prinzinger, R., Dietz, V., 1995. Qualitative course of embryonic O_2 consumption in altricial and precocial birds. Res. Physiol. 100, 289–294.

Prinzinger, R., Lübben, I., Schuchmann, K.L., 1989. Energy metabolism and body temperature in 13 sun-birds species (Nectariniidae). Comp. Biochem. Physiol. A 92, 393–402.

Prinzinger, R., Misovic, A., Schleucher, E., 1993. Energieumsatz und Körpertemperatur bei der Zwergwachtel *Coturnix chinensis* und beim Bindenlaufhühnchen *Turnix suscitator*. J. F. Ornithol. 134, 79–84.

Prinzinger, R., Preßmar, A., Schleucher, E., 1991. Body temperature in birds. Comp. Biochem. Physiol. 99A, 499–506.

Prinzinger, R., Schmidt, M., Diets, V., 1995. Embryogeny of oxygen consumption in 13 altricial and precocial birds. Rep. Physiol. 100, 283–287.

Puigserver, P., Spiegelman, P.M., 2003. Peroxisome proliferator-activated receptor-γ coactivator 1α (PGC-1α): transcriptional coactivator and metabolic regulator. Endocr. Rev. 24, 78–90.

Richards, M.P., Proszkowiec-Weglarz, M., 2007. Mechanisms regulating feed intake, energy expenditure, and body weight in poultry. Poult. Sci. 86, 1478–1490.

Ro, J., Williams, J.B., 2010. Respiratory and cutaneous water loss of temperate zone passerine birds. Comp. Biochem. Physiol. 156A, 237–246.

Romijn, C., Lockhorst, W., 1955. Chemical heat regulation in the chick embryo. Poult. Sci. 34, 649–654.

Ruben, J.A., 1995. The evolution of endothermy in mammals and birds: from physiology to fossils. Annu. Rev. Physiol. 57, 69–95.

Saarela, S., Klapper, B., Heldmaier, H., 1995. Daily rhythm of oxygen consumption and thermoregulatory responses in some European winter- or summer-acclimatized finches at different ambient temperatures. J. Comp. Physiol. B 165, 366–376.

Saito, N., Grossmann, R., 1998. Effect of short-term dehydration on plasma osmolality, levels of arginine vasotocin and its hypothalamic gene expression in the laying hen. Comp. Biochem. Physiol. 121A, 235–239.

Schmidt, A., Alard, F., Handrich, Y., 2006. Changes in body temperatures in king penguins at sea: the result of fine adjustments in peripheral heat loss. Am. J. Physiol. Regul. Intgr. Comp. Physiol. 291, R608–R618.

Shaffer, S.A., 2011. A review on seabird energetics using doubly labeled water method. Comp. Biochem. Physiol. A 158, 315–322.

Shinder, D., Rusal, M., Giloh, M., Yahav, S., 2009. The effect of repetitive acute cold exposures at the latest phase of embryogenesis of broilers on cold resistance during life span. Poult. Sci. 88, 636–646.

Shmidt-Nielsen, K., 1975. Respiration in air. In: Shmidt-Nielsen, K. (Ed.), Animal Physiology Adaptation and Environment. Cambridge University Press, New York, pp. 27–79.

Silva, J.E., 2006. Thermogenic mechanisms and their hormonal regulation. Physiol. Rev. 86, 435–464.

Simon, E., Notel, P., 1990. Temperature dependence of thermal and nonthermal regulation: hypothalamus thermo- and osmoregulation in the duck. In: Bligh, J., Voigt, K. (Eds.), Thermoreception and Thermoregulation. Springer, Heidelberg.

Smith, E., Morowitz, J., 2004. Universality in intermediary metabolism. Proc. Natl. Acad. Sci. USA 101, 13168–13173.

Southwick, E.E., 1973. Remote sensing of body temperature in a captive 25 g bird. Condor 75, 464–466.

Stewart, M., Webster, J.R., Schaefer, A.L., Cook, N.J., Scott, S.L., 2005. Infrared thermography as a non-invasive tool to study animal welfare. Anim. Welf. 14, 319–325.

Sturkie, P.D., 1967. Cardiovascular effects of acclimatization to heat and cold in chickens. J. Appl. Physiol. 22, 13–15.

Swann, G.S., Brake, J., 1990. Effect of dry-bulb temperature, relative humidity and egg shell conductance during days seventeen to twenty one of incubation on egg weight loss and chick weight. Poult. Sci. 69, 545–553.

Swanson, D.L., Weinacht, D.P., 1997. Seasonal effects on metabolism and thermoregulation in northern bobwhite. Condor 99, 478–489.

Tazawa, H., Rahn, H., 1987. Temperature and metabolism of chick embryos and hatchlings after prolonged cooling. J. Exp. Zool. (Suppl. 1), 105–109.

Tazawa, H., Wakayama, H., Turner, J.S., Paganelli, C.V., 1988. Metabolic compensation for gradual cooling in developing chick embryos. Comp. Biochem. Physiol. 89A, 125–129.

Tazawa, H., Suzuki, Y., Musashi, H., 1989. Simultaneous acquisition of ECG, BCG, and blood pressure from chick embryos in the egg. J. Appl. Physiol. 67, 478–483.

Tazawa, H., Moriya, K., Tamura, A., Komor, T., Akiyama, R., 2001. Ontogenetic study on thermoregulation. J. Therm. Biol. 26, 281–286.

Teunissen, L.P.J., Daanen, H.A.M., 2011. Infrared thermal imaging of the inner canthus of the eye as an estimator of body core temperature. J. Med. Eng. Technol. 35, 134–138.

Tieleman, B.I., Williams, J.B., 2000. The adjustment of avian metabolic rates and water fluxes to desert environments. Physiol. Biochem. Zool. 73, 461–479.

Tieleman, B.I., Williams, J.B., Michaeli, G., Pinshow, B., 1999. The role of the nasal passages in the water economy of crested larks and desert larks. Physiol. Zool. 72, 219–227.

Tieleman, B.I., Williams, J.B., Buschur, M.E., 2002. Physiological adjustments to arid and mesic environments in Larks (*Alaudidae*). Physiol. Biochem. Zool. 75, 305–313.

Tzschentke, B., Basta, D., Nichelmann, M., 2001. Epigenetic temperature adaptation in birds: peculiarities and similarities in comparison to acclimation. News Biomed. Sci. 1, 26–31.

Tzschentke, B., Basta, D., Janke, O., Maier, I., 2004. Characteristics of early development of body functions and epigenetic adaptation to the environment in poultry: focus on development of central nervous mechanisms. Avian Poult. Biol. Rev. 15, 107–118.

Uni, Z., Gal-Garber, O., Geyra, A., Sklan, D., Yahav, S., 2001. Changes in growth and function of chick small intestine epithelium due to early thermal conditioning. Poult. Sci. 80, 438–445.

Van Brecht, A., Hens, H., Lemaire, J.L., Aerts, J.M., Degraeve, P., Berckmans, D., 2005. Quantification of the heat exchange of chicken eggs. Poult. Sci. 84, 353–361.

Versteegh, M.A., Helm, B., Dingemanse, N.J., Tieleman, B.I., 2008. Repeatability and individual correlates of basal metabolic rate and total evaporative water loss in birds: a case study in European stonechats. Comp. Biochem. Physiol. A. 150, 452–457.

Vidal-Puig, A., Solanes, G., Grujic, D., Flier, J.S., Lowell, B.B., 1997. UCP3: an uncoupling protein homologue expressed preferentially and abundantly in skeletal muscle and brown adipose tissue. Biochem. Biophys. Res. Commun. 235, 79–82.

Walter, I., Seebacher, F., 2009. Endothermy in birds: underlying molecular mechanisms. J. Exp. Biol. 212, 2328–2336.

Weathers, W.W., Koenig, W.D., Stanback, M.T., 1990. Breeding energetic and thermal ecology of the acorn woodpecker in central coastal California. Condor 92, 341–359.

Webb, D.R., 1987. Thermal tolerance of avian embryos: a review. Condor 89, 874–898.

Weber, J.M., 2009. The physiology of long distance migration: extending the limits of endurance metabolism. J. Exp. Biol. 212, 593–597.

Wechselberger, M., Wright, C.L., Bishop, G.A., Boulant, J.A., 2006. Ionic channels and conductance-based models for hypothalamic neuronal thermosensitivity. Am. J. Physiol. Regul. Integr. Comp. Physiol. 291, R518–R529.

West, G.C., 1965. Shivering and heat production in wild birds. Physiol. Zool. 38, 111–120.

Whittow, G.C., 1976. Regulation of body temperature. In: Sturkie, P.D (Ed.), Avian Physiology, third ed. Springer Verlag, New York, pp. 146–173.

Whittow, G.C., 1986. Regulation of body temperature. In: Sturkie, P.D. (Ed.), Avian Physiology, fourth ed. Springer Verlag, New York, pp. 221–252.

Whittow, G.C., Sturkie, P.D., Stein Jr, G., 1964. Cardiovascular changes associated with thermal polypnea in the chicken. Am. J. Physiol. 207, 1349–1353.

Whittow, G., Tazawa, H., 1991. The early development of thermoregulation in birds. Physiol. Zool. 64, 1371–1390.

Williams, J.B., 1996. A phylogenetic perspective of evaporative water loss in birds. Auk 113, 457–472.

Williams, J.B., du Plessis, M.A., Siegfried, W.R., 1991. Green woodhoopoes (*Phoeniculus purpureus*) and obligate cavity roosting provide a test of the thermoregulatory insufficiency hypothesis. Auk 108, 285–293.

Williams, J.B., Tieleman, I., 2002. Ecological and evolutionary physiology of desert birds: a progress report. Integr. Comp. Physiol. 42, 68–75.

Wilson, H.R., 1991. Physiological requirements of the developing embryo: temperature and turning. In: Tullet, S.G. (Ed.), Avian Incubation. Butterworth-Heinemann, Oxford, pp. 145–156.

Wojciechowski, M.S., Pinshow, B., 2009. Heterothermy in small, migrating passerine birds during stopover: use of hypothermia at rest accelerates fuel accumulation. J. Exp. Biol. 212, 3068–3075.

Wolf, B.O., Walsberg, G.E., 1996. Respiratory and cutaneous evaporative water loss at high environmental temperatures in a small bird. J. Exp. Biol. 199, 451–457.

Yahav, S., 2000. Domestic fowl—strategies to confront environmental conditions. Poult. Avi. Biol. Rev. 11, 81–95.

Yahav, S., 2009. Alleviating heat stress in domestic fowl—different strategies. Worlds Poult. Sci. J. 65, 719–732.

Yahav, S., Giloh, M., 2012. Infrared thermography—applications in poultry biological research. In: Prakash, R.V. (Ed.), Infrared Thermography. Intech Publications, pp. 93–116.

Yahav, S., Goldfeld, S., Plavnik, I., Hurwitz, S., 1995. Physiological responses of chickens and turkeys to relative humidity during exposure to high ambient temperature. J. Therm. Biol. 20, 245–253.

Yahav, S., Straschnow, A., Plavnik, I., Hurwitz, S., 1997. Blood system response of chickens to changes in environmental temperature. Poult. Sci. 76, 627–633.

Yahav, S., Straschnow, A., Luger, D., Shinder, D., Tanny, J., Cohen, S., 2004. Ventilation, sensible heat loss, broiler energy, and water balance under harsh environmental conditions. Poult. Sci. 83, 253–258.

Yahav, S., Shinder, D., Tanny, J., Cohen, S., 2005. Sensible heat loss—the broiler's paradox. World's Poult. Sci. J. 61, 419–435.

Yahav, S., Shinder, D., Ruzal, M., Gilo, M., Piestun, Y., 2009. Controlling body temperature—the opportunities for highly productive domestic fowl. In: Cisneros, A.B., Goins, B.L. (Eds.), Body Temperature Control. NovaScience Publishers, New York, pp. 65–98.

Zhou, W., 2000. Physiological significance of the change in blood viscosity of broiler chickens under high ambient temperature. Jpn. Poult. Sci. 37, 201–211.

Zimerman, I., Piestun, Y., Yahav, S., 2013. The effect of thermal manipulations during embryogenesis on performance and FCR of turkeys raised up to marketing age. In "Turkeys Production" European Working Group No 10. Berlin (abstract).

Chapter 38

Avian Molting

Alistair Dawson
NERC Centre for Ecology & Hydrology, Midlothian, Edinburgh, UK

38.1 INTRODUCTION

Good-quality plumage is essential for the survival of birds. However, feathers wear out and need to be replaced normally on an annual basis. This process of molt is one of the key life-history stages of birds' annual cycles. Despite the critical importance of molt and research over many decades, we know remarkably little about its physiological control. Voitkevich wrote a 335-page treatise on the subject (translated from Russian and published in 1966). Payne wrote later in a review (1972): "The physiological bases of the control of the time and rate of molting and the integration of molt into the annual cycle are of considerable interest, and their discovery remains a challenge to the comparative physiologist"—a challenge that remains (Bridge, 2011; Kuenzel, 2003).

38.2 ANATOMICAL AND ECOLOGICAL CONSIDERATIONS

38.2.1 Feathers

Feathers are the characteristic feature of birds, with birds being defined as feathered vertebrate animals. Birds evolved from reptiles, and there is evidence that dinosaurs had feathers (Zelenitsky et al., 2012) and that the ontogeny of their feathers resembled that of modern birds (Xu et al., 2010).

The plumage of birds—the collective name for the covering of feathers—serves a number of functions. It provides insulation and waterproofing, in an analogous way to pelage in the other group of homeotherms, the mammals. Uniquely and crucially, it also enables most birds to fly by providing surfaces that give aerodynamic lift on the wings and tail. The contoured body shape reduces air resistance during flight; in waterbirds, plumage provides buoyancy. The color of plumage can provide camouflage or patterns that are used for sexual or aggressive displays.

To provide these functions, the plumage is made up of a variety of specialized feather types: flight, contour, semiplume, down, bristle, and filoplume (Ginn and Melville, 1983). All feathers are constructed largely of keratin and are essentially two-dimensional with a tapering central shaft, the base of which is embedded in the feather socket or follicle (Voitkevich, 1966). In the contour feathers, which include the flight feathers, the base of the shaft is hollow, cylindrical, and devoid of side branches. The rest of the shaft (rachis) is solid and bears rows of parallel side branches (barbs). Each barb has two rows of smaller side branches (barbules). There are two forms of barbule: those that point towards the tip of the feather are laterally flattened with hooked projections, and the others which lack the hooked projections. The hooks on one barbule link onto the adjacent barbule. This linkage system forms a two-dimensional lattice, the feather vane, which is light but strong. Other forms of feather differ in that they do not have the hooked projections on the barbules so that the barbules remain unlocked and downy. The color of feathers can be provided in two ways: (1) the feathers can contain pigments, of which there are three types, melanins, carotenoids and porphyrins; and (2) the structure of the feather can provide either iridescent or noniridescent color.

The feathers are distributed into seven or eight distinct tracts (pterylae) with areas between them (apteria) that are devoid of feathers or have only sparse coverage. In some species, such as penguins, the apteria are few and small so that feather coverage is almost even over the whole body. Even in species with significant apteria, these are not obvious externally because feathers in adjacent pterylae completely overly them. The total number of feathers on a bird varies widely, with large birds tending to have more. However, feather density also varies, which may be related to insulation requirements; penguins have particularly high densities, and, within species, feather density can be higher during winter.

Sturkie's Avian Physiology.

Each feather is formed from a cone-shaped structure deep in the feather follicle. Development of the complex feather structure may be the result of changing topology of stem cells within the follicle (Yue et al., 2005). The first generation of feathers forms at different rates in different species. In many species, particularly precocial species such as domestic poultry, the first feathers develop before hatching, and there are subsequent molts before birds are fully grown. In most altricial species, birds hatch naked and their juvenile plumage develops during the nestling period (Samson, 1976). There may, or may not, be a postfledging molt into adult plumage.

The importance of good quality plumage means that birds spend a significant portion of each day in feather care. The bill is used to preen—to clean the feathers, restore their structure, and apply preen oil from the uropygial gland. Nevertheless, feathers are dead structures; they degrade with time and cannot be repaired. They become bleached through the effects of UV radiation, abraded through contact with adjacent feathers and possibly vegetation in the bird's habitat; they may break and they will probably degrade as a result of infestation with ectoparasites. Because good-quality plumage is essential for survival, feathers need to be replaced on a regular basis. This renewal process is molt. During molt, there is also regeneration of other epidermal structures, particularly the bill and claws.

The molting process is the shedding of old feathers and the development of new ones in their place. It is thought that development of a new feather pushes out the old feather. However, if an old feather is pulled out, it is often replaced by a new one, suggesting the reverse—that it is the shedding of the old feather that stimulates development of a new feather.

38.2.2 Molt

All birds molt at least once during the year. The plumage that birds have during the nonbreeding and overwintering period is called the basic plumage. The molt that results in this plumage is therefore termed the prebasic molt; this often, but not always, occurs after breeding and hence is sometimes called the postnuptial molt (Ginn and Melville, 1983; Jenni and Winkler, 1994). The prebasic molt is the subject of this chapter. Some species of birds additionally undergo a prenuptial (or prealternate) molt into the breeding, or alternate, plumage. This molt involves some of the head and/or body plumage and occurs during spring in advance of breeding and brightens the appearance of the bird or emphasizes a "badge".

During the prebasic molt, all feathers are eventually replaced (except in some large birds), but this happens in a strict bilaterally symmetrical sequence so that the various functions of plumage are not normally compromised. Molt often starts with the shedding and replacement of the proximal primary feather, followed by the next primary and so on to the end of the wing. Normally, only a few primaries are in the process of molt at any one time so that flight performance is not compromised. If birds are under time constraint to finish molt rapidly, this is achieved by molting more feathers at once, rather than increasing the growth rate of each feather. The two wings are usually at exactly the same stage of molt. After primary feather molt has started, molt of the other feather tracts proceeds, again symmetrically. Molt of the other tracts is normally completed before the end of the primary molt. As a general rule, larger birds tend to take longer to molt than smaller birds. Rohwer et al. (2009) showed that, due to allometric constraints, large birds may not have time to molt all of their flight feathers sequentially each year and so must adopt one of two strategies. Some large waterbirds, such as ducks, swans, grebes, pelicans, and auks, molt all of their flight feathers simultaneously. During this period, they cannot fly and complete the process on lakes, estuaries, or at sea in order to minimize their vulnerability to predators. Other large birds—those that depend on flight to feed—molt only some of their flight feathers each year, and require 2 or even 3 years to complete molt. Some species have an additional full molt each year (Prys-Jones, 1991; Underhill et al., 1992) and some have partial repeated molts (Bridge et al., 2007).

In the U.S. poultry industry, induced molting is used to increase or rejuvenate egg production (Bell, 2003; Berry, 2003; Webster, 2003). Induced molting and ovarian regression are achieved by feeding a nutritionally deficient diet and reducing photoperiod. The minimum photoperiod is 8 h of day light and water is available *ad libitum*. Subsequent gonadal recrudescence is achieved by restoring a nutritionally replete diet and stimulatory photoperiod. This practice is not permitted in the European Union.

In many species, the plumage is sexually dimorphic, often with the male having the brighter showy plumage. In species that molt only once per year, the prebasic molt, the sexual dimorphism is under genetic control (Owens and Short, 1995). Castration of males or ovariectomy of females has no effect on the plumage. Similarly, treatment with exogenous testosterone or estrogen is without effect. In those species in which the males undergo a prenuptial molt in spring from a dull winter plumage into bright breeding plumage, it is frequently assumed that this is testosterone-dependent. This is not the case. Unlike mammals, in birds it is the males that are homogametic, so the showy plumage is the default state. Estrogens block the subsequent production of the bright male plumage. Castration of males has no effect on this plumage, but ovariectomy of females causes them to develop the bright male plumage. Thus, showy male plumage is not the result of masculinization by testosterone (Owens and Short, 1995).

38.2.3 Molt as a Life History Stage

Birds, as other animals, live in conditions that change on an annual cycle; to maximize fitness, they need to retain phenotypic flexibility. They switch from one life history stage to the next in a one-way sequence. Molt is important for birds' survival; birds that have been experimentally prevented from molting by treatment with testosterone (Nolan et al., 1992) or that have had to molt rapidly so that they had poor quality plumage (Nilsson and Svensson, 1996) survive less well over winter. Therefore, all birds have annual cycles in which molt features as a life history stage. In nonmigratory birds, there may be just three basic life history stages: winter/nonbreeding, breeding, and molt (Wingfield, 2008). In migratory species, there are two additional stages: a vernal migration (which has to precede breeding) and an autumn migration (which follows breeding).

Breeding and molt are both energetically demanding and take time (Murphy and King, 1992), and molt impairs flight performance (Swaddle and Witter, 1997; Tucker, 1991). Therefore, the prebasic molt is usually, but not always, delayed until breeding has finished (Hemborg, 1999; Hemborg and Merila, 1999; Morton, 1992). Even in equatorial or neotropical species, prebasic molt may closely follow breeding (Dittami and Gwinner, 1985; Moore et al., 2005; Wikelski et al., 2003). Consequently, in these situations, breeding birds start to molt later than nonbreeders; females often start to molt later than males, and birds with extended or late breeding molt later than those that curtailed breeding activity earlier (Morton and Morton, 1990; Newton, 1966; Newton and Rothery, 2000). Some temperate and high-latitude species are opportunistic breeders—they can breed at widely differing times in different years depending on the availability of food. Nevertheless, molt occurs at a comparatively fixed time of year and breeding does not occur during this period (Hahn, 1998). Some species, however, do overlap breeding and molt. This is more common in larger species because the time required for rearing young, as well as the time required to replace flight feathers, increases with body size. For example, in some raptors, the female starts to molt while she is incubating and being fed by her mate, so she does not need to fly (Young et al., 2009). It is also common in species with long breeding seasons and in low-latitude opportunistic breeders. In such cases, the rate of molt is slow during breeding and few flight feathers are replaced at a time, presumably to minimize the energetic and flight performance costs of molting on reproduction (Echeverry-Galvis and Hau, 2012; Hahn et al., 1992).

Migration is also energy demanding and, of course, molt impairs flight efficiency. Therefore, molt and migration tend also to be temporally separated, but there are a variety of strategies to achieve this (Barta et al., 2008). Some species molt immediately after breeding on the breeding grounds, whereas others migrate to their wintering grounds before molt is initiated (Kjellén, 1994). Some species molt after breeding before migration and again after migration on the wintering grounds, and others interrupt migration in order to molt. The scheduling of molt is the result of a tradeoff between having a high feather quality during breeding versus during the nonbreeding period (Holmgren and Hedenstrom, 1995) and the comparative availability of food on the breeding or wintering grounds (Barta et al., 2008).

Because life history stages tend to be temporally separate, excess time taken over one will incur constraints on the time available for subsequent stages. In particular, birds which breed late or for longer have less time available for the prebasic molt. Data on the duration of molt in free-living birds are rare, probably because of the difficulty of retrapping the same individuals over the period of molt. Nevertheless, several studies have shown that birds that start to molt late will molt more rapidly (Bojarinova et al., 1999; Morton and Morton, 1990; Newton, 1966). During a normal molt, new feather mass is accumulated at a fairly constant rate (Dawson, 2003; Seel, 1976; Underhill and Joubert, 1995). When birds molt more rapidly, they achieve this by growing more feathers simultaneously (Dawson, 2004). As a result, feather quality is compromised (Dawson et al., 2000), which leads to a subsequent decrease in survival. The timing of the end of breeding and the start of molt is a tradeoff between reproduction and individual survival.

38.3 ENVIRONMENTAL AND PHYSIOLOGICAL CONTROL

Physiology has to control how an old feather is shed, how a new feather starts to grow, and the appropriate development of that feather. It has to control the sequence of molt, when molt of each feather tract starts and how it progresses along that tract, and control the bilateral symmetry of the progress of molt. It has to integrate environmental cues, photoperiodic and nonphotoperiodic, so that molt starts at the appropriate time and progresses at the optimum rate. We have very little understanding of the process(es) involved.

38.3.1 Photoperiodic and Nonphotoperiodic Control

In free-living birds within the tropics, the timing of prebasic molt may be dependent on the timing of breeding. For example, in equatorial stonechats (*Saxicola torquata axillaris*), molt follows breeding, but the cues that control the time of breeding are unclear (Dittami and Gwinner, 1985). Two equatorial populations of rufous-collared sparrows (*Zonotrichia capensis*), at the same latitude and separated by just 25 km, breed seasonally but out of phase with each other and also molt out of phase (Moore et al., 2005). However, for birds outside the tropics, the timing of prebasic molt within the annual cycle, like other life history stages, is largely controlled by the annual cycle in photoperiod. The timing of the start of molt, the subsequent rate of molt, and hence the duration of molt, are largely under photoperiodic control. Experimental manipulation of photoperiod for captive birds gives insight into how this may operate.

38.3.1.1 Start of Molt

The start of molt is usually closely associated with the end of breeding and regression of the reproductive system, which in turn is dependent on the development of absolute or relative (Nicholls et al., 1988). In general, birds do not molt under short photoperiods—birds need to experience long photoperiods. In some species, those that become absolutely photorefractory, molt starts at some time after exposure to long photoperiods, and in such cases the time until the start of molt decreases as photoperiod increases. For example, white-crowned sparrows (*Zonotrichia leucophrys gambelii*) moved from a short mid-winter photoperiod to 20h of light per day (20L:4D) started to molt sooner than birds transferred to 16L:8D (Chilgren, 1978). Starlings (*Sturnus vulgaris*) moved from 8L:16D to 18L:6D started to molt sooner than birds moved to 13L:11D (Dawson, 1994). However, there are differences in the length of photoperiod needed to induce molt. White-crowned sparrows moved to 12L:12D do not molt (Chilgren, 1978; Donham et al., 1983; Farner et al., 1980). Starlings moved to 11L:13D never molt, but birds moved to 11.5L:12.5D (Dawson, 2007) or 12L:12D do molt and then may undergo repeated circannual cycles of gonadal maturation, regression, and molt. Such repeated cycles have been shown in a number of species held on constant neutral photoperiods (Dolnik and Gavrilov, 1980; Guyomarc'h and Guyomarc'h, 1995; Gwinner, 2003). Starlings moved to 13L:11D molt once but not thereafter (Gwinner et al., 1989). However, if starlings are moved after completion of molt from 13L:11D to 18L:6D, another molt is induced (Dawson, 1994).

In species that become relatively photorefractory, such as Japanese and European quail (*Coturnix coturnix*), birds need to experience a decrease in photoperiod after a period of exposure to long photoperiods to induce molt to start (Boswell et al., 1993; Robinson and Follett, 1982). There are also intermediate situations, such as house sparrow *Passer domesticus* (Dawson, 1991, 1998; Hahn and Ball, 1995) song sparrows *Melospiza melodia* (Wingfield, 1993), and Cardueline finches (Pereyra et al., 2005), where birds will start to molt spontaneously at some time after transfer to a constant long photoperiod, but in which a decrease in photoperiod will advance the onset of molt.

38.3.1.2 Rate of Molt

Under different constant long photoperiods, although molt starts sooner under longer photoperiods, the subsequent rate of molt is similar (Dawson, 1994). However, changes in the photoperiod after the start of molt affect the rate of molt. An increase in photoperiod slows the rate of molt, a decrease to a shorter photoperiod increases the rate (Dawson, 1994, 1998; Kobayashi, 1953a), but a decrease to a short photoperiod can stop further progress of molt. Most species of birds in temperate latitudes molt after the summer solstice; individuals molting later in the year will be subject to shorter days and a more rapid decrease in photoperiod during their molt. Presumably this accounts for the shorter molt durations seen in late breeding birds (see Section 38.2.3).

38.3.1.3 Natural Photoperiods

All of the above experimental data were obtained from birds transferred from one fixed photoperiod to another. This situation would never be experienced by free-living birds. In such birds, molt normally occurs after the breeding season. In birds with short breeding seasons early in the year, molt starts before the summer solstice when photoperiod is still increasing. In birds with longer breeding seasons, molt starts during decreasing photoperiods. How is the timing of molt controlled by natural changes in photoperiod? One way to examine this is to look at the timing of molt within populations of the same species at different latitudes. Miller (1960) showed that the end of the period of active spermatogenesis, which is associated with the start of molt, within species and subspecies of the genus *Zonotrichia* ended at the same time at all latitudes. Two independent studies investigated the timing of molt in California quail (*Lophortyx californicus*). Genelly (1955) studied free-living quail on San Pablo Ridge, California (38°N) and Anthony (1970) studied them along the Snake River, southeastern Washington (47°N). In both studies, molt started at the same time: mid-June for males and mid-July for females. In the white-crowned sparrows of the Pacific seaboard (*Zonotrichia leucophrys nuttalli* and *Zonotrichia leucophrys pugetensis*), males begin molting earlier than females by as much as 2weeks; for both sexes, the date on which molt begins is independent of latitude. However, molt duration decreases northward by an average of 2.6days per degree of latitude between the southernmost (35.2°N, molt duration 83days) and northernmost (48.9°N, 47days) limits of the breeding range (Mewaldt and King, 1978). In common starlings, molt starts in early June in birds at 37°N and at 52°N (Dawson, 2013 Rothery et al., 2001) and blackbirds *Turdus merula* in the north of England start and finish molt at the same time as birds in southern England (Dawson, 2013). In all of these studies, birds at different latitudes will have experienced different changes in photoperiod; both absolute photoperiod and the rate of change in photoperiod will have differed, and so neither can directly control the timing of molt.

The conclusion that molt starts at the same time at different latitudes is confirmed by studies in which captive birds are kept under photoperiods simulating annual changes at different latitudes. In starlings on photoperiods simulating 9°N or 52°N, molt occurred at the same time (Dawson, 2007). In siskins (*Carduelis spinus*) exposed to natural photoperiods simulating 40°N, 55°N, and 70°N, which encompass the natural breeding range of this species, molt occurred at the same time in all three groups (Newton and Dawson, 2011). Similarly, greenfinches (*Carduelis chloris*) held on photoperiods simulating natural cycles at 20°N and 60°N showed testicular regression and molt at the same time (Figure 38.1). These data suggest that birds are not using absolute photoperiod, or rate of change in photoperiod, to time gonadal regression and molt. They suggest that birds are using photoperiod in some more subtle way. They also suggest that the earlier start to molt in birds transferred to longer constant photoperiods is an artifact of the experimental design.

38.3.1.4 Nonphotoperiodic Environmental Control

Although the annual cycle in photoperiod is the major environmental cue used to time the start of molt, other supplementary information can be important (Dawson, 2008). Because the start of molt is closely associated with the end of breeding and gonadal regression, any factors affecting this will also affect molt. Unfavorable conditions that delay breeding, or favorable conditions that allow birds to raise an extra brood, will consequently delay the start of molt. Lower temperature delays the timing of regression and the end of breeding, resulting in a later start to molt, with the consequence that the subsequent molt occurs more rapidly (Dawson, 2005; Visser et al., 2011; Figure 38.1).

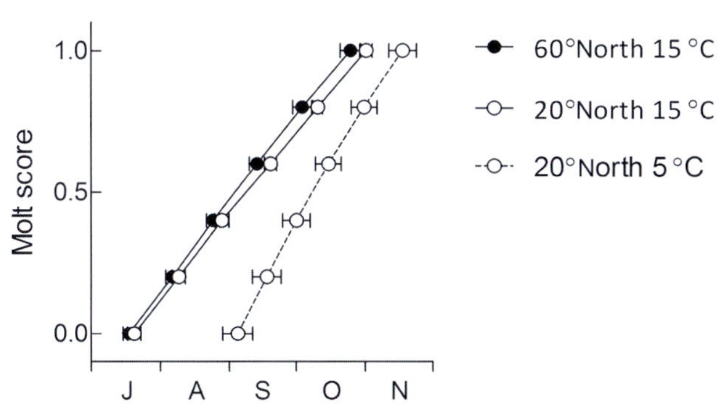

FIGURE 38.1 Molt in greenfinches (*Carduelis chloris*). Three groups of captive greenfinches were kept under photoperiodic regimes simulating latitudes of 60°N or 20°N. Two groups, one at each latitude, were held at a constant temperature of 15 °C and the other, at 5 °C. Molt is expressed as a proportion of total new primary feather mass (Dawson and Newton, 2004). The groups simulating 60°N and 20°N, at the same temperature, both started to molt at the same time (during July) and had the same duration of molt. The group at the lower temperature, 5 °C, started molt considerably later than the group at 15 °C and 20°N, and then molt proceeded more rapidly.

38.3.2 Physiological Control

There must be physiological processes that translate photoperiodic information into control of the whole process of molt: shedding of a feather, growth of the replacement feather, the sequence of this process along each feather tract, lateral symmetry in the progress of molt, timing of the start of molt, and subsequent rate of molt. The first studies on the endocrine control of molting were done nine decades ago, yet to date we still have little understanding of these processes. There is often a close temporal correlation between the end of breeding and the start of molt. This may mean that both share a common physiological trigger or, possibly, that molt is a comparatively passive event that is otherwise inhibited by an active reproductive state. Evidence for the latter view is that a complete molt can be triggered in a chronically reproductively inactive bird in the absence of a reproductive period. At least three hormone systems have been shown to play some role—thyroid hormones, steroid hormones, and prolactin—and there may be interactive effects between these.

38.3.2.1 Evidence for the Involvement of Thyroid Hormones

There is clear and long-standing evidence that thyroid hormones are in some way involved in molt (Torrey and Horning, 1922). Removal of the thyroid glands before transfer to a long photoperiod, which would normally induce a molt, prevents molt. Acute treatment with exogenous thyroid hormones can induce molt at times when a molt would not naturally occur (Himeno and Tanabe, 1957). However, molt is closely associated with gonadal regression and photorefractoriness, and thyroid hormones have long been known to be involved in these processes (Benoit, 1936; Boulakoud and Goldsmith, 1991; Chaturvedi and Meier, 1989; Dawson, 1989; Follett and Nicholls, 1985; Goldsmith and Nicholls, 1984; Lien and Siopes, 1993; Reinert and Wilson, 1996; Wilson and Reinert, 1993; Woitkewitsch, 1940). Consequently, a specific role for thyroid hormones in molt, rather than molt being a consequence of reproductive changes, cannot be substantiated. Exogenous thyroid hormones can to some extent act to mimic the effects of a longer photoperiod. For example, starlings held on a short photoperiod (8L:16D) and treated with exogenous thyroxine show enhanced gonadal maturation but do not molt. Starlings held on a photoperiod of 11L:13D and treated with thyroxine become photorefractory and molt as if photoperiod was greater than 12L:12D. Thyroidectomizing birds just before transferring them to long photoperiods does not apparently affect gonadal maturation but does prevent the subsequent onset of photorefractoriness and molt, suggesting a specific role for thyroid hormones in one or both of these processes. However, the full effects of thyroidectomy take some time. In chronically thyroidectomized starlings, all photoperiodic responses become suppressed: photoperiodically-induced gonadal maturation, the onset of photorefractoriness, and molt. In other words, thyroidectomized birds become photoperiodically blind (Dawson, 1993).

Several studies have measured changes in concentrations of thyroxine and triiodothyronine following change in photoperiod and during the annual cycle. In general, thyroxine concentrations tend to be higher under long photoperiods but with no consistent result in relation to molt. Scanes et al. (1979) and Lien and Siopes (1993) found that thyroxine was maximal during molt in turkeys (*Meleagris gallopavo*). Bentley et al. (1997) investigated whether the long-photoperiod-induced increase in plasma thyroxine is necessary for the induction of photorefractoriness and molt. Photosensitive starlings were thyroidectomized, given thyroxine in their drinking water at concentrations that resulted in plasma thyroxine at short-photoperiod physiological concentrations or lower, and then transferred to long photoperiods. The group with short-photoperiod concentrations of plasma thyroxine became photorefractory and molted at the same time as intact controls transferred to long photoperiods. It would appear, therefore, that thyroid hormones need to be present and have a passive role in all photoperiodically-controlled events rather than a direct causal role in molt.

Not only do thyroid hormones need to be present for molt to start, they are also needed for normal feather formation. Thyroid hormones are involved in somatic growth and development. Birds thyroidectomized soon after hatching fail to grow successfully, their plumage is slow to develop, and feathers grow abnormally (Dawson et al., 1987, 1994; Voitkevich, 1966). The barbs on normal contour feathers have highly asymmetric barbules (see Section 38.2.1). The barbs of feathers from thyroidectomized birds have long smooth symmetrical barbules, so adjacent barbs do not link together and the feathers appear more like down. This feather structure is characteristic of adult ratites, such as ostriches and emus. The neotenous condition of ratites (de Beer, 1956) may be related to hypothyroidism (Dawson et al., 1996). Because renewal of feathers during molt is essentially the same as growth of first feathers, this too can be considered a developmental process. Interestingly, if thyroidectomized birds are transferred to a long photoperiod, they do not molt. If they are treated with an acute dose of thyroxine, this does trigger a molt to start. Although the exogenous thyroxine will rapidly be cleared from the circulation, molt progresses to completion. However, in the absence of thyroxine, the feathers grown develop the downy structure described above (Dawson et al., 1994).

38.3.2.2 Evidence for the Involvement of Gonadal Steroids

Because birds aim to start molt as soon as breeding has finished, but at the same time avoid breeding/molt overlap as much as possible, it is intuitive to think that a reproductive hormone could act to inhibit the start of molt. And, indeed, that is true. As with thyroid hormones, the first indication that gonadal steroid hormones may influence molt dates back many decades (Greenwood and Burns, 1940; Kobayashi, 1954). Treatment with implants of exogenous testosterone prevents or delays the onset of molt (Dawson, 1994, 2004; Nolan et al., 1992; Schleussner et al., 1985). If the implants are then removed during the period that molt would have naturally progressed, molt starts and then proceeds more rapidly, in a way analogous to birds in which the start of molt is delayed by late breeding (Morton and Morton, 1990; Newton, 1966). If the implant is removed after the normal duration of molt, when photoperiod has decreased, then molt does not start. If testosterone is implanted after molt has started, this stops the progress of molt; if the implant is removed, molt either resumes from where it was stopped or a new molt starts. Thus, testosterone inhibits molt during the period of normal photoperiodically timed molt; a decrease in testosterone is not the stimulus that triggers molt. Castrated birds also molt at the normal time of year, strongly suggesting that changes in circulating gonadal steroid concentrations are not implicit in the control of the timing of molt (Dawson and Goldsmith, 1984; Dawson and Sharp, 2010; Greenwood and Burns, 1940).

38.3.2.3 Evidence for the Involvement of Prolactin

As with thyroid hormones and gonadal steroid hormones, the first suggestion that prolactin may play a role in molt goes back several decades (Juhn and Harris, 1958; Kobayashi, 1953b). There is considerable circumstantial evidence suggesting that prolactin has a role. Several studies have investigated seasonal cycles in prolactin secretion in free-living birds: rooks (*Corvus frugilegus*; Lincoln et al., 1980), starlings (Dawson and Goldsmith, 1982), white-crowned sparrows (Hiatt et al., 1987), song sparrows (Wingfield and Goldsmith, 1990), great tits (*Parus major*; Silverin et al., 1997), white-winged crossbills (*Loxia leucoptera*; Deviche and Sharp, 2001). Other studies have investigated captive birds under natural changes in photoperiod: starlings (Dawson and Goldsmith, 1984), grey partridge (*Perdix perdix*; Sharp et al., 1986), European quail (Boswell et al., 1996), and turkeys (Lien and Siopes, 1993). In each of these, peak prolactin concentrations were found at or just before the start of molt, or when molt is known to occur if it was not recorded in the study. In starlings held on a simulated natural cycle in photoperiod at different temperatures, birds at the higher temperature started to molt before those at the lower temperature, but in each case, peak prolactin coincided with the start of molt (Dawson, 2005).

Prolactin secretion is photoperiodically controlled. In starlings under experimental photoperiodic manipulations, prolactin only increases to a peak under photoperiods that are also long enough to induce a molt (i.e., equal to or greater than 12L:12D), and then peak prolactin concentrations coincide fairly closely with the start of molt (Dawson and Goldsmith, 1983). Under constant photoperiods of 12L:12D, starlings undergo repeated circannual cycles of gonadal maturation and molt; in this condition, prolactin also cycles with peak values coinciding with the onset of molt (Dawson, 1997). Importantly, when starlings were actively immunized against vasoactive intestinal peptide (the prolactin-releasing hormone in birds) and transferred to a long photoperiod, the increase in normal prolactin was prevented and these birds did not molt (Dawson and Sharp, 1998).

In these studies on free-living and captive experimental birds, high prolactin concentrations and the start of molt both followed breeding (or gonadal maturation and regression) and so it is impossible to differentiate a direct role for prolactin from a temporal coincidence. However, Dawson (2006) exposed starlings to photo-schedules that challenged the normally close relationship between gonadal regression and molt. Photorefractory birds on moderately long photoperiods that had completed molt were moved to longer photoperiods or briefly to short photoperiods and then back to the longer photoperiod. This induced renewed prolactin secretion and another molt, without renewed gonadal maturation. Therefore, there was no consistent relationship with gonadal regression and the start of molt (molt could be triggered in the absence of a gonadal cycle), nor was there a consistent relationship between photoperiod (or the increase in photoperiod) and the timing of the start of molt. However, there was always an association between the start of molt and prolactin. In all cases where molt was induced, there had been an earlier increase in prolactin, strongly suggesting a causal role for prolactin. However, the timing of molt was related to the time of peak prolactin, not the magnitude of that peak.

The temporal coincidence between the time of peak prolactin and the start of molt, and the absence of molt in the absence of prolactin, suggests that high prolactin concentrations could be instrumental in initiating molt. However, there are problems with this conclusion. Although prolactin secretion is photoperiodically controlled, secretion increases much more in incubating birds and highest values occur during incubation (Dawson and Goldsmith, 1985; Goldsmith, 1982; Hiatt et al., 1987; Wingfield and Goldsmith, 1990). Yet, incubating birds do not start to molt during or soon after incubating their clutch, and multibrooded birds certainly do not start to molt soon after incubating their first clutch. In fact, the reverse is often true—breeding activity can delay the onset of molt (see Section 38.2.3).

However, the fact that peak prolactin coincides with the start of molt also means that molt starts at or about the time that prolactin starts to decrease, and molt progresses during the period of decreasing prolactin. So, it may not be the case that high concentrations of prolactin stimulate molt; the reverse may be true. High concentrations of prolactin may inhibit molt, so that molt only starts when prolactin begins to decrease. This would still result in a close correlation between peak prolactin and the start of molt and may be a mechanism that minimizes breeding/molt overlap, but at the same time ensures that molt starts as soon as breeding activity has finished. The pattern of molt in mute swans (*Cygnus olor*) offers a model to test this. As in many species, molt is delayed in breeding birds compared to nonbreeders. In addition, in successfully breeding swans, males initiate molt 4–6 weeks after their mate (Czapulak, 2002; Heinroth, 1911; Scott, 1972), whereas in nonbreeding flocks males and females molt at about the same time (Coleman et al., 2002). In breeding pairs that lose their young, both sexes molt at the same time (McCleery et al., 2007). It is possible that males with dependent young delay their molt until the female has well-grown feathers so that at least one of the parents has fully feathered wings to protect the young. The changes in prolactin correspond to these different molt patterns. The seasonal decrease in prolactin in nonbreeding birds of both sexes starts at the end of May and is associated with the initiation of molt 4 weeks later. The decrease in plasma prolactin in breeding females is more pronounced, as a consequence of increased prolactin secretion associated with incubation behavior (only females incubate), but also starts at the end of May, and is associated with the onset of molt 6 weeks later. In breeding males, plasma prolactin increases at the end of May when they start to care for their newly hatched cygnets. Correspondingly, prolactin began to decrease 3–5 weeks later in males than in females. These males start to molt in mid-August, at least 4 weeks later than females (Dawson et al., 2009).

This supports the view that molt is related to decreasing plasma prolactin. Prolactin secretion is tightly controlled by photoperiod, and is further stimulated by incubation behavior. This would explain why molt only occurs under long photoperiods and only starts after an increase in photoperiod. It would also explain why once molt has started, a decrease in photoperiod accelerates the rate of molt. In breeding birds it would explain why the start of molt tends to be delayed until the end of breeding, and why in late breeding birds molt then progresses more rapidly (because photoperiod would be shortened and decreasing more rapidly later in the year). However, the evidence to support this is still only correlational. There is no direct evidence showing that exogenous prolactin controls molt, except for contradictory early papers that show that prolactin inhibits molt (Kobayashi, 1953b) or stimulates molt (Juhn and Harris, 1958). It is difficult to understand how a decrease in a hormone concentration can have an effect, but this does seem to be the case since a complete absence of photoperiodically-induced prolactin secretion prevents molt (Dawson and Sharp, 1998).

38.3.2.4 Interrelationships between Thyroid, Gonadal Steroids, and Prolactin

The evidence for the involvement of prolactin is strong. Nevertheless, both thyroid hormones and gonadal steroids also have effects. Are there interrelationships that can rationalize this? Thyroid hormones are involved in some way in the perception of photoperiod. Thyroidectomized birds are photoperiodically blind and exogenous thyroid hormones can act as a long-photoperiod mimic. So thyroidectomized birds show no photoperiodically-induced increase in prolactin secretion and exogenous thyroxine stimulates prolactin secretion. Therefore, the involvement of thyroid hormones in molt can be explained through effects on prolactin. Evidence suggests that exogenous gonadal steroids can enhance prolactin secretion, so the effects of gonadal steroids may also be mediated by prolactin.

38.3.2.5 GnRH Agonist

The GnRH system of birds shows a remarkable degree of seasonal plasticity (Stevenson et al., 2009). The onset of molt coincides with gonadal regression and downregulation of GnRH gene expression and/or a decrease in GnRH secretion. Treatment of domestic hens with leuprolide, a GnRH agonist, induces gonadal regression and molt (Attia et al., 1994; Burke and Attia, 1994; Dickerman and Bahr, 1989). GnRH agonists cause ovarian regression by downregulating pituitary GnRH receptors followed by reduced gonadotropin secretion. The mechanism by which this treatment induces molt is unknown. A decrease in circulating gonadal steroid hormones (at least in progesterone) occurs, but any effect on prolactin is unknown.

38.4 CONCLUSIONS

Molt is critically important to birds. There is a lot of information on the timing of molt in different species and its relationship with other aspects of the annual cycle. There is a lot of information on the sequence of feather loss and renewal along the various feather tracts. There is quite a lot of information on how photoperiod controls the timing of the start of molt and the subsequent rate of molt. However, we know very little about the physiological control. There is correlational data implicating three types of hormone: thyroid hormones, gonadal steroid hormones, and prolactin. Of these, prolactin seems to be the best candidate for a role. Prolactin is a multifunctional hormone throughout vertebrates, but many of these roles can be related to the epidermis, including its eponymous role in milk production, in both mammals and birds (specifically pigeons; Riddle et al., 1932). Prolactin is involved in broody and parental behavior and so could clearly play a linking role between the end of breeding and the start of molt. Prolactin has also been linked to the control of molt in mammals. However, it is difficult to comprehend how a circulating hormone could control the sequence of feather loss and renewal—molt starting with the shedding of the proximal primary feather on each wing, and then progressing along the feathers of each tract in turn. Could receptors in adjacent follicles have different threshold triggers? It may be more plausible that a paracrine system is important rather than an endocrine system. Feather follicles are innervated, so perhaps the autonomic nervous system is involved (Kuenzel, 2003). Clearly many questions remain to be answered on the physiological control of molt.

REFERENCES

Anthony, R., 1970. Ecology and reproduction of California quail in southeastern Washington. Condor 72, 276–287.

Attia, Y.A., Burke, W.H., Yamani, K.A., 1994. Response of broiler breeder hens to forced molting by hormonal and dietary manipulations. Poult. Sci. 73, 245–258.

Barta, Z., McNamara, J.M., Houston, A.I., Weber, T.P., Hedenstrom, A., Fero, O., 2008. Optimal moult strategies in migratory birds. Philos. Trans. R. Soc. Lond., B, Biol. Sci. 363, 211–229.

Bell, D.D., 2003. Historical and current molting practices in the U.S. table egg industry. Poult. Sci. 82, 965–970.

Benoit, J., 1936. Role de la thyroide dans la gonado-stimulation par lumiere artificielle chez le canard domestique. C. R. Seances Soc. Biol. Fil. 123, 243–246.

Bentley, G.E., Goldsmith, A.R., Dawson, A., Glennie, L.M., Talbot, R.T., Sharp, P.J., 1997. Photorefractoriness in European starlings (*Sturnus vulgaris*) is not dependent upon the long-day-induced rise in plasma thyroxine. Gen. Comp. Endocrinol. 107, 428–438.

Berry, W.D., 2003. The physiology of induced molting. Poult. Sci. 82, 971–980.

Bojarinova, J.G., Lehikoinen, E., Eeva, T., 1999. Dependence of postjuvenile moult on hatching date, condition and sex in the Great Tit. J. Avian Biol. 30, 437–446.

Boswell, T., Hall, M.R., Goldsmith, A.R., 1993. Annual cycles of migratory fattening, reproduction and moult in European quail (*Coturnix coturnix*). J. Zool. Lond. 231, 627–644.

Boswell, T., Sharp, P.J., Hall, M.R., Goldsmith, A.R., 1996. Migratory fat deposition in European quail: a role for prolactin? J. Endocrinol. 146, 71–79.

Boulakoud, M.S., Goldsmith, A.R., 1991. Thyroxine treatment induces changes in hypothalamic gonadotrophin-releasing hormone characteristic of photorefractoriness in starlings. Gen. Comp. Endocrinol. 82, 78–85.

Bridge, E.S., 2011. Mind the gaps: what's missing in our understanding of feather molt. Condor 113, 1–4.

Bridge, E.S., Voelker, G., Thompson, C.W., Jones, A.W., Baker, A.J., 2007. Effects of size and migratory behavior on the evolution of wing molt in terns (Sternae): a phylogenetic-comparative study. Auk 124, 841–856.

Burke, W.H., Attia, Y.A., 1994. Molting single comb white leghorns with the use of the lupron depot(R) formulation of leuprolide acetate. Poult. Sci. 73, 1226–1232.

Chaturvedi, C.M., Meier, A.H., 1989. Thyroid involvement in the relative photorefractoriness of Japanese quail, *Coturnix coturnix japonica*. J. Exp. Zool. 250, 63–66.

Chilgren, J.D., 1978. Effects of photoperiod and temperature on postnuptial molt in captive white-crowned sparrows. Condor 80, 222–229.

Coleman, J.T., Spray, C.J., Percival, S.M., Rickeard, A.T., Yeoman, P., 2002. The dynamics of a flock of mute swans at Berwick-upon-Tweed with particular reference to the effects of age, sex, social status and body condition on molt. Waterbirds 25, 346–351.

Czapulak, A., 2002. Timing of primary molt in breeding mute swans. Waterbirds 25, 258–267.

Dawson, A., 1989. The involvement of thyroxine and daylength in the development of photorefractoriness in European starlings. J. Exp. Zool. 249, 68–75.

Dawson, A., 1991. Photoperiodic control of testicular regression and moult in male house sparrows, *Passer domesticus*. Ibis 133, 312–316.

Dawson, A., 1993. Thyroidectomy progressively renders the reproductive system of starlings, *Sturnus vulgaris*, unresponsive to changes in daylength. J. Endocrinol. 139, 51–55.

Dawson, A., 1994. The effects of daylength and testosterone on the initiation and progress of moult in starlings *Sturnus vulgaris*. Ibis 136, 335–340.

Dawson, A., 1997. Plasma luteinizing hormone and prolactin during circannual rhythms of gonadal maturation and molt in male and female European starlings. J. Biol. Rhythms 12, 371–377.

Dawson, A., 1998. Photoperiodic control of the termination of breeding and the induction of moult in house sparrows *Passer domesticus*. Ibis 140, 35–40.

Dawson, A., 2003. A detailed analysis of primary feather moult in the common starling *Sturnus vulgaris*–new feather mass increases at a constant rate. Ibis 145, E69–E76.

Dawson, A., 2004. The effects of delaying the start of moult on the duration of moult, primary feather growth rates and feather mass in common starlings *Sturnus vulgaris*. Ibis 146, 493–500.

Dawson, A., 2005. The effect of temperature on photoperiodically regulated gonadal maturation, regression and moult in starlings–potential consequences of climate change. Funct. Ecol. 19, 995–1000.

Dawson, A., 2006. Control of molt in birds: association with prolactin and gonadal regression in starlings. Gen. Comp. Endocrinol. 147, 314–322.

Dawson, A., 2007. Seasonality in a temperate zone bird can be entrained by near equatorial photoperiods. Proc. R. Soc. Lond., B, Biol. Sci. 274, 721–725.

Dawson, A., 2008. Control of the annual cycle in birds: endocrine constraints and plasticity in response to ecological variability. Philos. Trans. R. Soc. Lond., B, Biol. Sci. 363, 1621–1633.

Dawson, A., 2013. The effect of latitude on photoperiodic control of gonadal maturation, regression and molt in birds. Gen. Comp. Endocrinol. 190, 129–133.

Dawson, A., Deeming, D.C., Dick, A.C.K., Sharp, P.J., 1996. Plasma thyroxine concentrations in farmed ostriches in relation to age, body weight, and growth hormone. Gen. Comp. Endocrinol. 103, 308–315.

Dawson, A., Goldsmith, A.R., 1982. Prolactin and gonadotrophin secretion in wild starlings (*Sturnus vulgaris*) during the annual cycle and in relation to nesting, incubation, and rearing young. Gen. Comp. Endocrinol. 48, 213–221.

Dawson, A., Goldsmith, A.R., 1983. Plasma prolactin and gonadotrophins during gonadal development and the onset of photorefractoriness in male and female starlings (*Sturnus vulgaris*) on artificial photoperiods. J. Endocrinol. 97, 253–260.

Dawson, A., Goldsmith, A.R., 1984. Effects of gonadectomy on seasonal changes in plasma LH and prolactin concentrations in male and female starlings (*Sturnus vulgaris*). J. Endocrinol. 100, 213–218.

Dawson, A., Goldsmith, A.R., 1985. Modulation of gonadotrophin and prolactin secretion by daylength and breeding behaviour in free-living starlings, *Sturnus vulgaris*. J. Zool. Lond. 206, 241–252.

Dawson, A., Hinsley, S.A., Ferns, P.N., Bonser, R.H.C., Eccleston, L., 2000. Rate of moult affects feather quality: a mechanism linking current reproductive effort to future survival. Proc. R. Soc. Lond., B, Biol. Sci. 267, 2093–2098.

Dawson, A., McNaughton, F.J., Goldsmith, A.R., Degen, A.A., 1994. Ratite-like neoteny induced by neonatal thyroidectomy of European starlings, *Sturnus vulgaris*. J. Zool. Lond. 232, 633–639.

Dawson, A., Newton, I., 2004. Use and validation of a molt score index corrected for primary feather mass. Auk 121, 372–379.

Dawson, A., Perrins, C.M., Sharp, P.J., Wheeler, D., Groves, S., 2009. The involvement of prolactin in avian molt: the effects of gender and breeding success on the timing of molt in mute swans (*Cygnus olor*). Gen. Comp. Endocrinol..

Dawson, A., Sharp, P.J., 1998. The role of prolactin in the development of reproductive photorefractoriness and postnuptial molt in the European starling (*Sturnus vulgaris*). Endocrinology 139, 485–490.

Dawson, A., Sharp, P.J., 2010. Seasonal changes in concentrations of plasma LH and prolactin associated with the advance in the development of photorefractoriness and molt by high temperature in the starling. Gen. Comp. Endocrinol. 167, 122–127.

Dawson, A., Williams, T.D., Nicholls, T.J., 1987. Thyroidectomy of nestling starlings appears to cause neotenous sexual maturation. J. Endocrinol. 112, R5–R6.

de Beer, G., 1956. The evolution of ratites. Bull. Br. Mus. 4, 58–70.

Deviche, P., Sharp, P.J., 2001. Reproductive endocrinology of a free-living, opportunistically breeding passerine (white-winged crossbill, *Loxia leucoptera*). Gen. Comp. Endocrinol. 123, 268–279.

Dickerman, R.W., Bahr, J.M., 1989. Molt induced by gonadotropin-releasing hormone agonist as a model for studying endocrine mechanisms of molting in laying hens. Poult. Sci. 68, 1402–1408.

Dittami, J.P., Gwinner, E., 1985. Annual cycles in the African stonechat *Saxicola torquata axillaris* and their relationship to environmental factors. J. Zool. Lond., A 207, 357–370.

Dolnik, V.R., Gavrilov, V.M., 1980. Photoperiodic control of the molt cycle in the chaffinch (*Fringilla coelebs*). Auk 97, 50–62.

Donham, R.S., Moore, M.C., Farner, D.S., 1983. Physiological basis of repeated testicular cycles on twelve hour days (12L:12D), in white-crowned sparrows, *Zonotrichia leucophrys gambelii*. Physiol. Zool. 56, 302–307.

Echeverry-Galvis, M.A., Hau, M., 2012. Molt-breeding overlap alters molt dynamics and behavior in zebra finches, *Taeniopygia guttata castanotis*. J. Exp. Biol. 215, 1957–1964.

Farner, D.S., Donham, R.S., Moore, M.C., Lewis, R.A., 1980. The temporal relationship between the cycle of testicular development and molt in the white-crowned sparrow, *Zoonotrichia leucophrys gambelii*. Auk 97, 63–75.

Follett, B.K., Nicholls, T.J., 1985. Influences of thyroidectomy and thyroxine replacement on photoperiodically controlled reproduction in quail. J. Endocrinol. 107, 211–221.

Genelly, R.E., 1955. Annual cycle in a population of California quail. Condor 57, 263–285.

Ginn, H.B., Melville, D.S., 1983. Moult in Birds. BTO, Tring, U.K.

Goldsmith, A.R., 1982. Plasma concentrations of prolactin during incubation and parental feeding throughout repeated breeding cycles in canaries (*Serinus canarius*). J. Endocrinol. 94, 51–59.

Goldsmith, A.R., Nicholls, T.J., 1984. Thyroidectomy prevents the development of photorefractoriness and the associated rise in plasma prolactin in starlings. Gen. Comp. Endocrinol. 54, 256–263.

Greenwood, A.W., Burns, M., 1940. The problem of the moult in the castrated Brown Leghorn fowl. Q. J. Exp. Physiol. 30, 163–171.

Guyomarc'h, C., Guyomarc'h, J.-C., 1995. Moulting cycles in European quail (*Coturnix coturnix coturnix*) under constant photoperiodic conditions. Biol. Rhythm Res. 26, 292–305.

Gwinner, E., 2003. Circannual rhythms in birds. Curr. Opin. Neurobiol. 13, 770–778.

Gwinner, E., Gänshirt, G., Dittami, J.P., 1989. Starling circannual systems: are they arrested in long photoperiods? J. Comp. Physiol. A 165, 35–39.

Hahn, T.P., 1998. Reproductive seasonality in an opportunistic breeder, the red crossbill, *Loxia curvirostra*. Ecol. 79, 2365–2375.

Hahn, T.P., Ball, G.F., 1995. Changes in brain GnRH associated with photorefractoriness in House sparrows (*Passer domesticus*). Gen. Comp. Endocrinol. 99, 349–363.

Hahn, T.P., Swingle, J., Wingfield, J.C., Ramenofsky, M., 1992. Adjustments of the prebasic molt schedule in birds. Ornis Scand. 23, 314–321.

Heinroth, O., 1911. Beiträge zur biologie, namentilich ethologie und psychologie der Anatiden. Proc. Int. Ornithol. Congr. 5, 598–702.

Hemborg, C., 1999. Sexual differences in moult-breeding overlap and female reproductive costs in pied-flycatchers, *Ficedula hypoleuca*. J. Anim. Ecol. 68, 429–436.

Hemborg, C., Merila, J., 1999. Reproductive investment and moult-breeding overlap in the collared flycatcher *Ficedula albicollis*: an experimental approach. Ann. Zool. Fenn. 36, 1–9.

Hiatt, E.S., Goldsmith, A.R., Farner, D.S., 1987. Plasma levels of prolactin and gonadotrophins during the reproductive cycle of white-crowned sparrows (*Zonotrichia leucophrys*). Auk 104, 208–217.

Himeno, K., Tanabe, Y., 1957. Mechanism of molting in hen. Poult. Sci. 36, 1186–1193.

Holmgren, N., Hedenstrom, A., 1995. The scheduling of molt in migratory birds. Evol. Ecol. 9, 354–368.

Jenni, L., Winkler, R., 1994. Moult and Ageing of European Passerines. Academic Press, London.

Juhn, M., Harris, P., 1958. Moult of capon feathering with prolactin. Proc. Soc. Exp. Biol. Med. 98, 667–672.

Kjellén, N., 1994. Moult in relation to migration in birds – a review. Ornis Svecica. 4, 1–24.

Kobayashi, H., 1953a. Acceleration of moulting in the canary by reducing the daily-light period. Annot. Zool. Jpn. 26, 156–161.

Kobayashi, H., 1953b. Studies on moulting in the pigeon. VII. Inhibitory effect of lactogen on moulting. Jpn. J. Zool. 11, 21–26.

Kobayashi, H., 1954. Studies on moulting in the pigeon. VIII. Effects of sex steroids on moulting and thyroid gland. Annot. Zool. Jpn. 27, 22–26.

Kuenzel, W.J., 2003. Neurobiology of molt in avian species. Poult. Sci. 82, 981–991.

Lien, R.J., Siopes, T.D., 1993. The relationship of plasma thyroid hormone and prolactin concentrations to egg laying, incubation behavior, and molting by female turkeys exposed to a one-year natural daylength cycle. Gen. Comp. Endocrinol. 90, 205–213.

Lincoln, G.A., Racey, P.A., Sharp, P.J., Klandorf, H., 1980. Endocrine changes associated with spring and autumn sexuality of the rook, *Corvus frugilegus*. J. Zool. Lond. 190, 137–153.

McCleery, R.H., Perrins, C.M., Wheeler, D., Groves, S., 2007. The effect of breeding status on the timing of moult in mute swans *Cygnus olor*. Ibis 149, 86–90.

Mewaldt, L.R., King, J.R., 1978. Latitudinal variation of postnuptial molt in Pacific coast white-crowned sparrows. Auk 95, 168–179.

Miller, A.H., 1960. Adaptation of breeding schedule to latitude. 12, 513–522.

Moore, I.T., Bonier, F., Wingfield, J.C., 2005. Reproductive asynchrony and population divergence between two tropical bird populations. 16, 755–762.

Morton, G.A., Morton, M.L., 1990. Dynamics of postnuptial molt in free-living mountain white-crowned sparrows. Condor 92, 813–828.

Morton, M.L., 1992. Control of postnuptial molt in the mountain white-crowned sparrow: a perspective from field data. Ornis Scand. 23, 322–327.

Murphy, M.E., King, J.R., 1992. Energy and nutrient use during molt by white crowned sparrows *Zonotrichia leucophrys gambelii*. Ornis Scand. 23, 304–313.

Newton, I., 1966. The moult of the bullfinch *Pyrrhula pyrrhula*. Ibis 108, 41–67.

Newton, I., Dawson, A., 2011. Seasonal changes in moult, body mass and reproductive condition in siskins *Carduelis spinus* exposed to daylength regimes simulating different latitudes. J. Avian Biol. 42, 22–28.

Newton, I., Rothery, P., 2000. Timing and duration of moult in the bullfinch *Pyrrhula pyrrhula*: an appraisal of different analytical procedures. Ibis 142, 65–74.

Nicholls, T.J., Goldsmith, A.R., Dawson, A., 1988. Photorefractoriness in birds and comparison with mammals. Physiol. Rev. 68, 133–176.

Nilsson, J.-Å., Svensson, E., 1996. The cost of reproduction: a new link between current reproductive effort and future reproductive success. Proc. R. Soc. Lond., B, Biol. Sci. 263, 711–714.

Nolan, V., Ketterson, E.D., Ziegenfus, C., Cullen, D.P., Chandler, C.R., 1992. Testosterone and avian life histories–effects of experimentally elevated testosterone on prebasic molt and survival in male dark-eyed juncos. Condor 94, 364–370.

Owens, I.P.F., Short, R.V., 1995. Hormonal basis of sexual dimorphism in birds: implications for new theories of sexual selection. Trends Ecol. Evol. 10, 44–47.

Payne, R.B., 1972. Mechanisms and control of molt. In: Farner, D.S., King, J.R. (Eds.), Avian Biology. Academic Press, New York, pp. 104–155.

Pereyra, M.E., Sharbaugh, S.M., Hahn, T.P., 2005. Interspecific variation in photo-induced GnRH plasticity among nomadic cardueline finches. Brain Behav. Evol. 66, 35–49.

Prys-Jones, R.M., 1991. The occurrence of biannual primary molt in passerines. Bull. Br. Ornithol. Club 111, 150–152.

Reinert, B.D., Wilson, F.E., 1996. Thyroid dysfunction and thyroxine-dependent programming of photoinduced ovarian growth in American tree sparrows (*Spizella arborea*). Gen. Comp. Endocrinol. 103, 71–81.

Riddle, O., Bates, R.W., Dykshorn, S.W., 1932. Prolactin, a new and third hormone of the anterior pituitary. Anat. Rec. 54, 25–26.

Robinson, J.E., Follett, B.K., 1982. Photoperiodism in Japanese quail: the termination of seasonal breeding by photorefractoriness. Proc. R. Soc. Lond., B, Biol. Sci. 215, 95–116.

Rohwer, S., Ricklefs, R.E., Rohwer, V.G., Copple, M.M., 2009. Allometry of the duration of flight feather molt in birds. PLoS Biol. 7, e1000132.

Rothery, P., Wyllie, I., Newton, I., Dawson, A., Osborn, D., 2001. The timing and duration of moult in adult starlings *Sturnus vulgaris* in east-central England. Ibis 143, 435–441.

Samson, F.B., 1976. Pterylosis and molt in Cassin's finch. Condor 78, 505–511.

Scanes, C.G., Sharp, P.J., Harvey, S., Godden, P.M.M., Chadwick, A., Newcomer, W.S., 1979. Variations in plasma prolactin, thyroid hormones, gonadal steroids and growth hormones in turkeys during the induction of egg laying and moult by different photoperiods. Br. Poult. Sci. 20, 143–148.

Schleussner, G., Dittami, J.P., Gwinner, E., 1985. Testosterone implants affect molt in male European starlings, *Sturnus vulgaris*. Physiol. Zool. 58, 597–604.

Scott, P., 1972. The Swans. Wildfowl Trust, Slimbridge.

Seel, D.C., 1976. Moult in five species of Corvidae in Britain. Ibis 118, 491–527.

Sharp, P.J., Massa, R., Bottoni, L., Lucini, V., Lea, R.W., 1986. Photoperiodic and endocrine control of seasonal breeding in Grey partridge (*Perdix perdix*). J. Exp. Zool. 209, 187–200.

Silverin, B., Kikuchi, M., Ishii, S., 1997. Seasonal changes in follicle-stimulating hormone in free-living Great Tits. Gen. Comp. Endocrinol. 108, 366–373.

Stevenson, T.J., Lynch, K.S., Lamba, P., Ball, G.F., Bernard, D.J., 2009. Cloning of gonadotropin-releasing hormone I complementary DNAs in songbirds facilitates dissection of mechanisms mediating seasonal changes in reproduction. Endocrinology 150, 1826–1833.

Swaddle, J.P., Witter, M.S., 1997. The effects of molt on the flight performance, body mass, and behavior of European starlings (*Sturnus vulgaris*): an experimental approach. Can. J. Zool. 75, 1135–1146.

Torrey, H.B., Horning, B., 1922. The effects of thyroid feeding on the molting process and feather structure of the domestic fowl. Proc. Soc. Exp. Biol. Med. 19, 275.

Tucker, V.A., 1991. The effect of molting on the gliding performance of a Harris Hawk (*Parabuteo unicinctus*). Auk 108, 108–113.

Underhill, L.G., Joubert, A., 1995. Relative masses of primary feathers. Ringing Migration 16, 109–116.

Underhill, L.G., Prys-Jones, R.P., Dowsett, R.J., Herroelen, P., Johnson, D.N., Lawn, M.R., Norman, S.C., Pearson, D.J., Tree, J., 1992. The biannual primary moult of willow warblers *Phylloscopus trochilus* in Europe and Africa. Ibis 134, 286–297.

Visser, M.E., Schaper, S.V., Holleman, L.J.M., Dawson, A., Sharp, P.J., Gienapp, P., Caro, S.P., 2011. Genetic variation in cue sensitivity involved in avian timing of reproduction. Funct. Ecol. 25, 868–877.

Voitkevich, A.A., 1966. The Feathers and Plumage of Birds. Sidgwick and Jackson, London.

Webster, A.B., 2003. Physiology and behavior of the hen during induced molt. Poult. Sci. 82, 992–1002.

Wikelski, M., Hau, M., Robinson, W.D., Wingfield, J.C., 2003. Reproductive seasonality of seven neotropical passerine species. Condor 105, 683–695.

Wilson, F.E., Reinert, B.D., 1993. The thyroid and photoperiodic control of seasonal reproduction in American tree sparrows (*Spizella arborea*). J. Comp. Physiol. B 163, 563–573.

Wingfield, J.C., 1993. Control of testicular cycles in the song sparrow, *Melospiza melodia*: interaction of photoperiod and an endogenous program? Gen. Comp. Endocrinol. 92, 388–401.

Wingfield, J.C., 2008. Organization of vertebrate annual cycles: implications for control mechanisms. Philos. Trans. R. Soc. Lond., B, Biol. Sci. 363, 425–441.

Wingfield, J.C., Goldsmith, A.R., 1990. Plasma levels of prolactin and gonadal steroids in relation to multiple-brooding and renesting in free-living populations of the song sparrow, Melospiza melodia. Horm. Behav. 24, 89–103.

Woitkewitsch, A.A., 1940. Dependence of seasonal periodicity in gonadal changes on the thyroid gland in *Sturnus vulgaris*L.. C. R. (Doklady) Acad. Sci. URSS 27, 741–745.

Xu, X., Zheng, X., You, H., 2010. Exceptional dinosaur fossils show ontogenetic development of early feathers. Nature 464, 1338–1341.

Young, G.R., Dawson, A., Newton, I., Walker, L., 2009. The timing of gonadal development and moult in three raptors with different breeding seasons: effects of gender, age and body condition. Ibis 151, 654–666.

Yue, Z.C., Jiang, T.X., Widelitz, R.B., Chuong, C.M., 2005. Mapping stem cell activities in the feather follicle. Nature 438, 1026–1029.

Zelenitsky, D.K., Therrien, F., Erickson, G.M., DeBuhr, C.L., Kobayashi, Y., Eberth, D.A., Hadfield, F., 2012. Feathered non-avian dinosaurs from North America provide insight into wing origins and that the ontogeny of their feathers was similar to that of modern birds. Science 338, 510–514.

Chapter 39

Flight

C.M. Bishop
School of Biological Sciences, Bangor University, Bangor, Gwynedd, UK

P.J. Butler
School of Biosciences, University of Birmingham, Edgbaston, Birmingham, UK

39.1 INTRODUCTION

Birds are probably best known for the evolution of their intricate feathers and the subsequent development of wings along the forelimbs, eventually leading to flapping flight. Some of their Theropod dinosaur ancestors even experimented with feathered hindlimbs, which may have assisted their early attempts at gliding, or provided additional lift generation, but the vast majority of extinct and extant birds have two independent locomotor systems, the legs and the wings. Powered flight has evolved in four unrelated groups: the insects, the pterosaurs, the birds, and the bats (Bomphrey, 2012), and in most cases the wings are entirely specialized for aerial (or rarely, aquatic) flapping locomotion. All these animals have had to overcome the initial problem that weight support in air while traveling slowly requires specialized adaptations and is energetically expensive (Pennycuick, 1968).

As a result of these difficulties, many of the descendants of feathered dinosaurs probably never gained the ability to fly, while the reverse evolution of flying birds back to flightlessness may not have been uncommon among ancestral birds (e.g., *Hesperornis*), and is not only a feature of the modern day ratites (e.g., ostriches and emus) and penguins but has evolved on a number of occasions in rails, geese, ducks, grebes, ibises (McNab, 1994), and Galápagos cormorant *Phalacrocorax harrisi*. As these authors point out, flightlessness in terrestrial birds is always associated with a reduction in the mass of the pectoral muscles with a consequent reduction in basal energy consumption, which may be related to overall energy conservation. Conversely, a complete reliance on flight as the only means of locomotion is seen in just a few families, such as the swifts, swallows, hummingbirds, and sunbirds. Hummingbirds exhibit an exceptional form of flight in that they are the only birds known routinely to use prolonged periods of "stationary" or hovering flight and are even capable of flying backwards, due to their unique wing kinematics (Warrick et al., 2009). However, most birds have the best of both worlds, the ability to walk, run, or swim using the legs and the ability to fly in air (or rarely, even water) using the wings and tail.

Various ideas have been put forward to indicate how even a primitive wing design might have assisted early proto-birds to survive (Dial, 2003; Hutchinson and Allen, 2009), but, even once airborne, forward flapping flight is, for birds of similar mass, energetically more costly per unit of time than running or swimming at the surface (Butler, 1991). However, during flight over long distances, birds are able to maintain relatively high velocities, which means that the energy required for a given mass to travel a given distance (i.e., the cost of transport) is considerably less than the energy cost of transport during walking, running, or surface swimming (Tucker, 1970; Schmidt-Nielsen, 1972). Thus, a number of species of birds are able to migrate over long distances in a relatively short time period (Gill et al., 2009; Klaassen et al., 2011). Flight also enables birds to travel relatively long distances on a daily basis, to and from their nests or roosting areas (Jouventin and Weimerskirch, 1990).

The high rate of energy consumption required for flapping flight to be sustained over a prolonged period of time can only be maintained by "oxidative" or "aerobic" metabolism of fuel substrates, and places large demands on the respiratory, cardiovascular, and muscular systems of birds. Around 18% of the world's bird species undergo long-distance migration (Sekercioglu, 2007), and in temperate latitudes this is closer to 50%, despite the added problem of the birds initially carrying a relatively large amount of fuel (Alerstam and Lindström, 1990; Hedenström and Alerstam, 1992; Witter and Cuthill, 1993). At the same time they have to control body temperature and evaporative water loss in the face of the large

Sturkie's Avian Physiology.

amount of heat that is generated by the intense muscular activity. The energy cost of spring migration may be so high as to influence the subsequent energy dynamics of reproduction (Bromley and Jarvis, 1993). The postbreeding autumn migration follows a period of relatively sedentary behaviors such as incubation, brooding, and molt for the adults, during which there is evidence, for some species at least, that the main flight muscles atrophy (Mainguy and Thomas, 1985; Piersma, 1988; Bishop et al., 1996; Portugal et al., 2009). In addition, the energy and time taken for birds to molt and replace their flight feathers may be a major consideration in regulating the length of the breeding season and even the maximum size of flying birds. Thus, for postbreeding and postmolting adults, as well as for the season's fledglings, the flight muscles and supporting systems have to achieve an adequate level of aerobic fitness and functionality in preparation for the forthcoming migratory flights. Those species of birds that fly at high altitudes during their migration also have the combined problems of engaging in this energetically costly activity in a severely hypoxic and cold environment and at a reduced air density (Butler, 2010; Scott, 2011).

At the other extreme from sustained flight is the explosive "burst" flapping flight of many species of Galliformes and Tinamiformes birds, which they utilize to escape terrestrial predators or to roost in trees. These birds rely on "anaerobic" metabolism and are unable to sustain such activity for more than a few minutes and may be rendered incapable of flying at all after repeated bursts of activity (Marden, 1994). During takeoff, or when sprinting to maneuver or avoid predators, even the most aerobic of avian species, the Trochiliformes or hummingbird, can briefly support "burst" flight performance by high levels of creatine kinase in their muscles, supporting rapid phosphocreatine regeneration of adenosine triphosphate (ATP) (Chai et al., 1997). In between these two extremes of aerial locomotion are birds such as vultures, storks, and albatrosses which are able to remain airborne by soaring and gliding and expend a fraction of the energy required for forward flapping flight during foraging trips and migration. In addition, small species of birds appear to use different "gaits" or flight modes, such as flap-bounding and flap-gliding (Tobalske and Dial, 1994; Tobalske et al., 2009), in order to maintain the wingbeat frequency at the most energy-efficient rate (Rayner, 1985a).

This chapter will focus primarily on the energetics of bird flight, including long-distance migration; the function, physiology, and biochemistry of the flight muscles, including physical fitness in preparation for migration; deposition of fuel stores in preparation for migration and their use during migration; respiratory and cardiovascular adjustments associated with flight, including temperature control and water loss; and flight at high altitudes. However, it is not easy to obtain experimental data from a flying animal, and it should be remembered that it is doubtful if any one experimental method can adequately give all of the necessary information required for a full analysis of the physiological responses to flight. Thus, a number of techniques used to study the flight of birds will be discussed. To begin with, however, there is a short introduction to the significance of animal size and scale.

39.2 SCALING

It is important to be able to assess the significance of a particular characteristic or adaptation and compare it with those of other individuals and/or species, even though these other animals may vary greatly in size or mass. The study of the consequences of changes in shape or mass of different animals is called scaling (Schmidt Nielsen, 1984), and it is frequently used in comparative biology, due to the many orders of magnitude differences that occur in body mass both within and between animal groups. If two bodies are the same shape or "geometrically similar" but of different volume (V) and mass (M) (i.e., their various characteristics, such as length and surface area, are directly proportional to their difference in mass), then these bodies are said to scale isometrically with regard to each other. If we consider two cubes of different size, then each side will have a length (L), and the area will be proportional to the length squared ($A \alpha L^2$), while the volume or mass (assuming a constant density) will be proportional to the length cubed ($V \alpha L^3$). Thus, area will be proportional to volume or mass to the two-thirds power ($A \alpha V^{2/3}$), and any length will be proportional to volume or mass to the one-third power ($L \alpha V^{1/3}$). Because body mass (M_b) is a relatively easy variable to measure in most animals and one that represents the large size range seen in animals, it is most frequently used as the independent variable in scaling. Thus:

$$L \, \alpha \, M_b^{1/3} \text{ or } M_b^{0.33} \text{and} A \, \alpha \, M_b^{2/3} \text{ or } M_b^{0.67}$$

Pennycuick (1982) studied the flight and morphometric parameters of 11 species of Procellariiformes (e.g., petrels and albatrosses), ranging from 0.03 to 9 kg, and found that wing span (b) scaled in proportion to $M_b^{0.37}$ (Figure 39.1(A)) and wing area (S) in proportion to $M_b^{0.627}$ (Figure 39.1(B)). These results indicate that there is a slight tendency for larger birds to have relatively longer wings and smaller wing areas, but that overall they are geometrically very similar. However, Figure 39.1(D) shows that, if the outlines of each species are traced and then displayed so that they all have the same wing span, as the birds become smaller the shape of their wings changes in a systematic way (Pennycuick, 1992). This can be assessed by calculating the dimensionless variable called aspect ratio (Λ), which is the wing span divided by the mean wing width (or chord) and is equivalent to b^2/S. Figure 39.1(C) shows that Λ scales in proportion to $M_b^{0.116}$ and is highly significantly different from the expected isometric value of $M_b^{0.0}$. Thus, this feature of birds

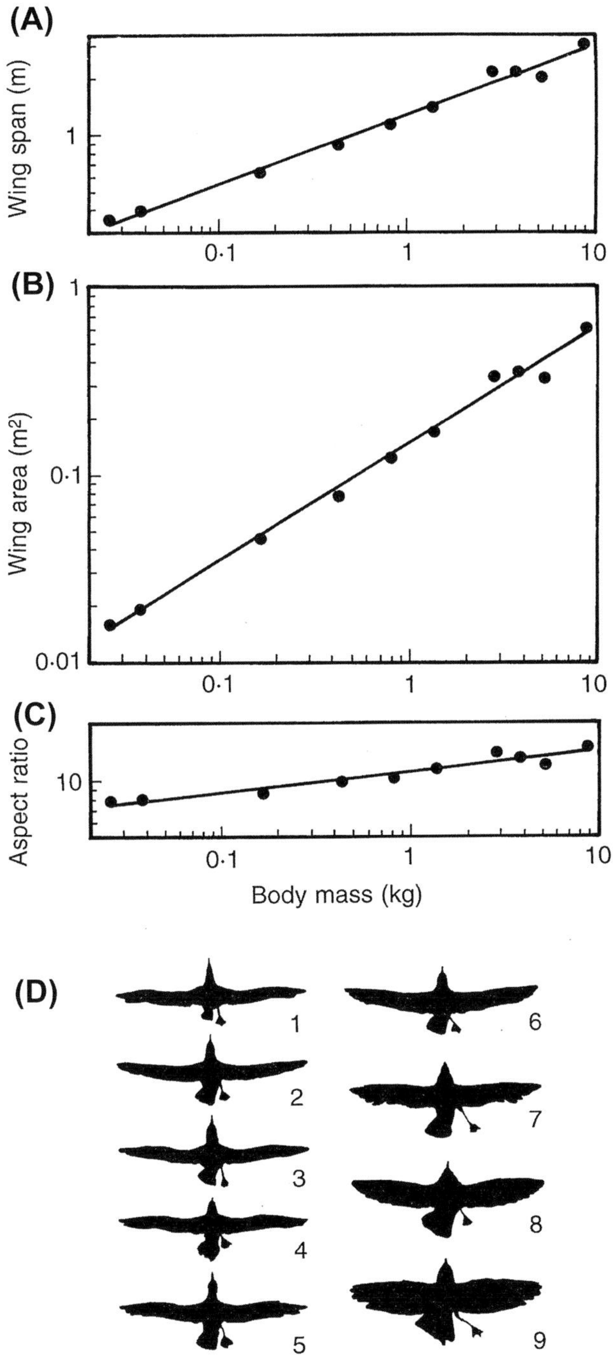

FIGURE 39.1 (A) Wing span, *b;* (B) wing area, *S;* (C) aspect ratio, Λ, of 11 species of Procellariiform birds, plotted against body mass, M_b (*From Pennycuick (1992), by permission of Oxford University Press.*). (D) Wing tracings from nine of these species, with the smaller species enlarged such that all have the same wing span. 1, wandering albatross, *Diomedia exulans*; 2, black-browed albatross, *D. melanophrys*; 3, grey-headed albatross, *D. chysostoma*; 4, light-mantled sooty albatross, *Phoebetria palpebrata*; 5, giant petrel, *Macronectes* sp.; 6, white-chinned petrel, *Procellaria aequinoctialis*; 7, cape pigeon, *Daption capensis*; 8, dove prion, *Pachyptila desolata*; 9,Wilson's storm petrel, *Oceanites oceanicus*. The wing span of the largest species, *D. exulans*, is 3.03 m and that of the smallest, *O. oceanicus*, is 0.393 m. *From Pennycuick (1982); Figure 39.4, The Royal Society.*

scales non-isometrically or allometrically and the study of such relationships is often called "allometry".

Such analyses can also be used to test predictions regarding the scaling of animal energetics. For example, as "geometrically similar" animals increase in mass, their surface area should decline proportional to $M_b^{0.67}$ and so should the rate at which heat will consequently be lost to the environment. Thus, it might be expected that basal metabolic rate (BMR) should also decline in proportion to $M_b^{0.67}$. Until recently, most studies of vertebrates had shown that BMR actually scales allometrically, at around $M_b^{0.72-0.75}$, i.e., slightly greater than predicted from the surface area law. However, at least with respect to the analysis of BMR in birds and mammals, it has become quite a controversial subject, with some authors suggesting that there is no rigid scaling exponent and that it varies with relative exercise intensity (Bishop, 1999) or body temperature. Other studies have suggested that it may also vary between taxonomic groupings or with ecological adaptations. Despite these difficulties, allometric analysis has been used to address a wide variety of issues in biology (Schmidt-Nielsen, 1984), and the scaling of metabolic and biomechanical parameters associated with bird flight will be considered throughout this chapter.

39.3 ENERGETICS OF BIRD FLIGHT

One of the most useful physiological variables is the rate at which a bird expends chemical energy (i.e., power input (P_i)) during flight. The P_i of both resting and exercising animals are important considerations in many ecological and physiological studies of animal behavior and evolution, particularly as it ultimately determines how much food must be foraged from the environment. However, the chemical energy available from the anaerobic or aerobic catabolism of the various fuel molecules must be transformed by the myofibrillar proteins of the flight muscles into the mechanical power output (P_o) required to fly, while the remainder is released as heat. The exact quantity of P_o produced from a given P_i will, thus, depend on the efficiency with which the relevant locomotor muscles can convert the chemical energy into mechanical power. The heat produced can be used to regulate body temperature, but it is almost invariably "lost" to the environment, while the biomechanical P_o results in various forms of locomotory behaviors. Some of the various techniques used to study the biology of flight will be discussed, primarily with respect to the estimation of P_i and P_o during flight.

The SI unit of power is the watt (W) which is equivalent to 1 J per second (J/sec). An estimate of P_i can be obtained by converting the rate of oxygen consumption ($\dot{V}O_2$, mL/sec) to W. The conversion of $\dot{V}O_2$ to P_i depends on the metabolic substrate; for metabolism of pure carbohydrate, with a respiratory quotient (RQ) of 1, 1 mL O_2/sec = 21.1 W; whereas for pure fat metabolism (with an RQ of 0.71), 1 mL O_2/sec = 19.6 W (Lusk, 1919; Brobeck and DuBois, 1980). Pure protein metabolism in birds would yield an RQ of 0.74 and 1 mL O_2/sec = 18.4 W (Schmidt-Nielsen, 1997). During short-duration exercise, it is likely that carbohydrate oxidation will dominate; for example, short hovering flights during foraging in hummingbirds (Suarez et al., 1990) and soon after takeoff in pigeons (Butler et al., 1977), but during longer bouts of exercise it is likely that fat oxidation will become dominant (Rothe et al., 1987). However, this may be influenced by the feeding regime and by the season: well-fed and summer birds start at a higher RQ and take longer to reach an RQ of approximately 0.7 (Nachtigall, 1995). Thus, if RQ is unknown, a compromise for an RQ of 0.8, of 1 mL O_2/sec = 20.1 W, is often used for calculations of P_i during aerobic activity (Schmidt-Nielsen, 1997).

39.3.1 Techniques Used to Study the Mechanical Power Output Required for Flight

39.3.1.1 Aerodynamic and Biomechanical Models

The use of aerodynamic and biomechanical models provides a theoretical framework for understanding the physiological adaptations and energetics of birds and the likely constraints acting on different species. It is theoretically possible to estimate the P_i of a particular bird, using estimates for P_o based on aerodynamic models (Rayner, 1979; Pennycuick, 1989), provided the efficiency by which the flight muscles are able to transduce chemical energy into mechanical energy is known or can be estimated. However, while Pennycuick (2008) has assumed that flight muscle efficiency is around 23% for all bird species, other authors have pointed out that the overall efficiency appears to scale positively with increasing body mass for many types of invertebrate and vertebrate locomotion, including flight (Alexander, 2005; Bishop, 2005; Askew et al., 2010). The P_o required for different modes of flight is primarily determined by the overall mass, wing kinematics, and detailed morphology of the bird under investigation (Pennycuick, 1968), but it is also modified by the prevailing environmental conditions such as air density, strength of gravity, local air thermals, and deflected currents from nearby surfaces (Pennycuick, 1989, 2008).

During forward flapping flight, a bird has to generate sufficient mechanical power to overcome the downward acceleration due to gravity acting on its own mass (induced power) and the frictional drag of the airflow over its body (parasite power) and wings (profile power), as well as inertial power to accelerate and decelerate the wings during each beat. Inertial power requirements are usually ignored for medium to fast flight speeds, but may be important during hovering and slow flight (Norberg, 1990). A number of aerodynamic models (Pennycuick, 1969; Tucker, 1973; Greenewalt, 1975; Rayner, 1979) have indicated that these components of the overall P_o of the flight muscles should vary with flight speed (U). Calculations indicate that at least parasite power should increase with speed, whereas induced power is high at low speeds and declines at higher speeds.

Thus, overall, P_o should vary with speed in a U-shaped fashion (Figure 39.2). This means that there is a speed at which P_o and, presumably also P_i, are at a minimum (U_{min}) and at which the bird should fly in order to maximize its flying time. However, this is not the speed at which the bird should fly in order to maximize the distance flown for a given amount of fuel consumption (i.e., the speed at which the energy cost of transport is lowest). This is the maximum range speed (U_{mr}, Figure 39.2) and is, theoretically,

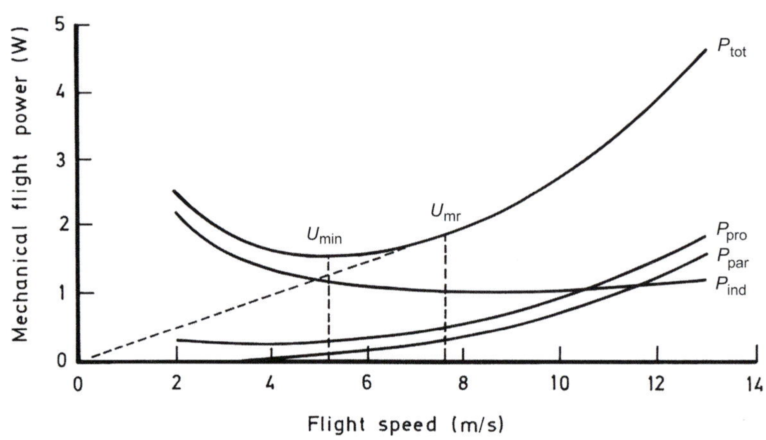

FIGURE 39.2 Calculated mechanical power output (P_{tot}) and its separate components (induced power, P_{ind}; parasite power, P_{par}; profile power, P_{pro}) for the European kestrel, *Falco tinnunculus* (mass 0.21 kg) flying at different speeds. Also shown are the minimum power speed (U_{min}) and maximum range speed (U_{mr}). *From Rayner (1993) by the American Mathematical Society.*

the speed at which a long-distance migrant should fly (Pennycuick, 1969), particularly if the supply of energy is limited for any reason (Alerstam and Lindström, 1990). Hedenström and Alerstam (1995) contend that birds should adapt their flight speed differently when migrating or transporting food compared with when they are foraging and this may depend on whether the bird is time or energy limited (Hedenström, 2008).

Measurement of drag components and the overall body drag coefficient of living birds is relatively difficult (Pennycuick et al., 1988) and the appropriate default values to be used in aerodynamic calculations for different groups of birds is uncertain (Pennycuick, 2008). Data from a teal (*Anas crecca*) and a thrush nightingale (*Luscinia luscinia*) flying in a wind tunnel indicate that the drag coefficients for these species may be only 20% of what was previously assumed (Pennycuick et al., 1996b). Subsequently, body drag estimates based on the speed of dives in wild birds supported the original, rather higher values. More recently, Hedenström and Rosén (2003) have indicated that at least for small passerine species of bird, a confounding variable may be the underestimation of the frontal area of the body. Thus, small birds may have a higher body drag coefficient than larger birds (Pennycuick, 2008). A phenomenon that could reduce induced drag is "ground effect", which is the result of a bird flying close above a plane surface, be it water (e.g., petrels) or land (e.g., certain raptors). It may also be important during takeoff for heavy birds such as swans and vultures. Rayner (1991) has developed a theory for ground effect under fixed wing conditions. While acknowledging the limitations of the model, particularly with respect to its neglect of flapping wings, Rayner (1991) concludes that there are improvements in performance for flight in ground effect, provided the flight speed is not too low. Conversely, attaching tracking devices and data loggers to the backs of birds will disturb the normal airflow over the body and base of the wings and increase overall drag (Obrecht et al., 1988; Bowlin et al., 2010). This may have negative effects on their behavior, energetics, and survival (Gessaman and Nagy, 1988; Barron et al., 2010).

The interspecies scaling of the biomechanical power required to hold the wings out and to support the mass of the body during gliding flight is discussed by Pennycuick (1989). Given the limited amount of data available, he tentatively concludes that the power required should scale with respect to body mass between $M_b^{0.67}$ and $M_b^{0.83}$, i.e., that the power required is approximately a fixed multiple of the aerobic basal metabolic rate in birds of different mass. This is consistent with the limited amount of data on the cost of gliding in birds (see below).

Pennycuick (1968, 1969, 1989) has also discussed the factors that may limit the flapping flight performance of different species of birds. He hypothesizes that, for geometrically similar birds, the P_o required for flight should scale proportional to $M_b^{1.17}$ and predicts that the P_o available from the flight muscles should scale proportional to $M_b^{0.67}$. It might be expected, therefore, that in order to maintain an equivalent flight performance, larger birds should develop relatively larger flight muscles than smaller birds. In fact, the mass of the flight muscles appears to scale in direct proportion to body mass (Greenewalt, 1962; Marden, 1987; Rayner, 1988; Bishop and Butler, 1995), which means that larger birds should have reduced flight performance in terms of rate of climb, speed of takeoff, ability to hover, and so on. In addition, Bishop and Butler (1995) point out that, as the maximum heart rate of vertebrates scales negatively with body mass, maximum sustained flight performance will actually be limited by the negative scaling of maximum cardiac output (Bishop, 1997; Bishop and Spivey, 2013). To some extent, this will be ameliorated by the observation that the aerobic scope of birds and mammals (the ratio between their maximum metabolic rate and their BMR) scales positively with body mass (Bishop, 1999). Nevertheless, because the mass exponents for the P_o required and the P_i available appear to be so different, there should be a clear upper limit to the size of birds that can perform flapping flight (Figure 39.3), and this would appear to be at approximately 12–20 kg (Pennycuick, 1968). Pennycuick (1968, 1969, 1989) suggests that the range of flight performance of any particular bird will depend on the relation between the P_o/U curve and the maximum P_o available from the flight muscles. He distinguishes two categories of maximum power: (1) absolute maximum power (P_{max}), which includes the aerobic (sustained) and anaerobic (burst) capacity of the flight muscles; and (2) maximum sustainable power (P_{ms}), which only includes the aerobic capacity of the muscles and is limited by the rate of delivery of oxygen and metabolite

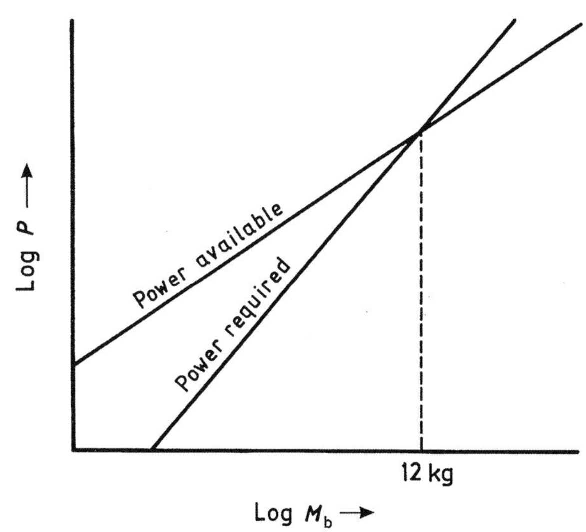

FIGURE 39.3 Double logarithmic plot of power required and power available (P) against body mass (M_b) for geometrically similar flying animals of different size. *From Pennycuick (1968), Company of Biologists, Ltd.*

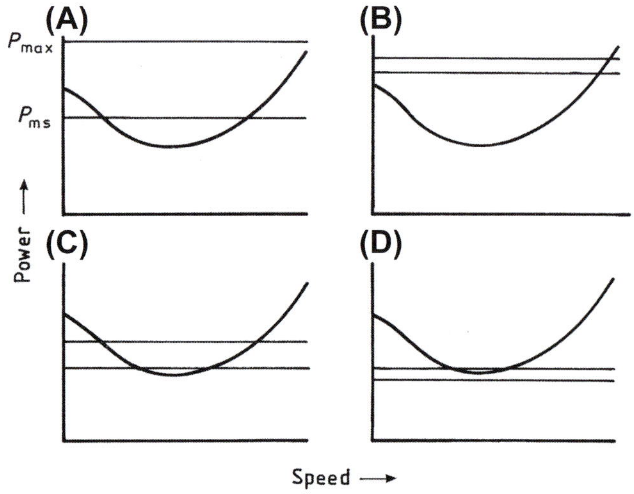

FIGURE 39.4 Power curves (see Figure 39.1) for (A) pigeon, *Columba livia*, (B) hummingbird spp., (C) white-backed vulture, *Gyps africanus*, and (D) California condor, *Gymnogyps californianus* (not to scale). The upper horizontal line represents maximum power available (P_{max}), and the lower line represents the maximum sustainable power (P_{ms}). *Modified from Pennycuick (1968), Company of Biologists, Ltd.*

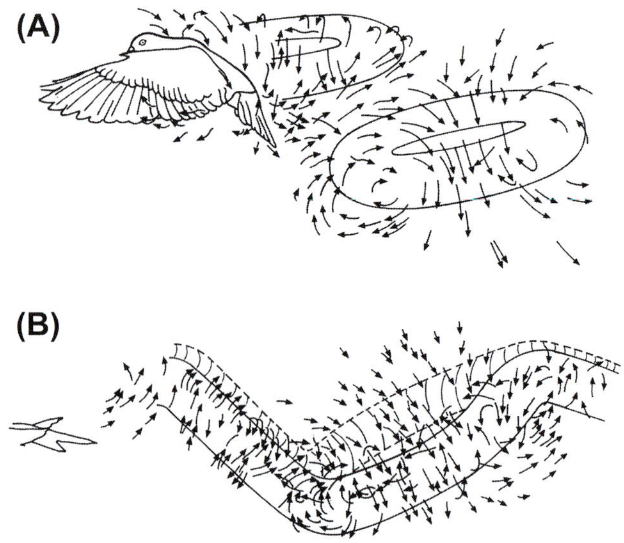

FIGURE 39.5 Reconstruction of the air movements and the position of the vortex core in the wake of (A) a pigeon, *Columba livia*, and (B) a kestrel, *Falco tinnunculus,* flying through a cloud of neutrally buoyant, helium-filled soap bubbles. *From Rayner (1988).*

flux through the cardiorespiratory system. In hummingbirds, both P_{ms} and P_{max} lie above the power required for hovering (Figure 39.4), as these birds can hover aerobically (Chai and Dudley, 1995). In contrast, for the white-backed vulture (*Pseudogyps africanus*), both P_{ms} and P_{max} lie between the minimum power required for flight and that required for hovering, as they cannot hover, whereas they can fly horizontally. Birds such as pigeons are intermediate between these two, possessing a P_{max} sufficient to hover for brief periods using anaerobic metabolism.

39.3.1.2 Airflow Visualization and Direct Force Measurements

Although the mathematical models for estimating aerodynamic P_o have been available for many years, it is only very recently that developments in technology have enabled measurements to be made that could attempt to test the accuracy of the models by obtaining measurements of the aerodynamic forces produced by the wings. Some ingenious early experiments performed by Spedding et al. (1984) and Spedding (1986) enabled the wake of pigeons (*Columba livia*) and jackdaws (*Corvus monedula*) flying in a 4-m long cage to be visualized as they flew slowly through a cloud of neutrally buoyant helium bubbles. Although the data obtained supported the hypothesis (Rayner, 1979) that the wake is composed of a chain of small vortex rings (Figure 39.5(A)), indicating that the upstroke is inactive, the energy (induced power) calculated from the rings was only 60% and 35% (for the pigeon and jackdaw, respectively) of those predicted from the model. The most likely explanation of this "deficit" was that the flights were of too short a duration and that, therefore, the birds were not flying at a constant speed as they flew through the cloud of bubbles. In a further study, Spedding (1987) flew a kestrel (*Falco tinnunculus*) through helium bubbles along a 36 m long corridor at a "medium" speed of approximately 7 m/s. In this case, the wake consisted of a pair of continuous, undulating trailing vortices (Figure 39.5(B)), indicating that the upstroke of the wing was aerodynamically active. As a result, a more simple wake model can be used to calculate induced power, which is very close to that calculated from the model of Pennycuick (1975), based on classical aerodynamic theory. Using wing-beat kinematics to infer lift production, Tobalske and Dial (1996) concluded that black-billed magpies (*Pica pica*) use a vortex-ring gait at all speeds, whereas pigeons use a vortex-ring gait at 6–8 m/s, a transitional gait at 10 m/s, and a continuous-vortex gait at higher speeds.

Recent work, utilizing sophisticated particle image velocimetry (PIV) equipment to monitor the wake of birds flying in wind tunnels, has found the apparently "missing" induced velocity component and demonstrated full weight support is present in the wake vortices (e.g., for thrush nightingales *L. luscinia*; Spedding et al., 2003). There is also a partial confirmation of the original vortex-ring and continuous-vortex gait idea but only at the extremes of the velocity range. In reality, the results for most flight speeds indicate a more complex and somewhat intermediate set of wakes, with strong and continuous wingtip vortices and, at least in the case of swifts (*Apus apus*), additional vortices shed at the root (or base) of the wing (Henningsson et al., 2011). Once again, hummingbirds demonstrate their unusual aerodynamic approach, with a bound circulation that is not shed as a vortex at the end of the wingbeat

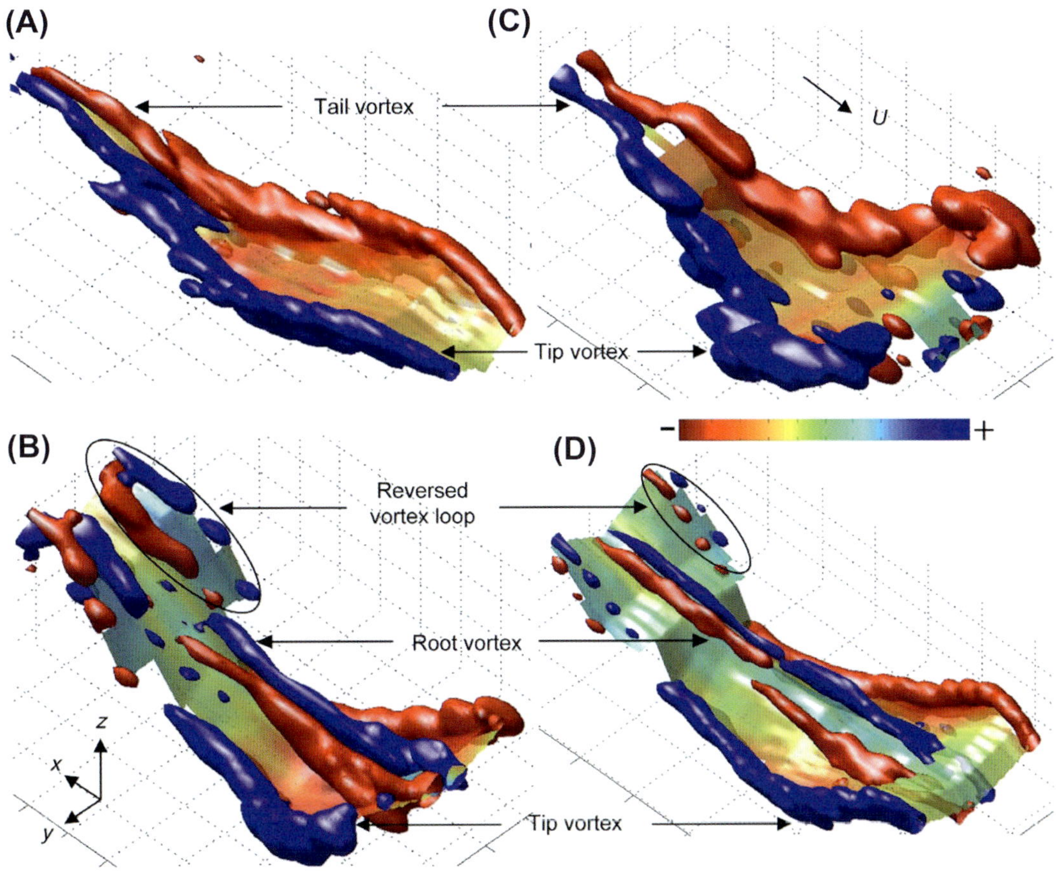

FIGURE 39.6 Reconstruction of the wake topology of a single flap cycle of two species of bird (A & C) and two species of bat (B & D) flying at 7 m/s. (A) Pied flycatcher, *Ficedula hypoleuca*. (B) Pallas' long-tongued bat, *Glossophaga soricina*. (C) blackcap, *Sylvia atricapilla*. (D) Lesser long-nosed bat, *Leptonycteris yerbabuenae*. The color-coded surface shows downwash (see color bar) as the animals fly from right to left. Bats produce root vortix at the base of the wing during the downstroke but have a slightly active upstroke that produces a reversed vortex loop due to generating thrust but negative lift. Birds have a largely inactive upstroke by retracting the wings towards the body and leaving a visible tail vortex. *From Muijres et al. (2012b).*

but persists, so that there is a near-continuous production of lift throughout the wing cycle (Warrick et al., 2009). As a result, some 25% of the weight support is generated by the aerodynamically "active" upstroke of the wing during hovering. While some other birds may hover for very limited periods, they retract the wing on the upstroke so that it almost folds up while sweeping close to the body and is, therefore, aerodynamically "inactive". To compensate for this, the pied flycatcher (*Ficedula hypoleuca*) can produce 23% of its weight support from a combination of it body and tail surfaces (Muijres et al., 2012a). Comparisons of the wake structures of birds and bats (Figure 39.6) show that birds have a higher aerodynamic efficiency (including a high lift to drag ratio) than bats, and this might contribute to the observation that birds tend to migrate more frequently and over longer distances than bats (Muijres et al., 2012b).

The first empirical attempt to measure directly the biomechanical force produced by the muscles of a bird during flight used a calibrated strain gauge, recording from the humerus of a starling (*Sturnus vulgaris*) and a pigeon (*C. livia*) and yielded estimates for the biomechanical P_o generated by the pectoralis muscle (Biewener et al., 1992; Dial and Biewener, 1993). A more complete series of measurements was achieved with black-billed magpies (*Pica hudsonia*), with flight speeds ranging from 0 to 14 m/s (Dial et al., 1997), followed by further studies on cockatiels (*Nymphicus hollandicus*) and turtle doves (*Streptopelia risoria*) over the next few years (Tobalske et al., 2003). For the magpie, only hovering resulted in a significantly higher value of P_o, compared with speeds of 2–14 m/s, yielding a so-called L-shaped curve, and a similarly rather flat profile or even J-shaped curve has been reported for many species when measuring P_i (see Section 39.3.3.2). However, for the cockatiel and the dove, changes in biomechanical P_o during flight showed the occurrence of significant U-shaped power curves with a change in velocity (as predicted by aerodynamic theory), with pectoralis mass-specific P_o ranging from 74 to 231 W/kg for the cockatiel (Figure 39.7). These authors point out that considerable flexibility in the precise shape and positioning of the power curve probably exist between avian species, which may have different flight styles and can functionally morph their wings and tail

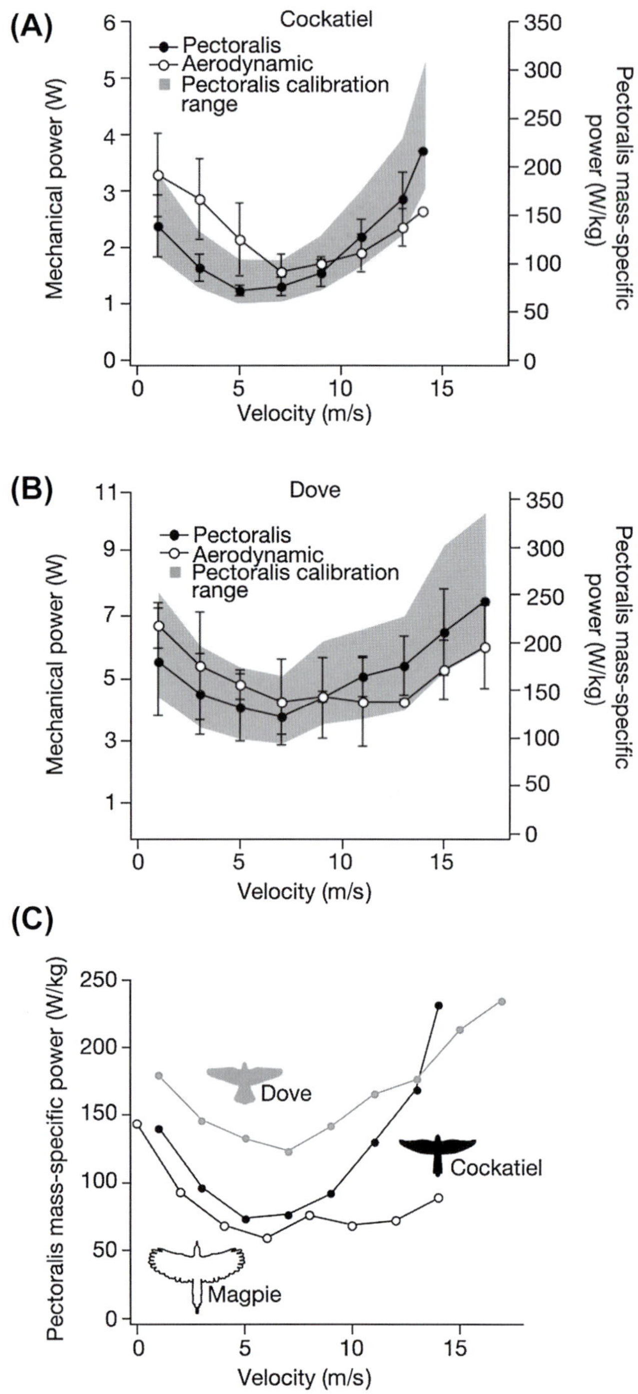

FIGURE 39.7 Mean pectoralis muscle and aerodynamic power output as a function of flight velocity in five cockatiels, *Nymphicus hollandicus* (A) and three ringed turtle-doves, *Streptopelia risoria* (B). The shaded region indicates the range of pectoralis power obtained when "low-power" versus "high-power" aerodynamic coefficients are input to the aerodynamic model. (C) Comparative mass-specific pectoralis power as a function of flight velocity in cockatiels, doves, and magpies, *Pica hudsonia*. *From Tobalske et al. (2003), Nature Publishing Group.*

profiles during flight and, thus, affect their relative P_o (Dial et al., 1997; Tobalske et al., 2003). Of course, this is a complex technique and the authors acknowledge that there may have been some limitations in the calibration of the strain gauges, which would have led to an underestimation of the maximal force and P_o.

Usherwood placed differential pressure sensors along and across the wings of Canada geese, *Branta canadensis*, and pigeons (Usherwood et al., 2005), in order to measure the net aerodynamic forces acting on the wings of birds during takeoff and slow forward flight. This technique was used to determine the lift and drag values for the wings and revealed a muscle mass-specific P_o of 273 W/kg for the pigeon. This value is much higher than that determined by Tobalske et al. (2003) for cockatiels and doves flying at a similar speed (80 and 150 W/kg muscle) but similar to those recorded during maximum fast flights in the wind tunnel and from the similarly sized American crow (*Corvus brachyrhynchos*) during vertical flight (Jackson and Dial, 2011). Askew and Marsh (2001) developed an *in vitro* method for measuring muscle force production in the blue-breasted quail (*Coturnix chinensis*), using typical *in vivo* length and activity patterns, to record average power outputs of 350 W/kg. Of course, these higher values for *in vitro* P_o may not accurately represent sustainable aerobic performance *in vivo*. A similar approach was taken by Askew and Ellerby (2007) to show that *in vitro* P_o from budgerigars (*Melopsittacus undulatus*) and zebra finches (*Taenopygia guttata*) does follow a classical U-shape.

39.3.2 Techniques Used to Measure the Power Input Required for Flight

39.3.2.1 Mass Loss

Some of the earliest measurements of P_i were estimated from the mass loss recorded during a long, nonstop flight, with the assumption that fat constituted by far the major part of this loss and that the net loss of water was negligible (Nisbet et al., 1963). Berger and Hart (1974) criticized this method, as they believe that total water loss may exceed the production of metabolic water, particularly at high environmental temperatures (cf. Nachtigall, 1995), leading to significant dehydration. Even so, Jehl (1994) has used this method to estimate the energy expenditure of black necked grebes (*Podiceps nigricollis*) forced to land during migration as a result of bad weather. The data were inconclusive, mainly because of uncertainty over the time the birds were flying. Another problem with this method is that there is evidence to indicate that migrating birds also catabolize protein (e.g., muscle and intestine) during flight (Piersma and Jukema, 1990; Jenni-Eiermann and Jenni, 1991; Battley et al., 2000), and, whereas fat has an energy density of 39.3 kJ/g, the value for muscle (which is composed of approximately 70% water) is only 5.4 kJ/g (cf. Schmidt-Nielsen, 1997).

Masman and Klaassen (1987) developed a more complex method of using changes in body mass to estimate flight costs. They studied the energy budgets of trained kestrels (*F. tinnunculus*) performing directional flights in the laboratory. Energy expenditure during flight was calculated by monitoring daily metabolizable energy intake, oxygen consumption, and carbon dioxide production during rest, and time spent flying per day. A similar approach was applied to the study of a single thrush nightingale (*L. luscinia*) flying for very long periods (seven different 12 h flight sessions) in a wind tunnel (Kvist et al., 1998). However, in general, estimates of energy consumption based on changes in body mass may be subject to considerable error as they are very sensitive to the values of protein and/or carbohydrate utilized in the calculation.

39.3.2.2 Doubly Labeled Water

The most commonly used method to determine the field metabolic rate (FMR) of air-breathing vertebrates is the doubly labeled water (DLW) method, and it has been used extensively with free-ranging birds (Hails, 1979; Ellis, 1984; Tatner and Bryant, 1989). This method was first proposed as a means of determining CO_2 production by Lifson and co-workers over 40 years ago (Lifson et al., 1949, 1955). It involves the injection of a mixture of ^2H_2O (or ^3H_2O) and $H_2{}^{18}O$ into the animal, allowing these isotopes to equilibrate with the body water pool(s), taking a body fluid sample after injection, and releasing the animal into the field. After a number of days, the animal is recaptured and another body fluid sample is taken. The difference between the rate of loss of the different isotopes between the two samples is assumed to be equivalent to the rate of CO_2 production ($\dot{V}co_2$), from which the metabolic rate and P_i of the animal can be estimated. The maximum time between these sampling points (the maximum duration of the experiment) varies according to the initial enrichment (the greater this is, the longer the time), the mass of the animal (for the same relative enrichment, the greater the mass, the longer the time), and, of course, the level of activity of the animal. The many assumptions that are made when using this technique have been thoroughly discussed by Nagy (1980), Speakman (1990). The relative pros and cons of the DLW technique in comparison with those of using measures of heart rate (see Section 39.3.2.3.1) have been reviewed by Butler et al. (2004).

Although DLW is a relatively simple method to use in the field, it has three major limitations. Perhaps most importantly for field work is the fact that the duration of an experiment is limited by the turnover rate of the ^{18}O, and thus it is necessary to recapture the animal after a sufficient delay following injection of the isotopes to enable a reasonable decline in the enrichment of ^{18}O, but before the enrichment is too low (Nagy, 1983). The second problem is that the technique only gives an average value for energy expenditure between the two sampling points and this is usually over a 24 h daily cycle of activity (Speakman and Racey, 1988). When DLW is used to determine the metabolic cost of a specific activity (e.g., flying), then, as with the body mass method, it is necessary to have accurate information on the duration of the flying period(s) between the initial and final sampling times. Ideally, the respiratory quotient (RQ) also needs to be known to convert the calculated $\dot{V}co_2$ to an estimate of $\dot{V}o_2$ and/or P_i (see above), but this will often introduce further errors. Finally, the method only provides estimates of $\dot{V}co_2$ and yet the results are invariably treated as being absolute. In other words, there is a degree of uncertainty around the individual values obtained, but this is usually ignored. If statistical analyses are performed on means of the individual results, this can lead to type I errors (i.e., a false positive).

Recent validation studies with captive birds exercising at different levels of activity (but not flying) over a period of 72 h have indicated that, provided mean data are used from a number of birds, the DLW method can give estimates of metabolic rate that are within a few percent of those measured by respirometry (Nolet et al., 1992; Bevan et al., 1994, 1995a), although they do have a greater variance around the mean. The only validation study on a flying bird was that of Ward et al. (2004), in which they used three independent techniques—DLW, mask respirometry, and heat transfer modeling—to measure the P_i of the European starling (*S. vulgaris*) flying in a wind tunnel between 6 and 14 m/s. Heat transfer modeling and mask respirometry yielded comparable results and a linear increase in P_i with increasing velocity, but the DLW results had a greatly increased scatter and variance between individual flight measurements.

39.3.2.3 Telemetry and Data Logging

Much progress has been made in the use of biotelemetry and data logging devices for studying the physiology of free-ranging animals. Some of the earlier studies involved attaching relatively large radio transmitters to the back of birds and recording data during flights of only a few seconds duration (Hart and Roy, 1966; Berger et al., 1970a,b). Of course, any externally mounted object is likely to influence the flight performance of a bird, not only as a result of the added mass but also by increasing the aerodynamic drag of the body (Bowlin et al., 2010). These effects may be minimal for a relatively small transmitter that the bird can preen under its contour feathers (Obrecht et al., 1988). However, in some circumstances, the effects can be quite substantial, particularly for high-performance birds like homing pigeons (Gessaman and Nagy, 1988). Butler and Woakes (1980) overcame the problem of increasing body drag by implanting relatively small (<10 g) radiotransmitters into the abdominal cavity of barnacle geese (*Branta leucopsis*) and were able to record heart rate (mean, 512 beats/min) and respiratory frequency (mean, 99 breaths/min) from two birds (mean mass, 1.6 kg) flying behind an open-topped vehicle. It is now possible to implant data loggers (weighing less than 30 g) that will store basic physiological information (heart rate, body temperature, etc.) over periods of many days. The data can then be downloaded after the animal is recaptured (Woakes et al., 1995; Bevan et al., 1995b).

39.3.2.3.1 Heart Rate

Heartbeat frequency (f_H) can be accurately measured and used as a proxy for metabolic energy consumption. Thus, if properly calibrated and validated, and if mean data are used from a number of animals, f_H may be used as an indicator of the rate of oxygen uptake in free-living birds (Butler et al., 2004). By storing f_H in implantable data loggers, Bevan et al. (1995b) estimated the cost of gliding flight, while having minimal effect on the behavior of the birds. However, the relationships between f_H and $\dot{V}O_2$ that are obtained during walking and running, or during swimming, may not be identical to those obtained when the birds are flying (Gessaman, 1980; Nolet et al., 1992; Ward et al., 2002). Ideally, it is recommended that f_H during flight should be calibrated during flight in a wind tunnel, and values recorded in the field should not exceed the range of f_H measured during calibration. The study by Ward et al. (2002) remains the only report of a direct calibration of heart rate against mask respirometry of a bird flying in a wind tunnel (Figure 39.8), in which there is a reasonable variation in both the measures of heart rate and the rate of oxygen consumption. The calibration relationship for both barnacle and bar-headed geese is different between running and flying but it is not known how typical this may be of other species. It is likely that it is the running relationship that is unusual in these species, given that geese are not thought to be particularly reliant on running in the wild.

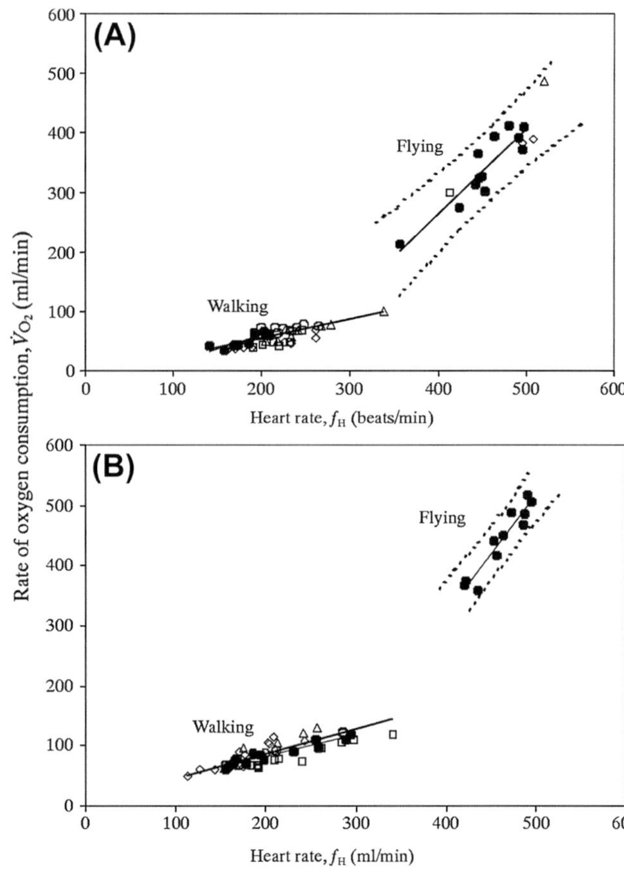

FIGURE 39.8 Linear relationships between mean rate of oxygen consumption ($\dot{V}O_2$) and mean heart rate (f_H) of (A) barnacle geese, *Branta leucopsis*, and (B) bar-headed geese, *Anser indicus*, walking on a treadmill and flying in a wind tunnel. The filled circles show data from a single bird of each species from which most data on the relationships between f_H and $\dot{V}O_2$ were obtained during flight (N=12 flights by barnacle goose B-B and 11 flights by bar-headed goose BH-O). The open symbols show data from other birds. The solid lines show the relationships between f_H and $\dot{V}O_2$ during walking and flight by barnacle goose B–B (walking, $\dot{V}O_2 = 0.47 f_H - 35.9$, r^2=0.78; flight, $\dot{V}O_2 = 1.42 f_H - 304$, r^2=0.82) and bar-headed goose BH-O (walking, $\dot{V}O_2 = 0.35 f_H + 12$, r^2=0.88; flight, $\dot{V}O_2 = 1.97 f_H - 467.5$, r^2=0.90). The broken lines show the 95% prediction intervals. *From Ward et al. (2002). Company of Biologists, Ltd.*

39.3.2.3.2 Accelerometry

In recent years, accelerometers have been mounted on the body of animals to provide high-resolution measures of acceleration (or changes in velocity). During steady state locomotion, the average acceleration, taken over a sufficient number of limb cycles (or flapping in the case of flight), is equivalent to the "static" component of acceleration due to gravity. The average of the individual raw acceleration data points minus the static acceleration for a given time period, gives a time-averaged "dynamic body acceleration" or DBA (Qasem et al., 2011). In multidimensional acceleration studies, summing the data taken from the different axes before taking the final average is called overall dynamic body acceleration (ODBA), and taking the vectoral length before averaging is termed VeDBA (Qasem et al., 2011). Both these measures have been shown to be well correlated with rates of oxygen consumption in human subjects and a number of other mammal and bird species while running on treadmills (Qasem et al., 2011).

Although no detailed and direct correlation of DBA and energy output has been conducted on birds during flight, ODBA was shown to increase during ascending and decrease during descending flights by a captive Griffon vulture *Gyps fulvus*. Recently, it has been shown that VeDBA is quite well correlated with the average daily energy expenditure of breeding thick-billed murres (*Uria lomvia*), as determined by measurements of DLW. However, as already mentioned, DLW only gives an estimate of $\dot{V}O_2$. In addition, Spivey and Bishop (2013) describe a mathematical model to help to interpret data collected by accelerometers mounted on the body of birds during flight. By assuming that the body motion of the bird during steady forward flight approximates to a sinusoidal waveform, they showed that the root mean square of acceleration (aRMS), another form of time-averaged DBA (which is closely related to VeDBA), can be naturally derived from the mathematics. The biomechanical power visible to the body-mounted accelerometer (P_{body}) can be interpreted in the SI units for mass-specific power (watts = $m^2 \times s^{-3}$) by a number of possible terms:

P_{body} is proportional to $(g \times a\text{RMS})/f_w$

P_{body} is proportional to $(a\text{RMS})^2/f_w$

Where g is the acceleration due to gravity (m/s^2) and f_w is wingbeat frequency (1/s). Uncertainties regarding the use of accelerometry center around the alignment of the sensor with the axis of movement of the body, the fact that the cost of limb movements may be partially "hidden" from a body-mounted sensor, as may changes in body mass itself, and that rotations of the body not directly linked to movement costs may exaggerate the apparent energy required. However, with more studies, it is likely to become an important technique in assessing animal movement and energetics.

Measurements of acceleration can also be used to determine relative body orientation and kinematics, in particular that of wingbeat frequency. Sato et al. (2008) looked at the foraging loads of chick rearing European shags (*Phalacrocorax aristotelis*) and the results suggested that changes in body mass of an individual could be reliably determined by measuring the wing flap frequency. Sato et al. (2009) studied the flapping frequency of five species of procellariiformes, during takeoff and during forward flight. They showed that during takeoff the wingbeat frequency was higher and scaled with $M_b^{-0.3}$, while during forward flight it scaled as $M_b^{-0.18}$. By extrapolating these two relationships until they intercepted, it was suggested that the maximum limits to practical soaring flight might be around 41 kg for a bird or pterosaur with a wing span of around 5.1 m (Figure 39.9).

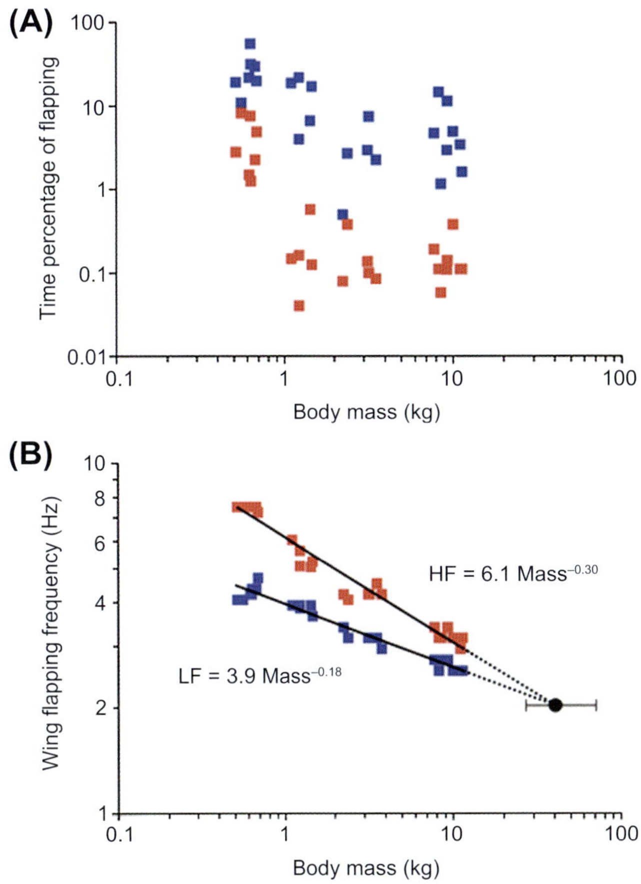

FIGURE 39.9 (A) The relationship between body mass and the percentage of time spent flapping relatively slowly (blue squares) and relatively quickly (red squares) during the foraging flights of five species of albatrosses and petrels. (B) The relationship between body mass and wingbeat frequency for the same species. Regression lines were calculated for high (red squares) and low (blue squares) frequencies and extrapolated for larger animals (dashed lines). The two lines intersect at a body mass of 41 kg (5.1 m wing span), with a 95% CI (26–75 kg). *From Sato et al. (2009).*

39.3.2.4 Wind Tunnel

P_i is most easily estimated indirectly, and under laboratory conditions, by directly measuring the $\dot{V}O_2$ of birds flying in a wind tunnel and then converting that to P_i (see above). The increased use of wind tunnels during the past 30 years has greatly added to our knowledge and understanding of the physiology of birds during flight (Butler and Woakes, 1990; Norberg, 1990). Perhaps the main advantage of being able to use a wind tunnel is that it is possible to make physiological measurements that would be virtually impossible by other means (e.g., repeated blood samples, blood pressure and flow measurements, direct measurements of force production). However, it must also be borne in mind that, as with almost any other method for studying a living animal, the wind tunnel itself can influence the behavior of the animal flying in it and hence affect the data obtained when compared with those from free-flying birds (Butler et al., 1977; Rayner, 1994). In order to obtain physiological data from a bird flying in a wind tunnel, it is often necessary to attach recording equipment to the bird, which will have some effect on the measurements. For example, gas exchange ($\dot{V}O_2$ and $\dot{V}CO_2$) is usually determined by means of a lightweight mask that covers the beak and nose and through which air is drawn via an attached tube (Tucker, 1968b, 1972; Bernstein et al., 1973; Butler et al., 1977; Gessaman, 1980; Hudson and Bernstein, 1983; Rothe et al., 1987; Ward et al., 2001, 2002; Peters et al., 2005). The mask and tubing will influence the recorded data because of their mass, but more so because of the extra drag that they impose. $\dot{V}O_2$ can also be measured in a similar way for hummingbirds while feeding at artificial flower-heads (Wells, 1993a, 1993b; Clark and Dudley, 2010). The only studies in which gas exchange has been determined in a wind tunnel without the use of a mask are those by Tucker (1966) and

Torre-Bueno and Larochelle (1978). These authors used air-tight, closed-circuit wind tunnels and measured the changing concentration of gases in the tunnel during the flight periods.

An aerodynamic analysis by Rayner (1994) suggests that the P_o required by a bird to fly is reduced when it flies in a wind tunnel with a closed section (unless the tunnel is at least 2.5 times the wing span of the bird and it flies at or near the center of the tunnel) and increased when it flies in a tunnel with an open section (again, the effect is related to the relative size of the tunnel), compared with the situation during free flight. However, at the moment there are few reliable experimental data to support these conclusions. In this context, it is interesting to note that the value for $\dot{V}O_2$ obtained by Butler et al. (1977) for pigeons (mean mass 0.442kg) flying in a relatively large wind tunnel at 10m/sec (182mL O_2/minkg, adjusted for 10% effect of drag of mask and tubing; Tucker, 1972) is similar to that obtained by LeFebvre (1964) using the DLW method for free-flying pigeons (199mL O_2/minkg, mean mass 0.384kg). However, it is somewhat less than the mean minimum value obtained by Rothe et al. (1987) for pigeons (mean mass 0.330kg) flying in a relatively small wind tunnel (248.5mL O_2/minkg, also adjusted for drag of mask and tubing, but for a 22% effect; Rothe et al., 1987). According to Rayner's analysis, aerodynamic factors alone would lead to $\dot{V}O_2$ being lower for the pigeons flown by Rothe et al. (1987). Rayner (1994) also concludes that, in closed wind tunnels, the P_o-versus-speed curve (see above) will be flatter than that for free-flying birds.

Contrary to the above, other studies have suggested that data on the energy cost of flight obtained from birds flying in wind tunnels are 30% to 50% higher than those obtained from some of the other methods discussed above, such as mass loss and DLW (Masman and Klaassen, 1987; Rayner, 1990). However, the former authors included in their wind tunnel data measurements from hovering hummingbirds and from other experiments that most definitely did not involve the use of a wind tunnel (Teal, 1969; Berger et al., 1970a). Rayner (1990) used the same dataset of 71 birds as Masman and Klaassen (1987), but supplemented it with 40 more values from more recent publications. Unfortunately, he did not give the source of these additional data, so it is not possible to check and if necessary reanalyze them. However, taking the nine values by the DLW technique and used by Masman and Klaassen (1987) and reanalyzing them using the method of reduced major axis (RMA; Sokal and Rohlf, 1981; Rayner, 1985b), P_i (W) during forward flapping flight is calculated as $P_i = 69.5\ M_b^{0.87}$, $r^2 = 0.83$, $n = 9$ M_b = body mass in kg). Taking the seven values for birds that did actually fly in wind tunnels from the list of Masman and Klaassen (1987) (i.e., numbers 44, 54, 57, 63, 64, 67, 68; note, P_i for bird 44 should be 4.08W and not 40.8W) gives $P_i = 58.8\ M_b^{0.76}$, $r^2 = 0.98$. Neither the slopes nor the elevations of these regressions are significantly different from one another. If the minimum values are taken from the seven studies of birds flying in wind tunnels, and if the drag of the mask and tubing is taken into account, then the equation is $P_i = 52.6\ M_b^{0.74}$, $r^2 = 0.95$. Thus, there is no justification for the conclusions made by Masman and Klaassen (1987) and Rayner (1990). It should be noted that Norberg (1996) made similar erroneous inclusions in her "wind tunnel" data (see Table 7.3) to those of Masman and Klaassen (1987).

39.3.2.5 Modeling of Cardiovascular Function

Some of the shortcomings of using the aerodynamic models to estimate P_o and then, subsequently, converting them to estimates of P_i (by assuming a value for efficiency) could possibly be reduced by devising a model to estimate total P_i directly. This was highlighted by the studies of Ward et al. (2001, 2004), who pointed out that starlings may not have a constant value of muscle efficiency throughout their speed range and that conversion of P_o to P_i was sensitive to the value of efficiency. Bishop and Butler (1995) attempted a modeling approach to estimating P_i by assuming that the mass of an animal's heart could be used as a basis for estimating cardiac stroke volume and calculating the animal's aerobic capacity. The role of the various components of the cardiovascular system in presenting O_2 to (and removing CO_2 from) the exercising muscles can best be described by a form of Fick's formula for convection:

$$\dot{V}O_2 = f_H \times V_S \times (C_aO_2 - C_{\bar{v}}O_2)$$

where f_H is heartbeat frequency (beats/min); V_S is cardiac stroke volume (mL); and $(C_aO_2 - C_{\bar{v}}O_2)$ is the difference between the oxygen content in arterial and mixed venous blood (mL of O_2 per mL of blood). In theory, each of these variables can scale independently with respect to body mass such that:

$$(\dot{V}O_2)\ M_b^z = (f_H)\ M_b^w \times (V_s)\ M_b^x \times (C_aO_2 - C_{\bar{v}}O_2)\ M_b^y.$$

There have only been two studies in which all four variables of the Fick equation have been measured in a bird during forward flapping flight (Butler et al., 1977; Peters et al., 2005). Bishop and Butler (1995) substituted each of the values from the Butler et al. (1977) study into the above equation, based on a series of general allometric equations ($y = a \cdot x^b$) for the scaling of f_H, V_s, and $(C_aO_2 - C_{\bar{v}}O_2)$, respectively. For birds flying close to their minimum power speeds, the individual components of the Fick equation were estimated to scale as follows: $f_H = 574\ M_b^{-0.19}$, $V_s = 3.48\ M_b^{0.96}$, $(C_aO_2 - C_{\bar{v}}O_2) = 0.083\ M_b^{0.00}$. The estimated scaling of V_s during flight was based on the assumption that V_s should be almost directly proportional to the mass of the heart (M_h), or $V_s = 0.3\ M_h^{1.05}$, while f_H during flight and M_h may be readily determined experimentally.

Bishop and Butler (1995) concluded that, for birds with a high aerobic capacity (such as pigeons), $\dot{V}O_2$ (mL/min) $= 166\ M_b^{0.77}$, for birds flying close to their minimum power speeds. This compares favorably with the relationship $(\dot{V}O_2 = 160\ M_b^{0.74})$ calculated by Butler (1991) and reanalyzed using RMA for seven species of birds during forward flapping flight in a wind tunnel. Bishop (1997) used a similar approach to estimate the maximum $\dot{V}O_2$ of birds during flight, again using M_h as the basis of the calculation. Although there is a need to obtain more data on the cardiovascular changes that occur in birds of different body mass and during different modes of flight, it would appear that the use of M_h to estimate the aerobic capacity of different species of birds may have more general applicability than the use of the allometric relationship obtained from a few species of birds flying in a wind tunnel. Bishop and Spivey (2013) did a meta-analysis of data from 24 species of birds and mammals that were said to be performing their "primary" modes of locomotion (running, swimming, or flying, respectively) and concluded that there was an overall allometric trend across all species for $\dot{V}O_2$ to be approximately proportional to heart rate squared (f_H^2). Individuals and different species may vary in how closely they followed this quadratic heart rate exponent, but knowledge of both heart mass and body mass gave the best predictive relationship for rate of oxygen consumption (Figure 39.10).

39.3.3 Empirical Data Concerning the Power Input during Flight

An individual flight by a free-ranging bird will include at least one takeoff and a period of climbing, probably several maneuvers (turns, short periods of "burst" flying, changes of gait) as well as various periods of gliding, soaring, bounding, and flapping flight. The power requirements associated with all of these various types of flight will be different and may be provided by either anaerobic or aerobic metabolic pathways or by a combination of the two.

Species belonging to the Galliformes (e.g., pheasants, grouse, and quail) routinely engage in a powerful "burst" type of flight during takeoff and short-duration flights. Their muscles use intracellular stores of muscle glycogen utilizing predominantly anaerobic metabolism, which leads to the accumulation of lactic acid and to a metabolic acidosis. Fatigue soon develops and, subsequently, the lactate is either oxidized as fuel by tissues such as the heart or reconverted to glycogen by the liver. Thus, a large proportion of the total cost of these explosive flights is "repaid" while the bird is resting on the ground following the flight (Bishop and Butler, 1995) and an assessment of the true cost of such explosive flights, if based on $\dot{V}O_2$, would have to include the post-flight period of recovery. In other words, measurements of $\dot{V}O_2$ can only be used for accurate estimates of P_i if the activity is sustainable for relatively long periods of time (i.e., if the bird is metabolizing aerobically).

39.3.3.1 Gliding and Soaring Flight

Previous study by Pennycuick has hypothesized that the cost of holding the wings outstretched when gliding and soaring should scale at around 1.5 times the BMR. In general, as BMR decreases as body size increases, while at the same time the speed of flight increases, then it would be expected that larger birds should benefit more than smaller species

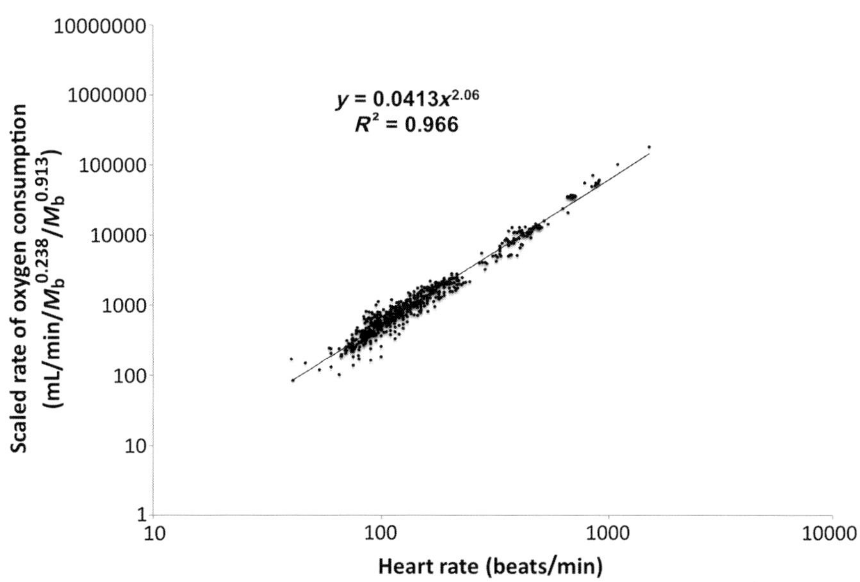

FIGURE 39.10 The relationship between heart rate (beats/min) and the rate of "rescaled" oxygen consumption $\left(\text{mL/min}/M_b^{0.328}/M_b^{0.913}\right)$ for 12 species of birds and 12 species of mammals, after taking into account differences in body mass (M_b) and heart mass (M_h) between species. Each species was flying, running or swimming, depending on what was considered their "primary mode" of locomotion, i.e., the form of locomotion that would yield the highest rate of oxygen consumption. *From Bishop and Spivey (2013). Elsevier publishing company.*

from the strategic use of gliding. In reality, most empirical studies have recorded slightly higher costs for gliding. In windy conditions, albatrosses exclusively use soaring and gliding flight and even in near calm conditions, they only beat their wings occasionally (Alerstam et al., 1993). However, it should be remembered that gliding birds can actively adjust the shape and span of their wings, including the sweep angle that they make to the body, in order to adjust air speed and the aerodynamic efficiency of the wing (Rosén and Hedenström, 2001). Using the DLW method, Adams et al. (1986), studying the wandering albatross (*Diomedia exulans*), and Costa and Prince (1987), studying the gray-headed albatross (*Diomedea chrysostoma*), estimated that the metabolic cost of flight is 3.0 and 3.2 times the predicted basal metabolic rate (BMR), respectively (49.7 W for birds of 8.4 kg mean mass, 36.3 W for birds of 3.7 kg mean mass). These estimates were based on the assumption that, when on water, the metabolic rates of the birds were close to those when they were at rest on land and their mean activity budgets were similar to those obtained from other studies on the same species. However, for herring gulls (*Larus argentatus*) gliding in a wind tunnel, P_i (calculated from measured $\dot{V}o_2$) was at 13.9 W for birds of 0.91 kg mean mass, only 2.1 times the resting value (Baudinette and Schmidt-Nielsen, 1974).

In a study on black-browed albatrosses (*Diomedea melanophrys*), Bevan et al. (1995b) used the f_H method to determine the field metabolic rate (FMR) and saltwater switches to determine when the birds were on the water. It was, therefore, possible for these authors to estimate the metabolic rate of their birds when on water and this was 2.6 times that recorded from the birds during incubation and only 1.9 times the predicted BMR; that is, somewhat lower than was assumed by Adams et al. (1986) and Costa and Prince (1987). Interestingly, the average metabolic cost of flying was 21.7 W for birds of 3.5 kg mean mass, 60% of that estimated for the similar sized gray-headed albatross by Costa and Prince (1987) and not significantly different from the value obtained when the birds were on the water. Other studies on sea birds, in which the DLW method has been used to estimate FMR, have used saltwater switches in order to determine when the bird was airborne, but the species studied (Sulidae) use a combination of gliding and flapping flight, so it was not possible to determine the energetic cost of either mode (Birt-Friesen et al., 1989; Ballance, 1995). Similar limitations apply to other studies in which the assumption is that, when not on the nest, the birds are continuously flying (Flint and Nagy, 1984; Obst et al., 1987).

The study by Bevan et al. (1995b) illustrates the importance of accurately determining the behavior of the animals under study and the advantages of using a method for determining FMR that allows the metabolic cost of the separate behaviors to be estimated. By using satellite transmitters and the logging of f_H data, Bevan et al. (1995b) were able to determine the metabolic costs of different sections of a foraging trip of an individual black-browed albatross (Figure 39.11). During section 2 of the trip, it covered a distance of 80.1 km

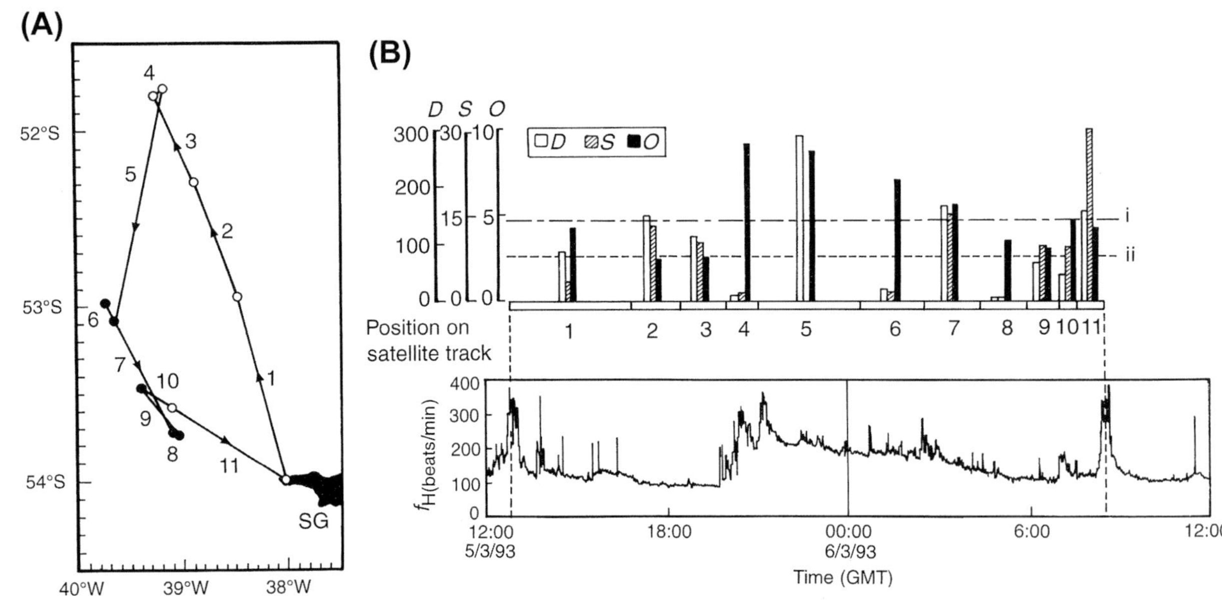

FIGURE 39.11 (A) Flight track of a single foraging trip from South Georgia (SG) by a black-browed albatross, *Diomedea melanophrys*, equipped with a satellite transmitter and a heart rate data logger. Circles represent the position of each satellite uplink, with the solid circles showing positions during the hours of darkness. The arrows show the direction traveled between each successive position, each section being numbered consecutively. (B) Top graph: distance traveled (*D*, km), ground speed (*S*, m/s), and estimated energy expenditure (*O*, W/kg) for each section of the trip depicted in a. The horizontal bars represent the time between each satellite uplink. Mean ground speed (horizontal broken line, i) is from Alerstam et al. (1993) and the incubating metabolic rate (horizontal broken line, ii) is that measured while birds were on the nest incubating (W/kg). Bottom graph: heart rate (f_H) trace over the same period. The vertical dashed lines joining the two graphs show when the bird left and returned to the colony. *Modified from Bevan et al. (1995b). The Royal Society.*

in 1.7 h, a mean ground speed of 13.1 m/s. This is very similar to the mean speed of 12.9 m/s that Alerstam et al. (1993) report for a visually tracked black-browed albatross. This suggests that the bird was flying continually during section 2. Metabolic rate was low during this period (2.4 W/kg) and barely greater than that during incubation (2.2 W/kg). The bird was probably searching for patches of prey and being carried by the wind (Alerstam et al., 1993). On the other hand, during section 4 of the trip, the straight-line distance traveled was only 6.3 km at a mean ground speed of 1.6 m/s. This suggests that the bird was relatively stationary on the water and/or was flying within a restricted area. It exhibited its highest metabolic rate (8.8 W/kg) during this period, indicating that the bird had engaged in some form of relatively strenuous activity, such as more active flight and/or feeding.

As gliding is thought to be more common in larger species of birds, it has seldom been quantified in smaller species, although some small species of birds, such as the Hirundines (martins and swallows) and the Apodidae (swifts) regularly glide. However, Sapir et al. (2010) showed that, like in the study of black-browed albatrosses (Bevan et al., 1995b), the heart rate of resting European bee-eaters was similar to that recorded while gliding and soaring, either on migration or at stopover sites (Figure 39.12). They suggest that not only was gliding metabolic rate only 1.9 times BMR but that the air speed of soaring was sufficiently fast (10.6 m/s) as to make gliding very economic. Assuming an f_H^2 relationship with oxygen consumption, Bishop and Spivey (2013) suggest that gliding flight may be at least five times less costly than typical flapping flight for the bee-eaters. Swifts (*A. apus*) have been predicted to glide most efficiently at velocities of 8–10 m/s (Lentink et al., 2007), which fits nicely with their behavior of climbing high above the ground and gliding slowly at around 10 m/s while sleeping on the wing. Similarly, magnificent frigate birds (*Fregata magnificens*) spend almost all of their time in slow soaring flight, constantly ascending and descending on thermals during the day and night, while looking for infrequent opportunities to feed (Weimerskirch et al., 2003).

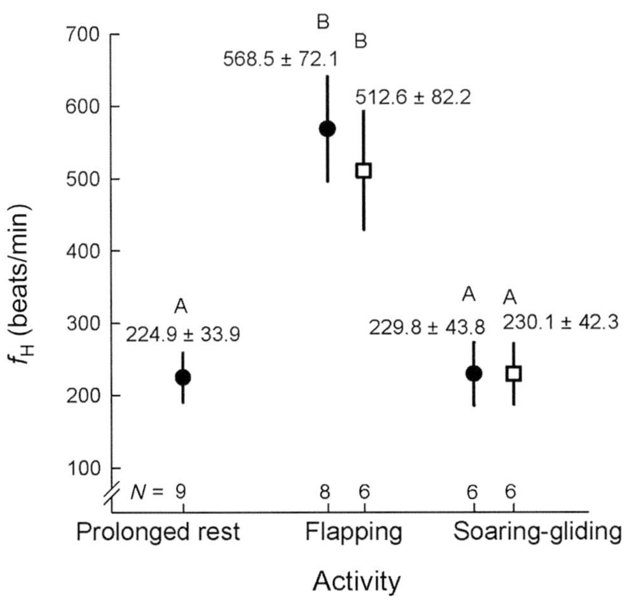

FIGURE 39.12 Heartbeat frequency (f_H) of European bee-eaters, *Merops apiaster*, measured in the field in relation to their behavior. The data is presented as the mean ± standard deviation of f_H for bee-eaters resting for prolonged duration or in different flight modes recorded, during stopovers (filled circles) and migratory cross-country flights (open squares). Different letters above the bars indicate groups that differed statistically. N = number of birds whose f_H was recorded during each activity. *From Sapir et al. (2010).*

During gliding flight it is important to minimize body (or parasite) drag and to maintain a high lift to drag ratio. Even turning the head from side to side can significantly increase the effective body drag. Indeed, because birds usually have laterally placed eyes with relatively low stereoscopic fields of view, they often prefer to see things either in one eye or the other. In the case of a fast-diving peregrine falcon (*Falco peregrinus*), it approaches its prey by flying along a spiral path so that it can keep its head facing forward on its body. Moving the head to one side or the other can increase parasite drag by around 50%.

39.3.3.2 Forward Flapping Flight

Some of the most comprehensive data we have on the overall energy consumed by the body (P_i) and the biomechanical power output by the muscles (P_o) during forward flapping flight are those obtained from birds flying in wind tunnels. In particular, the use of wind tunnels has made it possible to estimate both P_o and P_i at different flight speeds, although the details of these relationships appear to vary between species. The majority of studies of P_i were determined by measuring the rate of oxygen consumption while the birds wore a plastic respirometry mask (Figure 39.13). Results from the laughing gulls, fish crows, and white-necked ravens have shown a J-shaped curve (Tucker, 1972; Bernstein et al., 1973; Hudson and Bernstein, 1983), while the starling, barnacle, and bar-headed goose have a relatively flat "curve" (Ward et al., 2001, 2002). Only the budgerigar (*M. undulatus*), cockatiel, pigeon, and hummingbird have shown some kind of U-shaped curve for P_i (Tucker, 1966; Rothe et al., 1987; Bundle et al., 2007; Clark and Dudley, 2010), similar to that predicted by aerodynamic theory for mechanical power output (Pennycuick, 1969; Tucker, 1973; Greenewalt, 1975; Rayner, 1979). Arguably, the J-shaped P_i curves may not be so different from a U-shaped curve, as very low air speeds were not obtained with these species. Some early work on the P_i of hummingbirds, by Berger (1985), indicated that they might also exhibit a J-shaped curve, but a recent study of Anna's and Allen's hummingbirds (*Calypte anna* and *Selasphorus sasin*, respectively),

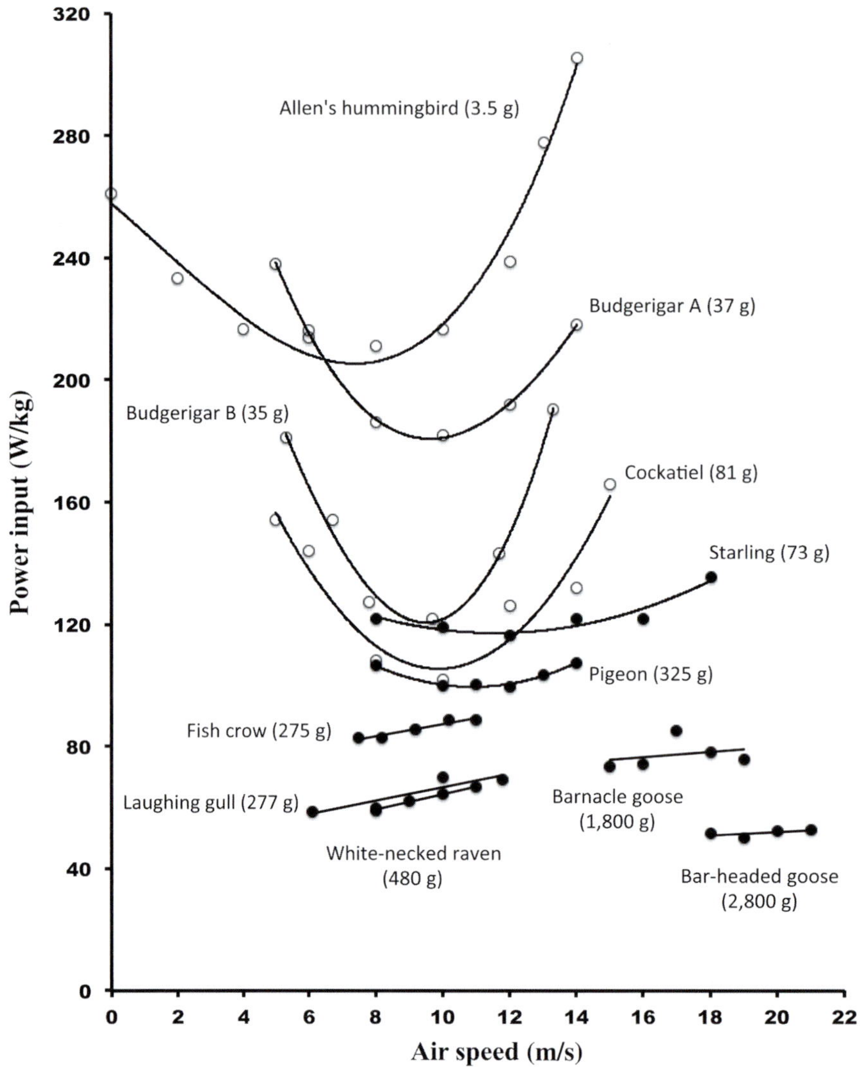

FIGURE 39.13 Power input (W/kg), converted from measurements of rates of oxygen consumption, measured at different air speeds for 10 species of birds during horizontal flapping flight in a wind tunnel. Data taken from Allen's hummingbird, *Selasphorus sasin* (Clark and Dudley, 2010); cockatiel, *Nymphicus hollandicus*, and budgerigar, *Melopsittacus undulatus* A (Bundle et al., 2007); budgerigar B; European starling, *Sturnus vulgaris* (Ward et al., 2001); pigeon, *Columba livia* (Rothe et al., 1987); fish crow, *Corvus ossifragus* (Berstein, Thomas and Schmidt-Nielsen, 1973); laughing gull, *Leucophaeus atricilla* (Tucker, 1972); white-necked raven, *Corvus albicollis* (Hudson and Bernstein, 1983); barnacle goose, *Branta leucopsis*, and bar-headed goose, *Anser indicus* (Ward et al., 2002).

flying between 0 and 14 m/s, indicates that there is a distinct P_i minima at intermediate speeds. Energetic costs during hovering and faster speeds were increased by around 30–40% (Clark and Dudley, 2010), while even relatively slow but backward flights showed a reduction of around 20% over that of hovering (Sapir and Dudley, 2012). The depth of the curve appeared to be slightly increased in birds with higher wing loadings, which was also more typical of males.

Figure 39.14 illustrates an allometric plot of the available data for the minimum P_i of seven species of birds during forward flapping flight in wind tunnels (Tucker, 1968b, 1972; Bernstein et al., 1973; Butler et al., 1977; Torre-ueno and Larochelle, 1978; Gessaman, 1980; Hudson and Bernstein, 1983), and it demonstrates two interesting features. First, on average, the minimum P_i (after adjusting for the drag of the mask and tubing, where necessary) of the seven species (mass range, 35–480 g) is 9.2 times the BMR calculated for nonpasserine birds (Prinzinger and Hänssler, 1980), and second, it is 2.2 times the P_i of mammals of similar mass (up to 900 g) running at their maximum sustainable speed (Pasquis et al., 1970). It can be seen from Figure 39.10 that the minimum metabolic rate during flight of the hummingbird, *Colibri coruscans* (Berger, 1985), is close (within 20%) to that predicted from the extrapolated regression line for the other seven species of birds. If the value from the hummingbird data is included in the regression analysis, the equation becomes $48.5\ M_b^{0.69}$, $r^2=0.98$,

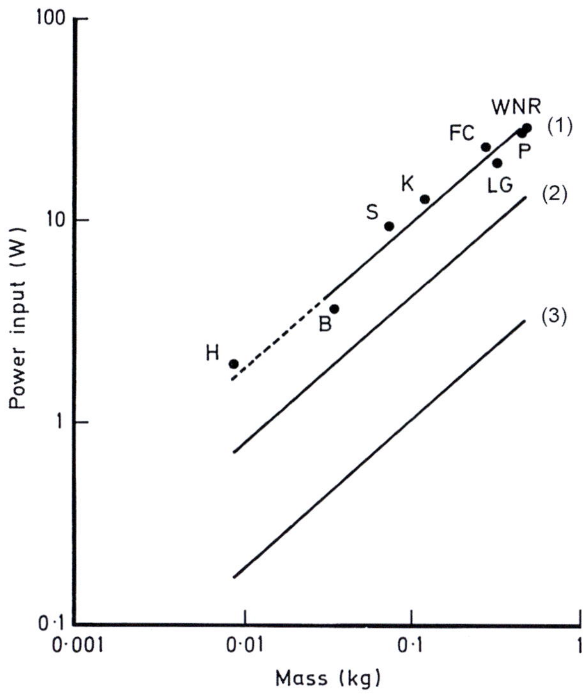

FIGURE 39.14 The relationships between power input (P_i) and body mass, M_b, using least-squares regression analysis for (1) seven species of birds during forward flapping flight in a wind tunnel; B, budgerigar, *Melopsittacus undulatus*; S, starling, *Sturnus vulgaris*, K, American kestrel, *Falco sparverius*; LG, laughing gull, *Larus atricilla*; FC, fish crow, *Corvus ossifragus*; P, pigeon, *Columba livia*; WNR, and white-necked raven, *Corvus cryptoleucus*. Minimum values of power input (P_{im}) were used to construct this curve and $P_{im}=50.7\ M_b^{0.72}$. (2) Small mammals during maximum sustainable exercise where $P_i=22.6\ M_b^{0.73}$ *(data from Pasquis et al. (1970))*. (3) Resting nonpasserine birds where $P_i=5.5\ M_b^{0.73}$ *(data from Prinzinger and Hanssler (1980))*. *(From Butler and Woakes (1985).)* Also included in (1), the minimum P_i of the hummingbird *Colibri coruscans* during forward flapping flight in a wind tunnel (H). *From Berger (1985).*

using least-squares regression analysis. The mean P_i of a 1.8 kg barnacle goose (*B. leucopsis*) flying in a wind tunnel was 102 W, while that of a 2.8 kg bar-headed goose (*Anser indicus*) was 135 W (Ward et al., 2002). These results are reasonably similar to the values of 81 and 113 W obtained for the predicted minimum P_i, determined from an extrapolation of the allometric equation derived from the seven species of birds flying in a wind tunnel (Butler, 1991), recalculated using RMA regression (see above). In addition, using the cardiovascular modeling approach of Bishop and Butler (1995), Butler et al. (submitted-a) estimated the minimum $\dot{V}O_2$ of premigratory barnacle geese (average $M_b=2.3$ kg) to be around 99 W. The relatively small errors between experimental measurements and these various predictions for flight costs, whether extrapolated towards smaller or larger species using allometric equations, or applying cardiovascular modeling approaches (Bishop and Butler, 1995; Bishop, 1997), suggests that the measurement of P_i is reasonably reliable across different studies.

39.3.3.3 Hovering Flight

As discussed above, hovering flight should be relatively costly, at least compared to intermediate air speeds. While a number of studies show that this is the case for muscle P_o, very few species of birds can sustain hovering for long enough so that P_i can be determined (usually by measuring the rate of oxygen consumption). Both budgerigars (Tucker, 1966; Bundle et al., 2007) and cockatiels (Bundle et al., 2007; Morris et al., 2010) show distinct U-shaped P_i curves, but the slowest speeds that they were willing to fly at while wearing a face-mask were 5 m/s. However, it does seem likely that slower velocities, including hovering, would have increased their flight costs even further. Only hummingbirds have been studied in which $\dot{V}O_2$for both hovering and forward flapping flight have been determined for the same bird (Berger, 1985; Clark and Dudley, 2010; Sapir and Dudley, 2012). The more recent studies indicate that there is a small, but significant, difference in P_i between the two forms of flight (Figure 39.13). This is consistent with the fact that the values for $\dot{V}O_2$ from six species of hovering hummingbirds lie close to the extrapolated line obtained from larger birds during forward flapping flight, even though the allometric relationships for the two groups are substantially different ($\dot{V}O_2=160\ M_b^{0.74}$ for birds during forward flapping flight and $463\ M_b^{0.91}$ for hovering hummingbirds; Butler (1991), and reanalyzed using RMA). If data from three more species of hummingbirds are added to those used by Butler (1991) in his analysis (those from Wolf and Hainsworth (1971) and Wells (1993a)), the relationship for the nine species of hummingbirds, using RMA analysis, becomes: $\dot{V}O_2=449\ M_b^{0.90}$, $r^2=0.94$. It is interesting to note that the analysis by Bishop and Butler (1995) indicates that minimum $\dot{V}O_2$ during forward flapping flight of hummingbirds should scale $314\ M_b^{0.90}$.

39.3.3.4 Scaling of Flight Muscle Efficiency and Elastic Energy Storage

The efficiency with which the flight muscles can produce mechanical power from chemical energy is of great physiological significance as it determines the overall energy required by the animal during exercise. Excess heat must be lost to the environment in order to prevent hyperthermia. However, in order to estimate efficiency it is necessary to measure both the P_i and the P_o of the flight muscle. As pointed out by Alexander (2005), determination of P_o generally involves calculations that add a good deal of uncertainty. Thus, it seems sensible when reporting on the mechanochemical conversion efficiency of flight muscles to refer to their "apparent" efficiency calculated when specifically combining particular values for aerobic P_i and P_o (Bishop, 2005). Pennycuick (1975) recommended that the value of the mechanochemical conversion efficiency of avian flight muscles should be around 0.23 and treated as a constant. This was based on the determination of partial efficiency (i.e., the change in P_o/change in P_i) during the flight of budgerigars, laughing gulls, and fish crow (Tucker, 1972; Bernstein et al., 1973), in which partial efficiency ranged from 0.19 to 0.30. However, this runs counter to the more general observation that locomotory muscles for various vertebrate (Alexander, 2005) and invertebrate (Askew et al., 2010) groups exhibit "apparent" muscle conversion efficiencies that scale positively with M_b and negatively with respect to contraction frequency. Theoretically, it should not be possible for muscle fibers to operate with an overall efficiency greater than 28% (Rall, 1985), and in practice, muscle fibers of larger mammals appear to operate with maximal efficiencies of between 20% and 25% (Taylor, 1994).

By using detailed kinematic and morphological analyses, Wells (1993a) calculated an estimate for P_o (100 W/kg) during hovering of the broad-tailed hummingbird (*Selasphorus platycercus*) based on the aerodynamic model of Ellington (1984). $\dot{V}o_2$ was measured as 50 mL/g h, which gave an estimated efficiency of 9–11% (assuming perfect elastic storage of energy). By adding weights to the birds, Wells (1993b) determined that maximum P_o of the pectoral muscles for this species during hovering was around 117 W/kg (assuming perfect elastic storage of energy) and that the efficiency remained approximately 10%. Using the novel and elegant method of flying ruby-throated hummingbirds (*Archilochus colcubris*) in an increasing atmosphere of He/O_2, Chai and Dudley (1995) were able to reduce the density of the medium from the normal air value of 1.23 kg/m^3 to the point (on average, 0.54 kg/m^3) at which the birds were no longer able to hover for more than a few seconds. At this density, the P_o of the highly oxidative pectoralis muscles was calculated to be around 133 W/kg (assuming perfect elastic storage of energy), $\dot{V}o_2$ was measured as 62 mL/g h, and the efficiency was estimated to be around 10% regardless of the intensity of the flight "effort". The above calculations include an allowance of 10% for the cost of cardiorespiratory function (Tucker, 1973; Pennycuick, 1989), but there is no allowance for BMR. However, taking this into account (Krüger et al., 1982) only increases efficiency by approximately 1% in both studies.

Biewener et al. (1992) and Dial and Biewener (1993) used calibrated strain recordings from the humerus of flying birds to measure the force and estimate the power generated by the pectoralis muscles. In the first of these studies, mean P_o for three starlings (*S. vulgaris*) between 70 and 73 g mass and flying in a wind tunnel (for unstated duration) at an estimated speed of 13.7 m/s was 1.12 W. Torre-Bueno and Larochelle (1978) reported mean P_i to be 8.9 W for three starlings of mean mass 73 g and flying in a wind tunnel for 30 min over a range of speeds from 8 to 18 m/s. Thus, after subtracting BMR (Prinzinger and Hänssler, 1980) and allowing 10% for the cost of cardiorespiratory function, the estimated efficiency of the flight muscles for this species is 15.4%. In the second study, mean P_o for four pigeons between 301 and 314 g and flying at approximately 8 m/s along a 47 m long corridor, was 10.5 W/kg. Dial and Biewener (1993) used a P_i value of 106 W/kg (cf. Rothe et al., 1987). Thus, estimated flight muscle efficiency is approximately 11.2%. However, Rothe et al. (1987) estimated that their measurements of P_i were elevated by between 15% and 30% by the presence of the mask and tubing. If a mean overestimate of 22% is assumed, the actual value of P_i would be 87 W/kg, which would give an efficiency of 13.8%. If the value of P_i, after adjustment for the drag on the mask and tube, is taken from Butler et al. (1977) for pigeons of mean mass 442 g and flying at 10 m/s (61 W/kg), a flight muscle efficiency power of 19.9% is obtained. The more recent study of Ward et al. (2001) estimated that flight muscle efficiency in two starlings flying in a wind tunnel varied between 0.13 and 0.23 depending on the bird, the flight speed, and the aerodynamic model used to calculate P_o.

Bishop (2005) estimated the maximum $\dot{V}o_2$ for 15 species of birds studied during migratory climbing flights by Hedenström and Alerstam (1992). The estimated P_i was based on the cardiovascular modeling approach (Bishop and Butler, 1995; Bishop, 1997) and was then compared to the estimated aerodynamic P_o based on the model of Pennycuick (1989). This analysis suggested that the "apparent" mechanochemical efficiency of the avian flight muscles scaled allometrically with body mass, as 0.3 $M_b^{0.14}$ (Figure 39.15). Thus, taken together, these various studies appear to indicate that there may be an increase in overall flight efficiency with an increase in body mass (cf. Rayner, 1990). However, this is a slightly controversial subject due to the potential errors

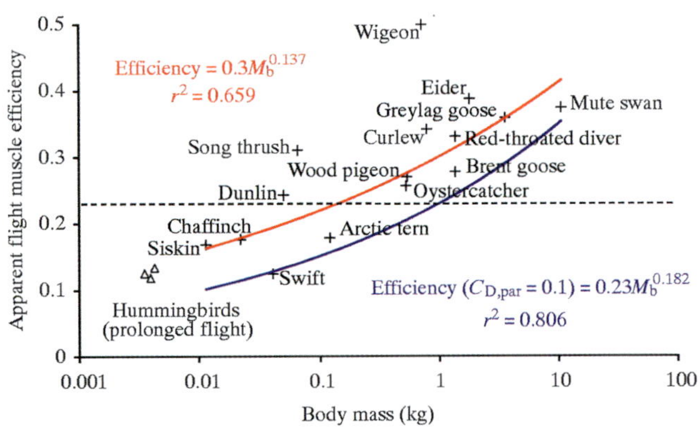

FIGURE 39.15 Estimated "apparent" flight muscle mechanochemical efficiency plotted against body mass (kg) for 15 bird species during migratory climbing flights over Sweden (see Hedenström and Alerstam, 1992). The wigeon, *Anas penelope*, is left out of the analysis as an outlier. The red regression line is based on the estimated sustainable biomechanical power output, calculated using default values for the parasite body drag coefficient ($C_{D,par}$) ranging from 0.25 to 0.4 (Pennycuick, 1989). The blue regression line is based on a value for $C_{D,par}$ of 0.1. Estimates of efficiency for three species of hummingbirds (**Δ**) are also plotted. *From Bishop (2005), Company of Biologists, Ltd.*

in calculating both P_i and P_o and the fact that the larger species of birds appeared to have efficiencies of over 0.3. As mentioned previously, the calculation of profile and parasite drag values in some aerodynamic models is uncertain, both with regard to differences in body mass between species and with regard to changes in air speed. Typically, there is also a tendency to ignore the inertial costs of accelerating and decelerating the wings of birds by assuming that the energy can be recovered by doing either useful aerodynamic work at the end of the wing stroke or by using elastic storage of wing kinetic energy to reaccelerate the wing in the following beat.

Wells (1993a) and Chai and Dudley (1995) concluded that hummingbirds are able to store the "inertial energy" required to accelerate the wing through each beat in an elastic component of their muscles. This assumption greatly reduces the value for the calculation of the power required to fly, especially for small species of birds that have high wingbeat frequencies. Bird flight muscles lack any associated specialized elastic components, such as long tendons, so any elastic component is likely to be within the cross-bridges of the muscle (Alexander and Bennett-Clark, 1977). Dial et al. (1987) measured the electrical activity of the pectoralis muscles of the pigeon during flight and demonstrated that the pectoralis muscles are active before the end of the upstroke of the wings. They conclude that the cross-bridges of the flight muscles are active during the final phase of the wing stroke and that the muscle could act like a spring (Goslow and Dial, 1990).

As pointed out by Spedding (1994), if zero elastic storage is assumed (cf. Weis-Fogh, 1972), the calculation of efficiency during hummingbird flight is around 25% and that this is still within the theoretical possibilities for vertebrate muscle (Weis-Fogh and Alexander, 1995; Pennycuick, 1992). The aerodynamic models of Pennycuick (1975, 1989) and Rayner (1979) also do not include the costs of accelerating and decelerating the wing during each cycle, as it is argued that the inertial costs are recovered aerodynamically and may, in any case, be relatively small during forward flight at medium to fast speeds. However, Van den Berg and Rayner (1995) suggest that it may be necessary to consider this cost. Van den Berg and Rayner (1995) estimated that the inertial power requirement should scale as approximately 5.8 $M_b^{0.80}$ (allowing for a 25% reduction due to wing flexion on the upstroke). Thus, in the light of the reduced estimates of flight costs of Pennycuick et al. (1996b), the relative importance of the inertial power requirements may have to be increased. Whatever the merits of each side of the debate, it seems possible that perfect elastic storage may prove to be unrealistic. Thus, any increment of the inertial costs that must be met by the flight muscles will increase the estimate of the "apparent" efficiency of the flight muscles. If "near" zero elastic storage is correct, then the inertial costs would dominate the total cost of hovering flight in hummingbirds (Spedding, 1994).

39.4 THE FLIGHT MUSCLES OF BIRDS

39.4.1 Flight Muscle Morphology and Fiber Types

The flight muscles of birds are structurally similar to the striated muscles of other vertebrates and consist of large numbers of long fibers, or cells, aligned essentially in parallel (Figure 39.16(A)). Each fiber can have a distinct biochemical character. At its simplest, each fiber can be specialize either for aerobic or anaerobic energy consumption, they can vary in diameter and cross-sectional area, and are supplied with various metabolites and oxygen via a network of capillaries (Figure 39.16(B)). Different species of birds have different amounts of flight muscle with respect to body mass, termed the flight muscle: mass ratio by Marden (1987), and these muscles are composed of four main cell or fiber types (cf. Rosser and George, 1986a): so-called slow oxidative fibers (SO), fast oxidative glycolytic fibers (FOG), fast glycolytic fibers (FG), and fibers that are intermediate (I) in their oxidative ability between the FOG and FG fibers.

Table 39.1 lists the major characteristics of these different fiber types. Essentially, the SO fibers have relatively slow rates of shortening, but are relatively efficient in producing force during isometric contractions, utilize oxidative metabolic pathways in the biosynthesis of energy rich ATP, and are resistant to fatigue. FOG fibers are also capable of oxidative metabolism and resistance to fatigue, but can also use carbohydrates as a fuel during anaerobic metabolism, have relatively faster rates of shortening, and are relatively efficient in producing force while shortening in length. FG fibers are susceptible to fatigue and are restricted to anaerobic metabolism of carbohydrates, but produce more power per unit mass of muscle than SO and FOG fibers. One reason for the latter is that they contain a lower volume fraction of mitochondria and associated structures and therefore have room for more myofibrilar proteins.

The function of SO fibers is considered primarily to be of advantage in postural muscles (e.g., in the back and neck muscles) and also in the legs during rest, due to their economic production of force during isometric contractions. So far, SO fibers have only been found in the deep layers of the pectoralis muscles of a limited number of species of birds that are especially adapted for gliding and soaring modes of flight, that is, species belonging to the avian orders of the Procellariformes, Pelecaniformes, Ciconiiformes, and Gruiformes (Rosser and George, 1986a; Rosser et al., 1994). Thus, Rosser et al. (1994) found that the deep "belly" of the pectoralis muscle of the American white pelican (*Pelecanus erythrorhynchos*) is composed exclusively of SO fibers, and other authors have reported SO fibers in the deep layers of the pectoralis muscles of the turkey vulture *Cathartes aura* (Rosser and George, 1986b), red-tailed hawk *Buteo jamaicensis* (Rosser and George, 1986a), double-crested

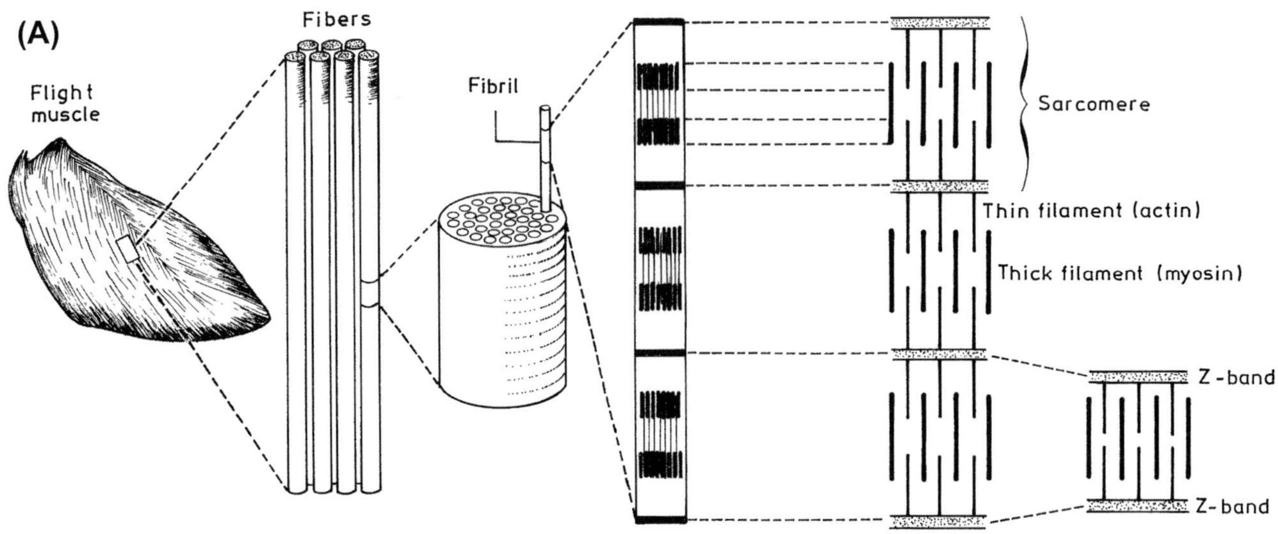

FIGURE 39.16 (A) Schematic illustration of the structure of a bird's pectoral muscle. The whole muscle is composed of many different fibers or cells, each of which consists of many fibrils and each fibril contains many thin actin filaments interspersed between thick myosin filaments. These filaments are organized into sarcomeres, which shorten during contraction as the actin and myosin filaments slide past each other. (*Modified from Schmidt-Nielsen (1997).*) (B) Light micrograph of a 1 μm thick transverse section of pectoralis muscle of the pigeon *Columba livia*. Fast oxidative glycolytic fibers (dark and small) and fast glycolytic fibers (pale and large) can be clearly identified. Capillaries are empty following vascular perfusion fixation. Bar = 40 μm. *From Mathieu-Costello (1991), Copyright © 1991 Wiley-Liss Inc., a division of John Wiley & Sons, Inc.*

cormorant *Phalacrocorax auritus* (Rosser and George, 1986a), and the domestic chicken *Gallus gallus* (Matsuda et al., 1983; Rosser and George, 1986a). However, in general, the occurrence of SO fibers is rare in birds, and they make up only a small percentage of the pectoralis muscle when they do occur.

Thus, while SO fibers may have some role in the thermal soaring flight of the pelican and the red-tailed hawk, or perhaps in the "outstretched" wing posture of the cormorant while drying its feathers, it would appear that most species of birds can manage without this specialized fiber type in their flight muscles (cf. Talesara and Goldspink, 1978; Rosser and George, 1986a).

The most abundant type of fiber to be found in the flight muscles of birds capable of prolonged flapping flight is the FOG (George and Berger, 1966; Rosser and George, 1986a). In general, the flight muscles of small birds are entirely composed of FOG fibers while larger birds have a mixture of different fiber types. However, there is a large degree of interspecies variation. George and Berger (1966)

TABLE 39.1 Simplified Characteristics of Muscle Fiber Types

Property	Slow Oxidative SO (Type I)	Fast Oxidative Glycolytic FOG (Type IIA)	Fast Glycolytic FG (Type IIB)
Speed of contraction	Slow	Medium–fast	Medium–very fast
Aerobic capacity	High	High	Low
Anaerobic capacity	Low	Moderate	High
Capillary supply	Good	Good	Poor
Triglyceride stores	High	Medium–high	Low
Glycogen stores	Medium	Medium–high	Medium–high
Fatigue resistance	High	High	Low
Cross-sectional area	Small	Small–medium	Medium–large

classified the pectoralis muscle of birds into six groups, with each fiber type listed in order of relative numerical abundance: (1) fowl type (FG, I, FOG); (2) duck type (FOG, FG, I); (3) pigeon type (FOG, FG); (4) kite type (mainly I); (5) starling type (FOG, I); and (6) sparrow type (FOG). While most species can be broadly assigned to one of these groups, there are many exceptions. Clearly, each individual species is adapted to cope with its own unique requirements based on its behavioral ecology, its detailed morphology, and the environmental conditions in which it lives.

Even more importantly, the histochemical classification of muscle fiber types can be misleading from a functional perspective. In reality, the fast-twitch fiber types are arbitrary classifications within a continuum of oxidative to glycolytic potentials. In addition, the term "fast" is also arbitrary, as there is likely to be a continuum for the rates at which the flight muscles are optimized to contract, due to the scaling of wingbeat frequency with body mass (Pennycuick, 1990, 1996). A small hummingbird may beat its wings at up to 50 Hz, while a large swan may only have a wingbeat of around 2.4 Hz. Therefore, there are likely to be many associated differences in the molecular characteristics of the contractile proteins from different species of birds. As a consequence, the FOG fibers of a hummingbird are not biochemically identical to the FOG fibers of a swan. It is probable that, even within the muscle of an individual bird, the FOG fibers of the deep layers may have slightly different characteristics to the FOG fibers from the more peripheral layers. Rosser and George (1986a) suggest that, given sufficient detailed knowledge about the protein isoforms of muscle, it is conceivable that almost all muscle fibers could be described as a unique subtype.

Despite the difficulties outlined above, it is still possible to draw some interesting conclusions about the functioning of avian flight muscles by quantifying their fiber type compositions. The Galliformes, such as the domestic chicken and the ruffed grouse (*Bonasa umbellus*), have a majority of FG fibers in their pectoralis muscles (Rosser and George, 1986a). This is matched by their relatively small heart masses and a low capacity to supply the muscles with oxygen (Bishop and Butler, 1995; Bishop, 1997). These birds are restricted to anaerobic flights of short duration. Typically, galliform birds live close to the edges of forests, have relatively short wings and relatively large flight muscles, and escape predation by being adapted for rapid takeoff followed by short flights into dense vegetation or into the branches of trees. The South American ecological equivalent of the Old World Galliformes, the tinamous, also have very small hearts and large flight muscles (Hartman, 1961) and live on the forest floor.

The Galliformes contrast with their closest relatives (Sibley and Alquist, 1990), the Anseriformes (the ducks, geese, and swans). Although of similar body size, the Anseriformes have relatively longer wings and live in open, uncluttered habitats. Their pectoralis muscles contain a high proportion of FOG fibers interspersed among the FG fibers, and their relative heart mass is also much larger (Magnan, 1922; Hartman, 1961). Thus, their wing morphology determines that they have a slightly more efficient form of flight (Rayner, 1988; Pennycuick, 1989), and their muscle physiology and cardiovascular adaptations determine that they are able to prolong these flights over relatively long distances. Similar adaptations are seen in the flight muscles of the doves and pigeons (Columbiformes).

The aerobic nature of the flight muscles of some relatively large species of birds, such as the bustards and swans, has an important consequence for the interpretation of the size limitations to flight performance in birds. It is necessary to distinguish between "burst" flight performance and "prolonged" or aerobic flight performance (Marden, 1994) and to compare flight muscles that are adapted for similar types of flight. The data of Marden (1987) and Pennycuick et al. (1989) on the maximum lifting ability of flying animals during takeoff refer only to "burst" flight performance, and bird species that have predominantly anaerobic FG flight muscle fibers will be at an advantage during takeoff compared to species that have predominantly aerobic FOG muscle fibers.

Caldow and Furness (1993) studied the histochemical adaptations of two ecologically similar seabirds, the great skua (*Catharacta skua*) and the herring gull (*L. argentatus*). They found that both species had FOG fibers exclusively in their pectoralis and supracoracoideus flight muscles, but that the mass-specific activity of both the oxidative and glycolytic enzymes (see below) were higher in the muscles of the great skua and concluded that this adaptation enabled the skua (Stercorariidae) to be a more effective aerial kleptoparasite than the herring gull (Lariidae). The mass of the pectoralis muscles is relatively small in the herring gull (12% of body mass), but it has a moderately large heart mass (0.9% of body mass), which is consistent with the fact that its flight muscles consist entirely of FOG fibers.

39.4.2 Biochemistry of the Flight Muscles

As discussed by Marsh (1981), the sustained, long-distance flights performed by many migrating birds require two interacting adaptions: the ability to store sufficient fuel reserves for the flight and the ability to maintain supplies of oxygen and fuel to the working tissues. Due to the very high energy density of lipids (39.3 kJ/g), compared to pure carbohydrates (17.6 kJ/g) and proteins (17.8 kJ/g), the energy for migratory flights is predominantly stored as fat in adipocytes. Thus, adaptations for the efficient oxidation of fatty acids and the potential for a high aerobic capacity are some of the most important features of most avian flight muscles. The relatively high mass-specific energy output required from the flight muscles necessitates high capillary and mitochondrial densities and high levels of mass-specific activities for the associated mitochondrial enzymes.

Hummingbirds are able to sustain the highest mass-specific metabolic rates of any vertebrate (Lasiewski, 1963), and Table 39.2 shows data from Suarez et al. (1986) for the maximum activities of various catabolic enzymes in hummingbird flight muscle. The following enzymes are of particular interest: citrate synthase (CS), as an indicator of general aerobic capacity; 3-hydroxyacyl-CoA dehydrogenase (HAD) and carnitine palmitoyl transferase (CPT), as indicators of fatty acid utilization; phosphofructokinase (PFK), as an indicator of general glycolytic flux; pyruvate kinase (PK), as an indicator of anaerobic glycolysis; and hexokinase (HEX) and glycogen phosphorylase (GPHOS), as indicators of plasma glucose and intracellular glycogen utilization, respectively. As expected, the flight muscles of hummingbirds have some of the highest catabolic enzyme activities ever measured. In particular, mass-specific CS

TABLE 39.2 Maximum Enzyme Activities in the Flight Muscle and Heart of the Rufous Hummingbird Selophorus Rufus[1]

Enzyme	Flight Muscle	*n*	Heart	*n*
Glycogen phosphorylase	31.22 ± 2.5	6	Not measured	
Hexokinase	9.18 ± 0.31	4	10.08 ± 1.9	4
Phosphofructokinase	109.8 ± 13	6	Unstable	
Pyruvate kinase	672.4 ± 27	6	507.3 ± 25	6
Lactate dehydrogenase	230.3 ± 23	6	357.4 30	6
Carnitine palmitoyl transferase	4.42 ± 0.46	6	2.83 ± 0.72	4
3-Hydroxyacyl CoA	97.10 ± 13	6	68.51 ± 10	4
Glutamate-oxaloacetate transaminase	1388 ± 70	5	576.4 ± 29	5
Glutamate-pyruvate transaminase	75.97 ± 6.0	5	16.31 ± 2.2	5
Citrate synthase	343.3 ± 8.8	6	190.3 ± 4.8	4
Creatine kinase	2848 ± 337	5	348.9 ± 59	5
Malate dehydrogenase	3525 ± 331	6	2024 ± 191	6
α-Glycerophosphate dehydrogenase	9.37 ± 2.1	6	8.10 ± 2.2	6

[1]*Values are expressed in μmol/min g wet wt and presented as means SE; n, number of birds (from Suarez et al., 1986).*

and the HEX activities are high and even higher (at 448.4 and 18.4 μmol/min g, respectively) in a subsequent study with a modified extraction procedure (Suarez et al., 1990).

These two studies (Suarez et al., 1986, 1990) indicate that while it is possible for hummingbirds to meet almost all their flight requirements by oxidizing lipid fuel, it is also possible for them to meet the requirements of hovering flight utilizing carbohydrate fuels. The latter possibility makes sense given that hummingbirds feed mainly on sucrose-rich nectar, and so during the day they would have a good supply of pure carbohydrate direct from the environment. However, for long-term storage, carbohydrates are stored as glycogen, which consists of at least 75% water, thus giving it a relatively poor energy density (4.4 kJ/g). It would not be possible, therefore, to store enough carbohydrates to fuel long-distance flight. In fact, it is unlikely that stored carbohydrate could support hovering flight for around 5 min (Suarez et al., 1990). However in the short-term, it is more efficient to oxidize plasma glucose directly to support flight, due to the 16% lower net yield for ATP resulting from the cost of synthesizing fatty acids from glucose for long-term storage. This led these authors to suggest that premigratory hummingbirds should behave as "carbohydrate maximizers" by keeping their foraging flights well under 5 min duration. This would provide efficient foraging while allowing excess glucose to be stored as fat. However, while the availability of plasma glucose may not be limiting for hummingbirds, the majority of bird species may place a higher emphasis on carbohydrate sparing. Bishop et al. (1995) found that HEX activity was relatively low in the pectoralis muscles of adult premigratory barnacle geese, and similar results have been reported for the pigeon and domestic chicken (Crabtree and Newsholme, 1972) and for the migratory semipalmated sandpiper, *Calidris pusilla* (Driedzic et al., 1993).

Many studies have shown that prior to long-distance flights, such as those performed during seasonal migrations, birds increase in body mass due primarily to the laying down of fat stores (Blem, 1976). This is usually correlated with hypertrophy of the pectoralis muscle in order to increase the power available for takeoff and forward flapping flight and species-specific changes in catabolic enzyme activities. While Marsh (1981) found no change in the premigratory activity of CS in the pectoralis muscle of the catbird (*Dumatella carolinensis*), he reported a significant increase in the level of HAD (indicating an increase in the potential for fatty acid oxidation). A similar increase in the ability to oxidize fatty acids was found in the semipalmated sandpiper, but the level of CS actually decreased (Driedzic et al. 1993). However, some species of small migratory passerine birds do appear to show an increase in the premigratory activity of both CS and HAD (Lundofgren and Kiessling, 1985, 1986), while CS activity in the premigratory barnacle goose was around 30% greater than that of postmolting birds approximately 4 weeks earlier in the season (Bishop et al., 1995).

Avian muscle capillarity and diffusion distances have been studied in a variety of different tissues, but no significant allometric trends with respect to different body masses were found for any of the variables measured (Snyder, 1990). However, Snyder (1990) did find a small but significant difference in the capillary to fiber ratio (C/F) between the aerobic fiber types (1.8 caps/fiber) and the glycolytic fibers (1.4 caps/fiber). Capillary density and diffusion distances are primarily determined by the cross-sectional area of the individual fibers, and the author argues that there may be little selective advantage in having a C/F greater than 2.0. Studies on avian pectoralis muscles have generally resulted in similar conclusions. Mathieu-Costello et al. (1992) found that the C/F of the rufous hummingbird (*Selaphorus rufus*) is only 1.55 caps/fiber compared to that of the pigeon, which is 2.0 caps/fiber (Mathieu-Costello, 1991). However, fiber diameter and the cross-sectional area are much smaller in the hummingbirds than in pigeons, and therefore the capillary density per unit area is much higher in hummingbirds. Mathieu-Costello et al. (1992) consider that one of the most important determinants of the rate of oxygen flux in aerobic fibers is the size of the capillary-to-fiber interface (i.e., the capillary surface per fiber surface area). However, in order to measure this feature accurately, it is necessary to standardize with respect to both the sarcomere length and the degree of tortuosity of the capillaries.

39.4.3 Neurophysiology and Muscle Function

Birds are capable of synchronizing the extension of the elbow and wrist joints of the wing during flapping flight, using automated coordinating mechanisms involving skeletal and muscular adaptations (Vazquez, 1994). However, the power and thrust generated during the wingbeat cycle, and the complex kinematics of the wing during the flapping flight of different species, require an active neuromuscular control mechanism and specialized adaptations of the flight muscles.

The detailed movements of the wing during flight actually involve a large number of different muscles (Dial et al., 1991; Dial, 1992a) and their differential (Dial, 1992a; Tobalske and Dial, 1994; Tobalske, 1995; Tobalske et al., 1997) or regional (Boggs and Dial, 1993) activation. For most studies, however, it is sufficient to focus on the functioning of the largest flight muscles, pectoralis and supracoracoideus. Dial (1992b) showed that birds were unable to take off or perform controlled landings without the use of the muscles of the forearm, but that they could sustain level flapping flight following an assisted takeoff.

The largest flight muscle, the pectoralis major, can be divided into two anatomical parts (Figure 39.17), the sternobrachialis (SB: which is superficial and lies along the sternum) and the thoracobrachialis (TB: which forms a deep layer lying along the sternum), and these are separated by the membrana intermuscularis (Dial et al., 1987). The SB is primarily innervated by the rostral nerve branch and the TB by the caudal nerve branch of the brachial ventral cord. In the pigeon, the fibers within the SB and TB have different orientations (Dial et al., 1987), while the SB has a lower percentage of FOG fibers and relatively more FG fibers, and the TB is primarily made up of FOG fibers (Kaplan and Goslow, 1989). Thus, the histochemical analysis and the neuroanatomy support the view that the pectoralis major is made up of at least two functional subunits, each with a potential for independent action on the wing during flight (Dial et al., 1987, 1988; Kaplan and Goslow, 1989; Goslow and Dial, 1990). Dial et al. (1987) showed that both the SB and TB muscles of the pigeon were maximally activated during takeoff or during large-amplitude wingbeats. However, during slow flight, the SB was relatively more important, and during low-amplitude wingbeats the activity of the TB could be almost zero. Interestingly, Dial et al. (1987) also indicate that there is a gradual reduction of large-amplitude

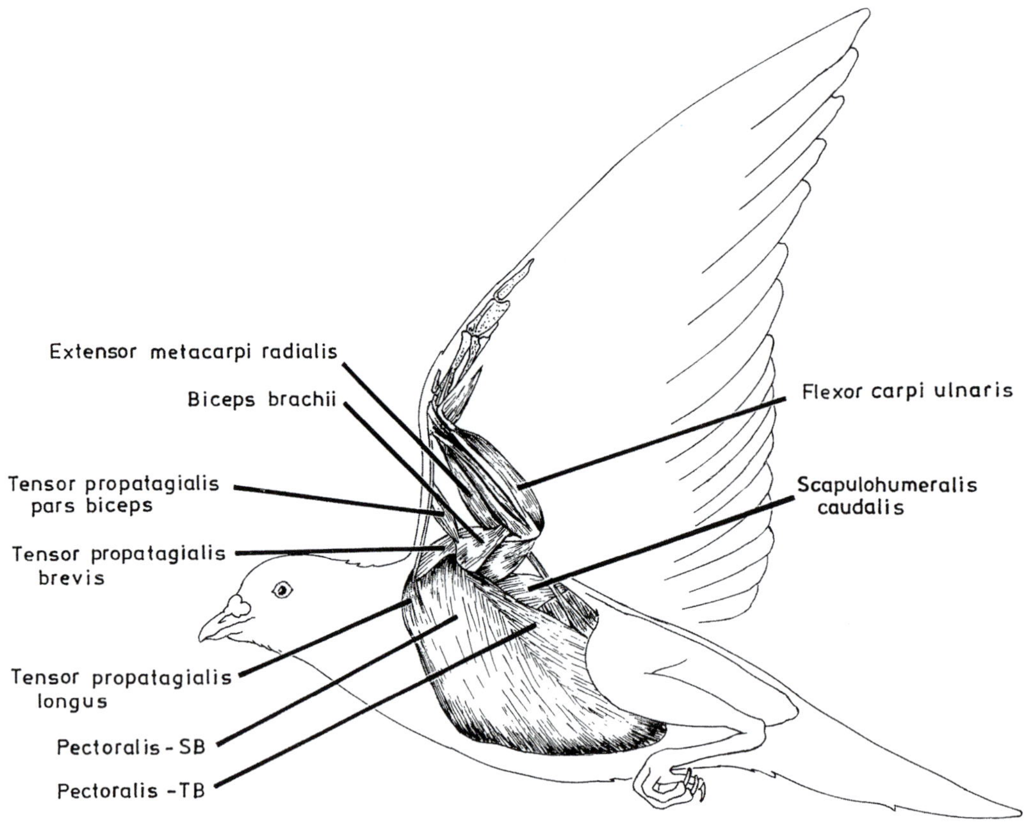

FIGURE 39.17 Illustration of the superficial flight muscles of the wings of a pigeon, *Columba livia*. The supracoracoideus muscles (not shown) are primarily used to power the upstroke and lie directly under the larger pectoralis muscles, which primarily power the downstroke. *Modified from Dial (1992a) Copyright © 1992 John Wiley & Sons, Inc. Reprinted by permission of Wiley-Liss, Inc., a subsidiary of John Wiley & Sons, Inc.*

electromyogram (EMG) spikes during the transition from takeoff to level flight and that they reappear during landing. This supports the hypothesis that FG fibers are derecruited during level flight.

Tobalske et al. (1997) studied the neuromuscular control and wing kinematics of the black-billed magpie, both in the wild and while flying in a wind tunnel, and showed that this species has a particularly complex flight style, consisting of alternating high- and low-amplitude wingbeats and occasional brief glides. This wide study found that the pectoralis consisted of FOG and I fiber types, as in the woodpecker (see below), and that the I fiber types were probably only recruited during the high-amplitude wingbeats. EMG recordings were made from six different wing muscles, and all showed that the relative intensity of the EMG signal exhibited a U-shaped output with respect to flight speed. Minimum intensity was recorded around 4.4 m/sec and the highest were recorded during hovering and the top speed of 13.4 m/sec. The pectoralis and the biceps brachii muscles showed strong activity during the end of the upstroke and the beginning of the downstroke, while the supracoracoideus, the humerotriceps, the scapulotriceps, and the scapulohumeralis caudalis were important during the end of the downstroke and the beginning of the upstroke.

A more detailed study of the EMG activity of 17 different muscles in the shoulder and forelimb of the pigeon during five different modes of flight was conducted by Dial (1992a). He concluded that the temporal pattern of activity did not vary much between different modes of flight, but that the intensity of the EMG signal was very important in determining the role of the different muscles during different modes of flight (Figure 39.18).

Tobalske (1996) studied the muscle composition and wing morphology of six different species of woodpecker, ranging in mass from 27 to 263 g, and related morphology to the scaling of intermittent flight behavior. Intermittent flight consists of regular alternation between periods of flapping and nonflapping. During the nonflapping phases, the bird's wings are either held folded at its side (flap-bounding flight) or held fully extended (flap-gliding flight). Biomechanical and physiological analyses suggest that intermittent flight is energetically efficient relative to continuous flapping (Rayner, 1985a), particularly for small species that have high wingbeat frequencies that are relatively "fixed" (Rayner, 1985a). Only FOG and I fiber types were found in these six species of woodpecker, with small species having almost exclusively FOG fibers in their pectoralis muscles, while species >100 g tended to have significant amounts of I fibers. All six species were capable of

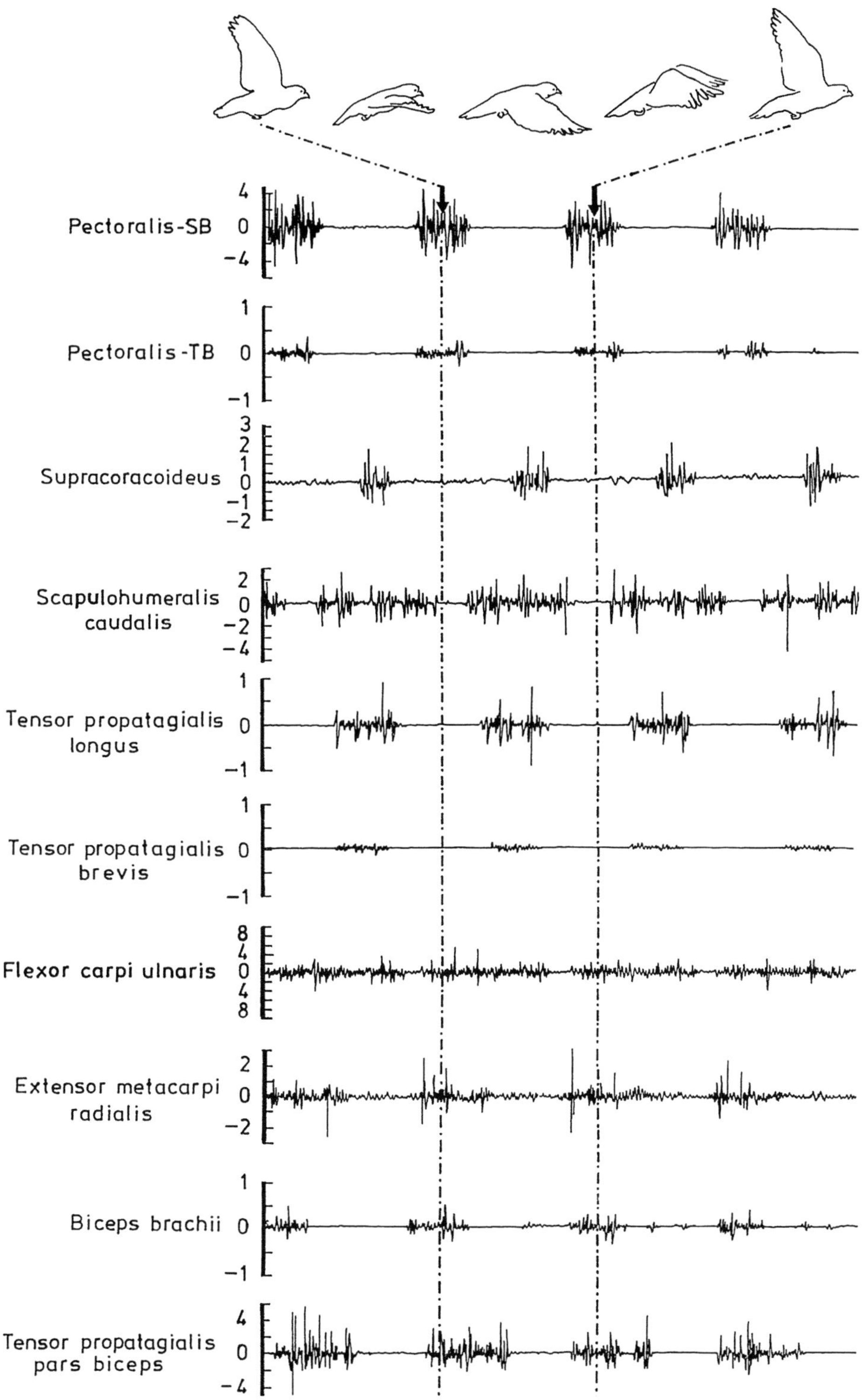

FIGURE 39.18 Electromyogram (EMG) signals recorded during level flapping flight from the various muscles of a pigeon, *Columba livia,* illustrated in Figure 39.13. EMG activity is presented with reference to the phase of the wingbeat cycle, with each downstroke indicated by the vertical dashed line. *Modified from Dial (1992a), Copyright © 1992 John Wiley & Sons, Inc. Reprinted by permission of Wiley-Liss, Inc., a subsidiary of John Wiley & Sons, Inc.*

exhibiting flap-bounding behavior, although the percentage of time spent flapping increased with increasing body mass. The empirical observations of Tobalske (1996) greatly exceed the theoretical considerations (Rayner, 1985a) that the upper limit for flap-bounding flight should be around 100 g of body mass.

However, the heterogeneous scaling of the fiber type composition of the pectoralis is likely to be one reason why the theoretical figures underestimate bird flight performance (Tobalske, 1996). FG and I fibers have the potential to produce more force per unit area than FOG fibers and therefore can produce more power for a given degree of fiber shortening and frequency. Thus, Tobalske (1996) argued that the positive scaling of the percentage and cross-sectional diameter of I fibers in the pectoralis of woodpeckers may be a direct result of the positive scaling of the power required for flight compared to the power available from the flight muscles (cf. Pennycuick, 1989; Marden, 1987; Ellington, 1991). Interestingly, among woodpeckers, the scaling of relative heart muscle mass is negative relative to increasing body mass (Hartman, 1961), which matches the decline in FOG content of the flight muscles. Thus, while large woodpecker species may be able to meet the power required to perform flap-bounding flight, it is likely that they will not be able to prolong these types of flight as easily as the smaller species, or perhaps that they might utilize a higher fraction of gliding flight.

Lewis' woodpecker (*Melanerpes lewis*) was found to be able to perform both flap-bounding and flap-gliding flight, contrary to theoretical predictions that it should not perform the former (Tobalske, 1996), and this behavior was not well correlated with the morphology of the wing. In addition, Tobalske and Dial (1994) found that the 35 g budgerigar was also capable of performing flap-gliding behavior when flying slowly, but switched to flap-bounding behavior during fast flight, despite the observation that they only have a single fiber type (FOG) in their pectoralis muscles (Rosser and George, 1986a). Thus, the hypothesis that the FOG fibers of small birds should not be used for gliding as it is inefficient appears to be undermined by the budgerigars, with the pectoralis being active during the gliding phases, but inactive during the bound. Tobalske and Dial (1994) predicted that all other species that show intermittent flight should also be facultative flap-gliders and flap-bounders and that the choice of flight mode is simply dependent on flight speed.

39.5 DEVELOPMENT OF LOCOMOTOR MUSCLES AND PREPARATION FOR FLIGHT

The development of the locomotor and cardiac muscles, together with other morphological features, have been studied in a Svalbard population of migratory barnacle geese, from hatch until it migrates to its wintering grounds in southern Scotland at the age of approximately 12 weeks (Bishop et al., 1996). Up until the age of 7 weeks, the goslings are unable to fly and spend their time walking and foraging. At the age of 5 weeks, the relative mass of the leg muscles is approximately 13% of body mass, while that of the pectoral muscles is only 3.5% of body mass. During the first 5 weeks of their life, there is a strong relationship between mass of the ventricles and mass of the leg muscles (Figure 39.19(B)). During the following 2 weeks, the flight muscles

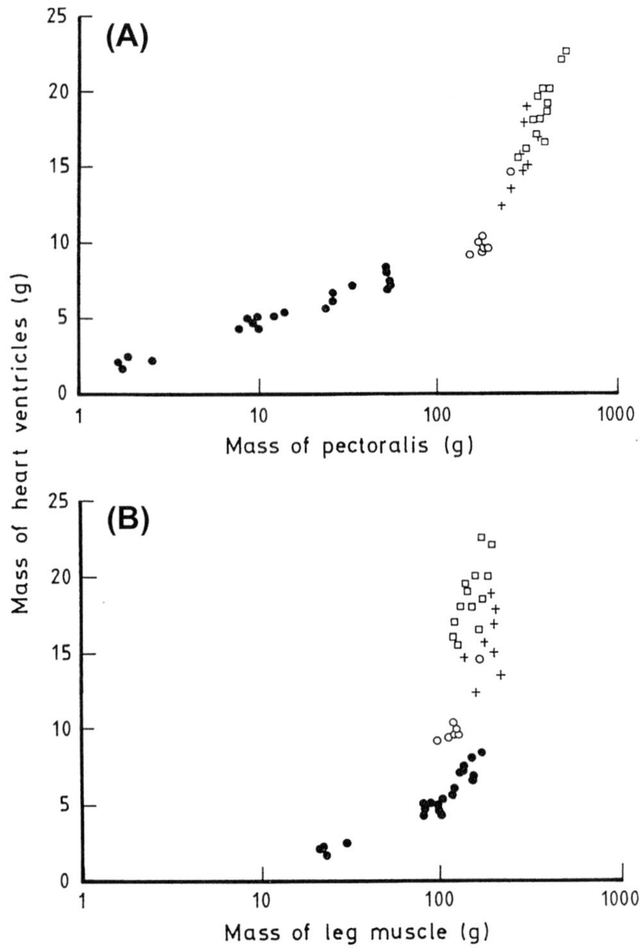

FIGURE 39.19 Scatter diagram of the masses (g) of the heart ventricles plotted against (A) the mass of the pectoral muscles and (B) the mass of the total leg muscles of wild barnacle geese, *Branta leucopsis*. Goslings of 1–5 weeks of age (●), fledgling goslings of 7 weeks of age (○),adult geese captured 7 weeks after the population hatch date (+), and premigratory adult geese (□). *Modified from Bishop et al. (1996), with permission.*

continue to grow in an exponential manner and by the time the goslings are 7 weeks old, these muscles are an impressive 14% of body mass, whereas the relative mass of the leg muscles has declined to approximately 9% of body mass (Figure 39.15). From the age of 7 weeks, the mass of the ventricles is proportional to the mass of the pectoral muscles (Figure 39.19(A)). At 5 weeks of age, ventricular mass is 5.2% of total mass of the leg muscles and at 7 weeks of age, it is 5.5% of mass of the pectoral muscles. Thus, as might be expected, the output capacity of the heart is closely matched to the oxygen requirements of the dominant set of locomotory muscles.

Choi et al. (1993) studied the skeletal muscle growth and development of thermogenesis in European starlings (*S. vulgaris*), the northern bobwhite (*Colinus virginianus*), and Japanese quails (*Coturnix japonica*). The body mass of the altricial starling developed more rapidly than that of the precocial quail species, although the percentage of the

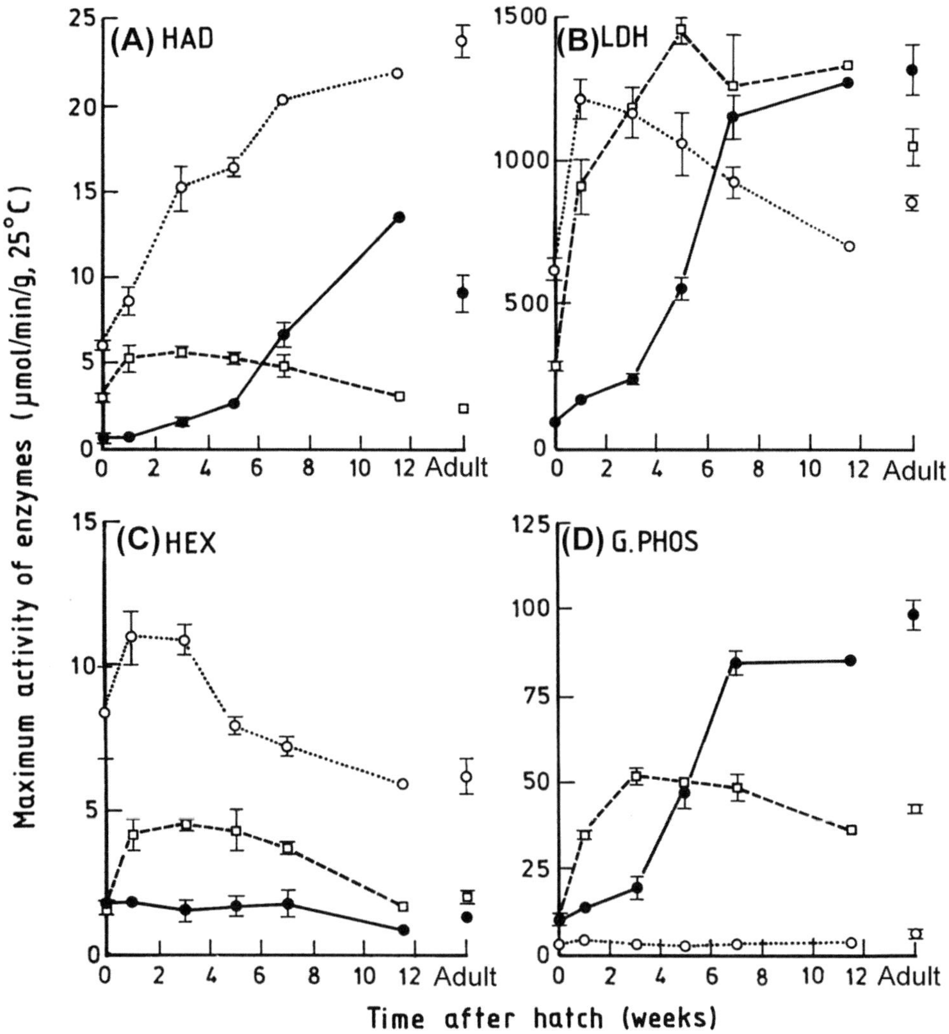

FIGURE 39.20 Mean ± SE of maximum activities of catabolic enzymes (mol/min g wet wt at 25 °C) in pectoral muscle (●——●), heart muscle (○⋯⋯○), and semimembranosus muscle (□------□) plotted against age of wild barnacle geese, *Branta leucopsis*. (A) 3-hydroxyacyl-CoA dehydrogenase (HAD), (B) lactate dehydrogenase (LDH), (C) hexokinase (HEX), and (D) glycogen phosphorylase (G.PHOS). $N=2$–8, except for a single individual at 11.5 weeks of age. Adult geese and 11.5 week old gosling are premigratory (see text). *Modified from Bishop et al. (1995) with permission.*

body mass that was accounted for by muscle mass was quite similar. The main difference between these groups was in the expression of muscle mass-specific enzyme activities. In particular, the activity of CS did not exceed 50 and activity 30 μmol/min g in the pectoralis muscles of *Coturnix* and *Colinus* quail, respectively, throughout development. In the starling pectoralis muscle, CS activity increased rapidly, and linearly, during development and reached 142 μmol/min g around fledging at 16 of age and approximately 230 μmol/min g in adult starlings. In bank swallows (*Riparia riparia*), CS activity increased from around 20 μmol/min g at 2 days of age and reached around 150 μmol/min g after 10 days (Marsh and Wickler, 1982).

A similar study of the barnacle goose found an even more exaggerated developmental pattern (Bishop et al., 1995), with CS expression showing almost no change in activity from hatch (12 μmol/min g) through to 5 weeks of age (20 μmol/min g). Subsequently, the activity increased rapidly to 75 μmol/min g at 30 μmol/min g in the pectoralis muscles of *Coturnix* 7 weeks of age and reached around 100 μmol/min g in the premigratory goslings and adults (see Figure 39.33). The mass-specific activity of four different mitochondrial enzymes showed broadly similar patterns of activity to each other during development (Bishop et al., 1995) and are typified by the results obtained for HAD activity in pectoralis, cardiac, and semimembranosus leg muscle (Figure 39.20(A)). Throughout most of development the cardiac muscle had the highest relative capacity for aerobic metabolic flux and fatty acid oxidation. HAD activity in the semimembranosus leg muscle was intermediate in value prior to fledging at 7 weeks (when the birds first begin to fly), but subsequently declined, while CS activity in the pectoralis flight muscle was initially very low, but reached a peak in the premigratory birds (see Figure 39.33).

The developmental profiles of various glycolytic enzymes, lactate dehydrogenase (LDH, Figure 39.20(B)), hexokinase (HEX, Figure 39.20(C)), and glycogen phosphorylase (GPHOS, Figure 39.20(D)), were more tissue-specific compared to those of the mitochondrial enzymes. Development of LDH and GPHOS activities in the pectoralis muscle show a similar exponential increase as that for CS, but values for HEX remain very low at all ages.

Overall, the results suggest that the relative use of circulating plasma glucose by the pectoralis muscle during flight is low, while intramuscular stores of glycogen are of primary importance during short-burst activity and of intermediate importance during longer-term activity, and that fatty acid oxidation is the main fuel during long-distance flight (Bishop et al., 1995). In general, moderate activity levels were detected in the semimembranosus muscle for all the enzymes that were measured, suggesting that the leg muscles are capable of utilizing all fuel types in almost equal amounts. However, the cardiac muscle had extremely low levels of GPHOS, indicating that it can only utilize aerobic pathways and that metabolic fuel must be primarily supplied by blood. LDH values were relatively high in the heart, while pyruvate kinase levels were very low (data not shown), indicating that the heart is capable of oxidizing circulating lactate as a fuel.

39.6 METABOLIC SUBSTRATE TRANSPORT

The use of fatty acids as the primary fuel during long-distance flights is an important adaptation, as the energy density of fat is much greater than that of other stored fuels. As pointed out earlier, the energy density of pure carbohydrate is 17.6 kJ/g; however, the storage of glycogen in cells involves between 3 and 5 g of water for each gram of glycogen, giving an effective energy density of approximately 4.4 kJ/g compared with that for lipids of 39.3 kJ/g (Schmidt-Nielsen, 1997). Protein may also be used to provide energy, but tissues such as muscle also consist of 70–80% water, so that the energy density of stored protein *in vivo* is around 5.4 kJ/g. Thus, during a long flight, it is likely that the majority of energy is provided by fatty acids. However, fatty acid transport in the plasma (from the adipocytes to the working muscles or the liver) may be a rate-limiting starvation the major source of glucose is via protein step during intense exercise (Newsholme and Leech, 1983; Butler, 1991) as it is insoluble in water and is transported by a carrier protein, albumin. In fact, mammals do not increase the rate of fatty acid metabolism as the intensity of exercise increases. This is achieved predominantly by an increase in carbohydrate metabolism (see Figure 39.21), mainly by the utilization of glycogen stored in the muscles. The rate of fatty acid metabolism may actually decrease with increased intensity of exercise. However, as migrating birds use predominantly FFA as their energy source, and as their rate of

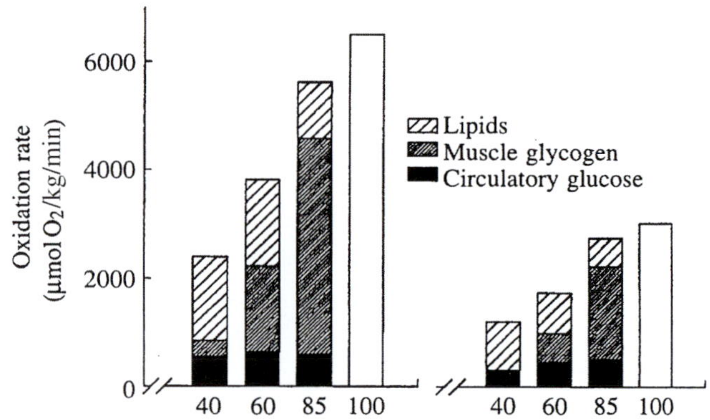

FIGURE 39.21 Substrate oxidation rates at different intensities of exercise in two species of mammals with different aerobic capacities. The large difference in the maximum rates of oxygen consumption of these two species is mainly the result of the larger rate at which glycogen is supplied from stores within the locomotor muscles themselves. *From Weber et al. (1996), Company of Biologists, Ltd.*

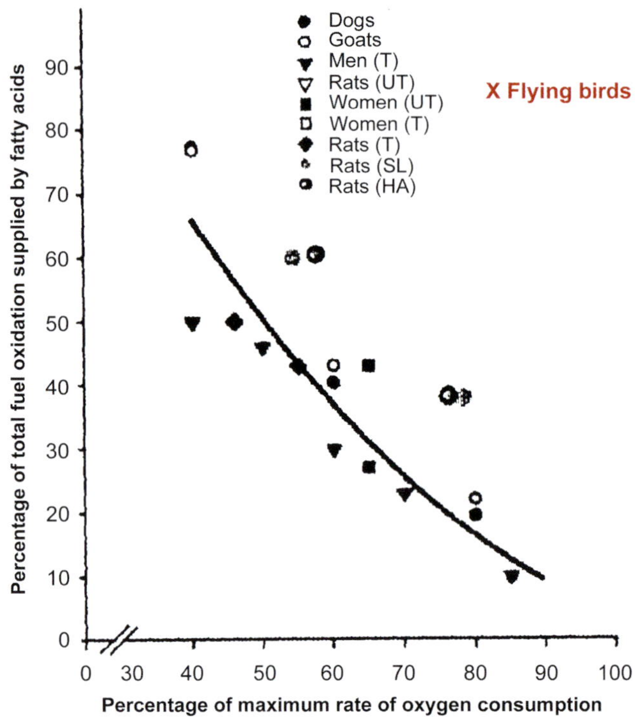

FIGURE 39.22 Percentage supply of fatty acid substrate to the muscles of mammals during increasing exercise intensity. X indicates an estimated value for flying birds. *From McCLelland (2004).*

aerobic metabolic rate is at least twice that of running mammals of similar size, the rate at which migrating birds need to transport FFA may be up to 20 times greater than that of running mammals (Figure 39.22).

A mechanism has been proposed for how small birds could deliver the high rates of fatty acid required by the flight muscles without increasing the concentration of albumin in the blood. Jenni-Eiermann and Jenni (1992) measured various metabolite concentrations in the plasma of three species

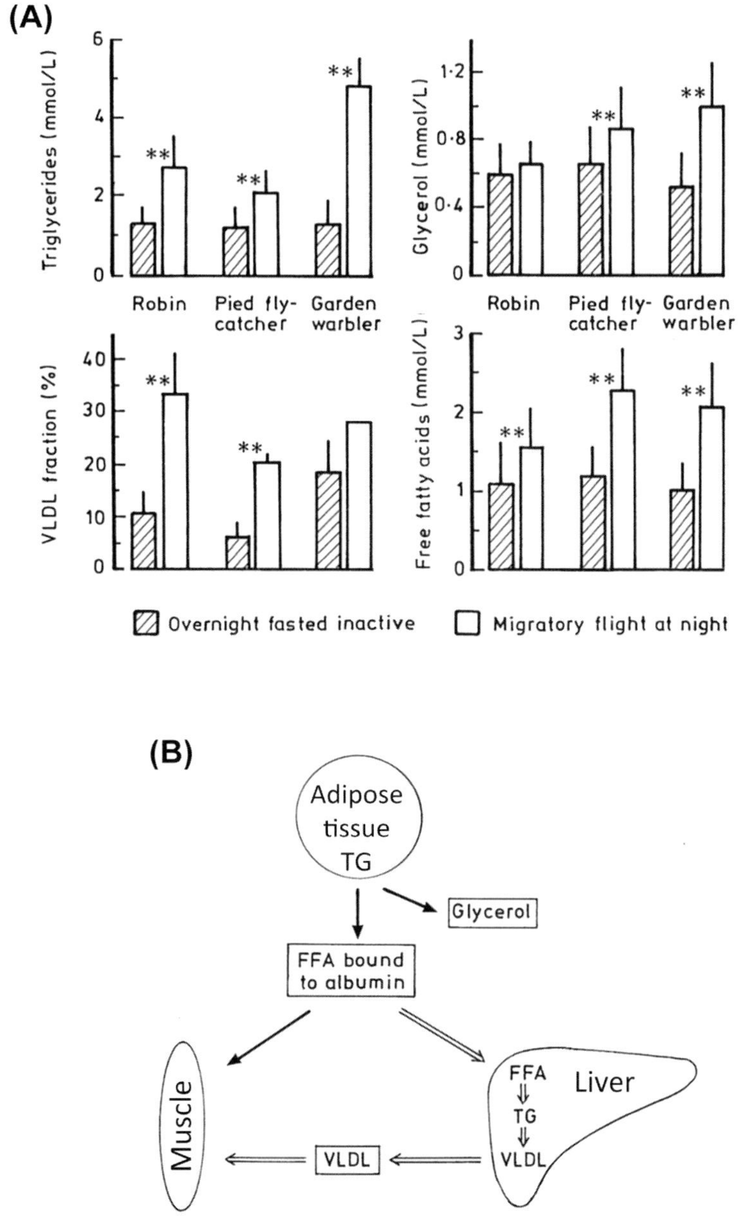

FIGURE 39.23 (A) Means ± SD of various fat metabolite concentrations and very low density lipoproteins (VLDL) in the plasma of three species of birds during overnight fasting (cross-hatched) or during nighttime migratory flights (clear). $^{**}p<0.01$, Mann–Whitney *U*-test. (B) Diagram illustrating the proposed pathway for fatty acid transport to the flight muscles of small birds. Free fatty acids (FFAs) are released from the triglyceride (TG) stores of the adipose tissue and either taken up directly by the muscles or converted to TG by the liver and transported to the muscles as VLDL. The activity of the latter pathway (double arrows) is suggested to increase in small birds during flight. *Modified from Jenni-Eiermann and Jenni (1992), with permission from The University of Chicago Press. © University of Chicago Press.*

of small night-migrating passerines (Figure 39.23(A)): the robin (*Erithacus rubecula*), garden warbler (*Sylvia borin*), and pied flycatcher (*F. hypoleuca*). Their results indicated that during long-distance flights, plasma levels of triglyceride, glycerol, free-fatty acids (FFAs), and very low density lipoproteins (VLDL) all increased significantly above values at rest.

The rise in triglycerides and VLDL was unexpected, but could provide a mechanism for increasing the FFA supply to the flight muscles. FFAs are re-esterified to triglyceride in the liver and subsequently delivered to the blood as VLDL. Thus, the removal of FFAs from the bloodstream by the liver as well as by the flight muscles will increase the rate at which albumin is able to transport FFAs from the adipocytes. Secondly, the VLDL can be hydrolyzed by lipoprotein lipase in the endothelium of the flight muscle capillaries, thus allowing the uptake of the released FFAs (Figure 39.23(B)). Apart from increasing the availability of FFAs to the flight muscles, this mechanism has the additional advantage that the overall protein (albumin) concentration of the blood is kept relatively low and prevents unacceptable increase in colloid osmotic pressure. Also, since the proportion of triglycerides in VLDL is 60–70% (compared to only 3% in albumin), there is a dramatic increase in the rate of fatty acid transport by the circulatory system of these birds.

In addition, the rate of uptake of the triglycerides via the VLDL can be controlled by the rate of lipoprotein lipase activity at the site of demand. During the migratory period, these birds accumulate fat in both their flight muscles and the adipocytes. Therefore, overnight-fasted birds were able to preferentially metabolize intramuscular stores of triglycerides, allowing a lower rate of utilization of adipocyte fatty acids and efficient sparing of protein and carbohydrate stores (Jenni-Eiermann and Jenni, 1996). It appears that this process is not as important in other, larger species (for example, Schwilch et al., 1996) and, in the barnacle goose, the maximum concentration of fatty acid-binding protein (considered to act as an intracellular carrier for fatty acids) in the pectoral muscle occurs at the same time as the maximum activity of HAD, just before the autumn migration (Pelsers et al., in press). So, it is still not clear how all species of migratory birds are able to transport FFAs in their blood at a sufficiently high rate to meet their high aerobic metabolic demands. Figure 39.21 shows that dogs have a greater aerobic capacity than pigmy goats, and McClelland et al. (1994) found that their albumin can bind 50% more FFAs than goat albumin. This adaptation of dogs does not, however, increase the proportion of their rate of oxidative metabolism fueled by FFAs. Transport in the blood is not the only problem, FFAs also have to cross the plasma membrane of the muscle cells and the outer and inner membranes of the mitochondria.

Up until the early 1980s it was thought that the uptake of long-chain fatty acids into cells was by passive diffusion. However, evidence now suggests that there are specific fatty acid transport proteins both in the cell membrane and in the cytosol of cells. It is possible, therefore, that there is both passive and protein-mediated transport of fatty acids into cells and into their organelles, such as the mitochondria. Figure 39.24 shows how the fatty acids are thought to cross the plasma membrane in mammals. Once dissociated from the albumin in the plasma, the fatty acids may cross the plasma membrane by passive diffusion and/or membrane-bound fatty acid binding proteins ($FABP_{pm}$) and fatty acid translocase (FAT, or CD36) may be involved in moving the fatty acids into the cytosol. Whichever is the case, once in the cytosol, the fatty acids are converted to fatty acyl-CoA esters either by binding to cytosolic or heart-type fatty acid binding proteins ($FABP_c$, or H-FABP), or directly by acyl-CoA synthetase (ACS). They are then bound to acyl-CoA binding protein (ACBP) and translocated to sites of usage within the cell. Alternatively, fatty acid transport

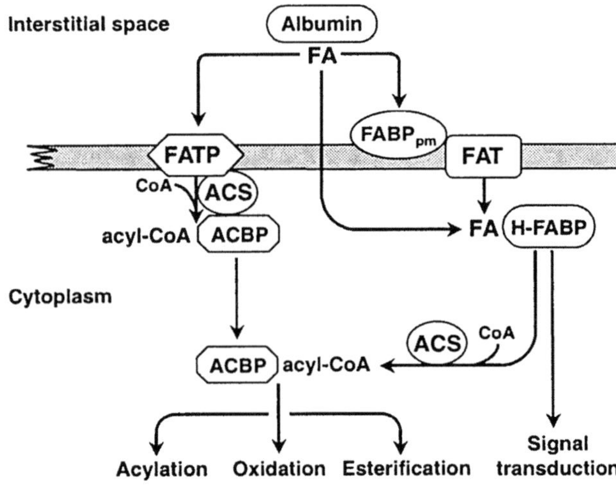

FIGURE 39.24 Long chain fatty acids (FA) are transported in the plasma of mammals bound to albumin. Following their dissociation from the albumin, their movement across a cell membrane into the cytoplasm is either: (1) by passive diffusion, or by membrane-bound fatty acid binding proteins ($FABP_{pm}$) and fatty acid translocase (FAT/CD36), or by a combination of the two; (2) by fatty acid transport protein (FATP), which is also located in the plasma membrane. Once inside the cytoplasm, the fatty acids are converted to fatty acyl-CoA esters by acyl-CoA synthetase (ACS), either via binding to a binding protein called heart-type fatty acid binding protein (H-FABP), or directly. Finally they bind to acyl-CoA binding protein (ACBP). *From Glatz et al. (2003).*

protein (FATP), which is also located in the plasma membrane, may transport the fatty acids into the cytosol, but as FATP is linked to acyl-CoA synthetase, the fatty acids are immediately converted to fatty acyl-CoA esters and bound to acyl-CoA binding protein. Recent evidence suggests that the membrane-bound transport proteins play the major role in fatty acid transport across the plasma membrane (Bonen et al., 2007; Schwenk et al., 2010) Cytosolic fatty acid binding protein acts as a "sink" for fatty acids, thereby increasing their solubility in the cytosol, their rate of removal from the inner surface of the plasma membrane, and their rate of intracellular diffusion.

Figure 39.25 shows how the FFAs enter the mitochondria. Firstly they are converted to their acyl carnitine esters by the enzyme carnitine acyl transferase-1 (CAT 1), which is present on the outer mitochondrial membrane. The acyl carnitine then crosses the inner mitochondrial membrane by the acyl carnitine/carnitine translocase system. Once inside the mitochondria, carnitine acyl transferase-2 (CAT 2), which is located on the inner membrane, regenerates acyl-CoA and free carnitine. This transport process could be the main rate-limiting step in the use of fatty acids by active muscles, with either carnitine or carnitine acyl transferase-1 being the important factors.

Studies on the flight muscles of birds indicate that they have greater concentrations of these fatty acid transporters than mammalian muscles. For example, the flight muscles of migratory western sandpipers (*Calidris mauri*) have

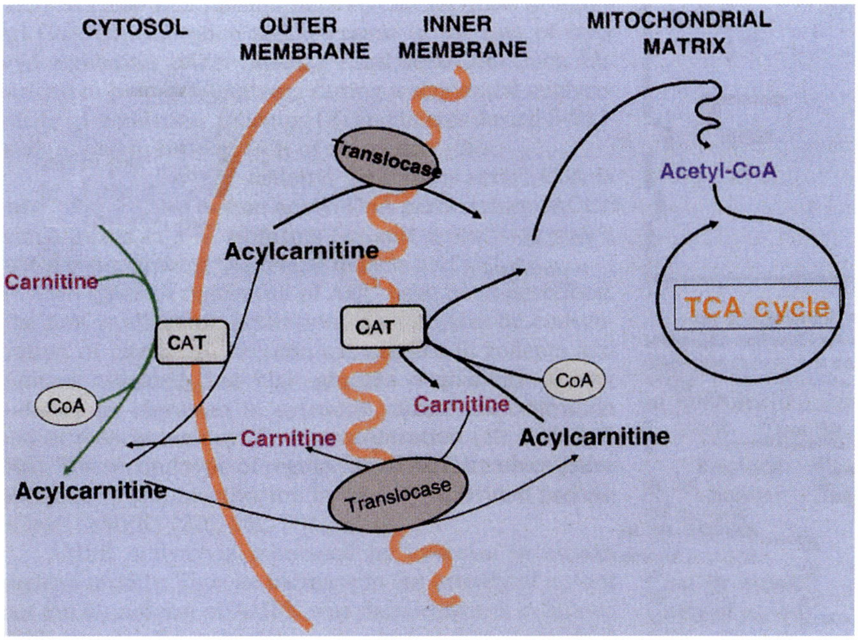

FIGURE 39.25 Transport of long-chain fatty acids (FA) into mitochondria of mammals. Acyl-CoA esters are unable to cross the inner membrane of the mitochondria so they are firstly converted to their acyl carnitine esters by carnitine acyl transferase 1 (CAT 1) which is present on the outer membrane. Acyl carnitine crosses the inner membrane via the translocase system and, once inside the mitochondrion, acyl-CoA and free carnitine are re-formed by CAT 2, which is located on the inner membrane. *From Kiens (2006).*

approximately 10 times greater concentrations of H-FABP than a typical mammalian muscle and approximately 14% of total cytosolic protein is H-FABP in sandpipers. This is the highest value reported for any vertebrate skeletal muscle (Guglielmo et al., 1998). Also, the concentration of H-FABP is 70% greater during the migratory period than during winter (Guglielmo et al., 2002). The presence of FAT/CD36 and $FABP_{pm}$ has been demonstrated in the flight muscle of white-throated sparrows (*Zonotrichia albicollis*), and the genes for these are strongly upregulated during the migratory season (McFarlan et al., 2009). These data indicate that at least three of the muscle fatty acid transport proteins are important in the transport of fatty acids during migration.

There has been much attention recently on the possible significance of a number of polyunsaturated fatty acids (PUFAs) in the diet of premigratory birds, although the evidence for some of these PUFAs having any influence on flight performance is somewhat inconclusive (Guglielmo, 2010). What we do know is that in white-throated sparrows (*Z. albicollis*) there is an increase in the PUFA 18:2n6 in triglyceride stores of muscles and adipose tissue during migration (Klaiman et al., 2009). Also, in European starlings (*S. vulgaris*) there is a 13% reduction in rate of energy expenditure during 6 h flights when they are fed a diet high in 18:2n-6 fatty acids compared with when they are fed a diet high in monounstaturated fatty acids (MUFAs; McWilliams and Pierce, 2006 as quoted by Guglielmo, 2010). The exact details of the mechanisms responsible for this significant saving in rate of energy expenditure are not clear. Two possibilities discussed by Price (2010) are the fuel hypothesis and the phospholipid hypothesis. The fuel hypothesis is based on the facts that unsaturated fatty acids are more soluble and have greater diffusion rates in water than saturated fatty acids. It is postulated, therefore, they should be transported more rapidly from the fat stores to the mitochondria, along the pathways outlined above. The phospholipid hypothesis proposes that, as fatty acids in the diet are incorporated in the phospholipids in membranes of cells and mitochondria, any alteration in composition of mitochondrial membranes affects the various metabolic processes in which they are involved and, therefore, overall performance of the whole organism. For example, in one of the research work done in 2012, the phospholipid hypothesis was developed by highlighting the evidence that fatty acid composition of mitochondrial membranes can alter the leakage of protons in intact mitochondrial preparations and that this proton leak can account for as much as 20% of basal metabolic rate in rats. In that research, it was also pointed out that a by-product of oxidative metabolism is the production of reactive oxygen species (ROS) and that the proton leak may modulate the production of ROS. In a series of experiments on European starlings, Gerson (2012) demonstrated that birds fed on a PUFA-rich diet and flown for approximately 3 h had lower rates of ROS production than those fed on a MUFA-rich diet. The disruption of mitochondrial function by ROS could result in less efficient production of ATP, which could account for the higher flight costs previously shown in European starlings fed a MUFA diet and flown in a wind tunnel. Gerson (2012) proposed, therefore, that there was reduced oxidative damage and impact on mitochondrial functions during long-duration flight in the PUFA-fed birds. Although protein stores are less energy dense than fatty acid adipocyte stores (see above), in reality it may not be possible entirely to exclude their use during periods of starvation and/or during prolonged periods of exercise. The brain and other nervous tissues are extremely dependent on glucose metabolism for the provision of energy (Newsholme and Leech, 1983). As carbohydrate stores are very limited in most animals, during starvation the major source of glucose is via protein (amino acid) degradation to various metabolic intermediates (called oxoacids), most of which can then enter the gluconeogenesis pathway in the liver. In addition, glycerol that is produced following the hydrolysis of triglycerides can also be converted to glucose. A small number of amino acids (leucine and lysine) are degraded directly to acetyl-CoA and so can contribute to ATP synthesis via the citric acid cycle (CAC) or be converted to fat, as acetyl-CoA cannot enter the gluconeogenesis pathway. Alternatively, excess amino acid production can provide intermediates (e.g., 2-oxalogluterate, succinyl-CoA, fumarate, and oxaloacetate) for the CAC if these should be in short supply or be converted to pyruvate and then to acetyl-CoA either to contribute to ATP synthesis or to be converted to fat. Certainly, there are a number of studies that show that protein, in the form of muscle and the intestinal tract, is catabolized during long periods of flight (Jehl, 1997) or during "simulated" flights by starvation (Biebach, in press).

Protein catabolism during migratory flight may not always be for the reasons discussed above. When catabolized, proteins produce more water per unit of energy released (0.155 g H_2O/kJ) than fats (0.029 g H_2O/kJ). Gerson and Guglielmo (2011a,b) have demonstrated that when flying for a mean duration of 2.7±0.3 h in conditions of low relative humidity (13%), Swainson's thrushes (*Catharus ustlatus*) have a lean mass (protein) loss of 3.55±0.91 mg/min, whereas when flying in high relative humidity (80%) air, there was no relationship between lean mass loss and flight duration. The greater rate of loss of lean mass when flying in relatively dry conditions led to a 21.7±4.9% increase in endogenous water production. These data demonstrate that when flying in dry conditions, the rate of water loss induces greater metabolic water production by increasing protein catabolism. However, for those species that do not complete their migration in one nonstop flight, relatively high rates of reductions in flight muscle and the digestive system may result in reduced ability to fly and to process food during the first day of stopover.

39.7 THE CARDIOVASCULAR SYSTEM

A major role of the cardiovascular system is to deliver metabolic substrates and oxygen to the tissues undergoing aerobic metabolism and to remove the waste products of that metabolism. Thus, the cardiovascular system in general, and the heart in particular, must be capable of meeting the demands of the flight muscles during sustained, flapping flight.

TABLE 39.3 Mean Values of Oxygen Uptake and Cardiovascular Variables Measured in the Pigeon *Columba livia* and Emu *Dromaius novaehollandiae*[1]

	Pigeon (0.442 kg)		Emu (37.5 kg)	
	Rest	Flying	Rest	Flying
Oxygen uptake (mL/min STPD)[2]	9.0	88.4	156.7	1807
Heart rate (beats/min)	115	670	45.8	180
Cardiac stroke volume (mL)	1.44 (1.14)	1.58	57	102.7
Oxygen content of arterial blood (vol%)	15.1	13.7	15.2	15.2
Oxygen content of mixed venous blood (vol%)	10.5	5.4	9.0	4.6

[1]*Measurements taken at rest and after 6 min of steady level flight in a wind tunnel at a speed of 10 m/s for the pigeon and after 20 min running on a treadmill at a 6-degree incline and a speed of 1.33 m/s for the emu (from Butler, 1991). Value for cardiac stroke volume in parentheses is recalculated from Butler et al. (1977), as discussed by Bishop and Butler (1995).*
[2]*STPD, Standard temperature and pressure, dry.*

39.7.1 Cardiovascular Adjustments during Flight

The pigeon (*C. livia*) is the only species in which all but one of the variables of the Fick equation (see above) have been determined during flight (Butler et al., 1977; Peters et al., 2005). In the study of Butler et al. (1977), $\dot{V}o_2$ in the pigeons was 10 times the resting value (Table 39.3) when flying in a wind tunnel at 10 m/sec. The respiratory system maintained C_aO_2 at slightly below the resting value, but $C_{\bar{v}}O_2$ was halved, giving a 1.8-fold increase in $(C_aO_2 - C_{\bar{v}}O_2)$. There was no significant change in V_s, so the major factor in transporting the extra oxygen to the muscles was the 6-fold increase in f_H. Unfortunately, all of these variables were not measured simultaneously on the same birds, and Bishop and Butler (1995) have argued that the original calculation of V_s at rest may have been inappropriate. This being the case, there would have been a 1.4-fold increase in V_s during flight. As mean arterial blood pressure did not change, total peripheral resistance must have declined by the same proportion as cardiac output (V_b, which is $f_H \times V_s$) increased.

In general, birds have larger hearts and lower resting heart rates than those of mammals of similar body mass (Lasiewski and Calder, 1971; Grubb, 1983). The higher V_b in birds is an important factor in their attaining a higher maximum $\dot{V}O_2$ ($\dot{V}O_{2,max}$) during flight than similar-sized mammals do when running. Bats have larger hearts and a higher blood oxygen-carrying capacity than other mammals of similar size (Jurgens et al.,1981), and their $\dot{V}O_2$ during flight is similar to that of birds of similar mass (Butler, 1991). Among birds, those that are not capable of sustained flight, such as Galliformes, tend to have slightly lower hematocrit (Hct, packed cell volume) and hemoglobin concentration (Hb) than those of good fliers (Balasch et al., 1973), and weaker fliers have a slightly lower Hct than stronger fliers (Carpenter, 1975). Wild barnacle geese migrating from their breeding grounds in the high Arctic to their wintering grounds in southern Scotland have surprisingly low heart rates during flight (see earlier), which means either that the energy cost of migratory flight is lower than was previously thought (maybe as a result of formation flight, which will be discussed later) and/or that ($C_aO_2 - C_{\bar{v}}O_2$) is greater in migrating geese than was assumed from the data obtained from pigeons flying in a wind tunnel (see Butler et al., submitted-a).

39.7.2 The Cardiac Muscles

Studies of exercising animals have indicated that both the locomotor and cardiac musculature are dynamic structures that can vary in mass seasonally and in direct response to demand (Hickson et al., 1983; Marsh, 1984; Dreidzic et al., 1993; Bishop et al., 1996). Thus, it might be anticipated that where additional energetic costs occur seasonally (e.g., due to migratory fattening or the development of secondary sexual characteristics), then the relevant cardiac and locomotor musculature might also be regulated seasonally. Bishop et al., (1996) showed that both cardiac and pectoralis muscles hypertrophy during the premigratory fattening period of the barnacle goose.

Bishop and Butler (1995) suggest that heart mass (M_h) can be used to model the availability of oxygen to the flight muscles of birds. Assuming that the adaptations of the flight muscles are appropriately matched to the cardiovascular system, then the relative M_h of birds of a similar body mass should be a good indicator of the relative aerobic power input available to the flight muscles. Bishop (1997) calculated that the $\dot{V}O_{2,max}$ of birds should scale with respect to body mass as 230 $M_b^{0.82}$ mL/min^{-1}, and that maximum aerobic power input (aerobic $P_{i,max}$) should scale as 11 $M_h^{0.82}$ (W). Thus, for studies of the prolonged aerobic flight performance of birds, the estimates of $\dot{V}O_{2,max}$ or $P_{i,max}$ based on relative M_h should be of more practical value than the use of general scaling equations based on body mass alone.

The mean values for M_h from selected avian families are plotted in Figure 39.26(A) (data from Hartman, 1961) and Figure 39.26(B) (data from Magnan, 1922) and indicate the interfamily adaptive diversity in aerobic capacity. The very small M_h of all three species of tinamou indicate that the members of this family have the lowest aerobic ability of all bird species. Other relatively sedentary forest birds with small M_h include the guans, motmots, puffbirds, and antbirds. This contrasts with the relatively large M_h of other bird families that live predominantly in forest, such as the parrots, trogons, kingfishers, and hummingbirds. Interestingly, Figure 39.18(B) shows that the M_h values of the bustard family (Otidae) appear to range above the general allometric trend. The relative M_h of the 0.83 kg little bustard and the 8.95 kg great bustard are 1.8 and 1.4%, respectively. These relative M_h values are even larger than those of the Anatidae, which typically range from 0.8% to 1.1%. Two other observations appear to substantiate this finding. Crile and Quiring (1940) gave the following values for the relative heart masses of the African Kori bustard; a 5.54 kg female Kori bustard (1.1%) and a 10 kg male Kori bustard (1.0%). In addition, Stickland (1977) shows that while 100% of the area of the pectoralis muscle of the African helmeted guinea-fowl is made up of "white" anaerobic muscle fibers (i.e., FG fibers), 82% of the area between the pectoralis muscle of the similarly sized white-bellied bustard is made up of "red" aerobic fibers (i.e., FOG fibers). This author also comments on the fact that the European species of bustards undergo local migrations, which is a further indication that they are capable of a "prolonged" mode of flight.

There are also some interesting intrafamily differences in aerobic capacity indicated in the M_h data of Hartman (1961) and Magnan (1922). The different genera of the Columbidae (pigeons and doves) have relative M_h means ranging from 1.29% to 0.57%, while among the genera of the Falconidae (falcons) these range from 1.14% to 0.6%. In both these families, the genera with the relatively small hearts occur predominantly in tropical forests and are likely to be relatively sedentary. As intrafamily pectoralis muscle mass is fairly constant in these examples, it is reasonable to assume that the relatively sedentary species are more dependent on anaerobic metabolism to support their flight activity and that their flight muscles consist of a relatively greater proportion of FG fiber types.

39.8 THE RESPIRATORY SYSTEM

The respiratory system is not only concerned with gas exchange (i.e., the supply of oxygen to and the removal of carbon dioxide from the circulatory system and the metabolizing tissues) but also with the control of body temperature and evaporative water loss.

39.8.1 Ventilatory Adjustments during Flight and Ventilatory/Locomotor Coupling

As for the cardiovascular system, there is a variation of Fick's formula for convection that describes the relationship between $\dot{V}O_2$ and the various components of the respiratory system:

$$\dot{V}O_2 = f_{resp} \times V_T \times (C_IO_2 - C_EO_2),$$

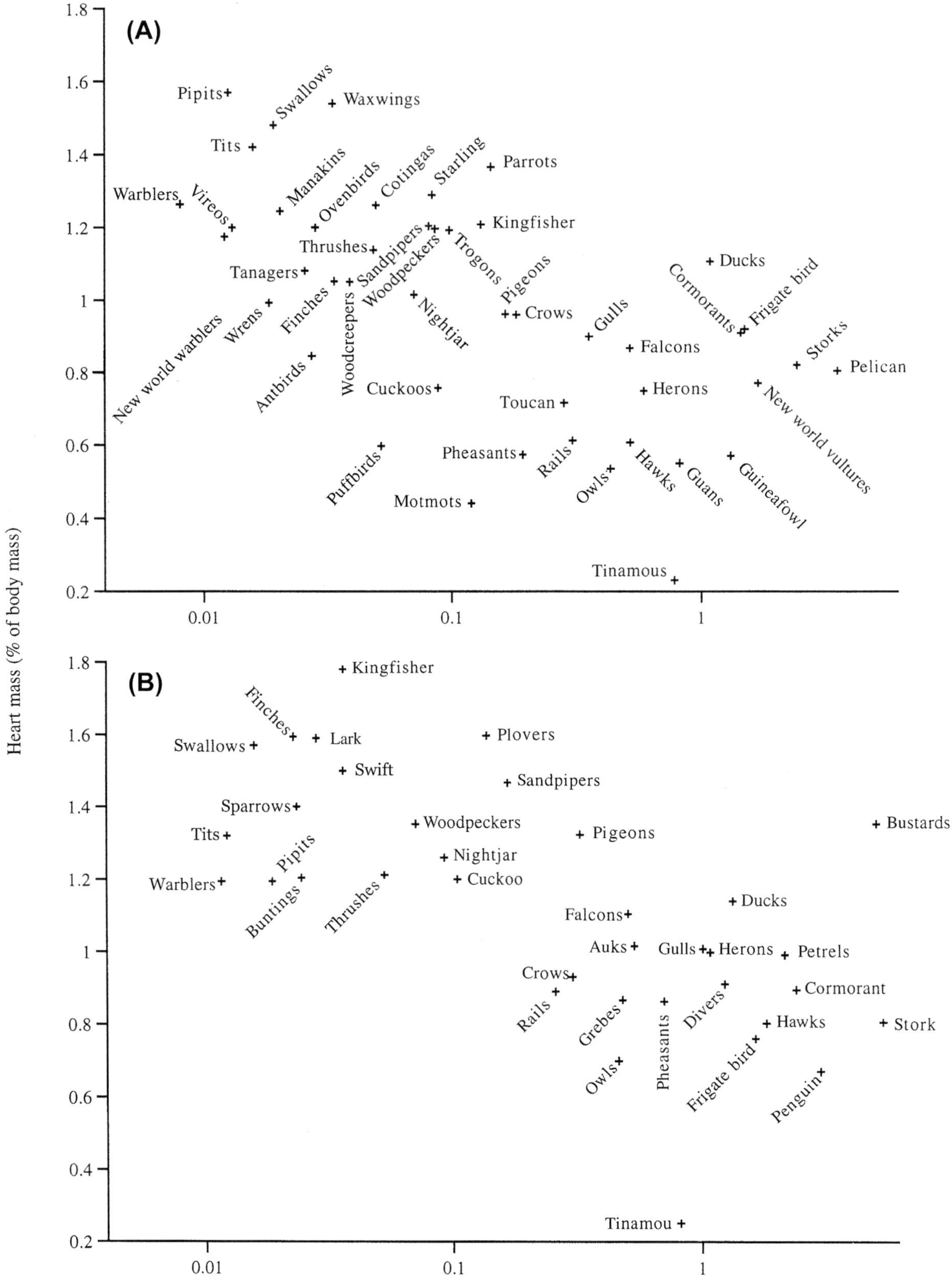

FIGURE 39.26 Mean values for cardiac muscle mass as a percentage of body mass (%), plotted against body mass (kg) for different families of birds. *(A) Data from Hartman (1961); (B) data from Magnan (1922).*

TABLE 39.4 Mean Values of Respiratory Frequency, Tidal Volume, Minute Ventilation Volume, Oxygen Extraction, and Oxygen Uptake[1]

		Rest					Flight				
	Mass (kg)	f_{resp}	V_T	$\dot{V}_I$	O_{2ext}	$\dot{V}O_2$	f_{resp}	V_T	$\dot{V}_I$	O_{2ext}	$\dot{V}O_2$
Hummingbird (>20 °C) (*Amazilia fimbriata fluviatilis*)[2]	0.006	–	–	–	–	–	280	0.63	0.18	0.135	4.1
Hummingbird (36 °C) (*Colibri coruscans*)[3]	0.008	–	–	–	–	–	330	0.38	0.12	0.24	5.0
Budgerigar (18–20 °C) (*Melopsittacus undulatus*)[4]	0.035	–	–	0.047	0.27	2.62	199	1.15	0.232	0.26	10.9
Starling (10–14 °C) (*Sturnus vulgaris*)[4]	0.073	92	0.67	0.061	0.28	3.16	180	2.8	0.504	0.31	28.1
Black-billed magpie (*Pica pica*)[5]	0.165	52.4	2.95	0.154	–	–	162	6.1	0.953	–	–
Fish crow (12–22 °C) (*Corvus ossifragus*)[4]	0.275	27.3	8.2	0.223	0.19	8.5	120	14.9	1.79	0.19	68.0
White-necked raven (14–22 °C) (*Corvus cryptoleucus*)[4]	0.48	32.5	10.5	0.34	0.24	17.0	140	10.7	1.40	0.29	84.9

[1]*Abbreviations: f_{resp}, respiratory frequency (1/min); V_T, tidal volume, (mL); $\dot{V}_I$, minute ventilation volume (1/min BTPS, except for fish crow where it is 1/min STPD); $O_{2\,ext}$, oxygen extraction; $\dot{V}O_2$, rate of oxygen uptake (mL O_2/min STPD) during rest and while hovering in two species of hummingbirds and while flying in a wind tunnel for five other species of birds. The values for $\dot{V}O_2$ during flight are the minima that have been recorded and have been corrected for the drag and mass of mask, etc., where necessary (see Butler et al. (1977) for further details). BTPS, body temperature and pressure, saturated; STPD, standard temperature and pressure, dry.*
[2]*Berger and Hart (1972).*
[3]*Berger (1978).*
[4]*Butler and Woakes (1990).*
[5]*Boggs et al. (1997a).*

where f_{resp} is respiratory frequency (breaths/min), V_T is respiratory tidal volume (mL), and ($C_IO_2-C_EO_2$) is the difference in the oxygen content in the inspired and expired gas (milliliter of O_2 per mL of gas). Data on ventilation are given in Table 39.4 for five species of birds at rest and during forward flapping flight in a wind tunnel. At relatively low ambient temperatures (<23 °C), minute ventilation volume V_I (=$f_{resp}\times V_T$) increases by a similar proportion as $\dot{V}o_2$ and the proportion of oxygen extracted from the inspired gas ($O_{2,ext}$ effectively equal to $C_IO_2-C_EO_2/C_IO_2$) during flight is similar to that in resting birds. The relative contributions of f_{resp} and V_T to the increase in V_I during flight vary between species. In white-necked ravens (*Corvus leucocephalus*), V_T does not change at all, whereas in fish crows and black-billed magpies it doubles and in starlings there is a fourfold increase. Thus, in the first three species, f_{resp} makes the greater contribution, but in the fourth, volume predominates. Respiratory frequency and V_T in the fish crow (Bernstein, 1976) and f_{resp} in the barnacle goose (Butler and Woakes, 1980) are independent of flight speed. This could mean that, like the starling and the fish crow, Vo_2 of the barnacle goose is also largely independent of flight speed. However, in the budgerigar, both $\dot{V}o_2$ and f_{resp} change with speed in a U-shaped fashion (Tucker, 1968b).

Despite the apparent matching between the increases in V_I and $\dot{V}o_2$ during flight in the four species listed in Table 39.4, there does appear to be an increase in effective lung ventilation above that required by metabolic rate (hyperventilation) during flight in starlings, as indicated by a decrease in the partial pressure of CO_2 (PCO_2, hypocapnia) in the anterior and posterior airsacs (Torre Bueno, 1978a). It can also be seen from Table 39.4 that $O_{2,ext}$ for the hovering hummingbird, *C coruscans*, is similar to those for the four birds during forward flapping flight, whereas that for *Amazilia fimbriata fluviatilis* is 50% or less.

Ever since Marey's (1890) pioneering studies on bird flight, it has been known that f_{resp} may be coordinated with the beating of the wings, and Marey himself suggested that the flapping of the wings during flight might have some impact on the airsacs. In crows (*C. brachyrhynchos*) and pigeons there is a 1:1 correspondence between f_{resp} and the frequency of wingbeating (f_{wb}), whereas ratios as high as 5:1 (f_{wb}:f_{resp}) have been reported for the black duck *Anas rubripes*, quail *Coturnix coturnix*, and ring-necked pheasant *Phasianus colchicus* (Hart and Roy, 1966; Berger et al., 1970b). However, as mentioned above, the flights reported by these authors were of only a few seconds duration, and it was concluded that coordination was not obligatory. During flights of up to 10-min duration, pigeons in a wind tunnel showed a very close relationship between these two activities. Wingbeats occurred in bursts and, although f_{resp} was often either slower or faster than f_{wb} between bursts of wingbeating, there was always close coordination between the two when the wings were flapping (Figure 39.27(A)). Almost immediately upon landing, the pigeons panted at a frequency identical to the mean resonant frequency of the respiratory system (i.e., 10 Hz, Kampe and Crawford, 1993). During flapping flight, however, f_{resp} was slower, at approximately 7 Hz (i.e. the same as f_{wb}).

With a 1:1 correspondence, it is possible to imagine how contractions of the flight muscles could assist respiratory airflow, but with higher ratios this is not always so obvious. Nevertheless, a correspondence of 3:1 has been found in free-flying barnacle (1.6 kg) and Canada (3.8 kg) geese (Butler and Woakes, 1980; Funk et al., 1993), and it is clear that the wingbeat is tightly locked to fixed phases of the respiratory cycle during flights of relatively long duration (Figure 39.27(B)). These phase relationships can be maintained, even following transient changes in one of the activities (Butler and Woakes, 1980). As Tucker (1968b) comments, "It is hard to believe that the contractions of the flight muscles have no influence in ventilation…." More recent studies have attempted to investigate this intriguing possibility.

Banzett et al. (1992) recorded respiratory airflow and the timing of each wingbeat for three starlings flying at 11 m/s for up to 5 min in a wind tunnel. Triggering on wingbeat and using the technique of ensemble averaging, these authors found that there was usually a 3:1 ratio between f_{wb} and f_{resp}, but that the effect of wingbeat on V_T ranged from 3% to 11%. They concluded, therefore, that wingbeat and breathing in starlings are essentially mechanically independent. A problem with ensemble averaging is that it examines the systems when they are out of synchrony; that is, when there would be a low influence of locomotion on ventilation. On the other hand, Jenkins et al. (1988) took high-speed X-ray cine films of starlings flying in a wind tunnel and suggested that the lateral movement of the furcula during the downstroke of the wing and the recoil during the upstroke may facilitate inflation and deflation, respectively, of the clavicular airsac. They also found, however, that the sternum is moved in an expiratory direction during the downstroke and in an inspiratory direction during the upstroke. As the action of the sternum will influence the more posterior airsacs, these authors proposed that the combined action of the movements of the furcula and of the sternum, which are caused by the beating wings, is to produce a secondary respiratory cycling mechanism between the airsacs and the lungs that performs independently of (the slower) inhalation and exhalation.

Boggs et al. (1997a,b) recorded pressures in, and simultaneous cineradiographic images of, the anterior and posterior

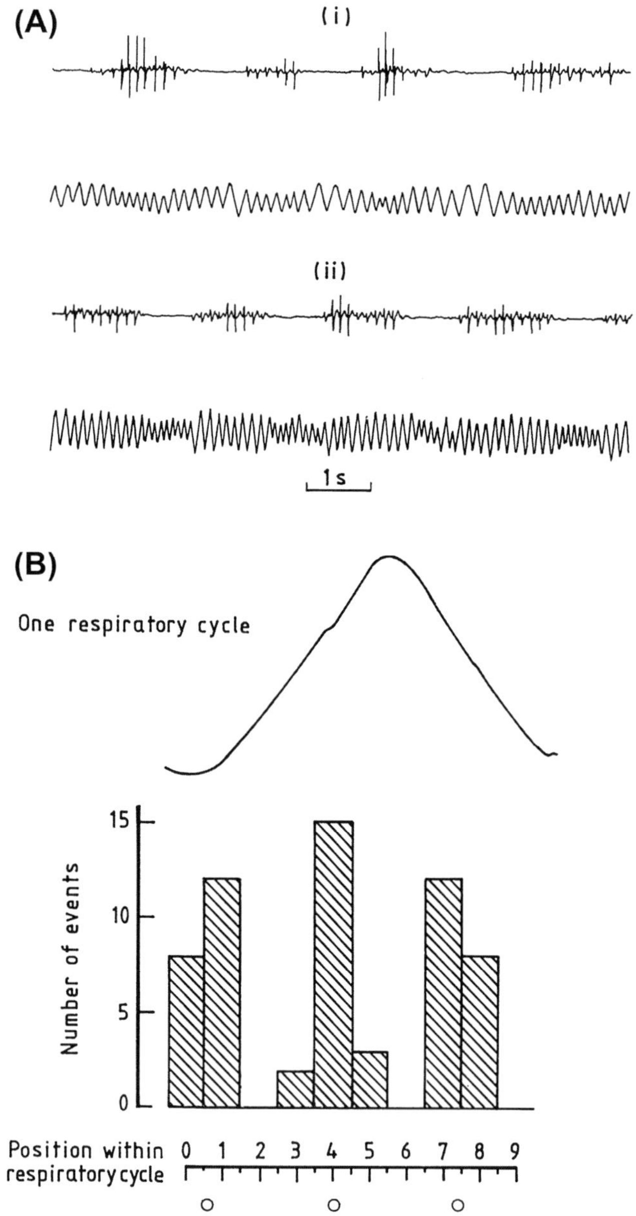

FIGURE 39.27 (A) Traces from a pigeon, *Columba livia* (0.45 kg), flying at a speed of 10 m/s in a wind tunnel, showing changes in respiratory frequency associated with periods of wingbeating and with periods of gliding (i) at the beginning of a flight (1 min after takeoff) when respiratory frequency decreased during the gliding period and (ii) later in the flight (6 min after takeoff) when respiratory frequency increased during the gliding period. In each series, the traces from top to bottom are electromyogram from the pectoralis muscle, respiratory movements (up to trace, inspiration). (*From Butler et al. (1977), Company of Biologists, Ltd.*) (B) Histogram showing the positions during the respiratory cycle at which the wings were fully elevated (called "events") during the flight of a barnacle goose, *Branta leucopsis,* trained to fly behind a truck. Data are from 20 respiratory cycles, which were divided into 10 equal parts. Below the histogram is plotted the mean position of each group of events (O). Above the histogram is a trace of one of f_{resp}, the respiratory cycles (inspiration, up). *From Butler and Woakes (1980), Company of Biologists, Ltd.*

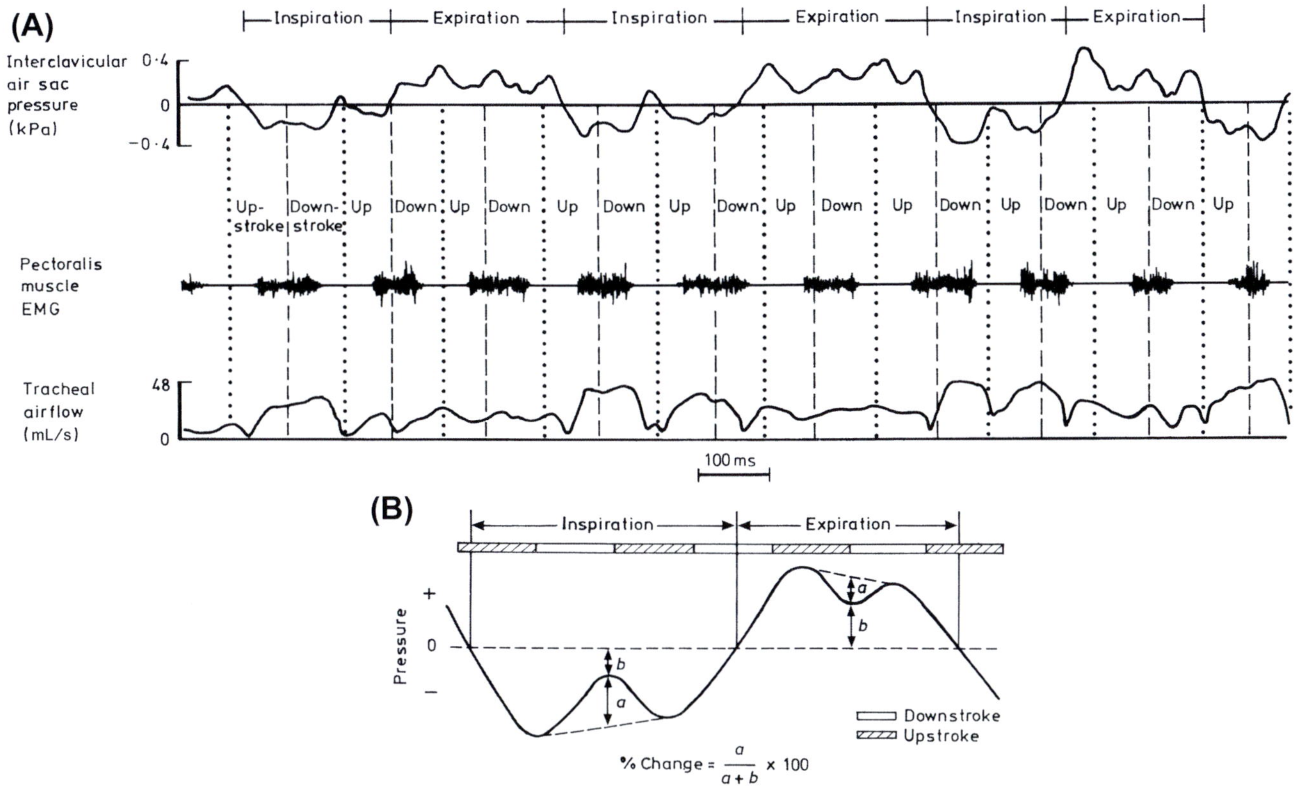

FIGURE 39.28 (A) The relationships between interclavicular airsac pressure, tracheal airflow, and electromyogram (EMG) of a pectoral muscle during flight of a black-billed magpie, *Pica pica,* in a wind tunnel. As the flow signal is not directional, inspiration is taken as that period during which pressure is subatmospheric (below 0) and expiration is that period during which pressure is above atmospheric (above 0). The EMG of the pectoral muscle indicates the upstrokes and downstrokes of the wings. When a downstroke occurs during inspiration, the subatmospheric pressure is driven up toward or above zero and when upstroke occurs during expiration, the supraatmospheric pressure is reduced toward zero (*from Boggs (1997); with permission*). (B) Diagram to indicate how the average changes in airsac pressure shown in (A) were quantified in terms of the percentage change in pressure caused during inspiration or expiration by downstroke or upstroke of the wings, respectively. *From Boggs et al. (1997a), Company of Biologists, Ltd.*

airsacs, airflow in the trachea, and EMGs of the pectoral muscles in black-billed magpies during short (10–20 s) flights in a wind tunnel (Figure 39.28(A)). Although they found similar patterns of movements of the furcula and sternum in the magpie during flight as Jenkins et al. (1988) described for the starling, they did not find that the pressure changes in the anterior and posterior airsacs are consistent with the internal, secondary cycling between the airsacs and the lungs, as postulated by Jenkins et al. (1988). However, the downstrokes and upstrokes of the wings do have compressive and expansive effects, respectively, on the thoracoabdominal cavity, most probably by way of inertial mechanisms. It was only possible to quantify the effect of these influences when the movements of the wings were opposing the action of the respiratory muscles (Figure 39.28(B)). Thus, the average change in airsac pressure caused when a downstroke occurred during inspiration was 94%, whereas the average change caused by an upstroke occurring during expiration was 41%. The corresponding average changes in flow and volume were 75% and 23%, and 35% and 11%, respectively. The conclusion is, therefore, that when the effects of the wings and the respiratory muscles are acting together (i.e., when downstroke occurs during expiration and upstroke during inspiration), ventilation of the lung is substantially enhanced. With the normal 3:1 ratio between f_{wb} and f_{resp}, the pattern of phasic coordination means that there are two upstrokes during inspiration (cf. Figure 39.27(B)) and two downstrokes during expiration, thus giving a net assistance to inspiration during the former and a net assistance to expiration during the latter. This phasic coordination is not disrupted when the birds breathe 5% CO_2 during flight. When they spontaneously switch to a 2:1 ratio, they shorten inspiratory time to ensure that upstroke occurs during most of inspiration and that downstroke corresponds with the transition to expiration (Boggs et al., 1997b). Both of these phenomena indicate that such coordination is probably of fundamental functional significance.

Studies with decerebrate Canada geese (Funk et al., 1992a,b) indicate that, in the absence of peripheral feedback from the flapping wings, there is predominantly a 1:1 ratio between f_{resp} and f_{wb} and that peripheral feedback is required to create the patterns of coordination seen in free-flying birds.

39.8.2 Temperature Control

At least 70% of the total energy expended during forward flapping flight is wasted as heat, and this heat load has to be dissipated in some way. One possible route is by evaporative heat loss via the respiratory system. The hyperventilation that occurs in starlings during forward flapping flight (Torre Bueno, 1978a) could be the result of the increase in body temperature (T_b), by at least 2 °C, that occurs in all birds so far studied, even at low (0 °C) ambient temperatures (T_a) (Torre-Bueno, 1976; Hudson and Bernstein, 1981; Hirth et al., 1987). However, when the usual elevation in T_b during running is prevented in ducks and domestic fowl, there are still signs of hyperventilation (Kiley et al., 1982; Brackenbury and Gleeson, 1983). In ducks, there was an increase in lactic acid and hence slight acidosis that could have stimulated ventilation, and in the fowl, hyperventilation seemed only to occur at higher workloads. Thus it appears as if factors other than hyperthermia contribute to the overall hyperventilation in birds during exercise.

At relatively low T_a (<23 °C), T_b does not change with varying T_a at constant speed in starlings, white-necked ravens, and pigeons flying in a wind tunnel (Torre-Bueno, 1976; Hudson and Bernstein, 1981; Hirth et al., 1987). Above a T_a of 23 °C, however, T_b does increase with increasing T_a in the same three species. In the raven, V_I increases with increasing T_b so that, in this species and in the fish crow flying at constant speed, V_I progressively rises above that required by the metabolic demands of the bird as T_a increases above about 23 °C. Thus, $O_{2,ext}$ falls from a value of 0.19 at a T_a of 20 °C to 0.13 at a T_a of 25 °C in the flying fish crow (Figure 39.29).

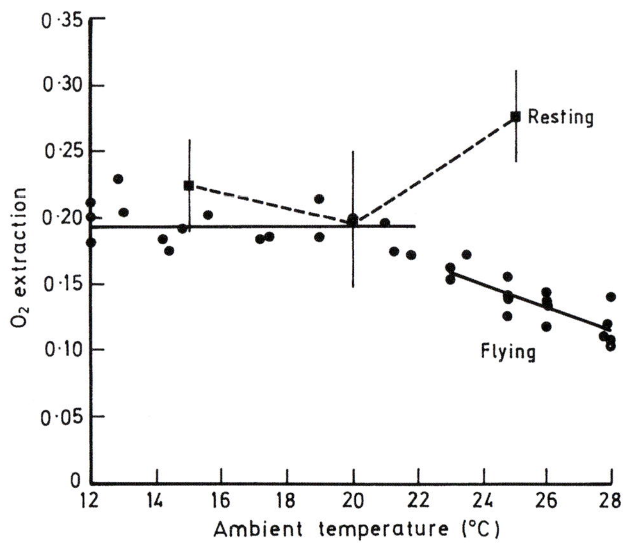

FIGURE 39.29 The relationship between ambient temperature and steady-state oxygen extraction of fish crows, *Corvus ossifragus,* while at rest and while flying in a wind tunnel. *Reprinted from Bernstein (1976), Copyright 1976 with permission from Elsevier Science–NL, Sara Burgerhartstraat 25, 1005 KV Amsterdam, The Netherlands.*

The increased hyperventilation at higher T_a would tend to cause further hypocapnia (seeTorre-Bueno, 1978a) and thus alkalosis. It has been demonstrated that the white-necked raven resorts to so-called compound ventilation when flying at high T_a. A high-frequency, shallow ventilatory component is superimposed upon a deeper, lower-frequency component and may serve to reduce hypocapnia during thermal panting (Hudson and Bernstein, 1978).

The physiological significance of the hyperthermia during flapping flight, even at low T_a, is uncertain, although Torre-Bueno (1976) concluded that birds adjust their insulation to allow a certain increase in T_b during flight in order to improve muscle efficiency and to increase maximal work output. Hudson and Bernstein (1981) point out that, assuming an overall efficiency of 25%, the almost 3 °C increase in T_b in white-necked ravens while flying at a speed of 10 m/s reflects the storage up to half of the metabolic heat produced during a flight of 5 min duration. At lower overall efficiencies, more heat would be produced, so the proportion stored under the above conditions would be less. A similar situation exists for the calculations of the proportion of metabolic heat lost by respiratory evaporation (see below).

Whatever the significance of the increase in core temperature during flapping flight, brain temperature is maintained at a lower level. In fact, even in birds under resting, thermoneutral conditions, there is approximately a 1 °C difference between brain and body temperatures (Bernstein et al., 1979a). This difference is at least maintained as kestrels (*Falco sparverius*) become hyperthermic during flapping flight and may even increase (Figure 39.30). The structure that appears to be largely responsible for this phenomenon is the *rete mirabile ophthalmicum* (RMO) (Kilgore et al., 1979; Bernstein et al., 1979a,b). It is closely associated with the circulatory system of the eye and warm arterial blood from the body is thought to be cooled by the countercurrent exchange with venous blood returning from the relatively cool beak, evaporative surfaces of the upper respiratory tract, and eye (Midtgård, 1983).

Hyperventilation during flight no doubt serves a thermoregulatory function, and as T_a increases, a greater proportion of total heat loss during flight is by respiratory evaporation. However, even at a T_a of 30 °C, this proportion is only approximately 20% of total heat loss for the budgerigar and fish crow (Tucker, 1968b; Bernstein, 1976) and 30% for the white-necked raven and pigeon (Hudson and Bernstein, 1981; Biesel and Nachtigall, 1987). In the hummingbird *Amazilia fimbriata*, maximum respiratory heat loss of 40% of the total occurred at a T_a of 35 °C (Berger and Hart, 1972). So, most metabolic heat must be dissipated by means other than respiratory evaporation. Indeed, herring gulls can lose up to 80% of total heat production through their webbed feet during flight (Baudinette et al., 1976). In the pigeon, the value is probably less, but nonetheless significant, at 50–65% (Martineau and Larochelle, 1988).

39.8.3 Respiratory Water Loss

During flapping flight at T_a above 18 °C for the budgerigar and 7.5 °C for the pigeon, water is lost at a greater rate than it is produced metabolically (i.e., the birds are dehydrating; Tucker, 1968b; Biesel and Nachtigall, 1987). The starling is in water balance only at T_a below 7 °C (Figure 39.31; Torre-Bueno, 1978b). However, because of the uncertainty regarding the rate of metabolic water production during flight, the author points out that the critical temperature lies between 0 and 12 °C. Only below a T_a of 0 °C was the hovering hummingbird *A. fimbriata* in water balance (Berger and Hart, 1972). Torre-Bueno (1978b) suggested that during migrations, birds ascend to altitudes where the air is cool enough to enable a greater proportion of heat to be dissipated by nonevaporative means, thus keeping them in water balance.

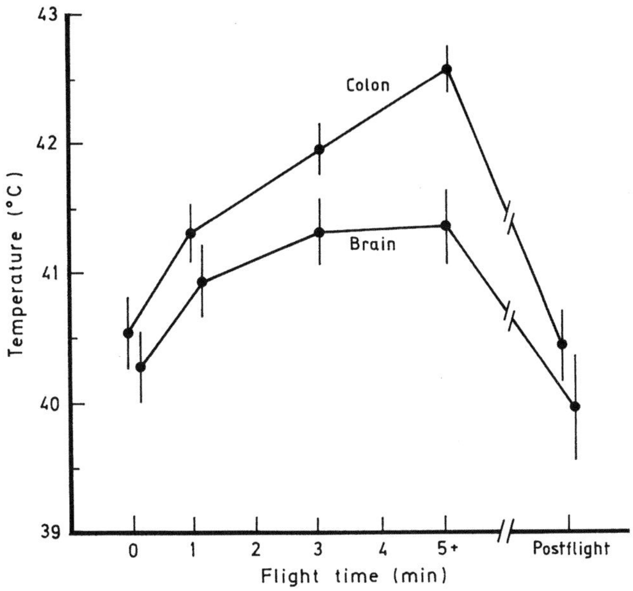

FIGURE 39.30 Mean (±2 SE) temperatures in the colon and brain of American kestrels, *Falco sparverius*, before, during, and after flying at a speed of 10 m/s and at an ambient temperature of 23 °C. Data at 5+ min were obtained as mean values over the period between 5 and 15 min after the onset of flight and represent steady-state values. Postflight values are those obtained after reestablishment of steady-state following each flight. *From Bernstein et al. (1979a) with permission.*

Some of the larger species of birds do indeed appear to migrate at high altitudes, in excess of 8000 m (Swan, 1970; Stewart, 1978). However, Carmi et al. (1992) used program one from Pennycuick (1989) to calculate P_i and concluded that smaller species might be better off staying below 1000 m because of the greater ventilation (and hence increased water loss) required by the reduced air density and partial pressure of oxygen (PO_2) associated with higher altitudes. These authors also emphasize the importance of $O_{2,ext}$ and the temperature of the exhaled. air (T_{exp}) in water conservation during flight. The greater the $O_{2,ext}$, the lower the V_I for a given $\dot{V}O_2$, and the lower the T_{exp}, the less water will be lost in expired air, provided the birds have an effective countercurrent heat exchanger in their respiratory passages (Schmidt-Nielsen et al., 1970). Thus, it should be advantageous for small migrants crossing the Sahara to fly at low altitude at night and to rest during the day. However, a study of the behavior of nocturnal migrants in a desert area indicates that only tailwind speed is closely related to the altitude at which the birds fly (Bruderer and Liechti, 1995). Thus, in autumn, that the birds make use of the northerly winds at relatively low altitudes, despite the implication for water balance, may not be surprising, but

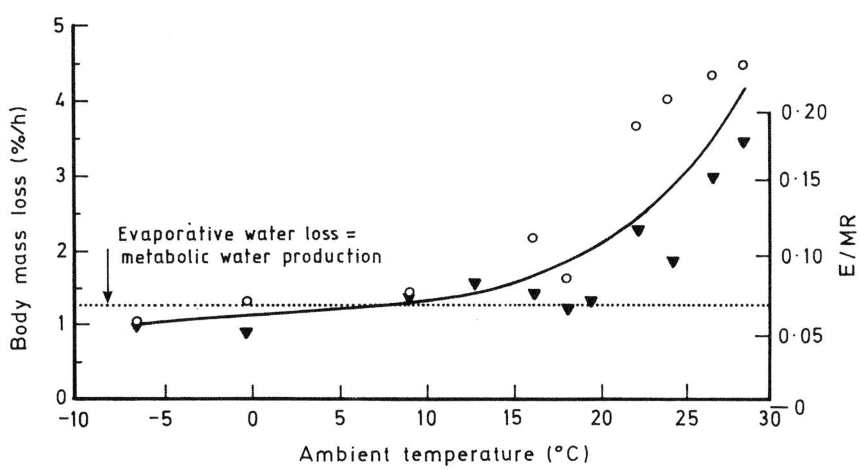

FIGURE 39.31 Data showing the rate of loss of body mass and the ratio of the rate of evaporative cooling to the metabolic rate (E/MR) as a function of ambient temperature in two starlings, *Sturnus vulgaris*, flying in a wind tunnel at between 9 and 14 m/s. *From Torre-Bueno (1978b), Company of Biologists, Ltd.*

how nocturnal migrants are able to detect wind direction is a fascinating question.

Carmi et al. (1993) also demonstrated that when at rest or during flight, birds have an ability to maintain plasma volume, despite significant reductions in body mass that are largely the result of water loss. This is of obvious significance during flight, particularly during long-distance migratory flights, when a reduction in plasma volume would result in an increase in viscosity of the blood, which in turn would increase the workload of the heart, resulting in an inadequate cardiac output and supply of oxygen to the metabolizing tissues.

39.9 MIGRATION AND LONG-DISTANCE FLIGHT PERFORMANCE

39.9.1 Preparation for Migration

As discussed by Fry and Ferguson-Lees (1972), it is a reasonable expectation that there will be correlated physiological adaptations associated with the extra power required to fly due to the laying down of fat as fuel for long-distance flights. Their work indicated that early in the fattening process of the yellow wagtail (*Motacilla flava*), the flight muscles showed a small hypertrophy. Other studies have also found muscle hypertrophy during fattening for both the pectoralis muscles (Marsh, 1984; Driedzic et al., 1993; Bishop et al., 1996; Jehl, 1997) and the cardiac muscles of birds (Driedzic et al., 1993; Bishop et al., 1996; Jehl, 1997). Thus, as protein as well as lipid is metabolized during long-duration flights, it is also replaced during stopovers, so the increase in body mass before a migration and during stopovers should be described as "fueling" rather than "fattening" (Piersma, 1998).

There has to be a close relationship between the amounts of fuel and flight muscle that are deposited before migration, as the greater the fuel load, the greater the mass of flight muscle required to carry that load. A study by Piersma (1998) indicates that the proportions of fat-free tissue and fat that are deposited before long-distance (>1500 km) migration of shore birds vary between similar-sized species depending on the duration and nature of their impending flights. Just before departure, the pectoral muscles and heart tend to hypertrophy whereas the stomach, intestine and liver tend to atrophy. Figure 39.32 indicates that the greater fat deposition and atrophy of the visceral organs (such as the liver) are most pronounced in those species that have the longest migrations and especially in those that have little or no opportunity for an emergency landing (e.g., by flying over an ocean). Of particular interest are the two subspecies of bar-tailed godwit. Although the distances of their migrations are vastly different (4500 km versus 11,000 km), they both have high fat fractions and low liver to pectoral muscle mass ratios. What they have in common is their inability to make emergency landings during their migrations.

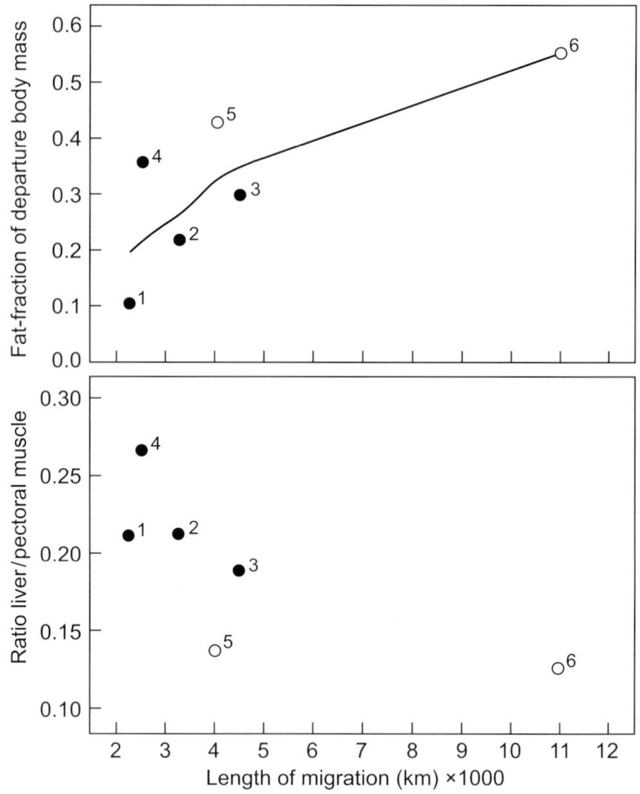

FIGURE 39.32 Fat loads and relative organ sizes before departure of six species of long-distance migrants. Top. The amount of fat as a fraction of body mass in relation to the distance of the migration to be undertaken. Bottom. The ratio of liver mass to pectoral muscle mass in relation to the distance of migration. Closed circles indicate species that would be able to make stopovers during their migration and open circles indicate subspecies that would not. (1) Golden plover, *Pluvialis apricaria;* (2) Ruff, *Philomachus pugnax;* (3) Bar-tailed godwit, *Limosa lapponica lapponica*; (4) Red knot, *Calidris canutus islandica;* (5) Bristle-thighed curlew, *Numenius tahitiensis;* (6) Bar-tailed godwit, *Limosa lapponica baueri*. *From Piersma (1998).*

Bishop et al. (1998) compared the development of a captive population of barnacle geese with that of the wild migratory population in order to investigate to what extent some of the migratory specializations of the cardiac and locomotory muscles might be determined by developmental processes and to what extent they might be modulated by differences in relative levels of activity. Postflight increases in the masses of the pectoralis muscles of both wild and captive geese tend to show an appropriate amount of hypertrophy in response to changes in body mass (Figure 39.33(A) and (B)). Regression equations were calculated for both the wild adult premigratory geese $\left(\text{pectoralis mass} = 0.5\ M_b^{0.86},\ r^2 = 0.93\right)$ and captive adult geese $\left(\text{pectoralis mass} = 0.3\ M_b^{0.92}\right)$. The body mass exponents (or slopes) of the two regression equations are not significantly different between the two populations, although the coefficients are slightly smaller in the captive geese. Thus, approximately 92% of the pectoralis muscle mass of

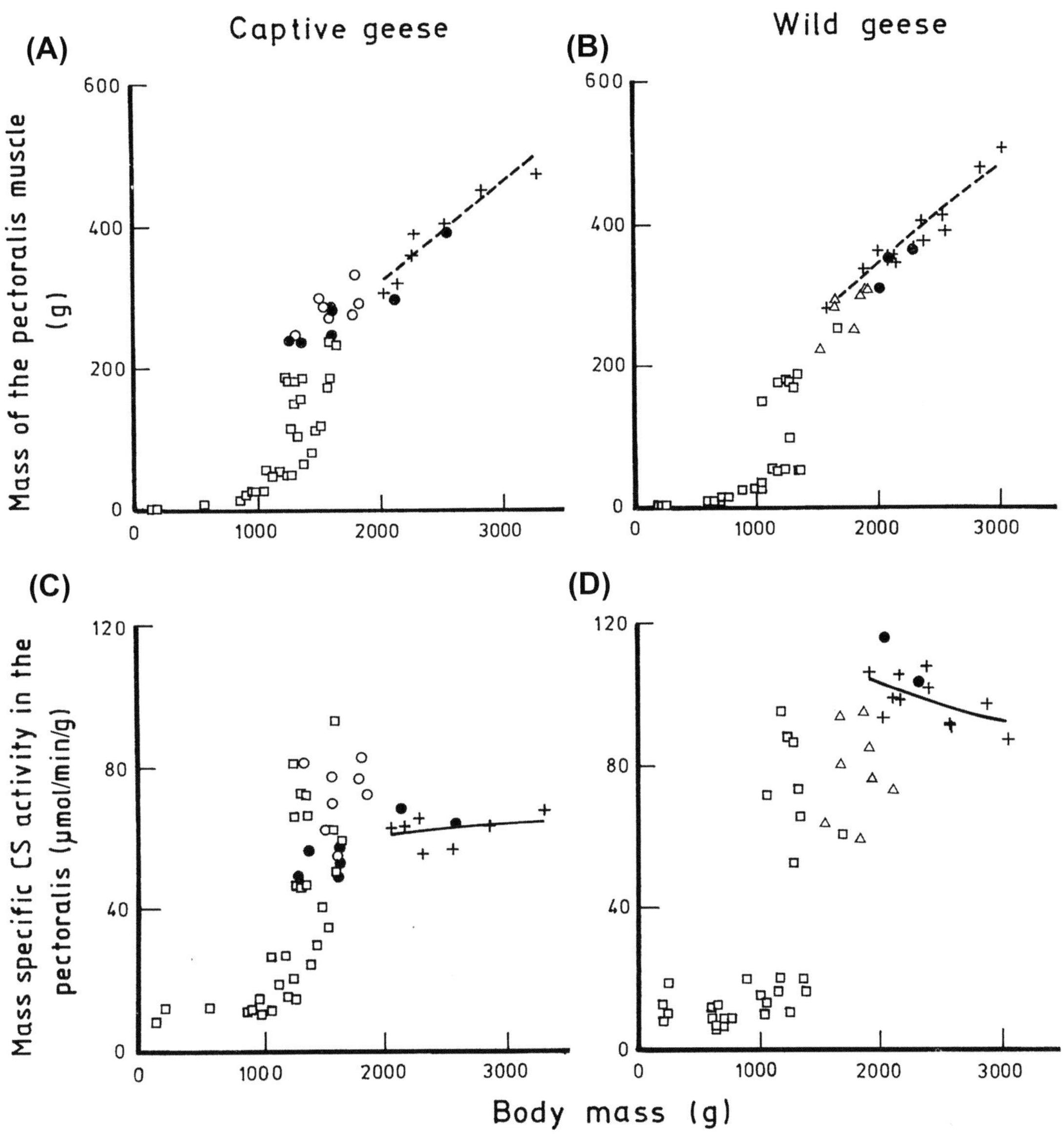

FIGURE 39.33 Scatter diagram of the masses of the pectoral muscles (g) and the mass-specific activity of citrate synthase (CS) in the pectoral muscles (μmol/min g wet wt) of captive (A and C) and wild (B and D) barnacle geese, *Branta leucopsis*, plotted against body mass (g). Goslings of 1 to 7 weeks of age (□), goslings of 12–20 weeks of age (●), adults at 7 weeks after the population hatch date, (Δ) and premigratory adults >10 weeks post–hatch (+). *From Bishop et al. (1998), with permission from The University of Chicago Press. © University of Chicago Press.*

premigratory barnacle geese appears to be almost independent of the experience of flight *per se*. Indeed, wild adult barnacle geese showed no increase in the amount of time they spent flying before their autumn and spring migrations (Portugal et al., 2012), and Portugal et al. (2009) found that hypertrophy of the pectoralis muscles of captive barnacle geese actually began even before their body mass increased.

Gaunt et al. (1990) studied the flight muscle changes of the eared grebe (*P. nigricollis*) during a staging period at Mono Lake, California, where the food resources were not limiting. Following a postbreeding molt at the lake, the birds become flightless and their flight muscles show a slight atrophy, but due to an abundance of food the birds put on a large amount of fat and body mass greatly increases. Just prior to departure for the winter grounds, the birds metabolize much of the fat, and it is at this time that the flight muscles hypertrophy. In this example, flight muscle atrophy occurs when the birds have plenty of food, while subsequent hypertrophy occurs while the birds are actually reducing in body mass. Marsh and Storer (1981) had originally suggested that the correlation between flight muscle mass and body mass in Cooper's hawk (*Accipiter cooperii*) was a natural analog of "power" training during flight. Subsequent studies have found no obvious mechanistic link between hypertrophy of the flight muscle and body mass, although the authors did associate increased wing flapping with the flight muscle hypertrophy. Dietz et al. (1999) found that two subspecies of red knots (*Calidris canutus*), which were kept under a constant photoperiod, showed similar increases in pectoral muscle hypertrophy (44%) during premigratory fattening, and that this represented between 29% and 39% of the total mass increase. This suggested that there was no direct photoperiodic control, and that the mechanism was likely to be endogenous, possibly with some kind of circannual regulation. It still remains an outside possibility that the relatively infrequent bouts of wing-flapping and occasional takeoffs exhibited by captive birds might provide a sufficient mechanistic link by which the birds are able to maintain the flight muscle to body mass relationship. However, as mentioned above, Portugal et al. (2012) showed that wild migratory barnacle geese did not show any change in the time spent flying during a 3 week premigratory period, thus, ruling out any overt evidence for flight training prior to the actual migration.

Figure 39.33(C) and (D) show the change in maximum mass-specific CS activity in the same barnacle geese (Bishop et al., 1998) for the data on pectoralis muscle mass (Figure 39.33(A) and (B)). There is little difference between the two populations up to 7 weeks of age. Specific activities of CS in the pectoralis muscles of the two wild 11.5 week old juveniles are, at 103 and 116 μmol of substrate/min g, outside the range of that for the 12 week old captive birds (55–59 μmol of substrate/min g) and the captive adult birds (56–58 μmol of substrate/min g) but similar to the range of wild adult birds (92–106 μmol/min g). The main conclusion from these results is that the captive adults have mass-specific values of activity for CS (mean = 62.1 ± 1.7 μmol/min g) that do not scale with body mass $\left(CS = 20\, M_b^{0.14},\ r^2 = 0.11\right)$ and are substantially below those of the wild premigratory birds (mean = 98.6 ± 1.9 μmol/min g, $p < 0.0001$). Results from captive goslings aged between 12 and 20 weeks also appear to be closely associated with the regression line resulting from the data from captive adult geese.

Peak values for the activity of CS in the pectoralis muscles of wild barnacle geese are found in the premigratory birds, although there is a tendency for the mass-specific activity to decline with increasing body mass $\left(CS = 862\, M_b^{-0.28},\ r^2 = 0.33\right)$. Thus, the activity of CS in the pectoralis of adult captive geese, and captive goslings over 11 weeks of age, is only around 60% of that measured in wild geese. In addition, CS activity in the pectoralis muscles of a group of postmolting, wild, adult geese was similar to that of long-term captive birds. Thus, it is suggested that the rise in the activity of CS in the premigratory birds may be a reaction to the increase in flight activity *per se,* while the rise in CS activity during development of the goslings up to fledging could be primarily under endogenous control. In addition, heart ventricular mass is reduced in captive and postmolt wild adults compared to that in wild premigratory birds, and is qualitatively similar to the reduction shown in the aerobic capacity of the pectoralis muscle. It is suggested that this reduction in ventricular mass is also likely to be a direct result of the lower activity levels experienced by captive birds (Bishop et al., 1998).

One of the factors that may be important in the development of the aerobic capability of the pectoral muscles of barnacle geese is thyroxine (T_4). Circulating levels of T_4 show a similar developmental profile as mass and mass-specific activity of CS of the pectoral muscles (Bishop, 1997). While artificially accelerating the increase in circulating T_4 did not have a significant effect on either relative mass or on the mass-specific CS activity of these muscles (Deaton et al., 1997), hypothyroidism during development, induced by administration of the drug methimazole, did result in the retardation of their growth, mass-specific CS activity, fractional volume of mitochondria, and capillarity (Deaton et al., 1998). There was no such effect on the mass of the muscles of the leg.

If the mass-specific CS values for the pectoralis muscles of premigratory geese are generalized to the whole muscle, then the largest birds will have the lowest mass-specific aerobic capacity despite requiring the highest relative power outputs (Pennycuick, 1989). Consequently, during sustained flight, the larger birds may have to fly nearer to their minimum power speeds than smaller geese. This conclusion is also supported by the data on the heart ventricular mass (Bishop et al., 1996), which indicates that there is a tendency for relative heart mass to decline slightly with increasing body mass. Thus, given that maximum oxygen consumption is likely to be closely correlated with heart mass (Bishop and Butler, 1995; Bishop, 1997, 1999, 2005), it is suggested that during premigratory fattening, P_i available is likely to scale relative to body mass with an exponent value considerably lower than that of the P_o exponent theoretically required $\left(M_b^{1.59}\right)$ for flight performance to be maintained at a similar level according to aerodynamic theory (Rayner, 1990). However, a number of studies have questioned whether the allometric exponent for the scaling of P_i with increasing body mass is similar to that predicted for P_o. Tucker (1972) measured the rate of oxygen consumption, from a single gull flying in a wind tunnel at 10.8 m/s, on 13 different occasions and with body mass naturally varying from 0.328 to 0.420 kg. He determined that the cost of carrying the extra body mass was quite small, with a body mass exponent of only $M_b^{0.325}$ but 95% confidence limits of 0.05–0.6. A very similar result was reported by Kvist et al. (1998) using the DLW technique to determine the energy expenditure of red knots flying in a wind tunnel on different days while they naturally put on body mass during the autumn. They calculated a body mass exponent of only $M_b^{0.325}$, with 95% confidence limits of 0.08–0.62. Subsequently, quite low exponents of $M_b^{0.55}$ and $M_b^{0.58}$ were also calculated for flying rose-colored starlings, *Sturnus roseus*, and barn swallows, *Hirundo rustica*, also using

DLW (Engel et al., 2006; Schmidt-Wellenburg et al., 2007). Kvist et al. (1998) suggest that birds may be able to change the mechanochemical conversion efficiency of their flight muscles as they fuel up for the migration. However, it might be expected that birds should maximize muscle efficiency at all times.

39.9.2 Migratory Behavior

When migrating, it might be more appropriate for the bird to minimize the time of migration (T_{min}), rather than the energy consumed, particularly if it is important for it to arrive at its destination before most of its competitors (Alerstam and Lindstrom, 1990). In this case, the optimal flight speed will be greater than U_{mr}. In an analysis of data in the literature for 48 species during migration, Welham (1992) concluded that lighter species do, indeed, fly faster than U_{mr}, and heavier species tend to fly slower. However, in an even more comprehensive study of 138 species (Alerstam et al., 2007), this apparent allometric compression of the expected speed range could be partially explained through the lack of geometric similarity with increases in body mass (e.g., aspect ratio of the wings scales positively). Flight speed was found to be strongly influenced by wing loading and phylogenetic relationships, but some of the variation could not be fully accounted for.

It is now possible to track individual birds of many different species on their migratory flights, some of which cover many thousands of kilometers over land and sea. One of the fastest flights recorded is that of great snipe (*Gallinago media*), despite the fact that it is mostly conducted over land (Klaassen et al., 2011). This species made nonstop flights from Sweden to Central Africa and back again, covering between 4000 and 6800 km at estimated air speeds of 16–26 m/s. Bar-tailed godwit (*Limosa lapponica*) makes an even longer nonstop flight, largely over the Central Pacific Ocean, while traveling between Alaska and New Zealand (Gill et al., 2009). They averaged 10,153 km in 7.8 days, equivalent to an air speed of around 15 m/s. Other species of wading birds have also been shown to make impressive and rapid migratory flights (Minton et al., 2011; Niles et al., 2012; Johnson et al., 2012) in which they have to greatly increase their body mass before departure (Gudmundsson et al., 1991). More specialized pelagic species of seabird, such as those within the orders Procellariiformes and Charadriiformes, regularly make quite long foraging trips during the breeding season. In the winter months they frequently disperse around the oceans of the world on massive migrations, such as that of the sooty shearwater, *Puffinus griseus* (Figure 39.34; Shaffer et al., 2006) and Arctic tern, *Sterna paradisaea* (Egevang et al., 2010).

It has been suggested that some species of birds, such as geese, may be able to reduce induced drag by flying in V-formation (Lissaman and Scholenberger, 1970; Hummel, 1995). This would entail their maintaining positions so that the tips of the wings on the inside of the V are close to the centers of the trailing vortices from the wings on the outside of the bird ahead (Hainsworth, 1987). Assuming a fixed wing at maximum span, Hainsworth (1987) found considerable variability in wing tip spacing (WTS) of between

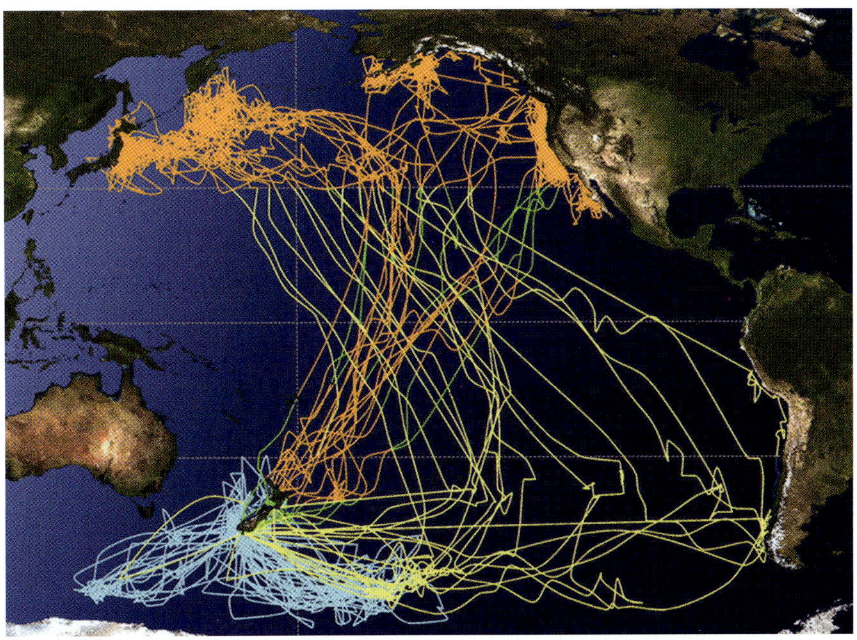

FIGURE 39.34 Geolocation tracks of sooty shearwater, *Puffinus griseus*, flights originating from breeding colonies in New Zealand. Interpolated geolocation tracks of 19 sooty shearwaters during breeding (light blue) and subsequent migration pathways (yellow, start of migration and northward transit; orange, wintering grounds and southward transit). *From Shaffer et al. (2006).*

−128 and 289 cm, both between and within formations of Canada geese (*B. canadensis*), but found that the median WTS for 55 geese (−19.8 cm) was close to the theoretical optimum (−16 cm) and corresponded to an estimated saving in induced power of 36%. Even so, in order to achieve such savings with beating wings, there would have to be a high degree of synchrony in wingbeat frequencies, as there is little variation in depth (the distance between adjacent birds' body centers, parallel to the flight path) between individuals (Hainsworth, 1988). This author found close synchrony (a difference of <0.1 beat/s) to be present in 48% of the birds he studied and, because of the calm conditions at the time, concluded that this probably represents an upper limit. Indeed, even when not taking wingbeat frequency into account, Cutts and Speakman (1994) calculated mean saving in induced power for 54 skeins of pink-footed geese (*Anser brachyrhynchus*) to be only 14% and concluded that the relationship between depth and WTS supported the communication hypothesis (Gould and Heppner, 1974). This contends that the birds position themselves so as to avoid collisions and to maintain flock unity. Although not on migration, Weimerskirch et al. (2001) studied the flight performance and heart rate of great white pelicans (*Pelecanus onocrotalus*) while they were flying alone and when they were flying in V-formation. Mean wingbeat frequency was reduced by up to 50% for pelicans flying behind the first two birds in formation, as a result of an increase in the time spent gliding. Heart rate was reduced by 14% for pelicans in formation compared to birds flying alone. If we apply the allometric trend that $\dot{V}O_2$ is proportional to f_H^2 (Bishop and Spivey, 2013), then this would suggest that the mean energetic saving for flying in V-formation was around 27% relative to unaccompanied flight.

39.10 FLIGHT AT HIGH ALTITUDE

The majority of avian migratory flights occur below 1 km above the ground (Liechti and Schaller, 1999; Berthold, 1993); thus migrations where birds climb to higher altitudes to select for favorable wind directions, or where traditional flyways route birds over large mountain barriers, are of particular interest. Early studies indicated that most small passerines migrating at night fly below 2 km above sea-level (Lack, 1960; Nisbet, 1963), although it was demonstrated by Bruderer and Liechti (1995) that wind direction may cause the majority of birds crossing the Negev desert in the south of Israel to fly above 1.8 km in spring, but with 90% below 3.5 km. Occasionally, when a low-level jet offered tailwinds that were particularly strong, birds flying over Israel reached heights between 5 km up to almost 9 km (Liechti and Schaller, 1999), and with ground speeds >25 m/s and up to 50 m/s. Even more rarely, some accidental observations of various species of birds have been reported at similarly extreme high altitudes (>6 km). A flock of 30 swans (probably whooper, *Cygnus cygnus*) was located by radar off the west coast of Scotland at an altitude of 8–8.5 km, where the temperature was −48 °C (Stewart, 1978; Elkins, 1979), but it is unclear how these birds managed to reach such high altitudes (Pennycuick et al., 1996a). Bar-headed geese, *A. indicus,* have also been reported as flying at altitudes of well over 8 km (where PO_2 is around 6.7 kPa, approximately one-third of the sea-level value) during their migration across the Himalayas (Swan, 1961). However, these anecdotal observations have never been substantiated by any scientific evidence, while the highest ever directly recorded flight for a bar-headed goose was 7290 m (Hawkes et al., 2012), using a GPS satellite tag. Thus, birds can certainly fly at altitudes at which non-acclimatized mammals find walking difficult (Tucker, 1968a). Unfortunately, there have been few physiological studies on birds flying at high altitude, real or simulated, although Tucker (1968a) did train a budgerigar wearing a $\dot{V}O_2$ mask to fly in a hypobaric wind tunnel for 30 s at the equivalent of 6.1 km. Two particular aspects of flying at high altitude (low air temperature and evaporative water loss) have already been mentioned. Other aspects are the effect of the decrease in air density and in the partial pressure of oxygen with increasing altitude, which will have both physiological and biomechanical influences on flight performance. Pennycuick (1975) argues that, as birds fly higher, they must fly faster in order to obtain sufficient lift from the wings. However, as speed increases with height, so does the power required to maintain it. Also, because the rate of gas exchange will be affected by the value for PO_2, there will be a particular altitude at which the bird can only just obtain O_2 at a rate sufficient to maintain even its minimum power speed, at which point it will not be able to climb any higher under its own power.

Gudmundsson et al. (1995) studied the migration of brent geese (*Branta bernicla*) from Iceland to Greenland and tracked them by satellite as they flew up and across the Greenland ice cap. These authors found that the brent geese (fitted with 60 g satellite transmitters) were not able to fly continuously over the Greenland ice cap and had to make frequent and long rests during the climb. Gudmundsson et al. (1995) concluded that this evidence supported the biomechanical predictions of Pennycuick (1989) that these birds were up against the limit of the biomechanical power available from the flight muscles due to negative scaling of wingbeat frequency with respect to body mass. However, the analyses of Bishop and Butler (1995) and Bishop (1997, 2005) suggest that the limit to the sustained climbing performance of birds, such as brent geese, is more likely to be related to the biomechanical performance of the heart. As geese have both FG and FOG fiber types in their flight muscles (Rosser and George, 1986a), the net effect is that during "prolonged" activity the mass of the flight muscle is effectively reduced to that of the "aerobic" FOG fibers and the P_o available becomes proportional to the ability of the cardiovascular and respiratory system to supply the working (aerobic) muscle tissue. With a small additional contribution from the FG fibers, the brent geese would have the potential to fly up the steeper slopes using anaerobic power, but the birds would require regular stops (as was observed). The empirical data of Gudmundsson et al. (1995) indicate that it is not the biomechanical power available from the whole flight muscles *per se* that is limiting, but the power available from the mass of the FOG fibers, which is likely to be closely matched to the ability of the cardiovascular system to perfuse the flight muscle with blood. Pennycuick et al. (2011) also tracked brent geese flying over the Greenland ice cap and arrived at the same conclusion.

Due to its iconic migration over the Himalayan mountains and across the Tibetan-Qinghai plateau (Figure 39.35), the bar-headed goose has been the subject of many studies concerned with the physiological adaptations for flying at high altitude (reviewed in Butler, 2010; Scott, 2011). Unlike Pekin ducks, bar-headed geese do not increase their Hct and Hb when exposed to simulated high altitude (Black and Tenney, 1980). This means that there is no increase in the viscosity of the blood, thus preventing a possible reduction in its circulation. This is more than counterbalanced by the higher affinity for oxygen (low P_{50}) of the Hb of the goose (P_{50} for blood of the bar-headed goose at pH 7.5 is approximately 5 kPa compared with 7.5 kPa for the duck), which allows the maintenance of a higher C_aO_2 (and hence $C_aO_2 - C_{\bar{v}}O_2$) at high altitudes than in the duck.

A comparison between the bar-headed goose and the plain-dwelling graylag goose, *Anser anser*, indicates that their Hb differs by only four amino acid substitutions, one of which (α^{119} Pro→Ala) may, by altering the contact between the α1 and β1 chains, confers a small increase in the O_2 affinity of the Hb from the bar-headed goose (Perutz, 1983), and this is then amplified by the interaction with inositol pentaphosphate (Rollema and Bauer, 1979). The complementary substitution (β^{55} Leu→Ser) at the same position may have a similar effect on the P_{50} of the Hb of the Andean goose *Cloephaga melanoptera* (Weber et al., 1993). Temporary movements to high altitude in some other species may

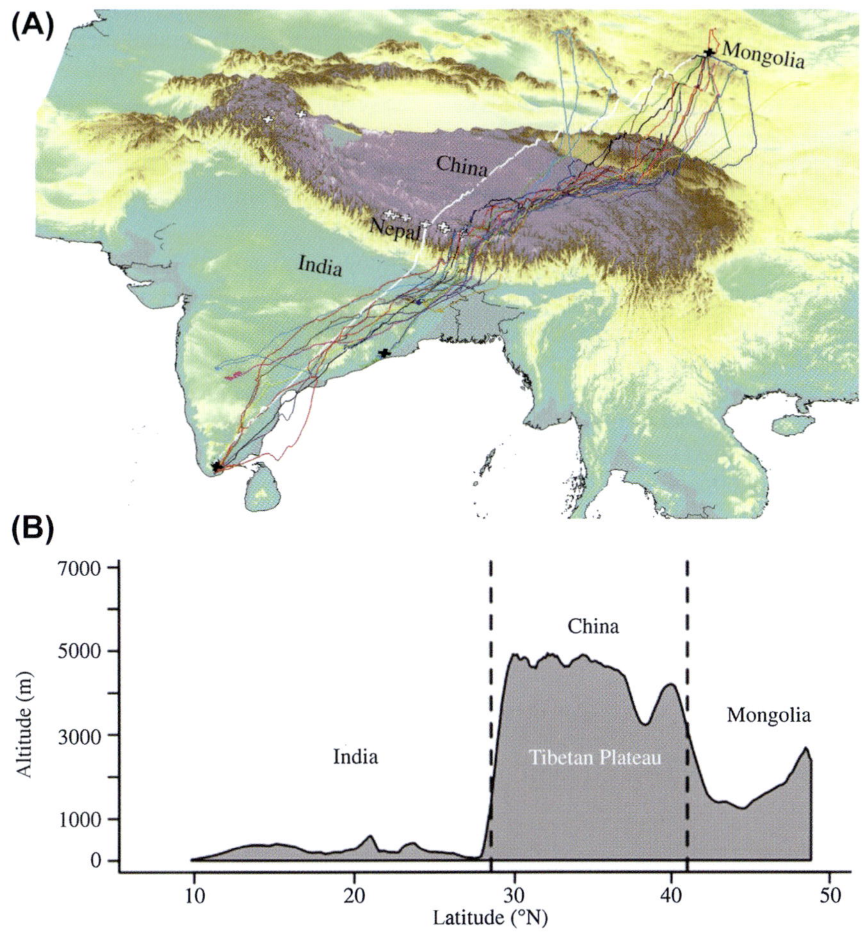

FIGURE 39.35 Satellite tracks of migrating bar-headed geese, *Anser indicus*: (A) three-dimensional map showing release locations (black crosses) of bar-headed geese in India ($n=2$ sites) and Mongolia ($n=1$ site). Colored lines represent 16 individual geese, and colored background shading indicates elevation. Solid thick white line shows the great circle route. White crosses show locations of the mountains peaks over 8000 m in elevation. (B) Cross-section of land elevation underlying the route of the northwards migration, based on the arithmetic mean of all flights. *From Hawkes et al. (2013).*

have different effects. The natural annual movements of wild quail from 200 to 1200 m in Catalonia is accompanied by an increase in Hb of 24% but a decrease in the affinity of Hb for O_2 (P_{50} from 3.6 to 4.2 kPa, Prats et al., 1996). Thus, there is an increase in O_2-carrying capacity of the blood and an enhancement of its ability to release O_2 to the tissues.

It is clear from the data of Black and Tenney (1980) that the respiratory system of bar-headed geese is able to maintain a very small difference between the PO_2 of inspired air (P_IO_2) and that in the arterial blood (P_aO_2), when at high altitude. For example, when at sea level, $P_IO_2-P_aO_2$ in bar-headed geese is approximately 7 kPa, whereas at a simulated altitude of 10.67 km it is only 0.5 kPa. The large increase in respiratory minute volume $(\dot{V}_I)$ that is required to maintain such a small difference between P_IO_2 and P_aO_2 also causes a decline in P_aCO_2 and associated increase in arterial pH (pHa); the bird becomes hypocapnic and alkalotic. Scott and Milsom (2007) showed that the increase in $\dot{V}_I$ in bar-headed geese was the result of a larger increase in tidal volume and a correspondingly lower increase in respiratory frequency than that of the lowland-living graylag goose (*A. anser*) or Pekin duck.

In a number of mammals (dog, monkey, rat, human) the hypocapnia induced by hyperventilation causes a reduction in cerebral blood flow. This is not the case in ducks and bar-headed geese (Grubb et al., 1977; Faraci and Fedde, 1986). Also, hypoxia causes a greater increase in cerebral blood flow in ducks than it does in dogs, rats, and man (Grubb et al., 1978). At high altitude, of course, both of these factors occur together (hypocapnic hypoxia) and under these conditions hypocapnia appears to attenuate the increase in cerebral blood flow caused by hypoxia (Grubb et al., 1979). However, these authors found that blood flow is similar at a given O_2 content in both normocapnic and hypocapnic ducks. This is because during hypocapnia and alkalosis, there is a leftward shift of the O_2 equilibrium curve, as a result of the Bohr effect, so that a given O_2 content is achieved at a lower PO_2. In fact, in bar-headed geese, the alkalosis during severe hypoxia is greater than that in Pekin ducks, which together with the higher affinity of their Hb for O_2, means that at a given (low) P_aO_2, C_aO_2 is much (at least two times) greater in the geese (Faraci et al., 1984a). In addition, the pulmonary vasoconstrictor response to hypoxia is smaller in the bar-headed goose than in other birds and mammals (Faraci et al., 1984b). Thus, similar (or even greater) O_2 deliveries to the brain and heart can be achieved in hypoxic bar-headed geese at lower cerebral and coronary blood flows than in hypoxic ducks. Indeed, this species of goose is able to maintain, or even to increase, its perfusion of all tissues during severe hypocapnic hypoxia (Faraci et al., 1985).

Delivery of O_2 to the locomotory muscles are also influenced by the architecture of the capillaries. Capillary density and capillary:fiber ratio are greater in the skeletal muscle (gastrocnemius) of bar-headed geese than in that of Canada geese (Snyder et al., 1984), and this is a genetic adaptation that is not dependent on exposure to low atmospheric oxygen during development. Perhaps surprisingly, there was only a 24% increase in the capillary:fiber ration of the bar-headed goose flight muscle compared with that of the barnacle goose and no significant difference in capillary density (Scott et al., 2009). All of these factors are, no doubt, very important features when the bar-headed geese need to fly at higher altitudes. Scott and Milsom (2006) suggested that adaptations to decrease the value for hemoglobin P_{50} and to increase tissue diffusion capacity might be the most beneficial for high-flying birds.

It has been suggested that the more effective lungs of birds, compared with those of mammals, may contribute significantly to the ability of birds better to tolerate high altitude (Scheid, 1985). However, Shams and Scheid (1989) concluded that, although the parabronchial lung of birds does confer an advantage at a PO_2 equivalent to that at the top of Mt. Everest, at the highest simulated altitude (11.58 km) tolerated by their ducks, there was no advantage at all. The authors concluded that the major difference between birds and mammals at these extreme altitudes is the ability of the former to tolerate lower P_aCO_2, thus enabling the respiratory system to maintain P_aO_2 at as high a level as possible. In a subsequent paper, the same authors (Shams and Scheid, 1993) reported that when hypoxia is accompanied by the appropriate hypobaria (reduced atmospheric pressure), V_I and P_aO_2 are slightly, but significantly greater. Although the authors have no explanation for these differences, they would allow a bird at the elevation of Mt. Everest to gain another 700 m in height before the "increase" in P_aO_2 was eliminated. Gas exchange across the *rete mirabile ophthalmicum* is thought to enhancing the supply of oxygen to the brain (Bernstein et al., 1984), and improvement in gas transport as a result of acclimation to high altitude (Weinstein et al., 1985) may be another way in which birds are adapted for survival and activity at high altitudes.

Despite all of these apparent adaptations to life at high altitude, as mentioned at the beginning of this section, there is very little accurate information with regard to the heights that these birds typically fly at during their migrations. Hawkes et al. (2011) tracked bar-headed geese flying North out of India and climbing up and through the Himalayan mountains, on their way to breed in China and Mongolia (Figure 39.36). The birds typically climbed between 4 and 6 km in 7–8 h in a single day, with regular sustained climb rates of between 0.2 and 0.6 m/s. Over the whole migratory period, the data suggested that bar-headed geese normally minimize flight altitude by traveling through the mountain valleys and only flying as high as the terrain dictates (Hawkes et al., 2012). Maximum altitudes reached were 7290 and 6540 m, for southbound and northbound geese, respectively, but with 95% of locations recording <5489 m.

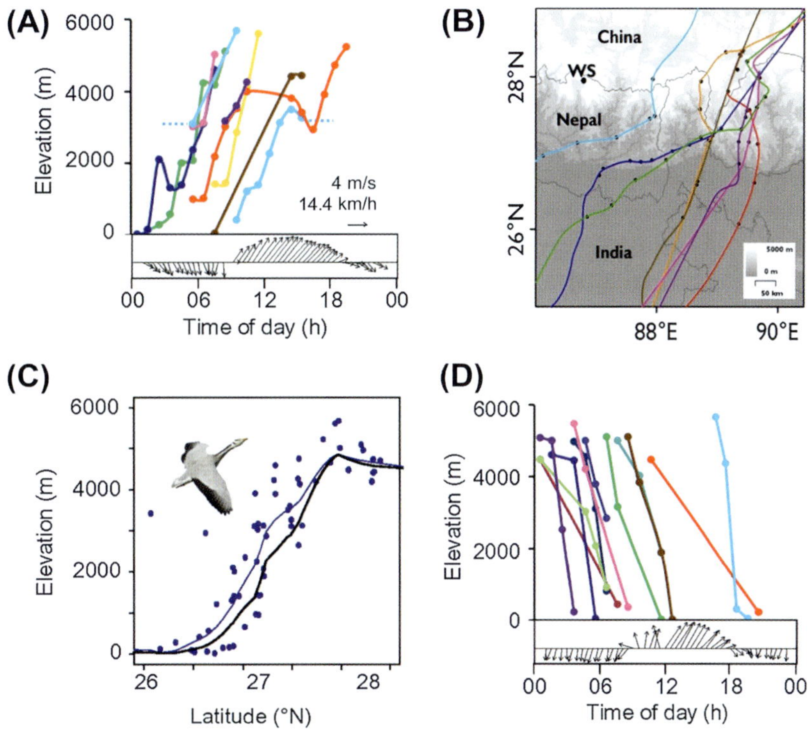

FIGURE 39.36 Timing of migrations with 30-min average wind speed and direction from the Nepal Climate Observatory at Pyramid station for bar-headed geese, *Anser indicus*, migrating (A) northward ($n=8$) and (D) southward ($n=12$) over the Himalayas. Arrows show cardinal direction (north pointing up to 0°) in which the wind was blowing and arrow length (indicated in (A)) is proportional to wind speed in (A and D). (B) Map showing the northward migration routes; weather station (WS) location is indicated. (C) Elevation of the mean northward track across the Himalayas (for all crossing locations from all eight geese); blue circles show individual data points and blue line shows Lowess smoother for mean ground elevation under the track (black line). *From Hawkes et al. (2011).*

Blum (1980) reports seeing bar-headed geese flying over the summit ridge of Annapurna I at >7622 m and took photos of them. Clearly, it is possible that occasional rare flights to very high altitudes could occur, but it is not clear if these flights are only possible if assisted by favorable wind conditions in the mountains or whether they are sustained entirely by the various physiological adaptations for flight at high altitude.

Experiments on bar-headed geese running on a treadmill under hypoxic conditions, similar to those at the top of Mt. Everest, indicated possible severe limitations to oxygen uptake (Fedde et al., 1989). Under normoxic conditions, running at 0.6 m/s and at a 2-degree incline caused a doubling of $\dot{V}o_2$ and of cardiac output. Under the severe hypoxic conditions of only 7% atmospheric O_2, there was a decline in resting $\dot{V}o_2$ and, surprisingly, $\dot{V}o_2$ did not increase significantly during the subsequent 6 min of exercise. Cardiac output also apparently remained unchanged during exercise, although there was a significant reduction in V_S. The authors suggest that there may be a hypoxic depression of cardiac contractility, but clearly this cannot occur when the birds are flying at high altitude. It may be that the birds used in this study were stressed by the surgical techniques associated with the cannulations and so on (Woakes and Butler, 1986). Certainly, their resting heart rate was almost three times higher than that of resting barnacle geese (Butler and Woakes, 1980; Ward et al., 2002). A similar study of bar-headed geese (Hawkes et al. submitted) obtained similar blood gas and metabolite values to those of Fedde et al. (1989) but significantly lower resting heart rates during normoxia and hypoxia. They also used accelerometry to indicate that the cost of locomotion and found that the requirement for oxygen was not different between running in hypoxia and in normoxia. Finally, it was noted that, while lactate levels had risen significantly in the geese running in hypoxia (Fedde et al., 1989; Hawkes et al., submitted), it could only account for a maximum of 10% of the locomotory costs after 6 min of running, which was incompatible with the 50% reduction in $\dot{V}o_2$ measured by Fedde et al. (1989) in severely hypoxic geese. Put together, these results suggest that there was a problem with the experimental determination of $\dot{V}o_2$ by Fedde et al. (1989) under hypoxic conditions during running, and that the maximum running speeds of bar-headed geese are not significantly affected by severe hypoxia (Hawkes et al., submitted).

These recent studies of bar-headed geese in the wild (Takekawa et al., 2009; Hawkes et al., 2011, 2012) are beginning to reveal the true nature of their migration across

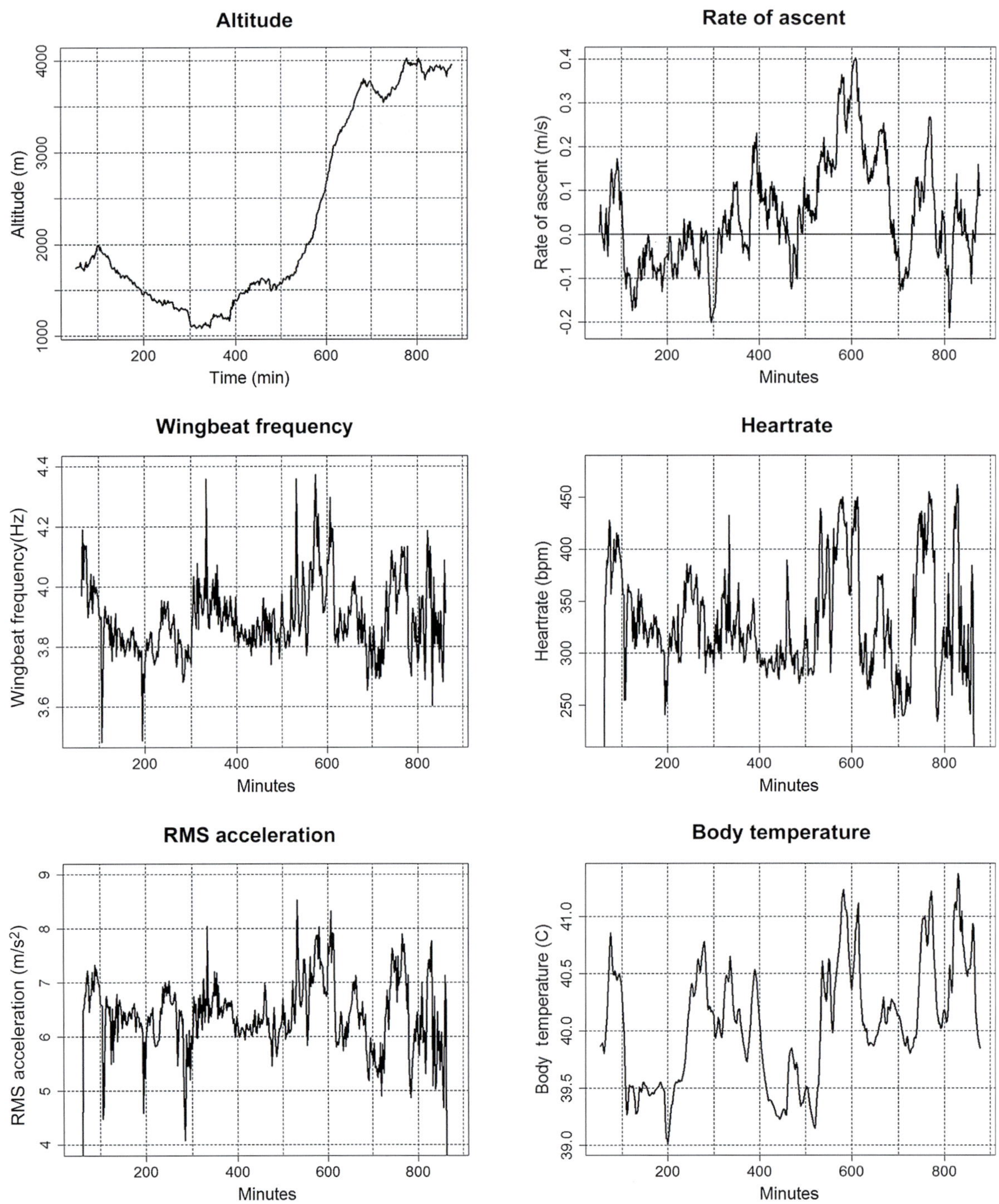

FIGURE 39.37 Simultaneous recording of heart rate, body temperature, wingbeat frequency, and root mean squared ventral-dorsal acceleration, from a wild bar-headed goose, *Anser indicus*, on its autumnal, southward migration from Mongolia up onto the edge of the Tibetan-Qinghai Plateau. All four variables are very well correlated with each other and, in particular, body temperatures fluctuate between 1 and 2 °C in direct association with changes in flight costs, as the geese alter their rate of ascent and descent during flight. *From Bishop, C.M., Spivey, R.J., Hawkes, L.A., Butler P.J., Unpublished data.*

the Himalayan Mountains and Tibetan-Qinghai plateau. Technological approaches are allowing the simultaneous recording of altitude, heart rate, and accelerometry during their migratory flights (Figure 39.37; Bishop, Hawkes, Butler et al. unpublished data). These data indicate that even a comparatively gentle climb rate of around 0.1 m/s, between an altitude of 3600 to 4000 m, required a heart rate of well over 400 beats/min, compared to a maximum heart rate of around 500 beats/min (Ward et al., 2002). However, overall the mean heart rates are remarkably low and indicate that there is still much to discover with regard to the maximum flight performance of these high-flying birds.

ACKNOWLEDGMENTS

The authors wish to thank BBSRC, NERC, The Royal Society, and the Leverhulme Trust for financial support for their recent and past work on this topic.

REFERENCES

Adams, N.J., Brown, C.R., Nagy, K.A., 1986. Energy expenditure of free-ranging wandering albatrosses *Diomedea exulans*. Physiol. Zool. 59, 583–591.

Alerstam, T., Lindström, Å., 1990. Optimal bird migration: the relative importance of time, energy, and safety. In: Gwinner, E. (Ed.), Bird Migration. Springer-Verlag, Berlin/Heidelberg/New York, pp. 331–351.

Alerstam, T., Gudmundsson, G.A., Larsson, B., 1993. Flight tracks and speeds of Antarctic and Atlantic seabirds: radar and optical measurements. Philos. Trans. R. Soc. Lond., B 340, 55–67.

Alerstam, T., Rosén, M., Bäckman, J., Ericson, P.G.P., Hellgren, O., 2007. Flight speeds among bird species: allometric and phylogenetic effects. PLoS Biol. 5, e197.

Alexander, R.M., 2005. Models and the scaling of energy costs for locomotion. J. Exp. Biol. 208, 1645–1652.

Alexander, R. McN., Bennett-Clark, H.C., 1977. Storage of elastic strain energy in muscle and other tissues. Nature 265, 114–117.

Askew, G.N., Ellerby, D.J., 2007. The mechanical power requirements of avian flight. Biol. Lett. 3, 445–448.

Askew, G.N., Marsh, R.L., 2001. The mechanical power output of the pectoralis muscle of blue-breasted quail (*Coturnix chinensis*): the in vivo length cycle and its implications for muscle performance. J. Exp. Biol. 204, 3587–3600.

Askew, G.N., Tregear, R.T., Ellington, C.P., 2010. The scaling of myofibrillar actomyosin ATPase activity in apid bee flight muscle in relation to hovering flight energetics. J. Exp. Biol. 213, 1195–1206.

Balasch, J., Palomeque, J., Palacios, L., Musquera, S., Jimenez, M., 1973. Hematological values of some great flying and aquatic-diving birds. Comp. Biochem. Physiol. A 49, 137–145.

Ballance, L.T., 1995. Flight energetics of free-ranging red-footed boobies (*Sula sula*). Physiol. Zool. 68, 887–914.

Banzett, R.B., Nations, C.S., Wang, N., Butler, J.P., Lehr, J.L., 1992. Mechanical independence of wingbeat and breathing in starlings. Respir. Physiol. 89, 27–36.

Barron, D.G., Brawn, J.D., Weatherhead, P.J., 2010. Meta-analysis of transmitter effects on avian behavior and ecology. Methods Ecol. Evol. 1, 180–187.

Battley, P.F., Piersma, T., Dietz, M.W., Tang, S., Dekinga, A., Hulsman, K., 2000. Empirical evidence for differential organ reductions during trans-oceanic bird flight. Proc. R. Soc. B 267, 191–195.

Baudinette, R.V., Schmidt-Nielsen, K., 1974. Energy cost of gliding flight in herring gulls. Nature 248, 83–84.

Baudinette, R.V., Loveridge, J.P., Wilson, K.J., Mills, C.D., Schmidt-Nielsen, K., 1976. Heat loss from feet of herring gulls at rest and during flight. Am. J. Physiol. 230, 920–924.

Berger, M., 1978. Ventilation in the humming birds *Colibri coruscans* during altitude hovering. In: Piiper, J. (Ed.), Respiratory Function in Birds, Adult and Embryonic. Springer-Verlag, Berlin/Heidelberg/New York, pp. 85–88.

Berger, M., 1985. Sauerstoffverbrauch von Kolibris (*Colibri corus-* und *C. thalassinus*) beim Horizontalflug. In: Nachtigall, W. (Ed.), BIONA Report 3. Gustav Fischer, Stuttgart, pp. 307–314.

Berger, M., Hart, J.S., 1972. Die Atmung beim Kolibri *Amazilia fimbriata* wa¨hrend des Schwirrfluges bei verschiedenen Umgebungstemperaturen. J. Comp. Physiol. 81, 363–380.

Berger, M., Hart, J.S., 1974. Physiology and energetics of flight. In: Farner, D.S., King, J.R. (Eds.), Avian Biology, vol. IV. Academic Press, New York and London, pp. 415–477.

Berger, M., Hart, J.S., Roy, O.Z., 1970a. Respiration, oxygen consumption and heart rate in some birds during rest and flight. Z. Vgl. Physiol. 66, 201–214.

Berger, M., Roy, O.Z., Hart, J.S., 1970b. The co-ordination between respiration and wing beats. Z. Vgl. Physiol. 66, 190–200.

Bernstein, M.H., 1976. Ventilation and respiratory evaporation in the flying crow, *Corvus ossifragus*. Respir. Physiol. 26, 371–382.

Bernstein, M.H., Curtis, M.B., Hudson, D.M., 1979a. Independence of brain and body temperatures in flying American kestrels, *Falco sparverius*. Am. J. Physiol. 237, 58–62.

Bernstein, M.H., Duran, H.L., Pinshow, B., 1984. Extrapulmonary gas exchange enhances brain oxygen in pigeons. Science 226, 564–566.

Bernstein, M.H., Sandoval, I., Curtis, M.B., Hudson, D.M., 1979b. Brain temperature in pigeons: effects of anterior respiratory bypass. J. Comp. Physiol. 129, 115–118.

Bernstein, M.H., Thomas, S.P., Schmidt-Nielsen, K., 1973. Power input during flight of the fish crow, *Corvus ossifragus*. J. Exp. Biol. 58, 401–410.

Berthold, P., 1993. Bird Migration: A General Survey. Oxford University Press, Oxford. pp. 239.

Bevan, R.M., Butler, P.J., Woakes, A.J., Prince, P.A., 1995b. The energy expenditure of free-ranging back-browed albatrosses. Philos. Trans. R. Soc. Lond., B 350, 119–131.

Bevan, R.M., Woakes, A.J., Butler, P.J., 1994. The use of heart rate to estimate oxygen consumption of free-ranging black-browed albatrosses *Diomedea melanophrys*. J. Exp. Biol. 193, 119–137.

Bevan, R.M., Woakes, A.J., Butler, P.J., Croxall, J.P., 1995a. Heart rate and oxygen consumption of exercising gentoo penguins. Physiol. Zool. 68, 855–877.

Biebach, H. Phenotypic plasticity in the digestive system and the breast muscle in trans-Sahara migrating passerines. J. Avian Biol., in press.

Biesel, W., Nachtigall, W., 1987. Pigeon flight in a wind tunnel. IV. Thermoregulation and water homeostasis. J. Comp. Physiol. B 157, 117–128.

Biewener, A.A., Dial, K.P., Goslow Jr., G.E., 1992. Pectoralis muscle force and power output during flight in the starling. J. Exp. Biol. 164, 1–18.

Birt-Friesen, V.L., Montevecchi, W.A., Cairns, D.K., Macko, S.A., 1989. Activity-specific metabolic rates of free-living northern gannets and other seabirds. Ecology 70, 357–367.

Bishop, C.M., 1997. Heart mass and the maximum cardiac output of birds and mammals: implications for estimating the maximum aerobic power input of flying animals. Philos. Trans. R. Soc. Lond., B 352, 447–456.

Bishop, C.M., 1999. Maximum oxygen consumption and aerobic scope of birds and mammals: getting to the heart of the matter. Proc. R. Soc. 266, 2275–2281.

Bishop, C.M., 2005. Circulatory variables and the flight performance of birds. J. Exp. Biol. 208, 1695–1708.

Bishop, C.M., Butler, P.J., 1995. Physiological modelling of oxygen consumption in birds during flight. J. Exp. Biol. 198, 2153–2163.

Bishop, C.M., Spivey, R.J., 2013. Integration of exercise response and allometric scaling in endotherms. J. Theor. Biol. 323, 11–19.

Bishop, C.M., Butler, P.J., Egginton, S., El Haj, A.J., Gabrielsen, G.W., 1995. Development of metabolic enzyme activity in locomotor and cardiac muscles of the migratory barnacle goose. Am.J. Physiol. 269, R64–R72.

Bishop, C.M., Butler, P.J., El Haj, A.J., Egginton, S., Loonen, M.J.J.E., 1996. The morphological development of the locomotor and cardiac muscles of the migratory barnacle goose (*Branta leucopsis*). J. Zool. London 239, 1–15.

Bishop, C.M., Butler, P.J., El Haj, A.J., Egginton, S., 1998. Comparative development of captive and migratory populations. Physiol. Zool. 71, 198–207.

Black, C.P., Tenney, S.M., 1980. Oxygen transport during progressive hypoxia in high-altitude and sea-level waterfowl. Respir. Physiol. 39, 217–239.

Blem, C.R., 1976. Patterns of lipid storage and utilization in birds. Am. Zool. 16, 671–684.

Blum, A., 1980. Annapurna: A Woman's Place. Sierra Club Books, USA. pp. 247.

Boggs, D.F., 1997. Coordinated control of respiratory pattern during locomotion in birds. Am. Zool. 37, 41–53.

Boggs, D.F., Dial, K.P., 1993. Neuromuscular organization and regional EMG activity of the pectoralis in the pigeon. J. Morphol. 218, 43–57.

Boggs, D.F., Jenkins Jr., F.A., Dial, K.P., 1997a. The effects of wingbeat cycle on respiration in black-billed magpies *Pica pica*. J. Exp. Biol. 200, 1403–1412.

Boggs, D.F., Seveyka, J.J., Kilgore Jr., D.L., Dial, K.P., 1997b. Coordination of respiratory cycles with wingbeat cycles in black-billed magpie (*Pica pica*). J. Exp. Biol. 200, 1413–1420.

Bonen, A., Chabowski, A., Luiken, J.J.F.P., Glatz, J.F.C., 2007. Mechanisms and regulation of protein-mediated cellular fatty acid uptake: molecular, biochemical and physiological evidence. Physiol 22, 15–28.

Bomphrey, R.J., 2012. Advances in animal flight aerodynamics through flow measurement. Evol. Biol. 39, 1–11.

Bowlin, M.S., Henningsson, P., Muigres, F.T., Vleugels, R.H.E., Liechti, F., Hedenstrom, A., 2010. The effects of geolocator drag and weight on the flight ranges of small migrants. Methods Ecol. Evol. 1, 398–402.

Brackenbury, J.H., Gleeson, M., 1983. Effects of PCO_2 on respiratory pattern during thermal and exercise hyperventilation in domestic fowl. Respir. Physiol. 54, 109–119.

Brobeck, J.R., DuBois, A.B., 1980. Energy exchange. In: Mountcastle, V.B. (Ed.), Medical Physiology, vol. 2. pp. 1351–1365. St Louis.

Bromley, R.G., Jarvis, R.L., 1993. The energetics of migration and reproduction of dusky Canada geese. Condor 95, 193–210.

Bruderer, B., Liechti, F., 1995. Variation in density and height distribution of nocturnal migration in the south of Israel. Isr. J. Zool. 41, 477–487.

Bundle, M.W., Hansen, K.S., Dial, K.P., 2007. Does the metabolic rate-flight speed relationship vary among geometrically similar birds of different mass? J. Exp. Biol. 210, 1075–1083.

Butler, P.J., 1991. Exercise in birds. J. Exp. Biol. 160, 233–262.

Butler, P.J., 2010. High fliers: the physiology of bar-headed geese. Comp. Biochem. Physiol. A 156, 325–329.

Butler, P.J., Woakes, A.J., 1980. Heart rate, respiratory frequency and wing beat frequency of free flying barnacle geese *Branta leucopsis*. J. Exp. Biol. 85, 213–226.

Butler, P.J., Woakes, A.J., 1985. Exercise in normally ventilating and apnoeic birds. In: Gilles, R. (Ed.), Circulation, Respiration and Metabolism: Current Comparative Approaches. Springer-Verlag, Berlin, pp. 39–55.

Butler, P.J., Woakes, A.J., 1990. The physiology of bird flight. In: Gwinner, E. (Ed.), Bird Migration: Physiology and Ecophysiology. Springer-Verlag, Berlin/Heidelberg, pp. 300–318.

Butler, P.J., West, N.H., Jones, D.R., 1977. Respiratory and cardiovascular responses of the pigeon to sustained, level flight a wind-tunnel. J. Exp. Biol. 71, 7–26.

Butler, P.J., Woakes, A.J., Bishop, C.M. Behaviour and physiology of Svalbard barnacle geese, *Branta leucopsis*, during their autumn migration. J. Avian Biol., submitted-a.

Butler, P.J., Green, J.A., Boyd, I.L., Speakman, J.R., 2004. Measuring metabolic rate in the field: the pros and cons of the doubly labelled water and heart rate methods. Funct. Ecol. 18, 168–183.

Caldow, R.W.G., Furness, R.W., 1993. A histochemical comparison of fibre types in the M. pectoralis and M. supracoracoideus of the great skua *Catharacta skua* and the herring gull *Larus argentatus* with reference to kleptoparasitic capabilities. J. Zool. London 229, 91–103.

Carmi, N., Pinshow, B., Horowitz, M., Bernstein, M.H., 1993. Birds conserve plasma volume during thermal and flight-incurred dehydration. Physiol. Zool. 66, 829–846.

Carmi, N., Pinshow, B., Porter, W.P., Jaeger, J., 1992. Water and energy limitations on flight duration in small migrating birds. Auk 109, 268–276.

Carpenter, F.L., 1975. Bird hematocrits: effects of high altitude and strength of flight. Comp. Biochem. Physiol. A 50, 415–417.

Chai, P., Dudley, R., 1995. Limits to vertebrate locomotor energetics suggested by hummingbirds hovering in heliox. Nature 377, 722–725.

Chai, P., Chen, J.S.C., Dudley, R., 1997. Transient hovering performance of hummingbirds under conditions of maximal loading. J. Exp. Biol. 200, 921–929.

Choi, I., Ricklefs, R.E., Shea, R.E., 1993. Skeletal muscle growth, enzyme activities and the development of thermogenesis: a comparison between altricial and precocial birds. Physiol. Zool. 66, 455–473.

Clark, C.J., Dudley, R., 2010. Hovering and forward flight energetics in Anna's and Allen's hummingbirds. Physiol. Biochem. Zool. 83, 654–662.

Costa, D.P., Prince, P.A., 1987. Foraging energetics of grey-headed albatrosses *Diomedea chrysostoma* at Bird Island, South Georgia. Ibis 129, 149–158.

Crabtree, B., Newsholme, E.A., 1972. The activities of phosphorylase, hexokinase, phosphofructokinase, lactate dehydrogenase and glycerol 3-phosphate dehydrogenase in muscles from vertebrate and invertebrates. Biochem. J. 126, 49–58.

Crile, G., Quiring, D.P., 1940. A record of the body weight and certain organ and gland-weights of 3690 animals. Ohio J. Sci. XL, 219–259.

Cutts, C.J., Speakman, J.R., 1994. Energy savings in formation flight of pink-footed geese. J. Exp. Biol. 189, 251–261.

Deaton, K.E., Bishop, C.M., Butler, P.J., 1997. The effect of thyroid hormones on the aerobic development of locomotor and cardiac muscles in the barnacle goose. J. Comp. Physiol. 167, 319–327.

Deaton, K.E., Bishop, C.M., Butler, P.J., 1998. Tissue-specific effects of hypothyroidism on postnatal muscle development in the goose. J. Exp. Biol. 201, 827–836.

Dial, K.P., 1992a. Activity patterns of the wing muscles of the pigeon (*Columba livia*) during different modes of flight. J. Exp. Zool. 262, 357–373.

Dial, K.P., 1992b. Avian forelimb muscles and nonsteady flight: can birds fly without using the muscles in their wings? Auk 109, 874–885.

Dial, K.P., 2003. Wing-assisted incline running and the evolution of flight. Science 299, 402–404.

Dial, K.P., Biewener, A.A., 1993. Pectoralis muscle force and power output during different modes of flight in pigeons (*Columba livia*). J. Exp. Biol. 176, 31–54.

Dial, K.P., Biewener, A.A., Tobalske, B.W., Warrick, D.R., 1997. Mechanical power output of bird flight. Nature 390, 67–70.

Dial, K.P., Kaplan, S.R., Goslow, G.E.J., Jenkins, F.A.J., 1987. Structure and neural control of the pectoralis in pigeons: implications for flight mechanics. Anat. Rec. 218, 284–287.

Dial, K.P., Kaplan, S.R., Goslow, G.E.J., 1988. A functional analysis of the primary upstroke and downstroke muscles in the domestic pigeon (*Columba livia*) during flight. J. Exp. Biol. 134, 1–16.

Dial, K.P., Goslow, G.E.J., Jenkins, F.A.J., 1991. The functional anatomy of the shoulder in the european starling (*Sturnus vulgaris*). J. Morphol. 207, 327–344.

Dietz, M.W., Piersma, T., Dekinga, A., 1999. Body-building without power training: endogenously regulated pectoral muscle hypertrophy in confined shorebirds. J. Exp. Biol. 202, 2831–2837.

Driedzic, W.R., Crowe, H.L., Hicklin, P.W., Sephton, D.H., 1993. Adaptations in pectoralis muscle, heart mass, and energy metabolism during premigratory fattening in semipalmated sandpipers (*Calidris pusilla*). Can. J. Zool. 71, 1602–1608.

Egevang, C., Stenhouse, I.J., Phillips, R.A., Petersen, A., Fox, J.W., Silk, J.R.D., 2010. Tracking of Arctic terns *Sterna paradisaea* reveals longest animal migration. Proc. Natl. Acad. Sci. 107, 2078–2081.

Elkins, N., 1979. High altitude flight by swans. Br. Birds 72, 238–239.

Ellington, C.P., 1984. The aerodynamics of hovering insect flight. VI. Lift and power requirements. Philos. Trans. Soc. Lond., B 305, 145–181.

Ellington, C.P., 1991. Limitations on animal flight performance. J. Exp. Biol. 160, 71–91.

Ellis, H.I., 1984. Energetics of free-ranging seabirds. In: Whittow, C.G., Rahn, H. (Eds.), Seabird Energetics. Plenum, New York, pp. 203–234. 261, pp. 73–79.

Engel, S., Biebach, H., Visser, G.H., 2006. Metabolic costs of avian flight in relation to flight velocity: a study in rose coloured starlings (*Sturnus roseus*). J. Comp. Physiol. B 176, 415–427.

Faraci, F.M., Fedde, M.R., 1986. Regional circulatory responses to hypocapnia and hypercapnia in bar-headed geese. Am. J. Physiol. 250, R499–R504.

Faraci, F.M., Kilgore Jr., D.L., Fedde, M.R., 1984a. Oxygen delivery to the heart and brain during hypoxia: Pekin duck vs. bar-headed goose. Am. J. Physiol. 247, R69–R75.

Faraci, F.M., Kilgore Jr., D.L., Fedde, M.R., 1984b. Attenuated pulmonary pressor response to hypoxia in bar-headed geese. Am. J. Physiol. 247, R402–R403.

Faraci, F.M., Kilgore Jr., D.L., Fedde, M.R., 1985. Blood flow distribution during hypocapnic hypoxia in Pekin ducks and bar-headed geese. Respir. Physiol. 61, 21–30.

Fedde, M.R., Orr, J.A., Shams, H., Scheid, P., 1989. Cardiopulmonary function in exercising bar-headed geese during normoxia and hypoxia. Respir. Physiol. 77, 239–262.

Flint, E.N., Nagy, K.A., 1984. Flight energetics of free-livingsooty terns. Auk 101, 288–294.

Fry, C.H., Ferguson-Lees, I.F., 1972. Flight muscle hypertrophy and ecophysiological variation of yellow wagtail (*Motacilla flava*) races at Lake Chad. J. Zool. London 167, 293–306.

Funk, G.D., Milsom, W.K., Steeves, J.D., 1992a. Coordination of wingbeat and respiration in the Canada goose. I. Passive wing flapping. J. Appl. Physiol. 73, 1014–1024.

Funk, G.D., Sholomenko, G.N., Valenzuela, I.J., Steeves, J.D., Milsom, W.K., 1993. Coordination of wing beat and respiration in Canada geese during free flight. J. Exp. Biol. 175, 317–323.

Funk, G.D., Steeves, J.D., Milsom, W.K., 1992b. Coordination of wingbeat and respiration in birds. II. "Fictive" flight. J. Appl. Physiol. 73, 1025–1033.

Gaunt, A.S., Hikida, R.S., Jehl Jr., J.R., Fenbert, L., 1990. Rapid atrophy and hypertrophy of an avian flight muscle. Auk 107, 649–659.

George, J.C., Berger, A.J., 1966. Avian Myology. Academic.Press, London/New York.

Gerson, A.R., Guglielmo, C.G., 2011a. Flight at low ambient humidity increases protein catabolism in migratory birds. Science 333, 1434–1436.

Gerson, A.R., Guglielmo, C.G., 2011b. House sparrows (*Passer domesticus*) increase protein catabolism in response to water restriction. Am. J. Physiol. 300, R925–R930.

Gessaman, J.A., 1980. An evaluation of heart rate as an indirect measure of daily energy metabolism of the American kestrel. Comp. Biochem. Physiol. A 65, 273–289.

Gessaman, J.A., Nagy, K.A., 1988. Transmitter loads affect the flight speed and metabolism of homing pigeons. Condor 90, 662–668.

Gill, R.E., Tibbitts, T.L., Douglas, D.C., Handel, C.M., Mulcahy, D.M., Gottschalck, J.C., Warnock, N., McCaffery, B.J., Battley, P.F., Piersma, T., 2009. Extreme endurance flights by landbirds crossing the Pacific Ocean: ecological corridor rather than barrier? Proc. R. Soc. B 276, 447–457.

Glatz, J.F.C., Schaap, F.G., Binas, B., Bonen, A., van der Vusse, G.J., Luiken, J.J.F.P., 2003. Cytoplasmic fatty acid-binding protein facilitates fatty acid utilization by skeletal muscle. Acta Physiol. Scand. 178, 367–371.

Goslow, G.E.J., Dial, K.P., 1990. Active stretch-shorten contractions of the M. Pectoralis in the European starling (*Sturnus vulgaris*): evidence from electromyography and contractile properties. Neth. J. Zool. 40, 106–114.

Gould, L.L., Heppner, F., 1974. The vee formation of Canada geese. Auk 91, 494–506.

Greenewalt, C.H., 1962. Dimensional relationships for flying animals. Smithson. Misc. Collect. 144, 1–46.

Greenewalt, C.H., 1975. The flight of birds. Trans. Am. Philos. Soc. 65, 1–67.

Grubb, B.R., 1983. Allometric relations of cardiovascular function in birds. Am. J. Physiol. 245, H567–H572.

Grubb, B., Mills, C.B., Colacino, J.M., Schmidt-Nielsen, K., 1977. Effect of arterial carbon dioxide on cerebral blood flow in ducks. Am. J. Physiol. 232, H596–H601.

Grubb, B., Colacino, J.M., Schmidt-Nielsen, K., 1978. Cerebral blood flow in birds: effect of hypoxia. Am. J. Physiol. 234, H230–H234.

Grubb, B., Jones, J.H., Schmidt-Nielsen, K., 1979. Avian cerebral blood flow: influence of the Bohr effect on oxygen supply. Am. J. Physiol. 236, H744–H749.

Gudmundsson, G.A., Benvenuti, S., Alterstam, T., Papi, F., Lilliendahl, K., Åkesson, S., 1995. Examining the limits of flight and orientation performance: satellite tracking of brent geese migrating across the Greenland ice-cap. Proc. R. Soc. Lond., B 261, 73–79.

Gudmundsson, G.A., Lindström, Å., Alerstam, T., 1991. Optimal fat loads and long-distance flights by migrating knots *Calidris canutus*, sanderlings *C. alba* and turnstones *Arenaria interpres*. Ibis 133, 140–152.

Guglielmo, C.G., 2010. Move that fatty acid: fuel selection and transport in migratory birds and bats. Integr. Comp. Biol. 50, 336–345.

Guglielmo, C.G., Haunerland, N.H., Williams, T.D., 1998. Fatty acid binding protein, a major protein in the flight muscle of migrating western sandpipers. Comp. Biochem. Physiol. B 119, 549–555.

Guglielmo, C.G., Haunerland, N.H., Hochachka, P.W., Williams, T.D., 2002. Seasonal dynamics of flight muscle fatty acid binding protein and catabolic enzymes in a migratory bird. Am. J. Physiol. 282, R1405–R1413.

Hails, C.J., 1979. A comparison of flight energetics in hirundines and other birds. Comp. Biochem. Physiol. A 63, 581–585.

Hainsworth, F.R., 1987. Precision and dynamics of positioning by Canada geese flying on formation. J. Exp. Biol. 128, 445–462.

Hainsworth, F.R., 1988. Wing movements and positioning for aerodynamic benefit by Canada geese flying in formation. Can. J. Zool. 67, 585–589.

Hart, J.S., Roy, O.Z., 1966. Respiratory and cardiac responses to flight in pigeons. Physiol. Zool. 39, 291–306.

Hartman, F.A., 1961. Locomotor mechanisms of birds. Smithson. Misc. Collect. 143, 1–91.

Hawkes, L.A., Balachandran, S., Batbayar, N., Butler, P.J., Chua, B., Douglas, D.C., Frappell, P.B., Hou, Y., Milsom, W.K., Newman, S.H., Prosser, D.J., Sathiyaselvam, P., Scott, G.R., Takekawa, J.K., Natsagdorj, T., Wikelski, M., Witt, M.J., Yan, B., Bishop, C.M., 2013. The paradox of extreme high-altitude migration in bar-headed geese *Anser indicus*. Proc. R. Soc. B. http://dx.doi.org/10.1098/rspb.2012.2114.

Hawkes, L.A., Balachandran, S., Batbayar, N., Butler, P.J., Frappell, P.B., Milsom, W.K., Natsagdorj, T., Newman, S.H., Scott, G.R., Sathiyaselvam, P., Takekawa, J.K., Wikelski, M., Bishop, C.M., 2011. The trans-Himalayan flights of bar-headed geese (*Anser indicus*). Proc. Natl. Acad. Sci. 108, 9516–9519.

Hedenström, A., 2008. Adaptations to migration in birds: behavioural strategies, morphology and scaling effects. Philos. Trans. R. Soc. B 363, 287–299.

Hedenström, A., Alerstam, T., 1992. Climbing performance of migrating birds as a basis for estimating limits for fuel-carrying capacity and muscle work. J. Exp. Biol. 164, 19–38.

Hedenström, A., Alerstam, T., 1995. Optimal flight speed of birds. Philos. Trans. R. Soc. Lond., B 348, 471–487.

Hedenström, A., Rosen, M., 2003. Body frontal area in passerine birds. J. Avian Biol. 34, 159–162.

Henningsson, P., Muijres, F.T., Hedenström, A., 2011. Time-resolved vortex wake of a common swift flying over a range of flight speeds. J. R. Soc., Interface 8, 807–816.

Hickson, R.C., Galassi, T.M., Dougherty, K.A., 1983. Repeated development and regression of exercise-induced cardiac hypertrophy in rats. J. Appl. Physiol. 54, 794–797.

Hirth, K.D., Biesel, W., Nachtigall, W., 1987. Pigeon flight in a wind tunnel. III. Regulation of body temperature. J. Comp. Physiol. B 157, 111–116.

Hudson, D.M., Bernstein, M.H., 1978. Respiratory ventilation during steady-state flight in the white-necked raven, *Corvus cryptoleucus*. Fed. Proc. 37, 472.

Hudson, D.M., Bernstein, M.H., 1981. Temperature regulation and heat balance in flying white-necked ravens, *Corvus cryptoleucus*. J. Exp. Biol. 90, 267–281.

Hudson, D.M., Bernstein, M.H., 1983. Gas exchange and energy cost of flight in the white-necked raven, *Corvus cryptoleucus*. J. Exp. Biol. 103, 121–130.

Hummel, D., 1995. Formation flight as an energy-saving mechanism. Isr. J. Zool. 41, 261–278.

Hutchinson, J.R., Allen, V., 2009. The evolutionary continuum of limb function from early theropods to birds. Naturwissenschaften 96, 423–448.

Jackson, B.E., Dial, K.P., 2011. Scaling of mechanical power output during burst escape flight in the Corvidae. J. Exp. Biol. 214, 452–461.

Jehl Jr., J.R., 1994. Field estimates of energetics in migrating and downed black-necked grebes. J. Avian Biol. 25, 63–68.

Jehl Jr., J.R., 1997. Cyclical changes in body composition in the annual cycle and migration of the Eared Grebe, *Podiceps nigricollis*. J. Avian Biol. 28, 132–142.

Jenkins Jr., F.A., Dial, K.P., Goslow Jr., G.E., 1988. A cineradiographic analysis of bird flight: the wishbone in starlings is a spring. Science 241, 1495–1498.

Jenni-Eiermann, S., Jenni, L., 1991. Metabolic responses to flight and fasting in night-migrating passerines. J. Comp. Physiol. B 161, 465–474.

Jenni-Eiermann, S., Jenni, L., 1992. High plasma triglyceride levels in small birds during migratory flight: a new pathway for fuel supply during endurance locomotion at very high mass-specific metabolic rates? Physiol. Zool. 65, 112–123.

Jenni-Eiermann, S., Jenni, L., 1996. Metabolic differences between the postbreeding, moulting and migratory periods in feeding and fasting passerine birds. Funct. Ecol. 10, 62–72.

Johnson, O.W., Fielding, L., Fisher, J.P., Gold, R.S., Goodwill, R.H., Bruner, A.E., Furey, J.F., Brusseau, P.A., Brusseau, N.H., Johnson, P.M., Jukema, J., Prince, L.L., Tenney, M.J., Fox, J.W., 2012. New insight concerning transoceanic migratory pathways of Pacific golden-plovers (*Pluvialis fulva*): the Japan stopover and other linkages as revealed by geolocators. Wader Study Group Bull. 119, 1–8.

Jouventin, P., Weimerskirch, H., 1990. Satellite tracking of Wandering albatrosses. Nature 343, 746–748.

Jurgens, K.D., Bartels, H., Bartels, R., 1981. Blood oxygen transport and organ weights of small bats and small nonflying mammals. Respir. Physiol. 45, 243–260.

Kampe, G., Crawford Jr., E.C., 1993. Oscillatory mechanics of the respiratory system of pigeons. Respir. Physiol. 18, 188–193.

Kaplan, S.R., Goslow, G.E.J., 1989. Neuromuscular organization of the pectoralis of the pigeon (*Columbia livia*): implications for motor control. Anat. Rec. 224, 426–430.

Kiens, B., 2006. Skeletal muscle lipid metabolism in exercise and insulin resistance. Physiol. Rev. 86, 205–243.

Kiley, J.P., Kuhlmann, W.D., Fedde, M.R., 1982. Ventilatory and blood gas adjustments in exercising isothermic ducks. J. Comp. Physiol. 147, 107–112.

Kilgore Jr., D.L., Boggs, D.F., Birchard, G.F., 1979. Role of the rete mirable ophthalmicum in maintaining the body-to-brain temperature difference in pigeons. J. Comp. Physiol. 129, 119–122.

Klaassen, R.H.G., Alerstam, T., Carlsson, P., Fox, J.W., Lindström, Å., 2011. Great flights by great snipes: long and fast non-stop migration over benign habitats. Biol. Lett. 7, 833–835.

Klaiman, J.M., Price, E.R., Guglielmo, C.G., 2009. Fatty acid composition of pectoralis muscle membrane, intramuscular fat stores and adipose tissue of migrant and wintering white-throated sparrows (*Zonotricia albicollis*). J. Exp. Biol. 212, 3865–3872.

Krüger, K., Prinzinger, R., Schuchmann, K.-L., 1982. Torpor and metabolism in hummingbirds. Comp. Biochem. Physiol. A 73, 679–689.

Kvist, A., Klaassen, M., Lindström, Å., 1998. Energy expenditure in relation to flight speed: what is the power of mass loss rate estimates? J. Avian Biol. 29, 485–498.

Lack, D., 1960. Migration across the north sea studied by radar. Part 2. The spring departure 1956–59. Ibis 102, 27–59.

Lasiewski, R.C., 1963. Oxygen consumption of torpid, resting, active and flying hummingbirds. Physiol. Zool. 36, 122–140.

Lasiewski, R.C., Calder Jr., W.A., 1971. A preliminary allometric analysis of respiratory variables in resting birds. Respir. Physiol. 11, 152–166.

LeFebvre, E.A., 1964. The use of D2O18 for measuring energy metabolism in *Columba livia* at rest and in flight. Auk 81, 403–416.

Liechti, F., Schaller, E., 1999. The use of low-level jets by migrating birds. Naturwissenschaften 86, 549–551.

Lifson, N., Gordon, G.B., McClintock, R., 1955. Measurement of total carbon dioxide production by means of D^2O^{18}. J. Appl. Physiol. 7, 704–710.

Lifson, N., Gordon, G.B., Visscher, M.B., Nier, A.O., 1949. The fate of utilized molecular oxygen and the source of the oxygen of respiratory carbon dioxide, studied with the aid of heavy oxygen. J. Biol. Chem. 180, 803–811.

Lissaman, P.B.S., Shollenberger, C.A., 1970. Formation flight of birds. Science 168, 1003–1005.

Lundgren, B.O., Kiessling, K.-H., 1985. Seasonal variation in catabolic enzyme activities in breast muscle of some migratory birds. Oecologia 66, 468–471.

Lundgren, B.O., Kiessling, K.-H., 1986. Catabolic enzyme activities in the pectoralis muscle of premigratory and migratory juvenile reed warblers *Acrocephalus scirpaceus*. Oecologia 68, 529–532.

Lusk, G., 1919. The Elements of the Science of Nutrition. W. B. Saunders, Philadelphia/London.

Magnan, A., 1922. Les caracteristiques des oiseaux suivant le mode de vol. Ann. Sci. Nat. 5, 125–334.

Mainguy, S.K., Thomas, V.G., 1985. Comparisons of body reserve buildup and use in several groups of Canada geese. Can. J. Zool. 63, 1765–1772.

Marden, J.H., 1987. Maximum lift production during takeoff in flying animals. J. Exp. Biol. 130, 235–258.

Marden, J.H., 1994. From damselflies to pterosaurs: how burst and sustainable flight performance scale with size. Am. J. Physiol. 266, R1077–R1084.

Marey, E.J., 1890. Le Vol des Oiseaux. Masson, Paris.

Martineau, L., Larochelle, J., 1988. The cooling power of pigeon legs. J. Exp. Biol. 136, 193–208.

Marsh, R.L., 1981. Catabolic enzyme activities in relation to premigratory fattening and muscle hypertrophy in the gray catbird *Dumetella carolinensis*. J. Comp. Physiol. 141, 417–423.

Marsh, R.L., 1984. Adaptations of the gray catbird *Dumatella carolinensis* to long-distance migration: flight muscle hypertrophy associated with elevated body mass. Physiol. Zool. 57, 105–117.

Marsh, R.L., Storer, R.W., 1981. Correlation of flight muscle size and body mass in Cooper's Hawks: a natural analogue of power training. J. Exp. Biol. 91, 363–368.

Marsh, R.L., Wickler, S.J., 1982. The role of muscle development in the transition to endothermy in nestling bank swallows *Riparia riparia*. J. Comp. Physiol. 149, 99–105.

Masman, D., Klaassen, M., 1987. Energy expenditure during free flight in trained and free-living Eurasian kestrels (*Falco tinnunculus*). Auk 104, 603–616.

Mathieu-Costello, O., 1991. Morphometric analysis of capillary geoometry in pigeon pectoralis muscle. Am. J. Anat. 191, 74–84.

Mathieu-Costello, O., Suarez, R.K., Hochachka, P.W., 1992. Capillary-to-fiber geometry and mitochondrial density in hummingbird flight muscle. Respir. Physiol. 89, 113–132.

Matsuda, R., Bandman, E., Strohman, R.C., 1983. Regional differences in the expression of myosin light-chains and tropomyosin subunits during development of chicken breast muscle. Dev. Biol. 95, 484–491.

McClelland, G.B., 2004. Fat to the fire: the regulation of lipid oxidation with exercise and environmental stress. Comp. Biochem. Physiol. B 139, 443–460.

McClelland, G., Zwingelsein, G., Taylor, C.R., Weber, J.-M., 1994. Increased capacity for circulatory fatty acid transport in a highly aerobic mammal. Am. J. Physiol. 266, R1280–R1286.

McFarlan, J.T., Bonen, A., Guglielmo, C.G., 2009. Seasonal upregulation of fatty acid transporters in flight muscles of migratory white-throated sparrows (*Zonotrichia albicollis*). J. Exp. Biol. 212, 2934–2940.

McNab, B.K., 1994. Energy conservation and the evolution of flightlessness in birds. Am. Nat. 144, 628–642.

Midtgård, U., 1983. Scaling of the brain and the eye cooling system in birds: a morphometric analysis of the *Rete ophthalmicum*. J. Exp. Zool. 225, 197–207.

Minton, C., Gosbell, K., Johns, P., Christie, M., Klaasen, M., Hassell, C., Boyle, A., Jessop, R., Fox, J., 2011. Geolocator studies on ruddy turnstones *Arenaria interpres* and greater sandplovers *Charadrius leschenaultii* in the East Asian-Australasia flyway reveal widely different migration strategies. Wader Study Group Bull. 118, 87–96.

Morris, C.R., Nelson, F.E., Askew, G.N., 2010. The metabolic power requirements of flight and estimations of flight muscle efficiency in the cockatiel (*Nymphicus hollandicus*). J. Exp. Biol. 213, 2788–2796.

Muijres, F.T., Bowlin, M.S., Johansson, L.C., Hedenstrom, A., 2012a. Vortex wake, downwash distribution, aerodynamic performance and wingbeat kinematics in slow-flying pied flycatchers. J. R. Soc., Interface 9, 292–303.

Muijres, F.T., Johansson, L.C., Bowlin, M.S., Winter, Y., Hedenstrom, A., 2012b. Comparing aerodynamic efficiency in birds and bats suggests better flight performance in birds. PLoS One 7, e37335.

Nachtigall, W., 1995. Impositions of energy balance in prolonged flight: wind tunnel measurements with "model birds.". Isr. J. Zool. 41, 279–295.

Nagy, K.A., 1980. CO_2 production in animals: analysis of potential errors in the doubly labeled water method. Am. J. Physiol. 238, R466–R473.

Nagy, K.A., 1983. The Doubly Labeled Water ($^3HH^{18}O$) Method: A Guide to its Use. University of California at Los Angeles. Publication No. 12-1417, 45 pages.

Newsholme, E.A., Leech, A.R., 1983. Biochemistry for the Medical Sciences. Wiley, Chichester/New York.

Niles, L.J., Burger, J., Porter, R.R., Dey, A.D., Koch, S., Harrington, B., Laquinto, K., Boarman, M., 2012. Migration pathways, migration speeds and non-breeding areas used by northern hemisphere wintering red knots *Calidris canutus* of the subspecies *rufa*. Wader Study Group Bull. 119, 1–9.

Nisbet, I.C.T., 1963. Measurements with radar of the height of nocturnal migration over Cape Cod, Massachusetts. Bird Band. 34, 57–67.

Nisbet, I.C.T., Drury Jr., W.H., Baird, J., 1963. Weight loss during migration, Part I: deposition and consumption of fat by the blackpoll warbler *Dendroica striata*. Bird Band. 34, 107–159.

Nolet, B.A., Butler, P.J., Masman, D., Woakes, A.J., 1992. Estimation of daily energy expenditure from heart rate and doubly labeled water in exercising geese. Physiol. Zool. 65, 1188–1216.

Norberg, U.M., 1990. Vertebrate Flight. Mechanics, Physiology, Morphology, Ecology and Evolution. Springer-Verlag, Berlin/Heidelberg.

Norberg, U.M., 1996. Energetics of flight. In: Carey, C. (Ed.), Avian Energetics andNutritional Ecology. Chapman & Hall, New York, pp. 199–249.

Obrecht, H.H., Pennycuick, C.J., Fuller, M.R., 1988. Wind tunnel experiments to assess the effect of back-mounted radio transmitters on bird body drag. J. Exp. Biol. 135, 265–273.

Obst, B.S., Nagy, K.A., Ricklefs, R.E., 1987. Energy utilization by Wilson's storm-petrel (*Oceanites oceanicus*). Physiol. Zool. 60, 200–210.

Pasquis, P., Lacaisse, A., Dejours, P., 1970. Maximal oxygen uptake in four species of small mammals. Respir. Physiol. 9, 298–309.

Pelsers, M.A.L., Bishop, C.M., Butler, P.J., Glatz, J.F.C. Fatty acid-binding protein in heart and skeletal muscles of the migratory barnacle goose throughout development. Am. J. Physiol., in press.

Pennycuick, C.J., 1968. Power requirements for horizontal flight in the pigeon *Columba livia*. J. Exp. Biol. 49, 527–555.

Pennycuick, C.J., 1969. The mechanics of bird migration. Ibis 111, 525–556.

Pennycuick, C.J., 1975. In: Farner, D.S., King, J.R., Parkes, K.C. (Eds.), Mechanics of Flight, vol. V. Academic Press, New York, pp. 1–75.

Pennycuick, C.J., 1982. The flight of petrels and albatrosses (Procellariiformes), observed in South Georgia and its vicinity. Philos. Trans. R. Soc. Lond., B 300, 75–106.

Pennycuick, C.J., 1988. Empirical estimates of body drag of large waterfowl and raptors. J. Exp. Biol. 135, 253–264.

Pennycuick, C.J., 1989. Bird Flight Performance: A Practical Calculation Manual. Oxford Univ. Press, Oxford.

Pennycuick, C.J., 1990. Predicting wingbeat frequency and wavelength of birds. J. Exp. Biol. 150, 171–185.

Pennycuick, C.J., 1992. Newton Rules Biology: A Physical Approach to Biological Problems. Oxford Univ. Press, Oxford/New York/Tokyo.

Pennycuick, C.J., 1996. Stress and strain in the flight muscles as constraints on the evolution of flying animals. J. Biomech. 29, 577–581.

Pennycuick, C.J., 2008. Modelling the Flying Bird. Theoretical Ecology Series. Academic Press, Elsevier Inc. USA. pp. 480.

Pennycuick, C.M., Einarsson, O., Bradbury, T.A.M., Owen, M., 1996a. Migrating whooper swans (*Cygnus cygnus*): satellite tracks and flight performance calculations. J. Avian Biol. 27, 118–134.

Pennycuick, C.J., Fuller, M.R., McAllister, L., 1989. Climbing performance of Harris Hawks (*Parabuteo unicinctus*) with added load: implications for muscle mechanics and for radiotracking. Science 177, 222–228. J. Exp. Biol. 142, 17–29.

Pennycuick, C.J., Klaassen, M., Dvist, A., Lindström, Å., 1996b. Wingbeat frequency and the body drag anomaly: wind-tunnel observations on a thrush nightingale (*Luscinia luscinia*) and a teal (*Anas crecca*). J. Exp. Biol. 199, 2757–2765.

Perutz, M.F., 1983. Species adaptation in a protein molecule. Mol. Biol. Evol. 1, 1–28.

Peters, G.W., Steiner, D.A., Rigoni, J.A., Mascilli, A.D., Schnepp, R.W., Thomas, S.P., 2005. Cardiorespiratory adjustments of homing pigeons to steady wind tunnel flight. J. Exp. Biol. 208, 3109–3120.

Piersma, T., 1988. Breast muscle atrophy and constraints on foraging during the flightless period of wing moulting great crested grebes. Ardea 76, 96–106.

Piersma, T., 1998. Phenotypic flexibility during migration: optimization of organ sizes contingent on the risks and rewards of fueling and flight? J. Avian Biol. 29, 511–520.

Piersma, T., Jukema, J., 1990. Budgeting the flight of a long-distance migrant: changes in nutrient reserve levels of bar-tailed godwits at successive spring staging sites. Ardea 78, 315–337.

Portugal, S.J., Green, J.A., White, C.R., Guillemette, M., Butler, P.J., 2012. Wild geese do not increase flight behavior prior to migration. Biol. Lett. 8, 469–472.

Portugal, S.J., Thorpe, S.K.S., Green, J.A., Myatt, J.P., Butler, P.J., 2009. Testing the use/disuse hypothesis: pectoral and leg muscle changes in captive barnacle geese *Branta leucopsis* during wing moult. J. Exp. Biol. 212, 2403–2410.

Prats, M.-T., Palacios, L., Gallego, S., Riera, M., 1996. Blood oxygen transport properties during migration to higher altitude of wild quail, *Coturnix coturnix coturnix*. Physiol. Zool. 69, 912–929.

Price, E.R., 2010. Dietary lipid composition and avian migratory flight performance: development of a theoretical framework for avian fat storage. Comp. Biochem. Physiol. A 157, 297–309.

Prinzinger, R., Hänssler, I., 1980. Metabolism–weight relationship in some small non-passerine birds. Experientia 36, 1299–1300.

Qasem, L., Cardew, A., Wilson, A., Griffiths, I., Halsey, L.G., Shepard, E.L.C., Gleiss, A.C., Wilson, R., 2011. Tri-axial dynamic acceleration as a proxy for animal energy expenditure; should we be summing values or calculating the vector?. PLoS One 7, e31187.

Rall, J.A., 1985. Energetic aspects of skeletal muscle contraction: implications of fiber types. Exerc. Sport Sci. Rev. 13, 33–74.

Rayner, J.M.V., 1979. A new approach to animal flight mechanics. J. Exp. Biol. 80, 17–54.

Rayner, J.M.V., 1985a. Bounding and undulating flight in birds. J. Theor. Biol. 117, 47–77.

Rayner, J.M.V., 1985b. Linear relations in biomechanics: the statistics of scaling functions. J. Zool. London 206, 415–439.

Rayner, J.M.V., 1988. Form and function in avian flight. In: Johnston, R.F. (Ed.), Current Ornithology, vol. 5. Plenum, New York, pp. 1–66.

Rayner, J.M.V., 1990. The mechanics of flight and bird performance. In: Gwinner, E. (Ed.), Bird Migration. Springer-Verlag, Berlin/Heidelberg, pp. 283–299.

Rayner, J.M.V., 1991. On the aerodynamics of animal flight in ground effect. Philos. Trans. R. Soc. Lond., B 334, 119–128.

Rayner, J.M.V., 1993. On aerodynamics and the energetics of vertebrate flapping flight. Contemp. Math. 141, 351–399.

Rayner, J.M.V., 1994. Aerodynamic corrections for the flight of birds and bats in wind tunnels. J. Zool. London 234, 537–563.

Rollema, H.S., Bauer, C., 1979. The interaction of inositol pentaphosphate with the hemoglobins of highland and lowland geese. J. Biol. Chem. 254, 12038–12043.

Rosén, M., Hedenström, A., 2001. Gliding flight in a jackdaw: a wind tunnel study. J. Exp. Biol. 204, 1153–1166.

Rosser, B.W.C., George, J.C., 1986a. The avian pectoralis: histochemical characterization and distribution of muscle fiber types. Can. J. Zool. 64, 1174–1185.

Rosser, B.W.C., George, J.C., 1986b. Slow muscle fibres in the pectoralis of the turkey vulture (*Cathartes aura*): an adaptation for soaring flight. Zool. Anz. 217, 252–258.

Rosser, B.W.C., Waldbillig, D.M., Wick, M., Bandman, E., 1994. Muscle fiber types in the pectoralis of the white pelican, a soaring bird. Acta Zool. 75, 329–336.

Rothe, H.J., Biesel, W., Nachtigall, W., 1987. Pigeon flight in a wind tunnel. II. Gas exchange and power requirements. J. Comp. Physiol. B 157, 99–109.

Sapir, N., Dudley, R., 2012. Backward flight in hummingbirds employs unique kinematic adjustments and entails low metabolic cost. J. Exp. Biol. 215, 3603–3611.

Sapir, N., Wikelski, M., McCue, M.D., Pinshow, B., Nathan, R., 2010. Flight modes in migrating European bee-eaters: heart rate may indicate low metabolic rate during soaring and gliding. PLoS One 5, e13956.

Sato, K., Daunt, F., Watanuki, Y., Takahashi, A., Wanless, S., 2008. A new method to quantify prey acquisition in diving seabirds using wing stroke frequency. J. Exp. Biol. 211, 58–65.

Sato, K., Sakamoto, K.Q., Watanuki, Y., Takahashi, A., Katsumata, N., Bost, C.A., Weimerskirch, H., 2009. Scaling of soaring seabirds and implications for flight abilities of giant pterosaurs. PLoS One 4, e5400.

Scheid, P., 1985. Significance of lung structure for performance at high altitude. In: Ilyichev, V.D., Gavrilov, V.M. (Eds.), ACTA XVII, Congressus Internationalis Ornithologici, Moscow, pp. 976–977.

Schmidt-Nielsen, K., 1972. Locomotion: energy cost of swimming, flying, and running. Science 177, 222–228.

Schmidt-Nielsen, K., 1984. Scaling: Why is Animal Size So Important? Cambridge Univ. Press.

Schmidt-Nielsen, K., 1997. Animal Physiology: Adaptation and Environment, fifth ed. Cambridge Univ. Press.

Schmidt-Nielsen, K., Hainsworth, F.R., Murrish, D.E., 1970. Counter-current heat exchange in the respiration passages: effect on water and heat balance. Respir. Physiol. 9, 263–276.

Schmidt-Wellenburg, C.A., Biebach, H., Daan, S., Visser, G.H., 2007. Energy expenditure and wing beat frequency in relation to body mass in free flying barn swallows (*Hirundo rustica*). J. Comp. Physiol. B 177, 327–337.

Schwenk, R.W., Holloway, G.P., Luiken, J.J.F.P., Bonen, A., Glatz, J.F.C., 2010. Fatty acid transport across the cell membrane: regulation by fatty acid transporters. Prostaglandins Leukot. Essent. Fatty Acids 82, 149–154.

Schwilch, R., Jenni, L., Jenni-Eiermann, S., 1996. Metabolic responses of homing pigeons to flight and subsequent recovery. J. Comp. Physiol. 166, 77–87.

Scott, G.R., 2011. Elevated performance: the unique physiology of birds that fly at high altitudes. J. Exp. Biol. 214, 2455–2462.

Scott, G.R., Milsom, W.K., 2006. Flying high: a theoretical analysis of the factors limiting exercise performance in birds at altitude. Respir. Physiol. Neurobiol. 154, 284–301.

Scott, G.R., Egginton, S., Richards, J.G., Milsom, W.K., 2009. Evolution of muscle phenotype for extreme high altitude flight in the bar-headed goose. Proc. R. Soc. B 276, 3645–3653.

Sekercioglu, C.H., 2007. Conservation ecology: area trumps mobility in fragment bird extinctions. Curr. Biol. 17, R283–R286.

Shaffer, S.A., Tremblay, Y., Weimerskirch, H., Scott, D., Thompson, D.R., Sagar, P.M., Moller, H., Taylor, G.A., Foley, D.G., Block, B.A., Costa, D.P., 2006. Migratory shearwaters integrate oceanic resources across the Pacific Ocean in an endless summer. Proc. Natl. Acad. Sci. 103, 12799–12802.

Shams, H., Scheid, P., 1989. Efficiency of parabronchial gas exchange in deep hypoxia: measurements in the resting duck. Respir. Physiol. 77, 135–146.

Shams, H., Scheid, P., 1993. Effects of hypobaria on parabronchial gas exchange in normoxic and hypoxic ducks. Respir. Physiol. 91, 155–163.

Sibley, C.G., Ahlquist, J.E., 1990. Phylogeny and Classification of Birds: A Study in Molecular Evolution. Yale Univ. Press, Haven/London.

Snyder, G.K., 1990. Capillarity and diffusion distances in skeletal-muscles in birds. J. Comp. Physiol. B 160, 583–591.

Snyder, G.K., Byers, R.L., Kayar, S.R., 1984. Effects of hypoxia on tissue capillarity in geese. Respir. Physiol. 58, 151–160.

Sokal, R.R., Rohlf, F.J., 1981. Biometry: The Principles and Practice of Statistics in Biological Research. W. H. Freeman, New York.

Speakman, J.R., 1990. Principles, problems and a paradox with the measurement of energy expenditure of free-living subjects using double-labelled water. Stat. Med. 9, 1365–1380.

Speakman, J.R., Racey, P.A., 1988. Consequences of non-steady-state CO_2 production for accuracy of the doubly labeled water technique: the importance of recapture interval. Comp. Biochem. Physiol. A 90, 337–340.

Spedding, G.R., 1986. The wake of a jackdaw (*Corvus monedula*) in slow flight. J. Exp. Biol. 125, 287–307.

Spedding, G.R., 1987. The wake of a kestrel (*Falco tinnunculus*) in flapping flight. J. Exp. Biol. 127, 59–78.

Spedding, G.R., 1994. On the significance of unsteady effects in the aerodynamic performance of flying animals. Contemp. Math. 141, 401–419.

Spedding, G.R., Rayner, J.M.V., Pennycuick, C.J., 1984. Momentum and energy in the wake of a pigeon (*Columba livia*) in slow flight. J. Exp. Biol. 111, 81–102.

Spedding, G.R., Rosén, M., Hedenström, A., 2003. A family of vortex wakes generated by a thrush nightingale in free flight in a wind tunnel over its entire natural range of flight speeds. J. Exp. Biol. 206, 2313–2344.

Spivey, R.J., Bishop, C.M., 2013. Interpretation of body-mounted accelerometry in flying animals and estimation of biomechanical power. J. R. Soc., Interface 10, 20130404.

Stewart, A.G., 1978. Swans flying at 8,000 metres. Br. Birds 71, 459–460.

Stickland, N.C., 1977. Succinic dehydrogenase distribution in the pectoralis muscle of several East African birds. Acta Zool. 58, 41–44.

Suarez, R.K., Brown, G.S., Hochachka, P.W., 1986. Metabolic sources of energy for hummingbird flight. Am. J. Physiol. 251, 537–542.

Suarez, R.K., Lighton, J.R.B., Moyes, C.D., Brown, G.S., Gass, C.L., Hochachka, P.W., 1990. Fuel selection in rufous hummingbirds: ecological implications of metabolic biochemistry. Proc. Natl. Acad. Sci. U.S.A. 87, 9207–9210.

Swan, L.W., 1961. The ecology of the high Himalayas. Sci. Am. 205, 68–78.

Swan, L.W., 1970. Goose of the Himalayas. Nat. Hist. 79, 68–74.

Takekawa, J.Y., Heath, S.R., Douglas, D.C., Perry, W.M., Javed, S., Newman, S.H., Suwal, R.N., Rahmani, A.R., Choudhury, B.C., Prosser, D.J., Yan, B., Hou, Y., Batbayar, N., Natsagdorj, T., Bishop, C.M., Butler, P.J., Frappell, P.B., Milsom, W.K., Scott, G.R., Hawkes, L.A., Wikelski, M., 2009. Geographic variation in bar-headed geese *Anser indicus*: connectivity of wintering areas and breeding grounds across a broad front. Wildfowl 59, 102–125.

Talesara, G.L., Goldspink, G., 1978. A combined histochemical and biochemical study of myofibrillar ATPase in pectoral, leg and cardiac muscle of several species of bird. Histochem. J. 10, 695–710.

Tatner, P., Bryant, D.M., 1989. Doubly-labeled water technique for measuring energy expenditure. In: Bridges, C.R., Butler, P.J. (Eds.), Techniques in Comparative Respiratory Physiology: An Experimental Approach. Cambridge Univ. Press.

Taylor, C.R., 1994. Relating mechanics and energetics during exercise. Adv. Vet. Sci. Comp. Med. A 38, 181–215.

Teal, J.M., 1969. Direct measurement of CO_2 production during flight in small birds. Zoologica 54, 17–24.

Tobalske, B.W., 1995. Neuromuscular control and kinematics of intermittent flight in the European starling (*Sturnus vulgaris*). J. Exp. Biol. 198, 1259–1273.

Tobalske, B.W., 1996. Scaling of muscle composition, wing morphology, and intermittent flight behavior in woodpeckers. Auk 113, 151–177.

Tobalske, B.W., Dial, K.P., 1994. Neuromuscular control and kinematics of intermittent flight in budgerigars (*Melopsittacus undulatus*). J. Exp. Biol. 187, 1–18.

Tobalske, B.W., Dial, K.P., 1996. Flight kinematics of black- billed magpies and pigeons over a wide range of speeds. J. Exp. Biol. 199, 263–280.

Tobalske, B.W., Hearn, J.W.D., Warrick, D.R., 2009. Aerodynamics of intermittent bounds in flying birds. Exp. Fluids 46, 963–973.

Tobakske, B.W., Hedrick, T.L., Dial, K.P., Biewener, A.A., 2003. Comparative power curves in bird flight. Nature 421, 363–366.

Tobalske, B.W., Olson, N.E., Dial, K.P., 1997. Flight style of the black-billed magpie: variation in wing kinematics, neuromuscular control and muscle composition. J. Exp. Zool. 279, 313–329.

Torre-Bueno, J.R., 1976. Temperature regulation and heat dissipation during flight in birds. J. Exp. Biol. 65, 471–482.

Torre-Bueno, J.R., 1978a. Respiration during flight in birds. In: Piiper, J. (Ed.), Respiratory Function in Birds, Adult and Embryonic. Springer-Verlag, Berlin/Heidelberg/New York, pp. 89–94.

Torre-Bueno, J.R., 1978b. Evaporative cooling and water balance during flight in birds. J. Exp. Biol. 75, 231–236.

Torre-Bueno, J.R., Larochelle, J., 1978. The metabolic cost of flight in unrestrained birds. J. Exp. Biol. 75, 223–229.

Tucker, V.A., 1966. Oxygen consumption of a flying bird. Science 154, 150–151.

Tucker, V.A., 1968a. Respiratory physiology of house sparrows in relation to high-altitude flight. J. Exp. Biol. 48, 55–66.

Tucker, V.A., 1968b. Respiratory exchange and evaporative water loss in the flying budgerigar. J. Exp. Biol. 48, 67–87.

Tucker, V.A., 1970. Energetic cost of locomotion in animals. Comp. Biochem. Physiol. 34, 841–846.

Tucker, V.A., 1972. Metabolism during flight in the laughing gull, *Larus atricilla*. Am. J. Physiol. 222, 237–245.

Tucker, V.A., 1973. Bird metabolism during flight: evaluation of a theory. J. Exp. Biol. 58, 689–709.

Usherwood, J.R., Hedrick, T.L., McGowan, C.P., Biewener, A.A., 2005. Dynamic pressure maps for wings and tails of pigeons in slow, flapping flight, and their energetic implications. J. Exp. Biol. 208, 355–369.

Van den Berg, C., Rayner, J.M.V., 1995. The moment of inertia of bird wings and the inertial power requirement for flapping flight. J. Exp. Biol. 198, 1655–1664.

Vazquez, R.J., 1994. The automating skeletal and muscular mechanisms of the avian wing (Aves). Zoomorphol 114, 59–71.

Ward, S., Bishop, C.M., Woakes, A.J., Butler, P.J., 2002. Heart rate and the rate of oxygen consumption of flying and walking barnacle geese (*Branta leucopsis*) and bar-headed geese (*Anser indicus*). J. Exp. Biol. 205, 3347–3356.

Ward, S., Möller, U., Rayner, J.M.V., Jackson, D.M., Bilo, D., Nachtigall, W., Speakman, J.R., 2001. Metabolic power, mechanical power and efficiency during wind tunnel flight by the European starling *Sturnus vulgaris*. J. Exp. Biol. 204, 3311–3322.

Ward, S., Möller, U., Rayner, J.M.V., Jackson, D.M., Nachtigall, W., Speakman, J.R., 2004. Metabolic power of European starlings *Sturnus vulgaris* during flight in a wind tunnel, estimated heat transfer modelling, doubly labelled water and mask respirometry. J. Exp. Biol. 207, 4291–4298.

Warrick, D.R., Tobalske, B.W., Powers, D.R., 2009. Lift production in the hovering hummingbird. Proc. R. Soc. B 276, 3747–3752.

Weber, R.E., Jessen, T.-H., Malte, H., Tame, J., 1993. Mutant hemoglobins (α119-Ala and β55-Ser): functions related to high-altitude respiration in geese. J. Appl. Physiol. 75 (6), 2646–2655.

Weber, J.-M., Roberts, T.J., Vock, R., Weibel, E.R., Taylor, C.R., 1996. Design of the oxygen and substrate pathways III: partitioning energy provision from carbohydrates. J. Exp. Biol. 199, 1659–1666.

Weimerskirch, H., Chastel, O., Barbraud, C., Tostain, O., 2003. Frigatebirds ride high on thermals. Nature 421, 333–334.

Weimerskirch, H., Martin, J., Clerquin, Y., Alexandre, P., Jiraskova, S., 2001. Energy saving in flight formation. Nature 413, 697–698.

Weinstein, Y., Bernstein, M.H., Bickler, P.E., Gonzales, D.V., Samaniego, F.C., Escobedo, M.A., 1985. Blood respiratory properties in pigeons at high altitudes: effects of acclimation. Am. J. Physiol. 249, R765–R775.

Weis-Fogh, T., 1972. Energetics of hovering flight in hummingbirds and in *Drosophila*. J. Exp. Biol. 56, 79–104.

Weis-Fogh, T., Alexander, R.McN., 1995. The moment of inertia of bird wings and the inertial power requirement for flapping flight. J. Exp. Biol. 198, 1655–1664.

Welham, C.V., 1992. Flight speeds of migrating birds: a test of maximum range speed predictions from three aerodynamic equations. Behav. Ecol. 5, 1–8.

Wells, D.J., 1993a. Muscle performance in hovering hummingbirds. J. Exp. Biol. 178, 39–57.

Wells, D.J., 1993b. Ecological correlates of hovering flight of hummingbirds. J. Exp. Biol. 178, 59–70.

Witter, M.S., Cuthill, I.C., 1993. The ecological costs of avian fat storage. Philos. Trans. R. Soc. Lond., B 340, 73–92.

Woakes, A.J., Butler, P.J., 1986. Respiratory, circulatory and metabolic adjustments during swimming in the tufted duck, *Aythya fuligula*. J. Exp. Biol. 120, 215–231.

Woakes, A.J., Butler, P.J., Bevan, R.M., 1995. Implantable data logging system for heart rate and body temperature: its application to the estimation of field metabolic rates in Antarctic predators. Med. Biol. Eng. Comput. 33, 145–151.

Wolf, L.L., Hainsworth, F.R., 1971. Time and energy budgets of territorial hummingbirds. Ecology 52, 980–988.

Chapter 40

Physiological Challenges of Migration

Eldon J. Braun

Department of Physiology, College of Medicine, University of Arizona, Tucson, AZ, USA

40.1 GENERAL CONCEPTS

There are a number of theories suggesting why birds migrate and the evolutionary origin of this phenomenon, but the fundamental premise is that this type of mass movement over long distances at predictable times of the year does occur (Salewski and Bruderer, 2007). This very complex phenomenon must be supported by a metabolism that is equally complex. Theories vary a great deal as to the way birds move. Do birds migrate towards breeding territories that provide necessary resources for reproduction and rearing of young, then return to "home" areas to avoid harsh climates that occur with annual cycles (Zink, 2011)? It has been suggested that birds gradually expand ranges toward more favorable territories. As this expansion continues or grows with changing seasons, the farthest points of the range become less favorable and birds return to home territory where conditions are more favorable (Zink, 2011).

40.2 EVOLUTION OF MIGRATION

Most avian species may have had their origin in tropical regions or were forced to the tropics by glaciation events (Klaassen et al., 2012). Such events may have may have markedly increased competition for resources in southern regions and forced species to gradually expand to more favorable regions to ensure breeding success and the rearing of offspring. These activities may have lead to mass movements of species in bidirectional movements (i.e., patterns of migration).

40.3 COST OF MIGRATION

No matter the evolutionary source of migratory patterns, they do exist. These long-distance—and at times, uninterrupted—flights must be supported by the metabolism of substrates stored in the bodies of birds. Through a process of evolution, the metabolic fuel of choice of all migratory birds appears to be lipids or fatty acids, which can be stored within the body without the accompaniment of water (Guglielmo, 2010). The source of this high-energy metabolite substrate is variable, as is the manner in which it is stored in the body of birds. This storage is dependent on the body mass of the species and the distance of the migratory route. Very small birds (body mass <25 g) accumulate lipid reserves prior to departure, but in some cases have to "come down" to refuel (Jenni and Jenni-Eiermann, 1998). However, other small species sustain migratory trips without refueling. Good examples are hummingbirds, which may have a body mass of 8 g before departure and arrive at their destination with body masses of less than 5 g (Calder, 1987). Small European garden warblers with body masses of approximately 20 g accumulate lipids prior to departure for sub-Saharan Africa, but they do not sustain flight for the entire trip (Klaassen and Biebach, 1994). They stop to refuel two or three times before their final destination is reached. The refueling is of the lipid stores. These birds, as all avian species, do not metabolize carbohydrates (glucose) as a metabolic fuel, as is suggested by a lack of change in plasma glucose concentration during the migration. Either the glucose is not metabolized or it is constantly replenished as it is used. Parenthetically, all birds have high plasma glucose concentrations compared to other vertebrates, particularly mammals and reptiles, when compared on the basis of body mass. It is not clear why this is the case. The high levels of plasma glucose may serve functions unrelated to being consumed as metabolic fuel.

Although the metabolism of lipids is the most important metabolic fuel for birds undergoing sustained flight, there is increasing evidence that the breakdown of body protein also serves as an important resource during migration. Not only can protein be catabolized for energy, but the metabolism of it produces much more water than the metabolism of lipid. The catabolism of protein produces five times more water per unit of energy produced than does the metabolism of lipid (Jenni and Jenni-Eiermann, 1998). In a study using garden warblers (*Sylvia borin*) as an experimental animal, Bauchinger et al. (2005) examined the mass changes of 13 organs during spring migration. In preparation for the journey across the Sahara, some organ masses increased 1.5 times. After the flight, most organs (including the heart and flight muscles) lost mass. The liver, spleen, kidney, and the gastrointestinal tract decreased in mass by 50%. During refueling stopovers, the mass of the organs was restored. Interestingly, the testes mass increased fourfold, probably in preparation for reproduction. Because there are no special tissues to store protein, the change in mass of the normal organs facilitates the storage of protein as a supplemental energy and water source. Therefore, protein is a multifunctional substrate that can facilitate long-distance flight.

Using computer simulations, field data, weather, and global positioning system tracking data, Pennycuick et al. (2011) estimated fuel reserves for three species of geese along several migratory routes: from Iceland to Arctic Canada, from Scotland to Greenland, and from Scotland to northern Norway. The major result of the study (other than employing a combination of technologies) was the prediction that the geese would arrive on their breeding grounds with sufficient energy reserves that could be used for breeding activities.

Another observation that strongly supports the use of lipid storage and reserves for long-distance flight is that of several species of plovers that use the Bay of Fundy as a stopover on their way to their wintering grounds in northeastern South America, a journey that takes about 72 h nonstop (Turcotte et al., 2013). The bay is a rich source of the invertebrate *Corophium volutator*, commonly known as the mud shrimp. The body mass of these small mud shrimp is 43% lipids (Napolitano and Ackman, 1989). The birds remain in the bay area for 10–20 days to rest and feed on the shrimp and markedly increase their body mass primarily as result of lipid storage.

The reliance on lipids as metabolic fuel presents problems. Their lack of aqueous solubility means that they not only have to be transported from fat depots to cells, but they must cross cell membranes to be metabolized. They are carried in the plasma by serum albumin, and evidence suggests that they are transported into cells by a protein FAT/CD36 (fatty acid translocating protein) not by simple diffusion. Once inside cells, they are bound to the fatty acid binding protein (FABP), which serves as a sink to facilitate their continued entry into cells. The role of FABP has been demonstrated in heart muscle of western sandpipers (*Calidris mauri*), where its concentration about 10-fold greater than in mammalian heart muscle and its concentration 70% higher during migration than in winter resident populations of sandpipers (Guglielmo et al., 1998).

The role of protein, not only as an energy source but also as a water source, was demonstrated in studies on Swainson's thrush (*Catharus ustulatus*). In controlled wind tunnel experiments, birds flown in dry air lost 3.55 mg/min lean body mass, whereas birds flown in moist air did not lose lean body mass (Gerson and Guglielmo, 2011). This study suggested lean body mass (muscle) was metabolized to produce water lost to the dry air.

The emerging pattern appears to be that lipids are the primary energy source for the demands of long-distance flight, with protein serving as the energy reserve and a source of water to prevent marked changes in osmolality of total body water. Lipids have a negative feature of low aqueous solubility; however, this can be viewed as positive because, as an energy storage source, lipids do not require water. In an overall balance, lipids supply approximately 90% of the energy for flight and protein supplies approximately 10%. The third metabolic substrate carbohydrate (glucose) appears to play no role as a source of energy for long-distance flight. The physiological role of the high plasma glucose concentrations is undetermined at this time.

Adaptations of birds to fly at high elevations during migration are exemplified by bar-headed geese (*Anser indicus*), which fly over the Himalayas as they move from their wintering grounds in the low lands of India to their breeding areas on the high-altitude plateaus of central Asia (Koppen et al., 2010). The flight to the breeding grounds last about 8 h, but the return flight is accomplished in about one-half that time. Although there are limited data in the literature on body mass changes or the metabolic fuel consumed, the principal adaptations exhibited by the bar-headed geese appear to be at the level of the muscle microcirculation and the mitochondria (Scott et al., 2011). The capillary density appears to be higher in both cardiac and skeletal muscle (to reduce diffusion distance) and the nature of cytochrome oxidase is altered compared to geese whose lifespan is at low altitudes.

The mass movement of large numbers of birds of a given species (migration) is an interesting and complex area of study; the evolutionary bases for it are not completely understood. Nonetheless, the phenomenon does occur and birds have evolved strategies to use primarily lipids as metabolic substrates to support this endurance activity.

REFERENCES

Bauchinger, U., Wohlmann, A., Biebach, H., 2005. Flexible remodeling of organ size during spring migration of the garden warbler (*Sylvia borin*). Zoology 108, 97–106.

Calder, W.A., 1987. Southbound through Colorado: migration of Rufous Hummingbirds. Natl. Geogr. Res. 3, 40–51.

Gerson, A.R., Guglielmo, C.G., 2011. Endurance flight at low ambient humidity increases protein catabolism in migratory birds. Science 333, 1434–1436.

Guglielmo, C., Haunerland, N.H., Williams, T.D., 1998. Fatty acid binding protein, a major protein in the flight muscle of migrating western sandpipers. Comp. Biochem. Physiol. 119B, 549–555.

Guglielmo, C., 2010. Move that fatty acid: fuel selection and transport in migratory birds and bats. Integr. Comp. Biol. 50, 336–345.

Jenni, L., Jenni-Eiermann, S., 1998. Fuel supply and metabolic constraints in migrating birds. J. Avian Biol. 129, 521–528.

Klaassen, M., Biebach, H., 1994. Energetics of fattening and starvation in the long-distance migratory garden warbler, *Sylvia borin*, during the migratory phase. J. Comp. Physiol. B 164, 362–371.
Klaassen, M., Hoye, B.J., Nolet, B.A., Buttemer, W.A., 2012. Ecophysiology of avian migration in the face of current global hazards. Philos. Trans. R. Soc. Lond., B Biol. Sci. 367, 1719–1732.
Koppen, U., Yakovlev, A.P., Barth, R., Kaatz, M., Berthold, P., 2010. Seasonal migrations of four individual bar-headed geese *Anser indicus* from Kyrgyzstan followed by satellite telemetry. J. Ornithol. 151, 703–712.
Napolitano, G.E., Ackman, V., 1989. Lipids and hydrocarbons in *Corophium volutator* from Minas Basin, Nova Scotia. Mar. Biol. 100, 333–338.
Pennycuick, C.J., Griffin, L.R., Colhoun, K., Angwin, V., 2011. A trial of a non-statistical computer program for monitoring fuel reserves, response to wind and other details from GPS tracks of migrating geese. J. Ornithol. 152, 87–99.
Salewski, V., Bruderer, B., 2007. The evolution of bird migration–a synthesis. Naturwissenschaften 4, 268–779.
Scott, G.R., Schulte, P.M., Egginton, S., Scott, A.L., Richards, J.G., Milsom, W.K., 2011. Molecular evolution of cytochrome C oxidase underlies high-altitude adaptation in the bar-headed goose. Mol. Biol. Evol. 28, 351–363.
Turcotte, Y., Lamarre, J.-F., Bêty, J., 2013. Staging ecology of semipalmated plover (*Charadrius semipalmatus*) and semipalmated sandpiper (*Calidris pusilla*) juveniles in the St. Lawrence river Estuary during fall migration. Can. J. Zool. 91, 802–809.
Zink, R., 2011. The evolution of avian migration. Biol. J. Linn. Soc. 104, 237–250.

Chapter 41

Actions of Toxicants and Endocrine-Disrupting Chemicals in Birds

Mary Ann Ottinger

Department of Biology and Biochemistry, University of Houston, Houston, TX, USA, Department of Animal and Avian Sciences, University of Maryland, College Park, MD, USA

Meredith Bohannon, Leah Carpenter and Tiffany Carro

Department of Animal and Avian Sciences, University of Maryland, College Park, MD, USA

Johanna R. Rochester

The Endocrine Disruption Exchange, Paonia, CO, USA

Karen M. Dean

University of Lethbridge, Lethbridge, Canada

41.1 INTRODUCTION

The use of chemicals has been critical in crop production, pest control, health and beauty products, and in industrial systems crucial in our global economies. Plastics and industrially produced compounds have been essential to our lifestyle and are key factors in the efficiency of production systems, packaging, and storage. Not surprisingly, as we learn more about the chemical components used in many of these processes, in storage and production, and for control of pests and our environment, adverse effects have come to light. One of the earliest findings indicative of potential adverse effects of industrial chemicals and pesticides was the thinning of eggs in bald eagles in response to exposure to dichlorodiphenyltrichloroethane (DDT), which dramatically illustrates the potency of these chemicals in wild birds. Since that time, there have been a series of workshops aimed at assessing the actions and risks associated with exposure to environmental chemicals across vertebrate and invertebrate species. Workshops, including the Wingspread Meetings and research publications, have documented impacts of EDCs and toxicants in laboratory studies and in wild populations of avian species (Berg et al., 1999; Bowerman et al., 2007; Brunstrom et al., 1991; Cohen-Barnhouse et al., 2011; Corbitt et al., 2007; Custer et al., 2010a,b; Custer and Read, 2006; Hotchkiss et al., 2008; Kavlock et al., 2005; Rattner, 2009; Rattner et al., 2000, 2010, 2001; Safe et al., 1985; Levengood et al., 2007). In general terms, the potential for deleterious results have been supported by observations of endocrine-related effects to organisms in areas in which dioxin, PCB, or other chemicals occurred. Furthermore, the exposure of wildlife to environmentally relevant levels is often at sublethal concentrations, making the determination of effects of EDCs challenging, especially when attempting to separate other interacting factors in the animal's environment. Additional variables may include environmental conditions, food availability, disease, or confounds such as simultaneous exposure to several chemicals. It is therefore important to develop reliable and sensitive measures that are appropriate for the endocrine or organ system that are the targets of the EDCs. This chapter will consider toxicants that affect avian species, with emphasis on environmental chemicals that attack endocrine systems through similarities in structure and/or function to hormones and growth factors. Background will be provided on some of the early studies to circumscribe the potential for adverse impacts of these toxicants and underpin the testing paradigms required for the commercial use of these chemicals. The impacts of these environmental chemicals will be considered in terms of broad categories of compounds, based on their characteristics and targets. Finally, mechanisms of action will be described in the context of the variety of avian species, their habitats and reproductive strategies, and characteristics that make them unique among vertebrates.

41.2 ENDOCRINE-DISRUPTING CHEMICALS: UTILITIES AND HAZARDS?

A number of factors must be considered when evaluating the cost/benefit for the use of environmental chemicals. The productivity of our agricultural systems is outstanding, in part due to the use of chemical treatments that limit pests and make applications, storage, and preparation more efficient. The use of chemicals has dramatically increased as our knowledge of chemistry and biochemistry has enabled the development of designer chemicals for specific applications. However, many of these chemicals are effective because they specifically interact with endocrine-related processes or other physiological functions in pest species. For example, an effective pesticide for gypsy moths is tebufenozide, which is an ecdysone inhibitor that acts on the timing of molting in larvae. Because these compounds were developed specifically to act on physiological targets in insects, coincident effects on other vertebrate classes are generally regarded as minimal risk compared to benefits.

The U.S. Environmental Protection Agency (EPA) formed the Endocrine Disruptor Screening and Testing Advisory Committee (EDSTAC), a committee of scientists and stakeholders that was charged by the EPA to provide recommendations on how to implement its Endocrine Disruptor Screening Program (EDSP). The EDSP is critical for assessing the potential for EDCs to impact vertebrate species, with the added benefit that identifying EDCs with adverse effects in vertebrates and banning them from use would then prevent human exposure. Upon recommendations from the EDSTAC (Harvey and Everett, 2006), the EPA expanded the EDSP using the Administrator's discretionary authority to include the androgen and thyroid hormonal systems as well as the reproductive endocrine systems of wildlife (see http://www.epa.gov/endo/pubs/edspoverview/edstac.htm). Following broader international concerns and the creation of similar programs in other countries, the Organisation for Economic Co-operation and Development (OECD) established the Endocrine Disrupters Testing and Assessment (EDTA) Task Force in 1998 within its Test Guidelines Programme, with the charge of developing an internationally harmonized testing strategy for the screening and testing of EDCs.

To address possible effects in birds, the Tier 2 Japanese quail (*Coturnix japonica*) Avian Toxicity Test was developed to detect potential disruption of the endocrine system by EDCs; this and other tests will be reviewed later in this chapter. One of the primary challenges, given the diversity of chemicals considered potential EDCs is to ascertain risk across vertebrate classes. Because birds have a broad range of life histories and unique developmental characteristics, there are differential sensitivities across avian species with changing lifetime vulnerability. Further, migratory birds have enormous metabolic demands, requiring highly functional metabolic and thyroid endocrine systems. As such, EDCs that impact metabolic and in particular the thyroid systems pose risk to migratory avian species. Furthermore, sexual differentiation of reproductive and other endocrine systems is unique to birds, being driven by primarily estradiol-mediated mechanisms. Developmentally, songbirds and other passerine species are altricial, while ducks, galliforms, and ground-dwelling species are precocial. This developmental difference appears to relate to the extent and timing of sensitivity to EDCs, because precocial species are fully functional and relatively independent at hatch. They undergo sexually differentiation *in ovo;* similarly other physiological systems are also well developed at hatch as compared to altricial birds that require intensive parental care. Finally, the potential for adverse effects of EDCs on long- or short-lived birds is unknown, especially in the context that long-lived birds tend to have small clutches and raise fewer chicks per year compared to short-lived birds (Holmes and Ottinger, 2003; Ottinger et al., 2009a; Ottinger, 2005).

Historically, there are many studies in wild birds that document cases of exposures with higher concentrations of toxicants originating from chemical applications and industrial spills. Often these occurrences were associated with high mortality, thereby prompting federal and regional responses aimed at protecting human and wildlife health. However, less is known about nonlethal impacts of EDCs on wild birds and the potential for adverse effects translating even to the level of a risk to the population. The deleterious effects of DDT was a recognized adverse effect of a pesticide acting as an EDC in bald eagles, with endocrine actions on females that resulted in eggshell thinning (Lundholm, 1997). As a consequence, embryonic eagles failed to develop to hatch. In addition, there are extensively documented cases, such as incidences of avian impacts from the Great Lakes, PCBs released into the Housatonic and Hudson Rivers, and high concentrations of munitions and other contaminants that indicate impacts on birds (Best et al., 2010; Bowerman et al., 2007; Custer et al., 2010a; Darnerud, 2003; Franceschini et al., 2008). Data collected from some extraordinary datasets also exist or are emerging from long-term monitoring and research programs, including the Great Lakes region; many of these studies have considered physiological and reproductive impacts of exposure to chemicals found in those areas. Laboratory studies have provided additional insights into potential mechanisms of action of EDCs as well as periods of sensitivity throughout the life-cycle. Information from many of these studies will be discussed below in order to put the proposed two-generation test into the context of testing in avian species. Furthermore, it is important to consider unique characteristics of the avian physiology with emphasis on sexual differentiation, hormonal modulation of endocrine and behavioral components of reproduction, and functional impacts of EDCs on neuroendocrine regulatory systems, especially considering potential adverse outcomes.

41.3 LIFE-CYCLE OF EDCS IN THE ENVIRONMENT

There are many sources of EDCs in the environment and there are excellent resources available to search for information and distribution about many of these compounds. For example, the United States Geological Service has a Contaminant Exposure and Effects—Terrestrial Vertebrates Database that provides over 20,000 searchable records that provide information on taxonomy and chronology of ecotoxicological exposure and effects as well as geographic location of contaminants (see http://www.pwrc.usgs.gov/contaminants-online/pages/ceetv/ceetvintro.htm). In considering the use and fate of environmental chemicals, it is useful to consider the life-cycle of a chemical or compound (Figure 41.1). The fate of a compound in the environment relates to both the movement in soil, water, and air as well as the metabolism of the parent or original chemical. Some chemicals enter directly into the air as a consequence of the manufacturing process, as in the case of airborne chemicals through smokestacks or steam. Moreover, chemicals in the environment become metabolized with exposure to air, sun, heat, and microbial metabolism. Metabolites of parent compounds are often bioactive and in some cases impact different physiological targets with even greater toxicity than the parent compound. Alternatively and more frequently, chemicals are manufactured for a specific purpose such as in the case of DDT, which as an effective insecticide became ubiquitous in the environment with wide use as a pesticide and in the battle against malaria-carrying mosquitos (Asawasinsopon et al., 2006; Borga et al., 2007; Braune et al., 2005; Rattner, 2009). As may be seen in Figure 41.2, DDT applied to fields as an insecticide then migrates into the water via rain events and run-off. In the rivers and streams, DDT becomes part of the food chain by ingestion into invertebrates, which are eaten by fish and eventually by organisms higher on the food chain. This brings to attention the effects. In the DDT example, effects were observed in bald eagles and more recently in herons at Lake Michigan on eggshell thinning (Lundholm, 1997; Rattner, 2009). The life-cycle of DDT also includes metabolism into DDE (dichlorodiphenyldichloroethylene), an antiandrogenic compound in vertebrates (Bowerman et al., 2007; Quinn et al., 2008). Despite years of banning, significant concentrations of DDT and its metabolites are found in agricultural areas and in wildlife living and breeding in proximity to areas that had wide use of DDT as an insecticide (Weseloh et al., 1990).

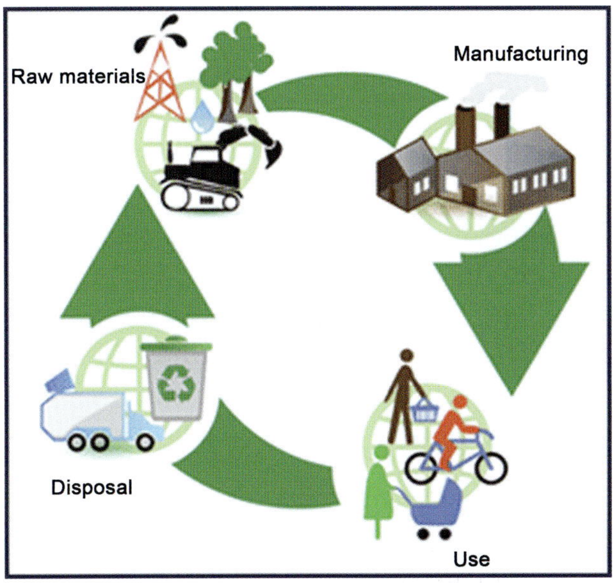

FIGURE 41.1 **Diagrammatic representation of the life-cycle of compounds;** once manufactured, they are used and then disposed allowing reentry into the environment. *From Environmental Protection Agency "Risk Management Sustainability Technology", http://www.epa.gov/nrmrl/std/lifecycle.html.*

In the case of industrial compounds such as lubricants, plastics, flame retardants, and nanofibers, there may be purposeful or inadvertent release into the environment, especially at the time of usage or disposal. This is the case with the release of polychlorinated biphenyls (PCBs) into the environment, which occurred both as a consequence of the manufacturing practices for electrical transformers containing Aroclors® and affected the Housatonic River in western Massachusetts (Custer and Read, 2006; Custer et al., 2012a; Barnthouse et al., 2003). Similarly, General Electric's Hudson Falls and Fort Edward plants discharged PCBs into the upper Hudson River from 1947 until 1952 (http://www.darrp.noaa.gov/northeast/hudson/index.html). This, in combination with leakage of these compounds from the manufacturing process for capacitors, resulted in long-term contamination of the river and continued persistence of PCBs in the environment (Erickson et al., 2005; Butcher and Garvey, 2004; Cho et al., 2002; Foley, 1992; Man et al., 2011). Not only do PCBs have a very long half-life in the environment, they also remain chemically stable in the sediment and become available when the sediment is disturbed. Further, measurements in tree swallow and sandpiper eggs revealed that there were two mixtures present in the Hudson River environs, reflecting both released congeners and other congeners associated with metabolism (Echols et al., 2004; Nichols et al., 2004). As such, there are dynamics in

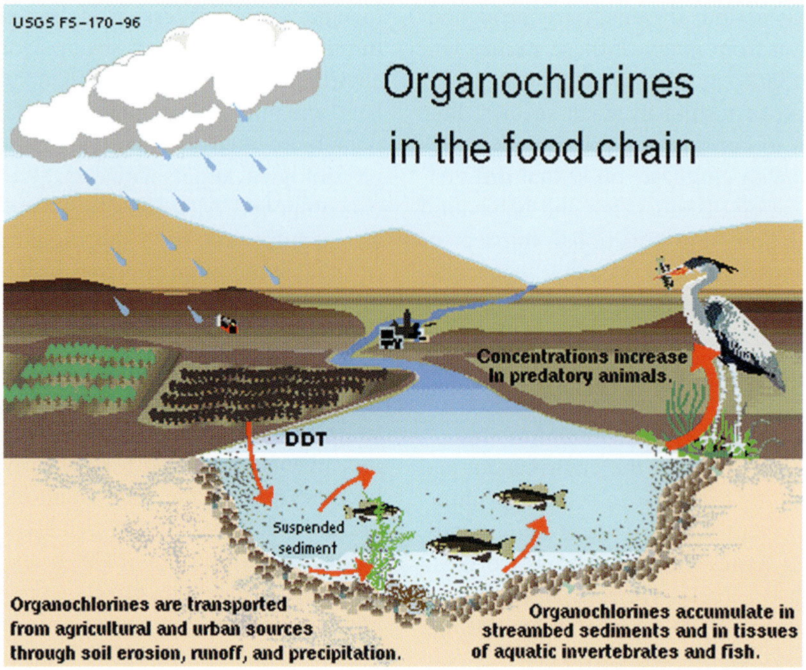

FIGURE 41.2 The life-cycle of DDT in the environment, from application on fields, transport to aquatic environments, and uptake by invertebrates and fish that are eaten by vertebrates higher on the food chain.

the environment including rainfall events as well as climate change that will impact the movement and potential for exposure across living organisms.

41.4 CLASSES OF EDCs

As the potential risk from EDCs became more apparent, the Environmental Protection Agency formed a federal advisory committee, the EDSTAC in 1996 and charged the members to carefully consider the issue of EDCs and to recommend appropriate regulatory measures (see http://www.epa.gov/endo/pubs/edspoverview/edstac.htm). Over two years, the EDSTAC members considered the complexities of potential and recognized EDCs, and risks to vertebrate and invertebrate organisms, and recommended a series of testing paradigms to be developed and verified (see http://www.epa.gov/endo/pubs/edspoverview/finalrpt.htm for the EDSTAC final report). These tests would include screening and testing, utilizing cell-based screens and bioassays sensitive to endocrine disruption of the reproductive and thyroid endocrine systems. The tests were separated into Tier 1 containing *in vitro* high throughput screens to allow rapid assessment and identification of endocrine active compounds and *in vivo* specific assays to target known EDC mechanisms of action (Table 41.1). The mechanisms of action considered in these Tier 1 tests focus on estrogenic, androgenic, or thyroid active compounds (Table 41.2). These Tier 1 assays are intended to precede more extensive Tier 2 assays that would serve as an in-depth test of risk for selected compounds that may be endocrine disrupting over the life-cycle. Tier 2 assays are multigenerational, including stages over the life-cycle and as such intended to capture periods of vulnerability to EDC effects. With the exception of the two-generation rat reproduction test, testing protocols are undergoing technical validation and refinement. Other two-generation tests under consideration include the medaka multigeneration test, Japanese quail two-generation test, mysid two-generation toxicity test, and larval amphibian growth and development test. All tests are being subjected to review by expert panels (FIFRA Science Advisory Panels) and have strengths and weaknesses that are being evaluated for reliable detection of EDCs of concern for regulatory oversight (Touart, 2004).

41.4.1 Categorizing Potential Endocrine Disruptors According to Structure and Function

Clearly, there are many factors to be considered in order to understand the passage of chemicals through the environment and to assess the potential for exposure to living organisms. Moreover, individuals are generally exposed to multiple environmental chemicals concurrently, and all living organisms carry some burden within their body. In order to deal with the potential for an estimated 80,000+ chemicals to be endocrine active, there have been attempts to organize potential and proven EDCs into classes of chemicals; chemicals have been grouped by mechanism of action, physiological target, and relative activity including capacity for toxicity. These groupings are based on structure–function relationships, and when available, on experimental findings in the

TABLE 41.1 List of Assays Proposed by the Endocrine Screening and Testing Advisory Committee

Assays Recommended for Consideration for the Tier 1 Screening Battery	
Assay	**Reason for Inclusion**
Estrogen receptor binding or transcriptional activation assay	An *in vitro* test to detect chemicals that may affect the endocrine system by binding to the estrogen receptor.
Estrogen receptor transcriptional activation assay	An *in vitro* test to detect chemicals that may affect the endocrine system by binding to the estrogen receptor.
Androgen receptor binding assay	An *in vitro* test to detect chemicals that may affect the endocrine system by binding to the androgen receptor.
In vitro steroidogenesis assay	An *in vitro* test to detect chemicals that interfere with the synthesis of the sex steroid hormones.
Placental aromatase assay	An assay to detect interference with aromatase.
Uterotrophic assay	An *in vivo* assay to detect estrogenic chemicals.
Hershberger assay	An *in vivo* assay to detect androgenic and antiandrogenic chemicals.
Pubertal male assay	An *in vivo* assay to detect chemicals that act on androgen or through the hypothalamic-pituitary-gonadal (HPG) axis that controls the estrogen and androgen hormone systems. It is also enhanced to detect chemicals that interfere with the thyroid system. This assay could in part substitute for the female pubertal assay.
Pubertal female assay	An *in vivo* assay to detect chemicals that act on estrogen or through the HPG axis that controls the estrogen and androgen hormone systems. It is also enhanced to detect chemicals that interfere with the thyroid system.
Amphibian metamorphosis assay	An *in vivo* assay for detection of chemicals that interfere with the thyroid hormone system.
Fish screening assay	An *in vivo* assay for detection chemicals that interfere with the HPG axes.

From EPA publication October 21, 2009; Tier 1 Screening Battery; http://www.epa.gov/endo/pubs/assayvalidation/tier1battery.htm.

literature for a specific or similar compound (Le Page et al., 2011; Newbold et al., 2008; Safe et al., 1998; Watanabe et al., 2002; Zacharewski, 1998; Foster, 1998; Kavlock and Ankley, 1996). A great deal of emphasis has gone into understanding the actions of estrogen active or xenoestrogens, because much of the initial recognition of endocrine-disrupting activities emerged from estrogenic effects of environmental chemicals. However, it has become clear that many compounds and mixtures are androgen active, either having androgen-like activity or antiandrogenic actions. In addition, many compounds including polybrominated biphenyls (PBDEs) and PCBs have thyroid-axis activities, with several physiological targets. Finally, there is increasing recognition of a wide range of other actions due to EDCs, which include impacts on the stress axis, growth factors modulating development of organs and organ systems, epigenetic actions associated with alterations at the genetic level, and lifetime effects on immune function and metabolism. At this time, most EDCs are considered as estrogenic, androgen active, or thyroid modulators. There are data emerging that also show EDC effects via other mechanisms on the stress axis and growth factors, and exerting epigenetic effects. Therefore, it will be critical to continue to modify testing methods to detect endocrine disruption as more data become available that provide insights into the varied actions of these compounds.

Analysis of the chemical structure of a compound provides valuable insight into the potential actions and physiological targets. In the case of PCBs, there is increasing toxicity associated with the number and molecular placement of chlorine atoms. Extremely toxic compounds such as dioxins provide a “yardstick” for comparison of toxic effects from other toxicants. As pointed out earlier, it is challenging to discern frank toxicity from endocrine disruption. Therefore, it has become important to utilize approaches, such as Adverse Outcomes Pathway (AOP; see below) that describe the suite of effects associated with exposure to a compound or mixture. Because the presumed targets of EDCs have been the reproductive axis and more specifically via interaction with steroid hormone

TABLE 41.2 The Mechanism Considered in the Selection of Screening Tests; Tests Focus on Known and Potential Mechanisms of Action for Compounds Having Endocrine Disrupting Activity

Complementary Modes of Action among Screening Assays in the EDSP Tier 1 Battery								
	Modes of Action							
	Receptor Binding				Steroidogenesis			
Screening Assays	E[2]	Anti-E	A[2]	Anti-A	E[2]	A[2]	HPG[3] Axis	HPT[3] Axis
In vitro								
ER binding[1]	■	■[4]						
ERα transcriptional activation	■							
AR binding[1]			■	■				
Steroidogenesis H295R					■	■		
Aromatase recombinant					■			
In vivo								
Uterotrophic	■							
Hershberger			■	■		■		
Pubertal male			■	■		■	■	■
Pubertal female	■	■[4]			■		■	■
Amphibian metamorphosis								■
Fish short-term reproduction (male & female)	■	■[4]	■	■	■	■	■	

[1]*Estrogen and androgen receptor binding.*
[2]*Estrogen and androgen.*
[3]*Hypothalamic-pituitary-gonadal or -thyroidal axis.*

[4]*Assays are expected to detect anti-estrogens, but this was not established during the validation process since no estrogen receptor antagonists were tested.*
From http://www.epa.gov/endo/pubs/assayvalidation/tier1battery.htm.

receptors, many of the strategies for high-throughput screening have utilized cell-based systems that detect interactions with these steroids (Fisher, 2004; Hartig et al., 2002; Kase et al., 2009; Kusk et al., 2011; D'Ursi et al., 2005). However, these *in vitro* approaches must be integrated with information about physiological and toxicological effects. As a consequence, computer-based predictive models have been developed that provide a rapid method for ascertaining potential endocrine activity based on structure–function and known mechanisms of action (Wambaugh et al., 2013; Knudsen et al., 2013; Kavlock et al., 2012; Martin et al., 2009; Dix et al., 2007). As more information becomes available, these approaches provide a powerful technique for assessing the potential for emerging chemicals to be endocrine disruptors.

41.4.2 General Mechanisms of Action of EDCs in Vertebrates

In addition to the development of extensive databases and mechanism-based models, a number of investigations have explored the mechanisms of endocrine-disrupting activity (Stoker et al., 2000; Zacharewski, 1998; Watanabe et al., 2002; Le Page et al., 2011; Janer et al., 2005; Gore, 2010; Harvey and Everett, 2006). Findings have revealed that EDCs can be endocrine active; however, the primary targets are not as clear. A number of EDCs that are estrogen mimics interact with estradiol receptors and/or with estrogen-responsive systems as agonists or antagonists (Watson et al., 2013; Hall and Korach, 2002; Marino et al., 2012). Conversely, androgen-active compounds are now gaining more attention, with DDE, trenbolone, and other EDCs acting as potent EDCs on endocrine systems (Hartig et al., 2007; Manikkam et al., 2013; Soto and Sonnenschein, 2010; Guerrero-Bosagna and Skinner, 2009; Prins, 2008; Skinner, 2007; Maffini et al., 2006; Bigsby et al., 1999; Santti et al., 1998). Many of the assays developed have taken advantage of the mechanisms revealed by these studies, for example the vitellogenic effects of EDCs in fish (Wang et al., 2011; Larkin et al., 2002). It is also important to recognize that there are a number of additional effects of EDCs, some of

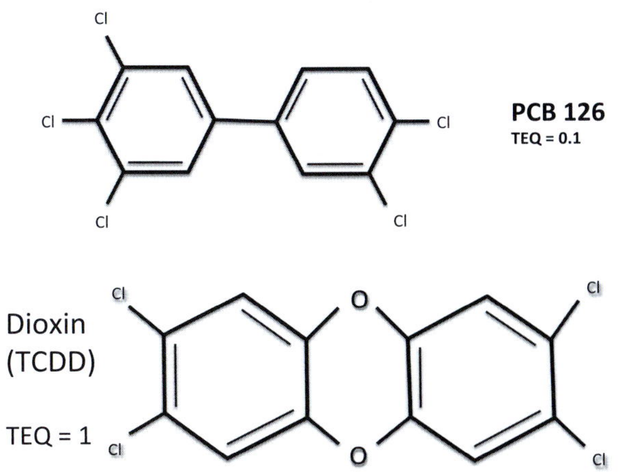

FIGURE 41.3 Structures of polychlorinated biphenyl 126 (PCB 126) with a toxic equivalency quotient (TEQ) of 0.1 compared to dioxin with a TEQ of 1.

which may be due to direct toxicological effects of these compounds in addition to endocrine effects (Fucic et al., 2012; Carro et al., 2013a,b; Quinn et al., 2007b). In some cases, induction of cancer in reproductive organs has also been linked to EDC exposure, especially during vulnerable life stages such as embryonic development (Ohlson and Hardell, 2000; Fucic et al., 2012).

There is an extensive literature now available that clearly demonstrates the interaction of EDCs with steroid hormone receptors, particularly those specific for estrogens, androgens, and thyroid hormones (Kusk et al., 2011; Zhang and Trudeau, 2006; Masuyama et al., 2002). This literature show similar mechanisms of action in birds compared to other classes of organisms, including mammals. However, the outcomes from exposures may differ widely, as will be reviewed in later sections of this chapter. Interactions with receptors depend on the structural characteristics of the EDC. In the case of polychlorinated biphenyls (PCBs), the toxicity and relative effect relates to the number and location of chlorine molecules on the ring structure (Figure 41.3) (Fischer et al., 1998; Hennig et al., 2002). Primary effects of EDCs rely on the ability to interact with steroid or thyroid hormone receptors, which then exert competitive inhibition or antagonistic effects or in some cases may be agonistic and stimulate the system. The up or down-regulation of cellular processes via interaction with these receptors then has the capacity to alter a host of physiological functions that are modulated through steroid hormone or thyroid hormones. Although less is known about the potential for EDCs to act on the stress axis, a stressed individual would be predicted to be more vulnerable to other insults such as those from EDC exposure (Fairbrother et al., 2004; Fernie et al., 2005a; Grasman, 2002; Grasman et al., 1998; Franceschini et al., 2008). Data are emerging that demonstrate effects of EDCs on multiple growth and differentiation factors, which in turn affect the ontogeny of that organ system as well as impact function in the mature individual (Thackaberry et al., 2005; Walker and Catron, 2000; Head et al., 2008). For example, effects of EDCs on the ryanodine receptor have been shown to impact calcium signaling, which in turn alters neural function, neurodevelopment, and neurodegenerative processes (Pessah et al., 2010). EDCs have the potential to affect immune function both short- and long-term in individuals and can be especially potent if there are other environmental stressors (Schug et al., 2011; Ottinger et al., 2005a; Quinn et al., 2007b; Lavoie et al., 2010). At a more cellular level, dioxins and other EDCs have been shown to affect oxidative damage processes, with potential to impact mitochondrial function (Pereira et al., 2013).

41.4.3 Predicting Risk: Adverse Outcomes Pathways

A number of approaches have been developed to conceptualize the myriad effects associated with endocrine disruption, with the overall intent to span molecular and cellular effects to the whole organism and eventual population-level impacts (Figure 41.4). A comprehensive approach has been developed and termed Adverse Outcomes Pathways (AOP). This approach has been conceptualized as well as applied to aquatic species (Lalone et al., 2013; Villeneuve et al., 2013; Ankley et al., 2010; Currie et al., 2005). Subsequent studies will utilize this and other approaches to establish probable outcomes with exposure to EDCs and model the potential for population level effects. In addition, it will be critical to integrate data collected from these and future studies to further refine the mechanisms that are responsible for both short- and long-term impacts upon individuals and potentially for wildlife and human populations (Ottinger et al., 2013; Villeneuve et al., 2013; Zawatski and Lee, 2013).

41.4.4 Predicting Impact: Toxic Equivalency Factor and Toxic Equivalence Quotient

There is a large body of literature on the role of the aryl hydrocarbon receptor (AhR) in response to exposure to toxicants. The signaling pathway showing activation of the AhR by dioxin or dioxin-like compounds involves interaction of an EDC with the AhR, which then binds to Hsp 90 followed by translocation into the nucleus. After dimerization, the aryl hydrocarbon nuclear translocator complex binds to the xenobiotic response element on the DNA with transcription of mRNA for translation into CYP proteins (Figure 41.5). As this is a well-characterized response to dioxin-like compounds, an accepted method for assessing the potential activity of a toxicant utilizes the relative activation of the AhR toxicological measure for environmental chemicals, the toxic equivalency factor (TEF), and the

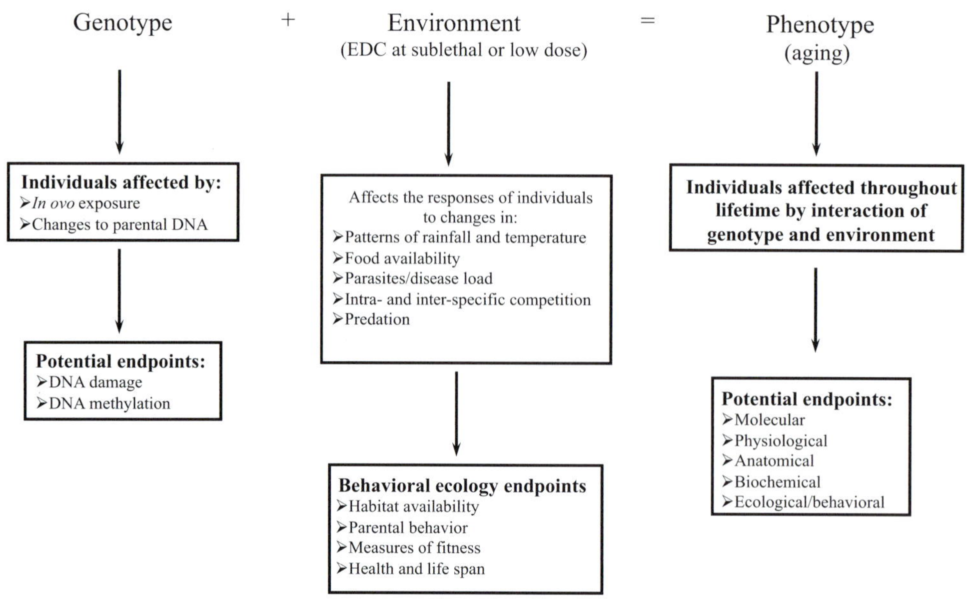

FIGURE 41.4 Conceptual range of effects of EDCs range from effects on individuals relative to altered gene expression.

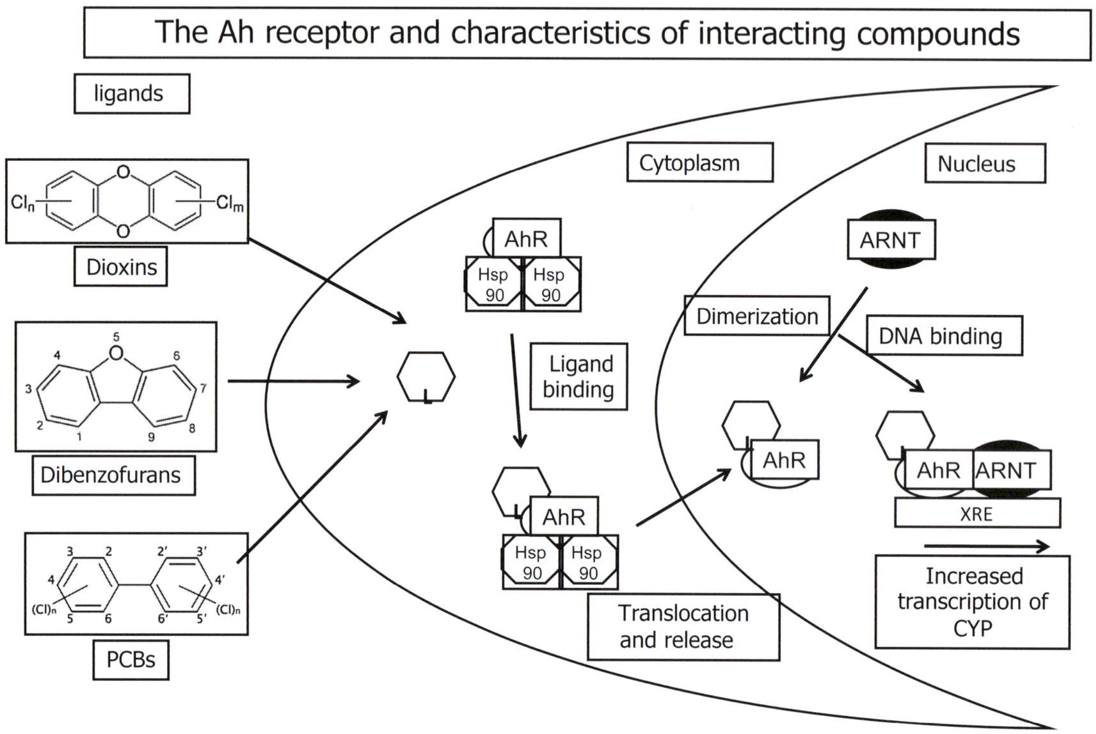

FIGURE 41.5 Aryl hydrocarbon receptor (AhR) signaling pathway is comprised of interaction of EDCs with affinity to AhR, translocation of the complex to the nucleus, dimerization, and binding of the aryl hydrocarbon nuclear translocator complex to the xenobiotic response element on the DNA, ultimately leading to increased transcription of CYP mRNA for translation to CYP enzymes.

added toxicity potential for a mixture or compound that is termed the toxic equivalency quotient (TEQ). This measure utilizes the literature and other information derived from functional similarities to compare toxicity against dioxin (2,3,7,8-tetrachlorodibenzo-*p*-dioxin (TCDD)) (Van Den Berg et al., 2013). In birds as well as other vertebrates, a range of adverse physiological outcomes have been documented with exposure to dioxin (Hervé et al., 2010; Hsu et al., 2007; Kuil et al., 1998; Watanabe et al., 2009). There has been a series of studies that have focused on the relative sensitivity and response to a range of EDCs using activation of the AhR as a relative measure among avian species (Head et al., 2008). These studies have demonstrated a wide variation in the vulnerability of avian species to toxicants, with molecular differences at the level of the AhR structure and function (Farmahin et al., 2012; Karchner et al., 2006; Manning et al., 2012; Zhang et al., 2013). It is clear that as in mammals, EDCs can exert wide-ranging effects on endocrine processes in birds, with the potential for lifetime effects that will impact fitness of the individual and ultimately have a potential population-level impact.

41.4.5 Why are Birds Unique?

As will be discussed in more detail below, avian species have specific attributes that distinguish them from other vertebrate classes (Table 41.3). These characteristics underpin observed differences in vulnerability to EDCs and contribute to the unique responses observed in birds compared to other vertebrate classes. A distinctive feature of avian biology is the higher body temperature, which is 105 °F. As in many other vertebrate classes, there is a large range of reproductive strategies and habitats utilized across avian species. In addition, many birds migrate seasonally often over long distances. In seasonally breeding birds, the gonads regress and reproductive function ceases; thereby allowing reducing weight and concentrating energy utilization on migration. There are documented seasonal changes in metabolism that also relate to energetics, exposure to environmental chemicals, and endocrine modulators (Broggi et al., 2003; Hallanger et al., 2011; Hurlbert and Haskell, 2003; Majewski et al., 2005; Wilson and Reinert, 1999). There is a differential pattern in the development of the thyroid and metabolic regulatory systems in altricial and precocial birds; precocial birds have a relatively functional hypothalamic–pituitary–gonadal axis at hatch (Mcnabb, 2007; Mcnabb and Fox, 2003; Wada et al., 2009; Webb and Mcnabb, 2008; Wilson and Reinert, 1999). Further, both seasonality and exposure to environmental contaminants alter thyroid hormone system responses (Mcnabb, 2005; Nost et al., 2012; Ross et al., 2011; Webb and Mcnabb, 2008; Wilson and Reinert, 1999). Birds also differ from other vertebrate classes in the endocrine and neural mechanisms involved in sexual differentiation. Moreover, mature females have only one functional ovary. Many of these characteristics, including breeding characteristics, neural mechanisms, and lifetime impacts, will be discussed below.

TABLE 41.3 Unique Characteristics of Avian Species that Influence Response to EDCs

- Growth and metabolism
  - High body temperature (105 °F) with high metabolic rate
  - Rapid growth rate
  - Seasonal metabolic changes associated with migration
  - Seasonal shift in mobilization of lipid reserves
  - Differential developmental patterns in altricial versus precocial birds
  - Thyroid system critical for pre-migratory fattening
- Reproductive system
  - Female has one functional ovary
  - Altered gonadal differentiation results in ovotestes
- Hormones and behavior
  - Males adversely impacted by xenoestrogens and androgenic compounds
  - Females less sensitive to xenoestrogens
- Sexual differentiation
  - Males are the homogametic sex (ZZ); females (ZW)
  - HPG axis relies on relative exposure to estradiol and testosterone
    - Males experience relatively higher exposure to testosterone
    - Females experience relatively higher exposure to estradiol
  - Song system differentiation occurs in passerines with potential role of steroids
- Lifespan
  - Long-lived birds produce few offspring annually over many years.

41.4.6 Discerning EDC Impacts in Field Birds

Much of the documentation for effects of toxicants that affect wild populations of birds has been associated with spills and critical exposures that resulted in the death of a number of individuals. Conversely, it is often difficult to detect nonlethal effects in wild populations due to the lack data for individuals and the difficulty in tracking birds, especially through migration. There are large datasets on wild birds from the USGS (United States Geological Service) Bird Banding lab and from annual surveys species and individual counts. However, there are few consolidated reports that begin to bring together potential long-term effects of exposure to environmental chemicals, especially those that are endocrine disruptors, because the effects may be subtle and occur over an individual's lifetime. As such, it is a challenge to develop reliable and sensitive measurement endpoints that can serve as indices of exposure in birds. Inherent in the selection of endpoints to be used as bioindicators is having an understanding of the association of an adverse impact on the individual and risk to populations as a whole. Exposure to steroid hormones or EDCs disrupt embryonic development including organ development, sexual differentiation, reproduction and immune function, reproductive behavior, and metabolic processes (Best et al., 2010; Berg et al., 1999; Darnerud, 2003; Ottinger et al., 2005a,b; Fernie et al., 2005a; Franceschini et al., 2008; Fox and Grasman, 1999). Biomedical and epidemiological data reveal clear linkages between EDC exposure, particularly early events and later disease or even epigenetic effects. In addition to biomedical data, field studies have correlated EDC exposure to immunosuppression in a number of taxa demonstrating that the effects of these compounds have conserved mechanisms and impact across a range of species, including invertebrates. Laboratory studies have been successful in establishing clear effects of various classes of EDCs on a number of physiological systems (Berg et al., 1998; Halldin et al., 2005). Because many early EDCs of concern such as methoxychlor, dioxin, and some polychlorinated biphenyls (PCBs) have estrogenic activity, attention focused on reproductive impact of these compounds in comparison to estradiol. However, at any one time, wildlife species are actually exposed to many different types of EDCs (e.g., androgenic, thyroid active) as well as other toxicants often with potentially different mechanisms of action (e.g., metals) (Levengood et al., 2007; Hotchkiss et al., 2008). As a result, it has been difficult to develop comprehensive risk-assessment models. Furthermore as mentioned above, some EDCs such as DDE are androgen receptor (AR) blockers, whereas other compounds such as trenbolone are androgenic (Mura et al., 2009; Panzica et al., 2007; Quinn et al., 2007a; Quinn et al., 2008). As in mammals, EDCs are physiologically active on a variety of pathways, including not only steroid-sensitive pathways but also pathways involving thyroid, metabolic, stress, and immune systems (Head et al., 2008; Fairbrother et al., 2004; Fernie et al., 2005a; Fox and Grasman, 1999; Mcnabb and Fox, 2003; Kavlock and Ankley, 1996; Knudsen et al., 2009). Therefore, use of the AOP (Hotchkiss et al., 2008) is an important tool for conceptualizing effects of EDCs including fundamental actions and molecular targets underpinning effects on the organism and population-level impacts.

Much of the available documentation on wild birds is from cases of highly concentrated exposures associated with toxicants originating from chemical applications and industrial spills. Often these occurrences are associated with high mortality; thereby prompting federal and regional responses aimed at protecting human and wildlife health. However, less is known about nonlethal impacts of EDCs on wild birds and the potential for adverse effects translating even to the level of a risk to the population. There are extensively documented cases, such as incidences of avian impacts from the Great Lakes, PCBs released into the Housatonic and Hudson Rivers, and high concentrations of munitions and other contaminants that indicate impacts on birds (Custer et al., 2010a, 2012a,b; Custer and Read, 2006; Custer et al., 2010d; Franceschini et al., 2008; Levengood et al., 2007). Data collected from some extraordinary datasets contain long-term monitoring and research programs, including the Great Lakes region and many of these studies have considered physiological and reproductive impacts of exposure to chemicals found in those areas. Laboratory studies provide the controlled conditions to filter out some of the myriad confounding factors challenging field birds, thereby allowing us to discern potential mechanisms of action of EDCs as well as periods of sensitivity throughout the life-cycle. Information from both types of studies is critical to conduct informed risk analysis for specific chemicals. Furthermore, it is important to consider unique characteristics of the avian physiology with emphasis on sexual differentiation, hormonal modulation of endocrine and behavioral components of reproduction, and functional impacts of EDCs on neuroendocrine regulatory systems, especially considering potential adverse outcomes.

Further, EDCs in the environment seldom often appear in combination and exert their actions the background of endogenous steroids. As such, EDCs may interfere with endogenous processes critical for sexual differentiation, maturation, breeding, and parental behavior. It is against this backdrop that exogenous EDCs act, with variations in specific mechanism(s) of action and targets for a particular compound. Although much of the emphasis of EDC studies has been on the reproductive axis, it is clear that the thyroid, immune, and stress axes are also potential targets of endocrine disruption. Toxicological effects are also observed with many of the EDCs, and some measures of activation of the liver detoxification processes, such as ethoxyresorufin-*O*-deethylase (EROD) activity, are excellent indicators of exposure. Elegant studies have also shown that subtle differences in the sequence of the aryl hydrocarbon receptor contribute to observed variations in the sensitivity of wild birds species to EDCs (Head et al., 2008; Head and Kennedy, 2007a,b). Finally, behavior is a sensitive measure, and more work is needed to establish methods to assess behavioral impairment related to EDC exposure. It is known from laboratory studies that EDCs do impair male sexual behavior in male quail and that diminished behavioral responses often parallel effects on other physiological responses (Mura et al., 2009; Ottinger et al., 2013; Panzica et al., 2007). Behavioral measures also provide useful indicators of effects, including sensory deficits, stress responses, and impaired reproductive function. Finally, behavioral measures may prove to be very revealing in songbirds in which the neural circuitry is steroid dependent and in distinguishing EDC effects on altricial versus precocial birds.

41.4.6.1 Altricial versus Precocial Birds: Developmental Patterns and Vulnerable Stages

Avian species rely on genetically based sexual differentiation with well-characterized developmental events occurring during ontogeny for both precocial and altricial birds. The genetic basis for gender is reversed from that of mammals in that the male is the homozygotic sex (ZZ) and the female is the heterozygotic (ZW). Two developmental strategies occur in birds; altricial chicks require extensive parental care whereas precocial chicks are well developed at hatch. The song system is steroid dependent, with elements that are sexually differentiated primarily post-hatch, and remains plastic throughout the life-cycle, showing seasonal regression and hormone-induced seasonal neurogenesis (Adkins-Regan and Watson, 1990; Adkins-Regan et al., 1990; Adkins-Regan and Ascenzi, 1990; Wade and Arnold, 2004; Fusani and Gahr, 2006; Ball and Balthazart, 2010). This neuroplasticity contributes to an apparent lifelong vulnerability yet potential resilience to EDC exposure (Iwaniuk et al., 2006; Rochester et al., 2010; Rochester and Millam, 2009). Adverse effects of EDCs have also been observed in finches and the dark-eyed junco (Hoogesteijn et al., 2008; Satre et al., 2009).

Precocial birds have often been the subject for toxicological studies and regulatory testing, such as Japanese quail (*C. japonica*), which are well developed at hatch, with sexual differentiation of the reproductive endocrine system and functional competency of other physiological systems already relatively complete (Adkins-Regan, 2009; Balthazart et al., 2009; Ottinger et al., 2005b; Ottinger and Dean, 2011). These studies have shown that the male Japanese quail is exquisitely sensitive to exogenous steroids, with both estradiol and androgen treatments resulting in impaired male reproductive behavior. As such, estradiol is very useful as a positive control and as a means of comparison for the relative activity of EDCs with varying impacts on endocrine, neural, and behavioral components of reproduction (Berg et al., 2001; Halldin et al., 1999; Berg et al., 1998; Adkins-Regan and Watson, 1990; Panzica et al., 2007). Japanese quail have a 17 day incubation period, and during development, each level of the reproductive (hypothalamic–pitutitary–gonadal; HPG) axis develops and begins to function as early as embryonic day 5 (gonadal differentiation), embryonic days 12–14 (hypothalamus and pituitary gland), and later in embryonic development (accessory structures). Gonadal function is initiated as early as embryonic day 5, with differential patterns of circulating steroid hormones in males and females (Figure 41.6). Estradiol is a critical element in sexual differentiation in quail. During embryonic development, females have relative high concentrations of estradiol/androgen whereas males have a relative low estradiol/androgen ratio. Moreover, there are sex-related differences in the patterns of plasma steroid levels during embryonic development. In female embryos, plasma E_2 rose until hatch and decreased post-hatch (Ottinger et al., 2008, 2005a). In males, plasma androgen peaked at embryonic days 14–17 (E14–E17), during the 17 day incubation period, and declined post-hatch. In addition, yolk steroid hormone content reflects embryonic steroid hormone levels. These steroid hormones are present and available to the embryo during sexual differentiation of multiple organ and endocrine systems throughout incubation. Once the embryonic gonads and adrenal glands begin to produce steroid hormones, circulating concentrations rise to make the gender-related pattern in changes more evident, especially during the last half of embryonic development. Both genders experience increased steroid hormones after embryonic days 10–12 in males and embryonic days 10–16 in females. Sexual differentiation of endocrine and behavioral components of reproduction occur during this time, organizing the HPG axis and gender-specific behaviors (Adkins-Regan and Watson, 1990; Ottinger et al., 2005a). There is a large literature that links steroid hormones, brain regions that modulate reproductive behavior, onset of reproduction during maturation, and adult reproductive function. The primary hypothalamic hormone that regulates the HPG axis is gonadotropin releasing hormone-I (GnRH-I), which is produced by the cell bodies located in the preoptic-septal region of the hypothalamus. Hypothalamic GnRH-I levels rise between embryonic days 10–15; followed by a sharp drop that appears associated with activation of the function of the HPG axis and feedback regulation of GnRH-I (Li et al., 1991). Both the endocrine and behavioral components of reproduction in Japanese quail are sexually differentiated during embryonic development.

In concert with the changing steroid hormones during embryonic development, the adrenal axis and the thyroid axis develop and achieve function. Circulating concentrations of adrenal and thyroid hormones rise during embryogenesis to peak at about the time of hatching in precocial birds, with a slight delay in this pattern for altricial birds. Additionally, the steroid hormones circulating during embryonic development also impact the immune system, with circulating testosterone in males inducing regression in the bursal tissue in males (Grasman, 2010; Lavoie et al., 2007; Lavoie and Grasman, 2007). Clear effects of EDCs have been observed for the immune system, especially for

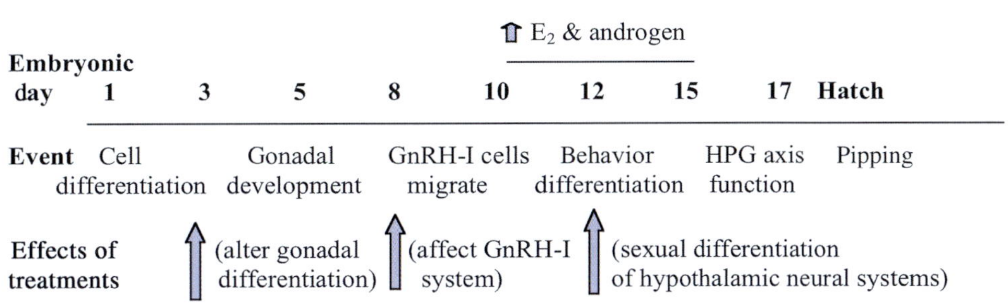

FIGURE 41.6 **Stages in embryonic development.** Diagram shows timing of developmental events during embryogenesis with indications of endogenous endocrine change and potential periods of vulnerability.

the response of the bursa, which is exquisitely sensitive to steroid hormones during embryonic development. Additionally, it has been well documented that the thyroid axis is also a primary target for many EDCs, including PCBs, flame retardants, and other compounds. Effects of EDCs exerted during embryonic development may also impair the thyroid axis, which is critical for hatching as well as overall metabolic function (Mcnabb and Fox, 2003).

Maternal deposition of steroids and EDCs into the egg. A primary mode of exposure to the avian embryo is from maternal deposition of steroids and EDCs into the egg. There is increasing recognition of the important role of maternally deposited steroid hormones, as well as corticosterone and thyroid hormones (Almasi et al., 2012a,b; Hayward and Wingfield, 2004). Mounting data point to both direct and epigenetic effects of maternally deposited hormones, which vary according to the condition of the female, health, reproductive status, and stressors. These steroids can affect viability as well as developmental characteristics of the chicks (Lipar et al., 1999; Schwabl, 1993,1996a,b). Deposition of EDCs is layered on top of these endogenously produced hormones from the hen. Moreover, most maternally transferred compounds and EDCs are lipid soluble, thereby deposited primarily into the egg yolk; whereas, water soluble compounds such as atrazine may be more evenly distributed throughout the egg compartments. There is significant deposition of EDCs into the eggs of field birds (Custer et al., 2010a,b,c,d). Similarly, laboratory studies have demonstrated that EDCs, including exogenous estradiol, methoxychlor, and soy phytoestrogens (see Figure 41.7 for phytoestrogens), all readily transfer from the hen into the egg and partition in the egg compartment according to lipid solubility (Lin et al., 2004; Ottinger et al., 2005b).

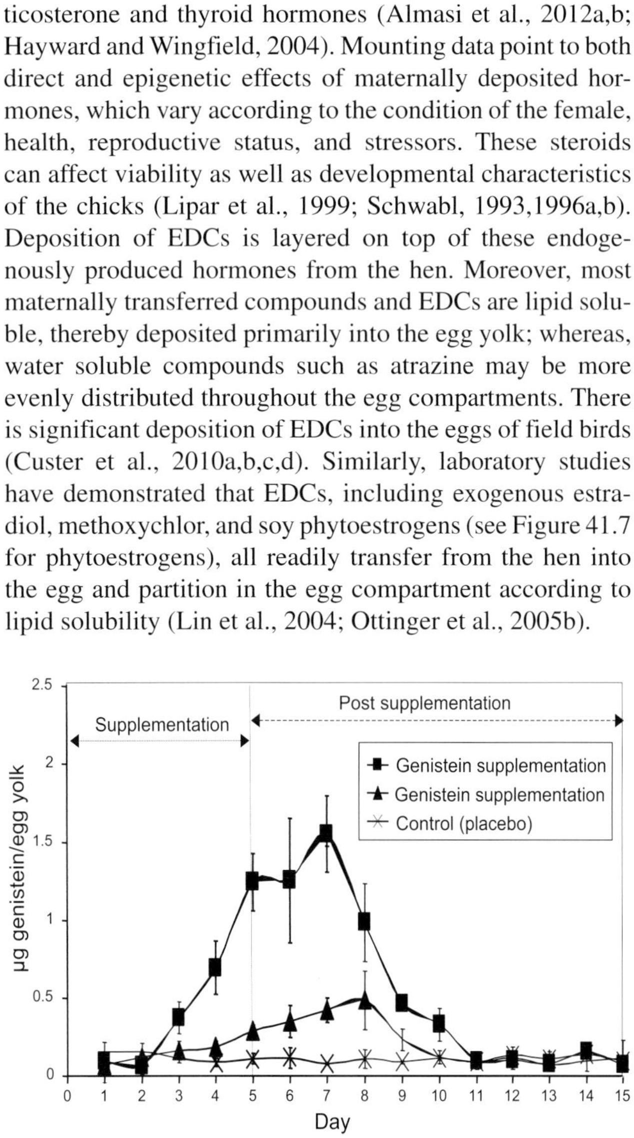

FIGURE 41.7 Maternal deposition of the phytoestrogen, genistein by the quail hen into eggs. Concentrations of Genistein in Japanese Quail egg yolks supplemented with genistein (aglycone), genistin (glucoside) or placebo capsules. Data points represent average of 4 replicates for treatment groups and 2 replicates for the control group. *Modified from Lin et al. (2004).*

Post-hatch growth and maturation are periods that are vulnerable to interference from EDCs. Steroid hormones continue to decline from the late embryonic and post-hatch levels associated with sexual differentiation to remain at low quiescent concentrations until the onset of sexual maturation. During sexual maturation, reproductive endocrine and behavioral responses are initiated earlier in males, so that they mature by 6–8 weeks of age (Ottinger and Brinkley, 1979, 1978). Females begin to mature slightly later than males, with onset of egg production by 8–10 weeks of age. This dimorphism is reflected in hypothalamic GnRH-I levels, which increase in males earlier than females (Ottinger et al., 2004). Organizational effects of steroids in quail are also reflected morphologically in the sexually dimorphic preoptic region of the quail, which is larger in males and testosterone sensitive in the adult; high levels of aromatase metabolize testosterone to estradiol, which is essential for sexual behavior. Because this area in known to be important in modulating courtship and mating behavior, a number of studies have focused on the neural systems contained in this area in Japanese quail; this area is potentially vulnerable to the effects of EDCs (Panzica et al., 2007). In addition, the preoptic-septal region (POA-SL) contains many of the gonadotropin releasing hormone (GnRH) cell bodies; these cells project to the median eminence of the hypothalamus, making the POA-SL critical in the regulation of both endocrine and behavioral components of reproduction. All these characteristics of the HPG axis and sexually dimorphic neuroendocrine systems are vulnerable to EDC exposure, especially during organization of the reproductive axis during embryonic sexual differentiation.

41.4.7 Developing Testing Paradigms to Reveal Endocrine Disruption in Birds

41.4.7.1 Investigating EDCs in an Avian Model: The Japanese Quail Two-Generation Test

Laboratory tests provide important information about EDC actions and impacts in birds and have used several experimental approaches. Most studies on the actions of steroid hormones have used egg injection, with treatment with known concentrations in order to understand the role of steroids in ontogeny and sexual differentiation. Many of the studies on EDC effects have used a similar approach, and these studies have been conducted in domestic chick embryos, mallard duck, and Japanese quail. The U.S. EPA oversees the safety of chemicals in use by industry, agricultural practices, recreational application, and individual residences. The Japanese quail two-generation test (JQTT) is under consideration to provide a multigenerational test to assess the impacts of EDCs on all life stages, both separately and as a continuum. Current testing for registration of chemicals may require Ecological Effects Studies, including

Avian Acute Oral Toxicity, Avian Subacute Dietary Toxicity, Avian Reproduction, or Simulated or Actual Field Testing (http://www.epa.gov/oppefed1/ecorisk_ders/toera_analysis_eco.htm) to ascertain safety for birds. These tests are conducted on a variety of species according to the test and are conducted in tandem with tests in other vertebrate and invertebrate species.

As discussed in the introduction to this chapter, EDCs may be difficult to detect, especially their nonlethal effects that have more subtle and long-term impacts on organisms. As a result, a series of testing protocols has been developed to address potential effects on vertebrates and invertebrates over the life-cycle. These two-generation testing protocols consider a range of measurement endpoints to assess effects and potential risks to selected organisms. In birds, there is consideration of a multigeneration test using Japanese quail, which contains each primary phase in the life-cycle with isolated exposure at that phase as well as cumulative exposure over the multiple life stages. Initial exposure in the current testing protocol occurs in proven breeders followed by continuous exposure to F1 offspring to evaluation of their F2 offspring as the final step in the testing protocol. As will be described below, there are a number of measures considered to assess health, toxicity effects, and endocrine disruption. The general list of measurement endpoints include measures of health and overt toxicity (food consumption, body weights, secondary sex characteristics, activity, chick health, growth and maturation, lethargy, morbidity or other clinical signs of toxicity, mortality); reproductive measures (sexual differentiation, sexual maturation, egg production, egg quality, fertility, embryo death, hatchability, chick survival and condition, sex ratio, reproductive behavior and reproductive hormones, testes histology and sperm counts, cloacal gland size and foam); endocrine and physiological endpoints (weight and histology of liver, testes, thyroid, adrenals, gonads, and brain, fecal steroid hormones, serum hormones and vitellogenin, glandular thyroid hormone). The reasoning for use of these endpoints in assessing EDCs is briefly discussed below to provide a more complete overview of the response of the endocrine and physiological systems to EDCs.

41.4.8 Pertinent Endpoints for Assessing Potential Endocrine Disruption

The selection of endpoints depends on the dosing regimen (i.e., amount of compound administered and the timing of administration), projected dose–response, the window of exposure (i.e., life-stage), projected mechanisms of action, potential endocrine activity or induction of vitellogenin or other biochemical responses, and mode of exposure; these are all critical variables in evaluating potential EDCs. As more information becomes available on EDC actions, it will be possible to use the structure proposed in the framework offered by the AOP to elucidate both the targeted physiological systems vulnerable to a chemical/chemical class and the potential for significant impacts to avian species. This approach is not currently integrated into the testing paradigm, but may prove useful for providing structure to interpreting the observations and findings. For example, the developing embryo is much more sensitive to the effects of EDCs than the adult, and developmental defects (i.e., permanent central and peripheral disruption) often persist into maturation and adult life stages. Conversely, adult exposure does not generally have as much impact or effects appear to be more transient, especially in precocial species. Advances in our understanding of molecular mechanisms have revealed that EDCs often exert epigenetic effects. This has changed our understanding of both the activational and organizational effects of EDCs, allowing for the possibility that exposure can cause lasting effects in multiple generations (Skinner et al., 2011; Hochberg et al., 2011). Therefore, it is extremely important to understand the physiology of the model (i.e., Japanese quail) and the species-specific endocrine mechanisms behind the selection of endpoints.

41.4.8.1 Survival

Survival is a basic endpoint for acute and chronic toxicity testing of compounds. Determining the median lethal dose that will kill 50% of the population being tested (LD50) is an important value for traditional toxicology studies. In dietary studies, survival must be monitored in order to assess health effects. In the context of endocrine disruption studies, it is unlikely that the doses will approach lethal concentrations because circulating concentrations of endogenous hormones are in the pico- or femtograms. Nonetheless, sublethal treatments are important in ascertaining potential risk from endocrine disruption. For example, 50 μg/kg/day of BPA has been determined to be a safe exposure by classic toxicology testing (Myers et al., 2009). However doses 100 to 1000 times lower than that have been shown to induce feminization of the gonads in Japanese quail, with no changes in survivability among groups (Oshima et al., 2012). Monitoring survival provides toxicological information and this important measure has not been generally reported in basic studies focused on understanding mechanisms. As such, the literature has a critical gap that ignores potential individual variability in response and the range of responses across a population including survivor effects.

41.4.8.2 Food Consumption and Body Weight

Food consumption is a basic measure of the general health; changes in food intake can indicate acute or chronic toxic effects of a compound. Additionally, if the route of chemical exposure is through the feed, it is important to monitor the consumption carefully to determine the exact exposure. It is also important to measure the stability of treatment compounds. Reduced food consumption has been associated with EDC exposure in Japanese quail (Yamashita et al., 2011). Similarly, body weight is closely linked with food intake and overall health and can be indicative of toxicological effects. For example, perchlorate, a thyroid hormone system disruptor, significantly reduced the growth (measured in body mass) of zebra finches dosed for the first two weeks of life (Rainwater et al., 2008). These birds also exhibited altered begging and fledging behaviors indicating that other endpoints may be linked to altered body weight, depending on the mechanism of disruption.

41.4.8.3 Accessory Sex Characteristics and Sexual Maturation

A number of secondary sex characteristics are gender-specific and provide excellent hormone-dependent response indices. Plumage differs in males and females; these sexual dimorphisms first appear as early as three weeks of age. Ultimately, the male has a rust-hued chest, which is even in color; females have a buff-colored chest with small dark areas in a dappled pattern. In general, feather color, especially in more brightly colored species such as bluebirds, depends on dietary carotenoids as well as being related to overall health. In addition, males often have sexually dimorphic characteristics that are testosterone dependent. In Japanese quail, the cloacal gland produces a glycoproteinaceous secretion, termed cloacal gland foam, that is critical for sperm transfer to the female (Ottinger and Brinkley, 1979). Measurements of the cloacal gland area have provided a reliable correlated measure for the sexual maturation and reproductive status of an individual male, and the foam produced can be estimated on a scale of 1–5 or a similar ordinal subjective measurement scale. Since Japanese quail are ground dwellers and males establish territories during breeding season, cloacal gland foam provides a marker delineating their territory; the foam has been called the "topping on the dropping" (Schleidt and Shalter, 1973). Monitoring the effects of EDCs on sexual maturation provides insight into the mechanisms of action. In males, monitoring for presence of cloacal gland foam in males provides an initial indicator of testicular activity and production of androgens. Further, depending on an EDC's action(s), there may be differential effects in males and females with males maturing slightly earlier than females. By 5–6 weeks of age, testicular activity initiates and begins to produce increasing concentrations of testosterone, leading to increasing expression of reproductive endocrine function, courtship and mating behavior, whereas females initiate egg production by 9–10 weeks of age.

41.4.8.4 Behavioral Indicators

Although the female Japanese quail exhibits a range of sex-specific behaviors, these behaviors are not readily quantifiable. Males and females both exhibit aggressive behavior, which can be quantified (Ramenofsky, 1985). Male courtship and mating behavior provide excellent bioindicators of reproductive competency. Sexual behaviors are steroid-hormone dependent; meaning that there must be sufficient circulating concentrations of androgens to enable the expression of male sexual behavior. Moreover, male sexual behavior is an exquisitely sensitive endpoint for embryonic exposure to estrogenic and highly active androgen-active compounds. A three minute test is conducted in which a receptive female (control diet female that is from the same population; non-experimental) is introduced into the male's home cage. Latency to mount, mount attempts, and cloacal contacts by the male are recorded. An individual may be tested on three successive days if it is important to determine the effects of experience on behavioral responses. Interpretation of the outcomes of behavioral testing relates to the potential mechanisms of action of compounds under testing. More specifically, the female quail requires exposure to estradiol during embryonic development and is masculinized by exposure to androgens between embryonic days 12–18. Conversely, males are demasculinized as evidenced by impaired male sexual behavior as adults, if they have been exposed to either estrogens or androgens during this critical period. As such, the male Japanese quail is exquisitely sensitive to exposure to EDCs during sexual differentiation.

41.4.8.5 Egg Production, Shell Quality, Fertility, and Embryo Viability

Egg production and onset of laying are important measures of reproductive function in birds, and are easily monitored by checking cages daily and marking eggs. EDC exposure has related to reduced egg production as well as delayed onset of egg production by acting on the gonads, gametes, or sperm-host glands, potentially resulting in decreased fertilization success. Similarly, the fertilized egg may not develop properly due to interference from an EDC and the embryo may not survive beyond the first few hours or days of development. During this time, the uptake of yolk (white and yellow yolk) is minimal and as such reflects primarily a toxic effect of a compound or specific effects impairing the function of homeobox genes and disruption of cellular lineages contributing to the formation of organ systems, such as in the case of heart effects (Carro et al., 2013b). Eggshell thickness, strength, and gross abnormalities can affect embryo viability. As early as the 1960s, exposure to DDT and DDE was linked to eggshell thinning and population decline in many wild bird species (Rattner, 2009).

Later embryonic effects and lethality are more likely to relate to disruption of essential physiological systems, such as impacting the thyroid endocrine axis and other systems essential for the development and synchronization of endocrine and other physiological systems. It has also been shown that a combination of early and late embryo mortality is a stronger indicator of toxicity that either alone. Moreover, there are distinct strain differences in Japanese quail, and there is an extensive literature that deals with a variety of physiological responses in Japanese quail that have been selected for multiple generations (Blohowiak et al., 1984; Bursian et al., 1983; Marks, 1996; Marks and Siegel, 1980). The Japanese quail chick begins to pip at ED15 in preparation for hatching at ED17. Once pipped, the chick is able to vocalize and often there is synchronizing in hatching of chicks in proximity. Pipping is followed by the chick pecking a hole, similar to an escape hatch, and requires a great deal of energy. The hatching success of chicks has been shown to be highly sensitive to chemical perturbation, and thus hatchability is an important endpoint in EDC studies.

41.4.8.6 *Histopathology*

Many organ systems are highly sensitive to hormonal perturbations, particularly if exposed during development. This includes reproductive organ (i.e., testes/ovaries/oviduct) weight, gross disruption of organ morphology, and microscopic disruption (i.e., disrupted follicular function or spermatogenesis). Thyroid glands, liver, bursa, and other organs should be examined, depending on the projected target(s) for the EDC. Because sexual differentiation of the avian female involves regression of one of the two primordial ovaries with its accompanying oviduct, the presence/regression of the right ovary and oviduct provides initial information about appropriate sexual differentiation of the female reproductive axis. There are sperm-host glands in the oviduct of the avian female, in which sperm may reside for some weeks; however, there are no data available for a potential impact of EDCs on these glands. The ovary in the ovulating female showed multiple follicles in varied stages of recruitment and maturation. Histological analyses will provide information as to the status of ovarian function, and linking ovarian morphology to these components of ovarian function provides insight into potential EDC targets and actions. Early exposure to EDCs in males can impact gonadal development in a genetic male; however, there is less information about this type of long-term impact. In some studies, EDC exposure resulted in transient presence of ovotestes, which are found in the hatchling but generally do not persist into the adult. Liver histology can provide information on any abnormalities associated with toxic effects. In addition, the activation of detoxification enzymes are excellent markers for initiation of these enzyme systems; conduct of the ethoxyresorufin-*O*-deethylase (EROD) assay on liver subsamples can provide supporting evidence for histopathology findings.

There is a link between embryonic exposure to PCBs and cardiac malformations, and kidney histology can provide information about toxicological impacts of EDCs due to the critical role of the kidneys in detoxification and elimination of water-soluble toxicants. Pathophysiological measurements of the bursa provide a powerful tool to assess endocrine disruption affecting the immune system. The bursa of Fabricius is exquisitely sensitive to steroid hormones, especially during development, with testosterone having an immunosuppressive effect (Ottinger et al., 2005b). As shown in Table 41.4, embryonic exposure to trenbolone acetate resulted in persistent effects on bursal morphology with fewer, smaller follicles (Quinn et al., 2007b). Conversely, estradiol treatment resulted in larger bursal size (Table 41.5); there was evidence of histopathology in treated individuals (Quinn et al., 2009). Many of these studies have also found a consistent relationship between bursal morphology and immune response. As such, bursal histopathology appears to be a reliable measure of impact upon the immune system, especially assessing impacts during embryonic treatment and the persistence of these effects into adulthood.

41.4.8.7 *Neuroendocrine Systems Regulating Reproduction, Metabolism, and Stress*

Embryonic exposures to steroids or compounds that are estrogen- or androgen-active result in impaired behavior in male behavior in adult quail (Ottinger et al., 2005a; Panzica et al., 2007; Ottinger et al., 2009b, 2008). Further, it has been shown that specific neural systems, including neurotransmitters (norepinephrine, dopamine, serotonergic) and neuropeptides (vasotocinergic), are modified in response to embryonic exposure to steroid hormones and to EDCs. Assessing the effects of embryonic exposure to EDCs on neurotransmitters and neuropeptides can provide information about direct effects of these compounds on regulatory systems that modulate reproductive, thyroid, and adrenal endocrine axes. Gonadotropin-releasing hormone (GnRH-I) neurons are located in the preoptic/lateral septal region of the hypothalamus. Although redundant, this system may be vulnerable to endocrine disruption (Ottinger et al., 2009b). Impact to neurotransmitter systems such as acetylcholine esterase (ACHase) has been used as a measure of toxicant exposure (Rattner et al., 1986).

Songbirds and other altricial birds also show impacts of EDC exposure on avian brain morphology (Iwaniuk et al., 2006; Millam et al., 2001). Further, there is a large literature about the impacts on steroid hormones on the sexual differentiation and function of the nuclei that direct and modulate song control in songbirds (Wade et al., 2004; Grisham et al., 2008; Arnold and Itoh, 2011; Wade and Arnold, 2004; Gilbert et al., 2007). Because of the neuroplasticity in songbirds, EDC effects may be more transient in songbirds;

TABLE 41.4 Summary of Significant ($p < 0.05$) Effects of Trenbolone Acetate and p,p-DDE on Immune, Reproductive, and Behavioral Measures

Endpoint Group	Endpoint	Trenbolone Acetate (Androgen)	p,p-DDE (Anti-androgen)
Immune measures	Bursa-body weight index	Decreased	Increased
	Bursal follicle number	Decreased in hatchlings and adults	Decreased in chicks
	Bursal follicle size	Increased at 0.05 μg/g, decreased at 50 μg/g	NA
	Spleen-body weight index	NA	NA
	Humoral response to chukar red blood cells	NA	NA
	Cell-mediated response to phytohemagglutinin	NA	NA
	Hatchling plasma immunoglobulin G	Increased at only 0.05 and 0.5 μg/g	NA
	Hatchling total leukocyte counts	NA	Increased in chicks
	Hatchling differential leukocyte counts	Increased heterophil : lymphocyte ratio	NA
Physiological reproductive measures	Gonad-body weight indices	NA	NA
	Testes morphology	NA	NA
	Ovarian follicle count	NA	NA
	Proctodeal foam gland weight	Decreased	NA
	Onset of puberty	Prolonged in males	Shortened in females
	Sperm penetration of perivitelline layer	NA	NA
Male copulatory behavior	Number of attempts to mount female	Decreased	Decreased
	Number of successful copulations	Decreased	NA
	Time to initial mount attempt	NA	NA
	Time to achieve first successful copulation	NA	Increased

"NA" denotes measures that were not affected by treatments.
Adapted from Quinn and Ottinger (2006). Reprinted with permission, The Journal of Poultry Science 43, 1–11, Japan Poultry Science Association, Tsukuba, Ibaraki, Japan.

however, there are insufficient data to ascertain if this is the case. Nonetheless, there are clearly documentations of EDC effects on brain regions that modulate singing and other behaviors that are steroid dependent in both precocial and altricial birds.

In addition, metabolic systems have been investigated in detail in birds, especially for domestic poultry. The thyroid system in birds has proven vulnerable to specific types of endocrine disruption in which this endocrine system is a target. Specifically, polychlorinated biphenyls, flame retardants, and other classes of EDCs have been implicated in having thyroid system actions that impact birds, both in the laboratory and in field birds (Scanes and Mcnabb, 2003; Mcnabb and Fox, 2003; Mcnabb, 2005; Chen et al., 2008; Webb and Mcnabb, 2008; Fernie et al., 2005b). It is not clear if histological methods provide a sensitive method for revealing these EDC effects; however, impaired production and release of thyroid hormones would be observed in abnormal thyroid gland follicles size and distribution and in differences in circulating thyroxine (T4) and the more bioactive form, triiodothyronine (T3) (please refer to Chapter 24). Finally, the hypothalamic–pituitary–adrenal axis is another endocrine system potentially responsive to EDC challenges. Measuring corticosterone in feathers may be an interesting technique to determine if long-term stress has occurred using a noninvasive technique. Seasonality and migration also complicate responses by birds to EDCs. Birds often exhibit seasonal patterns in reproduction, and species that migrate often show complete regression of the reproductive axis with cessation of function followed by restimulation of the reproductive system, or seasonal recrudescence for the next breeding cycle. Initial stimulation of reproductive function in birds relies on environmental triggers, termed zeitgebers, and in the case of temperate-zone species this environmental cue is often photoperiod. Birds are generally long day breeders due to their relatively short

TABLE 41.5 Summary of Responses to Estradiol or Trenbolone with Day of Egg Injection from the Studies Reviewed Above

	Hatchling Effects Summary[1]			
Endpoint	Estradiol ED4	Estradiol ED11	Trenbolone ED4	Trenbolone ED11
Embryo mortality	(↑)	–	(↑)	(↑)
Hatchability	(↓)	–	(↓)	(↓)
Body mass	–	–	–	–
Male: androgen	–	(↑↓)	(↓)	(↓)
Female: estradiol	–	(↑)	(↑↓)	(↓)
Bursa mass	(↓)	–	(↓)	(↑)
Aromatase	(↑)	(↑)	–	–
Norepinephrine	–	(↑)	–	–
Dopamine	–	(↑)	–	–
Serotonin	–	–	–	–
5 HT	–	–	–	–
GnRH-I	–	–	–	–

[1](↑) or (↓) means effect/trend was not significant, (↑↓) means a nonmonotonic dose response was observed.

incubation and relatively quick maturation times for their offspring. As such, most photoperiodic birds become reproductive with day length longer than 12h. There is a very large literature on this subject that deals with the neural mechanisms, response via the eyes and pineal gland, and the hypothalamic response to regulate reproductive function; review of this literature is beyond the scope of this chapter, but it is important to recognize seasonal variations in energy demands due to reproduction, migration, and environmental challenges, especially as EDCs impact critical endocrine systems involved in these seasonal and adaptive responses.

41.5 CONCLUSIONS

Estradiol and estrogen-active compounds are potent disruptors early in embryonic development in birds. Effects in females may be due to direct effects on the ovary as well as on other physiological systems. Males are exquisitely sensitive to impacts from xenoestrogens, with impaired endocrine and behavioral components of reproduction, altered immune responses, and modifications in other physiological systems. Androgen-active compounds and EDCs that impact thyroid system function alter reproductive, metabolic, and immune responses. Further, EDC impacts on stress responses place individuals at risk from multiple environmental challenges. It is clear that early exposure has greater overall impacts on precocial species, which have relatively developed and functional physiological systems at hatch. Altricial birds follow another developmental pattern and as such may differ in the timing and relative vulnerability to EDC exposure. Nonetheless, there are distinct and documented effects of EDCs in birds that must be considered for effective conservation of field populations of birds and to protect these populations from adverse effects from these environmental compounds.

ACKNOWLEDGMENT

Research from the Ottinger lab supported by EPA grants #R826134010 (Star Grant) and R-82877801; Battelle contract for EPA-EDSTAC validation studies, NRI #92-37203 and NSF #9817024; MAES, University of Maryland, College Park; Fish and Wildlife Service and Hudson River NRDA Trustees. The conclusions and opinions presented here are those of the authors, they do not represent the official position of any of the funding agencies, the Hudson River Trustees, or the United States.

REFERENCES

Adkins-Regan, E., 2009. Hormones and sexual differentiation of avian social behavior. Dev. Neurosci. 31, 342–350.

Adkins-Regan, E., Abdelnabi, M., Mobarak, M., Ottinger, M.A., 1990. Sex steroid levels in developing and adult male and female zebra finches (*Poephila guttata*). Gen. Comp. Endocrinol. 78, 93–109.

Adkins-Regan, E., Ascenzi, M., 1990. Sexual differentiation of behavior in the zebra finch: effect of early gonadectomy or androgen treatment. Horm. Behav. 24, 114–127.

Adkins-Regan, E., Watson, J.T., 1990. Sexual dimorphism in the avian brain is not limited to the song system of songbirds: a morphometric analysis of the brain of the quail (*Coturnix japonica*). Brain Res. 514, 320–326.

Almasi, B., Rettenbacher, S., Muller, C., Brill, S., Wagner, H., Jenni, L., 2012a. Maternal corticosterone is transferred into the egg yolk. Gen. Comp. Endocrinol. 178, 139–144.

Almasi, B., Roulin, A., Korner-Nievergelt, F., Jenni-Eiermann, S., Jenni, L., 2012b. Coloration signals the ability to cope with elevated stress hormones: effects of corticosterone on growth of barn owls are associated with melanism. J. Evol. Biol. 25, 1189–1199.

Ankley, G.T., Bennett, R.S., Erickson, R.J., Hoff, D.J., Hornung, M.W., Johnson, R.D., Mount, D.R., Nichols, J.W., Russom, C.L., Schmieder, P.K., Serrrano, J.A., Tietge, J.E., Villeneuve, D.L., 2010. Adverse outcome pathways: a conceptual framework to support ecotoxicology research and risk assessment. Environ. Toxicol. Chem. 29, 730–741.

Arnold, A.P., Itoh, Y., 2011. Factors causing sex differences in birds. Avian Biol. Res. 4, 44–51.

Asawasinsopon, R., Prapamontol, T., Prakobvitayakit, O., Vaneesorn, Y., Mangklabruks, A., Hock, B., 2006. Plasma levels of DDT and their association with reproductive hormones in adult men from northern Thailand. Sci. Total Environ. 355, 98–105.

Ball, G.F., Balthazart, J., 2010. Seasonal and hormonal modulation of neurotransmitter systems in the song control circuit. J. Chem. Neuroanat. 39, 82–95.

Balthazart, J., Taziaux, M., Holloway, K., Ball, G.F., Cornil, C.A., 2009. Behavioral effects of brain-derived estrogens in birds. In: Vaudry, H., Roubos, E.W., Coast, G.M., Vallarino, M. (Eds.), Trends in Comparative Endocrinology and Neurobiology.

Barnthouse, L.W., Glaser, D., Young, J., 2003. Effects of historic PCB exposures on the reproductive success of the Hudson River striped bass population. Environ. Sci. Technol. 37, 223–228.

Berg, C., Halldin, K., Brunstrom, B., 2001. Effects of bisphenol A and tetrabromobisphenol A on sex organ development in quail and chicken embryos. Environ. Toxicol. Chem. 20, 2836–2840.

Berg, C., Halldin, K., Brunstrom, B., Brandt, I., 1998. Methods for studying xenoestrogenic effects in birds. Toxicol. Lett. 102–103, 671–676.

Berg, C., Halldin, K., Fridolfsson, A.K., Brandt, I., Brunstrom, B., 1999. The avian egg as a test system for endocrine disrupters: effects of diethylstilbestrol and ethynylestradiol on sex organ development. Sci. Total Environ. 233, 57–66.

Best, D.A., Elliott, K.H., Bowerman, W.W., Shieldcastle, M., Postupalsky, S., Kubiak, T.J., Tillitt, D.E., Elliott, J.E., 2010. Productivity, embryo and eggshell characteristics, and contaminants in bald eagles from the Great Lakes, USA, 1986 to 2000. Environ. Toxicol. Chem. 29, 1581–1592.

Bigsby, R., Chapin, R.E., Daston, G.P., Davis, B.J., Gorski, J., Gray, L.E., Howdeshell, K.L., Zoeller, R.T., Vom Saal, F.S., 1999. Evaluating the effects of endocrine disruptors on endocrine function during development. Environ. Health Perspect. 107 (Suppl. 4), 613–618.

Blohowiak, C.C., Dunnington, E.A., Marks, H.L., Siegel, P.B., 1984. Body size, reproductive behavior, and fertility in three genetic lines of Japanese quail. Poult. Sci. 63, 847–854.

Borga, K., Hop, H., Skaare, J.U., Wolkers, H., Gabrielsen, G.W., 2007. Selective bioaccumulation of chlorinated pesticides and metabolites in Arctic seabirds. Environ. Pollut. 145, 545–553.

Bowerman, W.W., Bryan Jr., A.L., Robinette, J.R., Wing, J.M., Wiley, F.E., Murugasan, S., 2007. Concentrations of p,p′-DDE in plasma of nestling wood storks from Georgia. Chemosphere 68, 1506–1510.

Braune, B.M., Outridge, P.M., Fisk, A.T., Muir, D.C.G., Helm, P.A., Hobbs, K., Hoekstra, P.F., Kuzyk, Z.A., Kwan, M., Letcher, R.J., Lockhart, W.L., Norstrom, R.J., Stern, G.A., Stirling, I., 2005. Persistent organic pollutants and mercury in marine biota of the Canadian Arctic: an overview of spatial and temporal trends. Sci. Total Environ. 351–352, 4–56.

Broggi, J., Koivula, K., Lahti, K., Orell, M., 2003. Seasonality in daily body mass variation in a hoarding boreal passerine. Oecologia 137, 627–633.

Brunstrom, B., Hakansson, H., Lundberg, K., 1991. Effects of a technical PCB preparation and fractions thereof on ethoxyresorufin O-deethylase activity, vitamin-A levels and thymic development in the mink (Mustela-vison). Pharmacol. Toxicol. 69, 421–426.

Bursian, S.J., Polin, D., Olson, B.A., Shull, L.R., Marks, H.L., Siegel, H.S., 1983. Microsomal enzyme induction, egg production, and reproduction in three lines of Japanese quail fed polybrominated biphenyls. J. Toxicol. Environ. Health 12, 291–307.

Butcher, J.B., Garvey, E.A., 2004. PCB loading from sediment in the Hudson River: congener signature analysis of pathways. Environ. Sci. Technol. 38, 3232–3238.

Carro, T., Dean, K., Ottinger, M.A., 2013a. Effects of an environmentally relevant polychlorinated biphenyl (PCB) mixture on embryonic survival and cardiac development in the domestic chicken. Environ. Toxicol. Chem. 32, 1325–1331.

Carro, T., Taneyhill, L.A., Ann Ottinger, M., 2013b. The effects of an environmentally relevant 58-congener polychlorinated biphenyl (PCB) mixture on cardiac development in the chick embryo. Environ. Toxicol. Chem. 32, 1317–1324.

Chen, Y., Sible, J.C., Mcnabb, F.M.A., 2008. Effects of maternal exposure to ammonium perchlorate on thyroid function and the expression of thyroid-responsive genes in Japanese quail embryos. Gen. Comp. Endocrinol. 159, 196–207.

Cho, Y.C., Sokol, R.C., Rhee, G.Y., 2002. Kinetics of polychlorinated biphenyl dechlorination by Hudson River, New York, USA, sediment microorganisms. Environ. Toxicol. Chem. 21, 715–719.

Cohen-Barnhouse, A.M., Zwiernik, M.J., Link, J.E., Fitzgerald, S.D., Kennedy, S.W., Giesy, J.P., Wiseman, S., Jones, P.D., Newsted, J.L., Kay, D., Bursian, S.J., 2011. Developmental and posthatch effects of in ovo exposure to 2,3,7,8-TCDD, 2,3,4,7,8-PECDF, and 2,3,7,8-TCDF in Japanese quail (*Coturnix japonica*), common pheasant (*Phasianus colchicus*), and white leghorn chicken (*Gallus gallus domesticus*) embryos. Environ. Toxicol. Chem. 30, 1659–1668.

Corbitt, C., Satre, D., Adamson, L.A., Cobbs, G.A., Bentley, G.E., 2007. Dietary phytoestrogens and photoperiodic response in a male songbird, the Dark-eyed Junco (*Junco hyemalis*). Gen. Comp. Endocrinol. 154, 16–21.

Currie, R.A., Orphanides, G., Moggs, J.G., 2005. Mapping molecular responses to xenoestrogens through gene ontology and pathway analysis of toxicogenomic data. Reprod. Toxicol. 20, 433–440.

Custer, C.M., Custer, T.W., Dummer, P.M., 2010a. Patterns of organic contaminants in eggs of an insectivorous, an omnivorous, and a piscivorous bird nesting on the Hudson River, New York, USA. Environ. Toxicol. Chem. 29, 2286–2296.

Custer, C.M., Gray, B.R., Custer, T.W., 2010b. Effects of egg order on organic and inorganic element concentrations and egg characteristics in tree swallows, *Tachycineta bicolor*. Environ. Toxicol. Chem. 29, 909–921.

Custer, T.W., Custer, C.M., Gray, B.R., 2010c. Polychlorinated biphenyls, dioxins, furans, and organochlorine pesticides in belted kingfisher eggs from the upper Hudson River basin, New York, USA. Environ. Toxicol. Chem. 29, 99–110.

Custer, T.W., Custer, C.M., Gray, B.R., 2010d. Polychlorinated biphenyls, dioxins, furans, and organochlorine pesticides in spotted sandpiper eggs from the upper Hudson River basin, New York. Ecotoxicology 19, 391–404.

Custer, C.M., Custer, T.W., Hines, J.E., 2012a. Adult tree swallow survival on the polychlorinated biphenyl-contaminated Hudson River, New York, USA, between 2006 and 2010. Environ. Toxicol. Chem. 31, 1788–1792.

Custer, C.M., Custer, T.W., Schoenfuss, H.L., Poganski, B.H., Solem, L., 2012b. Exposure and effects of perfluoroalkyl compounds on tree swallows nesting at Lake Johanna in east central Minnesota, USA. Reprod. Toxicol. 33, 556–562.

Custer, C.M., Read, L.B., 2006. Polychlorinated biphenyl congener patterns in tree swallows (*Tachycineta bicolor*) nesting in the Housatonic River watershed, western Massachusetts, USA, using a novel statistical approach. Environ. Pollut. 142, 235–245.

D'Ursi, P., Salvi, E., Fossa, P., Milanesi, L., Rovida, E., 2005. Modelling the interaction of steroid receptors with endocrine disrupting chemicals. BMC Bioinform. 6 (Suppl. 4), S10.

Darnerud, P.O., 2003. Toxic effects of brominated flame retardants in man and in wildlife. Environ. Int. 29, 841–853.

Dix, D.J., Houck, K.A., Martin, M.T., Richard, A.M., Setzer, R.W., Kavlock, R.J., 2007. The ToxCast program for prioritizing toxicity testing of environmental chemicals. Toxicol. Sci. 95, 5–12.

Echols, K.R., Tillitt, D.E., Nichols, J.W., Secord, A.L., Mccarty, J.P., 2004. Accumulation of PCB congeners in nestling tree swallows (*Tachycineta bicolor*) on the Hudson River, New York. Environ. Sci. Technol. 38, 6240–6246.

Erickson, M.J., Turner, C.L., Thibodeaux, L.J., 2005. Field observation and modeling of dissolved fraction sediment-water exchange coefficients for PCBs in the Hudson River. Environ. Sci. Technol. 39, 549–556.

Fairbrother, A., Smits, J., Grasman, K.A., 2004. Avian immunotoxicology. J. Toxicol. Environ. Health B Crit. Rev. 7, 105–137.

Farmahin, R., Wu, D., Crump, D., Herve, J.C., Jones, S.P., Hahn, M.E., Karchner, S.I., Giesy, J.P., Bursian, S.J., Zwiernik, M.J., Kennedy, S.W., 2012. Sequence and in vitro function of chicken, ring-necked pheasant, and Japanese quail AHR1 predict in vivo sensitivity to dioxins. Environ. Sci. Technol. 46, 2967–2975.

Fernie, K.J., Mayne, G., Shutt, J.L., Pekarik, C., Grasman, K.A., Letcher, R.J., Drouillard, K., 2005a. Evidence of immunomodulation in nestling American kestrels (*Falco sparverius*) exposed to environmentally relevant PBDEs. Environ. Pollut. 138, 485–493.

Fernie, K.J., Shutt, J.L., Mayne, G., Hoffman, D., Letcher, R.J., Drouillard, K.G., Ritchie, I.J., 2005b. Exposure to polybrominated diphenyl ethers (PBDEs): changes in thyroid, vitamin A, glutathione homeostasis, and oxidative stress in American kestrels (*Falco sparverius*). Toxicol. Sci. 88, 375–383.

Fischer, L.J., Seegal, R.F., Ganey, P.E., Pessah, I.N., Kodavanti, P.R., 1998. Symposium overview: toxicity of non-coplanar PCBs. Toxicol. Sci. 41, 49–61.

Fisher, J.S., 2004. Are all EDC effects mediated via steroid hormone receptors? Toxicology 205, 33–41.

Foley, R.E., 1992. Organochlorine residues in New York waterfowl harvested by hunters in 1983-1984. Environ. Monit. Assess. 21, 37–48.

Foster, W.G., 1998. Endocrine disruptors and development of the reproductive system in the fetus and children: is there cause for concern? Can. J. Public Health 89 (Suppl. 1), S37–S41. S52, S41–S46.

Fox, L.L., Grasman, K.A., 1999. Effects of PCB 126 on primary immune organ development in chicken embryos. J. Toxicol. Environ. Health A 58, 233–244.

Franceschini, M.D., Custer, C.M., Custer, T.W., Reed, J.M., Romero, L.M., 2008. Corticosterone stress response in tree swallows nesting near polychlorinated biphenyl- and dioxin-contaminated rivers. Environ. Toxicol. Chem. 27, 2326–2331.

Fucic, A., Gamulin, M., Ferencic, Z., Katic, J., Krayer Von Krauss, M., Bartonova, A., Merlo, D.F., 2012. Environmental exposure to xenoestrogens and oestrogen related cancers: reproductive system, breast, lung, kidney, pancreas, and brain. Environ. Health 11 (Suppl. 1), S8.

Fusani, L., Gahr, M., 2006. Hormonal influence on song structure and organization: the role of estrogen. Neuroscience 138, 939–946.

Gilbert, L., Bulmer, E., Arnold, K.E., Graves, J.A., 2007. Yolk androgens and embryo sex: maternal effects or confounding factors? Horm. Behav. 51, 231–238.

Gore, A.C., 2010. Neuroendocrine targets of endocrine disruptors. Hormones (Athens) 9, 16–27.

Grasman, K.A., 2002. Assessing immunological function in toxicological studies of avian wildlife. Integr. Comp. Biol. 42, 34–42.

Grasman, K.A., 2010. In vivo functional tests for assessing immunotoxicity in birds. Methods Mol. Biol. 598, 387–398.

Grasman, K.A., Scanlon, P.F., Fox, G.A., 1998. Reproductive and physiological effects of environmental contaminants in fish-eating birds of the Great Lakes: a review of historical trends. Environ. Monit. Assess. 53, 117–145.

Grisham, W., Lee, J., Park, S.H., Mankowski, J.L., Arnold, A.P., 2008. A dose–response study of estradiol's effects on the developing zebra finch song system. Neurosci. Lett. 445, 158–161.

Guerrero-Bosagna, C.M., Skinner, M.K., 2009. Epigenetic transgenerational effects of endocrine disruptors on male reproduction. Semin. Reprod. Med. 27, 403–408.

Hall, J.M., Korach, K.S., 2002. Analysis of the molecular mechanisms of human estrogen receptors alpha and beta reveals differential specificity in target promoter regulation by xenoestrogens. J. Biol. Chem. 277, 44455–44461.

Hallanger, I.G., Warner, N.A., Ruus, A., Evenset, A., Christensen, G., Herzke, D., Gabrielsen, G.W., Borga, K., 2011. Seasonality in contaminant accumulation in Arctic marine pelagic food webs using trophic magnification factor as a measure of bioaccumulation. Environ. Toxicol. Chem. 30, 1026–1035.

Halldin, K., Axelsson, J., Brunström, B., 2005. Effects of endocrine modulators on sexual differentiation and reproductive function in male Japanese quail. Brain Res. Bull. 65, 211–218.

Halldin, K., Berg, C., Brandt, I., Brunstrom, B., 1999. Sexual behavior in Japanese quail as a test end point for endocrine disruption: effects of in ovo exposure to ethinylestradiol and diethylstilbestrol. Environ. Health Perspect. 107, 861–866.

Hartig, P.C., Bobseine, K.L., Britt, B.H., Cardon, M.C., Lambright, C.R., Wilson, V.S., Gray Jr., L.E., 2002. Development of two androgen receptor assays using adenoviral transduction of MMTV-luc reporter and/or hAR for endocrine screening. Toxicol. Sci. 66, 82–90.

Hartig, P.C., Cardon, M.C., Lambright, C.R., Bobseine, K.L., Gray Jr., L.E., Wilson, V.S., 2007. Substitution of synthetic chimpanzee androgen receptor for human androgen receptor in competitive binding and transcriptional activation assays for EDC screening. Toxicol. Lett. 174, 89–97.

Harvey, P.W., Everett, D.J., 2006. Regulation of endocrine-disrupting chemicals: critical overview and deficiencies in toxicology and risk assessment for human health. Best Pract. Res. Clin. Endocrinol. Metab. 20, 145–165.

Hayward, L.S., Wingfield, J.C., 2004. Maternal corticosterone is transferred to avian yolk and may alter offspring growth and adult phenotype. Gen. Comp. Endocrinol. 135, 365–371.

Head, J.A., Hahn, M.E., Kennedy, S.W., 2008. Key amino acids in the aryl hydrocarbon receptor predict dioxin sensitivity in avian species. Environ. Sci. Technol. 42, 7535–7541.

Head, J.A., Kennedy, S.W., 2007a. Differential expression, induction, and stability of CYP1A4 and CYP1A5 mRNA in chicken and herring gull embryo hepatocytes. Comp. Biochem. Physiol. C Toxicol. Pharmacol. 145, 617–624.

Head, J.A., Kennedy, S.W., 2007b. Same-sample analysis of ethoxyresorufin-O-deethylase activity and cytochrome P4501A mRNA abundance in chicken embryo hepatocytes. Anal. Biochem. 360, 294–302.

Hennig, B., Meerarani, P., Slim, R., Toborek, M., Daugherty, A., Silverstone, A.E., Robertson, L.W., 2002. Proinflammatory properties of coplanar PCBs: in vitro and in vivo evidence. Toxicol. Appl. Pharmacol. 181, 174–183.

Hervé, J.C., Crump, D., Giesy, J.P., Zwiernik, M.J., Bursian, S.J., Kennedy, S.W., 2010. Ethoxyresorufin O-deethylase induction by TCDD, PeCDF and TCDF in ring-necked pheasant and Japanese quail hepatocytes: time-dependent effects on concentration–response curves. Toxicol. In Vitro 24, 1301–1305.

Hochberg, Z., Feil, R., Constancia, M., Fraga, M., Junien, C., Carel, J.C., Boileau, P., Le Bouc, Y., Deal, C.L., Lillycrop, K., Scharfmann, R., Sheppard, A., Skinner, M., Szyf, M., Waterland, R.A., Waxman, D.J., Whitelaw, E., Ong, K., Albertsson-Wikland, K., 2011. Child health, developmental plasticity, and epigenetic programming. Endocr. Rev. 32, 159–224.

Holmes, D.J., Ottinger, M.A., 2003. Birds as long-lived animal models for the study of aging. Exp. Gerontol. 38, 1365–1375.

Hoogesteijn, A.L., Kollias, G.V., Quimby, F.W., De Caprio, A.P., Winkler, D.W., Devoogd, T.J., 2008. Development of a brain nucleus involved in song production in zebra finches (*Taeniopygia guttata*) is disrupted by Aroclor 1248. Environ. Toxicol. Chem. 27, 2071–2075.

Hotchkiss, A.K., Rider, C.V., Blystone, C.R., Wilson, V.S., Hartig, P.C., Ankley, G.T., Foster, P.M., Gray, C.L., Gray, L.E., 2008. Fifteen years after "Wingspread" – environmental endocrine disrupters and human and wildlife health: where we are today and where we need to go. Toxicol. Sci. 105, 235–259.

Hsu, P.C., Pan, M.H., Li, L.A., Chen, C.J., Tsai, S.S., Guo, Y.L., 2007. Exposure in utero to 2,2′,3,3′,4,6′-hexachlorobiphenyl (PCB 132) impairs sperm function and alters testicular apoptosis-related gene expression in rat offspring. Toxicol. Appl. Pharmacol. 221, 68–75.

Hurlbert, A.H., Haskell, J.P., 2003. The effect of energy and seasonality on avian species richness and community composition. Am. Nat. 161, 83–97.

Iwaniuk, A.N., Koperski, D.T., Cheng, K.M., Elliott, J.E., Smith, L.K., Wilson, L.K., Wylie, D.R., 2006. The effects of environmental exposure to DDT on the brain of a songbird: changes in structures associated with mating and song. Behav. Brain Res. 173, 1–10.

Janer, G., Sternberg, R.M., Leblanc, G.A., Porte, C., 2005. Testosterone conjugating activities in invertebrates: are they targets for endocrine disruptors? Aquat. Toxicol. 71, 273–282.

Karchner, S.I., Franks, D.G., Kennedy, S.W., Hahn, M.E., 2006. The molecular basis for differential dioxin sensitivity in birds: role of the aryl hydrocarbon receptor. Proc. Natl. Acad. Sci. U.S.A. 103, 6252–6257.

Kase, R., Hansen, P.D., Fischer, B., Manz, W., Heininger, P., Reifferscheid, G., 2009. Integral assessment of estrogenic potentials in sediment-associated samples: part 2: study of estrogen and anti-estrogen receptor-binding potentials of sediment-associated chemicals under different salinity conditions using the salinity-adapted enzyme-linked receptor assay. Environ. Sci. Pollut. Res. Int. 16, 54–64.

Kavlock, R., Ankley, G.T., Collette, T., Francis, E., Hammerstrom, K., Fowle, J., Tilson, H., Toth, G., Schmieder, P., Veith, G.D., Weber, E., Wolf, D.C., Young, D., 2005. Computational toxicology: framework, partnerships, and program development: September 29–30, 2003, Research Triangle Park, North Carolina. Reprod. Toxicol. 19, 265–280.

Kavlock, R., Chandler, K., Houck, K., Hunter, S., Judson, R., Kleinstreuer, N., Knudsen, T., Martin, M., Padilla, S., Reif, D., Richard, A., Rotroff, D., Sipes, N., Dix, D., 2012. Update on EPA's ToxCast program: providing high throughput decision support tools for chemical risk management. Chem. Res. Toxicol. 25, 1287–1302.

Kavlock, R.J., Ankley, G.T., 1996. A perspective on the risk assessment process for endocrine-disruptive effects on wildlife and human health. Risk Anal. 16, 731–739.

Knudsen, T., Martin, M., Chandler, K., Kleinstreuer, N., Judson, R., Sipes, N., 2013. Predictive models and computational toxicology. Methods Mol. Biol. 947, 343–374.

Knudsen, T.B., Martin, M.T., Kavlock, R.J., Judson, R.S., Dix, D.J., Singh, A.V., 2009. Profiling the activity of environmental chemicals in prenatal developmental toxicity studies using the U.S. EPA's ToxRefDB. Reprod. Toxicol. 28, 209–219.

Kuil, C.W., Brouwer, A., Van Der Saag, P.T., Van Der Burg, B., 1998. Interference between progesterone and dioxin signal transduction pathways. Different mechanisms are involved in repression by the progesterone receptor A and B isoforms. J. Biol. Chem. 273, 8829–8834.

Kusk, K.O., Kruger, T., Long, M., Taxvig, C., Lykkesfeldt, A.E., Frederiksen, H., Andersson, A.M., Andersen, H.R., Hansen, K.M., Nellemann, C., Bonefeld-Jorgensen, E.C., 2011. Endocrine potency of wastewater: contents of endocrine disrupting chemicals and effects measured by in vivo and in vitro assays. Environ. Toxicol. Chem. 30, 413–426.

Lalone, C.A., Villeneuve, D.L., Burgoon, L.D., Russom, C.L., Helgen, H.W., Berninger, J.P., Tietge, J.E., Severson, M.N., Cavallin, J.E., Ankley, G.T., 2013. Molecular target sequence similarity as a basis for species extrapolation to assess the ecological risk of chemicals with known modes of action. Aquat. Toxicol. 144–145, 141–154.

Larkin, P., Folmar, L.C., Hemmer, M.J., Poston, A.J., Lee, H.S., Denslow, N.D., 2002. Array technology as a tool to monitor exposure of fish to xenoestrogens. Mar. Environ. Res. 54, 395–399.

Lavoie, E.T., Grasman, K.A., 2007. Effects of in ovo exposure to PCBs 126 and 77 on mortality, deformities and post-hatch immune function in chickens. J. Toxicol. Environ. Health A 70, 547–558.

Lavoie, E.T., Wiley, F., Grasman, K.A., Tillitt, D.E., Sikarskie, J.G., Bowerman, W.W., 2007. Effect of in ovo exposure to an organochlorine mixture extracted from double crested cormorant eggs (*Phalacrocorax auritus*) and PCB 126 on immune function of juvenile chickens. Arch. Environ. Contam. Toxicol. 53, 655–661.

Lavoie, R.A., Champoux, L., Rail, J.-F., Lean, D.R.S., 2010. Organochlorines, brominated flame retardants and mercury levels in six seabird species from the Gulf of St. Lawrence (Canada): relationships with feeding ecology, migration and molt. Environ. Pollut. 158, 2189–2199.

Le Page, Y., Vosges, M., Servili, A., Brion, F., Kah, O., 2011. Neuroendocrine effects of endocrine disruptors in teleost fish. J. Toxicol. Environ. Health B Crit. Rev. 14, 370–386.

Levengood, J.M., Wiedenmann, L., Custer, T.W., Schaeffer, D.J., Matson, C.W., Melancon, M.J., Hoffman, D.J., Scott, J.W., Talbott, J.L., Bordson, G.O., Bickham, J.W., Rattner, B.A., Golden, N.H., 2007. Contaminant exposure and biomarker response in embryos of black-crowned night-herons (*Nycticorax nycticorax*) nesting near Lake Calumet, Illinois. J. Great Lakes Res. 33, 791–805.

Li, Q.C., Alston-Mills, B., Ottinger, M.A., 1991. Avian LHRH during embryonic development: measurement by competitive ELISA with a monoclonal antibody. Gen. Comp. Endocrinol. 82, 444–450.

Lin, F., Wu, J., Abdelnabi, M.A., Ottinger, M.A., Giusti, M.M., 2004. Effects of dose and glycosylation on the transfer of genistein into the eggs of the Japanese quail (*Coturnix japonica*). J. Agric. Food Chem. 52, 2397–2403.

Lipar, J.L., Ketterson, E.D., Nolan Jr., V., Casto, J.M., 1999. Egg yolk layers vary in the concentration of steroid hormones in two avian species. Gen. Comp. Endocrinol. 115, 220–227.

Lundholm, C.D., 1997. DDE-induced eggshell thinning in birds: effects of p,p′-DDE on the calcium and prostaglandin metabolism of the eggshell gland. Comp. Biochem. Physiol. C Pharmacol. Toxicol. Endocrinol. 118, 113–128.

Maffini, M.V., Rubin, B.S., Sonnenschein, C., Soto, A.M., 2006. Endocrine disruptors and reproductive health: the case of bisphenol-A. Mol. Cell. Endocrinol. 254–255, 179–186.

Majewski, P., Adamska, I., Pawlak, J., Baranska, A., Skwarlo-Sonta, K., 2005. Seasonality of pineal gland activity and immune functions in chickens. J. Pineal Res. 39, 66–72.

Man, Y.B., Chow, K.L., Wang, H.S., Lau, K.Y., Sun, X.L., Wu, S.C., Cheung, K.C., Chung, S.S., Wong, M.H., 2011. Health risk assessment of organochlorine pesticides with emphasis on DDTs and HCHs in abandoned agricultural soils. J. Environ. Monit. 13, 2250–2259.

Manikkam, M., Tracey, R., Guerrero-Bosagna, C., Skinner, M.K., 2013. Plastics derived endocrine disruptors (BPA, DEHP and DBP) induce epigenetic transgenerational inheritance of obesity, reproductive disease and sperm epimutations. PLoS One 8, e55387.

Manning, G.E., Farmahin, R., Crump, D., Jones, S.P., Klein, J., Konstantinov, A., Potter, D., Kennedy, S.W., 2012. A luciferase reporter gene assay and aryl hydrocarbon receptor 1 genotype predict the LD50 of polychlorinated biphenyls in avian species. Toxicol. Appl. Pharmacol. 263, 390–401.

Marino, M., Pellegrini, M., La Rosa, P., Acconcia, F., 2012. Susceptibility of estrogen receptor rapid responses to xenoestrogens: physiological outcomes. Steroids 77, 910–917.

Marks, H.L., 1996. Long-term selection for body weight in Japanese quail under different environments. Poult. Sci. 75, 1198–1203.

Marks, H.L., Siegel, H.S., 1980. Divergent selection in Japanese quail for the plasma cholesterol response to ACTH. Poult. Sci. 59, 1700–1705.

Martin, M.T., Judson, R.S., Reif, D.M., Kavlock, R.J., Dix, D.J., 2009. Profiling chemicals based on chronic toxicity results from the U.S. EPA ToxRef Database. Environ. Health Perspect. 117, 392–399.

Masuyama, H., Inoshita, H., Hiramatsu, Y., Kudo, T., 2002. Ligands have various potential effects on the degradation of pregnane X receptor by proteasome. Endocrinology 143, 55–61.

Mcnabb, F.M., 2007. The hypothalamic-pituitary-thyroid (HPT) axis in birds and its role in bird development and reproduction. Crit. Rev. Toxicol. 37, 163–193.

Mcnabb, F.M., Fox, G.A., 2003. Avian thyroid development in chemically contaminated environments: is there evidence of alterations in thyroid function and development? Evol. Dev. 5, 76–82.

Mcnabb, F.M.A., 2005. Biomarkers for the assessment of avian thyroid disruption by chemical contaminants. Avian Poult. Biol. Rev. 16, 3–10.

Millam, J.R., Craig-Veit, C.B., Quaglino, A.E., Erichsen, A.L., Famula, T.R., Fry, D.M., 2001. Posthatch oral estrogen exposure impairs adult reproductive performance of zebra finch in a sex-specific manner. Horm. Behav. 40, 542–549.

Mura, E., Barale, C., Quinn Jr., M.J., Panzica, G., Ottinger, M.A., Viglietti-Panzica, C., 2009. Organizational effects of DDE on brain vasotocin system in male Japanese quail. Neurotoxicology 30, 479–484.

Myers, J.P., Vom Saal, F.S., Akingbemi, B.T., Arizono, K., Belcher, S., Colborn, T., Chahoud, I., Crain, D.A., Farabollini, F., Guillette Jr., L.J., Hassold, T., Ho, S.M., Hunt, P.A., Iguchi, T., Jobling, S., Kanno, J., Laufer, H., Marcus, M., Mclachlan, J.A., Nadal, A., Oehlmann, J., Olea, N., Palanza, P., Parmigiani, S., Rubin, B.S., Schoenfelder, G., Sonnenschein, C., Soto, A.M., Talsness, C.E., Taylor, J.A., Vandenberg, L.N., Vandenbergh, J.G., Vogel, S., Watson, C.S., Welshons, W.V., Zoeller, R.T., 2009. Why public health agencies cannot depend on good laboratory practices as a criterion for selecting data: the case of bisphenol A. Environ. Health Perspect. 117, 309–315.

Newbold, R.R., Padilla-Banks, E., Jefferson, W.N., Heindel, J.J., 2008. Effects of endocrine disruptors on obesity. Int. J. Androl. 31, 201–208.

Nichols, J.W., Echols, K.R., Tillitt, D.E., Secord, A.L., Mccarty, J.P., 2004. Bioenergetics-based modeling of individual PCB congeners in nestling tree swallows from two contaminated sites on the upper Hudson River, New York. Environ. Sci. Technol. 38, 6234–6239.

Nost, T.H., Helgason, L.B., Harju, M., Heimstad, E.S., Gabrielsen, G.W., Jenssen, B.M., 2012. Halogenated organic contaminants and their correlations with circulating thyroid hormones in developing Arctic seabirds. Sci. Total Environ. 414, 248–256.

Ohlson, C.G., Hardell, L., 2000. Testicular cancer and occupational exposures with a focus on xenoestrogens in polyvinyl chloride plastics. Chemosphere 40, 1277–1282.

Oshima, A., Yamashita, R., Nakamura, K., Wada, M., Shibuya, K., 2012. In ovo exposure to nonylphenol and bisphenol A resulted in dose-independent feminization of male gonads in Japanese quail (*Coturnix japonica*) embryos. Environ. Toxicol. Chem. 31, 1091–1097.

Ottinger, M.A., 2005. Avians Species Comparison Study—A Protocol Development Study for the Avian 2-Generation Tier II Assay. . Report to the Endocrine Disruptor Methods Validation Advisory Committee. College Park, MD.

Ottinger, M.A., Abdelnabi, M., Li, Q., Chen, K., Thompson, N., Harada, N., Viglietti-Panzica, C., Panzica, G.C., 2004. The Japanese quail: a model for studying reproductive aging of hypothalamic systems. Exp. Gerontol. 39, 1679–1693.

Ottinger, M.A., Brinkley, H.J., 1978. Testosterone and sex-related behavior and morphology: relationship during maturation and in the adult Japanese quail. Horm. Behav. 11, 175–182.

Ottinger, M.A., Brinkley, H.J., 1979. Testosterone and sex related physical characteristics during the maturation of the male Japanese quail (coturnix coturnix japonica). Biol. Reprod. 20, 905–909.

Ottinger, M.A., Carro, T., Bohannon, M., Baltos, L., Marcell, A.M., Mckernan, M., Dean, K.M., Lavoie, E., Abdelnabi, M., 2013. Assessing effects of environmental chemicals on neuroendocrine systems: potential mechanisms and functional outcomes. Gen. Comp. Endocrinol. 190, 194–202.

Ottinger, M.A., Dean, K.M., 2011. Neuroendocrine impacts of endocrine-disrupting chemicals in birds: life stage and species sensitivities. J. Toxicol. Environ. Health B Crit. Rev. 14, 413–422.

Ottinger, M.A., Lavoie, E., Thompson, N., Barton, A., Whitehouse, K., Barton, M., Abdelnabi, M., Quinn Jr., M., Panzica, G., Viglietti-Panzica, C., 2008. Neuroendocrine and behavioral effects of embryonic exposure to endocrine disrupting chemicals in birds. Brain Res. Rev. 57, 376–385.

Ottinger, M.A., Lavoie, E.T., Abdelnabi, M., Quinn, M.J., Marcell, A., Dean, K., 2009a. An overview of dioxin-like compounds, PCB, and pesticide exposures associated with sexual differentiation of neuroendocrine systems, fluctuating asymmetry, and behavioral effects in birds. J. Environ. Sci. Health C Environ. Carcinog. Ecotoxicol. Rev. 27, 286–300.

Ottinger, M.A., Lavoie, E.T., Thompson, N., Bohannon, M., Dean, K., Quinn Jr., M.J., 2009b. Is the gonadotropin releasing hormone system vulnerable to endocrine disruption in birds? Gen. Comp. Endocrinol. 163, 104–108.

Ottinger, M.A., Quinn Jr., M.J., Lavoie, E., Abdelnabi, M.A., Thompson, N., Hazelton, J.L., Wu, J.M., Beavers, J., Jaber, M., 2005a. Consequences of endocrine disrupting chemicals on reproductive endocrine function in birds: establishing reliable end points of exposure. Domest. Anim. Endocrinol. 29, 411–419.

Ottinger, M.A., Wu, J.M., Hazelton, J.L., Abdelnabi, M.A., Thompson, N., Quinn Jr., M.L., Donoghue, D., Schenck, F., Ruscio, M., Beavers, J., Jaber, M., 2005b. Assessing the consequences of the pesticide methoxychlor: neuroendocrine and behavioral measures as indicators of biological impact of an estrogenic environmental chemical. Brain Res. Bull. 65, 199–209.

Panzica, G.C., Viglietti-Panzica, C., Mura, E., Quinn Jr., M.J., Lavoie, E., Palanza, P., Ottinger, M.A., 2007. Effects of xenoestrogens on the differentiation of behaviorally-relevant neural circuits. Front. Neuroendocrinol. 28, 179–200.

Pereira, S.P., Pereira, G.C., Pereira, C.V., Carvalho, F.S., Cordeiro, M.H., Mota, P.C., Ramalho-Santos, J., Moreno, A.J., Oliveira, P.J., 2013. Dioxin-induced acute cardiac mitochondrial oxidative damage and increased activity of ATP-sensitive potassium channels in Wistar rats. Environ. Pollut. 180, 281–290.

Pessah, I.N., Cherednichenko, G., Lein, P.J., 2010. Minding the calcium store: ryanodine receptor activation as a convergent mechanism of PCB toxicity. Pharmacol. Ther. 125, 260–285.

Prins, G.S., 2008. Endocrine disruptors and prostate cancer risk. Endocr. Relat. Cancer 15, 649–656.

Quinn Jr., M.J., Ottinger, M.A., 2006. Embryonic effects of androgen active endocrine disrupting chemicals on avian immune and reproductive systems. J. Poult. Sci. 43, 1–11.

Quinn Jr., M.J., Lavoie, E.T., Ottinger, M.A., 2007a. Reproductive toxicity of trenbolone acetate in embryonically exposed Japanese quail. Chemosphere 66, 1191–1196.

Quinn Jr., M.J., Mckernan, M., Lavoie, E.T., Ottinger, M.A., 2007b. Immunotoxicity of trenbolone acetate in Japanese quail. J. Toxicol. Environ. Health A 70, 88–93.

Quinn Jr., M.J., Summitt, C.L., Ottinger, M.A., 2008. Consequences of in ovo exposure to p,p′-DDE on reproductive development and function in Japanese quail. Horm. Behav. 53, 249–253.

Quinn Jr., M.J., Mckernan, M., Lavoie, E.T., Ottinger, M.A., 2009. Effects of estradiol on the development of the bursa of Fabricius in Japanese quail. J. Exp. Zool. A Ecol. Genet. Physiol. 311, 91–95.

Rainwater, T.R., Wood, M.B., Millam, J.R., Hooper, M.J., 2008. Effects of perchlorate on growth and behavior of a granivorous passerine, the zebra finch (*Taeniopygia guttata*). Arch. Environ. Contam. Toxicol. 54, 516–524.

Ramenofsky, M., 1985. Acute changes in plasma steroids and agonistic behavior in male Japanese quail. Gen. Comp. Endocrinol. 60, 116–128.

Rattner, B.A., 2009. History of wildlife toxicology. Ecotoxicology 18, 773–783.

Rattner, B.A., Clarke, R.N., Ottinger, M.A., 1986. Depression of plasma luteinizing hormone concentration in quail by the anticholinesterase insecticide parathion. Comp. Biochem. Physiol. C 83, 451–453.

Rattner, B.A., Hoffman, D.J., Melancon, M.J., Olsen, G.H., Schmidt, S.R., Parsons, K.C., 2000. Organochlorine and metal contaminant exposure and effects in hatching black-crowned night herons (*Nycticorax nycticorax*) in Delaware Bay. Arch. Environ. Contam. Toxicol. 39, 38–45.

Rattner, B.A., Horak, K.E., Warner, S.E., Johnston, J.J., 2010. Acute toxicity of diphacinone in Northern bobwhite: effects on survival and blood clotting. Ecotoxicol. Environ. Saf. 73, 1159–1164.

Rattner, B.A., Mcgowan, P.C., Hatfield, J.S., Hong, C.S., Chu, S.G., 2001. Organochlorine contaminant exposure and reproductive success of black-crowned night-herons (*Nycticorax nycticorax*) nesting in Baltimore Harbor, Maryland. Arch. Environ. Contam. Toxicol. 41, 73–82.

Rochester, J.R., Forstmeier, W., Millam, J.R., 2010. Post-hatch oral estrogen in zebra finches (*Taeniopygia guttata*): is infertility due to disrupted testes morphology or reduced copulatory behavior? Physiol. Behav. 101, 13–21.

Rochester, J.R., Millam, J.R., 2009. Phytoestrogens and avian reproduction: exploring the evolution and function of phytoestrogens and possible role of plant compounds in the breeding ecology of wild birds. Comp. Biochem. Physiol. A Mol. Integr. Physiol. 154, 279–288.

Ross, A.W., Helfer, G., Russell, L., Darras, V.M., Morgan, P.J., 2011. Thyroid hormone signalling genes are regulated by photoperiod in the hypothalamus of F344 rats. PLoS One 6.

Safe, S., Bandiera, S., Sawyer, T., Robertson, L., Safe, L., Parkinson, A., Thomas, P.E., Ryan, D.E., Reik, L.M., Levin, W., et al., 1985. PCBs: structure-function relationships and mechanism of action. Environ. Health Perspect. 60, 47–56.

Safe, S., Wang, F., Porter, W., Duan, R., Mcdougal, A., 1998. Ah receptor agonists as endocrine disruptors: antiestrogenic activity and mechanisms. Toxicol. Lett. 102–103, 343–347.

Santti, R., Makela, S., Strauss, L., Korkman, J., Kostian, M.L., 1998. Phytoestrogens: potential endocrine disruptors in males. Toxicol. Ind. Health 14, 223–237.

Satre, D., Reichert, M., Corbitt, C., 2009. Effects of vinclozolin, an anti-androgen, on affiliative behavior in the Dark-eyed Junco, *Junco hyemalis*. Environ. Res. 109, 400–404.

Scanes, C.G., Mcnabb, F.M.A., 2003. Avian models for research in toxicology and endocrine disruption. Avian Poult. Biol. Rev. 14, 21–52.

Schleidt, W.M., Shalter, M.D., 1973. Stereotypy of a fixed action pattern during ontogeny in *Coturnix coturnix coturnix*. Z. Tierpsychol. 33, 35–37.

Schug, T.T., Janesick, A., Blumberg, B., Heindel, J.J., 2011. Endocrine disrupting chemicals and disease susceptibility. J. Steroid Biochem. Mol. Biol. 127, 204–215.

Schwabl, H., 1993. Yolk is a source of maternal testosterone for developing birds. Proc. Natl. Acad. Sci. U.S.A. 90, 11446–11450.

Schwabl, H., 1996a. Environment modifies the testosterone levels of a female bird and its eggs. J. Exp. Zool. 276, 157–163.

Schwabl, H., 1996b. Maternal testosterone in the avian egg enhances postnatal growth. Comp. Biochem. Physiol. A Physiol. 114, 271–276.

Skinner, M.K., 2007. Endocrine disruptors and epigenetic transgenerational disease etiology. Pediatr. Res. 61, 48R–50R.

Skinner, M.K., Manikkam, M., Guerrero-Bosagna, C., 2011. Epigenetic transgenerational actions of endocrine disruptors. Reprod. Toxicol. 31, 337–343.

Soto, A.M., Sonnenschein, C., 2010. Environmental causes of cancer: endocrine disruptors as carcinogens. Nat. Rev. Endocrinol. 6, 363–370.

Stoker, T.E., Parks, L.G., Gray, L.E., Cooper, R.L., 2000. Endocrine-disrupting chemicals: prepubertal exposures and effects on sexual maturation and thyroid function in the male rat. A focus on the EDSTAC recommendations. Endocrine Disrupter Screening and Testing Advisory Committee. Crit. Rev. Toxicol. 30, 197–252.

Thackaberry, E.A., Nunez, B.A., Ivnitski-Steele, I.D., Friggins, M., Walker, M.K., 2005. Effect of 2,3,7,8-tetrachlorodibenzo-p-dioxin on murine heart development: alteration in fetal and postnatal cardiac growth, and postnatal cardiac chronotropy. Toxicol. Sci. 88, 242–249.

Touart, L.W., 2004. Factors considered in using birds for evaluating endocrine-disrupting chemicals. ILAR J. 45, 462–468.

Van Den Berg, M., Denison, M.S., Birnbaum, L.S., Devito, M.J., Fiedler, H., Falandysz, J., Rose, M., Schrenk, D., Safe, S., Tohyama, C., Tritscher, A., Tysklind, M., Peterson, R.E., 2013. Polybrominated dibenzo-p-dioxins, dibenzofurans, and biphenyls: inclusion in the toxicity equivalency factor concept for dioxin-like compounds. Toxicol. Sci. 133, 197–208.

Villeneuve, D., Volz, D.C., Embry, M.R., Ankley, G.T., Belanger, S.E., Leonard, M., Schirmer, K., Tanguay, R., Truong, L., Wehmas, L., 2013. Investigating alternatives to the fish early-life stage test: a strategy for discovering and annotating adverse outcome pathways for early fish development. Environ. Toxicol. Chem.

Wada, H., Cristol, D.A., Mcnabb, F.M.A., Hopkins, W.A., 2009. Suppressed adrenocortical responses and thyroid hormone levels in birds near a mercury-contaminated river. Environ. Sci. Technol. 43, 6031–6038.

Wade, J., Arnold, A.P., 2004. Sexual differentiation of the zebra finch song system. Ann. N.Y. Acad. Sci. 1016, 540–559.

Wade, J., Peabody, C., Coussens, P., Tempelman, R.J., Clayton, D.F., Liu, L., Arnold, A.P., Agate, R., 2004. A cDNA microarray from the telencephalon of juvenile male and female zebra finches. J. Neurosci. Methods 138, 199–206.

Walker, M.K., Catron, T.F., 2000. Characterization of cardiotoxicity induced by 2,3,7, 8-tetrachlorodibenzo-p-dioxin and related chemicals during early chick embryo development. Toxicol. Appl. Pharmacol. 167, 210–221.

Wambaugh, J.F., Setzer, R.W., Reif, D.M., Gangwal, S., Mitchell-Blackwood, J., Arnot, J.A., Joliet, O., Frame, A., Rabinowitz, J., Knudsen, T.B., Judson, R.S., Egeghy, P., Vallero, D., Cohen Hubal, E.A., 2013. High-throughput models for exposure-based chemical prioritization in the ExpoCast project. Environ. Sci. Technol. 47, 8479–8488.

Wang, J., Shi, X., Du, Y., Zhou, B., 2011. Effects of xenoestrogens on the expression of vitellogenin (vtg) and cytochrome P450 aromatase (cyp19a and b) genes in zebrafish (*Danio rerio*) larvae. J. Environ. Sci. Health A Tox. Hazard Subst. Environ. Eng. 46, 960–967.

Watanabe, H., Iguchi, T., Morohashi, K., 2002. Endocrine disruptors and nuclear receptors. Nihon Rinsho 60, 397–403.

Watanabe, M.X., Jones, S.P., Iwata, H., Kim, E.-Y., Kennedy, S.W., 2009. Effects of co-exposure to 2,3,7,8-tetrachlorodibenzo-p-dioxin and perfluorooctane sulfonate or perfluorooctanoic acid on expression of cytochrome P450 isoforms in chicken (*Gallus gallus*) embryo hepatocyte cultures. Comp. Biochem. Physiol. C Toxicol. Pharmacol. 149, 605–612.

Watson, C.S., hu, G., Paulucci-Holthauzen, A.A., 2013. Rapid actions of xenoestrogens disrupt normal estrogenic signaling. Steroids.

Webb, C.M., Mcnabb, F.M.A., 2008. Polychlorinated biphenyl effects on avian hepatic enzyme induction and thyroid function. Gen. Comp. Endocrinol. 155, 650–657.

Weseloh, D.V., Mineau, P., Struger, J., 1990. Geographical distribution of contaminants and productivity measures of herring gulls in the Great Lakes: Lake Erie and connecting channels 1978/79. Sci. Total Environ. 91, 141–159.

Wilson, F.E., Reinert, B.D., 1999. Long days and thyroxine program american tree sparrows for seasonality: evidence for temporal flexibility of the breeding season of euthyroid females. Gen. Comp. Endocrinol. 113, 136–145.

Yamashita, R., Oshima, A., Hasegawa-Baba, Y., Wada, M., Shibuya, K., 2011. Endocrine disrupting effects of low dose 17 beta-estradiol (E2) on the Japanese quail (*Coturnix japonica*) were detected by modified one-generation reproduction study. J. Toxicol. Sci. 36, 43–54.

Zacharewski, T., 1998. Identification and assessment of endocrine disruptors: limitations of in vivo and in vitro assays. Environ. Health Perspect. 106 (Suppl. 2), 577–582.

Zawatski, W., Lee, M.M., 2013. Male pubertal development: are endocrine-disrupting compounds shifting the norms? J. Endocrinol. 218, R1–R12.

Zhang, D., Trudeau, V.L., 2006. Integration of membrane and nuclear estrogen receptor signaling. Comp. Biochem. Physiol. A Mol. Integr. Physiol. 144, 306–315.

Zhang, R., Manning, G.E., Farmahin, R., Crump, D., Zhang, X., Kennedy, S.W., 2013. Relative potencies of aroclor mixtures derived from avian in vitro bioassays: comparisons with calculated toxic equivalents. Environ. Sci. Technol. 47, 8852–8861.

Index

Note: Page Numbers followed by *f* indicate figures; *t*, tables; *b*, boxes.

D

F

G

H

N

O

P

Q

R

S

U

V

W

Y